Springer Praxis Books

Astronomy and Planetary Sciences

Textbooks and monographs published in this series from 2013 onward are written for advanced undergraduate students in astronomy and the planetary sciences and advanced amateur astronomers. The editors insist on good readability and encourage new approaches to teaching astronomy and planetary sciences. Books published before 2013 serve a spectrum of readership: Some are at advanced amateur to advanced undergraduate level. Others are targeted at PhD students and researchers. Topics covered in the series include

- Astronomical telescopes and instrumentation
- Astronomical techniques, software and data
- Astrophysics, Astrochemistry, Astrobiology
- Solar system science (excluding the Earth sciences proper) and exoplanets
- Stellar physics and black hole astrophysics
- Galactic astronomy
- Extragalactic astronomy and cosmology.

The books are well illustrated with line diagrams and photographs throughout, with targeted use of colour for scientific interpretation and understanding. Many feature worked examples or problems and solutions.

Kanaris Tsinganos

A Century of Cosmic Discoveries

Heliophysics and Space Physics, Astrophysics and Cosmology

PRAXIS

Springer

Kanaris Tsinganos
Department of Physics, Section
of Astronomy and Astrophysics
University of Athens
Zografos, Greece

ISSN 2945-7475 ISSN 2945-7483 (electronic)
Springer Praxis Books
ISSN 2366-0082 ISSN 2366-0090 (electronic)
Astronomy and Planetary Sciences
ISBN 978-3-032-06281-9 ISBN 978-3-032-06282-6 (eBook)
https://doi.org/10.1007/978-3-032-06282-6

Cover illustration:
Herbig-Haro 211 (HH 211), a bipolar jet travelling through interstellar space at supersonic speeds. At roughly 1,000 light-years away from Earth in the constellation Perseus, the object is one of the youngest and nearest protostellar outflows, making it an ideal target for the NASA/ESA/CSA James Webb Space Telescope. These Herbig-Haro objects are luminous regions surrounding newborn stars, and are formed when stellar winds or jets of gas spewing from such newborn stars form shock waves, as they collide with nearby gas and dust at high speeds. This spectacular image of HH 211 reveals an outflow from a *Class 0 protostar*, an infantile analogue to our Sun, when it was no more than a few tens of thousands of years old and with a mass only 8% of the present-day Sun (it will eventually grow into a full growth star like the Sun).
Infrared imaging is powerful in studying newborn stars and their outflows, because such stars are invariably still embedded within the gas from the molecular cloud in which they formed. The infrared emission of the star's outflows penetrates the obscuring gas and dust, making an HH object like HH 211 ideal for observation with Webb's sensitive infrared instruments. Molecules excited by the turbulent conditions, including H_2, CO and SiO, emit infrared light that Webb can collect to map out the structure of the outflows. The image showcases a series of bow shocks to the lower-left and upper-right, as well as the narrow bipolar jet that powers them in unprecedented detail. The inner jet is seen to "wiggle" with mirror symmetry on either side of the central protostar. This is in agreement with observations on smaller scales and suggests that the protostar may in fact be an unresolved binary star.
Credit: ESA/Webb, NASA, CSA, T. Ray (Dublin Institute for Advanced Studies)

This Springer imprint is published by the registered company Springer Nature Switzerland AG
The registered company address is: Gewerbestrasse 11, 6330 Cham, Switzerland

If disposing of this product, please recycle the paper.

The Muses Urania and Calliope by Simon Vouet and Studio (French, ca. 1634, Samuel H. Kress Collection, oil on Wood, 31 7/16 inches by 49 3/16 inches, National Gallery of Art, Washington D.C. and Wikipedia Commons Public Domain). Platon describes Urania and Calliope as the Muses "who are chiefly concerned with heaven and thought, they are divine as well as human, they have the sweetest utterance, providing inspiration about heaven and knowledge"

Foreword

The *Nature* of beings and the *Reasons* of beings
(Η Φύση των όντων και οι Λόγοι των όντων)

Astrophysics has a charm that no other science has; both, with what it shows and proves and with what it hides and hints at; both, for the ordinary person and for the specialized researcher. Impressive words, concepts and terms such as elusive solar neutrinos, exoplanets, extraterrestrial life, white dwarfs, red giants, neutron stars, rapidly rotating pulsars, galaxies, black holes, quasars, gravitational lenses, Big Bang, accelerating expansion of the Universe, dark matter and energy, curved space-time, etc., stimulate our imagination, arouse our thoughts, unveil our existential quest and incite our admiration and inspiration.

This wonderful volume compiled by Prof. Kanaris Tsinganos – researcher and professor for half a century – whom I met during my graduate studies at the Center for Astrophysics at Harvard University – deals with all major astrophysical achievements of the last century, including 20 astrophysics related Nobel prizes in Physics. Topics, which literally fascinate, propel human thought into areas that are extremely difficult to access. When reading this book, you feel as if you have put your eye on the telescopes and are immersed in the view of our magnificent world. You see the achievements of the past and you foresee and imagine those of the future. You often feel as if you are touching the distant, the incomprehensible, the ungraspable, time and future.

Impressive images, descriptions and observations that reach us even from billions of light years away, miraculous achievements, inspiring facts are presented in this so to speak Encyclopedia of modern Astrophysics. However, along with the admiration of our technology and ingenious devices, along with the enjoyment of the beauty of the world and the grandeur of Physics, all these create a bitter feeling of scientific inadequacy and inability to provide answers to the great questions and *reasons* for our existence.

We live in a Universe characterized by extraordinary physical phenomena, by harmonious symmetries, astonishing disharmonies and inconceivable irregularities; a vast and majestic world, a world inaccessible through our senses, yet accessible through science! Yet, we may see everything, but the deepest secrets remain hidden.

Along with the Universe, we admire man, who thrown in the middle of nowhere, observes it and studies it. As a physical being, he is so small compared to the vast Universe: smaller than a gene, more minimal than a particle, less than dust. Biologically and physically insignificant! Yet, he is a thinking Universe himself, who by studying beauty can discover reality, feel the truth, contemplate the *Truth*. An expanding Universe himself! However, as knowledge increases, so does the realm of the unknown. And, if you are not deceived by your fake greatness, the more you learn, the more you experience what is beyond your capacity, what is beyond knowledge and understanding. You begin to discern faintly everywhere, like a cosmical microwave background radiation, the *reason* and its *cause*, the *mystery* of its *truth*. The other amazing Universe!

When you cannot bear with the challenge, when you cannot handle the unfathomable magnitude of space and the multitude of cosmic entities, when you are overwhelmed by the lack of answers and perplexed by the questions of purpose, reasons, causes and whys, and you are not content with knowing only the "how and what", then, either you confess in despair your weakness, your ignorance and your smallness, or you complacently push away your concerns and questioning. However, you can open yourself up to another world a little beyond physics, logic and your senses, considerably larger than the Universe, much more convincing than your scientific theories. Although you are small for what you think and know, yet you are very important. You feel your value, the greatness of being human. This creates within you a sense of mystery and of the reality of the spiritual world.

Nature may captivate the senses with what it shows, but its great beauty lies in what it hides. So, the question is how much it leads to the *truth*, because the satisfaction of knowing the *nature* (φύσις) of beings can cover the thirst for the *truth* of the reasons (λόγοι) of beings. That is, the discovery of the beauty of the natural world can hide the truth of existence, which is turn, conceals a greater beauty.

In the book *«Truth and Beauty»* of the truly genius theoretical astrophysicist Subrahmanyan Chandrasekhar[1], one may discern an admirable beauty emerging from symmetries and asymmetries, regularities and irregularities, from magnitudes inconceivable to the human mind, from bright flashes that alternate with cold yet meaningful darkness, from secrets that only a human genius could approach, from unyielding constants, which, however, define the harmony of the world and the maintenance of its existence and functioning. In addition, through the beauty of the natural world, Chandrasekhar discerns the absolute *truth* to be identical with its reality. But

[1] Subrahmanyan Chandrasekhar: Truth and Beauty: Aesthetics and Motivations in Science, University of Chicago Press, 1987.

shouldn't *truth* explain physical reality and therefore transcend it? The scientific approach, no matter how advanced it is, probably has limits. *Truth* has no limits.

Without any doubt, our times are fascinating in many ways. Science has exponentially increased knowledge, and revolutionary technological applications have so permeated life that they end up creating a world as inconceivable as the Universe and a human being perhaps quite different from *homo sapiens*. Discoveries such as the double helix of DNA or its recombination, the mapping of the human genome and the possibility of its editing, the technologization of the biological microcosm and the mechanisms of reproduction, interventions at the nanoscale, technologies to enhance humans at the cognitive level, robotic rehabilitation, artificial intelligence, all of these inevitably lead us to oscillate between admiration for the human spirit and fear for the consequences of human technologies.

Recently, theoretical physicist David Gross (Physics Nobel 2004) commenting in an interesting interview at Scientific American on the consequences of nuclear technology, stated: "what now concerns me more is the survival of humanity than conducting the actual research"[2]. At the same wavelength, Geoffrey Hinton, the Nobel Prize-winning computer scientist widely known as the "Godfather of AI", warns that artificial intelligence poses profound short-term risks to society and a 10% to 20% chance of causing human extinction within the next three decades.

How much is scientific knowledge worth, if it is unable to ensure the survival of humanity? How can technological achievements help when they also pose a threat? How much wisdom does Artificial Intelligence have? How necessary is it? What is the value of knowledge, if it is not capable of explaining the reasons for the beauty of this world, the reasons of the beings (λόγους των όντων)?

The achievements of Astrophysics of the last century presented in this unique volume are breathtaking, but they do not cause fear. They create awe. The reason being that the Universe cannot be technologically processed. Human beings have great potential to study it, to discover it and to examine it, but no capabilities to change it or to fix it.

Moreover, humans do not only possess a body and a mind, intelligence, sensation and logic; they are not an advanced thinking biological machine. They have wider horizons, which science cannot approach, which technology cannot affect. The latter can only examine human phenotypical identity, which is wonderful. But it is not the whole of the truth. It is only a very small part of it. Humans are something much more, which is approached through another way of seeing and knowing[3].

This book surpasses all natural laws and transcends human logic: It makes the dark universe look bright, turns its coldness into warmth and fills its emptiness. Universe "radiates" wisdom, harmony and incredible beauty. It proves that the insignificant human being is much greater and continually expanding. It bestows upon him new dimensions. It honors him with value and grandeur.

---

[2] Interview of David Gross to Lee Billings: Nuclear war may keep humanity from finding a 'theory of everything,' top physicist says, Scientific American, 28.4.2026.

[3] Dionysius the Aeropagite, On Mystical Theology, ch. 2 To Timothy.

Prof. Kanaris Tsinganos explains the nature of beings in a wonderful way and inspires you to search for the reason of beings.

"*The heavens, indeed, declare the glory of God*"!

Spata, Greece

Rev. Dr. Nikolaos Chatzinikolaou
Metropolitan of Mesogaia
and Lavreotiki

Prologue

The writing of this volume started in 2019, when the International Astronomical Union (IAU) celebrated the 100th anniversary since its foundation (1919). To commemorate this year-long milestone, celebrations were held around the world to raise public awareness for the splendid cosmical discoveries of the previous century. These centennial celebrations for IAU100 stimulated worldwide interest in astronomy and space science, reaching out to the global astronomical community, national science organizations, societies, policy-makers, students, families, and the general public. The author of this book, in his capacity as the national outreach coordinator in his country, organized several activities in the framework of IAU100 and all these events and experience inspired the writing of the present volume.

But it may be appropriate and worth to begin this volume by briefly mentioning the discoveries by some pioneering classical astronomers, who left their indelible traces on the evolution of early astronomy, such as: *Aristarchos* of Samos (310–230 BC), the systematic proposer of the heliocentric system, the so-called Copernicus of antiquity, *Eratosthenes* of Cyrene (276–194 BC), who measured with admirable accuracy the circumference of the Earth and *Hipparchos* of Rhodes (190–120 BC), the prince of observational astronomers of antiquity, a classical astronomer and the founder of trigonometry. We shall also mention very briefly some other astronomers and cosmic thinkers.

Aristarchos (Fig. 1) first noted that during a lunar eclipse, as the Moon travels through the Earth's shadow, the Moon is smaller than the Earth's shadow; therefore, he argued, the Moon must have of smaller size than Earth. Furthermore, by observing the movement of the Moon through the Earth's shadow during such a lunar eclipse, he judged that the diameter of the Earth is about three times the diameter of the Moon. Today we know that the actual ratio of the diameters of Earth/Moon is ≈ 3.67.

Additionally, he also observed and argued that at the time of the first and last quarters of the Moon, the triangle Sun-Moon-Earth (SME in Fig. 2) – with vertices at the Sun (S), the Moon (M), and the Earth (E) – is orthogonal at the position of the Moon. Thence, he estimated that the angle of the two celestial bodies from the center of the Earth, SEM $\approx 87°$ (the true value is $89.853°$) and therefore the angle ESM $\approx 3°$. By using correctly trigonometry [since $1/\cos(87°) = \mathrm{ES/EM} \approx 19$], he

Fig. 1 *Aristarchos* of Samos (310–230 BC) [Credit: webpage of Dr Gerti Tashko]

Fig. 2 *Aristarchos'* trigonometry to measure the ratio of the solar to lunar distance from Earth

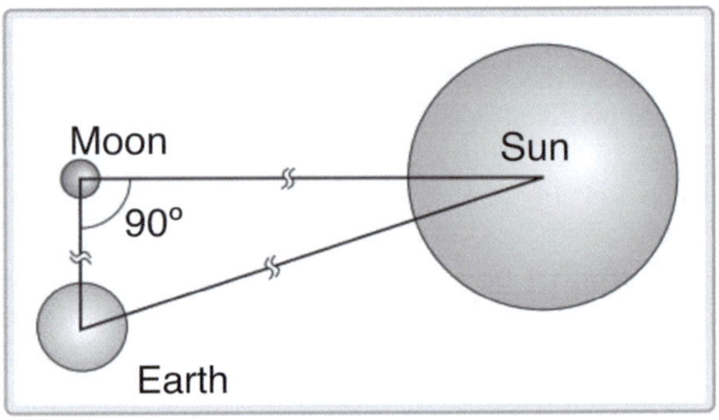

concluded that the Sun was 19 times farther from Earth than the Moon is (actually the Sun is about $1/\cos(89.853°) = ES/EM \approx 390$ times farther, with the difference due to the error in calculating the values of the observation angle SEM).

Aristarchos also noted that the Sun should have almost the same apparent angular diameter as the Moon, in order to have total solar eclipses. From this fact he concluded that the ratio of their diameters should be equal to the ratio of their distances from the Earth. So, he concluded that the Sun's diameter was about 19 times greater than the diameter of the Moon (i.e., the solar diameter was 6.33 times the diameter of the Earth), which is computationally reasonable. Therefore, he argued that *it is not possible for such a large Sun to revolve around a much smaller Earth*! This reasoning led him to the *heliocentric* Universe. *Aristarchos* was the first natural philosopher after the *Pythagoreans*, who proposed the *heliocentric model* of the Solar System, placing the Sun, not the Earth, at the center of the known Universe. But his ideas about Astronomy were not accepted during his time. About 1800 years later (1543),

Fig. 3 *Eratosthenes* (c. 276 BC–c. 194 BC), together with Aristarchos and Theocritos (left) and Ptolemy (right) in a painting on the façade of the University of Athens' central building, by Carl Rahl and Edward Lebietsk

Copernicus relying on the theories of *Aristarchos* and the *Pythagoreans* – as he *explicitly* wrote in the introduction of his work *De revolutionibus orbium coelestium* (1543) – reintroduced the *heliocentric* system as we know it today – although the original reference to Aristarchos was visibly deleted by someone in the original manuscript later on! In fact, Archimedes (287–c. 212 BC) in his mathematical treatise *Psammites (Sand Reckoner)* writes: And Aristarchos of Samos ... supposed that the stars and the Sun remain stationary, while the Earth revolves around the Sun in a circular circumference....

Eratosthenes (Fig. 3) was a mathematician, astronomer, and the father of geography. He was born in Cyrene, an ancient Greek colony and later a Roman city in northeastern Libya. *Eratosthenes* was a famous man of learning and this led the king of Egypt Ptolemy III Euergetes to place him as the director of the great Library of Alexandria (Egypt) in the year 245 BC. *Eratosthenes*, then 30 years old, accepted Ptolemy's invitation and moved to Alexandria, where he lived for the rest of his life, as the chief Librarian, expanding the library's holdings, tutoring also the children of Ptolemy, including Ptolemy IV Philopator, who became the fourth Ptolemaic pharaoh. *Eratosthenes* made several important contributions to mathematics and science, and around 255 BC, he invented the *armillary sphere*. He was a contemporary and friend of *Archimedes*. He is mostly known for the calculation of the circumference and radius of the Earth around 240 BC, with high accuracy, as illustrated in Fig. 4.

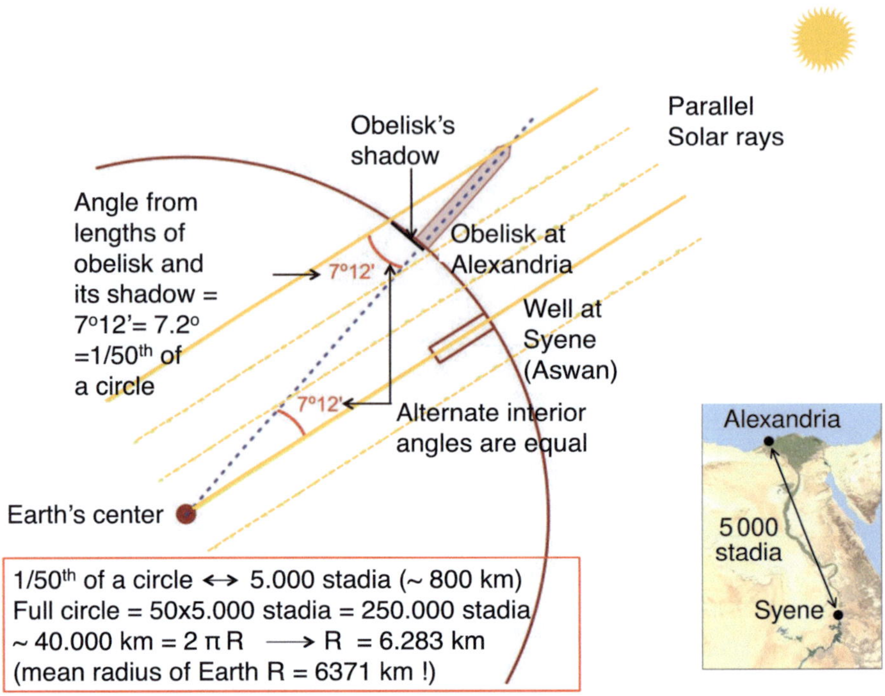

Fig. 4 Illustration of the calculation of the radius of the Earth by *Eratosthenes* (ca. 240 BC)

As director of the Library of Alexandria, *Eratosthenes* was aware of observations indicating that in Syene, in the tropic of Cancer, now the city of southern Egypt Aswan (23.5° N), at that line of latitude, at noon on the summer solstice (June 20 or 21), the Sun is directly overhead, when the Sun reaches its highest position in the sky, so that objects cast no shadow and the light entered the wells to their very depths. The tropic of Cancer is the northernmost latitude on the Earth's northern hemisphere where the sun can appear directly overhead. This meant that the sun's rays fell perpendicular to the earth in that area. In Alexandria, on that same day, the objects did cast shadows. That is, those same sunbeams did not fall perpendicularly, but at a certain angle (Fig. 4). *Eratosthenes* already held the view that the Earth was spherical, and this difference in the angle of incidence of the solar rays reinforced his belief. He also believed that the sun was far enough away to consider that its rays would reach the Earth practically, parallel to each other. The distance between Alexandria and Syene was about 5000 stadia (about 800 km), with the *stadion* a measure used at the time and estimated to be equivalent to about 158 meters. From trigonometry, *Eratosthenes* was able to measure the angle at which the sun's rays hit Alexandria to be approximately 7.2°, an angle which equals 1/50 the full circumference of a circle, 360° (Fig. 4): from that he inferred that the circumference C of the Earth should be about 50 times the distance between Alexandria and Syene, that is 50 x 5000 = 250,000 stadia, or, about 40,000 km. By that time Archimedes had already made an

Fig. 5 *Hipparchos* of Rhodes (190–120 BC). [Courtesy of Wikipedia Commons and a 19th century engraving based on an engraved amethyst from the Poniatowski gem collection.]

approximate calculation for the value of $\pi \cong 22/7$ and the relationship between the length of circumference C of a circle and its radius R was known to be $C = 2\pi R$. Thus, he calculated the radius of the Earth, with a great accuracy of the order of 1%.

Hipparchos (Fig. 5) calculated with admirable accuracy the length of the solar tropical year to be 365.25 days - 1/300 of a day = 365.24667 (i.e., 365 days, 5 h, 55 min, 12 s, where the true value is 365 days, 5 h, 48 min, 46 s, or 365.2422 days); also, the precession of the equinoxes (the gradual shift in the orientation of Earth's axis of rotation in a cycle of approximately 26,000 years, 46″ annually, very close to the value 50.26″ accepted today; *Hipparchos* estimated the Earth–Moon mean distance by using the visually identical sizes of the solar and lunar disks and observations of the Earth's shadow during lunar eclipses, he found a relationship between the lunar and solar distances that enabled him to calculate that the Moon's mean distance from Earth is approximately 63 times Earth's radius.). He introduced the stellar photometric magnitudes we use today and produced *the first catalog of 1022 stars* grouped by constellations, with their exact position (right ascension and declination), as well as the luminosity of each one – with the stars being divided into six classes, all the brightest ones being assigned to the 1st magnitude, and all those only just visible to the naked eye to the 6th – a catalog that was later included in Ptolemy's *Almagest. Hipparchos* perfected earlier astronomical instruments, such as the *Dioptra* – with which he estimated the apparent diameter of the Sun and the Moon, their distance and their actual size – the *Gnomon*, the *Heliotropion*, the *Heliorologion*, the *Cathetion*, the *Hourglass*, the *Rings* and the *Armillary sphere* (or a *spherical astrolabe*), an early astronomical device for representing the great circles of the heavens, including in the most elaborate instruments the horizon, meridian, equator, tropics, polar circles, and an ecliptic hoop. This sphere is a skeleton celestial globe, with circles divided into degrees for angular measurements. He directed the Library (Museum) of Alexandria for 20 years, where he studied all the information gathered in the Great Library of Alexandria by previous researchers. *Hipparchos* processed a large volume of observations and devised methods for solving problems of spherical astronomy, developing the idea of mathematical determination of stellar positions. He is regarded as the possible maker of the *Antikythera Mechanism*, this first astronomical calculator, a replica of which is exhibited at the Museum of the National Observatory of Athens (Fig. 6).

Other important classical astronomers include: *Thales of Militos* (ca. 626/623–ca. 548/545 BC), one of the *Seven Sages*, who according to Herodotos, had predicted a year in advance (!) the exact date of a solar eclipse that occurred in 585 BC and it coincided with a battle in Asia minor between the Lydians and the Medes. *Thales* calculated the timings of the solstices and equinoxes, and is the namesake of *Thales's geometry theorem* (if A, B, and C are distinct points on a circle where the line AC is a diameter, the angle ∠ ABC is a right angle). He is the first philosopher in the classical tradition, breaking from the prior use of *mythology* to explain the world and instead using *natural philosophy*, i.e., the first to have engaged in mathematics, science, and deductive reasoning. *Anaximander*, a student of Thales and a teacher of *Pythagoras*, a prominent member of the *Ionian school*, realized that the northern sky seems to turn around the North star, which led him to the concept of the *Celestial sphere*

Fig. 6 Replica in 3:1 scale of the *Antikythera* Mechanism (first - second century BC) placed by the author at the Museum of the National Observatory of Athens (NOA), during his capacity as the director of NOA (2011–2016)

around Earth. And, as the sky seems to vary with latitude, he also considered that Earth's surface may be curved as well. The most famous successors of the tradition initiated by *Thales* were *Platon* and *Aristotle*: while much thought continued to rely on intuition, the lasting legacy of their work was that offered non-supernatural explanations for the normal operations of the Universe. *Pythagoras* (ca. 570–ca. 495 BC) established the first Pythagorean community in the ancient Greek colony of Kroton, in modern Calabria (Italy) circa 530 BC and was already well known in ancient times for his *Pythagorean theorem* (in a right-angled triangle the square of the hypotenuse is equal to the sum of the squares of the other two sides). *Philolaos* (ca. 470–ca. 385 BC), one of the most prominent figures of *Pythagoreanism*, preceded *Aristarchos* and *Copernicus* in moving the Earth from the center of the Cosmos and making it a planet. According to Aristotle (4th century BC) on the Pythagorean astronomical system, the Earth is one of the stars, creating night and day by its circular motion about the center.

Meton was an astronomer who lived in Athens in the 5th century BC, a contemporary of *Pericles*, best known for calculations involving the eponymous 19-year *Metonic cycle*, which he introduced in 432 BC into the *lunisolar Attic calendar*. According to this *Metonic calendar*, 19 solar years ($19 \times 365.2422 = 6939.6018$ days, i.e., 6939 days, 14.5 h) and 235 lunar months ($235 \times 29.53 = 6939.6884$ days, i.e., 6939 days, 16.5 h) are very nearly of the same duration (the difference is only about 2 h), such that, a given day of a lunar month will often occur on the same day of the solar year as it did 19 years previously. *Callippos* a century after *Meton* expanded on the work of *Meton*, proposing what is now called the *Callippic cycle* which runs for 76 years, or four *Metonic cycles*. The world's oldest known astronomical calculator, the *Antikythera Mechanism* (Fig. 6), an ancient Greek hand-powered orrery, model of the solar system predicting solar and lunar eclipses and the location of the planets, regarded as the oldest known example of an analogue computer performs calculations based on both the *Metonic* and *Callipic calendar cycles*, with separate dials for each. *Archimedes* (ca. 287–ca. 212 BC), or *Hipparchos* are possibly the constructors of this *Mechanism*.

The foundations of *Meton's observatory*, the *Heliotropion*, are still visible today just behind the podium of the *Pnyx* in Athens, wherein was the ancient Athenian parliament. *Meton* found the dates of equinoxes and solstices by observing sunrise from this observatory: from that point of observation, during the summer solstice, sunrise was in line with the local hill of Mount Lycabetos, while six months later, during the winter solstice, sunrise occurs over the high brow of Mount Hymettos in the southeast. So, from *Meton's observatory* the Sun appears to move along a 60° arc between these two points on the horizon every six months, with the bisector of the observatory's solstitial arc in the same line with the view of *Acropolis*! These topological features are important because the summer solstice was the point in time from which the Athenians measured the start of their calendar years; *Heraclides Ponticos* (ca. 390 BC–ca. 310 BC) proposed that the apparent daily motion of the stars was created by the rotation of the Earth on its axis once a day; *Archimedes* (ca. 287–ca. 212 BC) who is also a possible constructor of the *Antikythera Mechanism*; *Apollonios of Perga* (ca. 262 BC–ca. 190 BC), defined the terms of the conic sections: ellipse, parabola, and hyperbola as we use them today and the hypothesis of eccentric orbits used to explain the apparent orbits of the planets; *Claudius Ptolemy* (ca. 100–ca. 170 AD), an Alexandrian mathematician and astronomer, who composed an astronomical treatise known as the *Almagest*, the only surviving comprehensive ancient treatise on astronomy, for over a thousand years the authoritative text on astronomy (Fig. 7).

After this quick reference to the early contributions of several classical thinkers, we move to the main body of this volume which describes, by chronological order and in each subfield of Astronomy (Sun, Planetary System, Star/Planet Formation, Exoplanets, Astrophysics, Cosmology), the most important discoveries which took place since the beginning of the 20th century and the establishment of the International Astronomical Union (1919). From the outset, it was encouraging to us the fact that such a volume has not been published so far in this subject in the international bibliography, i.e., a book covering *all* these subfields of Astronomy, from the Sun and the Heliosphere to Cosmology.

Fig. 7 The well known fresco by Raphael, named the *School of Athens*, painted between 1509 and 1511, now at the Stanze di Raffaello in the Apostolic Palace in the Vatican City. The elder Platon and the younger Aristotle are at the center. Pythagoras is represented by his theory of mathematical harmony on the left, Euclid by the perfection of Geometry on the right – the elaborate perspective design of the architectural setting seems to embody both those main schools. Pythagoras (writing in a book), Parmenides holding at the feet of Pythagoras a panel on which is inscribed Pythagoras' harmony of the spheres, the Alexandrian woman philosopher Hypatia is standing next to Parmenides, Anaximander leaning towards Pythagoras is at the lower left, Heraklitos (sitting alone) in front of Pythagoras, while the philosopher of Cordoba Averroës identified by his turban is behind Pythagoras; Euclid is at the lower right (bent over a panel), next to Ptolemy (holding a sphere), etc. [Courtesy of Wikipedia Commons]

In designing the organization of the material of this book, we confronted several questions. *First*, a difficult decision we faced to take was the selection of the *most important discoveries* in each year in this vast period of more than a century, in this rather wide field. In other words, the problem was how to choose those outstanding discoveries in the wide range of the general field of Astronomy, and discuss them under the constraint of a *reasonable length*, for a book that can be equally read by specialists working in all those diverse subfields of Astronomy, as well as by nonspecialists. This was not an easy decision, but a subjective and risky choice to make. At the end, we decided to select only those *fewer* discoveries which we thought as the most outstanding, neglecting inevitably some others, which according to other writers could be equally important. For discoveries at the beginning of the last century, this was an easier choice, since the elapsed time highlighted and refined the game changers. However, for the recent decades, the situation is different and this has been a matter of subjective and risky choice. For this we ask for the understanding of our readers, if some of the recent discoveries are not included in the following tables and throughout the book. Perhaps, in a more extended writing they should be presented.

A *second* important question was that an undertaking like the one of putting together the material of this book, presupposes that the writer has the needed expertise and feels comfortable in describing the latest evolutions in all those subfields of Astronomy. Fortunately, the present author started his career as a solar physicist – Ph.D. post doctoral at the University of Chicago supervised by Eugene Parker and at the same time enjoyed fruitful discussions on topics of plasma astrophysics with the Nobel prize winner S. Chandrasekhar – before continuing later on as a Research associate in astrophysics at the Harvard Smithsonian Center for Astrophysics, collaborating with Robert Rosner and the rich environment of the CfA group. In other words, we feel blessed that we lived for several years in the same corridor, wherein we had next to our office, on one side the office of Gene Parker (the unquestioned top heliospheric physicist) and on the other the office of Chandra (the unquestioned top astrophysicist). Subsequently, after those early inspiring and fertile years, our personal research has been oscillating back and forth between heliophysics and astrophysics, trying to transfer the experience gained from understanding our nearest star, the Sun, to the rest of astrophysics and vice versa. Hence, it was only after such research experience of half a century in the wider area of plasma astrophysics, that we have undertaken this difficult job of composing the present book. It is left to our readers to judge if it was worth of it.

A *third*, equally important choice we made in composing this volume, is to avoid as much as possible mathematical formulae, or any other technical material. Instead, the emphasis is on the new physics ideas and breakthroughs, as well as to key observations in the various branches of astronomy, which opened new ways of thinking and established new areas of study. Recalling the advice of Gene Parker, the essence of astrophysics is in physics and not in mathematics: *when your math becomes complicated stop and think again about your physics*, was his usual advice. This makes the material of the various astronomy discoveries presented in this volume, accessible even to those readers who are not familiar and versatile with mathematical techniques.

Then, talking about the prospective readers of this volume, the question arises who they might be. First, we believe that this volume will be of interest to practicing researchers in any particular subfield of astronomy, who are interested to follow what has happened over the last century on the research front of the other neighboring subfields; the same volume will be also of interest to undergraduate and graduate students, in the framework of their introductory or advanced courses in heliophysics, space physics, astrophysics and cosmology, in order to get a global view of the various fields of astronomy, especially before they choose the particular subfield in which the subject of their Ph.D. thesis will belong; additionally, this same volume will be also of interest to University teachers and researchers in the broad field of the physical sciences, as this writing provides a *holistic* physical picture of what has happened in heliophysics, space physics, astrophysics and cosmology in the last more than 100 years; finally, the present book will be useful to the educated general public who want to glean the important advances in science and technology during the last century, and till today.

Note that there are scattered in the text, many short narratives that describe in a smooth way some of the most impressive features of the discoveries, in a separate box and a light blue background. These are comprehensible descriptions that any reader may easily remember. For example, throughout the book there are answers to simple questions like:

- How does our Sun produces its radiant energy and what *nuclear reactions* burn for us in its core?
- At what rate the Sun produces the *elusive neutrinos* in its central regions and how many billions bombard us, passing through our head every second?
- How severe are *particle collisions* in solar/stellar cores?
- What kind of *music* is the Sun playing for us in its interior and why do we say that the Sun produces sound waves from millions of ringing *solar cosmic bells?*
- How devastating and violent solar flares can be and how unlucky are planets around wild *flare stars*?
- How expensive can be solar blackouts, why do we say that when the Sun *roars*, artificial satellites *tremble* and why *Space Weather* causes several troubles on Earth and hence it should be monitored?
- What is the effect of solar variability on *Earth's climate*?
- Could be the disappearance of the *Anasazi Indians* due to a severe drought during the warm period of the medieval *grand maximum* of solar activity?
- Is the *rise of magnetic flux tubes* from the solar interior similar to the rise of *air bubbles* in an aquarium, or a swimming pool?
- Which was the *fierce dispute* of the two giants of science A. Eddington and S. Chandrasekhar?
- *What is the mystery* of the superhot solar *corona and* which are its building blocks?
- Are tangential magnetic field discontinuities formed *spontaneously* in magnetized astrophysical plasmas?

- Why was delayed the publication of Parker's famous *solar wind* theory?
- What are *magnetic switchbacks* and *solar campfires*?
- Who were the *human computers* at the Jet Propulsion Laboratory in California, performing the complex mathematical calculations required to plan the early satellite orbits?
- Did we discover large amounts of water vapor in *Mercury*'s outer atmosphere, as well as *water ice* and carbonaceous organic compounds in permanently shaded craters at the poles of this near the Sun planet?
- How low is the atmospheric pressure in the planet Mars and why do we call *Mars*, apart from being a *desert* planet, a *silent* planet?
- Is *water*, detected today in the form of ice, not only at the poles, under the frozen dry ice of carbon dioxide, but also underground, a few centimeters below the Martian surface?
- Why Mars did loose its *magnetosphere* and the protection it gave to the planet from the solar wind?
- Why are sunsets *red* on our *blue* planet Earth, and *blue* in the *red* planet Mars?
- What is the *longest journey* of a human spacecraft in the Universe?
- Do all giant planets have *auroras*?
- How many *active volcanoes* are in our planetary system and which satellite has most of them?
- Is there *microbial life* in the oceans beneath the icy surfaces of the moons of Jupiter's Europa, Saturn's Enceladus, and Uranus' Miranda?
- How does *Titan's atmosphere* resemble Earth's primitive atmosphere and what kind of lakes and rivers are flowing on its surface?
- Where are Voyager 1 and Voyager 2 today, what are they telling us about the *shape of the heliospheric* bubble and when will they reach the nearest star?
- Which is the *silent killer* gas on Earth, that informs us on how life begins in space?
- Can we distinguish the two lights of a car on the Moon with the ESO/VLT in Chile?
- Which is the closest to us cosmic factory in our galaxy for the production of *newborns stars*, most of them surrounded by accretion disks?
- Do we observe water in *newborn planetary systems* today?
- Do we observe *weird exoplanets today* wherein rains down from the sky iron, red rubies, blue sapphires (corundum), molten glass, or, snows titanium oxide?
- Are observed exoplanets in *perpendicular orbits* around young stars?
- Are *biosignatures*, or extreme *supersonic winds* observed in the atmospheres of exoplanets?
- Did we discover *water vapor* in the habitable-zone of exoplanets?
- Do we observe huge quantities of *whiskey* and complex organic molecules in nebulae in the interstellar medium?
- Is there an energy integral and criterion distinguishing the formation of *uncollimated winds from collimated jets*?
- How the *ex machina Deus* intervenes in collapsing protostellar disks to give birth to stars, inside dark nebulae?

- Why it had been so difficult a task to discover small planets orbiting other stars till today?
- What is *Fermi's paradox*, and why after all, although we have discovered thousands of distant exoplanets in the last 31 years to date, we are surrounded by *cosmic silence* and astronomical *loneliness*?

or,

- When and how *white dwarfs* and *pulsars*, spinning hundreds of times per second were discovered, or, how was the existence of the exotic *black holes* theoretically proposed by a physicist who was serving on the front of the first world war?
- How were discovered the mysterious *quasars* at the distant edges of our observable Universe?
- What role do they play *gravitational lenses* in our efforts to understand the distribution of dark matter in the universe?
- How do we detect *supermassive black holes* at the center of our Galaxy and those in many other galaxies?
- What is the infinitesimally small amplitude of the *gravitational waves* we detect?
- How were the *precious metals* gold and platinum in our jewelry created when two neutron stars collided?
- How was photographed a *supermassive black hole* at the center of a distant galaxy?
- How did a *mule driver* made a significant contribution to one of Cosmology's greatest discoveries?
- How was created matter in the Universe *in just a few minutes* after the Big Bang?
- How do galaxies form a *spidery cosmic web*?
- What is cosmological *inflation*?
- What is the cause of the *accelerated expansion* of the Universe?
- How is *dark matter* detected?
- How is the existence of *dark energy* in the Universe deduced?
- Is the Universe spatially *finite* or *infinite*?
- Why does the solar thermonuclear reactor have a *carefully tuned efficiency*?
- Why there is a *fine-tuning* of the fundamental constants in our Universe?
- Is this just a coincidence, Divine Providence, or something else?

The purpose of this original volume is to present a *synopsis* of the splendid and exciting discoveries in heliophysics and space physics, astrophysics and cosmology during the period 1919–2026, that have answered these and many other similar cosmic questions, presenting thus a *holistic* picture of *our wonderful Universe*, via the *discoveries during the period 1919–2026*, something useful to students, teachers, researchers and the educated general public.

We start first with *our amazingly active Sun* (Chap. 1), presenting a century of great discoveries which led to the understanding of our life-giving star, the *Rosetta Stone* of Astrophysics. Then, the impressive in situ exploration of our home planetary system follows, from the dawn of the Space Age (1957), to the robotic exploration of Mars and the other planets and their satellites, to the exit of the Voyagers from the Heliosphere (Chap. 2). The theoretical understanding and key observations of the *spectacular star formation* ranks highly among the *marvelous* achievements of the 20th century astronomy and is next discussed (Chap. 3). The relatively more recent discoveries of Earth's *exotic* exoplanet relatives follow (Chap. 4), which echo and influence Earth's glaring loneliness in Space. Chapters 5 and 6 recount the golden age of exciting discoveries that literally turned astrophysics and cosmology upside down in the last century: unexpected discoveries of rapidly rotating pulsars, enigmatic black holes, red giants and white dwarfs, dark matter and dark energy and the observation of the spatial distribution of galaxies in a spidery cosmic web with concentrations in clusters, superclusters and filaments are just a few examples. They also describe the paradoxically *accelerating* expansion of the Universe, the diffuse cosmic microwave radiation, gravitational waves detected by merging neutron stars and black holes, the relativity of space-time, the possible existence of multi-Universes, and more.

All these wonderful discoveries are discussed in this *same volume*, such as experts in one subfield of astronomy, e.g., in astrophysics and cosmology can follow what has happened in another subfield of Universe studies, e.g., in heliophysics and vice versa.

Chronologically, we have surprising and exciting astronomical discoveries in the last century, such as:

- the discovery of extremely strong magnetic fields in the spots that appear periodically on the photosphere of the Sun (1909);
- the discovery of tiny stars with a radius similar to that of our Earth, but with a mass similar to the Sun, which means they have a mass density of several tons per cubic centimeter, and the artistic nickname white dwarfs (1910);
- the observation of extremely well focused jets of superheated plasma erupting at speeds approaching the speed of light from the cores of active galaxies and enigmatic Quasars (1918, 1970),
- the detection of radio-wave emission from the Milky Way, where electrons as they spiral around magnetic lines emit synchrotron radiation (1931);
- the discovery of the existence of a mysterious and unknown matter – the so-called *dark matter* – which dominates ordinary matter that exists around us (1933);
- the fierce dispute of two giants of science (A. Eddington and S. Chandrasekhar) on the existence of black holes (1933);
- the recognition that the remnants of massive stars left by supernova explosions are expected to contain rotating neutron stars (pulsars) associated with the acceleration of cosmic rays (1934);
- the revelation of the mystery of the superhot solar corona (1942);
- the astonishing prediction of the nucleosynthesis of the light chemical elements in the early Universe, where the chemical elements hydrogen and helium, which together constitute more than 99.9% of the (baryonic) mass of the Universe, were created in just the first few minutes of its life (1948);
- the start of the space race and the discovery of the van Allen radiation belts (1958);
- the theoretical prediction of the Solar Wind, together with the emergence of cosmical MHD by of E. N. Parker (1958);
- the discovery of the solar dynamo, the electric generator of the Sun (1961);

- the discovery of helioseismology, the *music* which deciphers the secrets of the solar and stellar interiors (1962);
- the observation of mysterious quasars at the edges of our known Universe which emit enormous amounts of energy from a relatively small volume and therefore have a star-like image in a telescope (1963);
- the observation of the oldest light in the Universe, the diffuse cosmic microwave radiation, which is also the most surprising landmark discovery in the history of Cosmology that definitively verified the Big Bang theory (1965);
- the first landing on the Moon (1969);
- the observation of highly magnetized and rapidly rotating neutron stars emitting beams of electromagnetic radiation from their magnetic poles, the famous beacons of space (pulsars) (1967) and recently their extremes, the magnetars (1992);
- the observation of extremely high-energy γ-ray bursts, which constitute the most powerful explosions in the Universe and are of isotropic extragalactic origin (1967);
- the first detection of cosmic X-ray sources emitted by million-degree superheated gases, by the first satellites launched into orbit outside the terrestrial atmosphere (1970);
- the discovery of coronal mass ejections, elevating from the sun the mass of a mountain range and creating the agitated space weather (1971);
- the discovery of the solar coronal loops, the ubiquitous magnetic arches of the solar atmosphere, the building blocks of the solar corona (1973);
- the start of the longest journey of a spacecraft in the Universe by the Voyagers 1 and 2, (1977);
- the discovery of the colorful and violent volcanos on Jupiter's satellite Io, with nearly 400 active volcanos (1979);
- the introduction of the epoch of cosmological inflation in the very early moments of the Universe to explain the isotropy of the diffuse cosmic microwave radiation and the flatness of the Universe (1980);
- the realization that the rise of the magnetic field from the solar interior is similar to the rise of air bubbles in an aquarium (1980);
- the detailed mapping of the cosmic microwave diffuse radiation of the Universe and its anisotropies with the COBE satellite (1989);
- the observation of the existence of supermassive black holes at the centers of many galaxies (1990) but also of a supermassive black hole at the center of the Galaxy (1995);
- the observation of the first exoplanet around a star similar to our Sun by the radial velocity method, a discovery that led to the birth of the field of exoplanet research (1995);
- the impressive discoveries by the solar observatories: SOHO (1996), STEREO (2006) and SDO (2010);
- the fundamental observation of the accelerating expansion of the Universe, a stunning observation that struck the scientific community as a "lightning in the air", because until the end of the last century, we thought that the expansion of the Universe should naturally be slowing down due to gravity (1998);

- the solution of the solar neutrino problem (1999);
- the detailed observation of the anisotropy of the diffuse cosmic microwave radiation by the WMAP and Planck space missions and from this fact the deduced Euclidean geometry of the Universe (2001, 2009);
- the discovery of hydrocarbon lakes and a murky, orange-tinged atmosphere with methane showers on Titan (2005);
- the discovery of jets of water from an underground ocean in Saturn's mysterious moon Enceladus (2005);
- the inventory of the total matter that the Universe contains, with the surprising result that ordinary matter is only about 5%, dark matter about 26% and the rest, about 69% consists of dark energy of unknown nature (2010);
- the discovery of molecular hydrogen and complex organic molecules in the ocean of Enceladus (2009);
- the epic exit of Voyager 1 and 2 from the Heliosphere (2012);
- the observation of kilonova explosions formed during the merger of two neutron stars and the subsequent cosmic alchemy and synthesis of the noble metals gold and platinum (2017);
- the detection of the first gravitational waves (2015) and the announcement of the first photograph of a black hole (2019);
- the remarkable observations by the Parker Solar Probe and the Solar Orbiter of the so-called campfires and switchbacks, which probably decipher the long standing problem of the coronal heating via nanoflares (1920);
- the launching of the *James Webb Space Telescope* which opened a new era in the observation of the Universe at the penetrating infrared wavelengths (2021);
- the discovery of CO, SO_2, CO_2 and Na besides water vapor, and also biosignatures (DMS) in the atmospheres of exoplanets, as well water in the early stages of a cool planet-forming protoplanetary disk (2024);
- the *launching of the ESA twin Proba-3 spacecrafts in December 2024*, beginning effectively the era of *formation flying* spacecrafts.

The following tables provide a brief chronology of a restricted selection of those astronomical discoveries, from the early 20th century to the present. And, the following chapters of the book provide a more detailed presentation of those splendid discoveries in solar and space physics, astrophysics, and cosmology during the more than last hundred years.

May 2026

Kanaris Tsinganos
University of Athens
Department of Physics, Section
of Astronomy and Astrophysics
Zografos, Greece

Acknowledgements My thanks go to the president of the Academy of Athens (2024) Dr. Stamatios (Tom) Krimigis, for his support in writing this volume. Academician S. Krimizis is a leading and internationally acclaimed pioneer of the exploration of Space, with many hundreds of research publications in the international literature, while some of the scientific instruments he has built have visited or crossed many bodies of our planetary system and still travel today in the interstellar space, like the two Voyager spacecrafts.

I thank the Rev. Dr. Nikolaos Chatzinikolaou, the Metropolitan of Mesogaia and Lavreotiki, for discussions and also for his writing the Foreword in the book, a piece of extraordinary written text that I read myself holding my breath. I had the good fortune to meet Nikolaos Chatzinikolaou, during his postgraduate studies in the Department of Astronomy of Harvard University, while I was doing my post-doctoral research at the Harvard-Smithsonian Center for Astrophysics and teaching at the Physics Department of the same University. Nikolaos Chatzinikolaou later completed his Ph.D. thesis on Biomedical Engineering in a joint program of Harvard – MIT, circa 1985. The general philosophical and metaphysical questions that emerge from the discoveries described in this book constitute the invisible link of mutual respect, which unites us undiminished, despite the passage of so many decades that have intervened since then.

I am also grateful to the distinguished researchers Thanasis Economou of the University of Chicago and Angelos Vourlidas of Johns Hopkins University, who provided me with material from the exploration of our planetary system and our Sun, respectively.

I should also mention with gratitude all my scientific collaborators, with which we have co-published hundreds of publications in authoritative scientific journals. First, the late professor of the University of Chicago, Eugene Parker, supervisor of my Ph.D. thesis. Then, Prof. Robert Rosner, my colleague at Harvard University, now Professor at the University of Chicago, former president of the American Physical Society (2023), and Boon Chye Low, Senior Scientist at the High Altitude Observatory, NCAR. I also acknowledge the research interaction with my research associates and Ph.D. students, some of whom are currently gifted academic teachers at various Universities: Christophe Sauty, Professor at the University of Montpellier and Observatoire de Paris, Drs Veronique Cayatte, Zakaria Meliani and Loic Chantry (LUTH, Observatoire de Paris), Joao Lima, Professor at the University of Porto, Vassilis Archontis and Spiros Patsourakos, Professors at the University of Ioannina, Ioannis Contopoulos and Costis Gontikakis Heads of Research at the Academy of Athens, Odysseas Dionatos, Researcher at the Institute of Astrophysics of the University of Vienna, Titos Matsakos and Petros Syntelis, Researchers at UK companies, Drs. Matthias Stute and Jose Garcia, Researchers in German Universities, and Gordon Petrie, Researcher at the National Solar Observatory. Also, my collaborators at the University of Torino, Profs. Edo Trussoni, Silvano Massaglia, Gianluigi Bodo, Paola Rossi and Andrea Mignone, Prof. Sergey Bogovalov of the National Research Nuclear University (MEPHI), Russia and Prof. Tom Ray of the Dublin Institute for Advanced Studies. Finally, at the National and Kapodistrian University of Athens, professors Antonis Paschalis, Nektarios Vlahakis, Despina Hatzidimitriou, while at the National Observatory of Athens the Director of the Geodynamics Institute Vasilis Karastathis and the Head Directors George Balasis, Alceste Bonanos and Manolis Xilouris. Finally, Gordon Petrie, Researcher at the National Solar Observatory, and Iakovos Ouranos, physicist/artist at Crete University Press.

I am deeply indebted to several international organizations and many individual researchers who are individually acknowledged in the book and they allowed me to include in this volume several images: NASA, ESA, ESO, EHT, MPIfR, DOT, A&A, ApJ/AAS/IOP, MNRAS, Wikipedia and Wikimedia Commons. To all of them I have given the due credit in the caption accompanying the corresponding image/figure. Indeed, without their courtesy and kind permission to include all those figures/images, this volume would be definitely less appealing and less complete.

Regarding the citations to important papers, my apologies are due to colleagues that I did not mention their own contributions: referring to all important papers during the more that a century period covered by this volume, it would require a very lengthy Bibliography!

The recent years were for me an intense period, reviewing the exciting discoveries of heliophysics, astrophysics and cosmology since the establishment of the International Astronomical Union (IAU) and till today, additionally to my own research. The understanding shown by my family, Dr Georgia Vlastou[4], Dr Christos Tsigkanos, Dr Anthony Tsigkanos (both shown in Fig. 1.19 at an early age(!), and Eugene Tsigkanos (researcher on AI, born in 1995, during the Fodele mtg honoring Eugene Parker, Fig. 1.36), was important during this long-time engagement with this volume, for which I am grateful to them.

In closing, I also feel grateful to all studious listeners of my lectures, who constantly urged me to write this volume, especially after the year 2019, in which the IAU appointed me a national representative for the centennial since its establishment.

[4] Tsinganos K, Vlastou G (1988) Helicoidal Astrophysical Outflows. In Mass Outflows from Stars and Galactic Nuclei. In: Proceedings of the 2nd Torino Workshop, held in Torino, Italy, May 4–8, 1987, Dordrecht: Kluwer, 1988, Luciana Bianchi, and Roberto Gilmozzi (eds) Astrophysics and Space Science Library, vol 142, p 239

Urania, the muse of Astronomy, Engraving by J. Zucchi after Angelica Kauffman (1781), published by J. Boydell of London. [Copyright: The Trustees of the British Museum, reprinted with permission from the British Museum]

Competing Interests The author has no competing interests to declare that are relevant to the content of this manuscript.

Important Discoveries/Innovations

Discovery	Year	Researchers	Image
Discovery of *strong magnetic fields in Sunspots*	1908	George Hale, Mount Wilson Solar Observatory, USA	
Discovery of *White Dwarfs*	1910	Williamina Fleming, Edward Pickering, Henry Norris Russell, Cambridge, USA	

(continued)

(continued)

Discovery	Year	Researchers	Image
Measurement of stellar *masses* and *diameters*	1912	Henry Norris Russell, Harlow Shapley, Princeton, USA	
Measurement of *redshifts* in the spectroscopic lines of «*nebulae*»	1912	Vesto Slipher, Lowell Observatory, AZ, USA	
Period/luminosity relation for Cepheid variable stars	1908–1912	Henrietta Leavitt, Harvard College Observatory, Cambridge, MA, USA	

(continued)

(continued)

Discovery	Year	Researchers	Image
Hertzsprung-Russel diagramm	1914	Ejnar Hertzsprung, Denmark, Henry Norris Russel, USA	
Mass-luminosity relation for stars	1915	Jacob Halm, Cape of Good Hope Observatory, Ejnar Hertzsprung, Potsdam	
General Theory of Relativity (*Special Theory of Relativity*)	1915 (1905)	Albert Einstein, Germany (**Nobel prize in physics 1921**)	

(continued)

(continued)

Discovery	Year	Researchers	Image
First *static model of the Universe*	1917	Albert Einstein, Germany	
Shapley Model of the Galaxy by using spherical clusters	1918	Harlow Shapley, Director, Harvard College Observatory 1921–1952 USA	
Observation of first jet from giant elliptical galaxy M87	1918	Heber Curtis, Lick Observatory, California, USA	

(continued)

(continued)

Discovery	Year	Researchers	Image
Observation of gravitational bending of light rays by the Sun	1919	Arthur Eddington, Cambridge, UK	
The Shapley–Curtis great Debate on the true distance of «nebulae»	1920	Harlow Shapley, Harvard Observatory, USA, Heber Curtis, Lick Observatory, USA	
Thermonuclear fusion as the energy source of the Sun	1920	Arthur Eddington, Cambridge, UK	

(continued)

(continued)

Discovery	Year	Researchers	Image
Saha ionization equation of a gas	1921	Megh Nad Saha, India	
Diameters of red giants by using interferometry, experimental measurement of speed of light (1887)	1921	Albert Michelson, University of Chicago, USA (1st American **Nobel prize in physics 1907**)	
Photoexcitation – photoionization of interstellar medium	1921	Henry Russel, USA, 1921, D. Menzel, USA, 1924, B. Stromgren, Denmark (1939)	

(continued)

(continued)

Discovery	Year	Researchers	Image
Friedmann's model of the Universe	1922	Alexander Friedman (1922, 1924), USSR	
Distribution of stars in the Galaxy and galactic disk density	1922	Jacob Kapteyn, The Netherlands	
Ionization states of ions in stellar atmospheres	1923	Ralph Fowler, UK [introduced P. Dirac, J. Lennard-Jones, G. Birkhoff in quantum theory], Edward Milne, UK (1923, 1924)	

(continued)

(continued)

Discovery	Year	Researchers	Image
Theory of stellar structure and evolution (*internal constitution of stars*)	1924	Arthur Eddington, Cambridge, UK (1916–1924)	
Spiral «nebulae» *are extragalactic systems*	1925	Knut Lundmark, Sweden, Edwin Hubble, USA (1920, 1925)	
Rotation of the Galaxy	1925	Bertil Lindbland, Sweden	

(continued)

(continued)

Discovery	Year	Researchers	Image
Chemical composition of the stars	1925	Cecilia Payne-Gaposhkin, Harvard, USA	
Expansion of the Universe – «The Primeval Atom»	1927	Georges Lemaitre, Belgium	
Differential rotation of Milky Way	1927	Jan Oort, Leiden, The Netherlands	

(continued)

(continued)

Discovery	Year	Researchers	Image
Identification of lines of *nonexistent Nebulium* as forbidden lines of O^{2+}	1927	Ira Sprague Bowen, USA – first Director of the Palomar Observatory	
Theory of *White Dwarfs* and their maximum mass	1929	S. Chandrasekhar, University of Chicago, USA (**Nobel prize in physics 1985**), Wilhelm Anderson, Estonia	
Tunnel effect in solar thermonuclear reactions	1929	Fritz Houtermans, Germany, Ph.D. with James Frank, Robert Atkinson, USA	

(continued)

(continued)

Discovery	Year	Researchers	Image
Hubble, the recession and speed-distance relationship of «nebulae»	1929	Edwin Hubble, Carnegie Institutions, Mount Wilson Observatory, California, USA	
Discovery of extragalactic BL *Lac type objects*	1929	Cuno Hoffmeister Sonnenbeg Observatory Germany (1929), J. M. MacLeod, J. M. Andrew (1968), National Research Council, Ottawa, Canada	
Discovery of *interstellar absorption (extinction)* by dust particles	1930	Robert Trumpler (1930), USA, John Plaskett, John Pearce (1933), Canada, Alfred Roy (1939), USA	

(continued)

(continued)

Discovery	Year	Researchers	Image
Theoretical prediction of existence of *dark matter* in clusters of galaxies	1933	Fritz Zwicky, Switzerland/USA	
Discovery of *radio emission* from our Milky Way galaxy	1933	Karl Jansky, Bell Labs, NJ, USA	
Development of *large-scale structures in the Universe*. The problem of primordial fluctuations	1933	Georges Lemaitre (1933) Belgium, Richard Tolman (1934) USA, Evgeny Lifschitz (1946) USSR	

(continued)

(continued)

Discovery	Year	Researchers	Image
Supernovae – Neutron Stars – Cosmic Rays	1934	Walter Baade Mount Wilson Observatory (USA), Fritz Zwicky, ETH Switzerland	
Discovery *of the Robertson-Walker metric*	1935	Howard Robertson (Pasadena, USA), Arthur Walker (Liverpool, UK)	
Discovery of *the p-p cycle* in solar thermonuclear reactions	1936	Hans Bethe, Univerrsity of Chicago, (**Nobel prize in physics 1967**) Robert Atkinson, USA	

(continued)

(continued)

Discovery	Year	Researchers	Image
Observation of *the first Radio Source* Cygnus A	1939	Grote Reber, Illinois, USA, Stanley Hey, UK (1946)	
Discovery of *upper bound in the mass of neutron stars* (Oppenheimer–Tolman–Volkoff limit)	1939	Robert Oppenheimer, George Volkoff, Richard Tolman, Los Alamos and University of California, Berkeley, USA	
Discovery of *magnetohydrodynamic (Alfvén) waves*	1942	Hannes Alfven, Royal Institute of Technology, Stockholm, Sweden (**Nobel prize in physics 1970**)	

(continued)

(continued)

Discovery	Year	Researchers	Image
Active Galactic Nuclei and Seyfert Galaxies	1943	Carl Seyfert, Mount Wilson Observatory, USA	
Nucleosynthesis of light elements and diffuse cosmic microwave radiation	1948	Ralph Alpher, George Gamow, George Washington University, USA	
Detection of *gravitational redshift* of light from white dwarfs	1954	Daniel Popper, University of California, Los Angeles, USA	

(continued)

(continued)

Discovery	Year	Researchers	Image
Launch of *Sputnik*, the first artificial satellite of Earth	1957	USSR Academy of Sciences, Soviet Union	
First magnetohydrodynamic model of *solar prominences*	1957	Rudolf Kippenhahn, A. Schluter, Max Planck Institute fur Physik, Gottingen, Germany	
Thermonuclear fusion reactions in stellar interiors	1957	William Alfred Fowler, Margaret Burbidge, Goefrey Burbidge, Fred Hoyle, Caltech, USA (**Nobel prize in physics 1983**)	

(continued)

(continued)

Discovery	Year	Researchers	Image
Discovery of the *Solar Wind*	1958	Eugene Parker, University of Chicago, USA	
Discovery of the *Van Allen radiation belts*	1958	James Van Allen, University of Iowa, USA	
Construction of the first accurate models of *the stellar internal structure*	1960	Icko Iben, University of IL, USA, Louis Henyey, Berkeley, USA, Rudolf Kippenhahn, MPIA, Germany	

(continued)

(continued)

Discovery	Year	Researchers	Image
Determination of Hubble's constant and the *age of the Universe*	1960	Allan Sandage, assistant to Hubble's work after 1954, USA	
A *3rd integral for stellar dynamics* in galaxies	1960	George Contopoulos, Aristotle University of Thessaloniki/University of Athens, Greece	
Evolutionary *Hayashi track* in the Hertzsprung–Russel diagram	1961	Chushiro Hayashi, Kyoto University, Japan	

(continued)

(continued)

Discovery	Year	Researchers	Image
Babcock-Leighton *solar/ stellar dynamo model*	1961	Horace W. Babcock and Robert Leighton, Caltech, USA	
Development of radio interferometry and *discovery of pulsars*	1962	Martin Ryle, Antony Hewish Cambridge, UK (**Nobel prize in physics 1974**)	
Discovery of first *X-ray sources* and *X-ray background*	1962	Riccardo Giacconi et al., Cambridge, USA (**Nobel Prize in physics 2002**)	

(continued)

(continued)

Discovery	Year	Researchers	Image
Discovery of *solar 5-minute oscillations*	1962	Robert Leighton et al., Caltech, USA	
Discovery of *Quasars*	1963	Maarten Schmidt, California Institute of Technology, USA	
Discovery *of molecular lines in the interstellar medium* via radio waves	1963	S. Weinreb, A. H. Barrett, M. L. Meeks, J. C. Henry, Lincoln Lab. M.I.T., USA	

(continued)

(continued)

Discovery	Year	Researchers	Image
Adsorption model as a source energy of Quasars and AGN	1964	Edwin Salpeter, Cornell University and Goddard Institute, New York, USA	
Discovery of *the diffuse cosmic microwave radiation*	1965	Arno Penzias and Robert Wilson, Bell Labs, New York, USA (**Nobel prize in physics 1978**)	
Discovery of *Lyman clouds* as cosmological distance indicators	1965	James Gunn, Bruce Peterson, Princeton University Observatory, USA	

(continued)

(continued)

Discovery	Year	Researchers	Image
The problem of the *baryonic-antibaryonic matter asymmetry*	1965	Yakov Zel'dovich, Moscow State University and Sternberg Astronomical Institute, USSR, Wagoner Stanford University, USA	
Detection in *the infrared of sources in the Orion Nebula*	1966	Gerry Neugebauer, Eric Becklin, Caltech, California, USA	
Superluminal motion in radiogalaxies and quasars	1966	Martin Rees, University of Cambridge, UK	

(continued)

(continued)

Discovery	Year	Researchers	Image
Development of *interferometry* via an array of radio telescopes – the Very Long Baseline Interferometry	1967	Kenneth Kellermann and Marshall Cohen (NRAO in the USA) and Norman Broten, Thomas Legg, and John Galt (Algonquin Radio Observatory and Dominion Radio Astrophysical Observatory in Canada)	
Sakharov's Baryogenesis rules	1967	Andrei Sakharov, Moscow State, USSR Academy of Sciences, USSR (**Nobel Peace prize 1975**)	
Discovery *of γ-ray flares* from the Vela satellites	1967	Ray Klebesadel et al., Los Alamos National Laboratory, USA	

(continued)

(continued)

Discovery	Year	Researchers	Image
Detection of galactic emission of γ-rays *with OSO-III and SAS-2*	1968	George Clark, M.I.T., G. P. Garmire, Caltech, W. L. Kraushaar, University of Wisconsin, USA	
Thermal history of young Universe, era of *recombination and reionization*	1968	Jim Peebles, Princeton, USA, Yakov Zel'dovich, Sternberg Institute, USSR (**Nobel Prize in physics 2019**)	
Discovery of *Pulsars*, as rotating neutron stars	1968	Anthony Hewish, Jocelyn Bell, Cavendish Laboratory, University of Cambridge, UK (**Nobel Prize in physics 1974**)	

(continued)

(continued)

Discovery	Year	Researchers	Image
Start of the experiment for the detection of *solar neutrinos*	1968	Raymond Davis, Homestake AZ, John Bahcall, Princeton, USA (**Nobel prize in physics 2002**)	
Cosmological problems of the *cosmic horizon and flatness*	1969	Robert Dicke (1961), Peebles and Dicke (1969), Princeton University, USA	
Exponential damping of the curve of *supernova light* due to radioactive decays	1969	Stirling Colgate, Chester McKee New Mexico Institute of Mining and Technology, Socorro, NM, USA	

(continued)

(continued)

Discovery	Year	Researchers	Image
Penrose and Hawking theorems for gravitational singularities	1969	Roger Penrose, University of Oxford, St. Hawking, University of Cambridge, UK	
Discovery of *the double pulsar* and indirectly of gravitational wave radiation	1974	Joseph Taylor, Alan Hulse, University of Massachusetts Amherst, MA, USA (**Nobel prize in physics 1993**)	
Discovery of *Sgr A*, a very compact radio source at the center of our Milky Way galaxy*	1974	Bruce Balick, Robert Brown, National Radio Astronomy Observatory (NRAO), University of Washington, USA	

(continued)

(continued)

Discovery	Year	Researchers	Image
Viking 1 and Viking 2 spacecrafts on Mars	1975	NASA, Viking Program to Mars, Orbiter and Lander USA	
Multiple discoveries of *Voyager 1 and 2, launched in 1977*	1977	Stamatis Krimigis, P. I. Low Energy Charged Particles, JHU/APL, USA	
Discovery of First *Gravitational Lens* (Twin Quasar QSO 0957+561)	1979	Dennis Walsh, Bob Carswell and Ray Weymann with Kitt Peak National Observatory 2.1 meter telescope, USA	

(continued)

(continued)

Discovery	Year	Researchers	Image
Concept of *cosmological inflation*	1979	Alan Guth, Cornell University, Demosthenes Kazanas, GSFC, USA	
Hydromagnetic outflows from disks and the *production of radio-jets*	1982	Roger Blandford and David Payne, Caltech, USA	
First discoveries in *Astroseismology*	1983	Jørgen Christensen-Dalsgaard, Aarhus Universiteit, Aarhus, Denmark	

(continued)

(continued)

Discovery	Year	Researchers	Image
Discovery *of diffuse cosmic background radiation with COBE*	1989	George Mather (GSFC), George Smoot (MIT), (**Nobel prize in physics 2006**)	
Launching and first discoveries with the *Hubble Space Telescope*	1990	NASA	
Launching of the *ULYSSES* spacecraft – first observation of the *Polar Solar Wind*	1992	ESA and NASA	
Use of self-similarity to model magnetohydrodynamic astrophysical flows. Discovery of a *criterion for the transition of winds to jets*	1994	Kanaris Tsinganos and Nektarios Vlahakis, University of Athens, Greece, Christophe Sauty, Observatoire de Paris, France	

(continued)

(continued)

Discovery	Year	Researchers	Image
Launching of the SoHO spacecraft	1995	ESA and NASA	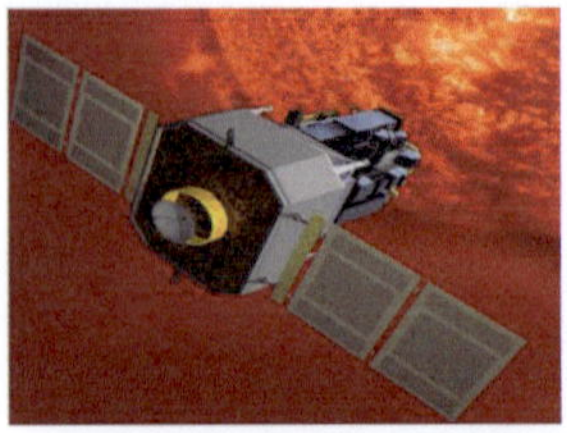
Observation of *first exoplanet* around a solar-type star	1995	Michel Mayor, Didier Queloz, University of Geneva, Switzerland, (**Nobel prize in physics 2019**)	
Discovery of a black hole at the center of our Milky Way galaxy	1995	Andrea Ghez. Caltech, USA, Reihard Genzel, MPIfEP, Germany (**Nobel prize in physics 2020**)	
Solar neutrino oscillations and *final resolution of the solar neutrino problem*	1996 1999	T. Kajita,Super-kamiokande, Japan, A. McDonald, Sudbury Neutrino Observatory, Ontario, Canada (**Nobel prize in physics 2015**)	
Launching of the Cassini–Huygens mission: Exploring Saturn and Titan	1997	NASA/ESA/ASI (2004–2017)	

(continued)

(continued)

Discovery	Year	Researchers	Image
First rover landing on Mars: *NASA Mars Pathfinder mission*	1997	Thanasis Economou, P. I. of Alpha Proton X-Ray Spetroscopy, University of Chicago, USA	
Discovery of *the accelerated expansion of the Universe*	1998	Adam Rees (J. Hopkins University), Saul Permutter (Berkeley), USA Brian Schmidt (Australian National University). (**Nobel prize in physics 2011**)	 From left: Adam Riess, Saul Perlmutter and Brian Schmidt shared the Nobel Prize in physics awarded Tuesday.
Launching of the 4 ESA/ NASA Cluster mission satellites to observe the terrestrial magnetosphere	2000–2025	@	
Chandra spacecraft *WMAP* spacecraft *Integral* spacecraft	1999 2001 2002	NASA X-ray Mission, NASA Microwave Probe, ESA γ-ray mission	
Discoveries of the Martian landscape with the Rovers *Spirit and Opportunity*	2003	NASA Mars Exploration Program	
Discoveries of *Mercury topography* with the *MESSENGER* spaceship	2004	NASA (2004–2015)	

(continued)

(continued)

Discovery	Year	Researchers	Image
NASA's New Frontiers program: *New Horizons, Juno, OSIRIS-Rex*	2006 2011 2016	NASA	
Stereoscopic observations of the Sun with the twin *STEREO* spaceships	2006	Angelos Vourlidas, SECCHI Project Scientist, GSFC, USA	
Discoveries with the Herschel (infrared) and Planck (microwaves) ESA space telescope	2009	ESA	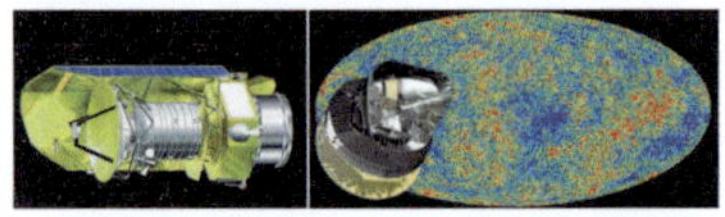
Observation of the Sun in many wavelengths with the *Solar Dynamics Observatory* (*SDO*)	2010	NASA's Living with a Star (LWS) Program, USA	
Mars Rover *Curiosity*	2011	NASA, Mars Science Laboratory	
First direct detection of *gravitational waves*	2015	Barry Barish, Rainer Weiss, Kip Thorn, LIGO Collaboration, USA, (**Nobel prize in physics 2017**)	
Mars Interior Exploration using Seismic Investigations, Geodesy and Heat Transport (*InSight*)	2018	NASA	

(continued)

(continued)

Discovery	Year	Researchers	Image
Launching and Observations of the Sun by the NASA/Parker Solar Probe, discovery of "magnetic switchbacks"	2018	NASA	 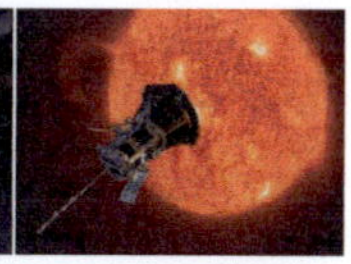
First ever photo taken of silhouette of *black hole* in the galaxy M87	2019	Event Horizon Telescope,international collaboration	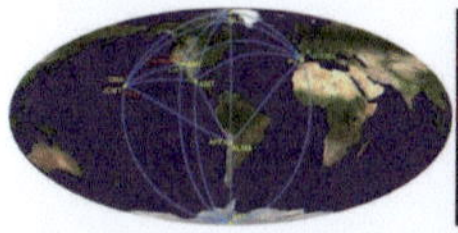 
First ever discovery of water vapor beyond our solar system, in the atmosphere of exoplanet K2-18b in the habitable-zone by the Hubble ST	2019	NASA	 
Launching and Observations of the Sun by the ESA/Solar Orbiter, discovery of "campfires"	2020	ESA	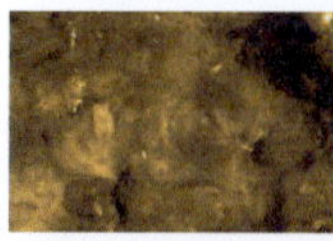 
An exoplanet wherein it rains down droplets of iron, discovered by ESO's VLT	2020	ESO	 
Launching and beginning of the era of the *James Webb Space Telescope*	2021	NASA/ESA/CSA	 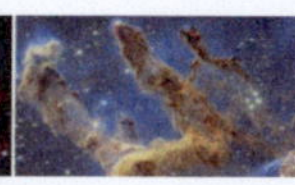 
Discovery of H_2O, CO, SO_2, CO_2, Na, K and photochemistry in the atmosphere of exoplanet WASP 39b by the *James Webb Space Telescope*	2022	NASA/ESA/CSA	 
Detection by the James Webb ST of *biosignature* of a molecule produced by life [dimethyl sulfide DMS] in K2-18b	2023	NASA/ESA/CSA	 
Discovery of water in the cool planet-forming disk of HL Tauri by ALMA	2024	ESO/ALMA	 

(continued)

(continued)

Discovery	Year	Researchers	Image
Launching and first observations of the twin Proba 3 spacecrafts, beginning of the era of *Formation Flying* spacecrafts	2024 2025	ESA	

Atlas, holding the Earth, a 2nd century Roman copy of a Hellenistic work. [Image credit: su concessione del Minister Culture - Museo Archeologico Nazional di Napoli, Italia - Courtesy of the Minister of Culture - National Archeological Museum of Napoli, Italy]

Contents

About the Author

Kanaris Tsinganos is a Professor of theoretical astrophysics at the University of Athens. With a PhD from the University of Chicago under Eugene Parker, the author has lectured at the Universities of Chicago and Harvard, Paris, Louvain, St.Andrews, Turin, Porto, Crete and Athens. He has held several distinguished positions including a directorship of the historic National Observatory of Athens, a presidentship of the Hellenic Astronomical Society, a national representation to ESA/SPC, etc. He has published over 300 research papers and books, including University textbooks on theoretical mechanics and plasma astrophysics, while he has also organized many international conferences and advanced study institutes.

His ongoing research topics are in the wide fields of theoretical astrophysics and magnetohydrodynamics with applications to our Sun, star formation and the study of plasma jets from newborn stars, AGN, quasars and black holes and participates as a PI in space missions, such as Solar Orbiter and Proba 3.

Chapter 1
The Amazingly Variable and Active Sun

But if you long to sing, my heart,
do not wish to see, apart from the Sun,
another star shining during the day
with more life-giving warmth in the desert ether...

Pindar, 522–443 BC,
ancient Greek lyric poet from the city of Thebes,
[1st of Olympian Odes, 1.3–1.6]

The appearance of sunspots is an instructive instance of the regular irregularity
and irregular regularity, characterizing solar phenomena...
R. Carrington
(1826–1875)

If it were not for its variable magnetic field,
the Sun would have been a rather uninteresting star...

E.N. Parker
(1928–2022)

Out of the trillions of stars in our own Milky Way galaxy and those in the other trillions of galaxies – in total about 10^{24} stars - it is only our Sun that is resolved in great detail and also affects our technology in space via Space Weather and our lives here on Earth…

K. Tsinganos, *A Century of Cosmic Discoveries*, Springer Praxis Books,
https://doi.org/10.1007/978-3-032-06282-6_1

Our Unique Sun, the Life Supporting and Life-Threatening Star

Our most vivid encounter with Astronomy occurs in a moonless summer night, when we gaze the sky decorated with thousands of stars, or – if we are lucky – when we see through a telescope. Almost all of these points of light belong to the Milky Way galaxy. The majority of these stars, (90–92)%, are main sequence stars. Our galaxy is estimated to contain about 24 billion yellow G-type stars, representing only about 6% of the galaxy's estimated 400 billion stars. These stars, which include our Sun, are also referred to as *sun-like* stars. Next in number come the orange K-type stars, about 13%. The remaining large majority of 73% of stars are red dwarfs. Finally, the very small minority of a few percent are giant and supergiant stars. Hence, with the number of galaxies estimated to be of the order of trillions, in the Universe we may have 10^{22}–10^{23} *solar-like* stars (Fig. 1.1).

And yet, out of these huge number of *sun-like* stars in the Universe, the Sun is the only G-type star that we resolve, model, and study in great detail, and at the

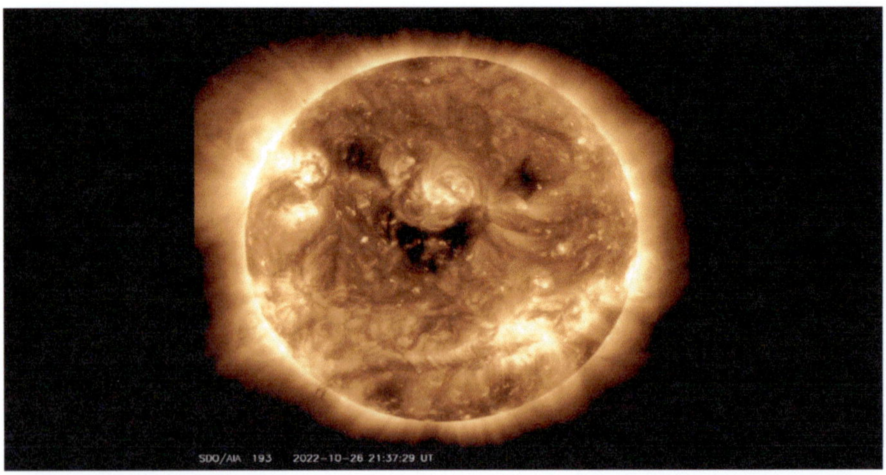

Fig. 1.1 NASA's *Solar Dynamics Observatory* caught the Sun *smiling*, on October 26, 2022 but warned us about some rough *Space Weather* ahead, as the Sun slowly approached its 25th maximum, in July 2025, with a peak of 115 *sunspots*. Note that **25** maxima have occurred in the **265** years that have elapsed since solar cycle **1**, in 1755–1766. This 25th cycle is expected to continue until about 2030. *Flares* and ejections during a rough *Space Weather* impact radio communications, electric power grids, navigation signals, and pose risks to spacecrafts and astronauts. In the image, the Sun is seen in the extreme ultraviolet light of 193 Angstroms, emitted from a 1.25 million degree coronal plasma of 11 times ionized iron (Fe XII). The dark patches on the Sun are *coronal holes*, regions where fast *solar wind* gushes out into space (*Image credit* NASA/SDO/AIA)

same time, the only star that affects positively – and sometimes negatively – life on Earth!
Hence, due to its proximity, our Sun provides a unique opportunity to perform high-resolution observations of its atmosphere throughout the whole electromagnetic spectrum. In addition, we may theoretically study most of the fascinating physical phenomena taking place on the solar corona and their impact on the solar heliosphere. Many of these phenomena can be appropriately well studied within the framework of magnetohydrodynamics, the standard theory of plasma astrophysics.

Since its creation in the depths of time, Earth has tirelessly completed billions of complete revolutions around its parent star, the Sun, which is after all the only essential source of energy for our planet. The Sun is ultimately a huge sphere of hydrogen gas, where the weight of the overlying layers is supported by the heat pressure produced inside it by thermonuclear reactions. Thanks to this thermonuclear energy, we and every other form of life on Earth, owe our existence. The solar energy reaches the Earth in the form of electromagnetic radiation. Solar radiation is the most important factor shaping the Earth's climate, due to its uneven distribution on the terrestrial surface. It participates to all physical and chemical processes, and phenomena that occur on our planet. The uneven heating of the Earth contributes to the creation of winds, the circulation of the atmosphere, the formation of clouds, rain, rivers, and the photosynthesis of plants. In the solar radiation is due the blue color of the sky, the rainbow, the dawn, and twilight, etc.

At the same time, however, an important byproduct of the solar energy – perhaps a side effect and collateral negative consequence we could to say – is also the fact that the production of this solar energy continuously supplies the dangerous solar activity, where colossal explosions, hurling into space a mass as great as a great mountain of ionized superhot gas (plasma) of the solar atmosphere and transforming interplanetary space into an environment inhospitable to life. Murderous flares when they reach the Earth, "whip" its magnetic field, creating magnetic storms, destroying pipelines, affecting communications and melting transformers, while if there are astronauts in their path they are in mortal danger. At the same time, such solar explosions and energetic particles can destroy sensitive and expensive high-tech instruments on modern spacecraft, some costing hundreds of millions of Euros. Fortunately, Earth's magnetosphere and atmosphere protect us from these cataclysmic solar phenomena. Last but not least, these eruptions affect our climate. Similar phenomena, some on a much larger scale, take place in more distant stars, for which the Sun is the *rosetta* stone.

Then, in the context of the need to understand the function of the Sun and the protection of the Earth, this fragile blue dot in Space, many exciting discoveries have been made in the last hundred years. Specifically, regarding the fiery solar sphere and its superheated atmosphere we may count critical discoveries whose consequences "touch" the Earth, such as: the observation of very strong magnetic fields in the dark sunspots that appear periodically on the solar photosphere; the solution to the famous solar neutrino problem, wherein neutrinos, those enigmatic elementary particles of almost zero mass, produced by thermonuclear reactions in the solar core, when they reach Earth, pierce each one of us at a rate of billions every second; the further deciphering of the conditions that prevail in the solar interior using helioseismology; the observation that the photosphere of the Sun moves up and down with a period of 5 min, wherein the solar orchestra plays a wonderful music, through which the Sun and the other stars reveal to us their well-hidden secrets beneath their photospheres; the description of the 11-year magnetic solar cycle, wherein the dynamo mechanism produces the reversal of the solar magnetic field; the observation of magnetic flux tubes on the solar photosphere, which rise from the solar interior through magnetic buoyancy, like air bubbles in an aquarium; the discovery of solar active regions where opposite magnetic polarities, emerging from the solar interior create a panspermia of impressive solar magnetic fireworks, such as: photospheric faculae, that shine like fireflies in the lower solar atmosphere; the giant prominences, the killer flares and the spectacular coronal plasma ejections; the small chromospheric jets (spicules) and the larger coronal jets which make the solar chromosphere at the rim of the disk resemble a "burning meadow", with hundreds of thousands of flames; the flashes on the lines of the chromospheric fligree, a luminous spider's web spread over the chromosphere; the discovery that the building blocks of the solar atmosphere are the coronal loops, the ubiquitous magnetic "arches"; the discovery of eruptive prominences, those huge suspended unstable curtains of plasma that are cooler and denser than the surrounding solar corona and are ejected into space; the theoretical prediction of the solar wind, where millions of tons of solar mass are released and flood interplanetary space every second; the discovery by the SoHO observatory that the solar corona is more active than we thought, changing from minute to minute; the 3D imaging with the STEREO observatory of the evolution of solar wind density inhomogeneities due to coronal mass ejections, as these start at the base of the solar atmosphere and propagate to interplanetary space to Earth, thus shaping space weather; the detailed and impressive data of the Solar Dynamics Observatory that have helped us to understand solar variability and how it affects Earth and space weather; the strong observational evidence that solar activity affects Earth's weather and climate, mainly through the ultraviolet component of the electromagnetic spectrum, which gave rise to the new science of solar activity forecasting, the *Space Weather*.

We discuss all these splendid discoveries in detail in this chapter.

1755–1766: *Sunspots*, the Discovery of Their Periodical Appearance and Solar Cycle 1

First of all, we should recall that since nothing in this vain material world is perfect, the same applies to the surface of the *photosphere*, which bears on its face many black spots, the *sunspots*. These spots are visible on the solar disk with diameters that can be several times larger than the dimensions of the Earth (Fig. 1.4). Historically, the first reference to observing *sunspots* with the naked eye is attributed to the Athenian Greek philosopher and naturalist *Theophrastus*, a student of Aristotle at the Academy of Athens, around 350 BC, who refers to the relevant observations of the Athenian astronomer *Meton*,[1] a contemporary of Pericles (ca. 495–429 BC). For example, we can read in *Theophrastus'* surviving fragments of his treatise *De Signis Tespestatum (On Weather signs): If the Sun when it rises has* ***a black mark****, or if it rises out of clouds, it is a sign of rain.*

The first modern observation of *sunspots*, however, had to wait for the discovery of the telescope and its use in astronomy, i.e., the year 1610. During this period, parallel observations of the Sun were made by Johann Fabricius, Christoph Scheiner, Galileo Galilei (Fig. 1.2) and Johann Kepler, who turning their telescopes toward the Sun for the first time, noticed the sunspots. These spots were dark and distinct formations of the solar *photosphere*, several thousand kilometers in diameter. In fact, it has been explained that the real rivalry between Christoph Scheiner and Galileo Galilei over who was the first to observe *sunspots* with a telescope, led Galileo Galilei to the *Inquisition*, but the real instigator was the Jesuit monk Christoph Scheiner.

Further, Galileo recorded their motion on the solar disk, thus providing evidence that the Sun rotates with a period of about 27 days. Galileo also realized that although the spots were black spots on the Sun, their actual brightness exceeded considerably the brightness of the Moon.

The observations that followed until 1750 had demonstrated the main characteristics of the spots and the separation of the central shadow from the surrounding shadow. In 1769 Alexander Wilson noticed that spots located near the edge of the solar disk appear more condensed, i.e., smaller in diameter (Wilson and Maskelyne 1774). He studied and explained this phenomenon considering that the two main parts of the spot, the *umbra* and the *penumbra*, are located at a greater depth than the level of the *photosphere*, so at the rim of the disk there is an optical obstruction that causes this phenomenon (the so-called Wilson effect).

In 1843 another great discovery was added to the theory of *sunspots*, when after 17 years of continuous observation of the Sun Heinrich Schwabe concluded that the appearance of *sunspots* has a maximum occurrence period of *about* 11 years. The pattern of this periodicity was confirmed by Richard Carrington who added that the appearance of the spots at the beginning of the cycle occurs near the geographical width of 30° and as the duration of the cycle progresses their appearance converges toward the equator.

From direct solar observations of the last at least 400 years, it is inferred that during this period, the appearance of *sunspots* shows, in terms of their number and position, an approximately eleven-year periodicity. This 11-year cycle of solar activity that the German astronomer Samuel Heinrich Schwabe (1789–1875, Fig. 1.3a) had first

---

[1] The foundations of Meton's observatory in Athens, *the Heliotropium*, are still visible today just behind the podium of the *Pnyx*, one of the 7 hills of Athens, the ancient Athenian parliament. Its base is very close to the Dorides telescope of the National Observatory of Athens, whose historical facilities are located on the adjacent hill of the *Nymphs* just across *Acropolis*.

Fig. 1.2 (*left*) Aristotle (left part of fresco) with his pupils Theophrastus – (middle of fresco) and Straton (right part of fresco), at the portico of the University of Athens painted by Carl Rahl, ca. 1888. (*right*) Portrait of Galileo Galilei (ca. 1640) by the artist Justus Sustermans (1597–1681) (courtesy of Wikipedia Commons). (*bottom*) About 2,000 years later after Theophrastus, Galileo made those drawings of *direct* observations of sunspots on June 28, 1613 and published them in "Istoria e Dimostrazioni Intorno Alle Macchie Solari e Loro Accidenti" (History and Demonstrations Concerning Sunspots and their Properties) (*Image credit* Associazione Nazionale delle Case d'Aste Italiane)

Fig. 1.3 (*left*) Samuel Heinrich Schwabe (1789–1875) (*Image courtesy* of Wikipedia Commons). (*right*) Rudolf Wolf (1816–1893) (Image from the collection of the ETH-Bibliothek, published on Wikimedia Commons as part of a cooperation with Wikimedia CH)

identified, by systematically counting the daily number of *sunspots* in the period 1826–1843. Next, the Swiss astronomer Rudolf Wolf (1816–1893, Fig. 1.3b), organized and systematized these and other observations of *sunspots* and solar activity, going back in time to the beginning of the seventeenth century, i.e., essentially to the time when Galileo pointed his telescope to the Sun. Thus, after the first systematic observations of spots by Schwabe, Wolf defined in 1849 the index R of the average annual number of spots, that bears his name (Wolf number), which is still in use today (Fig. 1.6).

Based on Wolf's observations, the cycle of the period 1755–1766 is called cycle **1**, and the cycle we are going through today (2025) is cycle **25** (2020–2031), Fig. 1.5. It is noteworthy that during the period of 70 years between 1645 and 1715 a very small number of spots were recorded. This period was named the *Maunder minimum*, in honor of Edward Walter Maunder, who particularly studied the strange absence of spots during this period (Fig. 1.6).

At the same time, other similar extended solar minima have also been indirectly detected in the past, in earlier periods lasting several decades to thousands of years by the method of radio dating of the carbon-14 isotope, but also from *carrots* extracted from deep glaciers, where the relative abundance of the beryllium-10 isotope is measured, as we explain next. There were periods in which *sunspots* were absent (Fig. 1.6): A.D. 1654 to 1714 (*Maunder minimum*), 1416 to 1534 (*Spörer minimum*) and 1282 to 1342 (*Wolf minimum*). A less precisely defined minimum occurred near A.D. 1040 (*Oort minimum*), Fig. 1.6.

The distribution of sunspot area as a function of latitude is shown in Fig. 1.5, in the so-called *Butterfly Diagram*. As it was already noted by Carrington (1858) and

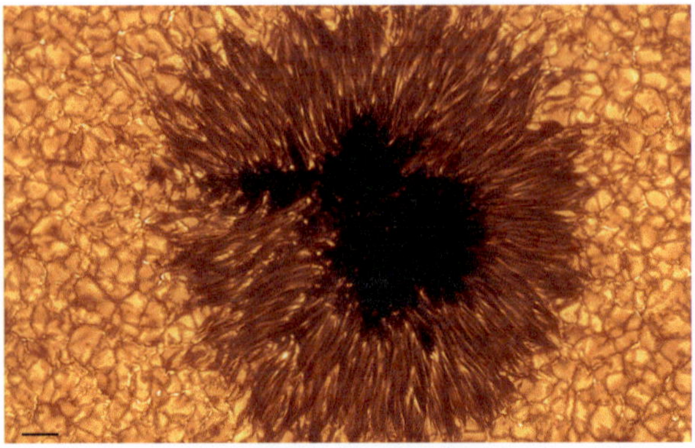

Fig. 1.4 The image shows the main *sunspot* in active region AR 12585 observed on 5-Sep-2016 with The Swedish 1-m Solar Telescope (SST)/CHROMIS on La Palma (Spain), on the neutral iron line, at 395.4 nm. The central dark area, the *umbra*, has the strongest and most vertical magnetic fields. The dark *umbra* is surrounded by the filamentary *penumbra* where magnetic fields are more horizontal. Many photospheric *faculae* are visible around the spot. The black line in the lower left measures 2500 km on the Sun (*Image credit* The Swedish Solar Telescope, La Palma, Spain. Observation by Luc Rouppe van der Voort and Shahin Jafarzadeh (University of Oslo, Norway). Data reduction was made by Jaime de la Cruz Rodríguez (ISP/Stockholm University, Sweden))

refined later by Gustav Spörer, *sunspots* and active regions appear in two bands at the start of a solar cycle, around 30° to 45° latitude on the *photosphere*. As the cycle progresses, they appear at lower and lower latitudes, until they average 15° at solar maximum. This average latitude then continues to drift lower, down to about 7° and then while the old sunspot cycle fades, active regions of the new cycle start appearing at high latitudes. The colors of the butterfly wings are related to how much of the solar surface the strips cover in that latitude (black/red/yellow). Note that the cycles overlap, e.g., often at minimum there are *sunspots* at the equator and at the same time there are *sunspots* at high latitudes from the next cycle.

Figure 1.6 (top) shows the number of *sunspots* during the last 4 centuries, with the eleven-year cycle evident and the periods in which *sunspots* were absent, as we discuss in the following. For a review on the solar cycle, see Hathaway (2015).

Let us first discuss briefly a few facts about *cosmic rays*, the anti-correlation of their penetration into the solar system with solar activity, and then how the two isotopes, carbon-14 and beryllium-10, are produced from cosmic rays. It is indeed a very well-documented fact that during periods of low solar activity, we have greater penetration of *cosmic rays* into the atmosphere, so there is an anti-correlation between the level of solar activity/number of *sunspots* and the *cosmic rays* entering the our solar system.

Cosmic rays are high-energy particles of cosmic origin, mainly coming from outside the solar system, even from the most distant galaxies. Galactic *cosmic rays* are mainly composed of protons (85%), helium nuclei (12%), electrons (2%) and other heavy nuclei (only 1%). When a cosmic ray particle enters the upper layers of the Earth's atmosphere, it interacts with the various atoms, creating a shower

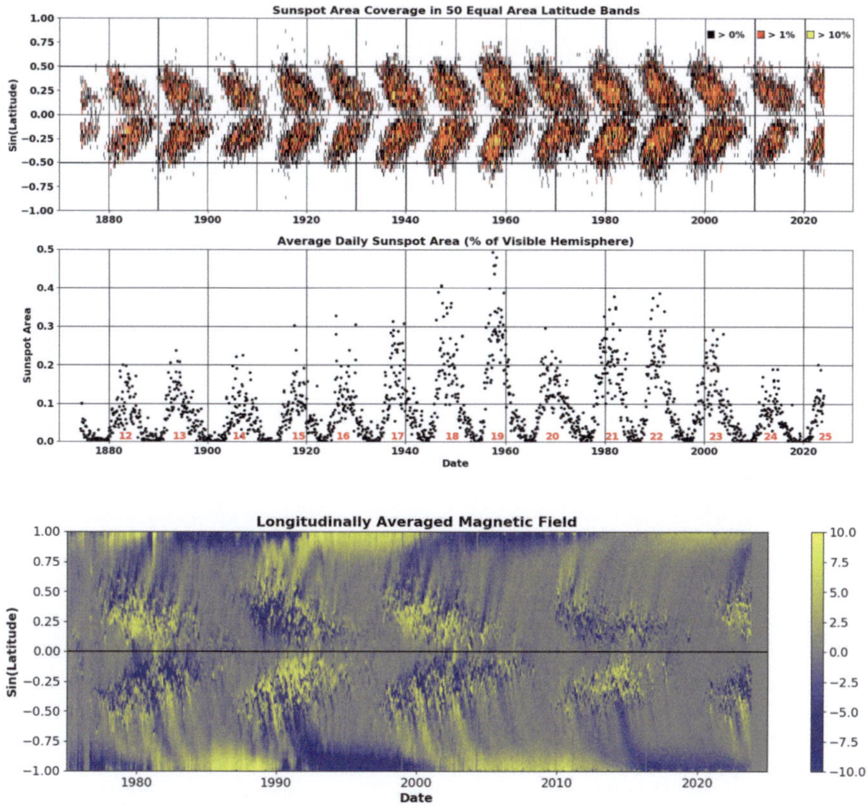

Fig. 1.5 (*top*) *Sunspot Area Butterfly Diagram*, showing the distribution of sunspot area as a function of latitude since 1874, with the number of *sunspots* starting in the year 1874 (solar cycle **11**) till now. (*middle*) Average daily *sunspot* area as percentage of the visible hemisphere. (*bottom*) *Magnetic Butterfly Diagram*, showing the distribution of the surface radial component of the solar magnetic field (longitudinally averaged) over the last four solar cycles. The poles have opposite polarities that switch from one cycle to the next near the time of solar maximum. The *butterfly* signature of *sunspots* is clearly visible at lower latitudes (*Image credits* Diagrams constructed by the solar group at NASA Marshall Space Flight Center, also available in Wikipedia Commons)

of interactions, before it finally reaches the ground. In this way, one cosmic ray particle can create millions of other secondary particles. The overall products of this shower of interactions are of three categories: photons and electrons, hadrons (protons, neutrons, and pions) and muons. The relative ratios of the three components of the avalanche of these chains of interactions depend on the height and therefore on the density of the atmosphere.

Carbon-14 is a radioactive isotope of the stable element carbon-12. It is continuously generated in nature by the interaction of neutrons with nitrogen-14 in the Earth's atmosphere. The neutrons required for this reaction are produced by *cosmic rays*, as they interact with the atmosphere. Subsequently, the carbon-14 atoms created in

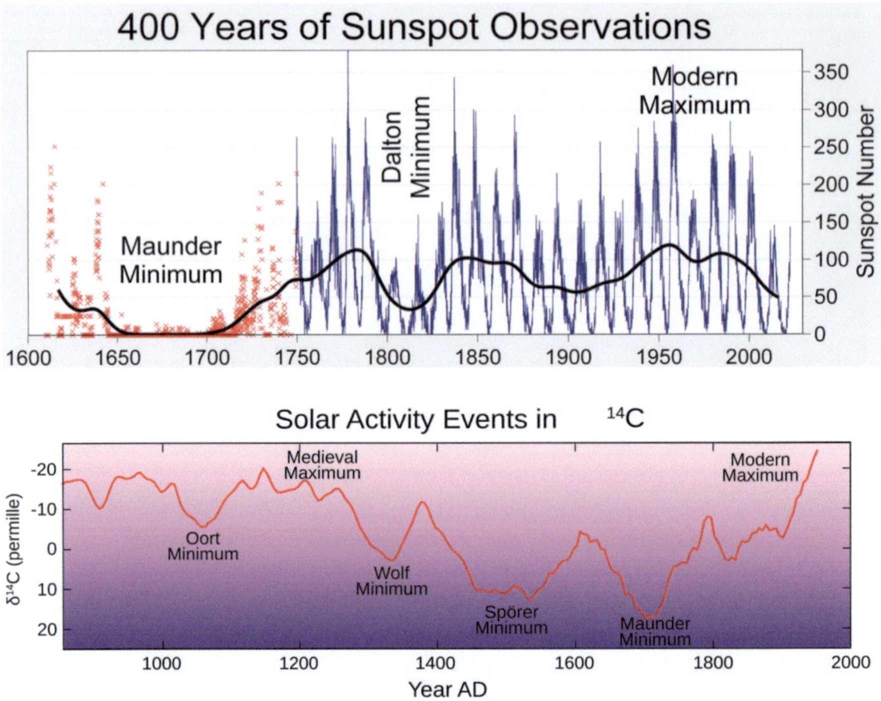

Fig. 1.6 (*top*) Detailed recording of the number of *sunspots* during the last 4+ centuries, with the eleven-year cycle evident, from 1700 onwards. The figure summarizes *sunspot* number observations since ca. 1749, when began the extensive recording of solar sunspot activity (solar cycle **1**). Continuous monthly averages of *sunspot* activity have been available and are shown here as reported by the Solar Influences Data Analysis Center, World Data Center for the *Sunspot Index*, at the *Royal Observatory of Belgium*. Figure is based on an average of measurements from many different observatories around the world. Prior to 1749, only sporadic observations of *sunspots* are available. These data were compiled and placed on consistent monthly framework by Hoyt and Schatten (1998a, b). Figure was prepared by Robert A. Rohde and is part of the Global Warming Art project (*Source* Wikipedia Commons). (*bottom*) The various minima and maxima of solar activity over the past millennium, using carbon-14 radio dating of tree rings. This figure was produced by Leland McInnes and is licensed under the GFDL (*Courtesy* of Wikipedia Commons)

this way by *cosmic rays* combine with oxygen atoms to form carbon dioxide, which is absorbed by plants. In this way, carbon-14 is eventually incorporated into the plant fibers and trees via *photosynthesis*. Carbon-14 dating is based on the fact that when an organism dies, or a tree is cut down, the radioactive carbon-14 ceases to be replenished by food intake. Then, instead to be found in a fixed proportion within the material or the organism, as well as in the environment, it begins to decrease, decaying with the emission of beta particles, and the carbon-14 nuclei begin to decay with a half-life of 5,730 years. The concentration of carbon-14 in the sample can be measured at any moment. Thus, by measuring the current concentration of carbon-14 atoms in a sample, measures the age of the sample since death, or, from the moment carbon dioxide intake through food, photosynthesis and respiration has stopped.

Beryllium-10 is a radioactive isotope of the stable beryllium-9. It is created in the Earth's atmosphere mainly through nuclear fusion caused by neutrons and protons produced by *cosmic rays* as they interact with atmospheric nitrogen and oxygen. Beryllium-10 has a half-life of 1.39 million years and decays via β-decay into stable Boron-10. The beryllium then precipitates with rain and snow and passes into the glaciers.

In glacier drilling, the so-called *ice carrots* are extracted, with a diameter of a few centimeters, from great depths. An *ice carrot* is a deep sample from a glacier of a high mountain or arctic region (e.g., Greenland). Since the glacier is formed by an increasing accumulation of annual layers of snow, the lowest layers are older than the upper ones, and an *ice carrot* contains ice formed over a sequence of many years. *Carrots* are extracted with hand augers (for shallow holes), or with motor drills. They can thus reach depths of more than 3 km and contain ice up to 800,000 years old.

In plots of the Wolf number R for the observation of *sunspots*, as counted by the use of the telescope from 1610 until today, with the percentages of carbon-14 and beryllium-10, it is found a complete anti-correlation. During periods of low solar activity with small average numbers of sunspot appearances, the percentages of these two radioisotopes are increased, due to greater penetration of *cosmic rays* into the planetary system and consequent greater production of the radioisotopes carbon-14 and beryllium-10. Hence, increased ^{14}C and ^{10}Be production during times of increased cosmic radiation serves as a proxy for solar activity.

Figure 1.7 show the observed recorded number of *sunspots* during the last 4 centuries from 1600 onwards, with an eleven-year cycle and the various minima and maxima of solar activity over the last millennium, using the method of carbon-14 radio dating in tree rings. The production of the beryllium-10 and carbon-14 isotopes, as we mentioned, is sensitive to the bombardment of the upper atmosphere with *cosmic rays* (Greenland Ice Core Project, GRIP, Steinhilber et al. 2012). The flux of *cosmic rays* reaching Earth decreases during periods of high solar activity. The relative percentages of these two isotopes of beryllium and carbon, in relation to the other isotopes of these elements, which determined by drilling in research drilling programs in ice fields in Greenland, Antarctica, and the Arctic, but also tree rings show to be significantly increased during the times of the *Maunder, Spörer, Wolf* and *Oort minima* and well before them (Figs. 1.6 and 1.7). Such changes significantly affect the earth's climate and it has been observed that the growth of trees, which is of course linked to rainfall, changes with the eleven-year cycle, as shown by the rings of the trunks of centuries-old trees.

The presence of grand minima in solar activity on the long-term scale has been discovered, using the C-14 radioisotope data in tree rings. Specifically, a list of 27 grand minima have been identified in the quantitative solar-activity reconstruction of the last 11,000 years (Usoskin et al. 2007). Thus, besides the *Maunder minimum* centered in 1680 with a duration of 80 years, we have the Sporer *minimum* centered in 1470 with a duration of 160 years, the *Wolf minimum* centered in 1305 with a duration of 700 years, etc. The cumulative duration of the grand minima is about

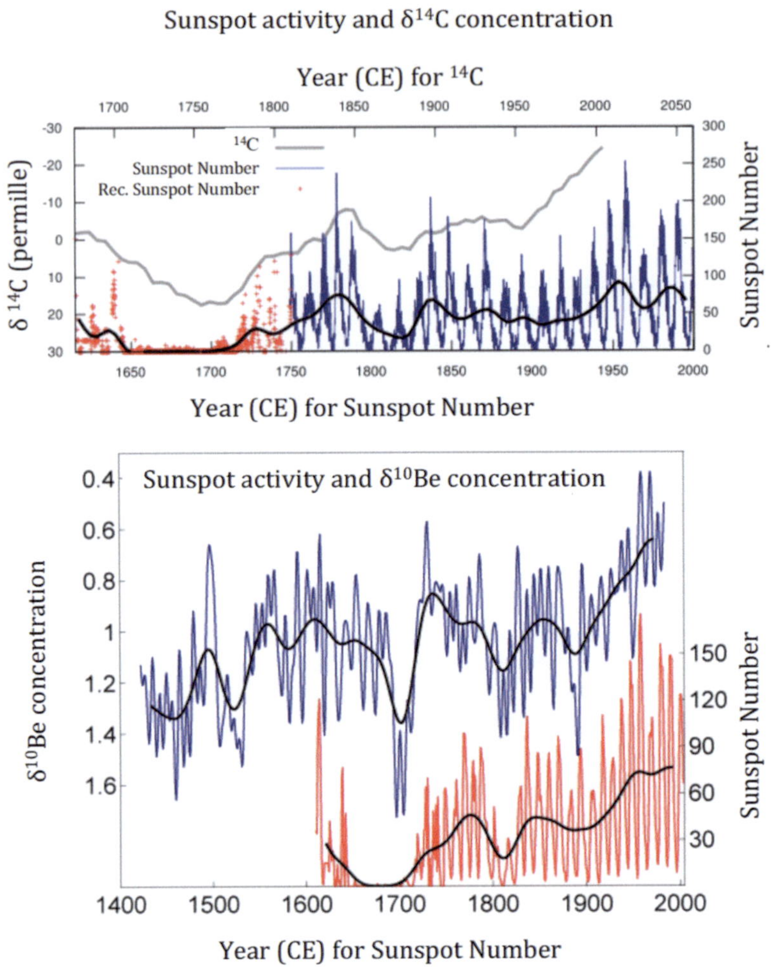

Fig. 1.7 Anticorrelation of sunspot occurrence with carbon-14 (*top*) and beryllium-10 (*bottom*) percentages. Sunspot data in red (in top) is reconstructed from group sunspot numbers. There is a roughly 60 year delay between changes in sunspot activity and ^{14}C; in this chart the time scale for ^{14}C has been separated (labeled at the top of the chart) and offset by 60 years. The concentration of ^{14}C from 1950 onwards has been distorted by nuclear testing. The carbon and beryllium scales are inverted so that the minima of the mean curve of these concentrations indicate a higher concentration of carbon-14 and beryllium-10. The bottom figure shows two different proxies of solar activity during the last several hundred years. In red is shown the Group Sunspot Number as reconstructed from historical observations by Hoyt and Schatten (1998a, b). In blue is shown the beryllium-10 concentration [10^4 atoms/(gram of ice)] as measured in an annually layered ice core from Dye-3, Greenland (Beer et al. 1994) (*Picture credits* top image was produced by Leland McInnes using gnuplot and Inkscape and is licensed under the GFDL. All data are from publicly available sources. Data Sources (gray) Carbon-14 data: http://www.radiocarbon.org/IntCal04%20files/intcal04.14c (blue) Sunspot data: http://sidc.oma.be/DATA/monthssn.dat (red) Reconstructed Sunspot Data: ftp://ftp.ngdc.noaa.gov/STP/SOLAR_DATA/SUNSPOT_NUMBERS/GROUP_SUNSPOT_NUMBERS/monthrg.dat Bottom image was produced by Robert A. Rohde from previously published data, is also licensed under the GFDL and is incorporated into the Global Warming Art project. Both images are courtesy of Wikipedia Commons)

1900 years, in this period of 11,000 years, indicating that the Sun in its present evolutionary stage spends ~1/6 (17%) of its time in a quiet state, corresponding to grand minima. Studying these grand minima has led to the conclusion that their occurrence is defined by stochastic/chaotic processes (Usoskin et al. 2007). For a reviews on the history of solar activity over millennia, see Usoskin (2008, 2023).

Sunspot numbers and solar activity in the last 11,400 years have been reproduced by adopting and extending the correlation of solar activity Earth's climate, using the methods of *dendroclimatology*, based on the carbon-14 dating method. *Dendroclimatology* is the science of determining climate from studying the characteristics of tree trunks, such as the thickness of their rings. Tree rings are thicker when there are climatic conditions that favor the growth of the tree. Using rings from old trees, researchers have calculated the terrestrial climate hundreds to thousands of years ago. In this way it has been calculated in the distant past, locally and globally, Earth's climate and subsequently the number of *sunspots* and the anticipated solar activity.

1859: The Strongest Solar Flare Ever Recorded!

Richard Christopher Carrington (Fig. 1.8) was a prominent English amateur astronomer, the second son of a wealthy brewer. He studied theology at Trinity College, Cambridge, because initially his wealthy parents intended him to be a clergyman. However, eventually his love for astronomy prevailed, when he attended the lectures of the director of the Cambridge observatory and Plumian Professor, James Challis. Carrington was an interesting personality. He began as an observational astronomer, but subsequently devote his life in the search for an understanding of the Sun.

Fig. 1.8 Richard Carrington (1826–1875) (*Picture credit* The High Altitude Observatory/NCAR)

Let us then move to a historic day in Carrington's private solar observatory, shortly after 11 o'clock on the morning of September 1, 1859. There, almost every day, his telescope projected the image of the Sun onto a white screen disc. In those days the *sunspots* were unusually numerous, because the Sun was at that period approaching the maximum of its 10th cycle (1855–1867) – a solar cycle which lasted 11.3 years, beginning in December 1855 and ending in March 1867. That day, Carrington noticed something strange: suddenly, as noon approached, the Sun suddenly became brighter. Two arcs of blinding light appeared above a system of *sunspots*, which then grew and took the shape of a loop. The sight was impressive and accompanied by an unusual behavior of the telegraph's systems: the wires and telegraph paper caught fire, while telegraph operators were electrocuted. Even when the power batteries were disconnected, the inductive currents created in the powerful *aurora* continued to allow the messages to be transmitted. Just before the sunrise of the next day, September 2, 1859, the sky over almost the entire Earth was lit up by a red, green and blue *aurora*, visible even in the tropics, from Cuba and Jamaica to El Salvador and Hawaii. In the USA, on September 1, and 2, 1859, the *aurora* borealis at night was so bright in the Rocky Mountains that coal miners there woke up early and began preparing their breakfast in the middle of the night, thinking that dawn had come! At the same time in Europe it was announced that the light from the *aurora* over Rome it was enough to read a newspaper in the middle of the night! This was, in short, the so-called *Carrington episode of 1859* and its consequences on Earth. Although initial assumptions were based on the fact that it was a meteorite that hit the surface of the Sun, this was the first and strongest solar flare ever recorded!

Afterward (1852), at the University of Durham, Carrington built his own observatory at Redhill, Surrey and equipped it with suitable telescopes (1853). Later (1857), presented his observations to the Royal Astronomical Society (RAS), in a list of "*3,735 Circumpolar Stars observed at Redhill in the years 1854, 1855, and 1856, and reduced to Mean Positions for 1855*". Alongside his astronomical observations, he also began to look into the original sketches of observed *sunspots* that were available in the RAS archives. Realizing the lack of systematic solar observations, he decided to place this subject at a first priority in his research and thus devoted his daytime to observations of the Sun, while at night he continued observations of stars. He was particularly concerned with solar rotation, for which there was then great ambiguity and confusion.

Leaving photography to be perfected by the subsequent generations, he chose the method of direct observation of the Sun. Specifically, he projected the solar image that produced his 4.5 inch (11.5 cm) equatorial telescope, at such a distance from the telescope eyepiece, such as to form a 12–14 inch (30.5–35.5 cm) diameter disk on a screen. He had also attached to the focus of the telescope two gold threads, perpendicular to each other, and inclined at 45° on each side of the meridian. Thus, as the inverted solar image crossed the observation screen, the observer recorded the moments in time when the solar limb passed from the *filaments*, so that their heliocentric position accurately recorded various formations on the solar disk. In this way, over a period of 7–8 years, he recorded 5,290 observations of 954 different formations, for many of which he accurately traced their position and shape. Unfortunately, the death of his father in 1858 forced him to stop the pursuit of solar observations and to deal with the family brewery. In the meantime, however, he published his observations on a RAS volume entitled: "*Observations of the Spots on the Sun from November 9, 1853, to March 24, 1861, made at Redhill (London, 1863)*". These data revolutionized solar physics in terms of the rotation of the Sun and the rotation of the *sunspots* at various heliographic latitudes. In particular, Carrington gave the following *empirical* formula for the diurnal rotation of *sunspots* located at heliographic latitude φ: $865' \pm 165 \times \sin 7/4\ (\varphi - 1°)$, although he did not provide any theoretical explanation for this formula. This describes the so-called *differential rotation* of the Sun. Since then, we know that the Sun's synodic period at the equator is about 25 days, while it increases with the heliographic latitude such that at the poles is about 36 days, although we still do not understand the reason why there exists precisely this rotation. Also, Carrington determined the inclination of the solar equator with respect to the plane of the ecliptic, (7° 15′) and the longitude of the ascending node (73° 40′), both for 1850; he also noticed a peculiarity in the appearance of *sunspots* during the minimum of 1856, wherein he noted:

- *an instructive instance of the regular irregularity and irregular regularity, characterizing solar phenomena*!

As the minimum approached, the active regions descended toward the equator and finally they disappeared there; soon after, two fresh new centers appeared spontaneously at relatively high heliographic latitudes, and then they also began to move toward the equatorial latitudes. Wolf had already spotted a similar behavior in Böhm's observations in the period 1833–6, while the same was observed by Spörer and Secchi and the same was repeated in 1867. If we cannot absolutely say that Carrington discovered the eleven-year solar activity cycle, his observations of *sunspots* led to the enumeration of solar cycles by his name. The period of the solar rotation is now conventionally taken to be 27.2753 days and to each revolution of the Sun around its axis, on this assumption, corresponds a number, the Carrington rotation number, starting from November 9, 1853, when Carrington began his observations. This date is defined as the 1st rotation. The heliographic angular length of a solar structure refers to an angular distance in relation to the central meridian, i.e., that defined by the Sun-Earth line. The *Carinngton heliographic latitude* of the same structure refers to a randomly chosen fixed point reporting a hypothetical rotation of the Sun as a solid body, as originally defined by Carrington.

In summary, Carrington discovered: (i) the first unexpected observation of a great flare (1859), (ii) the differential rotation of the Sun, (iii) the migration of *sunspots* toward the equator, during the solar activity circle, around the same time as Gustav Spörer in Germany (who discovered the so-called Spörer's law) and (iv) the determination of the axis of rotation of the Sun with unprecedented accuracy for his time.

Flares, *the Ultimate Solar Menace!*

Solar *flares* (Figs. 1.9, 1.10 and 1.11) occur in the upper *chromosphere* and *corona*, heating the local plasma to tens of millions of degrees Kelvin, accelerating electrons, protons and heavier ions to high velocities, some close to the speed of light. In addition, they produce electromagnetic radiation at all wavelengths of the electromagnetic spectrum, from long radio waves, to visible, ultraviolet and short wavelengths of the soft and hard X- and γ-rays. *Flares* occur over a period of minutes, and then they turn off within an hour or two, i.e., these powerful *fireworks* last less than an hour (Priest 1982, 2014) (Figs. 1.9, 1.10 and 1.11).

Solar *flares* are considered the most violent and powerful explosions in the entire solar system. *Loops* dozens of times larger than the size of the Earth, filled with billions of tons of superheated plasma are ejected from the *photosphere* to the interplanetary space. In this process, energy of the order of 10^{27}–10^{32} ergs is released in about 10 min. Specifically, in solar *flares*, within a few minutes, the energy which is released from a relatively *small* area of the solar *corona* is released from the entire solar disk and would be equivalent to the eruption of 10 million volcanoes. In other words, the amount of energy released in a big flare is equivalent to millions of 100-megaton hydrogen bombs[2] exploding at the same time!

In a typical flare, a shock wave is propagated from the region of the flare, accompanied by a large mass of plasma ejected from this region of the *corona*. During *flares*, the accelerated solar particles that have large amounts of energy can reach our planet within even 10 min, almost as fast as sunlight. These are particles endemic to the solar atmosphere, i.e., nitrogen, carbon, magnesium, neon, iron and silicon

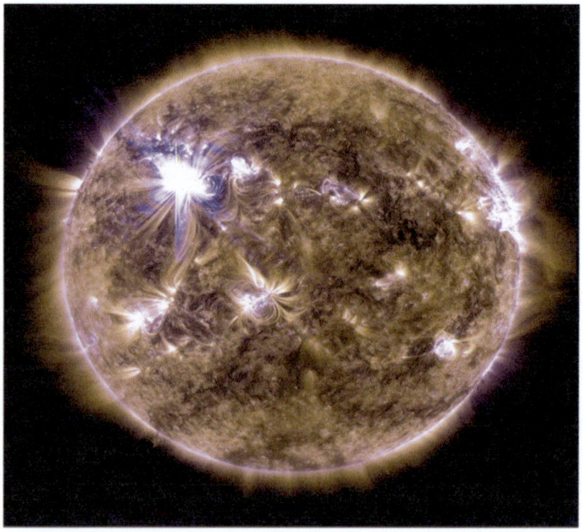

Fig. 1.9 An X5.4 class solar flare flashes near the upper left edge of the Sun on March 07, 2012. This image was captured by NASA's Solar Dynamics Observatory and shows a blend of light from the 171 and 131 Angstrom wavelengths of the UV spectrum (*Image credit* NASA/SDO)

[2] Hydrogen bombs of more than 50 megatons have been detonated.

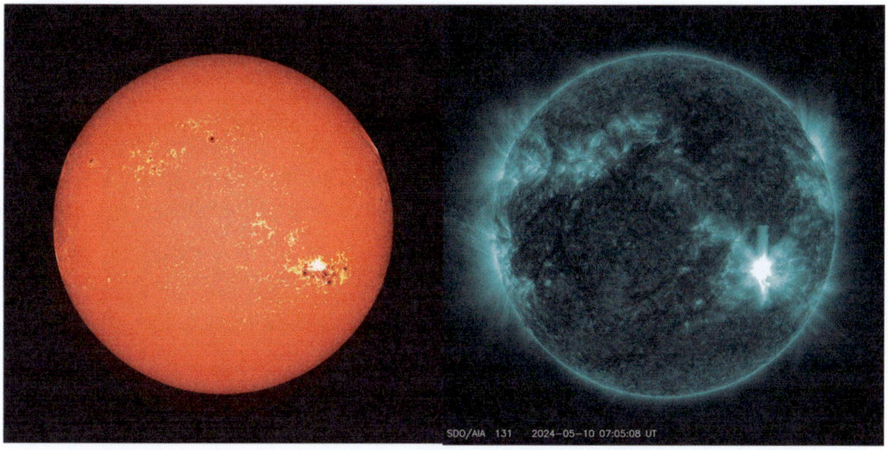

Fig. 1.10 (*left*) NASA's Solar Dynamics Observatory captured this image of an X3.9 flare on May 10, 2024. The image shows a blend of AIA 1600 Angstrom light and HMI Continuum (the continuum intensity of the solar spectrum in the region of the Fe I absorption line at 6173 Å on the surface of the sun). (*right*) SDO image of this X3.9 flare on May 10, 2024. The image shows 131 Angstrom light, a subset of extreme ultraviolet light (*Image credit* NASA/SDO)

ions. These particles constitute the high-energy solar *cosmic rays*, or otherwise, the so-called solar energetic particles (SEP).

The largest solar flare ever recorded since satellites started to measure them in 1976 occurred on Nov 4, 2003, with a magnitude X45! If the companion particles had been aimed at Earth, the damage to satellites and power grids could be very serious. The radiation of X-rays in the atmosphere was equivalent to that of 5,000 Suns. The accompanying coronal mass ejection reached a speed 2,300 km per second, or 8.2 million kilometers per hour.

An X-ray flare passes from the following phases:

(1) *Pre-impulsive* (gradual rise), observed only in some *flares*
(2) *Impulsive* (rise) phase, with bursty hard X-rays (HXR), fast but smoothly rising soft X-rays (SXR)
(3) *Gradual* (decay) phase – little to no HXR, gradual decline in SXR.

Devastating and Murderous *Flares*!

Often, but not always, as happened in the *Carrington episode*, flares are accompanied by coronal mass ejections (CMEs, e.g., Fig. 1.47) where billions of tons of particles are ejected into interplanetary space. The particular CME of 1-9-1859 was directed directly toward the Earth and produced the unusual phenomena of the following day which we mentioned. Thus, the 1859 flare produced the largest geomagnetic storm in the following about 160 years of available observations. Even if one looks for traces of such events in ice cores

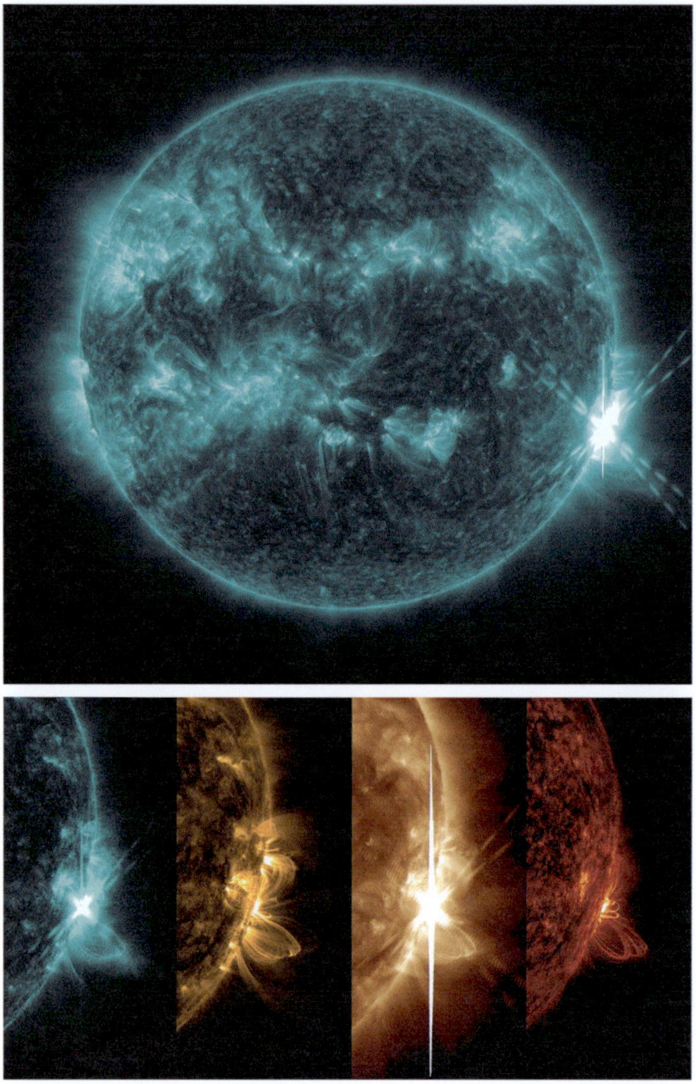

Fig. 1.11 May 2024 has proven to be a particularly stormy month for our Sun. During the first full week of May, a barrage of large solar *flares* and *Coronal Mass Ejections* (CMEs) launched clouds of charged particles and magnetic fields toward Earth, creating the strongest solar storm to reach Earth in two decades – and possibly one of the strongest displays of *auroras* on record in the past 500 years. The Sun has produced more than **100 M-class** and **16 X-class *flares*** since May 1, and a series of Earth-directed CMEs – some of which merged into one powerful CME and a historic G5 extreme geomagnetic storm (i.e., the most powerful type of geomagnetic storm) on May 10 and 11. NASA's Solar Dynamics Observatory (SDO) observed nine X-class solar *flares* erupting from the Sun, including the largest in this solar cycle to date on May 14 that peaked at X8.7 (*above*), showing a subset of extreme ultraviolet light that highlights the extremely hot material in *flares* and which is colorized in teal. (*below*) The same flare is shown (from left) in 131, 171, 193 and 304 Angstroms (*Image credit* NASA/SDO)

from the Arctic or Antarctic, Carrington's flare remains the strongest event for at least the last 500 years. But, even the mildest solar flares create magnetospheric storms that can cause significant problems on Earth. For example, on August 4, 1972, a geomagnetic storm followed by a powerful flare knocked out telephone lines in Illinois, USA. Because of this event, the American telecommunications company AT&T was forced to change the electrical power design of its transatlantic ship cables. Another powerful flare on March 13, 1989 knocked out an entire hydroelectric plant in Quebec and left 6 million local residents without power for about 9 h. Also, induced currents from the intense aurora melted transformers in New Jersey, USA. In December 2005, X-rays from another flare caused disruptions to satellite communications and the GPS network. This interruption lasted only ten minutes, but surely no one would want to be passenger of a plane, which during this time would try to land using the GPS system, as it could send the plane off course…

The X-rays emitted during a flare and incident on the Earth's ionosphere, the ionized layer of the atmosphere located at a height of about 60 km, cause geomagnetic disturbances and storms recorded by magnetometers. This recording was made for the *Carrington episode* by a tape recorder at Kew Gardens, the Royal Botanic Garden in south-west London. Of course, this could not be estimated at that time, before the discovery of X-rays by Roentgen (1895) and the ionosphere by Heaviside (1902).

The frequency of solar *flares* varies, from several per day, when the Sun is particularly active, to less than once each week, when the Sun is quiet. In other words, solar *flares* become more frequent and stronger near solar maximum, the peak of the 11-year cycle of solar activity. Large *flares* are less frequent than smaller ones. So during solar maximum we have the highest frequency of *flares*, with many of them recorded and monitored by various terrestrial and satellite solar observatories. The smaller *flares* are known as *A-type flares*. These are followed by the B, C and M *flares*, while the strongest of all are the *X-type flares*. Similar to the Richter scale for earthquakes, each letter represents an increase in flare intensity by a factor of 10. Thus, an X-type flare is 10 times stronger than an M-type, etc. Within each type there is a subdivision denoted on a scale of 1 to 9. For example, the 24 February 2011 X 2.2 type flare was the strongest of the last cycle No. 24 and lasted for about 90 min. It was powerful enough and lifted above the *chromosphere* a huge loop of plasma, as massive as a mountain range!

The 11-year solar cycle is pretty predictable, but no one knows when a flare similar to that of 1859 will occur again. What is certain, however, is that next time we will not only worry about the telegraph. With humanity relying more and more on electronic systems, damage to power grids, mobile networks and multi-output satellites, strong *flares* could cost tens of billions of dollars. Additionally, so will be at risk the crew of the International Space Station from this solar menace. The only positive difference is that this time there will be warnings from spacecrafts: the SOHO, Hinode, SDO, etc. spacecrafts are vigilantly pointed to the Sun!

Flares stimulate *Space Weather*, inducing on Earth an increased energetic proton flow, geomagnetic storms and intense *aurora borealis*. It is estimated that an enormous solar storm could short-out telecom satellites, radio communications, and power grids, leading to trillions of dollars in damages!

First of all, the high-energy protons ejected from the site of solar *flares* reach the Earth within about 10–30 min. Some of them they spiral along the lines of the earth's magnetic field reaching the upper layers of the atmosphere. On their way to space, they often destroy spacecraft microchips, or charge and short circuit other sensitive instruments. Do not forget that they are particularly dangerous for astronauts who happen to be in space, but also passengers on supersonic flights near the polar regions. For example, the high-intensity energetic solar protons of the March 13, 1989 flare, which as reported put out of service the hydroelectric plant of Quebec, they could have killed an astronaut if he was there at that time exposed on the Moon or in space: with a single blow he can receive as much radiation the human body can withstand.

The understanding of the phenomenon of solar *flares* is expected to lead to the understanding of other similar, but often much stronger, cosmic phenomena that occur far away from us, within our Galaxy. Such phenomena occur in flare stars, or UV Ceti-type stars. Of the billions of stars in our Galaxy, there is also a *lowly* category of stars, much fainter than the Sun (with corresponding temperatures from 2,500° to 4,000 °K) of low mass (from 0.1 to 0.6 solar masses) located at the lower end of the Main Sequence, the red dwarf stars. Often these stars emit emission lines of hydrogen and calcium, which shows that they have chromospheric activity. These flare stars have been observed at distances of up to 1,000 light years from Earth. Their systematic observation started by Alfred Joy & Milton Humason in 1949 from the observatory of Mount Wilson. These astronomers then noticed that in a star located in the constellation of Cetus (UV Ceti) – which since then has been the prototype of the class of stellar *flares* – the stellar brightness increases dramatically at unpredictable periods of time for minutes to hours, particularly in X-rays, by four orders of magnitude, with the temperature also rising to 10,000°, returning to the calm state in less than a day. So the fact that, in these unimpressive stars – in terms of brightness – their brightness increases sharply over a period of minutes to days across the electromagnetic spectrum, from X-rays to radio waves, makes them a special and interesting object of observation.

Solar *flares* are caused by local disturbances of the solar magnetic field. Specifically, they are another case of a solar phenomenon wherein magnetic reconnection plays a key role and is considered to be its cause. In addition, the solar *photosphere* is covered by millions of small magnetic areas with dimensions from 50 to 1,000 km. These small magnetic dipoles are stirred and agitated by *photosphere* granulation, so that regions of opposite magnetic lines often come into contact. The result of this is a magnetic field higher up in the *corona* in a state of almost continuous magnetic reconnection, responding to the movements of the photospheric *magnetic carpet*. As E. N. Parker has suggested, the energy released by these magnetic reconnections is a natural candidate and the cause of coronal heating, in the form of a series of very small *flares* (nanoflares), each of which individually yields a limited amount of

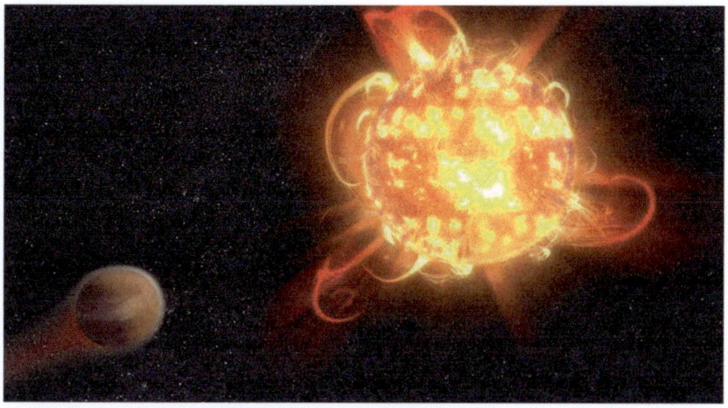

Fig. 1.12 Stellar flares could threaten life on red dwarf planets (*Image credit* NASA/ESA/D. Player (STScI))

energy, but all together ensure the required plasma heating, which creates the coronal temperature of millions of degrees.

We may distinguish three characteristic stages or phases in a solar flare. In the precursor (pre-) phase (1), we have a release of magnetic energy, which is accompanied by the emission of sufficient intensity soft X-rays (SXR). In the second explosive phase (2), the exponentially increasing energy accelerates protons and electrons to energies exceeding 1 MeV. During this phase, radio waves, hard X-Rays (HXR) and γ-rays are emitted. In the third phase (3) a gradual decrease in soft X-rays (SXR) is observed. The duration of these phases can be quite short, a few seconds, or, an hour or two.

Unlucky Planets Around Wild **Flare Stars***!*

The category of variable stars of the RS Canum Venaticorum type, abbreviated RS CVn – Canes Venatici, Latin for 'hunting dogs', the two favorite hounds of Bootes, Asterion and Chara, which accompany him, while he goes out to hunt the bears – are also known for their *flares*, but these *flares* are rather caused by the companion star in a double system.

On April 23, 2014, NASA's Swift satellite detected the strongest, hottest, and longest-lasting sequence of stellar *flares* ever, that we have seen in a red dwarf, in the binary system DG Canum Venaticorum, located about 60 light years away in the constellation of the Canes. The flare came from one member of the binary system, where two red dwarfs, about one-third the mass of the Sun and 30 million years old, orbit at a distance three times the distance of the Earth from the Sun. The initial burst from this series of *flares*, if it had occurred in the Sun and was observed from Earth, would be up to 10,000 times more powerful than the largest solar flare ever recorded

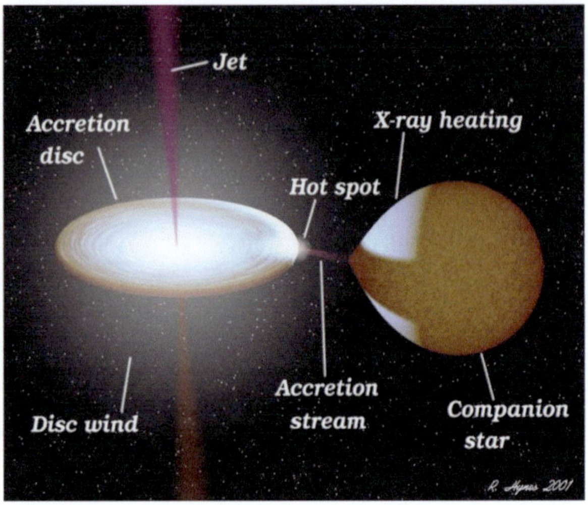

Fig. 1.13 Flares are seen in X-rays and γ-rays in accretion disks that form in X-ray emitting binary systems that contain a stellar black hole, such as Low-Mass X-Ray Binaries (LMXRB) and high-mass double systems (High-Mass X-Ray Binaries, HMXRB) (*Image credit* NASA/R. Hynes)

on Earth (X28) and on the scale of the *flares* it would be of the type X100,000. But how is it possible that such a small star with a mass equal to a third the mass of the Sun, to produce such powerful *flares*? The answer might be, that it is so because DG Canum Venaticorum rotates very fast, 30 times faster than the Sunits age is "only" 30 million years – resulting in strong magnetic fields. When the Sun was younger at such an age and was similarly spinning very fast, it is expected that he too would produce such superflares. Unfortunately, or fortunately, today the Sun does not allow us to observe such dangerous superflares!

The violent release of stored magnetic energy, in magnetic *flares* is also predicted to be a fairly common phenomenon in the magnetized atmospheres (coronas) of astrophysical disks that form in binary stellar systems containing a stellar black hole, Fig. 1.13.

If the current understanding of solar *flares* is correct, *flares* occur due to the kinematic distortion of magnetic flux tubes by subphotospheric fluxes. Similar bursts are therefore expected to occur in magnetic flux tubes in accretion disks when they are deformed by sufficiently strong movements of the plasma. Indeed, such magnetic effects may be responsible for the observed bursts, highly variable X- ray emission, and abrupt outbursts emission of synchrotron radiation in the radio part of the electromagnetic spectrum from binary stellar systems containing accretion disks around protostars, galactic sources with *black holes*, but also from nuclei of active galaxies.

These disks are strongly magnetized and subject to magneto-rotational instability: an initial weak magnetic field within the disk is strengthened by the dynamo mechanism and then emerges into the outer atmosphere due to magnetic buoyancy. This magnetic field in the disk *corona* is strongly inhomogeneous and we expect energy release in localized active regions, as in the solar *corona*. There, magnetic reconnection is the dominant mechanism that can explain the observed variable emission from compact astronomical sources. On the other hand, the hard component of the radiation in X-rays in the spectra of galactic sources containing *black holes* and Seyfert galaxies is likely due to multiple Compton scattering (Comptonization) of the low-energy (infrared) photon population from a population of hot (high-energy) electrons located in a area surrounding the disk itself. For a review on stellar flares, see Kowalski (2024).

1859: The Emergence of *Space Weather* and the Solar–Terrestrial Interaction

Interplanetary space contains charged particles, photons, electric and magnetic fields, dust and *cosmic rays*. Although the density of these particles is very low, compared to the atmospheric density on the surface of the Earth, it is nevertheless important to affect spaceships, the International Space Station and the various other artificial satellites, and also astronauts, who might be exposed there. Especially, when stormy phenomena of solar origin occur in interplanetary space, their effects appear more prominently as changes of the state and intensity of the geomagnetic field and also as the splendid and impressive *aurora* fireworks in the Earth's polar regions. Hence, a high-intensity solar activity causes in the interplanetary space an increased flow of energetic protons and on Earth geomagnetic disturbances in the form of *magnetic storms* and *magnetospheric substorms*, i.e., large changes of the terrestrial magnetic field and breathtaking *aurora borealis*. All these disturbances in space and the near environment of the Earth and the phenomena that accompany them, make up what we call *Space Weather*.

Although *Space Weather* is a fairly recent term, there is a rich history of similar terms being used initially, in the middle to late 1800s. As is summarized in the article by Cade and Chan-Park (2015), the first uses of various terms related to *Space Weather* are *Solar meteorology* [for *sunspots* and the conditions of the Sun's atmosphere, in Herschel (1847)], *magnetic weather* [for disturbances in the terrestrial magnetic field, in Phillips (1851)], and *cosmic meteorology* [for Solar-terrestrial interaction by Donati (1872), actually the first to assign a name to this science we now call *Space Weather*). All these happened before and after the Carrington event (1859). Later on, the term *Electrical weather* was also used [for the dynamics of the ionosphere and resulting magnetic disturbances, by Ewing (1955)], *Space weather* [for the weather of interstellar space, the motions and composition of the vast clouds of matter in the voids between stars in Science News Letter, Society for Science and the Public (1959)]. Hence, the actual first appearance of *Space Weather* can be attributed to the publication Science News Letter (1959) and was possibly coined by the editor at the time, Watson Davis. Later, the term *Space Weather* is attributed to Thomas Gold (1959). Finally, during the 1960s, *Space Weather* becomes more and more popular, as in New York Times articles (1966), or Hahn (1968) with the similar terms *Solar weather* and *magnetic weather* eventually merging into the discipline of *Space Weather*.

As previously mentioned, the geomagnetic field protects and shields the terrestrial space environment from the high-speed solar wind gusts that reach 2–3 million kilometers per hour, but also dangerous *coronal mass ejections* (CMEs). Furthermore, when there is a suitable orientation of the interplanetary solar magnetic field carried by the *solar wind* toward the planets, so that it is antiparallel to the geomagnetic field at the sunward side of the magnetosphere (see Fig. 1.14), then occurs the physical process of magnetic reconnection: two magnetic field lines with opposite orientation annihilate and neutralize each other. This occurs at the so-called magnetopause distance, which is calculated by equating the dynamical pressure of the incoming *solar wind* with the magnetic pressure of the resisting terrestrial magnetic field. For typical values of the density and velocity of the *solar wind* at the distance of the Earth (say, 7 protons per cubic centimeter and 400 km per second, respectively), the radius of the magnetopause is approx. 9–10 times the radius of the Earth, i.e., about 60,000 km above the surface of the Earth. In Fig. 1.14 this is indicated by the dark gray area in which the terrestrial magnetic field has been compressed by the *solar wind*, with a similar area in the magnetotail. Hence, from these regions, where the geomagnetic field reconnects, Earth's magnetic shielding is lost and then solar particles and available magnetic energy enter in the magnetosphere. The magnetic energy originates from and corresponds to the interplanetary magnetic field lines, which due to magnetic reconnection merge with the geomagnetic field and they are added to the magnetosphere. But this addition of mass and energy to the magnetosphere makes it extremely unstable, as happens in any physical system. The magnetosphere then looks for ways to *get rid of* this extra energy which has been added to the system. The specific ways of energy discharge of the magnetosphere are the two explosive phenomena, we call *magnetic storms* and *magnetospheric substorms*.

The name *magnetic storms* was given to this phenomenon for the first time (1806) by the polymath Prussian Alexander von Humboldt (1769–1859), geographer, explorer, geologist, naturalist, etc., i.e., a true scientist of his time. Humboldt after several travels to Siberia, persuaded the Czar of Russia to organize a network of magnetometers in Russia and combine these measurements with others in the British Empire, from Toronto to Tasmania. This network showed that magnetic storms were of global in nature, with a sharp drop in the magnetic field to 50–300 nanotesla within 6–24 h, followed by slow return to the initial state of 30–60,000 nanotesla lasting 1–4 days. The name of *magnetic storms* was introduced by Humboldt in his five-volume encyclopedia *Cosmos* (also introducing for the first time the use of this word to the scientific community), through which he communicated to the wider scientific community the sunspot observations by Heinrich Schwabe. The Dst index ("D" for Disturbance = disturbance, "st" for storm = storm) gives an average value of geomagnetic disturbances measured by stations on the equator. It is also a measure of the magnitude of the ring current flowing in the equator and the energy of the magnetic storm (Fig. 1.14). The Dst index is a measure of the decrease in the horizontal component of the Earth's magnetic field near the magnetic equator due to increases in the magnetospheric ring current. Values less than −50 nanotesla (nT) indicate high geomagnetic activity. Estimates for the peak Dst index of the Carrington event superstorm of 1859 are between −800 and −1750 nT. The May

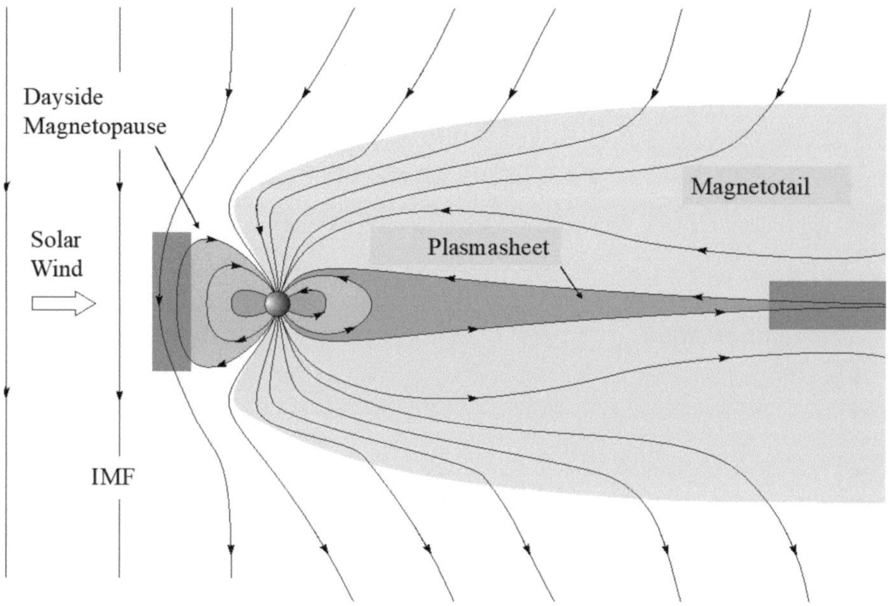

Fig. 1.14 Schematic representation of magnetic reconnection in the Earth's magnetosphere, where on the sunward side of the Earth, the solar magnetic field carried by the solar wind (Interplanetary Magnetic Field, IMF) has an opposite direction to the terrestrial magnetic field (dark gray rectangle). Similar magnetic reconnection conditions also exist in the magnetotail, where we also have magnetic reconnection (dark gray rectangle) (Image adapted from NASA/GSFC, MMS)

2024 solar storms reached a negative peak Dst index of −412 nT at 03:00 UTC on 11 May, 2024 (Fig. 1.15).

The ***second*** geomagnetic disturbance, *magnetospheric substorms*, which occur a few times a day on the night side of the magnetosphere, last 1–3 h and are associated with the appearance of the *aurora*. In contrast, *magnetic storms* occur less frequently, perhaps once per month, they last for days and are not always easy to predict, but they affect the magnetosphere as a whole. *Magnetospheric substorms* do not seem to require for their initiation a strong *external* cause, such as do magnetic storms, which are caused by solar energetic particles, or, fast gusts of the *solar wind*, or, strong *coronal mass ejections* (CMEs), or, a co-rotating interaction region (CIR). These bursts of *solar wind* and CMEs/CIR disturb the outer part of the Earth's magnetic field, which undergoes a complex oscillation, generating associated electric currents in the near-Earth space environment, which in turn generate additional magnetic field variations: all these constitute a *magnetic storm*. The *magnetospheric substorms* on the other hand, appear to be caused by instabilities in the magnetotail of the Earth's magnetosphere, which occur when the solar magnetic field reconnects with the oppositely oriented Earth's magnetosphere. During *magnetic storms* the ring current, which flows in the space around the Earth, increases because of the magnetic trapping of extra charged particles, which enter the Earth's magnetosphere from the

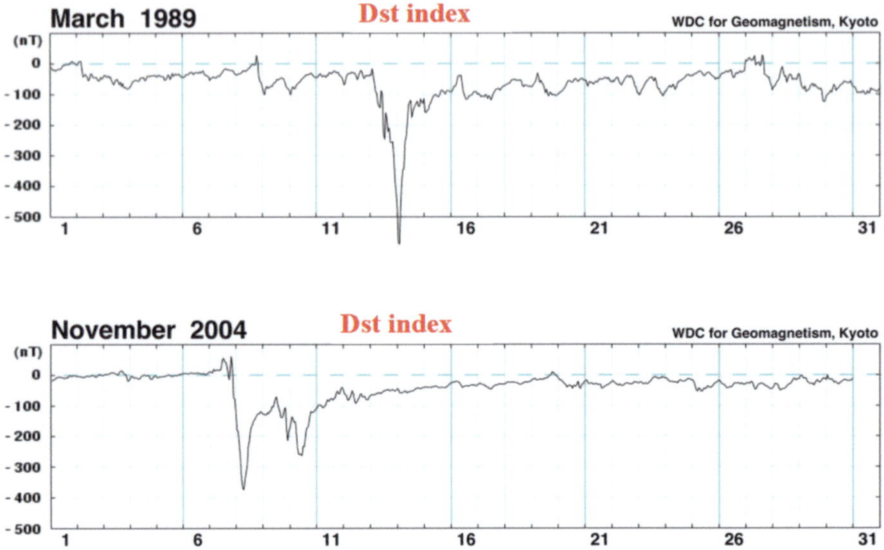

Fig. 1.15 The *Dst index* gives an average value of the geomagnetic disturbances from stations distributed over the equator. It also gives us an idea of the intensity of the ring current and its value is also proportional to the energy of the *magnetic storm*. Two strong storms have been selected from the database of the World Data Center for Geomagnetism, Kyoto: the famous storm of March 13, 1989 where Dst ~ –600 Nanotesla and the strong storm of November 7, 2004, where Dst ~ –400 Nanotesla (1,000 nanotesla corresponds to 0.01 Gauss) (Figures provided courtesy of Dr Shun Imajo at the World Data Center for Geomagnetism, Kyoto, Japan)

solar wind through magnetic reconnection. The central feature of a magnetic storm is its *global* nature, i.e., the reduction of the horizontal component of the magnetic field everywhere on Earth. The common feature of both types of storms is the consumption of energy input through the process of magnetic reconnection, energy which is stored in the magnetotail. Also noteworthy is the power of these magnetic storms, which may exceed 1,000 gigawatts. [For comparison, a typical power plant has a power of 1 gigawatt, [the installed electrical power in Greece is a few tens of gigawatts].

The ***third*** geomagnetic disturbance is the *aurora borealis* and *aurora australis* (Fig. 1.17). While usually the *aurora* is limited around the two polar regions of the Earth, during geomagnetic storm is diameter expands significantly toward the equatorial latitudes. As early as the eighteenth century, there was observed a correlation of *aurora* appearances toward equatorial latitudes and geomagnetic disturbances. As deduced from the observed annual frequency of *aurora* appearance in equatorial latitudes, it follows that the *aurora* is more intense during the period of the solar maximum than during the period of solar minimum. The observed correlation between the intensity/extent of *aurora* borealis and geomagnetic disturbances and on the other hand, of the number of *sunspots*, suggests a strong relationship between solar and geomagnetic activity, i.e., there is evidence of some form of solar-terrestrial coupling. When high-energy protons are ejected from sites of solar *flares*, they reach

Earth quickly, within minutes, and some spiral along Earth's magnetic field lines, slamming into Earth's upper atmosphere, at altitudes around 100 km, at speeds of up to 70 million km/h: Earth's magnetic field redirects the particles toward the north and south poles.

The *green* [O I] light at 5577 Å represents the observed emission transitions which dominate the *aurora* at relatively low altitudes around 100 to 300 km, and is emitted from *forbidden* transitions, by oxygen atoms relaxing from an ^1S state to a ^1D state. This is a relatively slow process, which on average takes 0.74 s (Fig. 1.16). This rather slow transition won't usually happen at the kind of air pressure we see at ground level, because the excited atom will have lost energy by bumping into another atom before it has a chance to send out a green photon. But in the atmosphere's upper heights, where there is lower air pressure and therefore fewer oxygen molecules, they have more time before bumping into one another and therefore have a chance to release a photon. Nevertheless, even after emitting this green photon, the oxygen atom finds itself in yet another excited state with no allowed relaxation. It stays there for 133.8 s seconds (Fig. 1.16). The only escape is via another forbidden transition, from the ^1D to the ^3P state – which emits a rosy red photon (6300/6364 Å). Those red *aurora*s are comparatively less frequent and are usually associated with a rather intense solar activity. They occur when solar energetic particles react with oxygen at higher altitudes, around 300 to 400 km, so they are seen in lower geographical altitudes. At this height, oxygen is less concentrated and is excited at a longer wavelength, making reds visible. Red *aurora*s are often seen at the higher edges of the display (Fig. 1.17).

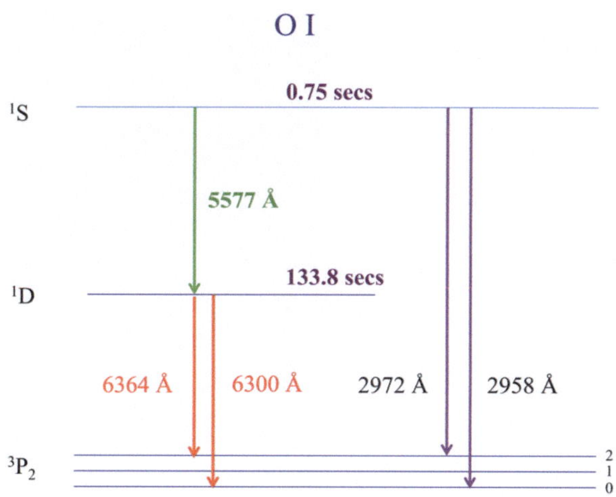

Fig. 1.16 Allowed and forbidded transitions of [O I]. The green and red lines represent the observed [O I] *forbidden* emission transitions in the water-poor comet C/2016 R2 [Raghuram et al. (2020), Fig. 1]. The values in parentheses represent the radiative lifetime of the excited state, and are taken from Wiese and Fuhr (2009) and Wiese et al. (1996) (The image is adapted from the paper Raghuram et al. (2020), with the kind permission of the authors)

But what are the specific practical consequences when the geomagnetic field is disturbed? First of all, the electromagnetic components of various modern technological systems, are affected. Already in the middle of the nineteenth century, anomalies had been observed in the operation of the telegraph, which had begun to be used in Europe and the USA in the first half of the nineteenth century. During the Carrington incident on the 1st of September 1859, the pioneer US telegraph entrepreneur Cyrus Field, who had just set up his transatlantic cable from Newfoundland to Ireland, wrote a message on the telegraph that he had never seen before in his fifteen years of experience in this object the impressive consequences of an *aurora* similar to that event; the *aurora* borealis was visible even in tropical latitudes; after disconnecting the batteries of the telegraph lines, everyone was talking about the fact that the telegraph was not led by the electrical batteries. The electrical hurricane on the night of the observations had reached its maximum. The conclusions about that storm were written in newspapers and quickly spread all over the globe being discussed ecstatically at length.

Today, problems are created in principle in the electric power supply lines, as well as in the telecommunications networks, or, in various satellite systems. Problems in the lines of electric power transport are due to induction currents flowing into the ground, after sudden changes in the geomagnetic field that occur during magnetic storms and substorms, according to Faraday's law of induction. In other words, *Space Weather* disturbances lead to abrupt changes in the Earth's magnetic field

Fig. 1.17 (*top*) A green, nearly round *aurora australis* curls over the southern hemisphere, in a photo taken with NASA's IMAGE satellite from its orbit more than 1,000 km over Earth in September 11th, 2005, four days after a record-setting solar flare sent plasma flying toward the Earth (*Credit* NASA/GSFC). (*bottom*) A red *aurora borealis* seen from the city of Leptokaria in the Greek Pieria district over the sky of the northern city of Greece, Thessaloniki in Macedonia, in November 6th, 2023 (*Credit* Greek Reporter)

to the extent that strong induction currents develop in all pipelines that are several kilometers long, such as cables of electricity transport and oil and gas pipelines on Earth. This phenomenon is of course more intense near the Earth's magnetic poles and results, due to the large change in currents, in the melting of some high-voltage transformers, up to even total interruption of electricity distribution in the affected areas.

As far as telecommunication networks and various military systems are concerned, the problems that arise are due to malfunctions, or losses, of the relevant supporting satellites, mainly due to the resulting destruction of the control boards from the energetic solar electrons and ions. Additionally, these high-energy solar protons can destroy spacecraft microchips, or charge and short circuit their other sensitive instruments. At the same time, astronauts are particularly at risk when they might be outside the spacecraft, as are passengers on commercial air flights near the polar regions. There, the crews and, secondarily, the passengers are subjected to extensive hazardous radiation exposure.

Expensive Solar Blackouts…

During extreme events of strong magnetic storms and substorms, induced currents can damage, by overheating, large power grid transformers and cause blackouts. The most famous incident occurred on March 13, 1989, in Montreal, Canada, where 6 million residents were left without electricity for 9 h due to such a powerful geomagnetic storm. That blackout costed more than $500 million. It has been estimated that if this same event were to occur in the densely populated Northeastern US, it would cost at least $5 billion. Similar smaller events of strong currents on telegraph lines on the same date, as intense *aurora* borealis, happened on March 19, 1847, on October 27 and 28, 1848, etc.

Geomagnetic storms also cause the occurrence of corresponding ionospheric storms, which make radio broadcasts difficult for communications of ships, submarines, airplanes, military installations, radar, intercontinental radio broadcasts, GPS, etc. The metal conductors are corroded by inductive currents despite the use of rectifiers of voltages applied to cancel the induced currents due to *Space Weather*. In particular, solar *flares* emit powerful electromagnetic radiation and strong X- and γ-rays. The X-rays further ionize the ionosphere and change the propagation conditions of electromagnetic radiation in it. That is why long-distance telecommunication conditions are affected and often drastically changed, because the composition and height of the Earth's ionosphere changes, where short electromagnetic waves are reflected, which in normal conditions are transmitted over the horizon. During intense changes in the ionosphere, which are caused by the intense emission of ultraviolet radiation and X-rays emitted during solar *flares*, long-distance telecommunications and short-wave radio, which rely on ionospheric reflection propagation, are disrupted. Propagation conditions at various other frequencies, such as HF, VHF,

VLF, LF, and MF, are similarly altered. For a review on the solar perspective of Space Weather, see Schwenn (2006a, b).

When the Sun *Roars*, Artificial Satellites *Tremble* …

As of early 2026, there are approximately 14,000 to 15,000 active satellites in Earth's orbit, largely driven by the growth of large commercial constellations. The majority of these are located in Low Earth Orbits. SpaceX's Starlink alone account for over 10,000 active satellites. At any time energetic ions and electrons in interplanetary space can create vital problems in their electronic systems, since solar energetic particles (protons and heavier ions of relatively high energy) affect their micro-electronics. The high flux of high-energy electrons and protons that come to Earth as a result of the solar CMEs, destroy the electronics of artificial satellites, possibly from discharges created when the spacecraft is overcharged. For example, a common problem is the strong charging of satellites, usually negative, of the order of 500 Volts. If this charge is homogeneous, it simply acts on counting charged particles that potentially perform various scientific instruments of the satellite (e.g., repelling electrons), while if this charge is different in different parts of the satellite, the induced electric currents can have destructive consequences in satellite instruments. In addition, charged particles with high energy destroy the photocells of the satellites, or, even, they can penetrate electrical circuits and cause damage or give false signals.

As previously mentioned, on March 13, 1989, a powerful magnetic storm hit Earth during the maximum of the solar activity (cycle 22). This magnetic storm on Earth was preceded by a powerful X15 class solar flare on March 6, 1989, which was followed by a large CME from the same solar active region on March 9, 1989. The high-energy charged particles of the CME arrived to Earth after 3.5 days. The storm coincided with a spectacular aurora borealis visible as far away as Texas and Florida in the US. During this *Space Weather* episode contact was lost with many satellites in polar orbit. For example, communication was lost with operational satellites of the American Geostationary Operational Environmental Satellite System (GOES), which is managed by the National Oceanic and Atmospheric Administration (NOAA) to provide valuable meteorological information and weather forecast data, track of cyclones, as well as for meteorological research.

Space Weather makes it difficult to connect with spaceships and has a temporary but sometimes even permanent loss of them. There have been cases where artificial Earth satellites have been destroyed by significant *Space Weather* disturbances. What happens is that geomagnetic storms as well as increased emission of solar ultraviolet radiation heat the Earth's atmosphere, causing it to expand and swell, so as the atmosphere rises, the satellites' orbits change. Low earth orbit (LEO), causing friction in their motion. These frictions sometimes lead to their gradual decline. NASA estimates that the International Space Station *descends* more than 300 m during periods of intense solar activity. The relevant US agency NO-RAD (North American Aerospace Defense Command) had to re-locate hundreds of space objects and record their new position after the great storm of March 13, 1989. Thus, the "useful life" of many low earth orbit (LEO) satellites is possible to be reduced even to half of the originally foreseen.

Space Weather Causes Several Troubles on Earth…

A magnetic storm on May 20, 1998 placed out of service temporarily the telecommunications satellite Galaxy IV, which is in a geostationary orbit 99°W and which lost contact with 80% of the PanAmSat company's customers for direct calls (pagers) with beepers. What had happened was failure of the central processor of the satellite's memory due to a hit by a solar energetic particle. The backup central processor had also experienced a similar problem and thus PanAmSat was forced to declare the satellite permanently lost on 20 May 1998. Here are some of the immediate consequences: many fuel selling stations with an automatic payment system (i.e., gas stations with automatic pumps) which were subscribers to the company could not verify the balances of their customers and thus they were unable to proceed with the sale of petrol. Major news agencies such as Reuters were affected, disrupting their new programs. The CBS television network changed providers to continue its program. All 600 stations of the American National Public Radio (NPR) were out of service, along with many other radio stations. Eventually the company moved another satellite into Galaxy IV's orbital position to temporarily cover its loss, before launching its new Galaxy VI satellite into orbit.

The premature fall of the Skylab space laboratory in July 1979, after staying in orbit for 24 weeks in 1973 and 1974, is due to higher than expected solar activity during the maximum of the solar cycle 21 (1974–1985). During the great storm on March 13–14, 1989, again near the maximum of the next solar cycle 22 (1986–1997), the NASA SMM (Solar Maximum Mission) spacecraft "lost" a large amount of altitude due to atmospheric friction – it was reported that it was like "hitting a wall".

Space Weather also has a transient effect on the operation of several geographical systems, such as the geographic positioning system (GPS) that we use to find our position on Earth and which work using many spacecrafts, the European Global Navigation Satellite System (GNSS) which the EU is trying to complete by means of the European GNSS company Galileo, costinh around €10 billion. At this time (05/2024), of the 30 satellites that Galileo will eventually have at its disposal, 26 have already been put into orbit. The rest 4 is under construction. In full operation, this system will have an accuracy of 1 m for the general public and 1 cm, for special encrypted applications.

The space conditions throughout the heliosphere, and especially in the interplanetary space between the Sun and the Earth, temporarily affect the intensity of *cosmic rays* on Earth. *Coronal mass ejections* create the phenomena Forbush on *cosmic rays*, wherein dips in cosmic ray intensity are the heliospheric phenomenon that concerns changes in the density and anisotropy of galactic *cosmic rays*, caused by large-scale propagating perturbations of the *solar wind*.

Disturbed *Space Weather* creates *aurora*e in regions around the magnetic poles of other planets with bipolar magnetic fields and therefore similar phenomena occur on other planets as well.

It is believed that solar activity affects the weather and climate on Earth, through the ultraviolet component of the electromagnetic spectrum. This is probably a result of the change in the so-called solar constant, i.e., the power of the Sun, by a percentage of 0.2%, which although it seems small, in the end it is not negligible at all. However, the mechanisms of the effect of the change of solar radiation on Earth's weather and climate are not yet fully understood. Finally, it is speculated, although also not convincingly proven, that the geomagnetic storms affect human health too.

In conclusion, the effects of magnetic storms and substorms are the creation of very strong currents that flow between the Earth and of interplanetary space. Those currents, such as the Birkeland currents, are associated with the occurrence of the *aurora* and are directed to the polar funnel from the magnetotail of the Earth's magnetosphere and vice versa. During magnetic storms, the currents spread equatorward, so the *aurora* borealis reaches the low latitudes, the more densely populated areas of our planet. Magnetic storms and substorms induce electric currents in electric power transmission lines which overload and sometimes cause a disruption of electricity transmission, thus creating a widespread electrical blackout in some large areas. For watching all these phenomena, an extensive network of satellites has been created, many of which are in geosynchronous orbit, with an apogee 6.6 radii of the Earth, observing solar phenomena, many of which are responsible for significant disturbances on the magnetosphere. A coronal mass ejection creates shock waves which reach Earth in 5–6 days and create the magnetic storms. For this reason, it is necessary to constantly monitoring the Sun, which is solely responsible for all these magnetic phenomena on Earth. Also, satellite streams at the Lagrange distance L1 at four times the Earth-Moon distance continuously monitor the Sun and can predict upcoming magnetic storms and high-energy charged particles that reach the Earth in about an hour and are dangerous for astronauts in space. The systematic understanding of *Space Weather* and the possibility of timely and reliable forecasting is an important priority of our time. This effort can be compared with the birth of operational meteorology at the beginnings of the twentieth century. The practical benefits are also comparable: successful *Space Weather* forecasts will help protect the technologies that are affected by *Space Weather*, and which year by year increase in number and complexity. Predictions will also be of vital importance for future interplanetary travel and colonization of other planets. It is therefore clear that continuous monitoring of *Space Weather* is of key importance and is a particularly critical priority in our electronic space age.

Effects of Solar Variability on Earth's Climate

There is strong observational evidence that solar activity does affect Earth's weather and climate, mainly through its ultraviolet component of the solar light. Recent measurements have shown that the *total* solar radiation varies by up to 0.2% during the solar cycle, with a decrease at the minimum and an increase at the maximum. Of particular importance, however, is the solar radiation that reaches Earth in the far ultraviolet part of the spectrum, which has an increased intensity during the years around solar maximum. In a narrow interval of wavelengths in the far ultraviolet, the variation of solar radiation with the solar cycle is not only limited to the overall general small variation of 0.2%, but varies by a factor of 10 or more. This change is expected to affect the chemical and thermal structure of the upper atmosphere. Although the exact mechanisms of the effect of solar irradiance on Earth's weather and climate have not yet been fully understood, the subject is too critical for Earth to avoid discussing. Thus, in the following we shall list the observational data that support a connection of the number of *sunspots* and, by extension, solar activity with the Earth's climate.

An interesting *coincidence* is that the period of the *Maunder minimum* coincided with a particularly cold period in Western Europe and North America. Specifically, the period from the middle of the sixteenth century up to the middle of the nineteenth century, it has been characterized as the so-called *little ice age*, although it is not a period with real glacial conditions. During this period there were observed noticeably lower temperatures than those of the present era, with exceptionally extreme weather phenomena: the Thames River in London froze, the lake in the city of Ioannina (Greece), the canals in the Netherlands, and also other areas which today are not covered by ice today (Figs. 1.18 and 1.19).

This cold period was preceded by a period of unusually high temperatures, between ~950 and ~1250 in North Atlantic. The relative proportions of the isotopes of carbon-14 and beryllium-10 are very decreased in the period 1100–1250 AD, possibly due to an unusually high solar activity. This period is associated with a warm period in Medieval Europe and also with a period of drought in North America, which probably brought about the extinction of the culture of the Indian tribe Anasazi in the second half of the thirteenth century.

These data led us to consider a possible association of Earth's climate with the number of *sunspots* and the level of solar activity, as documented by Eddy (1976).

Therefore, the observations show that the level of solar activity does affect the Earth's climate, although the exact mechanism of the effect of this interaction has not yet been clarified and is a topical and interesting issue of research topic.

It should be noted that in the present time period, the Sun is likely to enter a new *Maunder-type minimum*. The new solar cycle 25 (2020–) is projected to be one of the weakest in a period of more than 50 years. At the same time, there are indications that the magnetic fields of *sunspots* have a decreasing tendency.

The number of *sunspots* has been observed to be related to the intensity of solar radiation. In 1979, accurate satellite measurements became available of the absolute

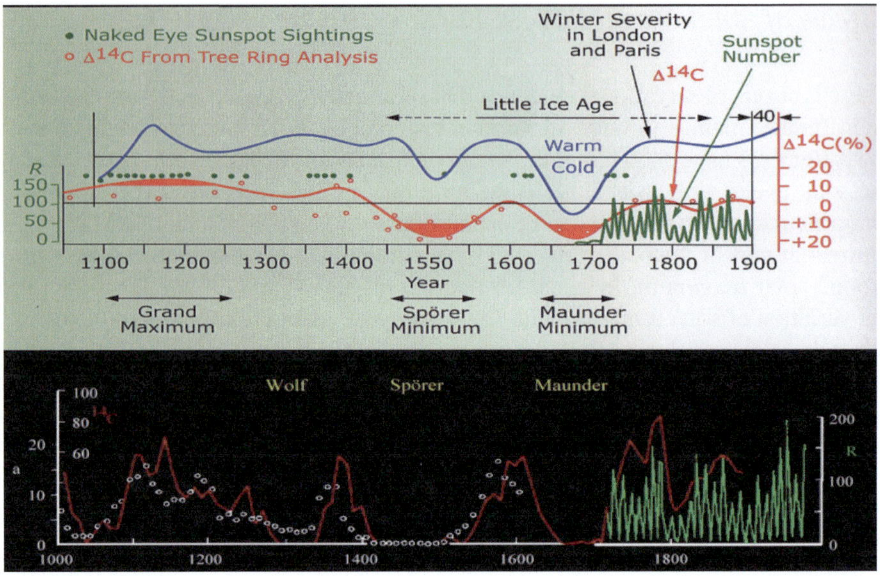

Fig. 1.18 (*top*) Winter temperatures in NW Europe are correlated with solar activity. Note the coincidence of the *Little Ice Age* with the *Maunder Minimum* in sunspots. (*bottom*) Sunspot number (green line) and the concentration of Carbon-14 in tree rings (red) as proxies of solar activity versus glacier fluctuations (white circles) (*Credits* The top graph is from Carpenter et al. (2004) and is reproduced here with the kind permission of Ken Carpenter. The second graph shows the solar cycles (green line), the Carbon-14 proxy from tree rings (red) and the length of glaciers (white circles) and is taken from the site https://klimaatgek.nl/wordpress/de-zon/] of Rob de Vos)

flux of solar radiation. Because *sunspots* are darker than the average solar *photosphere*, more *sunspots* would be expected to be accompanied by less solar radiation and a reduced solar constant. However, exactly the opposite happens: overall, when the Sun has more and larger *sunspots*, the overall solar brightness increases. The change caused by the sunspot cycle in the solar constant is small, of the order of 0.1% from maximum to minimum (or, 1.3 Watts per square meter from the value of 1,366 Watts per square meter, the average solar constant). Additionally, longer term changes in this basic cycle are known: e.g., from 1900 to the 1960s the number of spots in the maximum of solar activity was increasing, while from 1990 until today there is a continuous decreasing trend.

What we have definitely found is that when solar activity declines over the long term, glacier areas go down toward mid latitudes and the climate cools. During periods of high long-term solar activity, the glaciers retreat toward the terrestrial poles and the climate is warming. In conclusion, the variable long-term solar activity affects the Earth’s climate.

Fig. 1.19 Winter landscape with ice scating and bird trap, a painting attributed to the Flemish painter Pieter Bruegel the Elder (1526–1569). The painting was in the possession of Archduke Leopold of Austria and is now located at the Royal Museum of Fine Arts of Belgium, Brussels. It shows a village scene where people are skating on a frozen river (Courtesy of Wikimedia Commons)

The Disappearance of the *Anasazi Indians* Due to a Severe Drought During the Warm Period of the Medieval *Grand Maximum* of Solar Activity

According to historians, the *Anasazi Indians* were the builders of the charming *cliff pueblos*, which are scattered in the area where the four American states of Arizona, New Mexico, Colorado and Utah join at right angles. The *Anasazi* appeared in the area of these four states at the time of the birth of Christ and for a millennium after that their civilization flourished and left us their original houses on the steep sides of the canyons that we see today. These houses with stone walls, located below of overhanging rocks, to protect them from the changing weather conditions, were made of sandstone and mud and are the world's longest-lived residential structures. These cave dwellings were accessible only by climbing and thus constituted an interesting defense system at the same time, against attacks by other neighboring tribes. In the US *Mesa Verde National Park*, Colorado I had the opportunity to visit many of the approximately 150 rooms which occupy a deep recess below the rim of a canyon.

The *Anasazi Indians* had a pioneer culture, building roads in their villages and irrigation systems to water their crops. At the same time, they were diligent and systematic in the construction of their dwellings. This can be seen from the fact that the woods they used in many of these dwellings, did not come from

the immediate area, but they were transported from forests that were more than eighty kilometers away. It is puzzling how the *Anasazi* transported such large tree trunks from so far away.
Why the *Anasazi Indians* abruptly abandoned these impressive homes around 1275–1300 A.D. is still a mystery to researchers. The popular speculation is that a period of increased solar activity during the medieval *grand maximum* (Fig. 1.18) led to a warm period of prolonged drought, which shortened the grain growing season around the middle of the thirteenth century. This caused a shortage of food when their population had already grown quite large. The resulting difficulties may have led to tensions and conflicts. There is also a speculation regarding the increased presence of other nomadic tribes in the area which pushed them away. Finally, the people of the *Pueblos* of the area *Mesa Verde* decided to migrate south, where the rains were more abundant. Today, the *Hopi Indians* are considered their descendants.

1906: The Solar Radiation and *Convection Zones*

During its production in the solar core, solar energy consists of γ-ray photons, with energies in the range of MeV. These photons then follow a tortuous path through the solar interior. Their final destination is, after crossing the hundreds of thousands of kilometers of the dense plasma in the solar interior, to reach the solar *photosphere*. In this long and episodic journey of photons through the radiation zone, which extends for hundreds of thousands of km, the plasma electrons absorb the passing photons and re-emit them at lower energies. Thus, they eventually reach the solar *photosphere* as many more photons of visible light, i.e., with an energy much lower than the initial one, in the range of electron-volts (eV). From the *photosphere*, they travel freely in space, some reaching Earth. In the path of photons through the solar radiation zone, the plasma electrons do not only absorb the passing photons, but re-emit them in random directions: this random course of the emitted photons is likened to the walk of a drunk! Thus, as these *drunk* lower-energy photons are traveling slowly through the radiation zone, gradually finding their way toward their exodus from the solar interior, i.e., toward the solar *photosphere*, about 200,000 km above the radiation zone, the temperature of plasma gradually decreases significantly so that the electrons are captured by the bare nuclei, forming atoms. At the same time the gas is thermally unstable, as explained below.

First, it is worth to bear in mind that the temperature of the solar interior decreases from about 15 million degrees in the core to a few thousand degrees in the *photosphere*. As we approach the *photosphere*, this temperature decreases at a faster rate. Specifically, while at a distance of 500,000 km from the solar center the temperature has decreased from the 15 million in the core by only a factor of 10 to about 1.5 million degrees, in the last 200,000 km of the 700,000 km radius solar sphere it undergoes a drastic reduction to about 5,000°. That is, in the last 30% of solar radius we have a particularly dramatic decrease of the temperature. This has an important consequence to the thermal stability of the gas of the solar interior, which we explain in the following.

Let us consider a small gas cell inside which the fluid is in equilibrium with its environment, i.e., it has the same temperature, the same pressure and the same density as the external environment. Suppose that this bubble is initially at some position A, for example at a distance of 500,000 km from the center. Let us next assume that this small bubble of fluid moves radially outward from location A to a nearby new location B. For the simplicity of the discussion, let's assume that the new radius is 500,000 + 1 km from the core. *Suppose that in this new position B the pressure inside the bubble comes into balance with the pressure of the external environment.* However, this external temperature decreases independently of the internal temperature inside the bubble, at some given rate determined by the initial and boundary conditions that prevail inside the Sun, the value of the opacity of the gas, the value of the total brightness of the Sun, etc. On the other hand, suppose that as the cell moves from position A to position B, it does not exchange heat with the environment, i.e., it follows what we call an *adiabatic* change, which means that the walls of the cell are impermeable to heat. Then, as the cell rises from position A to position B within the Sun's gravity field, it expands and increases its volume, so its density decreases, and by extension its internal temperature also decreases, i.e., it becomes colder than it was initially. This happens because an decrease in density in the adiabatic change implies a corresponding decrease in temperature, since in the adiabatic change the density is proportional to a power of the temperature. There are then two possibilities for the evolution of the displacement of the cell, from position A to position B, depending on whether the internal density and temperature of the cell at B are lower or higher than the external density and temperature at B.

If the internal temperature of the bubble is lower than the temperature of the external environment, the bubble finds itself in the new location B to be denser than the environment, within the solar gravity field. This is because the product of density and temperature is the same at position B inside and outside the cell, due to the assumption we made that the pressure (which is the product of density and temperature) at position B is the same inside and outside the alveolus. For example, if the internal temperature of the bubble is 1/3 of the external temperature, then, because temperature–pressure product, which is the same inside and outside the bubble, the inner density will be three times the outer density! Thus, lower temperature inside the cell at B than outside means higher density inside the cell at B than outside it. Therefore, if the cell is released at B, it will be heavier than the environment and thus returns to A, since gravity will act as a restoring force and the fluid will be driven to oscillate around position A. That is, the plasma is stable to small perturbations. This occurs when the outer temperature decreases slowly with distance from the core and determines the extent of the radiation zone, where the temperature decreases slowly with the distance from the core (sub-adiabatic temperature decrease).

On the other hand, let's see what happens when the internal temperature of the cell at B is *higher* than the ambient temperature. This occurs when the ambient temperature decreases very rapidly, relative to the adiabatic decrease of the temperature inside the cell. As discussed above, this means that the cell will have a lower density at B than the environment, i.e., it will find itself lighter than the environment. Therefore, if the bubble is released at B, because of buoyancy within the Sun's gravitational field, it will continue to move from B toward the *photosphere*, causing convection currents, just as a bubble of gas from the bottom of a swimming pool moves by buoyancy toward the surface. That is, the plasma is unstable to small perturbations. This happens when the external temperature decreases rapidly with distance from the core and determines the extent of the so-called *convection zone*, where the temperature decreases rapidly with distance from the core (superadiabatic temperature decrease).

That is, in the region of the *convection zone*, atoms absorb heat energy coming from the core and the upper boundary of the radiation zone and then they transfer it to the outer *photosphere*, just as it happens when we heat water over fire: warm upward currents transport to the upper layers the provided heat, while colder downward currents fill their space, and these again are heated, and so on. This criterion was discovered by Karl Schwarzschild (1906).

When the photons finally reach the *photosphere* they are free to escape into space. So while it takes hundreds of thousands of years to travel the 700 thousand kilometers of the solar sphere, in just 8 min they cross the 150 million kilometers of the distance from the Sun to the Earth. In other words, while it will take hundreds of thousands of years on Earth to realize a possible *strike* or even exhaustion of the fuel in the thermonuclear factory of the solar core, an explosion in our solar *photosphere* is noticeable within just a few minutes. In other stars, of different mass from the Sun, the radiation and *convection zone*s have different order and depth, depending on their mass (Fig. 1.20).

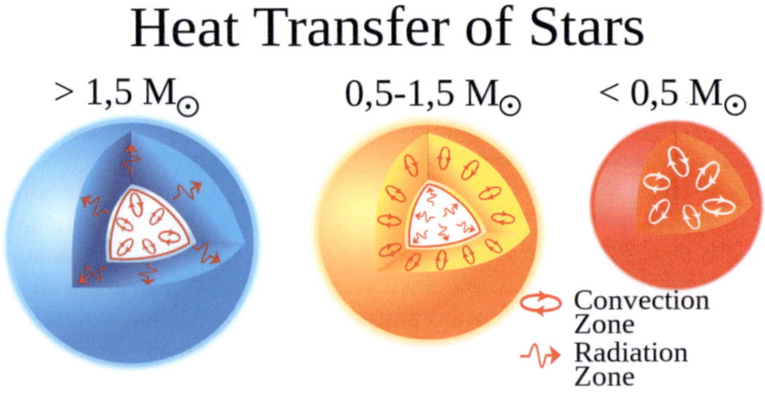

Fig. 1.20 The different transport mechanisms in high-mass, intermediate-mass and low-mass stars (From Wikimedia Commons)

1908: First Discovery of Strong Magnetic Fields in the Sun and of Their Dominant Role in *Sunspots* and the Solar Atmosphere

The Sun is a glowing ball of gas, mainly hydrogen. When on a hot summer noon the Sun is high in the sky and we can't even catch a glimpse of the dazzling solar disk – we should never look directly at the Sun, because our eyes are in danger from its intense ultraviolet radiation! – perhaps it doesn't cross our mind that on its *photosphere*, take place incredible explosions and terrifying phenomena that affect Earth, Space and the entire planetary system. As it is said, the Sun surrounds Earth with *love* and *hate at the same time.* With love, because every form of life on Earth depends on the life-giving solar radiation. And with hatred, because from its atmosphere emanate titanic bursts of energy, high-energy charged particles, and huge quantities of superhot mass, all pointing toward Earth, the planets and the interplanetary Space. But what we don't realize when we view the Sun, is that behind the continuous unrest of the solar *photosphere* and surrounding atmosphere is hidden the solar magnetic field. But really, how does the solar magnetic field arise?

Because of the high solar temperature, there are no neutral atoms, but the hydrogen gas that dominates the Sun has decayed into positively charged protons and negatively charged electrons. Thus, at these high temperatures, we have the state of a *plasma,* i.e., protons and electrons, a fluid with high electrical conductivity similar to that of copper wires in an electrical conductor. Consequently, the flow of electric charges, protons and electrons creates electric currents. However, because electric currents create magnetic fields, magnetic fields are everywhere in the interior and the atmosphere of the Sun. These magnetic fields interact with the electric currents and thus magnetic forces act on the solar plasma. Hence, *the prime mover* of solar activity and the creator of the *solar fireworks* which directly affect our Earth is the solar magnetic field. The strongest forms of the magnetic field appear in Sunspots. We shall discuss later on the wide range of explosive solar phenomena which are due to the solar magnetic field.

With the development of spectroscopy, the study of the Sun advanced significantly in areas that could explain the creation process of spots. In 1814 Joseph Fraunhofer discovered absorption lines of the chemical element of the Sun and a little later Gustav Kirchhof realized that these also revealed the chemical composition of the solar atmosphere.

The first solar physicist to spot it existence of magnetic fields in the Sun is the American astronomer George Ellery Hale (1868–1938, Fig. 1.21). Hale is particularly well known for the discovery of magnetic fields at *sunspots*, but also for the opposite magnetic polarity of *sunspots* in the two solar hemispheres (Hale's law). George Hale upgraded the solar spectrograph, the instrument that analyzes the solar spectrum lines and managed to take pictures of the solar surface in a single spectral line. Thus, in 1908 taking advantage of the *Zeeman effect*, i.e., the breaking of the spectral lines of one gas into components within a magnetic field, he was able to measure the intensity of magnetic fields in the sunspot region.

Fig. 1.21 George Ellery Hale (1868–1938)

Specifically, the presence of magnetic fields in some area where the plasma radiates, makes it possible to calculate the magnetic fields from the splitting of the spectral lines. Thus, the magnetic field within of *sunspots* can be calculated in an accurate and direct way, because in the presence of the magnetic field the energy levels of atoms, ions and molecules are separated into more levels. This forces the emission or absorption spectral lines to be separated into more than one line, with the separation being proportional to the size of magnetic field. This is called the *Zeeman effect*, and the corresponding *breaking* of the spectral lines is called Zeeman splitting. Therefore, we can infer the presence of magnetic fields if we observe in the radiation spectrum the Zeeman splitting. Additionally, we can measure the strength of the magnetic field by calculating the magnitude of the line splitting due to this Zeeman effect. In this way, very strong magnetic fields have been measured within *sunspots*.

Let us see briefly next a few facts about the magnetic forces which are acting on an ionized gas, i.e., a plasma. According to physics, in every conductor wherein flows an electric current and is located within a magnetic field experiences a magnetic force per unit volume, which is proportional to the product of the surface density of the electric current and the magnetic field, the so-called Lorentz force in Physics, in honor of the Dutch physicist Hendrik Antoon Lorentz (1853–1928).[3] According to Ampere's law of electromagnetism, the total surface density of electric currents is related to the spatial variation of magnetic fields. Consequently, the magnetic force exerted on an electrical conductor can ultimately be expressed only as a function of the magnetic field it creates and its spatial changes. From this expression, it follows a

[3] Pieter Zeeman shared the 1902 Nobel Prize in Physics with Hendrik Lorentz for discovering the *Zeeman effect* in 1896, which showed that spectral lines split into multiple components under a magnetic field.

magnetic pressure in all directions of space and a magnetic tension along the magnetic field lines. If the magnetic field extends up to a boundary, then the magnetic pressure is directed outward in the surface of this boundary. Thus, the magnetic force per unit area, i.e., the magnetic pressure, behaves like the thermal pressure of a gas and expresses the force per unit area exerted on the surface of the boundary. The only difference is that the magnetic pressure is carried by the magnet-field and not from the kinetic energy of the gas molecules! On the other hand, the magnetic tension is similar to the force exerted by a circular rubber band toward its center when we stretch it, increasing its radius.

The existence of magnetic pressure is also responsible for the dark appearance of *sunspots*, the most distinct formation of the solar *photosphere*. First, within a sunspot, the Zeeman effect, reveals the existence of strong magnetic fields, 2,000–4,000 Gauss at their center, about 1,000 Gauss at their periphery, thousands of times stronger than the average solar magnetic field. This magnetic field strength is similar to what we have in modern hospital MRI imaging machines (computed tomography, CT scan). This very strong magnetic field exerts a pressure on the outer gas, the sum of which together with the pressure of the inner gas must balance the pressure of the outer gas. Therefore, the plasma inside the tube has a reduced pressure compared to the gas surrounding the tube externally. If the gas density inside and outside the tube is similar, the outside-inside pressure difference requires a lower temperature inside the spot, relative to the outside environment. Specifically, while the external temperature of the *photosphere* is about 5,780 °K (about 5,500 °C), the internal temperature of the spots is about 4,500 °K (4,200 °C). Then, because the intensity of the emitted radiation depends on the fourth power of the temperature, the intensity of the radiation coming from the spots is equal to 1/2 to 1/4 of the radiation coming from the surrounding *photosphere*.

The end result is that the spots appear dark – like a candle flame placed in front of a lighted electric bulb it looks dark.

The real generative cause however of its colder temperature of plasma within the *sunspots* is the existence of strong magnetic fields at the *convection zone*, just below the spot: these strong fields hinder heat transfer. That is, because the magnitude of the magnetic field there is so strong, the "magnetic lines" behave as rigid "wires" which cannot be "bent" by the eddy currents carrying the heat coming from the deepest layers of the solar interior. This results in spots being cooler than the surrounding *photosphere* (Fig. 1.23).

Another important property of the magnetic field is that when a tube of magnetic flux is inside a gravity field, appears the so-called *magnetic buoyancy*. To understand this, let us consider a magnetic tube inside the Sun (Figs. 1.22 and 1.23). The total pressure within the magnetic flux tube is equal to the sum of the pressure of the gas contained in the tube and the magnetic pressure. Outside the tube we only have the gas pressure which is proportional to the external density and the external temperature. The pressure of the gas inside the tube is proportional to the internal density and the internal temperature. We can assume, without loss of generality, that thermal

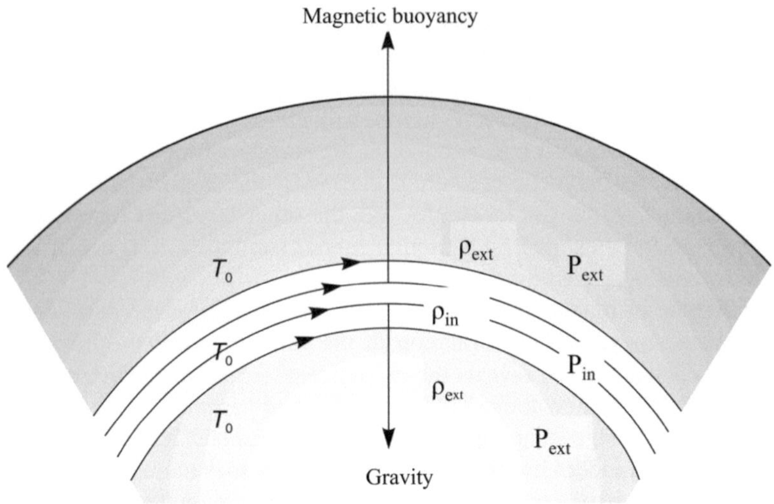

Fig. 1.22 Magnetic buoyancy in a magnetic flux tube located in the interior of the Sun, where gravity inside and outside the tube is in the radial direction, pressure and density are (P_{in} and ρ_{in}) and (P_{ext} and ρ_{ext}), respectively, while the temperature, To is homogeneous, i.e., the same inside and outside the tube. (*Source Plasma Astrophysics*, K. Tsinganos, p. 152 (2017))

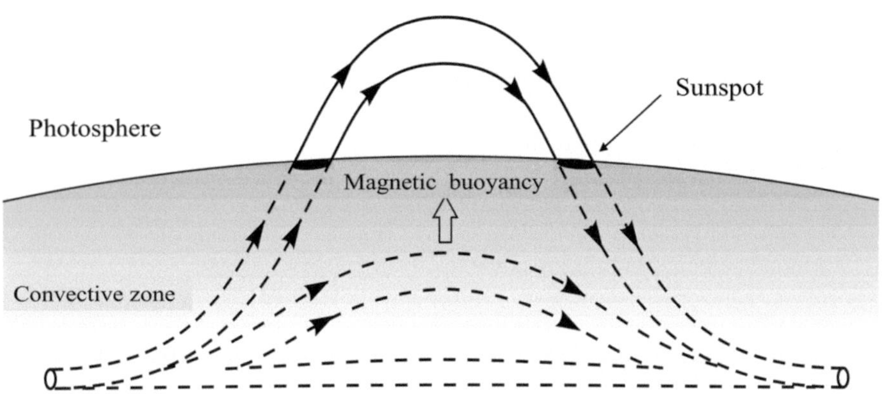

Fig. 1.23 A tube of magnetic flux created through the dynamo mechanism in the deepest part of the Sun's *convection zone*, emerges due to magnetic buoyancy in the *photosphere*, creating a bipolar magnetic region. (*Source Plasma Astrophysics,* K. Tsinganos, p. 152, (2017))

equilibrium has been restored in the system, so that the temperature inside and outside the tube is approximately the same. Then, equating the total pressure on the walls of the tube, we get that the internal density of the gas is lower than the external. Thus, the magnetic tube is lighter than the environment and is therefore *buoyant* (Fig. 1.23) rising due to this fact in the solar *photosphere*, just as a bubble of air rises to the surface of an aquarium, or a swimming pool.

1979: Aerodynamic Forces on Thin Magnetic Flux Tubes Emerging in the Solar Photosphere

Another known observational fact is that the magnetic field of the solar *photosphere* is concentrated in many, small diameter, strong and independent magnetic flux tubes, with radius of a few hundred of kilometers and intensity of the magnetic field within them approximately 1,500 Gauss. It has also been suggested that large *sunspots* are not a single homogeneous magnetic flux tube, but consist of a loose conglomeration of many such small magnetic flux tubes (Parker 1979a, b).

Therefore, the observational evidence is in the direction that, for some reason, the solar magnetic field prefers to appear in small diameter tubes, avoiding for some reason the larger diameter ones. On the other hand, the theory of the production of the magnetic field in the solar/stellar interior, has concluded that the solar dynamo that creates the solar magnetic field, operates at the lower depths of the convective zone, producing large-diameter magnetic tubes, mainly in the azimuthal direction. The Rayleigh–Taylor instability–, discussed in the following, of the upper layer of

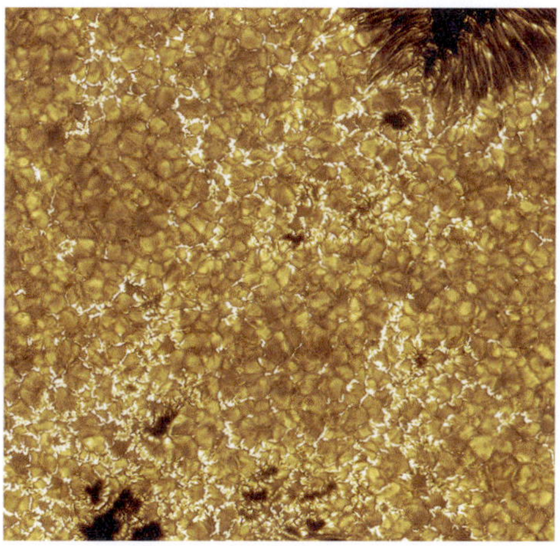

Fig. 1.24 In close-up, the solar photosphere consists of bright *granules* surrounded by dark rims within which glow *photospheric faculae* and *bright points,* like tiny *fireflies. Granules* are created by convective cells, wherein hot gas rises from their center and after cooling sinks in their periphery. The *bright points* are not related to *sunspots* that come and go with the solar magnetic cycle, but are constantly observed in the *photosphere*. They glow because of the strong magnetic field and increased magnetic pressure that they have, wherein the density of the gas within them is reduced due to the presence of the magnetic field and thus makes them transparent by opening a window toward the deeper and hotter layers of the Sun. This image shows high-resolution solar *granulation* with magnetic fields at the periphery of this granulation, forming a bright photospheric network, where we see deeper in the sun, because of the concentrated magnetic field (*Image credit* Observer and image processing: Vasco Henriques, Wavelength: 396.47 nm Ca II H wing, 23 May 2010, Institute for Solar Physics, Sweden, Swedish 1-m Solar Telescope, Spanish Observatorio del Roque de los Muchachos of the Instituto de Astrofísica de Canarias, La Palma, Spain)

such horizontal magnetic fields within a stratified atmosphere, triggers and fuels the upward escape of the magnetic field from the solar interior and its rapid emergence rate in the solar *photosphere* (with an emergence rate of about 10 magnetic flux tubes per day).

Therefore, an interesting question is what happens when a magnetic tube is within the solar *convection zone* and in particular what is the connection between the creation of the magnetic field at the bottom of the convective zone and its emergence at the solar *photosphere*. A motivation for this study is the open question of stability of an independent magnetic tube within a velocity field in the solar gravitational field.

This open question can be studied in principle in the context of hydrodynamics where we can consider the similar problem of the stability of rising bubbles in a liquid, where fragmentation occurs of those bubbles that have a relatively large diameter, greater than approx. 0.5 cm. This problem was studied as part of the author's Ph.D. thesis under the supervision of prof. Eugene Parker at the University of Chicago, and the results of this study are explained in the next paragraph. For a review of solar magnetic fields, see also Priest (2014).

1980: The Rise of Magnetic Flux Tubes from the Solar Interior Is Similar to the Rise of Air Bubbles in an Aquarium, or a Swimming Pool!

In one of the author's first research papers, we studied the analogy of the hydrodynamic instability of gas bubbles in an aquarium or a swimming pool and magnetic flux tubes in the Sun, as they move within a gravity field, because of magnetic buoyancy. In this study, we had put special emphasis on the simultaneous appearance of the Rayleigh–Taylor and *Kelvin–Helmholtz* hydrodynamic instabilities. These instabilities are capable of causing a final fragmentation of the gas bubbles, or the magnetic flux tubes, when there is unavailable a stabilizing mechanism, such as surface tension on gas bubbles, or, an azimuthal magnetic field in the case of magnetic flux tubes.

The results of the instability analysis were first applied to gas bubbles rising from the base of an aquarium (Fig. 1.25b, c). The theoretical results were found to be in excellent agreement with the experimental data: the instabilities break up the bubbles with a large initial radii, into smaller radii bubbles, which are stable, because the surface tension on the small bubbles with a radius of less than about 0.5 cm, is capable to stabilize them against these two hydrodynamic instabilities. Next, the theoretical results were applied to magnetic flux tubes in the solar *convective zone*. Then, as happens with gas bubbles, it was found that magnetic flux tubes are unstable, unless they are wound by relatively strong azimuthal magnetic fields. Magnetic flux tubes have a longitudinal magnetic field component along the axis of the tube and additionally a second azimuthal magnetic field component that wraps around them. Tubes are unstable when the azimuthal component is very strong relative to the longitudinal, in another instability, the kink instability: when we twist too much a telephone cord into a helix, then it kinks folding around its axis. At the other extreme, tubes with no azimuthal component at all are unstable to *Rayleigh–Taylor* and *Kelvin–Helmholtz– instabilities.*

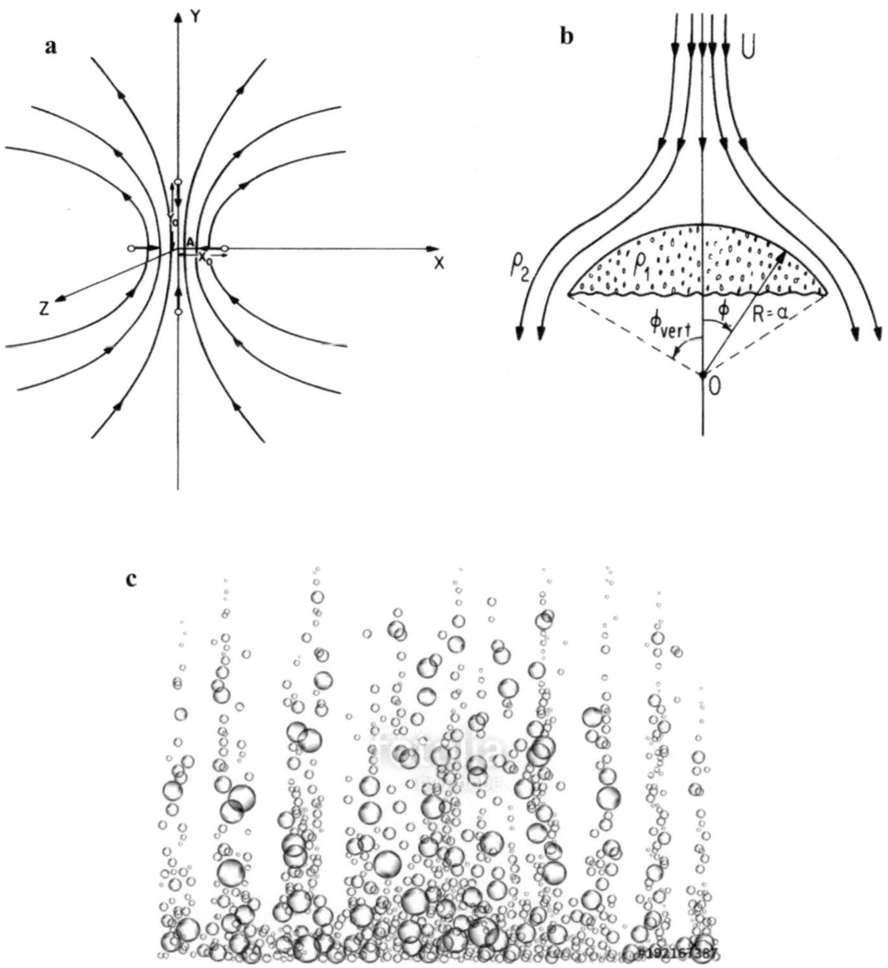

Fig. 1.25 **a**, **b** (*left*) The AZ-axis is a stable equilibrium direction for a tube extending in the Z-direction, normal to the XY-plane of a converging ideal flow, as far as the aerodynamic fluid force is concerned, Tsinganos (1979). (*right*) A gas bubble of radius R and internal density ρ_1 smaller than the external density ρ_2 of the surrounding gas moves, due to buoyancy, through the surrounding external gas with velocity U. Equivalently, we can consider that the external gas moves around the bubble with a velocity of the same magnitude U. The end result is that the bubble breaks in smaller ones, as is actually observed to happen in an aquarium, where air is channeled into its base, Fig. 1.25c, Tsinganos (1980). **c** In an aquarium, where air is channeled into its base, bubbles break in smaller ones. The rise of magnetic flux tubes from the solar interior is similar to the rise of such air bubbles in an aquarium, or a swimming pool! (Tsinganos 1980)

As we all know, olive oil is less dense than water, so it always rises and stands above the water in a glass (as in a candlestick). The opposite distribution with water on top of oil is unstable to the so-called *Rayleigh–Taylor instability*. The *Kelvin–Helmholtz– instability* on the other hand, occurs when there is a velocity shear in a single continuous fluid, or, where there is a velocity difference at the interface of two fluids. An example is the creation of small waves when a wind blows over sea water. In addition to sea water, the Kelvin–Helmholtz– instability also occurs in atmospheric clouds, in the solar atmosphere, *Saturn*'s rings and *Jupiter*'s *Red Spot*. For typical values of the twisting of solar magnetic flux tubes, we found an upper limit of a few hundred kilometers on the diameter of stable magnetic flux tubes in the *photosphere* (Tsinganos 1980). That is, observations show that the emerging magnetic flux tubes in the *photosphere* that have survived their ascent through the *convective zone*, do not have larger diameters than a few hundred kilometers, as observed! Therefore, the instability of large magnetic tubes appears to be the main reason for their absence from the solar *photosphere*. The only case of existence of large-diameter magnetic flux tubes, as in *sunspots*, is understood to be the result of the *clustering* of small-diameter magnetic tubes to form the observed large-diameter *sunspots*, as proposed by Parker (1979a, b).

1933: Stellar Evolution and the Fierce Dispute of Two Giants of Science (A. Eddington and S. Chandrasekhar), at the Gates of the Astrophysical Hell (*Black Holes*)!

One of the most prominent figures of astrophysics, mathematics, philosophy and a master of the popularization of science in the early twentieth century, a scientist to whom we often refer in this book, was Sir Arthur Eddington (Fig. 1.26). He was a pioneering researcher, particularly in the theoretical study of the internal structure of the Sun and other stars.

Fig. 1.26 Sir Arthur S. Eddington (1888–1944) (From Wikipedia)

During Eddington's time (1920) the subject of the energy source of the Sun and the other stars was an unsolved mystery. In this scientifically murky landscape, Eddington formulated the first radical model of the solar interior, wherein he proposed that it is the radiation pressure that balances solar gravity. The prevailing view held until then was that the energy available due to gravitational contraction provided the corresponding thermal pressure, which was able to balance the gravity of the Sun, or any other star. However, this hypothesis led to the conclusion that, due to conservation of angular momentum, such bodies they had to rotate at a high rotational speed, which was not observed in the Sun or other stars, such as, for example, the variable stars Cepheids. Instead of this mechanism of resistance to gravity, Eddington proposed (1920) that there were taking place *thermonuclear* reactions, which are producing the needed solar energy, and specifically the proton–proton cycle, via the mass deficit of this reaction, according to Einstein's principle of mass–energy equivalence ($E = mc^2$). With this mechanism the Sun and other stars could survive, on the one hand resisting gravity and being in a state of equilibrium during their lives in the *Main Sequence* and, on the other hand, to maintain their observed luminosities. Thus, he concluded, the core temperature of the Sun should be a few million degrees.

Long before the detailed energy source of the stars was described in 1938 by the German–American physicist Hans Bethe (Nobel Prize in Physics, 1967), Eddington had aptly described the situation at the heart of stars as follows.

The Severity of Particle Collisions in Stellar Cores

Within a cubic centimeter of the stellar core are contained more than a trillion trillion (10^{24}) atoms, about twice as many free electrons, and about 20 billion trillion (10^{18}), X-rays. These X-rays travel at the speed of light, while the electrons at 16,000 kilometers per second. Most atoms are simply protons (i.e., hydrogen nuclei) running at about 500 kilometers per second. There exist also in the stellar core heavier atoms moving with slower speeds, about 60 kilometers per second. With those speeds you can well imagine the frequency and severity of the collisions that follow [from Eddington (1927), Stars and Atoms. Oxford, Clarendon Press].

Eddington also dealt with Einstein's General Theory of Relativity (GTR), trying to spread GTR to the English-speaking world. Also, as we explain later in this book, in 1919 he led a scientific expedition to the islet of Principe, in the Gulf of Guinea, East Africa, for the observation of a total solar eclipse that was visible there on May 29, 1919. During this eclipse he made the first verification of the GTR, observing the angular deviation (curvature) of stellar light rays, when they pass very close, almost tangentially, to the solar disc. Eddington was director of the Cambridge Observatory already from 1914, professor at the University of Cambridge and in May 1914, he was elected a fellow of the Royal Society from which he was awarded the prestigious Royal Medal (1928). His book *The Internal Constitution of the Stars* (1920) was the universal reference book for astrophysicists. In conclusion, Eddington was very highly regarded internationally, one of the leading astrophysicists of his time.

One of Eddington's Ph.D. students had been the Indian–American astrophysicist Subrahmanyan Chandrasekhar (also known as Chandra, Fig. 1.27), who was awarded the Nobel Prize in Physics (1983) with William A. Fowler for "*…theoretical studies on natural processes which are important in understanding the structure and evolution of stars*". Chandrasekhar's theoretical studies of stellar evolution led to the modern models of the evolutionary stages of high-mass stars and *black holes* (Fig. 1.28). The Chandrasekhar limit of 1.44 solar masses was named after him, while NASA's Chandra X-ray space telescope was named in his honor. However, remains proverbial the conflict of the *teacher* (Eddington) with the *student* (Chandra), on one of the most important topics – which preoccupied at that time mainstream astrophysics, the internal constitution and the final stages of stellar evolution. It is interesting to follow some details of this conflict, several of which were told to the author by Chandra himself at the University of Chicago (1977–1981), and also later, during his several visits to the University of Crete (1985–1990), Fig. 1.29.

Chandrasekhar was born in Lahore, India (1910). He did his undergraduate studies at the Presidency College, Madras, from which graduated with a degree in Physics (1930). Because of his excellent undergraduate performance, he received a scholarship to continue his postgraduate studies at Trinity College, of the University of Cambridge (UK). With these already important parchments and many ambitions he left for the best University that existed at the time, the University of Cambridge, in the United Kingdom. He arrived in London in July 1930. There he completed his doctoral studies (1935) with Eddington as his supervisor. However, Eddington never accepted the results of his student's thesis regarding the existence of *black holes* as

Fig. 1.27 Subrahmanyan Chandrasekhar (1910–1995)

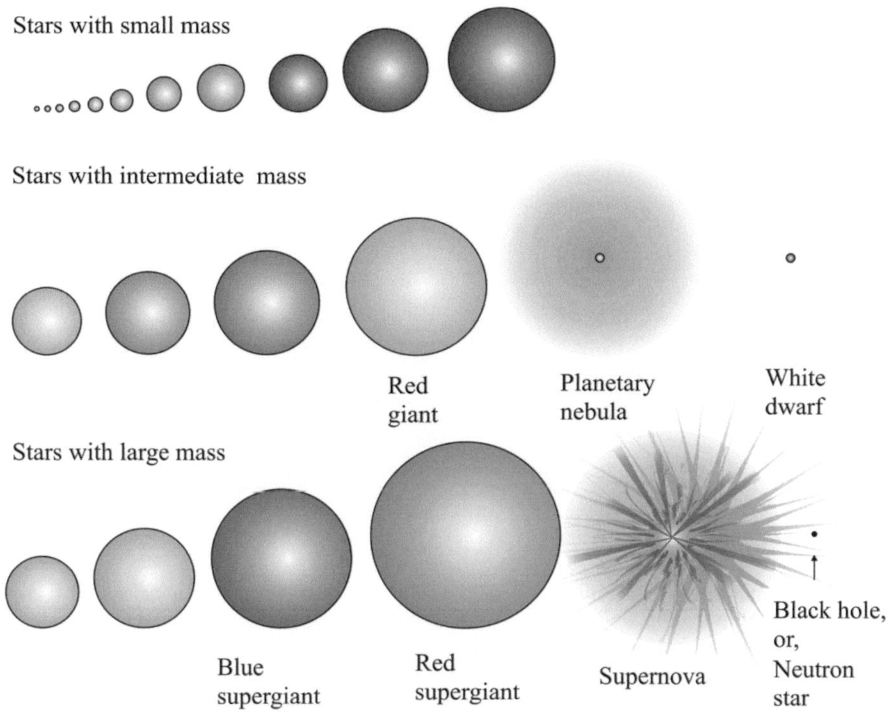

Fig. 1.28 The various possibilities of stellar evolution, according to the mass of the stars. (above) Stars lower in mass than the Sun live nearly an eternal life. (middle) Stars with a mass roughly equal to the solar one, they have fuel for billions of years and at the end of their lives end up as stellar ashes, i.e., *white dwarf*s. (bottom) Stars with a mass greater than the Sun's, after a few million years of life, explode as *supernovae*, leaving behind a *neutron star,* or a *black hole*

the final stage of massive stars, which forced Chandrasekhar to leave the UK and continue his academic career in the USA.

The following interesting fact in connection with Chandrasekhar's development is also noteworthy. During a long journey from India to Europe in 1921, Indian physicist CV Raman wondered why the sky is blue. The answer he gave to this question led to the discovery of the well-known *Raman effect* in Physics[4] which gave him the *Nobel prize* (1930) for light scattering and the Raman effect. The same year (1930) at which Raman became the first Indian to be awarded a *Nobel Prize* in Physics, his nephew, Subrahmanyan Chandrasekhar, made the same steamship journey to London to attend Trinity College. Likewise his uncle, the 20-year-old Chandrasekhar spent

---

[4] In the Raman effect (1928), a light ray is scattered as it passes through a transparent medium, in different incident directions and at different frequencies. This phenomenon is due to the fact that some of the molecules of the medium absorb or provide energy to the incident photons, as a result of which, after their scattering, they appear with a lower or higher frequency, respectively. With the technique of *Raman spectroscopy*, are determined the molecules of the different characteristic groups of organic chemicals.

Fig. 1.29 Chandrasekhar stood out not only as a leading scientist and teacher, but also for his love, simplicity and human friendliness. In this photo (1990), during one of Chandra's many visits to Heraklion, Crete (Greece) from left, the author (at his home) with his two children Christos and Anthony, Chandra with his wife Lalitha, and Chandra's late research associate and my beloved undergraduate classmate at the Un. of Thessaloniki and later distinguished colleague at the University of Crete, Vassilis Xanthopoulos. Chandra and Vassilis were both admirable scientists and kind persons! (From author's photo archive, picture taken by Georgia Vlastou)

hours on the ship's deck, gazing up at the night sky and wondering about the ultimate fate of all those stars. Three years later at Cambridge, he arrived at the conclusion that he had understood the complete theory that explained the final stages of stellar evolution. Unfortunately, however, the scientific community, led by his supervising professor, solemnly rejected the conclusions of his thesis as *unnatural*. Hence, his results were soon forgotten. *But not for very long*! His theory came back in the 50s, and was accepted by the international scientific community, when it was verified by observational data. Then, Chandrasekhar's theory was included in astrophysics textbooks, and after 30 more years, gave himself the *Nobel Prize in physics* (1983), as happened to his uncle of C. V. Raman, but 50 years since Chandra's original studies (!).

Chandrasekhar was 24 years old when he completed his thesis at Cambridge. He presented his results at the annual meeting of the Royal Astronomical Society in London on January 11, 1935, before a distinguished audience, that included his thesis supervisor (Eddington). The limit of 1.44 solar of masses he had discovered is the maximum possible mass a *white dwarf* can have. This limit results from the quantum mechanical study of a relativistic gas of *fermions*, subatomic particles with half-integer spin. Two or more *fermions* cannot be in exactly the same quantum state due to the so-called *Pauli exclusion principle*. Consequently, even at zero temperature, *fermions* occupy high energy levels as a result of which pressure will develop which will oppose the concentration of *fermions* in the same state. In a large concentration of *fermions*, such as it happens for example inside a *white dwarf*, the electron gas will exert an outward force due to the above pressure of the *fermions*, which will resist the inward gravitational force. However, above a critical mass value, the gravitational pull overcomes the pressure of the star's degenerate electrons. Then, the negatively charged electrons cancel out each other with the positively charged protons turning them into neutrons. Then, the system collapses into a *neutron star* or *black hole*. This is precisely the value of the critical mass, the so-called *Chandrasekhar limit,* which Chandra had discovered in his Ph.D.

In summary, then, the theory concluded that at the end of the life of a star, its final fate depends on whether its mass is less or greater than about 1.44 times the solar mass. Stars with a lower mass than this, when they run out of fuel, collapse into the dense *white dwarf* state. But stars greater than this critical mass continue to collapse under their own weight, becoming smaller and denser. These objects are now the celebrated *black holes* and the marginal mass above which they appear *the Chandrasekhar limit* (Fig. 1.28).

However, the prevailing opinion then was that all stars, regardless of the value of their final mass, end up becoming *dwarf* stars. At a conference in Paris where Chandra presented his theory, Eddington had arranged to be the next speaker after him. And, Eddington's presentation focused on describing Chandrasekhar's theory of a critical mass as an *absurd theory*. He claimed that somehow (in a way we do not know) nature would not allow a star to continue infinitely losing energy by radiation as it continues to contract into the final stages of its evolution. Because simply such a result would be *illogical.* He used the extraordinary abilities he had to outline his arguments, to crush the previous young speaker (Chandrasekhar)!

The international prestige and influence of Eddington was such that they did not even allow physicists, who believed Chandrasekhar might be right, to publicly stand by him. Chandrasekhar himself told me: *Eddington ridiculed me. I was desperate. I didn't know if I should even continue my career*.

The Eddington-Chandrasekhar disagreement continued for years with the publication of papers and discussions at scientific conferences. When the two crossed swords at a conference in Paris in 1935, Eddington again found himself on the attack and Chandra on the defense. Four years later, again in Paris, similarly the same situation. After all, ironically, Eddington claimed that there has been no observational data in the meantime to determine whether Chandra's theory is correct, or his own, in which massive *white dwarf*s do not collapse. Only the distinguished observational astronomer Gerard Kuiper, an expert on *white dwarf*s, intervened to report that observational evidence supporting Chandra's theory had just been found! At the end of this conference, Eddington and Chandra both had a brief tête-à-tête encounter:

- *I'm sorry if I've upset you*, Eddington told Chandra.
- Chandra then asked him if his scientific opinion on the well-known subject has changed.
- *No*, Eddington snapped back.

– *Then why are you sorry*? was Chandra's equally curt reply, who then ostentatiously walked away.

In 1936 Chandra left Cambridge and fled to the University of Chicago where, summarizing his theory, he wrote the book: "An Introduction to the study of stellar structure (1938), published by the University of Chicago Press". After that he stopped dealing with the specific subject, and started the study of other problems: the scattering of light, the study of plasma behavior in magnetic fields and the theory of Magnetohydrodynamics, the hydrodynamic stability of rotating bodies, and finally GTR and the study of *black holes*. Meanwhile, at that time in the laboratories of the Un. of Chicago researchers were working on the development of the hydrogen bomb. A phenomenon that they quickly realized was happening in conditions that were similar to those in an exploding star. The same force that broke up a star into a supernova could be used on Earth to create the conditions of the *Biblical Apocalypse* created during a hydrogen bomb explosion. Research efforts were completed in 1966 at *Livermore National Laboratory* in California, when researchers combined the computational codes of astrophysics and the hydrogen bomb. Then finally, the scientific community was convinced that a star could collapse into a black hole.

At the age of 36 Chandrasekhar became a *distinguished professor* at the Department of Physics of the University of Chicago. He also took over as editor-in-chief of the prestigious *Astrophysical Journal*, from 1952 to 1971, while he was honored with many medals, including the Bruce Medal of the Astronomical Society of the Pacific, the Gold Medal of the Royal Astronomical Society (1953), the Rumford Medal of the American Academy of Arts and Sciences (1957), the Royal Medal of the Royal Society, London (1962), the National Medal of Science, USA (1967) and the Draper Medal of the American National Academy of Science (1971).

In 1972 a powerful X-ray source in the Cygnus constellation, later named Cygnus X-1, at a distance of 57,427 trillion kilometers (about 6,070 light years), was identified as the first *black hole*. Other similar observations followed. So, 40 years after the original theoretical prediction, Chandra was vindicated observationally and Eddington was proved wrong. Chandra won the *Nobel Prize* in 1983 for his seminal studies on *white dwarf*s. But, the emotional pain of his traumatic student experience at Cambridge University tortured him until his death in 1995, as I had personally witnessed.

1990: Truth and Beauty, by S. Chandrasekhar

A really nice popular science book written by Chandra is entitled *Truth and Beauty: Aesthetics and Motivations in Science*, University of Chicago Press, 1st edition (October 15, 1990). This book has been translated in several languages, including the Greek one by Chandra's Ph.D. student, later professor at the University of Thessaloniki, Sotiris Persidis. I mention in passing that in 1990 at the University of Crete, in Heraklion, Professor S. Persidis saved the lives of many of us, professors and students, by sacrificing his health, in his attempt to

fend off the raiding into the classroom of an armed psychologically troubled person, who finally murdered, for incomprehensible reasons, the distinguished professors Vassilis Xanthopoulos and Stefanos Pneumatikos, when they were teaching an innovative postgraduate course (*Matthematica*).
Prof. S. Persidis writes in the preface of the above book *Truth and Beauty*: "The period of the human race has been identified with increasing knowledge and improving culture. More than politicians and great leaders, scientists have contributed irreversible processes to our development and brought our lives closer to natural laws, closer to God, discovering and applying the rules of the Universe. Science and its daughter technology shape our days and our destiny. S. Chandrasekhar served science for more than 70 years, with consistency and faith, with hard work, with humility and devotion, as a monk serves in a monastery. In his career he sometimes pondered deeply into the very soul of Science, its aesthetics and motives, its relationship with human nature. These speeches [in this book] present his findings and the contribution of other great scientists on the same subject. They are a lasting monument of *Truth and Beauty*. It is a great honor for me to present in the Greek language the writings of Chandrasekhar, my teacher and an outstanding scientist and man".

Chandrasekhar: A Dedicated Academic Teacher

Chandrasekhar, during the longest part of his academic career worked at the University of Chicago, part of the teaching staff of the University, from 1937 until his death in 1995. For many years he also worked at the Yerkes Observatory. Chandra was a dedicated academic teacher with a special love for teaching physics.
In this connection, Chandra told me the following story. During a (harsh!) Chicago winter, he drove each week a dangerous distance over 100 km from the *Yerkes observatory*, where he lived, until the University, to teach a course. "Do you know how many students I had in that class?" he told me. "Just 2, Yang and Lee, who later received the *Nobel Prize* in Physics (1957) for their *parity violation* theory… So, it seems that it was probably worth my trip from Yerkes to Chicago every week!".

Chandra was the author of many books, such as: An Introduction to the Study of Stellar Structure (1939), Principles of Stellar Dynamics (1942), Stochastic Problems in Physics and Astronomy (1943), Radiative Transfer (1950), Hydrodynamic and Hydromagnetic Stability (1961), Ellipsoidal Figures of Equilibrium (1969), The Mathematical Theory of *Black Holes* (1983), Eddington: The Most Distinguished

Astrophysicist of his Time (1983), Newton's Principia for the Common Reader (1992), etc.

1942: Formulation of the *Mystery* of the Superhot Solar *Corona*

The *solar corona* is the upper layer of the atmosphere of the Sun. Similarly there exist *stellar coronae* in other similar stars, or, *accretion disk coronae* in various accretion disks. The astrophysical coronae consist of fully ionized gas, mainly free protons and electrons, i.e., the gas is in the form of plasma. The solar *corona* extends for millions of kilometers away from the solar *photosphere*, deep into interplanetary space. However, the solar *corona* is clearly visible from the surface of the Earth only during a total eclipse of the Sun, as well as with a *coronagraph*. In a total solar eclipse, the *corona* is easily visible with the naked eye. The name of the *corona* derives from the royal *crown* and appears in a total solar eclipse, to surround the *king* of the solar system, the Sun, similarly to the crown that adorns the head of kings.

The first description of the solar *corona* was given by the Byzantine historian and chronicler Leo the Deacon, who vividly describes a total solar eclipse, which took place in Constantinople on December 22, 968 AD, wherein appeared the *chromosphere* and the solar *corona*[5,6,7]:

> "*...at the fourth hour of the day... the Earth was covered with darkness and all the bright stars appeared. And it was possible to observe the sun's disk, without light, unlit, while appeared around this dark disk, a thin circular zone with a faint and sickly moonlight...*".

[5] According to the ancient Greek historian *Herodotos*, the eclipse of *Thales* was the earliest recorded instance of a predicted solar eclipse, possibly occurring on May 28, 585 B.C. Again according to *Herodotus*, the appearance of this eclipse was interpreted as an *omen*, interrupting a battle in a long-standing war between the Medes and the Lydians. American writer Isaac Asimov described this battle as the earliest historical event whose date is known with precision to the day, and called that prediction *the birth of science*.

[6] The astronomer and explorer George Davidson made a scientific trip to the *Chilkat Valley* of Alaska. He told the *Chilkat Indians* that he was anxious to observe a total eclipse of the Sun that was predicted to occur the following day, August 7, 1869. This prediction was considered to have saved Davidson's expedition from an attack, similarly to the prediction by *Columbus* of the total lunar eclipse of Feb. 29, 1504 that saved him and his crew from the *Arawak Indians* in Jamaica.

[7] During the total eclipse on August, 7, 1869, American astronomers William Harkness and Charles Young observed a weak spectral emission line from the solar *corona* in the green region of the spectrum, which was visible for a couple of minutes (Fig. 1.30, left). However, despite an enormous amount of work, both at the telescope during subsequent eclipses and in the laboratory, this emission line could not be attributed to any known chemical element.

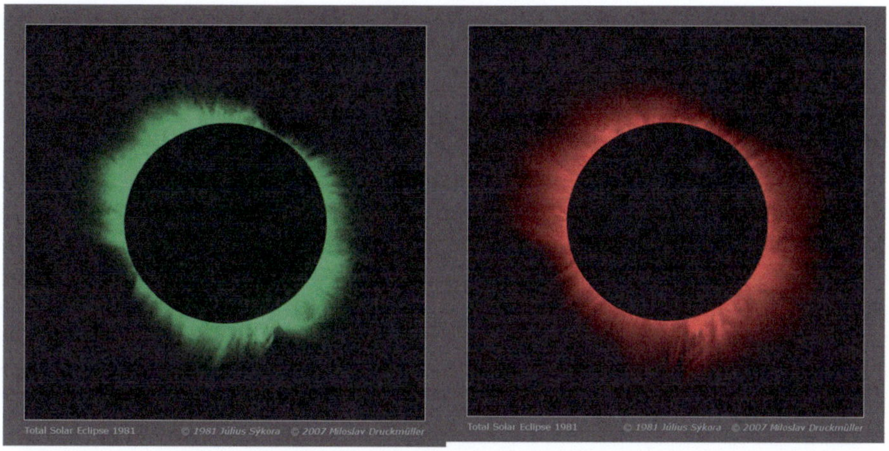

Fig. 1.30 Two figures from the total solar eclipse of 31.07.1981, Bratsk, Siberia, Russia: (*left*) This image was taken through a green narrow-band filter on the wavelength of the most intense coronal spectral line. This prominent spectral line was discovered by Harkness and Young during the total solar eclipse of August 7, 1869 visible from eastern Russia, Alaska, across Canada, and the northeastern United States and for the following 70 years the physical origin of this line was frightening generations of astronomers because they were not able to identify the element responsible for the line. This line in fact corresponds to 13× ionized iron, Fe XIV (530.3 nm, 1.8×10^6 K) and since an iron atom (Fe) has 26 electrons, in Fe XIV the iron atom has lost half of its electrons. (*right*) Another prominent coronal line is the red line 637.4 nm of 9× ionized iron, Fe X. This radiation visualizes relatively cooler coronal structures of about 800,000 K. The red color of the image is precisely computed, so the color for the human vision is the same as the color of the 637.4 nm light. Compare also with Fig. 1.33 (© 1981 Július Sýkora, © 2007 Miloslav Druckmüller, reproduced with the kind permission of M. Druckmüller)

The solar *corona* plays a basic role in the *solar wind* phenomenon, which could not have been understood as a concept, before we realized the nature and properties of the solar *corona*. The density of the *corona* is extremely low, of the order of one hundred million protons per cubic centimeter. Although this may sound like a large number of protons contained within each cubic centimeter of coronal gas (as much as a small bubble), if compare this to the density of the earth's atmosphere around us, it is a hundred billion times thinner! We may add that just one out of a million *photospheric* photons is scattered by the *corona*, giving it a rather small brightness, in relation to the Sun's *photosphere*. Nevertheless, the coronal plasma is extremely warm, with temperatures of a few million degrees. A result of these high temperatures is the ultrasonic expansion of the *coronal* plasma toward the interplanetary space, through the so-called *coronal holes* in the form of the *solar wind*, reaching speeds up to 3 million kilometers per hour.

About 99% of the light coming from the *corona* is the effect of *Thomson scattering* of *photospheric* light by free electrons. About 1% of the coronal light is emitted by excited gas atoms of the *corona* itself. Fraunhofer photospheric lines can be recognized in the spectrum of light from the *corona*, even though they are quite broad, because of the very high thermal speed of the coronal electrons.

In the following, let's take things in order, to see how we arrived to the fact that the *corona* consists of a superheated gas with a temperature of millions of degrees.

As the techniques of spectroscopy advanced, during a total eclipse in the year 1869 in Siberia, a coronal green emission spectral line was first detected. As the years passed, more emission lines were detected and their wavelengths were determined.

The invention by Bernard Lyot (1939) of the *coronagraph* was an important step that led to the discovery and accurate determination of several emission lines in the inner part of the *corona*. The problem that arose then was that none of these spectral emission lines could be attributed to some known chemical element. Then, a new hypothetical element, *Coronium*, was invented, with the expectation that this new element would soon be discovered in the laboratory, repeating the remarkable success of Jansson and Lockyer with the chemical element helium, which was first detected through its spectrum in the Sun in 1868. The helium was finally detected in minerals on Earth by William Ramsey in 1895. However, the periodic table was already full and there was no place for *Coronium*!

Then followed the inventive laboratory works of researchers of the electromagnetic spectrum, such as those of Grotrian (1933, 1934) and Edlén (1936, 1942). At the same time, the experimental determination by Lyot (1939) of the various coronal lines and the corresponding widths of these lines provided the definitive answer to this basic astrophysical question. By using the isoelectronic series, e.g., Li I, Be II, B III, CIV, NV, etc., it became evident that the coronal emission lines correspond to forbidden transitions of multiply ionized atoms, such as Fe X, Fe XI, Ca XII, Ca XIII, etc., which require temperatures of millions of degrees, or even higher. Today we know that such spectral lines are due, for example, to iron atoms with a high degree of ionization, such as Fe XIV, wherein we have these iron atoms 13 times ionized, a fact indicating a temperature greater than one million degrees (1.8 million degrees).

The measured widths of these spectral lines confirmed this surprising result, of the very high temperature of the *corona*. The observed *coronal scale height* gives an average molecular weight comparable to that of ionized hydrogen. The *corona*, then, is simply the extended outer atmosphere of the Sun, with the same abundance of elements that we encounter in the *photosphere*, and extends deep into space because of its incredibly high temperature. To give a specific example, the scale height for the pressure, i.e., the distance over which the pressure varies appreciably, for a temperature of 2 million degrees in the low part of *corona* is about as long as one solar radius (Fig. 1.30).

Therefore, it follows the surprising result that the Sun's atmosphere has a temperature of millions of degrees Celsius, heated from the *photosphere* which has a temperature of only a few thousand degrees! Nevertheless this conclusion, first of all seems to a first view violate the second law of thermodynamics, something that had been pointed out since the nineteenth century by Lord Kelvin (1824–1907) when he wondered: *how is it possible at all, a cold layer to heat another layer that already has a higher temperature*? If this is so, we have a violation of the famous second law of thermodynamics, which states that heat always flows spontaneously from hot to cold bodies and never from cold to hot, without the expense of some energy. The situation is analogous to being able to boil water at a temperature of 100 degrees on a kitchen electric stove that has a temperature of only 20 °C. And in the case of the Sun the temperature difference is far greater! (Figs. 1.30–1.33)

Therefore, this scientific discovery of a superheated *corona* was a benchmark discovery in astrophysics that revealed a phenomenon that had never been imagined before and today we know that this is true for most stars, and even for galaxies and galaxy clusters. At these high temperatures, the *corona* is the main source of X-ray emission from a star. Indeed, X-rays were detected from the Sun for the first time by Burnight in 1949. Furthermore, as we shall see below, the *corona* is connected directly with the solar corpuscular radiation, i.e., the *solar wind*.

Fig. 1.31 (top) The solar corona during the total solar eclipse that was visible from the Greek island of Kastellorizo on March 29, 2006 (*Source* Section of Astrophysics, Astronomy & Mechanics, Department of Physics, National and Kapodistrian University of Athens). (*bottom*) White-light eclipse of the March 29, 2006 corona made by a 1600 mm telescope at Libya and SOHO EIT He II (30.4 nm). Superposition of disk images in the EUV with eclipse images allow features on the Earth facing half of the Sun that are footprints for coronal structures to be so identified and allow those features to be followed from their bases in the *photosphere* or *chromosphere* up through the corona (© 2006 Miloslav Druckmüller, Peter Aniol, ESA/NASA, Figure reprinted with the kind permission of M. Druckmuller; see also Pasachoff (2007))

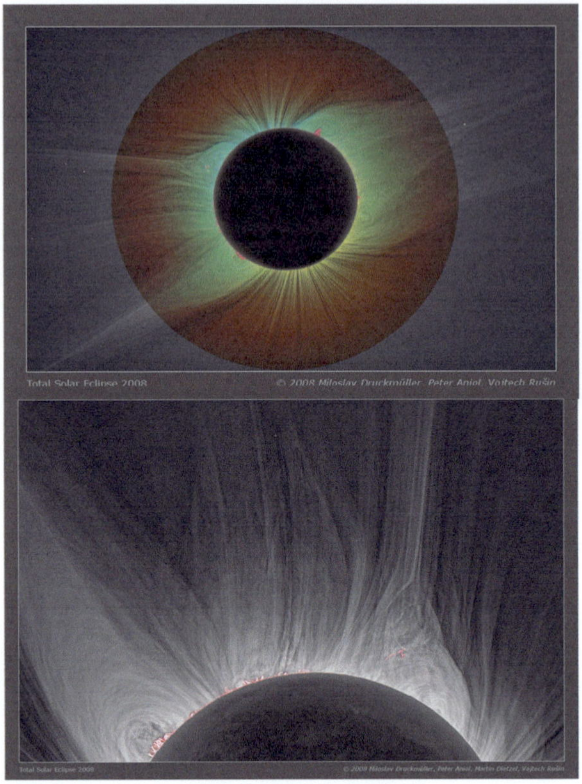

Fig. 1.32 (top) The total solar eclipse of August 1, 2008, visible in Mongolia (Bor Udzuur), with K-corona filters showing the continuous corona, 529.1 nm and the green line of iron Fe XIV, 530.3 nm, via a composition of 23 images. (*bottom*) Processed composite of 25 images of the white-light corona above the eastern solar limb. Note the extent of the cavity associated with the *prominences* at left (© 2008 Miloslav Druckmüller, Peter Aniol, Vojtech Rušin, reproduced with the kind permission of M. Druckmüller, for more details see also Pasachoff et al. (2009))

The obvious next question was which heat source is responsible for such a high temperature of the *corona*? The classic answer would be that falling of meteorites in the solar surface maintain the overall brightness of the Sun and the absorption of interstellar gas can create the *corona*. However, the antisolar orientation of the comet tails precludes such absorption of interstellar gas.

The answer to this still open question of how the *corona* is heated is fundamental to astrophysics and also concerns the superheated outer atmospheric layers of most stars. There, some form of mechanical disturbance, e.g., convective phenomena below the visible surface of the Sun (or an accretion disk, or an expanding shell in other cases), impart mechanical energy to the fluid and magnetic field. Some part of the mechanical work is consumed in creating *acoustic waves*, *internal gravity waves*, *Alfven waves* or other *magnetohydrodynamic waves*, and some other part is converted into free magnetic energy. Such waves are free to propagate up to the *chromosphere* and the lower *corona*, wherein they dissipate due to viscosity and thermal conduction, thereby converting their mechanical energy into heat. In addition, the free magnetic energy in decaying dipole magnetic fields is largely consumed in *nanoflares* generated as a result of internal discontinuities produced by the field topology in static equilibrium (Parker 1983a, b, 1988, 1994; Tsinganos 1982), or, from the collision of

Fig. 1.33 A true-color image of the solar corona in which the white light (continuum) in attenuated approximately 400×, from the total solar eclipse of 20.04.2023 (Island in archipelago of Lowendal Islands and Exmouth, Australia), a composition of 225 eclipse images of Fe X (637.4 nm, red, 800,000 K), Fe XIV (530.3 nm, green, 1.8×10^6 K) and continuum (white light image). Charged particles are tracers of magnetic field lines and they are the only tools available to visualize the *shape* of magnetic field lines in the solar *corona*, thus yielding an image of their distribution in the plane of the sky. Free coronal electrons, scattering the *photospheric* light toward the observer, are everywhere in the *corona* and provide a visualization of the solar magnetic field in the white-light (continuum). Fe X and Fe XIV populate different magnetic structures with different temperatures. The composition presented in this image provides valuable information about the structure of the solar magnetic field and at the same time visualizes the distribution of Fe ions (© 2023 Pavel Štarha, Matěj Štarha, Jana Hoderová, Shadia Habbal, Miloslav Druckmüller, image processing by Miloslav Druckmüller, reproduced with the kind permission of M. Druckmüller)

one dipole magnetic field with another, as in the following Fig. 1.65, etc. and Tsuneta (1997) indeed discovered small-scale fluctuations (*nanoflares*) in the X-rays from the active *corona* observed by the Japanese satellite *Yohkoh*. Recent observations by the SOHO spacecraft have shown that the *photosphere* is covered by many active small-scale magnetic fields that have high density, fluctuate and radiate, apparently due to small explosions from fast magnetic reconnections. The heat flux required to maintain the temperature of the quiet *corona* is calculated to be a small percentage of the characteristic flow of the kinetic energy in the *photosphere*. Therefore, the solar *corona* can be heated either through the deposition of wave energy, or through the consumption of the energy of strong electric currents. But which of these two candidate energy sources heats the *corona* has not yet been clarified. The first *Solar Orbiter* observations (July 2020) argue in favor of heating the solar *corona* with many small *flares* (*nanoflares*), according to Parker's theory.

The Hitherto Unsolved Problem (Mystery) of Heating the Solar *Corona*

The *chromosphere*, the crimson thin ring of fire that surrounds the solar disk and is revealed during total solar eclipses, has a temperature of about 10,000 degrees, and lies between the cooler *photosphere* and the warmer solar *corona*. Among others, the *chromosphere* is also the source of changes in emission in the ultraviolet part of the solar spectrum, which can contribute and affect the Earth's climate. But how the temperature increases as we move away from the cooler photosphere toward the hot corona, through the *chromosphere*, is an unanswered question that has plagued us, theoretical solar physicists for many years. That is, the situation is similar to be near a big fire, but, as we move away from this fire (the *photosphere*), the temperature increases (in the *chromosphere* and *corona*)!

1958: Beginning of the Legacy of E.N. Parker (Theoretical Prediction of the Solar Wind, Theorem on the Non-equilibrium of Non-invariant Fields, the Spontaneous Formation of Tangential Discontinuities, etc.)

1958: Parker's Prediction of the Solar Wind

At the dawn of the *Space Age*, in the late fifties, a conceptual revolution in astronomy was the prediction by Eugene Parker that an extremely hot solar *corona* cannot be confined by solar gravity, but expands supersonically into interplanetary space to fill the whole solar system and affect all planets. Similar reasoning applies to the coronae of stars and accretion disks.

Actually, the basic physical explanation of the solar wind phenomenon is *extremely simple* and it is based on the following facts:

1. The solar corona has a *high* temperature T $\sim 10^6$–10^7 K, in contrast to the *low* temperature of the interstellar space, T $\sim$ 100 K.

2. The plasma has a *high thermal conductivity* κ. Thus, if heat transfer occurs only by thermal conduction in a static corona, then the heat flow obeys Fourier's law:

$$J_Q = -\kappa(T)\vec{\nabla}T$$

For a high-temperature plasma the thermal conductivity coefficient is:

$$\kappa(T) \approx 2 \times 10^{-14} T^{5/2} \frac{\text{cal}}{\text{cm s K}},$$

and therefore for the temperatures of the solar corona ($T \sim 10^6$ K) we have $\kappa \approx 20$ cal/cm s K, a value higher even for the thermal conductivity coefficient of excellent conductors like copper, or silver wherein $\kappa \approx 1$ cal/cm s K.

3. Due to the high thermal conductivity in a static corona ($\vec{V} = 0$), the heat flow satisfies the energy conservation equation:

$$\vec{\nabla}[\kappa(T)\vec{\nabla}T] = 0 \Rightarrow T = \frac{T_0}{R^{2/7}}$$

where R is given in units of the solar radius r_0 and T_0 is the temperature at the base of the corona. At the boundary of the heliosphere, $r \approx 122$ AU, $R \approx 2.6 \times 10^4 \Rightarrow T = 10^6/[(2.6)^{2/7} \times (10)^{8/7}] \approx 10^5$ K $\gg 100$ K ~ T_{ISM}. Therefore a static atmosphere cannot provide the correct corona-interstellar space matching. Therefore, expansion of the solar atmosphere would be the inevitable result.

4. Bernoulli's law, according to which the sum of thermal, kinetic and gravitational energy is constant, in the case of the solar wind gives,

$$\frac{\gamma}{\gamma-1}\frac{2kT_0}{m} + \underbrace{\frac{V_0^2}{2}}_{0} - \frac{GM_\odot}{r_0} = \underbrace{\frac{\gamma}{\gamma-1}\frac{2kT_\infty}{m}}_{0} + \frac{V_\infty^2}{2} - \underbrace{\frac{GM_\odot}{r_\infty}}_{0},$$

wherein the initial flow speed V_0 and the final temperature T_∞ are negligible, say approximately zero, at large radial distances r_∞, such that the corresponding three terms in the above equation are negligible. Then the remaining nonzero two terms in the LHS of the above equation are equal to the asymptotic kinetic energy term in the RHS. Hence, the corona can expand in such a way that it acquires a large and approximately constant velocity over large distances. From this equation it appears that the asymptotic (finite) velocity of the solar wind V_∞ comes from the conversion of thermal energy at the base (reduced by the gravitational energy), into kinetic energy.

By using the basic equations of magnetohydrodynamics, Parker submitted for publication a seminal paper (1958d, 1963) concluding that the hot solar *corona* should expand producing a supersonic flow of particles.

The first approach of Parker to describe coronal expansion was rather simplified (Parker 1958d). First, he used hydrodynamics, ignoring magnetic phenomena. Additionally, the fluid was considered homogeneous so that no distinction was made between electrons and protons. A time-independent state of the phenomenon was considered ($\partial/\partial t = 0$) with the absence of internal friction phenomena of the fluid (negligible viscosity). The gas assumed to expand in a radial and spherically symmetric way and rotation phenomena were neglected. Finally, because of the very high conductivity of the corona, the expansion was assumed isothermal. Figure 1.34b is the first plot of the expansion speed (solar wind) as a function of the dimensionless radial distance for various values of the uniform coronal temperature T_0. Note that the wind speed increases with the radial distance to infinity at large r, since there is a continuous deposition of energy in the gas because of the isothermal hypothesis.

a

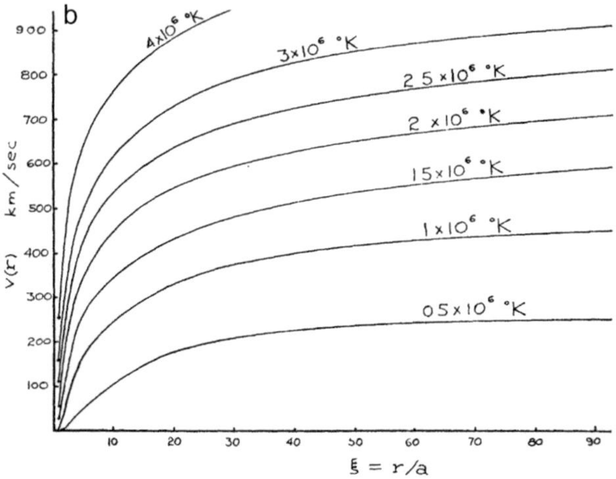

Fig. 1.34 **a** Eugene Parker (1927–2022) **b** Spherically symmetric hydrodynamic expansion velocity v(r) of an isothermal solar corona with temperature T_0 plotted as a function of r/a, where a is the radius of the corona taken to be 10^{11} cm. This is the first figure in Parker's first and original paper for the Solar Wind (Parker 1958d). **c**. Topology of the solutions for the dimensionless flow speed $M_0(R) = V(R)/V_{s0}$ in units of the sound speed V_{s0} at the base $r = r_0$, of a polytropic solar wind model, $P = \rho^{\gamma}$ with the dimensionless pressure P and density ρ normalized in units of their base values P_0 and density ρ_0. M is the Mach number $M(R) = V(R)/V_s(R)$ in units of the sound speed $V_s(R)$. All these branches correspond to different values of the dimensionless conserved *Bernoulli* energy E, in units of V_{s0}^2, the sum of the kinetic, thermal and gravitational energies. For a critical value of $E = E_c$, is obtained the single wind-type transonic solution with finite asymptotic speed $M_{0,\infty}$ and zero pressure $P_\infty = 0$ and density $\rho_\infty = 0$, while $T_\infty = 0$. For $E > E_c$, correspond the *breeze* solutions with finite P_∞, ρ_∞ and T_∞. **d**. The *heliospheric current sheet*, the largest plasma structure in the solar system (Photo courtesy of NASA). **e** The most prominent solar physicist and astrophysicist of his time, Eugene N. Parker, circa 1977 (Photo courtesy of the University of Chicago). **f** A sketch of a unidirectional hydromagnetic field (**B**, **v**) in the z-direction, $\mathbf{B} = B_0(x, y)\hat{z}(x, y)\hat{z}$, $\mathbf{v} = v_0(x, y)\hat{z}$ extending in the infinite space $z > 0$. (**b**) The same field after a perturbation has been applied at the level $z = 0$ manipulating the footprints of the field and stream lines in an arbitrary manner. (From Tsinganos 1982)

c

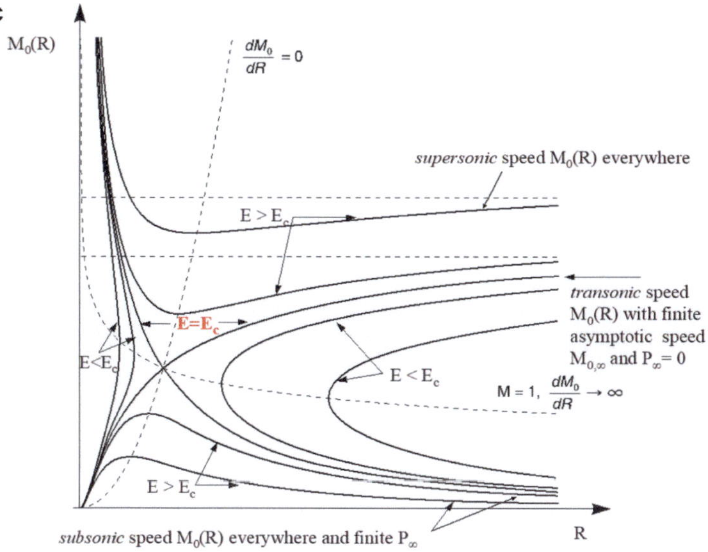

d

Fig. 1.34 (continued)

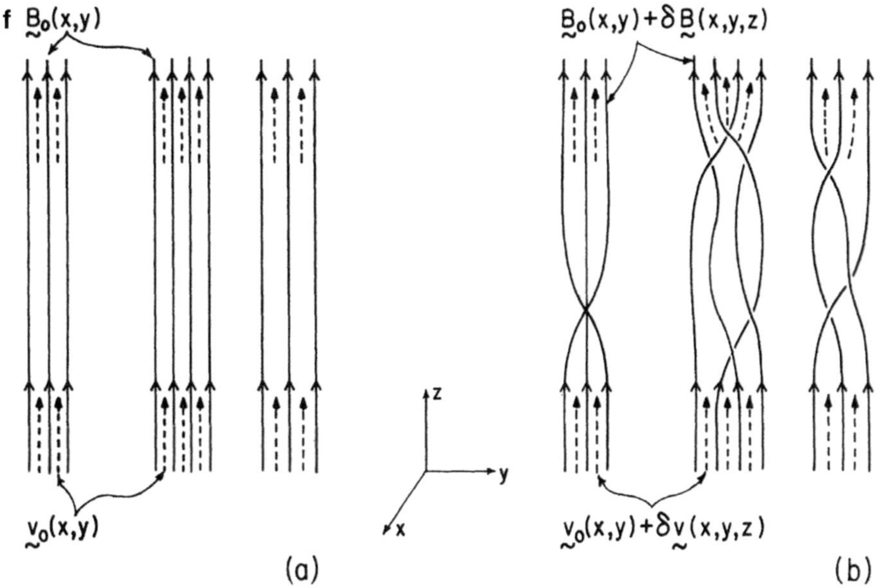

Fig. 1.34 (continued)

Figure 1.34c, is a plot of the solution topologies of the coronal expansion (dimensionless flow speed $M_0(R) = V(R)/V_{s0}$ in units of the sound speed V_{s0} at the base $r = r_0$), for a polytropic coronal gas with the dimensionless pressure P and density ρ related via a polytropic equation, $P = \rho^\gamma$. All those branches correspond to different values of the dimensionless conserved Bernoulli energy E. Note that for a critical value of the Bernoulli energy $E = E_c$, a single transonic solution is obtained with *finite* asymptotic speed $M_{0,\infty}$, zero pressure $P_\infty = 0$ and density $\rho_\infty = 0$ at infinity where $T_\infty = 0$. For values of the Bernoulli energy greater than the critical value of the Bernoulli energy ($E > E_c$) the so-called "breeze" solutions result which have finite pressure P_∞, density ρ_∞ and temperature T_∞ at infinity. The various branches Fig. 1.34c are obtained by integrating the Mach number equation,

$$\frac{1}{M_o^2}\frac{dM_o^2}{dR} = \frac{2}{R}\frac{2 - \frac{\lambda}{R}\frac{1}{\rho^{\gamma-1}}}{\frac{M_o^2}{\rho^{\gamma-1}} - 1} = \frac{2}{R}\frac{2 - \frac{\lambda}{R}\frac{M^2}{M_o^2}}{M^2 - 1}$$

together with the mass conservation equation,

$$\rho M_o R^2 = \mu$$

where

$$\lambda = \frac{GM_\odot}{r_0 V_{so}^2}, \quad V_{so}^2 = \frac{d\overline{P}}{d\overline{\rho}} = \gamma K \overline{\rho}_o^{\gamma-1} = \gamma \frac{k\overline{T}_o}{\overline{m}},$$

where the barred quantities have dimensions and the unbarred quantities are dimensionless and $\bar{m}$ is the mean molecular weight.

Additionally, Parker correctly predicted the twisted shape that the rotating Sun's magnetic field would take as the *solar wind* carried it to the outer solar system, which is now called the *Parker spiral* (Fig. 1.34d).

The heliospheric current sheet HCS within the solar system (Fig. 1.34d) is a spirally undulating surface – a spirally wavy shape similar to the dress of a spinning and swaying ballerina. In the HCS the polarity of the interplanetary magnetic field changes from north to south. This surface, which has a thickness of about 10,000 km, is located near the solar equator.

Because of the change of direction of the radial magnetic field across the HCS, the related azimuthal electric current I_φ flowing through the equatorial plane from the Sun to the Earth at a distance of one AU can be calculated from Ampere's law to be: $I_\varphi = 3.336 \times 10^{-3}$ $Am^{-1} \times 1.5 \times 10^{11}$ m $\simeq 5 \times 10^8$ Ampere. For distances considerably greater than the Earth-Sun distance, the solar magnetic field is approximately azimuthal since $\Omega_\odot r \gg V_{SW}$. In this case, the magnitude of the azimuthal magnetic field varies significantly near the solar equator where $B_\varphi = B_\varphi(\theta)$. The total radial current passing through a circumference of radius r at distance r can be also calculated from Ampere's law to be about $I_r \approx 3.9 \times 10^9$ Ampere. This outward flow of equatorial radiation is accompanied by an inflow of current from the polar regions, so that overall the Sun remains electrically neutral. At intermediate distances, the current has an azimuthal and radial component, that is, like the magnetic field, follows a spiral shape, being perpendicular to the magnetic lines. For the solar wind sources and their variations, see Schwenn (2006a, b).

Thus, Eugene Parker, as a young researcher at the University of Chicago (1955–1957), developed a simple theory for the *solar wind* phenomenon, wherein about a million tons of hot gas of the solar atmosphere are poured out from the Sun every second and flood interplanetary space. The *solar wind* originates deep within the extended solar atmosphere, beginning to blow relatively *slowly,* at only a few thousand kilometers per hour, to end up like a wild *tornado*, as it overtakes the Earth, moving at speeds of a few millions of kilometers per hour.

We can reasonably hypothesize that the supersonic *solar wind* has existed and propagated since our solar system was formed several billion years ago, feeding the interplanetary space with dynamical plasma and the corresponding magnetic field. It seems that the initial dense protosolar wind and the strong primordial magnetic field, removed the initial high angular momentum of the young Sun in the first 300 million years of its life, slowing down the rapid initial rotation of the solar equator to the slow rotation of about 25 days that we observe today.

Kepler - Newton - Parker

From a more general perspective, one may think in the following way. Kepler was the first to discover *observationally* the physical laws governing the motion of the planets around the Sun (1609). In 1687, 78 years later, Newton was able to *theoretically* explain Kepler's laws of planetary motions, by using his theory of classical mechanics and the law of the gravitational attraction of the Sun on each planet, in the well known two-body problem. As it happens in most of physics and astrophysics, observation/experiment was ahead of theory. Nevertheless, both, Kepler and Newton assumed an empty interplanetary medium for the planets to move around the Sun, without a *solar wind*. In 1957, i.e., 348 yearsafter Kepler and 270 years after Newton, it was left to Parker to discover, *theoretically* this time, that the interplanetary medium is not empty at all, but it is flood with the supersonic *solar wind* that travels at millions of kilometers per hour across the planets. And, this is one of the few cases of physical phenomena where theory was ahead of observations. Nevertheless, Kepler, Newton and Parker were lucky, because such fundamental physical laws and phenomena can be discovered only once!

It is also remarkable that Parker's theory was largely ignored until 1959, when Gringauz flew *hemispherical ion traps* on the Soviet Lunik 1, 2, 3 and Venera 1 missions to measure the total electric charge deposited by the arriving ions. He reported that the signal fluctuated as the spacecraft spun around its axis, suggesting that an ion flow was entering the instrument whenever it faced the Sun. Analysis of the data however failed to find the necessary signal. In 1961 Bridge & Rossi with the MIT team obtained more detailed observations with an elaborate ion trap on NASA's Explorer 10, a spacecraft designed to explore the nightside tail of the magnetosphere which however was often immersed in the *solar wind*. In 1962, when *Mariner* II the first probe to travel beyond Earth's magnetic field, flew toward *Venus*, not only detected a continuously flowing *solar wind*, but also observed in it fast and slow streams, approximately repeating at 27 day intervals, suggesting that their sources rotated with the Sun (Neugebauer and Snyder 1966). This spacecraft observed *in situ* the supersonic *solar wind* and Parker was thus vindicated.

However, before those observations, Parker's conclusion was roundly criticized. This is in accordance to Parker's saying: *I have never made an important proposition, wherein there was not a crowd saying, Ain't so, can't possibly be so.* or, as he said in an interview at the *Uchicago News,*

> If you do something new or innovative, expect trouble. But think critically about it because if you're wrong, you want to be the first one to know that.

Hence, finally Parker's theory of the *solar wind* prevailed, despite the severe criticism it received in the beginning, as we explain in the following. Parker's theory of the solar wind (Parker 1958b, d). remains even today *one of the most beautiful chapters of modern astrophysics*, both for its simplicity and also for the plausible and wonderful way of explaining a not-so-simple natural phenomenon (for a review of an extended MHD modeling of the steady solar corona and the solar wind, see Gombosi et al. 2018). Today, Parker's theory has been extended and generalized by his Ph.D. students, to collimated MHD outflows in astrophysical *jets*, as we discuss in the following Chaps. 3 and 5 (e.g., Sauty and Tsinganos 1994; Chantry et al. 2022).

An (Anekdotal) Background About the Publication of the Solar Wind *Theory...*

The following story is very interesting about the background of the publication of the theory of the solar wind, which Eugene Parker himself once told to the author. This also gives a light note to the formal description and assumptions of the theory of the solar coronal expansion.

E. N. Parker's *first* paper presenting the solar coronal expansion and the concept of the solar wind (Parker 1958d) in the context of an isothermal corona did not contain the term *solar wind*, for the simple reason that Parker hadn't thought of that name, yet. On the other hand, several of his following papers, which however were eventually published before this first paper, did use the name *solar wind* to emphasize the hydrodynamic nature of the phenomenon. It is interesting to see how this time reversal came about.

Parker's *first* paper was received for publication to the *Astrophysical Journal* on January 2, 1958 while its writing was completed toward the end of 1957. The next publication on the solar wind had as its subject the modulation of the intensity of the galactic cosmic radiation by the outward moving magnetic solar wind anomalies (Parker 1958a) and was received for publication by the journal *Physical Review* on February 19, 1958. This was the first paper to use the term *solar wind* and was published in the June 15th, 1958 issue of the *Physical Review.* This paper was followed by the study of the collision of the solar wind with the geomagnetic field, which was received by the journal *Physics of Fluids* on April

25, it was revised on May 16, according to some suggestions of the reviewer and was finally published in the May–June 1958 issue of the *Physics of Fluids* (Parker 1958b). Parker's interest then turned to the structure of collisionless shock waves. This paper (Parker 1958c) was received by the *Astrophysical Journal* on July 29, 1958, and was published in the issue of January 1959.
Sometime in the late spring of 1958, the reviewer's report of the first submitted paper on the solar wind was mailed to Parker, recommending to the author to become somewhat more familiar with the subject, before attempting to write such a scientific article on the solar cospuscular radiation. There was *none specific objection* to the arguments, or the calculations of the submitted work. Parker pointed out to the editorial director of the journal the absence of any valid and substantial objective argument, by the reviewer of this article, against the theory. The editorial director of the journal, prof. S. Chandrasekhar (1983 *Nobel Prize in Physics*), sent the paper to a *second* reviewer. Sometime in the summer of 1958, the second reviewer responded that the paper was giving erroneous conclusions and suggested that this paper should not be published, again without any specific criticism, except that the author was apparently unfamiliar with the bibliography. Once again, Parker's reply to the editorial director pointed out the fact that there was no substantial criticism and no valid and objective argument from this reviewer either, nor that any specific error had been identified. Apparently, the idea was roundly criticized.
As Parker himself told UChicago News in 2018: The first reviewer on the paper said, 'Well, I would suggest that Parker go to the library and read up on the subject before he tries to write a paper about it, because this is *utter nonsense*!'.
A few days later, Professor Chandrasekhar went to Parker's office shaking at his hand this paper and told him something like this:

- *Parker, do you really want to publish this paper*?

Parker replied that he saw no reason why the paper should not be published, on the basis of lack of any substantial criticism. Then, Chandrasekhar said:

- *I sent this to two distinguished scientists, specializing in this scientific field, but both decided that the work is misleading and far from having the required scientific level!*

Parker replied that the problem with the referees was that, besides that clearly they did not like the work itself, they had nothing more specific to say.
Prof. Chandrasekhar was left silent for a moment and then he said:

- *Ok Parker, I'll publish it.*

And so the paper without the term solar wind finally appeared in the November 1958 issue of the *Astrophysical Journal* (Parker 1958c), essentially after papers written later that already used the term solar wind (Parker 1958a, b). This interesting detail explains the reversal of the name of the solar wind which did not appear in the first paper (Parker 1958d), but it did appear in the later submitted papers of 1958.

Several times Parker wondered what would have happened if someone else was the journal's editorial director. Considering the robust claim of the two "*distinguished reviewers*" that the paper had no particular scientific value, its rejection would be the most likely outcome. Parker would probably be forced to submit the work to another less prestigious journal, e.g., in a journal that does not use such *distinguished* (!?) scientists as reviewers, resulting, of course, in further delays...

Coronal Heating in Thin Electrical Current Sheets and the Characteristic Time of Coronal Heating

In the following we shall calculate the characteristic heating time of the solar corona via the Joule mechanism, as well as the spatial scale of the electric current sheets where the electrical energy is consumed.

When in a plasma of finite electrical conductivity σ we have Joule heating, the rate of increase of its thermal energy per unit volume of the plasma d $\mathcal{V}$ is,

$$\frac{\mathrm{d}E}{\mathrm{d}\mathcal{V}t} \simeq \frac{E}{\mathcal{V}\cdot\tau} = \frac{I^2 R}{S\cdot\ell} = \frac{S^2 J^2}{S\cdot\ell}\cdot\frac{1}{\sigma}\frac{\ell}{S} = \frac{J^2}{\sigma}\Big(\frac{\mathrm{erg}}{\mathrm{s\,cm^3}}\Big).$$

From Ampere's law we have an approximate relationship between the current density J and the spatial scale of change of the magnetic field, ℓ, J = cB/4π ℓ. Therefore the volumetric rate of Joule heating is,

$$\frac{J^2}{\sigma} = \frac{c^2}{4\pi\sigma}\frac{B^2}{4\pi}\frac{1}{\ell^2} = \frac{\eta}{\ell^2}\frac{B^2}{4\pi}, \quad \eta = \frac{c^2}{4\pi\sigma}$$

In a completely ionized gas, the classical Spitzer electrical conductivity σ and the corresponding diffusion coefficient η (Spitzer and Härm 1953) are given by the expressions,

$$\sigma \simeq 10^7 T^{3/2}, \quad \eta = 10^{13} T^{-3/2},$$

In CGS units. Thus, the characteristic heating time is,

$$\tau = \frac{E/\mathcal{V}}{\eta\frac{B^2}{4\pi}}\ell^2.$$

For a typical plasma temperature of the order of millions of degrees, such as in the solar corona where T $\sim 10^6$ K, the diffusion coefficient is $\eta \simeq 10^3 - 10^4$, in CGS units. For a typical solar magnetic field of the order of hundreds of Gauss, B $\simeq$ 100 Gauss we have,

$$\tau = \frac{(E/\mathcal{V})}{10^3\cdot 10^3}\ell^2.$$

Because in a gas of N atoms and mean temperature T, its total thermal energy is E = NkT, or, E/$\mathcal{V}$ = nkT, where n is the number density and k is Boltzmann's constant, $k = 1.38 \times 10^{-16}$ ergs/K.

High in the tenuous solar corona the electron number density is $n \simeq 10^8$ per cm^3. However, in quiescent and active region solar coronal loops the electron number density is higher, typically around 10^9–10^{10} cm^{-3}, and it can be significantly higher (by a factor of 10 or more) in flaring loops, say, 10^{10}–10^{11} cm^{-3}. These loops can vary in temperature from 10^5 K (cool loops) to a few 10^7 K (flaring loops). Thus, we have $E/\mathcal{V} \sim nkT \sim 10^{10} \times 1.38 \times 10^{-16} \times 10^7 \simeq 14$ erg/cm^3,

$$\Rightarrow \tau = \frac{14}{10^6}\ell^2 \simeq 10^{-5}\ell^2.$$

This is a relationship between the characteristic heating time τ and the characteristic length over which thermal energy is consumed in a typical volume $\mathcal{V}$ of a heated solar plasma.

On the other hand, from observations of the emission of radiation in solar active regions, the flux F of energetic losses of the corona, mainly through X-ray and ultraviolet (UV) radiation, can be estimated. Specifically, the flux of these energy losses F also coincides with the calculation of the solar radiation flux at the distance of the Earth, so that for the known value of the solar luminosity $L \simeq 10^{33}$ erg/s and the Earth-Sun distance $r \simeq 10^{13}$ cm, we have $F = L/4\pi r^2$,

$$F \simeq 10^7 \mathrm{erg/cm^2 s}.$$

However, since we can express the radiation flux F as a function of the energy per unit volume E/$\mathcal{V}$, the scale height Λ and the characteristic heating time τ,

$$F \simeq \frac{E}{\mathcal{V}}\frac{\Lambda}{\tau},$$

where $\Lambda \simeq 10^{10}$ cm is the scale height of the solar atmosphere, we can calculate the characteristic heating time of the solar corona τ,

$$\tau \simeq \frac{E}{\mathcal{V}}\frac{\Lambda}{F} \simeq \frac{10 \cdot 10^{10}}{10^7} \simeq 10^4\,\mathrm{s} \simeq 2\,\mathrm{h}.$$

In other words, the solar corona is maintained hot from whatever mechanism from hour to hour and this heating source needs to be operating continuously in the short period of hours. Otherwise, the solar corona cannot maintain its high temperature and radiate its energy flux F!

Thus, combining with the previous relationship of the characteristic heating time τ and the spatial scale of change of the magnetic field, ℓ, we have

$$\ell = \sqrt{10^5 \tau} \simeq 10^5\,\mathrm{cm} \simeq 1\,\mathrm{km}.$$

That is, the heating takes place through very thin electrical current sheets. Within these very thin current sheets, the electric current density is high enough so that despite the high value of the electrical conductivity σ, the ratio J^2/σ is ultimately significant, that is, we have substantial and significant heating of the plasma.

It is worth noting that the current spatial resolution of observations in the solar corona is of the order of 100 km. These thin current sheets correspond to a change in the polarity of the magnetic field on the scale ℓ, within which the opposing magnetic fields reconnect.

These simple but basic considerations led Parker to propose that the solar corona is heated via the very thin electric current sheets, the *nanoflare model* for *coronal heating,* an issue that we discuss in the following.

1972: Parker's Theorem on the Non-equilibrium of Non-invariant Fields and the Spontaneous Formation of Tangential Discontinuities

Plasma confinement in stable equilibrium states constitutes a fundamental and still unresolved question in plasma astrophysics and thermonuclear fusion research. In 1972 Parker focused on the topology of the lines of the magnetic force, and in particular on the non-equilibrium of non-invariant fields along some large-scale direction (Parker 1972; Tsinganos 1982). In the so-called *Parker theorem*, he proposed that if an equilibrium magnetic field is subjected to an arbitrary, small perturbation, then, under ideal plasma dynamics, the resulting magnetic field will in general not relax toward a smooth equilibrium, but rather, toward a state containing *tangential discontinuities.* And as such a singular state is approached, dissipation must eventually become important, leading to the onset of rapid magnetic reconnection (Pontin and Horning 2020; Pontin and Priest 2022) and energy dissipation. In other words, the formal theory of the *spontaneous* formation of tangential discontinuities leads to the result that magnetic fields subject to continuous deformation at very large magnetic or Lundquist numbers tend toward internal surfaces of tangential discontinuities, across which rapid reconnection continues until the field topology is reduced to such simple form that discontinuities are no longer an essential part of the static equilibrium. In the presence of continuing shuffling of the footpoints of the field – as is the case in the Sun – the reconnection never ceases (Low 1987, 2010). This principle seems to be the essential effect providing the dissipation of magnetic free energy that creates the X-ray *corona* of the Sun and similar stars. That is to say, it looks as though *spontaneous discontinuities* are the basis for much of X-ray astronomy (Parker 1979a, b, 1994).

This topological dissipation mechanism remains a key ingredient to the *nanoflare model* for *coronal heating.* That is, the prediction by Parker (1988) that the solar X-ray *corona* might be heated by a myriad of tiny *nanoflares*, miniature brightenings resembling solar *flares*, which would occur all over the surface of the Sun. In fact, observations of the Sun with high temporal and spatial resolution in UV and X-rays show that the emission from small isolated magnetic bipoles is intermittent and impulsive, while the steadier emission from larger bipoles appears as the sum of many individual impulses (Antolin et al. 2021). Parker refered to this basic unit of impulsive energy release as a *nanoflare*. The observations suggest, then, that the active X-ray *corona* of the Sun is to be understood as a swarm of such *nanoflares*. This interpretation suggests that the X-ray *corona* is created by the dissipation at the many tangential discontinuities arising spontaneously in the bipolar fields of the active regions of the Sun as a consequence of random continuous motion of the footpoints of the field in the photospheric convection. The quantitative characteristics of the process are inferred from the observed coronal heat input. This has become the leading theory to explain the coronal heating problem, and the recently observed *campfires* by the *Solar Orbiter* we discuss in the following may be the long-sought Parker's *nanoflares* (Fig. 1.77), related possibly to the so-called magnetic *switchbacks* of the magnetic field observed by the PSP at several solar radii, as we discuss in the following (Fig. 1.41).

Some of Parker's Other Seminal Contributions

The *generation of magnetic fields* in turbulent fluids, including the hydromagnetic dynamo problem, in planetary, solar, stellar and galactic environments were first studied by Parker (1955). A rotating sphere of conducting fluid can regenerate a dipole magnetic field. It is only sufficient for the angular velocity of rotation to vary with distance from the axis of rotation (Ω-effect) and for cyclonic fluid motions to be present (α-effect). The nonuniform rotation generates a toroidal field from the poloidal dipole field; the cyclones generate, from the toroidal field, via the Coriolis force, *loops* of flux in the meridional plane which coalesce to finally amplify the dipole field (Fig. 1.42). This is in a few words the classical theory of the α- and Ω-dynamo, which remains since then literally as the «α» and the «Ω» (the first and last letters in the Greek alphabet) for other dynamo theories later on. It is worth to note that the rotating sphere can be the liquid core of the Earth producing with such a dynamo mechanism the geomagnetic dipole field.

On the subject of the *magnetic buoyancy*, Parker published two seminal papers. In the first, Parker (1955) has shown that a horizontal magnetic flux tube in an electrically conducting atmosphere is buoyant and will tend to rise, in a direction opposite to gravity. This magnetic buoyancy is large enough to bring an occasional strand of magnetic flux from the general solar toroidal field up into the *photosphere*, by assuming field densities of a few hundred gauss at the bottom of the convective zone. Identifying the intersection of such ropes with the *photosphere* as the source of *sunspots*, several general characteristics of the spots are deduced, e.g., their east–west orientation, bipolarity, appearance only in low latitudes, migration, reversal of polarity, etc. And, secondly, Parker (1966) emphasized the importance of the *magnetic Rayleigh–Taylor buoyancy instability* in the Galactic disk, including the effects of cosmic ray and pressure, which explains the formation of interstellar cloud complexes. This so-called *Parker instability* of a galactic gas layer with a horizontal magnetic field supported against gravity by magnetic fields and *cosmic rays*, wherein the plasma falls down along the bent magnetic field lines is thought to be dynamically important for galaxy evolution, possibly promoting molecular cloud formation and the galactic dynamo.

Magnetic reconnection refers to the breaking and reconnecting of oppositely directed magnetic field lines in a plasma. In this process, magnetic field energy is converted to plasma kinetic and thermal energy. The first theoretical framework of magnetic reconnection was laid down by Sweet and Parker at a conference in 1956. Sweet (1958) pointed out that by pushing two plasmas with oppositely directed magnetic fields together, resistive diffusion is able to occur on a length scale much shorter than a typical equilibrium length scale. Parker was in attendance at this conference and developed scaling relations for this model during his return travel, i.e., in a semi-quantitative manner. He pointed out that two oppositely directed sunspot fields with scales of about 10^4 km could be merged by Sweet's mechanism, if shoved firmly together, in about two weeks, in contrast to their normal interdiffusion time which would be of the order of 600 years! Parker (1957) proposed that Sweet's mechanism gives a means of altering quickly the configuration of magnetic fields in ionized gases, allowing a stable field to go over into an unstable configuration, subsequently converting much of the magnetic energy into kinetic energy of the fluid. Thus, he suggested that Sweet's mechanism may be of considerable astrophysical importance, wherein reconnection is at the heart of many spectacular events in our solar system (see also Cargill et al. 2010).

The *shape of the solar wind bubble within the interstellar medium*, where our planetary system lives in and which has been known as the *Heliosphere*, is a problem which today still remains controversial. For long we compared this shape to a comet, with a round *nose* on one side and a long tail extending in the opposite direction. But in 2015 new computational models and data from Voyager 1, came to a different conclusion, i.e., that the solar wind bubble is actually shaped more like a *crescent*. That regenerated the discussion about the shape of the Heliosphere. It has been shown that the depletion of the pickup ions, cools the Heliosphere downstream of the termination shock, *deflating* it and leading to a narrower heliospheric sheet and a smaller and rounder shape, in agreement with energetic neutral atom observations by the Cassini spacecraft (Dialynas et al. 2017; Krimigis et al. 2019; Opher et al. 2020). By timing particles echoing off the boundary of the heliosphere and correlating them with ions measured by the twin Voyager spacecraft, Cassini data concluded that the Heliosphere is actually very nearly round and symmetrical: neither a comet nor a crescent or a croissant, but more like a deflated *beach ball*. All these new evolvements, bring us back six decades ago, when Parker (1961) made the first rough calculation of the shape of the cavity formed by a stellar wind in a large-scale interstellar magnetic field, showing that this cavity shape and dimensions for a stationary flow depend on the pressure at infinity. As the pressure increases, the cavity changes from a circular cylinder, through an increasingly bulbous shape of finite radius, to a sphere. He discussed observations of the *solar wind* with a termination shock (TS) at a radial distance in the interval 10–10^3 AU and that of the M giant α Herculis wherein the TS may extend many parsecs, comparing thus the gross dynamical features of the stellar-wind regions associated with G class main sequence and M giant stars for various interstellar environments. Note that the *solar wind* bubble acts as our solar system's shield against *cosmic rays* and the rest of the galaxy. This propagation of *cosmic rays* in the Heliosphere is generally described by a transport equation which is a special case of the general Fokker–Planck equation (Parker 1965). Hence, understanding the *solar wind* and the shape of the cavity the *solar wind* creates in the ISM is critical for understanding how these influence the rate of galactic *cosmic rays* which penetrate our solar system and are thus key ingredients in the recipe for life on Earth. Also, recent studies of exoplanets should compare our *Heliosphere* with *Astrospheres* around other stars. And, to understand the potential habitability of exoplanets, we need to understand if our *solar wind* cavity more closely resembles the relatively shortened astrosphere of BZ Cam, the long astrosphere of Mira, or has another totally different shape. Effectively, Parker started that comparison six decades ago, even when there were no observations of exoplanets at that time!

In a long series of papers entitled *Sunspots and the physics of magnetic flux tubes* starting in (1979) – with the author contributing couple of papers in this series – the basic properties and interactions of isolated twisted magnetic flux tubes, were studied. An analysis of the dynamical stability of a large flux tube suggests that the magnetic field of a sunspot must divide into many separate tubes within the first 1000 km below the surface. Buoyancy of the Wilson depression at the visible surface and probably also a downdraft beneath the sunspot hold the separate tubes in a loose cluster. Thus, Parker proposed that *sunspots* are a conglameration of isolated flux tubes, wherein hydrodynamic effects suggest that the separate flux tubes beneath the *sunspots* exert significant attractive forces on each other. In particular, in the strongly stratified convective zone of the Sun, a downdraft herds flux tubes together into a cluster, while an updraft disperses them. To account for the observed strong cohesion of the cluster of flux tubes that make up a sunspot, Parker proposed a downdraft of the order 2 km s^{-1} through the cluster of separate tubes beneath the sunspot. Furthermore, the mutual hydrodynamic forces between neighboring tubes parallel were calculated; tubes in a uniform flow are attracted or repelled, depending on whether they are side by side, or one ahead of the other, respectively. We have also shown that magnetic flux tubes in a fluid velocity field are intrinsically unstable, unless twisted. With the known dynamical limits on the degree of twisting, an upper limit on the diameter of the order of a few hundred kilometers is found for the stability of individual flux tubes in the *photosphere*. Observations show that tubes, except when clustered to form *sunspots*, are generally not larger than this amount. Therefore, it is suggested that the hydrodynamic instability of larger tubes is a prime cause of their absence from the solar *photosphere* (Fig. 1.25b, Tsinganos 1980).

The most stringent, mass-independent upper limit on the flux of magnetic monopoles is based upon the survival of the galactic magnetic fields, the so-called *Parker limit* (Turner et al. 1982), wherein it was explained that magnetic monopoles would short circuit the galactic magnetic field, just as electrons short-out electric fields in the environment of any highly conducting matter. The quantum theory of magnetic monopoles (MMs) and their charge g started with a paper by Paul Dirac in 1931 (the *Dirac quantization condition*). In this paper, Dirac showed that if any

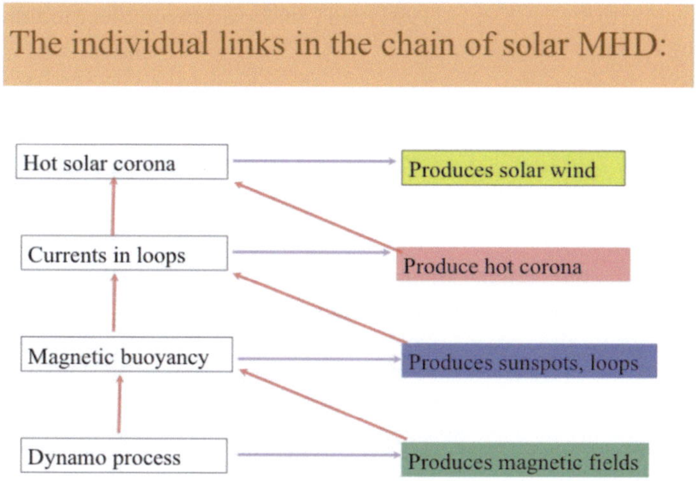

Fig. 1.35 Schematic diagram of solar MHD, as E. N. Parker viewed and established it, via his systematic research output. A similar comment appears in a tribute to E. N. Parker by Parker's former Ph.D. student, prof. Arnab Choudhuri

magnetic monopoles exist in the Universe, then all electric charges in the Universe must be quantized and the relation between the elementary electric charge e and the magnetic charge g is $g = n\, g_D = n\, (hc/2e) = (137/2)ne$, where n = 1, 2, 3,... and g_D is the unit of the magnetic charge. Grand unified and superstring theories, predict the existence of magnetic monopoles with a mass $M\_m \gtrsim 10^{16} \div 10^{17}$ GeV c^{-2}. However, there are no definite theoretical predictions of the *abundance* of cosmic MMs. A bound largely used is the so-called *Parker limit*, $F \lesssim 10^{-15}$ cm^{-2} s^{-1} sr^{-1} obtained by requiring that MMs do not short circuit the galactic magnetic field faster than the dynamo mechanism can regenerate it. By taking into account typical coherence lengths ($\ell \sim 1$ kpc) of the galactic magnetic fields, the limit becomes $F \lesssim m_{17}$ 10^{-15} cm^{-2} s^{-1} sr^{-1}, for $m_{17} \geq 1$, with $m_{17} = M\_m/10^{17}$ GeV c^{-2}. Similar considerations applied to the survival of an early seed of the galactic magnetic field yield a more stringent bound, the *Extended Parker Bound* (EPB): $F_{EPB} \lesssim m_{17}$ 10^{-16} cm^{-2} s^{-1} sr^{-1}.

Parker's name is littered across astrophysics and it is almost impossible to describe here all his scientific contributions. E. N. Parker published more than 420 scientific papers, in the great majority of which he is the sole author[8] and which have received more than 27,000 citations. He has also published the following monographs that are reference books in the field.

Interplanetary Dynamical Process (1963), Cosmical Magnetic Fields: Their Origin and their Activity (1979), Spontaneous Current Sheets in Magnetic Fields,

[8] To this rule, there are very few exceptions, e.g., Bernoulli Forces on an Elliptical Cylinder by a Nonuniform Flow, E. N. Parker and K. Tsinganos, Physics of Fluids, 22(10), 1847 – 1851 (1979).

with Application to Stellar X-rays (1994) and Conversations on Electric and Magnetic Fields in the Cosmos (2007).

1985: The Crete Meeting, a Review of the MHD Approach to Describe Cosmical Phenomena

By the end of the 20th century, modern observations including those recently with the Hubble Space Telescope, have revealed that the Universe is replete with analogous outflows from all kinds of objects, ranging from the rich varieties of stars to the galaxies. Parker initiated the use of the magnetohydrodynamic (MHD) approach to describe all such cosmical phenomena. Since then, MHD is increasingly used in several key problems of solar plasmas, such as, magnetoconvection and magnetic field generation, sunspots and prominences, MHD equilibria in plasma loops and arcades, magnetic non-equilibrium and coronal heating, coronal mass ejections and the acceleration of the solar wind. On the other hand, the same MHD approach has also increasingly been used in several central and puzzling aspects of more distant astrophysical plasmas, such as, stellar winds across the main sequence, the dynamics of the interstellar medium and the galactic center, collimated outflows from young stellar objects and accretion disks, molecular outflows and planetary nebulae, jets associated with enigmatic binaries and symbiotic stars, relativistic flows associated with superluminal micro quasars in our own galaxy, astrophysical *jets* from galaxies, or far away active galactic nuclei and quasars probably fueled by supermassive *black holes*. However, although the underlying similarities in the physics of these phenomena are rather striking, nevertheless quite often researchers in one field are unaware of the methods employed in the study of the other.

To emphasize the similarities of astrophysical outflows, as they are described via Parker's MHD theory, we organized an Advanced Study Institute on Solar and Astrophysical Magnetohydrodynamic Flows in June 11–23, 1995 at Fodele, the birth place of the Cretan painter D. Theotokopoulos (El Greco), in Crete, Greece. This conference was dedicated to Prof. Parker who has been the inspiration for many of us to study the role of magnetism in the Cosmos. Hundreds of researchers, experts in MHD and participants from 25 countries gathered to honor Eugene Parker on the occasion of his retirement from the University of Chicago, after more than 50 years in Solar physics, Heliophysics and Astrophysics. What attracted them was also the scientific program which was naturally dedicated to Eugene Parker's interests. Thus, appropriately, special reviews were presented on topics Parker has immensely contributed, such as, new developments in the Parker instability in the interstellar medium, the spontaneous formation of electric current sheets in magnetized plasmas, magnetic reconnection, sunspot modeling, coronal heating, stellar winds and also a generalization of his model of the *solar wind* by our group via *exact self- similar* solutions of the MHD equations.

The presentations during this conference,[9] included a discussion of the generation, storage and eruption of magnetic fields in our Sun deep in its convective layers, as prototypes of the fields of other stars, galaxies and accretion disks. The erupted through the *photosphere* magnetic field forms intense flux tubes and *sunspots* and thus it siphon flows and Alfven waves along such magnetic flux tubes. The impressive high-resolution pictures of the Sun in X- rays by modern satellites were reviewed and indeed show a solar atmosphere constituted of many steady and transient magnetic *loops* like those expected to emerge from deep within the solar convective zone. The X-ray data also show clear evidence of intense heating due to magnetic reconnection. Thus, Parker's spontaneous formation of current sheets in astrophysical magnetic fields and the various MHD processes in the solar *corona* (*flares*, *coronal mass ejections*, etc.) are today observed everywhere. Parker's ideas have been tested by *numerical simulations* of solar and astrophysical MHD flows (magnetic *loops*, *flares*, solar X-ray *jets* as well as the nonlinear evolution of the Parker instability and magnetically accelerated astrophysical *jets*). Observational data from the south and north solar polar pass of the Ulysses spacecraft are showing, for example, that the *solar wind* is indeed a rather meridionally anisotropic outflow with the proton speed at large heliolatitudes almost twice its equatorial value. Stellar winds from solar – as well as non-solar-type stars were also reviewed in the conference and a discussion of some general implications of solar and stellar activity, such as the profound effects of the variation of solar activity to the terrestrial climate (Figs. 1.7, 1.18 and 1.19).

Like Chandrasekhar, Gene Parker was not only a great physicist but at the same time he remained humble and approachable: till today I recall that whenever I was anxiously knocking his door to report on my research results, Gene would willingly interrupt his work to chat with me. His favorite advice was:

> when your math becomes complicated stop and think again about your physics...

Brief biographical elements of E. N. Parker

Prof. Eugene Newman Parker (1927–2022) has been considered by many, as the most prominent solar physicist and astrophysicist of his time. He was born in Houghton, Michigan, on June 10, 1927. He received a BS in physics from Michigan State University (1948) and a Ph.D. in physics from the California Institute of Technology (1951), supervised by Howard P. Robertson. Subsequently, Parker was an Instructor at the Department of Mathematics of the University of Utah (1951–1953) and then a research associate with Walter M.

[9] Published in the Volume, *Solar and Astrophysical Magnetohydrodynamic Flows*, K. Tsinganos (ed.), Kluwer Academic Publishers, Mathematical and Physical Sciences, Vol. 461, pages 745 (1986).

Fig. 1.36 Participants of the 15/day conference in Fodele, Crete (birthplace of D. Theotokopoulos – El Greco), organized by the author in honor of Eugene Parker (1995). With the arrow in the last row (!) is E. N. Parker

Fig. 1.37 Gene Parker (center) surrounded by four of his many former Ph.D. students. (*from left to right*) Eugene Levy, Tom Bogdan, Boon Chye Low and Kanaris Tsinganos, during Parker's 60th birthday (1987). (Photo from the author's photo archive)

Fig. 1.38 (from left) Eugene Parker, the author, Robert Rosner and Thanassis Economou (Senior scientist at the Enrico Fermi Institute of the University of Chicago), during a mtg at the Un. of Chicago (2017), dedicated to the 60th birthday of prof. R. Rosner (president of the American Physical Society in 2023) and professor at Harvard University when the author was a Research Associate at the Harvard–Smithsonian Center for Astrophysics and a Lecturer at Harvard University (1981–1985). (From author's photo archive)

Elsasser in the Department of Physics. Then, Parker moved to the University of Chicago in June 1955 as a research associate with John A. Simpson. In 1957, Parker was appointed an assistant professor in the Department of Physics and the Enrico Fermi Institute, and professor in 1962. He was also appointed professor in the Department of Astronomy and Astrophysics (1967). He was elected to the National Academy of Sciences in 1967. Parker retired from the University of Chicago in 1995, but continued to publish books and articles.

Parker received numerous scientific awards over the years, beginning with the Space Science Award from the American Institute of Aeronautics and Astronautics (1964) and including among others the Henrijk Arctowski Medal of the National Academy of Science (1967), the Hale Award of the American Astronomical Society (1978), the United States National Medal of Science (1989), the William Bowie Medal of the American Geophysical Union (1990), the Gold Medal of the Royal Astronomical Society (1992) and the Bruce Medal of the Astronomical Society of the Pacific (1997). Parker was the recipient

Fig. 1.39 Four generations of researchers on MHD Flows. E.N. Parker (gen 1, with the hat), the author (gen 2, PhD student with Parker, left), N. Vlahakis and C. Sauty (gen 3, in the middle, PhD students with the author), and Z. Meliani (gen 4, PhD student with C. Sauty, on the right) in front of the National Observatory of Athens (September 15th, 2003)

of the 2003 Kyoto Prize in basic science and also the 2020 Crafoord Prize in Astronomy, which is awarded every three years by the Royal Swedish Academy of Science and consists of a gold medal and a sum of about $600,000. The Academy recognized Gene Parker for his pioneering and fundamental studies of the *solar wind* and the role of magnetic fields from the solar to stellar and to galactic scales.

NASA honored E. N. Parker's scientific contributions in 2018 by naming a spacecraft after him, the *Parker Solar Probe* (PSP) spacecraft – the first time NASA had named a spacecraft after a living person – which has been exploring the Sun since 2018, aiming to approach gradually to a distance of about 10 solar radii, where the temperature exceeds 3.5 million degrees. In August 2018, at the age of 91, he became the first person to witness the successful launch of the namesake spacecraft, which has since provided unprecedented views of the Sun.

At the COSPAR General Assembly in Athens (July 2022), Parker was leading the SOC of the session that the author organized on *Exploring the cradle of the solar wind with Parker Solar Probe/Solar Orbiter/Proba-3: what do we really know about the inner solar corona?* Parker's last publication was entitled *Exploring the innermost solar atmosphere* (Nature Astronomy, 2020).

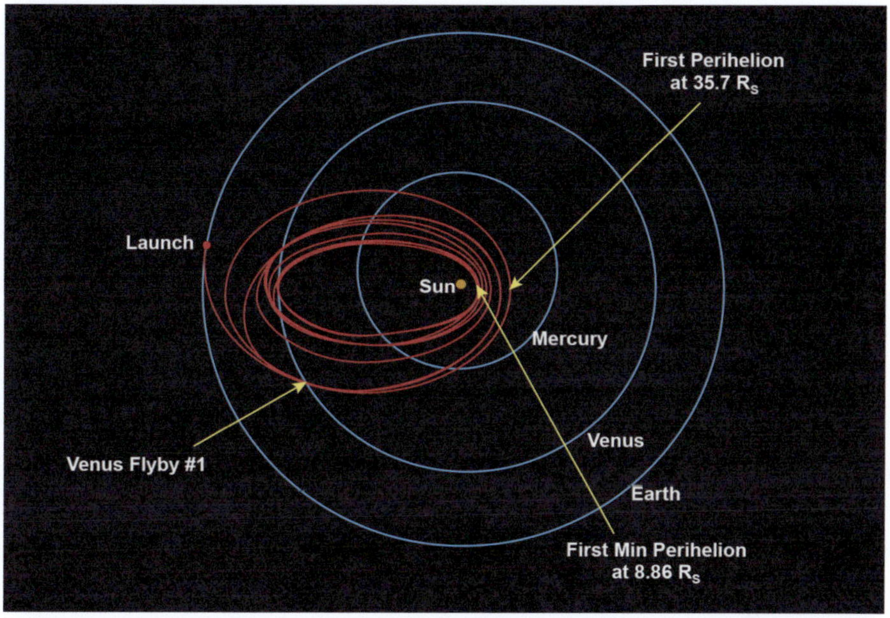

Fig. 1.40 The Parker Solar Probe was launched on August 12, 2018. The first close pass took place on November 6, 2018, roughly three months after the August launch. On September 15, 2025, completed its 25th close approach wherein the probe has completed over 20 orbits, successfully flying through the corona, continuing to set records for the closest distance to the Sun and highest speed. On December 24, 2024, NASA's Parker Solar Probe achieved a historic milestone, passing within 6.1 million km of the Sun's surface, the closest-ever approach to the Sun by a human-made object. During this 23rd perihelion, the probe reached top speeds of approx. 692,000 km/h, diving into the solar corona to study the solar wind, magnetic fields, and the coronal structure (Illustration by NASA, JHU/APL).

1959–2024: 65 Years of Observing the *Solar Wind* – the *Parker Solar Probe*

The first in situ measurement of the speed and density of the *solar wind* took place by the Soviet spacecraft Luna 1 (1959), which had as destination the Moon and also the American spacecraft *Mariner II* (1962) *en route* to *Venus*. Both data verified Parker's theory. A close exploration of the Sun, the solar *corona* and wind is now under way with the *Parker Solar Probe* (PSP, Fig. 1.40), which after 7 close flybys of *Venus* over the next 7 years (Fig. 1.40), it will gradually decrease the perihelion of its orbit, to approach the Sun to about 9 solar radii. Figure 1.41 APL shows Eugene Parker, during his October 2017 visit to the spacecraft that bears his name, NASA's *Parker Solar Probe* (PSP). Engineers in the clean room at the Johns Hopkins Applied Physics Laboratory in Laurel, Maryland, where the PSP was designed and built, explained to him the various spacecraft instruments, that will collect the data as the spacecraft

Fig. 1.41 From the visit of E. N. Parker, to Johns Hopkins University, Applied Physics Laboratory (JHU/APL), where the spacecraft that bears his name, NASA's Parker Solar Probe (PSP) was built. Behind Parker is Dr. T. Krimigis, Director of the Space Department (1991–2004) and the engineers in the clean room of the JHU/APL, Laurel, Maryland, displaying the spacecraft's instruments. (*Source* Johns Hopkins University Applied Physics Laboratory, Ed Whitman)

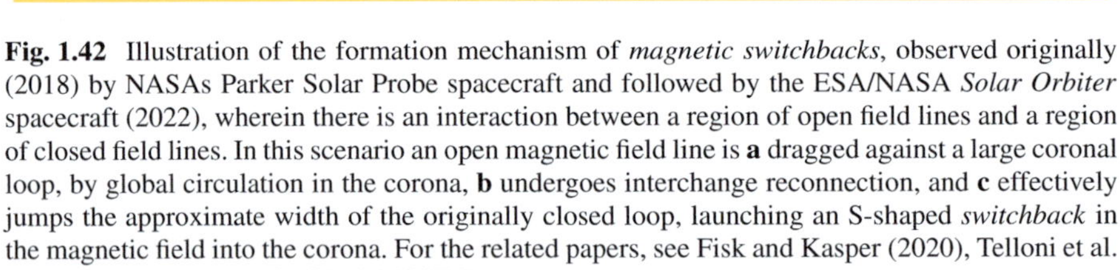

Fig. 1.42 Illustration of the formation mechanism of *magnetic switchbacks*, observed originally (2018) by NASAs Parker Solar Probe spacecraft and followed by the ESA/NASA *Solar Orbiter* spacecraft (2022), wherein there is an interaction between a region of open field lines and a region of closed field lines. In this scenario an open magnetic field line is **a** dragged against a large coronal loop, by global circulation in the corona, **b** undergoes interchange reconnection, and **c** effectively jumps the approximate width of the originally closed loop, launching an S-shaped *switchback* in the magnetic field into the corona. For the related papers, see Fisk and Kasper (2020), Telloni et al. (2022) (*Image Credit* ESA/NASA/GSFC)

will pass through the superheated atmosphere of the Sun, passing very close to it, where no other spacecraft has been, until now.

One of the recent findings of the PSP is the *magnetic switchbacks* (Fig. 1.42), sudden, large radial deflections of the *solar wind* magnetic field, widely revealed in interplanetary space by the *Parker Solar Probe*. According to NASA, during *Parker Solar Probe*'s 6th flyby of the Sun, it found that these flips of the *solar wind* magnetic field are aligned with magnetic *funnels,* which litter the Sun's surface. These funnels emerge from between supergranules. The *switchbacks*' formation mechanism and sources are still unresolved, although there are several candidate mechanisms, such as interchange reconnection (Fig. 1.42), Alfvénic turbulence, shear-driven Kelvin–Helmholtz instabilities, and geometrical effects related to the Parker spiral.

1961: Discovering the Solar Dynamo (the Electric Generator of the Sun)

The creation of the magnetic fields inside the Earth, the Sun and of other stars is a basic and particularly interesting problem of modern physics. Today we believe that these cosmic magnetic fields are produced by the *dynamo* mechanism. The first dynamo was built in 1831 by the English chemist and physicist Faraday (1791–1867). The best known and simplest dynamo is the *dynamo* of bicycles, where the kinetic energy of the bicycle wheel is converted into electricity and the electric current illuminates the lamp (light) of the bicycle. In our cars, the dynamo is, together with the starter and the battery, among the most important electrical components of the car.

The dynamo converts kinetic energy (the motion of the crankshaft via a belt) to electric energy. Hence, the basic parts of a dynamo are the *rotor* and the *stator*. The rotor is essentially an electromagnet that, when rotated, produces an alternating electric current, following the law of the electromagnetic induction. A voltage rectifier follows that converts the produced current to DC. This final electric current is consumed from our car's electrical systems, while part of it is also stored in the battery. On a larger scale, dynamo is found in power plants, where the mechanical energy which supplies the dynamo comes from wind in wind turbines, waterfalls in hydroelectric plants, steam in locomotives, etc. In hydroelectric plants, the impact of water sets in motion a turbine, which is connected to a frame in an external magnetic field. The rotation of this frame in the external magnetic, produces the electric voltage.

The modern understanding of the mechanism for the appearance of *sunspots* began with George Ellery Hale (1908), who associated *sunspots* with solar magnetic fields. Hale suggested that the sunspot cycle lasts 22 years, a period covering 2 reversals of the dipolar solar field. Horace W. Babcock proposed (1961) a qualitative model for the solar dynamo. Indeed, on a macroscopic scale, it is observed that the Sun has an oscillating magnetic field which is characterized by a periodicity of 22 years. This oscillation is known as the Babcock–Leighton dynamo–cycle and basically consists of an oscillatory transfer of energy between the poloidal and toroidal components of the magnetic field.

The three components of the dynamo mechanism also exist in the interiors of the Earth and the other stars and thus the conditions for the generation of the magnetic field are met. The magnetic field after its generation it is quickly transported due to *magnetic buoyancy* to the solar *photosphere*, wherein it creates *sunspots*, magnetic

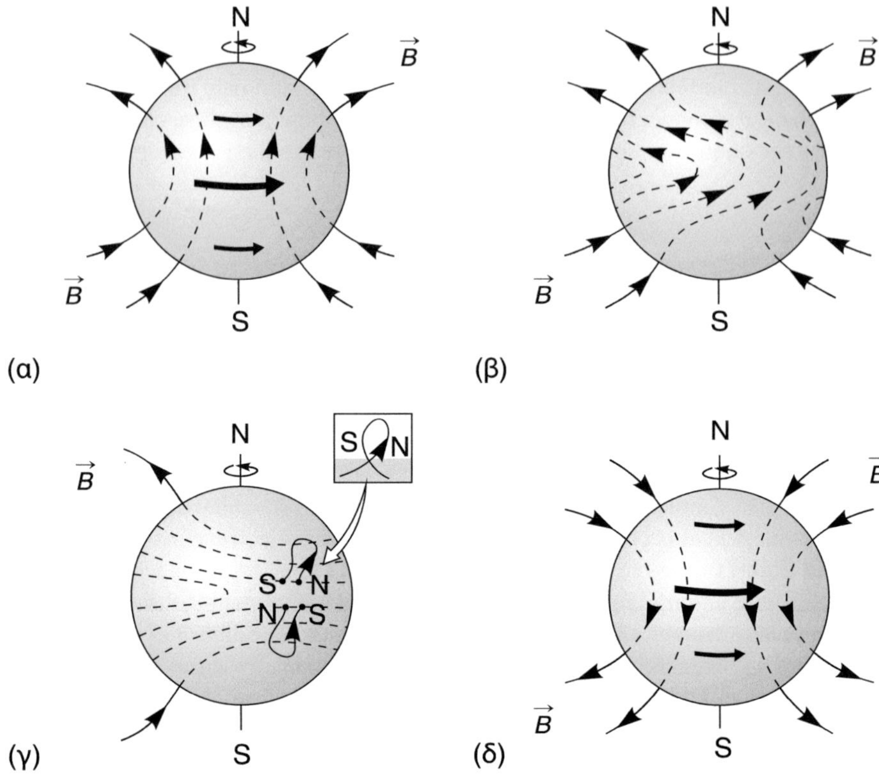

Fig. 1.43 Schematic representation of the creation of the solar magnetic field by the dynamo mechanism, illustrating the main processes at work in the solar dynamo model. The Ω-effect (top) transforms an initial primary poloidal field into a toroidal field by means of the differential rotation. Hence, the initial polar magnetic field (α) creates a toroidal magnetic field (β) within the convective zone due to the solar differential rotation. The generation of a poloidal field is next followed by the α-effect (γ) and/or by the Babcock–Leighton mechanism (bottom). In the α-effect case, the toroidal field at the base of the convection zone is subject to cyclonic turbulence. Secondary small-scale poloidal fields are thereby created, and produce on average a new, large-scale, poloidal field (δ). Alternatively, in the Babcock–Leighton mechanism, the primary process for poloidal field regeneration is the formation of sunspots at the solar surface from the rise of buoyant toroidal magnetic flux tubes from the base of the convection zone. The magnetic fields of those sunspots nearest to the equator in each hemisphere diffuse and reconnect, while the field due to those sunspots closer to the poles has a polarity opposite to the current one, which initiates a polarity reversal. The newly formed polar magnetic flux is transported by the meridional flow to the deeper layers of the convection zone, thereby creating a new large-scale poloidal field (Adapted from Parker (1955))

flux tubes and magnetic *loops*, while when opposite fields come into contact they create all the solar magnetic fireworks we will discuss later on.

Hence, the necessary components for the operation of the dynamo are the availability (a) of an initial magnetic field and (b) the supply of mechanical energy. However, we may also have a supply of mechanical energy in a fluid that "*boils*", as happens in the solar interior, due to the existence of anodic and cathodic currents.

Additionally, in the Sun there is also the differential rotational motion of the fluid which also plays an important role, transforming an original poloidal magnetic field into a toroidal one, as we will see below. That is, inside the Sun there exist the necessary components for the operation of the dynamo: some initial magnetic field, the existence of anodic and cathodic currents and the differential rotational motion of the plasma.

According to Fig. 1.43, we can distinguish the following phases of the dynamo that lead to a reversal of the solar magnetic field within 11 years:

1. Starting within the convective zone, 2–3 years before the start of the cycle, there exists an initial poloidal magnetic field, with its magnetic lines emerging at the high heliographic latitudes of the solar poles. In the solar interior, these magnetic lines are confined to a depth at the bottom of the convective zone, immediately below the *photosphere*. During the solar minimum, the toroidal magnetic field has its minimum value while the bipolar field a maximum value. This the time wherein, the number of *sunspots* is minimal.
2. Due to the solar differential rotation, where the equatorial regions rotate faster than the polar regions and the magnetic lines are frozen in the plasma (Tsinganos 2017), after several rotations, the magnetic field lines wrap several times around the solar center, with opposite direction in the two hemispheres. At the same time, the intensity of the magnetic field increases, because the density of the dynamic lines increases. Kinetic energy from the differential rotation is converted into magnetic energy and the field is strengthened from an initial value of the order of a few Gauss to values of a few hundreds Gauss, within about 3 years. The amplification of the toroidal magnetic field is done by the so-called Ω-mechanism, which is basically due to the freezing of the magnetic lines in the plasma.
3. Then, due to the convective cells which transfer thermal energy through the transport of mass in the convective zone and the action of magnetic buoyancy, the strong toroidal magnetic field created in the previous stage emerges in the *photosphere* from the convective zone.
4. During the ascent to the *photosphere* of the magnetic *loops*, Coriolis forces are exerted on them, creating meridian magnetic *loops* (mechanism-α). Due to the magnetic buoyancy, these *loops* emerge in the *photosphere* as poloidal regions of opposite orientation in the two hemispheres, as shown in Fig. 1.22. Sunspot production is maximum at this stage.
5. The created meridional magnetic *loops* after their emergence in the *photosphere*, reconnect with the pre-existing meridional magnetic field and finally, create a new polar magnetic field of opposite polarity wrt the initial poloidal polarity when the cycle started, as in Fig. 1.43. During this phase the appearance of *sunspots* decreases continuously.

Therefore, we can summarize the magnetic solar cycle as follows. At the beginning of the cycle, the magnetic field is roughly dipolar. As the Sun rotates, the solar equator rotates faster than the poles, due to the differential rotation, so the magnetic field lines which are frozen in the solar plasma – the charged particles are spinning around the magnetic lines – they lengthen in the direction of the equator like elastic strings. When strengthened in this way the azimuthal magnetic field emerges in the *photosphere* due to magnetic buoyancyas bubbles rise to the surface of a swimming pool – wherein they create *sunspots*, magnetic flux tubes, magnetic *loops* and a host of other magnetic phenomena. The Ω-effect converts poloidal magnetic field to toroidal by the differential rotation. The α-effect twists the toroidal field by the Coriolis force, as it rises within the convective zone (Choudhuri and Gilman 1987; Choudhuri et al. 1995, 2012; Hazra et al. 2014).

In Fig. 1.43 are illustrated these three physical processes which contribute in the operation of solar dynamo. First, the Ω-effect (α) transforms a primary poloidal field into a toroidal field by means of the differential rotation. Next, the poloidal field regeneration is accomplished either by the α-effect (γ) and/or by the Babcock–Leighton mechanism. In the α-effect case, the toroidal field at the base of the *convection zone* is subject to cyclonic turbulence. Secondary small-scale poloidal fields are thereby created, and produce on average a new, large-scale, poloidal field. In the Babcock–Leighton mechanism, the primary process for poloidal field regeneration is the formation of *sunspots* at the solar surface from the rise of buoyant toroidal magnetic flux tubes from the base of the *convection zone*. The magnetic fields of those *sunspots* nearest to the equator in each hemisphere diffuse and reconnect, while the field due to those *sunspots* closer to the poles has a polarity opposite to the current one, which initiates a polarity reversal (δ). The newly formed polar magnetic flux is transported by the meridional flow to the deeper layers of the *convection zone*, thereby creating a new large-scale poloidal field.

A Simple Example of the Creation of a Magnetic Field via the Dynamo Mechanism

Suppose we have a weak initial magnetic field $\vec{B} = B_0\hat{y}$ in a plasma of infinite electrical conductivity σ. Suppose we also impose a velocity field which acts from time $t = 0$ onwards,

$$\vec{V} = V_0 e^{-\frac{y^2}{y_0^2}}\hat{x}.$$

Then,

$$\vec{V} \times \vec{B} = V_0 B_0 e^{-\frac{y^2}{y_0^2}}\hat{z},$$

and,

$$\vec{\nabla} \times (\vec{V} \times \vec{B}) = \hat{x} V_0 B_0 \frac{\partial}{\partial y}\left[e^{-\frac{y^2}{y_0^2}}\right] = -\hat{x} V_0 B_0 \frac{2y}{y_0^2} e^{-\frac{y^2}{y_0^2}} = \frac{\partial \vec{B}}{\partial t}.$$

Therefore, a new magnetic field is created in the $\hat{x}$ direction with a magnitude that linearly increases with time,

$$B_x = -\frac{2V_0 B_0 y t}{y_0^2} e^{-\frac{y^2}{y_0^2}}.$$

Thus, from an initial weak magnetic field in the $\hat{y}$ direction, a strong magnetic field is produced – mainly near the $\hat{y}$ axis – in the orthogonal $\hat{x}$ direction.

1962: Helioseismology (Listening to the Music and Thus Deciphering the Secrets of the Solar Interior)

Sir A. Eddington was one of the most influential English astrophysicists of the early twentieth century. He was one of the few that had understood, believed and embraced the General Theory of Relativity (GTR) that Einstein had then just proposed. Thus, he had made the first famous verification of the GTR, observing during a total solar eclipse in Africa on May 29, 1919, the bending of stellar rays, when they pass very close, almost tangentially, from the solar disk, as we discuss later on in this volume.

In 1927 Cambridge University Press published A. Eddington's book *The internal constitution of the stars*, where he mentions: "... at first sight it would seem that the deep interior of the Sun and stars is less accessible to scientific investigation than any other region of the Universe. Our telescopes may probe farther and farther into the depths of space; but how can we ever obtain certain knowledge of that which is hidden behind substantial barriers? What appliance can pierce through the outer layers of a star and test the conditions within?".

Eddington goes on to state that ultimately the only possible way to examine what is happening inside the Sun and other stars is by using basic physics and thus only through it to determine the structure of a star. Ultimately, this is indeed the most important and sure way of progress in astrophysics. However, with the emergence of the field of helioseismology, he was later proven wrong in his first claim, because there are also observatories and other ways to *penetrate* under the *photosphere* and *examine* the interior of the Sun!

Thus, one of the largest observational discoveries of solar physics took place in 1962. Then, Caltech's physics professor Robert Leighton (Fig. 1.44), and his doctoral students Robert Noyes and George Simon, first noticed that the solar *photosphere boils* with a period of 5 min and a wavelength 1,000–5,000 km.

This is detected using the Doppler effect: when the gas emitting photospheric line moves outward (i.e., toward us), we see its spectral emission line shifted toward the blue part of the spectrum, while when falling toward the *photosphere*, the wavelength of the emission line is red-shifted.

Ulrich later (1970) elaborated theoretically on these observations by Leighton and collaborators and showed that these 5-min oscillations are acoustic waves trapped

Fig. 1.44 Robert Leighton (1919–1997)

below the *photosphere*. The power of these sounds in a wave vector versus frequency diagram is significant only at certain characteristic eigenfrequencies around 3 mHz, several octaves below the human audible frequencies (16–20,000 Hz). Today, thousands of such eigenfrequencies of the Sun's oscillation have been observed with great precision by various international collaborations.

This discovery launched the relatively new field of helioseismology, Fig. 1.45 (and asteroseismology) through which we can today calculate with amazing precision the structure, dynamics and properties of the solar and stellar interiors, Christensen-Dalsgaard (2021). These are oscillations that have characteristic frequencies and wavelengths which are determined by the physical conditions prevailing inside the Sun and the stars. This finding, along with the corresponding theoretical analysis, made it possible to use the oscillations as a diagnostic method for the interior of the Sun. Thus, we can determine critical physical parameters of the solar interior, such as the speed of sound, the dependence of density and temperature on solar depth, the internal rotation speed, the depth of the *convection zone*, the relative abundance of the chemical element helium, the ratio of the abundance of heavy elements to hydrogen in the transport zone, etc.

It is noteworthy that helioseismology has brought about a number of scientific revolutions, the most important of which is that the flux of neutrinos coming from the solar core is very accurately predicted by the standard solar model, and to solve the solar neutrino problem, nothing should change in the solar model, but instead we should change some understanding of particle physics. Thus, by allowing a finite mass of the three neutrinos, arose the oscillations between the three forms of neutrinos.

Helioseismology also made it possible to calculate precisely the gravitational potential of the Sun (a calculation of the quadrupole and higher-order moments) in agreement with the General Theory of Relativity. The first even calculations of the internal solar rotation as a function of depth and heliographic latitude yielded a solidly rotating inner and a differentially rotating overlying *convection zone*. The shell between these two regions, also known as the *tachocline*, plays a critical role in the operation of the solar dynamo: although it coincides with the base of the *convection zone*, it is a separate part of the solar interior.

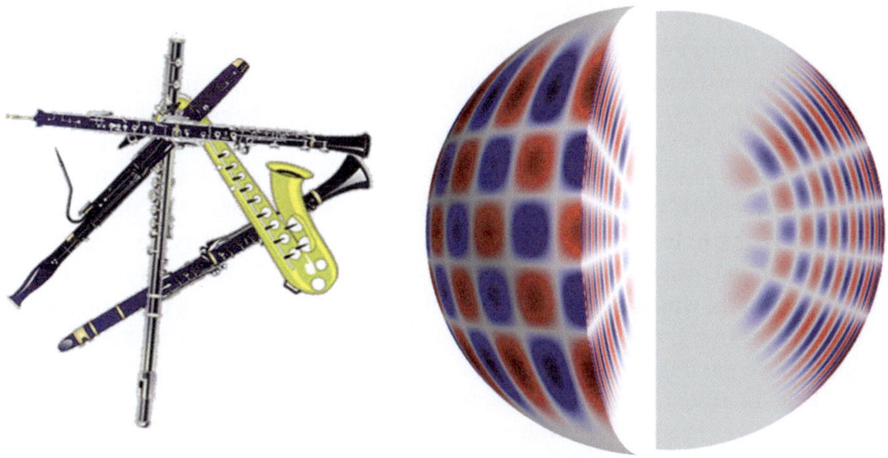

Fig. 1.45 Air trapped in a musician's pipe instrument sounds in different characteristic tones of music. Likewise, gas trapped in the solar interior turns the Sun into a giant musical instrument. Although they cannot be "heard" by the human ear, the characteristic tones of acoustic solar waves appear in an "up-and-down" pattern of oscillations in the solar photosphere. The particular mode of vibration in the image has n = 14 radial maxima–minima (14 radial nodes), l = 20 meridional maxima–minima (20 meridional nodes) and m = 16 azimuthal maxima–minima (equal azimuthal nodes). The interior shows the radial displacement computed using a standard solar model. Note that the increase in the speed of sound as these waves approach the dolar center causes a corresponding increase in the acoustic wavelength (Courtesy of Wikipedia Commons, licensed under the Creative Commons Attribution-Share Alike 4.0 International license. Right part of the figure is adapted with permission also from J. Christensen-Dalsgaard)

Turbulent motions in the solar convective zone is the stimulus for the production within the Sun of waves similar to earthquakes propagating inside the Earth Figs. 1.46 and 1.47. The produced longitudinal and transverse waves at the epicenter of an earthquake propagate inside the Earth, where the usual variation in the speed of sound is to increase with depth. Therefore, they are refracted and reach the surface, where they can be recorded by seismic instruments. In the Sun, the propagating waves can be common standing acoustic waves (pressure, or p-modes), where the restoring force is the increase/decrease of the gas pressure, just as it happens with the propagation of our voice in the air which surrounds us. They have frequencies greater than 1 mHz, mostly in the interval 2–4 mHz, width around 1,000 km and bring to the *photosphere* movements at speeds of around 500 m per second. These are the five-minute oscillations of the solar *photosphere* with an angular range from zero degrees (purely radial displacements) up to a few hundred degrees.

The second possibility is to have gravity waves (gravity, or g-modes), where the restoring force is buoyancy and therefore indirectly gravity, from where they also borrow their name. They have lower frequencies (0–0.4 mHz) and are located within the radiative zone of the solar interior. They are not detected in the *photosphere*. Finally, we may have gravitational waves nearby the *photosphere* (f-modes, or surface waves) analogous to water waves appearing at the air-sea interface. Thus, the Sun acts as a loudspeaker allowing the propagation of only those disturbances which have the appropriate wavelength, creating standing waves and where some frequencies are amplified via a constructive interference. These waves are transmitted till the outer *photosphere* and are observed at almost all observations of the Sun (Fig. 1.46).

According to physics, in every kind of wave there is a characteristic functional relation between the cyclic frequency (ω) and the wavenumber (k), the so-called dispersion relation. In the case of the Sun with spherical symmetry, to describe the *eigenoscillations*, instead of the wave of number k, three integers corresponding to the radial order n, which expresses the number of wave nodes from the center to the surface, the degree l, which expresses how many wavelengths are contained in the circumference passing through the poles, and the azimuth number m.

But how do acoustic waves propagate within the interior of the Sun? First, when sound propagates in a medium where the temperature (and the speed of sound which is proportional to the square root of the temperature) increases in the direction of propagation, as in the case of a disturbance in Fig. 1.47, then the refracted ray moves away from the perpendicular at the point of incidence. This causes the acoustic beam

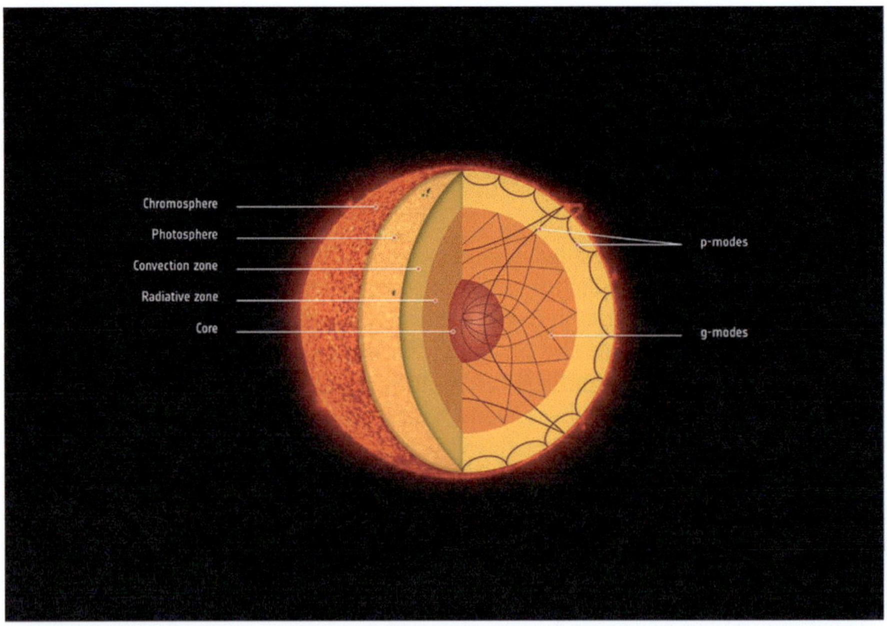

Fig. 1.46 Schematic diagram of the propagation of acoustic waves (p) and gravity waves (g) within the solar interior, where standing acoustic waves (pressure, or p-modes) are formed with frequencies 2–4 mHz, i.e., with periods around 5 min wherein the restoring force is gas pressure fluctuations (as with the propagation of our voice) and gravity waves (gravity, or, g-modes) at lower frequencies (0–0.4 mHz), with periods around 50 min, wherein the restoring force is buoyancy due to the existence of gravity. While p-mode oscillations are able to travel from the surface all the way to the core, g-modes are unable to escape the radiative zone (*Credit* ESA)

to form a concave-up curve, as in Fig. 1.47. We have the same phenomenon with the propagation of acoustic sound waves through the Earth's interior. And there the speed of sound increases with distance, with the result that the waves are refracted and the acoustic ray traces a curve with the concave parts upwards.

In the solar interior, temperature increases with depth and so does the speed of sound. This results in a perturbation which, starting from the *photosphere* follows a trajectory which is concave upward, reaches a lower point and then returns to the *photosphere*, depending on the value of the radial number l. For different values of l, following a similar trajectory acoustic ray reaches a lower point before returning to the *photosphere*. This means that for each disturbance and each particular mode of oscillation (i.e., for a given value of l), we will observe in the *photosphere* an oscillation that will have an observed value of frequency (e.g., 3 mHz) and also an observed wavelength (e.g., $\lambda = 1{,}000$ km). The next step is to reproduce this mode of oscillation with some particular function of the variation of the temperature (and sound speed) with the depth of the *convection zone*. This way we can finally find the law of increasing temperature with depth within the *convection zone*, if we manage to reproduce the data of the observations of oscillations in the *photosphere*.

With the modern methods of helioseismology we have the possibility to even calculate how much hydrogen has been burned to date in the solar core! In practical matters, because the acoustic solar oscillations change their frequency by 1 tenth of a millionth during the 11/year solar cycle, this gives us a new possibility of calculating the exact duration of the solar cycle which affects *Space Weather*, through *flares*, etc.

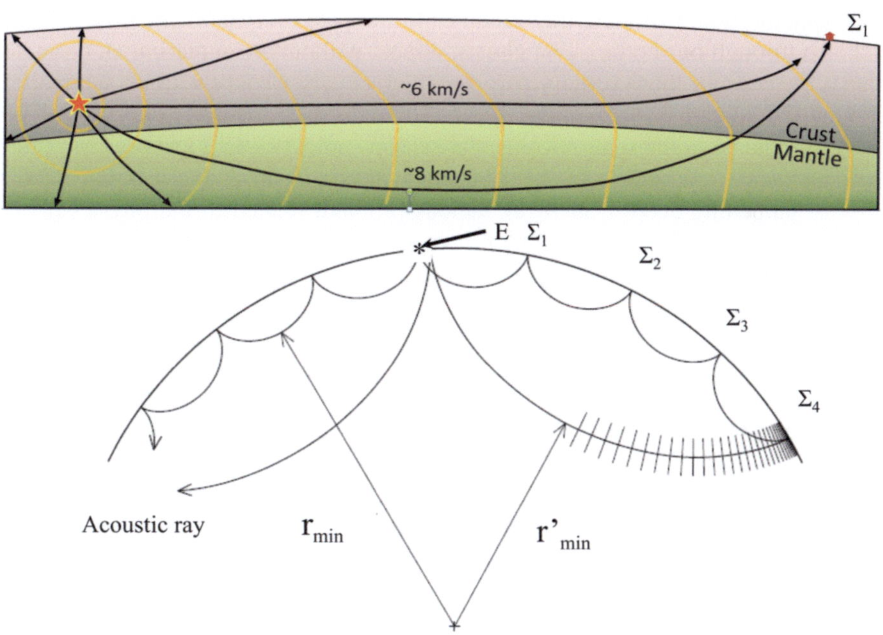

Fig. 1.47 Schematic analogy of seismic waves (above) with solar acoustic waves

1968: The Famous Solar Neutrino Problem

Neutrinos are the smallest in mass elementary particles. At the same time they are also the most numerous particles in the Universe, after photons. Despite the fact that neutrinos interact very weakly and rarely with our known matter, their total number is so large, that they significantly affect the overall structure and especially the evolution of the Universe. Neutrinos are created during the interactions of cosmic radiation with the atoms that make up the Earth's atmosphere, while others are produced during thermonuclear reactions in the center of the Sun and stars. The interaction of neutrinos with matter is very weak and so almost nothing can stop them. That's why neutrinos are nature's most elusive and elusive particles!

Trillions of neutrinos pass through our bodies every second, while reading this text. And yet, we don't realize them. Specifically, about one hundred billion neutrinos pass through one square centimeter of our body. And it is also worth remembering that a sunny summer day, when we are sunbathing on the beach, we are bombarded with hundreds of millions more photons, i.e. 10^{18} photons. Both neutrinos and photons come from the Sun's core, i.e., the deep solar interior. Neutrinos reach the solar *photosphere* from the center of the Sun in about 2 s, while photons in tens to hundreds of thousands of years and after being transformed by invisibles high-energy photons to low-energy visible photons.

Solar neutrinos, those ghostlike particles produced in the nuclear reactions occurring in the core of the Sun and providing the Sun's energy, are for us the key informants, playing the role of spies for the thermonuclear reactions which occur in the deep solar interior. Billions of these solar neutrinos pass every moment through Earth and also through our bodies, like ghosts, almost without encountering an obstacle. This is so, because these solitary particles interact very weakly with known matter, through the so-called *weak interactions*. Although they transport only 2% of the energy produced in the solar nucleus, they escape quickly – in 2 s they emerge into the *photosphere* coming from the solar coreand convey to us the *mandates* about what is happening deep in there. A famous problem related to the physics of the solar interior, as well as the interior of most of the other stars, is the problem having to do with the *identity* of the solar neutrinos.

In the first place, our knowledge for what is happening in the solar nucleus comes from the collaboration of astrophysicists – who build models of the solar and stellar structure – with the nuclear physicists – those who know the details of the thermonuclear reactions. Ultimately, the results of this theoretical modeling collaboration should be compared with the experimental data of neutrino detection in large underground mines, where the background noise is minimized. From the known production rate of energy in the Sun, we can calculate the rate of production of the neutrinos released in the solar core and then calculate their flux, i.e., the number which arrives here on Earth each second per unit area. Thus, we estimate that about 100 billion neutrinos should arrive each second in every square centimeter here on Earth. However, these elementary particles, of almost zero mass, are detected here on Earth with about *half* this flux. What happens to the other half of the flux? To explain these experimental data, there was a great disagreement and shifting of responsibilities between astrophysicists and nuclear physicists!

As we discussed earlier, according to the standard model of weak interactions in particle physics, neutrinos come in three *flavors*: muon neutrinos (ν_μ), electron neutrinos (ν_e) and neutrino – τ, (ν_τ). In 1965 an ambitious experiment started in an abandoned gold mine, in Homestake, South Dakota, 1.5 km underground, by Raymond Davis and his colleagues Fig. 1.68, to measure the flux of neutrino coming from the Sun. These neutrinos are produced by the nuclear reactions that occur deep in the core of the Sun, producing neutrinos with electronic *flavor* (ν_e). The original purpose was to make a direct measurement of the nuclear reactions taking place in the core of the Sun, so that to check the model of the solar interior. For this purpose, Raymond Davis' detector consisted of a huge tank filled with about 600 tons of the chemical solvent tetrachloroethane (C_2Cl_4), which in the industry is called *perc* and is used to clean delicate clothes that cannot withstand the harsh treatment of a washing machine. Specifically, he used the reaction $^{37}\mathrm{Cl} + \nu_e \rightarrow {}^{37}\mathrm{Ar} + e$, wherein the capture of an electron neutrino by the nucleus of the chemical element Chlorine produces the nucleus of the chemical element Argon, together with an electron. At regular intervals, the presence of isotope-37 atoms of the colorless noble gas Argon (^{37}Ar) which were produced in the liquid was checked. The relatively few produced nuclei of the gas ^{37}Ar are unstable and decay with a half-life of about 35 days. By a laborious process, Raymond Davis, after waiting several weeks for sufficient atoms of Argon to be formed, emptied the tank by using a Helium gas, which entrained the argon nuclei and after many complex stages, he obtained the final number of Argon atoms. Finally, by measuring the radioactive decay of Argon atoms with a Geiger counter, Davis calculated the number of the high-energy neutrinos that came from the Sun and interacted with ^{37}Cl, according to the previous reaction to form those ^{37}Ar atoms.

From the very first results of the experiment, it was clear that the measured flux of neutrinos was considerably lower, about 33–50% of the theoretically expected value. In the years that followed, the performance of the detector was improved and also the theories about the nuclear reactions inside the Sun were consolidated. Nevertheless, the flux measured by the experiment was smaller compared to the one predicted by the solar interior models. R. Davis's experiment continued for 30 years. In fact, for the careful, painstaking and systematic efforts made by both Raymond Davis and Masatoshi Koshiba, running a similar detector at the Kamioka Observatory in Japan (1989), the two were awarded the *Nobel Prize* in Physics (2002).

More recent experiments (1991–1997) using detectors which used the chemical element Gallium (Ga), such as the GALLEX experiment at the Italian Grand Sasso National Laboratory under the Grand Sasso mountain, 120 km from Rome, Italy, and the SAGE (Soviet–American Gallium Experiment) experiment at the Neutrino Observatory on the Baksan River in the Caucasus Mountains in Russia, measured the flux of lower-energy neutrinos compared to those in the experiment of Davis. However, their results were also consistent with a flux smaller than that predicted by the standard solar interior model. Hence, the results of all those different experiments gave a neutrino flux at the Earth that fell short of what the standard solar interior model predicted!

1976: Could be a Black Hole at the Solar Center?

The University of Chicago (UoC) is one of the most historic American Universities. In 1907, UoC professor Albert A. Michelson was the *first* American scientist to be awarded a *Nobel Prize*, for measuring the speed of light. On December 2, 1942, the first human-made self-sustaining nuclear chain reaction was initiated during an experiment led by the Nobel laureate (1938) Enrico Fermi and his collaborators in the laboratories of the University of Chicago: the Atomic Age began inside an enormous tent on a squash court under the stands of the University of Chicago's Stagg Field. In 1980, the Swedish Academy also honored UoC prof. Jim Cronin's discovery that certain basic interactions at the microscopic level are not invariant to time reversal, which allowed us to understand the phenomenon of the creation of matter (and not anti-matter!) in the primordial Universe. And, in 1983 fellow UoC professor S. Chandrasekhar was awarded the *Nobel Prize* for his studies on the structure and evolution of stars and his theoretical research on *white dwarf*s and *black holes*.

In 1975, right after I graduated from the Dept. of Physics of the Aristotle University of Thessaloniki (1974) and after completing my military service in the Greek-Turkish border, I was accepted with a scholarship by the Physics Department of the Univ. of Chicago to continue my graduate studies there. The weight of the tradition in the UoC is felt by its students, since until 1975, already 18 members of the Dept. Physics of the UoC and its partner institutes had received *Nobel Prize*s – today this number is 29. Therefore, studying in such an environment automatically meant increased obligations. One of them was the test of the graduate physics (PhD qualifying exams), the famous qualifying exams of the UoC, a prerequisite to allow Ph.D. candidates to go on to obtain a Ph.D. At the end of these exams, the graduate students present a research paper to a committee of UoC professors.

The research paper I had chosen was about trying to solve the then open solar neutrino problem. Specifically, a way to justify the observed lower solar neutrino flux was to reduce the rate of nuclear reactions at the solar core. How would this be done? The idea was simple: if there was a black hole at the center of the Sun with a mass much less than Earth's mass, the core would produce neutrinos at a lower rate, in agreement with experiment. Although the idea was really simple, the calculations were complicated, but feasible. They were based on a hypothesis, that of the existence of a hidden black hole at the center of the Sun, which could only be verified by its results. I presented these calculations to a committee of professors for my UoC Master of Science degree in Physics (1976).

1971: Coronal Mass Ejections (Elevating from the Sun the Mass of a Mountain Range)

Related to solar *jets*, of which they are the large-scale sister phenomena, but also to *flares* and explosive *prominences*, which often they follow, are the impressive *Coronal Mass Ejections* (CMEs, Figs. 1.48 and 1.49). In CMEs, within a few minutes, billions of tons of coronal solar plasma (as much as the mass of a mountain range!) are thrown into interplanetary space at a speed of 250–3,000 km per second, or, about 1–10 million kilometers per hour. The fastest CME reach Earth in 15–18 h, while the slowest ones they travel for a few days.

The average velocity of all CMEs throughout solar cycle 23 (1996–2007) was about 490 km per second, a value slightly greater than the average speed of the *solar wind*, which was 466 km per second. Usually there are three phases in the kinematic evolution of CMEs: initially, a slow rise (with a speed $V < 80$ km per second) within a few tens of minutes, a second phase of sharp acceleration of the order of 100–500 km per second at distances of 1.4–1.5 $R_{\odot}$ and a third phase of CME propagation with constant or decreasing speed. The average mass of CMEs is 3.5×10^{14} gr and their average kinetic energy is 2.9×10^{29} erg. In distances $R < 7\ R_{\odot}$, mass and energy increase monotonically, and then almost are stabilized.

CMEs contain hot plasma (mainly electrons and protons) as well as magnetic fields (1–10 Gauss). For comparison, the largest solar *flares* emit mainly electromagnetic radiation. The frequency of occurrence of CMEs has been observed to follow the solar activity cycle in phase and intensity wherein the relationship connecting the number of CMEs to the number of spots is approximately linear. When the Sun is at its maximum activity, we can have up to 10 coronal plasma spurts per day, while when it is at its minimum we usually have about one such eruption per day (Fig. 1.50).

CMEs vary in size and sometimes appear as narrow structures like *jets*, and sometimes like explosions coming from the entire Sun. Specifically, in the event that their

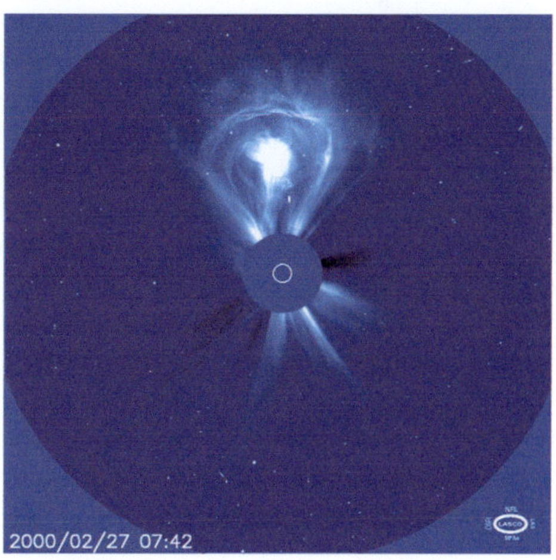

Fig. 1.48 Observations of an impressive CME (shaped like an electric lamp), observed by SOHO on 2000/02/27, with a three components structure: the leading loop, the cavity and the brilliant core (*Source* NASA)

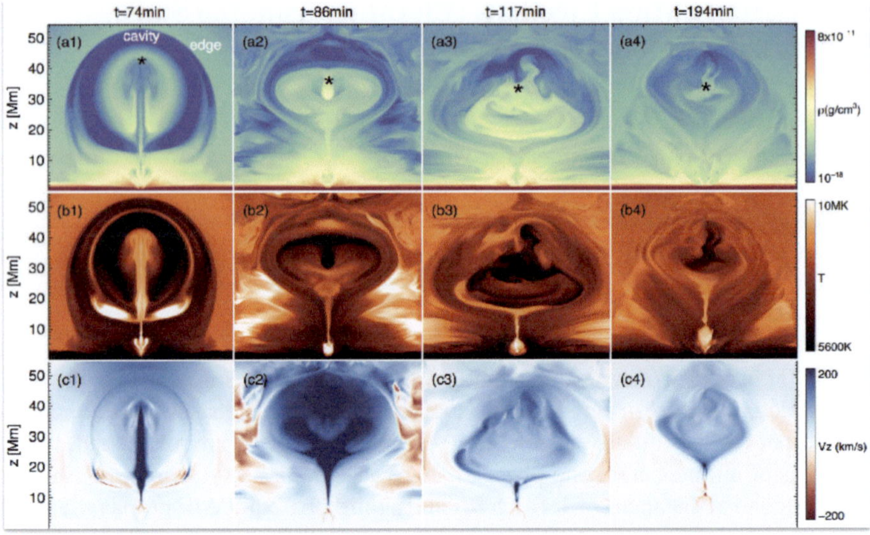

Fig. 1.49 Numerical simulation of the *iterative* formation of micro coronal mass ejections, through the dynamic emergence of a strongly twisted magnetic field from the solar interior, due to magnetic buoyancy. The first row gives the temperature, the second the density and the third the velocity, for 4 repeated explosions corresponding to the 4 columns of the image (From Syntelis et al. 2017)

angular width is 360°, they are called *Halo CMEs*. Actually these CMEs come from a central region of the Sun and are directed toward the detector, with the result that they are observed as if they are coming from the entire solar surface. Figure 1.51 shows the currently most accepted model of SEPs, which is consistent with observations.

Coronary plasma ejections (CMEs) vary in appearance. They usually have a three-part structure, the compressed plasma radiating and preceding a twisted magnetic tube, followed by a cavity enclosing a bright filament or projection (see Fig. 1.51).

Coronal mass ejections are very often accompanied by solar *flares* and it is assumed that there is a close correlation between the two phenomena (Fig. 1.49, Syntelis et al. 2016), although there are some coronal plasma *jets* which do not appear to be accompanied by *flares*. However, this can occur either because the region of the flare is behind the solar disk, or, because the outpouring of energy is such that the event is not characterized as flare. On the other hand, there are indeed many *flares* which do not appear to be associated with coronary mass ejections at all.

CMEs are also associated with *filaments* and erapting *prominences*. Part of the erupting prominence becomes the brilliant CME core and the rest of the prominence falls toward the solar disk. This strong correlation is crucial for modeling coronal plasma ejections. It is considered that when *flares* and *coronal mass ejections* are correlated, they are different sides of the same magnetic explosion.

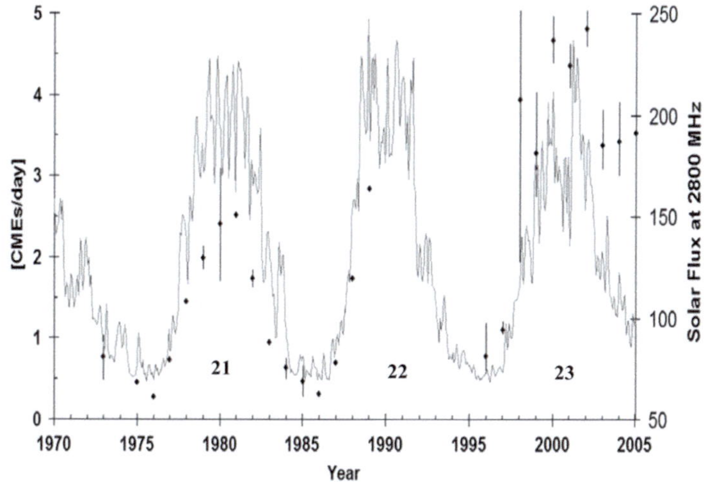

Fig. 1.50 Rate of CMEs per day during cycles 21, 22 and 23. The small cycles correspond to the solar flux at 2800 MHz. The highest CME rate occurs during the maximum of solar activity. (*Source* Cremades and St Cyr (2007) reprinted with the kind permission of the author Hebe Cremades)

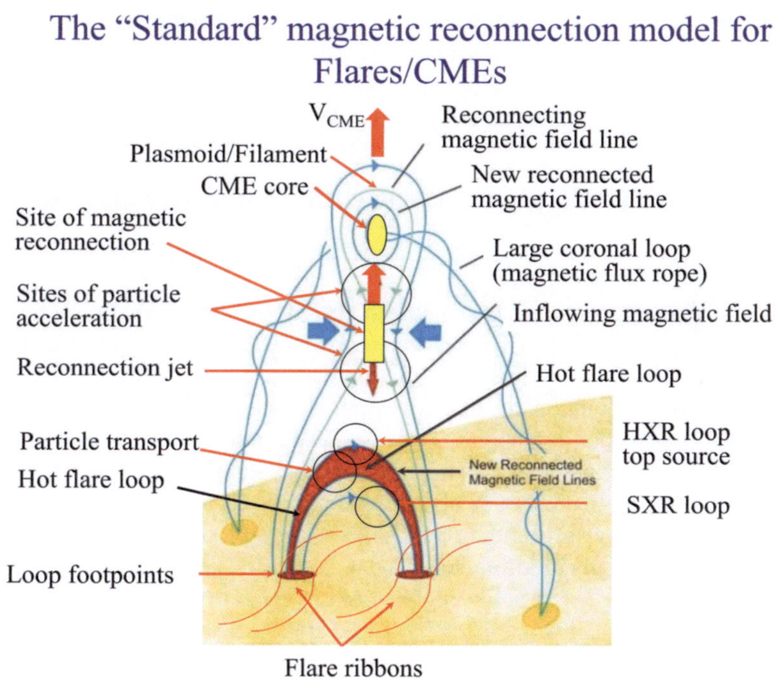

Fig. 1.51 Schematic representation of the standard CSHKP Flare/CME model, in which the reconnection of the magnetic lines of an extended coronal magnetic loop in the solar atmosphere leads to the generation of a flare and a CME. (Figure adapted from Shibata et al. (1995) and Christe et al. (2017))

CMEs affect significantly *Space Weather* and Earth. When they reach Earth, they disrupt the terrestrial magnetosphere and upper atmosphere, Patsourakos et al. (2016).

The twin STEREO spacecraft provided us with a very useful set of observations of CMEs, (Vourlidas et al. 2012).

The standard Flare/CME model is shown in Fig. 1.51. The CSHKP acronym refers to four key theoretical papers that appeared over the time frame 1964–1976: Carmichael, Sturrock, Hirayama, Kopp and Pneuman (Carmichael 1964; Sturrock 1966; Hirayama 1974; Kopp and Pneuman 1976). See also, Shibata et al. (1995), Christe et al. (2017), Pucci et al. (2020). For a recent review on solar flares as magnetohydrodynamic processes, see Shibata and Magara (2011).

Solar energetic particle (SEP) events are closely related to both *flares* and *coronal mass ejections* (CMEs), with the time-correlation presented in Anastasiadis et al. (2019): the fluxes of SEP range from MeV to GeV, with the flare energies from gamma-rays to radio waves. In this figure, the onset of the solar flare is indicated with a vertical red line with the resulting protons and the CME following. The gray-shaded area depicts the time interval between the start time of the onset of the solar flare and the CME start and the SEP event at MeV energies. For the sources of solar energetic particles see Vlahos et al. (2019). For solar flares, coronal mass ejections and solar energetic particle event characteristics, see Papaioannou et al. (2016), while for predicting flares and solar energetic particle events via the FORSPEF tool, Anastasiadis et al. (2017, 2019).

1972: Solar Active Regions, Areas with Opposite Magnetic Polarities

Solar active regions are very clearly visible in the red chromospheric line of hydrogen (Hα). The dimensions of active regions can reach 150,000 km and their orientation is in the east–west direction, but with a slight inclination to the equator. Active regions are divided into two components that have opposite magnetic polarities: the leading (relative to the solar rotation) and the following. These two areas are separated from the neutral line of the magnetic field, i.e., where the magnetic field is horizontal. In the *corona*, the magnetic field of an active region traces coronal *loops*, which contain plasma hotter from the surrounding gas, have X-ray emission and extend in height several tens of thousands of kilometers.

First, the appearance of an active region looks as a disturbance in the photospheric granulation pattern, wherein magnetic flux tubes emerge from the solar interior. Often they are seen as arch *filaments* (systems similar to *filaments* in appearance, connecting opposite polarities of newly emerging flux regions), or fibrils, (dark regions in the Hα line, which have an almost parallel arrangement and their legs are in regions of opposite magnetic polarity). They show upward plasma motions at their top and downward plasma motions at their feet. This suggests that they represent tubes of emerging magnetic flux that rise, while at the same time the plasma they contain escapes from their ends.

When the emergence of the magnetic flux is completed at a particular point of the solar *photosphere*, the active region may remain fixed for a long period of time that can reach three months. The dissolution of the active region occurs gradually due to photospheric movements and magnetic reconnection that destroys the concentrated magnetic flux. At their maximum development active regions manifest themselves in the *photosphere* as a group of *sunspots* comprising large spots and small pores.

Pores (Fig. 1.52) are small dark areas arising from the emergence of the magnetic flux in the *photosphere*. They look as small *sunspots*, but without a shadow around them and have a strong magnetic field of about 1,000 Gauss while their size is about 2,500 km, 2–3 times larger than the size of the individual granules. From the development or the coalescence of pores, *sunspots* seem to be created.

Solar active regions started to be enumerated from January 5, 1972. As of June 14, 2002, 10,000 had been counted. Since then they are referred to by the last digits only, i.e., solar active region AR0030 corresponding to AR10030.

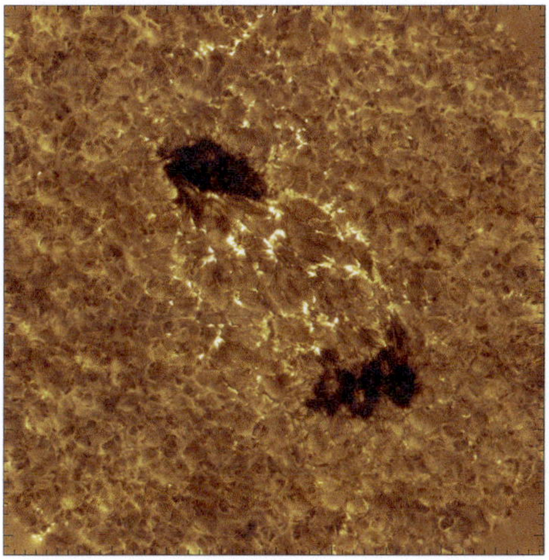

Fig. 1.52 Two pores through CRISP (CRisp Imaging Spectro Polarimeter) and the web-like and approximately hexagonal chromospheric network (bright white areas) between the two pores (black areas), small premature sunspots, of relatively smaller size (approximately 1,000 km), without penumbra. Observations performed on the 22nd of Apr. 2008, at the wavelength of Fe I 630.2 nm by Tomas Hillberg, with image processing by Mats Löfdahl (*Credit* Tomas Hillberg, Institute for Solar Physics (ISP), 1-m Swedish Solar Telescope, Spanish Observatorio del Roque de los Muchachos of the Instituto de Astrofísica de Canarias, La Palma, Canary Islands, Spain)

1973: Discovery of *Coronal Holes*

Coronal holes were noticed already in 1957 by the Swiss astronomer M. Waldmeier, director of the Zurich Observatory. They are regions which have an occasional dark appearance with abnormally low density and temperature. Synoptic observations from the Orbiting Solar Observatory series (OSO-4, OSO-6 and OSO-7, 1967, 1969 and 1971, respectively) established that *coronal holes* are indeed 3-dimensional solar objects. However, the highly successful Skylab mission (1973–1979), that carried an X-ray telescope, revealed *coronal holes* and for the first time, providing thus the first systematic observation of *coronal holes* which obtained specific interest because they are associated with high-speed *solar wind* streams: all large near-equatorial *coronal holes* seen during the Skylab period were associated with high-velocity *solar wind* streams observed at 1 AU.

Today, we understand a coronal hole as a region of a relatively cool and less dense plasma in the solar *corona*. Compared to the *corona*'s usual closed magnetic field that arches between regions of opposite magnetic polarity, the open magnetic field of a coronal hole allows *solar wind* to escape into space at a much quicker rate. This results in decreased temperature and density of the plasma at the site of a coronal hole, as well as an increased speed in the average *solar wind* measured in interplanetary space. If streams of high-speed *solar wind* from *coronal holes* encounter Earth, they can cause noticeable displays of *aurora*e. Near solar minimum, when solar activity, such as *coronal mass ejections* is less frequent, high-speed *solar wind* streams are the main cause of geomagnetic storms and associated *aurora*e. Many impressive *coronal holes* are observed via the AIA instruments of the SDO observatory, wherein in extreme ultraviolet, appear as relatively dark patches in the Sun's *corona* (Figs. 1.1, 1.53, and 1.54).

The size, frequency and location on the solar disk of *coronal holes* are related to the phase of the solar cycle. First, *coronal holes* tend to be larger and more frequent near the solar poles, where the magnetic field lines are more open and less tightly packed. Furthermore, the size and frequency of *coronal holes* are linked to the solar cycle, which follows an 11-year pattern of magnetic and sunspot activity. As the Sun heads toward solar maximum, *coronal holes* move closer and closer to the Sun's poles. During solar maxima, the number of *coronal holes* decreases until the magnetic fields on the Sun reverse and afterward, new *coronal holes* appear near the new poles. The *coronal holes* then increase in size and number, extending farther from the poles as the Sun moves toward a solar minimum again. During solar minima, *coronal holes* are the primary sources of *Space Weather* disturbances, including *aurora*e. Typically, geomagnetic storms originating from *coronal holes* have a gradual commencement of hours and are not as severe as storms caused by *coronal mass ejections* (CMEs), which usually have a sudden onset. Because *coronal holes* can last for several solar rotations, i.e. several months, predicting the recurrence of this type of disturbance is often possible significantly farther in advance, than for CME-related disturbances.

When a coronal hole is facing Earth, we can expect to experience an increased geomagnetic activity within 2–3 days of the hole rotating onto the Earth facing side

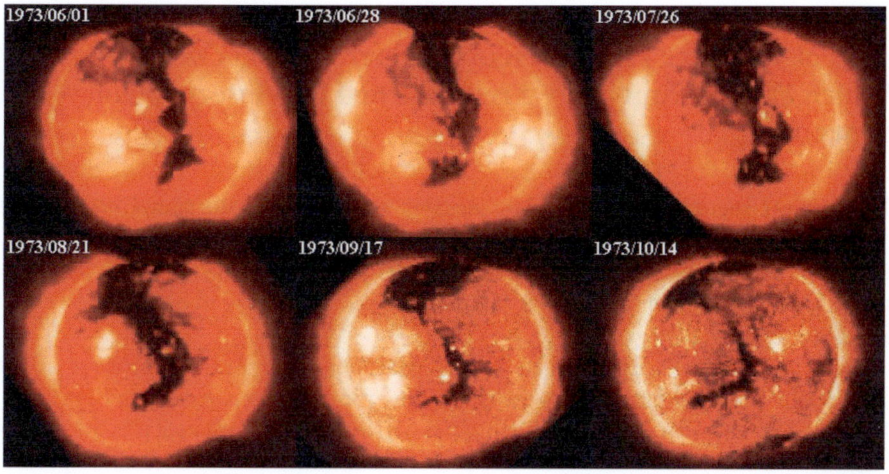

Fig. 1.53 "The Italian Boot" coronal hole observed with the X-ray telescope onboard Sky lab. Coronal holes were observed to rotate fairly rigidly and maintain their shape through several 27-day solar rotations in spite of the variations in rotation rate of the solar surface (*Credit* NASA/MSFC)

of the Sun. Increased geomagnetic activity is another term for increased potential of seeing the northern lights. Due to the increase in the *solar wind* speed, more charged particles within the *solar wind* will enter the atmosphere of the Earth, they will then react with oxygen and nitrogen molecules in the atmosphere and create the *aurora* borealis.

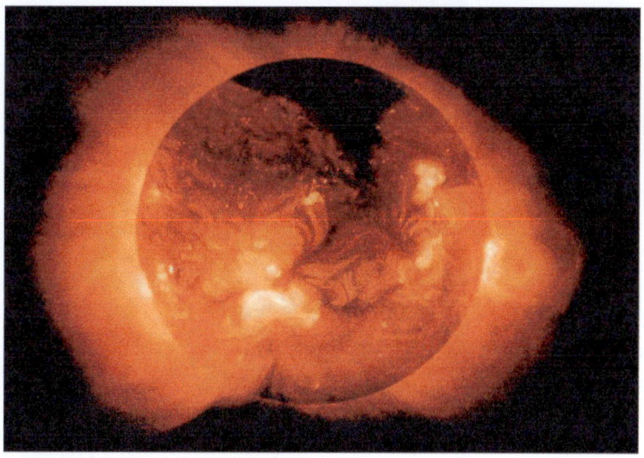

Fig. 1.54 Image of the solar corona observed in soft X-rays by the Yohkoh spacecraft on May 8, 1992. The coronal hole appears dark due to lower emission. The bright features are active regions with closed magnetic configuration (*Credit* Yohkoh data courtesy of the NASA-supported Yohkoh Legacy data Archive at Montana State University)

1973: *Prominences* (Huge Floating Unstable *Curtains* of Cooler/Denser Plasma Than the Surrounding *Corona*, Which When Erupt They Are Launched into Interplanetary Space)

The earliest record of solar *prominences* is found in the Russian Chronicle of Novgorod, where, during the total solar eclipse of May 1, 1185 is described as follows: "*… and the Sun was darkened and became like the Moon while its horns shone with a glow that looked like red-hot coals. It was scary to us to observe this divine sign…*"

Solar *prominences* are thin *curtains* of cold, dense plasma which hover in the strong solar gravitational field, in the hotter and thinner solar *corona* (see Figs. 1.55, 1.56, 1.57(left), and 1.58). Specifically, the plasma in the prominence is 100 times brighter and 100 times denser than the plasma in the surrounding *corona*. A prominence forms within a day or so, but may survive in the *corona* for many weeks and months, spanning hundreds of thousands of kilometers above the solar *photosphere*, i.e., at lengths reaching the solar radius. The reddish material of the protrusions is plasma composed of ionized hydrogen and helium. This plasma moves along twisted magnetic lines that have emerged in the solar *corona*. Coming from the solar interior where they are produced by the dynamo mechanism. When they explode due to

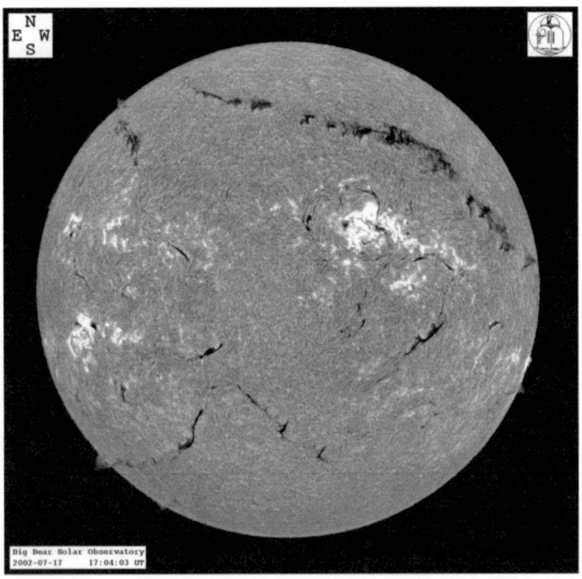

Fig. 1.55 An extraordinarily long, nearly straight, filament during solar cycle 23 (Jul 15, 2002) in Hα, one of the longest *filaments* that have ever observed on the Sun, that developed a maximum length of 160 heliographic degrees (*Credit* Michael Anderson and Sara F. Martin (2005), BBSO. Figure reproduced with the kind permission of Sara Martin)

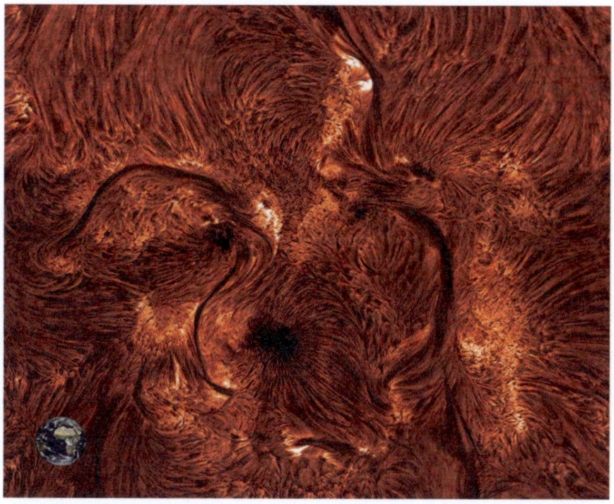

Fig. 1.56 The active region AR10786 in the red line of hydrogen Hα, on 8 July 2005, with the Dutch Open Telescope. The image area covers 182 × 133 arcsecs. Because an opening angle of 1 arcsec corresponds on the Sun to 727 km, in this image we see 132,314 × 96,691 km of the solar chromosphere. There are also a few sunspots (4) which are dark regions in the Hα line. Many *fibrils* emerge from these spots and connect different areas, like iron filings over a magnet. The white areas around the spots are *chromospheric plages*, where many magnetic tubes converge. The large elongated black structures are *filaments*, i.e., projections onto the solar disk of *prominences* and terminate in dipole regions that emerge through the photosphere (*Source* Dutch Open Telescope, Observatorio del Roque de Los Muchachos, Canary island of La Palma, Spain, prof. Dr. Robert J. Rutten and Wikimedia Commons)

instabilities, they develop into spectacular coronal plasma eruptions, spewing plasma at 600–2,000 km per second.

Projected onto the solar disk *prominences* appear dark elongated *filaments*, while above the *sunspots* they form huge *loops* of plasma. These huge *curtains* of dense plasma that float in the solar *corona* are supported by magnetic forces, opposed by the strong solar gravity.

Theoretical models of *prominences* aim to describe the natural conditions within them and in their immediate vicinity, as well as the natural conditions that lead to their formation and preservation, and also their final destruction. To the plasma trapped within the *prominences*, the force of the magnetic field, the force of gravity and the force due to the plasma pressure difference are in equilibrium, wherein these three forces are balanced. For a recent review on solar prominences, see Vial and Engvold (2015) and Gibson (2018).

A giant erupting prominence was photographed on July 31, 1992 UT, by the Japanese Norikura Solar Observatory in the red line Ha, at 656.3 nm. Using simultaneous observations in the visible, X-rays (by the Yohkoh satellite) and radio waves (Nobeyama Radioheliograph), the evolution and morphology of this explosive prominence was calculated in detail. Thus, the bulge began to move away at 0:43 am (UT) at a speed of 30 km per second. At 1:37 am (UT), the height of its loop overhang

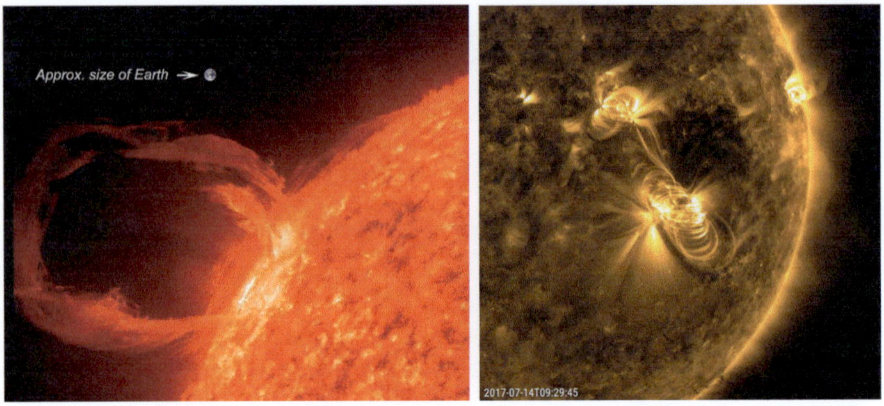

Fig. 1.57 (*left*) The large exploding solar prominence of March 30, 2010, viewed from outside the solar disk. When the prominence projects onto the solar disk we have a solar *filament*, as in the previous figure. The image combines two wavelengths, 171 and 304 Angstroms, where multiple *loops* are visible. (*right*) A medium-sized *flare* (M2) and a coronal plasma jet erupted from the active region AR9077 of the Sun on July 14, 2000 (Bastille day solar flare and coronal mass ejection). The appearance of erupting *loops* is common after flares. The *flare* lasted almost two hours, a rather quite long duration. Coronagraphs on the SOHO spacecraft show a substantial cloud of charged particles blasting into space just after the blast. The coils arcing over this active region are particles spiraling along magnetic field lines, which were reorganizing themselves after the magnetic field was disrupted by the blast. The image is in the extreme ultraviolet light (*Credit* Solar Dynamics Observatory, NASA)

reached 590,000 km and its speed 220 km the second. At its upper end the ledge had a helical structure, something indicating the existence of magnetic fields. Beneath the *prominence*, the formation of magnetic *loops* was observed in X-rays. Then the ledge went off into the space.

Often *filaments* suspended in the solar atmosphere are fired by neighboring flares which activate various instabilities, so that eventually they break and their plasma is ejected into space. When the charged particles reach Earth, they create a dynamic and impressive aurora. One such case is shown in Fig. 1.58 for the explosive prominence of August 31, 2012.

1974: Photospheric and Chromospheric Networks (Luminous *Spider Webs* Spread Over the *Photosphere* and *Chromosphere*)

Although the most visible feature of the solar *photosphere* is the *sunspots*, most of the magnetic flux emerges in small magnetic structures, *the magnetic flux tubes*. Recent high-resolution observations of the photospheric granulation pattern highlight the fact that these magnetic tubes are scattered throughout the *photosphere*, but are especially

Fig. 1.58 An explosive solar prominence at August 31, 2012, viewed from outside the solar disk, with the Solar Dynamics Observatory (SDO), at four different wavelengths. From top left and clockwise: 335 Angstroms (radiation from iron-16 (Fe XVI) atoms in a hot active region with temperature T ~ 2,500,000 K), 171 Angstroms (radiation of iron-9 (Fe IX) atoms of the quiescent corona and coronal loops temperature T ~ 600,000 K), 304 Angstroms (radiation of helium-2 (He II) atoms of the upper transition zone and chromosphere, temperature T ~ 50,000 K) and 131 Angstroms (radiation of atoms of iron-20 (Fe XX) and iron-23 (Fe XXIII) temperature T > 10,000,000 K). Because each wavelength corresponds to a different temperature of the emission plasma, we can track the plasma motions at different temperatures. The coronal mass ejection, or CME, traveled at over 900 miles per second. The CME did not travel directly toward Earth, but did connect with Earth's magnetic environment, the magnetosphere, with a glancing blow causing aurora to appear on the night of Monday, September 3 (*Credit* NASA/GSFC/SDO)

located on the boundaries of large cells, i.e. at the edges of the supergranulation and also in the dark areas of the granules where the gas sinks after exposure to the *photosphere* and is cooling there by radiation (Fig. 1.24). These magnetic tubes are concentrated on the chromospheric network and their magnetic field is perpendicular to the *photosphere*, while within this chromospheric network there does not seem to be any magnetic flux. The magnetic field within these small tubes is vertical and the magnetic pressure inside them causes the pressure of the gas inside them to fall short of the external gas pressure and thus be swept by the external currents of granulation and supergranulation toward the boundaries of the transport cells.

The *chromosphere*, is dominated by jet-like, dynamic structures, called *mottles* when found in quiet regions, fibrils when found in active regions and *spicule*s when observed at the solar limb. When the light emission of the *chromosphere* is observed in the red line of hydrogen Hα (wavelength 656 nm) and the emission lines in the ultraviolet of the element calcium (Ca II H and K – from calcium atoms that have lost an electron, 396.9 and 393.4 nm, respectively) the solar disk appears to be covered by a luminous spider web. This is the so-called chromospheric network, i.e., a beautiful light arrangement that looks like a giant spider's web, or a wire light

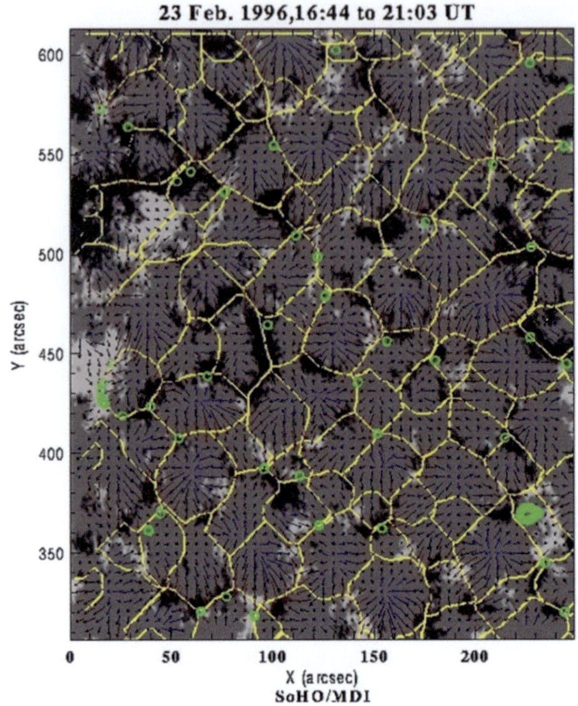

Fig. 1.59 A SOHO/MDI magnetogram, where the boundaries of supergranule cells are marked in yellow and arrows indicate the supergranular flow pattern. From Tsiropoula et al. (2012) (*Credits* ESA/NASA/ SoHO/MDI

network, spread over the solar *chromosphere*. The chromospheric network follows the supergranulation cells and is due to the concentration of magnetic flux at the boundaries of these cells, due to the movement of the cooling fluid toward them (Tsiropoula et al. 2012) (see Fig. 1.59).

On the bright borders of the chromospheric network appear *bright points* called *filigree*, or *bright points*. These *bright points* have a short life span of only a few minutes, i.e., they also flicker like the photospheric and chromospheric plages, like fireflies. Interestingly, recent observations of ultrahigh analysis showed that each bright point of the chromospheric fligree is resolved into more *bright points* below it, in the *photosphere*, in the dark zones between the granules (see Fig. 1.24).

Sound Waves Produced from Million of *Solar Cosmic Bells*...

The solar interior is constantly pulsating and the harmonious sounds (*chimes)* that ring there from millions of *bells* are propagating through the magnetic field/waveguides that act as channels transmitting the produced sounds. These magnetic channels increase their cross section as they extend into the chromosphere, wherein the acoustic waves are converted into *shock waves*, due to the large reduction in gas density, to finally deposit their energy in the chromosphere. In other words, the magnetic field acts like the gate that opens and closes the doors of the solar interior to allow sound waves to escape from the solar interior to the chromosphere!

Chromospheric Jets *(a* Burning Meadow, *with Hundreds of Thousands Flames...)*

When the *chromosphere* is observed at the solar limb on the line Hα, it appears as a *burning meadow* of many small and narrow *jets*of plasma. At any given moment, the solar *chromosphere*, is filled with up to 10 million of the so-called chromospheric *jets* (*spicules*) when observed in the solar limb (Fig. 1.60). These same *jets* when observed on the disk are called *fibrils* or *mottles* (Fig. 1.61a) and appear to connect *sunspots* with brighter chromospheric *plages*, (Fig. 1.61). These fireworks of plasma ejection from the *photosphere* to the solar *corona* reach heights of 3,000 to 10,000 km and last only a few minutes (usually around 15 min). Each of these formations reach speeds of 20 to 30 km per second, i.e., over 100,000 km per hour, before collapsing, disappearing and reappearing in a matter of minutes. They have a diameter about 500 km and trace the magnetic field. They are impressive and were observed for the first time by the priest Angelo Secchi in the Observatory of the College of Rome (1877).

Recent observations with a combination of ground-based telescopes, such as the Swedish Solar Telescope on La Palma, observations from specially configured aircrafts, observations from SoHO and TRACE, as well as numerical simulations, conclude that the magnetic field behaves like a waveguide channeling the wave energy from the solar interior. Thus, allows sound waves to propagate along thin, fountain-like *jets*the chromospheric *jets* (*spicules* on the solar limb) *fibrils* and *mottl*es on the disk – and finally deposit their energy in the solar *chromosphere*.

Specifically, the upper left part of Fig. 1.61 shows solar active region 10380, in June 2004, which is dominated by those dark elongated structures, the small chromospheric *spicules*. On the left side of the image there is a pair of spots (in black) with a strong magnetic field, as well as dark elongated fibrils connecting them. In the center and on the left, below the spot and *filaments* appears the brighter area (in white) corresponding to *plages* (or *chromospheric torches*) that also have strong magnetic fields.

As is explained in the site of this observation: "imagine a pipe as wide as a state and as long as half the Earth. Now imagine that this pipe is filled with hot gas moving with 50,000–100,00 km per hour. Further imagine that this pipe is not made of metal but of transparent magnetic field. You are envisioning just one of hundreds of thousands of young *spicules* on the active Sun. Pictured in Fig. 1.61, is perhaps the highest resolution image yet of these enigmatic solar flux tubes. *Spicules* dot the above frame of solar active region AR10380 in Fig. 1.61a that crossed the Sun in

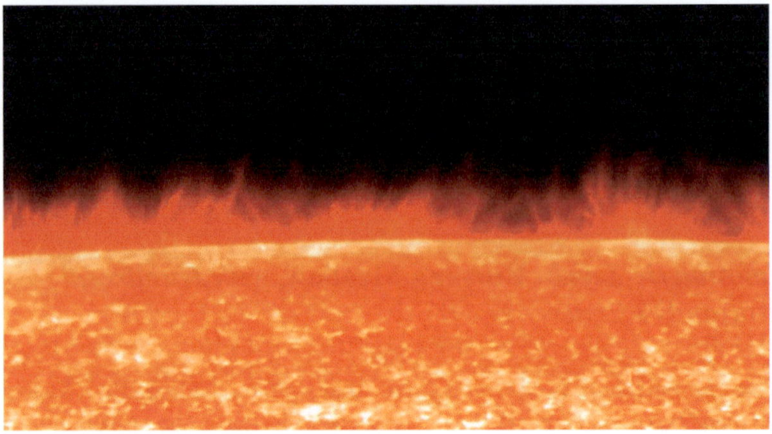

Fig. 1.60 The chromosphere at the edge of the solar disk in the Ha line looks like a *burning meadow*. These millions of elongated plasma *jets*, the so-called chromospheric *jets* (*spicules*), cover the sun's surface like whiskers. Those fireworks of plasma ejection from the chromosphere to the solar corona, reach heights of 3,000 to 10,000 km, usually last about 10–15 min and have speeds of 20 to 30 km per second, i.e., over 100,000 km per hour, before collapsing, disappearing and reappearing again in a matter of minutes. They have diameters of about 500 km and trace the magnetic field. More in the text (*Credit* NASA/Goddard Space Flight Center/JAXA/Hinode)

June 2004, but are particularly evident as a carpet of dark tubes on the right. Time-sequenced images have recently shown that *spicules* last a few minutes, starting out as tall tubes of rapidly rising gas but eventually fading as the gas peaks and falls back down to the Sun. These images also indicate that the ultimate cause of spicules is sound-like waves that flow over the Sun's surface but leak into the Sun's atmosphere.

The upper right and the two lower parts of Fig. 1.61 show solar active region 10675, in September 29, 2004. The dark spot is an isolated sunspot, relatively small in diameter and of regular shape. In particularly in the lower right (Fig. 1.61d), many *fibrils* are seen to emerge from the spot and connect different magnetic regions, like iron filings over a magnet. These large elongated black structures are *filaments*, ending in dipole regions, which emerge from the photosphere. In these black areas, we have a reversal of the magnetic polarity, i.e., the magnetic lines if were directed toward the outside of the Sun, in the *filaments*, return to the Sun again. The dark *filaments* are unstable and the source of many explosive phenomena. In the central dark shadow of the spots can be seen some *bright spots*. Outside of the central shadow (*umbra*), the Hα line does not *see* at such low photospheric depths but only at the highest chromospheric levels. At these heights they are seen the chromospheric *fibrils* (Fig. 1.60), elongated *jets* that magnetically connect the spot with distant bright areas, the *chromospheric torches* (*plages*), where many magnetic tubes are gathered together.

The upper left part of Fig. 1.61 looks like the part of a sunflower. In the specific color of the red light emitted by hydrogen, as featured here, some regions of the solar *chromosphere* may resemble a rose. The petals dominating the frame are actually

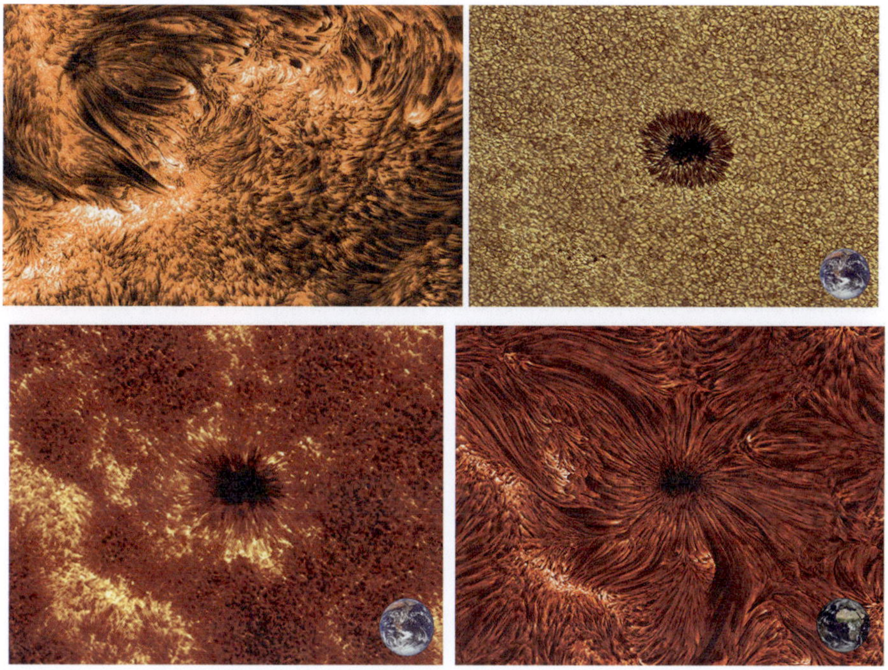

Fig. 1.61 (*top, left*) On June 16, 2004, the Swedish 1-m Solar Telescope, photographed AR10380 near the solar limb, in the bright red emission of the chromospheric Hα spectral line (656 nm), 130 km above the Sun. The entire region covers 65,000 × 43,000 km on the Sun (*Image credit* SST, Royal Swedish Academy of Sciences, LMSAL). (*top right and bottom left and right*). Image of active region AR10675 on September 29, 2004, at successive higher levels in the chromosphere in the red hydrogen line Hα, with the Dutch Open Telescope (DOT). The image of Earth gives the relative scale. The colors are artificial. (*top right*) The photospheric surface as it normally appears, is covered with *granules*. The large dark sunspot has a clear dark umbra in the center surrounded by a lighter *penumbra*. (*bottom left*) The Sun in light predominantly emitted a few hundred kilometers above the photosphere. At this height, the continent-sized bubbling *granules* appear reversed, and long lines of constant magnetic fields begin to appear. The last image shows the Sun at a few thousand kilometers into the *chromosphere*. Here magnetic field lines can be clearly followed outward from the sunspot to distant regions (*Image Credit* Dutch Open Telescope, Sterrekundig Instituut Utrecht, prof. Dr. Robert J. Rutten)

magnetically confined tubes of hot plasma, the *fibrils*, some of which extend longer than the diameter of the Earth. In the central region many of these *fibrils* are seen end-on, while the surrounding regions are typically populated with curved *fibrils*. When seen over the Sun's edge, these huge plasma tubes are called *spicules* (Fig. 1.60), and when they occur in passive regions they are termed *mottles*. This active region survived for several more days before the complex and tumultuous magnetic field poking through the Sun's surface evolved yet again.

Sunspots are normally seen flat on the Sun. A digital metamorphosis, however, shows a sunspot as it appears at increasing heights, effectively in three dimensions (Fig. 1.61b–d). The above false-colored image sequence of the solar active region AR

10675 was taken in three very specific colors that effectively isolate different layers above the solar *photosphere*. The first image (Fig. 1.61b) shows the Sun's photosphere as it normally appears, covered with *granules* (G band, 430.5 nm). The large dark sunspot has a clear dark *umbra* in the center surrounded by a lighter *penumbra*. The image appearing in Fig. 1.61c (lower left) shows the Sun in light predominantly emitted a few hundred kilometers above the photosphere, in the CaII H line (396.8 nm). At this height, the continent sized bubbling *granules* appear reversed, and long lines of constant magnetic fields begin to appear. The last image (Fig. 1.61d) shows the Sun at heights a few thousand kilometers into the *chromosphere*. Here magnetic field lines can be clearly followed outward from the sunspot to distant regions. The sunspot in the center of this image is single, rather small, and regularly shaped. In its dark *umbra* it shows bright *umbral dots* located in the deep *photosphere*. Outside the *umbra*, Hα does not sample such deep layers but instead samples the high *chromosphere*. It is filled with *fibrils* which outline magnetic connections from the spot to distant locations with bright *plage*.

1975: Solar Coronal *Loops* (The Ubiquitous *Magnetic Arches* of the Solar Atmosphere, the Building Blocks of the Solar *Corona*)

Coronal loops dominate the observations of the solar *corona* and are its basic magnetic building blocks (Fig. 1.62). For a review on solar coronal loops, see Reale (2014). These *loops* are emitting strongly and are therefore easily observable by space telescopes, such as the Einstein Observatory, Yohkoh, Soho, TRACE (Handy et al. 1999), SDO, etc. They are real laboratories of plasma physics for understanding phenomena such as plasma oscillations and wave activity, *nanoflares*, plasma heating and radiation, etc. Also, coronal *loops* are particularly important for understanding the still unsolved problem of solar *corona* heating.

Magnetic *loops* emerge from the solar interior because of magnetic buoyancy and are filled with hot plasma that strongly emits in the UV and X-rays, since the plasma trapped in these structures has temperatures over 1 million degrees Kelvin. The legs of the *loops* are firmly anchored to the *photosphere* and then extend through the *chromosphere* high in the *corona*.

Unlike the *solar wind* and *coronal holes* where the magnetic lines are open, in the *loops* we have closed magnetic lines. Often the plasma of the *loops* moves from one leg to the other through a *siphonic flow* due to a pressure difference between them. Coronal *loops* can be either short-lived as they can only last a few seconds or minutes (in the case of *flares*), or, survive for hours and less often for days. The longest *loops* are usually known as quiescent *loops*, where there is an equilibrium between the heating of the loop and its energy losses. For a model of the quiescent, inhomogeneous solar *corona* based upon the concept of loop structures as the basic structural element of the *corona*, see Rosner et al. (1978), while for a MHD theoretical

Fig. 1.62 (*Top left and right*) Solar loops observed by the Transition Region and Coronal Explorer (TRACE), on November 6, 1999. The images are not in the visible, but in the extreme ultraviolet (EUV) at a wavelength of 17.1 nm (171 Å, Angstroms). This section of the spectrum allows us to see features of the Sun's atmosphere that have temperatures around 1 million degrees Celsius (*Source* NASA/TRACE, ESA (D. Müller)). (*Bottom left*) SDO/AIA cascading loops on Jan. 15–16, 2012 following a solar flare eruption, in extreme ultraviolet light. The loops were "illuminated" by a neighboring flare, which filled the loops with heated plasma at a temperature of about 1 million degrees (emission at 17.1 nm, of ionized iron atoms that have lost their 8 outer electrons, FeIX) and thus created the emission of their radiation. Each of these loops can fit a few Earth-like planets inside. (*Bottom Right*). The changing appearance of the corona from solar minimum to maximum: the solar atmosphere is full of solar loops, especially during the maximum of solar activity, which are very well distinguished here in the ultraviolet part of the spectrum. These images are taken by AIA/SDO in EUV toward the end of the latest solar minimum activity period in May 2010 (left half) and during the current solar maximum period in December 2014 (right half). The three-color red–green–blue image channels are composed of observations made in three AIA channels: 171, 193, and 211 Å, respectively, corresponding to their most dominant emission lines of Fe IX, Fe XII, and Fe XIV with formation temperatures of ~0.7, 1.2, and 2.0 MK. The image has been processed using multiscale Gaussian normalization (*Source* NASA/SDO)

study of coronal *loops*, see Petrie et al. (2002), Gontikakis et al. (2005), Petrie et al. (2005), Syntelis et al. (2012).

The appearance of *coronal loops* is a function of the solar cycle and their number is proportional to the number of *sunspots* and, more generally, to *solar activity* (Morgan and Druckmüller 2014). They often appear near *sunspots* and are due to the intense magnetic activity in their regions, often precursors to solar *flares* and coronal plasma *jets*. In the *loops* the plasma can cool creating dark *filaments* above the solar surface, or, solar erupting *prominences* above the limb of the Sun (Figs. 1.55, 1.56, 1.57, and 1.58).

1980: Discovery of the (Invisible) Internal Rotation of the Sun and Other Stars

It has been observed that the outer solar layers rotate differentially, with the equator having a rotation period of 24.47 days and the polar regions rotating much slower, at 36 days (Fig. 1.63). Differential rotation is the phenomenon in which different parts of a rotating object move at different angular velocities. In Astronomy, the observation of differential rotation in a celestial body suggests that the particular body is not solid. This differential rotation of the Sun, plays an important role in producing its magnetic field with the *dynamo* mechanism. For a review on the internal solar rotation, see Howe (2009).

As early as 1630, Christoph Scheiner had observed that the Sun had different rotation periods at the equator and the poles, in good agreement with today's values for this rotation.

However, determining the rotation of the solar interior could not be done observationally until the advent of helioseismology, since there were no ways to penetrate below the solar *photosphere*.

Helioseismology determines the internal rotation of the Sun with great precision. In particular, helioseismological methods can measure and the rotational speed of the gas inside it. This is possible because the motion of the gas causes a separation of the frequencies of the acoustic waves due to the Doppler effect: an acoustic wave propagating in the direction of motion has a different frequency than an acoustic wave moving in the opposite direction. Thus, the rotation removes the degeneracy of the azimuthal order m that we would have if the Sun did not rotate. The results are shown in Figs. 1.63 and 1.64.

The conclusion is that the differential rotation observed in the *photosphere*, where the solar equatorial regions rotate more rapidly of those at greater heliographic latitudes, it decreases with depth in the interior and practically disappears at the base of the *convection zone*, 28.7% of the distance to the solar center (Fig. 1.64). A rotation frequency 320 nHz corresponds to a period of about 36 days, while a rotation frequency of 460 nHz corresponds to a period of about 25 days.

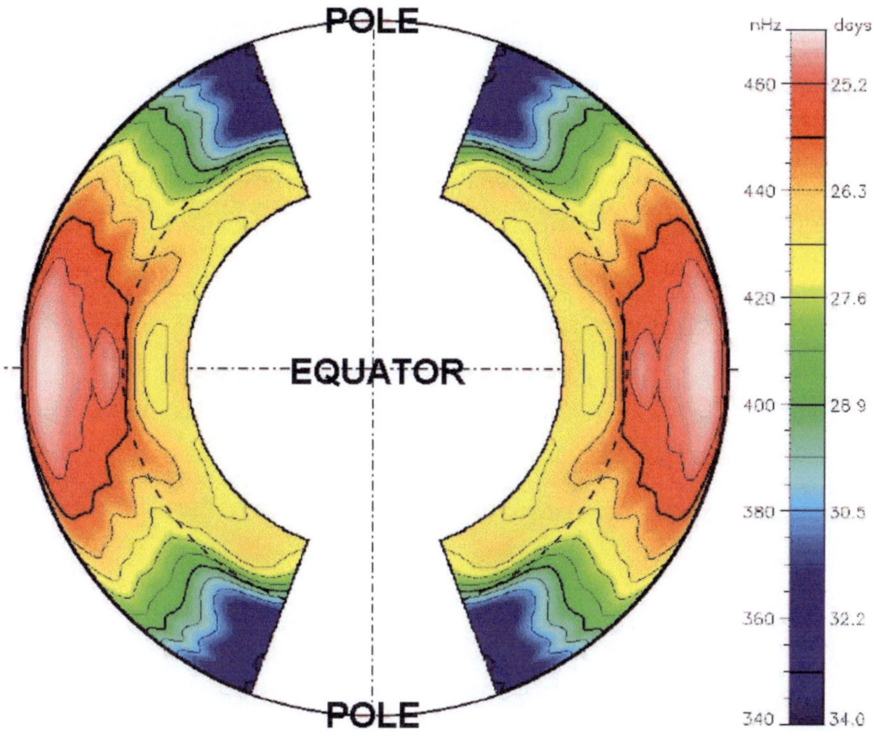

Fig. 1.63 Internal solar rotation, showing the location of the tachocline where rigid rotation gives way to differential rotation throughout the convection zone. The color scales are for solar differential rotation frequency (nHz) and the corresponding rotation period (day) (*Credit* NASA/Marshall Space Flight Center)

The rotation in the inner solar layers, at heliographic latitudes 0 (solar equator), 30°, 45°, 60° and 75° has been calculated from 144 days of continuous observation data, with the Michelson Doppler Imager (MDI), at the SOlar and Heliospheric Observatory (SOHO). Below the convective zone, the rotational speed changes significantly, and these large changes in speed in the small thickness of this of the layer, may be the source of solar magnetism in the *solar dynamo*, i.e. the mechanism for producing the solar magnetic field. Analysis of over 5 years of observations of solar acoustic eigenfrequencies with the Global Oscillations Network Group (GONG) and SOHO's MDI instrument, infers the rotation rate of the deep solar layers. The rotation there has an approximately constant angular velocity and a period of 27–28 days at the radiation zone, below the base of the *convection zone*, 0.713–0.25 of the solar radius. The acoustic waves do not reach the solar center where solar energy is produced.

With the methodology we discussed earlier, helioseismology and asteroseismology allow us to infer from the observations of the oscillations in the *photosphere*, a model for the temperature variation with the depth within the convective zone, and then the dependence density with solar depth, the internal rotation speed, the depth of the transport zone, etc.

Finally, Eddington (1882–1944) could not even imagine the revolution that helioseismology and asteroseismology would bring after 1962.

It is truly amazing that we can penetrate and understand what is going on inside distant stars, just by the analysis of the observations of the oscillations in their *photosphere*.

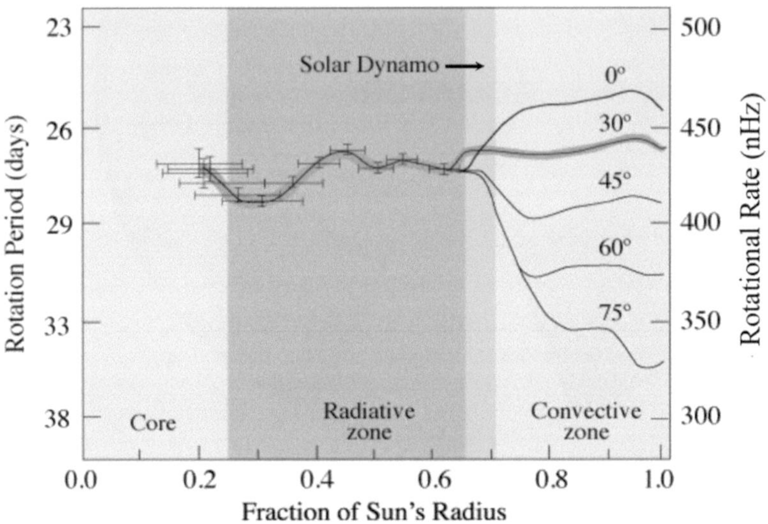

Fig. 1.64 Solar rotation in the convective zone, at heliographic latitudes 0 (solar equator), 30°, 45°, 60° and 75° from 144 days of continuous observation data, with the Michelson Doppler Imager (MDI), aboard the Solar and Heliospheric Observatory (SOHO). Analysis of over 5 years of observations of solar acoustic Eigen frequencies with the Global Oscillations Network Group (GONG) and SOHO's MDI instrument, infers the rotation rate of the deep solar layers, with the uncertainty lines shown in the diagram. The above figure is adapted from Thomson et al. (2003), which was adapted from K. Lang's book Essential Astrophysics (2013), with the MDI data of solar rotation in the convective zone by Alexander G. Kosovichev and the GOLF/MDI data for the radiation zone by Sebastien Couvidat, Rafael García, and Sylvaine Turck-Chièze. Data were acquired by GONG instruments operated by NISP/NSO/AURA/NSF with a contribution from NOAA (Image courtesy of GONG and Alexander G. Kosovichev)

1991: Solar Coronal *Jets* (the Giant Counterparts on the Sun of the US National Park's Geyser in Yellowstone, Old Faithful)

One of the attractions of the Yellowstone National Park in the US state of Wyoming are its famous geysers. The most famous of these is the geysers, is *the Old Faithful*, as it was called by the first explorers who visited it around 1870. In it, the jetof steam reaches 32–56 m high and lasts 1–5 min, while from the year 2000 onwards it has been relatively predictable, with its release occurring every 44–125 min – enough time for visitors to gather every hour on the special platform, waiting for the next release...

The solar coronal *jets* (Fig. 1.65) are a giant analog of these terrestrial thermojets. Only they reach a height of hundreds of thousands of kilometers and last longer than the terrestrial geysers. These are impressive eruptions of solar plasma, relatively small in diameter, visible mainly in the ultraviolet part of the electromagnetic spectrum and X-rays (Fig. 1.65). There are several types of solar coronal *jets* that have been observed at different wavelengths, such as in X-rays, far ultraviolet (EUV), or the Hα line of hydrogen. Recent solar space missions, such as the Einstein Observatory (1978), Yohkoh (1991), TRACE (1998), Hinode (2006), STEREO (2006), and SDO (2010), as well as high-resolution ground-based solar telescopes, such as the Dutch Open Telescope (DOT) at La Palma have today provided a relatively detailed picture of the solar coronal *jets*. A statistical study of about 100 *jets* observed with the Soft X-ray Telescope carried by Japan's Yohkoh Solar Space Observatory provided the following characteristics: most *jets* are combined and coexist with small *flares* occurring in their legs, extend from 10,000–400,000 km in length, have a diameter of 10,000–40,000 km, and apparent plasma velocities 150–800 km per second, an average speed of 200 km per second, while their lifespan extends to a few hours at most. High-resolution observations reveal that they generally have a helical structure, a feature that strongly hints that magnetic mechanisms are involved and play an important role in their initiation and development.

In the work Moschou et al. (2013), we analyzed observations of 6 *jets* detected with the Solar Dynamics Observatory in the far ultraviolet (EUV) part of the spectrum, and specifically at the wavelength of 304 Angstroms, emitted by atoms of the ionized element helium (HeII) in the *chromosphere*, at a plasma temperature of 50,000 °K.

The length of the *jets* ranged from 63,000–190,000 km, while their lifetime was between 20–50 min. Some of these *jets* had a helical appearance. The analyzed sample contained 4 *jets* of 25 June 2010, 2 of which are near the solar equator and 2 of 30 June 2010 with the aim of analyzing their evolution and basic physical properties. All these events evolved in the time intervals of a few tens of minutes, they reached heights of 10–30% of the solar radius above the solar *photosphere*, while some of them showed repeatability.

In another paper in the journal Astronomy & Astrophysics ((Archontis et al. 2010), Fig. 1.jetreconnection), we presented detailed numerical simulations and we arrived at a model where, after the emergence of magnetic flux tubes from the *photosphere* due to magnetic buoyancy, and since they encounter an opposite magnetic field in the *photosphere*, these two magnetic fields interact with the mechanism of magnetic reconnection and create the jet, according to the simple description of Fig. 1.65. Also, our observations have given us some re-susceptibility to the appearance and emission of these plasma *jets*, such as we have shown in our previous work.

Thus, in our papers we have proposed that solar *jets* are the result of the dynamic interaction, through magnetic reconnection, of a dipole magnetic field emerging in the *photosphere*, with a pre-existing dipole magnetic field in the *photosphere*. Magnetic reconnection is a process in which two opposing magnetic lines of different magnetic subsystems that come into contact, such as systems 1 and 2 in Fig. 1.65, reconnect and convert their free magnetic energy into kinetic and thermal energy of the plasma. This natural mechanism of magnetic reconnection of opposite magnetic fields leads to the ejection of *jets* high up in the solar atmosphere, while sometimes powerful *jets* even reach interplanetary space.

1992–2009: Exploring the Polar *Solar Wind* with the Spaceship ULYSSES

The Ulysses spaceship, a joint project of NASA and the European Space Agency (ESA), was launched on the 6th of October 1990, with the aim of studying the solar environment in the polar solar regions. After sixteen months of interplanetary cruise, Ulysses reached *Jupiter* in 1992 and used its gravity assist to climb vertically out of the ecliptic plane, the first spacecraft to explore interplanetary regions outside

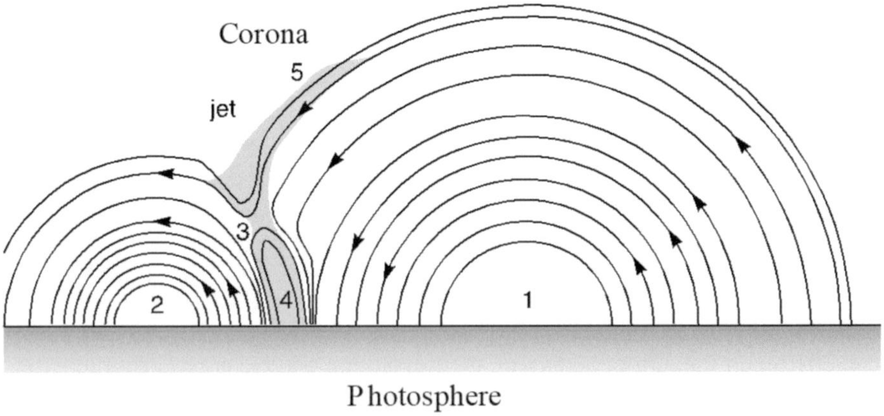

Fig. 1.65 Diagram of the production mechanism of a solar coronal jet, according to observations from the Solar Dynamics Observatory at 304 Å and the study in the paper, Moschou et al. (2013). The emerging magnetic flux of region (1) reconnects with the pre-existing in the photosphere, but in the opposite direction, magnetic flux of region (2), resulting in magnetic reconnection in region (3) and the launch of the solar jet (5, gray region) (*Source* Gontikakis et al. (2010))

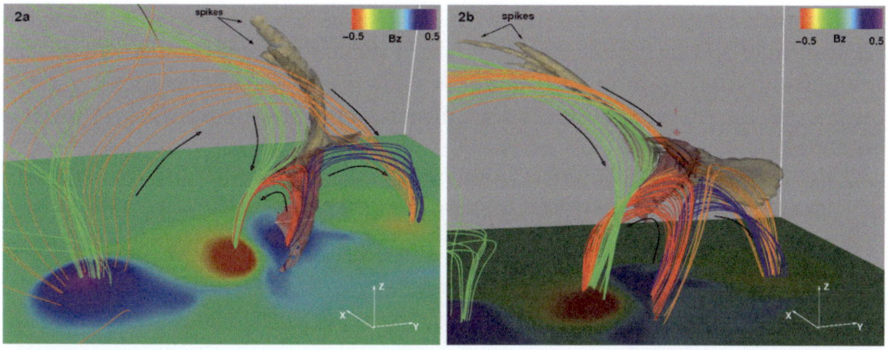

Fig. 1.66 Three-dimensional visualization of the formation of coronal solar jets through numerical simulation of the dynamic emergence of a magnetic field from the solar interior and its interaction with another pre-existing photospheric magnetic field. The two magnetic dipoles of the emerging field (green lines) and the pre-existing field (blue lines) interact and produce via the magnetic reconnection the red and orange lines, while the yellow/brown sheets are the isovelocity surfaces of the forming jets. Shown are the magnetic lines at two consecutive times, t = 144 (left) and t = 184 (right) as well as the magnetogram (blue and red polarities) at a given z-height. Remarkable is the formation of two elongated jets moving sideways up and to the left. The arrows (black) show the direction of the magnetic field (From the article Archontis et al. (2010))

the ecliptic plane. The spacecraft entered an elliptical, heliocentric orbit inclined at 80.2° to the solar equator, with an orbital period of 6.2 years, aphelion at ~5.4 AU from the Sun and a perihelion distance of ~1.34 AU. Ulysses measured the *solar wind* (SW) speed for a full orbit around the Sun during the 1992–1998 solar minimum period (Fig. 1.69). In particular, Ulysses observed that the speed of the SW in the polar regions, about 800 km per second, are twice the slow speed of the SW in the equatorial regions, about 400 km per second. Ulysses, after completing three full orbits around the Sun in its 18-year journey of 8.85 billion kilometers, was decommissioned in 2009 (Fig. 1.68).

1996–present: Discoveries with the Solar and Heliospheric Observatory (SOHO)

The Solar and Heliospheric Observatory (SOHO) is a project of international collaboration between ESA and NASA to study the Sun from its deep core to the outer *corona* and the *solar wind*. SOHO was launched on December 2, 1995. It orbits the Sun together with the Earth every 365 days, located at the Lagrange point L1 between the Earth and the Sun, so that to observe the Sun continuously. The SOHO spacecraft was built in Europe by an industry team under overall management by ESA. Of SOHO's nearly $1-billion cost, NASA picked up $477 million and ESA paid the rest. Weighing in at 1,875 kg, the satellite stretches about 8 m across with its solar panels extended.

The twelve instruments on board SOHO were provided by European and American scientists. Nine of the international instrument consortia are led by European

Principal Investigators (PI's) and three by PI's from the US. Large engineering teams and more than 200 co-investigators from many institutions supported the PI's in the development of the instruments and in the preparation of their operations and data analysis. NASA was responsible for the launch and is now responsible for mission operations.

Some notable instruments are the extreme ultraviolet EIT telescope and the LASCO coronagraphs. Three of its instruments (GOLF, MDI, VIRGO) are used

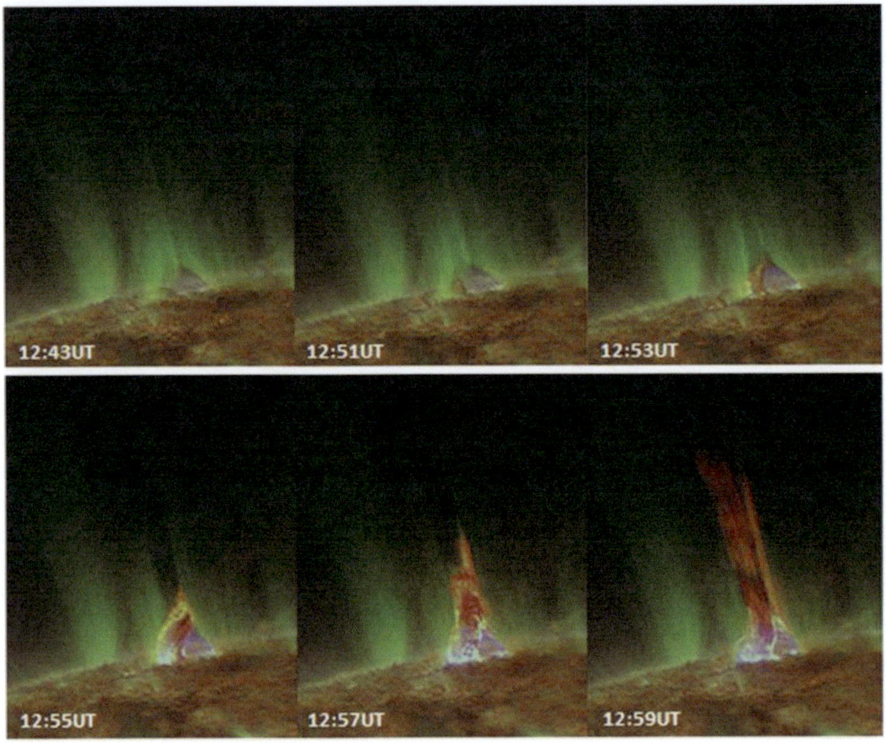

Fig. 1.67 Solar Dynamics Observatory (SDO) observation of a 30 January 2015 coronal jet at the limb of the solar disk, 16 arcminutes long, detectable in the far ultraviolet part of the electromagnetic spectrum (EUV) from 12:55 to 14:00 UT. It was not observed in X-rays, which came mainly from the nearby NOAA active regions 2268 and 2277. The latter produced an M2-type flare at 12:16UT, while the former increased X-ray emission at 12:16UT. 56UT. This coronal jet consisted of relatively *cold* plasma, as its emission is reduced by fitting filters to the detector that detect higher temperatures. The jet was detectable in the Hα line, with a temperature of about 10,000 degrees. From images on the SDO/AIA 304 Angstrom line, the velocity of the jet is estimated to be about 280–410 km/s. This speed is high enough that some of the jet's plasma escaped from the Sun into space. This is monitored with the coronographs of SOHO, although these observations are difficult due to the decrease in density and brightness of the jet as it moves away from the Sun. Another part of this jet returns and falls on the lower layers of the solar corona. The mechanism that we believe produces these jets is explained in the Fig. 1.64, based on our theoretical studies, Archontis et al. (2010) (*Source* SDO/AIA/Solar Influences Data Analysis Center)

Fig. 1.68 The underground tank where Brookhaven National Laboratory's solar neutrino experiment took place. The neutrino interactions took place in this massive tank, 1.5 km underground in the Homestake gold mine, in South Dakota (*Credit* U.S. Department of Energy)

to study the solar interior with helioseismology. Very important results came from the combined observations of the coronagraphs and the EIT telescope on the evolution of *Coronal Mass Ejections* (CME) away from the solar disk and their sources on the solar disk, linking for the first time solar activity on the disk and away from the disk. SoHO discovered that the solar *corona* is more active than we thought, changing from minute to minute. In the low *corona* we have the first indications of CMEs, especially in the extreme ultraviolet (EUV). SoHO established the CMEsas the main interplanetary driver of *Space Weather* and especially the halo-CMEs, as responsible for geomagnetic disturbances. To this day, it gives us valuable information about the outbursts of *Space Weather*, warning up to 3 days in advance of upcoming geomagnetic disturbances with observations at 4 temperatures: 60,000°–80,000° (EUV, HII, 304 Å), 0.9 million degrees (Fe IX/X, 171 Å), 1.5 million degrees (Fe XII, 195 Å) and 2 million degrees (Fe XV, 284 Å).

SOHO is the longest-lived Sun-watching satellite to date. Numerous mission extensions have enabled the spacecraft to observe two 11-year solar cycles (Fig. 1.70) and to discover thousands of comets. SOHO was meant to operate until 1998, but it was so successful that ESA and NASA decided to prolong its life several times and endorsed several mission extensions. It celebrated its 25th launch anniversary in 2020 and its mission is ongoing.

1998 and 1999: Solution of the Solar Neutrino Problem

For the solution of the solar neutrino problem, besides the Raymond Davis detector at the Homestake of North Dacota (Fig. 1.68), other detectors were also constructed. One of them is the Super-Kamiokande (SK) detector in Japan, located at a depth of 2 km below the ground. This detector measures muon neutrinos (ν_μ), τ-neutrinos (ν_τ) as well as electron neutrinos (ν_e). Another detector, the Sudbury Neutrino Observatory (SNO) detector in Canada (Fig. 1.71), measures only electron neutrinos (ν_e).

The Super-Kamiokande (SK) detector in Japan contains 50,000 tons of pure water and is surrounded by 11,000 sensitive photodetectors that detect flashes of Cerenkov light radiation which are produced by the motion of electrons which are emitted in the reactions involving neutrinos. Also in this experiment, the counted number of nuclear reactions was smaller compared to the theoretically predicted one.

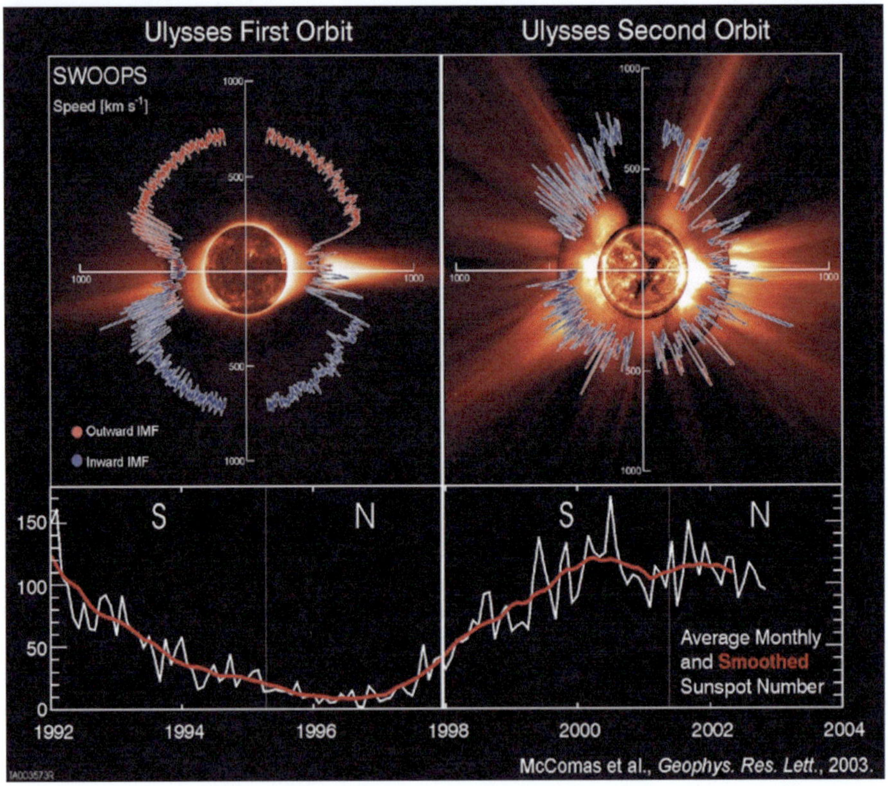

Fig. 1.69 Polar plots of solar wind speed as a function of latitude for Ulysses' first two orbits (McComas et al. 2003). The solar wind speed data was obtained by the SWOOPS instrument (Solar Wind Observations Over the Poles of the Sun). The bottom panel shows the sunspot number over the period 1992–2003. The first orbit occurred through the solar cycle declining phase and minimum while the second orbit spanned solar maximum. Both solar wind plots are plotted over solar images characteristic of solar minimum (17 August 1996) and maximum (7 December 2000). The solar images are composites of (from the center out): the Solar and Heliospheric Observatory (SOHO) Extreme ultraviolet Imaging Telescope (Fe XII at 195 Å), the Mauna Loa K-coronameter (700–950 nm), and the SOHO C2 Large Angle Spectrometric *Coronagraph* (white light). The increased polar velocities of the SW relative to the equatorial ones remind that the Sun in its "youth" (i.e. a few billion years ago) was ejecting from its pole jets (see our related article in Astronomy & Astrophysics, Sauty et al. (2005)) (*Credit* D. McComas and ESA/NASA, reprinted with the kind permission of D. McComas)

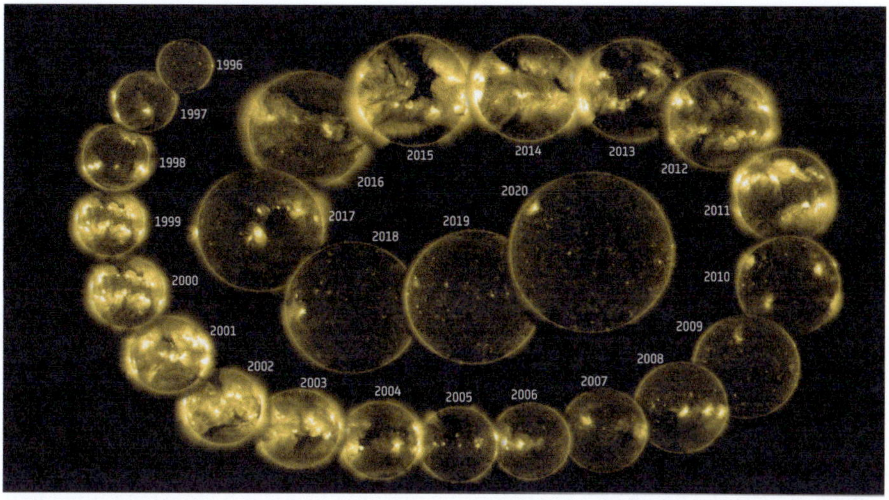

Fig. 1.70 The ESA/NASA Solar and Heliospheric Observatory (SOHO) has been observing the Sun for 28 years. In that time, SOHO has observed two complete of the Sun's 11-year sunspot cycles (23rd & 24th), while in the last few years observed the 25th solar cycle as well, as solar activity waxes and wanes. This montage of 25 images captured by the spacecraft's Extreme Ultraviolet Imaging Telescope provides a snapshot of the changing face of our Sun. The individual images show gas with a temperature of about two million degrees Celsius in the Sun's atmosphere. The brightest images occur around the time of solar maximum, when the Sun's magnetic field is strongest and highly dynamic, changing its configuration and releasing energy into space (*Credit* SOHO (ESA&NASA))

Fig. 1.71 Outside view of the Sudbury Neutrino Observatory (SNO) detector (Courtesy of SNO, also reprinted in Wikipedia Commons)

The same SK detector, however, could also measure muon neutrinos (ν_μ) created in the Earth's atmosphere by Cosmic Radiation. The result was that more muon neutrinos (ν_μ) were detected coming *from above* than *from below*, i.e., those that had crossed the entire Earth, where they managed to transform from ν_μ to ν_e and ν_τ. This showed that the muon neutrinos (ν_μ), as they traveled and interacted more with matter coming from *below the Earth* had time to change their identity. On the other hand, the muon neutrinos (ν_μ) that came directly *from above* and had interacted with matter only in a few tens of kilometers of the atmosphere, they did not have the time to transform to another flavor!

Thus, researchers at the Super-Kamiokande neutrino detector in Japan had shown that neutrinos originating from reactions in the atmosphere and within of Earth oscillate between two states, as they make their way to the detector. This measurement provided the first evidence of neutrino oscillation (1998).

At the same time, solar electron neutrinos (ν_e) were detected one year later (1999) and by another detector, at the Sudbury Neutrino Observatory (SNO), in Ontario, Canada, where at a depth of 2 km below the ground there was a large tank filled with 1,000 tons of heavy water, which is surrounded by 9,500 photomultipliers (Fig. 1.71). Heavy water is like usual water that instead of two atoms of hydrogen and one atom of oxygen has two atoms of an isotope of hydrogen, deuterium, and one atom of oxygen. The deuterium nucleus contains a proton and a neutron, unlike hydrogen nucleus containing only one proton. Heavy water is a colorless and odorless liquid that is mixed with ordinary water in a ratio of 1 to 6,760 parts of water.

SNO measures electron neutrinos (ν_e), from its telltale flashes of light caused when these neutrinos travel through heavy water. The SNO team could not detect muon neutrinos (ν_μ) and τ-neutrinos (ν_τ) because these do not break up the molecules of heavy water. When an electron neutrino ν_e collides with a deuterium nucleus in heavy water it turns the deuterium nucleus neutron into a proton, also ejecting an energetic electron, which then emits the characteristic Cherenkov radiation. Of the 60 billion solar electron neutrinos (ν_e) that reach Earth per second in every square centimeter, the SNO recorded only 1/3 of this theoretically expected number of e-neutrinos (ν_e) that it should *see*. It should be noted that this SNO experiment is extremely accurate and took about 10 years of preparation to be completed.

In conclusion, the prevailing explanation that has been given so far is that neutrinos, because in theoretical physics have an *identity crisis*, they change character in their flight from the Sun to the Earth, and half of them reach here as muon neutrinos (ν_μ), although they all started as electron neutrinos (ν_e).

The 2015 *Nobel Prize* in Physics was awarded to Takaaki Kajita, of the SK detector in Japan, and Arthur B. McDonald, of the SNO in Canada, for solving the solar neutrino problem and discovering neutrino oscillations. We should note here that the transformation of neutrinos requires that they have mass.

For particle physics this was a historic discovery because these experiments led to the fundamental conclusion that neutrinos, which until then were thought to have zero mass, must indeed have mass, even if it is very small. This discovery changed our understanding of the structure of the innermost components of matter and may also be crucial for understanding the structure of the Universe.

The *Standard Model* of theoretical physics, providing the mathematical description of elementary particles of matter and their interactions, and combines the electromagnetic with the weak and strong interactions and classifies all elementary particles, has been extremely successful in the last decades after 1970, in interpreting all experimental data. However, it requires neutrinos to have zero mass, like photons, which means that it can no longer be the full theory of the microcosm. Possibly, the *Standard Model* is only part of a larger theory that governs the microcosm of elementary particles and also describes our understanding for the creation of the Universe.

2002: Discovery of Pulsations in Other Stars (*Asteroseismology*)

After the detection of solar oscillations, it became possible to detect corresponding sound waves from other stars, beyond our Sun. Similarly with *Helioseismology*, in *Asteroseismology* the stellar oscillations are the visible manifestations of standing waves inside the stars. Main Sequence stars and subgiants that have *convection zones* exhibit solar-type oscillations that are basically acoustic and are caused by the turbulent motion in their *convection zones*. The dominant periods of these oscillations are of the order of a few minutes and cause fluctuations in the stellar brightness of the order of one millionth.

The frequencies of the oscillations depend on the particular structure of the stars and provide detailed information for calculating their mass, radius and age with an accuracy that cannot be achieved by other ways. Most recently, thanks to the European space telescope COROT (COnvection, Rotation and planetary Transits), which was launched in 2006 and was operational until 2013, and the corresponding NASA Keplertelescope, which was launched in 2009 and was operational until 2018, we *heard* the sounds of hundreds of stars.

It is interesting to note here that *asteroseismology* was not the main goal of the COROT and Kepler space missions: they were interested for a *space hunt* of exoplanets that have the characteristics of Earth in terms of size and distance from the parent star. But because both those missions sought to detect the minimal fluctuations of the brightness caused by a planet passing in front of the parent star, they were able to record imperceptible fluctuations of the order of one millionth in stellar brightness. Therefore, in theory, they could detect stellar acoustic waves. As is gratifyingly reported, asteroseismologists enjoyed a free ride on these two space telescopes!

There have already been interesting results from the observations, by COROT and Kepler, especially for red giants. They form when solar-type stars run out of their hydrogen fuel in their cores and suffer a fuel shortage crisis, resulting paradoxically in a stellar inflation, wherein the stellar radius reaches the size of the Earth's orbit around the Sun. We wanted to distinguish the following two possibilities: the first phase, wherein the giant star has exhausted hydrogen in its core, but continues to burn hydrogen in a peripheral shell around the dense core, a little bigger than Earth and a later phase, where the giant has started to burn the chemical element helium in its core. Knowing exactly when each phase happens, will give us the possibility to understand the ages of the giant stars, how fast they evolve, as well as the amounts of gas and heavy nuclei that scatter in space in each evolutionary phase.

In this way, it was measured, for the first time, the rotation frequency of the core of these giant stars and it was calculated that they rotate ten times faster than their surface! This result verifies the standard model of the formation of red giants – i.e., solar-type stars, which ejected their outer layers into space, while the cores are shrinking. Due to conservation of angular momentum of basic physics (the phenomenon of the ballerina rotating faster when she closes her arms), the contracting core must rotate faster from the outer layers, as indeed observed.

Thus, the spectral type G7 red giant ξ in the constellation Hydra, at a distance of 130 light years, with a luminosity 58 times higher than the solar, a radius ten times that of the Sun, a mass three times that of the Sun, and an age 510 million years, oscillates at much lower frequencies, 0.05–0.140 mHz, i.e., its photospheric oscillations have a period of 2–5.5 h, (Frandsen et al. 2002; Bedding and Kjeldsen 2003), Fig. 1.72.

The yellow-white F5IV star Procyon, in a binary system that is also close to us, at a distance of about 11.5 l.y, with an age of almost 2 billion years, a surface temperature of 6,650 K, a mass one and a half times the mass of the Sun and a diameter of 1.86 times greater than the solar one, also shows solar-like oscillations, with a period of about 20–25 min, Fig. 1.72 (Hekker et al. 2007; Huber et al. 2011).

Photospheric oscillations have also been observed in the triple system of α Centauri which is also the closest multiple stellar system to our solar system and consists of two solar-type stars, α Centauri A and B, which constitute the double star system α Centauri AB, and the nearest to us proxima (α Centauri C). The oscillations of the A member have a frequency of 2.25 mHz and a period of 7.4 min (Fig. 1.72), while the period of the oscillations of the B member is about 4 min (Bouchy and Carrier 2002). All these different oscillation periods result from differences in the properties of the stellar structure, i.e., radius, mass, photospheric temperature and age.

The Cosmic Music of Pulsating Stars…

In his book *Stars and Atoms* (Oxford University Press, 1926), A.S. Eddington writes (p. 89): "It is not surprising that these pulsating stars should be regarded with special interest. Ordinary stars must be viewed respectfully like the objects in glass cases in museums; our fingers are itching to pinch them and test their resilience. Pulsating stars are like those fascinating models in the Science Museum provided with a button which can be pressed to set the machinery in motion. To be able to see the machinery of a star throbbing with activity is most instructive for the development of our knowledge!".

The emergence of the field of helioseismology a few decades later showed that there are indeed observation instruments and ways to *penetrate* through the photosphere and *examine* the invisible interior of the Sun and other stars. The pulsating solar photosphere is the additional *button* that we can press to set the mechanism in motion. By analyzing the frequencies, helioseismology captures the *songs* of the Sun as a cosmic music, from a band that plays in sync, with a loud speaker the very *body* of our Sun. It is a wonderful music of the solar orchestra, through which the Sun and the other stars reveal to us the well-hidden secrets beneath their photospheres!

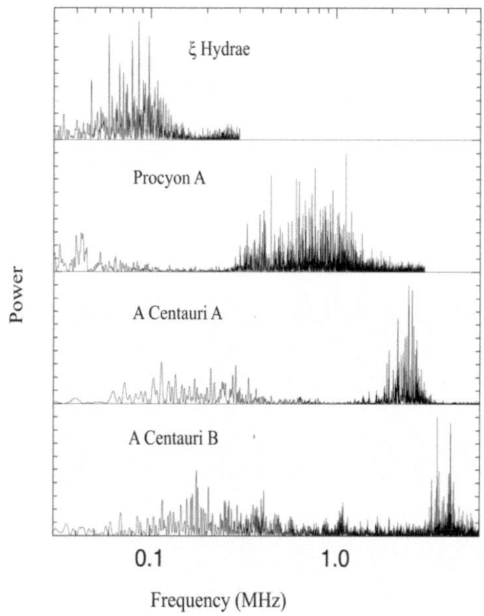

Fig. 1.72 Power spectra of time series observed in radial velocity for the four stars ξ Hydrae (G7 III; Frandsen et al. 2002), Procyon (F5 IV; Arentoft et al. 2008), α Centauri A (G2 V; Bedding et al. 2004) and α Centauri B (K2 V; Kjeldsen et al. 2005). The oscillation periods for the four stars are: around 3–4 hfor ξ Hydrae, 20–25 min for Procyon, 7 min for α Cen A and 4 min for α Cen B. These differences, and those in the detailed structure of the power spectra, reflect the differences in stellar properties (radius, mass, surface temperature and age). Note that the vertical scales are normalized – the actual amplitudes decrease from top to bottom in the figure (Figure reproduced from Kjeldsen et al. (2008) with the kind permission of the authors)

2006 to Present: The Innovative Solar TErrestrial RElations Observatory (STEREO)

STEREO, launched by NASA in 2006, consists of 2 spacecrafts, A and B, which view the Sun from 2 different angles. The angle that the Sun sees each spacecraft and the Earth increases by about 22° every year. STEREO carries five instruments whose observations are particularly useful for calculating the three-dimensional distribution of density and compression in the shock wave that precedes the propagation front of *Coronal Mass Eruptions*, as observed by the *coronagraphers* (Kwon and Vourlidas 2018). Also very useful for *Space Weather* monitoring are the observations of propagating *solar wind* density inhomogeneities throughout interplanetary space, from the solar *photosphere* to Earth (Fig. 1.73), see also DeForest et al. (2013). Principal investigator of the Sun Earth Connection Coronal and Heliospheric Investigation (SECCHI) instruments of STEREOA is the solar physicist Angelos Vourlidas (John Hopkins University, Applied Physics Lab). For predicting the geoeffective properties of coronal mass ejections, see Vourlidas et al. (2019) and for a multi-viewpoint coronal mass ejection catalog based on STEREOCOR2 observations, Vourlidas et al. (2017).

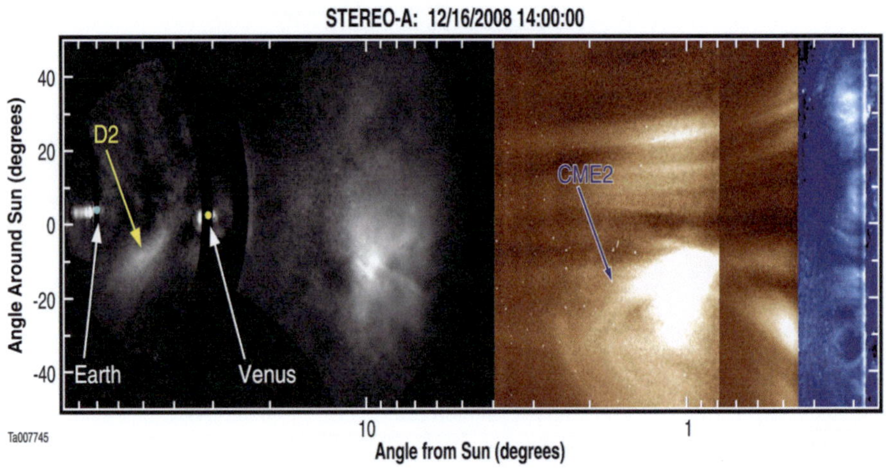

Fig. 1.73 One of the most important achievements of STEREOis the three-dimensional visualization of the evolution of inhomogeneities of the density of the *solar wind* due to *coronal mass ejections*, as these start from the base of the solar atmosphere and propagate through interplanetary space to Earth, thus shaping *Space Weather* (Reprinted from Howard et al. (2012), with the kind permission of C.E. Deforest)

2010 to Present: The Impressive Observations of the *Solar Dynamics Observatory* (SDO)

The *Solar Dynamics Observatory* (SDO) was placed by NASA in 2010 in a geosynchronous orbit at a height of 35,786 km above Earth, i.e., it is always above the same point on the Earth's surface. The main scientific goal of the SDOis to give us an

Fig. 1.74 This is how the SDO instruments *see* and *depict* the photosphere and external atmosphere of the Sun, at various wavelengths and temperatures. They are the *Atmospheric Imaging Assembly* (AIA), imaging the Sun in 10 wavelengths, every 10 s, the *Helioseismic and Magnetic Imager* (HMI) and the *Extreme Ultraviolet Variability Experiment* (EVE). (*Clockwise from left*) The HMI Dopplergram which maps the velocities observed on the solar photosphere. The HMI Magnetogram mapping the photosphere's magnetic field, in black the magnetic fieldlines directed away from Earth and with white those coming toward the earth. 1700 Å: Ultraviolet (UV) continuum from the photosphere and chromosphere at 4500 K. 4500 Å: Continuous white light from the photosphere at 6000 K. 1600 Å: Radiation emitted by carbon-4 (C IV) atoms of the transition zone at temperature T ~ 10,000 K. 304 Å: Radiation emitted by helium-2 (He II) atoms of the upper transition zone and the chromosphere at temperature T ~ 50,000 K. 171 Å: Radiation emitted by iron-9 atoms (Fe IX) of the quiescent corona and coronal loops at temperature T ~ 600,000 K. 193 Å: Radiation emitted by iron-12 atoms (Fe XII) of the hot corona at temperature T ~ 1,000,000 K and iron-24 (Fe XXIV) of the hottest plasma during flares at T ~ 20,000,000 K. 211 Å: Radiation emitted by iron-14 atoms (Fe XIV) of hot active regions in 2,000,000 K. 335 Å: Radiation emitted by hotter active iron-16 (Fe XVI) atoms, with temperature ranges T ~ 2,500,000 K. 094 Å: Radiation emitted by iron-18 (Fe XVIII) atoms at T ~ 6,000,000 K during flares. 131 Å: Radiation emitted by atoms iron-20 (Fe XX) and iron-23 (Fe XXIII) at a temperature greater than 10,000,000 K, plasma during flares (*Source* NASA/SDO)

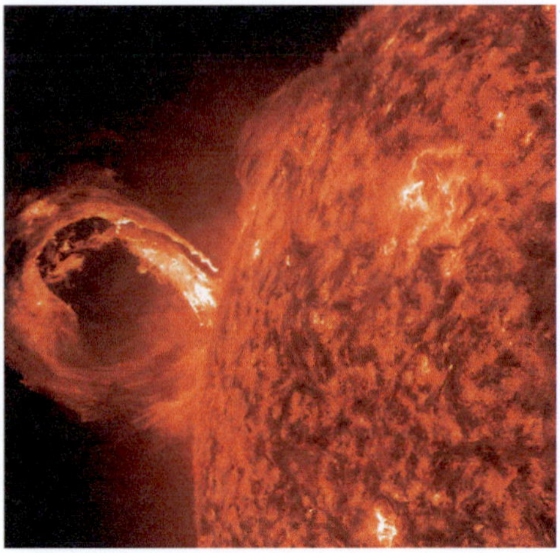

Fig. 1.75 A *Coronal Mass Ejection* (CME), carrying over a billion tons of particles reaching a speed over a million miles per hour, associated with a solar flare, blew out from just around the edge of the Sun on May 1, 2013 in a glorious roiling wave. Viewing the sun in the extreme ultraviolet wavelength of 304 angstroms, SDO provided a beautiful view of the initial arc as it left the solar surface. SOHO's C2 and C3 *coronagraphs* showed a large, bright, circular cloud of particles heading out into space. The STEREO spacecraft, from their different perspectives in space, observed the flare (*Credit* NASA/SDO)

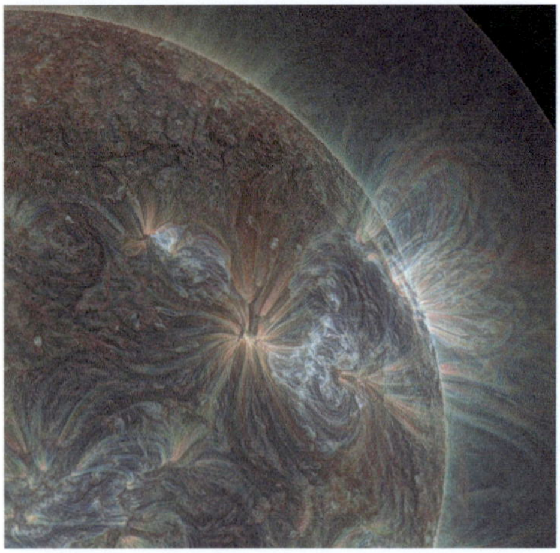

Fig. 1.76 The solar atmosphere is full of solar *loops*, especially during the maximum of solar activity (here on Jan. 17, 2013) which are very well distinguished in the ultraviolet part of the spectrum. In this image we see the Sun through a UV lens, in a combination of three different wavelengths taken by NASA's *Solar Dynamics Observatory*. The light has been artificially colorized in red, green and blue (*Source* NASA/SDO and Nathalia Alzate)

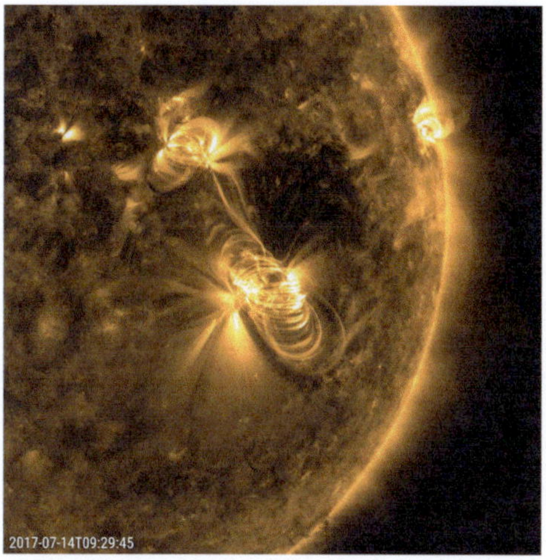

Fig. 1.77 The large active region of July 14, 2017. A medium-size (M2) flare erupted from this active region, with quite long duration – almost 2 h – accompanied by a *coronal mass ejection*. *Coronagraphs* on the SOHO spacecraft show a substantial cloud of charged particles blasting into space just after the blast. The coils arcing over this active region are particles spiraling along magnetic field lines, which were reorganizing themselves after the magnetic field was disrupted by the blast. Images were taken in a wavelength of extreme ultraviolet light (*Credit* NASA/GSFC/Solar Dynamics Observatory)

understanding of solar variability and how it affects our Earth. One of its 3 instruments is the *Atmospheric Imaging Assembly* (AIA), which observes the solar disk and *corona* at 10 wavelengths every 10 s, at temperatures ranging from 20,000 to and over 20 million degrees K, scanning the solar atmosphere at many different altitudes (Fig. 1.74). At the same time, the *Helioseismic and Magnetic Imager* (HMI) studies the origin of the solar variability by observing motions on the solar *photosphere* at a wavelength of 6.173 Å, with a resolution of 1 arcsecond (about 700 km above the Sun) to analyze solar oscillations as well as the polarities of the magnetic fields, taking a terabyte of data every day. HMI is a successor to the *Michelson Doppler Imager* on the *Solar and Heliospheric Observatory*. The *Extreme Ultraviolet Variability Experiment* (EVE) measures with high precision the solar luminosity, in the extreme ultraviolet (EUV), affecting terrestrial the climate and near interplanetary space (Figs. 1.75, 1.76 and 1.77).

2020 to Present: *Solar Orbiter* (Observing the Sun Outside the Plane of the Ecliptic)

Solar Orbiter (SolO) is the first mission of the *European Space Agency's* (ESA) *Cosmic Vision* program for the period 2015–2025, which is complementary to the *Parker Solar Probe*. With a total cost of 1.5 billion Euros and a weight of 1,800 kg vessel, built by Airbus, SolO was launched in 10-2-2020 with a American Atlas V rocket. SolO will approach the Sun at a distance of about 42 million kilometers to observe the solar atmosphere with unprecedented clarity and discretion, while at the same time carrying out in situ measurements of the solar plasma very close to the Sun and especially the inhospitable regions, hitherto inaccessible, reaching 33° outside the plane of the ecliptic, thus making the first observations of the solar poles. The operation of the mission's sensitive instruments in these exceptional conditions that prevail very close to the solar *furnace* is a technological and scientific challenge in itself, as special techniques have been developed for the protection and insulation of the instruments in these particularly inhospitable near solar environment (Fig. 1.78).

On SolO there are the following ten instruments: (1) The Energetic Particle Detector (EPD, Spain), a detector of charged particles, measuring electrons, protons, and heavy ions with high temporal resolution over a wide energy range, from suprathermal energies up to several hundreds of MeV/nucleons, (2) The Magnetometer (MAG, UK), a high-resolution magnetometer measuring the solar magnetic field, (3) the Radio & Plasma Waves experiment (RPW, France), a high-resolution detector of waves and fields, designed to measure *in situ* magnetic and electric fields and waves up to a few hundreds of kHz, also observing solar radio emissions up to 16 MHz. RPW is the only instrument on Solar Orbiter that makes both in situ and remote sensing measurements (4) the Solar Wind Analyser (SWA, UK), an analyzer of protons, electrons and heavy ions, designed to measure the 3D velocity distribution functions of *solar wind* protons and alpha particles with high time cadence, energy and angular resolution (5) the METIS Coronograph (Italy), a *coronagraph* for observing the *corona* outside the disc, in the visible and ultraviolet part of the spectrum, (6) The Extreme Ultraviolet Imager (EUI, Belgium), a suite of remote-sensing telescopes which will image at high-resolution structures in the solar atmosphere from the *chromosphere* to the *corona*, (7) the Polarimetric & Helioseismic Imager (PHI, Germany), a diffraction limited, wavelength tunable, quasi-monochromatic, polarization sensitive imager, providing maps of the magnetic vector and of the line-of-sight velocity in the solar *photosphere*; it will thus probe the deepest layers of the Sun (including the solar interior using helioseismology) of all the instruments on *Solar Orbiter*, (8) the Heliospheric Imager (SoloHI, USA) a wide-field camera that is designed to image the *solar wind* in interplanetary space, far from the Sun itself, observing the *corona* outside the disc, (9) the Spectrometer/Telescope for Imaging X-rays (STIX, Switzerland), an X-ray imaging spectrometer, whose purpose is to study the extremely hot solar plasma and the high-energy electrons accelerated during a solar flare, (10) the Spectral Imaging of the Coronal Environment (SPICE), a facility instrument on the *Solar Orbiter* mission which is an imaging spectrograph

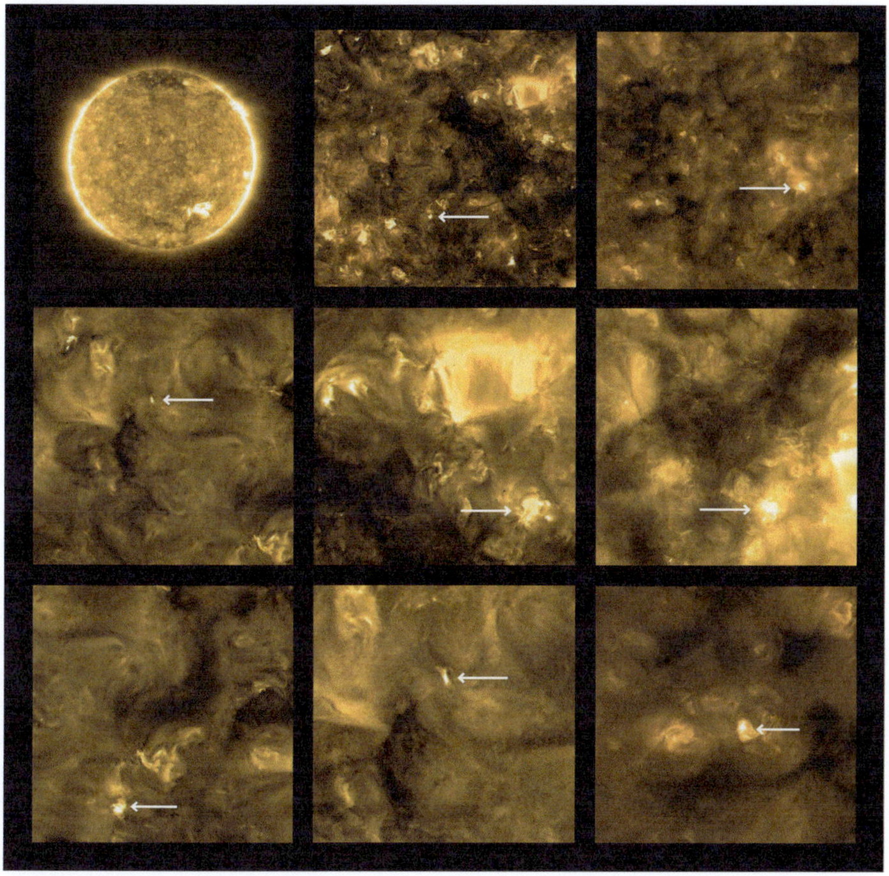

Fig. 1.78 *Campfires* seen on the Sun by Solar Orbiter. They may help explain why the Sun's atmosphere is 300 times hotter than its surface (*Image Credit* Solar Orbiter/EUI Team (ESA & NASA); CSL, IAS, MPS, PMOD/WRC, ROB, UCL/MSSL)

designed to make observations at extreme ultraviolet wavelengths, 70.4–79.0 nm and 97.3–104.9 nm for the spectral observation of the coronal environment.

SolO is the first mission with multiple Greek participations, since Greece became a full member of ESA (2005), with participation in the experiments (1)–(3)–(5)–(9). SolO survived marginalization within ESA thanks also to our efforts that culminated in the organization by the author of the 2nd *Solar Orbiter* Workshop in Athens (October 2006).

The first images from *Solar Orbiter*, have revealed omnipresent miniature solar, dubbed *campfires*, near the surface of the Sun, which may be Parker's *nanoflares* (Fig. 1.78).

2024–Present: Proba 3 (Ground-Breaking Space Solar *Coronagraph*)

ESA's *Proba-3* (*PR*oject for *O*n*B*oard *A*utonomy) space mission is dedicated to the observation of the solar *corona* with similar conditions to those of total solar eclipsesbut with the extended duration of several hours in each orbit – and at the same time demonstrating the technologies and techniques of high-precision satellite formation flight. Achieving high-precision formation spaceflight opens a new era for science and applications, so Proba-3 marks the next step in long-form formation flight accuracy. The twin spacecrafts of the *Proba 3* mission are separated by a fixed distance of about 150 m – with a millimeter precision – orbit the Earth *locked,* thus forming a giant *Coronagraph.* One spaceship carries the telescope (*Coronagraph spacecraft*), detectors, electronics, etc. which observes the solar *corona hidden* inside the shadow of the solar *photosphere*, which is created by the second spaceship (*Occulter spacecraft*), which is used as a curtain. The *Proba 3* spacecraft were launched in a highly eccentric elliptical orbit on December 8, 2024, with a perigee of 600 km and an apogee of 60,530 km (about 10 Earth radii) and an orbital period of 19.6 h. Near the apogee of their orbit and for 6 h intervals, the two spacecraft are perfectly aligned among themselves and with the Sun, making possible observations of the solar *corona* (Figs. 1.79, 1.80, 1.81, 1.82, 1.83, 1.84, 1.85 and 1.86).

The ASPIICS (Association de Satellites Pour l'Imagerie et l'Interferometrie de la Couronne Solaire) is the scientific instrument of the space mission *Proba 3*. The distance of m between the twin spacecrafts allows observations of the *corona* from its base, minimizing the scattered light of the *photosphere* and diffraction phenomena caused by the intervening disc. The main scientific goals of the program are the understanding on the one hand of the physical processes that occur in the quiet solar *corona* and on the other hand of the physical processes that lead to the powerful coronal plasma ejections (CMEs) which result in the creation of unstable *Space Weather*, with all its consequences for Earth and our infrastructure in Space.

The Greek research team of the Department of Informatics & Telecommunications of the University of Athens, has undertaken the construction of special innovative data compression algorithms; the wider local scientific team includes researchers from the University and Academy of Athens, as well as distinguished Greek researchers from foreign institutions, coordinated by the author, wherein the Greek involvement started as early as in 2008 with the participation in the ESA/Startiger preliminary Proba 3 studies in the LAM in Marseilles/France.

Fig. 1.79 The twin Proba-3/ESA spacecraft are observing the Solar Corona with conditions similar to those of total solar eclipses. One of the achievements of the mission, encompassing a large consortium of over 29 companies from 17 countries, is the demonstration of the viability of *high-precision formation flying* by satellites in space (*Credit* ESA)

Fig. 1.80 First light from Proba 3: the Sun's inner corona appears greenish in this image taken on 23 May 2025 by the ASPIICS coronagraph aboard Proba-3, ESA's formation-flying mission capable of creating artificial total solar eclipses in orbit. The twin Proba-3/ESA spacecraft are observing the Solar Corona with conditions similar to those of total solar eclipses. This image, captured in the visible light spectrum, shows the solar corona similarly to how a human eye would see it during an eclipse through a green filter. The hair-like structures were revealed using a specialized image processing algorithm (*Credit* ESA/Proba-3/ASPIICS/WOW algorithm)

Fig. 1.81 First light from Proba 3: the ASPIICS coronagraph aboard ESA's formation-flying Proba-3 mission is able to observe the Sun's corona in the gap between the fields of view of solar extreme ultraviolet imagers and conventional coronagraphs, making it uniquely suited for studies of the inner solar corona. This image is a combination of observations made on 23 May 2025 by three different European instruments aboard different missions: the Sun's disk (artificially colored in yellow), as captured by an extreme ultraviolet telescope (SWAP) aboard Proba-2; the outer corona (in red) observed by the LASCO C2 coronagraph aboard SOHO; and the inner corona (in green), imaged in detail by Proba-3's ASPIICS coronagraph, filling the gap (*Credit* ESA/NASA/Proba-2/Proba-3/SOHO/SWAP/ASPIICS/LASCO C2/WOW algorithm)

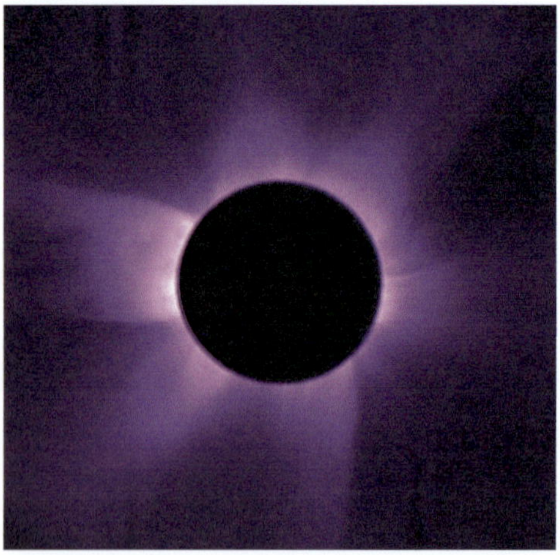

Fig. 1.82 First light from Proba 3: the Sun's inner corona, colored artificially to appear violet, in an image taken on 23 May 2025 by the ASPIICS coronagraph aboard Proba-3. This image shows the corona in polarized white light, captured using a special technique which allows the separation of the hot corona's polarized light from the light scattered by interplanetary dust (*Credit* ESA/Proba-3/ASPIICS)

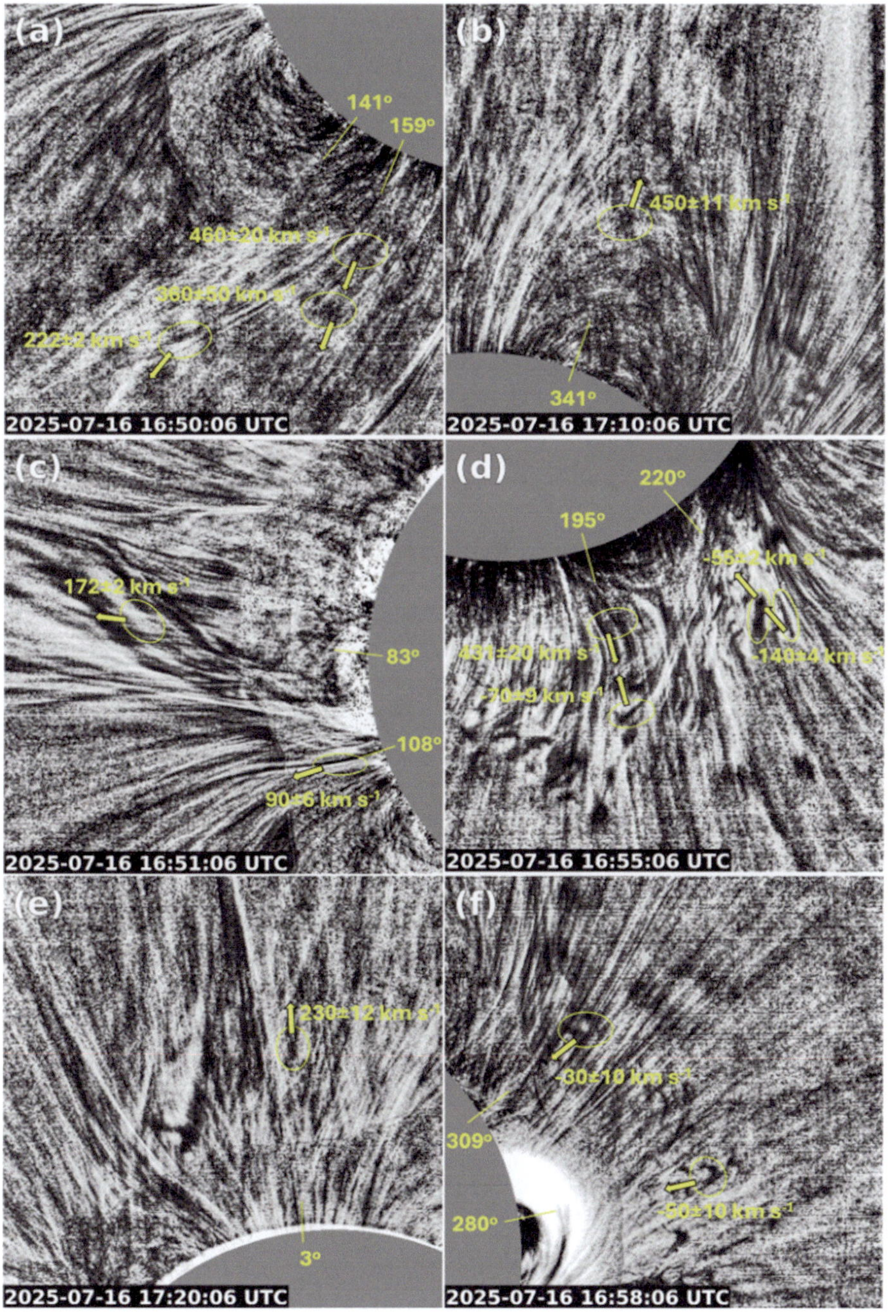

Fig. 1.83 Contrast-enhanced difference images with examples of small-scale features. Position angles of speed measurements are marked in yellow, and the tracked features are encircled. Arrows indicate propagation direction, and speeds are calculated from the quadratic fits of the height-time plots at the moment of the image (Zhukov et al. 2026a)

Fig. 1.84 A CME observed by Proba-3/ASPIICS on 16 June 2025. (Left panel) He I 5876 Å image of the corresponding erupting prominence taken at 07:27:41 UTC. (Right panel) Fe XIV 5303 Å image of the CME taken at 07:38:47 UTC, with the image taken at 07:28:58 UTC subtracted. The CME bright front, cavity, and the erupting prominence can be seen. A radial filter was applied to enhance the visibility of the weak bright front (Zhukov et al. 2026b)

Fig. 1.85 A coronal jet observed by Proba-3/ASPIICS in the Fe XIV (5303 Å) passband on 16 June 2025 at 09:48:47 UTC. A previous image taken at 09:38:47 UTC was subtracted. The jet first appeared in the ASPIICS field of view at 09:17 UTC and was visible for at least 50 minutes. The jet width was around 2°, a typical value for these phenomena (Zhukov et al. 2026c)

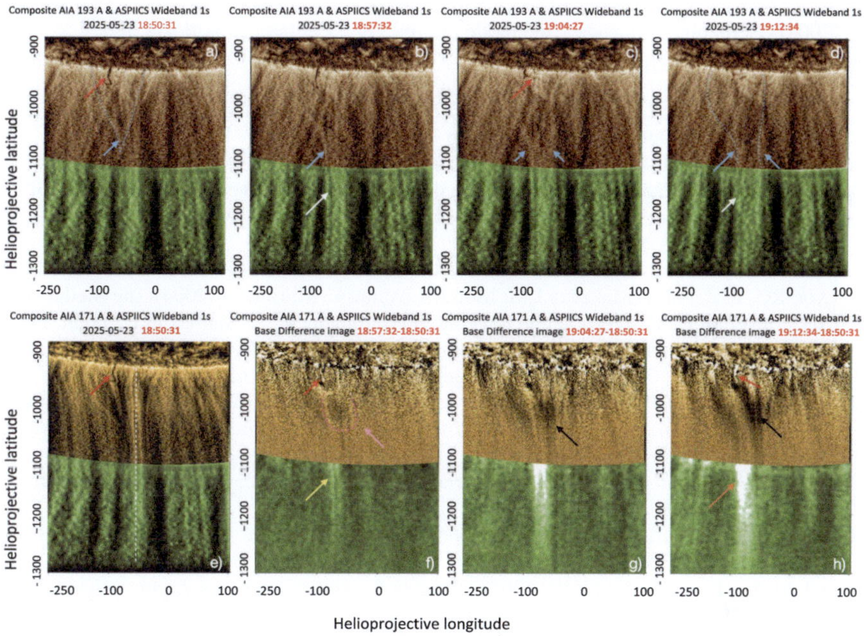

Fig. 1.86 Interaction of a solar filament, a coronal jet and a coronal plume, observed on 2025 May 23 with Proba-3/ASPIICS and SDO/AIA (**a–d**). Composite images of AIA 193 Å and the ASPIICS Wideband channel at 18:50:00, 18:57:32, 19:04:27 and 19:12:24 UT, respectively. **e** Similar to panel (a), but replacing AIA 193 Å with AIA 171 Å. (**f–h**) Composite of AIA 171 Å and ASPIICS Wideband base-difference images at the same time steps as panels (b-d). The coloured arrows indicate different features: red for solar filament; white for the plume expansion observed in ASPIICS; pink for smaller compact loops arching over filaments observed in AIA 171 Å; blue for streamer-like structures observed in AIA 193 Å; yellow for the jet front with two thread-like structures; and orange for the single thread structure at the jet's lower height (Ngampoopun et al. 2026)

References

Anastasiadis A, Papaioannou A, Sandberg I, Georgoulis M, Tziotziou K, Kouloumvakos A, Jiggens P (2017) Predicting flares and solar energetic particle events: the FORSPEF tool. Sol Phys 292(9):134. https://doi.org/10.1007/s11207-017-1163-7

Anastasiadis A, Lario D, Papaioannou A, Kouloumvakos A, Vourlidas A (2019) Solar energetic particles in the inner heliosphere: status and open questions. Philos Trans R Soc A Math Phys Eng Sci 377(2148):20180100

Antolin P, Pagano P, Testa P, Petralia AP, Reale (2021) Reconnection nanojets in the solar corona. Nat Astron 5:54–62

Archontis V, Tsinganos K, Gontikakis C (2010) Recurrent solar jets in active regions. Astron Astrophys 512:L2–L6

Arentoft T, Kjeldsen H, Bedding TR et al (2008) A multi-site campaign to measure solar-like oscillations in procyon. I. Observations, data reduction and slow variations. ApJ. arXiv:0807.3794

Bedding, Kjeldsen (2003) Solar-like oscillations. Publ Astron Soc Austr 20:203

Bedding TR, Kjeldsen H, Butler RP, McCarthy C, Marcy GW, O'Toole SJ, Tinney CG, Wright JT (2004) Oscillation frequencies and mode lifetimes in α Centauri A. ApJ 614:380
Beer J, Baumgartner St, Dittrich-Hannen B, Hauenstein J, Kubik P, Lukasczyk Ch, Mende W, Stellmacher R, Suter M (1994) Solar variability traced by cosmogenic isotopes. In: Pap JM, Fröhlich C, Hudson HS, Solanki SK (eds) The sun as a variable star: solar and stellar irradiance variations. Cambridge University Press, pp 291–300
Bouchy F, Carrier F (2002) The acoustic spectrum of alpha Cen A. Astron Astrophys 390:205–212
Cade III, William B, Chan-Park C (2015) The origin of "space weather". Space Weather 13(2)
Carmichael H (1964) In: Hess WN (ed) Proceedings of AAS-NASA Symposium on the Physics of Solar Flares. NASA-SP 50, p 451
Carpenter KG, Schrijver CJ, Allen RJ, Brown A, Chenette D, Danchi WC, Karovska M, Kilston S, Lyon RG, Marzouk J, Mazzuca LM, Moe RV, Walter F, Murphy N (2004) The Stellar Imager (SI): a revolutionary large-baseline imaging interferometer at the Sun-Earth L2 point. In: New frontiers in stellar interferometry, proceedings, vol 5491. Event: SPIE Astronomical Telescopes + Instrumentation, Glasgow, United Kingdom. https://doi.org/10.1117/12.550572
Chantry L, Cayatte V, Sauty C, Vlahakis N, Tsinganos K (2022) Double flows anchored in a Kerr black hole horizon-I. Meridionally self-similar MHD models with loading terms. Monthly Notices R Astron Soc 515(3):3796–3817
Choudhuri AR, Gilman PA (1987) The influence of the coriolis force on flux tubes rising through the solar convection zone. Astrophys J 316:788–800
Choudhuri AR, Schussler M, Dikpati M (1995) The solar dynamo with meridional circulation. Astron Astrophys 303:L29
Choudhuri AP, Hazra G, Karak BB (2012) Origin of grand minima in sunspot cycles. Phys Rev Lett 109:171103
Christe S, Krucker S, Glesener L, Shih A, Saint-Hilaire P, Caspi A, Allred J, Battaglia M, Chen B, Drake J et al (2017) Exploring impulsive solar magnetic energy release and particle acceleration with focused hard X-ray imaging spectroscopy. arXiv:1701.00792
Christensen-Dalsgaard J (2021) Solar structure and evolution. Living Rev Sol Phys 19, article number 1 (2022), 18, article number 2 (2021)
Cremades H, St Cyr OC (2007) Coronal mass ejections: solar cycle aspects. Adv Space Res 40(7):1042–1048
DeForest CE, Howard TA, McComas DJ (2013) Tracking coronal features from the low corona to Earth: a quantitative analysis of the 2008 December 12 coronal mass ejection. Astrophys J 769:43. https://doi.org/10.1088/0004-637X/769/1/43
Dialynas K, Krimigis SM, Mitchell DG, Decker RB, Roelof EC (2017) The bubble-like shape of the heliosphere observed by Voyager and Cassini. Nat Astron 1, id. 0115
Donati GB (1872) I fenomeni luminosi della grande aurora polare. In: Memorie del R (ed) Osservatorio di Firenze ad Arcetri, vol 1. Le Monnier, Florence, p 30
Eddy JA (1976) The Maunder minimum: the reign of Louis XIV appears to have been a time of real anomaly in the behavior of the sun. Science 192(4245):1189–1202
Eddy JA (1977a) The case of the missing sunspots. Sci Am 236:80–88
Eddy JA (1977b) Climate and the changing sun. Clim Change 1:173–190
Edlen B (1936) Mg I - Ähnliche Spektren der Elemente Titan bis Cobalt, Ti XI, V XII, Cr XIII, Mn XIV, Fe XV und Co XVI. Zeitschrift für Physik 103:536–541
Edlen B, Swings P (1942) Term analysis of the third spectrum of Iron (Fe III). Astrophys J 95:532
Ewing A (1955) International look at Earth. Sci News Lett 67:42
Fisk LA, Kasper JC (2020) Global circulation of the open magnetic flux of the sun. Astrophys J Lett 894:L4
Frandsen S, Carrier F, Aerts C, Stello D, Maas T, Burnet M, Bruntt H, Teixeira TC, de Medeiros JR, Bouchy F, Kjeldsen H, Pijpers F, Christensen-Dalsgaard J (2002) Detection of solar-like oscillations in the G7 giant star ξ Hydra. Astron Astrophys 394:L5–L8
Gibson S (2018) Solar prominences: theory and models. Living Rev Sol Phys 15, article number 7
Gold T (1959) Plasma and magnetic fields in the solar system. J Geophys Res 64:1665–1674

Gontikakis C, Petrie GJD, Dara HC, Tsinganos K (2005) A solar active region loop compared with a 2D MHD model. Astron Astrophys 434(3):1155–1163

Gontikakis C, Archontis V, Tsinganos K (2010) Advances in hellenic astronomy during the IYA09. In: Tsinganos K, Hatzidimitriou D, Matsakos T (eds) Study of a solar active region jet. ASP conference series, vol 424

Grotrian W (1933) Ergebnisse der Potsdamer Expedition zur Beobachtung der Sonnenfinsternis am 9. Mai 1929 in Takengon (Nordsumatra) 8. Mitteilung. Über den Intensitätsverlauf und das Intensitätsverhältnis der Koronalinien. Mit 12 Abbildungen, Zeitschrift für Astrophysik 7:26

Grotrian W (1934) Über das Fraunhofersche Spektrum der Sonnenkorona. Mit 10 Abbildungen. Zeitschrift für Astrophysik 8:124

Hahn W (1968) Space weather forecasting as a part of the new Environmental Science Services Administration (first appearance in peer-reviewed literature)

Haigh JD (2007) The sun and the earth's climate. Living Rev Sol Phys 4(2)

Handy BN et al (1999) The transition region and coronal explorer. Sol Phys 187(2):229–260. https://doi.org/10.1023/A:1005166902804

Hathaway DH (2015) The solar cycle. Living Rev Solar Phys 12. https://doi.org/10.1007/lrsp-2015-4

Hazra G, Karak BB, Choudhuri AR (2014) Is a deep one-cell meridional circulation essential for the flux transport solar dynamo? Astrophys J 782:93 (12pp)

Heaviside O (1902) Telegraphy. In Encyclopaedia Britannica, p. 214. Edinburgh and London: Adam and Charles Black.

Hekker S et al (2007) Oscillations in Procyon A: First results from a multi-site campaign. arXiv: 0710.3772v4

Herschel JFW (1847) Results of astronomical observations made during the years 1834, 5, 6, 7, 8, at the Cape of Good Hope. Smith, Elder, and Co., London

Hirayama T (1974) Theoretical model of flares and prominences: I: evaporating flare model. Sol Phys 34:323

Howard TA, DeForest CE, Reinard AA (2012) White-light observations of solar wind transients and comparison with auxiliary data sets. Astrophys J 754:10 pp, article id. 102

Howe R (2009) Solar interior rotation and its variation. Living Rev Sol Phys 6, article number 1

Hoyt DV, Schatten KH (1998a) Group sunspot numbers: a new solar activity reconstruction, part 1. Sol Phys 179:189–219

Hoyt DV, Schatten KH (1998b) Group sunspot numbers: a new solar activity reconstruction, part 2. Sol Phys 181(2):491–512

Huber D et al (2011) Solar-like oscillations and activity in Procyon: a comparison of the 2007 MOST and ground-based radial velocity campaigns. arXiv:1102.2894v2

Kjeldsen H, Bedding TR, Butler RP, Christensen-Dalsgaard J, Kiss LL, McCarthy C, Marcy GW, Tinney CG, Wright JT (2005) Solar-like oscillations in α Centauri B. ApJ 635:1281

Kjeldsen H, Bedding TR, Christensen-Dalsgaard J (2008) Measurements of stellar properties through asteroseismology: a tool for planet transit studies. arXiv:0807.0508v1

Kopp RA, Pneuman GW (1976) Magnetic reconnection in the corona and the loop prominence phenomenon. Sol Phys 50:85

Kowalski AF (2024) Stellar flares. Living Rev Sol Phys 21:1. https://doi.org/10.1007/s41116-024-00039-4

Krimigis SM, Decker RB, Roelof EC, Hill ME, Bostrom CO, Dialynas K, Gloeckler G, Hamilton DC, Keath EP, Lanzerotti LJ (2019) Energetic charged particle measurements from Voyager 2 at the heliopause and beyond. Nat Astron 3:997–1006

Kwon R-Y, Vourlidas A (2018) The density compression ratio of shock fronts associated with coronal mass ejections. J Space Weather Space Clim 8:14 pp, id.A08

Low BC (1987) Electric current sheet formation in a magnetic field induced by continuous magnetic footpoint displacements. Astrophys J 323:358–3671

Low BC (2010) The Parker magnetostatic theorem. Astrophys J 718:717–723

Lyot MB (1939) The study of the solar corona and prominences without eclipses (George Darwin Lecture, delivered by M. Bernard Lyot, Assoc. R.A.S., on 1939 May 12). Monthly Notices R Astron Soc 99:580–594

McComas DJ, Elliott HA, Schwadron NA, Gosling JT, Skoug RM, Goldstein BE (2003) The three-dimensional solar wind around solar maximum. GRL 30(10)

Miroshnichenko LI (ed) (2003) Radiation hazard in space. Astrophys Space Sci Libr 297. https://doi.org/10.1007/978-94-017-0301-7

Morgan H, Druckmüller M (2014) Multi-scale Gaussian normalization for solar image processing, in Coronal magnetometry. In: Tomczyk S, Zhang J, Bastian T (eds) Springer, pp 323–333

Moschou SP, Tsinganos K, Vourlidas A, Archontis V (2013) SDO observations of solar jets. Sol Phys 284(2):427–438

Ngampoopun N, Peter H, Chitta P, Dolliou A, Inhester B, Zhukov A, Dolla L, Shestov S, Bourgoigne B, Fineschi S, Gunar S, Lamy P, Mierla M, Rudawy P, Tsinganos K (2026) Fine-Scale Dynamics of a Coronal Plume Driven by a Coronal Jet. Observations from the Proba-3/ASPIICS coronagraph. Astron Astrophys (submitted)

Neugebauer M, Snyder CW (1966) Mariner 2 observations of the solar wind: 1. Average properties. JGR 71:4469–4484

Opher M, Loeb A, Drake J, Toth G (2020) A small and round heliosphere suggested by magnetohydrodynamic modelling of pick-up ions. Nat Astron 4:675–683

Papaioannou A, Sandberg I, Anastasiadis A, Kouloumvakos A, Georgoulis MK, Tziotziou K, Tsiropoula G, Jiggens P, Hilgers A (2016) Solar flares, coronal mass ejections and solar energetic particle event characteristics. J Space Weather Space Clim 6:A42. https://doi.org/10.1051/swsc/2016035

Parker EN (1955) Hydromagnetic dynamo models. Astrophys J 122:293

Parker EN (1957) Sweet's mechanism for merging magnetic fields in conducting fluids. J Geophys Res 62(4):509–520

Parker EN (1958a) Dynamics of the interplanetary gas and magnetic fields. Astrophys J 128:664

Parker EN (1958b) Interaction of the solar wind with the geomagnetic field. Phys Fluids 1:171–187

Parker EN (1958c) Cosmic-ray modulation by the solar wind. Phys Rev 110(6):1445–1449

Parker EN (1958d) Plasma dynamic determination of shock thickness in an ionized gas. Astrophys J 129:217

Parker EN (1961) The stellar wind regions. Astrophys J 134(1):20–27

Parker EN (1963) Interplanetary dynamical processes. Wiley Interscience, New York

Parker EN (1965) The passage of energetic particles through interplanetary space. Planet Space Sci 13(1):9–49

Parker EN (1966) The dynamical state of the interstellar gas and field. Astrophys J 145:811

Parker EN (1972) Topological dissipation and the small-scale fields in turbulent gases. Astrophys J 174:499

Parker EN (1979a) Cosmical magnetic fields. Oxford University Press, Oxford

Parker EN (1979b) Sunspots and the physics of magnetic flux tubes. I. The general nature of the sunspot. Astrophys J 230:905–913

Parker EN (1983a) Magnetic neutral sheets in evolving fields I. General theory. Astrophys J 264:635–647

Parker EN (1983b) Magnetic neutral sheets in evolving fields II. Formation of the Solar Corona. Astrophys J 264:642

Parker EN (1988) Nanoflares and the Solar X-Ray Corona. Astrophys J 330:474

Parker EN (1994) Spontaneous current sheets in magnetic fields: with applications to stellar x-rays, International series in astronomy and astrophysics, vol 1. Oxford University Press, New York

Parker EN, Tsinganos K (1979) Bernoulli forces on an elliptical cylinder by a nonuniform flow. Phys Fluids 22(10):1847–1851

Pasachoff JM, Rušin V, Druckmüller M, Aniol P, Saniga M, Minarovjech M (2009) The 2008 August 1 eclipse solar-minimum corona unraveled. Astrophys J 702:1297–1308

Patsourakos S, Georgoulis MK, Vourlidas A, Nindos A, Sarris T, Anagnostopoulos G, Anastasiadis A, Chintzoglou G, Daglis IA, Gontikakis C, Hatzigeorgiu N, Iliopoulos AC, Katsavrias C, Kouloumvakos A, Moraitis K, Nieves-Chinchilla T, Pavlos G, Sarafopoulos D, Syntelis P, Tsironis C, Tziotziou K, Vogiatzis II, Balasis G, Georgiou M, Karakatsanis LP, Malandraki OE, Papadimitriou C, Odstrcil D, Pavlos EG, Podlachikova O, Sandberg I, Turner DL, Xenakis MN, Sarris E, Tsinganos K (2016) The major geoeffective solar eruptions of 2012 March 7: comprehensive sun-to-earth analysis. Astrophys J 817:21 pp

Petrie GJD, Vlahakis N, Tsinganos K (2002) Systematic construction of exact 2-D MHD equilibria with steady, compressible flow in Cartesian geometry and uniform gravity. Astron Astrophys 382:1081–1092

Petrie GJD, Tsinganos K, Neukirch T (2005) Steady 2D prominence-like solutions of the MHD equations with field-aligned compressible flow. Astron Astrophys 429:1081–1092

Phillips J (1851) On isoclinical magnetic lines in Yorkshire. In: Report of the twentieth meeting of the British Association for the advancement of science, vol II notices and abstracts of communications to the British Association for the Advancement of Science. John Murray, London, p 14

Pontin DI, Horning G (2020) The Parker problem: existence of smooth force-free fields and coronal heating. Living Rev Sol Phys 17, article number 5

Pontin DI, Priest ER (2022) Magnetic reconnection: MHD theory and modelling. Living Rev Sol Phys 19, article number 1

Priest ER (2014) A life of fun playing with solar magnetic fields (special historical review). Sol Phys 289(10):3579–3615

Priest ER (1982) Solar magnetohydrodynamics. D. Reidel, Holland

Pucci F, Velli M, Shi C, Singh KAP, Tenerani A, Alladio F, Ambrosino F, Buratti P, Fox W, Jara-Almonte J, Ji H, Yamada M, Yoo J, Okamura S, Ergun R, Hoilijoki S, Schwartz S (2020) Onset of fast magnetic reconnection and particle energization in laboratory and space plasmas. J Plasma Phys 86(6)

Raghuram S, Hutsemékers D, Opitom C, Jehin E, Bhardwaj A, Manfroid J (2020) Forbidden atomic carbon, nitrogen, and oxygen emission lines in the water-poor comet C/2016 R2 (Pan-STARRS). A&A 635:A108

Reale F (2014) Coronal loops: observations and modeling of confined plasma. Living Rev Sol Phys 11, article number 4 (2014)

Rosner R, Tucker WH, Vaiana GS (1978) Dynamics of the Quiescent Solar Corona. Astrophys J 220:643–665

Sauty C, Tsinganos K (1994) Nonradial and nonpolytropic astrophysical outflows. III. The transition from jets to winds. Astron Astrophys 287:893–926

Sauty C, Lima JJG, Iro N, Tsinganos K (2005) Nonradial and nonpolytropic astrophysical outflows. VII. Fitting ULYSSES solar wind data during minimum. Astron Astrophys 432(2):687–698. https://doi.org/10.1051/0004-6361:20041606

Schwenn R (2006a) Space weather: the solar perspective. Living Rev Sol Phys 3:2. https://doi.org/10.12942/lrsp-2006-2

Schwenn R (2006b) Solar wind sources and their variations over the solar cycle. Space Sci Rev 124(1–4):51–76

Shibata K, Magara T (2011) Solar Flares: magnetohydrodynamic processes. Living Rev Sol Phys 8, article number 6

Shibata K, Masuda S, Shimojo M, Hara H, Yokoyama T, Tsuneta S, Kosugi T, Ogawara Y (1995) Hot-plasma ejections associated with compact-loop solar flares. Astrophys J Lett 451:L83

Spitzer L Jr, Härm R (1953) Transport phenomena in a completely ionized gas. Phys Rev 89(5):977–981

Steinhilber F et al (2012) 9.400 years of cosmic radiation and solar activity from ice cores and tree rings. Proc Natl Acad Sci USA 109(16):5967–5971

Sturrock PA (1966) Model of the high-energy phase of solar flares. Nature 211:695

Sweet PA (1958) The neutral point theory of solar flares. In: Lehnert B (ed) IAU symposium 6, electromagnetic phenomena in cosmical physics, vol 123. Kluwer, Dordrecht

Syntelis P, Gontikakis C, Georgoulis MK, Alissandrakis CE, Tsinganos K (2012) Study of the three-dimensional shape and dynamics of coronal loops observed by Hinode/EIS. Sol Phys 280:475–489

Syntelis P, Gontikakis C, Patsourakos S, Tsinganos K (2016) The spectroscopic imprint of the pre-eruptive configuration resulting into two major coronal mass ejections. Astron Astrophys 588:14 pp

Syntelis P, Archontis V, Tsinganos K (2017) Recurrent CME-like eruptions in emerging flux regions. I. On the mechanism of the eruption. Astrophys J 850:15 pp

Telloni D, Zank GP, Stangalini M, Downs C, Liang H, Nakanotani M, Andretta V, Antonucci E, Sorriso-Valvo L, Adhikari L, Zhao L, Marino R, Susino R, Grimani K, Fabi M, D'Amicis R, Perrone D, Bruno R, Carbone F, Mancuso S, Romoli M, Da Deppo V, Fineschi S, Heinzel P, Moses JD, Naletto G, Nicolini G, Spadaro D, Teriaca L, Frassati F, Jerse G, Landini F, Pancrazzi M, Russano G, Sasso C, Biondo R, Burtovoi A, Capuano GE, Casini C, Casti M, Chioetto P, De Leo Y, Giarrusso M, Liberatore A, Berghman D, Auchère F, Cuadrado RA, Chitta LP, Harra L, Kraaikamp E, Long DM, Mandal S, Parenti S, Pelouze G, Peter H, Rodriguez L, Schühle U, Schwanitz C, Smith PJ, Verbeeck C, Zhukov AN (2022) Observation of a magnetic switchback in the Solar Corona. ApJL 936 L25

Thompson MJ, Christensen-Dalsgaard J, Miesch MS, Toomre J (2003) The internal rotation of the sun. Annu Rev Astron Astrophys 41:599–643

Tsinganos K (1982) MHD Equilibrium. IV. Non-Equilibrium of Non-Symmetric Hydromagnetic Topologies. Astrophys J 259:832–834

Tsinganos K (2017) Plasma astrophysics, Athens

Tsinganos KC (1979) Sunspots and the physics of magnetic flux tubes. IV. Aerodynamic lift on a thin cylinder in convective flows. Astrophys J Part 1 231:260–269

Tsinganos KC (1980) Sunspots and the physics of magnetic flux tubes. X On the hydrodynamic instability of buoyant fields. Astrophys J Part 1 239:746–760

Tsiropoula G, Tziotziou K, Kontogiannis I, Madjarska MS, Doyle JG, Suematsu Y (2012) Solar fine-scale structures. I. Spicules and other small-scale, jet-like events at the chromospheric level: observations and physical parameters. Space Sci Rev. arXiv:1207.3956v1

Turner MS, Parker EN, Bogdan TJ (1982) Magnetic monopoles and the survival of galactic magnetic fields. Phys Rev D 26:1296

Ulrich RK (1970) The five-minute oscillations on the solar surface. Astrophys J 162:993

Usoskin IG (2008) A history of solar activity over millennia. Living Rev Sol Phys 5:3

Usoskin IG, Solanki SK, Kovaltsov GA (2007) Grand minima and maxima of solar activity: new observational constraints. A&A 471:301–309. arXiv:0706.0385

Usoskin IG (2023) A history of solar activity over millennia. 20, article number 2 Korolev Sergei

Vial JC, Engvold O (2015) Solar prominences, astrophysics and space science library, vol 415. Springer, Cham.https://doi.org/10.1007/978-3-319-10416-4

Vlahos L, Anastasiadis A, Papaioannou A, Kouloumvakos A, Isliker H (2019) Sources of solar energetic particles (2019), Philos Trans R Soc Lond Ser A 377(2148):20180095. https://doi.org/10.1098/rsta.2018.0095. arXiv:1903.08200 [astro-ph.HE]

Vourlidas A, Syntelis P, Tsinganos K (2012) Uncovering the birth of a coronal mass ejection from two-viewpoint SECCHI observations. Sol Phys 280:509–523

Vourlidas A, Balmaceda LA, Stenborg G, Dal Lago A (2017) Multi-viewpoint coronal mass ejection catalog based on STEREO COR2 observations. Astrophys J 838(2):141. https://doi.org/10.3847/1538-4357/aa67f0

Vourlidas A, Patsourakos S, Savani NP (2019) Predicting the geoeffective properties of coronal mass ejections: current status, open issues and path forward. Philos Trans R Soc Lond Ser A 377(2148):20180096. https://doi.org/10.1098/rsta.2018.0096

Wiese WL, Fuhr JR, Deters TM (1996) Atomic transition probabilities of carbon, nitrogen, and oxygen: a critical data compilation. American Chemical Society, Washington, D. C.

Wiese WL, Fuhr JR (2009) Accurate atomic transition probabilities for hydrogen, helium, and lithium. J Phys Chem Ref Data 38:565

Wilson A, Maskelyne M (1774) Observations on the solar spots. By Alexander Wilson, M. D. Professor of Practical Astronomy in the University of Glasgow. Communicated by the Rev. Nevil Maskelyne, Astronomer Royal. Philos Trans 64:1–30

Zhukov AN, Patel BD, Debrandere A, Dolla L, Mierla M, Sestov S, Bourgoignie B, Jean C, Nicula B, Talpeanu, DC, Zontou Z, Fineschi S, Gunar S, Lamy P, Peter H, Rudawy P, Tsinganos K, Abbo L, Aime C, Andretta V, Auchere F, Berghmans D, Besliu-Ionescu D, Gibson SE, Giordano S, Heinzel P, Inchester P, Magdalenic J, Marque C, Rodriguez L, Steslicki M, Zangrilli L, Galano D, Rougeot R, Versluys J, Thizy C (2026a) Ubiquitous small scale dynamics in the slow solar wind formation region observed by Proba-3/ASPIICS. ApJL vol. 999:L41

Zhukov AN, Thizy C, Galano D, Bourgoignie B, Dolla L, Jean C, Nicula B, Shestov S, Galy C, Rougeot R, Versluys J, Zender J, Lamy P, Fineschi S, Gunar S, Inhester B, Mierla M, Rudawy P, Tsinganos K, Koutchmy S, Howard S, Peter H, Vives S, Abbo L, Aime C, Aleksiejuk K, Baran J, Bak-Steslicka U, Bemporad A, Berghmans D, Besliu-Ionescu D, Buckley S, Buiu O, Capobianco G, Cimoch I, DHuys E, Dziezyc M, Fleury-Frenette K, Gibson SE, Giordano S, Golub L, Grochowski K, Heinzel P, Hermans A, Jacobs J, Jejcic S, Kranitis N, Landini F, Loreggia D, Magdalenic J, Maia D, Marque C, Melich R, Morawski M, Mosdorf M, Noce V, Orleanski P, Paschalis A, Peresty R, Rodriguez L, Seaton DB, Short L, Simar J-F, Steslicki M, Sorensen R, Terrasa G, Van Vooren N, Verstringe F, Zangrilli L (2026b) The ASPIICS solar coronagraph aboard the Proba-3 formation flying mission. Scientific objectives and instrument design. Astron Astrophys (submitted) arXiv:2509.00253v1

Zhukov AN, Dolla L, Mierla M, Patel BD, Shestov S, Bourgoignie B, Debrabandere A, Jean C, Nicula B, Talpeanu D-C, Zontou Z, Fineschi S, Gunár S, Lamy P, Peter H, Rudawy P, Tsinganos K, Abbo L, Aime C, Auchère F, Berghmans D, Besliu-Ionescu D, Gibson SE, Giordano S, Heinzel P, Inhester B, Magdalenic J, Marqué C, Rodriguez L, Steslicki M, Zangrilli L, Galano D, Rougeot R, Versluys J, Thizy C (2026c) Observing the solar corona from a formation-flying mission. First results of Proba-3/ASPIICS./ Astron Astrophys (submitted) arXiv:2511.01679

Further Reading List

For relatively recent and also classical reviews on the subject of this chapter, in addition to the references given in the text, the interested reader may consult the following articles:

Al-Haddad N, Lugaz N (2025) The magnetic field structure of coronal mass ejections: a more realistic representation. Space Sci Rev, Springer Nature 221:12

Antonucci E, Harra L, Susino R, Telloni D (2020) Observations of the solar corona from space. Space Sci Rev 216:117. https://doi.org/10.1007/s11214-020-00743-1

Biswas A, Karak BB, Usoskin I et al (2023) Long-term modulation of solar cycles. Space Sci Rev 219:19

Cargill P, Parnell C, Browning P, De Moortel I, Hood A (2010) Magnetic reconnection in the solar atmosphere: From proposal to paradigm. Astron & Geophys 51(3):3.31

Charbonneau P (2020) Dynamo models of the solar cycle. Living Rev Sol Phys 17:4

Drake JF, Antiochos SK (2025) Magnetic reconnection in solar flares and the near-sun solar wind. Space Sci Rev, Springer Nature, 221, article number 27

Eddington, AS (1927) Stars and Atoms. Oxford, Clarendon Press

Friedman H (1967) Ultraviolet and X rays from the sun. Annu Rev Astron Astroph 1:59–96

Goldberg L (1967) Ultraviolet and X rays from the sun. Annu Rev Astron Astroph 5:279–324

Harra L, Barczynski K, Auchère F et al (2025) The dynamics of the extreme ultraviolet quiet sun and coronal holes in the solar orbiter era. Space Sci Rev 221:50

Neupert WM (1969) X rays from the sun. Annu Rev Astron Astroph 7:121–148

Ramos A, Cheung MCM, Chifu I et al (2023) Machine learning in solar physics. Living Rev Sol Phys 20:4

Raouafi NE, Matteini L, Squire J et al (2023) Parker solar probe: four years of discoveries at solar cycle minimum. Space Sci Rev 219:8

Shestov S, Zhukov AN, Rougeot R, Aime C, Bourgoignie B, Dolla L, Fineschi S, Gunár S, Lamy P, Mierla M, Peter H, Rudawy P, Tsinganos K (2026b) Diffraction in ASPIICS: observations and modeling. Astron Astrophys (submitted)

Wang CG, Branduardi-Raymont, Escoubet CP, Forsyth C (2025) Solar wind magnetosphere ionosphere link explorer (SMILE): science and mission overview. Space Sci Rev, Springer Nature 221, article number 9

West MJ, Seaton DB, Wexler DB et al (2023) Defining the middle corona. Sol Phys 298:78

Chapter 2
Our Privileged Solar Planetary System

Nothing is more precious than water
(Pindar, 522–443 BC, ancient Greek lyric poet from Thebes, 1st Olympian Ode, 1.1).

1957: The Dawn of the Space Race

The dawn of the *Space Age* appeared on October 4, 1957 and continuous today with the Voyager spacecrafts traveling outside the Heliosphere (Figs. 2.1 and 2.2). It was then when the first artificial satellite of the Earth, Sputnik was placed into orbit by the Soviet Union (Fig. 2.1, left). The launch was made with an R-7 8K72 Vostok rocket. Sergei Korolev (1906–1966), at the time, was considered the *father* of rocket science in the Soviet Union.[1]

Sputnik was a metallic sphere 58 cm in diameter and weighing 86 kg, equipped with four antennas. It orbited Earth for 21 days every 96 min, with a perigee of 215 km and an apogee of 939 km above the Earth's surface. *Sputnik*'s speed was about 29,000 km per hour, which means that it could travel the driving distance between the cities Athens and Thessaloniki (500 km, about 5 hours) in less than 1.5 min. Sputnik carried no scientific instruments. Its launch sparked political, military, and technological developments and the start of the space race.

The Americans responded immediately to the launching of Sputnik by the Soviets with the launch of *Explorer* 1 (Fig. 2.1, right), the first American satellite, on January 31, 1958, and the establishment of NASA on 29/07/58 by a decision of the President of the USA Dwight D. Eisenhower. *Explorer* 1 did carry scientific instruments and made the first discovery in space: the *Van Allen* radiation belts around Earth. Then, on April 12, 1961, Yuri Gagarin began his historic journey inside *Vostok* 1, becoming the first person to travel to space and enter in orbit around Earth.

---

[1] Korolev, who had Greek roots from his mother, previously (1938) had been deported to a concentration camp for dissidents in Siberia, Kolyma, a Gulag-type archipelago.

K. Tsinganos, *A Century of Cosmic Discoveries*, Springer Praxis Books,
https://doi.org/10.1007/978-3-032-06282-6_2

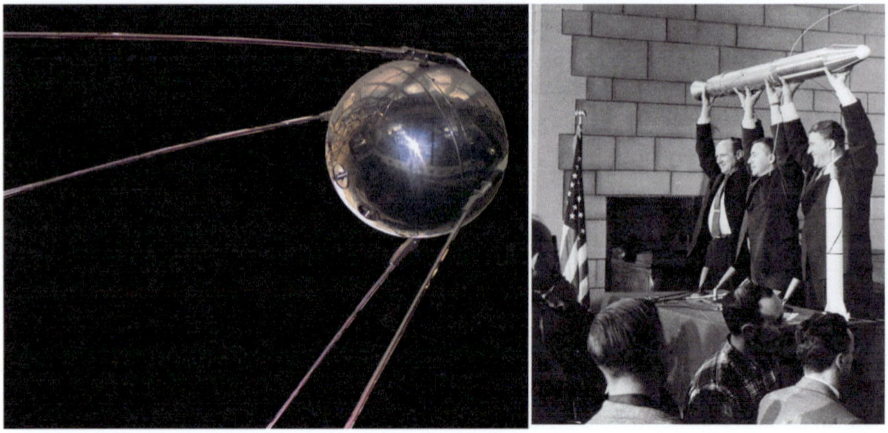

Fig. 2.1 Earth's first artificial satellite, Sputnik (left). In the right, Wernher von Braun, James Van Allen, and James Pickering (from right) hold a replica of America's first satellite, *Explorer*. (*Source* NASA)

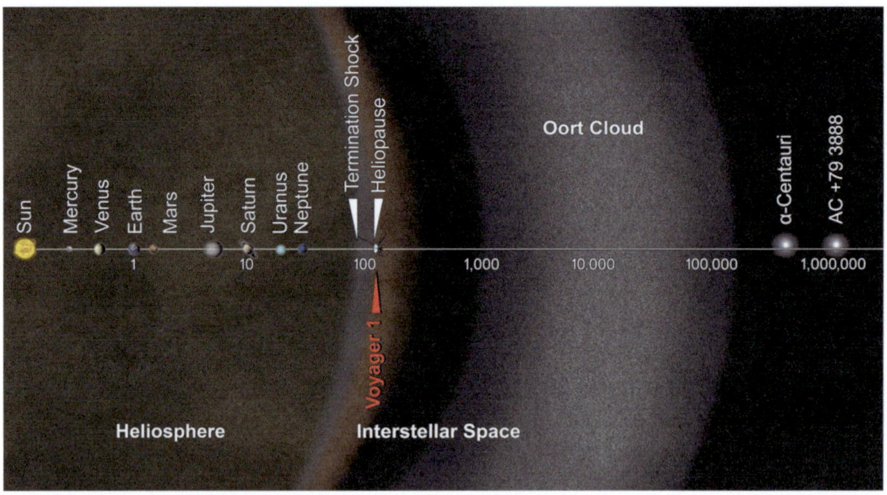

Fig. 2.2 NASA's *Voyager* 1, humankind's most distant spacecraft, was around 172 AU from Earth as of March 2026; its radio signals took about 24 h to reach Earth and it continues to send data on interstellar plasma, magnetic fields, and particles, with active instruments expected to function into the 2030s. It entered interstellar space, on August 25, 2012. *Voyager* 1 traveling with a speed more than 61,000 km/h, it will take about 300 years to reach the inner edge of the *Oort* Cloud at about 1000 AU and possibly about 30,000 years to fly beyond the outer edge of the Oort Cloud. Alpha Centauri at 253,000 AU (about 4 light years) is currently the closest star to our solar system. In 40,000 years, *Voyager* 1 will be closer to the red dwarf star AC +79 3888 (Gliese 445) than to our own Sun. AC + 79 3888 is actually traveling faster toward *Voyager* 1 than the spacecraft is traveling toward it. The Voyager 1 probe will pass within 1.6 light years of Gliese 445 in about 40,000 years. (*Credit* NASA/JPL-Caltech)

Fig. 2.3 From the visit of the first Russian cosmonaut Yuri Gagarin to the National Observatory of Athens (12/2/1962). (Photograph kindly provided by N. Matsopoulos/NOA)

As part of a triumphant world tour, Yuri Gagarin visited Greece on February 12, 1962. He received an enthusiastic welcoming by the crowd on his way from the airport towards the center of Athens. He met with Prime Minister Konstantinos Karamanlis, was admitted to the Academy of Athens, declared a honorary citizen of the city, and visited the Acropolis and the National Observatory of Athens (Fig. 2.3).

1958: Discovery of the *van Allen* Radiation Belts

Our Earth is surrounded by two huge doughnut-shaped zones of highly energetic charged particles trapped at high altitudes in the terrestrial magnetic field. They are filled with charged particles of deadly high energy, the *killer electrons* as they are called, which are trapped in our planet's magnetic field and from whose the smaller one is inside the bigger one. Discovered by the first satellite (Explorer 1) when it went out of the thermosphere (90–1000 km) is the region where the International Space 36 Station is today (about 400 km) and where there are many other artificial satellites. These zones were named after *James van Allen*, who had built the Geiger counter that first detected them (Fig. 2.4).

The inner region is centered approximately at 3,000 km above the terrestrial surface. The outer region of maximum density is centered at an altitude of about 15,000–20,000 km, though some estimates place it as far above the surface as six Earth radii. The strongest part of the inner belt is at a height of about 3,200 km above the ground and consists of electrons and protons. The strongest region of the second zone is located around 16,000 km above the Earth's equator and contains mainly high-energy electrons with energies of 0.1–10 MeV, which means that the speed of these electrons almost reaches the speed of light. These zones are very dangerous for satellites and manned missions. Under normal conditions, they remain well above the International Space Station and satellites moving in low earth orbit ($R \leq 1{,}000$ km).

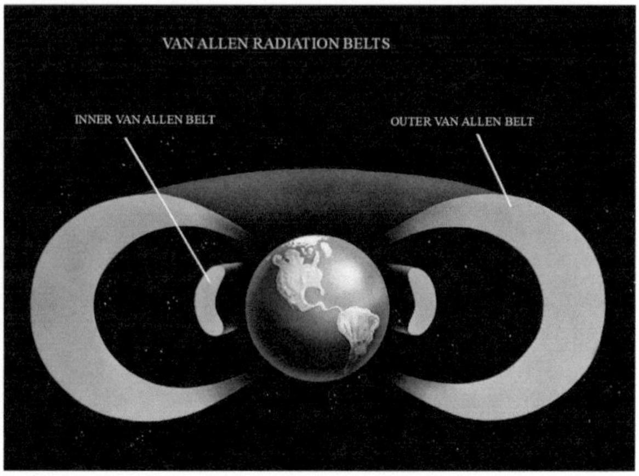

Fig. 2.4 Early schematic of the *Van Allen Belts'* structure, created after the first American satellite discovered their existence in 1958. The inner Van Allen Belt extends at altitudes (0.2–2) $R_\oplus \approx$1,000–6,000 km, where $R_\oplus$ is radius of the Earth, $R_\oplus$ = 6,371 km, and the outer at heights (4–5) $R_\oplus$ above the equator. (*Credit* NASA's Goddard Space Flight Center/Historic image of *Van Allen* Belts courtesy of NASA's Langley Research Center(*Source* NASA)

But when the belts deform and expand during solar flares and subsequent magnetic storms, they can cause damage to satellites and problems in radio communications.

1960: The Initial Competition Between the USA and the Soviet Union for the Exploration of the Planet *Mars*

In the former Soviet Union, encouraged by the success of the *Sputnik* 1 launching, they tried to start the adventure of the exploration of the red planet *Mars* with *MARS 1 M No. 1* and *MARS 1 M No. 2* on October 10 and 14, 1960. Unfortunately, however, both of these spacecraft exploded during launch. Then, they persevere and come back after 2 years with *Mars 2MV-4 No. 1* 10/24/1962 (dissolved in Earth orbit), *Mars 1* (11/1/1962, lost en route), *Mars 2MV-3 No. 1* (4/11/1962, abandoned in Earth orbit), and *Zond 2* (30/11/1964, lost contact while in Earth orbit). In this period (1964), in the Martian exploration entered the Americans with *Mariner 3* (5/11/1964, which however failed to open its cover of the payload after launch) and *Mariner* 4 (11/28/1964), the first successful mission in which participated via instrument construction Tom Krimigis. *Mariner 4* succeeded to send back the first photographs of the planet, as it was equipped with the first digital camera ever constructed. Next, followed *Mariner* 6 (2/25/1969) and *Mariner* 7 (3/27/1969), which also sent back interesting photos. Then, the Soviets returned with *Mars 1969A* (27/3/1969) and *Mars* 1969B (2/4/1969), which however exploded during launch. The Americans returned again with *Mariner* 8 (8/5/1971, which however exploded during launch) and *Mariner 9* on 5/30/1971.

Mariner 9 (launched on May 30, 1971) was the first major American success, since it was the first spacecraft to orbit *Mars*. The Soviets then attempted to land on *Mars* with *Cosmos 419* (5/10/1971, but it was destroyed in Earth's orbit), *Mars 2* on 5/19/1971 (it was crashed on landing), *Mars 3* (5/28/1971, with a pre-landing failure). Despite these failures, however, the Soviets persevere with the *Mars 4* mission (21/7/1973, which however failed to meet *Mars*), *Mars 5* (25/7/1973, this time succeeded in orbiting the planet), *Mars 6* (5/8/1973, which ultimately failed to land), and *Mars 7* (9/8/1971, again with landing failure).

The American turn comes next, with the successful *Viking* 1 (20/8/1975) and *Viking* 2 (9/9/1975) which landed normally, and they sent Earth interesting data. *Viking* 1 landed in Chryse Planitia on July 20, 1976, and *Viking 2* in Utopia Planitia on September 3, 1976. The Soviets persisted and tried again – but failed – by sending the 2 *Phobos* 1 spacecraft (7/7 /1988, which lost contact en route) and *Phobos* 2 (12/7/1988, again loss of communication, near the *Mars* satellite *Phobos*). At this point, the Soviets drop out of the race, and only two Russian missions followed. The *Mars 96* (16/11/1996) exploded on launch. Much later, the *Phobos-Grunt* mission launched on 8/11/2011 and scheduled to arrive in 2013 at the *Mars* moon *Phobos*, where it was supposed to collect samples of its soil (grunt in Russian) which would be ejected with a capsule, to reach Earth in 2014. The 14.6-tonne *Phobos-Grunt* was the largest interplanetary craft ever launched. When it failed to fire its engine to enter *Mars*

Fig. 2.5 *Viking* 1 and the first image transmitted from the surface of *Mars* by *Viking* 1 on July 20, 1976. Rocks in front of one leg of the spacecraft are visible to the right. (*Credit* NASA/JPL)

Fig. 2.6 The American spaceship *Viking* 1 was the first spaceship to land in *Chryse Planitia* of *Mars* on July 20, 1976. It was in operation for about 6 years, until November 11 1982

orbit, it remained stuck in low Earth orbit until January 15, 2012, when it finally crashed into the Pacific Ocean, west of Chile.

It should be added here that in the preparation of *Mars 96* (originally called *Mars 94*), we also participated as a research group of the University of Crete together with Dr. Thanasis Economou E.N. who belonged to the University of Chicago (Laboratory for Astrophysics and Space Research, E. Fermi Institute).

1960 to Present: The Robotic Exploration of the Planet *Mars*

Mars is known from mythology as the god of war and is called the *red* planet because of the red color on its surface, something which is due to the presence of iron trioxide. To this day, many space vehicles have been sent to explore the Martian landscape and atmosphere. As a consequence, today we know about *Mars* more facts, than for any other planet in our solar system.

In a Martian exploration *family photo* one may put 40+ missions to *Mars* (regardless of their success or not), plus the last and most recent *Insight* mission which landed on *Mars* on November 26, 2018. The overwhelming majority of these missions are US-flagged, 6 former USSR-flagged, 2 Russian-flagged, 1 belonging to ESA, and 1 to Japan. In addition, in the Martian terrain there are already 8 American spacecraft today, the *Viking* 1 and 2 (Fig. 2.6), *Phoenix*, as well as the 5 NASA robots: the small *Pathfinder* (Fig. 2.13), the twins *Spirit* + *Opportunity* (Fig. 2.14), the nuclear-powered Curiosity, and the geological laboratory *Insight*.

1962 to Present: Exploration of the Planet *Venus*

Venus is the brightest object in the night sky, after the Moon. It is the well-known *Eosphorus* (the one that brings the dawn), the morning star, early in the morning, before sunrise, or *Hesperus* (Fig. 2.7), the equivalent for Roman Vesper (the one that brings the evening), the *evening star*, shortly after sunset. *Venus* is the first celestial body to become visible in the sky at the evening and is the last one to disappear from the sky at sunrise. Nevertheless, *Venus* is the least explored planet to date. For what reason? Because of *Venus*' dense, hot, and toxic atmosphere! In the exploration of the outer planets, NASA's epic space missions *Voyager* 1 and 2 (1977–), Cassini (2007–2017), and New Horizons (2006–) played an important role and to a lesser extent Gallileo (1989) and Juno (2011). This atmosphere consists mainly of carbon dioxide, while the pressure on its surface reaches 92 atmospheres! Specifically, the composition of the *Venusian* atmosphere is 96.5% carbon dioxide and 3.5% nitrogen. The main component of the clouds above the carbon dioxide layer is sulfuric acid, the familiar caustic *vitriol* that causes burns when it gets on the skin. Carbon dioxide and sulfuric acid combined in the atmosphere of *Venus* maintain an intense *greenhouse effect*.

Venus rotates slowly in retrograde, i.e., from east to west. The result of this *reverse* slow rotation is that the length of the *Venusian* solar day, that is, the time interval from one sunrise to the next is 116.75 Earth days. To an observer on the surface of *Venus*, the Sun would rise in the west and set in the east, although *Venus*'s opaque clouds prevent observing the Sun from the planet's surface! Even so anyway, one

Fig. 2.7 *Hesperus* the *Evening Star*, as the personification of the evening with his arrows/ promise of love, rises as the Sun sets, by Anton Raphael Mengs (1765)

hemisphere remains in total darkness for 58 days, while the other receives all solar radiation by another 58 days. This results in the appearance in the upper atmosphere of fierce winds. These, in combination with the high density of the atmosphere and the strong greenhouse effect, distribute evenly the temperature, so that there is no large temperature difference between the two hemispheres.

Venus is the most accessible planet in the solar system for space missions from Earth, since a trip to this planet requires, with current technology, less than four months. That's why it was a target of many unmanned exploratory space missions mainly by the USA and the Soviet Union. But interest waned when the extreme conditions prevailing on its surface were found. The first probes to land on *Venus* survived only a few hours before being destroyed by the incredibly high temperatures they encountered. Previously, we had only a few brief *glimpses* and believed that under the dense clouds there might have been oceans and possibly life, something that seems to be quite the opposite from what really happens. Sulfur compounds abound in the clouds of *Venus*. The corrosive chemistry and the dense, moving atmosphere cause significant surface weathering and erosion.

The first successful mission to *Venus* was the American *Mariner* 2 mission, which sent the first close-up pictures of the planet in 1962. This was followed by the Soviet series of spacecrafts *Venera* (1961–1984), that achieved the first landing of a spacecraft on another planet with *Venera* 3 (1966), followed the next year by *Venera* 4 and *Mariner* 5. But the first actual landing on the planet was with *Venera* 7 (1970), which sent us 23 min of data from the surface of *Venus*. It was indeed the first spacecraft to successfully land on another planet. Before this mission, all other missions, either observed the planet from their orbit, or, they crashed by the high pressure of the planet's atmosphere. The Soviets continued to send the Venera spacecraft to *Venus* until 1983, analyzing the soil and atmosphere and taking pictures of the surface, starting with the *Venera* 9 (1975).

The next step in the exploration of the planet was made with NASA's spacecraft *Magellan* (1989), which systematically mapped for four years with its radar most of the planet (98%), revealing features of the surface, mechanisms that shape the surface, as well as the planet's magnetic field. Likewise, the Galileo spacecraft en route to *Jupiter* in 1990 mapped the surface by penetrating with its infrared radar the atmosphere of *Venus*. The first European mission to *Venus* was the *Venus Express* (2006), which mainly studied the atmosphere of *Venus* at various wavelengths, as well as its climate and surface temperature. Atmospheric lightning, which was suspected to occur, was finally confirmed in 2007 by the *Venus Express*. Although on Earth, *Jupiter*,and *Saturn*, lightning is associated with water clouds, on *Venus* it is associated with clouds of sulfuric acid!

Figure 2.8 shows a three-dimensional representation of the surface of *Venus* around *Maat Mons* mountain, which is 8 km high from the middle surface. The mountain is named in honor of *Maat*, the Egyptian goddess of truth, order, harmony, morality, and justice. The simulated de-colorings are based on color images captured by the Soviet *Venera* 13 and 14 probes.

Fig. 2.8 Maat Mons, rising 8 km above the mean Venus*ian* surface, is displayed in this three-dimensional perspective view of the surface of *Venus*. Lava flows extend for hundreds of kilometers across the fractured plains shown in the foreground, to the base of *Maat Mons*. The data come from the *Magellan* spacecraft's synthetic aperture radar, combined with radar altimetry to develop a three-dimensional map of the surface. (*inset*) Image of *Maat*, Egyptian goddess and the personification of truth, cosmic balance, and justice. Her ostrich feather represents truth. The image was produced at the JPL Multimission Image Processing Laboratory. (*Image Credit* NASA/JPL)

1969: Man's First Landing on the Moon

The pinnacle of space exploration was the American space program *Apollo* (1969–1972), leading to the first landing of man on the Moon with *Apollo* 11, on July 19, 1969. The commitment of President Kennedy (JFK), "to landing a man on the Moon by the end of this decade and return him safely to Earth" (speech of JFK to Congress, May 25, 1961), achieved this goal with six missions of the program (*Apollo* 11, 12, 14, 15, 16, and 17).

The *Apollo* 11 spacecraft carried into orbit around the Moon the three astronauts Neil Armstrong, Buzz Aldrin, and Michael Collins. The first two landed with the lunar module Eagle in the Sea of Tranquility (Mare Tranquillitatis) on July 20, 1969. Armstrong and Aldrin remained for 21.5 h on the surface of the Moon before returning to the spaceship Columbia – which remained in lunar orbit at an altitude of about 100 km orbiting the Moon every hour or so – taking with them 21.5 kg of lunar matter that they had collected.

Armstrong's first steps on the Moon (Fig. 2.9) were broadcast televised around the world and were accompanied by his historic phrase *a small step for a man, one giant leap for humankind*.

Apollo 11 completed the first phase of the space race and essentially satisfied the national goal set in 1961 by US President John F. Kennedy, that before the end of the 60 s, they should sent a man on the Moon who will return safely back to Earth.

Fig. 2.9 Astronaut Edwin E. Aldrin, Jr., the lunar module pilot of the first lunar landing mission, on July 20, 1969, during an *Apollo* 11 Extravehicular Activity (EVA) on the surface of the Moon. (*Source* NASA)

Note is passing here that many of the complex mathematical calculations required to plan the orbits of satellites at the American Jet Propulsion Laboratory (JPL) in California were performed by JPL's human computer team, as in Fig. 2.10.

1974: Exploration of the Planet *Mercury*

Mercury is the smallest planet in our solar system by size, which barely exceeds that of the Moon. It is the closest planet to the Sun and the distance separating this planet from the Sun is relatively small (0.39 AU): If one could stand on its heated surface, the Sun would appear three times larger than it appears on Earth. During the day, its surface temperature exceeds a scorching 427°, but because it has no atmosphere, during the night it can fall at –193°.

The first telescopic observations of *Mercury* were made by Galileo at the beginning of the seventeenth century. In 1639 Giovanni Zupi discovered telescopically that the planet had orbital phases similar to those of *Venus* and the Moon, which proved that *Mercury* revolves around the Sun. In the 1880s, Giovanni Schiaparelli mapped the planet with great precision and calculated its rotation period to be 88 days. The attempt to map the surface of *Mercury* was continued by Evgenios Antoniadis at the Observatoire de Paris with a dioptric telescope of 83 cm (*The Grand Lunette*). In 1934 he published a book that included maps with his observations. Many of *Mercury*'s surface features take their names from Antoniadis' map.

Fig. 2.10 During the 1940s and 1950s, the American Jet Propulsion Laboratory (JPL) in California used the term "*computers*" for humans, not for the computing machines we use today. The photograph's all-female JPL's computer team, many of whose members had just graduated from high school, performed the complex mathematical calculations required to plan the satellites' orbits. This practice continued the successful tradition that the Harvard observatory had established (1910–1920) and is thoroughly described elsewhere in this volume. (*Credit* Courtesy NASA/JPL-Caltech) (NASA/JPL-Caltech/JPL photo number P-163)

Mercury is, however, the planet that has been least explored, because of its proximity to the Sun. A spaceship accelerates as it approaches the Sun and has to slow down to orbit the planet, which entails a significant consumption of fuel. This is opposite of what happens to get a spacecraft into orbit around one of the outer planets, beyond Earth's orbit. The lack of an atmosphere on *Mercury* also doesn't help, as a parachute cannot be used.

The first spacecraft to pass by *Mercury* was *Mariner* 10, which visited the planet three times over a one-year period, from March 1974 to March 1975. *Mariner* 10 photographed nearly half of *Mercury*'s surface, sending us 10,000 pictures that revealed a surface very similar to that of the Moon, truly *pockmarked* by the countless craters caused by the fall of asteroids and comets. In fact, one of the largest and most impressive surface features that can be observed on *Mercury* is the *Caloris Planitia*, with a diameter of about 1,500 km. This feature was created by the impact of a giant asteroid when our solar system was still in its infancy.

Mercury's surface during the day is literally sizzling by the nearby massive Sun at temperatures that melt even metals, while at night it is freezing cold due to the lack of an atmosphere. The surface of Mercury is heavily cratered due to impacts

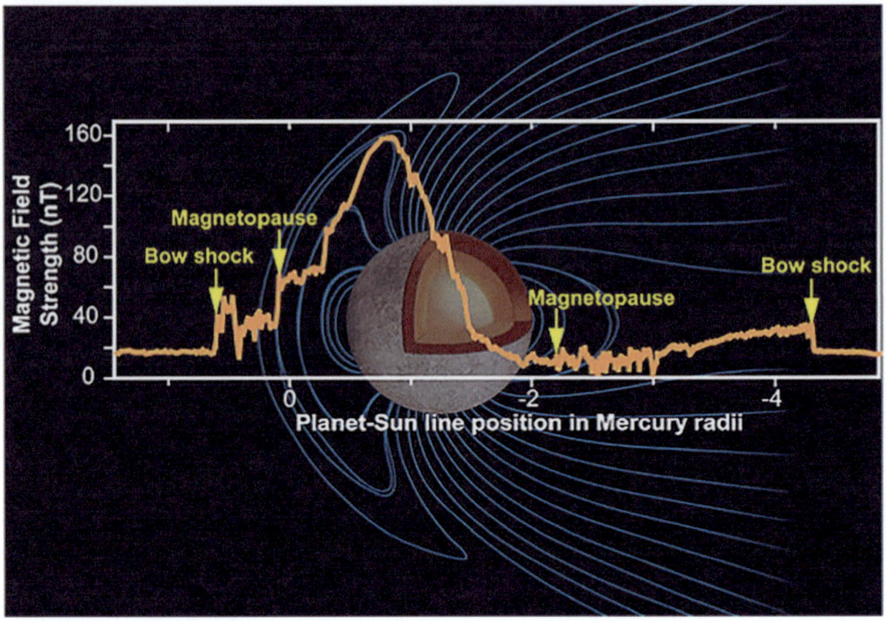

Fig. 2.11 The magnetosphere and the magnitude of *Mercury*'s weak magnetic field, according to data from *MESSENGER* (2008). (*Source* NASA/Johns Hopkins University, Applied Physics Laboratory/Carnegie Institution of Washington)

from meteorites. The *Mariner* 10 spacecraft provided the first close-up images of *Mercury*'s surface, showing huge cliffs criss-crossing the surface of the planet.

During the first close approach, *Mariner* 10's instruments also detected a magnetic field of about 0.0033 Gauss (330 nT)[2] at the equator, much to the astonishment of the planetary geologists, because *Mercury*'s self-rotation is too slow (59 days) to create an appreciable field with the dynamo mechanism, Fig. 2.11.

The second spacecraft to visit *Mercury* was NASA's mission *MESSENGER* (MERcury Surface, Space Environment, Geochemistry, and Ranging), which launched in 2004 and arrived at *Mercury* in 2008. There, the spacecraft made three close passes mapping 98% of the planet's surface. Mapping was completed in 2012, when it entered in orbit around the planet. The mission was then terminated due to lack of fuel (2015). Among the discoveries of *MESSENGER* were the unexpectedly high concentrations of magnesium and calcium which were found on *Mercury*'s dark side, and the fact that the axis of the magnetic dipole in *Mercury* does not pass through its center, but north of the center of the planet. *MESSENGER* also discovered large amounts of water vapor in *Mercury*'s outer atmosphere, as well as water ice and carbonaceous organic compounds in permanently shaded craters at the north pole of the planet. *MESSENGER* built the most detailed and accurate map of *Mercury* that we have to date, Fig. 2.12.

[2] 1 nanoTesla (nT) = 10^{-5} Gauss.

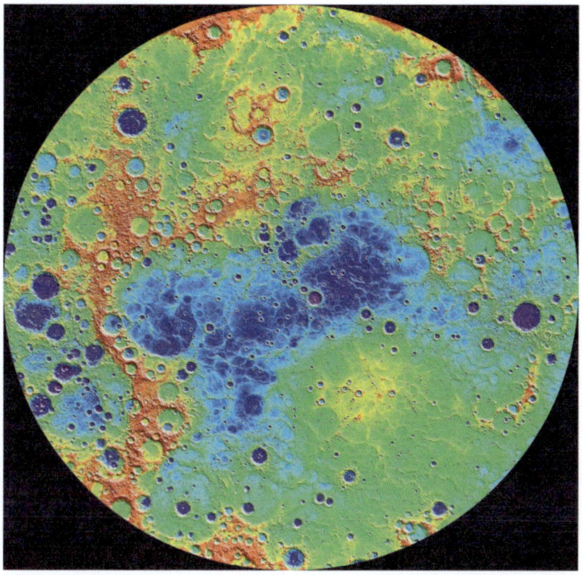

Fig. 2.12 High-precision mapping of the topography of Mercury's northern hemisphere with *MESSENGER*'s Mercury Laser Altimeter (MLA) instrument. Lowest areas are shown in purple and highest in red. The difference in elevation between the lowest and highest regions shown here is roughly 10 km. Among the prominent features visible here are the smooth northern volcanic plains and the enigmatic northern rise. The low-lying craters near the North pole host radar-bright materials, thought to be water ice. *MESSENGER* collected over 250,000 photographs of the surface of Mercury. Measurements from *MESSENGER*'s MLA instrument during the spacecraft's greater than four-year orbital mission have mapped the topography of Mercury's northern hemisphere in great detail. The view shown here is an interpolated shaded relief map of these data. (*Source* NASA/Johns Hopkins University Applied Physics Laboratory/Carnegie Institution of Washington)

However, there are still several unanswered questions related to *Mercury*. For example, why such a small planet has a magnetic field while *Venus*, *Mars* and the Moon do not? Does *Mercury*'s magnetized environment have radiation belts and magnetic storms like Earth? How does its magnetic field interact with the solar wind? How is *Mercury*'s exosphere formed and why the planet does not have an ionosphere? Why spectroscopic observations do not reveal the presence of iron, when this element is supposed to be the main component of *Mercury*'s interior? Is *Mercury*'s core liquid or solid? Is *Mercury* tectonically active today?

The next mission, *Beppi-Colombo*, a collaboration between ESA and Japan's JAXA, will try to answer these and other questions. The *Beppi-Colombo* spacecraft was launched on October 20, 2018, with an arrival at Mercury planned for November 2026, after a flyby of Earth, and two flybys of Venus. It will complete six flybys of Mercury. Its many instruments are expected to provide answers to the previous questions.

Fig. 2.13 First photo of the 11.5 kg Sojourner (Pathfinder) rover, just after it *walked* on the Martian surface, July 4, 1997, equipped with the Alpha particle X-ray Spectrometer (APXS) developed by Dr. Th. Economou at the University of Chicago. APXS analyzed the chemical composition of the soil samples from scattered α-particles (Rutherford scattering) and X-ray fluorescence radiation, when irradiated with α-particles and X-rays from radioactive sources. (*Source* NASA)

1997 to Present: the Robotic Exploration of *Mars* by NASA

After 2011, the Russians completely abandon the race to explore *Mars*. Thence, the Americans continued on their own (i.e., proceeding without a rival!) and they sent the *Mars Observer* (September 25, 1992), but communication with the spacecraft was lost when it was already close to *Mars*. The *Mars Global Surveyor* (7/11/1996) was placed in Martian orbit and sent back data from there. On 4/12/1996 left for *Mars* the *Mars Pathfinder* mission which landed the first small robot, the cute little sojourner rover[3] (Fig. 2.13). At 4/7/1998, the Japanese tried their own luck with *Nozema*, but this spacecraft did not even manage to meet the planet. On 4/12/1996 left for *Mars* the *Mars Climate Orbiter mission*, which however crashed because the ground software was giving commands in the wrong units (pound.secs instead of Newton.secs); this was due to a miscommunication between NASA and the company that had built it (Lockheed Martin)!

[3] Dr. Thanasis Economou has been building instruments for interplanetary spacecraft since the mid-1960s. He has developed tools and techniques that enabled several of NASA's famous missions to be carried out successfully, such as the Alpha Proton X-ray Spectrometer that successfully performed the first chemical analysis of Martian rocks aboard the Mars Pathfinder rover in 1997 and cosmic dust instruments that flew on several other NASA missions to comets, asteroids and outer solar system planets. His name is linked with most of NASA's 50 year history of lunar and planetary exploration. He is the author of more than 200 publications and has received many prestigious awards: the NASA award for the APXS on Surveyor and the Pathfinder Mission, the National Air and Space Museum 1998 Achievement Trophy for the Pathfinder Team, the Group award for the Sojourner rover development, the NASA Group and individual awards for the Mars Rover mission Spirit and Opportunity, and the NASA award for the STARDUST mission.

On 3/1/1999, the *Mars Polar Lander* took off for a landing on *Mars*, but unfortunately crashed when tried to land on the planet. On 7/3/2001 the *Mars Odyssey mission* was launched, whose name was inspired by the film *2001 a Space Odyssey* of Arthur Clarke and Stanley Kubrick, and which still remains in orbit around *Mars*. On 2/6/2003 the European Space Agency (ESA) sent *Mars Express* to *Mars*, which entered in an orbit around *Mars*, but unfortunately lost communication with the small *Beagle* lander it carried, immediately after it detached from the mothership. On 10/6/2003 and 7/7/2003 left for *Mars* the twin *Mars* Exploration Rovers, Spirit and *Opportunity* which landed successfully on Mars (Fig. 2.14). *Opportunity* continued to send us impressive data from exploration of the surface of *Mars*, till June 10, 2018, when it was abandoned. Its twin *Spirit* was stuck on a shaky ground in 2009, and despite the persistent efforts of NASA did not manage to get off the ground, so finally in 2010 it was abandoned. At 12/8/2005 the *Mars Reconnaissance Orbiter* departed and is on a *Mars* orbit, until today. On 4/8/2007 the *Phoenix* was launched for *Mars*, landing in the northern polar regions of the planet in search of water. On 26/11/2011 the *Mars Science Laboratory* was launched together with the rover *Curiosity* which landed in the Gale Crater, possibly an ancient dried up lake. On 11/18/2013 *Mars* Atmosphere and Volatile Evolution mission (*MAVEN*) was launched with the aim of studying the Martian atmosphere in orbit around *Mars* at an altitude of 170 km. Finally, on 5/5/2018 departed from Earth and landed on *Mars* (26/11/2018) the *Insight* spacecraft (Interior Exploration using Seismic Investigations, Geodesy, and Heat Transport), the 41st mission to *Mars*, which was designed to study the deep interior of the planet.

Note that the planet *Mars* is popular with us earthlings. The reasons for this are – apart from the many science fiction movies – its relative proximity on our own planet, the hypotheses for the existence of life on its surface, as well as the relative ease, in relation to the other planet near us *Venus*, with which a human colony could be established on the Martian surface. The launch window to *Mars*, i.e., the period

Fig. 2.14 Rover *Spirit* which landed on the Martian surface on January 3, 2004. The 185 kg rover covered in total about 8 km through March 22, 2010. A composite image is shown on how the Spirit rover probably looks, stuck in Gusev Crater. (*Source* NASA)

which a spacecraft can be launched, so as to make the trip to the planet in the shortest possible time, opens every two years or so, due to the motions of Earth and *Mars* around the Sun. In the course of Theoretical Mechanics at the Physics Department of the University of Athens, we calculate this window as well as the time duration of an economical trip to *Mars* with the so-called Hohmann transfer orbit, which is 258 days (Modern Theoretical Mechanics, Tsinganos 2021, Chap. 12, p. 274).

2012 to Present: Extraordinary Images from the Robotic Exploration of *Mars*

It's worth mentioning that these missions push technology to its limits, especially robotics, but at the same time have a rather high cost. For example, the cost of the *Curiosity* mission was not negligible at all, since it reached 2.5 billion dollars. One of the reasons for this high cost was the larger size of the rover, which is equal to that of a small car, but also the nuclear batteries for its energy needs, in contrast to the previous missions, where, for example, the *Spirit* and *Opportunit*y cost *only* 400 million dollars, using solar photocells, or, the small *Pathfinder* which costed 175 million dollars. Good things come at a cost in this life, but I believe that these missions *get their money's worth*, because as they say, for every dollar invested in space there is a financial return of about 10 dollars.

The picture we all have for our neighbor planet *Mars* is what its surface looks similar with the arid rocky or desert regions of our Earth. In the landscape of the planet, a reddish color dominates everywhere, due to the color of oxidized iron, i.e., of rust.

The first major exploration of *Mars* was made by the two NASA spacecraft *Viking* 1 and 2 which, as we already mentioned, landed on the Martian surface in August and September of 1976. The next phase in the exploration of *Mars* began in the 1990s, with the *Mars Global Surveyor* missions and NASA's *Mars Pathfinder*, which studied the planet. The first mission is from an orbit and the second is from the surface, respectively. After a few failed attempts to reach Mars, in 2005 the twin six-wheeled, nuclear-powered robots Spirit and Opportunity reached Mars. As mentioned above, the *Spirit* got stuck somewhere on the surface where the ground was loose, while *Opportunity* continued to study its surface till 2028, and has found indications of the existence, in the distant past, of liquid water on the surface of the planet (Figs. 2.15, 2.17 and 2.18).

Mars does not appear at first glance to be hospitable to life, despite some similarities with Earth, such as its 24/h day, or its 4 seasons year (in the northern hemisphere spring lasts 7 months, summer 6, autumn 5.3, and winter 4). First of all, Mars is smaller than Earth (with a radius equal to with 38% of Earth's) and an astronaut who weighs 68 kg on Earth, on *Mars* it would weigh only 26 kg! This also created its big problem, i.e., *Mars* lost its atmosphere and what is left today is an atmosphere 100 times thinner than Earth's, creating a pressure of only 0.6% of Earth's atmospheric

Fig. 2.15 (above) *Victoria Crater*, as viewed by the Mars Reconnaissance Orbiter HiRISE camera on October 3, 2006. At its center, are the remains of an ancient dried out lake. The 730 m diameter and 70 m deep Victoria crater was visited by the Opportunity rover. (*Credit* JPL-Caltech/ Uni. of Arizona/NASA). (*below*) The *beach* Duck Bay, of the Martian crater Victoria which was explored by the rover opportunity. (*Source* NASA)

pressure. Because of the thin atmosphere, its speed of sound is small. So the sounds do not spread very far, just a few tens of meters. That's why *Mars*, apart from being a desert, is also a silent planet!

The composition of the Martian atmosphere is very different from that of Earth's and it consists basically of 96% of the colorless, odorless and tasteless, denser than air, greenhouse gas, carbon dioxide (CO_2), while the rest, only a few percent are nitrogen, methane, carbon monoxide, etc. CO_2 freezes at very low temperatures creating dry ice.

The loss of the Martian atmosphere appears to have occurred 3.7–4.2 billion years ago in its distant past. Then, due to the loss of the Martian atmosphere, *Mars* cooled down its core as a result of which it lost both its magnetosphere and the protection it gave to the planet from the solar wind, which then literally *swept away* its remaining atmosphere. The average surface temperature of –63° is not particularly attractive, although during the summer on *Mars*, the robot Spirit recorded heat waves with + 35°.

Despite these negative features, the landscape on *Mars* seems to be not very different from the landscape on Earth (Fig. 2.16). Thus, we see dry rivers and lake beds

Fig. 2.16 Two rocky and rugged areas, at *Cape St. Vincent* of the Martian Crater *Victoria*, explored by the *Opportunity* rover. In the top image, the layer on top of the promontory is composed of mudstone, but then further down into the crater, the ground abruptly turns into solid rock. This transition is characterized by a bright area of rock, which can be seen everywhere throughout the mouth of the crater. It is believed that it was rather the surface of Mars before the meteorite impact formed Victoria Crater. At the distant layer is the horizon of Mars. (*Source* NASA/JPL/Cornell)

(e.g., *Gale Crater*), rocks that form only when water is present, other sedimentary rocks, and cliffs similar to those on Earth (as in the *Cape St. Vincent* image of the *Victoria Crater*, Fig. 2.16a) and dried beaches (as in the image of *Duck Bay* in the *Victoria Crater*, Fig. 2.15b).

But, as T. Krimigis notes in his popular science book *Journey to the Solar System*, all these are the remains of a lost globe, which remains dry for billions of years, since *Mars* did not take care to save what was required to keep its water! Thus, without water, *Mars* turned into the frozen reddish desert we see today.

Water, however, exists today in the form of ice, not only at the poles, under the frozen dry ice of carbon dioxide, but also underground, a few centimeters below the surface, where hydrates have been detected. At the same time, pictures show seasonal streams, which probably are formed in the summer, since the water remains liquid in the low temperatures that prevail on *Mars* due to the salts it contains, as they have

Fig. 2.17 Images of wind-sculpted ripples by the *Curiosity rover* are evident in this view of the top surface of Martian *sand dunes* in the Martian Gale Crater area, Bagnold Dune Field. (*Source* NASA/JPL-Caltech/MSSS)

shown images from the *Mars Reconnaissance Orbiter*, which has seen seasonal dark narrow streaks in crater slopes (Fig. 2.18).

Thus, in these areas of the subsoil there may have been microorganisms that possibly live there in the subsoil of the planet, even today. The question whether there is today – or has ever been in the past – biological activity on *Mars* doesn't seem to have been answered yet. NASA's *Insight* mission which landed on *Mars* (26/11/2018) investigated the Martian subsoil, before reaching the end of its mission after searching for and studying the interior of *Mars*. Not only did the robot study the planet's interior, but it was also sent to the planet to measure the seismic activity on it. It was launched on November 26, 2018, and after four years, the robot has just sent its final message to Earth.

Fig. 2.18 Dark narrow streaks, called *recurring slope lineae*, emanate from the walls of Garni Crater on Mars, in this view constructed from observations by the High-Resolution Imaging Science Experiment (HiRISE) camera on NASA's *Mars Reconnaissance Orbiter*. The dark streaks here are up to few hundred meters long. They are hypothesized to be formed by flow of briny liquid water on Mars. (*Credit* NASA)

Fig. 2.19 Red sunset on the blue planet (Earth, left) and blue sunset in the red planet (Mars, right). The difference is due to the fact that on Earth we have *Rayleigh scattering* of sunlight, wherein the blue wavelengths are scattered (blue sky) while the red ones penetrate the atmospheric depths, while on Mars there is a lot of dust in the atmosphere and so we have *Mie scattering* of light. On Mars, this was the first sunset observed in color by the rover *Curiosity*. The color has been calibrated and white-balanced to remove camera artifacts. *Mastcam*, the camera that was taking color images and color video footage of the Martian terrain, sees color much the way the human eye does, although it's a little less sensitive to blue. (*Credit* NASA/JPL-Caltech/MSSS/Texas A&M Univ)

1974–2024: Other Space Missions

A series of missions to explore space have followed, targeting all planets of our solar system, many of their satellites, and also some asteroids and comets. Thus, as we mention, *Mercury* was explored by *Mariner* 10 (1974–1975) and *MESSENGER*

(2008–2012). *Venus* was observed by American, Soviet, and European missions, such as the seven successful *Mariner* missions (1962–1973), the *Magellan* mission (1989), the Soviet *Venera* series of missions (1961–1984), and the European *Venus Express* mission (2006), respectively. The ongoing exploration of *Mars*, in addition to the spacecrafts which landed on the planet, such as *Viking* 1 and 2 (1976) and *Phoenix* (2008), also includes several American robotic missions with the *Pathfinder* rovers (1997), the twin Mars Exploration Rovers, *Spirit and Opportunity* (2004), as well as *Curiosity* (2012) and *Insight* (2018), which we have already discussed. In the exploration of the outer planets, NASA's epic space missions *Voyager* 1 and 2 (1977–), *Cassini* (2007–2017), and New *Horizons* (*2006–*) played an important role and to a lesser extent *Gallileo* (1989) and *Juno* (2011). *Cassini/Huygens* (2005), landed on *Saturn*'s large moon Titan. *NEAR/Shoemaker* (Near Earth Asteroid Rendezvous) was the first spacecraft to orbit and land on an asteroid, becoming the first human-made object to orbit an asteroid (Eros, NEAR Shoemaker, 1996–2001). Remarkably, this spacecraft, although it was not designed as a lander, survived touchdown on the asteroid and returned valuable data for about two weeks. The European Space Agency's *Rosetta* (2004–2016) was the first mission designed to orbit and land on a comet (67P/Churyumov–Gerasimenko, 2004) with its comet lander *Philae*. Also, ESA's *Rosetta* mission has returned the first close-up images of the asteroid *Lutitia* (2010) showing that it is most probably a primitive survivor from the violent birth of the Solar System – Lutetia has been revealed as a battered world of many craters. *Dawn* (2007–2018) was NASA's first truly interplanetary spaceship. The mission featured extended stays at two very different extraterrestrial bodies: giant asteroid *Vesta* and dwarf planet Ceres, the two most massive bodies in the main asteroid belt between *Mars* and *Jupiter*. Both these small worlds reside in the debris-strewn main asteroid belt. *Lucy* (2021) is a NASA space probe on a twelve-year journey to eight different asteroids, as well as six *Jupiter* Trojans – asteroids that share *Jupiter*'s orbit around the Sun, orbiting either ahead of or behind the planet. The mission is named after the *Lucy hominin fossils*, because study of the Trojans could reveal the *fossils of planet formation*, materials that clumped together in the early history of the solar system to form planets and other bodies.

Psyche is a NASA Discovery Program space mission launched on October 13, 2023, to explore the origin of planetary cores by orbiting and studying the metallic asteroid 16 Psyche beginning in 2029 (Bell et al. 2025). *Deep Impact* (2005–2013) was a NASA space probe to study the interior composition of the comet Tempel 1 (9P/Tempel), by releasing an impactor into the comet, by releasing an impactor into the comet. *MAVEN* (Mars Atmosphere and Volatile Evolution, 2013–) is a NASA spacecraft orbiting *Mars* to study the loss of that planet's atmospheric gases to space, providing insight into the history of the planet's climate and water. The *Gravity Recovery and Interior Laboratory* (GRAIL, 2011–) was an American lunar science mission in NASA's Discovery program which used high-quality gravitational field mapping of the Moon to determine its interior structure. The Stardast mission (1999–) mission was aimed to collect dust samples from the coma of comet Wild 2, as well as

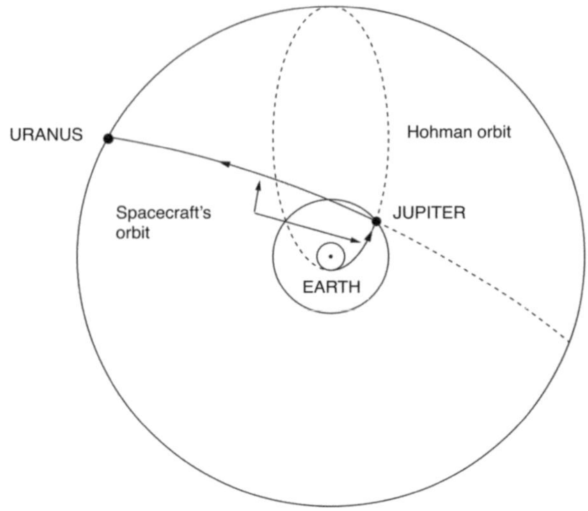

Fig. 2.20 Complete elliptic transfer Hohman orbit of a spacecraft to *Uranus* is shown by dots. The spacecraft initially is placed in an orbit around the Earth, and then its orbit is changed when it approaches *Jupiter*, before it finally reaches the planet *Uranus* (continuous line)

samples of cosmic dust, and return them to Earth for analysis. It was the first sample return mission of its kind.

1977 to Present: Space Billiard with the Gravity Assist of Spacecrafts

As is well known, in the table tennis game (aka pig pong), when we hit hard the ball with the racket, the ball leaves toward the other side with a higher speed than when it arrives to our racket. But what is less known is that the same phenomenon is used to send spacecrafts to distant planets, sooner than they would get there with the initial velocity given to them when they were launched from Earth (Figs. 2.20 and 2.21). For example, a spacecraft launched from Earth and headed toward the distant planet *Uranus*, at a distance $\alpha_{Ura} = 19.2$ AU following the economic in fuel *Hohmann transfer orbit*[4] (see the dotted elliptical curve in Fig. 2.20), needs a time equal to half the period T of this elliptical transfer orbit to get there. But according to Kepler's harmonic law relating the period T to the semi-major axis α of an elliptical orbit, i.e., $T = \alpha^{3/2}$, where T is in years and α in AU, $\alpha = (\alpha_{Ura} + 1)/2 = 10.5$ AU (Fig. 2.20). Thus, we have $T \approx 32$ years, and therefore, the time τ from Earth to *Uranus* is $\tau = T/2 = 16$ years.

However, since this is a rather long time, another smart and shorter way to send the spacecraft to a distant planet, such as the planet *Uranus*, is the one which uses the idea of gravity assist. Specifically, this is achieved by exploiting the gravitational field

[4] The *Hohmann transfer orbit* of a spacecraft from Earth to a distant planet is that elliptical orbit that is tangential to the orbits of Earth and the planet around the Sun, as in Fig. 2.20.

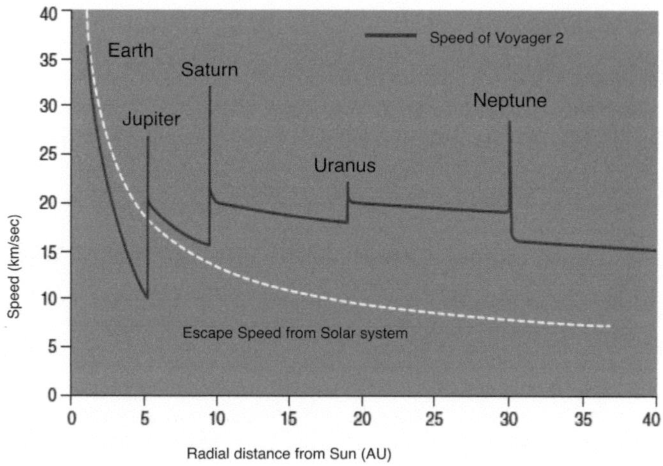

Fig. 2.21 When a spacecraft is launched from Earth toward the outermost planets Jupiter, *Saturn*, Uranus, and *Neptune* and is properly scattered by the gravitational field of these planets, the magnitude of the spacecraft's speed increases. This was the case with the Voyager 1 & 2, a situation which however requires an alignment of the outer planets, something that occurs once every 175 years. (*Source* T. Krimigis)

of an intermediate planet, such as, for example, the planet *Jupiter*. The spacecraft is appropriately placed in an orbit that brings it close to *Jupiter* as it moves around the Sun and behind *Jupiter*. Then, *Jupiter* drags the spacecraft along the direction of its movement, changing the direction and magnitude of the spacecraft's speed.

1977 to Present: the Longest Journey of a Spacecraft in the Universe

In 1977, a rare alignment of the outer planets *Jupiter*, *Saturn*, *Uranus,* and *Neptune* occurred – something that only happens once every 175 years – and allowed a single spacecraft to visit all of these giant planets and their satellites in succession. This ambitious project was undertaken by NASA's unmanned Voyager 2 spacecraft, which was launched on 08/20/1977. Together with its twin *Voyager* 1 (launched 09/05/1977), they became the protagonists of the longest journey, from a human perspective, in the Universe to date (Fig. 2.22). More than four decades after their launch, these two well-traveled human objects have already crossed the borders of the solar system. In

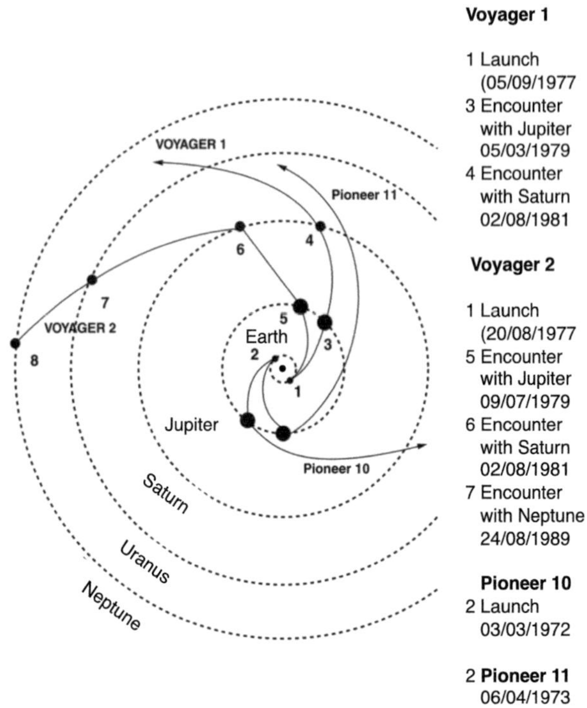

Fig. 2.22 Schematic diagram of the positions of the outer planets *Jupiter, Saturn, Uranus, and Neptune* in their orbits around the Sun, as well as the orbits of the Voyager 1 & 2, and Pioneer 10 & 11 spacecrafts that explored them

2012, *Voyager* 1 became the first spacecraft to enter interstellar space, while *Voyager* 2 is the only spacecraft to pass by all four outer planets. Today, the two *Voyager*s are still communicating to Earth, with signals traveling more than twenty hours to arrive on Earth, revealing to us what is happening in deep Space.

Voyager 1 began its journey on September 5, 1977. After a 35-year journey, on August 25, 2012, it crossed the *heliopause* at a distance of 122 AU and entered interstellar space (ISM), the first human object ever to reach there (Fig. 2.22).

2004, 2012, 2014: The Auroras and Great Dark Spots of the Giant Planets

From Space, planetary *auroras* – also known as *polar lights* – appear as bright rings in the polar regions of the planets where the magnetic field is stronger. *Auroras* are created when charged particles are trapped by the magnetic field of the planet. These charged particles are then accelerated to high energies and reach the upper layers of the planetary atmosphere. There, collisions with the molecules and atoms of the planet's atmosphere create flashes of visible, ultraviolet, and infrared light.

Figure 2.23 shows a Hubble Space Telescope photograph of the bright *aurora* of *Jupiter*. In this picture, for the first time, NASA's *Juno* spacecraft detected electrons

Fig. 2.23 Photo of the Jupiter's brilliant **aurora**, with the Hubble Space Telescope (2014). In the lower right part, the great red spot can be seen. (*Source* NASA, ESA, and J. Nichols (University of Leicester))

being ejected toward the upper atmosphere of Jupiter from an energy potential of 400,000 Volts. This corresponds to a huge energy, which then creates the impressive *aurora* of the planet *Jupiter*. Of course, these giant energies appear rarely, but the question is what exactly is their origin. For *Jupiter*, we know that the charged particles that create the *aurora* come from its moon Io, whose volcanoes eject large amounts of sulfur and oxygen into Space, thus loading the Jovian magnetic field with particles.

Figure 2.24 shows the variability of the *aurora* of *Saturn*, from a combination of observations by the Hubble Space Telescope in the ultraviolet and photos by the *Cassini* spacecraft. The *aurora* here appears blue because it glows in ultraviolet light. But for an observer on *Saturn*, the *aurora* would appear red due to flashes of its atmospheric hydrogen. In the terrestrial aurora, however, the charged particles when colliding with atmospheric nitrogen and oxygen atoms emit green red and blue flashes.

In Fig. 2.24, observations also reveal the dynamical character of *Saturn*'s *aurora*. Specifically, these images show that the *aurora* on *Saturn* changes from day to day, just as it does on Earth. But while on Earth the *aurora* changes within a few minutes and lasts only a few hours, on *Saturn* the *aurora* remains bright for many days. Additionally, observations show that *Saturn*'s *aurora* appears to be generated by the Solar Wind (SW). The great luminosity in this *aurora* on January 28, 2004, was due to the arrival at *Saturn* of a major SW disturbance. And, when the *aurora* becomes brighter, the bright ring at the pole becomes of smaller diameter.

Until recently, the best-studied planetary *auroras* were those of *Jupiter* and *Saturn*, while the *aurora* of their next neighbor, the icy giant planet *Uranus*, had not been observed and studied. Hubble has observed *auroras* on *Uranus* on various occasions:

Fig. 2.24 Variability in the aurora of Saturn, with a combination of observations of the Hubble Space Telescope in the UV and photos from the *Cassini* spacecraft. The strong brightening of the aurora corresponded with the arrival to Saturn of a large disturbance in the solar wind (*Source* NASA/ESA/JPL)

In 2011, when the telescope became the first to image the phenomenon from the vicinity of Earth, then again in 2012 and 2014, taking extra data beyond visible light. In 2011 we had the first related observations of the polar lights of *Uranus* with *Voyager 2* and the Hubble space telescope. A major disturbance of the SW affected the planet *Uranus* and created the appearance of an impressive *aurora*.

We should mention that the planet *Uranus* is a rather weird planet, with the orientation of its magnetic field being completely unusual: The magnetic poles do not coincide at all with the geographic ones, but the magnetic axis is inclined at an angle of 59° with respect to the axis of rotation. Also, this rotation axis is not vertical, but almost parallel to the plane of the planet's orbit around the Sun. In addition, the axis of the magnetic field doesn't even pass through the center of the planet (Fig. 2.25).

It has been possible to use the ultraviolet observing capabilities of the Hubble Space Telescope with the Space Telescope Imaging Spectrograph (STIS), to monitor the propagation of strong interplanetary shocks in the solar wind, from the Sun to the *Uranus*, to trace their effect on the planet's magnetic field and the final creation of *aurora* in *Uranus* (Lamy et al. 2012). In doing so, Hubble detected the most intense *aurora* which was ever observed in *Uranus*. Furthermore, by following the time evolution of this *aurora*, it was discovered that the auroral regions rotate with the planet. Thus, the lost poles of *Uranus*' magnetic field were located: They had been lost since their initial detection by *Voyager* 2 in 1986, due to uncertainties in the observations and also because of the smooth surface of the planet which has no fixed features.

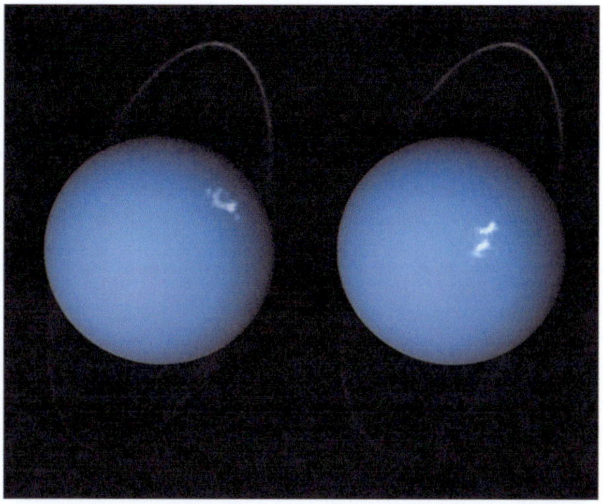

Fig. 2.25 Composite image of rare observations of intense polar aurorae on Uranus, using the Hubble Space Telescope on the 24th and 2nd of November 2014 [© ESA/Hubble & NASA, L. Lamy/Observatoire de Paris]

New measurements (Lamy et al. 2025) using an innovative method to track the rotational motion of Uranus' *aurorae* have revealed that Uranus completes a full rotation in 17 h, 14 min, and 52–28 s longer than the estimate obtained by NASA's Voyager 2 during its 1986 flyby.

For the first time in 2025, the NASA/ESA/CSA James Webb Space Telescope has captured bright *auroral* activity on the planet Neptune (Melin et al. 2025). In addition to the visible glow in the Fig. 2.26, the spectrum from Webb also found an extremely prominent emission line signifying the presence of the trihydrogen cation (H3+), which can be created in *auroras*. The trihydrogen cation is a positively charged ion consisting of three hydrogen nuclei sharing two electrons. It is one of the most abundant triatomic ions in the universe and plays a significant role in interstellar chemistry.

The auroral activity seen on Neptune is noticeably different from what we are accustomed to seeing here on Earth, or even Jupiter or Saturn. Instead of being confined to the planet's northern and southern poles, Neptune's *auroras* are located at the planet's geographic *mid*-latitudes. This is due to the strange nature of Neptune's magnetic field, originally discovered by NASA's Voyager 2 in 1989, which is tilted by 47° from the planet's rotation axis. Since auroral activity is based where the magnetic fields converge into the planet's atmosphere, Neptune's *auroras* are far from its rotational poles.

From the Webb observations, the temperature of the top of Neptune's atmosphere is measured (Melin et al. 2025) for the first time since Voyager 2's flyby. The results hint at why Neptune's auroras remained hidden from astronomers for so long: Neptune's upper atmosphere has cooled by several hundreds of degrees.

The Solar System's outermost gas giant shows a nearly uniform blue hue created by small amounts of methane drifting in a thick atmosphere of nearly colorless hydrogen and helium. Dark spots do appear, however, that are anti-cyclones: large high-pressure systems that swirl in Neptune's cold cloud tops. Two dark spots are

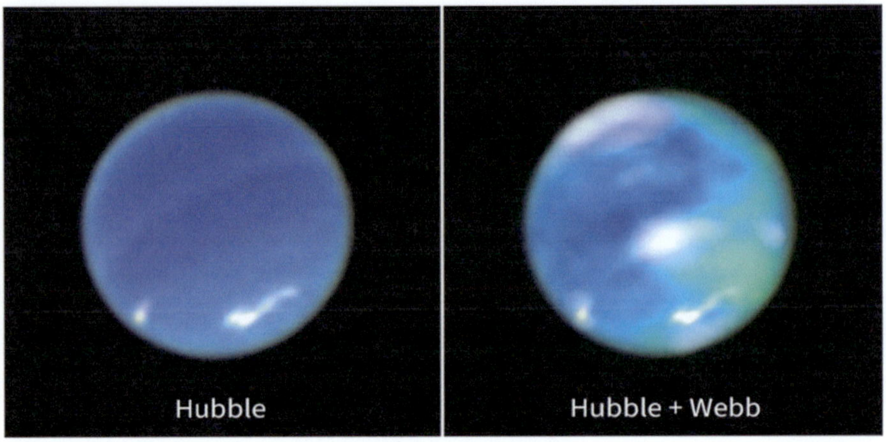

Fig. 2.26 At the left, an enhanced-color image of Neptune from NASA's Hubble Space Telescope. At the right, that image is combined with data from NASA's James Webb Space Telescope. (*Credit* NASA, ESA, CSA, STScI, Heidi Hammel (AURA), Henrik Melin (Northumbria University), Leigh Fletcher (University of Leicester), Stefanie Milam (NASA-GSFC))

visible in Fig. 2.27 taken by Voyager 2 in 1989: an Earth-sized Great Dark Spot (GDS) located on the far left, and Dark Spot 2 located near the bottom. A bright cloud dubbed *Scooter* accompanies the Great Dark Spot. Computer simulations (Stratman et al. 2001) indicate that scooters are methane clouds that might commonly be found near dark spots.

The GDS is the most impressive atmospheric formation of the planet *Neptune*. It appears in the southern hemisphere of the planet and is accompanied by large bright clouds. These clouds seem to persist throughout the time *Voyager* was passing from this area. In the northern hemisphere there were also some white clouds similar to those in the South.

Like Jupiter's red spot, the GDSs are *anticyclonic storms*. However, their interiors are relatively cloud-free, and unlike Jupiter's spot, which has lasted for hundreds of years, their lifetimes appear to be shorter, forming and dissipating once every few years or so. Based on observations taken with Voyager 2 and since then with the Hubble Space Telescope, Neptune appears to spend somewhat more than half its time with a GDS. Little is known about the origins, movement, and disappearance of the dark spots observed on the planet since 1989.

Fig. 2.27 This photograph of Neptune was reconstructed from two images taken by Voyager 2's narrow-angle camera, through the green and clear filters. The image shows three of the features that Voyager 2 has been photographing during recent weeks. At the north (top) is the Great Dark Spot, accompanied by bright, white clouds that undergo rapid changes in appearance. To the south of the Great Dark Spot is the bright feature nicknamed *Scooter*. Still farther south is the feature called Dark Spot 2, which has a bright core. Each feature moves eastward at a different speed, so it is only occasionally that they appear close to each other, such as at the time this picture was taken. (*Image Credit* NASA/JPL)

1977 to Present: Discoveries of the Voyager 1 & 2 Missions

The most notable achievements of the *Voyager 1 & 2* missions (Fig. 2.28) include:

- The discovery that Io, *Jupiter*'s closest moon, has many active volcanoes, which are activated by tidal heating of its interior (Fig. 2.29).
- The first detailed images of the *rings* of *Jupiter*, *Uranus,* and *Neptune* and the discovery of the complex structure and details of *Saturn*'s rings, such as various gaps, new narrow ringlets, density waves, and transient spokes (Fig. 2.24).
- The first high-resolution images of the four giant planets of our solar system (Figs. 2.23, 2.24, 2.25, 2.26 and 2.27) and their largest satellites, and furthermore the discovery of dozens of new satellites.
- The discovery that the planets *Uranus* and *Neptune* have magnetic fields, which differ from those of the other planets in that their axis has a large inclination to the axis of rotation of the planet. The *Voyagers* also gave us the first data on the magnetospheres surrounding these planets, such as their dimensions, densities, composition, and plasma waves. They also observed that aurora activity occurs in the upper atmospheres of the giant planets *Jupiter*, *Saturn, Uranus,* and *Neptune* (Figs. 2.23, 2.24, 2.25, and 2.26).

Fig. 2.28 (*top*) Artist's concept of the Voyager spacecrafts in flight. Their main characteristics are: weight 825 kg, power 465 (today 242) Watts, radio transmitter power 22.4 Watts (as a small refrigerator lamp), antenna diameter 3.7 m, fuel 22 kg. (*bottom*) Artist's concept of the Cassini spacecraft during its Saturn Orbit Insertion (SOI) maneuver. The SOI maneuver, lasting about 90 min, allowed Cassini to be captured by Saturn's gravity in a 5-month orbit. Cassini's proximity to the planet after this maneuver offered a unique opportunity to observe Saturn and its rings with extremely high resolution. Cassini arrived at Saturn on July 1, 2004, and the spacecraft's distance from Saturn was about 0.3 Saturn radii (20,000 km). (*Source* NASA/ESA, A. Simon (GSFC), MH Wong (University of California, Berkeley), and the OPAL Team)

- The first detailed measurements of the atmospheres of *Saturn*'s moon Titan (Fig. 2.31) and *Neptune*'s moon *Triton*.
- The discovery that *Neptune* radiates 2.6 times more energy than it receives from the Sun. The source of this energy still remains unknown.
- Measurements of the composition, wind speed, temperature distribution, and pressure of the planetary atmospheres. *Voyager 2* in particular showed that winds rage in *Neptune*'s atmosphere at speeds reaching up to 2,000 km per hour, as well as the existence of a huge cyclonic system, the great dark spot.

- The *Voyagers* dramatically improved our knowledge of the masses, dimensions, shapes, and gravitational fields of all the giant planets and many of their satellites.
- Both spacecrafts carried gold-plated disks on which selected sounds and images of Earth have been recorded, intending to present the wide range of biodiversity and culture of our planet. These records could be read by other advanced civilizations that might encounter the spacecrafts on their long journey (Figs. 2.2 and 2.22).
- *Voyager 1*, as it drifted away, took a "family photo" of the planets, including Earth as a "faint blue gem," which helps us realize where we are in space and how small and fragile our planet is in the vastness of the Universe (Fig. 2.35).

1979: *Io* (the Colorful and Violent Volcanic Moon of *Jupiter*, Unique in Our Solar System, with Nearly 400 Active Volcanos)

In mythology, *Io* had the bad luck to place herself between Zeus and Hera, resulting in a stressed life, so that eventually Zeus was forced to transform her into a cow to save her from the anger of jealous Hera. Similarly, the nearest satellite of *Jupiter* which has the same name lives a "stressed" life because of the tidal forces of the planet *Jupiter* and its other large satellites, *Europa*, *Ganymede,* and *Callisto*, which pull it at constantly different angles, with the result that sometimes Io is stretched and others it is compressed. As *Io* rotates, the gravity of *Jupiter* and its moons periodically weakens and strengthens, so that the surface of *Io* constantly rises and falls. These pressures and stresses heat the interior of *Io* in a process called *tidal heating*. This effect is so strong on *Io*, so that it can strongly heat its interior and melt the rocks; the lava that is thus created together with sulfurous gases is thrown to the satellite's surface in violent volcanic eruptions, thus creating its many volcanoes (Fig. 2.29). Hence, the torturous life that *Io* lives makes this satellite to have the uniqueness of the most geologically active body in our solar system!

Thus, the smell of the atmosphere on *Io* is like that of a spoiled egg due to various sulfur compounds; this, together with the enormous radiation *Io* receives from the neighboring Jupiter, make it a not-so-attractive tourist destination for us earthlings!

In 2007, the *New Horizons* mission, as it flew past *Io*, recorded a truly monstrous explosion: volcanic materials were ejected into *Io*'s atmosphere, 320 km above its surface: the Pele volcano, the first active volcano outside Earth, reaches a height of about 300 km, while the fall area of lava covers an area the size of France.

It has been estimated that we have on *Io* continuously (at any time) 12–15 active volcanoes, according to observations by *Voyager 1 & 2, Ulysses, Galileo, Cassini, and New Horizons* flyby, over a period of 28 years.

Io has a metallic core (probably of iron and iron sulfate) with a diameter of about 900 km, while newer information states that this satellite consists of two layers. Above the core there is a mantle with partially fluid rocks and a thin crust. On the surface, it has lakes of lava, composed of liquid sulfur, at a temperature of 1,100°.

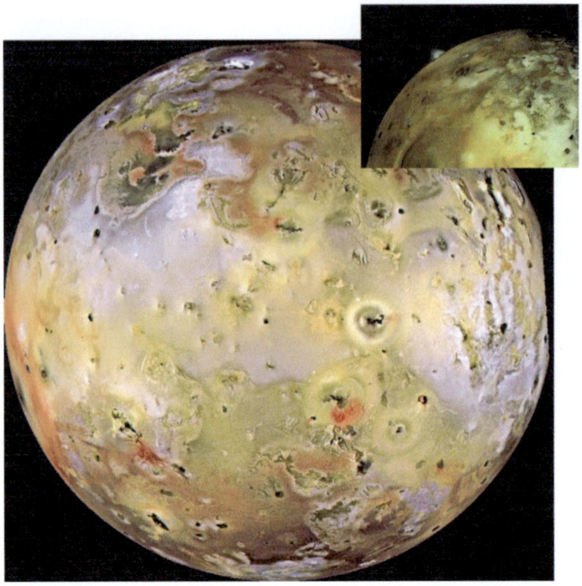

Fig. 2.29 Highest resolution image of the landscape of *Io*, the most volcanic body in our solar system, by NASA's *Galileo* spacecraft. The smallest features that can be distinguished are 2.5 km in size. Rugged mountains several kilometers high are visible, with layers of material forming plateaus and many irregular depressions, the volcanic calderas. Many of the dark features correspond to hot spots, probably active ones with lava flows. There are no impact crater-like features, as volcanoes cover the surface with new deposits of material faster than comets or asteroids could create such large impact craters. The image is centered on the side of *Io* facing opposite to Jupiter. The North Pole is at the top [*Source* NASA/JPL/University of Arizona]. (top right) Active volcano in *Io*. *Voyager* 1 photographed this volcano on March 4, 1979. (*Source* NASA)

The volcanoes on the colorful but tortured *Io* were discovered on 1979 by the *Voyager 1* spacecraft, and its surface was mapped by *Galileo* (Fig. 2.29). It is estimated that *Io* produces 100 times more lava each year than all volcanoes combined on Earth. Due to the reduced gravity in *Io*, some of the approximately 400 active volcanoes shoot their incandescent materials high, at hundreds of kilometers, while flues of sulfur dioxide are released from the interior and, rising, freeze and fall back into the ground like beautiful colored snow that covers *Io* at a rate of 10 million tons per year.

Thus, the volcanic activity on *Io* colors its surface with various red, yellow, white, green, and black shades, as clearly seen in Fig. 2.29. It is due to the same intense activity and the great height which some of its mountains reach to higher than Mount Everest here on Earth. Names of its regions are: *Bosphorus, Colchis, Caucasus, Bactria, Argos, Nemea, Mycenae, Lernea, Prometheus, Tarsus, Iopolis*, etc.

1990: The Search for Water in Our Planetary System

The main discovery from the systematic exploration of our planetary system in recent decades is that *water exists everywhere*: on *Mars*, in the moons of *Jupiter* and *Saturn*, in asteroids and comets, even in the distant Pluto.

NASA's policy since the 1990s has been to *follow the water*, according to the decision of its respective subcommittee. Specifically:

- From spectroscopic observations we know that the dwarf planet (asteroid). Demeter is composed largely of water ice, which in quantity is perhaps more than the whole of fresh water which exists on Earth!
- *Mars* is not the dry planet we thought after all. Except for the existence of abundant water in the form of ice at its poles, during its summer months, at low latitudes, traces were detected from running water and especially hydrates, caused only from the presence of salty water. As we discussed earlier (Fig. 2.18), the images released by NASA's *Mars Reconnaissance Orbiter* (MRO) satellite capture long, narrow dark formations on the surface of *Mars* and record the changes that marked in them according to the alternation of the seasons. According to the MRO images, there are seasonal changes in crater slopes where the tracks are five meters wide and 100 m long and change during the hottest months, while they "fade out" when the conditions are cooler.
- On *Saturn*'s satellite *Enceladus*, *Cassini* flying below the South pole at a height of several tens of kilometers detected fountains of ice crystals that are ejected at heights of 25 km from the frozen surface! Just like *Jupiter*'s *Europa*, *Enceladus* ejects jets from several km-long cracks in its frozen surface of its south pole (Fig. 2.30b). There are indications of a similar situation in *Jupiter*'s satellite Ganymede.

1997: Search for Water on *Jupiter's* Satellite Europa

Europa, the second closest moon of *Jupiter*, which with a diameter of 3 122 km is slightly smaller than the Moon and the smallest of the 4 moons of *Jupiter* that Galileo had observed, despite having temperatures that never exceed minus 160 °C, seems that it may be hospitable to life after all! This surprising fact emerged from the data that sent the Galileo spacecraft, which was launched in 1989 and reached *Jupiter* in 1995, while in December 1997 made its closest flyby of Europa, flying within only 400 km from its surface. So Galileo's instruments recorded for five minutes at Europa's atmosphere unusual indications, both for its magnetic field, and the plasma density of particles, which are estimated to have originated from a geyser of water vapor that reached a height of hundreds of kilometers.

What we initially knew about Europa is that it has a smooth frozen surface, 15–25 km thick, with few craters but with hundreds of lines and grooves, i.e., the ice appears to have broken into many individual plates. What is hidden under them? The data

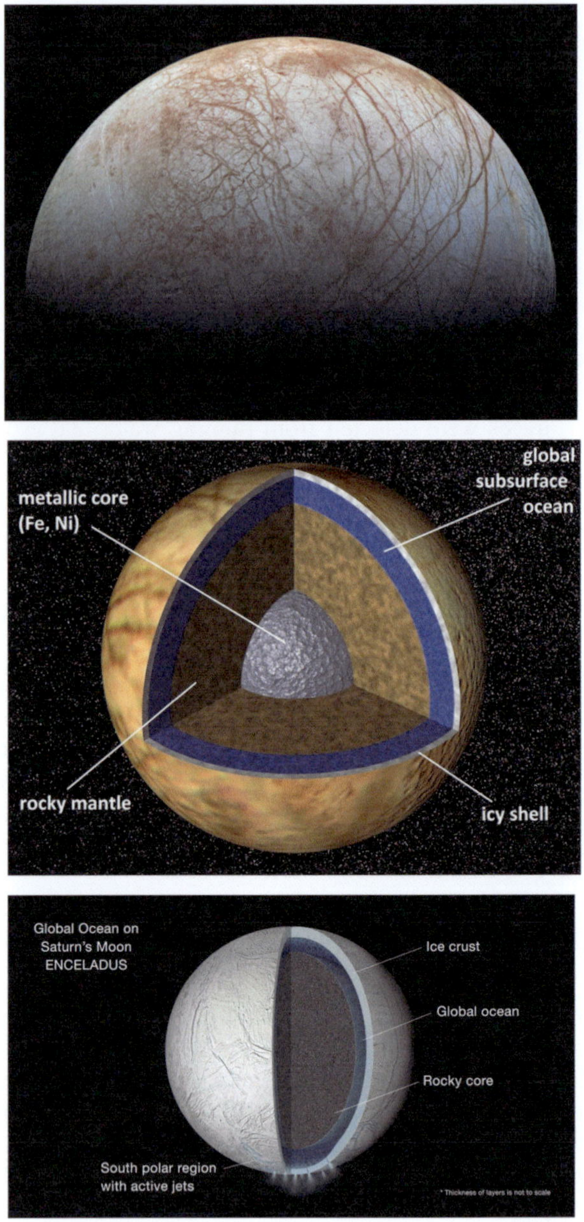

Fig. 2.30 (top) Reprocessed color view of Jupiter's moon Europa made from images taken by NASA's Galileo spacecraft in the late 1990s. (*Credit* NASA/JPL-Caltech/ SETI Institute). (*middle*) Annotated cross-section of the Moon-sized *Europa*, one of the satellites of Jupiter, 2nd in distance from Jupiter. Beneath *Europa*'s icy crust appears to be an underground ocean. (*bottom*) Annotated cross-section of *Enceladus*, the mysterious moon of Saturn, with a subsurface ocean between the rocky interior and the icy crust. The giant exotic geysers recently photographed by Cassini, named Alexandria, Cairo, Baghdad, and Damascus, spew material at 2,500 km/h and reach a height of hundreds of kilometers above the small *Enceladus*, which is only 500 km in diameter. They probably contain water droplets. This discovery indicates that beneath the icy surface of this moon lies a subterranean ocean. (*Source* NASA/JPL-Caltech)

from the measurements of Galileo between 1995 and 2003 and also from the Europa Clipper mission, lead us to believe that in Europa, beneath its thick frozen surface, there is a vast ocean (Roberts et al. 2023, Fig. 2.30a).

The question then arises: is there life in Europa's underground salty ocean? According to one estimate, Europa's subsurface ocean, 60–150 km deep, holds twice as much water as all oceans combined on Earth! Because it disturbs the magnetic field of Jupiter, it is deduced that this water is conductive, meaning that it contains salt. This salty ocean is kept liquid by the heat released when Europa's interior is shaped by the successive kneading and successive compressions that bring the tidal forces of Jupiter and its other neighboring moons. Within this vast ocean, conditions are likely to exist which are friendly to life development. As it seems, there are the three prerequisites for the development of some forms of life: (a) abundant liquid water, (b) organic compounds, (c) energy, from the warm interior. So we might have hydro-thermal activity in the dark depths of Europa, where might be ecosystems of anaerobic organisms that feed on methane or hydrogen sulfide. A truly exotic possibility of developing life forms!

2005: *Saturn's* Mysterious Moon *Enceladus*, with Jets of Water from an Underground Ocean

We have a similar situation to *Europa* on *Saturn*'s moon, *Enceladus*, regarding the existence of an underground ocean. This small satellite *Enceladus* – seven times smaller than the Moon – one of the 82(!) moons of *Saturn*, is located in the region of its rings. Its surface is frozen, but what *Cassini* discovered while traversing space (2005–2017) is that in the night side of *Enceladus*' south pole, at a height of several tens of km, there are fountains of ice crystals that shoot to heights of 25 km from the icy surface of *Enceladus*, where *Cassini* passed! Just like in *Europa*, *Enceladus* ejects these jets from several km-long cracks in the icy surface of its south pole. Data from the *Cassini* mission confirmed that under a crust of ice 30–40 km thick there is a salty ocean about 10 km deep, that is, as deep as the deepest point of the earth's oceans. So, at the south pole of this moon there are fissures through which heat escapes from the center of *Enceladus*. As the warm water rises to the surface, it breaks the ice and it shoots out jets of ice crystals (Fig. 2.30c)!

According to Joshua Colwell of the University of Central Florida, who collaborated on the study with NASA and University of Colorado researchers, there are only three locations in the solar system where liquid water is known or suspected to exist near the surface: Earth, *Jupiter*'s satellite *Europa*, and *Saturn*'s satellite *Enceladus*.

2009: Molecular Hydrogen and Complex Organic Molecules in the Ocean of *Enceladus*

An extremely interesting result discovered by *Cassini* is that molecular hydrogen was detected within these jets, possibly a product of biological activity (Fig. 2.30c). Hydrogen is a source of chemical energy which favors the growth and survival of microbes in Earth's oceans (earthworms have been found near geothermal resources on the ocean floor) and something similar could be happening in Engeladus. Also, complex organic molecules more than ten times heavier than methane were detected, probably originating in chemical reactions between the rocky core of the satellite and the warm water of its subsurface ocean. Thus, *Enceladus* with the discovery of complex molecules emerging from its liquid ocean of water, it appears to be the only body in the solar system beyond Earth known to simultaneously satisfy all the basic requirements for life as we know it. It is therefore a particularly fertile environment for the development of life. It would be very interesting then if in the future a spaceship passes again through the jets of *Enceladus*, collecting more data, which will allow us to understand how these organic molecules are formed and to what extent these may have a biological origin, i.e., they may be related to some microorganisms.

2024: *Miranda*, an Ocean of Water Beneath its Surface?

Miranda is the smallest and innermost of *Uranus*'s five round moons (Oberon, Titania, Ariel, Umbriel) and orbits the planet around once every 1.4 Earth days. A new study (2024) suggests that *Miranda* might be hiding an ocean of water beneath its surface. The few images Voyager 2 captured in 1986 show Miranda's southern hemisphere (the only part that we've seen) which is a grooved terrain quartered off by rough scarps and cratered areas, like squares on a quilt. These bizarre structures are the result of tidal forces and heating within the moon. Hence, *Miranda* might be a similar case to *Enceladus*: it's comparable in size and composition to *Enceladus*, and may be actively releasing material into space, possibly through plumes. If true, it would make this distant moon one of only a few places in our solar system which could potentially support life.

2005: Hydrocarbon Lakes and a Murky, Orange-Tinged Atmosphere with Methane Showers on *Titan*

Titan, *Saturn*'s largest moon and the second largest moon in our solar system after *Jupiter*'s moon Ganymede (larger even than the planet *Mercury*), is the only body in our solar system except of the Earth, which is confirmed to have large liquid reserves in seas and lakes on its surface, but not from water but from hydrocarbons, that is, liquefied natural gas (e.g., methane). The basic ingredient that drives much of Titan's chemistry is methane, which gets split apart by sunlight or energetic electrons from Saturn's magnetosphere, and then recombines with other molecules to make substances like ethane (C_2H_6) along with more complex carbon-bearing molecules (Sotin et al. 2021).

Most of Titan's lakes and seas are located in the northern hemisphere and are likely replenished by an occasional rain of methane and ethane. Its dense atmosphere, denser than Earth's, due to the nitrogen and hydrocarbons it contains, has a characteristic cloudy orange color: it is an intriguing world cloaked in a yellowish, smoggy haze.

As on Earth we have the water cycle, so on *Titan* there is the cycle of methane, i.e., methane evaporation, clouds, and rain. But while its atmosphere, like the Earth's, contains mostly nitrogen, in a ratio of more than 95% (on Earth we have 78% nitrogen), it has little oxygen (which on Earth we have 21%) and instead contains plenty methane. The atmosphere of frigid Titan has a methane (CH_4) cycle, which evaporates from the surface and rises into the atmosphere, where it condenses to form methane clouds. Occasionally it falls as a chilly, oily rain onto a solid surface where water ice is hard as rocks. On Earth the lowest layer of the atmosphere, the *troposphere*, extends up to an altitude of about 12 km. However, on Titan, whose lower gravity allows the atmospheric layers to expand, the *troposphere* extends up to about 45 km.

Due to the great distance from the Sun, the temperature there is very low, but methane exists in liquid, gaseous, and solid form, like water on Earth, covering about 2% of *Titan*'s surface with liquid hydrocarbons. The triple point of methane (where solids, gases, and its liquid form) is –182 °C. Because on the surface of the lakes of *Titan* the temperature is –180 °C, methane on *Titan* plays the role that water has on Earth. There are three large seas near its north pole (the largest is the Kraken mare, as big as the Caspian sea, reaching a depth of 300 m, with the second largest the Lygeian sea, about 400 km wide and the third, the Punga mare, about 380 km in diameter), which are surrounded by dozens of smaller lakes, while in its southern hemisphere there is only Lake Ontario Lacus. From the radar of the *Cassini* spacecraft, which passed by the satellite in 2013, we learned that large canyons, about a kilometer wide, 240–570 m deep and steep slopes of more than 40°, are filled with methane and other hydrocarbons, before ending up in large methane lakes. The densest network of eight canyons ends into the 400 km long Vid Flumina river, which in turn empties into the methane lake Ligeia (Fig. 2.31). This is an extensive network of rivers ending in the area of *Titan*'s second largest lake, the *Lygean Sea*, which was named after the clear-voiced Siren of Lygea.[5] In some canyons, the level of liquid hydrocarbons is at the same level as Ligeia, while in others tens of meters higher.

In 2006, the *Saturn* system explorer *Cassini* sent to *Titan* the Huygens robotic spacecraft, which entered *Titan*'s atmosphere (1/14/2005) and landed, sending back

[5] The other Sirens were *Aglaope* (with a brilliant face), *Thelxiepia* (charming in speech), *Parthenope* (with a virginal face), *Nicosia* (with a white substance), *Molpi* (with a melodious voice), and *Thelxinoe* (with a beautiful mind).

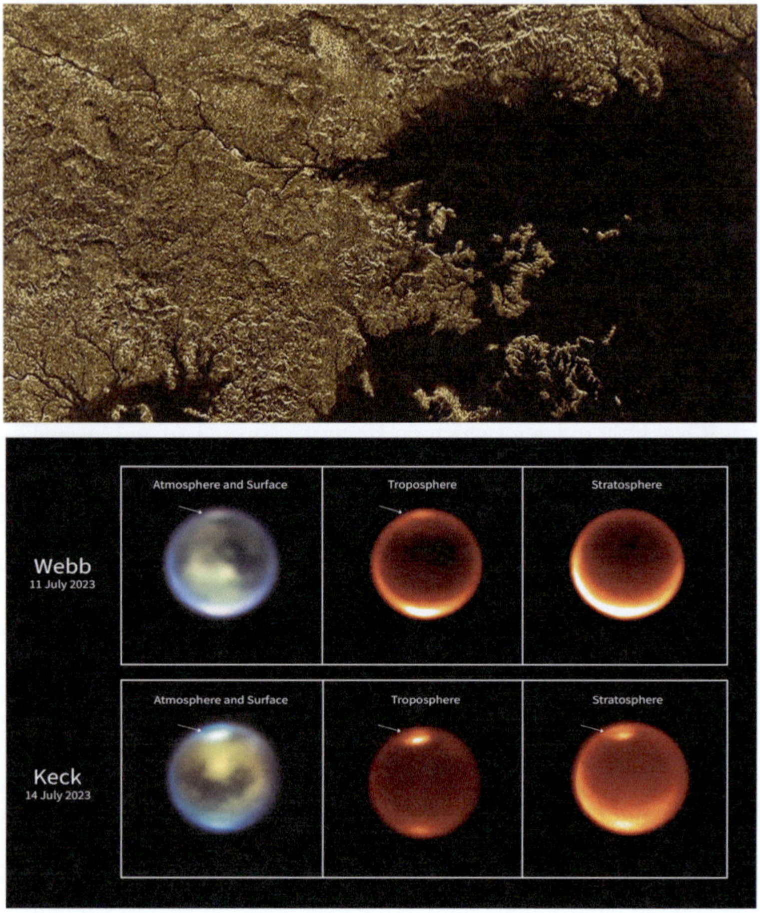

Fig. 2.31 *(top)* The gorge of the river Vid Flamina (center and slightly left) that is as long as the Nile and tributaries of, through which methane flows, before finally emptying into the Lygean lagoon, near the north pole of Saturn's moon *Titan*. The lower altitude areas are dark and correspond to liquid methane. (*Credit* NASA/JPL/ESA/ASI). *(bottom)* Recent images of Titan taken by the James Webb Space Telescope on July 11, 2023 (*top row*), and the ground-based W.M. Keck Observatories on July 14, 2023 (*bottom row*). They show methane clouds (denoted by the white arrows) appearing at different altitudes in Titan's northern hemisphere. On the left side are representative-color images from both telescopes. In the Webb image, light at 1.4 μm is colored blue, 1.5 μm is green, and 2.0 μm red (all in the NIR). In the Keck image light at 2.13 μm is colored blue, 2.12 μm is green, and 2.06 μm is red. The leftmost images, labeled "atmosphere and surface," show a mottled globe of brown and yellow with a hazy blue edge. In the middle column of the bottom panel, which labeled "troposphere" are shown single-wavelength images taken by Webb and Keck at 2.12 μm. This wavelength, sensitive to emission from Titan's lower troposphere, shows a dark orange globe with a brighter edge. The only features are bright spots near the top and bottom. The rightmost images are labeled "stratosphere" and also show a dark orange globe with a brighter edge [emission at 1.64 μm (Webb) and 2.17 μm (Keck)], which favor higher altitudes, in Titan's upper troposphere and stratosphere. It demonstrates that the clouds are seen at higher altitudes on July 14 than earlier on July 11, indicative of upward motion! (*Credit* NASA, ESA, CSA, STScI, Keck Observatory)

to Earth a wealth of impressive information. It was the first historical landing on a satellite of the major planets.

We believe that *Titan* is in a situation similar to that where the Earth was in its infancy, before plants appeared to enrich the atmosphere with oxygen. It should be noted that hydrocarbons are the basis for amino acids, which in turn are necessary for the creation of life.

The *Cassini* spacecraft spent 13 years orbiting its environment *Saturn*. Its measurements revolutionized our understanding of the *Saturn* system, including its moons, the radiation belts, the rings, but also in our understanding of where life can develop in our planetary system. For example, *Saturn* is known for its famous rings, of which Cassini sent us excellent pictures. In Dialynas et al. (2018) it is reported a new radiation belt on *Saturn*. When Cassini finally *ran out* of fuel, the control center on Earth guided it into an orbit in the upper atmosphere of *Saturn*, where it performed a *death dive* and was finally destroyed, on September 15, 2017, in order to avoid falling to one of *Saturn's* many satellites and infect it with Earth microbes!

Recently (2025) the James Webb Space telescope and the Keck II telescope found evidence of cloud convection on Titan.

The next ambitious exploration program of *Titan* will be done with the robot *Dragon fly*.

2015: *New Horizons* (Flying Above the Icy *Heart* of the Outermost Dwarf Planet Pluto)

After the exploration of the systems of the giant gas planets *Jupiter* and *Saturn* and the giant icy planets *Uranus* and *Neptune*, by the *Voyager 2* spacecraft and the *Saturn* system by *Voyager 1* and *Cassini*, the last remaining target to explore was the dwarf planet Pluto (Fig. 2.32) and the Kuiper belt of the solar system which contains the remnants of the creation of the planets.

This was undertaken by the *New Horizons* mission, led by T. Krimigis (Fig. 2.33), a mission which costed about 700 million dollars. The weight of the spacecraft was 478 kg and was launched on an Atlas-5 rocket on January 19, 2006. The *New Horizons* mission departed in 2006 and reached Pluto in 9.5 years, in 2015, at a distance of about 40 AU. This is a relatively short duration and was achieved thanks to the gravity assist it received during the passage from *Jupiter*, which increased the speed of the spacecraft to 53,000 km per hour. However, because of this great speed, the spacecraft was able to study Pluto for only 24 h, sending nevertheless to Earth a unique material with the secrets of this planet, from a distance about 12,000 km above its surface. At that time, Pluto was at a distance of 5 billion km, i.e., an electromagnetic signal to come to Earth was travelling for about 5 h. These photos showed us that one side of the planet's surface is dominated by a giant bright heart-shaped plain, 1,600 km wide. This consists of frozen nitrogen and is surrounded by high mountains 3 km high, consisting of hard water ice, since the temperature there is about –230 °C. These

Fig. 2.32 *Pluto*'s image taken by *New Horizons* on July 14, 2015, from a distance of 35,445 km. The striking features on *Pluto* are clearly visible, including the bright expanse of *Pluto*'s icy, nitrogen-and-methane rich "heart," Sputnik Planitia. (*Source* NASA/HST and NASA/New Horizons)

mountains seem to float on sheets of frozen nitrogen. There are almost no craters on the surface of Pluto, an indication that there is some geological activity that renews the surface and which possibly is caused by energy production within the planet's interior from the decay of radioactive elements.

A similar picture of the planet's satellite, Charon was obtained. *New Horizons*' brief flyby of *Pluto* and five known of its moons – at increasing distance from the planet: *Charon, Styx, Nix*, *Kerberos, and Hydra* – gave us a first "taste" of the *Kuiper Belt*, in which many bodies are found, from dwarf planets to space "icebergs."

Analyzing the evidence that has emerged from the mission of *New Horizons*, many of the planet's first secrets began to be deciphered, including that *Pluto*'s surface is covered with a layer of frozen nitrogen with traces of methane, which is sublimated when the planet is located closer to the Sun, thus forming a thin and rudimentary atmosphere around this celestial body. However, due to *Pluto*'s eccentric orbit, when it moves away from the Sun, its atmosphere gradually freezes and falls on its surface like snow. Therefore, on the surface of *Pluto* there are nitrogen ice, methane ice, and water ice. Its mountains reach a height of 3 km and are mainly made of hard water ice.

New Horizons also sent back evidence of several deep cracks on the surface of *Pluto*, the largest of which is 322 km long. The top of the fissure is approximately 4 km higher than its base – more than twice the height of the walls of the Grand Canyon of the USA.

Pluto, which takes 248 Earth years to complete one orbit around the Sun, has active geological activity, possibly fueled by the heat generated by the release of radioactive elements from rocks in its subsurface. Perhaps because of this, its surface ices appear very soft, something unique to our solar system.

Fig. 2.33 Dr Stamatis (Tom) Krimigis (born in Vrontados, Chios, Greece, studied physics at the University of Minnesota and at the University of Iowa with PhD advisor James Van Allen)[6] has played an important role in the development of new scientific instruments for space and the promotion of robotics in space exploration. He is the only scientist in the world, who has led or has participated in space physics experiments that have been explored all nine planets of our solar system. He was the principal investigator on five NASA missions, including the two Voyagers and the Cassini missions to Saturn. Specifically, he designed, developed, sent into space, and analyzed the data from 23 scientific instruments, used in various American and European space missions, such as the Low-Energy Charged Particle (LECP) instrument on the two Voyagers. Already, the LECP data from *Voyager 1* have been instrumental in confirming that the spacecraft has left our solar system for the first time, moving now to outer space. He has been involved with the *Voyagers* 1 and 2 missions since 1971. (*Source* Johns Hopkins University/Applied Physics Laboratory)

On July 15, 2015, a date in which *New Horizons* explored *Pluto*, it was exactly 50 years since July 15, 1965, when the *Mariner* 4 spacecraft reached *Mars* and sent us the first photos of this planet. Thus, in these 50 years, the exploration of the nine planets and other bodies of our planetary system was completed.

[6] T. Krimigis participated in establishing NASA's Discovery Program of low-cost planetary missions, as well as the New Frontiers program, for which the APL-built New Horizons to Pluto was the first mission. He is also a co-investigator on the Parker Solar Probe, co-investigator for LAN/HI-SCALE on Ulysses solar polar orbiter, EPIC on GEOTAIL, EDP for Galileo mission, TRD on Mariner 3, and for the LECR on Mariner 4. T. Krimigis has also worked on the Cassini, Advanced Composition Explorer, the Mariner 5, MESSENGER and New Horizons programs.

In 1986, T. Krimigis briefed Ronald Reagan on the AMPTE mission. He also met with Mikhail Gorbachev in 1987 and George W. Bush in 1990 to discuss space exploration.

In 2016, T. Krimigis received NASA's highest service honor, the NASA Distinguished Public Service Medal, for his lifelong efforts to advance space exploration and science. In 1999, the International Astronomical Union named the asteroid 8323 Krimigis in his honor. He is Head Emeritus of the Space Department at the Johns Hopkins Applied Physics Laboratory in Laurel, Maryland.

Presently, T. Krimigis holds the Chair of Science of Space at the Academy of Athens, Greece

Meanwhile, the *New Horizons* spacecraft continues its journey, heading toward the Kuiper belt (Fig. 2.2), now being at a distance more than five billion kilometers from Earth. The spacecraft's new target is an unusually shaped celestial body, 2014MU69, named Arrokoth, 35 km long, which consists of 2 stuck planetesimals, 22 km in diameter and 14 km in diameter, joined in their major dimension.

2012: *Voyager* 1 and 2's Epic Exit from the Heliosphere

As we mention, the *Voyager* 1 spacecraft began its journey on September 5, 1977. After almost 35 years, on August 25, 2012, it crossed the *Heliopause* and entered interstellar space (Fig. 2.34). This passage of the *Heliosphere* was made, according to NASA, when the spacecraft was at a distance of 122 AU (AU) and its signals from there reached Earth after about 17 h. It will take about 300 years for *Voyager* 1 to reach the inner edge of the Oort cloud and possibly about 30,000 years to fly beyond it (Fig. 2.2). It is the first human construct ever to enter interstellar space!

In June, 2025, some 48 years since its launch, the *Voyager 1* spacecraft is about 167 AU from the Sun, and its electromagnetic signals reach Earth after 23 h and 6 min. *Voyager 2*, on the other hand, is currently about 140 AU from the Sun, and its electromagnetic signals reach Earth after about 19 h.

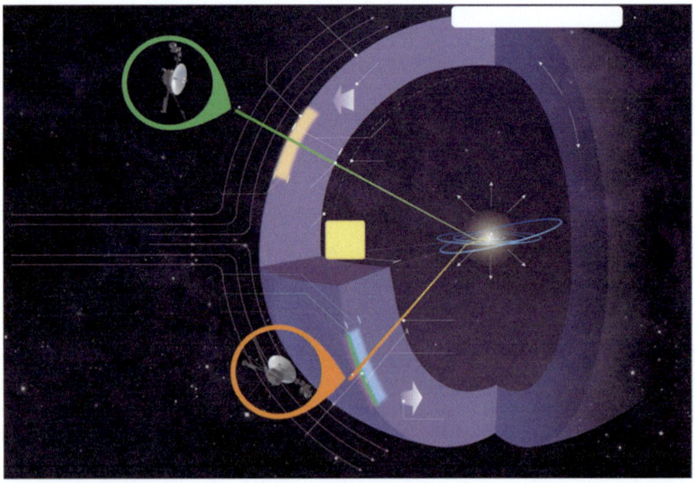

Fig. 2.34 Global heliosphere summarizing the findings of Voyager 1 (V1) and Voyager 2 (V2). A sectioned view of the global heliosphere depicts the regions traversed by the 2 spacecraft, such as the boundaries of the terminal shock (TS) and the Heliopause (HP), with V1 north and V2 south of the ecliptic plane. The areas highlighted in the figure are discussed in the publication Krimigis et al. (2019). The angle θ is the meridional angular distance between V1 and V2. (*Source* Image kindly provided by Krimigis et al. (2019))

Of the 10 instruments on *Voyager 1* and *Voyager 2*, 4 are operational: the Low-Energy Charged Particles (LECPs) detector, the Cosmic Ray Subsystem (CRS), the wave detection instrument (Plasma Wave Subsystem – PWS), and the Magnetometer. The remaining six instruments are no longer functional. *Voyagers 1* and *2, as they* move away from the Sun at about 61,200 km/h and 55,347 km/h, respectively, have enough electricity and fuel to sustain their current suite of scientific instruments for a few more years.

1990: A Bright Blue Gem, from the Far Away Distance of *Pluto*

On February 14, 1990, *Voyager 1* was at the far distance of 6.5 billion km, in the sunless realm of *Pluto*, at the entrance of Hades. There, the outermost planet of our solar system, *Pluto* accompanies its 5 moons, which have the names of mythical monsters: the ferryman of Hades Charon, the great river of the underworld Styx, the daughter of Chaos Night, the wild dog with 50 heads Cerberus, and the many-headed Hydra. Darkness and frost reign there!

Before *Voyager 1* leaves the Heliosphere for good and heads toward the unknown interstellar space, casted one last nostalgic glance behind it. From this far distance, its camera captured in the vastness of Space a *tiny bright blue gem*, mingling alone in a vast ocean of the other planets and stars. It is our Earth, with its sun-drenched mountain peaks, crystal-clear shores, and verdant plains with the amazing biodiversity of the plant and animal kingdoms and all of us, the roughly 8 billion of citizens! Light and warmth reign here!

In the distant photo of our Earth in Fig. 2.35, our planet appears to be hovering helpless and fragile, alone in Space!

Fig. 2.35 Earth as viewed from NASA's Cassini spacecraft from 1,445 million km (9.6 AU) away. (*Image Credit* NASA/JPL-Caltech/Space Science Institute)

1977 to Present: the *Voyagers*, World Travelers

Voyager 1 is moving away at a speed of 17 km per second, or, about 61,000 km per hour, or, 3.6 AU per year. *Voyager* 1 is headed to the star AC + 79 3888 (Gliese 445), a spectral type M red dwarf, with a mass and radius equal to 1/3 of those of the Sun, a member of the Giraffe constellation, located near the Polaris star and which moves toward the constellation Ophiuchus (Fig. 2.2). Today Gliese 445 is 17.6 light years from the Sun. However, since it is rapidly approaching the Sun, after a few tens of thousands of years it will be at a distance of about 3.45 light years from the Sun, closer than is today Proxima Centauri which is now 4.36 light years away, but will have moved away by then farther from the Sun than Gliese 445. So *Voyager 1* aims to pass very close to Gliese 445. Specifically in about 40,000 years *Voyager* 1 will be at a distance of 1.6 l.y. from our nearest then star, Gliese 445.

Also, in about 40,000 years, *Voyager 2* will pass 1.7 light years from the star Ross 248, and in about 296,000 years, it will pass within 4.3 light years from Sirius, the brightest star in the sky. We should have patience until then! After these stellar encounters, our cosmic travelers are destined – perhaps eternally – to wander in the Galaxy.

2012: Cheers, Whoever You Are. We Come in Peace, as Friends to Those Who Consider as Friends

With this greeting, recorded on a gilded copper disc embedded in the spaceships *Voyager 1* and *Voyager 2*, humanity will greet in ancient Greek, among other languages, some extraterrestrial life form, which perhaps the two spaceships may meet during their long journey in interstellar space, which began 40+ years ago (Figs. 2.2 and 2.22).

Thus, on the discs of the two *Voyagers* have been recorded 115 images, natural sounds, like the rustle of the wind, the sound of the rain, the chirping of birds, the call of whales and also music from our planet – from Mozart's Magic Hall to Chuck Berry's Johnny B. Goode – and also a radioactive source, which will enable those who may discover them to calculate the time elapsed since the launch of the two *Voyagers*, as well also Earth's position in our Solar System and the Milky Way.

Could either of these two spacecraft be spotted by an alien civilization, assuming of course that there is really one in our Galaxy? Unknown! This applies on the condition that the two Voyagers will manage to survive the inhospitable environment of interstellar space surrounding the Heliosphere. With this disc, one intelligent extraterrestrial life form will be informed about the diversity of life and the civilization on planet Earth.

Mankind's most distant and long-lived spacecrafts, *Voyager* 1 and 2, have completed more than 48 years of operation and exploration. Despite their huge distance, they continue to communicate with NASA every day, exploring even further distances into interstellar space.

Their story has influenced not only current and future generations of scientists and engineers, but also Earth's culture, including film, art, and music. Given that these spaceships could survive for billions of years and, who knows, they could one day in the future be the only remaining traces of human civilization.

References

Bell JF et al (2025) The psyche multispectral imager investigation: characterizing the geology, topography, and multispectral properties of a metal-rich world. Space Sci Rev 221(4):47

Dialynas K, Krimigis SM, Decker, Robert B, Mitchell DG (2019) Plasma pressures in the Heliosheath from Cassini ENA and voyager 2 measurements: Validation by the voyager 2 heliopause crossing. Geophys Res Lett 46(14):7911–7919

Krimigis SM, Decker RB, Roelof EC, Hill ME, Bostrom CO, Dialynas K, Gloeckler G, Hamilton DC, Keath EP, Lanzerotti LJ (2019) Energetic charged particle measurements from Voyager 2 at the heliopause and beyond. Nat Astron 3:997–1006

Lamy L, Prangé R, Hansen KC et al (2012) Earth-based detection of Uranus' aurorae. Geophys Res Lett 39(7). https://doi.org/10.1029/2012GL051312

Lamy L, Prangé R, Berthier J et al (2025) A new rotation period and longitude system for Uranus. Nat Astron. https://doi.org/10.1038/s41550-025-02492-z

Melin H, Moore L, Fletcher LN et al (2025) Discovery of and infrared aurorae at Neptune with JWST. Nat Astron. https://doi.org/10.1038/s41550-025-02507-9

Roberts JH et al (2023) Exploring the interior of Europa with the Europa Clipper. Space Sci Rev 219(6):46

Sotin C, Kalousová K, Tobie G (2021) Titan's interior structure and dynamics after the Cassini-Huygens mission. Annual Review of Earth and Planetary Sciences, 49 (1), pp. 579–607

Stratman PW, Showman AP, Dowling TE, Sromovsky LA (2001) EPIC simulations of bright companions to neptune's great dark spots. Icarus 151(2):275–285

Tsinganos K (2021) Modern theoretical mechanics with 200 solved problems and exercises. Unibooks

Chapter 3
Spectacular Star and Planet Formation Inside Dark Nebulae

Protostellar jets are the *ex machina Deus* in star formation! The term originates in Greek tragedy. At the climax of the tragedy, when the plot of the drama seems to be leading to a dead end, an actor suddenly appears on the scene, with the help of some mechanism, impersonating some deity or divine person, to solve the impasse.

Similarly, jets from newborn stars by stealing angular momentum they leave gravity free to create the star. For that reason, we might not be here today if jets did not accompany star formation, since the natural law of conservation of the disk angular momentum would not allow the collapse of the material for the formation of the protosun and subsequently of Earth and the other planets from the remaining material.

(Among the conclusions of the research of the author's group on MHD winds and jets).

Cosmic Star Formation

Star formation, in contrast to the subsequent mature life and inevitable final death of stars, occurs relatively fast and within small dimensions, which are not easily detectable in the large astronomical distances at which they are located. Furthermore, newborn stars and their planets are difficult to be detected because they are dim and hidden within the dark parent nebulae wherein they were created and from which eventually they emerge. But when stars enter their adult phase of life, they are very bright and easily detectable, even at great galactic distances. Also, when they end their lives in the form of red supergiants and planetary nebulae, or, when they explode as spectacular supernovae, they are detectable from afar. Hence, only their birth, in the dark matrix of opaque nebulae, goes relatively unnoticed. For this reason, the first part of the *trilogy: early birth – mature life – final death* of stars is the last one that has become accessible, observable, and understandable to us, in the last decades, thanks to advances in telescope related technology and the general progress of modern astrophysics. A key role, for example, was played by observations after 1990 with

K. Tsinganos, *A Century of Cosmic Discoveries*, Springer Praxis Books,
https://doi.org/10.1007/978-3-032-06282-6_3

Fig. 3.1 Composite image of HD 163296 shows two jets detected by MUSE, superimposed to the disk detected by ALMA at 1.2 mm (Andrews et al. 2018; Isella et al. (2018). The central star in this image of HD 163296 is still forming, but seen already surrounded by a rotating disk and outward moving jets . The disk is shown in radio waves taken by ALMA , and shows gaps likely created by the gravity of very-young planets. The jets, shown in visible light taken by the Very Large Telescope, expel fast-moving gas – mostly hydrogen – from the disk center. The system spans hundreds of AU. Details of these new observations are being interpreted to bolster conjectures that the jets are generated and shaped, by magnetic fields in the rotating disk. (*Image Credit*Visible: VLT/MUSE (ESO), Xie et al. (2020); Radio: ALMA Atacama Large Millimeter Array (ALMA) (ESO/NAOJ/NRAO))

the Hubble and Spitzer space telescopes of star-forming regions in molecular nebulae of the Milky Way and now with the James Webb Space Telescope, Fig. 3.2a,b,c).

The theoretical understanding of the formation and evolution of stars ranks highly among the marvelous achievements of the twentieth-century astronomy (Ray 1998, 2001). This area of theoretical astrophysics has managed to explain the formation of a star, starting from interstellar matter having a density of about 1 molecule per cubic centimeter and reaching a density increase by 24 orders of magnitude, i.e., to a density of about 10^{24} molecules/cm^3, the stellar density. This increase in density by 24 orders of magnitude corresponds to a shrinking of distances by a factor of 100 million. In the following, we pick up Ariadne's thread from the beginning, trying to answer the question of star and planet formation in the Universe from the point of view of observations and then discuss the physical processes. We shall briefly discuss how and by what natural processes some "*dark and spidery corners of the sky*" can come to life as a physical system and give a self-sustaining parent star, which will then give its light to those roaming around planets, some of which may then be teeming with life and to understand their existence and their place in the rest of the Universe.

Figure 3.2b, c shows the Sculptor and Sombrero nearby starburst Galaxies. The Sculptor Galaxy in the constellation *Sculptor* is one of the closest (D ~ 11 Mly) massive ($M_{star} \sim 4.4 \times 10^{10}\ M_{\odot}$) star-forming galaxies to the Milky Way, Congiu et al.

a

b

JAMES WEBB SPACE TELESCOPE
SOMBRERO GALAXY | M104
8,500 LIGHT-YEARS
N
E
MIRI Filters
F1130W
F1280W

◀**Fig. 3.2** **a** (*top*) The dazzling young star cluster NGC 346, in the southern constellation of Tucana, home to more than 2500 newborn stars, a prolific star factory in the Small Magellanic Cloud (SMC), at infrared, optical, and ultraviolet wavelengths with the Hubble Space Telescope. This open star cluster is located near the center of the brightest H II region in the SMC, N66. The cluster's most massive stars, which are many times more massive than our Sun, blaze with an intense blue light in this image, behaving like stellar sculptors, carving out a bubble from the nebula N66. Those NGC 346's hot, massive stars produce intense radiation and fierce stellar winds that pummel the billowing gas of their birthplace and begin to disperse the surrounding nebula. The glowing pink nebula and snakelike dark clouds are the remnant of the birth site of the stars in the cluster. This nebula N66, is the brightest example of an H II region in the SMC, which is set aglow by ultraviolet light from the hot young stars. The presence of the brilliant nebula indicates the young age of the star cluster, as an H II region shines only as long as the stars that power it, a mere few million years for the massive stars pictured here. The cluster is still engaged in high-mass star formation and its star formation rate is estimated at $(4 \pm 1) \times 10^{-3}\ M_{\odot}\mathrm{yr}^{-1}$. Note that the SMC is less rich in elements heavier than helium than the Milky Way, making conditions in the SMC similar to what existed in the early Universe. (Image credit: ESA/Hubble and NASA, A. Nota, P. Massey, E. Sabbi, C. Murray, M. Zamani (ESA/Hubble)). **b** (*middle*) This image shows the Sculptor Galaxy with ESO's VLT in specific wavelengths of light released by hydrogen, nitrogen, sulfur and oxygen. These elements exist in gas form all over the galaxy, but the mechanisms causing this gas to glow can vary throughout the galaxy. We can see the full structure of the galaxy disk, with the prominent bar and complex dust filaments that follow the distribution of the spiral arms of the galaxy. The pink light represents gas excited by the radiation of newborn stars, while the cone of whiter light at the center is caused by an outflow of gas from the black hole at the galaxy's core. (*Credit* ESO/Congiu et al. (2025)). (*bottom*) The Sombrero galaxy (M104), is shown in a new mid-infrared image from the NASA/ESA/CSA James Webb Space Telescope. A smooth inner disk and details of the galaxy's outer ring are providing insights into how the dust, an essential building block for astronomical objects in the Universe, is distributed. The clumpy nature of the dust, where MIRI detects carbon-containing molecules of polycyclic aromatic hydrocarbons, can indicate the presence of young star-forming regions. These rings of the Sombrero galaxy produce less than one solar mass of stars per year, in comparison to the Milky Way's roughly two solar masses a year. (*Image credit* NASA, ESA, CSA, STScI)

(2025). Its star formation rate has been estimated to be between ~4.9 and ~6.5 $M_{\odot}$ yr^{-1}. Almost a third of it ($\sim 2\ M_{\odot}\ \mathrm{yr}^{-1}$) is produced by 10 giant molecular clouds distributed in a starburst ring ($\sim$ 500 pc in diameter) around its nucleus, making this object the closest massive starburst to our Galaxy. In addition, the starburst is responsible for launching a powerful multi-phase that is ejecting large amounts of gas (14–39 $M_{\odot}\ \mathrm{yr}^{-1}$ for the molecular component only, from the center of the galaxy into its circumgalactic medium. Part of this gas seems to remain in the gravitational well of the galaxy, where it cools down to the neutral phase and moves toward the outskirts of the disk, potentially reaccreting and fueling future star formation, Congiu et al. (2025).

The Sombrero galaxy (M104), at a distance of about 31 million light years from Earth in the constellation *Virgo*, strung out from the southern edge of the *Virgo Supercluster* is shown in a new mid-infrared image from the NASA/ESA/CSA James Webb Space Telescope (Fig. 3.2b). The clumpy nature of the dust, where MIRI detects carbon-containing molecules of polycyclic aromatic hydrocarbons, can indicate the presence of young star-forming regions. These rings of the Sombrero galaxy produce less than one solar mass of stars per year, in comparison to the Milky Way's roughly two solar masses a year.

The cosmic star formation rate density – star formation occurring per unit time and volume at a given epoch – peaked approximately 3.5 billion years after the Big Bang, at $z \approx 1.9$, and declined exponentially at later times, with an e-folding timescale of 3.9 billion yrs. Half of the stellar mass observed today was formed before a redshift $z = 1.3$. About 25% formed before the peak of the cosmic star formation rate density, and another 25% formed after $z = 0.7$. Less than ~1% of today's stars formed during the epoch of reionization (Madau and Dickinson 2014).

1784: Observations on the Genesis of Stars and Protoplanets from the ISM

In 1610, when Galileo pointed the first then available telescope to the sky, he immediately noticed that the Milky Way is made up of stars. At the end of this century (1700), a systematic observation of stars took place when William Herschel built several telescopes. In the early 1800s, the first theories of how stars formed were proposed, noticeably, Kant's and Laplace's theories of the role of the parent proto-nebula. In the late 1800s, theories of how stars shine were proposed – such as Kelvin's and Helmholtz's theory of the production of stellar energy through gravitational contraction. At the beginning of the 20th century, prominent astrophysicists of that era, such as Eddington, Jeans, Chandrasekhar and Einstein, dealt with the theoretical question of the internal structure of stars. However, only at the end of the 29th century took place the systematic *observation* of the formation of stars, as many ground and space telescopes became available.

So let's take a comprehensive look at what astronomical observation has given us in recent decades, regarding star formation. The well observed and relatively nearby to Earth *Orion* nebula is perhaps the most representative star formation region. Along with the dark *Horse's Head* Nebula, they appear very nicely in images with a filter permeable to the Hα red emission line of the hydrogen atom. Observations of the nebula in the infrared with the Spitzer space observatory revealed about 1,000 newborn stars, a few million years old (Fig. 3.15). At the same time, huge disks of gas (Figs. 3.1 and 3.16), hundreds to thousands of AU in size, appear swirling around most newborn stars providing a strong indication that star formation is a fairly common process in our galaxy.

Star formation occurs when a giant molecular nebula, hundreds of light years across, collapses under its own weight; this cloud contains 90% hydrogen atoms, a little less than 10% helium atoms, and a very small percentage of the heavier elements, as well as water, ammonia, organic compounds, etc. Due to the law of conservation of angular momentum, as the nebula collapses, it rotates faster and faster. Then, like an initially spherical piece of pizza dough spun in the air by a chef expands and takes the shape of a flat pizza, so the rapidly spinning nebula turns into a lenticular disk. At the center of the disk is the protostar, which only after a few million years and continuous accretion of matter from its disk acquires in its center the right conditions (high temperature and density) suitable for the initiation of thermonuclear reactions.

Our knowledge of the various phases of the lives of stars, from their birth until their death are based on theoretical models, the predictions of which have been generally confirmed by the observational data. According to the prevailing models, stars are created by the gravitational collapse of interstellar clouds composed mainly of hydrogen. The cloud begins to contract, increasing its density. With the increase in pressure and temperature inside it, when the temperature reaches some millions of degrees, the thermonuclear fusion reactions of hydrogen into helium begin to take place in the core of the protostar. The released energy causes a dramatic increase in the internal pressure of the protostar, which compensates for its initial gravitational collapse. Thus, the initial contraction stops and a state of dynamical equilibrium is created. Then, we say that a new star has been born!

In the last decades, there have been published many papers with observations, modeling and laboratory experiments of protostellar outflows, jets and their associated accretion disks. The interested reader may see, e.g., the proceeding volumes of various conferences organized by the four-year EC Marie Curie Research Training Network JETSET (*Jet Simulations, Experiments and Theory*, JETSET) which focused on plasma jet studies, particularly in the context of star formation. JETSET aimed to integrate astrophysical observations, theoretical modeling, laboratory experiments, and advanced computational techniques, creating a synergy among diverse European academic institutions. The network's objectives included the development of sophisticated theoretical jet models, high-resolution observational techniques, diagnostic tools for data comparison, and experimental validation, thereby bridging the gap between application-focused practitioners and theorists in the field. Proceedings of various JETSET conferences have been published by Springer Nature as follows:

- *Jets from young stars I. Models and Constraints,* Ferreira et al. (2007).
- *Jets from young stars II. Clues from High angular resolutions observations,* Bacciotti et al. (2008).
- *Jets from young stars III. Numerical MHD and instabilities,* Massaglia et al. (2008).
- *Jets from young stars IV. From models to observations and experiments,* Gracia and Ferreira (2010).
- *Jets from young stars V. High performance computing and applications,* Gracia et al. (2009a).
- *Protostellar Jets in Context,* Tsinganos et al. (2009).
- *Jet Simulations experiments and Theory. Ten years after Jetset: What is next?* Sauty (2019).

1784: Holes in the Sky

On a clear night in 1784, the German-born British astronomer, telescope maker and musician William Herschel (Fig. 3.3) observed with his telescope a region of the sky in the Scorpius constellation. Next to him stood his sister Carolina, who was taking

Fig. 3.3 Sir William Herschel (1738–1822). Portrait by Lemuel Francis Abbott (1785)

notes of the observations. Some moment, in the still night, Herschel turns abruptly and says surprised to Carolina:

Hier ist wahrhaftig ein Loch im Himmel!

(Here is clearly visible a hole in the sky!)

At that moment the dark or molecular nebulae, as we call them today, had been discovered! Herschel published this observation in the journal *Philosophical Transactions*, but did not pay much attention to his discovery, because he thought he had discovered nothing more than *a hole in the sky*. Instead, Carolina attached more importance to this observation and continued them with the help of her nephew John Herschel, creating the first catalog of dark nebulae.

1902: Gravitational Jeans Instability

Assume such a giant cloud of molecular hydrogen, in which no star has yet formed and is in equilibrium in some region of the spirals of our Milky Way galaxy. We also assume that this cloud is initially located there in dynamic equilibrium and at some point in time the nebula is somehow disturbed. For example, the well-known explosions of nearby supernovas can act as sources of such disturbances, which with the strong shock waves they produce can *shake* the gas enough to cause its collapse, as shall see in the following. Also, the same result caused by hot O- and B-type stars, which emit quite strong winds, that can compress and disrupt the interstellar cloud. After all, such young stars may have already been formed and exist within the same cloud, causing thus its further collapse. A third source of disturbances is the AGB stars of the Asymptotic Giant Branch. These are low-to intermediate-mass (0.6–10 $M_{\odot}$) stars in the asymptotic giant branch phase, a region of the Hertzsprung–Russell diagram populated by evolved cool and luminous stars. This is a phase of

Fig. 3.4 Sir James Hopwood Jeans (1877–1846)

stellar evolution that pass all-stars toward the end in their life. During this phase, stars usually have a variable long-period luminosity and lose mass at a high rate in the form of strong stellar winds. They also produce thermal pulses with even greater mass loss that can lead to ejected large circumstellar mass shells. With this process a star can loose 50–70% of its mass in the AGB phase.

Therefore, due to such external disturbances, concentrations and dilutions are created within the nebula. Suppose that the distance between them is smaller than a characteristic value, which we will call the Jeans length (λ_J) – in honor of the English physicist and astronomer James Jeans (Fig. 3.4), who initially studied this phenomenon (Jeans 1902). Then, each density inhomogeneity is surrounded by many others similar inhomogeneities, so that the enhanced inward gravitational force is compensated by the outward gravitational forces of the neighboring densities. In this case, the disturbance cannot lead to collapse of this cloud. On the other hand, assume that the distance between these density enhancements and dilutions is greater than the characteristic Jeans value, λ_J. Then, because the neighboring dense regions are far enough from each other, the inward gravitational force is not compensated by the outward gravitational forces of the neighboring density inhomogeneities. In this case, the external disturbance leads to further collapse of the particular region. We then say that the nebular gas is unstable to the *Jeans gravitational instability*. Thus, the Jeans gravitational instability can lead to further fragmentation of the original cloud under certain pre-conditions. The product of the cloud density and its volume $(4\pi/3)\lambda_J{}^3$ defines the *critical Jeans mass*, M_J, which depends on the temperature and density of the cloud. When the mass of a cloud is much smaller than the characteristic Jeans mass, M_J, the cloud is gravitationally stable, i.e., it does not collapse due to its own gravity. For standard densities and temperatures of dense and cold molecular nebulae, the Jeans mass M_J is a few solar masses. Thus, molecular nebulae are not homogeneous at all. They are "*clumpy,*" that is, they are filled with condensation nuclei, i.e., regions which are considerably denser and cooler than their environment. These regions are the dark clouds, the Bok globules that we discussed earlier and which are believed to be the sites of star formation through further gravitational collapse.

1919: Discovery of the Dark Molecular Nebulae

The fact that dark nebulae were not just holes in the distribution of stars in the sky, but dense areas that prevented us from seeing the stars behind them, was only understood at the beginning of the twentieth century by the astronomers Edward Barnard (1857–1923, Lick Observatory, USA, Fig. 3.5), Max Wolf (1863–1932, Heidelberg Observatory, Germany), etc. The Lick Observatory, where Barnard made his observations, belongs to the University of California. It is located on the top of Mount Hamilton, in the "*Devil's Range*", east and a short distance from the city of San Jose, in California (USA). In 1919, Barnard with observations from the Lick Observatory published the first catalog of dark nebulae (Barnard Catalog of Dark Markings in the Sky, or Barnard Catalog) which contained 182 dark nebulae, while a newer version in 1927 contained 369.

Later (1947), Bart Bok pointed out that the dense regions within the dark nebulae, the *Bok globules* as they are called today, they were at a stage of collapse and they could form stars. Representative such nearby dark nebulae are Fig. 3.6.

- The Snake nebula (Barnard 72) in the Ophiuchus constellation, at a distance of about 650 l.y. in Ophiuchus (Fig. 3.6). Laocoon was, according to mythology,

Fig. 3.5 (from left) Bart Bok (1906–1983, Harvard College Observatory), Max Wolf (1863–1932, Heidelberg Observatory) and Edward Emerson Barnard (1857–1923, Lick Observatory). [From Wikipedia]

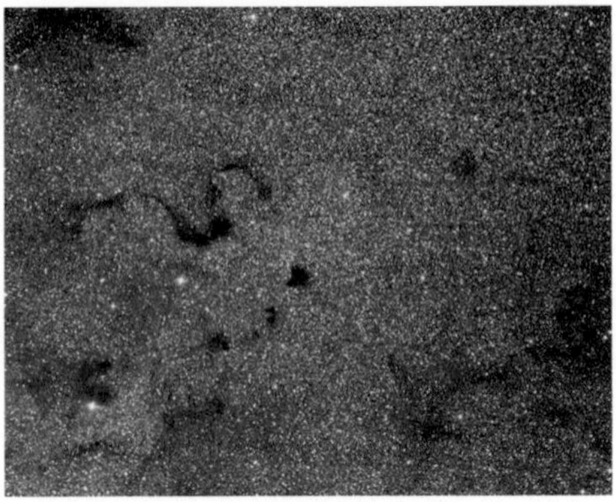

Fig. 3.6 Snake Nebula (Barnard 72) in the Ophiuchus constellation, about 6500 light years away. (*Source* Wikipedia)

the priest of Apollo,[1] who warned Trojans not to put the Trojan horse within their walls. Neptune sent then some sea reptiles to shut him up. The Laocoon complex has inspired many artists whose works adorn many museums. The famous complex of Laocoon's statue, which was placed in the palace of the Roman emperor Titus and now is exhibited at the Vatican Museum, depicts the Trojan (Thymbraian) priest of Apollo, Laocoon, and his sons, Antiphantes and Thymbraeus, being attacked by sea serpents, a work attributed to the Rhodian sculptors Agesander, Athanadorus and Polydorus.

- The "Duck nebula" (Duck nebula, Barnard 163), in the constellation of Cepheus, king of Ethiopia and father of Andromeda, in the distance about 2,500 l.y. (Fig. 3.14).
- The "E" Nebula (Barnard 142 and 143) in the constellation Aquila, a pair of dark nebulae about 2,000 AU away, with a diameter equal to half the apparent diameter of the Full Moon. The name Eagle is due to the mythological bird of Zeus. In this constellation, the ancient Greeks imagined that they saw an eagle flying and carrying Ganymede with its claws to Zeus to serve him as a cupbearer.

[1] Laocoon was the son of the king of Troy Priam, who became famous for warning the Trojans, in vain, against accepting the Trojan horse from the Greeks when they were besieging Troy. Virgil describes this event in *The Aeneid* (Book II, 48–49) and devotes Laocoon the famous line "Equo ne credite, Teucri ! Quidquid id est, timeo Danaos et dona ferentes," i.e., "Do not trust the horse, Trojans. Whatever it is, I fear the Greeks even bearing gifts." This gave birth to the famous line, "Beware of Greeks bearing gifts." The Trojans, however, cast his thoughts aside and saw the horse as a gift of Apollo and worthy of worship. In a fit of rage, Laocoon thrust his spear into the side of the horse. Even though it hit some Greek soldiers hiding in the belly of the horse, no one heard the moans and finally, they brought the horse in. This eventually led to Troy's defeat. Poseidon, who was supporting the Greeks was angered by Laocoon, and sent sea serpents to strangle him and his sons. This name was given in 1978 to the asteroid 3240 Laocoon, a member of the Trojan group of asteroids.

Fig. 3.7 In the center of the Eagle Nebula (Messier 16 or NGC 6611), are the so-called *Pillars of Creation*, left in the visible and middle in the infrared taken by the Hubble Space Telescope in 2014, and right in the near-infrared taken by the James Webb Space Telescope in 2022. The colors are: green for emission from hydrogen and nitrogen atoms, red for emission from ionized sulfur atoms [SII], and cyan for emission from doubly ionized oxygen atoms [O IIV]. In this view, bright stars and baby stars are revealed being formed within the pillars themselves. The ghostly outlines of the pillars seem much more delicate, and are silhouetted against an eerie blue haze (*Source* NASA, ESA, the Hubble Heritage Team and the James Webb Space Telescope)

1920: The Pillars of Creation…

In the Eagle Nebula, about 7,000 light years from Earth, are seen the so-called *pillars of creation*[2] with the Hubble Space Telescope (Fig. 3.7, left in the visible and middle/right in the infrared/near infrared). The longer wavelengths in the infrared penetrate dust and the gas of the nebula. The columns consist mainly of molecular hydrogen and dust and are corroded by the hot ultraviolet radiation of neighboring hot stars. The left pillar is about 4 light years high. They belong in the Eagle emission nebula (M16), which spans an area of 70 by 55 light years in the constellation Serpens (the serpent of Asclepius). The young open cluster of stars in the constellation Serpens Eagle nebula was discovered by the Swiss astronomer Jean-Philippe Loys de Chéseaux (1745) and the pillars in it by Duncan (1920) on a plate made with the Mount Wilson Observatory 60-inch telescope.

The small bumps at the top of each column (the little *horns* on their heads!) are slightly larger than our planetary system and are visible as the shadow of *Evaporating Gaseous Globules* (EGGs). EGGs shield and protect the gas behind them from intense UV radiation and strong winds of neighboring new stars, which slowly but surely erode these columns of gas and dust. EGGs are incubators and hatcheries of new stars that are revealed when the *cocoon* or the *shell* of the EGGs is eroded.

---

[2] The name is based on a phrase used by Charles Spurgeon in his 1857 sermon *The Condescension of Christ,* when he was a pastor of the congregation of the New Park Street Chapel (the Metropolitan Tabernacle) in London for 38 years.

1970: Discovery of Carbon Monoxide Emission from Giant Molecular Nebulae

Molecular clouds (MCs) are the densest and coldest phase of interstellar gas found between the stars of galaxies, an accumulation of interstellar gas and dust. Their inwards internal gravity within some of their dense cores can eventually overcome the outward thermal pressure and when it also overcomes the barrier of the magnetic forces may cause the further collapse of the cloud forming concentrations of newborn stars and protoplanetary disks that are swirling around them. The various constituent molecules of MCs are important in this process because they help to cool the gas down to a few degrees above absolute zero. This cooling removes from the pressure of the gas the ability to support the equilibrium of the nebula against gravity. Microscopic dust particles, less than a micron in size, also play an important role because they block heat from radiation of neighboring stars, something which also promotes cooling. Stars form under almost the same conditions everywhere in the Universe, where there is cold and dense gas and where self-gravity dominates all other forces.

In 1964, Robert Wilson and Arno Penzias had carried out one of the greatest discoveries in Cosmology, the cosmic microwave background radiation (CMB), the aftermath of the Big Bang, as we discuss in Chap. 6. For this discovery, later (1978) they received the Nobel Prize in Physics, as discuss in this Chap. 6. The CMB radiation corresponds to blackbody emission with temperature of about 3° above absolute zero, with the maximum emission at about 1 mm. Familiar as they were with the detection of such millimeter radio waves, it was expected and just a matter of time before they *fell upon* the emission of interstellar carbon monoxide, which also happens to be at approximately the same wavelength of 2–3 mm. This is because the temperature of molecular clouds is correspondingly cool, about 10° above absolute zero. Indeed, a few years later (1970), the same researchers in collaboration with Keith Jefferts, all at Bell Telephone Laboratories in the United States, made the first observations of molecular nebulae by discovering the emission of interstellar monoxide carbon (CO) near the Orion nebula, but also in other galactic sources, at the spectral frequency of 115 GHz, which corresponds to a wavelength of 2.6 mm.

The research team led by Robert Wilson, who made the first detection of a rotational spectral line of CO in the Orion molecular nebula, also detected the same line in eight other galactic regions of star formation (Wilson et al. 1970): the W3 region in its Cassiopeia constellation (at a distance of about 6,200 light years), the sources Sgr A and Sgr B2 of the galactic center, the ω nebula (M17) in Sagittarius, (at a distance of about 5,000–6,000 light years, a region which contains about 30,000 solar masses), the region W43 in the Aquila constellation (at a distance of about 20,000 light years), the region W44 in the same Aquila constellation (at a distance of about 10,000 light years, which also contains the remnant of supernova W44), the region W49 which activates a giant star with a luminosity millions of times the solar luminosity, as well as one of the largest regions of star formation, the W51 region, one of the largest "*star factories*" in the Milky Way galaxy. Carbon monoxide turned out to be easily detected throughout the solar neighborhood and inner regions of the

Milky Way as it is a tightly-bound molecule composed of relatively abundant atoms, Carbon and Oxygen, and it also radiates away its rotational energy rather quickly, being asymmetrical in structure. A far more abundant molecule in these dense cold regions is molecular Hydrogen (H_2), but that does not radiate nearly as well as CO because it is too symmetric. Thus, CO is the primary tracer of the clouds where stars form. Hence, these observations of CO have become since then the usual method of detecting cold molecular interstellar gas and this first detection of CO was the foundational observation of the field of millimeter wave astronomy.

These large, dense, and cold molecular clouds that constitute the birth regions of new stars and planets are among the most interesting regions of interstellar space. Their composition is mainly molecular hydrogen and helium, with small amounts of heavier gases. Molecular clouds with a mass exceeding 100,000 solar masses, the so-called giant molecular clouds (GMCs) are the largest inhabitants of galaxies, with dimensions of more than about 15 light years. GMCs contain enough dense gas and dust to form many stars in their densest parts. Compared to the mainly ionized hydrogen in other areas of the interstellar medium, less than 1% of all hydrogen in the Milky Way is bound in molecular clouds. The rest is in the form of atomic hydrogen which can become a molecular cloud at a later point in time. The size of these molecular clouds can be from a few light years up to 300 light years and their total mass can reach several million solar masses.

Molecular clouds are very cold, with temperatures that fluctuate from about 10 to 50° above absolute zero and therefore the hydrogen is primarily composed of molecular hydrogen (H_2). They usually do not emit their own light in the visible and thus appear dark in optical telescopes. In these cold and dense clouds, many atoms can combine to form molecules. Thus, many complex organic molecules have been detected, such as water, ammonia, alcohol, formaldehyde, etc. It is believed that these molecules passed from the solar protosolar cloud to the newborn planets and thus they relate to the emergence of life on Earth. GMCs can survive in equilibrium, for long periods of time, from 10 to 100 million years before being eroded by heat, radiation and the strong stellar winds of newborn stars conceived within their guts.

An average spiral galaxy, such as our own Milky Way galaxy, contains about 1,000–2,000 GMCs, as well as many smaller *dark nebulae*. Because GMCs are usually *opaque* in visible light, to observe them we need to use infrared radiation, which can penetrate even the densest cores where new stars are formed. Thus, infrared telescopes are being used to directly observe the star-forming regions within these GMCs, giving us valuable new information about the regions that are collapsing to form new stars. Through infrared spectroscopy we can calculate the temperature and density of a collapsing region, as well as the rate of collapse. We can also see which molecules are present within star-forming regions and determine their relative abundance.

The "Silent Killer" on Earth, Informs us How Life Begins in Space…

- In interstellar space, it is not necessary to detect directly molecular hydrogen H_2 in order to detect molecular clouds. More than 125 other types of molecules have now been identified in the interstellar medium and the carbon monoxide (CO) molecule is invaluable in locating these nebulae. It has been shown that for each CO molecule there are about 10,000 hydrogen molecules, which means we can detect H_2 through emission from CO. This emission from CO is the primary method for detecting molecular clouds.
- CO proved to be an easily detectable molecule in the solar neighborhood and the inner regions of the Milky Way galaxy. Of all the diatomic molecules it has the highest binding energy in its two constituent atoms. It is therefore a tightly-bound molecule that is additionally composed of the relatively abundant atoms, carbon and oxygen. Also, as it has an asymmetric structure, emits its rotational energy quite quickly. The much more abundant molecule H_2 in these dense cold regions does not radiate as well as CO, because it is symmetrical. Thus, CO has been the primary tracer of nebulae where new stars form.
- CO is a colorless, odorless, and tasteless but highly toxic gas for humans. This gas has been characterized as a *silent killer* because it is completely odorless and the initial symptoms of the poisoning it causes are often confused with those of a simple cold. It is a gas which easily accumulates in houses that are heated by burning organic materials (wood, or coal in braziers), in cases of poor maintenance of the combustion gas exhausts. Many cases of poisoning have also happen in closed garages, when the car engine is left running. So many people have lost their lives because of it.
- Hence ironically, it is this *silent killer* that gives us critical information about the nebulae where stars and planets are born and, in some of them, life may flourish…

1977: Identification of Jets from Protostars with Herbig-Haro Emission Nebulae

Argo (Argo Navis) was formerly a large constellation of the southern hemisphere, near the constellations of *Hydra* and the *Unicorn*. The name of the constellation came from the mythical Argo, the ship of the Argonautic expedition used by Jason and the Argonauts to travel to Colchis in search of the Golden Fleece. According to mythology, when the journey of the Argonautic expedition was over, the goddess Athena placed the ship in the heavens. But because it was an extensive constellation, in the nineteenth century it was divided into 4 smaller sections, which are the constellations officially recognized today by the IAU: *Vela* (Masts), *Puppis* (the Stern), *Carina* (the Keel) and *Pyxis* (the Compass).

In the constellation of *Vela* are the remnants of the *Vela* supernova, a supernova that exploded about 11,000–12,300 years ago – before the beginning of the *Mesolithic era* on Earth – at a distance of about 800 light years, leaving behind the *Vela* pulsar, a neutron star with a rotation period of 89, i.e., rotating 11 times every second. In this *Vela* constellation, however, there are not only the remnants of old or long-disappeared stars, but also young ones, in the dark *Bok globules*, relatively small

Fig. 3.8 *(top) The* Herbig-Haro object HH 47 emerges from a dark nebula on the left, where a newborn star is hiding. A high-mass part of the jet collides with the interstellar gas, creating the white bowshock on the right. The jet is at a distance of 1,400 l.y. in the Vela constellation and its length is about 30,000 AU. This image is a composite of separate exposures acquired by the WFPC2 instrument on HST. The color results from assigning different colors to each monochromatic (grayscale) image associated with an individual filter. In this case, the assigned colors are: Cyan: F656N (H-alpha), Orange: F673N ([S II]) (Image Credit: NASA, ESA, B. Reipurth (University of Hawaii) and P. Hartigan (Rice University)). *(bottom left)* Radio observations of HH 47/46 with ALMA (orange and green, at bottom right part of image) and optical observations with ESO's NTT. ALMA observations reveal an energetic jet moving away from us, but hidden at visible wavelengths by the dust and gas of the dark parent nebula. Diagonally on the left is the other visible part of the jet (pink and purple), which is heading toward us. (Image Credit: ESO/ALMA, ESO/NAOJ/NRAO)/ H. Arce). Acknowledgements: Bo Reipurth). *(bottom right)* Large-scale image of HH 47 with NTT, as it emerges from the giant dark parent dark nebula. (*Image Credit* ESO/Bo Reipurth)

dark nebulae containing dense cosmic dust and gas from which star formation may take place. In one of these dark *Bok globulae* of the Vela constellation and at a distance of about 1,470, is found the Herbig-Haro 46/47 system of young objects. Four emission nebulae (HH 46, HH 47A, HH 47C and HH 47D) and a jet, HH 47B, have been identified (Figs. 3.8). The system contains also a molecular outflow and two large shock waves on the two opposite sides of the source star. The total size of this system is about 10 light years, while the total length of the jet is about 30,000 astronomical units (AU).

HH 46/47 were discovered in 1977 by the American astronomer, Schwartz (1977), who according to the naming convention for Herbig-Haro objects (HH objects), named the two main nebulae detected there, HH 46 and HH 47, as they were the 46th and 47th HH objects that were at that time discovered. A jet, the first discovered near a protostar and other nebulae were soon spotted there. Before these observations it was not clear how Herbig-Haro objects form.

According to the models of that time, HH objects reflected the light of neighboring stars that were in dense nebulae and hence they were considered as reflection nebulae.[3] Based on spectral similarities between supernova remnants and HH objects, Schwartz theorized in 1975 that HH objects are produced by radiation coming from shock waves: stellar winds from T Tauri type stars collide with the surrounding gas causing shock waves and creating the emission of radiation, which then is reflected by the nebulae. But with the discovery of the jet in the HH 46/47 system, it became clear that the HH objects were not reflection nebulae but emission nebulae caused by the shock waves created by the jets emitted by these protostars themselves! Since 1977, the HH 46/47 system has been among the most studied objects, showing the effects of jet propagation around protostars and also the jet collision with the surrounding material.

The object HH 46/47 (Fig. 3.8) came to the fore again in 2013, as it was the observation target with the submillimeter wave radio symbolometer ALMA during its first early phase of operation. Until then, we could distinguish only one jet in the visible (top of Fig. 3.8) since the other was well hidden within the dark parent nebula, as seen in the bottom right of Fig. 3.8. In Fig. 3.8 (bottom left), the sought after jet is revealed by ALMA, which is well hidden in the dense gas and dus, invisible with conventional telescopes. The object HH 47 does appear to emerge from a dark nebula in the middle image of Fig. 3.8 where the newborn star is also hidden. Indeed, the image in the left part of the lower panel combines radio observations with the Atacama Large Millimeter Array (ALMA) and much shorter wavelength optical observations with ESO's New Technology Telescope (NTT). ALMA observations (orange and green, bottom right of lower left image) reveal an energetic jet moving away from us, but hidden at visible wavelengths by dust and the gas of the dark parent nebula. The left part of this image shows the other visible part of the jet (pink

[3] Reflection nebulae are interstellar nebulae of gas and dust that reflect the light of a nearby star or stars whose energy is insufficient to ionize the nebula's gas and create an emission nebula, but is sufficient to give sufficient scattering of light so that the dust becomes visible. So, the frequency spectrum displayed by reflection nebulae is similar to that of the stars that illuminate them.

and purple) which is directed toward us. In the lower right panel, a large-scale image of HH 47 is shown with the New Technology Telescope of ESO as it emerges from the giant parent dark nebula.

Such glowing, clumpy streams of material shown in these NASA/ESA Hubble Space Telescope images are the signposts of star birth. Ejected episodically by young stars like cannon salvos, the blobby material zips along at more than 700000 kms per hour. Those speedy jets are confined to narrow beams by the powerful stellar magnetic field. Called Herbig-Haro, or HH, objects, these outflows have indeed a bumpy ride through space. When those fast-moving blobs collide with slower-moving gas, bow shocks arise as the material heats up. Bow shocks are glowing waves of material similar to waves produced by the bow of a boat ploughing through water. In HH 47, at the top of Fig. 3.8, the blobs of material look like a string of cars on a crowded motorway, which ends in a chain-reaction accident. The smash-up creates the bow shock, in the right.

In the well observed case of HH 34 (top of Figs. 3.9), a grouping of merged bow shocks reveals regions that brighten and fade over time as the heated material cools where the shocks intersect. A highly collimated six-knot shock-excited jet of length about 8000 AU and heliocentric radial velocity about -60 km/s is found to extend toward HH 34 from a 0.5-solar-luminosity low-mass pre-main sequence star with an emission-line spectrum, red continuum, and NIR excess (Reipurth et al. 1986; Hartigan et al. 2011). The outburst of an extremely young star, in the earliest phase of formation, is captured in this spectacular image from the NASA/ESA Hubble Space Telescope. The colorful wisps, found in the lower right image, are painted onto the sky by a young star cocooned in the partially illuminated cloud of obscuring dust seen to the upper right. Pictured punching through the enshrouding dust is an extremely hot, blue jet of gas released by the young star. The jet resides approximately 1400 light years away near the Orion Nebula.

1998: The facility VLT in Chile (Distinguishing the Two Lights of a Car on the Moon)

Observations of star formation regions are made from the ground, with the 4 large octameter telescopes of the Very Large Telescope facility (VLT) (Fig. 3.10). This astronomical facility, operated since 1998 by the European Southern Observatory (ESO), the jewel of (ESO), is located in the Atacama desert, at Cerro Paranal of Northern Chile, at an altitude of 2,635 m. For comparison, the mythical mountain of Olympus, the highest mountain in Greece, has its highest peak, Mytikas, at 2,917 m– there lived the *Olympian gods*, according to the religion of the ancient Greeks. It is also the second highest mountain in the Balkans, and also in the entire region of Europe, from the Alps to the Caucasus. Such *Olympian* heights are chosen by astronomers to observe the distant Universe with clarity.

These VLT telescopes are technological marvels, the vanguard and the pride of astronomers. Each of them has a diameter of 8.2 m and can detect objects that

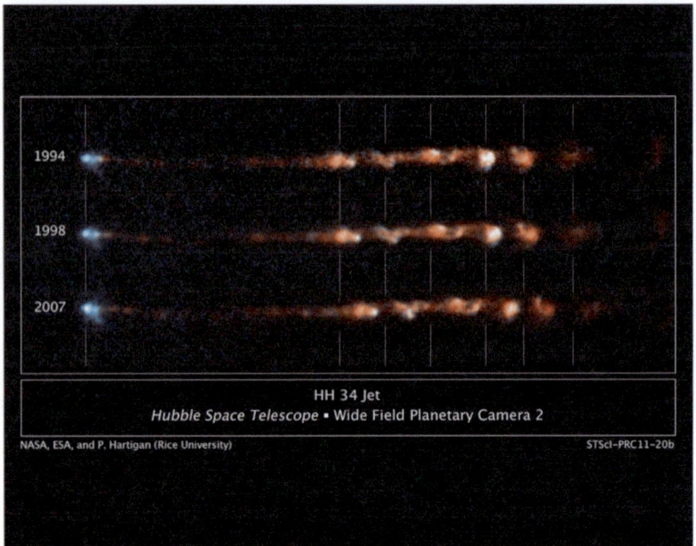

Fig. 3.9 *(top)* Changes in the jet emission of Herbig-Haro 34 (HH 34) with the Hubble Space Telescope's Wide Field Planetary Camera 2, 1994, 1998, and 2007. (*Image Credit* NASA, ESA and P. Hartigan, Rice University). (*bottom left*) Panoramic image in 3 pseudo-colors of the dipolar jet HH 34: green for emission from hydrogen atoms (Ha, 656 nm), red for emission from ionized sulfur atoms [SII, 673 nm] and cyan for emission from doubly ionized oxygen atoms [O IIV, 42 nm]. To the left of the jet appears an inexplicable "waterfall"-like object. (*Image Credit* ESO). (*bottom right*) The HH 34 jet's hot gas is ejected from the disk and protostar at speeds of up to (200–300) km/sec and collides with the cooler interstellar medium resulting in the creation of a bowshock, shown in the lower left of the image. (*Image Credit* ESA/Hubble and NASA)

are *billion times fainter those that the human eye can see.* The 4 telescopes can be combined using 4 Auxiliary Telescopes, each 1.8 m in diameter, achieving via optical interferometry in the so-called VLTI facility, an angular resolution 2 milliarcseconds, which is 25 times better than the 0.05 arcseconds achieved by

Fig. 3.10 The four telescopes, eight meters each, of the Very Large Telescope facility), the jewel of the European Southern Observatory (ESO), in the Atacama desert, in Cerro Paranal, Northern Chile, at an altitude of 2635 m. (*Source* ESO)

each individual telescope. This means that the VLTI can distinguish between the two lights of a car on the Moon.
The VLTI facility is the most productive ground-based astronomical facility in visible wavelengths. Among the ground-breaking VLTI observations is the first direct observation of an exoplanet (β Pictoris b), the tracking of the stellar orbits around the central black hole of the Milky Way (2020 Nobel Prize in Physics) and observations in the visible of the most distant GRBs (γ-ray bursts).

1998: Seeing More Light Through the Darkness…

Figure 3.11 shows a comparison of two pictures of the central area of the Bok globule known as Barnard 68 toward the southern constellation Ophiuchus. When light passes through the cloud, it is absorbed and scattered by the dust inside. The effect depends on the light color (wavelength). Hence, the background stars which appear redder than they really are. It is evident that the obscuration caused by the cloud diminishes dramatically with increasing wavelength. It is also proportional to the amount of obscuring material and is therefore largest for stars that are situated behind the cloud's center. The cloud is obviously in a temporary fine state of near-equilibrium, where the inward force of gravity caused by its mass more or less balances that of the outward pressure due to its temperature; a situation, however, that may not last long.

Barnard 68 is one of the nearest dark nebulae, located at the relatively close distance of 410 light years from Earth. Its diameter is about 12,500 AU, or as long

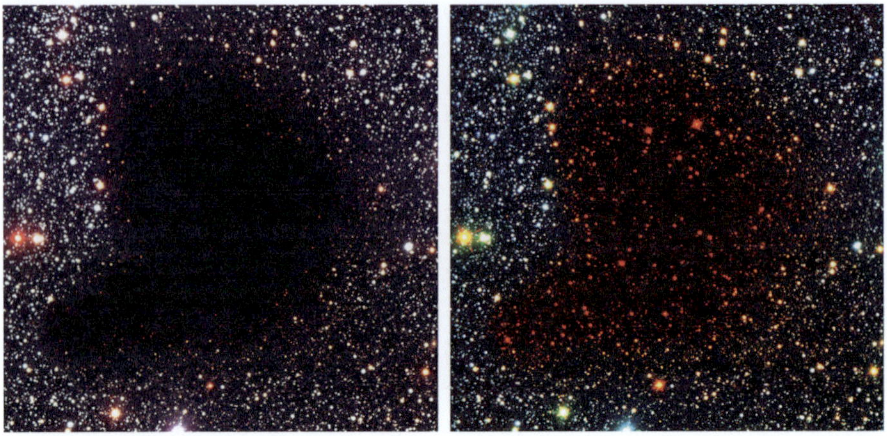

Fig. 3.11 Central region of the dark globule Barnard 68. Composite of visible and near-infrared light, with ESO's telescopes in La Silla, Chile and filters B (= 440 nm, cyan), V (= 550 nm, green), I (= 900 nm, red), and K (= 2,160 nm, red). The composition has been given false colors which they contain the visible in the B bands (440 nm – here in blue) and V (550 nm – here in green), as well as in band I (900 nm – here in red). These 3 optical filters were used on the FORS1 instrument on the VLT/ANTU 8.2-m telescope in the Atacama Desert on March 17, 1999. The image on the right was additionally obtained with the K filter (2,160 nm – here in red), with the SOFI probe at ESO's 3.58-m New Technology Telescope (NTT) at La Silla of Chile. The pictures are a color composite of visible and near-infrared on the left and a false-color composite based on a visible (here rendered as blue), a near-infrared (green) and an infrared (red) on the right. (*Image credit* ESO)

as the average dimension of the Oort cloud, which surrounds our planetary system and from where originate the long-period comets. Its temperature is only 16 degrees Kelvin (i.e., −257 °C) and the pressure at its outermost points 0.0025 nanoPascal, that is, 10 times higher than that of the interstellar medium (but 40,000 trillion times lower than the atmospheric pressure on the Earth's surface!). The total mass contained in this kidney shaped nebula is equal to approximately two solar masses and it measures about half a light year across. Behind this dark nebula "hide" about 3,700 stars, according to long-exposure photography with the 3.58-m NTT telescope at La Silla and the 8.2-m VLT telescope at Paranal of Chile. Of these, about 1000 can only be seen in infrared wavelengths. The nebula is in balance between the inward force of gravity and the outward force of internal pressure. However, the nebula is marginally in equilibrium and is actually on the verge of collapse. Today, we believe that Barnard 68 and its surrounding *cousins* (Barnard 69, 70, and 72) probably constitute the first phase of star formation of an isolated and low-density group of new stars of solar type.

But how did all these dark and dense nebulae come to be? An answer is that these form the durable cores of a larger dark nebula. So far, most of the big nebula has been eroded by the hot ultraviolet radiation and strong stellar winds of hot and massive stars, or by the *storms* of neighboring supernovae. However, observations show that bodies as massive as Barnard 68 can be in a state of *fragile* temporary equilibrium, surviving for some time before they finally start to fall apart.

2003: The Elephant's Trunk Nebula…

In Fig. 3.12, is shown with the Spitzer telescope an image of IC 1396A, the dark and dense nebula within IC 1396, which is informally called the *elephant's trunk* because of its characteristic shape. This dark nebula is a star formation region, as it contains many young (that is, less than 100,000 years old), stars that which were discovered in the infrared in 2003. Two larger stars (but still relatively young, about 2 million years old) are in a small circular cavity in the head of the globula. The winds from these young people stars have emptied the cavity. This image was derived from observations with filters in the Ha line (red), sulfur [SII] (cyan) as well as with filter I (orange).

The Spitzer space telescope image of IC 1396A (Fig. 3.12) reveals a brilliant stellar *nursery* inside the *elephant's trunk*. The nebula is compressed by the surrounding ionized gas. The Spitzer space telescope penetrated the dark nebula and revealed the genesis of new protostars or stellar embryos, as well as invisible stars. Spitzer's infrared photo is a composite of 4 infrared wavelengths, consisting of emission at 3.6 microns (cyan), 4.5 microns (green), 5.8 microns (orange), and 8.0 microns (red). The filamentous appearance of the nebula is the result of the corrosive effect of opposing physical processes. The strong winds from high-mass stars on the left of the image, create a circular corona that creates the *head* of the globule and a trailing

Fig. 3.12 NASA's Spitzer space telescope image of a glowing stellar nursery provides a spectacular contrast to the opaque cloud seen in visible light (inset). IC 1396A is seen with the Spitzer space telescope, at 4 infrared wavelengths: 3.6 microns (cyan), 4.5 microns (green), 5.8 microns (orange), and 8.0 microns (red). The image of IC 1396A (inset), a dark and dense nebula located within the giant emission nebula IC 1396 and called the *elephant's tusk*. Image from observations with filters on the Ha (red) line, sulfur [SII] (cyan) and I (orange). (*Image Credit* NASA/JPL-Caltech/W. Reach (SSC/Caltech))

tail of gas. A pair of young stars (LkHa 349a and LkHa 349c) that formed from the dense gas have *sculpted* a spherical cavity within the head of the nebula (something like an *eye* of the head…). Although one of them is significantly fainter than the other in the visible, at Spitzer's infrared wavelengths both they have the same brilliance. This implies the existence of a thick and dense dust disk around LkHa 349c. These circumstellar disks are the precursors to the formation of planetary systems. The disks are much thicker during the early stages of star formation, when there is still plenty of placental material (gas and dust) that will form the star.

2006: Star Formation in the Carina Nebula with the Hubble Space Telescope

The progress of astronomical observation gave us in the last decades of the twentieth century, for the first time, the possibility to penetrate into invisible nebulae, where there are stellar *nurseries* in which stellar *newborns* aged *only* a few million years, while some of them are predicted to live thousands of times longer. With an average human age of 1,000 months, a one-month-old newborn infant is about the age of 1/1,000 of the human age. Therefore, on the human scale stars millions of years old are infants, the corresponding human age of only a few weeks. These nascent young stars surrounded by dense disk-like cocoons of matter and shoot elaborate jets of plasma in interstellar space, they will become young Suns with planets orbiting them when they come of age. Probably some of these planets will obtain conditions suitable for the development of life. The study of genesis of stars therefore provides us with the possibility of following a similar process of creation of our own planetary system some two-billion years ago, when of course we did not exist!

The Hubble Space Telescope observed a region of new star formation, at a distance of about 7,500 l.y. in the Carina nebula. The image of Fig. 3.13 is a composite of data taken with two of Hubble's science instruments. Data taken with the Advanced Camera for Surveys (ACS) in 2006 isolated light emitted by hydrogen. More recent data, taken in 2008 with the Wide Field Planetary Camera 2 (WFPC2), isolated light emitted by sulfur and oxygen gas. To create a color composite, the data from the sulfur filter are represented by red, from the oxygen filter by blue, and from the hydrogen filter by green. Figure 3.13 is of impressive aesthetic beauty as well, hence its name *mystic mountain*.[4]

Hot young stars erode and sculpt the nebula into this fantastic landscape as they emit high-speed stellar winds and hot ultraviolet radiation. In this area there are many dense regions in which star formation has already begun in a dimension of a few light years.

[4] The name mystical mountain is the title of a motion picture that develops into a Himalayan mountain, starring a little boy searching for his father. It is also the name of an exotic rainforest in Jamaica. And now it also corresponds to an interstellar nebula...

Fig. 3.13 Region of star formation in the Carina Nebula. Many dense areas in which star formation has already begun can be seen. The dimension of the tall pillar starting from the center in the picture is about 3 l.y. These pillars are being assaulted from within, as infant stars buried inside them fire off jets of gas that can be seen streaming from towering peaks: from the tip of the "finger" upwards extends a bipolar jet (HH 901/902), while a second jet extends from the center to the left, almost diagonally to the upper left peak. Hubble's Wide Field Camera 3 observed this nebula on Feb. 1–2, 2010. The colors in this composite image correspond to the glow of oxygen (blue), hydrogen and nitrogen (green), and sulfur (red). (Image Credit: NASA, ESA, Mario Livio (STScI), Hubble twentieth Anniversary Team (STScI))

Low-density regions of the nebula appear hollow, while those with higher density resist erosion by creating dense and cold pillars. Within these dark pillars the creation of new stars continues. During the star formation process, a disk forms around the central condensation from which gas is adsorbed into the protostar. A small part of this disk gas is ejected perpendicular to the axis of the disk forming jets that propagate at a speed of hundreds of kilometers per second. As these jets propagate into the surrounding nebula, they create small bright radiating spots, so-called Herbig-Haro (HH) objects. Elongated gas formations are seen in the upper right part of the image as well and in the center.

2007: A Huge Duck that Lays Stars Instead of Eggs …

Figure 3.14 shows the dark nebula Barnard 163 (the *duck* nebula) in the constellation Cepheus. With an opening of the duck wings a few light years away, Barnard 163's interior is far colder than its exterior, so we have there conditions that allow to gas to condense and form stars. The reddish background is from the emission nebula IC 1396, which is on the right. IC 1396 is a giant emission nebula, 3° in diameter. The nebula is illuminated by a triple system of hot, massive O-type stars at the center of (HD 206267). This system emits a very fast stellar wind, with velocities of more

Fig. 3.14 Duck dark nebula (Barnard 163) in the constellation Cepheus, about 2,500 light years away, with a wing span measured in light years. It may look to some like a duck, but it lays stars instead of eggs. Barnard 163's insides are surely colder than its exterior, allowing conditions where gas can clump and eventually form stars. The red glow in the background results from IC 1396, a large emission nebula that houses the elephant's trunk nebula. (Credit & Copyright: T. Rector (U. Alaska Anchorage), H. Schweiker, WIYN, NOAO, AURA, NSF, reproduced with the kind permission of prof. T. Rector)

than 3,000 km/sec. The entire region of the IC 1396 nebula is ionized by the hot and massive star system, except for some dense globules that still resist the hot UV radiation of the central hot star system.

2010: A Stellar **Nursery** *with About a Thousand Newborns at the Tip of the Sword of the Mythical Hunter Orion*

NASA's Spitzer and Hubble satellite telescopes combined to give us a picture of the chaos they create inside their stellar nursery, hundreds of newborn stars, at a distance of about 1,500 light years, where the spectacular nebula M42 of Orion is located.[5]

[5] Note in passing, that the famous mythical hunter Orion, according to mythology, was the son of Poseidon, the God of the seas. He died in Delos from the sting of a scorpion sent by Artemis, because Orion had fled from her, as in love with Eos (Greek Hώ, Latin Aurora), the *rose-fingered dawn* (Odyssey, rhapsody E 121). She was the goddess and personification of the dawn, rising each morning from her home at the edge of the river Oceanus to deliver light and disperse the night. Since she had an affair with Ares (Mars), Eos was cursed by Ares's other lover Aphrodite to have an insatiable appetite for men. Orion was the third *conquest* of Eos (Fig. 3.15 (inset)). The first was Astraeus (personification of the starry sky) with whom he obtained the numerous stars, Hesperus

Fig. 3.15 Looking like a graceful watercolor painting, an impressive composite image in visible and infrared light of the Orion Nebula, with NASA's Hubble and Spitzer space telescopes, spans about 24 light years across, and shines with both emitted and reflected light from thousands of new stars. Shades of green in the Hubble image in the visible and ultraviolet light correspond to emission of the nebula's gas by hydrogen and sulfur atoms that have been heated, ionized and excited by the intense ultraviolet radiation of the *Trapezium*. Spitzer's image in the infrared comes from the emission of carbon-rich organic molecules in the nebula's gas that have also been excited by the *Trapezium* stars and appear as shades of red and orange. In combination, the two telescopes expose Orion's stars as small colorful dots of light. The Spitzer's red–orange dots are actually newborn stars hidden inside cocoons of dust and gas. Hubble shows less obscured stars in shades of green, while intervening field stars correspond to the small blue dots. (Image Credit: NASA/JPL-Caltech/ T. Megeath (University of Toledo) and M. Robberto (STScI)). The inset image of Eos is from a painting by Evelyn de Morgan (1895)

(Venus) and the 4 main winds, Argestes, North, Zephyrus, and South. The second was Tithonus, a most attractive youth, son of Laomedon, whom Eos abducted to her palaces on the shores of the Ocean, after he had been succeeded to be given by Zeus his immortality, forgetting however to ask for his eternal youth! Thus, although she fed him with ambrosia and clothed him with brilliant garments, the hairs on his head began to turn white and wrinkles furrowing his face, so that he became unrecognizable, retaining only his sweet voice. Then Eos asked Zeus and turned him into a tettiga (cicada). The name of Eos is given today to the journal of the American Geophysical Union (EOS/AGU). According to another version of the myth, Orion died stung by a scorpion sent by the Goddess Gaia, in order to protect her animal kingdom, because he had boasted that he would liberate the Earth of all wild animals. The constellation faces west, followed by two dogs, the constellations Canis Minor and Canis Major, which were fighting Taurus (Taurus constellation), who had kidnapped Europa, as the Scorpio (Scorpio constellation) rises menacingly in the east. From then, Orion and Scorpio, were placed in antidiametrical points on the sky, so that when one sets, the other rises, reminding how they are in a perpetual hunt with each other.

The characteristic shape of the quadrilateral, reminiscent of the mythical hunter, consists of 77 stars, the main four of which are:

(a) The red supergiant Betelgeuse (α-Ori), at a distance of about 650 l.y., one of the biggest and brightest stars in the sky – its diameter is so large that if it were in the center of the solar system, its surface would reach up to the orbit of Jupiter, *swallowing* all the inner planets and asteroids. Its photographic magnitude varies from 0 to + 1.6, which is the largest variability for a first-magnitude star. Its luminosity varies from 90,000 to 150,000 solar luminosities, while its mass is ten times that of the solar one. Although Betelgeuse is *only* 10 million years old, its large mass has led to its rapid evolution, and it is believed that within the next few millennia it will explode as a supernova. Recently (2020), Betelgeuse has become much fainter (magnitude 1.506) and *rumors* are spreading that it may become supernova (Type II) at any time! If this actually happens, it will be the first supernova in the Milky Wat to become visible from Earth after 1604 (Kepler's supernova, De Stella Nova in Pede Serpentarii, the newfound star at the foot of Ophiuchus). Betelgeuse left the main sequence about a million years ago and has been a red supergiant for 40,000 years already. It is a progenitor of a core-collapse supernova, since it has formed all the chemical elements until iron.
(b) The blue supergiant Rigel (β-Ori), whose mass is about 20-times the mass of the Sun, and its brightness is over 60,000 times greater than the solar one.
(c) The blue-white giant Bellatrix (in Latin it means *female warrior*, the star of the Amazon, γ-Ori) with a high temperature of 22,000° and a mass of about 9 solar masses.
(d) Also the blue supergiant Saiph (κ-Ori), with a mass of about 15 solar masses but with a strong wind through which it loses a solar mass every a million years.

Except for those big stars, in the heart of the Orion nebula there are 4 giant stars, which form the so-called *Trapezium* (table) of Orion. These monstrous stars are about 100,000 times brighter than our Sun, about 300,000 years old, while the energy they radiate heats the gases of the surrounding area of the nebula to 10,000 °C. This central set of stars can be identified as a yellow spot near the center of Fig. 3.15 and are primarily responsible for the impressive and phantasmagoric, (spectacular) appearance that this nebula has.

Shades of green in the Hubble visible image and UV correspond to emission of the nebula's gas by hydrogen and sulfur atoms which have been heated, ionized and excited by the intense ultraviolet radiation of the *Trapezium*. At the same time, Spitzer's image in the infrared corresponds to the emission of carbon-rich molecules of the nebula's gas, the so-called aromatic polycyclic hydrocarbons. These organic molecules have also been excited by the stars of the *Trapezium* and appear as shades of red and orange. On Earth, such molecules of aromatic polycyclic hydrocarbons, we have from the car exhaust evaporations and also from baked toast.

Combined, the two telescopes expose Orion's young stars as small colorful dots of light. The yellow-orange dots of Spitzer are actually newborn stars hidden within dust and gas cocoons. Hubble shows less hidden stars in shades of green, while the intervening field stars correspond in the small blue dots.

The constellation is visible in the winter in the southern part of the sky, in the northern hemisphere. The nebula can only be seen with binoculars, or with the help of a small telescope. The blue color in Fig. 3.15 corresponds to wavelengths of 430, 500, and 530 nm. Green at wavelengths of 600, 650, and 910 nm. Orange corresponds to wavelengths of 3,600 nm and red to 8,000 nm.

Stellar winds from clusters of young stars sculpt and erase various ridges, peaks and troughs in the nebula. For example, the big cavity around the right part of the image was created by the winds of the stars of the *Trapezium*, in image 3.15.

The Orion Nebula is the closest to us cosmic factory for the production of stars and is estimated to contain about 1,000 newborns stars, most of them surrounded by accretion disks (Fig. 3.16) and similarly in Taurus (Fig. 3.17). Their dimensions range from 200 to 2000 AU, i.e., they are planetary systems at their birth. Two

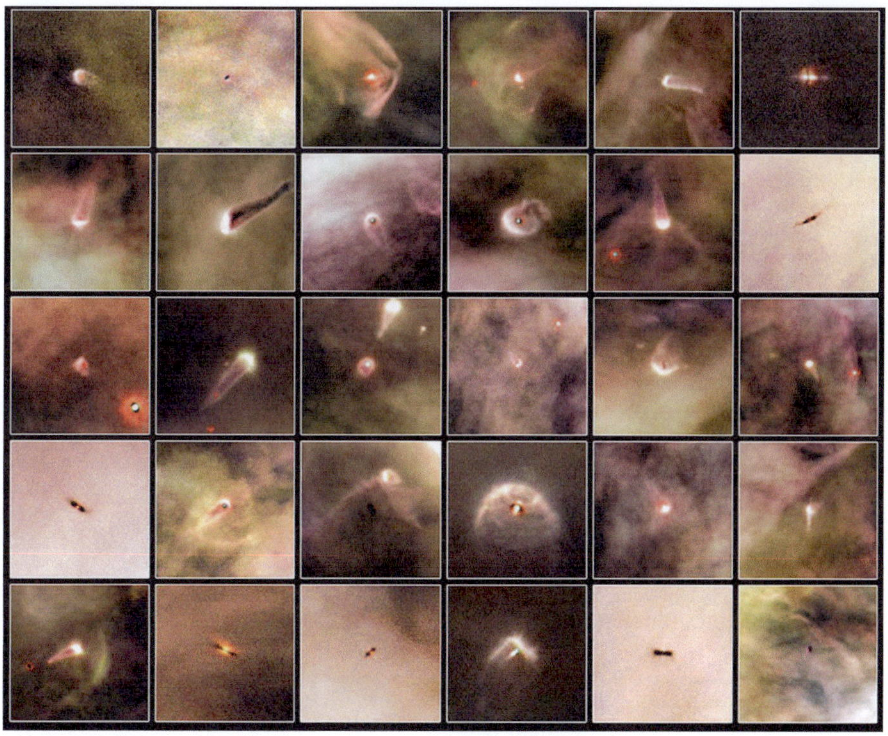

Fig. 3.16 Dozens of protoplanetary disks (*proplyds*) in Orion, can be seen in this image in the visible part of the spectrum with the Hubble Space Telescope. These condensations of gas and dust surround newborn stars and are nascent planetary systems. The central rapidly rotating regions of the disk are heated and form the protostar, while the rest of the peripheral material will subsequently form planets. The nearest disks to the Trapezium shine reflecting the light of the brightest giant star of the Trapezium θ^1-Orionis C ($M \approx 40\, M_\odot$, $R \approx 10R_\odot$, $L \approx 200{,}000\, L_\odot$). The more distant ones, however, receive less radiation and therefore appear as dark silhouettes in the illuminated nebula background (Image credit: NASA/ESA/HST and L. Ricci, ESO)

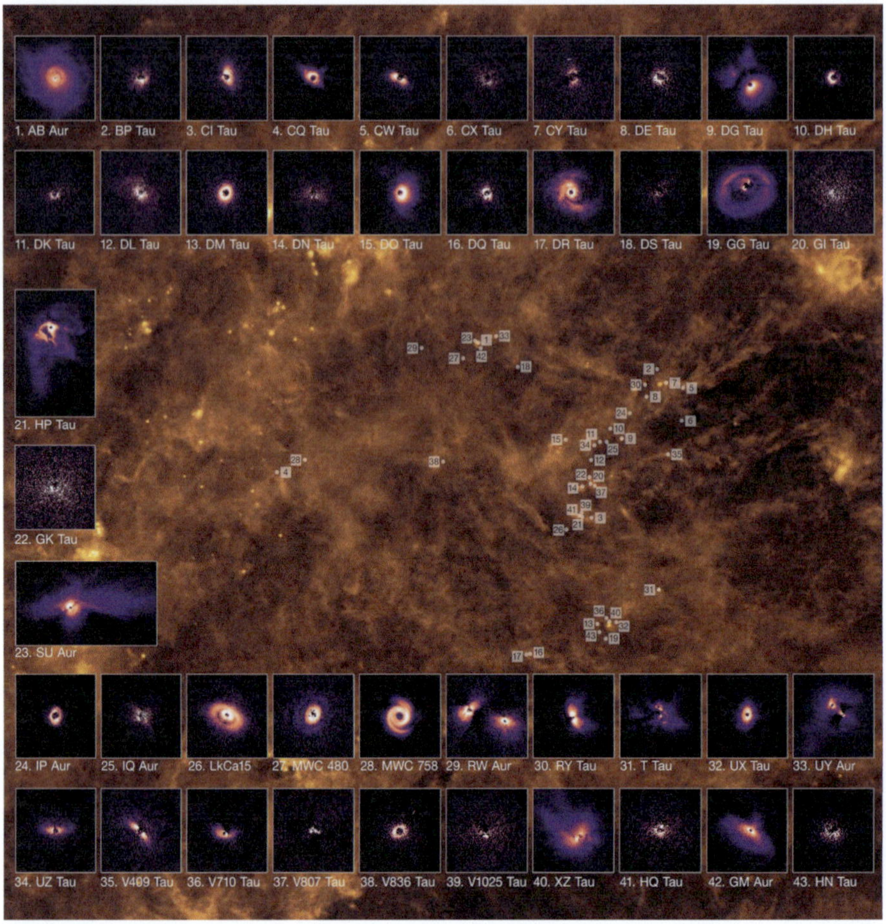

Fig. 3.17 Star and planet forming disks around young stars and their location within the gas-rich cloud of Taurus, about 600 light years from Earth. The stunning images of the disks were captured using the Spectro-Polarimetric High-contrast Exoplanet REsearch (SPHERE) instrument mounted on ESO's Very Large Telescope (VLT). In total, there have been observed 43 protostars in the Taurus region, all of which are pictured here (star and planet forming disks were only detected in 39 of these targets). The background image shows an infrared view of Taurus captured by the Infrared Astronomical Satellite. (Image Credit: ESO/A. Garufi et al.; IRAS)

different types of disks are seen: those that lie close to the brightest star in the cluster (θ^1-Orionis) and those farther away from it. The bright star heats up the gas in the nearby disks, causing them to shine brightly. The disks that are farther away do not receive enough of the energetic radiation from the star to set the gas ablaze; thus, they can only be detected as dark silhouettes against the background of the bright nebula, as the dust that surrounds these disks absorbs background visible light.

2015: Star Wars Episode VII – The Force Awakens …

Science fiction has always been a source of inspiration for researchers. Similarly, powerful motivation for great scientific discoveries have also occurred the spectacular discoveries with NASA's Hubble Space Telescope.

The force of a star awakens: at the right time, along with the release of the new episode of the film production "Star Wars, Episode VII: The Force Awakens," NASA's Hubble Space Telescope photographed Image 3.18, which contains the cosmic double-edged lasers that appear in this film.

In the center of the image of Fig. 3.18, partially obscured by a dark cloud, which with a little imagination it could be one of the main *Jedi* protagonists in the Star Wars movie, there is a newborn star hidden, which "fires" two antidiametrical light shots into space, to announce his birth to the Universe!

HH 24 consists of a protostar and the two jets that start from the center of the image. The protostar cannot be seen because it is surrounded by dense dust. Jets from

Fig. 3.18 *Cosmic lightsabre*: the system of HH 24 jets starting from the center of the image, consists from a protostar, which cannot be seen because it is surrounded by dense dust. Jets from other two newborn stars are also seen starting from the center, at about the same point in the image as HH 24 (bottom left inset). One of them, resembling a blue tunnel, runs through the image from the center to the top toward left, coming from an invisible protostar. A third jet appears to have tunneled through the dense and dark nebula toward the upper right of the image. (*Source* ESA/Hubble and NASA, D. Padgett (GSFC), T. Megeath (University of Toledo), and B. Reipurth (University of Hawaii))

at least two other newborn stars also appear to originate from the center at the same point in the image as HH 24 and can be seen in the lower left inset. One of them, resembling a blue tunnel, runs through the image from the center to the top above left, originating from the invisible protostar. A third jet appears to have tunneled through the dense and dark nebula toward the upper right of the image.

These cosmic double-edged swords are not found in some distant galaxy but within our own Galaxy. They are observed within a turbulent region of star formation known as the Orion B Molecular Nebula, about 1,350 light years from Earth.

When stars form in giant cold and dense molecular hydrogen nebulae, some of the surrounding material collapses under the influence of gravity to form a rotating lenticular disk around the forming star. Although planets will later form inside the disk, at an early stage the protostar is fed by the material of this disk by absorbing huge amounts of gas, just as "Jabba the Hutt" swallowed large amounts of food in the episodes of the star wars. Gas from the disk rains down on the protostar and floods it. At the same time, superheated material is ejected from the star in two opposite directions, along the path of minimum resistance – the axis of rotation of the star. Shock waves are created along of the jets, when they collide with an earlier flow and heat up there the gas in thousands of degrees. Also, as the jets collide on the road them with the surrounding gas and dust, open passages, as when we drop a high-velocity stream of water with a watering hose on a pile of sand. The fronts of these shock waves form bright concentrations, the Herbig-Haro (HH) objects we discussed before. The most prominent such HH object in this image 3.18 is HH 24.

Just to the right of the shrouded protostar, a pair of luminescent light sources can be seen. These are newborn stars that also faintly show us their own double-edged swords – one of which has tunneled into nebula toward the upper right corner of the image.

Hubble's visible and infrared observations of this nebula have revealed several Herbig-Haro objects there. Observations particularly in the infrared enable the telescope to penetrate the cocoon of gas and dust that contains newborn stars and give a clear picture of these objects. However, these will constitute the ideal goals of its upcoming James Webb Space Telescope NASA, which will have even greater infrared observing capabilities and penetrate deeper into the dust surrounding such new stars.

2015: Focused and Symmetric Twin Jets

There are many twin jets associated with protostars (e.g., Figs. 3.1, 3.18, and 3.19), on both sides of an accretion disk surrounding the central young star. Both jets are visible if they are not obscured by the dark cloud where the protostar is born. These jets are focused magnetocentrifugally, as explained later on and in many cases, like the ones in Figs. 3.19, they are impressively mirror symmetric.

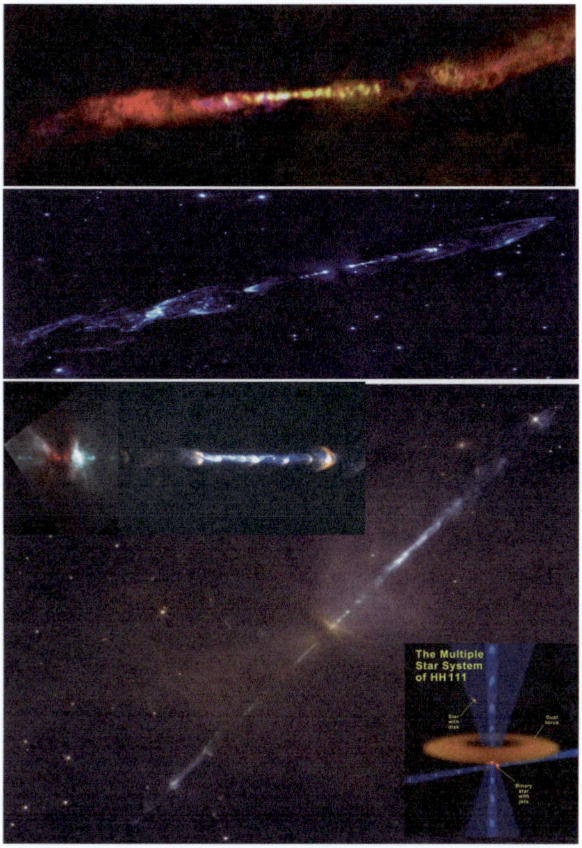

Fig. 3.19 (*top*) Twin massive jets of dense gas with an almost perfect mirror symmetry emanating from a single source at the center of this image, from the extremely young star known as CARMA-7, about 1400 light years from Earth observed in 2015 with ALMA. The larger area – a densely-packed stellar nursery known as the Serpens South Star Cluster – is home to around 30 other protostars, with many of them similar in size, mass and age. Each jet has a length about 16,000 AU (*Source* ESO ALMA/A. Plunkett). (*bottom*) The twin Herbig-Haro 111 jet system (HH 111), in the dark nebula L1617 of the Orion B molecular cloud, at a distance of about 1,400 light years from us, with the WFPC2 of the Hubble ST, is one of the best focused jets that extends to a length of about 12 l.y. from the star. The jet appears in visible light while the disk and protostars are in the infrared. These protostars are located near a huge torus (donut) of gas and dust from which were formed. The torus appears tilted and at the edge, the dark area near the left part of the image. The highly collimated six-knot shock-excited jet has a length of about 8,000 AU and radial velocity about 60 km/s. (Image Credit: ESA/Hubble and NASA, B. Nisini). (*bottom inset1, upper left*) The Herbig-Haro object HH 111, seen by the Hubble Space Telescope. In the Orion B molecular cloud, wherein the bright knots are moving outwards (*Source* NASA/ESA and B. Reipurth, CASA, Univ. of Colorado). (bottom inset2, lower right) The trio of newborn stars in HH 111, take part in a complex dance. Two are now closely embracing each other, while the third has parted from their company. The HST images in infrared light, combined with ground-based radio observations, reveal three young stars. These stars are located near a huge torus, or donut, of gas and dust from which they formed

2019: Observation of Two-Component Jet Outflows

A very compact and extremely high-velocity flow toward the Class 0 protostellar source MMS 5 has been observed with ALMA in the Orion Molecular Cloud 3 (OMC-3) star-forming region (Fig. 3.20, lower panel, Matsushita et al. 2019). Both high- and low-velocity outflows are observed around this protostar by molecular line emission. The high-velocity outflow is collimated and located at the root of the V-shaped outflow (Fig. 3.20, lower left). The CO outflow extends up to ~14,000 AU. Two scenarios have been proposed to explain the driving of both low- and high-velocity flows: (i) nested disk wind (Tomisaka et al. 2002; Machida et al. 2008) and (ii) entrainment scenarios (Arce et al. 2007 and references therein). In the former scenario, low- and high-velocity flows are directly driven by the inner and outer regions of the circumstellar disk, respectively. On the other hand, in the latter scenario, only the high-velocity flow is accelerated near the protostar and entrains the surrounding gas until the entrained gas reaches supersonic speed and creates the low-velocity outflow. The observations are more consistent with the nested wind scenario, a fast collimated wind around the system axis surrounded by a slower disk outflow. In the context of our model, Fig. 3.38, the jet has two components: a slow one from the disk and a fast one which appears to come from the invisible star. Hence, those observational data are consistent with a two-component outflow model (Sauty et al. 1999, 2002, 2011; Bogovalov and Tsinganos 1999, 2001; Tsinganos and Bogovalov 2000; Tsinganos et al. 2009; Matsakos et al. 2008, 2009, 2012). For more details, the interested reader may consult the proceedings of the series of conferences we mention in the beginning of this Chapter.

The similar idea of a two-component MHD outflow has been also applied to relativistic jets as well, with the obvious application to AGN radio jets (Bogovalov and Tsinganos, 2005). This two-component model consists of a relativistic wind-type outflow from the central source collimated magnetorotationally by a non-relativistic wind from a surrounding disk, as we discuss in more detail in Chap. 5.

2022: Star Formation in the Carina Nebula with the Webb Space Telescope

We have been now immersed already into the era of the James Webb Space Telescope (JWST), NASA's flagship observatory and successor to the Hubble Space Telescope. In Fig. 3.21, what looks much like craggy mountains on a moonlit evening is actually the edge of a nearby, young, star-forming region NGC 3324 in the Carina Nebula. Captured in infrared light by the Near-Infrared Camera (NIRCam) on the NASA/ESA/CSA James Webb Space Telescope, this image reveals previously obscured areas of star birth.

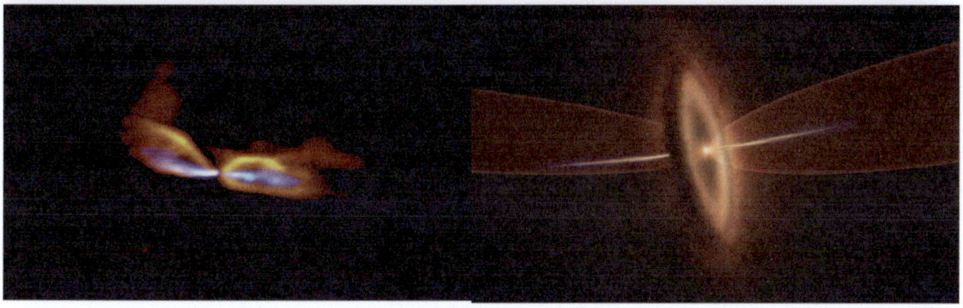

Fig. 3.20 (*top*) **ALMA** (Atacama Large Millimeter Array), a revolutionary array of radio astronomical telescopes, which consists of 66 giant antennas that observe at millimeter and submillimeter wavelengths (0.32–3.6 mm), achieving a resolution of 10 milliarcseconds. It is located on the majestic Chajnantor Plateau, at an altitude of 5,000 m at the Chilean Andes, where extremely low humidity conditions prevail. Its construction cost was about $1.5 billion. (Image *Source* ESO/Y. Beletsky). (*bottom left*) ALMA image of protostar **MMS5/OMC-3** in the center with jets shooting left and right. There is a two-component collimated outflow: a slow shown with orange, and fast jets shown in blue. The axes of the external slow outflow and the jets are not aligned (Credit: ALMA (ESO/NAOJ/NRAO), Matsushita et al. (2019). (*bottom right*) artist rendering of the system. (*Source* NAOJ)

Webb reveals emerging stellar nurseries and individual stars that are completely hidden in visible-light pictures. Because of Webb's sensitivity to infrared light, it can peer through cosmic dust to see these objects. Protostellar jets, which emerge clearly in this image, shoot out from some of these young stars. The youngest sources appear as red dots in the dark, dusty region of the cloud. Objects in the earliest, rapid phases of star formation are difficult to capture, but Webb's extreme sensitivity, spatial resolution, and imaging capability can chronicle these elusive events.

In MIRI's view, young stars and their dusty, planet-forming disks shine brightly in the mid-infrared, appearing pink and red. MIRI reveals structures that are embedded in the dust and uncovers the stellar sources of massive jets and outflows. With MIRI, the hot dust and soot-like material on the surface of the ridges glow, giving the appearance of jagged rocks.

Fig. 3.21 *Cosmic Cliffs*, a region at the edge of a gigantic, gaseous cavity within the NGC 3324 nebula, one of the first images taken by the James Webb Space Telescope, Near-Infrared Camera (NIRCam). (Image Credit: NASA, ESA, CSA, and STScI)

Called the *Cosmic Cliffs*, this Webb's seemingly three-dimensional picture resembling to a landscape of *mountains* and *valleys* speckled with glittering stars is actually the edge of a nearby, young, star-forming region, which looks like craggy mountains on a moonlit evening. In reality, it is the edge of a giant, gaseous cavity within NGC 3324, wherein the tallest *peaks* in the image are about 58 light years high! The cavernous area has been carved from the nebula by the intense ultraviolet radiation and stellar winds from extremely massive, hot, young stars located in the center of the bubble, above the area shown in this image. Located roughly 7,500 light years away, NGC 3324 was first catalogued by James Dunlop in 1826, residing in the constellation Carina, which is home to the Keyhole Nebula and the active, unstable supergiant star Eta Carinae. NGC 3324 is at the northwest corner of the Carina Nebula (NGC 3372). The entire Carina Nebula complex lies in the constellation Carina.

(1958, 1967, 1982, 1994): Using MHD to Theoretically Understand Protostellar Jets and Early Star Formation

In the following, we briefly describe how we understand quasi-analytically the appearance of the dichotomy of MHD outflows in the form of either cylindrical jets, or, radial winds.

1982: Basic Theoretical MHD Understanding of Magnetorotational Acceleration and Collimation of Disk Winds and Jets

The morphologies of collimated outflows have been studied, to a first approximation, in the framework of ideal stationary, or time-dependent magnetohydrodynamics (MHD). The equations of magnetohydrodynamics are a combination of:

[1] Maxwell's equations describing electromagnetism (e.g., describing the propagation of light),
[2] Newton's or Einstein's equations describing gravity (e.g., describing the motion of the Earth around the Sun, or, the expansion of the Universe), and
[3] the first law of thermodynamics which describes the conservation of energy (in an isolated system energy is neither destroyed nor is created from scratch, but simply transformed from one form to another).

MHD is technically a rather complex mathematical system of many (10) coupled and non-linear, partial differential equations containing mathematical singularities, at the so-called *critical points in 1-D, or, magnetosonic separatrix surfaces in 2-D* (Tsinganos et al. 1996; Vlahakis et al. 2000). Thus, our question is: how can we extract from this complex mathematical system of entangled differential equations the description of cosmic jets?

Of time-independent studies, after the pioneering 1-D (spherically symmetric) works of Parker (1958, 1963, 1965a, b, 2001), Weber and Davis (1967), Michel (1969), Blandford and Payne (1982) subsequently demonstrated that astrophysical jets may be accelerated magnetocentrifugally from Keplerian accretion disks, if the poloidal field lines are inclined by an angle of 60° or less, to the disk midplane (Fig. 3.23). This study introduced the *bead on a rotating rigid wire* picture (Fig. 3.24), although these solutions are limited by the fact that they contain singularities along the axis of the system and also terminate at finite heights above the disk (Tsinganos et al. 1996; Vlahakis et al. 2000).

Sakurai (1985) extended the Weber and Davis (1967) equatorial solution (Fig. 3.22c) to all space around the star by iterating numerically between the set of the *Bernoulli* and the *transfield* equations; thus, a polewards deflection of the poloidal field lines was found not only in an initially radial magnetic field geometry, but also in a split-monopole one appropriate to disk-winds (Sakurai 1987). Heyvaerts and Norman (1989) have shown analytically that the asymptotics of a particular field line in non-isothermal polytropic outflows is parabolic, if it does not enclose a net current to infinity, and, if a field line exists that does enclose a net current to infinity, then somewhere in the flow there exists a cylindrically collimated core. Later, Bogovalov (1995) showed analytically that there always exists a field line in the outflowing part of a rotating magnetosphere that encloses a finite total poloidal current and therefore the asymptotes of the outflow always contain a cylindrically collimated

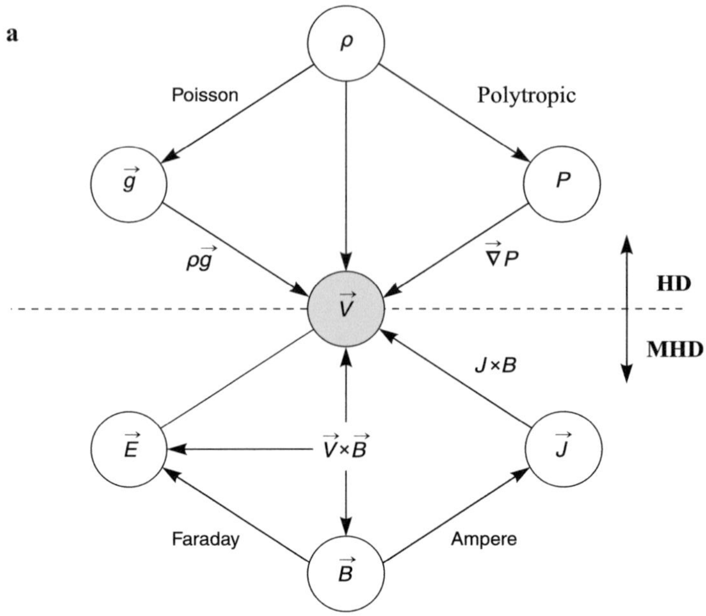

Fig. 3.22 **a** Schematic diagram for illustrating the Hydrodynamic (HD) and magnetohydrodynamic (MHD) interactions. The upper part refers to hydrodynamics and the lower to magnetohydrodynamics. **b** The basic magnetohydrodynamic equations. **c** A subtle point in the construction of wind-type MHD solutions is the requirement that the single physically accepted solution should cross some *critical points*, or, *magnetosonic separatrix surfaces,* a proxy of the boundary physical conditions at infinity. In the Parker thermally driven wind (Parker 1958, inset in the upper right), there is a single critical point wherein the outflow speed equals to the sound speed in the atmosphere. In the Weber and Davis (1967) magnetized and rotating outflow in the equatorial region, the physically accepted solution passes through the slow, Alfven and fast MHD critical points. With such superfast solution, causality is satisfied, since there is no way for the outflow at large distances to communicate and affect by the boundary conditions at the base of the atmosphere. **d** Graph of the equation $\varepsilon(u) = u^2 + 2\omega^2/u$. **e.** (*top*) Radial and azimuthal velocity and (*bottom*) radial accelerations due to magnetic, centrifugal, gravitational and pressure gradient forces, as a function of radius, as derived from a magnetized wind solution that fits the present-day, slow solar wind. **f.** (*top*) Radial and azimuthal velocity and (*bottom*) radial accelerations due to magnetic, centrifugal, gravitational and pressure gradient forces, as a function of radius, as derived from a magnetized WD wind solution that fits a solar-type young star of age 5×10^7 years

core. For self-consistent, time-dependent simulations of supersonic jets launched from magnetized accretion disks, using high-resolution numerical techniques, see also Zanni et al. (2007).

b

Gauss' law

$$\vec{\nabla} \cdot \vec{E} = 4\pi\delta \simeq 0$$

Absence of magnetic monopoles

$$\vec{\nabla} \cdot \vec{B} = 0$$

Faraday's law

$$\vec{\nabla} \times \vec{E} = -\frac{1}{c}\frac{\partial \vec{B}}{\partial t}$$

Ampere's law

$$\vec{\nabla} \times \vec{B} = \frac{4\pi}{c}\vec{J} + \frac{1}{c}\frac{\partial \vec{E}}{\partial t} \simeq \frac{4\pi}{c}\vec{J}$$

Ohm's law

$$\vec{J} = \underbrace{\sigma}_{\infty} \underbrace{\left(\vec{E} + \frac{\vec{V} \times \vec{B}}{c}\right)}_{0}$$

Newton's law

$$\rho\frac{\mathrm{d}\vec{V}}{\mathrm{d}t} = -\vec{\nabla}P + \frac{\vec{J} \times \vec{B}}{c} + \rho\vec{g}$$

with the notation:

- $\vec{E}(\vec{r}, t)$: Electric field in plasma
- $\vec{B}(\vec{r}, t)$: Magnetic field in plasma
- $\rho(\vec{r}, t)$: Mass density in plasma
- $\delta(\vec{r}, t)$: Charge density in plasma
- $\vec{J}(\vec{r}, t)$: Electric current density in plasma
- $\vec{V}(\vec{r}, t)$: Bulk flow speed of plasma
- σ: Electrical conductivity of plasma
- $\vec{P}(\vec{r}, t)$: Pressure in plasma
- $\vec{g}$: Gravitational field in plasma

c

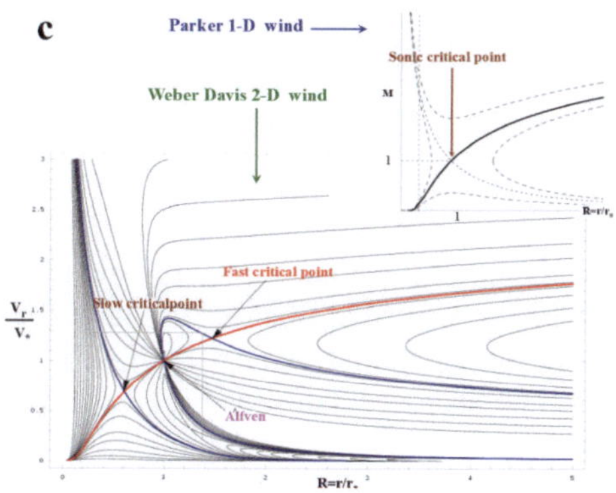

d

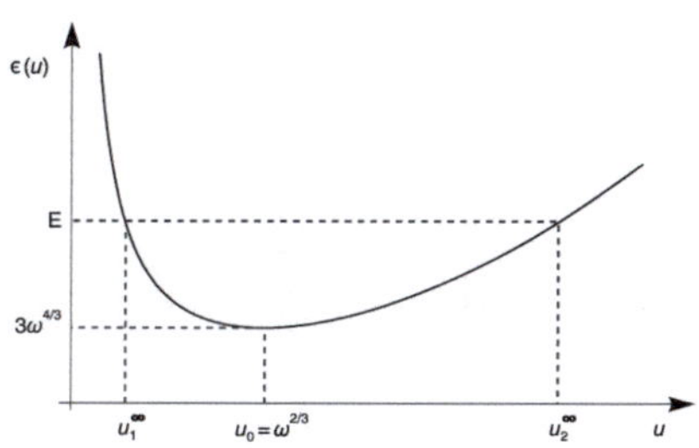

Fig. 3.22 (continued)

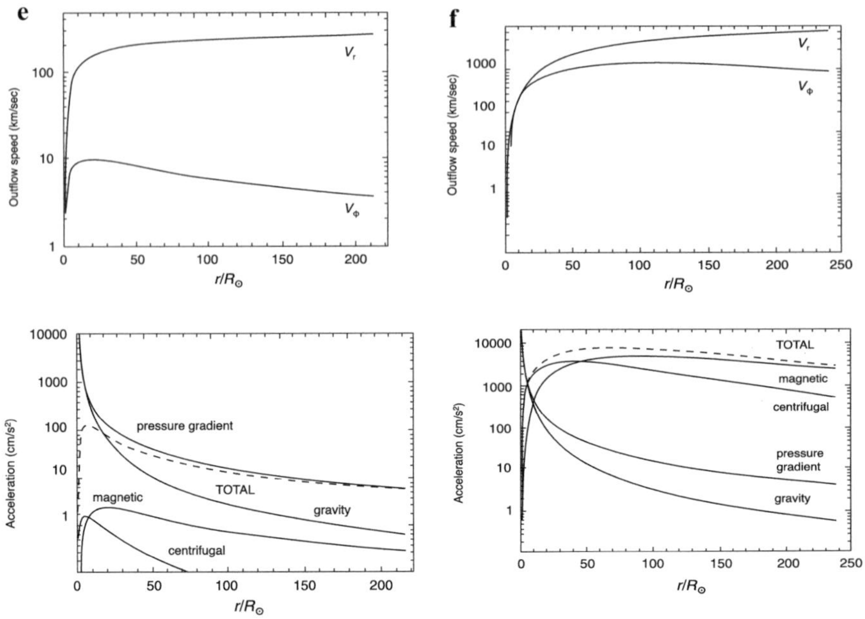

Fig. 3.22 (continued)

1967: Fast/Slow Magnetic Rotators and the Weber and Davis MHD Model for Angular Momentum Loss by a Rotating Magnetized Star

For the simplest description and study of magnetized stellar winds, the model of Weber and Davis (1967) can be used, which is systematically, albeit briefly, described below. In this model, the MHD equations in the equatorial plane $\theta = \pi/2$ of the rotating star are solved step by step using spherical coordinates (r, θ, φ). Important approximations are that we ignore the meridional components in the direction $\hat{\theta}$, as well as the dependence of the variables on the azimuthal coordinate φ. Thus, the five basic assumptions of the Weber and Davis model (hereinafter referred to as the WD model) for a magnetorotational wind are as follows:

1. Time-independent state of the system, $\partial/\partial \mathrm{t} = 0$.
2. Existence of symmetry around the rotation axis z, $\partial/\partial\phi = 0$.
3. Study of the problem only in the equatorial plane, $\theta = 90^{\circ}$.
4. Ignoring the meridional components of the fields ($\vec{V}$, $\vec{B}$), which are assumed to have only radial and azimuthal components that depend only on the radial distance r,

$$\vec{V} = \vec{V}_r(r) + \vec{V}_\varphi(r),$$

$$\vec{B} = \vec{B}_r(r) + \vec{B}_\varphi(r).$$

5. A polytropic relationship is assumed between pressure and density, $P = K\rho^\gamma$.

Also, the following characteristic physical quantities enter into this model:

- The Alfvén velocity in the radial direction,

$$V_{Ar} = \frac{B_r(r)}{\sqrt{4\pi\rho(r)}}.$$

- The Alfvén number M_r (corresponding to the Mach number in the hydrodynamic case) for magnetized fluids,

$$M_r = \frac{V_r(r)}{V_A(r)} \equiv M(r).$$

- The Alfvén radius $r_\star$, that is, the radial distance at which the velocity of the outflow $V_r(r)$ equals the Alfvén velocity V_{Ar}, $M(r_\star) = 1$.
- The Alfvén density, that is, the density of the plasma at the Alfvén distance $r_\star$, $\rho(r_\star) = \rho_\star$,
- The radial velocity at the radial Alfvén distance, $V_\star = V_A(r_\star)$,
- The dimensionless radial distance in units of the Alfvén radius, $R = r/r_\star$,
- The escape velocity from the star at distance $r_\star$,

$$V_{esc}(r_\star) = \sqrt{\frac{2G\mathcal{M}}{r_\star}},$$

where $\mathcal{M}$ is the mass of the star. From the previous physical quantities of the model, the following characteristic dimensionless constants can be defined, if we normalize the escape velocity, the speed of sound and the rotational velocity to the Alfvén distance $r_\star$ with the Alfvén velocity in the radial direction at the distance $r_\star$, V_*, as well as expressing the total energy in units of $V^2{}_*/2$. Thus we have,

- the dimensionless escape speed ν:

$$\nu = \frac{V_{esc}}{V_\star} = \sqrt{\frac{2G\mathcal{M}}{r_\star V_\star^2}},$$

- the dimensionless speed of sound s:

$$s = \sqrt{\frac{2\gamma}{\gamma - 1}\frac{V_s^2(r_\star)}{V_\star^2}} = \sqrt{\frac{2\gamma k\rho_\star^{\gamma-1}}{(\gamma - 1)V_\star^2}},$$

- the dimensionless rotational velocity at the Alfvén distance: $\omega = \Omega r_\star / V_\star$,
- the dimensionless total energy of the flow per unit mass: $\varepsilon = 2\mathrm{E}/\mathrm{V}^2{}_\star$

These characteristic quantities (v, s, ω, ε) fully characterize the WD model. In the following, we discuss the basic features of the WD model of a rotating magnetized star.

In a Parker hydrodynamic wind, the total energy E per unit mass is equal to the sum of the kinetic, thermal (h) and gravitational energy. In the case of magnetorotational wind, the kinetic energy of the azimuthal motion and the contribution of the electromagnetic energy must be added. Thus, the generalized Bernoulli integral is,

$$E = \underbrace{\frac{1}{2}V_r^2 - \frac{GM}{r} + h}_{\text{Hydrodynamic term (Parker)}} + \underbrace{\frac{1}{2}V_\phi^2 - \frac{rB_\phi \Omega}{\psi_A}}_{\text{Magneto}-\text{rotational term}} = \text{constant}$$

In dimensionless units, $\varepsilon = 2E/V_*^2 = \text{constant}$, the generalized Bernoulli integral can be written,

$$\varepsilon(M, R) = \frac{M^4}{R^4} - \frac{v^2}{R} + \frac{s^2}{M^{2(\gamma-1)}} + \frac{\omega^2}{R^2}\left(1 + \frac{(2M^2-1)(1-R^2)^2}{(M^2-1)^2}\right)$$

As the generalized Bernoulli integral $\varepsilon(M, R)$ is a constant, we may equate to zero its derivative with respect to R. This leads to the generalized Alfven number differential equation. Thence, by using the dimensionless constants of the model defined above, the equation of motion of the plasma in the radial direction can be written in the form,

$$\frac{dM}{dR} = \frac{N(R, M)}{D(R, M)},$$

where

$$N(R, M) = \left[\frac{4M^4}{R^4} - \frac{v^2}{R} + \frac{2\omega^2}{R^2}\right](M^2-1)^3 + \frac{2\omega^2}{R^2}(2M^2-1)(1-R^4)(M^2-1)$$

and,

$$D(R, M) = RM\left\{\left[\frac{4M^2}{R^4} - \frac{2s^2(\gamma-1)}{M^{2\gamma}}\right](M^2-1)^3 - \frac{4\omega^2}{R^2}M^2(1-R^2)^2\right\}.$$

For each value of the constant ε, the full topology of the solutions $M(R)$, can be constructed. However, the physically interesting solution needs to cross the critical points. To obtain, however, the single solution crossing all critical points, the set of the constants (v, s, ω, ε) needs to be appropriately tuned (Fig. 3.22c). One way to do that is to impose the condition $N(R, M) = 0$, $D(R, M) = 0$ at the slow, Alfven and fast critical points. The easiest critical point to determine is the Alfven critical point, i.e., the one at which the radial plasma flow velocity becomes equal to the Alfven velocity, which occurs at the Alfven distance $r_{\star}$. The Alfven critical point is therefore the point (1, 1) of the R–M plane, as follows from the definitions of M and R wherein obiously $N(R = 1, M = 1) = 0$, $D(R = 1, M = 1) = 0$. The difficulty of the Weber-Davis model lies in the determination of the slow and fast critical points. This is achieved through the numerical solution of the system of equations: $N(R, M)|_{\text{critical}} = 0$ and $D(R, M)|_{\text{critical}} = 0$, searching for the solution of this system at distances smaller than the Alfven distance for the slow critical point, and at distances larger than the Alfven distance for the fast one. The positions of these two critical points naturally depend on the values of the parameters s, v, ω of the model, and the polytropic index γ.

There are *two* ways to proceed in order to get the critical solution. In the *first*, we judiciously prescribe the set of the four parameters of the model (s, v, ω, γ) and then solve the system of the two equations $\partial\varepsilon/\partial R|_{(R_s,M_s)} = 0$ and $\partial\varepsilon/\partial \mathrm{M}|_{(R_s,M_s)} = 0$. This gives the values of the distance $R_s < 1$ and Alfven number $M_s < 1$ at the slow critical point. We repeat the same at the fast critical point, $\partial\varepsilon/\partial R|_{(R_f,M_f)} = 0$ and $\partial\varepsilon/\partial M|_{(R_f,M_f)} = 0$ which gives the values of the distance $R_f < 1$ and Alfven number $M_f < 1$ at the fast critical point. After determining the critical points, we substitute the coordinates of each point into the generalized Bernoulli integral, thus calculating two of its values, $\varepsilon_s = \varepsilon(R_s, M_s)$ and $\varepsilon_f = \varepsilon\left(R_f, M_f\right)$. Since the Bernoulli integral ε is a constant of motion with a specific value, ideally these two values should be equal. However, the precise determination of the critical points implies a small difference between these two values, $\left(\varepsilon_s, \varepsilon_f\right)$. That depends on the accuracy of our previous determination of the critical points. If this is satisfactory, we calculate the value of the energy to which the contour curve of the physical solution corresponds, which passes through all three critical points.

In the *second* method, we judiciously prescribe the set of the values of the coordinates of the critical points (R_s, M_s) and (R_f, M_f). Then we solve the system of the equations $\partial\varepsilon/\partial R|_{(R_s,M_s)} = 0$ and $\partial\varepsilon/\partial M|_{(R_s,M_s)} = 0$, which finally determines the set of the four parameters of the model (s, v, ω, γ). With either of these two methods we construct finally the complete topology of solutions, as shown in Fig. 3.22c.

For example, the case presented in Fig. 3.22e, corresponding, to a slow magnetic rotator like the Sun today, the calculated values of the four parameters of such a model are: (s, v, ω, γ) = (2.84215, 1.401256, 1.016858, 1.133537) and (R_s, M_s) = (0.608906, 0.505081), (R_f, M_f) = (1.468125, 1.622813). The two values ε_s and ε_f of the generalized Bernoulli integral are: (ε_s, ε_f) = (8.75723, 8.75725), which differ in the fifth decimal place, with a difference: $\Delta\varepsilon = \varepsilon_f - \varepsilon_s = 2 \times 10^{-5}$.

For the case shown in Fig. 3.22f corresponding to a fast magnetic rotator, like a young stellar object. The calculated values of the four parameters of such a model are: $(s, v, \omega, \gamma) = (2.01494, 2.50798, 2.327558, 1.195)$ and $(R_s, M_s) = (0.593922, 0.448758)$, $(R_f, M_f) = (3.78628, 5.08441)$. The corresponding two values of the Bernoulli integral are: $(\varepsilon_s, \varepsilon_f) = (11.4206, 11.4109)$, differing by 0.01.

In the following, we shall see how the above method of constructing a physically relevant solution in the 1-D WD model can be generalized to the case of the 2-D self-similar problem, wherein the critical points are generalized to the 2-D slow and fast magnetosonic surfaces, i.e., the slow and fast critical points of the 1-D WD model become the 2-D slow and fast magnetosonic separatrices. The speed at these separatrices does not correspond to the slow/fast speed but it is modified by the imposed symmetry of the corresponding radial or meridional self-similarity (Tsinganos 1996).

Furthermore, those solutions can be analytically studied at small ($R \to 0$), or large ($R \to \infty$) radial distances. At large radii, the generalized Bernoulli integral simplifies considerably, since in this case, as $R \to +\infty$, $M_\infty \to \infty$ and we have,

$$\varepsilon \simeq \frac{M^4}{R^4} + \frac{2\omega^2 R^2}{M^2}$$

In terms of the dimensionless poloidal speed $u = V_r/V_\star$, this can be rewritten as,

$$\varepsilon = u^2 + \frac{2\omega^2}{u} = \varepsilon(u).$$

The function $\varepsilon(u)$ exhibits an extremum where its derivative vanishes,

$$\left.\frac{d\varepsilon}{du}\right|_{u_o} = 0 \Rightarrow u_o = \omega^{2/3}, \quad \varepsilon(u_o) = 3\omega^{4/3} = \varepsilon^{\min}.$$

For each value of ε, two finite velocities at infinity arise, u_1^∞, u_2^∞, which are such that the corresponding Alfvén numbers at large distances tend to infinity, $\{M_1^\infty, M_2^\infty\} \to \infty$, while the corresponding densities,

$$\{\rho_1^\infty, \rho_2^\infty\} \propto \frac{1}{u^\infty R^2} \to 0,$$

at infinity go to zero. This case is interesting since it has the characteristics of a stellar wind.

The double root of the Bernoulli equation gives,

$$\begin{aligned} V_M &= V_\star \omega^{2/3} \Rightarrow V_M^3 = V_\star^3 \left(\frac{\Omega^2 r_\star^2}{V_\star^2} \right) = \frac{\Omega^2 B_\star^2 r_\star^4}{4\pi \rho_\star V_\star r_\star^2}, \\ \mathcal{F}_B &= B_r r^2, \quad \mathcal{F}_m = 4\pi \rho_\star V_\star r_\star^2, \\ V_M &= \left(\frac{\Omega^2 \mathcal{F}_B^2}{\mathcal{F}_m} \right)^{1/3}. \end{aligned}$$

The characteristic velocity V_M introduced by Michel (1969) is of central importance with regard to magnetic phenomena. As we discuss in the following, if $V_r \ll V_M$, magnetorotational phenomena can be well neglected in relation to thermal ones, and vice versa for $V_r \gg V_M$.

Slow and Fast Magnetic Rotators

The energy per unit mass of the fluid, which according to the generalized Bernoulli integral is a constant,

$$E = \frac{1}{2}V^2 + h - \frac{GM}{r} - \frac{rB_\phi \Omega}{\Psi_A} = \underbrace{\frac{1}{2}V_o^2 + h_o - \frac{GM}{r_o}}_{E_o} - \underbrace{\frac{r_o B_\phi^o \Omega}{\Psi_A}}_{\Omega L},$$

can be written as the sum of two terms,

$$E = E_0 + \Omega L$$

where E_0 is the energy of the thermally driven Parker wind and ΩL is the electromagnetic Poynting energy of the magnetic rotor. Depending on which of these two terms dominates, we have the following two cases:

1. $E_0 \gg \Omega L$: Weak (slow) magnetic rotator. In this case, we have a thermally driven Parker wind.
2. $E_0 \ll \Omega L$: strong (fast) magnetic rotator. In this case, we have a magnetorotationally driven wind.

Another way to distinguish weak from strong magnetic rotators can be done by comparing the Parker velocity (V_P) with the Michel velocity (V_M) so that:

- when $V_M \gg V_P$, we have a *strong* magnetic rotator, while
- when $V_M \ll V_P$, we have a *weak* magnetic rotator.

For today's Sun we have the following values of the two characteristic velocities V_M and V_P:

$$V_M = \left(\frac{\Omega^2 \mathcal{F}_B^2}{\mathcal{F}_m}\right)^{1/3} \simeq 60 - 90\,\text{km/sec} \ll V_P = 600\,\text{km/sec}.$$

So, today the Sun is a *slow magnetic rotator*. We expect that in the past the Sun was a fast magnetic rotator, like most young stars we observe today (Fig. 3.22e, f).

1982: Minimum Inclination of the Poloidal Fieldlines of a MHD Outflow with the Axis of Rotation

Let us then see how we understand theoretically the magnetic acceleration and subsequent focusing of a wind from an accretion disk (a *disk wind*) into a jet, in the framework of magnetohydrodynamics (Tsinganos 1981, 1982a; b, c). Initially, as the plasma swirls onto the accretion disk, it simultaneously may have a very small outward radial velocity, for example, initiated by some thermal pressure gradient (Fig. 3.23). Namely, the charged particles of the disk's ionized gas move circularly around the strong and inflexible meridional magnetic lines and at the same time they slide along them due to magnetorotational forces and the reduction of the outward pressure.

In disk winds, we can relatively easily prove the existence of a geometric criterion for the initial angle that the outflow must form with the axis of symmetry and rotation in the poloidal plane. In the inner parts of the accretion disk, the very rapid Keplerian rotation has a rotational kinetic energy considerably greater than the magnetic

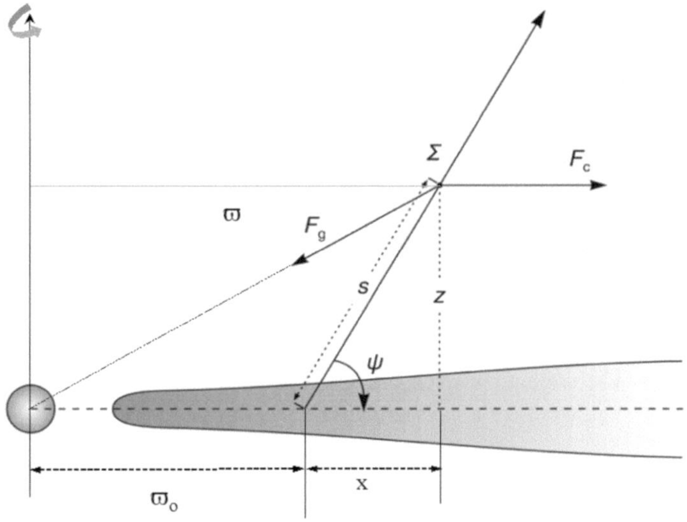

Fig. 3.23 Schematic representation of the various physical quantities involved in the calculation of the maximum required angle of the magnetic line with the plane of the accretion disk in order to have a magnetorotational acceleration of an outflow from an accretion disk. In the poloidal plane and cylindrical coordinates of the plasma volume element at $\Sigma(\varpi, z)$ the poloidal magnetic field line makes an angle ψ with the plane of the disk intersecting it at $z = 0$, at the cylindrical distance $\varpi = \varpi_o$. Gravitational F_g and centrifugal forces F_c act on this volume element Σ of the plasma at a distance s from the disk plane on the poloidal field/flow line, i.e., at the cylindrical coordinates $\varpi = \varpi_0 + x$ and z. The analysis can be analyically done for points Σ close to the plane of the disk, that is, for $x \ll \varpi_0$, $z \ll \varpi_o$, since we are interested in calculating for which initial values of the angle ψ of the magnetic line with the disk we have plasma acceleration and removal from the disk, that is, starting the jet (Tsinganos 2017)

energy. On the other hand, in the poloidal meridional $z-\varpi$ plane, the magnetic field is sufficiently strong, so as to drive the charged particles in Larmor motions around the magnetic lines, with their guiding center sliding along these lines,

$$\frac{1}{2}\rho V_\phi^2 \gg \frac{B_p^2}{8\pi} \gg \frac{1}{2}\rho V_p^2,$$

that is, on and very close to the disk, the Keplerian rotation velocity is much greater than the Alfvén velocity which in turn is much greater than the initial poloidal velocity. The simple analogous mechanical problem is the one where we have passed beads (the charged particles) around rigid wires (the lines of the poloidal magnetic field), while at the same time these wires rotate around the symmetry axis z with the Keplerian angular velocity Ω_K, with which we assume that the plasma on the accretion disk rotates. This Keplerian rotational angular frequency Ω_K is calculated by equating the attractive gravitational force from the central body of mass M on the plasma element with the centripetal force for particle motion in a circular orbit of cylindrical radius ϖ,

$$\Omega_K = \left(\frac{G\mathcal{M}}{\varpi_o^3}\right)^{1/2}.$$

The centrifugal force exerted on the rotating charged particles is,

$$F_c = \Omega_K^2 \varpi = -\frac{\partial}{\partial \varpi}\left(-\frac{\Omega_K^2 \varpi^2}{2}\right).$$

Therefore, in the system of rotating particles they see the centrifugal potential which, added to the attractive gravitational potential of the central body of mass M, gives the total potential $\Phi(\varpi, z)$ within which the plasma moves.

$$\Phi(\varpi, z) = -\frac{G\mathcal{M}}{\varpi_o}\left[\frac{1}{2}\left(\frac{\varpi}{\varpi_o}\right)^2 + \frac{\varpi_o}{\left(\varpi^2 + z^2\right)^{1/2}}\right].$$

Let us write the cylindrical distance of the plasma element as $\varpi = \varpi_0 + x$ where ϖ_0 is the cylindrical distance that the magnetic line meets the equatorial plane of the disk $z = 0$. Considering points Σ very close to the plane of the disk, as in Fig. 3.23, that is, for $x \ll \varpi_0$, $z \ll \varpi_0$ we can use the binomial Taylor expansion,

$$(1 + \varepsilon)^n \approx 1 + n\varepsilon + \frac{n(n-1)}{2}\varepsilon^2 + \ldots,$$

so that the potential $\Phi(x, z)$ felt by the particle is written,

$$\Phi(x, z) = -\frac{G\mathcal{M}}{\varpi_o}\left\{\frac{1}{2}\left(\frac{\varpi_o + x}{\varpi_o}\right)^2 + \frac{\varpi_o}{\left[(\varpi_o + x)^2 + z^2\right]^{1/2}}\right\}$$

$$= -\frac{G\mathcal{M}}{\varpi_o}\left\{\frac{1}{2} + \frac{\not{x}}{\not{\varpi_o}} + \frac{\not{x}^2}{\not{2}\,\not{\varpi_o^2}} + 1 - \frac{\not{x}}{\not{\varpi_o}} - \frac{\not{x}^2}{\not{2}\,\not{\varpi_o^2}} - \frac{z^2}{2\varpi_o^2} + \frac{3}{2}\frac{x^2}{\varpi_o^2} + \ldots\right\}$$

$$= -\frac{G\mathcal{M}}{\varpi_o}\left\{\frac{3}{2} + \frac{3x^2 - z^2}{2\varpi_o^2}\right\}.$$

Let s be the length on the poloidal magnetic line measured from the base of the outflow on the plane of the disk $z = 0$. Also, let ψ be the angle of this poloidal flux line with the plane of the disk,

$$z = s \sin\psi, x = s\cos\psi.$$

Substituting we get the potential as a function of s and ψ,

$$\Phi(s, \psi) = -\frac{G\mathcal{M}}{\varpi_o}\left\{\frac{3}{2} + s^2\frac{3\cos^2\psi - \sin^2\psi}{2\varpi_o^2}\right\}.$$

The force F_s exerted on the particles at a distance s for a given angle ψ is,

$$F_s = -\left.\frac{\partial\Phi}{\partial s}\right|_{\psi} = \frac{G\mathcal{M}s}{\varpi_o^3}\{3\cos^2\psi - \sin^2\psi\}.$$

Therefore, when $F_s > 0$ we have acceleration and outflow along the poloidal streamline,

$$F_s > 0 \Rightarrow 3\cos^2\psi > \sin^2\psi \Rightarrow \tan^2\psi < 3 \Rightarrow \psi < 60^{\circ}.$$

Acceleration and outflow along the poloidal flux line occur only when $\psi < 60°$, that is, when the inclination of the line with respect to the axis of symmetry and rotation is greater than 30°. In other words, the poloidal magnetic field and flux lines must be sufficiently inclined with respect to the axis of rotation so that the centrifuge neutralizes gravity and the plasma escapes the disk and accelerates to higher heights $z > 0$ by sliding along the magnetic lines. In other words, a condition for the initial acceleration of the plasma outwards at the disk level is that the meridional magnetic lines form an angle ψ with the plane of the accretion disk, which is less than 60° (Fig. 3.23).

1982: The "Bead on a Rotating Rigid Wire" Picture for Magnetorotational Acceleration and Flow Collimation

The magnetic field of the disk drives the plasma and gradually accelerates it along the meridional magnetic lines, in their first segment wherein they are *rigid* and *straight*, as explained in Figs. 3.23 and 3.24. This acceleration continues until the kinetic energy of the plasma gradually increases and reaches the energy of the meridional magnetic field, in the so-called *Alfven* distance. Beyond the *Alfven* distance, the magnetic field no longer drives the plasma, but the opposite happens: the magnetic field passively follows the rotational motion of the plasma on the disk, causing the magnetic lines to start spiraling around the z-axis, creating thus a strong azimuthal magnetic component, which then focuses by its tension the outflow toward the z-axis, creating thus a cylindrical jet!

In more detail, the mechanical analog is the following: in the first section of a meridian magnetic line which is near the accretion disk, the magnetic field lines being rooted on the disk, follow its rotation and rotate as a *rigid wire,* up to the Alfven distance dragging the plasma along with in this movement (Fig. 3.24).

After the Alfven distance, however, the plasma has been already accelerated by thermal pressure gradients to a high speed and thence no longer moves passively along the magnetic fieldlines – which had behaved up to this distance like *rigid wires* – but rather along deformable magnetic lines that now behave as *flexible threads.* Hence, after the Alfven distance, the plasma forces this second helically deformable part of the magnetic line to passively follow it in rotation, because now the magnetic lines have turned into *flexible threads* and the inertia of the plasma coils them around the axis of rotation creating thus a strong azimuthal magnetic field. This component of the azimuthal magnetic field exerts a magnetic tension toward the central axis of the system and an entanglement of the magnetic lines toward the z-axis, as illustrated in Fig. 3.24. A result of this tendency is the focusing of the initially approximately radial magnetics of lines to the rotational z-axis of the system, thus creating a cylindrical jet! Therefore, as the star tries to emerge through its gaseous womb, from the surrounding disk shoot like elaborate fireworks thin jets of plasma propagating intact for several light years, many thousands of times the Earth–Sun distance. According to this picture, jets are actually plasma of the accretion disk which is accelerated away from that disk, due to magnetorotational forces. This phenomenon, which is analogous to that of sliding beads on a rotating *stiff* wire, forming an angle greater than 30° with the axis of disk rotation, was studied by Blandford and Payne (1982). A more complete description of this phenomenon has been described and studied through analytical self-similarity solutions, as we discuss in the following (Tsinganos et al. 1996; Tsinganos 2017; Vlahakis and Tsinganos 1999; Vlahakis et al. 2000).

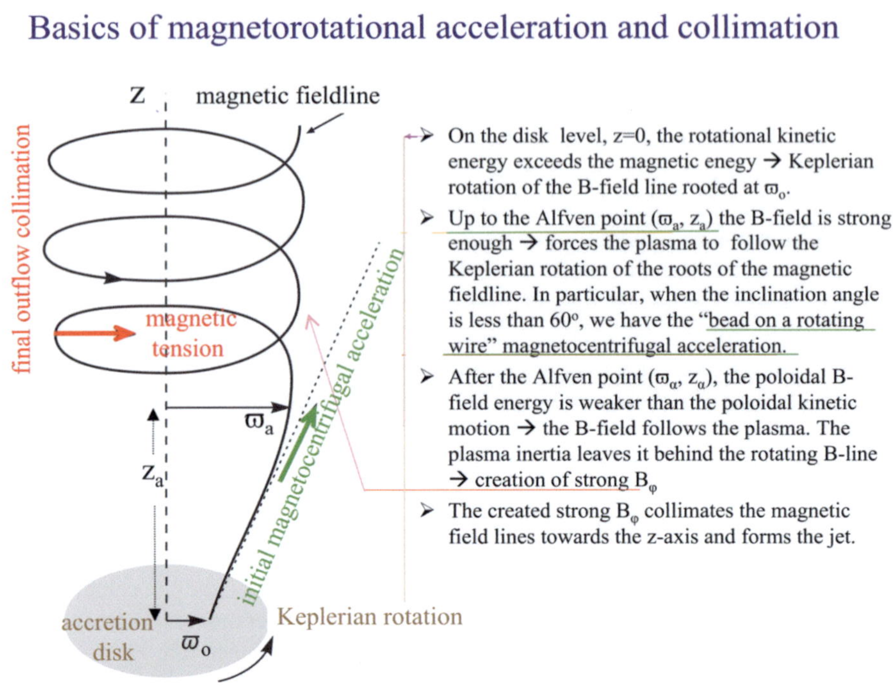

Fig. 3.24 *Bead on a rotating rigid wire picture*: initially, the disk's poloidal magnetic field is strong and the plasma's flow speed weak. Thus, the magnetic field which is rooted on the rotating disk behaves as a rotating rigid wire and drives the plasma which behaves as *a bead on a rotating wire*, accelerating magnetocentrifugally almost radially along the poloidal magnetic line until the plasma's kinetic energy equals to the energy of the poloidal magnetic field at the Alfvén distance. At larger distances, the magnitude of the magnetic field drops with distance as $1/r^2$ and no longer drives the fast-moving plasma but the opposite happens: the magnetic field now behaves like a non-rigid thread and passively follows the rotational motion of the plasma, causing the magnetic lines to start spiraling around the z-axis, creating a strong azimuthal component B_φ which then focuses the magnetic field toward the axis creating a cylindrical jet

1999: A Simple Numerical Demonstration of the Initial Magnetocentrifugal Acceleration and Final Collimation of Astrophysical Outflows

A simple demonstration of the magnetocentrifugal collimation of astrophysical outflows that illustrates Fig. 3.24 has been provided by Bogovalov and Tsinganos (1999). Initially ($t = 0$), a split-monopole magnetic field with a constant outflow speed V_0 along the magnetic field lines is assumed, i.e., along the lines of a radial magnetic field $B_r = B_0/R^2$, where B_0 is the magnetic field at the base of the atmosphere $R = r/r_0 = 1$, we have a plasma outflow with a constant speed V_0. The square of the Alfvén number $M(R)$ of the flow is,

$$M^2(R) = \frac{V_o^2}{V_A^2} = 4\pi\rho\frac{V_o^2}{B^2} = 4\pi\rho\frac{V_o^2}{B_o^2}R^4 = 4\pi\rho_a\frac{R_a^2}{R^2}\frac{V_o^2}{B_o^2}R^4,$$

$$M^2(R) = 4\pi\rho_a\frac{V_o^2}{B_o^2/R_a^4}\frac{R_a^2R^2}{R_a^4} = 4\pi\rho_a\frac{V_o^2}{B_a^2}\frac{R^2}{R_a^2} = \frac{R^2}{R_a^2},$$

because due to conservation of mass $\rho = \rho_a(R_a{}^2/R^2)$, while at the Alfvén distance $M = 1$, $4\pi\rho_a V_0{}^2 = B_\alpha{}^2$. That is, we finally have,

$$M = \frac{R}{R_a}.$$

Let us then assume that the base of the outflow rotates with some angular velocity Ω. From the MHD equations (Tsinganos 1981, 1982a) we have for the ratio of the azimuthal and radial magnetic field,

$$\frac{B_\varphi}{B_r} = -\frac{\Omega\varpi_a^2}{\varpi(B_r/\Psi_A)}\frac{\varpi^2/\varpi_a^2 - 1}{M^2 - 1},$$

where Ψ_A is the mass flux per unit magnetic flux. Let us assume that the flow remains approximately radial even after the start of rotation. Then, the cylindrical distance is $\varpi = R\sin\theta$, so that $M \approx \varpi/\varpi_a$ and for distances considerably larger than the Alfvén cylindrical distance $\varpi \gg \varpi_a$, $R \gg R_a$, $M^2 \approx \varpi^2/\varpi_a^2$, we have,

$$\frac{B_\varphi}{B_r} \approx -\frac{\Omega\varpi_a^2}{\varpi}\frac{\Psi_A}{B_r} = -\frac{\Omega\varpi_a^2}{\varpi}\frac{\Psi_A^2}{4\pi\rho}\frac{4\pi\rho}{\Psi_A B_r} = -\frac{\Omega\varpi_a^2}{\varpi}\frac{\Psi_A^2}{4\pi\rho}\frac{1}{V_o}.$$

But,

$$\frac{\Psi_A^2}{4\pi\rho} = \frac{4\pi\rho V_o^2}{B_r^2} = M^2 \approx \frac{\varpi^2}{\varpi_a^2},$$

and thus finally,

$$\frac{B_\varphi}{B_r} \approx -\frac{\Omega}{V_o}\varpi.$$

That is, the magnitude of the azimuthal magnetic field increases with cylindrical distance, relative to the radial magnetic field. Therefore, although in the axis of rotation the magnetic tension is negligible, as the ratio of the azimuthal magnetic field to the radial magnetic field increases with distance from the axis of rotation, after a certain cylindrical distance the azimuthal magnetic field prevails over the poloidal magnetic field B_r. The magnetic pressure and tension of this azimuthal magnetic field exerts a force toward the axis of symmetry. The next question is to see what force could balance this force directed toward the axis. Since the azimuthal velocity is,

$$V_\varphi = \frac{\Omega \varpi_a^2}{\varpi}\left[1 - \frac{\varpi^2/\varpi_a^2 - 1}{M^2 - 1}\right] \approx 0,$$

the centrifugal force is negligible. Another possibility is to have an increased thermal pressure in the central regions of the flow, which gives a gradient of thermal pressure opposite to the direction of the axis, i.e., radially outwards. However, in a gas in dynamic equilibrium in the gravitational field of the central body, the thermal pressure gradient decreases with the fourth power of the cylindrical distance, ϖ^4, since the density decreases with the square of the cylindrical distance, ϖ^2, due to conservation of mass, while the gravitational force drops with ϖ^2.

Because the magnetic tension and pressure fall off with the cube of the distance ϖ^3, the total magnetic force dominates over the thermal pressure gradient. Therefore, the magnetic forces alone are left, unopposed to focus the outflow, as in Fig. 3.24 creating the cylindrical geometry of the jets that we observe.

In Fig. 3.25, we see that as the base of the outflow starts rotating, in the initially radial magnetosphere, an MHD wave propagates outwards carrying the effect of the rotation and deflecting the field lines polewards by the Lorentz force. At times sufficiently long after the shock wave has reached the end boundary of the simulation ($t >> 1$), a final equilibrium state is reached (Fig. 3.26). In this stationary state, the poloidal magnetic field and the plasma density are increased along the axis because of the focusing of the field lines toward the pole. The disk's poloidal magnetic field drives the plasma and accelerates it almost radially along the poloidal magnetic lines until its kinetic energy equals the energy of the poloidal magnetic field at the Alfvén distance. At longer distances the magnetic field no longer drives the plasma but the opposite happens.

In Fig. 3.26, the magnetic lines have been drawn on a logarithmic scale in order to increase the focus is disturbed. Despite the enormous distances of 10^{10} stellar radii, the focusing of the magnetic lines is insignificant in the case of the slow magnetic rotator (left). In the right diagram, the central star is rotating at a speed 10 times larger than that of the Sun, as occurs in newborn stellar objects. In this case, it appears that the magnetic lines are significantly focused toward the axis of rotation of the central body, forming a jet.

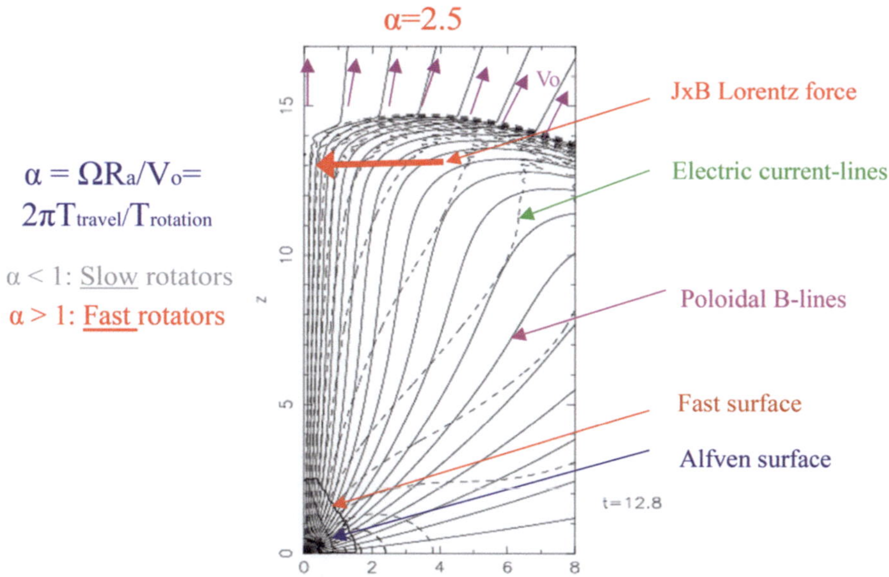

Fig. 3.25 Near-zone snapshot on the poloidal plane (Z, ϖ) from the simulation showing the change of shape of the poloidal magnetic field lines from an initially uniform-with-latitude radial monopole and before a stationary state is reached. This is a transition state, where after the start of rotation of the base of the outflow, the magnetic lines are focused toward the axis of rotation of the system by the Lorentz force $\mathbf{J} \times \mathbf{B}$. A magnetohydrodynamic wave carries the information of focusing toward the plasma which is at large distances from the central body. Distances are given in units of the Alfven radius R_a and time in units of the Alfven crossing time $t_a = R_a/V_a$. (*Source* Bogovalov and Tsinganos (1999))

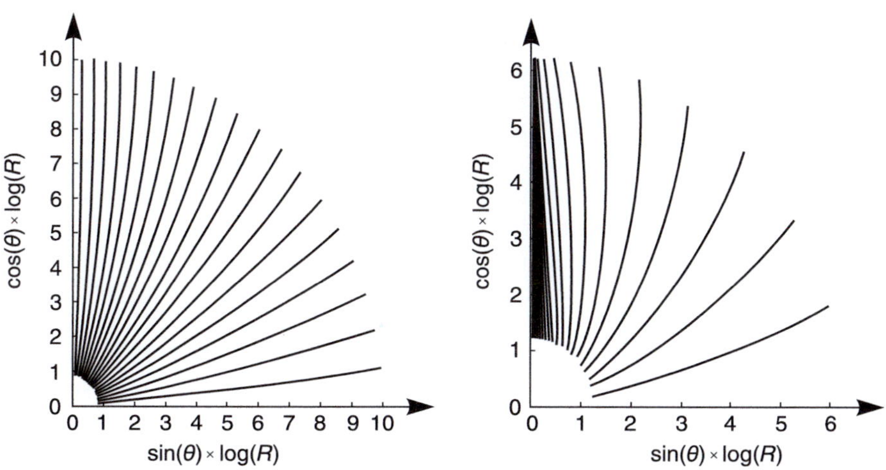

Fig. 3.26 Comparison of the focusing induced in the magnetic lines of an outflow by a slow magnetic rotator (left), with that of a fast magnetic rotator, producing a jet (right), according to the numerical simulations, as in Fig. 3.25. (*Source* Bogovalov and Tsinganos (1999))

1994: Use of Self-similarity for Constructing Analytical MHD Models of Astrophysical Outflows and Jets

In mathematics and physics, an object is said to be *self-similar* if it appears to have the same shape/form at any scale. Simple examples are shown in Fig. 3.27a, b.

Astrophysical jets posses such *self-similarity* in their morphology. This is the case with jets observed to be associated, either with star-forming regions, or, with compact objects in binary galactic systems (white dwarfs, neutron stars and black holes) and also relativistic jets from active galactic nuclei and quasars, which show this property of self-similarity, either when comparing these different classes of objects, or, when each jet is observed at various different scales (Tsinganos 1996; Tsinganos et al. 2009). For example, observations of extragalactic jets in the radio, visible, X- ray, etc., as in the cases of Cygnus and M87, show that these jets retain their *self-similar* appearance as we zoom in from their longer dimensions of millions of light years to the smallest we have been able to distinguish them, even of a few light years.

The realization of this simple fact was for our research group the trigger for a detailed analytical description of astrophysical jets with exact solutions of the equations of magnetohydrodynamics (Tsinganos 1981, 1982a, 2017) in the frame of *self-similarity.*

Rotating magnetized objects with hot coronae play a crucial role in accelerating and collimating astrophysical outflows on the solar, galactic, and extragalactic scales. As discussed above, the magnetohydrodynamical (MHD) approach is usually considered as the first zeroth order step for modeling such astrophysical outflows. However, the mathematical difficulties in the treatment of the MHD equations are of such a high level that several approximations are still unavoidable to obtain solutions useful for an interpretation of the corresponding observed phenomena. For example, in order to make some progress in the quasi analytical modeling of jets, we have assumed, additionally that the system we want to describe is *time-independent* and *axisymmetric* and the morphology of the jet is *self-similar*, as indicated by observations (Figs. 3.1, 3.18, and 3.19). Under these assumptions, it has been shown that there exist only *three* cases of *self-similarity* in Euclidean space, those characterized by *radial* and *meridional* self-similarity in a three-dimensional space and finally, *planar* self-similarity over a plane, Fig. 3.28. To each of these 3 cases of *self-similarity* corresponds a number of families of analytic solutions of the complete set of the MHD equations (Sauty and Tsinganos 1994; Sauty et al. 1999, 2002, 2004, 2005, 2006, 2007, 2011; Vlahakis and Tsinganos 1998, 1999; Vlahakis et al. 2000; Lima et al. 2001; Petrie et al. 2002, 2005). All these three possible general cases of *steady* and *axisymmetric* MHD flows in Euclidean space are summarized in Fig. 3.28. They correspond to stellar disk winds with *meridional* self-similarity, jets from protostars with a *radial* self-similarity and MHD flows in solar coronal loop with a *planar* self-similarity.

In particular, for MHD flows that originate in the near environment of a central gravitating astrophysical object, it has been shown by a systematic method that several general classes of exact, self-similar, axisymmetric magnetohydrodynamic (MHD) solutions can be constructed (Vlahakis and Tsinganos 1998). Technically

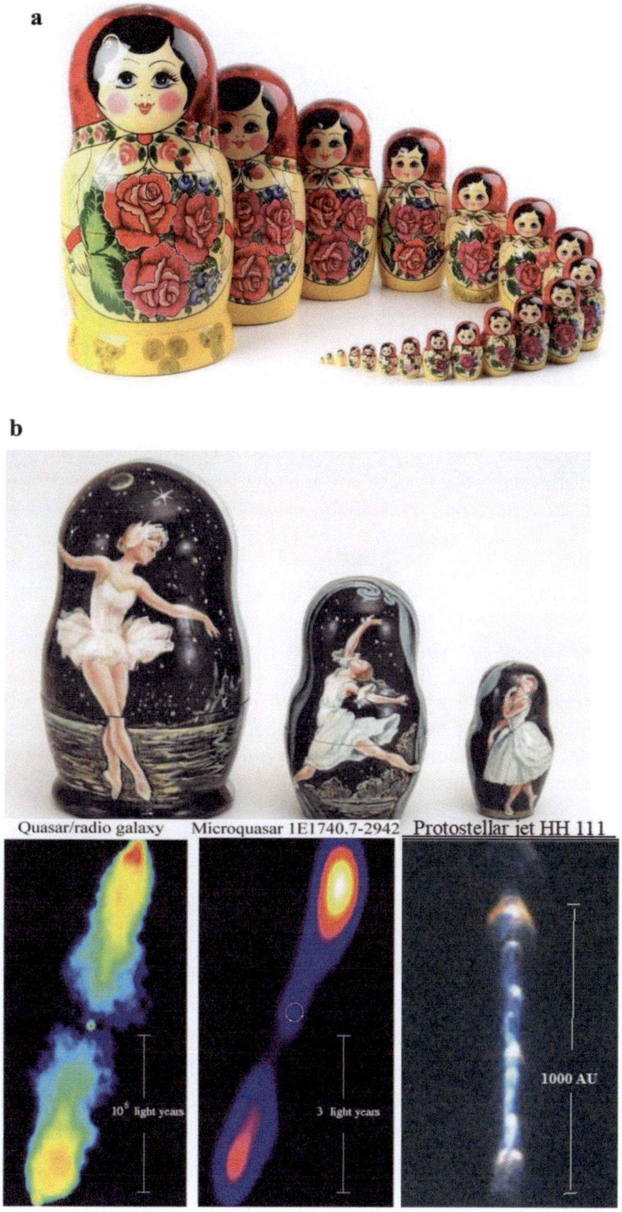

Fig. 3.27 An example of self-similarity is the traditional Russian nested wooden dolls (Russian Matryoshkas) with decreasing dimensions so that they *nest* exactly inside each other. (bottom) Illustration of *self-similarity* in the morphology of several classes of astrophysical jets, from those observed to be associated with star-forming regions (such as HH 111), to jets associated with compact objects in binary galactic systems (such as in the microquasar 1E1740.7-2942), to relativistic jets from active galactic nuclei and quasars

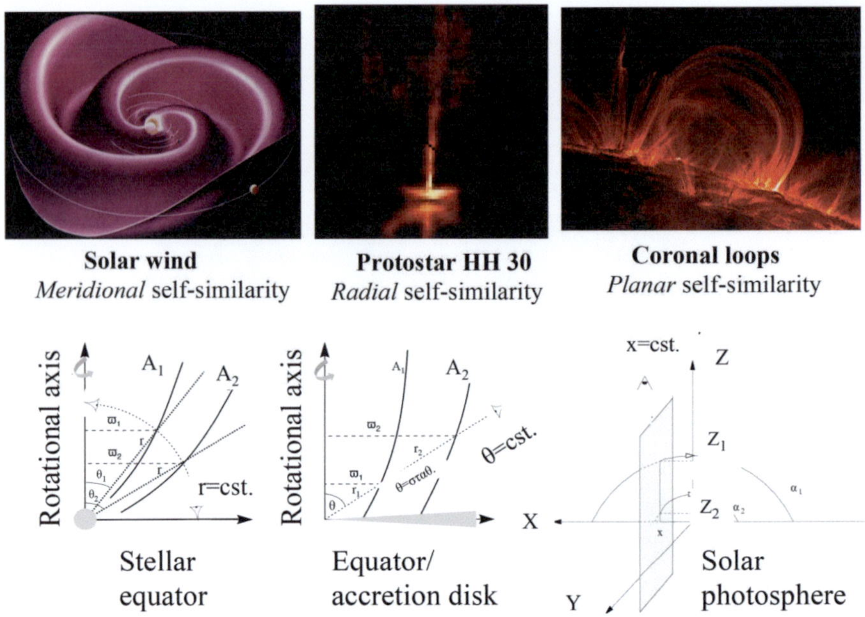

Fig. 3.28 Three cases of *self-similarity* in *steady* and *axisymmetric* MHD flows in Euclidean space. (from left) They correspond to a stellar wind (*meridional* self-similarity), a jet from the accretion disk of a protostar (*radial* self-similarity) and a solar coronal loop (*planar* self-similarity). In *meridional* self-similarity, we see exactly the same geometrical shape of any streamline A_1 or A_2 when viewning along a sphere of radius r, in r*adial* self-similarity, observing along a cone of aperture θ we see exactly the same shape for any streamline A_1 or A_2. Finally, in *planar* self-similarity, when a plane intersects the horizontal x-axis we see exactly the same shape for any streamline α_1 or α_2. (Credits: (top) NASA and Werner Heil, NASA and Alan Watson, Karl Stapelfeldt, John Krist, Chris Burrows, NASA/TRACE. (bottom) Sauty and Tsinganos (1994), Vlahakis and Tsinganos (1998) and Petrie et al. (2002))

what happens, is that the variables of the spherical distance r and the colatitude θ in an inertial coordinate system of spherical coordinates (r, θ, φ) centered at the central object, separate in those two cases, and instead of solving the (more complicated) governing non-linear MHD partial differential equations in r and θ, we end up to solve the (less complicated) ordinary differential equations (ODE) in θ alone (*radially* self-similar case), or r alone (*meridianally* self-similar case). Hence, this unifying scheme contains two large groups of exact MHD outflow models:

(I) *Radially* self-similar models with *conical* critical and Alfven surfaces, wherein the MHD non-linear partial differential equations (PDE) are transformed to ordinary differential equations (ODE) with derivatives depending only on the colatitude θ. Specifically, the magnetic flux function $A(r, \theta)$, which determines the poloidal magnetic field is separable in the poloidal variables (r, θ), $A(r, \theta) = f(\theta)r^x$ where x is a parameter. Additionally, the Mach numbers M are all assumed to be some unknown functions of θ, $M(\theta)$. With such a judicious choice of the radial dependence r in $A(r, \theta)$, the unknown functions $f(\theta)$ and $M(\theta)$ are determined from the final reduced ODE MHD equations.

(II) *Meridionally* self-similar models with *spherical* critical and Alfven surfaces, wherein the MHD PDE's are transformed to ODEs with derivatives depending now only on the spherical distance r. Specifically, the magnetic flux function $A(r, \theta)$, is separable in the poloidal variables (r, θ) in the form, $A(r, \theta) = f(r) \sin^{2\varepsilon}\theta$, where ε is a parameter. Additionally, the Mach numbers M are all now assumed to be some unknown functions of the spherical distance r (or, the dimensionless distance $R = r/r_0$). With such a new judicious choice of the angular dependence θ in $A(r, \theta)$, the unknown functions $f(r)$ and $M(r)$ are determined from the final reduced ODE MHD equations.

(III) For the sake of completeness, we add the remaining third possibility of *translationally* self-similar MHD flows in a constant gravitational field g along the dimensionless vertical direction z, as in the case of MHD flows along loops in the solar corona, where in this third case, the magnetic flux function A is separable in the variables (x, z) in the form, $A(x, z) = f(x)\, e^{-z}$. Additionally, the Mach numbers M are all now assumed to be some unknown functions of the horizontal variable x. With this new judicious choice of the dependence in dimensionless z in $A(x, z)$, the unknown functions $f(x)$ and $M(x)$ are determined from the final reduced ODE MHD equations (Petrie et al. 2002, 2005).

This classification includes known polytropic models, such as the classical Parker (1958) description of a stellar wind and the Blandford and Payne (1982) model of a disk wind. Besides the unification of all known cases under a common scheme, several new classes have emerge (Vlahakis and Tsinganos 1998) and some have been already analyzed in detail (Fig. 3.29).

1994: Magnetosonic Separatrices in Self-Similar MHD Outflows (the Analog of the Limit Cycles in Van Der Pol's Non-linear Differential Equation, or, the Event Horizon in General Relativity)

In *ideal steady and axisymmetric* MHD, by assuming a polytropic relationship between pressure and density $P = K\rho^{\gamma}$, the sound speed C_s and the two other characteristic flow velocities, the *slow* and *fast* magnetohydrodynamic velocities, V_{slow}, V_{fast}, can be defined to satisfy the biquartic equation,

$$V_p^4 - V_p^2\left(C_s^2 + V_A^2\right)_0 + C_s^2 V_{Ap}^2 = 0.$$

where V_A is the total Alfvén velocity and V_{Ap} is the Alfvén velocity in the poloidal plane.

Substituting the expressions for the azimuthal components B_φ and V_φ

Fig. 3.29 The *European self-similar MHD flows* group in front of the historic Meudon great refractor (the *Grande Lunette* in Meudon, 83-cm) at the Observatoire de Paris (2018). (from left) The author with his former PhD students: Christophe Sauty (prof. at the Observatoire de Paris, laboratory LUPM, Montpellier University), Nektarios Vlahakis (prof. at the Department of Physics of the University of Athens), Joao Lima (prof. at the Department of Physics and Astronomy, Faculty of Sciences of the University of Porto). At the center is our collaborator Dr Veronique Cayatte (Centre National de la Recherche Scientifique, laboratory LUTh, at the Observatoire de Paris). This group extended and unified the Parker's thermal solar wind model (1958) and the Blandford & Payne (1982) model of magnetocentrifugally driven winds applied to jets from accretion disks in AGN, as discussed in this chapter

$$V_\phi = \frac{L}{\varpi}\left[1 + \frac{1 - \frac{\Omega}{L}\varpi^2}{M^2 - 1}\right],$$

$$B_\phi = \Psi_A \frac{L}{\varpi}\frac{1 - \frac{\Omega}{L}\varpi^2}{M^2 - 1}.$$

into the generalized Bernoulli energy integral (Tsinganos 1981),

$$E = \frac{1}{2}V^2 - \frac{GM}{r} + \frac{\gamma K}{\gamma - 1}\rho^{\gamma-1} - \frac{\varpi \Omega B_\phi}{\Psi_A} = \sigma\tau\alpha\theta,$$

gives an algebraic equation expressing the conservation of momentum *along* a streamline A = constant (generalized Bernoulli integral). The conservation of momentum *perpendicular* to the streamlines is expressed by a second equation, the final *transfield* differential equation that needs to be solved in order to calculate the function that determines the poloidal magnetic field and flow, and which has the following general form,

$$\left[\frac{1-M^2}{\varpi^2}\right]\left[\nabla^2 A - \frac{\vec{\nabla}A\cdot\vec{\nabla}(\vec{\nabla}A)^2}{2(\vec{\nabla}A)^2}\frac{V_p^4}{V_p^4 - V_p^2\left(C_s^2 + V_A^2\right) + C_s^2 V_{Ap}^2}\right] = F_o$$

The system of these two equations determines the two unknowns, the density ρ and the magnetic flux function A. However, this second differential equation is of mixed type, elliptic/hyperbolic.

Hence, in order to self-consistently calculate the shape of the field/streamlines, the flow speeds and the other physical quantities, one has to solve this last PDE which involves derivatives of the spherical coordinates r and θ. This equation is of *mixed* elliptic/hyperbolic type: close to the disk, or central object the equation is of *elliptic* type. But as we move downstream along a poloidal streamline the equation changes to *hyperbolic* type. Further downstream, the equation switches back to *elliptic*, and further downstream changes finally again to a *hyperbolic* type (Fig. 3.30).

A perturbation of the initial (or boundary) conditions of an elliptic or parabolic differential equation is propagated, to essentially all points in the domain. However, the solutions of hyperbolic equations are *wave-like*: if a disturbance is made in the initial or boundary conditions of a hyperbolic differential equation, then not every point of space feels the disturbance at once. Relative to a fixed time coordinate, disturbances have a finite propagation speed. In the case of MHD, they are the characteristic speeds that MHD signals/waves propagate. These signals travel along the so-called *characteristics* of the equation. This feature qualitatively distinguishes drastically hyperbolic equations from elliptic and parabolic partial differential equations, as illustrated in Fig. 3.30.

Within each of the two hyperbolic regions there is a *separatrix surface*, the *slow magnetosonic separatrix surface* (SMSS) in Y_1 and the *fast magnetosonic separatrix surface* (FMSS) in Y_2. In Y_1 all curves of one of the two families of slow characteristics, $C^{-}{}_{\text{slow}}$, are asymptotically tangent to the SMSS. Correspondingly, in Y_2 all the curves of one of the two families of fast characteristics, $C^{-}{}_{\text{fast}}$, are asymptotically tangent to the FMSS. The two critical points of the MHD outflow are located on these separatrix surfaces which play the role of the *MHD event horizon*.

In particular, the *fast magnetosonic separatrix surface* (FMSS) plays the role of the real event horizon of the problem in the sense that if the poloidal velocity of the plasma exceeds the velocity on this FMSS, then no perturbation can affect the solution interior to the FMSS that is, between the base of the outflow and the FMSS. In other words, by imposing boundary conditions on the FMSS we produce a solution which extends correctly to infinity and no disturbance can return to the base and affect the initial state. Thus, our solution is very well defined. Examples of the FMSS and SMSS are shown in the following Figs. 3.32 and 3.34 and correspond to two cases of self-similar solutions describing focused MHD outflows.

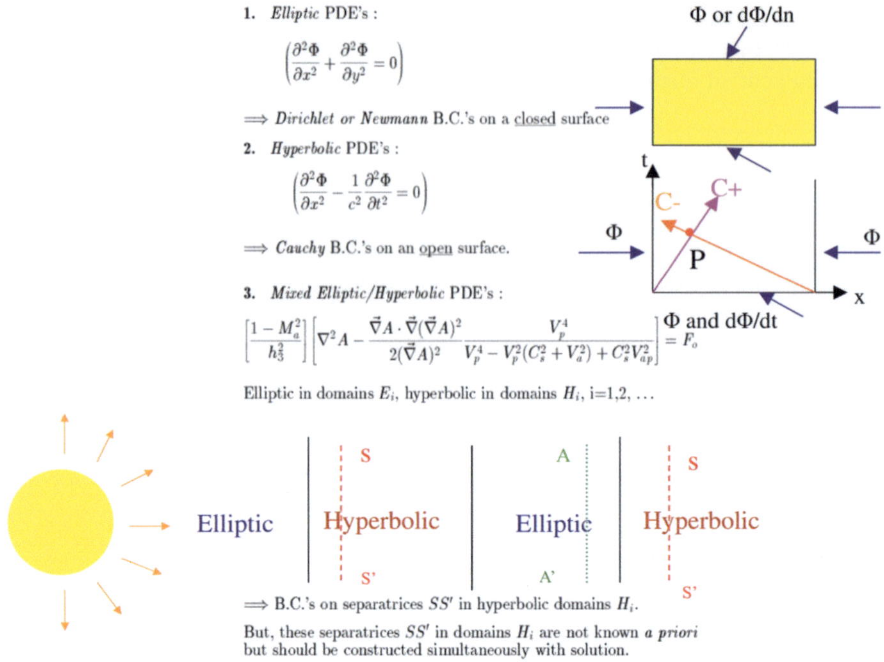

Fig. 3.30 Illustration of the mixed-type (elliptic/hyperbolic) MHD set of equations for *self-similar* outflows and the existence of the characteristics, including the *separatrices*. In hyperbolic regimes exist *separatrices* separating causally areas which cannot communicate with each other via an MHD signal. They are the analog of the limit cycles in Van der Pol's non-linear differential equation, or, the event horizon in relativity. The MHD critical points appear on these separatrices which do not coincide in general with the fast/slow MHD surfaces. To construct a correct solution we need to know the *limiting characteristics*, but this requires an a-priori knowledge of the solution we seek for.

In general, the SMSS and the FMSS do not coincide with the slow and fast MHD surfaces. Only in particularly symmetric or degenerate cases, the SMSS coincides with the slow MHD surface and the FMSS with the fast MHD surface. In the following examples illustrated in Figs. 3.32 and 3.34, indeed they do not coincide. Specifically, in the example of the self-similar MHD outflows, we have the symmetries of self-similarity and axial symmetry. In spherical coordinates (r, θ, ϕ), the direction of self-similarity $\hat{s}$ is either $\hat{s} \equiv \hat{\theta}$ or $\hat{s} \equiv \hat{r}$ while the direction of axial symmetry is $\hat{\varphi}$. Therefore, an MHD wave that preserves these two symmetries should propagate in the direction $\hat{\phi} \times \hat{s} // \hat{\chi}$ of the poloidal plane. In addition to the incompressible Alfvén wave that propagates along the magnetic field (B) with the Alfvén velocity V_A, the compressible slow/fast MHD waves propagate in the $\hat{\chi}$ direction with phase velocities $V_\chi \equiv V_{(\mathrm{slow},\,\chi)}$, or, $V_\chi \equiv V_{(\mathrm{fast},\,\chi)}$ which satisfy the biquadratic equation

$$V_\chi^4 - V_\chi^2\left(V_A^2 + C_s^2\right) + C_s^2 V_{A,\chi}^2 = 0.$$

where $V_\chi = V \cdot \hat{\chi}$. Therefore, on the surfaces on which this equation is satisfied, the MHD equations have X-type critical points. In the radially self-similar solutions $\hat{s} \equiv \hat{r}$ and thus $\hat{\phi} \times \hat{s}//\hat{\chi} = \hat{\theta}$, $V_\chi = V_\theta$. On the other hand, in the meridionally self-similar solutions $\hat{s} \equiv \theta$ and thus $\hat{\phi} \times \hat{s}//\hat{\chi} = \hat{r}$, $V_\chi = V_r$. That is, on the FMSS of the radially self-similar solutions, V_θ equals to the θ-component of the fast MHD speed, while on the FMSS of the meridionally self-similar solutions, V_r equals the r-component of the fast MHD speed.

1994: Modeling Radially Self-similar *Disk Winds*

A typical representative of a *radially* self-similar plasma outflows are magneto-rotational winds that start from a magnetized accretion disk, where the plasma swirls at high speeds, with the ultimate goal of collapsing in the center, feeding a central object, which may be a protostar, a neutron star, or, a black hole. The magnetic field of the disk accelerates and simultaneously focuses the plasma in two jets directed on either side of the disk, along the direction of symmetry of the magnetic field and its axis of rotation (Fig. 3.31).

A difficulty to construct analytically an MHD outflow solution is the passage from the so-called critical points (in 1-D), or critical surfaces (in 2-D). In the 1-D thermally driven outflow there exists only the sonic X-type critical point, while in the magnetized and rotating *equatorial* outflow (Weber and Davis 1967) there exist the classical *slow*, *Alfven* and *fast* critical points (Fig. 3.22c). In the case, however, of the

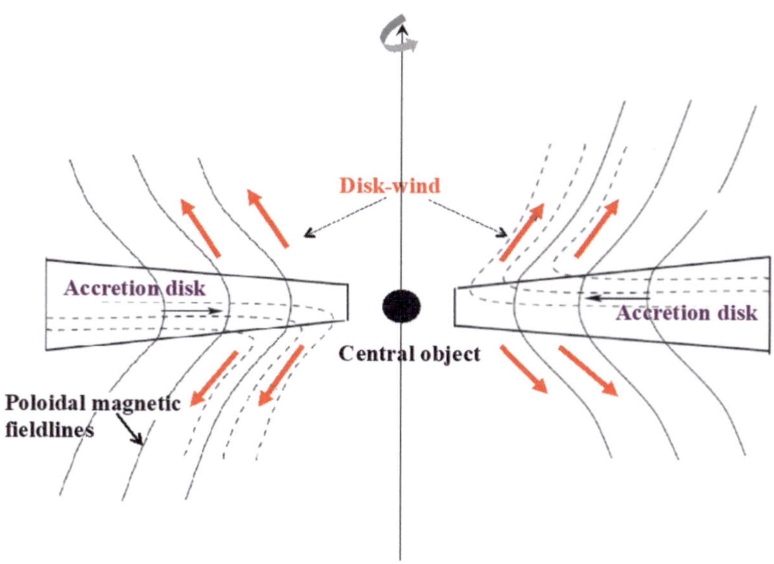

Fig. 3.31 Simple schematic drawing of a disk wind from the accretion disk around a central object

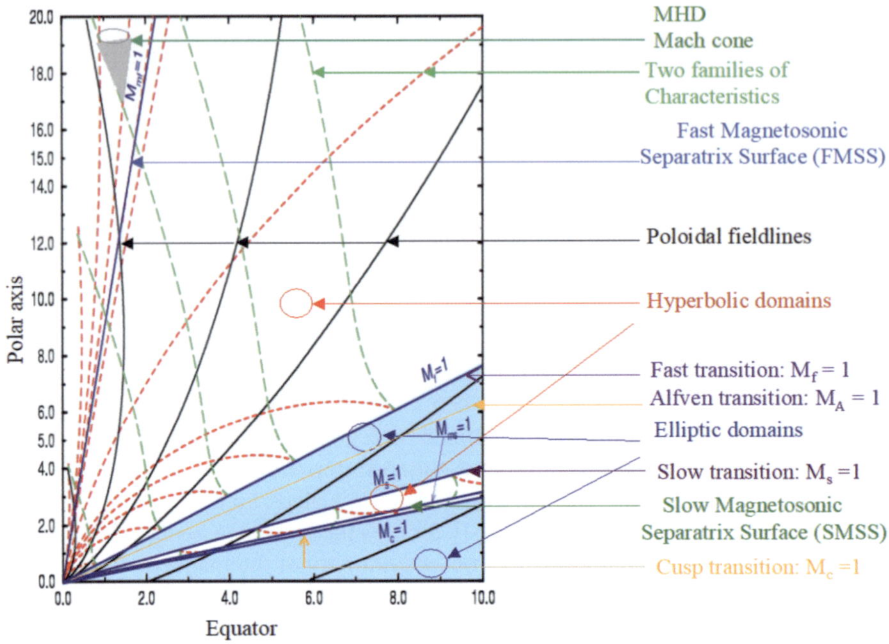

Fig. 3.32 Poloidal fieldlines (solid) and characteristics (dashed green and red) for a radially self-similar polytropic model (Tsinganos 1996; Vlahakis et al. 2000). The cones of the different transitions are shown, as well as the regions where equations are elliptical (blue shades). The red/blue circles indicate the hyperbolic/elliptical domains of the final equation for the Mach number $M(\theta)$

2-D *radially* self-similar disk winds, the situation is more complicated. In that case, as we discussed previously, a physically acceptable solution needs to cross the so-called *modified* (by self-similarity) critical surfaces (Tsinganos et al. 1996) which appear at the so-called slow and fast magnetosonic *separatrix surfaces* (SMSS, FMSS) which are not known a-priori, but they need to be determined with the solution (Figs. 3.29 and 3.32).

In fact, Bogovalov (1995, 1999) and Tsinganos et al. (1996) have emphasized that the singularities in MHD steady flows do not always coincide with the positions where the flow and the known magnetosonic velocities coincide, but with the limiting characteristics, i.e., the FMSS and the SMSS. In our case, the separatrices are found where V_θ is equal to either one of the triplet of the characteristic speeds $[V_{(s,\theta)}, V_{(A,\theta)}, V_{(f,\theta)}]$. This is so because in addition to the azimuthal direction $\hat{\varphi}$ due to the assumed axisymmetry, we have a second symmetry direction, which is the radial direction $\hat{r}$ because of the assumed radial self-similarity. Therefore, a compressible slow/fast MHD wave that preserves those two symmetries can only propagate along the $\hat{\theta}$-direction which is perpendicular to both the $\hat{\varphi}$- and $\hat{r}$-directions; the speed of propagation of such a wave satisfies exactly the quartic equation for the propagation of the MHD waves (for details see Tsinganos et al. 1996). Solutions of the meridionally self-similar relativistic MHD equations crossing the SMSS and FMSS have been also studied in Ceccobello et al. (2018).

Then, let us explain briefly what happens at those *modified* by self-similarity critical surfaces and the characteristics of the mathematical problem. In Fig. 3.32, we plot the characteristics of this problem in both hyperbolic regimes (white areas in the poloidal plane) of a *radially* self-similar MHD outflow. In order to self-consistently calculate the shape of the field/streamlines, the flow speeds and the other physical quantities, one has to solve a set of ordinary equations in the colatitude θ. For example, the final differential equation giving the Alfven number of the outflow, $dM(\theta)/d\theta$, is an equation of mixed elliptic/hyperbolic type: close to the disk (for small θ) the equation is of *elliptic* type without characteristics. But as we move downstream (for larger θ) along a poloidal streamline the equation changes to *hyperbolic* type, in which case, as it is known, exist *characteristics*. Further downstream, the equation switches back to *elliptic*, and further downstream changes finally again to *hyperbolic* type.

So, in each of the two *hyperbolic* regimes there exist two families of *characteristics*. These two families of characteristics in the first hyperbolic domain (i.e., closer to the disk) are bounded by the *cusp* and *slow* surfaces (i.e., where the poloidal outflow speed equals to the cusp and slow magnetosonic speeds), while in the super-fast region is bounded from below by the fast magnetosonic surface, i.e., where the poloidal speed of the outflow equals to the fast magnetosonic speed, Fig. 3.32. One family of these characteristics (red) is tangent to the so-called *slow magnetosonic separatrix surface* (SMSS) where the *modified* – by self-similarity – slow magnetosonic number equals to one, $M_{(\mathrm{mod},s)} = 1$, while the other (green) crosses it. Similarly, in the outer hyperbolic domain bounded by the cone where $M_f = 1$ one of the family of the characteristics (red) is tangent to the *fast magnetosonic separatrix surface* (FMSS) indicated by $M_{(\mathrm{mod},f)} = 1$ while the other (green) crosses it. It is crucial to note that these singularities appear not when the *total* poloidal flow speed, but instead where only its meridional component V_θ coincides with the corresponding meridional component of the slow $V_{(s,\theta)}$ and fast magnetosonic speed, $V_{(f,\theta)}$. These singularities are typical "X-type" critical points (critical transitions at these separatrices), and it is crucial to stress that those singularities appear not where the flow speed, but instead where its *meridional* component V_θ coincides with the meridional component of the slow/fast magnetosonic velocity, $V_{(s,\theta)}$ and $V_{(f,\theta)}$, respectively. It is also important to note that after the FMSS the MHD Mach cone is facing downstream, a fact that guarantees that causality is satisfied, since the outflow is not influenced by the base conditions, as it should.

Next, it is worth to briefly explain the numerical technique for the search of the physically acceptable solution, for some set of model parameters [for more details see again in Vlahakis et al. (2000)]. The main inherent difficulty of the problem is due to the fact that the positions of the *critical transitions* at the separatrices are not known *a-priori*, but need to be calculated simultaneously and self-consistently with the sought for solution. At these critical points (transitions) we do know, however, some relations between various functions; for example, at the Alfven surface the corresponding regularity condition should be satisfied. However, this knowledge alone is not practically enabling us to directly find a solution.

One way to proceed, in order to construct a physically accepted solution through all critical points, is to use the *shooting method* with successive iterations. As it is known, the essence of the *shooting method is to* start the integration from an angle $\theta = \theta_i$ and then to reach a singular point where, e.g., the denominator in the expression of the final MHD equations giving the slope of the Alfven Mach number $dM/d\theta$ vanishes, but not the numerator. We need then to go back and change some parameter and integrate again until the procedure converges, i.e., the denominator and the numerator vanish simultaneously. A similar procedure is followed to cross the other singularity. A rather key point is the most efficient choice of the starting position of integration. Although most of the previous studies integrated the equations by starting from the equator (Blandford and Payne 1982; Contopoulos and Lovelace 1996), it occurred to us that it is more convenient to integrate the equations by starting from the Alfven critical point, i.e., from the conical surface $\theta = \theta_*$ and move upstream (toward the base) and downstream (toward the external asymptotic region).

Furthermore, for the numerical integration, besides the model parameters, we need also to choose the value of the colatitude $\theta = \theta_*$ and the value there of the slope of the square of the Alfven number, $p_* = dM^2/d\theta|\theta_*$ together with the angle of expansion of the poloidal streamlines ($\theta = \theta_i$). Some of these quantities must be tuned to fulfill the singularity conditions at the three critical points. It turned out convenient for the assumed numerical technique to tune the value of one only of the model parameters (λ, controlling rotation of the flow) together with p_* for getting the critical solution.

Hence, we first prescribed the model parameters [γ (polytropic index), x (exponent of the radial distance r, λ (parameter controlling the rotational speed) and κ (parameter controlling the angular dependence of the gas pressure)], as well as the slope p_* of $dM^2/d\theta|\theta_*$, and the angles at the Alfven point θ_* and another model parameter emerging from the *regularity condition* at the Alfven point, such that the poloidal fieldlines avoid kinks. The integration can then start upstream from $\theta = \theta_*$ and the SMSS is encountered, but we cannot pass through it as, e.g., the denominator of $dM/d\theta$ vanishes there, but not the numerator. We integrate again with a different value of slope p_* until we find the opposite behavior around the slow magnetosonic singularity (the numerator vanishes but not the denominator). *Iteratively*, by fine-tuning the value of p_*, a solution is finally found which pass through the SMSS.

Then, we integrate downstream from the Alfven surface and the FMSS is encountered; but in general, it is not crossed. Changing the value of one of the model parameters (λ), we integrate upstream again tuning to a new value of p_* until the SMSS is crossed. Then, we integrate downstream toward the fast magnetosonic singularity and repeat all the procedure until we find the right values of p_* and λ that allow the crossing of the two singularities. At this point the complete solution is obtained by integrating, with the correct values for all the parameters, upstream to the base and downstream toward the asymptotic region. As described above, the procedure to obtain a critical solution is extremely lengthy and rather time consuming, but if followed in a systematic way yields the required final solution.

By following the above procedure, a physically acceptable solution is finally obtained with low velocity and high density at the base, and high speed and low-density asymptotically which smoothly cross the SMSS, the Alfven and FMSS singularities. Such solutions we have widely analyzed in our previous cited papers and are consistent with the observational data on collimated stellar jets. However, it is unescapable that also the fast magnetosonic singularity should be regularly crossed in order to have a steady structure causally disconnected from the asymptotic region, where the jet interacts with the environment (Bogovalov 1995).

Figure 3.33a, b is plotted the components of the outflow speed (in units of the initial z-component of the flow speed at the disk level) and respective Mach numbers, versus z for two typical sets of parameters of the model (solid and dashed lines, Vlahakis et al. 2000). Close to the disk level, the dimensionless escape speed is high, $V_{esc} \approx 440$, the initial rotational speed is lower, $V(\varphi, 0) \approx 110$ and of the order of the Keplerian speed, $V_{Kep} \approx 3V_{(\varphi,0)}$. The azimuthal speed V_φ after some increase in the region of corotation, approximately up to the Alfven critical point at $z \approx 4$ decays to zero transferring its corresponding kinetic energy to the poloidal motion. Thus, the z- and ϖ-components of the poloidal motion grow from their sub-slow and sub-escape values at the disk level where $V_z \approx 1$ to the high values obtained at the modified fast critical point where $V_Z \approx 10^3$ at $z \approx 10^4$. The poloidal speed exceeds the local escape speed around the Alfven transition.

Downstream of the Alfven transition the azimuthal component of the magnetic field grows to very high values in comparison to the poloidal component. At the modified fast critical point practically all the magnetic flux is in the azimuthal direction. For example, $B_\varphi/B_p \approx 1$ at the disk, while $B_\varphi/B_p \approx 60$ after the modified fast transition. The flow velocity is largely in the z-direction with very small components along the φ and cylindrical ϖ-directions. These results are in full agreement with the qualitative description of the basic magnetocentrifugal acceleration of outflows given in Fig. 3.24.

The behavior of the various components of the conserved total energy E versus z, are plotted in Fig. 3.33c, providing information on the different driving mechanisms that govern the dynamics of the outflow. Upstream of the SMSS and close to the base, most of the energy flux is electromagnetic plus some amount of enthalpy. The kinetic energy of the plasma is negligible. As the slow magnetosonic surface is approached, the kinetic energy sharply increases with a corresponding decrease of the thermal energy.

Downstream of the Alfven surface the Poynting flux rapidly decreases; the poloidal kinetic energy keeps increasing, becoming largely the main component of the energy flux at the position of the FMSS. There is some contribution to the acceleration of thermal origin up to the *modified slow* critical point after which the pressure drops to a rather constant value while the magnetic pressure maintains considerably higher values up to the Alfven transition.

The two solutions we have analyzed here correspond to efficient magnetic rotators in the terminology of Bogovalov and Tsinganos (1999), since the ratio of the corotational speed to the poloidal Alfven speed at the Alfven critical surface (the parameter α in their notation) has a value greater than unity ($\alpha \approx 2.13$).

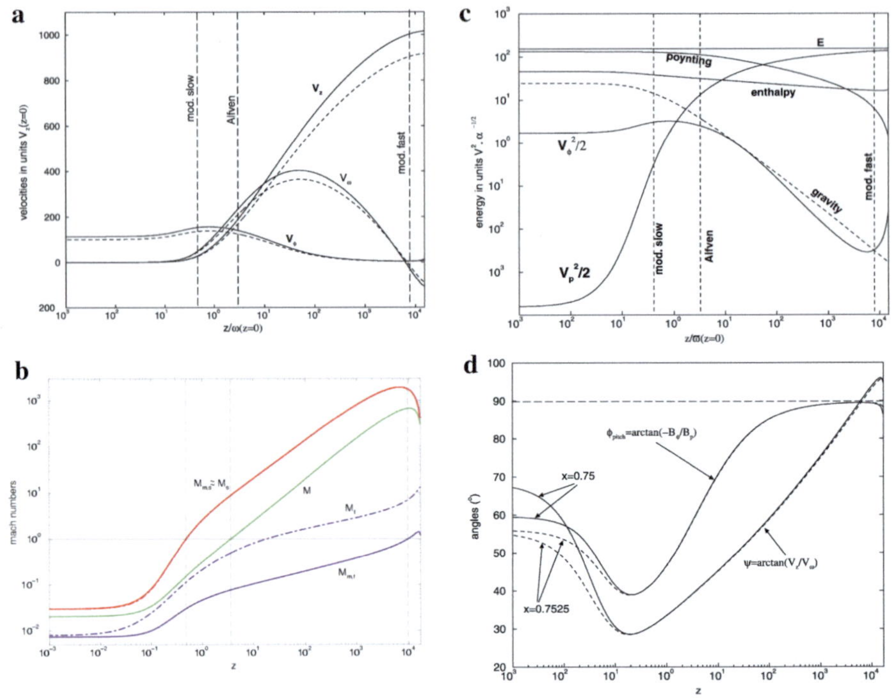

Fig. 3.33 a-d. Plots of the components of the dimensionless outflow speed (*top panel*), the corresponding Mach numbers (*second panel*), the various components of the conserved total energy E, versus the dimensionless height z given in units of the cylindrical distance at the disk level $z = 0\left[z/\varpi_{(z=0)} = z/\varpi_0\right]$ (*third panel*), and finally the ratios of the magnetic components B_φ/B_p and poloidal flow speeds V_z/V_ϖ, *(fourth panel)* via the angles φ and ψ of the corresponding ratios of the magnitudes of the components of the magnetic and flow fields. Note the increase with the altitude z above the disk of the z-component of the speed and also the transfer of the Poynting energy flux to the kinetic energy flux of the outflow, as the wind collimates. The critical points are indicated as the wind becomes superfast-modified at about $z = 10^4$. The plots are for two radially self-similar models (solid/dotted lines) analyzed in Vlahakis et al. (2000)

1994: Modeling Meridionally Self-similar *Winds and Jets*

An archetype example of *meridionally* self-similar plasma outflows are magnetized astrophysical winds that originate from the superheated atmosphere of a magnetized central object. The plasma is initially thermally accelerated at the base of the outflow and gradually acquires superfast MHD velocities. As in the case of *radially* self-similar outflows, a difficulty to construct analytically an MHD outflow solution is the passage from the so-called critical points (in 1-D), or critical surfaces (in 2-D). In the following, we discuss this issue of the critical surfaces through which a solution is required to pass.

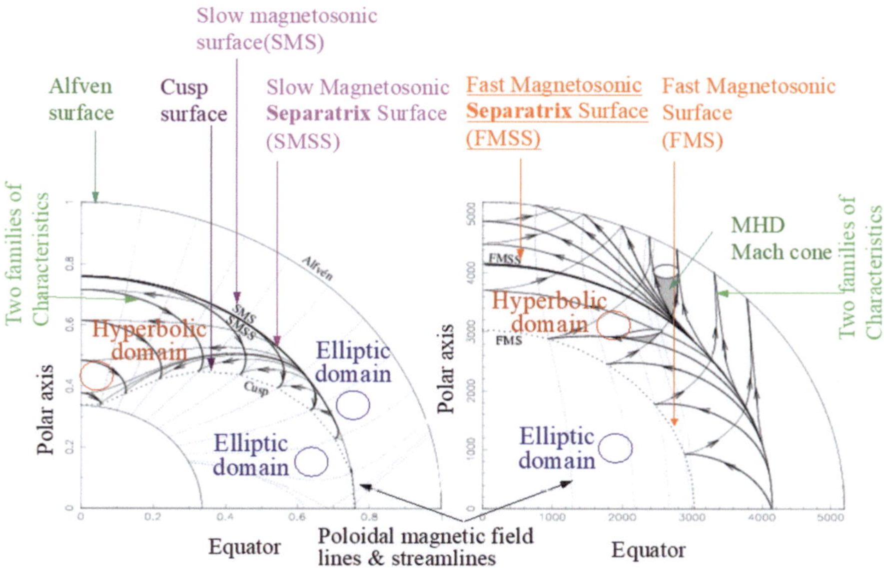

Fig. 3.34 Poloidal fieldlines (solid) and the two families of characteristics (black with arrows) for *meridionally* self-similar model of winds/jets, as in Sauty et al. (1999). The elliptical/hyperbolic domains are bounded by the *cusp* surface and the *slow magnetosonic surface* SMS (left) and the *fast magnetosonic surface* FMS (right). In the hyperbolic domain characteristics exist, one family of which is tangent to the *slow magnetosonic separatrix surface* (SMSS, left) and the *fast magnetosonic separatrix surface* (FMSS, right). The MHD Mach cone after the FMSS is facing downstream and causality is thus satisfied, since the outflow is not influenced by the base conditions, as it should. [From Sauty et al. (2004)]

In Fig. 3.34a, b, we plot the characteristics in both hyperbolic regimes of a *meridionally* self-similar outflow. Similarly to the previous *radially* self-similar case, in each of the two hyperbolic regimes there are two families of characteristics. Those two families of characteristics in the first hyperbolic domain closer to the central source are bounded by the *cusp* and *slow* surfaces, while in the superfast region are bounded from below by the *fast* magnetosonic surface, Fig. 3.34a, b. One family of these characteristics is tangent to the *slow magnetosonic separatrix surface* (SMSS) where the modified slow magnetosonic number equals to one, $M_{(\mathrm{mod},\, s)} = 1$, while the other crosses it. Similarly, in the outer hyperbolic domain bounded by the cone where $M_f = 1$ one of the family of the characteristics is tangent to the *fast magnetosonic separatrix surface* (FMSS) indicated by $M_{(\mathrm{mod}, f)} = 1$ while the other crosses it. It is crucial to note that these singularities appear not when the total pooidal flow speed, but instead where its meridional component V_r coincides with the meridional component $V_{(s,\, r)}$ of the slow and fast magnetosonic velocity, $V_{(f,\, r)}$. These singularities are typical "X-type" critical points, and it is crucial to stress that they appear not where the flow speed, but instead where its *meridional* component V_r coincides with the meridional component of the slow/fast magnetosonic velocity, $V_{(s,\, r)}$ and $V_{(f,\, r)}$, respectively (Tsinganos et al. 1996). As before in the case of *radially* self-similar outflows, the MHD Mach cone is facing downstream and causality is thus satisfied, since the outflow is not influenced by the base conditions, as it should.

By eliminating the angular dependent terms on θ in the MHD transfield equation for momentum balance across and along the fieldlines, via the explicit dependence on θ of the physical quantities dictated by the present *meridional* self-similarity, one finds three non-linear, ordinary differential equations for the pressure $\Pi(R)$, the expansion factor of the poloidal fieldlines $F(R)$ and the Alfven number $M(R)$, which as in the case of *radially* self-similarity, all have singularities at the separatrices. The mixed-type (elliptic/hyperbolic) of these equations and the existence of the characteristics, including the separatrices are shown in Fig. 3.34a, b.

As with the previous case of radial self-similarity, the singularities in the final equations do not always coincide with the positions where the flow and the magnetosonic velocities coincide, but with the limiting characteristics (separatrices), i.e., the *slow magnetosonic separatrix surface* (SMSS) and the *fast magnetosonic separatrix surface* (FMSS). In the present case, these separatrices are found where V_r is equal to either one of the triplet of the characteristic speeds $[V_{(s,\,r)}, V_{(A,\,r)}, V_{(f,\,r)}]$. This is so because in addition to the azimuthal direction $\hat{\varphi}$ due to the assumed axisymmetry, we have a second symmetry direction, which is the $\hat{\theta}$-direction, because of the assumed *meridional* self-similarity. Therefore, a compressible slow/fast MHD wave that preserves those two symmetries can only propagate along the $\hat{r}$-direction which is perpendicular to both the $\hat{\varphi}$- and $\hat{\theta}$-directions; the speed of propagation of such a wave satisfies exactly the quartic equation for the propagation of the MHD waves (for details see Tsinganos et al. 1996).

The numerical technique for obtaining a physically acceptable solution, for some set of model parameters, is similar to the previous case [for more details see Sauty and Tsinganos (1994)]. Before discussing the derived shape of streamlines and the speeds of the outflow, it is required to introduce an interesting integral which exists in this case of meridional self-similarity.

An Energy Integral and Criterion for the Formation of Uncollimated Winds and Collimated Jets

An important physical quantity in magnetized outflows is the so-called magnetic rotator energy EMR $= \Omega L$, where $\Omega(\alpha)$ is the corotation frequency, or angular velocity at the base of the flow for each poloidal streamline defined by $\alpha =$ constant, and $L(\alpha)$ is the total specific angular momentum carried by the flow and the magnetic field (Michel 1969; Belcher and McGregor 1976). The basal Poynting energy $E_{(\mathrm{Poynt.},\,o)}$, defined as the ratio of the Poynting flux density S_z per unit of mass flux density ρV_z, is roughly equal to this magnetic rotator energy ΩL, if the base radius of the jet $G(R)$ is much smaller than its Alfven radius ($G_o \ll 1$) and the Alfven number is also negligibly small ($M_o \ll 1$). Let E_o be the sum of the kinetic, gravitational and thermal energies per unit mass at the base of the outflow. Then, as in the WD model that we discussed before, the total available energy for the outflow at the base is $E \approx E_o + \Omega L$. Accordingly, we have an outflow from a fast magnetic rotator (FMR) when $E_o \ll \Omega L$ and an outflow from a slow magnetic rotator (SMR) in the opposite case of $E_o \gg \Omega L$.

In the present case of *meridionally* self-similar models, it is assumed that the thermal pressure has the form $P(R, \alpha) = \left(\rho_* V_*^2/2\right)\Pi(R)(1 + \kappa\alpha)$ similarly to the density $\rho(R, \alpha) = \rho_*(1 + \delta\alpha)/\mathrm{M}^2(\mathrm{R})$, where κ, δ are constants and ρ_* is the density at the Alfven transition where $V = V_*$ and $M = 1$. It turns out that a global integral (constant) ε exists, which is related to the variation across the fieldlines, of the specific energy which is left available to collimate the outflow, once the thermal content converted into kinetic energy and into balancing gravity has been subtracted (Sauty et al. 1999). For $\varepsilon > 0$ collimation is mainly provided by magnetic terms, while for $\varepsilon < 0$ the outflow is confined mainly by thermal pressure. Accordingly, we may define flows with positive or negative ε as *efficient* or *inefficient* magnetic rotators, respectively (EMR or IMR). Note that ε is not a mere generalization of the Bernoulli constant but rather a variation of the total energy per unit mass from one magnetic fieldline to the next. Therefore, it also contains information on the transfield force balance equation and the energies that control the shape of the flow tube, rather than only information on the various driving mechanisms.

In Figure 3.35a, b are plotted the Alfven number $M(R)$ and the cylindrical distance to the polar axis of a poloidal field and streamline, normalized to the Alfven cylindrical distance ϖ_a of the same line, $\mathrm{G(R)} = \varpi/\varpi_\mathrm{a}$, for a fixed value of the parameter λ which controls the rotation at the roots of the poloidal field and streamlines at the base, say $\lambda = 10$, for $\kappa = 0$ and several values of ε. Asymptotically, momentum balance in the cylindrical ϖ-direction requires that the outwards centrifugal force is completely balancing the sum of the inwards magnetic pincing force and the inwards gradient of the magnetic pressure associated with the increasing with the cylindrical distance ϖ strength of $\mathrm{B}_\varphi(\varpi)$. Note that for superAlfvenic outflows ($M > 1$, $G > 1$) and the following set of ε values (−10, −1, 0, 1, 10, 20, 70, 117) the corresponding values of $G_\infty = \varpi_\infty/\varpi_a$ are (∞, ∞, ∞, 14, 5.4, 3, 1.5, 1), with the last value $\varepsilon_{\max} = 2\lambda^2(2 - \sqrt{2}) = 117$ corresponding to ($G_\infty \to 1, M_\infty \to 1$), because the flow should remain superAlfvenic, so $M\infty > 1$.

Figure 3.37 is shown a more general ($\kappa \neq 0$) classification scheme, for the asymptotic properties of meridionally self-similar, superAlfvenic outflows. The terminal Alfven number, M_∞, the dimensionless asymptotic radius, G_∞ and asymptotic velocity of the jet, V_∞/V_* depend on three parameters (Π_∞, ε, κ), where Π_∞ denotes the terminal value of the spherically symmetric part of the pressure. Besides this terminal pressure Π_∞, the two other crucial parameters are:

(1) κ, on which depends the radial and longitudinal components of the thermal gas pressure gradient. The outflow can be either overpressured, in relation to the rotational axis of the system ($\kappa < 0$), or underpressured ($\kappa > 0$).
(2) ε, which measures the magnetorotational contribution to the collimation of the outflow. This divides the sources of outflows into two broad classes: those associated with *efficient magnetic rotators* (EMR) corresponding to positive values

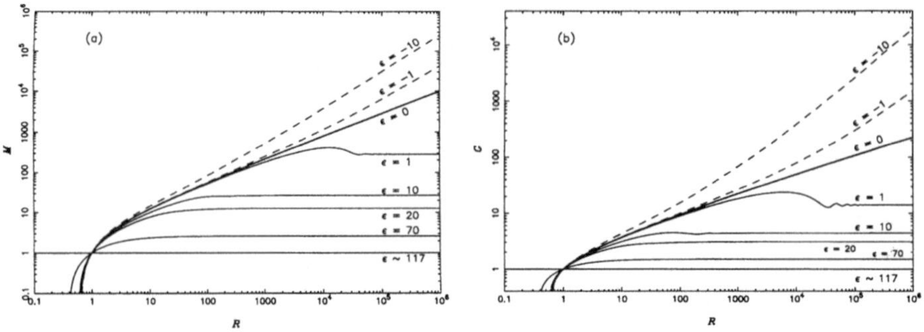

Fig. 3.35 Alfven number $M(R)$ (left) and the dimensionless radius of the jet G(R) (right) are plotted for various values of the energy integral ε from $\varepsilon = -1$ to $\varepsilon = 117$ in *meridionally* self-similar outflows, with spherically symmetric pressure, $\kappa = 0$. Oscillations in the radius of the jet are seen for $\varepsilon > 0$, while negative values of ε correspond to symptotically radial outflows (winds)

of ε wherein there is a strong magnetic contribution to collimation, and those associated with *inefficient magnetic rotators* (IMR) which have $\varepsilon < 0$, and they can be collimated only with the help of the gas pressure gradient, i.e., large values of $\kappa > 0$.

Despite this simple criterion, the asymptotic behavior of the outflow still depends on the value of each parameter taken separately as shown in the classification scheme of Fig. 3.37, as follows:

(a) *Underpressured at the polar axis outflows.* For $\kappa > 0$ the outflow always obtains cylindrical asymptotics. In the formed jet, collimation can be provided either by magnetic pinching or by the gas pressure, depending on the value of ε and the value of the asymptotic gas pressure Π_∞. For IMR ($\varepsilon < 0$) asymptotic magnetic confinement can be achieved only through the strong pressure gradients occurring between the base of the flow and infinity. If Π_∞ does not vanish, the jet radius has a maximum when moving from the magnetic to the thermal regime (by reducing ε).

(b) *Overpressured at the polar axis outflows.* For $\kappa < 0$ jet formation can be achieved only by magnetic pinching, and the jet is supported either by the centrifugal force or by the thermal pressure gradient. In such overpressured outflows, cylindrical configurations are attained only for values of $\varepsilon/2\lambda^2$ higher than some threshold depending on the pressure parameters, Π_∞ and κ. Below this positive threshold value the outflow reaches conical, or paraboloidal asymptotics if $\varepsilon > 0$ (EMR), or only purely conical asymptotics if $\varepsilon < 0$ (IMR).

In other words, we may distinguish two subcases:

(a) If the central body where the magnetic lines of the outflow are *rooted* rotates rapidly, and/or the magnetic field is strong (and we have a so-called *fast magnetic rotator*), and also the thermal pressure increases with the colatitude θ, then the magnetorotational and thermal pressure gradient forces focus the plasma toward the spin axis and thus eventually the outflow takes the form of a jet, such as the aligned outflows from T Tauri newborn stellar objects (jets associated with HH objects, Figs. 3.1, 3.18, and 3.19), or, the highly focused jets from radio quasars, discussed in Chap. 5. A similar situation we have at Fig. 3.35a (albeit with $\kappa = 0$). This corresponds to the upper right (yellow) and upper left (green) quadrant wherein either $[\varepsilon > 0, \kappa > 0]$ and the collimation is due to the magnetorotational forces assisted by thermal pressure, or to thermal pressure confinement $[\varepsilon < 0$ and $\kappa > 0]$.

(b) On the contrary, i.e., when the central body where the magnetic lines of the outflow are *rooted* rotates relatively slowly, and/or the magnetic field is weak

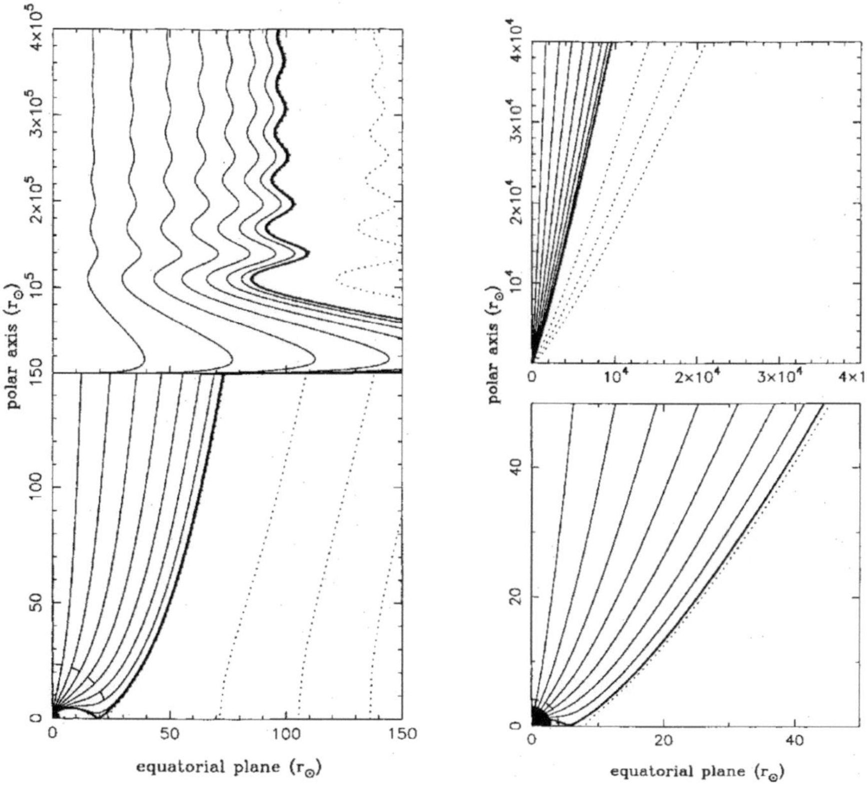

Fig. 3.36 Plot of the shape of the polar streamlines of the flow for two cases of meridionally self-similar solutions of the MHD equations corresponding to: (left), an asymptotically cylindrical outflow, with oscillations in the width of the jet, for $\varepsilon = 0.07$, $\lambda = 0.8$ and a spherically symmetric pressure, $\kappa = 0$, as is the case of a jet originating from the accretion disk of a protostar, or, a radio quasar. The stellar wind (thin solid) is separated from the disk wind (dotted) by the last line connected to the star by a thick solid line, with the dotted lines connected to the disk. (right) An asymptotically radial flow, for $\varepsilon = -2.54$, $\lambda = 1$, $\kappa = 0$, as in the case of the polar solar wind. The Alfven surface is plotted in both cases with a dashed line. In the lower panel is plotted, the initial acceleration and collimation region and on the upper the far region. (*Source* Sauty et al. (1994))

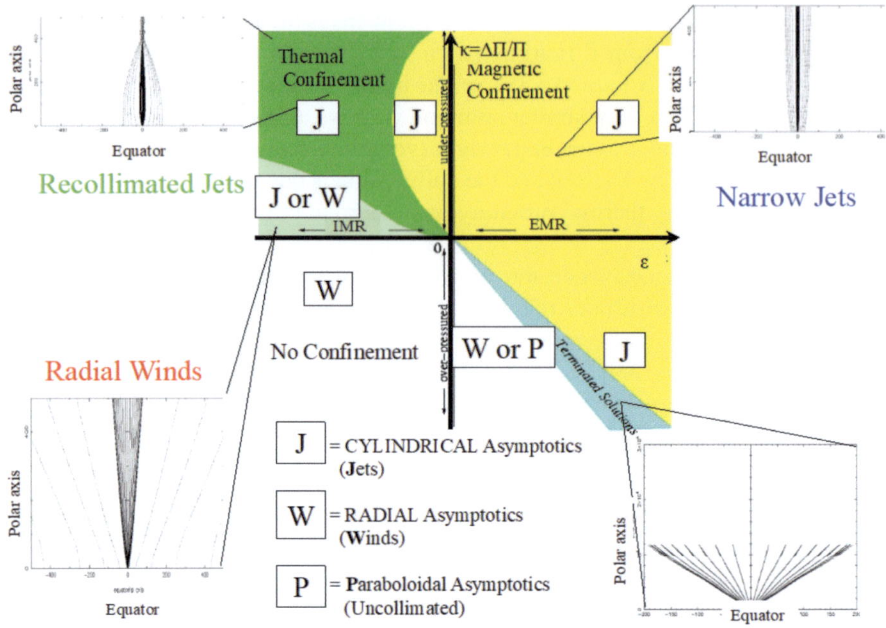

Fig. 3.37 Classification sketch of the various confinement regimes and asymptotical shapes of the meridionally self-similar MHD flows, in the plane $[\kappa/2\lambda^2, \varepsilon/2\lambda^2]$. The thermal pressure confined regime corresponds to the green area, while the magnetically confined regime corresponds to the yellow filled area. In the bluish small area we have terminated solutions. In the remaining (white) areas of the plane only uncollimated solutions (expanding radially, or paraboloidally) are found. [From Sauty et al. (1999)]

(and we have the so-called *slow magnetic rotator*), then the outflow asymptotically extends radially occupying all available space. Then we have what is called a wind, with the archetype example being the classical Parker Solar Wind, or, outflows from radio-quiet Seyfert galaxies. A similar situation we have in Fig. 3.36b (albeit with $\kappa = 0$).

Figure 3.38 shows three representative cases of the diagram 3.37, i.e., a radial wind for $\varepsilon = -0.5$, $\kappa = 0.001$, a magnetically collimated jet for $\varepsilon = 0$, $\kappa = 0.01$ and a recollimated jet with oscillations in its width for $\varepsilon = 0$, $\kappa = 0.01$.

During the evolutionary course of a medium-mass star like the Sun, when the protostellar material from the surrounding disk is adsorbed and the system shrinks, we have an increase in its rotation (angular velocity) such that the protostar rotates rapidly with a period of a few days (a phenomenon analogous to the increase of the spin of a skater when she/he closes her/his hands, Fig. 3.40). After a few tens of millions of years, the magnetized wind removes momentum from the central protostar, and thus we have the transition from a fast magnetic rotator to a slow magnetic rotator. This happens gradually in the evolutionary course of a medium-mass star like the Sun, when the star reaches the age of a few hundred million years.

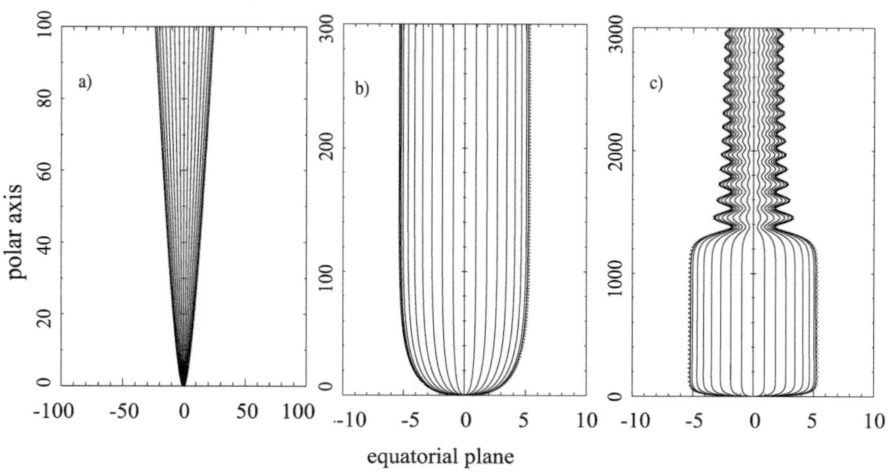

Fig. 3.38 Polar lines of a meridionally self-similar outflow in three cases of *meridionally* self-similar solutions of the MHD equations. They correspond to an asymptotically radial flow with $\varepsilon = -0.5$, $\kappa = 0.01$, as in the polar solar wind (*left, a*), an asymptotically cylindrical outflow, without oscillations with $\varepsilon = 0$, $\kappa = 0.01$ (*middle* b) and an asymptotically cylindrical outflow, with oscillations in the width of the jet, with $\varepsilon = 0$, $\kappa = 0.01$ (*right* c), as is the case of jets originating from the accretion disk of a protostar, or, a radio quasar. [From Sauty et al. (2002)]

Thus, in the case of a protostellar system, as we discuss in the following, the jets play the role of the *ex machina* Deus, stealing momentum from the disk so that the plasma is gradually driven into the central regions of the system, where increasing the density and temperature there, triggers the ignition of thermonuclear reactions and thus the main sequence life of star finally begins (Sauty and Tsinganos 1994).

A Criterion for the Transition from Jets to Winds, in Meridionally Self-similar MHD Outflows

The systematic studies of self-similar outflows led us to the discovery of an MHD *integral* (Sauty and Tsinganos 1994; Sauty et al. 1999, 2002), depending on the value of which we can decide whether the outflow will be asymptotically cylindrical (**jet**) or not (**wind**), as illustrated in Fig. 3.37. The outflow starts subsonically and subAlfvenically from the central gravitating source and its surrounding accretion disk and after crossing the MHD critical points, high values of the Alfven Mach number may be reached. Three broad types of solutions are found: (a) collimated jet-type outflows from efficient magnetic rotators where the outflow is confined by the magnetic hoop stress; (b) collimated outflows from inefficient magnetic rotators where the outflow is cylindrically confined by thermal pressure gradients; and (c) radially expanding wind-type outflows analogous to the solar wind. In most of the cases examined, cylindrically collimated (jet-type) outflows are naturally emerging with thermal and magnetic effects competing in the acceleration and the confinement of the jet.

In the special case of a spherically symmetric gas pressure ($\kappa = 0$), outflows from efficient magnetic rotators (EMR, $\varepsilon > 0$) become asymptotically cylindrically collimated, while outflows from inefficient magnetic rotators (IMR, $\varepsilon < 0$) become asymptotically radial or paraboloidal. Thus, by determining the analytical value of the integral ε at the base predefines the asymptotics of the outflow, as in Fig. 3.36. The transition from winds to jets results when there is a surplus of energy along any non-polar streamline as compared to the available energy along the polar streamline. In this case, the outflow collimates into a jet. Hence, the shape of the streamlines of the outflow from a rotating object switches over from cylindrical to conical asymptotics, as the central object looses angular momentum and passes from the stage of a pre-main sequence YSO to that of a main sequence star.

Based on this integral we can similarly classify the different classes of outflows from active galactic nuclei and quasars, as in Sauty et al. (2007) (see also Bogovalov and Tsinganos 1999, 2001, 2005; Tsinganos and Bogovalov 2000, 2002, 2005; Sauty et al. 2006; Tsinganos et al. 2009) (Fig. 3.40). More on models and constraints on jets from young stars can be found in Ferreira et al. (2007).

2014: MHD Numerical Modeling of Protostellar Jets and Formation of Shocks Along the Jet

There have been published several studies of time-dependent simulations of supersonic jets launched from magnetized accretion disks, using high-resolution numerical techniques as well as laboratory experiments. The interested reader may see, e.g., Zanni et al. (2007) and the proceedings of the JETSET conferences Massaglia et al. (2008), Gracia et al. (2009a, b). In the following, we review an indicative particular MHD model which combines as initial conditions two analytical outflows, a disk wind and a stellar outflow.

In the study of Teşileanu et al. (2014), two analytical MHD outflow solutions are combined as initial conditions, a stellar jet (Analytical Disk Outflow, ADO) and a stellar wind (Analytical Stellar Outflow, ASO). The numerical code PLUTO for computational astrophysics (Mignone et al. 2007) has been used to carry out the MHD simulations. These two analytical MHD outflow solutions are complementary to each other, with their numerical combination naturally addressing their shortcomings. The analytical solutions are normalized to each other, and then properly combined (upper panel of Fig. 3.39). The ADO solution is adopted from Vlahakis et al. (2000) and describes a magneto-centrifugally accelerated disk wind that successfully crosses all three magnetosonic critical surfaces. The ASO model is taken from Sauty et al. (2002) and is a pressure-driven outflow solution with a large lever arm capable of spinning down the protostar (see Mignone et al. (2007), for more details on implementing the ADO and ASO solutions).

Then, started the 2.5D (axisymmetric) numerical simulations for the two-component jet, finally generating synthetic emission maps, by adopting a sophisticated cooling function that follows the ionization and optically thin radiation losses of several ions. In particular, a velocity variability is applied at the base of the outflow, with faster velocity overtaking the slower velocity, in order to produce shocks and knots along the axis. As expected, this variability applied on the bottom boundary indeed produces shocks. These internal shocks locally heat the gas, and in turn the post-shock regions are susceptible to larger energy losses and strong emission. Moreover, the shocks propagate along the axis and locally heat the gas, and in turn the post-shock regions are susceptible to larger energy losses and strong emission. More specifically, the knots have an enhanced surface brightness that can be ten times higher than the bulk flow. The introduced variability has been intentionally chosen to be a few years, producing knot structures every few hundred AU's. This is the typical length scale observed in YSO jets, such as in the systems HH 1&2, HH 34, and HH 47, for which high angular resolution astronomical data are available (e.g., Hartigan et al. 2011).

Young stellar objects (YSOs) frequently show the collimated outflows to terminate in a hot spot, something usually interpreted as the bow shock originated by the head of a supersonic jet as it interacts with the external medium. In addition to this hot spot, some stellar jets present knots of emission along their axes, whose origin has been ascribed to Kelvin–Helmholtz (K–H) reflected modes evolved into shocks. In connection to this phenomenology, it has been analyzed the stability of an infinite cylindrical fluid beam undergoing radiative losses and it has been found that radiative cooling has a stabilizing effect on reflected modes, that critically depends on small variations of the temperature of the jet's matter. Also, together with K–H modes, a radiative jet can be made unstable by thermal instability (Ray, 1981; Massaglia et al. 1992).

The jet velocity is on average 380 km s^{-1}, whereas the speed of the shocks relative to the bulk flow is about 100 km s^{-1}. In general, the models correctly capture the typical values of observed YSO jets. Optically thin radiation losses and adaptive mesh refinement (AMR) are employed, that significantly increase the effective resolution.

The collisionally excited emission lines of observational interest treated in this work are the forbidden emission-line doublets of N II (6584 + 6548 Å), OI (6300 + 6363 Å), and SII (6717 + 6727 Å). The cooling function includes a network of a maximum of 29 ion species, selected appropriately to capture most of the radiative losses in YSO jets up to temperatures of 200,000 K. These ion species represent the first five ionization states (from I to V) of C, N, O, Ne, and S, as well as H I, H II, He I, and He II. Finally, 2D emission maps were calculated from the simulation output. In the present work, the first three ionization states were used, a choice that is adequate for the physical conditions developing in the areas of interest.

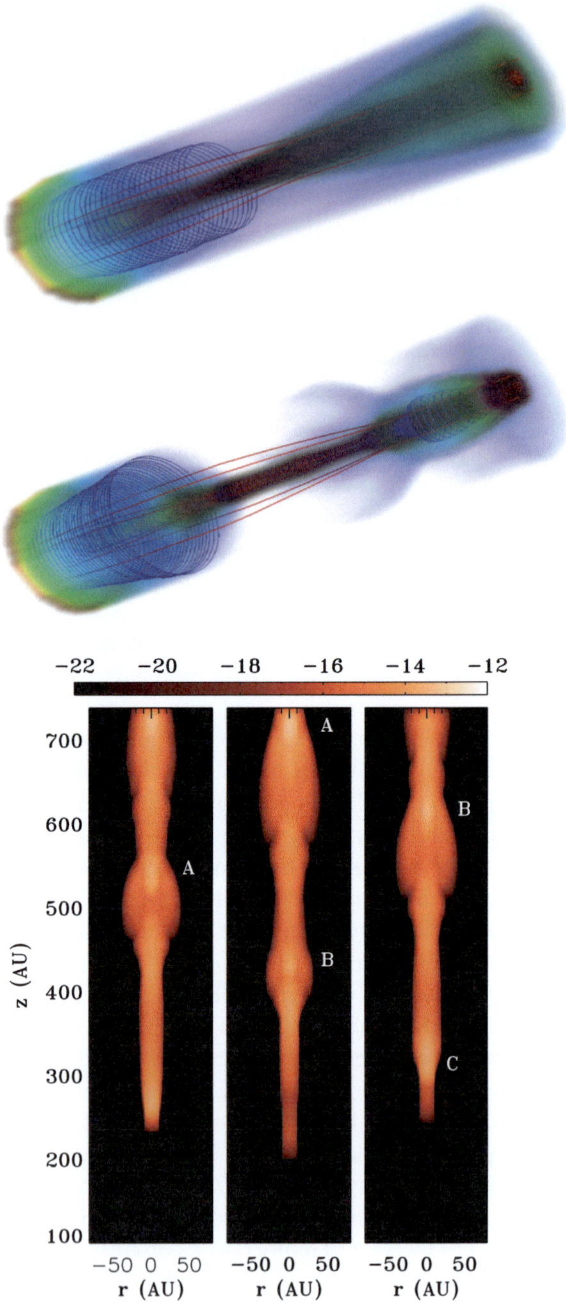

Fig. 3.39 (*top panel*) A typical magnetohydrodynamic model of the jets associated with protostars, such as of HH 34, HL Tau and RW Aur. Shown is the 3D representation of the density distribution of the initial conditions (*top*) and at a later time during evolution (*bottom*). Blue/red corresponds to low/high density, the thin blue circular lines denote the azimuthal magnetic field focusing the jet and the thick red lines the streamlines of the flow. (*bottom panel*) Analytical/numerical model, which is a combination of two-component analytical outflow solutions, a stellar jet and a disk wind and subsequently generating synthetic emission maps. Logarithmic maps of the surface brightness of the forbidden doublet [SII] (6717 + 6727 Å). The image is in units of erg cm^{-2} $arcsec^{-2}$ s^{-1} and has been projected with a declination angle of 80° with respect to the line of sight, convolved with a point spread function, for three time intervals, 16.8, 19.2, and 21.6 years in modeling the outflow of a magnetized jet from a protostar and disk system. The jet is propagating upward, and the 3 knots are labeled with the letters A, B, and C. (*Source* Tesileanu et al. (2014)]

The longitudinal velocity decreases with the jet radius, i.e., the flow consists of a fast jet close to the axis and a slow wind at the outer radii. During the first steps of the simulation, the cooling function lowers the temperature of the plasma, especially in the inner hot regions. Owing to the pressure drop, the collimating forces squeeze the jet, and in turn the jet radius is decreased (upper panel of Fig. 3.39).

The results of the modeling have been compared with the observations of the YSOs RW Aur, HH 30, and HL. The jet radius, as well as the opening angle, are found to be close to these typical YSO jets. The dense and hot inner part of the jet emits strongly, with the synthetic emission maps showing a well-collimated outflow that closely resembles real observations. The emission knots propagate along the axis and demonstrate enhanced emission because they are dense post-shock regions (Fig. 3.39). The predicted emission-lines match the observations of the YSO jets at high altitudes. The simulated jets have a width in the interval 40–60 AU, in good agreement with observations (Ray et al. 2001 and references therein). The variations in emission along the jet axis are due to the applied speed variability. The opening angle is also comparable with observations and it is a few degrees.

The presence of the less dense and slower surrounding wind cannot be seen in this emission line. Nevertheless, it fills up the rest of the computational domain and is considered to emanate from the outer disk radii, contributing to the mass and angular momentum loss rates.

2009: Protostellar Jets: The "*Ex Machina Deus*" in Star Formation…

As we may have noticed when the water swirls in our bathtub or sink, as it approaches the discharge siphon, its rotational speed increases at the expense of the radial incidence speed. Similarly in an astrophysical accretion disk, the plasma swirls around the central body with some initial angular velocity and the adsorbed material increases its rotational speed as it approaches toward the central body. It is, therefore, necessary to remove angular momentum at some rate, so that the plasma can also have a (usually small) radial velocity toward the center in order to reach the central body.

In other words, for a protostar to form, the disk material must get rid of the initial angular momentum it possesses and which is of course conserved in an isolated system, as is well-known. That is, angular momentum conservation is the enemy of star formation! Therefore, for a star to form, the protoplanetary nebula must get rid of a large percentage of its initial angular momentum. But how is this done?

With the formation of jets! Specifically, as the star tries to emerge from the dark cloud wherein typically starts its life, plasma jets are fired from the disk that propagate intact for large distances, many thousands of times the Earth–Sun distance. As illustrated in Fig. 3.40, the focused disk wind (the jet), removing only a small percentage of the accreted mass, may abduct almost all of its angular momentum, thus allowing the plasma to reach the central condensation. This is a very interesting result and shows how important is the role of focused magnetized disk winds. For this reason, we might not be here today if the "*angular momentum thieves (jets)*" did not accompany star formation!

Actually, a fairly easy way to efficiently remove angular momentum from an astronomical accretion disk is through the outflow of a focused jet, which removes a significant part of the angular momentum of the disk, without at the same time removing a relatively large percentage of its mass. This is an extremely interesting result and it shows how important is the role of the magnetized disk winds (jets), which although steal only a small percentage of mass from the spinning material of the disk, they do remove most of its angular momentum leaving the remaining free to collapse into the central condensation to form the protostar. Hence, jets from newborn stars play the role of an *ex machina Deus*[6] because stealing momentum they leave gravity free to make the star. For that reason, maybe we would not be here today if jets did not accompany star formation, since the natural law of conservation of the disk angular momentum would not allow the collapse of the material for the formation of the protosun and subsequently of the planets from the remaining material, (Fig. 3.40)[7]!

Extracting Angular Momentum From an Accretion Disk via an MHD Disk Wind

A related way to efficiently extract angular momentum from an accretion disk can be done when from the disk we have a plasma outflow which removes a significant part of the disk's angular momentum, without simultaneously removing necessarily a large percentage of its mass. Here is how this could be understood.

Suppose that a magnetized disk wind blows from a disk, removing in the disk wind a percentage f of its accreting angular momentum with $0 < f < 1$,

$$\dot{J}_w = f \dot{J}_a .$$

In this disk wind, the plasma is decoupled from the magnetic field when it is beyond the cylindrical Alfvén distance up to which it rotates like a solid body, while from the disk the plasma is released beyond the distance ϖ_0 where the magnetic line intersects the plane of symmetry of the disk, as we discussed in Fig. 3.24. That is, the angular momentum of an outflowing elementary mass δm_w of the disk

---

[6] The term originates in Greek tragedy. At the climax of the tragedy, when the plot of the drama seems to be leading to a dead end, an actor suddenly appears on the scene, with the help of some mechanism, impersonating some deity or divine person, to solve the impasse. For example, in "Medea" of Euripides, the god Helios listening to her at the request of Horus, sends a chariot drawn by dragons to carry his child-killing grand daughter Medea away from her husband Jason to the safety of Athens.

[7] Actually, our research group at the Un. of Athens has contributed to this interesting result.

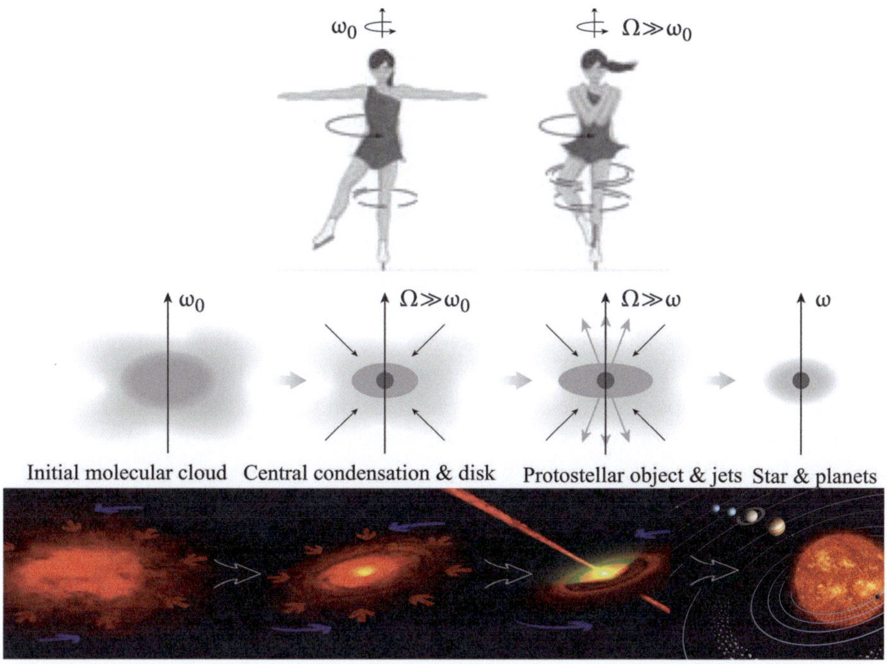

Fig. 3.40 Collapse stages of a dense molecular nebula: a roughly spherical nebula has an initial rotational frequency ω_o, as does the ice skater, when her arms are open. Then, when the nebula collapses to form a lenticular structure with a protostellar core in the center and a surrounding disk, its rotation frequency increases to $\Omega >> \omega_o$, just like the rotation of the ice skater when she closes her hands. Then, the collapse stops due to the high spin of the material, which as it shrinks rotates with an ever-increasing frequency Ω. At this stage, the jets appear, like the “*ex machina Deus*”, which steal momentum and reduce the spin of the star to $\omega << \Omega$ leaving the material free to collapse into the center and form the new star surrounded by its planets

wind, rotating at the Alfvén distance ϖ_* is $\delta m_w \Omega_K \varpi_*^2$ and if this mass is removed via the disk wind, the rate of removal of angular momentum via the disk wind is $(\delta m_w/\delta t)\Omega_K \varpi_*^2 = \dot{J}_a$. On the other hand, a ring of matter of mass δm_a rotating at a distance ϖ_0 with angular velocity Ω_K on the disk has angular momentum $\delta m_a \Omega_K \varpi_0^2$. Half of this angular momentum is carried away by one jet (with the other half by the other) and therefore the angular momentum of this ring is $\left[\delta m_a \Omega_K \varpi_0^2\right]/2$. Thus, the rate of change of the rotational momentum of the disk is $\left[(\delta m_a/\delta t)\Omega_K \varpi_0^2\right]/2 = \dot{J}_a$. Equating the fraction f of this rate of angular momentum that must be removed with that removed by the disk wind, we get,

$$\dot{J}_w = \dot{M}_w \Omega_K \varpi_a^2 = \frac{f}{2}\dot{M}_* \Omega_K \varpi_0^2,$$

from which it follows,

$$\frac{\dot{M}_w}{\dot{M}_a} = \frac{f}{2}\frac{\varpi_0^2}{\varpi_*^2}.$$

For a given streamline, the Alfvén arm ϖ_* is usually considerably larger than the corresponding rotation arm of the streamline line in the disk plane, ϖ_0, as we saw in the Weber and Davis model for magnetized winds. To give a specific example, let $\varpi_* = 7\varpi_0$. Then, from the previous equation it is obvious that the disk wind, by removing only a few percent of the adsorbed mass, can remove almost all of its angular momentum, thus allowing the plasma to accumulate on the central condensation. This is an extremely interesting result and shows how important the role of magnetized disk winds is, which, although they may remove only a small percentage of mass from the spinning disk material, remove almost all of its angular momentum, leaving the rest free to collapse into the central condensation to form the protostar.

For this reason, we might not be here today if the protosolar jets did not accompany the formation of the protosun, since the physical law of conservation of the disk's angular momentum would not allow the collapse of the material to form the protosun and subsequently the planets from its remaining material!

2024: Cheers! Webb Finds Ethanol and Other Icy Ingredients in IRAS 2A (10^{28} Bottles of Whiskey in Another Nebula in Space…)

Molecular clouds of hydrogen and dust, typically, have interstellar medium densities of 10–30 cm^{-3} and are in near-absolute cold, around 270° below zero, or three degrees above absolute zero, i.e., they have densities and temperatures which allow the formation of molecules, mainly molecular hydrogen. They constitute approximately 50% of the total interstellar gas in a galaxy, unlike other regions of the interstellar medium which contain mainly ionized gas.

As we discussed above, giant molecular clouds are the largest structures in galaxies: they have a mass of thousands to millions of solar masses and dimensions of 30–300 light years, i.e., they are capable of creating millions of stars each. Star formation, however, occurs in their densest, darkest, and coldest parts, where the temperature is extremely low, around 10 degrees Kelvin, i.e., 263 °C below zero. At these truly freezing temperatures, the main component of dark nebulae is molecular hydrogen, with a density of about 100–500 molecules per cubic centimeter, higher than the density of the interstellar material which is about one molecule per cubic centimeter. A typical galaxy, like our own Milky Way, or the Andromeda galaxy, contain 1,000–2,000 giant molecular clouds. Also, dark clouds contain dust, about 1% of their mass, which, nevertheless, protects the contained molecules from the hot ultraviolet radiation of neighboring hot stars. They also contain carbon monoxide (CO), ammonia (NH_3), amino acids, ethanol, poly-aromatic hydrocarbons, etc.

The NASA/ESA/CSA James Webb Space Telescope have recently discovered (Rocha et al. 2024) a variety of molecules, ranging from relatively simple ones like methane (CH_4, the simplest flammable hydrocarbon[8]) to complex compounds like acetic acid (CH_3CO_2H, a byproduct of fermentation, giving vinegar its characteristic odor) and ethanol (CH_3CH_2OH a clear, colorless liquid and an ingredient in alcoholic beverages like beer, wine or brandy), in early-stage protostars where planets have not yet formed. These are key ingredients for making potentially habitable worlds.

The presence of *complex organic molecules* (COMs) in the *solid phase* in protostars was first predicted decades ago from laboratory experiments, and tentative detections of these molecules have been made by other space telescopes. This includes Webb's Early Release Science program *Ice Age*, which discovered diverse ices in the darkest, coldest regions of a molecular cloud measured to date.

Now, with the unprecedented spectral resolution and sensitivity of Webb's Mid-InfraRed Instrument (MIRI), as part of the JOYS + (James Webb Observations of Young ProtoStars) program, (Rocha et al. 2024, Fig. 3.41) these COMs have been individually identified and confirmed to be present in the interstellar ices. This includes the robust detection of acetaldehyde (CH_3CHO a colorless, flammable liquid generated during alcohol metabolism, with pharmacological properties), ethanol (CH_3CH_2OH what we call in everyday life alcohol), methyl formate ($HCOOCH_3$ a flammable liquid hazardous substance used as an insecticide, and also used to manufacture various organic chemicals) and likely acetic acid (CH_3CO_2H the acid in vinegar), in *the solid phase*.

One of the long-standing questions in *astrochemistry*, concerns the origin of those COMs in space. In particular, if they are made in the gas phase or in ices. The detection of COMs in ices suggests that solid-phase chemical reactions on the surfaces of cold dust grains can build complex kinds of molecules (Rocha et al. 2024).

As several COMs, including those detected in the solid phase Webb, were previously detected in the warm gas phase, it is now believed that they originate from the sublimation of ices (sublimation is a change directly from a solid to a gas without becoming a liquid). Therefore, detecting COMs in ices makes us hopeful about developing an improved understanding of the origins of other even larger molecules in space.

We are also keen to explore to what extent these COMs are transported to planets at much later stages in the evolution of the protostar (Fig. 3.41). COMs in ices are transported more efficiently into planet-forming disks, than gas from clouds. These icy COMs can, therefore, be inherited by comets and asteroids which in turn may collide with planets in formation. In that scenario COMs can be delivered to those planets, potentially providing the ingredients for life to flourish.

---

[8] Although the concentration of methane in Earth's atmosphere is small (around 1.8 parts per million), it is an important greenhouse gas because it is such a potent heat absorber. The concentration of methane in our atmosphere has risen by about 150% since the start of the industrial revolution (ca. 1750), apparently largely due to human activities. Methane accounts for about 20% of the heating effects by all of the greenhouse gases combined. Both natural and human sources supply methane to Earth's atmosphere.

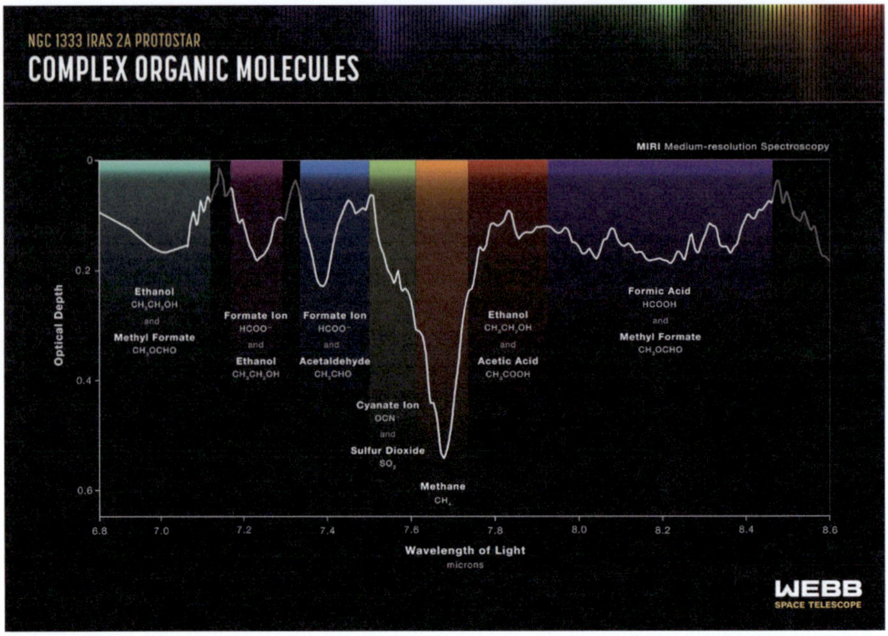

Fig. 3.41 James Webb Space Telescope has identified a wealth of complex organic molecules surrounding two protostars. This graphic shows the spectrum of one of the two protostars, IRAS 2A. It includes the fingerprints of acetaldehyde, ethanol, methylformate, and likely acetic acid, in the solid phase. (Image Credit: NASA, ESA, CSA, L. Hustak (STScI))

All of these molecules can become part of comets and asteroids and eventually new planetary systems when the icy material is transported inward to the planet-forming disks as the protostellar system evolves, as the second co-author of the paper Rocha et al. (2024), van Dishoeck, has emphasized. The science team in Rocha et al. (2024) also detected simpler molecules, including methane, formic acid (which makes the sting of ants painful), sulfur dioxide, and formaldehyde. Sulfur dioxide in particular allows the team to investigate the sulfur budget available in protostars. In addition, it is of prebiotic interest because existing research suggests that sulfur-containing compounds played an important role in driving metabolic reactions on the primitive Earth. Negative ions were also detected; they form part of salts that are crucial for developing further chemical complexity at higher temperatures. This indicates that the ices may be much more complex.

Of particular interest is that one of the sources investigated, IRAS 2A, is characterized as a low-mass protostar. IRAS 2A may therefore have similarities with the primordial stages of our own solar system. If that is the case, the chemical species identified in this source were likely present in the first stages of development of our solar system and were later delivered to the primitive Earth.

In a light note related to this issue, we mention that in another molecular cloud, the large molecular nebula B2 in Sagittarius, near the center of the Galaxy, an Irish researcher and my friend – Professor Tom Ray – estimated that it contains about 10^{28} bottles of whiskey!

Recently, it has been reported the first detection of a "*handed*" molecule, (propylene oxide) in interstellar space, near the center of our Galaxy in Sagittarius (Sgr) B2, a massive star-forming region. Propylene oxide is an acutely toxic and carcinogenic organic compound. This colorless volatile liquid with the molecular formula CH_3CHCH_2O, one of a class of the so-called *chiral* molecules – molecules that have an identical chemical composition, but right- and left-handed versions. These *chiral* molecules are essential for life and their discovery in deep space may help us understand why life on Earth relies on a certain handedness to perform key biological functions.

2024: Planet-Forming Disks in the Early Universe and Low Metallicity Nebulae Live Longer

A comprehensive summary of the various stages of star formation in our galaxy is schematically presented in Fig. 3.42 wherein the various stages of formation of a medium-mass star such as the Sun are schematically described, as well as their duration and observed emission of radiation at each stage. For more details, as well as the latest developments on the subject, the interested reader may find the proceedings of an international conference on star formation organized in the island of Rhodes in 2008 (Tsinganos et al. 2009).

Regarding the duration of the life of protoplanetary disks in low metallicity environments, such as in the early Universe, a new understanding has emerged recently thanks to observations by Hubble and the James Webb Space Telescope. First, in 2003, Hubble spotted a massive planet, 2.6 the mass Jupiter, around a very old star, almost as old as the Universe. This planet lies in the core of the ancient globular star cluster M4, located 5,600 light years away in the constellation of Scorpius. As known, globular clusters are deficient of heavier elements because they formed rather early in the Universe wherein heavier elements had not been cooked up in abundance in the nuclear furnaces of stars. So, it had been argued that globular clusters cannot contain planets. Then, this Hubble observation implied that some planet formation happened when our Universe was very young. And those types of planets had time to form and grow big inside their young disks, to grow even bigger than Jupiter. Yet, back then, stars possessed only small amounts of heavier elements that are the building blocks of planets. So, the question arises on how such planets could form.

A similar situation occurs in the nebula NGC 346 in the SMC, one of the most dynamic star-forming regions in nearby galaxies (Fig. 3.2a). NGC 346 is also more primeval in that it lacks a large percentage of the heavier elements that are forged in successive generations of stars through nuclear fusion. Therefore, its metallicity is similar to that of the early Universe, i.e., the cluster serves as a nearby proxy for studying stellar environments with similar conditions in the early, distant Universe.

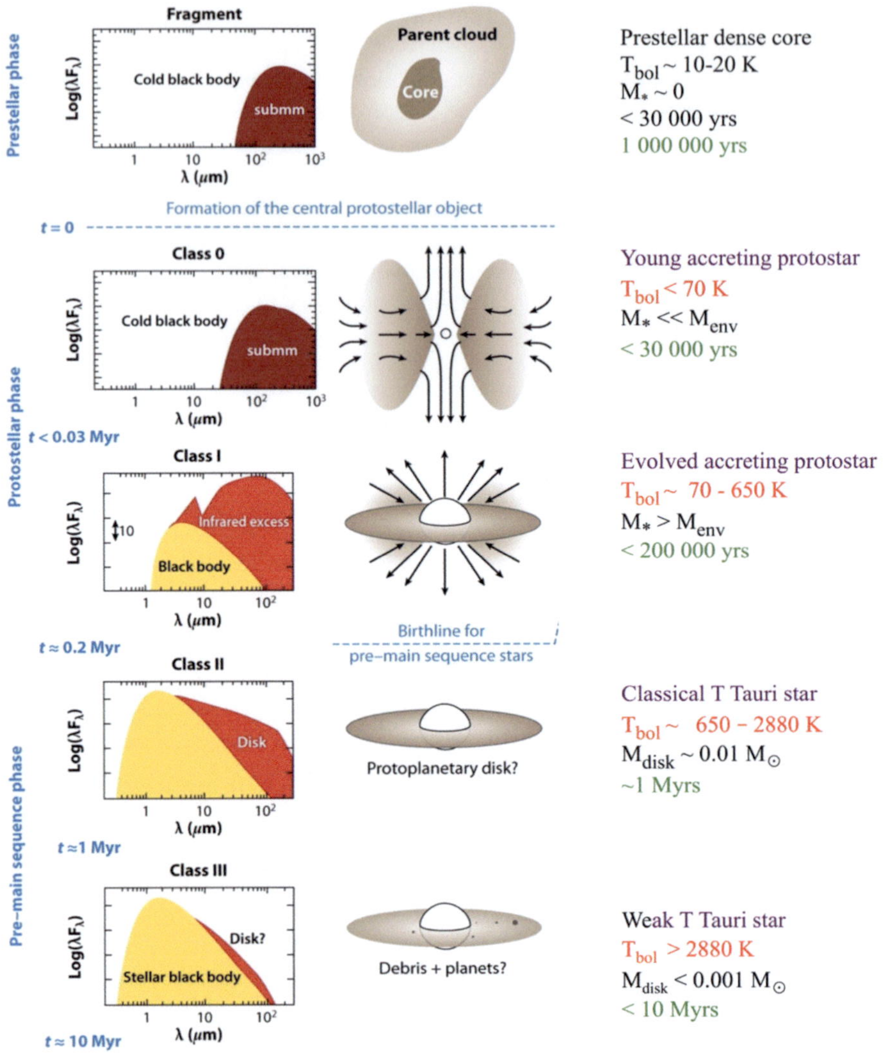

Fig. 3.42 Young stellar objects evolutionary stages/classes/parameters. (*left column*) Evolutionary stage of the star formation process from the protostellar dense core (first case on top, to young accreting/evolved protostar, Class 0/Class I), to the pre-main sequence Phase (Classical T Tauri/Weak T Tauri star, Class II/Class III. (*right column*) Temperature, circumstellar mass and timescale for each evolutionary stage. [Figure adapted from Lada (1987), Andre (2002) and Wikimedia Commons]

Hubble observations of NGC 346 from the mid-2000s revealed many stars about 20–30 million years old that seemed to still have planet-forming disks around them. However, this went against the conventional belief that such disks would dissipate after a few million years (Fig. 3.42).

In NGC 346 and in the early Universe, stars formed from mostly hydrogen and helium, and very few heavier elements such as carbon and iron, which came through the first generation of supernova explosions. And, to be able to blow away the disk, the star's light applies pressure to the gas in the disk. This process is more efficient when the gas contains elements heavier than hydrogen and helium. The massive star cluster NGC 346 has only about 10% of the heavier elements that are present in the chemical composition of our Sun. So, perhaps it simply takes longer for a star in this cluster to disperse its disk.

Hence, we see that in NGC 346 these new stars are indeed surrounded by disks and are still in the process of gobbling material, even at the relatively old age of 20 or 30 million years, something that implies that planets have more time to form and grow around these stars than in nearby star-forming regions in our own galaxy (Fig. 3.42).

There are two possibilities to explain this finding. In the *first*, to be able to blow away the disk, the star applies radiation pressure, which in order to be effective, elements heavier than hydrogen and helium would have to reside in the gas. With the massive star cluster NGC 346 only having about 10% of the heavier elements that are present in the chemical composition of our Sun, it simply takes longer for a star in this cluster to disperse its disk.

The *second* possibility is that, for a Sun-like star to form when there are few heavier elements, it would have to start from a larger cloud of gas. A bigger gas cloud will produce a bigger disk. So, there is more mass in the disk, and therefore it would take longer to blow the disk away, even if the radiation pressure were working in the same way.

References

André P (2002) Representation of the evolution of young stellar objects. Nearby protoclusters as laboratories for understanding star formation on galactic scales.Astrophys Space Sci 281(1):51–66

Andre P (2002) The initial conditions for protostellar collapse: observational constraints. In:Bouvier J, Zahn JP (eds) Star formation and the physics of young stars, Les Ulis, Fr.: EDP Sci, pp 1–38

Andrews SM, Huang J, Pérez LM et al (2018) The disk substructures at high angular resolution project (DSHARP). I. Motivation, sample, calibration, and overview. ApJ 869:L41

Arce HG, Shepherd D, Gueth F, Lee C-F, Bachiller R, Rosen A, Beuther H (2007) Molecular outflows in low- and high-mass star-forming regions. Protostars Planets V 245

Bacciotti J, Wheelan E, Testi L (eds) (2008) Jets from young stars II. Clues from High angular resolutions observations, vol 742. Springer, LNP

Belcher JW, MacGregor KB (1976) Magnetic acceleration of winds from solar-type stars. Astrophys J 210:498–507

Blandford RD, Payne DG (1982) Hydromagnetic flows from accretion disks and the production of radio jets. Mon Not R Astron Soc 199:883–903

Bogovalov S (1995) Formation of jets during the ejection of plasma by an axisymmetric rotator. Astron Lett 21(4):565–571

Bogovalov S, Tsinganos K (1999) On the magnetic acceleration and collimation of astrophysical outflows. MNRAS 305:211–224

Bogovalov S, Tsinganos K (2001) Magnetically collimated jets with high mass flux. MNRAS 325:249–256

Bogovalov S, Tsinganos K (2005) Shock formation at the magnetic collimation of relativistic jets. Mon Not R Astron Soc 357(47):918–928

Ceccobello C, Cavecchi Y, Heemskerk MHM, Markoff S, Polko P, Meier D (2018) A new method for extending solutions to the self-similar relativistic magnetohydrodynamics equations for black hole outflows. Mon Not R Astron Soc 473(4):4417–4435

Congiu E et al (2025) The MUSE view of the Sculptor galaxy: survey overview and the planetary nebulae luminosity function. Astron Astrophys

Duncan JC (1920) Bright nebulae and star clusters in Sagittarius and Scutum photographed with the 60-inch reflector. Astrophys J 51:4–12

Ferreira J, Dougados C, Wheelan E (eds) (2007) Jets from young stars I. In: Models and constraints, vol 723. Springer, LNP

Gracia J, de Colle F, Downes T (eds) (2009a) Jets from young stars V. In: High performance computing and applications, vol 791. Springer, LNP

Gracia J, Vlahakis N, Agudo I, Tsinganos K, Bogovalov SV (2009b) Synthetic synchrotron emission maps from MHD models for the jet of M87. Astrophys J 695:503–510

Gracia PJV, Ferreira JM (eds) (2010) Jets from young stars IV. From models to observations and experiments, vol 793. Springer, LNP

Hartigan P, Frank A, Foster JM, Wilde BH, Douglas M, Rosen PA, Coker RF, Blue BE, Hansen JF (2011) Fluid dynamics of stellar jets in real time: third epoch Hubble space telescope images of HH 1, HH 34, and HH 47. ApJ 736:29

Heyvaerts J, Norman CA (1989) The collimation of magnetized winds. Ap J 347:1055

Isella A, Huang J, Andrews SM et al (2018) The disk substructures at high angular resolution project (DSHARP). IX. A high-definition study of the HD 163296 planet-forming disk. ApJ 869:L49

Jeans JHI (1908) The stability of a spherical nebula. In: Philosophical transactions of the royal society of London. Series A, vol 199 (CXCI), pp 312–320. https://doi.org/10.1098/rsta.1902.0012

Lada CJ (1987) Star formation—from OB associations to protostars, star forming regions. In: Dordrecht DR (ed) Proceedings of the symposium, Tokyo, Japan, 11–15 Nov 1985 (A87-45601 20-90) Publishing Co., pp. 1–17

Lima JJG, Priest ER, Tsinganos K (2001) An analytical MHD wind model with latitudinal dependences obtained using separation of the variables. Astron Astrophys 371:240–249

Machida MN, Inutsuka S, Matsumoto T (2008) High-and low-velocity magnetized outflows in the star formation process in a gravitationally collapsing cloud. Ap J 676:1088–1108

Madau P, Dickinson M (2014) Cosmic star-formation history. Annu Rev Astron Astrophys 52:415–486. arXiv:1403.0007

Massaglia S, Bodo J, Mignone A, Rossi P (eds) (2008) Jets from young stars III. In: Numerical MHD and instabilities, vol 754. Springer, LNP

Massaglia S, Trussoni E, Bodo G, Rossi P, Ferrari A (1992) Radiative unstable modes in the jets of young stellar objects. Astron Astrophys 260(1–2):243–249

Matsakos T, Massaglia S, Trussoni E, Tsinganos K, Vlahakis N, Sauty C, Mignone A (2009) Two-component jet simulations. II. Combining analytical disk and stellar MHD outflow solutions. Astron Astrophys 502(1):217–229

Matsakos T, Tsinganos K, Vlahakis N, Massaglia S, Mignone A, Trussoni E (2008) Two-component jet simulations. I. Topological stability of analytical MHD outflow solutions. Astron Astrophys 477(2):521–533

Matsakos T, Vlahakis N, Tsinganos K, Karampelas K, Sauty C, Cayatte V, Matt SP, Massaglia S, Trussoni E, Mignone A (2012) Velocity asymmetries in young stellar object jets. Intrinsic Extrinsic Mech, A&A 545:A53

Matsushita Y, Takahashi S, Machida MN, Tomisaka K (2019) A very compact extremely high velocity flow toward MMS 5/OMC-3 revealed with ALMA. arXiv:1811.08060v1

Michel FC (1969) Relativistic stellar-wind torques. Astrophys J 158:727

Mignone A, Bodo G, Massaglia S, Matsakos T, Tesileanu O, Zanni C, Ferrari A (2007) PLUTO: a numerical code for computational astrophysics. ApJS 170:228

Parker EN (1958) Dynamics of the interplanetary gas and magnetic fields. Astrophys J 128:664

Parker EN (1963) Interplanetary dynamical processes. Wiley Intersience, New York

Parker EN (1965) The passage of energetic particles through interplanetary space, planet. Space Sci 13(1):9–49

Parker EN (1965a) On the existence of slow solutions in coronal hydrodynamics. Astrophys J 141:320–324

Parker EN (2001) A history of early work on the heliospheric magnetic field. J Geophys Res 106(A8):15797–15802

Petrie GJD, Tsinganos K, Neukirch T (2005) Steady 2D prominence-like solutions of the MHD equations with field-aligned compressible flow. Astron Astrophys 429:1081–1092

Petrie GJD, Vlahakis N, Tsinganos K (2002) Systematic construction of exact 2-D MHD equilibria with steady, compressible flow in Cartesian geometry and uniform gravity. Astron Astrophys 382:1081–1092

Ray TP (1981) Kelvin-Helmholtz instabilities in radio jets. Mon Not R Astron Soc 196:195–207

Ray TP (1998) Jets: a star formation perspective. In: Massaglia S, Bodo G (eds) Astrophysical jets: open problems. Amsterdam, The Netherlands: Gordon and Breach Science Publishers, p 173

Ray TP (2001) Setting the scene: a star formation perspective. In: Williams IP, Thomas N (eds) Solar and extra-solar planetary systems, Lecture notes in physics, vol 577, p 12

Reipurth B, Bally J, Graham JA, Lane AP, Zealey WJ (1986) The jet and energy source of HH 34. Astron Astrophys 164:51–66

Rocha WRM, van Dishoeck EF, Ressler ME, van Gelder ML, Slavicinska K, Brunken NGC, Linnartz H, Ray TP, Beuther H, Caratti o Garatti A, Geers V, Kavanagh PJ, Klaassen PD, Justtanont K, Chen Y, Francis L, Gieser C, Perotti G, Tychoniec Ł, Barsony M, Majumdar L, le Gouellec VJM, Chu LEU, Lew BWP, Henning T, Wright G (2024) JWST observations of young protoStars (JOYS+): detecting icy complex organic molecules and ions. A&A 683

Sakurai T (1985) Magnetic stellar winds: a 2-D generalization of the Weber-Davis model. A&A 152:121

Sakurai T (1987) Magnetically collimated winds from accretion disks. Astron Soc Jpn Publ 39(5):821–835. ISSN 0004-6264

Sauty C (2019) JET simulations experiments and theory. In: Ten years after Jetset: What is next? Astrophysics and space science proceedings, vol 55. Springer Nature Switzerland AG. ISBN 978-3-030-14127-1

Sauty C, Lima JJG, Iro N, Tsinganos K (2005) Nonradial and nonpolytropic astrophysical outflows. VII. Fitting ULYSSES solar wind data during minimum. Astron Astrophys 432(2):687–698. https://doi.org/10.1051/0004-6361:20041606

Sauty C, Lima JJG, Tsinganos K, Aibeo A, Meliani Z, Vlahakis N (2007) Solar wind and stellar jets, from newtonian to relativistic ones. AIP Conf Proc 895:87–96

Sauty C, Meliani Z, Lima JJG, Tsinganos K, Cayatte V, Globus N (2011) Nonradial and nonpolytropic astrophysical outflows IX. Modeling T Tauri jets with a low mass-accretion rate, Astron Astrophys 533:A46

Sauty C, Meliani Z, Trussoni E, Tsinganos K, Vlahakis N (2006) Relativistic jet modeling: application to AGN, Albert Einstein century international conference. AIP Conference Proceedings, vol 861, pp 736–742

Sauty C, Trussoni E, Tsinganos K (2002) Nonradial and nonpolytropic astrophysical outflows. V. Acceleration and collimation of self-similar winds. Astron Astrophys 389:1068–1085. arXiv: astro-ph/0205203

Sauty C, Trussoni E, Tsinganos K (2004) Nonradial and nonpolytropic astrophysical outflows VI. Overpressured winds and jets. A&A 421:797–809

Sauty C, Tsinganos K (1994) Nonradial and nonpolytropic astrophysical outflows. III. The transition from jets to winds. Astron Astrophys 287:893–926

Sauty C, Tsinganos K, Trussoni E (1999) Nonradial and nonpolytropic astrophysical outflows. IV. Magnetic or thermal collimation of winds into jets? Astron Astrophys 348:327–349

Schwartz RD (1977) Evidence of star formation triggered by expansion of the Gum Nebula. Ap J Lett 212:L25

Teşileanu O, Matsakos T, Massaglia S, Trussoni E, Mignone A, Vlahakis N, Tsinganos K, Stute M, Cayatte V, Sauty C, Stehlé C, Chièze J-P (2014) Young stellar object jet models: from theory to synthetic observations. Astron Astrophys 562(A117):9

Tomisaka K (2002) Collapse of rotating magnetized molecular cloud cores and mass outflows. Ap J 575:306

Tsinganos K (1981) MHD equilibrium. I. Exact solutions of the equations. Astrophys J 245:764–782

Tsinganos K (1982a) MHD equilibrium. II. General integrals of the equations with one ignorable coordinate. Astrophys J 253:775–790

Tsinganos K (1982b) MHD equilibrium. III. Helically symmetric field. Astrophys J 259:820–831

Tsinganos K (1982c) MHD equilibrium. IV. Non-equilibrium of non-symmetric hydromagnetic topologies. Astrophys J 259:832–834

Tsinganos K (1996) Solar and astrophysical magnetohydrodynamic flows, NATO ASI Series. Ser C Math Phys Sci 481. Dordrecht, Boston: Kluwer Academic Publishers

Tsinganos K, Bogovalov SV (2000) Magnetic collimation of the solar and stellar winds. Astron Astrophys 356:989–1002

Tsinganos K, Bogovalov SV (2002) Magnetic collimation of relativistic outflows in jets with a high mass flux. MNRAS 337:553–558

Tsinganos K, Bogovalov SV (2005) Magnetic collimation of the relativistic jet in M87. In: 2nd international symposium on high energy gamma-ray astronomy, AIP conference proceedings, vol 745, pp 148–159

Tsinganos K, Ray T, Stute M (2009) Protostellar jets in context. In: Astrophysics and space science proceedings series. Berlin: Springer

Tsinganos K, Sauty C, Surlantzis G, Trussoni E, Contopoulos J (1996) On the relation of limiting characteristics to critical surfaces in magnetohydrodynamic flows. MNRAS 283:811–820

Tsinganos K (2017) Plasma astrophysics. Unibooks

Vlahakis N, Tsinganos K (1998) Systematic construction of exact MHD models of astrophysical winds and jets. MNRAS 298(3):777–789

Vlahakis N, Tsinganos K (1999) A class of exact MHD models for astrophysical jets. MNRAS 307:279–292

Vlahakis N, Tsinganos K, Sauty C, Trussoni E (2000) A disk-wind model with correct crossing of all MHD critical surfaces. MNRAS 318(2):417–428

Weber EJ, Davis L (1967) The angular momentum of the solar wind. Astrophys J 148:217–227

Wilson RW, Jefferts KB, Penzias AA (1970) Carbon monoxide in the orion nebula. Astrophys J 161:L43

Xie C, Haffert SY, de Boer J, Kenworthy MA, Brinchmann J, Girard J, Snellen IAG, Keller CU (2020) Searching for proto-planets with MUSE, A&A 644:A149. arXiv:2011.08043

Zanni C, Ferrari A, Rosner R, Bodo G, Massaglia S (2007) MHD simulations of jet acceleration from Keplerian accretion disks the effects of disk resistivity. Astron Astrophys 469:811–828

Chapter 4
Exotic Exoplanets

A single ear of wheat in a great plain,
is as strange as a single world in an infinite space.
(Mitrodorus of Chios, fourth century BC)

In Space there are an infinite number of worlds like ours.
De l'infinito universo e mondi (On the Infinite Universe and Worlds),
A book by Giordano Bruno,
(1584) where he suggests the existence of an unlimited number of worlds (planets) orbiting other suns (stars), many of which could be inhabited by intelligent beings.

... with about 10^{12} galaxies, each one hosting about 10^{12} stars
and roughly each star surrounded by at least one planet,
we should have totally something of the order of 10^{24} planets in the Universe…

.... the light that our Earth reflects and shines is only one billionth
of the light that is emitted from the Sun.
And, if an alien observer tried to observe the diminution of sunlight
as the Earth passed in front of the solar disk,
they should be able to detect a decrease in solar brightness by a factor of 10,000....

... in order to detect a planet the size of the Earth,
i.e., with a mass approximately 300,000 times smaller than that of the Sun,
one should be able to detect a radial velocity equal to 30 km per second)/300,000 = 10 cm per second..

..Iron vapor is observed in the morning, while it rains down iron droplets from the sky in the evenings, toward the night side of the exotic exoplanet WASP-76 b!

K. Tsinganos, *A Century of Cosmic Discoveries*, Springer Praxis Books,
https://doi.org/10.1007/978-3-032-06282-6_4

1950: Fermi's Paradox

In 1950 Enrico Fermi pointed out the logical *paradox* that, till today, no sign of intelligent life has been detected elsewhere in the Universe. The *paradox* lies in the complete absence of any evidence for the existence of some extraterrestrial civilizations, despite the fact that, according to astronomical observations, there seem to be billions of other planets in our Milky Way galaxy alone. Specifically, with about 10^{12} galaxies, each one hosting about 10^{12} stars and roughly each star surrounded by at least one planet, we should have totally something of the order of 10^{24} planets in the Universe. As this number is huge, theoretically there is a high probability that life has developed on some percentage of those planets, and subsequently on a smaller percentage life could have evolved into an intelligent form, developing also a capacity for interstellar travel. Hence, based on the numbers involved, Earth should already have been visited by some extraterrestrial civilization!

There have been many attempts to explain the *paradox*, ranging from the hypothesis that intelligent extraterrestrial civilizations are extremely rare, to the belief that the vast distances within the Universe are a rather prohibitive barrier for a contact between civilizations based at different planets; and from the extreme case that there are no other intelligent beings and civilizations, to the other extreme case that they are already here on Earth but get unnoticed. For example, in response to this question for the absence of any evidence for extraterrestrial civilizations, Russian space scientist Konstantin Tsiolkovsky developed the *zoo theory*, in which extraterrestrials avoid communication with us and simply limit themselves to observe us, considering that humanity is not yet ready for a contact with an advanced extraterrestrial species.

However, we should realize that communication with an equally technologically advanced civilization in our galaxy requires many years. Assume for a moment that there exist right now about 50 million technologically advanced civilizations in our galaxy, the nearest of which should be at least a few tens of light years away. This means that a message this civilization could send to us will travel, with the speed of light, a few dozen years to get here. If we get the message with the SETI program, decode it and then respond by sending a corresponding message, this will need a total of several dozen years to reach this nearest alien civilization which initially sought to contact us, and this just for a simple greeting.

This is so, because the astronomical distances are *enormous*, so that even with a highly developed technology, cosmic communication is rather time consuming. However, as technology evolves and discoveries increase and as long as our civilization survives its self-destructive tendencies and the degradation of the environment we live, the Universe may surprise us at any time!

The first discovery of an exoplanet around a solar-type star was made in 1995 by Mayor and Queloz, who were awarded the Nobel Prize in Physics (2019) for their discovery. Since then, almost 6,000 exoplanets have been discovered, and this by exploring a relatively small region of the Milky Way, at a distance of less than 3,000–4,000 light years from Earth (Fig. 4.1). From this sample exploration, however, strong

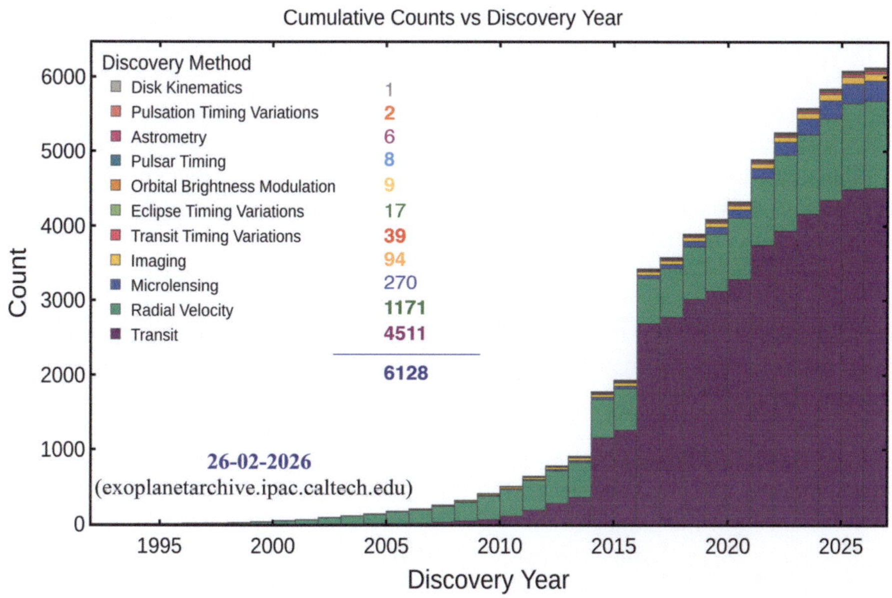

Fig. 4.1 Histogram of all exoplanets discovered by the various methods up to February 26th, 2026: 6128 confirmed exoplanets. (*Source* NASA Exoplanet Archive, Caltech)

evidence has been accumulated that the presence of planets in the Universe is ubiquitous and reaches or even surpasses the myriads of stars. Thus, our Milky Way galaxy, with hundreds of billions of stars, may eventually host a corresponding number of extrasolar planetary systems. Many of these planetary systems have already been observed to have up to eight planets, as our solar planetary system does. It is even interesting that some of these exoplanets are located in the so-called *habitable zone* around the parent star, that is, in a region where all three forms of water coexist: solid, liquid, and gas. And, as the basic prerequisites for the development of life seem to be abundant, who knows, what happened on Earth could have happened in those places. For now, however, Fermi's basic question remains unanswered about the *cosmic silence* that surrounds us, even though there seem to be so many myriads of planets.

Exoplanet observation has historically been dominated by space-based facilities, and the new infrared capabilities of JWST are now uncovering the atmospheres of exoplanets in an unprecedented way. The chemical signatures of various planets are actively probed and detected, with new chemical species being detected, including oxygen, carbon, and nitrogen molecules. This opens up new horizons in cosmical atmospheric chemistry in a wide range of planetary types, temperatures, and metallicities.

The Difficulty of Observing Exoplanets: The Impressive Numbers

To realize how difficult a task is to discover small planets orbiting other stars, it is important to realize that if an alien were to observe Earth from a distant star, this observer would see that the light that our little planet reflects and shines is only one billionth (10^{-9}) of the light that is emitted from the Sun. And, if this alien observer tried to observe the diminution of sunlight as the Earth passed in front of the solar disk, they should be able to detect a decrease in solar brightness by a factor of 10,000 (10^{-4}).
It is also worth noting that in order to detect a planet the size of the Earth, i.e., with a mass approximately 300,000 times smaller than that of the Sun, one should be able to detect a radial velocity equal to 30 km per second)/300,000 = 10 cm per second.
This speed arises because in a star and planetary system that both move around their common center of mass (CM), the stellar velocity around the CM is equal to the speed of the planet around the CM, times the ratio λ of the mass of the planet and the star. Since the speed of the Earth around the Sun is 30 km per second and the ratio $\lambda = 1/300{,}000$, the speed of the star around the CM is equal to 30/300,000 km per second = 10 cm per second. Such a low speed is just at the limits of our current observational techniques.

Methods for Observing Exoplanets

As we mentioned before and also in the following, the stars of our Milky Way galaxy are at enormous distances from Earth. And, exoplanets orbiting these distant stars have no light of their own, but glow faintly only by the reflected light of their parent star, and are billions of times fainter. Also, depending on their distances from Earth, the angular distance of exoplanets from their star is very small, only a few seconds of arc. This means that their detection is observationally a rather difficult task. Hence, the discovery of the first exoplanets had to wait for the development of the relevant technology and specifically the development of high-resolution spectrographs and high-precision photometry.

The main methods for observing exoplanets are by: (1) transits, (2) radial velocities, (3) microlensing, (4) direct imaging, (5) transit-timing variations, (6) eclipse timing variations, (7) pulsar timing, (8) orbital brightness modulations (9) astrometry, (10) pulsation timing variations, and (11) disk kinematics.

1992: Discovery of the *First* Exoplanetary System by Pulsar Timing

As *pulsars* rotate, they emit intense electromagnetic radiation that is detected on Earth as regular and precisely timed pulses, which are so regular that they are even more accurate than an atomic clock. Slight regular variations in the timing of the pulses indicate that the pulsar is moving back and forth, orbiting the center of mass of a planetary system with one or more planets. The orbit, as well as the mass of these planets, can be deduced by precisely measuring irregularities in the timing of the *pulsars*.

In 1992, three years earlier than the discovery of an exoplanet around a solar-type star, with the method of the radial velocities, Aleksander Wolszczan and Dale Frail announced the discovery of the first planetary system around a neutron star, the millisecond *pulsar* PSR 1257 + 12, by the pulsar timing method. The *pulsar* is located at a distance of 2,300 light years in the constellation Virgo and has three planets, A, B, C, at distances (major semi-axes of their orbits) of 0.19, 0.36, and 0.46 AU from the neutron star and masses of 0.02, 4.3, and 3.9 times the mass of the Earth. The exoplanet A, with a mass equal to about twice the mass of the Moon, is the lowest mass exoplanet discovered to date. The orbit as well as the mass of these exoplanets can be deduced by precisely measuring irregularities in the timing of the pulsars.

The neutron star has a mass equal to 1.4 times the solar mass and a radius equal to 10 km, typical quantities for neutron stars. Its surface temperature, however, is extremely high, 28,856 degrees Kelvin (K), so that its planets are literally *baked* as they move at those very close distances around the pulsar. The pulsar formed about 3 billion years ago, as a result of a merger of two white dwarfs and turning the system into a rapidly rotating neutron star with a period equal to about 6 ms. The existence of other planetary systems around neutron stars is an extremely rare event, and to date only 8 planets have been detected around pulsars by the pulsar timing method.

The method is so sensitive that it can detect planets as small as one-tenth the mass of Earth! However, there are two main drawbacks to the pulsar timing method: pulsars are relatively rare, and special circumstances are required for a planet to form around a pulsar. Therefore, it is unlikely that a large number of planets will be found this way. Additionally, life as we know it could not survive on these *pulsar* orbiting planets because the high-energy radiation emitted by the *pulsars* is extremely intense.

Surprising but Worth to Remember: The Milky Way Galaxy is Practically Empty *of Stars*

Our Milky Way galaxy contains several hundreds billion stars, with the Sun just a typical star. However, despite this large number, we can say that our Milky Way is practically *empty* of stars. Nevertheless, today we have indirectly detected many exoplanets, using either the effects of the periodic motion of the star around the common star-planet center of mass, or, by calculating the periodic diminution of the star's radiation due to the transit of the planet in front of the stellar disc.

To understand why we say that our Milky Way galaxy is practically *empty* of stars and how great the distances are between us and the other stars, let's build a 1:100 billion scale model of the Milky Way galaxy. Then distances equal to 1 cm in our model correspond to actual galactic distances equal to 1 million kilometers. In such a model, the solar diameter would be equal to 1.5 cm (i.e., as the bead of a rosary), while at a distance from the Sun equal to 1.5 m Earth would be with a diameter less than 1/10 of a millimeter, i.e., as big as the nose of a pin. At a distance of 7.5 m we would meet the largest of the planets *Jupiter* with a diameter of about 1.5 mm, i.e., the size of the head of a pin, while distant *Pluto* would be at a distance of about 60 m. The nearest star, *Proxima Centauri* would be 400 km away, while the bright *Sirius* would be about 800 km away. That is, according to this model, a star 1,000 light years away would be, according to our model, at a distance 90,000 km and the outskirts of our Milky Way galaxy would be at a distance of approximately 2 million kilometers, i.e., outside Earth, far in space. So, it is obvious that in the vast distances from the Sun to the other stars, it is extremely difficult to detect faint planets.

1995: Discovery of the *First* Exoplanet in the Milky Way Around a Solar-Type Star by the Method of the Radial Velocities

In 1994, Michel Mayor and his graduate student Didier Queloz at the Observatory of Geneva in Switzerland, after having completed the construction of a new spectrograph (ELODIE), at the Observatory of Haute-Provence in the south of France, started to use this instrument in order to measure accurately over a series of nights the radial velocities of 140 solar-like stars of spectral-type G and K, i.e., stars having a similar spectral type (yellow color) to our Sun. This study was initially aimed at the discovery of large planets around brown dwarf stars. At the end of 1994, Major and Queloz noticed that an ordinary star like our Sun, 51 Pegasi, at a distance of about 50 l.y., showed a periodic change in its radial velocity that could correspond to the orbit around the star of a planet smaller in dimensions than the planet Jupiter.

The observation of about 5,000 lines of light absorption from the atmosphere of this solar-type star, inferred a mean displacement of this spectrum corresponding to a speed of about 5 m per second (Fig. 4.2). This Doppler shift is due to the motion of the star around the common center of mass that the star shares with its planetary companion, which was calculated to have a mass M_* similar to the mass of Jupiter ($0.5\ MJ < M_* < 2\ MJ$) and to move around the parent star in a roughly circular orbit, at the fairly close distance of 0.05 AU, i.e., much smaller than the distance of Mercury from the Sun (0.39 AU).

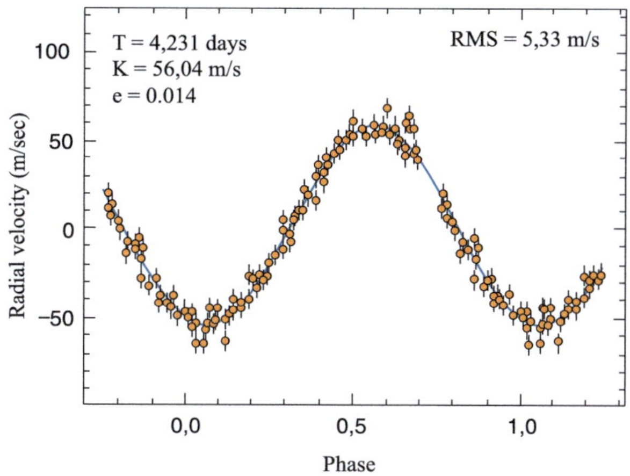

Fig. 4.2 Light curve of the star 51 Pegasi gives the radial velocity as a function of the phase of the orbital motion around the center of mass of this double system. The curve is a fit to spectroscopic observational data and corresponds to a circular motion of the star with a period of 4.2293 ± 0.0011 days. (Figure adapted from Mayor and Queloz (1996) and reproduced with the kind permission of D. Queloz. Figure appears also in Marcy et al. (1997))

The most remarkable result was that the orbital period of this planet around the star 51 Pegasi appeared to be only 4.2 days, contradicting theories of planetary system formation. Additional evidence allowed Major and Queloz to announce with certainty the discovery of the first extrasolar planet orbiting around a Main Sequence star on October 6, 1995 (Mayor and Queloz 1995).

On October 8, 2019, the Swedish Academy of Sciences announced the three scientists who shared the 2019 Nobel Prize in Physics, "*for their contribution to the understanding of the evolution of the Universe and the place of the Earth in it. In particular, Major and Queloz systematically explored the nearest to Earth regions of space in an attempt to locate unknown planets, eventually discovering an exoplanet orbiting a solar-type star, while James Peebles with his theoretical studies contributed to the understanding of evolution of the Universe after the Big Bang.*"

The Swedish Academy stated that these discoveries have permanently changed the way we perceive the Universe. They also added, in fact, that the existence of a large number of planned observations, aimed toward the exploration of exoplanets, may in the future provide an answer to the eternal question about the existence of extraterrestrial life, especially intelligent extraterrestrial life, in the Universe.

The radial velocity method is best for exoplanets with small orbital radius, or large mass, whereas it can only determine the minimum mass of exoplanets. According to the NASA exoplanet archive, as of February 26, 2026, 1171 planets have been discovered by the radial velocity method, out of a total 6128 exoplanets, i.e., about 19%.

1999: Discoveries of Exoplanets by the *Transit* Method

A second way of detecting large planets in orbit around stars uses the partial eclipse of the star's light, when it passes in front of the stellar disk. Figure 4.3a shows the light curve of the star HD 209458 (V376 Pegasi, Osiris) located at a distance of 153 l.y. in the constellation Pegasus, as measured by the Hubble Space Telescope's STIS camera, in April–May 2000.

This is the first exoplanet detected directly by the transit method (Brown et al. 2001, Fig. 4.3), i.e., by passing in front of the disk of the parent star. The planet has the dimensions of Jupiter and a period of 3.5247 days. The lower limit on the planet's mass M_p is $M_p \sin i = (0.69 \pm 0.05)M_J$ with an inclination angle i of the plane of the planetary orbit $i = 86{,}71° \pm 0.05°$, and a radius R_p of the planet $R_p = (1.347 \pm 0.060)\ R_J$, where R_J is the radius of the planet Jupiter.

It is impressive that this method is able to also detect rings or satellites around the planet, as well as to set some restrictions on the composition of their atmosphere. The radius of the planetary orbit is about 7 million kilometers, i.e., about 0.047 AU, or one-eighth the radius of the orbit of the planet Mercury. This small radius of the planetary orbit results in a year of only 3.5 days, with the planet experiencing a high temperature in this orbit, of about 1,000°.

The transit method is suitable for exoplanets with short periods, possessing the advantages of measuring various parameters of exoplanets simultaneously, but can confuse exoplanets and other celestial bodies with the same radius as the planets. Most known exoplanets have been discovered using the transit method.

Another example of observed exoplanet via the transit method is shown in Fig. 4.4. LHS 475 b is a rocky, Earth-sized exoplanet that orbits a red dwarf star roughly 41 light years away, in the constellation Octans. The planet is extremely close to the parent star, completing one orbit in two Earth-days. Confirmation of the presence of the planet was made possible by the James Webb data.

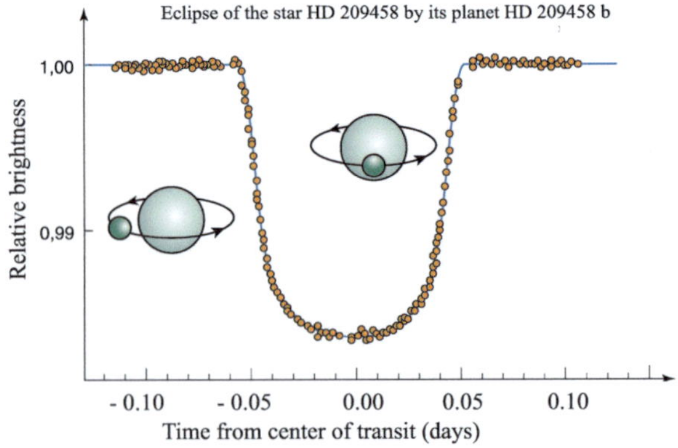

Fig. 4.3 Star HD209458 shows partial **eclipses** by a Jupiter-size planet passing in front of its disk. Within a few (3) hours, the brightness of the star decreases by 1.7%. (Figure adapted from Brown et al. (2001), and is based on observations with the NASA/ESA Hubble Space Telescope, with the kind permission of the coauthors)

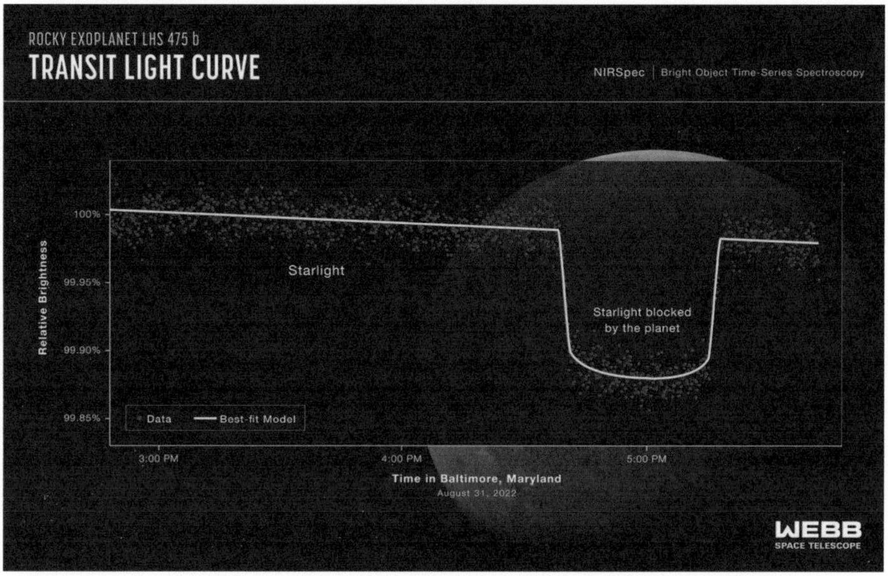

Fig. 4.4 Graphic shows the change in relative brightness of the star-planet system spanning three hours, in the rocky exoplanet LHS 475 b. The spectrum shows that the brightness of the system remains steady until the planet begins to transit the star. It then decreases, representing when the planet is directly in front of the star. The brightness increases again when the planet is no longer blocking the star, at which point it levels out. (*Credit* NASA, ESA, CSA, L. Hustak (STScI), K. Stevenson, J. Lustig-Yaeger, E. May (Johns Hopkins University Applied Physics Laboratory), G. Fu (Johns Hopkins University), and S. Moran (University of Arizona))

Specifically, with this method of transits, according to the NASA exoplanet archive, as of February 26, 2026, 4511 planets have been observed out of a total of 6128 exoplanets, i.e., about 74%.

In addition to the previous two very efficient methods of radial velocities and transits, other methods are also used to discover exoplanets, such as the method of gravitational microlensing, the direct imaging method, the method of pulsar timing and the method of astrometry, that we briefly discuss in the following.

2003: Discovery of Exoplanets by the Method of Gravitational Microlensing

Gravitational microlensing predicted by General Relativity is a phenomenon caused by the gravitational lens effect. It can be used to detect objects with the mass of a planet, regardless of the light. With this method, a star (lens) along with its planetary system can pass between Earth and another distant star. This method is able to detect multiple planets around stars. Microlensing works by waiting until an exoplanet passes in front of a background star.

The light from the distant star is bent as it passes in front of the star-lens. If the star-lens is accompanied by some planet, then a light spike will be added to the light curve of the curved light of the star-lens, something which is now detectable photometrically. This method requires the planet and the lens star to be intervened between the distant star and the observer. This method can detect small planets like Earth and is the only method that can detect planets in other galaxies as well. The microlensing method can find exoplanets that are extremely far from Earth, or even rogue planets, as well as their mass. However, it does not allow to observe exoplanets found by it twice. According to the NASA exoplanet archive, as of February 26, 2026 270 planets have been discovered by the microlensing method out of a total of 6128 exoplanets, i.e., about 4.4%.

The Optical Gravitational Lensing Experiment (OGLE) is a Polish astronomical project based at the University of Warsaw that runs a long-term variability sky survey (1992–present). Most of the observations have been made at the Las Campanas Observatory in Chile. Cooperating institutions include Princeton University and the Carnegie Institution.

2004: Discovery of Exoplanets by the Methods of Direct Imaging, Astrometry, Transit Time Variation, Eclipse Timing Variations, Orbital Brightness Modulation, and Disk Kinematics

Direct Imaging. Another method for discovering exoplanets is the method of *direct imaging* (e.g., Figures 4.5, 4.16, and 4.17), wherein the search of a planet is via a direct photography, particularly in the mid-infrared, with ground-based or space telescopes and the method of *astrometry*.

Out of about 6000 extrasolar planets detected through indirect means, 94 exoplanets have been directly imaged (e.g., Marois et al. 2008). Despite this relatively low yield of discoveries, direct imaging comprises a large fraction of the known exoplanet population amenable to atmospheric characterization, since the method provides photons from the planets themselves instead of just inferring their presence (e.g., Currie et al. 2011). Almost all of the directly imaged planets known thus far are young, self-luminous gas giants detected in thermal emission. They represent the extremes of planet formation: planets with masses of 2 M_J or more orbiting at moderate to wide separations (∼10–250 AU).

2M1207 bis a Jupiter-like planet, 5 times more massive than Jupiter, orbiting a brown dwarf at a distance 55 AU from the star, 230 light years from Earth, in the constellation of Hydra. It was imaged for the first time in 2004 by ESO'sVery Large Telescope. The imaging method offers a tool to directly observe exoplanets in the

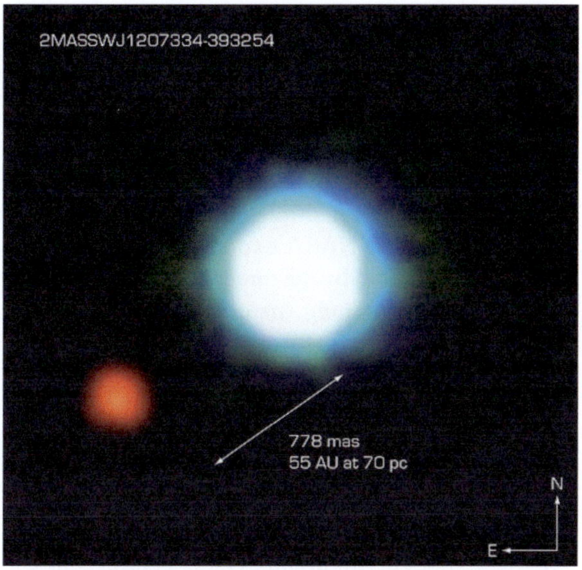

Fig. 4.5 This composite image shows an exoplanet (the reddish spot on the lower left), orbiting the brown dwarf 2M1207 (the blueish object on the center). The two objects are separated by less than one arc second in Earth's sky. The exoplanet orbits at a distance from the brown dwarf that is nearly twice as far as Neptune is from the Sun. The image was taken by the ESO's 8.2m VLT. 2M1207 b is the first exoplanet directly imaged and the first discovered orbiting a brown dwarf. (*Image Credit* ESO)

infrared band. It can detect exoplanets that are extremely far from their host star with relatively high temperatures or rogue planets.

Controversial has been the possible detection of a planet around the young – 440 million years old – brilliant blue giant Fomalhaut, with a surface temperature of 8,590°. Fomalhaut has a luminosity 16 times the solar and a radius about twice that of the Sun, only 25 light years from Earth, i.e., it is a relatively nearby star. Fomalhaut b has been observed from 2004 to 2014. Previously thought to be an exoplanet, it is now believed to be an expanding dust cloud.

Astrometry. With the method of *astrometry*, a small but repeated oscillation of the position of the star due to the existence of some invisible companion may be detected. Astrometry has been used to discover a handful number of exoplanets, mostly gas giants more massive than Jupiter. It is based on measuring a star's proper motion, and seeing how that position changes over time: a planet with a sufficiently large mass is able to gravitationally pull its host star, making its proper motion vary over large timescales. This causes a proper motion discrepancy between two observations (e.g., Hipparcos and Gaia). Today, the large ESO 8 m telescopes have achieved an accuracy of 20 microseconds of arc, which means that they can detect shifts in the position of stars corresponding to a few centimeters (the size of a ping-pong ball in the distance of the Moon). ESA's GAIA mission observed billions of stars in the Milky Way with a high precision, up to a few thousand light years away from Earth.

Transit-timing variation is a method for detecting exoplanets by observing variations in the timing of a transit. This provides an extremely sensitive method capable of detecting additional planets in the system with masses potentially as small as that of Earth. In tightly packed planetary systems, the gravitational pull of the planets among themselves causes one planet to accelerate and another planet to decelerate along its orbit. The acceleration causes the orbital period of each planet to change. Detecting this effect by measuring the change is known as transit-timing variations. The first significant detection of a non-transiting planet using transit-timing variations was carried out with NASA's Kepler telescope. The transiting planet Kepler-19 b shows transit-timing variation with an amplitude of 5 min and a period of about 300 days, indicating the presence of a second planet, Kepler-19 c, which has a period that is a near-rational multiple of the period of the transiting planet.

According to the NASA exoplanet archive, as of February 26, 2026, 94 planets have been discovered by the direct imaging method and 39 planets have been discovered by the *transit-timing variation* method, out of a total of 6128 exoplanets, i.e., about 1.5% and 0.6%, respectively.

Eclipse Timing Variations. When a binary star system is aligned such that from the Earth's point of view the stars pass in front of each other in their orbits, the system is called an *eclipsing binary* star system. The time of minimum light, when the star with the brighter surface is at least partially obscured by the disk of the other star, is called the primary eclipse, and approximately half an orbit later, the secondary eclipse occurs when the brighter surface area star obscures some portion of the other star. These times of minimum light, or central eclipses, constitute a time stamp on the system, much like the pulses from a pulsar, except that rather than a flash, they are a dip in brightness. If there is a planet in circumbinary orbit around the binary stars, the stars will be offset around a binary-planet center of mass. As the stars in the binary are displaced back and forth by the planet, the times of the eclipse minima will vary. The periodicity of this offset may be the most reliable way to detect extrasolar planets around close binary systems. With this method, planets are more easily detectable if they are more massive, orbit relatively closely around the system, and if the stars have low masses. According to the NASA exoplanet archive, as of February 26, 2026, 17 planets have been discovered by the eclipse timing variations, out of a total of 6128 exoplanets, i.e., about 0.3%.

Orbital Brightness Modulation Method. This method is applicable to large planets, and it is possible measure the starlight reflected by the planet itself, with nine planets already detected with this method.

Disk Kinematics. Planets in formation can be detected by the signatures they produce in their natal protoplanetary disks. The velocities of the gas in a protoplanetary disk can be observed, and their morphology can reveal the presence of planets. Planets perturb the gas velocities by imprinting strong variations from Keplerian motion. This method is now referred to as *disk kinematics*. Notable examples of protoplanetary disks around young stars with signatures of embedded planets include HD 97048, HD 163296 and HD 100546.

2015: Discovery of Planetary Systems at Their Infancy with ESO's Telescopes

In the highlands of the Andes in Chile, a place with low cloud cover, low humidity and zero light pollution are the telescopes of the European Observatory of the South, better known today as ESO. Thus, in La Silla (in Spanish meaning *saddle*, from the shape of the double peak) at an altitude of 2,400 m, north of Santiago, Chile, there are currently 9 telescopes installed, of which stands out ESO's3.6 m Telescope (in 1976, when it started operations it was comparable to the USA 5 m Large Telescope at the Palomar Mountains), equipped with the most sensitive spectrograph available today, HARPS (High Accuracy Radial Velocity Planet Searcher). This spectrograph allows the detection of the very small motion of the central star, which is due to the gravitational attraction of planets smaller than Jupiter, that revolve around the star. With this spectrograph, they have discovered approximately two thirds of the known low-mass exoplanets. Also in La Silla ESOoperates the 3.58-m New Technology Telescope(NTT), which in 1989 broke new ground for telescope engineering and design and was the first in the world to have a computer-controlled main mirror, technology developed at ESO. Specifically, the main mirror is flexible and its shape is actively adjusted by actuators during observations, using a reference star, to preserve the optimal image quality. The secondary mirror position is also actively controlled in three directions. This technology, developed by ESO, known as *active optics*, is now applied to all major modern telescopes, such as the Very Large Telescope at Cerro Paranal and the future Extremely Large Telescope. The design of the octagonal enclosure housing the NTTis another technological breakthrough. The telescope dome is relatively small, and is ventilated by a system of flaps that makes airflow smoothly across the mirror, reducing turbulence and leading to sharper images. La Silla is also home to the Belgian 0.60 m robotic telescope TRAPPIST(TRAnsiting Planets and Planetesimals Small Telescope–South), a collaboration between the University of Liège and the Geneva observatory, which specializes in the search for exoplanets. Recently, it discovered a planetary system with seven planets, of the TRAPPISTclass (TRAPPIST-1), which orbit around a very cool dwarf parent star.

At the top of Cerro Paranal of the Atacama desert of northern Chile at 2,635 m is installed the *Very Large Telescope* (VLT), the undisputed jewel of ESO, consisting of 4 telescopes each of 8.2 m, which have been designed so that to operate independently, but also connected to each other, as a larger telescope, Fig. 3.10. When these telescopes work together, the VLT gathers the equivalent light that a 16.4 m diameter telescope would gather, so that it has the exceptional resolution required to see astronauts on the surface of the Moon!

At an even higher point, on the Chajnantor plateau of the Atacama desert, at an altitude of 5,000 m, one of the driest and most arid places on Earth, is the large array of radio telescopes ALMA(Atacama Large Millimeter Array), an international collaboration between Europe, the USA, Canada, Japan, South Korea, Taiwan and Chile. With a total cost of $1.4 billion it is the most expensive ground-based telescope for the construction of which ESOhas played a central role. The array consists of 66 parabolic radio telescopes with a diameter of 12 and 7 m (Fig. 3.20a). ALMA's radio telescopes work interconnected so that they correspond to one giant radio telescope with an effective diameter of 14 km. This means that the resolution of ALMA is 10 ms of arc, i.e., 5 times better than the resolution of the Hubble telescope.

The observation that the state-of-the-art Atacama Large Millimeter/submillimeter Array (ALMA) located in Chile's Atacama Desert has ever made with the highest resolution, is that of the protoplanetary disk HL Tauri. Thanks to ALMA'sincreased resolution, structures within this disk are revealed that have never been observed before and show possible positions of planets at their birth, which may form in the dark voids of the system (Figs. 4.6 and 4.15). The radius of this disk is about 200 AU and the solar system, with the distance of the planet Neptunefrom the Sun at 40 astronomical units, would fit comfortably in the inner part of this disk.

A related study by Tamayo et al. (2015) is the first to suggest that gaps in protoplanetary disks are evidence of planetary formation because the gaps are separated by amounts consistent with what's called a *special resonant configuration.* In other words, these planets avoid violent collisions with each other by having specific orbital periods where they miss each other, similar to how Pluto has avoided Neptune for billions of years despite the two orbits cross one another.

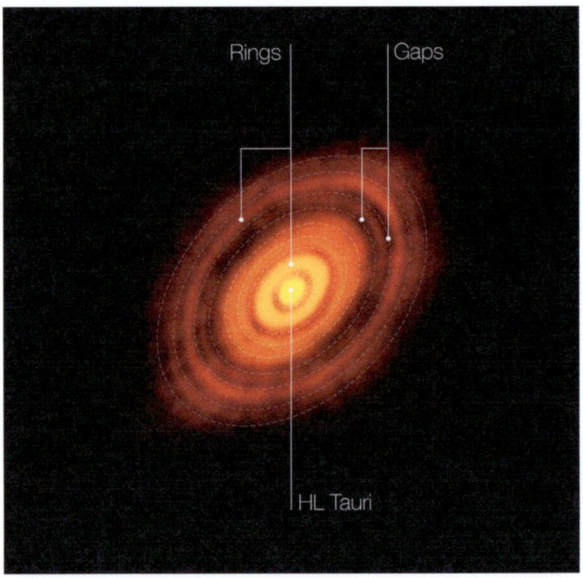

Fig. 4.6 This is the sharpest image ever taken by ALMA – sharper than is routinely achieved in visible light with the NASA/ESA Hubble Space Telescope, showing the protoplanetary disk surrounding the young starHL Tauri in the constellation Taurus at a distance of about 450 l.y. Its age is estimated to be approximately 100 million years. HL Tau is a planetary system in its infancy. The radius of the disk is about 200 astronomical units. (*Source* ESO/NAOJ/NRAO)

2017: Discovery of the Most Populous Extrasolar Planetary System

The star Kepler-90, is an F-type Main Sequence star in the constellation of Draco, similar in size to our Sun, which has the most populous in planets planetary system (Fig. 4.7). With a mass and radius 1.2 times the solar radius, a photospheric temperature of 6,080 °K and an age of about 2 billion years – as known the corresponding values for the Sun are 5,778° and 4.6 billion years – Kepler-90 is at a distance of 2,840 l.y. from us (855 pc). The name Kepler90 given to this star comes from the fact that this particular star was with number 90 in the KeplerSpace Telescope's catalog of stars that have confirmed the observation of planetary systems. Specifically, around Kepler-90, the Kepler space telescope discovered, with the method of transits, a planetary system with 8 planets (Fig. 4.7). With the letters corresponding to the year of discovery, these planets are Kepler-90 b, c, d, e, f, g, h, and i. Interestingly, the newly discovered Kepler-90ieighth planet was last discovered on December 14, 2017 – a sizzling hot, rocky planet that orbits its star once every 14.4 days – was found using machine learning from Google.

This planetary system has therefore the same number of planets as the *solar* planetary system and in addition has six rocky planets nearby to the star and two gas giants(Kepler g and h) further away. Kepler i, which is quite similar in dimensions to

Fig. 4.7 Kepler-90 is an extrasolar planetary system that has been observed by the Kepler spacecraft with eight planets, i.e., as many planets as our solar system has. Its six inner planets are rocky (like Earth) while the two outermost are gas giants, like Jupiter and Saturn. In this artistic representation, the relative sizes of the planets are real, but their distances from the parent star are not. The period of the closest planet is 7 days and the outermost 331 days. All planets are *locked* so that they have one hemisphere constantly facing the parent star, which means that this hemisphere always has day and the other always night. (*Image Credit* NASA/Ames Research Center/Wendy Stenzel)

Earth, is however located at a distance of 0.1 astronomical units from the parent star, where the temperature is about 436° (close to the average temperature of Venuswhich is 467°); this planet has a radius 1.32 times the radius of Earth and a mass equal to 2.5 times the mass of the Earth. Keplerg and h are located at distances close to the distance of the Earth from the Sun and have *years* of 210 and 331 days respectively, while their masses and dimensions are close to those of Jupiter. The planet which is further away, Kepler h, is orbiting the parent star with a period of 331 days, i.e., very close to the period of rotation of the Earth around the sun. The Kepler-90planetary system is the only extrasolar planetary system observed with Keplerhaving eight planets.

The inner planets, Kepler-90 b, c, i, d, and e hemisphere *locked* due to strong tidal forces (like the Moon from the Earth), so that one of their hemisphere sees permanently the parent star and the other hemisphere is eternally plunged in darkness.

Kepler notable for similarities it shares of the planetary system to that of the Solar System, in which rocky planets are nearer the star and gas giants farther away. The six inner planets range from super-Earths to mini-Neptunes in size. The two outermost planets are gas giants. The most distant known planet orbits its host star at about the same distance as Earth from the Sun.The Kepler is one of only two eight-planet candidate systems from Kepler, together with Kepler-385, and the second to be discovered after the Solar System.

2019: First Discovery of Water Vapor of an Exoplanet in the Habitable-Zone

For the first time, water vapor signatures were detected in the atmosphere of a planet, beyond our solar system, residing in the *habitable zone*, the region around a star wherein liquid water could potentially pool on the surface of a rocky planet by the NASA/ESA Hubble Space Telescope. The results revealed the molecular signature of water vapor, also indicating the presence of hydrogen and helium in the planet's atmosphere. This is K2-18 b, an exoplanet around a small red dwarf star about 110 light years away in the constellation Leo (Fig. 4.8).

K2-18 bis now the only super-Earth exoplanet known to host both *water* and temperatures that could support life. However, given the high level of activity of the parent red dwarf star, the conditions in K2-18 bmay be more hostile to life as we know it in Earth, since it is likely that the planet is exposed to more high-energy radiation. The planet has a mass about eight times the mass of the Earth. That means the surface gravity on this planet would be significantly higher than on our planet.

The analyzed data were captured by Hubble and analyzed as the host star's light filtered through K2-18 b's atmosphere. The results revealed the molecular signature of water vapor and also suggest the presence of hydrogen and helium in the planet's atmosphere. K2-18 b is one of hundreds of *super-Earths* – exoplanets with masses between those of Earth and Neptune – found by Kepler.

Fig. 4.8 This illustration shows what exoplanet K2-18 b could look like based on the available data. The exoplanet K2-18 c is shown between K2-18 b and its star. (*Image Credit* NASA, ESA, CSA, Joseph Olmsted (STScI))

New observations with NASA's James Webb Space Telescope into K2-18 b has revealed the presence of carbon-bearing molecules including methane and carbon dioxide.

The abundance of methane and carbon dioxide, and shortage of ammonia, support the hypothesis that there may be a water ocean underneath a hydrogen-rich atmosphere in K2-18 b. The James Webb Space Telescope discovered (2023) also biosignatures in the atmosphere of this Hycean exoplanet, as we discuss later.

2020: Exoplanets Where Rains Down from the Sky Droplets of Iron, or, Red Rubies and Blue Sapphires (Corundum), or, Molten Glass, or, Snows Titanium Oxide (Sunscreens), etc...

WASP-76 b is an interesting entry in the list of planets, whose environs seem to be taken straight out of science fiction: ESO'sVery Large Telescope (VLT) spotted this exoplanet wherein it rains iron! The exotic planet WASP-76 b (Fig. 4.9), which is about 640 light years away in the constellation Pisces, rotates with an orbital radius of about 5 million km around a hot star (type F), which has an atmospheric temperature more than 6,250°. The planet has a mass of about 0.9 Jupiters, taking 43 h to complete one orbit around the parent star at a distance 0.033 AU. The equilibrium temperature of WASP-76 b is estimated to be around 1,920 °C; however, the measured daytime temperature is higher, reaching approximately 2,227 $\pm$ 200 °C.

Not only does WASP-76 b has different day-night temperatures, but also distinct day-night chemistry, according to observations. Using the new ESPRESSO instrument at ESO'sVLT in Atacama, Chile, chemical changes have been detected for the

Fig. 4.9 Illustration showing a night-side view of the exoplanet WASP-76 b. This ultra-hot giant exoplanet has a day side where temperatures climb above 2400 °C, high enough to vaporize metals. Strong winds carry iron vapor to the cooler night side where it condenses into iron droplets. To the left of the image, we see the evening border of the exoplanet, where there is a transition from day to night. (*Source* ESO/M. Kornmesser/L. Calçada)

first time in the atmosphere of this giant planet which has twice the radius of Jupiter, but with high atmospheric temperatures. It was specifically identified a strong presence of iron vapors in the evenings that distinguish the planet's day side from its night side. In other words, iron vapor is observed in the morning, while it *rains iron* in the evenings, toward the night side of the exotic exoplanet!

This rather strange phenomenon of *iron rain* occurs because the planet turns only one of its faces – the day, or frontal hemisphere – toward the parent star, while its cooler night hemisphere remains plunged into eternal darkness. Like the Moon in its orbit around the Earth, similarly WASP-76 b is *locked* into always facing continuously the parent hot star on the same side; it takes that long to orbit around its spin axis, as to revolve around the parent star. During the day of the planet, it receives thousands of times more radiation from the parent star, than Earth receives from the Sun. The planetary atmosphere is so hot that molecules split into atoms and metals, such as iron, vaporize. The extreme temperature difference between the day and the night sides results in strong winds that carry iron vapors from the extremely hot 2,400 °C day, temperatures which are high enough to vaporize the metals, to the cooler night side, where temperatures drop at about 1,500 °C, at which the iron liquefies. Then, strong winds blowing with more than 18,000 km per hour carry the iron vapors to the colder night side, where they condense into iron raindrops!

Another *tidally locked* gas-giant hot exoplanet is HAT-P-7 b discovered in 2008, which also orbits an F-type star. Its mass is 1.84 Jupiters; it takes 2.2 days to complete one orbit around its star, at a distance 0.03676 AU. Due to the extreme heat that it receives from the parent star, the dayside temperature is predicted to be 2,360–2,600 °C, while night-side temperatures are 1,938–1,965 °C. HAT-P-7 b is also one of the darkest planets ever observed, with an albedo of less than 0.03 – meaning it absorbs more than 97% of the visible light that strikes it. Condensation models of HAT-P-7 b predict precipitation of Al_2O_3 (corundum, a very hard mineral known for its exceptional hardness, second only to diamond) on the night side of the planet's atmosphere. The clouds of the planet themselves are likely made up of *corundum*, the mineral which forms red rubies and blue sapphires!

HD 189733 b is a giant blue alien exoplanet where observational data suggest that *glass* rains down from the sky. This exoplanet is approximately 64.5 light years away from the Solar System. With a mass 11.2% higher than that of Jupiter and a radius 11.4% greater, HD 189733 b orbits its host star once every 2.2 days. This far-off blue planet may look like a friendly haven, but weather there is deadly. The planet's cobalt blue color may be the result of Rayleigh scattering and comes from a hazy, blow-torched atmosphere containing clouds laced with glass and the planet's weather includes raining molten glass! (Fig. 4.10).

Kepler-13A b is one of the hottest known planets, with a dayside temperature of nearly 2,760 °C. It orbits very close to the star Kepler-13A, which lies at a distance of 1,730 light years from Earth. On Kepler13A b, it *snows titanium oxide*, which is a relative of the UV-blocking component of the strongest *sunscreens* we use here on Earth. Powerful winds carry the titanium monoxide gas molecules in the dayside might around to the colder night time side, where it condenses into crystalline flakes, forms clouds, and precipitates as snow. The problem is that the sunscreen (titanium oxide) precipitation only happens on the planet's permanent night time side.

Fig. 4.10 Illustration of HD 189733 b, a huge gas giant that orbits very close to its host star HD 189733. The planet's atmosphere is scorching with a temperature of over 1000 °C, and it rains glass, sideways, in howling 7000 km/h winds. At a distance of 63 light years from us, this turbulent alien world is one of the nearest exoplanets to Earth that can be seen crossing the face of its star. By observing this planet before, during, and after it disappeared behind its host star during orbit, it was deduced that HD 189733 b has a deep, azure blue color – reminiscent of Earth's color as seen from space. (*Image Credit* NASA Goddard)

Fig. 4.11 This artist's impression shows the seething hot planet Kepler-13A b that circles very close to its host star, Kepler-13A. Seen in the background is the star's binary companion, Kepler 13B, and the third member of the multiple-star system is the orange dwarf star Kepler-13C (*Image Credit* NASA/ESA/G. Bacon (STScI))

According to NASA, any possible visitors to the exoplanet would need to bottle up some of that sunscreen, because they won't find it on the sizzling hot, daytime side, which always faces its host star (Fig. 4.11)!

Kepler-70 b (KOI-55) could well be another *circle of hell* planet with an average temperature hotter than the Sun's photosphere, an exoplanet orbiting a B-type subdwarf star which passed through the red giant phase, losing much of its mass, and is now in a helium-burning phase; the star is nearly 6 times as hot as the surface temperature of the Sun. The planet is located about 4000 light years away from our Solar System, has a mass equal to 0.44 the mass of Jupiter, taking 0.24 days to complete an orbit around the parent star, at a distance of 0.006 AU (by contrast, Mercury is nearly 70 times further from the Sun). It may once have been a gas giant, having lost its gaseous envelope owing to the ferocious heat of its nearby star; now only the core of the former planet remains, little more than a ball of molten rock. At about 6,800 C, this planet is one of the hottest planets discovered.

2022: Discovery of Water Vapor in the *Transmission Spectrum of the* Atmosphere of the Gas-Giant Exoplanet WASP-96 b by the J. Webb Space Telescope

While the Hubble Space Telescope has analyzed numerous exoplanet atmospheres over the past two decades, capturing the first clear detection of water in 2013, J. Webb's immediate and more detailed observation marks a giant leap forward in the quest to characterize potentially habitable planets beyond Earth. Thus, Webb revealed in exquisite detail the steamy atmosphere of the distant planet WASP-96 b, 1150 l.y. from Earth in the constellation Phoenix, a type of gas giant that has no direct analog in our Solar System, Fig. 4.12. With a mass less than half that of Jupiter and a diameter 1.2 times greater, WASP-96 b is much *puffier* than any planet orbiting our Sun. And with a temperature greater than 500 °C, it is significantly hotter. WASP-96 b orbits extremely close to its Sun-like star, just one-ninth of the distance between Mercury and the Sun, completing onc orbit cvcry 3½ Earth-days.

The spectrum of light shown in Fig. 4.12b – which contains information about the makeup of the planetary atmosphere 1,150 light years away – reveals the distinct signature of water. This *transmission spectrum* is made by comparing starlight filtered through a planet's atmosphere as it moves across the star, to the unfiltered starlight detected when the planet is beside the star. Each of the 141 data points (white circles) on this graph represents the amount of a specific wavelength of light that is blocked by the planet and absorbed by its atmosphere. The amount of starlight blocked ranges from about 13,600 parts per million (1.36%) to 14,700 parts per million (1.47%). To capture these data, Webb stared at the WASP-96 star system for 6 h 23 min, beginning about 2½ hours before the transit and ending about 1½ hours after the transit was complete. The transit itself lasted for just under 2½ hours. The curve includes a total of 280 individual brightness measurements – one every 1.4 min.

2022: Discovery of CO, CO_2, SO_2, and Na, Additionally to H_2O Vapor, in the Atmosphere of a Gas Exoplanet

WASP-39 b in the constellation Virgo, about 700 light years from Earth, is a *hot Jupiter* extrasolar planet discovered in 2011 by the WASP (Wide Angle Search for Planets) project, Fig. 4.13. WASP-39 b is notable for containing a substantial amount of water in its atmosphere and the first exoplanet found to contain carbon dioxide and sulfur dioxide in its atmosphere. WASP-39 b has a mass of about 0.28 times that of Jupiter and a radius about 1.27 times that of Jupiter. It is a *hot gas-giant planet* with a high temperature of 900 °C. The exoplanet orbits very close (7 million km) to WASP-39, its host star, every 4 days.

Figure 4.13 shows the transmission spectrum of WASP-39b, which is made by comparing starlight filtered through a planet's atmosphere as it moves in front of the star, to the unfiltered starlight detected when the planet is beside the star. Each of

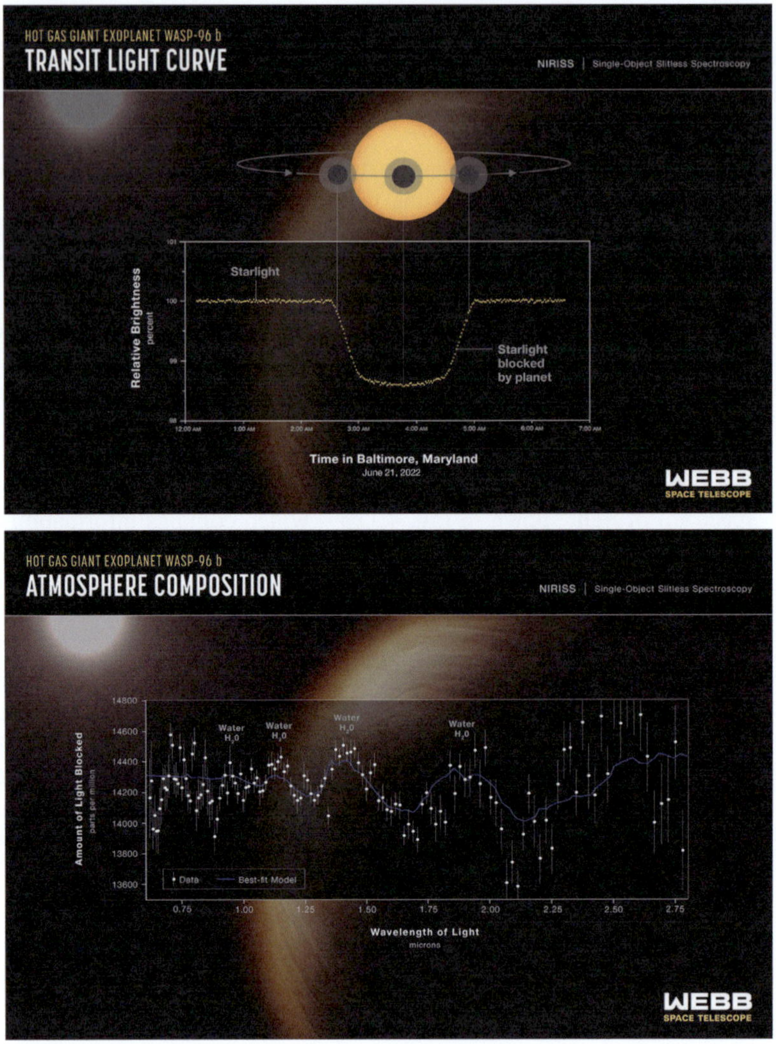

Fig. 4.12 (*top*) The light curve from Webb's Near-Infrared Imager and Slitless Spectrograph (NIRISS) showing the change in brightness of light from the WASP-96 star system over time as the planet transits the star. The transit itself lasted for just under 2½ hours. The curve includes a total of 280 individual brightness measurements – one every 1.4 min. The actual dimming caused by the planet is extremely small: the difference between the brightest and dimmest points is less than 1.5%. (*Credit* NASA). (*bottom*) Observation of the steamy atmosphere of WASP-96 b by the James Webb space telescope, showing also evidence for *clouds* and *haze*. The light curve of the transmission spectrum shows the overall dimming of starlight during the transit, revealing the brightness change of individual wavelengths of infrared light between 0.6 (red) and 2.8 microns (infrared), a part of the spectrum particularly sensitive to water. (*Image Credit* NASA, ESA, CSA, and STScI)

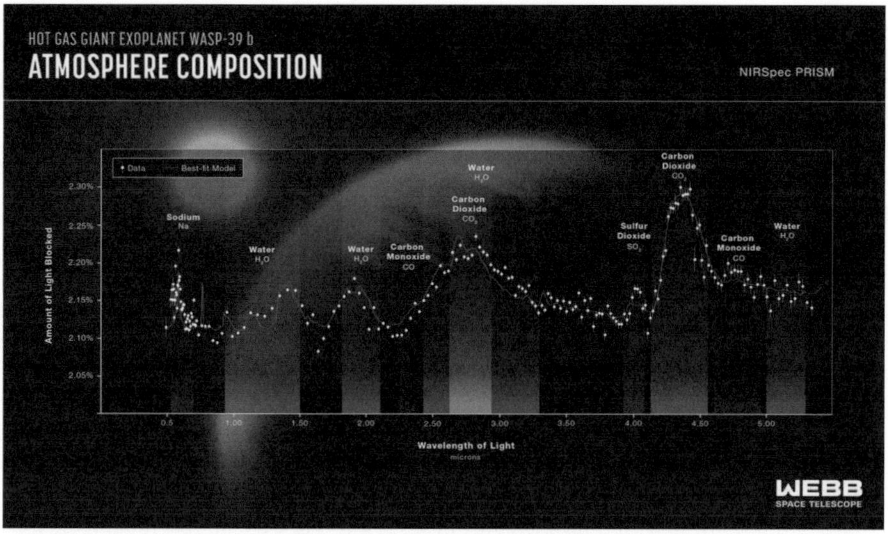

Fig. 4.13 Atmospheric composition of the hot gas-giant exoplanet WASP-39 b, as revealed by the James Webb Space Telescope. This graph displays data from Webb's NIRSpec instrument, indicating signatures of sodium (Na), potassium (K), water (H_2O), carbon monoxide (CO), sulfur dioxide (SO_2), and carbon dioxide (CO_2). (*Image Credit* NASA, ESA, CSA, J. Olmsted (STScI))

the data points (white circles) on these graphs represents the amount of a specific wavelength of light that is blocked by the planet and absorbed by its atmosphere. Wavelengths that are preferentially absorbed by the atmosphere appear as peaks in the transmission spectrum.

Figure 4.13 shows the atmospheric constituents detected by the Webb telescope which includes sodium (Na), potassium (K), water vapor (H_2O), carbon dioxide (CO_2), and carbon monoxide (CO) for WASP-39 b. Among the unprecedented revelations is the first detection in an exoplanet atmosphere of CO_2, a molecule produced from chemical reactions triggered by high-energy light from the planet's parent star. On Earth, the protective ozone layer in the upper atmosphere is created in a similar way.

2023: Discovery of Biosignatures in the Atmosphere of a *Hycean* Exoplanet

A hycean planet is a hypothetical type of planet, described as a hot, water-covered planet with a hydrogen atmosphere. The term *hycean* is a portmanteau of hydrogen + ocean. The presence of extraterrestrial liquid water makes Hyceanplanets promising candidates for planetary habitability. Density data imply that both rocky super-Earths and sub-Neptunes can fit this type, and it is thus expected that they will be common exoplanets.

K2-18 b is a rocky super-Earth exoplanet, 8.6 times as massive as the Earth with 2.6 the terrestrial radius, that orbits the cool dwarf star K2-18 in the habitable zone and lies 120 light years from Earth in the constellation Leo, Fig. 4.8. K2-18 b was initially discovered by the Kepler space telescope in 2015. Exoplanets such as K2-18 b, which have sizes between those of Earth and Neptune, are unlike anything in our solar system. This lack of equivalent nearby planets means that these "*sub-Neptunes*" are poorly understood, and the nature of their atmospheres is the subject of intense study. The suggestion that the sub-Neptune K2-18 b could be a *Hycean* exoplanet is intriguing, as some astronomers believe that these worlds are promising environments to search for evidence of life on exoplanets.

The first insight into the atmospheric properties of this habitable-zone exoplanet came from observations with NASA's Hubble Space Telescope. The following Webb observations (Fig. 4.14) revealed the abundance of methane and carbon dioxide, and shortage of ammonia, supporting thus the hypothesis that there may be a water ocean underneath a hydrogen-rich atmosphere in K2-18 b. JWST watches planets as they pass in front of (or transit) their stars. As a planet transits, some of the starlight passes through its atmosphere, during which some of the chemicals in the atmosphere leave behind a signature in the light. These initial Webb observations also provided a possible detection of a molecule called *dimethyl sulfide (DMS). On Earth, this is only produced by life.* The bulk of the DMS in Earth's atmosphere is emitted from phytoplankton in marine environments.

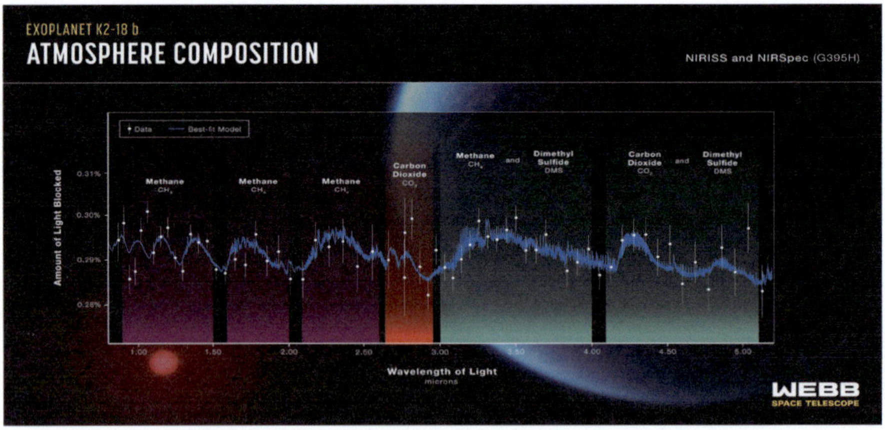

Fig. 4.14 Spectra of K2-18 b, obtained with Webb's Near-Infrared Imager and Slitless Spectrograph (NIRISS) and Near-Infrared Spectrograph (NIRSpec) displays an abundance of methane and carbon dioxide in the exoplanet's atmosphere, as well as a possible detection of a molecule called dimethyl sulfide (DMS), supposedly produced by life forms. (*Image Credit* NASA, ESA, CSA, Ralf Crawford (STScI), Joseph Olmsted (STScI))

Red dwarfs are stars smaller than our Sun that make up the most common stars in the Milky Way. Although they are not easily observable because of their low brightness, it is estimated that about *three-quarters* of the stars of our Milky Way are red dwarfs. The planets of the category of *super-Earths* (with a mass that is between the masses of Earth and the planet Neptune, $M_{\oplus} < M < 17\ M_{\oplus}$), are also the most common planets in our Galaxy. It is the first time worldwide that water has been detected in gaseous form in the atmosphere of a planet with a temperature similar to Earth's temperature. Water is either in liquid form on its surface, whence and evaporates, or exists only in its atmosphere. Which of the two cases applies is something that is currently impossible to determine.

Out of the several thousands exoplanets that have been confirmed to date, K2-18 b is two steps ahead, as far as it concerns the possibility of hosting life. However, due to the high level of activity in its parent star, it is possible that the exoplanet faces more hostile conditions from its environment, compared to Earth, as it is rather exposed to more dangerous stellar radiation. In other words, since the planet is very close to its star, it is bombarded with a great deal of high-energy radiation, which any organisms on its surface would have to be able to survive.

In this particular exoplanet, its spectroscopic data were studied in 2016 and 2017 by the Hubble Space Telescope. The results revealed water molecules in gaseous form, further indicating the presence of hydrogen and helium in the atmosphere of the planet.

2024: Discovery of Water in the HL Tauri Disk

Ring-shaped gaps are carved out in gas- and dust-rich disks by orbiting young planet-like bodies as they gather up material and grow. In a recent finding (2024), ALMA has been able to spatially resolve water in the cool planet-forming disk of HL Tau. A significant amount of water was found in the region where the gap in the HL Tauri disk exists (Figs. 4.6 and 4.15). The ALMA observations show a stringent lower limit of 3.7 Earth oceans of water vapor available within the inner 17 astronomical units of the system. This suggests that this water vapor could affect the chemical composition of planets forming in those regions.

2024: James Webb Images a Cold Exoplanet 12 Light Years Away

The direct imaging of exoplanets is particularly valuable for characterization, because we may directly collect light from the observed planet and compare its brightness at different wavelengths. Till today, only a few tens of exoplanets have been directly imaged by space-and ground-based observatories. Current direct imaging efforts are

Fig. 4.15 Water vapor in a disk around a young star exactly where planets may be forming (HL Tauri). In this image, the new observations by ALMA show the water vapor in shades of blue. Near the center of the disc, where the young star lives, the environment is hotter and the gas brighter. The red-hued rings are previous ALMA observations showing the distribution of dust around the star. (*Image Credit* ALMA(ESO/NAOJ/NRAO)/Facchini et al. (2024))

limited to enormous, bright planets. These worlds are typically super-Jupiters that are less than a hundred million years old – so young that they're still glowing from heat leftover from their formation, which makes them detectable in infrared light. Of the approximately 25 directly imaged planets to date, all are younger than 500 Myr, and all but six are younger than 100 Myr.

The last member of this group, known as Epsilon Indi A b, orbits the K-type star Epsilon Indi A (Eps Ind A), which is around the age of our Sun, but slightly cooler (Matthews et al. 2024). Eps Ind A b was observed by using the coronagraph on Webb's Mid-Infrared Instrument (MIRI) (Fig. 4.16). This planet is quite similar to Jupiter, albeit warmer ($T_J = -110\ °C < T_{Eps\ Ind\ A\ b} = 2\ °C$). Note that previously imaged exoplanets tend to be the youngest, hottest exoplanets that are still radiating much of the energy from when they first formed. As planets cool and contract over their lifetime, they become significantly fainter and therefore harder to image.

Epsilon Indi A b, as of 2024, is the nearest exoplanet to be directly imaged, the twelfth closest exoplanet to Earth known to date and the closest planet more massive than Jupiter. It orbits at around 28 AU (almost as far as Neptune from the Sun) with a period of around 174 years and a relatively high eccentricity of 0.4, and has a mass around 6 times that of Jupiter. It was directly imaged using the James Webb Space Telescope and is one of the coldest exoplanets to be directly detected.

Fig. 4.16 Direct imaging of the exoplanet Epsilon Indi A b with the coronagraph on the NASA/ESA/CSA James Webb Space Telescope's MIRI (Mid-Infrared Instrument). A star symbol marks the location of the host star Epsilon Indi A, whose light has been blocked by the coronagraph, resulting in the dark circle marked with a dashed white line. Blue scale-like features are visible in the background – light at 10.6 microns was assigned the color blue, while light at 15.5 microns was assigned the color orange. The exoplanet is visible on the left as a bright orange circle. (*Credit* ESA/Webb, NASA, CSA, STScI, E. Matthews (Max Planck Institute for Astronomy))

2025: J. Webb Infers the Chemistry of Exoplanet Atmospheres Through Direct Imaging

Of the 6,000 exoplanets discovered, few have been directly imaged, since even giant planets are many thousands of times fainter than their stars. By using Webb's Near-Infrared Camera (NIRCam) coronagraph, which blocks light from bright stars, otherwise hidden worlds are revealed. One such case is HR 8799 (Fig. 4.17a), a young system about 30 million years old, a fraction of our solar system's 4.6 billion years. This planetary system, is still *hot* from their tumultuous formation, with the planets within HR 8799 emitting large amounts of infrared light that give us valuable data on how they formed.

The observations indicate that the well-studied planets of HR 8799 are rich in *carbon dioxide gas* (CO_2). This provides strong evidence that the system's four giant planets formed much like Jupiter and Saturn, by slowly building solid cores that attract gas from within a protoplanetary disk.

The strong carbon dioxide features shown in Fig. 4.17b, indicate that there is a sizable fraction of heavier elements, like carbon, oxygen, and iron, in these planets' atmospheres, a result indicating that they formed via core accretion (Balmer et al. 2025).

a

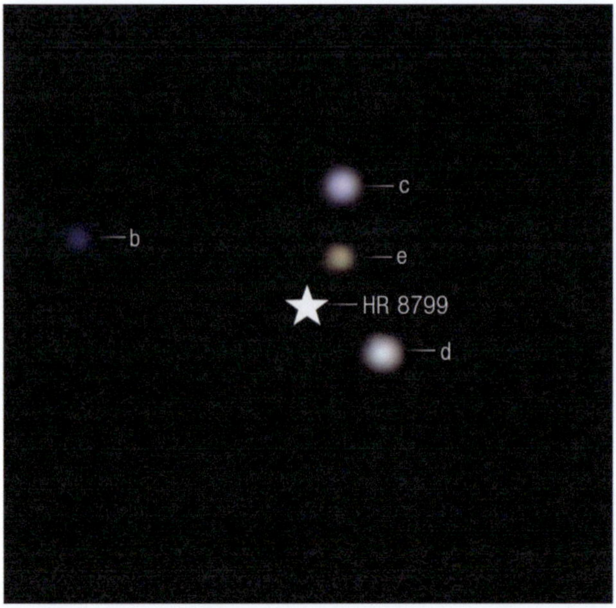

b

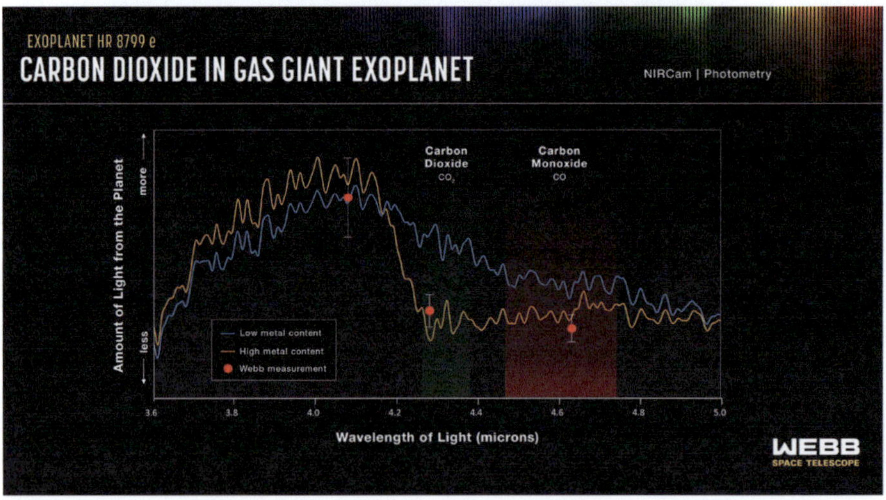

Fig. 4.17 a, b. (*top*) Image of the planetary system HR 8799. At the center of the image, there is a symbol representing the star HR 8799, which blocks the light from the host star. There are four exoplanets, which look like fuzzy dots, pictured in the image surrounding the star. Furthest from the star is a fuzzy, faint blue dot, labeled b, at the 10 o'clock position. At the one o'clock position, second furthest from the star is a blueish-white fuzzy dot labeled c. Just below that is an orange dot labeled e. At the four o'clock position, still nearby the star, is another fuzzy white dot labeled d. (*Credit* NASA, ESA, CSA, STScI, W. Balmer (JHU), L. Pueyo (STScI), M. Perrin (STScI)). (*bottom*) Spectrum of one of the planets in the HR 8799 system, HR 8799 e, which displays the amounts of near-infrared light detected from the planet by Webb at different wavelengths. The Webb data is consistent with a high metallicity planet. (*Credit* NASA, ESA, CSA, STScI, W. Balmer (JHU), L. Pueyo (STScI), M. Perrin (STScI))

2025: Gas-Giant Exoplanet with Extreme Supersonic Winds

WASP-127 b is a giant exoplanet, located 520 light years away, about 30% larger than Jupiter but only 16% of its mass, making it one of the *puffiest* planets with a low density. It is very close to its host sun-like star rotating at a distance ~5% of an AU, completing one orbit in about 4 days. One side of the planet always faces the star (dayside), the other remains dark (night side). By using the ESO'sVLT in Chile, it was discovered that in the planet's atmosphere, mainly hydrogen and helium with traces of carbon monoxide and water, extremely powerful supersonic winds pummeling the equator of WASP-127 band reaching speeds up 9 km per second – 33 000 km/h – making up the fastest jet stream of its kind ever measured on a planet (Fig. 4.18). Specifically, it was found that part of the atmosphere of this planet is moving toward us at a high velocity while another part is moving away from us at the same speed, which is nearly six times the speed at which the planet rotates. In comparison, the fastest wind ever measured in the Solar System was found on Neptune, moving at 0.5 km per second (1800 km/h). Further it was reported in Nortmann et al. (2025) that the poles are cooler than the rest of the planet. There is also a slight temperature difference between the morning and evening sides of WASP-127b, showing that the planet has complex weather patterns just like Earth and other planets of our own System.

Fig. 4.18 An artist's visualization of WASP-127 b, with its newly discovered supersonic jet winds that move around the planet's equator, with a speed of 33,000 km/h, the fastest jet stream of its kind ever measured in the Universe. By tracking the speed of molecules in the atmosphere with the CRIRES + instrument on ESO's Very Large Telescope, it was found that one side of the planet's atmosphere is moving toward us and the other away from us, indicating that there is a powerful wind current going around the planet. (*Credit* ESO/L. Calçada)

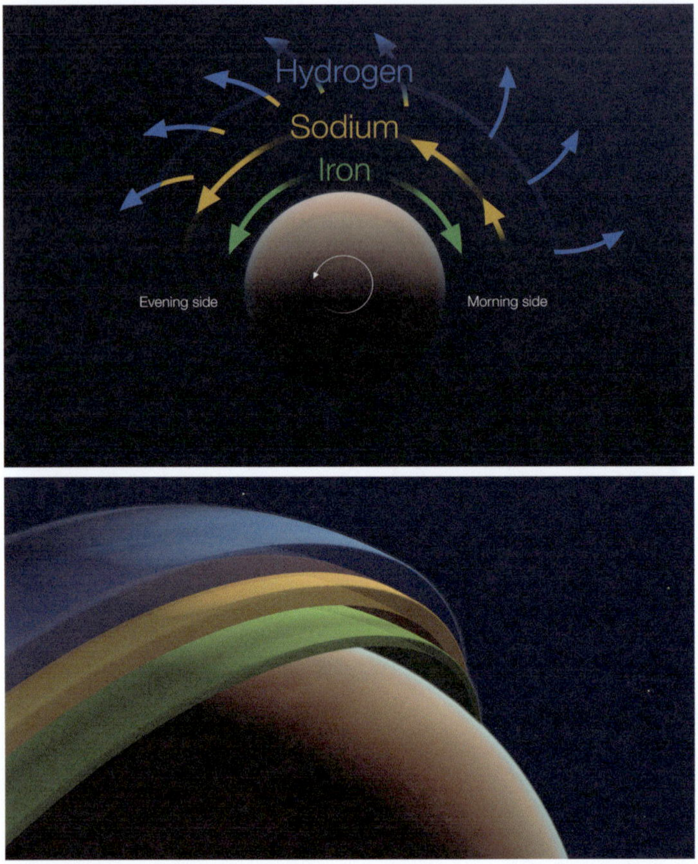

Fig. 4.19 Map of the atmosphere's 3D structure in WASP 121 b with winds in the deep, mid and shallow layers of the planet's atmosphere (*Credit* ESO)

2025: The 3D Structure of the Atmosphere of the Exoplanet Tylos

Figure 4.19 shows the structure and motion of the atmosphere of the exoplanet Tylos[1] (WASP-121 b). The exoplanet is shown from above in the top figure, looking at one of its poles. The exoplanet rotates counter-clockwise, in such a way that it always shows the same side to its parent star, so it's always day on one half of the planet and night on the other. The transition between night and day is the *morning* side while the *evening* side represents the transition between day and night; its morning side is to the right and its evening side to the left.

[1] Tylos (ancient Greek: Τύλος) was the Greek exonym of ancient Bahrain in the classical era, during which the island was a center of maritime trade and pearling.

As the planet crosses in front of its host star, atoms in the planet's atmosphere absorb specific colors/wavelengths of the star's light, which can be measured with the ESPRESSO spectrograph on ESO'sVery Large Telescope. From these data, we can reconstruct the composition and velocity of different layers in the atmosphere, as in Fig. 4.19.

Specifically, the deepest layer is a wind of iron that blows away from the point of the planet where the star is directly overhead. The lower layer is characterized by the presence of iron, which is in a gaseous state due to the temperature of 2,700° on the side of the planet that always has day. Strong winds blowing through this layer carry superhot material to the dark side, which remains relatively cool at 1,400 °C. Above this layer there is a very fast jet of sodium that moves faster than the planet rotates. Sodium is an element that does not exist in the Earth's atmosphere. This jet actually accelerates as it moves from the morning side to the evening side of the planet. Finally, there is an upper layer of hydrogen wind blowing outwards. This hydrogen layer overlaps with the sodium jet below (Seidel et al. 2025).

Tylos is a gaseous, giant exoplanet located some 900 light years away in the constellation Puppis. It's an ultra-hot Jupiter, a gas-giant orbiting its host star so closely that a year there lasts only about 30 Earth hours. Moreover, one side of the planet is scorching, as it is always facing the star, while the other side is much cooler. Interestingly, the observations also revealed the presence of titanium just below the jet stream, as highlighted in Prinoth et al. (2025).

2025: Planet in a Perpendicular Orbit Around a Pair of Young Brown Dwarfs

The European Southern Observatory's Very Large Telescope (VLT) discovered with radial velocities for the first time a *polar planet* orbiting a stellar pair (Fig. 4.20), i.e., a planet that orbits at an angle of 90° around a rare pair of peculiar stars (as in the fictional *Star Wars* world *Tatooine*). The new exoplanet, named 2M1510(AB) b, orbits a pair of young brown dwarfs ($\sim$30 M_{jup}), objects bigger than gas-giant planets but too small to be proper stars, which form an eclipsing binary. There are so far only 16 known circumbinary exoplanets, all of which lie in the same orbital plane as the host binary. There have previously been hints that planets on polar, orbits around binary stars could exist. In theory, these orbits are stable, and planet-forming disks on polar orbits around stellar pairs have been detected. However, until now, we lacked direct evidence that these polar planets do exist (Fig. 4.20).

2M1510 is a double-lined eclipsing binary composed of two equal-mass brown dwarfs of mass 0.0331 and 0.0332 $M_\odot$ respectively. The binary has an orbital period of 20.9 days, an eccentricity of 0.36, and there is also a visual brown dwarf tertiary companion, at a projected separation of $\sim$ 250 AU from the central and an orbital period of $\sim 11,000$ years, from the eclipsing pair (Baycroft et al. 2025). This distant

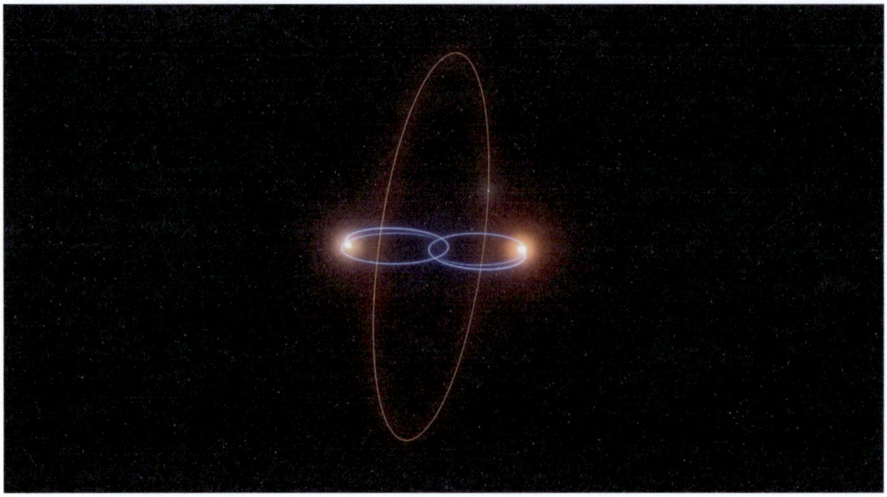

Fig. 4.20 Configuration of the system 2M1510 (*Credit* ESO/L. Calçada.)

tertiary companion cannot be the cause of the observed apsidal precession of 2M1510 (AB).

2025: Total Exoplanet Discoveries

The year 2026, when this book was completed, 31 years since the first discovery of an exoplanet around a solar-type star (1995), the almost daily discovery of new exoplanets seems to have become a routine. On February 26, 2026, a related NASA account listed 6,128 confirmed exoplanets, in more than 4,000 planetary systems, with about 1000 systems having more than one planet (Fig. 4.1). Most of these planets were discovered by the Kepler space telescope.

Specifically, planets have been identified:

- to orbit around solar-type stars of spectral-type G, i.e., *yellow* stars, such as around the star Kepler 69 (2013), with several of these systems having planets with sizes roughly equal to the terrestrial sizes and revolving within the *habitable zone* of stars like the Sun, most of which were discovered by the Kepler mission (2016), e.g., Kepler 296 f (2014);
- around the nearest star to Earth, the red dwarf Proxima Centauri at a distance 4.244 l.y., which however lives in a particularly inhospitable environment, since the parent star does emit strong UV radiation;
- around red dwarfs, some of which are prone to releasing potentially deadly flares and others having variability thought to be caused by large starspots moving in and out of view as the star rotates (Gliese 876, 2005), while some others do not

emit large amounts of dangerous radiation, such as Ross 128 b, at a distance of around 11 light years from Earth, the nearest exoplanet around a quiet red dwarf, which is considered one of the best candidates for habitability (2017);
- around brown dwarfs, some of which are observed via direct imaging, as 2M1207 b (2004);
- around giant orange stars (i Draconis, 2001);
- around solitary pulsars (PSR B1257 + 12, 1992);
- around very old stars, 12.7 billion years old, in orbit around binary systems containing pulsars and white dwarfs (PSR B1620-26b, 2003);
- around distant stars, at a distance of up to 21,500 l.y. toward the center of the Milky Way (OGLE-2005-BLG-390Lb, 2006), in whose atmospheres has been detected sodium (HD 209458 b, 2001), water vapor and silicon dust clouds (HD 209458 b, HD 189733 b, 2007), or, methane (HD 189733 b, 2008);
- planets with large ice cores (Gliese 436 b, 2007);
- planets in planetary systems that have a pair of planets similar to the Jupiter-Saturn pair (OGLE-2006-BLG-109Lb 2008);
- small planets, of terrestrial dimensions, with conditions favorable for the development of life (COROT-7b, 2010, GLIESE 581 e, 2009);
- planet-oceans composed of 75% water and 25% of rock (GJ 1214 b, 2009), in whose atmosphere are observed superstorms of carbon monoxide, from the side of the warm day to the cold night side (HD 209458 b, 2010), with the size of Jupiter, and which are possibly habitable (PH2 b, 2013);
- planets with a long year, such as 704 days (Kepler 421 b, 2014);
- rocky planets, like our own Earth (HD 219434 b, 2015);
- planets with Earth-like atmospheres and moderate temperatures which allow water in a liquid form on their surface (Trappist-1b, Trappist-1c, 2017);
- a planetary a system with eight planets, just like our own solar system (Kepler-90, 2017);
- a planetary system with no large planets but only with 5 small planets, with radii ranging from 1.6 to 3.3 the terrestrial radius and periods 2.35–3.56–5.4–8.26–12.76 days (K2-138, 2018);
- rocky, gaseous, odd, and other exotic planets exhibiting each one, separately, special interest.

The following list contains the 12 *weirdest* exoplanets ever discovered:

- TrES-2 b – the *darkest* exoplanet (reflecting less than 1% of incident light).
- KELT-9 b – the *hottest* exoplanet (with dayside temperatures 4,327 °C).
- HR 5183 b – the planet with the *strangest orbit* ($e \simeq 0.87$).
- HD 189773 b – a world where *glass rains* down from the sky.
- Kepler-13A b – a world where *snows titanium oxide* and *sunscreens.*
- Wasp 76 b – an exoplanet where it rains down from the sky droplets of iron.
- K2-18 b – a *hycean* exoplanet with *biosignatures* in its atmosphere.
- TOI 849 b – a world *stripped bare*, the remnant core of a gas giant.
- WASP-12 b – *egg-shaped* planet *consumed by its star*, black as asphalt.
- HAT-P-7 b – an ultra-hot Jupiter, *dark as charcoal with a sapphire sky.*

- 55 Cancri e – a *diamond* planet, with *volcanic* activity.
- Rogue worlds: exoplanets on the loose.

References

Balmer WO (2025) JWST-TST high contrast: Living on the wedge, or, NIRCam bar coronagraphy reveals CO_2 in the HR 8799 and 51 Eri Exoplanets' Atmospheres, AJ 169 209

Baycroft TA, Sairam L, Triaud AHMJ, Correia ACM (2025) Science advances. https://doi.org/10.1126/sciadv.adu0627, arXiv:2504.12209v1

Brown TM, Charbonneau D, Gilliland RL, Noyes RW, Burrows A (2001) Hubble space telescope time-series photometry of the transiting planet of HD 209458. Astrophys J 552:699–709. arXiv: astro-ph/0101336

Currie T, Burrows A, Itoh Y, Matsumura S, Fukagawa M, Apai D, Madhusudhan N, Hinz PM, Rodigas TJ, Kasper M, Pyo T-S, Ogino S (2011) A combined Subaru/VLT/MMT 1–5 μm study of planets orbiting HR 8799: Implications for atmospheric properties, masses, and formation. Astrophys J 729(2):20, article id. 128

Facchini S, Testi L, Humphreys E, Donckt MV, Isella A, Wrzosek R, Baudry A, Gray MD, Richards AMS, Vlemmmings W (2024) Resolved ALMA observations of water in the inner astronomical units of the HL Tau disk. Nat Astron 8:587–595 (arXiv:2403.00647)

Marcy GW, Butler RP et al (1997) The Planet around 51 Pegasi, Astrophys J 481:926

Marois C, Macintosh B, Barman, Zuckerman B, Song I, Patience J, Lafrenière T, Doyon R (2008) Direct imaging of multiple planets orbiting the star HR 8799. Science 322(5906):1348

Matthews EC, Carter AL, Pathak P, Morley CV, Phillips MW, PM, Krishanth S, Feng F, Bonse MJ, Boogaard LA, Burt JA, Crossfield IJM, Douglas ES, Henning T, Hom J, Ko C-L, Kasper M, Lagrange A-M, Petit dit de la Roche D, Philipot F (2024) A temperate super-Jupiter imaged with JWST in the mid-infrared. Nature 633(8031):789–792

Mayor, M., and D. Queloz. 1995. A Jupiter-mass companion to a solar-type star. *Nature* 378 (6555): 355–359.

Nortmann L et al (2025) CRIRES+ transmission spectroscopy of WASP-127 b, Detection of the resolved signatures of a supersonic equatorial jet and cool poles in a hot planet. A&A, 693

Prinoth B et al (2025) Titanium chemistry of WASP-121 b with ESPRESSO in 4-UT mode, A&A 694

Seidel JV et al (2025) Vertical structure of an exoplanet's atmospheric jet stream, Nature. arXiv: 2502.12261

Tamayo D, Triaud AHMJ, Menou K, Rein H (2015) Dynamical stability of imaged planetary systems in formation: Application to HL Tau. Astrophys J 805(2), article id. 100, 15

Chapter 5
Marvellous Discoveries in Astrophysics During the Last 100 Years

Astrophysics is older than Physics, and its beginnings lie in the attempt to understand the harmonious simplicity of the movements of stars and planets…

The most amazing discovery of modern Astrophysics is that the stars are made of the same atoms that we find here on Earth,
and the same laws of Physics which apply in our laboratory apply also in the Universe…

The Feynman Lectures on Physics, Vol. I, Chap. 3, 3–4: Astronomy
[Richard Feynman, Theoretical Physicist, *Nobel Prize* in Physics (1965)

1920: The *Great Debate* on Spiral Nebulae and the Size of the Universe

In astrophysics, which actually began as a science at the beginning of the twentieth century, fascinating discoveries have taken place since then and until today, which have radically changed its landscape. First of all, Einstein's mass-energy equivalence relationship, $E = mc^2$, led to the solution of the basic problems of the production of energy inside the cores of stars and also the nuclear composition of most of the chemical elements of the periodic table. Subsequently, large ground-based and space telescopes, which were built and deployed to understand the Universe collected a wealth of spectroscopic and photometric astronomical data. The effort to understand them led to such cornerstone-discoveries in astrophysics, such as the Hertzsprung-Russell diagram (Fig. 5.5) and the realization that the Universe contains a multitude of galaxies, similar to our Milky Way galaxy (Fig. 5.3).

K. Tsinganos, *A Century of Cosmic Discoveries*, Springer Praxis Books,
https://doi.org/10.1007/978-3-032-06282-6_5

The *Great Debate*, also called the Shapley–Curtis Debate, was held on April 26, 1920 at the Smithsonian Museum of Natural History, between the astronomers Heber Curtis and Harlow Shapley (Fig. 5.2). In brief, the controversy concerned the scale and makeup of the universe, i.e., the nature of the so-called *spiral nebulae* and the size of the Universe. Shapley believed that these nebulae were relatively small and lay in the outskirts of the Milky Way galaxy, which then thought to be the entire Universe. He stated that "it seems to me that the evidence is opposed to the view that the spirals are individual galaxies comparable with our own. In fact, there appears as yet no reason for modifying the tentative hypothesis that the spirals are not composed of typical stars at all, but are truly nebulous objects". On the other hand, Curtis held the view that they were in fact independent galaxies, implying that they were exceedingly large and distant. In Curtis words, "the evidence at present

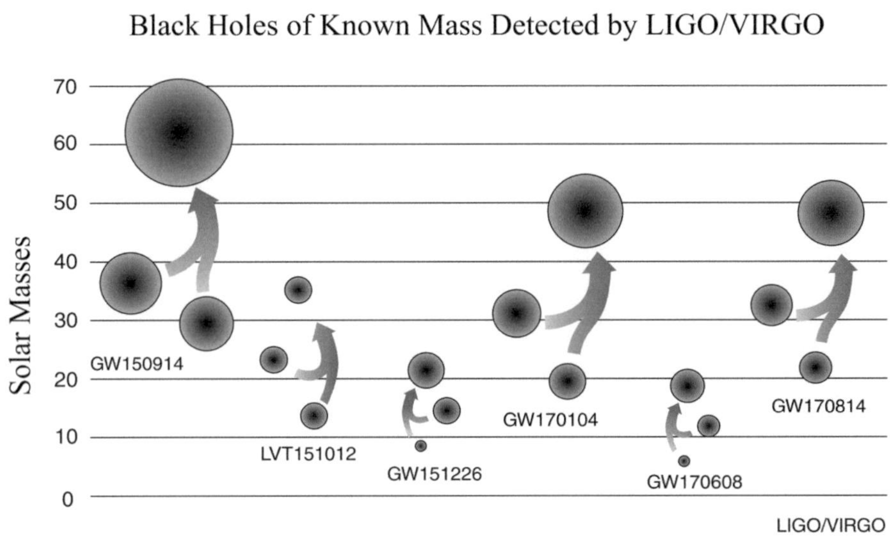

Fig. 5.1 The most important discovery in Astrophysics during the last hundred years is the detection of *gravitational waves* from the source GW150914 via the coalescence of two black holes. This schematic diagram shows the black holes detected as of November 2017 by the LIGO and VIRGO interferometers. These black holes emitted gravitational waves when they merged. The discovered new population of *black holes* has masses that are larger than what had been seen before with X-ray studies alone. At lower masses, gravitational wave sources such as GW170608 and GW151226 have masses comparable to those seen in X-ray emitting binaries. The three confirmed detections by LIGO (GW150914, GW151226, and GW170104), and one lower-confidence detection (LVT151012), point to a population of stellar-mass binary *black holes* that, which once merged, are larger than 20 solar masses – larger than what was known before. The sources GW150914, GW170104, and GW170814 appear to belong to a population of higher masses, which was not observed before the detection of gravitational waves. Also visible in the image is the source LVT151012, a LIGO candidate source, but which was too faint to finally be registered as a real detection. (*Image credit* LSC/LIGO/Caltech/Sonoma State (Aurore Simonnet))

Fig. 5.2 H.D. Curtis (left) and Harlow Shapley (right), the two protagonists of the *Great Debate* (1920) about the true distances of the *nebulae*, i.e., the galaxies as we call them today. (Credit Curtis photo Special Collections, UC Santa Cruz and Wikipedia Commons)

available points strongly to the conclusion that the spirals are individual galaxies, or island Universes, comparable with our own galaxy in dimension and in number of component units."

Hence, Shapley argued that the universe was comprised of a *single* galaxy, while Curtis proposed that the Universe contained *many* galaxies.

Later on (1929), Edwin Hubble showed that Andromeda was far outside the Milky Way by measuring Cepheid variable stars, proving thus that Curtis was correct. It is now known that the Milky Way is only one of as many as an estimated 200 billion to 2 trillion or more galaxies in the observable Universe, proving thus that Curtis was the accurate party in the debate.

An amazing progress has been taken place since then in observational astronomy. With the Hubble and James Webb Space Telescopes we have seen those spiral galaxies at huge distances from our Milky Way galaxy. Quite recently (2024), the NASA/ESA/CSA James Webb Space Telescope has released a collection of 19 face-on spiral galaxies in near- and mid-infrared light, shown in Fig. 5.3. Webb's near-infrared camera (NIRCam) have captured millions of stars in these images.

Older stars appear blue in Fig. 5.3, and are clustered at the galaxies' cores. The telescope's mid-infrared instrument (MIRI) observations highlight glowing dust, showing where it exists around and between stars – appearing in shades of red and orange. Stars that haven't yet fully formed and are encased in gas and dust appear bright red. Webb's high-resolution images are the first to show large, spherical shells in the gas and dust in such exquisite detail. These holes may have been created by stars that exploded and carved out giant regions in the interstellar material.

Fig. 5.3 A collection of 19 Webb images of face-on spiral galaxies are combined in this mosaic, which is at once overwhelming and awe-inspiring. Galaxies' spiral arms appear in shades of orange, and many of their centers have light blue hazes. Several galaxy cores are awash in pink-and-red diffraction spikes, clear signs that these galaxies may have central active supermassive black holes or central star clusters. (*Image Credit* NASA, ESA, CSA, STScI, J. Lee (STScI), T. Williams (Oxford), PHANGS Team, E. Wheatley (STScI))

These spiral galaxies are Webb's first big batch of contributions to the **P**hysics at **H**igh **A**ngular resolution in **N**earby **G**alaxie**S** (PHANGS) program, that includes existing images and data from the NASA/ESA Hubble Space Telescope and ESO's Very Large Telescope's Multi-Unit Spectroscopic Explorer (MUSE), and the Atacama Large Millimeter/submillimeter Array (ALMA). With those observations/images, researchers can now examine these galaxies in ultraviolet, visible, infrared, and radio light.

After the 2nd World War, with the development *of technology and space science*, new scientific branches emerged and began to be studied, such as radio astronomy and the exploitation of other than the visible components of the electromagnetic spectrum which revealed the so called *invisible Universe*. These new astronomical branches were ultraviolet, X-ray, and γ-ray astronomy, as well as infrared, radio, and millimeter-wave astronomy. By using such regions of the electromagnetic spectrum, new areas of astrophysics emerged and the respective discoveries came one after the other. And since the discoveries in these new areas of astrophysics had a rather wide impact on physics, they were honored with Nobel prizes: this is the case for the discovery of the theory of *white dwarfs* and their maximum mass (1929/Nobel 1985), the discovery of the proton-proton cycle for the solar thermonuclear reactions (1936/Nobel 1967), the discovery of magnetohydrodynamic Alfvén waves (1942/ Nobel 1970), the physical processes affecting the structure, evolution, and chain of thermonuclear fusion reactions inside stars (1957/Nobel 1983), the discovery of the aperture synthesis (radio interferometry) and the first pulsar (1962/Nobel 1974), the discovery of the diffuse cosmic microwave radiation(1965/Nobel 1978), the discovery of baryogenesis – the imbalance of matter (baryons) over antimatter (antibaryons) in the observed Universe – and Sakharov's conditions (1967/Nobel 1975), the first detection of cosmic neutrinos (1968/Nobel 2002), the observation of the double pulsar and indirectly of the gravitational waves (1974/Nobel 1993), solar neutrino oscillations and the solution of the solar neutrino problem (1996–1999/ Nobel 2005), the first direct detection of gravitational waves (2015/Nobel 2017) and more recently the first observation of an exoplanet orbiting a solar-type *main sequence* star (1995/Nobel 2019) and the discovery that black hole formation is a robust prediction of the general theory of relativity and for the discovery of a supermassive compact object at the center of our galaxy (2017/Nobel 2020).

A presentation of *all* discoveries in astrophysics is technically impossible to fit within this book. Therefore, in the following, we have selected only some of those discoveries which, in our opinion, changed the main course of astrophysics. We ask for an understanding from our readers and fellow astrophysicists, because we cannot mention the remaining here. In any case, several astrophysical discoveries, year by year, are mentioned in the tables preceding the first chapter of this book.

1910: The Hertzsprung-Russell Diagram (HR Diagram)

The Hertzsprung-Russell diagram (HR diagram, Fig. 5.5), was created around 1910 by Ejnar Hertzsprung and Henry Norris Russell, who independently discovered a relationship between stellar luminosity and surface temperature. This HR diagram was an important step in understanding stellar evolution and is a continuation of the systematic classification of stars by the Harvard Observatory research team. The diagram is similar to a human height-weight diagram, Fig. 5.4.

In Fig. 5.5, 22,000 stars from the catalog resulting from the Hipparcos astrometric mission of ESA (Hipparcos Catalogue) have been placed, as well as 1,000 stars from the Gliese catalog of nearby stars. All these thousands of stars are organized and located in certain areas of this HR chart. The *denser* and most *populated* region lies on the diagonal from the upper left corner (hot and bright stars) to the lower right

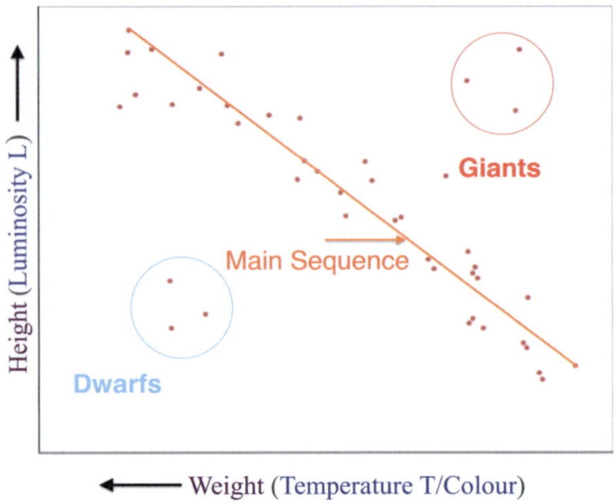

Fig. 5.4 Height/weight chart of a representative group of humans, or, stars. Most points are on the diagonal (main sequence), but there are some "exceptions" (circled), at the lower left and upper right

corner (cooler and fainter stars), the Main Sequence (MS), Fig. 5.5. In the lower left part are the white dwarfs and above the MS the subgiant, giant, and supergiant stars. The Sun lies in the MS at magnitude 1 (absolute magnitude 4.8) and color index BV = 0.66 (temperature 5,780° and spectral type G2V).

Henry Norris Russell, one of the most prominent American astrophysicists of the interwar period, visited the National Observatory of Athens (1937). His visit was recorded as one of the most memorable events in the relatively long history of the National Observatory of Athens, the oldest Research Institution in Greece, which was established in 1842 (Fig. 5.6a, b).

1910: Discovery of White Dwarfs

White dwarfs is the *artistic nickname (!)* of one of the three types of *stellar corpses*, the other two being *neutron stars* and *black holes* (Figs. 1.28 and 5.7). They are the remnants of the core of a low- or medium-mass star that remain after their death. In other words, *white dwarfs* are the end of the evolution sequence of relatively low-mass stars and are those stellar remnants that we have the best understanding of, since the gravitational compression has not yet reached the point of causing exotic phenomena, such as in the case of *neutron stars* and *black holes*, Fig. 5.7. The energy radiated by a *white dwarf* comes from the stored thermal energy of the progenitor star, since thermonuclear reactions no longer occur inside it to produce new energy. Stars whose mass does not exceed about 10 solar masses are driven to this final state. This means that about 97% of all stars in the Galaxy will end up as *white dwarf*s,

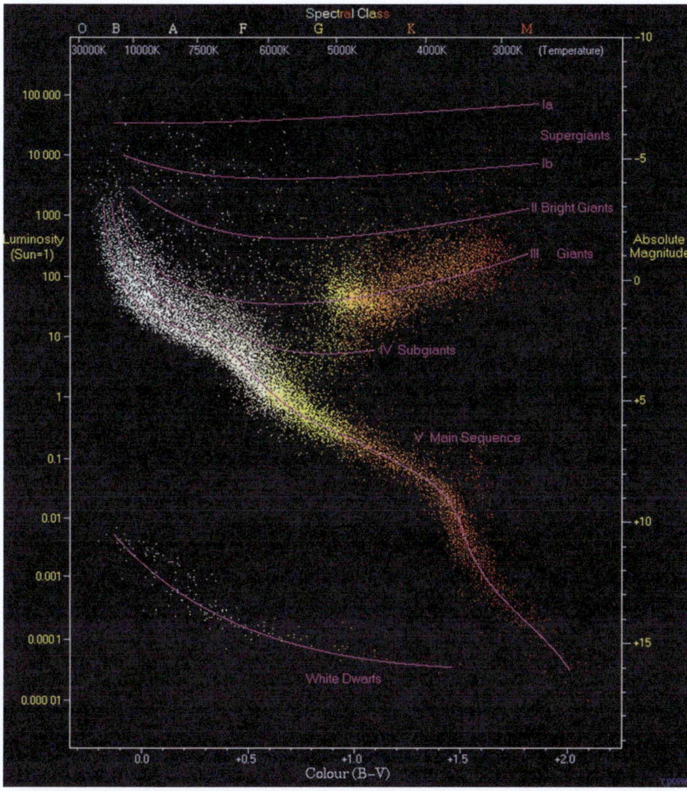

Fig. 5.5 The famous observational Hertzsprung-Russell diagram with 22,000 stars from the Hipparcos catalog and 1,000 stars from the Gliese catalog of nearby (up to 81.5 l.y.) stars. (*Source* http://www.atlasoftheuniverse.com/hr.html, Richard Powell and Wikimedia commons)

and this percentage includes our Sun, whose inner layers will turn into a *white dwarf* in about five billion years, surrounded by a planetary nebula.

A popular – for night sky watchers and amateur astronomers alike – example of a planetary nebula is the *ring nebula* in the constellation Lyra, south of its brightest star, Vega, which also marks the brightest corner of the Summer Triangle. This nebula is located at a distance of about 2,300 l.y. (light years). It is about 7,000 years old, and its current diameter is about 1 light year. Although the nebula is probably hourglass-shaped reflecting the fact of a focused plasma outflow, it appears annular because we see along the axis of symmetry that passes through its center.

Other planetary nebulae are listed in Fig. 5.8a, which includes 6 planetary nebulae observed with the Hubble Space Telescope.

IC 3568 (the "*Lemon Slice*" planetary nebula) is a relatively young planetary nebula in the constellation Camelopardalis (Giraffe), just 7.5° from the Polar Star, a few thousand light years distant from the Earth. Its central core has a diameter of 0.4 l.y.

(a)

(b)

Fig. 5.6 (*above*) From the visit of Henry Norris Russell (third from right) to the National Observatory of Athens, NOA (1937). Second from right is the Director of NOA in 1937, Captain George Hors and in the far left the Director of the Astronomical Institute S. Plakidis. (*below*) Henry Norris Russell (in the light-colored suit) at the entrance of the Headquarters of NOA (1937), with the Observatory's staff. To his right is the Director of NOA, George Hors and on the far right the Director of the Astronomical Institute, prof. S. Plakidis. (*Source* Album of the National Observatory of Athens, 2016)

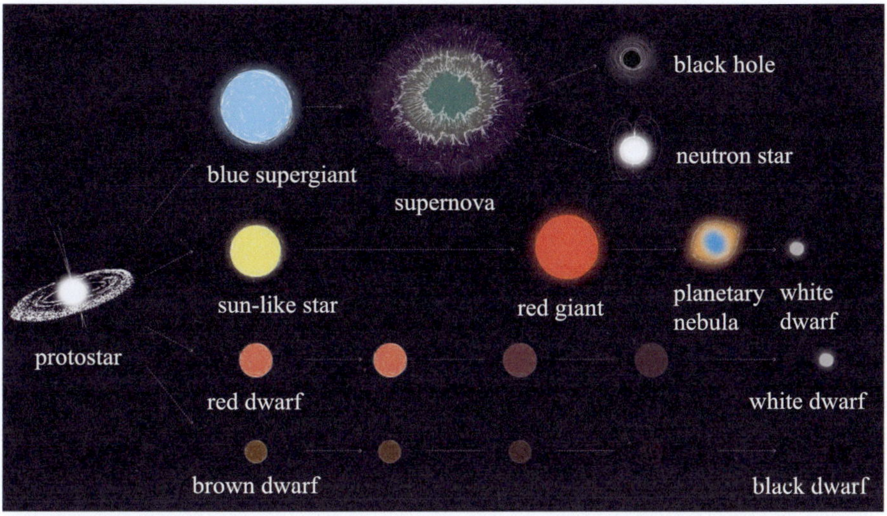

Fig. 5.7 Explanatory diagram of the evolutionary path of stars with different initial masses. (*Source* ESA)

(i.e., about 800 times the diameter of the solar system). It is known as one of the simplest nebulae (the ideal example for theorists to model!), with an almost perfectly spherical morphology consisting of a central hot star, a bright circular inner shell within a fainter outer circular environment. The central hot star, a red giant in the final stages of its life and evolution, has a temperature of 57,000 °K and is ionizing the outer Helium shell. The bright circular inner shell corresponds to the front of the shock wave which was created when a fast supersonic stellar wind travelling at a speed of about a thousand kilometers per second collided with a slow preceding and leading slow wind from the central star.

NGC 6826 (the *human eye* planetary nebula) is similar to IC 3568. It is located at a distance of 2,200 light years in Cygnus and its central core is 0.2 l.y. in diameter. The central hot star, a *white dwarf*, emits a fast stellar wind into a shell that has been created by an earlier and leading slow wind, and at their interface a shock wave is created that pushes away the gas in front. The surrounding faint shell contains about half of the original stellar mass. A characteristic of this nebula are the two red clusters on the left and right.

NGC 3918 (the *light blue* planetary nebula) was discovered by Sir John Herschel in 1834. It is located at a distance of 3,900 l.y. in the constellation Centaurus and is also 0.2 l.y. in diameter. At its center is a dying red giant star, from which the *white dwarf* will eventually emerge. It has a similar structure to the previous ones, a hot central star from which two colliding stellar winds emerge and create the two shells.

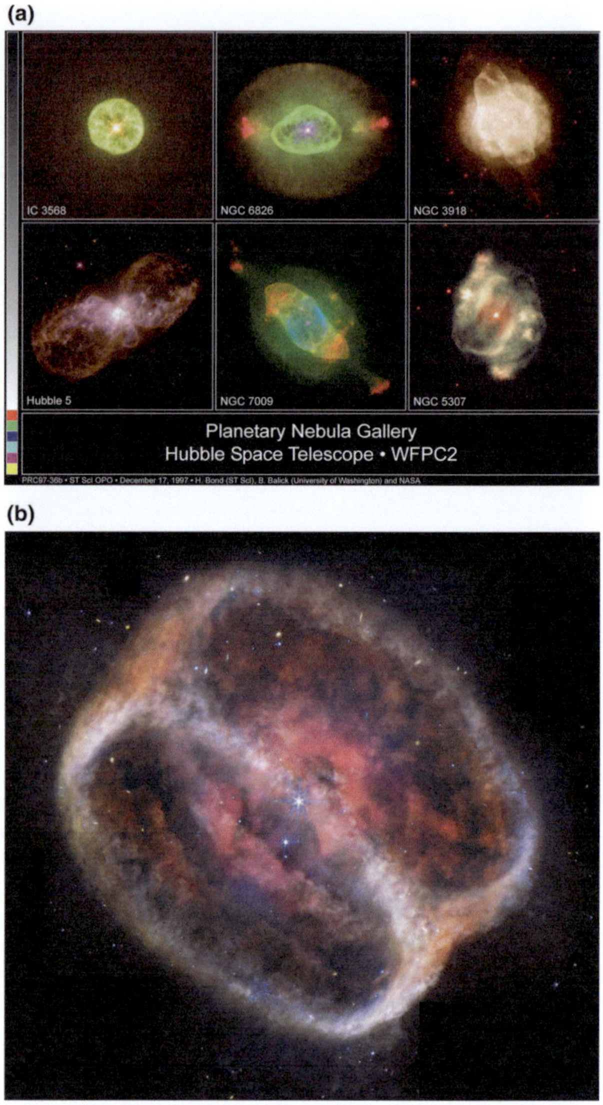

Fig. 5.8 **a** A relatively recent planetary nebulae gallery with the Hubble Space Telescope. (*Source* NASA/ESA). **b** In NGC 1514, two stars look like a single bright star, shine with bright purple diffraction spikes at the center of a large, diffuse cylinder of gas and dust that is tipped to the right. There are two large rings seen at a roughly 60° angle that appear joined at top left and bottom right. The edges are denser, and form shallow V-shapes that go inward. The rings appear orange at top left and bottom right, and are blue at bottom and center right. There is diffuse orange material around the body. The black background of space is speckled with tiny stars and galaxies mostly in blues and yellows. Areas Webb did not observe are along the top edges, a thin vertical near the nebula at top left, and at the bottom left and right corners. (*Credit* NASA, ESA, CSA, STScI, Michael Ressler (NASA-JPL), Dave Jones (IAC))

Hubble 5 (the *butterfly* planetary nebula) is an example of a butterfly-shaped bipolar nebula. The thermal pressure of the fast wind inflates the two lobes, as it happens in two balloons that have an internal heating.

NGC 7009 (the *Saturn* planetary nebula) was discovered by William Herschel in 1782 and resembles the planet *Saturn* and its rings. It is at a distance of about 1,400 l.y. in the constellation of Aquarius. It is similar to NGC 6826 in having a bright central star (*white dwarf*) and two *hooks* left and right which are connected to the central cavity via two jets.

NGC 5307 (the *spiral* planetary nebula) is also in the constellation Centaurus at a distance of about 10,000 l.y. and has a diameter of about 0.6 l.y. Every density increase on one side of the central star, it also appears on the opposite side.

NGC 1514 (the hourglass shape nebula in Taurus, about 1,500 light years from Earth). This planetary nebula has been studied since the late 1700s, with William Herschel noting in 1790 that NGC 1514 was the first deep sky object to appear genuinely cloudy, as he could not resolve what he saw into individual stars within a cluster, like other objects he cataloged. Figure 5.8b shows a considerably clearer view with the mid-infrared data from the James Webb Space Telescope, which makes it look like a can is being poured, but it's far more likely that NGC 1514 takes the shape of an *hourglass* with the ends lopped off, a pinched waist near top left and bottom right, where the dust is orange and drifts into shallow V-shapes. When this star was at its peak of losing material, the companion could have gotten very close, resulting in these unusual shapes. Instead of producing a sphere, this interaction might have instead formed rings. This scene has been forming for at least 4,000 years and will continue to change over many more millennia. At the center are two stars that appear as one in Webb's observation, and are set off with brilliant diffraction spikes. The stars follow a tight, elongated nine-year orbit and are draped in an arc of dust represented in orange. One of these stars, which used to be several times more massive than our Sun, took the lead role in producing this scene. Once the star's outer layers were exhausted, only its hot, compact core remained. As a white dwarf star, its winds both sped up and weakened, which might have swept up material into thin shells.

Planetary nebulae dissolve in about 10,000 years. The *white dwarf*, however, will survive for trillions of years, before finally being extinguished as a *black dwarf*. *Black dwarfs* are theoretical objects, predicted by models of stellar evolution. They are created when a *white dwarf* radiates all its energy and cools down to near absolute zero. However, the transition from a *white dwarf* to a *black dwarf* takes, according to the theory, a time longer than the present age of the Universe. Therefore, there should not yet be any *black dwarfs* in the Universe. Also, due to the fact that they do not radiate energy and heat, it is practically impossible to observe them.

White dwarfs were discovered in the mid-eighteenth century. The first *white dwarf* was discovered in the triple stellar system 40 Eridani, which contains the bright Main Sequence star 40 Eridani A, around which orbits a binary system consisting of the *white dwarf* 40 Eridani B and the red dwarf 40 Eridani C, Fig. 5.11. The pair 40 Eridani B/C was discovered by William Herschel (1783).

In 1844 Bessel discovered the *white dwarf* accompanying Sirius, the brightest star in our night sky, located 8.6 light years from Earth, Sirius B. This Sirius B is the nearest *white dwarf*, with a mass of 1.02 solar masses and a radius similar to the radius of Earth, while its surface temperature is 5,200 K. Because it has no internal heat supply, it will cool steadily for the next 2 billion years. It originates from the evolution of a 5 solar mass, spectral type B, Main Sequence star. The interior of Sirius B is a mixture of carbon and oxygen formed by the fusion of helium into its progenitor, while the outer layers contain lighter elements. Its outer surface consists almost exclusively of hydrogen.

The bright star Sirius has been known since ancient times and belongs to the constellation Canis Major. The Great Dog (Canis Major) is one of the two hounds of Orion, which helped the mythical hunter in his battle against Taurus, when he was trying to prevent the violent abduction of Europa by Jupiter in the form of a bull! For the ancient Greeks, the rising of Sirius *together* with the Sun (*eoa epitoli of Sirius, εώα επιτολή του Σείριου*), i.e., the rising of the brightest star of the night just before the rising of the Sun, signaled the arrival of the hot summer weather, which caused the canine burns, during these canine days. It happens once in a year, toward the end of July.

Sirius was often associated with drought and heat, since the Sun was in Leo when Sirius rose. The rising of Sirius together with the Sun was considered by the Egyptians as a harbinger of the beginning of the flood season of the Nile river and their beneficial effect on agricultural crops. The poet Hesiod, perhaps the first to refer to Homer's "*fruitful*" star, *the Dog star*, under the name of Sirius, advises his brother in his "*Works and Days*" (582–617), what to do during these days of the *Dog*.

The two stars are in orbit around their common center of mass, with a period of about 50 years and at an average distance of 20 AU. Therefore, although with the naked eye a single star is discernible, in fact it is a double star, consisting of a white Main Sequence star of spectral type A1V and a *white dwarf* of spectral type DA2.

The human weight on Sirius' White Dwarf…

Sirius B, although slightly smaller in size than Earth – with a diameter of about 12,000 km – has a mass approximately equal to the mass of the Sun. Thus, it is much denser than Earth, with a density of about 1 ton per cubic centimeter. Its gravitational field is about 400,000 times stronger than Earth's. So, an average human weighing 70 kg on Earth, if he/she went to the surface of Sirius B would weigh 28 million kg. Hence, it would be rather difficult to survive!

So, in 1844 the German mathematician and astronomer Friedrich Wilhelm Bessel, observing this star's proper motion – the angular change in position of a star across our line of sight, measured in arc seconds per year – suggested that Sirius is actually a *double* star (Fig. 5.9). Indeed, in 1862 the astronomer Alvan Graham Clark discovered a second star there, now referred to as Sirius B. Thus, the star visible from Earth is often called Sirius A. Sirius A has a mass twice that of the Sun, a radius of 1.2 million km, i.e., a radius of 1.7 times that of the Sun, and is 25 times brighter than the Sun. Sirius B was the first *white dwarf* to be discovered from its spectrum. This fact also meant that its progenitor star was more massive than Sirius A, as it evolved earlier beyond the Main Sequence. But the wavelength of the radiation emitted by Sirius B's photosphere is also greatly shifted toward the longer red wavelengths, because it has to *get out* of the strong gravitational field of the *white dwarf*.

This phenomenon, predicted by Einstein's General Theory of Relativity (1916), is observed in the radiation coming from dense, high-mass and compact stellar objects, where the strong gravitational field bends spacetime near their surface. The distance

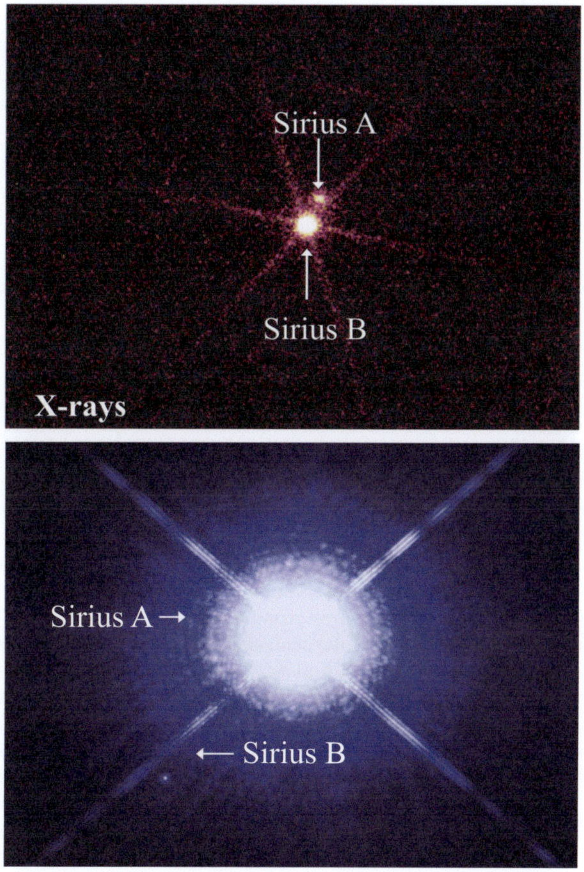

Fig. 5.9 Image of the Sirius binary system, with the Chandra X-ray Telescope (left) and the Hubble Space Telescope (right). Sirius A, in the visible, is the brightest star in our night sky, much brighter than Sirius B, but in X-rays it is Sirius B that is brighter. Sirius B, is a *white dwarf*. (*Source* NASA/SAO/CXC, HST)

between the nuclei of the carbon atoms inside Sirius B is about 10^{-10} cm, i.e., much smaller than the Bohr radius of an atom. Therefore, there is no room for atoms to exist inside a *white dwarf*, and the gas is completely degenerate, so that its description requires the use of *Quantum Physics*. There, the weight of the star is supported by the internal pressure of the gas of degenerate electrons.

Direct observations of Sirius B – and also of other *white dwarf*s, e.g., in Fig. 5.10 – is rather difficult because its luminosity is 10,000 times fainter than that of Sirius A. The surface temperature of Series A is 10,000°, but that of Sirius B is 25,200°. Because there is no internal heat source, Sirius B will continue to cool as its available heat will be radiated into space in about 2 billion years.

As mentioned earlier, a *white dwarf* is formed when a low/intermediate mass star completes the Main Sequence phase and passes into the Red Giant phase (Fig. 5.7). For Sirius B this happened when it was less than half its current age, i.e., about 120 million years ago. The progenitor star had a mass of about 5 solar masses and was a B-type star on the Main Sequence. The chemical composition of the *white dwarf* is

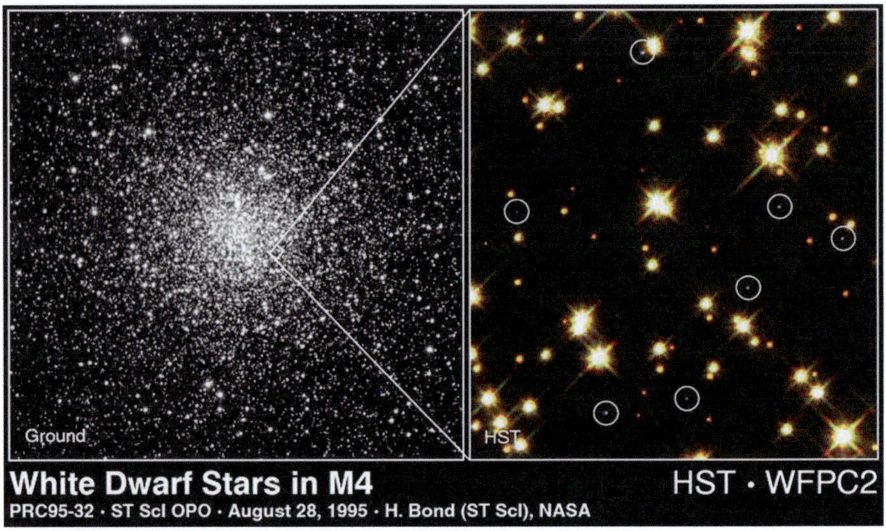

Fig. 5.10 *(left)*The globular cluster M4 is the closest stellar cluster to Earth (7,000 l.y.), with more than 100,000 stars. It was observed with the Hubble Space Telescope (HST) in a White Dwarf Survey. Red giant stars dominate this ground-based telescope image. The field is 47 l.y. in diameter. The small box (right/below of the center) is the area studied by the HST. In the HST image of the small region of the cluster, 0.63 l.y. in diameter, 7 White Dwarfs were observed among the populations of the much brighter yellow solar-type stars and cooler red dwarfs. In total, HST observed 75 White Dwarfs in this small region of M4, out of a total of about 40,000 White Dwarfs that the cluster is estimated to contain. This observation was very sensitive, since even the brightest White Dwarf at the distance of M4 is equivalent to a 100 watt light bulb at the distance of the Moon (384,000 km)! These results help astronomers to estimate the rate at which White Dwarfs are cooling, which is a prerequisite for calculating the age of the Galaxy and subsequently of the Universe. (SourceHarvey Richer (University of British Columbia, Vancouver, Canada), M. Bolte (University of California, Santa Cruz) and NASA/ESA NASA/HST

mainly carbon and oxygen, elements which were created by the fusion of the element helium in the progenitor star. Around this core of carbon and oxygen are arranged in shells the lighter elements, which are distributed in radial distance according to the their molecular weight, due to the strong gravity of the star. The outer atmosphere of Sirius B is almost pure hydrogen – the least massive element.

The Exotic Density of Matter of a *White Dwarf* …

Henry Norris Russell was quite concerned about the validity of the data, where 40 Eridani B occupied an unprecedented and unusual position, being a curious exception in his diagram. However, the Director of the Harvard Observatory, Edward C. Pickering, reminded him that *such exceptions are what sometimes "turn physics upside down"!* Really, this was a totally unexpected discovery.

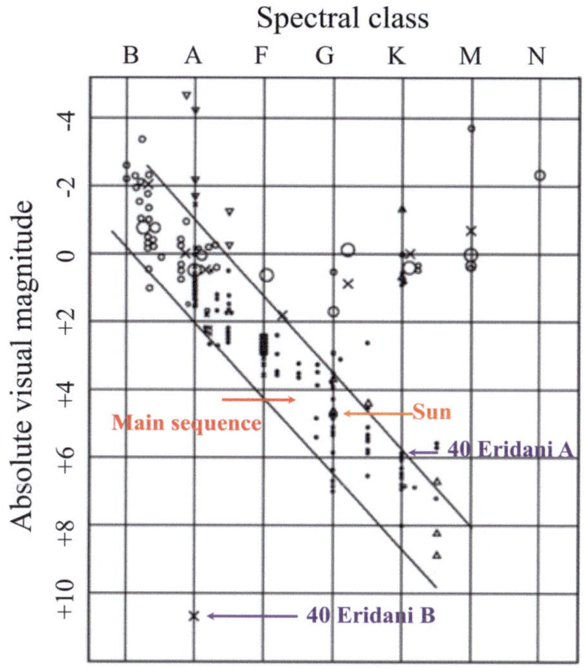

Fig. 5.11 The historical first diagram of resolved optical magnitude (vertical axis) and spectral Class (horizontal axis). The arrow marks the location of the White Dwarf, 40 Eridani B. *(Adapted from* Russell (1914))

As Eddington wrote later in his book *Stars and Atoms* (1926), the research community was surprised by this finding:
We learn about the stars by decoding the messages they send us by their light. Specifically, the message that Sirius' companion sends us, when decoded, tells us the following: "My composition is matter many thousands of times denser than anything else you have ever encountered on Earth. A ton of my material can fit in a matchbox. What can we respond to such a message? The response most of us initially gave in 1914 was, *Stop, don't talk nonsense*!"

In 1910, the team of Edward Charles Pickering, Director of the Harvard Astronomical Observatory and his group (Figs. 5.13, 5.14, 5.15 and 5.16) led by Williamina Fleming, (Fig. 5.13), were studying the spectra of several stars, and the results of this study were discussed with Henry Norris Russell, who used the method of parallaxes to calculate stellar distances. Based on these results, Henry Norris Russell formed the now classic Hertzsprung-Russell diagram of Fig. 5.5. In this study, the apparent surface temperature of the star was derived from the spectral lines and spectral type.

Then, with the star's distance known, its absolute brightness could be calculated from the apparent stellar brightness. Then, from the star's absolute luminosity and temperature, its surface area and ultimately its radius could be calculated. In double or triple systems, such as 40 Eridani and Sirius, the mass of the star can be calculated from the orbital data of each system. With this reasoning, Henry Norris Russell, Edward Charles Pickering and Williamina Fleming were led to the unexpected result that 40 Eridani B, although a very faint star, was of spectral type A, i.e., a white star. In the diagram of Fig. 5.11, 40 Eridani B stands out in the lower left as the only spectral type A star that is so faint.

Eddington wondered what would happen when this plasma cooled so that there was not enough energy available to keep the atoms ionized. The paradox was solved by R.H. Fowler (1926), applying the then new theory of *quantum mechanics*. Since electrons obey the Pauli exclusion principle, no two electrons can coexist in the same state and collectively follow the then (1926) newly discovered *Fermi–Dirac statistics*. At the temperature of absolute zero, the electrons cannot all occupy the lowest (ground) energy level, due to *Pauli's Exclusion Principle*, but they are distributed to the first available low energy levels above the ground, forming the Fermi energy sea. (Pauli's Exclusion Principle states that no two electrons in the same atom can have identical values for all four of their quantum numbers. In other words, (i) no more than two electrons can occupy the same orbital and (ii) two electrons in the same orbital must have opposite spins). This means that a *white dwarf*, even at the temperature of absolute zero can have high energy and pressure. This is the state of a *degenerate electron gas*. But this physics governing the interior of *white dwarf*s became known many decades after the discovery of the first *white dwarf* in the mid-eighteenth century, on the occasion of the Chandrasekhar-Eddington controversy that we discussed in Chap. 1.

An initially strange phenomenon which is a consequence of the degenerate state of matter inside these stars is that as their mass increases, their radius and volume decrease (Fig. 5.12)! This means that their density also increases and after a certain critical stellar mass, corresponding to a density of the order of *one ton per cubic centimeter*, the electrons become relativistic and lose their ability to counter balance the gravitational forces. This critical mass, also called the *Chandrasekhar limit* (M_{ch}), is an upper bound on the masses of *white dwarf*s. Stellar remnants with mass above this limit will either be *neutron stars* or *black holes*.

The white dwarf stars got the name "*dwarf*" from the fact that their radius does not exceed that of the Earth by much, although their mass is approximately equal to that of the Sun. The name "*white*" was given to them because of their high surface temperature, which in some cases can reach 100,000°. Despite the high temperatures that prevail inside them, the classical pressure of ideal gases is completely unable to contain the forces of gravity. It turns out that it is the pressure of the degenerate gas of electrons that provides the necessary counterbalance to gravity, as they are in a state of degeneracy due to the high density of matter inside them.

To calculate the stellar spectra, during the night observations, the astronomers at the Harvard College Observatory used photographic plates to record them. Then, during the day, the astronomical data was systematized and processed by a number of women, "*research assistants*," whom the Harvard Observatory, under the direction of Edward Charles Pickering (1877–1919), had hired for this purpose. These female research assistants, Figs. 5.13, 5.14 and 5.16, analyzed the spectra, calculated the stellar magnitudes and finally summarized and recorded their results in tables.

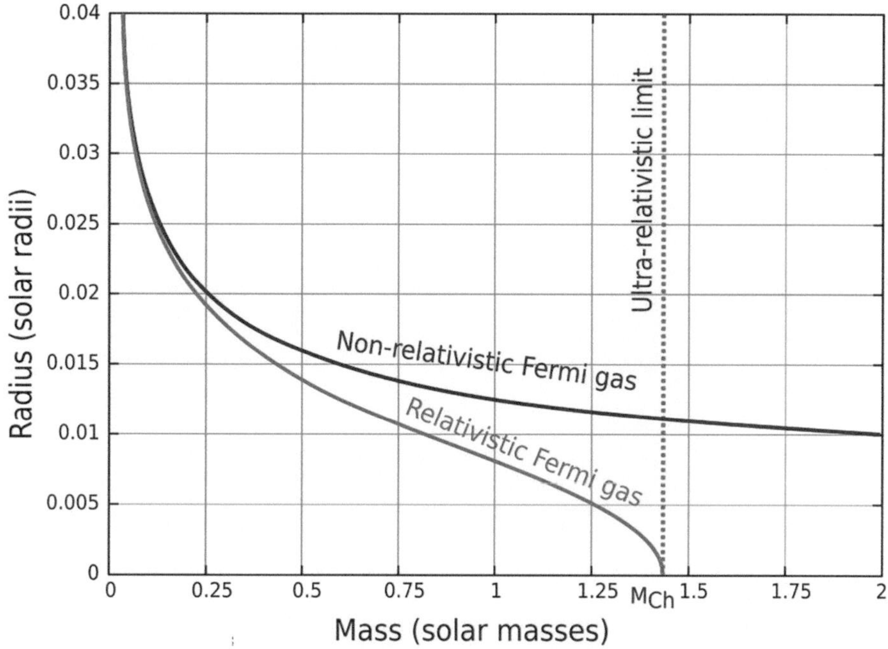

Fig. 5.12 The radius as a function of mass of a model *white dwarf* consisting of a Fermi electron gas. The mass M_{Ch} is the maximum mass that a *white dwarf* may have, the famous Chandrasekhar limit, $M_{Ch} = 1.4\ M_{\odot}$. Courtesy of Wikipedia

Fig. 5.13 Williamina PatonStevens Fleming (1857–1911), circa 1890 discovered the first White Dwarf. (*Source* Curator of Astronomical Photographs at Harvard College Observatory)

Fig. 5.14 The *Harvard Computers* at work, circa 1890. Shown are Henrietta Swan Leavitt (third from left, with magnifying glass), (1868–1921), Annie Jump Cannon (1863–1941), Williamina Fleming (standing center) (1857–1911) and Antonia Maury (1866–1952). (*Source* Harvard College Observatory and Wikimedia Commons)

Harvard was the first University to hire women into this type of computing research employment. They were called *Harvard Computers*. Their hourly pay was low (25 cents), half of men's earnings at the time. However, many of these women workers developed into outstanding and pioneering researchers who contributed first-line achievements to astronomy. For example, Mrs. Williamina Fleming, Fig. 5.13, made the first spectral identification of a *white dwarf*, 40 Eridani B, classified stars according to their hydrogen content, and was the first woman to become a member of the Royal Astronomical Society (1907) (Figs. 5.14 and 5.15).

Mrs. Annie Jump Cannon created the classic star classification scheme that is still valid today (Harvard Classification Scheme) for the classification of stars based on their temperature and spectral type where stars correspond to groups O, B, A, F, G, K, M (Oh, Be A Fine Girl, Kiss Me) according to their spectral characteristics. Mrs. Henrietta Swan Leavitt discovered the famous relationship between the period and the liminosity of the variable stars Cepheids, which is valuable to this day for calculating distances in the Universe. Mrs. Florence Cushman produced the 1918 edition of the Henry Draper catalog of the spectra of 222,000 stars, while Mrs. Antonia Maury processed many stellar spectra and produced an important catalog categorizing these results (1897). As a result of this work he observed the periodic doubling of some spectral lines in the spectrum of the star $\zeta 1$ of the big bear (Mizar A) which led to the identification of the first spectroscopic double system. Similarly, he identified the binary stellar system β of Lyra.

Fig. 5.15 E. Pickering and the Harvard-Computers, in front of Building C of the Harvard College Observatory, on May 13, 1913. (last row, left to right): Margaret Harwood, Mollie O'Reilly, Edward C. Pickering, Edith Gill, Annie Jump Cannon, Evelyn Leland (behind Cannon), Florence Cushman, Marion Whyte (behind Cushman), Grace Brooks. Front row: Arville Walker, unknown (probably Johanna Mackie), Alta Carpenter, Mabel Gill, Ida Woods. This photo, like the next one, is also interesting from a clothing point of view! (*Source* Harvard-Smithsonian Center for Astrophysics and Wikimedia Commons)

1915: Those Exotic Black Holes ...

Indeed, we all recognize that there is no other celestial body more exotic and mysterious, more alien and unreal, than *black holes*. But it is not widely known that these weird objects were discovered under the sounds of the cannon fire of the 1st World War, by a physicist suffering from a painful disease.

Einstein *formulated* but *he did not solve the* equations in his Theory of General Relativity. The *first exact solution* of the Einstein Field Equations of General Relativity (EFE) was found by the charismatic Austrian physicist Karl Schwarzschild, when he got his hands on the issue of the Proceedings of the Prussian Academy of Sciences of November 25, 1915, where Einstein's paper on General Relativity was published. He immediately informed Einstein, who was pleasantly *surprised* to learn that his equations were amenable to *exact* solutions, both because of their apparent complexity and because he himself had been unable to produce an exact solution.

Einstein had found only an approximate solution in his famous 1915 paper on Mercury's perihelion precession. In this paper, he used orthogonal coordinates to approximate the gravitational field around a spherically symmetric, non-rotating and uncharged gravitating body. Schwarzschild, in contrast, chose a more elegant polar coordinate system and was thus able to produce an accurate solution which he first

Fig. 5.16 The *Harvard Computers* outside the Harvard Observatory, circa 1890. (*Source* Harvard College Observatory)

Fig. 5.17 Karl Schwarzschild writing in his study at Potsdam. AIP Emilio Segre's Visual Archives, courtesy of Martin Schwarzschild.

communicated in a letter to Einstein on December 22, 1915, written while he was serving in the 1st World War on the Russian front. Schwarzchild concluded this letter to Einstein as follows (Figs. 5.16 and 5.17):

- "As you see, the war treats me quite *kindly,* and despite the cannonading around, it allows me to escape from all these petty troubles and take a pleasant walk in the land of your ideas".

Einstein immediately replied in December 1916 to Schwarzschild regarding this result:

- "I read your letter with great interest. I did not imagine that someone could formulate an exact solution to the problem in such a simple way. I really liked the mathematical treatment of the subject. Next Thursday I will present your solution at the Prussian Academy of Sciences, in Berlin, with a few words of explanation".

Schwarzschild's solution describes the gravitational field around a spherical mass, assuming that its electric charge and angular momentum, as well as the cosmological constant, are zero. This solution is a useful approach for describing slowly rotating astronomical objects, such as many stars and planets, including the Earth and the Sun.

The Schwarzschild black hole is characterized by a surrounding spherical boundary, called *the event horizon,* which lies at the so-called *Schwarzschild radius,* often called and the radius of the black hole,

$$R_{Sch} = \frac{2GM}{c^2}$$

where G is the gravitational constant, M is the mass of the central body, and c is the speed of light in vacuum. With the Schwarzschild metric the equations lead to a state of infinite gravity, a region in the Universe where physics as we know it ceases to apply!

In fact, however, in 1783 the English geologist John Mitchell,[1] *vicar* at the Thornhill Church in Yorkshire, had shown that if a star's mass was large enough, it could trap its own radiation, rendering the body invisible. Thus, a star of the same density as the Sun, but 500 times the radius, would have an escape velocity that would exceed the speed of light, and so light particles would not be able to escape the gravitational pull. Such a star should therefore appear black to the outside world and thus be invisible! Similarly, in 1796 the French physicist and philosopher Pierre-Simon Laplace published the same prediction in his famous work *The System of the World*. We had to wait until 1964 when it became known that actually there might exist such a dark star in the Cygnus constellation, with the "technical" name Cygnus X-1.

The next exact solution was discovered shortly afterwards (1916–1918) and involves a spherical, *charged* but non-rotating body and is the Reissner–Nordström metric. The problem for an uncharged but *rotating* black hole remained unsolved until 1963, when the corresponding exact solution, the Kerr metric, was discovered by Roy Kerr. The Kerr metric is a generalization to a rotating body of the Schwarzschild metric.

Schwarzchild died tragically in May 1916 at the age of 42, a few months after the publication of his famous work. The cause was a severe and painful autoimmune disease (Pemphigus), which affected mainly the Ashkenazi Jews, as he served in the German army, on the Russian front, during World War I.

We note in passing, that during his introductory speech at the Prussian Academy of Sciences (1913) Schwarzschild stated, among other things:

---

[1] John Michell (1724–1793) was an English natural philosopher and clergyman who provided pioneering insights into a wide range of scientific fields including astronomy, geology, optics, and gravitation. He is the first person known to have proposed the existence of *black holes*, in a paper for the Philosophical Transactions of the Royal Society of London, read on November 27, 1783 (Michell 1783). He invented an apparatus to measure the mass of the Earth, and explained how to manufacture an artificial magnet (Michell 1783).

Schwarzschild's Vision of the Whole

"...mathematics, physics and astronomy constitute a single body of knowledge, which, like Greek culture, must be regarded as an integral whole.
Only a vision of the whole, like that of a saint, a madman, or a mystic, will permit us to decipher the true organizing principles of the Universe."

A black hole[2] is that point in space where the gravitational forces are so strong that nothing – particles or electromagnetic radiation, such as light – can escape. The term "black hole", or, "black hole" is widespread and was adopted in 1967 by the American astronomer and theoretical physicist, John Wheeler. It is even said that in one of his classes, he wondered what would be the ideal name for these strange stars that "suck up" the light and look like holes in space. A student of his at the time jumped in and said, *Why not black holes*? Wheeler immediately adopted it, recognizing that it was a very *catchy* name. It does not, however, refer to an actual hole in the usual sense of the term, but to a region of space from which nothing can return.

In astrophysics, a black hole is that point in space where the core of a giant star which once existed, a core that contained more material than about two and a half solar masses and which, in the final phase of the star's evolution, lost the race against gravity, causing its material to collapse and compress even more than the material of a *neutron star*. The space in the neighborhood of this star is strongly curved and the dilation of time on its surface becomes infinite, that is, time does not flow at all, it "*freezes.*" Consequently, light is "trapped," its wavelength has become infinite, so the star must now appear completely dark.

If we compressed the Earth to a smaller and smaller radius, at some point the escape velocity from its surface would reach the speed of light. This would happen if Earth was compressed to a radius of 0.9 cm, that is, as much the size of a grape roga. This critical radius where the escape velocity reaches the speed of light is called the *Schwarzschild radius.*

Similarly, if the Sun were compressed to a radius of 3 km, i.e., about 4 millionths of its current radius, it would have turned into a black hole. Of course, there is no known process that could turn Earth, or the Sun, into a black hole. Around a Schwarzschild black hole, the nearest stable circular orbit occurs at a distance of 3 times the black hole's event horizon, i.e., $6GM/c^2$, Fig. 5.18. This is also the innermost radius of an accretion disc orbiting a Schwarzschild black hole. Inside of this radius: the "abyss"!

---

[2] Pierre-Simon Laplace: The gravitational attraction of a star with a diameter 250 times that of the Sun and comparable in density to the Earth would be so great no light could escape from its surface. The largest bodies in the Universe may thus be invisible by reason of their magnitude (Laplace 1799).

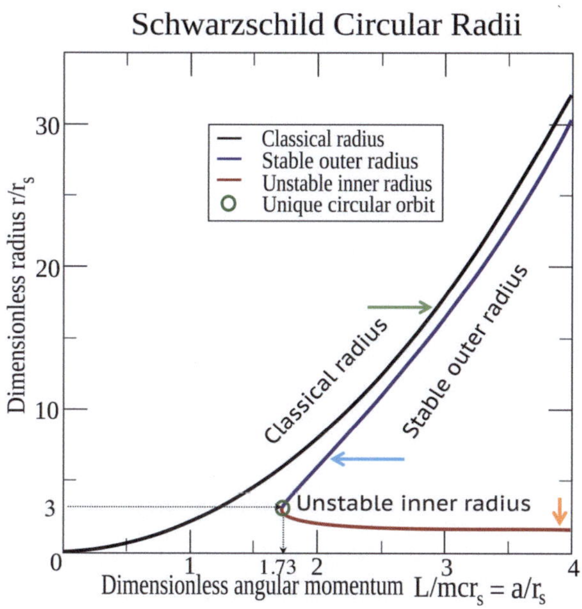

Fig. 5.18 The two-body Kepler problem in General Relativity refers to the determination of the motion and the gravitational field of two interacting material bodies, according to General Relativity. In the figure, the circular radii are given by using the Schwarzschild metric. The dimensionless radius r/r_S is given in terms of the dimensionless angular momentum L/mcr_S for circular orbits around a Schwarzchild black hole. To a given angular momentum L corresponds an unstable orbit (yellow curve) and a stable orbit (blue curve). The minimum stable orbit is at $3r_S$. This should be contrasted to the classical result wherein the orbit can be stable for any angular momentum L (green curve). (Figure adapted from Wikimedia Commons)

It is worth mentioning that although Einstein admired Schwarzschild's calculations, he considered them to be just a mathematical theory. In his opinion, nature should have a safety mechanism that does not allow such an anomaly to exist. But on April 10, 2019, the first-ever direct image of a black hole at the center of the giant elliptical galaxy M87 was published, as it was recorded by the Event Horizon Telescope in 2017 and discussed later in this Chapter (Fig. 5.19).

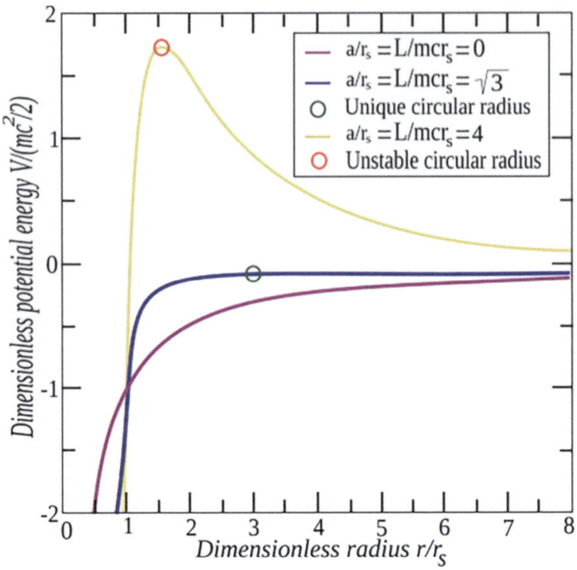

Fig. 5.19 Effective radial potential of the two-body Kepler problem in General Relativity for various angular momenta. At small radii, the energy drops precipitously, causing the particle to be pulled inexorably inwards to $r = 0$. However, when the normalized angular momentum $a/r_s = L/mcr_s$ equals the square root of three, a metastable circular orbit is possible at the radius highlighted with a green circle. At higher angular momenta, there is a significant centrifugal barrier (orange curve) and an unstable inner radius, highlighted in red. (*Source* Wikimedia Commons)

1915: Sir Arthur Eddington and the Inner Stellar Structure

Until the beginning of the twentieth century (1915), astronomers had no idea what was going on inside stars, simply because data on their interiors did not exist. Sir Arthur Eddington (Fig. 1.26) is credited with contributing to the rapid change in this scene. Specifically, he formulated the first radical model of the solar interior (1916), according to which the outward pressure of the central radiation could balance the inward gravity. Thus, he concluded, the core temperature of the Sun must be several million degrees.

Eddington contributed not only to the understanding of the internal stellar structure, but also to the understanding of stellar oscillations, the theory of relativity, cosmology, and the popularization of astronomy.

The Twofold Way of Understanding Stars and Atoms

The breadth of Eddington's research approach and thought is captured in the following sentence from his book *Stars and Atoms* (1927):

"I think we must think in a twofold way, since knowledge of the stars leads through the atom; and important knowledge of the atom has been reached through the stars."

1919: A Total Solar Eclipse Confirms the First of the Predictions of the General Theory of Relativity

Let us consider now the total solar eclipse of 1911. In that year, Albert Einstein had proposed in the GTR that light rays of stars, when they pass too close to the Sun, they curve. The bright rays deviate from straight propagation due to the curvature of space caused by the strong gravitational field of the Sun. It is like placing a bright object in a room from which we are separated by a wall and an opcn door at thc end of it, and we could see what is happening behind the opaque wall. According to Einstein, the curvature of space would reveal a world, which due to existence of the wall, it would not be directly visible to us!

Arthur Eddington, one of the few scientists at the time who understood the GTR, immediately understood what he had to do: during the first available total solar eclipse he could observe light from distant stars passing by (almost "scraping") the solar disk when the Moon covered the solar photosphere and the Sun's light disappeared for a few minutes. Thus, he could see and record the positions of stars that happened to be nearby the Sun in the sky. So, if Einstein was right, solar gravity would shift these stars in slightly different positions, compared to the positions they have in the night sky, at other times of the year, when the Sun is away from them. And the closer the star was to the Sun during the total eclipse, the greater its shift in the sky.

In the course of Modern Theoretical Mechanics at the Physics Department of the University of Athens, our students can calculate the deviation angle $\Delta\varphi$ (Fig. 5.20), which expresses the *curvature* of a light ray that passes almost tangentially across the surface of the Sun, making *Newtonian* calculations and applying the law of universal gravitation to the hyperbolic motion of photons in the gravitational field of the Sun, in the so-called two-body problem. Actually, this calculation was made for the first time in 1803 by the Bavarian astronomer Johann Georg von Soldner (1776–1833), who calculated the angle $\Delta\varphi$ that expresses the displacement of the positions of stars, $\Delta\varphi = 2GM/c^2R$ (in radians) where M and R are the mass and radius of the Sun respectively. Putting in this equation $R \approx 7 \times 10^8$ m and $M \approx 2 \times 10^{30}$ kg, it follows that: $\Delta\varphi \approx 0.42 \times 10^{-5}$ radians $= 0.87$ arcseconds (Soldner 1801–1804) (Fig. 5.20).

However, if we apply the GTR, as Karl Schwarzschild did in 1916, determining the shape of the spacetime geometry in the neighborhood of a spherical mass equal to the mass of the Sun by using the Einstein equations, it follows that the displacement will be **twice** the Newtonian one, approximately equal to 1.74 s of arc!

Hence, Eddington was the first to realize the possibility of measurement of the phenomenon of shifting the position of a star, which is just behind the Sun, during

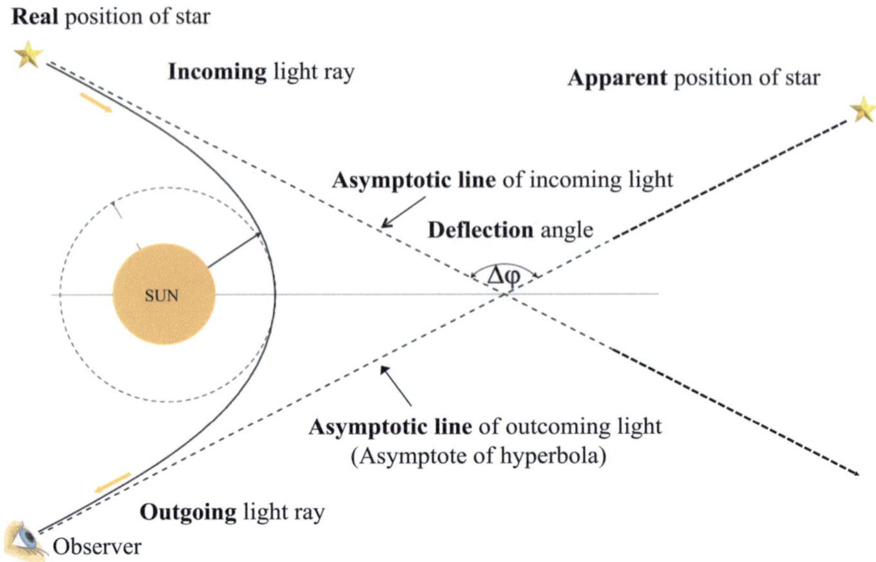

Fig. 5.20 Explanatory diagram of the curvature of the light rays of a star during the hyperbolic path they follow in passing through the Newtonian gravitational field of the Sun. The deviation of the position of the star in the sky is denoted by $\Delta\varphi$

a total solar eclipse, in order to confirm the first prediction of the GTR. Hence, he proposed to measure this phenomenon during the total solar eclipse on May 29, 1919. A donation of £1000 from the Royal Astronomer Sir Frank Dyson made this measurement possible. Two groups were then formed: the first, in which Eddington himself participated, went to the islet of Principe, in the Gulf of Guinea. The second went for the same purpose to Sobral, in Brazil.

Eddington's "*Moderation*"...

In the early years after the publication of Einstein's theory, Sir Arthur Eddington used his considerable authority to promote the work of his German colleague. Because the theory was so complex and difficult to understand, it was rumored that only *three* people in the world understood it.
There is the following anecdote about it:
American physicist Ludwik Silberstein, during one of Eddington's lectures, asked:
– "Professor Eddington, you should be one of the three people in the world who understand General Relativity."
Eddington paused, unable to answer. Then, Silberstein continued:
– "Don't be modest, Professor Eddington!".

Then, Eddington replied:
– "No, no! On the contrary, I'm trying to think who the third person is!...

The basic principle of the experiment they would attempt seemed simple. During the few minutes of the total solar eclipse, when the Moon completely hides the solar disk, the stellar field surrounding the Sun, appears suddenly. Using a telescope and photographic plates, astronomers take photographs of the eclipsed Sun and the surrounding field of stars. Then, by comparing these photographs, with photographs of the same stellar field taken without the presence of the Sun at its specific point sky, the angular deflection Δφ can be calculated, Figs. 5.20 and 5.21.

In practice, however, such an experiment is not really simple. An important complication arises because of various disturbances in the Earth's atmosphere. The

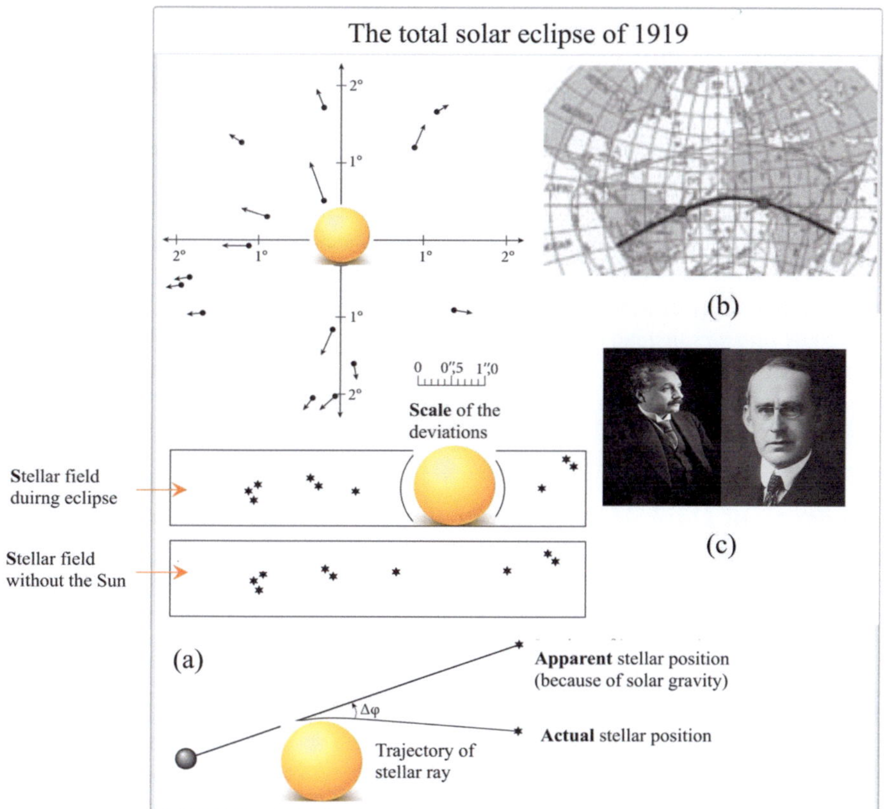

Fig. 5.21 (α) Illustrative diagram of the curvature of light rays as they passed almost tangentially to the solar disk, in the total solar eclipse of 1919. (β) The path of the Moon's shadow on the terrestrial surface during the total solar eclipse of 1919. (γ) Einstein and Eddington ca. 1919. (Adapted from NASA's cosmic times webpage)

stellar light passing through the terrestrial atmosphere is continuously refracted and deflected by only a few seconds of arc, a phenomenon which is responsible for the flickering brightness of the stellar light, i.e., when stars flicker when we look at them with the naked eye. But these deviations are random and they are altogether eliminated when we take the average of many images. So, one has to take as many photos as possible, i.e., as many stellar images as possible. And to achieve this, of course the sky has to be clear.

Hence, we can easily imagine Eddington's dismay when a terrible rainstorm broke out on the day of the eclipse. As the morning progressed Eddington began to be in despair. Luckily however, at the last moment, the weather started to improve and the rain stopped around noon. When the partial phase of the eclipse began, the Sun appeared through the clouds. Of the sixteen photographs taken through the remaining clouds, only two had reliably captured images of stars.

Comparison of these two plates with a comparison plate taken by the University of Oxford telescope before the mission gave a deflection of a light ray striking the surface of the Sun, $\Delta\varphi = 1.6 \pm 0.31$ s of arc, in agreement with Einstein's prediction of 1.74 s of arc! The mission at Sobral, he encountered better weather and managed to get 8 usable photographs, the analysis of which gave the value $\Delta\varphi = 1.93 \pm 0.12$ pts of arc. The results of the measurements, which were in favor of the General Theory of Relativity, were announced by Sir Frank Dyson on the 6th of November of 1919.

The accuracy of such observations was improved by the use of radio interferometry techniques in the microwave radiation from *quasars* in the late 1970s. Because these radio interferometry measurements are more precise compared to optical astronomy, and also because there doesn't even need the existence of a total solar eclipse for the calculation of the position of a microwave-emitting quasar, by observing microwaves emitted by *quasars* instead of visible light, it was shown with an error of only 1% that the result agreed with the GTR prediction. Thus, finally the GTR had successfully passed the experimental verification of the first of its predictions.

Discovery of *Dark Matter*, *Supernovae*, *Neutron Stars*, *Cosmic Rays* (1933–1934), Galactic Dynamics (1960)

1933: Astronomer Fritz Zwicky Discovers Dark Matter

In 1932, the Dutch astrophysicist de Sitter co-authored a research paper with Albert Einstein, in which they argued that there might be vast amounts of matter in the Universe that did not emit light and which today are known as *dark matter* (Einstein and de Sitter 1932).

Observationally, the enigma of the existence of *dark matter* was systematically laid bare in 1933, when Swiss astronomer Fritz Zwicky (Fig. 5.22) performed the first dynamical calculation of the mass of galaxy clusters and found that in the galaxy cluster of Coma Berenice the ratio of mass to brightness exceeds 400, compared to our own Milky Way where this ratio is only about 3. Specifically, Zwicky calculated the mass of each galaxy in the cluster based on its luminosity and summed all the galactic masses to calculate the total mass of the cluster. He then performed a second calculation, independent of the total visible mass which was based on measuring the atomic velocities of each galaxy within of the cluster. Much to his surprise, this second dynamical mass calculation gave a total mass of the cluster 400 times greater than mass based on the calculation of the total luminous mass of the galaxies, Fig. 5.75. Since then, all subsequent studies have verified Zwicky's original result (Zwicky 1933).

Dark Matter

By *dark matter* in astronomy and cosmology, we mean a hypothetical type of matter that contributes a large percentage to the total mass of the Universe. *Dark matter* cannot be observed directly with telescopes, because it neither emits nor absorbs light, or other electromagnetic radiation. Instead, its existence and properties are based on gravitational forces, effects on visible matter and radiation and on the large scale structure of the Universe. We believe that this matter

Fig. 5.22 Fritz Zwicky (1898–1974)

consists of hypothetical particles of unknown composition, which neither emit nor adequately reflect electromagnetic radiation, to become directly detectable by any known astronomical observation instrument.

According to the Planck research mission and based on the Standard of Cosmology, the total matter and energy content of the Universe is about 4.9% ordinary matter, 26.8% *dark matter* and 68.3% *dark energy* (Fig. 6.28). Therefore, *dark matter* is estimated to contribute about 84.5% to the total matter and 26.8% to the total content of the Universe.

Although experimental data were already available from the time Zwicky, it was only in the 70s that the solution of this *dark matter* enigma began to be studied systematically. This is because the existence of *dark matter* not only explained the stability of galaxy clusters, but also had far-reaching effects on the evolution and fate of the Universe itself.

An additional observational piece of evidence that argued for the existence of *dark matter* comes from the rotation curves of spiral galaxies. Spiral galaxies have a large population of stars which trace nearly circular orbits around the galactic center. As is also the case with the orbits of planets, it is expected that stars in galactic orbits of larger radii have lower orbital velocities, due to *Kepler's third law*. However, by conducting observations of the orbital velocities of stars at the outskirts of a large number of spirals galaxies, it was found that in no case this *Kepler's third law* was followed. Rather than decreasing the orbital velocities at large radii, they remained constant, suggesting that the mass contained in the large-radius orbit increases with distance. If taken into account that stars located near the periphery of a spiral galaxy, with orbital speeds of the order of 200 km per second, and if a galaxy consisted only of visible matter, these stars should all leave the galaxy in a short time, given that their orbital velocities are four times the escape speed from this galaxy. Since galaxies are not dissolved in this way, there must be some mass inside, which we don't take into account when adding up all the parts that we can see (Figs. 5.23 and 5.24).

1934: Supernovae, Neutron Stars, and Cosmic Rays Interrelated

In 1934, just two years after Chadwick's discovery of the neutron (1932), two eminent astrophysicists, Walter Baade and Fritz Zwicky published in the leading journals Physical Review and Proc. Natl. Acad. of Sci. USA, three important and far-reaching papers in astrophysics and cosmic rays, Baade and Zwicky (1934a, b, c).

With these papers, Baade and Zwicky introduced the then innovative terms *supernova* and *neutron star*, which they were connected with the origin of *cosmic rays*, thus creating one of the most amazing synthesis of ideas in Physics. Specifically,

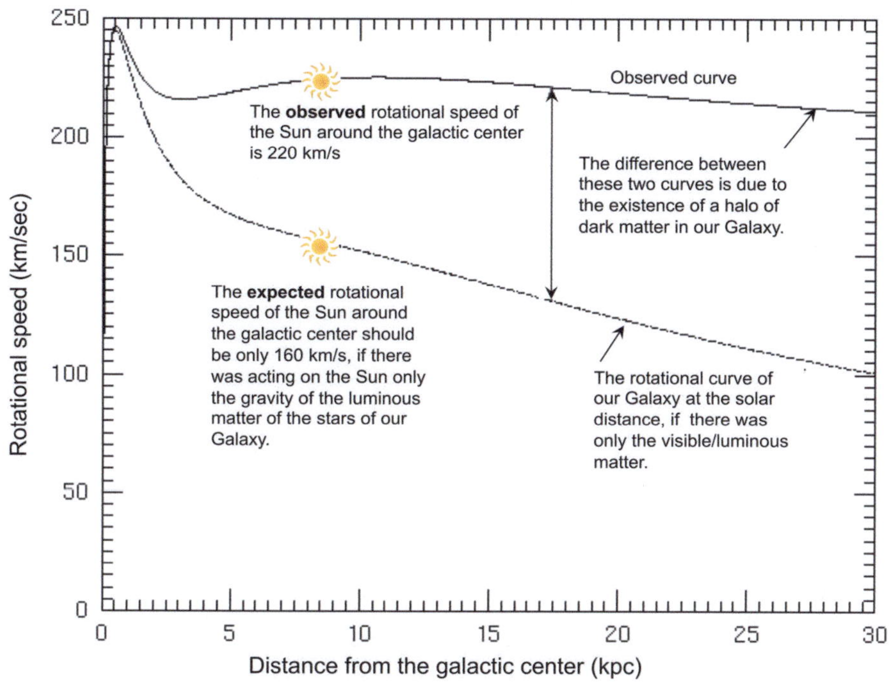

Fig. 5.23 The rotational speed of various star clusters and stars around the center of the Galaxy. The observed upper curve is due to the presence of *dark matter* in the halo of the Milky Way. This explains the high rotational speed of 220 km per second of the Sun around the galactic center. Based on this, the period of the Sun's orbit around the center of the Galaxy is about 250 million years. (*Source* K. Tsinganos, *Modern Theoretical Mechanics,* Unibooks Publications, Athens, p. 33 (2020))

they proposed that at the end of the life of massive stars, a colossal explosion takes place and the star turns into a *supernova*, with the byproduct of this explosion being very small and superdense stellar objects: during this explosion, and because of the enormous pressure created inside the star, conditions are right for the nuclear *inverse β-decay*, where the electrons interact and combine with protons to create neutrons and neutrinos. The neutrinos escape from the central stellar region carrying a significant amount of energy, leaving behind a very dense object consisting mainly of neutrons.

Baade and Zwicky named this new object a *neutron star*. Five years later (1939), the American theoretical physicist J. Robert Oppenheimer – who later was appointed head of the Manhattan program (1942–1944) at the University of Chicago, USA, for the construction of the first atomic bomb—in collaboration with the Canadian physicist George Volkof, calculated for the first time the expected size of these objects proposed by Baade and Zwicky. Then, according to quantum mechanics, *neutron stars* should have a very small diameter, of about 20 km, which nevertheless contains 1.4 times the solar mass. However, with such a small diameter of those *neutron stars*,

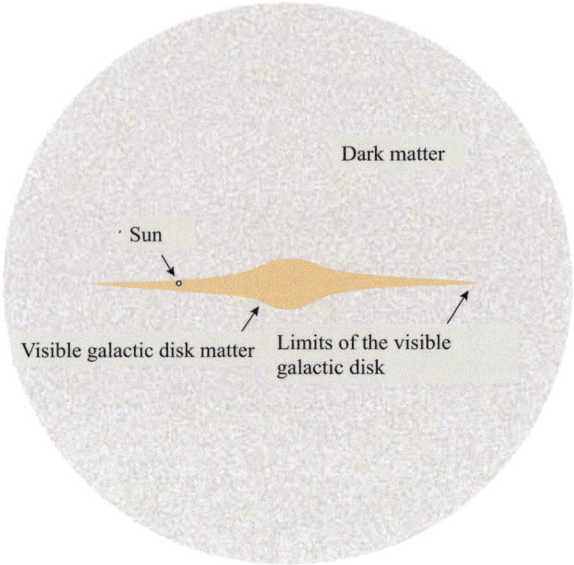

Fig. 5.24 Schematic image of our Milky Way, as it would appear from a distant point on the galactic plane. *Dark matter* forms a *halo* that extends well beyond the outer parts of the visible Milky Way. (*Source* K. Tsinganos, *Modern Theoretical Mechanics,* Unibooks Publications, Athens, p. 34 (2020))

it was concluded with dismay that they would be practically impossible to detect and thus verify Baade and Zwicky's theoretical predictions. This is because, on average, a supernova explodes every 50 years in a galaxy like ours, Fig. 5.25. The most recent supernova in our galaxy, SN 1604, was observed on October 9, 1604.

Fig. 5.25 Supernova SN 1994D, located in the outer reaches of the galaxy NGC 4526, is shown at the lower left part of the image of the galaxy. This supernova is 55 million light years away from Earth in the constellation Virgo. (*Image Credit* NASA/ESA, The Hubble Key Project Team and The High-Z Supernova Search Team)

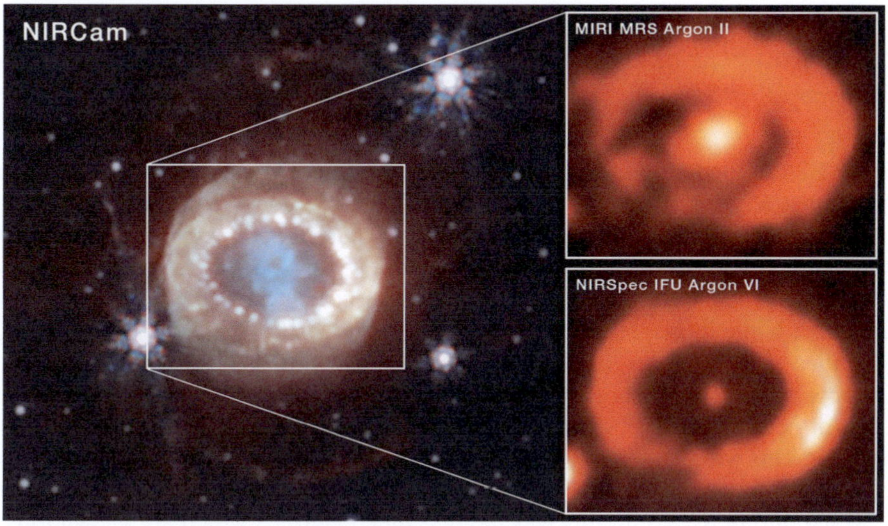

Fig. 5.26 (left): Webb's 2023 near-infrared camera (NIRCam) image of SN 1987A that highlights the object's central structure, expanding with several thousand km per second. The blue region is the densest part of the clumpy ejecta, containing heavy elements like carbon, oxygen, magnesium, and iron, as well as dust. The bright *ring of pearls* is the result of the collision of the ejecta with a ring of gas ejected about 20,000 years before the explosion. These are the locations of supernova shocks hitting more exterior material from the progenitor star. The outer ejecta is now illuminated by X-rays from the collision, while the inner ejecta is powered mainly by radioactivity and a putative compact object. (right): The top image features the data from Webb's MRS (Medium Resolution Spectrograph) mode of the MIRI instrument (Mid-InfraRed Instrument). The bottom image depicts data from Webb's NIRSpec (Near Infrared Spectrograph) at shorter wavelengths. Spectral analysis of the MIRI results showed a strong signal due to ionized argon from the center of the ejected material that surrounds the original site of SN 1987A. The NIRSpec data found even more heavily ionized chemical species, particularly five times ionized argon. Weak lines of ionized sulfur were also detected with MIRI. This indicated that there is a source of high-energy radiation in the center of the SN 1987A remnant, illuminating an almost point-like region in the center. The most likely source is believed to be a newly born *neutron star*. (*Credit* NASA, ESA, CSA, STScI, and C. Fransson (Stockholm University), M. Matsuura (Cardiff University), M. J. Barlow (University College London), P. J. Kavanagh (Maynooth University), J. Larsson (KTH Royal Institute of Technology))

Looking at this discovery after 85 years, the existence of *supernovae* in the context of astrophysics seems obviously of pivotal importance. During *supernovae*, the heavy elements of the periodic table are produced, while the shock waves they send to the interstellar medium trigger the birth of new stars, thus influencing the future of galaxies, while at the same time accelerate the charged particles of *cosmic rays*. Because *supernovae* have a high brightness, they shine at great distances in the Universe and are thus used to measure these great distances and the rate of expansion of the Universe. Most of them lead to the formation of *neutron stars* and *black holes*, Figs. 1.28 and 5.7.

2024: Evidence that the Supernova Remnant 1987A Has a **Neutron Star** *at Its Heart Found by the James Webb Space Telescope*

SN 1987A in the Large Magellanic Cloud, approximately 168,000 light years from Earth, was a type II core collapse supernova – i.e., the compact remains at its core are expected to have formed either a *neutron star* or a *black hole*. It was the first supernova that could be seen with the naked eye since Kepler's Supernova was observed in 1604. Light and neutrinos from the explosion reached Earth on February 23, 1987, while its brightness peaked in May of that year. The progenitor star was identified as a blue supergiant.

Approximately 2–3 h before the visible light from SN 1987A reached Earth, a burst of neutrinos was observed at three neutrino observatories lasting only a few seconds. This was likely due to neutrino emission, which occurs simultaneously with core collapse, but before visible light is emitted. Although only 25 neutrinos were detected during the event, this was the first time neutrinos which are known to be emitted from a supernova had been observed directly, a discovery that marked the beginning of *neutrino astronomy*. The observations were consistent with theoretical supernova models in which 99% of the energy of the collapse is radiated away in the form of neutrinos. The observations are also consistent with the models' estimates of a total neutrino count of 10^{58} with a total energy of 10^{53} ergs, i.e., a mean value of some dozens of MeV per neutrino (1 MeV = 1.6×10^{-6} ergs). The neutrino data allowed an upper bound on the electron neutrino rest mass, to be less than 16 eV/c^2 at 95% confidence level, i.e., 30,000 times smaller than the mass of an electron.

Since the supernova first became visible, an inconclusive search started for the collapsed core. The Hubble Space Telescope has taken images of the supernova regularly since August 1990 without a clear detection of a *neutron star*. The James Webb ST has found the best evidence yet for emission from a *neutron star* at the site of a recently observed *supernova*. Specifically, spectral analysis of the data from the MRS (Medium Resolution Spectrograph) mode of Webb's MIRI instrument showed a strong signal due to ionized argon from the center of the ejected material that surrounds the original site of SN 1987A. Furthermore, with the near infrared spectrograph (NIRSpec) at shorter wavelengths, more heavily ionized chemical species were found, particularly five times ionized argon (argon atoms that have lost five of their 18 electrons). Such ions require highly energetic photons to form, and those photons have to come from somewhere. To create these ions in the ejecta, there had to be a source of high-energy radiation in the centre of the SN 1987A remnant, with the best scenarios involving a newly born *neutron star*.

The Dazzling/Dizzling Numbers of the Energy Output of the Brightest Observed Supernova

Supernovae are *extremely bright objects*. The record holder is ASSASN-15lh (SN 2015L) that was observed in the summer of 2011. The total power of this explosion was calculated to be $(2.2 \pm 0.2) \times 10^{45}$ ergs per second, or, (2.2

$\pm$ 0.2) $\times$ 10^{38} Watts. This is as if the ASSASN-15lh illuminated the entire Universe with about 10^{36} lamps of 220 Watts each.
The radius of ASASSN-15lh, at peak brightness, was over 70,000 $R_{\odot}$, while its mass is 40 $M_{\odot}$. According to Krzysztof Stanek of Ohio State University, one of the principal investigators of ASASSN-15lh, if it was within our own galaxy, *it would shine brighter than the full moon*; there would be no night, and it would be easily seen during the day. Another way to see that huge power outburst of the energy radiated by ASSASN-15lh is via the Einstain mass to energy conversion, $E = mc^2$: it is equivalent to converting the entire mass of the Moon into pure energy within 30 s. In other words, it is as if this supernova *swallowed* a lunar mass every half a minute, converted this mass it has eaten into energy and subsequently radiated it! (Fig. 5.26).
The initial hypothesis was that ASASSN-15lh was the most extreme superluminous *supernova* so far seen, but it was recognized simultaneously as being unusual in several respects. Hence, other explanations have been discussed such as:

- *magnetars*, involving the conversion of rotational energy from a rapidly spinning neutron star into polar *jets* that heat surrounding material;
- a *quark-nova*, the hypothetical violent explosion resulting from the conversion of a neutron star to a quark star, produced by the neutron star remnant of the supernova, occuring a few days after the core collapse of a Wolf-Rayet star;
- a *tidal disruption event* of stars by a supermassive black hole;
- *gravitational lensing*. wherein unexpectedly bright visible objects can be produced by gravitational lensing of very distant objects by extremely massive objects closer to Earth.

This brightest observed supernova was a Type I explosion, powered by a submillisecond *magnetar* with field strength B $\sim$ few $\times$ 10^{13} Gauss. In extreme cases, such models might reach a peak luminosity of 2 $\times$ 10^{46} erg/s, radiating a total energy of up to 4 $\times$ 10^{52} erg. A *magnetar*-based model for the recent transient event, ASASSN-15lh, is presented in Sukhbold et al. (2016).
An additional comparison of this terrifying power is with the largest thermonuclear explosion ever produced on Earth, the 1 megaton H-bomb, 4.18 $\times$ 10^{15} Watts, released several billion trillion times *less* energy than this supernova produced in one second.
Our Sun produces energy at a rate equal to 3.8 $\times$ 10^{26} Watts. Then, the *supernova* ASSASN-15lh was about 580 billion times brighter than our Sun. The *supernova* explosion radiated every second as much energy as our Sun has produced in total over the last 18 millennia.
In the galaxy we live in, the Milky Way, all stars burn together and radiate a total of about 8 $\times$ 10^{36} Watts. Hence, for the *few days* that the *supernova* was glowing, before it finally went out, it was almost 30 times brighter than the

entire Milky Way. So in a brief flash, such a *supernova* is much brighter than the billions of stars in the entire galaxy.
A big power plant on Earth has a power output of the order of several billion (10^9) Watts, i.e., GWatts. Then, an estimate of the total electricity produced by all power plants around Earth is of the order of 8×10^{19} Joules per year: *In a single nanosecond, i.e., 1 billionth of a second, ASSASN-15lh radiated more energy than all the power plants on Earth could produce, at full power, for several billion of years.*
Last, but most important, *supernovae* are bad news for our planet, since they can wreck biospheres and flood planets with deadly radiation. Additionally, a recent study has added a new potential threat, that a special type of *supernova* can destroy a planet's ozone layer years after the initial explosion.

1960: Progress in Galactic Dynamics

In the area of galactic dynamics, some important discoveries have been made by the Athens research group on galactic dynamics (Contopoulos 2002, 2004):

- 1960; calculation of the *third integral*, which explains the shapes of stellar orbits in a galaxy;
- 1966: development of numerical programs for computing the higher order terms of the *third integral* and determination of a series expansion, which does not converge near stable periodic orbits, but converges near unstable periodic orbits, wherein there is a chaotic behavior and therefore it is useful for the study of chaotic trajectories. Another relevant contribution is the demonstration that overlapping resonances create *chaos;*
- 1971: proof that galactic spirals *follow* and do not *lead*;
- 1979: development of the non-linear theory of galactic spirals, as density waves;
- 1980: proof that galactic bars terminate at the *corotation distance*;
- 1985: proof that the most prominent galactic spirals in barless galaxies terminate near the 4/1 resonance. Also, proof that galactic spirals (beyond the corotation distance) consist of chaotic orbits and calculation of an analytical expression of these stellar orbits;
- 1990: the finding of chaotic behavior around two black holes, actually the first discovery of chaos in General Relativity;
- 1994: proof that the spectra of various dynamical systems are covariant and in the regions of chaotic orbits there are small regions of organized orbits;
- 2006: study of order and chaos in quantum mechanics (de Broglie – Bohm theory) and discovery of new cases of integrable systems.

1962–1970: X-ray and γ-ray Astronomy, γ-ray Bursts

1962: The Genesis of X-ray Astronomy

Since the terrestrial atmosphere absorbs X-rays that do not reach the ground, the discovery by Wilhelm Röntgen (1895, 1901 *Nobel Prize* in Physics) was not followed up in observational astronomy for the next 70 years. Thus, X-ray astronomy had to wait for 70 years before it became possible to use rockets that rose to heights greater than that of exploratory balloons. This happened on June 18, 1962 with the launch of the Aerobee 150 rocket. The payload of the 9.30 m long rocket was only 68 kg out of a total weight of 939 kg, while the maximum height reached was 270 km. The useful observation flight lasted only 5 min and 50 s. The rocket carried a small (primitive by today's standards) X-ray detector of the research team of the company American Science and Engineering, operating in Cambridge, Massachusetts and MIT. With this detector it discovered a very bright point source in the constellation Scorpius, Sco X-1, as well as a diffuse cosmic X-ray radiation. The experiment's research team included Riccardo Giacconi, Herbert Gursky, Frank Paolini, and Bruno Rossi.

Before 1962, the US Naval Research Laboratory (NRL) had attempted to send rockets to explore the upper atmosphere, something particularly useful for short-wave telecommunications. NRL's first successful flight of a V-2 rocket (1946) imaged the solar spectrum at ultraviolet wavelengths. In 1949 Herbert Friedman of NRL began studying solar X-ray emission and announced that the energy of solar X-rays was sufficient to explain all ionization of the E-layer of the ionosphere. This was followed by the launch of SkyLab, which had a multispectral probe for observing the Sun, which provided excellent X-ray and ultraviolet pictures of the Sun, Figs. 1.52 and 5.27. It also provided impressive photographs of the Earth, for example, the first image of the island of Crete from space. Skylab operated for 24 months from May 1973 to February 1974, during which time it was visited by three manned missions.

1970 Discovery of the First 300 Cosmic X-ray Sources by the UHURU Satellite

The UHURU satellite was launched in December 1970, off the coast of Kenya on its independence day, hence its name came from the word *uhuru* which means freedom in the local Swahili language. He detected in X-rays and in specifically in the energy range 2–20 keV, about 300 sources, among them some of the most *famous* today astronomical sources, Fig. 5.28. About half of the sources were in the Galactic plane of the sky, meaning they were of galactic origin. The remaining X-ray sources were evenly distributed across the sky.

Equally important was the discovery that there was X-ray emission from galaxy clusters. In particular, between the galaxies of the clusters there was plasma of low density, but with a temperature of millions of degrees, which actually came from

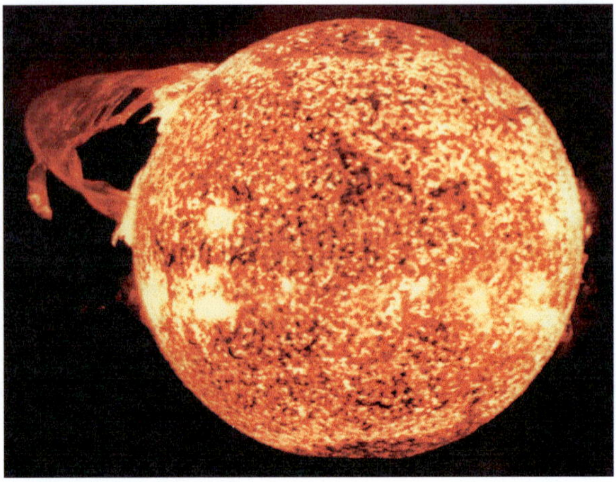

Fig. 5.27 Skylab photograph of the Sun, December 19, 1973. It shows one of the largest eruptive prominences in recorded history (as of 1998), taken by the third crew 19 December 1973 in the light of ionized helium, using the extreme-ultraviolet spectroheliograph of the U.S. Naval Research Laboratory. The twisted sheet of gas spans 588,000 km and seems to be unwinding itself. The darker areas at the top and bottom correspond to the solar poles. A spiral structure is evident, indicating the existence of magnetic fields. The Sun was then going through its 20th solar cycle (October 1964–March 1976). The solar poles are characterized by the absence of the supergranulation network and are darker than the central points of the disc, where there are many active regions. (*Source* NASA/Skylab)

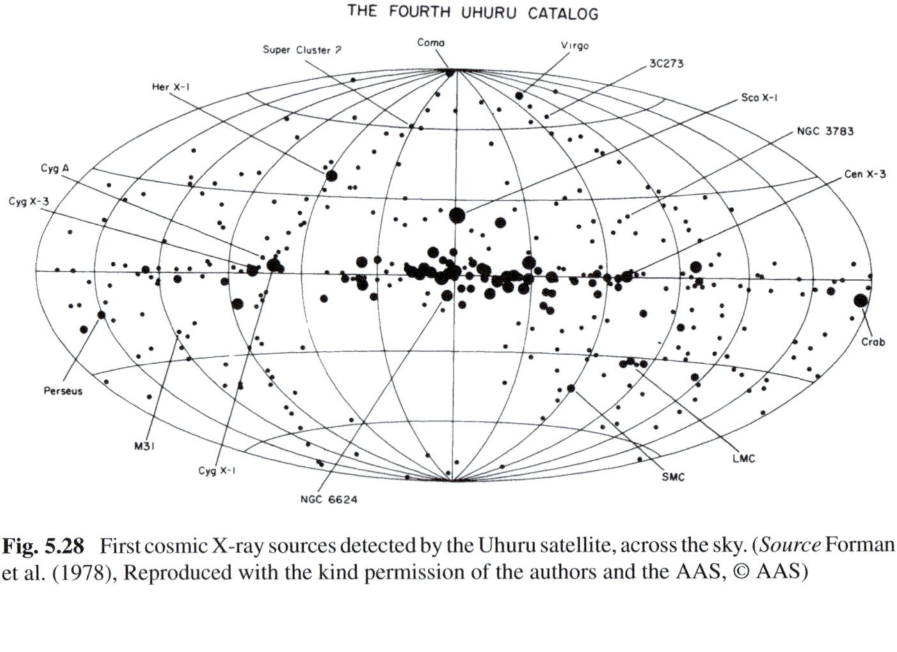

Fig. 5.28 First cosmic X-ray sources detected by the Uhuru satellite, across the sky. (*Source* Forman et al. (1978), Reproduced with the kind permission of the authors and the AAS, © AAS)

a mass greater than the total mass of the constituent galaxies of each cluster. This discovery opened new horizons in cosmological studies and the study of the early Universe.

2000 (1967): Discovery of γ-ray Bursts (GRBs)

Gamma ray bursts (GRBs) are giant cosmic explosions that produce γ-rays. They last from fractions of a second to several minutes. They have been observed to be associated with distant galaxies and occur at cosmological distances reaching as far back as the reionization epoch (400 million–1 billion years after the *Big Bang*). They must involve a relativistic increase in the intensity of the radiation which is emitted as a focused outflow in the form of a jet. After the initial flash, there is a longer aftershock that usually radiates at longer wavelengths, (X-rays, ultraviolet, optical, infrared, microwaves, and radio waves).

The first detection of γ-rays of cosmic origin was made by the Explorer 1 satellite, which also carried the 1st satellite γ-ray telescope (NASA 1958). Subsequently, the Orbiting Solar Observatory 3 (OSO-3) detected γ-rays of galactic origin, with energy greater than 100 MeV (1967). The American Vela 3 and 4 (1967) followed, while Small Astronomy Satellite 2 (SAS 2) carried an ionization chamber array and detected solitary γ-ray sources (NASA 1972). ESA's COS B satellite mapped the galactic plane (ESA 1975–1982) to be followed by the γ-ray space observatories, Compton (1991–2000), Integral (ESA 2002–) and Fermi (2008–).

The Most Powerful Explosions in the Universe Since the *Big Bang* …

Because the sources of γ-ray bursts (GRBs) are billions of light years away, to be visible from Earth means that their energy is enormous: indeed, in a few seconds they are estimated to generate and radiate as much energy as the entire Sun during all of its lifetime of about 10 billion years! At the same time, it appears that they are a rather rare astronomical phenomenon: only 1 γ-ray burst occurs in each galaxy every 1 million years! Of course there are billions of galaxies and that is why they are observed. Fortunately, however, all of these γ-ray bursts observed on Earth originate outside our Galaxy: just imagine that if a γ-ray burst were to occur in our Galaxy and hits the Earth, it could wipe out all traces of life from our planet!
It is interesting that the first detection of a γ-ray burst was made not by scientific, but by military artificial satellites. Specifically, the ban on nuclear weapons testing in the atmosphere in 1963 prompted the US Pentagon to launch a series of satellites, the *Vela* satellites, equipped with X-ray and γ-ray detectors, which were capable of recording γ-rays from all directions, thus *watching out* for any country illegally testing nuclear warheads on the Earth's surface, in the air or

in space. It is an irony that these satellites never detected a nuclear weapon explosion, but caught since their first operation (July 2, 1967) the distant γ-ray bursts emanating from the very beginnings of the Universe!

γ-ray bursts are momentary flashes of γ-rays at various random points in the sky Fig. 5.29. At first, it was not obvious whether they came from inside the galaxy, or if they were of extragalactic origin. Only when their population became quantitatively satisfactory, their isotropic distribution became evident, as in Fig. 5.29. Light curves of γ-ray flares observed by BATSE of the Compton GRO are shown in Fig. 5.32.

Specifically, thirty years after the first observation of γ-ray flares, one of the most important discoveries of the Compton Gamma Ray Observatory (Compton GRO) concerned the nature of these flares. Until then, most astronomers believed that γ-ray bursts came from *neutron stars* in our own Milky Way's halo, less than 200,000 light years away. Only a few astronomers, such as Bohdan Paczynski of Princeton University, believed that these sudden pulses of gamma rays reached us from tens of thousands of times greater distances, perhaps even from the edge of the visible Universe. Therefore, they betrayed the existence of phenomena hundreds of millions of times stronger, violent events in which energy equal to the total output of the Sun over the past billion years was released in a few seconds. The Compton GRO found that their distribution on the celestial sphere was isotropic, Fig. 5.29, that is, they came equally from all bisectors and did not show a concentration toward the hemisphere

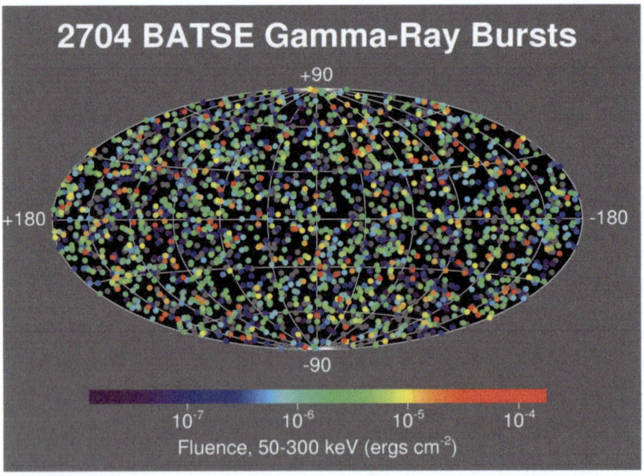

Fig. 5.29 The position of 2704 γ-ray bursts on a map of the sky (in galactic coordinates) observed by the Compton γ-ray observatory's (CGRO) Burst and Transient Source Experiment (BATSE) detector during its nine-year lifetime (1991–2000). This map clearly shows the isotropic distribution of GRBs throughout the celestial sphere. There is no concentration toward the galactic plane, or, toward the center region. Not a single GRB has been repeated at the same point in the sky. A similar distribution of 1405 bursts (1172 long duration and 227 short duration) has been observed by Fermi. (*Source* NASA/BATSE Gamma-Ray Astrophysics/NSSTC)

that included the center of our galaxy or the great Andromeda Galaxy (M31). This observation provided strong support for Paczynski's view. A first possible explanation for the generation of pulses of γ-ray bursts was that they are the traces of the death of binary *neutron star* systems in distant galaxies, with the two stars colliding with each other and creating a *black hole*.

More in detail, there are two observational varieties that are believed to result from two different types of progenitor stars. *Long-duration* GRBs are due to the death of a rare type of massive star that produces a Type Ibc *supernova* with powerful relativistic jets. *Short-duration* GRBs are due to mergers of binary *neutron stars*, which create a *central engine* with highly collimated bipolar relativistic outflows of ejected material similar to those thought to power long bursts. Short bursts have 90% of the total burst energy detected by the GBM within the first 2 s after the trigger (Fig. 5.30).

The Compton GRO's collaboration with the European X-ray detecting satellite BeppoSAX showed that γ-ray bursts (on 28 February 1997 and May 1997 in the constellations Orion and Camilopardalis, respectively) were also accompanied by the emission of X-rays. Since BeppoSAX's detectors could pinpoint the source's position on the sky with much greater precision, real alarm was sounded in both cases in the international astronomical community, and as quickly as possible the world's most powerful telescopes turned toward the directions they indicated the satellites. In the first case a faint point of light, a million times fainter than can be discerned by the naked eye, was detected, which disappeared before it was well studied. In the second case, however, a slightly brighter spot was detected a few hours after the γ-ray pulse, and astronomers managed to record its spectrum with the Keck II telescope in Hawaii three days later (May 11, 1997). Iron and magnesium lines in this spectrum were redshifted by a factor of 0.8, which corresponds to a distance of 4 billion light years!.

On December 25, 2010, a γ-ray flare was observed that lasted about 2 h, much longer than normal *flares* which last only a few seconds. Another such eruption was observed in December 2011 and lasted 7 h, the longest ever observed. The source of these massive flares is thought to be the destruction of stars with masses up to 1,000 times that of the Sun.

The extremely high energy of the observed γ-ray bursts appears to originate during *supernova* explosions, when a massive star collapses to form a *neutron star* or *black hole*. A subclass of GRBs (the *short* bursts) appear to originate from the *kilonova* explosions that accompany the merger of two *neutron stars,* Fig. 5.93, as we shall see later on, in the subject of gravitational wave emission.

A γ-ray burst can be initiated either by the merger of two *neutron stars*, or by the collapse of the core of a massive star. Both of these phenomena create a massive black hole, with an accretion disk around it. This disk-black hole system then ejects two *jets* of plasma moving at a speed very close to the speed of light. Formation of shock waves within these jets due to an inhomogeneous initial velocity creates the observed *flares*. The physical understanding of γ-ray *flares* is based on the most popular theoretical model of the *fireball*, which is described step by step in Fig. 5.31. On the left is the distance in centimeters of the progenitor binary system of the two

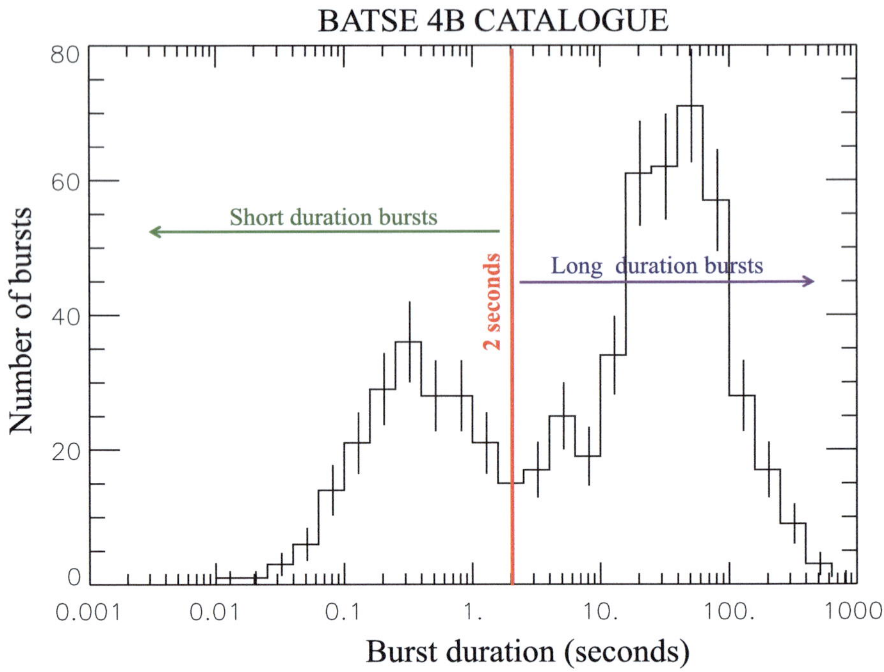

Fig. 5.30 The two populations of γ-ray bursts, after a count of their number as a function of their duration. According to data from the Burst and Transient Source Experiment (BATSE) instrument carried by the γ-ray Space Telescope Compton, we have the *short* γ-ray bursts, lasting less than 2 s, and the rest are the longer γ-ray bursts. Adapted from original figure at NASA/Goddard Space Flight Center

neutron stars, which with their merger give the central black hole, while on the right are described the various phases of the γ-ray *flare*.

γ-ray bursts have been statistically divided in two classes, *short* and *long* ones. The short GRBs were suggested to come from compact binary mergers and the *long* ones from the collapse of the cores of massive stars, greater that about 12 solar masses. In fact, in 2017 the coincident detection of gravitational waves from GW170817 and the *short* γ-ray burst GRB170817A confirmed the connection. The question arises however, if all *short* γ-ray bursts which have been detected are similar to GRB170817A, i.e., if not all *short* gamma-ray bursts come from compact binary mergers. In the following, we shall assume that the γ-ray flare originates from the atmosphere and the accretion disk in the immediate vicinity of the merger product of the two *neutron stars* that form finally the *black hole*.

The outflow starts at about 100 km from the center of the explosive fusion/collision of the two *neutron stars*: thermal or magnetic energy is converted into kinetic energy of the baryon components of the plasma, thus creating a *jet* of small angular opening, with a speed which gradually reaches very close to the speed of light. That is, in phase (1) we have the initial acceleration of the plasma which will have a short duration,

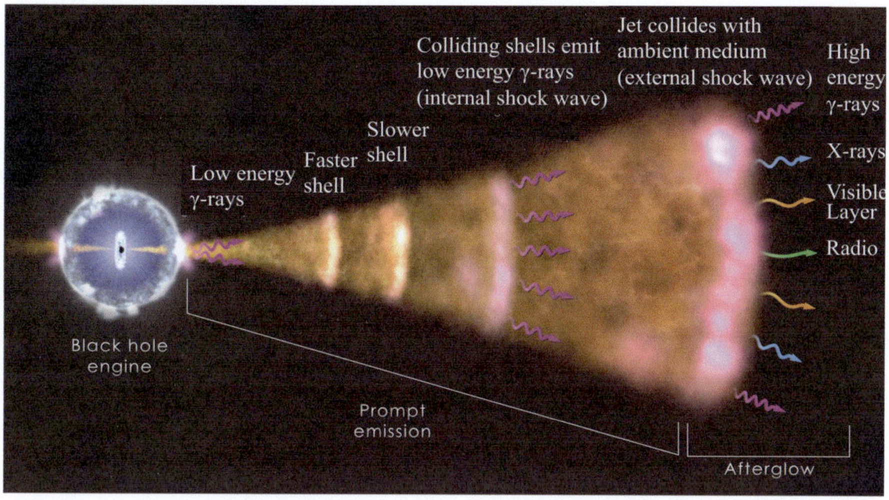

Fig. 5.31 Schematic of the *fireball shock model* of γ-ray bursts. The outflow from the central engine is accelerated and then it becomes transparent. The first stage of electromagnetic emission takes places when internal shocks occur in phase, after which the outflow propagates into the external medium, causing forward and reverse shocks producing electromagnetic radiation of lower energies. The *jet* becomes subrelativistic. The various phases of the flow across the cone of aperture θ are explained in more detail in the text. (*Credit* NASA/GSFC/ICRAR and Wikipedia)

less than 2 s, of a γ-ray flash from a source such as the product of the collision of two *neutron stars*. Such an origin appears to have had the gravitational wave GW170817 detected by ground-based and satellite telescopes across the electromagnetic spectrum, as we shall discuss in the following. This initial acceleration of the plasma that will then give the γ-ray flare takes place in a cone of small half-aperture θ.

In phase (2) which occurs at a distance of about 100 thousand kilometers, we have the end of phase (1) of the acceleration and there the velocity has now reached very close to the speed of light, so that the Lorentz factor γ of the outflow has already received its maximum value, equal to a few hundred.

At distance/phase (3) the plasma of the *jet*, having diluted significantly is now transparent in the so-called *photospheric radius*, about one solar radius. After this radial distance from the source, any available non-thermal energy is radiated in a semi-thermal emission. Then, at some distance from the source, the conditions in the accretion disc may change and give a faster outflow (4), so that the leading plasma in (5) is met by the faster outflow of matter (4) from the surroundings of the disk and the central *black hole*. Then we will inevitably have a collision of the faster outflow (4) with the slower leading outflow (5). This collision, because the outflows are supersonic, will create an internal *shock wave* (ISW), at an estimated distance of about 10 AU, which propagates through the *jet* plasma. Then the *jet* plasma is heated by this internal *shock wave* which becomes the source of the prompt γ-radiation we observe. Then, as the supersonic outflow moves further away from the central source, it reaches the interstellar medium where conditions are different from those prevailing

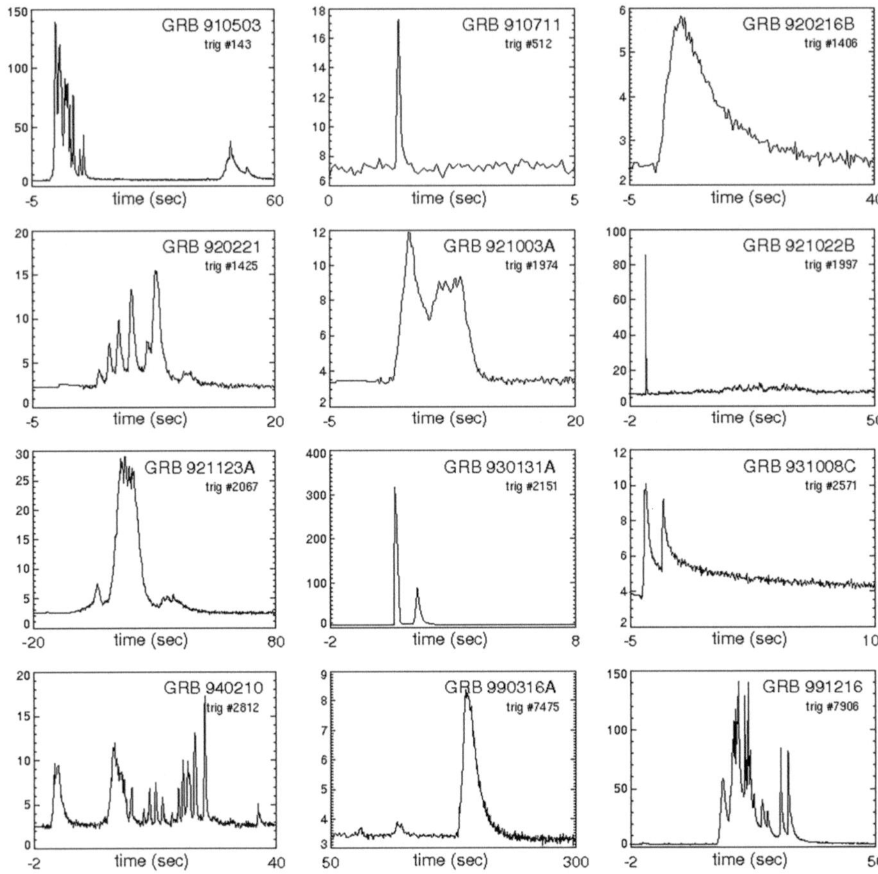

Fig. 5.32 Typical light curves of γ-ray bursts observed by BATSE. *Source* NASA/CGRO, Daniel Perley, and also Wikipedia Commons

within the outflow. As it happens in the *solar wind*, when the wind meets the thin and cold interstellar medium at the *heliopause*, a shock wave, (ESW), is inevitably created there, through which the velocity of the outflow is drastically reduced to non-relativistic values, to finally homogenize with the prevailing conditions of pressure, density and temperature of the interstellar medium (10).

Meanwhile, we have phases (7), (8), and (9) which characterize the behavior of the external shock wave (ESW) and begin with the deceleration of the *jet* as it collides with the interstellar medium: the phase (7) corresponds to a reverse shock wave (RSW), which is created based on the physical laws for this case and propagates backwards inside the *jet* toward the source. In phase (8) we again have, due to the physical laws that govern the creation of this external shock wave, the creation of a contact discontinuity. Next, in phase (9) we have the outwardly propagating shock wave (ESW) which is responsible for the emission of the afterglow radiation which we observe in X-rays, ultraviolet, visible, infrared, microwaves, and radio waves. Finally, phase (10) corresponds to a region where the velocity and the Lorentz factor of the outflow decrease dramatically, so that the outflow velocity becomes non-relativistic. This popular fireball model, which satisfactorily explains the observations of γ-ray bursts, is discussed in detail in various publications and monographs, e.g., see, Vedrenne and Atteia (2009) (Fig. 5.31).

Major Cataclysmic Extinctions of Living Species on Earth – And Perhaps in the Universe and Their Possible Connection to γ-ray Bursts

If a γ-ray burst occurs within a few thousand light years from Earth, it only takes 10 s for its effects to devastate our planet, while their damaging effects may last up to 15 years. Let's take a closer look at what could happen to Earth in this *nightmare* scenario.

During a γ-ray burst, the γ-rays break the bonds between the molecules of the elements oxygen and nitrogen, which thus turn into nitrogen monoxide – a gas that destroys ozone as it combines with oxygen – and nitrogen dioxide, a brown toxic gas. After a few weeks these nitrogen oxides will slowly destroy the ozone layer. As the ozone is destroyed, the amount of ultraviolet rays penetrating the Earth's surface will increase by 50%, which is enough to destroy the DNA of the living organisms. Winds will carry the gases around the planet, blanketing it in the brown gas of nitrogen dioxide. Hence, after a few months the whole planet will be covered by nitrogen dioxide. Although ozone will gradually begin to recover in the Earth's atmosphere (20% below the normal levels), UV radiation will still cause significant damage to living organisms on Earth. Ten years later, nitrous acids will be transported to lower altitudes, reacting with water to create nitric acid, such that acid rains will erupt everywhere. Although after about 15 years the atmosphere will try to gradually return to its normal state, the major damage will have already been done. Most of life will have disappeared, while a brown cloud will have covered the Sun and thus blocking sunlight and solar heat, causing a new ice age.

Studies have shown that the probability of a long-duration galactic γ-ray burst hitting Earth is 50% in the last 500 million years, 60% in the last 1 billion years, and 90% in the last 5 billion years (i.e., the lifetime of the solar system). Fortunately, very short-lived γ-ray bursts in the Galaxy, as well as extragalactic γ-ray bursts, do not appear to be as destructive to Earth.

A second conclusion is that γ-ray bursts may be a serious threat to life in other regions of our galaxy.

As is known, in the central galactic regions the stellar density is very high, because the stars are much closer to each other. This means that in these central regions the risk of stars being damaged by nearby γ-ray bursts is greater. In a region of a few thousand light years in radius in the central parts of our Milky Way Galaxy, where live 25% of the stars in the Milky Way, those stars face a 95% risk of being hit by a catastrophic γ-ray burst within one billion years. Therefore, life in the form that exists on Earth can survive with reasonable certainty only on the outskirts of the Milky Way, beyond about 32,000 AU from the Galactic Center!

My colleague and friend Tsvi Piran, an eminent professor of astrophysics at the University of Jerusalem, has studied the danger posed by γ-ray bursts to the Universe, in general. Specifically, because of the γ-ray *flares*, life – as we know it on Earth – could safely develop in only 10% of all galaxies. And such life could only have developed in the last 5 billion years. In the earlier era, galaxies were smaller in size and so γ-ray bursts were then much closer and frequent, having thus the potential to cause large-scale extinctions on planets that could potentially host and develop life.

This may be a partial explanation for the so-called *Fermi paradox*, or the «*great cosmic silence*» that we discussed in Chap. 4 of this book: why have we not encountered any traces of advanced civilization in the Universe so far? Our galaxy is much older than our planetary system, so there was enough time and space – the number of extrasolar systems with Earth-like conditions is huge – for life to develop elsewhere in our galaxy. So why haven't we encountered any advanced civilizations elsewhere in the galaxy until now? So, the answer to *Fermi's paradox* may lie in the fact that γ-ray bursts may have struck and wiped out life from many planets that began to harbor life! The most severe criticism of this answer is that «*we assume that on extrasolar planets there is life like we know it here on Earth*». However, one could certainly imagine very different forms of life that are resistant to the enormous radiation of γ-ray flares.

The results of these studies leave open the possibility that a galactic γ-ray burst may also be the cause of one of the five known large-scale mass extinctions that Earth appears to have suffered in the distant past during the *Paleozoic era* (510–250 million years ago) and the following *Mesozoic era* (250–60 million years ago):

- the first great extinction of the Late *Ordovician period* of the glaciers, where the first of the five mass extinctions of species occurred. This extinction occurred about **440** million years ago. It was probably the second most serious mass disaster that has ever taken place on our planet. Almost all species lived at that time in the seas and about 60% of them disappeared.
- the extinction toward the end of the *Devonian period*, which occurred about **375–359** million years ago and brought about significant environmental changes that caused a long-term extinction of species. It wiped out large groups of fish and stopped the growth of new coral reefs for the next 100 million years.
- the extinction at the end of the *Permian period*, which occurred **252** million years ago. This is the biggest mass extinction event of species. It affected Earth's ecology to the greatest extent, wiping out 97% of Earth's species forever.
- the extinction at the end of the *Triassic period*, 251–201 million years ago, which took place about **240** million years ago. At the beginning of the Triassic period, dinosaurs appeared and large amphibians and reptiles were the dominant animals on Earth. About 40% of the tetrapod amphibians, reptiles and mammals disappeared during this catastrophe.

- the destruction at the end of the *Cretaceous period*. Based on observational data, we now believe that an asteroid hit the Earth **66** million years ago and is responsible for the end of the «kingdom» of the dinosaurs.

Emergence of Radio Astronomy Before the 2nd World War, Active Galactic Nuclei (AGN) and *Quasars*

1931–1939: The Genesis of Radio Astronomy

Radio waves were discovered by the American physicist and radio engineer Karl Jansky in the laboratories of the telephone company Bell (Bell Labs). He detected for the first time in August 1931 radio waves coming from the center of our *Milky Way* galaxy. Jansky built the first radio antenna, which was used to detect astronomical sources of radio waves, the forefather of today's television *dishes*, Fig. 5.33a.

In particular, Jansky was looking for sources of radio noise that could affect wireless communication, a subject which interested the Bell company. The first such source he spotted was in the direction of the constellation *Sagittarius*, which is located at the center of our Galaxy. Jansky then proposed to Bell Labs the construction of a sensitive 30 m antenna, but his proposal was rejected on the grounds that it would not be practically useful for transatlantic telecommunications. Subsequently, he was transferred to another part of the company, despite the fact that his discovery appeared in the *New York Times* on May 5, 1933.

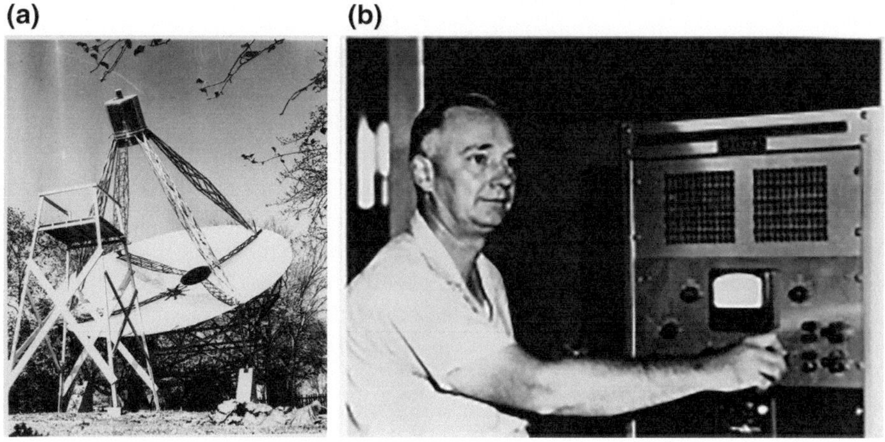

Fig. 5.33 (left) The 9-m antenna built by Reber (1940), was the first parabolic antenna and since then the prototype of radio telescopes built even after the 2nd World War. (right) Reber spent many hours each night *scanning* the sky with his radio telescope. (*Source* Wikimedia Commons)

The baton was then taken over from Jansky by another American, the amateur astronomer Grote Reber (1940), who remained for a decade, worldwide, the only radio astronomer. Reber, using an antenna of his own construction (Fig. 5.33a, costing $2,000), which he had placed in his backyard in Illinois, observed together with the radio emission of the Galaxy at Sagittarius, radio emission from the first astronomical radio source, in the constellation of Cygnus.

After this first observation of Cygnus A, it was proposed that the energy of the radio emission came from the collision of two galaxies. However, Jennison and Gupta (1953) discovered that something really strange was going on at Cygnus A: the radio emission itself was coming instead from a cosmic dumbbell, where the two *weights* – the two huge radio lobes containing high-energy particles – were about 300,000 l.y. apart from each other, while the visible galaxy, which actually was of much smaller extent, was lying in the middle! Therefore, it was not emission coming from two colliding galaxies, but emission from some mysterious explosive process, where high-energy particles were emitted from the central galaxy and moving around strong magnetic fields, produced the detected *synchrotron radiation* of the radio waves.

Cygnus can be seen at the zenith of the celestial dome during the summer period, as part of the popular *summer triangle*. In this constellation and at a distance of about 600 million light years lies the radio source Cygnus A.

Another remarkable fact about this source is that the variability of its emission at high energies is of the order of hours, or even minutes, i.e., too short a time span to be a galactic source.

Today, after detailed observations at various wavelengths of this most powerful radio source Cygnus A in the sky, we know that it has two radio jets, each about 150,000 l.y. in length, which originate at the center of this galaxy at Cygnus, where basks a giant *black hole* of mass about 2.5 billion solar masses, while the jets terminate in the *hot spots* of the two radio lobes.

Thus, since 1951, new components of the Universe have emerged due to *Radio Astronomy*. In principle, the radiation of most radio sources is *synchrotron radiation*, which is produced when charged bodies spiral around the lines of strong magnetic fields. But in order to explain the properties of this radio emission, three new components of the Universe had to be invoked and assumed: vast amounts of relativistic electrons, large-scale magnetic fields and huge masses ($\sim 10^6$ M_o) of relativistic plasma. All of this appears to be channeled, far enough away from the parent galaxy to maintain the thin diameter of the jets. Also, the theory of relativity is now essential to understand these astrophysical phenomena.

We must point out here the fact that the 2nd World War favored the development of radio astronomy, especially in England. Thus, by 1950 radio astronomers had already built sensitive radio telescopes and created extensive lists of radio sources. The best known catalog of cosmic radio sources was Cambridge's 3C. Many radio sources corresponded to distant known galaxies. But others do not. This naturally leads us to the next fundamental observational discovery.

All Radio Telescopes Have Collected Till Today About the Energy a Snowflake Has When It Hits the Ground

It is interesting to note that the energy of cosmic origin reaching the Earth in radio waves is extremely weak. In particular, the energy collected by all radio telescopes so far does not exceed that released by a snowflake when it travels through the atmosphere and reaches the ground!

1943: Discovery of Active Galactic Nuclei (AGN) and Quasars

An active galactic nucleus (AGN) is a compact region in the center of a galaxy which has a brightness much higher than usual, at least in some wavelengths of the electromagnetic spectrum. The characteristics of the emitted radiation indicate that it does not come from the usual brightness of the stars we know, for example, in terms of the emitted spectrum with the corresponding emission and absorption lines. This variation from the usual stellar luminosity is seen in the radio, microwave, infrared, optical, ultraviolet, X-ray, and gamma rays. So, when a galaxy *hosts* such an active galactic nucleus (AGN) it is called an *active galaxy*. The radiation of the active galactic nucleus is understood to be associated with the absorption of matter by a supermassive black hole at its center. The process of absorbing matter can convert the kinetic and potential energy of matter into radiation with great efficiency, assuming a supermassive *black hole* lurks at its center, Fig. 5.34.

The American astronomer Carl Seyfert discovered in 1943 that the light of certain spiral galaxies, which today bear his name (*Seyfert galaxies*), has peculiar characteristics, different from the usual characteristics of starlight, such as very broad emission lines. Among the earliest examples of AGN were the elliptical galaxy M87 (Figs. 5.39, 5.44a, b) and the radio galaxies Centaurus A and Cygnus A (Fig. 5.42). All *quasars*, such as the first discovered 3C 273 (Figs. 5.36 and 5.37) and 3C48, are the most extreme peculiar cases of AGN.

AGNs are located in distant galaxies, where extremely violent, energetic, and exciting phenomena take place, for the understanding of which modern physics is testing its limits. In radio galaxies we have the classic structure of two extended radio emission lobes at a distance of several million light years on either side of the central core (as in Hercules A Fig. 5.40), from which the radio lobes are fed with thin and often perfectly aligned beams of relativistic plasma. A relatively nearby example of an AGN is the giant elliptical galaxy M87 in the Virgo cluster. From the bright and compact central region of M87 emerges a plasma jet about 5,000 light years long in the visible, containing many equidistant luminous nodes, Figs. 5.39, 5.44a, b. Analysis of the radiation spectrum of the central region of the galaxy with the Hubble Space Telescope shows that around the core of the galaxy and in a few hundred light years away there is plasma in a disk-like distribution that rotates like the planets around the Sun. By calculating the rotational speed of this gas with the

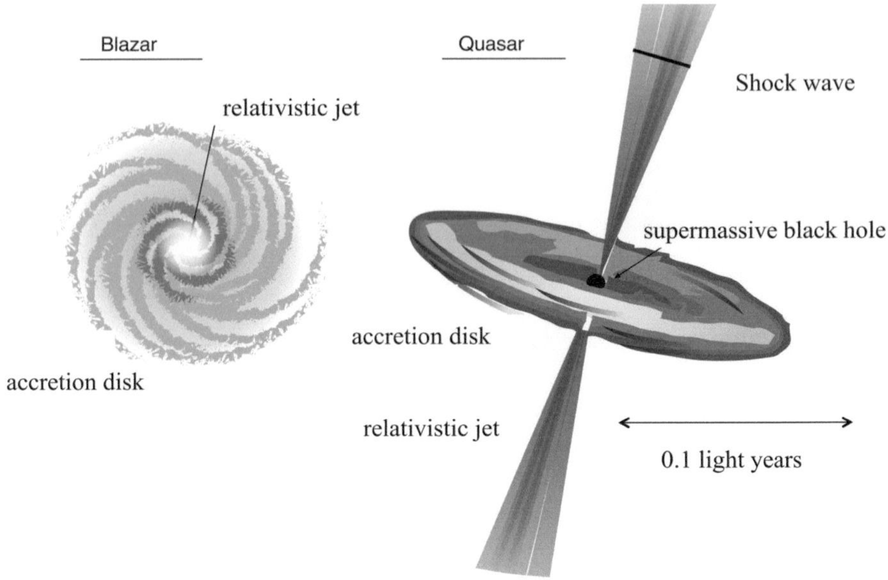

Fig. 5.34 Explanatory diagram of the *Quasar-Blazar* difference due to the different orientation of the jet axis with respect to the line of sight: in *blazars* the *jet* propagates at a very small angle to the line of sight. (From K. Tsinganos, Plasma Astrophysics, p. 409, 2015)

Doppler effect, we can *weigh* the mass of the central condensation, which is thus calculated to be about 6 billion solar masses, in the central region of a radius of about 200 light years. Such a concentration cannot be explained as an insanely dense star cluster and the only alternative explanation that has been given so far is that a supermassive black hole is lurking at the center of the disk (Figs. 5.39, 5.44a, b). See also Modern Theoretical Mechanics, K. Tsinganos, Unibooks, 2022, (p. 343).

Extreme and Unusual Characteristics of AGN

Some AGN, such as X-ray and γ-ray *quasars*, change their brightness significantly over a timescales of a few hours. Since a light signal travels in this time interval a distance that is approximately the diameter of our planetary system, these observations lead us to the conclusion that the regions emitting this variable radiation have dimensions smaller than our solar system. Also, the power radiated by AGN often exceeds a trillion solar luminosities. That is, within the volume of a planetary system fit and radiate dozens of galaxies like ours! There, the energy released each year is equivalent to the total conversion into energy of an amount of matter equal to the mass of our Sun. And this happens within such a relatively small area. Also, in some AGN plasma condensations seem to move faster than the speed of light, something strictly forbidden in the current view of modern physics. The prevailing explanation given so far is that these superluminal motions of the plasma in these long *jets* is actually a projection effect, an optical illusion: they are produced when the plasma is moving at a minimal angle to us, at speeds slightly less than that of light, as we discuss in the following. These characteristics of AGN make them not only the most powerful energy sources in nature, but also perhaps the most fascinating laboratories of physics.

1963: Discovery of the Exotic **Quasars,** *at the Edge of Our Known Universe*

The Universe, rich in natural phenomena, holds inexhaustible surprises for us. Rarely have theoretical astrophysicists been able to predict the highly unusual – for us tiny earthlings – phenomena occurring in the distant Universe. The prediction of the existence of the solar wind, that is, the flow of millions of tons of hot gas every second with which the solar atmosphere *bathes* the vast space, is perhaps one of the few cases where theory preceded observation. But in most other cases, we theoretical astrophysicists simply try to explain the provocative observations of observational astronomers after they have already been made. The discovery of *quasars* in 1963 is one such case. In these enigmatic cosmic objects occur unprecedented and unknown to us phenomena, such as, a blinding luminosity that is equivalent to that of hundreds of galaxies all together *crammed* into a relatively *small* area, as our planetary system, monstrous and voracious blacks holes of billions of solar masses that bathe in their center and *devour* many stars each year, relativistic jets of ionized gas that erupt from these black holes, forming brilliant hubs that often *seem* to move at speeds even exceeding and the speed of light (!), gravitational lenses, such as the "*Einstein's cross*", but also other extreme and exotic natural phenomena that occur in these peculiar astronomical environments, at the edges of our known Universe. But let's first see how these *quasars* were discovered.

In the first place, the precise radio location of radio galaxies was not easy, so radio astronomers devised various other clever workarounds. One of them was the occultation by the Moon of the radio source, whose current position in the sky is known with great precision and when it interferes with the direction of the radio source it cuts its radiation and thus gives us its exact position.

So, after measuring the precise coordinates of several strong radio sources, the focus for their explanation was found at the University of California Caltech, which operated the 1.1 m telescope on Mount Palomar. Observers focused their attention on 2 strong radio sources in the 3C catalog, 3C 273 (Figs. 5.36 and 5.37) and 3C 48 which appeared to coincide in the sky with two point blue stars. That's why they were called *quas-ars*, short for *quas(i-stell)ar radio source*. However, until then no stars had been discovered to emit radio waves, with the exception of the Sun, whose radio emission is however so weak that it would not be detectable at great extragalactic distances. Experienced observers obtained the spectra of these point sources with the world's largest then-available 1.1-m telescope at Caltech's Mount Palomar. To their great surprise, however, they found that their emission lines could not be matched with those of any chemical element on our familiar periodic table. Were they lines of some new chemical element? Did it come from some new and exotic kind of radio star within our Galaxy, or were they extragalactic objects? A great effort – not lacking in battles – began among observers to solve this *enigma.*

The scattered pieces of the puzzle were finally successfully assembled in 1963 by a young Dutch astrophysicist, Maarten Schmidt, Fig. 5.35, who having completed his thesis under the supervision of the famous astronomer Oort in 1956 at the Leiden Observatory, moved to Caltech in 1959. There, his clear mind noticed something very simple: that the 4 strongest emission lines of 3C273 had the same relative distance between them as the well-known but fundamental lines of the hydrogen atom. However, the wavelengths appearing in the spectrum of the radio source were shifted towards the red part of the spectrum by a significant percentage, about 11.8%! If this shift was due to the Doppler effect, this would mean that this object was moving away from us at a speed equal to 11.8% of the speed of light, i.e. 47,400 km per second, or, about 170 million kilometers per hour, located at a distance of 2.4 billion l.y.! The same lines in the spectrum of the radio galaxy 3C48 were redshifted even further, by 37% of the speed of light, i.e., 110,000 km per second, or about 400 million kilometers per hour, at a distance of 4,5 billion l.y., i.e., this radio source has been receding at about twice the speed of 3C273. Astrophysicists were speechless and dazzled, trying to understand how it is possible for these objects to be moving away from us at speeds of hundreds of millions of kilometers per hour. For example, when the light of Quasar 3C48 was emitted it was at a distance of 3.9 billion l.y. (light propagation time to Earth), meanwhile today, it has moved away from us and is already 4.5 billion light years away!

With further systematic observation and analysis, Schmidt and his colleagues showed that 3C48 and 3C273 were not exactly point sources, but each was surrounded by a faint halo. This halo indicated that some distant galaxy was there. According to the Hubble-Lemaitre expansion law, linking distance and speed in galaxies, these enigmatic objects were billions of light years away, meaning their light traveled billions of light years to reach us!

As is often the case in research, the solution of one puzzle leads to the appearance of others! Today we know of thousands of *quasars*. The farthest one detected so far is at a distance from which light traveled 12.9 billion years to reach Earth, that is, it began about 800 million years after the birth of the Universe in the *Big Bang*. It should be noted that the quasar 3C273's *jet*, first observed with the 64 m radio telescope of the Parkes Observatory in Australia, has a length greater than the radius of our Milky Way and is observed in the visible, X-rays, and radio waves, as it also exhibits the exotic phenomenon of *superluminal motion*, which we shall discuss and explain later on, i.e., its bright spots have an apparent speed of about 6 times the speed of light (Fig. 5.41) (Figs. 5.36 and 5.37).

Fig. 5.35 Maarten Schmidt (*Source* Caltech)

These Mysterious *Quasars*

Schmidt proved that *quasars* were not radio stars, but actually they were the most distant r*adiogalaxies* in the Universe. And, in order to have the observed large apparent luminosity, even though they are at these enormous distances, they must have an extremely large intrinsic luminosity. For example, 3C273 is over 4 trillion times more luminous than the Sun at visible wavelengths. The luminosity of 3C 273 is variable at nearly every wavelength from radio waves to gamma rays on timescales from a few days to decades. Furthermore, if the halo comes from the light of the surrounding galaxy, then its bright heart emits much stronger radiation. For example, if 3C 273 was at the distance of the star Pollux, i.e., at a distance of about 33 light years from Earth, it would appear as bright as the Sun! But then, what kind of phenomena is happening in the core of this galaxy so that it has thousands of the total luminosity of an ordinary galaxy (hundreds of billions the solar luminosity)?
The most dramatic changes in *quasar* brightness are equivalent to the energy released by hundreds of trillion Suns. To produce this much energy we would have to convert the total mass of about ten Earths into energy every minute.

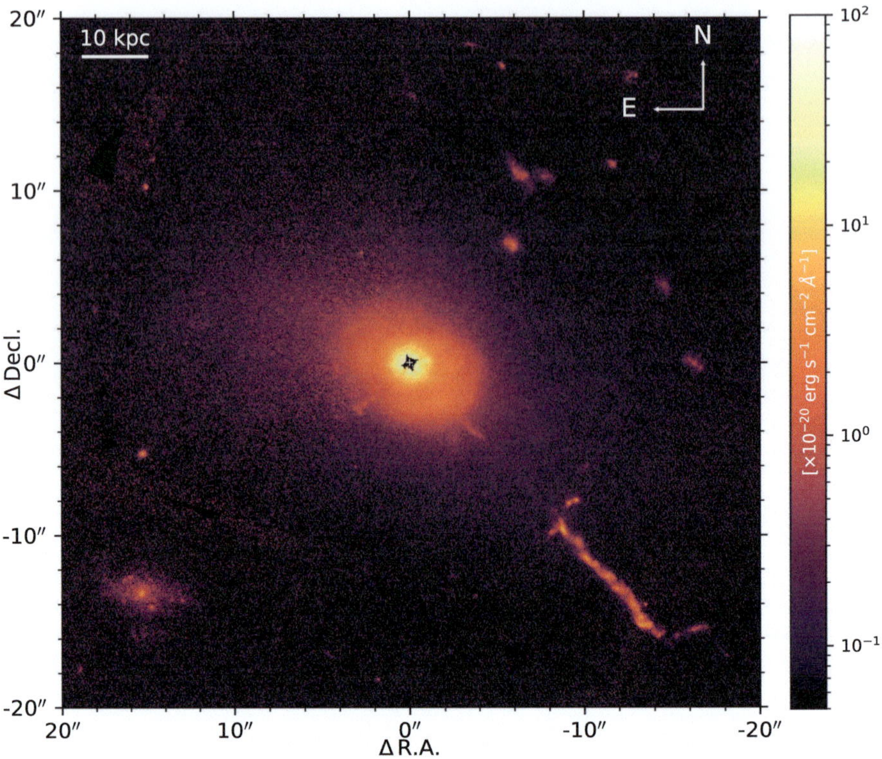

Fig. 5.36 The core and the *jet* of Quasar 3C 273 and surroundings in visible light seen with the HST/STIS coronagraph (Ren et al. 2024). (Permission to reproduce the image kindly provided by the author Bin Ren))

When and How the Enigmatic *Quasars* Formed and How Do They Shine (1964)?

We recall, that the discovery of *quasars* was the first major challenge to Fred Hoyle's theory of the *static Universe*, according to which the Universe was, is and will be forever the same, i.e., it is not changing. However, *quasars* have been mainly observed at very long distances, wherein, in the very distant past, they were about 100 times more numerous than they are today! *Therefore, the Universe is indeed changing.* That is why their discovery was a serious argument in favor of the *Big Bang* theory and a serious blow to Hoyle's *Static Universe* theory. But how did these enigmatic cosmic objects form?

A few million years after the formation of the first large stars, at the end of their short lives of millions of years, many of them exploded as *supernovae,* leaving behind as remnants *black holes.* But the Universe back in those early times was very dense and mergers were very frequent, creating thus larger *black*

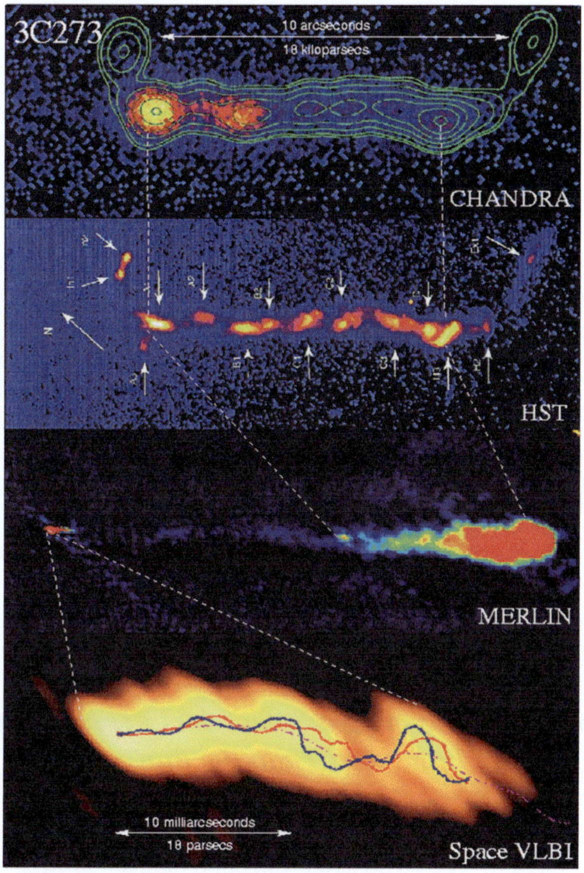

Fig. 5.37 This 4-panel image shows the *jet* blasting out of the core of the nearest bright *quasar* 3C273. The radio *jet* is seen to originate from the core and continue all the way to about 200,000 l.y., where the optical *jet* also arrives, although it does not appear to originate from the core. The scale on all three images is the same (about 10 arcseconds from one end of the *jet* to the other.) From top to bottom, the images are X-ray (Chandra), optical (HST), radio (MERLIN), and Space VLBI. (*Credit* X-ray: NASA/CXC, Optical: NASA/STScI, radio: Max Planck Institute for Radioastronomie, large scale radio: MERLIN and at a wavelength of 6 cm with the space VLBI mission VSOP (VLBI Space Observatory Programme), led by the Japanese Institute of Space and Astronautical Science.Permission to reproduce the image kindly provided by Andrei Lobanov, Thomas Muxlow and Robert Beswick. See also the first observations in Bahcall et al. (1995))

holes. At the same time, the gravitational attraction between the *protogalaxies* led them to further mergers, which gave high rates of star formation, as we observe today to happen when neighboring galaxies collide and merge. The black holes probably merged to form other supermassive black holes in the galactic centers with masses ranging from millions to billions of times the mass of the Sun and surrounded by accretion disks of gas. As the gas in the disc falls towards the central black hole, its gravitational energy is released in the form of electromagnetic radiation, across the electromagnetic spectrum. Hence, *quasars* are believed – and in many cases this is indeed confirmed – to be powered by the absorption of gas into supermassive black holes in the cores of distant galaxies, as proposed already in 1964 by Edwin Salpeter and Yakov Zel'dovich. Radiation cannot escape from within the horizon of a black hole. Therefore, the energy produced by a quasar comes from outside the black hole and is produced by enormous stresses and friction in the material closest to the black hole, as it swirls before falling inward.

The power radiated by *quasars* is enormous: the most powerful have luminosities thousands of times greater than a galaxy, similar to our own Milky Way. The huge luminosity of *quasars* comes from the release of the gravitational energy of the material adsorbed by the supermassive black holes and originating in the accretion discs surrounding them. Via this process, black holes can convert (6–32)% of an object's mass into energy, compared to only 0.7% of the mass conversion that occurs in the chain of the proton–proton nuclear fusion, during the energy production in stars like our Sun!

2024: A Black Hole Accreting a Solar Mass Per day, 500 Trillion Times More Luminous Than the Sun

Today, around a million *quasars* are known, but the quasar called J0529-4351, with a mass of 17 billion Suns (i.e., with a Schwarzchild radius 333 AU), eating just over a Sun per day, is the hungriest *black hole* ever seen! The black hole in this record-breaking quasar is growing in mass by the equivalent of one Sun per day, making it the fastest-growing black hole to date Wolf et al. (2024). The quasar, J0529-4351, is so far away from Earth (at redshift $z = 3.962$), that its light took over 12 billion years to reach us.

The plasma being pulled in toward this black hole, in the form of a hot disc, that measures seven light years in diameter (about 15,000 times the distance from the Sun to the orbit of *Neptune*) – perhaps the largest accretion disc in the Universe – emits so much energy that J0529-4351 is over 500 trillion times more luminous than the Sun (2×10^{48} ergs/s), the most luminous object known in the Universe to date!

Remarkably, this luminosity record-breaking quasar was overlooked until recently, hiding in plain sight, despite that this object showed up in images from the ESO Schmidt Southern Sky Survey dating back to 1980, but it was not recognized as a quasar until decades later.

The datasets from the search of *quasars* are so large, researchers often use machine-learning models to analyze them and distinguish *quasars* apart from other celestial objects. However, these models are trained on existing data, which limit the potential candidates to objects similar to those already known. If a new quasar is more luminous than any other previously observed, the program might reject it and classify it instead as a star not too distant from Earth! Thus, an automated analysis of data from the European Space Agency's Gaia satellite passed over J0529-4351 for being too bright to be a quasar, suggesting it to be a star instead! The researchers identified it as a distant quasar last year using observations from the Australian National University 2.3 m telescope at the Siding Spring Observatory in Australia. Discovering that it was the most luminous quasar ever observed, however, required a larger telescope and measurements from a more precise instrument. The X-shooter spectrograph on ESO's VLT in the Chilean Atacama Desert provided the crucial data (Fig. 5.38).

Fig. 5.38 The exceptional record-breaking *quasar* J0529-4351. This image shows the region of the sky in which the quasar is situated. Using ESO's Very Large Telescope (VLT) in Chile, this quasar has been found to be the most luminous object known in the Universe to date. This picture was created from images forming part of the Digitized Sky Survey 2, while the inset shows the location of the quasar in an image from the Dark Energy Survey. (*Credit* ESO/Digitized Sky Survey 2/Dark Energy Survey)

Extragalactic *Jets*

1918: Discovery of the First Extragalactic Plasma Jet

The first focused cosmic outflow was observed already more than a century ago (Curtis 1918) emerging from the giant elliptical galaxy (E0/1) M87, located near the center of the virgo cluster. Specifically, Curtis observed a *curious straight ray of light* extending from the center of M87, and clearly connected to its nucleus (Fig. 5.39).

This remarkable observation in 1918 marked the discovery of extragalactic jets.

However, the nature of the M87 *jet* remained unexplained for many decades, awaiting the advent of radio astronomy in the 1950s. At the time, the Virgo A radio source was one of the first radio sources observed. The strange straight visible jet in M87 was observed and studied by Baade and Minkowski (1954), who were also the first to introduce the term **jet**. Today, M87 is systematically observed from radio to X-rays, with indications that *jet* focusing already exists on scales of a few tens of Schwarzschild radii, for a black hole of a few billion solar masses, i.e., for distances less than about 1000 astronomical units, while the jet extends to distances of a few thousands of light years (Goddi et al. 2019, Fig. 5.44).

We shall return to this famous giant AGN later on in this chapter, and discuss the observed silhouette of a *black hole* which was photographed at its center.

Fig. 5.39 The *jet* of the galaxy M87 has a length of about 8,000 l.y. and in the image it points towards the left. (*Credits* NASA and the Wide Field/Planetary Camera Imaging Team, Tod R. Lauer (NOAO), Sandra M. Faber (CSC), C. Roger Lynds (NOAO))

1959: A Herculean Radio *Galaxy, with Its* Jets *Discovered in 2012 Extending from Its Central Black Hole to Lengths of Millions of Light Years Away*

Figure 5.40 shows the spectacular *jets* of the elliptical galaxy *Hercules A* in the constellation *Hercules*, in a composite of observations with the Hubble Space Telescope, the Chandra X-ray Space Telescope, and the Karl G. Jansky Very Large Array (VLA) radio telescope. in New Mexico, USA. These *jets* seem to be activated by a supermassive central *black hole* in the galactic nucleus.

At a distance of about 2 billion light years from Earth, this yellowish elliptical galaxy in the center of the image appears to be a quiet and ordinary galaxy in visible light. But let us have a look at its *Herculean* numbers: it is a giant galaxy with a mass about 1000 times the mass of our Milky Way galaxy – i.e., with four quadrillion solar masses (10^{15} M_o) – at the center of which lurks a supermassive *black hole* with mass 2.5 billion solar masses.

However, this seemingly quiet in the visible part of the spectrum galaxy is also one of the most powerful extragalactic radio sources, with the code 3C 348, observed already in 1959. It emits in radio waves billions more energy than our Sun does.

Radio observations with the VLA reveal two extremely long radio *jets* that extend on either side 1.5 million light years from its center, i.e., at distances much greater than the galaxy's visible-light galactic size. These *jets* consist of very high-energy subatomic particles and magnetic fields and are emitted from the neighborhood of the central black hole, traveling at nearly the speed of light. The outer parts of the

Fig. 5.40 Two spectacular radio *jets* 1.5 million light years long from end to end, energized and propagating in the gravitational field of a 2.5-billion-solar-mass central *black hole* at the core of the elliptical galaxy *Hercules A*. Synthesis of observations with the Wide Field Camera 3 of the Hubble space telescope (white), the Chandra X-ray telescope (purple) and the Karl G. Jansky Very Large Array (VLA) radio telescope in New Mexico (cyan). (*Source* visible, NASA/STScI, radio waves: NSF/NRAO/VLA, X-rays: NASA/CXC/SAO, https://arxiv.org/pdf/1711.07548, image reproduced with the kind permission of Bia Boccardi)

jets show ring-like condensations, which hints that there have been multiple bursts in the past from the neighborhood of the black hole at the center of the galaxy. The inner parts of the *jets* are not visible due to their extremely high speed, which causes relativistic effects that reduce the intensity of the radiation that reaches us. The galaxy is surrounded by a hot X-ray emitting halo, which is visible (purple) in this composite visible, X-ray, and radio image.

1970: Discovery of Superluminal Motion

One of the most impressive phenomena associated with *quasars*, is definitely *superluminal motion*, in which we see the motion of some clouds of plasma moving at *apparent* speeds greater than the speed of light in vacuum. Superluminal motion is nicely explained in *blazars* (Fig. 5.34), and is due to projection effects rather than a violation of relativity, according to which the maximum speed at which information can be transmitted is the speed of light. Basically it requires relativistic motion almost in the direction of the observer, Fig. 5.41.

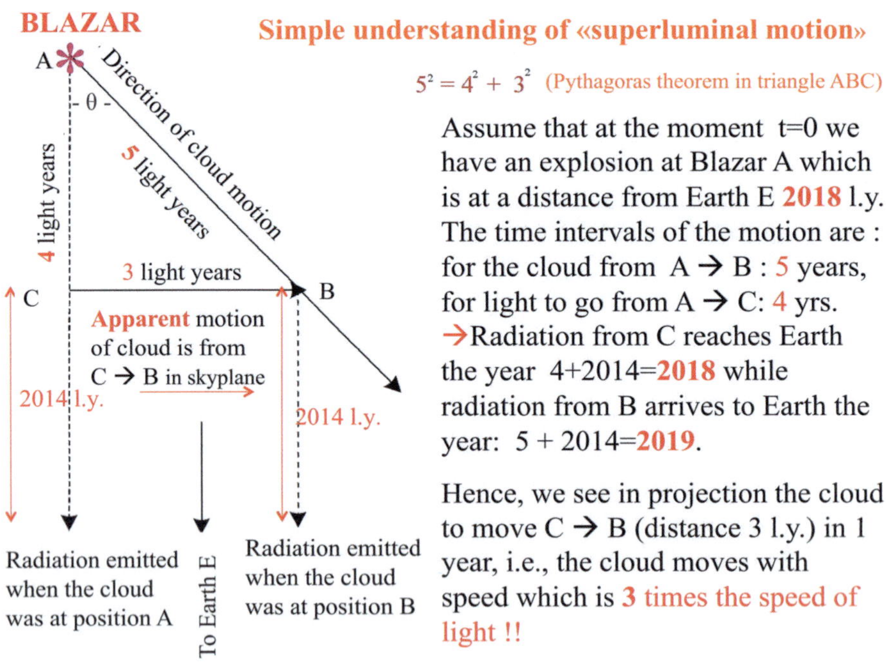

Fig. 5.41 Radio pictures of quasar 3C 273, taken in July 1977, March 1978, June 1979 and July 1980, showing *superluminal motion* of the inner jet with an apparent speed 9.6c (Pearson et al. 1981, 1982). The above schematic figure is an explanatory diagram of such observed motion in blazars, i.e., *quasars* that emit *jets* at a small angle θ with the direction we observe them. A more detailed explanation is in the text. (*Source* K. Tsinganos, Plasma Astrophysics, Unibooks, p. 445, 2015)

Almost all *quasars* are accompanied by relativistic *jets*. In some of them the jet moves at a small angle θ with respect to the direction we observe the *quasar*. Those *quasars* where the jet moves at such a small angle θ to the direction we observe the quasar are called *blazars* (Fig. 5.34). We also observe at many wavelengths, that the *jets* of *quasars* do not have a continuous distribution of intensity along them, but consist of many approximately equidistant *bright knots*. Various mechanisms have been proposed for the generation of those bright spots. Assuming that at the center of *quasars* lurks a supermassive black hole with mass of hundreds of millions to billions the solar mass, it absorbs material from the surrounding galaxy at a rate of a few solar masses per year. A fraction of this erratic adsorption leads to ejection and the creation of *jets*. When the fast plasma of these *jets* collide with a previous emission of slower plasma, shock waves are created where charged particles, electrons and positrons are accelerated, which then radiate in γ-rays, X-rays, etc., that we then observe.

By the 1960s, VLBI radio interferometry had developed to the extent that radio sources with angular resolution of milliarcseconds could be observed. Successive observations of the nucleus of *quasar* 3C 273 with the VLBI radio symbol meter for 4 consecutive years showed a density enhancement in the jet that during this time moved by 25 light years. Therefore, it moved at an apparent speed, in the plane of the sky, many times the speed of light. But this is forbidden, given that the speed of light is naturally the fastest speed with by which anything in the Universe can be propagated. So, here we have a *paradox*. How can we understand it? In what follows and in Fig. 5.41, a simple explanation of this *paradox* is given, namely that this is really a projection effect that requires movement at a speed close to the speed of light and in a direction making at a small angle to the line of sight.

So let's follow a simplified description of understanding this paradox. Suppose that at the time $t = 0$ the core A of the *blazar* is ignited, emitting a cloud of plasma on the *jet* line that forms a small angle θ with respect to the direction we observe it (Fig. 5.41). To simplify the discussion and without a loss of generality, let us assume that the signal reaches the Earth during the year 2018, i.e., the distance from point A in the blazar's core to the distance of the Earth is equal to 2018 l.y. Suppose that for the first 4 years the light from this flash travels from A to C and then travels the rest of the distance from C to Earth in the remaining 2014 years. At the same time, a plasma cloud (jet) is launched in the AB direction which is at a small angle θ to the straight line AC, with a speed, say 0.999c, i.e., practically equal to the speed of light. Then, in 5 years the plasma cloud moving along the line AB reaches B which is at a distance of 5 l.y. and from there a light signal is emitted that reaches from B to Earth in 2014 years, equal to the distance of this point from Earth. Hence, this signal reaches Earth in the year $5 + 2014 = 2019$. However, in reality we see the **projection** of the movement of the cloud on the plane of the sky, i.e., we see the movement of the cloud from C to B, which lasts: $2015–2014 = 1$ year. But according to the Pythagorean theorem in the right triangle ABC, the distance BC is equal to 3 light years, since $AC = 4$ l.y. and $AB = 5$ l.y. Therefore, we as observers conclude that the cloud moved from C to B (in projection) at a speed of 3c! This phenomenon of observing the motion of the emitted clouds from a blazar at a speed many times the speed of light is called *superluminal motion*. Hundreds of *blazars* have been observed in radio waves with the VLBA radio telescopes, wherein the apparent speed of the various emitting clouds exceeds the speed of light, from 1 to 20 times! (Fig. 5.42).

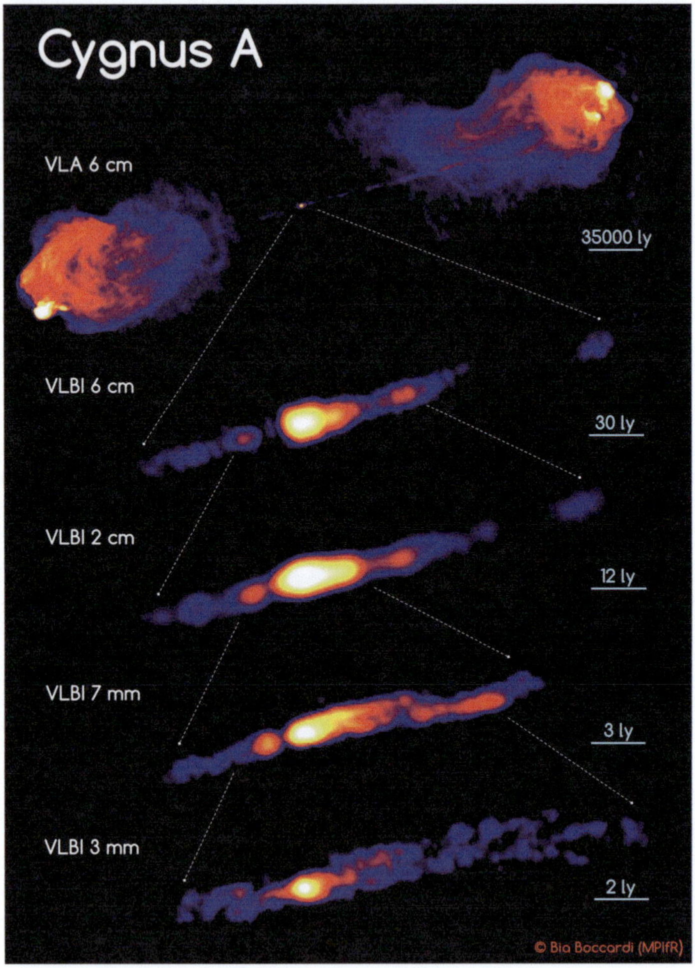

Fig. 5.42 The *jets* of the radio galaxy Cygnus A on scales from hundreds of kilo-parsecs imaged with the Very Large Array (Perley et al. 1984) to the sub-parsec probed with mm-VLBI (Boccardi et al. 2016b, a). The VLBI images are created after stacking several epochs. Data at 2 cm are from the MOJAVE survey (Lister et al. 2009). The self-similarity of the form of the two *jets* at scales differing by one million, 1:10^6, is obvious. (*Source* Radio observations of active galactic nuclei with mm-VLBI, Boccardi et al. (2017), arXiv e-print, arXiv:1711.07548, image reproduced with the kind permission of Bia Boccardi)

2015: Two-Component MHD Outflows with Shocks in Relativistic Jets

If the observed relativistic plasma outflows in AGN radio *jets* are magnetically collimated and a single-component model is adopted, consisting of a wind-type outflow from a central object (possibly an electron–positron plasma), then a problem arises with the inefficiency of magnetic self-collimation to collimate a sizeable portion of the mass and magnetic fluxes in the relativistic outflow from the central object. To solve this dilemma, we have applied the mechanism of magnetic collimation to a two-component model consisting of a relativistic wind-type outflow from a central source and a non-relativistic wind from a surrounding disc (Bogovalov and Tsinganos 2005). By employing a numerical code for a direct numerical solution of the steady-state problem in the zone of super-fast magnetized flow, which allows us to perform a determination of the flow with shocks, it is shown that in this two-component model it is possible to collimate into cylindrical *jets* all the mass and magnetic fluxes that are available from the central source. In addition, it is shown that the collimation of the plasma in this system is usually accompanied by the formation of oblique shock fronts. The non-relativistic disc-wind not only plays the role of the jet collimator, but it also induces the formation of shocks as it collides with the initially radial inner relativistic wind and also as the outflow is reflected by the system axis. Another interesting feature of this process of magnetic collimation is a sequence of damped oscillations in the width of the jet.

To understand the structure of the *jet* and how it is finally formed, it is necessary to consider the process of the collimation in some more detail. First, Fig. 5.43a shows schematically the interaction of the two components of the outflow which form the *jet* and the accompanying shock. The collimation of the relativistic outflow from the central source is performed by the tension of the toroidal magnetic field, which is generated by the rotating disc-wind. This wind which compresses the outflow from the central source in general induces the formation of the oblique shock (1) and a weak discontinuity (2). After the compression of the flow towards the axis, a reflection shock wave (3) may also be formed.

The results of the numerical simulation show in detail the gradual formation of the shock and the subsequent collimation of the outflow, when the interior relativistic flow is compressed by the external disk-wind, Fig. 5.43b–d. The shock is first formed at a distance $R \approx 100\ R_f$ from the source (Fig. 5.43b). Initially, close to the source the post-shock pressure is mainly provided by the pressure of the poloidal magnetic field, because there the poloidal magnetic field is still sufficiently strong. Thus, a layer of enhanced poloidal magnetic field is formed in the flow and this may be seen in Fig. 5.43a–c as a black ribbon. The further away from the source we move, the weaker the poloidal magnetic field becomes. Starting from some distance around $400\ R_f$, the poloidal magnetic field is no longer able to provide the needed pressure in the post-shock region. This role is now taken by the gas pressure, i.e., the total post-shock pressure becomes mainly of thermal origin. The shock front reaches the z-axis at a distance $z \approx 2000\ R_f$.

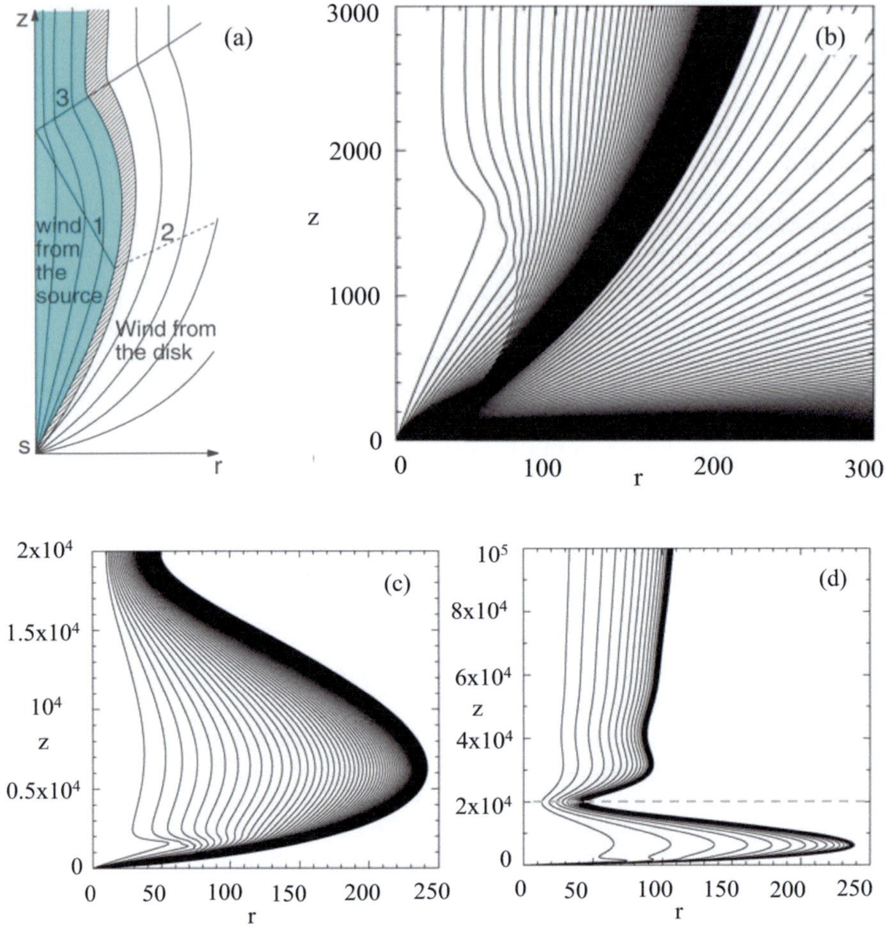

Fig. 5.43 **a** Sketch of the shock waves and singular surfaces expected to form in the general case of the collision of a two-component MHD outflow. The oblique shock front marked by "1" is formed at the collision of the two parts of the outflow: the exterior collimated disk wind and the still uncollimated interior relativistic outflow from the central object. An outgoing weak discontinuity from one end of this shock is marked by "2." The shock front marked by "3" is formed at the self-reflection of the collimated flow by the axis of rotation, in (**b**). In **c**, **d** the flow is shown at larger distances. Results of the numerical simulation up to z = 20,000 R_f are shown in (**b**) for the formation of the shock when the relativistic flow is compressed by the wind from the disc to form a jet with a minimum radius at z $\approx$ 20,000 R_f and an oscillatory radius further downstream, as it is evident in (**d**). (From Bogovalov and Tsinganos (2005))

At larger distances, R $\approx$ 20.000 R_f, the *jet* has been compressed to its minimum radius (Fig. 5.43c, d). At this stage the overpressure of the flow at the axis produces the reflecting shock wave (3) and this now turns the flow off the z-axis. The reflecting shock wave (3) is easily seen at relatively large distances from the axis, because the amplitude of the reflecting shock grows as the magnetic pressure decreases. Here

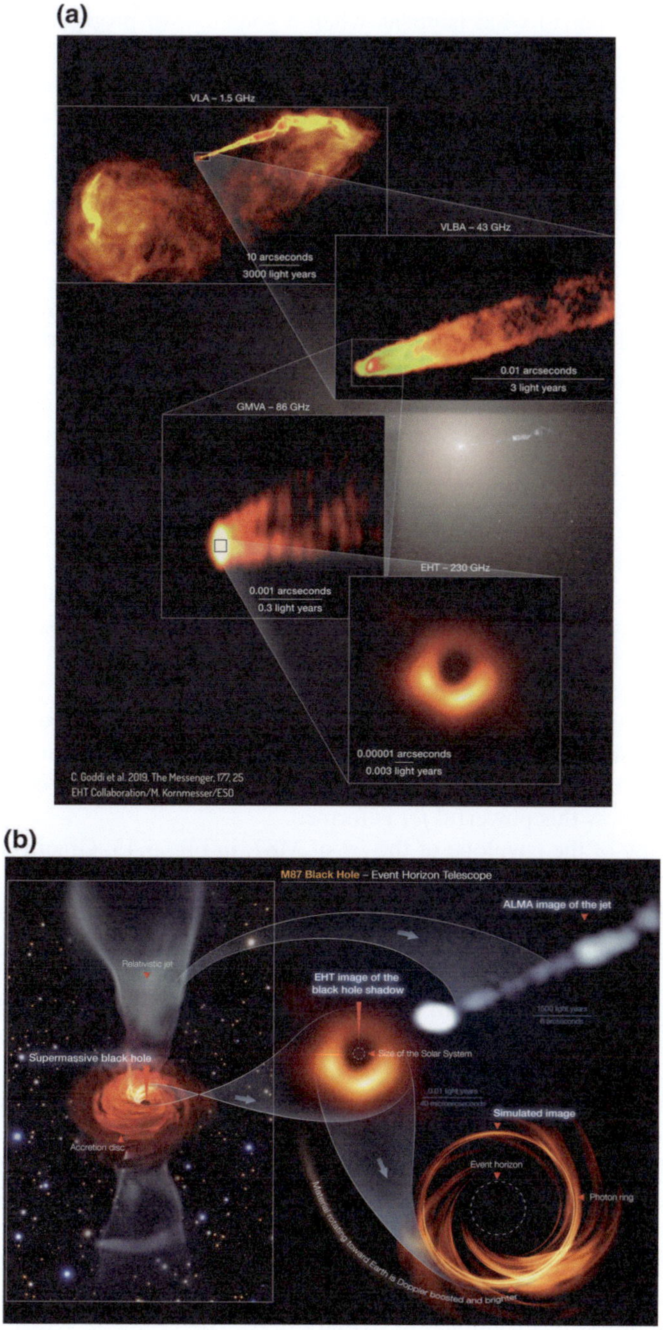

Fig. 5.44 **a**, **b** 20 years of radio observations of the jet of the giant elliptical galaxy M87 at an increasingly high resolution, finally reaching the central region of the galaxy, where the large Event Horizon Telescope (EHT) radio symbol meter revealed a supermassive black hole of 6.5 billion solar masses. (*Image Credits* Goddi et al. (2019), the Messenger 177.25, the EHT research collaboration, M. Kornmesser/ESO)

the effect is similar to what happens when a sound wave propagates in a region of decreasing density, wherein it is transformed into a shock wave (Zel'dovich and Raizer 2002). Indeed, at the be- ginning this distribution of the pressure creates a smooth motion of the plasma away from the z-axis. However, at larger distances where the total magnetic field decreases, this motion results in the formation of a reflecting shock wave, which corresponds to the predicted shock (3) in Fig. 5.43a.

Figure 5.44 are recent observations of the famous system *black hole* + *jet* in M87, which help us to understand the origin of the *jet* as close as possible to the *black hole* (Goddi et al. 2019). These new images show precisely for the first time, how the base of a *jet* connects with matter swirling around a supermassive *black hole*. Observations at different radio wavelengths using several radio-interferometers and the large EHT network can discern very small details in the region around M87's *black hole*. As matter orbits the *black hole*, it heats up and emits light. The black hole bends and captures some of this light, creating a ring-like structure around the *black hole* as seen from Earth. The darkness at the centre of the ring is the *black hole* shadow, which was first imaged by the Event Horizon Telescope (EHT) in 2017. Noteworthy is that the limb-brightened *jet* of M 87 has been observed down to the few Schwarzschild radii scale (Kim et al. 2018).

Actually, the *jet* and shadow of the supermassive black hole at the center of the M87 galaxy have been shown *together* for the first time. These observations were obtained with telescopes from the Global Millimeter VLBI Array (GMVA), the Atacama Large Millimeter/submillimeter Array (ALMA), and the Greenland Telescope forming a network of radio-telescopes around the globe working together as a virtual Earth-sized telescope. Such a large network can discern very small details in the region around M87's *black hole*. The observations provide the context needed to understand how this powerful *jet* is formed.

A modeling of the envelope of the *jet* of M87, in Fig. 5.45, Gracia et al. (2005) made a simplifying but crucial assumption in deriving the opening angle of their model for the *jet*. They identified the observed *jet* with the inner relativistic outflow of their two-component model. More specifically, the boundary of the *jet* was assumed to coincide with the shape of a specific fieldline Ψ_α, the one fieldline which separates the relativistic inner region from the non-relativistic outer disk-wind at the base, or the launching surface, of the outflow. So, the observed opening angle was strictly speaking fitted by the shape of a single magnetic fieldline. This model consists of two distinct zones: an inner relativistic outflow, which is identified with the observed *jet*, and an outer cold disk-wind. While the former does not selfcollimate efficiently due to its high effective inertia, as we discussed before, the latter fulfills all the conditions for efficient collimation by the magneto-centrifugal mechanism. Given the right balance between the effective inertia of the inner flow and the collimation efficiency of the outer disk wind, the relativistic flow is magnetically confined into a well collimated beam and matches the measurements of the opening angle of M87 over several orders of magnitude in spatial extent. The synthetic synchrotron maps reproduce the morphological structure of the *jet* of M87, i.e., center-bright profiles near the core and limb-bright profiles away from the core. At the same time, they also show a local increase of brightness at some distance along the axis, probably associated to a recollimation shock in the MHD model (Fig. 5.43).

In that connection, the optical knot HST-1 has received much attention. Most MHD models with acceptable fits to the opening angle distribution feature a shock at the distance 70–200 pc. This is particularly true for models with large initial opening angle α. The shock is due to the fact, that the *jet* is not in lateral force equilibrium at the launching surface and over-expands. Later, it is forced back toward the axis by magnetic hoop stress and produces a recollimation shock (Fig. 5.43, Bogovalov and Tsinganos 2005). In all cases the shock is visible as a bright spot in the synthetic maps. Its location coincides with the position of the optical knot HST-1. In addition, a best fitting model is consistent with a number of observational constraints such as the magnetic field in the knot HST-1, and the *jet*-to-counter *jet* brightness ratio (Fig. 5.46).

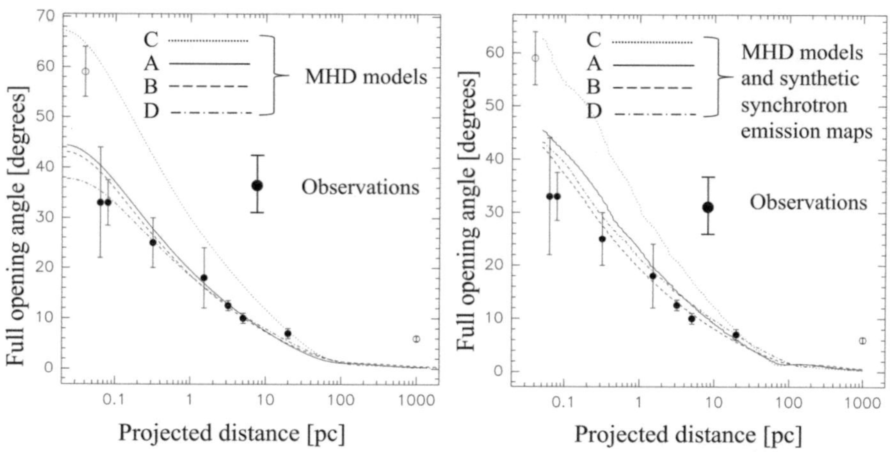

Fig. 5.45 Comparison of the opening angle calculated from the observational data for the *jet* of M87 and *MHD models/synthetic maps*: **a** *MHD models*: The lines show opening angle profiles as given by the separating field line Ψ_α for model A (solid line), i.e. the best model, and model D (dot-dashed), which best reproduces the data based on Ψ_α. The models B (dashed) and C (dotted) are shown for completeness. The data points marked by filled circles were taken into account in the fitting procedure. The innermost and outermost measurement (open circles) were disregarded. The data points marked by filled circles were taken into account in the fitting procedure. The opening angle close to the disk and black hole is about 40° at a projected distance less than 0.01 pc, in accordance with Fig. 5.44, while further away the jet becomes approximately cylindrical. **b** *synthetic maps*: The lines show opening angle profiles as given by the HWQM contour R_{map} of the overall best model A (solid line), and model B (dashed line) which fits the opening angle data best. Models C (dotted) and D (dot-dashed) are shown for comparison. Various symbols represent observational measurements. (*Source* Gracia et al. (2009))

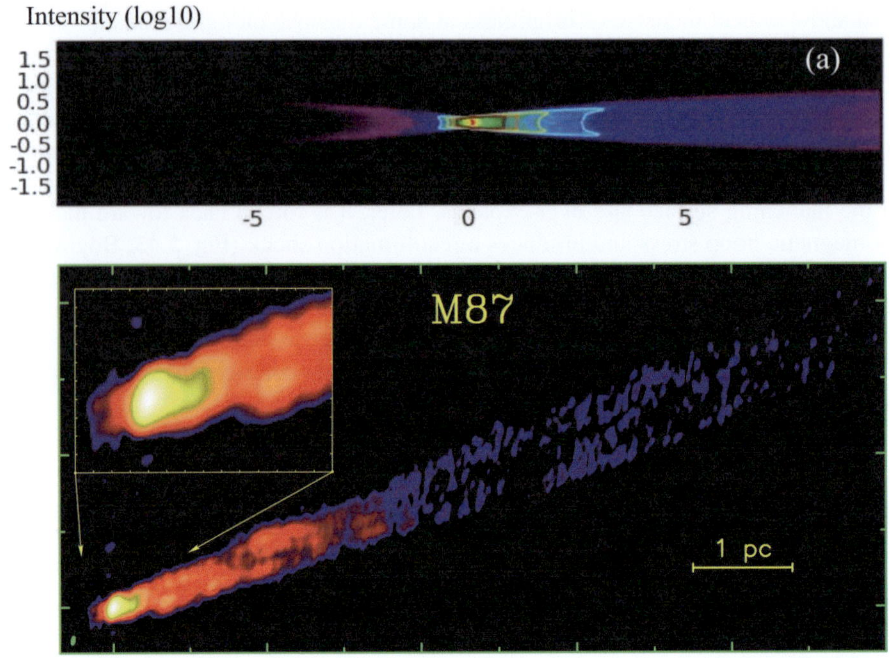

Fig. 5.46 Comparison and good agreement of a theoretical MHD model combined with synthetic synchrotron emission maps, for the jet of M87 (Gracia et al. 2009) and 2 cm VLBA images of the inner radio *jet* of M87 showing a limb-brightened structure and unambiguous evidence for a faint 3 mas long counter feature which also appears limb brightened (Kovalev et al. 2007). (The lower figure is reproduced with the kind permission of Yuri Kovalev])

2022: A Meridionally Self-similar MHD Model of a Relativistic Double Flow in Kerr Geometry, Applied to the M87 **Jet**

A meridionally self-similar MHD model of a relativistic *double flow* in Kerr geometry that incorporates mass, angular momentum and energy loading on magnetic field lines threading the *black hole* horizon has been presented in Chantry et al. (2022). This semi-analytical model allows the study of *double flow* solutions above the *black hole* pole; the solution passes through critical surfaces, including a spherical stagnation surface, wherein the loading of pairs occurs. The pair production is supposed to be due to two-photon interaction via gamma-ray emission from the accretion disk. The goal of this modelling is to describe MHD fields and flows close to the polar axis, where the total power is matter dominated resulting to a fast spine-*jet* propagating at large distances and a polar pair accretion onto the horizon (Fig. 5.47).

To perform the loading, two source terms are introduced, the mass injection rate and the 4-force exerted by the radiation at the origin of pairs, implying that mass, angular-momentum and energy fluxes along the field lines are not conserved. The following working assumptions at the stagnation surface were applied, in order to

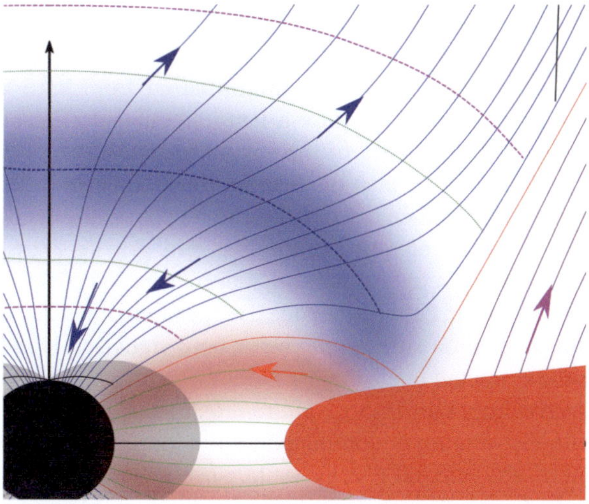

Fig. 5.47 Schematic representation of the inflow/outflow configuration in the poloidal plane of a meridionally self-similar MHD model for a relativistic double flow in Kerr geometry, applied to the *jet* of M87. The black sphere corresponds to the *black hole* horizon and the light grey shaded region to the ergosphere. The blue lines correspond to the poloidal magnetic field lines anchored on the *black hole* horizon and reaching infinity, while arrows indicate the direction of the flow velocity. The green lines correspond to those lines of the *black hole* magnetosphere which link the *disk* to the *black hole*. The magenta lines and arrow corresponds to the magnetic field lines anchored in the disk and reaching infinity. The blurred blue zone corresponds to the region where we expect that pair creation is efficient. The red zone corresponds to the disk and the blurred red zones correspond to the accretion columns. The dashed blue line corresponds to the stagnation surface, and the green dotted line to the inner slow-magneto-sonic transition, the magenta dashed line corresponds to the inner Alfvénic transition. The black line corresponds to the position of the inner and outer light cylinders. (From Chantry et al. 2022)

match the inflow and outflow solutions: continuity of the radial magnetic field component and the magnetic flux; continuity of the isorotation frequency and a meridional surface current, which creates a discontinuity of the current intensity along the polar axis. This meridional surface current, accompanied by a charge density on the stagnation surface, causes a discontinuity of the toroidal magnetic field component and the field line opening. It implies a contribution of the injected pairs to the electric current and then to the Poynting flux.

With those assumptions and applying three matching conditions for the inflow and outflow parameters, the inflow/outflow solution is obtained by solving a system of first order differential equations. Thus, the model allows to quantify the exchange between the *black hole* and the MHD fields, and the sources of energy and angular momentum due to pair loading. Three inflow/outflow solutions (M1, M2 and M3) were studied and applied to the case of the *black hole* mass of M87. For the magnetic field magnitude, the values observed by the EHT at a distance of a few Schwarzschild radii were used. The inflow model follows a scaling law similar to that in Zamaninasab et al. (2014) between the magnetic flux and the mass flux, assuming that the mass inflow rate is a fraction of the mass accretion rate. The efficiency is three times higher to create magnetic flux.

A range of mass injection rates by pair formation from hard disc photons is estimated. Solutions M2 and M3 require a large amount of injected mass and are located close to the upper boundary of this estimated range, while the M1 solution requires an injection rate higher than the maximum value by an order of magnitude. It could involve other mechanisms of injection to be physically relevant.

As expected the inflow acceleration is dominated by gravity. Pressure also plays a role, decelerating the inflow close to the stagnation layer and close to the *black hole* horizon and accelerating it in between. Inside the open field lines, the injected angular momentum on the stagnation surface can be of both signs and is of the same order of magnitude as the extracted one from the *black hole*. In terms of Noether's energy, most of the injected energy falls into the *black hole*. Inside of the open magnetic field lines the energetic distribution is dominated by the inertial term. The *black hole* is fed by mass, kinetic and internal energy. At larger colatitudes, the energetic budget is dominated by the Lense–Thirring effect for the inflow in M2 and by the Poynting flux for the inflow in M3. For this solution M3, the total Poynting power on the horizon, even if it is less than the inertial energy, represents a quarter of the total energy absorbed by the *black hole*. The pair injection contribution to the Poynting flux has been calculated for the M3 solution and is comparable to the Poynting flux extracted from the *black hole*.

The interesting solution (M3) of this modeling has an isorotation frequency equal to one half to that of the rotation frequency of the *black hole*; a high but reasonable injection mass rate and a reasonable total outflow power. The final Lorentz factor of the outflow is around $\gamma \approx 10$. The extracted Poynting power from the *black hole* is comparable to that corresponding to a force-free model with a monopole geometry, same isorotation frequency and same total magnetic flux. The outflow of this solution has a rather high magnetization just above the stagnation surface and a total power within the expected corresponding range for extragalactic *jets* (Fig. 5.48).

In Fig. 5.49a, a schematic diagram of a jet emerging from an accretion disk around an active galactic nucleus is shown, at the center of which there is possibly a high-mass black hole. The magnetic lines wind around the axis of the jet and thus the magnetic *tension* focuses the outflow. Some thickenings along the jet are also visible, the result of fluctuations in the initial velocity in the disk, but also instabilities during the interaction of the jet with the environment.

The study of the *jet* of the galaxy M87 has been, one of the favorite research subjects of the author's research team since the 1980s and several interesting results have emerged from our study (e.g., Ferrari et al. (1984), Meliani et al. (2004), Gracia et al. (2005, 2008, 2009), Chantry et al. (2018, 2022). The initial relativistic acceleration of the *jet* by the radiation of a hot disc ($T > 10^9$ K) results in *jet* plasma velocities approaching the speed of light at distances of the order of a light year from the disc (Ferrari et al. 1984, 1985, 1986; Ferrari and Tsinganos 1986). As the *jet* then propagates, bright nodes form along it, as a result of the generation of shock waves (Ferrari et al. 1986, Trussoni et al. 1988). Later, a method of magnetic focusing of the relativistic motion of the plasma along the M87 jet was studied (Tsinganos and

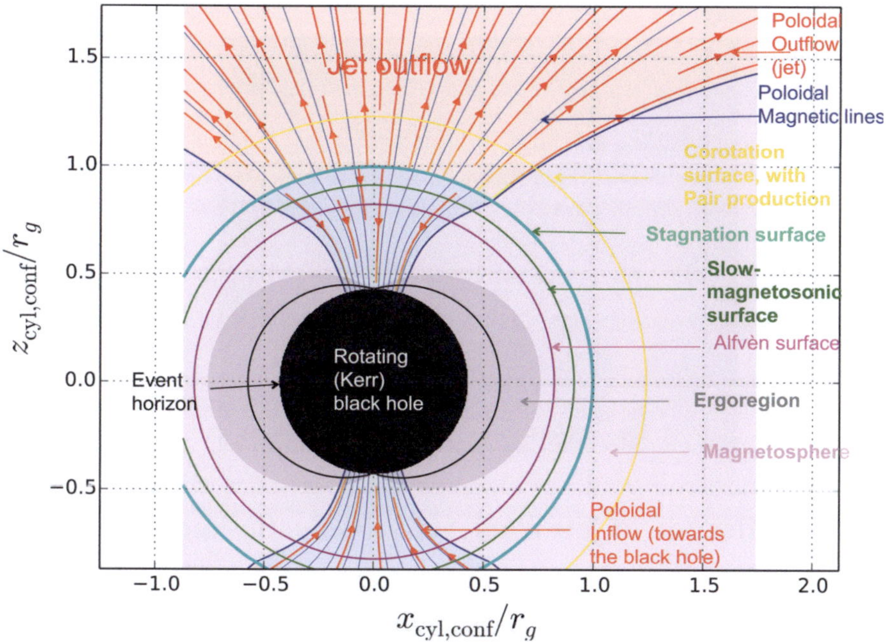

Fig. 5.48 Inner surfaces of the meridionally self-similar MHD model of a relativistic double flow in Kerr geometry, applied to the *jet* of M87. Above the marked cyan stagnation surface we have an outflow and inside this stagnation surface an inflow. The red arrows represent the mass flux and the thin blue lines, the poloidal magnetic field lines. The thick blue line marks the last open magnetic field line of the flow connected to the *black hole*. The yellow line represents the position of the corotation surface where $\Omega = \omega$. The cyan circle corresponds to the stagnation surface, the green ones correspond to the inner slow-magnetosonic surface and the magenta one to the inner Alfvén surface. The light cylinder surface is outside the scale, at about 100 Schwarzchild radii. The magnetosphere is represented in purple, the open line flow region in light red. The region where we expect the disk wind is in green (outside this scale), the ergoregion in light gray and the inner horizon region in black. (From Chantry et al. 2022)

Bogovalov 2005; Gracia et al. 2005). Next, we constructed a self-consistent magnetohydrodynamic model for the *jet* (Gracia et al. 2008, 2009). This model consists of two distinct zones and corresponding outflows: an inner zone of warm relativistic outflow, which we consider to correspond to the observed *jet*, and an outer cold disc wind. The first component is not effectively focused due to the high inertia of the relativistic plasma, but the second outer component is effectively focused by the magneto-centrifugal mechanism and then forces the inner *jet* component to focus as well. In the second part of this work we presented synthetic emission maps of the synchrotron radiation for the specific model, Gracia et al. (2009). The above two-band model can reproduce very well the morphological structure of M87's *jet* derived from radio/VLBA observations (Kovalev 2007). The *jet* angle decreases from an initial value of about 30 degrees very close to the source (0.3 l.y.) to a cylindrical outflow at a distance of 300 l.y. (Fig. 5.45). Also, the same model (Gracia et al. 2009) produces

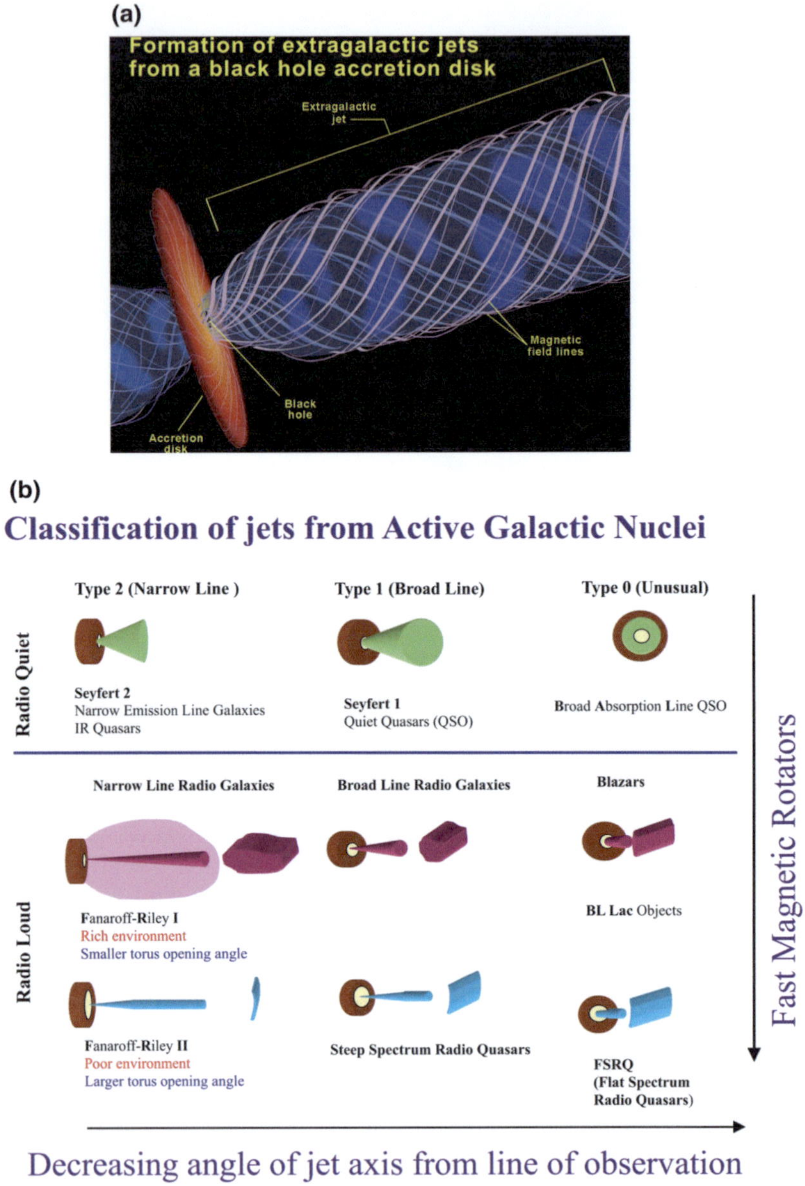

Fig. 5.49 **a** Artist's impression of the formation region of an accretion disk (red-yellow) surrounding a *black hole*, and its magnetic field lines twisting tightly to channel the outpouring subatomic particles into a narrow *jet*. The *jet* opens widely near the *black hole*, then is shaped into a narrower beam within about a light year from the *black hole*. (*Image Credit* NASA/ESA and Ann Feild (Space Telescope Science Institute)). **b** Classification of jets from AGN as a function of the angle between the jet axis and the viewing direction versus the efficiency of the AGN central magnetic rotator. From Sauty et al. (2007))

synthetic emission maps of the *jet* radiation with an increase in the brightness at the *jet* lip (limb brightening), but also a decrease in the intensity of the counter-*jet* by a factor equal to 13, which justifies its non-appearance in the observations.

Other questions concerning M87's jet, as well as most other extragalactic jets, is the origin of quasi-periodical bright knots along the jet and the stability of the jet. For example, the first studies of the effect of the Kelvin–Helmholtz (KH) instability in jets in the framework of the structure of radiogalaxies, Ferrari, Trussoni, Massaglia, Bodo and collaborators have shown that relativistic supersonic jets tend to be stable against the KH instability. Subsequently, a first comparison with observations of numerical simulations and the nonlinear evolution of the KH instability in different astrophysical contexts, were presented in Bodo et al. (1994). The 3D study of jet propagation as a crucial ingredient for understanding the dynamics that leads to FRI/FRII morphologies is presented in Massaglia et al. (2016).

Finally note that a classification of all known types of active galactic nuclei according to two parameters: (i) the angle between the line of sight and the *jet* axis and (ii) the magnetic rotator efficiency, as calculated from the magnetohydrodynamic integral ε of meridionally self-similar outflows has been presented in Sauty et al. (2007) (Fig. 5.49b).

Gravitational Lensing

1979: Observational Discovery of Gravitational Lensing

Associated with *quasars* are various exotic astrophysical phenomena, such as *superluminal motion*, which we just discussed and also gravitational lensing that we discuss in the following.

Gravitational lensing is a prediction of the General Theory of Relativity (GTR), and it was proposed by Einstein in 1936 as a direct consequence of the GTR. According to this theory, when the light of a distant light source, e.g., a quasar, passes through the strong gravitational field of a galaxy, or, a *black hole* (playing the role of the gravitational lense), the light rays are bent to create a circular ring, Figs. 5.52, 5.53, and 5.54 – when the group of lensing galaxies, or the quasar and the observer are well aligned – or, some other shape, when they are not, e.g., a cross, Fig. 5.51. An explanatory diagram of the gravitational lensing is shown in Fig. 5.50.

An interesting example of gravitational lensing has been observed by the Hubble Space Telescope. The galaxy – in the category of Luminous Red Galaxies – has an unusually large mass, containing about 100 times the mass of our Milky Way. In Fig. 5.52, however, the image has the shape of a ring, or rather a *horseshoe*, surrounding the red galaxy, which has and the greatest interest.

One of the most famous images and an excellent example of gravitational lensing is the *Einstein's cross*, observed with the Hubble Space Telescope (2012). This image is likely the sharpest image of the Einstein Cross ever made, Fig. 5.51. A *galaxy* with

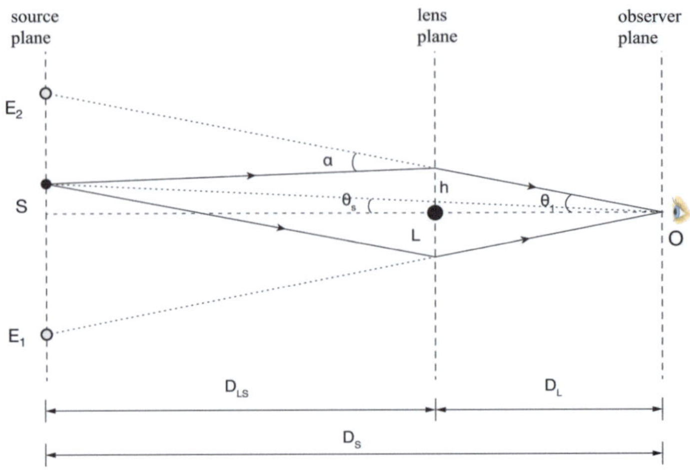

Fig. 5.50 Illustrative diagram of gravitational lensing: the source S (a distant quasar) is at a distance D_S from the observer O, the intermediate lens-galaxy L at a distance D_L, and the source S at a distance D_{LS} from the lens. The initial angular position of the source (without the lens effect) with respect the intervening lens is θ_s, but the observer sees two images, E_1 and E_2 at angles θ_1 and θ_2. (Adapted from Wikimedia commons)

Fig. 5.51 An impressive example of an *Einstein Cross*, as seen by the Hubble Space Telescope. The strong gravity of the elliptical lens-galaxy in the center ZW2237+030 (Huchra's Lens) with z = 0.04, creates the 4 peripheral images of the distant Quasar Q2237+0305 (Einstein Cross) z = 1.7 in Pegasus, by which the light is bent. When the quasar emitted its light it was about 8 billion light years from Earth. These two bodies are aligned with Earth in order that this type of mirages are observed. (*Source* NASA/ESA/Hubble)

Fig. 5.52 Pictured above, is the result of the gravity of a *luminous red galaxy* (LRG 3-757), to gravitationally distort the light from a much more distant *blue* galaxy to this cosmic *horseshoe*. Such light bending results in two discernible images of the distant galaxy. In this case, the lens alignment is so precise that the background galaxy is distorted into a nearly complete ring. LRG 3-757 was discovered in 2007 in data from the Sloan Digital Sky Survey, and this image is a follow-up observation taken with the Hubble Space Telescope's Wide Field Camera 3. Strong gravitational lenses like LRG 3-757 allow the determination of the mass and dark matter content of the foreground galaxy lenses. (*Source* ESA/Hubble and NASA)

five nuclei is really one galaxy surrounded by a mirage of four images of a distant quasar. The galaxy lies 400 million light years away; the quasar about 8 billion light years away.

The blue rings in Figs. 5.52 and 5.53 are actually imags of distant galaxies that have been enlarged and distorted into an almost circular ring by the strong gravitational pull of the giant, intermediate to us, bright red galaxy. For this circular distortion to occur in a distant galaxy and to form such an *Einstein ring* as it is often called, a very good alignment is required between the two lines, the one connecting the Earth to the lens galaxy and the one connecting the lens galaxy with the distant quasar (i.e., for small angles θ_s in Fig. 5.50).

The *cosmic ring* in Fig. 5.52 is one of the best examples of *Einstein rings*. At the same time, it gives us a tantalizing opportunity to observe the early Universe: the blue quasar's redshift – a measure of how much the wavelength of the light it emits has been stretched by the expansion of the Universe – is about $z = 2.4$. This means that we observe the Universe as it was about 3 billion years after the *Big Bang*, while today 13.8 billion years have already passed since the *Big Bang*. This Einstein ring was identified in 2007 through data from a systematic wide-field observation, the Sloan Digital Sky Survey. This image was taken with the Hubble Space Telescope's Wide Field Camera 3 in the visible and infrared. More impressive gravitational lenses are shown in Figs. 5.53 and 5.54.

Fig. 5.53 The blue circular arcs in these 8 Hubble Space Telescope images are giant distant galaxies, 2–4 billion light years from Earth. Their light is bent by the intervention of other, Earth-aligned, galaxies. Since such a lensing effect was generally predicted in some detail by Albert Einstein over 70 years ago, rings like this are now known as *Einstein Rings*. About 130 gravitational lenses have been observed to date and they are a way to calculate dark matter in intervening elliptical galaxies. (*Source* NASA, ESA, and SLACS *Survey team* A. Bolton (Harvard/Smithsonian), S. Burles (MIT), L. Koopmans (Kapteyn), T. Treu (UCSB), and L. Moustakas (JPL/Caltech))

Fig. 5.54 ALMA, in which ESO is a partner, has revealed an extremely distant galaxy that looks surprisingly like our Milky Way. The gravitationally lensed distant galaxy SPT0418-47 appears as a near-perfect ring of light around the nearby galaxy, because of their almost exact alignment. (*Credit* ALMA (ESO/NAOJ/NRAO), Rizzo et al.)

Lensed Circular View of the Early Galaxy SPT0418-47 with ALMA

SPT0418-47 (Fig. 5.54) is a gravitationally lensed, high-redshift ($z = 4.22$), dusty, star-forming galaxy, discovered with the South Pole Telescope (2013). The galaxy is so far away that its light has taken more than 12 billion years to reach us (light travel distance), or, 24 billion light years (present comoving distance), so we see it as it was when the Universe was just 1.4 billion years old, i.e., when the Universe was just 10% of its current age. This extremely distant galaxy and therefore very young galaxy, looks surprisingly like our Milky Way. It is also surprisingly unchaotic, contradicting theories that all galaxies in the early Universe were turbulent and unstable. While the galaxy, does not appear to have spiral arms, it has at least two features typical of our Milky Way: a rotating disc and a bulge, the large group of stars packed tightly around the galactic center. This is the first time a bulge has been seen this early in the history of the Universe, making SPT0418-47 *the most distant Milky Way look-alike*. Observations with NASA's James Webb Space Telescope (JWST) have revealed the presence of a companion galaxy, which may indicate that SPT0418-47 is a merging system of galaxies. In June 2023 it was announced the detection of organic molecules in SPT0418-47 using the James Webb telescope.

2009: Many Gravitational Lenses in the Field of the Galaxy Cluster Abell 370

NASA's Hubble Space Telescope captured this image in Fig. 5.55, of the galaxy cluster Abell 370 in visible and near-infrared light. Abell 370 contains several hundred galaxies tied together by their mutual pull of gravity. It was one of the first clusters in which we observed gravitational lensing. Entangled among the galaxies are arcs of blue light. These are actually distorted images of remote galaxies behind the cluster. These far-flung galaxies are too faint for Hubble to see directly. Instead, the gravity of the cluster acts as a huge lens in space, magnifying and stretching images of background galaxies like a funhouse mirror. Abell 370 was the last of six galaxy clusters imaged by the Frontier Fields project, which used the combined power of natural gravitational lenses caused by massive galaxy clusters and Hubble's ability to create long-exposure deep field images to see galaxies that would normally be far beyond the reach of telescopes. One of the most distinctive galaxies magnified by the cluster is nicknamed *the Dragon* (lower left of center). It is an extended feature that probably consists of several duplicated images of a single background spiral galaxy stretched along an arc. Abell 370 is located approximately 4 billion light years away in the constellation Cetus, the Sea Monster.

Fig. 5.55 Abell 370 appears to include several arcs of light, including the largest ever discovered, 30″ long. These arcs or deformations are mirages caused by gravitational lensing of distant galaxies by the massive galaxy cluster located between the observer and the magnified galaxies. Nearly a hundred distant galaxies behind this cluster have multiple images caused by the lensing effect. (*Credits* NASA, ESA, and J. Lotz and the HFF Team (STScI))

2022: Multiple Light Paths of a Single, Lensed Supernova in Abell 370

Three different moments in a far-off supernova explosion were captured in a single snapshot by the NASA/ESA Hubble Space Telescope in this group of galaxies. The progenitor star exploded more than 11 billion years ago, when the Universe was less than a fifth of its current age of 13.8 billion years. This is the first detailed look at a supernova at such an early time in the Universe's evolution. This observation was possible via gravitational lensing, and the light took three different paths (all in one picture!) through the cosmic lens of the massive galaxy cluster Abell 370, located nearly 5 billion light years away from Earth (at redshift $z = 0.375$), in the constellation of Cetus. Its core is made up of several hundred galaxies bending and magnifying the light from the more distant supernova located behind the cluster (Fig. 5.56). The three paths were of three different lengths, so when the light arrived at Hubble (on the same day in December 2010), the supernova appeared at three different stages of evolution. The Hubble also captured the fading supernova's rapid change of color, which indicates its changing temperature. The bluer the color, the hotter the supernova is. The earliest phase captured appears blue. As the supernova cooled its light turned redder.

Fig. 5.56 The top box shows a portion of Abell 370. The immense gravity of this galaxy cluster acted as a cosmic lens, bending and magnifying the light from the more distant supernova located behind the cluster. The box-within-the-box marks the area where the distant supernova was multiply lensed. The bottom image is a magnified version of this area with the light paths marked for the three images of the supernova. The right side of the bottom image shows the distant galaxy in which the supernova exploded. The lines show how the light travelled through the gravitational lens, with some of the light taking longer routes across *valleys* of warped space. (*Credit* NASA, ESA, CSA, Ivo Labbe (Swinburne), Rachel Bezanson (University of Pittsburgh), *Image Processing* Alyssa Pagan (STScI))

2023: Abell 2744 (Pandora's Cluster) with the James Webb Space Telescope, a Huge Gravitational Lens

Abell 2744 (nicknamed Pandora's Cluster) is a giant pile-up of smaller galaxy clusters, perhaps the result of the collision of several separate galaxy clusters. This megacluster is about 4 million light years across, has the mass of 4 trillion suns and is located at a distance of about 4 billion light years away containing several hundred galaxies and might be a pile-up of at least four smaller galaxy clusters. The galaxies in the cluster make up less than 5% of its mass, while the gas is around 20% and is so hot that it shines only in X-rays. The distribution of invisible *dark matter* makes up around 75% of the cluster's mass.

This object has been nicknamed Pandora's[3] Cluster because of the many different and strange phenomena that were unleashed by a huge collision that occurred over a period of 350 million years.

The cluster is so massive that its powerful gravity bends the light from galaxies far behind it, making the background objects appear larger and brighter via gravitational lensing. In other words, Abell 2744's strong gravitational field acts as a lens, brightening and magnifying the light of nearly 3,000 distant background galaxies. These lensed sources appear red in the image of Fig. 5.57b, and often as elongated arcs distorted by the gravitational lens. Many of these are galaxies from the early Universe, with their contents magnified and stretched out for astronomers to study.

In Fig. 5.57, it is shown a crowded galaxy field on a black background, with one large star dominating the image just right of center. Three areas are concentrated with larger white hazy blobs on the left, lower right, and upper right above the single star. Scattered between these areas are many smaller sources of light; some also have a hazy white glow, while many other are red or orange.

To Err Is Human: Even the Eminent Astrophysicists, Eddington, and Einstein Made Some Wrong Scientific Predictions!

As we have seen, Eddington (Fig. 5.58a), the greatest astrophysicist of the early twentieth century, the first to observe the phenomenon of the bending of light rays as they pass near the Sun, was wrong in his prediction that the final stage of evolution of high mass stars cannot be *neutron stars* or, *black holes.* As we mention previously, he never accepted the research findings of his PhD student S. Chandrasekhar, who predicted *black holes,* the superdense final stages of a star's evolution.

It Was His Big Mistake!

Einstein (Fig. 5.58b) had made the first unpublished calculations of the gravitational bending of light, as early as 1912. However, in his first relevant

[3] According to Greek mythology, in the very beginning Earth was free from toil and misery, the land was covered with flowers and Earth was inhabited only by men, who had been created by *Prometheus.* He made them of clay and modeled them after the gods, something however which angered *Zeus.* When *Prometheus* offended *Zeus* again by stealing fire from heaven to give to man, Zeus exacted revenge and ordered *Hephaestus*, to create *Pandora*, the first woman. The gods gave her many traits including beauty, curiosity, charm, and cleverness. Hence her name *Pandora*, meaning "*all gifted*," or, alternately, *a gift to all.* However, before putting *Pandora* on Earth, *Zeus* handed her a beautiful box saying, *This is my own special gift to you. Don't ever open it.* As *Zeus* anticipated, *Pandora*'s curiosity got the best of her, and she opened the box, ending earthly paradise. From the small chest flew troubles and woes – sorrow, disease, vice, violence, greed, madness, old age, death – to plague humankind forever. Nevertheless, *Zeus* did not realize that *hope* had been secretly added to the box by *Promethesus.* When *Pandora* opened the box and released trouble and woe into the world, hope was there to help people survive. (Above description accompanies the painting *Pandora (inset)*, by French painter Odilon Redon, 1840–1916, 1910/1912 oil on canvas, 143.5 × 62.9 cm, at the National Gallery of Art, Washington DC, USA).

Fig. 5.57 **a** The painting *Pandora* by French painter Odilon Redon, 1840–1916, 1910/1912, oil on canvas, 1430.5 × 62.9 cm, at the National Gallery of Art, Washington DC, USA **b** An estimated 50,000 sources of near-infrared light are represented in this image from the NASA/ESA/CSA James Webb Space Telescope. Their light has travelled through various distances to reach the telescope's detectors, representing the vastness of space in a single image. A foreground star in our own galaxy, to the right of the image center, displays Webb's distinctive diffraction spikes. Bright white sources surrounded by a hazy glow are the galaxies of Pandora's Cluster, a conglomeration of already-massive clusters of galaxies coming together to form a *mega cluster*. The concentration of mass is so great that the fabric of spacetime is warped by gravity, creating a natural, super-magnifying glass, a *gravitational lens* that can be used to see very distant sources of light beyond the cluster that would otherwise be undetectable, even to Webb. (*Credit* NASA, ESA, CSA, I. Labbe (Swinburne University of Technology), R. Bezanson (University of Pittsburgh), A. Pagan (STScI))

publication in 1936 in the journal *Science*, considering the phenomenon of the bending of light rays from a star, he predicted that the effect of gravitational lenses would *not* be observable:

> Of course, we do not hope to observe this phenomenon immediately. In principle, it is highly unlikely that the line connecting the observer to the lens coincides with the line connecting the lens to the source. Second, the angle θ_1 (Fig. 5.50) is so small that it is beyond the resolution of our telescopes. (A. Einstein, Science, vol 84, p. 506, 1936)

And yet today, more than 100 gravitational lenses are observed!

Additionally, Einstein insisted on a static Universe and invented the cosmological constant to achieve that goal, forcing thus the equations to fit his incorrect assumptions, and missing the expanding Universe. When it became clear that the Universe wasn't actually static, but was expanding instead, Einstein abandoned the cosmological constant, calling it the *biggest blunder* of his life.

Furthermore, as we shall discuss later on, Einstein predicted the existence of gravitational waves (1916), as a consequence of his theory of General Relativity. But he calculated that they create ripples in spacetime of such infinitesimal width that they are probably impossible to observe. Yet, they are indeed observed today!

All These Were His Rather Serious Scientific Mistakes!

Errare humanum est, sed in errare perseverare diabolicum. (To err is human, but to persist in error is diabolical, **Seneca**).

The falling is human, to repeat the same error again *is Luciferian, but to repent is divine!* (**St. John Chrysostom**).

*Black Hole*s at the Centers of Distant Galaxies and Our Own Milky Way

1990: Supermassive Black Holes *at the Centers of Many Galaxies*

With the launch of the Hubble Space Telescope (1990), its observations led to the conclusion that almost every large galaxy, including our Milky Way, has a supermassive *black hole* at its center, with some typical examples already shown. To date, however, we must admit that we do not fully understand the formation of these very massive, hundreds of thousands to billions of solar masses – *supermassive* as they are called – *black hole*s. By analogy with stellar *black hole*s, which we understand to be the result of the collapse of massive stars, it has been suggested that these supermassive *black hole*s were similarly formed by the collapse of giant gas clouds during the early phase of galaxy formation. Another theory suggests that as a stellar-mass *black hole* absorbs matter, over millions of years, its size increases to that of

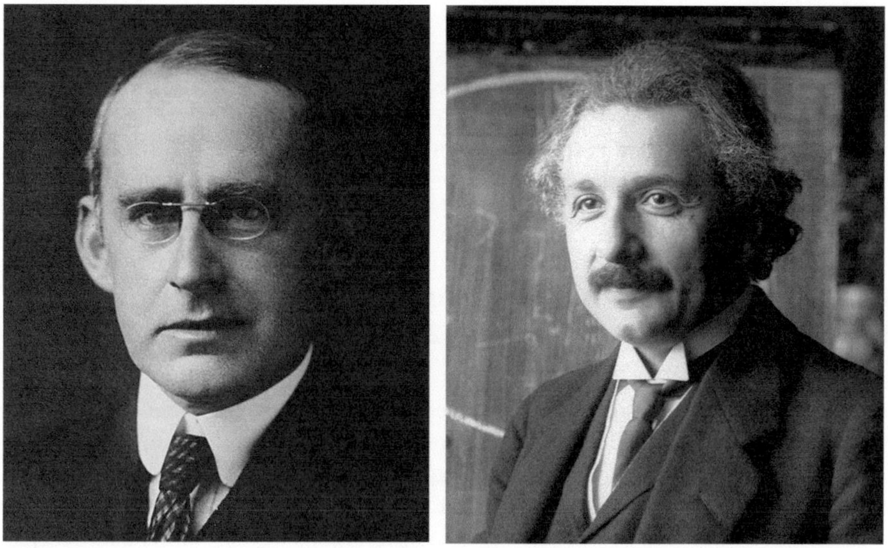

Fig. 5.58 (*left*) Arthur Eddington (from the collection of the Library of Congress) and Albert Einstein (*right*) during a lecture in Vienna in 1921 (age 42) (by the portraitist Ferdinand Schmutzer (1870–1928))

a supermassive *black hole*. Finally, a third theoretical proposal calls for a cluster of stellar *black holes* to gradually merge, forming a supermassive *black hole*.

Whatever is the mechanism of their creation, however, we all agree that the accretion of matter onto a supermassive *black hole* must be the ultimate cause of the observed wide spectrum of activity in AGN and *quasars*, including the creation of *jets* from them. In 1994, the Hubble's Faint Object Spectrograph, observing the ionized matter rotating around the center of the large elliptical galaxy M87, at a distance of about 55 million l.y. detected through the Doppler effect rotational velocities of $\pm$ 500 km per second. This fact is compatible with the existence of a massive *black hole* at the center of M87 of a few billion solar masses (6.5×10^9 $M_\odot$). The jet emerges from a disk of superheated gas (plasma) swirling around this black hole and is focused by the plasma's helical magnetic fields, accelerating to speeds approaching the speed of light. Also, radio data with the Very Long Baseline Array (VLBA), with the observation of the radio emission from the center of the galaxy M106, Miyoshi et al. (1995), lead to the conclusion that the radio emission originates from rotating gas clouds around a central mass of millions of solar masses (3.6×10^7 M_o), which is concentrated in a radius of about 0.4 light years. This study concluded that a solar-sized cluster of *black holes* orbiting such a concentration of mass could not survive without repeated collisions, ultimately resulting in the formation of a single supermassive *black hole* as the only final product. The Andromeda galaxy (M31), contains about 1 trillion stars and its center hosts a black hole of 100–200 million solar masses. The Seyfert spiral galaxy NGC 4151, at a distance of 45 million AU, hosts one of the brightest X-ray AGN and a central black hole of about 50 million solar masses and a well collimated jet. The center of the quasar 3C273 at a distance of 2.5 billion l.y. yet one of the closest quasars to Earth, hosts a supermassive black hole which has a mass of 300 million solar masses (with a Schwarzschild radius of about 20 AU) and a jet. Finally, on April 10, 2019, the Event Horizon Telescope research collaboration released the first *photograph* of the silouette of a supermassive *black hole* at the center of the galaxy M87, as we discuss leter on in this section.

1995: Discovery of a **Black Hole** *at the Center of Our Milky Way Galaxy (Physics Nobel Prize 2020)*

Already in 1931, Karl Jansky had detected radio emission from the center of our Milky Way Galaxy. In 1974 Balick and Brown using the American radio telescopes of the National Radio Astronomy Observatory (NRAO) discovered a bright and very compact radio source in the center of the Galaxy, Sgr A*, at a distance of about 27,000 light years, near the border of the constellations of Sagittarius and Scorpio. After 1995, systematic observation of the orbits of 50 stars around Sgr A* began with the 3.5 m New Technology Telescope and the 8.2 mVery Large Telescope (VLT) of the European Southern Observatory (ESO) in Chile, by the Max Planck Institute for Extraterrestrial Physics research team and the UCLA research group in California.

The UCLA research group in California has been monitoring the orbits of thousands of stars orbiting the galactic center in recent years. In particular, the orbit of two stars has been systematically observed: the star S0-2 with an orbital period of 11.56 years, and the star S0-102, with a period of 11.5 years, Figs. 5.59 and 5.60. At its closest approach to the galactic center the star S0-2 is only 17 light hours from the galactic center, i.e. about 4 times the distance of the planet Neptune from the Sun. In the period 1995–2016, the star S0-2 with a mass of about 14 solar masses ($M_\odot$) completed a fully elliptical orbit with a semi-major axis a = 970 astronomical units (AU), a period of about 16 years, and a perigee distance of 120 AU ~ 400 Schwarzschild radii. From this orbital data we can calculate the central mass contained in this stellar orbit and the M_{BH} mass of Sgr A*, $M_{BH} \sim 4.31 \times 10^6\ M_\odot$, at a distance from Earth of about 27,000 light years. For comparison, the Schwarzschild radius of this *black hole* is about 0.08 astronomical units (AU) ~ 17 solar radii. Another star, S0-120 follows an elliptical orbit with a period T ~ 11.6 years and closest distance from the center equal to 260 AU ~ 3250 Schwarzschild radii. The infrared adaptive optics method was used for these observations.

Thus, the observations have led to the conclusion that, beyond all doubt, the Sagittarius A* radio source at the Galactic center is a massive *black hole* with a mass of about 4 million solar masses, and a corresponding Schwarzschild radius of about 17 solar radii. For comparison, the orbit of the planet Mercury has a radius equal to about 83 solar radii.

These observations are valuable since, due to its proximity in comparison to other distant galaxies, the galactic center allows us direct observations of stellar orbits and thus an understanding of the physics of massive *black hole*s and their role in the formation and evolution of galaxies (Fig. 5.61).

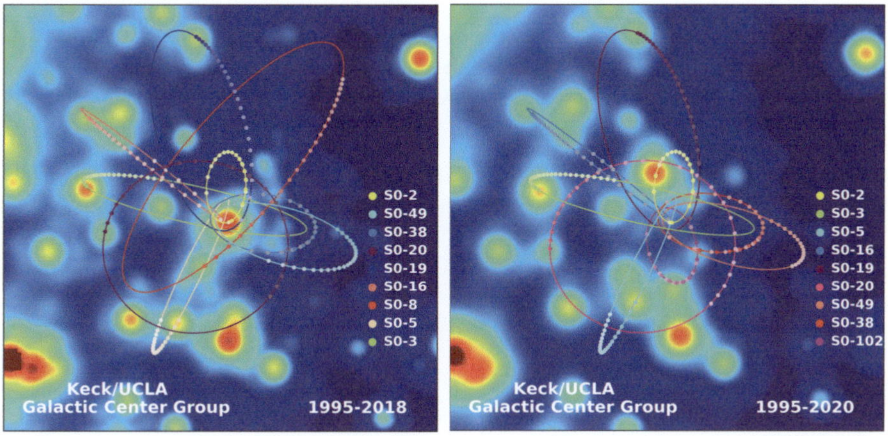

Fig. 5.59 Graphical representation of the orbits of 9 stars around the Galactic center for the periods 1995–2018 and 1995–2020, superimposed on the observed 1 × 1 arcsecond angular diameter image. In the left figure S0-8 is shown, while in the right figure, instead of S0-8, S0-102 is included. (*Source* UCLA Galactic Center Group – WM Keck Observatory Laser Team, reproduced with the kind permission of Dr Shoko Sakai)

Black Holes and the Dark Secret of Our Milky Way Galaxy …

Three scientists shared the Nobel Prize in Physics of 2020 for their discoveries about one of the most exotic objects in the Universe, black holes. In January 1965, ten years after the death of Einstein, Roger Penrose (University of Oxford, UK) proved, using clever mathematical methods, that black holes are a direct consequence of the General Theory of Relativity (GTR). It is noteworthy that Einstein believed that in reality such monsters do no actually exist, i.e., objects that adsorb everything in their neighborhood and nothing can escape them, not even light. Nevertheless, it seems that black holes hide at their center an singularity, where all known laws of physics cease to apply. Penrose's seminal article (Penrose 1965) is still considered to this day to be the most important contribution to GTR. Furthermore, Penrose's *singularity theorem*, contributed to the renaissance of the theory. It is also noteworthy that this same year 1965 we had the discovery and interpretation of the Cosmic Background Radiation and the first modern *singularity theorem*, 50 years after Einstein proposed field's equations (1915).

At the observational front, Reinhard Genzel and Andrea Ghez, each leading a group of astronomers (Max Planck Institute for Extraterrestrial Physics, Garching, Germany and California Institute of Technology, Pasadena, USA), focused since the early 1990s on a region in the center of our galaxy called Sagittarius A*. Using the world's largest telescopes (ESO's 4 eight meter VLT telescopes in Chile and the two 10-m Keck telescopes in Hawaii), they developed new methods to see through the vast and dense clouds of interstellar

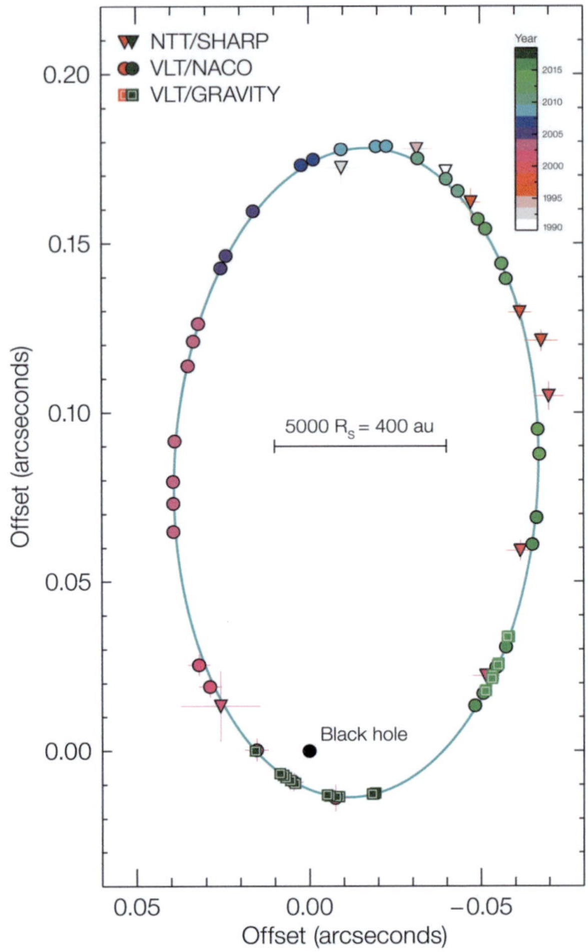

Fig. 5.60 Diagram showing the motion of the star S0-2 around the supermassive black hole at the center of the Milky Way. It was compiled from observations with the ESO telescopes and instruments over a period of more than 25 years. The star takes 16 years to complete one orbit and was very close to the black hole in May 2018. Note that the sizes of the black hole and the star are not to scale. (*Credit* ESO/MPE/GRAVITY Collaboration)

gas and dust at the center of our galaxy. Pushing the limits of technology, they refined new techniques to compensate for image distortions caused by Earth's atmosphere and, using pioneering instruments, in a long-term survey, mapped with increasing precision the orbits of the brightest stars closest to the Milky Way center. The two teams' independent observations agree that an extremely heavy, invisible object pulls on its neighboring stars, causing them to spin around it at speeds of thousands of kilometers per second, speeds which are truly dizzying. It is estimated that about four million solar masses are *crammed* into an area much smaller than our inner solar system. Their ground-breaking work has given us the most convincing evidence for the existence of a supermassive *black hole* at the center of our Milky Way galaxy.

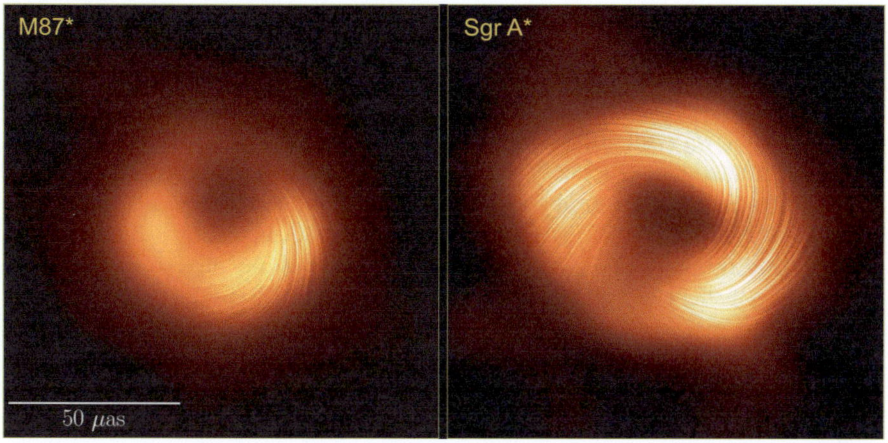

Fig. 5.61 The supermassive black holes in M87 and Sagittarius A, are shown here in polarized light, indicating that these beasts have similar magnetic field structures. This significant comparison suggests that the physical processes that govern how a black hole feeds and launches a jet may be universal features amongst supermassive black holes. The scale in M87 shows the apparent size on the sky for the image, in micro-arcseconds. *A finger held at arm's length measures 1° on the sky; an angle of one micro-arcsecond is 3.6 billion times smaller than that. In context, the images of these black holes have an apparent size similar to that of a donut, or a credit card, on the surface of the Moon.* (*Credit* EHT Collaboration)

2019: First Photograph of the Silhouette of a Supermassive Galactic Black Hole

As it has been anticipated for more than a century (1918), at the center of the giant elliptical galaxy M87 – in the Virgo cluster and at a distance of about 55 million light years from Earth – lurks a huge *black hole*, with an estimated mass of about 6.5 billion solar masses.

A Galaxy that "Fits" Inside the Size of a Small Planetary System …

The radius of the galaxy M87 is about 500,000 light years, but the horizon of its central black hole of 6.5 billion solar masses has a (Schwarzschild) radius of only R_{Sch} ~ 130 AU ~ 0.002 l.y. ~ 20 billion km. That is, the 500,000 light year radius of the galaxy M87 is about 250 million times larger than the radius of the black hole it contains at its center!
In other words, if the 6.5 billion solar mass black hole at the center of M87 were the size of a small ping-pong ball, the M87 galaxy, with a total mass of about 2.4 trillion solar masses, would have a radius of 2,500 km (about as long as the distance between Athens and Paris)!

Thence, there is a huge concentration of mass – 6.5 billion solar masses, about the size of a small galaxy – confined to the (relatively) tiny size of a planetary system (130 astronomical units).
In addition, a very hot, million-degree disk of gas and dust has been observed swirling around the center of M87. In its inner regions, the rotational speed of the disc is close to the speed of light. The absorption disc extends outward to about 25,000 AU, i.e., about 0.4 light years, but its inner radius is only a few Schwarzschild radii, (2–3) r_{Sch}, Fig. 5.62a. The gas of the disc superheats as it approaches the horizon of the black hole. But before it finally sinks into the abyss of the black hole, from where of course there is no return, its radiation forms a luminous ring, due to the gravitational curvature of the light rays. The final image is the one shown in Figs. 5.62a and 5.63.

But how can we understand the famous picture 5.63 *of the BH* silhouette*?*

First of all, the radius of the central dark region in Fig. 5.63 does not coincide exactly with the radius of the *black hole*'s (BH) event horizon, i.e., the Schwarzschild radius (r_{Sch} ~ 130 AU). *Why*; This is because there is a radius greater than the radius of the *black hole*'s event horizon within which the particle trajectories around the BH are *unstable.* This is the radius of the so-called *last stable orbit*, which for a non-rotating *black hole* is three times the Schwarzschild radius, $3r_{Sch}$, Figs. 5.62a and 5.63b. Matter orbiting inside this radius falls into the *black hole* and is lost forever.

But is there anything else that can rotate around the BH at radii smaller than 3 Schwarzschild radii? The answer is *yes*! Photons, having zero mass, can rotate outside the so-called *photon sphere*, which has a radius one and a half times the Schwarzschild radius (1.5 r_{Sch}, Fig. 5.62a). In other words, outside this distance the light is not absorbed by the BH. It is in that region where, for example, one can see the back of his/her head (!), as the photons there follow extremely curved trajectories. But this *photon sphere* is unstable, in the sense that from there the photons will either spiral towards the center of the *black hole*, or spiral away from it, Fig. 5.62b.

Finally, what does the radius of the black area in the picture 5.63 correspond to? Is it the radius of the *black hole* horizon, r_{Sch}, the radius of the photon sphere, 1.5 r_{Sch}, or, is it the radius of the last stable orbit (Fig. 5.64a)? *None of them*! Matters are complicated by the fact that light rays approaching the *black hole* region are bent and follow curved paths, as shown in Fig. 5.62b.

So now that we know how all these elements play a role in a *black hole*, we can answer the question:

In summary, what exactly do we see in Fig. 5.63 *(see also explanatory Fig.* 5.64*)?*

As Einstein's General Theory of Relativity predicts, the *black hole* will curve the space around it and change the path which the light travels. So, imagine that we send the parallel light rays of a small diameter beam from far away (infinity) toward the *black hole*. If this stream of photons were directed toward the center of the *black hole*, the BH would absorb the light and nothing would return. Then imagine that

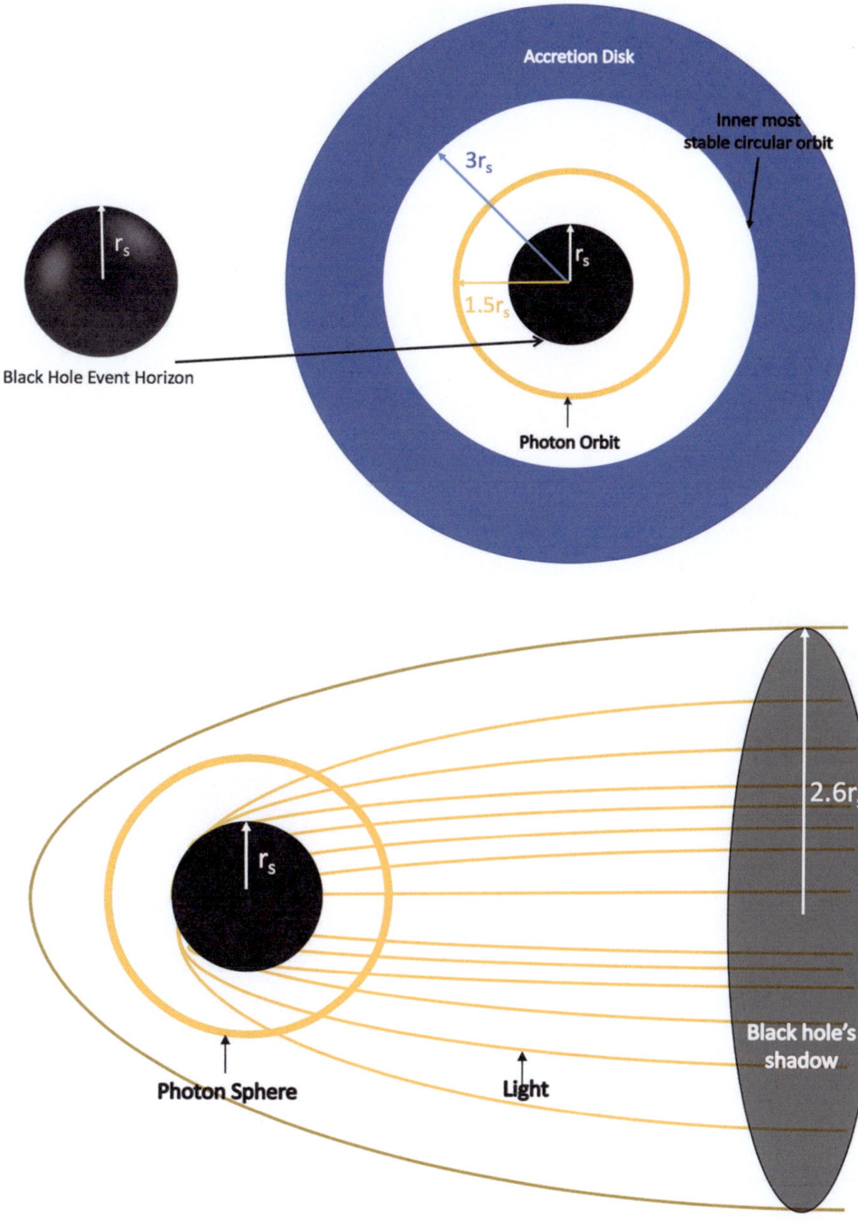

Fig. 5.62 (upper part) The spherical black hole (BH) horizon (r_S), the photon sphere ($1.5r_S$) and the orbit of the smallest stable radius ($3r_S$) around a non-rotating Schwarzschild-type black hole. (lower part) The radius of the circular region of the shadow of a black hole is about 2.6 Schwarzschild radii, assuming zero rotation of the BH. In this case, the light would be bent as seen from the penultimate darkest light ray and we could *catch* with our telescopes (at *infinity*), reproducing thus the image of the EHT (Adapted from image of the EHT Collaboration)

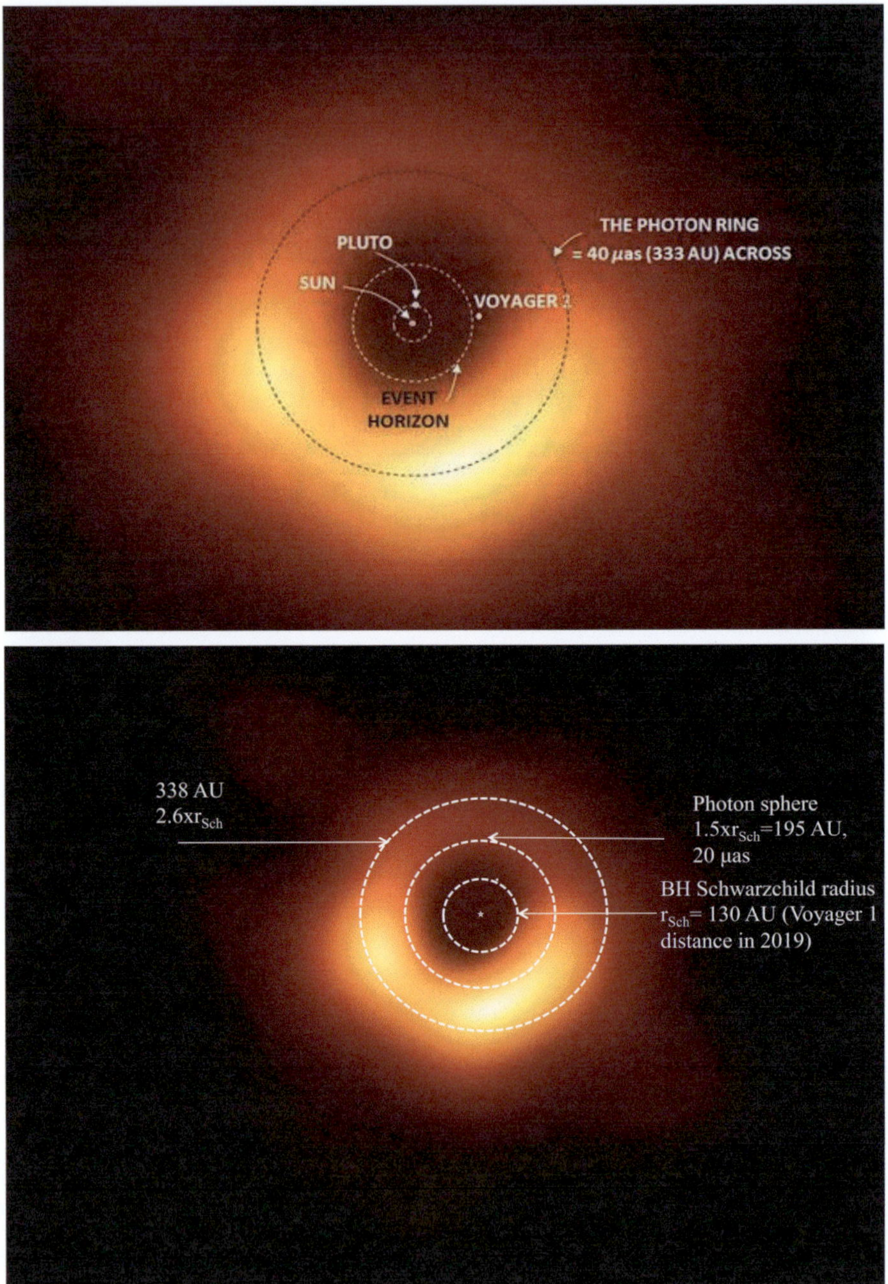

Fig. 5.63 The dark silhouette of the black hole at the center of the elliptical galaxy M87, with the Event Horizon Telescope. The event horizon of this black hole has a Schwarzschild radius r_{Sch} = 130 AU, about the distance the Voyager 1 spacecraft was in 2019. The dark silhouette is about 2.6 times the diameter of the black hole's horizon, i.e., about 40 billion kilometers, assuming zero rotation of the BH. The absorption disk extends out to 25,000 AU, ~0.4 light years, but its inner radius is a few Schwarzschild radii, (2–3) r_{Sch}. *Source* The EHT Collaboration, ESO

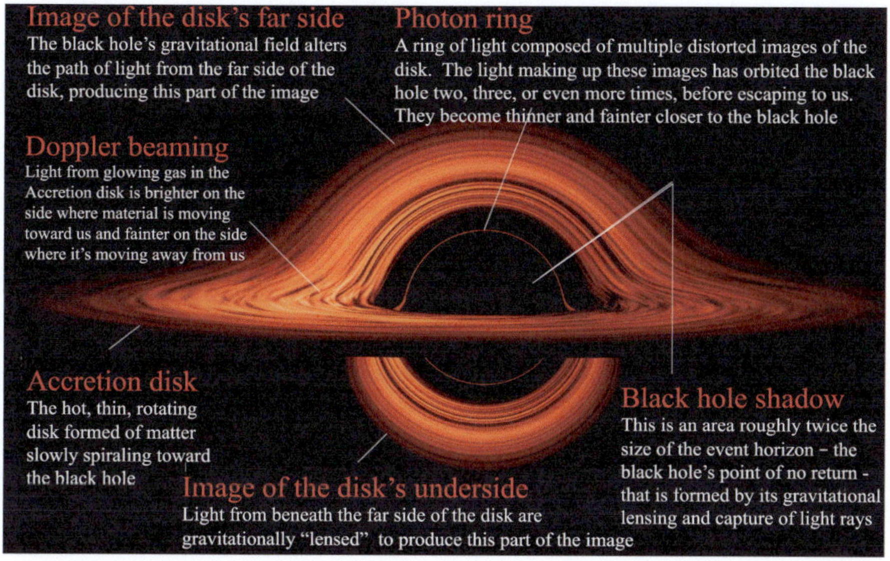

Fig. 5.64 This image explains various aspects of the black hole visualization. (*Image credit* NASA's Goddard Space Flight Center/Jeremy Schnittman)

instead, our light beam had a diameter slightly larger than the diameter of the *black hole*'s horizon. Then the light would be lost back into the *black hole* as it was bent by the BH, Fig. 5.62b. Only when the photons finally pass *right* past the circumference of the photon sphere (1.5 r_{Sch}) where they are bent by the BH field, could the light escape the gravitational pull of the *black hole* and return to the source (at infinity). It turns out that in this case, if the light beam coming from infinity has a radius of 2.6 r_{Sch}, the photons after making a U-turn will return to infinity again, as shown in Fig. 5.62b. Only then we will really start to *see* the silhouette of the *black hole*, as a dark disc of radius about 2.6 R_{Sch}. Then, it remains to answer one more question:

Why does half of the ring in images 5.63, 5.64 *appear brighter than the other half?*

This is due to the relativistic phenomenon *Doppler beaming* and is due to the fact that we do not observe the *black hole* at the center of M87 in a line parallel to the axis of the accretion disk (i.e., *face on*). If we were indeed observing M87 along a line perpendicular to the axis of the accretion disk (i.e., *edge on*), then the light from the upper part of the back part of the disk is bent before coming towards us. Similarly, the light from the lower part of the back part of the disk is curved before coming towards us. The same applies to the light coming from the top and bottom of the front part of the disk, which eventually give us a bright ring around the *black hole*. The angle between the observation direction and the disk axis which also coincides with the *jet* direction is $i > 17°$. The light coming from the part of the disk approaching us is amplified due to the increase in radiation intensity due to *Doppler beaming*. That explains why half of the ring in image 5.63 is brighter than the other.

In other words, viewed from the side, the disk looks brighter on the left than it does on the right. Glowing gas on the left side of the disk moves toward us so fast that the effects of Einstein's relativity give it a boost in brightness; the opposite happens on the right side, where gas moving away us becomes slightly dimmer. This asymmetry disappears when we see the disk exactly face on because, from that perspective, none of the material is moving along our line of sight.

Hence, seen nearly edgewise, the turbulent disk of gas churning around the *black hole* takes on a double-humped appearance. The *black hole*'s extreme gravity alters the paths of light coming from different parts of the disk, producing the warped image.

The silhouette of the *black hole* seen in Fig. 5.63 is the best we can see today emanating from the *black hole* itself. The boundary of the BH – the event horizon from which the EHT (Event Horizon Telescope) takes its name – is about 2.5 times smaller than the shadow we see in Fig. 5.63 and is less than 40 billion kilometers in diameter. Although this dimension may sound large, this ring has an angular diameter, as seen from Earth, of only about 40 μs of arc – equivalent to the angle at which we would see a credit card on the surface of the Moon!

What Is Actually This Silhouette of the *Black Hole* at the Center of M87?

In summary, the black hole image consists of an asymmetrically bright ring, which is formed by the bending of light rays caused by the black hole's strong gravitational field. It is recalled that in the environment of black holes light bends strongly and so, for example, one can see his/her back. The image is asymmetric due to light amplification by the part of the absorption disc moving in the direction of the observer and is in full agreement with the predictions of General Relativity. This image is the strongest evidence to date for the existence of massive *black holes* and opens a new window into the study of *black holes*, their *event horizons* and the theory of Gravity in general.
The observations were made with the *Event Horizon Telescope* (EHT), a combination of 8 ground-based radio telescopes that is equivalent to a radio telescope which has the equivalent size of the Earth's in diameter. EHT operating within the framework of Very Long Baseline Interferometry (VLBI) and at a wavelength of 1.3 mm, achieves a resolution of 20 μarcsec. This resolution is sufficient to enable one to read a newspaper in New York from a cafe in Paris, or to see a ping-pong ball on the surface of the Moon.
Although the 8 radio telescopes are not connected to each other, their observations are synchronized through individual clocks– hydrogen MASERS – which record with great precision the time when their observations are made. Each telescope produces a huge amount of data – about 0.5 petabytes per day – which is stored on liquid helium-cooled hard drives.

1 petabyte equals one thousand trillion bytes. It is estimated that the capacity of the human brain to store data in its memory is equivalent to about 2.5 petabytes of binary data. The LHC at CERN stores about 15 petabytes annually.
These disks are then transferred to specialized computers – known as *correlators* – at the Max Planck Institute for Radio Astronomy in Bonn and MIT's Haystack Observatory in Massachusetts, where the data are combined. Finally, the data are transformed into real images, using innovative algorithmic techniques developed by EHT's research team, which now numbers over 200 researchers.
The EHT can't merely take a picture of a black hole. The silhouette is about 40 microarcseconds wide, the size of a hydrogen atom seen at arm's length. To resolve that, you need a planet-size radio telescope – which of course doesn't exist. So instead, scientists piece together data taken at different observatories spread across the globe, carefully reconstructing the shadow from sparse information using complex computer algorithms. It's as though you were given the height, ear size, and skin texture of an elephant and then had to figure out exactly what the creature actually looks like!
The 8 radio telescopes of the Event Horizon Telescope Collaboration are: SPT (South Pole Telescope), ALMA, APEX (Atacama Pathfinder Experiment), JCMT (James Clerk Maxwell Telescope), SMT (Submillimeter Telescope), LMT (Large Millimeter Telescope), SMA (Submillimeter Array) and PV (Pico Valetta IRAM 30 m telescope).

2022: Comparison of the Sizes of Two **Black Holes***: M87 and Sagittarius A**

Figure 5.65 shows the scale of Sgr A^* in comparison with both M87 and other elements of the Solar System such as the orbits of Pluto and Mercury. Also displayed is the Sun's diameter and the current location of the Voyager 1 space probe, the furthest spacecraft from Earth. M87, which lies 55 million light years away, contains one of the largest black holes known. While Sgr A^*, 27,000 light years away, contains a black hole with a mass roughly four million times the solar mass, i.e., the M87 black hole is more than 1000 times more massive. Because of their relative distances from Earth, both black holes appear of the same size in the sky.

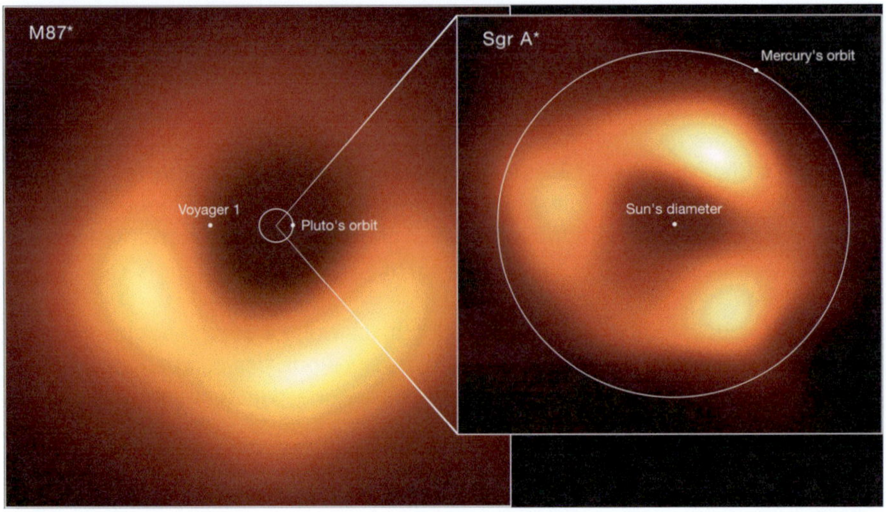

Fig. 5.65 Size comparison of the two black holes imaged by the Event Horizon Telescope (EHT) Collaboration: M87, at the heart of the galaxy Messier 87, and Sagittarius A* (Sgr A*), at the center of the Milky Way. *Credit* EHT collaboration, Lia Medeiros, ESO

2022: Most Distant Detection of a Black Hole *Swallowing a Star and Emitting Towards Us a Relativistic* Jet

In February 2022 the Zwicky Transient Facility collaboration observed a truly rare cosmic *lunch* of a supermassive *black hole* devouring a nearby star and expelling the leftovers in a relativistic *jet*, emitting multiwavelength radiation in this process (Fig. 5.66). Named AT2022cmc, this discovery was made while sieving through the ZTF survey data using a novel method designed to alert astronomers for such rare events in near real-time. The prospect of witnessing this rare phenomenon prompted observers to trigger in total 21 telescopes from across the globe to observe this mystery source in more detail, including ESO's VLT, which quickly observed this new event. Immediately, the VLT data placed the source at an unprecedented distance for such events, i.e., the light produced from AT2022cmc began its journey when the Universe was about one third of its current age ($z \sim 1.2$), such that AT2022cmc is at the cosmological distance of 8.5 billion light years away. A wide variety of light, from high energy γ- and X-rays (NuSTAR hard X-ray), as well as soft X-ray observations obtained by NICER, Swift, and XMM-Newton to radio and millimeter waves, was collected by these 21 telescopes around the world, the unique light curve of which transitioned into a luminous plateau within days (Andreoni et al. 2022). By modeling the spectral energy distribution evolution from radio to hard X-ray with NuSTAR, it is found that the sub-millimeter/radio emission originates from external shocks at large distances $\gtrsim 10^{17}$ cm from the *black hole*, the UV/optical light comes from a thermal envelope with radius $\sim 10^{15}$ cm, and the X-ray emission is consistent with

Fig. 5.66 Artist's impression illustrating how it might look when a star approaches (dangerously!) too close to a black hole; the star is squeezed by the intense gravitational pull of the black hole, such that the star's material gets pulled in and swirls around the black hole forming a disc. In rare cases, such as this one, *jets* of matter and radiation are shot out from the poles of the *black hole*. In the case of the AT2022cmc event, evidence of the *jets* was detected by various telescopes including the VLT, which determined this was the most distant example of such an event. (*Credit* ESO/M. Kornmesser)

synchrotron radiation powered by energy dissipation at intermediate radii within the (likely magnetically dominated) *jet* (Yao et al. 2023).

Theoretically, what we expect (with a morphology somehow similar to the fireball model of γ-ray bursts, Fig. 5.31) is the following: when a star crosses the tidal radius of a supermassive *black hole*, it will be spectacularly ripped apart, emitting a relativistic *jet* and an accompanying burst of radiation along the *black hole*'s spin axis. If this relativistic *jet* is pointing at us, it makes the event much brighter than it would otherwise appear, due to Doppler beaming. The *jet* is visible then over a broad span of the electromagnetic spectrum: IR/optical, radio/millimeter (from the material in the shell formed between the reverse and optical shock at about 0.1 pc from the BH), X- and γ-rays beamed from the *jet* itself, as well optical/UV from the BH disk which extends about 100 AU around the BH. So, the observations by ESO's VLT and the other 20 telescopes concluded that a supermassive *black hole* in a distant galaxy had indeed devoured a star. The VLT determined it to be the furthest example of such an event to have ever been observed. Because the *jet* is pointing almost toward us, this is also the first time it has been discovered with visible light, providing a new way of detecting these Tidal Disruption Events (TDEs). Approximately 1% of such TDEs cause *jets* of plasma and radiation to be ejected from the poles of the rotating *black hole*. Only a few tens of such tidal disruption events have now been identified in optical wavelengths, but the exact origin of the strong optical emission remains still inconclusive (Fig. 5.66).

2024: Hubble Finds Strong Evidence for an Intermediate-Mass **Black Hole** *in Ω Centauri*

*Black hole*s have been found over a wide range of masses, from stellar remnants with masses of 5–150 solar masses ($M_\odot$), to those found at the centres of galaxies with $M > 10^5\ M_\odot$. However, only a few debated candidate *black hole*s exist between 150 $M_\odot$ and $10^5\ M_\odot$.

The global cluster ω Centauri is visible from Earth with the naked eye in the southern hemisphere, and it is 17,700 light years away, lying just above the plane of the Milky Way and appearing almost as large as the full Moon when seen from a dark rural area. The global cluster ω Centauri was first listed in Ptolemy's catalogue nearly two thousand years ago as a single star. Edmond Halley reported it as a nebula in 1677, and in the 1830s John Herschel was the first to recognise it as a globular cluster.

The global cluster ω Centauri consists of roughly 10 million stars that are gravitationally bound. An enormous catalogue of the velocities for 1.4 million stars has been created over the years, by studying over 500 Hubble images of the cluster. This extensive catalogue, is the largest catalogue of motions for any star cluster to date. In a recent study Häberle et al. (2024), observations have been reported of seven fast-moving stars in the central 3 arcsec (0.08 pc) of ω Centauri, the most massive globular cluster of the Milky Way (Fig. 5.67). The velocities of the fast-moving stars are significantly higher than the expected central escape velocity of the star cluster, so their presence can be explained only by being bound to a massive *black hole*. From the velocities alone, a firm lower limit of the *black hole* mass of about 8,200 $M_\odot$ is infered, making this a good case for an intermediate-mass *black hole* in the local universe.

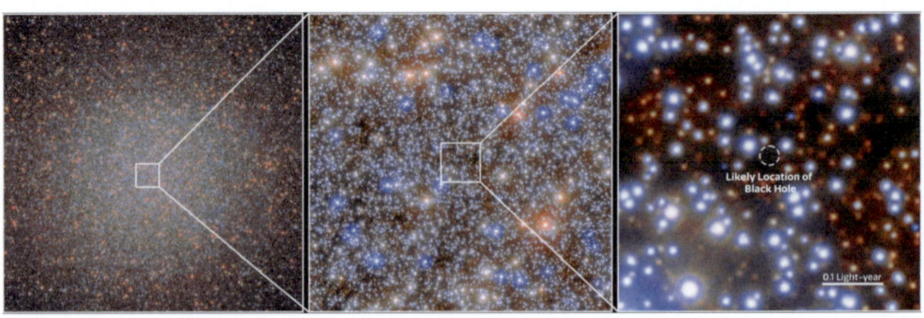

Fig. 5.67 The first image shows the global cluster ω Centauri, appearing as a highly dense and numerous collection of shining stars. The second image shows the details of the central region of this cluster, with a closer view of the individual stars. The third image shows the location of the IMBH candidate in the cluster. (*Image Credit* ESA/Hubble & NASA, M. Häberle (MPIA))

Pulsars

1967: Discovery of the First **Pulsar,** *a Beacon in Space*

In 1966 Anthony Hewish started at the Mullard Radio Astronomy Observatory, which was operated by the Cavendish Laboratory research group of the University of Cambridge (UK), the construction of an array of 4,100 dipoles in phase synchronization, spread on an area of 16 acres. Hewish's original goal was to study the high-frequency *flickers*, i.e., *flashes* of radio sources caused by the solar wind and the interplanetary medium. However, the ultimate goal of this Interplanetary Scintillation Array (IPS) arrangement was the discovery of distant *quasars* by studying the intervening medium using interstellar scintillation, which allows to distinguish point sources from extended ones. Specifically, these flashes of point-like *quasar sources* come from the diffraction of radio waves as they pass through the turbulent and inhomogeneous interplanetary material and the *solar wind*. Because these point sources of *quasars shimmer* more than the non-point and extended remaining radio sources, this IPS array technique was a clever way to discover new enigmatic *quasars*. The phenomenon is similar to how we distinguish the planets from the rest of the stars on a clear night: the point-like stars are *flickering* because of the atmosphere, while the planets do not. This IPS Array, or Pulsar Array operated at 3.7 m, 81.5 MHz and was also used to monitor *Space Weather*. Indicative of how dense this IPS arrangement was with many dipoles, is that sheep were used to clear the grass between the two poles, because the grass clearing machines did not have enough room to roll between them. The cost of constructing the IPS device, around €15,000, was relatively low and the first data were analyzed in July 1967 by Anthony Hewish's Irish graduate student Jocelyn Bell. This radio telescope scanned the entire sky every 4 days, producing about 30 m of recording paper per day, which Jocelyn Bell monitored systematically and daily.

Then, one late night in August 1967 – as known, graduate students in astrophysics often work late, after the midnight! – Jocelyn Bell was studying the records coming from the IPS radio station, in the context of the search for some new and unknown *quasars*, which they had already (1963) been discovered couple of years ago. It was then that she *stumbled upon* an unusual signal of the recording radio telescope. Bell was startled, surprised, and left almost speechless by the very *regularity* of the radio pulses, which were coming from deep space, at a rate of one pulse every about one second. Not knowing the exact origin of these regular pulses, humorously gave them the temporary name *Little Green Men* 1 (LGM-1). But the LGM signal ultimately had nothing to do with extraterrestrial civilizations. They were pulses emitted by a pulsar located at a distance of about 980 l.y., with a period of exactly 1.3373 s, each pulse lasting just for 0.04 s, while the pulsar's age is about 16 million years. After a few years this source, which is now known as pulsar PSR B1919+21 in the constellation Vulpecula, between the stars Vega (in the constellation of Lyra) and Altair (in the constellation of the Eagle) has been identified as a rapidly rotating *neutron star*, the first to be discovered in history, Fig. 5.68. Its code name arieses from its celestial coordinates, i.e., *right ascention* 19 h and 19 min (19 h 19 m) and *declination* 21° (+21°).

Thus, the IPS array that had started with a completely different goal – the discovery of *quasars* – *unexpectedly* led to the discovery of the first pulsar, something that often happens in research! For this discovery, Jocelyn Bell received a doctorate from the University of Cambridge with great praise, while the supervisor of her doctoral dissertation, Professor Anthony Hewish, shared the 1974 Nobel Prize in Physics with astronomer Martin Ryle, for the construction of IPS. Ryle had developed the pioneering technique of radio interferometric synthesis (aperture synthesis) in radio astronomy, where the appropriate combination of several radio telescopes leads to an equivalently larger radio telescope, whose resolution is greatly increased, depending only on the distance of the radio telescopes among themselves.

As Jocelyn Bell reports, after the discovery in the summer of 1967, many astronomers began flocking to her Observatory to see the data directly. Thus, to

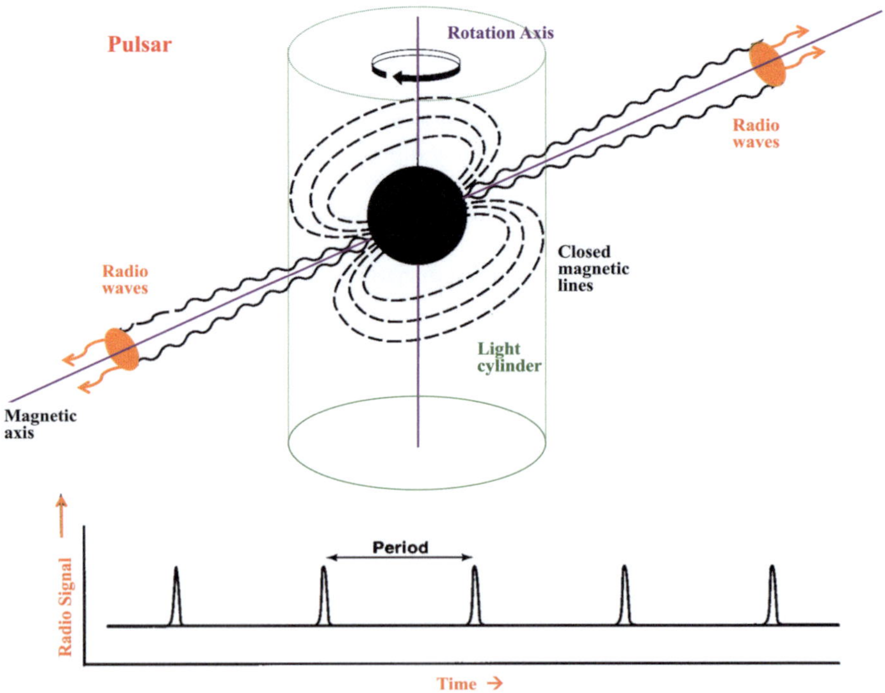

Fig. 5.68 Schematic diagram of a simple *pulsar* model: a rapidly rotating *neutron star* emits two very well-focused thin beams of radio waves from its polar regions in the direction of its strongest magnetic field. At the bottom of the image is a schematic diagram of the periodic radio pulses produced by a *pulsar*, as detected by a ground-based radio receiver. (Adapted from the Nobel Prize Lecture of Russell Hulse, *The discovery of the binary pulsar*, December 8, 1993)

satisfy the growing interest of all these visitors, she had to spend much of her time tracking down that strange source, as well as searching for other similar signals. In fact, she was not at all happy with that situation, because as she said, *some stupid little green men chose her antenna and her frequency to communicate with her..!*

1968: Models of Pulsars as Rotating Neutron Stars

After the discovery of the first pulsar, similar signals were discovered coming from other parts of the sky; they were called rotating *neutron stars*, or *pulsars*, Fig. 5.68. Today we know of many thousands of *radio pulsars* and *neutron star* systems, and their study occupies a large portion of the astronomical community. All these strange objects spin dozens or even hundreds of times every second. This is amazing for a body that has a diameter of about 10 km and a mass greater than the mass of our Sun. These are, in other words, objects that, although they have the enormous mass of a star, have the dimension of a city, rotating like perfect clocks with frequencies up to milliseconds! This also means that the rotational speeds at their equator are close to the speed of light.

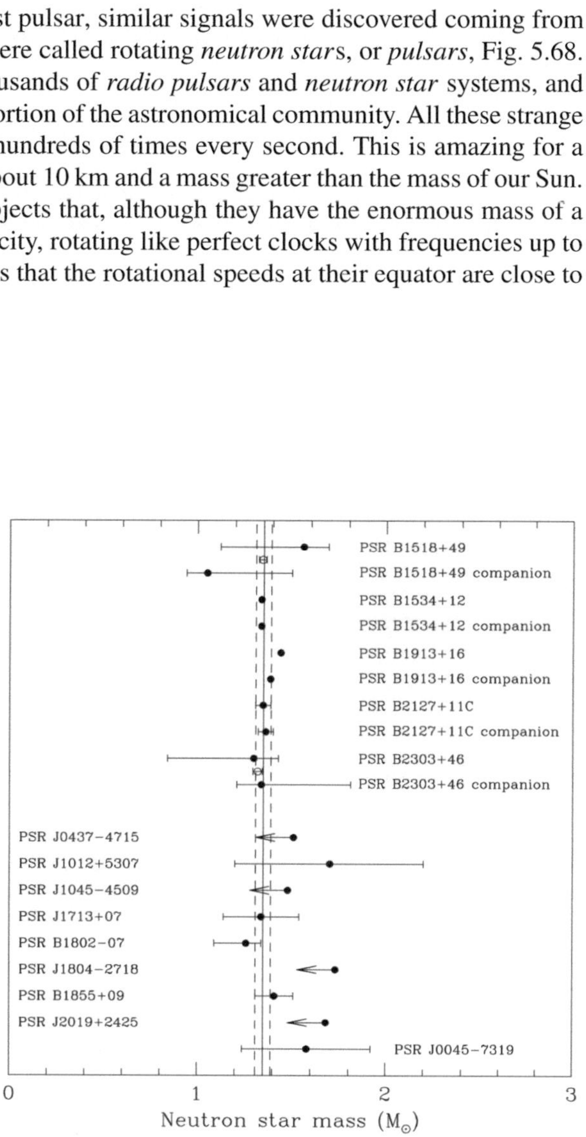

Fig. 5.69 Neutron star masses from observations of radio pulsar systems. Vertical lines are drawn at m = 1.35 ± 0.04 $M_{\odot}$. (From Thorsett and Chakrabarty (1999), with the kind permission of the authors)

Pulsars Are Giant Spinning Atomic Nuclei!

The physical conditions that prevail inside a neutron star are truly *unimaginable.* A small globule of the material of such a star, one millimeter in diameter (i.e., the size of the head of a pin), would weigh about 500,000 tons, that is, as much as five large aircraft carriers of 100,000 tons each!
The weight of the entire population of the Earth, if it was put on one side of a weighting scale would balance the weight of a small dice of the material of such a star, with an edge of one centimeter wide. This enormous density of *neutron star* material creates truly unimaginable gravitational forces.
These stars are essentially giant atomic nuclei spinning around their equator at speeds very close to the speed of light!

With these data, the next goal was to understand the physics and the way *pulsars* emit their radiation. It had to be explained exactly what process made them rival our best individual clocks in absolute accuracy. Three possible motions could create the precise clockwork of *pulsars*: oscillations of the size of their radius, rotation around another star, or, rotation around an axis. In the first case, the star could pulsate precisely, sending a signal with each pulse it makes, as is the case with pulsating *Cepheids* in the visible part of the electromagnetic spectrum. In the second case, the radio source could be orbiting a larger star, which cuts off its radiation whenever its orbit brings it behind this large companion. Finally, in the third case, the star could rotate and radiate like a beacon.

In the first place, a pulsating radial motion of the star itself cannot be accepted: the period of radial oscillations of *neutron star*s was found to be longer than one second and therefore did not agree with the periods of the first *pulsars* which were much shorter.

In Fig. 5.69 are shown the neutron star masses from observations of radio pulsar systems. All error bars indicate central 68% confidence limits, except upper limits are one-sided 95% confidence limits. Five double neutron star systems are shown at the top of the diagram. In two cases, the average neutron star mass in a system is known with much better accuracy than the individual masses; these average masses are indicated with open circles. Eight neutron-star—white-dwarf binaries are shown in the center of the diagram, and one neutron-star–mainsequence-star binary is shown at bottom. Vertical lines are drawn at $m = 1.35 \pm 0.04\ M_{\odot}$. Note that the most recent measurements of neutron-star masses are presented in Ozel and Freire (2016).

Next, the second case of a *neutron star* orbiting another companion was also ruled out, because with a period of less than 2 s the distance separating the two stars should be less than the sum of their two stellar radii.

Hence, there only remains the third solution of the rotating beacon, Fig. 5.68. The discovery of the pulsar in the Crab nebula in 1968 gave the final support to the rotating *neutron star* model for *pulsars*. This model was proposed by Thomas Gold of Cornell University. The Crab Nebula is the visible remnant of a supernova explosion recorded by Chinese astronomers in 1054 AD. This nebula has long been the focus of research because its luminosity has been difficult to be explained after so many years. So, there was the question of what was the source of its energy supply. It had been suggested that a mysterious star, which is actually visible at its center, played this role. However, this star did not belong to any known category. Along with the discovery of the first *pulsar* in 1967, Thomas Gold had proposed a year later (1968) that the nebula's power source was a *neutron star* with a strong magnetic field. After the discovery of *pulsars*, pulsations were observed from the mysterious star at the center of this Crab nebula, thus identified as a *pulsar*. The Crab nebula *pulsar* had a period of 33 ms, which was actually too short to be explained by other models of pulsar emission. Furthermore, the Crab nebula *pulsar* is so named because it is located at the center of the synonymous nebula, in agreement with Baade and Zwicky's 1934 prediction that supernovae leave a *neutron star* at the center of the remnant of the explosion. Thus, Gold initially identified the previously unknown *pulsars* with the *neutron stars* that had been theoretically studied by the American theoretical physicist J. Robert Oppenheimer, in collaboration with the Canadian physicist George Volkof. According to Gold, as a star's radius collapses and shrinks, its rotation rate and magnetic field increase, due to the conservation of the star's angular momentum and magnetic flux. Thus, the conservation of rotational angular momentum (spin) and magnetic flux during stellar collapse leads to the very fast rotation of *neutron stars* and at the same time to the high values of their magnetic fields, of the order of trillions of Gauss (for comparison, the Earth's magnetic field is about 0.5 Gauss and of the Sun about 50 Gauss). The characteristic magnetic field of the Crab pulsar is about 7.6×10^{12} G at its magnetic poles.

But there is another observational property of *pulsars* which was detected right from the beginning in the Crab nebula, and which gave the final clue to the solution of the puzzle. In particular, the period of *pulsars* has been observed to increase at a low but measurable rate: for the *pulsar* in the Crab nebula the observed increase in period was 36 billionths of a second each day. This feature does not arise in a model of an eclipsing binary system, where due to energy loss we expect the two stars to approach each other, shortening the orbital period around their common center of mass. Conversely, in a rotating star a decrease in period is expected. So the problem was finally solved: *pulsars* are indeed rotating beacons, Fig. 5.68.

There are two types of *pulsars*, *ordinary pulsars* and *millisecond pulsars*. Both categories can be understood by assuming that ordinary *pulsars* were originally members of a binary star system, along with a companion, Fig. 5.70. There, matter was transferred from the companion onto the *pulsar*. By this process of adding angular momentum, the *pulsar* increased its spin and became one of the millisecond *pulsars*. Later, when the companion ran out of fuel it turned into a *white dwarf*, *neutron star*, or *black hole*, depending on its size. If the companion remained in orbit with the *pulsar*, we have the formation of a double pair of millisecond *pulsars*. The fastest rotating *neutron star* known today is PSR J1748-2446ad, with 1,122 revolutions per second!

The youngest *pulsars* are found inside supernova remnants, which is exactly where we expect *neutron star*s to be born. Therefore, the physical explanation arises

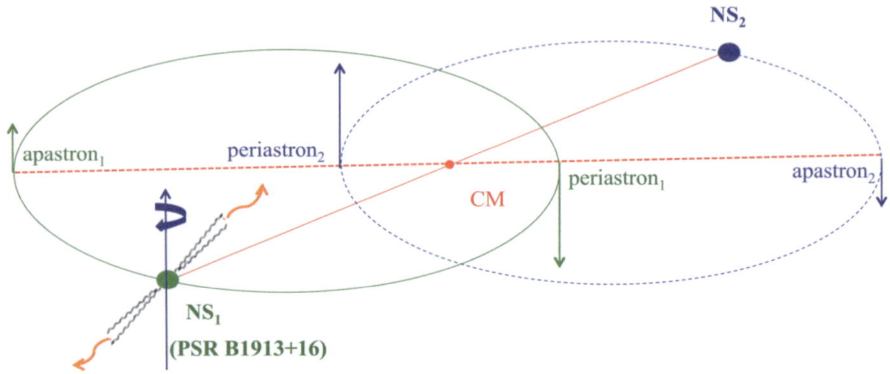

Fig. 5.70 Schematic diagram showing the motion of the neutron star NS_1: PSRB1913+16 with mass 1.438 $M_\odot$ orbiting around the common center of mass CM with another neutron star NS_2 of mass 1.390 $M_\odot$. (Adapted from the Nobel Prize Lecture of Russell Hulse, *The discovery of the binary pulsar*, December 8, 1993)

that *pulsars* are rapidly rotating *neutron star*s that emit radio waves in the direction of their magnetic axis. But this does not mean that all rotating *neutron star*s are necessarily detectable *pulsars*!

Beacons in Space

So, as a strongly magnetized star rotates rapidly, large amounts of the electrons or positrons trapped in its magnetosphere slide along the magnetic lines, directed from the equatorial regions to the region of the magnetic poles where the magnetic lines also become denser. From these magnetic poles of the star where the magnetic lines are practically open, they finally manage to escape. Because the magnetic axis usually does not coincide with the axis of rotation of the star, the escape of particles into space is detected only whenever this magnetic axis is directed in the observer's line. The escaping positrons and electrons, due to the existence of the magnetic field, lose energy in the form of microwaves and are scattered as jets of radio waves throughout space. That is, these particles are dragged along during the rotational movement of neutron stars and due to the enormous accelerations they acquire, they emit *synchrotron* radiation that has the shape of a thin cone that is tangent to their orbits, Fig. 5.68.

The radio emission beams from some of them may not *illuminate* Earth and therefore they are not detected. Also, other *neutron star*s may have been *pulsars* in the past, but now the emission process (which is not fully understood anyway) may no longer be at work, or, may have faded and become undetectable. An additional limitation for *neutron star*s to emit radio pulses is that they also have a sufficiently strong magnetic field. Therefore, only 1% of *pulsars* are expected to emit and send their focused radiation to Earth in the right direction for us to observe their radio pulse, out of a total of about 100,000 *pulsars* estimated to live in our Galaxy. So, we expect only a few thousand *pulsars* to be observable, at most. To date, several hundred such objects have already been discovered.

1974: Discovery of the Double Pulsar PSR B1913+16

In 1974 Joseph Hooton Taylor, Jr., prof. at the University of Massachusetts (Amherst) and his graduate student Russell Alan Hulse, using the 305 m radio telescope in Arecibo, Puerto Rico, discovered the first pulsar in a binary system, PSR B1913+16. In this system, two *neutron star*s, one of which is a pulsar, each with a mass of about 1.4 times the solar mass rotate around their common center of mass, Fig. 5.70. The pulsar rotates on its axis 17 times per second, i.e., its period of rotation is 59 ms.

By systematically monitoring the radio pulses of this system over a period of time, Hulse and Taylor noticed that they arrived sometimes earlier and sometimes later, and these changes were repeated every 7.75 h. Based on this they concluded that it was a double star system. The pulsar was one member of the pair, while the second star, later identified as a *neutron star*, was the other member, although no radio pulses from it reached Earth, perhaps because its rotating *beacon* did not *sweep* the Earth. The distance of the pair from Earth is 21,000 light years. Since the arrival difference of the *pulsars* is 3 s it follows that the perihelion of the pulsar's orbit is 3 light seconds. Thus, the distance between the two members of the pair ranges from 746,000 to 3,153,000 km, i.e., from 1.1 to 4.8 times the radius of the Sun. The orbit has an eccentricity $e = 0.6171334$ and an inclination of 45° to the plane of the sky.

The speed of the rotational movement ranges from 110 to 450 km/s, while the phenomenon of the precession of the orbit is also observed, Fig. 5.71.

1983: Change in the Orbital Period of the Double Pulsar Due to the Emission of Gravitational Waves, According to the Theory of General Relativity

In the double pulsar, some phenomena are observed which only the General Theory of Relativity (GTR) can explain. The first is the precession of the apastron (i.e., the furthest distance of the star's elliptical orbit to the system's center of mass, around which each *neutron star* revolves, Fig. 5.71). It is known that one of the most important predictions of the GTR is the precession of the perihelion of Mercury's orbit around the Sun: 43.2 arc seconds per century, which remained unexplained in the framework of Newtonian physics. In fact, in all planets, and not only Mercury, the perihelion (and in general the periastron) of a planet's orbit around the Sun

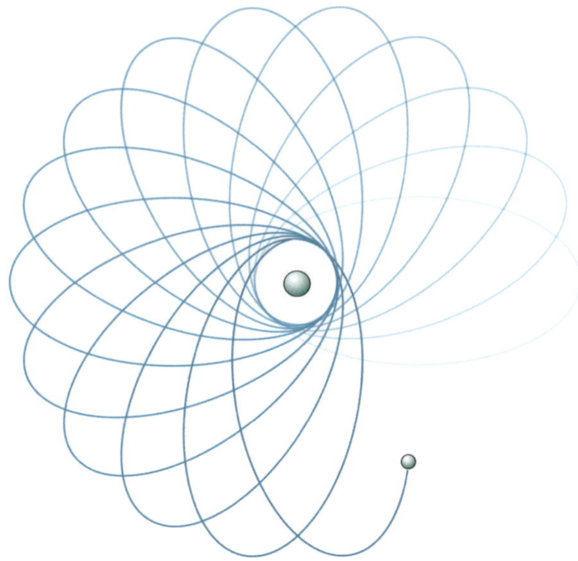

Fig. 5.71 Graphical representation of the precession of the neutron star's orbit as it moves within the gravitational field created by another neutron star

exhibits a precession which is inexplicable within the Newtonian theory of gravitation (Fig. 5.71).

However, for the closest planet to the Sun, Mercury, the phenomenon is more intense, because this planet is located close to the Sun, that is, in areas where the

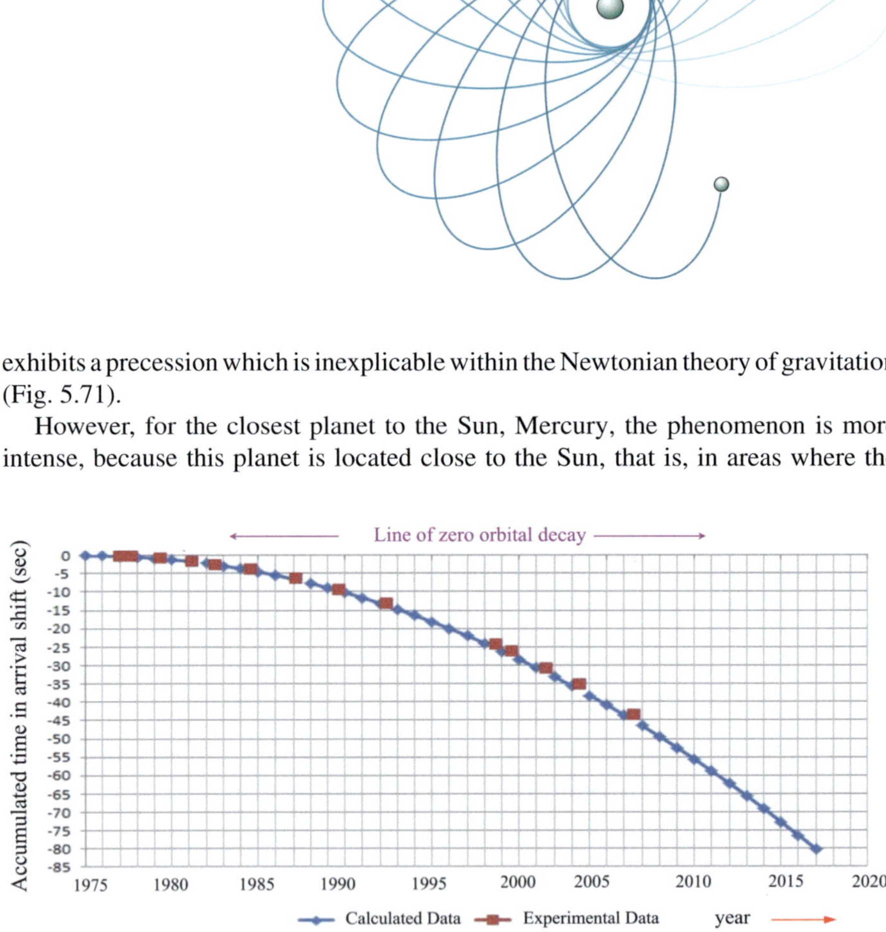

Fig. 5.72 Comparison between the observed accumulated shift in *periastron* arrival time (red points, Weisberg, Nice and Taylor 2010) and the predicted shift according to general relativity (blue curve, Peters and Matthews 1963). The slow decrease in the real system's period is due to the emission of gravitational radiation and causes the parabolic shape of the graph. The excellent agreement between theory and observations provided conclusive – albeit indirect – proof of gravitational waves and led to the 1993 Nobel prize for Hulse and Taylor. (Adapted from Wikipedia Commons and Hamilton and Cho (2017))

gravitational field is stronger and its rotational speed is quite high. The GTR explained precisely this precession. In the double pulsar, the precession of the periastron of the orbit is much greater: 4.2 degrees/year. In other words, the periastron of the double pulsar's orbit shifts every day by the same angle as the perihelion of Mercury's orbit shifts in a century! In other words, the precession angle of the periastron of the double pulsar is about 36,000 times greater than the precession angle of Mercury's perihelion. This large difference in the value of the perihelion angle of the double pulsar's orbit is due, in part, to the double pulsar's spin in its orbit: the period of the pulsar's orbit is only 7.75 h, while that of Mercury is about in 88 days, i.e. the pulsar completes approximately 272 complete rotations within a Mercury year of 88 days. A second factor is the greater orbital speed of the double pulsar in the periastron, which is 450 km per second, i.e., 6.5 times faster than that of Mercury, which is only 70 km per second.

After studying the orbital decay, the natural question is when the *pulsar* and its companion would coalesce. To solve for the collision time, the orbital radius decay equation can be set up as a differential equation, which after integration yields, coalescence in about 140 million years! (Hamilton and Ye Hie Cho 2018).

Duration of the Deadly Embrace of the Two *Neutron Stars* in PSR B1913+16

The second important result that emerged from the observations made by J. Taylor and P. Hulse, which only GTR can explain, is the emission of gravitational waves due to the movement of the two masses of the system. Observations of PSR B1913+16 over the past 35 years have shown that the *pulsar's* orbit is gradually shrinking. This is understood as a result of the *pulsar* losing energy due to the emission of gravitational waves, as described by Einstein's GTR, Fig. 5.72. Thus, the emission of gravitational waves causes the orbit to shrink, with the *pulsar* reaching the *periastron* a little earlier. The decrease in the *pulsar* period is 76.5 μs per year. Although this change is minimal, it nevertheless corresponds very precisely to the value predicted by the Theory of General Relativity. This means that the duration of the deadly embrace of the two neutron stars will last for about 140 million years!

1982: Discovery of Neutron Star *Spin-Down and Discovery of Millisecond Pulsars*

The oldest *neutron stars* observed to date rotate with a period of a few seconds. Over time, these spinning *neutron stars* slow down their rotation, as the electromagnetic fields absorb energy from the spin and re-emit it. This increase in period is called spin-down. The rate at which a *neutron star* slows its rotation is usually constant and very small. If the period of rotation is p, the rate of deceleration of rotation $\dot{\mathbf{p}}$ is a dimensionless quantity, with dimensions of seconds per second. This amount of **ṗ** of *pulsars* varies from 10^{-22} to 10^{-9} s per second, with the observed shorter-period (or faster-rotating) *pulsars* usually having the smallest **ṗ.** Over time and as the pulsar's rotation slows down (i.e. as p increases) eventually the rotation rate becomes too small, the period p becomes long, so the pulsar can no longer be detected.

The p – **ṗ** diagram shown in Fig. 5.73, is very useful for tracking the lifetime of *pulsars*. It contains an enormous amount of information about the entire currently known population of *pulsars* and their properties. Using these parameters, one can estimate the pulsar's age, its magnetic field strength, and its rotation period, which has a lower limit of about 1.4 ms for the fastest rotating pulsar observed today, i.e., rotation once every 1.4 ms, or, 716 rotations per second. So far the slowest pulsar, in the constellation Cassiopeia, about 5,200 light years from Earth, rotates once every 23.5 s. The p – **ṗ** diagram encodes an enormous amount of information about all *pulsars* and their properties and is of similar pivotal importance to the Hertzsprung-Russell diagram for the rest of the stars.

In 1982, Don Backer of the University of California, Berkeley and his colleagues studied a mysterious source that was sending out a radio emission with a highly polarized radio spectrum. Backer et al. (1982) found that this source was indeed a *pulsar*, but that the pulses were emitted at such short intervals that observers did not have the time to separate them. This source, now named PSR B1937+21, has a period of just 1.56 ms. In other words, this *neutron star* is spinning faster than 641 revolutions per second, or, 38,462 times per minute. As we mentioned above, the fastest rotating *neutron star* known today is PSR J1748-2446ad, at 714 revolutions per second, or 42,960 times per minute! The speed at which in the *neutron star* gravitational field, it would not be able to hold its material and it would be torn apart by centrifugal forces, is about 3,000 revolutions per second. So these rapidly rotating *neutron star*s are close to this breakup limit.

For PSR B1937+21, its extremely short period, however, was not the only surprising feature. Observers also measured that the decrease in the pulsar's period was extremely small. Given such a small period, and a very small decrease in period (**ṗ**), PSR B1937+21 is placed in a previously empty part of the p – **ṗ** diagram, in its lower left region, Fig. 5.73. Knowing now some characteristics of millisecond *pulsars*, astronomers designed their own field studies (surveys) to specifically look for these millisecond *pulsars*. Soon, many other such sources were discovered in the following years. Thus, since 1982 the diagram p – **ṗ** acquired many such *occupants*, Fig. 5.73.

But what mechanism moves the *pulsars* to the lower left part of the p – **ṗ** diagram as shown by the arrows in Fig. 5.73? A clear hint comes from the fact that *pulsars* with short period and small **ṗ** are old. But even more important is the observation that the vast majority of millisecond *pulsars* are observed to be members of binary star systems.

Very soon after the discovery of PSR B1937+21 it was proposed that millisecond *pulsars* were previously dead *pulsars* that had been recycled by transport of matter from a companion star.

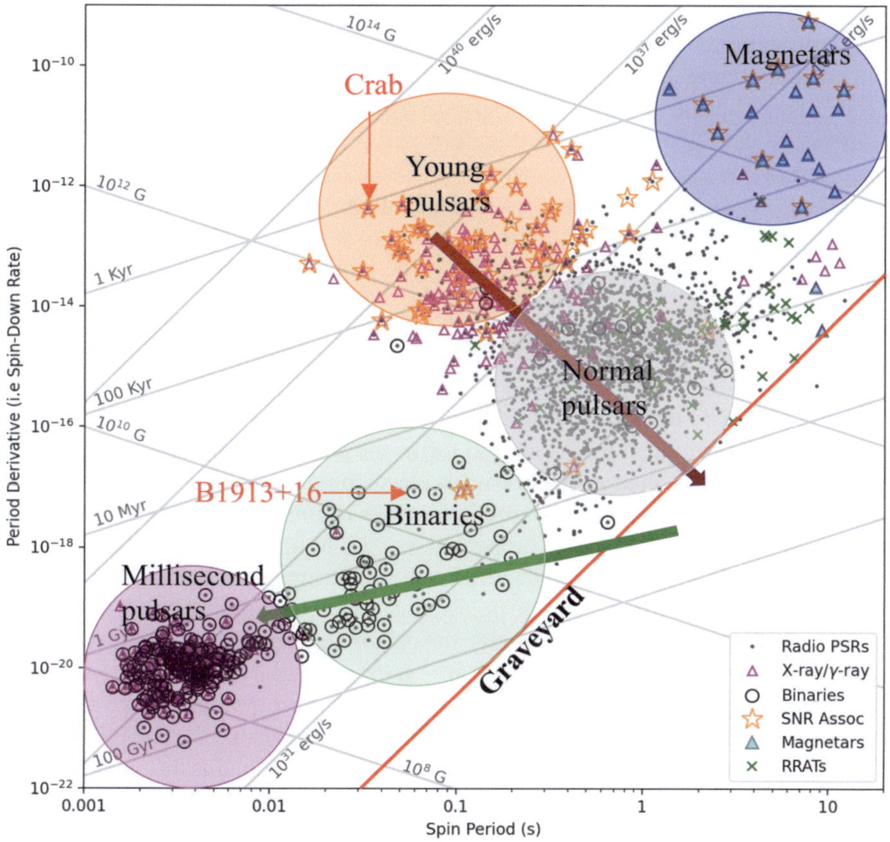

Fig. 5.73 The observational p − ṗ diagram for pulsars, containing about 2700 radio *pulsars* in our galaxy. Shown as dots are the hundreds of pulsars discovered since 1999 by the wide-field Parkes multibeam survey. Diagonally from lower right to upper left, the lines with the values of the magnetic field with values from 10^{10} to 10^{14} Gauss are shown. The diagonal lines from bottom left to top right show the age of the *pulsars*, with values from 100,000 years to 10 billion years. The two arrows show the path of recycling *pulsars* in binary systems. Bottom left are the millisecond *pulsars*. More explanations in the text. (Adapted from an original diagram with the kind permission of NRAO/Scott Ransom and Jim Condon)

The Fastest Spinning Pulsar, with Over 700 Revolutions Every Second

The fastest rotating neutron star known to date is PSR J1748-446ad, with a period of 1.4 ms, a frequency of 716 revolutions per second, i.e., 42,960 revolutions per minute! This corresponds to a rotation speed at its equator of faster than 70,000 km per second, i.e., equal to about 24% of the speed of light. It is thus rotating very close to its breaking speed, where its all-powerful gravity would not be able to hold its material back against the enormous opposing

centrifugal forces. Indeed, if its radius were 16 km, it would be close to such a breaking point. Thus, these millisecond *pulsars* are approaching their limits, i.e., a break up due to their rotation speed, despite their very strong gravitational forces of cohesion. These are therefore very exotic objects at the limits of physics.
PSR J1748-2446ad was observed in a 10-billion-year-old cluster of old stars called Terzan 5, located 28,000 light years away in the center of the Milky Way. Terzan 5 contains 33 ms *pulsars* – the largest number in a star cluster. To date, more than 150 ms *pulsars* have been detected. To this day, this is the shortest periods of all known *pulsars*.

In this scenario, the *pulsar* progenitor is the most massive member of a binary star system. This progenitor evolves and forms the *pulsar* in a supernova explosion. It is possible for the binary system to survive such an explosion. The *pulsar* lives its normal life and eventually dies after a few tens of millions of years. Meanwhile, the progenitor of its lower-mass companion evolves until it reaches the giant phase, during which it transfers mass to the dead pulsar. During this mass transfer, angular momentum is also transferred to the pulsar, and thus it begins to rotate faster until it reaches a millisecond period. Eventually, the *pulsar* starts emitting radio waves again and appears in the lower left part of the $p - \dot{\mathbf{p}}$ diagram as a millisecond *pulsar*, or, a *recycled pulsar*! But the mass that falls into the *neutron star* during the recycling process *destroys* a part of the magnetic field that is responsible for the pulses, thus reducing the field strength by a factor of about 1000. This explains the small rotation rate of the pulsar. This evolutionary scenario is summarized in the $p - \dot{\mathbf{p}}$ diagram with the two large arrows shown in Fig. 5.73.

Oppenheimer and Volkov (1939) used quantum mechanics to predict a mass equal to 1.4 solar masses for a *neutron star*. The exact value of mass depends on the equation of state of matter of the star, describing the internal pressure and density of this object as a function of mass, determining its radius.

The equation of state of the star's matter is not known with certainty for such high densities, as those found in *neutron star*s. From observations of double *pulsars*, however, we can measure the mass for a number of *neutron star*s. Observations show an average mass of 1.35 solar masses, which is very close to the theoretical predictions.

In summary, the $p - \dot{p}$ diagram is useful for following the evolution of *pulsars*, playing a role similar to the Hertzsprung–Russell diagram for ordinary stars. It encodes a very large amount of information about the pulsar population and its properties in terms of the observed spin period p and $\dot{p}$, the time derivative of the spin period. For each pulsar, the $p - \dot{p}$ diagram contains the characteristic age τ, the minimum magnetic field strength, and the spin-down luminosity $-\dot{\mathbf{E}}$. Thus, in the diagram of Fig. 5.73 are shown the contour lines for τ, B and $-\dot{\mathbf{E}}$. Young *pulsars* in the upper middle of the diagram are often associated with supernova remnants (SNRs) and emit high-energy radiation. They slow down and evolve over millions of years to become one of the large population of *slow pulsars*, until their rate of rotation is too slow to power the radio emission mechanism, and they disappear from view. Rotating Radio Transients (RRATs) are *pulsars* that sporadically emit single pulses instead of continuous pulse trains. The millisecond *pulsars* in the lower left are mostly in binary systems and were *recycled* via accretion from their companion stars (Fig. 5.73).

1992: Magnetars

In order to explain the properties of some transient sources of gamma rays, now known as soft gamma repeaters), Robert Duncan and Christopher Thompson proposed that a type of *neutron star* with an extremely powerful magnetic field of about 10^{13}–10^{15} Gauss, exists, the so-called *magnetars*, which are the most powerful magnetic objects detected throughout the Universe. Specifically, they calculated that when the spin, temperature and magnetic field of a newly formed *neutron star* falls into the right ranges, a dynamo mechanism could act, converting heat and rotational energy into magnetic energy and increasing the magnetic field, normally an already enormous 10^{12} Gauss, to more than 10^{15} Gauss. These are tremendous magnetic fields, such that at a distance of halfway from Earth to the Moon, a magnetar could wipe information from the magnetic stripes of all credit cards on Earth!

Some quite remarkable things happen with such magnetic field strengths. X-ray photons readily split in two or merge. The vacuum itself is polarized, becoming strongly birefringent, like a calcite crystal. Atoms are deformed into long cylinders

Fig. 5.74 Artist's impression of a gamma-ray burst and supernova powered by a magnetar. (*Image credit* ESO and Jochen Greiner)

thinner than the quantum-relativistic de Broglie wavelength of an electron. In field strengths of about 10^9 Gauss, atomic orbitals deform into rod shapes. At 10^{14} Gauss, a hydrogen atom becomes 200 times as narrow as its normal diameter,[4,5] Such extremely strong magnetic-fields power the emission of high-energy electromagnetic radiation, particularly very strong and characteristic bursts of X-rays and γ-rays.

Till today, about 30 confirmed *magnetars* are known in the Milky Way. It is estimated that about 1 in 10 supernova explosions results in a *magnetar* rather than a more standard *neutron star* or *pulsar*. In 2008, ESO announced what it ascertained was the first optically active *magnetar*-candidate yet discovered, SWIFT J195509+261,406 and in 2015 observations again from ESO's La Silla and Paranal Observatory have for the first time demonstrated a link between a very long-lasting burst of gamma rays and the unusually bright supernova explosion GRB 111209A, from a very massive star, about 50 times the mass of the Sun. These observations show that this supernova was not driven by radioactive decay, as expected, but was instead powered by the decaying super-strong magnetic fields around a *magnetar*.

Like other *neutron stars*, *magnetars* are around 20 kms in diameter, and have a mass of about 1.4 solar masses. They are formed by the collapse of a star with a mass 10–25 times that of the Sun. The active life of a *magnetar* is short compared to other celestial bodies. Their strong magnetic fields decay after about 10,000 years, after which activity and strong X-ray emission cease. Given the number of *magnetars* observable today, one estimate puts the number of inactive *magnetars* in the Milky Way at 30 million or more.

A recently studied by the NASA/ESA Hubble Space Telescope magnetar, SGR 0501+4516 at about 15,000 light years away (Chrimes et al. (2025), does not seem to be born in the neighboring supernova HB9, which is located promisingly close, as previously thought. If SGR 0501+4516 was not born in the supernova remnant HB9, the magnetar must either be far older than its reported 20,000-year age, or it must have formed in another way. Magnetars may also be able to form through the merger of two lower-mass neutron stars or through a process called *accretion-induced collapse*. Accretion-induced collapse requires a binary star system containing a white dwarf, the crystallized core of a dead Sun-like star. If the white dwarf ensnares gas from its companion, it can grow too massive to support itself, leading to an explosion, or possibly the creation of a magnetar. Magnetars that form through accretion-induced collapse could provide an explanation for some of the mysterious cosmic signals called *fast radio bursts*.

[4] Scientific American Magazine Vol. 288 No. 2, February 2003, p. 93.

[5] According to NASA, if a magnetar flew by Earth at half the Moon's distance, its intense field would wipe out every credit card on our planet. If a human got within 1000 kms, the magnetar would become a proverbial sci-fi death-ray, ripping apart every atom inside the body.

Galactic Hierarchy

The Fellowships of Galaxies …

Observational studies have shown that galaxies in the Universe behave as social beings: they rarely appear isolated in the vastness of space. As if they were afraid of loneliness, they form small or larger *groups*! (Figs. 5.75, 5.80 and 5.81).

1978: Galactic Hierarchy: Groups, Clusters, and Superclusters

At a *first level* of organization, small *gatherings* of galaxies form galaxy *groups*, consisting of around 50 members, at most. Around our own Milky Way galaxy there is such a group consisting of about 50, or so, mostly *dwarf galaxies*. This group is called the *Local Group* and includes our own Milky Way, the Andromeda galaxy (M31), the Triangulum galaxy and other smaller galaxies, spanning an area of 10 million light years. The two largest galaxies in the Local Group are the Andromeda Galaxy, with about 1 trillion stars, and our own Milky Way, with 200–400 billion stars. Both are barred spirals with hundreds of thousands of light years in diameter.

At a *second level* of organization, larger *gatherings* of galaxies form – like swarms of bees – *galaxy clusters*, which include from a few dozen up to a few thousand of galaxies, with total typical masses ranging from 10^{14} to 10^{15} solar masses. They typically have diameters from 3 to 15 million light years, while the observed spread of velocities for the individual galaxies is about 800–1000 km/s. X-ray observations have revealed in these *clusters* the presence of large amounts of intergalactic gas radiating by thermal bremsstrahlung, with temperatures between 10 and 100 million degrees Kelvin. The total mass of this intergalactic gas is about twice the mass in galaxies. Thence, in a typical cluster less that about 5% of the total mass is in the galaxies themselves, 10% is in hot X-ray emitting gas, and the rest is *dark matter*, almost *90%*! Notable galaxy clusters in the relatively nearby Universe include the *Virgo Cluster*, the *Coma Cluster*, the *Fornax Cluster* and the *Hercules Cluster*.

The nearest *galaxy cluster* is the *Virgo cluster* of galaxies containing about 2000 members, with a total mass of about 1000 trillion (10^{15}) solar masses. In the center of this cluster, at a distance of over 55 million light years, *sits round* majestically, like the king of the *cluster*, the famous giant elliptical galaxy M87. In its center M87 is engulfing a supermassive *black hole* of billions of solar masses, which was photographed recently (2017–2019), as we discuss later. The *Virgo Cluster* is at the core of a much larger cosmic structure, called the *Local supercluster, or the Virgo supercluster*, consisting of at least 100 groups and galaxy clusters, including the *Local Group*.

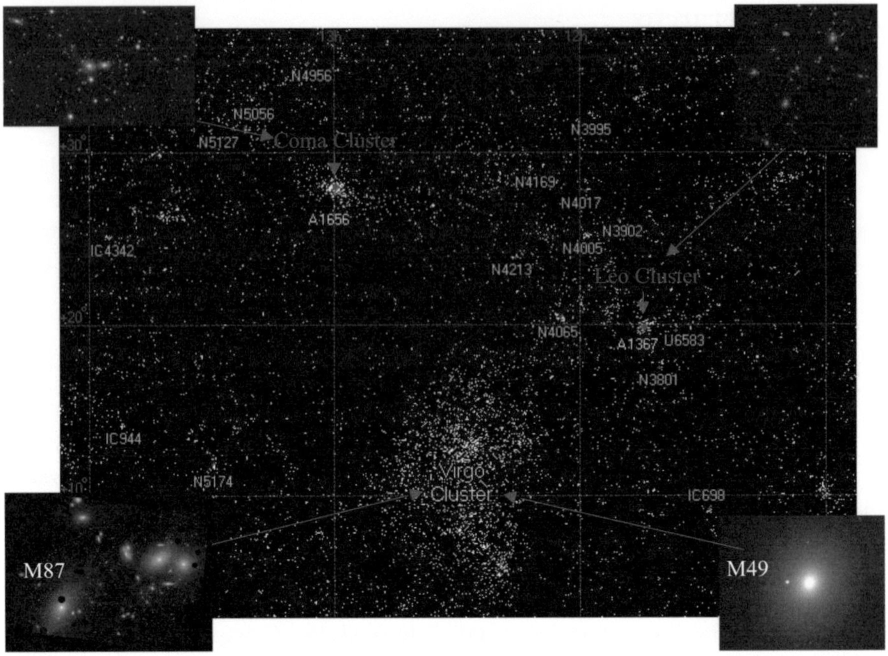

Fig. 5.75 The supercluster of galaxies in the constellation *Coma Berenice*, the *Coma supercluster*, or *Local Supercluster* (LS), a relatively nearby supercluster of galaxies that includes the 2 galaxy *clusters Coma* (Abell 1656) and *Leo* (Abell 1367), shown as insets in the upper part of the figure. It is roughly spherical, about 20 million light years in diameter, and contains more than 3,000 galaxies and galaxy groups (e.g., the N4065 group of galaxies). Located about 300 million light years from Earth, it is at the center of the *GreatCosmic Wall* and part of the *Coma Cosmic Filament*. The *Coma supercluster,* is the closest galaxy supercluster to our own Virgo *supercluster*. The *Coma* contains 90% dark matter. The *Virgo supercluster* is a large gathering of about 100,000 galaxies that contains the *Virgo cluster* and the *Local Group* of galaxies, which contains our own Galaxy and that of Andromeda. At least 300 *galaxy groups* and *clusters* lie within its 110 million light year diameter. Two elliptical galaxies, M87 and M49 are shown as insets in the lower part of the figure. The LS has a total mass of $M \approx 10^{17}$ M_o and a mass/luminosity ratio of about 300. The *Coma* and *Virgo superclusters* are two of the approximately 10 million galactic *superclusters* in the observed Universe. The *Virgo SC* is located on the galactic *cosmic filament*, between the constellations of Pisces and Cetus. (Adapted from the Atlas of the Universe by Richard Powell and Wikipedia Commons)

The *Coma cluster* contains also more than 1000 galaxies and is only a few degrees from the north galactic pole of the sky. In this cluster abound both dwarf and giant elliptical galaxies and it is one of the first places in the Universe where observed gravitational anomalies were considered as indicative of the existence of *dark matter* in the Cosmos. In 1933 Fritz Zwicky showed that the galaxies in the *Coma cluster* were moving too fast so that the cluster could not be held together alone by the visible matter of its galaxies. Although the idea of *dark matter* was not accepted for another fifty years, Zwicky noted that these galaxies should be held together by some strange *dark matter*. The gravitational mass of galaxies in the cluster is at least 400 times its luminous mass, which means that most of the matter is *dark*. Specifically, about 90% of the mass of the *Coma cluster* is believed to be *dark matter*. However, the distribution of *dark matter* throughout the cluster is not fully understood (Fig. 5.75).

Another particularly interesting galaxy cluster is *El Gordo* (ACT CL J0102-4915) (Fig. 5.76) the largest distant galaxy cluster observed at a distance of about 7 billion light years from Earth. The cluster is called *El Gordo, the fat* in Spanish, because of its gigantic mass, and as of 2014 it still holds the record of the largest distant *galaxy cluster* ever discovered, with a total mass of 3 quadrillion solar masses, $M \cong 10^{15}\ M_o$. *El Gordo* consists of two separate subclusters colliding at several million kilometers per hour. Its discovery was first announced in 2012 by the Chandra X-ray space observatory and ground-based optical telescopes. *El Gordo* was thus determined to be the most massive and hottest cluster, as it emits more X-rays than any other known galaxy cluster at its distance. Figure 5.76 shows a composite X-ray (Chandra) and optical (Hubble) image of El Gordo. Newer Hubble study found that most of the cluster's mass is invisible, as *dark matter*.

The location of *dark matter* is mapped in Fig. 5.76 in blue. Since *dark matter* does not emit any radiation, we can instead measure precisely how its gravity distorts images of galaxies which are behind and in the background. This allows us to estimate the total mass of the cluster. Chandra X-ray data are shown in pink and have been combined with the optical data from Hubble showing the individual galaxies in the cluster as well as stars in the field of view. The X-ray image of *El Gordo* has a cometary appearance. Together with data in the visible, this shows that *El Gordo* consists, in fact, by two colliding subclusters of galaxies.

Fig. 5.76 Two subclusters of galaxies colliding at several million kilometers per hour. X-ray (Chandra, in pink) and optical (Hubble, white) composite image of the cluster ACT-CL J0102-4915 (El Gordo). Most of the cluster's mass is invisible, as *dark matter* and its location is mapped in blue. *Dark matter* has been precisely measured by the distortion in images of back and background galaxies, allowing us to estimate the total mass of the cluster. (*Source* NASA/ESA/Hubble/Chandra)

This and other characteristics make *El Gordo* similar to another well-known cluster, the *Bullet Cluster*, where a smaller cluster of galaxies has passed like a bullet through an extremely large *galaxy cluster*, which is located at a distance of about 4 billion light years from Earth. Specifically, X-ray data by the Chandra X-ray Observatory, show a rapidly moving galaxy cluster with a shock wave trailing behind to have hit another cluster at high speed. When the galaxies collided, the temperature reached 160 million degrees and X-rays were emitted in great intensity, claiming the name of the *hottest* known galactic cluster for the *Bullet Cluster*. The third component, *the dark matter*, was detected indirectly by the gravitational lensing of background objects. The hot gas of the two colliding components, seen in X-rays, represents most of the baryonic, or *ordinary* matter in the cluster pair. Also in the *bullet cluster*, the data show that the baryonic matter, composed mostly of hot, luminous X-ray gas, has been disrupted by the putative *dark matter*. This hot gas in each subcluster has been slowed down by the collision, but not the *dark matter*. The gases of the intracluster medium interact electromagnetically, causing the gases of both clusters to slow down much more than the stars.

At a *third level* of organization, superclusters of galaxies, consisting of groups, clusters and even individual galaxies, resembling countries containing many states. Although until 1980 galaxy clusters were observed to be the largest structures in the Universe, in 1980 even larger cosmic structures were discovered, i.e., clusters of clusters, or *superclusters* of galaxies. Subsequently, these superclusters form the largest structures of *cosmic filaments* and *cosmic walls* that span distances of hundreds of millions to billions of light years.

The nearest supercluster of galaxies is the *Virgo supercluster* (Fig. 5.75), or *Local Supercluster* (LS). It is a large gathering of about 100,000 galaxies, which includes the *Virgo cluster* and the *Local Group* of galaxies, which contains our own galaxy and that of Andromeda, Fig. 5.75. At least 100 galaxy groups and clusters lie within its diameter of about 110 million light years. The LS has a total mass $M \cong 10^{15}$ M_0 and a mass/luminosity ratio of about 300, a figure that is consistent with results obtained for other superclusters. This is one of the main arguments in favor of the presence of large amounts of *dark matter* in the Universe. If *dark matter* did not exist, much smaller luminosity mass ratios should be expected. *The Virgo supercluster* is located on a galactic *cosmic filament* in the constellation Pisces-Cetus and is one of the approximately 10 million *galactic superclusters* in the observable Universe.

Another galaxy supercluster is in the constellation Coma Berenices, the *Coma Berenices supercluster*, also a relatively nearby supercluster of galaxies that includes *Coma* (Abell 1656) and *Leo* (Abell 1367), Fig. 5.75. It is roughly spherical, with a diameter of approximately 20 million light years and contains more than 3000 galaxies. Located at an average distance from Earth of about 300 million light years, it is located at the center of the *Great Wall* and is part of the *Coma Filament*. The *Coma supercluster* is the nearest supercluster of galaxies to our own *Laniakea supercluster*. Being one of the first superclusters to be discovered, it helped us understand the large-scale structure of the Universe.

1978: Discovery of Large Cosmic Voids and Filaments in the Distribution of Galaxies (the Bubbles of the Cosmic Web)

A Complex Network of Cosmic Highways …

The omnipresent gravity in the Universe that governs the (known) baryonic as well as the (unknown) dark matter, has woven on very large spatial scales a cosmic veil, the *cosmic web*, consisting of gas which resembles many packed soap bubbles, or, a tangled spider's web, or, an intricate network of cosmic highways!

On the great road map of the Universe, the busy hubs of galactic *superclusters* are interconnected by great plasma highways that enclose the deserts of the *cosmic voids*. These interstellar highways are known as *cosmic filaments* and can stretch for hundreds of millions of light years, filled only with dust, gas, and busy electrons that *drive* very close to the universal speed limit (the light speed)!

As observations of space mapping with the help of telescopes, but also numerical simulations on supercomputers show, the structures on a large cosmic scale form a giant *spider web*, Fig. 5.77. This vast network forms huge *bubbles*, which contain very few galaxies inside, while on their surface extend filamentous structures of galaxies, galaxy clusters and galaxy *superclusters*. These filamentous structures connect the galactic clusters and *superclusters*, which appear scattered as nodes of accumulated mass. But even though initial computer simulations first revealed this large-scale structure of the *cosmic web* decades ago, it was very hard to imagine that anything like this could be observed anytime soon. It's easier to see light coming from galaxies and galaxy clusters, but the thin gas bridges from one galaxy cluster to another had largely escaped detection. Of course, we knew that the *cosmic web* must exist because it had already been indirectly detected. For example, if we observe through the *cosmic web* a bright source behind it at a great distance, the light from this source will show a *forest* of hydrogen absorption lines, from all the *filaments* that blocked the path of light from the distant source. But, there aren't many bright, distant sources out there. And the gas between galaxies is so thin that it doesn't even emit detectable light on its own. Unless this intergalactic gas is magnified by galactic lenses, or, galaxies with explosive formation of newborn stars, which can light up their surroundings, thus irradiating the thin hydrogen gas that surrounds them (Fig. 5.78).

Cosmic voids were first discovered in 1978 in a pioneering study by Stephen Gregory and Laird Thompson of the Kitt Peak National American Observatory (Gregory and Thomson 1978). In 1981, the *great void* was discovered in the direction of Bootes, at a distance of about 800 million light years, where a region of radius about 300 million light years is empty of galactic superclusters. *Cosmic voids* are vast volumes between *cosmic filaments*, the largest elongated structures in the Universe, that contain very few, if any, galaxies. They typically span a diameter of 30–300 million light years.

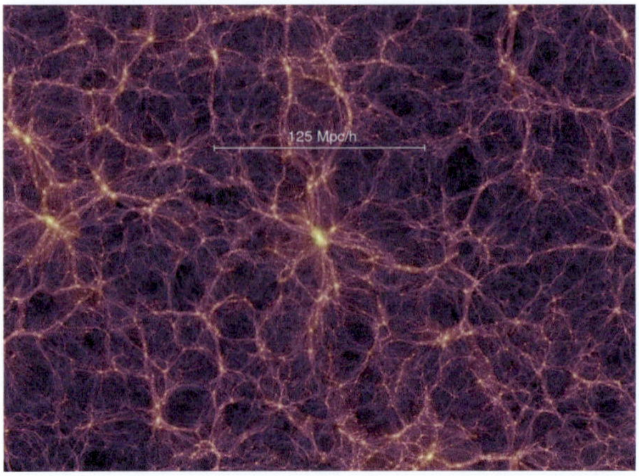

Fig. 5.77 *The cosmic network*: numerical simulations suggest that galaxies lie on massive *filaments* hundreds of millions of light years across. Computer simulations of the Universe have been in common use since the 1980s, and are now a vital tool in helping us interpret data from increasingly powerful telescopes. As we have discussed, amongst other things, simulations have helped establish the case for *dark matter* and *dark energy*, and have been key to creating a broad consensus around the idea that galaxies start small and grow over time through merging. The observed diversity of different galaxy sizes, shapes and colors can be reproduced in a statistical sense, but there is still considerable uncertainty around which causal processes give rise to this diversity. (*Image Credit* ESA, Volker Springel (Max Planck Institute for Astrophysics))

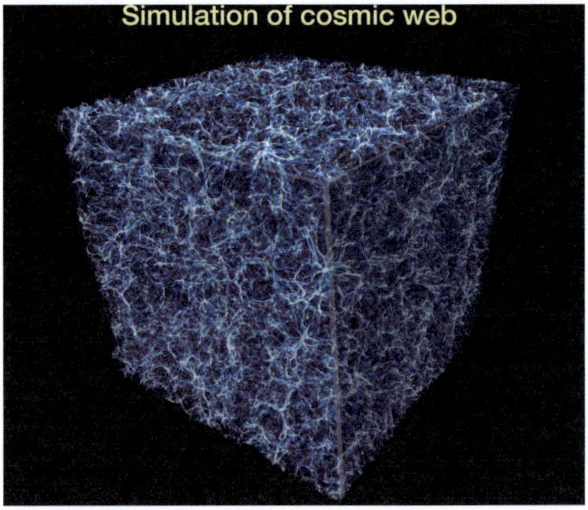

Fig. 5.78 Graphical depiction of a slice of the structure of the Universe, a type of spider web, the so-called *cosmic web*. These elongated filaments are mainly composed of *dark matter* found in the intergalactic space. (*Source* NASA/Hubble and E. Hallman (University of Colorado, Boulder))

By using the MUSE (Multi Unit Spectroscopic Explorer) instrument attached to the European Southern Observatory's Very Large Telescope in Chile (ESO/VLT), the research team of Hideki Umehata of RIKEN/University of Tokyo focused on a distant galaxy collection, known as SSA22, Fig. 5.79. From this massive protocluster of galaxies, light takes 12 billion years to reach Earth ($z = 3.02$). In SSA22 the galaxies themselves are bright with many newborn stars, *black hole* activity at their centers, or both. The light emitted by these galaxies illuminates the intergalactic gas, enhancing its emission. Two *filaments*, each about 3 million light years long, run horizontally through this MUSE image in Fig. 5.79. The white dots are galaxies with explosive formation of new stars, which were observed with the ALMA radio telescopes in Chile. There are other galaxies with *black holes* at their centers, imaged by the Chandra X-ray Space Telescope, which are not seen there. The total of 24 galaxies associated with this group are embedded in these filaments of hydrogen gas in Fig. 5.79.

Ironically, however, observations of the *cosmic web* will remain – for now – only possible for the distant, early Universe. The light emitted is the result of new stars forming in or around these regions, or, around the supermassive *black holes* at the centers of galaxies. So it is initially in the ultraviolet wavelengths, but is then red-shifted into the visible and infrared wavelengths as it passes through the expanding Universe. To observe the *cosmic web* at closer distances, access to ultraviolet wave-lengths is required, but these are blocked by the Earth's atmosphere. Observing the complete history of the Universe by monitoring gas emissions requires an ultraviolet (UV) type satellite in space, but this is currently difficult for economic reasons.

But how were these *cosmic voids* and the first galaxies really created? *Cosmic voids* are believed to have formed from baryonic acoustic oscillations that came from the *Big Bang*, and subsequently led to the collapse of compressed baryonic matter into sheets and subsequently into *filaments*. Starting from small anisotropies from quantum fluctuations in the early Universe, these anisotropies increased in scale with time. Regions of higher density collapsed more rapidly under the influence of gravity, eventually resulting in a large-scale foamy structure, the so-called *cosmic web*, of *cosmic voids* and *filaments* of hydrogen gas seen today. *Cosmic voids* found in high-density galactic environments are smaller than *cosmic voids* found in other low-density regions of the Universe. *Cosmic voids* also appear to be correlated with the observed temperature of *cosmic microwave radiation* (CMR): cooler regions of the CMR are associated with *cosmic voids* and warmer regions are associated with *filaments*.

Thus, large-scale structures owe their existence to the fact that the distribution of matter in the primordial Universe was not uniform, but showed microscopic variations in its density. These fluctuations were the cosmic *seeds*, from which the primary structures were born. Then, as the Universe continued to expand, the density fluctuations gradually increased under the influence of gravity, creating regions of even greater density, which eventually formed the first stars and the first small galaxies. In addition, where the gravitational pull prevailed over cosmic expansion, galaxies formed galactic clusters. But what was the role of *dark matter* in the primordial Universe in the formation of these structures?

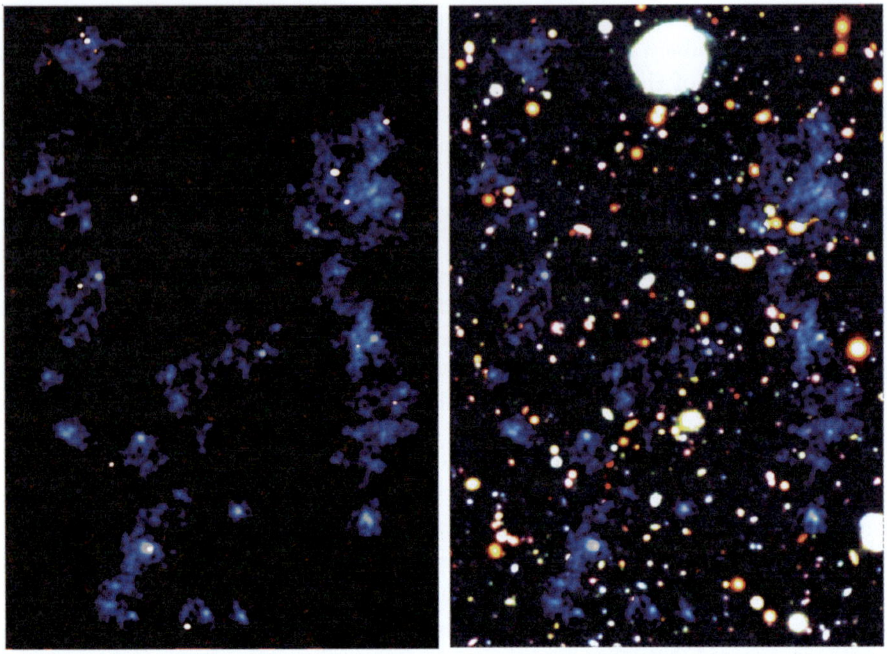

Fig. 5.79 (left) Map showing two *cosmic filaments*, each about 3 million light years long (blue) running from the top to the bottom of the image, detected using the MUSE instrument at the ESO Very Large Telescope in Chile. The white dots embedded within these filaments are very active star forming *starburst galaxies* which are being fed by the filaments, and which are detected using the Atacama Large Millimeter/submillimeter array There are additionally other galaxies with black holes at their centers, imaged by the Chandra X-ray space observatory, which are not seen here. The 24 galaxies associated with this cosmic filament are embedded in filaments of hydrogen gas. (right) Map showing the gas filaments (blue) running from the top to the bottom of the image, detected using the MUSE instrument at the Very Large Telescope. In colors, the optical light as seen by the Subaru telescope, coming from the galaxies that are visible in this region of the night sky, including galaxies that are physical associated with the filaments but also galaxies in the foreground and background of this structure, Umehata et al. (2019). (Images reproduced with the kind permission of Hideki Umehata)

Due to the enormous temperatures and densities that prevailed in the early stages of the Universe's evolution and the fact that matter was coupled to photons, radiation prevented through the pressure it exerted, the final gravitational collapse of these regions and the formation of the first stars and galaxies. *Dark matter*, however, which does not interact with photons, did not face this problem and began to widen the primordial fluctuations in its density long before the radiation was released from the matter. In the extreme, i.e., temperatures of this primordial epoch, because the radiation pressure prevailed over gravity, classical baryonic matter was unable to collapse into larger concentrations. Instead, *dark matter*, which does not interact with radiation, collapsed gravitationally, forming the primordial clumps of matter, an invisible spider web, which subsequently trapped with its gravitational attraction the visible matter, from which the first stars and the first small galaxies were formed. The first-generic gas that filled the young Universe first collapsed into extended sheets which then condensed into long filaments, separated by vast Voids.

We generally believe that the *filaments* in the early Universe fueled the formation of galaxies and supermassive *black hole*s in places where the *filaments* intersect, creating dense regions of matter. The hydrogen gas created in the *Big Bang* collapsed first into sheets and then into *filaments* that stretch long distances into space. In places where the *filaments* cross or meet, galaxies or superclusters of galaxies were formed (Fig. 5.79). At the same time, the *filaments* continued to fuel the growth of galaxies, supplying them with a steady stream of gas. Recent observations of the *cosmic web* reinforce the above cold *dark matter* theory of galaxy formation (Fig. 5.80).

The existence of *cosmic voids* is important in providing physical evidence for the distribution of *dark energy*. The coexistence of the largest known groups of *cosmic voids* and galaxies requires about 70% of the *dark energy* in the Universe today, according to the latest data from CMR. *Cosmic voids* act as bubbles in the Universe that are sensitive to background cosmological changes. This means that the evolution of the shape of the voids is partly the result of the expansion of the Universe. Since this acceleration is believed to be caused by *dark energy*, studying the changes in the

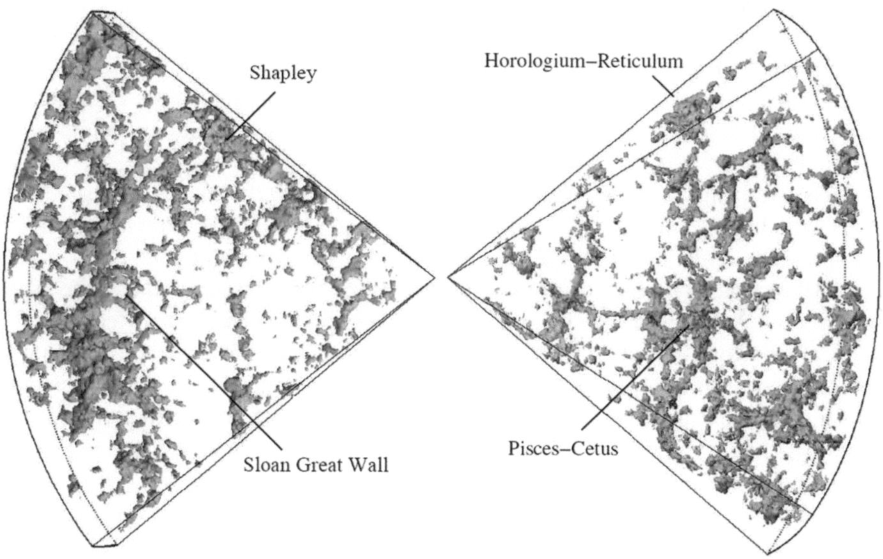

Fig. 5.80 Large-scale cosmic structures in two large slices of the distant Universe, about 2.5 billion light years away, via the Two-degree-Field Galaxy Redshift Survey. The Sloan *Great Wall*, the Shapley and Horologium-Reticulum *Superclusters*, and the Pisces-Cetus *Cosmic Filament* are visible. The Sloan *Great Wall* is one of the largest *Cosmic Filaments/Walls,* formed by a giant wall of galaxies, 1.38 billion light years long and located approximately one billion light years away. Observations are with the 3.9 m Anglo-Australian Telescope. (*Source* Willem Schaap (Kapteyn Institute, Un. Groningen) et al., 2dF Galaxy Redshift Survey, Australian Astronomical Observatory)

shape of voids over time can be used to place constraints on the standard cosmological model of cold *dark matter* (ΛCDM) which we will discuss below.

The elaborate structure of the local Universe of galaxies is evident in Fig. 5.80, where several massive *superclusters* of galaxies, i.e., clusters of galaxies, are visible. Such are the *Shapley supercluster* of galaxies and the supercluster of galaxies in the Horologium-Reticulum which contains about 30,000 giant galaxies and 300,000 dwarf galaxies, which form about 5,000 galaxy groups. The total mass of the supercluster has been calculated to be 10^{17} solar masses. Also shown in Fig. 5.80 is the Pisces-Cetus *cosmic filament*, which contains the *Virgo supercluster* and the *Local Group* of galaxies to which our Milky Way belongs. The *Sloan great wall* (Geller and Huchra 1989; Colless 2001; Gott et al. 2005; van de Weygaert and Schaap 2009) is located in the region of the constellations Corvus, Hydra, Centaurus. It was discovered through the *2dF Galaxy Redshift Survey,* a topographical survey of galaxies carried out by the Australian Astronomical Observatory with the 3.9 m Anglo-Australian Telescope from 1997–2002.

2019: Detection of Magnetic Fields in *Cosmic Filaments*

Like an ethereal cosmic spider web, the *cosmic filaments* of gas form a complex, interconnected structure that connects clusters of galaxies together (Fig. 5.78). However, just like the delicate threads of the spider web are almost invisible, the *cosmic web* is dim and hard to detect. Numerical simulations indeed predicted the existence of such a *cosmic web*, but we had not observed the network spanning between clusters of galaxies so far.

Recently, the first detailed picture of the light emitted by the gas of the *cosmic filaments*, which stretch for millions light years, has been obtained. There is a strange and of impressive length radio bridge of length about 10 million light years, connecting two colliding galaxy clusters, Abell 0399 and Abell 0401 observed using the Low-Frequency Array (LOFAR). The array of LOFAR radio telescopes is perfectly suited to detect very long-wavelengths, low-frequency radio waves from space and so was used to observe this particular pair of galaxy clusters because other earlier observations had already detected the *filament* connecting them, a part of the *cosmic web* that fills a large portion of the space around them. This particular pair was very interesting because both galaxy clusters had been found to have magnetic fields. As it is known, electrons that move very fast within a magnetic field, produce the *synchrotron radiation*, which is detected as radio wave emission. So, the next question was whether this *filament* between the two colliding galaxy clusters had a magnetic field. The best way to answer this question was to investigate whether there was radio emission between the two clusters. This was precisely detected when, with the help of LOFAR, the *cosmic filament* between the two galaxy clusters was observed: an emission bridge was revealed by radio waves spanning 10 million light years and connecting the two clusters, Govoni et al. (2019a). The magnetic fields run between these two galaxy clusters Abell 0399 and Abell 0401, which are beginning to merge, about 1 billion light years from Earth. Numerical simulations indicate that

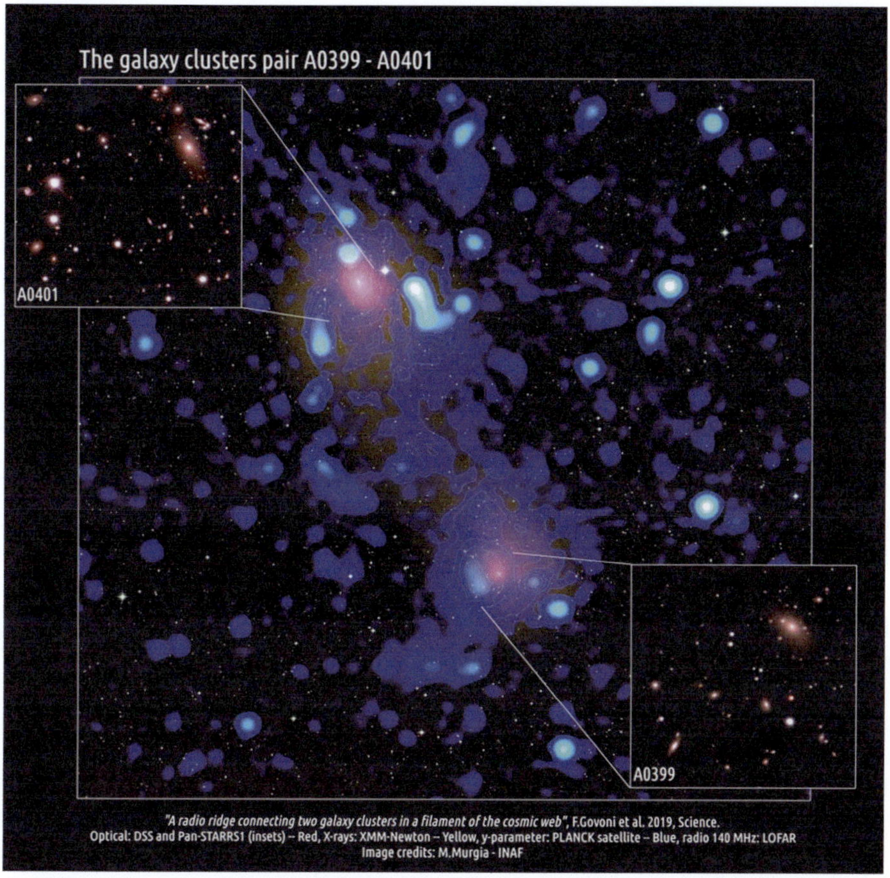

Fig. 5.81 Two distant galaxy clusters, Abell A0401 (upper left) and Abell A0399 (lower right), are connected by magnetic fields containing high-energy electrons. X-rays from the cluster cores € purple, and radio emission from the magnetic bridge forms the blue-toned smear in between. (Image reproduced courtesy of Matteo Murgia and Federica Govoni (INAF), Dss, Pan-Starrs1, XMM-Newton, Planck Satellite, Lofar, Astron)

weak shock waves from the early stages of this galaxy cluster merger can't accelerate normal electrons in the gaseous filament enough to generate the synchrotron emission observed. Instead, the filament must already have contained high-energy electrons that are being reaccelerated by merger shock waves.

Figure 5.81 is based on the related work Govoni et al. (2019b), of the collaboration of DSS and Pan-STARRS1 (optical), XMM-Newton (X-rays), PLANCK satellite (yparameter), LOFAR (radio). The system is at approximately 1 billion light years from Earth while the 2 galaxy clusters are about 10 million light years apart in projection. The cores of the two galaxy clusters are permeated by a high-temperature plasma that emits in X-rays (red tones), while observations in microwaves show a tenuous filament of matter connecting the two clusters. The low-frequency image in the radio waves reveals several bright discrete sources associated to individual galaxies and two diffuse radio halos towards the centers of the two clusters of galaxies. A striking ridge of radio emission is visible along the filament connecting Abell 0399 and Abell 0401 revealing the presence of a vast magnetic field illuminated by a population of high-energy electrons.

This is the first time, when magnetic fields between two galaxy clusters are observed, a result suggesting that some of the largest scale structures in the universe are indeed magnetized.

The next question that remains to be considered is whether the existence of magnetized *cosmic filaments* is a common phenomenon in the *cosmic web*. Fortunately, an additional observational tool will soon be available for this research, the *Square Kilometer Array* (SKA), which is expected to be the most powerful radio telescope in the world. One of the first goals of SKA is the detection of magnetic fields within the *cosmic web* itself.

2019: Discovery by the Hubble Space Telescope of Faint Light Inside the Galaxy Cluster Abell S1063 Showing the Distribution of *Dark Matter*

The *galaxy cluster* Abell S1063 is located at a distance of about 4 billion light years, in the constellation of the bird Crane (Grus, Flamingo) and consists of hundreds of galaxies, Fig. 5.82. This cluster has the truly gigantic mass of hundreds of trillions of suns. There, the strong sensitivity and resolution of Hubble, captured a faint blue emission among the myriad galaxies of Abell S1063, called the *inner cluster light*. The stars that emit this inner light they are *orphans*, having long since abandoned their parent galaxies. It is probably the result of the violent collision of the cluster's galaxies and now they wander among them as strays. Result of the collision of galaxies is also the intense X-ray emission from the cluster.

Everything Eventually May Be Useful...

These *unruly* and *uncommunicative* stars of the Abell S1063 galaxy cluster prefer to live solitary lives, rather than being tamed by a galaxy's particular gravitational field and thus living a *normal social life*. In other words, the dynamics, motion and detailed behavior of these *lonely stars* is no longer affected by the gravitational field of the particular parent galaxy they came from, but by the overall gravity of the cluster. So the interesting thing about this faint blue light seen by Hubble is that astronomers have correlated it with a map of mass distribution in the general gravitational field of the cluster, which is a good indication of how the invisible dark mass is distributed throughout the cluster. In fact, this method is better than X-ray emission in determining the dark matter of the cluster. Here, then, is a practical utility of these lonely and *elderly stars*.

Fig. 5.82 A *cluster of colliding galaxies,* Abell S1063, observed with the Hubble Space Telescope contains hundreds of galaxies, with a total of about 10^{15} stars, at a distance of about 4 billion light years from Earth (z = 0.351). The faint bluish emission, the *inner light of the cluster, comes from* stars *wandering* among the galaxies and is the result of the violent collision of other galaxies in the cluster. The collision of galaxies also causes a strong emission of X-rays from the cluster. Many arcs are seen around the cluster, effects of *gravitational lensing,* as light from distant galaxies is deflected, distorted and amplified as it passes through the space around the cluster. (*Source* NASA, ESA, and M. Montes (University of New South Wales))

Therefore, everyone ultimately needs the *social* stars of a galaxy, as they reveal to us the distribution of its light mass, but also we need the *uncommunicative ones*, as they reveal to us the distribution of the dark mass on the large scale of the cluster!

Galaxy collisions are a frequent phenomenon in the Universe. In a scale model where the galaxies are the size of a basketball, they are only a few meters apart. Besides, our Milky Way too is on a collision course with its neighboring Andromeda galaxy – the fatal collision that is predicted to occur in about 4.5 billion years. It should be noted that during the collision of two galaxies, the individual stars of the two galaxies do not collide with each other, simply because the galaxies are *empty* of stars, resulting in one *passing through the other*. But other phenomena occur, such as the creation of *orphan* stars discussed above, which escape the potential of the parent galaxy, or, concentrations of gas which strong *astrogenesis* phenomena will

Fig. 5.83 A ghost? ! No, a Hubble Space Telescope (6/19/2019) observation of two galaxies of equal size in a collision phase. At a distance of 704 million light years from Earth, this system is cataloged as Arp-Madore 2026-424. The blue areas around the perimeter trace areas of intense *astrogenesis*. This case foreshadows what will happen to our Milky Way galaxy when it collides with the Andromeda galaxy in about 4.5 billion years. *Source* NASA, ESA, J. Dalcanton, B.P. Williams and M. Durbin (University of Washington)

begin. In the Hubble Space Telescope Image – a ghost-like image – two galaxies of equal size 704 million light years from Earth have already collided, Fig. 5.83. After all, this is what the blue areas around the perimeter and made up of new blue stars betray.

2019: Spectacular Galaxy Clashes

In a relatively recent image from the Hubble Space Telescope, Fig. 5.83, it can be seen the collision of two galaxies of equal size, which looks like the face of a *ghost*. This system is 704 million years away light from Earth, and is named Arp-Madore 2026-424 in the Arp Madore (Catalogue of Southern Peculiar Galaxies and Associations). The observation was made with Hubble on June 19, 2019 in visible light by the Advanced Camera for Surveys.

Although galaxy collisions are common – especially in the early Universe – most do not happen head-on, like the one that created the Arp-Madore 2026-424 system. Such a violent collision gives the system its impressive ring structure, but only for a short time. The collision has pulled out and distorted the discs gas, dust, and stars that the two galaxies originally had, forming a ring, where a strong rate of new star formation is observed and gives the *nose* and *face* features to this system.

Ring galaxies are rare, and only a few hundred are found in our wider cosmic neighborhood. Such galaxies are created when the collision is in the right orientation, so that their mutual action creates this ring. Soon after the two galaxies will have merged completely, obliterating their episodic past.

The side-by-side appearance of the two central stellar bulges of primordial galaxies seen here is also unusual. Given the fact that the two bulges forming the *eyes* appear to be of the same size, we can infer that the two original galaxies involved in this collision were of about equal size. This is something rare because on more common collisions, the smaller galaxies are swallowed up by the their older neighbors. Astronomer Halton Arp published a catalog of 338 unusual collapsing galaxies in 1966. He later collaborated with astronomer Barry Madore to expand this list of unusual galactic collisions in the southern sky. So to date we have several thousand such galaxy collisions in this 1987 wide-field survey.

2024: Collision of Galaxies IC 2163 and NGC 2207, via Webb and Hubble

Two other galaxies in collision in a mix of mid-infrared light from the NASA/ESA/CSA James Webb Space Telescope, and visible and ultraviolet light from the NASA/ESA Hubble Space Telescope (Fig. 5.84). The pair grazed one another millions of years ago. The smaller spiral on the left, catalogued as IC 2163, passed behind NGC 2207, the larger spiral galaxy at right. Both have increased star formation rates. Combined, they are estimated to form the equivalent of two dozen new stars that are the size of the Sun annually. Our Milky Way galaxy forms the equivalent of two or three new Sun-like stars per year. Both galaxies have hosted seven known *supernovae* – a high number compared to an average of one every 50 years in the Milky Way – each of which may have cleared space in their arms, rearranging gas and dust that later cooled, and allowed many new stars to form.

Fig. 5.84 Two spiral galaxies take up almost the entire view and appear to be overlapping. The galaxy at left, IC 2163, is smaller and more compact than the galaxy at right, NGC 2207. The black background of space is filled with foreground stars and extremely distant galaxies. (*Credit* NASA, ESA, CSA, STScI)

Over many millions of years, the galaxies may swing by one another repeatedly. It's possible that their cores and arms will meld, leaving behind completely reshaped arms, and an even brighter, cyclops-like *eye* at the core. Star formation will also slow down once their stores of gas and dust deplete, and the scene will calm.

2024: Interacting Galaxies Arp 142, via NIRCam and MIRI of the James Web Space Telescope

Another pair of colliding galaxies, known jointly as Arp 142, located at 326 million light years from Earth in the constellation Hydra, made their first pass between 25 and 75 million years ago – causing *fireworks*, or new star formation, in one of them, called the *Penguin* (Fig. 5.85). In the most extreme cases, mergers can cause galaxies to form thousands of new stars per year for a few million years. For the *Penguin*, it has been shown that about 100–200 stars have formed per year. By comparison, our Milky Way galaxy, which is not interacting with a galaxy of the same size, forms roughly six to seven new stars per year.

This gravitational shimmy also remade the *Penguin's* appearance. Its coiled spiral arms unwound, and gas and dust were pulled in an array of directions, like it was releasing confetti. It is rare for individual stars to collide when galaxies interact (space is vast!), but the galaxies' mingling disrupts their stars' orbits. Today, the *Penguin*'s galactic center looks like an eye set within a head, and the galaxy has prominent

Fig. 5.85 Two interacting galaxies known as Arp 142. At left is NGC 2937, nicknamed *the Egg* for its appearance. At right is NGC 2936, nicknamed *the Penguin* for its appearance. The latter's beak-like region points toward and above the *Egg*. (*Credit* NASA, ESA, CSA, STScI)

star trails that take the shape of a beak, backbone, and fanned-out tail. A faint, but prominent dust lane extends from its beak down to its tail.

Despite the *Penguin* appearing far larger than the *Egg,* these galaxies have approximately the same mass. This is one reason why the smaller-looking *Egg* hasn't yet merged with the Penguin. If it was less massive, it may have merged earlier. The oval *Egg* is filled with old stars, and little gas and dust, which is why it isn't sending out *streamers* or *tidal tails* of its own and instead has maintained a compact oval shape.

2024: Encounter of the Milky Way with the LMC Blew Away Most of the LMC's Gaseous Halo

The dwarf galaxy of the Large Magellanic Cloud (LMC), with 10 percent the mass of the Milky Way, looms large in the southern nighttime sky at 20 times the apparent diameter of the full Moon. And yet, this is the first time now that we have been able to measure the size of the LMC's halo by analyzing Hubble's ultraviolet observations from the Mikulski Archive for Space Telescopes at STScI. Specifically, data from Hubble's Cosmic Origins Spectrograph (COS) were used to detect the presence of the halo gas by the way it absorbs certain colors of light from 28 bright background *quasars*. With COS, Mishra et al. (2024) measured the velocity of the gas around the LMC, which allowed them to determine the size of the halo.

Specifically, in the study Mishra et al. (2024), it has been found that an extremely small gaseous halo has remained in the LMC – about 50,000 light years across, i.e., around 10 times smaller than the halos of other galaxies that have the same mass with the LMC – after its collision with the Milky Way. Actually, the LMC has just completed its closest approach to the much more massive Milky Way. This passage has blown away most of the spherical halo of gas that surrounds the LMC. Still, the LMC has held onto enough of its gas to keep forming new stars. The LMC still retains a compact, stubby halo of gas – something that it wouldn't have been able to hold onto gravitationally had it been less massive. A smaller galaxy wouldn't have survived such an encounter. But its compactness tells the story of its encounter with the Milky Way.

Most of the LMC's halo was blown away by *ram-pressure stripping*. The dense environment of the Milky Way pushed back against the incoming LMC, creating a wake of gas trailing the dwarf galaxy – like the tail of a comet (Fig. 5.86a). What we're seeing now is the compact leftover of this collision. Because of its mass and proximity to the Milky Way, the LMC is a unique astrophysics laboratory. Seeing the LMC's interplay with our galaxy help us to understand what happened in the early Universe, when galaxies were closer together. It also shows just how messy and complicated the process of galaxy interaction is.

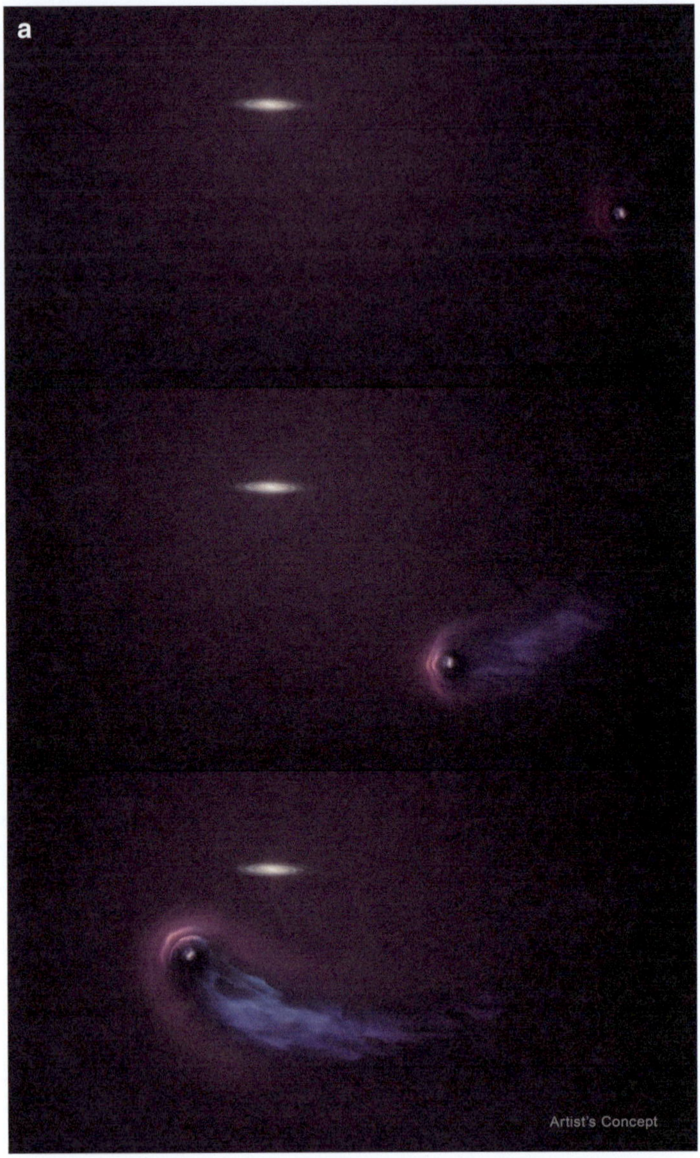

2025: Should We Worry About a Milky Way—Andromeda Collision?

As far back as 1912, a collision of our Milky Way galaxy containing almost 300 billion stars, with the neighboring large galaxy of Andromeda (M31) – then thought to be only a nebula – containing almost one trillion stars, both members of the *Local Group* of galaxies, was predicted to occur in about 4.5 billion years, with the formation of the new *Milkdromeda* galaxy. Specifically, by measuring the sideways

◄**Fig. 5.86** **a** This artist's concept illustrates the large magellanic cloud's (LMC's) encounter with the Milky Way galaxy's gaseous halo. In the top panel, at the middle of the right side, the LMC begins crashing through our galaxy's much more massive halo. The bright purple bow shock represents the leading edge of the LMC's halo, which is being compressed as the Milky Way's halo pushes back against the incoming LMC. In the middle panel, part of the halo is being stripped and blown back into a streaming tail of gas that eventually will rain into the Milky Way. The bottom panel shows the progression of this interaction, as the LMC's comet-like tail becomes more defined. A compact LMC halo remains. The bottom panel shows the tail becoming longer and more defined as the now tadpole-like object curves below the spiral galaxy and sweeps toward the upper left. The LMC is just past its closest approach to the Milky Way and is moving outward into deep space again. (*Image Credit* NASA, ESA, R. Crawford (STScI)). **b** These three images of galaxy interactions illustrate three possible encounter scenarios between our Milky Way galaxy and the neighboring Andromeda galaxy. (*Top left*) Scenario 1 is similar to the interacting galaxies M81 (*bottom*), a grand design spiral galaxy about 12 million light years away in the constellation Ursa Major and M82 (the Cigar galaxy, *top*). The Cigar galaxy experiences gravitational interactions with its galactic neighbor, M81 and NGC 3077, located off the edges of this image, causing it to have an extraordinarily high rate of star formation – a starburst galaxy. (*top right*) Scenario 2 is similar to the case of the pair of interacting galaxies NGC 6786 (right) and UGC 11,415 (left) imaged by the Hubble Space Telescope, wherein NGC 6786 being the larger galaxy, is likely that it will absorb UGC 11,415 in the future. (*bottom*): Scenario 3, NGC 520 (the Flying Ghost), a pair of colliding spiral galaxies about 105 million light years away in the constellation Pisces. Simulations indicate that this object consists of two galactic disks that began interacting about 300 million years ago. The system is still in an early stage of its merger. (*Credit* NASA, ESA, STScI, DSS, Till Sawala (University of Helsinki); *Image Processing* Joseph DePasquale (STScI)) **c** This artist's impression shows a "cosmic joust" of two galaxies: the galaxy on the right which hosts a quasar at its core, is piercing the other galaxy on the left with a lance (its quasar's beam). As the radiation of the quasar interacts with the galaxy on the left, it disrupts the clouds of gas and dust within, leaving behind only the smallest and densest regions. These regions are likely rendered incapable of star formation after the process. (*Image Credit* ESO and M. Kornmesser)

motion of Andromeda with the NASA/ESA Hubble Space Telescope it was found that it was so negligible that an eventual head-on collision with the Milky Way seemed almost certain. So, it was concluded that the Andromeda galaxy is threateningly approaching our own Milky Way galaxy, at a speed of about 110 km per second, i.e., almost 400,000 km per hour, as indicated by blueshift. A smashup between our own galaxy and Andromeda would trigger a firestorm of star birth, supernovae, and maybe toss our Sun into a different orbit. Simulations had suggested that it was inevitable.

Nevertheless, the probability of a collision of two particular stars from these two galaxies during such a collision is negligible, due to the vast distances between their stars. For example, the nearest star to our Sun, *Proxima Centauri*, is about 4.2 light years away from Earth, or, 30 million times the solar diameter. Thus, if the Sun were the size of a ping-pong ball (say, one centimeter in radius), this nearest star would be about 600 km away and our Milky Way galaxy, would have a diameter of about 30 million km. Although stars are closely packed at the centers of each galaxy, the average distance between them is still enormous and stars are trillion kilometers apart. Therefore, it is extremely unlikely that any two stars will collide when these two galaxies collide in the very distant future.

Fig. 5.86 (continued)

However, a key parameter for the galaxy collision is the *lateral* speed (proper motion) which is very difficult to measure with sufficient precision to draw reasonable conclusions. Hence, until 2012, it was not known whether the possible collision was definitely going to happen or not.

Recently, however, a new study (Sawala et al. 2025) found that, based on the latest observational data from *Hubble* as well as the *Gaia* space telescope, there is only a 50–50 chance of the two galaxies colliding within the next 10 billion years. This study also found that the presence of the large magellanic cloud and the effects of the orbits of Andromeda's large satellite galaxy, M33, could affect the trajectory of the Milky Way and make the collision less likely. The study considered 22 different variables that could affect the potential collision between our galaxy and our neighbor, and ran 100,000 Monte Carlo simulations stretching to 10 billion years into the future: the Milky Way and Andromeda alone would remain in the same plane as they orbit each other, but this doesn't mean they need to crash (panel 1 in Fig. 5.86a). They could still go past each other. Specifically, the extra mass of M33 pulls the Milky Way a little bit more toward it, but the LMC pulls the Milky Way off the orbital plane and away from Andromeda. It doesn't mean that the LMC will save us from that merger, but it makes it a bit less likely.

In about half of the simulations, the two main galaxies fly past each other separated by around half a million light years or less (five times the Milky Way's diameter), (panel 2 in Fig. 5.86b). They move outward but then come back and eventually merge in the far future (as in the lower panel 3 of Fig. 5.86b). The gradual decay of the orbit is caused by dynamical friction between the vast dark-matter halos that surrounds each galaxy at the beginning. In most of the other cases, the galaxies don't even come close enough for dynamical friction to work effectively. In this case, the two galaxies can continue their orbital waltz for a very long time.

The new result also still leaves a small chance of around 2% for a head-on collision between the galaxies in only 4–5 billion years. Thus, predicting the long-term future of galaxy interactions is highly uncertain, but the new findings challenge the previous consensus and suggest that the fate of the Milky Way remains an open question.

As remarked by the lead author of the study T. Sawala, considering that the warming Sun makes Earth uninhabitable in roughly 1 billion years, and the Sun itself will likely burn out in 5 billion years, a collision with Andromeda is the least of our cosmic worries.

(2025) *Cosmic Joust*: a Pair of Galaxies in a Deep-Space Battle

A major merger event is observed with the quasar J 012555.11−012925.00 at redshift $z = 2.6618$ and its companion (Fig. 5.86b, lower panel). The two galaxies, at a physical projected distance of about 5 kpc, and approaching each other at speed $\approx$550 km s^{-1}. The companion and quasar host galaxies are both massive, with stellar masses close to 10^{11} $M_{\odot}$, form stars, and contain a substantial molecular mass. Radiation from the quasar in one galaxy directly alters the gas properties in the other galaxy. The observation shows an ongoing major merger event, which is an important feeding mechanism for quasars. Specifically, the gas within this system is being transformed into a highly clumpy medium wherever exposed to the intense UV radiation powered by the actively accreting supermassive black hole. This quasar disrupts the molecular gas in a manner analogous to the impact of newborn stars on their parent molecular clouds. The observations infer that, wherever exposed to the quasar radiation, molecular gas is disrupted, leaving behind surviving dense clouds too small to give birth to new stars. In other words, the observations show that the radiation of the quasar dampens the wounded galaxy's ability to form new stars.

Combined observations from both the European Southern Observatory's Very Large Telescope (ESO's VLT) and the Atacama Large Millimeter/submillimeter Array (ALMA), reveal that these two galaxies are locked in a thrilling war (Balashev 2025). Over and over, they charge toward each other at speeds of 550 km/s on a violent collision course, only to land a glancing blow before retreating and winding up for another round, similarly to the medieval sport of *joust*. But these galactic knights aren't exactly chivalrous, and one has a very unfair advantage: it uses a quasar to pierce its opponent with a spear of radiation! This is the first time that we see how a quasar wounds a regular galaxy and its ability to form stars.

These mergers of galaxies are thought to bring huge amounts of gas to supermassive black holes residing in galaxy centers (Balashev 2025). In the cosmic joust, new reserves of fuel are brought within reach of the black hole powering the quasar. As the black hole feeds, the quasar can continue its damaging attack. The light from this *cosmic joust* has taken over 11 billion years to reach us, so we see it as it was when the Universe was only 18% of its current age.

Gravitational Waves

2015: First Direct Detection of Gravitational Waves

In physics, an *event* is characterized by its spatial coordinates (i.e., exactly *where* it happened) and time (i.e., *when* exactly it happened). For example, the instantaneous position of an observer, or, the Earth in space at a particular moment in time are such *events*. The combination of Euclidean space of three dimensions (length, width, height) with the time of a clock as an additional dimension leads to a four-dimensional continuum (a *manifold* in mathematics) which we call *spacetime*. Although the fourth dimension, that of time, is different from the other three that concern lengths in Euclidean space, it can be treated mathematically equivalently to the other three spatial dimensions. Spacetime is the *field* in which all physical events play and take place, i.e., the basic constituent elements of spacetime are events (Fig. 5.87).

According to the General Theory of Relativity (GTR), the local shape of spacetime (e.g., whether it is flat or curved) is a result of the density of matter and energy

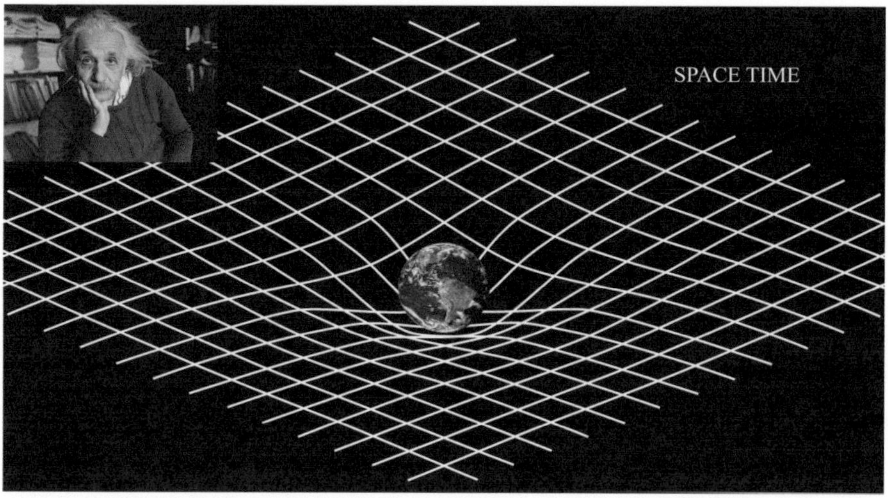

Fig. 5.87 Explanatory diagram of the curvature of spacetime from the concentration of matter and energy, according to Einstein's General Theory of Relativity

concentrated at that point. Einstein's GTR equations describe exactly this form of spacetime as a function of matter and energy density (Fig. 5.87).

Gravitational waves are wave changes of the local curvature of space in space–time, which propagate from the source in all its external directions. In cases where the concentration of matter/energy changes dramatically with time, for example, when two *neutron star*s or two *black hole*s collide, temporal oscillations occur in space, which propagate as waves throughout the Universe.

An analogy often used is that of a stone falling into the still water of a lake: concentric circles of disturbed water move around the point where the stone fell, spreading over long distances.

The Infinitesimal Amplitude of Gravitational Waves

Einstein predicted the existence of gravitational waves in 1916, as a consequence of his theory of General Relativity. But he estimated that they create *ripples* in spacetime of such infinitesimal amplitude that they are probably impossible to observe.

Indeed, the amplitude of a gravitational wave is extremely small, $1{:}10^{21}$. This means that each of the 4 km long arm of a modern interferometer will be distorted by 10^{-18} of a meter, i.e., a billionth of a billionth of a meter, or, one ten-thousandth of the radius of a proton. Also, if the arm of the interferometer was as long as the distance to the nearest star, i.e., 4 light years, then the width

of a gravitational wave would have been smaller than the thickness of a hair. And yet gravitational waves were detected!
Einstein was also wrong about the detectability of gravitational waves—but even Einstein made mistakes, as we discussed previously..

Gravitational waves carry energy in the form of gravitational radiation. It must be understood that gravitational waves are essentially due to the finite speed of propagation of physical interactions in the context of the GTR, in contrast to the Newtonian theory of gravity, wherein gravitational waves do not exist because there physical interactions propagate at infinite speed, i.e., the interaction is *instantaneous*. Gravitational waves can theoretically propagate at all frequencies. Practically, however, very low frequency gravitational waves are impossible to detect.

The first indirect evidence for the existence of gravitational radiation was the observation of the contraction of the orbit of the double pulsar PSR B1913+16 by the J. Taylor and P. Hulse discussed earlier. In this case the reduction of the energy of the system corresponded to the emission of gravitational radiation. But the first *direct* observation of the existence of gravitational radiation was made in September 2015 and announced in February 2016 by the research team of the two interferometers LIGO (Laser Interferometer Gravitational Wave Observatory) as the first detection of gravitational waves, coming from a pair of *black hole*s which merged. The two LIGO detectors at Hanford (Washington), and Livingston (Louisiana), (Fig. 5.89) each have a four-kilometer arm that form a right angle (the shape of the letter L). They started their operation in 2002, were upgraded in 2015 and in the same year they made the historic first discovery of gravitational waves from the Universe, Fig. 5.88.

***Whispers* of the Universe …**

The spectrum of gravitational waves is estimated to extend from 10^{-7} to 10^{11} Hz. Audible frequencies for humans range from approximately 20–20 kHz. Thus, because the frequencies of gravitational waves are frequencies that the human ear can hear, today we may say that we can hear for the first time the *whispers* of the Universe *talking* to us.
It should also be mentioned that gravitational radiation allows the observation of objects in the Universe that do not emit light, such as *black holes* and *dark matter*. At the same time, we can get closer observationally to the moment of the *Big Bang* itself, than using the electromagnetic radiation collected by the various conventional telescopes. This is because electromagnetic radiation collectors can observe what happened in the Universe only after 380,000 years from the time of the *Big Bang*.

The observation of the source GW150914 (Figs. 5.1 and 5.88) confirmed the existence of gravitational waves, a development hailed as *the greatest discovery*

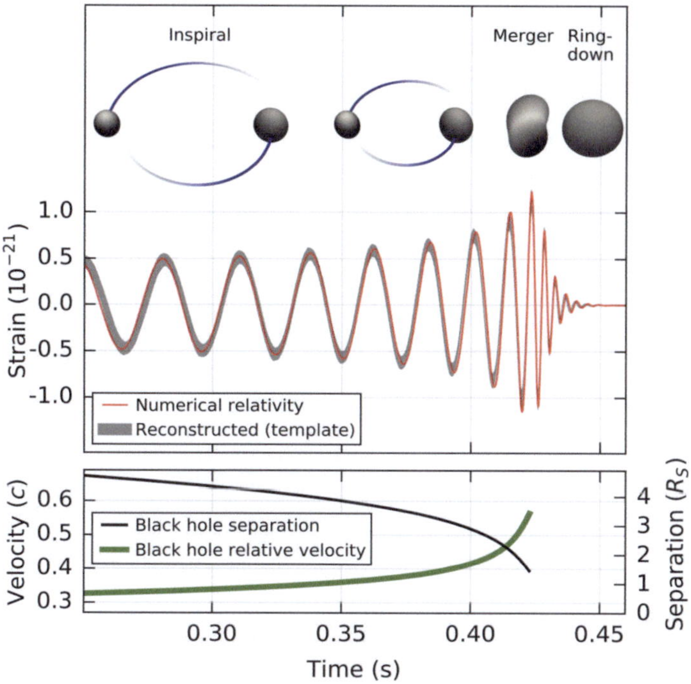

Fig. 5.88 Explanatory diagram of the spiral motion of the 2 black holes (BH) and the theoretical and observational pulse from the 2 LIGO interferometers at the source GW150914. Also shown is the velocity in units of the speed of light c and the distance of the 2 black holes in event horizon radii (R_{Sch}), before their final merger. (*Source* LIGO/Virgo Collaboration)

of the century to date. The damping of the waveform produced corresponds to the damping of the wave produced by a rotating Kerr *black hole* that is gradually brought to rest after the last dramatic fractions of a second before the merger of its two progenitors and its own birth. The exact description of the waveform detected by the two LIGO interferometers results from the numerical solution of the system of coupled differential equations of the Theory of General Relativity (GTR) and is in full agreement with the TGR predictions for the orbital contraction and final merger of a pair of rotating *black hole*s with masses around **36** and **29** solar masses ($M_{\odot}$) and emission of the subsequent *ringdown* waveform from the *black hole* that formed. The *black hole* that was finally formed has a mass of 62 solar masses, i.e., it was depleted by 3 $M_{\odot}$. The two stars had initial masses of 40 and 100 solar masses and began their life about 2 billion years after the *Big Bang*. This event took place at a distance of about 1.4 billion light years.

To date, many cases of gravitational waves have been detected, some from the merger of a pair of *black hole*s. After the upgrade of the two LIGO interferometers, now they are up and running, and hence these detectors will be able to detect *neutron star* mergers about a billion light years away, almost twice as far as today.

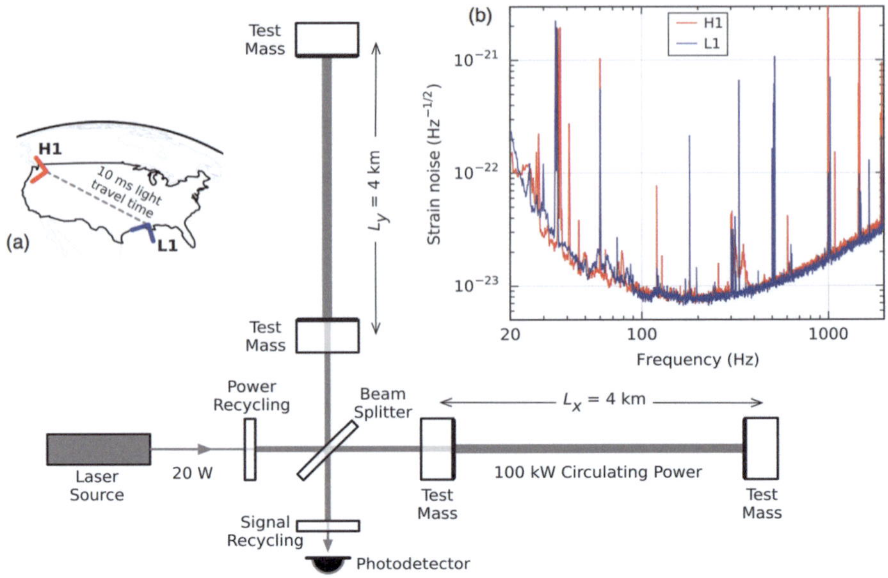

Fig. 5.89 Simplified diagram, not to scale, with arms along the x- and -y axes with lengths $L_x = L_y = 4$ km. Top left shows the position of LIGO at Hanford (H1) and Livingstone (L1) in a distance of about 3,000 km, with their arms in a different orientation. (*inset*) The instrument noise for each detector near the time of the signal detection. (*Credit* Publicly available image at https://www.ligo.caltech.edu/)

The Energy Produced by the Merger of 2 Supermassive *Black Holes* Exceeds the Energy Emitted by All Stars Combined in the Observable Universe …

Interestingly, during the 20 ms of the merger of the two black holes, the power of the gravitational wave radiation reached a maximum value of 3.6×10^{49} W, i.e., 50 times the total energy per second that all stars of the observable Universe shine together!

2017: Discovery and First Identification of a Gravitational Wave (GW170817) by the LIGO and VIRGO Interferometers and 70 Ground-Based and Space Telescopes Globally Observed Across the Electromagnetic Spectrum

2017 marks the beginning of a new era in Astrophysics. On the 17th of August of this year, for the first time, the *ripples* in spacetime, which we call gravitational waves, were directly detected, alongside the electromagnetic waves that came from the merger of two *neutron stars*. It is the first time that gravitational and electromagnetic waves have been detected simultaneously from the same source. This event was characterized as *the most important discovery of the year 2017*.

The detection was first performed by the LIGO–Virgo interferometers, Figs. 5.89 and 5.90 almost simultaneously, on August 17, 2017 and was immediately followed by the γ-ray detection of the space telescopes *Fermi* and *Integral*. The signal originated from the merger of two *neutron stars* – a source we now call GW170817. It was the first time LIGO–Virgo researchers had seen the merger of two *neutron stars*, and in the following five hours they had marked with great accuracy the very region of this source in the sky.

As we have already discussed, *neutron stars* are the densest stellar objects known in the Universe. They are formed when massive stars run out of their fuel and explode creating supernovae. When these superdense stars are members of a binary stellar system, moving around their common center of mass, they lose energy by emitting gravitational waves. As they lose energy in this way, their tight embrace causes them to around their common center of mass, before their final merger. So, the gravitational waves were detected for about 100 s. When the stars finally collided, they made sure to let us know about their dramatic end, giving us and the rest of the Universe a flash of γ-ray light that was detected about two seconds after the gravitational waves were detected. In the hours, days, and weeks following this collision, other forms of electromagnetic radiation were detected, including X-rays, ultraviolet, visible, infrared, and radio waves, Fig. 5.92. Two hundred days after the merger, observations

Fig. 5.90 The Virgo gravitational wave detector in Cascina, a small town near Pisa, home to the European Gravitational Observatory, an international collaboration of scientists from France, Italy, the Netherlands, Poland, and Hungary. Its name comes from a cluster of about 1,500 galaxies located in the constellation Virgo, about 54 million light years from Earth, one of the prime targets for gravitational wave detection. Each of its arms spans 3 km. (*Credit* The Virgo Collaboration)

combining radio telescopes in Europe, Africa, Asia, Oceania, and North America also confirmed the existence of a *jet* resulting from this violent collision (Fig. 5.94).

These coordinated observations provided valuable insight into what happens when two *neutron star*s collide and produce a new object we call a *kilonova*. These observations also gave us useful clues about how heavy precious metals, such as gold and platinum are produced in the Universe. They also helped us solve an old problem that has to do with the formation of high-energy, very short-duration γ-ray bursts. Finally, this observational event gave us a new and independent way of calculating the expansion of the Universe.

An Unprecedented Global Scientific Mobilization ...

At 8:41 am Eastern Daylight Time on August 17, 2017, the automatic gravitational pulse detection algorithm of one of the two LIGO interferometers *caught* a gravitational signal. Almost simultaneously, the Gamma-ray Burst Monitor on NASA's Fermi Gamma-ray Space Telescope also *caught* a γ-ray flash. The same happened with the second LIGO detector, the third Virgo interferometer and the *Integral* space telescope. These simultaneous observations meant that the 8.41 am event on August 17 was no coincidence. Immediately then an *alarm* was issued and the related information mechanism of many observatories and telescopes around the world was activated in order for them to observe it as well.

Robotic visible telescopes were then activated automatically in Chile and were able to pinpoint the source of GW170817 within hours. In the hours and days that followed, more than 70 telescopes searched for and eventually located the source GW170817 in a distant galaxy, about 130 million light years away, collecting a wealth of observational data in γ-rays, X-rays, the visible, the infrared, and the radio part of the electromagnetic spectrum.

Each interferometer consists of two perpendicular to each other tunnels 4 km long which form an L shape (Figs. 5.89 and 5.90). At their junction, a laser beam splits into two, sending the two sub-beams toward the ends of the two tunnels. There the light is reflected back in the direction it came from by two mirrors which are suspended by straps to avoid ground vibrations. When no gravitational signal passes through the detector, the two beams return to where they started at exactly the same time. But when a gravitational wave passes through the device, it alters the length of one of the two arms of the device, resulting in a very faint signal due to the formation of interference fringes.

The twin LIGO interferometers (Fig. 5.89) are funded by the US National Science Foundation (NSF) and were designed, built, and operated by Caltech and MIT. Additional financial support comes from Germany (Max Planck Society), UK (Science and Technology Facilities Council), and Australia (Australian Research Council), which are participating countries in the program. A total of about 1,200 scientists from 100 Universities and scientific centers around the world participate in this collaboration (LIGO Scientific Collaboration), which also includes the European GEO collaboration (GEO Collaboration) as well as the corresponding Australian collaboration (Australian collaboration OzGrav).

The Virgo interferometer (Fig. 5.90) is funded by the Istituto Nazionale di Fisica Nucleare (INFN) in Italy and the Center National de la Recherche Scientifque (CNRS) in France and operates under the European Gravitational Observatory (EGO). About 280 physicists and technicians from 20 European research groups are participating: 6 from the Center National de la Recherche Scientifque (CNRS) in France, 8 from the Istituto Nazionale di Fisica Nucleare (INFN) in Italy, 2 from the Netherlands, the MTA Wigner RCP team from Hungary, the Polish group POLGRAW, the University of Valencia in Spain and the European Gravitational Observatory, EGO, the laboratory operating the Virgo interferometer near Pisa, Italy funded by CNRS, INFN, and Nikhef. In other words, a total of approximately 1,500 scientists work closely together to operate the LIGO and Virgo interferometers as well as to process and understand the results of the collected data.

Today there is an admirable global scientific collaboration among the gravitational wave observatories to which the interferometers of Japan and India are planned to be added. And, with the participation and cooperation of many observatories of the electromagnetic spectrum around the world, a global effort has been organized which opens new paths in astronomy and the understanding of the Universe using gravitational wave detectors and astronomical probes. But let's look at the timeline of this great discovery in slow motion.

The 1 Meter 2 Hemisphere (1 m2H) Swope telescope (Chile) made the first detection of a gravitational wave of GW170817 in the visible, 10.86 h after the initial observation of this source with the LIGO/Virgo interferometers, Fig. 5.92.

This small (1-m) but fully equipped telescope was commissioned in 1971 by the Carnegie Institution at Cerro Las Campanas, Chile. The same Cerro Las Campanas observatory in Chile houses the two large 6.5 m Magellan telescopes of the Carnegie Institution. The telescope was named after a Carnegie astronomer, Henrietta Swope, a collaborator of the eminent astronomer Walter Baade and author of many important works in astrophysics. The first optical detection of the gravitational wave source GW170817 was made by the team at the University of California, Santa Cruz, led by Ryan Foley of UC Santa Cruz), by comparing the new image SSS17a of the lenticular galaxy NGC 4993 with earlier images taken by the Hubble Space Telescope. This galaxy NGC 4993 is located about 130 million light years away in the constellation Hydra. Source SSS17a was not present in earlier Hubble Space Telescope views of this galaxy, but appeared in the new Swope telescope observation at 8.14 pm, 10 h,and 22 min after the merger of the two *neutron star*s. This process was not at all easy because it was necessary to compare the images of many galaxies before and after 8/17/2017 within a wide area of the sky indicated by the first detection of the source by the LIGO interferometers.

Alongside the 1 Meter 2 Hemispheres Swope telescope, the effort to locate the source GW170817 also involved the small robotic panchromatic robotic optical monitoring and polarimetry telescopes (PROMPT) in the highlands of Chile: these telescopes survey galaxies up to 130 million AU, discovering about 10 supernovae per year, within a day of their initial explosion, in the so-called wide field survey DLT40.

The Hubble Space Telescope observed GW170817 from August 22 to 28, 2017, by which time the source had become extremely faint, Fig. 5.91 (Fig. 5.92).

In a schematic model of the physical and geometric parameters of the γ-ray burst that followed the merger of the two *neutron stars* in the source GW170817, Mooley et al. (2018), the event GW170817 contained a *jet* that has created a cocoon around itself through this *jet*'s interaction with the energetic debris on the spherical surface around it (the cocoon is an envelope of silk made by silkworms and the wax they secrete wraps around their body and as it hardens it forms the cocoon). The emergence of the shock wave from the cocoon probably generated the γ-ray emission and the other interaction of the cocoon with the interstellar medium producing the first (up to 2 months after the merger) emission in X-rays and radio waves. The relativistic core of the *jet* is inside a half-aperture cone $\theta_{jet} \leq 5°$. The Earth is at an angle $\theta_{obs} = 14°–28°$ from the *jet* axis. The event GW170817 most notably produced a brief γ-ray burst directed at this angle relative to Earth. The interstellar medium produced

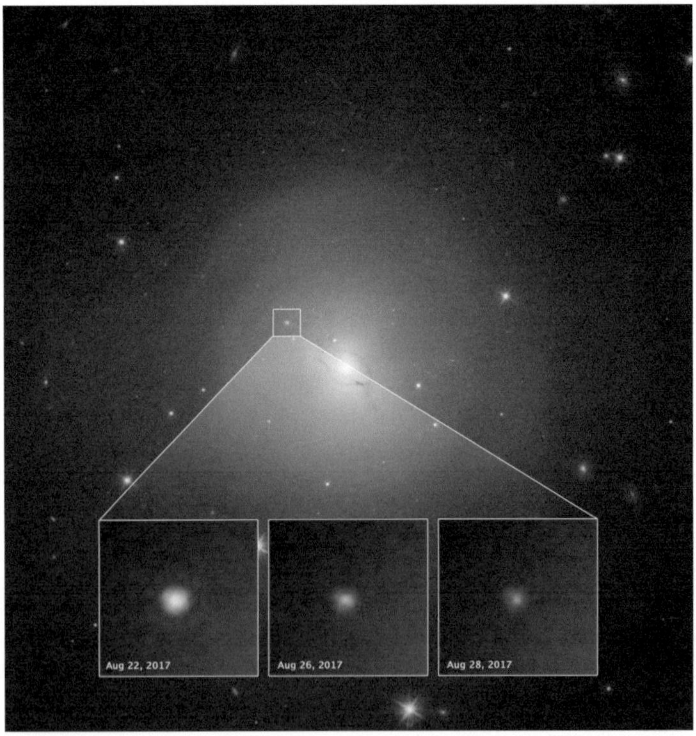

Fig. 5.91 The kilonova explosion was observed by the Hubble Space Telescope after 5 days, from August 22 to 28, 2017, when the source became very faint. (*Source* NASA, ESA, ESO)

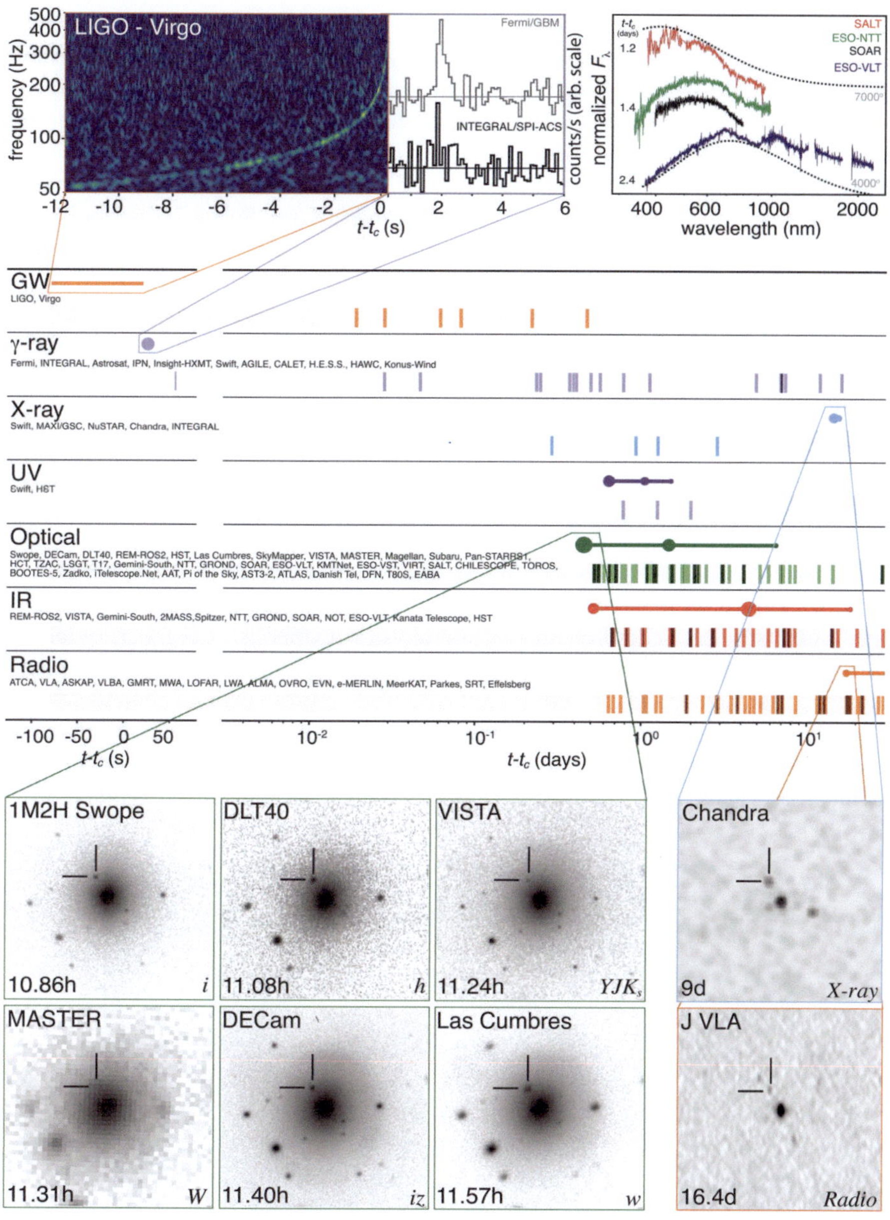

Fig. 5.92 Global identification of the source of gravitational waves GW170817. From above, with LIGO/Virgo, *Fermi* (GBM) and *Integral* in γ-rays, various optical telescopes (from 10.86 to 11.57 h), the *Hubble* Space Telescope (in 5 days), *Chandra* (in 9 days), and finally with *VLA* on radio waves (in 16 days). (*Source* arXiv e-prints (arXiv:1710.05833))

the delayed emission in X-rays and radio waves. Measurements with the VLBI array give a Lorentz factor for the *jet* $\gamma \approx 4$, 150 days after the merger (at the maximum of the radio light curve, when the core of the *jet* appeared). The total energy of the *jet*-cocoon system is 10^{49}–10^{50} ergs. The density (n) of the environment around the merger is 10^{-4}–5×10^{-3} cm^{-3} (Mooley et al. 2018), arXiv.org > astro-ph > arXiv:1806.09693, 2018).

The X-ray emission was detected by the Chandra X-ray Space Telescope 9 days after it was first detected by the LIGO-Virgo interferometers and visually identified by the 1m2H Swope and PROMPT telescopes. This appears to be related to the viewing angle of the *jet* created after the collision of the two *neutron stars* because it took a few days for the *jet* to *open up* and thus be observable from Earth. After 16 days it was also observed in radio waves by the Very Large Array (VLA).

2017: Discovery of Kilonova Explosions, Cosmic Alchemy, and the Creation of the Precious Metals

Most of us have seen and recognized the periodic table of the chemical elements since our first High School chemistry and physics textbooks. It is a list, in tabular form, of the chemical elements, in an arrangement organized based on the atomic number of each chemical element (i.e., the number of protons in its nucleus), which is equal to the number of electrons orbiting the nucleus, their electronic structure and repeated chemical properties. However, it may not be known that for most of these chemical elements in the periodic table we do not know their exact origin!

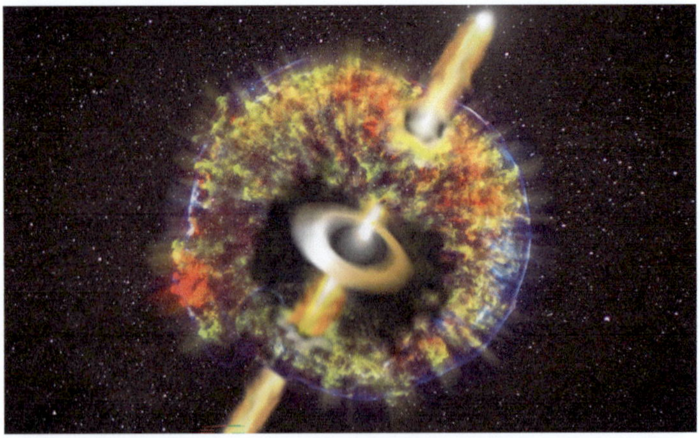

Fig. 5.93 Artist's illustration of the merger of two neutron stars, ejecting material into space with the observed jet penetrating the shell and propagating out of it. The jet is produced by the black hole, surrounded by a hot disc, which was formed after the merger. (*Source* O.S. Salafia, G. Ghirlanda, NASA/CXC/GSFC/B. Williams et al.)

What Is the Origin of Most of the Chemical Elements Around Us?

Out of the 118 chemical elements in the periodic table, 94 are found on Earth and the remaining 24 are made synthetically in large laboratories, but they are unstable. Of the 94 stable chemical elements, the first two, hydrogen and most of helium, were formed in the first 3 min after the Big Bang and are the most abundant chemical elements in the Universe: in the Sun, in every sample of 1,000 atoms we will find 921 hydrogen atoms, 28 helium atoms, and just one atom from the remaining elements. The chemical element immediately following helium on the periodic table is lithium, a percentage of which was also formed in the Big Bang. Subsequently, the next chemical elements of the periodic table, beryllium and boron which also occur in the Sun, are formed by collisions of cosmic rays with large nuclei.
The remaining 89 elements, we don't know exactly how our little planet got them. Certainly thcy did not form here on Earth: either they came from the primordial materials from which our planetary system was made, or they were transported by comets and asteroids that happened to collide with our planet. Therefore, an important question remains unanswered:
what Is the Origin of the Vast Majority of the Chemical Elements?

In order to manufacture new chemical elements, enormous amounts of energy, destructive pressures and extremely strong forces are required in principle which, overcoming the strong repulsion between the charged subatomic particles of the nucleus, can finally manage to *throw* them into the nucleus, where the tiny-range – but terrifyingly powerful – nuclear forces act, thus forming stable nuclei. These nuclei can then be suitably combined with electrons to form stable atoms.

Although the simplest elements hydrogen and helium formed 13.8 billion years ago, during the so-called *epoch of nucleosynthesis* after the *Big Bang*, the elements lithium, beryllium, and boron were produced by cosmic ray collisions and subsequent nuclear fissions of large nuclei. The next heavier elements carbon and oxygen were made much later, after the *Big Bang*, in the fiery stellar furnaces of high-mass stars, by the fusion of hydrogen into helium, helium into carbon, etc. This last process can also create the remaining chemical elements, up to iron (atomic number 26). But the creation of the heavier to iron chemical elements (with atomic number > 26) requires special conditions in which nuclei are bombarded by free neutrons, which when combined with these nuclei, produce the heaviest elements of the periodic table, namely those beyond iron.

It had been suggested that these special conditions exist in another cosmical phenomenon, *supernovae*, but this particular mechanism cannot exactly explain their observed cosmic ratio, i.e., the relative abundance they have in the Universe. Therefore, recent research has turned to the event of the dynamical merger of two *neutron stars*, where large amounts of neutron-rich matter are produced, which could be the first material and the source of the creation of heavy elements. But how and

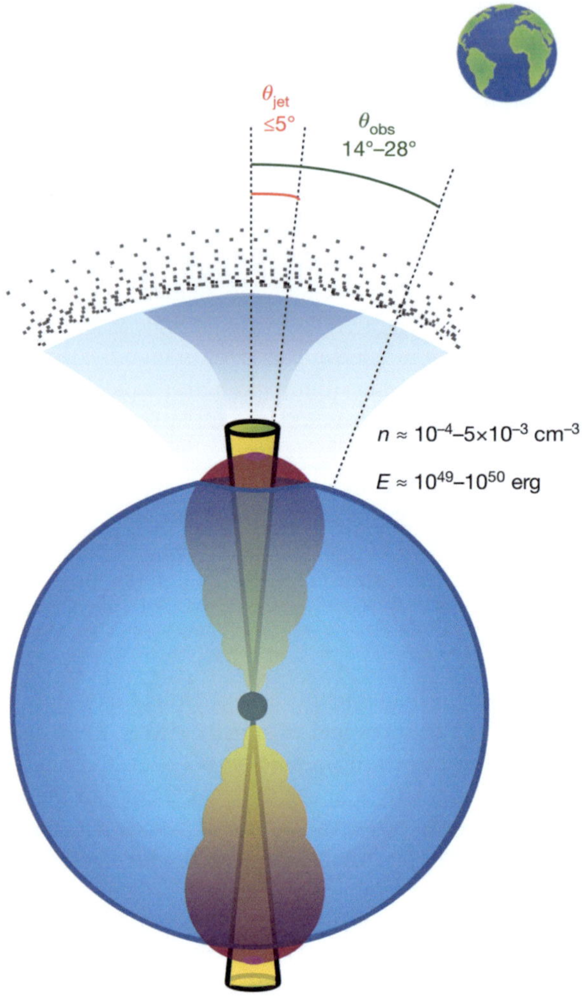

Fig. 5.94 Schematic of the physical and geometric parameters derived for GW170817. (Figure reproduced with the kind permission of K.P. Mooley)

where exactly this process occurs has been an open research question in modern astrophysics. So, after the discovery of the observation of gravitational waves from the source GW170817 where the collision of two *neutron star*s occurred, several researchers turned there in search of an answer to the question of the formation of the heavy elements. The first conjectures about the emission of radiation from these newly formed heavy elements when they form in an astrophysical environment date back to the 1990s. More recently, the luminosity of emitted radiation due to the heating by the radioactivity of these new nuclei, in the fragments of the collision of the two *neutron star*s was calculated in 2010 in great detail by a research team at the University of California, Berkeley. As the fragments of the collision of the two *neutron star*s are ejected expanding into the surrounding space, the emitted radioactivity by the radioactive chemical elements that are formed heats the fragments, causing them to radiate and glow. Accurate calculations have shown that the brightness of the radiation emitted by the merger of the two *neutron star*s is a thousand times greater than that of the novae in our galaxy. That is why they named these exotic bursts *kilonovae,* Fig. 5.93. Thus, *kilonovae* owe their name to the fact that they are a thousand times brighter than novae, although they are fainter than *supernovae*. The reason they have not been discovered until now is that they fade very quickly and astronomical observatories do not even have time to detect and observe them. During the explosion of a *kilonova*, only a small percentage of the matter, about 1%, is ejected, albeit with a high speed, about 30% of the speed of light. But many questions remained unanswered about how and where exactly these *kilonovae* appear, since the debris from the collision is a completely unusual and alien matter, a mixture of precious metals and radioactive debris.

On August 22, the space telescope began observing GW170817's *kilonova* in the visible and infrared, and particularly analyzing its near-infrared spectrum. This is particularly useful as it reveals the movement and composition of the expanding materials formed after the collision. The *kilonova*'s visible and near-infrared emission comes mainly from the heating caused by the radioactive materials that form in the neutron-rich fragments. Theoretical calculations showed that any presence of heavy metals in the fragments of the collision, would mean that these fragments would have a reddish color and very high opacity, i.e., they would be up to a hundred times more opaque than what we usually see in the debris that is created during *supernova* explosions. Therefore, the formation of the heavy elements would be betrayed by the characteristic red color of the fragments resulting from the collision of the two *neutron star*s. According to the Las Cumbres Observatory observations, the *kilonova* turned red and much fainter by a factor of 20 within days. The Las Cumbres Observatory Global Telescope (LCOGT) is a private Institute that operates a global network of 21 robotic astronomical survey telescopes covering many latitudes of the globe so that continuous monitoring of many astronomical sources is possible.

In the end, that's exactly what we noticed. Three days after the two *neutron star*s merged and exploded, the bright blueish emission from the lighter elements receded in the outer polar regions and gave way to red emission from the heavier elements. This red emission lasted for more than two weeks. The electromagnetic radiation detected after the collision of the two *neutron star*s revealed a *radioactive soup,* from which huge quantities of gold, platinum, and silver were produced. Also, it was discovered that neutron collisions are *universal factories* for the production of heavy elements, such as iodine, uranium, and bismuth.

As the data slowly came in and was analyzed, they began to piece together a final picture that was surprisingly anticipated by the theorists of the nucleosynthesis of the chemical elements.

The physical and geometrical parameters derived for the jet created during the event GW170817 can be calculated (Mooley et al. 2018). The jet drives a cocoon through interaction with the dynamical ejecta. The shock-breakout from the cocoon likely produced the γ-ray signal and the cocoon's interaction with the ISM produced the early-time (up to ~2 months post-merger) radio and X-ray emission. The relativistic core of the jet has a half-opening angle (θ_{jet}) of $\leq 5°$. The Earth is located 14°–28° away (the viewing angle, θ_{obs}) from the core of the jet. GW170817 most likely gave rise to a short γ-ray burst pointing at such an angle away from the Earth. The interaction between the jet and the ISM produced the late-time radio and X-ray emission. VLBI measurement suggest that the Lorentz factor of the jet 150 days post-merger (i.e., at the peak of the radio light curve, when the core of the jet came into view) is $\Gamma \approx 4$ (Fig. 5.94).

$10^{30}$$ Worth of Gold Produced in the Collision of 2 *Neutron Stars* …

One of the fundamental questions of astrophysics is to understand the cosmic origin and evolution of the chemical elements. For example, gold or platinum, are formed by the capture of neutrons on pre-existing seed-nuclei. However, although the fundamental principles of such neutron captures have been understood since a long time, the astrophysical explosions in which this actually happens has been a matter of fierce debates for more than half a century.

Indeed, during the first two nights of observations during the event GW170817, the color of the fusion product of the two *neutron star*s was blue, which agreed with the theory of composition in the outer layers of the merger, of lighter precious metals such as silver. In the following days, however, the color of the emission changed to red, irrefutable evidence that the heaviest, most precious metals, such as gold, platinum, and uranium, had formed in the inner, warmer layers of the fragments. It was truly amazing how well the observations agreed with theoretical predictions, in this first case of live observation of the merger of two *neutron star*s. It should be noted that the construction of a complete theoretical picture of this phenomenon involves a wide range of fields of physics, such as the GTR, hydrodynamics, nuclear and atomic physics. The successful combination of all these fields is truly a triumph of theoretical astrophysics!

References

Andreoni et al (2022) A very luminous jet from the disruption of a star by a massive black hole. Nat Astron 612:430–434

Baade W, Minkowski R (1954) On the Identification of Radio Sources. Astrophys J 119:215

Baade W, Zwicky F (1934a) Remarks on supernovae and cosmic rays. Phys Rev 46:76–77

Baade W, Zwicky F (1934b) On supernovae. Proc Natl Acad Sci USA 20(5):254–259

Baade W, Zwicky F (1934c) Cosmic rays from supernovae. Proc Natl Acad Sci USA 20(5):259–263

Backer DC, Kulkarni SR, Heiles C, Davis MM, Goss WM (1982) A millisecond pulsar. Nature 300(5893):615–618

Bahcall JN, Kirhakos S, Schneider DP, Davis RJ, Muxlow TWB, Garrington ST, Conway RG, Unwin SC (1995) Hubble space telescope and MERLIN observations of the jet in 3C 273. Astrophys J Lett 452:L91. arXiv:astro-ph/9509028

Balashev S, Noterdaeme P, Gupta N, Krogager J-K, Combes F, Lopez S, Petitjean P, Omont A, Srianand R, Cuellar R (2025) Quasar radiation transforms the gas in a merging companion galaxy. Nature. eprint arXiv:2505.15766

Boccardi B, Krichbaum TP, Bach U, Bremer M, Zensus JA (2016a) First 3 mm-VLBI imaging of the two-sided jet in Cygnus A. Zooming Radio observations of active galactic nuclei with mm-VLBI 37 into the launching region. A&A 588:L9

Boccardi B, Krichbaum TP, Bach U, Mertens F, Ros E, Alef W, Zensus JA (2016b) The stratified two-sided jet of Cygnus A. Acceleration and collimation. A&A 585:A33

Boccardi B, Krichbaum TP, Ros ′E, Zensus JA (2017) Radio observations of active galactic nuclei with mm-VLBI. Astron Astrophys Rev 25:4. [arXiv:1711.07548v1 [astro-ph.HE]

Bodo G, Massaglia S, Ferrari A, Trussoni E (1994) Kelvin-Helmholtz instability of hydrodynamic supersonic jets. Astron Astrophys 283:655

Bogovalov S, Tsinganos K (2005) Shock formation at the magnetic collimation of relativistic jets. Mon Not Roy Astron Soc 357(47):918–928

Chantry L, Cayatte V, Sauty C, Vlahakis N, Tsinganos K (2018) Nonradial and nonpolytropic astrophysical outflows X. Relativistic MHD rotating spine jets in Kerr metric. Astron Astrophys 612:A63

Chantry L, Cayatte V, Sauty C, Vlahakis N, Tsinganos K (2022) Double flows anchored in a Kerr black hole horizon—I. Meridionally self-similar MHD models with loading terms. Mon Not Roy Astron Soc 515(3):3796–3817

Chrimes AA et al (2025) The infrared counterpart and proper motion of magnetar SGR 0501+4516. A&A 696

Colless M et al (2001) The 2dF galaxy redshift survey: spectra and redshifts. Mon Not Roy Astron Soc 328:1039–1063

Contopoulos G (2004) Adventures in order and chaos, a scientific autobiography. Springer

Contopoulos G (2004) Adventures in order and chaos, a scientific autobiography. Springer

Curtis HD (1918) Descriptions of 762 nebulae and clusters photographed with the crossley reflector. Publ Lick Obs 13:9–42

Einstein A, de Sitter W (1932) On the relation between the expansion and the mean density of the universe. Contributions from the mount wilson observatory, vol 3, pp 51–52; Proc Natl Acad Sci 18:213–214

Ferrari A, Trussoni E, Rosner R, Tsinganos K (1986) On wind-type flows in astrophysical jets. II—Propagation outside the nucleus and the case of M87. Astrophys J 300:577–590

Ferrari A, Habbal SR, Rosner R, Tsinganos K (1984) A wind-type model for the generation of astrophysical jets. Astrophys J Lett 277:L35–L39

Ferrari A, Trussoni E, Rosner R, Tsinganos K (1985) On wind-type flows in astrophysical jets. I. The initial relativistic acceleration. Astrophys J 297:397–418

Ferrari A, Tsinganos K (1986) Wind-type flows and astrophysical jets. Can J Phys 64(4):456–462

Forman W, Jones C, Cominsky L, Julien P, Murray S, Peters G (1978) The fourth Uhuru catalog of X-ray sources. ApJS 38:357

Geller MJ, Huchra JP (1989) Mapping the universe. Science 246(4932):897–903

Goddi C, Crew G, Impellizzeri V, Martí-Vidal I, Matthews LD, Messias H, Rottmann H, Alef W, Blackburn L, Bronzwaer T, Chan CK, Davelaar J, Deane R, Dexter J, Doeleman S, Falcke H, Fish VL, Fraga-Encinas R, Fromm CM, Herrero-Illana R, Issaoun S, James D, Janssen M, Kramer M, Krichbaum TP, De Laurentis M, Liuzzo E, Mizuno Y, Moscibrodzka M, Natarajan I, Porth O, Rezzolla L, Rygl K, Roelofs F, Ros E, Roy AL, Shao L, van Langevelde HJ, van Bemmel I, Tilanus R, Torne P, Wielgus M, Younsi Z, Zensus JA (2019) Event horizon telescope collaboration. The Messenger 177:25–35

Gott JR III et al (2005) A map of the universe. Astrophys J 624(2):463–484. arXiv:astro-ph/0310571

Govoni F, Orrù E, Bonafede AM, Paladino IR, Vazza F, Murgia M, Vacca V, Giovannini G, Feretti L (2019a) A radio ridge connecting two galaxy clusters in a filament of the cosmic web. Science 364(6444):981–984. arXiv: 1906.07584

Govoni F, Orrù E, Bonafede A, Iacobelli M, Paladino R, Vazza F, Murgia M, Vacca V, Giovannini G, Feretti L, Loi F, Bernardi G, Ferrari C, Pizzo RF, Gheller C, Manti S, Brüggen M, Brunetti G, Cassano R, de Gasperin F, Enßlin TA, Hoeft M, Horellou C, Junklewitz H, Röttgering HJA, Scaife AMM, Shimwell TW, van Weeren RJ, Wise M (2019b) A radio ridge connecting two galaxy clusters in a filament of the cosmic web. Science 364(6444):981–984. arXiv:1906.07584

Gracia J, Tsinganos K, Bogovalov SV (2005) Magnetic collimation of the relativistic jet in M 87. Astron Astrophys 442:L7–L10

Gracia J, Bogovalov SV, Tsinganos K (2008) MHD models and synthetic synchrotron maps for the jet of M87. Int J Mod Phys D 17 (10):1687–1693

Gracia J, de Colle F, Downes T (eds) (2009) Jets from young stars V. High performance computing and applications. Springer, LNP 791

Gregory SA, Thompson LA (1978) The Coma/A1367 supercluster and its environs. Astrophys J 222:784–799

Häberle M, Neumayer N, Seth A, Bellini A, Libralato M, Baumgardt H, Whitaker M, Dumont A, Alfaro-Cuello M, Anderson J, Clontz C, Kacharov N, Kamann S, Feldmeier-Krause A, Milone A, Nitschai M, Pechetti R, van de Ven G (2024) Fast-moving stars around an intermediate-mass black hole in ω Centauri. Nature 631 (8020):285–288

Hamilton K, Cho YH (2017) J Res Progress 1. Howard Community College

Hamilton K, Cho YH (2018) PSR1913+16: the binary pulsar, mathematical reproduction and extension. Jv Res Progress 1

Herschel W (1783) On the Diameter and Magnitude of the Georgium Sidus; With a Description of the Dark and Lucid Disk and Periphery Micrometers. Philos Trans R Soc Lond 73:4–14

Kim JY, Krichbaum TP, Lu R-S, Ros E, Bach U, Bremer M, de Vicente P, Lindqvist M, Zensus JA (2018) The limb-brightened jet of M 87 down to 7 Schwarzschild radii scale. Astron Astrophys 616:188

Kovalev YY, Lister ML, Homan DC, Kellermann KI (2007) The inner jet of the radio galaxy M87. Astrophys J 668:L27–L30. [astro-ph:0708.2695v2]

Laplace PS (1799) in allgemeine geographische ephemeriden

Lister ML, Aller HD, Aller MF, Cohen MH, Homan DC, Kadler M, Kellermann KI, Kovalev YY, Ros E, Savolainen T, Zensus JA, Vermeulen RC (2009) MOJAVE: monitoring of jets in active galactic nuclei with VLBA experiments. V. Multi-epoch VLBA images. AJ 137:3718–3729

Massaglia S, Bodo G, Rossi P, Capetti S, Mignone A (2016) Making Faranoff-Riley I radio sources-I. Numerical hydrodynamic 3D simulations of low-power jets. Astron Astrophys 596:A12

Michell J (1783) On the means of discovering the distance, magnitude, &c. of the fixed stars, in consequence of the diminution of the velocity of their light, in case such a diminution should be found to take place in any of them, and such other data should be procured from observations, as would be farther necessary for that purpose. By the Rev. John Michell, B. D. F. R. S. In a Letter to Henry Cavendish, Esq. F. R. S. and A. S. Philos Trans Roy Soc Lond 74:35–57

Mishra S, Fox AJ, Krishnarao D, Lucchini S, D'Onghia E, Cashman FH, Barger KA, Lehner N, Tumlinson J (2024) The truncated circumgalactic medium of the large magellanic cloud. ApJ. arXiv:2410.11960v1

Miyoshi M, Moran J, Hernnstein J, Greenhill L, Nakal N, Diamond P, Inoue M (1995) Evidence for a black hole from high velocities in a sub-parsec region of NGC4258. Nature 373:127–129

Mooley KP, Deller AT, Gottlieb O, Nakar E, Hallinan G, Bourke S, Frail DA, Horesh A, Corsi A, Hotokezaka K (2018) Superluminal motion of a relativistic jet in the neutron-star merger GW170817. Nature 561(7723):355–359. arXiv:1806.09693v2 (astro-ph.HE) 30 Oct 2018

Oppenheimer JR, Volkoff GM (1939) On massive neutron cores. Phys Rev 55:374

Özel F, Freire P (2016) Masses, radii, and the equation of state of neutron stars. Annual Rev Astron Astrophys 54:401–440

Penrose R (1965) Gravitational collapse and space-time singularities. Phys Rev Lett 14:57

Pearson TJ, Unwin SC, Cohen MH et al (1981) Superlumina expansion of quasar 3C273. Nature 290

Pearson TJ, Unwin SC, Cohen MH, Linfield RP, Readhead ACS, Seielstad GA, Simon RS, Walker RC (1982) Superluminal expansion of 3C 273. In: Heeschen DS, Wade CM (eds) Extragalactic radio sources, IAU, vol 97. Springer, pp 355–356

Peters PC, Mathews J (1963) Gravitational radiation from point masses in a keplerian orbit. Phys Rev 131:435

Perley RA, Dreher JW, Cowan JJ (1984) The jet and filaments in Cygnus A. ApJ 285:L35–L38

Reber W (1940) Cosmic static. Astrophys J 91:621–624

Ren BB, Fogarty K, Debes JH, Meyer ET, Mo Y, Mawet D, Perrin MD, Ogle PM, Sahlmann J (2024) 3C 273 host galaxy with hubble space telescope coronagraphy. Astron Astrophys [arXiv:2402.09505v1]

Russell HN (1914) Relations between the spectra and other characteristics of the stars. Popul Astron 22:275

Sauty C, Lima JJG, Tsinganos K, Aibeo A, Meliani Z, Vlahakis N (2007) Solar wind and stellar jets, from newtonian to relativistic ones. AIP Conf Proc 895:87–96

Sawala T, Delhomelle J, Deason AJ et al (2025) No certainty of a Milky Way–Andromeda collision. Nat Astron

Soldner JG (1801–1804) On the deflection of a light ray from its rectilinear motion, by the attraction of a celestial body at which it nearly passes by. Berl Astronomisches Jahrbuch 161–172

Sukhbold T, Woosley SE (2016) The most luminous supernovae. Astrophys J Lett 820(2):5

Thorsett SE, Chakrabarty D (1999) Neutron star mass measurements. I. Radio pulsars. Astrophys J 512:288. arXiv:astro-ph/9803260

Trussoni E, Ferrari A, Rosner R, Tsinganos K (1988) On wind-type flows in astrophysical jets. III. Temporal evolution of perturbations and the formation of shocks. Astrophys J 325:417

Tsinganos K (2022) A Tribute to prof. Eugene Parker, BAAS. https://baas.aas.org/pub/2022i039/release/1

Tsinganos K, Bogovalov SV (2005) Magnetic collimation of the relativistic jet in M87. In: 2nd international symposium on high energy Gamma-Ray astronomy. AIP conference proceedings, vol 745, pp 148–159

Umehata H, Fumagalli M, Smail I, Swinbank AM, Cantalupo SC, Ivison RJ, Steidel CC, Shapley AE, Vernet J, Yamada T, Tamura Y, Kubo NK, Kajisawa M, Hatsukade B, Kohno K (2019) Gas filaments of the cosmic web located around active galaxies in a proto-cluster. Science 366(6461):97–100. arXiv:1910.01324

van de Weygaert R, Schaap W (2009) The cosmic web: geometric analysis. In: Martínez VJ, Saar E, Martínez-González E, Pons-Bordería MJ (eds) Data analysis in cosmology, vol 665. Lecture notes in physics. Springer, Berlin, pp 291–413

Vedrenne G, Atteia J (2009) Gamma-ray bursts. Springer, Berlin

Weisberg JM, Nice DJ, Taylor JH (2010) Timing measurements of the relativistic binary pulsar PSR B1913+16. Astrophys J 722:1030–1034

Wolf C, Lai S, Onken CA, Amrutha N, Bian F, Hon WJ, Tisserand P, Webster RL (2024) The accretion of a solar mass per day by a 17-billion solar mass black hole. Nat Astron https://doi.org/10.1038/s41550-024-02195-x

Yao Y et al (2023) The on-axis jetted tidal disruption event AT2022cmc: x-ray observations and broadband spectral modeling. arXiv:2308.09834v1 [astro-ph.HE]

Zamaninasab M, Clausen-Brown E, Savolainen T, Tchekhovskoy A (2014) Dynamically important magnetic fields near accreting supermassive black holes. Nature 510:126

Zel'dovich YB, Raizer YP (2002) Physics of shock waves and high-temperature hydrodynamic phenomena. In: Dover books on physics

Zwicky F (1933) Die Rotverschieb ung von extragalaktischen Nebeln (The redshift of extragalactic nebulae). Helvetica Physica Acta 6:110–127

Additional Review Articles

Abell GO (1965) Clustering of galaxies. Annu Rev Astron Astroph 3:1–22
Bahcall NA (1974) Clusters of galaxies. Annu Rev Astron Astroph 15:505–540
Baym G, Pethick C (1975) Neutron stars. Annu Rev Astron Astroph 25:27–77
Baym G, Pethick C (1979) Physics of neutron stars. Annu Rev Astron Astroph 17:415–443
Cameron AGW (1970) Neutron stars. Annu Rev Astron Astroph 8:179–208
Fabian AC (1994) Cooling flows in clusters of galaxies 32:277–318
Fich M, Tremaine S, (1991) The mass of the galaxy 29:409–445
Kravtsov AV, Borgani S (2012) Formation of galaxy clusters. Annu Rev Astron Astroph 50:353–409
Langer N (2012) Presupernova evolution of massive single and binary stars. Annu Rev Astron Astroph 50:107–164
Marchant P, Bodensteiner J (2014) The evolution of massive binary stars. Annu Rev Astron Astroph 62:21–61
McCarthy PJ (1993) High redshift radio galaxies. Annu Rev Astron Astroph 31:639–688
Yakovlev DG, Pethick C (2004) Neutron star cooling. Annu Rev Astron Astroph 42:169–210

Chapter 6
Fundamental Discoveries in Cosmology in the Last 100 Years

Observations of the anisotropy of the diffuse cosmic microwave radiation by WMAP at scales of approximately one degree in diameter have shown that:

(a) the Universe is flat,
(b) its density is very close to the critical density,
(c) the Universe
(d) we observe only a part of it
(e) besides the mysterious dark matter and dark energy, the ordinary matter that makes up everything we see – from stars and galaxies to planets and humans – amounts to only about 5% percent in the Universe.

Observations of the anisotropy of the diffuse *cosmic microwave radiation* by WMAP.

It is believed that during the infinitesimally small period of inflation the dimension of the Universe increased dramatically by at least a factor of $10^{26} \approx e^{60}$. *An expansion by a factor of 10^{26} is equivalent to expanding an object the size of a hair (0.1 mm) to a dimension about the diameter of our Galaxy (about 300,000 light years).*

1962: Cosmology Becomes an Exact Science After COBE/WMAP/Planck

Cosmology is the branch of *Astrophysics* that examines how and when the Universe was created, its structure on a large scale, the stages of its evolution, up to and including its eventual fate. Today, *Cosmology* has gradually become a *precise observational science*, having first passed through many evolutionary stages. In the beginning, ancient humans created the *mythological Cosmogony*, inventing myths behind every cosmological phenomenon, trying to answer existential questions about the surrounding world. With the genesis and development of *Philosophy* and as a result of the intellect of the *Ionian* natural philosophers and thinkers, *mythological Cosmogony* was replaced by *scientific Cosmology*, which uses the methods of physics. This

K. Tsinganos, *A Century of Cosmic Discoveries*, Springer Praxis Books,
https://doi.org/10.1007/978-3-032-06282-6_6

Cosmology was transmitted to Europe during the Middle Ages and the Renaissance, after the fall of *Constantinople* to the Ottomans, and evolved into its present form, especially after the development of spectroscopy and the use of terrestrial and satellite telescopes.

The main milestones of Cosmology during the last century include: the introduction of the *General Theory of Relativity* (GTR) which formed the backbone framework of the physical description of the Universe, the discovery of the expansion of the Universe and its subsequent creation through the *Big Bang* which created an ever-expanding space, the realization of the existence of dark matter and dark energy, of unknown texture acting on large scales, the rediscovery of diffuse cosmic microwave radiation and its anisotropies, which carry the echo and the oldest light of the Universe that reaches us today and originates hundreds of thousands of years after the *Big Bang*·the composition of the nuclei of the light elements and the physics of the first minutes of the Universe, where theories and experimental results of the physics of elementary particles are applied,·the introduction of the theory of inflation and the discovery of the accelerating expansion of the Universe, and the discovery of supercluster structures that formed after primordial inhomogeneities around concentrations of dark matter.

Specifically, Gamow suggested (1948) that the Universe is flooded by a diffuse cosmic microwave radiation (CMR) of current temperature about 270 degrees below zero, which was accidentally detected by Penzias & Wilson (1965). Measurements by instruments on the NASA COBE spacecraft (1989–1993) provided double support for the *Big Bang* theory, since the CMR has a near-perfect blackbody spectrum as well as very small *anisotropies*. Since then, Cosmology has became an exact science. COBE was followed by two more advanced spacecraft, WMAP (2001–2010) and Planck (2009–2013).

Nevertheless, the outstanding discovery in Cosmology during the last century is definitely the *expansion of the Universe,* which is expressed by the Hubble-Lemaitre law (Lemaitre 1927; Hubble 1929). In particular, observations of galaxies show that they are all moving away from us, the nearby ones at small speeds and the more distant ones at enormous speeds. So, it seems like everyone wants to avoid us! But any observer in any other galaxy would see the same effect, i.e., that all other galaxies would have been also moving away. However, galaxies do not move independently of the space of the Universe, but the space between them somehow *lips out*. This is what we mean when we say that the Universe is expanding. We must also clarify that by referring to the expansion of the Universe, we are not implying that the Universe is expanding into some pre-existing space, but that space itself (as well as time) was created during the singular moment of *Big Bang* and since then the relative distance between any two points grows. In this sense, the Universe has neither a center, nor is there anything outside of it. A simple way to visualize the expansion of 3-D space in 2-D is to liken it to the two-dimensional elastic surface of a balloon on which we have painted galaxy symbols. We see that the more we inflate the balloon, the further away each galaxy symbol moves from each other. By reversing time, a natural consequence of the expansion of the Universe is the *Big Bang* itself, which did not happen in one specific location and spread out from there, but happened *suddenly*, *everywhere.* According to the classic *Big Bang* model, the Universe was created at a given point in time in the past, and since then it has been continuously expanding and cooling.

As with the preceding chapter on astrophysics, a presentation of *all* discoveries in cosmology is technically impossible to fit within this single volume. Therefore, in the following, we have selected only the most important of those discoveries which, in our opinion, changed the main course of the field of Cosmology (Figs. 6.1 and 6.2).

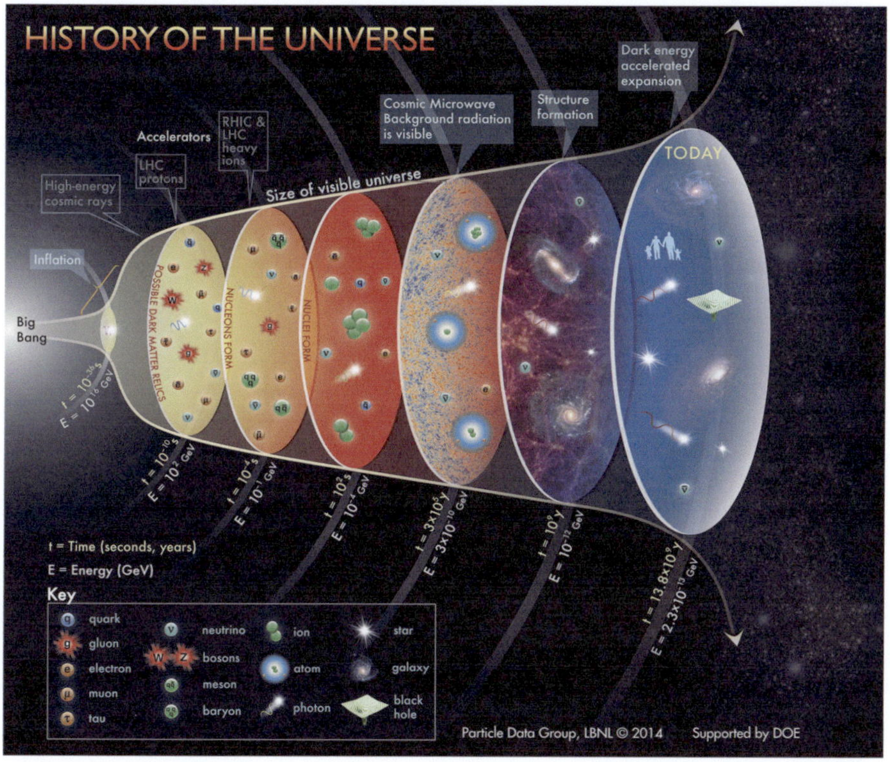

Fig. 6.1 Timeline of the evolution of the Universe, from the original *Big Bang* (BB) to the present day. *Inflation* occurs around 10^{-36} secs after the BB, the *particle era* is expected to have taken place around a few microseconds, the *cosmic microwave background radiation* becomes visible at 380.000 years after the BB, while the structure formation occurs around/before a billion years after the BB. (Figure courtesy CERN Particle Data Group, LBNL 2014)

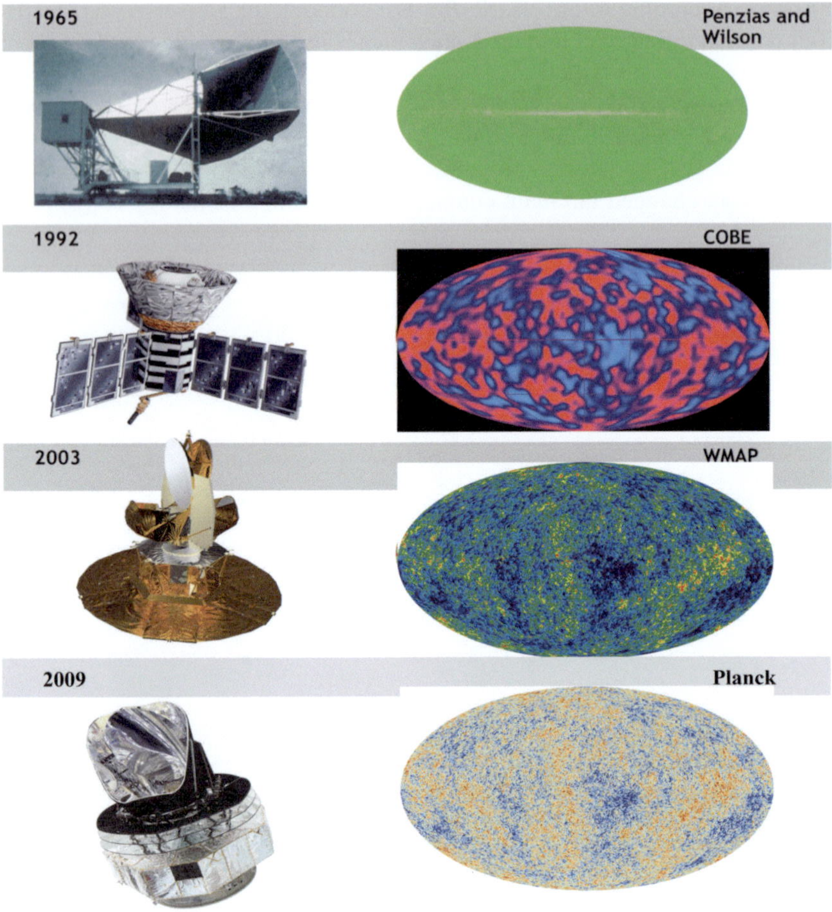

Fig. 6.2 The first discovery of the *diffuse cosmic microwave radiation* by Penzias & Wilson with the Bell Labs antenna (1962), was followed by the COBE (1989–1993), WMAP (2001–2010) and Planck (2009–2013) space missions, which measured the anisotropies of the diffuse cosmic microwave radiation, with increasing *precision* (*Image Sources* NASA/ESA)

1915–1930: Early Discoveries in Cosmology

1915: Albert Einstein's General Theory of Relativity

The expansion of the Universe is considered as the most important discovery of the last century in Cosmology and is regarded as the cornerstone of the *Big Bang* theory. However, this discovery did not come only from observations, but also emerged theoretically, by considering and solving the equations of Einstein's General Theory of Relativity (GTR), an extension of Newton's theory of gravity. *General Relativity*, also known as the *General Theory of Relativity* (GTR) and Einstein's Theory of Gravity, is the geometric theory of gravitation published by Albert Einstein in 1915 and is the current description of gravitation in modern physics.

Albert Einstein proposed and published the two interrelated theories, which combined together are known as the *Theory of Relativity.* He published the theory of special relativity in 1905 and the theory of general relativity in 1915 (Fig. 6.3).

The GTR describes the gravitational interaction via the curvature of spacetime brought about by the presence of mass and energy. Einstein's equations are a generalization of Poisson's equation which relates the density of a matter distribution to the potential it creates.

Einstein's equations of the *General Theory Relativity* are expressed mathematically in terms of the so-called *tensors* (from *tensus* in Latin, meaning stretch or tension), which can be thought of as generalized vectors, describing linear relationships between scalar numbers, vectors, and other tensors. In mathematics, a tensor is an algebraic object which may be visualized as a (potentially multidimensional) orthogonal array of numbers. The important feature of *tensors* is that they are *independent of the choice of a particular coordinate system*, i.e., *tensors* are defined independently of any basis, although they are often referred to their components in a basis related to a particular coordinate system; those components form an array, which can be thought of as a high-dimensional matrix, e.g., as an array A having **n** rows and **m** columns. The rank, or degree, or order of a *tensor* is the number of indices needed to name and separate an element of that array, e.g., A_{nm}. The familiar vector can be represented as a one-dimensional array and is a *tensor* of rank 1. Numerical quantities are simple numbers and thus *tensors* of rank zero. A simple example of

Fig. 6.3 Albert Einstein (1879–1955)

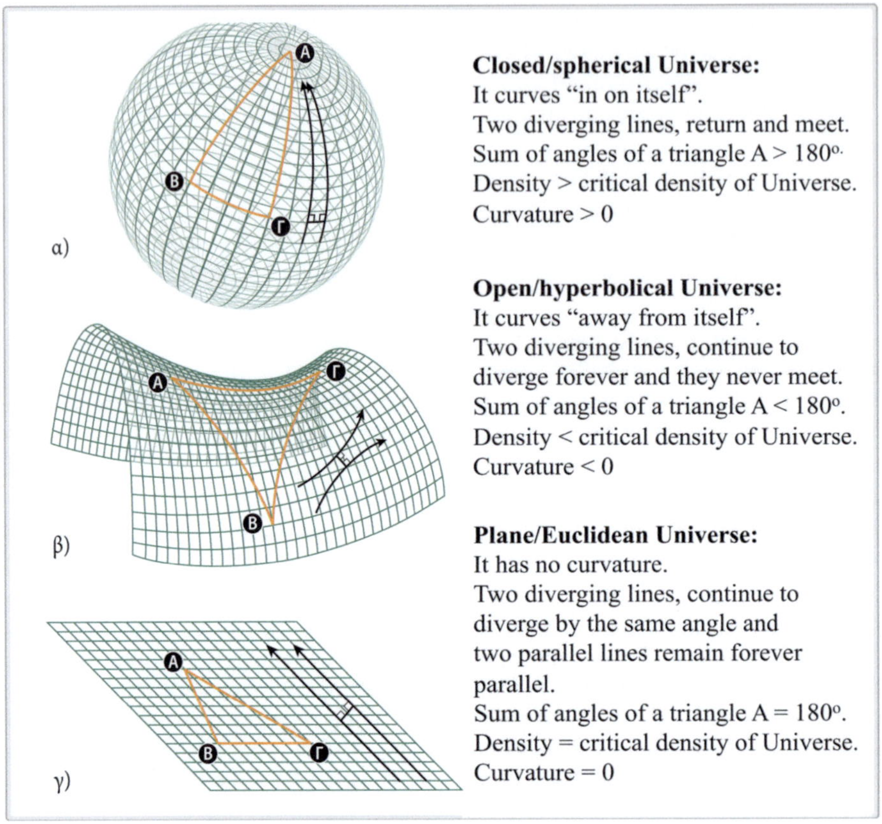

Fig. 6.4 Since it is difficult to visualize 3D curved spaces, we use often 2D curved spaces by analogy. Thus, the three possibilities for the spatial geometry of the Universe are: **a** *spherical* – two lines that are initially parallel, intersect, **b** *hyperbolic* – two lines that are initially parallel, diverge and they never intersect, **c** *planar* – two lines that are initially parallel, they remain parallel throughout and they never intersect. The sum A of the angles of a spherical triangle in case **a** is A > 180°, in case **b** A < 180° and in case **c** A = 180°. In case (c) of Euclidean space, the volume of a sphere of radius R equals to $V_o = 4\pi R^3/3$, in case (a) of spherical geometry, this volume V is smaller ($V < V_o$) and in case (b) of hyperbolic geometry it is greater than the Euclidean volume ($V > V_o$). In case (**a**) of spherical geometry, the volume of the Universe is finite without the Universe having an end: any two-dimensional being moving along a geodesic line will return to where it started. In cases (**b**) and (**c**) of hyperbolic and planar geometry, the volume of the Universe is infinite and the Universe has no ends

a *tensor* is *stress*, which is defined as the force per unit area (pressure, shear), on two directions, one normal to the area and the other perpendicular to it, consisting of nine components σ_{ij}, with $i = 1, 2, 3$, or, $i = x, y, z$ that completely define the state of stress at a point inside a material in a deformed configuration, for example, an elastic cube.

Tullio Levi–Civita and Gregorio Ricci popularized *tensors* in 1900 – continuing the earlier work of Bernhard Riemann, Elwin Bruno Christoffel, and others – as part of the absolute *differential calculus*. The concept enabled an alternative formulation of the intrinsic *differential geometry* of a manifold in the form of the Riemann curvature *tensor*. Einstein used *tensors* to formulate the theory of *General Relativity*, which is formulated completely in the language of *tensors*, because they provide a concise mathematical framework for solving problems of gravity. Einstein had learned about them – with great difficulty (!) – from the geometer Marcel Grossmann. In fact, Levi–Civita had initiated a correspondence with Einstein to correct Einstein's mistakes in his use of *tensor* analysis. The correspondence lasted 1915–17, and was characterized by mutual respect.

Einstein to Levi–Civita:

I admire the elegance of your method of computation; it must feel really nice to ride through these fields upon the horse of true mathematics, while the rest of us have to make our way tiringly on foot.

After the publication of the *Special Theory of Relativity* (1905), Einstein presented his famous *field equations (Einstein field equations, EFE)* in a series of lectures before the Prussian Academy of Sciences (Einstein 1919). It should be noted that his last lecture caused an uproar in the scientific world, as Einstein essentially presented a new theory that replaced Isaac Newton's theory of gravity. According to this theory, gravity is not the result of a force, but is rather due to the curvature of spacetime, which is caused by the mass and energy contained in this spacetime:

$$\mathrm{R}_{\mu\nu} - \frac{1}{2}\mathrm{R}\mathrm{g}_{\mu\nu} + \Lambda g_{\mu\nu} = \frac{8\pi G}{C^4}\mathrm{T}_{\mu\nu}, \tag{6.1}$$

where $\mathrm{g}_{\mu\nu}$ is the so-called spacetime *metric*, $\mathrm{R}_{\mu\nu}$ is Ricci's curvature *tensor* depending on this metric $\mathrm{g}_{\mu\nu}$ and $\mathrm{T}_{\mu\nu}$ another *tensor* expressing the energy and momentum distribution. The scalar constants R and Λ are the one-dimensional curvature and the famous *cosmological constant*, respectively. The indices of the *tensors* take values (μ, ν) = 0, 1, 2, 3. The metric *tensor* $\mathrm{g}_{\mu\nu}$ expresses the magnitude of the distance between two neighboring points of the four-dimensional spacetime, of a given curvature. The Ricci *tensor* $\mathrm{R}_{\mu\nu}$ defines how the curvature changes from point to point in spacetime. It is produced by a suitable combination of the metric *tensor* $\mathrm{g}_{\mu\nu}$ and its second derivatives with respect to spacetime coordinates.

The EFE can then be interpreted as a set of equations dictating how stress–energy–momentum determines the curvature of spacetime.

Einstein's Field Equations of General Relativity

Simply put, the previous *tensor* Eq. (6.1) says that the curvature of space-time (left- hand member) depends on and is shaped by the distribution of the amount of matter and energy it contains (right-hand member). In other words, analogously to the way that electromagnetic fields are related to the distribution of charges and currents via Maxwell's equations, the EFE relate the spacetime geometry to the distribution of mass–energy, momentum and stress, that is, they determine the metric *tensor* of spacetime for a given arrangement of stress–energy–momentum in the spacetime. The solutions of the EFE give the components of the *metric tensor*. The inertial trajectories of particles and radiation (*geodesics*) in the resulting geometry are then calculated using the geodesic equation.

Note that the Einstein field Eqs. (6.1) are not just one equation, but rather a suite of 16 different non-linear partial differential equations, one for each of the 4×4 combinations of the indices $(\mu,\nu) = 0, 1, 2, 3$.. Because of symmetry, they reduce to 10 field equations in which must be added the continuity equation that ensures the evolution of the *energy tensor* $T_{\mu\nu}$ (+4 equations), as well as the so-called *equation of state* that connects density to pressure. This results in a total of at least 14 equations for 20 unknowns, the 10 independent components of the metric $g_{\mu\nu}$ and the 10 independent components of the *energy tensor* $T_{\mu\nu}$.

Mathematically, Einstein's equations are a complex mathematical system of coupled, non-linear differential equations and are therefore very difficult to solve exactly. The solution of the above field equations gives the *metric tensor* $g_{\mu\nu}$ and thus the distance between two neighboring points of the four-dimensional spacetime is finally calculated. In the case of Cosmology, Einstein's equations give the scale factor, which gives the expansion of the coordinate system and from it a measure of the expansion of the Universe. What Einstein wrote about his theory is typical:

The Logical Completeness of the *General Theory of Relativity*

"The main appeal of the theory lies in *its logical completeness.* If even one of its conclusions turns out to be wrong, the theory must be abandoned. Any modification of it without destroying its entire structure is rather impossible" [Einstein (1919), "What is the theory of Relativity?", *German History in Documents and Images.*

In particular, Einstein proposed 3 tests of the GTR: (1) the *precession* of the perihelion of Mercury's orbit, (2) the *curvature of* stellar light rays as they pass near the photosphere of the Sun, and (3) the gravitational *redshift* of *white dwarf* light. All these predictions have today been verified! We can add (4) the detection *of gravitational waves,* but for these Einstein was not optimistic that they could be detected, as we discussed in the previous Chap. 5 on astrophysics.

However, it is truly *amazing* that a system of mathematical equations, however complex it is – such as the equations of General Relativity – can describe a fully self-consistent model of the Universe that agrees with observations. It is perhaps the pinnacle of the greatness of the humble, small, and finite man to be able to describe in the language of mathematics and physics and to understand the infinite and shocking greatness of the creation and evolution of the Universe!

Exact solutions to the field equations of *General Relativity* can only be found by assuming simplifying assumptions, such as various symmetries. Special classes of exact solutions are most often studied as they shape many gravitational phenomena, such as rotating black holes, or, an expanding Universe. Further simplification is achieved by approximating the actual curved spacetime as a small perturbation of the Euclidean spacetime, i.e., by linearizing the equations of *General Relativity*. These linearized equations are used to study phenomena such as gravitational waves.

1917: Einstein's Homogeneous/Isotropic, Static, Spherical Universe

After the publication of the GTR, Einstein felt that he then had the possibility to describe a fully self-consistent model of the Universe. Thus, in 1917 he applied his theory to the Universe as a whole, opening the field of *relativistic cosmology*. At that time (1917), the expansion of the Universe was not known and most astronomers and cosmologists believed in an eternal, static, and unchanging Universe. After all, this was a very convenient point of view, because if the Universe existed forever and ever the same and unchanging, there would be no need to ask various critical and perhaps uncomfortable cosmological questions, such as when it was created and what/who created it! So influenced by this surrounding scientific atmosphere, in 1917 Einstein tried to explain the existence of such a static Universe in the framework of the corresponding equations of the GTR. For this purpose, he accepted three assumptions in his cosmological model.

1. Cosmological principle: according to this first hypothesis, the Universe on a large scale is *homogeneous* and *isotropic*, that is, the same on average at any distance and in any direction we observe it. This is consistent with observations of the spatial and angular distribution of galaxies, where statistically the same number of galaxies per unit solid angle is indeed observed, in any direction, as well as the same number of galaxies per unit volume, anywhere in the Universe.[1]

[1] In the recent past, astronomers have claimed that the Universe is expanding non-uniformly. In a recent publication by Migkas et al., (2010, 2021), using data from the Chandra (NASA) and XMM-Newton (ESA) space telescopes to

2. Closed Universe: according to the second hypothesis, this *homogeneous* and *isotropic* Universe has a *closed*, *spherical* spatial geometry, Fig. 6.4a. Then, the total volume of a three-dimensional region of space with uniform positive curvature it would have to be finite, without ends, to agree with the first hypothesis.
3. Static Universe: according to this third hypothesis, the Universe is static, that is, its large-scale properties do not change with time.

Thus, according to his scientific atmosphere at that time (1917), Einstein considered a *closed* Universe with a *spherical* geometry, Fig. 6.4a. To obtain such a Universe, he introduced the *cosmological constant* Λ, with a chosen value, $\Lambda_o = 4\pi G\rho/c^2$ and a positive curvature $k = \Lambda_o/R_o^2$. In this way, the solutions of his equations indeed corresponded to a *static* Universe.

The philosophical appeal of the idea that the Universe on average is not alone *homogeneous* and *isotropic* in space, but also unchanging in time, it was so widespread in the scientific community that leading astrophysicists and cosmologists at that time, such as Herman Bondi, Fred Hoyle and Tomas Gold, called it the *perfect cosmological principle* in the so-called *steady state model*. As we shall discuss below, for the Universe to be in a state of constant density, due to expansion, the cosmology of steady state had to be introduced.

1922: Discovery of the First Solutions of Einstein's Equations: Friedmann's Models of the Universe

The first theoretical models of a *homogeneous* and *isotropic* Universe were discovered by the Soviet Physicist and Mathematician Alexander Friedmann, (Fig. 6.5). In his first article, *On the Curvature of Space, Zeitschrif für Physik, Vol. 10, pp. 377–386, 1922*, he went one step further than Einstein, solving the equations of the *General Theory of Relativity* (GTR), for the first time. Abandoning the model of the *static* Universe, he proposed a *dynamic* Universe, where space and its curvature had a time dependence. The new evidence that emerged was that the equations of the GTR had many solutions and not just one, as Einstein had assumed. Also, the models concluded that there was no reason for a *static* Universe to exist, unless cosmological constancy were adopted. Friedmann further published his models in the journal Zeitschrif for Physik, Vol. 21, pp. 326–332, (1924). Specifically, for an isotropic and homogeneous Universe, it was calculated by Friedmann, but also independently by Lemaître, Robertson and Walker, in their separate papers, an interesting *metric* $g_{\mu\nu}$ of this space, which has since been named after them, the Friedmann-Lemaître-Robertson-Walker metric (FLRW *metric*). With this *metric*, the above first solutions

---

many galaxy clusters and using a new technique, concluded that different parts of the Universe appear to be expanding at different rates. If this is verified, it may mean that dark energy is unevenly distributed in the Universe. The *Euclid* satellite telescope (Euclid, ESA, 2022) which will study billions of galaxies, could perhaps illuminate the enigma.

Fig. 6.5 Alexander Friedmann (1888–1925)

of Einstein's equations of the GTR are obtained, for *an isotropic* and *homogeneous* Universe.

Unfortunately, Friedmann died early, at the age of only 37 years, due to misdiagnosed typhoid fever. Also, unfortunately for him, this happened on his return from the Crimea, where he spent his honeymoon (he had allegedly contracted the bacteria on return from his honeymoon in Crimea, when he ate an unwashed pear bought at a railway station!). Thence, Cosmology lost one of its pioneer researchers. One of Friedmann's well known students was George Gamow, who discovered the synthesis of the nuclei of the light elements that occurred in the very first minutes after the *Big Bang*, and proposed the existence of the diffuse *cosmic microwave radiation* (1948), as we discuss later on in this book.

Without a *cosmological constant*, the simplest and most symmetric solution to Einstein's equations in vacuum is the flat, 4-dimensional Minkowski spacetime of special relativity. In 1917, the Dutch astronomer Willem de Sitter found an analogous solution, when the *cosmological constant* Λ is different from zero. If Λ is positive, the solution is called "De Sitter space" and Λ is its vacuum energy, which curves spacetime. In the de Sitter space, an empty from matter Universe is expanding exponentially, Fig. 6.6, which is the basis for the inflationary model of the Universe. De Sitter's model could be stable only if it contained no matter. De Sitter hoped the model might somehow be adjusted to describe the real universe, provided the density of matter was close enough to zero (de Sitter 1916, 1917). Einstein initially rejected the de Sitter solution because it was contrary to his intuition, since it meant that spacetime could be curved even without the presence of matter. After some period, however, he was forced to accept this idea.

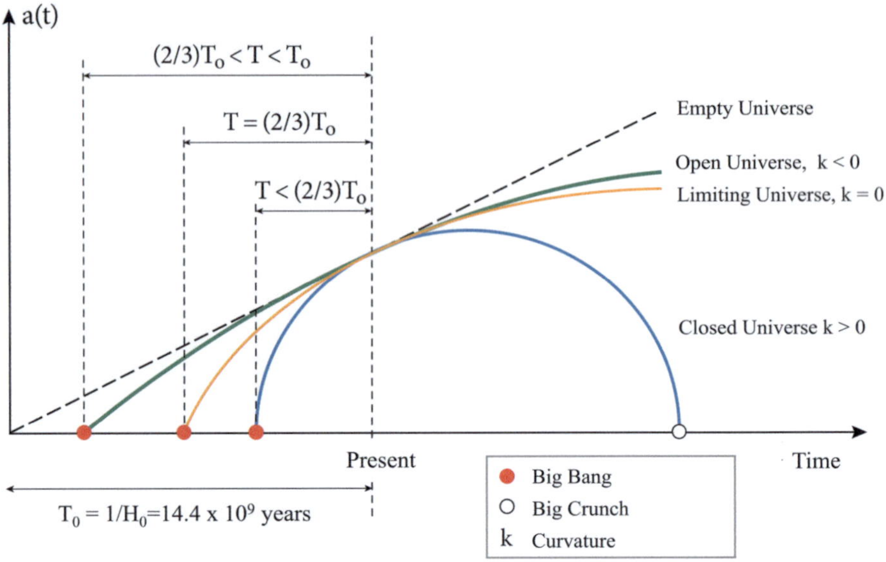

Fig. 6.6 Time evolution of the cosmic scale factor a(t) from Friedmann's original solutions of Einstein's equations of the GTR. The age of the Universe T depends on the curvature k of spacetime, assuming a *homogeneous* and *isotropic* Universe. In these models, the scale factor a(t), a measure of the distance of galaxies, is plotted as a function of time t. All models start with a *Big Bang*. The *closed* Universe model *of positive* curvature and *spherical* geometry results in the *Big Crash.* In the *limiting* and *open* Universe models with *Euclidean* and *hyperbolic* geometry, respectively, the expansion continues to infinity. In the following, we shall see that the presence of dark matter and dark energy modifies these models and corresponds to a model with Euclidean geometry ($k = 0$) but accelerated expansion and an age of about 13.8 billion years. The upper limit on the Hubble age of the Universe $T_0 = 1/H_0 \approx 14.4$ billion years (*the Hubble time*) corresponding to a value of the Hubble constant $H_0 \sim 67.8$ km/s/Mpc

1927: Independent Discovery of the Expansion of the Universe by Georges Lemaître

Georges Lemaître, the Belgian Roman Catholic priest and professor of physics at the University of Louvain[2] in Belgium is the first person to discover the expansion of the Universe. Subsequently, by reversing time, he was led to suggest that the Universe inevitably had a beginning, in the so-called *Big Bang*.

Lemaître began his scientific career with a strong mathematical background. His doctorate in 1920 was on functions of several real variables (*l'Approximation des fonctions de plusieurs variables reales*). Then, after ordination as a Roman Catholic priest (1923), he moved to the University of Cambridge, UK, where he was introduced

[2] The Katholieke Universiteit Louvain (Catholic University of Leuven) was founded in 1425 by Pope Martin V and is the oldest Catholic University in the world and the oldest University in Belgium. In the year 1996 I had the opportunity to teach there a course on *Plasma Astrophysics*, as an invited visiting professor.

to the then open and challenging subject of Cosmology by the most recognized astrophysicist and cosmologist of that time, Sir Arthur Eddington. In the following year (1924), he moved to the USA, at the Harvard College Observatory, in Cambridge (Massachusetts), where he collaborated with the director of the Observatory, Harlow Shapley, on the great research topic of that time, which was none other than: *what is the size of the Universe*? Shapley was already famous from his work on the nature of *spiral nebulae* – known today as spiral galaxies – and whether they are located within the Milky Way, or, as Curtis argued in the *great debate*, they are separate *island Universes*, far away and outside our own galaxy. This topic and the related discussion formed the beginning of the extragalactic astronomy. Also at Cambridge's Massachusetts Institute of Technology, Lemaître attended its Ph.D. program.

In 1925, Lemaître returned to his homeland, full of ideas, having visited the two world centers of Astrophysics and Cosmology, the University of Cambridge in the UK and the Harvard Observatory in USA, and having collaborated there with the leading international researchers Eddington and Shapley. The Catholic University of Louvain gave him a part-time Lecturer position. Then he began to write his paper, which was published in 1927 and it was to bring him great international recognition. In it Lemaître discovered, independently of Friedmann (1922, 1924), solutions of the Einstein equations for a homogeneously expanding Universe, (Figs. 6.7, 6.8, 6.9 and 6.10).

He then proved that Einstein's solution for a static Universe, with the help of the *cosmological constant*, is unstable to small perturbations of this *cosmological constant*. In the next step, he used the already published Doppler drift velocities of 42 galaxies and their corresponding distances and showed that these observed

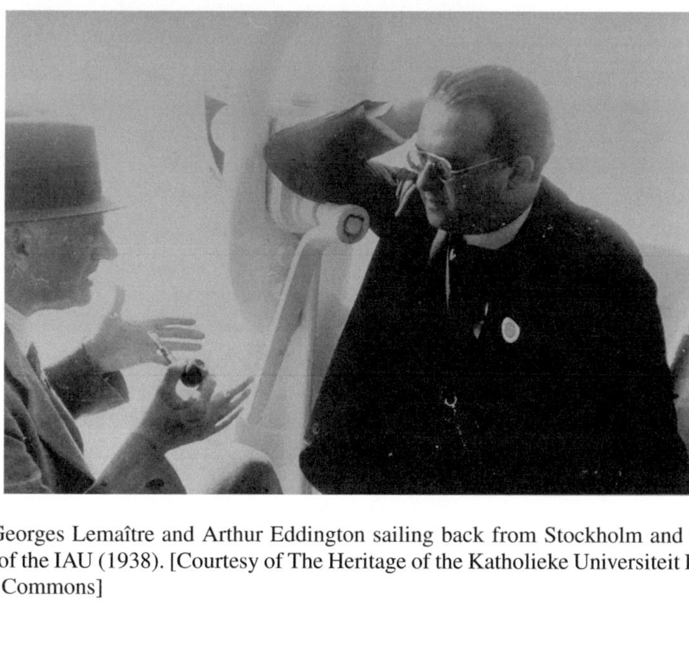

Fig. 6.7 Georges Lemaître and Arthur Eddington sailing back from Stockholm and the General Assembly of the IAU (1938). [Courtesy of The Heritage of the Katholieke Universiteit Louvain and Wikipedia Commons]

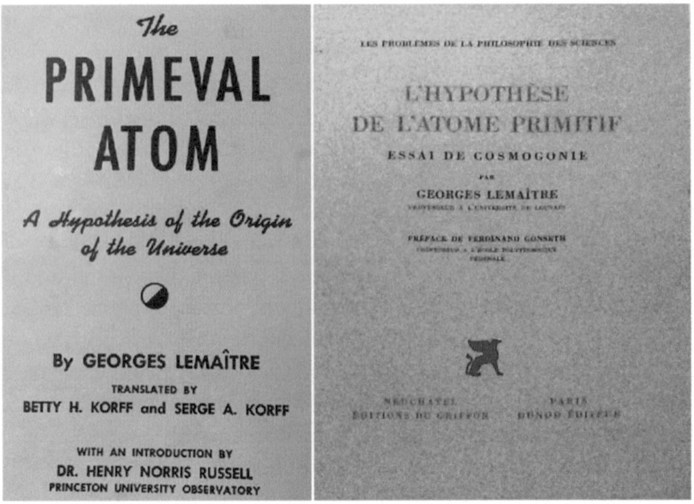

Fig. 6.8 (*left*) First English edition of G. Lemaitre's book, The PRIMEVAL ATOM (1950), which is still available by Amazon. (*right*) *L'Hypothèse de l' Atome primitif* (*The Primeval Atom – an Essay on Cosmogony*) (1946) [Image Credit: Collection of the Musée L, Ottignies-Louvain-la-Neuve; Belgium and Wikipedia Commons]

Fig. 6.9 (*from left*) Robert Millikan, Georges Lemaître and Albert Einstein after Lemaître's lecture at the California Institute of Technology in January 1933. [Courtesy of Wikipedia Commons]

velocities V are nothing but the result of the expansion of the Universe, the velocity being proportional to the distance D, i.e., the relation V = HD, where H is a constant, the inverse of which approximates the age of the Universe! He published this work in the journal *Annales de la Société Scientifique de Bruxelles* (*Annals of the Scientific Society of Brussels*) under the title *Un Univers homogène de masse constante et de*

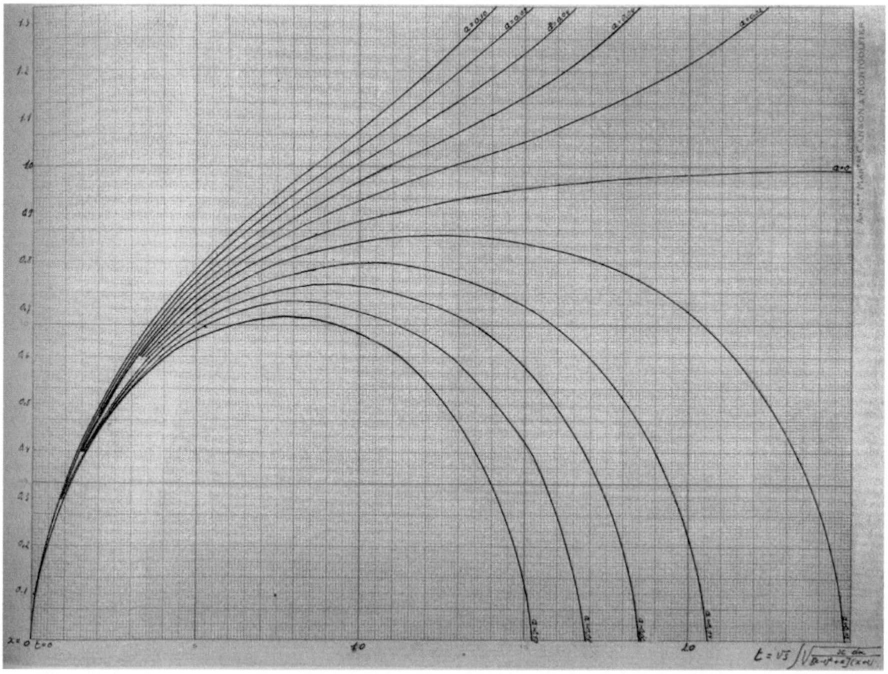

Fig. 6.10 Lemaître's unpublished manuscript diagram found in his own diary, together with the original text of his article (1927). It represents the time evolution of the scale factor a(t), i.e., the function of time a(t) according to which the size of a homogeneous and isotropic Universe changes. The curves are plotted with the *cosmological constant* as a parameter, for all possible models of the Universe that have positive curvature. All models start with a singularity at the origin of the axes (x = 0, t = 0). For large values of the *cosmological constant*, an open Universe results. Recent data on the accelerated expansion of the Universe agree with Lemaître's solution for a positive curvature and an accelerated expansion (upper curve). [Image Source: Manuscript, Archives Georges Lemaître, Katholieke Universiteit Louvain–Georges Lemaître manuscripts, Catholic University of Leuven, 1927]

rayon croissant rendant compte de la vitesse radiale des nébuleuses extragalactices– A homogeneous Universe of constant mass and growing radius accounting for the radial velocity of extragalactic nebulae, Vol. 47, p. 49, 1927).

Thus, Lemaître was the first to calculate the value of the constant H in the expansion law of the Universe, and from this the age of the Universe, already two years before Hubble (1929). Unfortunately, when his original article was translated into English and published [Lemaître, G. 1931, MNRAS, 91, 483], it was curiously altered. It was specifically omitted the most critical result, Eq. (24) of the original French text, which calculated the age of the Universe. In any case, this result led him to propose the *Big Bang* theory, *the hypothesis of the Primitive Atom* (Fig. 6.8, Lemaître 1945).

Hoyle and the Naming of the *Big Bang*

Fred Hoyle, a committed atheist and an advocate of the rival theory of *continuous creation*, ironically referred to Lemaître as the *Big Bang man.* Indeed, he was the first to use the expression *Big Bang* in 1948, in a BBC radio interview, despite not believing it!
Lemaître's theory was received with great skepticism by the scientific community. Einstein, though he couldn't find anything wrong with the math of Lemaître's theory, refused to accept that the Universe was expanding: *Vos calculus sont correct, mais votre physique est abominable, i.e., your calculations are right, but your physics is unacceptable.* Similarly, Eddington found Lemaître's idea *unpleasant*.
Lemaître's ideas raised many consequential questions, to answer which physics and astrophysics joined forces: e.g., what was really this unheard *Primitive Atom* and why this exploded? Lemaître pondered these questions for some time, and also he proposed that there must exist the diffuse cosmic microwave radiation, the remnants of the titanic explosion of the primordial atom. At the same time, he was also interested in the philosophical extensions of his theory, which were indeed many and very important.

Additionally, other researchers dealt with the theory of the *Big Bang* (BB). For several years, the scientific community was divided on them who accepted the *Big Bang* theory and those who supported the theory of *continuous creation*, according to which the Universe was and will be forever the same and unchanging. This scientific controversy ended when Arno Penzias and Robert Wilson (1965) discovered the diffuse cosmic microwave radiation, which Lemaître and other theoretical cosmologists have identified as the remnants of the *Big Bang*, which occurred billions of years ago.

Einstein met several times and collaborated with Georges Lemaître of the Catholic University of Leuven. In 1934, Lemaître received from King Leopold III the Francqui Prize, the highest Belgian scientific distinction. Proponents were Albert Einstein, Charles de la Vallée Poussin (Belgian Mathematician) and Alexandre de Hemptinne (Belgian physicist and chemist). The members of the international jury were Eddington, Langevin, Téophile de Donder and Marcel Dehalu. In the same year, he also received the Mendel Medal of Villanova University. He was also honored with many others awards and distinctions.

Georges Lemaître died on June 20, 1966, ironically, shortly after the discovery of the diffuse *cosmic microwave radiation*, which confirmed his theory of the origin of the Universe!

Lemaître: The Universe was created by a *Creator*

The eminent Canadian astronomer and cosmologist Sidney van den Bergh (vice-president of the International Astronomical Union 1972–1982) writes in his article (Sidney van den Bergh, The Curious Case of Lemaitre's Equation No 24, Journal of the Royal Astronomical Society of Canada, Vol. 105, No. 4, p. 151, 2011): *In August 1961, Abbé Georges Lemaître mentioned to me that, as a devout Christian, he was leaning toward the idea that the Universe was created by a Creator. Therefore, the discovery of the theoretical proof and observational verification of the expansion of the Universe should have given him an indescribable inner satisfaction.*

In 1951, Pope Pius XIII declared that Lemaître's theory provided a scientific validation of the Bible. However, Lemaître countered this remark by the Pope, stating that his theory was neutral and there was neither connection nor contradiction between his religion and his theory and advised the Pope to avoid such connections. Lemaître was a devout Catholic, but he opposed mixing science with religion, physics with metaphysics, although he saw that the two fields were not in conflict at all!

This episode matches well with the personality of Msgr. Lemaître, since the Belgian cosmologist had already on a previous occasion also "corrected" Albert Einstein, pointing out that the general relativity field equations, which described the Universe in its entirety and which the German physicist believed should only be applied to a *static* Universe, actually better described an expanding Universe. Einstein did not initially accept Lemaître's point of view and introduced a parameter ad hoc into the equations (the term Λ) precisely to "balance" this expansion, having to change his mind later and confess to having made the biggest mistake of his life on that occasion. As it has been characteristically written by Giuseppe Tanzella-Nitti, professor of Fundamental Theology at the Pontifical University of the Holy Cross, Rome: "who is this breezy, confident Monsignor-cosmologist who is able, within the span of a few short years, to correct the father of Relativity and a Pontiff of the Catholic Church in one fell swoop?"

The renaming of the expansion law of the Universe.

In 2018, I had the opportunity to participate in the XXX General Assembly of the International Astronomical Union (IAU) in Vienna. There, after an open vote of all its members, the IAU decided to rename the law relating the expansion speed V of the galaxies in the Universe, in proportion to their corresponding distance D, $V = HD$, to the Hubble-Lemaître law. Thus, an injustice was restored, even after 90 years!

1929: The Hubble Velocity-Distance Relation

In 1912, the American astronomer Vesto Slipher at the Lowell Observatory in Arizona observed the redshift of the spectral lines of galaxies, the so-called *nebulae* as the galaxies were then called (Slipher 1917). By 1925, he had shown that most exhibited redshifts in their spectral lines. He interpreted this as meaning that they were in fact moving away relative to us so their spectrum was shifted to longer (i.e., redder wavelengths). Taken together, the evidence eventually convinced astronomers that *spiral nebulae* such as Andromeda were in fact separate *island Universes* of billions of stars like our own Milky Way but more distant (Fig. 6.11).

Edwin Hubble, aided by Milton Humason, extended the work of Slipher by using the larger *Hooker telescope*, by taking long exposures of the spectra of faint galaxies. Then, Hubble calculated the distances of these galaxies (Fig. 6.12), by using the variable stars *Cepheids*, by means of their period-luminosity relation discovered by Henrietta Leavitt of the Harvard Observatory (1912). Pulsating stars of this type were observed for the first time in 1784, but they were given special attention after Miss

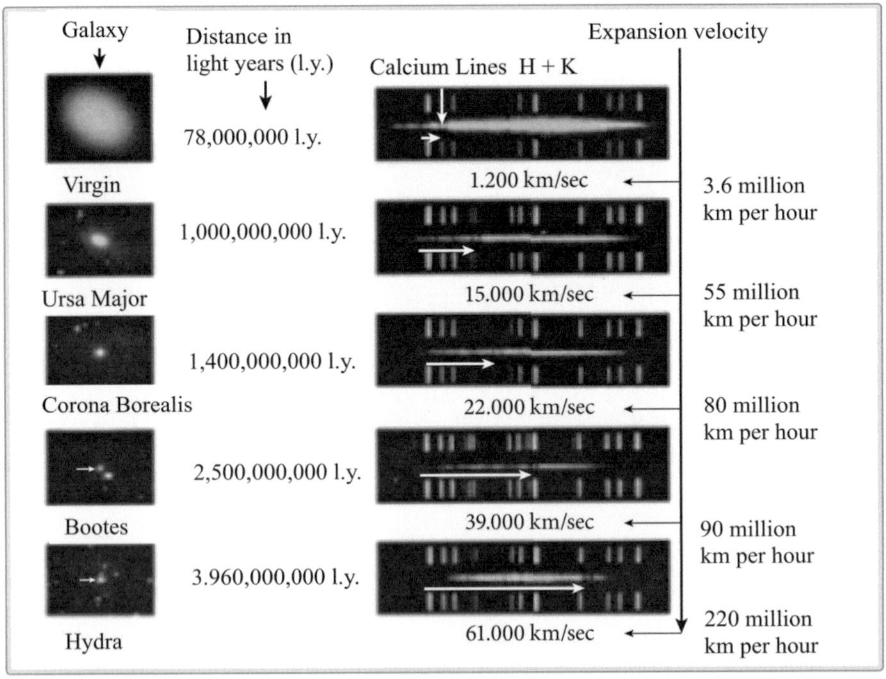

Fig. 6.11 Calculated values of the redshift of the spectral lines and thus the recession velocity for 5 extragalactic nebulae at the respective distances. The arrows in the spectra on the right show the redshift of the calcium H + K lines. Most of the spectra were obtained with the 100-inch telescope at Mount Wilson. The distances here assume expansion of the Universe at a rate of 50 km s^{-1} Mpc^{-1}. (*Credit* Adapted from the original, courtesy of the Palomar Observatory, California Institute of Technology)

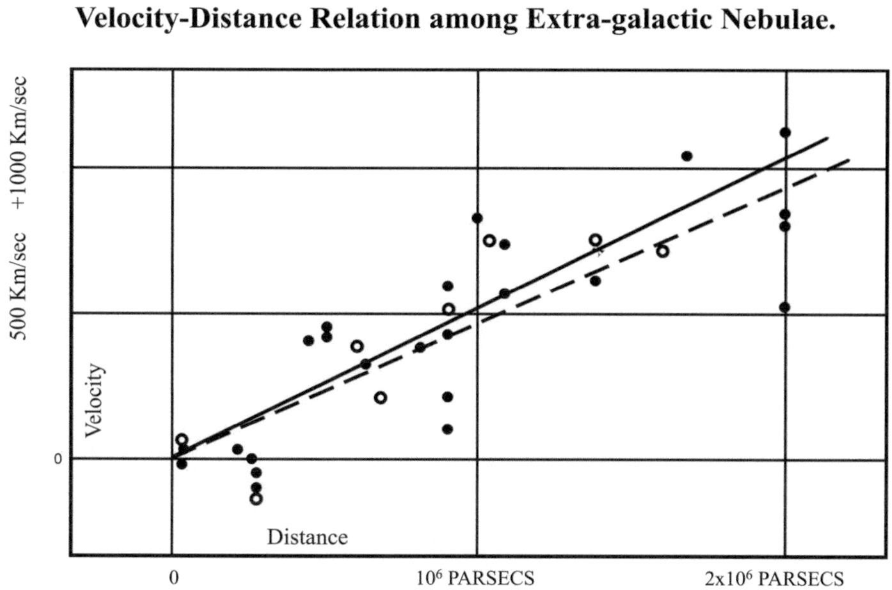

Fig. 6.12 Diagram of Edwin Hubble's first historic publication. Presented on January 17, 1929, at the US National Academy of Sciences. It shows the linear relationship between the escape velocity and the corresponding distance of 24 nebulae (galaxies) for which their distance could be calculated. The vertical axis of the diagram gives the solar rotation-corrected escape velocities as a function of the distances on the horizontal axis, calculated from the variable stars within each nebula. The solid circles and solid line correspond to the results taking into account the solar rotation and using each nebula separately. The empty circles and dashed line are from combining the nebulae into groups. Recoil velocities were calculated by Vesto Melvin Slipher, Lowell Observatory, AZ and distances by Hubble and Humason. (*Source* Hubble, PNAS (1929))

Leavitt's observation that there was a relation between their luminosity and their period.

In particular, Miss Leavitt observed that the more luminous were these stars, the longer was their period. She reached this result from the observation of stars of this type in the *Small Magellanic Cloud*. This meant that, because those stars were roughly at the same distance, the relation which connected their apparent magnitude with their period, it should also connect the absolute magnitudes with their period. Hence, the relationship of apparent magnitude – period was actually a relation of absolute magnitude – period. In other words, if we could measure the period of such a star we could determine its absolute magnitude. Then, by measuring the apparent magnitude we should be able to determine the corresponding distance of the star! This was a really important discovery, which opened the way for the measurement of distances in the Universe, since such stars existed in all galaxies!

Hence, combining Hubble's measurements of galaxy distances D with Slipher's measurements of the expansion velocities V derived from the redshifts of the galaxy spectra, Edwin Hubble and Humason at the Mount Wilson Observatory in California, discovered (1929) that these two quantities V and D are *proportional* to each other, **V = HD**, where H is some proportionality constant. This proportional relationship is now known as the *Hubble-Lemaître law* and is the cornerstone of the modern model of the expanding Universe.

Note that in the original Hubble diagram, there were only 24 galaxies. Hubble's contribution lies in calculating their distances up to the Virgo cluster, using three different methods for calculating distances: variable *Cepheids*, brighter stars in galaxies, and the mean luminosities in the Virgo supercluster.

Later followed other publications by Hubble (1936), Humason (1936), Hubble and Humason (1931, 1934), of more distant galaxies, at distances much greater than a few million light years that Hubble's original publication (1929), Fig. 6.13.

The Mount Wilson Observatory in California was the world's most important observatory during the first half of the twentieth century. There were the biggest telescopes and there the biggest discoveries were made in astronomy, which changed our view of the Universe. Edwin Hubble proved that the mysterious nebulae, about which there was a long scientific debate about whether they are inside or outside the Milky Way, were galaxies like our own Galaxy (1923). Then Hubble, with an excellent collaboration with Humason discovered that the texture of the vast dimensions of the Universe was expanding: space itself was *overflowing* and getting bigger and bigger (1929). This amazing effect, reversing time, led to the theory of the *Big Bang*, according to the previous works of Lemaître (1927). *Thus began modern Cosmology*!

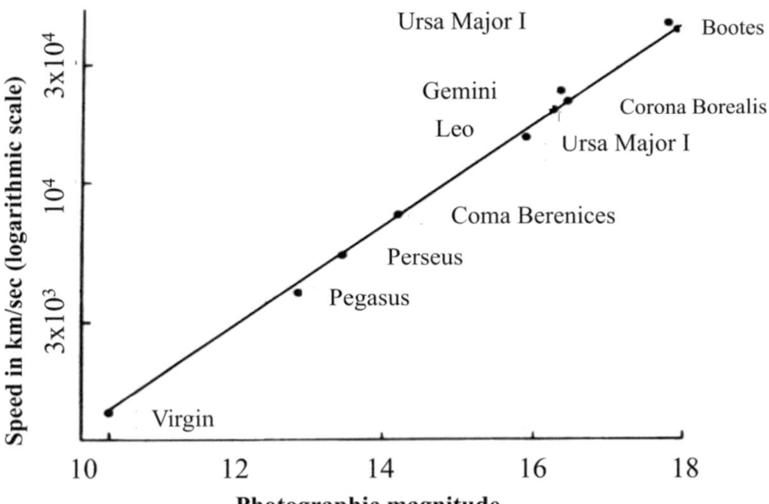

Fig. 6.13 Hubble in his 1936 diagram calculated spectroscopically the velocity as a function of distance in superclusters of galaxies in which he took the fifth brightest member of each cluster. This diagram extended the initial observations of 1929 shown in the previous diagram of Fig. 6.12 and which were limited to the relatively close distances of a few million light years, while those of the present diagram reached billions of light years, like the distance of Bootes. The magnitudes were photographic, i.e., they gave the brightness of a galaxy as depicted on a photographic film

Subsequently, Fritz Zwicky observed the first indications of the existence of dark matter (Zwicky 1933). Nicholson discovered Jupiter's moons Lysithea and Carmis (1938). Walter Baade discovered the two stellar populations I and II (1952).

A *Mule Driver* Contributes to One of Cosmology's Greatest Discoveries...

The observations of the spectral displacement of galaxies were made by Milton *Humason*, whose relevant biography is interesting. First of all, he dropped out of school at an early age, so his education stopped at the age of 14. Because he liked mountains, he asked to work as a mule driver, one of those hired to transport the building materials and equipment to the Mount Wilson Observatory in California – only for the telescope's main mirror, due to its large weight, one of the first primitive trucks of the time was used, Fig. 6.14 (lower panel). Curious about the material he had helped to carry up in the mountain, he volunteered as a night helper there. In 1917, *Humason* was appointed as a cleaning staff at the same observatory. His skill and quiet character soon made him the best assistant to the astronomers. Recognizing his talent, observatory director George Hale appointed him in 1919 as a member of the observatory's scientific staff. This was still unusual for the time, as *Humason* not only did not have a graduate degree, he did not even have a high school diploma. Soon, however, *Humason's* progress proved Hale's judgment correct, as he made very important observational discoveries. He gained a reputation as a particularly careful observer, taking photographs and difficult spectrograms of faint galaxies. These played an important role in the development of Cosmology, when he helped Edwin Hubble to formulate the law of the expansion of the Universe. The two collaborated harmoniously for many years, Fig. 6.14 (top panel).

1948–1965: The Final Prevalence of the *Big Bang* Theory

1948: Tremors in the Edifice of the Steady State Theory

The *steady state theory* (also called the theory of *continuous creation*) argued that, given the observed expansion of the Universe, for its density to remain constant, matter should have to be created from nothing. One of its proponents, Fred Hoyle, calculated that *the creation from nothing of one atom of hydrogen per cubic meter every one million years is capable of keeping the density of the Universe constant*.

Unlike the *Big Bang* theory, the *steady state theory* has no point far back in time when a *creation event* occurred causing the Universe to come into being. To Fred Hoyle, who was a committed atheist, this was a particularly attractive feature of the theory!

Fig. 6.14 (*top*). Excellent collaboration of Edwin Hubble (1889–1953) and Milton Humason (1891–1972) led to the discovery of the expansion of the Universe. (*top middle*) The historic 100 inch (2.5 m) Hooker telescope. It started operations in 1917 and was the world's largest reflecting optical telescope until 1949. (lower panel). The mirror of the 2.5 m Hooker telescope is transported on a Mack truck in 1917 up the dangerous road to Mount Wilson to be installed on the telescope at the 1750 m summit. *Source Wikimedia Commons* (*Photos Credit* Michael Richmond)

Although new matter is created from nothing, this model succeeds to keep the *total* amount of energy contained in the Universe constant. This is because the inertial energy of the created matter is provided by the gravitational energy of the expansion of the Universe. This can be achieved because the gravitational potential energy has a negative sign and thus balances the positive energy of the created matter. In other

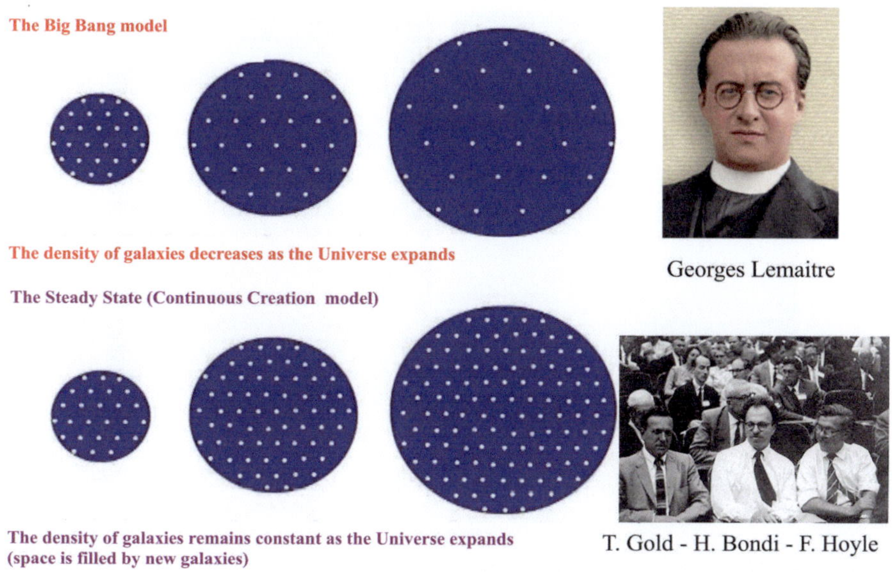

Fig. 6.15 Until 1948, there were two conflicting theories for the evolution of the Universe: the *Big Bang* theory and the *continuous creation (steady state) theory,* with the latter having the lead in acceptance by the scientific community. However, Lemaître's publication of a summary of his theory in the book *The Primeval Atom*, along with the theoretical prediction by Gamow and his doctoral student Alpher of the nucleosynthesis of the nuclei of the light elements and the existence of the diffuse cosmic microwave radiation, led to the definitive prevalence of the *Big Bang theory and the disappearance from the foreground of the* theory of *continuous creation*

words, this model manages to be compatible with the fundamental conservation principles of physics!

However, the theory could not be reconciled with some "disturbing" observations. For example, it predicted:

- The formation of the chemical element helium *only inside stars*, something which is, however, insufficient to justify the large proportional amount of this element in the Universe,
- the observation of *quasars* only at great distances (and thus their existence only in the distant past),
- the non-slowing down of the expansion of the Universe, as predicted by the theory.
- Nevertheless, the final nail in the coffin of the Steady State theory was the discovery in 1965 of the *cosmic microwave background radiation.*

These are indeed particularly *unpleasant* observational data, which an ambitious theory should satisfy, even if this theory is proposed by some famous astrophysicists [(Fig. 6.15 (lower right panel)].

Fig. 6.16 Ralph Alpher (1921–2007) (left), Ph.D. student of George Gamow (1904–1968), (right)

Stephen Hawking on the Steady State Theory[3]

The Steady State theory was what Karl Popper would call a *good scientific theory*: it made definite predictions, which could be tested by observation, and possibly falsified. Unfortunately for the theory, they were falsified…

1948: The Theoretical Prediction of the Nuclear Fusion of Light Nuclei and the Diffuse Cosmic Microwave Radiation

As already mentioned, until 1948 there were two *conflicting* theories about the evolution of the Universe: the Big Bang theory and the *continuous creation*, or, *steady state* theory. In 1945, Lemaître published a summary of his theory in the book *The Primeval Atom: An Essay on Cosmogony*, Fig. 6.8, albeit he still failed to convince the scientific community of the correctness of the *Big Bang* theory. Two years earlier, the theory of *continuous creation*, which was supported by Thomas Gold in the USA and Hermann Bondi and Fred Hoyle in the UK, seemed to be gaining ground in acceptance of the authoritative scientific community.

The further development and bringing to the fore of the *Big Bang* theory, was dealt with systematically and promoted by George Gamow – a student of Alexander Friedmann – as early as the 1940s. Gamow had seceded from the Soviet Union

[3] http://www.hawking.org.uk/the-beginning-of-time.html.

and was teaching at George Washington University in the US. In particular, Gamow assigned there to his graduate student Ralph Alpher, as his doctoral thesis topic, the numerical solution of a system of differential equations describing the detailed process of composition of the elements, according to the hot *Big Bang* model. The results of Alpher's Ph.D. thesis were published in 1948, as the *αβγ paper*, as it became known from the initials of the 3 authors (Alpher, R.A., Bethe, H., Gamow, G., The Origin of Chemical Elements, Physical Review, vol. 73, Issue 7, pp. 803–804 (1948).

Big Bang Nucleosynthesis

In the *αβγ paper*, the authors proposed that if the young Universe was very hot, then the genesis of the first light nuclei (deuterium, helium, lithium) could be explained during the first 3 min, a time where the temperature of the Universe was about one billion degrees. The important result of this *αβγ work* is that it explained the relative abundances of the isotopes of the elements hydrogen and helium, which together constitute more than 99.9% of the baryonic mass of the Universe. Indeed, in the solar system, the ratio of the elements to the number of nuclei is as follows: for every 1000 atoms, 921 are hydrogen atoms, 78 are helium and just one atom from the rest of the periodic table! The corresponding ratio by mass is about 75% hydrogen, 25% helium, and a much smaller percentage of the remaining elements. After hydrogen and helium, the most abundant chemical elements in the solar system are oxygen-16, carbon-12, nitrogen-14, and neon-20.

The main features of the so-called *Big Bang* nucleosynthesis that suggested this famous *work αβγ* are the following:

The neutron-proton ratio was already set during the first seconds after the *Big Bang* by the physics of the Standard Model of elementary particle interactions. The Universe was already homogeneous and radiation dominated. Nuclear fusion occurred between about 10 and 20 min after the *Big Bang*. This corresponds to the temperature range where the Universe was cold enough so that the deuterium survives and does not decay, but sufficiently hot and dense so that the fusion reactions proceed at a considerable rate. This nucleosynthesis was widespread, covering the entire observable Universe.

The basic parameter that allows the calculation of the results of the *Big Bang* nucleosynthesis is the ratio η of the number of baryons to photons, an extremely small number, $\eta \cong 6 \times 10^{-10}$, Fig. 6.17. This number also comes from the *Standard Model* of the interactions of elementary particles in high energy physics. This small parameter η controls the rate at which nucleons collide and react. Given the value of this parameter η, it is possible to calculate the abundances of the elements when the process of the *Big Bang* nucleosynthesis was completed. Although the value of the baryon number and photons ratio η is important for determining the final cosmic abundance of the elements, its exact value makes very little difference to the final result. Therefore, without significant changes in the *Big Bang* theory, nuclear fusion will eventually lead to cosmic abundances with approximately 75% hydrogen-1, about 25% helium-4, about 0.01% deuterium and helium-3, very small amounts of lithium-10 and even smaller proportions of heavier elements. The fact that the observed abundances of the chemical elements in the Universe, Fig. 6.17, are generally consistent with those numbers derived from the theoretical analysis of the nucleosynthesis of the lighter chemical elements is considered a very strong indication for the *Big Bang* theory!

Therefore, *nuclear fusion* during the first 3 min or so of the Universe's evolution is responsible, to a large extent, for the observed proportions of the elements hydrogen, deuterium, helium-3, and helium-4, in the Universe, Fig. 6.17. The nuclei of these

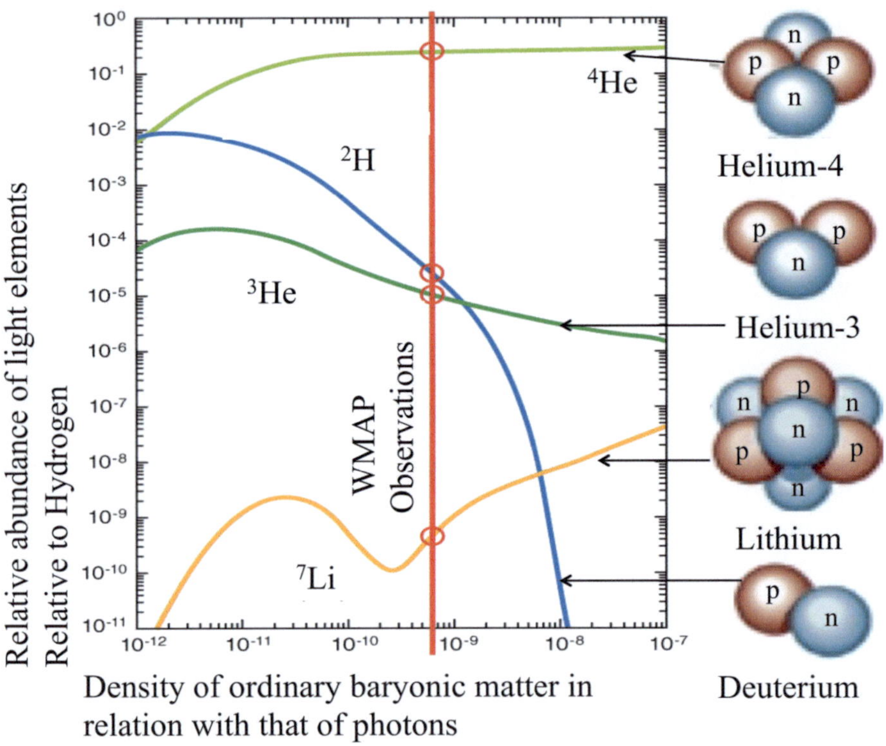

Fig. 6.17 Diagram of the chemical composition of the young Universe as a function of the ratio of baryons to photons. The abundances of helium-4, deuterium, helium-3, and lithium-7 are shown as a function of the ratio of baryons to photons, according to the *Big Bang* theory. Baryonic matter is currently observed to be about 4.6% of the energy density of the Universe. The energy density of the young Universe was dominated by the radiation of the hot black body. By the time the Universe was seconds old, the temperature had dropped to about $T \cong 10$ trillion degrees and neutrons were converting to protons and electrons, or, combining with protons to form deuterium nuclei, more of which then combined to form helium nuclei during the first 3 min of the Universe's life. Interestingly, WMAP observations of the baryon-to-photon ratio in the young Universe (red line) are in excellent agreement with the *Big Bang* model. (*Image Source* NASA/WMAP science team)

light elements, together with some amounts of lithium-7 and beryllium-7, are believed to have formed when the Universe was between about 100 and 300 s old, when the primordial quark-gluon plasma froze to form protons and neutrons. Due to the very short period of time during which the nuclear synthesis of these elements, before it stops due to the expansion and subsequent cooling of the Universe, no other element heavier than lithium would be able to form.

Consisting of only one proton and one neutron, deuterium is the simplest chemical element after hydrogen and thus the first and critical nucleus formed in the primordial Universe. This happened when the ratio of neutrons to protons in the Universe was about 1/2. It should be noted that the neutrons outside nuclei have a lifetime of only about 15 min. Nevertheless, other nuclear chain reactions followed that formed heavier nuclei, mainly helium-4, and much smaller amounts of other lighter nuclei, such as helium-3 and lithium-7, Figs. 6.17 and 6.18, as follows.

First, the neutron-proton ratio is a key element of *Big Bang* nuclear fusion and is determined by the physics of the Prototype Model, which was valid before the time of *Big Bang* nucleosynthesis. Explosion, essentially within the first second after the *Big Bang*. Neutrons can react with positrons or electron neutrinos for to create protons and other products. Much earlier than 1 s, these reactions were fast and kept the ratio of neutrons to protons close to 1:1. But as the temperature decreased, the balance shifted in favor of protons due to their slightly lower mass, so the ratio of neutrons to protons also decreased. These reactions continued until the ever-decreasing temperature and density of the Universe caused an excessive slowness of reactions. This happened when the temperature of the Universe T became equal to $T \cong 0.7$ MeV, or, 10 billion degrees (the so-called freezing temperature) and the age of the Universe was about 1 s. Then, the neutron-proton ratio was about 1/2. However, free neutrons are unstable with an average lifetime of 880 s. As neutrons continue to decay in the next few minutes before fusion with nuclei, the ratio of final neutrons to protons after the end of fusion becomes about 1/7.

Almost all neutrons that fuse, instead of decaying, eventually combine into helium-4 nuclei, due to the fact that helium-4 has the highest binding energy per nucleon among the light chemical elements. This process provides that eventually about 8% of all atoms should end up in helium-4, creating a helium-4 mass fraction of about 25%, which is consistent with the observations. Small traces of deuterium and helium-3 remained as it was insufficient time and density to react and form helium-4, Fig. 6.17. Later, at 380,000 years after the *Big Bang*, the Universe became transparent and the diffuse *cosmic microwave radiation* was released that we detect today, something we will discuss below.

In the same year (1948), Alpher in collaboration with Robert Herman proposed that, when the temperature of the Universe decreased due to expansion in some thousands of degrees, the negative electrons joined the positively charged ones protons forming neutral atoms, so the Universe became transparent to radiation. This allowed, for the first time, photons to propagate over long distances and penetrate it. But the most interesting thing was that Alpher and Hermann (1948) calculated that the *echo* of the *Big Bang* should be *heard* today, in the form of a *black body radiation* with a temperature of about 5 degrees. Again, however, their prediction did not come

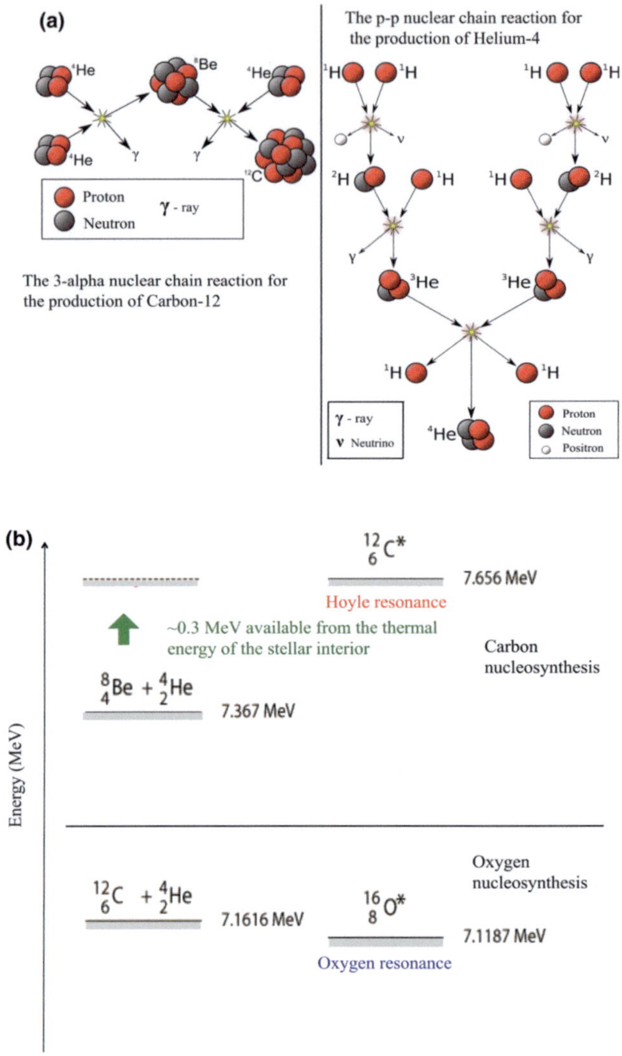

Fig. 6.18 **a**. (*right*) Proton-proton I chain reaction to produce helium nuclei ($_2He^4$) dominates the interior of solar-size stars. (*left*) The 3-α process is a series of nuclear fusion reactions, by which two helium-4 nuclei (α-particles) fuse to initially form a Beryllium nucleus ($_4Be^8$), which upon further fusion with an α-particle (via the *Hoyle resonance*) produces a carbon nucleus ($_6C^{12}$). [From Wikipedia Commons]. **b**. Illustration of the *Hoyle resonance*, wherein the fusion of two Beryllium and Helium nuclei produce an excited state of Carbon ($_6{}^{12}C^*$) in stellar interiors. Note that also an excited state of Oxygen is near but below the energy of the excited state of Carbon, with the result that it does not act to compete the formation of Carbon formation

true received with particular enthusiasm by the scientific community, despite the fact that their work was published in the most authoritative journal Nature (Alpher and Hermann 1948). Even when by 1956 they improved their calculations considerably, they again failed to arouse the interest of the scientific community, nor was there any attempt to detect the radiation at a *black body temperature* of a few degrees. Later, in 1964, Robert Dicke et al. in Princeton and Yakov Zel'dovich in Moscow theoretically rediscovered cosmic microwave radiation, a year before its observational discovery by Penzias and Wilson (1965). Experiment and observation (as always, the final judges in Physics), had not yet said their last word…

How the World Came into Being in Only 5 Min…

Alpher's doctoral thesis (1948) had made such an impression that it was attended by hundreds of people, including many journalists. This discovery had a great impact and on April 14, 1948, the American newspaper *Washington Post* had the front page headline *World began in 5 mins.* It was an unusual doctoral thesis.

From the Creation of Matter in the First Minutes, to the Much Later Creation of Humans in the Universe…

The fact that the chemical elements hydrogen and helium, which together make up more than 99.9% of the (baryonic) mass of the Universe, were created in the first few minutes in the life of the Universe is undoubtedly one of the most fascinating events in the history of science.
George Gamow had characteristically commented: *it took about an hour to create atoms, a few million years to create galaxies, but five billion years to create humans*!

1954: Hoyle's Prediction for the Creation of Carbon

A-particles are helium nuclei containing 2 protons and 2 neutrons ($_2He^4$), occupying the second place in the periodic table of chemicals elements. Hydrogen ($_1H^1$) occupy the first place in the same periodic table, consisting of a single proton at its core. The critical 3-α process is a nuclear process according to which carbon can be produced by fusing 3 nuclei of the element helium (Fig. 6.18a) and subsequently from the produced carbon, by further fusing with helium, oxygen can also furthermore be produced.

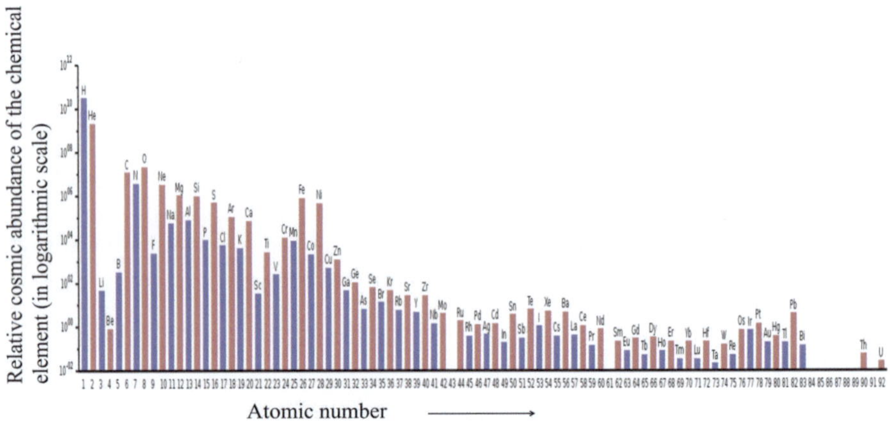

Fig. 6.19 Relative cosmic abundance of the chemical elements, in logarithmic scale. (From the Wikipedia)

However, this process is *inefficient* for the production of carbon and oxygen at the pressures and temperatures that occurred at the nucleosynthesis, during the first moments of the *Big Bang*. A consequence of this is that no significant amounts of carbon and oxygen were created with the 3-α process during the *Big Bang* epoch of nucleosynthesis. And yet, something must have happened so that the process 3-α proceeded, forming those crucial and life-useful elements carbon and oxygen, Figs. 6.18a, b and 6.19.

The carbon nucleus, contains 6 protons and 6 neutrons and its creation is a key step in the chain of nucleosynthesis – the process with which are produced the rest of the heavier elements inside the stars, e.g., oxygen and the other elements, Fig. 6.19. Physicists studying stellar fusion in the 1940s and 1950s believed that carbon-12 ($_6C^{12)}$) could be formed when two helium-4 nuclei fuse to produce a beryllium nucleus (beryllium-8) – which could then fuse with a third helium-4 nucleus. But there was a problem with this case. The energy of the beryllium and helium nuclei to be fused is significantly greater than that of the ground state of carbon-12. This means that the new particle is actually extremely unlikely to form in this way. *But how is the great abundance of carbon (and oxygen) in the Universe explained*?

To overcome this apparent contradiction, the British astronomer Fred Hoyle proposed in 1953 that the carbon-12 nucleus ($_6C^{12)}$) has an excited state that had never been observed before. The idea was that the carbon-12 nucleus would readily form in this excited state and then de-excite to its original state, emitting a well-defined amount of energy (7.6 MeV) upon de-excitation. The excited state of the carbon-12 nucleus was indeed observed three years later in experiments by researchers at the California Institute of Technology!

As Hoyle characteristically pointed out:
if this process did not take place within the stars I would not exist !

For the past 70 years, nuclear physicists have been trying to understand the nature of this *Hoyle resonance* (Dunbar et al. 1953), which is not predicted by standard nuclear models. Researchers have recently found that in its ground state, the carbon-12 nucleus consists of three aggregates of helium-4 nuclei arranged in a compact formation of an equilateral triangle, while in the Hoyle state the three clusters of helium-4 nuclei form the shape of an obtuse triangle. This more open configuration, the researchers explain, is a result of the higher energy of the system (Epelbaum et al. 2012).

Hoyle was working on the modeling of stellar nucleosynthesis (1950) and considered carbon synthesis in the light of the observed carbon abundance in the stars. Carbon could be formed by the fusion of three α-particles, but the probability is relatively so low that this would be too slow to explain the observed carbon abundance. In other words, when the production of carbon by this process was modeled, it still seemed to be too slow to account for the observed carbon abundance, and this led Hoyle to propose that carbon had a nuclear resonance in the neighborhood of 7.7 MeV, even though none had been observed at that time. In fact, Dunbar et al. (1953) did discover a resonance at 7.65 MeV which brought the modeled production rate into agreement with carbon observations.

Specifically, the formation of $_4Be^8$ from two α-particles is energetically favorable since they have almost the same energy. Then, the subsequent fusion of $_4Be^8$ with another α-particle ($_2He^4$) would create a $_6C^{12}$ nucleus,

$$_2He^4 +_2 He^4 \rightarrow_4 Be^8 - 0.092\ \mathrm{MeV},$$

$$_4Be^8 +_2 He^4 \rightarrow_6 C^{12} + 2\gamma(+7.367\ \mathrm{MeV})$$

The small needed energy of 0.3 MeV to reach the resonant carbon energy $(_6C^{12})^*$ is thermally provided at the stellar interiors. Hence, the nuclear resonance of carbon increases the rate of carbon production in stellar interiors. An oxygen resonance is near (at 7.1187 MeV) but below the carbon-helium combination, so it does not act to increase the rate of oxygen formation at the expense of carbon.

Implications of the Hoyle resonance

The dramatic implications of the *Hoyle resonance* in the modeling of the triple-α process are highlighted in a statement from Stephen Hawking and Leonard Mlodinow: *Such calculations show that a change of as little as 0.5% in the strength of the strong nuclear force, or 4% in the electric force, would destroy either nearly all carbon or all oxygen in every star, and hence the possibility of life as we know it.*

With further increases in temperature and density, the nucleosynthesis process produces nuclei of chemical elements up to nickel-56, which later decays to iron-56, Fig. 6.19. Heavier elements beyond Ni in the periodic table are produced mainly by neutron capture. The slow neutron capture, the so-called s-process (slow-process), produces about half the elements beyond iron. The other half of the elements are produced by rapid neutron capture, the rapid-process, which probably occurs in the cores of stars during violent supernova explosions, and also in the extremely high pressure and temperature conditions that they develop during the explosive mergers of neutron stars, the kilonova.

Some Interesting "*Coincidences*"..

Usually, the probability of the triple-α process for carbon formation is extremely small (only once in about 2420 times releases energy and changes into the stable base form of carbon-12). Two α-particles can fuse to Be-8, but Be-8 is extremely unstable and will not last unless there is enough He-4 (α-particles) around to almost immediately fuse with the Be-8 nucleus to form carbon.
However, the ground state of beryllium-8 (${}_4Be^8$) just «*happens*» (!) to have almost the same energy with that of the 2 alpha particles and so the coalescence/fusion reaction of the 2-alpha particles proceeds to form beryllium-8. Furthermore, in the second step, the nuclear reaction ${}_4Be^8 + {}_2He^4$ has almost exactly the same energy as that of an excited state of the nucleus of ${}_6C^{12}$. This *resonance* greatly increases the probability that an incoming α-particle nucleus will combine with a nucleus of beryllium-8 to form a nucleus of carbon-12 (${}_6C^{12)}$).
The existence of this *resonance* was predicted by Fred Hoyle before its actual observation in the laboratory, based on the physical necessity of the existence of this nucleus in order to form carbon inside stars. The prediction and the subsequent discovery of this energy resonance and process gave very important support to for Hoyle's hypothesis of stellar nucleosynthesis, which postulates that all chemical elements were originally formed inside stars from hydrogen, this truly primordial atom.

An Important Prediction of the *Anthropic Principle*

The *anthropic principle* was first proposed by the theoretical astrophysicist Brandon Carter during a symposium in Krakow (1973) on the occasion of the 500th anniversary of the birth of Copernicus. Carter formulated this principle in contrast to the *Copernicean principle*, according to which humans do not

occupy any special place in the Universe. As Carter noted (1973): Our Position May not Necessarily Be Central to the Universe, but It is Inevitably Privileged, to Some Extent.
In particular, he disagreed with the use of the *Copernicean principle* to justify the *perfect cosmological principle*, according to which all regions and times in the Universe are statistically identical. This last principle basically supports the *continuous creation theory*, which was shown to be invalid after the discovery of the diffuse *cosmic microwave radiation* in 1965, as discussed below.
The *anthropic principle* refers to the specific values that certain parameters happen to have in physics. That is, the philosophical dilemma that created the *anthropic principle* is that the constants of the Universe at the microcosmic, macrocosmic, and cosmological levels seem to be in extremely perfect harmony with each other in order for intelligent life to flourish and evolve in the Universe. This reflection on how beings with consciousness like us come to exist in the Universe is called the *anthropic principle*, and it takes several forms.
Further, Carter formulated the *weak anthropic principle*:

We should be prepared to consider that our position in the universe is necessarily privileged to a degree compatible with our existence as observers,

While its *strongest* form is:

The fundamental parameters of the universe must be such as to allow the creation of observers in the universe at some stage.

Paraphrasing Rene Descartes who wrote: Je pense, done je suis, i.e.,

I think, therefore I exist,

We may say, cogito ergo sum, cogito ergo mundus talis est

I think, therefore the world is as it is

It is extremely interesting that the *anthropic principle* was used by Hoyle to explain the fact that the nuclear resonance for the formation of carbon, Fig. 6.18, was finely tuned to create large quantities of the life-critical elements carbon and oxygen in the Universe.

Historic Discovery of the Diffuse *Cosmic Microwave Radiation*

1965: The Discovery of the Diffuse Cosmic Microwave Radiation

As it was previously discussed, until the 1960s, the scientific community remained very skeptical about the *Big Bang* model of Lemaître (1927, 1945), despite that it was proposed already 30 years ago. As a consequence, Gamow's (1948) subsequent prediction of the existence today of an isotropic microwave radiation, corresponding to the *black body emission* with a temperature of a few degrees Kelvin, as a consequence of the initial moments of the young hot Universe, was ignored. It is truly questionable and worthy of study that no one cared to even check observationally whether this very important idea, which was essentially at the core of the model of the *Big Bang* was right! The observational support for this truly important idea came completely by chance, like many great discoveries in Physics (see, for example, the discovery of X-rays by Roentgen), by the two researchers Penzias and Wilson, working at the laboratories of the Bell company in the USA Fig. 6.20.

In particular, in those laboratories of the American telephone company Bell, a large 6-m conical antenna had been constructed for the practical telecommunication purpose (which was subsequently abandoned) of communicating with satellites at centimeter wavelengths. With this antenna, the physicist Arno Penzias (1933–) and the radio astronomer Robert Wilson (1936–) in the process of preparation for their measurements, faced a *technical*, as they initially thought, problem. Their antenna was showing some mysterious low *noise* in the centimeter wavelength, coming from all directions in the sky, night and day and from which, despite their efforts, they could not get rid of. They investigated whether this static electrical noise was due to the proximity to the city of New York, or even to the *white dielectric substance* of the antenna (aka the cooing of two pigeons that had build their nest inside the antenna), Fig. 6.20.

After carefully checking all their electronics, they cleaned the antenna from this *white dielectric substance* and moved the two pigeons elsewhere; they noticed that the electrical *parasites* remained. Then, they concluded that this strange noise was coming from *somewhere outside*, perhaps from our galaxy.

Furthermore, after they were convinced that this microwave radiation had a physical origin, they sought help from their colleagues working in Cosmology at nearby Universities. Then, they were surprised to learn from Robert H. Dicke of the Univ. of Princeton that they were indeed lucky, because they were the first humans on Earth

Fig. 6.20 Arno Penzias and Robert Wilson on the gigantic 15-m Holmdel Horn Antenna at Bell Labs, with which they discovered – by chance! – cosmic microwave radiation. (*Credit* NASA)

to literally *hear* the echo of the Universe's titanic *Big Bang*. In fact, Robert Dicke, P. J. E. Peebles, P. G. Roll and D. T. Wilkinson interpret this radiation as a signature of the *Big Bang*, light that had started at the superheated temperature of a few thousand degrees, 380,000 years after the primeval moment of the fiery Creation, but now circulated extremely weakened in a temperature of only 2.74 °K, i.e., 269.5 °C below zero, due to the expansion of the Universe that followed the initial explosion! They named this new radiation Diffuse *Cosmic Microwave Radiation* (CMR), Penzias & Wilson (1965). Later, the COBE satellite (Cosmic Background Explorer) made more detailed measurements and calculated the temperature with greater precision to be at 2.725 °K, as we discuss later, Fig. 6.21.

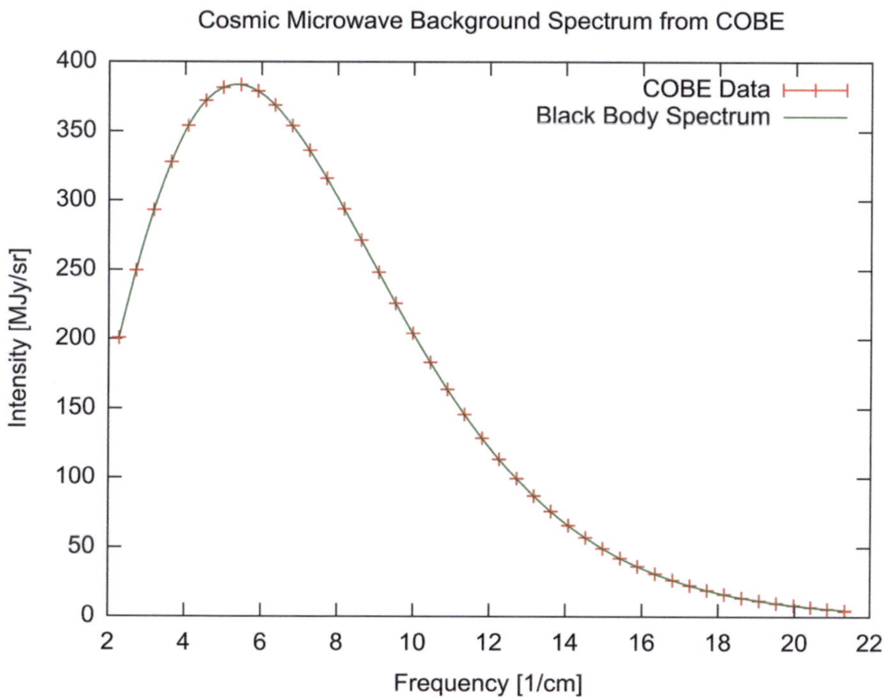

Fig. 6.21 Inverse wavelength distribution of the intensity of the diffuse cosmic microwave radiation, as measured by the Cosmic Background Explorer (COBE) satellite, launched in November 1989. The plot shows the intensity of the radiation as a function of the inverse of its wavelength. The points are from measurements by the FIRAS (Far-InfraRed Absolute Spectrophotometer) on COBE. The continuous curve is the theoretical spectral distribution of black body radiation, temperature T = 2.725 ± 0.001 °K. The perfect agreement of the observational data with the theoretical prediction is impressive. This is the most accurate blackbody spectrum ever observed in nature, since it is practically impossible to separate the observational data from the theoretical curve! (*Source* NASA/COBE, Mather et al. (1990))

Pesky *Parasites* Yielded the Most Important Observational Discovery in the History of Cosmology…

The electrical *noise* that concerned Penzias and Wilson was nothing more than the primordial cosmic radiation, the *echo* from the titanic creation of the Universe, after the Big Bang itself. That is, the annoying *parasites* that Penzias and Wilson were trying to get rid of were the most important observational discovery in the history of Cosmology, after the discovery of the expansion of the Universe. Thus, with this discovery, the coincidence of the theoretical model of the Big Bang of Lemaître with the observational data of Penzias and Wilson was so striking that the theory of the Big Bang triumphantly returned to the fore, after more than 30 years of rejection. It was now treated, not as a metaphysical hypothesis, but as a serious scientific theory. In 1978, Penzias and Wilson were awarded the Nobel Prize in Physics.

Accelerated Expansion of the Universe

1998: Discovery of the Accelerating Expansion of the Universe

By the end of the twentieth century there existed three fundamental discoveries on observational Cosmology, on which the *Big Bang* theory is based to this day:

- the observation of the expansion of the Universe itself [Lemaître (1927) and Hubble (1929)],
- the observed abundances of the light elements hydrogen, helium, and lithium, which perfectly agree with the nucleosynthesis in the young Universe, as predicted by the *Big Bang* theory [studies of Gamow and collaborators (1948)],
- the (accidental) discovery by Penzias and Wilson's of the diffuse *cosmic microwave background radiation*, the thermal *echo* of the superheated past of the primordial Universe.

These three observational discoveries, combined with the theoretical studies of Einstein, Friedmann, Lemaître, Gamow et al. led to the conclusion that if we reverse time in the evolution of the Universe, going back to when the galaxies were closer to each other and the Universe was younger and hotter, we shall reach a peculiar state, wherein the density and temperature of the Universe tend to infinity. This is the very moment of the *Big Bang* itself, when, about 14 billion years ago, space and time were created, together with all matter and energy content of the Universe. From that moment on, the Universe continued to expand at a slower rate due to the pull of gravity, which tends to brake its expansion, while its temperature has been constantly decreasing.

Type Ia Supernovae

However, more recently (1998), a fourth fundamental observational discovery appeared, which shaked the foundations of Cosmology, when two research groups presented their results. These research teams from Berkeley, Johns Hopkins University and the Australian National University aimed their ground-based and space-based telescopes toward the most distant from Earth *supernovae*. In particular, they focused on a special type of stellar explosions, *supernovae type Ia*. These are explosions of aging dense stars that are as massive as the Sun, but at the same time as small in size, as the Earth (known by the artistic name *white dwarfs*). Such supernova explosions temporarily shine, for only a few days, as bright as an entire galaxy. In total, these research groups observed 50 distant *supernovae Ia*, whose light was fainter than expected – i.e., they were at greater distances than expected – revealing that the Universe is expanding in an accelerated manner!

Type Ia supernova is a particular type of supernova that occurs in binary systems, wherein two stars orbit around each other, and in which one of them is a *white dwarf*. The other star can be anything from a giant star to a much smaller *white dwarf*. In such a binary system, the *white dwarf* gradually accretes mass from its binary companion, or merges with the second *white dwarf*. Regardless of the exact details of how the *supernova* ignites, it is generally accepted that a substantial fraction of the carbon and oxygen of the *white dwarf* fuses into heavier elements within a period of only a few seconds. The accompanying huge release of energy ($1–2 \times 10^{44}$ J) increases the internal temperature of the star to billions of degrees, and this energy is more than sufficient to unbind the star, which subsequently explodes violently ejecting its matter at speeds of the order of 5,000–20,000 km/sec, causing an extreme increase in its luminosity.

The interesting fact with these *Type Ia supernove* is that they shine at a fairly constant peak luminosity, because of the fixed critical mass at which the *white dwarf* will explode. The useful for us feature of this phenomenon is that because these *supernovae* have almost the same internal (absolute) brightness, we can measure the distance of these distant and extremely dim objects, by measuring their apparent brightness, i.e., their known peak luminosity allowing them to be used as *standard candles*. In other words, combining their theoretically known absolute luminosity with their observed luminosity, it gives the distance of their host galaxies from Earth. The distance can then be related to the cosmological *supernova* redshift, which is a measure of the expansion of the Universe since the *supernova* explosion occurred. The surprising finding was that the speeds at which distant galaxies are receding from us were observed to increase. Thus, as we studied the explosions of these distant supernovae, we concluded that the Universe is currently expanding at an accelerating rate, and probably in the future it will progressively lead to a state described by the de Sitter spacetime model.

This result fell to the scientific community as *lightning in the air*, because until the end of the last century, we believed that the expansion of the Universe should naturally slow down, because on these huge cosmic scales, the prevailing physical interaction is gravity, which as it is known is always attractive. That's why when the international research program *Supernova Cosmology Project* was launched in 1987 and in 1995 and the second research team, the *High-z Supernova Team* was added, the main goal of their research was to calculate the expected rate of deceleration of the expansion of the Universe, through the detection of selected *supernova* explosions. But the results of the two research groups, which were officially announced in January and February 1998, Fig. 6.22, instead of leading to the calculation of the expected rate of slowdown of the expansion of the Universe, as we all expected, led to exactly the opposite conclusion, namely, that the expansion of the Universe is *accelerating*!

Figure 6.22 displays distance vs. redshift for Type Ia *Supernovae*, generally referred to as Hubble diagrams, as it is similar, but not identical to the original diagram showing the expansion of the universe. The points are the *supernova* measurements assembled by two independent groups, *the High-z Supernova Search Team* (Riess et al. 1998) and the *Supernova Cosmology Project* (Perlmutter et al. 1998). The nearby *supernovae* have been measured over about one decade, while the distant objects have been observed in short campaigns of a few months length. The lines represent the expectations for a standard candle in the specific world models. They represent an Einstein-de Sitter model in which the Universe is completely dominated by matter $\Omega_M = 1.0$, $\Omega_\Lambda = 0.0$), an empty universe $\Omega_M = 0.0$, $\Omega_\Lambda = 0.0$) with nothing in it, and a universe dominated by the *cosmological constant* $\Omega_M = 0.0$, $\Omega_\Lambda = 1.0$). The location of the model lines is set by the data for the nearby *supernovae*. The dotted line is a model which is consistent with all current data and has $\Omega_M = 0.3$ and $\Omega_\Lambda = 0.7$.

In other words, the only way to explain the accelerated expansion of the Universe is that it is a result of the existence of the so-called *dark energy*, i.e., some kind of *repulsive antigravity*. Specifically, the observed expansion of the Universe seems to correspond in cosmological models with about 30% matter and 70% dark energy, i.e., $\Omega_M = \rho_M/\rho_0 = 0.3$ and $\Omega_\Lambda = \rho_\Lambda/\rho_0 = 0.7$ where ρ_M and ρ_Λ are the densities of matter and dark energy, respectively, and ρ_o is the critical density of the Universe. Of course, what exactly is the nature of this *dark energy* remains an enigma that no one has been able to solve so far – perhaps the greatest scientific puzzle of today. This accelerated expansion of the Universe appears to have started when the Universe passed into the era of *dark energy* dominance, about 5 billion years ago.

At this point, we need to clarify the *difference* between the *cosmological redshift* and the *ordinary Doppler redshift*. Laboratory experiments here on Earth have shown that every chemical element in the periodic table when excited emits (or absorbs) photons with a frequency and wavelength that exclusively characterize this, and only this, chemical element. These photons appear, either as emission lines, or, as absorption lines, in the spectrum of astronomical sources. By observing the emission or absorption spectral lines of the astronomical sources we establish the existence of each chemical element of the periodic table in the distant astronomical body. Even more so, from the relative intensity of the lines of different chemical elements we can

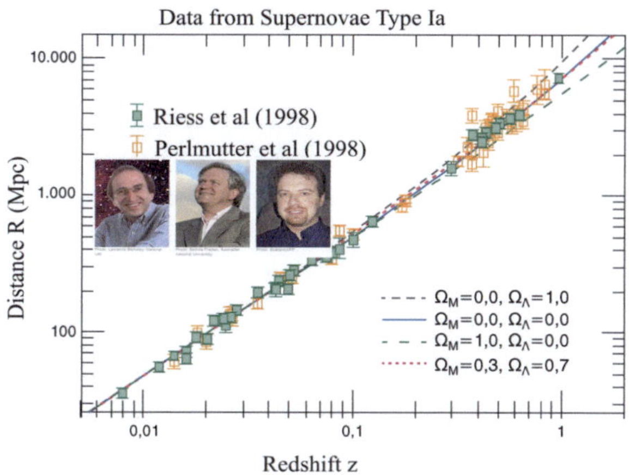

Fig. 6.22 Plot of the distance as a function of redshift z for *Type Ia supernovae* with data from the observation of the accelerated expansion of the Universe, from the two independent research groups: *the High-z Supernova Search Team* (Riess et al. 1998) and *the Supernova Cosmology Project* (Perlmutter et al. 1998, 1999). The nearby *supernovae* have already been observed for a decade, while the more distant ones during short observational surveys lasting a few months. It is noteworthy that there are many *type Ia supernovae* at redshift z = 0.6, which, with a value of the Hubble constant 71 km/s/Mpc, are at a distance of about 5 billion light years away. The lines correspond to the following models of the Universe: Einstein-de Sitter completely dominated by matter ($\Omega_M = 1.0$, $\Omega_\Lambda = 0.0$, green dashed line), an empty Universe ($\Omega_M = 0.0$, $\Omega_\Lambda = 0.0$, blue dashed line), and a Universe dominated by the cosmological constant ($\Omega_M = 0.0$, $\Omega_\Lambda = 1.0$, black dashed line). The best fit of these data corresponds to a fourth model Universe with about 30% matter and about 70% dark energy ($\Omega_M = 0.3$, $\Omega_\Lambda = 0.7$, red dashed line). [Adapted from original at Leibundgut and Sollerman (2001), ESO]. (inset) The Physics Nobel Prize winners (2011) for the observation of the accelerated Universe: S. Perlmutter of the *Supernova Cosmology Project* (Berkeley, *left*), A. Riess (Johns Hopkins University, Space Telescope Science Institute, *right)* and B. Schmidt (Australian National University, *center*)

also deduce the relative abundance of the elements, however, distant the astronomical body may be!

But when astronomers record astronomical spectra, they notice that for most astronomical sources, the spectral lines are all shifted to longer (redder) wavelengths. This shift of the spectral lines is called the *cosmological redshift*. The *cosmological redshift* z is caused solely by the expansion of the Universe and its value gives the source's velocity, or, its distance.

Although the *cosmological redshift* defined as $z = \Delta\lambda/\lambda_o = (\lambda_{obs}-\lambda_o)/\lambda_o$, with λ_o the natural/laboratory wavelength of the emitted or absorbed line, λ_{obs} the observed redshifted wavelength and $\Delta\lambda = \lambda_{obs}-\lambda_o$, when the source velocity with respect to the observer is V, while c is the speed of light, seems to cause the same effect with the Doppler effect, it is essentially rather different from it. In the Doppler effect, the wavelength of the emitted radiation depends on the movement of the source at the moment of emission of the photons. If the source moves closer to the observer, the wavelength decreases as it moves toward the blue part of the spectrum, while if the source moves away from the observer, the wavelength increases as it shifts toward the red part of the spectrum. At the *cosmological redshift*, the wavelength at which the radiation initially emitted increases as it *travels* through outer space. The *cosmological redshift* is caused by the expansion of space itself and not from the movement of the specific source!

For example, in a distant binary system, we are likely to measure simultaneously the Doppler shift and the cosmological redshift. The Doppler shift is caused by the movement of each source around the common center of mass of the system, depending on whether the source is moving toward the observer, or away from it, at the moment of the emission of the photons. The cosmological redshift depends on how far away the system was when the photons were emitted. The more distant the system, the more time the photons spend in thee expanding space and the larger is the measured cosmological redshift z, since as we mention the wavelength of the photons *stretches*, or *lips out* along with space.

Moving necessarily to a more quantitative and inevitably somewhat mathematical level, let us introduce a critical parameter that measures the relative expansion of the Universe, the so-called dimensionless cosmic *scale factor* a(t), which is a function of time t, and is counted since the birth of the Universe, such that if $t_o = 13.8$ billion years is the current age of the Universe, the current value of a(t) is $a(t_o) = 1$. This function, is also called the *scale factor* a(t) of Friedmann–Lemaître-Robertson–Walker (FLRW) and is a key component of the *metric* $g_{\mu\nu}$ of the spacetime contained in the Friedmann equations, $ds^2 = g_{\mu\nu}\, dx^\mu\, dx^\nu = -\, c^2dt^2 + a^2(t)d\sigma^2$, where $d\sigma^2$ is the spatial part of the *metric* $g_{\mu\nu}$, with $d\sigma^2 = dr^2/(1–kr^2) + r^2(d\theta^2 + \sin^2\theta d\varphi^2)$. The constant k is the spatial curvature of the *metric* that takes values (–1, 0, +1), which correspond to the three cases of the *open*, *flat* and *closed* Universe, respectively, Fig. 6.4.

Let $R(t_o) = R_o$ be the distance between two galaxies at time t_o, e.g., at the present, so that at a later (or earlier) time t their distance will be $R(t) = a(t)R_o$. The exact expression of the function a(t) is obtained by solving Einstein's *General Relativity* equations. In particular, by assuming that the Universe is *homogeneous and isotropic*, we have the classical Friedmann solutions, which give us the exact expression of the function a(t).

⇒ As light travels great distances and is redshifted, its wavelength may be shifted even by a factor of 10. So, light that starts out as ultraviolet may become infrared by the time it gets to us!

The Hubble parameter H is defined as the derivative of the logarithm a(t), i.e., $H = \dot{a}/a$, where $\dot{a}$ is the time derivative of the function a(t), da(t)/dt. Then, since $R(t) = a(t)R_o$ we will also have $\dot{R}(t) = R_o\dot{a}$, where similarly $\dot{R}(t)$ is the time derivative of the function R(t). Since $R_o = R(t)/a(t)$ we will have $\dot{R}(t) = R(t) \cdot \dot{a}/a(t) = H \cdot R(t)$ which is precisely Hubble's law for the expansion of the Universe.

Since the observational data give an accelerated expansion of the Universe, this means that the first derivative of a(t) increases with time, å(t) > 0. According to the FLRW metric used to describe an expanding Universe, if today we observe a distant galaxy with redshift $z \equiv \Delta\lambda/\lambda_o = (\lambda_{obs} - \lambda_o)/\lambda_o$, then it follows that $1/(1+z) = \lambda_o/\lambda_{obs}$. In other words, the redshifted wavelength λ_{obs} is larger than the original wavelength λ_o by the factor (1 + z), $\lambda_{obs} = \lambda_o\ (1+z)$.

For relatively close sources we have $z = \Delta\lambda/\lambda_o = V/c$, the familiar Doppler formula. Hence, according to the Friedmann-Lemaître-Robertson-Walker (FLRW) *metric*, which is used to describe an expanding Universe, if today we observe a distant galaxy at redshift z, then the scale factor of the Universe a(t) when the light started from that galaxy was $a(t) = 1/(1+z)$. *At large distances where the largest redshifts are observed, the previous expression for the redshift gives values of the expansion velocity that can be even greater than the speed of light*! We must note, however, that in these cases the upper limit for the speed of light set by the theory of Relativity is not violated, because nothing moves at that speed, but it's just that the distance between the source and the observer grows, as spacetime *lips out*! However, the previous relation is a rather complicated relation that requires, for the calculation of the redshift z, the knowledge of the general expansion history of the Universe. That's why we use the simpler expansion velocity V obtained by multiplying the source distance D by the Hubble parameter H, at this redshift z, $z \approx H{\cdot}D/V - 1$, since $z + 1 = 1/a = R_o/R(t) = [R_o \cdot H]/[R(t) \cdot H] = H{\cdot}D/V$.

By using the scale factor a(t), we find that the quantities of the energy and matter density vary differently with time. Thus, in the initial early stages of the *Big Bang*, most of the energy was in the form of radiation which dominated the other forms of energy and determined the expansion law of the Universe (*period of radiation).* Later, due to the drop in temperature brought about by the expansion, the roles of radiation and matter were reversed and the Universe passed into the age of matter (*period of matter*). Observations of *type Ia supernovae* show that about 5 billion years ago, the Universe entered into a new period dominated by *dark energy* (period of *dark energy*).

More quantitatively, the Universe has passed through the following three distinct epochs: (i) the radiation-dominated epoch, $z \geq 3{,}600$, (ii) the era of dominance of matter, $3{,}600 \geq z \geq 0.5$, and finally, (iii) the era where *dark energy* dominates, $z \leq 0.5$.

(i) Specifically, after the time of *cosmic inflation* (which we discuss later) and up to about 47,000 years from the initial time of the *Big Bang*, the relativistic dynamics of the components of the early Universe (basically photons and neutrinos) was determined by radiation. In such a Universe, where radiation dominates over matter, the time evolution of the scale factor a(t) with an FLRW *metric*, gives from the solution of the Friedmann equations, a scale factor that increases with the square root of time, $a(t) \cong t^{1/2}$.

(ii) Next, in the time period between 47,000 years from the *Big Bang* and up to 9.8 billion years after the initial explosion, the density of matter exceeded the density of radiation, as well as the density of the vacuum. In other words, at

47,000 years after the initial *Big Bang* (redshift $z = 3{,}600$), the energy density of matter exceeded the energy density of radiation, although the Universe continued to remain optically thick to radiation, up to the point in time of about 380,000 years from the *Big Bang* (redshift $z = 1{,}100$). It must be emphasized here that this second moment in time (which is close to the recombination time) where photons of the cosmic background radiation were last scattered is often mistakenly considered to mark the end of the radiation epoch. In such a Universe, where matter dominated radiation, the time evolution of the scale factor a(t) with an FLRW metric, gives from the solution of the Friedmann equations, a scale factor that increases with the cube root of the square of the time, $a(t) \cong t^{2/3}$.

(iii) Finally, the *third* phase of the Universe is the *dark energy* dominated era. The *cosmological constant* Λ, contained in Einstein's equations, can be thought of as the equivalent *mass* of vacuum/empty space, or, as the *dark energy*. Because the *cosmological constant* Λ increases as the volume of the Universe increases, the expansion pressure remains constant, regardless of the scale factor of the Universe, while the other terms of the energy density of the matter and the density of the radiation decrease with time. Therefore, as these energy densities of matter and radiation decrease to very low values, the *cosmological constant* (or, *dark energy*) gradually dominates the total energy density of the Universe. For a Universe where the *dark energy* dominates over matter and radiation, the time evolution of the scale factor a(t) with an FLRW metric, follows from the solution of the Friedmann equations. Specifically, it follows a scale factor that grows exponentially with time, $a(t) \cong \exp(Ht)$. The Hubble parameter H in this exponential is equal to $H = (8\pi G\rho/3)^{1/2} = (\Lambda c^2)^{1/2}$. This exponential dependence of the scale factor a(t) with time corresponds to a geometry of spacetime similar to a model Universe with a de Sitter metric, that we discussed earlier.

Altogether, the young Universe was dominated by radiation over matter, hence, we say that the young Universe lived through the age of *radiation*. But after about 47,000 years following the *Big Bang,* the density of matter exceeded the density of radiation and matter began to dominate, so the Universe entered in the era of *matter.* Finally, in the last 5 billion years it seems that the density of matter has decreased compared to the density of dark energy, which now dominates the Universe and since then and until today we live in the era of *dark energy*.

In closing this topic, we note that for this great discovery, namely that the expansion of the Universe is accelerating instead of slowing down due to gravity, the two respective research groups, led by S. Perlmutter of the *Supernova Cosmology Project* (Berkeley), A. Riess (Johns Hopkins University, Space Telescope Science Institute, right) and B. Schmidt (Australian National University, center) of the High-z *Supernova* Search Team, were awarded the Nobel Prize in Physics (2011) (Fig. 6.22).

What is the *Fate* of Our Universe?

The discovery of the accelerated expansion of the Universe has been something truly unexpectable, which changed fundamentally our thinking and seems to be responsible for accelerating the expansion of the Universe. However, its nature is a *mystery* and its understanding is one of the great challenges in modern physics. If this accelerated expansion continues, three theories can describe the possible fate and end of the Universe, Fig. 6.23.

According to the *first*, with the dark energy represented by the cosmological constant, the accelerated expansion of the Universe will gradually lead it to cooling and finally to a *thermal death*, when the temperature of the Universe will reach $T = 0$ (absolute zero), in the so-called *Big Freeze*: the space between clusters of galaxies will grow at an increasing rate, the galaxies will move further and further away from each other, the supply of gas needed for star formation will be exhausted. As existing stars run out of fuel and cease to shine, the Universe will slowly and inexorably (after many billion years) will *freeze* and eventually *extinguish*, turning into something utterly dark and cold. Actually, this *heat death* of the Universe, was formulated by Lord Kelvin (Thomson 1862) as a *reductio ad absurdum* argument, using thermodynamics to show the impossibility of an infinitely old Universe: if the Universe was eternal *(as postulated e.g.,* by the *Steady State theory)*, there must have been enough time for the stars to cool and warm their surroundings. Everywhere should therefore be the same temperature and there should either be no stars, or everything should be as hot as stars. The Universe should thus achieve, or asymptotically tend to, a *thermodynamic equilibrium*, which corresponds to a state where no thermodynamic free energy is left, and therefore no further work is possible.

According to the *second* theory, known as the, the Hubb*Big Rip*le constant goes to infinity within a finite time interval and the Universe could be ripped apart by dark energy if its influence steadily increases. The truth of this hypothesis relies on the type of dark energy present in our Universe. The type that could prove this hypothesis is a constantly increasing form of dark energy known as *phantom energy* (Caldwell 2003) – a hypothetical form of dark energy satisfying the equation of state $w = P/\rho c^2 < -1$, possessing additionally negative kinetic energy, and predicting expansion of the Universe in *excess* of that predicted by a cosmological constant – could cause the expansion of the Universe to accelerate very quickly. This acceleration necessarily passes the speed of light, causing more and more objects to leave our observable Universe faster than its expansion, as light and information emitted from distant stars and other cosmic sources cannot *catch up* with the expansion. As the observable Universe expands, objects will be unable to interact with each other via fundamental forces, and eventually, the expansion will prevent any action of forces between any particles, even within atoms, *ripping apart* the Universe, making distances between individual particles infinite. Eventually, there would be nothing left: no particles, no planetary systems, no clusters of galaxies, not even space itself. Hence, this hyper expansion will *stretch* the Universe, causing it to eventually *tear apart* and self-destruct after many billion years (Mack 2020).

Hence, it is interesting to note that if *dark energy* continues its *gallop* at its current rate, then two nightmarish final cosmic evolution scenarios looms on the horizon: at some point in the future everything will have moved so far from Earth that nothing will be visible on the sky. Earth – and every other planet – will live in total darkness. Even worse, atoms and subatomic particles may one day be dissolved from within, severing all bonds of matter, so that there will be no Earth and no Universe in material form: this is the ultimate cosmic catastrophe scenario. However, if *dark energy* fluctuates over time, so it might be *braked* in the future, then one can hope for a more optimistic future for humanity and our planet.

A *third* theory, based on *loop quantum cosmology* (*Agullo and Singh 2016*), a branch of *loop quantum gravity*, assumes that at some point the expansion stops and the Universe collapses. One of the main problems with the Big Bang theory is that there is a singularity of zero volume and infinite energy at the moment of the Big Bang. This is normally interpreted as a breakdown of physics as we know it. However, one expects *quantum effects* to become important and avoid the singularity at a point where the quantum effects of gravity become so strongly repulsive that the Universe rebounds back out, forming a new branch. This scenario is independent of the accelerated expansion and is called the *Big Bounce*. The main idea behind the quantum theory of a *Big Bounce* is that, as density approaches infinity, the behavior of quantum foam changes. Gravity will force the Universe to collapse and self-destruct through *a reverse Big Bang* which, although it will destroy the Universe, will simultaneously create the conditions for a new *Big Bang* to occur, from which a new Universe will emerge (Craig 2018). In *loop quantum cosmology*, the *Big Bounce* was first discovered for isotropic and homogeneous models by Ashtekar et al. (2006).

Recapitulating, as can be seen in Fig. 6.23, a *decelerating* Universe reaches its current size in the shortest amount of time. This Universe could eventually shrink and collapse in a *Big Crash*. A *limiting* Universe is older than the decelerating, but will not collapse, expanding at an ever-increasing rate of deceleration. A *hyperbolic* Universe is longer-lived and older than the decelerating and coasting Universes, because it takes longer time to reach its present size and continue to expand at a small rate of deceleration. An *accelerating* Universe, on the other hand, it is even older than all the previous ones. Its expansion rate increases due to a repulsive force that pushes the galaxies at increasingly greater distances, Fig. 6.23.

However, surely there will be newer developments that may change our views on the future of the Universe. For example, it has recently been proposed that the mysterious dark energy *swallows up* the equally mysterious dark matter, thus turning the Universe into a giant, empty, and completely barren world. Nevertheless, there is no reason for us to worry, since if any of these scenarios

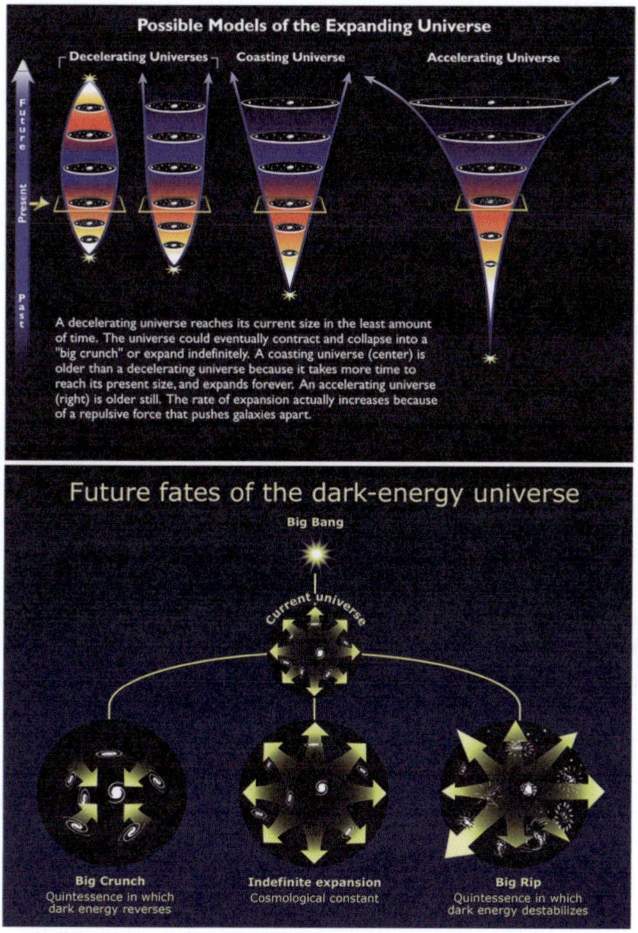

Fig. 6.23 (*top*) Various models for the evolution of the Universe. In the decelerating Universe there is enough matter to halt expansion, in the coasting Universe there is not enough dark matter to halt expansion, as is the case in the accelerating Universe. [Adapted from NASA & Jeffrey Newman (Univ. of California at Berkeley)]. (*bottom*) An illustration of the possible fates of an accelerating Universe. Dark energy could cause a *Big Rip* (the Universe expands violently and galaxies, stars, planets, and atoms come unglued), or, a *Big Crunch* (the Universe implodes or compresses). (Image courtesy of NASA/STScI/Ann Feild)

occurs, it will happen after many billions of years. It's probably more likely that we'll have destroyed ourselves as a species by then, much sooner than the Universe itself would take care of alone. Unless, in the meantime, we change our mindsets.

COBE-WMAP-Planck

1989: COBE, Beginnings of Exact *Observational Cosmology*

In Cosmology, the discovery of the primeval light in the Universe, by Penzias & Wilson (1965), the culmination of work initiated in the 1940s, was followed by the launch of the COBE (Cosmic Background Explorer, 1989–1993) satellite. COBE essentially marked a new era for observational Cosmology, delimiting its beginning as a precise science. It was followed by NASA's WMAP (Wilkinson Microwave Anisotropy Probe 2001–2010) and ESA's more technologically advanced satellite, Planck (2009–2013), Fig. 6.24.

The fundamental discovery of the diffuse *cosmic microwave radiation (*CMR), the first tangible evidence of the *Big Bang* theory, revealed an isotropic *cosmic microwave radiation*, that fills all space in the observable universe, with a frozen temperature of about 2.7 °K, i.e., about minus 270.3 °C. But how *precisely isotropic* was this CMR? If this temperature was really *isotropic*, regardless of the direction we observe it, then this means that it would have been also isotropic at the time of 380,000 years after the *Big Bang*, when the decoupling of matter from radiation occurred. In this case, however, at that time there would had been no inhomogeneities in the radiation field, neither in the matter distribution. This means that there would not had been the primordial seeds from which the galaxies formed. But such a conclusion contradicts the formation of the galaxies we observe today. Specifically, modern large-scale mapping of the sky shows clusters and superclusters of galaxies with dimensions reaching 300 million light years. Hence, how may we interpret the evolution of a perfectly homogeneous distribution of radiation and matter in the embryonic Universe, which today is dominated by superclusters of galaxies, in *great walls* scattered among *great voids*? Therefore, we should expect that today the temperature of the CMR should not be exactly isotropic, i.e., the same in all directions. COBE was launched the 1989, to answer exactly this question. An important theoretical work by Sachs & Wolfe had already shown that 2-dimensional anisotropies in the diffuse CMR with an angle larger than one degree (>1°) should exist today (Sachs and Wolfe 1967).

NASA's COBE satellite was the first mission to provide a first good mapping of the diffuse CMR of the Universe and its anisotropies. Specifically, COBE was equipped with three instruments:

Fig. 6.24 Detection of anisotropies in the *cosmic microwave radiation* by Penzias & Wilson (1965) and the COBE (1989–1993), WMAP (2003–2012) & Planck (2009–2013) spacecrafts. (*Image Credits* NASA, ESA)

(1) Far-InfraRed Absolute Spectrophotometer (FIRAS),
(2) Differential Microwave Radiometer (DMR), and
(3) Diffuse InfraRed Background Experiment (DIRBE).

with the corresponding science project investigators of these three instruments, George Mather, George Smoot and MG Hauser, respectively.

The FIRAS array performed frequency analysis covering a broad range of the spectrum, with wavelengths from 0.10 to 10 mm, mapping 98% of the sky until exhausted its liquid helium for the cooling of the electronics. The purpose of this experiment was to compare the spectrum of the CMR with the theoretical spectrum of a perfect *black body*. The processing of data already in the very first minutes of operation of the experiment, demonstrated an impressive coincidence of the experimental points with the theoretical curve corresponding to the emission of a *black body* with an absolute temperature of 2.72548 ± 0.00057 °K—270.725 °C below zero, Fig. 6.21. Strikingly, the deviation of the spectrum of the diffuse CMR, with respect to the theoretical prediction, as measured by FIRAS, was no more than 1%. The second instrument DMR (Differential Microwave Radiometer) of the COBE satellite found very faint variations of the temperature by direction, equal to 30 μK (30 millionths of a degree) and width $\Delta T/T = 5 \times 10^{-6}$ (Smoot et al. 1992), in agreement with the theoretical predictions (Wright et al. 1992). These fluctuations in temperature correspond to fluctuations in the density of matter that caused the formation of galaxies. The objective of the third instrument DIRBE was to measure the diffuse cosmic radiation in the infrared, as well as the observation of distant galaxies in the infrared.

The COBE mission was a great success. All three instruments of the mission functioned exceptionally well and contributed to cosmology, which now became, for the first time, an exact science. The predictions of the *Big Bang* model were impressively confirmed: the overall temperature of the radiation perfectly fitted a blackbody spectrum. Anisotropies were detected in the diffuse *cosmic microwave radiation* of the order of 1:100,000.

COBE Announces in 1992 that Big Bang is Alive and Healthy

The relevant announcement of the finding of the anisotropies of the cosmic microwave radiation by George Smoot on April 23, 1992, that *the Big Bang is alive and indeed very healthy*, was received worldwide with particular excitement, as well as a relief.

(2001, 2009): More Detailed Observations of the **Anisotropy** *and* **Polarization** *of the CMR, with the WMAP and Planck Space Missions*

Until the beginning of the twenty-firstcentury, cosmology differed from the other sciences in that, while there were many theories available, the necessary observational data to select the correct one were unavailable! In other words, the proper cosmic experiments that would verify the correct theoretical models were did not exist. However, things changed radically when in 2001 NASA launched the WMAP (Wilkinson Microwave Anisotropy Probe, 2009–2013) satellite, a collaboration between NASA and Princeton University and ESA followed immediately with an even more technologically advanced satellite, the European *Planck* mission (2009–2013). These two missions perfected the first revolutionary results, brought already by the COBE mission. The WMAP and Planck satellites orbited Earth in stable orbits around the L2 Lagrange point of the Earth-Sun system, 1.5 million km away from our planet. The main goal of these missions of NASA and ESA was to measure the variations in the cosmic microwave diffuse radiation – the oldest light recorded in the evolution of the Universe – with a precision defined by fundamental astrophysical limits, mapping, with the most accurate maps to date, the cosmic microwave diffuse radiation.

WMAP and Planck were equipped with sophisticated technology and had a specific purpose: to collect data related to the age, temperature, and expansion of the Universe. That is, to verify in great detail how the Universe has been expanding, contrary to the theory of *continuous creation*, as well as when the Universe began its expansion and how it has evolved so far. The measurements of the two satellites played a decisive role in establishing the generally accepted model of Cosmology today, the Λ *cold dark matter model* (Λ-CDM).

The agreement of the data of the WMAP and Planck missions with the predictions of the *Big Bang* scenario not only confirmed the existence of small inhomogeneities in the density of the Universe during the first *infant* phase of its evolution, before radiation was decoupled from matter, but it also allowed us to determine the nature of these microfluctuations of the density. In other words, it established that during the evolution of the early Universe, matter/energy were pulsed by acoustic waves of a fundamental wavelength (*the fundamental harmonic*) and its submultiples (*the higher harmonics*), which on the one hand formed the areas of condensation of matter, so that they later gave birth to galaxies, and on the other hand, they were imprinted in the diffuse CMB (Figs. 6.24 and 6.26).

CMR: Echo of the Primordial Cosmic Music[4]

The *cosmic microwave radiation* has *written* within it the *primordial cosmic music*, with its basic *note* and its higher harmonics, which vibrated in the still unformed young Universe and determined its subsequent structure. And human civilization can now *listen* to the *echo* of this *primordial cosmic music*. The *Pythagoreans* would be happy to see that their idea of *harmony of the planets* comes back to the foreground, not of course at the level of the planets, but as a *harmony* in the newborn Universe, from which sprang galaxies, stars, planets, and finally, the Pythagoreans themselves (Economou 2016).

[4] *Pythagoras* proposed that the Sun, Moon and planets all emit their own unique music based on their orbital revolutions, while *Platon* described astronomy and music as "*twinned* studies of sensual recognition: *astronomy for the eyes, music for the ears*: as the eyes, seem formed for studying astronomy, so do the ears seem formed for harmonious motions, and these seem to be twin sciences to one another, as also the Pythagoreans say".

Another important discovery of the WMAP mission was that the first generation of stars began to shine already only about 200 million years after the *Big Bang*, i.e., much earlier than expected. To calculate exactly when stars began to form, WMAP did not see directly the light of these first stars, but measured the polarization of the diffuse *cosmic microwave radiation* from the free electrons released during the re-ionization epoch. The polarization of the microwave radiation comes from two similar phenomena that occurred at different times: during its initial scattering (*recombination epoch*) and during the *re-ionization epoch*. In both cases, the radiation undergoes Thomson scattering, as explained below, resulting in some degree of polarization.

Thus, the diffuse *cosmic microwave radiation* is another source of information about the re-ionization epoch, Fig. 6.32, which we discuss in more detail below. So, as the matter of the Universe began to be re-ionized, immediately after the formation of the first bright and large mass stars, the released electrons affected the polarization of the light. A free electron can interact with an electromagnetic wave by the process of *Thomson scattering*. In this scattering, the electron absorbs the energy of the wave and is accelerated, emitting light that is polarized in the direction of the electron's motion. This phenomenon was intense during the era of *re-ionization*. In the following, as the Universe continued to expand, the density of the free electrons decreased, reducing their ability to polarize light.

Hence, the WMAP satellite was used to study the degree of polarization of the diffuse CMR. This degree of polarization gave an indication of the free electron density around the time of *re-ionization*, where higher polarization means higher free electron density. From these studies, it emerged that *re-ionization* began about 400 million years after the *Big Bang* and was completed at 400–500 million years later. This result is also in agreement with the observations of *quasars*, which formed about 900 million years after the *Big Bang*.

2001: Measuring the Geometry of the Universe

In the context of the Newtonian theory of gravity, but also in the context of the General Theory of Relativity, it can be calculated that the critical density of the Universe, which corresponds to a marginally flat Universe, is $\rho_o = 3H^2/8\pi G$. For a value of the Hubble constant $H_o \cong 70$ km/sec/Mpc, the critical density equals to $\rho_o \cong 1.0 \times 10^{-29}$ gr/cm^3, i.e., about 6 hydrogen atoms per cubic meter (m^3).

If the density of the Universe is greater than the critical value ρ_o ($\Omega = \rho/\rho_o > 1$), the Universe is *closed*, the geometry of space is *spherical* and the curvature *positive* ($k = +1$). As already mentioned, in such a *closed* Universe, the sum of the angles of a triangle is always greater than 180° (Fig. 6.4). In this case, the Universe is *finite without ends* (as is the surface of a sphere in a 2-dimensional space), so it is true that all space and mass of the Universe were originally concentrated in one point. The expansion rate of such a closed Universe goes to zero at finite time and then its contraction follows (Fig. 6.6). The evolution of the Universe is reversible, that is, after the Universe reaches its maximum dimensions, it will begin, under the influence of gravity, to contract and finally collapse toward a second singularity, similar to the one from which it came from. All galaxies will merge, radiation will become dominant and matter will disintegrate into its elementary forms. The final collapse – the *Great Crash* – will occur just like the original *Big Bang*, only now it will happen in reverse time, Fig. 6.23a. Although this is a simple scenario, the available observational data do not seem to agree with this.

If on the other hand, the density of the Universe ρ is less than the critical value ρ_o ($\Omega < 1$), the Universe is *open*, the geometry of space is *hyperbolic* and the space curvature is *negative* ($k = -1$). In this case, the Universe has a *hyperbolic* geometry, where the sum of the angles of a triangle is less than 180° (Fig. 6.4). If the density is less than the critical value, then the expansion rate is reducing with time, but it never becomes zero (Fig. 6.6). But an open Universe means that it is also infinite in extend, so at the moment of its creation, the Universe was not concentrated in one single point. In this case, we could say that the *Big Bang* happened *everywhere*.

Finally, if the density of the Universe equals to the critical density ρ_o ($\Omega = 1$), the space of the Universe is *flat*, of zero curvature ($k = 0$) and the Universe is marginally *open*. In a *flat* Universe, the sum of the angles of a triangle is exactly equal to 180°, as we know from Euclidean geometry. The *open* and *flat* Universes evolve in a non-reversible way, that is, they expand to infinity (Figs. 6.4 and 6.6).

The WMAP mission measured indeed the geometry of the Universe. In a *flat* Universe of Euclidean geometry, the brightest anisotropies (hottest regions) of diffuse *cosmic microwave radiation* would be about 1° in angular diameter. In a *closed* Universe with spherical geometry, the brightest anisotropies of the diffuse *cosmic microwave radiation* would be greater than 1 degree in diameter. Finally, in an *open* Universe with hyperbolic geometry, the brightest anisotropies of the diffuse *cosmic microwave radiation* would be *less* than 1° in diameter.

Already during the 1990s, preliminary ground measurements and balloon experiments, e.g., the BOOMERanG experiment, had reported that the highest power fluctuations occur at scales of approximately one degree in diameter. Thus, it was concluded that the Universe is *flat* to within an accuracy 15%. The WMAP satellite on board verified these results with an even greater precision, so that today we know that the Universe is *flat* to within 0.4%.

It should be noted that if the density of the Universe at some point in time was equal to the critical one, then it will continue to be equal to the critical one *forever*. The density corresponding to the visible mass of the Universe is approximately 25 times lower, while that corresponding to dark matter is approximately 5 times lower than the critical density.

2010: Correlation of the **Cosmic Microwave Radiation** *Anisotropy with the Critical Density of the Universe Ω*

The temperature variations $\Delta T/T_o$ of the diffuse *cosmic microwave radiation* are a function of the meridional and azimuthal angles θ and φ, respectively, on the celestial sphere. Developing this angular dependence of the temperature variations in *spherical harmonics,* the diagram in Fig. 6.25 results, where the size of the inhomogeneities of the cosmic microwave background radiation corresponds to different angular scales: small ℓ correspond to large angular scales and large ℓ to small scales. Further analysis of the angular dependence of these *spherical harmonics* shows that these functions correspond to ℓ wavelengths of oscillation, performed on a great cycle of the celestial sphere. The angle corresponding to a half-wavelength – that is, the interval between a maximum and its neighboring minimum – is $\theta = 180°/\ell$. This is the angular resolution that a microwave detector needs to have to be able to analyze the spectrum of the radiation up to the corresponding ℓ. Observations of the angular size distribution of inhomogeneities of the microwave background radiation according to the observations of WMAP (Spergel et al. 2007), Acbar (Kuo et al. 2004), Boomerang (Jones et al. 2005) and CBI & VSA (Rajguru et al. 2005) show a maximum at the angular scale of one degree, Fig. 6.25. At high resolution, at angles below 2 degrees, a lot of variation is evident. These fluctuations led to the formation of the first structures in the Universe.

Because the maximum of the observed angular distribution is associated with the normalized density of the Universe with the relation, $\Omega_{tot} = 1/[\theta_o]^{1/2}$, by setting $\theta_o = 1°$, it follows that $\Omega_{tot} = 1$, i.e., the density of the Universe equals the critical density ($\Omega = 1$). Therefore, the space of the Universe is *flat*, of zero curvature ($k = 0$). The Universe is marginally open and expands to infinity. And this was derived in a really simple way from observations. As we discuss, immediately below the best fit of these data is by a *cold dark matter model* (Λ-CDM) with about 30% matter and 70% dark energy.

In summary, analysis of the observations for the anisotropy of the diffuse *cosmic microwave radiation* by NASA's Wilkinson Microwave Anisotropy Probe (WMAP) mission have shown that the density of the Universe is very close to the critical density ($\Omega = 1$) and the Universe is *flat*. This means that the Universe is *infinite in extent*. But because it has a finite age, *we observe only a part of it.*

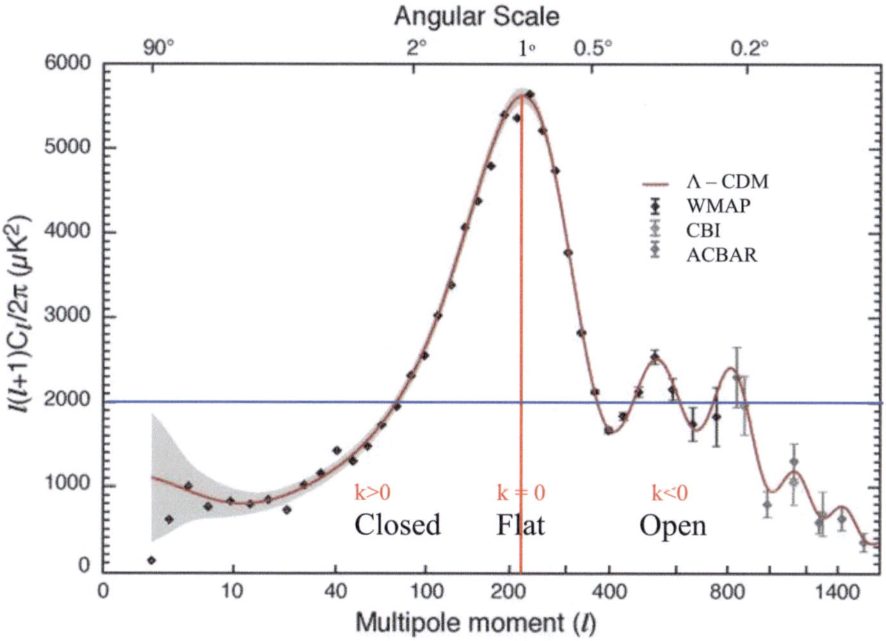

Fig. 6.25 Angular distribution of the magnitude of the *diffuse cosmic microwave radiation* inhomogeneities according to the Λ-CDM model, observations by WMAP (Spergel et al. 2007), Acbar (Kuo et al. 2004), Boomerang (Jones et al. 2005) and CBI & VSA (Rajguru et al. 2005). The first large peak is due to the *flat* geometry of the Universe, the second peak to its right to *baryonic matter* and the third to *dark matter*. See also Fig. 6.18 of Hinshaw et al. (2007). (*Credit* Adapted from a NASA/WMAP figure)

(2008, 2011): The Λ-cold Dark Matter Model and the Expansion of the Universe

The Λ-*cold dark matter model* (Λ-CDM) is a parameterization of the *Big Bang* cosmological model in which the Universe contains a *cosmological constant* denoted by Λ, as well as cold dark matter. It is often called the *Standard Big Bang model*, as it is the simplest model which is in a rather good agreement with the following observations of:

- The large-scale structure formation (galaxies),
- the existence and inhomogeneities in the diffuse *cosmic microwave background* radiation, leading to the flat geometry of the Universe as observed by WMAP and Planck,
- the accelerating expansion of the Universe, as observed by *supernovae* in distant galaxies.

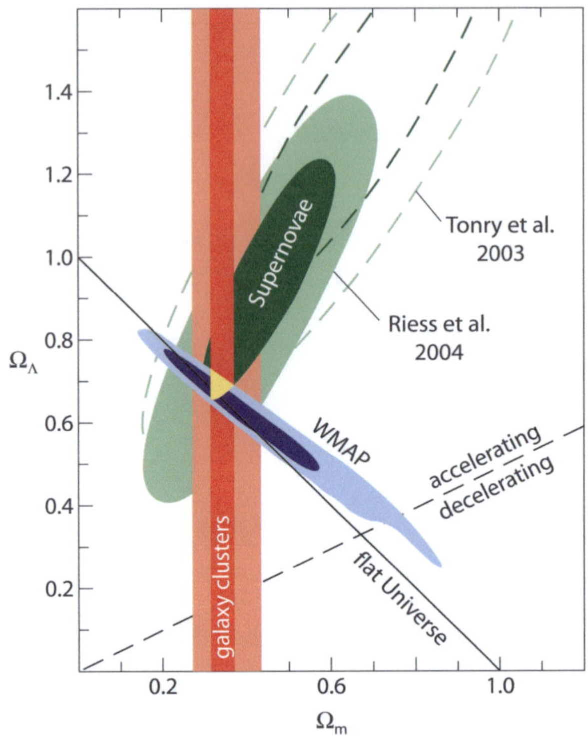

Fig. 6.26 Plane of the cosmological parameters of the normalized mass density Ω_m ($= \rho_m/\rho_o$) and *dark energy* density Ω_Λ ($= \rho_\Lambda/\rho_o$), with four independent observations. Colored areas represent constraints on the parameters from the observations, as follows: the values of (Ω_m, Ω_Λ) should be in the green area to agree with the type Ia SN observations, within the bluish area to agree with the CMB observations (WMAP), within the vertical red/brown area to agree with galaxy cluster data representing BAO constraints from the SDSS galaxy survey. Above the dashed line, the Universe is speeding up, on the solid line, the Universe expands forever, while above the Universe collapses. The solid line which starts at $(\Omega_m, \Omega_\Lambda) = (1, 0)$ and ends at $(\Omega_m, \Omega_\Lambda) = (0, 1)$ corresponds to a flat Universe wherein $\Omega_m + \Omega_\Lambda = 1$. The area in the upper left corner corresponds to a Universe without a *Big Bang* (the density of matter never becomes infinite). A Universe with $(\Omega_m, \Omega_\Lambda) = (1, 0)$ has been ruled out by the observations, Silvestri and Trodden (2009). The observations converge to the yellow triangle with coordinates approximately $\Omega_\Lambda \approx 69\%$ and $\Omega_M \approx 31\%$. (*Image Credit* ESO)

The *Big Bang* Standard Model is based on *General Relativity* as the correct theory of gravity on cosmological scales. It emerged in the late 1990s as a result of the synthesis of different cosmological observations and includes cosmological *inflation*. It is characterized as a model of cold dark matter (CDM), a hypothetical type of *dark matter*. Observations show that about 85% of the matter in the Universe is *dark matter*, with only a small fraction of ordinary baryonic matter that composes stars, planets, and living organisms. The composite name *cold* refers to the fact that dark matter moves slowly compared to the speed of light, while *dark* indicates that it interacts very weakly with ordinary matter and electromagnetic radiation. The nature of the *cold dark matter* (CDM) is currently unknown. The model also includes *dark energy*, due to the observational limitations we discuss immediately below.

Let us then see how the various percentages of the total matter in the Universe (Ω_{tot}) came about, in a diagram with axes the percentages of *baryonic and dark matter* (Ω_{M}) and *dark energy* (Ω_{Λ}), Fig. 6.26.

1. The Λ-CDM model makes *cold dark matter* responsible for the formation of all large-scale structures we observe today – galaxies and galaxy clusters. The data of the hierarchical formation of galactic structures in the Universe have a hierarchy from bottom-up (Eisenstein et al. 2005). Thus, the observations of dwarf galaxies play a crucial role in this theory. These seem to have first been created by small variations of the density in the early Universe and then followed the physical building blocks that formed the largest structures. That is, according to the Λ-CDM model, the formation of the first structures in the Universe – galaxies and galaxy clusters – develops hierarchically, with the small objects to collapse first under their own gravity and then to merge in subsequent successive phases to form structures of greater mass. By contrast, in the warm dark matter model, popular in the early 1980s, albeit less popular today, the structures do not form hierarchically, i.e., bottom-up, but form by fragmentation, i.e., top-down, with the largest super clusters forming first into planar structures and then by breaking them to form smaller structures, such as our Milky Way. In conclusion, the part of the plane that allows observations of structure formation is the red/brown area in the diagram of Fig. 6.26.
2. The data of the WMAP and Planck satellites on the inhomogeneities of the cosmic microwave background radiation also set constraints that are captured in the bluish area in the diagram of Fig. 6.26 (Dunkley et al. 2009).
3. The accelerated expansion of the Universe resulting from the observation of distant Type Ia *supernovae* (Fig. 6.22) discussed earlier, places additional observational constraints, which are schematically depicted by the area above the dotted line separating the area of acceleration/deceleration in the plot with axes the percentages of baryonic and dark matter (Ω_{M}) and dark energy (Ω_{Λ}), Fig. 6.26 (Kowalski et al. 2008).
4. As we discussed earlier, observations of the fluctuations of the temperature of the diffuse *cosmic microwave radiation* show that the total dimensionless density of the Universe, Ω_{tot}, the total sum of the density of baryonic matter in the form of atoms, etc., and of *dark matter*, (Ω_{M}) as well as *dark energy* (Ω_{Λ}) equals unity, $\Omega_{\text{tot}} = \Omega_{\text{M}} + \Omega_{\Lambda} = 1$. We therefore have the solid line on which $\Omega_{\text{M}} + \Omega_{\Lambda} = 1$, in the diagram of Fig. 6.26.

All these observational constraints intersect at the yellow triangle with coordinates approximately $\Omega_{\Lambda} \approx 68\%$ and $\Omega_{\text{M}} \approx 27\%$, Fig. 6.26 (Silvestri and Trodden 2009).

In the diagram of Fig. 6.26, it is also noted that if the *cosmological constant* were zero, the critical density would also mark the dividing line between eventual re-collapse of the Universe into a *Big Crash* and a perpetual expansion. In the Λ-CDM model with a positive value of the *cosmological constant* (as observed), the

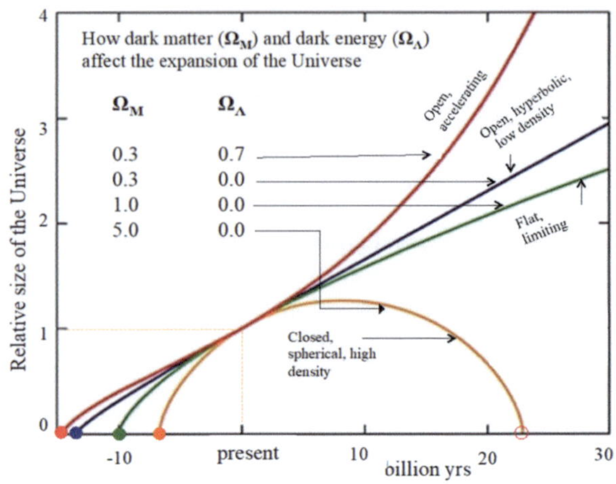

Fig. 6.27 Diagram showing the dependence of the geometry and the mode of expansion of the Universe, for several values of the dimensionless critical densities, matter density Ω_m ($= \rho_m/\rho_o$) and dark energy density Ω_Λ ($= \rho_\Lambda/\rho_o$) of the Universe. There is growing evidence that our universe is following the red curve. (*Image Credit* Adapted from NASA/WMAP)

Universe is predicted to expand forever, no matter if the total density is slightly above or below the critical density. However, there are other models where *dark energy* is not constant but actually it depends on the time (Fig. 6.27).

The Density and Other Parameters of the Universe

As we discussed above, from observations of the anisotropy of the *cosmic microwave radiation* conclude that the Universe is flat and its density appears to be close to the critical density. This is the density which is needed for gravity to just halt its expansion after an infinite time. From the Friedmann equations, the value for the critical density is $\rho_c = 3H_0^2/8\pi G$ where G is the gravitational constant and $H = H_0$ is the present value of the Hubble constant. The Hubble *constant* H_0 is not really a constant quantity, but has a different value at different cosmological times. The greater the value of H_o at a given time, the faster the Universe is expanding at that time. Gravity slowed the expansion of the Universe, so the early Universe was expanding faster than it is now. This means that the critical density was greater in earlier times. The critical density changes by the same factor that the actual density of the Universe changes during expansion. Thus, if the Universe starts out with a density greater than the critical one, then the density will always be greater than the critical one. If the Universe starts with a density less than the critical density, then its density will always be less than the critical density.

This density ρ_c, however, is the total density of both matter and energy. The value for H_0, as given by the European Space Agency's Planck Telescope, is $H_0 = 67.15$ km/s/Mpc. Therefore, the current value of the critical density is $\rho_c \cong 8.50093 \times 10^{-27}$ kg/m^3. For a proton mass $m_p = 1.67 \times 10^{-27}$ kg, it corresponds to about 5.0824 hydrogen atoms per cubic meter, on average. This density includes four significant

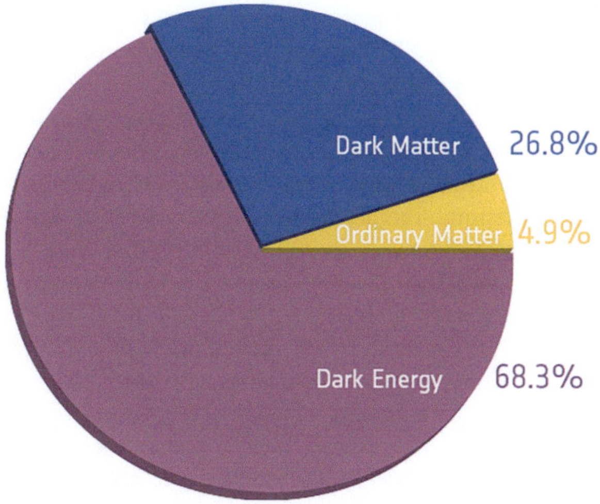

Fig. 6.28 The inventory of the *total* matter contained in the Universe, according to the Planck mission. In the ordinary matter, density is included 0.1% of neutrinos. (*Image Credit* ESA)

types of energy/mass: *ordinary matter* (4.8%), *neutrinos* (0.1%), *cold dark matter* (26.8%), and *dark energy* (68.3%) (Fig. 6.28).

Thus, observations have shown that ordinary matter accounts for 4.9% of this total density, while 22.8% of it is due to *dark matter*. The remaining 68.3% is due to dark energy, the mysterious energy field causing the Universe's expansion to accelerate. The nature of dark matter and dark energy are not fully understood. The density of ordinary matter, as measured by Planck, is 4.8% of the total critical density, or 4.08×10^{-28} kg/m^3. So, ordinary matter has a density corresponding to about one proton for every four cubic meters of volume. Note also that the value of the cosmological constant as measured by WMAP is $\Lambda = 1.8 \times 10^{-52}$ m^{-2} and the age of the Universe is 13.772 billion years.

2018: Inventory of Ordinary/Baryonic Matter in the Universe

The Universe, according to the latest data from the WMAP and Planck space missions, consists of about 5% ordinary baryonic matter, about 27% cold dark matter (CDM) which neither emits nor absorbs light and about 68% dark energy, in the form of the *cosmological constant*, which accelerates the expansion of the Universe (Fig. 6.28). Less than 1% accounts for by neutrinos. That is, *dark matter* is about 85% of the total in the Universe, while *dark matter* and *dark energy* together constitute about 95% of the total mass–energy content.

Dark matter is not known to interact with ordinary *baryonic matter* and *radiation* except through gravity, making it difficult to detect in the laboratory. The most prevalent explanation is that dark matter is some subatomic particle, such as either weakly interacting massive particles (WIMPs) or axions, undiscovered so far. The other main possibility is that dark matter is composed of primordial black holes.

Besides the mysterious *dark matter* and *dark energy*, *ordinary* matter that makes up everything we see – from stars and galaxies to planets and humans – amounts to only about 5% percent of the total mass–energy content of the Universe.

The amount of *dark matter* and *energy* in the Universe appears to play a crucial role in defining the geometry of space. In accordance with the Λ-CDM model, the age of the Universe is 13.772 ± 0.059 billion years.

If we take the expansion rate of the Universe (the Hubble constant) to be around 67 km/sec/Mpc, this leads to a *flat Euclidean geometry* with zero curvature. The data also support in many ways the inflationary evolution of the Universe, explaining especially the problem of flatness.

Most of the Universe's *ordinary matter*, or *baryons*, must be lurking in the filaments of the *cosmic web*, where matter has nevertheless very low density, and therefore it is rather difficult to observe. Using different techniques over the years, astronomers were able to locate a good chunk of this intergalactic material, mainly its cool component, known as the *Lyman-alpha forest*, which makes up about 28% of all baryons, as well as its warm component making up about 15% of all baryons, Fig. 6.29.

The Lyman-alpha transition corresponds to an electron transitioning between the ground state (n = 1) and the first excited state (n = 2). The Lyman-alpha spectral line has a rest wavelength of 1216 Å, in the ultraviolet portion of the electromagnetic spectrum. The *Lyman-alpha forest* was first discovered in 1970 by Roger Lynds in an observation of the *quasar* 4C 05.34, the farthest object observed to that date, wherein Lynds noted an unusually large number of absorption lines in its spectrum.

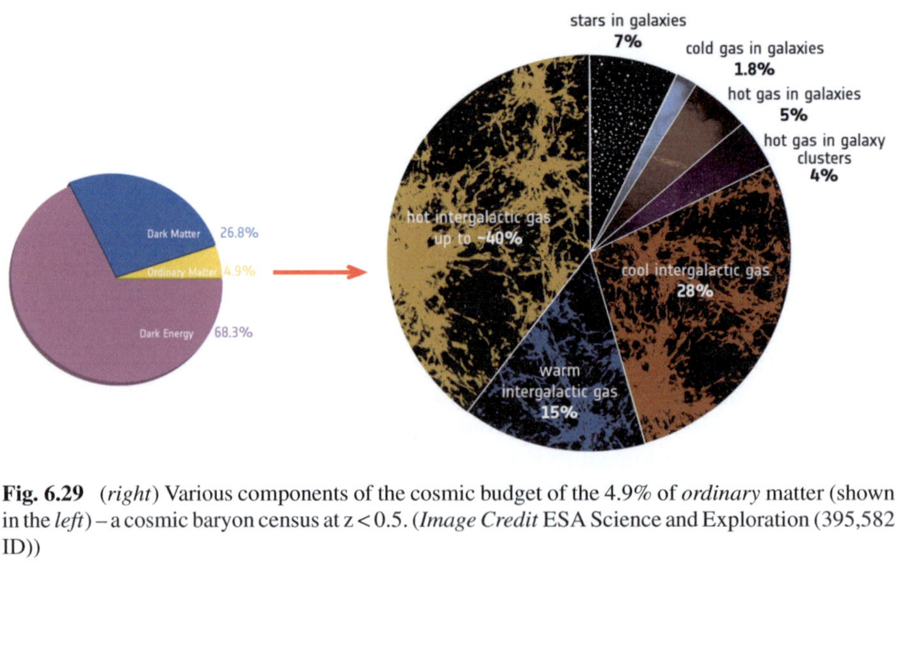

Fig. 6.29 (*right*) Various components of the cosmic budget of the 4.9% of *ordinary* matter (shown in the *left*) – a cosmic baryon census at z < 0.5. (*Image Credit* ESA Science and Exploration (395,582 ID))

He suggested that most of the absorption lines were all due to the same Lyman-alpha transition. As the light travels through multiple gas clouds with different redshifts, multiple absorption lines are formed.

In 2018, after two decades of observations, ESA's XMM-Newton space observatory has detected the warm-hot [$10^5 \leq$ T(K) < $10^{5.7}$ – $10^{5.7} \leq$ T(K) < $10^{2.2}$] component of the *intergalactic* material along the line of sight to a distant *quasar*. Specifically, by using XXM-Newton, as light from a quasar, residing 4 billion light years away, passes through the Universe, regions with hot baryons absorb some of the light, effectively blocking out specific wavelengths and thus leaving their mark on the remaining light from the *quasar* that we observe. These long observations found signatures of oxygen, indicating the presence of baryons. The amount of hot intergalactic gas detected in these observations amounts up to 40% of all baryons in the Universe (Nicastro et al. 2018, Fig. 6.29).

Part of the remaining of the 4.9% of ordinary matter is in stars in galaxies across the Universe, making up about 7% of all ordinary matter. The *cold interstellar gas* that permeates galaxies, the raw material to create stars, amounts to about 1.8% of the total, while the hot, diffuse gas in the haloes that encompass galaxies makes up roughly 5%, and the even hotter gas that fills galaxy clusters accounts for 4%, Fig. 6.29.

The Hubble Tension

One of modern Cosmology's pressing puzzles is the Hubble tension, a discrepancy between the value of the Hubble constant (H_0) from the local method (*distance ladder*) and the cosmic microwave background (CMB) method. Specifically, the *distance ladder method* gives a value around 73 km/s/Mpc, whereas the *CMB method* gives 67 km/s/Mpc. This difference is significant and definitely indicates a gap in our understanding the expansion rate of the Universe by the two methods (Fig. 6.30).

And, it is worth to note that Lemaitre's and Hubble's original estimates (in 1927 and 1929, respectively) were 650 km/s/Mpc and 500 km/s/Mpc, or, about more than seven times the value we think it is today. Generations of astronomers have improved upon those original methods and developed new ones, bringing down the Hubble constant to around 70 km/s/Mpc, but it seems that there's still a long way to go.

In more detail, recent estimates of the Hubble constant from CMB anisotropies by the Planck Collaboration, give a value around 67 km/s/Mpc ($H_0 = 66.93 \pm 0.62$ km s^{-1} Mpc^{-1}, Planck Collaboration Int. XLVI 2016). Also, Kozmanyan et al. (2029) by analyzing data from a sample of 61 galaxy clusters with redshifts up to $z <$ 0.5 observed with Planck and XMM-Newton found $H_0 = 67 \pm 3$ km s^{-1} Mpc^{-1}. Nevertheless, these values highlight the known discrepancy with the value derived by Riess et al. (2018a) using Cepheid-calibrated Type Ia supernovae (SNIa), giving a value around 73 ($H_0 = 73.48 \pm 1.66$ km s^{-1} Mpc^{-1}, or $H_0 = 73.52 \pm 1.62$ km s^{-1} Mpc^{-1} from Riess et al. 2018b). (Adam Riess of Johns Hopkins University and the team he leads, known as SH0ES, have turned this method into a fine art, deriving H_0

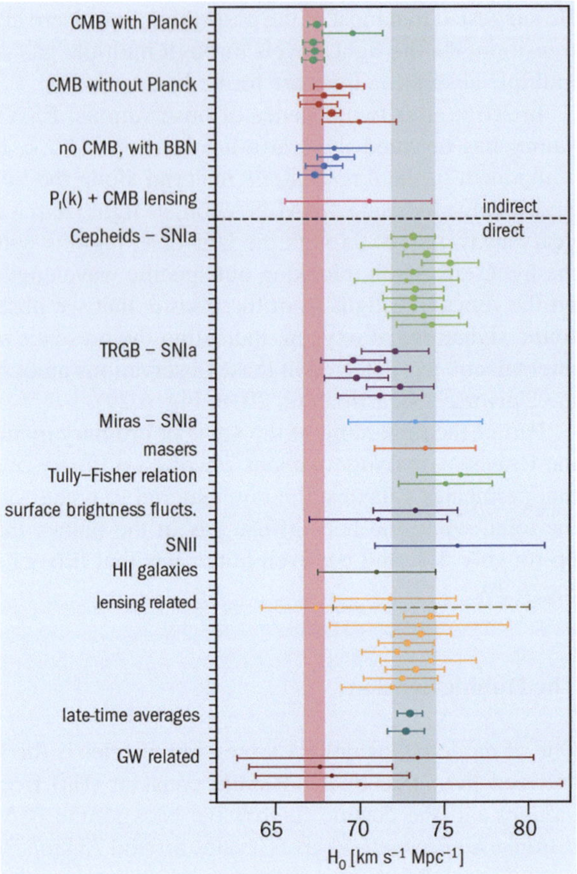

Fig. 6.30 Values of the Hubble constant from direct and indirect measurements by different missions, with the gray and pink bands showing the 68% confidence-level values from SH0ES and Planck, respectively. *Source* arXiv:2103.01183. (*Image Credit* CERN Courier)

by measuring hundreds of Cepheids in 37 nearby galaxies that are host to 42 Type Ia supernovae and then extending measurements out to 277 more distant supernovae. In 2022, using the Hubble Space Telescope, the team put H_0 at 73 *km/s/Mpc* with a 1.4% precision).

The *distance ladder* method uses nearby objects like Cepheid variable stars and Type 1a supernovas as candles, which are objects whose brightness is known to us. Since the distances to these objects and their redshifts are measured, this can be used to calculate the Hubble constant. On the other hand, methods using CMB rely on the *sound horizon,* a standard measure in cosmology, which is a measure of the maximum distance traveled by sound waves in the early Universe before the decoupling of light and matter. These approaches require specific assumptions about the ΛCDM model. In more detail, the ΛCDM cosmological model is a simple and elegant framework that has been found to provide a very good fit to almost all of the observational probes available until present. However, despite its great success, the model is based on the

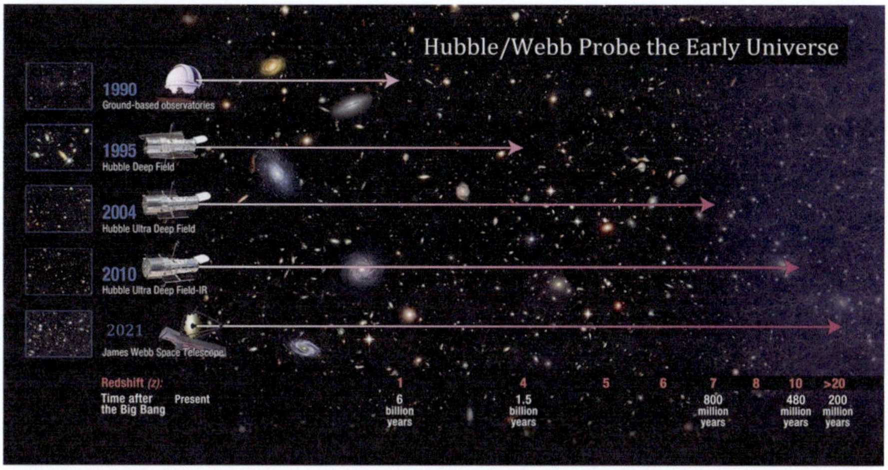

Fig. 6.31 Diagram showing how Hubble and Webb have revolutionized the study of the distant, early Universe. Before Hubble was launched (1990), ground-based telescopes were able to observe up to a redshift of around 1, about half way back through cosmic history. Hubble's latest instrument, Wide Field Camera 3 has identified a candidate galaxy at a redshift of 10 – around 96% of the way back to the Big Bang. The James Webb Space Telescope NASA/ESA/CSA is seeing further still to z $\approx$ 12.5. (*Image Credit* Adapted from a NASA, ESA original)

assumption of three basic ingredients: (i) CDM, (ii) a cosmological constant, and (iii) inflation whose underlying physics are largely unknown.

This has led efforts to modify early Universe physics to reduce the *sound horizon*, which increases the Hubble constant's value derived from CMB data, lowering the discrepancy. Krolewski et al (2025) introduced a new method for measuring the Hubble parameter from low-redshift large-scale observations that is independent of the comoving sound horizon. The method uses the baryon-to-photon ratio determined by the primordial deuterium abundance, together with big bang nucleosynthesis calculations and the present-day cosmic microwave background (CMB) temperature, to determine the physical baryon density $\Omega_b \boldsymbol{h}^2$. By including type Ia supernovae and uncalibrated BAO, they measure $H_0 = 67.1 + 6.3 - 5.3$ km s^{-1} Mpc^{-1}. As is expressed by Krolewski, "the Hubble tension between early-time, large-scale, and local, late-time measurements of the Universe's expansion rate has now reached the

5-sigma level,[5] i.e., there is a probability of less than 0.0000003 of occurring by chance"!

The *Observed* Early Universe

The First Stars and Galaxies We Observe in the Early Universe

Today, we are fortunate to have observations available from ground-based and space telescopes that penetrate ever greater cosmological distances, deep into space, so that we approach the outer limits of the visible Universe. In this process, the further we observe, the further back we *see* in time, Fig. 6.32.

2009: GOODS (The Great Observatories Origins Deep Survey)

GOODS unites extremely deep observations from NASA's Great Observatories, Spitzer, Hubble, and Chandra, ESA's Herschel and XMM-Newton, and from the most powerful ground-based facilities, to survey the distant universe to the faintest flux limits across the electromagnetic spectrum. GOODS data track the formation and evolution of galaxies across cosmic time and map the history of universal expansion using high-redshift *supernovae*. The survey covers a total of roughly 320 square arcminutes in two fields centered on the Hubble Deep Field North and the Chandra Deep Field South. It incorporates a Spitzer Space Telescope Legacy Program to carry out the deepest observations with that facility at 3.6–24 μm, a Hubble Space Telescope Treasury Program for deep high-resolution optical imaging, and a Herschel Space Observatory open time Key Program with the deepest far-infrared observations at 100–500 μm. The space-based observations are complemented by ground-based imaging and spectroscopy, including extensive commitments of ESO and NOAO observing time.

[5] 5-σ is considered *the gold standard* in particle physics because it guarantees an extremely low likelihood of a claim being false. A result that has a statistical significance of 5-σ means the almost certain likelihood that a bump in the data is caused by a new phenomenon, rather than a statistical fluctuation. In the case of the Higgs boson, physicists needed enough data for the statistical significance to pass the threshold of 5-σ. Only then could they announce the discovery of a Higgs-like particle. The 6-σ level is the highest quality level we usually try to achieve. However, it would be best to consider the sector, the context, the risk and the cost of getting to 6-σ level process quality. For example, in an airline or a large pharmaceutical company, they cannot afford issues with 4 customers per million, they would need to be higher than 6-σ level.

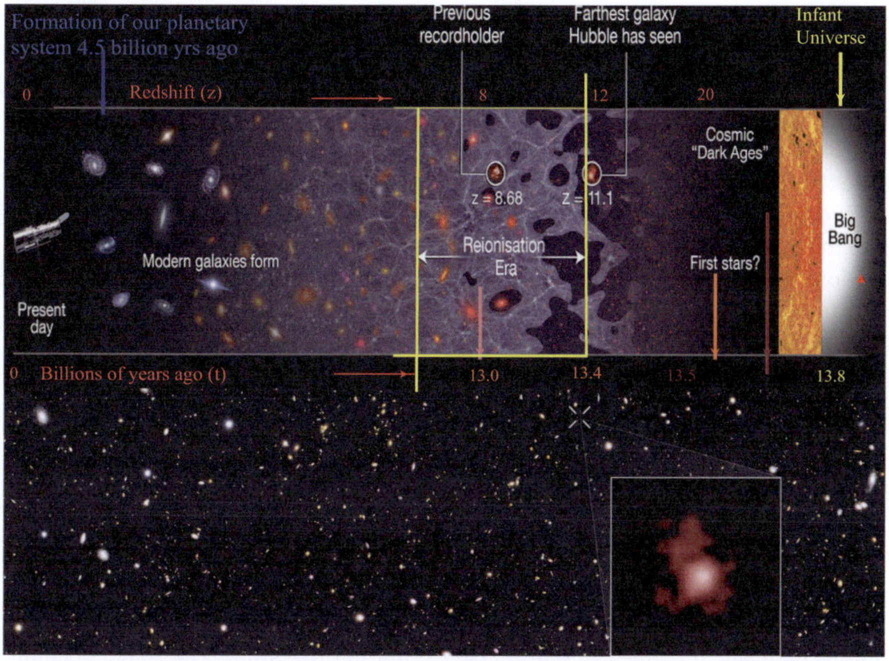

Fig. 6.32 The Hubble Space Telescope, observing the northern hemisphere field with the Great Observatories Origins Deep Survey (GOODS), has measured the distance to the farthest galaxy ever seen. The survey field contains tens of thousands of galaxies stretching far back into time. Galaxy GN-z11, shown in the inset, is seen as it was 13.4 billion years in the past, just 400 million years after the big bang, when the Universe was only three percent of its current age. The galaxy is ablaze with bright, young, blue stars, but looks red in this image because its light has been stretched to longer spectral wavelengths by the expansion of the universe. (*Image Credits* Adapted from original of NASA, ESA, P. Oesch (Yale University), G. Brammer (STScI), P. van Dokkum (Yale University), and G. Illingworth (University of California, Santa Cruz)

(2015–2023): Formation of the First Stars, Galaxies and **Quasars** *in the Early Universe (From t ≈ 200 Million Years → t ≈ 1 Billion Years After the Big Bang)*

Observations using the Hubble Space Telescope identified in 2015 a galaxy 13.4 billion light years away from us, with the code name GN-z11 [GOODS-N: Great Observatories Origins Deep Survey-North). This was till more recent observations by the James Webb Telescope, the oldest and most distant known galaxy identified in the observable universe, having the largest cosmological spectroscopic redshift observed. In early 2023 JWST observed with greater accuracy this galaxy and reported a definitive redshift of $z = 10.6034 \pm 0.0013$. Its light was emitted only 400 million years after the creation of the Universe in the *Big Bang* and belongs to the first generation of galaxies that formed and ever existed. Because it is in its infancy, GN-z11 may be

25 times smaller than our own galaxy and its total stellar mass is only 1% of that of our own galaxy, yet it is growing rapidly, creating new stars at a rate 20 times greater than our own Milky Way. With a stellar age estimated at 40 million years, it appears that the galaxy formed its stars relatively rapidly.

The time light has been traveling from this farthest galaxy is 13.4×10^9 years, since the light of this galaxy was emitted 13.4×10^9 light years ago, i.e., about 400 million light years after the *Big Bang*. The present *(proper)* distance, however, is $\approx$ 32 billion light years from us, because of the expansion of the Universe.

More recently (2023), NASA's James Webb Space Telescope (JWST) observed even more early galaxies, which are very unusual in many ways and were previously hidden beyond the grasp of all other telescopes until now. Two exceptionally such bright galaxies were discovered in the GLASS-JWST images (=Grism Lens-Amplified Survey from Space): GLASS-z12, 350 million years after the *Big Bang* (with a redshift of approximately $z \approx 12.5$) and a second distant galaxy named GLASS-z10 which existed approximately 450 million years after the *Big Bang* (with a redshift of approximately $z \approx 10.5$). GLASS-z12 and GLASS-z10 were spotted in the flanking field of the Hubble Frontier Fields cluster Abell 2744, a group of roughly 500 galaxies – also known as ACO 2744 and the Pandora's Box – located approximately 4 billion light years from Earth.

Before the formation of the first galaxies, the first stars had been formed during the interval between 200 and 250 million years after the *Big Bang*. The heavy elements were produced into those monster stars of pure hydrogen and helium. Without heavier elements, these stars had dozens or even hundreds of times the mass of our own Sun. These are the Population III stars, a new hypothetical division added in 1978, a hypothetical population of extremely massive, luminous and hot stars with virtually no *metals*, except possibly for intermixing ejecta from other nearby, early population III supernovae. These stars likely triggered the Universe's period of re-ionization. The oldest stars observed thus far have low metallicity and are known as population II, while our Sun is considered population I, a recent star with a relatively high 1.4% metallicity.

Subsequently, these massive stars also evolved very quickly, and their nucleosynthetic processes created the first 26 elements, up to iron in the periodic table. Many theoretical stellar models show that most high-mass population III stars rapidly exhausted their fuel and likely exploded in extremely energetic pair-instability supernovae. Those explosions would have thoroughly dispersed their material, ejecting metals into the interstellar medium, to be incorporated into the following generations of stars. Their destruction suggests that no galactic high-mass population III stars should be observable. However, some population III stars might be seen in high-redshift galaxies whose light originated during the earlier history of the Universe.

These first stars in the Universe subsequently formed the *primordial* galaxies. The constituents of those primordial galaxies were dark matter mixed with ordinary matter, i.e., mainly hydrogen. The gradual cooling of hydrogen gradually drove it into the centers of primordial galaxies, but the dark matter remained distributed over its original long distances. The dense regions of these hydrogen nebulae formed large and brilliant stars, hundreds of times more massive and luminous than the Sun. The ultraviolet radiation emitted by these early stars ionized the surrounding neutral hydrogen. Thus, as more stars formed, the regions of ionized gas merged and the interstellar material became more ionized. Thus, the Universe was led to era of re-ionization that we discuss next.

A *dark star* (Freese et al. 2007) is a hypothetical type of star that may have existed early in the Universe before conventional stars were able to form and thrive. *Dark stars* would be composed mostly of normal matter, like modern stars. However, a high concentration of *neutralino dark matter* – neutralino is a particle predicted by the popular *supersymmetric* extension to the Standard Model of particle physics and a favored candidate to comprise *dark matter* – within them would generate heat via annihilation reactions between the *dark-matter* particles, while glowing with infrared radiation. This heat would prevent such stars from collapsing into the relatively compact and dense sizes of modern stars and therefore prevent nuclear fusion among the *normal* matter atoms from being initiated.

Under this model, a *dark star* is predicted to be an enormous cloud of molecular hydrogen and helium ranging between 1 and 960 astronomical units in radius and with a surface temperature and luminosity low enough that the emitted radiation would be invisible to the naked eye. The spectral analysis of these objects suggested that they were between 500,000 and 1 million solar masses ($M_\odot$), as well as having a luminosity of billions of Suns ($L_\odot$); they would also likely be huge, possibly with radii surpassing 10,000 solar radii ($R_\odot$), far exceeding the size of the largest modern stars.

In the unlikely event that *dark stars* have endured to the modern era, they could be detectable by their emissions of gamma rays, neutrinos, and antimatter and would be associated with clouds of cold molecular hydrogen gas that normally would not harbor such energetic, extreme, and rare particles.

One of the most exciting and puzzling recent (2024) findings of the James Webb Space Telescope (JWST) has been the discovery of a large population of massive *black holes* within the first billion years after the *Big Bang*, wherein their large masses, are difficult to reconcile with the standard black hole formation scenarios. The JWST has also revealed that the interplay between these early *black holes* and their host galaxies was probably quite different than what has been observed at later cosmic times. The JWST is also finding a large population of *dual* black holes, which might be in the process of merging, something indicating that this might be an additional route for their early growth and also an early source of *gravitational waves*.

The Age of Re-ionization ($6 \lesssim z \lesssim 20$, $19\,K \lesssim T \lesssim 60\,K$)

After the first hundreds of millions of years since the *Big Bang*, the Universe passed through a critical point in its evolution, the epoch of *re-ionization*, during which the primordial neutral interstellar medium was ionized with the appearance of the first light sources that had formed immediately before. During this period, the primordial fluctuations in the density of the dominant *dark matter* began to collapse through their own gravity, forming halo structures. Ordinary baryonic matter also collapsed into these halos due to its own fluctuations in its density, so the first stars were formed which then emitted strong ultraviolet radiation. This UV radiation had the required high energy (13.6 electron volts) so that it could expel electrons from surrounding matter atoms creating the cosmic re-ionization. Figure 6.32 provides a good graphical illustration of the history and evolution of the Universe, placing the era of re-ionization in the overall picture.

Before the epoch of re-ionization and after about 380,000 years after the Big Bang, the matter of the Universe that turned into neutral atoms became invisible in a wide range of the electromagnetic spectrum. Any emission of short-wavelength radiation was quickly absorbed by the atomic gas. Thus, began a long period known as *dark ages*. During the Dark Ages, the temperature of the universe cooled from some 4000 K to about 60 K. At slow rates, the gravitational collapse of overdense regions, those same regions of the previous era, that we observe today as fluctuations in the diffuse *cosmic microwave radiation* (CMR), led to the formation of increasingly prominent structures in the neutral medium. So they finally started to emerge and form gradually the first stars, galaxies, and *quasars*. The exact mechanism and nature of this formation, little constrained by observation, is a matter of modern research of particular importance. We know what looked like the Universe during the diffuse CMR era and we also know how it is now, but how it evolved from one state to another, it eludes our full understanding to this day.

As the collapse of the first cosmic structures proceeded, temperature fluctuations developed. Gradually, the energetic radiation that began to be emitted by the first light sources caused a local heating and then the ionization of the primordial hydrogen in the Universe. It is very likely that this heating began within *bubbles* of ionized plasma that surrounded the most energetic sources (Fig. 6.33). As the bubbles increased and became numerous, they began to overlap so that more and more of the neutral intergalactic and interstellar medium was exposed to the hard ionizing ultraviolet radiation, which could propagate unhindered through the ionized regions. Then, the final phase of *re-ionization* of the Universe may have happened rather quickly. Finally, when the bulk of the Universe matter was ionized, light from primordial galaxies and *quasars* at many wavelengths escaped, revealing the distant Universe we see today with our optical and infrared telescopes.

Re-ionization was complete in about 1 billion years after *Big Bang*, which corresponds to a redshift of about $z = 2.5$ (Fig. 6.32). Before this time, i.e., at larger z, observations become increasingly difficult. In general, it is hoped to find isolated, very bright objects whose radiation of some form manages to reach us through the increasingly neutral intergalactic medium. Perhaps a useful probe of these early epochs is the neutral hydrogen 21 cm line, which is redshifted at frequencies below 200 MHz, or a wavelength of 150 cm. Sensitive emission and absorption observations of this line can deeply probe the reheating and *re-ionization epochs* and give us a detailed picture of the density, temperature, and velocity field of the material. We would be able to see not only the isolated bright objects and the material that happens to be in front of them, but also the large volumes of the Universe at any particular redshift. Such an observation would provide a wealth of data from which

Fig. 6.33 First stars in the Universe were unimaginably large – perhaps a million times the mass of our Sun. They would have helped clear the fog enshrouding the early universe before dying in *supernova* explosions, seen here in an artist's impression. *Image Credit* Adolf Schaller and NASA)

to derive the early history of the formation of large structures, but also the origin of the stars, galaxies, clusters, and *quasars* we see today (Fig. 6.34).

Fig. 6.34 Artist's concept of the first stars in the Universe turning on some 200 million years after the Big Bang. These first stars were made of almost pure hydrogen and helium. They and later generations of stars cooked up the heavier elements from these simple ones. (*Image Credit* NASA/ WMAP Science Team)

2024: Webb Reveals that Galaxy Mergers Are the Solution to Early Universe Mystery for the Observation of UV Light

New James Webb observations have found what lies in the local environment of galaxies in the very early Universe: small, dwarf faint galaxies surrounding the larger galaxies that show the *inexplicable* hydrogen emission. In conjunction with state-of-the-art simulations of galaxies in the early Universe, the observations have shown that the chaotic merging of these neighboring galaxies is the source of this hydrogen emission.

The very earliest galaxies were sites of vigorous and active star formation, and as such they were rich sources of Lyman-α emission. Those Lyman-alpha emitters are typically low mass galaxies of 10^8 to 10^{10} solar masses. They are typically young galaxies that are 200–600 million years old, and they have the highest specific star formation rate of any galaxies known.

However, during the epoch of *re-ionization*, an immense amount of neutral hydrogen gas surrounded these areas of active star formation, known as stellar nurseries. Furthermore, the space between galaxies was filled by more of this neutral gas than is the case today. The gas can very effectively absorb and scatter Lyman-α emission, so it has long been predicted that the abundant Lyman-α emission released in the very early Universe should not be observable today. However, examples of very early hydrogen emission have previously been observed. This has presented a mystery: how is it that this hydrogen emission – that should have long since been absorbed or scattered – is being observed?

Thus, it seems that the surroundings of these Lyman-α emitters appear to be a much busier place than we previously thought, filled with small, faint galaxies which were interacting and merging with one another, and Webb has revealed that galaxy mergers play an important role in explaining the mystery emission from the earliest galaxies, Fig. 6.35.

More specifically, Pandora's Cluster has been used as the gravitational lens to investigate the sources of the Universe's *period of re-ionization*. Gravitational lensing magnifies and distorts the appearance of distant galaxies, so they look very different from those in the foreground. The galaxy cluster *lens* is so massive that it warps the fabric of space itself, so much so that light from distant galaxies that passes through the warped space also takes on a warped appearance. The magnification effect allowed the team to study very distant sources of light beyond Abell 2744, revealing eight extremely faint galaxies that would otherwise be undetectable, even to Webb.

It was thus found that these faint galaxies are immense producers of ionizing radiation, at levels that are four times larger than what was previously assumed. This means that most of the photons that re-ionized the Universe likely came from these dwarf galaxies.

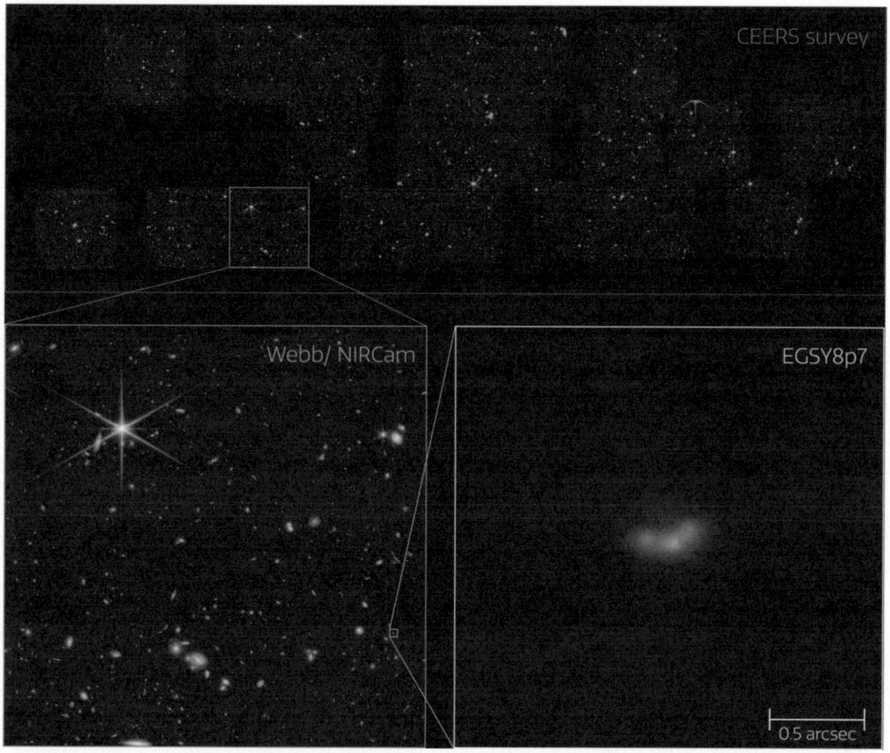

Fig. 6.35 Lyman-α emitting galaxy EGSY8p7 in the CEERS survey field (NIRCam image) is a distant galaxy in the constellation of Boötes, with a spectroscopic redshift of $z = 8.68$ (age 13.2 billion l.y., i.e., observed as it existed 570 million years after the Big Bang. The observations with Webb's high sensitivity NIRCam instrument were able to resolve, two companion fainter galaxies that surround the bright galaxy where previous observations saw only one larger galaxy in its place and from which the *inexplicable* hydrogen emission has been detected. (*Credit* ESA/Webb, NASA & CSA, S. Finkelstein (UT Austin), M. Bagley (UT Austin), R. Larson (UT Austin), A. Pagan (STScI), C. Witten, M. Zamani (ESA/Webb)

The Enigmatic Epoch of the Cosmic **Dark Ages** *(20 $\lesssim z \lesssim$ 1100, 60 K $\lesssim T \lesssim$ 4000 K)*

Even deeper in space and closer to the *Big Bang* in time, the Universe went through the era of the *cosmic dark ages* that was followed in time by the era of *recombination*[6] (Figs. 6.32 and 6.40).

[6] The epoch at which atoms formed for the first time, when the universe was at an age of 380,000 years with a temperature of about 3000 K is somewhat oxymoronically referred to as *recombination*, despite the fact that in the previous history of the Universe, electrons and nuclei had never before combined into atoms.

According to the Standard Cosmological model, as the Universe continued to expand after the *Big Bang*, matter dominated over radiation, becoming denser than radiation after 47,000 years. Later, when the initial hot radiation cooled to about 4,000°, the hitherto free electrons formed atoms combining with the nuclei of hydrogen and helium. Thus, the former state of ionization, the plasma state, ceased to exist. Photons no longer interacted with matter and the Universe became transparent and was ushered into the age of *recombination.* In other words, since plasma is opaque – photons can't travel far without bumping into something – and gas is not, this means that at the r*ecombination* era, photons were finally free to propagate. Those photons had been emitted all the time inside the plasma earlier and reflected the temperature of the plasma, so right after the *recombination* era, the universe would have been filled with a red/orange light from the afterglow of that plasma.

This later enabled the formation of the first stars and galaxies, as then the radiation pressure did not prevent the formation of stars in the first gravitational condensations of matter. But before this era of *recombination*, radiation pressure caused re-homogenization of any disturbances that could form stars. The Universe then had a radius of about several million light years, much larger than our galaxy, but much smaller than it is today (Fig. 6.40).

Thus, after the *recombination* era, the Universe was *transparent to radiation* and had cooled enough that photons could travel unimpeded for big distances. However, there were still no light sources, such as stars and galaxies. These were created later, when dense regions of gas were formed under the influence of gravity, a process, however, that takes a relatively long time, in an almost perfectly homogeneous medium. Therefore, it is estimated that star formation began a few hundred million years – 200 million years according to newer studies – after the time of *recombination.* So, the era that began immediately after 380,000 years from the great explosion, marks the beginning of the period of the *cosmic dark ages*.

Therefore, the Universe went through the *cosmic dark ages*, because at that time the stars and galaxies we see today had not been formed to shine, as yet. Most of the matter of the Universe in this period was *dark matter* with minimal amounts of baryonic matter, mainly deuterium hydrogen and helium.

During this time, the temperature of the Universe gradually decreased due to the cosmic expansion from a few thousand degrees Kelvin – during the era of *recombination* – to about 60 degrees Kelvin by the end of the dark ages, 150 million years later. At this time, there were only two photon sources: the primordial photons formed during the recombination epoch – when the first neutral hydrogen and helium atoms were formed – which we detect today as the diffuse *cosmic microwave radiation* (CMR) and photons periodically generated from neutral hydrogen atoms, at the 21 cm line in radio waves (frequency = 1,420 MHz). These photons are produced as follows. In the lowest energy, or ground state, the spins of the proton and electron are in opposite directions. When the atom is excited, either by absorbing a photon, or from collisions with other atoms, the electron absorbs a small amount energy, as a result, the spins of the two particles are the same direction. When the atom returns to its natural state, it loses the corresponding energy by emitting a photon with a wavelength of 21 cm, in the radio region of the electromagnetic spectrum.

The Empty Time Interval Between 2 Photos of the Young Universe...

The Standard Cosmological model describing how our Universe evolved tells us that about 380,000 years after the Big Bang, the Universe had cooled to about 4,000 degrees Kelvin, a temperature low enough for electrons and protons to bond and form neutral hydrogen atoms from the pre-existing gas of ions and electrons. The radius of the Universe was then about 42 million years. Electrons do not interact with neutral hydrogen (as they do with other atoms or molecules) and hence a hydrogen gas does not absorb photons efficiently. Thus, as the Universe was practically filled with neutral hydrogen – as is again the case today – it was transparent.

In contrast, when electrons and protons are free, the electrons mainly (and to a lesser extent the heavier protons) scatter the photons by *Thomson scattering* and thus a Universe full of ionized gas – as was the case until about 380,000 years since the Big Bang – is relatively opaque. The COBE, WMAP and Planck maps show us the Universe during the transparent stage, at the beginning of the *cosmic dark ages* of the Universe. This period of transparency ended when the Universe became ionized again in the *re-ionization era.*

We also have *snapshots* of an even more recent Universe: when galaxies were full of stars, about a billion years after the Big Bang – when the Universe became transparent again. Light from distant objects takes a long time to reach us, so we see these distant objects as they were many billions of years ago. So looking at these very distant objects, we are seeing light that has traveled for nearly 13 billion years – that is, we are seeing these objects as they were less than 1 billion years after the Big Bang.

However, what happened between these two *snapshots*, i.e., between the release of the diffuse *cosmic microwave radiation* 380,000 years after the Big Bang and the light emitted by these very distant galaxies almost 1 billion years later? When and how did this cosmic fog dissipate? What happened so that from a sea of particles without any structure arose a Universe lit by countless stars in the first young galaxies?

So, what today we are trying to understand is when and how the first stars and galaxies were born. Abraham Loeb of Harvard University puts it this way (Loeb 2002): *the problem we face is similar to having a photo album containing the first ultrasound image of an unborn baby and some photos of the same person as a teenager and as an adult. We are now searching for the lost pages of the cosmic photo album, which will show how the Universe has evolved during its infancy and how were made its building blocks: the various galaxies as well as our own Milky Way.*

The CMR photons, due to the continued expansion of the Universe underwent a wavelength shift from the visible to the infrared. So, from that period and until the formation of the first stars and galaxies, no photons were available in the visible part of the electromagnetic spectrum. That's why the Universe was really *black*, until a few hundred million years after the *Big Bang*.

The currently observed isotropy of the diffuse *cosmic microwave radiation* indicates that the original matter was uniformly distributed. And because there were no large luminous bodies to disturb the primordial cosmic soup, the radiation remained isotropic for millions of years after its creation. The only thing that happened was that the radiation was redshifted as the Universe was getting cold and dark.

We have no detailed observations for this era of cosmic dark ages of the Universe, except for the later epoch when the first stars and galaxies were formed. Exactly how long this age of cosmic dark ages lasted is the subject of modern research. However, at the end of this period, the first stars were formed and the Universe has entered the era of re-ionization we discussed earlier.

Deeper Era with Available Observational Data and the Epoch of *Recombination* [t $\approx$380,000 years, T $\approx$ 4,000 K, E $\gtrsim$ 0.4 eV]

Penetrating even deeper into the Universe, i.e., before the formation of the first stars and galaxies, we have presently no other observational data available, except from (i) the *CMR*, which spilled into space about 380,000 years after the *Big Bang* and (ii) the formation of nuclei with the observed cosmic abundances of the light nuclei.

Finally, at about $t \approx 380{,}000$ years after the *Big Bang* ($z \approx 1{,}100$), we have the *epoch of recombination*, the atomic era, where the temperature of the Universe dropped to about $T \approx 4{,}000$ degrees – the color of the *photosphere* of the Universe was reddish, close to the color of the light of the solar photosphere – with a maximum black body emission wavelength of about $\lambda_o \approx 7250$ Å,[7] while the size of the Universe was about 42 million light years, ($6000 \gtrsim z \gtrsim 1100$).

It is interesting to calculate how dense and opaque was the young Universe at that crucial time of its history, wherein we get its most ancient light coming from. The density of the baryonic matter today is of the order of 4.8% of the critical density ρ_c, i.e., $\Omega_b\rho_c \approx 0.048 \cdot (3H^2/8\pi G) \approx 4.7 \times 10^{-28}$ kg/m^3. This matter is mostly protons and neutrons, with about 88% of this mass density coming from protons. This means there were about 0.25 protons/m^3, and therefore the same number density of electrons/m^3. At the time of *recombination*, this would have been a factor of $(1100)^3$ higher, or roughly, 3.3×10^8 electrons per cubic meter. Multiplying by the Thomson scattering cross-section $\sigma_T \approx 2.652 \times 10^{-29}$ m^2 gives about 2×10^{-20} m^{-1}, which means that by taking the inverse, on average, a photon would be able to travel about 4.6×10^{19} m, i.e., about 4900 light years, a bit under 5000 light years, before scattering off from an electron. This is not *opaque* by normal standards, but for previous times wherein the mean free path was much less than 5000 years, it is a rather short period of time in cosmological terms, so *opaque* is a very reasonable approximation in that context.

[7] This is the soft glowing light from the sky when the sun is below the horizon, caused by the reflection of the sun's rays from the atmosphere. This *reddish twilight* is sometimes referred to poetically as the most beautiful color of the young Universe..

It was at this point that the Universe went from being totally opaque, to transparent. Note also that the CMBR contains photons emitted later than this period, with the oldest ones being around 380,000 years after the Big Bang and the last ones being around 480,000 years after the Big Bang, when recombination was completed, so it is not really a *surface* but it has a *depth* of about 100,000 light years.

Hence, when the mean free path of the photons because of Thomson scattering was greater than about 5000 light years, the nuclei could combine with electrons to create neutral atoms. As a result, photons no longer interacted as often with matter, the Universe became transparent – the *cosmic fog* was removed – and so the diffuse *CMBR* that flooded the Universe ($z \gtrsim 1100$) was created. Due to the expansion of the Universe we see today, this diffuse CMBR with a maximum blackbody emission wavelength redshifted by a factor of 1,100, i.e., $\lambda = \lambda_o(1 + z) = \lambda_o \times (1.100)$, i.e., approximately equal to one millimeter, in the microwave part of the electromagnetic spectrum.

In summary, the infant Universe was dominated by radiation over matter, and so the young Universe lived through the age of *radiation*. But after about 46,000 years ($z \approx 3{,}600$) following the Big Bang, the density of matter exceeded the density of radiation and began to dominate, with Universe entering the era of *matter*. Finally, in the last 5 billion years it seems that the density of matter has decreased compared to the density of *dark energy*, which now dominates the Universe and since then and until today we live in the era of *dark energy*.

It takes time for gravitation to collapse these atoms down to where they can form stars. Although the exact number isn't yet known, we can estimate it to last until the first stars form, say, 100 million years after the Big Bang. Namely, as we discussed previously, then followed the *dark ages* and within a few hundred million years, the first stars, galaxies, and quasars that we observe today were formed. The various eras of the infant Universe are shown in Fig. 6.1.

This is the last epoch on our path to the *Big Bang* that we have solid observational data. During the *era of radiation*, the Universe behaves like a giant super-luminous star, thousands of light years in diameter, in thermodynamic equilibrium, where photons interacted with matter and thus the Universe was opaque.

From the Blinding *Big Bang* to the First 3 Minutes

1967: The Infant Universe

In recent decades, particle physics has experienced a real boom. From an experimental point of view, experiments in large particle accelerators have confirmed the *Standard Model* (Glashow, Salam, Weinberg, Nobel Prize in Physics 1979) with 17 known elementary particles – 6 leptons, 6 quarks and 5 bosons. This model summarizes our knowledge and understanding of the fundamental building blocks of matter and a unifying way to describe three out of the four fundamental interactions via which these building blocks of matter interact, namely the weak, electromagnetic and the strong interaction, wherein gravity is excluded and is described separately by Einstein's General Theory of Relativity. Higher energies, however, well above several hundreds of GeV – the mass of the $W^{\pm}$ and Z bosons, the carriers of the electroweak force – remain out of the reach of any current or projected particle scattering experiment in today's accelerators, Fig. 6.1.

Hence, physicists realized that a unique *laboratory* for testing and applying ideas related to the various theories of the grand unification could be the Universe itself in the first infinitesimal moments of its evolution. The realization and the need to exploit this fact was a milestone in the subsequent development of ideas in theoretical physics and cosmology. Since then, the *microcosm* and the *macrocosm*, in other words the physics of elementary particles and cosmology, respectively, have been inextricably linked with each other. Thus, today theoretical studies have allowed astrophysicists and high-energy physicists to describe a self-consistent sequence of events that followed the *Big Bang*, from of the creation of time itself at $t = 0$, to the first minutes of the existence of the early Universe, as described in the following and schematically explained in Fig. 6.1.

In the following, we shall briefly summarize our knowledge about the first moments of the early Universe, for which nevertheless we do not have observational data, but only theoretical studies and results from the physics of elementary particles.

For a better understanding of the successive epochs of the infant Universe, we start from the moment of the *Big Bang* and work our way up to the first three minutes and the formation of the first nuclei of the light elements by the process of the Big Bang *nucleosynthesis*.

The Big Bang*: The Very First Moment (t = 0) of* Creation

In 1927 Georges Lemaître extended the idea of a continuously expanding Universe, where *the Universe was younger and smaller yesterday than it is today, until we had a day that had no yesterday*, i.e., in itself the first moment of Creation.

It is true that no one was present to watch in full the *Creation* of the Universe. At the same time, the temperature conditions, etc., which prevailed then are impossible to reproduce in our laboratories and accelerators. But what observational data dictate the existence of the initial *Big Bang*?

Undisputed Observational Data Supporting the Big Bang Theory

The *Standard Cosmological Model* of the *Big Bang* is supported by the following *solid* observational data, which we have already discussed:

- The *expansion* of the Universe itself,
- the diffuse *cosmic microwave radiation,*
- the *nucleosynthesis* of the light elements,
- the formation of galaxies and *large-scale structures* in the Universe,
- the appearance of *quasars*.

These *hard* observational data require us to accept that the Universe began its existence as an infinitely dense and hot *cosmic singularity*, during the so-called *Big Bang*, which marked the beginning of time and space.

Next, let's look in detail at the various epochs that characterized the evolution of the infant Universe, from the point of view and understanding we have today of high energy physics.

The Planck Epoch [0 $\lesssim$ t $\lesssim$ 10^{-44} Secs, T $\gtrsim$ 10^{32} K, E $\gtrsim$ 10^{19} GeV]

All matter and energy of the entire Universe is contained in a hot and dense gravitational singularity. During the Planck era, the Universe was so small that our laws of physics break down. To dive deeper back in time, we'll need a new scientific language.

The Planck epoch extends from this initial *singularity* at time t = 0, to the so-called *Planck time*,

$$t_p = \sqrt{\left(\hbar G/C^5\right)} \cong 5.4 \times 10^{-44} \text{ Seconds}$$

This specific value of the Planck time results from the combination of Heisenberg's quantum constant ($\hbar$), the gravitational constant (G), and the speed of propagation of light (c). It is very interesting that, in order to define the first time period of the Universe, the Planck time, the three main branches of physics, quantum physics of the microcosm, the theory of gravity of the macrocosm and electromagnetism which

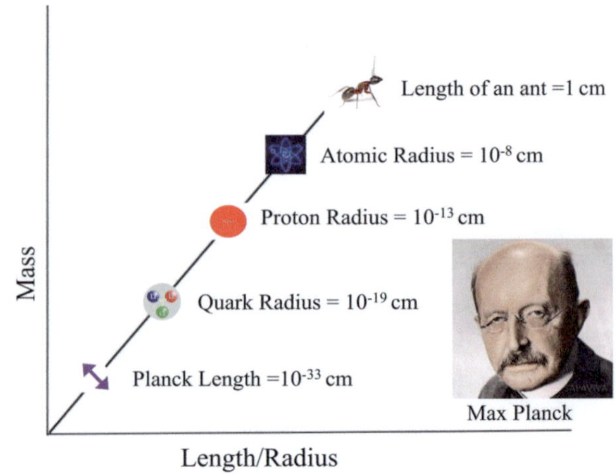

Fig. 6.36 The extremely small Planck length, relative to the size of a small ant and other subatomic lengths

describes the propagation of light, have contributed the best they have: their three main constants ħ, G and c, respectively!

This interval of about 10^{-44} s is an extremely short time: it is the infinitesimal fraction of a second that results when the second is divided by a huge, literally staggering number, the unit followed by 44 zeros. For comparison, the shortest period measured to date in the laboratory is 12 *attoseconds*, i.e., 12×10^{-18} s, 12 billionths of a *nanosecond*. For comparison, the time it takes for light to travel through the diameter of a hydrogen atom equals to 0.35 *attoseconds*. The *Planck time*, t_p, is a time interval about 26 orders of magnitude (10^{26}) smaller than the *attosecond*. Something truly unimaginable to our minds…[8]

The *Planck length*, L_p, Fig. 6.36, is the length traveled by light during a Planck time-interval and is equal to,

$$L_p = \sqrt{\left(\hbar G/c^3\right)} \cong 10^{-33}\ cm$$

Like time, the Planck length is also an extremely small spatial dimension, being about 10^{-20} times the diameter of a proton, i.e., about a hundred billion times smaller than the diameter of the proton. Here, too, we are talking about an unimaginably small dimension: if a proton were enlarged until it acquired the size of our solar system, then the Planck length would not be larger than the size of a virus, 0.025–0.25 μm, where one micrometer (μm) is equal to one millionth of a meter. At the age of the Planck time, the size of the Universe was about 10^{-33} cm, the Planck length. Thus, at the Planck instant of time, the size of the Universe was smaller than the quantum wavelength of the Universe, itself.

[8] The Nobel prize in Physics in 2023 was given to Pierre Agostini, Ferenc Krausz and Anne L'Huillier *for experimental methods that generate attosecond pulses of light for the study of electron dynamics in matter*.

Similarly, the *Planck mass* m_p is defined so that the gravitational potential energy between two similar such masses m_p at a distance r between them is equal to the energy of a photon of wavelength equal to this distance r,

$$m_p = \sqrt{(\hbar c/G)} = 2.2 \times 10^{-8}\,\text{kgrams} = 22\,\text{micrograms}$$

Thus, it follows that this most basic unit of mass in physics is not terribly smaller compared to masses on the biological scale: it is about equal to the mass of ten million bacteria, or, the mass of the smallest object we can discern with the naked eye – a grain of dust, for example.

The *Planck energy* E_p is defined as the combination of its quantum constant Heisenberg ($\hbar$) and the Planck time t_p.

$$E_p = \hbar/t_p = \sqrt{(\hbar c^5/Gk^2)} = 1.22\text{x}10^{19}\,\text{GeV}.$$

The Planck temperature T_p is obtained by equating the thermal energy kT_p with the rest energy of the Planck mass m_p.

$$T_p = \sqrt{(\hbar c^5/Gk^2)} = 1.41 \times 10^{32}\,\text{K},$$

where k is the Boltzmann constant, which relates the average kinetic energy of the particles of a gas to its temperature.

These characteristic quantities of the Planck length and time as well as the Planck temperature and mass were first introduced by Max Planck (1899).

Note that the atoms of matter around us are practically 99.999999999% empty space! If we subtract the space between the nuclei of all atoms and the distance of the classical electron radii where electrons move, for the entire population of Earth, then, the approximately 8 billion people inhabiting our planet would fit inside a small cube, like a cuboid of sugar.

Reaching the Limits of *Contemporary* Physics

In the Big Bang model, running time backwards and following the General Theory of Relativity, we reach the critical Planck length. At the Planck time, the mass density of the Universe is enormous and it is thought to approach the Planck density, c^5/hG^2, roughly 10^{94} g/cm^3, roughly, the mass of 100 billion galaxies squeezed into the size of an atomic nucleus. Under such extreme conditions, spacetime cannot be treated as a classical continuum and must be given a quantum interpretation. Earlier than this point, Einstein's equations are meaningless and the laws of physics breakdown, i.e., presently, we cannot project anything earlier than this time.

The 4 Basic Forces/Interactions of Nature

The various components of matter, as we know them today, are governed by the following 4 basic forces/interactions:

Table of basic interactions of nature and their range.

Name	Strength WRT strong	Range
Strong nuclear interactions	1	$< 10^{-15}$ m
Electromagnetic interactions	10^{-2}	infinite
Weak nuclear interactions	$< 10^{-7}$	$< 10^{-18}$ m
Gravitational interactions	10^{-38}	infinite

In the second column, we are rating the strength of these forces relative to the stronger one, the strong nuclear force. In such a comparison, the strength of the electromagnetic force would be roughly equal to 10^{-2}, i.e. 100 times smaller than the strong force. This ratio of the electromagnetic force to the strong force is equal to the so-called fine structure constant, $\alpha \approx e^2/\hbar c \approx 1/137$. Accordingly, the intensity of the weak nuclear force is rather weak at relatively large distances (macroscopic, atomic, and even nuclear distances) decreasing exponentially with distance. In radioactive decays the strength of the weak interaction is about 10^{-5} times the strength of the electromagnetic force and it can be as low as 10^{-13}, i.e., 10 trillion times weaker than the strong force, while that of gravity would be 10^{-38}, i.e. 100 trillion times weaker than the strong force.

Let us now come to the basic interactions that govern all physical phenomena. *First* of all, the *strong* nuclear force is an *attractive* force between protons and neutrons, holding them together, so that the nucleus does not break apart due to the electrical repulsion between protons. Similarly, the strong nuclear force holds quarks within protons, neutrons, and mesons. The strong nuclear force dominates in the microscopic atomic dimensions, but in order to act, nuclear particles must to be very close to each other, which means that this interaction is of very short range. The powerful nuclear forces participate in the nuclear fission that takes place in the reactors of nuclear plants and the thermonuclear fusion that occurs in the hydrogen bomb, and also the interior of the Sun and all the other stars.

Gluons, the massless quantum particles that govern the strong interaction, interact with quarks through a type of *charge* called *color*. This quantum *color* is analogous to the more familiar electric charge, but it exists in three types, which symbolically are named: ± red, ± green, ± blue and not in one only type like the electric charge. Therefore, the theory of *quantum chromodynamics*, which describes the interaction of quarks and baryons, describes a different type of force from the more familiar electrodynamics, with different rules of behavior.

Third in line is the weak nuclear force, which regulates and is related to the nuclear β-decay and our familiar radioactivity. As we discussed elsewhere in this book, the Sun is in pressure equilibrium and survives because of the thermonuclear fusion, the efficiency of which is only 0.007 (0.7%). This results in the Sun's metabolism being very low. Specifically, it is about 5,000 times lower than the human metabolism, which is 2,000 kcal/70 kgr/day. This justifies the solar longevity, ca. 10 billion years.

Fourth, let us turn to *gravity,* which is the most *mysterious* of all forces of Nature. It frustrated Einstein in his attempt to include it together with the other 3 forces in the General Theory of Relativity, and it also resists to this day, to be included in the Grand Unified Theory (GUT) of all forces, the so-called *Theory of Everything*. Due to the fact that it has a long-range, i.e., on the one hand it acts and *sees* at long distances, and on the other hand, it is *only attractive* – while the other forces either they are of a short range, or, they cancel each other out by being attractive and repulsive. Still, gravity remains dominant in the Universe, even though it is extremely weak. Gravity formed the huge and grand structures of the galaxies where the stars rotate inside them, such as our Sun, the planets around the stars and the satellites around the planets. This force both creates and destroys stars by creating black holes.

Going back to the infinitesimal duration of the *Planck era*, then all above interactions (forces) that we know today, i.e., the strong, electromagnetic, weak, and gravitational interactions, they were united and therefore the Universe had a complete *symmetry*. We don't know exactly what happened during this time period that *touches* the initial *Creation* at $t = 0$, since today's knowledge of physics is exhausted, reaching its limiting moments there. In order to be able to describe what exactly happened at this time, a theory is required for the so-called *quantum gravity*, which unfortunately we do not have available today.

Then let's briefly see what happened during the other epochs of the Universe, which followed the Planck epoch.

The Epoch of Grand Unification [10^{-44} Secs $\lesssim t \lesssim 10^{-36}$ Secs, 10^{32} K $\gtrsim T \gtrsim 10^{29}$ K, 10^{19} GeV $\gtrsim E \gtrsim 10^{16}$ GeV]

The *Grand Unification* epoch begins. At the post-Planck era, the Universe continued to expand and cool, successively passing through various lower transition temperatures, at which the various interaction/forces gradually separated from each other. These *phase changes* can be considered to be analogous to the phase changes that occur in water, during the condensation of water vapor and its liquefaction, as well as during the cooling of water and its solidification into ice. At certain temperatures/energies, for example at 100 and 0 °C for water, its molecules change behavior and structure and behave completely differently. At the Universe, a drastic change of behavior and structure occurs when the temperature/energy drop below a certain value, and the fields that determine the fundamental forces and the corresponding particles similarly to what happens when steam condenses to liquid, or liquid water condenses to ice. Because in the Universe the analogy occurs only at very high and extreme temperatures, these phase changes are not apparent to us. However, these phase changes in the fundamental interactions of the Universe are believed to be caused by a phenomenon occurring in the quantum fields involved, called *symmetry breaking*.

The success of symmetry has fueled hopes that we might ultimately understand the evolution of the universe from its beginning as a symmetry-breaking cascade, where at each step more and more of the initial symmetry is lost as the universe expands and cools down. In this view, the unsymmetrical world that we see around us is only the broken phase of a highly symmetrical theory at the origin of the universe, when forces, matter, and spacetime were unified into a single entity. However, this picture has so far been validated only up to energy scales accessible to the LHC, or equivalently distances down to 10^{-18} cm.

Simply put, as the Universe cools it has available some quantum fields which create the corresponding interactions/forces and the corresponding particles, to settle in lower energy levels, which are also distinguished by greater stability. When that happens, changes completely and the way they interact. Because forces and interactions arise as a result of these fields, the Universe can behave quite differently before and after such a phase change. For example, at a later time, a consequence of a phase change is that suddenly some particles that had no mass before, they gain mass after – i.e., start interacting differently with the Higgs field – and an initially single force, after a phase change manifests itself as two separate forces.

Assuming that at the ultrahigh energies of the Planck era there was a single unified interaction described by a universal unified theory (the *theory of everything*), the first phase change occurred when gravity separated from this universal super force. Of course, there does not exist yet solid evidence that such a global superpower ever existed, but many physicists believe that this may happen, by analogy with the gradual unification of the remaining other three forces. Subsequently, the era of Grand Unification ended, when a second phase change occurred, as the unified nuclear interaction – strong and electroweak – in turn splitted resulting in two separate interactions, manifesting themselves as the familiar strong and electroweak interactions, etc.

The energy above which the *Grand Unification* (GU) of the nuclear interactions occurs is about 10^{16} GeV. This energy corresponds to temperatures higher than 10^{29} K. Thus, in the infinitesimal time span of the life of the infant Universe, between the Planck time (10^{-44} s) and the time wherein inflation occurred (10^{-36} s), when the temperature of the Universe varied from 10^{32} to 10^{29} degrees, it is expected that there were only two different interactions in the Universe, the gravitational and the single strong/electroweak nuclear, with gravity having already splitted from the nuclear at the end of the Planck epoch. During the era of *Grand Unification*, physical characteristics such as mass, electrical charge, taste, and color charge were meaningless.

Experiments in accelerators have confirmed that at very high energies, the electromagnetic and weak interaction are unified into the single electromagnetic interaction. So, with example and guide/model the unification of the electric and magnetic forces from Maxwell's theory of electromagnetism in the nineteenth century, various theories have been proposed for such Grand Unified Theory (GUT) of the electroweak and the strong nuclear interaction. Of course, the physical implications and the mathematical structure of these new GUT theories are qualitatively different and rather more complicated compared to electromagnetism.

Although the unified strong nuclear and electroweak force has not been directly observed, there are several different GUT models that have been proposed to describe its existence. However, the main problem with grand unification theories is that there is no possibility of experimental verification, due to the ultra-high energies (10^{16} GeV) where this unification takes place. In any case, if such unification of these three interactions is indeed possible, raises the possibility that there was such an era of unification in the very early Universe, in which these three fundamental interactions were united. The new particles predicted by these Grand Unification Theories (GUT) models are expected to have extremely high masses around the 10^{16} GeV energy, which is far above any current or planned particle scattering experiment in accelerators. These extremely high energies and masses are beyond the reach of our particle accelerators. Therefore, the particles predicted by the GUT models will not be able to be verified in the near future. Instead, the predictions of those Grand Unification Theories could only be verified through indirect observations such as the half-time of the proton decay, the electric dipole moments of elementary particles, or, the properties of neutrinos. *It is worth to mention that some GUT models, such as the Pati-Salam model, predict the existence of magnetic monopoles*. The Grand Unified Force is mediated by (the hypothetical) X and Y bosons which allow early matter at this stage to fluctuate between baryon and lepton states. A wide array of exotic elementary particles result from the decay of these X and Y bosons, which include the $W^{\pm}$ and Z_o bosons and the Higgs bosons.

The first GUT theory was based on a simple Lie group of special symmetry [so-called SU(5)] and was proposed by Howard Georgi and Sheldon Glashow in 1974. The Georgi-Glashow model predicts a proton decay, something however that is not observed. It was preceded by a simple Lie algebra model by Abdus Salam and Jogesh Pati, who pioneered the effort to unify quantum interactions. The acronym GUT was first coined in 1978 by CERN researchers [John Ellis, Andrzej Buras, Mary K. Gaillard and Dimitris Nanopoulos] (Fig. 6.37).

The Extremely Short Epoch of the Big-Bang Inflation [10^{-36} Secs $\lesssim t \lesssim 10^{-32}$ Secs, 10^{29} K $\gtrsim T \gtrsim 10^{22}$ K, 10^{16} GeV $\gtrsim E \gtrsim$ 10^{9} GeV]

As we have already discussed, one problem with the original classical theory of the *Big Bang* was the observed isotropy of the diffuse cosmic microwave radiation. If the Universe originated from a point *singularity*, then microscopic perturbations – fragments of it would soon grow into huge inhomogeneities as the Universe expanded. But the microwave cosmic background radiation appears to be the same no matter which direction we observe.

One way that has been proposed to overcome this problem is by the theory of *cosmological inflation*. Specifically, around 10^{-36} s after the *Big Bang*, when the temperature of the Universe dropped to 10^{29} degrees, the Universe underwent an

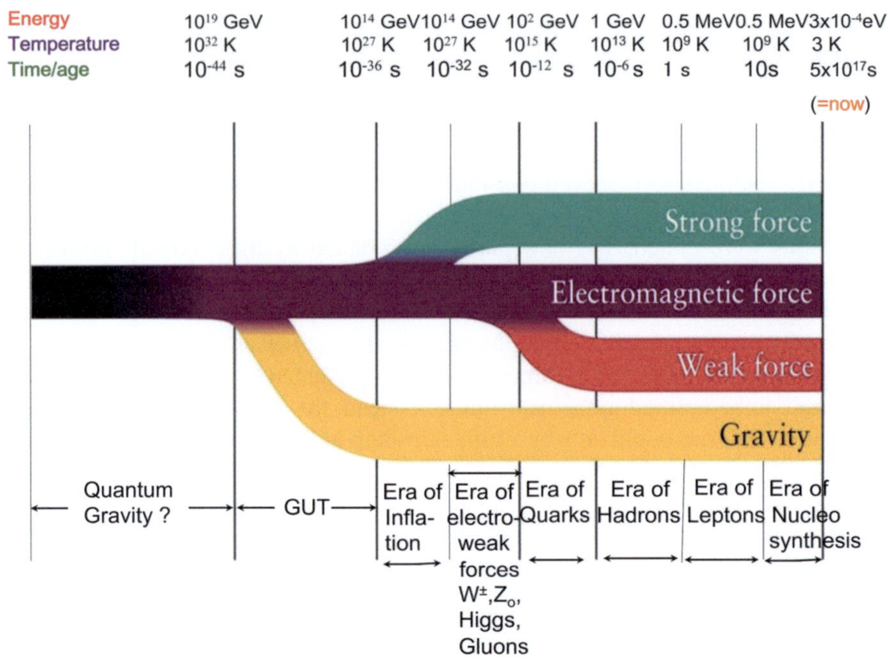

Fig. 6.37 The gradual unification of the 4 fundamental interactions (forces) of physics in the infant and young Universe, as time progresses backwards, heading toward the *Big Bang*. The timescales, temperatures and energies indicated on this diagram span an enormous range and are given correct only as orders of magnitude. One GeV is equivalent to about 10^{13}K. (Adapted from CERNCOURIER, 13 January 2017, https://cerncourier.com/a/gravity's-quantum-side/ article by Hermann Nicolai, Max Planck Institute for Gravitational Physics, Potsdam, Germany)

extremely rapid expansion lasting up to about 10^{-32} s. During this time, the Universe expanded exponentially, increasing its diameter by a factor of 10^{26} and its volume by a factor 10^{78}. Before the inflation phase, its components were in equilibrium with each other, so that they had the same temperature. Inflation ended at about 10^{-33} to 10^{-32} s and since then the Universe continued to expand normally. Inflationary expansion began when the strong and electroweak interactions due to the break of symmetry separated and the Universe passed into the state of *false vacuum*, as we discuss in the following.

In other words, at 10^{-36} s after the *Big Bang*, when the temperature of the Universe had decreased to about 10^{29} degrees, or, 10^{16} GeV, due to the symmetry breaking of the Grand Unified interaction (GUT) of the nuclear and electroweak interactions, the strong and electroweak interactions splitted and ceased to behave as a single interaction, according to the grand unification theories of fundamental interactions of theoretical physics, discussed earlier. It is possible that some part of this symmetry breaking process violated baryon number conservation and led to a small excess of matter over antimatter. This phase change is believed to have triggered the process of *cosmic inflation* that dominated the expansion of the Universe during the *inflationary era. The corresponding energy released during this symmetry breaking fueled the inflationary phase in which the volume of the Universe increased exponentially, by a factor of* $e^{34} = 10^{78}$*, in a tiny fraction of a second.*

The Phase of Inflation of the Universe ...

"When the age of the Universe was *one billionth of a billionth of a billionth of a billionth of a second*, it went through an extremely brief period of hyper-rapid expansion, swelling from the size of a proton to much larger than the size of an orange, expanding with a speed many-many times the speed of light. By assuming that the observable Universe had the size of the Planck length prior to inflation, you end up with a size of about 10^{-5} m after inflation. After that, the expansion slowed down by a significant amount. As unlikely as this may sound to you, it agrees with every observation astronomers have been able to make!" [Lemonick Michael and Nash Madeleine, *Cosmic Conundrum*, Time Magazine, 58, Nov 29, 2004].

Cosmological inflation is a theory, developed in the early 1980s, to describe the required exponential expansion of space in the young Universe and thus explain some unanswered cosmological questions of the classical Big Bang model. Today, many cosmologists believe that *inflation* explains why the Universe is isotropic, i.e., looks the same in all directions, and also why the cosmic microwave radiation (CMR) is also isotropic, that is, its intensity is practically the same in whichever direction of the sky we observe. Additionally, it explains why the Universe is *flat* and why magnetic monopoles are not observed. The theory also explains the origin of the large-scale structure of the Universe, where quantum fluctuations in a tiny region during *inflation*, when magnified on a cosmic scale, were the seeds that led to the creation of the large structures we observe today in the Universe. The inflationary period began when the age of the Universe was only 10^{-36} s and terminated when its age was between 10^{-33} and 10^{-32} s. After the era of *inflation*, the Universe continued to expand, but at a slower rate of expansion.

1980: Theory of the Cosmological Inflation

The *Big Bang* theory gave the impetus and basis to Cosmology to explain why the Universe is expanding, coming from an extremely dense and hot initial state, 13.8 billion years ago. However, subsequently some *annoying* questions appeared which the original *Big Bang* theory could not explain. For example, the extraordinary *flatness*, *homogeneity*, and *isotropy* of the Universe, which we describe next, could not be explained by the original *Big Bang* theory. Thus, arose the need for a theory, complementary to the original one, which would solve the above problems. This is the topic to discuss in the following.

Although we will return below to the complete chronology of the Universe, starting from the great explosion, in order to create the appropriate chronological framework of the presentation of cosmological inflation, it is sufficient to very briefly recapitulate here the first three *epochs* of the Universe:

⇒ Planck epoch, $0 < t \lesssim 10^{-44}$ secs [$T \gtrsim 10^{32}$ K, $E \gtrsim 10^{19}$ GeV].

This first era extends from the initial singularity, i.e., time $t = 0$, to the Planck time $t_P = 10^{-44}$ s. During this period, all interactions that we know today (strong, electromagnetic, weak and gravitational) were unified and the Universe had a complete symmetry, which we assume that it is described by the – today unavailable (!) – theory of *Quantum Gravity*. At about 10^{-44} s from the beginning we have the end of the Planck epoch, at which point gravity splits from the remaining 3 interactions.

⇒ Grand Unification epoch, $10^{-44} \lesssim t \lesssim 10^{-36}$ secs. (10^{32} K $\gtrsim T \gtrsim 10^{29}$ K, 10^{19} GeV $\gtrsim E \gtrsim 10^{16}$ GeV).

This epoch spans from the end of the Planck epoch – at about 10^{-44} s – to the time of about 10^{-36} s from the *Big Bang*. The Universe has cooled from about 10^{32} down to 10^{29} Kelvins and is in the new *Grand Unification epoch*. During this time, gravity remains separated from the remaining 3 *combined* interactions. The strong and electroweak interactions behave as a single interaction – something described by the theories of the *Grand Unification Theory* (GUT) of the fundamental interactions of Theoretical Physics.

⇒ Epoch of cosmological inflation, $10^{-36} \leq t \leq 10^{-32}$ secs (10^{29} K $\gtrsim T \gtrsim 10^{22}$ K, 10^{16} GeV $\gtrsim E \gtrsim 10^{9}$ GeV).

At about 10^{-36} s, the temperature of the Universe has decreased to 10^{29} degrees (10^{16} GeV). At 10^{-36} s we have the end of the *Grand Unification epoch*. Then comes the turn of the strong interaction to be splitted from the electromagnetic and the weak interaction, that these two at least stay together for a while still further, as electroweak interactions. Cosmic *inflation* expands space by a factor of the order of 10^{26} over a time of the order of 10^{-36} to 10^{-32} s. The Universe is supercooled from about 10^{29} down to 10^{22} Kelvins. The Universe is in its new era of *inflation*. We explain why/how this happened in the following.

In *Relativity*, applies the *principle of causality*, in which the cause precedes the final effect of this cause. Cosmological *inflation* is one of the most basic elements of the progress made in the original theory of the *Big Bang*, because it managed to eliminate the discrepancy between the law of expansion of the Universe and the *principle of causality* of *Relativity*, as we explain in the following. But let's see in more detail what are the main problems that *inflation* theory solved.

1. *Large-scale homogeneity and isotropy, or the horizon problem, i.e., to explain why the Universe and the CMR exhibit statistical homogeneity and isotropy.* The CMB we detect today comes from about 46 billion *comoving* light years away.[9] However, when the light was emitted the Universe was much younger (380,000 years old). In that time, light would have only reached as far as a circle of radius 380,000 light years. Two points at the sphere of radius 42.5 billion *comoving* light years away, would not have been able to communicate with each other, because their spheres of causality do not overlap. The simple analogous example is the density and temperature distribution of the air molecules in our room; the gas is homogeneous and isotropic, precisely because it is in *thermal equilibrium*. That is, the air molecules have enough time to interact among themselves and cancel out the impact of any inhomogeneities and anisotropies that their components might have, such as their temperature, density and speed distributions, for example, when we suddenly open a window of the room.
2. *Causality problem.* In the original *Big Bang* model, without the inflationary phase, the initial rapid expansion does not provide the required time in the early Universe to equilibrate any differences in the physical state of its components, as is the case in the example of the air in our room. In the original model of the *Big Bang* (without *inflation*), with only the presence of matter and radiation, some fairly distant regions of the *observable* Universe – such as the three regions shown schematically with the symbols of stars, galaxies, and black holes in Fig. 6.36 – have been unable to equilibrate their internal physical states in each one of the three, relative to the other two, because they move away from each other at a speed *faster* than the speed of light and thus lose their causal connection. In other words, a signal emitted by any one of them and heading to inform the others for its own internal state, does not have enough time to reach the other two, because the signal propagates at the speed of light, while they move away from each other at a speed greater than the speed of light!

So, normally they should not have the same internal thermal conditions, since they do not *know* what is happening in the others. Therefore, since they are not causally connected, the question that arises is: why should they be observed to have the same temperature, as required by the observed large-scale isotropy of the CMR? That is,

---

[9] The radius of the *observable* Universe is estimated to be about 46.5 billion light years (Lineweaver & Davis, 2005). The word *observable* refers to the physical limit created by the speed of light itself, wherein there is a maximum distance, the *particle horizon*, beyond which nothing can be detected, because the signals could not have reached us yet. This is the portion of the Universe that we can *observe*, since the end of the *inflationary epoch*, albeit the total actual size of the Universe is unknown. The *visible* Universe is different from the *observable* Universe, because it refers to signals coming from the end of the *recombination epoch* and thus it has a smaller radius, about 45.7 billion light years.

why should they be in thermal equilibrium, like the air molecules in our room, since they don't have the needed time to communicate with each other?

Consider the size of our *visible* Universe, along with the distance we can reach if we left, today, on a journey at the speed of light. The limit of the *visible* Universe is 46 billion light years, as that's the limit of how far away an object that emitted light that would just be reaching us today would be after expanding away from us for 13.8 billion years. Anything that occurs, right now, within a radius of 18 billion light years of us will eventually reach and affect us; anything beyond that point will not. Each year, another ~20 million stars/galaxies cross the threshold from being reachable to being unreachable.

3. *Horizon of observable Universe vs.* our *cosmic horizon*: To make the problem more quantitatively concrete, recall that observations show that the Universe is currently about 13.8 billion years old. Is correspondingly the radius of the observable Universe today 13.8 billion light years? The answer to this question is no! Because, meanwhile, in this period of 13.8 billion years, due to the expansion of space, the *comoving*[10] distance from Earth to a galaxy at the limits of the *observable* Universe has today reached a value of about 42.5 billion light years. So, the horizon of the *observable* Universe today is a sphere with a radius 42.5 billion light years, while *our cosmic horizon* is a sphere with a radius of only 13.8 billion light years. This is because until light comes to us here on Earth from this most distant galaxy, 13.8 billion years have passed since the beginnings of the Universe, during which time, this galaxy at the limits of the Universe has gradually moved even further away than the distance of 13.8 billion light years, at the new distance of 42.5 billion light years.

Using the FLRW metric and going back in time when the Universe was 380,000 years old after the *Big Bang*, we arrive at the time when the CMR photons were released and began to move toward all directions. According to the FLRW solution of the Einstein equations, the redshift factor z at that time had the value $z = 1091.64$. Therefore, the scale factor $a(t)$ at that time had the value $a(t) = 1/\,(1 + z) \cong a(t_o)/1{,}100$, relative to its current value of $a(t_o) = 1$, as discussed previously.

Note that the term *observable* does not refer to the capability of modern telescopes to detect light or other information from an object. It simply refers to the physical limit created by the finite speed of light itself; hence, there is a maximum distance (called the *particle horizon*) beyond which nothing can be detected, as the signals could not have reached us yet. Sometimes we distinguish between the *visible* universe, which includes only signals emitted since the recombination era (when hydrogen atoms were formed from protons and electrons and photons were emitted) – and the

[10] The *comoving* distance χ between two comoving galaxies, say our Galaxy and the Andromeda galaxy, is defined to be a fixed unchanging quantity independent of time, while the dynamic, changing distance between these two galaxies is the *proper* distance. *Comoving* and *proper* distances are numerically equal at the current age of the universe $t_o = 13.8$ billion light years, but they differ in the past and in the future; so, if the comoving distance between our Galaxy and the Andromeda galaxy is denoted by χ, the proper distance $d(t)$ at an arbitrary time t is simply given by $d(t) = a(t)\,\chi$, where $a(t)$ is the scale factor and $d(t_o) = a(t_o)\chi = \chi$, since $a(t_o) = 1$.

observable universe, which includes signals since the beginning of the cosmological expansion, i.e., the end of the inflationary epoch in modern cosmology. The horizon of the *observable* Universe, by analogy with the more familiar horizon caused by the curvature of Earth's surface, marks the boundary of the part of the Universe that an observer can see. Light (or other radiation) emitted by objects beyond this cosmological horizon in an accelerating universe never reaches the observer, because the space in between the observer and the object is expanding too rapidly. Every location in the Universe has its own *observable Universe*, which may or may not overlap with the one centered on Earth.

The current comoving distance to particles from which the CMR was emitted, which represents the radius of the visible universe, is about 45.7 billion light years. The comoving distance from Earth to the edge of the observable universe is about 42.5 billion light years in any direction. The observable Universe is thus a sphere with a diameter of about 93 billion light years.

As a consequence of the above, and since the Universe's expansion is *accelerating*, all currently observable objects, outside the *local supercluster*, will eventually appear to emit progressively redder and fainter light. For instance, objects with the current redshift z from 5 to 10 will remain observable for no more than 4–6 billion years. In addition, light emitted by objects currently situated beyond a certain comoving distance, currently about 62 billion light years, will never reach Earth (Loeb 2002).

So, the oldest CMR photons were emitted by individual atoms at the time when the Universe was $t = 380{,}000$ years old. These atoms today are in a distance 42.5 billion light years away from us, the radius of the *observable* Universe. Hence, going back in time, at the time of radiation-matter decoupling, the radius of the Universe was then about (42.5 billion l.y.)/$(1 + 1091) \approx$ (42.5 billion l.y.)/1100 $\approx$ 42 million light years (Fig. 6.38). Any mechanism, however, that could equalize any temperature differences in the Universe it ceased to act when it was 380,000 years old, a time when the cosmic horizon regions in the Universe had a radius of about 380,000 light years. Because from the law of the expansion of the Universe we know that these regions may have been apart about 42 million light years, we conclude then that these regions were located each outside the *individual* cosmic horizon of the other. In other words, the Universe we observe today came from a very tiny region, many orders of magnitude smaller than the cosmic horizon of the time of recombination, which allowed it to acquire thermal homogeneity, before the onset of its inflationary expansion (Fig. 6.38). Immediately after the beginning of the era of *inflation*, this region expanded rapidly, but maintaining its original uniformity and leading to a Universe that on larger scales is uniform, as we observe.

This is explained schematically in Fig. 6.38, where each of the three regions of stars, galaxies, and black holes had different temperatures inside their *atomic* horizon (small circles with dotted lines, each with a radius of about 380,000 l.y.), while the largest circle in which they contain has a radius of about 42 million light years (the outer circle with the solid line at the top left in Fig. 6.38. Therefore, there was a way to *communicate* with each other, so that their temperatures equalized with the temperature of the others. The result is that each of these three areas had a different temperature than the other. Therefore, in the present Universe (the *present* in

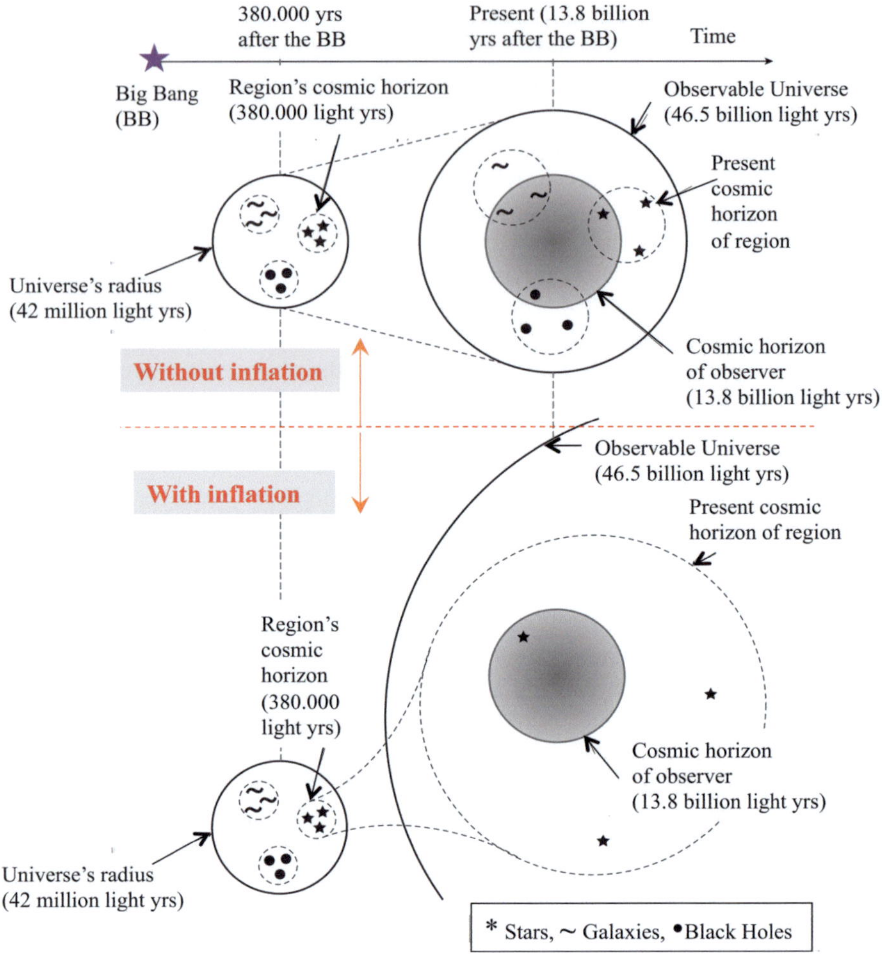

Fig. 6.38 Explanatory diagram of the change brought to the horizon problem by the idea of inflation. Without a loss of generality and for the sake of simplicity, we have chosen 3 regions in the Universe, each one containing, say, only individual stars, or only galaxies, or only black holes. Our *cosmic horizon* today is the gray sphere with radius 13.8 billion light years (l.y.)

Fig. 6.38, and the dark, gray region with a radius of 13.8 billion light years) the CMR that we observe, and the Universe as a whole, to show no homogeneity and isotropy at all. The situation becomes even worse if we go even further back in time, an infinitesimal fraction of a second after the *Big Bang*. It is estimated that the Universe then consisted of 10^{90} regions that did not communicate with each other. How did the Universe *manages* to acquire this high degree of isotropy and homogeneity that we observe today? All this is valid as long as there has not been an inflationary phase in the evolution of the Universe.

Thus, without *inflation* (top part of Fig. 6.38), at some point in the past the three regions of stars, galaxies, and black holes were each internally homogenized, but each had different internal physical states with respect to the other. Their cosmic horizon was 380,000 l.y. – so that we receive their *cosmic microwave radiation* – but the Universe then already had a radius of about 42 million l.y. (the outer circle with the continuous line at the top left). As we come to the present (top right), the expanding Universe has acquired the largest radius of 42.5 billion l.y. (the continuous outer line at the top right) and the observer at its own cosmic horizon (the gray area of radius 13.8 billion l.y.) should see different physical conditions, if observing toward these three different directions where the 3 regions coexist. In other words, it will not see that there is thermal isotropy, in contrast to the observations!

On the other hand, let's go to the inflationary, exponentially expanding, Universe (lower part of Fig. 6.38). An area, e.g., the one with the stars, has exponentially expanded its horizon (the dotted line at the bottom right), and the Universe, also expanding exponentially, has today acquired a much larger radius (the outer solid line, at the bottom right). However, the current cosmic limit of the observer remains at 13.8 billion l.y. (the gray area). But when the observer looks in these three directions, he/she will see the same internal physical states, because he/she will see only within the original area with the stars which is already homogenized, i.e., we have isotropy, as the observations show!

Some efforts had been made to solve the horizon problem already by Lemaitre, Tolman and Misne, but other problems arose. Thus the issue remained open until 1980, when its solution was proposed by Alan Guth, Andrei Linde, Paul Steinhardt and Demosthenes Kazanas.

In fact, it seems that basically the energy of the quantum vacuum is behind the *inflation* process. As it is known, a region of space within which there is no matter and no radiation (particles or photons), a rather constant state of physics, is called *vacuum*. This state of the lowest energy (i.e. the ground state) is sometimes called *true vacuum* and possibly corresponds to the empty regions of the Universe, as they are observed today. In contrast, an excited true vacuum is referred to as a *false vacuum*, or *pseudovacuum*, where the word *vacuum* indicates a state of the lowest possible energy density, and the word *false* is used to mean *temporary*. According to quantum physics, the *quantum vacuum* can contain energy, but corresponds to an unstable state. The energy of the *vacuum* can also be studied in terms of virtual particles, also known as fluctuations of the vacuum energy, and which are always created in the *pseudovacuum* in pairs of particles-antiparticle, which however soon decay and thus disappear, i.e., their existence is limited to time intervals determined by Heisenberg's uncertainty principle, so that the conservation of energy is not violated. Meanwhile, these pairs of virtual particles can interact with each other before they disappear, a process that is simply represented by Feynman diagrams. Although the *pseudovacuum* is a purely theoretical idea, and its properties depend to a large extent on the particular theory we are using, its existence is inferred in most Grand Unified Theories (GUT) of the fundamental interactions. An interesting feature of those various GUT's is that the energy of the pseudovacuum states (and the corresponding mass of this energy) is enormous, each cubic centimeter of space contains 10^{87}

Joules, when the energy that has an excited atom is only of the order of 10^{-18} Joules. So, for the excitation of the *true vacuum* to a *pseudovacuum* state, a huge energy is required. Therefore, we should not seek to find such *false vacuum* states in our Universe, because those exotic states were probably existing only during the infancy of our Universe.

The *false vacuum* arises naturally in any theory that contains scalar fields, i.e., fields that resemble electric or magnetic fields, except that they do not have direction. The Higgs fields of the Standard Model of particle physics, or the more speculative GUT are examples of such scalar fields.

The gravitational effect of the enormous energy of the excited quantum vacuum is strongly attractive because it corresponds to a very large mass. For example, the energy of one cubic centimeter of *pseudovacuum* has a mass of 10^{67} tons, an energy that is greater than the total mass of the observable Universe, about 10^{50} tons.

A special feature of such an *inflation* field is the fact that all observers, regardless of their speed, measure the same energy density of this field, in contrast to the electromagnetic field, in which a moving observer sees an energy density that it depends on his speed. This paradoxical property of the *inflation* field implies that it is characterized by negative pressure: a spherical volume of vacuum with non-zero energy density exerts a pressure toward its center, numerically equal to the energy density of the *inflation* field. This negative pressure is like the surface tension exerted on the surface of a gas bubble and balances the outward (positive) pressure of gas. The equilibrium of the positive energy density of matter and radiation with the negative energy density of the vacuum suggests that, despite the expansion of the Universe, it is possible for the Universe to maintain a constant energy density. The phenomenon is analogous to the phase change of water, from the solid to the liquid form. In a mixture of ice and water at zero temperature, providing external heat, the temperature of the mixture remains steady, until all the ice melts.

It is worth noting that the colossal energy and mass of the *pseudovacuum* and comparable colossal pressure, they compete for prevalence. The critical fact, however, is that the pressure is negative. Therefore, the pseudovacuum does not push outwards but suctions. A negative pressure produces a negative gravitational action, in other words, it behaves as an antigravity. Thus, the gravitational effect of the pseudovacuum involves a competition between the enormous effect of its energy and the enormous repulsive effect of its negative pressure. It turns out that the pressure eventually prevails, and the net result is the production of a gigantic repulsive force. This gigantic inflationary impulse caused the size of the Universe to double for every 10^{-34} s in its early moments.

As mentioned, the pseudovacuum is by its nature unstable and therefore it also tends to be *de-excited* to its ground state, i.e., to a true vacuum state. As a quantum process, it is subject to an unavoidable uncertainty and the random fluctuations of Heisenberg's uncertainty principle. This means that the de-excitation will not occur uniformly throughout all space, but there will be fluctuations. It is believed that this is the source of the variations observed by the WMAP satellite that mapped the variations of the microwave radiation in the sky.

Once the pseudo-vacuum is de-excited, the Universe continues its normal expansion. The energy that existed trapped in the pseudo-vacuum is released in the form of heat. At the end of cosmic *inflation*, the enormous expansion had cooled the Universe to a temperature very close to the absolute zero. With the end of inflation, however, the heat was released suddenly and reheats the Universe to the extremely high temperature of 10^{28} degrees. This vast reservoir of heat which was then created, survives till today, in the form of the CMR.

A by-product of the release of *pseudovacuum* energy is that the many virtual particles in it take up a small amount energy and develop into real particles. After further processes and changes, a remnant of these primordial particles arrived today to make up the approximately 10^{50} tons of matter we see in the *observable* Universe.

Assuming that the inflationary scenario is valid, then the basic structure and fundamental components of the Universe are determined by processes that were already completed at 10^{-32} s from the great explosion!

Just as in any other quantum field, the inflation field excitations are expected to be quantized. The quanta of the field of this are known as *infatons*. This term was first coined in the publication Nanopoulos et al. (1983). However, the nature of this field is not currently known (Fig. 6.39).

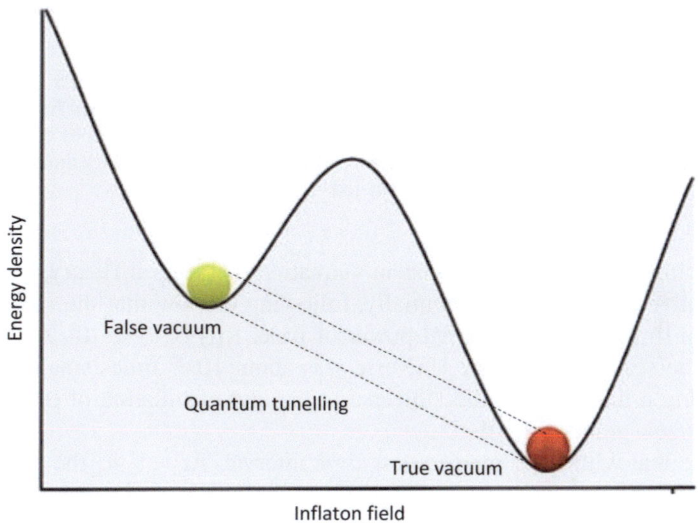

Fig. 6.39 Simplified model of inflation with the *true* and *false* vacuum energy levels. Initially, inflation is trapped in a *false* vacuum state where *bubbles* of negative pressure are formed which expand and collide resulting in occupying almost the entire volume of the Universe which thus ends up empty. From the local minimum of the *pseudovacuum, e*nergy is released when a transition is allowed to the global minimum through a quantum mechanical tunnel. The *inflation* phase begins when is allowed a slow transition from the *false* vacuum phase to the *true* vacuum phase, with oscillations in this lower energy state and conversion of the energy into particles of the Standard Model

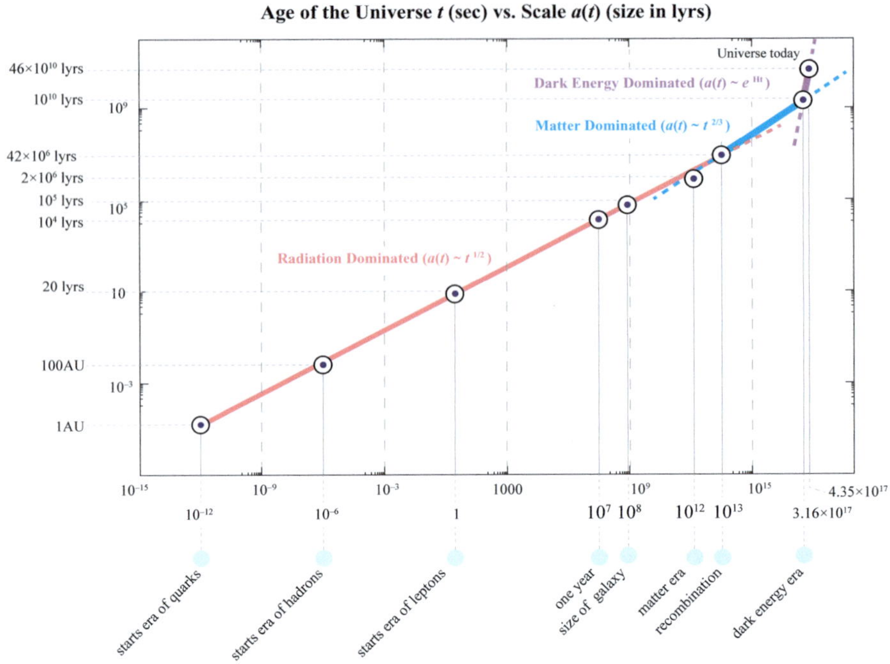

Fig. 6.40 Schematic logarithmic diagram of the age of the Universe (horizontal x-axis, in secs) versus the scale of the Universe a(t) (vertical y-axis, in light years/AU), obtained from the solution of the Friedmann (1924) equations. Nine characteristic time and scale milestones are marked. The red line corresponds to the radiation dominated era, the blue line to the matter dominated era and the violet line to the present dark energy dominated era

Combining this stability with Einstein's equations of General Theory, we conclude that the Universe expands exponentially, following the law that the radius of R(t) depends on time as an exponential power of time, $R(t) \propto e^{t} \propto 10^{t}$. At the end of *inflation*, the scale factor of the Universe was about 10^{-30} times smaller than it is today, giving a diameter for the Universe at the end of inflation of 0.88 mm, i.e., about the size a grain of sand!

Suppose that within a dimensionless time interval, $\Delta t = t_2 - t_1$ the radius of the Universe R(t) doubles, i.e., $R(t_2)/R(t_1) = 10^{\Delta t} = 2$. Then, Δt is equal to the logarithm of 2, $\Delta t \approx 0.3$. This is exactly the process of exponential increase of the radius of the Universe that is called *cosmological inflation*. Thus, cosmological inflation can bypass the *cosmic horizon* problem. Theories based on cosmological inflation assume that, after a time period corresponding to about N = 100 doublings, the vacuum collapses and its energy density is converted into the particles we know today (e.g., protons, neutrons, electrons) according to the well-known relationship $E = m.c^2$. Indeed, let us assume that in a dimensionless time interval $N\Delta t \approx 26$, the

Universe has expanded by a factor $10^{N\Delta t}$, i.e., $R(t_{end})/R(t_{start}) = 10^{N\Delta t}$. With N = 86 successive doublings of the diameter of the Universe, i.e., within a time interval $\Delta t = N \times 0.3 = 86 \times 0.3 \approx 22$. At the end of the inflationary process, which began at time equal to about 10^{-36} s of the age of the Universe, the final radius of the Universe will be 10^{26} times greater than the original!

The energy that created the exponential expansion during the period of inflation is thought to have come from the phase change that marked the end of the previous grand unification era, about 10^{-36} s after the *Big Bang*. One of the theoretical products of this phase change was the scalar field of *inflation*. When this scalar field settled into its lowest energy state throughout the Universe, enormous energy was released, which created the repulsive force which led to a very rapid expansion of space.

It is not known when the inflationary period ended, but it is assumed that this happened between 10^{-36} and 10^{-32} s after the *Big Bang*. The rapid expansion of space means that the elementary particles that remained from the era of the Grand Unification (GUT), they were distributed throughout the Universe. However, the enormous potential energy of the inflation field which was released at the end of the inflation epoch, it endowed the Universe with a dense and hot mixture of quarks, anti-quarks and gluons as the Universe entered into the following era of electroweak interactions.

To recap, going back in time, when the age of the Universe was about 10^{-36} s after the *Big Bang*, the dramatic inflation phase began. Then, in the later era of recombination and dissociation of the CMR from the first atoms, radiation was emitted from *small* homogenized areas, thus giving rise to the observed isotropy of the Universe and of the CMR. In this way, all points included within our current cosmic horizon, were included in the previous cosmic horizons, at every moment in the past and thus they were causally connected among themselves, in absolute agreement with the *General Theory of Relativity*.

⇒ What About the Dimensions of the Infant Universe Before and After the Era of Inflation?

In the first place, with the correct determination of the *dimensions* and *size* of the Universe, this question makes sense. According to the Standard Model of cosmology, the Universe is infinite, so it has no *size*. However, given that the *Big Bang* occurred about 13.8 billion years ago, we can define a size for the *observable* Universe. We could, for example, define as the size of the observable Universe, the distance a photon could travel since the *Big Bang*.

Consider, for example, a CM photon emitted as visible light about 380,000 years after the *Big Bang* and which is only now reaching our microwave detectors (the redshift of such a photon is z = 1089): this photon has traveled for 13.8 billion years, so it has traveled a distance of 13.8 billion light years. One could say that the current radius of the observable Universe is 13.8 billion light years. But, as we discussed earlier, during this time the Universe has expanded, so the current position of the atom that emitted this photon will now have moved to a distance of 42.5 billion light years. (In our time, the little temperature anisotropies of the magnitude of 10^{-5} of the CMR have already condensed into galaxies and stars at this new distance where is the atom that emitted the photon). Hence, this gives a diameter for today's observable Universe, which is equal to 93 billion light years.

Note that as time goes on, the size of the *observable* Universe is constantly growing. In fact, it will grow by much more than two (to convert radius to diameter) light years per year, due to the continuing (accelerating) expansion of the Universe. Also, note that we won't be able to use photons to explore the Universe earlier than 380,000 years after the *Big Bang*, because the Universe was opaque to photons at that time. However, in the future, we could use neutrino or gravitational wave telescopes to explore the earliest Universe.

⇒ *A Terrifying and Incredible Increase of the Size of the Infant Universe...*

Given the current size of the observable Universe, we may calculate its dimensions at some particular time in the past, according to the FLRW metric. According to Lineweaver (2005), at the end of the inflation era, the scale factor a(t) of the Universe was about 10^{-30} times smaller than it is today; this gives a dimension of the present observable Universe at the end of the inflation era, equal to 0.88 mm, about the size of a grain of sand, as we mention earlier. It is believed that during the period of inflation the dimension of the Universe increased by at least a factor of $10^{26} \approx e^{60}$. This is a dramatic increase in the dimensions of the Universe, which became 10^{26} times bigger. *An expansion by a factor of 10^{26} is equivalent to expanding an object the size of a hair (0.1 mm) to a dimension of about 300,000 light years, i.e., about the diameter of our Galaxy* ($10^{26} \times 0.1$ mm $= 10^{24}$ cm $\approx 3 \times 10^{18}$ cm x $3 \times 10^5 = 300.000$ light years).

So, going back in time, we find that the dimension of the Universe before the era of inflation was 8.8×10^{-30} m, a value many orders of magnitude smaller than the size of an individual atom Today, these regions have extended to a distance of about 93 billion light years in diameter, as we mention. The temperature of the Universe dropped by many orders of magnitude, from 10^{29} to 10^{22} degrees during the exponential expansion phase.

Inflation and Multiverses

Is there any direct observational confirmation of inflation? The answer is not yet. The BICEP3 (Background Imaging of Cosmic Extragalactic Polarization) experiment of the Harvard Smithsonian Center for Astrophysics research team is trying to detect gravitational wave signals from a small part of the sky, through the polarization of the CMR. In other words, they try to observe the polarization of the radiation that is the echo of the *Big Bang* in the microwave part of the electromagnetic spectrum. BICEP is based at the Amundsen–Scott South Pole Station.

The main objective of this experiment is to measure the polarization of the B oscillation modes of the diffuse Cosmic Microwave Radiation. Gravitational waves, the stamp of the violent expansion of the Universe during the first moments of its creation, at the time of 10^{-36} s after the *Big Bang*, cause wrinkles in space.

Such a picture of the *infant* Universe, apart from being able to provide information on the creation of galaxies, can confirm the theory of cosmological inflation and that our visible Universe is an infinitesimal part of a much larger world. Thus, a very interesting consequence of inflation, if it is ever (!) proven, is that our Universe is simply one of many *Multiverses*. When the Universe underwent a dramatic inflationary increase in its size, it may have created spacetime bubbles that fertilized other Universes, perhaps with different laws of physics.

The Cosmological Constant Problem

One of the first to systematically attempt to investigate the physical meaning of the *cosmological constant*, using fundamental ideas from quantum field theories, was the Soviet physicist Yakov Zel'dovich (1914–1987), who suggested in 1968 that the *cosmological constant* is mathematically equivalent to the quantum vacuum energy. In 2002, three of the architects of inflation theory, the physicists Alan Guth of MIT, Andrei Linde of Stanford, and Paul Steinhardt of Princeton shared the prestigious Dirac Prize for developing the concept of inflation in cosmology. In 2012, to Alan Guth and Andrei Linde was awarded the *Breakthrough Prize in Fundamental Physics* for the invention and the development of the concept of *inflation* in cosmology.

Although today the basic inflation model is accepted by most scientists, and a number
of inflation predictions have been confirmed from observations, however, a significant minority of scientists continue to disagree with this concept. Inflation is indeed an excellent theory that answers unsolved observational problems. Nevertheless, this theory too has also some *annoying* problems:

First, the detailed mechanism of elementary particle physics which is responsible for inflation is not known.

Second, and more important, there is a significant unsolved problem with inflation, the so-called *cosmological constant problem.* In Cosmology, the problem of the *cosmological constant* is due to the discrepancy between the observed value of the energy density of the quantum vacuum of the Universe, i.e., the low value of the *cosmological constant* $\Lambda = 10^{-52}$ m^{-2} resulting from observations and the enormous *theoretical* value of the fundamental zero-point energy of the quantum vacuum proposed by the quantum field theories. The difference is so dramatics, as to be said that the predicted value of the *cosmological constant* Λ is the worst theoretical prediction in all of the history of physics, or, the value of the *cosmological constant* constitutes physics' most embarrassing problem. Specifically, the calculated vacuum energy is huge, so that it should have expanded the Universe so forcefully and quickly that no stars and galaxies would have ever formed. Because that is clearly not the case, the vacuum energy in the Universe must be very small, about 120 orders of magnitude smaller than what quantum theory predicts…

According to quantum field theories, *each particle corresponds to a fluctuation of the corresponding quantum field*. Then, the concept of the classical vacuum, as a space that does not contain matter or energy, is differentiated. The vacuum is not really a completely empty space, since even if it does not contain any particles, the quantum fields corresponding to particles still exist. When these fields are properly excited, they create the corresponding particles. Therefore, the vacuum contains a non-zero amount of energy, the so-called *energy of the quantum vacuum*. The automatic appearance of particles as the result of energy fluctuations in the quantum vacuum does not violate the principle of conservation of energy, because it occurs within a time determined by the uncertainty principle of Heisenberg. Thus, the *cosmological constant* can be determined by the energy density of the quantum vacuum. But the theoretical value of the energy of the quantum vacuum resulting from quantum field theories is larger than the value of the *cosmological constant* by 40–120 orders of magnitude, i.e., it is 10^{40}–10^{120} times greater… This disagreement and state of affairs is described by physicists as the largest discrepancy between theory and experiment in all of science and the worst theoretical prediction in the history of physics!

Many solutions to the problem of the *cosmological constant* have been proposed, but none have been generally accepted. One proposal has to do with the *anthropic principle*: as string theory requires, perhaps we live in a Universe that belongs to a vast collection of *Multiverses* in each one of which there are different values of the vacuum energy. Such arguments of the anthropic principle that have been proposed in various forms at least as early as 1981, essentially propose that only low-energy regions of the quantum vacuum, such as the one we live in, are reasonably capable of supporting intelligent life.

The *cosmological constant* remains the best explanation for dark energy and the mysterious force causing the accelerating expansion of space. Another proposal is that dark energy isn't actually related to the *cosmological constant* or the vacuum energy. If somehow the *cosmological constant* is zero and the Universe's vacuum energy is somehow canceled out, dark energy might be the work of *quintessence*, some form of energy throughout space with a negative pressure. One version of *quintessence*, postulates an energy whose density increases with the age of the universe and results to an ultimate *big rip* wherein space is torn apart by runaway expansion until the distance between particles becomes infinite.

A consequence of the general theory of relativity is that the curvature of space k (Fig. 6.4) depends on the ratio of the density ρ to the critical density ρ_c, $\Omega = \rho/\rho_c$. A second question raised by the original *Big Bang* model is why the density of the Universe ρ is so close to the critical value of ρ_c, i.e., why does Ω take exactly the special value $\Omega = 1$? In other words, why is the Universe today observed to have a Euclidean (flat) geometry? The solutions of Friedmann and Lemaitre move easily above and below the critical value for very small density changes, from a closed, spherical Universe to an open hyperbolic Universe, so it is difficult to choose whether their particular model of the Universe is open or closed. Of all the possibilities that exist of a Universe with positive curvature (greater than critical density, $\Omega > 1$) up to a Universe with negative curvature (less than the critical density, $\Omega < 1$), the chosen state of the flat Universe $\Omega = 1$ is indeed a very *special* case. This preference for the flat Universe appears to have been even more carefully chosen as we get very close to the time of the *Big Bang*, because with time later it would increase. For example, if the density of the Universe was slightly greater than the critical one billion years after the *Big Bang*, it would have already collapsed today. Therefore, it appears that the density of the Universe is very carefully selected, already from times very close to the *Big Bang*.

The problem is that if today the parameter Ω is relatively close to unity, then the further back in time we go, the closer to unity this parameter should be. For example, one nanosecond – one billionth of a second – after the *Big Bang*, Ω should equal with 1, with an accuracy of 21 decimal places (Fig. 7.8), while in the Planck era, Ω should be equal to unity with an accuracy of 57 decimal places! But why is the density in the Universe so amazingly tuned to be so close to the critical density? In the context of the *Big Bang* Theory, such implausible coincidence of the Universe's density with the critical density such that $\Omega \cong 1$ is not explained. It is indeed a puzzle!

The theory of inflation predicts that the rapid increase in the radius of curvature of the Universe leads to the flat Universe we now observe. This is analogous to inflating a small balloon to the size of the Earth: the balloon gets flatter the more we inflate it. So, the rapid expansion of the Universe during the era of *inflation*, forces the geometry of space to effectively become flat, because in the context of its expansion any curvature in space quickly becomes undetectable. The balloon would still be curved, but the local geometry we actually see would be flat. The young Universe when it increased in radius by a large factor with the process of *inflation* became one almost *flat* Universe, like the one we now observe. Thus, the *cosmic inflation* also answers the question of the observed flatness of space.

The Problem of the Missing Magnetic Monopoles

The third cosmological question concerns *the problem of magnetic monopoles*: we should observe a large number of heavy, stable magnetic monopoles, i.e. magnetic particles with a single pole. This problem stems from grand unification theories (GUTs), which propose that at very high temperatures, the electromagnetic, strong, and weak forces are not independent fundamental forces but arise from the spontaneous symmetry breaking of a unified superforce. These theories predict that during the process of symmetry breaking, various stable and large-mass particles are produced, which nevertheless are not observed today. The most notorious of these particles are the magnetic monopole. These monopoles should have survived to this day. The special charm of the magnetic monopole lies in the fact that it has a huge mass, condensed into a tiny volume and a huge energy density, which is created by the Higgs field – just like the artificial pseudo-vacuum bubbles. However, they don't seem to exist! Why;

One solution to this puzzle could be that since the inflationary period that occurred later, at temperatures below those of the GUT symmetry breaking and monopole production, the result is that the density of magnetic monopoles has been *diluted* by several orders of magnitude and so we do not detect them today. In other words, even if there was produced a large number of monopoles in the early Universe, because the visible to us Universe has been produced from the huge expansion of a tiny volume, it is statistically impossible that there was even a small number of monopoles in this volume!

The Epoch of the Electroweak Interaction and the Primordial Hot Sea of Quark-Gluon Plasma [10^{-32} Secs $\lesssim t \lesssim 10^{-12}$ Secs, 10^{22} K $\gtrsim T \gtrsim 10^{15}$ K, 10^9 GeV $\gtrsim E \gtrsim 150$ GeV]

With the end of the era of *inflation*, the Universe was mostly made up of energy in the form of *photons*. The particles that were present at that time could not bind into larger stable particles because of the enormous energy densities of that time. The temperature was still too high for quarks to coalesce into hadrons, and a quark–gluon plasma persisted. We hypothesize that during this era, there was a collection of quarks and antiquarks together with their exchange particles (*gluons*), a situation described as a *hot sea of quarks*. During the period from 10^{-32} s to 10^{-12} s, when the temperature of the Universe dropped from 10^{22} to 10^{15} degrees, we have three *distinct* interactions, the *gravitational*, *strong*, and *electroweak*. In this period, the carriers of the electroweak force also appear, the $W^{\pm}$ and Z_o bosons, with finite mass around 100 GeV. This event may have occurred at about 10^{-32} s after the *Big Bang* when the potential energy of the field of inflation, *the inflation*, was released flooding the Universe with a dense and hot quark-gluon plasma. As we mention, particle interactions at this time were sufficiently high-energy to produce a large number of exotic particles, such as the stable $W^{\pm}$ and Z_o bosons, as well as the Higgs bosons.

Owing to a phenomenon known as *color confinement*, quarks are never found in isolation; they can be found only within hadrons, which include baryons, such as protons and neutrons and mesons, or, in quark–gluon plasmas.

This is the period that we might think of the Universe was in the form of a *primordial soup*, where every possible collision that we can imagine was taking place. As a result, no bound structures exist and no stable configurations. As the Universe continued expanding, the available energy for the interactions decreased, and at about 10^{-12} s, the $W^{\pm}$ and Z_o bosons ceased to be produced. The remaining $W^{\pm}$ and Z_o bosons decayed and the weak interaction converted into the short-range force of the following epoch *of the quarks*.

It is noted that the physics of the *electroweak era* is less hypothetical and much better understood than the physics of the earlier periods of the infant Universe, thanks to the theoretical studies of the unification of the electromagnetic and weak nuclear forces, notably by Glashow, Salam and Weinberg (1967). The existence of the $W^{\pm}$, Z_o and Higgs bosons has been fully verified today, as well as other predictions of the electroweak theory which have been tested experimentally. The basic theory that describes this era is the (unevolved) *Standard Model.*

In summary, at the end of the *electroweak period* (10^{-32} s to 10^{-12} s), the electromagnetic and weak forces undergo the final symmetry breaking, ending the era of unification of the *electroweak* interactions at about 10^{-12} s after the *Big Bang*.

The Quark Epoch [10^{-12} Secs $\lesssim t \lesssim 10^{-6}$ Secs, 10^{15} K $\gtrsim T \gtrsim 10^{13}$ K, 150 GeV $\gtrsim E \gtrsim$ 1 GeV]

The *quark epoch* is the period of evolution of the infant Universe, where the fundamental interactions of gravity, electromagnetism, strong and weak had taken their present forms, but the temperature of the Universe was still high enough to allow quarks to combine to form the bound states of hadrons, as the energies and densities were too great. The Universe is flooded with a *sea of quark-gluons*.

The era of quarks began at about 10^{-12} s after the *Big Bang*. This is the time when the previous era of the electroweak interaction, separated into the weak and electromagnetic interactions, ended. During the *quark era,* the Universe was flooded with a dense and hot quark-gluon plasma, which contained quarks, leptons and their antiparticles. However, the collisions between these particles were of extremely high energy and thus did they allow quarks to combine into the bound states of *mesons*, or *baryons*. These are the higher energies we can reach (TeV) today at the large Hadron Collider accelerator of CERN.

We classify the elementary particles into two categories, *hadrons* and *leptons*. Their difference lies in the fact that the former can participate in addition to the strong interaction, while leptons do not. *Hadrons* are divided into two subcategories: *baryons* which are *fermion* particles (with half-integer spin) and *mesons* which are bosons (particles with integral spin). *Hadrons* were so named from the Greek word αδρός (= strong), contrasted to *leptons*, named from the Greek word λεπτός (= thin).

In 1964, the American physicist Murray Gell-Mann proposed that protons and neutrons, also called *nucleons*, are not the basic building blocks of matter, but instead they consist of other smaller particles, which he called quarks.[11] *Quarks* are now considered as the fundamental elementary building blocks of matter, and are the ingredients of the *baryons* (consisting of a combination of three quarks) and *mesons* (consisting of a combination of two quarks). *Gluons* on the other hand, are elementary particles (bosons) with spin 1, zero rest mass and zero electric charge, which carry the

---

[11] In 1969 Gell-Mann was awarded the Nobel Prize in Physics for his contributions and discoveries on the classification of elementary particles and their interactions.

strong interaction (i.e., are the exchange particles of the strong interaction) via which quarks interact. They are named gluons (glue = something that glue) because they are the means by which the quarks stick to each other. This interaction is described by *quantum chromodynamics.*

When the temperature dropped below 10^{15} degrees, the four interactions, gravity, strong, electromagnetic, and weak, activated separately. Radiation-dominated matter and photons were very energetic, so that we had the formation of particle/antiparticle pairs, e.g., electrons and positrons, as long as the energy of the photons exceeded twice the rest mass of the pair's particles. These pairs were in thermal equilibrium with the radiation and could, after their formation, immediately transform back into radiation.

The *quark era* began about 10^{-12} s after the *Big Bang*. It ended when the infant Universe was about 10^{-6} secs old and the average energy of particle interactions fell below the binding energy of quarks into hadrons. The next period, when the quarks were bound into the hadrons, is known as the *epoch of hadrons*.

The Hadron Epoch and Matter–Antimatter Asymmetry [10^{-6} Secs $\lesssim t \lesssim 1$ s, 10^{13} K $\gtrsim T \gtrsim 10^{10}$ K, 1 GeV $\gtrsim E \gtrsim 1$ MeV]

At about 10^{-6} secs, the temperature of the Universe was about 10^{13} degrees, matter and antimatter were in thermal equilibrium with each other and the Universe was a cosmic soup of free quarks. The size of the Universe was the size of our solar system. Due to further expansion and cooling of the Universe, after these 10^{-6} secs, the quarks could no longer travel around free but could combine together in pairs, or triplets, forming *baryons.*

At the end of the *Hadron Epoch*, most hadrons and anti-hadrons experience annihilation reactions. The few survivors were matter, and hence this epoch is the epoch of *baryogenesis*, wherein antibaryons were annihilated. At the end, there are ~1 billion photons left over for every baryon, but no surviving antibaryons.

From this time on, an asymmetry was established between the numbers of baryons and antibaryons, which was due to the decay of a hypothetical X-boson created by an asymmetry in the electroweak interaction. X- and Y-bosons are elementary particles predicted by the Grand Unified theory (GUT) of Glashow and Giorgi, as analogs of the W- and Z-bosons. The X- and Y-bosons mediate in the quark-lepton interaction and allow the baryon number conservation to be violated, thus allowing the proton to decay. With the decay of the X-boson, for every one billion (10^9) antiparticles, one billion plus one ($10^9 + 1$) particles were created.

With a further drop in temperature, the stable states of protons and neutrons were formed from the quarks, but they annihilate from their antiparticles. For every proton in the Universe, there were 10^9 pairs of particles/antiparticles which decayed leaving behind one billion photons. So, because of this fortunate asymmetry in the number of baryons and antibaryons, one in a billion baryons survived, not having an antiparticle to annihilate. If this baryon-antibaryon asymmetry didn't exist between matter and antimatter, each particle would collide with its antiparticle, e.g., a proton with an antiproton, and thus they would annihilate. So, all that could be left would be a sea of photons. That is, if this asymmetry did not exist there would be no matter in the Universe today and we would not even be here!

The Lepton Epoch [1 s $\lesssim$ t $\lesssim$ 10 Secs, 10^{10} K $\geq$ T $\gtrsim$ 10^9 K, 1 MeV $\gtrsim$ E $\gtrsim$ 100 keV]

About a few seconds after the *Big Bang*, the Universe has a temperature of the order of a few billion degrees. The sphere of space that will become the observable universe is approximately 10 light years in radius at this time as, i.e., the distance of the nearest star (Proxima Centauri). Neutron/proton ratio freezes at approximately 1/2. However, the available temperature and corresponding energy was not enough to maintain electron–positron pairs. In other words, similarly with protons, only one in a billion of electrons survived. So, approximately a few seconds after the *Big Bang*, the temperature of the universe had fallen to the point where electron–positron pairs were gradually annihilated. The neutrinos, meanwhile, have lost enough energy and have not interacted with the rest of the particles since then, traveling vast distances without actually interacting, forming the cosmic neutrino background.

At the end of the lepton epoch, most Leptons and anti-leptons, such as electrons and positrons experience annihilation reactions. Antimatter was again underrepresented in the survivors, and hence this epoch is the epoch of *leptogenesis*, wherein antileptons were annihilated.

During the lepton epoch, in the interval from t $\approx$ 1 s to t $\approx$ 10 s, leptons are formed and dominated the mass of the Universe. Hence, overall the particle epoch in the Universe lasts for the time interval from t $\approx$ 10^{-12} to t $\approx$ 10 s after the *Big Bang*. The lepton era began about a few seconds after the *Big Bang*, when the majority of hadrons and anti-hadrons annihilated each other at the end of the epoch of hadrons. During the lepton epoch the temperature of the Universe was still high enough to create lepton/anti-lepton pairs, so that leptons and anti-leptons were in thermal balance. About 10 s after the *Big Bang*, the temperature of the Universe had collapsed to the point where no more lepton/anti-lepton pairs were created and most leptons and anti-leptons were annihilated with each other, leaving behind a small surplus of leptons. Then, the total energy of the Universe was dominated by photons as the evolution of the Universe is in the age of radiation.

Transition from the Radiation to the Matter Era at t = 47,000 Years (3000 $\gtrsim$ z $\gtrsim$ 0.5)

The *radiation age* is the period in the evolution of the infant Universe where photons dominated its energy. The *radiation era* began when most leptons and anti-leptons decayed at the end of the *lepton era*, about 10 s after the *Big Bang*.

Epoch of Nucleosynthesis [10 s $\lesssim$ t $\lesssim$ 10^3 s, 10^9 K $\gtrsim$ T $\gtrsim$ 10^7 K, 100 keV $\gtrsim$ E $\gtrsim$ 1 keV]

The atomic nuclei of the *light* elements were created in the process of Big Bang *nucleosynthesis* by nuclear fusion, that occurred during the first few minutes of the epoch of *radiation*, wherein helium, and trace amounts of deuterium, helium-3, and lithium-7 nuclei formed from protons and neutrons. At about 20 min, nuclear fusion ceased and normal matter consisted of 75% hydrogen nuclei and 25% helium nuclei, while free electrons begin scattering light. There weren't a lot of the heavier elements that life as we know it today depends on. You can't get oxygen without fusion inside a star. Our own Solar System is the result of several generations of supernovae that exploded, seeding our region with heavier and heavier elements.

Specifically, before the *recombination time* and the *dark ages* of the Universe that we discussed earlier and in the period before 47,000 years after the Big Bang, we have according to the Standard Cosmological model the absolute dominance of radiation over matter, the age of *radiation*. The Universe is a fully ionized plasma consisting of free electrons, protons, and neutrons. There are no atoms because the energies of the constituent particles are so high that the electrons with impacts are knocked off the atoms. The protons combine with other protons to create the hydrogen isotope deuterium. The temperature has dropped to about 1 million degrees, where deuterium is stable and combines with neutrons and protons to form helium nuclei consisting of 2 protons and 2 neutrons. The ratio of deuterium to helium isotopes created during the period is exactly what we observe today in the stars, an important observational fact that supports the *Big Bang* theory.

Photon Epoch [t $\lesssim$ 380.000 Years, 10^9 K $\gtrsim$ T $\gtrsim$ 4000 K, 100 keV $\gtrsim$ E $\gtrsim$ 0.4 eV]

At the end of this epoch, the spherical volume of space which will become the observable universe is about 300 light years in radius (Fig. 6.40), baryonic matter density is on the order of 4 g per m^3 (about 0.3% of sea level air density) – however, most energy at this time is in electromagnetic radiation. This lasts until the Universe reaches an age of about 380,000 years, and a temperature of merely ~ 4000 K. discussed above.

References

Alpher RA, Hermann R (1948) Evolution of the Universe. Nature 162(4124):774–775

Datasets (2008) Astrophys J 686:749–778

Dunbar DNF, Pixley RE, Wenzel WA, Whaling W (1953) Phys Rev 92:649

Dunkley J et al Five-Year Wilkinson Microwave Anisotropy Probe (WMAP) Observations

Einstein (1919) What is the theory of Relativity? In German history in documents and images, Volume 6. Weimar Germany (November 28, 1919), Source of original German text: Albert Einstein, Mein Weltbild (My Worldview). Querido Verlag, Amsterdam, pp 220–228

Eisenstein, D. J., et al. 2005. Detection of the baryon acoustic peak in the large-scale correlation function of SDSS luminous red galaxies. *Astrophysical J* 633: 560–574.

Epelbaum E, Krebs H, Lähde TA, Lee D, Meißner UG (2012) Structure and rotations of the Hoyle State, U.G. Phys Rev Lett 109:252501

Freese K, Gondolo P, Spolyar D (2007) The effect of dark matter on the first stars: A new phase of stellar evolution, astro-ph. arXiv:0709.2369

Friedmann A (1922) On the curvature of space. Zeitschrift für Physik 10:377–386

Friedmann A (1924) Über die Krümmung des Raumes. Zeitschrift für Physik 21:377–386

Gamow G (1948) The evolution of the Universe. Nature 162:680–682

Hubble E (1929) A relation between distance and radial velocity among extra-galactic Nebulae. In: Proceedings of the National Academy of Sciences of the United States of America, vol 15, Issue 3, pp 168–173

Hubble EP (1936) Realm of the Nebulae. Yale University Press, New Haven

Hubble, Humason ML (1931) The velocity-distance relation among extra-galactic Nebulae. Astrophys J 74:43

Hubble, Humason ML (1934) The velocity-distance relation for isolated extragalactic Nebulae. Proc Nat Acad Sci 20:264

Humason ML (1936) The apparent radial velocities of 100 extra-galactic nebulae. Astrophys J 83:10

Kowalski M et al Improved cosmological constraints from New, Old and Combined Supernova

Krolewski A, Percival WJ, Woodfinden A (2025) A new method to determine the H_0 from cosmological energy-density measurements. Phys Rev Lett 134:101002

Kuo CL, 13 coauthors (2004) High resolution observations of the CMB power spectrum with ACBAR. Astrophys J 600:32–51

Lemaître G (1931) A homogeneous universe of constant mass and increasing radius accounting for the radial velocity of extragalactic nebulae (a translation of his 1927 paper). Mon Not Roy Astron Soc 91:483–490

Lemaitre G (1927) L' Univers en expansion, Annales. Soc Sci Bruxelles, A 47:49

Lemaître G (1945) L' hypothèse de l' atome primitive. Act Soc Helv Sci

Likelihoods and Parameters from the WMAP data. 2009. *Astrophys J Suppl* 180: 306–329.

Lineweaver C, Davis TM (2005) Misconceptions about the Big Bang. Sci Am 292(3):36–45

Loeb A (2002) Long-term future of extragalactic astronomy. Phys Rev D 65(4):047301

Nanopoulos DV, Olive KA, Srednicki M (1983) After primordial inflation. Phys Lett B 127(1–2):30–34

Nicastro F, Kaastra J, Krongold Y et al Observations of the missing baryons in the warm–hot

Penzias AA, Wilson RW (1965) A measurement of excess antenna temperature at 4080 Mc/s. Astrophys J Lett 142:419–421

Perlmutter S et al (1999) Measurements of Ω and Λ from 42 high-redshift supernovae. Astrophys J 517:565–586. arXiv:astro-ph/9812133v1

Rajguru N, 34 more authors (2005) CMB observations from the CBI and VSA: A comparison of coincident maps and parameter estimation methods. Mon Not Roy Astron Soc 363:1125–1135

Riess, A. G., S. Casertano, W. Yuan, et al. 2018a. New parallaxes of galactic Cepheids from spatially scanning the hubble space telescope: implications for the Hubble constant. *The Astrophysical Journal* 855 (2): 136.

Riess AG, 19 coauthors (1998) Observational evidence from supernovae for an accelerating universe and a cosmological constant. Astrophys J 116(3):1009–1038

Riess AG, Casertano S, Yuan W et al (2018b) A 2.4% determination of the local value of the Hubble constant. Astrophys J 861:126

Sachs, R. K., and A. M. Wolfe. 1967. Perturbations of a cosmological model and angular variations of the microwave background. *The Astrophysical Journal* 147: 73.

Silvestri, A., and M. Trodden. 2009. Approaches to understanding cosmic acceleration. *Report Prog Phys.* 72: 096901.

de Sitter W (1917) On Einstein's theory of gravitation and its astronomical consequences. Third Paper. Mon Not Royal Astron Soc 78(1):3–28

de Sitter W (1916) On Einstein's theory of gravitation and its astronomical consequences. Second Paper. Mon Not Royal Astron Soc 77(2):155–184

de Sitter W (1916) On Einstein's theory of gravitation and its astronomical consequences. First Paper. Mon Not Royal Astron Soc 76(9):699–728

Slipher, V. M. 1917. Radial velocity observations of spiral nebulae. *The Observatory* 40: 304–306.

Smoot GF, 19 coauthors (1992) Structure in the COBE differential microwave radiometer first-year maps. Astrophys J Lett 396:L1

Spergel DN, 21 coauthors (2007) *T*hree-year wilkinson microwave anisotropy probe (wmap) observations: Implications for cosmology, Astrophys J Suppl 170:377

Wright EL, 20 coauthors (1992) Interpretation of the cosmic microwave background radiation anisotropy detected by the COBE differential microwave radiometer. Astrophys J Lett 396:L13

Zwicky, F. 1933. *Die Rotverschieb ung von extragalaktischen Nebeln* (The redshift of extragalactic nebulae). *Helvetica Physica Acta* 6: 110–127.

Chapter 7
Fine-Tuning of the Cosmic Parameters and Earth's Place in the Universe

Κάλλιστον Κόσμος,
ποίημα γαρ Θεού
[Cosmos is beautiful,
since it was created by God]

Thales the Milesian, 643-548 BC.

The heavens declare the glory of God and the skies proclaim the work of His hands.
Psalm 19:1-6

Our position may not be necessarily central to the Universe,
but it is inevitably privileged, to some extent.

Brandon Carter, theoretical astrophysicist, FRS (1973)

Concluding Remarks

In the previous chapters of the book at hand, starting from our base on this little blue jewel in Space, our Earth, we have wandered with surprise and admiration, throughout several of the far corners of our richly endowed with impressive phenomena, vast Universe. We watched the fiery genesis and first infinitesimal moments of the Universe and its subsequent rich evolution with many wonderful phenomena. At the same time, we discussed some of the great discoveries in heliospheric and space science, astrophysics, and cosmology, from the dawn of the last century to the present era. Hence, we obtained a taste of the many surprises that our Universe constantly has in store for us, with *the simplicity of its complexity* and the inexhaustible discoveries that are revealed one after the other.

K. Tsinganos, *A Century of Cosmic Discoveries*, Springer Praxis Books,
https://doi.org/10.1007/978-3-032-06282-6_7

Fig. 7.1 *God the Creator of the Universe*. The compass in this thirteenth-century manuscript is a symbol of God's act of *Creation*, since God has created the Universe after geometric and harmonic principles. God the Son creates the *Cosmos* by using a compass with one arm on the *«center»* of the Cosmos and the other on the *«periphery», while one of His foot lies outside of the Universe*. (*Source* Illustrated manuscript in the Austrian National Library, Vienna, ca. 1220–1230 AD, courtesy of Wikipedia Commons)

As *thinking inhabitants* of our Universe, we wish on the one hand to understand our position and relationship to the wider cosmic environment in which we live, and on the other, we often seek to delve into many of our own – perhaps with some metaphysical overtones – inner questions. This is of course especially true in the western cultural ecosystem, with the rich heritage of the ground breaking epistemological thinking of the ancient natural philosophers, that we started with. Since here, under a unique night sky decorated with stories and myths, or, walking in silence along the shores of the seas, similar fundamental questions were first raised 2–3 millennia ago by natural philosophers, and answers were sought, albeit without the assistance of the tools of today's science and technology. Similarly today, in some of these questions put forward first by the truth-seeking minds of those ancient thinkers, it would not be an exaggeration to say that space science, astrophysics, and cosmology continue to contribute catalytically today with their answers.

In the following we have chosen to list some specific scientific *facts* that may help us to answer fundamental questions about our Cosmos which indeed seems to be governed by harmonic principles. After all, the word *Cosmos* by itself means something that is *ornately arranged,* because the ancients believed that our Universe was perfectly harmonious and impeccably put in order. Hence, in this concluding chapter, we shall list some fundamental relevant scientific questions and facts, in the simplest possible way, so that the *thinking reader* may deduce by himself some sought-after useful answers.

The Physical Laws and the Values of the Cosmical Constants

Today we understand that the Universe is governed by some well-defined physical laws. At the same time, these laws of physics contain some constants, such as, the speed with which light propagates, the magnitude of the electric charge and the mass of the electron, the value of the gravitational constant of classical physics, and Planck's constant of quantum physics. The values of some of these constants come from the physical laws themselves, and the speed of light, for example, is derived from Maxwell's equations of electromagnetism. However, the magnitude of others seems to be orphaned and inexplicable. The physical laws would still hold just as well if these constants had different values, although many physical phenomena might appear radically different.

At the same time, the study of the Universe reveals the existence of some *universal numbers*, which constitute, together with the *physical laws* and the *natural constants*, the basis of its fundamental physical properties. It seems that every one of these universal numbers has a precisely *tuned* value that was required so that we evolved as *homo sapiens* and subsequently wonder how it was possible for intelligent life to flourish in our own corner – and possibly in others – of this Universe that hosts us. If some of these universal numbers were different, even to the smallest degree, it seems that galaxies, stars, planets, etc., with complex chemical elements and molecules and consequently life would never have been created, as at least we know on Earth today.

Therefore, all these *cosmic numbers* are critical to the formation of the basic building blocks of life. It is also interesting that the range of their possible values is actually very *narrow*. Outside and beyond that range in a possible fluctuation of their magnitudes, the existence of intelligent life seems likely to be rather thin.

As Stephen Hawking writes in his popular book *A Brief History of Time*, "The laws of physics, as we know them today, contain many fundamental numbers, such as the magnitude of the electric charge of the electron, or, the ratio of the proton–electron masses. We cannot, at least for now, predict the values of these numbers through our theory. We have to find them through observation. Perhaps one day we will discover a complete *unified theory of everything* that will predict all these numerical quantities. The remarkable fact, however, is that the values of these numbers seem to be rather precisely defined, so as to make possible the development of life."

The *Universality* and *Complementarity* of the Four Fundamental Interactions of Nature

Our walking down the street, the launch of a rocket and the orbital motion of a satellite around Earth as well as the annual motion of Earth around the Sun, or the faithful monthly movement of the Moon around Earth, the electric lighting of our homes, how a magnet sticks to our refrigerator, the production of light in the bowels of the Sun and the rest of the stars, the production of medical X-rays, the operation of nuclear plants as well as the destructive atomic weapons, *all* these physical phenomena are strictly determined by the *four* fundamental forces of nature. These interactions, which we feel and which permeate all the constituents of matter around us and also far beyond, in the entire Universe as we know it today, are in decreasing relative strength the following:

- the *strong nuclear force,* which is exerted between *quarks*, the elementary building blocks of matter and also the constituents of the nuclei, protons and neutrons; thus, it is the interaction subsequently responsible for the existence of the nuclei of atoms;
- the *electromagnetic force*, which is exerted between *electrically charged* particles and is responsible for the interaction of light with matter, for our electrical lighting, but also for determining the shape of our bodies;
- the *weak nuclear force* that acts in the depths of the atomic nuclei and is responsible for the *radioactivity* of certain elements, as well as the production of thermonuclear energy in the Sun and stars;
- the *gravitational force* which is exerted between any bodies having mass; it is responsible for the formation of planets, satellites, stars, galaxies and all concentrations of matter in the Universe.

To those four forces, we could add the existence of some newfound and hitherto with unknown details for their operation, interactions: those that govern *dark matter* and are related to the cosmic *dark energy*. However, from their naming it is evident that we currently do not have any understanding of what really they are, except only for some of their – indeed important – worldly effects.

The Universality and the Complementarity of the Fundamental Forces

The *universality* of those fundamental forces of nature, i.e., the fact that they govern all phenomena occurring on our little planet, Earth, as well as far away in the distant parts of the Universe, is an amazing fact. Also, there is *complementary* of the forces which work together without competition leading to the observed harmony of the World thereof. This *universality* and *complementarity* make us wonder because, in principle, things could not be exactly like that.

Why the *Fine-Structure Constant* Has the Value α ≈ 1/137?

The fine-structure constant **α** is a fundamental physical constant that characterizes the *strength of the electromagnetic interaction* between charged elementary particles. It is equal to the quotient of the square of the value of the electric charge of the electron (e) and the value of the product of the uncertainty constant of quantum mechanics (ħ)[1] and the speed of light (c), $\boldsymbol{\alpha} = e^2/\hbar c \approx 1/137$ (Fig. 7.2).

Actually the current estimate of the fine-structure constant is $\boldsymbol{\alpha} \approx 1/137.035999166 = 0.007\ 297\ 352\ 5693$ (Fan et al. 2023), with an uncertainty on the last two digits. The numerical value of the number **α** does not arise theoretically, and no one knows why it has this particular value. Currently, *it is one of the most mysterious numbers in the Universe*! It is determined **only experimentally** by the

Fig. 7.2 Bust of Arnold Sommerfeld (1868–1951) at the Ludwig-Maximilians-Universität Munich, at Theresienstrasse 37 by Theodor Georgii [Courtesy of Wikipedia Commons]

[1] This constant ħ appears in Heisenberg's *uncertainty principle*, which says that we cannot know both the position and speed of a particle, such as a photon or electron, with perfect accuracy.

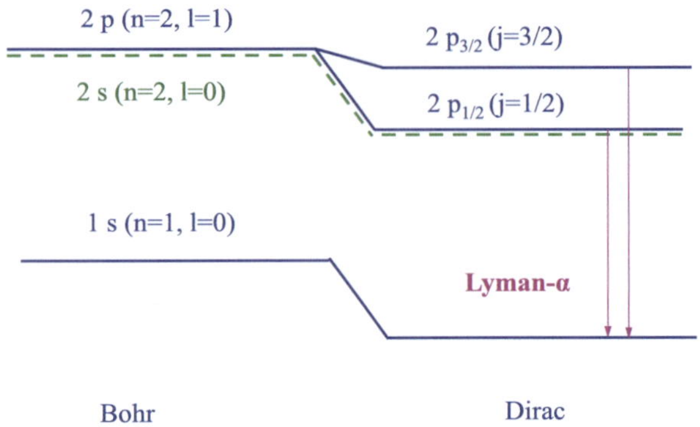

Fig. 7.3 Fine structure of the Lyman-α line of the hydrogen atom results from the coupling of the electron's orbital angular momentum (quantum number l) with its intrinsic angular momentum (spin, s). Thus, two transitions arise from the two energy levels [$2p_{3/2}$ (n=2, l=1, j=l+s=3/2] and $2p_{1/2}$ [(n=2, l=0, j=l+s=1/2] to the fundamental level 1s (n=1, l=0). The corresponding wavelengths 1215.674 and 1215.668 Angstroms (Å) are in the ultraviolet part of the electromagnetic spectrum. The resulting two spectral lines have a very small wavelength separation δλ of the order of $\boldsymbol{\alpha}^2$. It was first measured precisely for the hydrogen atom by Albert A. Michelson and Edward W. Morley in 1887, laying the basis for the theoretical treatment by Arnold Sommerfeld, introducing the fine-structure constant **α**. From this experimental fact originated the name fine structure of the spectrum and fine-structure constant **α**! [Courtesy of Wikipedia Commons]

separation Δλ of the wavelength λ of the corresponding two spectral lines of the hydrogen atom, for transitions of the outer electron (Fig. 7.3). This line separation corresponds to the two energy levels created by the coupling of the *orbital* angular momentum of the electron's motion around the nucleus (l) and the two values of the possible orientations (up/down) of its *spin* ($s = 1/2$ and $s = -1/2$), through relativistic corrections (Dirac) to the Schrödinger equation.

The fine-structure constant, **α**, was theoretically introduced by Sommerfeld (1916), Fig. 7.2, as the ratio of the speed of the electron in the first circular orbit of the relativistic Bohr atom to the speed of light in vacuum. Note also that the size of a hydrogen atom is a factor $\boldsymbol{\alpha}^{-2} \approx 20000$ times the size of the *classical electron radius*. Additionally, the number $Z = 137$, the inverse of the fine-structure constant, corresponds to the maximum possible positive charge of an atomic nucleus (i.e., one with atomic number $Z = 137$) that allows the existence of a stable orbit for electrons around the nucleus, according to the special theory of relativity and the Bohr atomic model.[2] Atoms with $Z > 137$ are theoretically not allowed to exist, as indeed it happens.

So far, most astrophysical experimental data agree that the value of the fine-structure constant, **α,** is truly time-invariant in the Universe and is therefore indeed a constant. For example observations of the redshift $z = 7.085$ quasar J1120 + 0641 have been used to search for variations of the fine-structure constant, **α**, over the

[2] For a relatively simple proof of this result, see Modern Theoretical Mechanics, by K. Tsinganos, Unibooks, Athens (2021).

redshift range 5.5–7.1, Wilczynska et al. (2020). These observations at $z = 7.1$ probe the physics of the Universe when it was only 0.8 billion years old.

It is noteworthy that the expression of the fine-structure constant $\boldsymbol{\alpha} = e^2/\hbar c \approx 1/137$ simultaneously involves *electricity* (due to the electric charge), *relativity* (due to the speed of light), and *quantum physics* (due to Heisenberg's uncertainty constant of quantum mechanics). Truly, a more wonderful and simple combination and representation of all areas of physics to make this mysterious dimensionless constant, I think could not exist!

Why Does the Fine-Structure Constant α Have its Particular Numerical Value?

Solid matter, and therefore life and intelligent beings, could not exist if the value of **α** was slightly different. If the value of this constant **α** had been slightly different, then stars would not have been able to sustain the nuclear reactions in their core that produce carbon, oxygen, etc. As we discussed in Chap. 6 in the section 1954: *Hoyle's prediction for the creation of carbon*, Fred Hoyle and others noted that the abundance of carbon in the Universe could be explained only if **α** had a specific value that made the nuclei of helium atoms far more likely to fuse to produce carbon nuclei than they otherwise would be. For example, increase the denominator of **α** from 137 to 138, and you decrease the value of **α** from about 0.00729 to about 0.00724, i.e., by 0.00005: this is enough, to prevent stars from creating carbon, thus halting the creation of life as we know it. As a result, there would not have been any carbon-based life forms in our Universe.

The strong force must be tuned to a precision of at least 0.5%, and the electromagnetic force to a precision of at least 4%, to prevent either carbon production or oxygen production from dropping significantly (Livio et al. 1989).

Feynman on the Fine Structure Constant α

In quantum electrodynamics (QED), **α** is directly related to the coupling constant determining the strength of the QED interaction between *electrons* and *photons*. **The theory does not predict its value!**. Therefore, **α** must be determined experimentally. In fact, **α** is one of the empirical parameters in the Standard Model of particle physics, whose value is not determined within the Standard Model. Richard Feynman, one of the originators and early developers of the theory of quantum electrodynamics, referred to the fine-structure constant in these terms: «There is a most profound and beautiful question associated with the observed coupling constant – the amplitude for a real electron to emit or absorb a real photon. It is a simple number that has been experimentally determined to be close to 137.03597 with an uncertainty of about 2 in the last decimal place. It has been a mystery ever since it was discovered more than fifty years ago, and all good theoretical physicists put this number up on their wall and worry about it. Immediately you would like to know where this number for the constant coupling comes from: is it related to π or perhaps to the base of natural logarithms e? Nobody knows. It's *one of the greatest damn mysteries of physics*: a magic number that comes to us with no understanding by humans. You might say the *hand of God* wrote that number, and *we don't know how He pushed His pencil.* We know what kind of a dance to do experimentally to measure this number very accurately, but we don't know what kind of dance to do on the computer to make this number come out – without putting it in secretly» (Feynman 1985)!

It is thus not obvious at all why this constant has this particular value. Why about 1/137 and not any other value? One might think the question is meaningless, but it is not. If the value of this constant had been slightly smaller or larger, even by as little as 4%, then stars would not have been able to sustain the nuclear reactions in their core that produce carbon, oxygen, etc. As a result, there would not have been any carbon-based lifeforms in our Universe. Therefore, the question why $\boldsymbol{\alpha} \approx 1/137$ is not completely irrelevant (Nath 2015).

At the beginning of the twenty-first century, many physicists including Stephen Hawking (1988) and Martin Rees (2001) started discussing the idea of *Multiverses*. In this context, the *fine-structure constant* was one of the universal constants leading to the idea that we might live in a rather *finely tuned* Universe.

Note also that the *anthropic principle* is another argument about why the fine-structure constant has the value it does: stable matter and therefore life and intelligent beings, could not exist if the value of $\boldsymbol{\alpha}$ was slightly different. One example is that if modern *grand unified theories* are correct, then $\boldsymbol{\alpha}$ needs to be between around 1/85 and 1/180 in order to have proton decay to be slow enough for life to be possible (Barrow 2001).

Some Remarkable Characteristics of the Strong Nuclear Force

We could relate the four interactions of nature according to their relative strength, starting with the pair of the strongest of them, i.e., the strong nuclear force in relation to the next in strength, the electromagnetic force. Thus, if we scaled the strength of the strong nuclear force with 1, then the strength of the electromagnetic force would be $\approx 10^{-2}$, i.e., about 100 times weaker. More precisely, this relative strength equals to the fine-structure constant $\boldsymbol{\alpha} \approx 1/137 \approx 7.3 \times 10^{-3}$, as we discussed above. The strong nuclear force (related to the nuclear energy), as its name suggests, is the most powerful of all forces of nature. As we mentioned, the strong nuclear force is about 137 times stronger than the next one, the electromagnetic force which holds electrons around atoms, 10^{13} times stronger than the weak nuclear force, which governs radioactive decay and 10^{38} times, i.e., one hundred trillion trillion trillion times stronger than gravity,[3] which keeps us pinned up on Earth and makes Earth to revolve around the Sun. As we have mentioned above, the particles that *glue* the elementary components of quark matter to each other are called *gluons*. The gluons, the *massless* quantum particles that mediate the strong interaction, interact with quarks and other gluons through a type of charge called *color*. This *quantum color* plays the role of the electric charge in electromagnetism. Gluons are the carriers of the strong force but they appear 8 independent color modes. In each hadron, however, the combination of colors of the component quarks is such that the hadron appears finally *colorless*! This is the way, the strong nuclear force assembles quarks into elementary particles, such as protons, neutrons, and hadrons.

The nuclei of atoms contain charged protons which repel each other electrically with enormous repulsive forces from each other, because they all have the same positive charge. The strong nuclear force is an attractive force between protons and neutrons that holds them together at the core. The nuclear force dominates the tiny atomic dimensions. But for it to work, the nuclear particles must be very close to each other, which means that their range is very short, i.e., about 1.5×10^{-15} m. The strong nuclear force therefore only works when subatomic particles come very close to each other. Specifically, they must be located at distances of 10^{-15} m between them, roughly the diameter of a proton.

For example, two protons are subject to two forces: the attractive nuclear force and the repulsive electrostatic force due to their electric charge. When these are very close to each other, the attractive nuclear force dominates, while when they are at a great distance from each other, the electrostatic force dominates. At an intermediate distance, of course, the two protons experience zero net force because the two opposing forces are equal. For the given electrical charges of the protons, the attractive force felt by two protons when they are deep inside the nucleus, with a range of action of the strong interactions of the order of 1.5×10^{-15} of the meter, is 137 times stronger of their mutual electric repulsion. That is why we say that the electromagnetic force is 137 times weaker than the strong nuclear force.

Nevertheless, the strong force also has some rather peculiar characteristics, because unlike the other fundamental forces, it weakens as subatomic particles get closer to each other! In fact, according to experiments in large accelerators, it has its maximum power when the particles are rather distant from each other. Acting like a *spring*, once the particles are within range, the massive charged bosons called *gluons* transmit the strong force between the quarks and thus keep them to stick together. A small fraction of the strong force called *residual strong force* acts between protons and neutrons. Although the protons in the nucleus repel each other because of their equal electric charge, this residual strong force can overcome the repulsion so that the particles remain bound within the nucleus of an atom. According to the theory of quantum chromodynamics, the strong nuclear force does not develop between nucleons, but rather between quarks of different colors. That is why it is also called *color interaction*. The strong nuclear force governs and controls the nuclear fission which takes place in nuclear plants and the atomic bomb.

---

[3] Assuming the respective forces between two protons.

If the Strong Nuclear Force Was 1–2% Stronger, Life on Earth Would not Exist.

If the strong nuclear force were (1–2)% stronger than it actually is and the other forces held their present strength, the *diproton* – one of the many *unstable* isotopes of the element helium which contains in its nucleus only two protons – **would be stable**! In this case, *diproton* formation would proceed one-way, instead of the formation of deuterium and then helium, carbon, etc., through the proton–proton chain and then the 3-α reaction, which we believe that takes place inside the Sun. Doing so, it would drastically change the physics of the stellar interiors and possibly preclude the existence of Earth-like life.
One of the necessary conditions for the emergence of life as we know it is the existence of hydrogen which forms together with Oxygen water (H2O). Additionally, the existence of the Sun itself that supports life and sustains its weight for several billion years burning hydrogen, would not exist if there was no hydrogen. In other words, the **stability of the *diproton* would short-circuit the slow fusion of hydrogen into deuterium**. After all, hydrogen could fuse so easily into a *diproton* and it would be possible that its entire quantity would had been already consumed in the first minutes after the Big Bang…

Why Are There *Exactly* Equal Numbers of Positive and Negative Electric Charges in the Universe, $Q_{tot} = 0$?

The electromagnetic force is an attractive or repulsive force between charged particles. It has a very long range and binds together the atoms, i.e., the negative electrons swirling around the positively charged nuclei. It is this force which prevents matter from collapsing under the force of gravity. For this reason, the electromagnetic force is responsible for some of the most common everyday phenomena, such as giving solids a given shape, determining the friction between the various bodies – the friction is due to the resistance of atoms or electrons to move away from their equilibrium position within the material – determining elasticity, etc. It is also the force that holds a book on a table – instead of gravity causing it to fall to the ground – something which is a consequence of the action of the electrons on the table to repel the electrons on the book's atoms.

The electric force between a proton and an electron at any distance between them is about 10^{39} times stronger than the gravitational force between them, so on the atomic scale gravity is negligible and essentially out of the game.

But matter in the Universe is *electrically neutral*. The number of positively charged particles is *exactly equal* to the number of negatively charged particles! Therefore, at large cosmic distances it is surpassed by the much weaker gravitational force. So, at cosmic distances gravity has the upper hand, despite being so weaker!

Hence, according to this fact, that the electric force between a proton and an electron is approximately 10^{39} times stronger than their gravitational force, it's really fortunate that at the Universe we have a strict electrical neutrality everywhere. Small deviations from electrical neutrality would have the result that stars and planets could not form, since the repulsive electrical forces would far outweigh the gravitational forces. Consequently, galaxies, stars, planets, etc., would not have formed.

Assume for a moment that the Earth and the Sun were not electrically neutral bodies, but they were charged with a very small electric charge. Then, a powerful force would develop between them – the electromagnetic force. This, if the bodies were charged with the same electrical charge, would create among them a great repulsion, without gravity being able to hold them, since it is so weak. If on the other hand the bodies were charged with the opposite electrical charge, they would be strongly attracted in relation to gravity. Hence, overall if there was a small charge difference in the Universe, either we would have no formation of stars and galaxies, or we would have a rapid formation of black holes.

The reason why the Universe is overall electrically neutral ($Q_{tot} = 0$) is not understood and is ultimately connected to the matter–antimatter asymmetry problem, also known as the *baryonic matter asymmetry*. After the *Big Bang*, equal amounts of matter and antimatter were created in the early Universe. But today, everything we see, from the smallest microorganisms on Earth to the largest stellar objects, they consist almost entirely of matter. In comparison, there doesn't seem to be enough antimatter around. Something must have happened to break the original matter–antimatter symmetry. One of the greatest challenges in physics is to understand exactly what happened in antimatter and disappeared, or, why we see today this *asymmetry* between matter and antimatter. In the previous chapter, we discussed as a possibility the decay of the X-boson predicted by the theories of the large unification (GUT) of Glashow and Giorgi. However, why the Universe we see today consists entirely of matter is one of the greatest mysteries of modern physics.

Antimatter particles have the same mass as their matter counterparts, but opposite properties, such as an opposite electric charge. The positively charged *positron*, for example, is the antiparticle of the negatively charged *electron*. Matter and antimatter particles are always produced as a pair and, if they come in contact, they dissipate leaving behind an equivalent amount of energy. During the first fractions of a second after the Big Bang, the ultrahot and superdense Universe was flooded with particle–antiparticle pairs. However, what we see today is only a small part of matter – about one particle per a billion pairs – which managed to survive annihilation with antimatter.

The *Right* Strength of the Electromagnetic Force Determines the *Right* Separation of Electrons from the Nucleus in the Atoms

If the electromagnetic force between the orbiting electrons and the protons of the nucleus was weaker, then there would not exist electrons orbiting at the reasonable (right) distance around the nucleus. If it was stronger, then the

atoms would not be able to form bonds with others atoms. Either way, there would be no molecules, which is a prerequisite for life to exist!

The Electric Repulsive Force Between Two Persons Who Are Electrically Charged with Only 1% of Their Total Electric Charge

As known, Richard Feynman (Fig. 7.4) was one of the most important theoretical physicists of the twentieth century. As we mention before, he founded quantum electrodynamics, the theory that accounts for the interaction of light and electromagnetic radiation with matter (Feynman 1985). Above all, however, Richard Feynman was a gifted teacher and a keen popularizer of physics, through both books and lectures of physics. From his popular book *The Feynman Lectures of Physics* we choose the following beautiful example. It turns out that this is ultimately impossible! The repulsive force between them, due to their electric charges of the same electrical sign which are known to repel each other electrically, would be enormous, i.e., it would equal to the weight of the entire Earth, capable of lifting the entire Earth! So, their handshake would be impossible.

Fig. 7.4 Richard Feynman (1918–1988), Nobel Prize in Physics (1965). [Courtesy of The Nobel Foundation and Wikipedia Commons]

The exact calculation is relatively simple. We first calculate how many atoms there are in an average person, taking into account that the composition of the human body is 62% (about 2/3) hydrogen atoms, 24% (about 1/4) oxygen atoms, 12% carbon atoms, 1.1% nitrogen atoms, and to a lesser extent the rest of the chemical elements. Then, Feynman identifies 1.404×10^{28} as the approximate number of electrons in a 70 kg person (for the calculation, see The Feynman Lectures of Physics, Chapter 1-1, Volume II). The electrical charge of 1% of them repels the same charge of the other person which stands at a distance of half a meter, giving a Coulomb repulsive force of about 10^{25} Newtons, or, 10^{24} kg, , i.e., as much as the weight of the entire Earth!

Why There Is a *Strict Electrical Neutrality* Everywhere in Our Universe?

The electromagnetic force is an attractive or repulsive force between electrically charged particles. It is long-range and holds the atoms, that is, the negative electrons orbiting the positive protons. Current estimates indicate that there are approximately 10^{80} atoms in the observable Universe. Let's use the same number for the number of electrons, since the vast majority of these are hydrogen atoms with one electron anyway.

So one of the most impressive facts in the Universe is that the number of positively charged particles is exactly equal to the number of negatively charged particles.

We don't know why there is such a strict electrical neutrality everywhere in the Universe. Did this happen by chance? Did it happen by simple coincidence? Was it set by Divine Providence; or, it occurred for some other unknown reason? The probability, however, that the ratio of the number of protons to the number of electrons in the Universe does not differ from unity, not even by one part in 10^{39}, is the same to the probability that a blinded person would draw a single Euro coin that we have hidden among billions of stacks of Euro coins, which we could conceivably stack from the Earth to the Moon!

If the ratio of the number of protons to the number of electrons in the Universe differed from unity by one part in 10^{39}, then the repulsive or attractive electric force would outweigh the attractive gravitational force. As a consequence, galaxies, stars, planets, and life itself would never have formed.

Why Does the Solar Thermonuclear Reactor Has a Carefully Tuned Efficiency $\varepsilon = 0.007$?

The *weak* interaction is responsible for *radioactivity*, but also for the operation of the giant thermonuclear plant in the bowels of the Sun and the other stars. The weak interaction participates in nuclear fission and nuclear fusion. Let's see briefly how it works.

First of all, the *Standard Model* of elementary particle physics provides a unifying framework for understanding the electromagnetic, weak and strong interactions. Therefore, it describes how an interaction proceeds when two particles – usually but not necessarily – with half-integer spin, with the *artistic* nickname *fermions* exchange the carrier particle of this particular force, which is a particle with an integral spin, and which also has the *artistic* nickname *boson.* The fermions involved in such exchanges can be either truly elementary (such as electrons or quarks), or, composite (such as protons, or neutrons), although at a deeper level, all weak interactions act ultimately among elementary particles.

In the weak interaction, therefore, the involved fermions can exchange not just one, but three different types of carriers of this particular force, the W^+, W^-, and Z_0 bosons. The masses of these bosons are much larger than the mass of the proton or the neutron, something which is consistent with the extremely small effective range of the weak force, which is limited to subatomic distances and is less than the proton dimension, say about 0.1% (10^{-18} m). At those distances around 10^{-18} m, the weak interaction has an intensity of a similar magnitude to the electromagnetic force, but this starts to decrease exponentially with increasing distance. Scaled up by just one and a half orders of magnitude, at distances of around 0.3×10^{-18}m, the weak interaction becomes 10,000 times weaker than that of the strong interaction. The characteristic action time of the weak interaction is 10^{-8}–10^{-13} s. Quarks, which make up more composite particles, such as neutrons and protons, come in six flavors: up, down, strange, charm, top, and bottom. The weak interaction has the unique characteristic that it allows the involved quarks to exchange their flavor with another *colleague particle (quark).* The exchange of these flavors is mediated by the interaction bosons. For example, let's take the so-called β^--decay. This particular reaction is a type of radioactive decay, where a heavy nucleus, with more neutrons than the number which is required for the nucleus to be stable, converts one of its neutrons into a proton, with the simultaneous emission of an electron and an electron antineutrino. Because the electrically neutral neutron consists of two down quarks (each with charge –1/3) and one up quark (with charge +2/3), changing the flavor of a down quark to that of an up quark creates a proton (two up quarks and a down quark, i.e., total charge

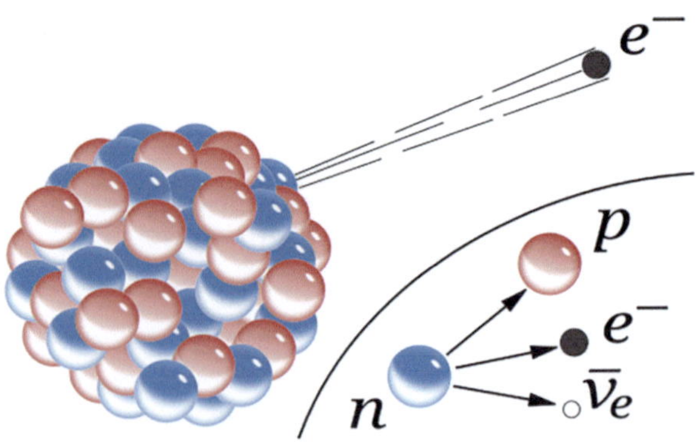

Fig. 7.5 Diagram showing the radioactive β^-- decay, due to the weak interaction, $n^0 \rightarrow p^+ + e^- + \bar{\nu}_e$ which transforms a neutron into a proton, an electron, and an electron antineutrino. [Image Courtesy of Wikipedia Commons]

+1). This phenomenon in which a quark can be transformed to another quark or a lepton to another lepton is operating exclusively within the framework of the weak interactions. These interactions are responsible for the fact that heavier quarks and leptons produce lighter quarks and leptons. For this reason all stable matter around us consists only of electrons and the two lightest types of up and down quarks.

For example, β^--decay makes radiocarbon dating possible, wherein carbon-14 (with 6 protons and 8 neutrons) decays via the weak interaction to nitrogen-14 (with 7 protons and 7 neutrons) with a half-life of 5,730 years. This nuclear β^--decay has been the first phenomenon observed to be due solely to the weak interaction. Yes, there is enough energy in an unstable nucleus to cause beta decay, which occurs when the nucleus has an unfavorable ratio of protons to neutrons. The energy available for this process, often called the Q-value, must be positive, meaning the initial nucleus has a higher mass-energy than the final products.

If the Weak Nuclear Force Was Weaker, Life on Earth as We Know It Could not Exist

What would happen if the weak nuclear force was, for example, weaker? One direct consequence would be that radioactive decay would be less efficient at heating the Earth's interior, wherein the Earth's magnetic field is produced and maintained. This would result in a weaker, or even an extinguished terrestrial magnetic field. As the Earth's magnetic field shields us from the harmful high-energy solar-charged particles, **life on Earth as we know it, could not exist.**

An extremely important example of a phenomenon involving the weak interaction involves the thermonuclear fusion of the element hydrogen into the chemical element helium, a process that powers our Sun. At the high temperatures of the solar core, four protons fuse to form a helium nucleus. The formed nucleus of the helium element consists of two protons and two neutrons. But its total mass is equal to 99.3% of the mass of the two initial protons and the two initial neutrons of which it is composed. The remaining 0.7% is released in the form of energy, mainly heat, during the fusion of the four protons. Although this seems like a small amount of mass, it is equal to 4.26 million metric tons of matter being converted to energy per second. Note in passing that thermonuclear fusion generates four times more energy per kilogram of fuel than nuclear fission (used in nuclear power plants) and nearly four million times more energy than burning oil or coal. Thus, hydrogen, which is the basic fuel inside the Sun, converts 0.7% of its mass into heat during a thermonuclear fusion, i.e., 7 thousandths of the mass of the reactants is converted into energy, or as we say, the efficiency of this reaction is $\boldsymbol{\varepsilon} = 0.007$.

What Would Happen if the Solar Thermonuclear Efficiency ε Had a Different Value?

As we mention, the Sun produces its energy through the thermonuclear fusion that occurs in its core, where 4 protons are converted into a helium nucleus and the mass difference is converted into energy which is radiated, a small part of which reaches us here on Earth. The efficiency of this reaction, governed by the weak nuclear forces, is 0.7%.

If the number **ε** of the solar thermonuclear efficiency was slightly smaller, say, **ε** = 0.006 (i.e., smaller by 1 part in one thousandth) instead of **ε** = 0.007, then the reaction would proceed much more slowly and the hydrogen would be less efficient as a fuel, so the Universe today would basically consist of only hydrogen. Not enough helium – and all the other chemical elements – would have formed inside stars, endowing the present Universe with the rich chemistry that is the basis of life.

On the other hand, if this reaction efficiency **ε** was slightly higher, for example **ε** = 0.008, fusion would proceed much faster. Then, there would be no hydrogen left from the Big Bang to fuel the Sun's core. Again, the solar system would not have formed and, by extension, there would have been no life.

The correct number is very precisely set between **ε** = 0.006 and **ε** = 0.008. All chemical elements even those that make up our bodies have been made in the fiery furnaces of the interior of stars that are now already extinguished and the leftovers of those fusion reactions have been scattered into space, before finally reaching us here. This would not have happened if the efficiency of the fusion reaction had not been set to the value that it now has (**ε** = 0.007). For a Universe with the complex chemistry and life, like the one we live and observe, the thermonuclear reactions should therefore proceed with an efficiency between 0.006 and 0.008. Therefore, without this correct value of this number **ε**, the Sun wouldn't exist as we know and also we wouldn't be here either.

Altogether then, **if the efficiency of this thermonuclear fusion reaction was 0.006, or 0.008, i.e., it different by 1 part in a thousand from the value of 0.007 with which currently operates inside the Sun, then the Sun would not exist and by extension, we would not be here without the life-giving solar energy** …

Additionally, not even the manufacturing of the nuclei of all chemical elements, such as those that make up our bodies, would have taken place in the fiery furnaces of the interior of the stars, if the performance of the above reaction had not been *exactly* what it is!

Why There Exists this *Unbearable Lightness* of Gravity in the Universe, with an Inconceivably Weak Strength, I = 10^{-38}?

Another universal number that logically seems to pop up to answer the question we asked at the beginning has to do with the incredibly weak strength of gravity, compared to the other forces, that determine everything in the Universe, at least as we know it today.

Gravity is the attraction between two objects that have mass, or energy. This is seen when we throw a stone from a bridge and is also responsible for the tides in the seas. Gravity could introduce itself as that force which keep people and everything pinned down to the surface of the Earth, despite the vertiginous speed of 107,000 km per hour with which we whirl around the Sun (we do not get dizzy because of this extremely high rotational speed, because of the enormous circular radius of the terrestrial orbit around thc Sun!). Similarly, as the Earth rotates around its axis with the lower speed at the equator of 1,670 km per hour, this only lightens our weight by 0.3% through the centrifugal force. However, despite the terrestrial rotations, Gravity is enough to hold the water vapor and the other precious gases of the atmosphere from escaping to space.

Let us think for a moment what would happen to Earth if, say, the Sun stopped gravitationally pulling the Earth: our stony planet will undergo what happens when a stone tied in one end of a rotating string is cut in the other one; similarly, Earth would also leave to travel lonely and lost in far space!

Similarly, in the scenario where the self-rotation of the Earth stopped, there would be some strange consequences. First, the terrestrial magnetic field would probably disappear because it is believed to be created in part by the Earth's self-rotation, as is the case with the solar magnetic field too. In addition to losing the colorful *aurora borealis*, Earth would be left bare and exposed to the Sun's wrath: as we discussed in Chap. 1, quite often the solar corona releases a blast of superhot gases (plasmas), consisting of high-energy charged solar particles; their radiation would kill everything in their path toward our planet, if the protective magnetosphere disappears. Also, when this happens, compasses will stop pointing north, birds won't know where to fly when they migrate, the solar wind would erode out atmosphere devoiding our planet of the life-giving air we breathe, etc. This would make life on Earth very difficult for human beings and other life forms. Earth could be rather similar to Mars, on whose surface we have seen features associated with volcanoes and moving tectonic plates. But these don't move any longer, since there is no protective magnetic field and only a very thin atmosphere remains today, which is relentlessly whipped by the solar wind and the hot charged solar particles.

According to its previous effects, gravity is probably the most intuitively perceived and familiar of all other fundamental forces of nature. But at the same time it is one of the most difficult to explain. Isaac Newton was the first to propose its existence and is said to have been inspired by watching an apple falling from an apple tree. In particular, he described gravity as the attraction between two objects. Centuries later, Albert Einstein proposed, through his theory of general relativity, that gravity is not really an attraction, or a force. On the contrary, it is a consequence of the *curvature* that the objects cause in the *space–time fabric*. Specifically, a massive spherical object behaves on space–time like a massive sphere placed in the middle of an elastic dimpling sheet: it affects the sheet by deforming it and also causing other, smaller objects to fall toward its position (Fig. 5.87).

Although gravity is the most familiar force to us, it is at the same time the most mysterious of all forces of nature. Gravity troubled Einstein in his attempt to include this interaction along with the other forces in the General Theory of Relativity and it resists to this day to be included in the Grand Unified Theory of forces, or a *Theory of Everything*. Actually, that is why there is no theory for *quantum gravity* to date.

However, if we had asked gravity itself, this force would boast first of all for its positive and creative role, i.e., that it is responsible for the creation of the Sun, the Earth, the Moon, and all the other planets and planetary moons. Perhaps gravity would hide from us that inside the Sun and the other stars has concluded only a *temporary truce* with the other forces. But what will happen when this truce ends? Then our low-mass star will dissolve into its components which gravity had put originally together, eventually forming a *white dwarf*. More massive stars will have a more spectacular end, finishing their lives via a supernova explosion, which will leave behind a stellar corpse. Depending on its mass, this will be a *black hole* or *a neutron star* (Fig. 1.28). The closest white dwarf, about the size of Earth, is in the double system of the brilliant Sirius, 8.6 light years away from us (Fig. 5.9). The nearest neutron stars to Earth are the neutron stars in a group called "the Magnificent Seven" a group of isolated young cooling neutron stars at a distance of 400–1600 light years away; the nearest to Earth (and perhaps dangerous!) black hole is about 3,000 light years away, a safe nevertheless distance from us.

But, the effects of gravity are not only local here, since it plays a catalytic role in the entire Universe. Today we observe gravity in action to form disks of primordial matter swirling around nascent stars, from which planetary systems, many like ours, will later emerge. Maybe some kind of life will develop in any of them? Who knows? After 1995 we have already observed a few thousands of such exoplanets. Within our Galaxy, but also in other galaxies, for every 3 stars, 2 are in double, or multiple star systems, i.e., they orbit around each other, again thanks to the attractive gravitational force between them. Gravity is still responsible for the rotation of the entire solar system around the center of the Galaxy at a speed of 720 km per hour, thus completing each such circular orbit in about 230 million years. Gravity also causes the approach of our Galaxy to the nearby galaxy of Andromeda, at speeds of about 400,000 km per hour; this attraction will necessarily bring them into a titanic cosmic embrace, after about 4 billion years.

Gravity would also boast that it is responsible for the large-scale panspermia in the Universe: gravity created not only the myriads of bright stars, but also the beautiful great structures of the Universe, the galaxies, in the first stages of the Universe's evolution, about 14 billion years ago. Its effect therefore extends throughout the Universe, that's why we say it has a wide range. On the grand cosmological scale, our hope was that gravity would slow down the frenzied expansion of the Universe caused by the *Big Bang*, reversing it and causing, in the far future, the *Great Crash*. Alas, however, our hopes today seem to be refuted by the documented presence of *dark energy*, which, acting as the mysterious *antigravity*, causes the accelerated expansion of space and thus the Universe seems destined to follow the nightmare scenario of eternal expansion and constant dilution, which will lead inevitably to his heat death.

If Gravity Was Slightly Stronger, *The Universe Would Consist Only of Black Holes*

Along with its two previous prestigious advantages (its creative *attractive* property and its *very long range*), gravity has also been endowed with a crucial *flaw* so to speak, i.e., that it is very, very weak – actually incredibly weak – compared to the other forces. That is, although gravity holds planets, satellites, stars, solar systems even galaxies together, it turns out to be the weakest of all the other fundamental forces, especially at the molecular and atomic scales. Let's think about it first of all in this simple way: how hard is it to pick up a ball off the ground or put our foot up or to jump up or to stick on the door of our refrigerator a small magnet, defying its weight? All this is done relatively easy, despite the fact that these energies have to compensate for the gravity of the entire Earth! But even at the molecular or atomic levels, gravity has almost no effect relative to the other fundamental forces. So how weak is gravity compared to the strongest of the other forces, the nuclear force? About 10^{38} times, i.e., the unit followed by 38 zeros, or, one hundred trillion trillion trillions! A number hard to fit to the human mind. And, as we shall see below, there is an important reason for this incredible "*humiliation*" – in relation to the other messmate Universal forces – which has been reserved in this deeply penetrating and omnipresent force of nature!

In other words, because gravity is endowed with (a) great range and the characteristic to be (b) only attractive, while the other forces are either short-range, or as attractive and simultaneously repulsive canceling each other out, gravity remains dominant in the Universe, even though it is hopelessly light and weak. So, what is the particular reason for gravity's feeble strength, the most enigmatic of all forces of Nature?

Why Gravity is so Weak?

It seems that one reason why gravity is so weak has to do with the long lifetimes of stars and the associated time for the development of life in their environment. With the value $\varepsilon = 0.007$ of the solar thermonuclear reaction efficiency, the Sun's metabolism has the rather small value of 2 erg/gr/sec, i.e., 5,000 times lower than the human metabolism (2,000 kcal/70kgr/day =

5,000 erg/gr/sec) justifying the solar longevity, about 10 billion years, which is enough time for life to evolve!

If Gravity Was Slightly *Weaker, or, Stronger*, The Sun and the Other Stars Would not be able to Shine, or, They Would Live a Shorter Life Span

If gravity was weaker, then inside the Sun and the other stars, their internal pressure, which always balances the weight of their overlying layers, would be lower. But a lower internal pressure would correspond to a lower thermal energy of the protons of the solar interior. In turn, these protons could not overcome the Coulomb barrier to initiate thermonuclear fusion reactions. The result would be that **the Sun and the other stars would not be able to shine**! Conversely, **if gravity were stronger** in the solar interior, the equilibrium pressure would be higher. Then, the thermonuclear fuel would burn at a faster rate and it would be exhausted much faster. As a result, the **stars would live a shorter life** span, before finally being given to a neighboring planet the opportunity in life to evolve and grow!

Recently, alternative theories have been developed to explain this particular aspect and behavior of gravity. One of them includes the Higgs boson which was discovered in 2012 and is believed to be directly related to the force of gravity. Specifically, the more mass the Higgs boson has, the stronger gravity would be in the Universe. Essentially, the question *why is gravity so weak* comes down to the question, *why does the Higgs boson does not have more mass*? New theories try to answer this question arguing that after the *Big Bang*, a large variety of different Higgs bosons were created, each with a different mass. The Higgs bosons with large masses were unstable, so they decayed into smaller ones. Thence, over time, all the heavier Higgs bosons were transformed into lighter and lighter ones, until only the lightest possible kind of boson survived. These extremely light Higgs bosons are what there were discovered in 2012 by the Large Hadron Collider at CERN.

What Would Happen if the Dimensions of Spacetime Were not (n, m) = 3 + 1

Very often, when we climb on a mountain, cross a street to enter our house, or go up the stairs, we take for granted that the dimensions of space can only be $n = 3$. And yet, in principle there could be only two ($n = 2$), or even four ($n = 4$) dimensions in the space around us. Perhaps in some other Universe, the dimensions of space–time are not $3 + 1 = 4$! How would physics be affected in such a case?

Consider for a moment the case where there would be four (4) spatial dimensions. Then, the law of universal gravitational attraction, or the law of electrostatic force between two electric charges (Coulomb's law), would give the corresponding force to vary inversely with the *cube* of the distance and not with the *square* of the distance. But here's what would happen in such a case: a small perturbation of a planet's orbit like the Earth moving around a star like the Sun, because of such a law gravity, would lead the planet, either to a rapid collapse in the warm embrace of its parent star, or in a desperate escape into distant and cold space. The same would happen to the orbit of an electron moving around a nucleus. This is in contrast to what happens with the inverse square law that applies in our three-dimensional space, where a planet that is slightly accelerated, or decelerated in its orbit, simply shifts to a slightly different orbit from the original one. It is worth noting that already in 1917, the Austrian theoretical physicist and mathematician Paul Ehrenfest (Fig. 7.6) had shown that if there were only one dimension for time, but more than three dimensions for space, the orbit of a planet around a central body, such as the Sun, would not be stable (Ehrenfest 1917).

It is worth noting that already in 1917, the Austrian theoretical physicist and mathematician Paul Ehrenfest (Fig. 7.6) had shown that if there were only one dimension for time, but more than three dimensions for space, the orbit of a planet around a central body, such as the Sun, is not stable (Ehrenfest 1917).

Fig. 7.6 Paul Ehrenfest (1880–1933) ca 1910 showed that there would be no atoms and no planetary systems if we did not live in a 3D space

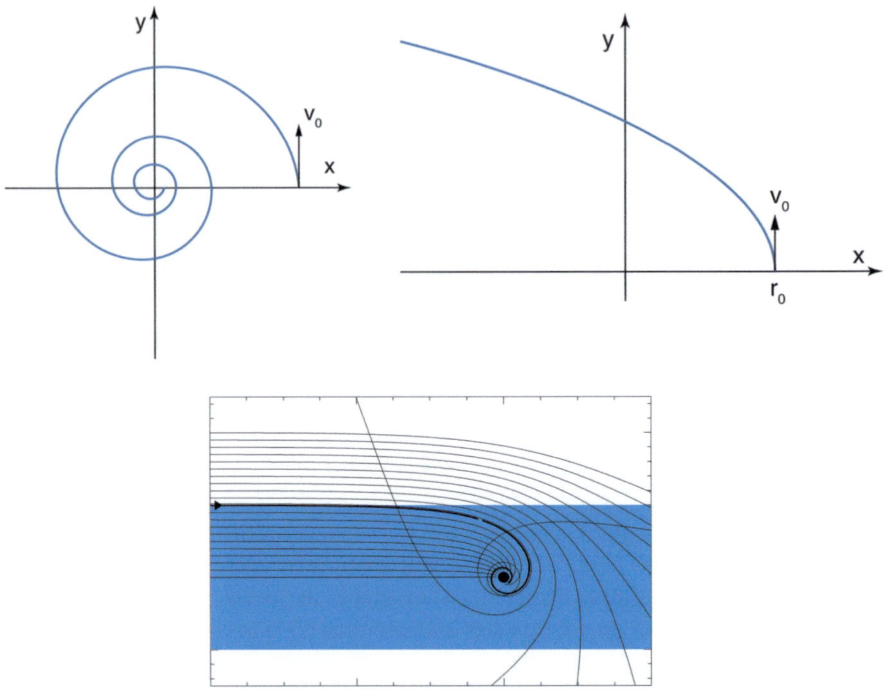

Fig. 7.7 (*top*) Motion of a particle under the influence of an attractive central force f (r) $\sim$ r^{-3}, which is the gravitational force corresponding to the two-body problem in four-dimensional space. The particle has an initial speed v_o at the position r_o of the plane (x, y). In the case of low angular momentum (left) the particle follows a spiral orbit and reaches the center of the central force at (x, y) = (0,0) at a finite time, as in the (*bottom*) figure, while in the case of high angular momentum (right), the particle goes to infinity along a parabolic orbit. (*Figure Credit* K. Tsinganos, Modern Theoretical Mechanics, UNIBOOKS I.K.E. (2021))

The problem is analogous to the movement of a body within the field of a central attractive force that varies inversely with the cube of the distance. In this case there are no elliptical orbit solutions, but the body either collapses to the center of the gravitational force following a spiral orbit, or, escapes to infinity following a parabolic orbit[4] (Figs. 7.7) The same applies for the orbit of a star around the center of the galaxy.

Consider for a moment the interesting two-body problem in four-dimensional space. If a swarm of light test particles are incident on a massive point particle (Fig. 7.7c), with a range of impact parameters, i.e., different angular momenta, we have two cases: those particles that start with a large impact parameter and angular momentum follow a parabolic orbit and escape to infinity (Fig. 7.7b), whereas those with smaller impact parameters and angular momentum spiral into a singular collision in a finite time (Fig. 7.7a). Hence, the light particles that approach the heavy one at

[4] K. Tsinganos, Modern Theoretical Mechanics, UNIBOOKS I.K.E. p. 199 (2021).

the center either escape to infinity or get sucked into a cataclysmic collision. There are no stable orbits, as in Figs. 7.7.

Ehrenfest also showed that if the number of dimensions of space is an even number, then different parts of a pulse move at different speeds. At about the same time (1922) Hermann Weyl had shown that Maxwell's equations for electromagnetism are valid only if we have three dimensions for space and one for time. The case with 1 spatial dimension and 3 temporal dimensions has the curious property that the speed of light in vacuum is a lower limit on the speed of matter, which consists of *tachyons*.

It has been furthermore argued (Tegmark 1997) that all worlds, except the one we live on, wherein (n, m) = 3 + 1 might correspond to *dead worlds*, devoid of observers and it is far from obvious that any other combination other than (n, m) = (3, 1) permits observers, since radical qualitative changes occur when n or m are altered.

With m > 1, or m < 1, the partial differential equations of physics would lack the hyperbolicity property that enables observers to make predictions, noting that hyperbolic equations allow well-posed initial-value problems. On the other hand, in a space with n > 3, there can be no traditional atoms and perhaps no stable structures. When n > 3, the two-body problem no longer has any stable orbits as solutions. A similar disaster occurs in quantum mechanics, where a study of the Schrodinger equation shows that the hydrogen atom has no bound states for n > 3 (Tangherlini 1963), i.e., the hydrogen atom has no ground state, but time-dependent states of arbitrarily negative energy.

The situation in general relativity is analogous. This means that such a world cannot contain any objects that are stable over time, and thus probably cannot contain stable observers too.

A space with n < 3 allows no gravitational force and may be too simple and barren to contain observers. It has also been argued (Whitrow 1955) that organisms would face insurmountable topological problems if n = 2: for instance, two nerves cannot cross. Another problem is the well-known fact that there is no gravitational force in general relativity when n < 3, Misner et al. (1973). The case n = 2, n = 1, and n = 0 offer vastly less complexity than the case n = 3, and therefore worlds with n < 3 are just too simple and barren to contain observers.

If Spacetime Did not Have 3 + 1 Dimensions, Intelligent Life Could not Emerge

There are many arguments that conclude that the space of a Universe wherein intelligent life can emerge capable of handling technology, as in the Universe we live in, can only be three-dimensional and time one-dimensional, to the exclusion of all other cases where the dimensions of space are not 3 and time is not only 1!

Why the Density Parameter of the Universe Seems to have been *Tuned* to The Precise Value $\Omega = 1$?

The parameter Ω of the density of the Universe, $\Omega = \rho/\rho_c$, is defined as the ratio of the actual mean density of matter and energy in the Universe ρ and the critical density ρ_c, i.e., the density which if the Universe had it would stop expanding only after infinite time (Figs. 6.6 and 6.27). To date, the critical density is estimated to be about 6 atoms of monatomic Hydrogen per cubic meter (equivalent to about 9.9 $\times$ 10^{-30} gr/cm^3, i.e., about 10^{-29} gr/cm^3), whereas the average density of ordinary baryonic matter in the Universe is believed to be 0.20–0.25 atoms per cubic meter (Fig. 6.28).

This cosmic number Ω counts for the total amount of matter and energy contained in our Universe: stars, galaxies, intergalactic gas, dark matter, and dark energy. In the absence of a cosmological constant, a Universe with $\Omega < 1$ is *open* and will expand forever, whereas a Universe with $\Omega > 1$ is *closed* and will eventually collapse. If $\Omega = 1$, then the Universe is *flat* and has the critical density. In other words, the flat Universe is one in which the amount of matter present is just sufficient to halt its expansion, but insufficient to re-collapse it (Figs. 6.6 and 6.27).

As we discussed previously, current observations and related research show that Ω is *very close to unity* (Figs. 6.25 and 6.26). However, it is still of ultimate importance of knowing whether Ω is slightly greater (less) than 1, or exactly equal to 1, as this also hides the ultimate fate of the Universe! In particular, if Ω is even slightly less than unity, the Universe is *open* and will continue to expand forever. If Ω is even slightly larger than unity, the Universe is *closed*, it will eventually brake its expansion, and then it will lead to a final collapse. If Ω is exactly equal to unity (which is indeed a remarkable case), then the Universe is *flat* and contains enough matter to eventually stop its expansion, but not enough to cause its collapse.

It is important to note that the density ρ used in the calculation of Ω is the *total* mass (and energy) density of the Universe. In other words, it is the sum of a number of different components, including both ordinary and dark matter, as well as of dark energy given by recent observations, for example with the Planck space mission. Thus, $\Omega = \Omega_B + \Omega_D + \Omega_\Lambda$, where Ω_B is the density parameter of ordinary baryonic matter, Ω_D is the density parameter of dark matter, and Ω_Λ is the density parameter for dark energy. The current analysis of the Planck Space Telescope observations conclude that we live in a Universe with $\Omega_\Lambda = 0.683$ i.e., dominated by dark energy, $\Omega_D = 0.268$ and $\Omega_B = 0.049$. Therefore, with today's precision of cosmological observations, it seems that we live in a *flat* ($\Omega = 1$) Universe (Fig. 7.8).

The Desnsity Parameter Ω

The meaning of the number Ω is that it expresses how important is the role that gravity plays compared to antigravity, i.e., the energy of expansion of the

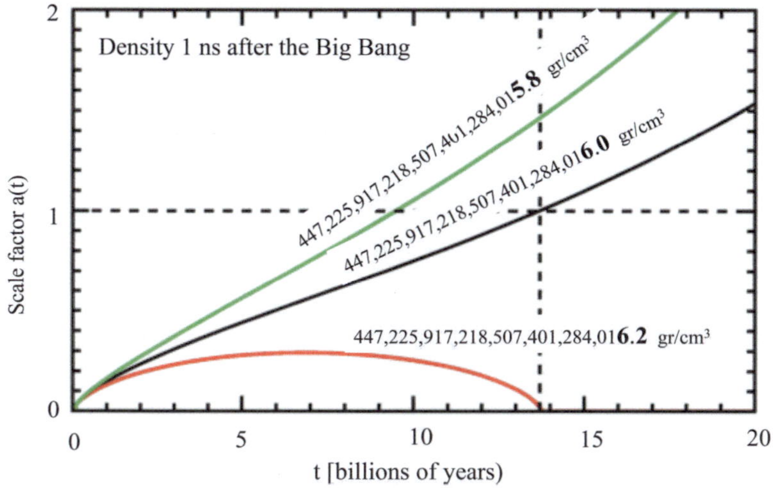

Fig. 7.8 Scale factor vs. time for 3 different models of the Universe: open (green curve), critical (black curve), and closed (red curve). Note that the density parameter of the Universe, Ω, at 1 billionth of a second (1 ns, 10^{-9} s) after the Big Bang differ from unity in the 21st decimal place in each model. (Adapted from prof. Ned Wright's Cosmology Tutorial at the UCLA with the author's kind permission)

Universe. If gravity and mass dominated, the Universe would have collapsed long ago. If, on the other hand, dark energy dominated, they wouldn't let neither galaxies nor stars to be formed. So, it seems that the initial expansion rate has been very carefully *tuned*. The density parameter Ω only *one second* after the Big Bang could not differ from unity more than one part in a million billion; otherwise the Universe would not be expanding today, about 13.8 billion years later (Rees 2001).

It is possible to plot the scale factor a(t) for three models with three different densities at a time 1 ns after the Big Bang (during the era of quarks, Figs. 6.37 and 6.40). In such a plot, the green curve in Fig. 6.27 would give a critical density case that matches the WMAP-based model, which has a density $\rho_{(t=1\ \mathrm{ns\ after\ BB})}$ = 447,225,917,218,507,401,284,016 gm/cm^3 at 1 ns after the Big Bang. Adding only 0.2 gm/cm^3 to this 447 sextillion[5] gr/cm^3 (i.e., for $\rho_{(t=1\ \mathrm{ns\ after\ BB})}$ = 447,225,917,218,507,401,284,016,2 gm/cm^3) would cause the *Big Crunch* to happen right now (13.8×10^9 yrs after the BB). On the other hand, taking away only 0.2 gm/

[5] One sextillion = 1,000,000,000,000,000,000,000 (one thousand million million million; 10^{21}. To help visualize this massive number 1 followed by 21 zeros, or expressed in scientific notation as 10^{21}, note that the Earth weighs approximately 5.98 sextillion metric tons, there are about 6 sextillion cups of water in the world's oceans and there are roughly 1 sextillion grains of sand on all of Earth's beaches combined!

Fig. 7.9 Einstein introduced the famous cosmological constant Λ (1917)

cm^3 (i.e., for $\rho_{(t=1\text{ ns after BB})}$ = 447,225,917,218,507,401,284,015,8 gm/cm^3) would give a model with a matter density Ω_M that is too low for our observations (Fig. 7.9).

Thus, the density 1 ns after the Big Bang was set to an accuracy of better than 1 part in 2235 sextillion (1 part in 2.235 x 10^{24}). Even earlier it was set to an accuracy better than 1 part in 10^{59} (https://astro.ucla.edu/~wright/A275.pdf). And, one second after the Big Bang could not have varied from unity by more than one part in a million billion (10^{15}), or the Universe would not still be expanding, 13.8 billion years on.

This small difference also reflects the essence of the problem of the flatness of the Universe that we have already discussed in the previous Chap. 6. Because the initial density of the Universe could take any value, it is extremely amazing to find it so well set up at the critical value. Hence, the *flatness problem* in Cosmology is the puzzle of explaining why the early Universe exhibits such fine-tuning in the value of this parameter!

If Ω Was not 1, the Universe Would not Contain Complex Structures Such As Galaxies, Stars, Planets, and Any Life Form

A very small deviation of Ω from unity in the early Universe would have enlarged during the billions of years of expansion to produce a current density very different from the critical one. In the case of the largest initial density ($\rho > \rho_c$) this would lead to a Universe so dense that it would stop expanding and collapse into the *Great Crash*. In the case of lower initial density ($\rho < \rho_c$), this would lead to a Universe that would expand so rapidly and become so dilute that it would soon appear essentially empty. Then gravity would not be strong enough to cause matter to collapse and form galaxies. In any case, the Universe would not contain complex structures such as galaxies, stars, planets, and any life form. So, conversely, the value $\Omega = 1$ seems to have been carefully tuned to be a prerequisite for the appearance of higher forms of conscious life in the Universe.

Furthermore, recent observations constrain Ω to be equal to unity with an error of less than 1%, i.e., the value of the term $|\Omega-1|$ nowadays is less than 0.01. That means, the value of the term $|\Omega-1|$ should have been less than 10^{-62} during the Planck epoch, while a billionth of a second (1 ns) after the Big Bang, as we discussed above, $|\Omega-1|$ was different at the 21st decimal place for the three cases of the open, critical and closed Universe shown in Fig. 6.27!

Why Does the Cosmological Constant Has the Low Value $\Lambda = 10^{-52}$ m^{-2}?

The cosmological constant Λ is the constant coefficient of the term $\Lambda g_{\mu\nu}$ that Albert Einstein temporarily added to his field Eq. (1) of general relativity in 1917 as a repulsive force required to keep the Universe in static balance, in the so-called *steady state model of the Universe*. In doing so, Einstein was influenced by the scientific accepted view/fashion of his time, which then wanted the Universe to be static, i.e., unchanging over time. However, when later Friedmann and Lemaitre showed that the solutions of Einstein's equations of General Relativity represented a changing and evolving in time Universe, Einstein faced with this news admitted that his introduction of the cosmological constant was "*his biggest blunder*", as he characteristically said.

However, decades later the observation of the accelerating expansion of the Universe brought the cosmological constant back to the fore. This is because in modern cosmology, among the possible causes of an accelerating Universe (such as enigmatic scalar fields, which can vary in time and space, quintessence, etc.) the cosmological constant is in most respects *the leading candidate* and *the simplest solution*. In other words, the cosmological constant is *a measure of the energy density of the vacuum of space*. Of course, although the vacuum may not contain matter, it is not really a true vacuum (!), because it can have an energy density due to the quantum fluctuations of the various fields it contains, as we discussed in the previous Chap. 6. This energy of the vacuum is added to the gravitational potential energy, but differs from it in the following important way: while the gradient of the gravitational potential energy gives the *attractive* gravitational force that slows down the expansion of the Universe, the gradient of the energy of the vacuum gives a new *repulsive* force, a cosmic *antigravity*, the existence of which we had not even suspected to exist previously and which determines the future of the Universe. This *antigravity* does not cause any observable phenomenon on a scale smaller than about 1 billion light years and is destined to overpower gravity and other forces of nature, as the Universe grows darker and emptier.

The expression of the cosmological constant depends on Hubble's constant H_0 and Ω_Λ the ratio between the energy density due to the cosmological constant and the critical density of the Universe, $\Lambda = 3(H_0/c)^2\ \Omega_\Lambda$. More specifically, according to the Λ-CDM model, the current *Standard Model* of cosmology which uses the Friedmann–Lemaître–Robertson–Walker metric, $\Omega_\Lambda = 0.6889 \pm 0.0056$ and the Hubble constant $H_0 = 67.66 \pm 0.42$ km/s/Mpc. Substituting we get the value of the cosmological constant, $\Lambda = 1.1056 \times 10^{-52}\ \mathrm{m}^{-2} = 2.888 \times 10^{-122}\ L_p{}^{-2}$. This value of the cosmological constant Λ is in excellent agreement with the latest measurements of the expansion of the Universe obtained by the High-Z Supernova Team and the Supernova Cosmology Project.

Our Existence Today Depends on the *Specific Numerical Value* of the Cosmological Constant Λ

The *specific numerical value* of the magnitude of the cosmological constant Λ seems to dictate, not only the future of the Universe itself, but also that we ourselves exist today! In the case where the value of Λ was greater, its effect would have stopped the formation of stars and galaxies, so that the cosmic evolution and the subsequent emergence of life would have been smothered before it could even begin!

As we discussed previously, the observation of the accelerated expansion of the Universe, the most important scientific discovery of 1998, led to the calculation of the specific *magnitude* of this constant $\Lambda = 10^{-52}\ \mathrm{m}^{-2}$. On the other hand, according to some theories, the magnitude of the vacuum energy density predicted by *quantum field theory* is about 120 orders of magnitude – i.e., 1 followed by 120 zeros – larger than the value implied by the cosmological observations of accelerating cosmic expansion, a discrepancy that has been called *the worst theoretical prediction in the history of physics (Hobson* et al. *2006)*! This issue is called the *cosmological constant problem*, and it is one of the greatest mysteries in science with many physicists believing *that the vacuum really holds the key to a full understanding of nature (*Davies 1985*)*. Nevertheless, fortunately for us, the value calculated for Λ according to observations is much less than *quantum field theories propose*.

Another theoretical approach that deals with the issue of the cosmological constant is that of *multiple Universes*, wherein an infinite number of parallel Universes may exist, albeit with different physical laws and/or values of the fundamentals constants. In this context, a possible explanation for the small but nonzero value of Λ is that we may live – and that is why we exist here – only in this particular of all those possible Universes, which has the potential to lead to some form of intelligent life. In other words, out of the infinite set of values of the vacuum energy which correspond to the infinite set of different parallel Universes, the formation of life-supporting structures would have been *suppressed* in all those Universes where the vacuum energy and Λ had very large values. More specifically, if the vacuum energy was negative, but with an absolute value significantly greater than it has in our observed Universe, say an absolute value greater by a factor of 10, holding constant all other physical quantities such as matter density, this would certainly lead to a closed Universe. But then, the lifetime of this closed Universe would be shorter than the age of our Universe, probably much shorter than is required for intelligent life to form and develop. On the other hand, a Universe with a large positive cosmological constant Λ would expand very quickly, thus preventing the formation of galaxies and stars. Only in the relatively rare case – which corresponds to our own Universe – where the vacuum energy gives the observed value of the cosmological constant Λ, this result is compatible with the development of life as we know it!

One possible explanation for the small but nonzero value of the cosmological constant was proposed by Weinberg (1987, 1992) following *the anthropic principle*. According to Weinberg, the present age of the Universe is not random. Instead, biological factors may have constrained the Universe to be more or less in a *golden age*, neither too young nor too old. If the Universe was *one-tenth* of the age it has today, there wouldn't have been enough time to build up appreciable metallicity abundances – for chemical elements other than hydrogen and helium – especially for carbon, through nuclear fusion reactions inside the stars. Back in this younger Universe, small rocky planets didn't exist yet. On the other hand, if the Universe was *10 times older* than it is today, most stars would have grown up enough, they would have moved away from the main sequence, and they would have become white dwarfs, except for the fainter red dwarfs. In such a case and time, the stable planetary systems would have already disappeared. With this reasoning, Weinberg explained the selected and tuned set of values of the dimensionless cosmic numbers which result from the constants of physics.

The Fine-Tuned Values of the Cosmic Numbers in the Universe

As we already mention previously, Richard Feynman (1985), the creator of the theory of quantum electrodynamics, wrote about the value of the fine-structure constant: *It's one of the greatest damn mysteries of physics: a magic number that comes to us with no understanding by man*! On the same subject, Paul M. Sutter[6] writes characteristically: *If you were to meet an alien from a distant star system, you'd have a pretty hard time communicating the value of the speed of light. Once you nailed down how we express our numbers, you would then have to define things like meters and seconds. But the fine-structure constant? You could just spit it out, and they*

[6] Paul M. Sutter is an astrophysicist at SUNY Stony Brook and the Flatiron Institute, host of "Ask a Spaceman" and "Space Radio," and author of "How to Die in Space."

Table 7.1 Seven cosmic numbers and their numerical values

• The fine-structure constant $\boldsymbol{\alpha}$	$\boldsymbol{\alpha} \approx 1/137$
• The Cosmic electrical neutrality $\mathbf{Q_{tot}}$	$\mathbf{Q_{tott}} = 0$
• The efficiency of solar thermonuclear reactor $\boldsymbol{\varepsilon}$	$\boldsymbol{\varepsilon} \approx 0.007$
• The ratio of gravitational to strong interactions **I**	$\mathbf{I} \approx 10^{-39}$
• The number of dimensions of space–time **(n, m)**	**(n, m)** = (3, 1)
• The density parameter of the Universe $\boldsymbol{\Omega}$	$\boldsymbol{\Omega} = 1$
• The cosmological constant $\boldsymbol{\Lambda}$	$\boldsymbol{\Lambda} = 10^{-52}$

would understand it (as long as they count numbers the same way as we do). Similar comments could be applied for the numerical values of all the 7 cosmic numbers in table.

As we already discussed, the ratio I of the gravitational force to the strong nuclear force, the so-called gravitational coupling constant $Gm_p^2/\hbar c \cong 6 \times 10^{-39}$, with m_p the mass of the proton, is an extremely small number and at the same time it is also extremely important. However, small magnitude of I does not come out of nowhere in physics.

The well-known theoretical physicist Paul Dirac,[7] whom I had the good fortune to meet at Fermi Lab in 1980,[8] hypothesized that this small number should be related to other numbers of similar size, characterizing the structure of the Universe (Dirac, 1937). However, most physicists today we believe that this dimensionless constant is provided by nature itself and cannot be calculated or related in any way to the others cosmic world numbers in Table 7.1 (see also, Dicke, 1961).

With the exemption of the cosmological constant, all these are *pure, dimensionless numbers.* And, there do not seem to be any correlations between those cosmic numbers, nor can we predict any of them through the values of the others. Each of them, however, seems to play a critical and distinct role in our Universe, while at the same time, only these few constants determine how the Universe evolves. After all, according to the German mathematician Leibniz, who pointed out that the real world should be the one who is *the simplest in assumptions and at the same time the richest in phenomena.*

[7] The Nobel Prize in Physics in 1933 was awarded jointly to Erwin Schrödinger and Paul Adrien Maurice Dirac, *for the discovery of new productive forms of atomic theory.*

[8] In 1980, during the completion of my Ph.D. studies at the University of Chicago, I had the good fortune to meet and discuss with Paul Dirac at Fermi Lab, IL (USA). This unique experience, together with those with the Un. of Chicago professors and Nobel laurates Jim Cronin and S. Chandrasekhar, as well as with my Ph.D. advisor Gene Parker, are indelibly etched in my memory and scientific life.

Why There Seems to be a *Fine-Tuning* of the Cosmic Parameters in Our Universe?

That there is some *fine-tuning* present in our Universe is undisputed by the scientific world in general. One could say that these seven numbers define a Universe which is properly regulated so that its structure of galaxies, stars, and planets are such that ultimately life emerges.
Is this just a *coincidence*, is it a *Divine Providence*, or *something else* altogether?

According to Rees (2001), there may be a finite or infinite set of other Universes in which some of these global cosmic numbers will be different. In this *Multiverse*, our Universe is a simple member. The idea of the *Multiverse* has been proposed to explain how our own Universe seems to be adapted to conscious life as we experience it. If there is an infinite number of Universes, each one with possibly different physical laws and/or different fundamental cosmic constants, then, some of these Universes would have the appropriate combination of laws and fundamental parameters for the organization of matter and elemental biological diversity, astronomical structures, stars and planets that can exist for a sufficient period of time so that life may emerge and evolve. Most of these Universes are *barren* and do not give birth to life, which could emerge and give rise to intelligent, self-aware beings, only in a Universe with the correct values of these cosmic numbers. Because gravity, despite being so weak, plays a dominant role in the Universe. How was this role of gravity properly regulated? Was this done for the reason that the Sun and the stars may live long? Did it happen to create the Earth and us at the right distance from the Sun and acquire the proper composition of the atmosphere so that life can eventually flourish? After all, everyone of the natural laws seems to be designed for the creation of life. Is this a coincidence? Are we so incredibly lucky to live in the luckiest Universe that could ever exist? Or, is this all part of a Grand Design? Do the simplicity, symmetry, harmony, and expediency of the natural laws testify to the presence of an omniscient Creator/Legislator? These are fundamental questions and their answers are beyond the scientific means.

The history of astronomy is really a continuous history of horizons that are sought after, approached but also constantly moving away. All of the discoveries of the last century which we have described in this volume very briefly prove this fact. However, of all the scientific discoveries made during the period of the 1919–2019 century, many of which we gathered in the pages of this book, perhaps the greatest discovery ever made is that the Universe as a whole, i.e., everything that exists and can come under our perception, can be studied in a logical way using observation, the tools of mathematics, and the methods of physics and astronomy.

From astronomy's point of view, we are ephemeral intelligent creatures who we inhabit a tiny planet, which orbits a life-giver Sun, one of the billions of similar stars that make up and revolve around the center of the Galaxy. And our Milky Way is a member of one of the numerous galaxy clusters that flood the inconceivably vast but finite cosmic space–time. Of course, we do not realize that the matter of this entire world of galaxies that we observe constitutes only a small part – about one twentieth – of the total content of the Universe. But at the same time, from a more general point of view, we little people it seems that we can gradually understand this Great Plan of the Universe in which we live, its immensity in space and time, but also the silent beauty of this World.

Quoting the words of Poincaré:

> "Astronomy is also useful in that it lifts humans out and beyond ourselves; it is useful because it is grand; that is what we should say. It shows how small and insignificant is the perishable body of man, but at the same time how magnificent and otherworldly is his spirit, since his intelligence can embrace the whole of this dazzling cosmic immensity, in which his body is only an obscure and insignificant point, and thus finally be able to enjoy the silent cosmic harmony. Thus we attain the consciousness of our power, and this is something which can not cost too dear, since this consciousness makes us mightier."

[Henri Poincaré, La Valeur de la Science, (the Value of Science), 1904].

Is Our Planet Earth Occupying a *Privileged* Position in the Universe?

Over the past centuries, astronomers have defined our place in the wider Universe. Space is filled with trillions of galaxies, grouped into vast clusters and even larger superclusters. Our own Milky Way galaxy contains a few hundred billion stars, with a wide variety of sizes, masses, and colors. The Sun is just an average star, orbited by a handful of tiny planets. One of those planets is the Earth – the cradle of humanity. This speck of cosmic dust is where all of society's history has played out.

In other words, from the point of view of the discoveries in heliophysics, astrophysics, and cosmology during the last century that we have discussed, we have realized that we inhabit a tiny rocky planet orbiting an average star, the Sun, which is just one of the hundreds of billions of stars that make up and orbit around the center of the Galaxy. And, our Milky Way is a member of one of the many clusters and superclusters of galaxies that flood the inconceivably vast but finite cosmic space–time. Therefore, it follows that the position of our Earth is not central to the Milky Way Galaxy, nor is our Galaxy central to the Universe.

And, as we already discussed in the preceding chapters, there is no *center* in the Universe wherein galaxies do not move apart in an empty infinite space. Instead, the expansion of the Universe is due to the *stretching* of this very space, which drags along all the galaxies. The Big Bang happened simultaneously in all space, as a result of which there is no specific center in the Universe, since the center is *everywhere.*

Hence, to answer the question: *does our Earth occupies a privileged position in the entire Universe*, we can only recall two facts. First, that the Sun is an inconspicuous star at the outskirts of one of the trillion galaxies in the Universe. And *second*, that our Earth, this small rocky world, just one of the eight planets orbiting our Sun, seems to occupy a privileged position within the solar planetary system. More on this is discussed in the following.

Universe with Selected Physical Constants

First of all, as we discussed in the previous sections of this Chapter, we seem to live in a Universe where a small set of the global constants of physics have selected values, such that life can exist and flourish in so many forms in our own Universe, in contrast, possibly, to other sterile Universes, where the natural laws and constants are not at all friendly to the development of life, at least in the form that exists on Earth. Similarly, we can consider the various conditions for the habitability of other extrasolar worlds within the Universe and in particular within our own Galaxy.

A Safe Galaxy

First, we must point out that we live in a *safe* galaxy. Due to the killer γ-ray bursts that we discussed at length in Chap. 5, life – as we know it on Earth – could safely develop in only about 10% of all galaxies. And, such a life could only have developed in the last 5 billion years. In earlier times, galaxies were smaller and those killer γ-ray bursts were then much closer to the planets. So, they had the potential to cause large-scale extinctions on planets which could potentially host and develop life.

Safe Location of Earth Within Our Milky Way Galaxy

But also within our galaxy, we live in its safe suburbs. As we discuss at length in Chap. 5, γ-ray bursts are a major threat to life in other more star-dense regions of our galaxy. In particular, in the central galactic regions, the stellar density is very high because there stars are much closer to each other. This means that in those central galactic areas the risk for a star to be hit by nearby γ-ray bursts is larger. In a region a few thousand light years from the center of our Milky Way galaxy, where 25% of the stars are located, the stars there face a mortal danger, with a 95% possibility of being struck by a catastrophic γ-ray burst, within one billion years. Therefore, life in the form that exists on Earth can survive and develop with great confidence only in the far reaches of the Milky Way, far beyond about 30,000 light years from the galactic center. We should also mention that in the center of the Milky Way, lurks a massive black hole (Nobel Prize in Physics 2020), which, due to its strongest gravity field and tidal forces, is dangerous for the neighboring stars. Similar is the situation in a globular cluster, where the huge densities of stars create environments hostile for the development of life. For example, excessive dangerous high energy radiation is emitted and potential planets undergo strong gravitational disturbances.

This may be a partial explanation, of the so-called *Fermi paradox* discussed in Chap. 4, for the *great cosmic silence*: why we have not encountered any advanced civilization in the Universe? Our Milky Way is much older than our solar planetary system. Thus, there was enough time and space – the number of extrasolar systems with Earth-like conditions is vast – for life to develop elsewhere in our galaxy. So why haven't we met someone superior civilization somewhere in the galaxy? An answer to the Fermi paradox can therefore be found in that γ-ray bursts *may have struck and wiped out* life from many planets, which initially hosted life in other, less secure, places of our galaxy.

Prerequisites for Life Development in a Cosmical Environment

This brings us to the question of various other factors that affect the habitability of a planet in any planetary system. As we discussed in Chap. 2, from a biological point of view, for some form of life to flourish on another alien world, at least as we know it on Earth – on some planet or one of its natural satellites – there are three basic prerequisites:

- availability of some energy source;
- existence of extensive areas with water in liquid form;
- abundance of carbon, nitrogen, sulfur, phosphorus, and other chemical elements for the formation of complex organic molecules.

In exoplanets and exosatellites, the energy for the emergence and development of life comes from the light of the parent star. But we must take into account that some classes of stars create a particularly inhospitable environment for life, through strong flares in dangerous X-rays and UV radiation. Also, their other astronomical features, such as, an elliptical orbit with a small eccentricity, a correct inclination of the axis of rotation with respect to the plane of the orbit, the existence of the protective shield of the magnetic field, the absence of tidal locking, etc., are conditions, *sine qua non*, for the development of life in them, as we discuss in more detail in the following. Appropriate chemical reactions are guided by proteins which absorb light in the case of photosynthesis. Carbon is the backbone of life on Earth because it can support an extraordinary variety of molecules used in biology. Liquid water is an excellent solvent through which chemical reactions take place of life on Earth. Other factors required for life on Earth include elements such as nitrogen, which is used to make proteins and DNA, among many other molecules. In the following, let's take in a systematic way the prerequisites (*sine qua non*) for the appearance of life on a planet.

The Planet in the Habitable Zone of the Parent Star

First of all, for a planet to be habitable, must be within the so-called *habitable zone* of its parent star. The habitable zone is a shell-shaped region surrounding a star within which liquid water could exist in the surface of the planet. Inside the habitable zone, the temperature is very high and liquid water evaporates. Beyond the outside border of the habitable zone, the temperature is low and liquid water freezes. The inner and outer radii of the habitable zone depend on the photospheric temperature of the star. In reddish stars, such as red dwarfs, the habitable zone is relatively close to the star, while in yellowish and blue stars, the habitable zone is farther away from the star.

Apart from the necessity of an energy source, liquid water is considered the most important ingredient for life, considering how crucial its role is to all life systems on Earth. This may reflect the well-known dependence of life on water: water is for life on Earth what blood is to the human organism, flowing in our veins and arteries.

Suitable Photospheric Temperature of the Parent Star

Because the existence of *liquid* water depends on the ambient temperature, one of the first factors affecting the habitable zone of a star is its photospheric temperature. The appropriate photospheric temperatures for a star that favor the habitability of planets range in the interval of 7,000 to 4,000 degrees. Stars of this category have characteristics which are considered friendly toward planetary habitability. Specifically:

They have lifetimes spanning at least a few billion years, thus providing a sufficient temporal window for life to develop. The brightest stars with temperatures greater than about 7,000 degrees typically live less than a billion years.

They emit enough UV radiation so that they have the potential to create suitable atmospheric conditions on their planets, such as an ozone layer – but not too much, so that an intense ionization caused by UV radiation that would destroy the emergence of life.

Water may exist in liquid form on the surface of planets orbiting stars, at distances where the so-called *tidal locking* does not occur, where the period of rotation of the planet around its axis, i.e., the length of the planet's day, lasts as long as its revolution around the star, i.e., the length of the planetary year. To such a tidally locked planet, one side is permanently facing the parent star and the other side is plunged in deep darkness. Such a situation causes high temperatures on the side facing the star and very low temperatures on the other side. On planets around red dwarfs, where the habitable zone is relatively close to the star, the planets may be tidally locked and therefore unsuitable for the development of life. Unless there is a large cloud cover of the planet, so that a large percentage of the incident radiation is reflected reducing the extreme temperatures of the illuminated side.

The above temperature range possibly corresponds to 5–10% of the stellar population in the Milky Way. The question if stars with lower luminosity – such as the red dwarfs – are able to support life on their planets, is one of the most important open questions today in the research for planetary habitability, since they make up the majority (70–90%) of the stellar population of the Milky Way.

Long Lifetime of the Parent Star

Red dwarfs have in principle a great advantage for the appearance of life on their planets over the other stars: they have a much longer life span. It took about 4.5 billion years to developed life on Earth, which as we know, will have suitable external conditions to continue for another 1 to 2.3 billion years. Red dwarfs, by contrast, can live for trillions of years, because the nuclear reactions in their interiors are extremely slow compared to those in the larger stars, which means life would have had more time to evolve and survive.

However, red dwarfs are considerably more variable and violent than their more stable, higher mass and temperature cousins. Often their photosphere is covered by huge spots that can reduce their brightness by up to 40%, with the spots lasting for months at a time, while giant flares also occurring in their atmospheres, oftentimes doubling their luminosity in a few minutes. Those changes in stellar brightness are very harmful to life, since they destroy complex organic molecules that could potentially form biological precursors.

Strong Planetary Magnetosphere

For a planet around a red dwarf to support life, it would need to have a sufficiently strong magnetic field to protect the planet inhabitants from deadly stellar flares. A tidally locked planet, however, spins rather slowly, so it cannot produce the required magnetic field with the dynamo action in the planetary interior.

Steady Radiation

The nonvariability of the radiation emitted by a parent star is another important factor for the habitability of planets. The Sun, in this respect, as well as in many others, is a relatively convenient star for the development of life in its habitable zone. The variability of the solar luminosity during the 11/year solar cycle, between the maximum and minimum of solar activity, does not exceed $\approx 0.1\%$. As we discussed in Chap. 1, there is strong evidence that even small changes in the solar luminosity had a significant impact on the Earth's climate in recent centuries. The *Little Ice Age* during the mid-second millennium, for example, is attributed to a relatively long-term reduction of the solar luminosity. Therefore, a star should not have large changes in its luminosity, in order not to endanger the habitability of its planets.

Approximate Circular Planetary Orbit

Another important factor that affects the habitability of a planet is the eccentricity e of the orbit – the difference between the maximum and minimum distance of the planet from its star, divided by the sum of these two distances. The eccentricity e describes the shape of the elliptical orbit, with the eccentricity of a circular orbit being $e = 0$. The greater the eccentricity, the greater the changes in temperature on the planet's surface, since the value of the eccentricity has important effects on exoplanets' climates. Although they might generally adapt, living organisms have a limit in which can withstand continuous changes, especially if these changes affect the melting and boiling point of the main material which is the basic solvent on the planet, as water is on Earth. If for example, the oceans of the Earth happened to boil and then to freeze over the course of a year, it is very hard to imagine how life could develop, as it did. The more complex the organisms, the greater their sensitivity to temperature changes. The terrestrial orbit is nearly circular, with an eccentricity less than $e = 0.02$. The data which have been collected as to the eccentricities of the extrasolar planets show that 90% of exoplanets have eccentricities which are greater than that of the Earth, with the average eccentricity being $e = 0.25$. This means that the large majority of exoplanets have highly eccentric orbits. Therefore, even if the average distance from their star remains within the habitable zone, due to those eccentric orbits they remain only for a short time within the zone habitable!

Short Duration of the Planetary day

An additional important factor for the habitability of a planet is its period of self-rotation, i.e., the length of its day. The planet should rotate fast enough, so that assuming that inside the planetary interior there is molten iron, the magnetic dynamo we discussed in Chap. 1 can create a sufficiently strong magnetic field on the planet, which then protects the planet from explosive phenomena and high-energy particles that emerge from the parent star. Also, when the planet rotates quickly, the lengths of the day and night will not be very long. But if the planet rotates slowly, the length of the day can last, compared to Earth, for whole months or even years, so the difference in temperature between the day of the planet and the night will be large, resulting in extreme physical conditions in its surface and atmosphere.

Mild Seasons on the Planet

The angle formed by the axis of rotation of a planet with the perpendicular to the plane of the orbit must be within some limits, so that there appear mild seasons on the planet. When this inclination is too small, there will be no seasons. Note that all dynamism of the terrestrial biosphere is due to this finite angle. On the other hand, if the inclination angle is large, it happens quite the opposite, and the change of seasons is extreme. For tidally locked planets, we have a limited area of open ocean centered at the substellar point and complete ice coverage on the night side.

And Our Earth?

The Earth seems to meet all the conditions discussed previously for the development of life. First of all, the Sun, our mother star is a mild yellow star, with only very small flares in its brightness. It is a star that lives on the safe outskirts of the Galaxy, avoiding the dangerous densely populated inner galactic regions. He has lived for the long span of 4.5 billion years, enough time to develop life. It spins relatively quickly around its axis, within 24 h, as a result of which it produces a magnetic field in the terrestrial interior that protects us from outbursts of solar activity, at the same time homogenizing the day and night atmospheric differences. The Earth's axis of rotation is inclined 23.4° to the perpendicular to the plane of its axis' orbit, as a result of which seasonal differences are created on its surface with a period of one year. Its orbit around the Sun is centered in the solar habitable zone, so that water, this wonderful and unique solvent, is found in its three forms, of solid, liquid, and gas. The Earth's orbit around the Sun is almost circular, with the result that the annual temperature changes on the surface of the planet are not extreme. All these together have resulted in an explosive increase in biodiversity, at a constant rate, except for some interruptions due to mass extinctions. Although it is estimated that more than 99% of the species of life that developed on Earth, i.e., more than 5 billion life species, that have existed on Earth today have disappeared, there are still around 10–14 million extant species, of which only about 1.2 million have studied by man, and the remaining 86% have not even been described…

On Earth we live today over 8 billion people, and we are the dominant species on our planet. Yet, what seems to differentiate us from the millions of the other species of the animal kingdom is the spark that Aristotle referred to as permeating us: «πάντες ἄνθρωποι τοῦ εἰδέναι ὀρέγονται φύσει», (all humans naturally desire knowledge). And this is so, because according to the Bible, the Creator created man in a unique way: «Let us make humankind in our image, after our likeness … and breathed into his nostrils the breath of life», that is why we try to overcome our smallness. And the Earth, indeed, is a precious blue gem floating in Space, in harmony with the wider diverse Universe, the *Cosmos*, which was so characterized by the ancient Greek thinkers and natural philosophers for its beauty and where everything seems to be *very good*.

At least we can, within the framework of reason and science, find out what kind of world is the one that surrounds us and what is our position in the *Cosmos*. Thus, we can really appreciate why our Earth is indeed a *precious blue gem* floating alone in Space, and therefore worth of our protection!

Hence, What is Our Place in the Universe? Where Do We Come From? What Are We? Where Are We Going?

This is the title of a painting created by the French artist Paul Gauguin (1848–1903) in Haiti (1897), Oil on canvas, 139 × 375 cm. Boston Museum of Fine Arts, Boston (MFA), Massachusetts, USA Paul Gauguin stated that the painting should be read from right to left, with the three large human groups depicting the questions posed in the title: the three women with the child represent the beginning of life. The middle one group symbolizes the daily life of adults. In the last group, an elderly woman nearing death appears to reconcile and renounce her thoughts, while at her feet, a strange white bird represents their futility reasons. The image in the background and to the left represents what Paul Gauguin described as the after life and soul.

This question posed by Gauguin is a rather difficult one to answer today, using all discoveries discussed in the previous chapters: why are we born into this world, who are we really, where do we end up, and where are we going? All in all, what is the destination of man? Using the terminology of mathematics, these questions may belong to another mathematical manifold, that of metaphysics.

For example, according to Christianity, the final destination and ultimate purpose of human existence is the process by which a person is transformed by divine grace to become what God is by nature (*Let us make man in our image and likeness*, Genesis 1:26).

References

Barrow JD (2001) Cosmology, life, and the anthropic principle. Ann N Y Acad Sci 950(1):139–153
Davies PCW (1985) Superforce: the search for a grand unified theory of nature. New York: Simon and Schuster, p 104
Dicke R (1961) Dirac's cosmology and Mach's principle. Nature 192:440–441
Ehrenfest P (1917) In what way does it become manifest in the fundamental laws of physics that space has three dimensions? In: Klein MJ (ed) Proceedings of the Royal Netherlands Academy of Arts and Sciences, vol 20, pp 200–209. Paul Ehrenfest—Collected Scientific Papers, Amsterdam, North Holland Publ. Co., pp 400–409
Fan X, MyersTG, Sukra BAD, Gabrielse G (2023) Measurement of the electron magnetic moment. Phys Rev Lett 130(7)
Feynman RP (2014) QED: The Strange Theory of Light and Matter. Princeton University Press, p 129
Livio M, Hollowell D, Weiss A, Truran JW (1989) The anthropic significance of the existence of an excited state of ^{12}C. Nature 340(6231):281–84
Misner CW, Thorne KS, Wheeler JA (1973) Gravitation, p 1205. San Francisco, CA: Freeman
Nath B (2015) The fine structure constant. Resonance 20:383–388. https://www.ias.ac.in/article/fulltext/reso/020/05/0383-0388
Rees M (2001) Just six numbers: the deep forces that shape the universe. Basic books
Tangherlini FR (1963) Einstein's Pseudo-tensor in n spatial dimensions for static systems with spherical symmetry. Nuovo Cimento 27:636

Tegmark M (1997) On the dimensionality of spacetime. Class Quantum Grav 14:L69–L75. arXiv: gr-qc/9702052

Whitrow CJ (1955) Why physical space has three dimensions? Br J Philos Sci 6:13–31

Wilczynska M et al (2020) Four direct measurements of the fine-structure constant 13 billion years ago. Sci Adv. arXiv:2003.07627

Bibliography

Abramenko V, Yurchyshyn V, Goode P, Kilcik A (2010) Statistical distribution of size and lifetime of bright points observed with the new solar telescope. Astrophys J Lett 725(1):L101–L105

Agullo I, Singh P (2016) Loop quantum cosmology: a brief review. Gen Relativ Quant Cosmol. arXiv:1612.01236v1

Akiyama K (2019) The event horizon telescope collaboration, first M87 event horizon telescope results. I. The shadow of the supermassive black hole. Astrophys J Lett 875:L1 (17pp)

Alpher RA, Bethe H, Gamow G (1948) The origin of chemical elements. Phys Rev 73(7):803–804 [the "αβγ" paper]

Albrecht A, Steinhardt PJ (1982) Cosmology for grand unified theories with radiatively induced symmetry breaking. Phys Rev Lett 48:1220

Anderson M, Martin SF (2005) Formation of an extraordinarily long filament channel. In: Sankarasubramanian K, Penn M, Pevtsov A (eds) Large-scale structures and their role in solar activity, ASP conference series, vol 346

Anderson L et al (2012) The clustering of galaxies in the SDSS-III Baryon Oscillation Spectroscopic Survey: baryon acoustic oscillations in the data release 9 spectroscopic galaxy sample. Monthly Not Roy Astron Soc 427:3435–3467.

Andre P (2002) The initial conditions for protostellar collapse: observational constraints. In: Bouvier J, Zahn JP (eds) Star formation and the physics of young stars, Les Ulis, Fr.: EDP Sci, pp 1–38

André P (2002) Representation of the evolution of young stellar objects. Nearby protoclusters as laboratories for understanding star formation on galactic scales. Astrophys Space Sci 281(1):51–66

Andrews SM, Huang J, Pérez LM et al (2018) The disk substructures at high angular resolution project (DSHARP). I. Motivation, sample, calibration, and overview. ApJ 869:L41

Andreoni et al (2022) A very luminous jet from the disruption of a star by a massive black hole. Nat Astron 612:430–434

Ashtekar A, Pawlowski T, Singh P (2006) Quantum nature of the Big Bang. Phys Rev Lett 96 (14):141301.

Baade W, Zwicky F (1934) Remarks on supernovae and cosmic rays. Phys Rev 46:76–77

Baade W, Zwicky F (1934) On supernovae. Proc Nat Acad Sci USA 20(5):254–259

Baade W, Zwicky F (1934) Cosmic rays from supernovae. Proc Nat Acad Sci USA 20(5):259–263

Babcock HW (1961) The topology of the sun's magnetic field and the 22-YEAR cycle. Astrophys J 133:572

Backer DC, Kulkarni SR, Heiles C, Davis MM, Goss WM (1982) A millisecond pulsar. Nature 300(5893):615–618

K. Tsinganos, *A Century of Cosmic Discoveries*, Springer Praxis Books,
https://doi.org/10.1007/978-3-032-06282-6

Bahcall JN, Kirhakos S, Schneider DP, Davis RJ, Muxlow TWB, Garrington ST, Conway RG, Unwin SC (1995) Hubble space telescope and MERLIN observations of the jet in 3C 273. Astrophys J Lett 452:L91 [arXiv:astro-ph/9509028]

Baker DN, Lanzerotti LJ (2016) Resource letter SW1: space weather. Am J Phys 84:166

Balashev S, Noterdaeme P, Gupta N, Krogager JK, Combes F, Lopez S, Petitjean P, Omont A, Srianand R, Cuellar R (2025) Quasar radiation transforms the gas in a merging companion galaxy. Nature. eprint arXiv:2505.15766

Balick B, Brown RL (1974) Intense sub-arcsecond structure in the galactic center. Astrophys J 194:265–270

Bally J, Odell CR, McCaughrean MJ (2000) Disks, microjets, windblown bubbles, and outflows in the Orion Nebula. Astron J 119:2919–2959

Bañados E, Mazzucchelli C, Momjian E, Eilers AC, Wang F, Schindler JT, Connor T, Andika IT, Barth AJ, Carilli C, Davies FB, Decarli R, Fan X, Farina EP, Hennawi JF, Pensabene A, Stern D, Venemans BP, Wenz L, Yang J (2021) The discovery of a highly accreting, radio-loud quasar at z = 6.82. Astrophys J 909(1)

Boccardi B, Krichbaum TP, Ros ´E, Zensus JA (2017) Radio observations of active galactic nuclei with mm-VLBI. Astron Astrophys Rev 25:4. [arXiv:1711.07548v1 [astro-ph.HE]

Boccardi B, Krichbaum TP, Bach U, Bremer M, Zensus JA (2016a) First 3 mm-VLBI imaging of the two-sided jet in Cygnus A. Zooming Radio observations of active galactic nuclei with mm-VLBI 37 into the launching region. A&A, 588:L9

Boccardi B, Krichbaum TP, Bach U, Mertens F, Ros E, Alef W, Zensus JA (2016b) The stratified two-sided jet of Cygnus A. Acceleration and collimation. A&A, 585:A33

Bodo G, Ferrari A, Massaglia S, Tsinganos K (1985) Accceleration and thermal instabilities in the jets of SS433. Astron Astrophys 149:246–252

Bodo G, Massaglia S, Ferrari A, Trussoni E (1994) Kelvin-Helmholtz instability of hydrodynamic supersonic jets. Astron Astrophys 283:655

Boisse I, 22 co-authors (2013) HAT-P-42b and HAT-P-43b, Two inflated transiting hot Jupiters from the HATNet survey. Astron Astrophys 558:A86

Byrne J (2015) Investigating the kinematics of coronal mass ejections with the automated CORIMP catalog. J Space Weather Space Clim 5(A19)14

van den Bergh S (2011) The curious case of Lemaitre's qquation No 24. J Roy Astron Soc Can 105(4):151

Caldwell RR, Kamionkowski M, Weinberg NN (2003) Phantom energy and cosmic doomsday. Phys Rev Lett 91(7)

Cargill P, Parnell C, Browning P, De Moortel I, Hood A (2010) Magnetic reconnection in the solar atmosphere: from proposal to paradigm. Astron Geophys 51(3):3.31

Carpenter KG (2005) The Stellar Imager/ A UV/Optical deep-space telescope to image stars and observe the Universe with 0.1 milli-arcsec angular resolution. https://hires.gsfc.nasa.gov/si/

Carter B (1974) Large number coincidences and the anthropic principle in cosmology, confrontation of cosmological theories with observational data. In: Proceedings of the symposium, krakow, Poland, 10–12 Sept 1973 (A75–21826 08–90) Dordrecht, D. Reidel Publishing Co., p 291–298

Chandrasekhar S (1990) Truth and beauty: aesthetics and motivations in science. University of Chicago Press

Chantry L, Cayatte V, Sauty C, Vlahakis N, Tsinganos K (2018) Nonradial and nonpolytropic astrophysical outflows X. Relativistic MHD rotating spine jets in Kerr metric. Astron Astrophys 612:A63

Choudhuri AP (2024) "Gene": a personal tribute to the life and science of Eugene Newman Parker. Rev Mod Plasma Phys 8:6

Chrimes AA et al (2025) The infrared counterpart and proper motion of magnetar SGR 696:0501+4516, A&A

Christensen-Dalsgaard J, Dappen W, Ajukov SV, Anderson ER, Antia HM, Basu S, Baturin VA, Berthomieu G, Chaboyer B, Chitre SM, Cox AN, Demarque P, Donatowicz J, Dziembowski WA, Gabriel M, Gough DO, Guenther DB, Guzik JA, Harvey JW, Hill F, Houdek G, Iglesias CA, Kosovichev AG, Leibacher JW, Morel P, Proffitt CR, Provost J, Reiter J, Rhodes EJ, Rogers FJ, Roxburgh IW, Thompson MJ, Ulrich RK (1996) The current state of solar modeling. Science 272(5266):1286–1292

Colless M et al (2001) The 2dF galaxy redshift survey: spectra and redshifts. Mon Not Roy Astron Soc 328:1039–1063

Contopoulos G (2002) Order and chaos in dynamical astronomy. Springer

Contopoulos G (2004) Adventures in order and chaos, a scientific autobiography. Springer

Craig DJ (2018) Was the big bang really a big bounce? Columbia Mag Columbia Univ

Curtis HD (1918) Descriptions of 762 nebulae and clusters photographed with the Crossley reflector. Publ Lick Obs 13:9–42

De Marchi G et al (2024) Protoplanetary disks around sun-like stars appear to live longer when the metallicity is low. Astrophys J 977:214

Di Valentino E, Mena O, Pan S, Visinelli L,Yang W, Melchiorri A, Mota DF, Riess AG, Silk J (2021) In the realm of the Hubble tension—a review of solutions. Class Quant Grav 38(15)

Dirac PAM (1937) The cosmological constants. Nature 139:323

Dicke RH, Peebles PJE, Roll PJ, Wilkinson DT (1965) Cosmic black-body radiation. Astrophys J Lett 142:414–419

Eddington AS (1926) The internal constitution of the stars. Cambridge University Press

Eddington AS (1927) Stars and atoms. Clarendon Press, Oxford

Einstein A (1905) Zur Elektrodynamik bewegter Körper (On the electrodynamics of moving bodies), Annalen der Physik 17, 891, which appeared in the book "The Principle of Relativity", published in 1923 by Methuen and Company, Ltd. of London. Most of the papers in that collection are English translations by W. Perrett and G.B. Jeffery from the German Das Relativatsprinzip (The principle of relativity), 4th ed., published by in 1922 by Tuebner. Another English translation of the paper On the Electrodynamics of Moving Bodies is by Megh Nad Saha (1920)

Einstein A, de Sitter W (1932) On the relation between the expansion and the mean density of the universe. In: Contributions from the Mount Wilson observatory, vol 3, pp 51–52, [Proceedings of the national academy of sciences, vol 18, pp. 213–214]

Economou EN (2016) From the quarks to the universe, a short physics course. Springer

Fermi E Paradox archived: https://web.archive.org/web/20070629174738/http://www.fas.org/sgp/othergov/doe/lanl/la-10311-ms.pdf

Ferrari A, Trussoni E, Zaninetti L (1978) Relativistic Kelvin-Helmholtz instabilities in extragalactic radio sources. Astron Astrophys 64:43

Ferrari A, Habbal SR, Rosner R, Tsinganos K (1984) A wind-type model for the generation of astrophysical jets. Astrophys J Lett 277:L35–L39

Ferrari A, Trussoni E, Rosner R, Tsinganos K (1985) On wind-type flows in astrophysical jets. I. Initial relativistic acceleration. Astrophys J 297:397–418

Ferrari A, Tsinganos K (1986) Wind-type flows and astrophysical jets. Can J Phys 64(4):456–462

Ferrari A, Trussoni E, Rosner R, Tsinganos K (1986) On wind-type flows in astrophysical jets. II—Propagation outside the nucleus and the case of M87. Astrophys J 300:577–590

Forman W, Jones C, Cominsky L, Julien P, Murray S, Peters G (1978) The fourth Uhuru catalog of X-ray sources. ApJS 38:357

Geller MJ, Huchra JP (1989) Mapping the universe. Science 246(4932):897–903

Gombosi TI, van der Holst B, Manchester WB, Sokolov IV (2018) Extended MHD modeling of the steady solar corona and the solar wind. Living Rev Solar Phys 15(4)

Gott JR III et al (2005) A map of the universe. Astrophys J 624(2):463–484. arXiv:astro-ph/0310571

Govoni F, Orrù E, Bonafede AM, Paladino IR, Vazza F, Murgia M, Vacca V, Giovannini G, Feretti L (2019) A radio ridge connecting two galaxy clusters in a filament of the cosmic web. Science 364(6444):981–984. arXiv: 1906.07584

Georgoulis M, Nindos A, Zhang H (2019) The source and engine of coronal mass ejections. Philos Trans Roy Soc A: Math Phys Eng Sci 377: 20180094

Goddi C, Crew G, Impellizzeri V, Martí-Vidal I, Matthews LD, Messias H, Rottmann H, Alef W, Blackburn L, Bronzwaer T, Chan CK, Davelaar J, Deane R, Dexter J, Doeleman S, Falcke H, Fish VL, Fraga-Encinas R, Fromm CM, Herrero-Illana R, Issaoun S, James D, Janssen M, Kramer M, Krichbaum TP, De Laurentis M, Liuzzo E, Mizuno Y, Moscibrodzka M, Natarajan I, Porth O, Rezzolla L, Rygl K, Roelofs F, Ros E, Roy AL, Shao L, van Langevelde HJ, van Bemmel I, Tilanus R, Torne P, Wielgus M, Younsi Z, Zensus JA (2019) Event horizon telescope collaboration. The Messenger 177:25–35

Golub L et al (1999) A new view of the solar corona from the transition region and coronal explorer (TRACE). Phys Plasmas 6 (5): 2205–2216

Goslar T (2003) 14C as an indicator of solar variability. PAGES News 11(2/3):12–14

Govoni F, Orrù E, Bonafede A, Iacobelli M, Paladino R, Vazza F, Murgia M, Vacca V, Giovannini G, Feretti L, Loi F, Bernardi G, Ferrari C, Pizzo RF, Gheller C, Manti S, Brüggen M, Brunetti G, Cassano R, de Gasperin F, Enßlin TA, Hoeft M, Horellou C, Junklewitz H, Röttgering HJA, Scaife AMM, Shimwell TW, van Weeren RJ, Wise M (2019) A radio ridge connecting two galaxy clusters in a filament of the cosmic web. Science 364(6444):981–984. arXiv:1906.07584

Gracia J, Tsinganos K, Bogovalov SV (2005) Magnetic collimation of the relativistic jet in M 87. Astron Astrophys 442:L7–L10

Gracia J, Bogovalov SV, Tsinganos K (2008) MHD models and synthetic synchrotron maps for the jet of M87. Int J Mod Phys D 17(10):1687–1693

Gregory SA, Thompson LA (1978) The Coma/A1367 supercluster and its environs. Astrophys J 222:784–799

Guth AH (1981) Inflationary universe: a possible solution to the horizon and flatness problems. Phys Rev D 23:347

Häberle M, Neumayer N, Seth A, Bellini A, Libralato M, Baumgardt H, Whitaker M, Dumont A, Alfaro-Cuello M, Anderson J, Clontz C, Kacharov N, Kamann S, Feldmeier-Krause A, Milone A, Nitschai M, Pechetti R, van de Ven G (2024) Fast-moving stars around an intermediate-mass black hole in ω Centauri. Nature 631(8020):285–288

Hale GE (1908) On the probable existence of a magnetic field in sun-spots. Astrophys J 28:315

Hamilton K, Cho YH (2017) J Res Prog 1. Howard Community College

Hamilton K, Cho YH (2018) PSR1913+16: the binary pulsar, mathematical reproduction and extension. J Res Progress 1

Hawking S (1988) A brief history of time. Bantam Dell Publishing Group

Hewish A (1986) The pulsar era. Q J Roy Astron Soc 27:548–558

Hewish A, Bell SJ, Pilkington JDH, Scott PF, Collins RA (1968–02) Observation of a rapidly pulsating radio source. Nature 217:709–713

Hinshow G et al (2007) Three-year Wilkinson Microwave Anisotropy Probe (WMAP) observations: temperature analysis. Astrophys J Suppl Ser 170 (2):288–334

Hobson MP, Efstathiou GP, Lasenby AN (2006) General relativity: an introduction for physicists. Cambridge University Press

Isella A, Huang J, Andrews SM et al (2018) The disk substructures at high angular resolution project (DSHARP). IX. A high-definition study of the HD 163296 planet-forming disk. ApJ 869:L49

Jones WC, 36 coauthors (2006) A measurement of the angular power spectrum of the CMB temperature anisotropy from the 2003 flight of BOOMERANG. Astrophys J 647:823–832

Kalas P, Graham JR, Fitzgerald MP, Clampin M (2013) STIS coronagraphic imaging of fomalhaut: main belt structure and the orbit of fomalhaut b. Astrophys J 775:56

Kazanas D (1980) Dynamics of the universe and spontaneous symmetry breaking. Astrophys J 241:L59-63

Kim JY, Krichbaum TP, Lu RS, Ros E, Bach U, Bremer M, de Vicente P, Lindqvist M, Zensus JA (2018) The limb-brightened jet of M 87 down to 7 Schwarzschild radii scale. Astron Astrophys 616:188

Konigl A, Choudhuri AR (1985) Force-free equilibria of magnetized jets. Astrophys J 289:173–187

Kozmanyan A, Bourdin H, Mazzotta P, Rasia E, Sereno M (2019) Deriving the Hubble constant using Planck and XMM-Newton observations of galaxy clusters. Astron Astrophys 621:A34

Kovalev YY, Lister ML, Homan DC, Kellermann KI (2007) The inner jet of the radio galaxy M87. Astrophys J 668: L27–L30 [astro-ph:0708.2695v2]

Krimigis SM, Mitchell DG, Roelof EC, Hsieh KC, McComas DJ (2009) Imaging the interaction of the heliosphere with the interstellar medium from Saturn with Cassini. Science 326(5955):971

Lada CJ (1987) Star formation—from OB associations to protostars, star forming regions. In: Dordrecht DR (ed) Proceedings of the symposium, Tokyo, Japan, 11–15 Nov 1985 (A87-45601 20-90) Publishing Co., pp. 1–17 (1987)

Lang KR (2013) Essential astrophysics. Springer, Heidelberg

Laplace PS (1799) In allgemeine geographische ephemeriden

Leibundgut B, Sollerman J (2001) A cosmological surprise: the universe accelerates. Europhys News 32(4)

Linde AD (1982) A new inflationary universe scenario: a possible solution of the horizon, flatness, homogeneity, isotropy and primordial monopole problems. Phys Lett B 108:389

Lineweaver CH (2005) Inflation and the cosmic microwave background. In: The new cosmology. In: Colless M (ed) Proceedings of the 16th international physics summer school. Held 3–14 February 2003 in Canberra, Australia (Anglo-Australian Observatory, Australia). Published by World Scientific Publishing Co. Pte. Ltd., ISBN #9789812702357, pp 31–65

Lineweaver C, Davis TM (2005) Misconceptions about the Big Bang. Sci Am 292(3):36–45

Lister ML, Aller HD, Aller MF, Cohen MH, Homan DC, Kadler M,, Kellermann KI, Kovalev YY, Ros E, Savolainen T, Zensus JA, Vermeulen RC (2009) MOJAVE: monitoring of jets in active galactic nuclei with VLBA experiments. V. Multi-Epoch VLBA Images AJ 137:3718–3729

Lobanov AP, Zensus JA (2001) A cosmic double helix in the archetypical quasar 3C273. Science 5, 294(5540):128–131. https://doi.org/10.1126/science.1063239]

Loeb A (2006) The dark ages of the universe. Sci Am 295 (5):46–53

Lorimer DR, Kramer M (2005) Handbook of pulsar astronomy. Cambridge University Press, Cambridge, UK

Low BC (1996) Solar activity and the corona. Solar Phys 167:217

Mack K (2020) Tearing apart the universe. Am Sci 108 (6):356

Madau P, Dickinson M (2014) Cosmic star formation history. Ann Rev Astron Astrophys 52:415–486. arXiv:1403.0007

Massaglia S, Bodo G, Rossi P, Capetti S, Mignone A (2016) Making Faranoff-Riley I radio sources-I. Numerical hydrodynamic 3D simulations of low-power jets. Astron Astrophys 596:A12

Mather JC, Cheng ES, Eplee RE, Isaacman RB, Meyer SS, Shafer RA, Weiss R, Wright EL, Bennett CL, Boggess NW, Dwek E, Gulkis S, Hauser MG, Janssen M, Kelsall T, Lubin PM, Moseley SH, Murdock TL, Silverberg RF, Smoot GF, Wilkinson DT (1990) A preliminary measurement of the cosmic microwave background spectrum by the cosmic background explorer (COBE) satellite. Astrophys J 354:L37–L40

Mayor M, Queloz D (1996) A search for substellar companions to solar-type stars via precise Doppler measurements: a first Jupiter mass companion detected, In: Pallavicini R, Dupree AK (eds) Proceedings of 9th Cambridge workshop, 3–6 October 1995, Florence, ASP, vol 109, pp 35–39

Meliani Z, Sauty C, Vlahakis N, Tsinganos K, Trussoni E (2006) Nonradial and nonpolytropic astrophysical outflows. VIII. A GRMHD generalization for relativistic jets, A&A 447:797–812

Michell J (1783) On the means of discovering the distance, magnitude, & c. of the fixed stars, in consequence of the diminution of the velocity of their light, in case such a diminution should be found to take place in any of them, and such other data should be procured from observations, as would be farther necessary for that purpose. By the Rev. John Michell BDFRS. In a Letter to Henry Cavendish, Esq. F. R. S. and A. S. Philos Trans Roy Soc Lond 74:35–57 (1783)

Miller AI (2005) Empire of the stars: obsession, friendship, and betrayal in the quest for black holes. In: Mariner books (2005)

Migkas K, Schellenberger G, Reiprich TH, Pacaud F, Ramos-Ceja ME, Lovisari L (2020) Probing cosmic isotropy with a new X-ray galaxy cluster sample through the LX−T scaling relation. Astron Astrophys 636:A15

Migkas K, Pacaud F, Schellenberger G, Erler J, Nguyen-Dang NT, Reiprich TH, Ramos-Ceja ME, Lovisari L (2021) Cosmological implications of the anisotropy of ten galaxy cluster scaling relations. Astron Astrophys 649:A151

Mishra S, Fox AJ, Krishnarao D, Lucchini S, D'Onghia E, Cashman FH, Barger KA, Lehner N, Tumlinson J (2024) The truncated circumgalactic medium of the large magellanic cloud. ApJ Ovt arXiv:2410.11960v1

Miyoshi M, Moran J, Hernnstein J, Greenhill L, Nakal N, Diamond P, Inoue M (1995) Evidence for a black hole from high velocities in a sub-parsec region of NGC4258. Nature 373:127–129

Mooley KP, Deller AT, Gottlieb O, Nakar E, Hallinan G, Bourke S, Frail DA, Horesh A, Corsi A, Hotokezaka K (2018) Superluminal motion of a relativistic jet in the neutron-star merger GW170817. Nature 561(7723):355–359. [arXiv:1806.09693v2 (astro-ph.HE) 30 Oct 2018]

Morgan H, Taroyan Y (2017) Global conditions in the solar corona from 2010 to 2017. Sci Adv 3

Mundt R, Buehrke T, Solf J, Ray TP, Raga AC (1990) Optical jets and outflows in the HL Tauri region. Astron Astrophys 232:37

New York Times (1966) Pioneer studies solar weather, New York

Oppenheimer JR, Volkoff GM (1939) On massive neutron cores. Phys Rev 55:374

Özel F, Freire P (2016) Masses, radii, and the equation of state of neutron stars. Annu Rev Astron Astrophys 54:401–440

Parker EN (1958) Dynamical instability in an anisotropic ionized gas of low density. Phys Rev 109:1874–1876

Parker EN (1959) Plasma dynamic determination of shock thickness in an ionized gas. Astrophys J 129:217

Parker EN (2020) Exploring the innermost solar atmosphere. Nat Astron 4:19–20

Pearson TJ, Unwin SC, Cohen MH et al (1981) Superlumina J expansion of quasar 3C273. Nature 290

Pearson TJ, Unwin SC, Cohen MH, Linfield RP, Readhead ACS, Seielstad GA, Simon RS, Walker RC (1982) Superluminal expansion of 3C 273. In: Heeschen DS, Wade CM (eds) Extragalactic radio sources, IAU, vol 97. Springer, pp 355–356

Perley RA, Dreher JW, Cowan JJ (1984) The jet and filaments in Cygnus A. ApJ 285:L35–L38

Perlmutter S (2003) Supernovae, dark energy and the accelerating universe. Phys Today 56(4):53

Peters PC, Mathews J (1963) Gravitational radiation from point masses in a Keplerian orbit. Phys Rev 131:435

Poggiali V, Mastrogiuseppe M, Hayes AG, Seu R, Birch SPD, Lorenz R, Grima C, Hofgartner JD (2016) Liquid-filled canyons on Titan. Geophys Res Lett 43:7887–7894

Penrose R (1965) Gravitational collapse and space-time singularities. Phys Rev Lett 14:57

Reichardt CL, 17 coauthors (2006) High resolution cmb power spectrum from the complete acbar data set. ApJ 647:823

Ren BB, Fogarty K, Debes JH, Meyer ET, Mo, Mawet YD, Perrin MD, Ogle PM, Sahlmann J (2024) 3C 273 host galaxy with Hubble space telescope coronagraphy. Astron Astrophys [arXiv:2402.09505v1]

Rivera-Ingraham A et al (2013) Herschel observations of the W3 GMC: clues to the formation of clusters of high-mass stars. Astrophys J 766:14

Rizzo F, Vegetti S, Powell D, Fraternali F, McKean JP, Stacey HR, White SDM (2020) A dynamically cold disk galaxy in the early universe. Nature 584:201–204. arXiv:2009.01251v1
Rodono M et al (1995) Starspot evolution, activity cycle and orbital period variation of the prototype active binary RS Canum Venaticorum. Astron Astrophys 301:75
Roussos E, Kollmann P, Krupp N, Kotova A, Regoli L, Paranicas C, Mitchell DG, Krimigis SM, Hamilton D, Brandt P, Carbary J, Christon S, Dialynas K, Dandouras I, Hill ME, Ip WH, Jones GH, Livi S, Mauk BH, Palmaerts B, Roelof EC, Rymer A, Sergis N, Smith HT (2018) A radiation belt of energetic protons located between Saturn and its rings. Science 362:6410
Russell HN (1914) Relations between the spectra and other characteristics of the stars. Popular Astron 22:275
Sanchez S, Fournier A, Pinheiro K, Aubert J (2013) A mean-field Babcock-Leighton solar dynamo model with long-term variability. arXiv:1307.4115
Sawala T, Delhomelle J, Deason AJ et al (2025) No certainty of a Milky Way–Andromeda collision. Nat Astron (2025)
Schwarzchild K (1916) Über das Gravitationsfeld eines Massenpunktes nach der Einsteinschen Theorie (On the gravitational field of a mass point according to Einstein's theory), Sitzungsberichte der Königlich Preussischen Akademie der Wissenschaften (Proceedings of the Royal Prussian Academy of Sciences), 189
Schwarzchild K (1916) Über das Gravitationsfeld einer Kugel aus inkompressibler Flüssigkeit nach der Einsteinschen Theorie (On the gravitational field of a sphere of incompressible fluid according to Einstein's theory), Sitzungsberichte der Königlich Preussischen Akademie der Wissenschaften zu Berlin (Meeting reports of the Royal Prussian Academy of Sciences in Berlin), 424
Schmidt M (1963) 3C 273: a star-like object with a large red-shift. Nature 197:1040
Shimizu T, Tsuneta S (1997) Deep survey of solar nanoflares with Yohkoh. Astrophys J 486:1045–1057
Society for science and the public. 1959. Chart space weather. Sci News Lett 76:229.
Soldner JG (1801–1804) On the deflection of a light ray from its rectilinear motion, by the attraction of a celestial body at which it nearly passes by. Berl Astron Jahrb 161–172
Springel V, Frenk CS, White SDM (2006) The large-scale structure of the Universe. Nature 440(7088):1137–1144
Sukhbold T, Woosley SE (2016) The most luminous supernovae. Astrophys J Lett 820(2):5, Article id. L38
Stuiver M (1961) Variations in radiocarbon concentration and sunspot activity. J Geophys Res 66:273–276
Stuiver M, Braziunas TF (1989) Atmospheric 14C and century-scale solar oscillations. Nature 338:405–408
Stuiver M, Braziunas TF, Becker B, Kromer B (1991) Climatic, solar, oceanic, and geomagnetic influences on late-glacial and Holocene atmospheric 14C/12C change. Quat Res 35:1–24
Syntelis P, Archontis V, Gontikakis C, Tsinganos K (2015) Emergence of non-twisted magnetic fields in the sun: jets and atmospheric response. Astron Astrophys 584:15
Syntelis P, Archontis V, Tsinganos K (2019) Recurrent CME-like eruptions in emerging flux regions, II. Scaling of energy and collision of successive eruptions. Astrophys J 876:8
Temmer M (2021) Space weather. Living Rev Solar Phys 18(4)
Thomson W (Lord Kelvin) (1862) On the age of the sun's heat. Macmillan's Mag 5:388–393
Thorsett SE, Chakrabarty D (1999) Neutron star mass measurements. I. Radio pulsars. Astrophys J 512:288. arXiv:astro-ph/9803260
Trussoni E, Ferrari A, Rosner R, Tsinganos K (1988) On wind-type flows in astrophysical jets. III. Temporal evolution of perturbations and the formation of shocks. Astrophys J 325:417
Trussoni E, Tsinganos K, Sauty C (1997) Analytical studies of collimated astrophysical winds. IV. Rotating collimated MHD flows. Astron Astrophys 325:1099–1114
Tsinganos K, Low BC (1989) Steady hydromagnetic flows in open magnetic fields. II. Global flows with static zones. Astrophys J 342:1028–1048

Tsinganos, Trussoni E (1990a) Analytical studies of collimated winds I. Topologies of 2-D hydrodynamic solutions. Astron Astrophys 231:270—276

Tsinganos, Trussoni E (1990b) Nonlinear MHD models of astrophysical winds. New York Acad Sci Ann (ISSN 0077-8923) 617:118–125

Tsinganos, Trussoni E (1991) Analytical studies of collimated astrophysical winds. II. Topologies of 2-D helicoidal MHD solutions. Astron Astrophys 249:156

Tsinganos K, Sauty C (1992a) Nonradial and nonpolytropic astrophysical outflows. I. Hydrodynamic solutions with flaring streamlines. Astron Astrophys 255:405

Tsinganos K, Sauty C (1992b) Nonradial and nonpolytropic astrophysical outflows. II. Topologies of MHD solutions with flaring streamlines. Astron Astrophys 257:790–806

Tsinganos K (2022) A tribute to prof. Eugene Parker, BAAS. https://baas.aas.org/pub/2022i039/release/1

Umehata H, Fumagalli M, Smail I, Swinbank AM, Cantalupo SC, Ivison RJ, Steidel CC, Shapley AE, Vernet J, Yamada T, Tamura Y, Kubo NK, Kajisawa M, Hatsukade B, Kohno K (2019) Gas filaments of the cosmic web located around active galaxies in a proto-cluster. Science 366(6461):97–100. arXiv:1910.01324

Usoskin IG, Mursula K, Kovaltsov GA (2003) The lost sunspot cycle: reanalysis of sunspot statistics. Astron Astrophys 403:743–748

Usoskin IG (2013) A history of solar activity over Millennia. arXiv:0810.3972v3

van de Weygaert R, Schaap W (2009) The cosmic web: geometric analysis. In: Martínez VJ, Saar E, Martínez-González E, Pons-Bordería MJ (eds) Data analysis in cosmology, vol 665. Lecture notes in physics, pp 291–413. Springer, Berlin

Vedrenne G, Atteia J (2009) Gamma-ray bursts. Springer, Berlin

Villata M, Tsinganos K (1993) Helically symmetric MHD equilibrium solutions. Phys Fluids B 7:2153–2164

Vlahakis N, Tsinganos K (1997) On the topological stability of astrophysical jets. MNRAS 292:591–600

Waldmeir M (1957) Die Sonnenkorona, 2, Verlag, Birkhauser Basel

Weinberg S (1987) Anthropic bound on the cosmological constant. Phys Rev Lett 59 (22):2607–2610

Weinberg S (1992) Dreams of a final theory: the scientist's search for the ultimate laws of nature. Pantheon Books, New York

Weisberg JM, Taylor JH, Fowler LA (1981) Gravitational waves from an orbiting pulsar. Sci Am 74

Weisberg JM, Nice DJ, Taylor JH (2010) Timing measurements of the relativistic binary pulsar PSR B1913+16. Astrophys J 722:1030–1034

Weisberg JM, Huang Y (2016) Relativistic measurements from timing the binary pulsar PSR B1913+16. Astrophys J 829(1):10

Wilson A, Maskelyne M (1774) Observations on the solar spots. By Alexander Wilson, M. D. Professor of Practical Astronomy in the University of Glasgow. Communicated by the Rev. Nevil Maskelyne. Astron Roy Philos Trans 64:1–30

Wolf C, Lai S, Onken CA, Amrutha N, Bian F, Hon WJ, Tisserand P, Webster RL (2024) The accretion of a solar mass per day by a 17-billion solar mass black hole. Nat Astron. https://doi.org/10.1038/s41550-024-02195-x

Xie C, Haffert SY, de Boer J, Kenworthy MA, Brinchmann J, Girard J, Snellen IAG, Keller CU (2020) Searching for proto-planets with MUSE, A&A 644:A149. arXiv:2011.08043

Yao Y et al (2023) The on-axis jetted tidal disruption event AT2022cmc: X-ray observations and broadband spectral modeling. arXiv:2308.09834v1 [astro-ph.HE]

Zamaninasab M, Clausen-Brown E, Savolainen T, Tchekhovskoy A (2014) Dynamically important magnetic fields near accreting supermassive black holes. Nature 51:126

Zel'dovich YB, Yu PR (2002) Physics of shock waves and high-temperature hydrodynamic phenomena. In: Dover books on physics

Index

K. Tsinganos, *A Century of Cosmic Discoveries*, Springer Praxis Books,
https://doi.org/10.1007/978-3-032-06282-6

D

N

O

P

Q

R

T

U

V

MIX
Papier aus verantwortungsvollen Quellen
Paper from responsible sources
FSC® C105338
FSC
www.fsc.org

If you have any concerns about our products,
you can contact us on
ProductSafety@springernature.com

In case Publisher is established outside the EU,
the EU authorized representative is:
Springer Nature Customer Service Center GmbH
Europaplatz 3, 69115 Heidelberg, Germany

Printed by Libri Plureos GmbH
in Hamburg, Germany

Kinase Inhibitor for Cancer Therapy

Kinase Inhibitor for Cancer Therapy

Editor

Francesca Musumeci

MDPI

Basel • Beijing • Wuhan • Barcelona • Belgrade • Novi Sad • Cluj • Manchester

Editor
Francesca Musumeci
Department of Pharmacy
University of Genoa
Genoa
Italy

Editorial Office
MDPI
St. Alban-Anlage 66
4052 Basel, Switzerland

This is a reprint of articles from the Special Issue published online in the open access journal *Pharmaceutics* (ISSN 1999-4923) (available at: www.mdpi.com/journal/pharmaceutics/special_issues/Kinase_Inhibitor_Cancer).

For citation purposes, cite each article independently as indicated on the article page online and as indicated below:

Lastname, A.A.; Lastname, B.B. Article Title. *Journal Name* **Year**, *Volume Number*, Page Range.

ISBN 978-3-7258-1290-5 (Hbk)
ISBN 978-3-7258-1289-9 (PDF)
doi.org/10.3390/books978-3-7258-1289-9

Contents

pharmaceutics

Editorial

Unlocking Potential and Limits of Kinase Inhibitors: The Highway to Enhanced Cancer Targeted Therapy

Francesca Musumeci * and Silvia Schenone

Department of Pharmacy, University of Genoa, Viale Benedetto XV, 3, 16132 Genoa, Italy; silvia.schenone@unige.it
* Correspondence: francesca.musumeci@unige.it

Kinases are a family of enzymes comprising over five hundred members, which, when overexpressed or hyperactivated, are implicated in the pathogenesis of numerous hematological and solid cancers. This evidence has spurred extensive research into molecules with targeted anti-cancer activity and led to the approval of the first monoclonal antibody, trastuzumab, in 1998 and the first small-molecule kinase inhibitor, imatinib, in 2001. Since then, the search for new kinase inhibitors has attracted the attention of scientists, as they can provide a powerful approach to mitigating the side effects associated with traditional antineoplastic agents and enabling convenient oral administration in the case of small-molecule inhibitors. The impact of targeted cancer therapy is underscored by the approval of over 70 kinase inhibitors, with approximately 25 new compounds emerging in the past five years [1,2]. Traditionally, small-molecule kinase inhibitors are classified into six types: type I inhibitors, which target the ATP-binding pocket of active kinases; type II inhibitors, which target the ATP-binding pocket of inactive kinases; type III inhibitors, allosteric inhibitors that bind a site adjacent to the ATP-binding pocket; type IV inhibitors, allosteric inhibitors that bind a site distant from the ATP-binding pocket; type V inhibitors, bivalent inhibitors; and type VI inhibitors, covalent inhibitors. Most currently approved kinase inhibitors belong to types I and II. Some examples of other types of kinase inhibitors include trametinib (a type III allosteric inhibitor targeting MEK1/2), asciminib (a type IV allosteric inhibitor targeting Bcr-Abl), and afatinib (a type VI covalent inhibitor targeting EGFR).

The fervent interest in kinase inhibitors from the academic and industrial world points toward a promising expansion of this field in the coming decades. This enthusiasm is driven by the untapped potential of targeting kinases alongside the pressing need to address current limitations.

Notably, despite significant progress, only approximately 20% of the kinome has been targeted to date, indicating vast opportunities for discovering novel treatment options once the roles of additional kinases are elucidated. Furthermore, ongoing research points out allosteric inhibition as a powerful approach to reduce adverse effects and fight resistance to the drug due to mutations in the active site of the enzyme. In this context, developing novel types of allosteric inhibitors, such as compounds targeting pseudokinase domains or extracellular domains, is a promising new route [2,3].

Regarding the current limitations of kinase inhibitors, major issues can be identified in the onset of mutations that confer drug resistance and the general suboptimal pharmacokinetic profiles exhibited by small-molecule kinase inhibitors.

Drug resistance poses a significant challenge to the long-term efficacy of kinase inhibitors, as treatment with most approved drugs is eventually interrupted following prolonged use. Resistance can arise through various mechanisms, with gene mutations being the most prevalent cause. Mutations in the catalytic domain frequently confer resistance to type I and II kinase inhibitors. This evidence encouraged, and is still encouraging, the search for new efficacious approaches [4].

First, the most intuitive strategy plans the design and synthesis of compounds that are also active against resistant isoforms. The approval of ponatinib is a pertinent example,

Citation: Musumeci, F.; Schenone, S. Unlocking Potential and Limits of Kinase Inhibitors: The Highway to Enhanced Cancer Targeted Therapy. *Pharmaceutics* **2024**, *16*, 625. https://doi.org/10.3390/pharmaceutics16050625

Received: 12 April 2024
Revised: 29 April 2024
Accepted: 30 April 2024
Published: 7 May 2024

as it is a multitargeted covalent kinase inhibitor that is able to inhibit the T315I Bcr-Abl gatekeeper mutation.

Second, as previously mentioned, the development of allosteric inhibitors constitutes another exciting opportunity to bypass resistance related to mutations in the catalytic domain. In particular, combining an ATP-competitive inhibitor and an allosteric inhibitor often led to improved efficacy compared to the single-agent regimen [3].

Third, the development of antibody–drug conjugates (ADCs) has been highlighted. An ADC comprises a monoclonal antibody and cytotoxic agent attached by a linker. The rationale behind the design of this class of compounds lies in the synergy between the highly specific targeting ability of monoclonal antibodies and the potent cytotoxic activity of the attached agents. Gemtuzumab ozogamicin (Mylotarg®) was the first ADC approved by the FDA in 2000. Since then, about 15 ADCs have been released. Notably, Fam-trastuzumab deruxtecan (Enhertu®) is an ADC targeting the HER kinase and incorporates the topoisomerase I inhibitor DXd as its cytotoxic agent. It was approved in 2019 as a second-line treatment for patients with unresectable or metastatic HER2-positive breast cancer [5].

Fourth, a recent approach gaining attention is synthetic lethality, which exploits the concept that in certain tumors, the expression of two or more genes is altered, and targeting both synthetic lethal factors enables the selective killing of cancer cells. The most relevant example of synthetic lethality is the use of the poly(ADP-ribose) polymerase inhibitor olaparib in the treatment of breast cancer associated with BRCA1/2 mutations [6]. However, kinase inhibitors also present an opportunity to leverage the concept of synthetic lethality. For instance, cyclin-dependent kinases (CDKs) are being investigated as synthetic lethal players when combined with specific oncogenes, such as MYC, TP53, and RAS [7].

At the same time, the poor suboptimal pharmacokinetic profile generally associated with kinase inhibitors has prompted the search for efficient drug delivery systems. In this frame, nanoformulations offer an effective tool to improve the absorption, distribution, metabolism, and excretion (ADME) properties and, consequently, the bioavailability of kinase inhibitors. Nanocarriers include liposomes, micelles, gold and polymeric nanoparticles, bovine serum albumin, and metal–organic frameworks. Notably, the nanoparticle surface can be functionalized with specific ligands, such as antibodies or peptides, capable of recognizing particular proteins on the surface of cancer cells, thus facilitating active targeting. This approach enables the selective release of the drug to the neoplastic cells, thereby enhancing therapeutic efficacy while minimizing off-target effects. Moreover, nanodelivery represents an effective strategy for overcoming resistance mechanisms [8].

The Special Issue features a comprehensive overview of the latest developments in kinase inhibitor research across various fronts, including in silico studies, the synthesis and identification of new compounds, drug delivery methods, overcoming resistance mechanisms, the potential to target specific kinases or use selected inhibitors, and biological and pharmacokinetic evaluations. Herein, an overview of the most innovative articles is reported. Čermáková et al. (contribution 1) provided deep insight into the ability of Bruton's tyrosine kinase inhibitor zanubrutinib to modulate cancer resistance by inhibiting anthracycline metabolism and efflux, suggesting a new combination therapy including both the drugs. Sunoqrot et al. (contribution 2) proposed vitamin E TPGS-poloxamer nanoparticles to deliver the new PI3Kα inhibitor R19 in breast cancer cell lines. The research carried out by the authors showed that the formulation possessed enhanced activity compared to R19 alone, good cancer cell selectivity, and high biocompatibility, paving the way for clinical translation. Poggialini et al. (contribution 3) performed a study to identify new candidates for glioblastoma (GBM) treatment since this tumor still has a poor prognosis. Starting from the in-house library of pyrazolo[3,4-*d*]pyrimidines, they performed extensive research, including enzymatic assays of Src/Bcr-Abl kinases, in vitro tests on four GBM cell lines, and a deep ADME evaluation that led to the discovery of a compound (namely derivative **5** of the paper) as a suitable candidate for in vivo evaluation. Sabetta et al. (contribution 4) analyzed the effect of dasatinib in 2D monolayer cultures of prostate cancer and glioblastoma cell lines and the corresponding 3D spheroids and

3D bioprinting models. Three-dimensional cell models better resemble tumor complexity associated with drug resistance than two-dimensional models and, for this reason, are gaining increasing attention in preclinical studies. In this work, the authors took a step forward in this field, identifying that the 3D bioprinted model utilizing an alginate–gelatin hydrogel was endowed with improved feasibility, reproducibility, and scalability than the classical 3D spheroid model.

Overall, this Special Issue offers a panoramic view of the current state of the field. Each contribution showcases the multifaceted approaches researchers are employing to unlock the power of kinase inhibition for cancer treatment. These studies contribute valuable data to the scientific community and pave the way for translating the findings into clinical applications.

Conflicts of Interest: The authors declare no conflicts of interest.

List of Contributions

1. Cermáková, L.; Hofman, J.; Laštovičková, L.; Havlíčková, L.; Špringrová, I.; Novotná, E.; Wsól, V. Bruton's Tyrosine Kinase Inhibitor Zanubrutinib Effectively Modulates Cancer Resistance by Inhibiting Anthracycline Metabolism and Efflux. *Pharmaceutics* **2022**, *14*, 1994. https://doi.org/10.3390/PHARMACEUTICS14101994.
2. Sunoqrot, S.; Aliyeh, S.; Abusulieh, S.; Sabbah, D. Vitamin E TPGS-Poloxamer Nanoparticles Entrapping a Novel PI3Kα Inhibitor Potentiate Its Activity against Breast Cancer Cell Lines. *Pharmaceutics* **2022**, *14*, 1977. https://doi.org/10.3390/PHARMACEUTICS14091977.
3. Poggialini, F.; Vagaggini, C.; Brai, A.; Pasqualini, C.; Crespan, E.; Maga, G.; Perini, C.; Cabella, N.; Botta, L.; Musumeci, F.; et al. Biological Evaluation and In Vitro Characterization of ADME Profile of In-House Pyrazolo[3,4-*d*]pyrimidines as Dual Tyrosine Kinase Inhibitors Active against Glioblastoma Multiforme. *Pharmaceutics* **2023**, *15*, 453. https://doi.org/10.3390/PHARMACEUTICS15020453.
4. Sabetta, S.; Vecchiotti, D.; Clementi, L.; Di Vito Nolfi, M.; Zazzeroni, F.; Angelucci, A. Comparative Analysis of Dasatinib Effect between 2D and 3D Tumor Cell Cultures. *Pharmaceutics* **2023**, *15*, 472. https://doi.org/10.3390/PHARMACEUTICS15020372.

References

1. Attwood, M.M.; Fabbro, D.; Sokolov, A.V.; Knapp, S.; Schiöth, H.B. Trends in Kinase Drug Discovery: Targets, Indications and Inhibitor Design. *Nat. Rev. Drug Discov.* **2021**, *20*, 839–861. [CrossRef] [PubMed]
2. Roskoski, R. Properties of FDA-Approved Small Molecule Protein Kinase Inhibitors: A 2024 Update. *Pharmacol. Res.* **2024**, *200*, 107059. [CrossRef] [PubMed]
3. Lu, X.; Smaill, J.B.; Ding, K. New Promise and Opportunities for Allosteric Kinase Inhibitors. *Angew Chem. Int. Ed. Engl.* **2020**, *59*, 13764–13776. [CrossRef] [PubMed]
4. Zhong, L.; Li, Y.; Xiong, L.; Wang, W.; Wu, M.; Yuan, T.; Yang, W.; Tian, C.; Miao, Z.; Wang, T.; et al. Small Molecules in Targeted Cancer Therapy: Advances, Challenges, and Future Perspectives. *Signal Transduct. Target. Ther.* **2021**, *6*. [CrossRef]
5. Fu, Z.; Li, S.; Han, S.; Shi, C.; Zhang, Y. Antibody Drug Conjugate: The "Biological Missile" for Targeted Cancer Therapy. *Signal Transduct. Target. Ther.* **2022**, *7*, 93. [CrossRef] [PubMed]
6. Li, S.; Topatana, W.; Juengpanich, S.; Cao, J.; Hu, J.; Zhang, B.; Ma, D.; Cai, X.; Chen, M. Development of Synthetic Lethality in Cancer: Molecular and Cellular Classification. *Signal Transduct. Target. Ther.* **2020**, *5*, 241. [CrossRef] [PubMed]
7. Li, K.; You, J.; Wu, Q.; Meng, W.; He, Q.; Yang, B.; Zhu, C.; Cao, J. Cyclin-Dependent Kinases-Based Synthetic Lethality: Evidence, Concept, and Strategy. *Acta Pharm. Sin. B* **2021**, *11*, 2738–2748. [CrossRef] [PubMed]
8. Xu, W.; Ye, C.; Qing, X.; Liu, S.; Lv, X.; Wang, W.; Dong, X.; Zhang, Y. Multi-Target Tyrosine Kinase Inhibitor Nanoparticle Delivery Systems for Cancer Therapy. *Mater. Today Bio* **2022**, *16*, 100358. [CrossRef]

pharmaceutics

MDPI

Review

Targeting EGFR/PI3K/AKT/mTOR Signaling in Hepatocellular Carcinoma

Jieun Bang †, Mihyeon Jun †, Soyun Lee †, Hyuk Moon and Simon Weonsang Ro *

Department of Genetics and Biotechnology, College of Life Sciences, Kyung Hee University, Yongin-si 17104, Republic of Korea; qkdwldms612@khu.ac.kr (J.B.); mihyeonjun@khu.ac.kr (M.J.); lsoyun15@khu.ac.kr (S.L.); hmoon@khu.ac.kr (H.M.)

* Correspondence: simonro@khu.ac.kr; Tel.: +82-31-201-5640; Fax: +82-31-204-8116

† These authors contributed equally to this work.

Abstract: Hepatocellular carcinoma (HCC) poses a significant global health concern, with its incidence steadily increasing. The development of HCC is a multifaceted, multi-step process involving alterations in various signaling cascades. In recent years, significant progress has been made in understanding the molecular signaling pathways that play central roles in hepatocarcinogenesis. In particular, the EGFR/PI3K/AKT/mTOR signaling pathway in HCC has garnered renewed attention from both basic and clinical researchers. Preclinical studies in vitro and in vivo have shown the effectiveness of targeting the key components of this signaling pathway in human HCC cells. Thus, targeting these signaling pathways with small molecule inhibitors holds promise as a potential therapeutic option for patients with HCC. In this review, we explore recent advancements in understanding the role of the EGFR/PI3K/AKT/mTOR signaling pathway in HCC and assess the effectiveness of targeting this signaling cascade as a potential strategy for HCC therapy based on preclinical studies.

Keywords: hepatocellular carcinoma; EGFR/PI3K/AKT/mTOR signaling; animal models; targeted therapy

Citation: Bang, J.; Jun, M.; Lee, S.; Moon, H.; Ro, S.W. Targeting EGFR/PI3K/AKT/mTOR Signaling in Hepatocellular Carcinoma. *Pharmaceutics* **2023**, *15*, 2130. https://doi.org/10.3390/pharmaceutics15082130

Academic Editors: Fabiana Quaglia and Xiaowei Zeng

Received: 25 July 2023
Revised: 7 August 2023
Accepted: 10 August 2023
Published: 14 August 2023

1. Introduction

Liver cancer presents a significant global health challenge due to its rising incidence and mortality rates. According to the World Health Organization, liver cancer accounted for approximately 800,000 deaths, positioning it as the fourth leading cause of cancer-related mortality [1]. Liver cancer comprises a heterogeneous group of malignant tumors with varying histologic characteristics and poor prognoses. Among these, hepatocellular carcinoma (HCC) represents the most prevalent primary liver cancer, accounting for approximately 80% of cases, followed by cholangiocarcinoma (CC), which contributes to 10–20% of primary liver cancers [2,3].

Existing treatment modalities for HCC offer limited success, with a considerably low 5-year survival rate. Surgical resection or local ablation therapy is typically favored for early-stage HCC; however, tumor recurrence occurs in approximately 70% of patients within 5 years [1,4]. For advanced HCC, systemic therapy is recommended as the standard treatment option, but the overall prognosis has been unsatisfactory [1,2].

HCC is a heterogeneous tumor with diverse phenotypic and genetic characteristics, and its tumorigenesis involves a range of molecular mechanisms [5,6]. Extensive research on the molecular pathogenesis of HCC has identified several critical signaling pathways involved in tumor initiation, promotion, and metastasis. These pathways include the mitogen-activated protein kinase/extracellular signal-regulated kinase (MAPK/ERK), Wnt/β-catenin, Hedgehog (HH), Hippo-YAP/TAZ, and phosphatidylinositol 3-kinase /AKT/mammalian target of rapamycin (PI3K/AKT/mTOR) signaling pathways. Recently, we have extensively reviewed the roles of MAPK/ERK, Wnt/β-catenin, HH, and Hippo-YAP/TAZ signaling in HCC development together with preclinical and clinical studies targeting signaling pathways in HCC [7,8].

In this review, our aim is to provide an overview of the tumor-promoting effects exerted by the PI3K/AKT/mTOR signaling pathway in hepatocellular carcinoma (HCC), which is primarily triggered by the epidermal growth factor receptor (EGFR). Additionally, we discuss the results from recent preclinical and clinical studies that target the EGFR/PI3K/AKT/mTOR signaling cascade as a potential therapeutic approach for treating HCC.

2. Role of EGFR/PI3K/AKT/mTOR Signaling Pathway in HCC

2.1. Overview of EGFR/PI3K/AKT/mTOR Signaling Pathway

Epidermal Growth Factor Receptor (EGFR) is a cell surface receptor that belongs to the ErbB family of receptor tyrosine kinases. EGFR is activated by binding specific ligands, such as the epidermal growth factor (EGF) or transforming growth factor-alpha (TGF-α). Ligand binding induces receptor dimerization, leading to the activation of the intrinsic kinase activity of EGFR. The catalytic domain of EGFR kinases carries out a transphosphorylation process by which an array of tyrosine residues at the c-terminal cytoplasmic domain of EGFR are phosphorylated by its dimerized partner. Phosphotyrosine (pY) at the cytoplasmic domain of an activated EGFR acts as a docking site for proteins containing Src homology 2 (SH2) domains, such as Phosphatidylinositol 3-kinase (PI3K). The SH2 domain of PI3K binds to phosphorylated tyrosine residues on EGFR, bringing PI3K in close proximity to the plasma membrane. Upon recruitment, PI3K converts phosphatidylinositol-4,5-disphosphate (PIP2) in the plasma membrane into phosphatidylinositol-3,4,5-trisphosphate (PIP3) [9,10]. This conversion is carried out by the lipid kinase activity of the catalytic subunit of PI3K. The conversion of PIP2 to PIP3 can be antagonized by lipid phosphatase PTEN that converts PIP3 into PIP2 [11]. PIP3 serves as a docking site for proteins containing pleckstrin homology (PH) domains, such as the serine/threonine kinase AKT (also known as protein kinase B). Thus, the presence of PH domains in AKT allows the serine/threonine kinase to be recruited to the plasma membrane where AKT is phosphorylated at two critical residues: Thr308 by phosphoinositide-dependent kinase 1 (PDK1) and Ser473 by the mammalian target of rapamycin complex 2 (mTORC2). These phosphorylation events lead to the full activation of AKT [12]. Fully activated AKT phosphorylates regulates numerous downstream effectors, including proteins involved in cell survival, protein synthesis, metabolism, and cell-cycle progression. One of the key targets of activated AKT is the mammalian target of rapamycin (mTOR), which exists in two complexes: mTORC1 and mTORC2 (Figure 1).

The mammalian target of rapamycin (mTOR) is a serine/ threonine protein kinase that regulates tumor growth, proliferation, metabolism, cell growth, and immunity [13]. mTOR has two complexes, which are mTORC1 and mTORC2. mTORC1 consists of mTOR, which is a regulator-associated protein of mTOR (Raptor), the DEP domain-containing mTOR interacting protein (DEPTOR), mammalian lethal with SEC13 protein 8 (mLST8), and the proline-rich AKT substrate 40 (PRAS40). mTORC2 contains the rapamycin-insensitive companion of mTOR (Rictor), DEPTOR, mLST8, Rictor-1 (Protor-1), Protor-2, and mammalian stress-activated protein kinase-interacting protein 1 (mSin1). mTORC1 is involved in cell growth and proliferation through nutritional sensing, and mTORC2 regulates the rebuilding of cytoskeletons and cell survival [14]. To activate mTORC1, RHEB, a small GTPase, is required. However, RHEB is usually inactivated by TSC1/2, a GTPase-activating protein (GAP) inhibitor [15]. The activated AKT can inhibit TSC1/2 and let RHEB be activated, resulting in the activation of mTORC1 and its downstream targets. Active mTORC1 increases protein synthesis by phosphorylating the eukaryotic translation inhibition factor 4E binding protein1 (4EBP1) and ribosomal proteins S6 kinase 1/2 (S6K1/2). Considering that these cellular processes are closely involved in carcinogenesis, it is no surprise that the dysregulation of the mTOR pathway has been observed in multiple human solid tumors [16].

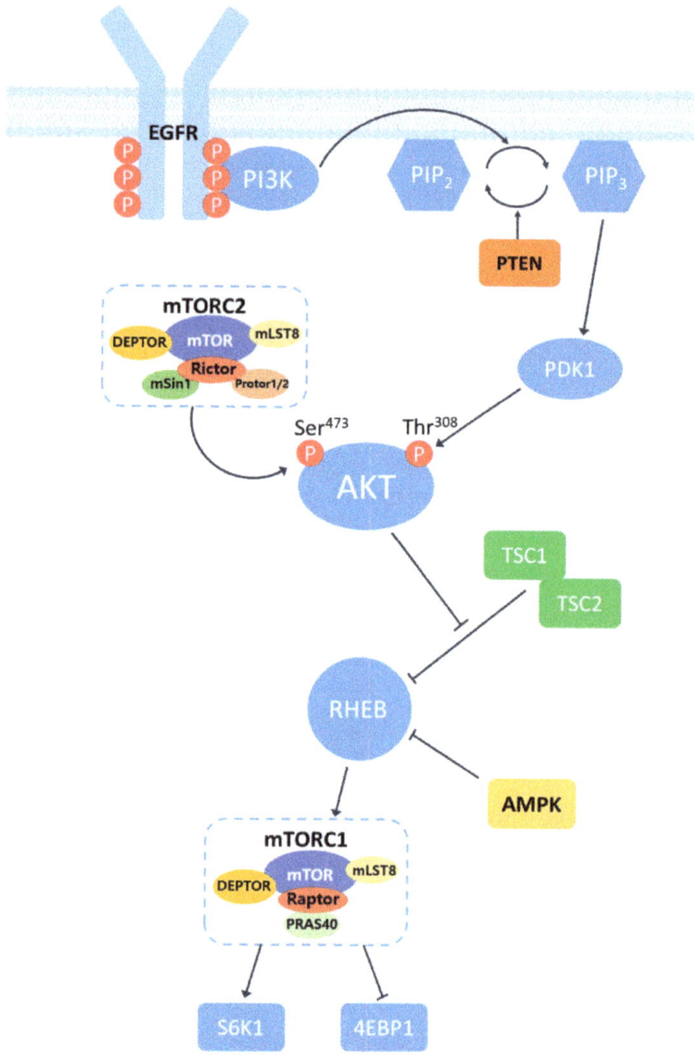

Figure 1. Schematic illustration of EGFR/PI3K/AKT/mTOR signaling pathway. Receptor dimerization induces the phosphorylation of tyrosine residues at the c-terminal cytoplasmic domain of EGFR. PI3K binds to phosphorylated tyrosine residues on EGFR and converts phosphatidylinositol-4,5-disphosphate (PIP2) in the plasma membrane into phosphatidylinositol-3,4,5-trisphosphate (PIP3), leading to the recruitment and subsequent phosphorylation of AKT by PDK1 and mTORC2. Phosphorylated AKT, in turn, leads to the activation of mTORC1 by regulating the activity of RHEB. Active mTOR induces the phosphorylation and activation of S6K1 and 4EBP1.

2.2. *Activation of EGFR/PI3K/AKT/mTOR Signaling Pathway in HCC*

The EGFR/PI3K/AKT/mTOR signaling pathway is a major pathway in diverse types of cancers. This pathway actively regulates various aspects of cellular processes, including cell proliferation, survival, migration, and metabolism. It also plays a pivotal role in regulating the tumor microenvironment via angiogenesis and the recruitment of inflammatory cells [17]. Also, the EGFR/PI3K/AKT/mTOR signaling is significantly involved in therapy response and metastasis [18,19]. Considering the multifaceted roles in carcinogenesis, it is no surprise that the signaling pathway is commonly upregulated in a variety of human cancers [20].

In HCC, the EGFR/PI3K/AKT/mTOR pathway is abnormally activated in approximately 50% of cases, and this dysregulated activation is involved in various cellular processes, including cell proliferation, tumor cell differentiation, autophagy, metabolism, and the epithelial–mesenchymal transition (EMT) [14]. EGFR overexpression occurs in 68% of human HCC cases and significantly correlates with metastasis, poor patient survival, and aggressive tumors [21]. Moreover, EGFR ligands are overexpressed in human liver cancer cells and tumor tissues [22]. In higher stages of HCC, this signaling pathway is associated with vascular invasion, poor differentiation, and reduced survival rates [23]. Additionally, it has been reported that 40% of HCC patients who underwent orthotopic liver transplantation exhibited elevated mTOR activity [24].

PTEN suppresses AKT activation by converting PIP3 to PIP2. The loss of tumor suppressor PTEN and reduced expression of PTEN are primarily observed in the majority of patients with HCC in the early stages [25,26]. Somatic mutations in PTEN have been reported in HCC tissues [27,28]. Moreover, Fujiwara et al. observed 12 allelic losses of PTEN in 37 patients with HCC [29]. Loss-of-function mutations in TSC1/2 were found in approximately 20% of patients with HCC, which serves as another major suppressor of the EGFR/PI3K/AKT/mTOR signaling pathway [30].

3. In Vitro Studies Investigating EGFR/PI3K/AKT/mTOR Signaling in HCC Cell Lines

3.1. Tumorigenic Roles of EGFR/PI3K/AKT/mTOR Signaling in HCC Cells

The activation of the EGFR/PI3K/AKT/mTOR pathway has been observed in various cancer types. Although ligand binding to EGFR is the natural process, which induces the dimerization of the receptors and subsequent phosphorylation of the cytoplasmic tails, the overexpression of EGFR alone can lead to the enhanced formation of its dimerization and activation of the downstream signaling pathway in the absence of its ligands. EGFR overexpression can be achieved through various mechanisms [31]. In addition to gene amplification and the epigenetic upregulation of the EGFR gene, changes in positive or negative regulators of EGFR can also affect its abundance in cancer cells. For example, NT5DC2 suppresses the ubiquitination of EGFR, preventing its ubiquitin-mediated proteasome degradation and leading to increased EGFR levels. In HCC cell lines, such as MHCC97H and PLC/RLF/5, the upregulated NT5DC2-induced overexpression of EGFR and the activation of the downstream PI3K/AKT/mTOR signaling pathway occurs [32]. Similarly, Song et al. reported that 14-3-3σ can interact with EGFR and stabilize the receptor, prolonging the activation of EGFR signaling in Huh7 and HepG2 cells [33]. Tropomodulin 3 (TMOD3), a member of the pointed-end capping protein family, is significantly upregulated in HCCs and is correlated with poor survival in patients with HCC. In various HCC cell lines, including Huh-7 and Hep3B, TMOD3 was found to facilitate the phosphorylation of the cytoplasmic tail of EGFR, triggering the activation of the downstream PI3K/AKT/mTOR signaling cascade [34].

In addition to the EGF/EGFR-mediated activation of the PI3K/AKT/mTOR signaling pathway, various molecules can contribute to eliciting this pathway. Recently, DEAD/DEAH box helicase 11 (DDX11) and Apelin (APLN) were identified as activators of the PI3K/AKT/mTOR signaling pathway in HCC. The overexpression and/or knockdown of these molecules substantially altered the PI3K/AKT/mTOR signaling in HCC cell lines and significantly affected the tumorigenic potentials of HCC cells [35,36]. Alpha-fetoprotein (AFP), which is strongly correlated with the aggressiveness of HCC, is a serum biomarker routinely used for the diagnosis and prognosis of HCC. Recently, Wang et al. investigated the role of AFP in HCC using two HCC cell lines. They reported that AFP interacted with PTEN, a negative regulator of AKT, and activated the PI3K/AKT/mTOR signaling pathway [37]. The role of claudin-6 (CLDN6) was investigated in HepG2 cells, where CLDN6 was found to activate the EGFR/PI3K/AKT/mTOR signaling pathway. The knockdown of CLDN6 led to decreases in the proliferation, migration, and invasion of HepG2 cells [38]. MicroRNAs (miRNAs) can also contribute to the activation of the EGFR/PI3K/AKT/mTOR signaling pathway. For example, miR-494-3p is correlated with aggressive clinicopathological characteristics and poor prognosis in HCC patients. Lin et al. showed that miR-494-3p bound to the 3′UTR of PTEN mRNA and repressed its translation in two HCC cell lines, SMMC7721 and HCCLM3 [39]. Notably, the suppression of PTEN expression by the ectopic expression of miR-494-3p led to the activation of the PI3K/AKT/mTOR pathway and enhanced the metastasis potentials of HCC cells [39]. Another in vitro study using the PLC/PRF/5 HCC cell line showed that the Mac-2-binding protein glycan isomer (M2BPGi) activated mTOR and exerted tumor-promoting effects on HCC [40].

3.2. *Anti-Tumor Effects of Targeting EGFR/PI3K/AKT/mTOR Pathway in HCC Cells*

Given that the EGFR/PI3K/AKT/mTOR signaling pathway exerts strong tumor-promoting effects on HCC, it is reasonable to attempt to inhibit this pathway for the treatment of HCC. EGFR is a direct target of miRNA-133b. The overexpression of miRNA-133b significantly suppressed EGFR protein expression and led to decreased activities of PI3K, AKT, and mTOR in the HepG2 cells [41]. Of note, the inhibition of the EGFR/PI3K/AKT/mTOR pathway via the overexpression of miRNA-133b induced the activation of caspase-3/-8 and apoptotic cell deaths in HepG2 cells. Methyltransferase like 14 (METTL14) can destabilize EGFR mRNA via an N6-methyladenosine (m6A) RNA methylation. METTL14 is significantly downregulated in HCC and is associated with the poor prognosis of HCC patients [42]. Similarly, an m6A-binding protein, YTH-Domain Family Member 2 (YTHDF2), binds to m6A sites in 3′UTR of EGFR mRNA and promotes the degradation of EGFR mRNA in HCC cells. In HCC cells such as HEP3B and SMMC7721, the overexpression of YTHDF2 suppressed cell proliferation via destabilizing EGFR mRNA and, thus, acting as a tumor suppressor [43]. In summary, inhibiting EGFR by the overexpression of endogenous EGFR suppressors such as miRNA-133b, METTL14, and YTHDF2 led to the subsequent inactivation of the downstream PI3K/AKT/mTOR signaling pathway and effectively reduced the malignancy of HCC cells by inducing cell apoptosis and suppressing cell proliferation.

In line with the observations in HCC cells characterized by the overexpression of endogenous suppressors of EGFR, the pharmacological inhibition of EGFR in HCC cell lines elicited similar effects. Specifically, it led to the suppression of cell proliferation and the induction of apoptosis in HCC cells through the downregulation of PI3K, p-AKT, and p-mTOR. For instance, the treatment of HepG2 cells with GW2974, an EGFR inhibitor, induced the attenuation of the downstream PI3K/AKT/mTOR signaling pathway, resulting in decreased cell proliferation and increased apoptosis in HCC cells [41]. Moreover, the inhibition of EGFR using erlotinib demonstrated inhibitory effects on the migratory capabilities and cell proliferation of HCC cells [32,44]. Similarly, the administration of EGFR inhibitors AG1478 and Gefitinib led to a reduction in cell proliferation, decreased invasion, and enhanced apoptosis in HCC cells [38,45].

In addition to targeting EGFR, the inhibition of its downstream effectors, such as PI3K, AKT, and/or mTOR, exerted similar tumor-suppressing effects on HCC cells (Table 1). Its treatment with LY294002 or Wortmannin, potent inhibitors of PI3K, downregulated the phosphorylated levels of AKT and induced apoptosis in HCC cells such as MHCC97 and Huh7 [46]. The treatment of HCC cells with a PI3K inhibitor 740Y-P showed similar effects [47]. These results consistently show that the inhibition of PI3K leads to the suppression of tumor growth and the induction of apoptosis. Cell cycle arrest was also observed in HCC cells when they were treated with inhibitors of PI3K. For example, the treatment of Huh7 and HepG2 cells with copanlisib induced cell-cycle arrest via the downregulation of CDK4/6 and cyclin D1, although this treatment had a minor effect on apoptosis [48].

Table 1. The inhibition of the EGFR/PI3K/AKT/mTOR pathway in HCC cell lines.

Drug	Target	HCC Cell Line	Phenotype	Reference
erlotinib	EGFR	MHCC97H, PLC/RLF/5	reduced cell proliferation	[32]
		Huh7	reduced migration	[44]
AG1478	EGFR	HepG2	reduced cell proliferation and invasion	[38]
gefitinib	EGFR	Huh7	induced apoptosis	[45]
GW2974	EGFR	HepG2	reduced cell proliferation, induced apoptosis	[41]
copanlisib	PI3K	Huh7, HepG2	induced apoptosis, inhibited cell growth, inducing cell cycle arrest	[45,48]
LY294002	PI3K	MHCC97-H	induced apoptosis	[49]
		Huh7, Mahlavu	suppressed cell growth, induced apoptosis	[46]
wortmannin	PI3K	Huh7, Mahlavu	suppressed cell growth, induced apoptosis	[46]
		HepG2	induced apoptosis	[46]
740Y-P	PI3K	SMMC-7721, MHCC-97H	reduced cell proliferation, induced apoptosis	[47]
MK2206	AKT	Hep3B, Huh7, PLC/RLF/5	growth-inhibitory, induced apoptosis	[50]
		HepG2	anti-proliferative	[51]
AKT inhibitor VIII	AKT	Huh7, Mahlavu	suppressed cell growth, induced apoptosis	[46]
rapamycin	mTOR	Hep3B	prevent enrichment of CSCs	[52]
RAD001 (everolimus)	mTOR	Hep3B, Huh7, PLC/RLF/5	growth-inhibitory, induced apoptosis	[50]
BEZ235	PI3K/mTOR	Hep3B, Huh7, PLC/RLF/5	growth-inhibitory, induced apoptosis	[50]
lenvatinib	AKT/mTOR	Hep3B, HepG2	reduced cell proliferation, migration	[53]

MK2206 effectively interacts with the pleckstrin–homology (PH) domain of AKT and hinders its recruitment in the plasma membrane, inhibiting PDK1 binding and the subsequent activation of AKT [54]. MK2206 has shown strong potency in inhibiting AKT [55]. Similar to the findings in HCC cells treated with PI3K inhibitors, treatment with MK2206 also induced growth inhibition and apoptosis in HCC cells [50,51]. The AKT Inhibitor VIII, which also blocks the activity of AKT through binding to the pleckstrin homology (PH) domain in AKT, suppressed cell growth and induced apoptosis in HCC cells [46].

Recent studies have indicated that mTOR plays a critical role in maintaining stemness-related functions in cancer stem cells (CSCs), and the inhibition of mTOR leads to the sensitization of CSCs to radiation therapy in breast cancer [56]. In line with these findings, treatment with rapamycin, the prototypic mTOR inhibitor, significantly reduces the frequency of CD133+/EpCAM+ cells in Hep3B and Huh7, which are widely considered liver cancer stem cell populations [52]. RAD001, also known as everolimus, is an inhibitor of mTOR. Its binding to the FK506-binding protein12 (FKBP12) allows the RAD001–FKBP12 complex to interact with mTOR, which subsequently inhibits S6K1 and 4EBP1 phosphorylation by mTOR. Treatment with RAD001 resulted in the induction of apoptosis [50], as well as a decrease in cell proliferation in diverse HCC cell lines, including Hep3B, Huh7, and HepG2 [50,51].

The PI3K/mTOR dual inhibitor BEZ235 induced growth inhibition and apoptosis in HCC cells [50]. Similarly, the treatment of HCC cells with Lenvatinib, which targets both AKT and mTOR, exhibited inhibitory effects on cell proliferation and migration [53]. In summary, targeting the EGFR/PI3K/AKT/mTOR signaling pathway via the inhibition of EGFR, PI3K, AKT, and/or mTOR has consistently shown anti-tumor effects on HCC cells in vitro (Table 1).

3.3. Targeting EGFR/PI3K/AKT/mTOR Signaling on Sorafenib-Resistant HCC Cells

Sorafenib is the first-line systemic therapeutic for patients with advanced HCC, which inhibits both receptor tyrosine kinases (RTKs) and RAF [57]. However, the development of resistance to drugs and disease progression are nearly inevitable during the course of treatment [58]. Sorafenib resistance appears to be associated with the activation of PI3K/AKT/mTOR signaling [59]. Therefore, the combination of sorafenib and an inhibitor of the EGFR/PI3K/AKT/mTOR signaling pathway has been proposed as an effective therapeutic approach [45,51,60].

Copanlisib, a PI3K inhibitor, down-regulates downstream targets of AKT, leading to cell-cycle arrest and the suppression of cell proliferation, although it has a minimal effect on apoptosis [48,61]. Copanlisib counteracted sorafenib-induced AKT phosphorylation and synergistically enhanced anti-tumor effects on HCC when combined with sorafenib [48]. Likewise, combined treatment with sorafenib and capsaicin, an inhibitor of the PI3K/AKT/mTOR signaling pathway, also showed enhanced anti-tumor effects in Hep3B and HuH7 cells [62].

Lenvatinib is another first-line treatment for HCC which also inhibits RTKs. The combination of lenvatinib and copanlisib effectively suppressed the phosphorylation of AKT, which had been induced by treatment with lenvatinib. Copanlisib enhanced pro-apoptotic effects on HCC cell lines that were resistant to Lenvatinib [45]. Moreover, the sequential treatment of Huh7 cells with rapamycin, an mTOR inhibitor, following treatment with sorafenib substantially increased the sensitivity of HCC cells to sorafenib and decreased the frequency of liver cancer stem cell (CSC)-like cells, which are considered primary cells resistant to chemotherapy [52].

In summary, the combination of sorafenib or lenvatinib with agents targeting PI3K/AKT /mTOR signaling can enhance the anticancer activity of RTK inhibitors and is expected to overcome the emergence of therapy-resistant cells.

4. Animal Studies Investigating EGFR/PI3K/AKT/mTOR Signaling in HCC

4.1. Animal Models for HCC Induced by Activated EGFR/PI3K/AKT/mTOR Signaling

Studies have demonstrated the induction of hepatocarcinogenesis through the modification of genes involved in the PI3K/AKT/mTOR signaling pathway (Table 2). PTEN functions as a negative regulator of PI3K/AKT/mTOR signaling because it counteracts PI3K-mediated AKT activation (Figure 1). For the inactivation of PTEN, specifically in the liver, genetically engineered mice were used that carried two *Pten* alleles flanked by loxP sites as well as the *Cre* transgene under the promoter of albumin. These mice (referred to as "AlbCre; $Pten^{fl/fl}$ mice") were prone to the development of HCC as well as intrahepatic cholangiocarcinoma (ICC). Tumors from the mice exhibited significant increases in the phosphorylated level of AKT, confirming the activation of the PI3K/AKT/mTOR signaling in *Pten*-deleted livers [63].

Table 2. Mouse models for HCC induced by activated EGFR/PI3K/AKT/mTOR signaling.

Gene	Mouse Model	Phenotype/Tumor Type	Ref.
Pten	Alb-Cre; $Pten^{fl/fl}$	ballooning, steatosis/ICC, HCC	[63]
	sgPten sgPten/c-Met	lipid accumulation/no tumor lipid accumulation/HCA, HCC	[64]

Table 2. *Cont.*

Gene	Mouse Model	Phenotype/Tumor Type	Ref.
AKT	HA-myr-AKT1 HA-myr-AKT1/V5-c-Met	hepatic steatosis, proliferation/HCC rapid liver tumor growth	[65]
	myr-AKT1 myr-AKT1/N-RasG12V	hepatocyte proliferation/HCA (12 w), HCC (6 m) spotty and pale color/nodular lesions (~4 w)	[66]
Tsc1/Tsc2	Alb-Cre; *Tsc1*$^{fl/fl}$	dysplastic foci, nodules, hepatomas/HCC	[67]
	Alb-Cre; *Tsc1*$^{fl/fl}$	moderate tumor incidence rate/HCC	[68]
	Alb-Cre; *Tsc1*$^{fl/fl}$ Alb-Cre; *Tsc1*$^{fl/fl}$/*Pten*$^{fl/fl}$	no steatosis/HCC, ICC hepatomegaly (early), large tumor (14 w)/HCC	[63]

ICC, intrahepatic cholangiocarcinoma; HCA, hepatocellular adenoma; HCC, hepatocellular carcinoma.

Recently, simple liver-specific transgenesis has been developed that allows the liver in adult mice to be transfected with DNA. This methodology, called hydrodynamic tail vein injection or simply HTVI, employs the application of physical force through the rapid injection of a large volume of DNA solution into the lateral tail vein. This process generates increased pressure within the vena cava, propelling the DNA solution into the large hepatic vein and eventually reaching the hepatic tissue and hepatocytes [69,70]. For genome editing in the liver using the CRISPR/Cas system, plasmids containing Cas9 and sgRNA are delivered to the liver via the HTVI method [71]. Using CRISPR-based gene editing combined with the HTVI method, the *Pten* gene was ablated in a subset of hepatocytes in murine livers [64]. In this setting, however, the tumor failed to develop in mice (referred to as "sgPten mice"). The discrepancy in these results between sgPten and AlbCre; *Pten*$^{fl/fl}$ mice were thought to be partially due to the lower frequency of hepatocytes having undergone the loss of PTEN in sgPten mice compared with that in AlbCre; *Pten*$^{fl/fl}$ mice. Moreover, in the case of the latter, the *Pten* gene was deleted during early embryogenesis, as opposed to sgPten mice, in which the deletion was induced in adult livers. Considering massive cell divisions during embryonic and fetal development, the deletion of *Pten* in embryos is expected to create a more favorable tissue environment for hepatocarcinogenesis [72]. Although the tumor failed to develop in the sgPten mice, concomitant c-Met overexpression in the liver led to the formation of hepatic adenomas (HCA) as well as HCC [64].

Plasmids delivered to the liver via HTVI remain as episomes in the nucleus and, thus, are prone to degradation. To overcome this limitation, the *Sleeping Beauty* transposon system is often coupled with HTVI, which allows genes of interest to be integrated into the genome. The HTVI of transposons encoding an activated form of AKT (myr-AKT1) alone induced HCC after 24 weeks post-HTVI [65,66]. Notably, the simultaneous expression of myr-AKT1 and c-Met significantly accelerated HCC development, causing complete lethality by 8 weeks post-HTVI [65]. Likewise, the co-expression of myr-AKT1 and N-RasG12V led to the rapid emergence of HCC [66]. The studies indicate that oncogenic AKT synergistically cooperated with another oncogene in the development of HCC.

TSC1 is an upstream inhibitor of mTOR which suppresses an mTOR activator, RHEB (Figure 1). The liver-specific knockout of *Tsc1* using a similar method, as described in AlbCre; *Pten*$^{fl/fl}$ mice, resulted in the development of HCC by the age of 9–10 months old [67]. Similar experiments using AlbCre; *Tsc1*$^{fl/f}$ mice conducted by other groups also showed HCC development with a minor presence of ICC by 40 weeks of age. Tumors from AlbCre; *Tsc1*$^{fl/f}$ mice consistently exhibited the activation of mTOR signaling [63,68]. Of note, the concomitant deletion of both *Tsc1* and *Pten* in the liver gave rise to the rapid induction of HCC, usually around the age of 14 weeks, showing the simultaneous deletion of the two major negative regulators of PI3K/AKT/mTOR signaling strongly enhanced the signaling pathway and carcinogenesis in the liver [63].

4.2. Preclinical Animal Studies Targeting EGFR/PI3K/AKT/mTOR Signaling in HCC

Since the activation of the EGFR/PI3K/AKT/mTOR signaling pathway significantly contributes to HCC development, various inhibitors of this signaling pathway have been administered to murine models of HCC to investigate their effects on HCC in vivo (Table 3). MUC15 is a member of the high-molecular-weight glycoprotein family of Mucins. It has been shown to bind to EGFR and induce EGFR degradation, subsequently suppressing the downstream PI3K/AKT/mTOR signaling pathway [73]. In xenograft models of HCC, mice transplanted with HCCLM3 cells overexpressing MUC15 displayed fewer and smaller tumors, leading to increased survival rates compared to the control mice transplanted with HCC cells overexpressing the green fluorescent protein (GFP): an inert protein. Additionally, the degradation of EGFR by MUC15 resulted in fewer lung metastases in the MUC15 group compared to the control [73]. RHEB, as an activator of mTOR, was also targeted in a xenograft model of HCC. When human HCC cells, which were manipulated to express short hairpin RNA downregulating RHEB (RHEB-shRNA), were transplanted to immune-deficient mice, they grew more slowly than tumor cells expressing the control of shRNA in vivo [74].

Table 3. HCC suppression in xenograft models by targeting EGFR/PI3K/AKT/mTOR signaling.

Agent	Target	Administration Route	HCC Cell Line Transplanted	Phenotype	Ref.
MUC15	EGFR	NA	HCCLM3	delayed tumor formation, higher survival rate	[73]
RHEB-shRNA	RHEB	NA	SMMC-7721	decrease in tumor growth	[74]
DZW-310	PI3K	OA	Hep3B	decrease in tumor growth	[75]
MK2206	AKT	IP	Hep3B, Huh7	decrease in tumor growth	[50]
ZJQ-24	AKT/mTOR	OA	HepG2	Tumor cell death/reduced tumor growth	[76]
RAD001	mTOR	OA	Patient-derived HCCs	decrease in tumor growth	[77]
rapamycin	mTOR	OA	HepG2	no effects	[78]
rapamycin + bevacizumab	mTOR + VEGF	OA + IP	HepG2	decreased tumor size/increased survival	[78]
metformin	mTOR	IP	Bel-7402	decrease in tumor growth	[79]
miRNA-199a-3p	mTOR	IV	Huh7	decreased tumor growth/increased survival	[80]
miRNA-99a	mTOR	IT	SMMC-LTNM	reduction in tumor size	[81]

NA, not applicable; OA, oral administration; IP, intraperitoneal; IV, intravenous; IT, intertumoral.

The pharmacological inhibition of PI3K, AKT, or mTOR has exerted similar anti-cancer effects in xenograft models of HCC. Recently, a novel small-molecule inhibitor of PI3K, DZW-310, was developed. The administration of the drug to a Hep3B xenograft tumor model resulted in a significant decrease in tumor growth [75]. As observed in in vitro studies using HCC cells (see above in Section 3.2), administration with an AKT inhibitor, MK2206, led to the growth inhibition of tumors in xenograft models transplanted with Hep3B and Huh7 [50]. A novel dual inhibitor was developed that simultaneously inhibited AKT and mTOR. The compound, ZJQ-24, suppressed tumor growth in HepG2 xenograft mice in a dose-dependent manner [76].

Rapamycin (sirolimus) and its derivatives, RAD001 (also known as everolimus), function as highly selective allosteric inhibitors of mTORC1. They bind tightly to its primary target, the FK506-binding protein12 (FKBP12), which allows the complex to associate with the FKBP-rapamycin binding domain (FRB domain) of mTOR, which is located proximal to the active site of mTOR kinase. The formation of the FKBP12–rapamycin–mTOR complex restricts the access of mTOR substrates and, thus, inhibits the phosphorylation and activation of S6K1 and 4EBP1 [82,83]. In patient-derived HCC xenograft models, treatment with RAD001 led to a significant reduction in tumor growth in a dose-dependent manner [77]. Molecular analysis confirmed significant decreases in the phosphorylated levels of S6K1 and 4EBP1, while the level of phosphorylated mTOR was not altered. RAD001-induced growth suppression was associated with the inhibition of cell proliferation via the downregulation of Cdk-6, Cdk-2, Cdk-4, cdc-25C, cyclin B1 and c-Myc [77]. Treatment with rapamycin alone in the HepG2 xenograft model, however, showed no significant benefit to animal survival compared to the control mice treated with the vehicle [78]. Instead, combination therapy with rapamycin and bevacizumab, a monoclonal antibody targeting the vascular endothelial growth factor (VEGF), led to reduced tumor sizes as well as prolonged survival in xenograft mice. Although the discrepancy in the results between RAD001 and rapamycin treatments was not clear, it was speculated that xenograft mice transplanted with HCC directly derived from patients might better represent the characteristics of human HCC compared to those transplanted with HepG2, which had been long maintained in the tissue culture in vitro. Metformin inhibits mTOR via the activation of AMPK: a strong suppressor of mTOR. It is noteworthy that treatment with metformin also retarded tumor development in HCC xenograft models transplanted with Bel-7402 cells [79].

Recently, there has been increasing interest in applying the RNA interference (RNAi) approach to HCC therapy [84–86]. A variety of short interfering RNAs (siRNAs), as well as miRNAs have been tested for anti-cancer effects in HCC cells in vitro, targeting the major components of the EGFR/PI3K/AKT/mTOR signaling pathway. For example, miRNAs targeting EGFR, such as miRNA-137, miRNA-302b, and miRNA-486-3p, have consistently shown anti-proliferative effects on HCC cells, including HepG2, Huh7, and SMMC-7721 [87–89]. Likewise, PI3K and AKT were downregulated in HepG2 cells using miRNA-124 and miRNA-149, respectively, which led to decreases in the proliferation and migration of HCC cells [90,91]. In particular, attempts have been made to inhibit mTOR using miRNA in vivo. The intravenous injection of miRNA-199a-3p encapsulated in nanoparticles efficiently reduced the growth of tumors in HCC xenograft mice [80]. In this study, target delivery to the liver was achieved by conjugating lactobionic acid (LA) with nanoparticles, the receptor of which was overexpressed in HCC cells. In another study, the intratumoral injection of cholesterol-conjugated miR-99a, which targets mTOR, led to a reduction in tumor mass in xenograft mice bearing HCC [81]. With the recent improvements in the stability of RNA, nanoparticle technology, and targeted delivery, RNAi-based therapy is expected to arise as a promising approach to target the EGFR/PI3K/AKT/mTOR signaling specifically and effectively in HCC in vivo.

5. Clinical Studies Targeting EGFR/PI3K/AKT/mTOR Signaling

Given the significant roles of the EGFR/PI3K/AKT/mTOR pathway in human cancer, there has been a strong emphasis on investigating its clinical potential by targeting the major components of the EGFR/PI3K/AKT/mTOR signaling pathway using specific small-molecule inhibitors. For instance, copanlisib, an inhibitor specific to PI3K, was administered to patients with cancers carrying an activating mutation in PI3K (NCT02465060) [92]. In a similar vein, a phase II clinical trial evaluated the efficacy of the allosteric AKT inhibitor MK-2206 (NCT01239355) [93]. Both clinical studies, however, were prematurely terminated due to discouraging results, including various adverse side effects (Table 4).

Various drugs have been employed to inhibit mTOR in clinical studies of HCC [94]. In the phase I/II study, patients with advanced HCC were given everolimus at 10 mg/day as a single agent. The treatment failed to consistently show effectiveness, as only two patients (8%) were progression-free at 24 weeks (NCT00516165). HCC patients treated with the mTOR inhibitor sirolimus (rapamycin) exhibited significantly longer overall survival (OS) compared to the control group [95]. A phase II clinical study demonstrated that sirolimus extended the survival of patients with HCC after liver transplantation (NCT01374750). Onatasertib (CC-223), an orally administered mTOR inhibitor, has been found to induce mitochondrial dysfunction in HCC cell lines [96]. In an open-label phase II trial (NCT03591965), researchers explored the use of onatasertib in subjects with hepatitis B virus (HBV)-positive HCC who had previously undergone at least one line of therapy. Another completed phase I/II study investigated AZD8055, a novel ATP-competitive mTOR kinase inhibitor, evaluating its safety, tolerability, pharmacokinetics, and preliminary efficacy [97]. Overall, although some clinical trials have shown that mTOR inhibitors can provide a survival benefit in patients with advanced HCC, especially when they were intolerant to sorafenib (a standard first-line therapy for advanced HCC), not all trials have demonstrated a significant improvement in overall survival with the treatment of an mTOR inhibitor.

Table 4. Clinical trials targeting the EGFR/PI3K/Akt/mTOR signaling pathway in HCC.

Agent	Target	Phase	Clinical Outcomes	Adverse Events (Side Effects)	NCT #
Copanlisib	pan-class I PI3K	II	discontinuation due to disease progression	hyperglycemia (63%), fatigue (40%), diarrhea (37%), hypertension (33%), nausea (33%), maculopapular rash (30%)	02465060
MK-2206	AKT	II	early termination for discouraging results	anemia (73.33%), hyperglycemia (60.00%), hypoalbuminemia (46.67%), hyperbilirubinaemia (13.33%)	01239355
everolimus (RAD001)	mTOR	I/II	only 2 patients (8%) responded to treatment	hyperglycemia (42.86%), diarrhea (39.29%), hyponatremia (32.14%)	00516165
sirolimus (rapamycin)	mTOR	II	MOS of 21.1 m vs. 14.1 m for control (survival benefit over control)	dyslipidaemia, oral mucositis, diarrhea	01374750
onatasertib (CC-223)	mTOR	II	preliminary antitumor activity	diarrhea (60.38%), hyperglycaemia (60.38%), thrombocytopenia (30.19%), hyperbilirubinaemia (11.32%)	03591965
AZD8055	mTOR	I/II	not yet determined	increased aspartate aminotransferase (22%), fatigue (16%)	00999882
rapamycin + bevacizumab	mTOR + VEGF	I	no survival benefit over bevacizumab-only treatment	hyperglycaemia (83%), thrombocytopenia (75%), fatigue (46%), mucositis (46%), anorexia (42%), Diarrhea (33%)	00467194
erlotinib + bevacizumab	EGFR + VEGF	II	no improvement over sorafenib-treated group	increased alkaline phosphatase (38.30%), hypoalbuminemia (29.79%)	00881751
temsirolimus + sorafenib	mTOR + RTK	II	no survival benefit over sorafenib-only group	hypophosphatemia (60.71%), diarrhea (28.57%), anemia (10.71%)	01687673

MOS, mean overall survival; VEGF, vascular endothelial growth factor; RTK, receptor tyrosine kinase.

The combinatory inhibition of mTOR and VEGF using rapamycin and bevacizumab significantly reduced tumor growth in xenograft models of HCC [78] (Table 3). A phase I clinical trial explored the combination of rapamycin with bevacizumab at the maximum tolerated dose in patients with HCC, demonstrating a complete response in one case and stable disease states in the majority, although this combinatory treatment did not show significant improvements in overall survival compared to patients treated with bevacizumab only (NCT00467194). Phase I/II clinical trials are ongoing to further investigate the efficacy of the combination therapy in HCC [94]. Erlotinib is an EGFR inhibitor that effectively suppresses the phosphorylation of its downstream effectors, such as AKT and mTOR. The combined inhibition of EGFR and VEGF using erlotinib and bevacizumab showed similar anti-cancer effects in patients with advanced HCC (NCT00881751) [98].

In an attempt to improve outcomes, a single-arm phase II trial combined temsirolimus and sorafenib to exploit mTOR inhibition along with sorafenib's effects on HCC [99]. However, it failed to achieve an improvement in overall survival compared to the control group, who were treated with sorafenib only (NCT01687673). Despite this setback, ongoing research continues to explore drug combinations, and several clinical studies are underway [100,101].

It is noteworthy that while mTOR inhibitors, as well as other therapeutics targeting the EGFR/PI3K/AKT/mTOR pathway, have shown some promise in preclinical studies, their efficacy in patients with advanced HCC remains uncertain. These findings highlight the complexity of targeting the EGFR/PI3K/AKT/mTOR pathway in HCC and underscore the need for further investigation to identify more effective treatment approaches.

6. Perspectives and Conclusions

A comprehensive understanding of the molecular pathway leading to carcinogenesis is crucial for predicting patient responses to targeted therapies, significantly impacting clinical decision-making. The development of HCC is a complex process involving various alterations in multiple signaling cascades [5,6]. Among the various oncogenic signals, the EGFR/PI3K/AKT/mTOR pathway stands out as it is activated in over 50% of HCC cases, making it a crucial target for patients with this condition. Research has emphasized the pivotal role of the EGFR/PI3K/AKT/mTOR cascade in the development of HCC. Promisingly, preclinical studies using human HCC cells have consistently demonstrated the effectiveness of targeting the key components of this signaling pathway. Such interventions have been shown to suppress the proliferation of tumor cells in vitro and inhibit the growth of HCC in vivo, offering hope for potential therapeutic approaches.

While preclinical studies have shown promise in targeting the PI3K/AKT/mTOR pathway, clinical trials have not consistently demonstrated a significant improvement in patient outcomes. The overall response rate and survival benefits observed in clinical studies targeting the EGFR/PI3K/AKT/mTOR signaling pathway have been modest or disappointing. These less-than-satisfactory outcomes could be attributed, in part, to the significant toxicities and adverse effects caused by PI3K/AKT/mTOR inhibitors, negatively impacting patients' quality of life and leading to treatment discontinuation. Notably, adverse effects routinely observed in administering these inhibitors include decreased liver functions, as determined by increased alkaline phosphatase (ALP) and aspartate aminotransferase (AST) levels, hypoalbuminemia, thrombocytopenia, and hyperbilirubinemia (Table 4). The side effects on the liver exerted by PI3K/AKT/mTOR inhibitors are especially detrimental to HCC patients who already have pre-existing liver damage.

In this sense, successful therapy for HCC should enable the efficient targeting of the EGFR/PI3K/AKT/mTOR pathway with minimal toxicities and adverse effects. One promising approach is the targeted delivery of drugs to HCC using polymeric nanoparticles (PNPs) [86,102]. In other words, for HCC-targeted delivery, inhibitors of EGFR/PI3K/AKT/mTOR signaling are encapsulated in PNPs that are conjugated with targeting ligands specific to HCC cells. For example, the asialoglycoprotein receptors (ASGP-R) and Glypican-3 (GPC3) are overexpressed in HCC and are considered highly specific markers for HCC [103,104]. Conjugating monoclonal antibodies or ligands that specifically recognize ASGP-R or GPC3 with PNPs are expected to allow the encapsulated drugs to be specifically delivered to HCC, minimizing various side effects and drug-related toxicities commonly found in the systemic delivery of these inhibitors for EGFR/PI3K/AKT/mTOR signaling. Thus, there is hope for the future as new drugs and therapeutic approaches are anticipated to be developed.

Funding: This work was supported by a National Research Foundation of Korea (NRF) grant, 2019R1A2C2009518 (awarded to S.W.R.) which was funded by the Korea government (MSIT).

Institutional Review Board Statement: Not applicable.

Informed Consent Statement: Not applicable.

Data Availability Statement: Not applicable.

Conflicts of Interest: The authors declare no conflict of interest.

References

1. Llovet, J.M.; Kelley, R.K.; Villanueva, A.; Singal, A.G.; Pikarsky, E.; Roayaie, S.; Lencioni, R.; Koike, K.; Zucman-Rossi, J.; Finn, R.S. Hepatocellular carcinoma. *Nat. Rev. Dis. Prim.* **2021**, *7*, 6. [CrossRef]
2. Singal, A.G.; Lampertico, P.; Nahon, P. Epidemiology and surveillance for hepatocellular carcinoma: New trends. *J. Hepatol.* **2020**, *72*, 250–261. [CrossRef]
3. Banales, J.M.; Marin, J.J.G.; Lamarca, A.; Rodrigues, P.M.; Khan, S.A.; Roberts, L.R.; Cardinale, V.; Carpino, G.; Andersen, J.B.; Braconi, C.; et al. Cholangiocarcinoma 2020: The next horizon in mechanisms and management. *Nat. Rev. Gastroenterol. Hepatol.* **2020**, *17*, 557–588. [CrossRef]
4. Villanueva, A. Hepatocellular Carcinoma. *N. Engl. J. Med.* **2019**, *380*, 1450–1462. [CrossRef]
5. Llovet, J.M.; Bruix, J. Molecular targeted therapies in hepatocellular carcinoma. *Hepatology* **2008**, *48*, 1312–1327. [CrossRef]
6. Dimri, M.; Satyanarayana, A. Molecular Signaling Pathways and Therapeutic Targets in Hepatocellular Carcinoma. *Cancers* **2020**, *12*, 491. [CrossRef]
7. Moon, H.; Ro, S.W. MAPK/ERK Signaling Pathway in Hepatocellular Carcinoma. *Cancers* **2021**, *13*, 3026. [CrossRef]
8. Park, H.; Park, H.; Baek, J.; Moon, H.; Ro, S.W. Target Therapy for Hepatocellular Carcinoma: Beyond Receptor Tyrosine Kinase Inhibitors and Immune Checkpoint Inhibitors. *Biology* **2022**, *11*, 585. [CrossRef]
9. Fruman, D.A.; Rommel, C. PI3K and cancer: Lessons, challenges and opportunities. *Nat. Rev. Drug Discov.* **2014**, *13*, 140–156. [CrossRef]
10. Osaki, M.; Oshimura, M.; Ito, H. PI3K-Akt pathway: Its functions and alterations in human cancer. *Apoptosis* **2004**, *9*, 667–676. [CrossRef]
11. Thorpe, L.M.; Yuzugullu, H.; Zhao, J.J. PI3K in cancer: Divergent roles of isoforms, modes of activation and therapeutic targeting. *Nat. Rev. Cancer* **2015**, *15*, 7–24. [CrossRef]
12. Gao, Y.; Moten, A.; Lin, H.K. Akt: A new activation mechanism. *Cell Res.* **2014**, *24*, 785–786. [CrossRef]
13. Saxton, R.A.; Sabatini, D.M. mTOR Signaling in Growth, Metabolism, and Disease. *Cell* **2017**, *169*, 361–371. [CrossRef]
14. Sun, E.J.; Wankell, M.; Palamuthusingam, P.; McFarlane, C.; Hebbard, L. Targeting the PI3K/Akt/mTOR Pathway in Hepatocellular Carcinoma. *Biomedicines* **2021**, *9*, 1639. [CrossRef]
15. Ma, X.M.; Blenis, J. Molecular mechanisms of mTOR-mediated translational control. *Nat. Rev. Mol. Cell Biol.* **2009**, *10*, 307–318. [CrossRef]
16. Laplante, M.; Sabatini, D.M. mTOR signaling in growth control and disease. *Cell* **2012**, *149*, 274–293. [CrossRef]
17. He, Y.; Sun, M.M.; Zhang, G.G.; Yang, J.; Chen, K.S.; Xu, W.W.; Li, B. Targeting PI3K/Akt signal transduction for cancer therapy. *Signal Transduct. Target. Ther.* **2021**, *6*, 425. [CrossRef]
18. Liu, R.; Chen, Y.; Liu, G.; Li, C.; Song, Y.; Cao, Z.; Li, W.; Hu, J.; Lu, C.; Liu, Y. PI3K/AKT pathway as a key link modulates the multidrug resistance of cancers. *Cell Death Dis.* **2020**, *11*, 797. [CrossRef]
19. Peng, Y.; Wang, Y.; Zhou, C.; Mei, W.; Zeng, C. PI3K/Akt/mTOR Pathway and Its Role in Cancer Therapeutics: Are We Making Headway? *Front. Oncol.* **2022**, *12*, 819128. [CrossRef]
20. Yu, L.; Wei, J.; Liu, P. Attacking the PI3K/Akt/mTOR signaling pathway for targeted therapeutic treatment in human cancer. *Semin. Cancer Biol.* **2022**, *85*, 69–94. [CrossRef]
21. Harada, K.; Shiota, G.; Kawasaki, H. Transforming growth factor-alpha and epidermal growth factor receptor in chronic liver disease and hepatocellular carcinoma. *Liver* **1999**, *19*, 318–325. [CrossRef]
22. Ito, Y.; Takeda, T.; Sakon, M.; Tsujimoto, M.; Higashiyama, S.; Noda, K.; Miyoshi, E.; Monden, M.; Matsuura, N. Expression and clinical significance of erb-B receptor family in hepatocellular carcinoma. *Br. J. Cancer* **2001**, *84*, 1377–1383. [CrossRef]
23. Zhou, L.; Huang, Y.; Li, J.; Wang, Z. The mTOR pathway is associated with the poor prognosis of human hepatocellular carcinoma. *Med. Oncol.* **2010**, *27*, 255–261. [CrossRef]
24. Sieghart, W.; Fuereder, T.; Schmid, K.; Cejka, D.; Werzowa, J.; Wrba, F.; Wang, X.; Gruber, D.; Rasoul-Rockenschaub, S.; Peck-Radosavljevic, M.; et al. Mammalian target of rapamycin pathway activity in hepatocellular carcinomas of patients undergoing liver transplantation. *Transplantation* **2007**, *83*, 425–432. [CrossRef]
25. Villanueva, A.; Chiang, D.Y.; Newell, P.; Peix, J.; Thung, S.; Alsinet, C.; Tovar, V.; Roayaie, S.; Minguez, B.; Sole, M.; et al. Pivotal role of mTOR signaling in hepatocellular carcinoma. *Gastroenterology* **2008**, *135*, 1972–1983.e11. [CrossRef]
26. Chen, D.; Li, Z.; Cheng, Q.; Wang, Y.; Qian, L.; Gao, J.; Zhu, J.Y. Genetic alterations and expression of PTEN and its relationship with cancer stem cell markers to investigate pathogenesis and to evaluate prognosis in hepatocellular carcinoma. *J. Clin. Pathol.* **2019**, *72*, 588–596. [CrossRef]
27. Kawamura, N.; Nagai, H.; Bando, K.; Koyama, M.; Matsumoto, S.; Tajiri, T.; Onda, M.; Fujimoto, J.; Ueki, T.; Konishi, N.; et al. PTEN/MMAC1 mutations in hepatocellular carcinomas: Somatic inactivation of both alleles in tumors. *Jpn. J. Cancer Res.* **1999**, *90*, 413–418. [CrossRef]
28. Yao, Y.J.; Ping, X.L.; Zhang, H.; Chen, F.F.; Lee, P.K.; Ahsan, H.; Chen, C.J.; Lee, P.H.; Peacocke, M.; Santella, R.M.; et al. PTEN/MMAC1 mutations in hepatocellular carcinomas. *Oncogene* **1999**, *18*, 3181–3185. [CrossRef]
29. Fujiwara, Y.; Hoon, D.S.; Yamada, T.; Umeshita, K.; Gotoh, M.; Sakon, M.; Nishisho, I.; Monden, M. PTEN/MMAC1 mutation and frequent loss of heterozygosity identified in chromosome 10q in a subset of hepatocellular carcinomas. *Jpn. J. Cancer Res.* **2000**, *91*, 287–292. [CrossRef]

30. Ho, D.W.H.; Chan, L.K.; Chiu, Y.T.; Xu, I.M.J.; Poon, R.T.P.; Cheung, T.T.; Tang, C.N.; Tang, V.W.L.; Lo, I.L.O.; Lam, P.W.Y.; et al. TSC1/2 mutations define a molecular subset of HCC with aggressive behaviour and treatment implication. *Gut* **2017**, *66*, 1496–1506. [CrossRef]
31. Wang, C.; Liao, Y.; He, W.; Zhang, H.; Zuo, D.; Liu, W.; Yang, Z.; Qiu, J.; Yuan, Y.; Li, K.; et al. Elafin promotes tumour metastasis and attenuates the anti-metastatic effects of erlotinib via binding to EGFR in hepatocellular carcinoma. *J. Exp. Clin. Cancer Res.* **2021**, *40*, 113. [CrossRef]
32. Li, K.S.; Zhu, X.D.; Liu, H.D.; Zhang, S.Z.; Li, X.L.; Xiao, N.; Liu, X.F.; Xu, B.; Lei, M.; Zhang, Y.Y.; et al. NT5DC2 promotes tumor cell proliferation by stabilizing EGFR in hepatocellular carcinoma. *Cell Death Dis.* **2020**, *11*, 335. [CrossRef]
33. Song, J.; Liu, Y.; Liu, F.; Zhang, L.; Li, G.; Yuan, C.; Yu, C.; Lu, X.; Liu, Q.; Chen, X.; et al. The 14-3-3σ protein promotes HCC anoikis resistance by inhibiting EGFR degradation and thereby activating the EGFR-dependent ERK1/2 signaling pathway. *Theranostics* **2021**, *11*, 996–1015. [CrossRef]
34. Zheng, H.; Yang, Y.; Hong, Y.G.; Wang, M.C.; Yuan, S.X.; Wang, Z.G.; Bi, F.R.; Hao, L.Q.; Yan, H.L.; Zhou, W.P. Tropomodulin 3 modulates EGFR-PI3K-AKT signaling to drive hepatocellular carcinoma metastasis. *Mol. Carcinog.* **2019**, *58*, 1897–1907. [CrossRef]
35. Chen, H.; Wong, C.C.; Liu, D.; Go, M.Y.Y.; Wu, B.; Peng, S.; Kuang, M.; Wong, N.; Yu, J. APLN promotes hepatocellular carcinoma through activating PI3K/Akt pathway and is a druggable target. *Theranostics* **2019**, *9*, 5246–5260. [CrossRef]
36. Yu, Y.; Zhao, D.; Li, K.; Cai, Y.; Xu, P.; Li, R.; Li, J.; Chen, X.; Chen, P.; Cui, G. E2F1 mediated DDX11 transcriptional activation promotes hepatocellular carcinoma progression through PI3K/AKT/mTOR pathway. *Cell Death Dis.* **2020**, *11*, 273. [CrossRef]
37. Wang, S.; Zhu, M.; Wang, Q.; Hou, Y.; Li, L.; Weng, H.; Zhao, Y.; Chen, D.; Guo, J.; Ding, H.; et al. Publisher Correction: Alpha-fetoprotein inhibits autophagy to promote malignant behaviour in hepatocellular carcinoma cells by activating PI3K/AKT/mTOR signalling. *Cell Death Dis.* **2019**, *10*, 214. [CrossRef]
38. Huang, L.; Zhao, C.; Sun, K.; Yang, D.; Yan, L.; Luo, D.; He, J.; Hu, X.; Wang, R.; Shen, X.; et al. Downregulation of CLDN6 inhibits cell proliferation, migration, and invasion via regulating EGFR/AKT/mTOR signalling pathway in hepatocellular carcinoma. *Cell Biochem. Funct.* **2020**, *38*, 541–548. [CrossRef]
39. Lin, H.; Huang, Z.P.; Liu, J.; Qiu, Y.; Tao, Y.P.; Wang, M.C.; Yao, H.; Hou, K.Z.; Gu, F.M.; Xu, X.F. MiR-494-3p promotes PI3K/AKT pathway hyperactivation and human hepatocellular carcinoma progression by targeting PTEN. *Sci. Rep.* **2018**, *8*, 10461. [CrossRef]
40. Dolgormaa, G.; Harimoto, N.; Ishii, N.; Yamanaka, T.; Hagiwara, K.; Tsukagoshi, M.; Igarashi, T.; Watanabe, A.; Kubo, N.; Araki, K.; et al. Mac-2-binding protein glycan isomer enhances the aggressiveness of hepatocellular carcinoma by activating mTOR signaling. *Br. J. Cancer* **2020**, *123*, 1145–1153. [CrossRef]
41. Wang, X.; Zeng, J.; Wang, L.; Zhang, X.; Liu, Z.; Zhang, H.; Dong, J. Overexpression of microRNA-133b is associated with the increased survival of patients with hepatocellular carcinoma after curative hepatectomy: Involvement of the EGFR/PI3K/Akt/mTOR signaling pathway. *Oncol. Rep.* **2017**, *38*, 141–150. [CrossRef] [PubMed]
42. Shi, Y.; Zhuang, Y.; Zhang, J.; Chen, M.; Wu, S. METTL14 Inhibits Hepatocellular Carcinoma Metastasis Through Regulating EGFR/PI3K/AKT Signaling Pathway in an m6A-Dependent Manner. *Cancer Manag. Res.* **2020**, *12*, 13173–13184. [CrossRef] [PubMed]
43. Zhong, L.; Liao, D.; Zhang, M.; Zeng, C.; Li, X.; Zhang, R.; Ma, H.; Kang, T. YTHDF2 suppresses cell proliferation and growth via destabilizing the EGFR mRNA in hepatocellular carcinoma. *Cancer Lett.* **2019**, *442*, 252–261. [CrossRef]
44. Xu, Y.; Xu, H.; Li, M.; Wu, H.; Guo, Y.; Chen, J.; Shan, J.; Chen, X.; Shen, J.; Ma, Q.; et al. KIAA1199 promotes sorafenib tolerance and the metastasis of hepatocellular carcinoma by activating the EGF/EGFR-dependent epithelial-mesenchymal transition program. *Cancer Lett.* **2019**, *454*, 78–89. [CrossRef] [PubMed]
45. Sun, D.; Liu, J.; Wang, Y.; Dong, J. Co-administration of MDR1 and BCRP or EGFR/PI3K inhibitors overcomes lenvatinib resistance in hepatocellular carcinoma. *Front. Oncol.* **2022**, *12*, 944537. [CrossRef]
46. Buontempo, F.; Ersahin, T.; Missiroli, S.; Senturk, S.; Etro, D.; Ozturk, M.; Capitani, S.; Cetin-Atalay, R.; Neri, M.L. Inhibition of Akt signaling in hepatoma cells induces apoptotic cell death independent of Akt activation status. *Investig. New Drugs* **2011**, *29*, 1303–1313. [CrossRef]
47. Gong, C.; Ai, J.; Fan, Y.; Gao, J.; Liu, W.; Feng, Q.; Liao, W.; Wu, L. NCAPG Promotes The Proliferation Of Hepatocellular Carcinoma Through PI3K/AKT Signaling. *Onco Targets Ther.* **2019**, *12*, 8537–8552. [CrossRef]
48. Ye, L.; Mayerle, J.; Ziesch, A.; Reiter, F.P.; Gerbes, A.L.; De Toni, E.N. The PI3K inhibitor copanlisib synergizes with sorafenib to induce cell death in hepatocellular carcinoma. *Cell Death Discov.* **2019**, *5*, 86. [CrossRef]
49. Song, L.; Luo, Y.; Li, S.; Hong, M.; Wang, Q.; Chi, X.; Yang, C. ISL Induces Apoptosis and Autophagy in Hepatocellular Carcinoma via Downregulation of PI3K/AKT/mTOR Pathway in vivo and in vitro. *Drug Des. Dev. Ther.* **2020**, *14*, 4363–4376. [CrossRef]
50. Ou, D.L.; Lee, B.S.; Lin, L.I.; Liou, J.Y.; Liao, S.C.; Hsu, C.; Cheng, A.L. Vertical blockade of the IGFR-PI3K/Akt/mTOR pathway for the treatment of hepatocellular carcinoma: The role of survivin. *Mol. Cancer* **2014**, *13*, 2. [CrossRef]
51. Grabinski, N.; Ewald, F.; Hofmann, B.T.; Staufer, K.; Schumacher, U.; Nashan, B.; Jücker, M. Combined targeting of AKT and mTOR synergistically inhibits proliferation of hepatocellular carcinoma cells. *Mol. Cancer* **2012**, *11*, 85. [CrossRef]
52. Kahraman, D.C.; Kahraman, T.; Cetin-Atalay, R. Targeting PI3K/Akt/mTOR Pathway Identifies Differential Expression and Functional Role of IL8 in Liver Cancer Stem Cell Enrichment. *Mol. Cancer Ther.* **2019**, *18*, 2146–2157. [CrossRef] [PubMed]
53. Zhao, Z.; Song, J.; Zhang, D.; Wu, F.; Tu, J.; Ji, J. Oxysophocarpine suppresses FGFR1-overexpressed hepatocellular carcinoma growth and sensitizes the therapeutic effect of lenvatinib. *Life Sci.* **2021**, *264*, 118642. [CrossRef] [PubMed]

54. Cheng, Y.; Zhang, Y.; Zhang, L.; Ren, X.; Huber-Keener, K.J.; Liu, X.; Zhou, L.; Liao, J.; Keihack, H.; Yan, L.; et al. MK-2206, a novel allosteric inhibitor of Akt, synergizes with gefitinib against malignant glioma via modulating both autophagy and apoptosis. *Mol. Cancer Ther.* **2012**, *11*, 154–164. [CrossRef]
55. Hirai, H.; Sootome, H.; Nakatsuru, Y.; Miyama, K.; Taguchi, S.; Tsujioka, K.; Ueno, Y.; Hatch, H.; Majumder, P.K.; Pan, B.S.; et al. MK-2206, an allosteric Akt inhibitor, enhances antitumor efficacy by standard chemotherapeutic agents or molecular targeted drugs in vitro and in vivo. *Mol. Cancer Ther.* **2010**, *9*, 1956–1967. [CrossRef]
56. Lai, Y.; Yu, X.; Lin, X.; He, S. Inhibition of mTOR sensitizes breast cancer stem cells to radiation-induced repression of self-renewal through the regulation of MnSOD and Akt. *Int. J. Mol. Med.* **2016**, *37*, 369–377. [CrossRef] [PubMed]
57. Llovet, J.M.; Hernandez-Gea, V. Hepatocellular carcinoma: Reasons for phase III failure and novel perspectives on trial design. *Clin. Cancer Res.* **2014**, *20*, 2072–2079. [CrossRef] [PubMed]
58. Niu, L.; Liu, L.; Yang, S.; Ren, J.; Lai, P.B.S.; Chen, G.G. New insights into sorafenib resistance in hepatocellular carcinoma: Responsible mechanisms and promising strategies. *Biochim. Biophys. Acta Rev. Cancer* **2017**, *1868*, 564–570. [CrossRef]
59. Samarin, J.; Laketa, V.; Malz, M.; Roessler, S.; Stein, I.; Horwitz, E.; Singer, S.; Dimou, E.; Cigliano, A.; Bissinger, M.; et al. PI3K/AKT/mTOR-dependent stabilization of oncogenic far-upstream element binding proteins in hepatocellular carcinoma cells. *Hepatology* **2016**, *63*, 813–826. [CrossRef]
60. Gedaly, R.; Angulo, P.; Chen, C.; Creasy, K.T.; Spear, B.T.; Hundley, J.; Daily, M.F.; Shah, M.; Evers, B.M. The role of PI3K/mTOR inhibition in combination with sorafenib in hepatocellular carcinoma treatment. *Anticancer Res.* **2012**, *32*, 2531–2536.
61. Schneider, P.; Schön, M.; Pletz, N.; Seitz, C.S.; Liu, N.; Ziegelbauer, K.; Zachmann, K.; Emmert, S.; Schön, M.P. The novel PI3 kinase inhibitor, BAY 80-6946, impairs melanoma growth in vivo and in vitro. *Exp. Dermatol.* **2014**, *23*, 579–584. [CrossRef] [PubMed]
62. Dai, N.; Ye, R.; He, Q.; Guo, P.; Chen, H.; Zhang, Q. Capsaicin and sorafenib combination treatment exerts synergistic anti-hepatocellular carcinoma activity by suppressing EGFR and PI3K/Akt/mTOR signaling. *Oncol. Rep.* **2018**, *40*, 3235–3248. [CrossRef] [PubMed]
63. Kenerson, H.L.; Yeh, M.M.; Kazami, M.; Jiang, X.; Riehle, K.J.; McIntyre, R.L.; Park, J.O.; Kwon, S.; Campbell, J.S.; Yeung, R.S. Akt and mTORC1 have different roles during liver tumorigenesis in mice. *Gastroenterology* **2013**, *144*, 1055–1065. [CrossRef] [PubMed]
64. Xu, Z.; Hu, J.; Cao, H.; Pilo, M.G.; Cigliano, A.; Shao, Z.; Xu, M.; Ribback, S.; Dombrowski, F.; Calvisi, D.F.; et al. Loss of Pten synergizes with c-Met to promote hepatocellular carcinoma development via mTORC2 pathway. *Exp. Mol. Med.* **2018**, *50*, e417. [CrossRef] [PubMed]
65. Hu, J.; Che, L.; Li, L.; Pilo, M.G.; Cigliano, A.; Ribback, S.; Li, X.; Latte, G.; Mela, M.; Evert, M.; et al. Co-activation of AKT and c-Met triggers rapid hepatocellular carcinoma development via the mTORC1/FASN pathway in mice. *Sci. Rep.* **2016**, *6*, 20484. [CrossRef]
66. Ho, C.; Wang, C.; Mattu, S.; Destefanis, G.; Ladu, S.; Delogu, S.; Armbruster, J.; Fan, L.; Lee, S.A.; Jiang, L.; et al. AKT (v-akt murine thymoma viral oncogene homolog 1) and N-Ras (neuroblastoma ras viral oncogene homolog) coactivation in the mouse liver promotes rapid carcinogenesis by way of mTOR (mammalian target of rapamycin complex 1), FOXM1 (forkhead box M1)/SKP2, and c-Myc pathways. *Hepatology* **2012**, *55*, 833–845. [CrossRef]
67. Menon, S.; Yecies, J.L.; Zhang, H.H.; Howell, J.J.; Nicholatos, J.; Harputlugil, E.; Bronson, R.T.; Kwiatkowski, D.J.; Manning, B.D. Chronic activation of mTOR complex 1 is sufficient to cause hepatocellular carcinoma in mice. *Sci. Signal.* **2012**, *5*, ra24. [CrossRef]
68. Luo, Y.D.; Fang, L.; Yu, H.Q.; Zhang, J.; Lin, X.T.; Liu, X.Y.; Wu, D.; Li, G.X.; Huang, D.; Zhang, Y.J.; et al. p53 haploinsufficiency and increased mTOR signalling define a subset of aggressive hepatocellular carcinoma. *J. Hepatol.* **2021**, *74*, 96–108. [CrossRef]
69. Chen, X.; Calvisi, D.F. Hydrodynamic transfection for generation of novel mouse models for liver cancer research. *Am. J. Pathol.* **2014**, *184*, 912–923. [CrossRef]
70. Ju, H.L.; Han, K.H.; Lee, J.D.; Ro, S.W. Transgenic mouse models generated by hydrodynamic transfection for genetic studies of liver cancer and preclinical testing of anti-cancer therapy. *Int. J. Cancer* **2016**, *138*, 1601–1608. [CrossRef]
71. Cho, K.; Ro, S.W.; Seo, S.H.; Jeon, Y.; Moon, H.; Kim, D.Y.; Kim, S.U. Genetically Engineered Mouse Models for Liver Cancer. *Cancers* **2019**, *12*, 14. [CrossRef]
72. Moon, H.; Park, H.; Chae, M.J.; Choi, H.J.; Kim, D.Y.; Ro, S.W. Activated TAZ induces liver cancer in collaboration with EGFR/HER2 signaling pathways. *BMC Cancer* **2022**, *22*, 423. [CrossRef]
73. Wang, R.Y.; Chen, L.; Chen, H.Y.; Hu, L.; Li, L.; Sun, H.Y.; Jiang, F.; Zhao, J.; Liu, G.M.; Tang, J.; et al. MUC15 inhibits dimerization of EGFR and PI3K-AKT signaling and is associated with aggressive hepatocellular carcinomas in patients. *Gastroenterology* **2013**, *145*, 1436–1448.e12. [CrossRef]
74. Liu, F.; Pan, Z.; Zhang, J.; Ni, J.; Wang, C.; Wang, Z.; Gu, F.; Dong, W.; Zhou, W.; Liu, H. Overexpression of RHEB is associated with metastasis and poor prognosis in hepatocellular carcinoma. *Oncol. Lett.* **2018**, *15*, 3838–3845. [CrossRef]
75. Wu, Y.; Xu, X.; Liu, M.; Qin, X.; Wu, Q.; Ding, H.; Zhao, Q. DZW-310, a novel phosphoinositide 3-kinase inhibitor, attenuates the angiogenesis and growth of hepatocellular carcinoma cells via PI3K/AKT/mTOR axis. *Biochem. Pharmacol.* **2022**, *201*, 115093. [CrossRef]
76. Liu, J.; Liu, Y.; Zhang, J.; Liu, D.; Bao, Y.; Chen, T.; Tang, T.; Lin, J.; Luo, Y.; Jin, Y.; et al. Indole hydrazide compound ZJQ-24 inhibits angiogenesis and induces apoptosis cell death through abrogation of AKT/mTOR pathway in hepatocellular carcinoma. *Cell Death Dis.* **2020**, *11*, 926. [CrossRef]
77. Huynh, H.; Chow, K.H.; Soo, K.C.; Toh, H.C.; Choo, S.P.; Foo, K.F.; Poon, D.; Ngo, V.C.; Tran, E. RAD001 (everolimus) inhibits tumour growth in xenograft models of human hepatocellular carcinoma. *J. Cell. Mol. Med.* **2009**, *13*, 1371–1380. [CrossRef]

78. Ong, L.C.; Song, I.C.; Jin, Y.; Kee, I.H.; Siew, E.; Yu, S.; Thng, C.H.; Huynh, H.; Chow, P.K. Effective inhibition of xenografts of hepatocellular carcinoma (HepG2) by rapamycin and bevacizumab in an intrahepatic model. *Mol. Imaging Biol.* **2009**, *11*, 334–342. [CrossRef]
79. Ling, S.; Song, L.; Fan, N.; Feng, T.; Liu, L.; Yang, X.; Wang, M.; Li, Y.; Tian, Y.; Zhao, F.; et al. Combination of metformin and sorafenib suppresses proliferation and induces autophagy of hepatocellular carcinoma via targeting the mTOR pathway. *Int. J. Oncol.* **2017**, *50*, 297–309. [CrossRef]
80. Varshney, A.; Panda, J.J.; Singh, A.K.; Yadav, N.; Bihari, C.; Biswas, S.; Sarin, S.K.; Chauhan, V.S. Targeted delivery of microRNA-199a-3p using self-assembled dipeptide nanoparticles efficiently reduces hepatocellular carcinoma in mice. *Hepatology* **2018**, *67*, 1392–1407. [CrossRef]
81. Li, D.; Liu, X.; Lin, L.; Hou, J.; Li, N.; Wang, C.; Wang, P.; Zhang, Q.; Zhang, P.; Zhou, W.; et al. MicroRNA-99a inhibits hepatocellular carcinoma growth and correlates with prognosis of patients with hepatocellular carcinoma. *J. Biol. Chem.* **2011**, *286*, 36677–36685. [CrossRef]
82. Hausch, F.; Kozany, C.; Theodoropoulou, M.; Fabian, A.K. FKBPs and the Akt/mTOR pathway. *Cell Cycle* **2013**, *12*, 2366–2370. [CrossRef]
83. Yip, C.K.; Murata, K.; Walz, T.; Sabatini, D.M.; Kang, S.A. Structure of the human mTOR complex I and its implications for rapamycin inhibition. *Mol. Cell* **2010**, *38*, 768–774. [CrossRef]
84. Perrone, F.; Craparo, E.F.; Cemazar, M.; Kamensek, U.; Drago, S.E.; Dapas, B.; Scaggiante, B.; Zanconati, F.; Bonazza, D.; Grassi, M.; et al. Targeted delivery of siRNAs against hepatocellular carcinoma-related genes by a galactosylated polyaspartamide copolymer. *J. Control. Release* **2021**, *330*, 1132–1151. [CrossRef]
85. Scarabel, L.; Perrone, F.; Garziera, M.; Farra, R.; Grassi, M.; Musiani, F.; Russo Spena, C.; Salis, B.; De Stefano, L.; Toffoli, G.; et al. Strategies to optimize siRNA delivery to hepatocellular carcinoma cells. *Expert. Opin. Drug Deliv.* **2017**, *14*, 797–810. [CrossRef]
86. Farra, R.; Musiani, F.; Perrone, F.; Čemažar, M.; Kamenšek, U.; Tonon, F.; Abrami, M.; Ručigaj, A.; Grassi, M.; Pozzato, G.; et al. Polymer-Mediated Delivery of siRNAs to Hepatocellular Carcinoma: Variables Affecting Specificity and Effectiveness. *Molecules* **2018**, *23*, 777. [CrossRef]
87. Ji, L.; Lin, Z.; Wan, Z.; Xia, S.; Jiang, S.; Cen, D.; Cai, L.; Xu, J.; Cai, X. miR-486-3p mediates hepatocellular carcinoma sorafenib resistance by targeting FGFR4 and EGFR. *Cell Death Dis.* **2020**, *11*, 250. [CrossRef]
88. Shi, T.; Fujita, K.; Gong, J.; Nakahara, M.; Iwama, H.; Liu, S.; Yoneyama, H.; Morishita, A.; Nomura, T.; Tani, J.; et al. Aspirin inhibits hepatocellular carcinoma cell proliferation in vitro and in vivo via inducing cell cycle arrest and apoptosis. *Oncol. Rep.* **2020**, *44*, 457–468. [CrossRef]
89. Wang, L.; Yao, J.; Shi, X.; Hu, L.; Li, Z.; Song, T.; Huang, C. MicroRNA-302b suppresses cell proliferation by targeting EGFR in human hepatocellular carcinoma SMMC-7721 cells. *BMC Cancer* **2013**, *13*, 448. [CrossRef]
90. Lang, Q.; Ling, C. MiR-124 suppresses cell proliferation in hepatocellular carcinoma by targeting PIK3CA. *Biochem. Biophys. Res. Commun.* **2012**, *426*, 247–252. [CrossRef]
91. Zhang, Y.; Guo, X.; Xiong, L.; Yu, L.; Li, Z.; Guo, Q.; Li, Z.; Li, B.; Lin, N. Comprehensive analysis of microRNA-regulated protein interaction network reveals the tumor suppressive role of microRNA-149 in human hepatocellular carcinoma via targeting AKT-mTOR pathway. *Mol. Cancer* **2014**, *13*, 253. [CrossRef] [PubMed]
92. Damodaran, S.; Zhao, F.; Deming, D.A.; Mitchell, E.P.; Wright, J.J.; Gray, R.J.; Wang, V.; McShane, L.M.; Rubinstein, L.V.; Patton, D.R.; et al. Phase II Study of Copanlisib in Patients With Tumors With PIK3CA Mutations: Results From the NCI-MATCH ECOG-ACRIN Trial (EAY131) Subprotocol Z1F. *J. Clin. Oncol.* **2022**, *40*, 1552–1561. [CrossRef] [PubMed]
93. Ahn, D.H.; Li, J.; Wei, L.; Doyle, A.; Marshall, J.L.; Schaaf, L.J.; Phelps, M.A.; Villalona-Calero, M.A.; Bekaii-Saab, T. Results of an abbreviated phase-II study with the Akt Inhibitor MK-2206 in Patients with Advanced Biliary Cancer. *Sci. Rep.* **2015**, *5*, 12122. [CrossRef]
94. Matter, M.S.; Decaens, T.; Andersen, J.B.; Thorgeirsson, S.S. Targeting the mTOR pathway in hepatocellular carcinoma: Current state and future trends. *J. Hepatol.* **2014**, *60*, 855–865. [CrossRef]
95. Lee, K.W.; Kim, S.H.; Yoon, K.C.; Lee, J.M.; Cho, J.H.; Hong, S.K.; Yi, N.J.; Han, S.S.; Park, S.J.; Suh, K.S. Sirolimus Prolongs Survival after Living Donor Liver Transplantation for Hepatocellular Carcinoma Beyond Milan Criteria: A Prospective, Randomised, Open-Label, Multicentre Phase 2 Trial. *J. Clin. Med.* **2020**, *9*, 3264. [CrossRef]
96. Tian, L.Y.; Smit, D.J.; Jücker, M. The Role of PI3K/AKT/mTOR Signaling in Hepatocellular Carcinoma Metabolism. *Int. J. Mol. Sci.* **2023**, *24*, 2652. [CrossRef]
97. Cervello, M.; McCubrey, J.A.; Cusimano, A.; Lampiasi, N.; Azzolina, A.; Montalto, G. Targeted therapy for hepatocellular carcinoma: Novel agents on the horizon. *Oncotarget* **2012**, *3*, 236–260. [CrossRef]
98. Thomas, M.B.; Garrett-Mayer, E.; Anis, M.; Anderton, K.; Bentz, T.; Edwards, A.; Brisendine, A.; Weiss, G.; Siegel, A.B.; Bendell, J.; et al. A Randomized Phase II Open-Label Multi-Institution Study of the Combination of Bevacizumab and Erlotinib Compared to Sorafenib in the First-Line Treatment of Patients with Advanced Hepatocellular Carcinoma. *Oncology* **2018**, *94*, 329–339. [CrossRef]
99. Kelley, R.K.; Joseph, N.M.; Nimeiri, H.S.; Hwang, J.; Kulik, L.M.; Ngo, Z.; Behr, S.C.; Onodera, C.; Zhang, K.; Bocobo, A.G.; et al. Phase II Trial of the Combination of Temsirolimus and Sorafenib in Advanced Hepatocellular Carcinoma with Tumor Mutation Profiling. *Liver Cancer* **2021**, *10*, 561–571. [CrossRef]
100. Chan, S.L.; Yeo, W. Development of systemic therapy for hepatocellular carcinoma at 2013: Updates and insights. *World J. Gastroenterol.* **2014**, *20*, 3135–3145. [CrossRef]

101. Lu, X.; Paliogiannis, P.; Calvisi, D.F.; Chen, X. Role of the Mammalian Target of Rapamycin Pathway in Liver Cancer: From Molecular Genetics to Targeted Therapies. *Hepatology* **2021**, *73* (Suppl. S1), 49–61. [CrossRef]
102. Avramović, N.; Mandić, B.; Savić-Radojević, A.; Simić, T. Polymeric Nanocarriers of Drug Delivery Systems in Cancer Therapy. *Pharmaceutics* **2020**, *12*, 298. [CrossRef]
103. Mu, H.; Lin, K.X.; Zhao, H.; Xing, S.; Li, C.; Liu, F.; Lu, H.Z.; Zhang, Z.; Sun, Y.L.; Yan, X.Y.; et al. Identification of biomarkers for hepatocellular carcinoma by semiquantitative immunocytochemistry. *World J. Gastroenterol.* **2014**, *20*, 5826–5838. [CrossRef] [PubMed]
104. Baumhoer, D.; Tornillo, L.; Stadlmann, S.; Roncalli, M.; Diamantis, E.K.; Terracciano, L.M. Glypican 3 expression in human nonneoplastic, preneoplastic, and neoplastic tissues: A tissue microarray analysis of 4,387 tissue samples. *Am. J. Clin. Pathol.* **2008**, *129*, 899–906. [CrossRef] [PubMed]

pharmaceutics

MDPI

Article

Synthesis and Anticancer Activity of Novel Dual Inhibitors of Human Protein Kinases CK2 and PIM-1 †

Patrycja Wińska [1,*], Monika Wielechowska [1], Mirosława Koronkiewicz [2] and Paweł Borowiecki [1,*]

1 Faculty of Chemistry, Warsaw University of Technology, 00-664 Warsaw, Poland; mwielechowska@ch.pw.edu.pl
2 Department of Biomedical Research, National Medicines Institute, 00-725 Warsaw, Poland; m.koronkiewicz@nil.gov.pl
* Correspondence: patrycja.winska@pw.edu.pl (P.W.); pawel.borowiecki@pw.edu.pl (P.B.); Tel.: +48-222345573 (P.W.); +48-222347677 (P.B.)

† Tribute to Zygmunt Antoni Kazimierczuk (1945–2021).

Abstract: CK2 and PIM-1 are serine/threonine kinases involved in the regulation of many essential processes, such as proliferation, differentiation, and apoptosis. Inhibition of CK2 and PIM-1 kinase activity has been shown to significantly reduce the viability of cancer cells by inducing apoptosis. A series of novel amino alcohol derivatives of parental DMAT were designed and synthesized as potent dual CK2/PIM-1 inhibitors. Concomitantly with the inhibition studies toward recombinant CK2 and PIM-1, the influence of the obtained compounds on the viability of three human carcinoma cell lines, i.e., acute lymphoblastic leukemia (CCRF-CEM), human chronic myelogenous leukemia (K-562), and breast cancer (MCF-7), as well as non-cancerous cells (Vero), was evaluated using an MTT assay. Induction of apoptosis and cell cycle progression after treatment with the most active compound and a lead compound were studied by flow-cytometry-based assay. Additionally, autophagy induction in K-562 cells and intracellular inhibition of CK2 and PIM-1 in all the tested cell lines were evaluated by qualitative/quantitative fluorescence-based assay and Western blot method, respectively. Among the newly developed inhibitors, 1,1,1-trifluoro-3-[(4,5,6,7-tetrabromo-1*H*-benzimidazol-2-yl)amino]propan-2-ol demonstrates the highest selectivity and the most prominent proapoptotic properties towards the studied cancer cells, especially towards acute lymphoblastic leukemia, in addition to inducing autophagy in K-562 cells.

Keywords: 4,5,6,7-tetrabromo-*N*,*N*-dimethyl-1*H*-benzimidazol-2-amine (DMAT) derivatives; protein kinase CK2; protein kinase PIM-1; dual protein kinase inhibitors; antitumor activity; apoptosis; autophagy

Citation: Wińska, P.; Wielechowska, M.; Koronkiewicz, M.; Borowiecki, P. Synthesis and Anticancer Activity of Novel Dual Inhibitors of Human Protein Kinases CK2 and PIM-1. *Pharmaceutics* **2023**, *15*, 1991. https://doi.org/10.3390/pharmaceutics15071991

Academic Editors: Francesca Musumeci and Anna Carbone

Received: 26 May 2023
Revised: 5 July 2023
Accepted: 17 July 2023
Published: 20 July 2023

1. Introduction

Protein kinase CK2 (an acronym derived from the old misnomer "casein kinase 2") and PIM-1 (proviral integration of Moloney virus-1) are protein serine/threonine kinases that have been implicated in cell growth and proliferation [1,2]. CK2 phosphorylates over 300 substrates and has multiple roles in the cell cycle, including in cell growth, proliferation, and survival [3,4]. The PIM kinase family contains three strongly evolutionary conservative isoforms: PIM-1, PIM-2, and PIM-3 [5]. Although CK2 and PIM kinases belong to different branches of the kinome [6], they exhibit structural and functional similarities. Unlike other eukaryotic kinases, both CK2 and PIM-1 demonstrate constitutive activity associated with a lack of phosphorylation sites in the activation loop [7,8]. Overexpression of these kinases occurs in many types of cancers, including leukemia, breast, and prostate cancer [9–12]. The increased catalytic activity of PIM-1 and CK2 kinases has been shown to enhance cell resistance to chemotherapy, inter alia, by inactivating BAD protein, a member of the Bcl-2 family [13], and interacting with the Myc factor [14]. It has been reported that the phosphorylation of BAD by CK2 and PIM kinases inhibits its proapoptotic functions in leukemia and prostate cancer, contributing to increased cell survival [13,15]. Moreover,

PIM-1 overexpression combined with Myc leads to the development of the advanced form of prostate carcinoma [16]. The data show that overexpression of CK2 and PIM-1 kinases is a poor predictor in many cancers, such as prostate cancer and breast cancer, among others [17–19], and the reduction in CK2 and PIM-1 activity by chemical or molecular methods induces apoptosis in tumor cells [20–22].

Recent studies have shown that the contribution of CK2 and PIM-1 kinases to the regulation of transcription, differentiation, or signaling of DNA damage/repair systems is achieved by regulating survival pathways and hypoxia [23]. The activity of both CK2 and PIM-1 has been demonstrated to be elevated in hypoxia, which increases HIF-1 transcriptional activity (hypoxia-induced factor), whereas under normoxia, this factor is degraded [24]. In contrast, in hypoxia, its active form impacts the expression of many genes, stimulating angiogenesis and tumor cell resistance to chemotherapy [25–27]. Moreover, both kinases activate the transcriptional factor NF-κB, which is observed in transformed cells [28]. Furthermore, tumor transformation of lymphocytes with PIM-2 involvement depends on the activation of NF-κB [29]. In addition, PIM-1 is essential in activating the NF-κB pathway, which allows for the survival of prostate cancer cells treated with docetaxel, whereas PIM-1 knockdown or expression of a dominant negative protein sensitizes cells to the cytotoxic effects of docetaxel [30].

Over the last three decades, a plethora of potent and cell-permeable ATP-competitive inhibitors for CK2 kinase has been developed [4]. Among the most prominent examples are the polyhalogenated compounds depicted in Figure 1.

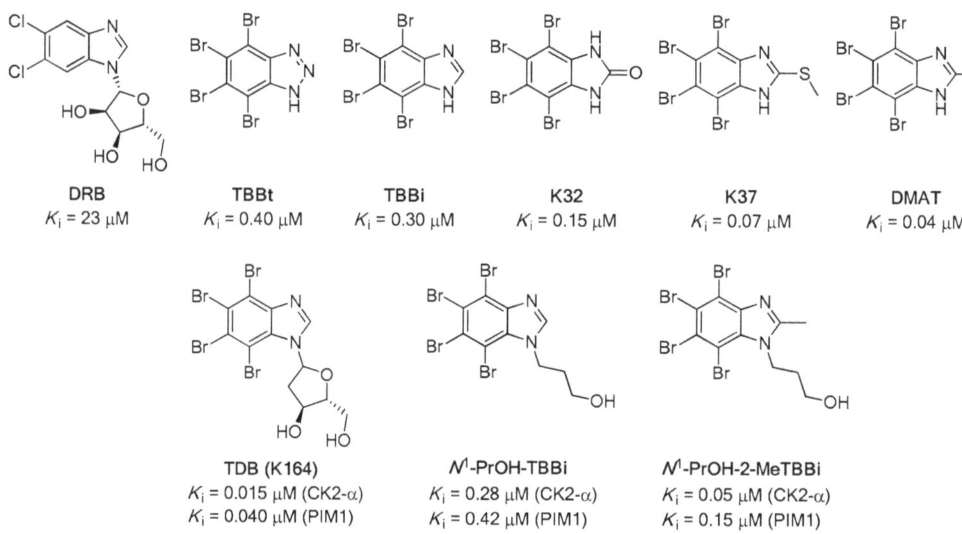

Figure 1. Chemical structures of the representative examples of CK2 inhibitors (5,6-dichloro-1-*β*-D-ribofuranosylbenzimidazole (DRB), 4,5,6,7-tetrabromo-1*H*-benzotriazole (TBBt), 4,5,6,7-tetrabromo-1*H*-benzimidazole (TBBi), 4,5,6,7-tetrabromo-2,3-dihydro-1*H*-1,3-benzodiazol-2-one (K32), 4,5,6,7-tetrabromo-2,3-dihydro-1*H*-1,3-benzodiazol-2-one (K37), and 4,5,6,7-tetrabromo-*N*,*N*-dimethyl-1*H*-benzimidazol-2-amine (DMAT)), and dual CK2/PIM-1 inhibitors (1-(*β*-D-2′-deoxyribofuranosyl)-4,5,6,7-tetrabromo-1*H*-benzimidazole (TDB, also termed K164), 3-(4,5,6,7-tetrabromo-1*H*-1,3-benzodiazol-1-yl)propan-1-ol (N^1-PrOH-TBBi), and 3-(4,5,6,7-tetrabromo-2-methyl-1*H*-1,3-benzodiazol-1-yl)propan-1-ol (N^1-PrOH-TBBi)).

Essential chemical features necessary for efficient inhibitory activity toward CK2 have already been well-identified. Moreover, most of the structural characteristics of halogenated compounds were proven beneficial for the inhibition of the catalytic activity of PIM-1 [31,32].

Since complex signaling pathways and multiple targets are often involved in cancers, synergistically inhibiting multiple targets may be a more effective therapeutic strategy than a convenient single-target approach [33]. To date, quite a few multitarget CK2 inhibitors have been disclosed, which have the potential to become clinical candidates [31]. In this regard, the real breakthrough in the field of developing cell-permeable dual inhibitors of protein kinases CK2 and PIM-1 was achieved by discovering 1-(β-D-2′-deoxyribofuranosyl)-4,5,6,7-tetrabromo-1*H*-benzimidazole (TDB, also termed K164) [34,35]. These studies have shown that the use of dual inhibitors of CK2 and PIM-1 is beneficial for the reduction in cell proliferation and induction of apoptosis in cancer cell lines, i.e., cervical cancer (HeLa) and chronic myeloid leukemia (CML) [35,36]. The most recent studies demonstrated that inhibitor K164 is a promising compound that can be considered a potential active agent in targeted therapy in selected types of breast cancer [37].

In our previous studies [38,39], we evaluated the influence of the structure of the TBBi-alkanol side chain. We found that introducing hydroxyl groups into the aliphatic substituent attached to the hydrophobic TBBi scaffold is critical for the efficient inhibition of CK2. Although this structural fragment is located towards the outer part of the CK2 ATP-binding site, it generates additional polar interactions with a catalytic cavity, which increase the affinity toward the kinase receptor. In view of these findings, we also obtained a novel potent dual CK2/PIM-1 inhibitor, namely 3-(4,5,6,7-tetrabromo-2-methyl-1*H*-1,3-benzodiazol-1-yl)propan-1-ol (N^1-PrOH-TBBi), which, after functionalization into its corresponding acetyl prodrug, turned out to be a very promising anticancer agent toward breast cancer cell lines [40]. Therefore, we expect that further structural modification of aliphatic substituents possessing efficient hydrogen bond-forming moieties installed at the appropriate distance from the TBBi core structure will enhance the affinity and selectivity of the designed 2-alkanol-TBBi derivatives toward titled kinases.

In this work, our ultimate goal was to design and synthesize novel amino alcohol derivatives of parental DMAT as potent dual CK2/PIM-1 inhibitors exhibiting promising anticancer activity. In this regard, intending to potentiate a DMAT-lead structure, we obtained a series of its model analogs differing in the structural topology of aliphatic amino alcohol substituents. Given that previous studies showed that breast cancer, as well as leukemic cells, is sensitive to dual CK2/PIM-1 inhibitor treatment [35–37], two leukemic cell lines, i.e., CCRF-CEM and K-562, and breast cancer cells, i.e., MCF-7, were used in the present study.

2. Materials and Methods

2.1. Chemistry

2.1.1. General Procedure for the Synthesis of 2-Bromo-1*H*-benzimidazole (**2**)

Compound **2** was synthesized according to the procedure reported by Ellingboe et al. [41]. Br_2 (12 mL, 0.24 mol) was added dropwise to a cooled (water bath, 5–10 °C) and mechanically stirred mixture of 2-mercapto-1*H*-benzimidazole (**1**, 10.0 g, 66.58 mmol), 48% aqueous HBr (10 mL), and glacial AcOH (100 mL) over 25 min. The mixture warmed slightly (40–45 °C) during the addition, and additional glacial AcOH (50 mL) was added to aid in the stirring of the thick mixture. After the addition was complete, stirring was continued at room temperature for 4 h. Afterwards, H_2O (200 mL) was added, and the resulting solution was cooled in an ice bath (0–5 °C). The pH was adjusted to 4 with solid NaOH (ca. 40 g), and the precipitate was collected by filtration to afford a crude product, which was further purified by recrystallization from acetone to afford 2-bromo-1*H*-benzimidazole (**2**, 6.93 g, 35.17 mmol, 53% yield) as a white solid. Mp 194–196 °C (acetone) [41] 190–192 °C (acetone); R_f [CH_2Cl_2/MeOH (99:1, v/v)] 0.24; ^{1}H NMR (500 MHz, DMSO-d_6): δ 13.18 (br. s, 1H), 7.51 (s, 2H), 7.19 (m, 2H), 2.07 (s, 1H); ^{13}C NMR (126 MHz, DMSO-d_6): δ 139.7, 127.2, 122.2, 114.6; MS (ESI-TOF) *m/z*: $[M + H]^+$ Calcd for $C_7H_6BrN_2^+$ *m/z*: 196.9709, Found 196.9697; FTMS (ESI-TOF) *m/z*: $[M + H]^+$ Calcd for $C_7H_6BrN_2^+$ *m/z*: 196.97089 and 198.96884, Found 196.97125 and 198.96854; GC [200–260 (10 °C/min)]: t_R = 3.16 min.

2.1.2. General Procedure for the Synthesis of 2,4,5,6,7-Pentabromo-1*H*-benzimidazole (**3**)

Compound **3** was synthesized according to the procedure reported by Andrzejewska et al. [42]. Br_2 (8 mL, 160 mmol) was added portionwise to a stirred and refluxed suspension of 2-bromo-1*H*-benzimidazole (**2**, 1.5 g, 7.6 mmol) in H_2O (60 mL) within 6 h. The reflux was continued for 24 h under irradiation of purple LEDs (390 nm) (for details, see Figure S1 in the Supplementary Materials). Afterward, the reaction mixture was cooled, and the orange precipitate was filtered off under suction. The collected solid was dissolved in MeOH/25% $NH_{3aq.}$ (80 mL, 3:1, v/v) and treated with charcoal (1.3 g) and Celite® 545 (3.5 g). After filtering the solids under suction, the pale-yellow solution of the permeate was brought to pH 4–5 with conc. AcOH (150 mL), and the formed precipitate was recrystallized from a mixture of MeOH/H_2O (50 mL, 1:1, v/v) to afford 2,4,5,6,7-pentabromo-1*H*-benzimidazole (**3**, 1.65 g, 3.22 mmol, 42%) as a yellowish solid.

Yield, 42% (1.65 g); yellowish solid; R_f [hexane/AcOEt (50:50, v/v)] 0.15; ^{1}H NMR (500 MHz, DMSO-d_6): δ Not found; ^{13}C NMR (126 MHz, DMSO-d_6 + 2 drops of 2M $HCl_{aq.}$ and registered for 16 h): δ 142.7, 138.9, 138.0, 131.3, 121.0, 120.9, 110.3; FTMS (ESI-TOF) *m/z*: $[M + H]^+$ Calcd for $C_7H_2Br_5N_2^+$ *m/z*: 512.60884, Found 512.60913.

2.1.3. General Procedure for the Synthesis of Dual CK2/PIM-1 Inhibitors—TBBi Amino Alcohol Derivatives (**4–11**)

Compounds **4–11** were synthesized according to the procedure reported by Kazimierczuk and Pinna et al. [43].

Method A: A mixture of 2,4,5,6,7-pentabromo-1*H*-benzimidazole (**3**, 150 mg, 0.30 mmol) and the respective amino alcohol (4.9 equiv) in anhydrous EtOH (4.5 mL) was heated in an ace pressure tube (bushing type, back seal, V = 15 mL, L × O.D. 10.2 cm × 25.4 mm, Sigma Aldrich (Darmstadt, Germany): Z181064) at 110–115 °C for 72 h. Afterward, the reaction mixture was cooled to room temperature, and the volatiles were evaporated under reduced pressure using a rotavap. The oil residue was purified by a column chromatography using a sequential mixture of hexane/AcOEt (500 mL, 50:50 v/v) and $CHCl_3$/MeOH (500 mL, 90:10 v/v) as an eluent to afford the desired solid-state products (**4**, *rac*-**6**, **8**, *rac*-**9**, **10**, and **11**).

Method B: A mixture of 2,4,5,6,7-pentabromo-1*H*-benzimidazole (**3**, 500 mg, 0.98 mmol) and racemic 2-hydroxypropylamine (1.10 g, 14.63 mmol, 1.13 mL) in anhydrous EtOH (7.5 mL) was heated in an ace pressure tube (bushing type, back seal, V = 15 mL, L × O.D. 10.2 cm × 25.4 mm, Sigma Aldrich (Darmstadt, Germany): Z181064) at 110–115 °C for 72 h. Afterward, the reaction mixture was cooled to room temperature, and the volatiles were evaporated under reduced pressure using a rotavap. The oil residue was purified by a column chromatography using a sequential mixture of hexane/AcOEt (500 mL, 50:50 v/v) and $CHCl_3$/MeOH (500 mL, 90:10 v/v) as an eluent to afford the desired product, 1-[(4,5,6,7-tetrabromo-1*H*-benzimidazol-2-yl)amino]propan-2-ol (*rac*-**5**, 247 mg, 0.49 mmol, 50%), as a white solid.

Method C: A mixture of 2,4,5,6,7-pentabromo-1*H*-benzimidazole (**3**, 50 mg, 98 μmol) and optically active (*S*)- or (*R*)-1-aminopropan-2-ol (110 mg, 1.46 mmol, 113 μL) in dry $PhCH_3$ (1 mL) was heated in an ace pressure tube (bushing type, back seal, V = 15 mL, L × O.D. 10.2 cm × 25.4 mm, Sigma Aldrich (Darmstadt, Germany): Z181064) at 100 °C for 48 h. Afterward, the reaction mixture was cooled to room temperature, and the volatiles were evaporated under reduced pressure using a rotavap. The oil residue was purified by a column chromatography using $CHCl_3$/MeOH (95:5 v/v) as an eluent to afford the desired enantiomerically pure (2*S*)-1-[(4,5,6,7-tetrabromo-1*H*-benzimidazol-2-yl)amino]propan-2-ol [(*S*)-**5**, 46 mg, 91 μmol, 93% yield, >99% ee] or (2*R*)-1-[(4,5,6,7-tetrabromo-1*H*-benzimidazol-2-yl)amino]propan-2-ol [(*R*)-**5**, 42 mg, 83 μmol, 85% yield, >99% ee] as a white solid.

4,5,6,7-Tetrabromo-N,N-dimethyl-1H-benzimidazol-2-amine (DMAT, **4***)*

Synthesized according to *Method A* (Section 2.3). Yield, 29% (41 mg); white solid; R_f [$CHCl_3$/MeOH (95:5 v/v)] 0.82; ^{1}H NMR (500 MHz, DMSO-d_6): δ 11.44 (br. s, 1H, *NH*), 3.13 (s, 6H, CH_3); ^{13}C NMR (126 MHz, DMSO-d_6): δ 158.1 ($CNCH_3$), 38.3 (CH_3), the rest of the peaks were not detected; FTMS (ESI-TOF) *m/z*: $[M + H]^+$ Calcd for $C_9H_8Br_4N_3^+$ *m/z*: 477.74053, Found 477.74015.

*1-[(4,5,6,7-Tetrabromo-1H-benzimidazol-2-yl)amino]propan-2-ol (rac-***5***)*

Synthesized according to *Method B* (Section 2.3). Yield, 50% (247 mg); white solid; R_f [hexane/AcOEt (50:50, *v*/*v*)] 0.22 or R_f [$CHCl_3$/MeOH (90:10, *v*/*v*)] 0.60 or R_f [$CHCl_3$/MeOH (95:5 *v*/*v*)] 0.54; ^{1}H NMR (500 MHz, acetone-d_6): δ 11.04 (br. s, 1H), 6.48 (br. s, 1H), 4.70 (br. s, 1H), 4.07–3.92 (m, 1H), 3.56 (ddd, J = 13.6, 6.5, 3.5 Hz, 1H), 3.33 (ddd, J = 13.5, 7.3, 5.1 Hz, 1H), 1.19 (d, J = 6.3 Hz, 3H); ^{1}H NMR (500 MHz, DMSO-d_6): δ 11.41 (br. s, 1H), 6.62 (br. s, 1H), 4.95 (br. s, 1H), 3.83 (d, J = 5.9 Hz, 1H), 3.38 (ddd, J = 13.2, 6.5, 4.4 Hz, 1H), 3.26–3.18 (m, 1H), 1.10 (d, J = 6.2 Hz, 3H); ^{13}C NMR (126 MHz, DMSO-d_6 + 2 drops of 2M $HCl_{aq.}$ and registered for 16 h): δ 154.8, 153.2, 131.5, 130.5, 120.8, 118.0, 106.4, 103.7, 65.2, 50.1, 20.8; FTMS (ESI-TOF) *m/z*: $[M + H]^+$ Calcd for $C_{10}H_{10}Br_4N_3O^+$ *m/z*: 507.75110, Found 507.75086; HPLC [*n*-hexane-2-PrOH (95:5, *v*/*v*); f = 0.8 mL/min; λ = 225 nm; T = 25 °C (Chiralpak AD-H)]: t_R = 19.105 (*S*-isomer) and 21.524 min (*R*-isomer).

*1,1,1-Trifluoro-3-[(4,5,6,7-tetrabromo-1H-benzimidazol-2-yl)amino]propan-2-ol (rac-***6***)*

Synthesized according to *Method A* (Section 2.3). Yield, 28% (46 mg); beige solid; R_f [$CHCl_3$/MeOH (95:5 *v*/*v*)] 0.55; ^{1}H NMR (500 MHz, DMSO-d_6): δ 11.69 (br. s, 1H), 6.76 (t, J = 5.9 Hz, 1H), 6.66 (br. s, 1H), 4.39–4.27 (m, 1H), 3.70 (ddd, J = 13.8, 6.1, 4.0 Hz, 1H), 3.49 (ddd, J = 13.9, 8.1, 5.8 Hz, 1H); ^{19}F NMR (470 MHz, CD_3CN): δ –77.26 (d, J = 7.4 Hz, 3F).; FTMS (ESI-TOF) *m/z*: $[M + H]^+$ Calcd for $C_{10}H_7Br_4F_3N_3O^+$ *m/z*: 561.72283, Found 561.72251.

*3-[(4,5,6,7-Tetrabromo-1H-benzimidazol-2-yl)amino]propan-1-ol (***7***)*

Synthesized according to *Method A* (Section 2.3). Yield, 44% (67 mg); light green solid; R_f [$CHCl_3$/MeOH (95:5 *v*/*v*)] 0.32; ^{1}H NMR (500 MHz, DMSO-d_6): δ 11.59 (br. s, 1H), 6.77 (t, J = 5.4 Hz, 1H), 3.50 (t, J = 6.1 Hz, 4H), 3.43 (q, J = 6.6 Hz, 2H), 1.71 (p, J = 6.5 Hz, 2H); FTMS (ESI-TOF) *m/z*: $[M + H]^+$ Calcd for $C_{10}H_{10}Br_4N_3O^+$ *m/z*: 507.75110, Found 507.75116.

*2-[Methyl(4,5,6,7-tetrabromo-1H-benzimidazol-2-yl)amino]ethanol (***8***)*

Synthesized according to *Method A* (Section 2.3). Yield, 37% (56 mg); white solid; R_f [$CHCl_3$/MeOH (95:5 *v*/*v*)] 0.60; ^{1}H NMR (500 MHz, DMSO-d_6): δ 11.44 (br. s, 1H), 4.89 (br. s, 1H), 3.67–3.59 (m, 4H), 3.17 (s, 3H); FTMS (ESI-TOF) *m/z*: $[M + H]^+$ Calcd for $C_{10}H_{10}Br_4N_3O^+$ *m/z*: 507.75110, Found 507.75116.

*1-(Diethylamino)-3-[(4,5,6,7-tetrabromo-1H-benzimidazol-2-yl)amino]propan-2-ol (rac-***9***)*

Synthesized according to *Method A* (Section 2.3). Yield, 64% (112 mg); white solid; R_f [$CHCl_3$/MeOH (95:5 *v*/*v*)] 0.48; ^{1}H NMR (500 MHz, DMSO-d_6): δ 8.22 (br. s, 1H), 5.32 (dt, J = 11.1, 5.5 Hz, 1H), 5.04 (ddd, J = 13.3, 5.8, 4.4 Hz, 1H), 4.85–4.76 (m, 1H), 4.24–3.80 (m, 12H); FTMS (ESI-TOF) *m/z*: $[M + H]^+$ Calcd for $C_{14}H_{19}Br_4N_4O^+$ *m/z*: 578.82459, Found 578.82454.

*2-[(4,5,6,7-Tetrabromo-1H-benzimidazol-2-yl)amino]propane-1,3-diol (***10***)*

Synthesized according to *Method A* (Section 2.3). Yield, 13% (20 mg); white solid; R_f [$CHCl_3$/MeOH (95:5 *v*/*v*)] 0.10; ^{1}H NMR (500 MHz, DMSO-d_6): δ 6.39 (d, J = 7.7 Hz, 1H), 3.88–3.79 (m, 1H), 3.55 (qd, J = 10.7, 5.4 Hz, 4H); FTMS (ESI-TOF) *m/z*: $[M + H]^+$ Calcd for $C_{10}H_{10}Br_4N_3O_2^+$ *m/z*: 523.74601, Found 523.74597.

*3-[(4,5,6,7-Tetrabromo-1H-benzimidazol-2-yl)amino]propane-1,2-diol (rac-***11***)*

Synthesized according to *Method A* (Section 2.3). Yield, 38% (60 mg); white solid; R_f [$CHCl_3$/MeOH (95:5 *v*/*v*)] 0.13; ^{1}H NMR (500 MHz, DMSO-d_6): δ 6.61 (t, J = 5.3 Hz, 1H), 4.09 (d, J = 4.7 Hz, 1H), 3.70–3.60 (m, 1H), 3.51 (ddd, J = 13.4, 6.3, 4.6 Hz, 1H), 3.43–3.36 (m, 1H), 3.17 (d, J = 2.6 Hz, 2H); FTMS (ESI-TOF) *m/z*: $[M + H]^+$ Calcd for $C_{10}H_{10}Br_4N_3O_2^+$ *m/z*: 523.74601, Found 523.74584.

2.2. *Biological Evaluation*

2.2.1. Reagents and Antibodies

Dimethyl sulfoxide (DMSO), a molecular-biology-grade solvent used for all stocks of the chemical agents, was obtained from Carl Roth (Karlsruhe, Germany). All reagents used in flow cytometry analysis were purchased from BD Biosciences Pharmingen (San Diego, CA, USA). Information about antibodies is provided in Supplementary Materials (Section S3.1).

2.2.2. Cloning, Expression, and Purification of Human CK2α, holoCK2, and PIM-1

CK2α, holoCK2, and PIM-1 were obtained according to Borowiecki [44] and Chojnacki [45]. The protein concentration in the final solution was 12.68 mg/mL for CK2α, 1.61 mg/mL for holoCK2, and 3.0 mg/mL for PIM-1 (determined by the Bradford method using bovine serum albumin as a standard) [46].

2.2.3. Inhibition of Recombinant CK2 and PIM-1

The obtained compounds were tested for their inhibitory activity toward human CK2α, human CK2 holoenzyme, and PIM-1 using a P81 filter isotopic assay as described previously [39]. IC_{50} values were determined for the tested compounds at eight concentrations in the range of 0.005 to 400 μM. The experimental data were fitted to the sigmoidal dose–response equation, i.e., (variable slope) $Y = Bottom + (Top-Bottom)/(1 + 10^{((LogIC50-X)*HillSlope)}$, in GraphPad Prism (Prism 9, v. 9.0.1).

2.2.4. Cell Culture and Agent Treatment

An acute lymphoblastic leukemia ALL cell line (named CCRF-CEM) was purchased from the European Collection of Authenticated Cell Cultures (ECACC), whereas MCF-7 (hormone-dependent breast adenocarcinoma), K-562 (human chronic myelogenous leukemia), and Vero cells (*Cercopithecus aethiops* kidney) were purchased from the American Type Culture Collection (ATCC, Manassas, VA, USA). For more details, see Supplementary Materials (Section S3.2).

2.2.5. 3-(4,5-Dimethylthiazol-2-yl)-2,5-diphenyltetrazolium Bromide (MTT)-Based Viability Assay

After incubation with the tested compounds, an MTT test was performed as described previously [39]. Optical densities were measured at 570 nm using a BioTek microplate reader (BioTek Instruments, Inc., Winooski, VT, USA). All measurements were carried out in a minimum of three biological replicates.

2.2.6. Detection of Apoptosis by Annexin V/propidium Iodide (PI) Labeling

MCF-7 cells were seeded in 6-well plates at 1.2×10^5 cells/well, whereas CCRF-CEM and K-562 cells were seeded in 24-well plates at a density of 2×10^5/mL. Cells were treated with the tested compounds used in 5 μM and 10 μM concentrations. Then, the plates were incubated for 48 h. After exposure to the examined compounds, the cells were collected and centrifugated at 200× *g* at 4 °C for 5 min, washed twice in cold phosphate-buffered saline (PBS), and subsequently suspended in binding buffer at 1×10^6 cells/mL. Subsequently, 100-μL aliquots of the cell suspension were labeled according to the instructions of the respective manufacturer's kit. Briefly, annexin V-fluorescein isothiocyanate and propidium iodide (BD Biosciences, Pharmingen, San Diego, CA, USA) were added to the cell suspension, and the mixture was vortexed, then incubated for 15 min at RT in the dark. A cold binding buffer (400 μL) was then added, and the cells were vortexed again and kept on ice. Flow cytometric measurements were performed within 1 h after labeling. Viable, necrotic, early, and late apoptotic cells were detected by flow cytometry using a BD FACSCanto II flow cytometer and analyzed using BD FACSDiva operating software (BD Biosciences, San Jose, CA, USA).

2.2.7. Mitochondrial Membrane Potential (ΔΨm) Assay

Mitochondrial membrane potential was assessed by flow cytometry using 5,5′,6,6′-tetrachloro-1,1′,3,3′-tetraethylbenzimidazolocarbocyanine iodide (JC-1; Sigma-Aldrich, St. Louis, MO, USA). JC-1 undergoes potential-dependent accumulation in mitochondria. In healthy cells, the dye accumulates in mitochondria, forming aggregates with red fluorescence (FL-2 channel), whereas in dead and apoptotic cells, the dye remains in the cytoplasm in a monomeric form and emits green fluorescence (FL-1 channel). Cells were harvested by centrifugation 48 h post treatment, suspended in 1 mL of complete culture medium at approximately 1 × 10^6 cells/mL, and incubated with 2.5 μL of JC-1 solution in DMSO (1 mg/mL) for 15 min at 37 °C in the dark. The stained cells were then washed with cold PBS, suspended in 400 μL of PBS, and examined by flow cytometry.

2.2.8. Microscopic Examination

K-562 cells were seeded in 6-well plates at 2.5 × 10^5/mL and subjected to the tested compounds. After 48 h of incubation, cells were centrifugated at 200× *g* at 4 °C for 5 min, subsequently suspended in PBS containing 2 μg/mL AO and Hoechst 33342, and incubated at 37 °C for 20 min in the dark. Subsequently, cells were centrifugated at 200× *g* at 4 °C for 5 min. Then, 5 μL of the sample was mounted on a glass slide and covered with a coverslip and examined under a Nikon ECLIPSE Y-TV55 fluorescent microscope.

2.2.9. Fluorescence Intensity Assay

K-562 cells were seeded in 6-well plates at 2.5 × 10^5/mL and subjected to the tested compounds. After 48 h of incubation, cells were centrifugated at 200× *g* at 4 °C for 5 min, subsequently suspended in PBS containing 2 μg/mL AO and Hoechst 33342, and incubated at 37 °C for 20 min in the dark. Subsequently, cells were centrifugated at 200× *g* at 4 °C for 5 min, and 150 μL of the sample was measured on dark plates (Sarstedt). The fluorescence intensity of AO at Ex. 502 nm/Em. 520–524 nm (aggregated or DNA complexed form), Ex. 457 nm/Em. 630–644 nm (aggregated or RNA complexed form), Ex. 540 nm/Em. 640–660 nm (red-stained lysosomes), and Ex. 488 nm/Em. 540–550 nm (yellowish stained lysosomes) and that of Hoechst (nuclei, Ex. 361 nm/Em. 497 nm) in K-562 cells was measured with a Synergy H4 Hybrid Multi-Mode Microplate Reader (BioTek Instruments, Inc., Winooski, VT, USA). The data were normalized to the intensity of Hoechst (nuclei, Ex. 361 nm/Em. 497 nm) in comparison to control cells serving as the reference point, showing 100%.

2.2.10. Detection of Cell Cycle Progression by Flow Cytometry

MCF-7, CCRF-CEM, and K-562 cells were cultured in 6-well plates and treated with the tested compounds for 48 h. After exposure to the compounds, the cells were collected and washed with cold PBS and fixed at −20 °C in 70% ethanol for at least 24 h. Subsequently, cells were washed in PBS and stained with 50 μg/mL PI (propidium iodide) and 100 μg/mL RNase solution in PBS supplemented with 0.1% *v*/*v* Triton X-100 for 30 min in the dark at RT. Cellular DNA content was determined by flow cytometry employing a BD FACSCanto II flow cytometer (BD Biosciences, San Jose, CA, USA). The obtained DNA histograms were analyzed using MacCycle software (Phoenix Flow Systems, San Diego, CA, USA) for evaluation of the distribution of the cells in different phases of the cell cycle.

2.2.11. Western Blotting

All the procedures are described in detail in Supplementary Materials (Section S3.3). The protein concentration was determined using a Bradford assay [46].

2.2.12. Densitometry

For densitometry, immunoblots were scanned using G Box Chemi (Syngene, Cambridge, UK), and the density of each lane of phosphorylated and total protein was quantified using GeneSys software (Syngene, Cambridge, UK). Phosphorylated protein densities were normalized to GAPDH densities, assuming 1 for untreated cells; then, they were converted to a percentage of the appropriate control.

2.2.13. Statistical Evaluation

Results are represented as mean ± s.e.m. of at least three independent experiments. Statistical analysis was performed using GraphPad Prism 5.0 software (GraphPad Software Inc., San Diego, CA, USA). Significance was determined using a one-way ANOVA analysis.

The statistical significance of differences is indicated in figures by asterisks as follows: * $p \leq 0.05$, ** $p \leq 0.01$, and *** $p \leq 0.001$.

2.3. Molecular Docking

2.3.1. Molecular Docking Preparation

Molecular docking studies to establish favorable ligand binding geometries for both studied inhibitors, namely 1,1,1-trifluoro-3-[(4,5,6,7-tetrabromo-1*H*-benzimidazol-2-yl)amino]propan-2-ol (*rac*-**6**) and 3-[(4,5,6,7-tetrabromo-1*H*-benzimidazol-2-yl)amino]propane-1,2-diol (*rac*-**11**), were performed using AutoDock Vina v. 1.1.2 (http://autodock.scripps.edu/; accessed on 21 October 2021) [47]. First, ligands *rac*-**6** and *rac*-**11** in non-ionizable form were prepared with ChemAxon MarvinSketch v. 14.9.1.0 (http://www.chemaxon.com/marvin/; accessed on 9 September 2014). The initial geometries of the ligands with the minimum energy conformation ($E_{calc.}$ = −226.758 kJ/mol for *rac*-**6** and $E_{calc.}$ = −171.375 kJ/mol for *rac*-**11**) were optimized in Avogadro v. 1.2.0. (http://avogadro.cc/; accessed on 15 June 2016) using General Amber Force Field (GAFF) [48] and/or MMFF94 force field with 500 steps and the steepest descent algorithm. The visualization of the optimized geometries was performed using POV-Ray for Windows v. 3.7.0.msvc10.win64 licensed under the terms of the GNU Affero General Public License (AGPL3) (Figure 2). Afterward, the Gasteiger partial charges were calculated with AutoDock Tools v. 1.5.6 (ADT, S3 http://mgltools.scripps.edu/; accessed on 29 October 2022). In contrast, all torsion angles for each ligand were considered flexible, and all the possible rotatable bonds and non-polar hydrogens were determined. The final 'ligand' files were saved as PDBQT files (.pdbqt format) and were ready for the docking procedure disclosed below in Section 2.3.2. Molecular Docking Procedure.

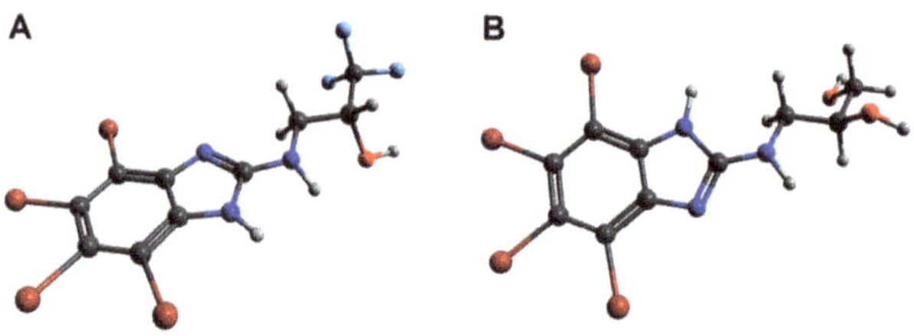

Figure 2. The geometries of 1,1,1-trifluoro-3-[(4,5,6,7-tetrabromo-1*H*-benzimidazol-2-yl)amino]propan-2-ol (*rac*-**6**, (**A**)) and 3-[(4,5,6,7-tetrabromo-1*H*-benzimidazol-2-yl)amino]propane-1,2-diol (*rac*-**11**, (**B**)) optimized in Avogadro v. 1.2.0. The figures rendered using POV-Ray v. 3.7.0 molecular visualization software. Nitrogen atoms are presented with blue color, oxygen atoms with red color, bromine atoms with burgundy (maroon) color, and fluorine atoms with light blue color, whereas hydrogen atoms are expressed as light grey balls.

The crystal structures of human protein kinases, namely CK2-α (PDB code: 4KWP) [35] of the highest available resolution (1.25 Å) and PIM-1 (PDB code: 4DTK) [49] with a resolution of 1.86 Å, were downloaded from the PDB database (http://www.rcsb.org/pdb/). The crude target proteins were prepared using the UCSF Chimera v. 1.11.2 package (http://www.cgl.ucsf.edu/chimera/; accessed on 2 December 2016) [50] after removing all nonstandard molecules, including 4,5,6,7-tetrabromo-1-(2-deoxy-beta-*D*-erythro-pentofuranosyl)-1*H*-benzimidazole (EXX), sulfate ion (SO_4), 1,2-ethanediol (EDO), triethylene glycol (PGE), di(hydroxyethyl)ether (PEG), and dimethyl sulfoxide (DMS) in the case of 4KWP, as well as (5*Z*)-5-{2-[(3*R*)-3-aminopiperidin-1-yl]-3-(propan-2-yloxy)benzylidene}-1,3-thiazolidine-2,4-dione (7LI), EDO, and SO_4 in the case of 4DTK. All the procedures are described in detail in Supplementary Materials (Section S5.1).

2.3.2. Molecular Docking Procedure

Docking was performed using standard protocols as described in Supplementary Materials (Section S5.2). For validation of the docking calculations, two prominent kinase inhibitors crystalized with 4KWP and 4DTK, i.e., 4,5,6,7-tetrabromo-1-(2-deoxy-beta-*D*-erythro-pentofuranosyl)-1*H*-benzimidazole (EXX) for CK2-α and (5*Z*)-5-{2-[(3*R*)-3-aminopiperidin-1-yl]-3-(propan-2-yloxy)benzylidene}-1,3-thiazolidine-2,4-dione (7LI) for PIM-1, were docked as control ligands. The docking modes of each studied ligand (i.e., 1,1,1-trifluoro-3-[(4,5,6,7-tetrabromo-1*H*-benzimidazol-2-yl)amino]propan-2-ol (*rac*-**6**) and 3-[(4,5,6,7-tetrabromo-1*H*-benzimidazol-2-yl)amino]propane-1,2-diol (*rac*-**11**)) were clustered and ranked based on a mutual ligand–protein affinity expressed as absolute free binding energies (ΔG_{calc} (kcal/mol)), as well as the values of root mean square deviation (rmsd) in both modes with respect to the rmsd lower bound (l.b.) and the rmsd upper bound (u.b.). The rmsd values were computed with reference to the input structure submitted to docking simulations. For CK2-α (PDB code: 4KWP), the used random seed amounted to +1342461868 for *rac*-**6** and −2037069392 for *rac*-**11**, whereas for PIM-1 (PDB code: 4DTK), the used random seed amounted to +956047904 for *rac*-**6** and −769683352 for *rac*-**11**. The results of docking are collected in Table S3. The optimized binding poses of *rac*-**6** and *rac*-**11** in hypothetical complexes with CK2-α and PIM-1 were visualized using PyMOL Molecular Graphics System software, v. 1.3, Schrödinger, LLC (https://www.pymol.org/; accessed on 13 October 2011).

3. Results

3.1. Chemical Synthesis

Taking DMAT (**4**) as the lead compound, a series of dual CK2/PIM-1 inhibitors were designed and synthesized in analogy to the methods already reported in the literature [41,42]. The general synthetic route is described in Scheme 1. Briefly, in the first step, a commercially available 2-mercapto-1*H*-benzimidazole (**1**) was brominated with bromine (Br_2) diluted in a mixture of 48% aqueous HBr and glacial acetic acid to afford 2-bromo-1*H*-benzimidazole (**2**) in 53% yield. Next, an exhaustive bromination of the resulting **2** with Br_2 in boiling water performed under irradiation of purple LEDs (390 nm) for 24 h afforded the desired 2,4,5,6,7-pentabromo-1*H*-benzimidazole (**3**) in 42% yield. Finally, the aminolysis of the key intermediate **3** was carried out using the appropriate amino alcohol in anhydrous ethanol in a pressure glass tube reactor at elevated temperatures (110–115 °C) to afford DMAT (**4**) and its derivatives (**5**–**11**) in the yield range of 13–64%.

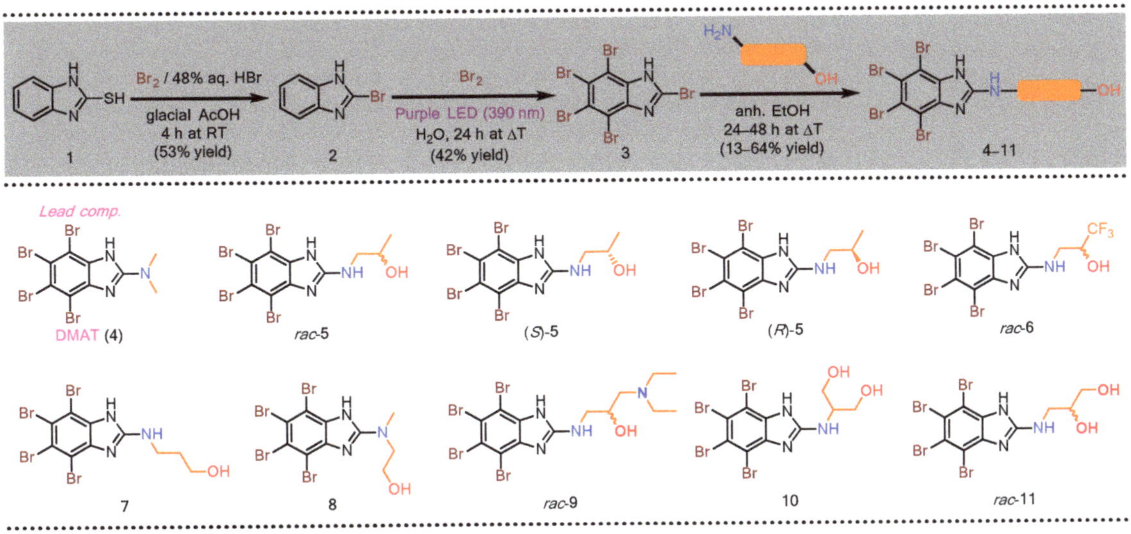

Scheme 1. Synthesis of amino alcohol-like DMAT derivatives (**5**–**11**) as potential dual CK2/PIM-1 inhibitors.

Due to the strong impact of the stereochemistry of xenobiotics on their biological activity in vivo, there is an urgent need to evaluate the single enantiomers of designed inhibitors toward target proteins, as well as cancer cell lines. Such an evaluation is incredibly valid especially when one of the optical isomers from the pair acts as the eutomer, while its counterpart behaves as the distomer. We envisioned that if the eudysmic ratio (ER) is high in the case of the developed chiral compounds (i.e., *rac*-**5**, *rac*-**6**, and *rac*-**11**), then the inhibitory activity of each enantiomer will significantly differ toward target kinases. Therefore, we found it pivotal for biological studies to elaborate on a highly efficient and stereoselective synthetic method for the preparation of both enantiomers of one of the chiral products (*rac*-**5**). This task was accomplished by employing commercially available, optically pure (*S*)- and/or (*R*)-1-aminopropan-2-ol (>99% ee) as chiral building blocks. Unfortunately, conducting the reaction with (*S*)-1-aminopropan-2-ol using a standard protocol afforded (*S*)-**5** with only 39% ee. To our delight, a detailed screening of the reaction conditions, including the selection of organic solvent, as well as the evaluation of the effect of the reaction time and temperature on the stereochemical outcome, led to obtaining enantiomerically pure antipodes (*S*)-**5** and (*R*)-**5** (>99% ee) in the 85–93% yield range (for details, see Table S1 in Supplementary Materials). Among the solvents used, only toluene ($PhCH_3$) guaranteed that the enantiomers of **5** were isolated without undesired racemization. Other tested solvents (i.e., 1,4-dioxane, DMF, and CH_3CN) achieved inferior results in terms of enantiomeric purity (26–83% ee). The considerable erosion in % ee-values with respect to the desired products ((*S*)-**5** and (*R*)-**5**) obtained from the reactions performed in aprotic polar solvents is interesting; however, the explanation for the racemization phenomenon is complex and definitely exceeds the scope of these studies.

All the resulting products (**4–11**) were characterized by single-proton nuclear magnetic resonance (^{1}H-NMR) spectroscopy, high-resolution mass spectrometry (HR-MS), and high-performance liquid chromatography (HPLC). A lack of ^{13}C-NMR spectra is typical for polyhalogenobenzimidazoles, since recording the narrow signals corresponding to quaternary carbon atoms is highly challenging for these compounds due to the electronic features of the TBB-ring possessing a N-H tautomeric proton. On the other hand, this problem is not observed in all the cases when the tautomeric proton is replaced by any substituent [45]. A series of modifications of the NMR experimental conditions implemented to overcome this drawback, including the significant extension of the time of analysis, changes in the values of relaxation times, application of high magnetic fields, and the use of lower temperatures to achieve slow exchange conditions, failed to obtain spectra with all the predicted signals. Interestingly, only the treatment of the samples with hydrochloric acid to avoid the existence of 1,3-tautomeric equilibrium by the protonation of nitrogen atom present in the imidazole ring allowed the ^{13}C-NMR spectra to be recorded with the appropriate signals. Nevertheless, these results are subject to the risk of error, as no correction for the substituent effect was performed. For details, see copies of the recorded spectra appended in Supplementary Materials.

3.2. Biological Evaluation

3.2.1. Inhibition of Recombinant CK2 and PIM-1

Inhibition of the human CK2 catalytic subunit (CK2α), CK2 holoenzyme (CK2α2β2), and PIM-1 by the newly obtained compounds was evaluated using a radiometric assay (Table 1, Figure S2). The synthetic peptide RRRADDSDDDDD was used as the substrate of CK2, and peptide ARKRRRHPSGPPTA was used as the substrate of PIM-1. The values of the inhibition constant (K_i) were calculated using the Cheng–Prusoff equation: $K_i = IC_{50}/(1 + [S]/Km)$ [51].

Table 1. Inhibition of CK2 and PIM-1 by compounds **4–11**.

Cpd.	K_i (μM)		
	CK2α	CK2α₂β₂	PIM-1
4	0.223	0.056	0.052
rac-**5**	0.296	0.127	0.064
(*S*)-**5**	0.271	0.167	0.099
(*R*)-**5**	0.394	0.104	0.089
rac-**6**	0.294	0.112	0.067
7	0.156	0.114	0.082
8	0.256	0.072	0.071
rac-**9**	1.671	0.605	0.267
10	0.139	0.124	0.093
rac-**11**	0.151	0.089	0.073

All the tested compounds were efficient inhibitors of both recombinant forms of CK2 and PIM-1, with K_i values in the range of 0.56–0.605 μM for CK2α2β2 and 0.052–0.267 μM for PIM-1. None of the newly obtained inhibitors inhibited CK2α2β2 or PIM-1 more strongly than the parent compound, DMAT (**4**); however, compounds **7**, **10**, and *rac*-**11** demonstrated lower K_i values for CK2α than for DMAT (**4**), i.e., 0.156 μM, 0.139 μM, and 0.151 μM, respectively. Among the newly obtained compounds, the *rac*-**11** derivative was the most efficient inhibitor of CK2α2β2, with K_i values equal to 0.089 μM, whereas *rac*-**5** was the most potent inhibitor of PIM-1, with K_i = 0.064 μM. The effectiveness of *rac*-**11** towards CK2 kinase can be attributed to the presence of two hydroxyl groups in its structure, which can additionally interact with structural water molecules accommodated in the enzyme's active site. On the contrary, the weakest inhibitor of CK2, compound *rac*-**9** (K_i = 0.605 μM), has the largest substituent when compared to other studied inhibitors and can therefore potentially undergo clashes with amino acid residues present in the binding pocket of the enzyme.

3.2.2. Cytotoxic Effect of DMAT Derivatives **4–11** toward Cancer Cell Lines: CCRF-CEM, K-562, MCF-7, and Non-Cancerous Vero Cells

To test the cytotoxicity of the DMAT derivatives, we treated the CCRF-CEM, K-562, MCF-7, and Vero cells with the newly synthesized compounds in the concentration range of 1.575–100 μM for 48 h. The representative plots demonstrating sigmoidal dose–response curves for compounds **4–11** are shown in Figure 3. The IC_{50} values describing the half-maximal effective concentration of each tested compound were calculated and are summarized in Table 2. Selectivity is an important feature of compounds demonstrating anticancer properties; therefore, the selectivity indices (SI) were calculated, as presented in Table 2. Using an MTT viability assay, we demonstrated that all tested compounds significantly decreased the viability of the studied cells, with IC_{50} values ranging from 9.66 μM to 41.53 μM. Among the newly obtained DMAT derivatives, *rac*-**6** and *rac*-**9** were the most cytotoxic toward the tested tumor cells (Table 2). The viability of CCRF-CEM and MCF-7 was most strongly reduced by *rac*-**6**, with IC_{50} values of 11.83 μM (for CCRF-CEM) and 9.66 μM (for MCF-7), whereas the viability of K-562 was the most strongly decreased by *rac*-**9**, with an IC_{50} value of 11.61 μM. Interestingly, *rac*-**6** demonstrated higher SI values than *rac*-**9**, with the highest value of 2.11 obtained for MCF-7. Considering the obtained results in terms of the lowest IC_{50} and the highest SI values, subsequent biological studies were devoted to *rac*-**6** and **4** (DMAT) as lead compounds.

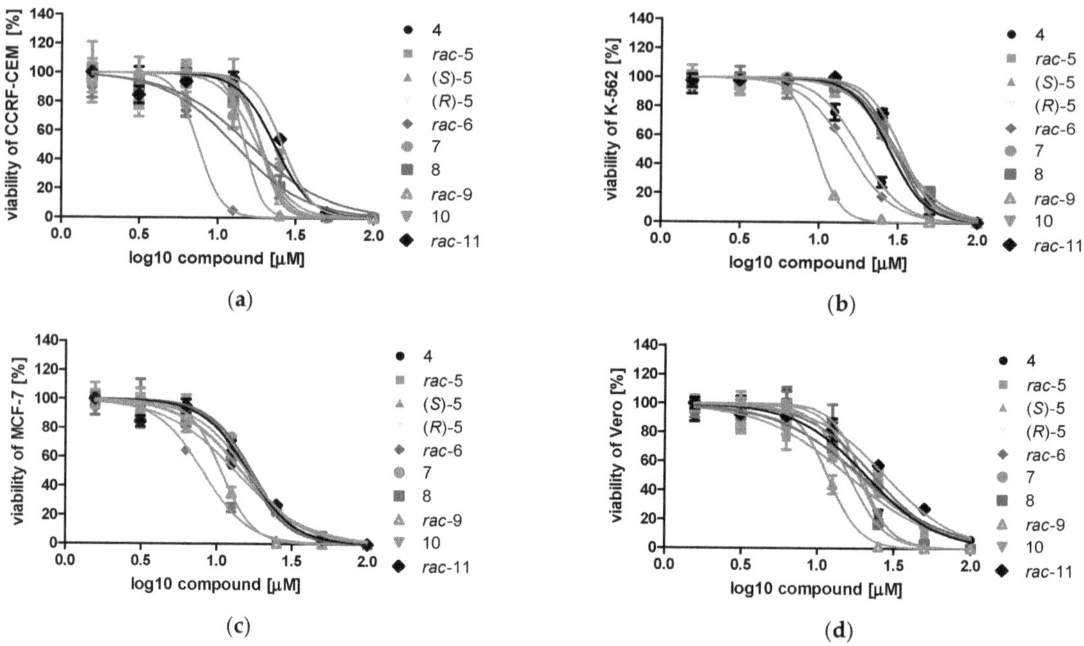

Figure 3. Sigmoidal dose–response curves for compounds **4–11** determined for CCRF-CEM (**a**), K-562 (**b**), MCF-7 (**c**), and Vero cells (**d**) after 48 h of treatment. Plots were generated by GraphPad Prism after fitting the MTT data to the sigmoidal dose–response equation: Y = 100/(1 + 10^((LogIC50 − X) ∗ HillSlope)).

Table 2. Viability of CCRF-CEM (acute lymphoblastic leukemia), K-562 (chronic myelogenous leukemia), MCF-7 (breast carcinoma), and non-cancerous Vero cells after treatment with the tested compounds. The IC_{50} values were calculated using an MTT-based assay and the following equation: Y = 100/(1 + 10^((LogIC50 − X) ∗ HillSlope)).

Cpd.	**Cell Line**				**SI ***			**log*P* ****
	CCRF-CEM	**K-562**	**MCF-7**	**Vero**	**CCRF-CEM**	**K-562**	**MCF-7**	
4	18.94 ± 0.88	22.54 ± 6.09	14.68 ± 0.27	18.72 ± 0.33	0.99	0.83	1.27	5.39
rac-**5**	21.70 ± 0.86	38.70 ± 8.08	16.86 ± 0.30	22.58 ± 0.13	1.04	0.58	1.34	4.41
(*S*)-**5**	23.21± 4.01	39.28 ± 9.05	16.55 ± 0.30	25.20 ± 0.54	1.11	0.64	1.52	4.41
(*R*)-**5**	22.31 ± 0.10	39.59 ± 8.11	16.09 ± 0.19	28.30 ± 0.43	1.27	0.71	1.76	4.41
rac-**6**	11.83 ± 0.40	17.74 ± 3.54	9.66 ± 0.31	20.38 ± 0.41	1.72	1.15	2.11	5.05
7	25.64 ± 0.54	33.70 ± 4.92	18.21 ± 0.30	29.46 ± 0.12	1.15	0.87	1.62	4.19
8	17.37 ± 0.27	40.02 ± 7.45	15.11 ± 0.39	16.07 ± 0.27	0.92	0.40	1.06	4.88
rac-**9**	13.59 ± 0.84	11.61 ± 2.85	10.80 ± 0.23	12.01 ± 0.21	0.88	1.03	1.11	4.74
10	24.47 ± 0.21	38.06 ± 5.56	16.65 ± 0.23	25.19 ± 0.10	1.03	0.66	1.51	3.55
rac-**11**	33.80 ± 0.26	41.49 ± 8.45	17.20 ± 0.39	41.53 ± 0.10	1.23	1.00	2.41	3.55

* SI (selectivity index) was calculated with the following formula: $x = \frac{IC50\ Vero}{IC50\ cancer\ cells}$. ** Logarithm of the partition coefficient of a given inhibitor between n-octanol and water calculated using ChemBioDraw Ultra 13.0 software (PerkinElmer Informatics Indications, Waltham, MA, USA).

3.2.3. Induction of Apoptosis in CCRF-CEM, K-562, and MCF-7 Cells

In order to evaluate the proapoptotic properties of *rac*-6, we analyzed annexin V-binding to phosphatidylserine using flow cytometry. The results are shown in Figure 4 and Table S2. The obtained results indicate that *rac*-6 induced apoptosis in CCRF-CEM at both concentrations, with the highest proportion of 47% of cells in early and late apoptosis after treatment with 10 µM conc. The examined racemate, *rac*-6, also induced apoptosis in MCF-7 cells at 10 µM conc. (statistically significant result), whereas K-562 cells were the least sensitive to *rac*-6 (no significant apoptosis observed). However, the percentage of total apoptotic K-562 cells was higher after treatment with 10 µM conc. of *rac*-6 than after treatment with parental compound 4 (DMAT), with values of 15.5% and 8.7%, respectively. We also observed the unusual presence of two different pools of unstained cells (considered alive) that were especially visible on cytograms obtained for K-562 cells treated with 10 µM conc. of *rac*-6 (Figure 4c). The results obtained for this line utilizing flow cytometry suggest a mechanism of death induction other than apoptosis for *rac*-6.

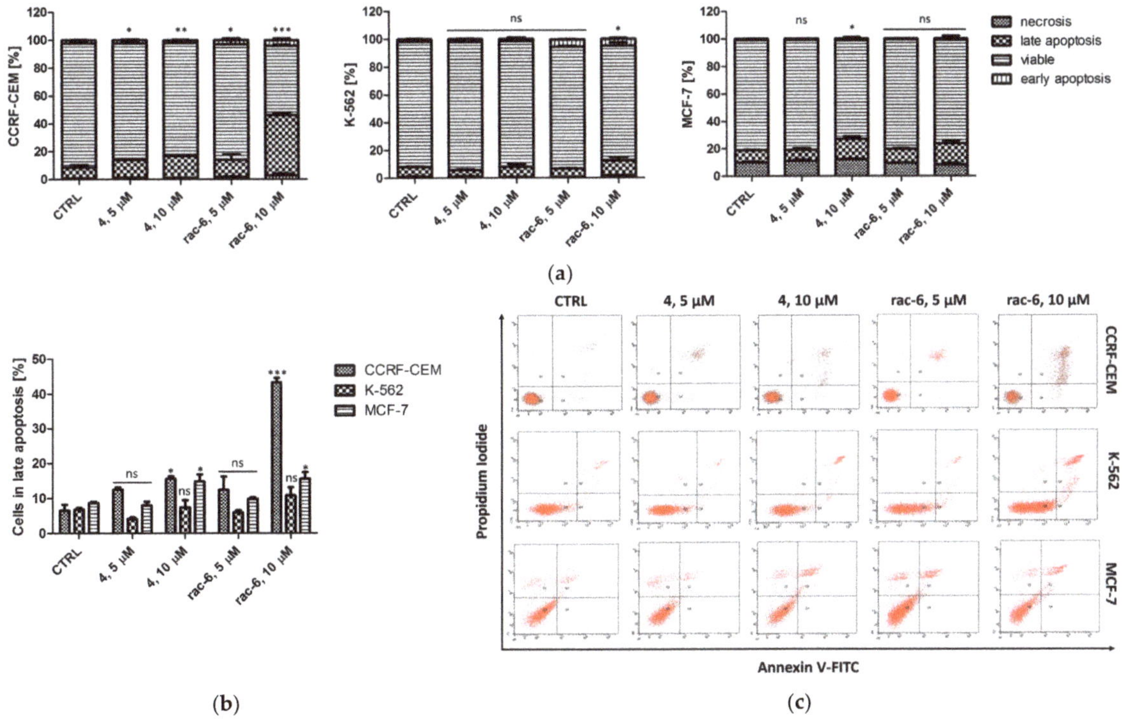

Figure 4. Induction of apoptosis in CCRF-CEM, K-562, and MCF-7 cells. The data were determined by a FACSCanto II flow cytometer after 48 h of treatment with compounds **4** and *rac*-**6**. Cells were stained with annexin V-FITC and PI (propidium iodide). (**a**) Mean and standard deviation (SD) of necrosis, as well as viable, early, and late apoptosis, as percent from three independent experiments each. (**b**) Percentage of late apoptosis in CCRF-CEM, K-562, and MCF-7 cells. (**c**) Representative cytograms. The data for viable cells (**a**) and in late apoptosis (**b**) were analyzed by Dunnett's multiple comparison test as follows: * $p < 0.05$, ** $p < 0.01$, and *** $p < 0.001$ relative to control; ns—not significant.

3.2.4. Mitochondrial Membrane Potential (ΔΨm) in CCRF-CEM

Regarding the significant proapoptotic effect of *rac*-6 on CCRF-CEM cells, in the next step of our investigation, we measured mitochondrial membrane potential (ΔΨm) in leukemia cells treated with compounds 4 and *rac*-6 for 48 h (Figure 5). Incubation with the tested compounds caused mitochondrial membrane depolarization in a dose-dependent manner (as evidenced by the shift from a red to green fluorescence ratio). The most significant effect, i.e., 45% of cells with reduced ΔΨm, was obtained in cells treated with 10 µM conc. of *rac*-6. The results correlate well with annexin binding studies of *rac*-6-treated CCRF-CEM cells, suggesting the intrinsic nature of the apoptotic pathway.

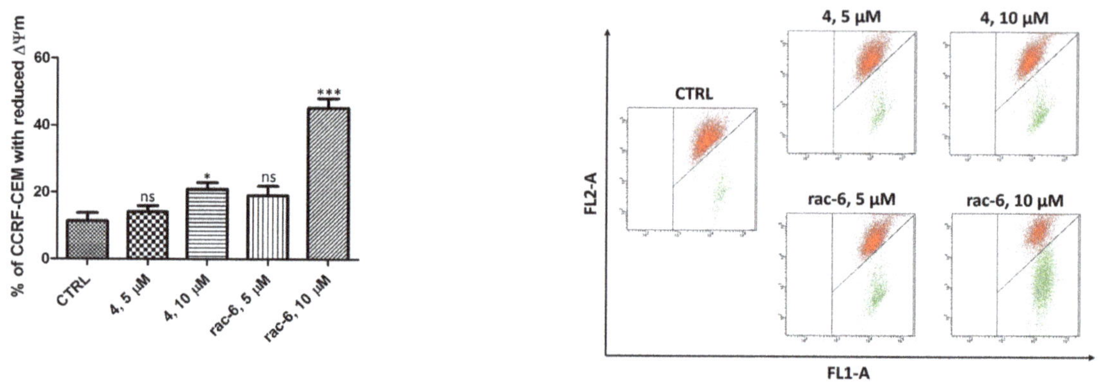

(a) (b)

Figure 5. Mitochondrial membrane potential (ΔΨm) in CCRF-CEM treated for 48 h with **4** and *rac*-**6**. (**a**) The percentage of CCRF-CEM cells with reduced ΔΨm. (**b**) Representative flow cytograms demonstrating changes in cells induced by 48 h of culturing with **4** or *rac*-**6**. The cells were stained with JC-1 dye. The cells in the lower-right region of panel (**b**) show increased green fluorescence (apoptotic cells). The data were analyzed by Dunnett's multiple comparison test as follows: * $p < 0.05$, *** $p < 0.001$ relative to control; ns—not significant.

3.2.5. Detection of Autophagy in K-562 Cells

In order to conclude the mechanism of death of K-562 cells treated with the tested active agents, microscopic observations and a fluorescence intensity assay were performed after staining the cells with an acridine orange (AO). The employed dye is a cell-permeable fluorophore that can be protonated and trapped in acidic vesicular organelles (AVOs). Its metachromatic shift to red fluorescence is concentration-dependent, and therefore, acridine orange fluoresces red in AVOs, such as autolysosomes, can be detected quantitatively [52]. In addition, OA has the ability to intercalate between the nitrogenous bases of the nucleic acids present in cells and their labeling. DNA or RNA is digested during the autophagy process, and the resulting damage makes it easier to bind with OA molecules [52].

After 48 h of treatment with 4 and *rac*-6, K-562 cells, were subjected to staining with Hoechst (nuclei staining) and AO, followed by fluorescent microscopy and a fluorescence intensity measurement. As shown in Figure 6, control cells primarily displayed green fluorescence, with minimal red fluorescence, indicating a lack of acidic vesicular organelles (AVOs). Both 4- and *rac*-6-treated cells showed yellowish-stained lysosomes and a fold-increase in red fluorescent AVOs compared to the controls (Figure 6). Interestingly, cells treated with *rac*-6 demonstrated reduced green fluorescence (DNA and cytosol staining) in comparison to control cells (especially at higher concentrations of the compound). Since the obtained microscopic data were inconclusive, we additionally measured fluorescence intensity of AO at four different wavelengths, i.e., Ex. 502 nm/Em. 520–524 nm (aggregated or DNA complexed form), Ex. 457 nm/Em. 630–644 nm (aggregated or RNA complexed form), Ex. 540 nm/Em. 640–660 nm (red-stained lysosomes), and Ex. 488 nm/Em. 540–550 nm (yellowish stained lysosomes) (Figure 6b). The obtained quantitative data indicated that the fluorescence of K-562 cells treated with the tested compounds was increased in a dose-dependent manner for aggregated or DNA/RNA-complexed AO, as well as for yellowish-stained lysosomes, with the highest percentage of lysosomes reaching 324% of the control in a 15 µM conc. of the *rac*-6-treated cells.

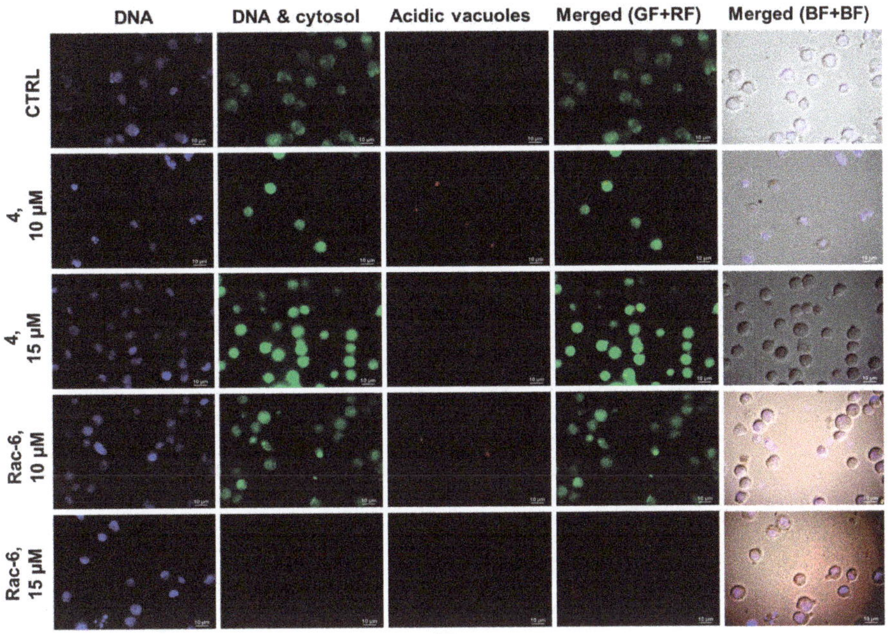

(a)

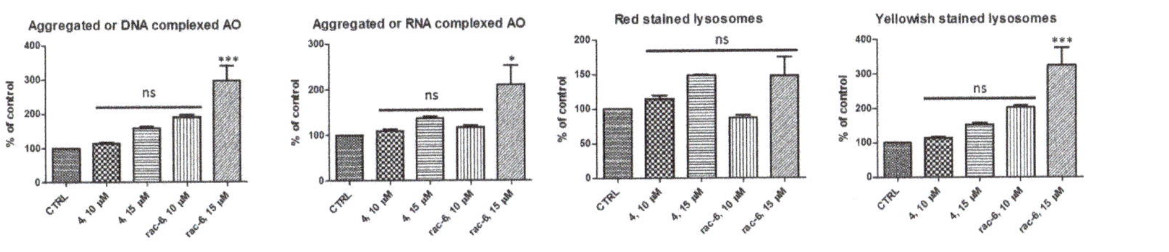

(b)

Figure 6. Induction of autophagy in K-562 cells. (**a**) Visualization of intracellular autophagic vacuoles in K-562 cells: fluorescence microscopy of Hoechst (blue-stained nuclei, BF) and AO-stained K-562 cells (green fluorescence (GF) and red fluorescence (RF)) treated for 48 h with **4** and *rac*-**6** (using a fluorescent microscope: ECLIPSE Y-TV55, Nikon, ×600, magnification). (**b**) The fluorescence intensity of AO in K-562 cells measured with a Synergy H4 Hybrid Multi-Mode Microplate Reader (Biotek) as described in the Methods section. The data were normalized to the intensity of Hoechst (nuclei, Ex. 361 nm/Em. 497 nm). The data were analyzed by Dunnett's multiple comparison test as follows: * $p < 0.05$, *** $p < 0.001$ relative to control; ns—not significant.

Furthermore, the only statistically significant results were obtained for 15 µM conc. of the *rac*-6-treated cells. Moreover, the red fluorescence was reduced in the case of 10 µM conc. of the *rac*-6-treated cells. On the contrary, the red fluorescence was increased in 4-treated cells (both concentrations) and 15 µM conc. of the *rac*-6-treated cells up to 148% of the control, although the results in these cases were not statistically significant.

Because proton pump-driven lysosomal acidity generates a significant pH gradient, resulting in an efficient concentration of AO within the lysosome organelles and because the effectiveness of the AO concentration process is sufficient to create intralysosomal concentrations, leading to precipitation of the AO into aggregated granules, the obtained results indicate that some of the treated cells can, in fact, undergo autophagy, especially after treatment with *rac*-6. The results obtained for an aggregated form of AO correspond to the increase in the yellowish-stained lysosomes and red-stained lysosomes, with the exception of 10 μM conc. of the *rac*-6-treated cells.

3.2.6. The Effect of 4 and *rac*-6 on Cell Cycle Progression in CCRF-CEM, K-562, and MCF-7 Cells

Since CK2 controls cell cycle progression and, consequently, its inhibition can affect the distribution of cells in the individual phases of the cell cycle, we tested the cell cycle progression in CCRF-CEM, K-562, and MCF-7 cells after treatment with 4 and *rac*-6. Representative plots DNA histograms with the calculations of cell percentages in each phase of the cell cycle are depicted in Figure 7. The results indicate differences in the mode of action of the two tested compounds, i.e., 4 led to S-phase arrest in all the tested cell lines, with a maximum of 57.7% of K-562 cells in that phase after treatment with 15 μM conc., whereas *rac*-6 induced G1-phase arrest in CCRF-CEM and MCF-7 cells, with a maximum accumulation of 49% of leukemia cells in that phase after using 10 μM conc. of this compound. Interestingly, *rac*-6 induced S-phase arrest only in K-562 cells, with a maximum of 56.1% of cells in that phase after 15 μM conc. of this compound. Furthermore, MCF-7 cells were observed in the sub-G1 phase (Figure 7d). Accumulation of cells in sub-G1 indicates DNA degradation, as often seen in apoptotic cells, confirming the results obtained with annexin V and PI staining.

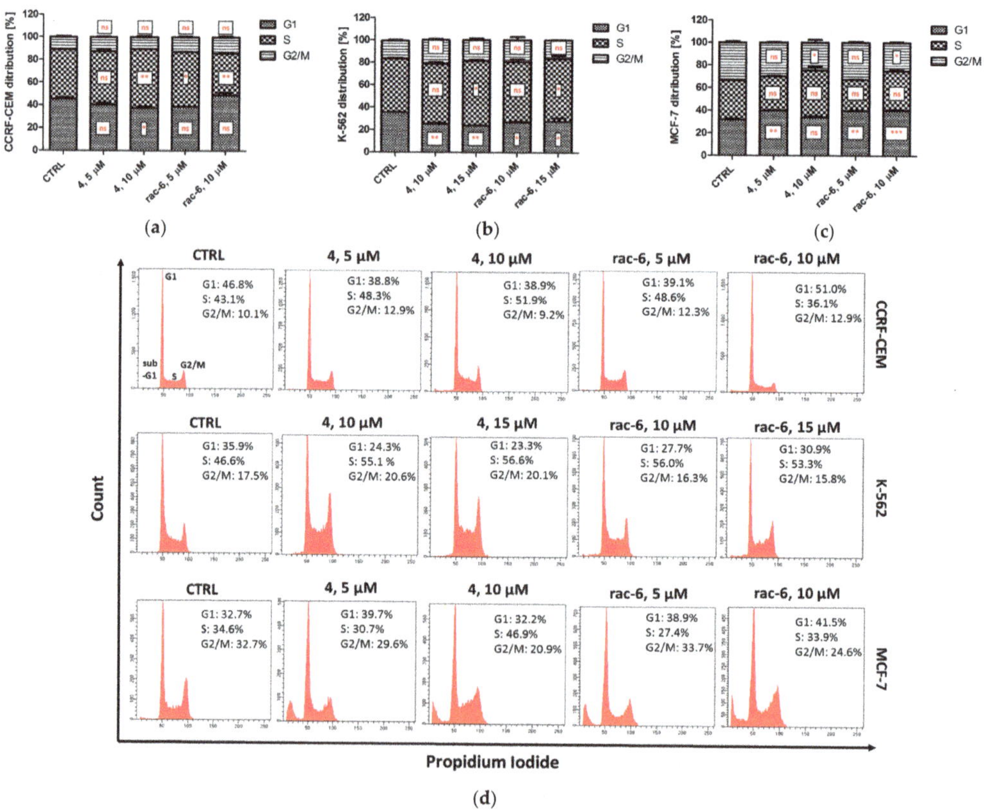

Figure 7. Cell cycle progression in CCRF-CEM, K-562, and MCF-7 cells. The data were determined by a FACSCanto II flow cytometer after 48 h of treatment with compounds **4** and *rac*-**6**. (**a**–**c**) The cell cycle distribution profiles determined by flow cytometry. (**d**) Representative DNA histograms of cell lines. The data were analyzed by Dunnett's multiple comparison test as follows: * $p < 0.05$, ** $p < 0.01$, and *** $p < 0.001$ relative to control; ns—not significant.

3.2.7. Intracellular Inhibition of Protein Kinase CK2 and PIM-1 in CCRF-CEM, K-562, and MCF-7 Cells

To confirm the intracellular inhibition of CK2 and PIM-1 by compounds 4 and *rac*-6 in the tested cell lines, we evaluated site-specific phosphorylation of Ser529 in NF-κBp65 (nuclear factor kappa-light-chain-enhancer of activated B-cells) and Ser112 in BAD after 48 h of treatment (Figure 8). The data obtained for p65 confirmed the intracellular inhibition of CK2 by the tested compounds in all the studied cell lines in a dose-dependent manner. Both tested compounds inhibited CK2-mediated phosphorylation of p65 to a similar extent, with the relative level of p65-P in the range of 0.18–0.47 μM in CCRF-CEM, 0.26–0.44 μM in K-562, and 0.19–0.73 μM in MCF-7. However, the most potent inhibition was observed in CCRF-CEM cells after treatment with 10 μM conc. of *rac*-6. Otherwise, Western blot data for BAD (a marker of PIM-1 activity) indicated inhibition of PIM-1 only in MCF-7 cells, with the most significant reduction in phosphorylated BAD in cells treated with 10 μM conc. of *rac*-6 (0.49 of control) (Figure 8b). Unexpectedly, the relative level of BAD-P in the drug-treated CCRF-CEM and K-562 cells was even higher than in control cells.

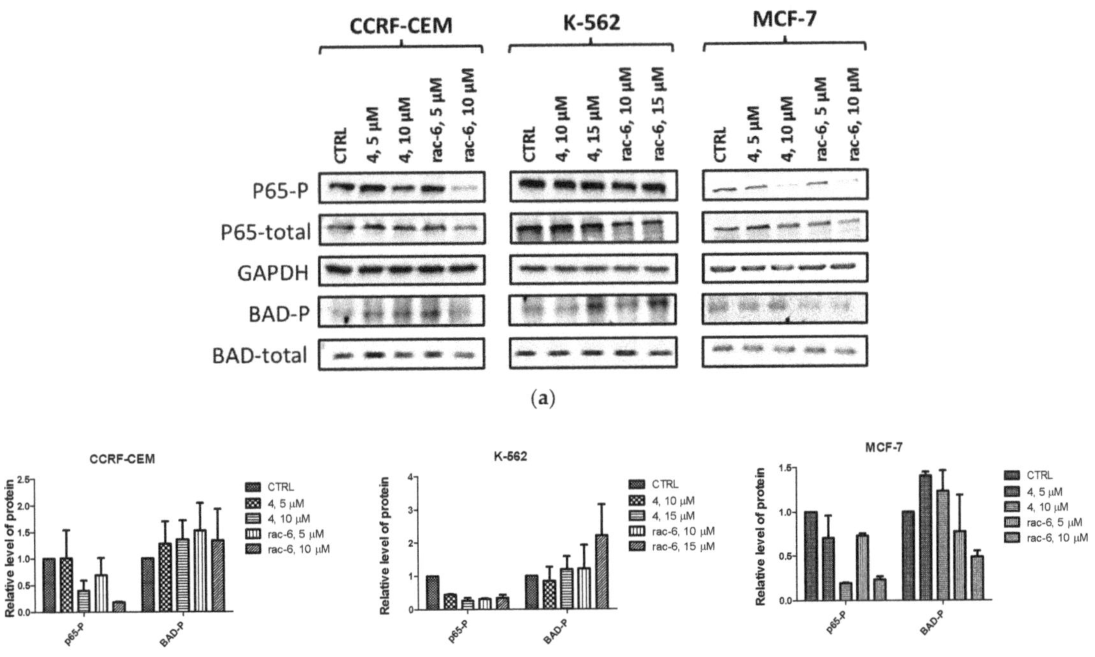

(a)

(b)

Figure 8. Intracellular inhibition of protein kinases CK2 and PIM-1 in CCRF-CEM, K-562, and MCF-7 cell lines after 48 h treatment with compounds **4** and *rac*-**6**. (**a**) Representative Western blots of p65-P-Ser 529 (nuclear factor kappa-light-chain-enhancer of activated B-cells), p65 total, BAD-P-Ser112, BAD total, and GAPDH in the tested cells. (**b**) Densitometric quantifications for each tested protein were calculated with untreated cells (CTRL) serving as the reference point, showing relative protein levels. The preparation of cell extracts and protein detection is described in the Materials and Methods Section.

3.3. Molecular Docking

In order to rationalize the results of in vitro enzymatic assays and cellular activity in cancer cell lines of the most potent dual CK2-α/PIM-1 inhibitors with the highest selectivity factor (SI), comprehensive in silico enzyme–substrate docking calculations were performed. For this purpose, the respective TBBi derivatives (*rac*-6 and *rac*-11)were docked with receptor molecules prepared based on the crystal structures of both titled kinases, CK2-α (PDB code: 4KWP) [35] and PIM-1 (PDB code: 4DTK) [49], which were retrieved from the Protein Data Bank (PDB; https://www.rcsb.org/). As a result of molecular docking experiments, nine of the most energetically favorable binding modes for the ligand–protein complexes for ligands *rac*-6 and *rac*-11 and target CK2-α and PIM-1 proteins were generated. The results of their binding affinity energies expressed as ΔG_{calc} (kcal/mol) are presented in Table S3 (Supplementary Materials). A visualization of the representative docking poses of *rac*-6 and *rac*-11 to CK2-α and PIM-1 with close contacts with amino acid residues located in the related ATP-binding sites of both studied kinases is presented in Figure 9.

Inspection of the productive pose of *rac*-6 in CK2-α (Figure 9A–C) showed that this inhibitor forms strong 2.1–2.2 Å-long hydrogen bonds with crystal waters (HOH-744 and HOH-954) present in the catalytic cavity. At the same time, the 4,5,6,7-tetrabromo-benzimidazole moiety of *rac*-6 is located deep inside the ATP-binding pocket, thus exhibiting strong hydrophobic interactions with Val53, Val66, and Val116. In contrast, *rac*-11 establishes hydrogen bonding between the backbone carbonyl oxygen of His160 residue and the hydrogen atom of the secondary hydroxyl moiety present in the propane-1,2-diol substituent (Figure 9D–F). Moreover, the CK2-α-*rac*-11 complex is stabilized through an additional water-mediated hydrogen bond between the imidazole-NH moiety and the oxygen atom of the bridging conserved water molecule (HOH-744). In this case, the ATP-binding pocket, which is surrounded by hydrophobic amino acids, was also occupied by the TBBi scaffold, providing alkyl CH–CH van der Waals (vdW) and π–alkyl interactions with Val53, Leu45, Val66, Glu114, Val116, Met163, and Ile174 residues. It is worth noting that docking simulations revealed that both TBBi inhibitors bind to CK2α, exploiting two different poses. In this context, *rac*-6 is anchored to the hinge region through Br1 and Br2 closer to Leu45, Val53, and Val66 residues (1.9–2.7 Å-long distance), while *rac*-11 is oriented through Br2 and Br3 closer to Glu114, Val116, Asn118, and Met163 residues (2.4–3.4 Å-long distance). As a consequence of the second orientation, ligand *rac*-11 can avoid unfavorable repulsion and steric clashes of the tailed hydrophilic propane-1,2-diol chain and hydrophobic residues located more profoundly in the ATP-binding pocket. The second plausible explanation for this orientation is that the additional primary hydroxyl group present in the amino alcohol substituent introduces a steric hindrance due to the proximity of Met163, which pushes *rac*-11 out of the hinge region. Interestingly, although *rac*-11 slightly protrudes from the ATP-binding site as compared to *rac*-6, the drop in the inhibitory potency of this compound was not detected during in vitro kinase activity assays.

In turn, complexes of PIM-1 and the selected top-scoring binding modes of *rac*-6 (Figure 9G–I) and *rac*-11 (Figure 9J–L) revealed that both inhibitors are accommodated inside the ATP-binding cleft with similar poses, explaining the marginal differences in inhibitory potency towards target protein in vitro. The only difference among the studied PIM-1 complexes concerns the feature of the H-bonding network between the inhibitors' amino alcohol skeletons and interacting amino acid residues, as well as crystal water. The docking of *rac*-6 with PIM-1 protein shows that the secondary hydroxyl group of the alkyl substituent formed H-bond interactions with PIM-1 via amine groups of Lys67 or Asp186, whereas the TBBi-NH moiety interacted via crystal water (HOH-680). On the other hand, the secondary hydroxyl functionality of *rac*-11 interacted only with Lys67, while the TBBi-NH moiety was H-bounded with HOH-680. Interestingly, when comparing both inhibitors complexed with target kinases, one can see that in the case of CK2-α, it is the proton of imidazole-NH moiety that forms H-bonds with crystal water. In contrast, in the case of PIM-1, both inhibitors' TBBi-NH scaffolds are engaged in the formation of H-bonding interaction with the crystal water.

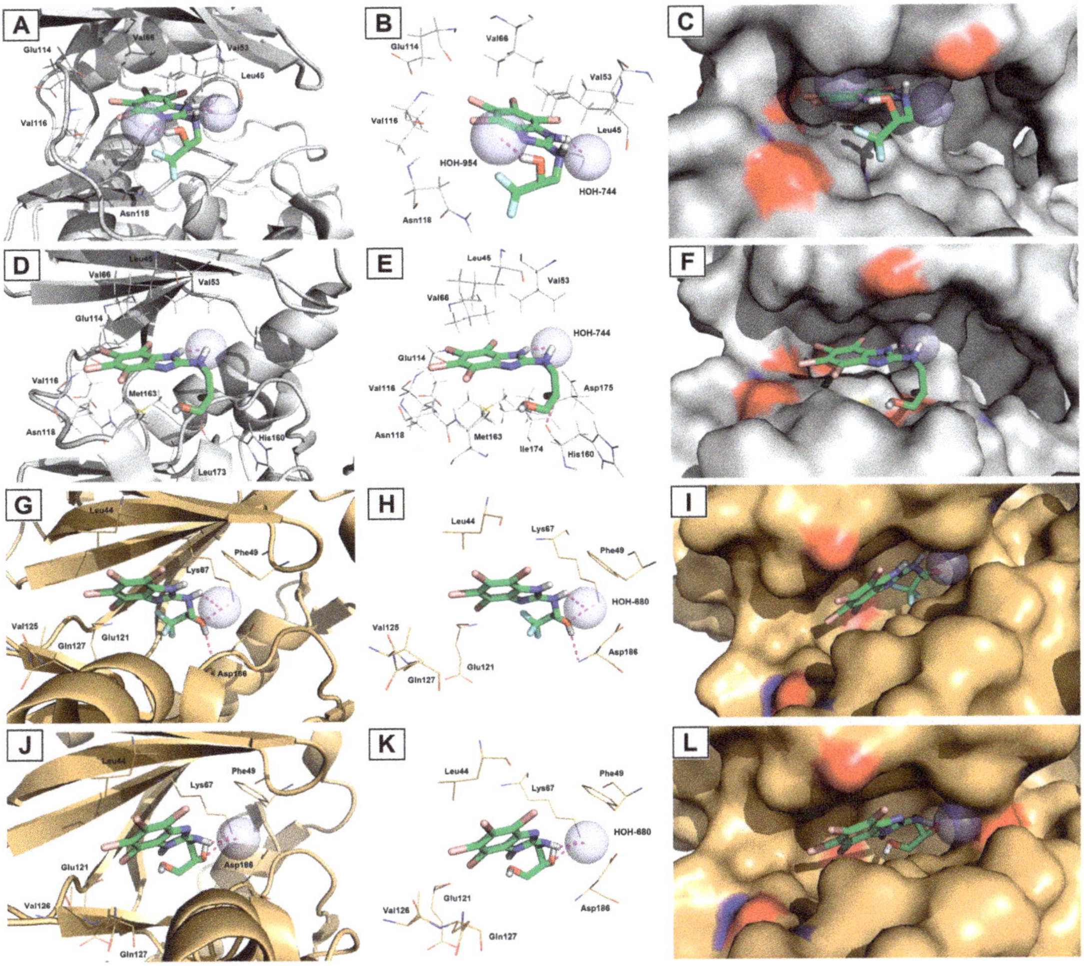

Figure 9. Representative three-dimensional (3D) binding modes of 1,1,1-trifluoro-3-[(4,5,6,7-tetrabromo-1*H*-benzimidazol-2-yl)amino]propan-2-ol (*rac*-**6**) (**A–C**,**G–I**) and 3-[(4,5,6,7-tetrabromo-1*H*-benzimidazol-2-yl)amino]propane-1,2-diol (*rac*-**11**) (**D–F**,**J–L**) with human protein kinases, CK2-α (PDB code: 4KWP; (**A–F**)) and PIM-1 (PDB code: 4DTK; (**G–L**)) with close contacts with residues in the ATP-binding site. All nine *panels* (**A–L**) show inhibition of protein kinases CK2-α and PIM-1 by blocking the substrate-binding site through direct interaction of the ligands with critical amino acids of the receptors in the catalytic cavity. Ligands are shown as *sticks*. The overall receptor structures are shown as a *cartoon* diagram (left column) or *surface* representation (right column), where CK2-α is light grey and PIM-1 is gold. The most significant amino acid residues contributing to the stabilization of the ligand molecules in the complex with CK2-α and PIM-1 by polar interactions and by CH–CH van der Waals (vdW) or π−alkyl interactions are shown as *lines*. Nitrogen atoms are presented with blue color, oxygen atoms with red color, fluorine atoms with light blue color, and bromine atoms with burgundy (maroon) color, whereas the hydrogen atoms (attached to nitrogen and/or oxygen atoms of the ligand molecules) are represented by grey color. Only polar hydrogens are shown for clarity. The bridging conserved water molecules (crystal waters) are presented as light blue *spheres*. The formation of intermolecular hydrogen bonds is represented by magenta dashed lines.

Notably, no halogen bonding was detected among the docked ligand *rac*-6, suggesting that the fluorine atoms of the -CF_3 group are less detrimental for polar interactions within CK2-α and PIM-1 receptor molecules. Therefore, it is more likely that the organofluorine moiety in the studied lead compound *rac*-6 is responsible for enhanced membrane permeation and higher bioavailability (due to the greater lipophilicity of bioisosteric fluorine compared to hydrogen), as well as improved metabolic stability (due to resistance of the C–F bond toward intracellular biotransformations compared to the C–H bond). Both aforementioned features can modulate the performance and pharmacokinetics of this compound in vivo but have no significant influence on the increased binding affinity of fluorinated derivatives to target kinases. Nevertheless, such a phenomenon concerning the physicochemical properties and pharmacological behavior of TBBi-based derivatives demands a deeper understanding.

4. Discussion

A series of novel amino alcohol derivatives (5–11) of parental DMAT (4) were designed and synthesized as potent dual CK2/PIM-1 inhibitors. All the compounds were obtained in up to 14% total yield after a three-step synthetic procedure, following a well-known route elaborated by Andrzejewska et al. [42], which includes: (i) bromination of 2-mercapto-1H-benzimidazole (1) with Br_2 and 48% aqueous HBr in glacial AcOH, (ii) an exhaustive bromination of 2-bromo-1H-benzimidazole (2) with Br_2 in boiling water and modified using additional LED (390 nm) irradiation to intensify the process, and (iii) aminolysis of the resulting key intermediate, namely 2,4,5,6,7-pentabromo-1H-benzimidazole (3), with commercial amino alcohols different than those used in the literature. Because the designed compounds included chiral small molecules, we also extended our synthetic efforts toward the preparation of the corresponding (R)- and (S)-enantiomers using commercially available chiral building blocks as starting materials. This strategy was applied to evaluate the structure–activity relationship with respect to the absolute configuration of stereogenic centers present in one of the studied DMAT analogs: (2S)- or (2R)-1-[(4,5,6,7-tetrabromo-1H-benzimidazol-2-yl)amino]propan-2-ol [(S)-5 or (R)-5].

Although the results of in vitro kinetics studies revealed that all inhibitors decrease the catalytic activity of recombinant CK2- and PIM-1 enzymes to a similar extent as a parental DMAT, the cytotoxicity and selectivity of the tested compounds were even better than in the case of DMAT. Moreover, the cytotoxic activity of the tested compounds corresponds better with their lipophilicity than their inhibitory activity. This correlation is in agreement with our previous findings demonstrating that the most promising compound characterized by a higher log*P* value can penetrate cell membranes more efficiently, therefore exhibiting improved intracellular inhibition of CK2 [40]. The present results were supported by molecular docking studies for the most selective compounds, i.e., *rac*-6 and *rac*-11. The in silico results demonstrated that both compounds are accommodated inside the ATP-binding cleft with similar poses, explaining the marginal differences in inhibitory potency towards target protein in vitro. These findings confirm that efficient inhibitors of recombinant CK2 should possess (nearby hydrophobic TBBi scaffold) a polar carboxyl or hydroxyl moiety, which are prone to forming strong hydrogen bonds similar to the phosphate groups present in the ATP physiological substrate.

Because cytotoxicity against tumor cells and selectivity are important features of compounds demonstrating anticancer properties, further investigations of biological action of the most promising compounds, i.e., *rac*-6 and 4, were performed. We observed that among tested cell lines, K-562, a Philadelphia (Ph) chromosome-positive (BCR-ABL-positive) leukemia cell line derived from chronic myeloid leukemia (CML) in blast crisis [53], was the most resistant toward the tested compounds. Previous studies demonstrated that CK2 expression was increased in leukemia cells from CML patients in blast crisis as compared to healthy peripheral blood mononuclear cells and showed that Bcr-Abl in K-562 cells physically interacts with CK2α, affecting its activity [54]. The therapeutic resistance of K-562 cells can also be partially correlated with an increased metabolic flux towards the Warburg phenotype, which promotes survival and proliferation [55]. It has been demonstrated that hexokinase-II (HK-II) is expressed predominantly in cancer cells, which promotes the Warburg metabolic phenotype and protects the cancer cells from drug-induced apoptosis. It was proven that K-562 cells have multifold higher levels of HK-II, glucose uptake, and endogenous ROS with respect to normal peripheral blood mononuclear cells [55].

The resistance of K-562 cells to the tested compounds correlates well with the lack of apoptosis in these cells after their treatment with *rac*-6 and 4. Moreover, the poor proapoptotic activity of the tested compounds towards K-562 cells also correlates with the lack of efficient intracellular inhibition of CK2 or PIM-1 protein kinases after 4- and *rac*-6-treatment. Considering that CK2 has been demonstrated as an essential mediator of BCR-ABL oncogenic signals and the BCR-ABL/CK2 complex is responsible for mediating BCR-ABL-induced cell proliferation and survival [56], the lack of efficient intracellular inhibition of CK2 by the tested compounds is in agreement with the poor proapoptotic activity of this treatment. We also observed that the phospho-BAD (Ser112) level in K-562 cells increased significantly after treatment with both inhibitors, which supports cell survival and prevents the occurrence of apoptosis. Moreover, it was demonstrated that the inactivation of BAD by PIM-1-mediated phosphorylation of Ser112 can affect BCL-2, an apoptotic cell death suppressor, which consequently leads to cell survival [57]. A similarly increased level of phospho-BAD (Ser112) was observed in TNBC MDA-MB-231 cells after treatment with a dual CK2/PIM-1 inhibitor, 1-(β-D-2′-deoxyribofuranosyl)-4,5,6,7-tetrabromo-1*H*-benzimidazole, named K164 (TDB) [37]. Interestingly, the same inhibitor reduced the catalytic activity of both CK2 and PIM-1 kinases, as manifested by decreased phosphorylated levels of Akt and BAD in K-562 cells, and induced apoptosis in this cell line [36].

Our present studies show that both tested compounds are prone to induce autophagy in the K-562 cell line. Although autophagy is a physiological cellular process that leads to the degradation and recycling of damaged cellular components [58], depending on the degree of activation, it can lead to death. In cancer, autophagy exhibits contradictory behavior, and depending on the cell type, it may be an important factor for the induction of cell death or tumor progression [59]. Previous studies have demonstrated that a well-established inhibitor of tyrosine kinase, imatinib, induces autophagy in the K-562 cell line and in primary cultures of patients with CML [60]. In turn, novel JAK inhibitor ruxolitinib (INCB018424) reported by Lin et al. [61] is able to notably decrease the expression of AKT, mTOR, and STAT autophagy inhibitor genes in K-562 cells relative to the control cell line.

Among the tested cell lines, apoptosis was induced by *rac*-6 to the greatest extent in CCRF-CEM. Moreover, the proapoptotic activity of this compound was even better than the parental DMAT and corresponded to its strong ability to reduce ΔΨm, suggesting the intrinsic nature of the apoptotic pathway. The obtained results concerning CK2-mediated phosphorylation of NF-κB strongly correlate with the proapoptotic properties of *rac*-6, confirming the antiapoptotic role of this kinase. It was demonstrated that CK2 and its substrates protect cells from apoptosis by phosphorylating a wide range of proteins involved in the apoptotic response [62,63]. Interestingly, intracellular inhibition of CK2 in the *rac*-6-treated CCRF-CEM cells was not accompanied by inhibition of PIM-1 (increased level of phosphorylated BAD).

Among studied cell lines, the decreased level of (PIM-1)-mediated phosphorylation of BAD was observed only in MCF-7, and similarly to CCRF-CEM cells, CK2-mediated phosphorylation of p65 was reduced in MCF-7 cells treated with 4 and *rac*-6. The obtained results correlate with the ability of the tested compounds to induce apoptosis in breast cancer cells; however, its intensity is lower than in CCRF-CEM cells. The indicated differences in the proapoptotic efficacy of the tested inhibitors between CCRF-CEM and MCF-7 cell lines may be related to the caspase-3 deficiency in MCF-7 cells, which consequently undergo cell death due to a lack of typical apoptotic properties [64]. Interestingly, intracellular inhibition of CK2 after treatment with *rac*-6 detected in CCRF-CEM and MCF-7 cells correlates with G1-phase arrest in these cells, and it is opposite to S-phase arrest in

K-562 cells. The G1-arrest that occurred in CCRF-CEM and MCF-7 cells after inhibition of CK2 is in agreement with the literature data, showing cell cycle regulation by CK2. It was demonstrated that treatment of cells with a selective inhibitor of CK2, i.e., CX-4945, resulted in reduced phosphorylation of a key cell cycle inhibitor protein [p21 (T145)] and increased the stability and levels of total p21 and p27 [65]. These cell cycle inhibitors are responsible for stopping the cell cycle in the G1 phase, activation of repair mechanisms, and apoptosis of cells with damaged DNA.

5. Conclusions

All the tested compounds were found to be efficient dual inhibitors of both recombinant forms of CK2 and PIM-1, with activity comparable to that of the parent compound. We concluded that the cytotoxic activity of the tested compounds corresponds better with their lipophilicity than their inhibitory activity. This correlation confirms our previous findings demonstrating that the most promising compound characterized by a higher log*P* value can penetrate cell membranes more efficiently.

Among the newly developed amino alcohol derivatives of DMAT, 1,1,1-trifluoro-3-[(4,5,6,7-tetrabromo-1*H*-benzimidazol-2-yl)amino]propan-2-ol (*rac*-6) demonstrates the most promising anticancer properties. Its ability to induce apoptosis in breast cancer cells can be attributed to efficient intracellular inhibition of both CK2 and PIM-1 activity. Moreover, *rac*-6 demonstrates the best anticancer properties towards acute lymphoblastic leukemia cells, inducing apoptosis via an intrinsic apoptotic pathway. Interestingly, both studied inhibitors, i.e., DMAT and *rac*-6, are able to induce autophagy in the BCR-ABL-positive chronic myeloid leukemia (CML) cell line.

The obtained results also support the concept that CK2 kinase is a vital factor for breast cancer cell survival and an appealing molecular drug target in the development of novel antineoplastic agents. Taking into account the obtained data, as well as our previous results demonstrating a synergistic effect of TBBi derivatives and classical cytostatics [66–68], it may be valuable to test a combination of *rac*-6 with 5-fluorouracil or methotrexate against breast cancer and leukemic cells.

Supplementary Materials: The following supporting information can be downloaded at: https://www.mdpi.com/xxx/s1. Table S1: Screening conditions for the synthesis of (2*S*)-1-[(4,5,6,7-tetrabromo-1*H*-benzimidazol-2-yl)amino]propan-2-ol [(*S*)-5] using commercially available (*S*)-1-aminopropan-2-ol (>99% ee); Table S2: Induction of apoptosis in CCRF-CEM, K-562, and MCF-7 cells; Table S3: Docking scoring of the respective ligands (*rac*-6 and *rac*-11) complexed with human protein kinases. Next, the copies of HPLC chromatograms of inhibitors in their racemic and optically active form, as well as copies of NMR and FTMS spectra, can be found. Figure S1. Synthesis of 2,4,5,6,7-pentabromo-1H-benzimidazole (3) as a key precursor for novel dual CK2/PIM-1 inhibitors carried out under (A) convenient heating conditions, and (B–C) the influence of purple light (390 nm). Light-induced chemical reactions was conducted using self-prepared (made-in-home) device equipped with light-emitting diodes (LEDs). Figure S2. Sigmoidal dose-response curves. Inhibition of human CK2 catalytic subunit (CK2α), CK2 holoenzyme (CK2 $\alpha 2\beta 2$), and PIM-1 kinases by the newly synthesized compounds were evaluated using the radiometric assay. The synthetic peptide RRRADDSDDDDD was used as the substrate of CK2, and peptide ARKRRRHPSGPPTA as the substrate of PIM-1. The experimental data were fitted to sigmoidal dose-response (variable slope) Y = Bottom + (Top-Bottom)/(1 + 10^((LogIC50 − X) * HillSlope) equation in GraphPad Prism. Reference [69] is cited in the supplementary materials.

Author Contributions: Conceptualization, P.W. (biological studies) and P.B. (synthetic and computational studies); Methodology and Investigation, P.W. (biological studies with cells), M.W. (assays with recombinant enzymes), M.K. (flow cytometry analyses), and P.B. (synthesis and analysis of the organic compounds; molecular docking studies); Software, P.B.; Validation, P.W. and P.B.; Formal Analysis, P.W. and P.B.; Resources, P.W. and P.B.; Data Curation, P.W., M.K. and P.B.; Writing—Original Draft Preparation, P.W. and P.B.; Writing—Review and Editing, P.W. and P.B.; Visualization, P.W. and P.B.; Supervision, P.W. and P.B.; Project Administration, P.W. and P.B.; Funding Acquisition, P.W. and P.B. All authors have read and agreed to the published version of the manuscript.

Funding: This research was funded by the Warsaw University of Technology (WUT) and partially by the National Science Centre (NCN) of Poland within the framework of the SONATA 15 grant (No. 2019/35/D/ST4/01556). This work was also supported by Polish Ministry of Science and Higher Education (financial support for statutory activity of the National Medicines Institute).

Institutional Review Board Statement: Not applicable.

Informed Consent Statement: Not applicable.

Data Availability Statement: The data presented in this article are openly available.

Acknowledgments: The authors thank undergraduate students Kristina Fateyeva, Patrycja Puk, and Anna Gabor for assistance with the chemical and biological studies. P.B. is grateful to the IDUB project ('The Best of the Best' program) for providing individual support dedicated to young, talented WUT scientists.

Conflicts of Interest: All authors declare no financial/commercial conflict of interest.

References

1. Trembley, J.H.; Kren, B.T.; Afzal, M.; Scaria, G.A.; Klein, M.A.; Ahmed, K. Protein kinase CK2—Diverse roles in cancer cell biology and therapeutic promise. *Mol. Cell. Biochem.* **2023**, *478*, 899–926. [CrossRef] [PubMed]
2. Walhekar, V.; Bagul, C.; Kumar, D.; Muthal, A.; Achaiah, G.; Kulkarni, R. Topical advances in PIM kinases and their inhibitors: Medicinal chemistry perspectives. *Biochim. Biophys. Acta Rev. Cancer* **2022**, *1877*, 188725. [CrossRef]
3. Borgo, C.; Ruzzene, M. Role of protein kinases in antitumor drug resistance. *J. Exp. Clin. Cancer Res.* **2019**, *38*, 287. [CrossRef]
4. Atkinson, E.L.; Iegre, J.; Brear, P.D.; Zhabina, E.Z.; Hyvönen, M.; Spring, D.R. Downfalls of Chemical Probes Acting at the Kinase ATP-Site: CK2 as a Case Study. *Molecules* **2021**, *26*, 1977. [CrossRef]
5. Narlik-Grassow, M.; Blanco-Aparicio, C.; Carnero, A. The PIM family of serine/threonine kinases in cancer. *Med. Res. Rev.* **2014**, *34*, 136–159. [CrossRef] [PubMed]
6. Manning, G.; Whyte, D.B.; Martinez, R.; Hunter, T.; Sudarsanam, S. The protein kinase complement of the human genome. *Science* **2002**, *298*, 1912–1934. [CrossRef] [PubMed]
7. Qian, K.C.; Wang, L.; Hickey, E.R.; Studts, J.; Barringer, K.; Peng, C.; Kronkaitis, A.; Li, J.; White, A.; Mische, S.; et al. Structural basis of constitutive activity and a unique nucleotide binding mode of human Pim-1 kinase. *Biol. Chem.* **2005**, *280*, 6130–6137. [CrossRef]
8. Niefind, K.; Guerra, B.; Pinna, L.A.; Issinger, O.G.; Schomburg, D. Crystal structure of the catalytic subunit of protein kinase CK2 from Zea mays at 2.1 A resolution. *EMBO J.* **1998**, *17*, 2451–2462. [CrossRef]
9. Pucko, E.B.; Ostrowski, R.P. Inhibiting CK2 among Promising Therapeutic Strategies for Gliomas and Several Other Neoplasms Emanuela, B. *Pharmaceutics* **2022**, *14*, 331. [CrossRef]
10. Brault, L.; Gasser, C.; Bracher, F.; Huber, K.; Knapp, S.; Schwaller, J. PIM serine/threonine kinases in the pathogenesis and therapy of hematologic malignancies and solid cancers. *Haematologica* **2010**, *95*, 1004–1015. [CrossRef]
11. Valdman, A.; Fang, X.; Pang, S.T.; Ekman, P.; Egevad, L. Pim-1 expression in prostatic intraepithelial neoplasia and human prostate cancer. *Prostate* **2004**, *60*, 367–371. [CrossRef] [PubMed]
12. Tursynbay, Y.; Zhang, J.; Li, Z.; Tokay, T.; Zhumadilov, Z.; Wu, D.; Xie, Y. Pim-1 kinase as cancer drug target: An update (Review). *Biomed. Rep.* **2016**, *4*, 140–146. [CrossRef] [PubMed]
13. Yan, B.; Zemskova, M.; Holder, S.; Chin, V.; Kraft, A.; Koskinen, P.J.; Lilly, M.J. The PIM-2 kinase phosphorylates BAD on serine 112 and reverses BAD-induced cell death. *Biol. Chem.* **2003**, *278*, 45358–45367. [CrossRef] [PubMed]
14. Channavajhala, P.; Seldin, D.C. Functional interaction of protein kinase CK2 and c-Myc in lymphomagenesis. *Oncogene* **2002**, *21*, 5280–5288. [CrossRef] [PubMed]
15. Klumpp, S.; Maurer, A.; Zhu, Y.; Aichele, D.; Pinna, L.A.; Krieglstein, J. Protein kinase CK2 phosphorylates BAD at threonine-117. *J. Neurochem. Int.* **2004**, *45*, 747–752. [CrossRef]
16. Wang, J.; Kim, J.; Roh, M.; Franco, O.E.; Hayward, S.W.; Wills, M.L.; Abdulkadir, S.A. PIM-1 kinase synergizes with c-MYC to induce advanced prostate carcinoma. *Oncogene* **2010**, *29*, 2477–2487. [CrossRef]
17. Laramas, M.; Pasquier, D.; Filhol, O.; Ringeisen, F.; Descotes, J.L.; Cochet, C. Nuclear localization of protein kinase CK2 catalytic subunit (CK2alpha) is associated with poor prognostic factors in human prostate cancer. *Eur. J. Cancer* **2007**, *43*, 928–934. [CrossRef]
18. Cibull, T.L.; Jones, T.D.; Li, L.; Eble, J.N.; Baldridge, A.L.; Malott, S.R.; Luo, Y.; Cheng, L. Overexpression of Pim-1 during progression of prostatic adenocarcinoma. *J. Clin. Pathol.* **2006**, *59*, 285–288. [CrossRef]
19. Trembley, J.H.; Kren, B.T.; Abedin, M.J.; Vogel, R.I.; Cannon, C.M.; Unger, G.M.; Ahmed, K. CK2 Molecular Targeting-Tumor Cell-Specific Delivery of RNAi in Various Models of Cancer. *Pharmaceuticals* **2017**, *10*, 25. [CrossRef]
20. Ahmad, K.A.; Wang, G.; Slaton, J.; Unger, G.; Ahmed, K. Targeting CK2 for cancer therapy. *Anticancer Drugs* **2005**, *16*, 1037–1043. [CrossRef]

21. Hu, X.F.; Li, J.; Vandervalk, S.; Wang, Z.; Magnuson, N.S.; Xing, P.X.J. PIM-1-specific mAb suppresses human and mouse tumor growth by decreasing PIM-1 levels, reducing Akt phosphorylation, and activating apoptosis. *Clin. Invest.* **2009**, *119*, 362–375. [CrossRef] [PubMed]
22. Gowda, C.; Sachdev, M.; Muthusami, S.; Kapadia, M.; Petrovic-Dovat, L.; Hartman, M.; Ding, Y.; Song, C.; Payne, J.L.; Tan, B.H.; et al. Casein Kinase II (CK2) as a Therapeutic Target for Hematological Malignancies. *Curr. Pharm. Des.* **2017**, *23*, 95–107. [CrossRef] [PubMed]
23. Milani, M.; Venturini, S.; Bonardi, S.; Allevi, G.; Strina, C.; Cappelletti, M.R.; Corona, S.P.; Aguggini, S.; Bottini, A.; Berruti, A.; et al. Hypoxia-related biological markers as predictors of epirubicin-based treatment responsiveness and resistance in locally advanced breast cancer. *Oncotarget* **2017**, *8*, 78870–78881. [CrossRef] [PubMed]
24. Garziera, M.; Scarabel, L.; Toffoli, G. Hypoxic Modulation of HLA-G Expression through the Metabolic Sensor HIF-1 in Human Cancer Cells. *J. Immunol. Res.* **2017**, *2017*, 4587520. [CrossRef]
25. Chen, J.; Kobayashi, M.; Darmanin, S.; Qiao, Y.; Gully, C.; Zhao, R.; Yeung, S.C.; Lee, M.H. Pim-1 plays a pivotal role in hypoxia-induced chemoresistance. *Oncogene* **2009**, *28*, 2581–2592. [CrossRef]
26. Mottet, D.; Ruys, S.P.; Demazy, C.; Raes, M.; Michiels, C. Role for casein kinase 2 in the regulation of HIF-1 activity. *Int. J. Cancer.* **2005**, *117*, 764–774. [CrossRef]
27. Casillas, A.L.; Toth, R.K.; Sainz, A.G.; Singh, N.; Desai, A.A.; Kraft, A.S.; Warfel, N.A. Hypoxia-inducible PIM kinase expression promotes resistance to anti-angiogenic agents. *Clin. Cancer Res.* **2018**, *24*, 169–180. [CrossRef]
28. Landesman-Bollag, E.; Song, D.H.; Romieu-Mourez, R.; Sussman, D.J.; Cardiff, R.D.; Sonenshein, G.E.; Seldin, D.C. Protein kinase CK2: Signaling and tumorigenesis in the mammary gland. *Mol. Cell. Biochem* **2001**, *227*, 153–165. [CrossRef]
29. Hammerman, P.S.; Fox, C.J.; Cinalli, R.M.; Xu, A.; Wagner, J.D.; Lindsten, T.; Thompson, C.B. Lymphocyte transformation by Pim-2 is dependent on nuclear factor-kappaB activation. *Cancer Res.* **2004**, *64*, 8341–8348. [CrossRef]
30. Zemskova, M.; Sahakian, E.; Bashkirova, S.; Lilly, M.J. The PIM-1 kinase is a critical component of a survival pathway activated by docetaxel and promotes survival of docetaxel-treated prostate cancer cells. *Biol. Chem.* **2008**, *283*, 20635–20644. [CrossRef]
31. Sarno, S.; Papinutto, E.; Franchin, C.; Bain, J.; Elliott, M.; Meggio, F.; Kazimierczuk, Z.; Orzeszko, A.; Zanotti, G.; Battistutta, R.; et al. ATP site-directed inhibitors of protein kinase CK2: An update. *Curr. Top. Med. Chem.* **2011**, *11*, 1340–1351. [CrossRef] [PubMed]
32. Gingipalli, L.; Block, M.H.; Bao, L.; Cooke, E.; Dakin, L.A.; Denz, C.R.; Ferguson, A.D.; Johannes, J.W.; Larsen, N.A.; Dowling, J.E.; et al. Discovery of 2,6-disubstituted pyrazine derivatives as inhibitors of CK2 and PIM kinases. *Bioorg. Med. Chem. Lett.* **2018**, *28*, 1336–1341. [CrossRef] [PubMed]
33. Chen, Y.; Wang, Y.; Wang, J.; Zhou, Z.; Cao, S.; Zhang, J. Strategies of Targeting CK2 in Drug Discovery: Challenges, Opportunities, and Emerging Prospects. *J. Med. Chem.* **2023**, *66*, 2257–2281. [CrossRef] [PubMed]
34. Cozza, G.; Sarno, S.; Ruzzene, M.; Girardi, C.; Orzeszko, A.; Kazimierczuk, Z.; Zagotto, G.; Bonaiuto, E.; Di Paolo, M.L.; Pinna, L. Exploiting the repertoire of CK2 inhibitors to target DYRK and PIM kinases. *Biochim. Biophys. Acta* **2013**, *1834*, 1402–1409. [CrossRef] [PubMed]
35. Cozza, G.; Girardi, C.; Ranchio, A.; Lolli, G.; Sarno, S.; Orzeszko, A.; Kazimierczuk, Z.; Battistutta, R.; Ruzzene, M.; Pinna, L.A. Cell-permeable dual inhibitors of protein kinases CK2 and PIM-1: Structural features and pharmacological potential. *Cell. Mol. Life Sci.* **2014**, *71*, 3173–3185. [CrossRef]
36. Koronkiewicz, M.; Chilmonczyk, Z.; Kazimerczuk, Z.; Orzeszko, A. Deoxynucleosides with benzimidazoles as aglycone moiety are potent anticancer agents. *Eur. J. Pharmacol.* **2018**, *820*, 146–155. [CrossRef]
37. Koronkiewicz, M.; Kazimierczuk, Z.; Orzeszko, A. Antitumor activity of the protein kinase inhibitor 1-(β-D-2′-deoxyribofuranosyl)-4,5,6,7-tetrabromo- 1H-benzimidazole in breast cancer cell lines. *BMC Cancer* **2022**, *22*, 1069. [CrossRef]
38. Łukowska-Chojnacka, E.; Wińska, P.; Wielechowska, M.; Poprzeczko, M.; Bretner, M. Synthesis of novel polybrominated benzimidazole derivatives-potential CK2 inhibitors with anticancer and proapoptotic activity. *Bioorg. Med. Chem.* **2016**, *24*, 735–741. [CrossRef]
39. Chojnacki, K.; Wińska, P.; Skierka, K.; Wielechowska, M.; Bretner, M. Synthesis, in vitro antiproliferative activity and kinase profile of new benzimidazole and benzotriazole derivatives. *Bioorg. Chem.* **2017**, *72*, 1–10. [CrossRef]
40. Chojnacki, K.; Winska, P.; Karatsai, O.; Koronkiewicz, M.; Milner-Krawczyk, M.; Wielechowska, M.; Redowicz, M.J.; Bretner, M.; Borowiecki, P. Synthesis of Novel Acyl Derivatives of 3-(4,5,6,7-Tetrabromo-1H-benzimidazol-1-yl)propan-1-ols-Intracellular TBBi-Based CK2 Inhibitors with Proapoptotic Properties. *Int. J. Mol. Sci.* **2021**, *22*, 6261. [CrossRef]
41. Ellingboe, J.W.; Spinelli, W.; Winkley, M.W.; Nguyen, T.T.; Parsons, R.W.; Moubarak, I.F.; Kitzen, J.M.; Von Engen, D.; Bagli, J.F. Class III antiarrhythmic activity of novel substituted 4-[(methylsulfonyl)amino]benzamides and sulfonamides. *J. Med. Chem.* **1992**, *35*, 705–716. [CrossRef] [PubMed]
42. Andrzejewska, M.; Pagano, M.A.; Meggio, F.; Brunati, A.M.; Kazimierczuk, Z. Polyhalogenobenzimidazoles: Synthesis and their inhibitory activity against casein kinases. *Bioorg. Med. Chem.* **2003**, *11*, 3997–4002. [CrossRef] [PubMed]
43. Pagano, M.A.; Andrzejewska, M.; Ruzzene, M.; Sarno, S.; Cesaro, L.; Bain, J.; Elliott, M.; Meggio, F.; Kazimierczuk, Z.; Pinna, L.A. Optimization of protein kinase CK2 inhibitors derived from 4,5,6,7-tetrabromobenzimidazole. *J. Med. Chem.* **2004**, *47*, 6239–6247. [CrossRef] [PubMed]

44. Borowiecki, P.; Wawro, A.; Wińska, P.; Wielechowska, M.; Bretner, M. Synthesis of novel chiral TBBt derivatives with hydroxyl moiety. Studies on inhibition of human protein kinase CK2α and cytotoxicity properties. *Eur. J. Med. Chem.* **2014**, *84*, 364–374. [CrossRef] [PubMed]
45. Chojnacki, K.; Wińska, P.; Wielechowska, M.; Łukowska-Chojnacka, E.; Tölzer, C.; Niefind, K.; Bretner, M. Biological properties and structural study of new aminoalkyl derivatives of benzimidazole and benzotriazole, dual inhibitors of CK2 and PIM1 kinases. *Bioorg. Chem.* **2018**, *80*, 266–275. [CrossRef]
46. Bradford, M.M. A rapid and sensitive method for the quantitation of microgram quantities of protein utilizing the principle of protein-dye binding. *Anal. Biochem.* **1976**, *72*, 248–254. [CrossRef]
47. Trott, O.; Olson, A.J. AutoDock Vina: Improving the speed and accuracy of docking with a new scoring function, efficient optimization, and multithreading. *J. Comput. Chem.* **2010**, *31*, 455–461. [CrossRef]
48. Wang, J.; Wolf, R.M.; Caldwell, J.W.; Kollman, P.A.; Case, D.A. Development and testing of a general amber force field. *J. Comput. Chem.* **2004**, *25*, 1157–1174. [CrossRef]
49. Dakin, L.A.; Block, M.H.; Chen, H.; Code, E.; Dowling, J.E.; Feng, X.; Ferguson, A.D.; Green, I.; Hird, A.W.; Howard, T.; et al. Discovery of novel benzylidene-1,3-thiazolidine-2,4-diones as potent and selective inhibitors of the PIM-1, PIM-2, and PIM-3 protein kinases. *Bioorg. Med. Chem. Lett.* **2012**, *22*, 4599–4604. [CrossRef]
50. Pettersen, E.F.; Goddard, T.D.; Huang, C.C.; Couch, G.S.; Greenblatt, D.M.; Meng, E.C.; Ferrin, T.E. UCSF Chimera--a visualization system for exploratory research and analysis. *J. Comput. Chem.* **2004**, *25*, 1605–1612. [CrossRef]
51. Yung-Chi, C.; Prusoff, W.H. Relationship between the inhibition constant (Ki) and the concentration of inhibitor which causes 50 per cent inhibition (IC50) of an enzymatic reaction. *Biochem. Pharmacol.* **1973**, *22*, 3099–3108. [CrossRef] [PubMed]
52. Thomé, M.P.; Filippi-Chiela, E.C.; Villodre, E.V.; Migliavaca, C.B.; Onzi, G.R.; Felipe, K.F.; Lenz, G. Ratiometric analysis of Acridine Orange staining in the study of acidic organelles and autophagy. *J. Cell Sci.* **2016**, *129*, 4622–4632. [CrossRef] [PubMed]
53. Lozzio, C.B.; Lozzio, B.B. Human chronic myelogenous leukemia cell-line with positive Philadelphia-chromosome. *Blood* **1975**, *45*, 321–334. [CrossRef] [PubMed]
54. Heriche, J.K.; Chambaz, E.M. Protein kinase CK2α is a target for the Abl and Bcr-Abl tyrosine kinases. *Oncogene* **1998**, *17*, 13–18. [CrossRef]
55. Rai, Y.; Yadav, P.; Kumari, N.; Kalra, N.; Bhatt, A.N. Hexokinase II inhibition by 3-bromopyruvate sensitizes myeloid leukemic cells K-562 to anti-leukemic drug, daunorubicin. *Biosci Rep.* **2019**, *39*, BSR20190880. [CrossRef]
56. Morotti, A.; Carrà, G.; Panuzzo, C.; Crivellaro, S.; Taulli, R.; Guerrasio, A.; Saglio, G. Protein Kinase CK2: A Targetable BCR-ABL Partner in Philadelphia Positive Leukemias. *Adv. Hematol.* **2015**, *2015*, 612567. [CrossRef]
57. Hori, M.; Nogami, T.; Itabashi, M.; Yoshim, F.; Ono, H.; Koizumi, S. Expression of Bcl-2 in human breast cancer: Correlation between hormone receptor status, p53 protein accumulation and DNA strand breaks associated with apoptosis. *Pathol. Int.* **1997**, *47*, 757–762. [CrossRef]
58. Galluzzi, L.; Vitale, I.; Aaronson, S.A.; Abrams, J.M.; Adam, D.; Agostinis, P.; Alnemri, E.S.; Altucci, L.; Amelio, I.; Andrews, D.W.; et al. Molecular mechanisms of cell death: Recommendations of the Nomenclature Committee on Cell Death 2018. *Cell Death Differ.* **2018**, *25*, 486–541. [CrossRef]
59. Denton, D.; Kumar, S. Autophagy-dependent cell death. *Cell Death Differ.* **2018**, *26*, 605–616. [CrossRef]
60. Bellodi, C.; Lidonnici, M.; Hamilton, A.; Helgason, V.; Soliera, A.R.; Ronchetti, M.; Galavotti, S.; Young, K.W.; Selmi, T.; Yacobi, R.; et al. Targeting autophagy. *J. Clin. Investig.* **2009**, *119*, 1109–1123. [CrossRef]
61. Lin, Q.; Meloni, D.; Pan, Y.; Xia, M.; Rodgers, J.; Shepard, S.; Li, M.; Galya, L.; Metcalf, B.; Yue, T.-Y.; et al. Enantioselective synthesis of Janus kinase inhibitor INCB018424 via an organocatalytic aza-Michael reaction. *Org. Lett.* **2009**, *11*, 1999–2002. [CrossRef] [PubMed]
62. Izeradjene, K.; Douglas, L.; Delaney, A.; Houghton, J.A. Casein kinase ii (CK2) enhances death-inducing signaling complex (disc) activity in trail-induced apoptosis in human colon carcinoma cell lines. *Oncogene* **2005**, *24*, 2050–2058. [CrossRef] [PubMed]
63. Trembley, J.H.; Chen, Z.; Unger, G.; Slaton, J.; Kren, B.T.; Van Waes, C.; Ahmed, K. Emergence of protein kinase ck2 as a key target in cancer therapy. *BioFactors* **2010**, *36*, 187–195. [CrossRef] [PubMed]
64. Wang, S.; He, M.; Li, L.; Liang, Z.; Zou, Z.; Tao, A. Cell-in-cell death is not restricted by caspase-3 deficiency in mcf-7 cells. *J. Breast Cancer* **2016**, *19*, 231–241. [CrossRef] [PubMed]
65. Li, Y.; Dowbenko, D.; Lasky, L.A. AKT/PKB phosphorylation of p21Cip/WAF1 enhances protein stability of p21Cip/WAF1 and promotes cell survival. *J. Biol. Chem.* **2002**, *277*, 11352–11361. [CrossRef] [PubMed]
66. Wińska, P.; Skierka, K.; Łukowska-Chojnacka, E.; Koronkiewicz, M.; Cieśla, J.; Bretner, M. Effect of Simultaneous Inhibition of Protein Kinase CK2 and Thymidylate Synthase in Leukemia and Breast Cancer Cells. *Anticancer Res.* **2018**, *38*, 4617–4627. [CrossRef]
67. Wińska, P.; Widło, Ł.; Skierka, K.; Krzyśko, A.; Koronkiewicz, M.; Cieśla, J.M.; Cieśla, J.; Bretner, M. Simultaneous Inhibition of Protein Kinase CK2 and Dihydrofolate Reductase Results in Synergistic Effect on Acute Lymphoblastic Leukemia Cells. *Anticancer Res.* **2019**, *39*, 3531–3542. [CrossRef]

68. Wińska, P.; Karatsai, O.; Staniszewska, M.; Koronkiewicz, M.; Chojnacki, K.; Rędowicz, M.J. Synergistic Interactions of 5-Fluorouracil with Inhibitors of Protein Kinase CK2 Correlate with p38 MAPK Activation and FAK Inhibition in the Triple-Negative Breast Cancer Cell Line. *Int. J. Mol. Sci.* **2020**, *21*, 6234. [CrossRef]
69. Daina, A.; Blatter, M.-C.; Baillie Gerritsen, V.; Palagi, P.M.; Marek, D.; Xenarios, I.; Schwede, T.; Michielin, O.; Zoete, V. Drug design workshop: A web-based educational tool to introduce computer-aided drug design to the general public. *J. Chem. Educ.* **2017**, *94*, 335–344. [CrossRef]

pharmaceutics

MDPI

Article

An Aptamer against MNK1 for Non-Small Cell Lung Cancer Treatment

Rebeca Carrión-Marchante [1,†], Celia Pinto-Díez [2,†], José Ignacio Klett-Mingo [1], Esther Palacios [1], Miriam Barragán-Usero [1], M. Isabel Pérez-Morgado [1], Manuel Pascual-Mellado [1], Sonia Alcalá [3,4], Laura Ruiz-Cañas [3,4], Bruno Sainz, Jr. [3,4,5], Víctor M. González [1,*] and M. Elena Martín [1,*]

[1] Aptamer Group, Deparment Biochemistry-Research, IRYCIS—Hospital Universitario Ramón y Cajal, 28034 Madrid, Spain
[2] Aptus Biotech SL, 28035 Madrid, Spain
[3] Department of Cancer, Instituto de Investigaciones-Biomédicas "Alberto Sols" (IIBM), CSIC-UAM, 28034 Madrid, Spain
[4] Chronic Diseases and Cancer Area 3—Instituto Ramón y Cajal de Investigación Sanitaria (IRYCIS), 28034 Madrid, Spain
[5] Centro de Investigación Biomédica en Red, Área Cáncer—CIBERONC, ISCIII, 28029 Madrid, Spain
* Correspondence: victor.m.gonzalez@hrc.es (V.M.G.); m.elena.martin@hrc.es (M.E.M.); Tel.: +34-913-368-173 (V.M.G. & M.E.M.)
† These authors contributed equally to this work.

Abstract: Lung cancer is the leading cause of cancer-related death worldwide. Its late diagnosis and consequently poor survival make necessary the search for new therapeutic targets. The mitogen-activated protein kinase (MAPK)-interacting kinase 1 (MNK1) is overexpressed in lung cancer and correlates with poor overall survival in non-small cell lung cancer (NSCLC) patients. The previously identified and optimized aptamer from our laboratory against MNK1, apMNKQ2, showed promising results as an antitumor drug in breast cancer in vitro and in vivo. Thus, the present study shows the antitumor potential of apMNKQ2 in another type of cancer where MNK1 plays a significant role, such as NSCLC. The effect of apMNKQ2 in lung cancer was studied with viability, toxicity, clonogenic, migration, invasion, and in vivo efficacy assays. Our results show that apMNKQ2 arrests the cell cycle and reduces viability, colony formation, migration, invasion, and epithelial-mesenchymal transition (EMT) processes in NSCLC cells. In addition, apMNKQ2 reduces tumor growth in an A549-cell line NSCLC xenograft model. In summary, targeting MNK1 with a specific aptamer may provide an innovative strategy for lung cancer treatment.

Keywords: aptamer; MNK1; therapeutic target; non-small cell lung cancer; NSCLC; antitumor

Citation: Carrión-Marchante, R.; Pinto-Díez, C.; Klett-Mingo, J.I.; Palacios, E.; Barragán-Usero, M.; Pérez-Morgado, M.I.; Pascual-Mellado, M.; Alcalá, S.; Ruiz-Cañas, L.; Sainz, B., Jr.; et al. An Aptamer against MNK1 for Non-Small Cell Lung Cancer Treatment. *Pharmaceutics* **2023**, *15*, 1273. https://doi.org/10.3390/pharmaceutics15041273

Academic Editor: Francesca Musumeci

Received: 3 March 2023
Revised: 13 April 2023
Accepted: 14 April 2023
Published: 18 April 2023

1. Introduction

Lung cancer is the leading cause of cancer-related death worldwide and the second most commonly diagnosed tumor among all cancer types [1]. The 5-year survival rate is greater for early-stage lung cancer than for advanced-stage lung cancer. Lung cancer is usually diagnosed in advanced stages, or even when it metastasizes to other areas, which increases therapeutic failure, decreasing patient survival rates [2]. Therefore, the search for new therapies against lung cancer is a healthcare priority.

The mitogen-activated protein kinase (MAPK)-interacting kinases (MNKs) are serine/threonine kinases that are directly activated by an extracellular signal-regulated kinase (ERK) 1/2 or p38 MAP kinases [3,4]. In humans, MNKs comprise a group of four isoforms (MNK1a/b and MNK2a/b) from two genes by alternative splicing. The four isoforms are similar in their N-terminus but differ in their C-terminus. MNK1b was first described in our laboratory [5], and it lacks the MAPK binding site motif present in the C-terminal region of MNK1a. MNK1b has higher basal activity than MNK1a, being independent of

ERK1/2 and p38 MAPK activation [6]. MNKs phosphorylate several substrates including eukaryotic initiation factor 4E (eIF4E) [7] and 4G (eIF4G) [8], heterogeneous nuclear ribonucleoprotein A1 (hnRNPA1) [9], cPLA2 [10], Sprouty2 [11], and polypyrimidine-tract binding protein-associated splicing factor (PSF) [12], being eIF4E the best characterized.

The eIF4E improves the translation of proteins involved in proliferation, survival, invasion, and angiogenesis such as cyclin D1, c-Myc, Bcl-2, metalloproteases (MMPs), Snail, vascular endothelial growth factor (VEGF) and fibroblast growth factor (FGF) [13–17]. Phosphorylation of eIF4E on serine 209 by MNKs is associated with tumor progression and poor prognosis, with high levels of phosphorylated eIF4E (p-eIF4E) found in different types of cancer [18]. Similarly, overexpression of MNKs has been found in glioblastoma, lung cancer, hepatocellular carcinoma, ovarian cancer, and breast cancer [19–24], and high expression of MNK1 is correlated with poor prognosis in several types of cancers [25]. Of all MNK inhibitors developed thus far [25], some have been evaluated in clinical trials for cancer therapy such as BAY1143269 [26], eFT508 [27], or ETC-206 [28]. Likewise, in MNK1 and MNK2 knockout (KO) mice, while their development is normal, there is a delay in tumor progression in KO mice compared to wild-type mice [29]. Thus, since MNKs are not essential but overexpressed in many tumor types, they represent good therapeutic cancer targets, with theoretically no adverse effects in non-tumor cells.

Aptamers are single-strand nucleic acids (ssDNA or RNA) that bind to different targets by folding into a three-dimensional conformation. They are identified by the SELEX method [30,31], which represents an important tool for the discovery of novel biomarkers of both therapeutic and diagnostic interest. Aptamers offer several advantages over antibodies including their high stability, low immunogenicity, no batch-to-batch variability, small size, short generation time, and quick modification [32]. So far, Macugen® is the only aptamer approved by the FDA for the treatment of age-related macular degeneration [33]. However, several aptamers have entered into clinical trials [34]; some for cancer treatment such as the DNA aptamer AS1411 which targets nucleolin, inducing apoptosis [35], or NOX-A12, an RNA aptamer that inhibits tumor growth by binding to CXCL-12.

We selected an aptamer against MNK1 capable of inhibiting proliferation, migration, and colony formation in MDA-MB-231 breast cancer cells [36]. In order to decrease its size, we obtained four different sequences from the initial aptamer, where apMNKQ2 was the most efficient sequence at inhibiting the proliferation of breast cancer cell lines. Further studies showed that apMNKQ2 inhibits the tumorigenic and metastatic activity of breast cancer cells and reduces tumor growth and metastasis numbers in an in vivo model [37].

Thus, the objective of this work was to evaluate the potential of apMNKQ2 as an antitumor drug. For this purpose, we studied the effect of apMNKQ2 in another type of cancer where MNK1 is overexpressed, specifically lung cancer. Our results show that apMNKQ2 arrests the cell cycle and inhibits the proliferation, colony formation, migration, invasion, and adhesion of lung cancer cell lines, and reduces tumor growth in a mouse xenograft model of lung adenocarcinoma. These results highlight apMNKQ2 as a potential anticancer drug for lung cancer.

2. Materials and Methods

2.1. Materials

The ssDNA aptamers, apMNKQ2 (TGGGGTGGCGGCGGGGTGGGGTGGT) and p29 (GCGGTCGACTTAAATGTCCATCTCAAACT) were purchased from Biospring, Frankfurt am Main, Germany. The apMNKQ2 aptamer used in this study does not have any modification. The origin of the rest of the material is indicated in the text.

2.2. Cell Culture and Transfection

A549 (human lung adenocarcinoma cell line) and SW900 (human lung squamous cell carcinoma cell line) cells were authenticated in May of 2019, and H460 cells (human lung large cell carcinoma cell line) were authenticated in September of 2022, through the GenePrint® 10 System. SW900 and H460 were cultured in RPMI medium (PAA, Pasching, Austria) containing 10% fetal calf serum (Gibco, Grand Island, New York, NY, USA), 100 U/mL penicillin, 100 μg/mL streptomycin, and 25 μg/mL amphotericin (Sigma, St. Louis, MO, USA). A549 cells were cultured in Dulbecco's modified Eagle's medium (DMEM) (Biowest SAS, Nuaillé, France) with 10% fetal calf serum (Gibco, Grand Island, New York, NY, USA), 1% pyruvate, 100 U/mL penicillin, 100 μg/mL streptomycin, and 25 μg/mL amphotericin (Sigma, St. Louis, MO, USA). All cell lines were maintained at 37 °C in a humidified incubator with 5% CO_2. For cell transfection with aptamers, cells were seeded at different concentrations according to the assay and after 24 h, aptamers were transfected into cells using Lipofectamine^{TM} 2000 (Invitrogen, Carlsbad, CA, USA), according to the manufacturer's instructions. Cells were incubated at 37 °C in a humidified incubator with 5% CO_2 until the assay was carried out. Before transfection, aptamers were dissolved in selection buffer (20 mM Tris-HCl pH 7.4, 150 mM NaCl, 1 mM $MgCl_2$, and 5 mM KCl), denatured at 90 °C for 10 min and then cooled on ice for 10 min.

2.3. Protein and RNA Extraction

For protein extraction, cells were mechanically dissociated and washed once with cold buffer A (20 mM Tris-HCl pH 7.6, 1 mM ethylenediaminetetraacetic acid (EDTA), 1mM dithiothreitol (DTT), 1 mM benzamidine, 1 mM phenylmethylsulfonyl fluoride (PMSF), 10 mM sodium β-glycerophosphate, 10 mM sodium molybdate, 1 mM sodium orthovanadate, 120 mM potassium chloride (KCl), 10 μg/mL antipain, 1 μg/mL pepstatin A, and leupeptin). Then, cells were lysed in buffer A with 1% Tritón X-100 (volume ratio 1:2) and centrifugated for 10 min at 12,000× *g*. Supernatants were used to determine protein concentrations using the BCA kit (ThermoFisher Scientific, Waltham, MA, USA) and then they were stored at −80 °C until use.

For total RNA extraction, cells were trypsinized and centrifugated at 400× *g* for 5 min. Pellets were lysed in NucleoZOL (Macherey-Nagel, Düren, Germany) according to the manufacturer's instructions. RNA pellets were resuspended in 50 μL of RNase-free water, quantified, and stored at −80 °C until use.

2.4. Western Blot

Cell lysates were resolved in 7.5%, 10%, or 12% sodium dodecyl sulphate-polyacrilamide gels (SDS-PAGE) by electrophoresis and transferred onto polyvinylidene difluoride (PVDF) membranes. Membranes were incubated for 1h at room temperature with 5% non-fat milk in PBS and then overnight at 4 °C with monoclonal or polyclonal antibodies. After washing, membranes were incubated with the corresponding peroxidase-conjugated secondary antibodies for 1 h at room temperature and proteins were visualized using Clarity Western ECL Substrate (BioRad, Hercules, CA, USA) and a ChemiDoc^{TM} (BioRad, Hercules, CA, USA). Bands were quantified using ImageLab 6.1 software (BioRad, Hercules, CA, USA). PageRuler Plus Prestained Protein Ladder (ThermoFisher Scientific, Waltham, MA, USA) was the molecular weight marker used in all the experiments. β-actin (Sigma, St. Louis, MO, USA) antibody was used to control the homogeneity of loading. Antibodies used in this work can be found in Supplementary Table S1.

2.5. Cell Viability (MTT) and Cell Toxicity (LDH) Assays

To study cell viability, 3-(4,5-dimethylthiazol-2-yl)-2,5-diphenyl-2H-tetrazolium bromide (MTT) assays were performed. A549 and H460 cells were plated at 6×10^3 cells/well and SW900 at 10^4 cells/well in 96-well plates. After 16–24 h, cells were transfected with apMNKQ2 or the unspecific p29 aptamer as described above and incubated at 37 °C in a humidified incubator with 5% CO_2 for 48 h. Then, the medium was removed to add 100 μL/well of MTT (Sigma, St. Louis, MO, USA) at 1 mg/mL in culture medium. A549 and H460 cells were incubated for 1.5 h and the SW900 cell line for 3 h at 37 °C. Next, 100 μL/well of lysis buffer (10% SDS and 10 mM HCl) was added. After 24 h, absorbance was read at 540 nm in an Infinite F200 spectrophotometer (TECAN, Männedorf, Switzerland).

To determine cytotoxicity in cells, lactate dehydrogenase (LDH) assays were performed 48 h after cell transfection with aptamers. Supernatants were collected and the

Cytotoxicity Detection kit (LDH) (Roche, Madrid, Spain) reagent was added according to the manufacturer's instructions. After incubation for 30 min at room temperature, the reaction was stopped by adding 1M HCl and absorbance was read at 490 nm in an Infinite F200 spectrophotometer (TECAN, Männedorf, Switzerland). To calculate the percentage of cytotoxicity, two controls were used: the supernatant of untreated cells (LDHlow) and of lysed cells (LDHhigh) with 0.2% Tritón X-100. Results are expressed as:

$$Cytotoxicity\ (\%) = {\scriptstyle LDHtreated - LDHlow}/{\scriptstyle LDHhigh - LDHlow} \times 100$$

2.6. Trypan Blue Exclusion Test of Cell Viability

In order to determine the viable cells after transfection with apMNKQ2, cells were collected from the plates and 10 μL of cells were mixed with 10 μL of trypan blue and were counted on a TC20 counter (BioRad, Hercules, CA, USA). Those cells that excluded the dye were considered viable.

2.7. Cell Cycle Assay

Cells were plated at 5×10^5 cells/well in a 6-well plate and transfected with the aptamers. Cells were trypsinized 24 h post-transfection, and both attached and floating cells were collected, fixed with PFA 4%, washed twice with PBS, and stored at 4 °C. Next, cells were incubated with FxCycle Violet Ready Flow Reagent (Thermo Fisher Scientific, Waltham, MA, USA) for 30 min at 25 °C and DNA content was evaluated with a 4-laser Attune NxT Acoustic Cytometer (Thermo Fisher Scientific, Waltham, MA, USA). Results were analyzed using FCSalyzer 0.9.22 alpha Software.

2.8. Apoptosis Assay

Cells were plated at 5×10^5 cells/well in a 6-well plate, transfected with the aptamers during 24 h, resuspended, and stained with Annexin V Apoptosis detection kit (Canvax, Córdoba, Spain) as described previously [37] and analyzed with a 4-laser Attune NxT Acoustic Cytometer (Thermo Fisher Scientific, Waltham, MA, USA). Results were analyzed using FCSalyzer 0.9.22 alpha Software.

2.9. Clonogenic Assays

Cells were seeded at 5×10^4 cells/well in 24-well plates and 16–24 h later, they were transfected with aptamers. After 16–24 h, cells were collected and seeded at 10^3 cells/well in 6-well plates. Cells were incubated at 37 °C in a humidified incubator with 5% CO_2 during 5–8 days. Then, cells were fixed with 1 mL/well of ethanol for 10 min at room temperature and stained with Giemsa 0.2% (Sigma, St. Louis, MO, USA) for 30 min. Remains of Giemsa were removed with water, images of colonies were taken and colonies were counted using ImageJ Java 1.8.0_172 software (National Institutes of Health, Bethesda, MD, USA).

2.10. Migration Assays

Migration assays were performed using transwell insert chambers (Corning, New York, NY, USA). Cells were seeded at 5×10^4 cells/well in 24-well plates and transfected with aptamers 16–24 h later. After 16 h of serum-deprivation, cells were collected and added at 4×10^4 cells/well into the upper chamber in 300 μL of serum-free medium. In the lower chamber, 500 μL of complete medium (10% FSB) was added as a chemoattractant. After 24 h, both media were removed and cells in the insert were fixed for 2 min with 4% formaldehyde followed by 20 min with 100% methanol. Next, cells were stained with 30 μM Hoechst 33342 for 15 min and washed twice with PBS. At least five photographs were taken for each sample using a fluorescence microscope Olympus IX70 (Olympus Iberia, Barcelona, Sapin). Cells were counted using ImageJ Java 1.8.0_172 software (National Institutes of Health, Bethesda, MD, USA).

2.11. Cell Adhesion Assays

Plates were coated with 10 µg/mg type I collagen (Sigma, St. Louis, MO, USA) or BSA (Sigma, St. Louis, MO, USA) as a control for 1 h at 37 °C. Cells were seeded at 5×10^4 cells/well in 6-well plates and transfected with aptamers 16–24 h later. After 24 h, cells were collected, added to coated plates at 3×10^4 cells/well in serum-free medium, and allowed to bind at 37 °C for 1 h. Wells were washed twice with PBS to remove non-adhered cells and an MTT assay was performed with 1 h of incubation between the addition of the MTT reagent and lysis buffer. After 24 h, absorbance was read at 540 nm in an Infinite F200 spectrophotometer (TECAN, Männedorf, Switzerland).

2.12. Zymography

The MMP activity of cells was analyzed through zymography. Cells were transfected and after 18 h of serum-deprivation, supernatants were collected and concentrated using Centricon tubes with a 3 KDa cutoff (Merck, Rahway, NJ, USA). Samples were mixed with loading buffer without mercaptoethanol and then separated in 7.5% polyacrylamide gels with 1 mg/mL gelatin. Gels were then incubated for 30 min in 2.5% Triton X-100 at room temperature and overnight at 37 °C in developing buffer (50 mM Tris-HCl pH 7.5, 200 mM NaCl, 5 mM $CaCl_2$, and 0.02% Tween). Coomassie blue staining revealed the presence of MMP activity as clear bands against the blue background. Bands were quantified with ImageLab 6.1 software (BioRad, Hercules, CA, USA).

2.13. Aptamer Quantification

To quantify intracellular aptamers, cells were collected 4-, 24- and 48-h post-transfection, washed twice in PBS, lysed in 1 mL of H_2O for 5 min, vortexed, and boiled at 90 °C for 10 min. Lysates were centrifuged at 12,000× *g* for 10 min and the aptamer in the supernatant was quantified by quantitative PCR. Pellets were resuspended in 10 µL of 0.1 M NaOH to determine protein concentration by BCA as described above. Results are expressed in fmol of aptamer/µg of protein. To quantify aptamer in tumors and organs, 50–100 mg of tissue were lysed and homogenized in NucleoZOL reagent (Macherey-Nagel, Düren, Germany) following the manufacturer's instructions to isolate small RNA fraction, in which the aptamer is found. Results are expressed in fmol of aptamer/µg of tissue. Since apMNKQ2 is only 29 nucleotides in length, we designed a method that allows us to amplify this small-size aptamer in which apMNKQ2 acts as a primer on a template oligo-nucleotide called QR short (see Supplementary Table S2). After this reaction, both apMNKQ2 and QR short are elongated generating a 76-nucleotide template that can then be amplified with the appropriate primers (R3 and QF) (see Supplementary Table S2). Thus, in the absence of apMNKQ2, no amplification occurs, being the amplification proportional to the amount of apMNKQ2 present in the sample. The qPCR was performed using the AceQ qPCR SYBR® Green Master Mix-Vazyme (Quimigen, Madrid, Spain) according to the manufacturer's instructions in a StepOne Plus Real-Time PCR system (Applied Biosystem, Waltham, MA, USA). The reaction mixture consisted of 1× Mix FastGene® IC Green, 1 pmol QR short, 100 nM QF and R3 primers, and 1 µL of samples in a 10 µL/tube final volume. Aptamers were quantified using a standard curve (100 fmol–100 amol).

2.14. Quantification of mRNA

Total RNA was obtained as described above and was used to synthesize first strand cDNA using the SensiFAST™ cDNA Synthesis Kit (Bioline, Segovia, Spain) following the manufacturer's instructions. Products were used for qPCR amplification using the AceQ qPCR SYBR® Green Master Mix-Vazyme (Quimigen, Madrid, Spain) according to the manufacturer's protocol in a StepOne Plus Real-Time PCR system (Applied Biosystem, Waltham, MA, USA). Triplicate reactions of the targets and housekeeping genes were performed simultaneously for each cDNA template. Sequences of oligonucleotides are in Supplementary Table S2. To calculate the relative expression of each target gene, the $2^{-\Delta\Delta Ct}$

method was performed, using β-actin as a housekeeping normalization gene, where ΔCt is Ct_{target}-$C_{thousekeeping}$.

2.15. Tolerability Study

Non-tumor-bearing CD1 mice were used to determine the maximum tolerated dose (MTD) of apMNKQ2. The aptamer was administered once per day intraperitoneally (i.p.) for 7 consecutive days. We applied a modified "3 + 3" study design [38,39], using cohorts of 3 animals per dose and with the first cohort treated at a starting dose (1 mg/Kg). Subsequent cohorts were treated with increasing or decreasing doses according to the observed dose-limiting toxicities (DLTs) responses with guided subsequent doses, as described in [40]. DLT endpoints included weight loss (>20%), abnormal behavior, signs of physical discomfort, and/or death. Mice were evaluated daily and if any of the endpoints were met, animals were euthanized.

2.16. In Vivo Efficacy Assays

For in vivo experiments, mice were housed according to institutional guidelines, and all experimental procedures were performed in compliance with the institutional guidelines for the welfare of experimental animals approved by the Universidad Autónoma de Madrid Ethics Committee (CEI 60-1057-A068 and CEI 103-1958-A337) and La Comunidad de Madrid (PROEX 294/19) and in accordance with the guidelines for Ethical Conduct in the Care and Use of Animals as stated in The International Guiding Principles for Biomedical Research involving Animals, developed by the Council for International Organizations of Medical Sciences (CIOMS). Briefly, mice were housed according to the following guidelines: a 12-h light/12-h dark cycle, with no access during the dark cycle; temperatures of 65–75 °F (~18–23 °C) with 40–60% humidity; a standard diet with fat content ranging from 4–11%; sterilized water was accessible at all times; for handling, mice were manipulated gently and as little as possible; noises, vibrations, and odors were minimized to prevent stress and decreased breeding performance; and enrichment was always used per the facility's guidelines to help alleviate stress.

Female 9-week-old $Foxn1^{nu/nu*}$ (Janvier Labs, Le Genest-Saint-Isle, France) were injected subcutaneously in the dorsal flanks with 8×10^6 A549 cells resuspended in 100 μL Matrigel (Corning, New York, NY, USA) per injection. Once tumors were established, mice were randomized into 3 groups (5 mice per group) and injected intraperitoneally with vehicle control (selection buffer), 10 mg/kg apMNKQ2 or 25 mg/kg apMNKQ2. Aptamer injections were daily and tumor volumes were determined twice per week for 25 days using a manual caliper. At the time of sacrifice, tumors were excised and weighed. One half was frozen in liquid nitrogen for western blot and qPCR analysis and the other half was fixed in 4% PFA and processed for histologic analysis. At the end of the treatment, the TGI ratio (%) was calculated using the following formula: TGI (%) = [1 − (Volume of the treated group)/(volume of the control group)] × 100 (%).

2.17. Statistical Analyses

Results are presented as means ± standard error of the mean (SEM) unless stated otherwise. Pair-wise multiple comparisons were performed with one-way ANOVA (two-sided) with Turkey's test adjustment, as indicated in the figure legends. A student's *t*-test was used to determine differences between the means of groups. The p-values < 0.05 were considered statistically significant. All analyses were performed using GraphPad Prism version 8.0 (San Diego, CA, USA).

3. Results

3.1. apMNKQ2 Inhibits Cell Viability in NSCLC Cell Lines

In our previous work [37], four sequences were designed (apMNKQ1, apMNKQ2, apMNKQ3, and apMNKQ4) from the aptamer apMNK2F against MNK1b and tested in breast cancer cell lines. Here, to study the effect of the four sequences in the context of lung cancer, A549 and SW900 cell lines were transfected with the four aptamers at 250 nM, and cell viability was assessed 48 h later by measuring MTT activity. The nonspecific aptamer (p29) was used as a control. Results (Supplementary Figure S1) showed that apMNKQ2 had the greatest effect on MTT activity in both cell lines, producing a significant reduction in cell viability of 77% in A549 and 51% in SW900 cells.

Consequently, the effect of apMNKQ2 on cell proliferation and toxicity was evaluated in three cell lines (A549, SW900, and H460) representing the main types of non-small cell lung cancer (adenocarcinoma, squamous cell carcinoma, and large cell carcinoma, respectively). Cells were transfected with increasing concentrations (0–500 nM) of apMNKQ2 or the nonspecific control aptamer (p29). MTT activity was measured after 48 h and to study possible cell death by necrosis, LDH enzyme activity was simultaneously measured. Results showed that apMNKQ2 decreased MTT activity in all three cell lines in a concentration-dependent manner, with A549 cells being the most sensitive (Figure 1a). The nonspecific p29 aptamer had a slight effect on the cell lines (less than 20%). The IC50 values were 70 nM, 250 nM, and 100 nM for A549, SW900, and H460, respectively. This effect was not necrosis-mediated since there was little to no cytotoxicity (Figure 1b).

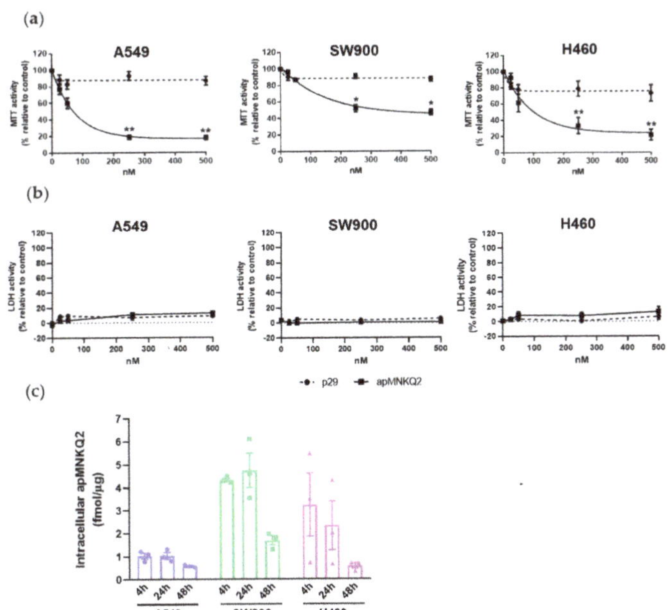

Figure 1. Aptamer effect on the viability of lung cancer cells. (**a**) Effect of apMNKQ2 on MTT and (**b**) LDH activity. A549 and H460 cells were seeded at 6×10^3 cells/well and SW900 cells were seeded at 10^4 cells/well in 96-well plates. After 48 h MTT and LDH assays were performed. Graphs represent the mean ± SEM of 6 and 3 independent experiments. * $p < 0.05$ and ** $p < 0.01$ relative to p29 control. (**c**) The half-life of apMNKQ2 in lung cancer cells. Cells were seeded at 5×10^5 cells/well in 6-well plates and transfected with the corresponding IC50 of apMNKQ2 16–24 h later. After 4, 24, and 48 h, cells were lysed and apMNKQ2 was quantified through qPCR. Results are expressed as fmol of aptamer/µg of protein. Bars represent mean ± SEM of 3 independent experiments. Solid blue circles (A549 cell line), green squares (SW900 cell line) and violet triangles (H460 cell line) represent one individual value.

When aptamers are used against intracellular targets it is important to determine their half-life inside the cell, which may be affected by their molecular nature or structure. To study the intracellular stability of apMNKQ2, cells were transfected according to their IC50.

After 4, 24, and 48 h, cells were lysed and apMNKQ2 was quantified by qPCR. As shown in Figure 1c, apMNKQ2 is stable in all three cell lines for at least 48 h.

3.2. *apMNKQ2 Induces Apoptosis, Cell Cycle Arrest, and Inhibits Colony Formation in Lung Cancer Cells*

To determine whether the decrease in cell viability was a consequence of apMNKQ2-induced apoptosis, we analyzed the apoptotic marker cleaved-PARP. Aptamer-transfected cells were lysed after 24 h of transfection (cell viability 24 h after transfection is shown in Supplementary Figure S2) and analyzed by western blot as described in the Material and Methods section (Section 2). Figure 2a shows that the cleavage of PARP by caspase 3 occurs in the three cell lines albeit to a different extent. A decrease in the uncleaved form of PARP was observed in A549 and SW900 cell lines, being statistically significant in A549 cells. In addition, we observed a significant increase in both the percentage of Annexin-V positive and propidium iodide (PI) negative cells (Figure 2b) and the percentage of the sub-G1 phase in the A549 cells (Supplementary Figure S2c), indicating that apMNKQ2 induces apoptosis in this cell line.

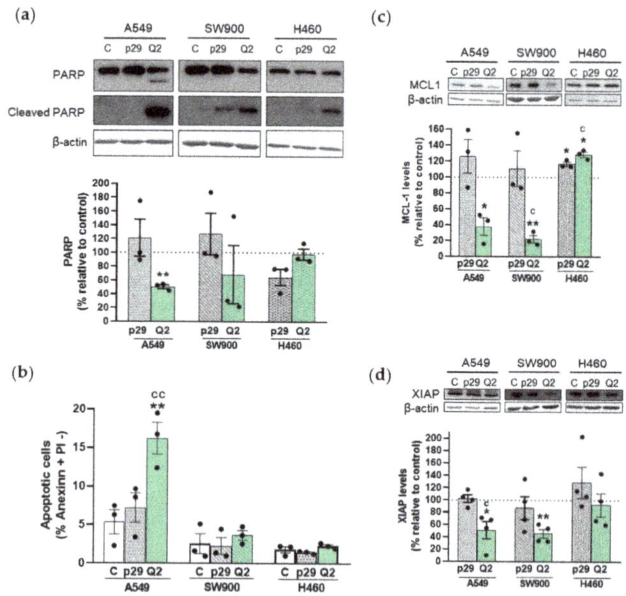

Figure 2. Apoptotic activity of apMNKQ2 on NSCLC cells. (**a**) On the top, lysates (20 μg) were resolved in 10% SDS-PAGE gels for PARP and cleaved PARP. On the bottom, densitometry analyses of total PARP in the three cell lines. Data were normalized to respective β-actin bands and expressed as the percentage relative to control (C) cells. The graphs represent the mean ± SEM of 3 independent experiments. (**b**) Quantification of early apoptotic cells by Annexin-V staining. The bars represent the mean percentage of Annexin-V positive and PI negative cells ± SEM of 3 independent experiments. (**c**,**d**) Densitometry analyses of XIAP and MCL-1. Both were normalized to respective β-actin bands and expressed as the percentage relative to control (C) cells. The graphs represent the mean ± SEM of 3–4 independent experiments. Lysates (20 μg) were resolved in 12% gels for XIAP and MCL-1 and western blot analysis was carried out using specific antibodies (see Supplementary Table S1). A representative blot is shown on the top of each graph. * $p < 0.05$ and ** $p < 0.01$ relative to control cells; ^{c} $p < 0.05$ and ^{cc} $p < 0.01$ relative to p29 control. Control is shown as C, p29 is the unspecific aptamer used as a negative control, and apMNKQ2 is shown as Q2. Each solid black circle represents one individual value.

We also analyzed XIAP and MCL1 levels since both are antiapoptotic proteins regulated by MNK1 [23,41,42]. Results show that apMNKQ2 significantly decreased both XIAP and MCL1 levels in A549 and SW900 cell lines, again confirming the proapoptotic effect of apMNKQ2 (Figure 2c,d). However, apMNKQ2 seemed to significantly increase the levels of MCL1 in H460 cells (Figure 2c).

Next, we determined the effect of the aptamers on cell cycle distribution. Cells were transfected and 24 h later analyzed by flow cytometry. Figure 3a shows that apMNKQ2 increased the percentage of cells in the G1 phase in A549 (6.6%) and H460 (7.5%) cells compared with control cells, indicating that apMNKQ2 induces an arrest in the G1 phase of the cell cycle. Surprisingly, apMNKQ2 induced an increase in the percentage of cells in the G2/M phase (3.8%) in SW900 cells.

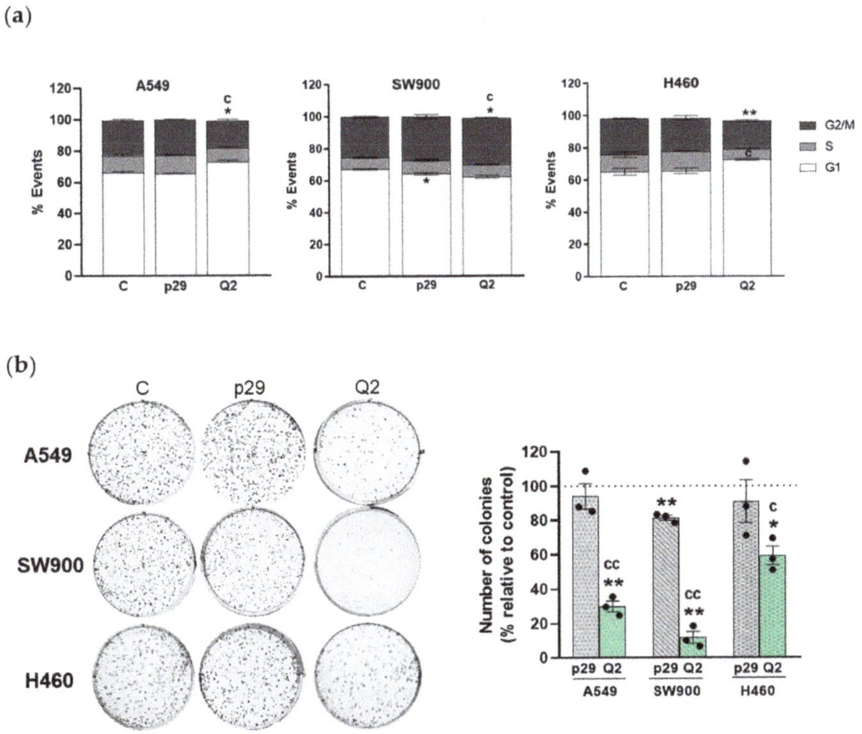

Figure 3. apMNKQ2 arrests cell cycle progression and colony formation in NSCLC cells. (**a**) Cells were transfected with apMNKQ2 at IC50 concentrations. After 24 h, cells were stained with PI and analyzed by flow cytometry. The bars represent the mean percentage of cells gated in each phase, G1 (white), S (grey), and G2/M (dark grey) ± SEM of 3 independent experiments. (**b**) Cells were transfected with aptamers at IC50 and after 24 h reseeded at 10^3 cells/well in 6-well plates. After 7–8 days, colonies were fixed, stained, and counted. (Left) shown are representative images. (Right) The bars represent the mean percentage of the number of colonies obtained ± SEM of 3 independent experiments relative to control (C). * $p < 0.05$ and ** $p < 0.01$ relative to control cells; ^{c} $p < 0.05$ and ^{cc} $p < 0.01$ relative to p29 control. Control is shown as C, p29 is the unspecific aptamer used as a negative control, and apMNKQ2 is shown as Q2. Each solid black circle represents one individual value.

Moreover, we studied the effect of apMNKQ2 on colony formation and observed that apMNKQ2 significantly reduced the clonogenic capacity of the three cell lines (Figure 3b). The unspecific p29 aptamer had a slight but significantly less effect compared to apMNKQ2.

3.3. apMNKQ2 Inhibits Migration, Invasion, Cell Adhesion, and Epithelial-Mesenchymal Transition (EMT) in Lung Cancer Cells

Metastatic dissemination requires cancer cells to detach from the primary tumor and colonize distant organs through pivotal steps such as cell migration, invasion, and adhesion. We have studied how apMNKQ2 affects all these cell features in order to determine its antimetastatic potential. We performed transwell migration assays to determine the migratory ability of lung cancer cells after being transfected with apMNKQ2. Results showed that the number of migrated cells was lower in the three cell lines following transfection with apMNKQ2, and statistically significant in both A549 and SW900 cells (Figure 4a).

The Invasiveness of cancer cell lines depends on the expression of proteins such as MMPs. Thus, we analyzed the invasive potential of the three cell lines by zymography in order to detect changes in the proteolytic activity of matrix metalloproteases MMP2 and MMP9 after transfection with apMNKQ2. As Figure 4b shows, apMNKQ2 decreased MMP9 and MMP2 activity in both A549 and SW900 cells, although the effect was only statistically significant for MMP9 in A549 cells. In H460 cells, however, apMNKQ2 seemed to increase MMP9 activity and did not produce changes in MMP2.

Next, we analyzed whether apMNKQ2 affected the ability of cells to adhere to extracellular matrix components such as type I collagen. For these assays, 24 h after transfection with aptamers, cells were reseeded in plates previously coated with type I collagen in order to measure adherence. Subsequently, MTT assays were performed to quantify the number of adherent cells. Results demonstrated that apMNKQ2 significantly reduced cell adhesion in A549 cells but did not have any effect on SW900 and H460 cells (Figure 4c).

Finally, we studied the effect of apMNKQ2 on epithelial-mesenchymal transition (EMT), in which cancer cells lose epithelial features and acquire a mesenchymal phenotype. For this purpose, cells were transfected with apMNKQ2 and the expression of the epithelial markers E-cadherin and occludin and the mesenchymal marker N-cadherin were analyzed by western blotting. E-cadherin and N-cadherin were only detected in A549 and SW900 cells while occludin was detected in the three cell lines (Supplementary Figure S3). Results showed that apMNKQ2 significantly reduced E-cadherin levels in both A549 and SW900 cells (Figure 4d) and increased occludin levels in H460 cells (Figure 4e). Moreover, apMNKQ2 produced a significant reduction in N-cadherin expression levels in SW900 cells (Figure 4f).

3.4. Effect of apMNKQ2 on MNK1 Isoforms

The expression levels of both MNK1 isoforms were analyzed as well as eIF4E phosphorylation after cell transfection with aptamers. Figure 5a shows that apMNKQ2 increased eIF4E phosphorylation in A549 cells while no changes were observed in SW900 and H460 cells. MNK1a was significantly reduced in the three cell lines after apMNKQ2 transfection (Figure 5b), but MNK1b did not change in any of the three cell lines (Figure 5c). The reduction of MNK1a observed in SW900 cells may be a consequence of a reduction in mRNA levels (Supplementary Figure S4).

3.5. apMNKQ2 Reduces Tumor Growth In Vivo

To investigate the antitumor efficacy of apMNKQ2, we first performed a tolerability study as described in the Materials and Methods section (Section 2). During the course of this experiment, no DLTs were observed at any of the doses tested, including the maximum feasible dose of 400 mg/kg. Thus, for practical purposes, doses of 10 and 25 mg/kg were used in the xenograft studies. Then, A549 cells were subcutaneously xenografted in athymic nude mice, and apMNKQ2 was administered intraperitoneally at 10 mg/kg and 25 mg/kg. Our preliminary results showed that treatment with apMNKQ2 produced a reduction in both tumor volume (Figure 6a) and weight (Figure 6b) reaching a tumor growth inhibition (TGI) of 18.2% and 28.5% for 10 mg/kg and 25 mg/kg of apMNKQ2, respectively. However, these results were not statistically significant, probably due to unexpectedly slow tumor growth and/or the necessity of a higher dose of the aptamer.

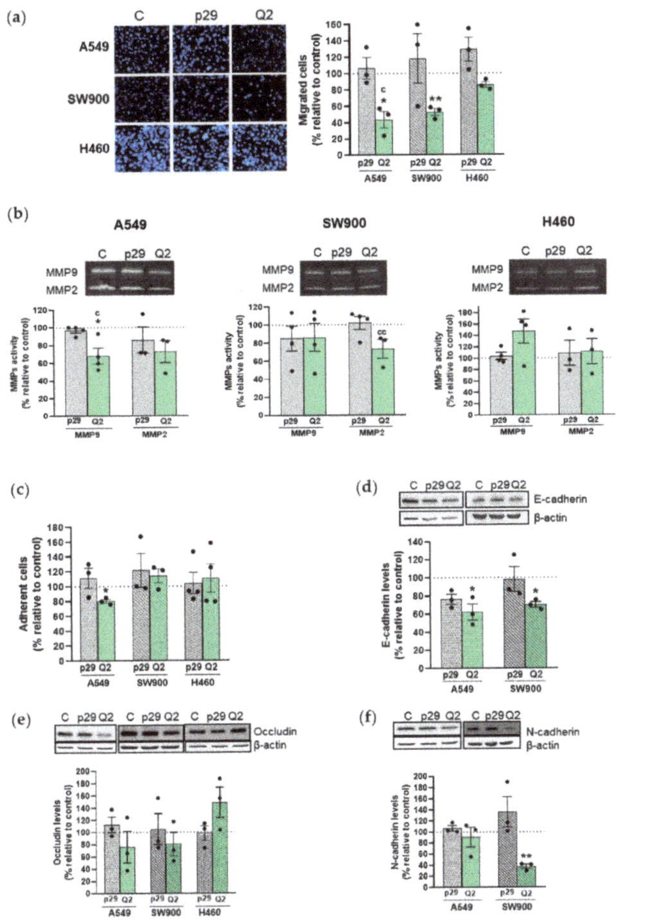

Figure 4. ApMNKQ2-mediated effects on migration, invasion, and EMT in NSCLC cells. (**a**) Cells were transfected with aptamers at IC50, maintained in serum-deprived medium for 16 h, and collected and reseeded at 4×10^4 cells/well in the upper chamber of a transwell. Representative images are shown in the left panel. The graphs represent the mean percentage of migrated cells ± SEM relative to the value of control (C) cells from 3 independent experiments. (**b**) Zymography was performed as described in Materials and Methods. The graphs represent the mean percentage of MMP activity ± SEM relative to the value of control (C) cells from 3–4 independent experiments. (**c**) Cells were transfected with aptamers at IC50 and reseeded in plates previously coated with type I collagen. After binding, cells were quantified by MTT assay. The graphs represent the mean percentage of adherent cells ± SEM relative to the value of control cells from 3 independent experiments. (**d–f**) Western blot and corresponding densitometry analyses of lysates (20 μg) that were subjected to SDS-PAGE (10%). Specific antibodies are shown in Supplementary Table S1. Actin detection was used as a loading control. A representative blot is shown at the top of each graph. Densitometric quantification of E-cadherin, occludin, and N-cadherin was normalized to respective β-actin bands, and levels are expressed as a percentage relative to control cells. The graphs represent mean percentage levels ± SEM of 3 independent experiments. * $p < 0.05$ and ** $p < 0.01$ relative to control cells; ^{c} $p < 0.05$ and ^{cc} $p < 0.01$ relative to p29 control. Control is shown as C, p29 is the unspecific aptamer used as a negative control, and apMNKQ2 is shown as Q2. Each solid black circle represents one individual value.

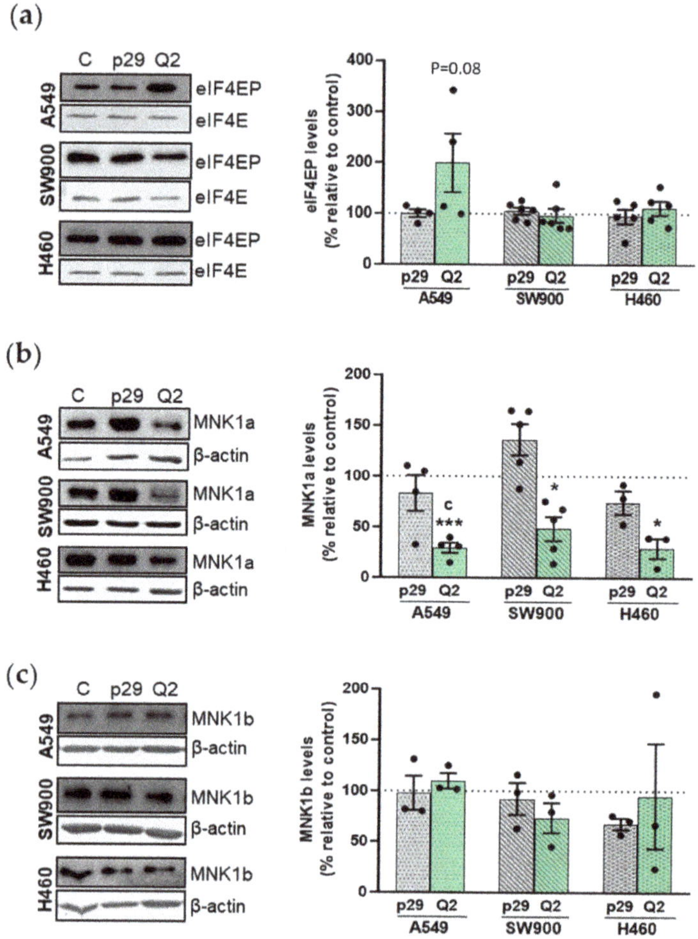

Figure 5. Effect of apMNKQ2 on eIF4E phosphorylation and MNK1 isoforms. Lysates (20 μg) were subjected to SDS-PAGE (12%) and western blotting was performed using specific antibodies (see Supplementary Table S1). Actin detection was used as a loading control. A representative blot and corresponding densitometry analyses are shown in each figure. (**a**) Quantification of eIF4EP was normalized to total eIF4E levels and expressed as a percentage relative to control (C) cells (**b**) and (**c**) quantification of MNK1a/b was normalized to β-actin levels and expressed as a percentage relative to control cells. The graphs represent mean values ± SEM of 3–5 independent experiments. * $p < 0.05$ and *** $p < 0.001$ relative to control cells; ^{c} $p < 0.05$ relative to p29 control. Control is shown as C, p29 is the unspecific aptamer used as a negative control, and apMNKQ2 is shown as Q2. Each solid black circle represents one individual value.

We analyzed aptamer uptake after intraperitoneal administration by qRT-PCR and demonstrate that apMNKQ2 reaches different organs, such as the lung or pancreas but not the brain and is cleared by the kidney and liver (Supplementary Figure S5). In tumors, apMNKQ2 reached the tumor in a dose-dependent manner (Figure 6c) and exerted a biological effect on the tumor, with no toxic macroscopic effects observed in other organs.

In order to determine the association between the anti-tumor effects observed in vivo and the inhibition of MNK1 downstream signaling, XIAP and MCL1 levels were analyzed from tumor samples by western blotting. Results showed a reduction of both proteins in

25 mg/kg apMNKQ2-treated tumor samples (Figure 6d). These results corroborate our in vitro findings of the apoptotic effect of apMNKQ2 in A549 cells.

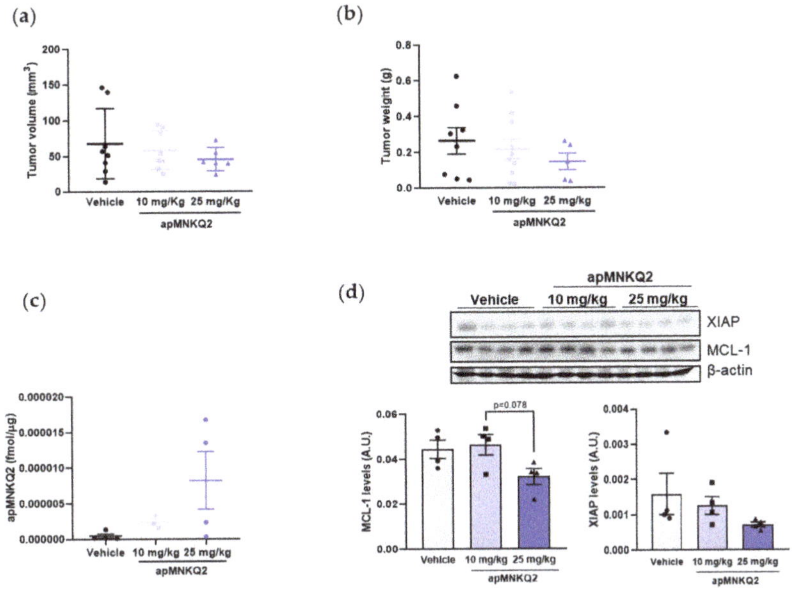

Figure 6. apMNKQ2 efficacy in an adenocarcinoma xenograft model. (**a**,**b**) Reduction in (**a**) volume and (**b**) weight of A549 xenografts following treatment with apMNKQ2 at 10 and 25 mg/kg daily. Mice were treated intraperitoneally with apMNKQ2 or vehicle control (selection buffer) for 4 weeks. Scatter plots represent the mean ± SEM from 6-8 tumors. Solid black circles (vehicle), blue squares (apMNKQ2 at 10 mg/Kg) or blue triangles (apMNKQ2 at 25 mg/Kg) represent one individual value. (**c**) Quantification of apMNKQ2 levels in tumors by qRT-PCR. RNA fraction, where apMNKQ2 is found, was obtained from tumor samples and used in qPCR analyses as described in Material and Methods. Results are expressed as pmol/µg of tissue. Scatter plot represents the mean ± SEM from 3–4 tumors. Each solid circle represents one individual value. (**d**) Proteins were obtained from tumor samples as described in Material and Methods and lysates (20 µg) were subjected to SDS-PAGE (12%). Western blotting was performed using specific antibodies (see Supplementary Table S1). Actin detection was used as a loading control. Blots are shown on the top of the graphs. The bars represent mean protein levels ± SEM, expressed as arbitrary units (A.U.) for 4 tumors from each treatment arm. Each solid black symbol represents one individual value.

4. Discussion

Lung cancer is considered one of the most invasive cancers and the fastest to metastasize, being the leading cause of cancer-related death worldwide and the third most frequently diagnosed in 2022 after colorectal and breast cancer [2]. To date, therapeutic strategies to treat lung cancer, such as chemotherapy, molecular targeted therapy, or immunotherapy, have been developed [43] and have improved the survival of lung cancer patients. Despite these advances, the prognosis for patients with lung cancer remains poor, highlighting the need for new therapeutic approaches.

MNKs are involved in several types of cancer [25] including NSCLC, where MNK1 overexpression correlates with poor overall patient survival [44]. This suggests that blocking the MNK/eIF4E pathway may be a good strategy to treat NSCLC.

Here we studied the antitumor potential of apMNKQ2 as an antitumor agent in NSCLC using in vitro and in vivo assays. The three cell lines used harbor KRAS mutations,

SW900 cells carry an inactivated mutation in the tumor suppressor gene TP53, and H460 cells also harbor PI3KCA mutations [45,46].

One of the most important characteristics of aptamers is their stability and half-life inside cells, and apMNKQ2 is stable inside lung cancer cells for at least 48 h, which may be an advantage for its clinical application in terms of decreasing the number of injections and improving patient comfort [47].

Inhibiting tumorigenic characteristics is an important aspect of the development of antitumor drugs. Consequently, we show that apMNKQ2 reduces the viability in A549 and SW900 cells via apoptosis induction, but to a less extent in the latter. While SW900 cells are not Annexin-V positive after transfection with apMNKQ2, PARP was cleaved. In addition, XIAP and MCL1 are antiapoptotic proteins that are overexpressed in several types of cancers and whose expression is regulated by MNK1 [23,41,42]. Moreover, MCL1 overexpression inhibits apoptosis and promotes cell survival in NSCLC [48]. The apMNKQ2 reduced both XIAP and MCL1 in A549 and SW900 cells both in vitro and in vivo (the latter only for A549) proving its effect through MNK1, and confirming the pro-apoptotic effect of the aptamer. However, despite the fact that other MNK inhibitors [26] promote the induction of apoptosis in H460 cells, apMNKQ2 did not induce it, which may explain the lack of effect on MCL1 and XIAP levels.

Some MNK inhibitors produce a cell cycle arrest in breast cancer cells [19,49]. For example, the MNK1 inhibitor BAY1143269 induces a G0/G1 arrest in NSCLC cells. This is consistent with our results in A549 and H640 cells in which a G1 phase arrest is produced when cells are transfected with apMNKQ2. Surprisingly, the aptamer induces a G2 phase arrest in SW900 cells. This difference among the three cell lines could be explained since both A549 and H460 cells contain a wild-type TP53, while SW900 cells carry an inactivated mutation in TP53.

MNKs also play an important role in metastasis [16], and several MNK inhibitors reduce cell migration and invasion [26,50–53]. Indeed, apMNKQ2 decreased migration and invasion of NSCLC cells. During metastasis, cells migrate and attach to extracellular matrix (ECM) components. apMNKQ2 was also able to reduce cell adhesion capacity in A549 cells. β1 integrin is involved in the adhesion of cells to the surroundings ECM and its inhibition reduces the adhesion of A549 cells by inhibiting the ERK1/2 signaling pathway [54]. This suggests that apMNKQ2 may reduce cell adhesion through MNK1 inhibition. The catalytic subunit p110α mediates β1 integrin-regulated activation of AKT [55], which could contribute to the tumorigenic properties of cells expressing constitutively active p110 α, as in H460 cells.

Similarly, apMNKQ2 alters cell migration and MMP activity only in A549 and SW900 cells. The PI3K/AKT/mTOR pathway plays an important role in lung cancer cell migration and invasion, specifically in the regulation of MMP2 and MMP9 [56]. These data suggest that the increased activity in the PI3K pathway in H460 cells could compensate for the effects of apMNKQ2.

Therefore, apMNKQ2 seems to exert more potent effects in A549 and SW900 cells than in H460 cells probably due to the PI3KCA mutation in the latter. It would be interesting to use apMNKQ2 in combination with inhibitors of the PI3K pathway to increase the sensitivity of H460 cells to apMNKQ2.

apMNKQ2 significantly reduced N-cadherin levels in SW900 cells, which is in accordance with other MNK inhibitors such as VNLG-152 that reverse EMT in prostate cancer [57]. The aptamer also increased the epithelial marker occludin in H460 cells, recovering part of the epithelial characteristics in these cells. Since Snail is a direct repressor of E-cadherin [58], the overexpression of Snail in A549 cells (data not shown) may be associated with low E-cadherin expression in these cells.

Some MNK inhibitors such as MNK-I1, BAY1143269, MNK-7g, eFT508 or novel retinamides inhibits eIF4E phosphorylation [26,51,59–61]. apMNKQ2 also inhibits eIF4E phosphorylation in breast cancer cells [37]; however, in NSCLC cells, apMNKQ2 did not affect eIF4E phosphorylation, which is in accordance with other MNK inhibitors such as apMNK2F and apMNK3R aptamers [36] or ferrocene analogs [62]. Since eIF4E phosphorylation is mainly mediated by MNK2 in breast cancer cells MDA-MB-231 [36], this could also occur in lung cancer cells. Moreover, MNK2b acts as a proto-oncogene increasing eIF4E phosphorylation [63] and in NSCLC, MNK2 is involved in tumorigenesis through eIF4E phosphorylation [21]. To answer this unknown, we propose to use of CRISPR/Cas, which is currently being developed in the laboratory.

Even though eIF4E is the best characterized substrate of MNK1, it is possible that eIF4E-independent effects, through other MNK substrates, may contribute to the observed MNK-mediated effects of apMNKQ2. Here we show that apMNKQ2 exerts its effects through proteins regulated by MNK1 such as XIAP and MCL1, but in the last few years, other proteins regulated by MNK1 have been described such as NODAL [64], NDRG1 [65] or ANGPTL-4 [66], which may be involved in the antitumoral mechanism of apMNKQ2.

We have studied different administration routes of apMNKQ2 in preliminary assays in mice, such as oral, intravenous, and intraperitoneal routes. The latter was the most effective in breast and pancreatic cancer tumor models and was therefore chosen for the in vivo experiments presented herein. We have shown that there is a downward trend in tumor growth after treatment with apMNKQ2. Although tumor growth with xenografted A549 cells was slower than expected, we have obtained promising results that can be improved by increasing the dose of the aptamer. It must be emphasized that in previous studies [37] we used a transfectant as a vehicle to administer apMNKQ2 into animals, while in this study apMNKQ2 was injected without the transfectant, which is an advantage since it allowed us to increase the aptamer dose. Importantly, in the absence of the transfectant, the aptamer still reached the tumor and was easily detectable by qRT-PCR and increased as a function of the dose. Indeed, further studies are still necessary to determine the antitumor mechanism and the full antitumor potential of apMNKQ2. Thus, apMNKQ2 is currently being tested in a patient-derived xenograft (PDX) of NSCLC. This will allow us to learn more about the therapeutic potential of the aptamer in a more physiologically relevant tumor model.

One of the most important points to take into account is the toxicity commonly observed with antitumor chemotherapeutics [67] or immunotherapy and molecular targeted therapy, even though the latter two are better tolerated, side effects can occur [68,69]. Toxicity regarding aptamers is very limited. For example, the aptamers NOX-A12 or ApTOLL have been administered to patients with no signs of toxicity [70]. Moreover, treatment with apMNKQ2 was well tolerated in breast cancer in vivo assays as indicated by no difference in body weight [37] indicating the low toxicity of apMNKQ2. Likewise, daily administration of apMNKQ2 in this study showed no obvious signs of toxicity in mice. Moreover, we performed a maximum tolerated dose (MTD) assay in CD1 mice and no toxicity was observed at any dose, including the highest dose tested (400 mg/kg, maximum feasible dose), indicating that the data obtained so far demonstrates low toxicity for apMNKQ2. However, future ADME Tox are still necessary and are currently planned.

In summary, we show that apMNKQ2 inhibits the tumorigenic and metastatic processes of lung cancer cells, and preliminary in vivo experiments indicate that apMNKQ2 can reduce tumor growth and induce apoptosis in an adenocarcinoma xenograft model. Further studies are still needed in order to elucidate the mechanism of action of apMNKQ2 as well as its pharmacokinetics and pharmacological security to expedite its possible use as a therapeutic tool in the different types of cancers where MNK1 plays an important biological role.

5. Patents

C.P-D, V.M.G. and M.E.M. declare that a patent application has been filed relating to this work.

Supplementary Materials: The following supporting information can be downloaded at: https://www.mdpi.com/article/10.3390/pharmaceutics15041273/s1, Figure S1: Effect of the four aptamers on MTT activity in A549 and SW900 cells; Figure S2: Aptamer effect on the viability of lung cancer cells 24 h after transfection; Figure S3: E-cadherin, N-cadherin, and occludin levels in NSCLC cells; Figure S4: ApMNKQ2 effect in MNK1a/b expression in NSCLC cells; Figure S5. Quantification of apMNKQ2 levels in several organs by qRT-PCR. Table S1: Antibodies used in this study; Table S2: Oligonucleotides used in this study.

Author Contributions: Conceptualization, V.M.G. and M.E.M.; investigation, R.C.-M., C.P.-D., J.I.K.-M., E.P., M.B.-U., M.I.P.-M., M.P.-M., L.R.-C. and S.A.; writing—original draft preparation, R.C.-M.; writing—review and editing, R.C.-M., C.P.-D., B.S.J., V.M.G. and M.E.M., Funding Acquisition, V.M.G. and M.E.M. All authors have read and agreed to the published version of the manuscript.

Funding: R.C.-M. was supported for predoctoral contracts (PEJD 2016-BMD-2145 and 2018-BMD-9201) from the Community of Madrid and grant RTC2019-07227-1. M.E.M. and V.M.G. are researchers from FIBio-HRC supported by Consejeria de Sanidad (CAM). This work was supported by grants RTC2019-07227-1 and PID2019-105417RB-I00, funded by MCIN/AEI/10.13039/501100011033 (Ministry of Economy and Competitiveness, Spain).

Institutional Review Board Statement: Not applicable.

Informed Consent Statement: Not applicable.

Data Availability Statement: Not applicable.

Acknowledgments: We thank Eloisa Jantus and Silvia Calabuig from Fundación para la Investigación del Hospital General Universitario de Valencia (FIHGUV) for kindly providing lung cancer cell lines used in this study.

Conflicts of Interest: The authors declare no conflict of interest.

References

1. Sung, H.; Ferlay, J.; Siegel, R.L.; Laversanne, M.; Soerjomataram, I.; Jemal, A.; Bray, F. Global Cancer Statistics 2020: GLOBOCAN Estimates of Incidence and Mortality Worldwide for 36 Cancers in 185 Countries. *CA Cancer J. Clin.* **2021**, *71*, 209–249. [CrossRef]
2. Li, C.; Wang, H.; Jiang, Y.; Fu, W.; Liu, X.; Zhong, R.; Cheng, B.; Zhu, F.; Xiang, Y.; He, J.; et al. Advances in lung cancer screening and early detection. *Cancer Biol. Med.* **2022**, *19*, 591–608. [CrossRef] [PubMed]
3. Fukunaga, R.; Hunter, T. MNK1, a new MAP kinase-activated protein kinase, isolated by a novel expression screening method for identifying protein kinase substrates. *EMBO J.* **1997**, *16*, 1921–1933. [CrossRef]
4. Waskiewicz, A.J.; Flynn, A.; Proud, C.G.; Cooper, J.A. Mitogen-activated protein kinases activate the serine/threonine kinases Mnk1 and Mnk2. *EMBO J.* **1997**, *16*, 1909–1920. [CrossRef]
5. O'Loghlen, A.; Gonzalez, V.M.; Pineiro, D.; Perez-Morgado, M.I.; Salinas, M.; Martin, M.E. Identification and molecular characterization of Mnk1b, a splice variant of human MAP kinase-interacting kinase Mnk1. *Exp. Cell Res.* **2004**, *299*, 343–355. [CrossRef]
6. O'Loghlen, A.; González, V.M.; Jurado, T.; Salinas, M.; Martín, M.E. Characterization of the activity of human MAP kinase-interacting kinase Mnk1b. *Biochim. Biophys. Acta* **2007**, *1773*, 1416–1427. [CrossRef]
7. Waskiewicz, A.J.; Johnson, J.C.; Penn, B.; Mahalingam, M.; Kimball, S.R.; Cooper, J.A. Phosphorylation of the cap-binding protein eukaryotic translation initiation factor 4E by protein kinase Mnk1 in vivo. *Mol. Cell Biol.* **1999**, *19*, 1871–1880. [CrossRef]
8. Shveygert, M.; Kaiser, C.; Bradrick, S.S.; Gromeier, M. Regulation of eukaryotic initiation factor 4E (eIF4E) phosphorylation by mitogen-activated protein kinase occurs through modulation of Mnk1-eIF4G interaction. *Mol. Cell Biol.* **2010**, *30*, 5160–5167. [CrossRef] [PubMed]
9. Buxade, M.; Parra, J.L.; Rousseau, S.; Shpiro, N.; Marquez, R.; Morrice, N.; Bain, J.; Espel, E.; Proud, C.G. The Mnks are novel components in the control of TNF alpha biosynthesis and phosphorylate and regulate hnRNP A1. *Immunity* **2005**, *23*, 177–189. [CrossRef]
10. Hefner, Y.; Borsch-Haubold, A.G.; Murakami, M.; Wilde, J.I.; Pasquet, S.; Schieltz, D.; Ghomashchi, F.; Yates, J.R., 3rd; Armstrong, C.G.; Paterson, A.; et al. Serine 727 phosphorylation and activation of cytosolic phospholipase A2 by MNK1-related protein kinases. *J. Biol. Chem.* **2000**, *275*, 37542–37551. [CrossRef] [PubMed]

11. DaSilva, J.; Xu, L.; Kim, H.J.; Miller, W.T.; Bar-Sagi, D. Regulation of sprouty stability by Mnk1-dependent phosphorylation. *Mol. Cell Biol.* **2006**, *26*, 1898–1907. [CrossRef] [PubMed]
12. Buxadé, M.; Morrice, N.; Krebs, D.L.; Proud, C.G. The PSF.p54nrb complex is a novel Mnk substrate that binds the mRNA for tumor necrosis factor alpha. *J. Biol. Chem.* **2008**, *283*, 57–65. [CrossRef] [PubMed]
13. Rousseau, D.; Kaspar, R.; Rosenwald, I.; Gehrke, L.; Sonenberg, N. Translation initiation of ornithine decarboxylase and nucleocytoplasmic transport of cyclin D1 mRNA are increased in cells overexpressing eukaryotic initiation factor 4E. *Proc. Natl. Acad. Sci. USA* **1996**, *93*, 1065–1070. [CrossRef]
14. Culjkovic, B.; Topisirovic, I.; Borden, K.L. Controlling gene expression through RNA regulons: The role of the eukaryotic translation initiation factor eIF4E. *Cell Cycle* **2007**, *6*, 65–69. [CrossRef]
15. Furic, L.; Rong, L.; Larsson, O.; Koumakpayi, I.H.; Yoshida, K.; Brueschke, A.; Petroulakis, E.; Robichaud, N.; Pollak, M.; Gaboury, L.A.; et al. eIF4E phosphorylation promotes tumorigenesis and is associated with prostate cancer progression. *Proc. Natl. Acad. Sci. USA* **2010**, *107*, 14134–14139. [CrossRef]
16. Robichaud, N.; del Rincon, S.V.; Huor, B.; Alain, T.; Petruccelli, L.A.; Hearnden, J.; Goncalves, C.; Grotegut, S.; Spruck, C.H.; Furic, L.; et al. Phosphorylation of eIF4E promotes EMT and metastasis via translational control of SNAIL and MMP-3. *Oncogene* **2015**, *34*, 2032–2042. [CrossRef]
17. Robichaud, N.; Hsu, B.E.; Istomine, R.; Alvarez, F.; Blagih, J.; Ma, E.H.; Morales, S.V.; Dai, D.L.; Li, G.; Souleimanova, M.; et al. Translational control in the tumor microenvironment promotes lung metastasis: Phosphorylation of eIF4E in neutrophils. *Proc. Natl. Acad. Sci. USA* **2018**, *115*, E2202–E2209. [CrossRef]
18. Yang, X.; Zhong, W.; Cao, R. Phosphorylation of the mRNA cap-binding protein eIF4E and cancer. *Cell Signal.* **2020**, *73*, 109689. [CrossRef] [PubMed]
19. Wheater, M.J.; Johnson, P.W.; Blaydes, J.P. The role of MNK proteins and eIF4E phosphorylation in breast cancer cell proliferation and survival. *Cancer Biol.* **2010**, *10*, 728–735. [CrossRef]
20. Grzmil, M.; Morin, P., Jr.; Lino, M.M.; Merlo, A.; Frank, S.; Wang, Y.; Moncayo, G.; Hemmings, B.A. MAP kinase-interacting kinase 1 regulates SMAD2-dependent TGF-β signaling pathway in human glioblastoma. *Cancer Res.* **2011**, *71*, 2392–2402. [CrossRef] [PubMed]
21. Guo, Z.; Peng, G.; Li, E.; Xi, S.; Zhang, Y.; Li, Y.; Lin, X.; Li, G.; Wu, Q.; He, J. MAP kinase-interacting serine/threonine kinase 2 promotes proliferation, metastasis, and predicts poor prognosis in non-small cell lung cancer. *Sci. Rep.* **2017**, *7*, 10612. [CrossRef] [PubMed]
22. Hou, S.; Du, P.; Wang, P.; Wang, C.; Liu, P.; Liu, H. Significance of MNK1 in prognostic prediction and chemotherapy development of epithelial ovarian cancer. *Clin. Transl. Oncol.* **2017**, *19*, 1107–1116. [CrossRef]
23. Pinto-Díez, C.; García-Recio, E.M.; Pérez-Morgado, M.I.; García-Hernández, M.; Sanz-Criado, L.; Sacristán, S.; Toledo-Lobo, M.V.; Pérez-Mies, B.; Esteban-Rodríguez, I.; Pascual, A.; et al. Increased expression of MNK1b, the spliced isoform of MNK1, predicts poor prognosis and is associated with triple-negative breast cancer. *Oncotarget* **2018**, *9*, 13501–13516. [CrossRef]
24. Wang, X.; Wang, Y.; Zhang, Q.; Zhuang, H.; Chen, B. MAP Kinase-Interacting Kinase 1 Promotes Proliferation and Invasion of Hepatocellular Carcinoma and Is an Unfavorable Prognostic Biomarker. *Med. Sci. Monit.* **2018**, *24*, 1759–1767. [CrossRef] [PubMed]
25. Pinto-Díez, C.; Ferreras-Martín, R.; Carrión-Marchante, R.; González, V.M.; Martín, M.E. Deeping in the Role of the MAP-Kinases Interacting Kinases (MNKs) in Cancer. *Int. J. Mol. Sci.* **2020**, *21*, 2967. [CrossRef]
26. Santag, S.; Siegel, F.; Wengner, A.M.; Lange, C.; Bömer, U.; Eis, K.; Pühler, F.; Lienau, P.; Bergemann, L.; Michels, M.; et al. BAY 1143269, a novel MNK1 inhibitor, targets oncogenic protein expression and shows potent anti-tumor activity. *Cancer Lett.* **2017**, *390*, 21–29. [CrossRef]
27. Reich, S.H.; Sprengeler, P.A.; Chiang, G.G.; Appleman, J.R.; Chen, J.; Clarine, J.; Eam, B.; Ernst, J.T.; Han, Q.; Goel, V.K.; et al. Structure-Based Design of Pyridone-Aminal eFT508 Targeting Dysregulated Translation by Selective Mitogen-Activated Protein Kinase Interacting Kinases 1 and 2 (MNK1/2) Inhibition. *J. Med. Chem.* **2018**, *61*, 3516–3540. [CrossRef]
28. Xu, W.; Kannan, S.; Verma, C.S.; Nacro, K. Update on the Development of MNK Inhibitors as Therapeutic Agents. *J. Med. Chem.* **2022**, *65*, 983–1007. [CrossRef]
29. Ueda, T.; Watanabe-Fukunaga, R.; Fukuyama, H.; Nagata, S.; Fukunaga, R. Mnk2 and Mnk1 are essential for constitutive and inducible phosphorylation of eukaryotic initiation factor 4E but not for cell growth or development. *Mol. Cell Biol.* **2004**, *24*, 6539–6549. [CrossRef] [PubMed]
30. Ellington, A.D.; Szostak, J.W. In vitro selection of RNA molecules that bind specific ligands. *Nature* **1990**, *346*, 818–822. [CrossRef]
31. Tuerk, C.; Gold, L. Systematic evolution of ligands by exponential enrichment: RNA ligands to bacteriophage T4 DNA polymerase. *Science* **1990**, *249*, 505–510. [CrossRef] [PubMed]
32. Zhou, J.; Rossi, J. Aptamers as targeted therapeutics: Current potential and challenges. *Nat. Rev. Drug Discov.* **2017**, *16*, 181–202. [CrossRef]
33. Ng, E.W.; Shima, D.T.; Calias, P.; Cunningham, E.T., Jr.; Guyer, D.R.; Adamis, A.P. Pegaptanib, a targeted anti-VEGF aptamer for ocular vascular disease. *Nat. Rev. Drug Discov.* **2006**, *5*, 123–132. [CrossRef]
34. Zhang, N.; Chen, Z.; Liu, D.; Jiang, H.; Zhang, Z.K.; Lu, A.; Zhang, B.T.; Yu, Y.; Zhang, G. Structural Biology for the Molecular Insight between Aptamers and Target Proteins. *Int. J. Mol. Sci.* **2021**, *22*, 4093. [CrossRef]
35. Fu, Z.; Xiang, J. Aptamers, the Nucleic Acid Antibodies, in Cancer Therapy. *Int. J. Mol. Sci.* **2020**, *21*, 2793. [CrossRef] [PubMed]

36. García-Recio, E.M.; Pinto-Díez, C.; Pérez-Morgado, M.I.; García-Hernández, M.; Fernández, G.; Martín, M.E.; González, V.M. Characterization of MNK1b DNA Aptamers That Inhibit Proliferation in MDA-MB231 Breast Cancer Cells. *Mol. Nucleic Acids* **2016**, *5*, e275. [CrossRef]
37. Pinto-Díez, C.; Ferreras-Martín, R.; Carrión-Marchante, R.; Klett-Mingo, J.I.; García-Hernández, M.; Pérez-Morgado, M.I.; Sacristán, S.; Barragán, M.; Seijo-Vila, M.; Tundidor, I.; et al. An optimized MNK1b aptamer, apMNKQ2, and its potential use as a therapeutic agent in breast cancer. *Mol. Ther. Nucleic Acids* **2022**, *30*, 553–568. [CrossRef]
38. Huang, X.; Biswas, S.; Oki, Y.; Issa, J.P.; Berry, D.A. A parallel phase I/II clinical trial design for combination therapies. *Biometrics* **2007**, *63*, 429–436. [CrossRef]
39. Le Tourneau, C.; Lee, J.J.; Siu, L.L. Dose escalation methods in phase I cancer clinical trials. *J. Natl. Cancer Inst.* **2009**, *101*, 708–720. [CrossRef]
40. Frank, K.J.; Mulero-Sanchez, A.; Berninger, A.; Ruiz-Canas, L.; Bosma, A.; Gorgulu, K.; Wu, N.; Diakopoulos, K.N.; Kaya-Aksoy, E.; Ruess, D.A.; et al. Extensive preclinical validation of combined RMC-4550 and LY3214996 supports clinical investigation for KRAS mutant pancreatic cancer. *Cell Rep. Med.* **2022**, *3*, 100815. [CrossRef] [PubMed]
41. Wendel, H.G.; Silva, R.L.; Malina, A.; Mills, J.R.; Zhu, H.; Ueda, T.; Watanabe-Fukunaga, R.; Fukunaga, R.; Teruya-Feldstein, J.; Pelletier, J.; et al. Dissecting eIF4E action in tumorigenesis. *Genes Dev.* **2007**, *21*, 3232–3237. [CrossRef] [PubMed]
42. Evans, M.K.; Brown, M.C.; Geradts, J.; Bao, X.; Robinson, T.J.; Jolly, M.K.; Vermeulen, P.B.; Palmer, G.M.; Gromeier, M.; Levine, H.; et al. XIAP Regulation by MNK Links MAPK and NFκB Signaling to Determine an Aggressive Breast Cancer Phenotype. *Cancer Res.* **2018**, *78*, 1726–1738. [CrossRef]
43. Alexander, M.; Kim, S.Y.; Cheng, H. Update 2020: Management of Non-Small Cell Lung Cancer. *Lung* **2020**, *198*, 897–907. [CrossRef] [PubMed]
44. Wen, Q.; Wang, W.; Luo, J.; Chu, S.; Chen, L.; Xu, L.; Zang, H.; Alnemah, M.M.; Ma, J.; Fan, S. CGP57380 enhances efficacy of RAD001 in non-small cell lung cancer through abrogating mTOR inhibition-induced phosphorylation of eIF4E and activating mitochondrial apoptotic pathway. *Oncotarget* **2016**, *7*, 27787–27801. [CrossRef]
45. Blanco, R.; Iwakawa, R.; Tang, M.; Kohno, T.; Angulo, B.; Pio, R.; Montuenga, L.M.; Minna, J.D.; Yokota, J.; Sanchez-Cespedes, M. A gene-alteration profile of human lung cancer cell lines. *Hum. Mutat.* **2009**, *30*, 1199–1206. [CrossRef] [PubMed]
46. Korrodi-Gregório, L.; Soto-Cerrato, V.; Vitorino, R.; Fardilha, M.; Pérez-Tomás, R. From Proteomic Analysis to Potential Therapeutic Targets: Functional Profile of Two Lung Cancer Cell Lines, A549 and SW900, Widely Studied in Pre-Clinical Research. *PLoS ONE* **2016**, *11*, e0165973. [CrossRef]
47. Zhang, Y.; Zhang, H.; Chan, D.W.H.; Ma, Y.; Lu, A.; Yu, S.; Zhang, B.; Zhang, G. Strategies for developing long-lasting therapeutic nucleic acid aptamer targeting circulating protein: The present and the future. *Front. Cell Dev. Biol.* **2022**, *10*, 1048148. [CrossRef]
48. Zhang, H.; Guttikonda, S.; Roberts, L.; Uziel, T.; Semizarov, D.; Elmore, S.W.; Leverson, J.D.; Lam, L.T. Mcl-1 is critical for survival in a subgroup of non-small-cell lung cancer cell lines. *Oncogene* **2011**, *30*, 1963–1968. [CrossRef] [PubMed]
49. Abdelaziz, A.M.; Basnet, S.K.C.; Islam, S.; Li, M.; Tadesse, S.; Albrecht, H.; Gerber, C.; Yu, M.; Wang, S. Synthesis and evaluation of 2′H-spiro[cyclohexane-1,3′-imidazo[1,5-a]pyridine]-1′,5′-dione derivatives as Mnk inhibitors. *Bioorg. Med. Chem. Lett.* **2019**, *29*, 2650–2654. [CrossRef] [PubMed]
50. Ramalingam, S.; Gediya, L.; Kwegyir-Afful, A.K.; Ramamurthy, V.P.; Purushottamachar, P.; Mbatia, H.; Njar, V.C. First MNKs degrading agents block phosphorylation of eIF4E, induce apoptosis, inhibit cell growth, migration and invasion in triple negative and Her2-overexpressing breast cancer cell lines. *Oncotarget* **2014**, *5*, 530–543. [CrossRef]
51. Beggs, J.E.; Tian, S.; Jones, G.G.; Xie, J.; Iadevaia, V.; Jenei, V.; Thomas, G.; Proud, C.G. The MAP kinase-interacting kinases regulate cell migration, vimentin expression and eIF4E/CYFIP1 binding. *Biochem. J.* **2015**, *467*, 63–76. [CrossRef]
52. Kwegyir-Afful, A.K.; Bruno, R.D.; Purushottamachar, P.; Murigi, F.N.; Njar, V.C. Galeterone and VNPT55 disrupt Mnk-eIF4E to inhibit prostate cancer cell migration and invasion. *Febs J.* **2016**, *283*, 3898–3918. [CrossRef] [PubMed]
53. Zhan, Y.; Guo, J.; Yang, W.; Goncalves, C.; Rzymski, T.; Dreas, A.; Żyłkiewicz, E.; Mikulski, M.; Brzózka, K.; Golas, A.; et al. MNK1/2 inhibition limits oncogenicity and metastasis of KIT-mutant melanoma. *J. Clin. Investig.* **2017**, *127*, 4179–4192. [CrossRef] [PubMed]
54. Guo, L.; Zhang, F.; Cai, Y.; Liu, T. Expression profiling of integrins in lung cancer cells. *Pathol. Res. Pr.* **2009**, *205*, 847–853. [CrossRef]
55. Zeller, K.S.; Idevall-Hagren, O.; Stefansson, A.; Velling, T.; Jackson, S.P.; Downward, J.; Tengholm, A.; Johansson, S. PI3-kinase p110α mediates β1 integrin-induced Akt activation and membrane protrusion during cell attachment and initial spreading. *Cell Signal.* **2010**, *22*, 1838–1848. [CrossRef] [PubMed]
56. Iksen; Pothongsrisit, S.; Pongrakhananon, V. Targeting the PI3K/AKT/mTOR Signaling Pathway in Lung Cancer: An Update Regarding Potential Drugs and Natural Products. *Molecules* **2021**, *26*, 4100. [CrossRef]
57. Ramamurthy, V.P.; Ramalingam, S.; Gediya, L.K.; Njar, V.C.O. The retinamide VNLG-152 inhibits f-AR/AR-V7 and MNK-eIF4E signaling pathways to suppress EMT and castration-resistant prostate cancer xenograft growth. *FEBS J.* **2018**, *285*, 1051–1063. [CrossRef] [PubMed]
58. Puisieux, A.; Brabletz, T.; Caramel, J. Oncogenic roles of EMT-inducing transcription factors. *Nat. Cell Biol.* **2014**, *16*, 488–494. [CrossRef]

59. Ramalingam, S.; Ramamurthy, V.P.; Gediya, L.K.; Murigi, F.N.; Purushottamachar, P.; Huang, W.; Choi, E.Y.; Zhang, Y.; Vasaitis, T.S.; Kane, M.A.; et al. The Novel Mnk1/2 Degrader and Apoptosis Inducer VNLG-152 Potently Inhibits TNBC Tumor Growth and Metastasis. *Cancers* **2019**, *11*, 299. [CrossRef] [PubMed]
60. Jin, X.; Merrett, J.; Tong, S.; Flower, B.; Xie, J.; Yu, R.; Tian, S.; Gao, L.; Zhao, J.; Wang, X.; et al. Design, synthesis and activity of Mnk1 and Mnk2 selective inhibitors containing thieno[2,3-d]pyrimidine scaffold. *Eur. J. Med. Chem.* **2019**, *162*, 735–751. [CrossRef]
61. Suarez, M.; Blyth, G.T.; Mina, A.A.; Kosciuczuk, E.M.; Dolniak, B.; Dinner, S.; Altman, J.K.; Eklund, E.A.; Saleiro, D.; Beauchamp, E.M.; et al. Inhibitory effects of Tomivosertib in acute myeloid leukemia. *Oncotarget* **2021**, *12*, 955–966. [CrossRef] [PubMed]
62. Sansook, S.; Lineham, E.; Hassell-Hart, S.; Tizzard, G.J.; Coles, S.J.; Spencer, J.; Morley, S.J. Probing the Anticancer Action of Novel Ferrocene Analogues of MNK Inhibitors. *Molecules* **2018**, *23*, 2126. [CrossRef]
63. Maimon, A.; Mogilevsky, M.; Shilo, A.; Golan-Gerstl, R.; Obiedat, A.; Ben-Hur, V.; Lebenthal-Loinger, I.; Stein, I.; Reich, R.; Beenstock, J.; et al. Mnk2 alternative splicing modulates the p38-MAPK pathway and impacts Ras-induced transformation. *Cell Rep.* **2014**, *7*, 501–513. [CrossRef]
64. Guo, Q.; Li, V.Z.; Nichol, J.N.; Huang, F.; Yang, W.; Preston, S.E.J.; Talat, Z.; Lefrère, H.; Yu, H.; Zhang, G.; et al. MNK1/NODAL Signaling Promotes Invasive Progression of Breast Ductal Carcinoma In Situ. *Cancer Res.* **2019**, *79*, 1646–1657. [CrossRef]
65. Tian, S.; Wang, X.; Proud, C.G. Oncogenic MNK signalling regulates the metastasis suppressor NDRG1. *Oncotarget* **2017**, *8*, 46121–46135. [CrossRef] [PubMed]
66. Yang, W.; Khoury, E.; Guo, Q.; Prabhu, S.A.; Emond, A.; Huang, F.; Gonçalves, C.; Zhan, Y.; Plourde, D.; Nichol, J.N.; et al. MNK1 signaling induces an ANGPTL4-mediated gene signature to drive melanoma progression. *Oncogene* **2020**, *39*, 3650–3665. [CrossRef]
67. Rossi, A.; Di Maio, M. Platinum-based chemotherapy in advanced non-small-cell lung cancer: Optimal number of treatment cycles. *Expert Rev. Anticancer* **2016**, *16*, 653–660. [CrossRef] [PubMed]
68. Barquín-García, A.; Molina-Cerrillo, J.; Garrido, P.; Garcia-Palos, D.; Carrato, A.; Alonso-Gordoa, T. New oncologic emergencies: What is there to know about inmunotherapy and its potential side effects? *Eur. J. Intern. Med.* **2019**, *66*, 1–8. [CrossRef]
69. Hsu, W.H.; Yang, J.C.; Mok, T.S.; Loong, H.H. Overview of current systemic management of EGFR-mutant NSCLC. *Ann. Oncol.* **2018**, *29*, i3–i9. [CrossRef]
70. Steurer, M.; Montillo, M.; Scarfò, L.; Mauro, F.R.; Andel, J.; Wildner, S.; Trentin, L.; Janssens, A.; Burgstaller, S.; Frömming, A.; et al. Olaptesed pegol (NOX-A12) with bendamustine and rituximab: A phase IIa study in patients with relapsed/refractory chronic lymphocytic leukemia. *Haematologica* **2019**, *104*, 2053–2060. [CrossRef]

pharmaceutics

MDPI

Review

Lights and Shadows on the Cancer Multi-Target Inhibitor Rigosertib (ON-01910.Na)

Ana Monfort-Vengut and Guillermo de Cárcer *

Cell Cycle and Cancer Biomarkers Group, Instituto de Investigaciones Biomédicas Alberto Sols (IIBM) CSIC-UAM, 28029 Madrid, Spain
* Correspondence: gdecarcer@iib.uam.es; Tel.: +34-91-585-4426

Abstract: Rigosertib (ON-01910.Na) is a small-molecule member of the novel synthetic benzyl-styryl-sulfonate family. It is currently in phase III clinical trials for several myelodysplastic syndromes and leukemias and is therefore close to clinical translation. The clinical progress of rigosertib has been hampered by a lack of understanding of its mechanism of action, as it is currently considered a multi-target inhibitor. Rigosertib was first described as an inhibitor of the mitotic master regulator Polo-like kinase 1 (Plk1). However, in recent years, some studies have shown that rigosertib may also interact with the PI3K/Akt pathway, act as a Ras–Raf binding mimetic (altering the Ras signaling pathway), as a microtubule destabilizing agent, or as an activator of a stress-induced phospho-regulatory circuit that ultimately hyperphosphorylates and inactivates Ras signaling effectors. Understanding the mechanism of action of rigosertib has potential clinical implications worth exploring, as it may help to tailor cancer therapies and improve patient outcomes.

Keywords: rigosertib; ON-01910.Na; cancer therapy; kinase allosteric inhibition; Plk1; Ras; JNK; PI3K/Akt; microtubule-destabilizing agents; stress-signaling pathways

Citation: Monfort-Vengut, A.; de Cárcer, G. Lights and Shadows on the Cancer Multi-Target Inhibitor Rigosertib (ON-01910.Na). *Pharmaceutics* **2023**, *15*, 1232. https://doi.org/10.3390/pharmaceutics15041232

Academic Editor: Francesca Musumeci

Received: 27 February 2023
Revised: 11 April 2023
Accepted: 11 April 2023
Published: 13 April 2023

1. Introduction

Rigosertib (ON-01910.Na) is a non-cyclooxygenase inhibitor sulfone metabolite with the ability to act as a non-ATP competitor kinase inhibitor. Originally described as an inhibitor for the master mitotic kinase Polo-like kinase (Plk1) [1], rigosertib has been reported to have broad activity against a number of different targets, including kinases such as PI3K/Akt or NF-kB complex proteins [2,3]. Interestingly, rigosertib has also been described as a Ras mimetic molecule, inhibiting Ras–Raf binding and, therefore, hampering Ras mitogenic signaling [4]. Recent data from CRISPR genome-wide screens describe rigosertib as a microtubule-depolymerizing agent that leads to cell-cycle alterations that are typical of other microtubule dynamic poisons [5]. It has also been proposed that the induced mitotic stress, upon rigosertib treatment, leads to oxidative stress resulting in a JNK-mediated inhibition of the Ras–MAPK pathway [6]. Overall, rigosertib is highly effective in killing tumoral cells in vitro and in vivo, with cells arrested in mitosis being the most prominent phenotype, although there is abundant data showing that rigosertib effectively stops cell proliferation through various combined effects.

Based on the promising in vitro data, rigosertib was subjected to clinical trials and showed good tolerability in patients with low levels of toxicity. As a result, rigosertib reached phase III trials for second-line treatment of patients with high-risk myelodysplastic syndrome (MDS) [7] and, in combination with gemcitabine, for patients with metastatic pancreatic cancer [8]. Unfortunately, these clinical trials were not successful, as the patients did not show a significant benefit over the standard of care, suggesting a need to find the right sensitivity biomarkers to achieve a therapeutic benefit. In summary, the multi-target activity of rigosertib, which is controversial, makes this drug an interesting molecule to study in order to deeply understand its mechanism of action and, subsequently, to find the biomarkers that can make rigosertib a profitable drug in the treatment of diseases.

This review collects data on the different targets and cell phenotypes associated with rigosertib response and attempts to clarify the targeting controversy raised in recent years. We also discuss some of the previously described mechanisms of resistance to rigosertib and, finally, summarize all the rigosertib clinical trials conducted to date. Overall, our aim is to help the growing number of researchers working with this small compound.

2. Development of Rigosertib (ON-01910.Na)

One of the main problems associated with kinase-inhibition-based strategies is the development of drug resistance due to the accumulation of mutations in the ATP binding site of the kinase [9]. The scientific community has focused its efforts on developing inhibitors that target regions outside the ATP binding site of the target kinases and may act as better drug candidates in cancer therapy.

Between 2002 and 2003, Reddy et al. described the synthesis of several small-molecule kinase inhibitors, based on styryl-benzyl-sulfones, that exhibited high antitumor activity and were not ATP-competitors, but allosterically inhibited substrate binding [10–12]. Since the cytotoxic activity depends on the nature and position of the styryl aromatic ring, the group synthesized a collection of molecules capable of targeting cancer cells while leaving non-malignant cells unaffected, thus lacking many side effects associated with commonly used chemotherapeutic agents. This advantage opened a therapeutic opportunity window for this collection of new small molecules against cancer.

One of the most analyzed styryl-benzyl-sulfones is rigosertib (ON-01910), commercially known as Estybon®. Rigosertib was derived from ON-01940 ((E)-2′,4′,6′-trimethoxystyryl-4-methoxy-3-aminobenzylsulfone) in a two-step synthetic reaction (Figure 1): first, ON-01940 is reacted with a methyl 2-bromo acetate in a mild basic sodium acetate media to generate ON-01500 ((E)-2-methoxy-5-(((2,4,6-trimethoxystyryl)sulfonyl)methyl)aniline). Second, ON-01500 is hydrolyzed with sodium hydroxide in ethanol and dichloromethane, followed by washing with methyl ethyl ketone to obtain ON-01910 (rigosertib, sodium (E)-2-{2-methoxy-5-[(2′,4′,6′-trimethoxystyrylsulfonyl)methyl]phenylamino}acetate [13]. After screening these compounds for activity against tumoral cell lines, the researchers found ON-01910 to be a potent inhibitor of the mitotic kinase Plk1, inducing mitotic arrest of tumor cells. It is worth noting that rigosertib also showed certain affinity to other kinases, such as PDGFR, Abl, Flt-1, CDK1, Plk2, Src, and Fyn, which share common features in their binding sites [1].

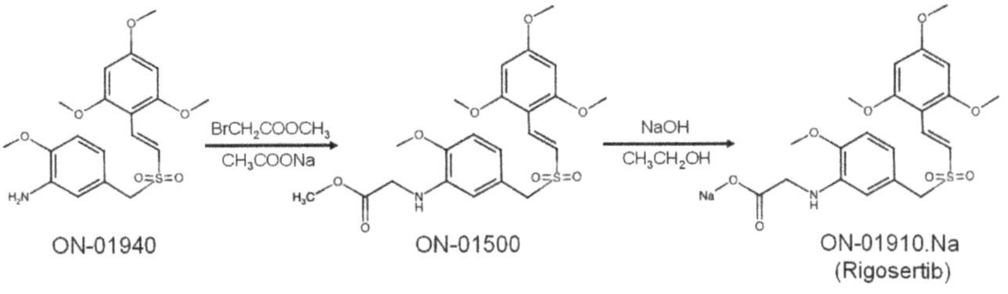

Figure 1. Scheme of development of the rigosertib molecule. Images generated using (chemaxon.com/marvin (accessed on 25 January 2023).

After rigosertib was first described, research began to flourish and many studies have provided new insights into this small molecule, describing new mechanisms and putative targets such as the PI3K–Akt pathway [14] the Ras–Raf signaling cascade [4], and microtubule dynamics [5] (Figure 2).

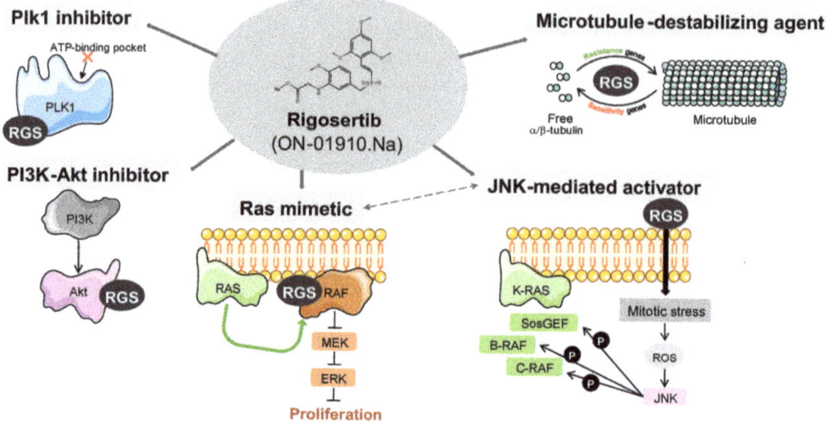

Figure 2. Diagram showing the described rigosertib targets and mechanisms of action. Rigosertib was first described as an allosteric Plk1 inhibitor [1]. Other data show that rigosertib can inhibit the PI3K–Akt signaling axis [2,3]. Rigosertib is also defined as a Ras–Raf binding competitor, altering the Ras–MAPK signaling cascade [4]. This Ras–Raf signaling impairment might be due to altering the response of stress kinases such as JNK [6]. Finally, using CRISPR genome-wide screens, rigosertib is also identified as a microtubule destabilizing agent [5].

3. Described Targets for Rigosertib

Over the years, many studies have been published in attempts to unravel the mechanism of action of rigosertib, to the point where it is now considered a multi-kinase inhibitor. In recent years, rigosertib has also been described as a non-kinase drug, further complicating the understanding of its mechanism of action. The main targets described are the following:

3.1. Rigosertib as a Plk1 Inhibitor

Rigosertib has been generally considered a Plk1 inhibitor. Pioneering phosphorylation experiments performed immediately after rigosertib synthesis, combining recombinant Plk1 and CDC25 or Casein as substrates, showed a very efficient rigosertib-mediated Plk1 inhibition with an IC_{50} of 9–10 nM [1]. In parallel, the chemical mechanism of action of rigosertib was defined as allosteric, since increasing concentrations of ATP in the kinase reaction did not alter the inhibition constant. Notably, the allosteric binding of rigosertib to Plk1 has not yet been demonstrated [15].

To further evaluate the specificity of rigosertib for Plk1, the authors evaluated the activity of a panel of 29 recombinant kinases and showed that rigosertib was most active against Plk1 and, therefore, likely to be its primary target. However, rigosertib also showed some activity against other kinases, such as PDGF receptor (PDGFR), Abl, Flt-1, CDK1, Plk2, and Src, when tested in recombinant kinase assays. Similarly, when the authors performed cell synchronization assays and tested for Plk1 activation, they showed that rigosertib inhibited Plk1 activity, although total levels of the kinase were not affected. This effect of rigosertib seemed to be specific for Plk1, since there is almost no alteration in CDK1 activity, although other kinases were not tested in these experiments. Since Plk1 is required for tumor cell proliferation, the authors showed that rigosertib induced apoptotic activity against 94 tumor cell lines with a GI_{50} between 50–200 nM, resulting in tumor cell death by mitotic arrest, demonstrating its potential efficacy. Notably, this cytotoxic effect was not observed in non-tumoral cells. More recently, in an effort to evaluate the efficacy of rigosertib in retinoblastoma-derived cells, Ma et al. showed suppression of Cdc25C phosphorylation as a readout for Plk1 activity and concomitant accumulation of CDK1 phosphorylation [16]. They also performed flow cytometry cell-cycle analysis and demonstrated that upon rigosertib treatment there was an arrest in the G2/M phase of the cell cycle, although it was not as strong as that identified in other reports.

Polo-like kinase 1 is a master regulator of cell division, controlling several important processes throughout the cell cycle, including centrosome maturation, spindle assembly, chromosome segregation, and, ultimately, cytokinesis [17]. The chemical inhibition or genetic depletion of Plk1 results in strong mitotic arrest and efficient cell death, making it a bona fide cancer target [18]. Importantly, Plk1 inhibition produces a highly recognizable phenotype, which is the presence of monopolar mitotic spindles [19,20] leading to impaired mitotic progression and subsequent mitotic arrest.

To confirm Plk1 as a substrate for rigosertib, confocal microscopy experiments were performed to examine the cell fate of treated cells. Rigosertib-treated HeLa cells exhibit mitotic spindle aberrations, such as multipolar spindles, chromosome mislocalization in metaphase cells, and centrosome fragmentation [1]. Although, as mentioned above, the chemical inhibition of Plk1 is known to induce mitotic arrest, the observed phenotypes in the HeLa-treated cells did not accurately reflect the classical Plk1 inhibition phenotype, which consists of monopolar spindle organization due to impaired centrosome duplication [21]. Indeed, other studies have shown that rigosertib-treated U2OS do not show alterations in centrosome maturation, a well-known function for Plk1, and present increased levels of gamma-tubulin loading at the centrosomes [22].

As the Plk1 inhibition phenotype is controversial, there are other possible explanations for the mitotic arrest observed with rigosertib treatment. Oussenko et al. described how rigosertib induces a prolonged phosphorylation of RanGAP1-SUMO1, thereby arresting cells in mitosis and ultimately inducing apoptosis in acute lymphoblastic leukemia (MOLT-3), prostate cancer (DU-145), and lymphoma cells (U937) [23]. As it is well known that RanGAP1 is phosphorylated at the onset of mitosis [24], it is difficult to determine whether RanGAP1-SUMO1 phosphorylation is directly due to rigosertib function or, on the other hand, a secondary effect of the mitotic arrest mediated by rigosertib. The authors claimed that RanGAP1-SUMO1 phosphorylation happens in short time after adding rigosertib, thus not having enough time to arrest in mitosis.

In summary, although there is still a tendency to use rigosertib as a Plk1 inhibitor [16,25–27], the Plk1 inhibition mechanism for rigosertib is controversial, as there is strong evidence that this may not be the real mechanism and some alternative mechanisms have been described for the mitotic arrest phenotype.

3.2. *Rigosertib as a PI3K–Akt Pathway Inhibitor*

Although most of the initial studies pointed to rigosertib as a Plk1 inhibitor, or more generally as a mitotic inhibitor [14,28–33], some reports additionally demonstrated that it disrupted other non-cell-cycle-related signaling pathways. Xu et al. hypothesized that since rigosertib was being used as a kinase inhibitor, it was worth studying the dysregulation of signaling transduction pathways after rigosertib treatment; they tested samples from myelodysplastic syndrome (MDS) patients and found 31 significant upregulated pathways and 14 downregulated pathways. The PI3K–Akt, MAPK, Jak-STAT, Wnt, and Notch signaling pathways were among the most dysregulated in MDS patients, which could point to apoptotic cell death regulation [14].

These phenotypes were also observed in cell lines derived from head and neck squamous cell carcinoma (HNSCC), where PI3K dysregulation is a common event. Authors demonstrated that rigosertib induced toxicity in HNSCC cell lines, inhibiting PI3K/Akt/mTOR pathways in a dose-dependent manner, which ultimately led to the inhibition of cell-cycle progression. It also induced oxidative stress that generated reactive oxygen species (ROS) and activated extracellular-regulated kinases (ERK1/2) and c-Jun NH2-terminal kinase (JNK). They demonstrated that rigosertib triggered the mitochondria translocation of the stress kinase ATF-2, which is also related to apoptosis induction [3]. In contrast, another study done in HNSCC found rigosertib to be an effective treatment, but by a completely different mechanism than PI3K–Akt inhibition [2]. Here, the authors did not observe any alterations in the PI3K pathway, such as the phosphorylation on Akt Ser-473 residue, or Ser-240/244 of the S6 ribosomal protein. On the contrary, a complete G2 arrest is observed, which fits with rigosertib being a mitotic inhibitor.

Overall, the possible inhibition of the PI3K–Akt axis by rigosertib seems to be cell-dependent and often accompanied by mitotic arrest. Therefore, it is difficult to interpret if both mechanisms are autonomous.

3.3. *Rigosertib as a Ras Mimetic and Ras–Raf–MEK Axis Inhbitor*

Up to this point, observations suggested that rigosertib's mechanism of action involved several alternatives and appeared to be dependent on cell type and genetic background. In 2016, the laboratory of E. Premkumar Reddy described rigosertib as the first Ras–RBD inhibitor, binding to the RBD (Ras-binding domain) of the Ras effectors and impeding Ras binding, therefore blocking the activation of Raf kinase activity and inhibiting Ras–Raf–MEK signaling [4]. The binding of rigosertib to the Raf–RBD domain was elegantly demonstrated by NMR studies, further showing that rigosertib does indeed impede Raf heterodimerization. Notably, the authors claimed that rigosertib specifically inhibits the Ser-388 residue phosphorylation of c-Raf that associates with Plk1 [34], and this may explain the relationship between rigosertib and Plk1 inhibition. Likewise, other studies showed that rigosertib treatment induces cell-cycle arrest and disrupts Ras signaling [35,36]. Finally, the authors also found that rigosertib can bind to the RBDs of Ral-GDS and PI3Ks, which might explain the inhibitory mechanism on PI3K and, therefore, its multi-target capacity.

On the other hand, Ritt et al. demonstrated that rigosertib leads to an inhibition of the Ras/Raf/MEK/ERK signaling axis, but in an indirect way. Rigosertib-induced mitotic arrest generates an oxidative stress that activates the stress cascade mediated by the cJun N-terminal kinases 1/2 (JNK1/2). This stress signaling hyperphosphorylates RasGEF, Sos1, and the Raf proteins, rendering them unresponsive to upstream signals [6]. Notably, this effect of rigosertib can be mimicked by other microtubule poisons such as taxol, suggesting that mitotic arrest is the main inducer of such a stress response, rather than a specific targeting of Ras–MEK signaling by rigosertib. In the same trend, Günther et al. provided more detail on the JNK-mediated mechanism of rigosertib, showing that rigosertib treatment activates p66Shc [37], which is a known effector of the JNK1/2 kinases [38]. Subsequently, p66Shc translocates to the mitochondria, where it induces ROS production [39].

Following the description of the Ras–Raf inhibitory activity of rigosertib, there has been a growing collection of articles exploiting this evidence. Rahmani et al. investigated the mechanism of rigosertib in colorectal cancer and confirmed that it inhibited cell proliferation and the cell cycle and that it was dependent on mutations in K-Ras [40], which also elevates ROS. More interestingly, they showed that rigosertib regulates the expression of CD31, an angiogenic marker, which opens an alternative for rigosertib as an antiangiogenic drug.

Rigosertib was also used as a canonical Ras–Raf signaling inhibitor with the intention to evaluate the impact of Ras signaling during the immune checkpoint blockade (ICB) in metastatic melanoma. The efficacy of rigosertib in melanoma is dependent on the activity of CD40, a receptor molecule of the TNF receptor family that is normally expressed on antigen-presenting cells [41]. Rigosertib Ras/Raf/PI3K inhibition promotes CD40 expression that induces immunogenic cell death and enhances the response to immune checkpoint inhibitors, thus conferring immunogenicity to rigosertib [42]. Whether this effect on CD40 is due to Ras–Raf signaling or to the dependence of CD40 function upon ROS induction [43,44] is not yet clear. Finally, in a non-cancer related study, Wang et al. studied lipopolysaccharide-induced sepsis, using rigosertib as a MEK1-ERK signaling inhibitor, concluding that it prevented the production of proinflammatory cytokines induced by LPS by disrupting the activation of the MEK1-ERK signaling axis [45].

Overall, although strong data show that rigosertib interferes with Ras–Raf–MEK signaling, the precise mechanism by which this effect is mediated is still unclear, and it is also unclear whether it is a direct molecular impact on Ras–Raf molecules or indirectly induced by cytotoxic stress mediated by ROS induction and stress kinase (JNK) signaling.

3.4. Rigosertib as a Tubulin Polymerization Destabilizer

A common cellular phenotype upon rigosertib treatment, observed by many different groups, is the G2/M arrest accompanied by mitotic aberrations. As mentioned above, this was originally assumed to be dependent on the Plk1 inhibition. However, there are examples providing evidence that rigosertib might be a microtubule dynamics inhibitor and this might be also responsible for the mentioned G2/M arrest.

Hyoda et al. showed that rigosertib induced the phosphorylation of histone H2AX and led to a DNA damage-induced G2/M arrest in HL-60, Jurkat, Ramos, and MDS-L cell lines. Rigosertib caused an abnormal localization of Aurora A kinase (AurKA), suggesting that it disrupts the spindle assembly machinery, and they finally concluded that the rigosertib effect had a possible involvement with genes related to microtubule kinetics [28]. More recently, the efficacy and mechanism of rigosertib have been tested in rhabdomyosarcoma (RMS) and neuroblastoma (NBs) cell lines. Here, the authors showed that the main cytotoxic effect is due to an efficient cell arrest in mitosis, mainly by interfering with mitotic spindle assembly, which leads to cell death [46,47]. In fact, the authors have shown a similar behavior of rigosertib with respect to other Vinca alkaloids, although it is better tolerated. Interestingly, these studies also provided evidence that rigosertib does not induce cell death by inhibiting the Ras pathway in either RMS or NB cells, demonstrating that the mechanism of action of rigosertib may be cell-type dependent.

An explanation for the rigosertib-mediated mitotic arrest was recently described by Jost et al., who claimed that rigosertib is indeed a microtubule-destabilizing agent [5]. Their study aimed to identify novel rigosertib response-associated biomarkers by performing a combined CRISPRi/a (inactivation/activation) genome-wide screen using the myeloid leukemia cell line K562. The authors demonstrated that rigosertib causes spindle and mitotic defects because it binds to $\alpha\beta$-tubulin heterodimers, in a region similar to that of to other microtubule destabilizers, such as colchicine. This binding of rigosertib to tubulin alters microtubule dynamics both in vitro and in vivo. Moreover, the authors used structural analysis to depict the residues involved in the binding of rigosertib to β-tubulin, with Lys-240 being critical for this interaction. Interestingly, the mutation of this Lys-240 to phenylalanine (L240F) confers resistance to rigosertib in K562 cells, confirming that the rigosertib effect is mainly mediated by altering microtubule dynamics and, subsequently, mitotic alterations.

An important aspect to consider in these results is that rigosertib is synthesized from ON-01500 (Figure 1), which is a microtubule poison. In fact, this feature opened a dispute when the Reddy group published a contrary article stating that the microtubule-destabilizing effect is most likely due to the presence of the late intermediate impurity ON-01500 in commercial rigosertib formulations, which has a potent tubulin depolymerizing activity [48]. NMR and mass spectrometry analysis determined that the commercial grade rigosertib contained at least 5% of ON-01500. This ON-01500 impurity may result from the degradation of rigosertib due to poor storage conditions, including elevated temperature, acidic pH, and exposure to intense light. In contrast, pharmaceutical-grade rigosertib directly supplied from Onconova Therapeutics (Onconova Therapeutics, Inc., Newtown, PA 18940, USA), which is free of the ON-01500 contaminant, does not exhibit in vitro tubulin-binding activity. More importantly, they also showed that different cell lines harboring the L240F mutations were sensitive to the pharmaceutical-grade rigosertib and failed to proliferate. The disagreement does not end here, as Jost et al. reevaluated the effect of commercial and clinical-grade rigosertib, showing indistinguishable phenotypes and agreeing with their previous findings [49]. Both rigosertib versions destabilize microtubules either in vivo or in vitro, and both show significantly reduced toxicity in cell lines expressing the L240F tubulin mutant. Interestingly, a very recent screen for identifying tubulin polymerization inhibitors pinpointed rigosertib as a major hit and demonstrated the capacity of rigosertib binding to tubulin by in silico docking analysis [50].

On the other hand, earlier studies made these conclusions even more controversial by showing that rigosertib-mediated effects do not depend on altered microtubule dynamics. Oussenko et al. not only described rigosertib-induced mitotic arrest as mentioned above, but also shed light on the microtubule destabilizer hypothesis by testing the ability of rigosertib and the inactive analog molecule ON-01911 to modify the tubulin polymerization. They revealed that clinical-grade rigosertib directly obtained from Onconova Therapeutics and control ON-01911 had a minor inhibition effect on tubulin polymerization, whereas nocodazole (another well-known microtubule poison) completely inhibited the process. This is indeed in agreement with the original article describing rigosertib as a Plk1 inhibitor, in which; they also established that rigosertib does not alter microtubule polymerization [1].

In summary, it is clear that rigosertib induces mitotic arrest and this appears to be a major cause of cell death. Whether this mitotic arrest is a direct effect of rigosertib due to changes in microtubule dynamics or other targets, such as Plk1, is still under debate. The source of the drug also appears to determine the outcome. While the clinical-grade rigosertib supplied by the proprietary pharmaceutical company Onconova Therapeutics appears to be more reliable than the commercial stocks, it is probably more important that the storage and handling conditions are precise to avoid degradation to synthesis intermediates, such as ON-01500, which has a strong affinity for microtubule heterodimers.

4. Rigosertib Clinical Trials for Cancer Therapy

Successful preclinical in vivo studies performed with rigosertib confirmed its safety profile with no evidence of myelotoxicity, neuropathy, or cardiotoxicity in xenograft mouse models, while successfully inhibiting liver, breast, and pancreatic cancers [1,51]. Rigosertib was, therefore, immediately included in a series of clinical trials to test its efficacy in a wide range of different cancers, including both solid tumors and hematological malignancies (Table 1).

Phase I trials suggested a maximum tolerated drug dose of 1800 mg/24 h in a three-day constant intravenous infusion rate, delivered every other week, with an acceptable toxicity profile [52–54]. The most common adverse events (AEs) included fatigue, gastrointestinal, and urinary symptoms. More severe grade 3 to 4 AEs were uncommon. Thrombocytopenia AE was also observed, yet to a lesser extent. Interestingly, drug-induced myelosuppression was also infrequent, thus providing a therapeutic advantage for myelodysplastic syndrome patients, given their compromised bone marrow function.

Rigosertib can be also administrated orally with a positive clinical activity [55]. Dose-limiting toxicity was established at 700 mg twice daily (b.i.d) upon severe grade 3 AE appearance, and the recommended dose was identified as 560 mg b.i.d. Overall, oral rigosertib was well tolerated. The most significant grade 2–3 toxicities included mainly urinary symptoms, abdominal pain and diarrhea, fatigue, hypotension, and anorexia. These positive toxicity and tolerability data encouraged clinicians to move forward into more advanced clinical trial phases.

Although some phase II tests were performed on some solid tumors, such as non-small cell lung cancer adenocarcinoma, and a wide variety of squamous cell carcinomas, they were predominately focused on hematological malignancies and, more concretely, on myelodysplastic syndromes (MDS) and myeloid leukemias. Similarly, phase III trials were mainly done in myelodysplastic syndromes, and only a recent phase III trial was performed on pancreatic adenocarcinoma patients (Table 1). It is also important to note that although most trials with rigosertib have been conducted as a monotherapy strategy (mainly phase I), advanced phase II and III studies have also been undertaken in combination with other chemotherapeutic agents in an effort to achieve a better response for those patients. Gemcitabine (DNA synthesis inhibitor) was combined with rigosertib in the phase III trial for pancreatic adenocarcinoma [8,56]. Azacitidine or decitabine (DNA methyltransferase inhibitors) were combined with rigosertib for MDS trials. Some solid tumor trials also combined rigosertib with either oxaliplatin (DNA synthesis inhibitor) or irinotecan (topoisomerase I inhibitor).

Despite the motivating number of clinical trials annotated in the U.S. National Library of Medicine website (ClinicalTrials.gov, accessed on 20 February 2023), very few of them provide results that indicate the efficacy of the drug. Data from phase I and II trials in MDS and myeloid leukemias were promising, with some patients showing a positive response to the treatment, with either a partial or complete bone marrow blast response and subsequent hematological improvement. This encouraged the initiation of a large-scale phase III trial comparing rigosertib administration to the standard best supportive care (BSC) for MDS (low dose cytarabine). Unfortunately, the results of the large trial were not positive [7]. The median overall survival (OS) in the rigosertib arm was 8.2 months and the median OS in the BSC arm was 5.9 months. In addition, no patients had a complete or partial response, although there is a modest increase in the bone marrow blast response with rigosertib, which was previously associated with improved survival.

In summary, rigosertib did not reach its primary objective of providing a survival benefit compared to the BSC-treated cohort, although certain subgroups appeared to benefit from treatment. Another interesting phase III trial is the combination of rigosertib with gemcitabine in a cohort of metastatic pancreatic adenocarcinoma, with 160 enrolled patients [8,56]. Unfortunately, despite the good data from preclinical studies using PDX models of pancreatic tumors [51], the combination treatment did not show an improvement in survival or response in pancreatic cancer patients and the authors stated that further studies with rigosertib in this type of cancer would not be performed.

In an attempt to understand the variability in sensitivity to rigosertib, and to shed light on what may be the best tumoral context for a rigosertib therapeutic strategy, we took advantage of the DepMap portal (depmap.org), which provides chemical and genetic cancer dependency information, allowing data visualization from a large scale of viability screens across hundreds of cell lines. We selected data generated by the Broad Institute CTD^2 Center originally published via the Cancer Therapeutics Response Portal (CTRP) [57,58] to plot rigosertib sensitivity (AUC, area under the curve) against different cancer lineages. We observed a specific sensitivity against rigosertib of hematological cancers cells (leukemias, myeloid and lymphoid) in comparison with solid cancer-derived cell lines (Figure 3). This probably explains why clinical trials have focused primarily on myeloid and lymphoid tumors, although in vitro data have never really translated into the clinical trials for these cancers.

Interestingly, and according to DepMap portal data, the most sensitive cell populations are those derived from neuroblastoma tumors. Indeed, the sensitivity of rigosertib to neuroblastoma models, both in vitro and in vivo, has been reported by others [46,47,59], demonstrating that rigosertib may be a good therapeutic alternative for this tumoral type. Perhaps the fact that neuroblastoma is a pediatric tumor is one of the reasons why its testing in clinical trials has been negatively affected. The data from the DepMap portal also confirm why the clinical trial that was focused on pancreatic cancer was completely unsuccessful. The cell lines derived from pancreatic adenocarcinoma are the most refractive to rigosertib treatment, among the 20 cohorts tested, with values equivalent to those for biliary tract and kidney tumors.

Table 1. Summary of rigosertib clinical trials.

Phase	NCT Number	Treatment	Disease	Participants	References
Phase I	NCT01048619	ON-01910.Na	Myelodysplastic Syndrome	36	[55]
	NCT01168011	ON-01910.Na	Solid Tumor	68	
	NCT01125891	Gemcitabine or ON-01910.Na	Malignant Neoplasms Solid Tumors	39	[60]

Table 1. *Cont.*

Phase	NCT Number	Treatment	Disease	Participants	References
Phase I	NCT00854646	ON-01910.Na	Acute Myelocytic Leukemia Acute Lymphocytic Leukemia Chronic Myelocytic Leukemia Chronic Lymphocytic Leukemia Myelodysplastic Syndromes	22	[61]
	NCT01538537	ON-01910.Na	Advanced Cancer Solid Tumors Cancer Neoplasms	29	[53]
	NCT01538563	ON-01910.Na	Solid Tumors Advanced Cancer Neoplasms	42	-
	NCT01165905	Gemcitabine or ON-01910.Na	Solid Tumor	10	-
	NCT00861783	Irinotecan + ON-01910.Na Oxaliplatin + ON-01910.Na	Hepatoma Advanced Solid Tumor	16	-
	NCT00861328	Irinotecan + ON-01910.Na Oxaliplatin + ON-01910.Na	Advanced Solid Tumors	18	-
	NCT00861510	ON-01910.Na	Lymphoma Mantle-cell Leukemia Lymphocytic Chronic B-Cell Leukemia Hairy Cell Waldenstrom Macroglobulinemia Multiple Myeloma	16	[62]
	NCT00533416	ON-01910.Na	Myelodysplastic Syndrome (MDS)	14	[63]
Phase I/II	NCT01167166	ON-01910.Na	Acute Myelocytic Leukemia Acute Lymphocytic Leukemia Myeloproliferative Disease Chronic Myeloid Leukemia	30	-
	NCT00854945	ON-01910.Na	Myelodysplastic Syndromes Acute Myeloid Leukemia	36	-
	NCT01926587	Azacitidine + ON-01910.Na	Myelodysplastic Syndromes Acute Myeloid Leukemia Chronic Myelomonocytic Leukemia	45	[64]
	NCT04263090	Nivolumab + ON-01910.Na	Non-small Cell Lung Cancer Adenocarcinoma Stage IV	20	-
	NCT03786237	ON-01910.Na	Epidermolysis Bullosa Dystrophica Squamous Cell Carcinoma	12	-
Phase II	NCT01807546	ON-01910.Na	Head and Neck Squamous Cell Carcinoma Anal Squamous Cell Carcinoma Lung Squamous Cell Carcinoma Cervical Squamous Cell Carcinoma Esophageal Squamous Cell Carcinoma Skin Squamous Cell Carcinoma Penile Squamous Cell Carcinoma	64	-
	NCT01904682	ON-01910.Na	Myelodysplastic Syndromes	45	-
	NCT01584531	ON-01910.Na	Myelodysplastic Syndromes MDS Trisomy 8	82	-
	NCT00906334	ON-01910.Na	Myelodysplastic Syndromes	14	[65]

Table 1. *Cont.*

Phase	NCT Number	Treatment	Disease	Participants	References
Phase III	NCT01241500	ON-01910.Na	Myelodysplastic Syndromes MDS RAEB Chronic Myelomonocytic Leukemia	299	-
	NCT01928537	ON-01910.Na	Myelodysplastic Syndromes Refractory Anemia with Excess Blasts Chronic Myelomonocytic Leukemia Cytopenia	67	-
	NCT02562443	Best supportive care (BSC) + ON-01910.Na Vs. Physician's choice (PC) + ON-01910.Na	Myelodysplastic Syndromes MDS Refractory Anemia with Excess Blasts RAEB	372	[7]
	NCT01360853	Gemcitabine + ON-01910.Na	Metastatic Pancreatic Adenocarcinoma	160	[8,56,66]

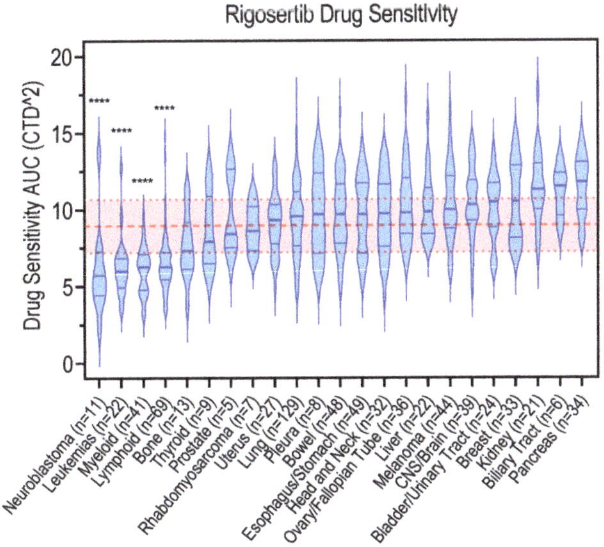

Figure 3. Data adapted from the DepMap Portal, retrieved from the Cancer Therapeutics Response Portal (CTRP), showing the sensitivity of cancer-cell lines of multiple origins to rigosertib. AUC reflects the area under the concentration-response curve. The lower the AUC, the more sensitive the cell line. Violin plots show median rigosertib sensitivity values (thick blue line). The thick dashed red line shows the average of all median AUCs. The shaded red area is defined by the standard deviation of the average median values. An unpaired t-test was performed, comparing the neuroblastoma, leukemias, myeloid, or lymphoid cohorts with the rest of all cohorts together and significance is indicated by $p < 0.0001$ (****). The number of cell lines (n) in each cohort is indicated in the X axis. Statistics were calculated using GraphPad Software version 9.5.1.

5. Genetic Biomarkers Associated with Rigosertib Response

In an effort to elucidate the mechanism of action of rigosertib and to identify therapeutic biomarkers, some researchers have recently begun to identify the genetic alteration that may determine the efficacy of the response to rigosertib.

Oussenko et al. performed cross-resistance experiments and showed that rigosertib did not induce an MDR1 (multi-drug resistance 1) or an MRP1 (multi-drug resistance-associated protein 1) response. Moreover, cells resistant to other drugs did not show cross-resistance to rigosertib, making it a favorable anticancer agent [23]; hence, rigosertib appears to escape the general resistance mechanisms described for other drugs. On the contrary, Sánchez-Burgos et al. described how deficiency of the FBXW7 gene leads to resistance to rigosertib treatment. Interestingly, FBWX7 deficiency also confers resistance to a wide range of other chemotherapeutic agents, making it a general drug-resistance mechanism, rather than rigosertib-specific [67].

Another described biomarker of rigosertib efficacy is the deubiquitinating enzyme USP28 [68]. USP28 destabilizes the Raf family proteins by a feedback-loop-mediated mechanism. When USP28 is downregulated, B-Raf is hyper stabilized and, therefore, enhances downstream MAPK signaling, promoting resistance to Raf inhibitor therapy strategies. As USP28 is frequently deleted in a proportion of melanoma patients, the authors suggested rigosertib as a potential therapeutic strategy for USP28-depleted cells, demonstrating an increased sensitivity of USP28-depleted cells to rigosertib. Surprisingly, rigosertib appears to kill USP28 null cells in an ERK independent manner, and with a strong mitotic arrest phenotype. Hence, there is still doubt as to whether this USP28-null synthetic lethality is due to an amplification of the Ras–Raf signaling pathway, or simply to an increase in the proliferation ratio and a strong dependence on mitosis. Indeed, the same authors observed that USP28-depleted cells are more sensitive to small molecules such as microtubule poisons (taxol, vincristine, nocodazole) or even Plk1 inhibitors such as BI-2536.

Focusing on myelodysplastic syndromes (MDS), a possible biomarker for rigosertib response is the presence of trisomy 8. Trisomy of chromosome 8 and monosomy of chromosome 7 are common alterations observed in MDS patients [69] and these cytogenetic features correlate with the upregulation of the oncogenes c-myc, Wilms Tumor 1 (WT1), and Cyclin D1, thus increasing the proliferation capacity of these cells [70]. Interestingly, rigosertib has a selective inhibitory effect in trisomy-8 MDS cells and, to a lesser extent, in monosomy-7. This was also seen in patients enrolled in the MDS clinical trial (NCT00533416) and occurred even when these chromosomal abnormalities were part of a more complex karyotype. [63]. Thus, trisomy-8/monosomy-7 and elevated Cyclin D1, can behave as bona fide biomarkers for rigosertib treatment, at least in MDS.

Seeking to identify relevant cellular biomarkers following rigosertib treatment, Jost et al. performed a chemical screen for rigosertib using a CRISPRi (inactivation) and a CRISPRa (activation) strategy and identified a signature of microtubule-related genes that antagonized or synergized with the drug [5]. Specifically, the authors determined that the overexpression of the KIF2C gene, which encodes the microtubule depolymerizing enzyme MCAK [71], sensitizes cells to rigosertib. At the same time, the knockdown of TACC3, a microtubule-binding protein involved in microtubule stability [72], sensitizes cells to rigosertib. Thus, genetic manipulations that destabilize microtubules lead to increased sensitivity to rigosertib, whereas genetic alterations that induce microtubule stabilization lead to a protective effect against rigosertib [5]. To validate this hypothesis, the authors knocked-down both genes involved in microtubule stability with CRISPR-Cas9 technology and confirmed in different tumoral cell lines that the downregulation of KIF2C and TACC3 dictate resistance or sensitivity to rigosertib, respectively. In addition, they confirmed that rigosertib destabilizes microtubules by docking into the colchicine-binding site of tubulin. These results, although requiring further validation, identify specific genes whose expression predicts response to rigosertib and may serve as biomarkers for rigosertib treatment in a clinical context.

Despite these initial efforts to identify biomarkers of response to rigosertib, there is still very little known, and further efforts are needed in order to find genetic signatures that might help to better stratify patients in future clinical trials.

6. Conclusions, Current Challenges, and Prospects

Rigosertib is considered a promising therapeutic strategy for cancer for several reasons. It is defined as a non-ATP competitor multi-kinase inhibitor, which makes it an attractive molecule, as it avoids the associated drug-resistance mechanisms that can arise with ATP-competitor molecules [9]. Data suggest that rigosertib does not present cross-resistance to other drugs and is not an MDR1 or MPR1 substrate, making rigosertib a good choice of treatment for multi-drug resistant cancers. Preclinical studies prompted rigosertib as a promising molecule for the treatment of cancer therapy because in vivo studies demonstrated its efficacy in killing cancer cells with little associated toxicity. This tolerable toxicity was also seen in many clinical trials performed; therefore, rigosertib was considered a promising therapeutic strategy. Unfortunately, advanced clinical trials turned out to not be as efficient as initially thought and there was very little improvement when compared to the standard therapies for several types of cancer. This indicates that we still lack the right biomarkers to better stratify the patients for a rigosertib-mediated therapy. In this regard, all efforts trying to define the drug target have been confusing and controversial.

Rigosertib's mechanism of action was originally suggested to be an allosteric inhibitor of the mitotic master regulator Plk1 [1]. Over the years, Plk1 specificity has been debated [22] and, in parallel, there was increasing evidence showing that rigosertib might alter other pathways. Indeed, it was described as a the PI3K–Akt axis inhibitor [2,3] or acting as a Ras–Raf binding mimetic leading to MEK signaling inhibition [4]. In the same vein, the inhibition of MEK signaling was attributed to the generation of mitotic stress [6], thus altering stress-signaling pathways. Finally, the mitotic arrest phenotype was recently explained by demonstrating that rigosertib is a microtubule-destabilizing drug [6]. This raised a strong controversy about the source quality of the drug, its stability, and its storage accuracy [48,49].

Because of the multiple mechanisms of action described, the scientific community has followed waves of studies, often viewing rigosertib as a unique inhibitor for Plk1, Ras, or PI3K/Akt, overlooking all other possible targets and side effects. Although some research groups have begun efforts to define the mechanisms of rigosertib, these studies are still at an early stage, and only a few genetic determinants have been described, with little clinical impact.

Therefore, a strong effort to elucidate rigosertib's mechanism of action and understand how rigosertib inhibits cancer cell growth is critical to better define the potential patients who could benefit from this inhibitor and, thus, advance clinical trials.

Author Contributions: Conceptualization, G.d.C.; data curation from DepMap portal and ClinicalTrials.gov web pages, A.M.-V. and G.d.C.; manuscript writing, A.M.-V. and G.d.C.; funding acquisition, G.d.C. All authors have read and agreed to the published version of the manuscript.

Funding: This research was funded by the Spanish Ministry of Science and Innovation MCIN/AEI/FEDER (doi:10.13039/501100011033), grants RTI2018-095496-B-I00 and PID2021-125705OB-I00, and by the Spanish Association Against Cancer (AECC) Scientific Foundation, grant LABAE16017DECA.

Institutional Review Board Statement: Not applicable.

Informed Consent Statement: Not applicable.

Data Availability Statement: Table 1 was generated retrieving data from the Clinical Trial web page (https://clinicaltrials.gov/ct2/home, accessed on 20 February 2023) of the U.S. National Library of Medicine. Figure 1 was generated using the chemical drawing tool Marvin JS (https://chemaxon.com/marvin, accessed on 25 January 2023). Figure 2 was drawn using templates from Servier Medical Art. Servier Medical Art by Servier is licensed under a Creative Commons Attribution 3.0 Unported License (https://creativecommons.org/licenses/by/3.0/, accessed on 15 January 2023). Figure 3 was generated using data from the DepMap Portal (https://depmap.org/portal/, accessed on 15 February 2023), retrieved from the Cancer Therapeutics Response Portal (CTRP).

Acknowledgments: G.d.C belongs to the Spanish National Research Council (CSIC)'s Cancer Hub (https://conexion-cancer.csic.es/).

Conflicts of Interest: The authors declare no conflict of interest.

References

1. Gumireddy, K.; Reddy, M.V.R.; Cosenza, S.C.; Nathan, R.B.; Baker, S.J.; Papathi, N.; Jiang, J.; Holland, J.; Reddy, E.P. ON01910, a Non-ATP-Competitive Small Molecule Inhibitor of Plk1, Is a Potent Anticancer Agent. *Cancer Cell* **2005**, *7*, 275–286. [CrossRef] [PubMed]
2. Anderson, R.T.; Keysar, S.B.; Bowles, D.W.; Glogowska, M.J.; Astling, D.P.; Morton, J.J.; Le, P.; Umpierrez, A.; Eagles-Soukup, J.; Gan, G.N.; et al. The Dual Pathway Inhibitor Rigosertib Is Effective in Direct Patient Tumor Xenografts of Head and Neck Squamous Cell Carcinomas. *Mol. Cancer Ther.* **2013**, *12*, 1994–2005. [CrossRef] [PubMed]
3. Prasad, A.; Khudaynazar, N.; Tantravahi, R.V.; Gillum, A.M.; Hoffman, B.S. ON 01910.Na (Rigosertib) Inhibits PI3K/Akt Pathway and Activates Oxidative Stress Signals in Head and Neck Cancer Cell Lines. *Oncotarget* **2016**, *7*, 79388–79400. [CrossRef] [PubMed]
4. Athuluri-Divakar, S.K.; Vasquez-Del Carpio, R.; Dutta, K.; Baker, S.J.; Cosenza, S.C.; Basu, I.; Gupta, Y.K.; Reddy, M.V.R.; Ueno, L.; Hart, J.R.; et al. A Small Molecule RAS-Mimetic Disrupts RAS Association with Effector Proteins to Block Signaling. *Cell* **2016**, *165*, 643–655. [CrossRef] [PubMed]
5. Jost, M.; Chen, Y.; Gilbert, L.A.; Horlbeck, M.A.; Krenning, L.; Menchon, G.; Rai, A.; Cho, M.Y.; Stern, J.J.; Prota, A.E.; et al. Combined CRISPRi/a-Based Chemical Genetic Screens Reveal That Rigosertib Is a Microtubule-Destabilizing Agent. *Mol. Cell* **2017**, *68*, 210–223.e6. [CrossRef] [PubMed]
6. Ritt, D.A.; Abreu-Blanco, M.T.; Bindu, L.; Durrant, D.E.; Zhou, M.; Specht, S.I.; Stephen, A.G.; Holderfield, M.; Morrison, D.K. Inhibition of Ras/Raf/MEK/ERK Pathway Signaling by a Stress-Induced Phospho-Regulatory Circuit. *Mol. Cell* **2016**, *64*, 875–887. [CrossRef] [PubMed]
7. Garcia-Manero, G.; Fenaux, P.; Al-Kali, A.; Baer, M.R.; Sekeres, M.A.; Roboz, G.J.; Gaidano, G.; Scott, B.L.; Greenberg, P.; Platzbecker, U.; et al. Rigosertib versus Best Supportive Care for Patients with High-Risk Myelodysplastic Syndromes after Failure of Hypomethylating Drugs (ONTIME): A Randomised, Controlled, Phase 3 Trial. *Lancet Oncol.* **2016**, *17*, 496–508. [CrossRef]
8. O'Neil, B.H.; Scott, A.J.; Ma, W.W.; Cohen, S.J.; Aisner, D.L.; Menter, A.R.; Tejani, M.A.; Cho, J.K.; Granfortuna, J.; Coveler, L.; et al. A Phase II/III Randomized Study to Compare the Efficacy and Safety of Rigosertib plus Gemcitabine versus Gemcitabine Alone in Patients with Previously Untreated Metastatic Pancreatic Cancer. *Ann. Oncol.* **2015**, *26*, 1923–1929. [CrossRef]
9. Sawyers, C.L. Opportunities and Challenges in the Development of Kinase Inhibitor Therapy for Cancer. *Genes Dev.* **2003**, *17*, 2998–3010. [CrossRef]
10. Reddy, E.; Reddy, M. Substituted Styryl Benzylsulfones for Treating Proliferative Disorders. U.S. Patent No. 6486210 B2, 22 November 2002.
11. Reddy, E.; Reddy, M. Preparation of α,β-Unsaturated Sulfones for Treating Proliferative Disorders. U.S. Patent No. 6541475, 7 March 2002.
12. Reddy, E.; Reddy, M. α,β-Unsaturated Sulfones for Treating Proliferative Disorders. U.S. Patent No. 6599932 B1, 29 July 2003.
13. Reddy, M.V.R.; Venkatapuram, P.; Mallireddigari, M.R.; Pallela, V.R.; Cosenza, S.C.; Robell, K.A.; Akula, B.; Hoffman, B.S.; Reddy, E.P. Discovery of a Clinical Stage Multi-Kinase Inhibitor Sodium (*E*)-2-{2-Methoxy-5-[(2′,4′,6′-Trimethoxystyrylsulfonyl)Methyl]Phenylamino}acetate (ON 01910.Na): Synthesis, Structure–Activity Relationship, and Biological Activity. *J. Med. Chem.* **2011**, *54*, 6254–6276. [CrossRef]
14. Xu, F.; He, Q.; Li, X.; Chang, C.-K.; Wu, L.-Y.; Zhang, Z.; Liu, L.; Shi, W.-H.; Zhu, Y.; Zhao, Y.-S.; et al. Rigosertib as a Selective Anti-Tumor Agent Can Ameliorate Multiple Dysregulated Signaling Transduction Pathways in High-Grade Myelodysplastic Syndrome. *Sci. Rep.* **2014**, *4*, 7310. [CrossRef] [PubMed]
15. Archambault, V.; Normandin, K. Several Inhibitors of the Plk1 Polo-Box Domain Turn out to Be Non-Specific Protein Alkylators. *Cell Cycle* **2017**, *16*, 1220–1224. [CrossRef] [PubMed]
16. Ma, H.; Nie, C.; Chen, Y.; Li, J.; Xie, Y.; Tang, Z.; Gao, Y.; Ai, S.; Mao, Y.; Sun, Q.; et al. Therapeutic Targeting PLK1 by ON-01910.Na Is Effective in Local Treatment of Retinoblastoma. *Oncol. Res. Featur. Preclin. Clin. Cancer Ther.* **2021**, *28*, 745–761. [CrossRef] [PubMed]
17. de Cárcer, G.; Manning, G.; Malumbres, M. From Plk1 to Plk5. *Cell Cycle* **2011**, *10*, 2255–2262. [CrossRef] [PubMed]
18. Chiappa, M.; Petrella, S.; Damia, G.; Broggini, M.; Guffanti, F.; Ricci, F. Present and Future Perspective on PLK1 Inhibition in Cancer Treatment. *Front. Oncol.* **2022**, *12*, 903016. [CrossRef]
19. Wachowicz, P.; Fernández-Miranda, G.; Marugán, C.; Escobar, B.; de Cárcer, G. Genetic Depletion of Polo-like Kinase 1 Leads to Embryonic Lethality Due to Mitotic Aberrancies. *BioEssays* **2016**, *38*, S96–S106. [CrossRef] [PubMed]
20. Lénárt, P.; Petronczki, M.; Steegmaier, M.; Di Fiore, B.; Lipp, J.J.; Hoffmann, M.; Rettig, W.J.; Kraut, N.; Peters, J.-M. The Small-Molecule Inhibitor BI 2536 Reveals Novel Insights into Mitotic Roles of Polo-like Kinase 1. *Curr. Biol.* **2007**, *17*, 304–315. [CrossRef] [PubMed]
21. Haren, L.; Stearns, T.; Lüders, J. Plk1-Dependent Recruitment of γ-Tubulin Complexes to Mitotic Centrosomes Involves Multiple PCM Components. *PLoS ONE* **2009**, *4*, e5976. [CrossRef]
22. Peters, U.; Cherian, J.; Kim, J.H.; Kwok, B.H.; Kapoor, T.M. Probing Cell-Division Phenotype Space and Polo-like Kinase Function Using Small Molecules. *Nat. Chem. Biol.* **2006**, *2*, 618–626. [CrossRef]
23. Oussenko, I.A.; Holland, J.F.; Reddy, E.P.; Ohnuma, T. Effect of ON 01910.Na, an Anticancer Mitotic Inhibitor, on Cell-Cycle Progression Correlates with RanGAP1 Hyperphosphorylation. *Cancer Res.* **2011**, *71*, 4968–4976. [CrossRef]

24. Swaminathan, S.; Kiendl, F.; Körner, R.; Lupetti, R.; Hengst, L.; Melchior, F. RanGAP1*SUMO1 Is Phosphorylated at the Onset of Mitosis and Remains Associated with RanBP2 upon NPC Disassembly. *J. Cell Biol.* **2004**, *164*, 965–971. [CrossRef] [PubMed]
25. Atanasova, V.S.; Pourreyron, C.; Farshchian, M.; Lawler, M.; Brown, C.A.; Watt, S.A.; Wright, S.; Warkala, M.; Guttmann-Gruber, C.; Hofbauer, J.P.; et al. Identification of Rigosertib for the Treatment of Recessive Dystrophic Epidermolysis Bullosa–Associated Squamous Cell Carcinoma. *Clin. Cancer Res.* **2019**, *25*, 3384–3391. [CrossRef] [PubMed]
26. Seo, Y.; Kang, Y.; Ham, Y.; Kim, M.-H.; Kim, S.-J.; Yoon, S.K.; Jang, S.K.; Park, J.B.; Cho, S.; Kim, J.H. PLK1-ELAVL1/HuR-MiR-122 Signaling Facilitates Hepatitis C Virus Proliferation. *Proc. Natl. Acad. Sci. USA* **2022**, *119*, e2214911119. [CrossRef] [PubMed]
27. Vulin, M.; Jehanno, C.; Sethi, A.; Correia, A.L.; Obradović, M.M.S.; Couto, J.P.; Coissieux, M.-M.; Diepenbruck, M.; Preca, B.-T.; Volkmann, K.; et al. A High-Throughput Drug Screen Reveals Means to Differentiate Triple-Negative Breast Cancer. *Oncogene* **2022**, *41*, 4459–4473. [CrossRef] [PubMed]
28. Hyoda, T.; Tsujioka, T.; Nakahara, T.; Suemori, S.; Okamoto, S.; Kataoka, M.; Tohyama, K. Rigosertib Induces Cell Death of a Myelodysplastic Syndrome-derived Cell Line by DNA Damage-induced G2/M Arrest. *Cancer Sci.* **2015**, *106*, 287–293. [CrossRef]
29. Lin, Y.G.; Shen, J.; Yoo, E.; Liu, R.; Yen, H.-Y.; Mehta, A.; Rajaei, A.; Yang, W.; Mhawech-Fauceglia, P.; DeMayo, F.J.; et al. Targeting the Glucose-Regulated Protein-78 Abrogates Pten-Null Driven AKT Activation and Endometrioid Tumorigenesis. *Oncogene* **2015**, *34*, 5418–5426. [CrossRef]
30. Liu, Z.; Wang, M.; Wang, H.; Fang, L.; Gou, S. Targeting RAS-RAF Pathway Significantly Improves Antitumor Activity of Rigosertib-Derived Platinum(IV) Complexes and Overcomes Cisplatin Resistance. *Eur. J. Med. Chem.* **2020**, *194*, 112269. [CrossRef]
31. Lu, T.; Laughton, C.A.; Wang, S.; Bradshaw, T.D. In Vitro Antitumor Mechanism of (*E*)-*N*-(2-Methoxy-5-(((2,4,6-Trimethoxystyryl)Sulfonyl)Methyl)Pyridin-3-Yl)Methanesulfonamide. *Mol. Pharmacol.* **2015**, *87*, 18–30. [CrossRef]
32. Malacrida, A.; Rigolio, R.; Celio, L.; Damian, S.; Cavaletti, G.; Mazzaferro, V.; Miloso, M. In Vitro Evaluation of Rigosertib Antitumoral and Radiosensitizing Effects against Human Cholangiocarcinoma Cells. *Int. J. Mol. Sci.* **2021**, *22*, 8230. [CrossRef]
33. Ruppenthal, S.; Kleiner, H.; Nolte, F.; Fabarius, A.; Hofmann, W.-K.; Nowak, D.; Seifarth, W. Increased Separase Activity and Occurrence of Centrosome Aberrations Concur with Transformation of MDS. *PLoS ONE* **2018**, *13*, e0191734. [CrossRef]
34. Mielgo, A.; Seguin, L.; Huang, M.; Camargo, M.F.; Anand, S.; Franovic, A.; Weis, S.M.; Advani, S.J.; Murphy, E.A.; Cheresh, D.A. A MEK-Independent Role for CRAF in Mitosis and Tumor Progression. *Nat. Med.* **2011**, *17*, 1641–1645. [CrossRef] [PubMed]
35. Dietrich, P.; Freese, K.; Mahli, A.; Thasler, W.E.; Hellerbrand, C.; Bosserhoff, A.K. Combined Effects of PLK1 and RAS in Hepatocellular Carcinoma Reveal Rigosertib as Promising Novel Therapeutic "Dual-Hit" Option. *Oncotarget* **2018**, *9*, 3605–3618. [CrossRef] [PubMed]
36. Urasaki, Y.; Fiscus, R.R.; Le, T.T. Detection of the Cell Cycle-Regulated Negative Feedback Phosphorylation of Mitogen-Activated Protein Kinases in Breast Carcinoma Using Nanofluidic Proteomics. *Sci. Rep.* **2018**, *8*, 9991. [CrossRef] [PubMed]
37. Galimov, E.R. The Role of P66shc in Oxidative Stress and Apoptosis. *Acta Nat.* **2010**, *2*, 44–51. [CrossRef]
38. Günther, J.K.; Nikolajevic, A.; Ebner, S.; Troppmair, J.; Khalid, S. Rigosertib-Activated JNK1/2 Eliminate Tumor Cells through P66Shc Activation. *Biology* **2020**, *9*, 99. [CrossRef]
39. Giorgio, M.; Migliaccio, E.; Orsini, F.; Paolucci, D.; Moroni, M.; Contursi, C.; Pelliccia, G.; Luzi, L.; Minucci, S.; Marcaccio, M.; et al. Electron Transfer between Cytochrome c and P66Shc Generates Reactive Oxygen Species That Trigger Mitochondrial Apoptosis. *Cell* **2005**, *122*, 221–233. [CrossRef]
40. Rahmani, F.; Hashemzehi, M.; Avan, A.; Barneh, F.; Asgharzadeh, F.; Moradi Marjaneh, R.; Soleimani, A.; Parizadeh, M.; Ferns, G.A.; Ghayour Mobarhan, M.; et al. Rigosertib Elicits Potent Anti-Tumor Responses in Colorectal Cancer by Inhibiting Ras Signaling Pathway. *Cell. Signal.* **2021**, *85*, 110069. [CrossRef]
41. Tong, A.W.; Stone, M.J. Prospects for CD40-Directed Experimental Therapy of Human Cancer. *Cancer Gene Ther.* **2003**, *10*, 1–13. [CrossRef] [PubMed]
42. Yan, C.; Saleh, N.; Yang, J.; Nebhan, C.A.; Vilgelm, A.E.; Reddy, E.P.; Roland, J.T.; Johnson, D.B.; Chen, S.-C.; Shattuck-Brandt, R.L.; et al. Novel Induction of CD40 Expression by Tumor Cells with RAS/RAF/PI3K Pathway Inhibition Augments Response to Checkpoint Blockade. *Mol. Cancer* **2021**, *20*, 85. [CrossRef]
43. Lee, J.R.; Koretzky, G.A. Production of Reactive Oxygen Intermediates Following CD40 Ligation Correlates with C-Jun N-Terminal Kinase Activation and IL-6 Secretion in Murine B Lymphocytes. *Eur. J. Immunol.* **1998**, *28*, 4188–4197. [CrossRef]
44. Liu, J.; Yoshida, Y.; Yamashita, U. Suppressive Effect of Reactive Oxygen Species on CD40-Induced B Cell Activation. *FEBS Lett.* **2007**, *581*, 5043–5049. [CrossRef] [PubMed]
45. Wang, Y.; Du, P.; Jiang, D. Rigosertib Inhibits MEK1–ERK Pathway and Alleviates Lipopolysaccharide-induced Sepsis. *Immun. Inflamm. Dis.* **2021**, *9*, 991–999. [CrossRef] [PubMed]
46. Kowalczyk, J.T.; Wan, X.; Hernandez, E.R.; Luo, R.; Lyons, G.C.; Wilson, K.M.; Gallardo, D.C.; Isanogle, K.A.; Robinson, C.M.; Mendoza, A.; et al. Rigosertib Induces Mitotic Arrest and Apoptosis in RAS-Mutated Rhabdomyosarcoma and Neuroblastoma. *Mol. Cancer Ther.* **2021**, *20*, 307–319. [CrossRef] [PubMed]
47. Radke, K.; Hansson, K.; Sjölund, J.; Wolska, M.; Karlsson, J.; Esfandyari, J.; Pietras, K.; Aaltonen, K.; Gisselsson, D.; Bexell, D. Anti-Tumor Effects of Rigosertib in High-Risk Neuroblastoma. *Transl. Oncol.* **2021**, *14*, 101149. [CrossRef] [PubMed]
48. Baker, S.J.; Cosenza, S.C.; Athuluri-Divakar, S.; Reddy, M.V.R.; Vasquez-Del Carpio, R.; Jain, R.; Aggarwal, A.K.; Reddy, E.P. A Contaminant Impurity, Not Rigosertib, Is a Tubulin Binding Agent. *Mol. Cell* **2020**, *79*, 180–190.e4. [CrossRef] [PubMed]

49. Jost, M.; Chen, Y.; Gilbert, L.A.; Horlbeck, M.A.; Krenning, L.; Menchon, G.; Rai, A.; Cho, M.Y.; Stern, J.J.; Prota, A.E.; et al. Pharmaceutical-Grade Rigosertib Is a Microtubule-Destabilizing Agent. *Mol. Cell* **2020**, *79*, 191–198.e3. [CrossRef] [PubMed]
50. Khachatryan, H.; Olszowy, B.; Barrero, C.A.; Gordon, J.; Perez-Leal, O. Identification of Inhibitors of Tubulin Polymerization Using a CRISPR-Edited Cell Line with Endogenous Fluorescent Tagging of β-Tubulin and Histone H1. *Biomolecules* **2023**, *13*, 249. [CrossRef]
51. Jimeno, A.; Chan, A.; Cusatis, G.; Zhang, X.; Wheelhouse, J.; Solomon, A.; Chan, F.; Zhao, M.; Cosenza, S.C.; Reddy, M.R.; et al. Evaluation of the Novel Mitotic Modulator ON 01910.Na in Pancreatic Cancer and Preclinical Development of an Ex Vivo Predictive Assay. *Oncogene* **2009**, *28*, 610–618. [CrossRef]
52. Navada, S.C.; Silverman, L.R. The Safety and Efficacy of Rigosertib in the Treatment of Myelodysplastic Syndromes. *Expert Rev. Anticancer Ther.* **2016**, *16*, 805–810. [CrossRef]
53. Ohnuma, T.; Lehrer, D.; Ren, C.; Cho, S.Y.; Maniar, M.; Silverman, L.; Sung, M.; Gretz, H.F.; Benisovich, V.; Navada, S.; et al. Phase 1 Study of Intravenous Rigosertib (ON 01910.Na), a Novel Benzyl Styryl Sulfone Structure Producing G2/M Arrest and Apoptosis, in Adult Patients with Advanced Cancer. *Am. J. Cancer Res.* **2013**, *3*, 323–338.
54. Silverman, L.R.; Greenberg, P.; Raza, A.; Olnes, M.J.; Holland, J.F.; Reddy, P.; Maniar, M.; Wilhelm, F. Clinical Activity and Safety of the Dual Pathway Inhibitor Rigosertib for Higher Risk Myelodysplastic Syndromes Following DNA Methyltransferase Inhibitor Therapy. *Hematol. Oncol.* **2015**, *33*, 57–66. [CrossRef] [PubMed]
55. Komrokji, R.S.; Raza, A.; Lancet, J.E.; Ren, C.; Taft, D.; Maniar, M.; Wilhelm, F.; List, A.F. Phase I Clinical Trial of Oral Rigosertib in Patients with Myelodysplastic Syndromes. *Br. J. Haematol.* **2013**, *162*, 517–524. [CrossRef] [PubMed]
56. Ottaiano, A.; Capozzi, M.; De Divitiis, C.; De Stefano, A.; Botti, G.; Avallone, A.; Tafuto, S. Gemcitabine Mono-Therapy versus Gemcitabine plus Targeted Therapy in Advanced Pancreatic Cancer: A Meta-Analysis of Randomized Phase III Trials. *Acta Oncol.* **2017**, *56*, 377–383. [CrossRef]
57. Rees, M.G.; Seashore-Ludlow, B.; Cheah, J.H.; Adams, D.J.; Price, E.V.; Gill, S.; Javaid, S.; Coletti, M.E.; Jones, V.L.; Bodycombe, N.E.; et al. Correlating Chemical Sensitivity and Basal Gene Expression Reveals Mechanism of Action. *Nat. Chem. Biol.* **2016**, *12*, 109–116. [CrossRef] [PubMed]
58. Seashore-Ludlow, B.; Rees, M.G.; Cheah, J.H.; Cokol, M.; Price, E.V.; Coletti, M.E.; Jones, V.; Bodycombe, N.E.; Soule, C.K.; Gould, J.; et al. Harnessing Connectivity in a Large-Scale Small-Molecule Sensitivity Dataset. *Cancer Discov.* **2015**, *5*, 1210–1223. [CrossRef] [PubMed]
59. Hansson, K.; Radke, K.; Aaltonen, K.; Saarela, J.; Mañas, A.; Sjölund, J.; Smith, E.M.; Pietras, K.; Påhlman, S.; Wennerberg, K.; et al. Therapeutic Targeting of KSP in Preclinical Models of High-Risk Neuroblastoma. *Sci. Transl. Med.* **2020**, *12*, eaba4434. [CrossRef] [PubMed]
60. Ma, W.W.; Messersmith, W.A.; Dy, G.K.; Weekes, C.D.; Whitworth, A.; Ren, C.; Maniar, M.; Wilhelm, F.; Eckhardt, S.G.; Adjei, A.A.; et al. Phase I Study of Rigosertib, an Inhibitor of the Phosphatidylinositol 3-Kinase and Polo-like Kinase 1 Pathways, Combined with Gemcitabine in Patients with Solid Tumors and Pancreatic Cancer. *Clin. Cancer Res.* **2012**, *18*, 2048–2055. [CrossRef] [PubMed]
61. Navada, S.C.; Fruchtman, S.M.; Odchimar-Reissig, R.; Demakos, E.P.; Petrone, M.E.; Zbyszewski, P.S.; Holland, J.F.; Silverman, L.R. A Phase 1/2 Study of Rigosertib in Patients with Myelodysplastic Syndromes (MDS) and MDS Progressed to Acute Myeloid Leukemia. *Leuk. Res.* **2018**, *64*, 10–16. [CrossRef]
62. Roschewski, M.; Farooqui, M.; Aue, G.; Wilhelm, F.; Wiestner, A. Phase I Study of ON 01910.Na (Rigosertib), a Multikinase PI3K Inhibitor in Relapsed/Refractory B-Cell Malignancies. *Leukemia* **2013**, *27*, 1920–1923. [CrossRef]
63. Olnes, M.J.; Shenoy, A.; Weinstein, B.; Pfannes, L.; Loeliger, K.; Tucker, Z.; Tian, X.; Kwak, M.; Wilhelm, F.; Yong, A.S.M.; et al. Directed Therapy for Patients with Myelodysplastic Syndromes (MDS) by Suppression of Cyclin D1 with ON 01910.Na. *Leuk. Res.* **2012**, *36*, 982–989. [CrossRef]
64. Navada, S.C.; Garcia-Manero, G.; OdchimarReissig, R.; Pemmaraju, N.; Alvarado, Y.; Ohanian, M.N.; John, R.B.; Demakos, E.P.; Zbyszewski, P.S.; Maniar, M.; et al. Rigosertib in Combination with Azacitidine in Patients with Myelodysplastic Syndromes or Acute Myeloid Leukemia: Results of a Phase 1 Study. *Leuk. Res.* **2020**, *94*, 106369. [CrossRef] [PubMed]
65. Seetharam, M.; Fan, A.C.; Tran, M.; Xu, L.; Renschler, J.P.; Felsher, D.W.; Sridhar, K.; Wilhelm, F.; Greenberg, P.L. Treatment of Higher Risk Myelodysplastic Syndrome Patients Unresponsive to Hypomethylating Agents with ON 01910.Na. *Leuk. Res.* **2012**, *36*, 98–103. [CrossRef] [PubMed]
66. Nazha, A.; Sekeres, M.A.; Komrokji, R.; Steensma, D.P.; Kantarjian, H.; Roboz, G.; Fenaux, P.; Prebet, T.; Azarnia, N.; Zbyszewski, P.S.; et al. Validation of a Post-Hypomethylating Agent Failure Prognostic Model in Myelodysplastic Syndromes Patients Treated in a Randomized Controlled Phase III Trial of Rigosertib vs. Best Supportive Care. *Blood Cancer J.* **2017**, *7*, 644. [CrossRef] [PubMed]
67. Sanchez-Burgos, L.; Navarro-González, B.; García-Martín, S.; Sirozh, O.; Mota-Pino, J.; Fueyo-Marcos, E.; Tejero, H.; Antón, M.E.; Murga, M.; Al-Shahrour, F.; et al. Activation of the Integrated Stress Response Is a Vulnerability for Multidrug-resistant FBXW7-deficient Cells. *EMBO Mol. Med.* **2022**, *14*, e15855. [CrossRef] [PubMed]
68. Saei, A.; Palafox, M.; Benoukraf, T.; Kumari, N.; Jaynes, P.W.; Iyengar, P.V.; Muñoz-Couselo, E.; Nuciforo, P.; Cortés, J.; Nötzel, C.; et al. Loss of USP28-Mediated BRAF Degradation Drives Resistance to RAF Cancer Therapies. *J. Exp. Med.* **2018**, *215*, 1913–1928. [CrossRef] [PubMed]

69. Haase, D.; Germing, U.; Schanz, J.; Pfeilstöcker, M.; Nösslinger, T.; Hildebrandt, B.; Kundgen, A.; Lübbert, M.; Kunzmann, R.; Giagounidis, A.A.N.; et al. New Insights into the Prognostic Impact of the Karyotype in MDS and Correlation with Subtypes: Evidence from a Core Dataset of 2124 Patients. *Blood* **2007**, *110*, 4385–4395. [CrossRef] [PubMed]
70. Schoch, C.; Kohlmann, A.; Dugas, M.; Kern, W.; Hiddemann, W.; Schnittger, S.; Haferlach, T. Genomic Gains and Losses Influence Expression Levels of Genes Located within the Affected Regions: A Study on Acute Myeloid Leukemias with Trisomy 8, 11, or 13, Monosomy 7, or Deletion 5q. *Leukemia* **2005**, *19*, 1224–1228. [CrossRef]
71. Tanenbaum, M.E.; Medema, R.; Akhmanova, A. Regulation of Localization and Activity of the Microtubule Depolymerase MCAK. *Bioarchitecture* **2011**, *1*, 80–87. [CrossRef] [PubMed]
72. Hood, F.E.; Royle, S.J. Pulling It Together. *Bioarchitecture* **2011**, *1*, 105–109. [CrossRef]

pharmaceutics

MDPI

Article

A Small Molecule Targeting the Intracellular Tyrosine Kinase Domain of ROR1 (KAN0441571C) Induced Significant Apoptosis of Non-Small Cell Lung Cancer (NSCLC) Cells

Amineh Ghaderi [1], Mohammad-Ali Okhovat [1,†], Jemina Lehto [2,†], Luigi De Petris [1,3], Ehsan Manouchehri Doulabi [1], Parviz Kokhaei [1,4], Wen Zhong [1], Georgios Z. Rassidakis [1], Elias Drakos [1,5], Ali Moshfegh [1,2], Johan Schultz [2], Thomas Olin [2], Anders Österborg [1,6], Håkan Mellstedt [1,*] and Mohammad Hojjat-Farsangi [1,*]

1 Department of Oncology-Pathology, BioClinicum, Karolinska University Hospital Solna, Karolinska Institutet, 171 64 Stockholm, Sweden
2 Kancera AB, Nanna Svartz Väg 4, 171 65 Solna, Sweden
3 Thoracic Oncology Center, Karolinska Comprehensive Cancer Center, 171 76 Solna, Sweden
4 Department of Immunology, Arak University of Medical Sciences, Arak 3848170001, Iran
5 Department of Pathology, Medical School, University of Crete, 700 13 Heraklion, Greece
6 Department of Hematology, Karolinska University Hospital Solna, 171 64 Solna, Sweden
* Correspondence: hakan.mellstedt@ki.se (H.M.); mohammad.hojat-farsangi@ki.se (M.H.-F.); Tel.: +46-735234706 (H.M.); +46-706589809 (M.H.-F.)
† These authors contributed equally to this work.

Abstract: The ROR1 receptor tyrosine kinase is expressed in embryonic tissues but is absent in normal adult tissues. ROR1 is of importance in oncogenesis and is overexpressed in several cancers, such as NSCLC. In this study, we evaluated ROR1 expression in NSCLC patients (N = 287) and the cytotoxic effects of a small molecule ROR1 inhibitor (KAN0441571C) in NSCLC cell lines. ROR1 expression in tumor cells was more frequent in non-squamous (87%) than in squamous (57%) carcinomas patients, while 21% of neuroendocrine tumors expressed ROR1 ($p = 0.0001$). A significantly higher proportion of p53 negative patients in the ROR1$^+$ group than in the p53 positive non-squamous NSCLC patients ($p = 0.03$) was noted. KAN0441571C dephosphorylated ROR1 and induced apoptosis (Annexin V/PI) in a time- and dose-dependent manner in five ROR1$^+$ NSCLC cell lines and was superior compared to erlotinib (EGFR inhibitor). Apoptosis was confirmed by the downregulation of MCL-1 and BCL-2, as well as PARP and caspase 3 cleavage. The non-canonical Wnt pathway was involved. The combination of KAN0441571C and erlotinib showed a synergistic apoptotic effect. KAN0441571C also inhibited proliferative (cell cycle analyses, colony formation assay) and migratory (scratch wound healing assay) functions. Targeting NSCLC cells by a combination of ROR1 and EGFR inhibitors may represent a novel promising approach for the treatment of NSCLC patients.

Keywords: ROR1; NSCLC; small molecules; KAN0441571C; erlotinib; ibrutinib; TTP

Citation: Ghaderi, A.; Okhovat, M.-A.; Lehto, J.; De Petris, L.; Manouchehri Doulabi, E.; Kokhaei, P.; Zhong, W.; Rassidakis, G.Z.; Drakos, E.; Moshfegh, A.; et al. A Small Molecule Targeting the Intracellular Tyrosine Kinase Domain of ROR1 (KAN0441571C) Induced Significant Apoptosis of Non-Small Cell Lung Cancer (NSCLC) Cells. *Pharmaceutics* **2023**, *15*, 1148. https://doi.org/10.3390/pharmaceutics15041148

Academic Editor: Gabriele Grassi

Received: 27 February 2023
Revised: 28 March 2023
Accepted: 31 March 2023
Published: 5 April 2023

1. Introduction

Lung cancer is one of the main causes of death in cancer patients [1], and non-small cell lung cancer (NSCLC) accounts for around 85% of lung cancer cases. Current treatments for NSCLC are not sufficiently effective and there is an urgent need for new precision medicines [2].

Receptor tyrosine kinases (RTKs) are phosphotransferase enzymes that are essential for intracellular signal transduction and that are important targets for cancer treatment [3]. ROR1 (receptor tyrosine kinase-like orphan receptor 1), also known as neurotrophic tyrosine kinase receptor-related 1 (NTRKR1), is a transmembrane tyrosine-kinase enzyme encoded by a gene on chromosome 1. ROR1 belongs to the ROR family (ROR1 and ROR2) and was discovered based on amino acid sequence homology to the Trk family of neurotrophin receptors [4]. ROR1 is highly expressed during embryonic development in central neurons, and respiratory, cardiac, and skeletal tissues, and is essential for the proliferation, differentiation, polarity, and migration of neurons [5]. ROR1 is downregulated after birth but maintains a low expression in a few normal tissues, such as adipocytes, stomach and duodenum, pancreatic islets, parathyroid glands, and early B-cells [6–8]. Wnt5a is a ligand for ROR receptor proteins [9].

ROR1 is, however, expressed in hematologic malignancies and solid tumors, such as chronic lymphocytic leukemia (CLL) [10,11], mantle cell lymphoma (MCL) [12], diffuse large B-cell lymphoma (DLBCL) [13–15], melanoma [16–22], and NSCLC [17,23]. ROR1 is essential for the expansion, survival, epithelial-to-mesenchymal transition (EMT), migration, and metastasis of malignant cells [15,16,24–26]. ROR1 is involved in sustaining the self-renewal of cancer stem cells, and is related to disease activity and resistance to chemotherapy [27–29]. Several key signaling pathways such as PI3K/AKT/mTOR and the canonical/non-canonical Wnt pathways, as well as transcription factors such as CREB and C-Jun, have been proposed to be associated with ROR1 activation [30–35]. In various tumors, such as CLL, breast, and gastric cancers, high expression of ROR1 was associated with advanced disease and short survival [31,36–40].

ROR1 targeting agents, such as siRNA, monoclonal antibodies (mAbs), chimeric antigen receptor-modified T cells (CAR-T), and small molecule inhibitors (SMI), induced significant tumor cell death in hematological malignancies and cancers of epithelial origin [31,36,37,41–45]. The first-in-class ROR1 inhibitor (KAN0439834), which binds to the intracellular tyrosine kinase (TK) domain of ROR1, induced apoptosis of CLL cells and inhibited tumor cell growth in NOD/SCID mice xenotransplanted with human CLL cells [46]. KAN0439834 dephosphorylated ROR1 and inhibited the PI3K/AKT/mTOR pathway [20]. The combination of the ROR1 inhibitor with ibrutinib or erlotinib showed additive effects on pancreatic tumor cell death, compared to treatment with either agent alone [20]. KAN0441571C is the second generation of a ROR1 tyrosine kinase inhibitor (TKI) that induced apoptosis and death of DLBCL cells and small cell carcinoma of the lung, and that acted synergistically with venetoclax on tumor cell apoptosis [13,47].

In this study, ROR1 expression was analyzed in NSCLC tumors from patients as well as in cell lines. The anti-tumor effects of KAN0441571C alone were evaluated in vitro and in combination with erlotinib (EGFR inhibitor) and ibrutinib (BTK inhibitor). Erlotinib is a standard drug for the treatment of lung cancer [48] and ibrutinib has been suggested to be evaluated in lung cancer treatment [49,50]. These three drug molecules represent compounds with different mechanisms of action (MOA). Our data indicate that KAN0441571C in combination with EGFR or BTK inhibitors had synergistic apoptotic effects on lung cancer cells.

2. Materials and Methods

2.1. Patients

In total, 287 surgically resected NSCLC tumor specimens were analyzed for ROR1 expression. The use of patient samples was in accordance with the Declaration of Helsinki and ethically approved by the national ethics committee (www.etikprovningsmyndigheten.se) (accessed on 3 August 2015). Diagnosis was based on the WHO classification [51–53]. Patients were diagnosed, registered, and treated at the Oncology Department of Karolinska University Hospital Solna, Stockholm, Sweden, according to Swedish national guidelines. Based on histology, patients were divided in 3 groups including non-squamous adenocarcinoma (n = 157), squamous (n = 106), and neuroendocrine (mainly carcinoids) (n = 24) tumors (Table S1). Of these, 118, 89, 40, and 9 patients were at stages IA, IB, II, and III-IV, respectively (Table S2).

2.2. Immunohistochemical (IHC) Assays

Immunohistochemistry (IHC) staining of ROR1 was performed as previously described [54], using an anti-ROR1 polyclonal antibody (Proteintech, Manchester, UK). ROR1 expression was scored as negative (0), weak (1), or strong (2) according to staining intensity [6]. Slides were blindly scored by 3 independent pathologists.

2.3. Lung Cancer Cell Lines

Five ROR1$^+$ NSCLC cell lines were used for in vitro experiments and obtained from ATCC: NCI-H1975 (EGFR mutated L858R/T790M), NCI-H23 (EGFR wild type (WT)), NCI-HCC827 (EGFR mutated E746-A750 del19), A549 (EGFR WT), and NCI-H1299 (EGFR WT) (Table S3).

2.4. ROR1, BTK and EGFR Inhibitors

KAN0441571C, an SMI targeting ROR1, was produced and developed by Kancera AB (Stockholm, Sweden) [13,46]. Erlotinib (EGFR inhibitor) and ibrutinib (BTK inhibitor) were purchased from Selleckchem (Rungsted, Denmark).

2.5. Flow Cytometry

Surface ROR1 expression was evaluated by flow cytometry, as previously described [46]. Briefly, cell lines (single-cell suspension) were harvested and washed in 100 µL phosphate buffered saline (PBS) and re-suspended in 100 µL cell staining buffer (BD Biosciences, San Jose, CA, USA). Next, 10^6 cells were incubated with an APC (allophycocyanin)-conjugated anti-ROR1 monoclonal antibody (Miltenyi Biotec, Bergisch Gladbach, Germany) and an isotype-matched antibody (Miltenyi Biotec) for twenty minutes at room temperature (RT). Cells were then washed with cell staining buffer (BD Biosciences) and were prepared for analysis by flow cytometry (Canto II, BD Biosciences). To analyze the data, the FlowJo program was used (Tree Star Inc., Ashland, OR, USA) [13].

2.6. Western Blot Analysis

Western blot was used to analyze the expression of proteins, as previously described [46]. Cells were lysed on ice for 1 h by lysis buffer [46]. Supernatants were collected, and a BCA Kit (ThermoFisher Scientific, Bartlett, IL, USA) was used to measure the protein concentration. Next, 10 to 20 µg of the protein lysate was mixed with loading dye and reducing agent, and then loaded onto 8 or 10% polyacrylamide gel (ThermoFisher Scientific). Proteins were separated by electrophoresis and transferred to a polyvinylidene fluoride membrane (Millipore Corporation, Bedford, MA, USA). Membranes were then blocked in 5% bovine serum albumin (BSA) (Santa Cruz Biotechnology, Heidelberg, Germany) or skimmed milk (Sigma-Aldrich, St. Louis, MO, USA). After blocking, membranes were incubated with the primary antibodies at 4 °C for overnight, washed and probed with a peroxidase (HRP)-conjugated mAb (Dako Cytomation, Glostrup, Denmark). Finally, membranes were washed and a chemiluminescence detection system (GE Healthcare, Uppsala, Sweden) was used for protein visualization. The following primary antibodies were used for protein staining: phospho (p) ROR1 (against amino acid residues Ser 652, Tyr 641, 646) [46], anti-ROR1 antibody (R&D Systems, Minneapolis, MN, USA), anti-EGFR and anti-pEGFR (Y1173) antibodies, anti-SRC and anti-p-SRC (Tyr 416) antibodies, anti-AKT and anti-p-AKT (Ser 473) antibodies, anti-mTOR and anti-p-mTOR (Ser 2448) antibodies, anti-cAMP response element-binding protein (CREB) and anti-p-CREB (Ser 133) antibodies, anti-cleaved poly ADP ribose polymerase (PARP) antibody, anti-B-cell lymphoma (BCL)-2 antibody, anti-BAX antibody, anti-myeloid cell leukemia (MCL)-1 antibody, anti-cleaved caspases 3, 8, 9 antibodies (Cell Signaling Technology, Danvers, MA, USA), anti-phosphoinositide 3-kinase (PI3K) p110δ and anti-p-PI3Kp110δ (Tyr 485) antibodies (Santa Cruz), and anti-β-actin antibody (Sigma-Aldrich). Image J 1.44p software (National Institute of Health, Bethesda, MD, USA) was used for the densitometric measurement of proteins. Ratios of phosphorylated protein/total protein were calculated.

2.7. MTT Cytotoxicity Assay

The MTT [3-(4,5-Dimethylthiazol-2-yl)-2,5-Diphenyltetrazolium Bromide] assay (Sigma-Aldrich) was applied to evaluate the cytotoxicity of KAN0441571C, ibrutinib, and erlotinib.

Briefly, 20,000 lung cancer cells were cultured in 96 well plates in 200 μL of RPMI-1640 (ThermoFisher Scientific) (triplicates) containing 10% FBS and the drugs (diluted in DMSO). The maximum concentration of DMSO in the medium was 1%. Cells were then incubated at 37 °C for several time points (24, 48, and 72 h), followed by adding 20 μL of 5 mg/mL of the MTT solution to each well, and then further incubated for 4 h at 37 °C. Next, 100 microliters of MTT stop solution (10% SDS in 0.01 M HCL) was then added to wells and incubated for 2–4 h at 37 °C. Cells treated with DMSO alone were used as the control. Optical density (OD) was defined using a plate reader at 570 nm. The effects of the drugs in vitro on the apoptosis (additive, synergistic, or antagonist effects) of the combined treatment of KAN0441571C with erlotinib or ibrutinib was evaluated by the Chou–Talalay method using the CompuSyn software (Combosyn Inc., New York, NY, USA) [55].

2.8. *Apoptosis Assay (Flow Cytometry)*

Lung cancer cells were incubated with the drugs and cultured in 6-well plates (10^6/well) at 37 °C for 24 h. After incubation, cells were prepared and stained for apoptosis assay as previously described [13,46].

2.9. *Immunofluorescence (IF) Assay*

The immunofluorescent (IF) assay was completed as previously described [46]. Briefly, cell lines were cultured on a sterile 8-well glass slide (BD Biosciences) for 24 hours to form a monolayer. Next, 4% formaldehyde was used to fix cells (15 min), which were then washed with PBS and blocked for 2 h in buffer containing 0.01% sodium azide, 2% bovine serum albumin, and 1% Tween 20 in PBS buffer. Cells were incubated with an anti-ROR1 antibody (2 μg/mL) (Sigma-Aldrich) and a non-specific antibody (mouse IgG) (eBioscience, San Diego, CA, USA) for 24 h at 4 °C, washed with PBS, and then treated with Alexa Flour 488-conjugated goat anti-mouse IgG (1:200) (ThermoFisher Scientific) for 1 h. After washing, the cell nuclei were stained with mounting media (VectaShield H-1000) containing 4′,6-diamidino-2-phenylindole (DAPI) (Vector Laboratories, Burlingame, CA, USA). A Zeiss Axioplan2 fluorescence microscope (Oberkochen, Germany) with a 63 X objective lens and the ZEN software (Carl Zeiss Microscopy, Munich, Germany) was used to take pictures.

2.10. *Colony Formation Assay*

Two-hundred cells per well were seeded in 6 well plates (BD Biosciences) and incubated at 37 °C overnight. KAN0441571C and DMSO alone (control) were added and the cells were incubated for 72 h at 37 °C. The medium was removed and the cells were washed with PBS. Fresh medium was added and cell colonies were maintained in culture until control cultures were confluent. The medium was then removed and the colonies were stained with 4% methylene blue/MeOH (Sigma-Aldrich) for 30 min. Finally, the cells were washed with water and the plates were left to dry. Colonies were counted and the relative colony count to DMSO control was calculated using the following formula: plating efficiency (%) = number of colonies (treated)/average number of colonies (DMSO) × 100. For each cell line, the assay was repeated 2–4 times.

2.11. *Scratch Wound Healing Assay (Migration Assay)*

Lung cancer cells were seeded in 96-well plates (6×10^4 cells/well) and incubated overnight at 37 °C. When cells reached >90% confluence, the IncuCyte® wound maker tool (Sartorius AG, Gottingen, Germany) was used to scratch all wells uniformly. Cell migration in the presence of KAN0441571C was assessed by measuring the migration of cells into the scraped wound area. The wound-healing process was captured after 0, 24, and 48 h by IncuCyte® S3 Live-Cell Analysis System (Sartorius). Three different parameters, namely wound width (μM), wound confluence (%), and relative wound density (%), were determined by scratch wound cell migration analysis using IncuCyte™ Software. For each cell line, the test was done in quadruplicate.

2.12. Cell Cycle Analysis

For cell cycle analysis, 100,000 A549 cells/well were seeded in 6-well plates and incubated at 37 °C overnight. KAN0441571C was added (500–5000 nM) in medium with 10% FBS and incubated for a further 24 h. DMSO 1% was used as the control. After incubation, the medium was removed and the wells were washed once with PBS. Adherent cells were detached by trypsin, transferred to tubes, and washed twice with PBS. Next, 700 µL of ice-cold 100% ethanol was added drop-wise while vortexing the tube. The tubes were incubated for 30 min to 1 h on ice and kept at 4 °C for 24 h. For propidium iodide (PI) staining, the cells were centrifuged at 1000 g at 4 °C for 10 min. The supernatant was discarded and the cells were washed with 250 µL cold PBS with 2% BSA. The samples were then centrifuged at 1000 g, at 4 °C for 10 min. The cells were re-suspended in 300 µL of staining buffer containing 40 µg/mL PI, 100 µg/mL RNase A, and 0.1% Triton X 100/PBS. The cells were transferred to FACS tubes, incubated for 20 min in RT in the dark, and then analyzed by flow cytometry (Navios, Beckman Coulter, Indianapolis, IN, USA). Data was analyzed with the Kaluza software (Navios, Beckman Coulter).

2.13. Statistical Methods

For comparison of ROR1 expression in the patient groups, the Kruskal–Wallis test was applied. The Mann–Whitney U test was used to check the association of variables among various groups (R version 3.3.2, The R Foundation for Statistical Computing, Vienna, Austria). The Chi-square test was utilized to find differences in the expression of ROR1. Time to progression (TTP) from diagnosis was demonstrated by the Kaplan–Meier method. Overall survival (OS) was determined from the time of diagnosis to death or last follow-up. Statistical significance was assessed by the log-rank test. For multivariable analyses, Cox regression models were used. Student's t-test and Mann–Whitney U test were utilized for comparison of EC_{50} values (GraphPad Software, Inc., La Jolla, CA, USA). p-values < 0.05 were considered significant. Asterisks represent p values of: *: 0.01 to 0.05, **: 0.001 to 0.01, ***: <0.001.

3. Results

3.1. ROR1 Expression in Subtypes of Lung Cancer and Cell Lines

ROR1 expression was significantly higher in non-squamous (136/157, 86%) and squamous (60/106, 57%) lung cancer patients compared to 5/24 (21%) patients with neuroendocrine tumors (p = 0.0001) (Table S1). No association between ROR1 expression and pathological disease stage was noted, irrespective of the histopathological subtype (Table S2). In adenocarcinoma, no statistical association between ROR1 expression and age, grading, smoking status, and overall survival was seen (Table S4). However, a trend to shorter TTP in $ROR1^+$ vs $ROR1^-$ patients was found (p = 0.09) (Figure S1). In addition, a significant higher proportion of p53 negative cases (63%) was seen in the $ROR1^+$ group as compared to p53 positive patients (25%) (p = 0.03) (Table S4).

In patients with squamous NSCLC, no statistically significant association was observed between ROR1 expression and age, stage, grading, smoking status, p53 expression, and overall survival.

Surface ROR1 expression was detected in all cell lines with the exception for NCI-H23 (Figure 1A). However, using Western blot and IF, ROR1 was shown to be expressed in all cell lines (Figure 1B and Figure S2). ROR1 was phosphorylated in all cell lines (Figure 1B).

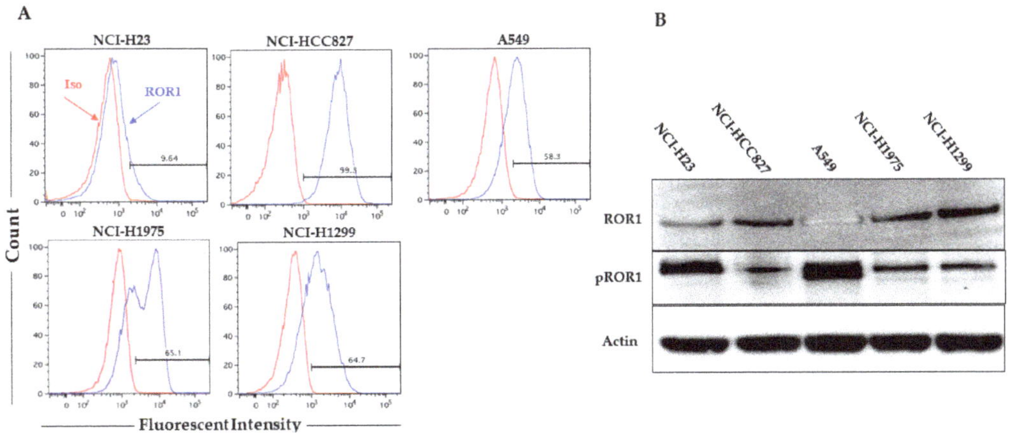

Figure 1. (**A**) Surface staining for ROR1 (flow cytometry) (red line: isotype control; blue line: anti-ROR1 antibody); (**B**) Western blot for ROR1 and pROR1 expression (130 KDa) in 5 lung cancer cell lines. Actin was used as internal control.

3.2. *Cytotoxicity of KAN0441571C, Ibrutinib and Erlotinib in Lung Cancer Cell Lines*

KAN0441571C induced a dose-dependent cytotoxicity (MTT) in all cell lines with the exception for A549. The cytotoxic effect of KAN0441571C varied between cell lines. In comparison to ibrutinib and erlotinib, the cytotoxic effect of KAN0441571C was similar or significantly better. Both erlotinib and ibrutinib induced the highest cytotoxic effect in HCC827 with EGFR Del19 (mutated E746-A750) compared to cell lines without EGFR Del19 or of the EGFR wild-type. However, in the HCC827 cell line, a dose-dependency within the used concentrations of erlotinib and ibrutinib could not been seen (compared to KAN0441571C in A549) (Figure 2). EC_{50} values for erlotinib and ibrutinib were >10,000 nM for all cell lines with the exception for NCI-HCC827, where EC_{50} was <1250 nM for both erlotinib and ibrutinib (Figure 2).

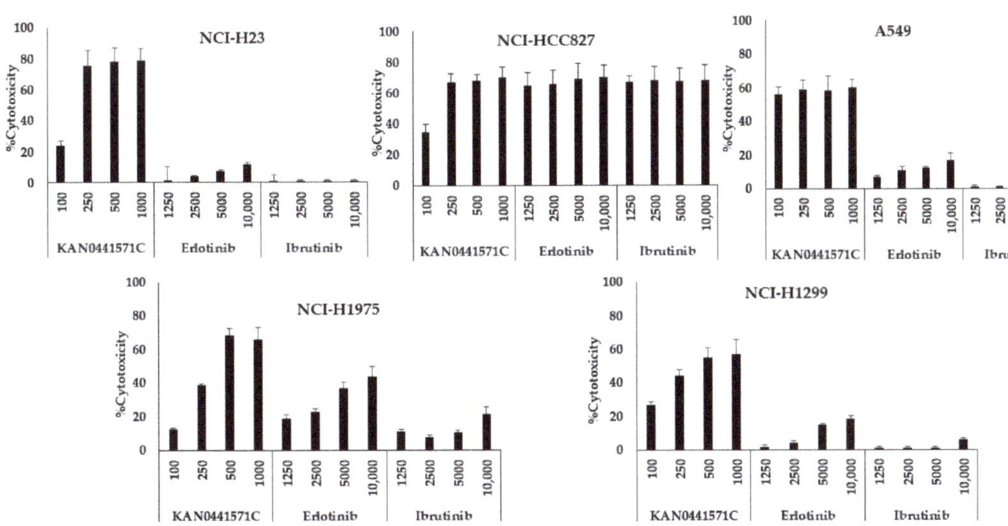

Figure 2. Cytotoxicity (MTT) (72 h) in 5 lung cancer cell lines incubated with various concentrations of KAN0441571C, erlotinib, or ibrutinib.

Apoptosis was confirmed by Annexin V/PI staining (Figure 3A). A dose-response relationship was noted for all cell lines with the exception for NCI-HCC827 (compared with MTT for KAN0441571C, which seemed to be the most resistant in the Annexin V/PI assay). EC_{50} values are also shown in Figure 3A. Apoptosis was accompanied by downregulation of BCL-2, and MCL-1, as well as by cleavage of PARP and caspases 3, 8, and 9, while the BAX protein was upregulated (Figure 3B,C). Caspase 8 and 9 cleavage indicates activation of both the extrinsic and intrinsic apoptotic pathways.

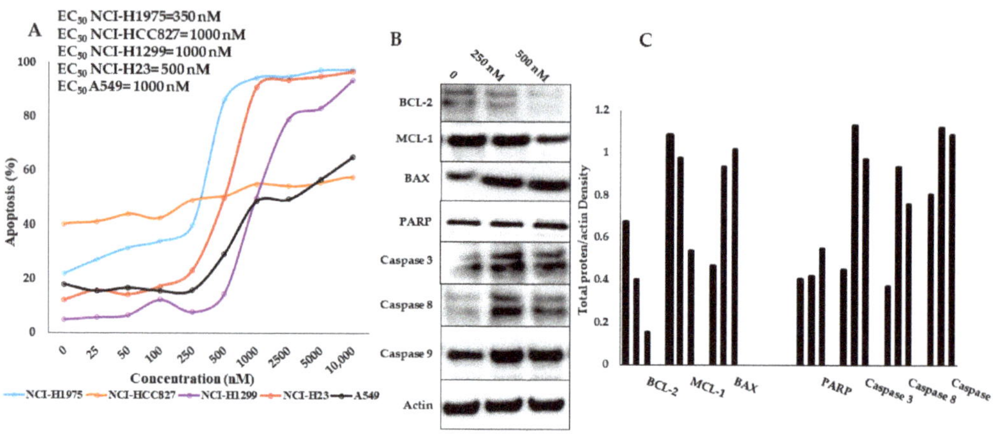

Figure 3. (**A**) Apoptosis (Annexin V/PI) of KAN0441571C in 5 lung cancer cell lines (EC_{50} values are shown to the upper left). (**B**) Western blots of apoptotic proteins BCL-2, MCL-1, BAX, PARP, and caspases-3, 8, and 9 after 24 h of incubation with KAN0441571C in NCI-H23 cells. One representative data of three individual experiments are shown. (**C**) Densitometric measurements of protein bands in Figure 3B expressed as total protein/actin intensity.

3.3. Colony Formation Assay

In a colony tumor cell formation assay (72 h), KAN0441571C reduced the ability of A549, NCI-HCC827, NCI-H23, and NCI-H1299 cells at concentrations ranging from 100 to 200 nM to form colonies, indicating that the drug compromised the clonogenic capacity of the cells (Figure 4) (NCI-H1975 was not included in this assay). Colonies were completely abolished at 200 nM of KAN0441571C in all cell lines. In H1299, only few colonies were seen at 125 nM. These findings indicate that inhibition of ROR1 may irreversible prevent NSCLC cells to recover after exposure to ROR1 inhibition.

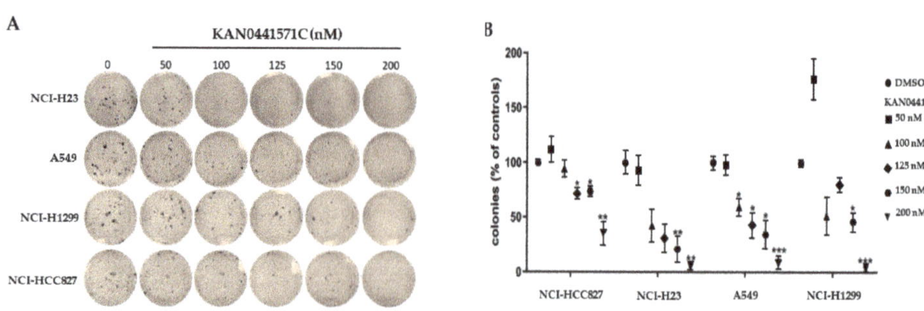

Figure 4. (**A**) Lung cancer cell lines (NCI-H23, A549, NCI-H1299, and NCI- HCC827) incubated with KAN0441571C (72 h) prevented formation of cell colonies in a dose-dependent manner (representative images). (**B**) Relative quantification of colonies (mean ± SEM) at various concentrations of KAN0441571C. * $p < 0.05$, ** $p < 0.01$, *** $p < 0.001$ (one-way ANOVA).

3.4. Migration of Lung Cancer Cells

The migratory ability of lung cancer cells was analyzed by the scratch wound healing assay (Figure 5 and Figure S3). After 48 h of incubation with KAN0441571C, untreated (DMSO alone) cells proliferated and migrated as expected. KAN0441571C inhibited migration in a dose- and time-dependent manner. At 1 µM, tumor cell mobility was completely abrogated.

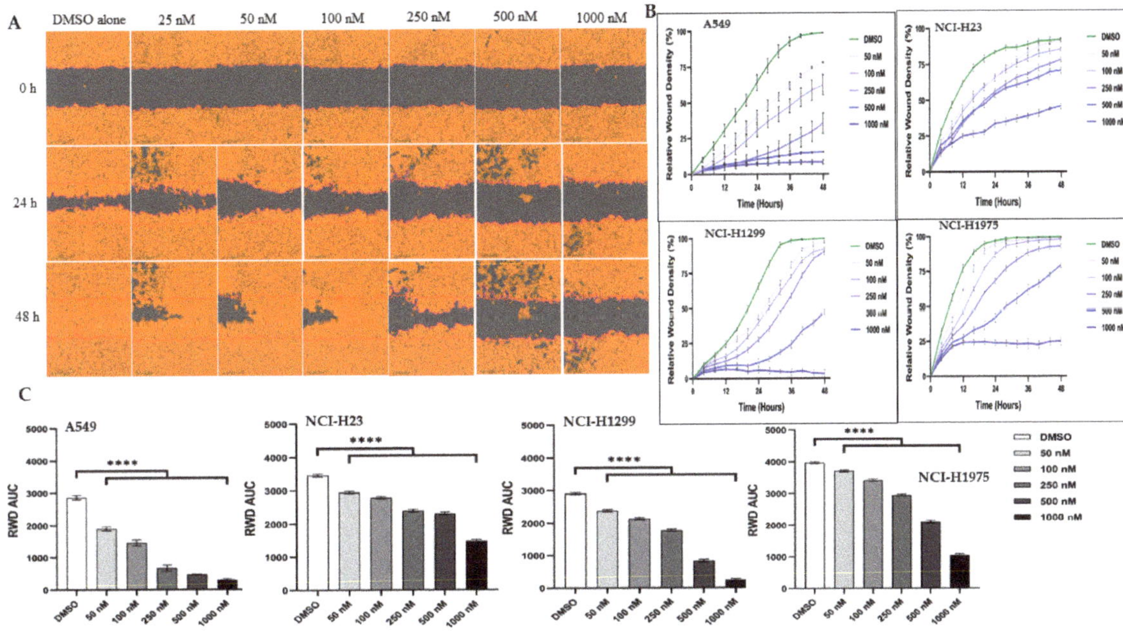

Figure 5. KAN0441571C inhibited migration of lung cancer cell lines in the scratch wound healing assay. (**A**) Migration of A549 was inhibited as shown by images acquired after 0, 24, and 48 h of incubation with different concentrations of KAN0441571C. Images of 1 representative experiment out of 4 for each cell line is shown. (**B**) Rate of wound closure (relative wound density) of lung cancer cells. The Incucyte Zoom Imager Software Analyzer was used to count the number of lung cancer cells during 48 h of incubation with of KAN0441571C (25–1000 nM). DMSO used as control. (**C**) Reduction (mean ± SEM) of cell migration (RWD AUC) after treatment with KAN0441571C (0–1000 nM) in 4 cell lines. **** $p < 0.0001$.

3.5. Cell Cycle Analysis

The cell cycle profile of the lung cancer cell line A549 cell was analyzed after treatment with KAN0441571C. A decrease in S phase cells, including the accumulation of G2/M phase cells and augmented cell death (sub-G1 cells), was noted with increasing concentrations of KAN0441571C in 24 h cell culture. A decrease in G1 cells was also seen at the highest concentration of KAN0441571C. Data indicate that replicating cells might be vulnerable to cell death at exposure to the ROR1 inhibitor KAN0441571C and/or that KAN0441571C induced a cytostatic effect in lung cancer cells preventing exit from G2/M phase and cell proliferation (Figure S4).

3.6. Effect of KAN0441571C on Signaling Molecules

As expected [56], KAN0441571C induced dephosphorylation of ROR1 in a dose-dependent manner, as well as EGFR, SRC, PI3K110δ/AKT/mTOR (non-canonical Wnt pathway), and CREB in the NCI-H23 lung cancer cell line (Figure S5).

3.7. *Effects on Tumor Cell Death of KAN0441571C Alone and in Combination with Ibrutinib or Erlotinib*

KAN0441571C induced a significant cell death (MTT) in all cell lines, at a concentration of 250 nM. For erlotinib and ibrutinib, significantly higher concentrations were required) 5000–10,000 nM) (Figure 6). The cytotoxic effect of KAN0441571C was significantly higher than that of erlotinib and ibrutinib with the exception of NCI-HCC827 (EGFR mutated E746-A750, Del19). When KAN0441571C was combined with ibrutinib or erlotinib, cytotoxicity increased significantly in 3 out of 5 cell lines.

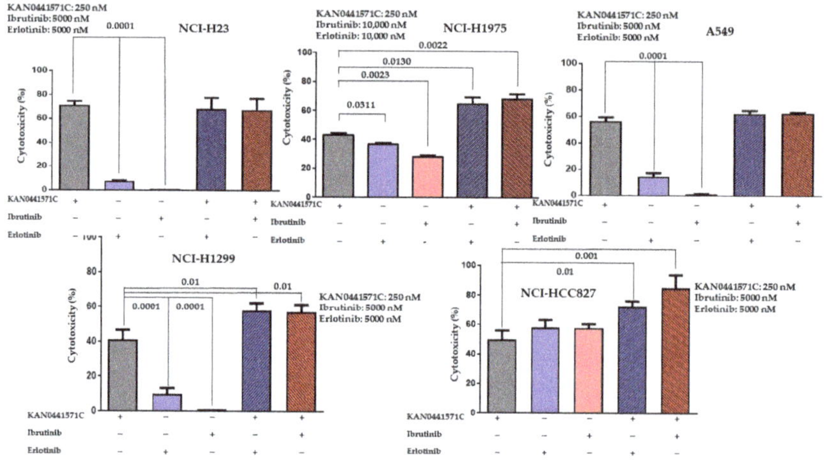

Figure 6. Cytotoxicity (MTT) (mean ± SEM) of five lung cancer cell lines incubated with KAN0441571C in combination with ibrutinib or erlotinib. Results of 3 independent experiments are shown. The concentration of the respective drug are shown for each cell line (−: No inhibitor added, +: inhibitor added).

To further assess the efficacy (antagonism, addition, or synergism) of the combinations of KAN0441571C/erlotinib and KAN0441571C/ibrutinib on apoptosis, the Chou–Talalay method was applied. The combinations had synergistic or additive apoptotic effect in all tested lung cancer cell lines (Figure 7 and Figure S6A,B).

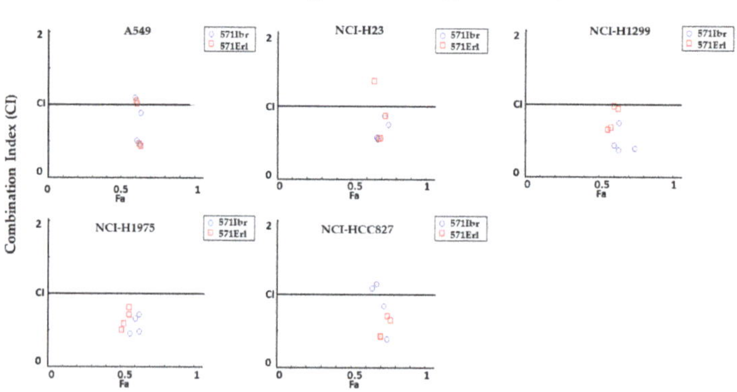

Figure 7. Combination index (CI) plots for apoptosis (Annexin V/PI) in lung cancer cell lines incubated (24 h) with a combination of KAN0441571C (250 nM) with ibrutinib (571Ibr) (blue symbols) or erlotinib (571Erl) (red symbols) (5000–10,000 nM, see Figure 6). Values across Fa in lung cancer cells are shown, where Fa is the % of cell death. CI < 1 synergistic effect; CI = 1 additive effect; CI > 1 antagonistic effect (Chou–Talalay method).

4. Discussion

In the present study, we analyzed ROR1 expression in primary tumors of patients with surgically resected NSCLC. In non-squamous and squamous carcinomas, 86% and 56%, respectively, expressed ROR1, while in neuroendocrine tumors only about 20% exhibited ROR1. No relationship between ROR1 expression and age, stage, tumor grading, smoking status, and overall survival was observed in adenocarcinoma or squamous NSCLC patients. However, an inverse relationship to p53 expression in non-squamous NSCLC ($p = 0.03$) was seen. Such a finding might be expected as low or no p53 expression is associated with a poor outcome in malignancies [57]. There was also a trend towards shorter TTP in ROR1$^+$ adenocarcinoma NSCLC patients compared to ROR1$^-$ patients.

KAN0441571C prevented ROR1 phosphorylation and inactivated both Wnt non-canonical pathway molecules and the transcription factor CREB. Apoptosis was induced in NSCLC cell lines both through the extrinsic and intrinsic pathways, including downregulation of pro-survival molecules (BCL-2 and MCL-1) and upregulation of the pro-apoptotic BAX protein. The ROR1 inhibitor was more effective than an EGFR inhibitor in inducing apoptosis. There seemed to be no relation between the response to KAN0441571C and mutations of EGFR, K-Ras, or p53. The combination of KAN0441571C with erlotinib or ibrutinib had synergistic apoptotic effects.

ROR1 is dysregulated during tumorigenesis and is constitutively expressed and phosphorylated in various tumors [13,29,31,39,58]. ROR1 has been shown to be a prognostic marker in several tumors [59]. A high level of ROR1 expression was noted to be associated with an aggressive and poor prognosis disease in, for example, CLL, DLBCL, MCL, ALL, as well as in ovarian, breast, pancreatic, gastric, colorectal, and lung cancers [15,18,21,32,59–63]. In meta-analyses, high ROR1 expression was found to relate to worse overall survival in hematologic malignancies and solid tumors [59,64]. The data may support an important role of ROR1 in cancer.

ROR1 is involved in a lot of functions in tumor cells, such as proliferation, survival, migration, stemness, epithelial to mesenchymal transition (EMT), chemotaxis, and drug resistance, through the planar cell polarity (PCP) activation and Ca2$^+$ dependent pathways of non-canonical Wnt signaling [25,26,65,66]. Binding of Wnt5a to ROR1 stimulates several signaling pathways, such as PI3K/AKT and RhoA/Rac1 GTPases, activating the transcriptional coactivator YAP/TAZ or polycomb complex protein BMI-1 to sustain stemness, metastatic ability, and drug resistance [66].

ROR1 has been described as an oncogenic molecule of interest for the development of targeted therapy for antibody-drug conjugates [67], chimeric antigen receptor T-cells (CAR-T) [68,69], mAbs [27], and bi-specific T-cell engager (BiTE) [70], as well as SMIs targeting the intracellular [20,56] or extracellular parts of ROR1 [42,43].

Targeting ROR1 by small molecules and mAbs has been shown to be an effective therapeutic approach in pre-clinical and clinical studies in various malignancies, such as CLL, MCL, lung, breast cancers, etc. [27,43,46,67,71]. The silencing of ROR1 significantly inhibited the proliferation of tumor cells in lung adenocarcinoma via the PI3K/AKT/mTOR signaling pathway [72,73]. NSCLC is still a major therapeutic challenge and there is a great need for new treatment alternatives.

In previous reports, we have described the effects of two small-molecule ROR1 inhibitors (KAN0439834 and KAN0441571C) targeting the cytoplasmic tyrosine kinase domain that dephosphorylated ROR1 and induced tumor cell apoptosis [13,20,46,74]. In the current study, we analyzed the effects of KAN0441571C in lung cancer cells. We also compared its tumor cell killing effect with erlotinib, which is one of the current drug options for the treatment of NSCLC patients with EGFR exon 19 deletions or exon 21 L858R substitution mutational status [75]. KAN0441571C was more effective than erlotinib in inducing apoptosis of lung cancer cells, irrespective of EGFR mutations. The combination of the two inhibitors had significantly improved the synergistic apoptotic effects, in the NCI-H1299, NCI-H1975, and NCI-HCC827 cell lines.

NKX2-1, a lineage-specific transcription factor, has been noted to be essential for the development of peripheral parts of the lung and in morphogenesis, and it is overexpressed in lung adenocarcinomas [24,76,77]. NKX2-1 induced ROR1 transcription and was critically involved in the maintenance of a balance between the pro-apoptotic p38 pathway and the pro-survival PI3K/AKT signaling pathway [16]. Furthermore, activated ROR1 binds to SRC and activates it. SRC activation inhibits PTEN activity and results in AKT activation. An association between EGFR mutations and NKX2-1 expression has been described in lung adenocarcinoma [78]. ROR1 has also been shown to form heterodimers with EGFR, promoting the maintenance of lung cancer cell survival [16,79]. High ROR1 expression in EGFR T790M NSCLC patients was related to an inferior progression-free survival in erlotinib-treated patients compared to those with low ROR1 expression, substantiating a role of ROR1 in the disease pathobiology and supporting that the inhibition of ROR1 might add to the therapeutic effect [79]. Overall, the data indicate a role of ROR1 expression in lung adenocarcinoma, which may be independent of ROR1 kinase activity [16].

ROR1 inhibition by KAN0441571C blocked the phosphorylation of ROR1 and EGFR. Dephosphorylation of EGFR may be due to the inhibition of the phosphorylation of ROR1, resulting in the inactivation of the ROR1-EGFR dimer and the suppression of survival signals [16,80].

In addition, to target the TK domain of ROR1 by KAN0441571C, ROR1 targeting SMIs against other extracellular and intracellular regions of ROR1 have been produced [42,43,46]. ARI-1 ((R)-5,7-bis (methoxymethoxy)-2-(4-methoxyphenyl) chroman-4-one) is a novel ROR1 inhibitor targeting the external CRD domain of ROR1, preventing NSCLC cell proliferation, migration, as well as inactivating the PI3K/AKT/mTOR signaling pathway [42]. ARI-1 decreased the growth of ROR1-expressing NSCLC in vitro and in vivo with no significant side effects in mice [42]. This inhibitor seemed to interrupt Wnt5a binding to ROR1 via CRD domain blockage.

Strictinin, isolated from *Myrothamnus flabellifolius*, targeted the intracellular region of ROR1 and induced apoptosis of triple-negative breast cancer cells (TNBC) through the intrinsic apoptotic pathway. Therefore, this inhibitor prevented ROR1 activation via inhibiting the binding to Wnt5a, dephosphorylating AKT and GSK3β, and blocking ROR1 signaling in a β-catenin-independent PI3K/AKT/GSK3ß signaling pathway [43].

In contrast to AR-1 and strictinin, KAN0441571C may not directly inhibit the binding of Wnt5a to ROR1, but instead might change the conformational structure of ROR1 [42,43], which thus prevents ROR1 from binding with Wnt5a. The treatment of human cells with KAN0441571C has also been shown to dephosphorylate SRC, which binds to phosphorylated intracellular regions of ROR1 [81].

In conclusion, ROR1 is expressed in NSCLC and seems to be of importance in the pathobiology of the disease. A ROR1 small molecule inhibitor (KAN0441571C) was highly effective in inducing apoptosis of NSCLC cells, and was superior to erlotinib and ibrutinib. The combination of KAN0441571C with erlotinib had significant synergistic cytotoxic effects. The development of new targeted drugs with other MOAs than those clinically available is warranted to improve the prognosis in NSCLC patients. A ROR1 inhibitor may be a new promising drug candidate.

Supplementary Materials: The following supporting information can be downloaded at: https://www.mdpi.com/article/10.3390/pharmaceutics15041148/s1, Figure S1: Time to progression (TTP) in non-squamous NSCLC patients with no ROR1 expression (red line) (n = 16) compared to those positive for ROR1 (green line) (n = 123) (p = 0.09). Figure S2: Immunofluorescent staining of lung cancer cell lines using anti-ROR1 antibodies (20×). Lung cancer cells were cultured on a sterile 8-well glass slide for 24 h, fixed, blocked in PBS containing BSA, and then incubated with an anti-ROR1 antibody for 24 h at 4 °C, and finally treated with Alexa Flour 488-conjugated goat anti-mouse IgG for 1 h. After washing, the cell nuclei were stained with mounting media containing DAPI. A Zeiss Axioplan2 fluorescence microscope was used. Figure S3: Migration of lung cancer cell lines (NCI-H1299 and NCI-H1975) in the wound scratch healing assay. Migration of lung cancer cell lines was inhibited by treatment with 50–1000 nM of KAN0441571C for 0, 24, and 72 h.

Figure S4: Cell cycle profiles of A549 lung cancer cells treated with KAN0441571C. KAN0441571C induced reduction of S phase cells and accumulation in the G2 phase with a simultaneous increased cell death. Representative histograms (A) and quantifications (B) of 3 independent experiments. Figure S5: Effect of KAN0441571C on ROR1-associated signaling molecules (Western blot) in the lung cancer cell line NCI-H23 (4 h). ROR1, EGFR, SRC, AKT, PI3K110δ, mTOR, and CREB were dephosphorylated. One of three representative experiments is shown. Figure S6A: Apoptosis (Annexin V/PI) in 5 lung cancer cell lines incubated with a combination of KAN0441571C (250 nM) with erlotinib (5000–10,000 nM). For each combination, data are shown as combination index plot (CI), isobologram, and dose reduction index (DRI). Combination index (CI) plot (Fa-CI plot): values across Fa in lung cancer cells, where Fa is the % of cell death. CI < 1 synergistic effect; CI = 1 additive effect; CI > 1 antagonistic effect. (Chou–Talalay) method). Figure S6B: Apoptosis (Annexin V/PI) in 5 lung cancer cell lines incubated with a combination of KAN0441571C (250 nM) with ibrutinib (5000–10,000 nM). For each combination, data are shown as combination index plot (CI), isobologram, and does reduction index (DRI). Combination index (CI) plot (Fa-CI-plot): values across Fa in lung cancer cell lines, where Fa is the % of cell death. CI < 1 synergistic effect; CI = 1 additive effect; CI > 1 antagonistic effect. (Chou–Talalay method). Table S1: Relation between ROR1 expression and subgroups of lung cancer patients. Table S2: ROR1 expression by pathological stage (7th TNM). Table S3: Characteristics of the ROR1 positive lung cancer cell lines. Table S4: Clinical characteristics of adenocarcinoma NSCLC patients ($n = 139$) in relation to ROR1 expression.

Author Contributions: A.G., M.-A.O., J.L., E.M.D., P.K. and W.Z. performed experiments, analyzed data and finalized the manuscript. J.S. and T.O. produced the small molecule and finalized the manuscript. L.D.P. and G.Z.R. provided clinical material, analyzed data and finalized the manuscript. E.D. and A.M. provided clinical material, analyzed data and finalized the manuscript. A.Ö., H.M. and M.H.-F. designed the study, performed experiments, supervised the study and wrote the manuscript. All authors have read and agreed to the published version of the manuscript.

Funding: This research was funded by Åke Olsson Foundation for hematology research, grant numbers 2017-00436; 2019-00396 (M.H.-F.). The Cancer and Allergy Foundation, grant numbers 2016/5, 2017/63, 2018/63, 2019/189, 2020/258 (A.Ö.); 2017/64, 2018/140, 2019/194, 2020/260 (H.M.). The Cancer Society in Stockholm, grant numbers 151313, 184203 (A.Ö.); 144142, 164122 (H.M.). The Swedish Cancer Society, grant numbers CAN 2015/408, CAN 2018/499 (A.Ö.). AFA Insurance, grant number 130054 (A.Ö.). The Karolinska Institutet Foundation (A.Ö., H.M.) and Region Stockholm, grant numbers 20150070, 20180030 (A.Ö.).

Institutional Review Board Statement: The study was conducted in accordance with Good Clinical Practice guidelines and the Declaration of Helsinki. All patients or their legally acceptable representative, or both (if possible), provided written informed consent. The study was authorized by the National Ethics Authority of Sweden (www.etikprovningsmyndigheten.se) (accessed on 3 August 2015) (approval code: 2016/2506-32 and date of approval: 20 December 2016).

Informed Consent Statement: Informed consent was obtained from all subjects involved in the study.

Data Availability Statement: Not applicable.

Acknowledgments: For excellent secretarial assistance we thank Leila Relander.

Conflicts of Interest: J.L. was employed by Kancera AB at the time of the project experiments and manuscript preparation, J.S. is employee and shareholder of Kancera AB. T.O. is founder, employee, and shareholder of Kancera AB. P.K. and M.H.F. are shareholders of Kancera AB. H.M. is founder and shareholder of Kancera AB and received research grant and honorarium from Kancera AB. The remaining authors declare no conflict of interest.

References

1. Sung, H.; Ferlay, J.; Siegel, R.L.; Laversanne, M.; Soerjomataram, I.; Jemal, A.; Bray, F. Global Cancer Statistics 2020: GLOBOCAN Estimates of Incidence and Mortality Worldwide for 36 Cancers in 185 Countries. *CA Cancer J. Clin.* **2021**, *71*, 209–249. [CrossRef] [PubMed]
2. Siegel, R.; DeSantis, C.; Virgo, K.; Stein, K.; Mariotto, A.; Smith, T.; Cooper, D.; Gansler, T.; Lerro, C.; Fedewa, Ś.; et al. Cancer treatment and survivorship statistics, 2012. *CA Cancer J. Clin.* **2012**, *62*, 220–241. [CrossRef] [PubMed]
3. Hojjat-Farsangi, M. Small-Molecule inhibitors of the receptor tyrosine kinases: Promising tools for targeted cancer therapies. *Int. J. Mol. Sci.* **2014**, *15*, 13768–13801. [CrossRef]

4. Masiakowski, P.; Carroll, R.D. A novel family of cell surface receptors with tyrosine kinase-like domain. *J. Biol. Chem.* **1992**, *267*, 26181–26190. [CrossRef] [PubMed]
5. Stricker, S.; Rauschenberger, V.; Schambony, A. ROR-Family Receptor Tyrosine Kinases. *Curr. Top. Dev. Biol.* **2017**, *123*, 105–142.
6. Balakrishnan, A.; Goodpaster, T.; Randolph-Habecker, J.; Hoffstrom, B.G.; Jalikis, F.G.; Koch, L.K.; Berger, C.; Kosasih, P.L.; Rajan, A.; Sommermeyer, D.; et al. Analysis of ROR1 Protein Expression in Human Cancer and Normal Tissues. *Clin. Cancer Res.* **2017**, *23*, 3061–3071. [CrossRef]
7. Hudecek, M.; Schmitt, T.M.; Baskar, S.; Lupo-Stanghellini, M.T.; Nishida, T.; Yamamoto, T.N.; Bleakley, M.; Turtle, C.J.; Chang, W.C.; Greisman, H.A.; et al. The B-cell tumor-associated antigen ROR1 can be targeted with T cells modified to express a ROR1-specific chimeric antigen receptor. *Blood* **2010**, *116*, 4532–4541. [CrossRef]
8. Yoda, A.; Oishi, I.; Minami, Y. Expression and function of the Ror-family receptor tyrosine kinases during development: Lessons from genetic analyses of nematodes, mice, and humans. *J. Recept. Signal. Transduct. Res.* **2003**, *23*, 1–15. [CrossRef]
9. Billiard, J.; Way, D.S.; Seestaller-Wehr, L.M.; Moran, R.A.; Mangine, A.; Bodine, P.V. The orphan receptor tyrosine kinase Ror2 modulates canonical Wnt signaling in osteoblastic cells. *Mol. Endocrinol.* **2005**, *19*, 90–101. [CrossRef]
10. Daneshmanesh, A.H.; Mikaelsson, E.; Jeddi-Tehrani, M.; Bayat, A.A.; Ghods, R.; Ostadkarampour, M.; Akhondi, M.; Lagercrantz, S.; Larsson, C.; Osterborg, A.; et al. Ror1, a cell surface receptor tyrosine kinase is expressed in chronic lymphocytic leukemia and may serve as a putative target for therapy. *Int. J. Cancer* **2008**, *123*, 1190–1195. [CrossRef]
11. Baskar, S.; Kwong, K.Y.; Hofer, T.; Levy, J.M.; Kennedy, M.G.; Lee, E.; Staudt, L.M.; Wilson, W.H.; Wiestner, A.; Rader, C. Unique cell surface expression of receptor tyrosine kinase ROR1 in human B-cell chronic lymphocytic leukemia. *Clin. Cancer Res.* **2008**, *14*, 396–404. [CrossRef]
12. Karvonen, H.; Chiron, D.; Niininen, W.; Ek, S.; Jerkeman, M.; Moradi, E.; Nykter, M.; Heckman, C.A.; Kallioniemi, O.; Murumagi, A.; et al. Crosstalk between ROR1 and BCR pathways defines novel treatment strategies in mantle cell lymphoma. *Blood Adv.* **2017**, *1*, 2257–2268. [CrossRef]
13. Ghaderi, A.; Daneshmanesh, A.H.; Moshfegh, A.; Kokhaei, P.; Vagberg, J.; Schultz, J.; Olin, T.; Harrysson, S.; Smedby, K.E.; Drakos, E.; et al. ROR1 Is Expressed in Diffuse Large B-Cell Lymphoma (DLBCL) and a Small Molecule Inhibitor of ROR1 (KAN0441571C) Induced Apoptosis of Lymphoma Cells. *Biomedicines* **2020**, *8*, 170. [CrossRef]
14. Lovat, F.; Gasparini, P.; Nigita, G.; Larkin, K.; Byrd, J.C.; Minden, M.D.; Andreeff, M.; Carter, B.Z.; Croce, C.M. Loss of expression of both miR-15/16 loci in CML transition to blast crisis. *Proc. Natl. Acad. Sci. USA* **2021**, *118*, e2101566118. [CrossRef] [PubMed]
15. Bicocca, V.T.; Chang, B.H.; Masouleh, B.K.; Muschen, M.; Loriaux, M.M.; Druker, B.J.; Tyner, J.W. Crosstalk between ROR1 and the Pre-B cell receptor promotes survival of t(1;19) acute lymphoblastic leukemia. *Cancer Cell* **2012**, *22*, 656–667. [CrossRef] [PubMed]
16. Yamaguchi, T.; Yanagisawa, K.; Sugiyama, R.; Hosono, Y.; Shimada, Y.; Arima, C.; Kato, S.; Tomida, S.; Suzuki, M.; Osada, H.; et al. NKX2-1/TITF1/TTF-1-Induced ROR1 is required to sustain EGFR survival signaling in lung adenocarcinoma. *Cancer Cell* **2012**, *21*, 348–361. [CrossRef] [PubMed]
17. Gentile, A.; Lazzari, L.; Benvenuti, S.; Trusolino, L.; Comoglio, P.M. Ror1 is a pseudokinase that is crucial for Met-driven tumorigenesis. *Cancer Res.* **2011**, *71*, 3132–3141. [CrossRef]
18. Avasarala, S.; Bikkavilli, R.K.; Van Scoyk, M.; Zhang, W.; Lapite, A.; Hostetter, L.; Byers, J.T.; Heasley, L.E.; Sohn, J.W.; Winn, R.A. Heterotrimeric G-protein, Galpha16, is a critical downstream effector of non-canonical Wnt signaling and a potent inhibitor of transformed cell growth in non small cell lung cancer. *PLoS ONE* **2013**, *8*, e76895. [CrossRef] [PubMed]
19. Zhang, R.; Wang, Z.; Yu, Q.; Shen, J.; He, W.; Zhou, D.; Yu, Q.; Fan, J.; Gao, S.; Duan, L. Atractylenolide II reverses the influence of lncRNA XIST/miR-30a-5p/ROR1 axis on chemo-resistance of colorectal cancer cells. *J. Cell. Mol. Med.* **2019**, *23*, 3151–3165. [CrossRef]
20. Daneshmanesh, A.H.; Hojjat-Farsangi, M.; Ghaderi, A.; Moshfegh, A.; Hansson, L.; Schultz, J.; Vagberg, J.; Bystrom, S.; Olsson, E.; Olin, T.; et al. A receptor tyrosine kinase ROR1 inhibitor (KAN0439834) induced significant apoptosis of pancreatic cells which was enhanced by erlotinib and ibrutinib. *PLoS ONE* **2018**, *13*, e0198038. [CrossRef] [PubMed]
21. Zhang, H.; Qiu, J.; Ye, C.; Yang, D.; Gao, L.; Su, Y.; Tang, X.; Xu, N.; Zhang, D.; Xiong, L.; et al. ROR1 expression correlated with poor clinical outcome in human ovarian cancer. *Sci. Rep.* **2014**, *4*, 5811. [CrossRef] [PubMed]
22. Fernandez, N.B.; Lorenzo, D.; Picco, M.E.; Barbero, G.; Dergan-Dylon, L.S.; Marks, M.P.; Garcia-Rivello, H.; Gimenez, L.; Labovsky, V.; Grumolato, L.; et al. ROR1 contributes to melanoma cell growth and migration by regulating N-cadherin via the PI3K/Akt pathway. *Mol carcinog* **2016**, *55*, 1772–1785. [CrossRef] [PubMed]
23. Schiavone, G.; Epistolio, S.; Martin, V.; Molinari, F.; Barizzi, J.; Mazzucchelli, L.; Frattini, M.; Wannesson, L. Functional and clinical significance of ROR1 in lung adenocarcinoma. *BMC Cancer* **2020**, *20*, 1085.
24. Ida, L.; Yamaguchi, T.; Yanagisawa, K.; Kajino, T.; Shimada, Y.; Suzuki, M.; Takahashi, T. Receptor tyrosine kinase-like orphan receptor 1, a target of NKX2-1/TTF-1 lineage-survival oncogene, inhibits apoptosis signal-regulating kinase 1-mediated pro-apoptotic signaling in lung adenocarcinoma. *Cancer Sci.* **2016**, *107*, 155–161. [CrossRef] [PubMed]
25. Cui, B.; Zhang, S.; Chen, L.; Yu, J.; Widhopf, G.F., 2nd; Fecteau, J.F.; Rassenti, L.Z.; Kipps, T.J. Targeting ROR1 inhibits epithelial-mesenchymal transition and metastasis. *Cancer Res.* **2013**, *73*, 3649–3660. [CrossRef] [PubMed]
26. Long, M.P.; Wang, H.L.; Luo, Y.B.; Yang, J.H. Targeting ROR1 inhibits epithelial to mesenchymal transition in human lung adenocarcinoma via mTOR signaling pathway. *Int. J. Clin. Exp. Pathol.* **2018**, *11*, 4759–4770.

27. Choi, M.Y.; Widhopf, G.F., 2nd; Ghia, E.M.; Kidwell, R.L.; Hasan, M.K.; Yu, J.; Rassenti, L.Z.; Chen, L.; Chen, Y.; Pittman, E.; et al. Phase I Trial: Cirmtuzumab Inhibits ROR1 Signaling and Stemness Signatures in Patients with Chronic Lymphocytic Leukemia. *Cell Stem Cell* **2018**, *22*, 951–959.e3. [CrossRef]
28. Jung, E.H.; Lee, H.N.; Han, G.Y.; Kim, M.J.; Kim, C.W. Targeting ROR1 inhibits the self-renewal and invasive ability of glioblastoma stem cells. *Cell. Biochem. Funct.* **2016**, *34*, 149–157. [CrossRef]
29. Zhang, S.; Cui, B.; Lai, H.; Liu, G.; Ghia, E.M.; Widhopf, G.F., 2nd; Rassenti, L.Z.; Chen, L.; Chen, Y.; Pittman, E.; et al. Ovarian cancer stem cells express ROR1, which can be targeted for anti-cancer-stem-cell therapy. *Proc. Natl. Acad. Sci. USA* **2014**, *111*, 17266–17271. [CrossRef]
30. Fukuda, T.; Chen, L.; Endo, T.; Tang, L.; Lu, D.; Castro, J.E.; Widhopf, G.F., 2nd; Rassenti, L.Z.; Cantwell, M.J.; Prussak, C.E.; et al. Antisera induced by infusions of autologous Ad-CD154-leukemia B cells identify ROR1 as an oncofetal antigen and receptor for Wnt5a. *Proc. Natl. Acad. Sci. USA* **2008**, *105*, 3047–3052. [CrossRef]
31. Zhang, S.; Chen, L.; Cui, B.; Chuang, H.Y.; Yu, J.; Wang-Rodriguez, J.; Tang, L.; Chen, G.; Basak, G.W.; Kipps, T.J. ROR1 is expressed in human breast cancer and associated with enhanced tumor-cell growth. *PLoS ONE* **2012**, *7*, e31127. [CrossRef]
32. Hojjat-Farsangi, M.; Khan, A.S.; Daneshmanesh, A.H.; Moshfegh, A.; Sandin, A.; Mansouri, L.; Palma, M.; Lundin, J.; Osterborg, A.; Mellstedt, H. The tyrosine kinase receptor ROR1 is constitutively phosphorylated in chronic lymphocytic leukemia (CLL) cells. *PLoS ONE* **2013**, *8*, e78339. [CrossRef] [PubMed]
33. Mikels, A.J.; Nusse, R. Purified Wnt5a protein activates or inhibits beta-catenin-TCF signaling depending on receptor context. *PLoS Biol.* **2006**, *4*, e115. [CrossRef] [PubMed]
34. Nomachi, A.; Nishita, M.; Inaba, D.; Enomoto, M.; Hamasaki, M.; Minami, Y. Receptor tyrosine kinase Ror2 mediates Wnt5a-induced polarized cell migration by activating c-Jun N-terminal kinase via actin-binding protein filamin A. *J. Biol. Chem.* **2008**, *283*, 27973–27981. [CrossRef] [PubMed]
35. Zhuo, W.; Kang, Y. Lnc-ing ROR1-HER3 and Hippo signalling in metastasis. *Nat. Cell. Biol.* **2017**, *19*, 81–83. [CrossRef]
36. Hojjat-Farsangi, M.; Moshfegh, A.; Daneshmanesh, A.H.; Khan, A.S.; Mikaelsson, E.; Osterborg, A.; Mellstedt, H. The receptor tyrosine kinase ROR1–an oncofetal antigen for targeted cancer therapy. *Semin. Cancer Biol.* **2014**, *29*, 21–31. [CrossRef] [PubMed]
37. Mao, Y.; Xu, L.; Wang, J.; Zhang, L.; Hou, N.; Xu, J.; Wang, L.; Yang, S.; Chen, Y.; Xiong, L.; et al. ROR1 associates unfavorable prognosis and promotes lymphoma growth in DLBCL by affecting PI3K/Akt/mTOR signaling pathway. *Biofactors* **2019**, *45*, 416–426. [CrossRef]
38. Cui, B.; Ghia, E.M.; Chen, L.; Rassenti, L.Z.; DeBoever, C.; Widhopf, G.F., 2nd; Yu, J.; Neuberg, D.S.; Wierda, W.G.; Rai, K.R.; et al. High-Level ROR1 associates with accelerated disease progression in chronic lymphocytic leukemia. *Blood* **2016**, *128*, 2931–2940. [CrossRef]
39. Daneshmanesh, A.H.; Porwit, A.; Hojjat-Farsangi, M.; Jeddi-Tehrani, M.; Tamm, K.P.; Grander, D.; Lehmann, S.; Norin, S.; Shokri, F.; Rabbani, H.; et al. Orphan receptor tyrosine kinases ROR1 and ROR2 in hematological malignancies. *Leuk. Lymphoma* **2013**, *54*, 843–850. [CrossRef]
40. Ikeda, T.; Nishita, M.; Hoshi, K.; Honda, T.; Kakeji, Y.; Minami, Y. Mesenchymal stem cell-derived CXCL16 promotes progression of gastric cancer cells by STAT3-mediated expression of Ror1. *Cancer Sci.* **2020**, *111*, 1254–1265. [CrossRef]
41. Choudhury, A.; Derkow, K.; Daneshmanesh, A.H.; Mikaelsson, E.; Kiaii, S.; Kokhaei, P.; Osterborg, A.; Mellstedt, H. Silencing of ROR1 and FMOD with siRNA results in apoptosis of CLL cells. *Br. J. Haematol.* **2010**, *151*, 327–335. [CrossRef]
42. Liu, X.; Pu, W.; He, H.; Fan, X.; Zheng, Y.; Zhou, J.K.; Ma, R.; He, J.; Zheng, Y.; Wu, K.; et al. Novel ROR1 inhibitor ARI-1 suppresses the development of non-small cell lung cancer. *Cancer Lett.* **2019**, *458*, 76–85. [CrossRef] [PubMed]
43. Fultang, N.; Illendula, A.; Chen, B.; Wu, C.; Jonnalagadda, S.; Baird, N.; Klase, Z.; Peethambaran, B. Strictinin, a novel ROR1-inhibitor, represses triple negative breast cancer survival and migration via modulation of PI3K/AKT/GSK3ss activity. *PLoS ONE* **2019**, *14*, e0217789. [CrossRef] [PubMed]
44. Srivastava, S.; Furlan, S.N.; Jaeger-Ruckstuhl, C.A.; Sarvothama, M.; Berger, C.; Smythe, K.S.; Garrison, S.M.; Specht, J.M.; Lee, S.M.; Amezquita, R.A.; et al. Immunogenic Chemotherapy Enhances Recruitment of CAR-T Cells to Lung Tumors and Improves Antitumor Efficacy when Combined with Checkpoint Blockade. *Cancer Cell.* **2021**, *39*, 193–208.e10. [CrossRef]
45. Khaledian, B.; Taguchi, A.; Shin-Ya, K.; Kondo-Ida, L.; Kagaya, N.; Suzuki, M.; Kajino, T.; Yamaguchi, T.; Shimada, Y.; Takahashi, T. Inhibition of heat shock protein 90 destabilizes receptor tyrosine kinase ROR1 in lung adenocarcinoma. *Cancer Sci.* **2021**, *112*, 1225–1234. [CrossRef] [PubMed]
46. Hojjat-Farsangi, M.; Daneshmanesh, A.H.; Khan, A.S.; Shetye, J.; Mozaffari, F.; Kharaziha, P.; Rathje, L.S.; Kokhaei, P.; Hansson, L.; Vagberg, J.; et al. First-in-Class oral small molecule inhibitor of the tyrosine kinase ROR1 (KAN0439834) induced significant apoptosis of chronic lymphocytic leukemia cells. *Leukemia* **2018**, *32*, 2291–2295. [CrossRef]
47. Wang, W.Z.; Shilo, K.; Amann, J.M.; Shulman, A.; Hojjat-Farsangi, M.; Mellstedt, H.; Schultz, J.; Croce, C.M.; Carbone, D.P. Predicting ROR1/BCL2 combination targeted therapy of small cell carcinoma of the lung. *Cell Death Dis.* **2021**, *12*, 577. [CrossRef] [PubMed]
48. Kumar, A.; Wakelee, H. Second- and third-line treatments in non-small cell lung cancer. *Curr. Treat. Options Oncol.* **2006**, *7*, 37–49. [CrossRef]
49. Hong, D.; Rasco, D.; Veeder, M.; Luke, J.J.; Chandler, J.; Balmanoukian, A.; George, T.J.; Munster, P.; Berlin, J.D.; Gutierrez, M.; et al. A Phase 1b/2 Study of the Bruton Tyrosine Kinase Inhibitor Ibrutinib and the PD-L1 Inhibitor Durvalumab in Patients with Pretreated Solid Tumors. *Oncology* **2019**, *97*, 102–111. [CrossRef]

50. Wang, A.; Yan, X.E.; Wu, H.; Wang, W.; Hu, C.; Chen, C.; Zhao, Z.; Zhao, P.; Li, X.; Wang, L.; et al. Ibrutinib targets mutant-EGFR kinase with a distinct binding conformation. *Oncotarget* **2016**, *7*, 69760–69769. [CrossRef]
51. Travis, W.D.; Brambilla, E.; Nicholson, A.G.; Yatabe, Y.; Austin, J.H.M.; Beasley, M.B.; Chirieac, L.R.; Dacic, S.; Duhig, E.; Flieder, D.B.; et al. The 2015 World Health Organization Classification of Lung Tumors: Impact of Genetic, Clinical and Radiologic Advances Since the 2004 Classification. *J. Thorac. Oncol.* **2015**, *10*, 1243–1260. [CrossRef]
52. Mengoli, M.C.; Longo, F.R.; Fraggetta, F.; Cavazza, A.; Dubini, A.; Ali, G.; Guddo, F.; Gilioli, E.; Bogina, G.; Nannini, N.; et al. The 2015 World Health Organization Classification of lung tumors: New entities since the 2004 Classification. *Pathologica* **2018**, *110*, 39–67.
53. Micke, P.; Mattsson, J.S.; Djureinovic, D.; Nodin, B.; Jirstrom, K.; Tran, L.; Jonsson, P.; Planck, M.; Botling, J.; Brunnstrom, H. The Impact of the Fourth Edition of the WHO Classification of Lung Tumours on Histological Classification of Resected Pulmonary NSCCs. *J. Thorac. Oncol.* **2016**, *11*, 862–872. [CrossRef]
54. Akyurek, N.; Drakos, E.; Giaslakiotis, K.; Knoblock, R.J.; Abruzzo, L.V.; Ning, Y.; Rassidakis, G.Z.; Medeiros, L.J. Differential expression of CKS-1B in typical and blastoid variants of mantle cell lymphoma. *Hum. Pathol.* **2010**, *41*, 1448–1455. [CrossRef]
55. Chou, T.C. Drug combination studies and their synergy quantification using the Chou-Talalay method. *Cancer Res.* **2010**, *70*, 440–446. [CrossRef]
56. Hojjat-Farsangi, M.; Moshfegh, A.; Schultz, J.; Norin, M.; Olin, T.; Osterborg, A.; Mellstedt, H. Targeting the Receptor Tyrosine Kinase ROR1 by Small Molecules. *Handb. Exp. Pharmacol.* **2021**, *269*, 75–99.
57. Kwan, K.; Castro-Sandoval, O.; Gaiddon, C.; Storr, T. Inhibition of p53 protein aggregation as a cancer treatment strategy. *Curr. Opin. Chem. Biol.* **2022**, *72*, 102230. [CrossRef] [PubMed]
58. Zhang, S.; Chen, L.; Wang-Rodriguez, J.; Zhang, L.; Cui, B.; Frankel, W.; Wu, R.; Kipps, T.J. The onco-embryonic antigen ROR1 is expressed by a variety of human cancers. *Am. J. Pathol.* **2012**, *181*, 1903–1910. [CrossRef] [PubMed]
59. Saleh, R.R.; Antras, J.F.; Peinado, P.; Perez-Segura, P.; Pandiella, A.; Amir, E.; Ocana, A. Prognostic value of receptor tyrosine kinase-like orphan receptor (ROR) family in cancer: A meta-analysis. *Cancer Treat. Rev.* **2019**, *77*, 11–19. [CrossRef] [PubMed]
60. Chien, H.P.; Ueng, S.H.; Chen, S.C.; Chang, Y.S.; Lin, Y.C.; Lo, Y.F.; Chang, H.K.; Chuang, W.Y.; Huang, Y.T.; Cheung, Y.C.; et al. Expression of ROR1 has prognostic significance in triple negative breast cancer. *Virchows Arch.* **2016**, *468*, 589–595. [CrossRef]
61. Zhou, J.K.; Zheng, Y.Z.; Liu, X.S.; Gou, Q.; Ma, R.; Guo, C.L.; Croce, C.M.; Liu, L.; Peng, Y. ROR1 expression as a biomarker for predicting prognosis in patients with colorectal cancer. *Oncotarget* **2017**, *8*, 32864–32872. [CrossRef] [PubMed]
62. Cetin, M.; Odabas, G.; Douglas, L.R.; Duriez, P.J.; Balcik-Ercin, P.; Yalim-Camci, I.; Sayan, A.E.; Yagci, T. ROR1 Expression and Its Functional Significance in Hepatocellular Carcinoma Cells. *Cells* **2019**, *8*, 210. [CrossRef]
63. Zheng, Y.Z.; Ma, R.; Zhou, J.K.; Guo, C.L.; Wang, Y.S.; Li, Z.G.; Liu, L.X.; Peng, Y. ROR1 is a novel prognostic biomarker in patients with lung adenocarcinoma. *Sci. Rep.* **2016**, *6*, 36447. [CrossRef]
64. Jeong, S.Y.; Lee, K.J.; Cha, J.; Park, S.Y.; Kim, H.S.; Kim, J.H.; Lee, J.J.; Kim, N.; Park, S.T. Meta-Analysis of Survival Effects of Receptor Tyrosine Kinase-like Orphan Receptor 1 (ROR1). *Medicina* **2022**, *58*, 1867. [CrossRef] [PubMed]
65. Katoh, M. Canonical and non-canonical WNT signaling in cancer stem cells and their niches: Cellular heterogeneity, omics reprogramming, targeted therapy and tumor plasticity (Review). *Int. J. Oncol.* **2017**, *51*, 1357–1369. [CrossRef]
66. Karvonen, H.; Barker, H.; Kaleva, L.; Niininen, W.; Ungureanu, D. Molecular Mechanisms Associated with ROR1-Mediated Drug Resistance: Crosstalk with Hippo-YAP/TAZ and BMI-1 Pathways. *Cells* **2019**, *8*, 812. [CrossRef]
67. Islam, S.S.; Uddin, M.; Noman, A.S.M.; Akter, H.; Dity, N.J.; Basiruzzman, M.; Uddin, F.; Ahsan, J.; Annoor, S.; Alaiya, A.A.; et al. Antibody-drug conjugate T-DM1 treatment for HER2+ breast cancer induces ROR1 and confers resistance through activation of Hippo transcriptional coactivator YAP1. *EBioMedicine* **2019**, *43*, 211–224. [CrossRef]
68. Huang, X.; Park, H.; Greene, J.; Pao, J.; Mulvey, E.; Zhou, S.X.; Albert, C.M.; Moy, F.; Sachdev, D.; Yee, D.; et al. IGF1R- and ROR1-Specific CAR T Cells as a Potential Therapy for High Risk Sarcomas. *PLoS ONE* **2015**, *10*, e0133152. [CrossRef] [PubMed]
69. Wallstabe, L.; Gottlich, C.; Nelke, L.C.; Kuhnemundt, J.; Schwarz, T.; Nerreter, T.; Einsele, H.; Walles, H.; Dandekar, G.; Nietzer, S.L.; et al. ROR1-CAR T cells are effective against lung and breast cancer in advanced microphysiologic 3D tumor models. *JCI Insight* **2019**, *4*, e126345. [CrossRef]
70. Gohil, S.H.; Paredes-Moscosso, S.R.; Harrasser, M.; Vezzalini, M.; Scarpa, A.; Morris, E.; Davidoff, A.M.; Sorio, C.; Nathwani, A.C.; Della Peruta, M. An ROR1 bi-specific T-cell engager provides effective targeting and cytotoxicity against a range of solid tumors. *Oncoimmunology* **2017**, *6*, e1326437. [CrossRef]
71. Daneshmanesh, A.H.; Hojjat-Farsangi, M.; Khan, A.S.; Jeddi-Tehrani, M.; Akhondi, M.M.; Bayat, A.A.; Ghods, R.; Mahmoudi, A.R.; Hadavi, R.; Osterborg, A.; et al. Monoclonal antibodies against ROR1 induce apoptosis of chronic lymphocytic leukemia (CLL) cells. *Leukemia* **2012**, *26*, 1348–1355. [CrossRef]
72. Liu, Y.; Yang, H.; Chen, T.; Luo, Y.; Xu, Z.; Li, Y.; Yang, J. Silencing of Receptor Tyrosine Kinase ROR1 Inhibits Tumor-Cell Proliferation via PI3K/AKT/mTOR Signaling Pathway in Lung Adenocarcinoma. *PLoS ONE* **2015**, *10*, e0127092. [CrossRef] [PubMed]
73. Zhou, Q.; Zhou, S.; Wang, H.; Li, Y.; Xiao, X.; Yang, J. Stable silencing of ROR1 regulates cell cycle, apoptosis, and autophagy in a lung adenocarcinoma cell line. *Int. J. Clin. Exp. Pathol.* **2020**, *13*, 1108–1120. [PubMed]
74. Ghaderi, A.; Zhong, W.; Okhovat, M.A.; Aschan, J.; Svensson, A.; Sander, B.; Schultz, J.; Olin, T.; Osterborg, A.; Hojjat-Farsangi, M.; et al. A ROR1 Small Molecule Inhibitor (KAN0441571C) Induced Significant Apoptosis of Mantle Cell Lymphoma (MCL) Cells. *Pharmaceutics* **2022**, *14*, 2238. [CrossRef]

75. Zhang, Y.X.; He, D.C.; Fang, W.F.; Kang, S.Y.; Chen, G.; Hong, S.D.; Sheng, J.; Zhan, J.H.; Chen, N.; Hu, Z.H.; et al. The Difference of Clinical Characteristics Between Patients With Exon 19 Deletion and Those With L858R Mutation in Nonsmall Cell Lung Cancer. *Medicine* **2015**, *94*, e1949. [CrossRef] [PubMed]
76. Yatabe, Y.; Mitsudomi, T.; Takahashi, T. TTF-1 expression in pulmonary adenocarcinomas. *Am. J. Surg. Pathol.* **2002**, *26*, 767–773. [CrossRef]
77. Liu, Z.; Yanagisawa, K.; Griesing, S.; Iwai, M.; Kano, K.; Hotta, N.; Kajino, T.; Suzuki, M.; Takahashi, T. TTF-1/NKX2-1 binds to DDB1 and confers replication stress resistance to lung adenocarcinomas. *Oncogene* **2017**, *36*, 3740–3748. [CrossRef]
78. Yatabe, Y.; Kosaka, T.; Takahashi, T.; Mitsudomi, T. EGFR mutation is specific for terminal respiratory unit type adenocarcinoma. *Am. J. Surg. Pathol.* **2005**, *29*, 633–639. [CrossRef]
79. Karachaliou, N.; Gimenez-Capitan, A.; Drozdowskyj, A.; Viteri, S.; Moran, T.; Carcereny, E.; Massuti, B.; Vergnenegre, A.; de Marinis, F.; Molina, M.A.; et al. ROR1 as a novel therapeutic target for EGFR-mutant non-small-cell lung cancer patients with the EGFR T790M mutation. *Transl. Lung Cancer Res.* **2014**, *3*, 122–130.
80. Wang, H.L.; Liu, Y.C.; Long, M.P.; Zheng, C.; Yang, J.H. Blocking ROR1 enhances the roles of erlotinib in lung adenocarcinoma cell lines. *Oncol. Lett.* **2019**, *18*, 2977–2984. [CrossRef]
81. Villarroel, A.; Del Valle-Perez, B.; Fuertes, G.; Curto, J.; Ontiveros, N.; Garcia de Herreros, A.; Dunach, M. Src and Fyn define a new signaling cascade activated by canonical and non-canonical Wnt ligands and required for gene transcription and cell invasion. *Cell. Mol. Life Sci.* **2019**, *77*, 919–935. [CrossRef] [PubMed]

pharmaceutics

MDPI

Review

Immunomodulatory Activity of the Tyrosine Kinase Inhibitor Dasatinib to Elicit NK Cytotoxicity against Cancer, HIV Infection and Aging

Andrea Rodríguez-Agustín [1,†], Víctor Casanova [1,†], Judith Grau-Expósito [1], Sonsoles Sánchez-Palomino [1,2], José Alcamí [2,3] and Núria Climent [1,2,*]

[1] HIV Unit, Hospital Clínic-IDIBAPS, University of Barcelona, 08036 Barcelona, Spain
[2] CIBER of Infectious Diseases (CIBERINFEC), 28029 Madrid, Spain
[3] AIDS Immunopathogenesis Unit, Instituto de Salud Carlos III (ISCIII), 28029 Madrid, Spain
* Correspondence: ncliment@recerca.clinic.cat; Tel.: +34-93-2275400 (ext. 3144); Fax: +34-93-2271775
† These authors contributed equally to this work.

Abstract: Tyrosine kinase inhibitors (TKIs) have been extensively used as a treatment for chronic myeloid leukemia (CML). Dasatinib is a broad-spectrum TKI with off-target effects that give it an immunomodulatory capacity resulting in increased innate immune responses against cancerous cells and viral infected cells. Several studies reported that dasatinib expanded memory-like natural killer (NK) cells and $\gamma\delta$ T cells that have been related with increased control of CML after treatment withdrawal. In the HIV infection setting, these innate cells are associated with virus control and protection, suggesting that dasatinib could have a potential role in improving both the CML and HIV outcomes. Moreover, dasatinib could also directly induce apoptosis of senescence cells, being a new potential senolytic drug. Here, we review in depth the current knowledge of virological and immunogenetic factors associated with the development of powerful cytotoxic responses associated with this drug. Besides, we will discuss the potential therapeutic role against CML, HIV infection and aging.

Keywords: dasatinib; tyrosine kinase inhibitors; CML; cancer; HIV-1; CMV; HIV functional cure; memory-like NK cells; $\gamma\delta$ T cells; anti-aging; senolytic

Citation: Rodríguez-Agustín, A.; Casanova, V.; Grau-Expósito, J.; Sánchez-Palomino, S.; Alcamí, J.; Climent, N. Immunomodulatory Activity of the Tyrosine Kinase Inhibitor Dasatinib to Elicit NK Cytotoxicity against Cancer, HIV Infection and Aging. *Pharmaceutics* **2023**, *15*, 917. https://doi.org/10.3390/pharmaceutics15030917

Academic Editor: Francesca Musumeci

Received: 18 February 2023
Revised: 7 March 2023
Accepted: 9 March 2023
Published: 11 March 2023

1. Introduction

1.1. Dasatinib

Dasatinib is a broad-spectrum tyrosine kinase inhibitor originally developed to treat chronic myeloid leukemia (CML) [1]. CML is a hematopoietic progenitor cell leukemia, in which overgrown myeloid cells accumulate in bone marrow and peripheral blood [2]. A translocation between chromosomes 9 and 22 results in an aberrant chromosome 22 (or Philadelphia (Ph) chromosome), generating the BCR-ABL oncogene. BCR-ABL constitutively activates tyrosine kinases (TK) that drives both Ph+ CML and Ph+ acute lymphoblastic leukemia (ALL) [3]. Treatment for CML was drastically improved with imatinib, one of the first TK Inhibitors (TKI) targeting BCR-ABL TK [4]. Furthermore, development of these small molecules led to the development of alternative inhibitors such as the second-generation drug dasatinib, which yield up to 20-300 times higher activity and has faster and deep molecular response (DMR) than imatinib [1].

Dasatinib has other off-target effects inhibiting other TK (Table 1) [5] with potential side effects such as hematological, pulmonary and gastrointestinal toxicity that could limit its clinical use [6]. Interestingly, off-target effects have been related to potential benefits such as an increased natural killer cell (NK)-mediated cytotoxic capacity against cancerous cells and viral infected cells [7] and an anti-aging capacity. In this review, we will discuss the potential immunomodulatory effects of dasatinib on NK cells and other innate cells and its therapeutic role against CML, HIV infection and aging (Figure 1).

Table 1. Tyrosine kinases (TKs) targeted by dasatinib and the cellular processes in which they are involved. Adapted from Araujo et al., 2010 [5]. Abbreviations: PDGFRβ, platelet-derived growth factor receptor beta; EPHA2, ephrin type-A receptor 2; EFNB, ephrin-B; M-CSF, macrophage colony-stimulating factor.

Tyrosine Kinase Target (s)	Pathways and Processes
SRC family (SRC, LCK, YES, FYN)	Oncogenic, invasive and bone-metastatic processes
BCR-ABL	Promotion of growth advantage of leukemic cells
c-KIT	Cell growth
PDGFRβ	Tumor growth capacity and cell survival
c-FMS	Macrophage behavior regulation by M-CSF
EPHA2 receptor	Interference with EFNB-dependent suppression of apoptosis/Cell behavior

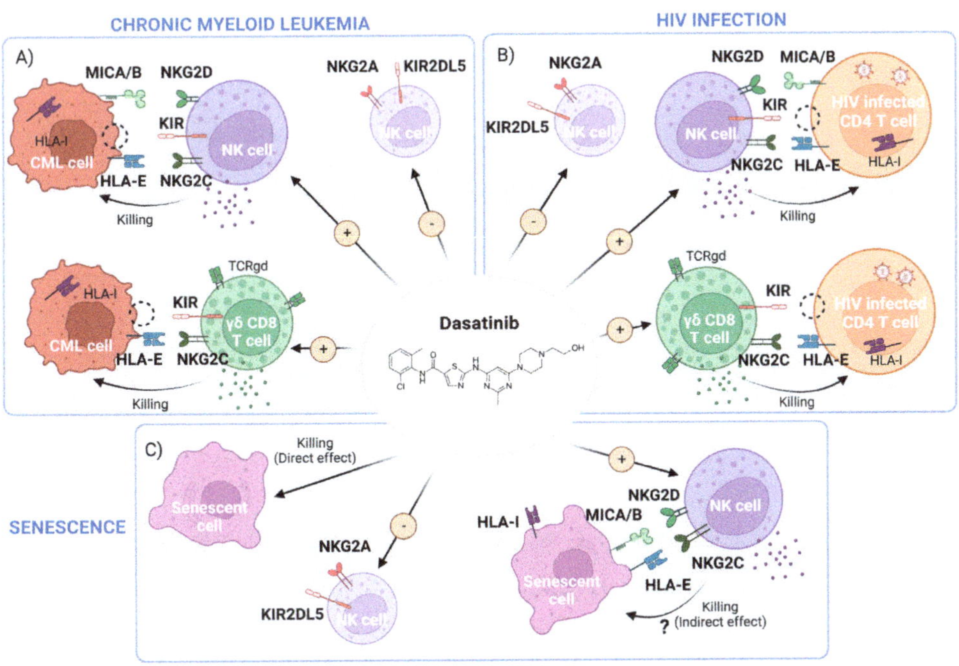

Figure 1. Scheme of dasatinib-mediated effects that may interfere with chronic myeloid leukemia (CML), HIV-1 infection and cellular senescence. (**A**) CML cell killing by the enhancement of memory natural killer (NK) cells and cytotoxic $CD8^+$ T cells expressing γδTCR and a reduction of inhibitory receptors driven by dasatinib. (**B**) HIV infected cells' clearance, carrying pro-viral DNA, by dasatinib activity through the potentiation of the above cell subpopulations. (**C**) Representation of both direct killing and a potential indirect effect by boosting NK and $CD8^+$ cells of senescent cells with dasatinib. Figure made with BioRender.com (accessed on 7 March 2023).

1.2. Natural Killer Cell Biology

Natural killer cells (NK cells), are cells from the innate immune system that show strong cytolytic function against stressed cells such as tumoral cells and virus-infected cells. NK activation state is determined by a balance of multiple activation and inhibition signals mediated by NK inhibitory and activating receptors that bind to NK ligands from other neighboring cells (Figure 2A). Most of these ligands are HLA class I molecules, such as HLA-A, B, C and HLA-E. Depending on the balance between these NK receptors (NKR) and their HLA ligands, NK cells will either be activated to kill the target cell or inhibited, allowing the target cell to survive. NK cells are an heterogeneous population harboring multiple subsets that differentially expressed NK receptors such as: killer cell immunoglobulin-like receptors (KIRs), natural killer group 2 such as NKG2A, NKG2C and NKG2D, and natural cytotoxicity receptors (NCRs), for example NKp30, NKp40 or NKp46 [8,9]. The main activating and inhibiting NK receptors and their respective target cell ligands are shown in Figure 2A. The NKG2C and NKG2D receptors are activating NK receptors that could detect abnormal or malignant cells by binding with HLA-E and the MIC A/B ligands, respectively. These interactions potentiate NK cell killing [10,11] (Figure 2A).

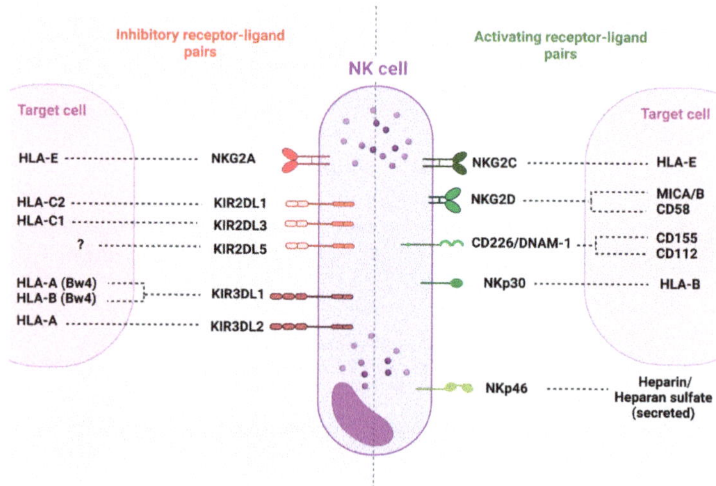

(A)

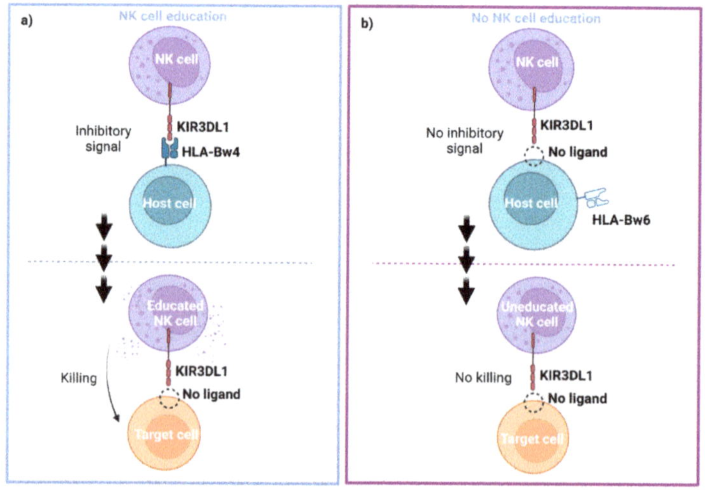

(B)

Figure 2. *Cont.*

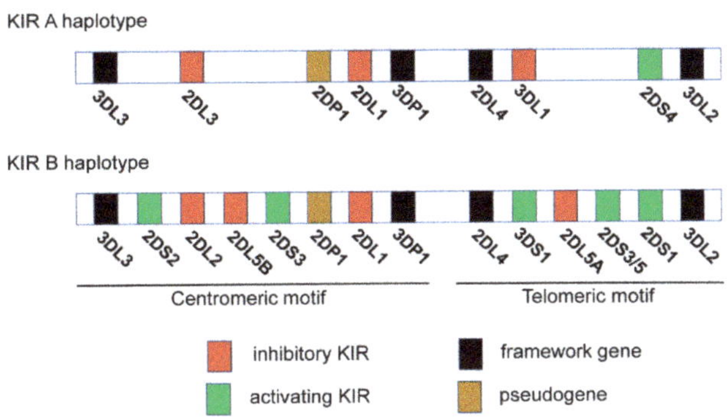

(C)

Figure 2. (**A**) Representation of NK cell inhibitory and activating receptor-ligand pairs. Adapted from [9]. Figure made with BioRender.com. (**B**) NK cell education process and "missing self" killing of target cells. (**a**) NK cells are recognized through KIR3DL1 receptor host cells expressing HLA-Bw4 and become educated to recognize its absence then in target cells. This process leads to an efficient target cell killing. (**b**) Absence of recognition of HLA-Bw4 by KIR3DL1 drives a deficient cell killing of target cells with "missing self" by miseducation of the NK cell. Adapted from [12]. Figure made with BioRender.com. (**C**) Genomic organization of KIR A and B haplotypes. Adapted from [12].

NK cells are activated by target cells that down regulate "self" HLA. This NK activation mechanism appears by a phenomenon known as education or licensing. Education requires the interaction between inhibitory NK cell receptors and their own HLA ligands [13,14]. Educated NK cells remain inactive by the interaction of NKRs inhibitor with autologous HLA-expressing neighboring cells. However, NK cells can also be activated by viral infected cells and abnormal or injured cells that usually have a lower HLA class I expression to evade cytotoxicity driven by conventional CD8+ T cells. NK cells can detect these anomalous target cells by the lack of inhibiting interaction (low o non-inhibitory HLA interactions). This phenomenon is called "missing self" detection and generally results in NK cell activation, degranulation and target cell death (Figure 2B). For NK cell activation, additional activating signal are necessary [15]. NK cells are activated when the ratio of inhibitory and activator NKR signaling favors activation [9] (Figure 2B). Many inhibitory KIRs from NK cells interact with HLA alleles. KIR and HLA proteins are both encoded by highly polymorphic alleles, enabling a wide diversity of receptor/ligand interactions. Education through inhibitory receptors sensitizes NK cells to detect "missing self" cells, while education through activating receptors inhibited NK cell cytotoxicity [16] (Figure 2B).

Despite the high level of KIR gene content variability, there are two main groups of KIR haplotypes, termed "A" and "B" (Figure 2C). The A haplotypes include inhibitory KIR genes coding for receptors such as KIR3DL1, which recognize HLA-A and HLA-B molecules that contain Bw4 epitopes. The B haplotypes include several activating KIR genes such as KIR3DS1 (Figure 2C). Then, the KIR genetic combination can be AA, AB or BB genotype. There are two possible established epitopes for HLA-B: Bw4 or Bw6. KIR3DL1 is highly polymorphic and, depending on the haplotype, NK cells can express: high, low, or null levels of KIR3DL1 [17]. In fact, KIR3DL1 and HLA-B genetic polymorphisms regulate NK cell function. Their stronger ligand affinity correlates with enhanced NK cell inhibition but also with increased education and, consequently, potentiates "missing self" NK capacity to detect and kill anomalous cells [18] (Figure 2B).

The inhibitory NKG2A receptor binds to HLA-E molecules. Signal peptides from the leader sequence of HLA-A, -B and -C proteins, which are codified by the first exon of the MHC gene, must bind to HLA-E in order to fold properly and reach the cell membrane and to become a ligand for the inhibitory NKG2A receptor. In addition, the anchor residue of the nonamer peptides that bind to HLA-E corresponds to residue -21 of the classical HLA class I leader sequence. There are two amino acid variations at position -21, a methionine (-21M) or a threonine (-21T) [19,20]. HLA-B antigens can have both -21M and -21T variants, depending on the HLA-B haplotypes that are associated with two NK cell education profiles. On the one hand, HLA-B -21M alleles contribute more effectively to NKG2A-mediated education, which suppresses NK cell activation when the target cell expresses stabilized levels of HLA-E. HLA-B -21T alleles mainly contribute to education through inhibitory KIRs, especially KIR3DL1 [15], as HLA-B -21T and HLA-Bw4 are genetically linked. Then, the HLA-Bw4 alleles are, in fact, HLA antigens encoded by -21T, able to interact and educate through KIR3DL1/HLA-Bw4, allowing the NK cell cytotoxicity upon "missing self" recognition [9,20] (Figure 2B).

2. Dasatinib-Mediated Immunomodulatory Effects in CML

2.1. Dasatinib Effect on NK Cells and Innate T Cells

Besides its direct impact on leukemic cells through inhibition of constitutive tyrosine kinase activity, dasatinib's off-target effects include: inhibition of proliferation and activation of T cells, and in vitro suppression of cytotoxic activity of NK cells [21,22] This is due to the potent inhibition of several off-target kinases of the Src family, such as Lck and Fyn in T cells [23]. TKI treatments can restore an anti-leukemic effector function, such as specific cytotoxic T lymphocytes' (CTL) responses against leukemia-associated antigen (LAA). Overall, this results in a reduction in leukemic cell load major molecular remission (MMR) and a better recovery [24,25]. Dasatinib has the potential to reduce Tregs and derived factors (sCTLA-4), especially in patients developing large granular lymphocytes (LGLs), lymphocytosis [26] and NK cell differentiation, promoting immune stimulation [27]. This is relevant because the proportion of Treg cells is abnormally elevated in CML individuals at diagnosis, compared to healthy controls. Furthermore, dasatinib also reduces myeloid-derived suppressor cells (MDSC) in CML patients [28].

Up to half of the patients receiving dasatinib treatment develop an LGL lymphocytosis, mainly composed of cytotoxic cells (NK cells and $CD8^+$ $\gamma\delta$ T cells) [29,30] but this also includes $CD57^+$ cytotoxic $CD4^+$ T cells with anti-leukemic properties [31]. This LGL expansion is associated with an improved anti-leukemic response in both CML and Ph+ ALL patients [29,32–35]. The $CD8^+$ T cell response in dasatinib-mediated LGL response includes TCR-$V\beta^+$ expansion of either oligoclonal or polyclonal origin with cells resembling healthy memory $CD8^+$ T cells [36]. Importantly, dasatinib may have an immunomodulatory role in non-conventional or innate T cells by increasing cell number, activation status and Th-1 polarization of innate $\alpha\beta$ iNKT-cells in treated CML patients [37]. In addition, it has been reported that dasatinib increased a novel innate subset termed innate $CD8^+$ T-cell-expressing IFNγ [37].

CML patients treated with dasatinib presented more classical NK cells ($CD3^-CD56^+$) and matured NK cells ($CD56^+CD57^+$) compared to imatinib- or nilotinib-treated patients [38]. These patients also presented lower expression of NK-inhibitory markers (KIR2DL5A, KIR2DL5B and KIR2DL5), which was associated with an MMR [39] and an increase in KIR2DL1 expression [8].

Dasatinib promoted NK-cell cytokine expression and cytotoxic activity towards the CML-derived cell line K562 [18,40], especially when KIR3DL1/HLA-Bw4 interactions were null, low or weak. Of note, Izumi et al. showed that blocking KIR3DL1/HLA-Bw4 binding with an anti-KIL3DL1 antibody potentiated NK cytotoxicity [18]. Furthermore, Shen et al. reported that this increased NK cytotoxic activity can be emulated in vitro using dasatinib treatment after IL-2/IL-15-mediated expansion of NK cells. In this setting, dasatinib also increases the percentage of $NKG2A^-CD57^+$ NK cells and the expression of activating receptors CD226 (DNAM-1), NKp46 and NKG2D [41], promoting their capacity to kill by degranulation of the CML cell line K562, not expressing HLA class I. Finally, dasatinib-mediated reduction of NKG2A also may boost NK cytotoxicity and improve MMR [18,42].

2.2. Dasatinib Increases Memory-like Natural Killer (NK) Subsets Displaying Activity against Both Leukemic and Cytomegalovirus (CMV) Infected Cells

There is discussion on whether development of LGLs with dasatinib is associated with previous immunity to CMV [32,43] or related to CMV viral load (VL) [7,43]. Ishiyama et al. reported that NK cells are the main component of LGLs in patients with CML treated with dasatinib, and NK cell expansion was highly associated with CMV-serostatus [7]. The authors performed a principal component analysis (PCA) with multiple markers on NK cells after dasatinib treatment and determined that NK cells from CMV^+ individuals had a CMV-associate phenotype, named memory-like NK cells, with highly differentiated NK phenotypes ($NKG2C^{high}NKG2A^{low}CD57^{high}LIR\text{-}1^{high}NKp30^{low}NKp46^{low}$). NK cells from CMV-uninfected individuals were negative for this CMV-related signature [7,44]. Remarkably, a higher grade of NK cell differentiation at CML diagnosis predicts both a greater expansion of CMV-related memory-like NK cells and a lower leukemic-cell load after dasatinib treatment in CMV^+CML patients [7,44]. Kadowaki et al. reported that a persistent, low-level CMV replication, often subclinical, triggered memory-like NK cell expansion. This suggests that CMV may trigger NK-cell expansion and both leukemia and dasatinib are enhancing factors that expand this NK subpopulation [45]. This hypothesis was reviewed by Climent et al. [46]. Importantly, this $NKG2C^+CD57^+$ subset of memory-like NK cells may represent NK cells with unique adaptive and editing properties [44]. Thus, the expansion of this subset could promote long-term memory and high cytotoxic activity against CML, even after dasatinib treatment interruption [45]. Recent studies have also observed an expansion of $CD56^{neg}$ NK cell populations exclusively in CMV^+ patients treated with dasatinib, and this increase parallels memory $NKG2C^+CD57^+$ NK cells [47]. These authors propose that $CD56^-$ NK cells may be an exhausted population induced by chronic activation through CMV reactivation but, paradoxically, are proposed as a hallmark of CML control because it predicts a better clinical outcome. Overall, these results suggest that dasatinib immunomodulatory effects on NK cell responses against CMV could also be relevant against malignancies or other viral infections [48], such as HIV infection [46].

2.3. CML Control and Therapeutic Treatment Interruption: Immunological Factors Involved in a Successful Treatment-Free Remission (TFR)

Once a DMR is observed, TKI treatment can be interrupted in selected patients with the aim of achieving a TFR [49]. Dasatinib and nilotinib treatments are associated with a stronger DMR, increasing the chances of longer TFR [49]. A better understanding of the factors that could predict longer TFR is of paramount importance [50–52]. Moreover, half of the individuals who stopped treatment were able to keep TFR during at least one year and developed high levels of NK cells [53–55] and neutrophils [56] but not T cells [53,57] indicating that these populations are key to keeping CML under control.

Regarding the implication of NK cells in TFR, individuals who controlled CML after imatinib therapy cessation showed higher NK cytotoxic function towards the target K562 cell line, lacking HLA class I. Furthermore, NKG2D gene polymorphisms [58] and the IFN-γ and TNF-α cytokine secretion by NK ($CD56^{dim}CD16^-$) cells correlated with the successful drug discontinuation and control of CML [57,59]. Increased mature ($CD57^+$) and cytotoxic ($CD16^+$ and $CD57^+$) NK cells, together with IFN therapy prior to TKI cessation, have been also shown to produce better CML outcomes after treatment interruption [60]. IFNα treatment also increased differentiated $NKG2C^+$ NK cells, increased NKp46 expression on the $CD56^{bright}/CD16^-$ NK cell subpopulation and modulated NK cell cytotoxicity [61].

TFR has also been evaluated in dasatinib stopping TKI trials [62]. The DADI study [53,63] demonstrated that high levels of NK cells ($CD56^+$), LGL NK cells ($CD56^+CD57^+$) and low levels of Treg ($CD25^+CD127^{low}$) preceding the dasatinib interruption were associated with longer TFR periods. Furthermore, a critical role of Treg inhibition by dasatinib, potentiating NK cell function, promotes a DMR [64]. Likely expanded NK cell functionality and lower Treg frequencies may decrease the probability of a worse outcome after dasatinib interruption and result in longer TFR [65].

Recent studies suggest that the features most consistently linked to longer TFR, independently of the type of TKI treatment stopped, are: (1) a high frequency of cytotoxic subsets such as NK, NKT and $CD8^{+}$ $\gamma\delta$ T cells [66,67]; (2) high level of NK-activating receptors such as NKG2D, NKp30 or NKG2C on NK [11,67,68] and NKT cells [67]; (3) enhanced expression of activation cytokines or granzyme B in NK cells after stimulating with HSP70; and (4) KIR homozygosis at haplotype AA, which includes KIR3DL1. These hallmarks may be useful as prognostic biomarkers of longer TFR [67]. Further clinical trials are needed to test these predicting biomarkers for TFR [69]. Moreover, the memory-like NK cells, characterized as $CD3^{-}CD56^{dim}CD57^{+}$ $NKG2A^{-}$ $NKG2C^{+}$, were increased in patients with TFR success [70].

Results concerning NKG2A expression on NK cells are controversial [11]. Several studies suggested that low levels of NKG2A expression in NK cells is associated with longer TFR [70] and better CML prognostic [42]. In stark contrast, Xu et al. reported that an elevated expression of NKG2A in NK cells, especially in the $CD56^{bright}$ subset, was a good prognostic biomarker for TFR [11], as also reported by Vigón et al. [67]. The inhibitor receptor NKG2A has a dual function on NK cells: firstly, it has a key function in the NK education process; second, after inhibition of NK cell activation it could send other signals to NK cells [71]. In fact, we recently found that high expression of NKG2A is present in NK cells able to kill cancerous cells such as reprogrammed cells or those that downregulate HLA-E. Reprogrammed cells express Yamanaka factors and this gives them the capacity of being pluripotent embryonic stem cells (e.g., teratoma-like cells). These educated NK cells are able to kill by the missing self-recognition [72]. NKG2A expression could not be per se detrimental to the function of all NK cell subsets. Altogether, the increased expression of NKG2A in the $CD56^{bright}$ subset could be interpreted as an increased killing capacity against target cells with lower expression of HLA-E, such as cancerous cells or CML cells [11].

2.4. NK Immunogenetics Associated with CML Control or TFR

The diversity of allotypes of KIR3DL1 and HLA-Bw4 is associated with the receptor/ligand avidity and the NK cytotoxic capacity [12,17,73,74]. KIR genotypes have high genomic variations or allotypes that are associated with NK cell cytotoxic activity against CML and extended TFR periods [75,76]. Some reports suggested that CML patients with KIR2DL5B and KIR2DL2 alleles reached higher DMR after TKIs, implying that some KIR alleles or a specific combination of KIR genes can modify NK cell activity against CML cells [77,78]. In fact, KIR AA haplotypes, which include many KIR inhibitors such as KIR3DL1, are associated with better outcomes in TKI-treated CML patients [77] and are also linked to patients with sustained TFR [8,75]. It is interesting that the A haplotype including KIR3DL1 is highly associated with NK education (Figure 2B,C). Interaction of KIR3DL1 and HLA-Bw4 could affect NK cell education and cytotoxicity. In fact, the KIR3DL1*005 allele was highly linked with DMR, suggesting the relevance of this specific KIR3DL1. Consistently, DMR was coupled with higher NK cell killing in vitro in a NK cell cytotoxic assay against the CML cell line K562 without HLA class I, indicating that these educated NK cells could contribute to eliminating CML cells in vivo (Figure 2B). However, verification is needed, and the specific A haplotype KIR gene with the greatest impact needs to be identified [12] (Figure 2B). It has been shown that haplotypes of KIRs and HLAs were linked to a better outcome in a Japanese cohort [18,76]; specifically, Ureshino et al. reported that TFR in patients with HLA-Bw4 was higher than in patients with HLA-Bw6 alleles [79]. Similarly, HLA-Bw4 has been related to HIV control [80]. Altogether these genetic associations suggest that NK cells from patients achieving DMR or TFR could be better educated by interaction between the inhibitor receptor KIR3DL1 and HLA-Bw4, allowing activation of NK cells against cancerous cells that downregulate HLA expression, triggering "missing-self activation" (Figure 2B). Clinical trials with large cohorts are needed in order to explore deeply the NK immunogenetic factors associated with CML control [12].

2.5. *Dasatinib as an Immunomodulator in Other Therapeutic Strategies against Cancer*

The immunomodulatory activity of dasatinib has also been evaluated in combination with other therapeutic strategies against advanced malignancies such (1) immunotherapy with immune checkpoint inhibitors, where dasatinib immunomodulatory capacity has been investigated in increased programmed cell death protein 1 (PD-1) and programmed cell death ligand 1 (PD-L1) (PD1-PDL1) immunotherapy [81,82]; or (2) with chimeric antigen receptor (CAR)-engineered T cells (CAR-T). In CAR-T therapies, T cells are ex vivo modified by adding a gene for a receptor that helps the T cells to target specific myeloid antigens. A combined CAR-T and TKI approach has also been evaluated in some studies to enhance antitumor immunity and demonstrated that dasatinib limits CAR-T cells' therapy side effects, such as the cytokine release syndrome (CRS) [83], and increases the anti-leukemia activity of CAR-T cells by decreasing cell exhaustion [84].

2.6. *Summary of the CML Section*

In brief, these findings support the idea that dasatinib contributes to better treatment response in CML patients through enhancement of the immune system, particularly via NK cell differentiation. CML patients with better outcomes could have done better due to genetic factors, such as AA alleles (homozygosis at KIR3DL1) and HLA-Bw4 associated with educated and highly cytotoxic NK cells able to detect malignant cells. Consequently, dasatinib could be useful, especially in the patients that do not have these protective features, to enhance cytotoxic activity of NK cells against CML cells by increasing memory-like NKG2C$^+$CD57$^+$ NK cells [70], $\gamma\delta$ T cells and other innate CD8$^+$ T cells [7,30,37,44]. In addition, these innate cells could express high levels of the activation receptor NKG2C, NKG2D, NKp46 or DNAM-1 and downregulate some inhibitory receptors such as NKG2A and KIR2DL5 [7,39,41,44] (Figure 1A).

3. Potential Use of Dasatinib in the Setting of HIV-1 Infection

Current antiretroviral therapy (ART) can prevent progression to AIDS, blocking new infections by interfering with the virus life cycle. ART efficiently decreases the plasma VL under the limit of detection of conventional techniques (50 copies/mL) but is unable to fully eliminate HIV from the body. Latently infected cells, established during the early stages of infection, harbor integrated forms of the virus that are responsible for viral rebound when ART is stopped. These viral reservoirs represent the major barrier to the complete eradication of HIV, because it is invisible to the immune system and inaccessible to treatment [85]. In recent years, several strategies have been proposed to eliminate these reservoirs, including the 'shock and kill' strategy, but none of them have demonstrated a significant decrease of the reservoir size [86]. Therefore, new strategies are needed to avoid the formation of the reservoir, but also its replenishment and maintenance, by additional mechanisms including immunotherapy and new immunomodulatory compounds [87].

3.1. *Effect of Dasatinib on HIV Infection and Reservoir: Direct Effect*

Dasatinib directly interferes with CD4$^+$ T-cell activation, which is the main HIV-1 cell target, by inhibiting the Src TK implicated in T cell receptor (TCR) signaling and the phosphorylation of SAM domain and HD domain-containing protein 1 (SAMHD1), a cell restriction factor. Here, we discuss the potential use of dasatinib as an adjuvant of the ART during HIV-1 infection. This strategy might reduce the viral reservoir and control infection [87–89]. Quite unexpectedly, CD4$^+$ T cells from CML individuals under dasatinib therapy were shown to be resistant to ex vivo HIV-1 infection [88,90]. Results from our own work showed that dasatinib preserved SAMHD1 activity, by maintaining its active conformation, which restricts HIV-1 replication [46,87–89]. This enzyme controls homeostatic balance of dNTPs. When active (unphosphorylated), SAMHD1 restricts HIV-1 replication by lowering dNTP to levels that do not allow an effective viral replication. In contrast, when SAMHD1 is inactive (phosphorylated), it increases intracellular dNTPs' availability, facilitating HIV-1 replication [46,87–89]. Moreover, Williams et al. described how the inhibition of cyclin-dependent kinases (CDKs) 1, 2, 4 and 6 in macrophages by dasatinib leads to blockade of HIV-1 infection by dephosphorylation of SAMHD1. This confers protection from the virus not only in the $CD4^+$ T cells but also in monocyte-derived macrophages [91].

Dasatinib targets multiple host non-receptor tyrosine kinases, in particular the ABL family of kinases such as ABL1 or ARG. It has been demonstrated that siRNA knockdown of ABL1 and ARG kinases inhibits HIV-1 at a step after reverse transcription during early infection of activated $CD4^+$ T cells, increasing the unintegrated two-long terminal repeat DNA circles (2-LTRcs) generated through the ligation of the cDNA ends by the host cell non-homologous DNA end-joining system. These results suggest preliminary efficacy of dasatinib treatment during acute HIV-1 infection [92]. Meanwhile, Bermejo et al. and Vigón et al. found that the frequency of proviral integration was strongly reduced in activated $CD4^+$ T cells from CML patients treated with dasatinib compared with those from untreated healthy donors [88,93]. Importantly, we collaborated to find that peripheral blood lymphocytes obtained from CML patients treated for at least 6 months with dasatinib and individuals that were in TFR for more than one year were resistant to HIV-1 infection [46,93]. This indicates that mechanisms other than SAMHD1 phosphorylation are implicated and suggests an immunomodulatory role of dasatinib.

Despite CML and HIV-1 infection not being commonly associated, several cases of people living with HIV(PLWH) who developed CML have been described. Studies performed on this special subset of HIV^+ individuals treated with ART and dasatinib showed that the frequency of latently infected cells was reduced more than 5-fold compared with PLWH only on ART. In addition, reactivation of proviruses in $CD4^+$ T cells isolated from PLWH with CML on ART and dasatinib was reduced 7-fold compared with individuals only on ART. Importantly, dasatinib treatment dramatically impaired SAMHD1 phosphorylation *both in vitro and in vivo* (30-fold and 21-fold, respectively) [94]. Dasatinib also blocks $CD4^+$ T cell proliferation induced by homeostatic cytokines such as IL-7 and IL-15, which are crucial for the stability of the viral reservoir. Innis et al. described the restriction of both homeostatic and antigen-driven proliferation in memory $CD4^+$ T cells from PLWH by dasatinib, consistent with promoting a smaller reservoir size [95]. In addition, the antimitotic properties associated with dasatinib can reduce the clonal expansion of infected cells, limiting the permanent filling of viral reservoirs [87].

3.2. Dasatinib's Indirect Effect in HIV-1 Infection: Potentiation of NK and $\gamma\delta$ $CD8^+$T Cell Responses

Vigón et al. suggested that potentiation of memory-like NK cells and a $\gamma\delta$ $CD8^+$ T cell responses in dasatinib-treated CML patients may contribute to the observed resistance of these individuals' PBMCs to HIV-1 infection. This mechanism would operate in parallel to the previously described phosphorylation of SAMHD1 [93]. Interestingly, this increased cytotoxic activity of memory-like NK cells and a $\gamma\delta$ $CD8^+$ T cell subsets might be effective both against HIV infection and CML. A similar cytotoxic cell signature has recently been described in a case of exceptional HIV long-term post-treatment control (>15 years, Barcelona patient). This control has been linked to a strong memory-like NK cell response, with $NKG2C^+CD57^+$ phenotype and $\gamma\delta$ cytotoxic $CD8^+$ T cells [96]. This NK cell phenotype ($NKG2C^+CD57^+$) was also found in long-term HIV elite controllers [97] and in individuals during primary infection who showed lower HIV RNA expression and more quickly reached an undetectable VL [98–100]. The above phenotype has been described together with a reduction of KIR2DL5, NKG2A (inhibitory receptors) and NKp30 in NK cells [93,97].

Taking all this into account, it could be considered that, despite immune exhaustion in PLWH, dasatinib is able to increase NK memory-like ($CD56^+CD57^+$) and $\gamma\delta$ $CD8^+$ T cells subsets and to reduce NKG2A expression in NK cells [42], contributing to HIV-1 infection control.

3.3. NK Cells' Role in HIV-Mediated Control and Functional Cure

Functional HIV cure, defined as control of viral replication in the absence of treatment, has been described in two scenarios: in those PLWH who maintain undetectable HIV VL (<50 copies/mL) in the absence of ART, known as "elite controllers" (EC), and those individuals who demonstrate sustained virologic suppression for months or years after treatment cessation, called "post-treatment controllers" (PTC). The definitions of these patients vary depending on the time elapsed without treatment, since in many of them a "loss of control" occurs over time, defined by an increase in VL, which makes it necessary to restart ART. The patients of greatest medical and biological interest are those who have controlled the infection for at least ten years in the absence of ART and will be called long-term or exceptional elite controllers (LTEC or EEC) and long-term post-treatment controllers (LTPTC). This rare subset of patients may serve as a realistic model of a functional cure for HIV-1. LTPTC have a NK-mediated HIV control [96]. Most LTPTC are patients treated in the acute infection phase in whom spontaneous control of the viral load is observed when the treatment is discontinued months or years later [101,102]. The percentage of LTPTC patients is difficult to estimate and although some authors put it at 10%, it is probably lower [101].

3.4. Immunogenetic Factors of HIV Control and Functional Cure

The most relevant genetic alleles that are associated with HIV control are the HLA-B allotypes, such as HLA-B*57 and HLA-B*27, both being members of the HLA-Bw4 epitope [103]. Interestingly, HLA-Bw4 homozygosis is associated with HIV control and protection [80,104–106]. HLA-A expression levels change from one allotype to another [107]. Decreased HLA-A expression correlates with HIV control, being seen in the EC patients that have reduced HLA-A levels [107]. In addition, -21T amino acid variation at the signal peptide HLA-B has been associated with being HLA-Bw4 [9,20]. For that reason, homozygosis at HLA-B -21T, likely associated with Bw4 homozygosis, contributes to HIV control [80]. Some KIR3DL1/Bw4 combinations also are associated with HIV control [9]. In this context, HLA alleles carrying HLA-B -21T, such as Bw4 and low-expression HLA-A allotypes, codify HLA molecules that reduce HLA-E levels -n HIV-infected $CD4^+$ T cells. Once KIR3DL1/HLA-Bw4-educated NK cells interact with HIV-infected target cells expressing low levels of HLA-E, activation of NK cells by lack of inhibitory signaling could be able to enhance cytotoxic function against HIV-infected cells, thus promoting HIV control [9,107] (Figure 3).

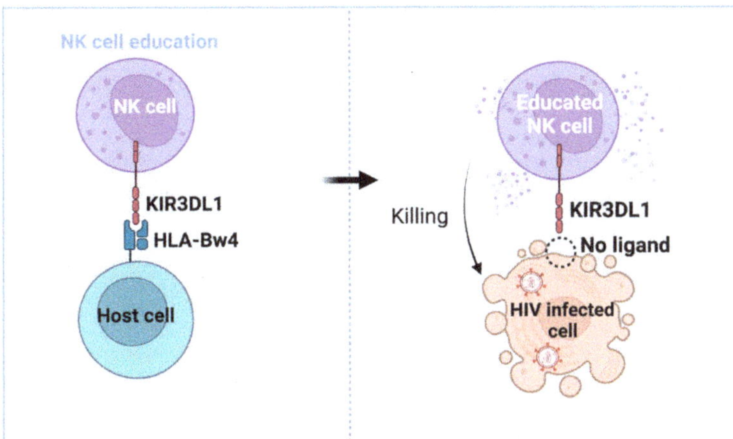

Figure 3. NK cell education outline in HIV context by KIR3DL1/HLA-Bw4, associated with HIV infection control. Adapted from [9]. Figure made with BioRender.com.

Consistently, in vitro experiments have suggested that KIR3DL1/HLA-Bw4 genotype combinations have higher educated NK cytotoxic responses than NK cells from Bw6 homozygotes against HLA-downregulated target CML cell lines (K562) [9,108–110] or HIV-infected $CD4^+$ T cells that decreased the levels of HLA-A, -B and -C by Nef and Vpu HIV proteins [111–114] emulating the missing-self NK cytotoxic activity (Figure 3) [4,114,115]. Moreover, a higher number of educated NK cells by KIR2DL1/HLA-C2 and KIR2DL3/HLA-C1 interactions can mediate the killing of K562 cell lines and HIV-infected $CD4^+$ T cells compared with uneducated NK cells [9,116,117].

HIV control in children Bw4 has been linked to reduced levels of HLA-A. In fact, levels of HLA-A in cells depends on the specific HLA-A alleles [105,107,118]. HLA-Bw4 mediates education of NK cells by its NK ligand, KIR3DL1. In fact, controller children reportedly had enriched Bw4 and co-expressed KIR3DL1, which educates these NK cells [9]. A deletion in the NKG2C gene is also associated with enhanced HIV infection, suggesting that this NK-activating receptor is a clue in the context of HIV protection [119].

Recently communicated data from several international congresses published by our group reported that LTPTC are enriched in alleles implicated in educated NK cells. Sáez-Cirión et al. found that LTPTC had certain education-related alleles such as homozygosis at HLA-Bw4 and its ligand, KIR3DL1, and also at HLAB*35, which is also an HLA-B -21 T allotype [120,121]. In fact, the HLA-B*35 allele is not a Bw4 epitope, but exceptionally is an HLA-B -21T [20,107]. These results suggest that HLA -21T (HLA B*35), Bw4 (also HLA -21T) and KIR3DL1 alleles have been associated with HIV LTPTC, suggesting an important role of NK cell education in NK-mediated HIV control [96,121]. These recent data point to a possible role of NK cells and, possibly, other innate T cells such as $\gamma\delta$ T cells, in LTPTC HIV control process [96,122].

Moreover, a recent communication indicated that EEC have a high content of homozygous Bw4 alleles. In fact, 75% of EEC have two Bw4 alleles, suggesting that a high amount Bw4 homozygosis could be related to HIV functional cure [123–129]. Consistently, another recent communication suggests that LTEC have a high percentage of $NKG2C^+CD57^+$ NK cells [97]. Due to similar features in EEC and LTEC, such as persistent HIV control for more than 10 years, we could hypothesize that the high frequency of memory-like NK cells found in LTEC could be related to favorable NK immunogenetic characteristics. Thus, these features could be linked to HIV long-term control and functional cure. Consistent with these data and as Sáez-Cirión communicated [120], our recently published results concerning the case report of the "Barcelona patient", reveal that this patient is a homozygote not only for the HLA haplotypes KIR3DL1/HLA-Bw4, but also for KIR2DL3/HLA-C1 (C*16) alleles. In fact, both genotypes are compatible with highly educated NK cells. Moreover, we found that this genotype is linked to an increased expansion of memory-like NK cells and $\gamma\delta$ $CD8^+$ T cell subpopulations associated with persistent HIV control and a low viral reservoir. These factors might be contributing to the observed HIV functional cure [96].

In summary, homozygosis at Bw4, and homozygosis at -21T and at KIR3DL1 genes is highly associated with persistent HIV control. These results may suggest that PLWH with these genetic features could educate NK cells by the KIR3DL1/HLA-Bw4 interaction having an enhanced capacity to mediate killing of HIV-infected $CD4^+$ T cells (Figure 3) than the ones with uneducated NK cells. This fact suggests that not only CTL but also NK cells could have an important role in HIV control and protection. Genetic studies of NK cell education alleles related to both NKG2A/HLA-E, KIR/HLA-B interactions and HLA-A and HLA-B -21 alleles could increase the knowledge of the NK immunogenetics associated with HIV functional cure. Larger clinical trials with cohorts of LTPTC, LTEC or EEC will be necessary to explore more deeply whether protective HLA and KIR allotypes are able to potentiate NK cells against HIV infected cells by KIR3DL1/HLA-Bw4 NK education [9].

3.5. Summary of the HIV Section

In conclusion, due to the fact that dasatinib is able to induce these innate NK and $\gamma\delta$ T cell subsets highly associated with long-term HIV control, the use of this drug could be a new strategy in the HIV functional cure field. Dasatinib could activate NK cells by up-regulation of activating receptors such as NKG2C, NKG2D, NKp46 and reducing the NKG2A/HLA-E interaction by reducing NKG2A expression in NK cells mediating HIV- infected cells killing (Figure 1B). Dasatinib could act against HIV-1 by maintaining the antiviral effect of SAMHD1 and by inhibiting $CD4^+$ T-cell activation and proliferation, reducing the establishment of viral reservoirs and limiting the replenishment and maintenance of infected cells. Additionally, the potentiation of subpopulations such as memory-like NK cells and $\gamma\delta$ $CD8^+$ T cells associated with low reservoirs, persistent HIV control and having high cytotoxic potential against HIV infected cells could be key against HIV-1 infection.

Clinical trials using dasatinib as a new immunomodulatory drug that potentiates innate cytotoxic cells could be relevant to increase the knowledge on the role of dasatinib as an NK-activating drug against HIV infection. In fact, our group has a recently approved pilot clinical trial (DASAHIVCURE NCT05527418) that will test the safety, tolerability, and antiretroviral activity of dasatinib in a cohort of recently asymptomatic HIV-infected individuals. This study will evaluate the capacity of dasatinib to inhibit HIV replication and to potentiate cytotoxic responses against HIV mediated by NK and $\gamma\delta$ T cells.

4. Dasatinib as a Senolytic Drug

4.1. Cellular Senescence Results in Increased Immune Surveillance

Cellular senescence is activated in response to a myriad of stressors or insults. Senescent cells display a series of features that result in the recruitment and activation of immune cells with the ultimate goal of removing damaged cells and restoring tissue homeostasis. Thus, activation of cellular senescence has beneficial effects in wound healing, tissue regeneration and blocking tumor progression [130–132]. However, accumulation of senescent cells during aging or due to different pathologies has deleterious consequences [133].

To target senescent cells, it is key to define characteristic biomarkers, taking into consideration that these will vary depending on cell type and the type of stress [134]. The upregulation of cell cycle inhibitors such as $p16^{INK4a}$, $p21^{CIP1}$ and p53, or the increased expression of BCL-2, an anti-apoptotic marker, define most senescent cell types. In addition, these cells present some metabolic changes, such as an augmented activity of lysosomal senescence-associated β-galactosidase (SA-β-gal) due to a mitochondrial expansion, and also secrete specific inflammatory factors named senescence-associated secretory phenotype or SASP [133].

Immunosurveillance mechanisms are key to detect and eliminate senescent cells. A growing body of evidence points towards NK cells and their activating receptors as key cells in removing senescent cells. Gasser et al. first reported that NKG2D ligands ULBP1, ULBP2 and ULBP3 were upregulated in human fibroblasts after ionizing radiation or inhibitors of DNA replication, stimuli that activate the ATM/ATR DNA damage response and are now known to induce cellular senescence [135].

Krizhanovsky's group [136] found that senescent cells accumulate in the liver of mice treated with CCl_4, a chemical known to induce liver fibrosis. Livers from p53 and INK4a/ARF double knockout animals showed up to a 50% increase in liver fibrosis markers after CCL4 treatment compared to WT animals, with little to no accumulation of senescent cells. While WT animals fully resolved liver fibrosis and eliminated senescent cells in 20 days after treatment, double KO animals were severely impaired in doing so, underscoring that inducing cell senescence is important for liver fibrosis resolution [136]. Through gene profiling of senescent hepatic stellate cells (HSCs), authors show that many genes upregulated in those cells mediate an increased NK cell function, including CD58, IL-8 and NKG2D ligands such as MICA, ULBP2. Depletion of NK cells in these mice models resulted in delayed fibrosis resolution and incomplete senescent cells' removal. Conversely, boosting NK cell function with Poly I:C (TL3-agonist) resulted in decreased fibrosis and reduced senescent cell persistence. Several studies have expanded this knowledge, showing that a wide variety of senescent cells express increased levels of NKG2D ligands, such as

MICA, ULBP1 and ULBP2, in response to DNA damage (etoposide), telomere shortening (replicative renascence) or H-RAS^{v12} overexpression (oncogene-induced senescence) [137]. Further, pharmacological blockade or gene knockdown (siRNA) of NKG2D interaction with MICA or ULBP2 ligands reduces NK-mediated cytotoxicity towards senescent cells by 80–90% in vitro [137], confirming the important role of NKG2D and NK cells in removing senescent cells.

Interestingly, a recent study shows that activated invariant natural killer T cells (iNKT) are able to target and eliminate senescent cells both in a human in vitro setting and in two different mouse in vivo models [138]. Authors report that beta-2-microglobulin (B2M) and CD1d are upregulated in senescent cells, forming a complex that activates iNKT cells [138].

It is widely known that senescent cells accumulate with chronological age and that immune function also declines over time. Recent studies show that perforin release and binding (a key effector mechanism of NK and other cytotoxic cells) in the immunological synapse declines with age [139]. This is further modeled in $prf^{-/-}$ *KO* mice, which lack perforin. These mice show an up to 2- to 4-fold increase in senescent cells across a variety of tissues, resulting in chronic inflammation and increased age-related disorders, showing that cytotoxic activity has an important role to avoid excessive senescent cell accumulation.

All of this body of evidence suggests that NK cells could be harnessed as a therapeutic option to remove senescent cells. To this end, [140] has already shown that infusion of autologous, in vitro expanded and activated NK cells (1×10^9) can reduce p16^{INK4a} and SA-β-gal in PBMCs up to 90 days after infusion.

4.2. NK Cells Can Remove Senescent Tumor Cells

In the cancer research setting, the contribution of cellular senescence to immune surveillance is demonstrated by reactivation of endogenous p53 in p53-deficient tumors [141]. Briefly, reactivating p53 in such tumors causes cell senescence, in vitro cell cycle arrest, and triggers an NK-mediated immune response that targets tumor cells in vivo, resulting in tumor regression. Further studies using the same p53 reactivation model in mice showed that the elimination of senescent tumor cells depend on NKG2D expression. In addition, p53 is key for senescent cells to secrete pro-inflammatory chemokines such as CCL2 that recruit NK cells to senescent cells [142]. A blocking antibody against CCL2 prevented NK cell recruitment to senescent tumors, reducing their elimination. This contribution of the SASP for an optimal NK cell function was also revealed in a model of therapy-induced tumor senescence, where targeting the p65 subunit of NF-κB reduced senescent SASP secretion and proper NK-cytotoxic function [143].

The fact that senescent cells are prone to be detected and cleared by the immune system prompted researchers to test inducing cell senescence in tumors to address whether malignant cells can be then eliminated by the immune system. New therapeutic avenues are being studied to sensitize cancer cells to NK-mediated killing. In this regard, the chemotherapeutic drug doxorubicin, which is also a senescence-inducing drug, is able to improve NK-mediated killing of the MCF7 breast cancer cell line, through upregulation of death receptors FASR [144].

4.3. Senescent Cells Can Avoid NK-Cell Recognition, Thwarting Immune Clearance

While activating NK ligands are clearly upregulated in senescent cells, these cells may also upregulate inhibitory signals, resulting in the inhibition of NK-cell function. In this regard, Pereira et al. elegantly showed that primary human dermal fibroblasts were made senescent by ionizing radiation upregulating HLA-E [145]. This results in NK cell and T CD8$^+$ cell inhibition as a result of NKG2A ligation in those cells. HLA-E upregulation seems greatly dependent on SASP mediators such as IL-6 and the activation of the P38 signaling pathway. Authors confirm that NKG2D is key for NK-mediated killing of senescent cells, as blocking NKG2D with a monoclonal antibody abrogates NK-mediated killing of senescent fibroblasts. Conversely, authors showed for the first time that blocking or inhibiting NKG2A with siRNA actually boosted NK-mediated senescent killing.

This underscores the key role of these receptors in senescent or tumor cell removal [146]. Additional mechanisms of NK-cell resistance include the SASP-mediated recruitment of $CCR2^+$ immature myeloid cells (iMC), which inhibit NK-cell inhibition in established tumors, thus further contributing to tumor growth [147,148].

4.4. Dasatinib as a Senolytic Therapy in Animal Models and Human Models

The senolytic effect of dasatinib was first reported by Zhu et al., after a transcript analysis of preadipocytes made senescent by ionizing radiation (10 Gy) [149]. Further analysis showed increased negative regulators of apoptosis and anti-apoptotic gene sets in senescent cells such as ephrin receptors, the BCL-2/BCL-XL family, P13K/AKT, HIF-1a and others. Out of a panel of 46 drug candidates targeting these pathways, dasatinib and quercetin (D+Q) showed promise in selectively eliminating senescent preadipocytes, although with much reduced effect in other senescent cell types [149]. Importantly, the broad-spectrum tyrosine kinase inhibition mediated by dasatinib interferes with the ephrin receptor family (EFNB), which are the most extended tyrosine kinase receptor family and mediate key antiapoptotic signals [149].

In Zhu et al.'s work, in vivo administration of D+Q reduced the senescent cell burden in chronologically aged, radiation-exposed, and progeroid $Ercc1^{-}/D$ mice, delaying age-related symptoms and pathologies [149]. A similar effect was observed by Xu et al., when they found a healthier, longer mouse lifespan as a result of a senolytic therapy with dasatinib [140]. Since this first observation, D or D+Q senolytic strategies have been successfully used in a wide range of mouse models of diseases such as lung fibrosis [150], renal fibrosis [151], T2 diabetes [152] vascular pathology and atherosclerosis [153]. In all these models, D+Q reduced the senescent cell burden and improved organ function and physical health of treated animals. In the mouse setting, intestinal inflammation was reduced after D+Q intervention [154]. These D+Q effects have also been addressed in zebrafish models of skin inflammation and fatty liver disease, also resulting in reduced inflammation after therapy [155].

There are two early phase clinical trials with published results on D+Q senolytic therapy. The first in-human open-label pilot study (NCT02874989) showed that intermittent D+Q treatment (100 mg D + 1250 mg Q over three consecutive days in three consecutive weeks) is feasible, well tolerated and can alleviate physical dysfunction in idiopathic pulmonary fibrosis [156]. The second study (NCT02848131) showed that a single, 3-day dose of D+Q significantly reduced senescent cell burden ($p16^{INK4a}$, $P21^{CIP}$ and SA-β-gal positive cells) and plasma pro-inflammatory SASP mediators in diabetic kidney disease [157].

Different clinical trials are currently ongoing comparing the D + Q senolytic strategy to other drugs such as Fisetin in a variety of conditions, ranging from Alzheimer disease (NCT04063124) to hematopoietic stem cell transplant survivors (who are at risk of premature aging) (NCT02652052) and skeletal muscle health in older individuals (NCT04313634). More clinical trials are under way [158], which will shed light on whether senolytic D+Q treatments can improve a number of conditions.

In addition to its senolytic effects, D+Q therapy has also been shown to reduce the secretion of pro-inflammatory molecules and SASP mediators. In this regard, D+Q reduced the secretion of pro-inflammatory cytokines in human adipose tissue explants from obese individuals [153]. In the HIV context, our own work shows that D+Q senolytic therapy is able to reduce senescence biomarkers in PBMCs derived from PLWH [159].

While NK function is key to remove senescent cells both in vitro and in vivo, we are not aware of studies addressing whether senolytic treatment (dasatinib or others) may impact NK cell-mediated senescent cell clearance. This would be valuable to better understand the beneficial effects of dasatinib senolytic treatments. Furthermore, while promising, D+Q senolytic therapy may not be effective in all in vitro models, underscoring the importance of the experimental setting and cell type chosen when evaluating D+Q effects [149,160,161].

Results of ongoing studies and clinical trials will give clues to demonstrate if D+Q remains effective against aging related pathologies and could be translated towards new effective and safe anti-aging therapies.

4.5. Summary of the Senescence Section

Dasatinib treatment is effective in removing senescent cells in a wide variety of conditions, as shown in human clinical models. In addition, more evidence points to NK cells being key in senescent cell clearance and immunosurveillance. Dasatinib has the potential to activate NK cells by downregulating inhibitory receptors such as NKG2A and increasing activating receptors such as NKG2C and NKG2D. Thus, dasatinib may potentiate the NK cytotoxic effect against senescent cells. More research is needed in the aging field to attempt to elucidate this open question and the interesting therapeutic avenue (Figure 1C).

5. Concluding Remarks and Perspectives

Dasatinib treatment in CML induced populations of NK memory-like cells, $\gamma\delta$ T cells and other innate T cells, which are associated with better prognosis of CML, a possible indefinite TFR and good cellular response. Knowledge on the precise mechanisms by which dasatinib treatment yields these immunomodulatory effects is important, as this may improve treatment interruption parameters in CML, but also help to fight against other malignancies (Figure 1A).

A future goal is to explore whether dasatinib could safely prevent HIV-1 replication in patients both directly inhibiting SAMHD1-phosphorilation and also indirectly promoting the expansion of NK ($NKG2C^+CD57^+$) cells, $\gamma\delta$ T cells and other innate T cells. These cell subsets show a highly effective anti-HIV response and seem to be expanded in CMV^+ individuals. Recently, our published data concerning the case of a functionally cured HIV patient suggest that there are genetic features associated with the expansion of these innate NK and $\gamma\delta$ T cell subsets that allow HIV persistent control. Consequently, strategies able to expand these subpopulations are important to potentiate NK cytotoxicity against HIV. This would be extremely valuable in PLWH without these genetic characteristics and may pave the way to achieve a possible functional cure. The use of dasatinib in combination with ART or new combined immunotherapies could protect $CD4^+$ T cells and macrophages from HIV infection and activate a powerful cytotoxic response that could promote the elimination of the viral reservoirs (Figure 1B) and also reduce HIV-associated inflammation and senescence.

Finally, dasatinib shows potential against inflammation and cellular senescence or aging (Figure 1C). While most senolytic effects are reported in synergy with quercetin, dasatinib on its own downregulates NKG2A and increases NKG2C and -D in NK cells. Thus, dasatinib could potentiate the already important role of NK cells in removing senescent cells, which may be an important contributor to the overall senolytic effects. More research is needed in the field to attempt to elucidate this open question.

Author Contributions: N.C., A.R.-A., V.C., J.G.-E., S.S.-P. and J.A. conceived and wrote the review manuscript. All authors have read and agreed to the published version of the manuscript.

Funding: This review was partially supported by grants from the "Programa de becas Gilead a la investigación Biomédica GLD21_00111"; "Fondo Europeo para el Desarrollo Regional" (FEDER); the SPANISH AIDS Research Network RD12/0017/0001, RD16/0025/0002 and RD16/0025/0014-ISCIII-FEDER (RIS); the "Fondo de Investigación Sanitaria" (FIS) PI04/0363, PI20/00676; HIVACAT Programme and the CERCA Programme/"Generalitat de Catalunya" SGR 615 and SGR 653. Our group received funding from "la Caixa" Foundation under agreement, as well as from CIBERINFEC from the "Instituto de Salud Carlos III", Madrid, Spain.

Acknowledgments: We would like to acknowledge all PLWH participants who made research possible. We would especially like to thank María José Maleno, Cristina Rovira and Carmen Hurtado for their technical support and advice.

Conflicts of Interest: The authors declare that the research was conducted in the absence of any commercial or financial relationships that could be construed as a potential conflict of interest.

References

1. Simoneau, C.-A. Treating Chronic Myeloid Leukemia: Improving Management through Understanding of the Patient Experience. *Clin. J. Oncol. Nurs.* **2013**, *17*, E13–E20. [CrossRef]
2. D'Antonio, J. Chronic Myelogenous Leukemia. *Clin. J. Oncol. Nurs.* **2007**, *9*, 535–538. [CrossRef]
3. Quintás-Cardama, A.; Kantarjian, H.; Cortes, J. Imatinib and Beyond–Exploring the Full Potential of Targeted Therapy for CML. *Nat. Rev. Clin. Oncol.* **2009**, *6*, 535–543. [CrossRef]
4. Thompson, P.A.; Kantarjian, H.M.; Cortes, J.E. Diagnosis and Treatment of Chronic Myeloid Leukemia in 2015. *Mayo Clin. Proc.* **2015**, *90*, 1440–1454. [CrossRef]
5. Araujo, J.; Logothetis, C. Dasatinib: A Potent SRC Inhibitor in Clinical Development for the Treatment of Solid Tumors. *Cancer Treat Rev* **2010**, *36*, 492–500. [CrossRef]
6. Cheng, F.; Xu, Q.; Li, Q.; Cui, Z.; Li, W.; Zeng, F. Adverse Reactions after Treatment with Dasatinib in Chronic Myeloid Leukemia: Characteristics, Potential Mechanisms, and Clinical Management Strategies. *Front Oncol* **2023**, *13*, 1113462. [CrossRef]
7. Ishiyama, K.; Kitawaki, T.; Sugimoto, N.; Sozu, T.; Anzai, N.; Okada, M.; Nohgawa, M.; Hatanaka, K.; Arima, N.; Ishikawa, T.; et al. Principal Component Analysis Uncovers Cytomegalovirus-Associated NK Cell Activation in Ph^+ Leukemia Patients Treated with Dasatinib. *Leukemia* **2017**, *31*, 268. [CrossRef]
8. Hsieh, Y.-C.; Kirschner, K.; Copland, M. Improving Outcomes in Chronic Myeloid Leukemia through Harnessing the Immunological Landscape. *Leukemia* **2021**, *35*, 1229–1242. [CrossRef] [PubMed]
9. Bernard, N.F.; Kant, S.; Kiani, Z.; Tremblay, C.; Dupuy, F.P. Natural Killer Cells in Antibody Independent and Antibody Dependent HIV Control. *Front. Immunol.* **2022**, *13*, 879124. [CrossRef] [PubMed]
10. Liu, H.; Wang, S.; Xin, J.; Wang, J.; Yao, C.; Zhang, Z. Role of NKG2D and Its Ligands in Cancer Immunotherapy. *Am. J. Cancer Res.* **2019**, *9*, 2064–2078. [PubMed]
11. Xu, Z.; Yin, J.; Sun, Q.; Hu, J.; Hong, M.; Qian, S.; Liu, W. The Prognostic Role of NKG2A Expression for Patients with Chronic Myeloid Leukemia after Treatment Discontinuation. *Leuk. Lymphoma* **2022**, *63*, 2616–2626. [CrossRef] [PubMed]
12. Shindo, T.; Ureshino, H.; Kojima, H.; Tanaka, H.; Kimura, S. Allelic Polymorphisms of KIRs and Antitumor Immunity against Chronic Myeloid Leukemia. *Immunol. Med.* **2021**, *44*, 61–68. [CrossRef] [PubMed]
13. Kim, S.; Poursine-Laurent, J.; Truscott, S.M.; Lybarger, L.; Song, Y.-J.; Yang, L.; French, A.R.; Sunwoo, J.B.; Lemieux, S.; Hansen, T.H.; et al. Licensing of Natural Killer Cells by Host Major Histocompatibility Complex Class I Molecules. *Nature* **2005**, *436*, 709–713. [CrossRef]
14. Boudreau, J.E.; Hsu, K.C. Natural Killer Cell Education and the Response to Infection and Cancer Therapy: Stay Tuned. *Trends Immunol.* **2018**, *39*, 222–239. [CrossRef]
15. Long, E.O.; Kim, H.S.; Liu, D.; Peterson, M.E.; Rajagopalan, S. Controlling Natural Killer Cell Responses: Integration of Signals for Activation and Inhibition. *Annu. Rev. Immunol.* **2013**, *31*, 227–258. [CrossRef] [PubMed]
16. Fauriat, C.; Ivarsson, M.A.; Ljunggren, H.-G.; Malmberg, K.-J.; Michaëlsson, J. Education of Human Natural Killer Cells by Activating Killer Cell Immunoglobulin-like Receptors. *Blood* **2010**, *115*, 1166–1174. [CrossRef]
17. Gardiner, C.M.; Guethlein, L.A.; Shilling, H.G.; Pando, M.; Carr, W.H.; Rajalingam, R.; Vilches, C.; Parham, P. Different NK Cell Surface Phenotypes Defined by the DX9 Antibody Are Due to KIR3DL1 Gene Polymorphism. *J. Immunol.* **2001**, *166*, 2992–3001. [CrossRef]
18. Izumi, K.; Shindo, T.; Ngo, H.T.; Nakayama-Hosoya, K.; Akahane, K.; Tamai, M.; Nguyen, T.T.T.; Kawana-Tachikawa, A.; Inukai, T.; Takaori-Kondo, A. KIR3DL1 Allotype-Dependent Modulation of NK Cell Immunity against Chronic Myeloid Leukemia. *Immunohorizons* **2021**, *5*, 687–702. [CrossRef]
19. Lee, N.; Goodlett, D.R.; Ishitani, A.; Marquardt, H.; Geraghty, D.E. HLA-E Surface Expression Depends on Binding of TAP-Dependent Peptides Derived from Certain HLA Class I Signal Sequences. *J. Immunol.* **1998**, *160*, 4951–4960. [CrossRef]
20. Horowitz, A.; Djaoud, Z.; Nemat-Gorgani, N.; Blokhuis, J.; Hilton, H.G.; Béziat, V.; Malmberg, K.-J.; Norman, P.J.; Guethlein, L.A.; Parham, P. Class I HLA Haplotypes Form Two Schools That Educate NK Cells in Different Ways. *Sci. Immunol.* **2016**, *1*, eaag1672. [CrossRef]
21. Marinelli Busilacchi, E.; Costantini, A.; Viola, N.; Costantini, B.; Olivieri, J.; Butini, L.; Mancini, G.; Scortechini, I.; Chiarucci, M.; Poiani, M.; et al. Immunomodulatory Effects of Tyrosine Kinase Inhibitor In Vitro and In Vivo Study. *Biol. Blood Marrow Transplant.* **2018**, *24*, 267–275. [CrossRef] [PubMed]
22. Damele, L.; Montaldo, E.; Moretta, L.; Vitale, C.; Mingari, M.C. Effect of Tyrosin Kinase Inhibitors on NK Cell and ILC3 Development and Function. *Front. Immunol.* **2018**, *9*, 2433. [CrossRef] [PubMed]
23. Giansanti, P.; Preisinger, C.; Huber, K.V.M.; Gridling, M.; Superti-Furga, G.; Bennett, K.L.; Heck, A.J.R. Evaluating the Promiscuous Nature of Tyrosine Kinase Inhibitors Assessed in A431 Epidermoid Carcinoma Cells by Both Chemical- and Phosphoproteomics. *ACS Chem. Biol.* **2014**, *9*, 1490–1498. [CrossRef] [PubMed]
24. Hughes, A.; Clarson, J.; Tang, C.; Vidovic, L.; White, D.L.; Hughes, T.P.; Yong, A.S.M. CML Patients with Deep Molecular Responses to TKI Have Restored Immune Effectors and Decreased PD-1 and Immune Suppressors. *Blood* **2017**, *129*, 1166–1176. [CrossRef] [PubMed]

25. Hughes, A.; Yong, A.S.M. Immune Effector Recovery in Chronic Myeloid Leukemia and Treatment-Free Remission. *Front. Immunol.* **2017**, *8*, 469. [CrossRef] [PubMed]
26. Nomura, S.; Ito, T.; Satake, A.; Ishii, K. Assessment of Soluble Cytotoxic T Lymphocyte-Associated Antigen-4, Transforming Growth Factor B1, and Platelet-Derived Microparticles during Dasatinib Therapy for Patients with Chronic Myelogenous Leukemia. *J. Blood Med.* **2019**, *10*, 1–8. [CrossRef]
27. Najima, Y.; Yoshida, C.; Iriyama, N.; Fujisawa, S.; Wakita, H.; Chiba, S.; Okamoto, S.; Kawakami, K.; Takezako, N.; Kumagai, T.; et al. Regulatory T Cell Inhibition by Dasatinib Is Associated with Natural Killer Cell Differentiation and a Favorable Molecular Response—The Final Results of the D-First Study. *Leuk. Res.* **2018**, *66*, 66–72. [CrossRef]
28. Christiansson, L.; Söderlund, S.; Mangsbo, S.; Hjorth-Hansen, H.; Höglund, M.; Markevärn, B.; Richter, J.; Stenke, L.; Mustjoki, S.; Loskog, A.; et al. The Tyrosine Kinase Inhibitors Imatinib and Dasatinib Reduce Myeloid Suppressor Cells and Release Effector Lymphocyte Responses. *Mol. Cancer Ther.* **2015**, *14*, 1181–1191. [CrossRef] [PubMed]
29. Mustjoki, S.; Ekblom, M.; Arstila, T.P.; Dybedal, I.; Epling-Burnette, P.K.; Guilhot, F.; Hjorth-Hansen, H.; Höglund, M.; Kovanen, P.; Laurinolli, T.; et al. Clonal Expansion of T/NK-Cells during Tyrosine Kinase Inhibitor Dasatinib Therapy. *Leukemia* **2009**, *23*, 1398–1405. [CrossRef] [PubMed]
30. Qiu, Z.-Y.; Xu, W.; Li, J.-Y. Large Granular Lymphocytosis during Dasatinib Therapy. *Cancer Biol. Ther.* **2014**, *15*, 247–255. [CrossRef]
31. Watanabe, N.; Takaku, T.; Takeda, K.; Shirane, S.; Toyota, T.; Koike, M.; Noguchi, M.; Hirano, T.; Fujiwara, H.; Komatsu, N. Dasatinib-Induced Anti-Leukemia Cellular Immunity through a Novel Subset of CD57 Positive Helper/Cytotoxic CD4 T Cells in Chronic Myelogenous Leukemia Patients. *Int. J. Hematol.* **2018**, *108*, 588–597. [CrossRef]
32. Kreutzman, A.; Juvonen, V.; Kairisto, V.; Ekblom, M.; Stenke, L.; Seggewiss, R.; Porkka, K.; Mustjoki, S. Mono/Oligoclonal T and NK Cells Are Common in Chronic Myeloid Leukemia Patients at Diagnosis and Expand during Dasatinib Therapy. *Blood* **2010**, *116*, 772–782. [CrossRef]
33. Mustjoki, S.; Auvinen, K.; Kreutzman, A.; Rousselot, P.; Hernesniemi, S.; Melo, T.; Lahesmaa-Korpinen, A.-M.; Hautaniemi, S.; Bouchet, S.; Molimard, M.; et al. Rapid Mobilization of Cytotoxic Lymphocytes Induced by Dasatinib Therapy. *Leukemia* **2013**, *27*, 914–924. [CrossRef]
34. Schiffer, C.A.; Cortes, J.E.; Hochhaus, A.; Saglio, G.; le Coutre, P.; Porkka, K.; Mustjoki, S.; Mohamed, H.; Shah, N.P. Lymphocytosis after Treatment with Dasatinib in Chronic Myeloid Leukemia: Effects on Response and Toxicity. *Cancer* **2016**, *122*, 1398–1407. [CrossRef]
35. Kim, D.H.; Kamel-Reid, S.; Chang, H.; Sutherland, R.; Jung, C.W.; Kim, H.-J.; Lee, J.-J.; Lipton, J.H. Natural Killer or Natural Killer/T Cell Lineage Large Granular Lymphocytosis Associated with Dasatinib Therapy for Philadelphia Chromosome Positive Leukemia. *Haematologica* **2009**, *94*, 135–139. [CrossRef]
36. Lissina, A.; McLaren, J.E.; Ilander, M.; Andersson, E.I.; Lewis, C.S.; Clement, M.; Herman, A.; Ladell, K.; Llewellyn-Lacey, S.; Miners, K.L.; et al. Divergent Roles for Antigenic Drive in the Aetiology of Primary versus Dasatinib-Associated CD8+ TCR-Vβ+ Expansions. *Sci. Rep.* **2018**, *8*, 2534. [CrossRef] [PubMed]
37. Barbarin, A.; Abdallah, M.; Lefèvre, L.; Piccirilli, N.; Cayssials, E.; Roy, L.; Gombert, J.-M.; Herbelin, A. Innate T-Aβ Lymphocytes as New Immunological Components of Anti-Tumoral "off-Target" Effects of the Tyrosine Kinase Inhibitor Dasatinib. *Sci. Rep.* **2020**, *10*, 3245. [CrossRef] [PubMed]
38. Hayashi, Y.; Nakamae, H.; Katayama, T.; Nakane, T.; Koh, H.; Nakamae, M.; Hirose, A.; Hagihara, K.; Terada, Y.; Nakao, Y.; et al. Different Immunoprofiles in Patients with Chronic Myeloid Leukemia Treated with Imatinib, Nilotinib or Dasatinib. *Leuk. Lymphoma* **2012**, *53*, 1084–1089. [CrossRef] [PubMed]
39. Kreutzman, A.; Jaatinen, T.; Greco, D.; Vakkila, E.; Richter, J.; Ekblom, M.; Hjorth-Hansen, H.; Stenke, L.; Melo, T.; Paquette, R.; et al. Killer-Cell Immunoglobulin-like Receptor Gene Profile Predicts Good Molecular Response to Dasatinib Therapy in Chronic Myeloid Leukemia. *Exp. Hematol.* **2012**, *40*, 906–913.e1. [CrossRef] [PubMed]
40. Hassold, N.; Seystahl, K.; Kempf, K.; Urlaub, D.; Zekl, M.; Einsele, H.; Watzl, C.; Wischhusen, J.; Seggewiss-Bernhardt, R. Enhancement of Natural Killer Cell Effector Functions against Selected Lymphoma and Leukemia Cell Lines by Dasatinib. *Int. J. Cancer* **2012**, *131*, E916–E927. [CrossRef]
41. Sheng, L.-X.; Wang, J.-P.; Lai, Y.-L.; Wu, H.; Sun, Y.-C.; Zhou, M.; Ouyang, G.-F.; Huang, H. Effects of Dasatinib on the Expansion, Subsets, Receptor Expression and Cytotoxic Function of NK Cells in Vitro. *Zhongguo Shi Yan Xue Ye Xue Za Zhi* **2020**, *28*, 1762–1768. [CrossRef] [PubMed]
42. Chang, M.-C.; Cheng, H.-I.; Hsu, K.; Hsu, Y.-N.; Kao, C.-W.; Chang, Y.-F.; Lim, K.-H.; Chen, C.G. NKG2A Down-Regulation by Dasatinib Enhances Natural Killer Cytotoxicity and Accelerates Effective Treatment Responses in Patients With Chronic Myeloid Leukemia. *Front. Immunol.* **2018**, *9*, 3152. [CrossRef] [PubMed]
43. Kreutzman, A.; Ladell, K.; Koechel, C.; Gostick, E.; Ekblom, M.; Stenke, L.; Melo, T.; Einsele, H.; Porkka, K.; Price, D.A.; et al. Expansion of Highly Differentiated CD8+ T-Cells or NK-Cells in Patients Treated with Dasatinib Is Associated with Cytomegalovirus Reactivation. *Leukemia* **2011**, *25*, 1587–1597. [CrossRef]
44. Rölle, A.; Brodin, P. Immune Adaptation to Environmental Influence: The Case of NK Cells and HCMV. *Trends Immunol.* **2016**, *37*, 233–243. [CrossRef]
45. Kadowaki, N.; Ishiyama, K.; Kitawaki, T. Cytomegalovirus Pulls Strings behind NK Cells. *Oncotarget* **2017**, *8*, 93297–93298. [CrossRef] [PubMed]

46. Climent, N.; Plana, M. Immunomodulatory Activity of Tyrosine Kinase Inhibitors to Elicit Cytotoxicity Against Cancer and Viral Infection. *Front. Pharmacol.* **2019**, *10*, 1232. [CrossRef] [PubMed]
47. Ishiyama, K.-I.; Kitawaki, T.; Otsuka, Y.; Takaori-Kondo, A.; Kadowaki, N. Programmed Cell Death 1-Expressing CD56-Negative Natural Killer (NK) Cell Expansion Is a Hallmark of Chronic NK Cell Activation during Dasatinib Treatment. *Cancer Sci.* **2021**, *112*, 523–536. [CrossRef] [PubMed]
48. Duerkop, B.A.; Hooper, L.V. Resident Viruses and Their Interactions with the Immune System. *Nat. Immunol.* **2013**, *14*, 654–659. [CrossRef]
49. Cortes, J.E.; Saglio, G.; Kantarjian, H.M.; Baccarani, M.; Mayer, J.; Boqué, C.; Shah, N.P.; Chuah, C.; Casanova, L.; Bradley-Garelik, B.; et al. Final 5-Year Study Results of DASISION: The Dasatinib Versus Imatinib Study in Treatment-Naïve Chronic Myeloid Leukemia Patients Trial. *J. Clin. Oncol.* **2016**, *34*, 2333–2340. [CrossRef] [PubMed]
50. Saußele, S.; Richter, J.; Hochhaus, A.; Mahon, F.-X. The Concept of Treatment-Free Remission in Chronic Myeloid Leukemia. *Leukemia* **2016**, *30*, 1638–1647. [CrossRef] [PubMed]
51. Cortes, J.; Rea, D.; Lipton, J.H. Treatment-Free Remission with First- and Second-Generation Tyrosine Kinase Inhibitors. *Am. J. Hematol.* **2019**, *94*, 346–357. [CrossRef] [PubMed]
52. Guru Murthy, G.S.; Atallah, E. Treatment-Free Remission in CML: The US Perspective. *Curr. Hematol. Malig. Rep.* **2019**, *14*, 56–61. [CrossRef]
53. Imagawa, J.; Tanaka, H.; Okada, M.; Nakamae, H.; Hino, M.; Murai, K.; Ishida, Y.; Kumagai, T.; Sato, S.; Ohashi, K.; et al. Discontinuation of Dasatinib in Patients with Chronic Myeloid Leukaemia Who Have Maintained Deep Molecular Response for Longer than 1 Year (DADI Trial): A Multicentre Phase 2 Trial. *Lancet Haematol.* **2015**, *2*, e528–e535. [CrossRef] [PubMed]
54. Mizoguchi, I.; Yoshimoto, T.; Katagiri, S.; Mizuguchi, J.; Tauchi, T.; Kimura, Y.; Inokuchi, K.; Ohyashiki, J.H.; Ohyashiki, K. Sustained Upregulation of Effector Natural Killer Cells in Chronic Myeloid Leukemia after Discontinuation of Imatinib. *Cancer Sci.* **2013**, *104*, 1146–1153. [CrossRef]
55. Rea, D.; Henry, G.; Khaznadar, Z.; Etienne, G.; Guilhot, F.; Nicolini, F.; Guilhot, J.; Rousselot, P.; Huguet, F.; Legros, L.; et al. Natural Killer-Cell Counts Are Associated with Molecular Relapse-Free Survival after Imatinib Discontinuation in Chronic Myeloid Leukemia: The IMMUNOSTIM Study. *Haematologica* **2017**, *102*, 1368–1377. [CrossRef] [PubMed]
56. Ureshino, H.; Kamachi, K.; Sano, H.; Okamoto, S.; Itamura, H.; Yoshimura, M.; Katsuya, H.; Ando, T.; Kimura, S. Higher Neutrophil Counts Are Associated with Successful Tyrosine Kinase Inhibitor Discontinuation in Patients with Chronic Myeloid Leukemia. *Hematology* **2022**, *27*, 1171–1175. [CrossRef] [PubMed]
57. Ilander, M.; Olsson-Strömberg, U.; Schlums, H.; Guilhot, J.; Brück, O.; Lähteenmäki, H.; Kasanen, T.; Koskenvesa, P.; Söderlund, S.; Höglund, M.; et al. Increased Proportion of Mature NK Cells Is Associated with Successful Imatinib Discontinuation in Chronic Myeloid Leukemia. *Leukemia* **2017**, *31*, 1108–1116. [CrossRef]
58. Hara, R.; Onizuka, M.; Matsusita, E.; Kikkawa, E.; Nakamura, Y.; Matsushita, H.; Ohgiya, D.; Murayama, H.; Machida, S.; Ohmachi, K.; et al. NKG2D Gene Polymorphisms Are Associated with Disease Control of Chronic Myeloid Leukemia by Dasatinib. *Int. J. Hematol.* **2017**, *106*, 666–674. [CrossRef] [PubMed]
59. Hughes, A.; Clarson, J.; White, D.L.; Ross, D.M.; Hughes, T.P.; Yong, A.S. Enhanced Natural Killer and Cytotoxic T Lymphocyte Responses, with Decreased Monocytic Myeloid Derived Suppressor Cells May Promote Treatment Free Remission in Chronic Myeloid Leukaemia Patients Following Tyrosine Kinase Inhibitor Cessation. *Blood* **2016**, *128*, 1122. [CrossRef]
60. Burchert, A.; Saussele, S.; Eigendorff, E.; Müller, M.C.; Sohlbach, K.; Inselmann, S.; Schütz, C.; Metzelder, S.K.; Ziermann, J.; Kostrewa, P.; et al. Interferon Alpha 2 Maintenance Therapy May Enable High Rates of Treatment Discontinuation in Chronic Myeloid Leukemia. *Leukemia* **2015**, *29*, 1331–1335. [CrossRef]
61. Puzzolo, M.C.; Breccia, M.; Mariglia, P.; Colafigli, G.; Pepe, S.; Scalzulli, E.; Mariggiò, E.; Latagliata, R.; Guarini, A.; Foà, R. Immunomodulatory Effects of IFNα on T and NK Cells in Chronic Myeloid Leukemia Patients in Deep Molecular Response Preparing for Treatment Discontinuation. *J. Clin. Med.* **2022**, *11*, 5594. [CrossRef]
62. Rea, D.; Nicolini, F.E.; Tulliez, M.; Guilhot, F.; Guilhot, J.; Guerci-Bresler, A.; Gardembas, M.; Coiteux, V.; Guillerm, G.; Legros, L.; et al. Discontinuation of Dasatinib or Nilotinib in Chronic Myeloid Leukemia: Interim Analysis of the STOP 2G-TKI Study. *Blood* **2017**, *129*, 846–854. [CrossRef] [PubMed]
63. Okada, M.; Imagawa, J.; Tanaka, H.; Nakamae, H.; Hino, M.; Murai, K.; Ishida, Y.; Kumagai, T.; Sato, S.; Ohashi, K.; et al. Final 3-Year Results of the Dasatinib Discontinuation Trial in Patients With Chronic Myeloid Leukemia Who Received Dasatinib as a Second-Line Treatment. *Clin. Lymphoma Myeloma Leuk.* **2018**, *18*, 353–360.e1. [CrossRef]
64. Yoshida, C.; Iriyama, N.; Najima, Y.; Fujisawa, S.; Wakita, H.; Chiba, S.; Okamoto, S.; Kawakami, K.; Takezako, N.; Kumagai, T.; et al. Association of Peripheral Regulatory T Cells with Achievement of Deep Molecular Response in Newly Diagnosed Chronic Phase Chronic Myeloid Leukemia Treated with Dasatinib - the Final Results of D-First Study. *Blood* **2016**, *128*, 1916. [CrossRef]
65. Takaku, T.; Iriyama, N.; Mitsumori, T.; Sato, E.; Gotoh, A.; Kirito, K.; Noguchi, M.; Koike, M.; Sakamoto, J.; Oba, K.; et al. Clinical Efficacy and Safety of First-Line Dasatinib Therapy and the Relevance of Velocity of BCR-ABL1 Transcript Decline for Achievement of Molecular Responses in Newly Diagnosed Chronic-Phase Chronic Myeloid Leukemia: Report from the Juntendo Yamanashi Cooperative Study Group. *Oncology* **2018**, *94*, 85–91. [CrossRef]
66. Cayssials, E.; Jacomet, F.; Piccirilli, N.; Lefèvre, L.; Roy, L.; Guilhot, F.; Chomel, J.-C.; Leleu, X.; Gombert, J.-M.; Herbelin, A.; et al. Sustained Treatment-Free Remission in Chronic Myeloid Leukaemia Is Associated with an Increased Frequency of Innate CD8(+) T-Cells. *Br. J. Haematol.* **2019**, *186*, 54–59. [CrossRef]

67. Vigón, L.; Luna, A.; Galán, M.; Rodríguez-Mora, S.; Fuertes, D.; Mateos, E.; Piris-Villaespesa, M.; Bautista, G.; San José, E.; Rivera-Torres, J.; et al. Identification of Immunological Parameters as Predictive Biomarkers of Relapse in Patients with Chronic Myeloid Leukemia on Treatment-Free Remission. *J. Clin. Med.* **2020**, *10*, 42. [CrossRef]
68. Irani, Y.D.; Hughes, A.; Clarson, J.; Kok, C.H.; Shanmuganathan, N.; White, D.L.; Yeung, D.T.; Ross, D.M.; Hughes, T.P.; Yong, A.S.M. Successful Treatment-Free Remission in Chronic Myeloid Leukaemia and Its Association with Reduced Immune Suppressors and Increased Natural Killer Cells. *Br. J. Haematol.* **2020**, *191*, 433–441. [CrossRef] [PubMed]
69. Stuckey, R.; López Rodríguez, J.F.; Gómez-Casares, M.T. Discontinuation of Tyrosine Kinase Inhibitors in Patients with Chronic Myeloid Leukemia: A Review of the Biological Factors Associated with Treatment-Free Remission. *Curr. Oncol. Rep.* **2022**, *24*, 415–426. [CrossRef] [PubMed]
70. Schlums, H.; Cichocki, F.; Tesi, B.; Theorell, J.; Beziat, V.; Holmes, T.D.; Han, H.; Chiang, S.C.C.; Foley, B.; Mattsson, K.; et al. Cytomegalovirus Infection Drives Adaptive Epigenetic Diversification of NK Cells with Altered Signaling and Effector Function. *Immunity* **2015**, *42*, 443–456. [CrossRef]
71. He, Y.; Tian, Z. NK Cell Education via Nonclassical MHC and Non-MHC Ligands. *Cell Mol. Immunol.* **2017**, *14*, 321–330. [CrossRef]
72. Melendez, E.; Chondronasiou, D.; Mosteiro, L.; Martínez de Villarreal, J.; Fernández-Alfara, M.; Lynch, C.J.; Grimm, D.; Real, F.X.; Alcamí, J.; Climent, N.; et al. Natural Killer Cells Act as an Extrinsic Barrier for in Vivo Reprogramming. *Development* **2022**, *149*, dev200361. [CrossRef] [PubMed]
73. Yawata, M.; Yawata, N.; Draghi, M.; Little, A.-M.; Partheniou, F.; Parham, P. Roles for HLA and KIR Polymorphisms in Natural Killer Cell Repertoire Selection and Modulation of Effector Function. *J. Exp. Med.* **2006**, *203*, 633–645. [CrossRef]
74. O'Connor, G.M.; Guinan, K.J.; Cunningham, R.T.; Middleton, D.; Parham, P.; Gardiner, C.M. Functional Polymorphism of the KIR3DL1/S1 Receptor on Human NK Cells. *J. Immunol.* **2007**, *178*, 235–241. [CrossRef] [PubMed]
75. Caocci, G.; Martino, B.; Greco, M.; Abruzzese, E.; Trawinska, M.M.; Lai, S.; Ragatzu, P.; Galimberti, S.; Baratè, C.; Mulas, O.; et al. Killer Immunoglobulin-like Receptors Can Predict TKI Treatment-Free Remission in Chronic Myeloid Leukemia Patients. *Exp. Hematol.* **2015**, *43*, 1015–1018.e1. [CrossRef]
76. Ureshino, H.; Shindo, T.; Kojima, H.; Kusunoki, Y.; Miyazaki, Y.; Tanaka, H.; Saji, H.; Kawaguchi, A.; Kimura, S. Allelic Polymorphisms of KIRs and HLAs Predict Favorable Responses to Tyrosine Kinase Inhibitors in CML. *Cancer Immunol. Res.* **2018**, *6*, 745–754. [CrossRef]
77. La Nasa, G.; Caocci, G.; Littera, R.; Atzeni, S.; Vacca, A.; Mulas, O.; Langiu, M.; Greco, M.; Orrù, S.; Orrù, N.; et al. Homozygosity for Killer Immunoglobin-like Receptor Haplotype A Predicts Complete Molecular Response to Treatment with Tyrosine Kinase Inhibitors in Chronic Myeloid Leukemia Patients. *Exp. Hematol.* **2013**, *41*, 424–431. [CrossRef]
78. Yeung, D.T.; Tang, C.; Vidovic, L.; White, D.L.; Branford, S.; Hughes, T.P.; Yong, A.S. KIR2DL5B Genotype Predicts Outcomes in CML Patients Treated with Response-Directed Sequential Imatinib/Nilotinib Strategy. *Blood* **2015**, *126*, 2720–2723. [CrossRef]
79. Ureshino, H.; Shindo, T.; Tanaka, H.; Saji, H.; Kimura, S. HLA Polymorphisms Are Associated with Treatment-Free Remission Following Discontinuation of Tyrosine Kinase Inhibitors in Chronic Myeloid Leukemia. *Mol. Cancer Ther.* **2021**, *20*, 142–149. [CrossRef]
80. Flores-Villanueva, P.O.; Yunis, E.J.; Delgado, J.C.; Vittinghoff, E.; Buchbinder, S.; Leung, J.Y.; Uglialoro, A.M.; Clavijo, O.P.; Rosenberg, E.S.; Kalams, S.A.; et al. Control of HIV-1 Viremia and Protection from AIDS Are Associated with HLA-Bw4 Homozygosity. *Proc. Natl. Acad. Sci. USA* **2001**, *98*, 5140–5145. [CrossRef] [PubMed]
81. Yu, G.-T.; Mao, L.; Wu, L.; Deng, W.-W.; Bu, L.-L.; Liu, J.-F.; Chen, L.; Yang, L.-L.; Wu, H.; Zhang, W.-F.; et al. Inhibition of SRC Family Kinases Facilitates Anti-CTLA4 Immunotherapy in Head and Neck Squamous Cell Carcinoma. *Cell Mol. Life Sci.* **2018**, *75*, 4223–4234. [CrossRef]
82. Tu, M.M.; Lee, F.Y.F.; Jones, R.T.; Kimball, A.K.; Saravia, E.; Graziano, R.F.; Coleman, B.; Menard, K.; Yan, J.; Michaud, E.; et al. Targeting DDR2 Enhances Tumor Response to Anti-PD-1 Immunotherapy. *Sci. Adv.* **2019**, *5*, eaav2437. [CrossRef]
83. Mestermann, K.; Giavridis, T.; Weber, J.; Rydzek, J.; Frenz, S.; Nerreter, T.; Mades, A.; Sadelain, M.; Einsele, H.; Hudecek, M. The Tyrosine Kinase Inhibitor Dasatinib Acts as a Pharmacologic on/off Switch for CAR T Cells. *Sci. Transl. Med.* **2019**, *11*, eaau5907. [CrossRef]
84. Zhang, H.; Hu, Y.; Shao, M.; Teng, X.; Jiang, P.; Wang, X.; Wang, H.; Cui, J.; Yu, J.; Liang, Z.; et al. Dasatinib Enhances Anti-Leukemia Efficacy of Chimeric Antigen Receptor T Cells by Inhibiting Cell Differentiation and Exhaustion. *J. Hematol. Oncol.* **2021**, *14*, 113. [CrossRef]
85. Board, N.L.; Moskovljevic, M.; Wu, F.; Siliciano, R.F.; Siliciano, J.D. Engaging Innate Immunity in HIV-1 Cure Strategies. *Nat. Rev. Immunol.* **2022**, *22*, 499–512. [CrossRef] [PubMed]
86. Rasmussen, T.A.; Tolstrup, M.; Søgaard, O.S. Reversal of Latency as Part of a Cure for HIV-1. *Trends Microbiol.* **2016**, *24*, 90–97. [CrossRef]
87. Coiras, M.; Ambrosioni, J.; Cervantes, F.; Miró, J.M.; Alcamí, J. Tyrosine Kinase Inhibitors: Potential Use and Safety Considerations in HIV-1 Infection. *Expert. Opin. Drug Saf.* **2017**, *16*, 547–559. [CrossRef] [PubMed]
88. Bermejo, M.; López-Huertas, M.R.; García-Pérez, J.; Climent, N.; Descours, B.; Ambrosioni, J.; Mateos, E.; Rodríguez-Mora, S.; Rus-Bercial, L.; Benkirane, M.; et al. Dasatinib Inhibits HIV-1 Replication through the Interference of SAMHD1 Phosphorylation in CD4+ T Cells. *Biochem. Pharmacol.* **2016**, *106*, 30–45. [CrossRef] [PubMed]

89. Coiras, M.; Bermejo, M.; Descours, B.; Mateos, E.; García-Pérez, J.; López-Huertas, M.-R.; Lederman, M.M.; Benkirane, M.; Alcamí, J. IL-7 Induces SAMHD1 Phosphorylation in CD4+ T Lymphocytes, Improving Early Steps of HIV-1 Life Cycle. *Cell Rep.* **2016**, *14*, 2100–2107. [CrossRef] [PubMed]
90. Bermejo, M.; Ambrosioni, J.; Bautista, G.; Climent, N.; Mateos, E.; Rovira, C.; Rodríguez-Mora, S.; López-Huertas, M.R.; García-Gutiérrez, V.; Steegmann, J.L.; et al. Evaluation of Resistance to HIV-1 Infection Ex Vivo of PBMCs Isolated from Patients with Chronic Myeloid Leukemia Treated with Different Tyrosine Kinase Inhibitors. *Biochem. Pharmacol.* **2018**, *156*, 248–264. [CrossRef]
91. Williams, E.S.C.P.; Szaniawski, M.A.; Martins, L.J.; Innis, E.A.; Alcamí, J.; Hanley, T.M.; Spivak, A.M.; Coiras, M.; Planelles, V. Dasatinib: Effects on the Macrophage Phospho Proteome with a Focus on SAMHD1 and HIV-1 Infection. *Clin. Res. HIV AIDS* **2022**, *8*, 1053.
92. McCarthy, S.D.S.; Leontyev, D.; Nicoletti, P.; Binnington, B.; Kozlowski, H.N.; Ostrowski, M.; Cochrane, A.; Branch, D.R.; Wong, R.W. Targeting ABL1 or ARG Tyrosine Kinases to Restrict HIV-1 Infection in Primary CD4+ T-Cells or in Humanized NSG Mice. *J. Acquir. Immune Defic. Syndr.* **2019**, *82*, 407–415. [CrossRef] [PubMed]
93. Vigón, L.; Rodríguez-Mora, S.; Luna, A.; Sandonís, V.; Mateos, E.; Bautista, G.; Steegmann, J.L.; Climent, N.; Plana, M.; Pérez-Romero, P.; et al. Cytotoxic Cell Populations Developed during Treatment with Tyrosine Kinase Inhibitors Protect Autologous CD4+ T Cells from HIV-1 Infection. *Biochem. Pharmacol.* **2020**, *182*, 114203. [CrossRef] [PubMed]
94. Vigón, L.; Martínez-Román, P.; Rodríguez-Mora, S.; Torres, M.; Puertas, M.C.; Mateos, E.; Salgado, M.; Navarro, A.; Sánchez-Conde, M.; Ambrosioni, J.; et al. Provirus Reactivation Is Impaired in HIV-1 Infected Individuals on Treatment with Dasatinib and Antiretroviral Therapy. *Biochem. Pharmacol.* **2021**, *192*, 114666. [CrossRef]
95. Innis, E.A.; Levinger, C.; Szaniawski, M.A.; Williams, E.S.C.P.; Alcamí, J.; Bosque, A.; Schiffer, J.T.; Coiras, M.; Spivak, A.M.; Planelles, V. Pharmacologic Control of Homeostatic and Antigen-Driven Proliferation to Target HIV-1 Persistence. *Biochem. Pharmacol.* **2021**, *194*, 114816. [CrossRef]
96. Climent, N.; Ambrosioni, J.; González, T.; Xufré, C.; Casadellà, M.; Noguera-Julian, M.; Paredes, R.; Plana, M.; Grau-Expósito, J.; Mallolas, J.; et al. Immunological and Virological Findings in a Patient with Exceptional Post-Treatment Control: A Case Report. *Lancet HIV* **2023**, *10*, e42–e51. [CrossRef]
97. Sánchez-Gaona, N.; Gallego, A.; Astorga, A.; Rallón, N.; Benito, J.M.; Falcó, V.; Genescà, M.; Buzón, M.J. Characterization of NK Cells in Elite Controllers Losig HIV Control. [CROI Abstract 307]. In Proceedings of the CROI 2022 Conference on Retroviruses and Opportunistic Infections, Virtual, 12–16 February 2022; p. 118.
98. Gondois-Rey, F.; Chéret, A.; Granjeaud, S.; Mallet, F.; Bidaut, G.; Lécuroux, C.; Ploquin, M.; Müller-Trutwin, M.; Rouzioux, C.; Avettand-Fenoël, V.; et al. NKG2C+ Memory-like NK Cells Contribute to the Control of HIV Viremia during Primary Infection: Optiprim-ANRS 147. *Clin. Transl. Immunol.* **2017**, *6*, e150. [CrossRef]
99. Flórez-Álvarez, L.; Hernandez, J.C.; Zapata, W. NK Cells in HIV-1 Infection: From Basic Science to Vaccine Strategies. *Front. Immunol.* **2018**, *9*, 2290. [CrossRef] [PubMed]
100. Peppa, D.; Pedroza-Pacheco, I.; Pellegrino, P.; Williams, I.; Maini, M.K.; Borrow, P. Adaptive Reconfiguration of Natural Killer Cells in HIV-1 Infection. *Front. Immunol.* **2018**, *9*, 474. [CrossRef]
101. Sáez-Cirión, A.; Bacchus, C.; Hocqueloux, L.; Avettand-Fenoel, V.; Girault, I.; Lecuroux, C.; Potard, V.; Versmisse, P.; Melard, A.; Prazuck, T.; et al. Post-Treatment HIV-1 Controllers with a Long-Term Virological Remission after the Interruption of Early Initiated Antiretroviral Therapy ANRS VISCONTI Study. *PLoS Pathog.* **2013**, *9*, e1003211. [CrossRef]
102. Namazi, G.; Fajnzylber, J.M.; Aga, E.; Bosch, R.J.; Acosta, E.P.; Sharaf, R.; Hartogensis, W.; Jacobson, J.M.; Connick, E.; Volberding, P.; et al. The Control of HIV After Antiretroviral Medication Pause (CHAMP) Study: Posttreatment Controllers Identified From 14 Clinical Studies. *J. Infect. Dis.* **2018**, *218*, 1954–1963. [CrossRef]
103. Carlson, J.M.; Listgarten, J.; Pfeifer, N.; Tan, V.; Kadie, C.; Walker, B.D.; Ndung'u, T.; Shapiro, R.; Frater, J.; Brumme, Z.L.; et al. Widespread Impact of HLA Restriction on Immune Control and Escape Pathways of HIV-1. *J. Virol.* **2012**, *86*, 5230–5243. [CrossRef] [PubMed]
104. Zhang, X.; Huang, X.; Xia, W.; Li, W.; Zhang, T.; Wu, H.; Xu, X.; Yan, H. HLA-B*44 Is Associated with a Lower Viral Set Point and Slow CD4 Decline in a Cohort of Chinese Homosexual Men Acutely Infected with HIV-1. *Clin. Vaccine Immunol.* **2013**, *20*, 1048–1054. [CrossRef]
105. Singh, K.K.; Qin, M.; Brummel, S.S.; Angelidou, K.; Trout, R.N.; Fenton, T.; Spector, S.A. Killer Cell Immunoglobulin-Like Receptor Alleles Alter HIV Disease in Children. *PLoS ONE* **2016**, *11*, e0151364. [CrossRef]
106. Maruthamuthu, S.; Rajalingam, R.; Pandian, K.; Madasamy, S.; Manoharan, M.; Pitchai, L.; Murugesan, A.; Mariakuttikan, J. Inhibitory Natural Killer Cell Receptor KIR3DL1 with Its Ligand Bw4 Constraints HIV-1 Disease among South Indians. *AIDS* **2018**, *32*, 2679–2688. [CrossRef]
107. Ramsuran, V.; Naranbhai, V.; Horowitz, A.; Qi, Y.; Martin, M.P.; Yuki, Y.; Gao, X.; Walker-Sperling, V.; Del Prete, G.Q.; Schneider, D.K.; et al. Elevated HLA-A Expression Impairs HIV Control through Inhibition of NKG2A-Expressing Cells. *Science* **2018**, *359*, 86–90. [CrossRef] [PubMed]
108. Boulet, S.; Song, R.; Kamya, P.; Bruneau, J.; Shoukry, N.H.; Tsoukas, C.M.; Bernard, N.F. HIV Protective KIR3DL1 and HLA-B Genotypes Influence NK Cell Function Following Stimulation with HLA-Devoid Cells. *J. Immunol.* **2010**, *184*, 2057–2064. [CrossRef]

109. Parsons, M.S.; Boulet, S.; Song, R.; Bruneau, J.; Shoukry, N.H.; Routy, J.-P.; Tsoukas, C.M.; Bernard, N.F. Mind the Gap: Lack of Association between KIR3DL1*004/HLA-Bw4-Induced Natural Killer Cell Function and Protection from HIV Infection. *J. Infect. Dis.* **2010**, *202* (Suppl. S3), S356–S360. [CrossRef] [PubMed]
110. Kamya, P.; Boulet, S.; Tsoukas, C.M.; Routy, J.-P.; Thomas, R.; Côté, P.; Boulassel, M.-R.; Baril, J.-G.; Kovacs, C.; Migueles, S.A.; et al. Receptor-Ligand Requirements for Increased NK Cell Polyfunctional Potential in Slow Progressors Infected with HIV-1 Coexpressing KIR3DL1*h/*y and HLA-B*57. *J. Virol.* **2011**, *85*, 5949–5960. [CrossRef]
111. Schwartz, O.; Maréchal, V.; Le Gall, S.; Lemonnier, F.; Heard, J.M. Endocytosis of Major Histocompatibility Complex Class I Molecules Is Induced by the HIV-1 Nef Protein. *Nat. Med.* **1996**, *2*, 338–342. [CrossRef]
112. Cohen, G.B.; Gandhi, R.T.; Davis, D.M.; Mandelboim, O.; Chen, B.K.; Strominger, J.L.; Baltimore, D. The Selective Downregulation of Class I Major Histocompatibility Complex Proteins by HIV-1 Protects HIV-Infected Cells from NK Cells. *Immunity* **1999**, *10*, 661–671. [CrossRef] [PubMed]
113. Bonaparte, M.I.; Barker, E. Killing of Human Immunodeficiency Virus-Infected Primary T-Cell Blasts by Autologous Natural Killer Cells Is Dependent on the Ability of the Virus to Alter the Expression of Major Histocompatibility Complex Class I Molecules. *Blood* **2004**, *104*, 2087–2094. [CrossRef]
114. Apps, R.; Del Prete, G.Q.; Chatterjee, P.; Lara, A.; Brumme, Z.L.; Brockman, M.A.; Neil, S.; Pickering, S.; Schneider, D.K.; Piechocka-Trocha, A.; et al. HIV-1 Vpu Mediates HLA-C Downregulation. *Cell Host Microbe* **2016**, *19*, 686–695. [CrossRef] [PubMed]
115. Song, R.; Lisovsky, I.; Lebouché, B.; Routy, J.-P.; Bruneau, J.; Bernard, N.F. HIV Protective KIR3DL1/S1-HLA-B Genotypes Influence NK Cell-Mediated Inhibition of HIV Replication in Autologous CD4 Targets. *PLoS Pathog* **2014**, *10*, e1003867. [CrossRef]
116. Kiani, Z.; Dupuy, F.P.; Bruneau, J.; Lebouché, B.; Retière, C.; Geraghty, D.E.; Bernard, N.F. The Education of NK Cells Determines Their Responsiveness to Autologous HIV-Infected CD4 T Cells. *J. Virol.* **2019**, *93*, e01185-19. [CrossRef]
117. Körner, C.; Simoneau, C.R.; Schommers, P.; Granoff, M.; Ziegler, M.; Hölzemer, A.; Lunemann, S.; Chukwukelu, J.; Corleis, B.; Naranbhai, V.; et al. HIV-1-Mediated Downmodulation of HLA-C Impacts Target Cell Recognition and Antiviral Activity of NK Cells. *Cell Host Microbe* **2017**, *22*, 111–119.e4. [CrossRef] [PubMed]
118. Vieira, V.A.; Adland, E.; Malone, D.F.G.; Martin, M.P.; Groll, A.; Ansari, M.A.; Garcia-Guerrero, M.C.; Puertas, M.C.; Muenchhoff, M.; Guash, C.F.; et al. An HLA-I Signature Favouring KIR-Educated Natural Killer Cells Mediates Immune Control of HIV in Children and Contrasts with the HLA-B-Restricted CD8+ T-Cell-Mediated Immune Control in Adults. *PLoS Pathog.* **2021**, *17*, e1010090. [CrossRef] [PubMed]
119. Thomas, R.; Low, H.Z.; Kniesch, K.; Jacobs, R.; Schmidt, R.E.; Witte, T. NKG2C Deletion Is a Risk Factor of HIV Infection. *AIDS Res. Hum. Retrovir.* **2012**, *28*, 844–851. [CrossRef]
120. Essat, A.; Scott-Algara, D.; Monceaux, V.; Avettand-Fenoel, V.; Didier, C.; Caillat-Zucman, S.; Orr, S.; Theodorou, I.; Goujard, C.; Boufassa, F.; et al. Association between Immunogenetic Factors and Post-Treatment Control of HIV-1 Infection. ANRS VISCONTI and PRIMO Studies. In Proceedings of the 22nd International AIDS Conference (AIDS 2018), Amsterdam, Netherlands, 23–27 July 2018. [CrossRef]
121. Sáez-Cirión, A. Mechanisms of Post-Treatment HIV Control: ANRS VISCONTI Study. In Proceedings of the IAS 2019, 10th International Conference on HIV Science, Mexico City, Mexico, 21 July 2019. [CrossRef]
122. Juno, J.A.; Kent, S.J. What Can Gamma Delta T Cells Contribute to an HIV Cure? *Front. Cell Infect. Microbiol.* **2020**, *10*, 233. [CrossRef] [PubMed]
123. Lambotte, O.; Boufassa, F.; Madec, Y.; Nguyen, A.; Goujard, C.; Meyer, L.; Rouzioux, C.; Venet, A.; Delfraissy, J.-F. SEROCO-HEMOCO Study Group HIV Controllers: A Homogeneous Group of HIV-1-Infected Patients with Spontaneous Control of Viral Replication. *Clin. Infect. Dis.* **2005**, *41*, 1053–1056. [CrossRef]
124. McLaren, P.J.; Ripke, S.; Pelak, K.; Weintrob, A.C.; Patsopoulos, N.A.; Jia, X.; Erlich, R.L.; Lennon, N.J.; Kadie, C.M.; Heckerman, D.; et al. Fine-Mapping Classical HLA Variation Associated with Durable Host Control of HIV-1 Infection in African Americans. *Hum. Mol. Genet.* **2012**, *21*, 4334–4347. [CrossRef]
125. Mendoza, D.; Johnson, S.A.; Peterson, B.A.; Natarajan, V.; Salgado, M.; Dewar, R.L.; Burbelo, P.D.; Doria-Rose, N.A.; Graf, E.H.; Greenwald, J.H.; et al. Comprehensive Analysis of Unique Cases with Extraordinary Control over HIV Replication. *Blood* **2012**, *119*, 4645–4655. [CrossRef] [PubMed]
126. Casado, C.; Galvez, C.; Pernas, M.; Tarancon-Diez, L.; Rodriguez, C.; Sanchez-Merino, V.; Vera, M.; Olivares, I.; De Pablo-Bernal, R.; Merino-Mansilla, A.; et al. Permanent Control of HIV-1 Pathogenesis in Exceptional Elite Controllers: A Model of Spontaneous Cure. *Sci. Rep.* **2020**, *10*, 1902. [CrossRef]
127. Jiang, C.; Lian, X.; Gao, C.; Sun, X.; Einkauf, K.B.; Chevalier, J.M.; Chen, S.M.Y.; Hua, S.; Rhee, B.; Chang, K.; et al. Distinct Viral Reservoirs in Individuals with Spontaneous Control of HIV-1. *Nature* **2020**, *585*, 261–267. [CrossRef]
128. Martinez-Picado, J. Exceptional HIV Elite Controllers. Towards an HIV Cure. In Proceedings of the AIDS 2022 Congress, Montreal, Canada, 28 July 2022; Abstract Supplement Abstracts from AIDS 2022—the 24th International AIDS Conference, Virtual, 29 July–2 August 2022. [CrossRef]
129. Turk, G.; Seiger, K.; Lian, X.; Sun, W.; Parsons, E.M.; Gao, C.; Rassadkina, Y.; Polo, M.L.; Czernikier, A.; Ghiglione, Y.; et al. A Possible Sterilizing Cure of HIV-1 Infection Without Stem Cell Transplantation. *Ann. Intern. Med.* **2022**, *175*, 95–100. [CrossRef] [PubMed]
130. Campisi, J. Aging, Cellular Senescence, and Cancer. *Annu. Rev. Physiol.* **2013**, *75*, 685–705. [CrossRef] [PubMed]

131. Demaria, M.; Ohtani, N.; Youssef, S.A.; Rodier, F.; Toussaint, W.; Mitchell, J.R.; Laberge, R.-M.; Vijg, J.; Van Steeg, H.; Dollé, M.E.T.; et al. An Essential Role for Senescent Cells in Optimal Wound Healing through Secretion of PDGF-AA. *Dev. Cell* **2014**, *31*, 722–733. [CrossRef]
132. Ritschka, B.; Storer, M.; Mas, A.; Heinzmann, F.; Ortells, M.C.; Morton, J.P.; Sansom, O.J.; Zender, L.; Keyes, W.M. The Senescence-Associated Secretory Phenotype Induces Cellular Plasticity and Tissue Regeneration. *Genes Dev.* **2017**, *31*, 172–183. [CrossRef]
133. Paramos-de-Carvalho, D.; Martins, I.; Cristóvão, A.M.; Dias, A.F.; Neves-Silva, D.; Pereira, T.; Chapela, D.; Farinho, A.; Jacinto, A.; Saúde, L. Targeting Senescent Cells Improves Functional Recovery after Spinal Cord Injury. *Cell Rep.* **2021**, *36*, 109334. [CrossRef]
134. Calcinotto, A.; Kohli, J.; Zagato, E.; Pellegrini, L.; Demaria, M.; Alimonti, A. Cellular Senescence: Aging, Cancer, and Injury. *Physiol. Rev.* **2019**, *99*, 1047–1078. [CrossRef] [PubMed]
135. Gasser, S.; Orsulic, S.; Brown, E.J.; Raulet, D.H. The DNA Damage Pathway Regulates Innate Immune System Ligands of the NKG2D Receptor. *Nature* **2005**, *436*, 1186–1190. [CrossRef]
136. Krizhanovsky, V.; Yon, M.; Dickins, R.A.; Hearn, S.; Simon, J.; Miething, C.; Yee, H.; Zender, L.; Lowe, S.W. Senescence of Activated Stellate Cells Limits Liver Fibrosis. *Cell* **2008**, *134*, 657–667. [CrossRef]
137. Sagiv, A.; Burton, D.G.A.; Moshayev, Z.; Vadai, E.; Wensveen, F.; Ben-Dor, S.; Golani, O.; Polic, B.; Krizhanovsky, V. NKG2D Ligands Mediate Immunosurveillance of Senescent Cells. *Aging* **2016**, *8*, 328–344. [CrossRef] [PubMed]
138. Arora, S.; Thompson, P.J.; Wang, Y.; Bhattacharyya, A.; Apostolopoulou, H.; Hatano, R.; Naikawadi, R.P.; Shah, A.; Wolters, P.J.; Koliwad, S.; et al. Invariant Natural Killer T Cells Coordinate Removal of Senescent Cells. *Med* **2021**, *2*, 938–950. [CrossRef]
139. Hazeldine, J.; Hampson, P.; Lord, J.M. Reduced Release and Binding of Perforin at the Immunological Synapse Underlies the Age-Related Decline in Natural Killer Cell Cytotoxicity. *Aging Cell* **2012**, *11*, 751–759. [CrossRef] [PubMed]
140. Chelyapov, N.; Nguyen, T.T.; Gonzalez, R. Autologous NK Cells Propagated and Activated Ex Vivo Decrease Senescence Markers in Human PBMCs. *Biochem. Biophys. Rep.* **2022**, *32*, 101380. [CrossRef] [PubMed]
141. Xue, W.; Zender, L.; Miething, C.; Dickins, R.A.; Hernando, E.; Krizhanovsky, V.; Cordon-Cardo, C.; Lowe, S.W. Senescence and Tumour Clearance Is Triggered by P53 Restoration in Murine Liver Carcinomas. *Nature* **2007**, *445*, 656–660. [CrossRef]
142. Iannello, A.; Raulet, D.H. Immune Surveillance of Unhealthy Cells by Natural Killer Cells. *Cold Spring Harb. Symp. Quant. Biol.* **2013**, *78*, 249–257. [CrossRef]
143. Ruscetti, M.; Leibold, J.; Bott, M.J.; Fennell, M.; Kulick, A.; Salgado, N.R.; Chen, C.-C.; Ho, Y.-J.; Sanchez-Rivera, F.J.; Feucht, J.; et al. NK Cell-Mediated Cytotoxicity Contributes to Tumor Control by a Cytostatic Drug Combination. *Science* **2018**, *362*, 1416–1422. [CrossRef] [PubMed]
144. Sawasdee, N.; Wattanapanitch, M.; Thongsin, N.; Phanthaphol, N.; Chiawpanit, C.; Thuwajit, C.; Yenchitsomanus, P.-T.; Panya, A. Doxorubicin Sensitizes Breast Cancer Cells to Natural Killer Cells in Connection with Increased Fas Receptors. *Int. J. Mol. Med.* **2022**, *49*, 40. [CrossRef]
145. Pereira, B.I.; Devine, O.P.; Vukmanovic-Stejic, M.; Chambers, E.S.; Subramanian, P.; Patel, N.; Virasami, A.; Sebire, N.J.; Kinsler, V.; Valdovinos, A.; et al. Senescent Cells Evade Immune Clearance via HLA-E-Mediated NK and CD8+ T Cell Inhibition. *Nat. Commun.* **2019**, *10*, 2387. [CrossRef]
146. Yang, H.; Rei, M.; Brackenridge, S.; Brenna, E.; Sun, H.; Abdulhaqq, S.; Liu, M.K.P.; Ma, W.; Kurupati, P.; Xu, X.; et al. HLA-E-Restricted, Gag-Specific CD8+ T Cells Can Suppress HIV-1 Infection, Offering Vaccine Opportunities. *Sci. Immunol.* **2021**, *6*, eabg1703. [CrossRef] [PubMed]
147. Di Mitri, D.; Toso, A.; Chen, J.J.; Sarti, M.; Pinton, S.; Jost, T.R.; D'Antuono, R.; Montani, E.; Garcia-Escudero, R.; Guccini, I.; et al. Tumour-Infiltrating Gr-1+ Myeloid Cells Antagonize Senescence in Cancer. *Nature* **2014**, *515*, 134–137. [CrossRef]
148. Eggert, T.; Wolter, K.; Ji, J.; Ma, C.; Yevsa, T.; Klotz, S.; Medina-Echeverz, J.; Longerich, T.; Forgues, M.; Reisinger, F.; et al. Distinct Functions of Senescence-Associated Immune Responses in Liver Tumor Surveillance and Tumor Progression. *Cancer Cell* **2016**, *30*, 533–547. [CrossRef] [PubMed]
149. Zhu, Y.; Tchkonia, T.; Pirtskhalava, T.; Gower, A.C.; Ding, H.; Giorgadze, N.; Palmer, A.K.; Ikeno, Y.; Hubbard, G.B.; Lenburg, M.; et al. The Achilles' Heel of Senescent Cells: From Transcriptome to Senolytic Drugs. *Aging Cell* **2015**, *14*, 644–658. [CrossRef]
150. Schafer, M.J.; White, T.A.; Iijima, K.; Haak, A.J.; Ligresti, G.; Atkinson, E.J.; Oberg, A.L.; Birch, J.; Salmonowicz, H.; Zhu, Y.; et al. Cellular Senescence Mediates Fibrotic Pulmonary Disease. *Nat. Commun.* **2017**, *8*, 14532. [CrossRef]
151. Li, C.; Shen, Y.; Huang, L.; Liu, C.; Wang, J. Senolytic Therapy Ameliorates Renal Fibrosis Postacute Kidney Injury by Alleviating Renal Senescence. *FASEB J.* **2021**, *35*, e21229. [CrossRef]
152. Aguayo-Mazzucato, C.; Andle, J.; Lee, T.B.; Midha, A.; Talemal, L.; Chipashvili, V.; Hollister-Lock, J.; van Deursen, J.; Weir, G.; Bonner-Weir, S. Acceleration of β Cell Aging Determines Diabetes and Senolysis Improves Disease Outcomes. *Cell Metab.* **2019**, *30*, 129–142.e4. [CrossRef]
153. Roos, C.M.; Zhang, B.; Palmer, A.K.; Ogrodnik, M.B.; Pirtskhalava, T.; Thalji, N.M.; Hagler, M.; Jurk, D.; Smith, L.A.; Casaclang-Verzosa, G.; et al. Chronic Senolytic Treatment Alleviates Established Vasomotor Dysfunction in Aged or Atherosclerotic Mice. *Aging Cell* **2016**, *15*, 973–977. [CrossRef] [PubMed]
154. Saccon, T.D.; Nagpal, R.; Yadav, H.; Cavalcante, M.B.; Nunes, A.D.d.C.; Schneider, A.; Gesing, A.; Hughes, B.; Yousefzadeh, M.; Tchkonia, T.; et al. Senolytic Combination of Dasatinib and Quercetin Alleviates Intestinal Senescence and Inflammation and Modulates the Gut Microbiome in Aged Mice. *J. Gerontol. A Biol. Sci. Med. Sci.* **2021**, *76*, 1895–1905. [CrossRef]

155. Hernández-Silva, D.; Cantón-Sandoval, J.; Martínez-Navarro, F.J.; Pérez-Sánchez, H.; de Oliveira, S.; Mulero, V.; Alcaraz-Pérez, F.; Cayuela, M.L. Senescence-Independent Anti-Inflammatory Activity of the Senolytic Drugs Dasatinib, Navitoclax, and Venetoclax in Zebrafish Models of Chronic Inflammation. *Int. J. Mol. Sci.* **2022**, *23*, 10468. [CrossRef] [PubMed]
156. Justice, J.N.; Nambiar, A.M.; Tchkonia, T.; LeBrasseur, N.K.; Pascual, R.; Hashmi, S.K.; Prata, L.; Masternak, M.M.; Kritchevsky, S.B.; Musi, N.; et al. Senolytics in Idiopathic Pulmonary Fibrosis: Results from a First-in-Human, Open-Label, Pilot Study. *EBioMedicine* **2019**, *40*, 554–563. [CrossRef] [PubMed]
157. Hickson, L.J.; Langhi Prata, L.G.P.; Bobart, S.A.; Evans, T.K.; Giorgadze, N.; Hashmi, S.K.; Herrmann, S.M.; Jensen, M.D.; Jia, Q.; Jordan, K.L.; et al. Senolytics Decrease Senescent Cells in Humans: Preliminary Report from a Clinical Trial of Dasatinib plus Quercetin in Individuals with Diabetic Kidney Disease. *EBioMedicine* **2019**, *47*, 446–456. [CrossRef] [PubMed]
158. Chaib, S.; Tchkonia, T.; Kirkland, J.L. Cellular Senescence and Senolytics: The Path to the Clinic. *Nat. Med.* **2022**, *28*, 1556–1568. [CrossRef]
159. Climent, N.; Casanova, V.; Maleno, M.J.; Rodríguez-Agustín, A.; Sánche-Palomino, S.; González, T.; Hurtado, C.; Martínez, E.; Mallolas, J.; Ambrosioni, J.; et al. HIV-Induced Cellular Senescence in PLWH Is Decreased by D+Q Senolytic Drugs. [CROI Abstract 223]. In Proceedings of the CROI 2022 Conference on Retroviruses and Opportunistic Infections, Virtual, 12–16 February 2022; p. 222.
160. Baar, M.P.; Brandt, R.M.C.; Putavet, D.A.; Klein, J.D.D.; Derks, K.W.J.; Bourgeois, B.R.M.; Stryeck, S.; Rijksen, Y.; van Willigenburg, H.; Feijtel, D.A.; et al. Targeted Apoptosis of Senescent Cells Restores Tissue Homeostasis in Response to Chemotoxicity and Aging. *Cell* **2017**, *169*, 132–147.e16. [CrossRef]
161. Kovacovicova, K.; Skolnaja, M.; Heinmaa, M.; Mistrik, M.; Pata, P.; Pata, I.; Bartek, J.; Vinciguerra, M. Senolytic Cocktail Dasatinib+Quercetin (D+Q) Does Not Enhance the Efficacy of Senescence-Inducing Chemotherapy in Liver Cancer. *Front. Oncol.* **2018**, *8*, 459. [CrossRef] [PubMed]

pharmaceutics

Review

Drugging Hijacked Kinase Pathways in Pediatric Oncology: Opportunities and Current Scenario

Marina Ferreira Candido [1], Mariana Medeiros [2], Luciana Chain Veronez [3], David Bastos [4], Karla Laissa Oliveira [4], Julia Alejandra Pezuk [5], Elvis Terci Valera [3] and María Sol Brassesco [5,*]

1 Department of Cell Biology, Ribeirão Preto Medical School, University of São Paulo, Ribeirão Preto 14049-900, SP, Brazil
2 Regional Blood Center, University of São Paulo, Ribeirão Preto 14049-900, SP, Brazil
3 Department of Pediatrics, Ribeirão Preto Medical School, University of São Paulo, Ribeirão Preto 14049-900, SP, Brazil
4 Department of Biology, Faculty of Philosophy, Sciences and Letters at Ribeirão Preto, University of São Paulo, Ribeirão Preto 14040-901, SP, Brazil
5 Departament of Biotechnology and Innovation, Anhanguera University of São Paulo, UNIAN/SP, São Paulo 04119-001, SP, Brazil
* Correspondence: solbrassesco@usp.br; Tel.: +55-16-3315-9144; Fax: +55-16-3315-4886

Abstract: Childhood cancer is considered rare, corresponding to ~3% of all malignant neoplasms in the human population. The World Health Organization (WHO) reports a universal occurrence of more than 15 cases per 100,000 inhabitants around the globe, and despite improvements in diagnosis, treatment and supportive care, one child dies of cancer every 3 min. Consequently, more efficient, selective and affordable therapeutics are still needed in order to improve outcomes and avoid long-term sequelae. Alterations in kinases' functionality is a trademark of cancer and the concept of exploiting them as drug targets has burgeoned in academia and in the pharmaceutical industry of the 21st century. Consequently, an increasing plethora of inhibitors has emerged. In the present study, the expression patterns of a selected group of kinases (including tyrosine receptors, members of the PI3K/AKT/mTOR and MAPK pathways, coordinators of cell cycle progression, and chromosome segregation) and their correlation with clinical outcomes in pediatric solid tumors were accessed through the R2: Genomics Analysis and Visualization Platform and by a thorough search of published literature. To further illustrate the importance of kinase dysregulation in the pathophysiology of pediatric cancer, we analyzed the vulnerability of different cancer cell lines against their inhibition through the Cancer Dependency Map portal, and performed a search for kinase-targeted compounds with approval and clinical applicability through the CanSAR knowledgebase. Finally, we provide a detailed literature review of a considerable set of small molecules that mitigate kinase activity under experimental testing and clinical trials for the treatment of pediatric tumors, while discuss critical challenges that must be overcome before translation into clinical options, including the absence of compounds designed specifically for childhood tumors which often show differential mutational burdens, intrinsic and acquired resistance, lack of selectivity and adverse effects on a growing organism.

Keywords: childhood cancer; kinases; chemical inhibitors; clinical trials

Citation: Candido, M.F.; Medeiros, M.; Veronez, L.C.; Bastos, D.; Oliveira, K.L.; Pezuk, J.A.; Valera, E.T.; Brassesco, M.S. Drugging Hijacked Kinase Pathways in Pediatric Oncology: Opportunities and Current Scenario. *Pharmaceutics* **2023**, *15*, 664. https://doi.org/10.3390/pharmaceutics15020664

Academic Editor: Francesca Musumeci

Received: 15 December 2022
Revised: 9 February 2023
Accepted: 10 February 2023
Published: 16 February 2023

1. Pediatric Cancer

The past two decades have witnessed tremendous advances in our understanding of cancer pathogenesis, with most neoplasms resulting from the accumulation of gains in function in proto-oncogenes and losses of tumor suppressors.

In the pediatric setting (between 0 and 19 years of age), cancer is defined as a group of several diseases that have in common the uncontrolled proliferation of abnormal cells that can occur in any region of the body. However, unlike adult tumors that are classified according to the primary site, the International Classification of Childhood Cancer (CICI) categorizes pediatric tumors into 12 main groups based on histological findings [1].

Of these, leukemias are the most frequent, representing 35% of all tumors that affect children and adolescents [2]. The second most common group is represented by lymphomas (20%), followed in descending order by tumors of the central (15%) and sympathetic (7%) nervous system, soft tissue sarcomas (6%), bone tumors (5%), kidney tumors (3%), germ cell tumors (3%), retinoblastoma (2%), carcinomas (2%), liver tumors (1%) and other rare pediatric neoplasms.

Several specific characteristics converge on the premise that childhood and adult cancer should be studied separately. First of all, most pediatric tumors have histological findings that resemble fetal tissues at different stages of development, being considered embryonic and carrying different levels of cell differentiation. Furthermore, the spectrum of tumors in the pediatric age group differs from that in adult patients. Medulloblastoma (MB), neuroblastoma (NB), rhabdomyosarcoma (RMS), Ewing's sarcoma (EWS), osteosarcoma (OS), retinoblastoma (RB) and Wilms' tumor (WT), which are the most frequent pediatric solid tumors, are rarely found in adulthood [3]

Moreover, latency periods are shorter in pediatric cancer, with several histologies presenting even shortly after birth. Such rapid proliferation in embryonic tissues relies mainly on genomic errors with lesser contribution of environmental factors [4]. Similarly, unlike what happens in adults, pediatric tumors are usually more aggressive with nonspecific signs and symptoms, confusing them with common childhood illnesses and making early diagnosis difficult.

Additionally, tumors in this age group show clear differences in their presentation, clinical course and response to treatment when compared to adult counterparts. Tumors of the EWS family, for example, in adults, in addition to presenting a more differentiated histology (PNET—primitive neuroectodermal tumor), manifest preferentially with greater volume at diagnosis, affecting soft tissues and with distant metastases, resulting in poorer survival. In children, EWS preferentially affects the bones of the extremities, have a smaller volume, and respond better to chemotherapy. Still, in this population, the chance of survival after 5 years (without metastasis at diagnosis) is 75%, significantly higher than that calculated for adults (50%) [5]. Similar observations have been reported for RMS. In a cohort of 1071 adults and 1529 children, for example, the survival rate in the first group was considerably lower (27% versus 61%), with tumors occurring in unfavorable locations and with rare histologies [6].

Another peculiarity of childhood cancer lies in the fact that specific histological subtypes and clinical behavior are also age-dependent, suggesting differential pathogenic mechanisms and underlying molecular alterations for tumor initiation and progression. The general incidence of acute lymphoblastic leukemia (ALL), for example, is highest in the 1–4-year-old age group, while the highest frequency of lymphomas occurs among adolescents (between 15 and 19 years old). Embryonic tumors (NB, WT, RB, etc.), on the other hand, share a descending incidence, which is highest early in life and almost dissipates after 5 years of age, while the incidence of bone sarcomas reaches a sharp peak at the time of the pubertal growth spurt [7].

Finally, with the methodologic refinements in the identification of genomic alterations, it has become increasingly evident that the spectra of mutations and the subsequent dysregulation of signaling pathways in pediatric neoplasms differ from those that occur predominantly in adult cancer. In fact, it has been stipulated that most pediatric tumors carry between 5 and 10 mutations; however, the average number of mutations in adult tumors varies between 33 and 66 (i.e., colon, breast or pancreas carcinomas) and increases up to 200 in tumors caused by mutagens (such as melanoma/ultraviolet radiation (UV) and lung cancer/smoking) and up to 1000 in tumors with defects in mismatch repair genes, as is the case of nonpolyposis colorectal cancer, among others [8]. In this regard, a recent integrative study based on whole-genome sequencing data from 24 tumor types (914 patients) showed that even though mutation frequencies (SNV and indels) vary between pediatric tumor types (from 0.02 to 0.49 per Mb), these were 14 times lower than in adult cancers [9]. Of note, a high prevalence of mutations affecting genes related to cancer predisposition syndrome are seen in children affected by different tumor types [10]. In addition, pediatric cancers are usually enriched by gene fusions, driving tumorigenesis and showing impact on both diagnostic and targeted-treatments [11].

In fact, this work identified 52 genes significantly mutated for childhood and juvenile tumors and 102 genes for adult tumors, 25 of which were shared by both groups. *TP53* was the most commonly mutated gene (4% of childhood tumors), followed by *KRAS*, *ATRX*, *NF1* and *RB1* (1–2% of tumors). In adult tumors, *TP53* was also the gene most affected by mutations, albeit tenfold more frequently. Furthermore, the burden of mutations increased with patient age except for tumors characterized by *kataegis* or *chromothripsis* events [9].

Moreover, the mutational identity may also vary. Glioblastoma (GBM) (grade IV astrocytoma), for example, is characterized by mutations in *PTEN* and epidermal growth factor receptor (*EGFR*) amplification in adults [12]; the pediatric counterpart more frequently presents mutations in the N-terminal tail of the histone variant 3.3 [13], in platelet-derived growth factor (*PDGF*) and its receptor (*PDGFR*) [14]. As previously described, gene fusions that are rare in adult tumors appear recurrently in pediatric tumors such as *BRAF/KIAA1549* in pilocytic astrocytoma (PA) [15], *C11orf95/RELA* [16] in supratentorial ependymoma (EPN), PAX3/FOXO1 and PAX7/FOXO1 in alveolar RMS [17], and variants involving the *EWS* gene (*EWS/FLI1*, *EWS/ERG* and others less frequent) in EWS [18]; the majority of these fusions involve transcription factors associated with the development/differentiation of the affected tissue.

More recently, with the establishment of cooperative study groups and progress in imaging associated with more accurate anatomopathological and molecular diagnosis, the mortality of children affected by cancer, especially those with leukemia and some types of solid tumors, has shown an important decline. Five-year overall survival rates increased from 56% in the 1970s to 77% in the following vicennial [19].

In accordance, today ALL represents the paradigm of curable cancer in children, with current overall survival rates exceeding 85% in most modern treatment protocols [20]. However, for some aggressive leukemia subtypes and certain solid tumor histologies, the persisting advances in biological characterization combined with new technologies in radiotherapy (RT), chemotherapy (CT) and supportive/rehabilitation care have resulted in marginal survival advantages, and cure rates have stagnated at around 70% [21]. Yet, in underdeveloped areas such as Eastern Europe, Africa and South America, these reductions in mortality have been less expressive, and a considerable portion of children with cancer fail to respond to traditional chemotherapy.

In this scenario, alternative rationally targeted pharmacological options are still needed to overcome clinical resistance, tumor progression, and prevent the adverse side effects of standard therapy.

2. Kinases as Cancer Drivers

The human genome encodes more than 500 protein kinases, enzymes responsible for turning protein functions "on" through the transference of γ-phosphate groups from ATP to one of their three amino acids with free hydroxyl groups: serine, threonine or tyrosine. Human protein kinases have been divided into nine classes which are further subdivided into families, and often subfamilies whose actions can alter up to 30% of all cell proteins [22]. The molecular shifts exerted by them through phosphorylation not only can affect the function of a given protein, but it can also stabilize it, localize it in a particular cellular compartment and modulate its association with other proteins [23].

Protein kinases may be triggered or deactivated in many ways, including cis- or autophosphorylation, binding with substrates or activator/inhibitor proteins. Once activated, they act as crucial regulators of many features of cell behavior and specialized functions by coupling reception of extracellular signals, intracellular signaling transduction and cellular responses [24].

Playing fundamental roles in cell division, survival and migration, their dysregulation is commonly associated with human malignancies and contributes to tumor initiation and all stages of cancer progression. In fact, innumerous mutations, translocations, and amplifications that result in constitutively overexpressed or active kinases have been demonstrated in many human cancers [23]. Mutations within the catalytic domain serve to stabilize the kinase in an active conformation and to destabilize cis-inhibitory interactions. Other domains can also be affected and elicit constitutive activity and hyperactive pathways as well. DNA translocations, on the other hand, predominantly create in frame gene fusions leading to chimeric proteins with novel/increased activity, leading to continued cancer cell growth and survival (Figure 1) [25,26]. Moreover, the dysregulation of different kinases has been repeatedly associated with tumor prognosis and categorized as a determinant of patient survival [27,28].

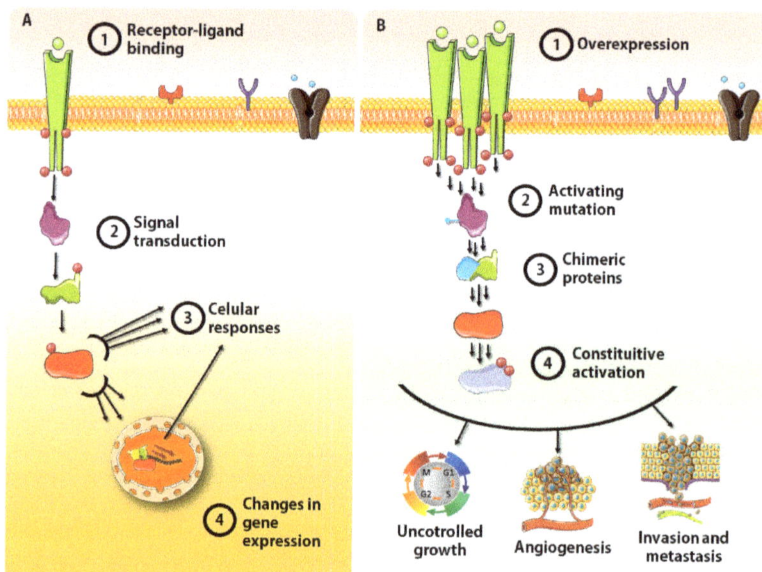

Figure 1. (**A**) Protein kinases may be triggered or deactivated in many ways, acting as key regulators of many features of cell behavior and specialized functions by coupling (1) reception of extracellular signals and (2) intracellular signaling transduction, leading to (3) direct cellular responses or (4) changes in gene expression. (**B**) In cancer, increased kinase activity may result from gene amplification (1,4), mutations that stabilize the kinase in an active conformation and destabilize cis-inhibitory interactions (2) and translocations that encode chimeric proteins with novel/increased activity (3). This figure was created using Servier Medical Art templates, which are licensed under a Creative Commons Attribution 3.0 Unported License; https://smart.servier.com (accessed on 14 December 2022).

Hence, the so-called "Kinome" (the complete set of protein kinases encoded in the human genome—about 2% of all genes) has become an attractive target for the treatment of a variety of tumors. Even so, despite the great diversity, over the years, it has become more apparent that only certain kinases are among the most frequently occurring drivers of human cancer, including tyrosine receptors (RTK) (i.e., FGFR, EGFR, VEGFR, RET, MET, ALK), members of the PI3K/AKT/mTOR and Mitogen-Activated Protein Kinase (MAPK) pathways, along with central coordinators of cell cycle progression (i.e., cyclin-dependent kinases) and chromosome segregation such as polo-like and aurora kinases [25].

In this way, the present study aimed to present evidence of the involvement of these kinases' dysregulation in the pathophysiology of pediatric tumors, their correlation with clinical outcomes and prospects of their inhibition through in silico analysis, along with an up-to-date revision of compound development and testing.

3. Protein Kinases in Pediatric Oncology and Their Association with Tumor Prognosis

As stated above, substrate reversible phosphorylation by protein kinases is nature's main molecular system for organizing cellular signal transduction and regulating cell metabolism, growth and differentiation. The phosphorylation state of a protein determines not only its function, subcellular distribution and stability, but also its interaction with other proteins or cellular components. Intrinsically, signaling pathways are remarkably complex and as our knowledge increases, it has become progressively evident that such molecular networks are not linear but contain modules of multi-protein complexes, many feedbacks, feedforwards and competing protein mechanisms that not only assemble at various intracellular compartments to process, integrate and transmit information that will ultimately specify a particular biological response, but also crosstalk with many other signaling pathways [29]. Thus, even though the kinases that were selected for this review will be treated separately, many, if not all, are directly or indirectly interconnected (Figure 2).

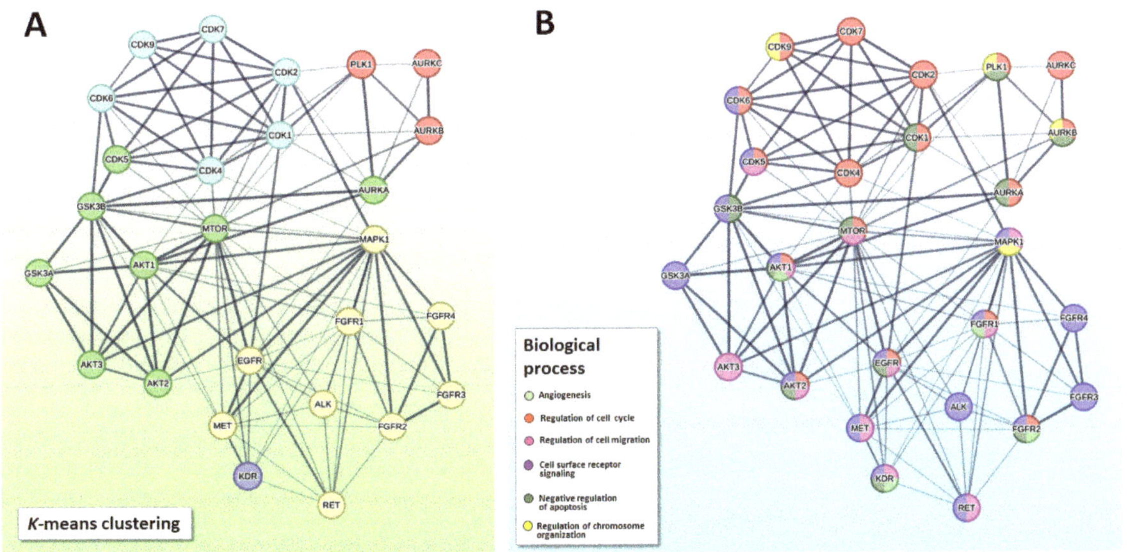

Figure 2. (**A**) Protein–protein interactions accessed through the software STRING v11.5 (available at https://string-db.org/ (accessed on 2 November 2022)). The parameters evaluated were text mining, experiments and databases. Network edges denote confidence and the minimum required interaction score was 0.700, considered high. (**A**) K-means clustering; (**B**) enrichment analysis for biological processes.

In this section, the different roles of protein kinases in oncogenic transformation and tumor prognosis in the pediatric setting were assessed by two different approaches: by a thorough search of published literature, and by a systematic search in publicly available data retrieved from expression arrays accessed through the R2: Genomics Analysis and Visualization Platform (http://r2.amc.nl (accessed on 15 October 2022)). For this, datasets were included if they met the following criteria: inclusion of pediatric samples (exclusively or in which adult variants could be omitted), normal counterparts and having information about clinical features of prognosis (Supplementary Figure S1). Datasets and probes are detailed in Supplementary Tables S1 and S2.

3.1. Published Evidence of Kinase Dysregulation in Pediatric Oncology

3.1.1. Receptor Tyrosine Kinases (RTK)

Humans express 58 receptor tyrosine kinases (RTK) which function as entry points for many extracellular signals and the recruitment of the intracellular signaling networks that orchestrate a particular response [30].

These cell surface receptors possess multi-domain identical architectures that are made up of an extracellular ligand-binding domain (which differs between subfamilies), a single transmembrane helix and an intracellular region that contains a juxtamembrane regulatory region (composed of 40–80 amino acids), a tyrosine kinase domain (TKD) and a carboxyl (C-) terminal tail [31].

Generally, RTK activation occurs upon binding of ligands (i.e., growth factors or cytokines) to their extracellular domains. This interaction results in RTK non-covalent dimerization/oligomerization, which juxtaposes the cytoplasmic TKDs and facilitates autophosphorylation in trans of tyrosine residues in the juxtamembrane regulatory region, inducing conformational changes that serve to stabilize the active state of the kinase. Then, a second phase of tyrosine autophosphorylation occurs on phosphotyrosines that recruit downstream signaling proteins that typically contain Src homology 2 (SH2) or phosphotyrosine-binding (PTB) domains. The recruitment of these adapter molecules then initiates a cascade of RTK-specific pathways that determine cell fate [22,31,32].

About 20 different RTKs classes or subfamilies have been described [33]. Their activity is tightly regulated in normal cells; however, constitutive kinase activity acquired through mutation, overexpression and/or autocrine/paracrine stimulation has been strongly associated with pathological disorders, neoplastic transformation and metastasis [34]. Dysregulation of the epidermal growth factor receptor (EGFR/ErbB), the receptor for insulin (IR), the platelet-derived growth factor receptor (PDGFR), the fibroblast growth factor receptor (FGFR), the vascular endothelial growth factor receptor (VEGFR) and the hepatocyte growth factor receptor (HGFR/MET), for example, results in uncontrolled activation of multiple downflow signal transduction pathways and provides a strong drive toward malignancy [35]. Some of these oncogenes are paradigms of certain tumor types, as is the case of the amplification of EGFR/ErbB in breast cancer or MET overexpression in non-small cell lung cancer (NSCLC) [36,37]. Nevertheless, information about RTKs' involvement in the pathophysiology of childhood cancer is less discernible, as illustrated below.

FGFR. The signaling cascades of fibroblast growth factor receptors (FGFR1, FGFR2, FGFR3 and FGFR4) play pivotal roles in the regulation of development and tissue repair and regeneration. These receptors are highly conserved and widely distributed and their dysregulation promotes tumor growth, survival and development of drug resistance, as well as the development of angiogenesis and immune evasion [38]. A recent pan-cancer next-generation sequencing profiling demonstrated that ~7% of cancers harbor gain-of-function FGFR aberrations, with varying frequencies between family members (FGFR1 > FGFR3 > FGFR2 > FGFR4) [39]. Gene amplifications or activating mutations have been observed in multiple cancer types, although they are most commonly detected in breast, lung, liver, stomach, uterus and bladder cancer [40]. In fact, most of the aberrations detected in the survey performed by Helsten et al. (2016) involved adult carcinomas, while the percentage of cases positive for *FGFR* aberrations in childhood tumors (NB and OS) was around 3% [39]. Additionally, *FGFR1* fusions (i.e., FGFR1-TACC1), TKD duplications and hotspot mutations (i.e., N5465K and K656E) are frequently observed in certain types of pediatric brain cancer, particularly dysembryoplastic neuroepithelial tumors (DNET) and PA [41]. Moreover, germline mutations in *FGFR1*, either complete or in mosaicism, may predispose low-grade central nervous system (CNS) tumors in children and adolescents [42,43].

Other research groups have also found correlations between altered expression of FGFRs and poor prognosis. *FGFR1* amplification, for example, was correlated with worse prognosis and poor response to chemotherapy in a large cohort of patients with OS [44]. Moreover, FGFR1 has been correlated with tumor development and lung metastasis in xenographic models, whereas its activation improved survival and radiation resistance, a phenotype which was reversed when FGFR1 was inhibited [45]. Similar to OS, *FGFR1* copy number gains are frequent in EWS, where patients with activating *FGFR1* mutations present higher incidence of metastatic disease [46]. In vitro, *FGFR1* suppression through interference RNA (RNAi) significantly reduced cell proliferation. Decreased xenograft tumor growth and 18F-fluorodeoxyglucose activity were also observed [47].

FGFR1 amplification and gene fusions have also been described in RMS with an active role cancer cell proliferation [48,49]. However, opposite expression patterns have been reported [50,51]. Nevertheless, it has already been shown that embryonal (ERMS) histologies present higher *FGFR1* expression levels compared to the alveolar forms (ARMS) [51].

Considering CNS tumors, upregulation of *FGFR1* has been associated with worse prognosis and shorter overall and recurrence-free survival in EPN and NB [52–55]. The analysis of FGFR1 protein abundance in human MB tissues, through an anonymized, validated MB, and cerebellum tissue microarray (TMA) and immunohistochemistry (IHC) found high levels of FGFR1 expression in 18% of the tumor tissues. In the case of gliomas, many mutations involving the *FGFR1* gene were described, one of which was associated with radioresistance [56–60]. Otherwise, in WT and MB, there is no published evidence.

Regarding FGFR2, information in the literature about its relevance in pediatric tumors is scarce. Besides correlations with higher tumor grade, radioresistance and poorer survival in gliomas [61–63], its phosphorylation (indicative of activation) was seen increased in NB samples (compared to normal tissues) and correlated with cisplatin resistance [62]. Alternatively, low expression of this kinase was observed MB [63] and in RMS when compared to normal myeloblasts [64]. Downregulation or undetectable expression of FGFR3 was also reported in RMS with no evidence of correlation with the clinical outcome [64]. However, a more recent study described a small population of FGFR3-positive cells as strongly tumorigenic with a stem cell-like phenotype [65].

FGFR3 is also downregulated in WT [66]; however, in pediatric CNS tumors, opposite *FGFR3* expression profiles are observed. In NB, high expression levels are associated with worse overall survival and event-free survival (EFS) [67]. In glioma, its upregulation was associated with increased patient age [52], a feature that denotes a more invasive phenotype in adult counterparts [59]. Moreover, *FGFR3* amplification [68] and fusions (*FGFR3-TACC3*) seem to play a role in tumor metabolism and tumor growth promotion in low-grade gliomas (LGG) [69–71]. Such correlation with poor prognosis was also observed in EPN in which *FGFR3* was associated with shorter overall survival and shorter time to tumor recurrence [52].

Moderate-to-high expression of *FGFR3* mRNA was also observed in 80% of samples from EWS family tumors [71]. Mutations in this gene were also reported in circulating tumor cell samples [72]. However, neither of these studies presented information about correlations with clinical outcomes. Of note, when FGFR3 is downregulated in OS cell lines (by long noncoding, microRNA or iRNA), there is a reduction in tumor growth and angiogenesis, reinforcing its relevance to this disease [73–76].

Regarding *FGFR4*, little information about its prognostic value has been published in the pediatric setting, with a few reports on glioma, MB, RMS and NB. The prognostic value of this kinase in the first group was initially evaluated by in 2019 by Jimenez-Pascual and Siebzehnrubl, who did not find any correlations between FGFR4 expression and clinical outcomes [59]. Nevertheless, a recent evaluation of transcriptomic glioma datasets from The Cancer Genome Atlas (TCGA) revealed a direct association of high FGFR4 expression and dismal prognosis, progressively upregulated in recurrent tumors. In addition, the contribution of FGFR4 to the malignant phenotype of a highly aggressive GBM subgroup was further validated by increased viability, adhesion, migration and clonogenicity in vitro, along with abolished xenograft formation in mice and reduced invasiveness in zebrafish xenotransplantation models [76].

In MB, high *FGFR4* expression levels were observed in HD-MBO3 cells and in a small cohort (n = 12) of primary MB tissues, even though there was no validation upon TMA [63]. Of note, a pilot study based on an independent blinded set of 112 samples showed that protein levels of FGFR4 in urine, together with cadherin-1 (CADH1) and fibrinogen beta chain (FIBB), could be used to discriminate MB patients from healthy control patients with acceptable accuracy. Moreover, the authors reported a positive correlation of urine FGFR4 detection with the age of affected patients [77].

Additionally, FGFR4 overexpression in RMS contributes to the failure of cells to complete normal skeletal muscle development, leading to constitutive signaling and unregulated growth in correlation with poor differentiation [78–81]. *FGFR4* mutations in childhood RMS (7–8% of tumors) are more frequently observed within the kinase domain. From those, N535K and V550E increase autophosphorylation of the receptor and promote proliferation and metastatic potential when expressed in vitro [79]. High expression levels of FGFR4 have also been associated with advanced stage and poor survival in RMS [82,83]. Furthermore, FGFR4 has been reported as a downstream target of PAX3 and PAX3–FOXO1 [81] and thus is commonly altered in fusion-positive RMS [83] and a key contributor to RMS invasion and metastasis [84].

Last but not least, a germline polymorphism in the *FGFR4* gene (rs351855) which results in the expression of an arginine at codon 388 (Arg388), rather than the more common glycine (Gly388), is frequently associated with decreased survival rates, treatment resistance and more aggressive disease in a variety of malignancies, and is associated with an increased prevalence of NB in children [85], and this association may be linked to differences in FGFR4 degradation rates [86]. It was also observed that cases with the *FGFR4* AA genotype were 2.5 times more likely to have tumors with *MYCN* amplification compared with those with AG and GG genotypes, although such association was not statistically significant [85].

EGFR. The epidermal growth factor receptor (EGFR) (also recognized as HER-1 or ERBB-1) is a transmembrane glycoprotein of the ERBB receptor tyrosine kinase superfamily. Overexpression and/or enhanced activity of EGFR activate the downstream pro-oncogenic signaling, including the RAS-RAF-MEK-ERK and AKT-PI3K-mTOR pathways. These consequently activate several biologic expressions that proceed human cancer progression [87].

Overexpression of EGFR has been reproducibly detected in a large number of tumor samples and found to act as a strong prognostic indicator in head and neck, ovarian, cervical, bladder and esophageal cancers, correlating to poorer survival rates [88,89]. In the pediatric setting, however, there are few reports about its prognostic relevance. In gliomas, for example, fewer molecular alterations in the EGFR gene (mutations and amplifications) are observed in children when compared to adult counterparts [90–95]. *EGFR* gene amplification/overexpression is a genetic hallmark in adult GBM (observed ~40% of tumors) [95], whereas this feature is only observed in 25% of pediatric cases; nevertheless, it is associated in a similar manner with higher proliferation and increased tumor grade [96–98].

Likewise, high-level amplification and EGFR overexpression correlate with shorter event-free survival and relapse in high-grade EPN, being considered an independent prognostic marker for intracranial forms [99–101]. In MB patients, high expression of HER-2, another member of the EGFR gene family, was also associated with limited survival and metastasis [102,103]. Moreover, this RTK often co-expresses with HER-4 (more than 50% of samples), suggesting that HER-2/HER-4 heterodimerization may be of particular biological significance in this disease [101].

HER-2 expression was also reported to be associated with the aggressive behavior of NB and to significantly reduce survival [103]. However, a later study demonstrated that EGFR and HER-2 positivity are more frequently found in favorable histological risk groups, including younger age ($\leq$18 months), localized disease, and favorable histological group [104]. Similar results were obtained by Izycka-Swieszewska et al. (2010), where

HER-2-negative cases were more often found in the metastatic tumor group, associated with increased mitotic index and higher KI67 expression. *MYCN* non-amplified tumors were more often HER-2-positive than amplified tumors [105]. In contrast, higher expression levels of HER-4 are more often found in patients with metastatic disease [104].

For other pediatric tumors, the biological relevance of EGFR family members remains to be clarified. In EWS, for example, while HER-2 is not considered an important prognostic factor, an association of HER-4 and metastasis was found [106]. In OS, *EGFR* expression is common [107], but correlations between EGFR or HER-2 expression and clinical prognosis have been controversial, as no treatment improvements are achieved when EGFR is inhibited in pre-clinical and clinical trials [106].

In RMS, RB and WT, no strong associations of EGFR expression with clinical data have been found [108–111], although the ERBB family seems to be important for the malignant phenotype of RMS: ERBB1 sustains cell proliferation and growth, ERBB2 regulates myoblast cell transformation and survival and ERBB3 induces myogenic differentiation. Additionally, activation of ERBB2 coupled with inactivation of p53 induces RMS in animal models [106].

VEGFR (KDR—kinase insert domain receptor). The vascular endothelial growth factor receptor (VEGFR) family consists of three members: VEGFR1, VEGFR2 and VEGFR3 [112]. These receptors are established players in the formation of new blood vessels and the maintenance and remodeling of existing ones, during development and in adult tissues [113]. As such, in neoplastic growth, VEGFRs play an essential role in tumor neovascularization, providing oxygen and nutrition, and they facilitate tumor cells to metastasize and spread to distant organs [114]. Regarding pediatric cancer, VEGFRs have already been quantitatively evaluated in various types of refractory brain tumors [115]. Both VEGFR1 and VEGFR2 were detected in anaplastic astrocytoma tumor cells, MB and EPN samples [116–120]. Moreover, these receptors are frequently mutated and highly expressed in gliomas, NB and OS; in all cases, there is a negative correlation with unfavorable prognosis, advanced tumor stage, metastasis and shorter overall survival [121–127].

RET. Under normal conditions, the RET ("rearranged during transfection") TKR pathway is activated by glial cell line-derived neurotrophic factor (GDNF) ligands that bind to coreceptors from the GDNF family receptor alphas (GFRαs), playing a major role during sympathetic and enteric nervous system development, where it signals toward proliferation, migration and differentiation. Apart from amplification, the constitutive activation of RET is caused by point mutations and gene rearrangements that drive malignancy in multiple tissues (i.e., papillary and medullary thyroid carcinomas and non-small cell lung carcinomas) [128–131].

RET rearrangements are also found in a high proportion of childhood papillary thyroid cancers [132–134]. Recently, it was observed that pediatric tumors (soft tissue sarcomas or medullary thyroid cancer) harboring either an RET-fused or RET-mutated pathogenic somatic alteration show clinical response to the RET inhibitor Selpercatinib [134].

RET mutations leading to dysfunctional ligand binding have also been described as the second most significant cancer-predisposing gene in the germline of patients with OS [135,136]. Moreover, RET is activated and can promote motility and colony formation in metastatic OS cells, contributing to the higher resistance of this tumor type to different chemotherapeutic agents [137–140]. Furthermore, NB cells and tumor samples demonstrated high RET expression levels [140], and its activation induces invasive spread NB in animal models [141].

c-MET. The mesenchymal–epithelial transition factor (c-MET), which is also known as hepatocyte growth factor receptor (HGFR), is an essential molecule for the survival and function of normal cells that promotes tissue remodeling and organ homeostasis [142]. MET's gain of function either via overexpression, amplification, aberrant splicing or mutations is associated with the constant activation of downstream classic signaling pathways that sustain rapid proliferation, promote cell migration, angiogenesis and survival of cancer cells [143]. Moreover, recent evidence indicates that MET signaling participates in the acquirement of mesenchymal phenotype, tumor plasticity and adaptive responses to metabolic stress, contributing to the recurrence and metastatic dissemination of cancer cells [144,145].

The c-MET gene was first identified in the human OS cell line (HOS) that had been treated with N-methyl-N′-nitro-N-nitrosoguanidine (MNNG) as a gene able to transform normal fibroblasts [146]. Since then, its involvement in cancer establishment and progression has been repeatedly described in a variety of common and high-risk pediatric solid tumors, including not only sarcomas, but also gliomas, MB, NB, WT and hepatoblastomas, among others [147]. Of note, infantile hemispheric gliomas were recently recognized to be driven by different RTKs, including somatic fusions and alterations involving ALK, ROS1, NTRK and c-MET [148]. Especially in anaplastic, diffuse and PA, c-MET levels often correlate with tumor grade [149].

Cytoplasmic c-MET immunoreactivity is also associated with poor clinical outcome, and tissues with overexpression often exhibit higher vascular proliferation and proliferative index [150–152]. Similar phenotypes have been observed in NB, where overexpression of this receptor promotes invasion and is associated with advanced metastatic stage [153,154].

Likewise, high levels of MET protein are associated with increased proliferative activity invasion and metastasis in WT [155,156], and represent a risk factor for invasion in RB [156].

In childhood sarcomas, several studies have pointed out c-MET as a promising biomarker capable of predicting poor prognosis. Forced expression of MET in primary osteoblasts induces transformation and is essential for the maintenance of the cancer phenotype [157], while loss-of-function approaches in OS cell lines (143B and U2OS) demonstrated that this oncogene promotes cell proliferation, migration and invasion, and inhibits cell apoptosis [158]. However, the study of genomic status of MET and other genes implied in ossification processes in a cohort of 91 children and teenagers showed that *MET* is mainly deleted, although the clinical subgroup with MET amplification presents worse outcomes [159].

In EWS, modest to high MET cytoplasmic/membranous expression is detected in the majority of tumor samples and is significantly correlated with a poor overall survival. However, there were no significant correlations between MET expression and clinical characteristics, including tumor stage, tumor location and age at diagnosis. The same group also detected genetic alterations that result in the formation of truncated MET proteins in 5% of patients and in two cell lines (ES-2 and ES-7) [160].

Finally, this RTK is overexpressed in RMS tumor samples [161–163] and cell lines, contributing to the metastatic and invasive features of this tumor type [163,164].

ALK. This RTK was first described in 1994, as a fusion partner in the t(2;5)(p23;q35) chromosomal translocation characteristic of anaplastic lymphoma from which takes its name [165]. In general, ALK activates multiple signaling cascades, such as the PI3K-AKT, CRKL-C3G, MEKK2/3-MEK5-ERK5, JAK-STAT and MAPK pathways, and its role in cancer may vary due to many factors, including not only its fusion partners (more than 30 described so far), but also the tumor type or its genetic background (its effects on NB, for example, are dependent on *MYCN* status) [166].

Next-generation sequencing has revealed the presence of several *ALK* mutations in pediatric cases with RMS, EWS, WT and OS [161,167–170]. Most mutations are located within the kinase domain and can be divided into three groups: ligand-independent mutations (F1174I, F1174S, F1174L and R1275Q), ligand-dependent mutations (D1091N, T1151M and A1234T) and a kinase-dead mutation (I1250T) [167].

Germline gain-of-function point mutations are observed in half of hereditary NB and in 9% of the sporadic forms [170], and correlate with high risk and poor prognosis [171,172], mainly because both the wildtype and mutant forms of ALK induce MYCN transcription and potentiate its oncogenic activity in this tumor type [173]. Other ALK-driven pediatric tumors include infantile hemispheric gliomas [149,174], inflammatory myofibroblastic tumors, renal cell carcinomas [167] and pediatric mesotheliomas [175].

ALK in-frame translocations have been described in EPN and EWS as detected by fluorescent in situ hybridization with the brea-apart of 5′ and 3′ probes [160,176]. Finally,

ALK expression is strongly associated with the WNT-activated MB subtype in which, differently from other pediatric tumors, it represents an independent indicator of good prognosis for medulloblastoma patients [177].

3.1.2. PI3K/AKT/mTOR Pathway

The phosphatidylinositol 3-kinase (PI3K)/Akt/mammalian target of rapamycin (mTOR) signaling pathway is among the best investigated in human biology, and is considered a key player in both physiological and pathological conditions [178].

The first step of activation of this pathway consists of the recognition of various growth factors and cytokines by RTKs localized at the cytoplasmic membrane. Then, these receptors dimerize and undergo autophosphorylation, activating GRB2 (Growth Factor Receptor Bound Protein 2) and SOS (Ras/Rac Guanine Nucleotide Exchange Factor). These activate Ras through the exchange of GDP with GTP, which then phosphorylates and activates the PI3K [179]. Active PI3K catalyzes the conversion of PIP2 (phosphatidylinositol 4,5-bisphosphate) to phosphatidylinositol-3,4,5-trisphosphate (PIP3), a second messenger that binds and recruits AKT to the cell membrane, which causes a conformational change in AKT and makes it more accessible to the PDK1-mediated phosphorylation of Thr308, followed by the phosphorylation at serine-473 by the mTOR2 complex [180]. This activation then induces a detachment of AKT from the inner surface of the plasma membrane, and the relocation to the nucleus where AKT isoforms phosphorylate and modulate the activity of several transcription factors. More than 100 different effectors have been described, including cyclin-dependent kinase inhibitor kip1 (p27kip1) through the FOXO family of Forkhead transcription factors, glycogen synthase kinase 3 (GSK3) and cell cycle stimulators, including cyclin D1 and c-Myc. AKT can regulate apoptosis by the inhibition of Fas ligand (FasL), BCL2-associated death promoter (BAD), BCL-2-interacting mediator of cell death (BIM) or BCL-2-associated X-protein (BAX), and by the degradation of p53 [181].

AKT also activates mTOR1, which has many different targets, including translation transcription factors that initiate transcription of genes associated with cell survival and growth and factors associated with hypoxia and angiogenesis (Figure 2) [182].

In cancer, this pathway can be dysregulated as a result of the activation of upstream oncoproteins including RTKs, by the loss or decreased level of its negative regulators such as the phosphatase PTEN (phosphate and tensin homolog deleted on chromosome 10), or directly through mutation and overexpression [183].

Undeniably, this pathway is activated in a wide variety of tumors (i.e., prostate, breast, lung and leukemia, among many others), leading to a profound disturbance of cell growth control, metabolic reprogramming and invasion/metastasis, as well as the suppression of autophagy and senescence [179,184–189]. Moreover, increasing evidence points to its critical participation in the maintenance of stemness in a variety of cancers, contributing directly to recurrence and chemoresistance [189].

In the pediatric setting, the PI3K/AKT/mTOR signaling axis has been described as abnormally activated in both hematologic and solid tumors, mainly as a consequence of chromosomal gains amplifying the AKT1 gene (described in rare cases of leukemia) [190–192] or the aberrant expression of PI3K isoforms [193]. Below, the involvement of this pathway's individual members in childhood tumors is explored.

PI3K. PI3K is a group of plasma membrane-associated lipid kinases, consisting of three subunits: p85 regulatory subunit, p55 regulatory subunit and p110 catalytic subunit. According to their structure and substrate specificity, these kinases are grouped into three categories (classes I, II and III) [194,195]. PIK3CA (phosphatidylinositol 3-kinase, catalytic, α-polypeptide), the gene encoding the p110α subunit, is frequently mutated in ~30% of common human cancers and has been studied most thoroughly [196,197]. Although numerous mutations in this gene have been described, most gain-of-function mutations cluster around two hotspots at exons 9 and 20 [197]; however, contrasting roles for mutations at each exon have been described depending on the tumor type [198,199]. PI3K amplifications have also been frequently described and correlated with aggressive phenotypes, chemoresistance and poor prognosis [200–203].

The prognostic power of PI3K alterations in childhood cancer has been less explored. In MB, however, although no mutations have been detected [196], the p110α isoform is typically overexpressed, promoting cell proliferation, chemoresistance and migration [204–206].

Dysregulation of PI3K signaling is also considered an important player in gliomagenesis, with key roles in regulating cell movement and thus contributing to the highly invasive phenotype of GBM. Compared with normal human astrocytes, overexpression of PI3K p110 catalytic subunits, p85 regulatory subunits and phosphorylated Akt (Ser473) was also detected in two pediatric GBM cell lines (GBM6840 and GBM2603) [206]. Likewise, overexpression of the catalytic p110δ and regulatory p85α isoforms was also detected in a panel of primary NB samples and cell lines with active roles in cell growth and survival. Especially, p110δ was correlated with *MYCN* amplification [207]. However, this gene is significantly lower in NB samples with loss of heterozygosity at 1p36 and associated with poor clinical outcome [208–210].

The regulatory domain of PI3K, p55, is upregulated in sarcoma stem-like cells and promotes invasion, migration and chemotherapy resistance [210]. In EWS, despite variable expression levels between samples, this positive regulator has an oncogenic role [211]. Moreover, p55 analysis on a human sarcoma TMA (that includes two EWS samples) performed by Yoon et al. demonstrated a 4.1-fold increase compared with normal tissues [210].

AKT (PKB). AKT or PKB (protein kinase B) is a serine/threonine kinase that functions as an important regulator of cell growth, survival and glucose metabolism. There are three isoforms of mammalian AKT which are encoded by different genes [212]. AKT1 and AKT2 are ubiquitously expressed and are mostly involved in regulating cellular survival and protein synthesis, involved in glucose transport through the insulin signaling pathway, respectively. The function of AKT3 is not yet fully understood and its expression is almost entirely limited to the nervous system tissue [213–215]. Nevertheless, it has been reported that despite the high similarity, AKT isoforms exert non-redundant specific effects under physiological and pathological conditions [215].

Gain-of-function mutations in all three AKT genes have been identified in ~40% of breast, colon, melanoma and ovarian cancers [216,217]. G49A mutations affecting the pleckstrin homology domain of AKT1, for instance, were identified in ~5% of breast, colorectal and ovarian cancers [217]; however, this mutation was not detected in any of the 100 cases of GBM or 75 cases of MB analyzed by Schüller et al. in 2008 [218]. In the pediatric population, MB samples show p-AKT, and cell lines have shown to be crucially dependent on PI3K/AKT pathway activation; however, the phenotype was attributed to PTEN inactivation as a result of the loss of heterozygosity of chromosome 10q or promoter [219].

In pediatric sarcomas, Akt1 has been shown to contribute to the maintenance of the undifferentiated state of myoblasts pointing towards Akt signaling as a critical RMS nodal point [220]. The AKT pathway is also considered to be an important mediating survival signal in EWS [221], Likewise, an increasing body of evidence has shown that this pathway is frequently hyperactivated in OS and contributes to disease initiation and development, including tumorigenesis, proliferation, invasion, cell cycle progression, inhibition of apoptosis, angiogenesis, metastasis and chemoresistance [222–224]. The AKT2 gene is significantly upregulated in chemoresistant OS cell lines [224] and tumor samples, being significantly associated with positive recurrence, the presence of metastasis, poor response to chemotherapy and shorter EFS and overall survival [224]. Moreover, the AKT3 isoform was evidently upregulated in OS tissues and positively associated with tumor size [225].

AKT2 also plays an important role in NB by regulating N-myc expression. Of note, attenuation of this AKT isoform impaired proliferation and anchorage-independent cell growth, and decreased the secretion of angiogenic factor VEGF and decreased the potential to metastasize to the liver in vivo, thus implicating AKT2 in multiple aspects of NB initiation and progression [226].

mTOR. mTOR is a serine/threonine protein kinase that forms the catalytic subunit of two structurally and functionally distinct protein complexes, known as mTOR Complex 1 (mTORC1) and 2 (mTORC2) [227]. mTORC1 consists of mTOR, Raptor, GβL (mammalian lethal with SEC13 protein 8) and domain-containing mTOR-interacting protein (DEPTOR), and plays active roles in integrating various signals that specify the availability of growth factors, nutrients and energy in order to endorse ribosomal biogenesis, protein translation during cell growth and the expression of metabolism-related genes, while inhibiting apoptosis and autophagy [228]. mTORC2, on the other hand, is composed of mTOR, Rictor, GβL, Sin1, PRR5/Protor-1 and DEPTOR, and regulates cytoskeletal dynamics, ion transport and promotes cell proliferation and survival through the activation of Akt [229–231].

mTOR is frequently improperly activated in human cancers and results in alteration of both mTORC1 and mTORC2 signaling pathways, leading to increased cell proliferation and decreased apoptosis. However, among 33 mTOR activating mutations identified in 2014 by Grabiner et al. [231], those that were functionally tested in vitro conferred varying degrees of pathway activation, and, most importantly, a few displayed some substrate preference towards the eukaryotic translation initiation factor 4E binding protein 1 (4EBP1) and ribosomal protein S6 kinase (S6K1), or towards AKT1, implying that such mutations had distinct effects on mTORC1 or mTORC2. Specifically, 4EBP1 activation by mTOR1 is a major contributor to accelerated cell proliferation or increased cell survival; the so-called eIF4E-sensitive mRNAs code for various cell cycle and apoptosis regulators, including cyclins D1 and D3, CDK2, MYC, PIM1, Bcl-2, Bcl-xL and VEGF, among others [232–234].

mTOR overactivation is observed in many childhood tumors, including EPN, MB and PA, high-risk NB, WT and RB, leading to worse prognosis and survival [105,235–241].

Constitutive activation of the mTOR pathway, predominantly through mTORC2, is observed in EWS, with active roles in metastasis formation [241–244]. The metastatic behavior of OS is also dependent on the PI3K/Akt/mTOR cascade, in which mTOR contributes to cellular transformation and poor cancer prognosis via its downstream effectors S6K1, 4EBP1 and eIF4E [245,246]. In RMS, lower disease-free or overall survival is also associated with the activation (phosphorylation) of multiple interconnected Akt/mTOR pathways [246]. Of note, rapamycin treatment can greatly reduce the growth of cell lines derived from these three sarcoma types [242].

GSK-3. The glycogen synthase kinase is a ubiquitously expressed serine/threonine kinase existing as GSK-3α and GSK-3β isoforms (encoded by separate genes), both of which are downstream effectors of AKT [247]. Differentially from other kinases, GSK3 is one of the few whose activity tends to be high in resting cells, and exposure of cells to growth factors, serum or insulin results in its catalytic inactivation [248].

The GSK3 kinases are pleiotropic, phosphorylate many proteins, and interact with multiple signaling pathways [249]. These kinases can modify the activity of transcription factors that have profound regulatory roles in cellular proliferation (such as p53 and NF-κB), transcription factors important for epithelial–mesenchymal transition (EMT) (i.e., Snail) and pro-apoptotic molecules including BCL2 and BAX [250]. Therefore, aberrant activity of GSK3s can result in many diseases and disorders and influence oncogenesis and metastasis [251]. However, since GSK3s are involved in a wide range of signal transduction cascades and a plethora of cellular functions [252,253], their roles in cancer establishment and maintenance can deviate from their chief tumor suppressor effects and also promote neoplastic transformation [254–257]. This dichotomy is also observed in the pediatric setting. Strong evidence provided by Wang et al. (2008) [257], for example, demonstrated that GSK-3 activity is essential for the maintenance of MLL-positive leukemias. MLL rearrangements are in >70% of infant leukemia, and irrespective of the translocation partner, they are associated with poor clinical outcomes [258–260]. Alternatively, as a key suppressor of the Wnt, Hedgehog and Notch pathways GSK3 has attracted much scrutiny. Within these pathways, this kinase is critical in regulating the turnover of the effectors β-catenin, c-Myc and c-Jun, targeting them for degradation/inactivation, and this inhibits proliferation and stem cell maintenance [250].

The literature about the prognostic value of GSK3A in pediatric cancer is scarce. No evidence was found in the literature about its involvement in EWS, OS, RMS, WT, RB, NB and EPN. However, its role in MB has been explored in vitro and in vivo, showing to be important for cell proliferation and tumor growth [260], a phenotype that seems to be similar in pediatric glioma [261].

On the other hand, the role of GSK3B is more extensively studied. In EWS, this gene can either promote or impair tumor growth and is associated with good prognosis [262–265]. Interestingly, in OS, the same gene acts as an oncogene [265], and is associated with worse response to neoadjuvant chemotherapy [266].

The oncogene status also occurs in alveolar RMS, where GSK3B is directly involved in regulating the transcriptional activity of PAX3/FKHR [267] at the same time that the chimeric protein enhances GSK3B activity, which in turn represses MYOGENIN, a member of the muscle regulatory factor family that orchestrates the terminal differentiation step of skeletal muscle cells [268]. GSK3B is also involved in the maintenance of undifferentiated phenotypes in ERMS [269].

GSK-3B is highly expressed in high-risk NB; however, its expression is not associated with clinical stage, survival or other clinicopathological parameters [270]. GSK-3B has also been involved in the protection of NB cells against chemotherapy by regulating NF-kB signaling [271]. In this regard, several authors have shown that GSK3 inhibitors are able to regulate MYCN mRNA levels and reduce NB cell viability through multiple mechanisms, including p53 and Wnt signaling, BDNF/TrkB/PI3K/Akt, suggesting that targeting this kinase might potentiate chemotherapy [271–274].

Alternatively, a predominantly tumor-suppressive role for GSK3B is observed in MB, in which its accumulation leads to the downregulation of GLI, the most important activator and driver of the SHH medulloblastoma subtype [274]. Constitutive phosphorylation leading to GSK3β activation improves cell survival and contributes to malignant transformation [275]. Dysregulated GSK3B also sustains the survival, immortalization, migration, invasion and maintenance of stem cells in glioma [276–278].

3.1.3. MAPK Pathway

The mitogen-activated protein kinases (MAPKs) comprise a group of serine–threonine protein kinases that control numerous cellular processes, including proliferation, differentiation, apoptosis, survival, inflammation and innate immunity [278]. In mammals, MAPKs include three main signaling axes, namely c-Jun NH2-terminal kinase (JNK), p38 MAPK and extracellular signal-regulated kinase (ERK), each of which exists in several isoforms [279].

This pathway mediates intracellular signaling triggered by extracellular stimuli such as growth factors and cytokines (ERK), or by intracellular stimuli such as genotoxic, osmotic, hypoxic, oxidative or endoplasmic reticulum (ER) stress (JNK and p38), for example [280,281].

The general cascade pattern includes initial activation of MAP4Ks (membrane-bond GTPases such as RAS, RHO, RAN, RAB and ARF) by RTKs, which phosphorylate intermediate MAP3Ks (i.e., RAF, MEKK). These then mediate phosphorylation and activation of MAP2Ks (MEK1/2—mitogen-activated protein/extracellular signal-regulated kinases, MKK4/7), followed by the positive phosphorylation of MAPKs (ERK1/2, p38 or JNK). Once activated, MAPKs phosphorylate diverse substrates, including transcription factors such as c-Jun, c-Myc, P53 and ATF2, thereby giving rise to the various cellular responses [282,283]. p38 MAPKs have also emerged as important modulators of gene expression by regulating chromatin modifiers and remodelers [281].

Thus, compromised MAPK signaling contributes to the pathology of a wide spectrum of human malignancies [282]. However, while the roles of JNK and p38 pathways are elusive [284–290], dysregulation of the RAS/RAF/MAPK(MEK)/ERK pathway explicitly drives the oncogenic process [291].

In fact, many of the cancer-associated mutations of components of MAPK signaling pathways have been found in RAS. Missense gain-of-function mutations in all three RAS genes (HRAS, KRAS and NRAS) are found in ~30% of all human cancers [292–294]. Other perturbations in GDP–GTP regulation, persistent receptor tyrosine kinase-mediated activation of GEFs, and miRNA deregulation are additional mechanisms of RAS activation in cancer and result in constant input signals with downstream kinases [291,295].

In this regard, the frequency of genomic alterations in the MAPK pathway as a whole parallels the direction of the signaling cascade: RAS > BRAF > MEK, and ERK mutations are exceptionally rare [296,297].

RAF. RAF has three isoforms (ARAF, BRAF and CRAF/RAF1), sharing a high similarity of domain organization. These cytoplasmic serine/threonine-specific protein kinases are essential effectors of the MAPK pathway through the association with activated RAS. This binding leads to their homo- or heterodimerization and activation with the phosphorylation of ERK.

Altered activation of RAF members results in increased proliferation in a broad range of human tumors [298,299]. The most common gain-of-function mutation in the members of the family occurs in BRAF codon 600, in which a valine is substituted for glutamic acid (BRAF-V600E). This point mutation is notably widespread in pilocytic astrocytoma (15%), melanomas (63%) and papillary thyroid carcinomas (more than 50%) [300,301]. Overexpression of full-length RAF or the truncated catalytic domain also leads to hyperactivated ERK signaling, resulting in increased malignant behavior [302].

ERK1/2. ERK1 and ERK2 are the prototypes of the eight isoforms of ERK and are activated by MAPK/ERK kinase (MEK) 1 or 2. Upon activation, ERK detaches from cytoplasmic anchoring proteins and translocates to the nucleus to exert its transcriptional regulation. Despite the well-recognized importance of ERK activation in cancer malignancy, mutations in these genes have rarely been reported as drivers in human cancers. Nonetheless, The Human Protein Atlas classifies them with enhanced expression compared to normal tissues.

The most compelling evidence of MAPK activity in cellular processes contributing to the development and progression of childhood tumors is represented by the duplication/rearrangement of *BRAF* at 7q34 leading to *KIAA1549:BRAF* fusion product, which is the most common molecular alteration in sporadic PA, occurring at the highest frequency in tumors of the posterior fossa [303–305]. ERK2 was identified as differentially expressed in tumor samples compared to normal tissues [306]. RAS/MAPK activation was associated with metastatic disease in MB [307,308]. Associations of ERK hyperexpression with distant metastasis and poor overall survival were also reported for childhood sarcomas, including RMS and OS [309–312]. For other pediatric tumors, activation of this pathway results from their interaction with dysregulated microRNAs [313,314].

3.1.4. Cell Cycle Kinases

The cell cycle is a complex and well-ordered series of irreversible events through which a cell duplicates its DNA and grows to produce two daughter cells with identical genomes [315,316]. Transitions from one state to the next are driven by many oscillating regulators that determine whether cells proceed through G1 into the S phase, and from G2 to M, each of which are characterized by distinct molecular features and functional outputs [317]. Central to this process are the cyclin-dependent kinases and other key regulators such as kinases from the Polo and Aurora families.

Cyclin-Dependent Kinases

Cyclin-dependent kinases (CDKs) comprise 13 key regulatory enzymes involved in cell proliferation through the regulation of cell cycle checkpoints and transcriptional events in response to extracellular and intracellular signals. These intracellular serine/threonine kinases, whose catalytic activities are regulated by interactions with the adaptor molecules cyclins and CDK inhibitors (CKIs), orchestrate the evolution through the sequential phases, including entry into the cell cycle from quiescence, the G1/S phase transition, DNA replication in the S phase, nuclear breakdown, chromosome condensation and segregation, and cytokinesis [318].

CDKs coordinate cell cycle regulation at different stages to ensure the coherence, integrity and maintenance of every step in a sequential manner. CDK1 and CDK2, for instance, are necessary to direct the transition from S to G2, but only CDK1 governs the G2/M transition and mitotic progression [319]. Other CDKs regulate the cell cycle indirectly by activating other members of the family (CDK7, CDK20) or transcription (CDK7, CDK8, CDK9, CDK12, CDK19) [320,321].

Changes in the expression and regulation of CDKs induce unscheduled proliferation and chromosomal instability, well-known hallmarks of cancer and tumor aggressiveness [322–327]. Amplification or mutation of genes encoding CDKs, cyclins or endogenous inhibitors of CDKs have been described in many solid cancer types, and are recurrent events in the development of breast cancer [328] and GBM [329], for example. Such alterations are also described as molecular drivers in childhood tumors [330–334].

CDK1. Cyclin-dependent kinase 1 (CDK1) is vital in governing cell division and transition from G2 to the M phase [335]. Its dysregulation is common in many tumors of diverse origins, leading to chromosomal instability via replication stress and enhanced proliferation of cells. A recent pan-cancer integrative analysis based on TCGA and GTEx databases performed by Liu et al. (2022) showed that CDK1 expression levels are increased in many tumor types when compared to normal tissues and are generally associated with poor clinical prognosis [336]. For example, CDK1 expression is positively and highly associated with advanced cancer stages in lung and endometrial cancer [337,338]. Similar results were reported for other tumors, such as breast [339–341]. Moreover, CDK1 expression is positively correlated with the expression of the stemness marker SOX2, indicating a direct action on tumor maintenance and chemoresistance [342,343].

Regarding pediatric tumors, this kinase is associated with lower overall survival and EFS rates in EPN [344,345], RMS [346,347] and NB [348]. In silico analyses have also demonstrated that CDK1 is differentially expressed in RB [349] and plays a key role in the development of OS, since its negative regulation or depletion leads to significant decreases in proliferation while inducing apoptosis [350–353].

Furthermore, according to the literature, a well-described relationship exists between CDK1 and EWS, WT and high-grade gliomas (HGG). Specifically, this kinase expression has been directly associated with tumor progression [354], being considered a hub gene for GBM [355].

CDK2. Cyclin-dependent kinase 2 (CDK2) drives the entry of cells into the S and M phases of the cell cycle. Except for a few exceptions (i.e., testis), the majority of normal tissues have low expression of this serine/threonine kinase [356], and its activity is not essential for normal development [357]. However, CDK2 has been associated with cancer progression and aggressiveness across several malignancies [339,358], contributing not only with genomic instability and under-replication of DNA in the late S phase [359] but also through interactions with other proteins in a wide range of biological processes such as DNA damage response, intracellular transport, protein degradation and signal transduction, among others. Differentially from other kinases, several investigations have demonstrated that CDK2 is not upregulated or amplified; instead, its dysregulation results from altered binding partners or alterations due to post-translational modifications [360]. In tumors with MYCN overexpression, as is the case of NB, interaction with CDK2 appears to be critical for senescence avoidance and immortalization [361], being associated with worse prognosis and considered a suitable therapeutic target in this tumor type [362,363]. CDK2 inhibitors effectively induced cell cycle arrest or apoptosis in MYC-driven MB [364].

The literature also shows overexpression of CDK2 in HGG compared to normal tissue and low-grade forms, with a direct association with worse prognosis due to immune cell infiltration [365]. Moreover, this kinase plays a central role in the development of NB. Even though there is no well-established relationship between this kinase and the development of this tumor type, Zhang et al. (2016) [366] demonstrated that the inactivation of TAZ (a biomarker of aggressiveness in RB through miR-125a-5p) inhibited proliferation and tumor formation by decreasing cyclin E and CDK2 expression [367]. Similarly, although there is no clear and explicit description in the literature about the relationship of this kinase with RMS, Knudsen et al. also discussed the relationship between sustained CDK2 levels in RD cells irrespective of the exposure of cells to differentiating culture media, explaining the inability of those cells to arrest growth and thus contributing to oncogenesis [368]. Moreover, there are reports of apoptosis induction in several sarcoma cells after CDK1 and CDK2 co-depletion [369,370]. Of note, in a microarray-based study, CDK2 was found to be overexpressed and associated with poor prognosis in EWS [371].

CDK4/6. Cyclin-dependent kinases 4 and 6 are highly homologous key components of the cell cycle to drive the passage from G1 to S phase. Upon interaction with any D-type cyclin (CCND1, CCND2 or CCND3), these interphase kinases phosphorylate Rb to release E2F from Rb and initiate the transcription of genes required for cell cycle progression. Besides proliferation, other roles of cyclin-D/CDK4/6 have been confirmed, including the regulation of senescence, apoptosis, migration/invasion and angiogenesis [372].

Consequently, the complex CCND/CDK4/6 shortens G1, and hence, its constitutive activation represents a driving force of tumorigenesis. These proteins are generally concurrently studied and, in many cases, they present themselves with similar patterns, being simultaneously dysregulated [373,374].

CDK4 was identified as a major risk factor for disease progression in Paget's disease [375] and its overexpression and/or hyperactivation is implicated in many types of human cancers [376–381]. Point mutations at the CDK4 locus (CDK4R24C) have also been reported [382].

Co-overexpression of both CCND1 and CDK4 is common in hepatoblastoma, a rare malignant liver tumor of childhood, and usually positively correlated with tumor recurrence [383]. In parallel, enhanced kinase activity of CDK6 has been associated with other childhood tumors [384]. This kinase plays an important role during hematopoiesis and is frequently altered in hematological malignancies of different immunophenotypes [385,386]. MLL-AF9 oncofusions in myeloid leukemia, for example, induce high CDK6 levels, acting as a blocker of myeloid differentiation and contributing to the maintenance of an immature phenotype [387]. This MLL fusion-driven activation of CDK6 (through MLL-AF4 and MLL-ENL) has also been described in infant leukemia [388].

Likewise, CDK4 and CDK6 have been described with similar frequencies in WT compared to normal mature kidneys, even though only CDK4 showed correlation with relapse [389]. However, a more recent study by Haruta et al. (2019) showed that WT samples with chromosome 12 trisomy does indeed show upregulation of this kinase, but that stronger expression is associated with better overall survival [390].

Activation of the CDK4/6 pathway is also a powerful driver of sarcomagenesis [391]. Amplification of 12q13-15 also occurs in OS, and a recent copy number analysis of pediatric high-grade OS detected a recurrent gain of chromosome 12q14.1 in ~25% of samples, which resulted in CDK4 overexpression. In vitro, higher expression of CDK4 was considered a predictive biomarker for resistance to cisplatin [392]. Indeed, elevated CDK4 expression is correlated with metastasis potential and poor prognosis in this tumor type [393–395]. Consistent with these findings, a recent study demonstrated that about 50% of OS samples present CDK4 somatic variants, 9.5% of which were identified as gain-of-function CNVs correlated with metastasis and death [396].

The inhibition of CDK4/6 also represents a promising precision medicine-guided therapy for other childhood sarcomas. A parcel of PAX3/PAX7-FOXO1-positive RMS tumors with amplification of the chromosomal region 12q13-q14, for example, also presents elevated CDK4 levels relative to non-amplified, fusion-negative forms [397,398]. In addition, in Brazilian cohorts, amplification or overrepresentation of CDK4 was evinced through qRT-PCR and immunoreactivity in both forms of RMS (ERMS and ARMS), along with several leiomyosarcoma samples [399,400]. Similarly, using a human TMA, Saab et al. (2007) demonstrated CDK4 expression in 82% of ARMS and 63% of ERMS tumors [401]. CDK6 was detected at high levels in six RMS-derived cell lines, reinforcing the prospects of its inhibition as a therapeutic opportunity [402]. In a similar manner, a shRNA-based screening demonstrated that CDK4 (together with CCND1) is required for survival and anchorage-independent growth in EWS [403].

More recently, a systematic evaluation of CDK4/6 as targets in a series 16 pediatric cancer types indicated that further preclinical evaluations are still needed to affirm the dependence of tumors on CDK4/6. Nevertheless, the results provided evidence for benefits in EWS, malignant peripheral nerve sheath tumors and MB [391]. Of note, within MB subgroups, CDK6 and CDK14 co-amplifications were identified in 20% samples from patients with relapsed group-4 MB [330].

Shubert et al. (2022) also pointed out that patients with atypical rhabdoid tumor/malignant rhabdoid tumor, NB or HGG may also benefit from anti-CDK4/6 therapy [391].

CDK4 and CDK6 are both highly expressed in NB compared to normal tissues [404]. Moreover, like CDK2, CDK4/CDK6 exert oncogenic roles in this tumor type, especially in MYC-amplified forms [405]. Additionally, when co-amplified with MDM2/FRS2, CDK4 and CDK6 are associated with poor prognosis and atypical clinical features, including poorly differentiated or undifferentiated histology and metastasis at diagnosis and at relapse [406].

In line with Schubert et al. (2022) [391], CDK4/6 upregulation also plays an important role in the pathogenesis and progression of high-grade gliomas with potential actionability [407–409]. However, the use of CDK4/6 inhibitors alone did not show satisfactory results, suggesting the use of combinatorial intervention [410].

CDK4 was likewise found overexpressed in EPN and associated with adverse outcomes [411]; accordingly, its inhibition restricted cell proliferation and reduced the expression of genes associated with the cell cycle and DNA repair (*CCNB1*, *TOP2A*, *CDK2*, *BRCA1* and *RAD51*), and induced morphological changes that culminated in cell death [412]. Considering EPN subgroups, De Almeida Magalhaes et al. (2020) showed that *CDK6* is overexpressed in ST-EPN-RELA tumors compared to other ST-EPN subgroups [413], even though others have suggested that the dysregulations of the p16-CDK4/6-pRB-E2F pathway might also compose the genetic background underlying the aggressive biology of posterior fossa EPN in infants less than 1 year old [414].

CDK5. The cyclin-dependent kinase 5 (CDK5) represents an unusual member of the family of cyclin-dependent kinases, which is activated upon binding to p35 and p39 proteins, which are not cyclins. Conversely, interactions with CCND1 or CCND can attenuate CDK5 activity [415]. CDK5 is expressed ubiquitously, but with higher activity in the nervous system, participating in neuron migration, neurite outgrowth and synaptogenesis. Nevertheless, increasing evidence points to a diverse array of functions in other tissues, ranging from cell proliferation to cytoskeleton remodeling and cell motility by regulating actin dynamics [416–418].

Apart from neurodegenerative disorders, amplification and increased expression of CDK5 have been described in multiple tumor types and are associated with worse prognosis and stemness [419–431]. Mutations located in key domains of CDK5 that influence its structure and post-translational modifications have also been described as contributors to tumorigenesis [432].

The participation of CDK5 in pediatric tumors is purported; however, the stimulation of cancer-related signaling pathways by this kinase remains obscure, and reports are scarce. CDK5 was found to be hyperactivated in NB and its inhibition resulted in cell cycle arrest and morphological differentiation [433,434]. Moreover, as a crucial regulator of neuronal signal transduction, CDK5 can be found differentially expressed in gliomas, progressively augmenting with tumor grade, suggesting an active role not only in tumorigenesis but in aggressiveness as well [424,435,436].

CDK5 also appears to be a central regulator of OS tumorigenesis, with high levels of expression being associated with low survival and increased angiogenesis [437,438]. Interestingly, CDK5 also plays a role in osteoblastic differentiation. Fu et al., for example, demonstrated that CDK5 inhibition promotes the expression of *Runx2, ALP, OCN* and *OPN* in mesenchymal stem cells, the mineralization of MC-3T3E1 cells and suppresses the migration of the OS cell line MG-63 [439]. Additionally, the CDK5/p35 complex strongly inhibits the Wnt/beta-catenin signaling pathway, also able to stimulate osteoblastic differentiation [440].

The WNT pathway defines a molecular subgroup of MB [441]; thus, it may be assumed that CDK5 might also contribute to this tumor malignancy. In fact, Cdk5 expression has been demonstrated in different MB cell lines and in a reduced cohort of patients; however, its deletion did not alter proliferation, reflecting the more favorable prognosis of MB with WNT activation [441]. Nevertheless, a role for CDK5 in tumor immune evasion through the regulation of PD-L1 was suggested [442].

CDK7/9. Cyclin-dependent kinase 7 (CDK7) and 9 (CDK9), apart from directing cell cycle progression, have critical roles in transcription initiation and elongation as regulators of the phosphorylation of the carboxy-terminal domain (CTD) of RNA polymerase II (CDK7 is a component of TFIIH, and CDK9, a subunit of pTEFb) [443–445]. CDK7/9 also controls many transcription factors, functioning to either promote their activity and/or regulate their turnover [445]. Recently, other uncovered transcription-associated functions have been revealed, including epigenetic modifications and mRNA-3′ termination [446–448].

CDK7/9 levels are elevated in several cancer types and are associated with clinical outcomes [449–455]. In many cases, they can indirectly impact gene expression profiles by aberrantly controlling the functioning of transcription factors that are critical in specific tumor types, as is the case of estrogen- or androgen receptor-mediated transcription in breast and prostate cancer, respectively [449,456,457]. MYCN-dependent transcription can also be affected, as demonstrated in NB cells, or contribute to histone-3 methylation in diffuse intrinsic pontine glioma (DIPG) [445,458].

With regard to other pediatric tumors, CDK7 has been shown to be upregulated in a panel of OS cell lines and tumor samples, being associated with worse prognosis and higher metastasis rates [459]. Accordingly, CDK7 knockdown in SJSA-1 cells reduced phosphorylation of the RNAPII CTD and reduced tumor volume and weight in xenograft models compared with tumors derived from wild-type cells [460]. Similarly, higher levels of expression of CDK9 have been associated with lower Huvos grade and lower survival rates, characterizing this kinase as a suitable therapeutic target, as determined through siRNA assays [461,462].

Descriptions about the relationship between CDK7 expression and EPN, RMS, EWS, WT, RB and NB are rare. Nevertheless, the use of THZ1 (CDK7 inhibitor) has exposed positive scenarios, considering that EWS cells are sensitive to this compound and that reduced EMT capacity of RB cells is observed after treatment [463,464]. In contrast, CDK9 kinase is widely expressed in RMS, where it impedes the physiological cellular differentiation [448,465–467]. Additionally, inhibition of CDK9 demonstrated a general disruption of transcription [465]. Similarly, this kinase is widely expressed in pediatric sarcomas, such as EWS, and its pharmacological inhibition (PHA-767491) enhanced the mithramycin-mediated suppression of the EWS-FLI1 transcriptional program, leading to a shift in the IC_{50} and striking regressions of mouse xenografts. Furthermore, this kinase is upregulated in NB, increasing with the degree of differentiation of the tumor [468]. Of note, Poon et al. (2020) demonstrated that CDK9 inhibitors are able to downregulate MYCN to varying degrees and to induce apoptosis, as detected by induction of poly (ADP-ribose) polymerase (PARP) cleavage [469].

Polo-Like Kinases

Polo-like kinases (PLKs) comprise a highly conserved multifunctional family of kinases of five members: PLK1, PLK2, PLK3, PLK4 and PLK5 [470–472]. These serine/threonine kinases are traditional controllers of cell cycle progression, with major roles in the formation of the mitotic spindle, chromatid separation, regulation of the anaphase-promoting complex, DNA damage response and cytokinesis [472,473]. Structurally, these proteins share an N-terminal highly conserved catalytic domain and a regulatory domain fundamental to the functionality and localization of PLKs, called Polo-Box (PBD) and located at the C-terminus [474].

PLKs are differentially expressed depending on the tissue and cell cycle phase [475]. Alterations in the expression of PLK genes have already been described in different types of cancer (breast, OS, leukemia, gliomas, among others) and have generally been correlated with dismal prognosis [476].

PLK1. PLK1 is the most studied member of the family. This protein plays key roles at different points of the cell cycle, especially during the progression of mitosis [477]. Nevertheless, other non-mitotic functions such as cell survival, genomic maintenance, cell fate and DNA damage control are also regulated by PLK1 through the interaction of with effector pathway components, including the oncogenes AKT, MYC, MDM2, B-catenin and the tumor suppressors P53, PRB, BRCA2 and PTEN [478,479].

A plethora of studies have firmly established the active role of this kinase in oncogenesis and its prognostic value along with its potentiality as a therapeutic target [480–486]. Childhood cancer is not an exception. Higher levels of PLK1 have been observed in a variety of cell lines, including EWS, OS, NB and RMS [483,484]. Moreover, this protein has been described as overexpressed in MB samples, where it is associated with higher recurrence and lower survival rates [487,488]. Furthermore, other studies have validated this positive correlation between PLK1 expression and higher cell proliferation in, mainly in undifferentiated tumors with the presence of massive choroidal invasion [205,476,489–491]. PLK1 overexpression is also present in unfavorable NB and associated with poor prognostic markers such as lower age at diagnosis and MYC amplification [492]. This interaction between PLK1 and MYC has also been observed in OS, in which the kinase contributes to MYC stabilization [493].

Aurora Kinases

Three members of the Aurora family of serine/threonine kinases have been identified in humans: Aurora kinase A (AURKA), Aurora kinase B (AURKB) and Aurora kinase C (AURKC). These kinases (named after the resemblance or their localization to the poles of the mitotic spindle to the way aurora borealis are observed at one of the poles of the earth) have pivotal parts in the execution of mitosis (AURKA and AURKB) and meiosis (AURKC), and even exerting conserved function, they cannot fully compensate for the loss of one another [494,495].

AURKA. Aurora kinase A is involved in the centrosome maturation process and promotes the transition from G2 to mitosis. AURKA levels increase along late S and G2 phases and reach a higher peak in mitosis, followed by proteasome-dependent degradation [496,497]. This kinase is early localized at the centrosome and regulates the progression of mitosis by phosphorylation of multiple substrates, promoting mitotic entry through the activation of Cyclin-B/CDK1 [498]. Moreover, AURKA progressively associates with the mitotic poles and the adjacent spindle microtubules, contributing to chromosome separation and bipolar spindle [499,500].

Among the three human aurora kinases, AURKA has been the family member most consistently associated with cancer. Amplification of the chromosomal region 20q13 where the AURKA gene is located is commonly observed in cancer cells [501–503]. Nevertheless, according to Mou et al., almost 90% of tumors present in the TCGA database show AURKA overexpression [504]. Indeed, high levels of AURKA expression can endorse abnormal cell cycle progression, resulting in genomic and chromosomal instabilities, which are hallmarks of highly proliferative tumors [505,506]. Thus, AURKA expression not only enhances proliferation, but may also influence other processes, including apoptosis evasion, EMT, drug resistance and metastasis [507–517].

Besides AURKA mitotic functions, other non-canonical and kinase-independent activities have been gradually discovered in cancer cells. After mitosis, most AURKA proteins degraded, but a remnant population may be still detected inside the nucleus, pointing to the possibility that the kinase could work as a transcriptional regulator [518]. In this regard, AURKA overexpression has been associated with the upregulation of stem cell markers such as SOX2 and NANOG, imposing participation in the maintenance of the self-renewal capacity of cancer stem cells (CSCs) [519].

In pediatric tumors, AURKA polymorphism rs8173 G > C has shown to decrease WT risk [520]. Conversely, AURKA plays an active role protecting MYCN from ubiquitinylation and proteolysis in NB, thus contributing to more aggressive phenotypes and poor survival probability [521,522]. Moreover, this role has also been described in RMS, where AURKA not only stabilizes MYC but also PAX3-FOXO1 [523]. Moreover, in this tumor type, AURK is overexpressed and considered a key factor in the observed aneuploidy and chromosomal instability [524]. AURKA is also closely related to the oncogenic process of EWS and is considered a chemotherapy resistance and may act as a potential biomarker for prognosis [525,526].

AURKA overexpression has also been associated with OS, as many cell lines are highly sensitive to its inhibition [527,528]. Similar patterns have been seen for WT, where AURKA inhibition impaired tumor growth and induced apoptosis both in vitro and in vivo; however, such effects were improved in RB1-deficient cell lines compared to those with *MYC* amplification [529].

Moreover, expression profiling of pediatric brain tumors has shown that AURKA was consistently and highly overexpressed (up to 106-fold) in tumor samples from all glioma grades and from patients varying from 4 months to 82 years old; however, mRNA expression showed only weak correlation with the Ki-67 labeling index, and significant associations with poor patient survival were only observed for GBM [530]. Overexpression of AURKA is also linked with survival in MB patients [531].

AURKB. The second member of the family, Aurora kinase B (AURKB) is one of the most intensively studied kinases because it provides catalytic activity to the chromosome passenger complex (CPC), formed by AURKB, INCEP (inner centromere protein), survivin and borealin (Cell 2002;13:3064–77). The CPC governs highly different processes, such as chromosome alignment, histone modification and cytokinesis [532–535]. Additionally, AURKB kinase activity is essential for faithful chromosome segregation and functions to correct any improper kinetochore attachment to the spindle [532,536–538]. Finally, Aurora kinase B (AURKB) is also essential in mitotic DNA damage response, protecting against DNA damage-induced chromosome segregation errors, including the control of abscission checkpoint and prevention of micronuclei formation [539]. Consequently, AURKB dysregulation results in aneuploidy and genomic instability and in turn promotes cell cycle progression and survival of cancer cells [540–543]. Indeed, expanding evidence at the gene, mRNA and protein levels supports a carcinogenic role of AURKB. Overexpression has been reported in clear cell renal cell carcinoma (RCC) and cervical carcinoma, among many others, with clear associations with clinicopathological parameters such as stage and tumor volume, chemoresistance, tumor progression and poorer survival [544–553].

In the pediatric setting, overexpression of AURKB was closely correlated with poor prognosis and carboplatin resistance in NB patients [552,554]. Similar profiles were observed for pediatric ALL and AML patients, especially in T-cell and E2A-PBX1-translocated ALL cases. Further in vitro assays demonstrated that AURKB is an essential protein for the proliferation and survival of acute leukemia cells [555].

The importance of Aurora kinases as potential therapeutic targets for childhood brain malignancies is highlighted by AURKB being highly and consistently overexpressed in the majority of high-grade gliomas, but despite reflecting the presence of aneuploidy, at least in EPN, it did not emerge as a prognostic factor [556,557]. For other tumor types, however, data about the prognostic value of AURKB are less explored and primarily rely on experimental assays using pharmacological inhibitors. In this context, OS, EWS and RB are included [558–560].

AURKC. Differentially from AURKA and AURKB, Aurora kinase C (AURKC) is limited to cells that undergo meiosis (sperm and oocyte). This kinase is located on human chromosome 19q13.43, regulated by promoter methylation and when expressed in germ cells, can undergo alternative splicing resulting in three protein variants [561–563]. As the major enzymatic component of the CPC during meiosis, it plays a specific role during human female meiosis and preimplantation embryo development [564].

A body of evidence shows that overexpression of AURKC in mitotic cells leads to centrosome amplification and multinucleation [565]. Its upregulation and other CPC components occur in cancer cells and may correlate with clinical characteristics [566,567]. In line with this, AURKC is overexpressed in tumors of the reproductive system and in breast and prostate cancer cell lines [568,569]. Nevertheless, varying degrees of CpG islands hypermethylation leading to lower AURKC mRNA levels have been described in WT, suggesting that this kinase might not be of importance in this childhood tumor [570]. Likewise, AURKC expression was not associated with survival or risk status in neuroblastoma patients [571]. In OS, knockout of AURKC displayed no changes in cell proliferation, migrated less and formed fewer colonies in soft agar compared to wild-type cells. Moreover, whole-transcriptome sequencing revealed over 400 differentially expressed genes which included genes encoding proteinaceous extracellular matrix components, suggesting that therapeutics targeting this aurora kinase isoform could decrease cancer cell metastasis and disease progression, the most limiting characteristic of survival [572].

3.2. *In Silico Analysis of Different Kinases Expression and Their Association with Clinical Prognosis*

According to our systematic in silico analysis of the selected group of kinases, comparisons of expression patterns between pediatric tumors and normal samples showed varying results, and despite what was expected from the data already published, few commonalities were found (Figure 3A,B, Supplementary Table S1). Considering CNS tumors, overexpression of *AKT1*, *AURKA*, *CDK2* and *CDK7* was observed in EPN and MB, but not in PA. Similarly, EPN showed higher levels of *CDK1*, *CDK4*, *EGFR*, *KDR* and *MET*, while for MB, the upregulation of *FGFR2*, *FGFR4* and *MAPK1* was highlighted. Conversely, EPN samples demonstrated low levels AKT3 and FGFR1, while CDK*1/4/6*, *FGFR1/3* and AURKB were downregulated in MB tissues. PA, on the other hand, exhibited high levels of only *ALK* and *FGFR1*, and downregulated genes included *AKT3*, *AURKB*, *CDK5*, *EGFR*, *FGFR2*, *FGFR4* and *MAPK1*.

Neuroblastoma was the tumor type with the more reduced number of hub genes. In this tumor type, CDK6 and FGFR2 were less expressed than in normal tissue, whereas *CDK7*, *MET*, *ALK* and *AURKB* stood out as upregulated in tumor samples. Concomitantly, RB showed high levels of *AKT3*, *ALK*, *AURKA*, *AURKB*, *CDK1*, *CDK2* and *FGFR2* genes and low levels of CDK6, CDK9, EGFR, KDR and MET.

Among sarcomas, RMS was the tumor type with the most altered kinase profile, including high levels of *AKT3*, *ALK*, *AURKA*, *AURKB*, *CDK4*, *CDK5*, *CDK6*, *CDK7*, *GSK3B*, *PLK1* and all *FGFR*. Then, *AKT2*, *EGFR*, *FGFR4* and *GSK3s* (A and B) showed higher expression in EWS, contrasting the low levels of *ALK*, *AURK*s (A and B), *CDK*s (1, 2, 4, 5 and 9) and *FGFRs* (1, 2 and 3). Finally, our analysis of OS samples demonstrated upregulation of the receptor genes *EGFR* and *FGFR* (1, 2 and 4), as well as *CDK9* and *GSK3A*. Alternatively, *CDK6/7*, *AURKA/B* and *AKT3* had low expression profiles.

Notwithstanding, further analysis showed that, in the same line as reported in the literature, the expression of most of the selected kinases is indeed associated with clinical features of worse prognosis, including associations with MYC amplification in NB and molecular subtypes in MB, and metastases in bone sarcomas (Supplementary Table S2).

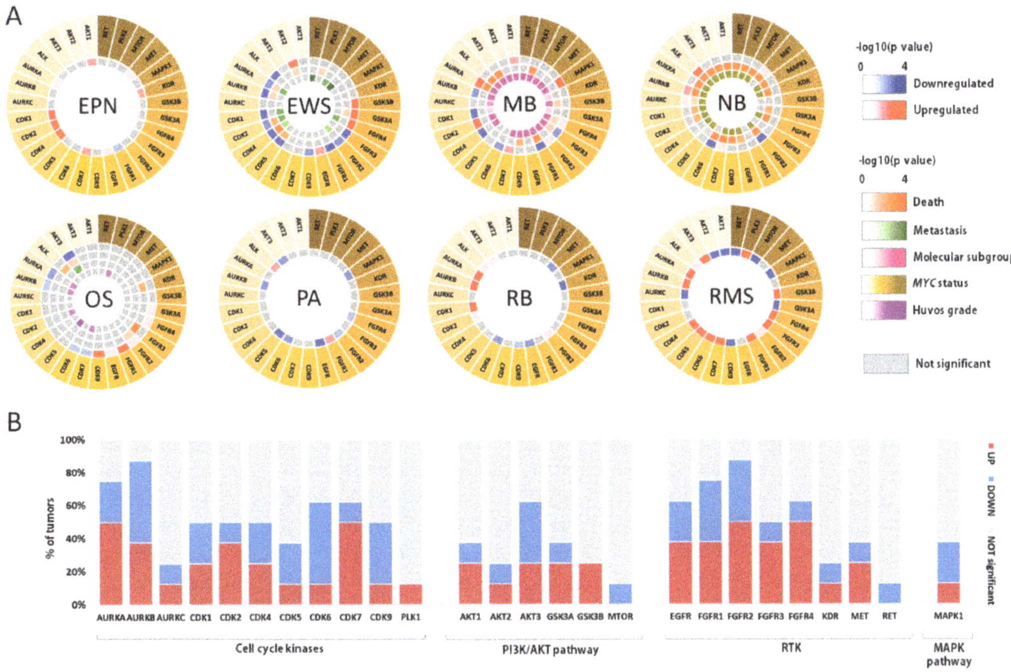

Figure 3. (**A**) Polar plots of differentially expressed kinases in pediatric tumors obtained through the analysis of available data on the R2: Genomics Analysis and Visualization Platform (http://r2.amc.nl (accessed on 15 October 2022)). Tumor abbreviations: EPN—ependymoma; EWS—Ewing sarcoma; MB—medulloblastoma; NB—neuroblastoma; OS—osteosarcoma; PA—pilocytic astrocytoma; RB—retinoblastoma; RMS—rhabdomyosarcoma. *p*-values are represented by differential coloring gradients. The external inner circle corresponds with "tumor versus normal tissue" results. The other concentric layers represent data related to associations with clinical features: metastasis, death, molecular subgroup (MB), MYC status (NB) and Huvos grade (OS). For actual *p*-values, refer to Supplementary Table S2. (**B**) Percentage of tumors with altered expression of each kinase. Few commonalities were found.

4. Kinases as Druggable Targets—Evidence and Limitations

The gradual advancements in genetics and biochemistry during the second half of the last century not only contributed to the better understanding of molecular events underneath signaling pathways in both natural and pathological settings, but also laid the foundation for the development of modern targeted agents. Perhaps the most expressive example of that trajectory involves chronic myeloid leukemia and the "Philadelphia chromosome". After its simple description (250 words) by Nowell and Hungerford, it took a decade to properly identify the chromosome pairs involved in the translocation [573]. It was only after the introduction of the G-bands by Marina Seabright that Janet D. Rowley from the University of Chicago that it was possible to identify the little chromosome as a result of the reciprocal translocation between chromosomes 9 and 22, specifically, t(9;22)(q34;q11) [574,575]. However, its molecular characterization only came to light between 1982 and 1984 [576–578], demonstrating the in-frame juxtaposition of the *ABL* oncogene (on chromosome 9) with the *BCR* gene (on chromosome 22), resulting in the hybrid *BCR/ABL* gene that gives rise to a chimeric protein with high tyrosine kinase activity and with a critical role in the development of leukemia [579]. Later, the discovery of this tumor-specific protein led to development of imatinib mesylate, providing an incredibly successful treatment that converted a fatal cancer into a manageable chronic condition, and pioneered an era of target-directed therapy [580].

More recently, the emergence of integrative laboratorial methods such as kinome-wide siRNA screens, next-generation sequencing (NGS) and phosphoproteomics have dramatically intensified the assortment of kinase inhibitors for the treatment of human cancers, currently accounting for about a quarter of all drug discovery research and development efforts.

Moreover, the increasing number of databases and analytical and visualization tools has facilitated advanced drug discovery not only by gathering information about the prognostic value of specific genes in oncology, but also it is now possible to access chemical structures and docking, affinities and structural features of approved small-molecule inhibitors in more easily, accessible and systematic ways, thus accelerating the discovery and optimizing screening to more direct translational assays.

In this regard, to further illustrate the importance of the selected kinases' dysregulation in the pathophysiology of pediatric cancer, other bioinformatic tools were used (Supplementary Figure S2). As a first step, we analyzed the vulnerability of different cancer cell lines against their inhibition through the Cancer Dependency Map portal (http://depmap.org accessed on 14 December 2022), a platform that provides information about how dependent different cell lines are on a specific gene depletion based on CRISPR and RNAi knockout experiments. The results are presented as a score generated by the platform itself: greater than zero (>0) indicates that the cell line is not dependent, less than zero (<0) indicates that the lineage is dependent and scores below -1 indicate that the analyzed gene is essential for the survival of the cell lineage.

Initial screening showed that all or most cell lines are dependent on the kinases analyzed, with comparable scores between adult or pediatric origins (Figure 4; Supplementary Table S3). Interestingly, a similar pattern occurs across the different tumors, irrespective of histology. Almost 100% of the cell lines are highly dependent on cell cycle kinases, especially AURKA/B AURKB, cyclin-dependent kinases CDK1 and CDK7/9, and PLK1. Cell lines were also highly dependent on mTOR. Conversely, cell lines were less vulnerable to the depletion of AURKC, AKT3, GSK3A and FGFR3, with more than 50% of cell lines presenting scores above 1 (in line with published data reviewed above).

Then, aiming to further exemplify information on individual kinase-targeted compounds and their analogs, we performed a search through the CanSAR knowledgebase (http://cansar.icr.ac.uk accessed on 14 December 2022), an integrative platform that compiles multidisciplinary data and provides useful drug discovery predictions. The analysis of predicted compounds for our selected group of kinases showed more than 55,000 potential compounds that are able to target RTK, with EGFR and VEGFR representing the most druggable ones. As shown in Figure 5, FDA-approved drugs are already identified for all the RTKs, and more than 20 additional drugs are being studied as novel clinical candidates.

For PI3K, although the CanSAR analysis revealed more than 20,000 potential compounds, only the mTOR inhibitors perhexiline and everolimus are described as FDA-approved drugs. Nonetheless, 15, 4 and 2 clinical candidates are being investigated as specific inhibitors for mTOR, AKT1 and GSK3B, respectively (Supplementary Figure S3). Approved drugs for MAPK are also scarce, including only sorafenib, regorafenib, dabrafenib and encorafenib, all of which target RAF1. For this specific kinase, CanSAR identified 4764 promising compounds and three clinical candidates (Supplementary Figure S3). Among the cell cycle kinases, CanSAR identified only approved drugs targeting CDK4 and CDK6. However, many clinical candidates (more than 15) targeting the other kinases of this group are being studied. According to the platform, cell cycle kinases are the second druggable category of kinases with the highest number of compounds described as potential specific drugs, totaling more than 48,000 (Supplementary Figure S3).

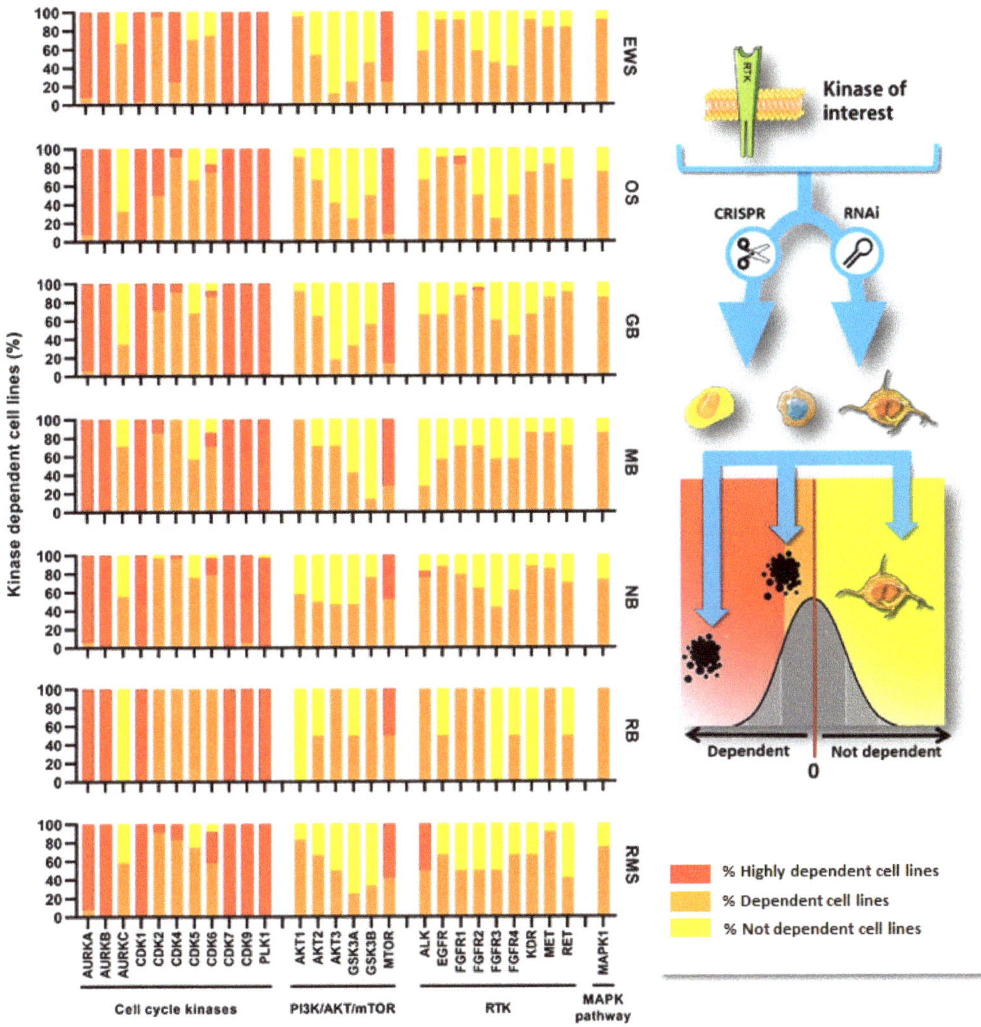

Figure 4. Percentage of pediatric cell lines dependent on the selected group of kinases. Dependency data were imported from the DepMap consortium (CRISPR (DepMap 22Q2 Public + Score, Chronos; https://depmap.org/ portal/ (accessed on 30 October 2022)) and classified as highly dependent, dependent or not dependent. The data were plotted on a histogram where it is possible to see the vulnerability of pediatric cell lines mainly to cell cycle kinases and PI3K/AKT/mTOR families. Cell lines selected included Ewing sarcoma (EWS), osteosarcoma (OS), glioma (GB), medulloblastoma (MB), neuroblastoma (NB), neuroblastoma (NB), retinoblastoma (RB) and rhabdomyosarcoma (RMS). Dependency scores for each cell line are shown in Supplementary Table S3.

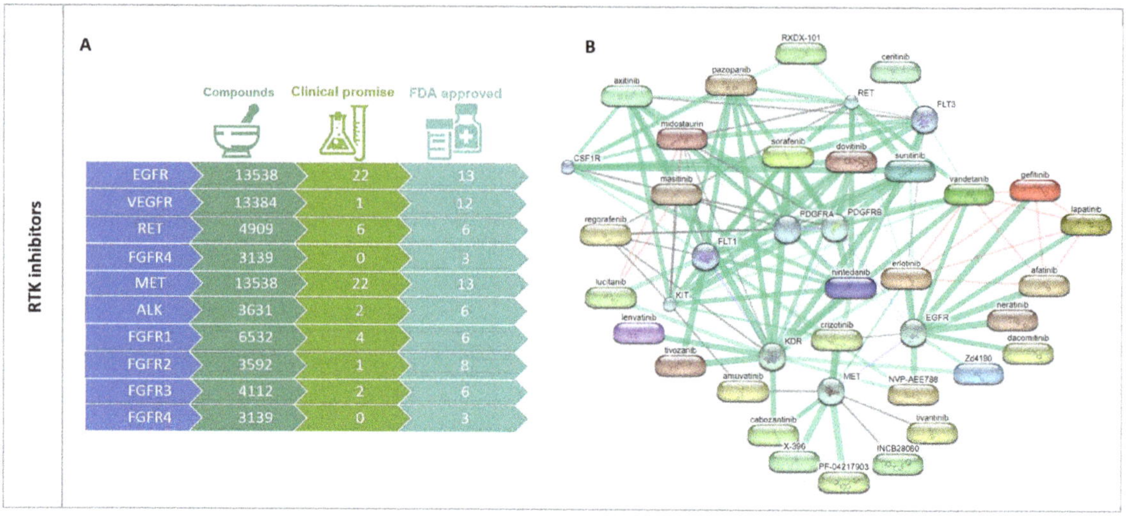

Figure 5. (**A**) Schematic illustrations of RTK druggability identified by the CanSAR database, including the total number of compounds with predicted interaction capacity with each kinase, as well as FDA-approved drugs and clinical candidates. (**B**) Interaction networks of RTK inhibitors and associated binding proteins according to STITCH (search tool for known and predicted interactions between chemicals and proteins available at http://stitch.embl.de (accessed on 1 November 2022)). Compounds are represented as pill-shaped nodes, while proteins are shown as spheres. Small nodes represent proteins of unknown 3D structures, while large nodes show proteins with known or predicted structures. Nodes that are associated with each other are linked by an edge: thicker lines represent stronger binding affinities. Networks were constructed considering a minimum required interaction score of 0.700, and based on associations reported in curated databases (gray lines), or on both databases and experimental/biochemical data (green lines). Purple lines represent functional links between proteins.

Further, for compilation of the preclinical results on kinase inhibitors, and considering that information in the literature often appears scattered and fragmented, the following section shows published evidence for (1) solid tumors that showed differential gene expression through in silico analysis (refer to Figure 3), (2) compounds considered "FDA approved" or with "clinical promise" and (3) compounds that have been assessed in vitro or in vivo before entering clinical trials. Thus, in the following subsections, experimental data on individual kinase inhibitors are detailed within each category. For further information, see Supplementary Table S4.

4.1. RTK Inhibitors

Erlotinib (Tarceva®). This compound is a quinazoline derivative that selectively and reversibly inhibits EGFR [581]. Erlotinib is an FDA-approved drug for the treatment of NSCLC and pancreatic cancer in combination with gemcitabine chemotherapy [582]. In the pediatric setting, this inhibitor has shown contrasting results. As monotherapy, Erlotinib was not efficient in reducing cell growth in a panel of NB cell lines, albeit effective indirect responses were obtained in xenograft tumors [583]. Similar results were obtained in sarcomas. This compound alone was ineffective in OS cells, in which the STAT3 cascade pathway has been pointed as the molecular mediator of both intrinsic and acquired resistance. In EWS, this compound alone or in combination did not inhibit growth of tumor xenografts and even led to a decrease in the therapeutic activity of cyclophosphamide when compared to single-agent activity [584,585]. Erlotinib had no effect on tumor progression in genetically engineered ARMS mouse models [586].

In contrast, satisfactory results were obtained in CNS tumors, where erlotinib therapy inhibited MB migration in vitro and successfully diminished the levels of phosphorylated EGFR in EPN models [587,588]. Treatment with this drug was also cytotoxic in Y79 and WERI RB cells in a dose-dependent manner, leading to cell cycle arrest and reduced migration, while oral administration dramatically reduced the growth of Y79 tumor grafts [589]. Regarding patients, phase I clinical studies have been developed in order to evaluate the acceptable tolerability profile in cases of brain and refractory solid tumors including RMS, soft tissue sarcomas, NB or germ cell tumor. Children appeared to tolerate erlotinib similarly to adult patients, and drug disposition was similar between these populations. The combination of temozolomide and erlotinib was well tolerated and it was also suggested in combination with radiotherapy [590,591].

Vandetanib (Caprelsa®). This is a multitargeted tyrosine kinase inhibitor with potent effects against VEGFR2/3, EGFR and RET [592,593]. This compound is approved to treat medullary thyroid cancer that cannot be removed by surgery and is locally advanced or has metastasized, and has demonstrated modest efficacy in patients with metastatic breast cancer [594,595]. Regarding pediatric tumors, vandetanib has been shown to inhibit the proliferation of NB cells mediated by the induction of G1-phase cell cycle arrest at lower concentrations and by apoptosis at higher concentrations. Migration and invasion were also markedly decreased compared with the control group [596,597]. Treatment also decreased (p)RET expression in five other NB cell lines and strongly impaired tumor growth in vivo in both MYCN/KI AlkR1279Q and MYCN/KI AlkF1178L mice, and was able to sensitize cisplatin-resistant NB subcutaneous tumor growth with less severe liver toxicity compared with high-dose cisplatin [598–600]. Moreover, vandetanib, in combination with 13-cis-retinoic acid, reduced tumor vascularity and induced apoptosis in NB xenografts [601]. Indeed, according to Craveiro et al. (2017), the narrow target spectrum of Vandetanib along with a favorable toxicity profile makes this drug ideal for multimodal treatment approaches. These authors tested this compound against SHH-TP53-mutated and MYC-amplified MB cell lines and found that it leads to a dose-dependent reduction in cell viability, interferes with clonogenicity and has pro-apoptotic effects after 48 h. Of note, combinations with GDC-0941 (clinically available PI3K inhibitor) and etoposide resulted in complete loss of cell viability [602]. The combination of vandetanib and celecoxib displayed a synergistic or additive antitumor effect on OS in vitro and in vivo [603]. However, combinations of gefitinib and vandetanib only inhibited the proliferation of EWS cell lines at very high concentrations (>1 μM vandetanib, >5 μM gefitinib), indicating the action on off-target effects [604].

Gefitinib (Iressa®). This drug, also known as ZD1839, is a member of the 4-anilinoquinazoline class of compounds that specifically and selectively inhibits EGFR [605]. In preclinical studies, gefitinib treatment was associated with growth inhibition and increased apoptosis in human cancer cell lines, and antitumor effects against xenografts of human tumors [606,607]. Gefitinib was shown to inhibit proliferation in juvenile PA primary cell cultures with an IC_{50} determined between 1.6 and 9.6 μM [608]. In addition, this compound was shown to inhibit invasion and metastasis of intratibial OS xenografts via inhibition of macrophage receptor interacting serine-threonine kinase 2 (RIPK2) [609]. Moreover, in children, Gefitinib has been tested for refractory solid tumors and CNS malignancies, showing similar pharmacokinetics as in adults [610].

Regorafenib (Stivarga®). This drug is an oral multikinase inhibitor that targets VEGFR1/3, FGFR and other receptor kinases [611]. This compound is already approved to treat metastatic cases of colorectal cancer and advanced hepatocellular carcinoma (HCC) previously treated with Sorafenib [610]. In vitro, regorafenib exhibits antiproliferative effects against a panel of 33 pediatric tumor cell lines, including MB (D341 Med, Med-Meb-8A), OS (IOR-OS-18), EWS (EW7, ORS, POE, SIM, STA-ET-1) NB and (SJ-NB-8, SK-N-BE(2), SH-SY5Y), with a mean half maximal growth inhibition of 12.5 μmol/L [612]. Particularly in the last, Regorafenib has shown to be effective against through the inhibition of RAS/MAPK, PI3K/Akt/mTOR and Fos/Jun pathways [612]. Similarly, regorafenib demonstrated antitumor activity in animals bearing subcutaneous RMS, EWS (STA-ET-1 and EW7) and NB (SJ-N-B8 and SK-N-AS) xenografts with tumor growth inhibition ranging from 73% to 93%. Moreover, when associated with radiation and irinotecan, it induced 100% regression in an MB patient-derived xenografts (PDX) model [610].

Dacomitinib (Vizimpro®). Also known as PF-00299804, this drug is an orally administered, second-generation, irreversible inhibitor of EGFR, HER2 and HER4, which has shown positive anticancer activities in some preclinical and clinical trials, being approved by the FDA for the treatment of metastatic NSCLC [613,614]. Besides its ATP-competitive action, dacomitinib covalently binds to Cys773 located in the ATP-binding cleft of EGFR, which irreversibly blocks ATP binding and inactivates the receptor [615]. For pediatric MB, dacomitinib has shown to block EGFR/HER signaling in DAOY cells and in orthotopic xenografts, extending median survival as a single agent; however, it was antagonistic when used in combination with standard frontline chemotherapy (4HPC, vincristine or cisplatin) [616].

Lapatinib (Tykerb®). This compound is an oral dual tyrosine kinase inhibitor that inhibits human EGFR and blocks the EGF receptor 2 (HER2) [617]. It was FDA-approved to treat HER2-positive advanced or metastatic breast cancer as monotherapy or in combination with other drugs [618]. Lapatinib has been tested in childhood solid tumors (including RMS, EWS and NB) and leukemia cells by the NCI-supported Pediatric Preclinical Testing Program (PPTP) [619]. In this study, among 23 cell lines, fifteen achieved at least 50% growth inhibition, and the median IC_{50} value for lapatinib against the entire cell line panel was 6.84 μM (range 2.08 μM to >10.0 μM). In vivo, however, lapatinib presented little activity against the 41 xenograft models of pediatric tumors [619,620].

Cetuximab (Erbitux®). This compound, available as Erbitux® (Merck Sereno), is a human–murine chimeric monoclonal antibody that competes to bind to the extracellular domain of EGFR and has been approved for the treatment of colorectal and head and neck cancer [620]. Information about preclinical use of this compound is scarce. However, a report showed that the proliferation of RMS cell lines was not influenced by this EGFR inhibitor [621]. However, later, it was shown that the combination of cetuximab and actinomycin D was highly effective in EGFR-positive RMS cells (RD and Rh30, of embryonal and alveolar origin, respectively), synergistically inhibiting cell growth and inducing apoptosis [622].

Sunitinib: Sold under the brand name Sutent®, this drug is a small-molecule multitarget inhibitor functioning on PDGFR, VEGFR, KIT, Flt-3 and RET [623–625]. In a preclinical study, this drug demonstrated limited growth inhibitory effects in a panel of 23 pediatric cell lines that included OS, ALL, EWS, RMS, MB, EPN, NR, GBM, WT and others [625]. However, in vivo, it presented growth inhibitory activity against pediatric solid xenograft models of EWS, RMS and NB [625]. A later study showed decreased cell proliferation and phosphorylation of VEGFRs NB cells after treatment with sunitinib, and tumor growth, angiogenesis and metastasis in tumor xenograft models [624]. Moreover, in combination with an mTOR inhibitor (rapamycin), it showed a synergic cytotoxic effect, which was more effective than the traditional chemotherapeutic agent cyclophosphamide [624].

Lenvatinib (Lenvima®). This drug is a synthetic, orally available type I tyrosine kinase inhibitor exhibiting powerful antiangiogenic activity currently used to treat certain types of thyroid cancer and potentially other tumor types [626]. Lenvatinib was initially reported in 2008 as a multitargeted RTK inhibitor of VEGFR1/2/3, but it also inhibits FGFR1–4, PDGFR-alpha and KIT [627–629]. Preclinical findings in sarcomas indicated that lenvatinib was able to inhibit tumor growth in xenografts obtained through direct implantation of patient tumor specimens in nude mice. The experiment showed positive results in 7 out of 10 xenografts accompanied by marked decrease in microvessel densities. However, in vitro, Lenvatinib did not show potent effects on tumor viability in OS-derived cell lines [629]. Others showed that the drug was able to inhibit tumor cell migration and invasion in U2OS cells [630]. Further, in a phase I/II study, lenvatinib as a single-agent reported a response rate of 7% and a median progression-free survival of 3 months in a cohort of 31 children and young adults with OS, although many patients had treatment-related adverse events of grade ≥ 3 [631]. In other pediatric tumors, the effects of lenvatinib remain to be investigated.

Pazopanib (Votrient®). This compound is an FDA-approved pan-VGFR inhibitor, even though it also targets PDGFR-α and -β, FGFR1/3, KIT as well as BRAF proteins [632]. In a pan-cancer study, pazopanib was unable to affect the viability of the any of the treated cell lines, which included SK-N-BE(2) (N-Myc amplified) and SH-SY5Y (non-N-Myc amplified) NB cell lines, the KHOS OS cell line, and the RMS cell lines RH30 and RD. However, in combination with topotecan, this compound showed significant antitumor activity in vitro and halted tumor growth in NB xenograft-bearing mice, but after 50 days, gradual growth was observed [633,634]. The combination of pazopanib with trametinib showed antitumor effects in vitro and in vivo against a panel of seven OS cell lines, in which treatment reduced proliferation and colony-forming capacity and increased the percentage of apoptotic and dead cells. In MNNG/HOS and KHOS xenograft models, both drugs induced a significant inhibition of tumor growth compared to the untreated controls [635]. The in vivo antitumor activity of pazopanib was also tested by the PPTP Program in a subset of sarcoma models that also included EWS and RMS. Although objective responses were not observed for any of the sarcoma xenografts studied, treatment prolonged survival [636]. Even with modest benefits, pazopanib has been approved for line treatment of metastatic non-adipocytic soft tissue sarcomas after the failure of standard chemotherapy. Its efficacy in patients with OS is limited to case reports [637]. One metastatic extraosseous EWS was also reported as successful after treatment with pazopanib [638].

Regarding CNS tumors, EPN cells showed to be sensitive to Pazopanib with a viability reduction of around 35% at 1 µmol/L [639]. Additionally, treatment with Pazopanib reduced the mobility of MB cell lines, inducing clumping of the actin microfilaments (which facilitated cell detachment), as detected by wound healing assays and Fluor-555-coupled phalloidin [640]. Further in vivo tests demonstrated delayed growth of group-3-MB cells transplanted into the cerebellum of mice and prolonged survival (by 10 days) of mice treated once daily by gavage with 60 mg/kg compared to untreated controls [641]. Alternatively, for patients with recurrent high-grade gliomas as part of phase I or II clinical trials, this drug has not been beneficial [642].

Cabozantinib (Cometriq®). This compound is an orally available multitarget tyrosine kinase inhibitor that inhibits VEGFR1/2/3, MET, KIT, RET, AXL and FLT3. FDA-approved since 2012, it is currently used to treat metastatic medullary thyroid cancer, RCC, HCC and differentiated thyroid cancer [643,644]. In preclinical studies, reports of its anticancer effects include the inhibition of metastasis, angiogenesis and tumor growth [645–647]. In vitro, cabozantinib has been shown to diminish the cell viability of EWS and OS cells in a dose-dependent manner [160]. Moreover, it also interferes with migration and the microenvironment by inducing the production of osteoprotegerin and causing a decrease in the synthesis of the RANK ligand by osteoblasts [648]. Positive effects on decreasing proliferation were also observed in MB with no differences between cell lines corresponding to different molecular subgroups [649]. Cabozantinib also exhibited anti-proliferative effects in NB cells and reduced cell migration in vitro and significantly inhibited tumor growth of orthotopic xenografts on a daily basis [650].

Nintedanib (Ofev®). This drug, commercially available under the brand names Ofev and Vargatefi, is an indolinone-derived inhibitor of multiple kinases including VEGFR, FGFR and PDGFR. Recently approved for the treatment of idiopathic pulmonary fibrosis and advanced non-small cell cancer of adenocarcinoma tumor histology, it exerts its antitumoral activity by reducing proliferation, migration and angiogenesis [651–653]. Considering pediatric tumors, nintedanib has been shown to inhibit growth in EWS (A673, CHP100) and OS (SaOS2) cell lines, with a key role in controlling OS lung metastatic growth by blocking the fibrogenic reprogramming of OS stem cells (OSCs) [654,655]. Growth inhibition was also observed in a panel of 13 RMS cells, with the PAX3–FOXO1 fusion-gene-positive ones more sensitive to treatment [656]. Moreover, there are reports of EPN cells being sensitive to nintedanib treatment, while this drug is able to extend the survival of mice bearing ST-RELA xenografts [657,658].

Midostaurin (Rydapt/Tauritmo®). Also known as PKC412 and CGP 41251, this small molecule acts as a multikinase inhibitor targeting PKCα/β/γ, Syk, Flk-1, Akt, PKA, c-Kit, c-Fgr, c-Src, FLT3, PDFRβ and VEGFR1/2. Presenting anticancer roles in vitro and in vivo, it is currently approved for the treatment of acute myeloid leukemia and advanced systemic mastocytosis [659,660]. Midostaurin has been shown to be an efficient anti-sarcoma agent. Indeed, it inhibited EWS cell proliferation in a dose- and time-dependent manner and decreased tumor growth in vivo [661,662]. Moreover, the combination of midostaurin with the cytokine oncostatin M has been shown to be efficient in reducing in vivo tumors, pointing to this combination as a potential adjuvant treatment for OS [663].

Axitinib (Inlyta®). Also known as AG-013736 this is an oral VGEFR1/3 and PDGFR inhibitor explored to control angiogenesis [664]. Currently, this compound is approved for treatment as monotherapy or in combination with other drugs for renal carcinoma, and is under phase I, II and III clinical trials for many other tumor types [664]. Pre-clinical studies in EPN showed that this drug inhibited PDGFRα and PDGFRβ, and reduced the expression of mitosis-related genes including *ASF1B*, *MKI67*, *HMGA1*, *BRCA2*, *ESPL1*, *TACC3*, *CDC25A*, *RAD51AP1*, *AURKA*, *BUB1B*, *CENPE* and *HELLS*. It also decreased proliferation resulting from cellular senescence [639]. Similar antiproliferative effects were observed in MB 2D and 3D cell cultures, without affecting normal brain cells. Of note, the compound efficiently crossed the blood–brain barrier (BBB), reducing growth rates of experimental brain tumors without acute toxicity in juvenile rats [649]. In GBM, the cytotoxic activity of Axitinib was also reported in vitro and in vivo, characterized by an anti-angiogenic effect and survival prolongation [665]. Moreover, combinations of axitinib and other therapeutic targets have been explored with satisfactory results [666]. Indeed, the combinatorial treatment of Axitinib and PLK4 inhibitor has shown to be beneficial in MB and RMS [667]. Combinations with etoposide or gemcitabine also showed favorable effects on preventing tumor progression in an orthotopic group-3-MB xenograft models [649,668]. Furthermore, in immunodeficient and immunocompetent orthotopic GBM models, axitinib + G47Δ-mIL12 resulted in an extensive decrease in vascularity, increased macrophage infiltration and significant tumor necrosis [669]. Such a antimetastatic effect was also observed in NB [670].

Ramucirumab (Cyramza®). This is a humanized monoclonal antibody that acts by binding to VEGFR-2, thus limiting angiogenesis and the proliferation and migration of human endothelial cells [671]. Preclinical studies in NB, RB, OS, RMS and EWS have also shown that ramacirumab enhances anti-tumor activity by abrogating endothelial cord formation, while in vivo, it has also induced tumor growth delay. However, modest or no response was observed in OS [672]. This compound was approved by the FDA in 2014 and indicated for the treatment of gastric cancer, NSCLC, colorectal cancer and HCC [673–675]. As a well-tolerated drug, its combinatorial use was also approved, even though its use in clinics is limited due to a lack of specific markers and high costs [676].

Alectinib (Alecensa®). Also known as CH5424802, this is an orally available selective ALK inhibitor already approved by the FDA for lung cancer treatment [677]. The compound is able to bind wild-type ALK and its fusions and its anticancer effects have been widely described. Noteworthy, it has shown acceptable results after treatment of intracranial EML4-ALK-positive tumors in rats with a high brain-to-plasma ratio, and permeability independent of P-glycoprotein transport [678]. Moreover, despite heterogeneous intratumoral distribution, alectinib delayed tumor growth in an NB mouse model, leading to increased survival, providing an option for future clinical treatment [679–681]. An interesting point in this regard is that Alectinib may improve sensitivity to chemotherapeutic since it increases the intracellular accumulation of ABCB1/ABCG2 substrates such as doxorubicin (DOX) and rhodamine [682]. Moreover, it has also shown effectiveness in combination with the histone deacetylase inhibitor vorinostat in NB harboring the ALK R1275Q mutation and after intensive radiotherapy for the treatment of a rare intraosseous RMS with FUS-TFCP2 fusion, evidencing the potential of this drug to treat extremely aggressive tumors [683,684].

AEE-788. This drug is an orally bioavailable bispecific EGFR/HER2 inhibitor that exerts significant anti-tumoral activities and radio-sensitizes EGFR-overexpressing cells [685]. By targeting this receptor, the compound efficiently reduced clonogenicity, proliferation and survival of EPN cells and prolonged the survival of tumor-bearing mice, probably due to the increase in apoptosis of endothelial cells (as shown by others in cutaneous cancer xenografts) [686,687]. AEE788 also inhibited cell proliferation and prevented epidermal growth factor- and neuregulin-induced HER1, HER2 and HER3 activation in chemosensitive and chemoresistant (cisplatin selected) MB cells in vitro and in vivo [688].

Crizotinib (Xalkori®). This drug is an orally available aminopyridine-based ATP-competitive inhibitor of ALK that has shown positive results against NSCLC [689]. In turn, in pediatric tumors, growth-suppressive activities have been reported in some tumor types, such as PA, EPN, EWS and MB [160,690,691]. This drug was also able to induce apoptosis and autophagy in a dose-dependent manner in RMS cells, reducing cell migration and invasion, as well [692]. However, despite these promising results, this compound lacks clinical significance in patients with FOXO1-rearranged ARMS [693]. Similarly, crizotinib responses in NB are variable and mostly dependent on the mutation variants present in the tumor, considerably limiting its applicability [694–696]. Moreover, the literature widely illustrates that despite initial effectiveness, the vast majority of tumors treated with this compound will develop resistance within a few years [697].

Capmatinib (Tabrecta®). This compound is an orally bioavailable inhibitor of c-MET [597]. The information about the effects of this compound in pediatric oncology is limited. There are reports of its action in pediatric HGG in which this compound appeared to be more efficient than crizotinib in terms of specificity, potency and brain availability, resulting in a higher cellular response compared to crizotinib treatment in vitro and in vivo [698]. Nevertheless, in a phase I dose escalation study that included EWS and OS patients, only mild responses were observed [699].

Tepotinib (Tepmetko®). This compound is a phenylmethyl-pyrimidine derivative developed to disrupt MET phosphorylation that received approval from the FDA and the Japanese Ministry of Health, Labour and Welfare for the treatment of patients with metastatic NSCLC harboring METex14 skipping alterations who progressed following platinum-based cancer therapy [700]. According to PubChem (CID 25171648), this compound has been investigated in the treatment of neuroblastoma.

PF-04217903. This compound is an ATP-competitive small-molecule inhibitor with 1000-fold selectivity for c-MET compared with more than 150 kinases, making it one of the most selective c-MET inhibitors described to date. In vitro, it inhibited tumor cell proliferation, survival and migration/invasion in cell lines where c-MET is activated by different mechanisms, including c-MET gene amplification, HGF/c-MET autocrine loop formation and c-MET overexpression [701]. In vivo, oral administration or subcutaneous minipump infusions led to a robust tumor growth inhibition at doses of 30 mg/kg with suitable tolerability. Reductions in microvessel density were also observed [701]. Considering pediatric tumors, similar results were obtained when two highly metastatic OS cell lines were injected by tail vein into immunodeficient mice. In this experiment, mice were treated with PF-04217903 (30 mg/kg) or vehicle control by gavage 5 days on and 2 days off for 30 days. Mice injected with MNNG-HOS cells (which has constitutively activated MET) treated with PF- 04217903 had a tenfold reduction in the number of metastatic nodules, while those with injected MG63.2-derived tumors (which have high levels of total and phospho-MET) had a 37% reduction in nodules compared to control mice [702]. This compound has also shown potential for the treatment of malignant peripheral nerve sheath tumors in NF1 patients [703].

Tivantinib. This compound, also known as ARQ 197, was described as an orally bioavailable small-molecule c-MET inhibitor with antitumor activity. Tivantinib inhibited cell viability with similar potency in both c-MET-addicted and nonaddicted adult carcinoma cells, pointing to alternative mechanisms of action [704]. Despite this, the failure of a pioneer a phase I clinical trial in pediatric tumors was attributed to the lack of selection for MET amplification during patient enrollment. In the study, which comprised 36 patients, including 4 glioma, 4 MB, 4 EPN, 4 EWS, 4 OS, 3 RMS, 2 WT and 2 NB, suboptimal responses were achieved when tivantinib was given with food to children with refractory solid tumors is 240 mg/m^2/dose. Moreover, while the drug was well tolerated, its pharmacokinetic profile was also variable, discouraging further investigation in this setting [705]. However, two of those patients (alveolar soft part sarcoma) who responded to tivantinib administration and were transitioned to a follow-up protocol experienced extended progression-free survival receiving 360 mg twice every day without adverse events [706].

Lorlatinib (Lorbrena®). This small molecule represents an orally available, ATP-competitive inhibitor developed by Johnson et al., and further investigated for the treatment of ALK-positive NSCLC [707]. Also named PF-06463922, the drug has shown minimal toxicity in adults and there has been much interest in its prospective use in NB treatment. In this regard, Infarnato et al. described higher potency of PF-06463922 across ALK variants in a panel of 10 NB cell lines, with IC_{50} values for inhibition of F1174L- and F1245C-mutated ALK significantly lower than those seen for its predecessor, crizotinib (0.2–10 nmol/L) [708]. Moreover, this compound at 10 mg/kg/day induced complete tumor regression in xenograft mouse models of NB, and in (PDX) harboring the crizotinib-resistant F1174L or F1245C mutations within 3 weeks [708]. Similar 10-fold lower IC_{50} values were obtained by Guan et al. (2016). In another group of cell lines, PF-06463922 inhibited growth, reduced levels of tyrosine 1278 (Y1278) phosphorylation on ALK, and induced apoptosis. Comparatively, treatment reduced tumor volume in subcutaneous and orthotopic xenograft models of NB, as well as in the Th-ALKF1174L/MYCN-driven transgenic NB mouse model [709]. PF-06463922 has also been tested sporadically in patients affected with NB. Two recent articles portray favorable responses in a 3-year-old boy with ALK-fusion-positive HGG and an adolescent with relapsed, refractory, metastatic ALK F1174L-mutated NB. The first, considering that the compound is able to cross the BBB, was treated through a nasogastric tube at a dose of 95 mg per square meter of body surface area once daily [710]. Histology after tumor resection showed a marked decrease in the proliferative index of the tumor and since the tumor was not seen on postsurgical MRI, therapy stopped. After 6 months, metastatic lesions were identified on cranial nerve VII and treatment was restarted at a dose of 95 mg per square meter administered by mouth once daily, achieving a near-complete response after 1 month [711]. In the second case, the patient had already shown no response to the first-generation ALK inhibitor crizotinib (240 mg/m2/dose given twice daily combined with the standard cytotoxic chemotherapy regimen). The tumor was reduced with continuous 95 mg/m2/dose lorlatinib and the only significant side effect observed was grade 2 hypercholesterolemia. However, differentially from the infant, she relapsed after 13 months of treatment and died from progressive disease 3 months later [712].

Ceritinib (Zykadia™). Formerly known as LDK378, it is an oral ALK inhibitor that also targets insulin-like growth factor receptor IGFR, insulin receptor and ROS1. This compound was approved by the FDA through an accelerated process to treat ALK-positive metastatic NSCLC [713]. Preclinical studies in the pediatric setting have indicated antiproliferative effects and improved inhibition (11-fold) compared to crizotinib [714]. However, in an exploratory study with a panel of NB cell lines, it was noted that inhibition occurs irrespective of ALK mutational status, and cell lines that carry other driver mutations (i.e., MYC amplification) are sensitive to treatment as well. The same authors further treated a child with ALK-I1171T high-risk NB that was not responding to conventional treatment due to an underlying congenital genetic condition, Fanconi anemia. Monotherapy with ceritinib was well tolerated and resulted in tumor shrinkage and complete clinical remission including all metastatic sites [715]. This compound can be given with food and penetrates the human brain, and thus presents itself as an option for the treatment of CNS tumors with ALK alterations such as EPN and MB [715–718]. However, in orthotopic PDX (from a 10-year-old boy with a multiple recurrent GBM), it was observed that even though ceritinib-treated mice lived longer, the drug had only a moderate effect [719]. Monotherapy was also inefficient in treating a 16-year-old patient with a long history of OS lung metastases, despite acceptable results in primary tumor cells of six other patients and the HOS cell line [720]. Similarly, Ceritinib treatment led to decreased cell proliferation, cell cycle arrest and apoptosis in a dose-dependent manner in a panel of RMS cell lines, all of which lack intrinsic ALK phosphorylation (PAX3-FOXO1-positive Rh30, Rh41 and -negative Rh18 and RD cell lines). The work showed that the compound affects the IGF1R signaling pathway without effects on the migratory ability of cells. Moreover, in subcutaneous Rh41 xenografts, a reduction in tumor growth was observed after approximately 2 weeks, albeit subsequent evaluation of tumor characteristics showed no difference in proliferation or vascularization between the treatment groups and controls [721]. Others also showed that even though LDK378 reduces cell viability and induces cell death in RMS cell lines at low micromolar concentrations irrespective of ALK expression levels or phosphorylation status, cells are far less sensitive compared with Karpas 299 non-Hodgkin's lymphoma cells carrying the NPM–ALK fusion gene [722].

Brigatinib (Alunbrig®). Originally named AP26113, this next-generation ALK inhibitor was first described in 2016 and is considered highly CNS-penetrant [723,724]. This compound was granted approval for the treatment of patients with metastatic ALK+ NSCLC and intolerance to crizotinib [725]. In an NB setting, preliminary indication of efficacy was observed after exposure of several NB cell lines, including CLB-BAR (MYCN amplification, ALK (Δ4-11) and amplified, ALK addicted), CLB-GE (MYCN amplification, ALK (F1174V) amplification, ALK addicted), IMR32 (MYCN amplification, WT ALK) and CLB-PE (MYCN amplified, WT ALK), in which treatment inhibited cell growth and ALK phosphorylation in a dose-dependent manner. However, while IC_{50} values varied between 75 and 100 nM in ALK-addicted cell lines, the compound was unable to inhibit growth of both non-ALK addicted NB cell lines, IMR32 and CLB-PE. The effects of brigatinib were further validated in vivo through two complementary models. The first used transgenic *Drosophila melanogaster* flies expressing two gain-of-function variants, F1174L and R1275Q, which disrupt the eye morphology, giving a "rough phenotype". The authors showed that larvae grown on food containing Brigatinib displayed a concentration-dependent improvement of the rough eye phenotype. Then, brigatinib was used as a single agent to treat BalbC/NUDE mice bearing ALK-addicted CLB-BAR xenografts. In this model, the compound also showed to be effective, with robust and potent anti-tumor activity [726].

Entrectinib (Rozlytrek®). This compound (also called RXDX-101, NMS-E628, NMS-01191372, Rozlytrek) is a selective, oral tyrosine pan-TRK, ALK and ROS1 inhibitor that has demonstrated preclinical efficacy in tumors with NTRK1/2/3, ALK and ROS1 alterations [727]. This inhibitor can pass through the BBB and has clinically proven to be effective against primary and metastatic brain diseases, with no adverse off-target activity [728].

Entrectinib also displays promising anti-tumor activity in NB, evinced by diminished Ki-67 and activation of caspase-3 in ALK wild-type, amplified or mutated cell lines [729]. In vivo growth inhibition and substantially reduced phosphorylation in TrkB-expressing NB xenografts were also observed after treatment as a single agent or in combination with irinotecan or temozolomide (TMZ), eliciting increased EFS when compared to controls [730]. Moreover, the ability of entrectinib to inhibit p-TrkB, p-PLCγ, p-Akt and p-Erk suggested that this compound may have improved efficacy compared to other targeted inhibitors previously evaluated in NB [172]. However, despite durable responses in pediatric patients with intracranial tumors or NB harboring NTRK1/2/3 or ROS1 fusions, its utility may be hampered by the appearance of acquired resistance in this tumor type [730,731].

X-396. This compound, also known as Ensartnib, is an aminopyridazine-based second-generation ALK/MET inhibitor that holds much clinical promise with increased potency as compared with crizotinib and other second-generation ALK inhibitors such as alectinib and ceritinib [732]. X-396 significantly reduced growth (by 40% at a 3 nM concentration) and ALK phosphorylation in SY5Y NB cells that harbor ALK-F1174L. Biochemical IC_{50} values for MET inhibition were 2-fold higher [733]. Ensartinib was significantly more effective than crizotinib at inhibiting the intracranial growth of the SH-SY5Y NB model harboring the F1174L mutation [732]. Furthermore, the activity of X-396 administered alone or in combination with liposomes carrying ALK-siRNAs (that are active irrespective of ALK gene mutational status) was later tested in a mouse model by Di Paolo et al. (2011). These authors corroborated previous in vitro data with a second NB cell line (LAN-5) and showed that in subcutaneous NB models, the compound acted in a dose-dependent manner, with adequate bioavailability, moderate half-life, high mean plasma and tumor concentrations. Moreover, against human NB orthotopic xenografts obtained by implanting of Luciferase stably transduced NB cells, SH-SY5Y-Luc and LAN-5-Luc, into the adrenal gland of nu/nu mice, significant dose-dependent anti-tumor activity was also observed, with even more reduced tumors and prolonged survival with the combination with the liposomal formulation [734].

Erdafitinib (Balversa™). This compound is an oral pan-FGFR inhibitor with quinoxaline structure [735]. Known as JNJ-42756493, this compound is already approved by the FDA for the treatment of advanced or metastatic urothelial carcinoma, and is now under clinical trials that also include childhood CNS tumors [736]. It inhibits FGFR1/2/3/4 with increasing IC_{50} values of 1.2, 2.5, 3.0 and 5.7 nM, respectively [737]. This compound inhibited proliferation on five different NB cell lines (SK-N-AS, SK-N-BE(2)-C, SK-N-DZ, SK-N-FI and SK-N-SH) as monotherapy, but showed variable synergistic, additive and antagonistic effects when combined with commonly used cytotoxic agents such as cisplatin, vincristine and doxorubicin [738]. Additionally, IC_{50} for FGFR4 inhibition by this compound on the A-204 RMS cell line was determined as 4.5 nM, while treatment of mice xenografts resulted in a 58% volume reduction after 21 days of treatment with daily doses of 30 mg/kg [735].

Erdafitinib has also been tested alone and in combination with cisplatin, vincristine and radiotherapy on the SHH-MB cell lines DAOY and UW228-3. Under all conditions, the cell lines showed dose-dependent decreases in viability and proliferation after 48 and 72 h [739].

Dovitinib. Also known as TKI258, this is a multi-targeted tyrosine kinase inhibitor with potent activity against FGFR1/3, VEGFR1/2/3 and to different extents, PDGFR-β, Flt3, c-Kit and CSF-1R, that showed promising results as an antitumoral and antiangiogenic compound [658]. This compound is already in clinical trials in adult patients [740]. However, in pediatric neoplasms, information about preclinical studies is limited. Preliminary results of Dovitinib in NB cells, which express high levels of FGFR, indicated anticancer-activity in this tumor type [741]. Similar results were reported for RMS, albeit it was demonstrated that this inhibitor is not as potent as other FGFR inhibitors (i.e., ponatinib) [656,742]. In addition, due to its ability to cross the BBB, this compound has been indicated as a suitable candidate for the treatment of CNS tumors. In this regard, in vitro, it reduced the capacity of EPN cells to re-adhere and proliferate in a dose-dependent manner [658]. In DIPG and GBM, however, true effects on viability were observed at high dovitinib concentrations (>400 nM) [743]. Of note, others showed that despite killing glioma cells in vitro (up to 55% of cells at the assay end point), the drug exerted minimal anti-tumoral effects in vivo, suggesting a microenvironment-mediated therapeutic resistance mechanism [744].

Masitinib (Masivet®). This compound, also known as AB1010, is an orally administered, novel, potent and selective phenyl aminothiazole-type tyrosine kinase inhibitor of KIT, used in the treatment of canine mast cell tumors acting as a blocker of mast cell degranulation, cytokine production and migration of bone marrow cells [745,746]. Masitinib is under clinical investigation in several human malignancies that harbor similar canine KIT mutations (i.e., gastro-intestinal stromal tumors, ovarian and prostate cancer). In fact, this inhibitor acts on several mutated forms of KIT, and other receptors, including PDGFR, FGFR3 and focal adhesion kinase (FAK) [747,748]. Noteworthy, a brain tumor xenograft model using pediatric GBM cells suggested that masitinib may potentiate the effects of TMZ, providing decreased tumor growth relative to either drug used as a monotherapy [749].

4.2. PI3K/AKT/mTOR Pathway Inhibitors

Everolimus (Afinitor®). Everolimus (Afinitor, Novartis) is an orally administered rapamycin derivative approved by the FDA and the European Medicines Agency for the treatment of RCC [749]. This compound reduces tumor cell proliferation and induces apoptosis and autophagy through the phosphorylation inhibition of mTOR [750,751]. Preclinically, the combination of everolimus with sorafenib yielded enhanced antiproliferative and proapoptotic effects, potentiated antiangiogenesis and reduced the metastatic potential of OS [751]. Prolonged exposure to everolimus also improved the CNS retention of dasatinib and extended the survival of mice bearing pediatric high-grade glioma tumors [752]. Comparatively, everolimus is synergistic with carboplatin in low-grade glioma models [753]. However, in the literature, there is significantly more information about clinical experience because, since its approval, everolimus has become widely accepted by the medical community where treatment options may be limited. One major clinical example involves subependymal giant cell astrocytomas (SEGA), tumors that are frequently diagnosed in patients with tuberous sclerosis complex (TSC) [754]. Loss of function of either TSC1 or TSC2 leads to downstream constitutional activation of the mTOR complex [755]. Besides surgical excision, patients with large or recurring SEGAs did not have robust treatment options, and Everolimus has been shown to induce tumor shrinkage and presents additional clinical benefits including seizure control [756,757].

Palomid-529. Also known as RES-529, this compound is a small-molecule drug dual novel inhibitor of mTOR complex 1 (mTORC1) and mTOR complex 2 (mTORC2). Palomid 529 likewise inhibits both VEGF-driven and bFGF-driven endothelial cell proliferation [758]. Due to its potential to penetrate the BBB without restriction by the ABCB1 and ABCG2 efflux transporters, the anti-glioma effects of this drug have been investigated [759–761]. In childhood cancer, a single report proved a potent inhibition of viability, cell cycle progression and proliferation of the OS cell line U2OS [762].

OSI-027. This compound is an orally bioavailable selective ATP competitive inhibitor of mTOR and off-targeted PI3Kα (100-fold selectivity for mTOR relative to PI3Kα) that has been studied in the treatment of many tumors [763]. OSI-027 is active in vitro against cell lines and primary cells of pediatric pre-T-ALL, with superior efficacy to rapalogs and in vitro synergy with a number of conventional cytotoxic agents [181]. Preliminary studies combining OSI-027 treatment with alpelisib demonstrate similar antineoplastic results inhibiting PI3K/mTOR signaling MB, EWS and RMS cell lines [764–767].

VS-5584. This dual inhibitor of mTORC1/2 and class I PI3-kinases has shown antitumor potential in a broad spectrum of tumor types in vitro and in vivo [768–770]. Noteworthy, evidence supports that this compound has an active role in reducing stem cell viability in multiple mouse xenograft models of human cancer (30-fold more potent compared to non-stem cells) [769]. Thus, the activity of VS-5584 was recently explored in OS, in which treatment dramatically suppressed growth and cell migration and synergized with CCT128930, an AKT2 inhibitor [771–773]. In the same way, this drug is cytotoxic, showing apoptosis induction and a robust limitation of the colony-forming ability in NB cell lines. Delay of tumor growth was also observed in mice subcutaneously inoculated with BE(2)-M17 cells and treated with VS-5584 (25 mg/kg, three times per week) for 2 weeks [773].

Dactolisib (BEZ235). This drug, also called BEZ235 or NVP-BEZ235, is a reversible PI3K/mTOR inhibitor belonging to the imidazoquinoline class already tested in a variety of cancers in preclinical studies. In sarcomas (EWS, OS and RMS), dactolisib showed promising results in vitro, such as a reduction in cell proliferation, G1 cell cycle arrest and decreased in cell migration [774,775]. Interestingly, it was also shown that BEZ235 elicits strong cytostatic effects in EWS cells and results in a global modulation of the transcriptome affecting other pathways related to splicing and metabolism. This drug also reduced the expression of EWS/FLI1 by 50%, reinforcing its potential for EWS treatment [773]. However, its capacity to induce apoptosis is uncertain. Mild results were obtained by Giorgi et al. (2018), and when OS cells were treated with a similar inhibitor range, U2-OS and MG63 presented no significant differences in apoptosis induction, although the drug was efficacious with either doxorubicin or vincristine [774–776]. In RB, GBM, and MB, decreased viability and proliferation in a dose-dependent pattern was observed in most cell lines [777–780]. When tested in vivo, dactolisib could reduce tumor volume, vascularity and metastasis and improve animal survival, especially when combined with other drugs, such as topotecan, carboplatin, vincristine or the SMO inhibitor LDE225 [773,781–783].

SF-1126. This is a pan and dual first-in-class soluble PI3K/mTORC inhibitor that exhibits antitumor and antiangiogenic activity against several malignancies [784,785]. In the literature, there are few reports of this inhibitor in pediatric preclinical models. SF-1126 promoted a decrease in cell viability of a panel of EWS cell lines and CD15+ stem cell population in SHH-driven MB [786,787]. Moreover, SF1126 has been shown to enhance the cytotoxicity of doxorubicin in NB cells, leading to p53-mediated activation of apoptosis [788]. Treatment of NB tumors with SF1126 also reduced MYC expression and inhibited growth in vivo, leading to tumor shrinkage and reduced neovascularization [789].

Triciribine. Triciribine is a pan-AKT 1 inhibitor with anticancer effects in various tumor types [790]. This compound has been shown to decrease the survival of SH-SY5Y NB cells in both 2D and 3D culture models, affecting the migratory abilities of their sphere-forming units [791]. Triciribine demonstrated activity in EWS cell lines as well, with a mean IC50 of 24 μM, with robust synergy when combined with dasatinib; however, it did not affect tumor growth in vivo [792]. Moreover, Smeester and colleagues (2020) tested this drug in OS and showed that ATK inhibition in HOS and SJSA-1 cell lines leads to decreased cell proliferation, migration and colony capacity, and increased apoptosis. Further assessment in an orthotopic OS model also demonstrated reduced tumor growth (volume and weight) and metastasis after triciribine 40 mg/kg three times weekly [793].

Sapanisertib. Also called MLN0128, INK-128 or TAK-228, this is an ATP-competitive mTOR inhibitor already tested for safety in an adult cohort [794]. In studies including pediatric models, this inhibitor has shown promising results in vitro, reducing cell viability and colony formation and inducing apoptosis in sarcoma cells (EWS, OS and RMS) without affecting human osteoblast and osteocyte cells (normal bone cells); effects were improved by combination with MK2206, an AKT-specific inhibitor [795,796]. Similarly, the inhibition of mTOR (oral gavage for 21 days-3 mg/kg twice daily 3 x/week in EWS and RMS, or 2.5 mg/kg, daily) in OS resulted in tumor volume reduction, without observable side effects [795,796]. Comparable results were obtained for brain tumors (MB, NB and GB), with reduced cell invasion at low concentrations [797,798]. Interestingly, it was also observed that sapanisertib promotes metabolic alterations, such as disrupting glutathione synthesis and reducing glucose and lactate (a common feature of cancer cells) [797,798]. When tested in murine models, it reduced tumor weight and size, improving animal survival [797–799]. However, in combination with trametinib (1 mg/kg; 3 x/week; p.o. + MAPK inhibitor; 1.5 mg/kg; 5 x/week; p.o.), despite showing antitumor effects (reduction in angiogenesis and improvement in animal survival), some adverse effects were observed, including weight loss and skin redness [800]. Finally, in RB models, sapanisertib inhibited growth and increased apoptosis, whereas it inhibited cell migration and angiogenesis [801].

LY-2090314. This drug belongs to the ATP-competitive class of GSK-3 inhibitors with limited activity against additional kinases. Preclinical data suggested partial anticancer activity as a single agent against solid-tumor-derived cancer cell lines in vitro and in xenograft models, although it seemed to potentiate platinum-based chemotherapy [802]. Only a single report of its activity against pediatric tumors was found in the literature. Kunnimalaiyaan et al. (2018) tested LY2090314 in a panel of NB cell lines with different genetic backgrounds: SH-SY-5Y (non-amplified MYCN or single-copy, wild-type TP53, F1174L ALK mutation), NGP (1p alteration, MYCN-amplified, wild-type ALK, TP53 mutated, MDM2-amplified) and SK-N-AS (1p deletion, MYCN single-copy, H168R TP53 mutation, wild-type ALK), and found that this GSK-3 inhibitor at nanomolar range promoted growth inhibition in a time- and dose-dependent manner irrespective of the cell line markers. Reduced growth resulted mainly due to apoptosis induction, evinced by a 2-fold increase in the expression of cleaved PARP and capsase-3/7 activity. Downregulation of survivin and cyclin 1 was also observed [803].

Tideglusib. This compound represents another GSK-3 inhibitor, although it acts in a non-ATP competitive manner. Evidence of its antineoplastic effects with a pediatric scope includes in vitro experiments in OS and NB cell lines. In both models, treatment showed a significant reduction in cell proliferation in a dose-dependent manner, cell cycle arrest, and apoptosis induction, even though micromolar concentrations are required to achieve comparable results to LY2090314 [804–806]. Nevertheless, inhibition of GSK-3 by Tideglusib importantly compromises stem cell characteristics of both cell types. In the OS, treatment decreases stem cell markers, including OCT4, CD133 and SOX2, while in NB, it decreases neurosphere self-renewal. In mice models, tideglusib treatment (10 or 20 mg/kg in OS- and NB-derived tumors, respectively) promoted a reduction in tumor growth with few side effects [804,806]. Of note, PDX-derived cell cultures of both variants of RMS (embryonal and alveolar) treatment with tideglusib substantially reduced β–catenin phosphorylation at 60 nM; however, tumor-bearing mice treated with 200 mg/kg of tideglusib daily by oral gavage did not benefit in terms of survival or myodifferentiation [807].

MK-2206. This compound is an orally bioavailable allosteric and non-ATP-competitive AKT inhibitor tested in several tumors [808]. In OS, for instance, this drug was able to induce cytotoxic effects both in vitro and in vivo [223,796,809,810]. Similarly, in NB cells, MK-2206 diminished cell viability and increased apoptosis in cells with high expression of FOXO3a [810,811]. In vivo, the drug promoted inhibition of tumor growth and increased animal survival, effects that were even improved by combination with etoposide [812]. Of note, EWS and RMS cells were not sensitive or had less sensitivity to AKT inhibition [809].

Ipatasertib. Also known as GDC-0068, this compound is an ATP-competitive pan-AKT inhibitor developed by Array BioPharma/Genentech Inc. Having similar activity against Akt-1 and Akt-3, it is effective against several tumor types [813]. So far, Choo and colleagues are the only group that has tested this compound in childhood sarcomas. The drug induced a reduction in cell viability; however, RMS cells were more sensitive to PI3K/AKT pathway inhibition than OS cells [814].

4.3. MAPK Pathway Inhibitors

Sorafenib (Nexavar®). Sorafenib is an inhibitor of VEGFR2/3, PDGFR, KIT, FGFR-1, RAF and RET, approved by the U.S. FDA for the treatment of unresectable HCC and advanced RCC [815]. In preclinical models of MB, this compound reduced cell viability and increased apoptosis in established cell lines and primary tumor cultures. Moreover, it induced cytoskeletal alterations that ended in impaired cell migration [640]. In vivo, sorafenib (100 μL of 10 μmol/L administered three times a week for five weeks) was able to reduce the volume of subcutaneous tumors [816]. Similar results were obtained NB, in vitro and in vivo. Of note, in this model, sorafenib also impaired angiogenesis and G1 cell cycle arrest [817,818]. Conversely, despite reducing initial viability in EPN and PA, growth-factor-driven rescue was also seen, reducing the potential of using sorafenib for treatment of these tumors [691]. Indeed, in pediatric patients with PA, sorafenib induced progressive tumor growth acceleration as a result of ERK upregulation, which resulted in premature termination of the study [819]. In bone sarcomas, dubious results were also observed. In OS, sorafenib treatment blocked cell proliferation and was able to reduce tumor growth in murine models [820–822]. However, other studies showed that sorafenib was only able to reduce tumor growth when combined with everolimus, or palbociclib, probably due to the capacity of sorafenib to induce mTORC activation [821,823]. At the same time, in EWS and RMS, sorafenib only showed efficacy when combined with doxorubicin or with ceritinib, respectively [823,824].

Regorafenib. Also called BAY 73–4506, this is a new-generation multi-tyrosine kinase inhibitor that already showed antitumor and antiangiogenic effects. This inhibitor diminished cell proliferation in a cell line panel from the Innovative Therapies for Children with Cancer (ITCC), which includes 5 MB, 7 EWS, 7 NB, 7 OS and 7 RMS cell lines [610]. Regorafenib also induced cell cycle arrest and promoted apoptosis in NB cells [612]. In vivo, 10 mg/kg/d or 30 mg/kg/d treatment resulted in tumor growth inhibition in RMS, EWS and NB orthotopic models and improved EFS in EWS, NB, OS and RMS [610,825,826].

4.4. Cyclin-Dependent Kinases Inhibitors

Milciclib. This is a second-generation ATP competitive pan-CDK inhibitor, developed by Tiziana Life Sciences, that also acts on TRKs (from tropomyosin receptor kinase A) [827,828]. In preclinical trials performed in different tumors, such as MB and gliomas, it showed promise given its ability to cross the BBB, even though there are reports of MDR transporters limiting the penetration into the brain [829–831]. Moreover, MYCN-amplified the Grp3-MB cell lines MB002, Sd425 and D283 and the *MYCN*-amplified NB cell line Kelly are particularly sensitive to MILCICLIB treatment, evinced by cell cycle arrest and massive apoptosis [829].

Terameprocol (CINelim™). This is a semi-synthetic inhibitor developed by Erimos Pharmaceuticals LLC from a plant lignan, showing antiviral and anti-cancer potential [829,832]. The drug is considered a global inhibitor of the transcription process, which in turn acts by preventing, for example, the synthesis and activation of survivin, by competing with the transcription factor Sp1 for specific Sp1 DNA-binding domains within gene-promoter regions during DNA synthesis [833]. To date, only a single in vitro study (that included the childhood GBM cell line SF188) showed that this inhibitor was able to reduce the proliferation capacity of the cells in a dose-dependent manner, and showed synergism with TMZ under simultaneous exposure for 48 h. Increased effects were also observed when combined with ionizing radiation. Moreover, as expected, this compound induced significant arrest in the G0/G1 phase, decreasing the mitotic index and almost killing all cells at 30 μM [833].

UCN. UCN-01, or 7-Hydroxystaurosporine, is a synthetic derivative of staurosporine with antineoplastic activity that acts on AKT, CDKs and calcium-dependent protein kinase C (in an ATP-competitive manner), and is able to act synergistically with others [834]. In experimental models described in the literature, UCN-01 showed promising results by inducing apoptosis in leukemic and colon cancer cells; moreover, according to the pediatric tumors included in this work, this inhibitor also showed potential against OS tumor cells, reducing viability, proliferation and migration [835,836]. Similar results were observed in a panel of NB cell lines (with genetic backgrounds differing in MYC, p53 and BCL2 statuses), where this inhibitor was the most effective compound in reducing cell proliferation (compared to BiCNU, docetaxel, flavopiridol, staurosporine) and induced apoptosis measured through both caspase activation and caspase-3 and PARP cleavage [837].

BMS-387032. Also called SNS-032, this is a small aminothiazole molecule that acts as an ATP-competitive cyclin CDK inhibitor, especially for CDK2/7/9. Preclinical studies have shown that as a cell cycle blocker, it causes cytotoxicity and prevents tumor cell growth in several models [838–840]. Regarding pediatric tumors, this inhibitor showed positive results in the OS cell line U2-OS, evoking downregulation of RNA polymerase II Ser2 phosphorylation and some degree of caspase activation at all doses tested [839]. Similarly, this CDK inhibitor showed encouraging results in a panel of 109 NB cell lines, consisting of 19 parental cell lines and 90 sublines with acquired resistance to 14 different anticancer drugs. Doses between 58.3 and 14,615 nM were able to reduce viability in a great proportion of cell lines and impaired the growth of the multidrug-resistant cisplatin-adapted UKF-NB-3 subline UKF-NB-3(r)CDDP(1000) injected into the right flank of NMRI:nu/nu mice. Of note, p53 status did not affect the response of NB cells; however, ABCB1 expression conferred resistance to this drug [841–843]. Other interesting results, albeit not in child-derived cell lines, were also published, including the inhibition of hypoxia-mediated GBM cell invasion and cell-mediated capillary formation of HUVEC cells when co-cultured with U87MG cells in the presence of the drug [844,845]. Nevertheless, further studies with this inhibitor were stopped due to its high toxicity and low selectivity [846–848].

Seliciclib (Roscovitine®). Formerly known as Roscovitine, CYC202 or R-roscovitine, this is a selective ATP-competitive pan-CDK inhibitor that blocks cell proliferation in almost all phases of the cell cycle. Seliciclib is a potent inhibitor of CDK9/cyclin T, CDK7/cyclin H, CDK2/cyclin E and CDK1/cyclin B. The negative influence of seliciclib on CDK7 and CDK9 also portrays a role for this inhibitor in modulating RNA polymerase II CTD phosphorylation [848]. Its antitumor activity has been explored in a wide spectrum of hematological and solid malignancies as a single agent and in combination with other cytotoxic agents [848–850]. Among pediatric tumors, Roscovitine showed promising results, in EWS, where it was able to reduce cell proliferation and induce caspase-dependent activation (half minimal dose 10 μmol/L) in a panel of six cell lines, while it slowed A4573-derived tumor growth in mice after intraperitoneal injection [851]. Similar results were found in OS, with reduced proliferation and migration at doses up to 90 μM [437,851]. Moreover, in NB, the drug resulted concentration-dependent cytotoxicity, both in vitro and in vivo, with doses between 10 and 200 μM [851–855]. Roscovitine also reduced MB viability, with IC50 values of around 25 μM [856]. Moreover, treatment of Pzp53med cells (derived from a mouse Ptc+/−/p53−/− tumor) with 10 nM roscovitine resulted in reduced levels of E2F1, FASN, Bmi1, cyclin D2, cdk2 and cdk4. Synergistic effects were also observed when combined with C75, an inhibitor of FASN [857].

Ribociclib (Kisqali®). Also known as Kisqali® (Novartis, Basel, Switzerland), this compound is a highly specific inhibitor of CDKs 4/6 that received FDA approval for use in the upfront treatment of hormone receptor-positive (HR^+), HER2-negative breast cancer in 2017 [858]. With respect to the pediatric tumors, this compound demonstrated adequate results in EWS, causing cell cycle arrest mainly in combination with IGF1R inhibitors [859]. Moreover, in NB, this drug was able to reduce proliferation in vitro and in vivo, with doses between 0 and 10,000 nmol/L [331,860,861]. Most importantly, ribociclib showed high CNS penetration (>10 nM) in vivo, suggesting prospects for its use in the treatment of brain tumors. In this regard, oral doses of ribociclib inhibited RB phosphorylation, downregulated E2F target genes (CNE2, CCNA2, MKI67, TOP2A and PLK1) and decreased proliferation in group-3-MB mouse and human orthotopic PDX. Additionally, the combination of ribociclib and gemcitabine slowed tumor progression and metastatic spread and increased survival, warranting further investigation [862].

Palbociclib (Ibrance®). Also known as PD-0332991 (Ibrance®, Pfizer, New York, USA), this compound represents an ATP-competitor with selective potency against CDK4/6, approved by the FDA in 2015 [859]. The effects of this inhibitor have been assessed in several childhood tumors. In primary EPN cells, for example, it was able to reduce proliferation at 0.5 μM, with G1 arrest and reduced expression of CDC6, MCM2, MAD2L1, CDK2, BRCA2 and RAD51 [863]. Similar results were observed in NB, where this inhibitor reduced proliferation, inhibited colony formation in a dose-dependent manner and affected cell differentiation, tumor progression and metastasis in a preclinical chick embryo model [863–866]. Palbociclib has also been shown to be a new option for targeted therapy in childhood sarcomas. Perez et al. (2015), by treating a panel of 10 low-passaged sarcoma cell lines generated directly from patient samples and two commercial cell lines of heterogeneous origin and different molecular karyotypes (including liposarcoma, leiomyosarcoma, EWS, RMS and myxofibrosarcoma), determined IC_{50} values ranging from 8 to 26 μM depending on their levels of CDK4 expression. Moreover, palbociclib was active in vivo against subcutaneously engrafted CDK4-expressing sarcomas, although responses were negative in tumors displaying low levels of CDK4 and high levels of p16ink4a [867]. Strong decreased cell proliferation and G0/G1-phase arrest with decreased S/G2 fractions were also observed in leiomyosarcomas by another group [868]. Most interestingly, this compound is capable of inhibiting growth in sarcomas with different translocation backgrounds. For example, Palbociclib (100 mg/kg) was able to reduce the volume of tumors originated from an EWS sample with CDKN2A/B loss and FUS-ERG fusion implanted in the right chest wall of nude mice [869]. Additionally, satisfactory results were obtained after the treatment of a child with a refractory pediatric sarcoma harboring paracentric inversion on the short arm of chromosome X, resulting in the fusion of the BCOR and CCNB3 genes [870]. Regarding assays in OS, this inhibitor reduced proliferation and migration with doses of 0.04, 0.16, 0.625, 2.5 and 10 μM [393]. Migration and invasion have also been hampered by palbociclib in glioma cell lines, both in vitro and in vivo, with doses ranging between 10 nM and 10 μM [871–873]. Moreover, this inhibitor showed significant therapeutic benefit in mice after intracranial transplant of genetically relevant murine or human astrocytoma cells expressing BRAFV600E, and extended survival of animals when combined with PLX4720 (PubChem CID24180719) [874]. Similar results were also obtained in a DIPG with PDGF-B overexpression and Ink4a-ARF loss. Palbociclib induced cell cycle arrest in vitro and in vivo. However, in models engineered for PDGF-B expression with p53 deletion, the results were disappointing. Regarding survival, Palbociclib treatment prolonged animal survival by 12%, which was further increased by combinations with a previous single dose of 10 Gy radiation therapy [875].

Abemaciclib (Verzenio®). This compound, under the name Verzenios® (Eli Lilly, Indianapolis, USA), is a highly selective CDKs 4/6 inhibitor that acts by competing for the ATP binding site. This inhibitor is the most different from its peers (palbociclibe and ribociclibe), being more lipophilic and able to cross the BBB and penetrate breast tissue [859]. In addition, this inhibitor has potent activity against recurrent ER+/HER2- breast cancers [876,877] However, its clinical adverse effects are not well described [878–880]. Preclinical studies in pediatric tumors indicate effectiveness against EP, NB, EWS and OS [412,880–882]. Moreover, in gliomas, this inhibitor has been shown to be efficient in reducing cell migration and invasion, as well [883,884]. Finally, combining abemaciclib with other inhibitors, one of them being trametinib (MEK inhibitor), synergistically reduced the survival of the RAS-mutant RMS cell line RD. However, when PDX-bearing mice were treated with that combination, they exhibited progressive disease compared to the RMS standard-of-care regimen (irinotecan + vincristine) [885].

AT-7519. This is a potent pan-CDK inhibitor, acting on CDK1/2/4/6/9. Preclinical studies have shown a reduction in cell proliferation and induction of cell death in many cell lines, regardless of tumor origin [886–888]. In addition, this inhibitor showed promise for the treatment of MYC-amplified NB, evinced by apoptosis induction in vitro and dose-dependent growth inhibition in PDX, with improved survival and tumor regression in 86% of patients 7 days of treatment initiation [889].

4.5. Polo-Like and Aurora Kinases Inhibitors

BI-2536. This is an ATP-competitor dihydropteridinone that has proved to be more than 1000 times more specific for PLK1 than for other kinases [471,890]. This compound has been tested in several tumor cells, although reports for pediatric tumors are more uncommon. Our group showed that this compound reduces proliferation in up to 64% of cases, causes G2/M arrest and induces apoptosis after 24 h of treatment in the SF188 cell line, and it exerts the strongest radiosensitizing effect among all the cell lines tested [891,892]. Anti-mitotic and sensitizing to ionizing radiation effects were also demonstrated by us in MB cells, even though the results were comparable to other PLK1 inhibitors [893]. Others also showed that this compound suppresses self-renewal of patient-derived primary cells with high PLK1 but not low PLK1 expression, and it did not affect the growth of normal neural stem cells. Finally, BI2536 extended survival in MB-bearing mice [487,488]. Sensitizing effects were also observed for hyperthermia in the RB cell lines Y79 and WERI-Rb-1 [894].

With IC_{50} lower than 100 nM, BI 3526 was also able to induce cell cycle arrest at the G2/M phase and cell apoptosis in NB cells [492,895,896]. It has recently been proposed that this drug induces cell death by regulating the expression of the minichromosome maintenance complex components 2 and 10, which are involved in DNA replication and have been associated with poor outcome in other tumors [897].

Perturbation of normal mitotic progression by BI 2536 nanomolar concentrations (10, 50 and 100 nmol/L) also significantly decreased cell proliferation and clonogenic capacity, inducing mitotic arrest and aneuploidy in OS cell lines, resulting in caspase-independent mitotic catastrophe followed by necrosis [898]. Conversely, in another set of OS cell lines, apoptosis induction was validated through PARP cleavage and caspase activation. Irrespective of this, BI 2536-treated xenograft mouse models presented significantly smaller tumors compared with controls [899]. Moreover, in RMS, PLK1 inhibition by BI 2536 led to elevated ubiquitination and rapid proteasomal degradation of the PAX3-FOXO1 chimeric oncoprotein in vitro, whereas it reduced PAX3-FOXO1-mediated gene expression and elicited tumor regression in a xenograft mouse model [900]. Moreover, in this tumor type, this drug presented high antiproliferative activity when combined with Eribulin, a microtubule-interfering drug [901].

NMS-1286937. Also known as Onvansertib or NMS-P937, this novel PLK1-specific inhibitor has shown high potency at low nanomolar concentrations on a large number of cell lines, both from solid and hematologic tumors; in addition, differentially from other PLK1 inhibitors that compulsorily need intravenous administration, this small molecule can be administered orally [902]. Considering pediatric tumors, onvasertib has shown promising results in OS and MB. In the former, this drug proved to be highly active in both drug-sensitive and drug-resistant cell lines, except for cell lines overexpressing the multiple-drug-resistant transporter ABCB1 [903]. Results were also very promising in group-3 MB, which is characterized by PLK1 overexpression. In the study, treatment of D341, D425 and D458 cell lines resulted in reduced colony formation, cell proliferation, stem cell renewal and G2/M arrest. The half-maximal inhibitory concentrations varied from 4.9 to 6 nM. Other cell lines within the SHH subgroup needed 27.94 nM for comparable results. Moreover, onvansertib acted as a radiosensitizer, and showed marked time- and dose-dependent growth arrest of neurospheres and patient-derived short-term cultures. Most notably, onvansertib dramatically improved the median survival of orthotopic PDX models from 68 to 95 days [904].

GSK-461364. This compound is a second-generation and potent ATP-competitive thiophene amide PLK1 inhibitor. The anti-mitotic effects of this compound have been demonstrated in several tumors; however, it has been observed that its activity can be hampered by the overexpression of the multidrug resistance pump ABCB1 [905]. Preclinical findings in the pediatric setting include reduced viability after treatment in NB, MB and OS, in all of which it diminished growth and caused cell cycle arrest with massive apoptosis at a low-dose nanomolar range [906,907]. This compound also demonstrated a synergistic cytotoxic effect with paclitaxel, even though combination with methotrexate, cisplatin, vinblastine or doxorubicin was not that effective [907]. Conversely, this PLK1 inhibitor has been shown to be an effective radiosensitizer [908]. In vivo, GSK461364 treatment (50 mg/kg body weight intraperitoneally administered) strongly delayed the establishment of high-risk NB tumors in nude mice (by 22 days) irrespective of MYC status of the cell lines used, and significantly increased survival time in the treated group [906].

Volasertib. This drug, also known as BI 6727, is also a dihydropteridinone derivative that induces a distinct prometaphase arrest phenotype (polo-arrest) and subsequent apoptosis. Regarding pediatric tumors, in 2014, the NCI-supported PPTP Program published initial results about the use of BI 6727 in pediatric tumors. The systematic work presented in vitro results on 24 cell lines that included 4 RMS, 4 EWS, 1 GBM, 4 NB and several leukemias, concluding that the compound was effective without histotype selectivity. Then, the responsiveness of solid tumor xenografts (that also included OS and WT) using a dose of 30 mg/kg for 3 weeks was also tested. Volasertib was able to induce regression in only a minority of the models tested; however, significant differences in EFS distribution compared to control in 59% of the evaluable xenografts were observed, with better results for OS, WT and NB (the only one that showed objective responses) [909]. The results for OS were later validated in vitro by our research group, where treatment with BI 6727 not only led to growth arrest, triggered apoptosis and radiosensitized cells, but also it seemed to be more efficient in sensitizing OS cells to standard cytotoxics compared with GSK461364 [908]. Anti-mitotic effects were also observed in MB cells [893].

Aurora kinase inhibitors. Over the past two decades, several small-molecule inhibitors of Aurora kinases have been developed, most of which primarily target Aurora B. Despite not being approved or with clinical promise by the CanSAR platform, some of these inhibitors have shown promising results. MLN8237, for instance, was evaluated against a panel of EWS (n = 11) and NB (n = 17) cell lines with acceptable results in vivo [910]. The drug also inhibited growth uniformly in the majority of the cell lines from the PPTP in vitro panel, with IC_{50} values ranging from 49 nM to 61 nM. In vivo, EFS was 80% higher in treated animals compared to controls, showing even more auspicious results than those obtained for Volasertib [911]. More recently another AURK inhibitor, designated as PHA-680626, disrupted the AURKA/N-Myc, presenting a new alternative for the treatment of high-risk NB [912,913]. Additionally, AMG-900 (pan-aurora inhibitor) blocked MB cell proliferation and increased apoptosis and acted synergistically with the histone deacetylase inhibitor SaHa [914].

5. Kinase Inhibitors in Clinical Trials

Clinical trials including children and using kinase inhibitors have been increasingly reported in recent decades. Among the 95 predicted compounds for our selected group of kinases retrieved through the CanSAR platform, for instance, 18 have already entered clinical trials (Figure 6A), most of which have focused on measuring cytotoxic effects on high-risk, refractory and recurrent tumors.

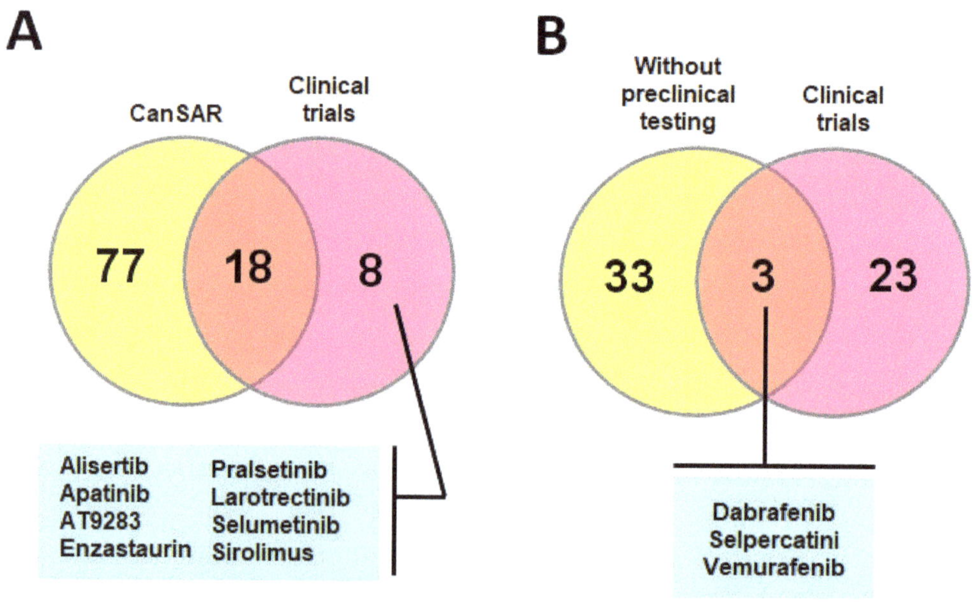

Figure 6. (**A**) Venn diagram showing the number of kinase inhibitors retrieved from the CanSAR platform versus those already tested in clinical trials. Around 30% of the compounds have been tested in patients. Of note, eight kinase inhibitors tested in clinical trials were not found in the CanSAR platform. (**B**) Venn diagram comparing kinase inhibitors found in the CanSAR platform without evidence of preclinical tests in pediatric cancer versus clinical trials. Three of the drugs are already tested in pediatric patients, without in vitro or in vivo evidence.

In order to gather an updated and comprehensive review on clinical data and outcome of pediatric tumors treated by these new TKIs compounds and already published to date, a PubMed search was performed (as per Oct 2022) using the following uniterms: ((cancer) AND (pediatric)) AND (kinase inhibitor). The following additional filters were set: clinical trial, meta-analysis, randomized controlled trial. In total, 233 articles were retrieved. Abstracts of the whole set of results were carefully read to exclude duplicated data/patients from the same clinical trial, articles where children and/or adolescents (<21 years of age) were not included or where clinical trials not testing kinase inhibitors were retrieved. A total of 53 articles met these inclusion/exclusion criteria, and were read and analyzed in full. Data on study design, study population, main clinical information and outcomes are summarized in Supplementary Table S5. The next part of this section does not review all studies on this subject, and also does not include clinical trials under development or in recruitment with unpublished data. Our main purpose was to gather and to discuss major sedimented data of clinical value and of clinical interest and applicability in this setting.

5.1. TRK—Tyrosine Receptor Kinases

5.1.1. EGFR and VEGFR

Abnormal or disrupted angiogenesis is considered to be one of the hallmarks of cancer, and an increasing interest in targeting EGFR and VEGFR pathways has been observed in clinical oncology [915]. Unfortunately, most of these studies are focused on the adult population, and experience with these drugs in pediatric cancer is less robust. Yet, many of these compounds were tested in clinical trials that included children and/or adolescents with cancer, with variable results.

Regarding VEGFR inhibitors, at least three major molecule subtypes have been described. Type I VEGFR inhibitors exert their action as competing molecules to ATP [916]. Some examples of type I VEGFR inhibitors include pazopanib, axitinib, sunitinib, ponatinib and others. Type II inhibitors bind to the inactive "DFG-out" conformation adjacent to the ATP-binding site. Some examples of type II inhibitors include sorafenib and lenvatinib. Type III inhibitors lead to an irreversible binding of kinases at specific sites [917]. In addition, many of these VEGFR inhibitor molecules are under clinical evaluation in association with different inhibitors, particularly immunotherapy [918]. In addition, VEGFR2 inhibitors of dual action against other tumor-associated biomarkers are gaining much attention lately [919].

Inaba et al. (2011) evaluated the combination of sorafenib, a potent multikinase inhibitor, in association with cytarabine and clofarabine to treat relapsed or refractory childhood leukemia [916]. A total of 12 patients (<21 years of age; 11 with acute myeloid leukemia (AML) and one with early T-cell precursor leukemia) entered this phase I study. Of note, complete remission (CR) was obtained in 6 out of 12 patients, CR without complete blood recovery in 2 cases and partial remission (PR) in 1 case. Dermatologic, gastrointestinal (GI), metabolic and infectious adverse events were observed, and were more pronounced in the sorafenib higher-dose stratum. Grade 3 hand–foot skin reactions and/or rash were dose-limiting toxicities (DLTs). Sorafenib was also evaluated in a phase II trial, in association with everolimus to treat patients with progressive and unresectable high-grade osteosarcoma who failed standard treatment [920]. Although some encouraging initial results with sorafenib were observed earlier in this setting, a larger phase II study by Grignani et al. has shown some minor activity for selected cases, and the trial did not reach the 6-month progression-free survival target in at least 50% of patients [920]. Sorafenib was also evaluated in a phase I study that included refractory or relapsed hepatic tumors (hepatoblastoma or HCC) in children. The drug was used in association with irinotecan [921]. Six patients were evaluable for tumor response: two patients survived with no evidence of disease (NOD), one patient was alive with disease (AWD) and two patients DOD upon publication date. Radiation therapy and/or metastasectomies were offered after study protocol based on individual clinical needs. Increased grade 3 or 4 transaminase levels or neutropenia were reported.

Axitinib, a VEGFR1, 2 and 3 inhibitor, was also evaluated in a phase I study that included refractory solid tumors, as part of a Children's Oncology Group (COG) trial and a pilot consortium trial ADVL1315. Nineteen patients were evaluated, with ages ranging from 9 to 17 years. Five patients achieved stable disease (SD), and a PR was observed in one case (an alveolar soft tissue sarcoma). The maximum tolerated dose (MTD) of axitinib was set at 2.4 $mg/m^2/dose$ [922]. Lenvatinib, a multiple oral tyrosine-kinase inhibitor against VEGFRs 1 to 3, RET, KIT, FGFRs and PDGFR-alpha were evaluated in a phase I/II pediatric and young adult trial for osteosarcomas [631]. The phase I study observed SD (some lasting for 23 weeks) as the best response obtained with Lenvatinib; the phase II study depicted two patients with partial response and thirteen children with SD. Although this single agent showed some activity in osteosarcoma, future studies will focus on the association of Lenvatinib with chemotherapy, or different molecules. Recently (September 2021), the FDA approved the use of cabozantinib for the treatment of patients (>12 years of age) with metastatic or locally advanced differentiated thyroid cancer (DTC), not amenable to receive iodine therapy, and who have failed different TKIs therapies. This approval was mainly achieved as a result of clinical findings of the COSMIC-311 study that observed prolonged progression-free survival in the group receiving the drug compared to the control (placebo) group.

Anti-EGFR therapy to treat pediatric malignancies has been less frequently evaluated in clinical trials. The Children's Oncology Group (COG) evaluated gefitinib, an oral EGFR tyrosine kinase inhibitor, in children with refractory solid tumors [923]. Twenty-five patients were enrolled, and although the drug was well tolerated, only one patient showed partial tumor response in this study cohort. Gefitinib was also evaluated in concomitance to radiotherapy in newly diagnosed children with brainstem gliomas (DIPGs). Forty-three eligible patients entered this study. Although the vast majority of patients experienced rapid and fatal tumor progression, three patients remained free of tumor progression for more than 36 months, pointing to a possible benefit of this approach for a small subset of DIPGs [924].

5.1.2. RET Inhibitors

Recently, RET-altered tumors were considered amenable to receive targeted therapy with RET inhibitors in a tissue-agnostic manner [925]. Vandetanib, a multi-TKI including RET inhibition, was evaluated in association with bortezomib in 22 patients (17 evaluable cases) with medullary thyroid cancer, with 27% showing partial responses [926]. Additionally, selpercatinib was evaluated in 42 patients with RET-fused tumors of different histologies other than lung and thyroid; durable antitumor activity across different tumor subtypes was observed, with only minor adverse effects [927]. Tissue-agnostic benefits of the use of RET inhibitors in patients with RET-fused tumors of different histologies were also confirmed with different drugs, such as Pralsetinib [928].

5.1.3. ALK Inhibitors

A consortium phase I study coordinated by the COG evaluated the use of crizotinib for childhood cancer with refractory solid tumors or anaplastic large-cell lymphomas (ALCL). Seventy-nine children (aged 6 years or older) and adolescents were enrolled; tumor responses were more pronounced among patients with tumors with activating ALK aberrations [929]. In addition, Fukano et al. investigated the role of alectinib in primary refractory ALCL, or after relapsing, in a phase II study that included both children and adults. Eight out of ten enrolled patients responded to alectinib, with minor adverse effects described [930].

Moreover, Entrectinib, a potent CNS-penetrant inhibitor of TRKA/B/C, ROS1 and ALK, was also evaluated to treat children and young adults with solid or primary CNS tumors harboring NTRK, ROS1 or ALK aberrations [730]. In this phase I/II trial, the objective response rate (ORR) was 57.7% among 43 response-evaluable patients. Entrectinib shows a suitable safety profile and is effective as an option to treat pediatric patients with solid tumors harboring NTRK1/2/3 or ROS1 fusions.

5.2. PI3K/AKT/mTOR Pathway

The use of mTOR inhibitors, particularly everolimus, may be considered an important hallmark to treat children (>3 years old) with subependymal giant cell astrocytomas (SEGAs) associated with tuberous sclerosis complex (TSC) and not amenable to surgical treatment. This indication is largely derived from a phase III study (EXIST-1) that evaluated 117 patients in a double-blind placebo controlled trial, showing 50% tumor reduction exclusively in the treatment arm. The most frequent adverse events were oral ulcer and pyrexia; however, there was no treatment discontinuation due to adverse events [931].

5.3. MAPK Pathway

Abnormal, disrupted or constitutively activated MAPK pathways are involved in many pediatric cancers, particularly in low-grade gliomas (LGG). Recently, BRAF V600E-mutated tumors in children were eligible for agnostic treatment with BRAF plus MEK inhibitors [932]. Patients were randomized to receive either dabrafenib plus trametinib or standard chemotherapy. Among 110 treated children, complete and partial responses were reached in 47% of patients receiving the targeted therapy versus 11% for patients receiving chemotherapy alone. Of note, the clinical benefit rate, defined as complete, partial and stable disease lasting for more than 24 weeks, was 86% for trametinib plus dabrafenib versus 46% for standard chemotherapy [933]. Monotherapy with the BRAF inhibitor Dabrafenib was also previously evaluated in refractory or relapsed BRAF V600-mutated LGGs in children in a phase I/II study [934]. Among 32 enrolled patients (aged 1 to <18 years), the ORR was 44% and the 1-year PFS was 85%. In addition, adverse events (AE) were described in 91% of the participants; the most frequent AEs were fatigue, skin rash, dry skin and fever.

Disrupted MAPK pathways are also observed in patients with NF-1, where germline pathogenic neurofibromin mutations lead to the abrogation of the repressive function of this protein, with consequent activation of the PI3-K/AKT and RAS/MAPK cell signaling [935]. Besides the augmented frequency of LGG in patients with NF-1, plexiform neurofibromas are also frequently diagnosed in these patients, sometimes with life-threatening clinical consequences. Selumetinib, an MEK inhibitor, was evaluated in patients with symptomatic and inoperable plexiform neurofibromas [936]. Fifty children were enrolled in this phase II study that showed sustained tumor reduction, associated with clinical benefits. Improvements were observed in reducing pain and recovering motor function. However, 5 out of 50 patients discontinued treatment due to adverse effects possibly related to selumetinib, and 6 patients experienced disease progression while receiving the medication.

5.4. Cell Cycle Kinases

Ribociclib, an oral CDK4/6 inhibitor with pre-clinical evidence of action in different types of pediatric cancer, was tested either alone or in combination with chemotherapy in phase I and I/II studies, respectively [331,937]. Stable disease was the best response observed for both trials. The most common AEs were hematologic, including neutropenia, anemia and lymphopenia. More recently, palbociclib, a different oral CDK4/6 inhibitor, was evaluated in a phase I study directed at children and adolescents with progressive brain tumors. MTD of palbociclib was set at 75 mg/m^2 (as monotherapy) for 21 days, followed by 7 days without medication. Neutropenia and thrombocytopenia were common AEs; no objective responses were observed among 35 enrolled patients [938].

In addition, different aurora-kinase inhibitors (AKIs) have undergone clinical trials in pediatric cancer. Thirty-seven patients were enrolled in a phase I COG study evaluating MLN8237, a selective AKI-A [939]. Myelosuppression, mucositis and hand–foot skin syndrome were common side effects. One PR and six prolonged SD were observed. AT9283, a different multitarget of AKIs A and B, was evaluated to treat pediatric patients with different types of solid tumors [940]. Of twenty-three evaluable patients, the authors described one confirmed PR and nine cases of disease stabilization after two courses of AT9283. More recently, Alisertib, a potent AKI-A, was evaluated in the pediatric population with both recurred/refractory solid tumors or leukemia (phase I). Five objective responses were reported, including two complete responses out of one-hundred and thirty-seven evaluable participants [941].

6. Final Remarks

Despite improvements, cancer is still responsible for 8% of all disease-related deaths in the pediatric setting [942]. Conventional chemotherapy not only is often ineffective but can also cause long-term complications that hamper the patient's quality of life.

Over the past two decades, precision oncology and the advent of innovator small-molecule drugs or immunotherapy have revolutionized the treatment of many adult cancers, such as CML, GBM and certain types of breast carcinomas [12,943–947]. The tidal increase in genomic, epigenomic, transcriptomic, proteomic and biochemical data has enormously improved our understanding of the specific molecular signatures of pediatric solid tumors, as well as allowing the sub-classification of some tumor types, as is the case of MB and EPN [948], and the application of corresponding specific therapies. Indeed, we are currently in a transition state where the broadly applied decades-old and not always conclusively curative cytotoxic drugs are being gradually substituted by targeted ones, and in the near future, molecular technology will steer diagnosis and personalized treatment [949].

With over 500 kinases in the human genome regulating key biological processes, many members of this molecular family have gained scientific limelight in oncology and academic pharma. Herein, we provided an in-depth review of published data on the roles of the dysregulation of a selected group of kinases in tumor pathophysiology and corroborated their importance as therapeutic candidates in the context of pediatric solid tumors.

By 2020, the FDA had approved 52 small-molecule therapeutics that target nearly 20 different protein kinases (half of them are multikinase inhibitors and the majority target RTKs) [950]. Other drugs targeting an additional 15–20 protein kinases remain in clinical trials worldwide. Nevertheless, a total of 40 target kinases represents only 10% of the kinase superfamily, and most of those kinase-directed drugs have not been tested in pediatric patients. Still, critical challenges must be overcome in experimental oncology and the translation into clinical options. These challenges include the following:

(1) **Selectivity:** Most inhibitors developed so far target the ATP-binding site, meaning that they may act on multiple targets simultaneously and open new opportunities for the treatment of different tumor histologies [951,952]. Imatinib, for example, which has led to a significant increase in CML survival rates by selectively targeting the tumor-specific protein BCR/ABL, was included for the treatment of gastrointestinal stromal tumors (GIST), which are characterized by KIT-activating mutations [953]. Nevertheless, most inhibitors discovered to date have faced several adversities limiting their clinical use. First of all, the high sequence similarity in the ATP-binding sites frequently results in poor selectivity (refer to Figure 5 and Supplementary Figure S3) that may lead to undesired side effects. Moreover, these small molecules must compete with high intracellular ATP levels, leading to differences in potency when measured in vivo by biochemical versus cellular assays. In fact, many compounds inhibit their enzymes at nanomolar concentrations when measured biochemically, but only inhibit tumor cell growth under 3-fold higher concentrations [954]. Nevertheless, the increasing number of recognized kinase-specific structural features has allowed the emergence of superior non-ATP competitive kinase inhibitors that target other allosteric sites, which mostly act by inducing a conformational shift in the target enzyme, depleting its function [951,955–957].

(2) **Adverse effects:** Imatinib and dasatinib, for instance, are both licensed for the treatment of children with CML. Despite its undeniable benefits, and with the spectrum of

side effects being comparable to what has been reported in adults (i.e., gastrointestinal toxicity, skin rash and muscle cramps), in a growing organism, imatinib treatment impairs longitudinal growth through the disturbance of osseous remodeling and inhibition of growth hormone secretion, which raises concerns about its lifelong use [958]. Moreover, despite the wealth of compounds that emerge on a daily basis, showing selectivity, potency and favorable pharmacological profiles, the probabilities for the translation into effective patient treatment for the great majority of them are extremely low. In this regard, less than 30% of the compounds approved or with clinical promise retrieved from the CanSAR platform have entered clinical trials in the pediatric setting (refer to Figure 6A). PLK1 inhibitors, for instance, despite the robust results obtained in vitro and in vivo, have demonstrated poor applicability due to severe hematological toxicity [959].

(3) **Mutational burden and lack of predictive biomarkers:** As stated before, the mutational identity may vary between adult and pediatric cancer, a feature that reflects in treatment response. Current treatments targeting ALK mutations in other cancers, for example, have not shown significant efficacy against NB. In this tumor, two hotspot mutations, at positions R1275Q and F1174L, occur in a high proportion of patients; tumors harboring the first are highly sensitive to crizotinib, while tumors bearing the second are resistant [960–962]. Moreover, as suggested by Bellantoni and Wagner (2021), childhood solid tumors may have fewer potentially targetable mutations, evinced by the inhibition of RTK for the treatment of OS, where it is necessary to target several relevant RTKs simultaneously to achieve desirable results [821]. Therefore, ground-breaking drugs for adult cancer may not be effective in the pediatric setting.
In parallel, inhibitors are not effective if the target is not essential to drive tumor growth or does not represent a prognostic factor, as is the case of ROCK kinases. Even though these proteins have gained popularity and progressively been researched as targets for the development of novel anti-cancer drugs due to their association with metastasis and poorer patient survival in adult tumors, the influence of both isoforms on the prognosis of childhood cancer remains controversial [963].

(4) **Intrinsic and acquired resistance:** Resistance to targeted therapies is considered a largely inevitable hurdle that has a substantial impact on patients. Refractoriness to chemotherapy due to acquired F1174S ALK mutation in NB has been reported [964]. Likewise, the location of EGFR mutations significantly changes the effectiveness of EGFR; several mutations conferring resistance to EGFR tyrosine kinase inhibitors (such as T790M, L833V, A839T, V851I, A871T and G873E) have been reported [938]. Other examples include inadequate response to imatinib due to BCR-ABL1 kinase domain mutations that impart varying degrees of drug insensitivity, observed as underlying mechanism in 5–10% of adults and children with CML, bypassing pathway activation [965,966]. Moreover, despite an initial benefit of the targeted drug in molecularly well-defined tumors, patients inevitably experience tumor progression due to the development of resistance (i.e., Crizotinib in ALK-rearranged NSCLC population and CNS relapses) [967].

(5) **Lack of compounds designed specifically for childhood tumors:** In general, few pediatric patients with cancer are enrolled in clinical trials. The perception that adult studies can be generalized to children with similar diseases is a major obstacle. Consequently, most treatments are based on modifications of previously approved regimes for the adult population, and many compounds enter clinical trials without preclinical testing in pediatric oncology (refer to Figure 6B), which is mandatory to obtain a more accurate interpretation of its possible therapeutic potential in a certain cancer entity [968]. Moreover, pediatric cancer is rare, and even among patients with the same cancer type, there is often broad heterogeneity in terms of prognosis, molecular features or pathology. Therefore, few institutions have sufficient patients and the chances of every potential agent or combination being tested are reduced. Even so, priorities for funding are typically assessed according to the "burden of

illness" for diseases, which is traditionally determined by disease frequency and mortality rate, leading to reluctance to distribute limited research funding to pediatric trials [969–971].

Regardless of the above challenges, kinase-based drug discovery has attained dramatic growth in the past 20 years. Although kinase inhibition represents a young therapeutic strategy compared with other traditional tactics, the FDA has approved a median of almost two small-molecule kinase inhibitors per year [952]. Thus, increasing numbers of targeted therapies are being tested for pediatric cancers, and many have shown undeniable success. Besides, the inhibition of kinases in normal cells can be clinically tolerated, presenting a therapeutic window that allows the softening of the acute side effects that generally lead to refusal and abandonment of treatment [972].

Moreover, as research advances, it has become clear that kinase inhibitors do not have to be absolutely selective. Crizotinib, for instance, was initially developed as an MET inhibitor, but later it was found to be even more efficient in cancers with ALK rearrangements. Additionally, molecularly targeted therapies are proving to be more effective in combination regimes to completely shut down the dysregulated pathway. As an example, it has been shown that everolimus improves CNS retention of vandetanib, dasatinib and sorafenib, which may have a great impact on the treatment of CNS tumors or brain metastases [751,752]. In the same vein, third- or fourth-generation inhibitors are being developed to avoid resistance and improve other biopharmaceutical properties such as brain penetration. These inhibitors, coupled with the increased ability to characterize tumors on molecular and genomic levels, will not only enable treatment refinement by identifying which patients may benefit most, but in the near future may conquer many diseases that are currently incurable.

Supplementary Materials: The following supporting information can be downloaded at: https://www.mdpi.com/article/10.3390/pharmaceutics15020664/s1, Figure S1: Flowchart depicting the identification of databases used in the present study, describing the number originally identified, included and excluded, and the reasons for exclusions; Figure S2: Flowchart representing kinases of interest analysis in cell line dependency score and the identification of inhibitors, preclinical studies and clinical test describing the number originally identified, included and excluded, and the reasons for exclusions; Figure S3: Schematic illustrations of kinases druggability identified by the CanSAR database in the other kinase families, including the total number of compounds with predicted interaction capacity with each kinase, as well as FDA-approved drugs, and clinical candidates. Interaction networks of kinase inhibitors and associated binding proteins according to STITCH ('search tool for interactions of chemicals'). Compounds are represented as pill-shaped nodes, while proteins are shown as spheres. Small nodes represent proteins of unknown 3D structures, while large nodes show proteins with known or predicted structures. Nodes that are associated to each other are linked by an edge: thicker lines represent stronger binding affinities. Networks were constructed considering a minimum required interaction score of 0.700, and based on associations reported in Curated Databases (gray lines), or on both Databases and Experimental/Biochemical Data (green lines). Purple lines represent functional links between proteins; Table S1: Differential expression of kinases (pediatric tumors vs normal tissue); Table S2: Knase expression and clinical features; Table S3: DepMap scores; Table S4: preclinical data; Table S5: Clinical trials.

Author Contributions: Conceptualization, funding acquisition, figures design, review and editing, M.S.B.; project administration M.F.C.; formal analysis, investigation, data curation, writing—original draft preparation M.F.C., M.M., L.C.V., D.B., K.L.O., J.A.P., E.T.V. and M.S.B. All authors have read and agreed to the published version of the manuscript.

Funding: This research was in part supported by Coordenação de Aperfeiçoamento de Pessoal de Nível Superior (CAPES—001) and CNPq—Bolsas de Produtividade em Pesquisa.

Institutional Review Board Statement: Not applicable.

Informed Consent Statement: Patient consent does not apply to this study.

Data Availability Statement: Not applicable.

Conflicts of Interest: The authors declare no conflict of interests.

References

1. Steliarova-Foucher, E.; Stiller, C.; Lacour, B.; Kaatsch, P. International Classification of Childhood Cancer, third edition. *Cancer* **2005**, *103*, 1457–1467. [CrossRef] [PubMed]
2. Instituto Nacional de Câncer (Brazil). *Coordenação de Prevenção e Vigilância and Sociedade Brasileira de Oncologia Pediátrica, Câncer na Criança e no Adolescente no Brasil: Dados dos Registros de Base Populacional e de Mortalidade*; Ministério da Saúde, Instituto Nacional de Câncer–INCA: Brasília, Brazil, 2008.
3. Downing, J.R.; Wilson, R.K.; Zhang, J.; Mardis, E.R.; Pui, C.-H.; Ding, L.; Ley, T.J.; E Evans, W. The Pediatric Cancer Genome Project. *Nat. Genet.* **2012**, *44*, 619–622. [CrossRef] [PubMed]
4. Toren, A.; Rechavi, G.; Ramot, B. Pediatric Cancer: Environmental and Genetic Aspects. *Pediatr. Hematol. Oncol.* **1996**, *13*, 319–331. [CrossRef] [PubMed]
5. Verma, V.; Denniston, K.A.; Lin, C.; Lin, C. A Comparison of Pediatric vs. Adult Patients with the Ewing Sarcoma Family of Tumors. *Front. Oncol.* **2017**, *7*, 82. [CrossRef]
6. Sultan, I.; Qaddoumi, I.; Yaser, S.; Rodriguez-Galindo, C.; Ferrari, A. Comparing Adult and Pediatric Rhabdomyosarcoma in the Surveillance, Epidemiology and End Results Program, 1973 to 2005: An Analysis of 2,600 Patients. *J. Clin. Oncol.* **2009**, *27*, 3391–3397. [CrossRef]
7. Spector, L.; Brown, M.B.; Wantman, E.; Letterie, G.S.; Toner, J.P.; Doody, K.; Ginsburg, E.; Williams, M.; Koch, L.; Schymura, M.J.; et al. Association of In Vitro Fertilization With Childhood Cancer in the United States. *JAMA Pediatr.* **2019**, *173*, e190392. [CrossRef]
8. Rahal, Z.; Abdulhai, F.; Kadara, H.; Saab, R. Genomics of adult and pediatric solid tumors. *Am. J. Cancer Res.* **2018**, *8*, 1356–1386.
9. Gröbner, S.N.; Worst, B.C.; Weischenfeldt, J.; Buchhalter, I.; Kleinheinz, K.; Rudneva, V.A.; Johann, P.D.; Balasubramanian, G.P.; Segura-Wang, M.; Brabetz, S.; et al. The landscape of genomic alterations across childhood cancers. *Nature* **2018**, *555*, 321–327. [CrossRef]
10. Sweet-Cordero, E.A.; Biegel, J.A. The genomic landscape of pediatric cancers: Implications for diagnosis and treatment. *Science* **2019**, *363*, 1170–1175. [CrossRef]
11. Vellichirammal, N.N.; Chaturvedi, N.K.; Joshi, S.S.; Coulter, D.W.; Guda, C. Fusion genes as biomarkers in pediatric cancers: A review of the current state and applicability in diagnostics and personalized therapy. *Cancer Lett.* **2020**, *499*, 24–38. [CrossRef]
12. The Cancer Genome Atlas Research Network. Comprehensive genomic characterization defines human glioblastoma genes and core pathways. *Nature* **2008**, *455*, 1061–1068. [CrossRef]
13. Schwartzentruber, J.; Korshunov, A.; Liu, X.-Y.; Jones, D.T.W.; Pfaff, E.; Jacob, K.; Sturm, D.; Fontebasso, A.M.; Khuong-Quang, D.-A.; Tönjes, M.; et al. Driver mutations in histone H3.3 and chromatin remodelling genes in paediatric glioblastoma. *Nature* **2012**, *482*, 226–231. [CrossRef]
14. Paugh, B.S.; Qu, C.; Jones, C.; Liu, Z.; Adamowicz-Brice, M.; Zhang, J.; Bax, D.A.; Coyle, B.; Barrow, J.; Hargrave, D.; et al. Integrated Molecular Genetic Profiling of Pediatric High-Grade Gliomas Reveals Key Differences With the Adult Disease. *J. Clin. Oncol.* **2010**, *28*, 3061–3068. [CrossRef]
15. Appay, R.; Fina, F.; Macagno, N.; Padovani, L.; Colin, C.; Barets, D.; Ordioni, J.; Scavarda, D.; Giangaspero, F.; Badiali, M.; et al. Duplications of KIAA1549 and BRAF screening by Droplet Digital PCR from formalin-fixed paraffin-embedded DNA is an accurate alternative for KIAA1549-BRAF fusion detection in pilocytic astrocytomas. *Mod. Pathol.* **2018**, *31*, 1490–1501. [CrossRef]
16. Fukuoka, K.; on behalf of the Japan Pediatric Molecular Neuro-Oncology Group (JPMNG); Kanemura, Y.; Shofuda, T.; Fukushima, S.; Yamashita, S.; Narushima, D.; Kato, M.; Honda-Kitahara, M.; Ichikawa, H.; et al. Significance of molecular classification of ependymomas: C11orf95-RELA fusion-negative supratentorial ependymomas are a heterogeneous group of tumors. *Acta Neuropathol. Commun.* **2018**, *6*, 134. [CrossRef]
17. Wachtel, M.; Schäfer, B.W. PAX3-FOXO1: Zooming in on an "undruggable" target. *Semin. Cancer Biol.* **2018**, *50*, 115–123. [CrossRef]
18. Giovannini, M.; Biegel, J.A.; Serra, M.; Wang, J.Y.; Wei, Y.H.; Nycum, L.; Emanuel, B.S.; Evans, G.A. EWS-erg and EWS-Fli1 fusion transcripts in Ewing's sarcoma and primitive neuroectodermal tumors with variant translocations. *J. Clin. Investig.* **1994**, *94*, 489–496. [CrossRef]
19. Jemal, A.; Murray, T.; Samuels, A.; Ghafoor, A.; Ward, E.; Thun, M.J. Cancer Statistics, 2003. *CA A Cancer J. Clin.* **2003**, *53*, 5–26. [CrossRef]
20. Pui, C.-H. Recent Research Advances in Childhood Acute Lymphoblastic Leukemia. *J. Formos. Med. Assoc.* **2010**, *109*, 777–787. [CrossRef]
21. Pezuk, J.A.; Valera, E.T.; Brassesco, M.S. PLK1 Inhibition: Prospective Role for the Treatment of Pediatric Tumors. *Curr. Drug Targets* **2016**, *17*, 1661–1672. [CrossRef]
22. Manning, G.; Whyte, D.B.; Martinez, R.; Hunter, T.; Sudarsanam, S. The Protein Kinase Complement of the Human Genome. *Science* **2002**, *298*, 1912–1934. [CrossRef] [PubMed]
23. Theivendren, P.; Kunjiappan, S.; Hegde, Y.M.; Vellaichamy, S.; Gopal, M.; Dhramalingam, S.R.; Kumar, S. Importance of Protein Kinase and Its Inhibitor: A Review. *Protein Kinases-Promis Targets Anticance. Drug Res. IntechOpen Ser. Biochem.* **2021**, *24*, 75–100. [CrossRef]

24. Alberts, B.; Johnson, A.; Lewis, J.; Raff, M.; Roberts, K.; Walter, P. *Bruce Alberts, Molecular Biology of the Cell*, 4th ed.; Garland Science: New York, NY, USA, 2002.
25. Turdo, A.; D'Accardo, C.; Glaviano, A.; Porcelli, G.; Colarossi, C.; Colarossi, L.; Mare, M.; Faldetta, N.; Modica, C.; Pistone, G.; et al. Targeting Phosphatases and Kinases: How to Checkmate Cancer. *Front. Cell Dev. Biol.* **2021**, *9*, 690306. [CrossRef] [PubMed]
26. Giamas, G.; Man, Y.L.; Hirner, H.; Bischof, J.; Kramer, K.; Khan, K.; Ahmed, S.S.L.; Stebbing, J.; Knippschild, U. Kinases as targets in the treatment of solid tumors. *Cell Signal.* **2010**, *22*, 984–1002. [CrossRef]
27. Armstrong, H.; Bording-Jorgensen, M.; Dijk, S.; Wine, E. The Complex Interplay between Chronic Inflammation, the Microbiome, and Cancer: Understanding Disease Progression and What We Can Do to Prevent It. *Cancers* **2018**, *10*, 83. [CrossRef]
28. Das, S.; Bhattacharya, B.; Das, B.; Sinha, B.; Jamatia, T.; Paul, K. Etiologic Role of Kinases in the Progression of Human Cancers and Its Targeting Strategies. *Indian J. Surg. Oncol.* **2019**, *12*, 34–45. [CrossRef]
29. McKay, M.M.; Morrison, D.K. Integrating signals from RTKs to ERK/MAPK. *Oncogene* **2007**, *26*, 3113–3121. [CrossRef]
30. Hubbard, S.R.; Miller, W.T. Receptor tyrosine kinases: Mechanisms of activation and signaling. *Curr. Opin. Cell Biol.* **2007**, *19*, 117–123. [CrossRef]
31. Lemmon, M.A.; Schlessinger, J. Cell Signaling by Receptor Tyrosine Kinases. *Cell* **2010**, *141*, 1117–1134. [CrossRef]
32. Ullrich, A.; Schlessinger, J. Signal transduction by receptors with tyrosine kinase activity. *Cell* **1990**, *61*, 203–212. [CrossRef]
33. Popovic, N.; Wilson, E. Cell Surface Receptors. In *Comprehensive Toxicology*; Elsevier: Amsterdam, The Netherlands, 2010; pp. 81–91. [CrossRef]
34. Paul, M.K.; Mukhopadhyay, A.K. Tyrosine kinase—Role and significance in Cancer. *Int. J. Med. Sci.* **2004**, *1*, 101–115. [CrossRef]
35. Schmidt-Arras, D.; Böhmer, F.-D. Mislocalisation of Activated Receptor Tyrosine Kinases—Challenges for Cancer Therapy. *Trends Mol. Med.* **2020**, *26*, 833–847. [CrossRef]
36. Bhargava, R.; Gerald, W.L.; Li, A.R.; Pan, Q.; Lal, P.; Ladanyi, M.; Chen, B. EGFR gene amplification in breast cancer: Correlation with epidermal growth factor receptor mRNA and protein expression and HER-2 status and absence of EGFR-activating mutations. *Mod. Pathol.* **2005**, *18*, 1027–1033. [CrossRef]
37. Drilon, A.; Cappuzzo, F.; Ou, S.-H.I.; Camidge, D.R. Targeting MET in Lung Cancer: Will Expectations Finally Be MET? *J. Thorac. Oncol.* **2017**, *12*, 15–26. [CrossRef]
38. Katoh, M. Fibroblast growth factor receptors as treatment targets in clinical oncology. *Nat. Rev. Clin. Oncol.* **2018**, *16*, 105–122. [CrossRef]
39. Helsten, T.; Elkin, S.; Arthur, E.; Tomson, B.N.; Carter, J.; Kurzrock, R. The FGFR Landscape in Cancer: Analysis of 4,853 Tumors by Next-Generation Sequencing. *Clin. Cancer Res.* **2016**, *22*, 259–267. [CrossRef]
40. Liu, G.; Chen, T.; Ding, Z.; Wang, Y.; Wei, Y.; Wei, X. Inhibition of FGF-FGFR and VEGF-VEGFR signalling in cancer treatment. *Cell Prolif.* **2021**, *54*, e13009. [CrossRef]
41. Ryall, S.; Tabori, U.; Hawkins, C. Pediatric low-grade glioma in the era of molecular diagnostics. *Acta Neuropathol. Commun.* **2020**, *8*, 30. [CrossRef]
42. Rivera, B.; Gayden, T.; Carrot-Zhang, J.; Nadaf, J.; Boshari, T.; Faury, D.; Zeinieh, M.; Blanc, R.; Burk, D.L.; Fahiminiya, S.; et al. Germline and somatic FGFR1 abnormalities in dysembryoplastic neuroepithelial tumors. *Acta Neuropathol.* **2016**, *131*, 847–863. [CrossRef]
43. Valera, E.T.; McConechy, M.K.; Gayden, T.; Rivera, B.; Jones, D.T.W.; Wittmann, A.; Han, H.; Bareke, E.; Nikbakht, H.; Mikael, L.; et al. Methylome analysis and whole-exome sequencing reveal that brain tumors associated with encephalocraniocutaneous lipomatosis are midline pilocytic astrocytomas. *Acta Neuropathol.* **2018**, *136*, 657–660. [CrossRef]
44. Amary, M.F.; Ye, H.; Berisha, F.; Khatri, B.; Forbes, G.; Lehovsky, K.; Frezza, A.M.; Behjati, S.; Tarpey, P.; Pillay, N.; et al. Fibroblastic growth factor receptor 1 amplification in osteosarcoma is associated with poor response to neo-adjuvant chemotherapy. *Cancer Med.* **2014**, *3*, 980–987. [CrossRef] [PubMed]
45. Kim, J.-A.; Berlow, N.E.; Lathara, M.; Bharathy, N.; Martin, L.R.; Purohit, R.; Cleary, M.M.; Liu, Q.; Michalek, J.E.; Srinivasa, G.; et al. Sensitization of osteosarcoma to irradiation by targeting nuclear FGFR1. *Biochem. Biophys. Res. Commun.* **2022**, *621*, 101–108. [CrossRef] [PubMed]
46. Ogura, K.; Elkrief, A.; Bowman, A.S.; Koche, R.P.; de Stanchina, E.; Benayed, R.; Mauguen, A.; Mattar, M.S.; Khodos, I.; Meyers, P.A.; et al. Prospective Clinical Genomic Profiling of Ewing Sarcoma: *ERF* and *FGFR1* Mutations as Recurrent Secondary Alterations of Potential Biologic and Therapeutic Relevance. *JCO Precis. Oncol.* **2022**, *6*, e2200048. [CrossRef] [PubMed]
47. Agelopoulos, K.; Richter, G.H.; Schmidt, E.; Dirksen, U.; von Heyking, K.; Moser, B.; Klein, H.-U.; Kontny, U.; Dugas, M.; Poos, K.; et al. Deep Sequencing in Conjunction with Expression and Functional Analyses Reveals Activation of FGFR1 in Ewing Sarcoma. *Clin. Cancer Res.* **2015**, *21*, 4935–4946. [CrossRef]
48. Rakheja, D.; Park, J.Y.; Yang, M.S.; Martinez, D.P.; Koduru, P.; Wilson, K.S.; Garcia, R.; Uddin, N. Rhabdomyosarcoma With Epithelioid Features And *NSD3::FOXO1* Fusion: Evidence For Reconsideration Of Previously Reported *FOXO1::FGFR1* Fusion. *Int. J. Surg. Pathol.* **2022**. [CrossRef]
49. Goldstein, M.; Meller, I.; Orr-Urtreger, A. FGFR1 over-expression in primary rhabdomyosarcoma tumors is associated with hypomethylation of a 5′ CpG Island and abnormal expression of theAKT1,NOG, andBMP4 genes. *Genes Chromosom. Cancer* **2007**, *46*, 1028–1038. [CrossRef]

50. Gasparini, P.; Fortunato, O.; De Cecco, L.; Casanova, M.; Iannó, M.F.; Carenzo, A.; Centonze, G.; Milione, M.; Collini, P.; Boeri, M.; et al. Age-Related Alterations in Immune Contexture Are Associated with Aggressiveness in Rhabdomyosarcoma. *Cancers* **2019**, *11*, 1380. [CrossRef]
51. Missiaglia, E.; Selfe, J.; Pritchard-Jones, K.; Kool, M.; Shipley, J.; Hamdi, M.; Williamson, D.; Schaaf, G.; Fang, C.; Koster, J.; et al. Genomic imbalances in rhabdomyosarcoma cell lines affect expression of genes frequently altered in primary tumors: An approach to identify candidate genes involved in tumor development. *Genes Chromosom. Cancer* **2009**, *48*, 455–467. [CrossRef]
52. Lehtinen, B.; Raita, A.; Kesseli, J.; Annala, M.; Nordfors, K.; Yli-Harja, O.; Zhang, W.; Visakorpi, T.; Nykter, M.; Haapasalo, H.; et al. Clinical association analysis of ependymomas and pilocytic astrocytomas reveals elevated FGFR3 and FGFR1 expression in aggressive ependymomas. *BMC Cancer* **2017**, *17*, 310. [CrossRef]
53. Cimmino, F.; Montella, A.; Tirelli, M.; Avitabile, M.; Lasorsa, V.A.; Visconte, F.; Cantalupo, S.; Maiorino, T.; De Angelis, B.; Morini, M.; et al. FGFR1 is a potential therapeutic target in neuroblastoma. *Cancer Cell Int.* **2022**, *22*, 174. [CrossRef]
54. Schmelz, K.; Toedling, J.; Huska, M.; Cwikla, M.C.; Kruetzfeldt, L.-M.; Proba, J.; Ambros, P.F.; Ambros, I.M.; Boral, S.; Lodrini, M.; et al. Spatial and temporal intratumour heterogeneity has potential consequences for single biopsy-based neuroblastoma treatment decisions. *Nat. Commun.* **2021**, *12*, 6804. [CrossRef] [PubMed]
55. Nobusawa, S.; Hirato, J.; Yokoo, H. Molecular genetics of ependymomas and pediatric diffuse gliomas: A short review. *Brain Tumor Pathol.* **2014**, *31*, 229–233. [CrossRef] [PubMed]
56. Park, S.-H.; Won, J.; Kim, S.-I.; Lee, Y.; Park, C.-K.; Kim, S.-K.; Choi, S.-H. Molecular Testing of Brain Tumor. *J. Pathol. Transl. Med.* **2017**, *51*, 205–223. [CrossRef] [PubMed]
57. Georgiou, V.; Gkretsi, V. The role of fibroblast growth factors and their receptors in gliomas: The mutations involved. *Rev. Neurosci.* **2018**, *30*, 543–554. [CrossRef]
58. Vega, J.E.V.; Brat, D.J. Incorporating Advances in Molecular Pathology Into Brain Tumor Diagnostics. *Adv. Anat. Pathol.* **2018**, *25*, 143–171. [CrossRef]
59. Jimenez-Pascual, A.; Siebzehnrubl, F.A. Fibroblast Growth Factor Receptor Functions in Glioblastoma. *Cells* **2019**, *8*, 715. [CrossRef]
60. Ohashi, R.; Matsuda, Y.; Ishiwata, T.; Naito, Z. Downregulation of fibroblast growth factor receptor 2 and its isoforms correlates with a high proliferation rate and poor prognosis in high-grade glioma. *Oncol. Rep.* **2014**, *32*, 1163–1169. [CrossRef]
61. Yan, Y.; Li, Z.; Zeng, S.; Wang, X.; Gong, Z.; Xu, Z. FGFR2-mediated phosphorylation of PTEN at tyrosine 240 contributes to the radioresistance of glioma. *J. Cell Commun. Signal.* **2019**, *13*, 279–280. [CrossRef]
62. Salm, F.; Cwiek, P.; Ghosal, A.; Buccarello, A.L.; Largey, F.; Wotzkow, C.; Höland, K.; Styp-Rekowska, B.; Djonov, V.; Zlobec, I.; et al. RNA interference screening identifies a novel role for autocrine fibroblast growth factor signaling in neuroblastoma chemoresistance. *Oncogene* **2012**, *32*, 3944–3953. [CrossRef]
63. Kumar, K.S.; Neve, A.; Stucklin, A.S.G.; Kuzan-Fischer, C.M.; Rushing, E.J.; Taylor, M.D.; Tripolitsioti, D.; Behrmann, L.; Kirschenbaum, D.; Grotzer, M.; et al. TGF-β Determines the Pro-migratory Potential of bFGF Signaling in Medulloblastoma. *Cell Rep.* **2018**, *23*, 3798–3812.e8. [CrossRef]
64. Vignovich, J.; Becker, D. Expression of BFGF and differential expression of FGF receptors in normal human myoblasts and rhabdomyosarcomas. *Int. J. Oncol.* **1993**, *2*, 637–642. [CrossRef]
65. Hirotsu, M.; Setoguchi, T.; Matsunoshita, Y.; Sasaki, H.; Nagao, H.; Gao, H.; Sugimura, K.; Komiya, S. Tumour formation by single fibroblast growth factor receptor 3-positive rhabdomyosarcoma-initiating cells. *Br. J. Cancer* **2009**, *101*, 2030–2037. [CrossRef]
66. Sahu, D.K.; Singh, N.; Das, M.; Rawat, J.; Gupta, D.K. Differential expression profiling of onco and tumor-suppressor genes from major-signaling pathways in Wilms' tumor. *Pediatr. Surg. Int.* **2022**, *38*, 1601–1617. [CrossRef]
67. Kostopoulou, O.N.; Holzhauser, S.; Lange, B.; Ohmayer, A.; Andonova, T.; Bersani, C.; Dalianis, T. Analyses of FGFR3 and PIK3CA mutations in neuroblastomas and the effects of the corresponding inhibitors on neuroblastoma cell lines. *Int. J. Oncol.* **2019**, *55*, 1372–1384. [CrossRef]
68. Ahrendsen, J.T.; Sinai, C.; Meredith, D.M.; Malinowski, S.W.; Cooney, T.M.; Bandopadhayay, P.; Ligon, K.L.; Alexandrescu, S. Molecular Alterations in Pediatric Low-Grade Gliomas That Led to Death. *J. Neuropathol. Exp. Neurol.* **2021**, *80*, 1052–1059. [CrossRef]
69. Johnson, A.; Severson, E.; Gay, L.; Vergilio, J.-A.; Elvin, J.; Suh, J.; Daniel, S.; Covert, M.; Frampton, G.M.; Hsu, S.; et al. Comprehensive Genomic Profiling of 282 Pediatric Low- and High-Grade Gliomas Reveals Genomic Drivers, Tumor Mutational Burden, and Hypermutation Signatures. *Oncologist* **2017**, *22*, 1478–1490. [CrossRef]
70. Frattini, V.; Pagnotta, S.M.; Tala; Fan, J.J.; Russo, M.V.; Lee, S.B.; Garofano, L.; Zhang, J.; Shi, P.; Lewis, G.; et al. A metabolic function of FGFR3-TACC3 gene fusions in cancer. *Nature* **2018**, *553*, 222–227. [CrossRef]
71. Kamura, S.; Matsumoto, Y.; Fukushi, J.-I.; Fujiwara, T.; Iida, K.; Okada, Y.; Iwamoto, Y. Basic fibroblast growth factor in the bone microenvironment enhances cell motility and invasion of Ewing's sarcoma family of tumours by activating the FGFR1–PI3K–Rac1 pathway. *Br. J. Cancer* **2010**, *103*, 370–381. [CrossRef]
72. Lee, S.; Lim, S.; Cho, D. Personalized genomic analysis based on circulating tumor cells of extra-skeletal Ewing sarcoma of the uterus: A case report of a 16-year-old Korean female. *Exp. Ther. Med.* **2018**, *16*, 1343–1349. [CrossRef]
73. Li, Z.; Dou, P.; Liu, T.; He, S. Application of Long Noncoding RNAs in Osteosarcoma: Biomarkers and Therapeutic Targets. *Cell. Physiol. Biochem.* **2017**, *42*, 1407–1419. [CrossRef]

74. Ren, T.; Qing, Y.; Dai, N.; Li, M.; Qian, C.; Yang, Y.; Cheng, Y.; Li, Z.; Zhang, S.; Zhong, Z.; et al. Apurinic/apyrimidinic endonuclease 1 induced upregulation of fibroblast growth factor 2 and its receptor 3 induces angiogenesis in human osteosarcoma cells. *Cancer Sci.* **2014**, *105*, 186–194. [CrossRef] [PubMed]
75. Bi, Y.; Jing, Y.; Cao, Y. Overexpression of miR-100 inhibits growth of osteosarcoma through FGFR3. *Tumor Biol.* **2015**, *36*, 8405–8411. [CrossRef]
76. Gabler, L.; Jaunecker, C.N.; Katz, S.; van Schoonhoven, S.; Englinger, B.; Pirker, C.; Mohr, T.; Vician, P.; Stojanovic, M.; Woitzuck, V.; et al. Fibroblast growth factor receptor 4 promotes glioblastoma progression: A central role of integrin-mediated cell invasiveness. *Acta Neuropathol. Commun.* **2022**, *10*, 65. [CrossRef] [PubMed]
77. Hao, X.; Guo, Z.; Sun, H.; Liu, X.; Zhang, Y.; Zhang, L.; Sun, W.; Tian, Y. Urinary protein biomarkers for pediatric medulloblastoma. *J. Proteom.* **2020**, *225*, 103832. [CrossRef] [PubMed]
78. El Demellawy, D.; McGowan-Jordan, J.; de Nanassy, J.; Chernetsova, E.; Nasr, A. Update on molecular findings in rhabdomyosarcoma. *Pathology* **2017**, *49*, 238–246. [CrossRef] [PubMed]
79. Taylor Vi, J.G.T.; Cheuk, A.T.; Tsang, P.S.; Chung, J.-Y.; Song, Y.K.; Desai, K.; Yu, Y.; Chen, Q.-R.; Shah, K.; Youngblood, V.; et al. Identification of FGFR4-activating mutations in human rhabdomyosarcomas that promote metastasis in xenotransplanted models. *J. Clin. Investig.* **2009**, *119*, 3395–3407. [CrossRef]
80. Wesche, J.; Haglund, K.; Haugsten, E.M. Fibroblast growth factors and their receptors in cancer. *Biochem. J.* **2011**, *437*, 199–213. [CrossRef]
81. Baird, K.; Davis, S.; Antonescu, C.R.; Harper, U.L.; Walker, R.L.; Chen, Y.; Glatfelter, A.A.; Duray, P.H.; Meltzer, P.S. Gene Expression Profiling of Human Sarcomas: Insights into Sarcoma Biology. *Cancer Res* **2005**, *65*, 9226–9235. [CrossRef]
82. Cao, L.; Yu, Y.; Bilke, S.; Walker, R.L.; Mayeenuddin, L.H.; Azorsa, D.O.; Yang, F.; Pineda, M.; Helman, L.J.; Meltzer, P.S. Genome-Wide Identification of PAX3-FKHR Binding Sites in Rhabdomyosarcoma Reveals Candidate Target Genes Important for Development and Cancer. *Cancer Res* **2010**, *70*, 6497–6508. [CrossRef]
83. Shern, J.F.; Chen, L.; Chmielecki, J.; Wei, J.S.; Patidar, R.; Rosenberg, M.; Ambrogio, L.; Auclair, D.; Wang, J.; Song, Y.K.; et al. Comprehensive genomic analysis of rhabdomyosarcoma reveals a landscape of alterations affecting a common genetic axis in fusion-positive and fusion-negative tumors. *Cancer Discov.* **2014**, *4*, 216–231. [CrossRef]
84. Ramadan, F.; Fahs, A.; Ghayad, S.E.; Saab, R. Signaling pathways in Rhabdomyosarcoma invasion and metastasis. *Cancer Metastasis Rev.* **2020**, *39*, 287–301. [CrossRef]
85. Whittle, S.B.; Reyes, S.; Du, M.; Gireud-Goss, M.; Zhang, L.; Woodfield, S.E.; Ittmann, M.; Scheurer, M.; Bean, A.J.; Zage, P.E. A Polymorphism in the FGFR4 Gene Is Associated With Risk of Neuroblastoma and Altered Receptor Degradation. *J. Pediatr. Hematol.* **2016**, *38*, 131–138. [CrossRef]
86. Sugiyama, N.; Varjosalo, M.; Meller, P.; Lohi, J.; Chan, K.M.; Zhou, Z.; Alitalo, K.; Taipale, J.; Keski-Oja, J.; Lehti, K. FGF receptor-4 (FGFR4) polymorphism acts as an activity switch of a membrane type 1 matrix metalloproteinase–FGFR4 complex. *Proc. Natl. Acad. Sci. USA* **2010**, *107*, 15786–15791. [CrossRef]
87. Hajjo, R.; Sweidan, K. Review on Epidermal Growth Factor Receptor (EGFR) Structure, Signaling Pathways, Interactions, and Recent Updates of EGFR Inhibitors. *Curr. Top. Med. Chem.* **2020**, *20*, 815–834. [CrossRef]
88. Nicholson, R.; Gee, J.; Harper, M. EGFR and cancer prognosis. *Eur. J. Cancer* **2001**, *37*, 9–15. [CrossRef]
89. Spano, J.-P.; Lagorce, C.; Atlan, D.; Milano, G.; Domont, J.; Benamouzig, R.; Attar, A.; Benichou, J.; Martin, A.; Morere, J.-F.; et al. Impact of EGFR expression on colorectal cancer patient prognosis and survival. *Ann. Oncol.* **2005**, *16*, 102–108. [CrossRef]
90. Zarghooni, M.; Bartels, U.; Lee, E.; Buczkowicz, P.; Morrison, A.; Huang, A.; Bouffet, E.; Hawkins, C. Whole-Genome Profiling of Pediatric Diffuse Intrinsic Pontine Gliomas Highlights Platelet-Derived Growth Factor Receptor α and Poly (ADP-ribose) Polymerase As Potential Therapeutic Targets. *J. Clin. Oncol.* **2010**, *28*, 1337–1344. [CrossRef]
91. Suri, V.; Das, P.; Jain, A.; Sharma, M.C.; Borkar, S.A.; Suri, A.; Gupta, D.; Sarkar, C. Pediatric glioblastomas: A histopathological and molecular genetic study. *Neuro-Oncology* **2009**, *11*, 274–280. [CrossRef]
92. Wong, A.J.; Bigner, S.H.; Bigner, D.D.; Kinzler, K.W.; Hamilton, S.R.; Vogelstein, B. Increased expression of the epidermal growth factor receptor gene in malignant gliomas is invariably associated with gene amplification. *Proc. Natl. Acad. Sci. USA* **1987**, *84*, 6899–6903. [CrossRef]
93. Kraus, J.A.; Felsberg, J.; Tonn, J.C.; Reifenberger, G.; Pietsch, T. Molecular genetic analysis of the TP53, PTEN, CDKN2A, EGFR, CDK4 and MDM2 tumour-associated genes in supratentorial primitive neuroectodermal tumours and glioblastomas of childhood. *Neuropathol. Appl. Neurobiol.* **2002**, *28*, 325–333. [CrossRef]
94. Pollack, I.F.; Hamilton, R.L.; James, C.D.; Finkelstein, S.D.; Burnham, J.; Yates, A.J.; Holmes, E.J.; Zhou, T.; Finlay, J.L. Rarity of*PTEN*deletions and*EGFR*amplification in malignant gliomas of childhood: Results from the Children's Cancer Group 945 cohort. *J. Neurosurgery Pediatr.* **2006**, *105*, 418–424. [CrossRef] [PubMed]
95. Hatanpaa, K.J.; Burma, S.; Zhao, D.; Habib, A.A. Epidermal Growth Factor Receptor in Glioma: Signal Transduction, Neuropathology, Imaging, and Radioresistance. *Neoplasia* **2010**, *12*, 675–684. [CrossRef] [PubMed]
96. Ganigi, P.; Santosh, V.; Anandh, B.; Chandramouli, B.; Kolluri, V.S. Expression of p53, EGFR, pRb and bcl-2 Proteins in Pediatric Glioblastoma Multiforme: A Study of 54 Patients. *Pediatr. Neurosurg.* **2005**, *41*, 292–299. [CrossRef] [PubMed]
97. Gilbertson, R.J.; Hill, D.A. ERBB1 is amplified and overexpressed in high-grade diffusely infiltrative pediatric brain stem glio-ma. *Clin. Cancer Res.* **2003**, *9 Pt 1*, 3620–3624. [PubMed]

98. Korshunov, A.; Golanov, A.; Timirgaz, V. Immunohistochemical markers for intracranial ependymoma recurrence: An analysis of 88 cases. *J. Neurol. Sci.* **2000**, *177*, 72–82. [CrossRef] [PubMed]
99. Mendrzyk, F.; Korshunov, A.; Benner, A.; Toedt, G.; Pfister, S.; Radlwimmer, B.; Lichter, P. Identification of Gains on 1q and Epidermal Growth Factor Receptor Overexpression as Independent Prognostic Markers in Intracranial Ependymoma. *Clin. Cancer Res.* **2006**, *12*, 2070–2079. [CrossRef]
100. Massimino, M.; Buttarelli, F.R.; Antonelli, M.; Gandola, L.; Modena, P.; Giangaspero, F. Intracranial ependymoma: Factors affecting outcome. *Futur. Oncol.* **2009**, *5*, 207–216. [CrossRef]
101. Gilbertson, R.J.; Perry, R.H.; Kelly, P.J.; Pearson, A.D.; Lunec, J. Prognostic significance of HER2 and HER4 coexpres-sion in childhood medulloblastoma. *Cancer Res.* **1997**, *57*, 3272–3280.
102. Bodey, B.; E Kaiser, H.; E Siegel, S. Epidermal growth factor receptor (EGFR) expression in childhood brain tumors. *Vivo* **2005**, *19*, 931–941.
103. Layfield, L.J.; Thompson, J.K.; Ms, R.K.D.; Kerns, B.-J. Prognostic indicators for neuroblastoma: Stage, grade, DNA ploidy, MIB-1-proliferation index, p53, HER-2/neu and EGFr–a survival study. *J. Surg. Oncol.* **1995**, *59*, 21–27. [CrossRef]
104. Izycka-Swieszewska, E.; Wozniak, A.; Drozynska, E.; Kot, J.; Grajkowska, W.; Klepacka, T.; Perek, D.; Koltan, S.; Bien, E.; Limon, J. Expression and significance of HER family receptors in neuroblastic tumors. *Clin. Exp. Metastasis* **2011**, *28*, 271–282. [CrossRef]
105. Izycka-Swieszewska, E.; Wozniak, A.; Kot, J.; Grajkowska, W.; Balcerska, A.; Perek, D.; Dembowska-Baginska, B.; Klepacka, T.; Drozynska, E. Prognostic significance of HER2 expression in neuroblastic tumors. *Mod. Pathol.* **2010**, *23*, 1261–1268. [CrossRef]
106. Wang, H.; Yang, Q.; Fu, Z.; Zuo, D.; Hua, Y.; Cai, Z. ErbB Receptors as Prognostic and Therapeutic Drug Targets in Bone and Soft Tissue Sarcomas. *Cancer Investig.* **2014**, *32*, 533–542. [CrossRef]
107. Wen, Y.H.; Koeppen, H.; Garcia, R.; Chiriboga, L.; Tarlow, B.D.; Peters, B.A.; Eigenbrot, C.; Yee, H.; Steiner, G.; Greco, M.A. Epidermal growth factor receptor in osteosarcoma: Expression and mutational analysis. *Hum. Pathol.* **2007**, *38*, 1184–1191. [CrossRef]
108. Liu, L.; Xiao, C.; Sun, Q. MiRNA-375 inhibits retinoblastoma progression through targeting ERBB2 and inhibiting MAPK1/MAPK3 signalling pathway. *Cutan. Ocul. Toxicol.* **2021**, *41*, 1–10. [CrossRef]
109. Vasei, M.; Modjtahedi, H.; Ale-Booyeh, O.; Mosallaei, A.; Kajbafzadeh, A.M.; Shahriari, M.; Ghaderi, A.A.; Soleymanpour, H.; Kosari, F.; Moch, H.; et al. Amplification and expression of EGFR and ERBB2 in Wilms tumor. *Cancer Genet. Cytogenet.* **2009**, *194*, 88–95. [CrossRef]
110. Little, S.E.; Bax, D.A.; Rodriguez-Pinilla, M.; Natrajan, R.; Messahel, B.; Pritchard-Jones, K.; Vujanic, G.M.; Reis-Filho, J.S.; Jones, C. Multifaceted Dysregulation of the Epidermal Growth Factor Receptor Pathway in Clear Cell Sarcoma of the Kidney. *Clin. Cancer Res.* **2007**, *13*, 4360–4364. [CrossRef]
111. Armistead, P.M.; Salganick, J.; Roh, J.S.; Steinert, D.M.; Patel, S.; Munsell, M.; El-Naggar, A.K.; Benjamin, R.S.; Zhang, W.; Trent, J.C. Expression of receptor tyrosine kinases and apoptotic molecules in rhabdomyosarcoma: Correlation with overall survival in 105 patients. *Cancer* **2007**, *110*, 2293–2303. [CrossRef]
112. Shibuya, M. Vascular endothelial growth factor and its receptor system: Physiological functions in angiogenesis and pathological roles in various diseases. *J. Biochem.* **2012**, *153*, 13–19. [CrossRef]
113. Risau, W. Mechanisms of angiogenesis. *Nature* **1997**, *386*, 671–674. [CrossRef]
114. Simons, M.; Gordon, E.; Claesson-Welsh, L. Mechanisms and regulation of endothelial VEGF receptor signalling. *Nat. Rev. Mol. Cell Biol.* **2016**, *17*, 611–625. [CrossRef] [PubMed]
115. Tamura, R.; Sato, M.; Morimoto, Y.; Ohara, K.; Kosugi, K.; Oishi, Y.; Kuranari, Y.; Murase, M.; Yoshida, K.; Toda, M. Quantitative assessment and clinical relevance of VEGFRs-positive tumor cells in refractory brain tumors. *Exp. Mol. Pathol.* **2020**, *114*, 104408. [CrossRef] [PubMed]
116. Slongo, M.L.; Molena, B.; Brunati, A.M.; Frasson, M.; Gardiman, M.; Carli, M.; Perilongo, G.; Rosolen, A.; Onisto, M. Functional VEGF and VEGF receptors are expressed in human medulloblastomas. *Neuro-Oncology* **2007**, *9*, 384–392. [CrossRef] [PubMed]
117. Gesundheit, B.; Klement, G.; Senger, C.; Kerbel, R.; Kieran, M.; Baruchel, S.; Becker, L. Differences in vasculature between pilocytic and anaplastic astrocytomas of childhood. *Med. Pediatr. Oncol.* **2003**, *41*, 516–526. [CrossRef]
118. Farschtschi, S.; Merker, V.L.; Wolf, D.; Schuhmann, M.; Blakeley, J.; Plotkin, S.R.; Hagel, C.; Mautner, V.F. Bevacizumab treatment for symptomatic spinal ependymomas in neurofibromatosis type 2. *Acta Neurol. Scand.* **2015**, *133*, 475–480. [CrossRef]
119. Virág, J.; Kenessey, I.; Haberler, C.; Piurko, V.; Bálint, K.; Döme, B.; Tímár, J.; Garami, M.; Hegedűs, B. Angiogenesis and Angiogenic Tyrosine Kinase Receptor Expression in Pediatric Brain Tumors. *Pathol. Oncol. Res.* **2013**, *20*, 417–426. [CrossRef]
120. Zhou, Q.; Yan, X.; Zhu, H.; Xin, Z.; Zhao, J.; Shen, W.; Yin, W.; Guo, Y.; Xu, H.; Zhao, M.; et al. Identification of three tumor antigens and immune subtypes for mRNA vaccine development in diffuse glioma. *Theranostics* **2021**, *11*, 9775–9790. [CrossRef]
121. Fakhari, M.; Pullirsch, D.; Abraham, D.; Paya, K.; Hofbauer, R.; Holzfeind, P.; Hofmann, M.; Aharinejad, S. Selective upregulation of vascular endothelial growth factor receptors neuropilin-1 and -2 in human neuroblastoma. *Cancer* **2001**, *94*, 258–263. [CrossRef]
122. Czapiewski, P.; Kunc, M.; Haybaeck, J. Genetic and molecular alterations in olfactory neuroblastoma: Implications for pathogenesis, prognosis and treatment. *Oncotarget* **2016**, *7*, 52584–52596. [CrossRef]
123. Behjati, S.; Tarpey, P.S.; Haase, K.; Ye, H.; Young, M.D.; Alexandrov, L.B.; Farndon, S.J.; Collord, G.; Wedge, D.C.; Martincorena, I.; et al. Recurrent mutation of IGF signalling genes and distinct patterns of genomic rearrangement in osteosarcoma. *Nat. Commun.* **2017**, *8*, 15936. [CrossRef]

124. Joseph, C.G.; Hwang, H.; Jiao, Y.; Wood, L.D.; Kinde, I.; Wu, J.; Mandahl, N.; Luo, J.; Hruban, R.H.; Diaz, L.; et al. Exomic analysis of myxoid liposarcomas, synovial sarcomas, and osteosarcomas. *Genes Chromosom. Cancer* **2013**, *53*, 15–24. [CrossRef]
125. Perry, J.A.; Kiezun, A.; Tonzi, P.; Van Allen, E.M.; Carter, S.L.; Baca, S.C.; Cowley, G.S.; Bhatt, A.S.; Rheinbay, E.; Pedamallu, C.S.; et al. Complementary genomic approaches highlight the PI3K/mTOR pathway as a common vulnerability in osteosarcoma. *Proc. Natl. Acad. Sci. USA* **2014**, *111*, E5564–E5573. [CrossRef]
126. Negri, G.L.; Grande, B.M.; Delaidelli, A.; El-Naggar, A.; Cochrane, D.; Lau, C.C.; Triche, T.J.; Moore, R.A.; Jones, S.J.M.; Montpetit, A.; et al. Integrative genomic analysis of matched primary and metastatic pediatric osteosarcoma. *J. Pathol.* **2019**, *249*, 319–331. [CrossRef]
127. Subbiah, V.; Cote, G.J. Advances in Targeting RET-Dependent Cancers. *Cancer Discov.* **2020**, *10*, 498–505. [CrossRef]
128. Takahashi, M.; Kawai, K.; Asai, N. Roles of the *RET* Proto-oncogene in Cancer and Development. *JMA J.* **2020**, *3*, 175–181. [CrossRef]
129. Li, A.Y.; McCusker, M.G.; Russo, A.; Scilla, K.A.; Gittens, A.; Arensmeyer, K.; Mehra, R.; Adamo, V.; Rolfo, C. RET fusions in solid tumors. *Cancer Treat. Rev.* **2019**, *81*, 101911. [CrossRef]
130. Adashek, J.J.; Desai, A.P.; Andreev-Drakhlin, A.Y.; Roszik, J.; Cote, G.J.; Subbiah, V. Hallmarks of RET and Co-occuring Genomic Alterations in *RET*-aberrant Cancers. *Mol. Cancer Ther.* **2021**, *20*, 1769–1776. [CrossRef]
131. Elisei, R.; Romei, C.; Vorontsova, T.; Cosci, B.; Veremeychik, V.; Kuchinskaya, E.; Basolo, F.; Demidchik, E.P.; Miccoli, P.; Pinchera, A.; et al. RET/PTC Rearrangements in Thyroid Nodules: Studies in Irradiated and Not Irradiated, Malignant and Benign Thyroid Lesions in Children and Adults1. *J. Clin. Endocrinol. Metab.* **2001**, *86*, 3211–3216. [CrossRef]
132. Rabes, H.M.; Klugbauer, S. Molecular genetics of childhood papillary thyroid carcinomas after irradiation: High prevalence of RET rearrangement. *Genes Environ. Cancer* **1998**, *154*, 248–264. [CrossRef]
133. Zimmerman, D. Thyroid neoplasia in children. *Curr. Opin. Pediatr.* **1997**, *9*, 413–418. [CrossRef]
134. Ortiz, M.V.; Gerdemann, U.; Raju, S.G.; Henry, D.; Smith, S.; Rothenberg, S.M.; Cox, M.C.; Proust, S.; Bender, J.G.; Frazier, A.L.; et al. Activity of the Highly Specific RET Inhibitor Selpercatinib (LOXO-292) in Pediatric Patients With Tumors Harboring *RET* Gene Alterations. *JCO Precis. Oncol.* **2020**, *4*, 341–347. [CrossRef] [PubMed]
135. Kovac, M.; Woolley, C.; Ribi, S.; Blattmann, C.; Roth, E.; Morini, M.; Kovacova, M.; Ameline, B.; Kulozik, A.; Bielack, S.; et al. Germline RET variants underlie a subset of paediatric osteosarcoma. *J. Med. Genet.* **2020**, *58*, 20–24. [CrossRef] [PubMed]
136. Greenfield, E.M.; Collier, C.D.; Getty, P.J. Receptor Tyrosine Kinases in Osteosarcoma: 2019 Update. *Adv. Exp. Med. Biol.* **2020**, *1258*, 141–155. [CrossRef] [PubMed]
137. Luo, J.; Xia, Y.; Yin, Y.; Luo, J.; Liu, M.; Zhang, C.; Zhao, Y.; Yang, L.; Kong, L. ATF4 destabilizes RET through nonclassical GRP78 inhibition to enhance chemosensitivity to bortezomib in human osteosarcoma. *Theranostics* **2019**, *9*, 6334–6353. [CrossRef] [PubMed]
138. Rettew, A.N.; Young, E.D.; Lev, D.C.; Kleinerman, E.S.; Abdulkarim, F.W.; Getty, P.J.; Greenfield, E.M. Multiple receptor tyrosine kinases promote the in vitro phenotype of metastatic human osteosarcoma cell lines. *Oncogenesis* **2012**, *1*, e34. [CrossRef]
139. Rettew, A.N.; Getty, P.J.; Greenfield, E.M. Receptor Tyrosine Kinases in Osteosarcoma: Not Just the Usual Suspects. *Curr. Adv. Osteosarcoma* **2014**, *804*, 47–66. [CrossRef]
140. Dabir, S.; Babakoohi, S.; Kluge, A.; Morrow, J.J.; Kresak, A.; Yang, M.; MacPherson, D.; Wildey, G.; Dowlati, A. RET Mutation and Expression in Small-Cell Lung Cancer. *J. Thorac. Oncol.* **2014**, *9*, 1316–1323. [CrossRef]
141. Cockburn, J.G.; Richardson, D.S.; Gujral, T.S.; Mulligan, L.M. RET-Mediated Cell Adhesion and Migration Require Multiple Integrin Subunits. *J. Clin. Endocrinol. Metab.* **2010**, *95*, E342–E346. [CrossRef]
142. Trusolino, L.; Bertotti, A.; Comoglio, P.M. MET signalling: Principles and functions in development, organ regeneration and cancer. *Nat. Rev. Mol. Cell Biol.* **2010**, *11*, 834–848. [CrossRef]
143. Park, K.C.; Richardson, D.R. The c-MET oncoprotein: Function, mechanisms of degradation and its targeting by novel anti-cancer agents. *Biochim. Biophys. Acta (BBA)-Gen. Subj.* **2020**, *1864*, 129650. [CrossRef]
144. Marona, P.; Górka, J.; Kotlinowski, J.; Majka, M.; Jura, J.; Miekus, K. C-Met as a Key Factor Responsible for Sustaining Undifferentiated Phenotype and Therapy Resistance in Renal Carcinomas. *Cells* **2019**, *8*, 272. [CrossRef]
145. Zambelli, A.; Biamonti, G.; Amato, A. HGF/c-Met Signalling in the Tumor Microenvironment. *TumorMicroenviron. Signal. Pathw. Part B* **2020**, *1270*, 31–44. [CrossRef]
146. Cooper, C.S.; Park, M.; Blair, D.G.; Tainsky, M.A.; Huebner, K.; Croce, C.M.; Vande Woude, G.F. Molecular cloning of a new transforming gene from a chemically transformed human cell line. *Nature* **1984**, *311*, 29–33. [CrossRef]
147. Grundy, M.; Narendran, A. The hepatocyte growth factor/mesenchymal epithelial transition factor axis in high-risk pediatric solid tumors and the anti-tumor activity of targeted therapeutic agents. *Front. Pediatr.* **2022**, *10*, 910268. [CrossRef]
148. Stucklin, A.S.G.; Ryall, S.; Fukuoka, K.; Zapotocky, M.; Lassaletta, A.; Li, C.; Bridge, T.; Kim, B.; Arnoldo, A.; Kowalski, P.E.; et al. Alterations in ALK/ROS1/NTRK/MET drive a group of infantile hemispheric gliomas. *Nat. Commun.* **2019**, *10*, 4343. [CrossRef]
149. Kunkel, P.; Müller, S.; Schirmacher, P.; Stavrou, D.; Fillbrandt, R.; Westphal, M.; Lamszus, K. Expression and localization of scatter factor/hepatocyte growth factor in human astrocytomas. *Neuro-Oncology* **2001**, *3*, 82–88. [CrossRef]
150. Li, Y.; Lal, B.; Kwon, S.; Fan, X.; Saldanha, U.; Reznik, T.E.; Kuchner, E.B.; Eberhart, C.; Laterra, J.; Abounader, R. The Scatter Factor/Hepatocyte Growth Factor: C-Met Pathway in Human Embryonal Central Nervous System Tumor Malignancy. *Cancer Res* **2005**, *65*, 9355–9362. [CrossRef]

151. Provençal, M.; Labbé, D.; Veitch, R.; Boivin, D.; Rivard, G.; Sartelet, H.; Robitaille, Y.; Gingras, D.; Béliveau, R. c-Met activation in medulloblastoma induces tissue factor expression and activity: Effects on cell migration. *Carcinog.* **2009**, *30*, 1089–1096. [CrossRef]
152. E Crosswell, H.; Dasgupta, A.; Alvarado, C.S.; Watt, T.; Christensen, J.G.; De, P.; Durden, D.L.; Findley, H.W. PHA665752, a small-molecule inhibitor of c-Met, inhibits hepatocyte growth factor-stimulated migration and proliferation of c-Met-positive neuroblastoma cells. *BMC Cancer* **2009**, *9*, 411. [CrossRef]
153. Hecht, M.; Papoutsi, M.; Tran, H.D.; Wilting, J.; Schweigerer, L. Hepatocyte Growth Factor/c-Met Signaling Promotes the Progression of Experimental Human Neuroblastomas. *Cancer Res* **2004**, *64*, 6109–6118. [CrossRef]
154. Alami, J.; Williams, B.R.G.; Yeger, H. Expression and localization of HGF and met in Wilms' tumours. *J. Pathol.* **2001**, *196*, 76–84. [CrossRef]
155. Cao, X.; Liu, D.-H.; Zhou, Y.; Yan, X.-M.; Yuan, L.-Q.; Pan, J.; Fu, M.-C.; Zhang, T.; Wang, J. Histone deacetylase 5 promotes Wilms' tumor cell proliferation through the upregulation of c-Met. *Mol. Med. Rep.* **2016**, *13*, 2745–2750. [CrossRef]
156. Nair, R.M.; Prabhu, V.; Manukonda, R.; Mishra, D.K.; Kaliki, S.; Vemuganti, G.K. Overexpression of metastasis-associated in colon cancer 1 in retinoblastoma. *Tumor Biol.* **2020**, *42*, 1010428320975973. [CrossRef]
157. Patanè, S.; Avnet, S.; Coltella, N.; Costa, B.; Sponza, S.; Olivero, M.; Vigna, E.; Naldini, L.; Baldini, N.; Ferracini, R.; et al. *MET* Overexpression Turns Human Primary Osteoblasts into Osteosarcomas. *Cancer Res* **2006**, *66*, 4750–4757. [CrossRef]
158. Chen, W.; Wu, S.; Huang, Y.; Zhang, T.; Dong, H.; Zheng, X.; Chen, T.; Gong, X.; Liu, G.; Zhao, X. A c-Met Inhibitor Suppresses Osteosarcoma Progression via the ERK1/2 Pathway in Human Osteosarcoma Cells. *OncoTargets Ther.* **2021**, *14*, 4791–4804. [CrossRef]
159. Entz-Werle, N.; Lavaux, T.; Metzger, N.; Stoetzel, C.; Lasthaus, C.; Marec, P.; Kalita, C.; Brugieres, L.; Pacquement, H.; Schmitt, C.; et al. Involvement of MET/TWIST/APC Combination or the Potential Role of Ossification Factors in Pediatric High-Grade Osteosarcoma Oncogenesis. *Neoplasia* **2007**, *9*, 678–688. [CrossRef]
160. Fleuren, E.D.; Roeffen, M.H.; Leenders, W.P.; Flucke, U.E.; Vlenterie, M.; Schreuder, H.W.; Boerman, O.C.; van der Graaf, W.T.; Versleijen-Jonkers, Y.M. Expression and clinical relevance of MET and ALK in Ewing sarcomas. *Int. J. Cancer* **2013**, *133*, 427–436. [CrossRef]
161. Yan, D.; Da Dong, X.; Chen, X.; Wang, L.; Lu, C.; Wang, J.; Qu, J.; Tu, L. MicroRNA-1/206 Targets c-Met and Inhibits Rhabdomyosarcoma Development. *J. Biol. Chem.* **2009**, *284*, 29596–29604. [CrossRef]
162. Du, J.; Wang, Y.; Meng, L.; Liu, Y.; Pang, Y.; Cui, W.; Zhang, L.; Li, Z.; Liu, Q.; Shang, H.; et al. c-MET expression potentially contributes to the poor prognosis of rhabdomyosarcoma. *Int. J. Clin. Exp. Pathol.* **2018**, *11*, 4083–4092.
163. Otabe, O.; Kikuchi, K.; Tsuchiya, K.; Katsumi, Y.; Yagyu, S.; Miyachi, M.; Iehara, T.; Hosoi, H. MET/ERK2 pathway regulates the motility of human alveolar rhabdomyosarcoma cells. *Oncol. Rep.* **2016**, *37*, 98–104. [CrossRef]
164. Taulli, R.; Scuoppo, C.; Bersani, F.; Accornero, P.; Forni, P.E.; Miretti, S.; Grinza, A.; Allegra, P.; Schmitt-Ney, M.; Crepaldi, T.; et al. Validation of Met as a Therapeutic Target in Alveolar and Embryonal Rhabdomyosarcoma. *Cancer Res* **2006**, *66*, 4742–4749. [CrossRef]
165. Morris, S.W.; Kirstein, M.N.; Valentine, M.B.; Dittmer, K.G.; Shapiro, D.N.; Saltman, D.L.; Look, A.T. Fusion of a Kinase Gene, *ALK*, to a Nucleolar Protein Gene, *NPM*, in Non-Hodgkin's Lymphoma. *Science* **1994**, *263*, 1281–1284. [CrossRef]
166. Hallberg, B.; Palmer, R. The role of the ALK receptor in cancer biology. *Ann. Oncol.* **2016**, *27*, iii4–iii15. [CrossRef]
167. Takita, J. The role of anaplastic lymphoma kinase in pediatric cancers. *Cancer Sci.* **2017**, *108*, 1913–1920. [CrossRef]
168. Peron, M.; Lovisa, F.; Poli, E.; Basso, G.; Bonvini, P. Understanding the Interplay between Expression, Mutation and Activity of ALK Receptor in Rhabdomyosarcoma Cells for Clinical Application of Small-Molecule Inhibitors. *PLoS ONE* **2015**, *10*, e0132330. [CrossRef]
169. Felkai, L.; Bánusz, R.; Kovalszky, I.; Sápi, Z.; Garami, M.; Papp, G.; Karászi, K.; Varga, E.; Csóka, M. The Presence of ALK Alterations and Clinical Relevance of Crizotinib Treatment in Pediatric Solid Tumors. *Pathol. Oncol. Res.* **2017**, *25*, 217–224. [CrossRef]
170. Aygun, N. Biological and Genetic Features of Neuroblastoma and Their Clinical Importance. *Curr. Pediatr. Rev.* **2018**, *14*, 73–90. [CrossRef]
171. Pastor, E.R.; Mousa, S.A. Current management of neuroblastoma and future direction. *Crit. Rev. Oncol.* **2019**, *138*, 38–43. [CrossRef]
172. Pacenta, H.L.; E Macy, M. Entrectinib and other ALK/TRK inhibitors for the treatment of neuroblastoma. *Drug Des. Dev. Ther.* **2018**, *12*, 3549–3561. [CrossRef]
173. Berry, T.; Luther, W.; Bhatnagar, N.; Jamin, Y.; Poon, E.; Sanda, T.; Pei, D.; Sharma, B.; Vetharoy, W.R.; Hallsworth, A.; et al. The ALKF1174L Mutation Potentiates the Oncogenic Activity of MYCN in Neuroblastoma. *Cancer Cell* **2012**, *22*, 117–130. [CrossRef]
174. Valera, E.T.; Neder, L.; Queiroz, R.G.; Santos, A.C.; Sousa, G.R.; Oliveira, R.S.; Santos, M.V.; Machado, H.R.; Tone, L.G. Perinatal complex low- and high-grade glial tumor harboring a novel *GIGYF2-ALK* fusion. *Pediatr. Blood Cancer* **2019**, *67*, e28015. [CrossRef]
175. Argani, P.; Lian, D.W.; Agaimy, A.; Metzler, M.; Wobker, S.E.; Matoso, A.; Epstein, J.I.; Sung, Y.-S.; Zhang, L.; Antonescu, C.R. Pediatric Mesothelioma With ALK Fusions. *Am. J. Surg. Pathol.* **2021**, *45*, 653–661. [CrossRef] [PubMed]
176. Olsen, T.K.; Panagopoulos, I.; Meling, T.R.; Micci, F.; Gorunova, L.; Thorsen, J.; Due-Tønnessen, B.; Scheie, D.; Lund-Iversen, M.; Krossnes, B.; et al. Fusion genes with*ALK*as recurrent partner in ependymoma-like gliomas: A new brain tumor entity? *Neuro-Oncology* **2015**, *17*, 1365–1373. [CrossRef] [PubMed]

177. Łastowska, M.; Trubicka, J.; Niemira, M.; Paczkowska-Abdulsalam, M.; Karkucińska-Więckowska, A.; Kaleta, M.; Drogosiewicz, M.; Tarasińska, M.; Perek-Polnik, M.; Krętowski, A.; et al. ALK Expression Is a Novel Marker for the WNT-activated Type of Pediatric Medulloblastoma and an Indicator of Good Prognosis for Patients. *Am. J. Surg. Pathol.* **2017**, *41*, 781–787. [CrossRef] [PubMed]
178. Porta, C.; Paglino, C.; Mosca, A. Targeting PI3K/Akt/mTOR Signaling in Cancer. *Front. Oncol.* **2014**, *4*, 64. [CrossRef]
179. Castellano, E.; Downward, J. RAS Interaction with PI3K: More than Just another Effector Pathway. *Genes Cancer* **2011**, *2*, 261–274. [CrossRef]
180. Revathidevi, S.; Munirajan, A.K. Akt in cancer: Mediator and more. *Semin. Cancer Biol.* **2019**, *59*, 80–91. [CrossRef]
181. Barrett, D.; Brown, V.I.; Grupp, S.A.; Teachey, D.T. Targeting the PI3K/AKT/mTOR Signaling Axis in Children with Hematologic Malignancies. *Pediatr. Drugs* **2012**, *14*, 299–316. [CrossRef]
182. Zoncu, R.; Efeyan, A.; Sabatini, D.M. mTOR: From growth signal integration to cancer, diabetes and ageing. *Nat. Rev. Mol. Cell Biol.* **2011**, *12*, 21–35. [CrossRef]
183. Aoki, M.; Fujishita, T. Oncogenic Roles of the PI3K/AKT/mTOR Axis. *Curr. Top. Microbiol. Immunol.* **2017**, *407*, 153–189. [CrossRef]
184. Shorning, B.; Dass, M.; Smalley, M.; Pearson, H. The PI3K-AKT-mTOR Pathway and Prostate Cancer: At the Crossroads of AR, MAPK, and WNT Signaling. *Int. J. Mol. Sci.* **2020**, *21*, 4507. [CrossRef]
185. Narayanankutty, A. PI3K/ Akt/ mTOR Pathway as a Therapeutic Target for Colorectal Cancer: A Review of Preclinical and Clinical Evidence. *Curr. Drug Targets* **2019**, *20*, 1217–1226. [CrossRef]
186. Fattahi, S.; Amjadi-Moheb, F.; Tabaripour, R.; Ashrafi, G.H.; Akhavan-Niaki, H. PI3K/AKT/mTOR signaling in gastric cancer: Epigenetics and beyond. *Life Sci.* **2020**, *262*, 118513. [CrossRef]
187. Bertacchini, J.; Heidari, N.; Mediani, L.; Capitani, S.; Shahjahani, M.; Ahmadzadeh, A.; Saki, N. Targeting PI3K/AKT/mTOR network for treatment of leukemia. *Cell. Mol. Life Sci.* **2015**, *72*, 2337–2347. [CrossRef]
188. Miricescu, D.; Totan, A.; Stanescu-Spinu, I.-I.; Badoiu, S.; Stefani, C.; Greabu, M. PI3K/AKT/mTOR Signaling Pathway in Breast Cancer: From Molecular Landscape to Clinical Aspects. *Int. J. Mol. Sci.* **2020**, *22*, 173. [CrossRef]
189. Gann, C.-N.; Morsli, N.; Chen, X.; Barrueco, J. Response to 'Dai W et al. Am J Cancer Res 2015;5(10):3270-3275′ from the makers of nintedanib. *Am. J. Cancer Res.* **2016**, *6*, 1547–1548.
190. Remke, M.; Pfister, S.; Kox, C.; Toedt, G.; Becker, N.; Benner, A.; Werft, W.; Breit, S.; Liu, S.; Engel, F.; et al. High-resolution genomic profiling of childhood T-ALL reveals frequent copy-number alterations affecting the TGF-β and PI3K-AKT pathways and deletions at 6q15-16.1 as a genomic marker for unfavorable early treatment response. *Blood* **2009**, *114*, 1053–1062. [CrossRef]
191. Armengol, G.; Canellas, A.; Álvarez, Y.; Bastida, P.; De Toledo, J.S.; Pérez-Iribarne, M.D.M.; Camós, M.; Tuset, E.; Estella, J.; Coll, M.D.; et al. Genetic changes including gene copy number alterations and their relation to prognosis in childhood acute myeloid leukemia. *Leuk. Lymphoma* **2009**, *51*, 114–124. [CrossRef]
192. Knobbe, C.B.; Reifenberger, G. Genetic Alterations and Aberrant Expression of Genes Related to the Phosphatidyl-lnositol-3′-Kinase/Protein Kinase B (Akt) Signal Transduction Pathway in Glioblastomas. *Brain Pathol.* **2006**, *13*, 507–518. [CrossRef]
193. Engelman, J.A.; Luo, J.; Cantley, L.C. The evolution of phosphatidylinositol 3-kinases as regulators of growth and metabolism. *Nat. Rev. Genet.* **2006**, *7*, 606–619. [CrossRef]
194. Okkenhaug, K.; Ahmadi, K.; White, S.; Timms, J.; Waterfield, M.D. Cellular Function of Phosphoinositide 3-Kinases: Implications for Development, Immunity, Homeostasis, and Cancer. *Annu. Rev. Cell Dev. Biol.* **2001**, *17*, 615–675. [CrossRef]
195. Zhao, L.; Vogt, P.K. Helical domain and kinase domain mutations in p110 of phosphatidylinositol 3-kinase induce gain of function by different mechanisms. *Proc. Natl. Acad. Sci. USA* **2008**, *105*, 2652–2657. [CrossRef]
196. Samuels, Y.; Wang, Z.; Bardelli, A.; Silliman, N.; Ptak, J.; Szabo, S.; Yan, H.; Gazdar, A.; Powell, S.M.; Riggins, G.J.; et al. High Frequency of Mutations of the *PIK3CA* Gene in Human Cancers. *Science* **2004**, *304*, 554. [CrossRef] [PubMed]
197. Jiang, W.; He, T.; Liu, S.; Zheng, Y.; Xiang, L.; Pei, X.; Wang, Z.; Yang, H. The PIK3CA E542K and E545K mutations promote glycolysis and proliferation via induction of the β-catenin/SIRT3 signaling pathway in cervical cancer. *J. Hematol. Oncol.* **2018**, *11*, 139. [CrossRef] [PubMed]
198. Ikenoue, T.; Kanai, F.; Hikiba, Y.; Obata, T.; Tanaka, Y.; Imamura, J.; Ohta, M.; Jazag, A.; Guleng, B.; Tateishi, K.; et al. Functional Analysis of PIK3CA Gene Mutations in Human Colorectal Cancer. *Cancer Res* **2005**, *65*, 4562–4567. [CrossRef] [PubMed]
199. Murat, C.; Braga, P.; Fortes, M.; Bronstein, M.; Correa-Giannella, M.L.; Giorgi, R. Mutation and genomic amplification of the PIK3CA proto-oncogene in pituitary adenomas. *Braz. J. Med. Biol. Res.* **2012**, *45*, 851–855. [CrossRef]
200. Shi, J.; Yao, D.; Liu, W.; Wang, N.; Lv, H.; Zhang, G.; Ji, M.; Xu, L.; He, N.; Shi, B.; et al. Highly frequent PIK3CA amplification is associated with poor prognosis in gastric cancer. *BMC Cancer* **2012**, *12*, 50. [CrossRef]
201. Holst, F.; Werner, H.M.; Mjøs, S.; Hoivik, E.A.; Kusonmano, K.; Wik, E.; Berg, A.; Birkeland, E.; Gibson, W.J.; Halle, M.K.; et al. *PIK3CA* Amplification Associates with Aggressive Phenotype but Not Markers of AKT-MTOR Signaling in Endometrial Carcinoma. *Clin. Cancer Res.* **2019**, *25*, 334–345. [CrossRef]
202. Huw, L.-Y.; O'Brien, C.; Pandita, A.; Mohan, S.; Spoerke, J.M.; Lu, S.; Wang, Y.; Hampton, G.M.; Wilson, T.R.; Lackner, M.R. Acquired PIK3CA amplification causes resistance to selective phosphoinositide 3-kinase inhibitors in breast cancer. *Oncogenesis* **2013**, *2*, e83. [CrossRef]

203. Salm, F.; Dimitrova, V.; Von Bueren, A.O.; Ćwiek, P.; Rehrauer, H.; Djonov, V.; Anderle, P.; Arcaro, A. The Phosphoinositide 3-Kinase p110α Isoform Regulates Leukemia Inhibitory Factor Receptor Expression via c-Myc and miR-125b to Promote Cell Proliferation in Medulloblastoma. *PLoS ONE* **2015**, *10*, e0123958. [CrossRef]
204. Guerreiro, A.S.; Fattet, S.; Fischer, B.; Shalaby, T.; Jackson, S.P.; Schoenwaelder, S.M.; Grotzer, M.A.; Delattre, O.; Arcaro, A. Targeting the PI3K p110α Isoform Inhibits Medulloblastoma Proliferation, Chemoresistance, and Migration. *Clin. Cancer Res.* **2008**, *14*, 6761–6769. [CrossRef]
205. Guerreiro, A.S.; Fattet, S.; Kulesza, D.W.; Atamer, A.; Elsing, A.N.; Shalaby, T.; Jackson, S.P.; Schoenwaelder, S.M.; Grotzer, M.A.; Delattre, O.; et al. A Sensitized RNA Interference Screen Identifies a Novel Role for the PI3K p110γ Isoform in Medulloblastoma Cell Proliferation and Chemoresistance. *Mol. Cancer Res.* **2011**, *9*, 925–935. [CrossRef]
206. Luk, S.K.; Piekorz, R.P.; Nürnberg, B.; To, S.-S.T. The catalytic phosphoinositol 3-kinase isoform p110δ is required for glioma cell migration and invasion. *Eur. J. Cancer* **2012**, *48*, 149–157. [CrossRef]
207. Boller, D.; Schramm, A.; Doepfner, K.T.; Shalaby, T.; von Bueren, A.O.; Eggert, A.; Grotzer, M.A.; Arcaro, A. Targeting the Phosphoinositide 3-Kinase Isoform p110δ Impairs Growth and Survival in Neuroblastoma Cells. *Clin. Cancer Res.* **2008**, *14*, 1172–1181. [CrossRef]
208. Fransson, S.; Martinsson, T.; Ejeskär, K. Neuroblastoma tumors with favorable and unfavorable outcomes: Significant differences in mRNA expression of genes mapped at 1p36.2. *Genes Chromosom. Cancer* **2006**, *46*, 45–52. [CrossRef]
209. Wang, Q.; Diskin, S.; Rappaport, E.; Attiyeh, E.; Mosse, Y.; Shue, D.; Seiser, E.; Jagannathan, J.; Shusterman, S.; Bansal, M.; et al. Integrative Genomics Identifies Distinct Molecular Classes of Neuroblastoma and Shows That Multiple Genes Are Targeted by Regional Alterations in DNA Copy Number. *Cancer Res* **2006**, *66*, 6050–6062. [CrossRef]
210. Yoon, C.; Lu, J.; Ryeom, S.W.; Simon, M.C.; Yoon, S.S. PIK3R3, part of the regulatory domain of PI3K, is upregulated in sarcoma stem-like cells and promotes invasion, migration, and chemotherapy resistance. *Cell Death Dis.* **2021**, *12*, 749. [CrossRef]
211. Staff, T.P.O. Correction: Variable expression of PIK3R3 and PTEN in Ewing sarcoma impacts oncogenic phenotypes. *PLoS ONE* **2015**, *10*, e0120830. [CrossRef]
212. Bellacosa, A.; Franke, T.F.; E Gonzalez-Portal, M.; Datta, K.; Taguchi, T.; Gardner, J.; Cheng, J.Q.; Testa, J.R.; Tsichlis, P.N. Structure, expression and chromosomal mapping of c-akt: Relationship to v-akt and its implications. *Oncogene* **1993**, *8*, 745–754.
213. Basu, A.; Lambring, C. Akt Isoforms: A Family Affair in Breast Cancer. *Cancers* **2021**, *13*, 3445. [CrossRef]
214. Meier, R.; Alessi, D.R.; Cron, P.; Andjelković, M.; Hemmings, B.A. Mitogenic Activation, Phosphorylation, and Nuclear Translocation of Protein Kinase Bβ. *J. Biol. Chem.* **1997**, *272*, 30491–30497. [CrossRef]
215. Hinz, N.; Jücker, M. Distinct functions of AKT isoforms in breast cancer: A comprehensive review. *Cell Commun. Signal.* **2019**, *17*, 1–29. [CrossRef]
216. Chen, H.; Zhou, L.; Wu, X.; Li, R.; Wen, J.; Sha, J.; Wen, X. The PI3K AKT pathway in the pathogenesis of prostate cancer. *Front. Biosci.* **2016**, *21*, 1084–1091. [CrossRef]
217. Carpten, J.D.; Faber, A.L.; Horn, C.; Donoho, G.P.; Briggs, S.L.; Robbins, C.M.; Hostetter, G.; Boguslawski, S.; Moses, T.Y.; Savage, S.; et al. A transforming mutation in the pleckstrin homology domain of AKT1 in cancer. *Nature* **2007**, *448*, 439–444. [CrossRef]
218. Schüller, U.; Ruiter, M.; Herms, J.; Kretzschmar, H.A.; Grasbon-Frodl, E. Absence of mutations in the AKT1 oncogene in glioblastomas and medulloblastomas. *Acta Neuropathol.* **2008**, *115*, 367–368. [CrossRef]
219. Hartmann, W.; Digon-Söntgerath, B.; Koch, A.; Waha, A.; Endl, E.; Dani, I.; Denkhaus, D.; Goodyer, C.G.; Sörensen, N.; Wiestler, O.D.; et al. Phosphatidylinositol 3′-Kinase/AKT Signaling Is Activated in Medulloblastoma Cell Proliferation and Is Associated with Reduced Expression of *PTEN*. *Clin. Cancer Res.* **2006**, *12*, 3019–3027. [CrossRef]
220. Granados, V.A.; Avirneni-Vadlamudi, U.; Dalal, P.; Scarborough, S.R.; Galindo, K.A.; Mahajan, P.; Galindo, R.L. Selective Targeting of Myoblast Fusogenic Signaling and Differentiation-Arrest Antagonizes Rhabdomyosarcoma Cells. *Cancer Res* **2019**, *79*, 4585–4591. [CrossRef] [PubMed]
221. Hotfilder, M.; Sondermann, P.; Senß, A.; van Valen, F.; Jürgens, H.; Vormoor, J. PI3K/AKT is involved in mediating survival signals that rescue Ewing tumour cells from fibroblast growth factor 2-induced cell death. *Br. J. Cancer* **2005**, *92*, 705–710. [CrossRef]
222. Ren, C.; Pan, R.; Hou, L.; Wu, H.; Sun, J.; Zhang, W.; Tian, X.; Chen, H. Suppression of CLEC3A inhibits osteosarcoma cell proliferation and promotes their chemosensitivity through the AKT1/mTOR/HIF1α signaling pathway. *Mol. Med. Rep.* **2020**, *21*, 1739–1748. [CrossRef]
223. Kuijjer, M.L.; Akker, B.E.W.M.V.D.; Hilhorst, R.; Mommersteeg, M.; Buddingh, E.; Serra, M.; Bürger, H.; Hogendoorn, P.C.W.; Cleton-Jansen, A.-M. Kinome and mRNA expression profiling of high-grade osteosarcoma cell lines implies Akt signaling as possible target for therapy. *BMC Med. Genom.* **2014**, *7*, 4. [CrossRef]
224. Zhu, Y.; Zhou, J.; Ji, Y.; Yu, B. Elevated expression of AKT2 correlates with disease severity and poor prognosis in human osteosarcoma. *Mol. Med. Rep.* **2014**, *10*, 737–742. [CrossRef] [PubMed]
225. Liu, W.; Zhou, Z.; Zhang, Q.; Rong, Y.; Li, L.; Luo, Y.; Wang, J.; Yin, G.; Lv, C.; Cai, W. Overexpression of miR-1258 inhibits cell proliferation by targeting AKT3 in osteosarcoma. *Biochem. Biophys. Res. Commun.* **2019**, *510*, 479–486. [CrossRef] [PubMed]
226. Qiao, J.; Lee, S.; Paul, P.; Qiao, L.; Taylor, C.J.; Schlegel, C.; Colon, N.C.; Chung, D.H. Akt2 Regulates Metastatic Potential in Neuroblastoma. *PLoS ONE* **2013**, *8*, e56382. [CrossRef] [PubMed]
227. Saxton, R.A.; Sabatini, D.M. mTOR Signaling in Growth, Metabolism, and Disease. *Cell* **2017**, *168*, 960–976, Erratum in *Cell* **2017**, *169*, 361–371. [CrossRef] [PubMed]
228. Hemmings, B.A.; Restuccia, D.F. PI3K-PKB/Akt Pathway. *Cold Spring Harb. Perspect. Biol.* **2012**, *4*, a011189. [CrossRef]

229. Zarogoulidis, P.; Lampaki, S.; Turner, J.F.; Huang, H.; Kakolyris, S.; Syrigos, K.; Zarogoulidis, K. mTOR pathway: A current, up-to-date mini-review (Review). *Oncol. Lett.* **2014**, *8*, 2367–2370. [CrossRef]
230. Zou, Z.; Tao, T.; Li, H.; Zhu, X. mTOR signaling pathway and mTOR inhibitors in cancer: Progress and challenges. *Cell Biosci.* **2020**, *10*, 31. [CrossRef]
231. Grabiner, B.C.; Nardi, V.; Birsoy, K.; Possemato, R.; Shen, K.; Sinha, S.; Jordan, A.; Beck, A.H.; Sabatini, D.M. A Diverse Array of Cancer-Associated *MTOR* Mutations Are Hyperactivating and Can Predict Rapamycin Sensitivity. *Cancer Discov.* **2014**, *4*, 554–563. [CrossRef]
232. Culjkovic, B.; Topisirovic, I.; Skrabanek, L.; Ruiz-Gutierrez, M.; Borden, K.L. eIF4E is a central node of an RNA regulon that governs cellular proliferation. *J. Cell Biol.* **2006**, *175*, 415–426. [CrossRef]
233. Mamane, Y.; Petroulakis, E.; Rong, L.; Yoshida, K.; Ler, L.W.; Sonenberg, N. eIF4E—From translation to transformation. *Oncogene* **2004**, *23*, 3172–3179. [CrossRef]
234. Wang, G.; Jia, Y.; Ye, Y.; Kang, E.; Chen, H.; Wang, J.; He, X. Identification of key methylation differentially expressed genes in posterior fossa ependymoma based on epigenomic and transcriptome analysis. *J. Transl. Med.* **2021**, *19*, 1–14. [CrossRef] [PubMed]
235. Machado, L.E.; Alvarenga, A.W.; Da Silva, F.F.; Roffé, M.; Begnami, M.D.; Torres, L.F.B.; Da Cunha, I.W.; Martins, V.R.; Hajj, G.N.M. Overexpression of mTOR and p(240–244)S6 in IDH1 Wild-Type Human Glioblastomas Is Predictive of Low Survival. *J. Histochem. Cytochem.* **2018**, *66*, 403–414. [CrossRef]
236. Shi, J.; Zhang, P.; Su, H.; Cai, L.; Zhao, L.; Zhou, H. Bioinformatics Analysis of Neuroblastoma miRNA Based on GEO Data. *Pharmacogenomics Pers. Med.* **2021**, *14*, 849–858. [CrossRef]
237. Pócza, T.; Sebestyén, A.; Turányi, E.; Krènacs, T.; Márk, A.; Sticz, T.B.; Jakab, Z.; Hauser, P. mTOR Pathway As a Potential Target In a Subset of Human Medulloblastoma. *Pathol. Oncol. Res.* **2014**, *20*, 893–900. [CrossRef]
238. Kaid, C.; Assoni, A.; Marçola, M.; Semedo-Kuriki, P.; Bortolin, R.H.; Carvalho, V.M.; Okamoto, O.K. Proteome and miRNome profiling of microvesicles derived from medulloblastoma cell lines with stem-like properties reveals biomarkers of poor prognosis. *Brain Res.* **2020**, *1730*, 146646. [CrossRef]
239. Chakraborty, S.; Khare, S.; Dorairaj, S.K.; Prabhakaran, V.C.; Prakash, D.R.; Kumar, A. Identification of genes associated with tumorigenesis of retinoblastoma by microarray analysis. *Genomics* **2007**, *90*, 344–353. [CrossRef]
240. Subbiah, V.; Brown, R.E.; Jiang, Y.; Buryanek, J.; Hayes-Jordan, A.; Kurzrock, R.; Anderson, P.M. Morphoproteomic Profiling of the Mammalian Target of Rapamycin (mTOR) Signaling Pathway in Desmoplastic Small Round Cell Tumor (EWS/WT1), Ewing's Sarcoma (EWS/FLI1) and Wilms' Tumor(WT1). *PLoS ONE* **2013**, *8*, e68985. [CrossRef]
241. Ahmed, A.A.; Sherman, A.K.; Pawel, B.R. Expression of therapeutic targets in Ewing sarcoma family tumors. *Hum. Pathol.* **2012**, *43*, 1077–1083. [CrossRef]
242. Dobashi, Y.; Suzuki, S.; Sato, E.; Hamada, Y.; Yanagawa, T.; Ooi, A. EGFR-dependent and independent activation of Akt/mTOR cascade in bone and soft tissue tumors. *Mod. Pathol.* **2009**, *22*, 1328–1340. [CrossRef]
243. Krishnan, K.; Bruce, B.; Hewitt, S.; Thomas, D.; Khanna, C.; Helman, L.J. Ezrin mediates growth and survival in Ewing's sarcoma through the AKT/mTOR, but not the MAPK, signaling pathway. *Clin. Exp. Metastasis* **2006**, *23*, 227–236. [CrossRef]
244. Hu, K.; Dai, H.-B.; Qiu, Z.-L. mTOR signaling in osteosarcoma: Oncogenesis and therapeutic aspects (Review). *Oncol. Rep.* **2016**, *36*, 1219–1225. [CrossRef] [PubMed]
245. Egas-Bejar, D.; Anderson, P.M.; Agarwal, R.; Corrales-Medina, F.; Devarajan, E.; Huh, W.W.; Brown, R.E.; Subbiah, V. Theranostic profiling for actionable aberrations in advanced high risk osteosarcoma with aggressive biology reveals high molecular diversity: The human fingerprint hypothesis. *Oncoscience* **2014**, *1*, 167–179. [CrossRef] [PubMed]
246. Petricoin, E.F.; Espina, V.; Araujo, R.P.; Midura, B.; Yeung, C.; Wan, X.; Eichler, G.S.; Johann, D.J.; Qualman, S.; Tsokos, M.; et al. Phosphoprotein Pathway Mapping: Akt/Mammalian Target of Rapamycin Activation Is Negatively Associated with Childhood Rhabdomyosarcoma Survival. *Cancer Res* **2007**, *67*, 3431–3440. [CrossRef]
247. Doble, B.W.; Woodgett, J.R. GSK-3: Tricks of the trade for a multi-tasking kinase. *J. Cell Sci.* **2003**, *116*, 1175–1186. [CrossRef]
248. Kockeritz, L.; Doble, B.; Patel, S.; Woodgett, J. Glycogen Synthase Kinase-3—An Overview of An Over-Achieving Protein Kinase. *Curr. Drug Targets* **2006**, *7*, 1377–1388. [CrossRef]
249. Sutherland, C. What Are the bona fideGSK3 Substrates? *Int. J. Alzheimer's Dis.* **2011**, *2011*, 505607. [CrossRef]
250. McCubrey, J.A.; Davis, N.M.; Abrams, S.L.; Montalto, G.; Cervello, M.; Basecke, J.; Libra, M.; Nicoletti, F.; Cocco, L.; Martelli, A.M.; et al. Diverse roles of GSK-3: Tumor promoter–tumor suppressor, target in cancer therapy. *Adv. Biol. Regul.* **2014**, *54*, 176–196. [CrossRef]
251. Patel, S.; Woodgett, J. Glycogen Synthase Kinase-3 and Cancer: Good Cop, Bad Cop? *Cancer Cell* **2008**, *14*, 351–353. [CrossRef]
252. Mancinelli, R.; Carpino, G.; Petrungaro, S.; Mammola, C.L.; Tomaipitinca, L.; Filippini, A.; Facchiano, A.; Ziparo, E.; Giampietri, C. Multifaceted Roles of GSK-3 in Cancer and Autophagy-Related Diseases. *Oxidative Med. Cell. Longev.* **2017**, *2017*, 1–14. [CrossRef]
253. Luo, J. Glycogen synthase kinase 3β (GSK3β) in tumorigenesis and cancer chemotherapy. *Cancer Lett.* **2008**, *273*, 194–200. [CrossRef]
254. Banerji, V.; Frumm, S.M.; Ross, K.N.; Li, L.S.; Schinzel, A.C.; Hahn, C.K.; Kakoza, R.M.; Chow, K.T.; Ross, L.; Alexe, G.; et al. The intersection of genetic and chemical genomic screens identifies GSK-3α as a target in human acute myeloid leukemia. *J. Clin. Investig.* **2012**, *122*, 935–947. [CrossRef] [PubMed]

255. Farago, M.; Dominguez, I.; Landesman-Bollag, E.; Xu, X.; Rosner, A.; Cardiff, R.D.; Seldin, D.C. Kinase-Inactive Glycogen Synthase Kinase 3β Promotes Wnt Signaling and Mammary Tumorigenesis. *Cancer Res* **2005**, *65*, 5792–5801. [CrossRef]
256. Ougolkov, A.V.; Fernandez-Zapico, M.E.; Savoy, D.N.; Urrutia, R.A.; Billadeau, D.D. Glycogen Synthase Kinase-3β Participates in Nuclear Factor κB–Mediated Gene Transcription and Cell Survival in Pancreatic Cancer Cells. *Cancer Res* **2005**, *65*, 2076–2081. [CrossRef] [PubMed]
257. Wang, Z.; Smith, K.S.; Murphy, M.; Piloto, O.; Somervaille, T.C.P.; Cleary, M.L. Glycogen synthase kinase 3 in MLL leukaemia maintenance and targeted therapy. *Nature* **2008**, *455*, 1205–1209. [CrossRef] [PubMed]
258. Brassesco, M.S.; Valera, E.T.; Meyer, C.; Marschalek, R.; Lopes, B.A.; Queiroz, R.G.D.P.; Calado, R.D.T.; Scrideli, C.; Tone, L.G. A new complex rearrangement in infant ALL: T(X;11;17)(p11.2;q23;q12). *Cancer Genet.* **2018**, *228–229*, 110–114. [CrossRef]
259. Meyer, C.; Burmeister, T.; Gröger, D.; Tsaur, G.; Fechina, L.; Renneville, A.; Sutton, R.; Venn, N.C.; Emerenciano, M.; Pombo-De-Oliveira, M.S.; et al. The MLL recombinome of acute leukemias in 2017. *Leukemia* **2018**, *32*, 273–284. [CrossRef]
260. Ocasio, J.K.; Bates, R.D.P.; Rapp, C.D.; Gershon, T.R. GSK-3 modulates SHH-driven proliferation in postnatal cerebellar neurogenesis and medulloblastoma. *Development* **2019**, *146*, dev177550. [CrossRef]
261. Silva-Evangelista, C.; Barret, E.; Ménez, V.; Merlevede, J.; Kergrohen, T.; Saccasyn, A.; Oberlin, E.; Puget, S.; Beccaria, K.; Grill, J.; et al. A kinome-wide shRNA screen uncovers vaccinia-related kinase 3 (VRK3) as an essential gene for diffuse intrinsic pontine glioma survival. *Oncogene* **2019**, *38*, 6479–6490. [CrossRef]
262. Lenz, J.E.; Riester, R.; Schleicher, S.B.; Handgretinger, R.; Boehme, K.A.; Traub, F. Interaction of arsenic trioxide and etoposide in Ewing sarcoma cell lines. *Oncol. Rep.* **2019**, *43*, 337–345. [CrossRef]
263. Machado, I.; López-Guerrero, J.A.; Navarro, S.; Alberghini, M.; Scotlandi, K.; Picci, P.; Llombart-Bosch, A. Epithelial cell adhesion molecules and epithelial mesenchymal transition (EMT) markers in Ewing's sarcoma family of tumors (ESFTs). Do they offer any prognostic significance? *Virchows Arch.* **2012**, *461*, 333–337. [CrossRef]
264. Ma, C.; Bower, K.A.; Chen, G.; Shi, X.; Ke, Z.-J.; Luo, J. Interaction between ERK and GSK3β Mediates Basic Fibroblast Growth Factor-induced Apoptosis in SK-N-MC Neuroblastoma Cells. *J. Biol. Chem.* **2008**, *283*, 9248–9256. [CrossRef]
265. Woodgett, J.R. Can a two-faced kinase be exploited for osteosarcoma? *Gynecol. Oncol.* **2012**, *104*, 722–723. [CrossRef]
266. Le Guellec, S.; Moyal, E.C.-J.; Filleron, T.; Delisle, M.-B.; Chevreau, C.; Rubie, H.; Castex, M.-P.; De Gauzy, J.S.; Bonnevialle, P.; Gomez-Brouchet, A. The β5/focal adhesion kinase/glycogen synthase kinase 3β integrin pathway in high-grade osteosarcoma: A protein expression profile predictive of response to neoadjuvant chemotherapy. *Hum. Pathol.* **2013**, *44*, 2149–2158. [CrossRef]
267. Zeng, F.-Y.; Dong, H.; Cui, J.; Liu, L.; Chen, T. Glycogen synthase kinase 3 regulates PAX3–FKHR-mediated cell proliferation in human alveolar rhabdomyosarcoma cells. *Biochem. Biophys. Res. Commun.* **2010**, *391*, 1049–1055. [CrossRef]
268. Dionyssiou, M.G.; Ehyai, S.; Avrutin, E.; Connor, M.K.; McDermott, J.C. Glycogen synthase kinase 3β represses MYOGENIN function in alveolar rhabdomyosarcoma. *Cell Death Dis.* **2014**, *5*, e1094. [CrossRef]
269. Belyea, B.; Kephart, J.G.; Blum, J.; Kirsch, D.G.; Linardic, C.M. Embryonic Signaling Pathways and Rhabdomyosarcoma: Contributions to Cancer Development and Opportunities for Therapeutic Targeting. *Sarcoma* **2012**, *2012*, 1–13. [CrossRef]
270. Ugolkov, A.V.; Bondarenko, G.I.; Dubrovskyi, O.; Berbegall, A.P.; Navarro, S.; Noguera, R.; O'Halloran, T.V.; Hendrix, M.J.; Giles, F.J.; Mazar, A.P. 9-ING-41, a small-molecule glycogen synthase kinase-3 inhibitor, is active in neuroblastoma. *Anti-Cancer Drugs* **2018**, *29*, 717–724. [CrossRef]
271. Li, Z.; Tan, F.; Thiele, C.J. Inactivation of glycogen synthase kinase-3β contributes to brain-derived neutrophic factor/TrkB-induced resistance to chemotherapy in neuroblastoma cells. *Mol. Cancer Ther.* **2007**, *6*, 3113–3121. [CrossRef]
272. Duffy, D.J.; Krstic, A.; Schwarzl, T.; Higgins, D.G.; Kolch, W. GSK3 Inhibitors Regulate *MYCN* mRNA Levels and Reduce Neuroblastoma Cell Viability through Multiple Mechanisms, Including p53 and Wnt Signaling. *Mol. Cancer Ther.* **2014**, *13*, 454–467. [CrossRef]
273. Dickey, A.; Schleicher, S.; Leahy, K.; Hu, R.; Hallahan, D.; Thotala, D.K. GSK-3β inhibition promotes cell death, apoptosis, and in vivo tumor growth delay in neuroblastoma Neuro-2A cell line. *J. Neuro-Oncology* **2010**, *104*, 145–153. [CrossRef]
274. Katoh, Y.; Katoh, M. Hedgehog Target Genes: Mechanisms of Carcinogenesis Induced by Aberrant Hedgehog Signaling Activation. *Curr. Mol. Med.* **2009**, *9*, 873–886. [CrossRef] [PubMed]
275. Urbanska, K.; Trojanek, J.; Del Valle, L.; Eldeen, M.B.; Hofmann, F.; Garcia-Echeverria, C.; Khalili, K.; Reiss, K. Inhibition of IGF-I receptor in anchorage-independence attenuates GSK-3β constitutive phosphorylation and compromises growth and survival of medulloblastoma cell lines. *Oncogene* **2006**, *26*, 2308–2317. [CrossRef] [PubMed]
276. Atkins, R.; Stylli, S.; Luwor, R.; Kaye, A.; Hovens, C. Glycogen synthase kinase-3β (GSK-3β) and its dysregulation in glioblastoma multiforme. *J. Clin. Neurosci.* **2013**, *20*, 1185–1192. [CrossRef] [PubMed]
277. Domoto, T.; Pyko, I.V.; Furuta, T.; Miyashita, K.; Uehara, M.; Shimasaki, T.; Nakada, M.; Minamoto, T. Glycogen synthase kinase-3β is a pivotal mediator of cancer invasion and resistance to therapy. *Cancer Sci.* **2016**, *107*, 1363–1372. [CrossRef] [PubMed]
278. Peti, W.; Page, R. Molecular basis of MAP kinase regulation. *Protein Sci.* **2013**, *22*, 1698–1710. [CrossRef]
279. Schaeffer, H.J.; Weber, M.J. Mitogen-Activated Protein Kinases: Specific Messages from Ubiquitous Messengers. *Mol. Cell. Biol.* **1999**, *19*, 2435–2444. [CrossRef]
280. Kim, E.K.; Choi, E.-J. Compromised MAPK signaling in human diseases: An update. *Arch. Toxicol.* **2015**, *89*, 867–882. [CrossRef]
281. Cuadrado, A.; Nebreda, A.R. Mechanisms and functions of p38 MAPK signalling. *Biochem. J.* **2010**, *429*, 403–417. [CrossRef]

282. Kim, E.K.; Choi, E.-J. Pathological roles of MAPK signaling pathways in human diseases. *Biochim. Biophys. Acta (BBA)-Mol. Basis Dis.* **2010**, *1802*, 396–405. [CrossRef]
283. Cicenas, J.; Zalyte, E.; Rimkus, A.; Dapkus, D.; Noreika, R.; Urbonavicius, S. JNK, p38, ERK, and SGK1 Inhibitors in Cancer. *Cancers* **2017**, *10*, 1. [CrossRef]
284. Mishima, K.; Inoue, K.; Hayashi, Y. Overexpression of extracellular-signal regulated kinases on oral squamous cell carcinoma. *Oral Oncol.* **2002**, *38*, 468–474. [CrossRef]
285. Kudaravalli, S.; Hollander, P.D.; Mani, S.A. Role of p38 MAP kinase in cancer stem cells and metastasis. *Oncogene* **2022**, *41*, 3177–3185. [CrossRef]
286. Tournier, C. The 2 Faces of JNK Signaling in Cancer. *Genes Cancer* **2013**, *4*, 397–400. [CrossRef]
287. Martínez-Limón, A.; Joaquin, M.; Caballero, M.; Posas, F.; de Nadal, E. The p38 Pathway: From Biology to Cancer Therapy. *Int. J. Mol. Sci.* **2020**, *21*, 1913. [CrossRef]
288. Pandey, P.; Raingeaud, J.; Kaneki, M.; Weichselbaum, R.; Davis, R.J.; Kufe, D.; Kharbanda, S. Activation of p38 Mitogen-activated Protein Kinase by c-Abl-dependent and -independent Mechanisms. *J. Biol. Chem.* **1996**, *271*, 23775–23779. [CrossRef]
289. Sui, X.; Kong, N.; Ye, L.; Han, W.; Zhou, J.; Zhang, Q.; He, C.; Pan, H. p38 and JNK MAPK pathways control the balance of apoptosis and autophagy in response to chemotherapeutic agents. *Cancer Lett.* **2014**, *344*, 174–179. [CrossRef]
290. Grossi, V.; Peserico, A.; Tezil, T.; Simone, C. p38α MAPK pathway: A key factor in colorectal cancer therapy and chemoresistance. *World J. Gastroenterol.* **2014**, *20*, 9744–9758. [CrossRef]
291. Masliah-Planchon, J.; Garinet, S.; Pasmant, E. RAS-MAPK pathway epigenetic activation in cancer: miRNAs in action. *Oncotarget* **2015**, *7*, 38892–38907. [CrossRef]
292. Schubbert, S.; Shannon, K.; Bollag, G. Hyperactive Ras in developmental disorders and cancer. *Nat. Rev. Cancer* **2007**, *7*, 295–308. [CrossRef]
293. Liu, F.; Yang, X.; Geng, M.; Huang, M. Targeting ERK, an Achilles' Heel of the MAPK pathway, in cancer therapy. *Acta Pharm. Sin. B* **2018**, *8*, 552–562. [CrossRef]
294. Malumbres, M.; Barbacid, M. RAS oncogenes: The first 30 years. *Nat. Rev. Cancer* **2003**, *3*, 459–465. [CrossRef] [PubMed]
295. Hobbs, G.A.; Der, C.J.; Rossman, K.L. RAS isoforms and mutations in cancer at a glance. *J. Cell Sci.* **2016**, *129*, 1287–1292. [CrossRef]
296. Yaeger, R.; Corcoran, R.B. Targeting Alterations in the RAF–MEK Pathway. *Cancer Discov.* **2019**, *9*, 329–341. [CrossRef] [PubMed]
297. Burotto, M.; Chiou, V.L.; Lee, J.-M.; Kohn, E.C. The MAPK pathway across different malignancies: A new perspective. *Cancer* **2014**, *120*, 3446–3456. [CrossRef] [PubMed]
298. Stefan, E.; Bister, K. MYC and RAF: Key Effectors in Cellular Signaling and Major Drivers in Human Cancer. *Poxviruses* **2017**, *407*, 117–151. [CrossRef]
299. Maurer, G.; Tarkowski, B.; Baccarini, M. Raf kinases in cancer–roles and therapeutic opportunities. *Oncogene* **2011**, *30*, 3477–3488. [CrossRef]
300. Lito, P.; Rosen, N.; Solit, D.B. Tumor adaptation and resistance to RAF inhibitors. *Nat. Med.* **2013**, *19*, 1401–1409. [CrossRef]
301. The Cancer Genome Atlas Network. Genomic Classification of Cutaneous Melanoma. *Cell* **2015**, *161*, 1681–1696. [CrossRef]
302. Samatar, A.A.; Poulikakos, P.I. Targeting RAS–ERK signalling in cancer: Promises and challenges. *Nat. Rev. Drug Discov.* **2014**, *13*, 928–942. [CrossRef]
303. Pfister, S.; Janzarik, W.G.; Remke, M.; Ernst, A.; Werft, W.; Becker, N.; Toedt, G.; Wittmann, A.; Kratz, C.; Olbrich, H.; et al. BRAF gene duplication constitutes a mechanism of MAPK pathway activation in low-grade astrocytomas. *J. Clin. Investig.* **2008**, *118*, 1739–1749. [CrossRef]
304. Bar, E.E.; Lin, A.; Tihan, T.; Burger, P.C.; Eberhart, C.G. Frequent Gains at Chromosome 7q34 Involving BRAF in Pilocytic Astrocytoma. *J. Neuropathol. Exp. Neurol.* **2008**, *67*, 878–887. [CrossRef]
305. Reis, G.F.; Bloomer, M.M.; Perry, A.; Phillips, J.J.; Grenert, J.P.; Karnezis, A.N.; Tihan, T. Pilocytic astrocytomas of the optic nerve and their relation to pilocytic astrocytomas elsewhere in the central nervous system. *Mod. Pathol.* **2013**, *26*, 1279–1287. [CrossRef]
306. Anagnostopoulos, A.K.; Dimas, K.S.; Papathanassiou, C.; Braoudaki, M.; Anastasiadou, E.; Vougas, K.; Karamolegou, K.; Kontos, H.; Prodromou, N.; Tzortzatou-Stathopoulou, F.; et al. Proteomics Studies of Childhood Pilocytic Astrocytoma. *J. Proteome Res.* **2011**, *10*, 2555–2565. [CrossRef]
307. MacDonald, T.J.; Brown, K.M.; LaFleur, B.; Peterson, K.; Lawlor, C.; Chen, Y.; Packer, R.J.; Cogen, P.; Stephan, D.A. Expression profiling of medulloblastoma: PDGFRA and the RAS/MAPK pathway as therapeutic targets for metastatic disease. *Nat. Genet.* **2001**, *29*, 143–152. [CrossRef]
308. Badodi, S.; Pomella, N.; Lim, Y.M.; Brandner, S.; Morrison, G.; Pollard, S.M.; Zhang, X.; Zabet, N.R.; Marino, S. Combination of BMI1 and MAPK/ERK inhibitors is effective in medulloblastoma. *Neuro-Oncology* **2022**, *24*, 1273–1285. [CrossRef]
309. Tsumura, H.; Yoshida, T.; Saito, H.; Imanaka-Yoshida, K.; Suzuki, N. Cooperation of oncogenic K-ras and p53 deficiency in pleomorphic rhabdomyosarcoma development in adult mice. *Oncogene* **2006**, *25*, 7673–7679. [CrossRef]
310. Na, K.Y.; Kim, Y.W.; Park, Y.-K. Mitogen-activated protein kinase pathway in osteosarcoma. *Pathology* **2012**, *44*, 540–546. [CrossRef]
311. Wu, J.; Zhang, C.; Chen, L. MiR-511 mimic transfection inhibits the proliferation, invasion of osteosarcoma cells and reduces metastatic osteosarcoma tumor burden in nude mice via targeting MAPK1. *Cancer Biomarkers* **2019**, *26*, 343–351. [CrossRef]
312. Guo, H.; Zhang, H.-Y.; Wang, S.-L.; Ye, L.; Yang, G.-H.; Bu, H. Smad4 and ERK2 stimulated by transforming growth factor beta1 in rhabdomyosarcoma. *Chin. Med. J.* **2007**, *120*, 515–521. [CrossRef]

313. Lynch, J.; Fay, J.; Meehan, M.; Bryan, K.; Watters, K.M.; Murphy, D.M.; Stallings, R.L. MiRNA-335 suppresses neuroblastoma cell invasiveness by direct targeting of multiple genes from the non-canonical TGF-β signalling pathway. *Carcinog.* **2012**, *33*, 976–985. [CrossRef]
314. Tabatabaei, S.N.; Derbali, R.M.; Yang, C.; Superstein, R.; Hamel, P.; Chain, J.L.; Hardy, P. Co-delivery of miR-181a and melphalan by lipid nanoparticles for treatment of seeded retinoblastoma. *J. Control. Release* **2019**, *298*, 177–185. [CrossRef] [PubMed]
315. Poon, R.Y.C. Cell cycle control: A system of interlinking oscillators. *Methods Mol. Biol.* **2016**, *1342*, 3–19. [PubMed]
316. Saka, Y.; Giuraniuc, C.V.; Ohkura, H. Accurate chromosome segregation by probabilistic self-organisation. *BMC Biol.* **2015**, *13*, 1–10. [CrossRef] [PubMed]
317. Gao, S.-W.; Liu, F. Novel insights into cell cycle regulation of cell fate determination. *J. Zhejiang Univ. B* **2019**, *20*, 467–475. [CrossRef]
318. Ding, L.; Cao, J.; Lin, W.; Chen, H.; Xiong, X.; Ao, H.; Yu, M.; Lin, J.; Cui, Q. The Roles of Cyclin-Dependent Kinases in Cell-Cycle Progression and Therapeutic Strategies in Human Breast Cancer. *Int. J. Mol. Sci.* **2020**, *21*, 1960. [CrossRef]
319. Pines, J. The cell cycle kinases. *Semin. Cancer Biol.* **1994**, *5*, 305–313.
320. Bruyère, C.; Meijer, L. Targeting cyclin-dependent kinases in anti-neoplastic therapy. *Curr. Opin. Cell Biol.* **2013**, *25*, 772–779. [CrossRef]
321. Chilà, R.; Guffanti, F.; Damia, G. Role and therapeutic potential of CDK12 in human cancers. *Cancer Treat. Rev.* **2016**, *50*, 83–88. [CrossRef]
322. Sherr, C.J. Cancer Cell Cycles. *Science* **1996**, *274*, 1672–1677. [CrossRef]
323. Hall, M.; Peters, G. Genetic Alterations of Cyclins, Cyclin-Dependent Kinases, and Cdk Inhibitors in Human Cancer. *Adv.Cancer Res.* **1996**, *68*, 67–108. [CrossRef]
324. Malumbres, M. Cyclins and related kinases in cancer cells. *J. BUON.* **2007**, *12* (Suppl. S1), S45–S52.
325. Malumbres, M.; Barbacid, M. Cell cycle, CDKs and cancer: A changing paradigm. *Nat. Rev. Cancer* **2009**, *9*, 153–166. [CrossRef]
326. Niwa, T.; Akaike, Y.; Watanabe, K.; Chibazakura, T. Hyperactivation of cyclin A-CDK induces centrosome overduplication and chromosome tetraploidization in mouse cells. *Biochem. Biophys. Res. Commun.* **2021**, *549*, 91–97. [CrossRef]
327. Viotto, D.; Russo, F.; Anania, I.; Segatto, I.; Vinciguerra, G.L.R.; Dall'Acqua, A.; Bomben, R.; Perin, T.; Cusan, M.; Schiappacassi, M.; et al. *CDKN1B* mutation and copy number variation are associated with tumor aggressiveness in luminal breast cancer. *J. Pathol.* **2020**, *253*, 234–245. [CrossRef]
328. Lu, Y.; Leow, A.; O Madu, C. The Role of Cyclin-Dependent Kinases on the Metastasis of Breast Cancer. *Nov. Approaches Cancer Study* **2020**, *4*, 377–385. [CrossRef]
329. Lam, Y.; E. di Tomaso, H.-K.; Ng, J.C.S.; Pang, M.F.; Roussel, N.M.; Hjelm, P. Expression of p19 INK4d, CDK4, CDK6 in glioblastoma multiforme. *Br. J. Neurosurg.* **2000**, *14*, 28–32. [CrossRef]
330. Richardson, S.; Hill, R.M.; Kui, C.; Lindsey, J.C.; Grabovksa, Y.; Keeling, C.; Pease, L.; Bashton, M.; Crosier, S.; Vinci, M.; et al. Emergence and maintenance of actionable genetic drivers at medulloblastoma relapse. *Neuro-Oncology* **2021**, *24*, 153–165. [CrossRef]
331. Wood, A.C.; Krytska, K.; Ryles, H.T.; Infarinato, N.R.; Sano, R.; Hansel, T.D.; Hart, L.S.; King, F.J.; Smith, T.R.; Ainscow, E.; et al. Dual *ALK* and *CDK4/6* Inhibition Demonstrates Synergy against Neuroblastoma. *Clin. Cancer Res.* **2017**, *23*, 2856–2868. [CrossRef]
332. Iolascon, A.; Faienza, M.F.; Coppola, B.; Rosolen, A.; Basso, G.; Ragione, F.D.; Schettini, F. Analysis of cyclin-dependent kinase inhibitor genes (CDKN2A, CDKN2B, andCDKN2C) in childhood rhabdomyosarcoma. *Genes Chromosomes Cancer* **1996**, *15*, 217–222. [CrossRef]
333. Komuro, H.; Valentine, M.B.; Rubnitz, J.E.; Saito, M.; Raimondi, S.C.; Carroll, A.J.; Yi, T.; Sherr, C.J.; Look, A.T. p27KIP1 Deletions in Childhood Acute Lymphoblastic Leukemia. *Neoplasia* **1999**, *1*, 253–261. [CrossRef]
334. Martinez-Soria, N.; McKenzie, L.; Draper, J.; Ptasinska, A.; Issa, H.; Potluri, S.; Blair, H.J.; Pickin, A.; Isa, A.; Chin, P.S.; et al. The Oncogenic Transcription Factor RUNX1/ETO Corrupts Cell Cycle Regulation to Drive Leukemic Transformation. *Cancer Cell* **2018**, *34*, 626–642.e8, Erratum in **2019**, *35*, P705. [CrossRef] [PubMed]
335. Diril, M.K.; Ratnacaram, C.K.; Padmakumar, V.C.; Du, T.; Wasser, M.; Coppola, V.; Tessarollo, L.; Kaldis, P. Cyclin-dependent kinase 1 (Cdk1) is essential for cell division and suppression of DNA re-replication but not for liver regeneration. *Proc. Natl. Acad. Sci. USA* **2012**, *109*, 3826–3831. [CrossRef] [PubMed]
336. Liu, X.; Wu, H.; Liu, Z. An Integrative Human Pan-Cancer Analysis of Cyclin-Dependent Kinase 1 (CDK1). *Cancers* **2022**, *14*, 2658. [CrossRef] [PubMed]
337. Ying, X.; Che, X.; Wang, J.; Zou, G.; Yu, Q.; Zhang, X. CDK1 serves as a novel therapeutic target for endometrioid endometrial cancer. *J. Cancer* **2021**, *12*, 2206–2215. [CrossRef]
338. Li, M.; He, F.; Zhang, Z.; Xiang, Z.; Hu, D. CDK1 serves as a potential prognostic biomarker and target for lung cancer. *J. Int. Med. Res.* **2020**, *48*. [CrossRef]
339. Kim, S.J.; Nakayama, S.; Miyoshi, Y.; Taguchi, T.; Tamaki, Y.; Matsushima, T.; Torikoshi, Y.; Tanaka, S.; Yoshida, T.; Ishihara, H.; et al. Determination of the specific activity of CDK1 and CDK2 as a novel prognostic indicator for early breast cancer. *Ann. Oncol.* **2007**, *19*, 68–72. [CrossRef]
340. Wang, C.; Xie, X.; Li, W.; Jiang, D. Expression of KIF2A, NDC80, CDK1, and CCNB1 in breast cancer patients: Their interaction and linkage with tumor features and prognosis. *J. Clin. Lab. Anal.* **2022**, *36*, e24647. [CrossRef]

341. Xing, Z.; Wang, X.; Liu, J.; Zhang, M.; Feng, K.; Wang, X. Expression and prognostic value of CDK1, CCNA2, and CCNB1 gene clusters in human breast cancer. *J. Int. Med. Res.* **2021**, *49*. [CrossRef]
342. Huang, Z.; Shen, G.; Gao, J. CDK1 promotes the stemness of lung cancer cells through interacting with Sox2. *Clin. Transl. Oncol.* **2021**, *23*, 1743–1751. [CrossRef]
343. Ravindran Menon, D.; Luo, Y.; Arcaroli, J.J.; Liu, S.; Krishnankutty, L.N.; Osborne, D.G.; Li, Y.; Samson, J.M.; Bagby, S.; Tan, A.C.; et al. CDK1 Interacts with Sox2 and Promotes Tumor Initiation in Human Melanoma. *Cancer Res.* **2018**, *78*, 6561–6574. [CrossRef]
344. Zhong, S.; Yan, Q.; Ge, J.; Dou, G.; Zhao, G. Identification of driver genes and key pathways of ependymoma. *Turk. Neurosurg.* **2018**. [CrossRef]
345. Pérez-Ramírez, M.; Hernández-Jiménez, A.J.; Guerrero-Guerrero, A.; Benadón-Darszon, E.; Pérezpeña-Díazconti, M.; Siordia-Reyes, A.G.; García-Méndez, A.; de León, F.C.-P.; Salamanca-Gómez, F.A.; García-Hernández, N. Genomics and epigenetics: A study of ependymomas in pediatric patients. *Clin. Neurol. Neurosurg.* **2016**, *144*, 53–58. [CrossRef]
346. Li, Q.; Zhang, L.; Jiang, J.; Zhang, Y.; Wang, X.; Zhang, Q.; Wang, Y.; Liu, C.; Li, F. CDK1 and CCNB1 as potential diagnostic markers of rhabdomyosarcoma: Validation following bioinformatics analysis. *BMC Med. Genom.* **2019**, *12*, 198. [CrossRef]
347. Lu, S.; Sun, C.; Chen, H.; Zhang, C.; Li, W.; Wu, L.; Zhu, J.; Sun, F.; Huang, J.; Wang, J.; et al. Bioinformatics Analysis and Validation Identify CDK1 and MAD2L1 as Prognostic Markers of Rhabdomyosarcoma. *Cancer Manag. Res.* **2020**, *12*, 12123–12136. [CrossRef]
348. Schwermer, M.; Lee, S.; Köster, J.; van Maerken, T.; Stephan, H.; Eggert, A.; Morik, K.; Schulte, J.H.; Schramm, A. Sensitivity to cdk1-inhibition is modulated by p53 status in preclinical models of embryonal tumors. *Oncotarget* **2015**, *6*, 15425–15435. [CrossRef]
349. Shi, K.; Zhu, X.; Wu, J.; Chen, Y.; Zhang, J.; Sun, X. Centromere protein E as a novel biomarker and potential therapeutic target for retinoblastoma. *Bioengineered* **2021**, *12*, 5950–5970. [CrossRef]
350. Liu, J.; Wu, S.; Xie, X.; Wang, Z.; Lei, Q. Identification of potential crucial genes and key pathways in osteosarcoma. *Hereditas* **2020**, *157*, 29. [CrossRef]
351. Liu, L.; Xu, Y.; Reiter, R.J. Melatonin inhibits the proliferation of human osteosarcoma cell line MG-63. *Bone* **2013**, *55*, 432–438. [CrossRef]
352. Zhang, J.; Zhu, X.; Li, H.; Li, B.; Sun, L.; Xie, T.; Zhu, T.; Zhou, H.; Ye, Z. Piperine inhibits proliferation of human osteosarcoma cells via G2/M phase arrest and metastasis by suppressing MMP-2/-9 expression. *Int. Immunopharmacol.* **2015**, *24*, 50–58. [CrossRef]
353. Cai, D.; Latham, V.M.; Zhang, X.; Shapiro, G.I. Combined Depletion of Cell Cycle and Transcriptional Cyclin-Dependent Kinase Activities Induces Apoptosis in Cancer Cells. *Cancer Res* **2006**, *66*, 9270–9280. [CrossRef]
354. Chen, H.; Huang, Q.; Zhai, D.; Dong, J.; Wang, A.; Lan, Q. CDK1 expression and effects of CDK1 silencing on the malig-nant phenotype of glioma cells. *Zhonghua Zhong Liu Za Zhi* **2007**, *29*, 484–488.
355. Zhou, Y.; Yang, L.; Zhang, X.; Chen, R.; Chen, X.; Tang, W.; Zhang, M. Identification of Potential Biomarkers in Glioblastoma through Bioinformatic Analysis and Evaluating Their Prognostic Value. *BioMed Res. Int.* **2019**, *2019*, 1–13. [CrossRef] [PubMed]
356. McCurdy, S.R.; Pacal, M.; Ahmad, M.; Bremner, R. A CDK2 activity signature predicts outcome in CDK2-low cancers. *Oncogene* **2016**, *36*, 2491–2502. [CrossRef] [PubMed]
357. Santamaría, D.; Barrière, C.; Cerqueira, A.; Hunt, S.; Tardy, C.; Newton, K.; Cáceres, J.F.; Dubus, P.; Malumbres, M.; Barbacid, M. Cdk1 is sufficient to drive the mammalian cell cycle. *Nature* **2007**, *448*, 811–815. [CrossRef] [PubMed]
358. Tadesse, S.; Caldon, E.C.; Tilley, W.; Wang, S. Cyclin-Dependent Kinase 2 Inhibitors in Cancer Therapy: An Update. *J. Med. Chem.* **2018**, *62*, 4233–4251. [CrossRef]
359. Teixeira, L.K.; Wang, X.; Li, Y.; Ekholm-Reed, S.; Wu, X.; Wang, P.; Reed, S.I. Cyclin E Deregulation Promotes Loss of Specific Genomic Regions. *Curr. Biol.* **2015**, *25*, 1327–1333. [CrossRef]
360. Tadesse, S.; Anshabo, A.T.; Portman, N.; Lim, E.; Tilley, W.; Caldon, C.E.; Wang, S. Targeting CDK2 in cancer: Challenges and opportunities for therapy. *Drug Discov. Today* **2019**, *25*, 406–413. [CrossRef]
361. Campaner, S.; Doni, M.; Hydbring, P.; Verrecchia, A.; Bianchi, L.; Sardella, D.; Schleker, T.; Perna, D.; Tronnersjö, S.; Murga, M.; et al. Cdk2 suppresses cellular senescence induced by the c-myc oncogene. *Nature* **2009**, *12*, 54–59. [CrossRef]
362. Chen, Z.; Wang, Z.; Pang, J.; Yu, Y.; Bieerkehazhi, S.; Lu, J.; Hu, T.; Zhao, Y.; Xu, X.; Zhang, H.; et al. Multiple CDK inhibitor dinaciclib suppresses neuroblastoma growth via inhibiting CDK2 and CDK9 activity. *Sci. Rep.* **2016**, *6*, 29090. [CrossRef]
363. Bo, L.; Wei, B.; Wang, Z.; Kong, D.; Gao, Z.; Miao, Z. Bioinformatics analysis of the CDK2 functions in neuroblastoma. *Mol. Med. Rep.* **2017**, *17*, 3951–3959. [CrossRef]
364. Bolin, S.; Borgenvik, A.; Persson, C.; Rosén, G.; Sundström, A.; Qi, J.; Bradner, J.E.; Weiss, W.A.; Cho, Y.-J.; Weishaupt, H.; et al. Abstract 2473: Combined BET-bromodomain and CDK2 inhibition in MYC-driven medulloblastoma. *Cancer Res* **2016**, *76*, 2473. [CrossRef]
365. Liu, H.; Weng, J. A comprehensive bioinformatic analysis of cyclin-dependent kinase 2 (CDK2) in glioma. *Gene* **2022**, *822*, 146325. [CrossRef]
366. Zhang, Y.; Xue, C.; Zhu, X.; Zhu, X.; Xian, H.; Huang, Z. Suppression of microRNA-125a-5p upregulates the TAZ-EGFR signaling pathway and promotes retinoblastoma proliferation. *Cell Signal.* **2016**, *28*, 850–860. [CrossRef]
367. Zhang, Y.; Xue, C.; Cui, H.; Huang, Z. High expression of TAZ indicates a poor prognosis in retinoblastoma. *Diagn. Pathol.* **2015**, *10*, 187. [CrossRef]
368. Knudsen, E.S.; Pazzagli, C.; Born, T.L.; Bertolaet, B.L.; Knudsen, K.; Arden, K.C.; Henry, R.R.; Feramisco, J.R. Elevated cyclins and cyclin-dependent kinase activity in the rhabdomyosarcoma cell line RD. *Cancer Res* **1998**, *58*, 2042–2049.

369. Fu, W.; Ma, L.; Chu, B.; Wang, X.; Bagui, T.K.; Bui, M.M.; Gemmer, J.; Altiok, S.; Letson, D.G.; Pledger, W.J. Abstract 3596: SCH727965, a cyclin-dependent kinases inhibitor, induces apoptosis in sarcoma cells through caspase 3- dependent pathway. *Cancer Res* **2011**, *71*, 3596. [CrossRef]
370. Musa, J.; Cidre-Aranaz, F.; Aynaud, M.-M.; Orth, M.F.; Knott, M.M.L.; Mirabeau, O.; Mazor, G.; Varon, M.; Hölting, T.L.B.; Grossetête, S.; et al. Cooperation of cancer drivers with regulatory germline variants shapes clinical outcomes. *Nat. Commun.* **2019**, *10*, 4128. [CrossRef]
371. Ohali, A.; Avigad, S.; Zaizov, R.; Ophir, R.; Horn-Saban, S.; Cohen, I.J.; Meller, I.; Kollender, Y.; Issakov, J.; Yaniv, I. Prediction of high risk Ewing's sarcoma by gene expression profiling. *Oncogene* **2004**, *23*, 8997–9006. [CrossRef]
372. Gao, X.; Leone, G.W.; Wang, H. Cyclin D-CDK4/6 functions in cancer. *Adv. Cancer Res.* **2020**, *148*, 147–169.
373. Dobashi, Y.; Goto, A.; Fukayama, M.; Abe, A.; Ooi, A. Overexpression of cdk4/cyclin D1, a possible mediator of apoptosis and an indicator of prognosis in human primary lung carcinoma. *Int. J. Cancer* **2004**, *110*, 532–541. [CrossRef]
374. Nadal, A.; Jares, P.; Pinyol, M.; Conde, L.; Romeu, C.; Fernández, P.L.; Campo, E.; Cardesa, A. Association of CDK4 and CCND1 mRNA overexpression in laryngeal squamous cell carcinomas occurs without CDK4 amplification. *Virchows Arch.* **2006**, *450*, 161–167. [CrossRef] [PubMed]
375. Hashimoto, H.; Kaku-Ito, Y.; Oda, Y.; Ito, T. CDK4: A Novel Therapeutic Target for Extramammary Paget's Disease. *Front. Oncol.* **2021**, *11*, 710378. [CrossRef] [PubMed]
376. Dong, Y.; Sui, L.; Sugimoto, K.; Tai, Y.; Tokuda, M. Cyclin D1-CDK4 complex, a possible critical factor for cell prolifera-tion and prognosis in laryngeal squamous cell carcinomas. *Int. J. Cancer* **2001**, *95*, 209–215. [CrossRef] [PubMed]
377. Chen, T.-J.; Lee, S.-W.; Lin, L.-C.; Lin, C.-Y.; Chang, K.-Y.; Li, C.-F. Cyclin-dependent kinase 4 overexpression is mostly independent of gene amplification and constitutes an independent prognosticator for nasopharyngeal carcinoma. *Tumor Biol.* **2014**, *35*, 7209–7216. [CrossRef]
378. Wu, A.; Wu, B.; Guo, J.; Luo, W.; Wu, D.; Yang, H.; Zhen, Y.; Yu, X.; Wang, H.; Zhou, Y.; et al. Elevated expression of CDK4 in lung cancer. *J. Transl. Med.* **2011**, *9*, 38. [CrossRef]
379. Lu, J.-W.; Lin, Y.-M.; Chang, J.-G.; Yeh, K.-T.; Chen, R.-M.; Tsai, J.J.P.; Su, W.-W.; Hu, R.-M. Clinical implications of deregulated CDK4 and Cyclin D1 expression in patients with human hepatocellular carcinoma. *Med. Oncol.* **2013**, *30*, 379. [CrossRef]
380. An, H.-X.; Beckmann, M.W.; Reifenberger, G.; Bender, H.G.; Niederacher, D. Gene Amplification and Overexpression of CDK4 in Sporadic Breast Carcinomas Is Associated with High Tumor Cell Proliferation. *Am. J. Pathol.* **1999**, *154*, 113–118. [CrossRef]
381. Yang, C.; Li, Z.; Bhatt, T.; Dickler, M.; Giri, D.; Scaltriti, M.; Baselga, J.; Rosen, N.; Chandarlapaty, S. Acquired CDK6 amplification promotes breast cancer resistance to CDK4/6 inhibitors and loss of ER signaling and dependence. *Oncogene* **2016**, *36*, 2255–2264. [CrossRef]
382. Wölfel, T.; Hauer, M.; Schneider, J.; Serrano, M.; Wölfel, C.; Klehmann-Hieb, E.; De Plaen, E.; Hankeln, T.; Büschenfelde, K.-H.M.Z.; Beach, D. A p16 ^{INK4a} -Insensitive CDK4 Mutant Targeted by Cytolytic T Lymphocytes in a Human Melanoma. *Science* **1995**, *269*, 1281–1284. [CrossRef]
383. Kim, H.; Ham, E.K.; Kim, Y.I.; Chi, J.G.; Lee, H.S.; Park, S.H.; Jung, Y.M.; Myung, N.K.; Lee, M.J.; Jang, J.-J. Overexpression of cyclin D1 and cdk4 in tumorigenesis of sporadic hepatoblastomas. *Cancer Lett.* **1998**, *131*, 177–183. [CrossRef]
384. Nebenfuehr, S.; Kollmann, K.; Sexl, V. The role of CDK6 in cancer. *Int. J. Cancer* **2020**, *147*, 2988–2995. [CrossRef]
385. Jena, N.; Sheng, J.; Hu, J.K.; Li, W.; Zhou, W.; Lee, G.; Tsichlis, N.; A Pathak, A.; Brown, N.; A Deshpande, A.; et al. CDK6-mediated repression of CD25 is required for induction and maintenance of Notch1-induced T-cell acute lymphoblastic leukemia. *Leukemia* **2015**, *30*, 1033–1043. [CrossRef]
386. Kollmann, K.; Sexl, V. CDK6 and p16INK4A in lymphoid malignancies. *Oncotarget* **2013**, *4*, 1858–1859. [CrossRef]
387. Placke, T.; Faber, K.; Nonami, A.; Putwain, S.L.; Salih, H.R.; Heidel, F.H.; Krämer, A.; Root, D.E.; Barbie, D.A.; Krivtsov, A.V.; et al. Requirement for CDK6 in MLL-rearranged acute myeloid leukemia. *Blood* **2014**, *124*, 13–23. [CrossRef]
388. Van der Linden, M.; Willekes, M.; van Roon, E.; Seslija, L.; Schneider, P.; Pieters, R.; Stam, R. MLL fusion-driven activation of*CDK6*potentiates proliferation in*MLL*-rearranged infant ALL. *Cell Cycle* **2014**, *13*, 834–844. [CrossRef]
389. Faussillon, M.; Monnier, L.; Junien, C.; Jeanpierre, C. Frequent overexpression of cyclin D2/cyclin-dependent kinase 4 in Wilms' tumor. *Cancer Lett.* **2005**, *221*, 67–75. [CrossRef]
390. Haruta, M.; Arai, Y.; Okita, H.; Tanaka, Y.; Takimoto, T.; Sugino, R.P.; Yamada, Y.; Kamijo, T.; Oue, T.; Fukuzawa, M.; et al. Combined Genetic and Chromosomal Characterization of Wilms Tumors Identifies Chromosome 12 Gain as a Potential New Marker Predicting a Favorable Outcome. *Neoplasia* **2018**, *21*, 117–131. [CrossRef]
391. Schubert, N.A.; Chen, C.Y.; Rodríguez, A.; Koster, J.; Dowless, M.; Pfister, S.M.; Shields, D.J.; Stancato, L.F.; Vassal, G.; Caron, H.N.; et al. Target actionability review to evaluate CDK4/6 as a therapeutic target in paediatric solid and brain tumours. *Eur. J. Cancer* **2022**, *170*, 196–208. [CrossRef]
392. Iwata, S.; Tatsumi, Y.; Yonemoto, T.; Araki, A.; Itami, M.; Kamoda, H.; Tsukanishi, T.; Hagiwara, Y.; Kinoshita, H.; Ishii, T.; et al. CDK4 overexpression is a predictive biomarker for resistance to conventional chemotherapy in patients with osteosarcoma. *Oncol. Rep.* **2021**, *46*, 1–11. [CrossRef]
393. Zhou, Y.; Shen, J.K.; Yu, Z.; Hornicek, F.J.; Kan, Q.; Duan, Z. Expression and therapeutic implications of cyclin-dependent kinase 4 (CDK4) in osteosarcoma. *Biochim. Biophys. Acta (BBA)-Mol. Basis Dis.* **2018**, *1864*, 1573–1582. [CrossRef]
394. Wunder, J.S.; Eppert, K.; Burrow, S.R.; Gogkoz, N.; Bell, R.S.; Andrulis, I.L. Co-amplification and overexpression of CDK4, SAS and MDM2 occurs frequently in human parosteal osteosarcomas. *Oncogene* **1999**, *18*, 783–788. [CrossRef] [PubMed]

395. Suehara, Y.; Alex, D.; Bowman, A.; Middha, S.; Zehir, A.; Chakravarty, D.; Wang, L.; Jour, G.; Nafa, K.; Hayashi, T.; et al. Clinical Genomic Sequencing of Pediatric and Adult Osteosarcoma Reveals Distinct Molecular Subsets with Potentially Targetable Alterations. *Clin. Cancer Res.* **2019**, *25*, 6346–6356. [CrossRef] [PubMed]
396. Guimarães, G.; Tesser-Gamba, F.; Petrilli, A.; Donato-Macedo, C.; Alves, M.; de Lima, F.; Garcia-Filho, R.; Oliveira, R.; Toledo, S. Molecular profiling of osteosarcoma in children and adolescents from different age groups using a next-generation sequencing panel. *Cancer Genet.* **2021**, *258–259*, 85–92. [CrossRef] [PubMed]
397. Hettmer, S.; Linardic, C.M.; Kelsey, A.; Rudzinski, E.R.; Vokuhl, C.; Selfe, J.; Ruhen, O.; Shern, J.F.; Khan, J.; Kovach, A.R.; et al. Molecular testing of rhabdomyosarcoma in clinical trials to improve risk stratification and outcome: A consensus view from European paediatric Soft tissue sarcoma Study Group, Children's Oncology Group and Cooperative Weichteilsarkom-Studiengruppe. *Eur. J. Cancer* **2022**, *172*, 367–386. [CrossRef]
398. Barr, F.G.; Duan, F.; Smith, L.M.; Gustafson, D.; Pitts, M.; Hammond, S.; Gastier-Foster, J.M. Genomic and clinical analyses of 2p24 and 12q13-q14 amplification in alveolar rhabdomyosarcoma: A report from the Children's Oncology Group. *Genes Chromosom. Cancer* **2009**, *48*, 661–672. [CrossRef]
399. de Andrade, C.R.; Jr, A.T.; Nishimoto, I.N.; Kowalski, L.P.; Lopes, M.A. Rhabdomyosarcoma of the head and neck: A clinicopathological and immunohistochemical analysis of 29 cases. *Braz. Dent. J.* **2010**, *21*, 68–73. [CrossRef]
400. Ragazzini, P.; Gamberi, G.; Pazzaglia, L.; Serra, M.; Magagnoli, G. Amplification of CDK4, MDM2, SAS and GLI genes in leiomyosarcoma, alveolar and embryonal rhabdomyosarcoma. *Histol. Histopathol.* **2004**, 401–411. [CrossRef]
401. Saab, R.; Bills, J.L.; Miceli, A.P.; Anderson, C.M.; Khoury, J.D.; Fry, D.W.; Navid, F.; Houghton, P.J.; Skapek, S.X. Pharmacologic inhibition of cyclin-dependent kinase 4/6 activity arrests proliferation in myoblasts and rhabdomyosarcoma-derived cells. *Mol. Cancer Ther.* **2006**, *5*, 1299–1308. [CrossRef]
402. Barghi, F.; Shannon, H.E.; Saadatzadeh, M.R.; Bailey, B.J.; Riyahi, N.; Bijangi-Vishehsaraei, K.; Just, M.; Ferguson, M.J.; Pandya, P.H.; Pollok, K.E. Precision Medicine Highlights Dysregulation of the CDK4/6 Cell Cycle Regulatory Pathway in Pediatric, Adolescents and Young Adult Sarcomas. *Cancers* **2022**, *14*, 3611. [CrossRef]
403. Kennedy, A.L.; Vallurupalli, M.; Chen, L.; Crompton, B.; Cowley, G.; Vazquez, F.; Weir, B.A.; Tsherniak, A.; Parasuraman, S.; Kim, S.; et al. Functional, chemical genomic, and super-enhancer screening identify sensitivity to cyclin D1/CDK4 pathway inhibition in Ewing sarcoma. *Oncotarget* **2015**, *6*, 30178–30193. [CrossRef]
404. Molenaar, J.J.; Ebus, M.E.; Koster, J.; van Sluis, P.; van Noesel, C.J.; Versteeg, R.; Caron, H.N. Cyclin D1 and CDK4 Activity Contribute to the Undifferentiated Phenotype in Neuroblastoma. *Cancer Res* **2008**, *68*, 2599–2609. [CrossRef]
405. Rader, J.; Russell, M.R.; Hart, L.S.; Nakazawa, M.S.; Belcastro, L.T.; Martinez, D.; Li, Y.; Carpenter, E.L.; Attiyeh, E.F.; Diskin, S.J.; et al. Dual CDK4/CDK6 Inhibition Induces Cell-Cycle Arrest and Senescence in Neuroblastoma. *Clin. Cancer Res.* **2013**, *19*, 6173–6182. [CrossRef]
406. Amoroso, L.; Ognibene, M.; Morini, M.; Conte, M.; Di Cataldo, A.; Tondo, A.; D'Angelo, P.; Castellano, A.; Garaventa, A.; Lasorsa, V.A.; et al. Genomic coamplification of *CDK4/MDM2/FRS2* is associated with very poor prognosis and atypical clinical features in neuroblastoma patients. *Genes Chromosom. Cancer* **2019**, *59*, 277–285. [CrossRef]
407. Huang, W.; Hao, Z.; Mao, F.; Guo, D. Small Molecule Inhibitors in Adult High-Grade Glioma: From the Past to the Future. *Front. Oncol.* **2022**, *12*, 911876. [CrossRef]
408. Liu, A.; Zhao, H.; Sun, B.; Han, X.; Zhou, D.; Cui, Z.; Ma, X.; Zhang, J.; Yuan, L. A predictive analysis approach for paediatric and adult high-grade glioma: miRNAs and network insight. *Ann. Transl. Med.* **2020**, *8*, 242. [CrossRef]
409. Rallis, K.S.; George, A.M.; Wozniak, A.M.; Bigogno, C.M.; Chow, B.; Hanrahan, J.G.; Sideris, M. Molecular Genetics and Targeted Therapies for Paediatric High-grade Glioma. *Cancer Genom.-Proteom.* **2022**, *19*, 390–414. [CrossRef]
410. Sepúlveda-Sánchez, J.M.; Gil-Gil, M.; Alonso-García, M.; Salgado, M.V.; Vicente, E.; Barroso, C.M.; Sánchez, R.; Durán, G.; Peñas, R.D.L.; Muñoz-Langa, J.; et al. Phase II Trial of Palbociclib in Recurrent Retinoblastoma-Positive Anaplastic Oligodendroglioma: A Study from the Spanish Group for Research in Neuro-Oncology (GEINO). *Target. Oncol.* **2020**, *15*, 613–622. [CrossRef]
411. Zangen, I.L.; Kneitz, S.; Monoranu, C.-M.; Rutkowski, S.; Hinkes, B.; Vince, G.H.; Huang, B.; Roggendorf, W. Ependymoma gene expression profiles associated with histological subtype, proliferation, and patient survival. *Acta Neuropathol.* **2007**, *113*, 325–337. [CrossRef]
412. Liang, M.-L.; Chen, C.-H.; Liu, Y.-R.; Huang, M.-H.; Lin, Y.-C.; Wong, T.-T.; Lin, S.-E.; Chu, S.-S.; Ding, Y.-H.; Hsieh, T.-H. Abemaciclib, A Selective CDK4/6 Inhibitor, Restricts the Growth of Pediatric Ependymomas. *Cancers* **2020**, *12*, 3597. [CrossRef]
413. Magalhães, T.D.A.; Cruzeiro, G.A.V.; de Sousa, G.R.; da Silva, K.R.; Lira, R.C.P.; Scrideli, C.A.; Tone, L.G.; Valera, E.T.; Borges, K.S. Notch pathway in ependymoma RELA-fused subgroup: Upregulation and association with cancer stem cells markers expression. *Cancer Gene Ther.* **2019**, *27*, 509–512. [CrossRef]
414. Lummus, S.C.; Donson, A.M.; Gowan, K.; Jones, K.L.; Vibhakar, R.; Foreman, N.K.; Kleinschmidt-DeMasters, B.K. *p16*Loss and E2F/cell cycle deregulation in infant posterior fossa ependymoma. *Pediatr. Blood Cancer* **2017**, *64*, e26656. [CrossRef] [PubMed]
415. Shupp, A.; Casimiro, M.C.; Pestell, R.G. Biological functions of CDK5 and potential CDK5 targeted clinical treatments. *Oncotarget* **2017**, *8*, 17373–17382. [CrossRef] [PubMed]
416. Sharma, S.; Sicinski, P. A kinase of many talents: Non-neuronal functions of CDK5 in development and disease. *Open Biol.* **2020**, *10*, 190287. [CrossRef] [PubMed]
417. Shah, K.; Rossie, S. Tale of the Good and the Bad Cdk5: Remodeling of the Actin Cytoskeleton in the Brain. *Mol. Neurobiol.* **2017**, *55*, 3426–3438. [CrossRef]

418. Liu, W.; Li, J.; Song, Y.-S.; Li, Y.; Jia, Y.-H.; Zhao, H.-D. Cdk5 links with DNA damage response and cancer. *Mol. Cancer* **2017**, *16*, 60. [CrossRef]
419. Liu, S.-L.; Wang, C.; Jiang, T.; Tan, L.; Xing, A.; Yu, J.-T. The Role of Cdk5 in Alzheimer's Disease. *Mol. Neurobiol.* **2015**, *53*, 4328–4342. [CrossRef]
420. Liang, Q.; Li, L.; Zhang, J.; Lei, Y.; Wang, L.; Liu, D.-X.; Feng, J.; Hou, P.; Yao, R.; Zhang, Y.; et al. CDK5 is essential for TGF-β1-induced epithelial-mesenchymal transition and breast cancer progression. *Sci. Rep.* **2013**, *3*, 2932. [CrossRef]
421. Zeng, J.; Xie, S.; Liu, Y.; Shen, C.; Song, X.; Zhou, G.-L.; Wang, C. CDK5 Functions as a Tumor Promoter in Human Lung Cancer. *J. Cancer* **2018**, *9*, 3950–3961. [CrossRef]
422. Lin, H.; Lin, T.-Y.; Juang, J.-L. Abl deregulates Cdk5 kinase activity and subcellular localization in Drosophila neurodegeneration. *Cell Death Differ.* **2006**, *14*, 607–615. [CrossRef]
423. de Porras, V.R.; Bystrup, S.; Heras, S.C.-D.L.; Musulén, E.; Palomero, L.; Alonso, M.H.; Nieto, R.; Arango, D.; Moreno, V.; Queralt, C.; et al. Tumor Expression of Cyclin-Dependent Kinase 5 (Cdk5) Is a Prognostic Biomarker and Predicts Outcome of Oxaliplatin-Treated Metastatic Colorectal Cancer Patients. *Cancers* **2019**, *11*, 1540. [CrossRef]
424. Zhou, Y.; Wang, X.; Lv, P.; Yu, H.; Jiang, X. CDK5 Knockdown inhibits proliferation and induces apoptosis and Cell Cycle Arrest in Human Glioblastoma. *J. Cancer* **2021**, *12*, 3958–3966. [CrossRef]
425. Lin, H.; Chen, M.-C.; Chiu, C.-Y.; Song, Y.-M.; Lin, S.-Y. Cdk5 Regulates STAT3 Activation and Cell Proliferation in Medullary Thyroid Carcinoma Cells. *J. Biol. Chem.* **2007**, *282*, 2776–2784. [CrossRef]
426. Oner, M.; Lin, E.; Chen, M.-C.; Hsu, F.-N.; Prince, G.M.S.H.; Chiu, K.-Y.; Teng, C.-L.J.; Yang, T.-Y.; Wang, H.-Y.; Yue, C.-H.; et al. Future Aspects of CDK5 in Prostate Cancer: From Pathogenesis to Therapeutic Implications. *Int. J. Mol. Sci.* **2019**, *20*, 3881. [CrossRef]
427. Huang, P.-H.; Chen, M.-C.; Peng, Y.-T.; Kao, W.-H.; Chang, C.-H.; Wang, Y.-C.; Lai, C.-H.; Hsieh, J.-T.; Wang, J.-H.; Lee, Y.-T.; et al. Cdk5 Directly Targets Nuclear p21CIP1 and Promotes Cancer Cell Growth. *Cancer Res* **2016**, *76*, 6888–6900. [CrossRef]
428. Sun, Y.-Q.; Xie, J.-W.; Xie, H.-T.; Chen, P.-C.; Zhang, X.-L.; Zheng, C.-H.; Li, P.; Wang, J.-B.; Lin, J.-X.; Cao, L.-L.; et al. Expression of CRM1 and CDK5 shows high prognostic accuracy for gastric cancer. *World J. Gastroenterol.* **2017**, *23*, 2012–2022. [CrossRef]
429. Kour, S.; Rana, S.; Contreras, J.I.; King, H.M.; Robb, C.M.; Sonawane, Y.A.; Bendjennat, M.; Crawford, A.J.; Barger, C.J.; Kizhake, S.; et al. CDK5 Inhibitor Downregulates Mcl-1 and Sensitizes Pancreatic Cancer Cell Lines to Navitoclax. *Mol. Pharmacol.* **2019**, *96*, 419–429. [CrossRef]
430. Mukherjee, S.; Tucker-Burden, C.; Kaissi, E.; Newsam, A.; Duggireddy, H.; Chau, M.; Zhang, C.; Diwedi, B.; Rupji, M.; Seby, S.; et al. CDK5 Inhibition Resolves PKA/cAMP-Independent Activation of CREB1 Signaling in Glioma Stem Cells. *Cell Rep.* **2018**, *23*, 1651–1664. [CrossRef]
431. Do, P.A.; Lee, C.H. The Role of CDK5 in Tumours and Tumour Microenvironments. *Cancers* **2020**, *13*, 101. [CrossRef]
432. Gao, G.-B.; Sun, Y.; Fang, R.-D.; Wang, Y.; Wang, Y.; He, Q.-Y. Post-translational modifications of CDK5 and their biological roles in cancer. *Mol. Biomed.* **2021**, *2*, 1–15. [CrossRef]
433. Peyressatre, M.; Laure, A.; Pellerano, M.; Boukhaddaoui, H.; Soussi, I.; Morris, M.C. Fluorescent Biosensor of CDK5 Kinase Activity in Glioblastoma Cell Extracts and Living Cells. *Biotechnol. J.* **2020**, *15*, e1900474. [CrossRef]
434. Binlateh, T.; Reudhabibadh, R.; Prommeenate, P.; Hutamekalin, P. Investigation of mechanisms underlying the inhibitory effects of metformin against proliferation and growth of neuroblastoma SH-SY5Y cells. *Toxicol. Vitr.* **2022**, *83*, 105410. [CrossRef] [PubMed]
435. Yushan, R.; Wenjie, C.; Suning, H.; Yiwu, D.; Tengfei, Z.; Madushi, W.M.; Feifei, L.; Changwen, Z.; Xin, W.; Roodrajeetsing, G.; et al. Insights into the clinical value of cyclin-dependent kinase 5 in glioma: A retrospective study. *World J. Surg. Oncol.* **2015**, *13*, 223. [CrossRef] [PubMed]
436. Catania, A.; Urban, S.; Yan, E.; Hao, C.; Barron, G.; Allalunis-Turner, J. Expression and localization of cyclin-dependent kinase 5 in apoptotic human glioma cells. *Neuro-Oncology* **2001**, *3*, 89–98. [CrossRef] [PubMed]
437. de Nigris, F.; Mancini, F.P.; Schiano, C.; Infante, T.; Zullo, A.; Minucci, P.B.; Al-Omran, M.; Giordano, A.; Napoli, C. Osteosarcoma cells induce endothelial cell proliferation during neo-angiogenesis. *J. Cell. Physiol.* **2012**, *228*, 846–852. [CrossRef] [PubMed]
438. Bao, H.-X.; Bi, Q.; Han, Y.; Zhao, C.; Zou, H. Potential mechanisms underlying CDK5 related Osteosarcoma progression. *Expert Opin. Ther. Targets* **2017**, *21*, 455–460. [CrossRef]
439. Fu, H.; Zhao, H.; Yang, Y.; Duan, K.; Guo, T. CDK5 Inhibitor Seliciclib Promotes Osteoblastic Differentiation of MSCs and Suppresses the Migration of MG-63 Osteosarcoma Cells. *BioRixv* **2020**. [CrossRef]
440. Saidak, Z.; Le Henaff, C.; Azzi, S.; Marty, C.; Da Nascimento, S.; Sonnet, P.; Marie, P.J. Wnt/β-Catenin Signaling Mediates Osteoblast Differentiation Triggered by Peptide-induced α5β1 Integrin Priming in Mesenchymal Skeletal Cells. *J. Biol. Chem.* **2015**, *290*, 6903–6912. [CrossRef]
441. Schwalbe, E.C.; Lindsey, J.C.; Nakjang, S.; Crosier, S.; Smith, A.J.; Hicks, D.; Rafiee, G.; Hill, R.M.; Iliasova, A.; Stone, T.; et al. Novel molecular subgroups for clinical classification and outcome prediction in childhood medulloblastoma: A cohort study. *Lancet Oncol.* **2017**, *18*, 958–971. [CrossRef]
442. Dorand, R.D.; Nthale, J.; Myers, J.T.; Barkauskas, D.S.; Avril, S.; Chirieleison, S.M.; Pareek, T.K.; Abbott, D.W.; Stearns, D.S.; Letterio, J.J.; et al. Cdk5 disruption attenuates tumor PD-L1 expression and promotes antitumor immunity. *Science* **2016**, *353*, 399–403. [CrossRef]
443. Sava, G.P.; Fan, H.; Coombes, R.C.; Buluwela, L.; Ali, S. CDK7 inhibitors as anticancer drugs. *Cancer Metastasis Rev.* **2020**, *39*, 805–823. [CrossRef]

444. Schachter, M.M.; Fisher, R.P. The CDK-activating kinase Cdk7: Taking yes for an answer. *Cell Cycle* **2013**, *12*, 3239–3240. [CrossRef]
445. Chipumuro, E.; Marco, E.; Christensen, C.L.; Kwiatkowski, N.; Zhang, T.; Hatheway, C.M.; Abraham, B.J.; Sharma, B.; Yeung, C.; Altabef, A.; et al. CDK7 Inhibition Suppresses Super-Enhancer-Linked Oncogenic Transcription in MYCN-Driven Cancer. *Cell* **2014**, *159*, 1126–1139. [CrossRef]
446. Fisher, R.P. Cdk7: A kinase at the core of transcription and in the crosshairs of cancer drug discovery. *Transcription* **2018**, *10*, 47–56. [CrossRef]
447. Bacon, C.W.; D'Orso, I. CDK9: A signaling hub for transcriptional control. *Transcription* **2018**, *10*, 57–75. [CrossRef]
448. Zhang, H.; Pandey, S.; Travers, M.; Sun, H.; Morton, G.; Madzo, J.; Chung, W.; Khowsathit, J.; Perez-Leal, O.; Barrero, C.A.; et al. Targeting CDK9 Reactivates Epigenetically Silenced Genes in Cancer. *Cell* **2018**, *175*, 1244–1258.e26. [CrossRef]
449. Patel, H.; Abduljabbar, R.; Lai, C.-F.; Periyasamy, M.; Harrod, A.; Gemma, C.; Steel, J.H.; Patel, N.; Busonero, C.; Jerjees, D.; et al. Expression of CDK7, Cyclin H, and MAT1 Is Elevated in Breast Cancer and Is Prognostic in Estrogen Receptor–Positive Breast Cancer. *Clin. Cancer Res.* **2016**, *22*, 5929–5938. [CrossRef]
450. Naseh, G.; Mohammadifard, M. Upregulation of cyclin-dependent kinase 7 and matrix metalloproteinase-14 expression contribute to metastatic properties of gastric cancer. *IUBMB Life* **2016**, *68*, 799–805. [CrossRef]
451. Jagomast, T.; Idel, C.; Klapper, L.; Kuppler, P.; Offermann, A.; Dreyer, E.; Bruchhage, K.-L.; Ribbat-Idel, J.; Perner, S. CDK7 Predicts Worse Outcome in Head and Neck Squamous-Cell Cancer. *Cancers* **2022**, *14*, 492. [CrossRef]
452. Kim, J.; Cho, Y.-J.; Ryu, J.-Y.; Hwang, I.; Han, H.D.; Ahn, H.J.; Kim, W.Y.; Cho, H.; Chung, J.-Y.; Hewitt, S.M.; et al. CDK7 is a reliable prognostic factor and novel therapeutic target in epithelial ovarian cancer. *Gynecol. Oncol.* **2020**, *156*, 211–221. [CrossRef]
453. Tang, L.; Zhu, C.; Jin, J.; Wang, X.; Yu, L.; Guan, X. Expression of CDK7 correlates with molecular subtypes and predicts clinical outcomes in breast cancer. *Transl. Cancer Res.* **2021**, *10*, 669–680. [CrossRef]
454. Kretz, A.-L.; Schaum, M.; Richter, J.; Kitzig, E.F.; Engler, C.C.; Leithäuser, F.; Henne-Bruns, D.; Knippschild, U.; Lemke, J. CDK9 is a prognostic marker and therapeutic target in pancreatic cancer. *Tumor Biol.* **2017**, *39*. [CrossRef] [PubMed]
455. Yang, W.; Liu, S.; Luo, Q.; Tan, X. Expression of CDK9 in endometrial cancer tissues and its effect on the proliferation of HEC-1B. *Open Life Sci.* **2021**, *16*, 1341–1346. [CrossRef] [PubMed]
456. Rasool, R.U.; Natesan, R.; Deng, Q.; Aras, S.; Lal, P.; Effron, S.S.; Mitchell-Velasquez, E.; Posimo, J.M.; Carskadon, S.; Baca, S.C.; et al. CDK7 Inhibition Suppresses Castration-Resistant Prostate Cancer through MED1 Inactivation. *Cancer Discov.* **2019**, *9*, 1538–1555. [CrossRef] [PubMed]
457. Gao, X.; Liang, J.; Wang, L.; Zhang, Z.; Yuan, P.; Wang, J.; Gao, Y.; Ma, F.; Calagua, C.; Ye, H.; et al. Phosphorylation of the androgen receptor at Ser81 is co-sustained by CDK1 and CDK9 and leads to AR-mediated transactivation in prostate cancer. *Mol. Oncol.* **2021**, *15*, 1901–1920. [CrossRef] [PubMed]
458. Nagaraja, S.; Vitanza, N.A.; Woo, P.J.; Taylor, K.R.; Liu, F.; Zhang, L.; Li, M.; Meng, W.; Ponnuswami, A.; Sun, W.; et al. Transcriptional Dependencies in Diffuse Intrinsic Pontine Glioma. *Cancer Cell* **2017**, *31*, 635–652.e6. [CrossRef]
459. Ma, H.; Dean, D.C.; Wei, R.; Hornicek, F.J.; Duan, Z. Cyclin-dependent kinase 7 (CDK7) is an emerging prognostic biomarker and therapeutic target in osteosarcoma. *Ther. Adv. Musculoskelet. Dis.* **2021**, *13*. [CrossRef]
460. Zhang, J.; Liu, W.; Zou, C.; Zhao, Z.; Lai, Y.; Shi, Z.; Xie, X.; Huang, G.; Wang, Y.; Zhang, X.; et al. Targeting Super-Enhancer–Associated Oncogenes in Osteosarcoma with THZ2, a Covalent CDK7 Inhibitor. *Clin. Cancer Res.* **2020**, *26*, 2681–2692. [CrossRef]
461. Ma, H.; Seebacher, N.A.; Hornicek, F.J.; Duan, Z. Cyclin-dependent kinase 9 (CDK9) is a novel prognostic marker and therapeutic target in osteosarcoma. *Ebiomedicine* **2018**, *39*, 182–193. [CrossRef]
462. Qin, J.-J. Is CDK9 a promising target for both primary and metastatic osteosarcoma? *Ebiomedicine* **2019**, *40*, 27–28. [CrossRef]
463. Iniguez, A.B.; Stolte, B.; Wang, E.J.; Conway, A.S.; Alexe, G.; Dharia, N.V.; Kwiatkowski, N.; Zhang, T.; Abraham, B.J.; Mora, J.; et al. EWS/FLI Confers Tumor Cell Synthetic Lethality to CDK12 Inhibition in Ewing Sarcoma. *Cancer Cell* **2018**, *33*, 202–216.e6. [CrossRef]
464. Ning, J.; Ma, X.; Long, C.; Mao, Y.; Kuang, X.; Huang, Z.; Fan, Y.; Zhang, H.; Xia, Q.; Wang, R.; et al. Anti-tumor Drug THZ1 Suppresses TGFβ2-mediated EMT in Lens Epithelial Cells via Notch and TGFβ/Smad Signaling Pathway. *J. Cancer* **2019**, *10*, 3778–3788. [CrossRef]
465. Cassandri, M.; Fioravanti, R.; Pomella, S.; Valente, S.; Rotili, D.; Del Baldo, G.; De Angelis, B.; Rota, R.; Mai, A. CDK9 as a Valuable Target in Cancer: From Natural Compounds Inhibitors to Current Treatment in Pediatric Soft Tissue Sarcomas. *Front. Pharmacol.* **2020**, *11*, 1230. [CrossRef]
466. Simone, C.; Giordano, A. Abrogation of signal-dependent activation of the cdk9/cyclin T2a complex in human RD rhabdomyosarcoma cells. *Cell Death Differ.* **2006**, *14*, 192–195. [CrossRef]
467. Richter, G.H.; Hensel, T.; Schmidt, O.; Saratov, V.; von Heyking, K.; Becker-Dettling, F.; Prexler, C.; Yen, H.-Y.; Steiger, K.; Fulda, S.; et al. Combined Inhibition of Epigenetic Readers and Transcription Initiation Targets the EWS-ETS Transcriptional Program in Ewing Sarcoma. *Cancers* **2020**, *12*, 304. [CrossRef]
468. De Falco, G.; Bellan, C.; D'Amuri, A.; Angeloni, G.; Leucci, E.; Giordano, A.; Leoncini, L. Cdk9 regulates neural differentiation and its expression correlates with the differentiation grade of neuroblastoma and PNET tumors. *Cancer Biol. Ther.* **2005**, *4*, 277–281. [CrossRef]
469. Poon, E.; Liang, T.; Jamin, Y.; Walz, S.; Kwok, C.; Hakkert, A.; Barker, K.; Urban, Z.; Thway, K.; Zeid, R.; et al. Orally bioavailable CDK9/2 inhibitor shows mechanism-based therapeutic potential in MYCN-driven neuroblastoma. *J. Clin. Investig.* **2020**, *130*, 5875–5892. [CrossRef]

470. Hamanaka, R.; Maloid, S.; Smith, M.R.; O'Connell, C.D.; Longo, D.L.; Ferris, D.K. Cloning and characterization of hu-man and murine homologues of the Drosophila polo serine-threonine kinase. *Cell Growth Differ.* **1994**, *5*, 249–257.
471. Johnson, E.F.; Stewart, K.; Woods, K.W.; Giranda, V.L.; Luo, Y. Pharmacological and Functional Comparison of the Polo-like Kinase Family: Insight into Inhibitor and Substrate Specificity. *Biochemistry* **2007**, *46*, 9551–9563. [CrossRef]
472. Cizmecioglu, O.; Warnke, S.; Arnold, M.; Duensing, S.; Hoffmann, I. Plk2 regulated centriole duplication is dependent on its localization to the centrosome and a functional polo-box domain. *Cell Cycle* **2008**, *7*, 3548–3555. [CrossRef]
473. Archambault, V.; Glover, D. Polo-like kinases: Conservation and divergence in their functions and regulation. *Nat. Rev. Mol. Cell Biol.* **2009**, *10*, 265–275. [CrossRef]
474. Glover, D.M.; Hagan, I.M.; Tavares, A. Polo-like kinases: A team that plays throughout mitosis. *Genes Dev.* **1998**, *12*, 3777–3787. [CrossRef]
475. Barr, F.; Sillјé, H.H.W.; Nigg, E. Polo-like kinases and the orchestration of cell division. *Nat. Rev. Mol. Cell Biol.* **2004**, *5*, 429–441. [CrossRef] [PubMed]
476. A Winkles, J.; Alberts, G.F. Differential regulation of polo-like kinase 1, 2, 3, and 4 gene expression in mammalian cells and tissues. *Oncogene* **2005**, *24*, 260–266. [CrossRef] [PubMed]
477. Iliaki, S.; Beyaert, R.; Afonina, I.S. Polo-like kinase 1 (PLK1) signaling in cancer and beyond. *Biochem. Pharmacol.* **2021**, *193*, 114747. [CrossRef] [PubMed]
478. Raab, C.A.; Raab, M.; Becker, S.; Strebhardt, K. Non-mitotic functions of polo-like kinases in cancer cells. *Biochim. Biophys. Acta (BBA)-Rev. Cancer* **2020**, *1875*, 188467. [CrossRef]
479. Cholewa, B.D.; Liu, X.; Ahmad, N. The Role of Polo-like Kinase 1 in Carcinogenesis: Cause or Consequence? *Cancer Res* **2013**, *73*, 6848–6855. [CrossRef]
480. Strebhardt, K. Multifaceted polo-like kinases: Drug targets and antitargets for cancer therapy. *Nat. Rev. Drug Discov.* **2010**, *9*, 643–660. [CrossRef]
481. Pellegrino, R.; Calvisi, D.F.; Ladu, S.; Ehemann, V.; Staniscia, T.; Evert, M.; Dombrowski, F.; Schirmacher, P.; Longerich, T. Oncogenic and tumor suppressive roles of polo-like kinases in human hepatocellular carcinoma. *Hepatology* **2009**, *51*, 857–868. [CrossRef]
482. Degenhardt, Y.; Lampkin, T. Targeting Polo-like kinase in cancer therapy. *Clin. Cancer Res. Off. J. Am. Assoc. Cancer Res.* **2010**, *16*, 384–389. [CrossRef]
483. Ando, K.; Ozaki, T.; Yamamoto, H.; Furuya, K.; Hosoda, M.; Hayashi, S.; Fukuzawa, M.; Nakagawara, A. Polo-like Kinase 1 (Plk1) Inhibits p53 Function by Physical Interaction and Phosphorylation. *J. Biol. Chem.* **2004**, *279*, 25549–25561. [CrossRef]
484. Li, S.; Li, H.; Cao, Y.; Geng, H.; Ren, F.; Li, K.; Dai, C.; Li, N. Integrated bioinformatics analysis reveals CDK1 and PLK1 as potential therapeutic targets of lung adenocarcinoma. *Medicine* **2021**, *100*, e26474. [CrossRef]
485. Li, H.; Wang, H.; Sun, Z.; Guo, Q.; Shi, H.; Jia, Y. The clinical and prognostic value of polo-like kinase 1 in lung squamous cell carcinoma patients: Immunohistochemical analysis. *Biosci. Rep.* **2017**, *37*. [CrossRef]
486. Ramani, P.; Nash, R.; Sowa-Avugrah, E.; A Rogers, C. High levels of polo-like kinase 1 and phosphorylated translationally controlled tumor protein indicate poor prognosis in neuroblastomas. *J. Neuro-Oncology* **2015**, *125*, 103–111. [CrossRef]
487. Harris, P.S.; Venkataraman, S.; Alimova, I.; Birks, D.K.; Donson, A.M.; Knipstein, J.; Dubuc, A.; Taylor, M.D.; Handler, M.H.; Foreman, N.K.; et al. Polo-like kinase 1 (PLK1) inhibition suppresses cell growth and enhances radiation sensitivity in medulloblastoma cells. *BMC Cancer* **2012**, *12*, 80. [CrossRef]
488. Triscott, J.; Lee, C.; Foster, C.; Manoranjan, B.; Pambid, M.R.; Berns, R.; Fotovati, A.; Venugopal, C.; O'Halloran, K.; Narendran, A.; et al. Personalizing the Treatment of Pediatric Medulloblastoma: Polo-like Kinase 1 as a Molecular Target in High-Risk Children. *Cancer Res* **2013**, *73*, 6734–6744. [CrossRef]
489. Pezuk, J.A.; Brassesco, M.S.; de Oliveira, R.S.; Machado, H.R.; Neder, L.; Scrideli, C.A.; Tone, L.G. PLK1-associated microRNAs are correlated with pediatric medulloblastoma prognosis. *Child's Nerv. Syst.* **2017**, *33*, 609–615. [CrossRef]
490. Ma, H.; Nie, C.; Chen, Y.; Li, J.; Xie, Y.; Tang, Z.; Gao, Y.; Ai, S.; Mao, Y.; Sun, Q.; et al. Therapeutic Targeting PLK1 by ON-01910.Na Is Effective in Local Treatment of Retinoblastoma. *Oncol. Res. Featur. Preclin. Clin. Cancer Ther.* **2021**, *28*, 745–761. [CrossRef]
491. Singh, L.; Pushker, N.; Sen, S.; Singh, M.K.; A Chauhan, F.; Kashyap, S. Prognostic significance of polo-like kinases in retinoblastoma: Correlation with patient outcome, clinical and histopathological parameters. *Clin. Exp. Ophthalmol.* **2015**, *43*, 550–557. [CrossRef]
492. Ackermann, S.; Goeser, F.; Schulte, J.H.; Schramm, A.; Ehemann, V.; Hero, B.; Eggert, A.; Berthold, F.; Fischer, M. Polo-Like Kinase 1 is a Therapeutic Target in High-Risk Neuroblastoma. *Clin. Cancer Res.* **2011**, *17*, 731–741. [CrossRef]
493. Mo, H.; He, J.; Yuan, Z.; Wu, Z.; Liu, B.; Lin, X.; Guan, J. PLK1 contributes to autophagy by regulating MYC stabilization in osteosarcoma cells. *OncoTargets Ther.* **2019**, *12*, 7527–7536. [CrossRef]
494. Mountzios, G.; Terpos, E.; Dimopoulos, M. Aurora kinases as targets for cancer therapy. *Cancer Treat. Rev.* **2008**, *34*, 175–182. [CrossRef] [PubMed]
495. Goldenson, B.; Crispino, J.D. The aurora kinases in cell cycle and leukemia. *Oncogene* **2014**, *34*, 537–545. [CrossRef] [PubMed]
496. Toji, S.; Yabuta, N.; Hosomi, T.; Nishihara, S.; Kobayashi, T.; Suzuki, S.; Tamai, K.; Nojima, H. The centrosomal protein Lats2 is a phosphorylation target of Aurora-A kinase. *Genes Cells* **2004**, *9*, 383–397. [CrossRef]
497. Naso, F.D.; Boi, D.; Ascanelli, C.; Pamfil, G.; Lindon, C.; Paiardini, A.; Guarguaglini, G. Nuclear localisation of Aurora-A: Its regulation and significance for Aurora-A functions in cancer. *Oncogene* **2021**, *40*, 3917–3928. [CrossRef] [PubMed]

498. Marumoto, T.; Honda, S.; Hara, T.; Nitta, M.; Hirota, T.; Kohmura, E.; Saya, H. Aurora-A Kinase Maintains the Fidelity of Early and Late Mitotic Events in HeLa Cells. *J. Biol. Chem.* **2003**, *278*, 51786–51795. [CrossRef]
499. Crane, R.; Gadea, B.; Littlepage, L.; Wu, H.; Ruderman, J.V. Aurora A, Meiosis and Mitosis. *Biol. Cell* **2004**, *96*, 215–229. [CrossRef]
500. Musacchio, A.; Hardwick, K.G. The spindle checkpoint: Structural insights into dynamic signalling. *Nat. Rev. Mol. Cell Biol.* **2002**, *3*, 731–741. [CrossRef]
501. Werner, M.; Mattis, A.; Aubele, M.; Cummings, M.; Zitzelsberger, H.; Hutzler, P.; Höfler, H. 20q13.2 Amplification in intraductal hyperplasia adjacent to in situ and invasive ductal carcinoma of the breast. *Virchows Arch.* **1999**, *435*, 469–472. [CrossRef]
502. Bui, V.-M.; Mettling, C.; Jou, J.; Sun, H.S. Genomic amplification of chromosome 20q13.33 is the early biomarker for the development of sporadic colorectal carcinoma. *BMC Med. Genom.* **2020**, *13*, 149. [CrossRef]
503. Tanner, M.M.; Tirkkonen, M.; Kallioniemi, A.; Holli, K.; Collins, C.; Kowbel, D.; Gray, J.W.; Kallioniemi, O.; Isola, J. Amplification of chromosomal region 20q13 in invasive breast cancer: Prognostic implications. *Clin. Cancer Res.* **1995**, *1*, 1455–1461.
504. Mou, P.K.; Yang, E.J.; Shi, C.; Ren, G.; Tao, S.; Shim, J.S. Aurora kinase A, a synthetic lethal target for precision cancer medicine. *Exp. Mol. Med.* **2021**, *53*, 835–847. [CrossRef]
505. Murga-Zamalloa, C.; Inamdar, K.V.; Wilcox, R.A. The role of aurora A and polo-like kinases in high-risk lymphomas. *Blood Adv.* **2019**, *3*, 1778–1787. [CrossRef]
506. Bast, R.C., Jr.; Hennessy, B.; Mills, G.B. The biology of ovarian cancer: New opportunities for translation. *Nat. Rev. Cancer* **2009**, *9*, 415–428. [CrossRef]
507. Sun, H.; Wang, H.; Wang, X.; Aoki, Y.; Wang, X.; Yang, Y.; Cheng, X.; Wang, Z.; Wang, X. Aurora-A/SOX8/FOXK1 signaling axis promotes chemoresistance via suppression of cell senescence and induction of glucose metabolism in ovarian cancer organoids and cells. *Theranostics* **2020**, *10*, 6928–6945. [CrossRef]
508. Wan, X.-B.; Long, Z.-J.; Yan, M.; Xu, J.; Xia, L.-P.; Liu, L.; Zhao, Y.; Huang, X.-F.; Wang, X.-R.; Zhu, X.-F.; et al. Inhibition of Aurora-A suppresses epithelial–mesenchymal transition and invasion by downregulating MAPK in nasopharyngeal carcinoma cells. *Carcinog.* **2008**, *29*, 1930–1937. [CrossRef]
509. Nguyen, T.T.B.; Silva, F.N.M.; Golemis, E.A. Aurora Kinases as Therapeutic Targets in Head and Neck Cancer. *Cancer J.* **2022**, *28*, 387–400. [CrossRef]
510. Do, T.-V.; Xiao, F.; E Bickel, L.; Klein-Szanto, A.J.; Pathak, H.B.; Hua, X.; Howe, C.; O'Brien, S.W.; Maglaty, M.; A Ecsedy, J.; et al. Aurora kinase A mediates epithelial ovarian cancer cell migration and adhesion. *Oncogene* **2013**, *33*, 539–549. [CrossRef]
511. Katayama, H.; Wang, J.; Treekitkarnmongkol, W.; Kawai, H.; Sasai, K.; Zhang, H.; Wang, H.; Adams, H.P.; Jiang, S.; Chakraborty, S.N.; et al. Aurora Kinase-A Inactivates DNA Damage-Induced Apoptosis and Spindle Assembly Checkpoint Response Functions of p73. *Cancer Cell* **2012**, *21*, 196–211. [CrossRef]
512. Anand, S.; Penrhyn-Lowe, S.; Venkitaraman, A.R. AURORA-A amplification overrides the mitotic spindle assembly checkpoint, inducing resistance to Taxol. *Cancer Cell* **2003**, *3*, 51–62. [CrossRef]
513. Cirak, Y.; Furuncuoglu, Y.; Yapicier, O.; Aksu, A.; Cubukcu, E. Aurora A overexpression in breast cancer patients induces taxane resistance and results in worse prognosis. *J. BUON.* **2016**, *20*, 1414–1419.
514. Reiter, R.; Gais, P.; Jütting, U.; Steuer-Vogt, M.K.; Pickhard, A.; Bink, K.; Rauser, S.; Lassmann, S.; Höfler, H.; Werner, M.; et al. Aurora Kinase A Messenger RNA Overexpression Is Correlated with Tumor Progression and Shortened Survival in Head and Neck Squamous Cell Carcinoma. *Clin. Cancer Res.* **2006**, *12*, 5136–5141. [CrossRef] [PubMed]
515. Noh, E.-M.; Lee, Y.-R.; Hong, O.-Y.; Jung, S.H.; Youn, H.J.; Kim, J.-S. Aurora kinases are essential for PKC-induced invasion and matrix metalloproteinase-9 expression in MCF-7 breast cancer cells. *Oncol. Rep.* **2015**, *34*, 803–810. [CrossRef] [PubMed]
516. Landen, C.N.; Lin, Y.G.; Immaneni, A.; Deavers, M.T.; Merritt, W.M.; Spannuth, W.A.; Bodurka, D.C.; Gershenson, D.M.; Brinkley, W.R.; Sood, A.K. Overexpression of the Centrosomal Protein Aurora-A Kinase is Associated with Poor Prognosis in Epithelial Ovarian Cancer Patients. *Clin. Cancer Res.* **2007**, *13*, 4098–4104. [CrossRef] [PubMed]
517. Tuncel, H.; Shimamoto, F.; Qi, H.K.; Aoki, E.; Jikihara, H.; Nakai, S.; Takata, T.; Tatsuka, M. Nuclear Aurora B and cytoplasmic Survivin expression is involved in lymph node metastasis of colorectal cancer. *Oncol. Lett.* **2012**, *3*, 1109–1114. [CrossRef]
518. Rannou, Y.; Troadec, M.-B.; Petretti, C.; Hans, F.; Dutertre, S.; Dimitrov, S.; Prigent, C. Localization of aurora A and aurora B kinases during interphase: Role of the N-terminal domain. *Cell Cycle* **2008**, *7*, 3012–3020. [CrossRef]
519. Xia, Z.; Wei, P.; Zhang, H.; Ding, Z.; Yang, L.; Huang, Z.; Zhang, N. AURKA Governs Self-Renewal Capacity in Glioma-Initiating Cells via Stabilization/Activation of β-catenin/Wnt Signaling. *Mol. Cancer Res.* **2013**, *11*, 1101–1111. [CrossRef]
520. Lu, T.; Li, L.; Zhu, J.; Liu, J.; Lin, A.; Fu, W.; Liu, G.; Xia, H.; Zhang, T.; He, J. AURKA rs8173 G>C Polymorphism Decreases Wilms Tumor Risk in Chinese Children. *J. Oncol.* **2019**, *2019*, 9074908–7. [CrossRef]
521. Otto, T.; Horn, S.; Brockmann, M.; Eilers, U.; Schüttrumpf, L.; Popov, N.; Kenney, A.M.; Schulte, J.H.; Beijersbergen, R.; Christiansen, H.; et al. Stabilization of N-Myc Is a Critical Function of Aurora A in Human Neuroblastoma. *Cancer Cell* **2009**, *15*, 67–78. [CrossRef]
522. Maris, J.M. Unholy Matrimony: Aurora A and N-Myc as Malignant Partners in Neuroblastoma. *Cancer Cell* **2009**, *15*, 5–6. [CrossRef]
523. Ommer, J.; Selfe, J.L.; Wachtel, M.; O'Brien, E.M.; Laubscher, D.; Roemmele, M.; Kasper, S.; Delattre, O.; Surdez, D.; Petts, G.; et al. Aurora A Kinase Inhibition Destabilizes PAX3-FOXO1 and MYCN and Synergizes with Navitoclax to Induce Rhabdomyosarcoma Cell Death. *Cancer Res* **2020**, *80*, 832–842. [CrossRef]

524. Goldstein, M.; Meller, I.; Issakov, J.; Orr-Urtreger, A. Novel Genes Implicated in Embryonal, Alveolar, and Pleomorphic Rhabdomyosarcoma: A Cytogenetic and Molecular Analysis of Primary Tumors. *Neoplasia* **2006**, *8*, 332–343. [CrossRef]
525. Zhao, R.; Li, Z.; Huang, Y.; Xiong, C.; Zhang, C.; Liang, H.; Xu, J.; Luo, X. A Novel Ferroptosis-Related Gene Signature for Prognosis Prediction in Ewing Sarcoma. *Anal. Cell Pathol.* **2022**, *2022*, 1–22. [CrossRef]
526. Liao, H.; Xie, X.; Xu, Y.; Huang, G. Identification of driver genes associated with chemotherapy resistance of Ewing's sarcoma. *OncoTargets Ther.* **2018**, *11*, 6947–6956. [CrossRef]
527. Huang, W.-T.; Liu, A.-G.; Cai, K.-T.; He, R.-Q.; Li, Z.; Wei, Q.-J.; Chen, M.-Y.; Huang, J.-Y.; Yan, W.-Y.; Zhou, H.; et al. Exploration and validation of downregulated microRNA-199a-3p, downstream messenger RNA targets and transcriptional regulation in osteosarcoma. *Am. J. Transl. Res.* **2019**, *11*, 7538–7554.
528. Zhu, X.; Mei, J.; Wang, Z. Aurora-A kinase: Potential tumor marker of osteosarcoma. *J. Cancer Res. Ther.* **2014**, *10*, 102–107. [CrossRef]
529. Yang, W.; Jiang, X.; Zhao, X.; Mao, P. Treatment of RB -deficient retinoblastoma with Aurora-A kinase inhibitor. *Kaohsiung J. Med. Sci.* **2021**, *38*, 244–252. [CrossRef]
530. Lehman, N.L.; O'Donnell, J.P.; Whiteley, L.J.; Stapp, R.T.; Lehman, T.D.; Roszka, K.M.; Schultz, L.R.; Williams, C.J.; Mikkelsen, T.; Brown, S.L.; et al. Aurora A is differentially expressed in gliomas, is associated with patient survival in glioblastoma and is a potential chemotherapeutic target in gliomas. *Cell Cycle* **2012**, *11*, 489–502. [CrossRef]
531. Liang, B.; Zhou, Y.; Jiao, J.; Xu, L.; Yan, Y.; Wu, Q.; Tong, X.; Yan, H. Integrated Analysis of Transcriptome Data Revealed AURKA and KIF20A as Critical Genes in Medulloblastoma Progression. *Front. Oncol.* **2022**, *12*. [CrossRef]
532. Vader, G.; Medema, R.; Lens, S.M. The chromosomal passenger complex: Guiding Aurora-B through mitosis. *J. Cell Biol.* **2006**, *173*, 833–837. [CrossRef]
533. Vagnarelli, P.; Earnshaw, W.C. Chromosomal passengers: The four-dimensional regulation of mitotic events. *Chromosoma* **2004**, *113*, 211–222. [CrossRef]
534. Adams, R.R.; Carmena, M.; Earnshaw, W.C. Chromosomal passengers and the (aurora) ABCs of mitosis. *Trends Cell Biol.* **2001**, *11*, 49–54. [CrossRef] [PubMed]
535. Minoshima, Y.; Kawashima, T.; Hirose, K.; Tonozuka, Y.; Kawajiri, A.; Bao, Y.C.; Deng, X.; Tatsuka, M.; Narumiya, S.; May, W.; et al. Phosphorylation by Aurora B Converts MgcRacGAP to a RhoGAP during Cytokinesis. *Dev. Cell* **2003**, *4*, 549–560. [CrossRef] [PubMed]
536. Hsu, J.-Y.; Sun, Z.-W.; Li, X.; Reuben, M.; Tatchell, K.; Bishop, D.K.; Grushcow, J.M.; Brame, C.J.; A Caldwell, J.; Hunt, D.F.; et al. Mitotic Phosphorylation of Histone H3 Is Governed by Ipl1/aurora Kinase and Glc7/PP1 Phosphatase in Budding Yeast and Nematodes. *Cell* **2000**, *102*, 279–291. [CrossRef] [PubMed]
537. Murnion, M.E.; Adams, R.R.; Callister, D.M.; Allis, C.D.; Earnshaw, W.C.; Swedlow, J.R. Chromatin-associated Protein Phosphatase 1 Regulates Aurora-B and Histone H3 Phosphorylation. *J. Biol. Chem.* **2001**, *276*, 26656–26665. [CrossRef] [PubMed]
538. Lan, W.; Zhang, X.; Kline-Smith, S.L.; E Rosasco, S.; A Barrett-Wilt, G.; Shabanowitz, J.; Hunt, D.F.; E Walczak, C.; Stukenberg, T. Aurora B Phosphorylates Centromeric MCAK and Regulates Its Localization and Microtubule Depolymerization Activity. *Curr. Biol.* **2004**, *14*, 273–286. [CrossRef]
539. Ma, H.T.; Poon, R.Y. Aurora kinases and DNA damage response. *Mutat. Res. Mol. Mech. Mutagen.* **2020**, *821*, 111716. [CrossRef]
540. González-Loyola, A.; Fernández-Miranda, G.; Trakala, M.; Partida, D.; Samejima, K.; Ogawa, H.; Cañamero, M.; de Martino, A.; Martínez-Ramírez, A.; de Cárcer, G.; et al. Aurora B Overexpression Causes Aneuploidy and p21 ^{Cip1} Repression during Tumor Development. *Mol. Cell. Biol.* **2015**, *35*, 3566–3578. [CrossRef]
541. Takeshita, M.; Koga, T.; Takayama, K.; Ijichi, K.; Yano, T.; Maehara, Y.; Nakanishi, Y.; Sueishi, K. Aurora-B overexpression is correlated with aneuploidy and poor prognosis in non-small cell lung cancer. *Lung Cancer* **2013**, *80*, 85–90. [CrossRef]
542. Porcelli, L.; Guida, G.; E Quatrale, A.; Cocco, T.; Sidella, L.; Maida, I.; Iacobazzi, R.M.; Ferretta, A.; A Stolfa, D.; Strippoli, S.; et al. Aurora kinase B inhibition reduces the proliferation of metastatic melanoma cells and enhances the response to chemotherapy. *J. Transl. Med.* **2015**, *13*, 26. [CrossRef]
543. Wang, C.; Chen, J.; Cao, W.; Sun, L.; Sun, H.; Liu, Y. Aurora-B and HDAC synergistically regulate survival and proliferation of lymphoma cell via AKT, mTOR and Notch pathways. *Eur. J. Pharmacol.* **2016**, *779*, 1–7. [CrossRef]
544. Wan, B.; Huang, Y.; Liu, B.; Lu, L.; Lv, C. *AURKB*: A promising biomarker in clear cell renal cell carcinoma. *Peerj* **2019**, *7*, e7718. [CrossRef]
545. Twu, N.-F.; Yuan, C.-C.; Yen, M.-S.; Lai, C.-R.; Chao, K.-C.; Wang, P.-H.; Wu, H.-H.; Chen, Y.-J. Expression of Aurora kinase A and B in normal and malignant cervical tissue: High Aurora A kinase expression in squamous cervical cancer. *Eur. J. Obstet. Gynecol. Reprod. Biol.* **2009**, *142*, 57–63. [CrossRef]
546. Pannone, G.; Hindi, S.; Santoro, A.; Sanguedolce, F.; Rubini, C.; Cincione, R.; De Maria, S.; Tortorella, S.; Rocchetti, R.; Cagiano, S.; et al. Aurora B Expression as a Prognostic Indicator and Possibile Therapeutic Target in Oral Squamous Cell Carcinoma. *Int. J. Immunopathol. Pharmacol.* **2011**, *24*, 79–88. [CrossRef]
547. Alafate, W.; Zuo, J.; Deng, Z.; Guo, X.; Wu, W.; Zhang, W.; Xie, W.; Wang, M.; Wang, J. Combined elevation of AURKB and UBE2C predicts severe outcomes and therapy resistance in glioma. *Pathol.-Res. Pract.* **2019**, *215*, 152557. [CrossRef] [PubMed]
548. Liu, M.; Li, Y.; Zhang, C.; Zhang, Q. Role of aurora kinase B in regulating resistance to paclitaxel in breast cancer cells. *Hum. Cell* **2022**, *35*, 678–693. [CrossRef]

549. Wang, Z.; Yu, Z.; Wang, G.-H.; Zhou, Y.-M.; Deng, J.-P.; Feng, Y.; Chen, J.-Q.; Tian, L. AURKB Promotes the Metastasis of Gastric Cancer, Possibly by Inducing EMT. *Cancer Manag. Res.* **2020**, *12*, 6947–6958. [CrossRef]
550. Nie, M.; Wang, Y.; Yu, Z.; Li, X.; Deng, Y.; Wang, Y.; Yang, D.; Li, Q.; Zeng, X.; Ju, J.; et al. AURKB promotes gastric cancer progression via activation of CCND1 expression. *Aging* **2020**, *12*, 1304–1321. [CrossRef]
551. Yuan, K.; Wu, M.; Lyu, S.; Li, Y. Identification of prognostic genes for early basal-like breast cancer with weighted gene co-expression network analysis. *Medicine* **2022**, *101*, e30581. [CrossRef]
552. Gao, X.; Jiang, A.; Shen, Y.; Lu, H.M.; Chen, R. Expression and clinical significance of AURKB gene in lung adenocarcinoma. *Medicine* **2021**, *100*, e26439. [CrossRef]
553. Yang, Y.; Sheng, Y.; Sun, D.; Sun, J.; Li, L.; Sun, L. AURKB promotes tumorigenesis and carboplatin resistance by regulating the ERK pathway in neuroblastoma cells. *Int. J. Neurosci.* **2021**, 1–11. [CrossRef]
554. Bogen, D.; Wei, J.S.; Azorsa, D.O.; Ormanoglu, P.; Buehler, E.; Guha, R.; Keller, J.M.; Griner, L.A.M.; Ferrer, M.; Song, Y.K.; et al. Aurora B kinase is a potent and selective target in MYCN-driven neuroblastoma. *Oncotarget* **2015**, *6*, 35247–35262. [CrossRef] [PubMed]
555. Hartsink-Segers, S.; Zwaan, C.; Exalto, C.; Luijendijk, M.W.J.; Calvert, V.; Petricoin, E.F.; Evans, W.; Reinhardt, D.; De Haas, V.; Hedtjärn, M.; et al. Aurora kinases in childhood acute leukemia: The promise of aurora B as therapeutic target. *Leukemia* **2012**, *27*, 560–568. [CrossRef] [PubMed]
556. Saletta, F.; Wadham, C.; Ziegler, D.; Marshall, G.M.; Haber, M.; McCowage, G.; Norris, M.D.; Byrne, J.A. Molecular profiling of childhood cancer: Biomarkers and novel therapies. *BBA Clin.* **2014**, *1*, 59–77. [CrossRef]
557. Gibson, S.E.; Zeng, W.F.; Weil, R.J.; Prayson, R.A. Aurora B Kinase Expression in Ependymal Neoplasms. *Appl. Immunohistochem. Mol. Morphol.* **2008**, *16*, 274–278. [CrossRef]
558. Wang, S.; Hwang, E.E.; Guha, R.; O'Neill, A.F.; Melong, N.; Veinotte, C.J.; Saur, A.C.; Wuerthele, K.; Shen, M.; McKnight, C.; et al. High-throughput Chemical Screening Identifies Focal Adhesion Kinase and Aurora Kinase B Inhibition as a Synergistic Treatment Combination in Ewing Sarcoma. *Clin. Cancer Res.* **2019**, *25*, 4552–4566. [CrossRef] [PubMed]
559. Zhao, Z.; Jin, G.; Yao, K.; Liu, K.; Liu, F.; Chen, H.; Wang, K.; Gorja, D.R.; Reddy, K.; Bode, A.M.; et al. Aurora B kinase as a novel molecular target for inhibition the growth of osteosarcoma. *Mol. Carcinog.* **2019**, *58*, 1056–1067. [CrossRef]
560. Borah, N.A.; Sradhanjali, S.; Barik, M.R.; Jha, A.; Tripathy, D.; Kaliki, S.; Rath, S.; Raghav, S.K.; Patnaik, S.; Mittal, R.; et al. Aurora Kinase B Expression, Its Regulation and Therapeutic Targeting in Human Retinoblastoma. *Investig. Opthalmology Vis. Sci.* **2021**, *62*, 16. [CrossRef]
561. Kimura, M.; Matsuda, Y.; Yoshioka, T.; Okano, Y. Cell Cycle-dependent Expression and Centrosome Localization of a Third Human Aurora/Ipl1-related Protein Kinase, AIK3. *J. Biol. Chem.* **1999**, *274*, 7334–7340. [CrossRef]
562. Fujii, S.; Srivastava, V.; Hegde, A.; Kondo, Y.; Shen, L.; Hoshino, K.; Gonzalez, Y.; Wang, J.; Sasai, K.; Ma, X.; et al. Regulation of AURKC expression by CpG island methylation in human cancer cells. *Tumor Biol.* **2015**, *36*, 8147–8158. [CrossRef]
563. Tseng, T.-C.; Chen, S.-H.; Hsu, Y.-P.P.; Tang, T.K. Protein Kinase Profile of Sperm and Eggs: Cloning and Characterization of Two Novel Testis-Specific Protein Kinases (AIE1, AIE2) Related to Yeast and Fly Chromosome Segregation Regulators. *DNA Cell Biol.* **1998**, *17*, 823–833. [CrossRef]
564. Santos, M.A.; van de Werken, C.; de Vries, M.; Jahr, H.; Vromans, M.J.; Laven, J.S.; Fauser, B.C.; Kops, G.J.; Lens, S.M.; Baart, E.B. A role for Aurora C in the chromosomal passenger complex during human preimplantation embryo development. *Hum. Reprod.* **2011**, *26*, 1868–1881. [CrossRef]
565. Khan, J.; Ezan, F.; Crémet, J.-Y.; Fautrel, A.; Gilot, D.; Lambert, M.; Benaud, C.; Troadec, M.-B.; Prigent, C. Overexpression of Active Aurora-C Kinase Results in Cell Transformation and Tumour Formation. *PLoS ONE* **2011**, *6*, e26512. [CrossRef]
566. Yan, X.; Cao, L.; Li, Q.; Wu, Y.; Zhang, H.; Saiyin, H.; Liu, X.; Zhang, X.; Shi, Q.; Yu, L. Aurora C is directly associated with Survivin and required for cytokinesis. *Genes Cells* **2005**, *10*, 617–626. [CrossRef]
567. Kobayashi, K.; Kiyomura, H. The theoretical analysis on the tooth movement (II). *Nihon Kyosei Shika Gakkai zasshi = J. Jpn. Orthod. Soc.* **1982**, *41*, 716–722.
568. Tsou, J.-H.; Chang, K.-C.; Chang-Liao, P.-Y.; Yang, S.-T.; Lee, C.-T.; Chen, Y.-P.; Lee, Y.-C.; Lin, B.-W.; Lee, J.-C.; Shen, M.-R.; et al. Aberrantly expressed AURKC enhances the transformation and tumourigenicity of epithelial cells. *J. Pathol.* **2011**, *225*, 243–254. [CrossRef]
569. Zekri, A.; Lesan, V.; Ghaffari, S.H.; Tabrizi, M.H.; Modarressi, M.H. Gene Amplification and Overexpression of Aurora-C in Breast and Prostate Cancer Cell Lines. *Oncol. Res. Featur. Preclin. Clin. Cancer Ther.* **2012**, *20*, 241–250. [CrossRef]
570. Seuánez, H.N.; Pereira, H.S.; Lima, S.C.S.; de Faria, P.S.; Cardoso, L.C.D.A. i RPS6KA4 i i MIR1237 i and i AURKC i promoter regions are differentially methylated in Wilms rsquo tumor. *Front. Biosci.* **2018**, *10*, 143–154. [CrossRef]
571. Hsieh, C.-H.; Cheung, C.H.Y.; Liu, Y.-L.; Hou, C.-L.; Hsu, C.-L.; Huang, C.-T.; Yang, T.-S.; Chen, S.-F.; Chen, C.-N.; Hsu, W.-M.; et al. Quantitative Proteomics of Th-MYCN Transgenic Mice Reveals Aurora Kinase Inhibitor Altered Metabolic Pathways and Enhanced ACADM To Suppress Neuroblastoma Progression. *J. Proteome Res.* **2019**, *18*, 3850–3866. [CrossRef]
572. Bejar, J.F.; DiSanza, Z.; Quartuccio, S.M. The oncogenic role of meiosis-specific Aurora kinase C in mitotic cells. *Exp. Cell Res.* **2021**, *407*, 112803. [CrossRef]
573. Nowell, P.C.; Hungerford, D.A. Chromosome studies on normal and leukemic human leukocytes. *J. Natl. Cancer Inst.* **1960**, *25*, 85–109.
574. Seabright, M. A rapid banding technique for human chromosomes. *Lancet* **1971**, *298*, 971–972. [CrossRef] [PubMed]

575. Rowley, J.D. A New Consistent Chromosomal Abnormality in Chronic Myelogenous Leukaemia identified by Quinacrine Fluorescence and Giemsa Staining. *Nature* **1973**, *243*, 290–293. [CrossRef] [PubMed]
576. Heisterkamp, N.; Stephenson, J.R.; Groffen, J.; Hansen, P.F.; de Klein, A.; Bartram, C.R.; Grosveld, G. Localization of the c-abl oncogene adjacent to a translocation break point in chronic myelocytic leukaemia. *Nature* **1983**, *306*, 239–242. [CrossRef] [PubMed]
577. de Klein, A.; van Kessel, A.G.; Grosveld, G.; Bartram, C.R.; Hagemeijer, A.; Bootsma, D.; Spurr, N.K.; Heisterkamp, N.; Groffen, J.; Stephenson, J.R. A cellular oncogene is translocated to the Philadelphia chromosome in chronic myelocytic leukaemia. *Nature* **1982**, *300*, 765–767. [CrossRef] [PubMed]
578. Groffen, J.; Stephenson, J.R.; Heisterkamp, N.; De Klein, A.; Bartram, C.R.; Grosveld, G. Philadelphia chromosomal breakpoints are clustered within a limited region, bcr, on chromosome 22. *Cell* **1984**, *36*, 93–99. [CrossRef]
579. Koschmieder, S.; Göttgens, B.; Zhang, P. Inducible chronic phase of myeloid leukemia with expansion of hematopoietic stem cells in a trans-genic model of BCR-ABL leukemogenesis. *Blood* **2005**, *105*, 324–334. [CrossRef]
580. Druker, B.J.; Talpaz, M.; Resta, D.J.; Peng, B.; Buchdunger, E.; Ford, J.M.; Lydon, N.B.; Kantarjian, H.; Capdeville, R.; Ohno-Jones, S.; et al. Efficacy and Safety of a Specific Inhibitor of the BCR-ABL Tyrosine Kinase in Chronic Myeloid Leukemia. *N. Engl. J. Med.* **2001**, *344*, 1031–1037. [CrossRef]
581. Festuccia, C.; Gravina, G.L.; Biordi, L.; D'Ascenzo, S.; Dolo, V.; Ficorella, C.; Ricevuto, E.; Tombolini, V. Effects of EGFR tyrosine kinase inhibitor erlotinib in prostate cancer cells in vitro. *Prostate* **2009**, *69*, 1529–1537. [CrossRef]
582. Abdelgalil, A.A.; Al-Kahtani, H.M.; Al-Jenoobi, F.I. Erlotinib. *Profiles DrugSubst. Excip. Relat. Methodol.* **2019**, *45*, 93–117. [CrossRef]
583. Ms, K.N.R.; Zweidler-McKay, P.A.; Van Roy, N.; Speleman, F.; Trevino, J.; Zage, P.E.; Hughes, D.P.M. Signaling of ERBB receptor tyrosine kinases promotes neuroblastoma growth in vitro and in vivo. *Cancer* **2010**, *116*, 3233–3243. [CrossRef]
584. Ji, X.L.; He, M. Sodium cantharidate targets STAT3 and abrogates EGFR inhibitor resistance in osteosarcoma. *Aging* **2019**, *11*, 5848–5863. [CrossRef]
585. Bandyopadhyay, A.; Favours, E.; Phelps, D.A.; Del Pozo, V.; Ghilu, S.; Kurmashev, D.; Michalek, J.; Trevino, A.; Guttridge, D.; London, C.; et al. Evaluation of patritumab with or without erlotinib in combination with standard cytotoxic agents against pediatric sarcoma xenograft models. *Pediatr. Blood Cancer* **2017**, *65*, e26870. [CrossRef]
586. Abraham, J.; Nelon, L.D.; Kubicek, C.B.; Kilcoyne, A.; Hampton, S.T.; Zarzabal, L.A.; Giles, F.J.; Michalek, J.E.; Rubin, B.P.; Keller, C. Preclinical Testing of Erlotinib in a Transgenic Alveolar Rhabdomyosarcoma Mouse Model. *Sarcoma* **2011**, *2011*, 1–5. [CrossRef]
587. Hernan, R.; Fasheh, R.; Calabrese, C.; Frank, A.J.; MacLean, K.H.; Allard, D.; Barraclough, R.; Gilbertson, R.J. ERBB2 up-regulates S100A4 and several other prometastatic genes in medulloblastoma. *Cancer Res* **2003**, *63*, 140–148.
588. Guan, S.; Shen, R.; Lafortune, T.; Tiao, N.; Houghton, P.; Yung, W.A.; Koul, D. Establishment and characterization of clinically relevant models of ependymoma: A true challenge for targeted therapy. *Neuro-Oncology* **2011**, *13*, 748–758. [CrossRef]
589. Shao, Y.; Yu, Y.; Zong, R.; Quyang, L.; He, H.; Zhou, Q.; Pei, C. Erlotinib has tumor inhibitory effect in human retinoblastoma cells. *Biomed. Pharmacother.* **2017**, *85*, 479–485. [CrossRef]
590. Jakacki, R.I.; Hamilton, M.; Gilbertson, R.J.; Blaney, S.M.; Tersak, J.; Krailo, M.D.; Ingle, A.M.; Voss, S.D.; Dancey, J.E.; Adamson, P.C. Pediatric Phase I and Pharmacokinetic Study of Erlotinib Followed by the Combination of Erlotinib and Temozolomide: A Children's Oncology Group Phase I Consortium Study. *J. Clin. Oncol.* **2008**, *26*, 4921–4927. [CrossRef]
591. Geoerger, B.; Hargrave, D.; Thomas, F.; Ndiaye, A.; Frappaz, D.; Andreiuolo, F.; Varlet, P.; Aerts, I.; Riccardi, R.; Jaspan, T.; et al. Innovative Therapies for Children with Cancer pediatric phase I study of erlotinib in brainstem glioma and relapsing/refractory brain tumors. *Neuro-Oncology* **2010**, *13*, 109–118. [CrossRef]
592. Frampton, J.E. Vandetanib. *Drugs* **2012**, *72*, 1423–1436. [CrossRef]
593. Karras, S.; Anagnostis, P.; E Krassas, G. Vandetanib for the treatment of thyroid cancer: An update. *Expert Opin. Drug Metab. Toxicol.* **2014**, *10*, 469–481. [CrossRef]
594. Hatem, R.; Labiod, D.; Château-Joubert, S.; de Plater, L.; El Botty, R.; Vacher, S.; Bonin, F.; Servely, J.-L.; Dieras, V.; Bièche, I.; et al. Vandetanib as a potential new treatment for estrogen receptor-negative breast cancers. *Int. J. Cancer* **2016**, *138*, 2510–2521. [CrossRef] [PubMed]
595. Valerio, L.; Bottici, V.; Matrone, A.; Piaggi, P.; Viola, D.; Cappagli, V.; Agate, L.; Molinaro, E.; Ciampi, R.; Tacito, A.; et al. Medullary thyroid cancer treated with vandetanib: Predictors of a longer and durable response. *Endocrine-Related Cancer* **2020**, *27*, 97–110. [CrossRef] [PubMed]
596. Ding, X.; Xiang, L.; Wang, N.; Zhao, Z.; Jin, X.; Sun, Y.; Duan, W.; Wang, S.; Jin, X. Vandetanib-induced inhibition of neuroblastoma cell migration and invasion is associated with downregulation of the SDF-1/CXCR4 axis and matrix metalloproteinase 14. *Oncol. Rep.* **2013**, *31*, 1165–1174. [CrossRef] [PubMed]
597. Li, H.; Li, C.-W.; Li, X.; Ding, Q.; Guo, L.; Liu, S.; Liu, C.; Lai, C.-C.; Hsu, J.-M.; Dong, Q.; et al. MET Inhibitors Promote Liver Tumor Evasion of the Immune Response by Stabilizing PDL1. *Gastroenterology* **2019**, *156*, 1849–1861.e7. [CrossRef]
598. Beaudry, P.; Nilsson, M.; Rioth, M.; Prox, D.; Poon, D.; Xu, L.; Zweidler-Mckay, P.; Ryan, A.; Folkman, J.; Ryeom, S.; et al. Potent antitumor effects of ZD6474 on neuroblastoma via dual targeting of tumor cells and tumor endothelium. *Mol. Cancer Ther.* **2008**, *7*, 418–424. [CrossRef]
599. Cazes, A.; Lopez-Delisle, L.; Tsarovina, K.; Pierre-Eugène, C.; De Preter, K.; Peuchmaur, M.; Nicolas, A.; Provost, C.; Louis-Brennetot, C.; Daveau, R.; et al. Activated Alk triggers prolonged neurogenesis and Ret upregulation providing a therapeutic target in ALK-mutated neuroblastoma. *Oncotarget* **2014**, *5*, 2688–2702. [CrossRef]

600. Li, C.; Yang, C.; Wei, G. Vandetanib inhibits cisplatin-resistant neuroblastoma tumor growth and invasion. *Oncol. Rep.* **2018**, *39*, 1757–1764. [CrossRef]
601. Zage, P.E.; Zeng, L.; Palla, S.; Fang, W.; Nilsson, M.B.; Heymach, J.V.; Zweidler-McKay, P.A. A novel therapeutic combination for neuroblastoma. *Cancer* **2010**, *116*, 2465–2475. [CrossRef]
602. Craveiro, R.B.; Ehrhardt, M.; Velz, J.; Olschewski, M.; Goetz, B.; Pietsch, T.; Dilloo, D. The anti-neoplastic activity of Vandetanib against high-risk medulloblastoma variants is profoundly enhanced by additional PI3K inhibition. *Oncotarget* **2017**, *8*, 46915–46927. [CrossRef]
603. Liu, J.; Wu, J.; Zhou, L.; Pan, C.; Zhou, Y.; Du, W.; Chen, J.-M.; Zhu, X.; Shen, J.; Chen, S.; et al. ZD6474, a new treatment strategy for human osteosarcoma, and its potential synergistic effect with celecoxib. *Oncotarget* **2015**, *6*, 21341–21352. [CrossRef]
604. Andersson, M.K.; Åman, P. Proliferation of Ewing sarcoma cell lines is suppressed by the receptor tyrosine kinase inhibitors gefitinib and vandetanib. *Cancer Cell Int.* **2008**, *8*, 1–6. [CrossRef]
605. Maloney, C.; Kallis, M.P.; Edelman, M.; Tzanavaris, C.; Lesser, M.; Soffer, S.Z.; Symons, M.; Steinberg, B.M. Gefitinib Inhibits Invasion and Metastasis of Osteosarcoma via Inhibition of Macrophage Receptor Interacting Serine-Threonine Kinase 2. *Mol. Cancer Ther.* **2020**, *19*, 1340–1350. [CrossRef]
606. Wakeling, A.E.; Guy, S.P.; Woodburn, J.R.; Ashton, S.E.; Curry, B.J.; Barker, A.J.; Gibson, K.H. ZD1839 (Iressa): An orally active inhibitor of epidermal growth factor signaling with potential for cancer therapy. *Cancer Res.* **2002**, *62*, 5749–5754.
607. Ciardiello, F.; Caputo, R.; Bianco, R.; Damiano, V.; Pomatico, G. Antitumor effect and potentiation of cytotoxic drugs activity in human cancer cells by ZD-1839 (Iressa), an epidermal growth factor receptor-selective tyrosine kinase inhibitor. *Clin Cancer Res* **2000**, *6*, 2053–2063.
608. Foreman, N.K.; Gore, L.; Wells, D.; Bs, J.S.; Heideman, R.; Donson, A.M. Gefitinib is effective against juvenile pilocytic astrocytoma in vitro. *Pediatr. Blood Cancer* **2005**, *47*, 293–298. [CrossRef]
609. Schaiquevich, P.; Panetta, J.C.; Throm, S.; Daw, N.C.; Geyer, J.R.; Furman, W.L.; Stewart, C.F. Population pharmacokinetic (PK) analysis of gefitinib in pediatric cancer patients. *J. Clin. Oncol.* **2008**, *26*, 2523. [CrossRef]
610. Daudigeos-Dubus, E.; Le Dret, L.; Lanvers-Kaminsky, C.; Bawa, O.; Opolon, P.; Vievard, A.; Villa, I.; Pagès, M.; Bosq, J.; Vassal, G.; et al. Regorafenib: Antitumor Activity upon Mono and Combination Therapy in Preclinical Pediatric Malignancy Models. *PLoS ONE* **2015**, *10*, e0142612. [CrossRef]
611. Ettrich, T.J.; Seufferlein, T. Regorafenib. In *Recent Results in Cancer Research*; Springer LLC: New York, NY, USA, 2018; Volume 211, pp. 45–56. [CrossRef]
612. Subramonian, D.; Phanhthilath, N.; Rinehardt, H.; Flynn, S.; Huo, Y.; Zhang, J.; Messer, K.; Mo, Q.; Huang, S.; Lesperance, J.; et al. Regorafenib is effective against neuroblastoma in vitro and in vivo and inhibits the RAS/MAPK, PI3K/Akt/mTOR and Fos/Jun pathways. *Br. J. Cancer* **2020**, *123*, 568–579. [CrossRef]
613. Carpenter, R.L.; Lo, H.-W. Dacomitinib, an emerging HER-targeted therapy for non-small cell lung cancer. *J. Thorac. Dis.* **2012**, *4*, 639–642. [CrossRef]
614. Shirley, M. Dacomitinib: First Global Approval. *Drugs* **2018**, *78*, 1947–1953. [CrossRef]
615. Popat, S.; Yap, T. Toward precision medicine with next-generation EGFR inhibitors in non-small-cell lung cancer. *Pharmacogenomics Pers. Med.* **2014**, *7*, 285–295. [CrossRef] [PubMed]
616. Endersby, R.; Whitehouse, J.; Hii, H.; Greenall, S.A.; Johns, T.G.; Gottardo, N.G. A Pre-Clinical Assessment of the Pan-ERBB Inhibitor Dacomitinib in Pediatric and Adult Brain Tumors. *Neoplasia* **2018**, *20*, 432–442. [CrossRef] [PubMed]
617. Moreira, C.; Kaklamani, V. Lapatinib and breast cancer: Current indications and outlook for the future. *Expert Rev. Anticancer. Ther.* **2010**, *10*, 1171–1182. [CrossRef] [PubMed]
618. Bouchalova, K.; Cizkova, M.; Cwiertka, K.; Trojanec, R.; Friedecký, D.; Hajduch, M. Lapatinib in breast cancer—the predictive significance of her1 (egfr), her2, pten and pik3ca genes and lapatinib plasma level assessment. *Biomed. Pap.* **2010**, *154*, 281–288. [CrossRef]
619. Gorlick, R.; Kolb, E.A.; Houghton, P.J.; Morton, C.L.; Phelps, D.; Schaiquevich, P.; Stewart, C.; Keir, S.T.; Lock, R.; Carol, H.; et al. Initial testing (stage 1) of lapatinib by the pediatric preclinical testing program. *Pediatr. Blood Cancer* **2009**, *53*, 594–598. [CrossRef]
620. Tebbutt, N.; Pedersen, M.W.; Johns, T.G. Targeting the ERBB family in cancer: Couples therapy. *Nat. Rev. Cancer* **2013**, *13*, 663–673. [CrossRef]
621. Herrmann, D.; Seitz, G.; Warmann, S.W.; Bonin, M.; Fuchs, J.; Armeanu-Ebinger, S. Cetuximab Promotes Immunotoxicity Against Rhabdomyosarcoma In Vitro. *J. Immunother.* **2010**, *33*, 279–286. [CrossRef]
622. Yamamoto, Y.; Fukuda, K.; Fuchimoto, Y.; Matsuzaki, Y.; Saikawa, Y.; Kitagawa, Y.; Morikawa, Y.; Kuroda, T. Cetuximab promotes anticancer drug toxicity in rhabdomyosarcomas with EGFR amplification in vitro. *Oncol. Rep.* **2013**, *30*, 1081–1086. [CrossRef]
623. O'Farrell, A.-M.; Abrams, T.J.; Yuen, H.A.; Ngai, T.J.; Louie, S.G.; Yee, K.W.H.; Wong, L.M.; Hong, W.; Lee, L.B.; Town, A.; et al. SU11248 is a novel FLT3 tyrosine kinase inhibitor with potent activity in vitro and in vivo. *Blood* **2003**, *101*, 3597–3605. [CrossRef]
624. Zhang, L.; Smith, K.M.; Chong, A.L.; Stempak, D.; Yeger, H.; Marrano, P.; Thorner, P.S.; Irwin, M.S.; Kaplan, D.R.; Baruchel, S. In Vivo Antitumor and Antimetastatic Activity of Sunitinib in Preclinical Neuroblastoma Mouse Model. *Neoplasia* **2009**, *11*, 426–435. [CrossRef]
625. Maris, J.M.; Bs, J.C.; Houghton, P.J.; Bs, C.L.M.; Kolb, E.A.; Lock, R.; Bs, M.T.; Reynolds, C.P.; Keir, S.T.; Wu, J.; et al. Initial testing (stage 1) of sunitinib by the pediatric preclinical testing program. *Pediatr. Blood Cancer* **2008**, *51*, 42–48. [CrossRef]
626. Hao, Z.; Wang, P. Lenvatinib in Management of Solid Tumors. *Oncologist* **2020**, *25*, e30. [CrossRef]

627. Matsui, J.; Yamamoto, Y.; Funahashi, Y.; Tsuruoka, A.; Watanabe, T.; Wakabayashi, T.; Uenaka, T.; Asada, M. E7080, a novel inhibitor that targets multiple kinases, has potent antitumor activities against stem cell factor producing human small cell lung cancer H146, based on angiogenesis inhibition. *Int. J. Cancer* **2007**, *122*, 664–671. [CrossRef]
628. Capozzi, M.; De Divitiis, C.; Ottaiano, A.; von Arx, C.; Scala, S.; Tatangelo, F.; Delrio, P.; Tafuto, S. Lenvatinib, a molecule with versatile application: From preclinical evidence to future development in anti-cancer treatment. *Cancer Manag. Res.* **2019**, *11*, 3847–3860. [CrossRef]
629. Bruheim, S.; Kristian, A.; Uenaka, T.; Suo, Z.; Tsuruoka, A.; Nesland, J.M.; Fodstad, A. Antitumour activity of oral E7080, a novel inhibitor of multiple tyrosine kinases, in human sarcoma xenografts. *Int. J. Cancer* **2011**, *129*, 742–750. [CrossRef]
630. Glen, H.; Mason, S.; Patel, H.; MacLeod, K.; Brunton, V.G. E7080, a multi-targeted tyrosine kinase inhibitor suppresses tumor cell migration and invasion. *BMC Cancer* **2011**, *11*, 309. [CrossRef]
631. Gaspar, N.; Campbell-Hewson, Q.; Melcon, S.G.; Locatelli, F.; Venkatramani, R.; Hecker-Nolting, S.; Gambart, M.; Bautista, F.; Thebaud, E.; Aerts, I.; et al. Phase I/II study of single-agent lenvatinib in children and adolescents with refractory or relapsed solid malignancies and young adults with osteosarcoma (ITCC-050). *ESMO Open* **2021**, *6*, 100250. [CrossRef]
632. Wilky, B.A.; Meyer, C.F.; Trent, J.C. Pazopanib in sarcomas. *Curr. Opin. Oncol.* **2013**, *25*, 373–378. [CrossRef]
633. Kumar, S.; Mokhtari, R.B.; Oliveira, I.D.; Islam, S.; Toledo, S.R.C.; Yeger, H.; Baruchel, S. Tumor Dynamics in Response to Antiangiogenic Therapy with Oral Metronomic Topotecan and Pazopanib in Neuroblastoma Xenografts. *Transl. Oncol.* **2013**, *6*, 493–503. [CrossRef]
634. Kumar, S.; Mokhtari, R.B.; Sheikh, R.; Wu, B.; Zhang, L.; Xu, P.; Man, S.; Oliveira, I.D.; Yeger, H.; Kerbel, R.S.; et al. Metronomic Oral Topotecan with Pazopanib Is an Active Antiangiogenic Regimen in Mouse Models of Aggressive Pediatric Solid Tumor. *Clin. Cancer Res.* **2011**, *17*, 5656–5667. [CrossRef]
635. Chiabotto, G.; Grignani, G.; Todorovic, M.; Martin, V.; Centomo, M.L.; Prola, E.; Giordano, G.; Merlini, A.; Miglio, U.; Berrino, E.; et al. Pazopanib and Trametinib as a Synergistic Strategy against Osteosarcoma: Preclinical Activity and Molecular Insights. *Cancers* **2020**, *12*, 1519. [CrossRef] [PubMed]
636. Keir, S.T.; Bs, C.L.M.; Wu, J.; Kurmasheva, R.T.; Houghton, P.J.; Smith, M.A. Initial testing of the multitargeted kinase inhibitor pazopanib by the pediatric preclinical testing program. *Pediatr. Blood Cancer* **2011**, *59*, 586–588. [CrossRef] [PubMed]
637. Aggerholm-Pedersen, N.; Rossen, P.; Rose, H.; Safwat, A. Pazopanib in the Treatment of Bone Sarcomas: Clinical Experience. *Transl. Oncol.* **2020**, *13*, 295–299. [CrossRef] [PubMed]
638. Mori, Y.; Kinoshita, S.; Kanamori, T.; Kataoka, H.; Joh, T.; Iida, S.; Takemoto, M.; Kondo, M.; Kuroda, J.; Komatsu, H. The Successful Treatment of Metastatic Extraosseous Ewing Sarcoma with Pazopanib. *Intern. Med.* **2018**, *57*, 2753–2757. [CrossRef] [PubMed]
639. Donson, A.M.; Amani, V.; Warner, E.A.; Griesinger, A.M.; Witt, D.A.; Levy, J.M.M.; Hoffman, L.M.; Hankinson, T.C.; Handler, M.H.; Vibhakar, R.; et al. Identification of FDA-Approved Oncology Drugs with Selective Potency in High-Risk Childhood Ependymoma. *Mol. Cancer Ther.* **2018**, *17*, 1984–1994. [CrossRef]
640. Schoen, L.F.; Craveiro, R.B.; Pietsch, T.; Moritz, T.; Troeger, A.; Jordans, S.; Dilloo, D. The PI3K inhibitor pictilisib and the multikinase inhibitors pazopanib and sorafenib have an impact on Rac1 level and migration of medulloblastoma in vitro. *J. Cell. Mol. Med.* **2022**, *26*, 5832–5845. [CrossRef]
641. Craveiro, R.B.; Ehrhardt, M.; Holst, M.I.; Pietsch, T.; Dilloo, D. In Comparative Analysis of Multi-Kinase Inhibitors for Targeted Medulloblastoma Therapy Pazopanib Exhibits Promising In Vitro and In Vivo Efficacy, 2014. Available online: www.impactjournals.com/oncotarget (accessed on 22 October 2022).
642. Scott, B.J.; Quant, E.C.; McNamara, M.B.; Ryg, P.A.; Batchelor, T.T.; Wen, P.Y. Bevacizumab salvage therapy following progression in high-grade glioma patients treated with VEGF receptor tyrosine kinase inhibitors. *Neuro-Oncology* **2010**, *12*, 603–607. [CrossRef]
643. Duke, E.S.; Barone, A.K.; Chatterjee, S.; Mishra-Kalyani, P.S.; Shen, Y.-L.; Isikwei, E.; Zhao, H.; Bi, Y.; Liu, J.; Rahman, N.A.; et al. FDA Approval Summary: Cabozantinib for Differentiated Thyroid Cancer. *Clin. Cancer Res.* **2022**, *28*, 4173–4177. [CrossRef]
644. Leavitt, J.; Copur, M.S. FDA Approved Uses of Cabozantinib. *Oncology* **2019**, *33*, 685004.
645. Yakes, F.M.; Chen, J.; Tan, J.; Yamaguchi, K.; Shi, Y.; Yu, P.; Qian, F.; Chu, F.; Bentzien, F.; Cancilla, B.; et al. Cabozantinib (XL184), a Novel MET and VEGFR2 Inhibitor, Simultaneously Suppresses Metastasis, Angiogenesis, and Tumor Growth. *Mol. Cancer Ther.* **2011**, *10*, 2298–2308. [CrossRef]
646. Bentzien, F.; Zuzow, M.; Heald, N.; Gibson, A.; Shi, Y.; Goon, L.; Yu, P.; Engst, S.; Zhang, W.; Huang, D.; et al. In Vitro and In Vivo Activity of Cabozantinib (XL184), an Inhibitor of RET, MET, and VEGFR2, in a Model of Medullary Thyroid Cancer. *Thyroid* **2013**, *23*, 1569–1577. [CrossRef]
647. Santoni, M.; Iacovelli, R.; Colonna, V.; Klinz, S.; Mauri, G.; Nuti, M. Antitumor effects of the multi-target tyrosine kinase inhibitor cabozantinib: A comprehensive review of the preclinical evidence. *Expert Rev. Anticancer. Ther.* **2021**, *21*, 1029–1054. [CrossRef]
648. Fioramonti, M.; Fausti, V.; Pantano, F.; Iuliani, M.; Ribelli, G.; Lotti, F.; Pignochino, Y.; Grignani, G.; Santini, D.; Tonini, G.; et al. Cabozantinib Affects Osteosarcoma Growth Through A Direct Effect On Tumor Cells and Modifications In Bone Microenvironment. *Sci. Rep.* **2018**, *8*, 4177. [CrossRef]
649. Pagnuzzi-Boncompagni, M.; Picco, V.; Vial, V.; Planas-Bielsa, V.; Vandenberghe, A.; Daubon, T.; Derieppe, M.-A.; Montemagno, C.; Durivault, J.; Grépin, R.; et al. Antiangiogenic Compound Axitinib Demonstrates Low Toxicity and Antitumoral Effects against Medulloblastoma. *Cancers* **2021**, *14*, 70. [CrossRef]

650. Daudigeos-Dubus, E.; Le Dret, L.; Bawa, O.; Opolon, P.; Vievard, A.; Villa, I.; Bosq, J.; Vassal, G.; Geoerger, B. Dual inhibition using cabozantinib overcomes HGF/MET signaling mediated resistance to pan-VEGFR inhibition in orthotopic and metastatic neuroblastoma tumors. *Int. J. Oncol.* **2016**, *50*, 203–211. [CrossRef]
651. Wind, S.; Schmid, U.; Freiwald, M.; Marzin, K.; Lotz, R.; Ebner, T.; Stopfer, P.; Dallinger, C. Clinical Pharmacokinetics and Pharmacodynamics of Nintedanib. *Clin. Pharmacokinet.* **2019**, *58*, 1131–1147. [CrossRef]
652. A Kharlamova, L. Study of the radiobiological reactions of Amoeba proteus. 2. The effect of the binuclear state on the radiosensitivity of amebae. *Radiobiologiia* **1972**, *12*, 934–937.
653. Lin, T.; Gong, L. Inhibition of lymphangiogenesis in vitro and in vivo by the multikinase inhibitor nintedanib. *Drug Des. Dev. Ther.* **2017**, *11*, 1147–1158. [CrossRef]
654. Patwardhan, P.P.; Musi, E.; Schwartz, G.K. Preclinical Evaluation of Nintedanib, a Triple Angiokinase Inhibitor, in Soft-tissue Sarcoma: Potential Therapeutic Implication for Synovial Sarcoma. *Mol. Cancer Ther.* **2018**, *17*, 2329–2340. [CrossRef]
655. Zhang, W.; Zhao, J.-M.; Lin, J.; Hu, C.-Z.; Zhang, W.-B.; Yang, W.-L.; Zhang, J.; Zhu, J. Adaptive Fibrogenic Reprogramming of Osteosarcoma Stem Cells Promotes Metastatic Growth. *Cell Rep.* **2018**, *24*, 1266–1277.e5. [CrossRef]
656. Milton, C.I.; Selfe, J.; Aladowicz, E.; Man, S.Y.K.; Bernauer, C.; Missiaglia, E.; Walters, Z.S.; Gatz, S.A.; Kelsey, A.; Generali, M.; et al. FGF7–FGFR2 autocrine signaling increases growth and chemoresistance of fusion-positive rhabdomyosarcomas. *Mol. Oncol.* **2021**, *16*, 1272–1289. [CrossRef]
657. Loetsch, D.; Gojo, J.; Kirchhofer, D.; Van Schoonhoven, S.; Pajtler, K.; Kool, M.; Haberler, C.; Czech, T.; Slavc, I.; Berger, W. OS5.2 FGFR a novel target in malignant pediatric ependymoma. *Neuro-Oncology* **2018**, *20*, iii224. [CrossRef]
658. Lötsch, D.; Kirchhofer, D.; Englinger, B.; Jiang, L.; Okonechnikov, K.; Senfter, D.; Laemmerer, A.; Gabler, L.; Pirker, C.; Donson, A.M.; et al. Targeting fibroblast growth factor receptors to combat aggressive ependymoma. *Acta Neuropathol.* **2021**, *142*, 339–360. [CrossRef] [PubMed]
659. Kasamon, Y.L.; Ko, C.-W.; Subramaniam, S.; Ma, L.; Yang, Y.; Nie, L.; Shord, S.; Przepiorka, D.; Farrell, A.T.; McKee, A.E.; et al. FDA Approval Summary: Midostaurin for the Treatment of Advanced Systemic Mastocytosis. *Oncologist* **2018**, *23*, 1511–1519. [CrossRef] [PubMed]
660. Midostaurin Gets FDA Nod for AML. *Cancer Discov.* **2017**, *7*, OF5. [CrossRef]
661. Kawamoto, T.; Akisue, T.; Kishimoto, K.; Hara, H.; Imabori, M.; Fujimoto, T.; Kurosaka, M.; Hitora, T.; Kawaguchi, Y.; Yamamoto, T. Inhibition of PKCalpha activation in human bone and soft tissue sarcoma cells by the selective PKC inhibitor PKC412. *Anticancer. Res.* **2008**, *28*, 825–832.
662. Boro, A.; Prêtre, K.; Rechfeld, F.; Thalhammer, V.; Oesch, S.; Wachtel, M.; Schäfer, B.W.; Niggli, F.K. Small-molecule screen identifies modulators of EWS/FLI1 target gene expression and cell survival in Ewing's sarcoma. *Int. J. Cancer* **2012**, *131*, 2153–2164. [CrossRef]
663. Brounais, B.; Chipoy, C.; Mori, K.; Charrier, C.; Battaglia, S.; Pilet, P.; Richards, C.D.; Heymann, D.; Rédini, F.; Blanchard, F. Oncostatin M Induces Bone Loss and Sensitizes Rat Osteosarcoma to the Antitumor Effect of Midostaurin *In vivo*. *Clin. Cancer Res.* **2008**, *14*, 5400–5409. [CrossRef]
664. Kelly, R.J.; Rixe, O. Axitinib (AG-013736). *Small Mol. Oncol.* **2009**, *184*, 33–44. [CrossRef]
665. Lu, L.; Saha, D.; Martuza, R.L.; Rabkin, S.; Wakimoto, H. Single agent efficacy of the VEGFR kinase inhibitor axitinib in preclinical models of glioblastoma. *J. Neuro-Oncology* **2014**, *121*, 91–100. [CrossRef]
666. Ehrhardt, M.; Craveiro, R.B.; Velz, J.; Olschewski, M.; Casati, A.; Schönberger, S.; Pietsch, T.; Dilloo, D. The FDA approved PI 3K inhibitor GDC -0941 enhances in vitro the anti-neoplastic efficacy of Axitinib against c-myc-amplified high-risk medulloblastoma. *J. Cell. Mol. Med.* **2018**, *22*, 2153–2161. [CrossRef]
667. Suri, A.; Bailey, A.W.; Tavares, M.T.; Gunosewoyo, H.; Dyer, C.P.; Grupenmacher, A.T.; Piper, D.R.; Horton, R.A.; Tomita, T.; Kozikowski, A.P.; et al. Evaluation of Protein Kinase Inhibitors with PLK4 Cross-Over Potential in a Pre-Clinical Model of Cancer. *Int. J. Mol. Sci.* **2019**, *20*, 2112. [CrossRef]
668. Schwinn, S.; Mokhtari, Z.; Thusek, S.; Schneider, T.; Sirén, A.-L.; Tiemeyer, N.; Caruana, I.; Miele, E.; Schlegel, P.G.; Beilhack, A.; et al. Cytotoxic effects and tolerability of gemcitabine and axitinib in a xenograft model for c-myc amplified medulloblastoma. *Sci. Rep.* **2021**, *11*, 1–15. [CrossRef]
669. Saha, D.; Wakimoto, H.; Peters, C.W.; Antoszczyk, S.J.; Rabkin, S.D.; Martuza, R.L. Combinatorial Effects of VEGFR Kinase Inhibitor Axitinib and Oncolytic Virotherapy in Mouse and Human Glioblastoma Stem-Like Cell Models. *Clin. Cancer Res.* **2018**, *24*, 3409–3422. [CrossRef]
670. Rössler, J.; Monnet, Y.; Farace, F.; Opolon, P.; Daudigeos-Dubus, E.; Bourredjem, A.; Vassal, G.; Geoerger, B. The selective VEGFR1-3 inhibitor axitinib (AG-013736) shows antitumor activity in human neuroblastoma xenografts. *Int. J. Cancer* **2010**, *128*, 2748–2758. [CrossRef]
671. Lu, D.; Jimenez, X.; Zhang, H.; Bohlen, P.; Witte, L.; Zhu, Z. Selection of high affinity human neutralizing antibodies to VEGFR2 from a large antibody phage display library for antiangiogenesis therapy. *Int. J. Cancer* **2001**, *97*, 393–399. [CrossRef]
672. Lowery, C.D.; Blosser, W.; Dowless, M.; Renschler, M.; Perez, L.V.; Stephens, J.; Pytowski, B.; Wasserstrom, H.; Stancato, L.F.; Falcon, B. Anti-VEGFR2 therapy delays growth of preclinical pediatric tumor models and enhances anti-tumor activity of chemotherapy. *Oncotarget* **2019**, *10*, 5523–5533. [CrossRef]
673. Casak, S.J.; Fashoyin-Aje, I.; Lemery, S.J.; Zhang, L.; Jin, R.; Li, H.; Zhao, L.; Zhao, H.; Zhang, H.; Chen, H.; et al. FDA Approval Summary: Ramucirumab for Gastric Cancer. *Clin. Cancer Res.* **2015**, *21*, 3372–3376. [CrossRef]

674. Syed, Y.Y. Ramucirumab: A Review in Hepatocellular Carcinoma. *Drugs* **2020**, *80*, 315–322. [CrossRef]
675. Tiwari, P. Ramucirumab: Boon or bane. *J. Egypt. Natl. Cancer Inst.* **2016**, *28*, 133–140. [CrossRef]
676. Debeuckelaere, C.; Murgioni, S.; Lonardi, S.; Girardi, N.; Alberti, G.; Fano, C.; Gallimberti, S.; Magro, C.; Ahcene-Djaballah, S.; Daniel, F.; et al. Ramucirumab: The long and winding road toward being an option for mCRC treatment. *Expert Opin. Biol. Ther.* **2019**, *19*, 399–409. [CrossRef] [PubMed]
677. Larkins, E.; Blumenthal, G.M.; Chen, H.; He, K.; Agarwal, R.; Gieser, G.; Stephens, O.; Zahalka, E.; Ringgold, K.; Helms, W.; et al. FDA Approval: Alectinib for the Treatment of Metastatic, ALK-Positive Non–Small Cell Lung Cancer Following Crizotinib. *Clin. Cancer Res.* **2016**, *22*, 5171–5176. [CrossRef] [PubMed]
678. Kodama, T.; Hasegawa, M.; Takanashi, K.; Sakurai, Y.; Kondoh, O.; Sakamoto, H. Antitumor activity of the selective ALK inhibitor alectinib in models of intracranial metastases. *Cancer Chemother. Pharmacol.* **2014**, *74*, 1023–1028. [CrossRef] [PubMed]
679. Ryu, S.; Hayashi, M.; Aikawa, H.; Okamoto, I.; Fujiwara, Y.; Hamada, A. Heterogeneous distribution of alectinib in neuroblastoma xenografts revealed by matrix-assisted laser desorption ionization mass spectrometry imaging: A pilot study. *Br. J. Pharmacol.* **2017**, *175*, 29–37. [CrossRef]
680. Chen, L.; Humphreys, A.; Turnbull, L.; Bellini, A.; Schleiermacher, G.; Salwen, H.; Cohn, S.L.; Bown, N.; Tweddle, D.A. Identification of different*ALK*mutations in a pair of neuroblastoma cell lines established at diagnosis and relapse. *Oncotarget* **2016**, *7*, 87301–87311. [CrossRef]
681. Lu, J.; Guan, S.; Zhao, Y.; Yu, Y.; Woodfield, S.E.; Zhang, H.; Yang, K.L.; Bieerkehazhi, S.; Qi, L.; Li, X.; et al. The second-generation ALK inhibitor alectinib effectively induces apoptosis in human neuroblastoma cells and inhibits tumor growth in a TH-MYCN transgenic neuroblastoma mouse model. *Cancer Lett.* **2017**, *400*, 61–68. [CrossRef]
682. Yang, K.; Chen, Y.; Wang, F. Correction: Alectinib (CH5424802) antagonizes ABCB1- and ABCG2-mediated multidrug resistance in vitro, in vivo and ex vivo. *Exp. Mol. Med.* **2020**, *52*, 989–990. [CrossRef]
683. Brunac, A.; Laprie, A.; Castex, M.; Laurent, C.; Le Loarer, F.; Karanian, M.; Le Guellec, S.; Guillemot, D.; Pierron, G.; Gomez-Brouchet, A. The combination of radiotherapy and ALK inhibitors is effective in the treatment of intraosseous rhabdomyosarcoma with *FUS-TFCP2* fusion transcript. *Pediatr. Blood Cancer* **2020**, *67*, e28185. [CrossRef]
684. Hagiwara, K.; Tokunaga, T.; Iida, H.; Nagai, H. Combined Inhibition of ALK and HDAC Induces Synergistic Cytotoxicity in Neuroblastoma Cell Lines. *Anticancer. Res.* **2019**, *39*, 3579–3584. [CrossRef]
685. Berezowska, S.; Diermeier-Daucher, S.; Brockhoff, G.; Busch, R.; Duyster, J.; Grosu, A.-L.; Schlegel, J. Effect of additional inhibition of human epidermal growth factor receptor 2 with the bispecific tyrosine kinase inhibitor AEE788 on the resistance to specific EGFR inhibition in glioma cells. *Int. J. Mol. Med.* **2010**, *26*, 713–721. [CrossRef]
686. Servidei, T.; Meco, D.; Trivieri, N.; Patriarca, V.; Vellone, V.G.; Zannoni, G.F.; Lamorte, G.; Pallini, R.; Riccardi, R. Effects of epidermal growth factor receptor blockade on ependymoma stem cells in vitro and in orthotopic mouse models. *Int. J. Cancer* **2012**, *131*, E791–E803. [CrossRef]
687. Park, Y.W.; Younes, M.N.; Jasser, S.A.; Yigitbasi, O.G.; Zhou, G.; Bucana, C.D.; Bekele, B.N.; Myers, J.N. AEE788, a Dual Tyrosine Kinase Receptor Inhibitor, Induces Endothelial Cell Apoptosis in Human Cutaneous Squamous Cell Carcinoma Xenografts in Nude Mice. *Clin. Cancer Res.* **2005**, *11*, 1963–1973. [CrossRef]
688. Meco, D.; Servidei, T.; Zannonit, G.F.; Martinelli, E.; Prisco, M.G.; de Waure, C.; Riccardi, R. Dual Inhibitor AEE78 Reduces Tumor Growth in Preclinical Models of Medulloblastoma. *Transl. Oncol.* **2010**, *3*, 326–335. [CrossRef]
689. Heigener, D.F.; Reck, M. Crizotinib. In *Recent Results in Cancer Research*; Springer LLC: New York, NY, USA, 2018; Volume 211, pp. 57–65. [CrossRef]
690. Zomerman, W.W.; Plasschaert, S.L.A.; Diks, S.H.; Lourens, H.-J.; Boer, T.M.-D.; Hoving, E.W.; Dunnen, W.F.A.D.; De Bont, E.S.J.M. Exogenous HGF Bypasses the Effects of ErbB Inhibition on Tumor Cell Viability in Medulloblastoma Cell Lines. *PLoS ONE* **2015**, *10*, e0141381. [CrossRef]
691. Sie, M.; Dunnen, W.F.A.D.; Lourens, H.J.; Boer, T.G.J.M.-D.; Scherpen, F.J.G.; Zomerman, W.W.; Kampen, K.; Hoving, E.W.; De Bont, E.S.J.M. Growth-Factor-Driven Rescue to Receptor Tyrosine Kinase (RTK) Inhibitors through Akt and Erk Phosphorylation in Pediatric Low Grade Astrocytoma and Ependymoma. *PLoS ONE* **2015**, *10*, e0122555. [CrossRef]
692. Megiorni, F.; McDowell, H.P.; Camero, S.; Mannarino, O.; Ceccarelli, S.; Paiano, M.; Losty, P.D.; Pizer, B.; Shukla, R.; Pizzuti, A.; et al. Crizotinib-induced antitumour activity in human alveolar rhabdomyosarcoma cells is not solely dependent on ALK and MET inhibition. *J. Exp. Clin. Cancer Res.* **2015**, *34*, 1–16. [CrossRef]
693. Schöffski, P.; Wozniak, A.; Leahy, M.G.; Aamdal, S.; Rutkowski, P.; Bauer, S.; Richter, S.; Grünwald, V.; Debiec-Rychter, M.; Sciot, R.; et al. The tyrosine kinase inhibitor crizotinib does not have clinically meaningful activity in heavily pre-treated patients with advanced alveolar rhabdomyosarcoma with FOXO rearrangement: European Organisation for Research and Treatment of Cancer phase 2 trial 90101 'CREATE'. *Eur. J. Cancer* **2018**, *94*, 156–167. [CrossRef]
694. Heuckmann, J.M.; Hölzel, M.; Sos, M.L.; Heynck, S.; Balke-Want, H.; Koker, M.; Peifer, M.; Weiss, J.; Lovly, C.M.; Grütter, C.; et al. ALK Mutations Conferring Differential Resistance to Structurally Diverse ALK Inhibitors. *Clin. Cancer Res.* **2011**, *17*, 7394–7401. [CrossRef]
695. George, R.E.; Sanda, T.; Hanna, M.; Fröhling, S.; Luther, W., II; Zhang, J.; Ahn, Y.; Zhou, W.; London, W.B.; McGrady, P.; et al. Activating mutations in ALK provide a therapeutic target in neuroblastoma. *Nature* **2008**, *455*, 975–978. [CrossRef]

696. Bresler, S.C.; Weiser, D.A.; Huwe, P.J.; Park, J.H.; Krytska, K.; Ryles, H.; Laudenslager, M.; Rappaport, E.F.; Wood, A.C.; McGrady, P.W.; et al. ALK Mutations Confer Differential Oncogenic Activation and Sensitivity to ALK Inhibition Therapy in Neuroblastoma. *Cancer Cell* **2014**, *26*, 682–694. [CrossRef]
697. Dagogo-Jack, I.; Shaw, A.T. Crizotinib resistance: Implications for therapeutic strategies. *Ann. Oncol.* **2016**, *27* (Suppl. S3), iii42–iii50. [CrossRef] [PubMed]
698. Zuckermann, M.; He, C.; Andrews, J.; Sloan-Henry, R.; Bianski, B.; Xie, J.; Wang, Y.; Twarog, N.; Onar-Thomas, A.; Ernst, K.; et al. MODL-06. Targeting c-MET in combination with radiation is effective in MET-fusion driven high-grade glioma. *Neuro-Oncology* **2022**, *24*, i169. [CrossRef]
699. Esaki, T.; Hirai, F.; Makiyama, A.; Seto, T.; Bando, H.; Naito, Y.; Yoh, K.; Ishihara, K.; Kakizume, T.; Natsume, K.; et al. Phase I dose-escalation study of capmatinib (INC 280) in Japanese patients with advanced solid tumors. *Cancer Sci.* **2019**, *110*, 1340–1351. [CrossRef] [PubMed]
700. Markham, A. Tepotinib: First Approval. *Drugs* **2020**, *80*, 829–833. [CrossRef]
701. Zou, H.Y.; Li, Q.; Lee, J.H.; Arango, M.E.; Burgess, K.; Qiu, M.; Engstrom, L.D.; Yamazaki, S.; Parker, M.; Timofeevski, S.; et al. Sensitivity of Selected Human Tumor Models to PF-04217903, a Novel Selective c-Met Kinase Inhibitor. *Mol. Cancer Ther.* **2012**, *11*, 1036–1047. [CrossRef]
702. Niswander, L.M.; Guenther, L.M.; Mendoza, A.; Khanna, C.; Christensen, J.G.; Helman, L.J.; Kim, S. Effect of modulation of MET with the small molecule inhibitor PF-04217903 on osteosarcoma metastasis in vivo. *J. Clin. Oncol.* **2010**, *28*, 9567. [CrossRef]
703. Lock, R.; Ingraham, R.; Maertens, O.; Miller, A.L.; Weledji, N.; Legius, E.; Konicek, B.M.; Yan, S.-C.B.; Graff, J.R.; Cichowski, K. Cotargeting MNK and MEK kinases induces the regression of NF1-mutant cancers. *J. Clin. Investig.* **2016**, *126*, 2181–2190. [CrossRef]
704. Katayama, R.; Aoyama, A.; Yamori, T.; Qi, J.; Oh-Hara, T.; Song, Y.; Engelman, J.A.; Fujita, N. Cytotoxic Activity of Tivantinib (ARQ 197) Is Not Due Solely to c-MET Inhibition. *Cancer Res* **2013**, *73*, 3087–3096. [CrossRef]
705. Geller, J.I.; Perentesis, J.P.; Liu, X.; Minard, C.G.; Kudgus, R.A.; Reid, J.M.; Fox, E.; Blaney, S.M.; Weigel, B.J. A phase 1 study of the c-Met inhibitor, tivantinib (ARQ197) in children with relapsed or refractory solid tumors: A Children's Oncology Group study phase 1 and pilot consortium trial (ADVL1111). *Pediatr. Blood Cancer* **2017**, *64*, e26565. [CrossRef]
706. Goldberg, J.M.; Gavcovich, T.; Saigal, G.; Goldman, J.W.; Rosen, L.S. Extended Progression-Free Survival in Two Patients With Alveolar Soft Part Sarcoma Exposed to Tivantinib. *J. Clin. Oncol.* **2014**, *32*, e114–e116. [CrossRef]
707. Johnson, T.W.; Richardson, P.F.; Bailey, S.; Brooun, A.; Burke, B.J.; Collins, M.R.; Cui, J.J.; Deal, J.G.; Deng, Y.-L.; Dinh, D.; et al. Discovery of (10*R*)-7-Amino-12-fluoro-2,10,16-trimethyl-15-oxo-10,15,16,17-tetrahydro-*2H*-8,4-(metheno)pyrazolo[4,3-*h*][]2,5,11]-benzoxadiazacyclotetradecine-3-carbonitrile (PF-06463922), a Macrocyclic Inhibitor of Anaplastic Lymphoma Kinase (ALK) and c-ros Oncogene 1 (ROS1) with Preclinical Brain Exposure and Broad-Spectrum Potency against ALK-Resistant Mutations. *J. Med. Chem.* **2014**, *57*, 4720–4744. [CrossRef]
708. Infarinato, N.R.; Park, J.H.; Krytska, K.; Ryles, H.T.; Sano, R.; Szigety, K.M.; Li, Y.; Zou, H.Y.; Lee, N.V.; Smeal, T.; et al. The ALK/ROS1 Inhibitor PF-06463922 Overcomes Primary Resistance to Crizotinib in ALK-Driven Neuroblastoma. *Cancer Discov.* **2016**, *6*, 96–107. [CrossRef]
709. Guan, J.; Tucker, E.R.; Wan, H.; Chand, D.; Danielson, L.S.; Ruuth, K.; El Wakil, A.; Witek, B.; Jamin, Y.; Umapathy, G.; et al. The ALK inhibitor PF-06463922 is effective as a single agent in neuroblastoma driven by expression of ALK and MYCN. *Dis. Model. Mech.* **2016**, *9*, 941–952. [CrossRef]
710. Collier, T.L.; Maresca, K.P.; Normandin, M.D.; Richardson, P.; McCarthy, T.J.; Liang, S.H.; Waterhouse, R.N.; Vasdev, N. Brain Penetration of the ROS1/ALK Inhibitor Lorlatinib Confirmed by PET. *Mol. Imaging* **2017**, *16*. [CrossRef] [PubMed]
711. Bagchi, A.; Orr, B.A.; Campagne, O.; Dhanda, S.; Nair, S.; Tran, Q.; Christensen, A.M.; Gajjar, A.; Furtado, L.V.; Vasilyeva, A.; et al. Lorlatinib in a Child with *ALK*-Fusion–Positive High-Grade Glioma. *New Engl. J. Med.* **2021**, *385*, 761–763. [CrossRef] [PubMed]
712. Liu, T.; Merguerian, M.D.; Rowe, S.P.; Pratilas, C.A.; Chen, A.R.; Ladle, B.H. Exceptional response to the ALK and ROS1 inhibitor lorlatinib and subsequent mechanism of resistance in relapsed *ALK* F1174L-mutated neuroblastoma. *Mol. Case Stud.* **2021**, *7*, a006064. [CrossRef] [PubMed]
713. Khozin, S.; Blumenthal, G.M.; Zhang, L.; Tang, S.; Brower, M.; Fox, E.; Helms, W.; Leong, R.; Song, P.; Pan, Y.; et al. FDA Approval: Ceritinib for the Treatment of Metastatic Anaplastic Lymphoma Kinase–Positive Non–Small Cell Lung Cancer. *Clin. Cancer Res.* **2015**, *21*, 2436–2439. [CrossRef] [PubMed]
714. Guan, J.; Fransson, S.; Siaw, J.T. Clinical response of the novel activating ALK-I1171T mutation in neuroblastoma to the ALK inhibitor ceritinib Running title: Clinical response to ceritinib in ALK-positive neuroblastoma. *Cold Spring Harb. Mol. Case Stud.* **2018**, *4*, a002550. [CrossRef]
715. Mittal, A.; Gupta, A.; Rastogi, S.; Barwad, A.; Sharma, S. Near-complete response to low-dose ceritinib in recurrent infantile inflammatory myofibroblastic tumour. *Ecancermedicalscience* **2021**, *15*. [CrossRef]
716. Russo, A.; Paret, C.; Lehmann, N.; Bender, H.; Wingerter, A.; Neu, M.A.; Alt, F.; Backes, N.; Roth, L.; Seidmann, L.; et al. Epen-29. individualized therapy of an anaplastic ependymoma pediatric patient with a notch1 germline mutation. *Neuro-Oncology* **2018**, *20*, i79. [CrossRef]
717. Russo, A.; Paret, C.; Alt, F.; Burhenne, J.; Fresnais, M.; Wagner, W.; Glaser, M.; Bender, H.; Huprich, S.; Harter, P.N.; et al. Ceritinib-Induced Regression of an Insulin-Like Growth Factor-Driven Neuroepithelial Brain Tumor. *Int. J. Mol. Sci.* **2019**, *20*, 4267. [CrossRef] [PubMed]

718. Lin, C.Y.; Erkek, S.; Tong, Y.; Yin, L.; Federation, A.J.; Zapatka, M.; Haldipur, P.; Kawauchi, D.; Risch, T.; Warnatz, H.-J.; et al. Active medulloblastoma enhancers reveal subgroup-specific cellular origins. *Nature* **2016**, *530*, 57–62. [CrossRef] [PubMed]
719. Tsoli, M.; Wadham, C.; Pinese, M.; Failes, T.; Joshi, S.; Mould, E.; Yin, J.X.; Gayevskiy, V.; Kumar, A.; Kaplan, W.; et al. Integration of genomics, high throughput drug screening, and personalized xenograft models as a novel precision medicine paradigm for high risk pediatric cancer. *Cancer Biol. Ther.* **2018**, *19*, 1078–1087. [CrossRef] [PubMed]
720. Beck, O.; Paret, C.; Russo, A.; Burhenne, J.; Fresnais, M.; Steimel, K.; Seidmann, L.; Wagner, D.-C.; Vewinger, N.; Lehmann, N.; et al. Safety and Activity of the Combination of Ceritinib and Dasatinib in Osteosarcoma. *Cancers* **2020**, *12*, 793. [CrossRef] [PubMed]
721. van Erp, A.E.M.; Hillebrandt-Roeffen, M.H.S.; van Houdt, L.; Fleuren, E.D.G.; van der Graaf, W.T.A.; Versleijen-Jonkers, Y.M.H. Targeting Anaplastic Lymphoma Kinase (ALK) in Rhabdomyosarcoma (RMS) with the Second-Generation ALK Inhibitor Ceritinib. *Target. Oncol.* **2017**, *12*, 815–826. [CrossRef] [PubMed]
722. Dolgikh, N.; Fulda, S. Rhabdomyosarcoma cells are susceptible to cell death by LDK378 alone or in combination with sorafenib independently of anaplastic lymphoma kinase status. *Anti-Cancer Drugs* **2017**, *28*, 1118–1125. [CrossRef]
723. Huang, W.-S.; Liu, S.; Zou, D.; Thomas, M.; Wang, Y.; Zhou, T.; Romero, J.; Kohlmann, A.; Li, F.; Qi, J.; et al. Discovery of Brigatinib (AP26113), a Phosphine Oxide-Containing, Potent, Orally Active Inhibitor of Anaplastic Lymphoma Kinase. *J. Med. Chem.* **2016**, *59*, 4948–4964. [CrossRef]
724. Mo, F.; Pellerino, A.; Soffietti, R.; Rudà, R. Blood–Brain Barrier in Brain Tumors: Biology and Clinical Relevance. *Int. J. Mol. Sci.* **2021**, *22*, 12654. [CrossRef]
725. Spencer, S.A.; Riley, A.C.; Matthew, A.; Di Pasqua, A.J. Brigatinib: Novel ALK Inhibitor for Non–Small-Cell Lung Cancer. *Ann. Pharmacother.* **2019**, *53*, 621–626. [CrossRef]
726. Siaw, J.T.; Wan, H.; Pfeifer, K.; Rivera, V.M.; Guan, J.; Palmer, R.H.; Hallberg, B. Brigatinib, an anaplastic lymphoma kinase inhibitor, abrogates activity and growth in ALK-positive neuroblastoma cells,*Drosophila*and mice. *Oncotarget* **2016**, *7*, 29011–29022. [CrossRef]
727. Drilon, A.; De Braud, F.G.; Siena, S.; Ou, S.-H.I.; Patel, M.; Ahn, M.-J.; Lee, J.; Bauer, T.M.; Farago, A.F.; Liu, S.V.; et al. Abstract CT007: Entrectinib, an oral pan-Trk, ROS1, and ALK inhibitor in TKI-naïve patients with advanced solid tumors harboring gene rearrangements: Updated phase I results. *Cancer Res* **2016**, *76*, CT007. [CrossRef]
728. Jiang, Q.; Li, M.; Li, H.; Chen, L. Entrectinib, a new multi-target inhibitor for cancer therapy. *Biomed. Pharmacother.* **2022**, *150*, 112974. [CrossRef]
729. Iyer, R.; Wehrmann, L.; Golden, R.L.; Naraparaju, K.; Croucher, J.L.; MacFarland, S.P.; Guan, P.; Kolla, V.; Wei, G.; Cam, N.; et al. Entrectinib is a potent inhibitor of Trk-driven neuroblastomas in a xenograft mouse model. *Cancer Lett.* **2016**, *372*, 179–186. [CrossRef]
730. Desai, A.V.; Robinson, G.W.; Gauvain, K.; Basu, E.M.; E Macy, M.; Maese, L.; Whipple, N.S.; Sabnis, A.J.; Foster, J.H.; Shusterman, S.; et al. Entrectinib in children and young adults with solid or primary CNS tumors harboring *NTRK*, *ROS1*, or *ALK* aberrations (STARTRK-NG). *Neuro-Oncology* **2022**, *24*, 1776–1789. [CrossRef]
731. MacFarland, S.P.; Naraparaju, K.; Iyer, R.; Guan, P.; Kolla, V.; Hu, Y.; Tan, K.; Brodeur, G.M. Mechanisms of Entrectinib Resistance in a Neuroblastoma Xenograft Model. *Mol. Cancer Ther.* **2020**, *19*, 920–926. [CrossRef]
732. Spitaleri, G.; Passaro, A.; de Marinis, F. Ensartinib (X-396) a novel drug for anaplastic lymphoma kinase-positive non-small cell lung cancer patients: We need smart trials to avoid wasting good bullets. *Chin. Clin. Oncol.* **2019**, *8*, S1. [CrossRef]
733. Lovly, C.M.; Heuckmann, J.M.; de Stanchina, E.; Chen, H.; Thomas, R.K.; Liang, C.; Pao, W. Insights into ALK-Driven Cancers Revealed through Development of Novel ALK Tyrosine Kinase Inhibitors. *Cancer Res* **2011**, *71*, 4920–4931. [CrossRef]
734. Di Paolo, D.; Yang, D.; Pastorino, F.; Emionite, L.; Cilli, M.; Daga, A.; Destafanis, E.; Di Fiore, A.; Piaggio, F.; Brignole, C.; et al. New therapeutic strategies in neuroblastoma: Combined targeting of a novel tyrosine kinase inhibitor and liposomal siRNAs against *ALK*. *Oncotarget* **2015**, *6*, 28774–28789. [CrossRef]
735. Perera, T.P.; Jovcheva, E.; Mevellec, L.; Vialard, J.; De Lange, D.; Verhulst, T.; Paulussen, C.; Van De Ven, K.; King, P.; Freyne, E.; et al. Discovery and Pharmacological Characterization of JNJ-42756493 (Erdafitinib), a Functionally Selective Small-Molecule FGFR Family Inhibitor. *Mol. Cancer Ther.* **2017**, *16*, 1010–1020. [CrossRef]
736. Luzzi, S.; Brambilla, I.; Mantelli, S.S.; Mosconi, M.; Foiadelli, T.; Savasta, S. Targeting the medulloblastoma: A molecular-based approach. *Acta Biomed.* **2020**, *91*, 79–100. [CrossRef]
737. Musumeci, F.; Cianciusi, A.; D'Agostino, I.; Grossi, G.; Carbone, A.; Schenone, S. Synthetic Heterocyclic Derivatives as Kinase Inhibitors Tested for the Treatment of Neuroblastoma. *Molecules* **2021**, *26*, 7069. [CrossRef] [PubMed]
738. Holzhauser, S.; Lukoseviciute, M.; Papachristofi, C.; Vasilopoulou, C.; Herold, N.; Wickström, M.; Kostopoulou, O.N.; Dalianis, T. Effects of PI3K and FGFR inhibitors alone and in combination, and with/without cytostatics in childhood neuroblastoma cell lines. *Int. J. Oncol.* **2021**, *58*, 211–225. [CrossRef] [PubMed]
739. Lukoseviciute, M.; Maier, H.; Poulou-Sidiropoulou, E.; Rosendahl, E.; Holzhauser, S.; Dalianis, T.; Kostopoulou, O.N. Targeting PI3K, FGFR, CDK4/6 Signaling Pathways Together With Cytostatics and Radiotherapy in Two Medulloblastoma Cell Lines. *Front. Oncol.* **2021**, *11*, 748657. [CrossRef] [PubMed]
740. Angevin, E.; Lopez-Martin, J.A.; Lin, C.-C.; Gschwend, J.E.; Harzstark, A.; Castellano, D.; Soria, J.-C.; Sen, P.; Chang, J.; Shi, M.; et al. Phase I Study of Dovitinib (TKI258), an Oral FGFR, VEGFR, and PDGFR Inhibitor, in Advanced or Metastatic Renal Cell Carcinoma. *Clin. Cancer Res.* **2013**, *19*, 1257–1268. [CrossRef]

741. Arnz, E.; Hertwig, F.; Krämer, A.; Ikram, F.; Roels, F.; Kocak, H.; Engesser, A.; Thomas, R.; Peifer, M.; Ackermann, S.; et al. Fibroblast growth factor receptors as therapeutic targets in neuroblastoma. *Klin. Pädiatrie* **2014**, *226*, A22. [CrossRef]
742. Li, S.Q.; Cheuk, A.T.; Shern, J.F.; Song, Y.K.; Hurd, L.; Liao, H.; Wei, J.S.; Khan, J. Targeting Wild-Type and Mutationally Activated FGFR4 in Rhabdomyosarcoma with the Inhibitor Ponatinib (AP24534). *PLoS ONE* **2013**, *8*, e76551. [CrossRef]
743. Schramm, K.; Iskar, M.; Statz, B.; Jäger, N.; Haag, D.; Słabicki, M.; Pfister, S.M.; Zapatka, M.; Gronych, J.; Jones, D.T.W.; et al. DECIPHER pooled shRNA library screen identifies PP2A and FGFR signaling as potential therapeutic targets for diffuse intrinsic pontine gliomas. *Neuro-Oncology* **2019**, *21*, 867–877. [CrossRef]
744. Yan, D.; Kowal, J.; Akkari, L.; Schuhmacher, A.J.; Huse, J.T.; West, B.L.; A Joyce, J. Inhibition of colony stimulating factor-1 receptor abrogates microenvironment-mediated therapeutic resistance in gliomas. *Oncogene* **2017**, *36*, 6049–6058. [CrossRef]
745. Georgin-Lavialle, S.; Lhermitte, L.; Suarez, F.; Yang, Y.; Létard, S.; Hanssens, K.; Feger, F.; Renand, A.; Brouze, C.; Canioni, D.; et al. Mast cell leukemia: Identification of a new c-Kit mutation, dup(501-502), and response to masitinib, a c-Kit tyrosine kinase inhibitor. *Eur. J. Haematol.* **2012**, *89*, 47–52. [CrossRef]
746. Dubreuil, P.; Letard, S.; Ciufolini, M.; Gros, L.; Humbert, M.; Castéran, N.; Borge, L.; Hajem, B.; Lermet, A.; Sippl, W.; et al. Masitinib (AB1010), a Potent and Selective Tyrosine Kinase Inhibitor Targeting KIT. *PLoS ONE* **2009**, *4*, e7258. [CrossRef]
747. Marech, I.; Patruno, R.; Zizzo, N.; Gadaleta, C.; Introna, M.; Zito, A.F.; Gadaleta, C.D.; Ranieri, G. Masitinib (AB1010), from canine tumor model to human clinical development: Where we are? *Crit. Rev. Oncol.* **2014**, *91*, 98–111. [CrossRef]
748. Fleming, T.; Cunningham, C.; Keir, S. *The Effect of Masitinib on Pediatric Glioblastoma*; Duke University: Durham, NC, USA, 2014.
749. Buti, S.; Leonetti, A.; Dallatomasina, A.; Bersanelli, M. Everolimus in the management of metastatic renal cell carcinoma: An evidence-based review of its place in therapy. *Core Évid.* **2016**, *11*, 23–36. [CrossRef]
750. Nashan, B. Review of the proliferation inhibitor everolimus. *Expert Opin. Investig. Drugs* **2002**, *11*, 1845–1857. [CrossRef]
751. Pignochino, Y.; Dell'Aglio, C.; Basiricò, M.; Capozzi, F.; Soster, M.; Marchiò, S.; Bruno, S.; Gammaitoni, L.; Sangiolo, D.; Torchiaro, E.; et al. The Combination of Sorafenib and Everolimus Abrogates mTORC1 and mTORC2 Upregulation in Osteosarcoma Preclinical Models. *Clin. Cancer Res.* **2013**, *19*, 2117–2131. [CrossRef]
752. Miklja, Z.; Yadav, V.N.; Cartaxo, R.T.; Siada, R.; Thomas, C.C.; Cummings, J.R.; Mullan, B.; Stallard, S.; Paul, A.; Bruzek, A.K.; et al. Everolimus improves the efficacy of dasatinib in PDGFRα-driven glioma. *J. Clin. Investig.* **2020**, *130*, 5313–5325. [CrossRef]
753. Poore, B.; Yuan, M.; Arnold, A.; Price, A.; Alt, J.; A Rubens, J.; Slusher, B.S.; Eberhart, C.G.; Raabe, E.H. Inhibition of mTORC1 in pediatric low-grade glioma depletes glutathione and therapeutically synergizes with carboplatin. *Neuro-Oncology* **2018**, *21*, 252–263. [CrossRef]
754. Salussolia, C.L.; Klonowska, K.; Kwiatkowski, D.J.; Sahin, M. Genetic Etiologies, Diagnosis, and Treatment of Tuberous Sclerosis Complex. *Annu. Rev. Genom. Hum. Genet.* **2019**, *20*, 217–240. [CrossRef]
755. Rosset, C.; Netto, C.B.O.; Ashton-Prolla, P. TSC1 and TSC2 gene mutations and their implications for treatment in Tuberous Sclerosis Complex: A review. *Genet. Mol. Biol.* **2017**, *40*, 69–79. [CrossRef]
756. Krueger, D.A.; Care, M.M.; Holland, K.; Agricola, K.; Tudor, C.; Mangeshkar, P.; Wilson, K.A.; Byars, A.; Sahmoud, T.; Franz, D.N. Everolimus for Subependymal Giant-Cell Astrocytomas in Tuberous Sclerosis. *N. Engl. J. Med.* **2010**, *363*, 1801–1811. [CrossRef]
757. Franz, D.N.; Lawson, J.A.; Yapici, Z.; Ikeda, H.; Polster, T.; Nabbout, R.; Curatolo, P.; de Vries, P.J.; Dlugos, D.J.; Voi, M.; et al. Everolimus for treatment-refractory seizures in TSC. *Neurol. Clin. Pract.* **2018**, *8*, 412–420. [CrossRef]
758. Xue, Q.; Hopkins, B.; Perruzzi, C.; Udayakumar, D.; Sherris, D.; Benjamin, L.E. Palomid 529, a Novel Small-Molecule Drug, Is a TORC1/TORC2 Inhibitor That Reduces Tumor Growth, Tumor Angiogenesis, and Vascular Permeability. *Cancer Res* **2008**, *68*, 9551–9557. [CrossRef] [PubMed]
759. Lin, F.; Buil, L.; Sherris, D.; Beijnen, J.H.; van Tellingen, O. Dual mTORC1 and mTORC2 inhibitor Palomid 529 penetrates the Blood-Brain Barrier without restriction by ABCB1 and ABCG2. *Int. J. Cancer* **2013**, *133*, 1222–1233. [CrossRef] [PubMed]
760. Gravina, G.L.; Mancini, A.; Colapietro, A.; Monache, S.D.; Sferra, R.; Pompili, S.; Vitale, F.; Martellucci, S.; Marampon, F.; Mattei, V.; et al. The Brain Penetrating and Dual TORC1/TORC2 Inhibitor, RES529, Elicits Anti-Glioma Activity and Enhances the Therapeutic Effects of Anti-Angiogenetic Compounds in Preclinical Murine Models. *Cancers* **2019**, *11*, 1604. [CrossRef] [PubMed]
761. Cerna, D.; Carter, D.; Flaherty, S.; Cao, L.; Sherris, D.; Yoo, S.S. Abstract 2506: Palomid 529, a PI3K/Akt/mTOR dual TORC1/2 inhibitor, is a radiosensitizer with effect in both subcutaneous and orthotopic U251 glioblastoma tumor xenograft models. *Cancer Res.* **2010**, *70*, 2506. [CrossRef]
762. Hu, X.; Wang, Z.; Chen, M.; Chen, X.; Liang, W. The anti-osteosarcoma cell activity by a mTORC1/2 dual inhibitor RES-529. *Biochem. Biophys. Res. Commun.* **2018**, *497*, 499–505. [CrossRef]
763. Bhagwat, S.V.; Gokhale, P.C.; Crew, A.P.; Cooke, A.; Yao, Y.; Mantis, C.; Kahler, J.; Workman, J.; Bittner, M.; Dudkin, L.; et al. Preclinical Characterization of OSI-027, a Potent and Selective Inhibitor of mTORC1 and mTORC2: Distinct from Rapamycin. *Mol. Cancer Ther.* **2011**, *10*, 1394–1406. [CrossRef]
764. Eckerdt, F.; Clymer, J.; Bell, J.B.; Beauchamp, E.M.; Blyth, G.T.; Goldman, S.; Platanias, L.C. Pharmacological mTOR targeting enhances the antineoplastic effects of selective PI3Kα inhibition in medulloblastoma. *Sci. Rep.* **2019**, *9*, 12822. [CrossRef]
765. Srivastava, R.K.; Guroji, P.; Jin, L.; Mukhtar, M.S.; Athar, M. Combined inhibition of BET bromodomain and mTORC1/2 provides therapeutic advantage for rhabdomyosarcoma by switching cell death mechanism. *Mol. Carcinog.* **2022**, *61*, 737–751. [CrossRef]
766. Eckerdt, F.; Beauchamp, E.; Bell, J.; Iqbal, A.; Su, B.; Fukunaga, R.; Lulla, R.R.; Goldman, S.; Platanias, L.C. Regulatory effects of a Mnk2-eIF4E feedback loop during mTORC1 targeting of human medulloblastoma cells. *Oncotarget* **2014**, *5*, 8442–8451. [CrossRef]

767. Clymer, J.; Eckerd, F.; Bell, J.; Lulla, R.; Goldman, S.; Platanias, L. MEDU-44. TARGETING SHH SIGNALING VIA PI3K/MTOR INHIBITION IN MEDULLOBLASTOMA AND EWING SARCOMA. *Neuro-Oncology* **2017**, *19*, iv47. [CrossRef]
768. Calimeri, T.; Ferreri, A.J.M. m-TOR inhibitors and their potential role in haematological malignancies. *Br. J. Haematol.* **2017**, *177*, 684–702. [CrossRef]
769. Kolev, V.N.; Wright, Q.G.; Vidal, C.M.; Ring, J.E.; Shapiro, I.M.; Ricono, J.; Weaver, D.T.; Padval, M.V.; Pachter, J.A.; Xu, Q. PI3K/mTOR Dual Inhibitor VS-5584 Preferentially Targets Cancer Stem Cells. *Cancer Res* **2015**, *75*, 446–455. [CrossRef]
770. Sun, J.-Y.; Hou, Y.-J.; Cui, H.-J.; Zhang, C.; Yang, M.-F.; Wang, F.-Z.; Sun, Z.; Fan, C.-D.; Sun, B.-L.; Oh, J.R. VS-5584 Inhibits Human Osteosarcoma Cells Growth by Induction of G1- phase Arrest through Regulating PI3K/mTOR and MAPK Pathways. *Curr. Cancer Drug Targets* **2020**, *20*, 616–623. [CrossRef]
771. Sun, J.-Y.; Hou, Y.-J.; Yin, Y.-B.; Wang, F.-Z.; Yang, M.-F.; Zhang, Y.-Y.; Fan, C.-D.; Sun, B.-L. CCT128930 induces G1-phase arrest and apoptosis and synergistically enhances the anticancer efficiency of VS5584 in human osteosarcoma cells. *Biomed. Pharmacother.* **2020**, *130*, 110544. [CrossRef] [PubMed]
772. Maira, S.-M.; Stauffer, F.; Brueggen, J.; Furet, P.; Schnell, C.; Fritsch, C.; Brachmann, S.; Chène, P.; De Pover, A.; Schoemaker, K.; et al. Identification and characterization of NVP-BEZ235, a new orally available dual phosphatidylinositol 3-kinase/mammalian target of rapamycin inhibitor with potent in vivo antitumor activity. *Mol. Cancer Ther.* **2008**, *7*, 1851–1863. [CrossRef]
773. Manara, M.C.; Nicoletti, G.; Zambelli, D.; Ventura, S.; Guerzoni, C.; Landuzzi, L.; Lollini, P.-L.; Maira, S.-M.; García-Echeverría, C.; Mercuri, M.; et al. NVP-BEZ235 as a New Therapeutic Option for Sarcomas. *Clin. Cancer Res.* **2010**, *16*, 530–540. [CrossRef]
774. Giorgi, C.; Boro, A.; Rechfeld, F.; Lopez-Garcia, L.A.; Gierisch, M.E.; Schäfer, B.W.; Niggli, F.K. PI3K/AKT signaling modulates transcriptional expression of EWS/FLI1 through specificity protein 1. *Oncotarget* **2015**, *6*, 28895–28910. [CrossRef]
775. Zhu, Y.-R.; Min, H.; Fang, J.-F.; Zhou, F.; Deng, X.-W.; Zhang, Y.-Q. Activity of the novel dual phosphatidylinositol 3-kinase/mammalian target of rapamycin inhibitor NVP-BEZ235 against osteosarcoma. *Cancer Biol. Ther.* **2015**, *16*, 602–609. [CrossRef]
776. Meng, W.; Wang, B.; Mao, W.; Wang, J.; Zhao, Y.; Li, Q.; Zhang, C.; Ma, J. Enhanced efficacy of histone deacetylase inhibitor panobinostat combined with dual PI3K/mTOR inhibitor BEZ235 against glioblastoma. *Nagoya J. Med. Sci.* **2019**, *81*, 93–102. [CrossRef]
777. Vazquez, N.; Lopez, A.; Cuello, V.; Persans, M.; Schuenzel, E.; Innis-Whitehouse, W.; Keniry, M. NVP-BEZ235 or JAKi Treatment leads to decreased survival of examined GBM and BBC cells. *Cancer Treat. Res. Commun.* **2021**, *27*, 100340. [CrossRef]
778. Xie, C.; Freeman, M.J.; Lu, H.; Wang, X.; Forster, C.L.; Sarver, A.L.; Hallstrom, T.C. Retinoblastoma cells activate the AKT pathway and are vulnerable to the PI3K/mTOR inhibitor NVP-BEZ235. *Oncotarget* **2017**, *8*, 38084–38098. [CrossRef] [PubMed]
779. Holzhauser, S.; Lukoseviciute, M.; Andonova, T.; Ursu, R.G.; Dalianis, T.; Wickström, M.; Kostopoulou, O.N. Targeting Fibroblast Growth Factor Receptor (FGFR) and Phosphoinositide 3-kinase (PI3K) Signaling Pathways in Medulloblastoma Cell Lines. *Anticancer. Res.* **2019**, *40*, 53–66. [CrossRef] [PubMed]
780. Gobin, B.; Battaglia, S.; Lanel, R.; Chesneau, J.; Amiaud, J.; Rédini, F.; Ory, B.; Heymann, D. NVP-BEZ235, a dual PI3K/mTOR inhibitor, inhibits osteosarcoma cell proliferation and tumor development in vivo with an improved survival rate. *Cancer Lett.* **2014**, *344*, 291–298. [CrossRef] [PubMed]
781. Chaturvedi, N.K.; Kling, M.J.; Griggs, C.N.; Kesherwani, V.; Shukla, M.; McIntyre, E.M.; Ray, S.; Liu, Y.; McGuire, T.R.; Sharp, J.G.; et al. A Novel Combination Approach Targeting an Enhanced Protein Synthesis Pathway in MYC-driven (Group 3) Medulloblastoma. *Mol. Cancer Ther.* **2020**, *19*, 1351–1362. [CrossRef] [PubMed]
782. Buonamici, S.; Williams, J.; Morrissey, M.; Wang, A.; Guo, R.; Vattay, A.; Hsiao, K.; Yuan, J.; Green, J.; Ospina, B.; et al. Interfering with Resistance to Smoothened Antagonists by Inhibition of the PI3K Pathway in Medulloblastoma. *Sci. Transl. Med.* **2010**, *2*, 51ra70. [CrossRef]
783. Garlich, J.R.; De, P.; Dey, N.; Su, J.D.; Peng, X.; Miller, A.; Murali, R.; Lu, Y.; Mills, G.B.; Kundra, V.; et al. A Vascular Targeted Pan Phosphoinositide 3-Kinase Inhibitor Prodrug, SF1126, with Antitumor and Antiangiogenic Activity. *Cancer Res* **2008**, *68*, 206–215. [CrossRef]
784. Mahadevan, D.; Chiorean, E.; Harris, W.; Von Hoff, D.; Stejskal-Barnett, A.; Qi, W.; Anthony, S.; Younger, A.; Rensvold, D.; Cordova, F.; et al. Phase I pharmacokinetic and pharmacodynamic study of the pan-PI3K/mTORC vascular targeted pro-drug SF1126 in patients with advanced solid tumours and B-cell malignancies. *Eur. J. Cancer* **2012**, *48*, 3319–3327. [CrossRef]
785. Singh, A.R.; Joshi, S.; Zulcic, M.; Alcaraz, M.; Garlich, J.R.; Morales, G.A.; Cho, Y.J.; Bao, L.; Levy, M.L.; Newbury, R.; et al. PI-3K Inhibitors Preferentially Target CD15+ Cancer Stem Cell Population in SHH Driven Medulloblastoma. *PLoS ONE* **2016**, *11*, e0150836. [CrossRef]
786. Goldin, A.N.; Singh, A.; Joshi, S.; Jamieson, C.; Durden, D.L.M. Augmented Antitumor Activity for Novel Dual PI3K/BDR4 Inhibitors, SF2523 and SF1126 in Ewing Sarcoma. *J. Pediatr. Hematol.* **2021**, *43*, e304–e311. [CrossRef]
787. Peirce, S.; Durden, D.; Garlich, J.; Findley, H. SF1126, a novel pan-PI3K inhibitor, inhibits activation of Mdm2 and in-creases sensitivity to doxorubicin in wild type p53 neuroblastoma cell lines. *Cancer Res.* **2007**, *67*, LB-294.
788. Erdreich-Epstein, A.; Singh, A.R.; Joshi, S.; Vega, F.M.; Guo, P.; Xu, J.; Groshen, S.; Ye, W.; Millard, M.; Campan, M.; et al. Association of high microvessel $\alpha v\beta 3$ and low PTEN with poor outcome in stage 3 neuroblastoma: Rationale for using first in class dual PI3K/BRD4 inhibitor, SF1126. *Oncotarget* **2016**, *8*, 52193–52210. [CrossRef]
789. Beljanski, V. Triciribine. In *xPharm: The Comprehensive Pharmacology Reference*; Elsevier: Amsterdam, The Netherlands, 2009; pp. 1–5. [CrossRef]

790. Bahmad, H.F.; Mouhieddine, T.H.; Chalhoub, R.M.; Assi, S.; Araji, T.; Chamaa, F.; Itani, M.M.; Nokkari, A.; Kobeissy, F.; Daoud, G.; et al. The Akt/mTOR pathway in cancer stem/progenitor cells is a potential therapeutic target for glioblastoma and neuroblastoma. *Oncotarget* **2018**, *9*, 33549–33561. [CrossRef]
791. Cubitt, C.L.; Menth, J.; Dawson, J.; Martinez, G.V.; Foroutan, P.; Morse, D.L.; Bui, M.M.; Letson, G.D.; Sullivan, D.M.; Reed, D.R. Rapid Screening of Novel Agents for Combination Therapy in Sarcomas. *Sarcoma* **2013**, *2013*, 1–12. [CrossRef]
792. Smeester, B.A.; Draper, G.M.; Slipek, N.J.; Larsson, A.T.; Stratton, N.; Pomeroy, E.J.; Becklin, K.L.; Yamamoto, K.; Williams, K.B.; Laoharawee, K.; et al. Implication of *ZNF217* in Accelerating Tumor Development and Therapeutically Targeting ZNF217-Induced PI3K–AKT Signaling for the Treatment of Metastatic Osteosarcoma. *Mol. Cancer Ther.* **2020**, *19*, 2528–2541. [CrossRef]
793. Voss, M.H.; Gordon, M.S.; Mita, M.; Rini, B.; Makker, V.; Macarulla, T.; Smith, D.C.; Cervantes, A.; Puzanov, I.; Pili, R.; et al. Phase 1 study of mTORC1/2 inhibitor sapanisertib (TAK-228) in advanced solid tumours, with an expansion phase in renal, endometrial or bladder cancer. *Br. J. Cancer* **2020**, *123*, 1590–1598. [CrossRef]
794. Slotkin, E.K.; Patwardhan, P.P.; Vasudeva, S.D.; de Stanchina, E.; Tap, W.D.; Schwartz, G.K. MLN0128, an ATP-Competitive mTOR Kinase Inhibitor with Potent In Vitro and In Vivo Antitumor Activity, as Potential Therapy for Bone and Soft-Tissue Sarcoma. *Mol. Cancer Ther.* **2015**, *14*, 395–406. [CrossRef]
795. Jiang, H.; Zeng, Z. Dual mTORC1/2 inhibition by INK-128 results in antitumor activity in preclinical models of osteosarcoma. *Biochem. Biophys. Res. Commun.* **2015**, *468*, 255–261. [CrossRef]
796. Maynard, R.E.; Poore, B.; Hanaford, A.R.; Pham, K.; James, M.; Alt, J.; Park, Y.; Slusher, B.S.; Tamayo, P.; Mesirov, J.; et al. TORC1/2 kinase inhibition depletes glutathione and synergizes with carboplatin to suppress the growth of MYC-driven medulloblastoma. *Cancer Lett.* **2021**, *504*, 137–145. [CrossRef]
797. Zhang, H.; Dou, J.; Yu, Y.; Zhao, Y.; Fan, Y.; Cheng, J.; Xu, X.; Liu, W.; Guan, S.; Chen, Z.; et al. mTOR ATP-competitive inhibitor INK128 inhibits neuroblastoma growth via blocking mTORC signaling. *Apoptosis* **2014**, *20*, 50–62. [CrossRef]
798. Miyahara, H.; Yadavilli, S.; Natsumeda, M.; Rubens, J.A.; Rodgers, L.; Kambhampati, M.; Taylor, I.C.; Kaur, H.; Asnaghi, L.; Eberhart, C.G.; et al. The dual mTOR kinase inhibitor TAK228 inhibits tumorigenicity and enhances radiosensitization in diffuse intrinsic pontine glioma. *Cancer Lett.* **2017**, *400*, 110–116. [CrossRef]
799. Arnold, A.; Yuan, M.; Price, A.; Harris, L.; Eberhart, C.G.; Raabe, E.H. Synergistic activity of mTORC1/2 kinase and MEK inhibitors suppresses pediatric low-grade glioma tumorigenicity and vascularity. *Neuro-Oncology* **2019**, *22*, 563–574. [CrossRef]
800. Tang, L.; Fu, Y.; Song, J.; Hu, T.; Li, K.; Li, Z. mTOR inhibition by TAK-228 is effective against growth, survival and angiogenesis in preclinical retinoblastoma models. *Pharmacol. Res. Perspect.* **2022**, *10*, e00930. [CrossRef] [PubMed]
801. Gray, J.E.; Infante, J.R.; Brail, L.H.; Simon, G.R.; Cooksey, J.F.; Jones, S.F.; Farrington, D.L.; Yeo, A.; Jackson, K.A.; Chow, K.H.; et al. A first-in-human phase I dose-escalation, pharmacokinetic, and pharmacodynamic evaluation of intravenous LY2090314, a glycogen synthase kinase 3 inhibitor, administered in combination with pemetrexed and carboplatin. *Investig. New Drugs* **2015**, *33*, 1187–1196. [CrossRef] [PubMed]
802. Kunnimalaiyaan, S.; Schwartz, V.K.; Jackson, I.A.; Gamblin, T.C.; Kunnimalaiyaan, M. Antiproliferative and apoptotic effect of LY2090314, a GSK-3 inhibitor, in neuroblastoma in vitro. *BMC Cancer* **2018**, *18*, 560. [CrossRef]
803. Wei, D.; Zhu, X.; Li, S.; Liu, G.; Wang, Y.; Wang, W.; Zhang, Q.; Jiang, S. Tideglusib suppresses stem-cell-like features and progression of osteosarcoma by inhibiting GSK-3β/NOTCH1 signaling. *Biochem. Biophys. Res. Commun.* **2021**, *554*, 206–213. [CrossRef]
804. Bahmad, H.F.; Chalhoub, R.M.; Harati, H.; Bou-Gharios, J.; Assi, S.; Ballout, F.; Monzer, A.; Msheik, H.; Araji, T.; Elajami, M.K.; et al. Tideglusib attenuates growth of neuroblastoma cancer stem/progenitor cells in vitro and in vivo by specifically targeting GSK-3β. *Pharmacol. Rep.* **2020**, *73*, 211–226. [CrossRef]
805. Mathuram, T.L.; Ravikumar, V.; Reece, L.M.; Karthik, S.; Sasikumar, C.S.; Cherian, K.M. Tideglusib induces apoptosis in human neuroblastoma IMR32 cells, provoking sub-G 0 /G 1 accumulation and ROS generation. *Environ. Toxicol. Pharmacol.* **2016**, *46*, 194–205. [CrossRef]
806. Bharathy, N.; Svalina, M.N.; Settelmeyer, T.P.; Cleary, M.M.; Berlow, N.E.; Airhart, S.D.; Xiang, S.; Keck, J.; Hayden, J.B.; Shern, J.F.; et al. Preclinical testing of the glycogen synthase kinase-3β inhibitor tideglusib for rhabdomyosarcoma. *Oncotarget* **2017**, *8*, 62976–62983. [CrossRef]
807. Hirai, H.; Sootome, H.; Nakatsuru, Y.; Miyama, K.; Taguchi, S.; Tsujioka, K.; Ueno, Y.; Hatch, H.; Majumder, P.K.; Pan, B.-S.; et al. MK-2206, an Allosteric Akt Inhibitor, Enhances Antitumor Efficacy by Standard Chemotherapeutic Agents or Molecular Targeted Drugs In vitro and In vivo. *Mol. Cancer Ther.* **2010**, *9*, 1956–1967. [CrossRef]
808. Gorlick, R.; Maris, J.M.; Houghton, P.J.; Lock, R.; Carol, H.; Kurmasheva, R.T.; Kolb, E.A.; Keir, S.T.; Reynolds, C.P.; Kang, M.H.; et al. Testing of the Akt/PKB inhibitor MK-2206 by the pediatric preclinical testing program. *Pediatr. Blood Cancer* **2011**, *59*, 518–524. [CrossRef]
809. Duan, L.; E Perez, R.; Hansen, M.; Gitelis, S.; Maki, C.G. Increasing cisplatin sensitivity by schedule-dependent inhibition of AKT and Chk1. *Cancer Biol. Ther.* **2014**, *15*, 1600–1612. [CrossRef]
810. Santo, E.E.; Stroeken, P.; Sluis, P.V.; Koster, J.; Versteeg, R.; Westerhout, E.M. FOXO3a Is a Major Target of Inactivation by PI3K/AKT Signaling in Aggressive Neuroblastoma. *Cancer Res* **2013**, *73*, 2189–2198. [CrossRef]
811. Qi, L.; Toyoda, H.; Xu, D.-Q.; Zhou, Y.; Sakurai, N.; Amano, K.; Kihira, K.; Hori, H.; Azuma, E.; Komada, Y. PDK1-mTOR signaling pathway inhibitors reduce cell proliferation in MK2206 resistant neuroblastoma cells. *Cancer Cell Int.* **2015**, *15*, 1–14. [CrossRef]

812. Li, Z.; Yan, S.; Attayan, N.; Ramalingam, S.; Thiele, C.J. Combination of an Allosteric Akt Inhibitor MK-2206 with Etoposide or Rapamycin Enhances the Antitumor Growth Effect in Neuroblastoma. *Clin. Cancer Res.* **2012**, *18*, 3603–3615. [CrossRef]
813. Kang, B.W.; Chau, I. Molecular target: Pan-AKT in gastric cancer. *ESMO Open* **2020**, *5*, e000728. [CrossRef]
814. Choo, F.; Odintsov, I.; Nusser, K.; Nicholson, K.S.; Davis, L.; Corless, C.L.; Stork, L.; Somwar, R.; Ladanyi, M.; Davis, J.L.; et al. Functional impact and targetability of *PI3KCA*, *GNAS*, and *PTEN* mutations in a spindle cell rhabdomyosarcoma with MYOD1 L122R mutation. *Mol. Case Stud.* **2022**, *8*, a006140. [CrossRef]
815. Abdelgalil, A.A.; Alkahtani, H.M.; Al-Jenoobi, F.I. Sorafenib. *Profiles Drug Subst. Excip. Relat. Methodol.* **2019**, *44*, 239–266. [CrossRef]
816. Yang, F.; Van Meter, T.E.; Buettner, R.; Hedvat, M.; Liang, W.; Kowolik, C.M.; Mepani, N.; Mirosevich, J.; Nam, S.; Chen, M.Y.; et al. Sorafenib inhibits signal transducer and activator of transcription 3 signaling associated with growth arrest and apoptosis of medulloblastomas. *Mol. Cancer Ther.* **2008**, *7*, 3519–3526. [CrossRef]
817. Chai, H.; Luo, A.Z.; Weerasinghe, P.; E Brown, R. Sorafenib downregulates ERK/Akt and STAT3 survival pathways and induces apoptosis in a human neuroblastoma cell line. *Int. J. Clin. Exp. Pathol.* **2010**, *3*, 408–415.
818. Kakodkar, N.C.; Peddinti, R.R.; Tian, Y.; Guerrero, L.J.; Chlenski, A.; Yang, Q.; Salwen, H.R.; Maitland, M.L.; Cohn, S.L. Sorafenib inhibits neuroblastoma cell proliferation and signaling, blocks angiogenesis, and impairs tumor growth. *Pediatr. Blood Cancer* **2011**, *59*, 642–647. [CrossRef]
819. Karajannis, M.A.; Legault, G.; Fisher, M.J.; Milla, S.S.; Cohen, K.J.; Wisoff, J.H.; Harter, D.H.; Goldberg, J.D.; Hochman, T.; Merkelson, A.; et al. Phase II study of sorafenib in children with recurrent or progressive low-grade astrocytomas. *Neuro-Oncology* **2014**, *16*, 1408–1416. [CrossRef] [PubMed]
820. Albarrán, V.; Villamayor, M.L.; Chamorro, J.; Rosero, D.I.; Pozas, J.; Román, M.S.; Calvo, J.C.; de Aguado, P.P.; Moreno, J.; Guerrero, P.; et al. Receptor Tyrosine Kinase Inhibitors for the Treatment of Recurrent and Unresectable Bone Sarcomas. *Int. J. Mol. Sci.* **2022**, *23*, 13784. [CrossRef] [PubMed]
821. Tian, Z.; Niu, X.; Yao, W. Receptor Tyrosine Kinases in Osteosarcoma Treatment: Which Is the Key Target? *Front. Oncol.* **2020**, *10*, 1642. [CrossRef] [PubMed]
822. Wu, C.-H.; Lin, K.-H.; Fu, B.-S.; Hsu, F.-T.; Tsai, J.-J.; Weng, M.-C.; Pan, P.-J. Sorafenib Induces Apoptosis and Inhibits NF-κB-mediated Anti-apoptotic and Metastatic Potential in Osteosarcoma Cells. *Anticancer. Res.* **2021**, *41*, 1251–1259. [CrossRef]
823. Dumont, S.; Yang, D.; Dumont, A.; Reynoso, D.; Blay, J.-Y.; Trent, J. Targeted polytherapy in small cell sarcoma and its association with doxorubicin. *Mol. Oncol.* **2014**, *8*, 1458–1468. [CrossRef]
824. Higuchi, T.; Igarashi, K.; Yamamoto, N.; Hayashi, K.; Kimura, H.; Miwa, S.; Bouvet, M.; Tsuchiya, H.; Hoffman, R.M. Osteosarcoma Patient-derived Orthotopic Xenograft (PDOX) Models Used to Identify Novel and Effective Therapeutics: A Review. *Anticancer. Res.* **2021**, *41*, 5865–5871. [CrossRef]
825. Chen, Z.; Zhao, Y.; Yu, Y.; Pang, J.C.; Woodfield, S.E.; Tao, L.; Guan, S.; Zhang, H.; Bieerkehazhi, S.; Shi, Y.; et al. Small molecule inhibitor regorafenib inhibits RET signaling in neuroblastoma cells and effectively suppresses tumor growth in vivo. *Oncotarget* **2017**, *8*, 104090–104103. [CrossRef]
826. Harrison, D.J.; Gill, J.D.; Roth, M.E.; Zhang, W.; Teicher, B.; Erickson, S.; Gatto, G.; Kurmasheva, R.T.; Houghton, P.J.; Smith, M.A.; et al. Initial in vivo testing of a multitarget kinase inhibitor, regorafenib, by the Pediatric Preclinical Testing Consortium. *Pediatr. Blood Cancer* **2020**, *67*, e28222. [CrossRef]
827. Jindal, A.; Thadi, A.; Shailubhai, K. Hepatocellular Carcinoma: Etiology and Current and Future Drugs. *J. Clin. Exp. Hepatol.* **2019**, *9*, 221–232. [CrossRef]
828. Aspeslagh, S.; Shailubhai, K.; Bahleda, R.; Gazzah, A.; Varga, A.; Hollebecque, A.; Massard, C.; Spreafico, A.; Reni, M.; Soria, J.-C. Phase I dose-escalation study of milciclib in combination with gemcitabine in patients with refractory solid tumors. *Cancer Chemother. Pharmacol.* **2017**, *79*, 1257–1265. [CrossRef]
829. Bolin, S.; Borgenvik, A.; Persson, C.U.; Sundström, A.; Qi, J.; Bradner, J.E.; Weiss, W.A.; Cho, Y.-J.; Weishaupt, H.; Swartling, F.J. Combined BET bromodomain and CDK2 inhibition in MYC-driven medulloblastoma. *Oncogene* **2018**, *37*, 2850–2862. [CrossRef]
830. Albanese, C.; Alzani, R.; Amboldi, N.; Degrassi, A.; Festuccia, C.; Fiorentini, F.; Gravina, G.L.; Mercurio, C.; Pastori, W.; Brasca, M.; et al. Anti-tumour efficacy on glioma models of PHA-848125, a multi-kinase inhibitor able to cross the blood-brain barrier. *Br. J. Pharmacol.* **2013**, *169*, 156–166. [CrossRef]
831. Martínez-Chávez, A.; Broeders, J.; Lebre, M.C.; Tibben, M.T.; Rosing, H.; Beijnen, J.H.; Schinkel, A.H. The role of drug efflux and uptake transporters ABCB1 (P-gp), ABCG2 (BCRP) and OATP1A/1B and of CYP3A4 in the pharmacokinetics of the CDK inhibitor milciclib. *Eur. J. Pharm. Sci.* **2021**, *159*, 105740. [CrossRef]
832. Smolewski, P. Terameprocol, a novel site-specific transcription inhibitor with anticancer activity. *IDrugs.* **2008**, *11*, 204–214.
833. Castro-Gamero, A.M.; Borges, K.S.; Moreno, D.A.; Suazo, V.K.; Fujinami, M.M.; Queiroz, R.D.P.G.; de Oliveira, H.F.; Carlotti, C.G.; Scrideli, C.; Tone, L.G. Tetra-O-methyl nordihydroguaiaretic acid, an inhibitor of Sp1-mediated survivin transcription, induces apoptosis and acts synergistically with chemo-radiotherapy in glioblastoma cells. *Investig. New Drugs* **2013**, *31*, 858–870. [CrossRef]
834. Akinaga, S.; Sugiyama, K.; Akiyama, T. UCN-01 (7-hydroxystaurosporine) and other indolocarbazole compounds: A new generation of anti-cancer agents for the new century? *Anti-Cancer Drug Des.* **2000**, *15*, 43–52.
835. Shao, R.-G.; Shimizu, T.; Pommier, Y. 7-Hydroxystaurosporine (UCN-01) Induces Apoptosis in Human Colon Carcinoma and Leukemia Cells Independently of p53. *Exp. Cell Res.* **1997**, *234*, 388–397. [CrossRef]

836. Lien, W.C.; Chen, T.Y.; Sheu, S.Y.; Lin, T.C.; Kang, F.C.; Yu, C.H.; Kuan, T.S.; Huang, B.M.; Wang, C.Y. 7-hydroxy-staurosporine, UCN-01, induces DNA damage response, and autophagy in human osteosarcoma U2-OS cells. *J. Cell. Biochem.* **2018**, *119*, 4729–4741. [CrossRef]
837. Shankar, S.L.; Krupski, M.; Parashar, B.; Okwuaka, C.; O'Guin, K.; Mani, S.; Shafit-Zagardo, B. UCN-01 alters phosphorylation of Akt and GSK3beta and induces apoptosis in six independent human neuroblastoma cell lines. *J. Neurochem.* **2004**, *90*, 702–711. [CrossRef]
838. Zhang, J.; Liu, S.; Ye, Q.; Pan, J. Transcriptional inhibition by CDK7/9 inhibitor SNS-032 abrogates oncogene addiction and reduces liver metastasis in uveal melanoma. *Mol. Cancer* **2019**, *18*, 1–17. [CrossRef]
839. Wu, Y.; Chen, C.; Sun, X.; Shi, X.; Jin, B.; Ding, K.; Yeung, S.-C.J.; Pan, J. Cyclin-Dependent Kinase 7/9 Inhibitor SNS-032 Abrogates FIP1-like-1 Platelet-Derived Growth Factor Receptor α and Bcr-Abl Oncogene Addiction in Malignant Hematologic Cells. *Clin. Cancer Res.* **2012**, *18*, 1966–1978. [CrossRef] [PubMed]
840. Xie, G.; Tang, H.; Wu, S.; Chen, J.; Liu, J.; Liao, C. The cyclin-dependent kinase inhibitor SNS-032 induces apoptosis in breast cancer cells via depletion of Mcl-1 and X-linked inhibitor of apoptosis protein and displays antitumor activity in vivo. *Int. J. Oncol.* **2014**, *45*, 804–812. [CrossRef] [PubMed]
841. Scrace, S.F.; Kierstan, P.; Borgognoni, J.; Denny, S.; Wayne, J.; Bentley, C.; Cansfield, A.D.; Jackson, P.S.; Lockie, A.M.; Curtin, N.J.; et al. Transient treatment with CDK inhibitors eliminates proliferative potential even when their abilities to evoke apoptosis and DNA damage are blocked. *Cell Cycle* **2008**, *7*, 3898–3907. [CrossRef] [PubMed]
842. Löschmann, N.; Michaelis, M.; Rothweiler, F.; Voges, Y.; Balónová, B.; Blight, B.A.; Jr, J.C. ABCB1 as predominant resistance mechanism in cells with acquired SNS-032 resistance. *Oncotarget* **2016**, *7*, 58051–58064. [CrossRef] [PubMed]
843. Löschmann, N.; Michaelis, M.; Rothweiler, F.; Zehner, R.; Cinatl, J.; Voges, Y.; Sharifi, M.; Riecken, K.; Meyer, J.; von Deimling, A.; et al. Testing of SNS-032 in a Panel of Human Neuroblastoma Cell Lines with Acquired Resistance to a Broad Range of Drugs. *Transl. Oncol.* **2013**, *6*, 685–IN18. [CrossRef]
844. Saha, D.; Ali, M.A.; Reis, A.; Ding, L.-H.; Story, M.D.; Habib, A.A.; Chattopadhyay, A. SNS-032 prevents hypoxia-mediated glioblastoma cell invasion by inhibiting hypoxia inducible factor-1α expression. *Int. J. Oncol.* **2009**, *34*, 1051–1060. [CrossRef]
845. Ali, M.A.; Choy, H.; Habib, A.A.; Saha, D. SNS-032 Prevents Tumor Cell-Induced Angiogenesis By Inhibiting Vascular Endothelial Growth Factor. *Neoplasia* **2007**, *9*, 370–381. [CrossRef]
846. Gaevyĭ, M.D.; Seĭf, M.N. Cerebral blood circulation during angiotensin-converting enzyme inhibition. *Fiziol Zh Im I M Sechenova* **1993**, *79*, 74–78.
847. Zhang, M.; Zhang, L.; Hei, R.; Li, X.; Cai, H.; Wu, X.; Zheng, Q.; Cai, C. CDK inhibitors in cancer therapy, an overview of recent development. *Am. J. Cancer Res.* **2021**, *11*, 1913–1935.
848. Aldoss, I.T.; Tashi, T.; Ganti, A.K. Seliciclib in malignancies. *Expert Opin. Investig. Drugs* **2009**, *18*, 1957–1965. [CrossRef]
849. Wu, T.; Qin, Z.; Tian, Y.; Wang, J.; Xu, C.; Li, Z.; Bian, J. Recent Developments in the Biology and Medicinal Chemistry of CDK9 Inhibitors: An Update. *J. Med. Chem.* **2020**, *63*, 13228–13257. [CrossRef]
850. Khalil, H.S.; Mitev, V.; Vlaykova, T.; Cavicchi, L.; Zhelev, N. Discovery and development of Seliciclib. How systems biology approaches can lead to better drug performance. *J. Biotechnol.* **2015**, *202*, 40–49. [CrossRef]
851. Tirado, O.M.; Mateo-Lozano, S.; Notario, V. Roscovitine Is an Effective Inducer of Apoptosis of Ewing's Sarcoma Family Tumor Cells In vitro and In vivo. *Cancer Res* **2005**, *65*, 9320–9327. [CrossRef]
852. Iurisci, I.; Filipski, E.; Reinhardt, J.; Bach, S.; Gianella-Borradori, A.; Iacobelli, S.; Meijer, L.; Lévi, F. Improved Tumor Control through Circadian Clock Induction by Seliciclib, a Cyclin-Dependent Kinase Inhibitor. *Cancer Res* **2006**, *66*, 10720–10728. [CrossRef]
853. Pizarro, J.G.; Folch, J.; Junyent, F.; Verdaguer, E.; Auladell, C.; Beas-Zarate, C.; Pallàs, M.; Camins, A. Antiapoptotic effects of roscovitine on camptothecin-induced DNA damage in neuroblastoma cells. *Apoptosis* **2011**, *16*, 536–550. [CrossRef]
854. Garrofé-Ochoa, X.; Cosialls, A.M.; Ribas, J.; Gil, J.; Boix, J. Transcriptional modulation of apoptosis regulators by roscovitine and related compounds. *Apoptosis* **2011**, *16*, 660–670. [CrossRef]
855. Chen, Y.; Tsai, Y.-H.; Tseng, S.-H. Inhibition of cyclin-dependent kinase 1–induced cell death in neuroblastoma cells through the microRNA-34a–MYCN–survivin pathway. *Surgery* **2012**, *153*, 4–16. [CrossRef]
856. Ribas, J.; Boix, J.; Meijer, L. (R)-Roscovitine (CYC202, Seliciclib) sensitizes SH-SY5Y neuroblastoma cells to nutlin-3-induced apoptosis. *Exp. Cell Res.* **2006**, *312*, 2394–2400. [CrossRef]
857. Krüger, K.; Geist, K.; Stuhldreier, F.; Schumacher, L.; Blümel, L.; Remke, M.; Wesselborg, S.; Stork, B.; Klöcker, N.; Bormann, S.; et al. Multiple DNA damage-dependent and DNA damage-independent stress responses define the outcome of ATR/Chk1 targeting in medulloblastoma cells. *Cancer Lett.* **2018**, *430*, 34–46. [CrossRef]
858. Bhatia, B.; Hsieh, M.; Kenney, A.M.; Nahlé, Z. Mitogenic Sonic hedgehog signaling drives E2F1-dependent lipogenesis in progenitor cells and medulloblastoma. *Oncogene* **2010**, *30*, 410–422. [CrossRef]
859. Braal, C.L.; Jongbloed, E.M.; Wilting, S.M.; Mathijssen, R.H.J.; Koolen, S.L.W.; Jager, A. Inhibiting CDK4/6 in Breast Cancer with Palbociclib, Ribociclib, and Abemaciclib: Similarities and Differences. *Drugs* **2020**, *81*, 317–331. [CrossRef] [PubMed]
860. Guenther, L.M.; Dharia, N.V.; Ross, L.; Conway, A.; Robichaud, A.L.; Catlett, J.L., II; Wechsler, C.S.; Frank, E.S.; Goodale, A.; Church, A.J.; et al. A Combination CDK4/6 and IGF1R Inhibitor Strategy for Ewing Sarcoma. *Clin. Cancer Res.* **2019**, *25*, 1343–1357. [CrossRef]

861. Martinez-Monleon, A.; Öberg, H.K.; Gaarder, J.; Berbegall, A.P.; Javanmardi, N.; Djos, A.; Ussowicz, M.; Taschner-Mandl, S.; Ambros, I.M.; Øra, I.; et al. Amplification of CDK4 and MDM2: A detailed study of a high-risk neuroblastoma subgroup. *Sci. Rep.* **2022**, *12*, 1–17. [CrossRef] [PubMed]
862. Guntner, A.S.; Peyrl, A.; Mayr, L.; Englinger, B.; Berger, W.; Slavc, I.; Buchberger, W.; Gojo, J. Cerebrospinal fluid penetration of targeted therapeutics in pediatric brain tumor patients. *Acta Neuropathol. Commun.* **2020**, *8*, 78. [CrossRef] [PubMed]
863. Liang, M.-L.; Hsieh, T.-H.; Liu, Y.-R.; Chen, Y.-W.; Lee, Y.-Y.; Chang, F.-C.; Lin, S.-C.; Huang, M.-C.; Ho, D.M.-T.; Wong, T.-T.; et al. Significance of cyclin D1 overexpression in progression and radio-resistance of pediatric ependymomas. *Oncotarget* **2017**, *9*, 2527–2542. [CrossRef]
864. D'Oto, A.; Fang, J.; Jin, H.; Xu, B.; Singh, S.; Mullasseril, A.; Jones, V.; Abu-Zaid, A.; von Buttlar, X.; Cooke, B.; et al. KDM6B promotes activation of the oncogenic CDK4/6-pRB-E2F pathway by maintaining enhancer activity in MYCN-amplified neuroblastoma. *Nat. Commun.* **2021**, *12*, 7204. [CrossRef]
865. Swadi, R.R.; Sampat, K.; Herrmann, A.; Losty, P.D.; See, V.; Moss, D.J. CDK inhibitors reduce cell proliferation and reverse hypoxia-induced metastasis of neuroblastoma tumours in a chick embryo model. *Sci. Rep.* **2019**, *9*, 9136. [CrossRef]
866. Rihani, A.; Vandesompele, J.; Speleman, F.; Van Maerken, T. Inhibition of CDK4/6 as a novel therapeutic option for neuroblastoma. *Cancer Cell Int.* **2015**, *15*, 76. [CrossRef]
867. Perez, M.; Muñoz-Galván, S.; Jiménez-García, M.P.; Marín, J.J.; Carnero, A. Efficacy of CDK4 inhibition against sarcomas depends on their levels of CDK4 and p16ink4 mRNA. *Oncotarget* **2015**, *6*, 40557–40574. [CrossRef]
868. Böhm, M.J.; Marienfeld, R.; Jäger, D.; Mellert, K.; von Witzleben, A.; Brüderlein, S.; Wittau, M.; von Baer, A.; Schultheiss, M.; Mayer-Steinacker, R.; et al. Analysis of the CDK4/6 Cell Cycle Pathway in Leiomyosarcomas as a Potential Target for Inhibition by Palbociclib. *Sarcoma* **2019**, *2019*, 1–10. [CrossRef]
869. Murakami, T.; Singh, A.S.; Kiyuna, T.; Dry, S.M.; Li, Y.; James, A.W.; Igarashi, K.; Kawaguchi, K.; DeLong, J.C.; Zhang, Y.; et al. Effective molecular targeting of CDK4/6 and IGF-1R in a rare *FUS-ERG* fusion *CDKN2A*-deletion doxorubicin-resistant Ewing's sarcoma patient-derived orthotopic xenograft (PDOX) nude-mouse model. *Oncotarget* **2016**, *7*, 47556–47564. [CrossRef]
870. Tramontana, T.F.; Marshall, M.S.; Helvie, A.E.; Schmitt, M.R.; Ivanovich, J.; Carter, J.L.; Renbarger, J.L.; Ferguson, M.J. Sustained Complete Response to Palbociclib in a Refractory Pediatric Sarcoma With *BCOR-CCNB3* Fusion and Germline *CDKN2B* Variant. *JCO Precis. Oncol.* **2020**, 466–471. [CrossRef]
871. Riess, C.; Koczan, D.; Schneider, B.; Linke, C.; del Moral, K.; Classen, C.F.; Maletzki, C. Cyclin-dependent kinase inhibitors exert distinct effects on patient-derived 2D and 3D glioblastoma cell culture models. *Cell Death Discov.* **2021**, *7*, 1–15. [CrossRef]
872. Sun, Y.; Sun, Y.; Yan, K.; Li, Z.; Xu, C.; Geng, Y.; Pan, C.; Chen, X.; Zhang, L.; Xi, Q. Potent anti-tumor efficacy of palbociclib in treatment-naïve H3.3K27M-mutant diffuse intrinsic pontine glioma. *Ebiomedicine* **2019**, *43*, 171–179. [CrossRef]
873. Morris-Hanon, O.; Marazita, M.C.; Romorini, L.; Isaja, L.; Fernandez-Espinosa, D.D.; Sevlever, G.E.; Scassa, M.E.; Videla-Richardson, G.A. Palbociclib Effectively Halts Proliferation but Fails to Induce Senescence in Patient-Derived Glioma Stem Cells. *Mol. Neurobiol.* **2019**, *56*, 7810–7821. [CrossRef]
874. Huillard, E.; Hashizume, R.; Phillips, J.J.; Griveau, A.; Ihrie, R.A.; Aoki, Y.; Nicolaides, T.; Perry, A.; Waldman, T.; McMahon, M.; et al. Cooperative interactions of BRAF ^{V600E} kinase and *CDKN2A* locus deficiency in pediatric malignant astrocytoma as a basis for rational therapy. *Proc. Natl. Acad. Sci. USA* **2012**, *109*, 8710–8715. [CrossRef]
875. Barton, K.L.; Misuraca, K.; Cordero, F.; Dobrikova, E.; Min, H.D.; Gromeier, M.; Kirsch, D.G.; Becher, O.J. PD-0332991, a CDK4/6 Inhibitor, Significantly Prolongs Survival in a Genetically Engineered Mouse Model of Brainstem Glioma. *PLoS ONE* **2013**, *8*, e77639. [CrossRef]
876. Lallena, M.J.; Boehnke, K.; Torres, R.; Hermoso, A.; Amat, J.; Calsina, B.; De Dios, A.; Buchanan, S.; Du, J.; Beckmann, R.P.; et al. Abstract 3101: In-vitro characterization of Abemaciclib pharmacology in ER+ breast cancer cell lines. *Cancer Res* **2015**, *75*, 3101. [CrossRef]
877. Torres-Guzmán, R.; Calsina, B.; Hermoso, A.; Baquero, C.; Alvarez, B.; Amat, J.; McNulty, A.M.; Gong, X.; Boehnke, K.; Du, J.; et al. Preclinical characterization of abemaciclib in hormone receptor positive breast cancer. *Oncotarget* **2017**, *8*, 69493–69507. [CrossRef]
878. Chong, Q.-Y.; Kok, Z.-H.; Bui, N.-L.; Xiang, X.; Wong, A.L.-A.; Yong, W.-P.; Sethi, G.; Lobie, P.E.; Wang, L.; Goh, B.-C. A unique CDK4/6 inhibitor: Current and future therapeutic strategies of abemaciclib. *Pharmacol. Res.* **2020**, *156*, 104686. [CrossRef]
879. Goetz, M.P.; Toi, M.; Campone, M.; Sohn, J.; Paluch-Shimon, S.; Huober, J.; Park, I.H.; Trédan, O.; Chen, S.-C.; Manso, L.; et al. MONARCH 3: Abemaciclib As Initial Therapy for Advanced Breast Cancer. *J. Clin. Oncol.* **2017**, *35*, 3638–3646. [CrossRef] [PubMed]
880. Dowless, M.S.; Lowery, C.D.; Shackleford, T.J.; Renschler, M.; Stephens, J.R.; Flack, R.; Blosser, W.; Gupta, S.; Stewart, J.; Webster, Y.; et al. Abemaciclib Is Active in Preclinical Models of Ewing Sarcoma via Multipronged Regulation of Cell Cycle, DNA Methylation, and Interferon Pathway Signaling. *Clin. Cancer Res.* **2018**, *24*, 6028–6039. [CrossRef] [PubMed]
881. Wang, D.; Bao, H. Abemaciclib is synergistic with doxorubicin in osteosarcoma pre-clinical models via inhibition of CDK4/6–Cyclin D–Rb pathway. *Cancer Chemother. Pharmacol.* **2021**, *89*, 31–40. [CrossRef] [PubMed]
882. Schubert, N.A.; Schild, L.; van Oirschot, S.; Keller, K.M.; Alles, L.K.; Vernooij, L.; Nulle, M.E.; Dolman, M.E.M.; van den Boogaard, M.L.; Molenaar, J.J. Combined targeting of the p53 and pRb pathway in neuroblastoma does not lead to synergistic responses. *Eur. J. Cancer* **2020**, *142*, 1–9. [CrossRef] [PubMed]
883. Cao, Y.; Li, X.; Kong, S.; Shang, S.; Qi, Y. CDK4/6 inhibition suppresses tumour growth and enhances the effect of temozolomide in glioma cells. *J. Cell. Mol. Med.* **2020**, *24*, 5135–5145. [CrossRef]

884. Mayr, L.; Guntner, A.S.; Madlener, S.; Schmook, M.T.; Peyrl, A.; Azizi, A.A.; Dieckmann, K.; Reisinger, D.; Stepien, N.M.; Schramm, K.; et al. Cerebrospinal Fluid Penetration and Combination Therapy of Entrectinib for Disseminated *ROS1/NTRK*-Fusion Positive Pediatric High-Grade Glioma. *J. Pers. Med.* **2020**, *10*, 290. [CrossRef]
885. Stewart, E.; McEvoy, J.; Wang, H.; Chen, X.; Honnell, V.; Ocarz, M.; Gordon, B.; Dapper, J.; Blankenship, K.; Yang, Y.; et al. Identification of Therapeutic Targets in Rhabdomyosarcoma through Integrated Genomic, Epigenomic, and Proteomic Analyses. *Cancer Cell* **2018**, *34*, 411–426.e19. [CrossRef]
886. Rana, S.; Kour, S.; Sonawane, Y.A.; Robb, C.M.; Contreras, J.I.; Kizhake, S.; Zahid, M.; Karpf, A.R.; Natarajan, A. Symbiotic prodrugs (SymProDs) dual targeting of NFkappaB and CDK. *Chem. Biol. Drug Des.* **2020**, *96*, 773–784. [CrossRef]
887. Squires, M.S.; Feltell, R.E.; Wallis, N.G.; Lewis, E.J.; Smith, D.-M.; Cross, D.M.; Lyons, J.F.; Thompson, N.T. Biological characterization of AT7519, a small-molecule inhibitor of cyclin-dependent kinases, in human tumor cell lines. *Mol. Cancer Ther.* **2009**, *8*, 324–332. [CrossRef]
888. Oghabi, M.; Safaroghli-Azar, A.; Pourbagheri-Sigaroodi, A.; Sayyadi, M.; Hamidpour, M.; Mohammadi, M.H.; Bashash, D. Anti-proliferative effects of a small molecule inhibitor of CDK AT7519 on chronic myeloid leukemia (CML) cells through halting the transition of cells from G2/M phase of the cell cycle. *Biocell* **2020**, *44*, 183–192. [CrossRef]
889. Dolman, M.E.M.; Poon, E.; Ebus, M.E.; Hartog, I.J.D.; van Noesel, C.J.; Jamin, Y.; Hallsworth, A.; Robinson, S.P.; Petrie, K.; Sparidans, R.W.; et al. Cyclin-Dependent Kinase Inhibitor AT7519 as a Potential Drug for MYCN-Dependent Neuroblastoma. *Clin. Cancer Res.* **2015**, *21*, 5100–5109. [CrossRef]
890. Steegmaier, M.; Hoffmann, M.; Baum, A.; Lénárt, P.; Petronczki, M.; Krššák, M.; Gürtler, U.; Garin-Chesa, P.; Lieb, S.; Quant, J.; et al. BI 2536, a Potent and Selective Inhibitor of Polo-like Kinase 1, Inhibits Tumor Growth In Vivo. *Curr. Biol.* **2007**, *17*, 316–322. [CrossRef]
891. Pezuk, J.A.; Brassesco, M.S.; Morales, A.G.; de Oliveira, J.C.; de Oliveira, H.F.; Scrideli, C.A.; Tone, L.G. Inhibition of Polo-Like Kinase 1 Induces Cell Cycle Arrest and Sensitizes Glioblastoma Cells to Ionizing Radiation. *Cancer Biotherapy Radiopharm.* **2013**, *28*, 516–522. [CrossRef]
892. A Pezuk, J.; Brassesco, M.S.; Morales, A.G.; de Oliveira, J.C.; Queiroz, R.G.D.P.; Machado, H.R.; Carlotti, C.G.; Neder, L.; A Scrideli, C.; Tone, L.G. Polo-like kinase 1 inhibition causes decreased proliferation by cell cycle arrest, leading to cell death in glioblastoma. *Cancer Gene Ther.* **2013**, *20*, 499–506. [CrossRef]
893. Pezuk, J.A.; Brassesco, M.S.; Ramos, P.M.M.; Scrideli, C.A.; Tone, L.G.; Pezuk, M.S.B.J.A. Polo-Like Kinase 1 Pharmacological Inhibition as Monotherapy or in Combination: Comparative Effects of Polo-Like Kinase 1 Inhibition in Medulloblastoma Cells. *Anti-Cancer Agents Med. Chem.* **2017**, *17*, 1278–1291. [CrossRef]
894. Yunoki, T.; Tabuchi, Y.; Hayashi, A.; Kondo, T. Inhibition of Polo-Like Kinase 1 Promotes Hyperthermia Sensitivity via Inactivation of Heat Shock Transcription Factor 1 in Human Retinoblastoma Cells. *Investig. Opthalmology Vis. Sci.* **2013**, *54*, 8353–8363. [CrossRef]
895. Li, Z.; Yang, C.; Li, X.; Du, X.; Tao, Y.; Ren, J.; Fang, F.; Xie, Y.; Li, M.; Qian, G.; et al. The dual role of BI 2536, a small-molecule inhibitor that targets PLK1, in induction of apoptosis and attenuation of autophagy in neuroblastoma cells. *J. Cancer* **2020**, *11*, 3274–3287. [CrossRef]
896. Grinshtein, N.; Datti, A.; Fujitani, M.; Uehling, D.; Prakesch, M.; Isaac, M.; Irwin, M.S.; Wrana, J.L.; Al-Awar, R.; Kaplan, D.R. Small Molecule Kinase Inhibitor Screen Identifies Polo-Like Kinase 1 as a Target for Neuroblastoma Tumor-Initiating Cells. *Cancer Res* **2011**, *71*, 1385–1395. [CrossRef]
897. Hsieh, C.-H.; Yeh, H.-N.; Huang, C.-T.; Wang, W.-H.; Hsu, W.-M.; Huang, H.-C.; Juan, H.-F. BI-2536 Promotes Neuroblastoma Cell Death via Minichromosome Maintenance Complex Components 2 and 10. *Pharmaceuticals* **2021**, *15*, 37. [CrossRef]
898. Morales, A.G.; Brassesco, M.S.; Pezuk, J.A.; Oliveira, J.C.; Montaldi, A.P.; Sakamoto-Hojo, E.T.; Scrideli, C.A.; Tone, L.G. BI 2536-mediated PLK1 inhibition suppresses HOS and MG-63 osteosarcoma cell line growth and clonogenicity. *Anti-Cancer Drugs* **2011**, *22*, 995–1001. [CrossRef]
899. Liu, X.; Choy, E.; Harmon, D.; Yang, S.; Yang, C.; Mankin, H.; Hornicek, F.J.; Duan, Z. Inhibition of polo-like kinase 1 leads to the suppression of osteosarcoma cell growth in vitro and in vivo. *Anti-Cancer Drugs* **2011**, *22*, 444–453. [CrossRef] [PubMed]
900. Thalhammer, V.; Lopez-Garcia, L.A.; Herrero-Martin, D.; Hecker, R.; Laubscher, D.; Gierisch, M.E.; Wachtel, M.; Bode, P.; Nanni, P.; Blank, B.; et al. PLK1 Phosphorylates PAX3-FOXO1, the Inhibition of Which Triggers Regression of Alveolar Rhabdomyosarcoma. *Cancer Res* **2015**, *75*, 98–110. [CrossRef] [PubMed]
901. Stehle, A.; Hugle, M.; Fulda, S. Eribulin synergizes with Polo-like kinase 1 inhibitors to induce apoptosis in rhabdomyosarcoma. *Cancer Lett.* **2015**, *365*, 37–46. [CrossRef] [PubMed]
902. Valsasina, B.; Beria, I.; Alli, C.; Alzani, R.; Avanzi, N.; Ballinari, D.; Cappella, P.; Caruso, M.; Casolaro, A.; Ciavolella, A.; et al. NMS-P937, an Orally Available, Specific Small-Molecule Polo-like Kinase 1 Inhibitor with Antitumor Activity in Solid and Hematologic Malignancies. *Mol. Cancer Ther.* **2012**, *11*, 1006–1016. [CrossRef]
903. Sero, V.; Tavanti, E.; Vella, S.; Hattinger, C.M.; Fanelli, M.; Michelacci, F.; Versteeg, R.; Valsasina, B.; Gudeman, B.; Picci, P.; et al. Targeting polo-like kinase 1 by NMS-P937 in osteosarcoma cell lines inhibits tumor cell growth and partially overcomes drug resistance. *Investig. New Drugs* **2014**, *32*, 1167–1180. [CrossRef]
904. Wang, D.; Veo, B.; Pierce, A.; Fosmire, S.; Madhavan, K.; Balakrishnan, I.; Donson, A.; Alimova, I.; Sullivan, K.D.; Joshi, M.; et al. A novel PLK1 inhibitor onvansertib effectively sensitizes MYC-driven medulloblastoma to radiotherapy. *Neuro-Oncology* **2021**, *24*, 414–426. [CrossRef]

905. Wu, C.-P.; Hsiao, S.-H.; Luo, S.-Y.; Tuo, W.-C.; Su, C.-Y.; Li, Y.-Q.; Huang, Y.-H.; Hsieh, C.-H. Overexpression of Human ABCB1 in Cancer Cells Leads to Reduced Activity of GSK461364, a Specific Inhibitor of Polo-like Kinase 1. *Mol. Pharm.* **2014**, *11*, 3727–3736. [CrossRef]
906. Pajtler, K.W.; Sadowski, N.; Ackermann, S.; Althoff, K.; Schönbeck, K.; Batzke, K.; Schäfers, S.; Odersky, A.; Heukamp, L.; Astrahantseff, K.; et al. The GSK461364 PLK1 inhibitor exhibits strong antitumoral activity in preclinical neuroblastoma models. *Oncotarget* **2016**, *8*, 6730–6741. [CrossRef]
907. Chou, Y.-S.; Yen, C.-C.; Chen, W.-M.; Lin, Y.-C.; Wen, Y.-S.; Ke, W.-T.; Wang, J.-Y.; Liu, C.-Y.; Yang, M.-H.; Chen, T.-H.; et al. Cytotoxic mechanism of PLK1 inhibitor GSK461364 against osteosarcoma: Mitotic arrest, apoptosis, cellular senescence, and synergistic effect with paclitaxel. *Int. J. Oncol.* **2016**, *48*, 1187–1194. [CrossRef]
908. Bogado, R.F.; Pezuk, J.A.; de Oliveira, H.F.; Tone, L.G.; Brassesco, M.S. BI 6727 and GSK461364 suppress growth and radiosensitize osteosarcoma cells, but show limited cytotoxic effects when combined with conventional treatments. *Anti-Cancer Drugs* **2015**, *26*, 56–63. [CrossRef]
909. Gorlick, R.; Kolb, E.A.; Keir, S.T.; Maris, J.M.; Reynolds, C.P.; Kang, M.H.; Carol, H.; Lock, R.; Billups, C.A.; Kurmasheva, R.T.; et al. Initial testing (stage 1) of the polo-like kinase inhibitor volasertib (BI 6727), by the Pediatric Preclinical Testing Program. *Pediatr. Blood Cancer* **2013**, *61*, 158–164. [CrossRef]
910. Carol, H.; Boehm, I.; Reynolds, C.P.; Kang, M.H.; Maris, J.M.; Morton, C.L.; Gorlick, R.; Kolb, E.A.; Keir, S.T.; Wu, J.; et al. Efficacy and pharmacokinetic/pharmacodynamic evaluation of the Aurora kinase A inhibitor MLN8237 against preclinical models of pediatric cancer. *Cancer Chemother. Pharmacol.* **2011**, *68*, 1291–1304. [CrossRef]
911. Maris, J.M.; Bs, C.L.M.; Gorlick, R.; Kolb, E.A.; Lock, R.; Carol, H.; Keir, S.T.; Reynolds, C.P.; Kang, M.H.; Wu, J.; et al. Initial testing of the aurora kinase a inhibitor MLN8237 by the Pediatric Preclinical Testing Program (PPTP). *Pediatr. Blood Cancer* **2010**, *55*, 26–34. [CrossRef]
912. Boi, D.; Souvalidou, F.; Capelli, D.; Polverino, F.; Marini, G.; Montanari, R.; Pochetti, G.; Tramonti, A.; Contestabile, R.; Trisciuoglio, D.; et al. PHA-680626 Is an Effective Inhibitor of the Interaction between Aurora-A and N-Myc. *Int. J. Mol. Sci.* **2021**, *22*, 13122. [CrossRef]
913. Brockmann, M.; Poon, E.; Berry, T.; Carstensen, A.; Deubzer, H.E.; Rycak, L.; Jamin, Y.; Thway, K.; Robinson, S.P.; Roels, F.; et al. Small Molecule Inhibitors of Aurora-A Induce Proteasomal Degradation of N-Myc in Childhood Neuroblastoma. *Cancer Cell* **2013**, *24*, 75–89. [CrossRef]
914. Geron, L.; Borges, K.S.; Andrade, A.F.; Suazo, V.K.; Scrideli, C.A.; Tone, L.G. Antitumour activity of AMG 900 alone or in combination with histone deacetylase inhibitor SaHa on medulloblastoma cell lines. *Neurol. Res.* **2015**, *37*, 703–711. [CrossRef]
915. Hanahan, D. Hallmarks of Cancer: New Dimensions. *Cancer Discov.* **2022**, *12*, 31–46. [CrossRef]
916. Grignani, G.; Palmerini, E.; Ferraresi, V.; D'Ambrosio, L.; Bertulli, R.; Asaftei, S.D.; Tamburini, A.; Pignochino, Y.; Sangiolo, D.; Marchesi, E.; et al. Sorafenib and everolimus for patients with unresectable high-grade osteosarcoma progressing after standard treatment: A non-randomised phase 2 clinical trial. *Lancet Oncol.* **2015**, *16*, 98–107. [CrossRef]
917. Liu, Y.; Li, Y.; Wang, Y.; Lin, C.; Zhang, D.; Chen, J.; Ouyang, L.; Wu, F.; Zhang, J.; Chen, L. Recent progress on vascular endothelial growth factor receptor inhibitors with dual targeting capabilities for tumor therapy. *J. Hematol. Oncol.* **2022**, *15*, 1–28. [CrossRef]
918. Chambers, A.; Kundranda, M.; Rao, S.; Niu, J. Anti-angiogenesis Revisited: Combination with Immunotherapy in Solid Tumors. *Lung Cancer* **1912**, *23*, 100. [CrossRef]
919. Inaba, H.; Rubnitz, J.E.; Coustan-Smith, E.; Li, L.; Furmanski, B.D.; Mascara, G.P.; Heym, K.M.; Christensen, R.; Onciu, M.; Shurtleff, S.A.; et al. Phase I Pharmacokinetic and Pharmacodynamic Study of the Multikinase Inhibitor Sorafenib in Combination With Clofarabine and Cytarabine in Pediatric Relapsed/Refractory Leukemia. *J. Clin. Oncol.* **2011**, *29*, 3293–3300. [CrossRef] [PubMed]
920. Grignani, G.; Palmerini, E.; Dileo, P.; Asaftei, S.D.; D'Ambrosio, L.; Pignochino, Y.; Mercuri, M.; Picci, P.; Fagioli, F.; Casali, P.G.; et al. A phase II trial of sorafenib in relapsed and unresectable high-grade osteosarcoma after failure of standard multimodal therapy: An Italian Sarcoma Group study. *Ann. Oncol.* **2011**, *23*, 508–516. [CrossRef] [PubMed]
921. Keino, D.; Yokosuka, T.; Hirose, A.; Sakurai, Y.; Nakamura, W.; Fujita, S.; Hayashi, A.; Miyagawa, N.; Iwasaki, F.; Hamanoue, S.; et al. Pilot study of the combination of sorafenib and fractionated irinotecan in pediatric relapse/refractory hepatic cancer (FINEX pilot study). *Pediatr. Blood Cancer* **2020**, *67*, e28655. [CrossRef] [PubMed]
922. Geller, J.I.; Fox, E.; Turpin, B.K.; Goldstein, S.L.; Liu, X.; Minard, C.G.; Kudgus, R.A.; Reid, J.M.; Berg, S.L.; Weigel, B.J. A study of axitinib, a VEGF receptor tyrosine kinase inhibitor, in children and adolescents with recurrent or refractory solid tumors: A Children's Oncology Group phase 1 and pilot consortium trial (ADVL1315). *Cancer* **2018**, *124*, 4548–4555. [CrossRef] [PubMed]
923. Daw, N.C.; Furman, W.L.; Stewart, C.F.; Iacono, L.C.; Krailo, M.; Bernstein, M.L.; Dancey, J.E.; Speights, R.A.; Blaney, S.M.; Croop, J.M.; et al. Phase I and Pharmacokinetic Study of Gefitinib in Children With Refractory Solid Tumors: A Children's Oncology Group Study. *J. Clin. Oncol.* **2005**, *23*, 6172–6180. [CrossRef]
924. Pollack, I.F.; Stewart, C.F.; Kocak, M.; Poussaint, T.Y.; Broniscer, A.; Banerjee, A.; Douglas, J.G.; Kun, L.E.; Boyett, J.M.; Geyer, J.R. A phase II study of gefitinib and irradiation in children with newly diagnosed brainstem gliomas: A report from the Pediatric Brain Tumor Consortium. *Neuro-Oncology* **2011**, *13*, 290–297. [CrossRef]
925. Subbiah, V.; Wolf, J.; Konda, B.; Kang, H.; Spira, A.; Weiss, J.; Takeda, M.; Ohe, Y.; Khan, S.; Ohashi, K.; et al. Tumour-agnostic efficacy and safety of selpercatinib in patients with RET fusion-positive solid tumours other than lung or thyroid tumours (LIBRETTO-001): A phase 1/2, open-label, basket trial. *Lancet Oncol.* **2022**, *23*, 1261–1273. [CrossRef]

926. Del Rivero, J.; Edgerly, M.; Ward, J.; Madan, R.A.; Balasubramaniam, S.; Fojo, T.; Gramza, A.W. Phase I/II Trial of Vandetanib and Bortezomib in Adults with Locally Advanced or Metastatic Medullary Thyroid Cancer. *Oncol.* **2018**, *24*, 16-e14. [CrossRef]
927. Subbiah, V.; Wolf, J.; Konda, B.; Kang, H.; Spira, A.I.; Weiss, J.; Takeda, M.; Ohe, Y.; Khan, S.A.; Ohashi, K.; et al. Tumor agnostic efficacy of selpercatinib in patients with *RET* fusion+ solid tumors: A global, multicenter, registrational trial update (LIBRETTO-001). *J. Clin. Oncol.* **2022**, *40*, 3094. [CrossRef]
928. Subbiah, V.; Cassier, P.A.; Siena, S.; Garralda, E.; Paz-Ares, L.; Garrido, P.; Nadal, E.; Vuky, J.; Lopes, G.; Kalemkerian, G.P.; et al. Pan-cancer efficacy of pralsetinib in patients with RET fusion–positive solid tumors from the phase 1/2 ARROW trial. *Nat. Med.* **2022**, *28*, 1640–1645. [CrossRef]
929. Mossé, Y.P.; Lim, M.S.; Voss, S.D.; Wilner, K.; Ruffner, K.; Laliberte, J.; Rolland, D.; Balis, F.M.; Maris, J.M.; Weigel, B.J.; et al. Safety and activity of crizotinib for paediatric patients with refractory solid tumours or anaplastic large-cell lymphoma: A Children's Oncology Group phase 1 consortium study. *Lancet Oncol.* **2013**, *14*, 472–480. [CrossRef]
930. Fukano, R.; Mori, T.; Sekimizu, M.; Choi, I.; Kada, A.; Saito, A.M.; Asada, R.; Takeuchi, K.; Terauchi, T.; Tateishi, U.; et al. Alectinib for relapsed or refractory anaplastic lymphoma kinase-positive anaplastic large cell lymphoma: An open-label phase II trial. *Cancer Sci.* **2020**, *111*, 4540–4547. [CrossRef]
931. Franz, D.N.; Belousova, E.; Sparagana, S.; Bebin, E.M.; Frost, M.; Kuperman, R.; Witt, O.; Kohrman, M.H.; Flamini, J.R.; Wu, J.Y.; et al. Efficacy and safety of everolimus for subependymal giant cell astrocytomas associated with tuberous sclerosis complex (EXIST-1): A multicentre, randomised, placebo-controlled phase 3 trial. *Lancet* **2013**, *381*, 125–132. [CrossRef]
932. Adashek, J.J.; Menta, A.K.; Reddy, N.K.; Desai, A.P.; Roszik, J.; Subbiah, V. Tissue-Agnostic Activity of BRAF plus MEK Inhibitor in BRAF V600–Mutant Tumors. *Mol. Cancer Ther.* **2022**, *21*, 871–878. [CrossRef]
933. Bouffet, E.; Hansford, J.; Garré, M.L.; Hara, J.; Plant-Fox, A.; Aerts, I.; Locatelli, F.; Van der Lugt, J.; Papusha, L.; Sahm, F.; et al. Primary analysis of a phase II trial of dabrafenib plus trametinib (dab + tram) in *BRAF* V600–mutant pediatric low-grade glioma (pLGG). *J. Clin. Oncol.* **2022**, *40*, LBA2002. [CrossRef]
934. Hargrave, D.R.; Bouffet, E.; Tabori, U.; Broniscer, A.; Cohen, K.J.; Hansford, J.R.; Geoerger, B.; Hingorani, P.; Dunkel, I.J.; Russo, M.W.; et al. Efficacy and Safety of Dabrafenib in Pediatric Patients with *BRAF* V600 Mutation–Positive Relapsed or Refractory Low-Grade Glioma: Results from a Phase I/IIa Study. *Clin. Cancer Res.* **2019**, *25*, 7303–7311. [CrossRef]
935. Lau, N.; Feldkamp, M.M.; Roncari, L.; Loehr, A.H.; Shannon, P.; Gutmann, D.H.; Guha, A. Loss of Neurofibromin Is Associated with Activation of RAS/MAPK and PI3-K/AKT Signaling in a Neurofibromatosis 1 Astrocytoma. *J. Neuropathol. Exp. Neurol.* **2000**, *59*, 759–767. [CrossRef]
936. Gross, A.M.; Wolters, P.L.; Dombi, E.; Baldwin, A.; Whitcomb, P.; Fisher, M.J.; Weiss, B.; Kim, A.; Bornhorst, M.; Shah, A.C.; et al. Selumetinib in Children with Inoperable Plexiform Neurofibromas. *N. Engl. J. Med.* **2020**, *382*, 1430–1442. [CrossRef]
937. Bautista, F.; Paoletti, X.; Rubino, J.; Brard, C.; Rezai, K.; Nebchi, S.; Andre, N.; Aerts, I.; De Carli, E.; van Eijkelenburg, N.; et al. Phase I or II Study of Ribociclib in Combination With Topotecan-Temozolomide or Everolimus in Children With Advanced Malignancies: Arms A and B of the AcSé-ESMART Trial. *J. Clin. Oncol.* **2021**, *39*, 3546–3560. [CrossRef]
938. Yang, Y.; Li, S.; Wang, Y.; Zhao, Y.; Li, Q. Protein tyrosine kinase inhibitor resistance in malignant tumors: Molecular mechanisms and future perspective. *Signal Transduct. Target. Ther.* **2022**, *7*, 1–36. [CrossRef]
939. Van Mater, D.; Gururangan, S.; Becher, O.; Campagne, O.; Leary, S.; Phillips, J.J.; Huang, J.; Lin, T.; Poussaint, T.Y.; Goldman, S.; et al. A phase I trial of the CDK 4/6 inhibitor palbociclib in pediatric patients with progressive brain tumors: A Pediatric Brain Tumor Consortium study (PBTC-042). *Pediatr. Blood Cancer* **2021**, *68*, e28879. [CrossRef] [PubMed]
940. Mossé, Y.P.; Lipsitz, E.; Fox, E.; Teachey, D.T.; Maris, J.M.; Weigel, B.; Adamson, P.C.; Ingle, M.A.; Ahern, C.H.; Blaney, S.M. Pediatric Phase I Trial and Pharmacokinetic Study of MLN8237, an Investigational Oral Selective Small-Molecule Inhibitor of Aurora Kinase A: A Children's Oncology Group Phase I Consortium Study. *Clin. Cancer Res.* **2012**, *18*, 6058–6064. [CrossRef] [PubMed]
941. Moreno, L.; Marshall, L.V.; Pearson, A.D.; Morland, B.; Elliott, M.; Campbell-Hewson, Q.; Makin, G.; Halford, S.E.; Acton, G.; Ross, P.; et al. A Phase I Trial of AT9283 (a Selective Inhibitor of Aurora Kinases) in Children and Adolescents with Solid Tumors: A Cancer Research UK Study. *Clin. Cancer Res.* **2015**, *21*, 267–273. [CrossRef] [PubMed]
942. Mossé, Y.P.; Fox, E.; Teachey, D.T.; Reid, J.M.; Safgren, S.L.; Carol, H.; Lock, R.B.; Houghton, P.J.; Smith, M.A.; Hall, D.; et al. A Phase II Study of Alisertib in Children with Recurrent/Refractory Solid Tumors or Leukemia: Children's Oncology Group Phase I and Pilot Consortium (ADVL0921). *Clin. Cancer Res.* **2019**, *25*, 3229–3238. [CrossRef] [PubMed]
943. Feliciano, S.V.M.; Santos, M.D.O.; Pombo-De-Oliveira, M.S. Incidência e Mortalidade por Câncer entre Crianças e Adolescentes: Uma Revisão Narrativa. *Rev. Bras. Cancerol.* **2019**, *64*, 389–396. [CrossRef]
944. Verhaak, R.G.W.; Hoadley, K.A.; Purdom, E.; Wang, V.; Wilkerson, M.D.; Miller, C.R.; Ding, L.; Golub, T.; Jill, P.; Alexe, G.; et al. Integrated Genomic Analysis Identifies Clinically Relevant Subtypes of Glioblastoma Characterized by Abnormalities in PDGFRA, IDH1, EGFR, and NF1. *Cancer Cell* **2010**, *17*, 98–110. [CrossRef]
945. Maximiano, S.; Magalhães, P.; Guerreiro, M.P.; Morgado, M. Trastuzumab in the Treatment of Breast Cancer. *Biodrugs* **2016**, *30*, 75–86. [CrossRef]
946. Lee, J.; Park, Y.H. Trastuzumab deruxtecan for HER2+ advanced breast cancer. *Futur. Oncol.* **2022**, *18*, 7–19. [CrossRef]
947. Hochhaus, A.; Baccarani, M.; Silver, R.T.; Schiffer, C.; Apperley, J.F.; Cervantes, F.; Clark, R.E.; Cortes, J.E.; Deininger, M.W.; Guilhot, F.; et al. European LeukemiaNet 2020 recommendations for treating chronic myeloid leukemia. *Leukemia* **2020**, *34*, 966–984. [CrossRef]

948. Louis, D.N.; Perry, A.; Wesseling, P.; Brat, D.J.; Cree, I.A.; Figarella-Branger, D.; Hawkins, C.; Ng, H.K.; Pfister, S.M.; Reifenberger, G.; et al. The 2021 WHO Classification of Tumors of the Central Nervous System: A summary. *Neuro-Oncology* **2021**, *23*, 1231–1251. [CrossRef]
949. Nørøxe, D.S.; Poulsen, H.S.; Lassen, U. Hallmarks of glioblastoma: A systematic review. *ESMO Open* **2016**, *1*, e000144. [CrossRef]
950. Roskoski, R., Jr. Properties of FDA-approved small molecule protein kinase inhibitors: A 2020 update. *Pharmacol. Res.* **2020**, *152*, 104609. [CrossRef]
951. Roskoski, R., Jr. Classification of small molecule protein kinase inhibitors based upon the structures of their drug-enzyme complexes. *Pharmacol. Res.* **2016**, *103*, 26–48. [CrossRef]
952. Wu, P.; Nielsen, T.E.; Clausen, M.H. FDA-approved small-molecule kinase inhibitors. *Trends Pharmacol. Sci.* **2015**, *36*, 422–439. [CrossRef]
953. Lopes, L.F.; Bacchi, C.E. Imatinib treatment for gastrointestinal stromal tumour (GIST). *J. Cell. Mol. Med.* **2009**, *14*, 42–50. [CrossRef]
954. Fischer, P.M. The Design of Drug Candidate Molecules as Selective Inhibitors of Therapeutically Relevant Protein Kinases. *Curr. Med. Chem.* **2004**, *11*, 1563–1583. [CrossRef]
955. A Bogoyevitch, M.; Fairlie, D. A new paradigm for protein kinase inhibition: Blocking phosphorylation without directly targeting ATP binding. *Drug Discov. Today* **2007**, *12*, 622–633. [CrossRef]
956. Gaumann, A.K.; Kiefer, F.; Alfer, J.; Lang, S.A.; Geissler, E.K.; Breier, G. Receptor tyrosine kinase inhibitors: Are they real tumor killers? *Int. J. Cancer* **2015**, *138*, 540–554. [CrossRef]
957. Roberti, M.; Bottegoni, G. Non-ATP Competitive Protein Kinase Inhibitors. *Curr. Med. Chem.* **2010**, *17*, 2804–2821. [CrossRef]
958. Suttorp, M.; Bornhäuser, M.; Metzler, M.; Millot, F.; Schleyer, E. Pharmacology and pharmacokinetics of imatinib in pediatric patients. *Expert Rev. Clin. Pharmacol.* **2017**, *11*, 219–231. [CrossRef]
959. Angel, M.C.-G.; Julia, A.P.; María, S.B.; Luiz, G.T. G2/M inhibitors as pharmacotherapeutic opportunities for glioblastoma: The old, the new, and the future. *Cancer Biol. Med.* **2018**, *15*, 354–374. [CrossRef] [PubMed]
960. Kwak, E.L.; Bang, Y.-J.; Camidge, D.R.; Shaw, A.T.; Solomon, B.; Maki, R.G.; Ou, S.-H.I.; Dezube, B.J.; Jänne, P.A.; Costa, D.B.; et al. Anaplastic Lymphoma Kinase Inhibition in Non–Small-Cell Lung Cancer. *N. Engl. J. Med.* **2010**, *363*, 1693–1703. [CrossRef] [PubMed]
961. De Brouwer, S.; De Preter, K.; Kumps, C.; Zabrocki, P.; Porcu, M.; Westerhout, E.M.; Lakeman, A.; Vandesompele, J.; Hoebeeck, J.; Van Maerken, T.; et al. Meta-analysis of Neuroblastomas Reveals a Skewed *ALK* Mutation Spectrum in Tumors with *MYCN* Amplification. *Clin. Cancer Res.* **2010**, *16*, 4353–4362. [CrossRef] [PubMed]
962. Bresler, S.C.; Wood, A.C.; Haglund, E.A.; Courtright, J.; Belcastro, L.T.; Plegaria, J.S.; Cole, K.; Toporovskaya, Y.; Zhao, H.; Carpenter, E.L.; et al. Differential Inhibitor Sensitivity of Anaplastic Lymphoma Kinase Variants Found in Neuroblastoma. *Sci. Transl. Med.* **2011**, *3*, 108ra114. [CrossRef]
963. de Sousa, G.R.; Vieira, G.M.; das Chagas, P.F.; Pezuk, J.A.; Brassesco, M.S. Should we keep rocking? Portraits from targeting Rho kinases in cancer. *Pharmacol. Res.* **2020**, *160*, 105093. [CrossRef]
964. Martinsson, T.; Eriksson, T.; Abrahamsson, J.; Caren, H.; Hansson, M.; Kogner, P.; Kamaraj, S.; Schönherr, C.; Weinmar, J.; Ruuth, K.; et al. Appearance of the Novel Activating F1174S ALK Mutation in Neuroblastoma Correlates with Aggressive Tumor Progression and Unresponsiveness to Therapy. *Cancer Res* **2011**, *71*, 98–105. [CrossRef]
965. Braun, T.P.; Eide, C.A.; Druker, B.J. Response and Resistance to BCR-ABL1-Targeted Therapies. *Cancer Cell* **2020**, *37*, 530–542. [CrossRef]
966. Katayama, R. Drug resistance in anaplastic lymphoma kinase-rearranged lung cancer. *Cancer Sci.* **2018**, *109*, 572–580. [CrossRef]
967. Rolfo, C.; Passiglia, F.; Castiglia, M.; E Raez, L.; Germonpre, P.; Gil-Bazo, I.; Zwaenepoel, K.; De Wilde, A.; Bronte, G.; Russo, A.; et al. ALK and crizotinib: After the honeymoon ... what else? Resistance mechanisms and new therapies to overcome it. *Transl. Lung Cancer Res.* **2014**, *3*, 250–261. [CrossRef]
968. Sharma, S.V.; Haber, D.A.; Settleman, J. Cell line-based platforms to evaluate the therapeutic efficacy of candidate anticancer agents. *Nat. Rev. Cancer* **2010**, *10*, 241–253. [CrossRef]
969. Massicotte, M.P.; Sofronas, M.; Deveber, G. Difficulties in performing clinical trials of antithrombotic therapy in neonates and children. *Thromb. Res.* **2006**, *118*, 153–163. [CrossRef]
970. Bond, M.C.; Pritchard, S. Understanding clinical trials in childhood cancer. *Paediatr. Child Health* **2006**, *11*, 148–150. [CrossRef]
971. Renfro, L.A.; Ji, L.; Piao, J.; Onar-Thomas, A.; Kairalla, J.A.; Alonzo, T.A. Trial Design Challenges and Approaches for Precision Oncology in Rare Tumors: Experiences of the Children's Oncology Group. *JCO Precis. Oncol.* **2019**, *1*, 1–13. [CrossRef]
972. Bavcar, S.; Argyle, D.J. Receptor tyrosine kinase inhibitors: Molecularly targeted drugs for veterinary cancer therapy. *Veter- Comp. Oncol.* **2012**, *10*, 163–173. [CrossRef]

pharmaceutics

Article

Biological Evaluation and In Vitro Characterization of ADME Profile of In-House Pyrazolo[3,4-*d*]pyrimidines as Dual Tyrosine Kinase Inhibitors Active against Glioblastoma Multiforme

Federica Poggialini [1], Chiara Vagaggini [1], Annalaura Brai [1], Claudia Pasqualini [1], Emmanuele Crespan [2], Giovanni Maga [2], Cecilia Perini [2], Noemi Cabella [2], Lorenzo Botta [3], Francesca Musumeci [4], Maria Frosini [5], Silvia Schenone [4] and Elena Dreassi [1,*]

1 Department of Biotechnology, Chemistry and Pharmacy (DBCF), University of Siena, 53100 Siena, Italy
2 Institute of Molecular Genetics (IGM), CNR "Luigi Luca Cavalli-Sforza", 27100 Pavia, Italy
3 Department of Ecological and Biological Sciences, University of Tuscia, Via S.C. De Lellis s.n.c., 01100 Viterbo, Italy
4 Department of Pharmacy, University of Genoa, 16132 Genoa, Italy
5 Department of Life Sciences, University of Siena, 53100 Siena, Italy
* Correspondence: elena.dreassi@unisi.it; Tel.: +39-0577-234321

Abstract: The therapeutic use of tyrosine kinase inhibitors (TKIs) represents one of the successful strategies for the treatment of glioblastoma (GBM). Pyrazolo[3,4-*d*]pyrimidines have already been reported as promising small molecules active as c-Src/Abl dual inhibitors. Herein, we present a series of pyrazolo[3,4-*d*]pyrimidine derivatives, selected from our in-house library, to identify a promising candidate active against GBM. The inhibitory activity against c-Src and Abl was investigated, and the antiproliferative profile against four GBM cell lines was studied. For the most active compounds endowed with antiproliferative efficacy in the low-micromolar range, the effects toward nontumoral, healthy cell lines (fibroblasts FIBRO 2-93 and keratinocytes HaCaT) was investigated. Lastly, the in silico and in vitro ADME properties of all compounds were also assessed. Among the tested compounds, the promising inhibitory activity against c-Src and Abl (K_i 3.14 μM and 0.44 μM, respectively), the irreversible, apoptotic-mediated death toward U-87, LN18, LN229, and DBTRG GBM cell lines (IC_{50} 6.8 μM, 10.8 μM, 6.9 μM, and 8.5 μM, respectively), the significant reduction in GBM cell migration, the safe profile toward FIBRO 2-93 and HaCaT healthy cell lines (CC_{50} 91.7 μM and 126.5 μM, respectively), the high metabolic stability, and the excellent passive permeability across gastrointestinal and blood–brain barriers led us to select compound **5** for further in vivo assays.

Keywords: glioblastoma; tyrosine kinase; c-Src/Abl dual inhibitors; pyrazolo[3,4-*d*]pyrimidines; ADME; antiproliferative activity

Citation: Poggialini, F.; Vagaggini, C.; Brai, A.; Pasqualini, C.; Crespan, E.; Maga, G.; Perini, C.; Cabella, N.; Botta, L.; Musumeci, F.; et al. Biological Evaluation and In Vitro Characterization of ADME Profile of In-House Pyrazolo[3,4-*d*]pyrimidines as Dual Tyrosine Kinase Inhibitors Active against Glioblastoma Multiforme. *Pharmaceutics* **2023**, *15*, 453. https://doi.org/10.3390/pharmaceutics15020453

Academic Editor: Yu-Fen Huang

Received: 28 October 2022
Revised: 19 January 2023
Accepted: 27 January 2023
Published: 30 January 2023

1. Introduction

According to the World Health Organization (WHO), glioblastoma multiforme (GBM) is classified as the most aggressive form of primary brain cancer (grade IV), associated with poor prognosis, high morbidity, and mortality [1]. Among GBM patients, the median survival time is typically no longer than 1 year from the diagnosis, and <3% of cases have a 5-year survival rate. Responsible for 18% of brain tumors, GBM is one of the most common primary tumors of the central nervous system in adults [2]. Histologically, GBM is typed by high microvascularity and necrotic areas, typical signs of the high proliferation, invasiveness, and angiogenesis of the tumor, demanding nutrients and oxygen for its development. The current standard treatments of GBM are represented by surgical resection, followed by a combination of radiotherapy and chemotherapy; despite these massive and intense treatments, the prognosis remains inauspicious [3]. Failure in GBM therapy is mostly represented by the lack of drugs able to hit effective molecular targets, overcome the blood–brain barrier, and escape from multidrug resistance mechanisms occurring during the therapy. Therefore, novel and more effective pharmacological approaches are urgently needed to improve the prognosis of GBM patients [4]. Genomic studies highlighted the aberrant activation of receptor tyrosine kinases (RTKs) over different types of cancers, including GBM. A common feature of GBM has been identified by transcriptional analyses in the downregulation of TP53, RB1, and RTKs/Ras/PI3K pathways, recognizing these as fundamental requirements for the pathogenesis of GBM. Moreover, the massive activation of RTKs leads to the downregulation of the downstream signaling cascade resulting in the activation of both ERK and MAPK pathways and non-receptor tyrosine kinases (TKs), such as Src and Abl. Src family kinases regulate cell proliferation, survival, and angiogenesis and play an essential role in cancer progression and diffusion [5]. Additionally, the upregulation of the Abelson (Abl) family members is a trademark of GBM, involved in tumorigenesis, proliferation, and migration [6,7]. Thus, targeted therapy toward activated Src and Abl might represent a favorable strategy to counteract the high proliferation rate and invasion of GBM. In the last few decades, great attention has been paid to the implication of TKs in glioma biology and development of novel small-molecule compounds able to target and inhibit these molecular targets. Although monotherapy with Src inhibitors, such as dasatinib and bosutinib, resulted unsatisfactory for recurrent GBM, TKs remain one of the most promising targets to arrest cancer invasion [4]. In this context, several pyrazolo[3,4-*d*]pyrimidines have been studied and developed by our research group as ATP-competitive TKIs. Several compounds showed potent antitumor activities, inducing apoptosis and downregulating cell proliferation, in several cancer cell lines, including solid tumors such as osteosarcoma [8], neuroblastoma [9], GBM [1,5,10], and mesothelioma [11], as well as hematological tumors such as leukemia and Burkitt lymphoma [12,13]. In vivo data, obtained from xenograft mouse models, supported the antitumor activity of these small molecules against neuroblastoma, leukemia, and GBM [14]. Compound **1** was previously reported by us as a potent c-Src/Abl dual inhibitor with favorable K_i values and micromolar efficacy against the U-87 GBM cell line [10]. The ADME profile was characterized by suboptimal aqueous solubility and good passive permeability across the gastrointestinal (GI) and blood–brain barrier (BBB) membranes, but metabolic stability was lower than 80%. With the purpose of identifying molecules with the right balance between efficacy against GBM and suitable pharmacokinetic properties, we selected a series of pyrazolo[3,4-*d*]pyrimidines from our internal library. We performed structure–activity relationship (SAR) studies along with biological and ADME evaluations to select a promising molecule active as a TKI against GBM. The compounds were characterized by heterogeneous chemical features at the N1, C4, and C6 positions. First, a cell-free assay was set up to determine the K_i values against c-Src and Abl, followed by the assessment of the antiproliferative activity against four GBM cell lines (U-87, LN18, LN229, and DBTRG). Furthermore, the cytotoxic profile was established on two nontumoral, healthy cell lines (fibroblasts FIBRO 2-93 and keratinocytes HaCaT) for the most active compounds of the series. Lastly, the in silico and in vitro ADME profile was investigated for all derivatives. On the basis of the biological and ADME studies, compound **5** resulted as the most promising pyrazolo[3,4-*d*]pyrimidine derivative of the panel. Its favorable inhibitory activity against c-Src/Abl, its promising antitumor efficacy against all GBM cell lines tested, and its safe cytotoxic profile led us to select compound **5** for further in vitro assays. Its ability to elicit an irreversible cytotoxic effect, arrest cellular migration, and induce apoptotic-mediated cell death was assessed. Moreover, the high metabolic stability and satisfactory passive permeability across the GI and BBB membranes suggest that compound **5**, with appropriate formulation studies to overcome solubility issues, could represent a strong candidate for future in vivo PK studies.

2. Materials and Methods

2.1. Drugs and Materials

Compounds herein reported were previously synthesized and reported by us [12,15–19]. The NMR characterizations, chemical names, and structures are presented in the Supplementary Materials (Chemistry chapter and Scheme S1). All reagents, solvents, and materials were purchased from Sigma Aldrich S.r.l. (Milan, Italy) and Carlo Erba Reagents S.r.l (Milan, Italy). Cell culture mediums, fetal bovine serum (FBS), L-glutamine, and penicillin–streptomycin were purchased from Euroclone S.p.A. (Milan, Italy). Human glioblastoma multiforme cell lines (U-87, LN18, LN229, and DBTRG) and the normal human dermal fibroblast (FIBRO 2-93) cell line were purchased from American Type Culture Collection (ATCC) (Manassas, VA, USA). Human keratinocytes (HaCaT) were kindly donated by Professor Federica Pessina from the Department of Molecular and Developmental Medicine (University of Siena, Italy). All the compounds were dissolved in dimethyl sulfoxide (DMSO) as a 20 mM stock solution immediately before use and diluted to final desired concentration with appropriate cell culture medium (see below).

2.2. Enzymatic Assay

Active recombinant kinases and specific peptide substrates were purchased from Promega (Madison, WI, USA). The reactions were carried out in accordance with the manufacturer's instructions, with minor modifications. The experiments were run using all substrates at least twice the concentration of apparent K_m. Src reactions were carried out with 500 μM Src-peptide (KVEKIGEGTYGVVYK), 100 μM ATP, and 0.00087% NP-40; Abl reactions were performed with 50 μM Abltide (EAI-YAAPFAKKK), 30 μM ATP, and 0.00087% NP-40, and T315I reactions were carried out with 100 μM Abltide, 60 μM ATP, and 0.0013% NP-40. All reactions were run out at 30 °C for 10 min with 10–50 ng of enzyme and 10 μL of 10% DMSO. Protein low-binding tubes were used to prevent peptide adsorption onto the plastic surface. To detect kinase activity, the ADP-Glo Kinase Assay (Promega, Madison, WI, USA) was used according to the manufacturer's protocol. Into white 384-well plates, reactions were stopped for 50 min at RT by adding 10 μL of ADP-Glo Reagent (Promega, Madison, WI, USA). After 30 min, 20 μL of Detection Reagent (Promega, Madison, WI, USA) was added, and the reaction was read using a GloMax Discover microplate reader (Promega, Madison, WI, USA). Data were plotted using GraphPad Prism 5.04 (GraphPad Software Inc., San Diego, CA, USA), and ID_{50} values were calculated according to Equation (1).

$$\nu = V/(1 + (I/ID_{50})), \quad (1)$$

where ν represents the measured reaction velocity, V is the apparent maximal velocity without the inhibitor, I is the inhibitor concentration, and ID_{50} is the 50% inhibitory dose. Compounds analyzed were considered to act as ATP-competitive inhibitors. Thus, K_i values were calculated using Equation (2).

$$K_i = (ID_{50}/(1 + KmATP/[SATP]))/(1 + Kmpep/[Spep]), \quad (2)$$

where K_i represents the affinity of the inhibitor for the enzyme, S is the ATP concentration, and Km is the affinity of ATP calculated according to Michaelis–Menten equation.

2.3. Cell Culture, Antiproliferative Effects, and Cytotoxicity on Cell-Based Assays

Cell-based assays were performed using four GBM cell lines, i.e., U-87, LN18, LN229, and DBTRG, cultured in Dulbecco's modified Eagle medium (DMEM) (U-87, LN18, and LN229) or Roswell Park Memorial Institute medium (RPMI 1640) (DBTRG). Both media were implemented with 10% fetal bovine serum (FBS), 2 mM L-glutamine, and 10,000 units/mL penicillin/streptomycin at 37 °C in a 5% CO_2 atmosphere. To evaluate the antiproliferative activity of the compounds reported as IC_{50} (i.e., drug concentration that caused 50% of cell growth inhibition), cells were seeded at a density of 1×10^4 in 96-well plates and, after 24 h of incubation, treated with increasing concentrations of compounds for an additional 72 h. The final concentration of DMSO used never exceeded 0.5% *v*/*v*, and appropriate controls were always performed in each assay. To assess the potential cytotoxic profile of the three most promising and active compounds toward nontumoral cells, reported as CC_{50} (i.e., concentration that reduced the proliferation of cells by 50%), further investigations were conducted using healthy fibroblasts FIBRO 2-93 and keratinocytes HaCaT. These were cultured in DMEM supplemented with 10% FBS, 2 mM L-glutamine, and 10,000 units/mL penicillin/streptomycin at 37 °C in a 5% CO_2 atmosphere. The protocol used to assess the cytotoxic effect on both healthy cell lines was the same described above for the antiproliferative assay. For both the antiproliferative and the cytotoxic assays, 3-(4,5-dimethylthiazol-2-yl)-2,5-diphenyltetrazolium bromide (MTT) was used to assess cell viability as previously described [20]. Briefly, 150 μL of MTT (5 mg/mL in FBS-free culture medium) was added to each well, and the plates incubated for 4 h in a 5% CO_2 atmosphere at 37 °C. Afterward, formazan violet crystals were solubilized in 150 μL of isopropanol, and the absorbance was quantified at 570 nm using a Multiskan SkyHigh Microplate Spectrophotometer (ThermoFisher, Waltham, MA, USA). Cell viability was expressed as a percentage of DMSO-treated cells (controls), taken as 100%. IC_{50} and CC_{50} values were calculated by fitting data to a nonlinear regression analysis (sigmoidal log concentration vs. normalized response curve; GraphPad Prism 5.04 software, GraphPad Software Inc., San Diego, CA, USA) using at least six points in a concentration range from 0% to 100% of the studied effects (total widest range 0.01–1000 μM). To support MTT data, cytotoxicity assays was always paralleled by a careful observation of cells under a phase-contrast light microscope, as described in USP 28 (United States Pharmacopeia edition 2005) as an alternative method [21–23]. This analysis was performed by an expert operator blind to the treatment, and exemplificative photos representing the effects of compound **5** on LN229 and DBTRG cells are reported in Figures S2 and S3 (Supplementary Materials).

2.4. *In Vitro* ADME Assays

2.4.1. HPLC/UV–MS Method

UV/LC–MS chromatographic analyses were carried out using an Agilent 1100 LC/MSD VL system (G1946C) purchased from Agilent Technologies (Palo Alto, CA, USA). Chromatographic separations were achieved at room temperature (RT) using a Phenomenex Kinetex C18-100 column (150 mm × 4.6 mm, 5 μm) and gradient elution consisting of a solution of solvent A (H_2O) and solvent B (ACN), both acidified with 0.1% *v*/*v* formic acid (FA). The analysis began with 5% B (t = 0–1 min), increased to 95% (t = 1–10 min), maintained at 95% (t = 10–19 min), and finally returned to 5% solvent A. The analyses were performed at a flow rate of 0.6 mL/min, UV detection was monitored at 254 nm, and spectra were acquired over the scan range of m/z 100–1000 in both positive and negative modes.

2.4.2. Parallel Artificial Membrane Permeability Assay (PAMPA)

The DMSO stock of each compound was diluted with phosphate buffer (PBS 25 mM, pH 7.4) to a final concentration of 500 μM in order to obtain the "donor". For GI and BBB permeability, filters were coated with 10 μL of 1% *w*/*v* dodecane phosphatidylcholine solution or 5 μL of 10% *w*/*v* $CHCl_3$/dodecane brain polar lipid solution. The donor (150 μL) was poured over the artificial membranes on the filter plate. DMSO/PBS (300 μL, 1:1 *v*/*v*) was then added to the acceptor wells. After assembling the sandwich plates, the experiments were run for 5 h at room temperature (RT). Lastly, the amount of compound passed though the artificial membranes was determined using the UV/LC–MS method described above. Apparent permeability (Papp) and membrane retention (MR%) were determined as previously described [24,25].

2.4.3. Metabolic Stability Assay

Each DMSO compound solution was incubated for 1 h at 37 °C in the presence of 500 μL of PBS (25 mM, pH 7.4), human liver microsomal (HLM) proteins (0.2 mg/mL), and 10 mM NADPH solution in $MgCl_2$ (48 mM). The reactions were stopped with 1.0 mL of cold ACN. The samples were centrifuged (4000 rpm for 10 min), and the supernatants were collected, dried under N_2 flow, and suspended in 100 μL of methanol (MeOH). Parent drugs and metabolites were determined using the previously described HPLC/UV–MS method. The percentage metabolic stability was calculated by comparing the metabolized and unmetabolized compounds.

2.5. *Reversible/Irreversible Effect on LN229 Cell Death*

To check if cytotoxic effects of compound **5** were reversible or irreversible, LN229 cells were seeded and treated as previously described. After 24 h of incubation with increasing concentration of compound **5** (0.1–1–10–100 μM), plates were washed with PBS and cultured in drug-free DMEM with 10% FBS for a further period of 48 h [20]. Afterward, the MTT assay was performed to assess cell viability as previously described.

2.6. *Cell-Cycle Analysis*

Flow cytometry was used to investigate the apoptotic effect of compound **5** on LN229 and DBTRG cells. These were seeded in a six-well plate (5×10^5 cells/well) and treated with compound **5** at the respective IC_{50} values concentrations for 72 h. Afterward, the cells were treated as previously reported [26]. Red fluorescence (DNA) was detected through a 563–607 nm bandpass filter (FL2 channel, 10^4 cells/sample) using an FACScan flow cytometer coupled with Cell Quest software v.3.0, (both purchase from BD Biosciences, San Jose, CA, USA), and the latter was used to calculate the percentage of cells in the different phases of the cell cycle.

2.7. *T-Scratch Assay*

To investigate the inhibitory effect of compound **5** on cell migration, LN229 cells (8×10^4 cells/well) were seeded in 24 well-plate such that they were at 90% confluence the following day, at which point a scratch was made in the cell layer using a sterile p200 pipet tip. After washing the cells with warm PBS to remove the detached cells, 500 μL of fresh DMEM containing 1% FBS and increasing concentrations (0.25–0.5–2.5–5 μM) of compound **5** were added to each well. The scratched area was observed and photographed at 10× magnification after 0–24–48–72 h (see Figure S5, Supplementary Materials). Finally, the size of the area covered by cells at selected timepoints was obtained using ImageJ software (National Institute of Health, Bethesda, MD, USA, 1.37v), and the percentage cellular migration was plotted as percentage wound closure using GraphPad Prism version 5.04 (GraphPad Software Inc., San Diego, CA, USA).

2.8. *Data Analysis*

Data are reported as the mean ± standard deviation (SD) of at least three independent experiments. Statistical analysis was performed using two-way ANOVA followed by Tukey's multiple comparison test (reversible/irreversible effect on LN229 cell death, T-scratch assay) and Student's *t*-test for unpaired samples (cell-cycle analysis) as appropriate.

3. Results and Discussion

3.1. *In Vitro Biological Evaluation*

3.1.1. Enzymatic Assay

GBM is a highly aggressive tumor with low expectations of recovery. Today, the standard treatment for GBM is surgical ablation, combined with and/or followed by radiotherapy and chemotherapy. The Src and Abl families play a crucial role in carcinogenesis, angiogenesis, and metastasis, and contribute to the development of the mechanisms of resistance, particularly in GBM, in which they are highly upregulated [5–7]. On the basis of these factors, the development of small molecules that act as dual Src/Abl inhibitors with improved pharmacokinetic properties, especially the ability to reach the target beyond the BBB, represents one successful strategy to attack GBM on several fronts while targeting the molecular mechanisms implicated in resistance. Thus, to investigate the structure–activity relationship (SAR) of the selected compounds, a cell-free enzymatic assay was used to assess the affinity toward c-Src and Abl, as shown in Table 1. As already reported [10], compound **1** inhibited both enzymes with K_i values in the micromolar and low-micromolar range (3 μM and 0.37 μM against c-Src and Abl, respectively). A wide-ranging SAR study on this class of derivatives suggested that the presence of a benzylamine moiety (**3**, **4**, **5**, **6**, **7**, **9**, and **11**) in the R_2 position resulted in more active derivatives than cyclopropylamine (**2**), aniline (**12**), or phenethylamine ones (**8** and **10**). Among the first group of compounds, with respect to parent compound **1**, an improvement in the inhibitory efficacy against c-Src was achieved thanks to the introduction of a fluorine atom in the ortho (**3**), para (**4**) [16], or meta (**11**) [19] position of the benzylamine group. These modifications led to a significant reduction in the inhibitory efficacy against Abl when compared to that of reference compound **1**. The introduction of a fluorine (**6**) [16] or a chlorine atom (**7**) [17] in the R position resulted in promising derivatives with K_i values in the low-micromolar range. Compound **5** did not show substantial changes in c-Src/Abl inhibition, despite having one more fluorine atom than **1**. Lastly, the introduction of a morpholine ethanethiol moiety (**9**) in R_3 provoked a favorable improvement in the inhibition of Abl, but no significant differences against c-Src [17].

Table 1. Chemical structures and enzymatic activity (μM) against c-Src and Abl of compounds **1–12**.

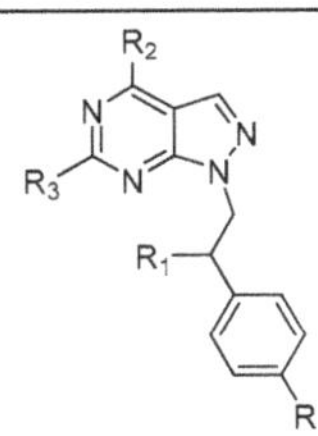

Compounds	R	R_1	R_2	R_3	Enzymatic Data c-Src K_i [a]	Abl K_i [a]
1 [b]	H	Cl	$NHCH_2C_6H_5$	H	3 *	0.37 *
2	H	Cl	NH-cyclopropyl	H	11.57 ± 1.08	8.01 ± 0.73
3 [c]	H	Cl	$NHCH_2C_6H_4$-*o*F	H	0.15 ± 0.02	0.73 ± 0.05 *
4 [c]	H	Cl	$NHCH_2C_6H_4$-*p*F	H	0.51 ± 0.08	2.00 ± 0.10 *
5 [c]	F	Cl	$NHCH_2C_6H_5$	H	3.14 ± 0.38	0.44 ± 0.06 *
6 [c]	F	Cl	$NHCH_2C_6H_4$-*o*F	H	0.07 ± 0.01	0.02 ± 0.01 *
7 [d]	Cl	Cl	$NHCH_2C_6H_4$-*p*F	H	0.80 ± 0.09	0.08 *
8 [e]	H	Cl	$NHCH_2CH_2C_6H_5$	SCH_3	0.70 *	7.30 *
9 [f]	H	Cl	$NHCH_2C_6H_5$	SCH_2CH_2-4-morpholino	2.90 ± 0.30 *	0.09 ± 0.01 *
10	Cl	Cl	$NHCH_2CH_2C_6H_5$	H	0.22 ± 0.03	0.31 ± 0.02
11 [e]	H	H	$NHCH_2C_6H_4$-*m*F	H	0.51 *	4.92 *
12 [b]	Br	Cl	NHC_6H_5	H	0.06 *	0.61 *

[a] Values are the means ± SD of three different experiments run in triplicate. * Compounds previously published. [b] Ref. [10]. [c] Ref. [16]. [d] Ref. [12]. [e] Ref. [19]. [f] Ref. [17].

While the cyclopropylamino derivative **2** showed K_i values approximately 10 μM against both enzymes, the anilino-substituted compound **12**, typed also by the presence of a bromine in R [10], was characterized by a relevant improvement in the inhibitory efficacy against c-Src. Among the phenethylamine derivatives, compound **8** possessed a promising activity against c-Src, but a lower activity against Abl, probably due to the presence of methanethiol group in R_3 position [19], while compound **10** substituted with a chlorine in R, showed a better profile, acting as a low-micromolar dual inhibitor (K_i values of 0.22 and

0.31 μM against c-Scr and Abl, respectively). The present enzymatic studies revealed a general improvement in the inhibitory efficacy against c-Src, with K_i values usually lower than the reference compound **1**. The low-micromolar inhibitory activity against Abl was confirmed and sometimes also improved; only a few exceptions resulted in potential TKI inhibition worse than compound **1**.

3.1.2. Cell Viability Assay

In light of the previous results, further SAR studies were conducted to investigate the antiproliferative activity of the pyrazolo[3,4-*d*]pyrimidine derivatives against four GBM cell lines (U-87, LN18, LN229, and DBTRG) (Table 2).

Table 2. Antiproliferative activity (μM) of compounds **1–12** against four GBM cell lines and cytotoxicity (μM) of the most promising ones on nontumoral, healthy FIBRO 2-93 and HaCaT cell lines.

Compounds	IC_{50} [a]				CC_{50} [b]	
	U-87	LN18	LN229	DBTRG	FIBRO 2-93	HaCaT
1 [c]	2.9 ± 1.1 *	3.6 ± 0.5	3.4 ± 0.8	4.3 ± 0.8	86.9 ± 4.2	88.7 ± 5.7
2	NA	49.4 ± 2.8	NA	NA	-	-
3	40.2 ± 3.6	27.5 ± 2.1	29.2 ± 2.6	15.1 ± 2.1	-	-
4	29.8 ± 2.5	9.0 ± 1.2	NA	68.5 ± 3.8	-	-
5	6.8 ± 0.3	10.8 ± 1.4	6.9 ± 0.5	8.5 ± 1.4	91.7 ± 5.3	126.5 ± 6.4
6	32.4 ± 2.7	11.7 ± 1.5	33.8 ± 2.3	20.1 ± 1.8	-	-
7	22.7 ± 2.3	3.9 ± 0.6	7.1 ± 1.1	14.6 ± 0.9	28.6 ± 3.7	37.8 ± 4.9
8	37.9 ± 3.9	5.9 ± 1.1	26.1 ± 2.8	31.2 ± 2.5	-	-
9	NA	2.9 ± 0.5	58.2 ± 3.4	59.4 ± 3.7	-	-
10	33.9 ± 2.9	6.2 ± 1.0	27.9 ± 2.1	46.5 ± 4.1	-	-
11	56.9 ± 3.3	34.1 ± 2.9	58.4 ± 3.0	69.2 ± 3.9	-	-
12 [c]	14.0 ± 1.9 *	NA	NA	NA	-	-

[a] IC_{50} = drug concentration that caused 50% of cell growth inhibition, [b] CC_{50} = concentration that reduced the proliferation of cells by 50%, both determined by measuring the cell viability with the colorimetric MTT assay. Values are the means ± SD of at least three different experiments run in triplicate. * Compounds previously published. [c] Ref. [10]. NA means not active (IC_{50} > 100 μM). Empty cells were not determined.

Compound **1**, previously reported to be active against U-87 cells in the low-micromolar range [10], confirmed its potent antitumor activity against the other GBM cell lines (LN18, LN229, and DBTRG). As shown in Table 2, the cyclopropylamine derivative **2** resulted the least active of the series with only a slight activity against LN18 cells, while compounds **3** and **4**, with a fluorine atom in the ortho and para positions of the benzylamine group, displayed promising IC_{50} values. Interestingly, the best results were obtained with compound **5** (characterized by an additional fluorine atom in the R position with respect to reference compound **1**), which showed low-micromolar activity against all GBM cells tested (concentration vs. cytotoxic effect curves relative to each cell line are reported in Figure S1, Supplementary Materials). Ortho and para fluorobenzylamino derivatives bearing a fluorine (**6**) or chlorine (**7**) atom in the R position presented a slight or significant improvement in anticancer efficacy compared to the corresponding unsubstituted compounds (**3** and **4**) on the N1 side-chain. To further investigate the SAR around the scaffold, substituents such as methanethiol and 4-morpholinoethanethiol moieties were introduced in R_3 position leading to compounds **8** and **9**, respectively. The phenethylamine derivative **8** demonstrated a mild ability to act as an antiproliferative agent, especially against LN18 cells, while the efficacy of compound **9** was reduced in the other GBM cell lines. Compound **10** followed the same trend as compound **8**, proving to be a valuable antiproliferative agent with encouraging IC_{50} values. Lastly, if compound **11** was a weak cytotoxic agent, the introduction of the aniline moiety in R_2 (**12**) produced a potent compound active against U-87 cells [10], but was ineffective against the other GBM cell lines. Taken together, the present results strongly suggest that our compounds are promising anticancer candidates toward GBM. Compound **1** and the two most encouraging derivatives (**5** and **7**), endowed with antiproliferative activity comparable to that of reference one, were further investigated for their safety profile by assessing their activity in two different nontumoral, immortalized healthy cells (FIBRO 2-93 and HaCaT). Except for compound **7** which affected the viability of both healthy cell lines (CC_{50} values of 28.6 and 37.8 μM on FIBRO 2-93 and HaCaT, respectively), compounds **1** and **5** were characterized by an optimal safety profile (CC_{50} ~87 μM for compound **1** and >90 μM for compound **5**, i.e., 10 times higher than their IC_{50}), thus suggesting a reassuring ability to distinguish between tumoral and healthy cells.

3.2. *In Vitro ADME Evaluation*

To assess the capability of the selected pyrazolo[3,4-*d*]pyrimidine derivatives to reach their target in vivo, the in silico and in vitro ADME properties were assessed for all compounds (Table 3). Our research group was efficiently involved in the attempt to overtake the suboptimal solubility of these compounds with prodrugs [10] or formulations such as albumin nanoparticles [27], cyclodextrins [28], and liposomes [27]. For this reason, we decided to analyze this property using a computational approach followed by proper in vitro investigations. The QikProp software confirmed the hydrophobicity of the derivatives. All compounds were characterized by suboptimal LogS values which could limit their in vivo bioavailability depending on the solvation rate. Nevertheless, the entire series of pyrazolo[3,4-*d*]pyrimidines had the optimal ability to cross the GI membrane and, most importantly, the BBB, as demonstrated by PAMPA assays. Similar to the reference compound **1**, previously reported to show high passive permeability across membranes, compounds **3**, **4**, **5**, **6**, **7**, **10**, and **11** were characterized by high apparent permeability values, never lower than 10 cm/s × 10^{-6}, and limited percentages of membrane retention.

Table 3. In silico and in vitro ADME properties of compounds **1–12**.

Compounds	QP LogS	GI Papp [a,b] (MR%) [c]	BBB Papp [a,b] (MR%) [c]	Metabolic Stability [a] (%)	Metabolite Formation (%) [d]
1 [e]	−5.715	11.08 *	16.5 *	78.3 *	M_1 = 14.5 * M_2 < 0.1 * M_3 = 2.6 * M_4 = 4.5 *
2	−4.862	6.46 (34.7)	4.96 (55.0)	96.3	M_1 = 2.1 M_2 = 1.6
3	−6.358	12.4 (7.3)	12.2 (4.8)	80.4	M_1 = 14.5 M_3 = 5.1
4	−6.784	11.2 (7.8)	10.6 (2.1)	92.2	M_1 = 5.9 M_2 = 1.0 M_3 = 0.9
5	−6.511	10.7 (9.2)	11.2 (5.6)	84.2	M_1 = 10.8 M_2 = 1.7 M_3 = 3.3
6	−6.955	11.4 (10.1)	12.4 (4.4)	69.9	M_1 = 21.9 M_2 = 3.0 M_3 = 5.2
7 [f]	−7.533	16.6 *	9.6 (4.4)	94.0 *	M_1 = 4 * M_3 = 2 *
8	−6.207	3.2 (42.7)	5.96 (18.3)	96.6	M_1 = 1.9 M_2 = 1.5

Table 3. *Cont.*

Compounds	QP LogS	GI Papp [a,b] (MR%) [c]	BBB Papp [a,b] (MR%) [c]	Metabolic Stability [a] (%)	Metabolite Formation (%) [d]
9 [g]	−5.6 *	9.1 *	2.6 (6.3)	95 *	M_1 = 5 *
10	−7.589	10.8 (17.2)	11.3 (4.8)	85.9	M_1 = 10.3 M_3 = 3.8
11	−5.973	10.1 (6.7)	10.8 (5.2)	93.0	M_1 = 0.4 M_2 = 6.7
12 [e]	−6.576	6.64 *	13.1 *	96.4 *	-

[a] Values represent the mean values of three independent experiments run in triplicate. [b] Apparent permeability (Papp) reported in cm/s × 10^{-6}. [c] Membrane retention %. [d] M_1 = M − HCl + O (−36 + 16); M_2 = M + OH (−16); M_3 = M—group bearing NH moiety in R_2; M_4 = M − Cl + OH (−35 + 17). * Compounds previously published. [e] Refs. [10,18]. [f] Ref. [12]. [g] Ref. [17]. Empty cells were not determined or not found.

The in silico and in vitro results confirmed the well-known suboptimal aqueous solubility characterizing the pyrazolo[3,4-*d*]pyrimidine derivatives, as well as the satisfactory ability to efficiently overcome membranes (especially the BBB) without remaining entrapped in the phospholipidic bilayer. Regarding the metabolic stability in the presence of HLMs, the principal metabolites were identified because of dechlorination (M_1 and M_4), oxidation (M_2), and the loss of the groups bearing the amine moiety in R_2 (M_3). The percentage of metabolite formation was calculated as previously reported [18]. With respect to reference compound **1**, most of the derivatives resulted in higher metabolic stability, around 90% or more (**2**, **4**, **7**, **8**, **9**, **11**, and **12**), when incubated in presence of HLMs. Compound **6** underwent massive metabolization with the formation of more than 20% of metabolite M_1, resulting definitively worse than the reference compound **1**. Lastly, compounds **3**, **5**, and **10** were characterized by improved metabolic stability of approximately 80–85%. As a general conclusion, metabolic stability studies highlighted the tendency of selected compounds to lead to the massive formation of metabolite M_1 (M − HCl + O), followed by the loss of the group bearing the NH moiety in R_2 (M_3).

3.3. Irreversible Cytotoxic Effect on LN229 Cell Death

Results so far obtained highlighted that compound **5** is a dual inhibitor, endowed with low-micromolar activity against four different GBM cell lines, an optimal safety profile toward healthy cells, and promising ADME properties, particularly a high BBB permeability that represents the Achilles heel of many GBM candidates. These characteristics strongly led us to further investigate its anticancer activity. In the case of cancer cells, it is mandatory to investigate whether a cell line may (or may not) be able to restart its proliferative activity upon drug treatment. It is crucial to demonstrate the presence of the so-called "point of no return", a limit line between cell injury and cell death, whereby surpassing this point drives irreversible damage. Thus, to better elucidate the biological effect and the mechanism of compound **5**-mediated cell death, further investigations were performed on the LN229 cell line. To check whether the growth inhibitory effect was reversible or irreversible, LN229 cells were initially treated with compound **5** for 24 h; after that, cells were carefully washed to remove the treatment and were cultured for a further 48 h in a complete drug-free medium. At the timepoint of 72 h, as described above, the MTT assay was used to assess cell viability. As reported in Figure 1, compound **5** caused an irreversible cytotoxic effect. This trend was emphasized at the concentrations of 10 and 100 μM with a further decrease in cell viability (~−30% for both concentrations) after 24 h of treatment, followed by 48 h of incubation without compound **5**.

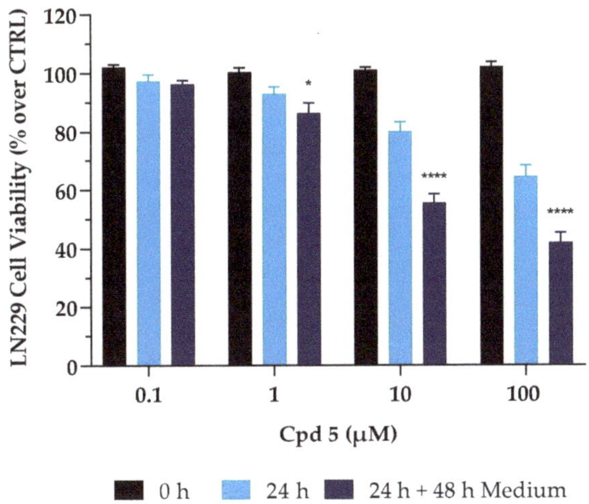

Figure 1. Irreversible cytotoxic effect of compound **5** (Cpd **5**) on LN229 GBM cell line. These were treated with the compound for 24 h (24 h, light-blue column) and then incubated for 48 h with fresh, drug-free medium (24 h + 48 h medium, dark-blue column). Afterward, cell viability was assessed using the MTT assay. Values are the means ± SD of $n = 3$ independent experiments run in triplicate. * $p < 0.05$, **** $p < 0.001$ vs. 24 h at the same concentration (ANOVA followed by Tukey's multiple comparison test).

Interestingly, it is important to note that, after 24 h of treatment, cytotoxic effects occurring at both 1 and 10 μM, although moderate (~−10% and −20%, respectively), were immediately irreversible, as cell viability was not restored after withdrawal of the drug. This might prove to be very useful to support a full cytotoxic effect achieved in the low μM range after 72 h of treatment.

3.4. Cell-Cycle Analysis

To further investigate whether compound **5**-mediated cell death involved some characteristics of apoptosis such as loss of DNA due to DNA fragmentation, as well as effects on cell-cycle progression, flow cytometry-mediated cell-cycle analyses were conducted by treating LN229 and DBTRG cells for 72 h with compound **5** at its IC_{50} concentrations. As reported in Figure 2a, a relevant accumulation of LN229 hypodiploid cells in the sub G0/G1 phase was observed (+32.1%, $p < 0.0001$ vs. control), which was accompanied by a proportional reduction in those in the G0/G1 phase (−32.4%, $p < 0.0001$ vs. control), suggesting that compound **5** induced apoptotic-mediated cell death. No significant differences were detected in the percentage of cells in the S and G2/M phases. As reported in Figure 2b, a significant accumulation of hypodiploid cells was also found in the sub G0/G1 area in DBTRG cells (+15.6%, $p < 0.0001$ vs. control) (see exemplificative histograms of cell-cycle analysis reported in Figure S4, Supplementary Materials), thus suggesting that, in the presence of compound **5**, both LN299 and DBTRG cells respond by initiating programmed cell death. This hypothesis was also supported by the analysis at phase-contrast microscopy, which highlighted significant morphological alterations, including a tendency to round up and detach from the culture plate, shrinkage, loss of contact with adjacent cells, membrane blebbing, and formation of apoptotic bodies (see Figures S2 and S3, Supplementary Materials). The increased number of cells with DNA fragmentation was accompanied by an arrest of DBTRG cells in the G2/M phase (+38.1%, $p < 0.0001$ vs. control).

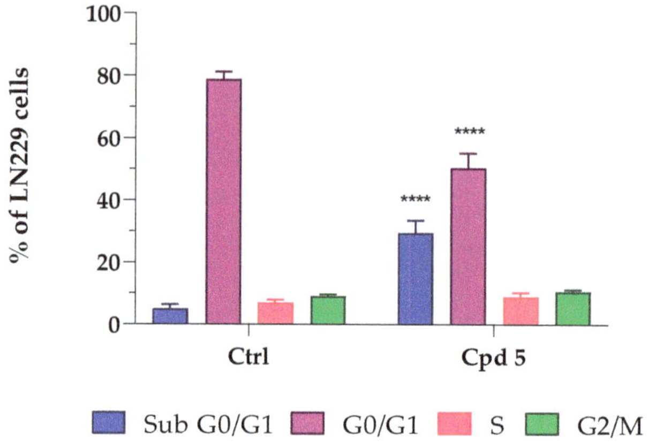

(a)

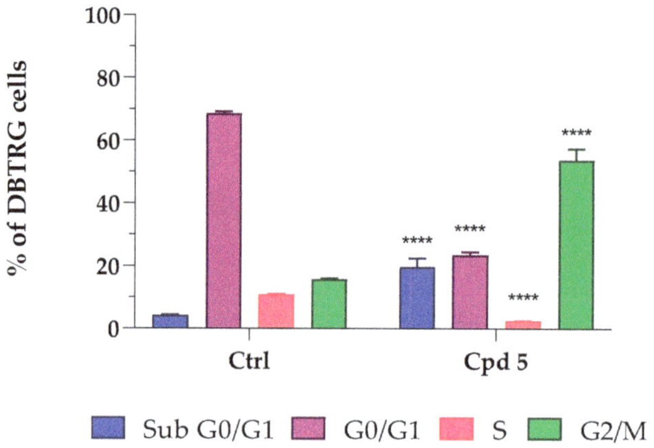

(b)

Figure 2. Compound **5**-mediated effects on the glioblastoma LN229 (**a**) and DBTRG (**b**) cell cycle. Cells were treated for 72 h at the compound **5** IC_{50} concentrations, i.e., 6.9 μM for LN229 and 8.5 μM for DBTRG. Percentages of cells in sub G0/G1 (Blue), G0/G1 (Purple), S (Pink), and G2/M (Green) phases are reported as the means ± SD of at least three independent experiments run in triplicate. **** $p < 0.0001$ vs. ctrl, same cell-cycle phase (unpaired Student *t*-test).

A cell-cycle completion is essential for proliferation, and cell apoptosis often results from cell-cycle arrest [29]. To complete the cell cycle, a successful G2/M phase is crucial, and uncontrolled cell proliferation might result from imbalanced cell-cycle regulation, which is a hallmark of cancer cells, including GBM [30,31]. Several anticancer drugs have been found to induce G2/M cell-cycle arrest and apoptosis in different GBM cell lines, effectively suppressing proliferation [32,33]. Moreover, the observation that, upon compound **5** treatment, G2/M phase arrest was observed in DBTRG, but not in LN299 cells can be explained by considering that these GBM cell lines possess a diverse genetic background and, in turn, possibly different mechanisms leading to tumor development [34], which might impact the cells' capacity to respond to drugs. Taken together, results from cell-cycle analysis, showing an increased population in cells with a reduced DNA content

(LN299 and DBTRG) and G2/M arrest (DBTRG), suggest the possibility of an apoptotic-mediated cell death caused by compound **5**. As types of cell death other than apoptosis might result in loss of DNA appearing in the sub G0/G1 region, the possibility that other mechanisms may drive compound **5** effects cannot be ruled out.

3.5. T-Scratch Assay

Glioblastoma is a malignant primary brain tumor strongly characterized by aggressive invasion of the surrounding cerebral tissues that obstacles surgery and target therapies. Although gliomas rarely metastasize out of the brain tissue, the massive infiltration of tumor cells along blood vessels, with matter tracts and the subarachnoid space is a common hallmark of this type of cancer [35,36]. Therefore, studying the effect of drugs on cell migration is of primary importance; to fulfill this task, we decided to use an easy, low-cost, and reproducible T-scratch assay. The overall ability of compound **5** to influence GBM cell migration was assessed in LN229 cells (Figure 3).

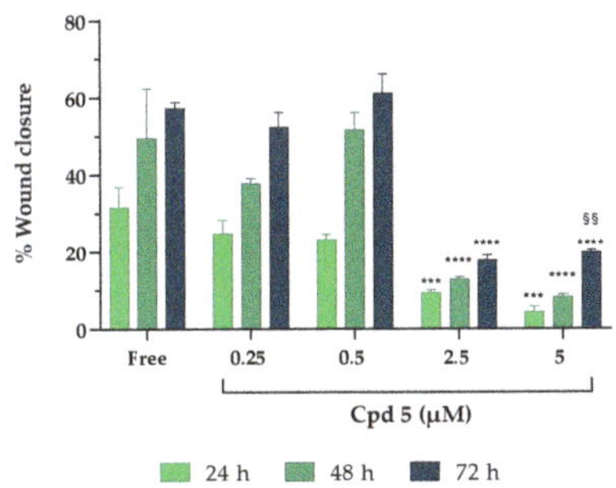

Figure 3. Effect of compound **5** on cancer cell migration. After scratching, LN229 cells were treated with increasing concentrations of compound **5** for 24–48–72 h. Results were obtained from $n = 3$ independent experiments run in triplicate; values are reported as the means ± SD. Two-way ANOVA followed by Tukey's multiple comparison test was performed to determine statistical significance. 2.5 μM and 5 μM: *** $p < 0.001$, **** $p < 0.0001$ vs. free at the same timepoints; 5 μM at 72 h: §§ $p < 0.01$ vs. the same concentration at 24 h.

While, in free samples and in the presence of the lowest concentrations of the inhibitor (0.25 and 0.5 μM), the closure of the monolayer significantly increased after 72 h (getting close to 60% of wound closure), at 2.5 and 5 μM, compound **5** suppressed healing, leaving more than 80% of the wound opened at all the timepoints analyzed, suggesting that compound **5** can inhibit the horizontal migration ability of LN229 cells already at the early stage (24 h) of drug stimulation and is sustained over time (72 h), albeit with decreased inhibition rates.

4. Conclusions

GBM continues to be a highly aggressive tumor with low expectations of recovery. Today, the standard treatment for GBM is surgical ablation, combined with and/or followed by radiotherapy and chemotherapy. However, such methods are not curative, and patients affected by GBM still have poor prognoses, primarily due to the onset of resistance, often caused by the presence of cancer stem cells [6]. Progressive and increasing interest in developing small-molecule kinase inhibitors has been recorded since the 1980s, with more than 30 new drugs approved by the FDA for the treatment of tumors [37]. In recent years, our research team has created a large library of pyrazolo[3,4-*d*]pyrimidines, capable of inhibiting tyrosine kinases, identified as potent antitumor agents, demonstrating a powerful action both in vitro and in vivo. Thus, moving from our in-house library, a panel of compounds characterized by the pyrazolo[3,4-*d*]pyrimidine scaffold and heterogeneously decorated at N1, C4, and C6 was selected for this study. Compounds were tested against c-Src and Abl, showing promising K_i values in the micromolar and low-micromolar range. Then, the antiproliferative activity of the compounds was assessed on four GBM cell lines (U-87, LN229, LN18, and DBTRG); for the most active derivatives, a safety profile was established with the healthy fibroblasts FIBRO 2-93 and keratinocytes HaCaT. The ability of the compounds to cross both GI and BBB membranes was assessed in vitro. The optimal passive permeability values suggest that several compounds could efficiently cross the BBB. Moreover, a complete ADME profile was studied for all compounds, revealing comparable solubility but higher metabolic stability with respect to parent compound **1**. Lastly, considering the in vitro biological and pharmacokinetic data, compound **5** was chosen to further characterize its biological effect on LN229 and DBTRG GBM cells. Compound **5** induced irreversible cytotoxicity, significantly reduced cell migration, and increased the number of cells with DNA fragmentation, possibly because of an apoptotic-mediated cell death. Overall, this study led to the identification of a new promising pyrazolo[3,4-*d*]pyrimidine derivative for GBM treatment. Although further investigations are needed to characterize its mechanism of action, as well as to evaluate the in vivo pharmacokinetic profile and efficacy, our present findings demonstrate that compound **5** could represent a new excellent approach to GBM treatment.

Supplementary Materials: The following supporting information can be downloaded at https://www.mdpi.com/article/10.3390/pharmaceutics15020453/s1, Chemistry; Scheme S1: Chemical Structures of tested compound; Figure S1: Effects of compound **5** on viability of four different GBM cell lines (U-87, LN18, LN229, DBTRG), as well as on two non-tumoral, healthy cells (the fibroblast FIBRO 2-93 cells and the keratinocytes HaCaT cells). Cell viability was expressed as percentage of that of DMSO-treated cells (controls), taken as 100%. IC_{50} and CC_{50} values were calculated by fitting data according to a non-linear regression analysis (sigmoidal log concentration vs. normalized response curve) and reported in Table 2. Each point represents mean $\pm$ SD (if not visible, SD was covered by the point); Figure S2: Morphological comparison performed at contrast phase microscopy (scale bar 180 µm) between controls, untreated LN229 cells (panel (a)) and those treated with compound **5** (6.9 µM, 72 h, panel (b)) in which significant morphological alterations, including tendency to round-up, shrinkage, loss of contact with adjacent cells, membrane blebbing and formation of apoptotic bodies are evident. Each photograph was representative of three independent observations; Figure S3: Morphological comparison performed at contrast phase microscopy (scale bar 180 µm) between controls, untreated DBTRG cells (panel (a)) and those treated with compound **5** (8.5 µM, 72h, panel (b)) in which significant morphological changes caused by the drug are evident. Compound **5** caused the cells to lose contact with adjacent cells, cell membrane blebbing and a general tendency to round-up. Each photograph was representative of three independent observations; Figure S4: Exemplificative cell cycle distribution histograms of the LN299 and DBTRG cell lines in control condition (CTRL) and after 72 h of treatment with compound **5** (6.9 µM and 8.5 µM, respectively). Following incubation, the DNA of the cells was stained with propidium iodide and its content was analysed by flow cytometry. The cells were normally distributed in the population of sub G0/G1, G0/G1, S, and G2/M of cell cycle phases in CTRL samples. After 72 h of treatment with compound **5**, an increase in sub G0/G1 cells in both cells lines was observed, while the same compound also caused an arrest the cell cycle in G2/M phase in DBTRG cells; Figure S5: Time-lapse images of a representative T-scratch assay to investigate the effect of compound **5** on GBM cancer cell migration. LN229 cells were seeded, scratched, and finally incubated with increasing concentrations of compound **5** (0.25–0.5–2.5–5 µM) for 24–48–72 h. Scale bar 100 µm.

Author Contributions: Conceptualization, F.P., C.V., A.B. and E.D.; methodology, F.P., C.V., A.B., C.P. (Claudia Pasqualini), G.M., M.F., S.S. and E.D.; formal analysis, F.P., C.V., A.B., E.C., F.M. and M.F.; investigation, F.P., C.V., A.B., C.P. (Claudia Pasqualini), E.C., F.M., L.B., C.P. (Cecilia Perini), N.C. and M.F.; resources, G.M., S.S., L.B., M.F. and E.D.; data curation, F.P., C.V., A.B., E.C., F.M. and M.F.; writing—original draft preparation, F.P.; writing—review and editing, C.V., A.B., C.P. (Claudia Pasqualini), E.C., G.M., L.B., F.M., M.F., S.S. and E.D.; visualization, F.P., C.V., A.B. and M.F.; supervision, G.M., E.C., L.B., M.F., S.S. and E.D.; project administration, E.D.; funding acquisition, E.D. All authors have read and agreed to the published version of the manuscript.

Funding: This research was funded by the TUSCAVIR.NET Bando Ricerca Salute 2018 CUP: B64I18013740002 funded by Tuscany Region, PI Elena Dreassi, the AIRC Foundation for Cancer Research in Italy (Fondazione AIRC per la Ricerca sul Cancro, AIRC): (i) Grant IG-2019, project code 23725, PI Silvia Schenone, (ii) Grant IG-2017, project code 20762, PI Giovanni Maga, (iii) Grant IG-2020, project code 24448, PI Emmanuele Crespan, and the Italian MIUR Project PRIN 2017 2017SA5837_004 (C.P. (Cecilia Perini) and G.M.).

Institutional Review Board Statement: Not applicable.

Informed Consent Statement: Not applicable.

Data Availability Statement: Data are reported in this article and the Supplementary Materials.

Acknowledgments: E.C. wishes to thank the Cariplo Foundation, project 2019-1836.

Conflicts of Interest: The authors declare no conflict of interest.

References

1. Fallacara, A.L.; Zamperini, C.; Podolski-Renić, A.; Dinić, J.; Stanković, T.; Stepanović, M.; Mancini, A.; Rango, E.; Iovenitti, G.; Molinari, A.; et al. A New Strategy for Glioblastoma Treatment: In Vitroand In Vivo Preclinical Characterization of Si306,a Pyrazolo [3,4-d]Pyrimidine DualSrc/P-Glycoprotein Inhibitor. *Cancers* **2019**, *11*, 848. [CrossRef]
2. Tamimi, A.F.; Juweid, M. Epidemiology and Outcome of Glioblastoma. *Glioblastoma* **2017**, 143–153. [CrossRef]
3. Ceccherini, E.; Indovina, P.; Zamperini, C.; Dreassi, E.; Casini, N.; Cutaia, O.; Forte, I.M.; Pentimalli, F.; Esposito, L.; Polito, M.S.; et al. SRC Family Kinase Inhibition through a New Pyrazolo[3,4-d]Pyrimidine Derivative as a Feasible Approach for Glioblastoma Treatment. *J. Cell. Biochem.* **2015**, *116*, 856–863. [CrossRef]
4. Molinari, A.; Iovenitti, G.; Mancini, A.; Gravina, G.L.; Chebbi, M.; Caruana, M.; Vignaroli, G.; Orofino, F.; Rango, E.; Angelucci, A.; et al. AuNP Pyrazolo[3,4-d]Pyrimidine Nanosystem in Combination with Radiotherapy against Glioblastoma. *ACS Med. Chem. Lett.* **2020**, *11*, 664–670. [CrossRef]
5. Cirotti, C.; Contadini, C.; Barilà, D. SRC Kinase in Glioblastoma: News from an Old Acquaintance. *Cancers* **2020**, *12*, 1558. [CrossRef]
6. Lamballe, F.; Toscano, S.; Conti, F.; Arechederra, M.; Baeza, N.; Figarella-Branger, D.; Helmbacher, F.; Maina, F. Coordination of Signalling Networks and Tumorigenic Properties by ABL in Glioblastoma Cells. *Oncotarget* **2016**, *7*, 74747–74767. [CrossRef]
7. Luttman, J.H.; Colemon, A.; Mayro, B.; Pendergast, A.M. Role of the ABL Tyrosine Kinases in the Epithelial–Mesenchymal Transition and the Metastatic Cascade. *Cell Commun. Signal.* **2021**, *19*, 59. [CrossRef]
8. Spreafico, A.; Schenone, S.; Serchi, T.; Orlandini, M.; Angelucci, A.; Magrini, D.; Bernardini, G.; Collodel, G.; Di Stefano, A.; Tintori, C.; et al. Antiproliferative and Proapoptotic Activities of New Pyrazolo[3,4-d]Pyrimidine Derivative Src Kinase Inhibitors in Human Osteosarcoma Cells. *FASEB J.* **2008**, *22*, 1560–1571. [CrossRef]
9. Molinari, A.; Fallacara, A.L.; Di Maria, S.; Zamperini, C.; Poggialini, F.; Musumeci, F.; Schenone, S.; Angelucci, A.; Colapietro, A.; Crespan, E.; et al. Efficient Optimization of Pyrazolo[3,4-d]Pyrimidines Derivatives as c-Src Kinase Inhibitors in Neuroblastoma Treatment. *Bioorg. Med. Chem. Lett.* **2018**, *28*, 3454–3457. [CrossRef]
10. Vignaroli, G.; Iovenitti, G.; Zamperini, C.; Coniglio, F.; Calandro, P.; Molinari, A.; Fallacara, A.L.; Sartucci, A.; Calgani, A.; Colecchia, D.; et al. Prodrugs of Pyrazolo[3,4-d]Pyrimidines: From Library Synthesis to Evaluation as Potential Anticancer Agents in an Orthotopic Glioblastoma Model. *J. Med. Chem.* **2017**, *60*, 6305–6320. [CrossRef]
11. Indovina, P.; Giorgi, F.; Rizzo, V.; Khadang, B.; Schenone, S.; Di Marzo, D.; Forte, I.M.; Tomei, V.; Mattioli, E.; D'urso, V.; et al. New Pyrazolo[3,4-d]Pyrimidine SRC Inhibitors Induce Apoptosis in Mesothelioma Cell Lines through P27 Nuclear Stabilization. *Oncogene* **2012**, *31*, 929–938. [CrossRef] [PubMed]
12. Radi, M.; Tintori, C.; Musumeci, F.; Brullo, C.; Zamperini, C.; Dreassi, E.; Fallacara, A.L.; Vignaroli, G.; Crespan, E.; Zanoli, S.; et al. Design, Synthesis, and Biological Evaluation of Pyrazolo[3,4- d]Pyrimidines Active in Vivo on the Bcr-Abl T315I Mutant. *J. Med. Chem.* **2013**, *56*, 5382–5394. [CrossRef] [PubMed]
13. Cozzi, M.; Giorgi, F.; Marcelli, E.; Pentimalli, F.; Forte, I.M.; Schenone, S.; D'urso, V.; De Falco, G.; Botta, M.; Giordano, A.; et al. Antitumor Activity of New Pyrazolo[3,4-d]Pyrimidine SRC Kinase Inhibitors in Burkitt Lymphoma Cell Lines and Its Enhancement by WEE1 Inhibition. *Cell Cycle* **2012**, *11*, 1029–1039. [CrossRef]
14. Schenone, S.; Bruno, O.; Bondavalli, F.; Ranise, A.; Mosti, L.; Menozzi, G.; Fossa, P.; Manetti, F.; Morbidelli, L.; Trincavelli, L.; et al. Synthesis of 1-(2-Chloro-2-Phenylethyl)-6-Methylthio-1H-Pyrazolo[3,4-d]Pyrimidines 4-Amino Substituted and Their Biological Evaluation. *Eur. J. Med. Chem.* **2004**, *39*, 153–160. [CrossRef] [PubMed]
15. Schenone, S.; Bruno, O.; Bondavalli, F.; Ranise, A.; Mosti, L.; Menozzi, G.; Fossa, P.; Donnini, S.; Santoro, A.; Ziche, M.; et al. Antiproliferative Activity of New 1-Aryl-4-Amino-1H-Pyrazolo[3,4-d]Pyrimidine Derivatives toward the Human Epidermoid Carcinoma A431 Cell Line. *Eur. J. Med. Chem.* **2004**, *39*, 939–946. [CrossRef] [PubMed]
16. Manetti, F.; Brullo, C.; Magnani, M.; Mosci, F.; Chelli, B.; Crespan, E.; Schenone, S.; Naldini, A.; Bruno, O.; Trincavelli, M.L.; et al. Structure-Based Optimization of Pyrazolo[3,4- d]Pyrimidines as Abl Inhibitors and Antiproliferative Agents toward Human Leukemia Cell Lines. *J. Med. Chem.* **2008**, *51*, 1252–1259. [CrossRef] [PubMed]

17. Radi, M.; Dreassi, E.; Brullo, C.; Crespan, E.; Tintori, C.; Bernardo, V.; Valoti, M.; Zamperini, C.; Daigl, H.; Musumeci, F.; et al. Design, Synthesis, Biological Activity, and ADME Properties of Pyrazolo[3,4- d]Pyrimidines Active in Hypoxic Human Leukemia Cells: A Lead Optimization Study. *J. Med. Chem.* **2011**, *54*, 2610–2626. [CrossRef] [PubMed]
18. Zamperini, C.; Dreassi, E.; Vignaroli, G.; Radi, M.; Dragoni, S.; Schenone, S.; Musumeci, F.; Valoti, M.; Antiochia, R.; Botta, M. CYP-Dependent Metabolism of Antitumor Pyrazolo[3,4-d]Pyrimidine Derivatives Is Characterized by an Oxidative Dechlorination Reaction. *Drug Metab. Pharmacokinet.* **2014**, *29*, 433–440. [CrossRef]
19. Radi, M.; Brullo, C.; Crespan, E.; Tintori, C.; Musumeci, F.; Biava, M.; Schenone, S.; Dreassi, E.; Zamperini, C.; Maga, G.; et al. Identification of Potent C-Src Inhibitors Strongly Affecting the Proliferation of Human Neuroblastoma Cells. *Bioorg. Med. Chem. Lett.* **2011**, *21*, 5928–5933. [CrossRef]
20. Chiaino, E.; Micucci, M.; Durante, M.; Budriesi, R.; Gotti, R.; Marzetti, C.; Chiarini, A.; Frosini, M. Apoptotic-Induced Effects of Acacia Catechu Willd. Extract in Human Colon Cancer Cells. *Int. J. Mol. Sci.* **2020**, *21*, 2102. [CrossRef]
21. Santulli, C.; Brizi, C.; Micucci, M.; del Genio, A.; de Cristofaro, A.; Bracco, F.; Pepe, G.L.; di Perna, I.; Budriesi, R.; Chiarini, A.; et al. Castanea Sativa Mill. Bark Extract Protects U-373 MG Cells and Rat Brain Slices Against Ischemia and Reperfusion Injury. *J. Cell. Biochem.* **2017**, *118*, 839–850. [CrossRef] [PubMed]
22. Durante, M.; Frosini, M.; Fusi, F.; Neri, A.; Sticozzi, C.; Saponara, S. In Vitro Vascular Toxicity of Tariquidar, a Potential Tool for in Vivo PET Studies. *Toxicol. Vitr.* **2017**, *44*, 241–247. [CrossRef] [PubMed]
23. Chiaino, E.; Micucci, M.; Budriesi, R.; Mattioli, L.B.; Marzetti, C.; Corsini, M.; Frosini, M. Hibiscus Flower and Olive Leaf Extracts Activate Apoptosis in SH-SY5Y Cells. *Antioxidants* **2021**, *10*, 1962. [CrossRef] [PubMed]
24. Tintori, C.; Brai, A.; Dasso Lang, M.C.; Deodato, D.; Greco, A.M.; Bizzarri, B.M.; Cascone, L.; Casian, A.; Zamperini, C.; Dreassi, E.; et al. Development and in Vitro Evaluation of a Microbicide Gel Formulation for a Novel Non-Nucleoside Reverse Transcriptase Inhibitor Belonging to the N -Dihydroalkyloxybenzyloxopyrimidines (N-DABOs) Family. *J. Med. Chem.* **2016**, *59*, 2747–2759. [CrossRef] [PubMed]
25. Brai, A.; Riva, V.; Saladini, F.; Zamperini, C.; Trivisani, C.I.; Garbelli, A.; Pennisi, C.; Giannini, A.; Boccuto, A.; Bugli, F.; et al. DDX3X Inhibitors, an Effective Way to Overcome HIV-1 Resistance Targeting Host Proteins. *Eur. J. Med. Chem.* **2020**, *200*, 112319. [CrossRef]
26. Chiaino, E.; Stella, R.; Peggion, C.; Micucci, M.; Budriesi, R.; Mattioli, L.B.; Marzetti, C.; Pessina, F.; Valoti, M.; Frosini, M. *Acacia catechu* Willd. Extract Protects Neuronal Cells from Oxidative Stress-Induced Damage. *Antioxidants* **2021**, *11*, 81. [CrossRef]
27. Vignaroli, G.; Calandro, P.; Zamperini, C.; Coniglio, F.; Iovenitti, G.; Tavanti, M.; Colecchia, D.; Dreassi, E.; Valoti, M.; Schenone, S.; et al. Improvement of Pyrazolo[3,4-d]Pyrimidines Pharmacokinetic Properties: Nanosystem Approaches for Drug Delivery. *Sci. Rep.* **2016**, *6*, 21509. [CrossRef]
28. Dreassi, E.; Zizzari, A.T.; Mori, M.; Filippi, I.; Belfiore, A.; Naldini, A.; Carraro, F.; Santucci, A.; Schenone, S.; Botta, M. 2-Hydroxypropyl-β-Cyclodextrin Strongly Improves Water Solubility and Anti-Proliferative Activity of Pyrazolo[3,4-d]Pyrimidines Src-Abl Dual Inhibitors. *Eur. J. Med. Chem.* **2010**, *45*, 5958–5964. [CrossRef]
29. Schwartz, G.K.; Shah, M.A. Targeting the Cell Cycle: A New Approach to Cancer Therapy. *J. Clin. Oncol.* **2005**, *23*, 9408–9421. [CrossRef]
30. Stewart, Z.A.; Westfall, M.D.; Pietenpol, J.A. Cell-Cycle Dysregulation and Anticancer Therapy. *Trends Pharmacol. Sci.* **2003**, *24*, 139–145. [CrossRef]
31. Hanahan, D.; Weinberg, R.A. Hallmarks of Cancer: The next Generation. *Cell* **2011**, *144*, 646–674. [CrossRef] [PubMed]
32. Xu, J.; Sun, S.; Zhang, W.; Dong, J.; Huang, C.; Wang, X.; Jia, M.; Yang, H.; Wang, Y.; Jiang, Y.; et al. Irigenin Inhibits Glioblastoma Progression through Suppressing YAP/β-Catenin Signaling. *Front. Pharmacol.* **2022**, *13*, 1027577. [CrossRef]
33. Othman, N.S.; Mohd Azman, D.K. Andrographolide Induces G2/M Cell Cycle Arrest and Apoptosis in Human Glioblastoma DBTRG-05MG Cell Line via ERK1/2 /c-Myc/P53 Signaling Pathway. *Molecules* **2022**, *27*, 6686. [CrossRef] [PubMed]
34. Demircan, T.; Yavuz, M.; Kaya, E.; Akgül, S.; Altuntaş, E. Cellular and Molecular Comparison of Glioblastoma Multiform Cell Lines. *Cureus* **2021**, *13*, 43. [CrossRef]
35. Hatoum, A.; Mohammed, R.; Zakieh, O. The Unique Invasiveness of Glioblastoma and Possible Drug Targets on Extracellular Matrix. *Cancer Manag. Res.* **2019**, *11*, 1843. [CrossRef] [PubMed]
36. Vollmann-Zwerenz, A.; Leidgens, V.; Feliciello, G.; Klein, C.A.; Hau, P. Tumor Cell Invasion in Glioblastoma. *Int. J. Mol. Sci.* **2020**, *21*, 1932. [CrossRef] [PubMed]
37. Bhullar, K.S.; Lagarón, N.O.; McGowan, E.M.; Parmar, I.; Jha, A.; Hubbard, B.P.; Rupasinghe, H.P.V. Kinase-Targeted Cancer Therapies: Progress, Challenges and Future Directions. *Mol. Cancer* **2018**, *17*, 48. [CrossRef]

pharmaceutics

MDPI

Article

Influence of Lenvatinib on the Functional Reprogramming of Peripheral Myeloid Cells in the Context of Non-Medullary Thyroid Carcinoma

Chunying Peng [1], Katrin Rabold [1], Mihai G. Netea [2,3], Martin Jaeger [2] and Romana T. Netea-Maier [1,*]

[1] Department of Internal Medicine, Division of Endocrinology, Radboud University Medical Center, 6525 GA Nijmegen, The Netherlands
[2] Department of Internal Medicine and Radboud Center for Infectious Diseases, Radboud University Medical Center, 6525 GA Nijmegen, The Netherlands
[3] Department of Immunology and Metabolism, Life and Medical Sciences Institute, University of Bonn, 53115 Bonn, Germany
* Correspondence: romana.netea-maier@radboudumc.nl; Tel.: +31-24-3614599

Abstract: Lenvatinib is a multitarget tyrosine kinase inhibitor (TKI) approved for the treatment of several types of cancers, including metastatic differentiated thyroid cancer (DTC). The intended targets include VEGFR 1–3, FGFR 1–4, PDGFRα, RET, and KIT signaling pathways, but drug resistance inevitably develops and a complete cure is very rare. Recent data has revealed that most of the TKIs have additional 'off-target' immunological effects, which might contribute to a protective antitumor immune response; however, human cellular data are lacking regarding Lenvatinib-mediated immunomodulation in DTC. Here, we investigated in ex vivo models the impact of Lenvatinib on the function of immune cells in healthy volunteers. We found that monocytes and macrophages were particularly susceptible to Lenvatinib, while neutrophiles and lymphocytes were less affected. In tumor-immune cell co-culture experiments, Lenvatinib exerted a broad inhibitory effect on the proinflammatory response in TC-induced macrophages. Interestingly, Lenvatinib-treated cells had decreased cellular M2 membrane markers, whereas they secreted a significantly higher level of the anti-inflammatory cytokine IL-10 upon LPS stimulation. In addition, prolonged exposure to Lenvatinib impaired macrophages survival and phenotypical differentiation, which was accompanied by remarkable morphological changes and suppressed cellular metabolic activity. These effects were mediated by myeloid cell-intrinsic mechanisms which are independent of Lenvatinib's on-target activity. Finally, using specific inhibitors, we argue that dual effects on p38 MAPK and Syk pathways are likely the underlying mechanism of the off-target immunological effects we observed in this study. Collectively, our data show the immunomodulatory properties of Lenvatinib on human monocytes. These insights could be harnessed for the future design of novel treatment strategies involving a combination of Lenvatinib with other immunotherapeutic agents.

Keywords: tyrosine kinase inhibitors; innate immunity; thyroid cancer; tumor-associated macrophages; cytokines

Citation: Peng, C.; Rabold, K.; Netea, M.G.; Jaeger, M.; Netea-Maier, R.T. Influence of Lenvatinib on the Functional Reprogramming of Peripheral Myeloid Cells in the Context of Non-Medullary Thyroid Carcinoma. *Pharmaceutics* **2023**, *15*, 412. https://doi.org/10.3390/pharmaceutics15020412

Academic Editor: Francesca Musumeci

Received: 15 December 2022
Revised: 21 January 2023
Accepted: 24 January 2023
Published: 26 January 2023

1. Introduction

Differentiated thyroid cancer (DTC) constitutes the majority of thyroid carcinomas (TC) worldwide. Generally, treatment with surgery and radioactive iodine are effective for most cases. However, up to 15% of patients develop radioiodine-refractory metastatic disease (RAI-refractory-DTC) and have poor long-term prognosis [1]. Currently, the only therapeutic option available for these patients is systemic molecular targeted therapy, which is represented by tyrosine kinase inhibitors (TKIs). Lenvatinib, a multitargeted TKI against VEGFR 1–3, FGFR 1–4, PDGFRα, RET, and KIT signaling networks, has been approved for the treatment of metastatic DTC since 2015. Despite positive results from a

SELECT trial, in a real-world setting, response to treatment differs among patients [2,3]. Oncological drug development is a time- and cost-consuming task, yet a high frequency of secondary mutations and paradoxical activation of alternative pathways often leads to drug resistance [3–5]. Therefore, innovative strategies to fully assess the antitumor activity of Lenvatinib are urgently needed.

Beyond targeting oncogenesis and angiogenesis processes, tumor immunotherapy remains a highly attractive yet not fully operational approach to treatment. Recent advances in immune checkpoint inhibitors (ICIs) offer a promising approach to achieve durable improvement by remobilizing the immune system against tumor proliferation and metastasis. However, the efficacy of ICIs is largely dependent on an immunogenic tumor microenvironment (TME), and, to date, ICI monotherapy has been proved effective only in a minority of solid cancers, as it largely blunted by the immunosuppressive factors with TME [6–8]. Myeloid cells constitute a highly plastic and heterogenous cell population in the tumor milieu and are often hijacked and reprogrammed by malignant cells to facilitate disease progression [9]. Notably, tumor-associated macrophages (TAMs) identified in advanced DTC display an immunosuppressive and tumor-promoting phenotype, and increased TAMs infiltration is correlated with poor prognosis [10–12]. This, in turn, makes them suitable targets for therapeutic intervention [13]. Interestingly, there has been emerging evidence indicating that most TKI therapies exert additional immunological effects via tumor intrinsic or extrinsic mechanisms [14–18]. Part of the immunomodulatory properties could be explained by their anti-angiogenesis effects via direct binding to the primary targets VEGFRs, which is also known as on-target activity. It should be noted that many FDA-approved TKIs interact with undesignated kinases/proteins expressed by healthy cells, termed as off-target effects. Interestingly, these off-target effects could lead to beneficial drug repurposing if the underlying mechanism could be characterized [19].

Recent results from different murine models revealed that an intact immune system was an essential prerequisite for the Lenvatinib-mediated tumor killing activity, and the combination with anti-PD-1 therapy significantly improved animal survival [20–22]. This is indicative of possible Lenvatinib-mediated immunomodulation, mainly involving the adaptive immune system. However, thus far, little has been known about the role of myeloid cells in these effects and how such a mechanism can be leveraged to improve clinical outcomes. Therefore, in this study, we set out to study whether Lenvatinib modulates the immune system in the context of TC and the potential mechanisms mediating these effects. This knowledge could contribute to the development of combination treatment strategies to enhance its effectivity and reduce toxicity. Our data show that ex vivo Lenvatinib treatment induced functional and metabolic reprogramming in circulating monocytes obtained from healthy volunteers, and this was mediated via dual inhibition of the p38 MAPK and Syk pathways.

2. Materials and Methods

2.1. Monocyte Isolation

Peripheral blood mononuclear cells (PBMCs) were isolated by density gradient centrifugation with Ficoll-Paque (GE healthcare, Diegem, Belgium) from buffy coats obtained from the Sanquin bloodbank, Nijmegen, The Netherlands. Percoll-monocyte enrichment was performed as previously described [23]. Briefly, PBMCs were layered on top of a hyper-osmotic Percoll solution, and the interphase layer was collected after centrifugation. Cells were resuspended in RPMI culture medium (Life Technologies, Carlsbad, CA, USA) supplemented with pyruvate (1 mM), glutamine (2 mM), gentamicin (50 µg/mL), HEPES (10 mM), glucose (5 mM), and 10% human pooled serum (HPS). In selected experiments, highly purified monocytes were obtained from PBMCs through negative selection using a human pan monocyte isolation kit, following the manufacturer's instruction (Miltenyi Biotec, Bergisch Gladbach, Germany).

2.2. Neutrophil Isolation and ROS Production Analysis

Neutrophil isolation and functional analysis were performed as previously described [24]. EDTA-anticoagulated peripheral blood was collected from healthy donors with informed consent. After density-gradient centrifugation, red blood cells were lysed with hypotonic buffer, and the purity of the remaining neutrophils was measured on a Sysmex XN-450 Hematology Analyzer (Sysmex Corporation, Kobe, Japan). Reactive oxygen species (ROS) assays were performed with freshly isolated neutrophils using a luminescence assay in quadruplicates. A total of 200,000 neutrophils per well were added into a 96-well white assay plate (Corning, Corning, NY, USA). Due to the relatively brief lifespan of neutrophils, ROS production was measured after 1 h preincubation of Lenvatinib. Cells were stimulated with 0.5 μg/mL of phorbol myristate acetate (PMA) or plasma-opsonized Zymosan A (Sigma-Aldrich, St. Louis, MO, USA) at a concentration of 1 mg/mL. Luminol (Sigma-Aldrich) was added with a final concentration of 0.1 mM, and chemiluminescence was measured at a wavelength of 425 nm every 142 s for 1 h.

2.3. Cell Culture, Viability Test, and Stimulation

Several Lenvatinib concentrations within the clinically achievable dose range (100 nM to 1 μM) were tested in our experimental setup with both PBMCs and monocytes. Cell viability was determined by flowcytometry using fixable viability dye (FVD) eFlour 780 (Invitrogen, Carlsbad, CA, USA). After 24 h of incubation with Lenvatinib, the cells were labelled with FVD for 30 min and stained with surface marker. $CD14^+$ viable cells were gated for analysis.

Cells stimulation was performed with 5×10^5 PBMCs/well in 96-well round-bottom plates or 1×10^5 Percoll-enriched monocytes/well in 96-well flat-bottom plates. After 24 h of preincubation with Lenvatinib/DMSO, the cells were stimulated with 1×10^6/mL of heat-killed Candida albicans for 7 days or 10 ng/mL of LPS for 24 h. Supernatant was collected and stored at −20 °C until measurement.

Three TC cell lines were used for co-culture assay; these were characterized by different pathological types and mutation status, namely, TPC-1 (RET/PTC rearrangement), BCPAP (BRAF V600E mutation), and FTC-133 (PTEN deficient). All cancer cell lines were grown in RPMI 1640 Dutch modification culture medium supplemented with 10% Fetal Bovine Serum. ThinCertTM cell culture inserts were placed in the 24-well plate to create the co-culture setting (Greiner Bio-One GmbH, Kremsmünster, Austria). Cell counts were determined with a Coulter particle counter (Beckman Coulter Inc., Brea, CA, USA). A total of 50,000 TPC-1 cells (or the other TC cell lines, as indicated) were added to the trans-well inserts (the upper chamber). The same number of monocytes were added into trans-well inserts as the control. After overnight attachment, TC cell lines were co-incubated with 5×10^5 monocytes in the lower chamber in 500 μL of medium containing 1 μM of Lenvatinib or DMSO vehicle control. In some experiments, tumor cells were pre-treated with Lenvatinib in serum-free medium for 6 h. Prior to co-culture, medium containing Lenvatinib was removed, cells were washed with PBS, and fresh medium without Lenvatinib was added into the upper and lower chambers; this was termed as the pre-treatment group. After 24 h of co-culture, the trans-well inserts were discarded, and monocytes were stimulated as previously described.

GM-CSF/M-CSF differentiated macrophages were generated by culturing Percoll-enriched monocytes in 10% HPS RPMI medium containing GM-CSF (10 ng/mL, Miltenyi) or M-CSF (50 ng/mL, Miltenyi) for 6 days. Cells were seeded with 2.5×10^6 per well in a 6-well plate for RNA, Western blot, and seahorse experiments; 5×10^5 cells per well in a 24-well plate for flowcytometry analysis. The cells were detached by gentle scraping and the concentration was adjusted based on the results of flowcytometry (Beckman Coulter Inc.).

2.4. Flow Cytometry Analysis

Adherent monocytes/macrophages were harvested by gentle scraping, followed by viability staining. The cells were preincubated at 4 °C for 10 min with staining buffer containing 2% HPS prior to mAb labelling to CD14, CD163, CD206, CD64, CD86, HLA-DR. Intracellular TNF-α was measured in macrophages stimulated for 6 h with LPS (10 ng/mL) in the presence of Golgi plug (BD Bioscience, Franklin Lakes, CA, USA). The cells were fixed and permeabilized (Invitrogen, Carlsbad, CA, USA) followed by incubation with mAb to TNF-α. For the T cell proliferation assay, PBMCs were loaded with CellTrace violet (Invitrogen, Carlsbad, CA, USA) per the manufacturer's instructions and cultured for 7 days. The CellTrace dilution signal was analyzed on gated $CD3^+$ cells. Data were acquired using Cytoflex (Beckman Coulter Inc.) and analyzed with FlowJo V10 software (Tree Star, Inc., Ashland, OR, USA). The results were presented as geometric mean fluorescence intensity (gMFI) or as the percentage of positive cell populations.

2.5. *Cytokine Measurements*

Cytokine (TNF-α, IL-6, IL-10, IL-17, IFN-γ, IL-22) concentrations in supernatant were determined using a commercial ELISA kit (R&D System, Minneapolis, MN, USA), following the instructions of the manufacturer.

2.6. *Metabolic Measurements*

GM-CSF differentiated macrophages were cultured as described above and detached on Day 6. A total of 50,000 cells/well were seeded onto overnight-calibrated cartridges in assay medium (DMEM supplemented with 1 mM of pyruvate, 2 mM of L-Glutamine, and 1 mM of glucose for the Mito Stress Test; DMEM supplemented with 1 mM of L-Glutamine for the Glyco Stress Test, pH = 7.4) and incubated for 1 h in a non-CO_2 corrected incubator at 37 °C. The oxygen consumption rate (OCR) and extracellular acidification rate (ECAR) were analyzed using the Mito Stress and Glyco Stress Test kit in an XFp analyzer (Seahorse bioscience, Copenhagen, Denmark). The Mito Stress Test was performed with sequential injections of mitochondrial respiration inhibitors at final concentrations of 1 μM of oligomycin, 1 μM of FCCP, 0.5 μM of Rotenone/Antimycin A. The Glyco Stress Test was performed with 2-DG (50 mM) in the presence of glucose (10 mM). Each condition was tested in 5 replicates. Lactate concentration was measured in serum-free medium using Amplex Red Lactate oxidase assay (Sigma-Aldrich).

2.7. *RT-qPCR*

Total RNA was extracted with the RNeasy Mini kit (Qiagen, Hilden, Germany) and reverse-transcribed into cDNA with the iScript™ kit (Bio-Rad, Hercules, CA, USA). A real-time PCR was performed with the SYBR Green assay (Applied Biosystems, Foster City, CA, USA). Relative quantification was performed using the $2^{-\Delta\Delta Ct}$ method, and RPL37A was used as the housekeeping gene.

2.8. *Western Blotting*

Monocytes or GM-CSF differentiated macrophages were cultured in 6-well plates and lysed with RIPA buffer for Western blot. Protein concentration was determined by a BCA assay. An amount of 20 μg of proteins was loaded onto 10-well pre-casted gradient gels (4–15%) (Bio-Rad). Protein was transferred using the Trans-Blot Turbo Transfer system (Bio-Rad), after which PVDF membrane was incubated overnight at 4 °C with monoclonal antibodies against VEGFR1 and FGFR1. β-actin were used to normalize the protein expression.

2.9. *Kinase Inhibition Experiments*

Kinase inhibitors SB 202190(p38 MAPK inhibitor), U0126 (ERK1/2 inhibitor), SP 600125 (JNK inhibitor), and R406 (Syk inhibitor) (Sigma-Aldrich) were dissolved in sterile DMSO and stored at −20 °C. Monocytes were treated with kinase inhibitors (0, 1 μM, 10 μM) with or without Lenvatinib (1 μM) for 4 h or 24 h before LPS stimulation. Cytokine production was determined as previously described.

2.10. Statistical Analysis

The statistical analysis was performed with Prism 9 (GraphPad Software Inc., La Jolla, CA, USA). Nonparametric Wilcoxon matched-pair tests were used to compare means. Data are shown as means ± SEM. * $p < 0.05$, ** $p < 0.01$, *** $p < 0.001$.

3. Results

3.1. Lenvatinib Modulates the Function of Monocytes

As a first step to evaluate whether Lenvatinib interferes with immune cell function, we performed a dose–response test on human circulating immune cells in vitro. The physiologically relevant dose range was determined based on pharmacokinetic data from Phase I/II clinical trials. Plasma maximum and trough concentrations were reported as 598.0 ng/mL (1401 nM) and 33.9ng/mL (79.4 nM), respectively [25,26]; hence, 100 nM to 1 μM was deemed as a clinically meaningful dose range in our study The immune response was determined by measuring the ROS production from neutrophils, and the secretion of monocyte-derived (TNF-α, IL-6) and lymphocyte-derived (IL-17, INF-γ, IL-22) cytokines upon stimulation (Figure 1A).

No significant changes in neutrophil ROS production were observed, either by pathogen-specific (zymosan) or non-specific (PMA) stimulation (Figure 1B). Moreover, no spontaneous cytokine secretion was detected in monocytes or PBMCs incubated with Lenvatinib for 24 h or 7 days. Interestingly, the production of TNF-a and IL-6 elicited by LPS was largely attenuated by Lenvatinib in a dose-dependent manner (Figure 1C). In contrast, no detectable effect on Lenvatinib on either T cell proliferation (Supplementary Figure S1) or IL-17 and IFN-γ production was observed after a 7-day Candida stimulation, except a slight increase in IL-22 at the highest Lenvatinib concentration (Figure 1D). Taken together, these data suggest that monocytes are particularly susceptible to Lenvatinib treatment, whereas lymphocytes and neutrophils are less affected.

3.2. Lenvatinib Partially Reverses the TC-Induced Monocyte Phenotype

Noting that Lenvatinib affects monocytes stimulated with exogenous ligands, we further investigated whether the functional rewiring of monocytes in the context of TC is also affected by Lenvatinib treatment. Therapeutic perturbation of the TC secretome by Lenvatinib might interfere with the functional and phenotypic differentiation of TAMs [27]. To test this hypothesis, Lenvatinib was added into the trans-well co-culture system, which aims to mimic the cell-to-cell communication within TME.

In these sets of experiments, we observed the spontaneous production of IL-6 and IL-8 from the TPC-1 cell line, whereas TNF-α and IL-10 were below the detection limit (data not shown). To exclude the possibility of any influence of tumor-derived cytokines on the interpretation of monocytes' immune response, TNF-a and IL-10 were used as the major readout of cytokine response. In agreement with previously published data [28], after 24 h of co-culture with TPC-1 cells, monocytes showed a higher expression of immunosuppressive surface markers, CD163 and CD206. Concomitantly, a pro-inflammatory cytokine response was observed, which was demonstrated as an elevated TNF-α and downregulated IL-10 level (Figure 2B). This points towards a mixed M1/M2 phenotype which resembles TAMs [29]. Having validated the efficacy of the co-culture model, we applied this method to evaluate the role of Lenvatinib in regulating TAMs induction. Interestingly, Lenvatinib significantly attenuated TNF-a production, while it increased IL-10 release. Furthermore, the expression of the M2 surface marker, CD206, was diminished by Lenvatinib, which was shown as a lowered percentage of $CD206^+$ $CD14^+$ cells. CD163 is generally expressed on almost all the $CD14^+$ monocytes at baseline; however, the signal intensity was attenuated by Lenvatinib (Figure 2D). In the meantime, the M1 markers (CD64, CD86, and HLA-DR) remained unchanged (Figure 2E). Flowcytometric cell discrimination and viability measurements revealed that cell viability was not affected after 24 h exposure to Lenvatinib within the clinically relevant dose range (100 nM to 1 μM) (Figure 2C).

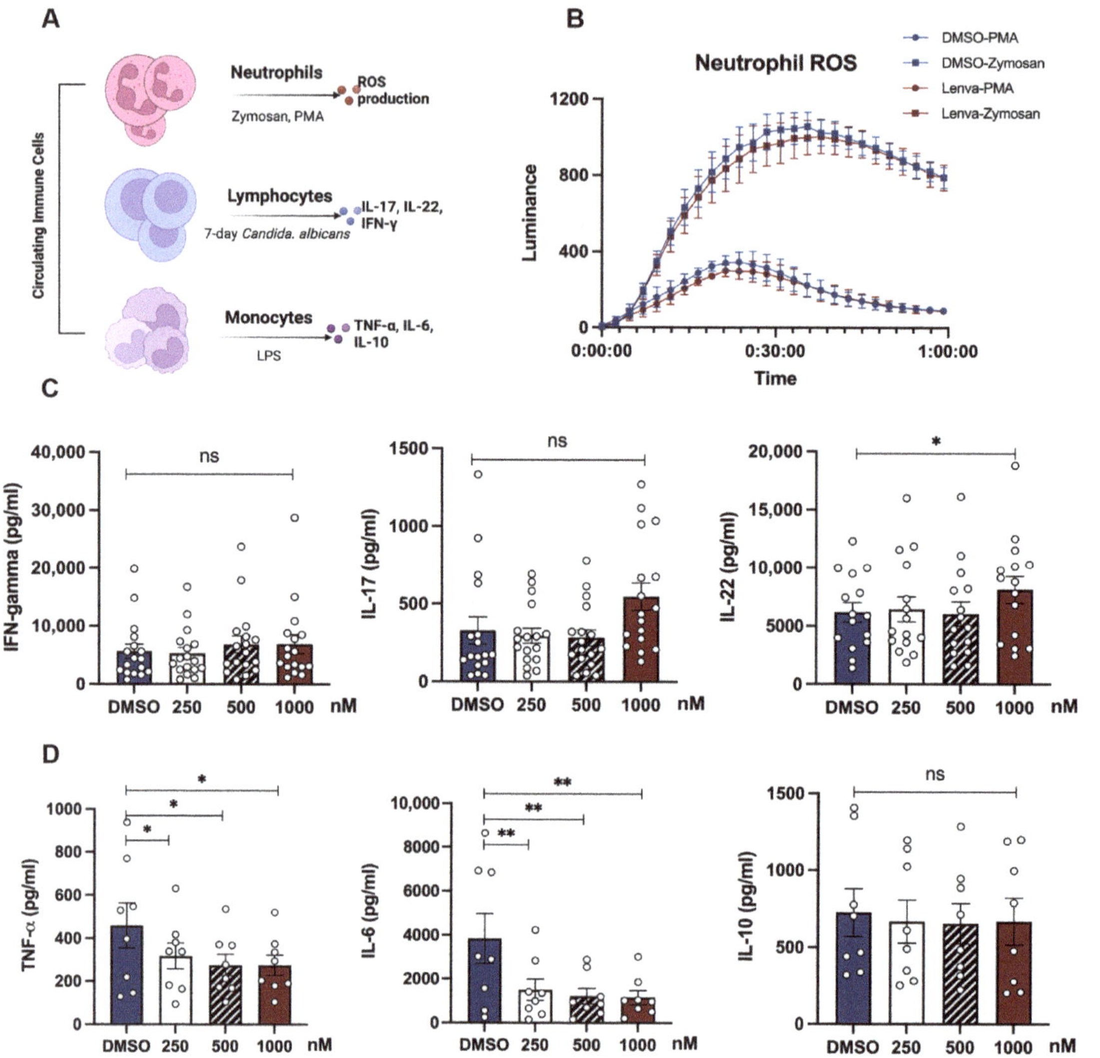

Figure 1. Lenvatinib modulates in vitro monocyte-derived cytokine secretion in dose–response experiments. (**A**) Schematic diagram of the dose–response test. Freshly isolated neutrophils were stimulated with opsonized zymosan or PMA for the measurement of ROS production, after brief exposure to Lenvatinib (1 μM) for 1 h. PBMCs from healthy donors were incubated with Lenvatinib with indicated concentrations for 24 h, after which cells were stimulated with LPS or Candida albicans. Cytokines levels in supernatants were determined by ELISA. (**B**) ROS production from neutrophils; data represent mean ± SME at each time point, n = 6. (**C**) Lymphocyte-derived cytokines (IFN-γ, IL-17, and IL-22) after 7-day Candida albicans stimulation, n = 15. (**D**) Monocyte-derived cytokines (IL-6, TNF-α, IL-10) after 24 h of LPS stimulation, n = 8. The bar graphs indicate mean ± SME; each dot represents one healthy donor. * $p < 0.05$, ** $p < 0.01$ by nonparametric Wilcoxon matched-pair tests; ns, not significant; PMA, phorbol myristate acetate; ROS, reactive oxygen species.

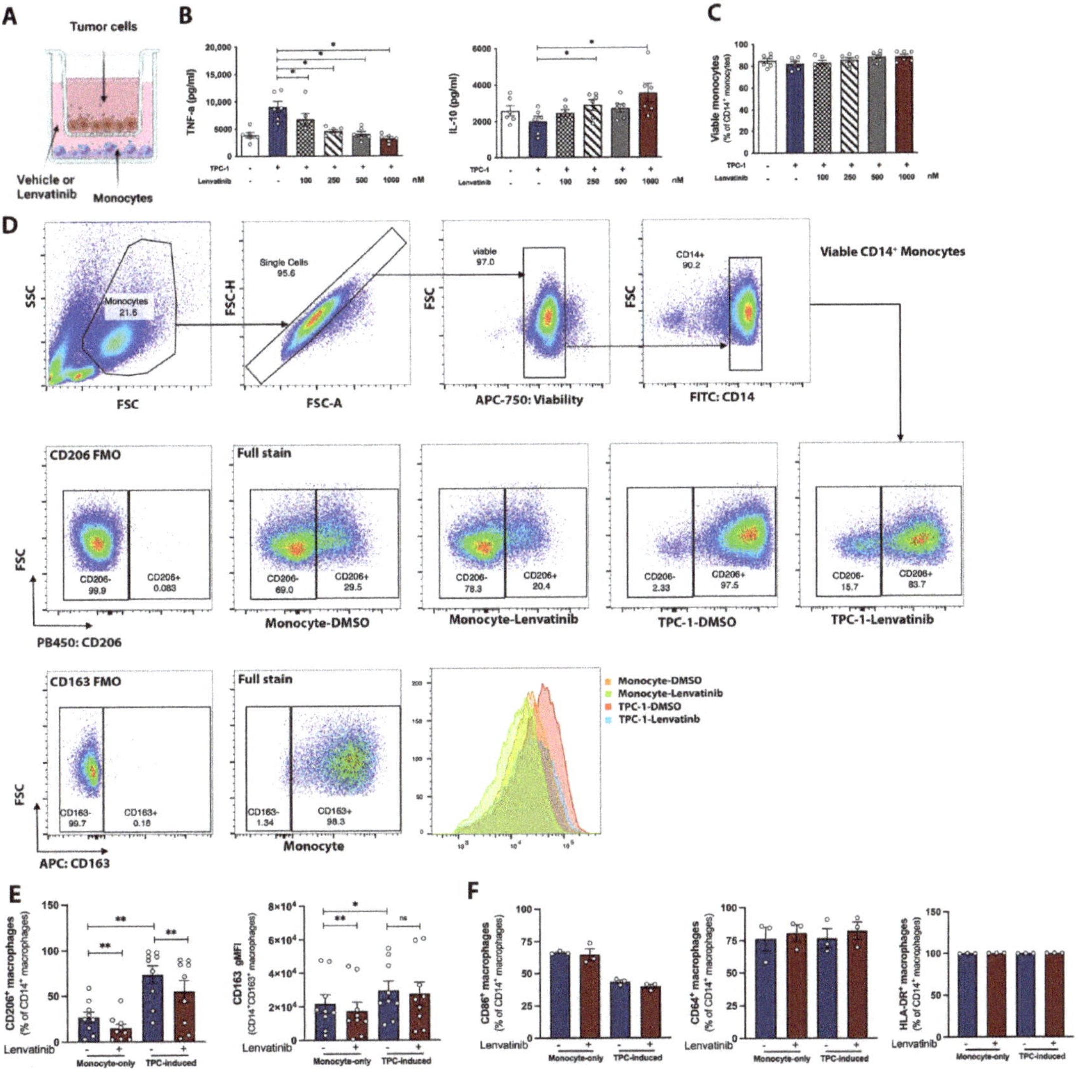

Figure 2. Lenvatinib interferes with monocytes' functional rewiring in co-culture experiments. (**A**) Schematic diagram of the co-culture experiments. (**B**) Cytokine secretion from TPC-1 induced monocytes treated with indicated concentrations of Lenvatinib; monocytes cultured in RPMI were included as the baseline control, n = 6. (**C**) Viability of monocytes after 24 h exposure to Lenvatinib; percentage of viable $CD14^+$ cells were gated for statistical analysis, n = 6. (**D**) Expression of M2 markers in TPC-1-induced monocytes. The upper panel shows the gating strategy to identify the viable $CD14^+$ monocytes. Identification of $CD206^+$ and $CD163^+$ subsets within viable $CD14^+$ monocytes was based on FMO controls. The lower panel shows the representative dot plot and histogram from one donor. (**E**) Quantification of the percentage of $CD206^+$ subsets within $CD14^+$ monocytes and fluorescence intensity of CD163 signal on $CD14^+CD163^+$ monocytes, n = 6. (**F**) Expression of M1 markers (CD86, CD64, and HLA-DR), n = 3. The bar graphs indicate mean ± SME; each dot represents one healthy donor. * $p < 0.05$, ** $p < 0.01$ by nonparametric Wilcoxon matched-pair tests. FMO, fluorescence minus one control; gMFI, geometric mean fluorescence intensity.

The fact that Lenvatinib concomitantly affected monocyte function raised the question regarding to what extent the alteration in TPC-1 secretome could explain the immunological effects we observed. To address this, we tested two additional TC cell lines, FTC-133 and BCPAP, in the co-culture setting. Previous reports have shown that the growth rate of TPC-1, which bears a Lenvatinib-targeted RTK mutation-RET rearrangement, is strongly inhibited by Lenvatinib, whereas FTC-133 and BCPAP, negative for Lenvatinib-targeted mutations, are not sensitive to the drug treatment [30]. Strikingly, similar changes in cytokine production caused by Lenvatinib were observed across all three co-cultures (Figure 3A). On the other hand, a TC-pre-treatment group was included as the control, where only TPC-1 cells had been pre-exposed to Lenvatinib for 6 h before starting the co-culture. Interestingly, incubation with Lenvatinib-pre-treated TC cells resulted in a similar (albeit less strong) effect on cytokine production, suggesting that Lenvatinib partially abolished the pro-inflammatory and tumor-promoting phenotype induced by TPC-1 (Figure 3B). These data demonstrated that the tumor cell-mediated mechanism alone was insufficient to explain the immunomodulatory nature of Lenvatinib in the co-culture experiments, arguing for the direct immunological effects of Lenvatinib on immune cells. These results were further supported by the similar results from GM-CSF- or M-CSF-primed monocytes in the absence of tumor cells (Figure 3C), which is suggestive of a more critical role of the monocyte-intrinsic mechanism underlying this effect.

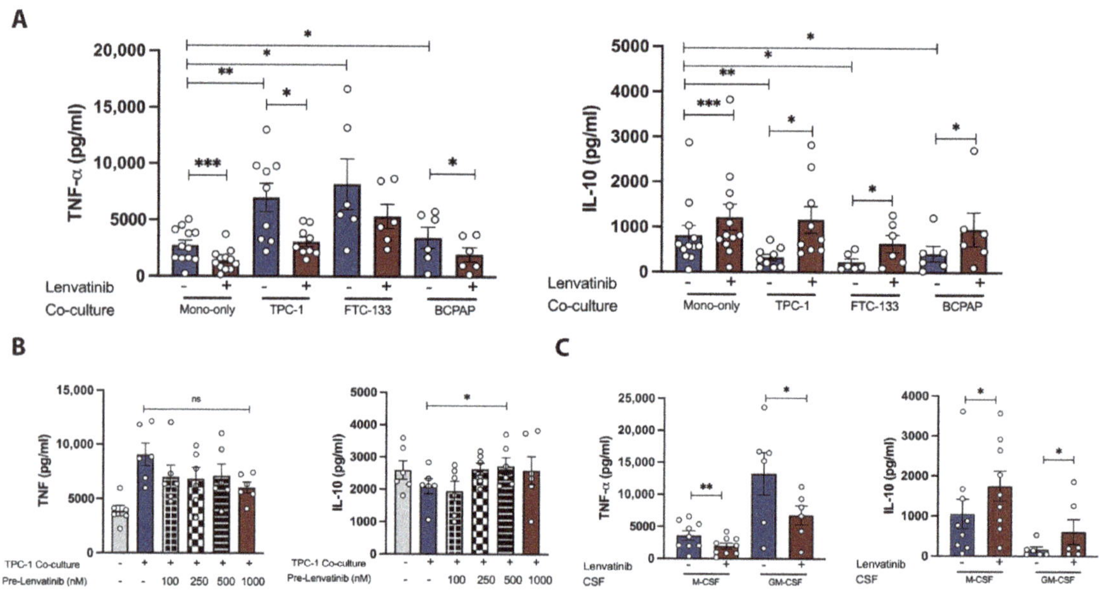

Figure 3. Similar patterns of cytokine production caused by Lenvatinib in different experimental settings. (**A**) In vitro generation of tumor-associated macrophages with 3 TC cell lines. (**B**) Co-culture experiments with TPC-1 cell line pretreated by Lenvatinib for 6 h. (**C**) Monocytes primed by GM-CSF and M-CSF for 24 h, n = 6. The bar graphs indicate mean ± SME; each dot represents one healthy donor. * $p < 0.05$, ** $p < 0.01$, *** $p < 0.001$ by nonparametric Wilcoxon matched-pair tests.

3.3. Prolonged Exposure to Lenvatinib Affects Macrophage Survival and Differentiation

We next investigated whether Lenvatinib could affect the survival and phenotypic differentiation of macrophages. To address this, monocytes were cultured with either GM-CSF (M1 differentiation) or M-CSF (M2 differentiation) for 6 days in the presence of Lenvatinib or a DMSO control. We found that long-term exposure to Lenvatinib impaired macrophage survival, with M-CSF-induced macrophages being more vulnerable to

Lenvatinib, resulting in approximately half of the cells being lost during the differentiation process (Figure 4A). More importantly, we observed drastic morphological changes in Lenvatinib-treated macrophages. As indicated in the flowcytometry FSC channel, the cells were significantly smaller (Figure 4B), suggesting the potential impairment of macrophage differentiation.

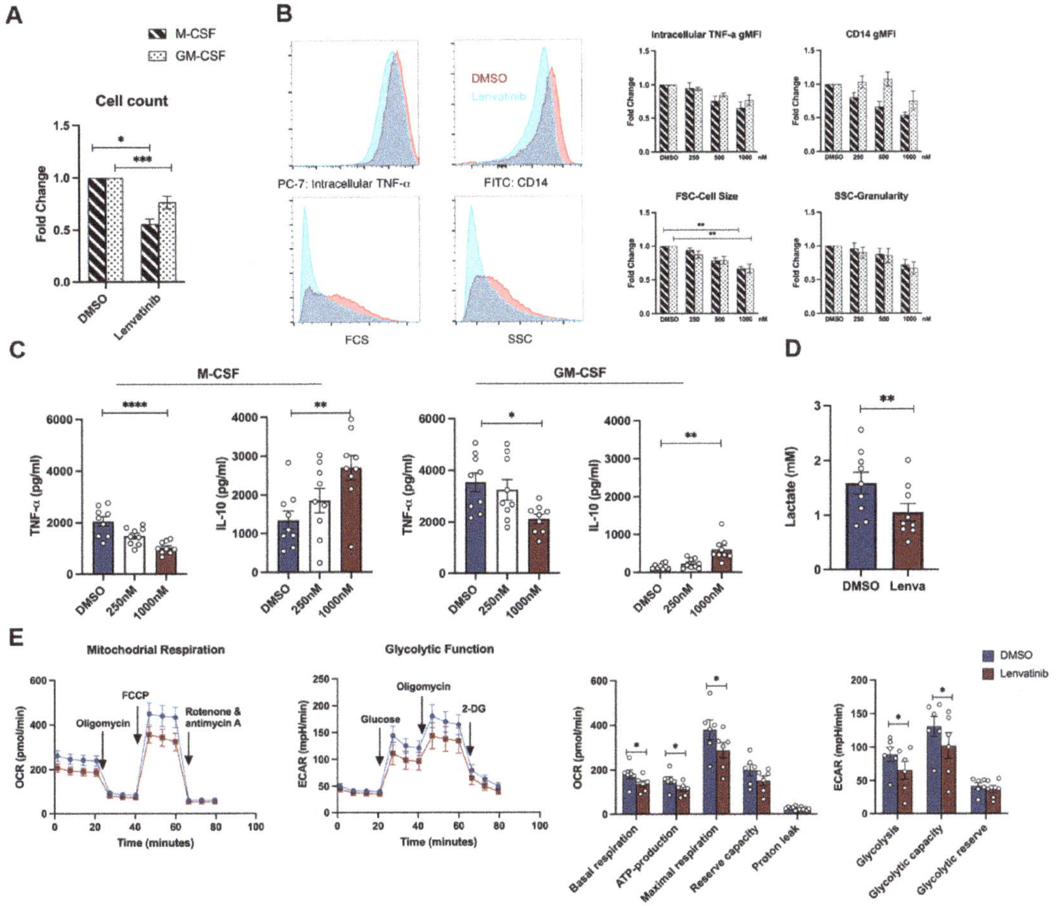

Figure 4. Morphological, functional, and metabolic changes in Lenvatinib-treated macrophages. (**A**) Overall macrophage survival determined by flowcytometry; the cell counts of macrophages were gated on FCS-A and SSC-A (n = 11). (**B**) Left panel indicates representative histogram of the parameters, including intracellular TNF, CD14 expression, cell size (FSC), and granularity (SSC), analyzed by flowcytometry from one donor. The right panel shows the corresponding quantification of gMFI, n = 6. (**C**) Cytokine production from GM-CSF/M-CSF primed macrophages, n = 9. (**D**) Lactate production from GM-CSF-induced macrophages after 24 h of stimulation with LPS, n = 6. (**E**) Analysis of ECAR and OCR from GM-CSF-polarized macrophages was performed with a seahorse XF analyzer. The left figure shows the OCR and ECAR profiles for macrophages exposed to Lenvatinib or DMSO vehicle control. The right figure shows mitochondrial bioenergetic and glycolytic parameters calculated from OCR and ECAR (n = 6). The bar graphs indicate mean ± SME; each dot represents one healthy donor. * $p < 0.05$, ** $p < 0.01$, *** $p < 0.001$, **** $p < 0.0001$ by nonparametric Wilcoxon matched-pair tests. FSC, forward scatter; ECAR, extracellular acidification rate; OCR, oxygen consumption rate; SSC, side scatter.

Cytokine production was determined after the adjustment of cell concentration, despite a prototypical anti- and pro-inflammatory cytokine production pattern upon LPS stimulation from GM-CSF/M-CSF primed macrophages. The modulation of TNF-a and IL-10 production induced by Lenvatinib was similar in both groups (Figure 4C). This was further supported by intracellular TNF-α level measurement (Figure 4B), recapitulating our observations from monocytes.

Metabolic reprogramming shapes a distinct immune response to stimulus [31]. To explore the cellular metabolic mechanisms that might contribute to the cytokine production pattern, we quantified the indicators of mitochondrial respiration and glycolytic capacity by oxygen consumption rate (OCR) and extracellular acidification rate (ECAR) using seahorse. Lenvatinib-treated macrophages displayed a trend towards suppressed metabolic activities, which is shown as lower basal and maximal respiration, lower glycolysis activity and glycolytic capacity, while the glycolytic reserve remains unchanged (Figure 4E). Consistent with these findings, the Lenvatinib-treated group produced lower amounts of lactate upon 24 h of LPS challenge (Figure 4D), which was indicative of general defects in both oxidative phosphorylation and glycolytic activities [32].

Taken together, these data demonstrate that Lenvatinib exerts multifaceted inhibitory effects on macrophage survival and maturation.

3.4. The Immunological Effects on Monocytes Are Independent of the Lenvatinib-Targeted Receptor Tyrosine Kinase (RTK) Pathways

To explore the underlying mechanism of Lenvatinib-mediated immunological effects, we evaluated the expression of Lenvatinib-targeted RTKs on myeloid cells. Mining into the publicly available transcriptomic database Human Protein Atlas [33], we found that most of the Lenvatinib targets were not detected in circulating myeloid cells, except VEGFR1 and FGFR1, with a relatively low level of mRNA expression in classical monocytes, shown as normalized transcript per million (nTPM) (Figure 5A). To further validate the expression of VEGFR1 and FGFR1, we performed RT-qPCR and Western blotting using samples harvested from buffy coat monocytes and 6-day differentiated macrophages. Despite comparatively consistent mRNA expression in both monocytes and macrophages (Figure 5B), the Western blots did not show any detectable VEGFR1 and FGFR1 expression on monocytes (Figure 5C). Therefore, it is hard to explain the Lenvatinib-mediated mediated immunomodulation by its on-target activity, pointing instead to potential off-target mechanisms.

To elucidate the potential alternative mechanisms of the action of Lenvatinib, we explored the ProteomeXchange database (dataset PXD005336), a public resource containing the target landscape of 243 clinically tested kinase drugs. The data were obtained via systemic kinase assays using four cancer cell lines [34]. Surprisingly, we found that Lenvatinib interacts with multiple unanticipated kinases (Supplementary Figure S2). Among all the additional targets, several mitogen-activated protein kinases (MAPKs) were of particular interest. The sequential induction of pro- and anti-inflammatory cytokine by LPS requires fine-tuned control via MAPKs, consisting of extracellular signal-regulated kinases1/2 (ERK1/2), JUN N-terminal kinase (JNK), and p38 MAPKs [35–37]. In addition to the MAPK family, the tyrosine kinase Syk regulates LPS-mediated TLR4 endocytosis [38]. If the TLR4 downstream kinases are responsible for Lenvatinib-induced immunomodulation, then the blockage of a certain kinase pathway with highly selective inhibitors could potentially abolish Lenvatinib's influence on monocytes. To test this hypothesis, we combined Lenvatinib with different inhibitors against p38 MAPK, ERK1/2, JNK, and Syk, respectively [39]. As expected, the inhibition of p38 MAPK, ERK1/2, and Syk alone was able to attenuate TNF-α release. More interestingly, the reduction in TNF-α production caused by Lenvatinib was completely abrogated by concomitant p38 MAPK inhibition, whereas Lenvatinib's effect persisted in the presence of the ERK1/2, JNK, and Syk inhibitors (Figure 6A–D). Despite this, p38 MAPK blockage alone was not able to recapitulate the IL-10 response induced by Lenvatinib. Strikingly, the Syk inhibitor, R406, resulted in enhanced IL-10 production by itself, and the dual treatment of R406 plus Lenvatinib did not further increase the IL-10 level [40]. Furthermore, we observed similar effects in GM-CSF/M-CSF primed monocytes (Supplementary Figure S3). Therefore, Lenvatinib's ability to block p38 MAPK and Syk appears to be the possible mechanism underlying its opposing effect on TNF-α and IL-10 production.

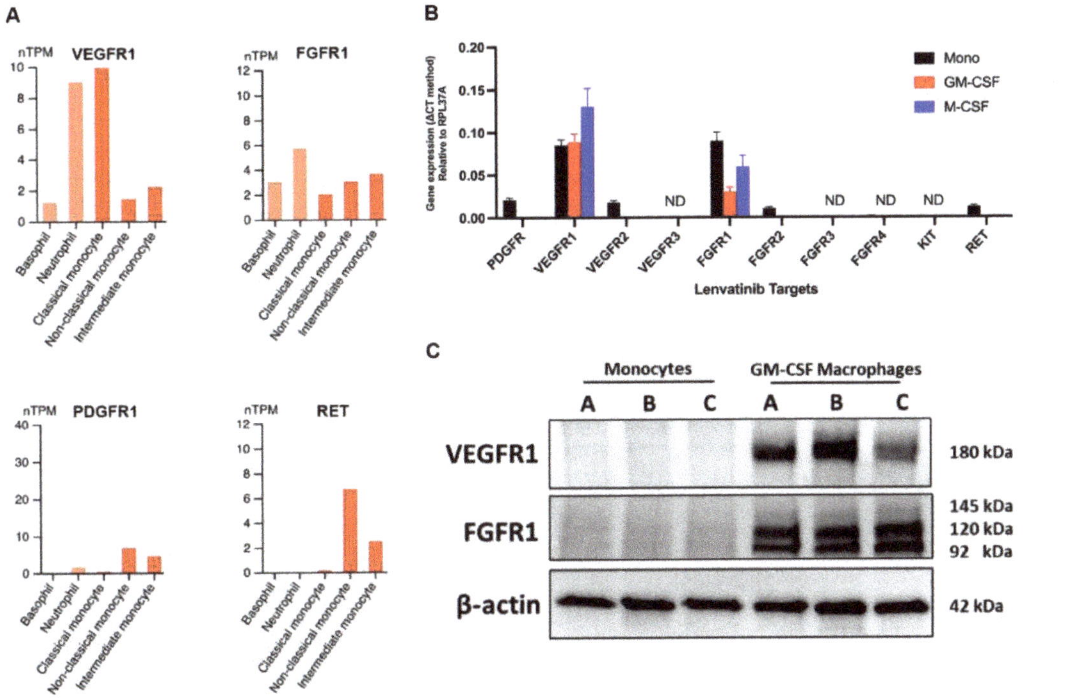

Figure 5. The expression of Lenvatinib-targeted RTKs on monocytes. (**A**) Normalized transcript expression levels of 4 detectable Lenvatinib-targeted RTKs (VEGFR1, FGFR1, PDGFR, RET) from the transcriptome signatures are visualized for different peripheral myeloid cell types from Monaco et al. The panels are screenshots from the Human Protein Atlas resource. (**B**) Quantified Lenvatinib-targeted RTKs mRNA expression by RT-qPCR. Transcript copy number was normalized to cellular control (RPL37A). n = 6. (**C**) Levels of VEGFR1 and FGFR1 in the cell lysate from buffy coat monocytes and GM-CSF polarized macrophages were analyzed by Western blot, and β-actin was used as the loading control. Three isoforms of FGFR1 were detected with molecular weights of 92, 120, and 145 kDa. n = 3. ND, not determined; nTPM, normalized transcripts per million.

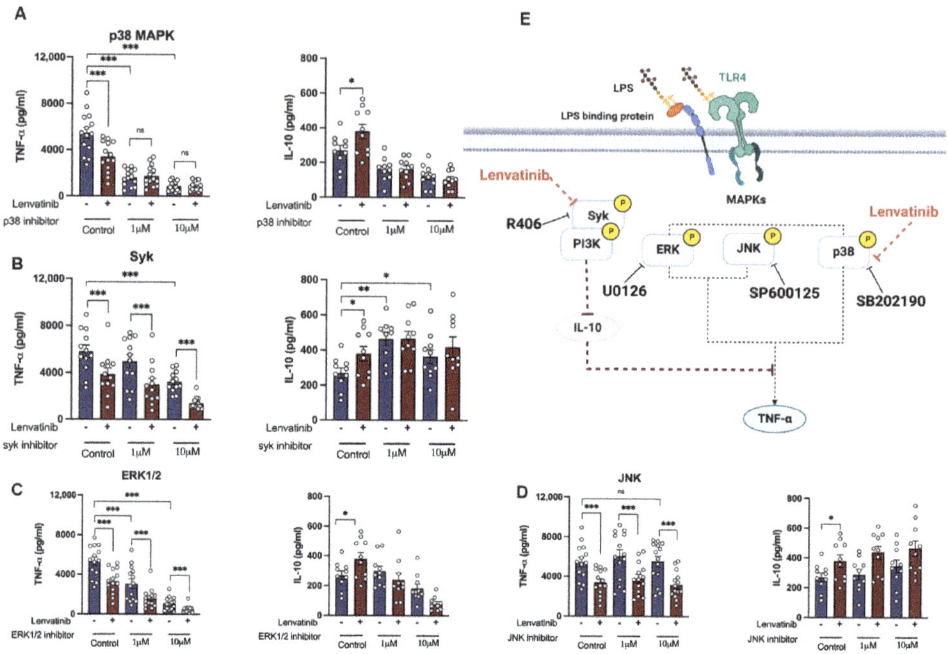

Figure 6. Effects of TLR4 downstream kinase inhibitors on Lenvatinib-mediated immunomodulation. (**A–D**) Monocytes were treated with selective inhibitors against kinases within the TLR4 signaling networks: SB 201019 as p38 MAPK inhibitor, R406 as Syk inhibitor, U0126 as ERK1/2 inhibitor, SP 600125 as JNK inhibitor. All the inhibitors were tested at two concentrations (1 μM and 10 μM), either used alone or in combination with Lenvatinib. Cytokine response was determined by ELISA after 24 h LPS stimulation. Kinase inhibitor single agent versus DMSO control was compared, and kinase inhibitor plus Lenvatinib versus kinase inhibitor alone was compared, n = 10 to 14. (**E**) Schematic diagram of the possible mechanism behind Lenvatinib's off-target effect on monocytes via dual inhibition of p38 and Syk pathways. The bar graphs indicate mean ± SME; each dot represents one healthy donor. * $p < 0.05$, ** $p < 0.01$, *** $p < 0.001$ by nonparametric Wilcoxon matched-pair tests. ERK, extracellular signal-regulated kinases; JNK, c-Jun N-terminal kinases; ns, not significant; p38 MAPK, p38 mitogen-activated protein kinases; Syk, spleen tyrosine kinase; TLR4, toll-like receptor 4.

4. Discussion

In the present study, we show that circulating monocytes are particularly vulnerable to ex vivo Lenvatinib treatment at the concentrations equivalent to pharmacological concentrations achievable during treatment, with the function of neutrophils and lymphocytes being less affected. Specifically, the induction of pro-inflammatory monocytic cytokine response was largely attenuated in the presence of Lenvatinib. Furthermore, the drug also interferes with the proinflammatory potential of TC-induced TAMs. Critically, these effects could not be explained by the drug impact on the TPC-1 tumor cells, but rather through direct effects on immune cells. The direct drug impact on monocytes is further supported by in vitro experiments of macrophage differentiation, showing effects of Lenvatinib on cell survival, morphology, cytokine response, and metabolism. Strikingly, we observed no detectable expression in the monocytes of Lenvatinib's classical targets. On the other hand, blockage of the p38 MAPK and Syk pathways partially diminished Lenvatinib's immunological effects. Thus, we propose that Lenvatinib directly impacts the functional status of human monocytes via non-canonical mechanisms.

Due to the importance of tumor–immune cell interaction, we initially hypothesized that Lenvatinib might interfere with TAMs-induction by an indirect tumor-instructed mechanism. Previous reports using murine models have shown that TKIs prompt tumor cells to release chemotactic factors (CXCL12 and HMGB-1) or induce immunogenic cell death, resulting in robust immune cell infiltration and tumor clearance [16,41]. Contrary to this, in our co-culture experiments, we did not observe significant differences between the settings with or without the presence of tumor cells. One of the possible explanations is that the tumor-killing capacity of Lenvatinib in vivo relies more on its anti-angiogenetic effects, which require the presence of TME. Supporting this, clinical trials revealed that tumor response to Lenvatinib was not correlated with targeted RTK mutation status [7,42]. Therefore, we were unable to fully recapitulate the tumor response to Lenvatinib with simple two-factor co-culture experiments. Future in vivo studies will be required to elucidate the possible tumor-mediated mechanism.

The direct influence of Lenvatinib on monocytes was interesting and unexpected. Recent advances in drug–target interaction led to the realization that the majority of previously established TKIs bind to additional kinases, including serine-threonine kinases that are not closely related to the function and sequence of the intended targets [34,43]. However, their immunological implications remain largely unexplored, and preclinical results reflect that the immunologic effects of individual TKIs are largely drug-specific and context-dependent [19,44,45].

Lenvatinib is a multitargeted kinase inhibitor with a particularly potent activity against the VEGFR2 via type V binding model [46]. The side effects of Lenvatinib, such as hypertension, are predominantly caused by its anti-angiogenetic activity [4], while no major immune-related side effects have been reported. It might be reasonable to postulate that Lenvatinib-mediated immunological effects are well tolerated by patients, yet they have been largely overlooked since the early stage of drug development [2]. No data on intratumoral Lenvatinib concentrations have been reported; thus, the working concentration in this study was based on plasma pharmacokinetic data. We selected 1 μM as the highest non-cytotoxic concentration and observed that monocyte activation status was significantly influenced at this concentration. In line with this, a recent study also demonstrated that, compared to other immune cells, monocytes were less resistant to in vitro TKIs treatment [47]. Similarly, a clinical observational study also reported decreased levels of circulating TNF-α and IL-6 in hepatocellular carcinoma (HCC) patients after Lenvatinib treatment [48].

As for the mode of action, it appears plausible that Lenvatinib might exert inhibitory activity through p38 MPAK and Syk kinase in monocytes. We found that the inhibition of p38 by SB202190 completely abolished the effect of Lenvatinib on TNF-α production. In line with our findings, it has been previously described that, in tumor cells, Lenvatinib directly binds with several components within the MAPK signaling pathway, including MAP4K2, MAP4K5, MAPK14, and MAPKAPK2 [49,50]. p38 MAPK plays a pivotal role in governing macrophage alternative activation and preventing apoptosis [51,52]. Indeed, we observed the selective downregulation of M2 cell surface markers on TC-induced monocytes and significant cell loss caused by Lenvatinib during macrophage differentiation. Additionally, a recent study investigating the immunological effects of Regorafenib, another TKI used for HCC, revealed that its synergizing efficacy with anti-PD1 therapy was associated with the suppression of p38 kinase phosphorylation, which reversed the M2-like polarization of murine bone marrow-derived macrophages (BMDMs) [45], thus supporting the off-target effects via p38MAPK blockage seen in our study.

A striking aspect of the Lenvatinib-treated monocyte is the persistent elevation of IL-10 production across all the experimental settings. The high production of IL-10 and TGF-β is a hallmark of TAMs [53]. However, seemingly counterintuitively, we did not observe the concomitant upregulation of other immunosuppressive markers in Lenvatinib-treated TC-induced monocytes, such as CD163 and CD206. The elevation of IL-10 seemed to be independent of the general immunosuppressive traits. As part of the host defense mechanism, the induction of IL-10 occurs simultaneously with proinflammatory cytokines via shared p38 MAPK and ERK pathways [54]. The opposing effects on IL-10 and TNF-a production indicate additional pathways influenced by Lenvatinib treatment.

Syk is a non-receptor tyrosine kinase transmitting signals of immunoreceptors, regulating the activation of innate immunity [55,56]. We found that Syk inhibitor, R406, was the only agent that could enhance IL-10 production upon LPS stimulation, and no synergizing effect with Lenvatinib was observed. It is tempting to speculate that the Syk pathway is also involved in Lenvatinib's off-target effects. Consistent with our findings, others have shown that R406 selectively blocks the induction of IL-6 and TNF-a, while restoring the IL-10 production in human monocytes [40]. However, this is at odds with previous reports showing that Syk functions as a negative-feedback mechanism to limit inflammation by promoting IL-10 production [56]. Nevertheless, the Syk inhibitor we used, R406, was proposed to be potent and selective against Syk, but it also binds to several other kinases, including Salt-inducible kinase 2 (SIK2) [34]. SIK2 is one of the most common additional targets of TKIs, as revealed by large-scale drug screening [34]. Due to its restriction on the formation of regulatory macrophages, the inhibition of SIK2 stimulates the production of IL-10 [57]. Importantly, drugs with an SIK2 affinity of greater than 500nM are capable of suppressing TNF-a and upregulating IL-10 in BMDM, recapitulating the cytokine response to Lenvatinib [34]. However, due to the complex crosstalk among signaling pathways and the model we used, it is not possible to make a definitive statement about the off-target effect of Lenvatinib on the p38 MAPK, Syk, or SIK2 pathways. Future work will be required to define the comprehensive drug-kinase activity in immune cells using high-throughput profiling techniques [58].

Clinical evidence highlights the beneficial role of Lenvatinib in combination with immunotherapy in metastatic renal cell carcinoma and endometrial cancer, mirrored by an increased objective response rate and prolonged progression-free survival [59,60]. Recently, clinical trials investigating the safety and efficacy of Lenvatinib plus Pembrolizumab (anti-PD-1 antibody) in advanced DTC (NCT02973997) and HCC (NCT03713593) have also been initiated. However, combination regimens generally lead to more severe side-effects than monotherapies. Therefore, attempts to incorporate TKIs into immunotherapy have been largely limited by an increased incidence of high-grade toxicities [61,62]. Thus, a deeper understanding of the neglected immune-cell intrinsic off-target effect is of great clinical significance. Our study provided preclinical evidence that Lenvatinib has a direct impact on the myeloid compartment of the immune system. Recent studies have demonstrated the deleterious role of inflammation in cancer survival, with proinflammatory mediators playing an important role in angiogenesis, as well as acting as survival factors for cancer cells [63]. Our data, demonstrating the anti-inflammatory effects of Lenvatinib, argue that this effect is likely to contribute to the therapeutic effect of this class of drugs, extending their potential therapeutic applicability in tumors characterized by a prominent inflammatory profile. Future work should characterize the immune landscape in patients receiving Lenvatinib and identify predictive biomarkers for sensitivity to immunotherapies.

Supplementary Materials: The following supporting information can be downloaded at: https://www.mdpi.com/article/10.3390/pharmaceutics15020412/s1, Figure S1: T cell proliferation in the presence of Lenvatinib; Figure S2: Target landscape of Lenvatinib in 4 cell lines; Figure S3. The impact of TLR4 downstream kinase inhibitors on Lenvatinib-mediated immunomodulation in GM-CSF/M-CSF-primed monocytes.

Author Contributions: Conceptualization, R.T.N.-M. and M.G.N.; methodology, C.P., K.R., and M.J.; validation, C.P.; formal analysis, C.P.; investigation, C.P., M.J., and K.R.; resources, R.T.N.-M., M.J., and M.G.N.; writing—original draft preparation, C.P.; writing—review and editing, R.T.N.-M., M.J., and M.G.N.; visualization, C.P.; supervision, R.T.N.-M.; project administration, R.T.N.-M.; funding acquisition, R.T.N.-M. All authors have read and agreed to the published version of the manuscript.

Funding: This investigator-initiated research was funded by an unrestricted research grant from EISAI Co. C.P. was supported by the China Scholarship Council, grant number 202006100049.

Institutional Review Board Statement: Not applicable.

Informed Consent Statement: Informed consent was obtained from all subjects involved in the study.

Data Availability Statement: The data presented in this study are available on request from the corresponding author. The data are not publicly available due to privacy issues.

Acknowledgments: The authors would like to thank Julia Brake for technical support in the ex vivo neutrophil experiments.

Conflicts of Interest: The authors declare no conflict of interest.

References

1. Cabanillas, M.E.; McFadden, D.G.; Durante, C. Thyroid cancer. *Lancet* **2016**, *388*, 2783–2795. [CrossRef] [PubMed]
2. Locati, L.D.; Piovesan, A.; Durante, C.; Bregni, M.; Castagna, M.G.; Zovato, S.; Giusti, M.; Ibrahim, T.; Puxeddu, E.; Fedele, G.; et al. Real-world efficacy and safety of lenvatinib: Data from a compassionate use in the treatment of radioactive iodine-refractory differentiated thyroid cancer patients in Italy. *Eur. J. Cancer* **2019**, *118*, 35–40. [CrossRef] [PubMed]
3. Schlumberger, M.; Tahara, M.; Wirth, L.J.; Robinson, B.; Brose, M.S.; Elisei, R.; Habra, M.A.; Newbold, K.; Shah, M.H.; Hoff, A.O.; et al. Lenvatinib versus Placebo in Radioiodine-Refractory Thyroid Cancer. *N. Engl. J. Med.* **2015**, *372*, 621–630. [CrossRef] [PubMed]
4. Capozzi, M.; De Divitiis, C.; Ottaiano, A.; von Arx, C.; Scala, S.; Tatangelo, F.; Delrio, P.; Tafuto, S. Lenvatinib, a molecule with versatile application: From preclinical evidence to future development in anti-cancer treatment. *Cancer Manag. Res.* **2019**, *11*, 3847–3860. [CrossRef]
5. Brose, M.S.; Robinson, B.; Sherman, S.I.; Krajewska, J.; Lin, C.-C.; Vaisman, F.; Hoff, A.O.; Hitre, E.; Bowles, D.W.; Hernando, J.; et al. Cabozantinib for radioiodine-refractory differentiated thyroid cancer (COSMIC-311): A randomised, double-blind, placebo-controlled, phase 3 trial. *Lancet Oncol.* **2021**, *22*, 1126–1138. [CrossRef]
6. Sharma, P.; Allison, J.P. The future of immune checkpoint therapy. *Science* **2015**, *348*, 56–61. [CrossRef]
7. French, J.D.; Bible, K.; Spitzweg, C.; Haugen, B.R.; Ryder, M. Leveraging the immune system to treat advanced thyroid cancers. *Lancet Diabetes Endocrinol.* **2017**, *5*, 469–481. [CrossRef]
8. O'Donnell, J.S.; Teng, M.W.L.; Smyth, M.J. Cancer immunoediting and resistance to T cell-based immunotherapy. *Nat. Rev. Clin. Oncol.* **2019**, *16*, 151–167. [CrossRef]
9. Cassetta, L.; Fragkogianni, S.; Sims, A.H.; Swierczak, A.; Forrester, L.M.; Zhang, H.; Soong, D.Y.H.; Cotechini, T.; Anur, P.; Lin, E.Y.; et al. Human Tumor-Associated Macrophage and Monocyte Transcriptional Landscapes Reveal Cancer-Specific Reprogramming, Biomarkers, and Therapeutic Targets. *Cancer Cell* **2019**, *35*, 588–602.e510. [CrossRef]
10. DeNardo, D.G.; Ruffell, B. Macrophages as regulators of tumour immunity and immunotherapy. *Nat. Rev. Immunol.* **2019**, *19*, 369–382. [CrossRef]
11. Ryder, M.; Ghossein, R.A.; Ricarte-Filho, J.C.M.; Knauf, J.A.; Fagin, J.A. Increased density of tumor-associated macrophages is associated with decreased survival in advanced thyroid cancer. *Endocr.-Relat. Cancer* **2008**, *15*, 1069–1074. [CrossRef]
12. Qing, W.; Fang, W.Y.; Ye, L.; Shen, L.Y.; Zhang, X.F.; Fei, X.C.; Chen, X.; Wang, W.Q.; Li, X.Y.; Xiao, J.C.; et al. Density of tumor-associated macrophages correlates with lymph node metastasis in papillary thyroid carcinoma. *Thyroid* **2012**, *22*, 905–910. [CrossRef]
13. Weber, R.; Fleming, V.; Hu, X.Y.; Nagibin, V.; Groth, C.; Altevogt, P.; Utikal, J.; Umansky, V. Myeloid-Derived Suppressor Cells Hinder the Anti-Cancer Activity of Immune Checkpoint Inhibitors. *Front. Immunol.* **2018**, *9*, 9. [CrossRef]
14. Hage, C.; Hoves, S.; Strauss, L.; Bissinger, S.; Prinz, Y.; Pöschinger, T.; Kiessling, F.; Ries, C.H. Sorafenib Induces Pyroptosis in Macrophages and Triggers Natural Killer Cell–Mediated Cytotoxicity Against Hepatocellular Carcinoma. *Hepatology* **2019**, *70*, 1280–1297. [CrossRef]
15. Petroni, G.; Buque, A.; Zitvogel, L.; Kroemer, G.; Galluzzi, L. Immunomodulation by targeted anticancer agents. *Cancer Cell* **2021**, *39*, 310–345. [CrossRef]
16. Liu, P.; Zhao, L.; Pol, J.; Levesque, S.; Petrazzuolo, A.; Pfirschke, C.; Engblom, C.; Rickelt, S.; Yamazaki, T.; Iribarren, K.; et al. Crizotinib-induced immunogenic cell death in non-small cell lung cancer. *Nat. Commun.* **2019**, *10*, 1486. [CrossRef]
17. Yi, C.; Chen, L.; Lin, Z.; Liu, L.; Shao, W.; Zhang, R.; Lin, J.; Zhang, J.; Zhu, W.; Jia, H.; et al. Lenvatinib Targets FGF Receptor 4 to Enhance Antitumor Immune Response of Anti–Programmed Cell Death-1 in HCC. *Hepatology* **2021**, *74*, 2544–2560. [CrossRef]
18. Peng, C.; Rabold, K.; Mulder, W.J.M.; Jaeger, M.; Netea-Maier, R.T. Kinase Inhibitors' Effects on Innate Immunity in Solid Cancers. *Cancers* **2021**, *13*, 5695. [CrossRef]
19. Zitvogel, L.; Rusakiewicz, S.; Routy, B.; Ayyoub, M.; Kroemer, G. Immunological off-target effects of imatinib. *Nat. Rev. Clin. Oncol.* **2016**, *13*, 431–446. [CrossRef]
20. Torrens, L.; Montironi, C.; Puigvehí, M.; Mesropian, A.; Leslie, J.; Haber, P.K.; Maeda, M.; Balaseviciute, U.; Willoughby, C.E.; Abril-Fornaguera, J.; et al. Immunomodulatory Effects of Lenvatinib Plus Anti–Programmed Cell Death Protein 1 in Mice and Rationale for Patient Enrichment in Hepatocellular Carcinoma. *Hepatology* **2021**, *74*, 2652–2669. [CrossRef]

21. Kimura, T.; Kato, Y.; Ozawa, Y.; Kodama, K.; Ito, J.; Ichikawa, K.; Yamada, K.; Hori, Y.; Tabata, K.; Takase, K.; et al. Immunomodulatory activity of lenvatinib contributes to antitumor activity in the Hepa1-6 hepatocellular carcinoma model. *Cancer Sci.* **2018**, *109*, 3993–4002. [CrossRef] [PubMed]
22. Kato, Y.; Tabata, K.; Kimura, T.; Yachie-Kinoshita, A.; Ozawa, Y.; Yamada, K.; Ito, J.; Tachino, S.; Hori, Y.; Matsuki, M.; et al. Lenvatinib plus anti-PD-1 antibody combination treatment activates CD8+ T cells through reduction of tumor-associated macrophage and activation of the interferon pathway. *PLoS ONE* **2019**, *14*, e0212513. [CrossRef] [PubMed]
23. Repnik, U.; Knezevic, M.; Jeras, M. Simple and cost-effective isolation of monocytes from buffy coats. *J. Immunol. Methods* **2003**, *278*, 283–292. [CrossRef] [PubMed]
24. Kuhns, D.B.; Priel, D.A.L.; Chu, J.; Zarember, K.A. Isolation and Functional Analysis of Human Neutrophils. *Curr. Protoc. Immunol.* **2015**, *111*, 7.23.21–27.23.16. [CrossRef]
25. Ikeda, K.; Kudo, M.; Kawazoe, S.; Osaki, Y.; Ikeda, M.; Okusaka, T.; Tamai, T.; Suzuki, T.; Hisai, T.; Hayato, S.; et al. Phase 2 study of lenvatinib in patients with advanced hepatocellular carcinoma. *J. Gastroenterol.* **2017**, *52*, 512–519. [CrossRef]
26. Boss, D.S.; Glen, H.; Beijnen, J.H.; Keesen, M.; Morrison, R.; Tait, B.; Copalu, W.; Mazur, A.; Wanders, J.; O'Brien, J.P.; et al. A phase I study of E7080, a multitargeted tyrosine kinase inhibitor, in patients with advanced solid tumours. *Br. J. Cancer* **2012**, *106*, 1598–1604. [CrossRef]
27. Bejarano, L.; Jordão, M.J.C.; Joyce, J.A. Therapeutic Targeting of the Tumor Microenvironment. *Cancer Discov.* **2021**, *11*, 933–959. [CrossRef]
28. Arts, R.J.; Plantinga, T.S.; Tuit, S.; Ulas, T.; Heinhuis, B.; Tesselaar, M.; Sloot, Y.; Adema, G.J.; Joosten, L.A.; Smit, J.W.; et al. Transcriptional and metabolic reprogramming induce an inflammatory phenotype in non-medullary thyroid carcinoma-induced macrophages. *Oncoimmunology* **2016**, *5*, e1229725. [CrossRef]
29. Christofides, A.; Strauss, L.; Yeo, A.; Cao, C.; Charest, A.; Boussiotis, V.A. The complex role of tumor-infiltrating macrophages. *Nat. Immunol.* **2022**, *23*, 1148–1156. [CrossRef]
30. Okamoto, K.; Kodama, K.; Takase, K.; Sugi, N.H.; Yamamoto, Y.; Iwata, M.; Tsuruoka, A. Antitumor activities of the targeted multi-tyrosine kinase inhibitor lenvatinib (E7080) against RET gene fusion-driven tumor models. *Cancer Lett.* **2013**, *340*, 97–103. [CrossRef]
31. Netea-Maier, R.T.; Smit, J.W.A.; Netea, M.G. Metabolic changes in tumor cells and tumor-associated macrophages: A mutual relationship. *Cancer Lett.* **2018**, *413*, 102–109. [CrossRef]
32. O'Neill, L.A.; Kishton, R.J.; Rathmell, J. A guide to immunometabolism for immunologists. *Nat. Rev. Immunol.* **2016**, *16*, 553–565. [CrossRef]
33. Monaco, G.; Lee, B.; Xu, W.; Mustafah, S.; Hwang, Y.Y.; Carré, C.; Burdin, N.; Visan, L.; Ceccarelli, M.; Poidinger, M.; et al. RNA-Seq Signatures Normalized by mRNA Abundance Allow Absolute Deconvolution of Human Immune Cell Types. *Cell Rep.* **2019**, *26*, 1627–1640.e1627. [CrossRef]
34. Klaeger, S.; Heinzlmeir, S.; Wilhelm, M.; Polzer, H.; Vick, B.; Koenig, P.-A.; Reinecke, M.; Ruprecht, B.; Petzoldt, S.; Meng, C.; et al. The target landscape of clinical kinase drugs. *Science* **2017**, *358*, eaan4368. [CrossRef]
35. Guha, M.; Mackman, N. LPS induction of gene expression in human monocytes. *Cell. Signal.* **2001**, *13*, 85–94. [CrossRef]
36. O'Neill, L.A.J.; Golenbock, D.; Bowie, A.G. The history of Toll-like receptors—redefining innate immunity. *Nat. Rev. Immunol.* **2013**, *13*, 453–460. [CrossRef]
37. Aksoy, E.; Taboubi, S.; Torres, D.; Delbauve, S.; Hachani, A.; Whitehead, M.A.; Pearce, W.P.; Berenjeno, I.M.; Nock, G.; Filloux, A.; et al. The p110δ isoform of the kinase PI(3)K controls the subcellular compartmentalization of TLR4 signaling and protects from endotoxic shock. *Nat. Immunol.* **2012**, *13*, 1045–1054. [CrossRef]
38. Zanoni, I.; Ostuni, R.; Marek, L.R.; Barresi, S.; Barbalat, R.; Barton, G.M.; Granucci, F.; Kagan, J.C. CD14 Controls the LPS-Induced Endocytosis of Toll-like Receptor 4. *Cell* **2011**, *147*, 868–880. [CrossRef]
39. Voss, O.H.; Murakami, Y.; Pena, M.Y.; Lee, H.N.; Tian, L.J.; Margulies, D.H.; Street, J.M.; Yuen, P.S.T.; Qi, C.F.; Krzewski, K.; et al. Lipopolysaccharide-Induced CD300b Receptor Binding to Toll-like Receptor 4 Alters Signaling to Drive Cytokine Responses that Enhance Septic Shock. *Immunity* **2016**, *44*, 1365–1378. [CrossRef]
40. Vogelpoel, L.T.C.; Hansen, I.S.; Rispens, T.; Muller, F.J.M.; van Capel, T.M.M.; Turina, M.C.; Vos, J.B.; Baeten, D.L.P.; Kapsenberg, M.L.; de Jong, E.C.; et al. Fc gamma receptor-TLR cross-talk elicits pro-inflammatory cytokine production by human M2 macrophages. *Nat. Commun.* **2014**, *5*, 5444. [CrossRef]
41. Patnaik, A.; Swanson, K.D.; Csizmadia, E.; Solanki, A.; Landon-Brace, N.; Gehring, M.P.; Helenius, K.; Olson, B.M.; Pyzer, A.R.; Wang, L.C.; et al. Cabozantinib Eradicates Advanced Murine Prostate Cancer by Activating Antitumor Innate Immunity. *Cancer Discov.* **2017**, *7*, 750–765. [CrossRef] [PubMed]
42. Schlumberger, M.; Jarzab, B.; Cabanillas, M.E.; Robinson, B.; Pacini, F.; Ball, D.W.; McCaffrey, J.; Newbold, K.; Allison, R.; Martins, R.G.; et al. A Phase II Trial of the Multitargeted Tyrosine Kinase Inhibitor Lenvatinib (E7080) in Advanced Medullary Thyroid Cancer. *Clin. Cancer Res.* **2016**, *22*, 44–53. [CrossRef] [PubMed]
43. Fabian, M.A.; Biggs, W.H.; Treiber, D.K.; Atteridge, C.E.; Azimioara, M.D.; Benedetti, M.G.; Carter, T.A.; Ciceri, P.; Edeen, P.T.; Floyd, M.; et al. A small molecule–kinase interaction map for clinical kinase inhibitors. *Nat. Biotechnol.* **2005**, *23*, 329–336. [CrossRef] [PubMed]

44. Heine, A.; Schilling, J.; Grünwald, B.; Krüger, A.; Gevensleben, H.; Held, S.A.E.; Garbi, N.; Kurts, C.; Brossart, P.; Knolle, P.; et al. The induction of human myeloid derived suppressor cells through hepatic stellate cells is dose-dependently inhibited by the tyrosine kinase inhibitors nilotinib, dasatinib and sorafenib, but not sunitinib. *Cancer Immunol. Immunother.* **2016**, *65*, 273–282. [CrossRef]
45. Ou, D.L.; Chen, C.W.; Hsu, C.L.; Chung, C.H.; Feng, Z.R.; Lee, B.S.; Cheng, A.L.; Yang, M.H.; Hsu, C. Regorafenib enhances antitumor immunity via inhibition of p38 kinase/Creb1/Klf4 axis in tumor-associated macrophages. *J. Immunother. Cancer* **2021**, *9*, 14. [CrossRef]
46. Okamoto, K.; Ikemori-Kawada, M.; Jestel, A.; von König, K.; Funahashi, Y.; Matsushima, T.; Tsuruoka, A.; Inoue, A.; Matsui, J. Distinct binding mode of multikinase inhibitor lenvatinib revealed by biochemical characterization. *ACS. Med. Chem. Lett.* **2015**, *6*, 89–94. [CrossRef]
47. Bellora, F.; Dondero, A.; Corrias, M.V.; Casu, B.; Regis, S.; Caliendo, F.; Moretta, A.; Cazzola, M.; Elena, C.; Vinti, L.; et al. Imatinib and Nilotinib Off-Target Effects on Human NK Cells, Monocytes, and M2 Macrophages. *J. Immunol.* **2017**, *199*, 1516–1525. [CrossRef]
48. Zhu, J.; Fang, P.; Wang, C.; Gu, M.; Pan, B.; Guo, W.; Yang, X.; Wang, B. The immunomodulatory activity of lenvatinib prompts the survival of patients with advanced hepatocellular carcinoma. *Cancer Med.* **2021**, *10*, 7977–7987. [CrossRef]
49. Foey, A.D.; Parry, S.L.; Williams, L.M.; Feldmann, M.; Foxwell, B.M.J.; Brennan, F.M. Regulation of Monocyte IL-10 Synthesis by Endogenous IL-1 and TNF-α: Role of the p38 and p42/44 Mitogen-Activated Protein Kinases. *J. Immunol.* **1998**, *160*, 920–928. [CrossRef]
50. Wu, C.H.; Hsu, F.T.; Chao, T.L.; Lee, Y.H.; Kuo, Y.C. Revealing the suppressive role of protein kinase C delta and p38 mitogen-activated protein kinase (MAPK)/NF-κB axis associates with lenvatinib-inhibited progression in hepatocellular carcinoma in vitro and in vivo. *Biomed Pharm.* **2022**, *145*, 112437. [CrossRef]
51. Jiménez-Garcia, L.; Herránz, S.; Luque, A.; Hortelano, S. Critical role of p38 MAPK in IL-4-induced alternative activation of peritoneal macrophages. *Eur. J. Immunol.* **2015**, *45*, 273–286. [CrossRef]
52. Seimon, T.A.; Wang, Y.; Han, S.; Senokuchi, T.; Schrijvers, D.M.; Kuriakose, G.; Tall, A.R.; Tabas, I.A. Macrophage deficiency of p38α MAPK promotes apoptosis and plaque necrosis in advanced atherosclerotic lesions in mice. *J. Clin. Investig.* **2009**, *119*, 886–898. [CrossRef]
53. Zhang, H.Y.; Li, R.C.; Cao, Y.F.; Gu, Y.; Lin, C.; Liu, X.; Lv, K.P.; He, X.D.; Fang, H.J.; Jin, K.F.; et al. Poor Clinical Outcomes and Immunoevasive Contexture in Intratumoral IL-10-Producing Macrophages Enriched Gastric Cancer Patients. *Ann. Surg.* **2022**, *275*, E626–E635. [CrossRef]
54. Saraiva, M.; O'Garra, A. The regulation of IL-10 production by immune cells. *Nat. Rev. Immunol.* **2010**, *10*, 170–181. [CrossRef]
55. Mocsai, A.; Ruland, J.; Tybulewicz, V.L.J. The SYK tyrosine kinase: A crucial player in diverse biological functions. *Nat. Rev. Immunol.* **2010**, *10*, 387–402. [CrossRef]
56. Lowell, C.A. Src-family and Syk Kinases in Activating and Inhibitory Pathways in Innate Immune Cells: Signaling Cross Talk. *Cold Spring Harb. Perspect. Biol.* **2011**, *3*, a002352. [CrossRef]
57. Clark, K.; MacKenzie, K.F.; Petkevicius, K.; Kristariyanto, Y.; Zhang, J.; Choi, H.G.; Peggie, M.; Plater, L.; Pedrioli, P.G.A.; McIver, E.; et al. Phosphorylation of CRTC3 by the salt-inducible kinases controls the interconversion of classically activated and regulatory macrophages. *Proc. Natl. Acad. Sci. USA* **2012**, *109*, 16986–16991. [CrossRef]
58. Ferguson, F.M.; Gray, N.S. Kinase inhibitors: The road ahead. *Nat. Rev. Drug Discov.* **2018**, *17*, 353–377. [CrossRef]
59. Makker, V.; Colombo, N.; Casado Herráez, A.; Santin, A.D.; Colomba, E.; Miller, D.S.; Fujiwara, K.; Pignata, S.; Baron-Hay, S.; Ray-Coquard, I.; et al. Lenvatinib plus Pembrolizumab for Advanced Endometrial Cancer. *N. Engl. J. Med.* **2022**, *386*, 437–448. [CrossRef]
60. Motzer, R.; Alekseev, B.; Rha, S.-Y.; Porta, C.; Eto, M.; Powles, T.; Grünwald, V.; Hutson, T.E.; Kopyltsov, E.; Méndez-Vidal, M.J.; et al. Lenvatinib plus Pembrolizumab or Everolimus for Advanced Renal Cell Carcinoma. *N. Engl. J. Med.* **2021**, *384*, 1289–1300. [CrossRef]
61. European Medicines Agency: EMA/621567/2021-Assessment report. Available online: https://www.ema.europa.eu/en/documents/variation-report/keytruda-h-c-003820-ii-0104-epar-assessment-report-variation_en.pdf (accessed on 21 January 2022).
62. European Medicines Agency: EMA/CHMP/618201/2021-Assessment report. Available online: https://www.ema.europa.eu/en/documents/variation-report/lenvima-h-c-003727-ii-0042-epar-assessment-report-variation_en.pdf (accessed on 21 January 2022).
63. Bottazzi, B.; Riboli, E.; Mantovani, A. Aging, inflammation and cancer. *Semin. Immunol.* **2018**, *40*, 74–82. [CrossRef] [PubMed]

pharmaceutics

Article

Comparative Analysis of Dasatinib Effect between 2D and 3D Tumor Cell Cultures

Samantha Sabetta, Davide Vecchiotti, Letizia Clementi, Mauro Di Vito Nolfi, Francesca Zazzeroni and Adriano Angelucci *

Department of Biotechnological and Applied Clinical Science, University of L'Aquila, 67100 L'Aquila, Italy
* Correspondence: adriano.angelucci@univaq.it

Abstract: Three-dimensional cell culture methods are able to confer new predictive relevance to in vitro tumor models. In particular, the 3D multicellular tumor spheroids model is considered to better resemble tumor complexity associated with drug resistance compared to the 2D monolayer model. Recent advances in 3D printing techniques and suitable biomaterials have offered new promises in developing 3D tissue cultures at increased reproducibility and with high-throughput characteristics. In our study, we compared the sensitivity to dasatinib treatment in two different cancer cell lines, prostate cancer cells DU145 and glioblastoma cells U87, cultured in the 3D spheroids model and in the 3D bioprinting model. DU145 and U87 cells were able to proliferate in 3D alginate/gelatin bioprinted structures for two weeks, forming spheroid aggregates. The treatment with dasatinib demonstrated that bioprinted cells were considerably more resistant to drug toxicity than corresponding cells cultured in monolayer, in a way that was comparable to behavior observed in the 3D spheroids model. Recovery and analysis of cells from 3D bioprinted structures led us to hypothesize that dasatinib resistance was dependent on a scarce penetrance of the drug, a phenomenon commonly reported also in spheroids. In conclusion, the 3D bioprinted model utilizing alginate/gelatin hydrogel was demonstrated to be a suitable model in drug screening when spheroid growth is required, offering advantages in feasibility, reproducibility, and scalability compared to the classical 3D spheroids model.

Keywords: 3D tissue culture; 3D tumor model; tumor spheroids; drug resistance; drug screening; preclinical study

Citation: Sabetta, S.; Vecchiotti, D.; Clementi, L.; Di Vito Nolfi, M.; Zazzeroni, F.; Angelucci, A. Comparative Analysis of Dasatinib Effect between 2D and 3D Tumor Cell Cultures. *Pharmaceutics* **2023**, *15*, 372. https://doi.org/10.3390/pharmaceutics15020372

Academic Editor: Sudip K. Das

Received: 15 December 2022
Revised: 14 January 2023
Accepted: 18 January 2023
Published: 21 January 2023

1. Introduction

Inhibition of key molecules in signaling pathways that regulate cancer cell proliferation and dissemination is one of the most promising strategies in modern cancer therapy. Specifically, tyrosine kinases represent a particularly important target because they mediate several biological processes, including cell growth and differentiation, metabolism, and apoptosis. Constitutive activation gives these enzymes an oncogene status, which directs the cell toward neoplastic transformation [1].

Src, a 60-kDa protein, is the most widely studied member of the Src family tyrosine kinases (SFKs), being frequently overexpressed in many tumors. SFKs are a large class of non-receptor kinases that control multiple signaling pathways in animal cells, the activation of which is necessary for mitogenic signaling of many growth factors, but also for the acquisition of the migratory and invasive phenotype. Indeed, oncogenic activation of SFKs has been shown to play an important role in solid tumors, promoting tumor growth and distant metastasis formation, and that is why they represent the most representative class of targeted proteins in cancer therapy [2]. This family consists of 11 highly homologous members: c-Src, Blk, Fgr, Fyn, Frk, Hck, Lck, Lyn, Yes, Yrk, and Srms [3,4]. Of the 11 members of SFKs, Src, Fyn, and Yes are implicated in tumorigenesis and metastasis formation [5]. In fact, Src, Fyn, and Yes, but also Frk, are widely expressed in a variety of tissues, while for the other members, protein expression is more restricted in tissues, with a prevalence in cells of hematopoietic origin [6].

Src activity is regulated, at the cellular level, by the phosphorylation state of two key tyrosine residues and by interactions with ligands mediated by its regulatory domains, which stabilize the kinase in active or inactive configuration. In the inactive configuration, the SH2 domain binds phosphorylated Tyr527 in the C-terminal domain, while the SH3 domain interacts with the linker domain on the backside of the catalytic domain, thus promoting a closed conformation that prevents interaction with substrates. In the active configuration, Tyr527 is dephosphorylated, and the SH2 and SH3 domains are released by intramolecular interactions and thus available for binding of heterologous molecular partners; this open conformation also allows Tyr416 to be self-phosphorylated.

In cancer cells, Src transmits signals that promote cell survival and mitogenesis; in addition, Src exerts a profound effect on cytoskeleton reorganization and adhesion systems that underlie cell migration and invasion. Src, therefore, not only promotes cancer cell growth but is also involved in the control of adhesiveness and migration, functioning as a key molecule that regulates signal transduction pathways triggered by various surface molecules, such as growth factor receptors and integrins [2,7]. Indeed, most evidence suggests that Src has a predominant role in the maintenance of the neoplastic phenotype and tumor progression, rather than tumor initiation or growth [8]. During tumor progression, Src activity becomes abnormally elevated, and because mutation activation or amplification of Src is very rare in human tumors, altered extrinsic control of Src phosphorylation by kinase or phosphatase may represent an important mechanism of Src upregulation.

Dasatinib (BMS-354825, Sprycel) is an orally available small molecule that in 2006 received FDA approval for the treatment of chronic myelogenous leukemia and Philadelphia-positive acute lymphoblastic leukemia. Dasatinib has a low inhibition specificity and its potential targets comprise Src in addition to BCR-Abl, c-kit, PDGFR, and other SFKs, including Lck, Fyn, and Yes, with IC50 < 1.0 nmol/L [9]. Preclinical data suggest that dasatinib could be effective also in solid cancer, and numerous results from phase I/II trials conducted in different cancer types have been published since 2009 [10]. Despite the encouraging preclinical results, the clinical validation of these data has been largely unproductive. In fact, although in vitro studies have demonstrated several gene signatures predictive of dasatinib response, these signatures have not yet permitted the definition of cohorts clinically sensitive to dasatinib [11]. In addition, available preclinical models have been shown to be largely inadequate in their predictive capacity, and new experimental strategies implementing microenvironment condition are needed. In fact, SFK activity is tightly dependent on ECM composition, and a changeable microenvironment could heavily impact SFK functions and thus treatment efficacy.

The tumor microenvironment is known to be complex, both in its content and dynamic nature, which is difficult to study using two-dimensional (2D) cell culture models. In contrast, three-dimensional (3D) cell models, such as spheroids, show peculiar molecular features that differ from monolayer cultures but are closer to the structural architecture of the tumor in vivo. Such models, in fact, can more accurately mimic both the structure of the malignant tissue and its microenvironment (physiological responses, secretion of soluble mediators, gene expression, patterns and mechanisms of drug resistance). Tumor cells within the spheroid reproduce the same concentric arrangement as tumors; in fact, they have a proliferative outer layer, a quiescent middle layer, and a hypoxic and necrotic core and have a growth pattern like solid tumors in the early non-vascularized stage. The use of 3D cell assays adds value to research and screening studies to identify potential anticancer compounds, bridging existing limitations between 2D cell cultures and animal models.

The theoretic superiority of 3D vs. 2D cultures has prompted the development of several protocols with the purpose to overcome the main weaknesses of 3D tissue cultures: technical complexity in performing and analyzing, high costs, reproducibility, and scalability. In particular, different approaches have been described to realize 3D sphere culture. Hanging drop protocol is a suspension culture method and it is one of the most popular approaches used for culture of 3D spheres, because it does not need particular or expensive instruments [12]. There are other suspension culture approaches for large-scale formation of spheres, each with their distinct advantages and disadvantages [13]. They include device-assisted culture, in order to force cell aggregation by movement or magnetic levitation, and gel embedding culture, able to mimic ECM microenvironment. In particular, the use of gels or scaffolds may guarantee a higher predictivity of the tissue models because cells organize a complex dynamic 3D architecture making contact also with ECM [14].

Three-dimensional bioprinting is an emerging technology that, using dedicated software and hardware to design 3D patterns and structures, aims to precisely produce an engineered structure or tissue that can be used for biological and pharmacological studies or directly in humans for regenerative medicine. It represents a relatively new approach that provides high reproducibility and precise control over automated fabricated constructs, potentially enabling high-throughput production of the desired 3D model. Bioprinting modalities include extrusion-based bioprinting [15], inkjet-based bioprinting [16], and laser-based bioprinting. Extrusion-based bioprinting is the most common and reliable system for 3D printing of biological tissues and is characterized by robotic delivery of a continuous stream of bioink, under pneumatic or motorized forces.

During the bioprinting process, a solution of a biomaterial or a mixture of several biomaterials in the form of a hydrogel, which usually encapsulates the desired cell types, called a bioink, is used to create tissue scaffolds with the final shape resembling the designed construct. Bioinks can be made alone from natural or synthetic biomaterials or from a combination of the two as hybrid materials. Hydrogel is the most widely used bioink, as it has properties like those of extracellular matrices and also allows the encapsulation of cells in a 3D environment that is both highly hydrated and mechanically stable. Sodium alginate is frequently used as hydrogel in 3D tissue culture thanks to its favorable physio-chemical characteristics, including gelling, viscous, and stabilizing properties, and the ability to retain water. Although sodium alginate is biocompatible and biodegradable, it is frequently mixed with other polymers of animal origin in order to enhance its gelling properties and the viability of encapsulated cells. The combination of sodium alginate and gelatin provides an excellent hydrogel for use as a substrate in 3D printing technology due to its biological properties, such as its biocompatibility, biodegradability, and non-toxicity. The easily modified mechanical properties of these materials can be adapted to living tissue, making them ideal environments for cell culture development. However, the chemical and mechanical cues provided by the specific hydrogel formulation as well as protocol used when performing 3D culture can affect the behavior of spheres and their resistance to drugs, requesting more confirmatory data [17,18].

In our study, we aimed to investigate the preclinical potentiality of 3D bioprinting technology in the field of experimental oncology. In particular, we verified the effect of dasatinib in 3D bioprinting models based on alginate/gelatin matrix utilizing two different tumor cell models, prostate cancer and glioblastoma, in comparison with the corresponding 3D spheroids models and 2D monolayer cultures.

2. Material and Methods

2.1. Cell Lines and 2D Monolayer Culture

Experiments were performed using U87 MG cell line (U87) human glioblastoma astrocytoma derived from a malignant glioma from a female patient by the explant technique, able to produce a malignant tumor consistent with glioblastoma in nude mice, and DU145 cell line human prostate adenocarcinoma isolated from the brain of a 69-year-old male. The U87 cells were cultured in DMEM high-glucose growth medium and DU145 cells in RPMI 1640 medium. For all cell lines, the growth medium was supplemented with 10% fetal bovine serum, 2 mM glutamine 100 IU/mL penicillin, and 100 μg/mL streptomycin. Cell lines were supplied by ECACC and underwent regular testing for mycoplasma by Hoechst DNA staining and PCR. Authentication procedures included species verification by DNA barcoding and identity verification by DNA profiling. U87 and DU145 cells when passed were initially plated at a density of 10^4 cells/cm^2, and incubated in 5% CO_2 at 37 °C. For protein extraction, we initially recovered the cells with the scraper and then centrifuged at 300× *g* for 10 min and then at 100× *g* for 5 min. Cells were processed for protein extraction with cell lysis buffer containing 0.1% Triton X-100, 10 mM Tris-HCl (pH 7.5), 150 mM NaCl, 5 mM EDTA, and supplemented with 1 mM Na_3VO_4 and 75 U of aprotinin (Sigma-Aldrich, St. Louis, MI, USA), and incubated for 20 min at 4 °C. Protein extracts were stored at −80 °C until use.

2.2. Preparation of Hydrogel and Bioprinting

The hydrogel used for bioprinting is composed of 2% alginate and 8% gelatin (bioink). The powders were initially exposed to UV light for 15 min and then dissolved in sterile DPBS on a magnetic stirrer at 50 °C under laminar flow. Once prepared, bioink was placed in sterile syringes and stored at 4 °C until use. Prior to the bioprinting procedure, bioink was equilibrated at 37 °C. Next, 20 × 10^6 cells were suspended in 400 μL of culture medium and then mixed with 3.7 mL of bioink to reach a final concentration of 5 × 10^6 cell/mL. Finally, a disposable cartridge was filled with cell-laden hydrogel and equilibrated at 29 °C for 30 min in a temperature-controlled printhead. Printing was carried out using BIO X™ bioprinter (Cellink, Gothenburg, Sweden) in a 12-well plate with a dispensing pressure of 50 kPa at a speed of 5 mm/s and setting the printbed temperature at 18 °C. The 3D structures were finally cross-linked with $CaCl_2$ for 10 min and for an additional 3 min with $BaCl_2$ before adding culture medium. For protein extraction, the 3D structures were quickly dissolved in a solution of EDTA 250 mM in PBS and cells were collected by centrifugation at 400× *g* for 10 min. The pellet was resuspended in PBS and centrifugated at 400× *g* for 5 min. The resulting pellet was dissolved in RIPA buffer for protein extraction.

2.3. Cell Viability and IC50 Analysis

DU145 and U87 cell lines were seeded at a density of 5 × 10^4 cells/mL in a 24 multiwell. After 24 h, the cells were treated with dasatinib by the serial dilution method for 72 h. At the end of 72 h, the growth medium was removed and substituted with fresh medium. Cell viability in both 2D and 3D cultures was assessed using PrestoBlue® (Thermo Fisher, Waltham, MA, USA) colorimetric viability assay, based on a ready-to-use, non-toxic, cell-permeable resazurin-based solution. It functions as a cell viability indicator by using the reducing power of living cells and quantitatively measuring the proliferation of cells. Cells were incubated with reagent in a 1:10 ratio for 2 h, in the dark. Then, a spectrophotometer reading was taken at wavelengths of 570 and 600 nm. For IC50 analysis, data collected in triplicate were elaborated with DRFit software (http://www.structuralchemistry.org/pcsb/drfit.php, accessed on 2 November 2022).

2.4. Western Blot

Protein extracts were centrifuged for 10 min at 300× *g* to eliminate nuclei and large debris. After protein dosage by Bradford Dye Reagent assay (Bio-Rad, Hercules, CA, USA), the same protein quantity for each sample was subjected to 10% sodium-dodecyl sulphate polyacrilamide gel electrophoresis (SDS-PAGE). Prestained protein molecular markers sharpmass VII (Euroclone, Milan, Italy) were loaded on a separate well for each gel. Then, at the end of the run, proteins were electrophoretically transferred from gel onto nitrocellulose membranes Amersham protran 0.2 μm (Cytiva Europe, Freiburg, Germany) for 90 min at 350 mA. Membranes were incubated for 1 h at RT with 10% nonfat dry milk in Tris-buffered saline (Bio-Rad) at pH 7.4 containing 20 mM Tris, 500 mM NaCl, and supplemented with 0.05% Tween 20 (Bio-Rad), and then probed for 1 h at RT with primary antibodies according to dilution suggested by the manufacturer: anti-Src (36D10), anti-Phospho-Src (Thr416) (E6G4R), and anti-GAPDH (0411) (all from Cell Signaling Technology, Danvers, MA, USA). Protein bands were visualized after 1 h of incubation with horseradish peroxidase (HRP)-conjugated anti-rabbit IgG or anti-mouse IgG (Cell Signaling Technology) at RT, and then with chemiluminescence reagents (Amersham, Buckinghamshire, UK).

Chemiluminescent signals were acquired by the Chemidoc XRS system and digitally analyzed for the determination of band molecular weight and density by Imagelab software (Bio-Rad).

2.5. 3D Multicellular Spheroid Model

For the formation of U87 and DU145 spheroids, we used the "hanging drops" method. The density chosen was 500,000 cells/mL, which were resuspended in 10 μL of growth medium with 25% FBS and 25 μL of collagen (10 μg/mL). This method involves the deposition of 1 μL of cells on the bottom of the lid of a Petri dish; inverting the dish forms a "hanging" drop, and the cells, by gravity, accumulate in the bottom of the drop, promoting cell aggregation into spheroids. The spheroids were kept in an incubator at 37 °C and 5% CO_2. After 30 min, on each drop, growth medium was added to prevent evaporation of the drop. Then, 6 mL of PBS was added to the bottom of the plate, with the purpose of keeping the moisture level high. After 24 h, the formed spheroids were transferred to a 96 round-bottom multiwell with 200 μL of DMEM growth medium for U87 and RPMI for DU145.

2.6. Spheroid Dissemination Assay

The dissemination assay was performed plating spheroids on collagen matrix. Briefly, a 24 multiwell was coated with 30 μL of collagen (4 mg/mL) for 24 h before transferring the spheroids. Once transferred, the spheroids were treated with dasatinib at selected concentrations and monitored by phase contrast microscopy up to 24 h. Digital images were taken at T0 (newly transferred spheroid), T8, and T24, and were analyzed by the public domain software ImageJ (https://imagej.nih.gov/ij/ accessed on 15 September 2022) for the semiquantitative evaluation of the invaded area.

2.7. Cell Migration Assay

For the migration assay, cells were first seeded at a density of 5×10^4 cells/mL in a 6-well multiwell and the next day were treated with dasatinib at selected concentrations for 24 h. Then, cells were detached from the culture plate by trypsin–EDTA and centrifuged at $300\times$ g for 10 min. Alternatively, cells were recovered from 3D bioprinted structures with a solution of EDTA 250 mM in PBS. Cells were first counted and then seeded in serum-free medium at a density of 3×10^4 cells/mL on top of the filter membrane of inserts that were placed in the 24-well multiwell. In each well was added 800 μL of complete growth medium (DMEM for U87 and RPMI for DU145), which serves as a chemoattractant. The planned incubation time was 2 h. The insert was removed from the multiwell plate, and was cleaned of the remaining cells that did not migrate from the top of the membrane. Then, 200 μL of filtered formalin was added to each insert for 10 min at room temperature, and then 200 μL of crystal violet (12 mM in a 20% solution of methanol in water) was added to each insert for 10 min at room temperature. Crystal violet was then removed, and washing with tap water was performed to remove excess dye. The cells were counted by phase contrast microscopy at $400\times$ magnification, in at least 5 fields per membrane, to obtain the mean number of migrated cells per field.

2.8. Statistical Analysis

All the statistical procedures were performed by GraphPad Prism Software Inc. (San Diego, CA, USA). Data are expressed as mean ± standard deviations (SD) of at least three independent experiments. The statistical significance between measure series was calculated with parametric Student t test and p values of less than 0.01.

3. Results

Tumor cell growth in bioprinted scaffold. Prostate cancer (DU145) and glioblastoma (U87) cells were resuspended in alginate/gelatin hydrogel at a density of 2×10^6 cells/mL. Cell suspension was bioprinted in a controlled environment onto each well of a 12-well plate following the digital model of a cuboid shape with four internal empty spaces (structure size: L 12 mm × W 12 mm × H 1.2 mm) (Figure 1a). Tumor cell growth was monitored by PrestoBlue® cell viability reagent and revealed a progressive increase in the number of viable cells that was similar in the two cell lines, with a doubling time of about 14 days (Figure 1b). When bioprinted structures were observed by phase contrast microscopy, it was evident that tumor cells tended to form spheroid aggregates with increasing diameter over time (Figure 1c).

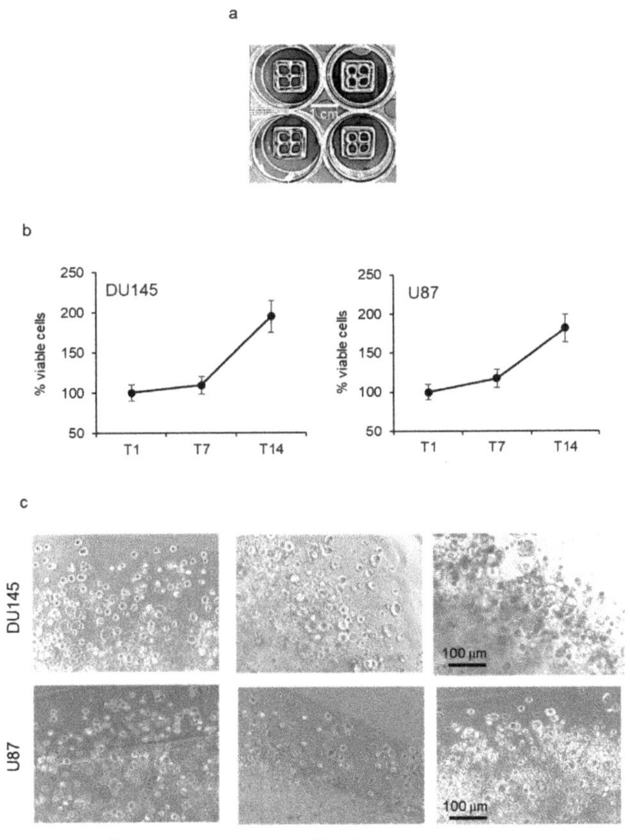

Figure 1. Tumor cell growth in bioprinted scaffold. (**a**) Exemplificative image of bioprinted algine/gelatin constructs printed in 12-well multiwell plate. (**b**) Mean percentage of viable DU145 and U87 cells grown in bioprinted scaffold and evaluated 1, 7, and 14 days (T1, T7, T14) after bioprinting. The mean values measured at T1 were adjusted at 100%. Each value represents the mean of five different bioprints (±standard deviation). (**c**) Representative images acquired by phase contrast microscopy (100× magnification) of bioprints containing DU145 (upper images) and U87 cells at days 1, 7, and 14 after printing.

Antiproliferative effect of dasatinib in 3D cell models. Bioprinted tumor cells and tumor spheroids, utilizing both DU145 and U87 cells, were subjected to increasing concentrations of dasatinib (1 µM, 10 µM, and 100 µM) for 72 h and tumor growth was monitored by PrestoBlue® cell viability reagent. Dasatinib determined a significant reduction in cell viability that was evident starting from 1 µM with similar trends in the two different cell models (Figure 2a,b). Dasatinib cytotoxicity was more marked in U87 cells compared to

DU145 cells. In addition, according to IC50 value, bioprinted cells demonstrated a higher sensitivity to dasatinib compared to spheroids, but both spheroids and bioprints were far less susceptible to dasatinib action compared to 2D cell culture treated in the same experimental conditions (Figure 2c).

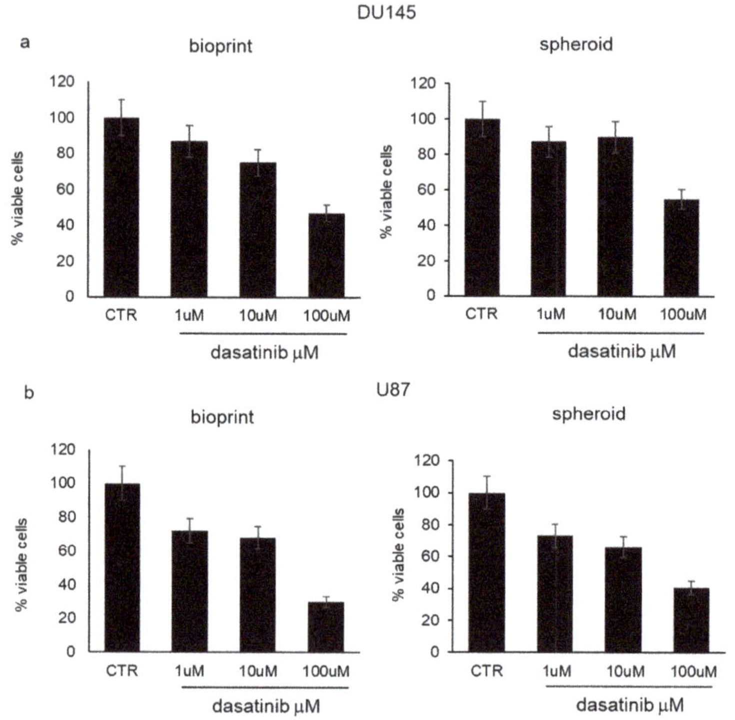

c

Cell line	2D	bioprint	spheroid
U87	8.0±0.6 μM	23±3 μM	42±4 μM
DU145	4.2±0.4 nM	89±7 μM	102±12 μM

Figure 2. Antiproliferative effect of dasatinib in 3D cell models. (**a**) Percentage of viable DU145 cells in bioprints (**left**) and in spheroids (**right**) 72 h after treatment with increasing concentrations of dasatinib (1, 10, and 100 μM). The mean values measured in untreated cells were adjusted at 100%. Each value represents the mean of five different measurements (±standard deviation). (**b**) Percentage of viable U87 cells in bioprints (**left**) and in spheroids (**right**) 72 h after treatment with increasing concentrations of dasatinib (1, 10, and 100 μM). The mean values measured in untreated cells were adjusted at 100%. Each value represents the mean of five different measurements (±standard deviation). (**c**) Mean IC50 values calculated for U87 and DU145 cells treated with dasatinib for 72 h. The IC50s were calculated in standard culture conditions (2D), in 3D bioprint model (bioprint), and in spheroids and were the result of three different experiments (±standard deviation).

Evaluation of Src activation. The expression of Src and of its active form, pSrc (Tyr416), was evaluated in 2D culture and in bioprints realized with DU145 and U87 cells. Expression levels of total Src, pSrc (Tyr416), and GAPDH were evaluated in total cell lysates by Western blot analysis in 2D cell cultures treated with increasing concentrations of dasatinib (0.1, 1, and 10 μM) for 24 h. In DU145 cells, a reduction in the active form of Src, but not in the total Src, was evident already from 0.1 μM dasatinib, while in U87 cells, the first effective concentration was 1 μM (Figure 3a). Expression levels of total Src, pSrc (Tyr416), and GAPDH were also evaluated in total cell lysates by Western blot analysis in bioprint models treated with 10 and 25 μM of dasatinib for 24 h. The results did not show an appreciable reduction in the active form of Src in the DU145 model, while the first effective concentration of dasatinib in U87 cells was 25 μM (Figure 2b).

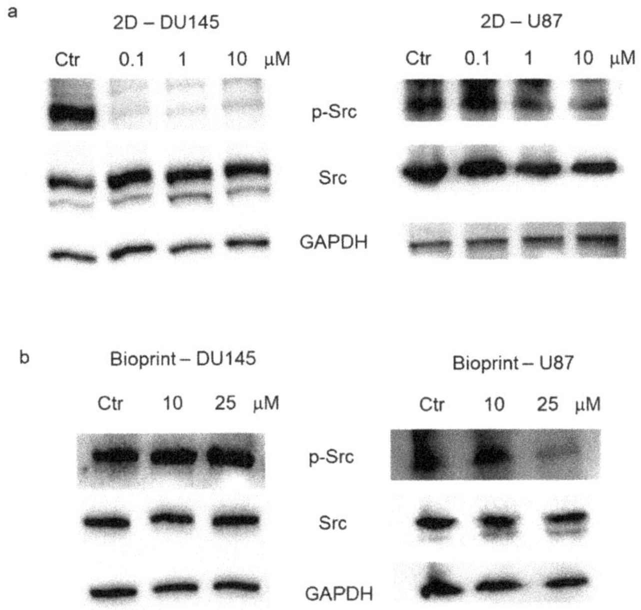

Figure 3. Evaluation of Src activation. (**a**) Western blot analysis of total cell lysates from DU145 (**left**) and U87 (**right**) cells cultured in standard conditions (2D). Cell lysates were recovered 24 h after dasatinib treatment with increasing concentrations (0.1, 1, and 10 μM) and analyzed for the expression of pSrc (Tyr416), total Src, and GAPDH (loading reference). (**b**) Western blot analysis of total cell lysates from DU145 (**left**) and U87 (**right**) cells cultured alginate/gelatin bioprinted scaffold (bioprint). Bioprints were treated for 24 h with 10 or 25 μM dasatinib and analyzed for the expression of pSrc (Tyr416), total Src, and GAPDH (loading reference).

Inhibition of cell migration by dasatinib. Two-dimensional cell culture and bioprint models were treated for 24 h with increasing concentrations of dasatinib (0.1, 1, and 10 μM), and then cells were recovered to evaluate their migration ability. The analysis of the number of cells able to cross the membrane of the transwell system demonstrated that in DU145 and U87 cells treated in standard culture conditions, dasatinib was effective in reducing cell migration at a concentration of 0.1 μM and 1 μM, respectively (Figure 4a,b). On the contrary, cells recovered from bioprints did not show a significant reduction in cell migration with respect to control cells for DU145 cells, and only at the concentration of 10 μM for U87 cells (Figure 4b).

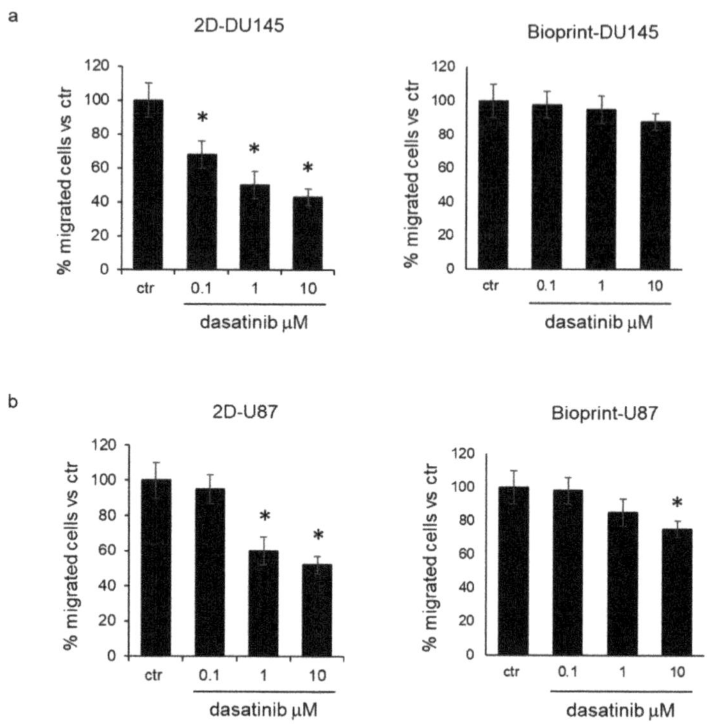

Figure 4. Inhibition of cell migration by dasatinib. (**a**) Percentage of migrated DU145 cells evaluated in transwell system utilizing cells from standard cell culture (2D, **left**) or from bioprint model (**right**) and treated for 24 h with increasing concentrations of dasatinib (0.1, 1, and 10 μM). The mean values measured in untreated cells were adjusted at 100%. Each value represents the mean of five different measurements (±standard deviation). * $p < 0.01$ vs. ctr, according to Student's t test. (**b**) Percentage of migrated U87 cells evaluated in transwell system utilizing cells from standard cell culture (2D, **left**) or from bioprint model (**right**) and treated for 24 h with increasing concentrations of dasatinib (0.1, 1, and 10 μM). The mean values measured in untreated cells were adjusted at 100%. Each value represents the mean of five different measurements (±standard deviation). * $p < 0.01$ vs. ctr, according to Student's t test.

Inhibition of cell dissemination in spheroids. In order to verify the effect of dasatinib on migration of cells grown in spheroids, DU145 and U87 cells were used to form spheroids for 48 h and then spheroids were transferred to collagen-coated plates, where half of them were treated with 10 μM dasatinib. Then, spheroids were monitored by phase contrast microscopy, and time-course digital images were taken. The evaluation of dissemination was performed by measuring the area occupied by cells adhered to collagen, subtracting the area of the spheroid. The analysis of the mean relative values of the covered area revealed that cells in the control spheroids were able to disseminate progressively onto collagen from the surface of the spheroid, and this phenomenon was significantly evident starting from 8 h after cell plating (Figure 5a,b). On the other hand, dasatinib-treated spheroid demonstrated a reduced capacity to disseminate, and a significant increase in mean disseminated area was visible only after 24 h but to a smaller extent with respect to the mean control area.

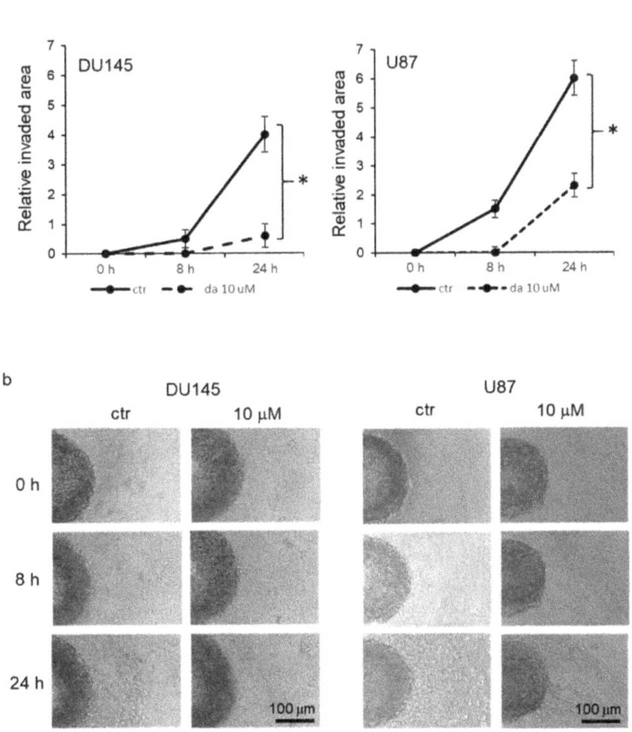

Figure 5. Inhibition of cell dissemination in spheroid. (**a**) Mean relative invaded area in DU145 and U87 spheroids evaluated 8 and 24 days after plating onto collagen-coated surface. The values represent the mean of five different measurements (±standard deviation) of spheroids treated or not with 10 μM dasatinib. * $p < 0.01$ between indicated points according to Student's *t* test. (**b**) Representative images acquired by phase contrast microscopy (100× magnification) of DU145 and U87 cells at 0, 8, and 24 h after plating in the presence or not of 10 μM dasatinib.

4. Discussion

Cancer 3D tissue cultures represent an experimental model able to potentially mimic in vivo growth more closely. In particular, they show distinct characteristics in terms of cellular phenotype, mass transport, and cell–cell interactions as compared with conventional 2D cell cultures [19]. These specific characteristics can significantly affect the sensitivity to antitumoral drugs, and for this reason, it is fundamental to acquire detailed knowledge about their response in order to plan their effective utilization in preclinical drug testing. Spheroids are considered an ideal model in order to mimic some important tumor features, such as structural organization and the gradients of oxygen, pH, and nutrients [20]. Indeed, cancer spheroids have been frequently considered in preclinical studies to evaluate tumor response to chemotherapy. Although different techniques have been tested for spheroid formation, several issues with applying this model at a preclinical level still remain, particularly reproducibility and high-throughput application [21]. Recently, 3D bioprinting has been recognized as a promising technology for creating a tissue-based platform with high reproducibility and scalability. A key feature in developing a 3D bioprint model useful in pharmaceutical applications is the choice of an appropriate printable biomaterial, commonly referred to as bioink, essential for determining cancer cell phenotypes and biophysical properties of the tissue. Among the natural polymers, sodium alginate plays a significant role in tissue engineering applications owing to its biocompatibility, bioavailability, low cost, and thixotropic property [22]. Gelatin is frequently added to alginate in order to enhance cytocompatibility and printability. In our study, we confirmed that DU145 and U87 cells survive in alginate/gelatin scaffold, showing a proliferation trend for at least 14 days. Cancer cells cultured in these conditions tended to grow as cell aggregates, resembling spheroids, and this is because the bioink does not represent an adhesive substrate for cells. When we compared 3D bioprinted tissue culture with 2D culture and spheroids in their susceptibility to dasatinib treatment, we ascertained that 3D bioprinted cells behaved more similarly to the spheroid model. In fact, both 3D bioprints and spheroids demonstrated higher resistance to dasatinib toxicity than 2D cell culture. The resistance was particularly evident in DU145 cells (about 20-fold higher IC50 than 2D) and slightly higher in spheroids than in bioprinted tissue cultures (about 2-fold higher IC50 in U87).

Spheroids frequently showed in the literature high resistance to most therapies, including chemotherapies and radiotherapies [23]. The acquisition of resistance in 3D models could be explained by different mechanisms. Besides biological features, including different interactions with surrounding cells or with the extracellular matrix and higher phenotypic heterogeneity, an important element to consider is the penetration of the drug into the spheroid. Large spheroids show higher drug resistance than small ones, and the penetration of drug is restricted to the outer layer. In our experiments, we utilized spheroids of >500 μm in diameter, in which drug resistance could be attributable just to their dimensions, as mentioned before [20]. Cells in bioprinted structures formed spheroids that after 12 days are heterogeneous in diameter and <500 μm; however, as indicated by the IC50, they express a drug resistance similar to that of the spheroids. However, also in this case, the reason could be attributable to a scarce penetrance of the drug. The analysis of the activation status of Src, the main target of dasatinib, showed a reduced ability of the drug to inhibit the formation of the phosphorylated Src. Thus, it is plausible that alginate/gelatin hydrogel generates a hurdle to the diffusion of dasatinib within the bioprinted structures. Resistance to dasatinib was also evident when cells were recovered from the bioprinted structure and subjected to migration assay. In fact, with respect to the expected reduction in migration demonstrated by cells treated in 2D culture, cells from bioprinted structures maintained a higher migration ability. The drug resistance was not apparently associated with a change in phenotype. In fact, the permanence of the susceptibility to dasatinib in 3D tissue culture was confirmed by the dissemination test performed with spheroids. In this test, dasatinib was effective in reducing the ability of cells to invade the surrounding microenvironment. Because the cells involved in the dissemination are mainly those present on the surface of the spheroid, this supports the hypothesis of the scarce penetrance of the drug within the 3D structure.

The application of bioprinting to drug screening in our experimental procedures offers different advantages with respect to spheroid formation by the hanging drop method that is performed manually. In fact, the hanging drop method requires a skilled operator, and its technical complications compromise its reproducibility and make it unsuitable for high-throughput screening. However, novel 3D tissue techniques emerge constantly. These include new multi-microwell platforms able to assure higher sphere formation yield, uniform sphere size, and scalability [24]. However, in these procedures, as in all suspension cell culture methods, a reliable non-adherent coating is critical to assure reproducibility in sphere formation and in their behavior during the prolonged culture. The use of hydrogel as embedding material may guarantee that cells can organize in a stable, complex, dynamic 3D architecture making contact also with extracellular matrices. Three-dimensional bioprinting does not require particular expertise and permits obtaining several uniform structures in a short time. Alginate/gelatin hydrogel supports 3D growth of cancer cells, and as we observed also in our experimental models, cells spontaneously formed spheroids without the need for a further repulsive procedure. In addition, the bioprinted structures, for their chemical and physical stability permitting a regular change in medium without affecting cell behavior, offer the possibility to perform frequent viability tests. At the same time, alginate/gelatin hydrogel can be dissolved with ion-chelating solution, freeing spheroids that once in suspension could be utilized for further analyses. With this study, we have verified that bioprinting offers an abundant cellular material that could be easily used to analyze protein expression.

Author Contributions: L.C. and S.S. performed experiments with 2D cell cultures and 3D spheroids, wrote the Materials and Methods section, and performed the bibliographic search; M.D.V.N. and S.S. performed experiments with 3D bioprinting models; D.V. supervised and developed protocols for 3D models; F.Z. critically revised the manuscript; A.A. and S.S. collected data for the preparation of figures and wrote the main manuscript text. All authors have read and agreed to the published version of the manuscript.

Funding: The study was funded by the Department of Biotechnological and Applied Clinical Sciences, University of L'Aquila, with intramural research programs.

Institutional Review Board Statement: Not applicable.

Informed Consent Statement: Not applicable.

Data Availability Statement: Most data generated or analyzed during this study are included in this article. The datasets and materials used and/or analyzed during the current study are available from the corresponding author on reasonable request.

Acknowledgments: Samantha Sabetta is supported by MIUR with a doctoral fellowship within the National Operational Program for Research and Innovation (Programma Operativo Nazionale Ricerca e Innovazione 2014-2020, ref: DOT13SR6G7).

Conflicts of Interest: The authors declare that they have no conflict of interest.

References

1. Angelucci, A. Targeting Tyrosine Kinases in Cancer: Lessons for an Effective Targeted Therapy in the Clinic. *Cancers* **2019**, *11*, 490. [CrossRef] [PubMed]
2. Yeatman, T.J. A renaissance for SRC. *Nat. Rev. Cancer* **2004**, *4*, 470–480. [CrossRef] [PubMed]
3. Kim, L.C.; Song, L.; Haura, E.B. Src kinases as therapeutic targets for cancer. *Nat. Rev. Clin. Oncol.* **2009**, *6*, 587–595. [CrossRef] [PubMed]
4. Rai, K. Personalized Cancer Therapy: YES1 Is the New Kid on the Block. *Cancer Res.* **2019**, *79*, 5702–5703. [CrossRef] [PubMed]
5. Stein, P.L.; Vogel, H.; Soriano, P. Combined deficiencies of Src, Fyn, and Yes tyrosine kinases in mutant mice. *Genes Dev.* **1994**, *8*, 1999–2007. [CrossRef] [PubMed]
6. Manning, G.; Whyte, D.B.; Martinez, R.; Hunter, T.; Sudarsanam, S. The protein kinase complement of the human genome. *Science* **2002**, *298*, 1912–1934. [CrossRef]
7. Frame, M.C. Src in cancer: Deregulation and consequences for cell behaviour. *Biochim. Biophys. Acta* **2002**, *1602*, 114–130. [CrossRef]
8. Summy, J.M.; Gallick, G.E. Src family kinases in tumor progression and metastasis. *Cancer Metastasis Rev.* **2003**, *22*, 337–358. [CrossRef]
9. Lombardo, L.J.; Lee, F.Y.; Chen, P.; Norris, D.; Barrish, J.C.; Behnia, K.; Castaneda, S.; Cornelius, L.A.; Das, J.; Doweyko, A.M.; et al. Discovery of N-(2-chloro-6-methyl- phenyl)-2-(6-(4-(2-hydroxyethyl)- piperazin-1-yl)-2-methylpyrimidin-4- ylamino)thiazole-5-carboxamide (BMS-354825), a dual Src/Abl kinase inhibitor with potent antitumor activity in preclinical assays. *J. Med. Chem.* **2004**, *47*, 6658–6661. [CrossRef]
10. Martellucci, S.; Clementi, L.; Sabetta, S.; Mattei, V.; Botta, L.; Angelucci, A. Src Family Kinases as Therapeutic Targets in Advanced Solid Tumors: What We Have Learned so Far. *Cancers* **2020**, *12*, 1448. [CrossRef]
11. Pusztai, L.; Moulder, S.; Altan, M.; Kwiatkowski, D.; Valero, V.; Ueno, N.; Esteva, F.; Avritscher, R.; Qi, Y.; Strauss, L.; et al. Gene signature-guided dasatinib therapy in metastatic breast cancer. *Clin. Cancer Res.* **2014**, *20*, 5265–5271. [CrossRef] [PubMed]
12. Kelm, J.; Timmins, N.; Brown, C.; Fussenegger, M.; Nielsen, L. Method for generation of homogeneous multicellular tumor spheroids applicable to a wide variety of cell types. *Biotechnol. Bioeng.* **2003**, *83*, 173–180. [CrossRef] [PubMed]
13. Lv, D.; Hu, Z.; Lu, L.; Lu, H.; Xu, X. Three-dimensional cell culture: A powerful tool in tumor research and drug discovery. *Oncol. Lett.* **2017**, *14*, 6999–7010. [CrossRef] [PubMed]
14. Loessner, D.; Stok, K.; Lutolf, M.; Hutmacher, D.; Clements, J.; Rizzi, S. Bioengineered 3D platform to explore cell-ECM interactions and drug resistance of epithelial ovarian cancer cells. *Biomaterials* **2010**, *31*, 8494–8506. [CrossRef] [PubMed]
15. Ozbolat, I.T.; Hospodiuk, M. Current advances and future perspectives in extrusion-based bioprinting. *Biomaterials* **2016**, *76*, 321–343. [CrossRef] [PubMed]
16. Gudapati, H.; Dey, M.; Ozbolat, I. A comprehensive review on droplet-based bioprinting: Past, present and future. *Biomaterials* **2016**, *102*, 20–42. [CrossRef]

17. Lan, S.; Starly, B. Alginate based 3D hydrogels as an in vitro co-culture model platform for the toxicity screening of new chemical entities. *Toxicol. Appl. Pharmacol.* **2011**, *256*, 62–72. [CrossRef]
18. Kimlin, L.C.; Casagrande, G.; Virador, V.M. In vitro three-dimensional (3D) models in cancer research: An update. *Mol. Carcinog.* **2013**, *52*, 167–182. [CrossRef]
19. Yamada, K.M.; Cukierman, E. Modeling tissue morphogenesis and cancer in 3D. *Cell* **2007**, *130*, 601–610. [CrossRef]
20. Thakuri, P.S.; Gupta, M.; Plaster, M.; Tavana, H. Quantitative Size-Based Analysis of Tumor Spheroids and Responses to Therapeutics. *Assay Drug Dev. Technol.* **2019**, *17*, 140–149. [CrossRef]
21. Han, S.J.; Kwon, S.; Kim, K.S. Challenges of applying multicellular tumor spheroids in preclinical phase. *Cancer Cell Int.* **2021**, *21*, 152. [CrossRef] [PubMed]
22. Fatimi, A.; Okoro, O.V.; Podstawczyk, D.; Siminska-Stanny, J.; Shavandi, A. Natural Hydrogel-Based Bio-Inks for 3D Bioprinting in Tissue Engineering: A Review. *Gels* **2022**, *8*, 179. [CrossRef] [PubMed]
23. Gong, X.; Lin, C.; Cheng, J.; Su, J.; Zhao, H.; Liu, T.; Wen, X.; Zhao, P. Generation of Multicellular Tumor Spheroids with Microwell-Based Agarose Scaffolds for Drug Testing. *PLoS ONE* **2015**, *10*, e0130348. [CrossRef] [PubMed]
24. Chen, Y.; Lou, X.; Zhang, Z.; Ingram, P.; Yoon, E. High-Throughput Cancer Cell Sphere Formation for Characterizing the Efficacy of Photo Dynamic Therapy in 3D Cell Cultures. *Sci. Rep.* **2015**, *5*, 12175. [CrossRef] [PubMed]

pharmaceutics

MDPI

Article

Micellar Curcumin Substantially Increases the Antineoplastic Activity of the Alkylphosphocholine Erufosine against TWIST1 Positive Cutaneous T Cell Lymphoma Cell Lines

Antonios G. X. Trochopoulos [1,†], Yana Ilieva [2,†], Alexander D. Kroumov [3], Lyudmila L. Dimitrova [2], Ivanka Pencheva-El Tibi [4], Stanislav Philipov [5], Martin R. Berger [6], Hristo M. Najdenski [2], Krassimira Yoncheva [7,*], Spiro M. Konstantinov [1] and Maya M. Zaharieva [2,*]

1 Department of Pharmacology, Pharmacotherapy and Toxicology, Faculty of Pharmacy, Medical University of Sofia, 2 Dunav Str., 1000 Sofia, Bulgaria
2 Department of Infectious Microbiology, The Stephan Angeloff Institute of Microbiology, Bulgarian Academy of Sciences, 26 Acad. G. Bonchev Str., 1113 Sofia, Bulgaria
3 Department of Applied Microbiology, The Stephan Angeloff Institute of Microbiology, Bulgarian Academy of Sciences, 26 Acad. G. Bonchev Str., 1113 Sofia, Bulgaria
4 Department of Pharmaceutical Chemistry, Faculty of Pharmacy, Medical University of Sofia, 2 Dunav Str., 1000 Sofia, Bulgaria
5 Department of Human Anatomy, Histology, General and Clinical Pathology and Forensic Medicine, Faculty of Medicine, University Hospital Lozenetz, Sofia University "St. Kliment Ohridski", 2 Kozyak Str, 1421 Sofia, Bulgaria
6 Unit of Toxicology and Chemotherapy, German Cancer Research Center, D-69120 Heidelberg, Germany
7 Department of Pharmaceutical Technology and Biopharmaceutics, Faculty of Pharmacy, Medical University of Sofia, 2 Dunav Str., 1000 Sofia, Bulgaria
* Correspondence: kyoncheva@pharmfac.mu-sofia.bg (K.Y.); zaharieva26@yahoo.com (M.M.Z.)
† These authors contributed equally to this work.

Abstract: Cutaneous T-cell lymphoma (CTCL) is a rare form of cancer with local as well as systemic manifestations. Concomitant bacterial infections increase morbidity and mortality rates due to impaired skin barrier and immune deficiency. In the current study, we demonstrated that the in vitro anti-lymphoma potential of erufosine is diminished by TWIST1 expression and micellar curcumin substantially increases its antineoplastic activity. Pharmacokinetic analysis showed that the micellar curcumin (MCRM) used in our study was characterized by low zeta potential, slow release of curcumin, and fast cell membrane penetration. The combination ratio 1:4 [erufosine:MCRM] achieved strong synergism by inhibiting cell proliferation and clonogenicity. The combined antiproliferative effects were calculated using the symbolic mathematical software MAPLE 15. The synergistic combination strongly decreased the expression of TWIST1 and protein kinase B/Akt as proven by western blotting. Significant reductions in NF-κB activation, induction of apoptosis, and altered glutathione levels were demonstrated by corresponding assays. In addition, the synergistic combination enhanced the anti-staphylococcal activity and prevented biofilm formation, as shown by crystal violet staining. Taken together, the above results show that the development of nanotechnological treatment modalities for CTCL, based on rational drug combinations exhibiting parallel antineoplastic and antibacterial effects, may prove efficacious.

Keywords: cutaneous T-cell lymphoma; TWIST1; erufosine; curcumin; synergy; nanotechnology

Citation: Trochopoulos, A.G.X.; Ilieva, Y.; Kroumov, A.D.; Dimitrova, L.L.; Pencheva-El Tibi, I.; Philipov, S.; Berger, M.R.; Najdenski, H.M.; Yoncheva, K.; Konstantinov, S.M.; et al. Micellar Curcumin Substantially Increases the Antineoplastic Activity of the Alkylphosphocholine Erufosine against TWIST1 Positive Cutaneous T Cell Lymphoma Cell Lines. *Pharmaceutics* **2022**, *14*, 2688. https://doi.org/10.3390/pharmaceutics14122688

Academic Editor: Tomáš Etrych

Received: 31 October 2022
Accepted: 27 November 2022
Published: 1 December 2022

1. Introduction

Cutaneous T-cell lymphomas (CTCL) represent a heterogeneous group of rare extranodal T-cell lymphoproliferative disorders (non-Hodgkin's lymphomas, NHLs) which primarily affect the skin by a clonal accumulation of skin-homing CD4+CD45RO+ helper/memory neoplastic T-lymphocytes [1,2]. CTCL progresses by involving the lymph nodes, blood, and visceral organs [3,4], and many patients develop relapsed/refractory disease with a potentially fatal prognosis [5]. Two of the most important subtypes of CTCL are the cutaneous mycosis fungoides (MF) [6] and the leukemic Sézary syndrome (SS) [7]. Both are characterized by a poor quality of life and may lead to seriously shortened overall survival, especially if an extracutaneous involvement is present. An additional important clinical problem associated with major morbidity and mortality rates is the high frequency of concomitant bacterial infections due to impaired barrier function of the skin and progressive immune deficiency [8]. Axelrod et al. found out that 396 of 478 documented microbial infections in CTCL patients were of bacterial origin, e.g., *Staphylococcus aureus*, and were intimately associated with the disease stage [9]. Other published data indicate that staphylococcal enterotoxins may promote the expansion of malignant T-cells [8].

The current therapy of CTCL is challenging, often empiric, and not typically based on specific molecular alterations due to limited insight into the genetic basis of the disease [10]. Recent research based on next-generation sequencing revealed potentially targetable oncogenic mutations in the nuclear factor κ B (NF-κB) and the Janus Kinase and Signal Transducer and Activator of Transcription (JAK-STAT) signaling pathways whose abnormal activation causes apoptosis resistance [11–13]. These cancer-promoting somatic mutations affect transcription factors such as TWIST1 (Twist-related protein 1), thereby altering the T-cell effector function and driving lymphomagenesis into proliferation [14,15]. The TWIST1 transcription regulator plays an essential role in cancer metastasis and is activated by a variety of signal transduction pathways, including protein kinase B (PKB/Akt), STAT3, mitogen-activated protein kinase (MAPK), Ras, and Wnt signaling. TWIST1 is thought to promote tumor progression in MF and SS via the p53 axes for cell G1/S cycle arrest with subsequent inhibition of the c-myc-induced apoptosis [16,17], which makes it an attractive molecular target in the personalized treatment approach.

The current therapeutic modalities of CTCL include skin-directed treatments (for patients with limited skin disease and favorable overall survival), retinoids and histone deacetylase inhibitors for advanced-stage disease, and classical cytoreductive chemotherapeutics for relapsed/refractory CTCL forms [7,18,19]. Extracorporeal photopheresis, which is characterized by an excellent side effect profile and moderate efficacy, is considered the first-line therapy for erythrodermic MF and SS [20,21]. Patients with significant nodal, visceral, or blood involvement are generally treated with biologic-response modifiers before escalating to systemic, single-agent chemotherapy. In highly-selected patients, allogeneic stem-cell transplantation may be considered, as this may be curative in some patients [22]. Systemic treatment for relapsed/refractory CTCL has historically relied on traditional chemotherapeutics, retinoids interferons, interleukins, phosphorylase inhibitors, histone deacetylase or proteasome inhibitors; however, responses are often short-lived [23]. Response rates of the clinically approved histone deacetylase inhibitors romidepsin and vorinostat are typically <35%. They can induce some durable responses in heavily pretreated patients and alleviate bothersome symptoms, such as pruritus. Failure to cure advanced SS and MF with large cell transformation peripheral T-cell lymphoma has resulted in a search for novel targeted agents, including antibodies and gene modulators [7], such as anti-CD30 antibody-drug conjugate brentuximab vedotin, anti-CCR4 antibody mogamulizumab, and the fusion protein immunotoxin A-dmDT390-bisFv(UCHT1) [5,18]. Nevertheless, none of these drugs were related to distinguished advances in CTCL therapy.

The drug erufosine (ERF) is a third-generation alkylphosphocholine (APC) with favorable pharmacokinetics and a broad spectrum of in vitro and in vivo antineoplastic activities [24–28], as well as antibacterial activity against pathogenic *Staphylococcus aureus* strains in clinically applicable concentrations [29]. In contrast to other APCs [30], ERF exhibits less pronounced cholinomimetic side effects [31] and can be given intravenously due to a lack of hemolytic properties [32]. The comprehensive mode of action of erufosine includes apoptotic and/or autophagy-mediated cell death in a dose-dependent manner, inhibition of the PKB/Akt-Rb and mTOR axes, and induction of G2/M cell cycle arrest through modulation of the cyclin-dependent kinase inhibitor p27Kip1 [28,33–43]. One of the most important advantages of ERF is the absence of bone marrow toxicity [38,44,45], which makes it an attractive candidate for combined chemotherapies. The progenitor of ERF, miltefosine, was tested in phase I-II studies for topical treatment (6% ointment) of cutaneous lymphomas and led to an overall response rate between 58 and 71% and response duration of 12 months without causing myelotoxicity [46–50]. However, more than 50% of the patients developed strong side effects such as erythema, scaling, skin atrophy, local desquamation, and pruritus [46–48], and the clinical trials were discontinued. These results raise the issue and give hope that erufosine could be a more suitable option for CTCL treatment, especially in rational synergistic combinations with anti-inflammatory compounds that could ameliorate possible adverse events without compromising the antineoplastic effect.

Curcumin (CRM) is the major active component of the spice turmeric (*Curcuma longa*, Zingiberaceae) [51] with pleiotropic pharmacological effects which has been used for centuries in the Indian traditional medicine as an anti-inflammatory and antimicrobial remedy [52]. The most important cellular target of curcumin is the pro-inflammatory mediator NF-κB which explains the chemoprotective, antiproliferative, anti-apoptotic, and anti-carcinogenic effects of the compound [53,54]. Curcumin is a potent inhibitor of lymphoblasts' proliferation in CTCL cell lines through modulation of the JAK/STAT and NF-κB signaling and induction of oxidative stress [55,56]. It is also a suitable candidate for combination therapies because of the low cytotoxicity on normal tissues. The cellular uptake of curcumin is higher in malignant cells than in normal; therefore, it is well tolerated in humans [57]. However, curcumin's clinical application is limited by the low water solubility and fast metabolism after absorption from the gastrointestinal tract leading to low bioavailability. As a consequence, only small amounts of curcumin are detectable in target tissues which can be overcome through incorporation into polymeric micellar systems suitable for cutaneous application [58–61].

Having in mind the published scientific data on the therapeutic approaches for CTCL and the gaps in this research area, we set ourselves the goal to evaluate in detail the in vitro pharmacological potential of basically new therapeutic modalities for targeted inhibition of signal transduction pathways involved in the carcinogenesis of CTCL such as TWIST1, PKB/Akt and NF-κB. The focus of the investigations falls on the antineoplastic activity and mode of action of rationally selected synergistic combinations between ERF and micellar curcumin (MCRM) in a panel of CTCL cell lines and the potential of such combinations to inhibit the growth and biofilm formation of pathogenic *Staphylococcus aureus* strains. Our specific aim was to demonstrate that (1) ERF is a suitable drug candidate for the treatment of CTCL, but the transcription factor TWIST 1 reduces the efficacy of ERF in TWIST1 expressing CTCL cell lines, and (2) combining ERF with MCRM will increase the antineoplastic effect of both substances, thus leading to a significant TWIST1 inhibition, deactivation of PKB/Akt and NF-κB, and suppression of *Staphylococcus aureus* biofilm formation.

2. Materials and Methods

2.1. Drugs and Chemicals

Curcumin (#C1386, Mw = 368.385 g/mol), absolute ethanol (#46139), methanol (#322415), methoxy poly(ethylene glycol)-block-poly(ε-caprolactone) (#900649), glacial acetic acid (#1005706 USP), crystal Violet (#C0775), Tris-HCl (#T5941), Sodium Dodecyl Sulfate (#L3771), glycerol (#G5516), DL-Dithiothreitol (#43815), skimmed milk powder (#70166), Tween 20® (#P1379), 5,5′-Dithiobis(2-nitrobenzoic acid) (#D8130), Hoechst 33342 (#14533) and 3-(4, 5-dimethylthiazolyl-2)-2, 5-diphenyltetrazolium bromide (#M2128, MTT dye) were purchased from Sigma® Life Science (Roedermark, Germany). Working solution of Gentamycin (40 mg/L) was prepared through dilution of commercially available stock (amp. 40 mg/mL, 1 mL, Sopharma®, Sofia, Bulgaria) in HPLC purified water. Erufosine (Mw = 503.74 g/mol) was kindly provided by Prof. Hans-Jörg Eibl [62] in the form of 20 mM stock solution in 0.9% NaCl and was stored at 4 °C.

2.2. Cell Lines and Cultivation Procedure

All three cell lines originated from the American Type Culture Collection (ATCC): HuT-78 (lymphoblasts, Sezary Syndrome, ATCC®TIB-161™), MJ (lymphoblasts, Mycosis fungoides, ATCC®CRL-8294™), HH (lymphoblasts, cutaneous T cell lymphoma; ATCC®CRL-2105™). Cell cultures were maintained at cell density 5×10^4–8×10^5 viable cells/mL under standard conditions (37 °C, 5% CO_2, humidified atmosphere, Panasonic CO_2 incubator, #MCO-18AC-PE, Osaka, Japan) according to the recommendations of ATCC for growth media and subculturing rate. The following buffers and media were used in the cell culturing procedures: (1) RPMI-1640 without Phenol Red (#RPMI-XRXA, Capricorn®, Düsseldorf, Germany), supplemented with 4 mM L-Glutamine (#G7513, Sigma® Life Science, Germany), 20% fetal bovine serum (#FBS-HI-12A, Capricorn®, Germany), 25 mM HEPES buffer solution (#HEP-B, Capricorn®, Germany) and 4.5 g/l D-(+)-glucose (#G8769, Sigma® Life Science, Germany); (2) IMDM (#IMDM-A, Capricorn®, Germany), supplemented with 4 mM L-Glutamine and 20% fetal bovine serum and (3) Dulbecco's Phosphate Buffered Saline (PBS, #D8537, Sigma® Life Science, Germany). For all experiments, cells were plated at cell density 3×10^5 viable cells/mL.

2.3. Bacterial Strains and Growth Conditions

The methicillin-resistant *Staphylococcus aureus* strain NBIMCC 8327 (MRSA, Bulgarian National Bank for Industrial Microorganisms and Cell Cultures, Sofia, Bulgaria) was used for the testing of combination effects and biofilm assay. The bacterial cultures were maintained at 37 °C under aerobic conditions using Trypticase Soy Broth (TSB, #LQ508) and Agar (TSA, #M1968) purchased from HiMedia®, Mumbai, India. For the biofilm inhibition assay Brain Heart Infusion Broth (BHI, # GM210, HiMedia®, India), supplemented with 2% D-Glucose (#G8769, Sigma® Life Science, Germany) was used.

2.4. Preparation and Characterization of Curcumin Loaded Micelles

The mPEG-PCL copolymer and CRM were dissolved in dioxane at a ratio 10:1 (*wt/wt*). The organic solution was gently stirred for 30 min (700 rpm), and after that purified water was added drop by drop. The resulting micellar dispersion was introduced into a dialysis membrane (MW = 6000–8000), which was further immersed in purified water. The outer aqueous medium was replaced 4 times. The size and zeta potential were determined by photon correlation spectroscopy and electrophoretic laser doppler velocimetry (Zetamaster analyzer, Malvern Instruments, Malvern, UK). Freshly prepared micellar dispersions were measured at 25 °C with a scattering angle of 90°. In vitro release of CRM from the micelles was examined by dialysis. Briefly, the micellar dispersion was poured into a dialysis membrane bag (MW = 6000–8000) that was further placed into 100 mL of phosphate buffer (pH = 7) containing 10% ethanol. Samples were withdrawn from the medium outside the dialysis bag, and the concentration of the released CRM was determined by UV-Vis spectrophotometry at $\lambda = 425$ nm.

2.5. Cell Viability Test

The cell viability was evaluated according to ISO10993-5, Annex C [63] (MTT dye reduction assay). Briefly, prior treatment cells were seeded in 96-well plates (3×10^5 cells/mL) under sterile conditions (Laminar Air Flow Telstar Bio II Advance, Terrassa, Spain), incubated for 24 h until entering the *log*-phase of the growth curve, and treated with erufosine (0–200 µM in serial twofold dilutions) for 24, 48 and 72 h. The schema for the combinations between erufosine and micellar curcumin followed the recommendations of Chou and Talalay for constant drug ratios [64]. All experiments were performed in triplicate, wherein every sample was repeated 4 times. The formazan intensity was measured on an Absorbance Microplate Reader EL-800 (Bio-Tek Instruments Inc., Winooski, VT, USA) at $\lambda = 550$ nm ($\lambda_{ref} = 690$ nm) against a blank solution (culture medium, MTT, and solvent).

2.6. Mathematical Modeling of Cytotoxic Effects after Single Drug Treatment

The calculation of the median single drug effects (median inhibitory concentrations, IC_{50}) was performed as published before [65]. Briefly, we coded a nonlinear regression procedure in MAPLE® software of symbolic mathematics based on weighted least squares statistical criterion as an objective function of the search. A numerical optimization algorithm was used to minimize the sum of weighted squares and to find the estimates of best-fitting parameter values. The median-dose model was applied to obtain the "IC_{50}" and "m", as presented in Equation (1):

$$\frac{F_a}{F_u} = \left(\frac{Dose}{D_m}\right)^m, \tag{1}$$

where F_a represents the affected fraction; F_u—the unaffected fraction $(1 - F_a) = F_u$; *Dose*—the applied drug concentration; D_m—the median-effect dose (in our case $D_m = IC_{50}$), and m—a slope of median-effect plot (for $m = 1$ the curve is hyperbolic; for $m > 1$—sigmoidal; for $m < 1$—negative (flat) sigmoidal). In addition, we performed response surface analysis (RSA) methodology in order to reveal the predictive power of the model as a function of the parameters "IC_{50}" and "m". The range of the value changes in the RSA 3D plot was determined based on the standard deviation of the "IC_{50}" and "m" values obtained during the statistical evaluation of the experimental data with the GraphPad Prism software.

2.7. Mathematical Modelling of Drug-Drug Interactions in Cell Lines

A computer program in the platform of the symbolic mathematical software MAPLE® was applied for calculation of the combination effects, and the simulation results were compared with calculations of CompuSyn (Paramus, NJ, USA) [64]. The simulations quantitatively evaluated the effects of action of two applied drugs MCRM and ERF (Dose A:Dose B). The combinations schema followed the recommendations of Chou and Martin [64]. Combination ratios 1:2 and 1:4 [ERF:MCRM] were applied. The mathematical equation of CI (combination index) was written as follows for earlier determined IC_{50} values of drugs A, B, and the combination AB:

$$CI_{(alpha = 0)} = D_1/Dx_1 + D_2/Dx_2 \tag{2}$$

$$CI_{(alpha = 1)} = D_1/Dx_1 + D_2/Dx_2 + alpha \times (D_1/Dx_1) \times (D_2/Dx_2), \tag{3}$$

where D_{x1} = dose of drug 1 only to obtain 50% cell inhibition; D_1 = dose of drug 1 to obtain 50% cell inhibition in combination with D_2; D_{x2} = dose of drug 2 only to obtain 50% cell killing; (D) 2 = dose of drug 2 to obtain 50% cell killing in combination with D_1; and the values of alpha = 0 for mutually exclusive or alpha = 1 for non-mutually exclusive actions of drugs. Only the results for mutually exclusive actions of drugs are presented in this study. The step-by-step, fully automated simulation procedure resulted in isobolograms for HuT-78 and MJ cell lines.

According to the theory included in the manuals of the software program CompuSyn [64], the meanings of CI can be classified in more detail as follows: CI > 1.3 shows antagonism; CI = 1.1 to 1.3 moderate antagonism; CI = 0.9 to 1.1 additive effect; CI = 0.8 to 0.9 slight synergism; CI = 0.6 to 0.8 moderate synergism; CI = 0.4 to 0.6 synergism; CI = 0.2 to 0.4 strong synergism. It should be noted that in Equations (2) and (3) the value of D_x (dose of a single drug) can be determined from the following equation:

$$D_x = D_m[F_a/(1 - F_a)]^{1/m}, \tag{4}$$

where F_a—stands for the affected fraction (has inhibition); F_u—means the unaffected fraction "u", $F_u = (1 - Fa)$ (no inhibition, control); D_x is the single dose of the drug; D_m means the dose giving the mean effect; hence, $Dm = IC_{50}$ (in our case) 50% inhibition; "m"—is the slope of the dose-effect curve.

2.8. Determination of Extracellular Curcumin Levels

The extracellular levels of micellar curcumin were determined in comparison to pure reference substance curcumin with a UV-spectrophotometric method using the following system: UV/VIS Spectrometer HP; Diode array detector; Analytical wavelength 426 nm ± 2 nm and operating software. The analytical calculations are based on Multicomponent analysis calculations. The following method options were used: multicomponent analysis (MCA), Beer's law calibration curve type, least squares fit (LSQ) algorithm, derivative order 0, polynomial degree 0, 1 smoothing point, 2 nm data interval, 426 nm analytical wavelength and temperature of 25 °C. The test was prepared in the following way: for reference solutions (a): accurately weighed masses curcumin RS were dissolved in ethanol to obtain a solution with the following concentrations: $C_1 = 1.00 \times 10^{-5}$ g/mL, $C_2 = 0.75 \times 10^{-5}$ g/mL, $C_3 = 0.65 \times 10^{-5}$ g/mL, $C_4 = 0.55 \times 10^{-5}$ g/mL, $C_5 = 0.50 \times 10^{-5}$ g/mL, $C_6 = 0.45 \times 10^{-5}$ g/mL, $C_7 = 0.30 \times 10^{-5}$ g/mL and $C_8 = 0.20 \times 10^{-5}$ g/mL; For the bioanalytical assay, test solution (b) was prepared as follows: to each sample aliquot containing curcumin were added 4.5 mL solvent mixture from acetonitrile and methanol (1:1, *v*/*v*), the samples were sonicated at 20 °C for 5 min and centrifuged (<5000 rpm, 5 min) for separating of the precipitates, the supernatant mixtures were filtered additionally and subjected to spectrophotometric determination against *blanks* prepared respectively.

2.9. Colony Forming Units (CFU) Assay

The clonogenicity survival assay was performed as published before [40]. Briefly, cells were treated as for MTT assay with selected combinations between ERF and MCRM and the responding single drug concentrations and incubated for 48 h. After that period of time, the cells in all treated groups were counted, and 3000 cells/mL from each group were seeded in semisolid medium (0.8% RPMI-methylcellulose, 40% fetal bovine serum), plated in 12-well plates (600 µL/well) in triplicate and cultured for 10 days. Colonies (≥20 cells in clusters) were counted using an inverted microscope (Boeco BIB-100, Hamburg, Germany).

2.10. Determination of Reduced Glutathione (GSH) Content

The spectrophotometric assay of Sedlak and Lindsay based on Ellman's reagent was used to determine the GSH levels after single and combined treatment [66]. Briefly, HuT-78 and MJ cells were seeded at a density of 0.3×10^6 cells/mL, incubated for 24 h, and treated with 20 µM ERF, 80 µM MCRM or the combination of both. After 48 h incubation, 5×10^5 cells/sample were washed with PBR, centrifuged (6000 rpm, 5 min), and lysed in 100 µL lysis buffer (0.5% triton-100 in 0.2 M EDTA) at 4°C (5 min). The proteins were precipitated with 20 µL 20% (*v*/*w*) trichloroacetic acid. The volume of each sample was adjusted to 200 µL with distilled water, followed by centrifugation at 14 5000 rpm (10 min). Supernatant aliquots were transferred in a 96-well plate (100 µL/well) and mixed with 160 µL Tris buffer (0.4 M, pH 8.9) and 4 µL Ellman's reagent (3.4 mg/mL in methanol). The resulting yellow product was measured at 405 nm on an Absorbance Microplate Reader EL-800 (Bio-Tek Instruments Inc., USA). The GSH levels were presented as a percentage of the untreated control.

2.11. Determination of Cytosolic Mono- and Oligonucleosomes

A photometric enzyme immunoassay (Cell Death ELISA kit, #11 544 675 001, Roche Applied Science, Penzberg, Germany) was used for quantitative in vitro determination of cytoplasmic histone-associated DNA fragments (mono- and oligonucleosomes). Briefly, HuT-78 and MJ cell lines were seeded at density of 0.3×10^6 cells/mL, cultivated for 24 h until entering the exponential growth phase, and exposed to 80 µM MCRM or 20 µM ERF or the combination thereof. After 72 h incubation period, cells were counted, and the immunoassay was performed according to the manufacturer's manual. The absorbance of the reaction product was measured at $\lambda = 405$ nm (490 reference wavelength) against a substrate solution blank.

2.12. Caspase-3 Activity Assay

HuT-78 and MJ cells were seeded in 6-well plates at a density of 0.3×10^6 cells/mL, incubated for 24 h, and treated thereafter with 20 μM ERF, 80 μM MCRM, or the combination of both for further 48 h. Cells were counted and 1×10^6 cells/sample were washed three times with PBS and frozen at −80 °C. The measurement of caspase-3 activation was performed with the Caspase-3 DEVD-R110 Fluorometric and Colorimetric Assay Kit (#30008, Biotium, Fremont, CA, USA) following the experimental protocol of the manufacturer. The absorbance was measured at $\lambda = 490$ nm on an Absorbance Microplate Reader EL-800 (Bio-Tek Instruments Inc., USA). As positive control, HL-60 cells (5×10^4) were used and incubated in hypertonic buffer (10 mM Tris, pH 7.4, 400 mM NaCl, 5 mM $CaCl_2$ and 10 mM $MgCl_2$) for 2 h at 37 °C.

2.13. Detection of Apoptosis with Hoechst Staining

HuT-78 and MJ cells were seeded and treated for the caspase-3 activity assay. Nuclear fragmentation was imaged by staining the cells with Hoechst 33342 (0.1 mg/mL final concentration) for 30 min according to the protocol of Chazotte [67]. Samples were examined under a Nikon TiU fluorescent microscope (UV filter, 200× magnification), and the images were acquired and processed using EZC1 software.

2.14. Western Blot Analysis for Protein Expression

The modulation of TWIST1 and related signal molecules, except for NF-κB, after treatment with erufosine or combinations between erufosine and micellar curcumin was tested by immunoblot analysis as published before [36]. Briefly, all three cell lines were seeded in 6-well plates at a density of 0.3×10^6 cells/mL and treated with 12.5, 25, or 50 μM of erufosine for 24 h or as for the caspase-3 assay. Thereafter, cells were washed in PBS and centrifuged for 5 min at 2000 rpm (Eppendorf® microcentrifuge, Hamburg, Germany). Cell pellets (2×10^6 cells) were lysed in a buffer containing 100 mM Tris-HCl with pH 8.0, 4% SDS, 20% Glycerol, 200 mM DTT, and complete protease inhibitor cocktail (#A7779 Applichem, Darmstadt, Germany). Lysates were boiled 10 min and centrifuged at 13,000 rpm for 10 min at 4 °C. Aliquots of 10 μL were taken from the lysates before adding DTT, diluted five-fold in distilled water and quantified for protein concentration with the Pierce BCA Protein Assay Kit (#23225, ThermoFisher Scientific, Waltham, MA, USA). The total protein lysates (50 mg) were subjected to electrophoresis (8% polyacrylamide SDS gels), and proteins were transferred onto PVDF membranes (#IPFL00005 Sigma–Aldrich). The specific antibody labeling was performed in Tris buffer saline supplemented with 0.1% Tween and 1% skimmed milk. The TWIST1 (sc-81417) and β-Actin antibodies (C-2, sc-8432) were purchased from Santa Cruz Biotechnology®, Inc. (Dallas, TX, USA), whereas p-Akt^{Ser473} (#4060) and p-Akt^{Thr308} (#13038) originated from Cell Signaling Technology® (Danvers, MA, USA). The antibodies were diluted according to the manufacturer's instructions. Immunoblots were developed using an HRP-conjugated anti-mouse m-IgGκ BP-HRP (sc-516102, Santa Cruz Biotechnology) or anti-rabbit IgG (#7074, Cell Signaling Technologies, USA) on a C-DiGit Blot Scanner (Li-Cor Biotechnology, Lincoln, NE, USA). The protein expression was normalized based on β-Actin levels by densitometric analysis of the digitized autographic images using the Quantity One 1-D Analysis software 4.6.6. (Bio-Rad, Hercules, CA, USA).

2.15. NF-κB p65 ELISA

HuT-78 and MJ cells were plated and treated for the caspase-3 activity assay. The NF-κB p65 activation after single and combined treatment was evaluated using the NF-κB p65 ELISA (#ADI-EKS-446, Enzo Life Sciences (ELS) AG, Lausen, Switzerland). The protocol was performed following the manufacturer's instructions.

2.16. Biofilm Formation Assay

The biofilm formation assay was performed according to the protocol of Stepanovic et al. [68]. Two-fold serial dilutions of the combinations between ERF and MCRM in concentrations ranging from 1.25/5 to 20/80 µM ERF/MCRM were prepared using BHI broth supplemented with 2% glucose (*w*/*v*). The samples were placed in 96-well polystyrene tissue culture plates at a final volume of 100 µL/well. An equivalent volume of MRSA bacterial inoculum (5×10^5 CFU/mL) was added to each well. Cells were incubated aerobically for 24 h at 37 °C under static conditions. The supernatant was discarded, and planktonic cells were removed three times by washing with PBS (250 µL/well). Cells attached in biofilm were fixed with methanol (200 µL/well, 15 min), air dried, and stained with 0.1% crystal violet (200 µL/well, 5 min). Excess stain was rinsed off with tap water and air dried. The biofilm formation was documented microscopically (40× magnification). Thereafter, the stained biofilm was re-solubilized in 160 µL of 33% acetic acid, and the OD was measured at $\lambda = 550$ nm. The biofilm inhibitory concentrations (BIC) were calculated with the GraphPad Prism software and presented in graphs by using a nonlinear regression model (dose-response inhibition, variable slope after normalization, and logarithmic transformation of the applied concentrations). The minimum biofilm inhibition concentration ($MBIC_{50}$) was defined as the concentration of the tested drug that led to 50% inhibition of the biofilm formation.

2.17. Statistical Evaluation

The experimental data were analyzed statistically with the GraphPad Prism software (Version 5.00, for Windows, GraphPad Software, La Jolla California, San Diego, CA, USA). Each experiment was performed in triplicate. Minimum of three samples for each concentration, the positive, negative, and untreated controls, were prepared. Data were presented as the mean ± SD. One-way and two-way analysis of variance and the two-independent sample Student's *t*-tests were applied to compare two or more groups. A value of $p < 0.05$ was considered statistically significant.

3. Results

3.1. Cytotoxic Effects of Erufosine on CTCL Cell Lines

The IC_{50} value of ERF for HH cells (8.6 µM) was nearly twofold lower than that for MJ cells (16 µM) and twice and a half lower than that determined for HuT-78 cells (approx. 19 µM). All coefficients of determination were higher than 0.95, which ensures the best curve fit. The response surface analysis (RSA) confirmed the reliability of the model (Figure 1). No cytotoxic effect (cell viability > 70%) was found for MCRM.

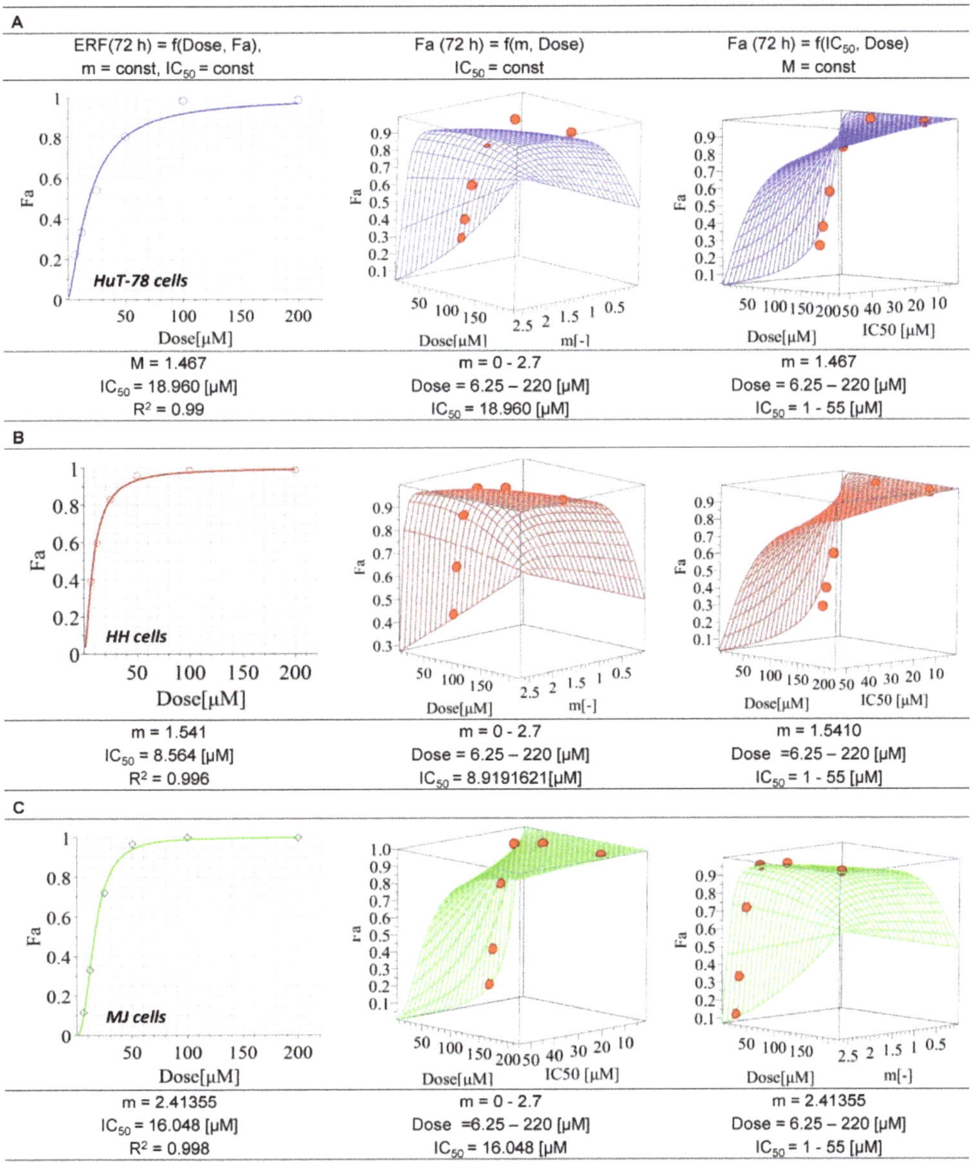

Figure 1. Median inhibitory concentrations of erufosine on the cell lines HuT-78, HH and MJ. **Legend:** ERF—erufosine; m—hillslope; R^2—coefficient of determination; IC_{50}—inhibitory concentration 50% (median inhibitory concentration); Fa—drug effect; (**A**)—Cell line HuT78 (Sézary Syndrome); (**B**)—Cell line HH (cutaneous T-cell lymphoma); (**C**)—Cell line MJ (Mycosis fungoides).

3.2. Production and Internalization Rate of MCRM

Due to the poor water solubility of curcumin, it was formulated in nano-sized micelles based on methoxy poly(ethylene glycol)-block-poly(ε-caprolactone) (mPEG-PCL) copolymer (Figure 2A). The mean diameter of the micelles loaded with curcumin was approximately 125 nm, and the zeta-potential was negative (−27 mV), providing colloidal stability of the resulting nanosystem. The encapsulation efficiency reached 80%, probably due to a high affinity between curcumin and the hydrophobic block of the selected copolymer (Figure 2B). This fact was related to the achievement of sustained release of curcumin (Figure 2C). In order to quantify the cell internalization of MCRM in comparison to CRM dissolved in ethanol (ECRM), spectrophotometric estimation of the curcumin content was performed in the cultivation medium up to 3 h after the start of treatment (Table 1). Rest concentrations for MCRM were substantially lower than that for ECRM, thus confirming the enhanced internalization of MCRM.

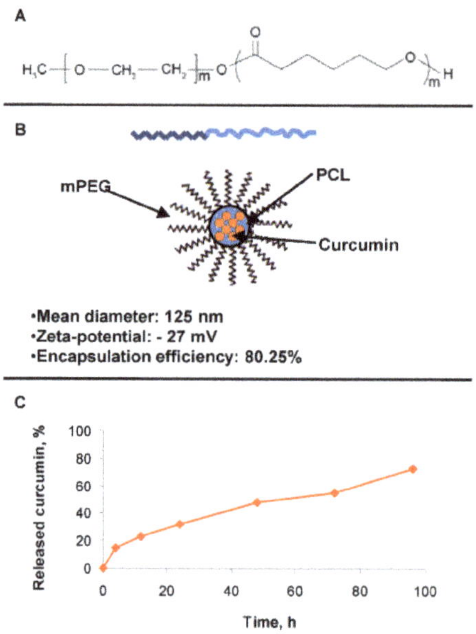

Figure 2. Structure of the micelles and achievement of sustained release with slightly pronounced burst release of curcumin. **Legend**: (**A**)—Copolymer carrier mPEG-PCL; (**B**)—curcumin loaded mPEG-PCL micelles and their physicochemical properties; (**C**)—curcumin release from the loaded mPEG-PCL micelles.

Table 1. Extracellular concentration of MCRM in comparison to pure curcumin after treatment of HuT-78 cells.

Drug	Incubation Time [h]	Absorption (AU) +/− SD	$A_{sample} - A_{blank}$ (AU)	Quantity of the Substance (mmol)
Control		0.05748 +/− 0.01	-	-
CRM	0	0.47531 +/− 0.01	0.41783	8.0290×10^{-5}
MCRM		0.25410 +/− 0.01	0.19662	3.3212×10^{-5}
Control		0.12833 +/− 0.02	-	-
CRM	1	0.38940 +/− 0.02	0.26107	4.4100×10^{-5}
MCRM		0.27864 +/− 0.02	0.15033	2.5394×10^{-5}
Control		0.10574 +/− 0.02	-	-
CRM	2	0.37878 +/− 0.02	0.27304	4.612×10^{-5}
MCRM		0.23807 +/− 0.02	0.13233	2.2354×10^{-5}
Control		0.15630 +/− 0.01	-	-
CRM	3	0.43492 +/− 0.01	0.27862	4.7064×10^{-5}
MCRM		0.23947 +/− 0.01	0.08317	1.4050×10^{-5}

Legend: AU—absorption units; CRM—pure curcumin dissolved in ethanol; MCRM—micellar curcumin.

3.3. Micellar Curcumin Potentiates the Antiproliferative and Anticlonogenic Effect of Erufosine in TWIST1 Expressing CTCL Cell Lines

Based on the calculated median inhibitory concentrations of ERF, we planned the combination treatment following the recommendations and the schema for constant combination ratio in the manual of Chou and Talalay [Manual of CompuSyn. Inc., Paramus, NJ, USA]. The following two combination ratios were investigated: 1:2 and 1:4 [ERF:MCRM]. The highest concentration of ERF used was near the IC_{50} value, and only the concentrations of MCRM were increased up to 80 μM as far as curcumin is less toxic than ERF to normal cells. Only the combination ratio of 1:4 led to synergistic (HuT-78) or additive (MJ) interactions. The results are presented in Figure 3A. The CI values of the combination 80 μM MCRM and 20 μM ERF for HuT-78 cells were lower than 0.9 for all three Fa observed (50, 75, and 90% inhibition of the cell viability), which is indicative of synergism ($CI_{Fa(0.5)} = 0.35$, $CI_{Fa(0.75)} = 0.2$, $CI_{Fa(0.9)} = 0.17$). For MJ cells, an additive effect was achieved by Fa0.9. The data from the CFU test (Figure 3B) revealed 100% CFU inhibition in HuT-78 cells for the combination [20 μM ERF/80 μM MCRM]. In MJ cells, treatment with 40 and 80 μM MCRM and the respective combinations with ERF led to complete CFU inhibition.

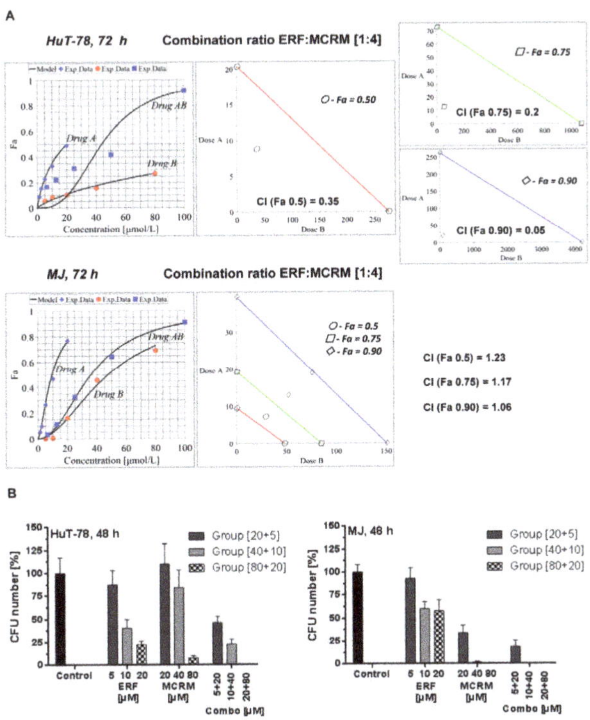

Figure 3. Synergistic and additive combinations between erufosine and micellar curcumin with an inhibitory effect on cell proliferation and clonogenicity of the cell lines HuT-78 and MJ. **Legend:** (**A**)–isobolograms for HuT-78 and MJ cells, calculated by MAPLE software from a MTT assay, where CI stands for Combination Index, Fa 0.5 is the median cytotoxic effect (denoted with points), Fa 0.75 is the effect killing 75% of the cells (denoted with squares) and Fa 0.9 is equal to 90% dead cells (denoted with rhombuses); (**B**)—clonogenic survival (CFU assay) of lymphoblasts from the TWIST1 expressing cell lines HuT-78 and MJ after treatment with combinations achieving Fa 0.9, including the single drug effects of the respective concentrations. Drug A stands for ERF, Drug B stands for MCRM and Drug AB stands for combination between ERF and MCRM.

The combination indexes for the combination ratio 1:2 from the MTT assay are as follows: (1) HuT-78—$CI_{Fa(0.5)}$ = 1.34, $CI_{Fa(0.75)}$ = 2.28, $CI_{Fa(0.9)}$ = 3.87; (2) MJ—$CI_{Fa(0.5)}$ = 1.57, $CI_{Fa(0.75)}$ = 1.25, $CI_{Fa(0.9)}$ = 1.01. An additive effect was achieved in MJ cells at Fa(0.9). All other variants led to an antagonistic effect.

3.4. MCRM Increases the Inhibitory Effect of ERF on TWIST1, PKB/Akt and NF-κB in TWIST1 Expressing CTCL Cell Lines

No expression of TWIST1 was detected in the cell line HH. HuT-78 and MJ cell lines express TWIST1 and ERF inhibited it in a dose-dependent manner (Figure 4A). ERF alone suppressed the expression of TWIST1 by 60% only at a concentration of 50 μM. In the presence of curcumin, a full TWIST1 inhibition was achieved by using more than a twofold lower concentration of ERF. The same effect was observed in the MJ cell line for single administration of ERF and the additive combination. A single administration of ERF (20 μM) on the MJ cell line inhibited both active forms of the protein kinase B. In the HuT-78 cell line, the dephosphorylation of PKB/Akt at Ser^{473} was not significant against the untreated control, while its dephosphorylation at Thr^{308} was more strongly expressed than the untreated control after treatment with 20 μM erufosine (value close to IC_{50}). The combination achieved complete inhibition of protein phosphorylation at both amino acid residues in the MJ cell line, whereas, for HuT-78, this effect was poorly expressed and was the same as after a single treatment with ERF. The NF-κB p65 activity was significantly inhibited (up to 60%) in both cell lines 24 h after administrating the combination.

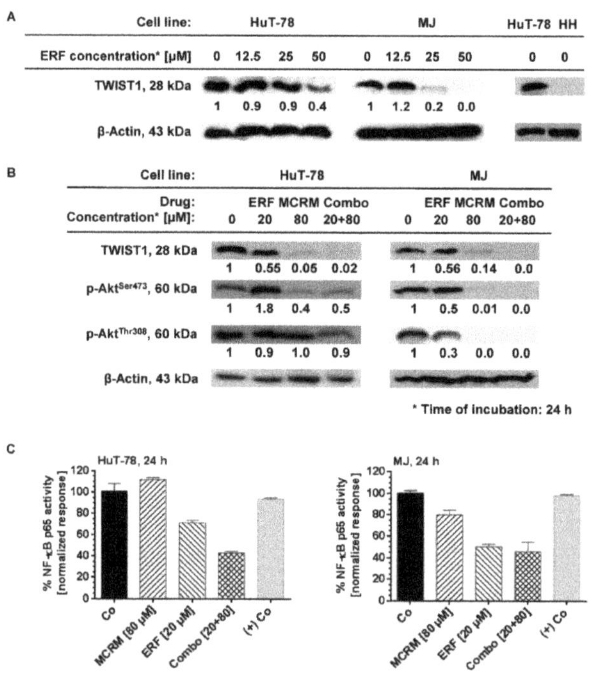

Figure 4. Micellar curcumin potentiates the inhibition effect of erufosine on NF-κB and TWIST1 in the CTCL cell lines HuT-78 and MJ. **Legend:** MCRM—micellar curcumin; ERF—erufosine; (+) Co—nuclear extract of TNFα-activated HeLa cells; Co—untreated control; Combo—combination between erufosine and micellar curcumin with the same concentrations as by the single treatment; (**A**)—expression of TWIST1 by western blot after exposure of the cell lines to erufosine; (**B**)—expression ot TWIST1, p-Akt^{Ser472} and p-Akt^{Thr308} by western blot after exposure of the cell lines to the combinations; (**C**)—NF-κB p65 activity assay of ERF, MCRM and the combinations combination.

3.5. MCRM Potentiates the Anti-Apoptotic Effect of ERF in TWIST1 Expressing CTCL Cell Lines

In both cell lines, the combinations between MCRM and ERF induced apoptosis and resulted in significant nuclear fragmentation in the treated cells to a greater degree than after single administration of the corresponding concentrations. The caspase-3 activity induced in MJ cells by the combinations was twice as strong as compared to that induced by MCRM or ERF separately (Figure 5A). The effect of the combination in HuT-78 resulted in a fourfold higher accumulation of mono- and oligonucleotides in the cytosol of the treated cells as compared to the untreated control (Figure 5C). The Hoechst microscope images (Figure 5D) revealed nuclear fragmentation at preserved nuclear membrane in HuT-78. In MJ cells, a breakdown of cellular structures and extracellular condensed nuclear fragments were observed.

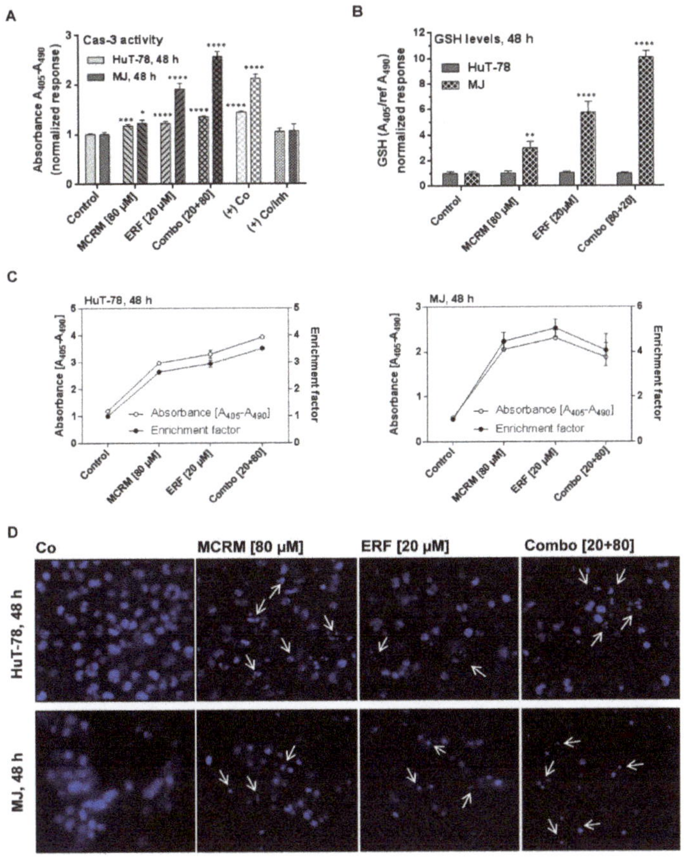

Figure 5. Synergistic interactions between micellar curcumin and erufosine in ratio 4:1 lead to apoptotic cell dead in HuT-78 and MJ cells. **Legend**: (+) Co—cells incubated with hypertonic buffer; (+) Co/Inh—cells incubated with hypertonic buffer in the presence of caspase 3 inhibitor; (**A**)—caspase 3 activity of samples treated with ERF, MCRM or combinations; (**B**)—GSH levels in the same samples, presented in part A; (**C**)—accumulation of mono- and oligonucleotides in the cellular cytosol in the same samples presented in parts A and B; (**D**)—nuclear fragmentation after Hoechst staining of the cells in the same samples presented in parts A, B and C (depicted with white arrows). Significant differences between groups in (**A**,**B**) are marked with asterisks: *—$p \leq 0.05$; **—$p \leq 0.01$; ***—$p \leq 0.001$, ****—$p \leq 0.0001$.

3.6. Effects of the Combinations between ERF and MCRM on Total GSH Levels in TWIST1 Expressing CTCL Cell Lines

Glutathione levels did not change at the 48th hour of the treatment in any of the HuT-78 samples, unlike the other cell line MJ, which showed a strong enhancement of total GSH after treatment with MCRM and the combination (Figure 5B). The result correlates with increased levels of caspase-3 in the same samples. Obviously, in MJ cells, glutathione levels were significantly enhanced—up to 10-fold higher than in the untreated control. A concentration-dependent manner of the effect was detected.

3.7. The Combination Ratio of 4:1 between MCRM and ERF Inhibits Biofilm Formation of MRSA

The synergistic concentration MCRM:ERF [4:1] significantly inhibited the biofilm formation of MRSA. The effect of the combination of concentrations 80 μM MCRM and 20 μM ERF is equal to that of ERF after a single application, demonstrating full biofilm eradication (Figure 6). In lower concentrations of both drugs, the effect on biofilm formation was antagonistic.

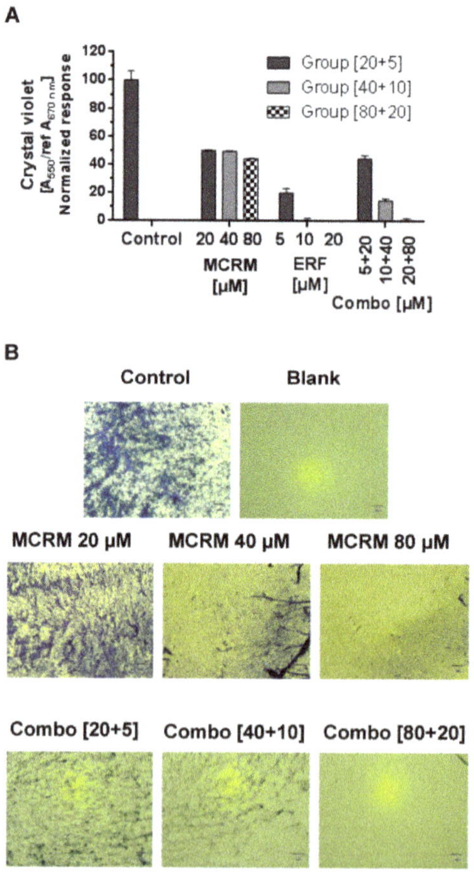

Figure 6. Effect of the synergistic combination [ratio 4:1] between micellar curcumin and erufosine on the biofilm formation of MRSA. **Legend:** ERF—erufosine; MCRM—micellar curcumin; (**A**)—crystal violet absorbance of MRSA biofilm; (**B**)—Morphological evaluation of MRSA biofilm (40× microscope magnification).

4. Discussion

In the present study, we compared and evaluated in detail for the first time, to our knowledge, the in vitro efficacy of the alkylphosphocholine ERF in combination with curcumin incorporated in micellar formulation with a prolonged release on T-cell lymphoma cell lines with and without expression of the oncogene TWIST1. The transcription factor TWIST1 is not detected in normal peripheral blood mononuclear cells. Any increase in its expression levels is abnormal and has been identified as a marker of metastasis and poor prognosis in patients with MF and SS [69]. Furthermore, it has been found that its expression increases in advanced MF/SS lesions and triggers chemotherapy resistance [16,70]. The underlying mechanisms are related to the inhibition of both p53 and Rb tumor-suppressor pathways and the up-regulation of protein kinase B (PKB/Akt) in malignant cells with increased TWIST1 levels [70,71]. In cancer cells, TWIST1 targets several enzymes from the DNA damage response pathway, thus neutralizing senescence and cell death [72]. Based on the accumulated knowledge about the mode of action of the APCs [40,73] and our previous studies on the effects of ERF on CTCL cell lines demonstrating clear inhibition of PKB/Akt [41], increase in the Rb expression levels in T-cells, modulation of the Rb-protein signaling pathway [40] and induction of apoptosis, we presumed that the expression levels of TWIST1 might affect the activity of ERF or vice versa. In our study, ERF showed different activity in the tested cell lines, with HuT-78 being the most resistant and HH being the most sensitive (Figure 1), as evidenced by the estimated IC50 values. The behavior of the RSA simulations, along with the dose-effect curves, confirmed the reliability of the chosen mathematical model in the MAPLE software. The deviation of the model from the experimental data (see points) for the selected 95% confidence interval was not very high, meaning that the values of all constants in the selected ranges of m and the IC50 parameters can be considered reliable and stable. Determination of the TWIST1 expression levels in all three cell lines through western blot analysis revealed that the protein was expressed in HuT-78 and MJ but not in the most sensitive cell line HH (Figure 4). Moreover, inhibition of TWIST1 occurred in MJ at a lower ERF concentration (25 μM) than in HuT-78 (50 μM). This finding could explain the differences in the sensitivity of the cell lines toward the drug.

According to published data, there is cross-talk between TWIST1 and PKB/Akt in the promotion of metastasis via the TGF-β signaling axes [71]. On the one hand, TWIST1 mediates the phosphorylation of PKB/Akt, thus contributing to uncontrolled cell proliferation and invasion. On the other hand, TWIST1 associates directly with the subunit RELA of the pro-inflammatory factor NF-κB to activate the transcriptional activity of the latter and to promote cell invasion through IL8 production [74]. ERF is a potent inhibitor of the PKB/Akt signaling pathway in leukemic and lymphoma cell lines [35–37,39,41], whereas the natural product curcumin is a well-known proteasome inhibitor of NF-κB [75–78]. Based on these facts, we built our hypothesis on the presumption that a combination between ERF and MCRM will achieve simultaneous down-regulation of PKB/Akt and NF-κB with subsequent inhibition of cell proliferation and induction of apoptosis. In addition, ERF and CRM possess in vitro antimicrobial activity against the pathogenic bacterial species *Staphylococcus aureus* [29], which was shown to accelerate the pathogenesis of CTCL [79,80]. That is why; we set the additional goal to study the MRSA anti-biofilm activity of the selected combinations, as far as it is a highly important mechanism providing survival of the microorganisms in conditions of external stress and chemotherapy. Taking into account both the poor curcumin solubility and its hydrolytic instability in a slightly alkaline medium, we assumed that polymeric micelles would be appropriate nano-carriers for cutaneous application. The hydrophobic nature of their micellar core enables the loading of active substances with poor water solubility. Moreover, the loading of unstable drugs into micelles would increase their stability against different degradation processes such as hydrolysis or oxidation [58–61]. Despite the advantages of the micelles, the studies of polymeric micelles intended for the cutaneous application of curcumin are limited to a few reports [81,82]. Therefore, in the present study, we developed curcumin-loaded nano-sized micelles based on the methoxy poly (ethylene glycol)-block-poly (ε-caprolactone) (mPEG-PCL) copolymer carrier, selected for its biocompatibility, biodegradation and suitable properties for loading of hydrophobic drugs (Figure 2). The nanoparticles were internalized into the treated cells significantly faster than pure CRM (Table 1), which is a prerequisite for the successful cytopenetration and the subsequent intracellular release of curcumin.

The tested interactions between ERF and MCRM in combination ratio 1:4 [ERF:MCRM] were proven to be synergistic for HuT-78 cells and additive for MJ cells (Figure 3A) in difference to the combination ratio 1:2, which led only to additive effects. As a final result, all CI values regarding the cell line HuT-78 were lower than 0.4 for the combination ratio 1:4, which indicates a strong synergism. The results clearly demonstrated that the applied strategy for experimental design and algorithms for calculation of isobolograms under particularly chosen combination ratios was robust and very reliable. The synergistic and additive effects were confirmed by the CFU assay (Figure 3B). The effect of the combination 1:4 [ERF:MCRM] in HuT-78 cells led to complete inhibition of cell clonogenicity, and therefore, this combination was subjected to a series of tests to study its effect on the expression of the transcription factor TWIST1, PKB/Akt phosphorylation, NF-κB activity, apoptosis induction, and glutathione reduction.

Undoubtedly, an important outcome was the complete suppression of the TWIST1 expression after the application of the synergistic combination [20 μM ERF:80 μM MCRM] on HuT-78 cells (Figure 4A) compared to the single application of erufosine that, although at a concentration of 50 μM, suppressed the TWIST1 expression by only 60% (Figure 4B). The achieved effect makes it feasible to reduce the effective dose of ERF and thus minimize the occurrence of adverse effects. The same effect was observed with the MJ cell line, except that ERF alone inhibited TWIST1 by 80% at a twofold lower concentration than in HuT-78 cells (25 μM vs. 50 μM), which may explain why the effect of the combination [20 μM ERF:80 μM MCRM] is additive and not synergistic (as in HuT-78 cells). Regarding PKB/Akt dephosphorylation, the combination inhibited almost fully the protein phosphorylation at Ser473 in both cell lines, which correlates with the results from the MTT and CFU assay, suggesting that proliferation inhibition occurs via inhibition of PKB/Akt. It also confirms that the lower sensitivity of HuT-78 cells to ERF is due to the weaker inhibition of this protein than in MJ cells. A single administration of ERF (20 μM) or MCRM (80 μM) achieved this effect to a greater extent in MJ cells than in HuT-78. In MJ cells, dephosphorylation at Thr308 was more pronounced than at Ser473, which is in line with previously published data [36]. In HuT-78, there was no significant difference in the levels of p-AktThr308 between the treated samples, whereas the levels of p-AktSer473 decreased significantly and to the same extent after exposure to MCRM or the combination. The activity of NF-κB p65 was also inhibited by up to 60% in both cell lines after administration of the combination (Figure 4C). The differences between the effects of the combination and the self-administered drugs were significant. The lack of effect of MCRM on NF-κB expression can be explained by the slow, gradual release of curcumin from the particular micelles during the incubation period—35% (28 μM) of the whole loaded dose, which is 80 μM. The released amount of CRM is obviously not sufficient to achieve the desired effect for this incubation period (24 h). In contrast, according to published data, 80 μM of pure CRM or CRM loaded in nanoparticles with rapid release significantly inhibits the activity of NF-κB [41,76,82,83]. In this regard, we could assume that not the gradual administration of low CRM doses over time but the fast release of higher doses in the first hours after treatment is crucial for the NF-κB inhibition. However, in combination with ERF, the low concentrations of CRM released from the micelles were sufficient to potentiate the effect of the inhibitory effect of the alkylphosphocholine on NF-κB, suggesting the unlocking of other mechanisms that deserve further investigation in a separate study. In the MJ cell line, the effect of ERF was the same as in the combination, which correlates with the determined additive effect.

The inhibition of NF-κB and PKB/Akt logically led to the induction of apoptosis in the treated cell lines. As visible in Figure 5, the combination 1:4 [ERF:MCRM] activated caspase-3, increased the fraction of mono- and oligonucleosomes in the cytosol of HuT-78 and MJ and led to nuclear fragmentation in both cell lines, indicating apoptosis (Figure 5C). Activation of caspase-3 and elevation of the reduced glutathione levels were more pronounced in MJ cells after treatment with the combination (Figure 5A,B). One possible explanation for this difference could be an earlier induction of apoptosis in HuT-78 cells, which could be proven in detail in a further study. As far as excess GSH in malignant cells correlates with increased metastasis and tumor progression [84,85], it could be speculated that the elevated GSH levels in the MJ cell line in our study could be one of the reasons why the best effect of the combination 1:4 [ERF:MCRM] on MJ cells is not synergistic but additive only.

Last but not least, the synergistic combination fully inhibited the formation of MRSA biofilm, which can bring additional benefits for patients with concomitant staphylococcal infections as this will further help reduce inflammation and limit the appearance of infected skin lesions. ERF alone is also a potent inhibitor of the MRSA biofilm, as published before [29], and the combination with MCRM did not reduce this effect.

5. Conclusions

The anti-lymphoma effect of ERF in CTCL cell lines was dependent on the TWIST1 expression. The combination between ERF and MCRM in a ratio of 1:4 achieves a synergistic or additive effect depending on the cell line, which was due to inhibition of TWIST1, dephosphorylation of PKB/AKT at Ser^{473} and/or Thr^{308}, and deactivation of the p65 subunit of NF-κB. Taken together, our experimental findings indicate that it is feasible to develop and investigate new micellar CRM formulations in rational combinations with the alkylphosphocholine ERF for future local and/or intravenous CTCL treatment perspectives. The low toxic profile of the two substances allows their combination to be used in a broader pharmacological regimen, with the combined antineoplastic, anti-inflammatory, and antibacterial activity assisting in overcoming primary and secondary drug resistance of CTCL-derived cells.

Author Contributions: Conceptualization, S.M.K., M.M.Z., K.Y., A.D.K., M.R.B. and H.M.N.; methodology, A.G.X.T., Y.I., I.P.-E.T., K.Y., M.M.Z. and A.D.K.; software, A.D.K.; formal analysis, A.G.X.T., Y.I., A.D.K., L.L.D., I.P.-E.T., K.Y. and M.M.Z.; investigation, A.G.X.T., Y.I., L.L.D., S.P., I.P.-E.T., K.Y. and M.M.Z.; resources, M.M.Z., H.M.N. and S.M.K.; writing—original draft preparation, M.M.Z., A.G.X.T., Y.I., A.D.K., I.P.-E.T. and K.Y.; writing—review and editing, S.M.K., M.R.B. and H.M.N.; visualization, M.M.Z. and S.M.K.; supervision, M.M.Z., H.M.N. and S.M.K.; project administration, M.M.Z. and S.M.K.; funding acquisition, S.M.K., M.M.Z., K.Y. and H.M.N. All authors have read and agreed to the published version of the manuscript.

Funding: This research was funded by the Bulgarian National Science Fund; grant number DN03/3 (16 December 2016).

Institutional Review Board Statement: Not applicable.

Informed Consent Statement: Not applicable.

Data Availability Statement: All raw data from the experiments are available from the authors.

Acknowledgments: Antonios G.X. Trochopoulos, was granted a research grant for 'Doctoral candidates and young academics and scientists' by the German Academic Exchange Service (DAAD), during the course of his doctoral studies. We would like to express our special gratitude to the Alexander von Humboldt Foundation for the Grant "Equipment subsidies" to Maya M. Zaharieva, used for conducting the cell cytotoxicity and clonogenicity tests and the biofilm assay. The authors are also thankful to Ts. Paunova-Krasteva for her help with the imaging of the Hoechst-stained preparations. Vasiliki Tsorbatzian, ESL/EFL teacher, proofread the article.

Conflicts of Interest: The authors declare no conflict of interest.

References

1. CheckOrphan. 2016. Available online: http://www.checkorphan.org/diseases/cutaneous-t-cell-lymphoma (accessed on 15 May 2021).
2. Rare Disease Registries in Europe. Orphanet Report Series: Rare Diseases Collection. 2016. Available online: www.orpha.net (accessed on 15 May 2021).
3. Bagherani, N.; Smoller, B.R. An overview of cutaneous T cell lymphomas. *F1000Research* **2016**, *5*, 1882. [CrossRef] [PubMed]
4. Cieza-Díaz, D.E.; Conde-Montero, E.; Menarguez-Palanca, J.; Longo-Imedio, I. Epidemiological and clinical features of patients diagnosed with cutaneous T-cell lymphomas in a spanish tertiary care hospital. *J. Eur. Acad. Dermatol. Venereol.* **2016**, *31*, e150–e153. [CrossRef] [PubMed]
5. Zinzani, P.L.; Bonthapally, V.; Huebner, D.; Lutes, R.; Chi, A.; Pileri, S. Panoptic clinical review of the current and future treatment of relapsed/refractory T-cell lymphomas: Cutaneous T-cell lymphomas. *Crit. Rev. Oncol.* **2016**, *99*, 228–240. [CrossRef] [PubMed]
6. Thestrup-Pedersen, K. Cutaneous T-Cell Lymphoma. A hypothesis on disease pathophysiology involving deficiency in DNA repair. *J. Eur. Acad. Dermatol. Venereol.* **2016**, *30*, 1682–1685. [CrossRef]
7. Duvic, M. Histone Deacetylase Inhibitors for Cutaneous T-Cell Lymphoma. *Dermatol. Clin.* **2015**, *33*, 757–764. [CrossRef]
8. Willerslev-Olsen, A.; Krejsgaard, T.; Lindahl, L.M.; Bonefeld, C.M.; Wasik, M.A.; Koralov, S.B.; Geisler, C.; Kilian, M.; Iversen, L.; Woetmann, A.; et al. Bacterial toxins fuel disease progression in cutaneous T-cell lymphoma. *Toxins* **2013**, *5*, 1402–1421. [CrossRef]
9. Axelrod, P.I.; Lorber, B.; Vonderheid, E.C. Infections complicating mycosis fungoides and Sézary syndrome. *JAMA* **1992**, *267*, 1354–1358. [CrossRef]
10. Brunner, P.M.; Jonak, C.; Knobler, R. Recent advances in understanding and managing cutaneous T-cell lymphomas. *F1000Research* **2020**, *9*, 331. [CrossRef]
11. Damsky, W.E.; Choi, J. Genetics of Cutaneous T Cell Lymphoma: From Bench to Bedside. *Curr. Treat. Options Oncol.* **2016**, *17*, 33. [CrossRef]
12. Nicolay, J.P.; Felcht, M.; Schledzewski, K.; Goerdt, S.; Geraud, C. Sezary syndrome: Old enigmas, new targets. *J. Dtsch. Dermatol. Ges.* **2016**, *14*, 256–264. [CrossRef]
13. Phyo, Z.; Shanbhag, S.; Rozati, S. Update on Biology of Cutaneous T-Cell Lymphoma. *Front. Oncol.* **2020**, *10*, 765. [CrossRef]
14. Khan, M.A.; Chen, H.-c.; Zhang, D.; Fu, J. Twist: A molecular target in cancer therapeutics. *Tumor Biol.* **2013**, *34*, 2497–2506. [CrossRef] [PubMed]
15. van Doorn, R.; Dijkman, R.; Vermeer, M.H.; Out-Luiting, J.J.; van der Raaij-Helmer, E.M.H.; Willemze, R.; Tensen, T.P. Aberrant Expression of the Tyrosine Kinase Receptor EphA4 and the Transcription Factor Twist in SΓ©zary Syndrome Identified by Gene Expression Analysis. *Cancer Res.* **2004**, *64*, 5578–5586. [CrossRef] [PubMed]
16. Goswami, M.; Duvic, M.; Dougherty, A.; Ni, X. Increased Twist expression in advanced stage of mycosis fungoides and Sézary syndrome. *J. Cutan. Pathol.* **2012**, *39*, 500–507. [CrossRef]
17. Kempf, W.; Kazakov, D.V.; Kerl, K. Cutaneous lymphomas: An update. Part 1: T-cell and natural killer/t-cell lymphomas and related conditions. *Am. J. Dermatopathol.* **2014**, *36*, 105–123. [CrossRef]
18. Spaccarelli, N.; Rook, A.H. The Use of Interferons in the Treatment of Cutaneous T-Cell Lymphoma. *Dermatol. Clin.* **2015**, *33*, 731–745. [CrossRef]
19. Bobrowicz, M.; Fassnacht, C.; Ignatova, D.; Chang, Y.-T.; Dimitriou, F.; Guenova, E. Pathogenesis and Therapy of Primary Cutaneous T-Cell Lymphoma: Collegium Internationale Allergologicum (CIA) Update 2020. *Int. Arch. Allergy Immunol.* **2020**, *181*, 733–745. [CrossRef] [PubMed]
20. Zic, J.A. Extracorporeal Photopheresis in the Treatment of Mycosis Fungoides and Sézary Syndrome. *Dermatol. Clin.* **2015**, *33*, 765–776. [CrossRef]
21. Vieyra-Garcia, P.A.; Wolf, P. Extracorporeal Photopheresis: A Case of Immunotherapy Ahead of Its Time. *Transfus. Med. Hemotherapy* **2020**, *47*, 226–235. [CrossRef]
22. Wilcox, R.A. Cutaneous T-cell lymphoma: 2016 update on diagnosis, risk-stratification, and management. *Am. J. Hematol.* **2016**, *91*, 151–165. [CrossRef]
23. Chung, C.G.; Poligone, B. Other Chemotherapeutic Agents in Cutaneous T-Cell Lymphoma. *Dermatol. Clin.* **2016**, *33*, 787–805. [CrossRef] [PubMed]
24. Eibl, H.; Kaufmann-Kolle, P. Medical Application of Synthetic Phospholipids as Liposomes and Drugs. *J. Liposome Res.* **1995**, *5*, 131–148. [CrossRef]
25. Kaufmann-Kolle, P.; Koetting, J.; Drevs, J.; Berger, M.; Unger, C.; Eibl, H. Intravenous application of alkylphosphocholines: Comparison of different homologues in lamellar structures. *J. Cancer Res. Clin. Oncol.* **1992**, *120*, R14.
26. Konstantinov, S.M.; Eibl, H.; Berger, M.R. BCR-ABL influences the antileukaemic efficacy of alkylphosphocholines. *Br. J. Haematol.* **1999**, *107*, 365–374. [CrossRef] [PubMed]
27. Georgieva, M.C.; Konstantinov, S.M.; Topashka-Ancheva, M.; Berger, M.R. Combination effects of alkylphosphocholines and gemcitabine in malignant and normal hematopoietic cells. *Cancer Lett.* **2002**, *182*, 163–174. [CrossRef] [PubMed]
28. Martelli, A.M.; Papa, V.; Tazzari, P.L.; Ricci, F.; Evangelisti, C.; Chiarini, F.; Grimaldi, C.; Cappellini, A.; Martinelli, G.; Ottaviani, E.; et al. Erucylphosphohomocholine, the first intravenously applicable alkylphosphocholine, is cytotoxic to acute myelogenous leukemia cells through JNK- and PP2A-dependent mechanisms. *Leukemia* **2010**, *24*, 687–698. [CrossRef] [PubMed]

29. Zaharieva, M.M.; Kroumov, A.D.; Dimitrova, L.; Tsvetkova, I.; Trochopoulos, A.; Konstantinov, S.M.; Berger, M.R.; Momchilova, M.; Yoncheva, K.; Najdenski, H.M. Micellar curcumin improves the antibacterial activity of the alkylphosphocholines erufosine and miltefosine against pathogenic *Staphyloccocus aureus* strains. *Biotechnol. Biotechnol. Equip.* **2019**, *33*, 38–53. [CrossRef]
30. Fleer, E.A.; Unger, C.; Kim, D.J.; Eibl, H. Metabolism of ether phospholipids and analogs in neoplastic cells. *Lipids* **1987**, *22*, 856–861. [CrossRef]
31. Fiegl, M.; Lindner, L.H.; Juergens, M.; Eibl, H.; Hiddemann, W.; Braess, J. Erufosine, a novel alkylphosphocholine, in acute myeloid leukemia: Single activity and combination with other antileukemic drugs. *Cancer Chemother. Pharmacol.* **2007**, *62*, 321–329. [CrossRef]
32. Berger, M.R.; Sobottka, S.B.; Konstantinov, S.; Eibl, H. Erucylphosphocholine is the prototype of i.v. injectable alkylphosphocholines. *Drugs Today* **1998**, *34*, 73–81.
33. Königs, S.K.; Pallasch, C.P.; Lindner, L.H.; Schwamb, J.; Schulz, A.; Brinker, R.; Claasen, J.; Veldurthy, A.; Eibl, H.; Hallek, M.; et al. Erufosine, a novel alkylphosphocholine, induces apoptosis in CLL through a caspase-dependent pathway. *Leuk. Res.* **2010**, *34*, 1064–1069. [CrossRef] [PubMed]
34. Yosifov, D.Y.; Konstantinov, S.M.; Berger, M.R. Erucylphospho-N,N,N-trimethylpropylammonium Shows Substantial Cytotoxicity in Multiple Myeloma Cells. *Ann. N.Y. Acad. Sci.* **2009**, *1171*, 350–358. [CrossRef] [PubMed]
35. Zaharieva, M.; Konstantinov, S.; Pilicheva, B.; Karaivanova, M.; Berger, M. Erufosine: A Membrane Targeting Antineoplastic Agent with Signal Transduction Modulating Effects. *Ann. N.Y. Acad. Sci.* **2007**, *1095*, 182–192. [CrossRef] [PubMed]
36. Kapoor, V.; Zaharieva, M.M.; Das, S.N.; Berger, M.R. Erufosine simultaneously induces apoptosis and autophagy by modulating the Akt-mTOR signaling pathway in oral squamous cell carcinoma. *Cancer Lett.* **2012**, *319*, 39–48. [CrossRef] [PubMed]
37. Dineva, I.K.; Zaharieva, M.M.; Konstantinov, S.M.; Eibl, H.; Berger, M.R. Erufosine suppresses breast cancer in vitro and in vivo for its activity on PI3K, c-Raf and Akt proteins. *J. Cancer Res. Clin. Oncol.* **2012**, *138*, 1909–1917. [CrossRef]
38. Yosifov, D.Y.; Todorov, P.T.; Zaharieva, M.; Georgiev, K.D.; Pilicheva, B.A.; Konstantinov, S.M.; Berger, M.R. Erucylphospho-N,N,N-trimethylpropylammonium (erufosine) is a potential antimyeloma drug devoid of myelotoxicity. *Cancer Chemother. Pharmacol.* **2010**, *67*, 13–25. [CrossRef]
39. Rudner, J.; Ruiner, C.-E.; Handrick, R.; Eibl, H.-J.; Belka, C.; Jendrossek, V. The Akt-inhibitor Erufosine induces apoptotic cell death in prostate cancer cells and increases the short term effects of ionizing radiation. *Radiat. Oncol.* **2010**, *5*, 108. [CrossRef]
40. Zaharieva, M.; Kirilov, M.; Chai, M.; Berger, S.M.; Konstantinov, S.; Berger, M.R. Reduced Expression of the Retinoblastoma Protein Shows That the Related Signaling Pathway Is Essential for Mediating the Antineoplastic Activity of Erufosine. *PLoS ONE* **2014**, *9*, e100950. [CrossRef]
41. Yosifov, D.Y.; Kaloyanov, K.A.; Guenova, M.L.; Prisadashka, K.; Balabanova, M.B.; Berger, M.R.; Konstantinov, S.M. Alkylphosphocholines and curcumin induce programmed cell death in cutaneous T-cell lymphoma cell lines. *Leuk. Res.* **2014**, *38*, 49–56. [CrossRef]
42. Chometon, G.; Cappuccini, F.; Raducanu, A.; Aumailley, M.; Jendrossek, V. The Membrane-targeted Alkylphosphocholine Erufosine Interferes with Survival Signals from the Extracellular Matrix. *Anti-Cancer Agents Med. Chem.* **2014**, *14*, 578–591. [CrossRef]
43. Ruiter, G.A.; Zerp, S.F.; Bartelink, H.; van Blitterswijk, W.J.; Verheij, M. Anti-cancer alkyl-lysophospholipids inhibit the phosphatidylinositol 3-kinase–Akt/PKB survival pathway. *Anti-Cancer Drugs* **2003**, *14*, 167–173. [CrossRef] [PubMed]
44. Konstantinov, S.M.; Topashka-Ancheva, M.; Benner, A.; Berger, M.R. Alkylphosphocholines: Effects on human leukemic cell lines and normal bone marrow cells. *Int. J. Cancer* **1998**, *77*, 778–786. [CrossRef]
45. Bagley, R.G.; Kurtzberg, L.; Rouleau, C.; Yao, M.; Teicher, B.A. Erufosine, an alkylphosphocholine, with differential toxicity to human cancer cells and bone marrow cells. *Cancer Chemother. Pharmacol.* **2011**, *68*, 1537–1546. [CrossRef] [PubMed]
46. Dummer, R.; Krasovec, M.; Röger, J.; Sindermann, H.; Burg, G. Topical administration of hexadecylphosphocholine in patients with cutaneous lymphomas: Results of a phase I/II study. *J. Am. Acad. Dermatol.* **1993**, *29*, 963–970. [CrossRef] [PubMed]
47. Dummer, R.; Vogt, T.; Hefner, H.; Sindermann, H.; Burg, G. Topical Application of Hexadecylphosphocholine in Patients with Cutaneous Lymphomas. *Prog. Exp. Tumor Res.* **1992**, *34*, 160–169. [CrossRef]
48. Dumontet, C.; Thomas, L.; Berard, F.; Gimonet, J.F.; Coiffier, B. A phase II trial of miltefosine in patients with cutaneous T-cell lymphoma. *Bull. Cancer* **2006**, *93*, E115–E118. [PubMed]
49. Pronk, L.; Planting, A.; Oosterom, R.; Drogendijk, T.; Stoter, G.; Verweij, J. Increases in leucocyte and platelet counts induced by the alkyl phospholipid hexadecylphosphocholine. *Eur. J. Cancer* **1994**, *30*, 1019–1022. [CrossRef] [PubMed]
50. Berkovic, D.; Bensch, M.; Bertram, J.; Wille, T.; Haase, D.; Binder, C.; Fleer, E. Effects of hexadecylphosphocholine on thrombocytopoiesis. *Eur. J. Cancer* **2001**, *37*, 503–511. [CrossRef]
51. Aggarwal, B.B.; Sung, B. Pharmacological basis for the role of curcumin in chronic diseases: An age-old spice with modern targets. *Trends Pharmacol. Sci.* **2009**, *30*, 85–94. [CrossRef]
52. Gupta, S.C.; Patchva, S.; Koh, W.; Aggarwal, B.B. Discovery of curcumin, a component of golden spice, and its miraculous biological activities. *Clin. Exp. Pharmacol. Physiol.* **2012**, *39*, 283–299. [CrossRef]
53. Duvoix, A.; Blasius, R.; Delhalle, S.; Schnekenburger, M.; Morceau, F.; Henry, E.; Dicato, M.; Diederich, M. Chemopreventive and therapeutic effects of curcumin. *Cancer Lett.* **2005**, *223*, 181–190. [CrossRef] [PubMed]

54. Syng-Ai, C.; Kumari, A.L.; Khar, A. Effect of curcumin on normal and tumor cells: Role of glutathione and bcl-2. *Mol. Cancer Ther.* **2004**, *3*, 1101–1108. [CrossRef] [PubMed]
55. Zhang, C.; Li, B.; Zhang, X.; Hazarika, P.; Aggarwal, B.B.; Duvic, M. Curcumin selectively induces apoptosis in cutaneous T-cell lymphoma cell lines and patients' PBMCs: Potential role for STAT-3 and NF-kappaB signaling. *J. Investig. Dermatol.* **2010**, *130*, 2110–2119. [CrossRef] [PubMed]
56. Khan, M.A.; Gahlot, S.; Majumdar, S. Oxidative Stress Induced by Curcumin Promotes the Death of Cutaneous T-cell Lymphoma (HuT-78) by Disrupting the Function of Several Molecular Targets. *Mol. Cancer Ther.* **2012**, *11*, 1873–1883. [CrossRef] [PubMed]
57. Kunwar, A.; Barik, A.; Mishra, B.; Rathinasamy, K.I.; Pandey, R.; Priyadarsini, K. Quantitative cellular uptake, localization and cytotoxicity of curcumin in normal and tumor cells. *Biochim. Biophys. Acta* **2008**, *1780*, 673–679. [CrossRef] [PubMed]
58. Anand, P.; Kunnumakkara, A.B.; Newman, R.A.; Aggarwal, B.B. Bioavailability of curcumin: Problems and promises. *Mol. Pharm.* **2007**, *4*, 807–818. [CrossRef]
59. Lapteva, M.; Mondon, K.; Möller, M.; Gurny, R.; Kalia, Y.N. Polymeric Micelle Nanocarriers for the Cutaneous Delivery of Tacrolimus: A Targeted Approach for the Treatment of Psoriasis. *Mol. Pharm.* **2014**, *11*, 2989–3001. [CrossRef]
60. Xue, B.; Wang, Y.; Tang, X.; Xie, P.; Wang, Y.; Luo, F.; Wu, C.; Qian, Z. Biodegradable Self-Assembled MPEG-PCL Micelles for Hydrophobic Oridonin Delivery In Vitro. *J. Biomed. Nanotechnol.* **2012**, *8*, 80–89. [CrossRef]
61. Lapteva, M.; Möller, M.; Gurny, R.; Kalia, Y.N. Self-assembled polymeric nanocarriers for the targeted delivery of retinoic acid to the hair follicle. *Nanoscale* **2015**, *7*, 18651–18662. [CrossRef]
62. Eibl, H.; Hilgard, C.; Unger, C. *Alkylphosphocholines: New Drugs in Cancer Therapy*; Eibl, H., Hilgard, C., Unger, C., Eds.; Karger: Basel, Switzerland, 1992; 173p.
63. *ISO 10993-5:2009*; Biological Evaluation of Medical Devices—Part 5: Tests for In Vitro Cytotoxicity. ISO: Geneva, Switzerland, 2017.
64. Chou, T.C.; Martin, N. (Eds.) *CompuSyn for Drug Combinations: PC Software and User's Guide: A Computer Program for Quantitation of Synergism and Antagonism in Drug Combinations, and the Determination of IC50and ED50 and LD50 Values*; ComboSyn Inc: Paramus, NJ, USA, 2005.
65. Zaharieva, M.; Trochopoulos, A.; Dimitrova, L.; Berger, M.; Najdenski, H.; Konstantinov, S.; Kroumov, A. New Insights in Routine Procedure for Mathematical Evaluation of in vitro Cytotoxicity Data from Cancer Cell Lines. *Int. J. Bioautomation* **2018**, *22*, 87–106. [CrossRef]
66. Sedlak, J.; Lindsay, R.H. Estimation of total, protein-bound, and nonprotein sulfhydryl groups in tissue with Ellman's reagent. *Anal. Biochem.* **1968**, *25*, 192–205. [CrossRef]
67. Chazotte, B. Labeling Nuclear DNA with Hoechst 33342. *Cold Spring Harb. Protoc.* **2011**, *2011*, pdb.prot5557. [CrossRef] [PubMed]
68. Stepanović, S.; Vuković, D.; Hola, V.; DI Bonaventura, G.D.; Djukić, S.; Ćirković, I.; Ruzicka, F. Quantification of biofilm in microtiter plates: Overview of testing conditions and practical recommendations for assessment of biofilm production by staphylococci. *APMIS* **2007**, *115*, 891–899. [CrossRef]
69. Wong, H.K. STAT Assays with a TWIST: Differentiating Sézary Syndrome from Erythrodermic Inflammatory Dermatitis. *J. Investig. Dermatol.* **2016**, *136*, 1313–1315. [CrossRef]
70. Merindol, N.; Riquet, A.; Szablewski, V.; Eliaou, J.-F.; Puisieux, A.; Bonnefoy, N. The emerging role of Twist proteins in hematopoietic cells and hematological malignancies. *Blood Cancer J.* **2014**, *4*, e206. [CrossRef] [PubMed]
71. Xue, G.; Restuccia, D.F.; Lan, Q.; Hynx, D.; Dirnhofer, S.; Hess, D.; Rüegg, C.; Hemmings, B.A. Akt/PKB-Mediated Phosphorylation of Twist1 Promotes Tumor Metastasis via Mediating Cross-Talk between PI3K/Akt and TGF-OI Signaling Axes. *Cancer Discov.* **2012**, *2*, 248–259. [CrossRef] [PubMed]
72. Zhao, Z.; Rahman, M.A.; Chen, Z.G.; Shin, D.M. Multiple biological functions of Twist1 in various cancers. *Oncotarget* **2017**, *8*, 20380–20393. [CrossRef] [PubMed]
73. Kaleağasıoğlu, F.; Zaharieva, M.; Konstantinov, S.M.; Berger, M.R. Alkylphospholipids are Signal Transduction Modulators with Potential for Anticancer Therapy. *Anti-Cancer Agents Med. Chem.* **2019**, *19*, 66–91. [CrossRef] [PubMed]
74. Li, S.; Kendall, S.E.; Raices, R.; Finlay, J.; Covarrubias, M.; Liu, Z.; Lowe, G.; Lin, Y.; Teh, Y.H.; Leigh, V.; et al. TWIST1 associates with NF-kB subunit RELA via carboxyl-terminal WR domain to promote cell autonomous invasion through IL8 production. *BMC Biol.* **2012**, *10*, 73. [CrossRef]
75. Ilieva, Y.; Kaloyanov, K.; Yosifov, D.; Robev, B.; Zhelezova, I.; Genova, M.; Mihova, A.; Balatzenko, G.; Zaharieva, M.; Berger, M.R.; et al. Antineoplastic potential of curcumin (cooperative study in Bulgaria and Germany). *Phytochem. Rev.* **2014**, *13*, 459–469. [CrossRef]
76. Aggarwal, B.B.; Kumar, A.; Bharti, A.C. Anticancer potential of curcumin: Preclinical and clinical studies. *Anticancer Res.* **2003**, *23*, 363–398. [PubMed]
77. Shishodia, S.; Amin, H.M.; Lai, R.; Aggarwal, B.B. Curcumin (diferuloylmethane) inhibits constitutive NF-kappaB activation, induces G1/S arrest, suppresses proliferation, and induces apoptosis in mantle cell lymphoma. *Biochem. Pharmacol.* **2005**, *70*, 700–713. [CrossRef] [PubMed]
78. Kunwar, A.; Kunwar, A.; Narang, H.; Priyadarsini, K.I.; Krishna, M.; Pandey, R.; Sainis, K.B. Delayed activation of PKCdelta and NFkappaB and higher radioprotection in splenic lymphocytes by copper (II)-Curcumin (1) complex as compared to curcumin. *J. Cell. Biochem.* **2007**, *102*, 1214–1224. [CrossRef] [PubMed]

79. Prasad, S.; Gupta, S.C.; Tyagi, A.K.; Aggarwal, B.B. Curcumin, a component of golden spice: From bedside to bench and back. *Biotechnol. Adv.* **2014**, *32*, 1053–1064. [CrossRef]
80. Obando, D.; Widmer, F.; Wright, L.C.; Sorrell, T.C.; Jolliffe, K.A. Synthesis, antifungal and antimicrobial activity of alkylphospholipids. *Bioorg. Med. Chem.* **2007**, *15*, 5158–5165. [CrossRef]
81. Gong, C.; Wu, Q.; Wang, Y.; Zhang, D.; Luo, F.; Zhao, X.; Wei, Y.; Qian, Z. A biodegradable hydrogel system containing curcumin encapsulated in micelles for cutaneous wound healing. *Biomaterials* **2013**, *34*, 6377–6387. [CrossRef]
82. Caon, T.; Mazzarino, L.; Simões, C.M.O.; Senna, E.L.; Silva, M.A.S. Lipid- and Polymer-Based Nanostructures for Cutaneous Delivery of Curcumin. *AAPS PharmSciTech* **2016**, *18*, 920–925. [CrossRef]
83. Trochopoulos, A.G.; Zaharieva, M.M.; Marinova, M.H.; Yoncheva, K.; Tibi, I.P.; Berger, M.R.; Konstantinov, S.M. Antineoplastic effect of a novel nanosized curcumin on cutaneous T cell lymphoma. *Oncol. Lett.* **2020**, *20*, 304. [CrossRef]
84. Kennedy, L.; Sandhu, J.K.; Harper, M.-E.; Cuperlovic-Culf, M. Role of Glutathione in Cancer: From Mechanisms to Therapies. *Biomolecules* **2020**, *10*, 1429. [CrossRef]
85. Bansal, A.; Celeste Simon, M. Glutathione metabolism in cancer progression and treatment resistance. *J. Cell Biol.* **2018**, *217*, 2291–2298. [CrossRef]

pharmaceutics

Article

From Serendipity to Rational Identification of the 5,6,7,8-Tetrahydrobenzo[4,5]thieno[2,3-*d*]pyrimidin-4(3*H*)-one Core as a New Chemotype of AKT1 Inhibitors for Acute Myeloid Leukemia

Andrea Astolfi [1,†], Francesca Milano [2,†], Deborah Palazzotti [1,†], Jose Brea [3], Maria Chiara Pismataro [1], Mariangela Morlando [1], Oriana Tabarrini [1], Maria Isabel Loza [3], Serena Massari [1], Maria Paola Martelli [2] and Maria Letizia Barreca [1,*]

1 Department of Pharmaceutical Sciences, "Department of Excellence 2018–2022", University of Perugia, 06123 Perugia, Italy
2 Hematology and Clinical Immunology, Department of Medicine and Surgery, University of Perugia, 06132 Perugia, Italy
3 CIMUS Research Center, University of Santiago de Compostela, 15782 Santiago de Compostela, Spain
* Correspondence: maria.barreca@unipg.it
† These authors contributed equally to this work.

Abstract: Acute myeloid leukemia (AML) is a heterogeneous hematopoietic malignancy whose prognosis is globally poor. In more than 60% of AML patients, the PI3K/AKTs/mTOR signaling pathway is aberrantly activated because of oncogenic driver alterations and further enhanced by chemotherapy as a mechanism of drug resistance. Against this backdrop, very recently we have started a multidisciplinary research project focused on AKT1 as a pharmacological target to identify novel anti-AML agents. Indeed, the serendipitous finding of the in-house compound **T187** as an AKT1 inhibitor has paved the way to the rational identification of new active small molecules, among which **T126** has emerged as the most interesting compound with $IC_{50} = 1.99 \pm 0.11$ μM, ligand efficiency of 0.35, and a clear effect at low micromolar concentrations on growth inhibition and induction of apoptosis in AML cells. The collected results together with preliminary SAR data strongly indicate that the 5,6,7,8-tetrahydrobenzo[4,5]thieno[2,3-*d*]pyrimidin-4(3*H*)-one derivative **T126** is worthy of future biological experiments and medicinal chemistry efforts aimed at developing a novel chemical class of AKT1 inhibitors as anti-AML agents.

Keywords: AKT1; kinase inhibitors; cancer; acute myeloid leukemia; rare disease; molecular modeling; molecular docking; similarity search; MD simulation; computer-aided drug discovery

Citation: Astolfi, A.; Milano, F.; Palazzotti, D.; Brea, J.; Pismataro, M.C.; Morlando, M.; Tabarrini, O.; Loza, M.I.; Massari, S.; Martelli, M.P.; et al. From Serendipity to Rational Identification of the 5,6,7,8-Tetrahydrobenzo[4,5]thieno[2,3-*d*]pyrimidin-4(3*H*)-one Core as a New Chemotype of AKT1 Inhibitors for Acute Myeloid Leukemia. *Pharmaceutics* **2022**, *14*, 2295. https://doi.org/10.3390/pharmaceutics14112295

Academic Editor: Francesca Musumeci

Received: 30 August 2022
Accepted: 22 October 2022
Published: 26 October 2022

1. Introduction

The AGC kinases AKT1-3 are key mediators of the PI3K/AKT/mTOR signaling pathway, which plays a crucial role in major cellular functions such as proliferation, migration, and antiapoptotic survival [1]. The main downstream targets of AKT include the mammalian target of rapamycin (mTOR) of the mTOR complex 1 (mTORC1), that through phosphorylation activates the p70 ribosomal S6 kinase (p70S6K) and inactivates the 4E-BP1 (an inhibitor of the elongation initiating factor 4E, eIF4E) ultimately promoting protein translation and cell growth [2]. Among the three AKT isoforms, AKT1 is expressed ubiquitously and a high rate of AKT1 amplification occurs in a wide range of human cancers [3] including acute myeloid leukemia (AML) [4,5], a rare but often fatal blood cancer. The major activators of the PI3K/AKT/mTOR pathways are upstream kinases, such as tyrosine kinase receptors (e.g., fms-like tyrosine kinase 3, FLT3) or K/NRAS frequently involved in gain-of-function gene mutations in AML [6]. Indeed, a constitutive activation of the AKT pathway and the phosphorylation of its targets such as mTOR, p70S6K, and 4E-BP1 are detected in about 60% of AML patients [7–9], and are mostly described as connected with a poorer prognosis [10–14]. The level of AKT1 expression and kinase activity is often associated with the degree of differentiation, hormone dependency, and aggressive behavior of cancer cells and/or with a less favorable prognosis [12–14]. Additionally, accumulating evidence indicates that the acquisition of resistance to chemotherapeutic drugs involves the activation of the PI3K/AKT/mTOR pathway [15], suggesting that AKT1 inhibitors could be useful for overcoming drug resistance and improving response [16,17]. Recently, AKT activation in AML has also been implicated as a resistance mechanism to the targeted therapy **selinexor**, a newly FDA-approved small molecule inhibitor of exportin 1 (XPO1) [18].

In the era of precision oncology, AKT1 thus represents an attractive target for the development of targeted therapies selectively hitting cancer cells characterized by an aberrant activation of the PI3K/AKT/mTOR signaling pathway. To date, no drugs against AKT1 are on the market, although some AKT1 inhibitors are currently in clinical trials for solid cancers [19,20], but not for AML, which remains an urgent medical need. Against this backdrop, we have very recently started a multidisciplinary research program aimed at rationally identifying new AKT1 inhibitors as anti-AML agents.

2. Materials and Methods

2.1. Computational Studies

2.1.1. Ligand Preparation

All compounds explored in molecular modelling studies (i.e., **T187**, the clinical candidates, and the selected UNIPG-lib compounds) were prepared using the LigPrep [21] to produce low energy 3D structures taking into account the ionization states, stereochemistry, the tautomerism, and ring conformations at the desired pH (7.5 ± 0.5).

2.1.2. Similarity Search

The BioSolveIT Feature Tree (FTrees) [22] software tool was used to perform fast 2D-similarity assessments. The similarity searches were performed using the following command line:

```
ftrees -i T187.sdf -l $library –minSimilarityThreshold X -o output.csv
```

where the -i option defines the query molecule, the -l option states the screened library, the –minSimilarityThreshold defines the similarity threshold below which molecules are discarded, and the -o option states the output file.

To compute the similarity between **T187** and the known AKT1 inhibitor clinical candidates, the compounds reported in Table S2 were used as the $library file and the -minSimilarityThreshold was set to 0.

To build the **T187**-like compounds' library, the prepared UNIPG-lib was used as the $library file and the –minSimilarityThreshold was set to 0.65.

2.1.3. Protein Preparation

The only two crystal structures of AKT1 co-crystallized with clinical candidates (i.e., **capivasertib** and **ipatasertib**) available from RCSB PDB [23] were PDB ID 4GV1 [24] and 4EKL [25], respectively. Considering the best resolution of the 4GV1 structure (1.49 Å vs. 2.00 Å for 4EKL), this protein conformation was used for further modeling studies. Firstly, the complex was prepared using Schrödinger's Protein Preparation Wizard (Schrödinger Release 2019-2) [26] in order to obtain satisfactory starting structures for the docking studies. The complex was pre-processed as follows: (i) hydrogen atoms were added and bond orders were assigned to amino acid residues and ligands, (ii) the missing side chains were filled, (iii) the protein was capped with acetyl (ACE) and N-methylamine (NMA) groups, (iv) all water molecules beyond 3 Å from heteroatoms were deleted, and (v) Epik (Schrödinger Release 2019-2) [27] was used to predict the ionization and tautomeric states for the ligands (pH = 7.0 ± 2). In addition, the structure presented some missed residues (loop 451–457) that were filled by using the homology modelling tool function implemented in the Prime package (Schrödinger Release 2019-2) [28]. Then, the H-bond network of the complex was optimized using PROPKA for the assignment of the residue protonation states (pH = 7.0). Finally, the complex was submitted to a restrained minimization (OPLS3 force field), which was stopped when the RMSD of heavy atoms reached 0.30 Å.

2.1.4. Molecular Docking

Docking studies were performed using Schrödinger's Glide software (Schrödinger Release 2019-2) [29]. In our protocol, the crystallographic position of **capivasertib** (PDB ID 4GV1) was used as reference to center the grid on the ATP-site. The docking space was defined as a cubic box (28 Å outerbox), with an inner cubic box (14 Å) defining the region where the centroid of the ligand had to be located. The centroid for the 4GV1 protein had coordinates of −20.00, 4.33, and 11.27 to the *x*-, *y*-, and *z*-axis, respectively. After grid preparation with the "receptor grid generation" tool, docking experiments were performed using the Glide XP (Extra Precision) protocol (Schrödinger Release 2019-2) [30], by generating three poses for each ligand. The generated poses were submitted to the MMGBSA minimization step using the "Refine Protein-Ligand Complex" tool in Prime (default settings—esidues at 5 Å within the ligand considered as flexible) and only the pose with the lower MM-GBSA ΔG of binding (MM-GBSA dG Bind) was retained.

The collected binding poses were then filtered for the presence of H-bond interaction(s) with the hinge residues Ala230 and/or Glu 228.

By using the above-described protocol, the initial hit, **T187,** was predicted to bind the catalytic site with a MM-GBSA dG Bind of −39.42 Kcal/mol. Based on this result, the **T187**-like compounds were docked and later filtered using a threshold of −40 Kcal/mol on the MM-GBSA dG Bind and then submitted to visual inspection.

The two co-crystallized advanced inhibitors were added to the ligand set for comparison. The best docking poses of **ipatasertib** (MM-GBSA dG Bind = −67.72 Kcal/mol) and **capivasertib** (MM-GBSA dG Bind = −62.42 Kcal/mol) agreed well with their experimental binding conformations (Figure S2).

2.1.5. Molecular Dynamics Simulations

Explicit solvent Molecular Dynamics (MD) simulations were performed by using the Desmond package v5.8 (Schrödinger Release 2019-2) [31] to investigate the stability of the MMGBGA binding modes generated for **T126** and **T125** in complex with the 4GV1 protein conformation. The OPLS3e forcefield was selected and the systems were solvated in an orthorhombic box (size = 20.0 × 20.0 × 20.0) exploiting the TIP3P water molecule mode. Na^+ and Cl^- counter ions at a concentration of 0.15 M were added to neutralize the system charge by means of the System Builder tool. The system setup step enabled the generation of the 4GV1-**T126** and 4GV1-**T125** complex systems consisting of 83,768 and 83,767 atoms, respectively. MD simulations were set up through the Molecular dynamics tool. The systems were relaxed before the simulation by using the protocol implemented in Desmond which comprises 5 equilibration steps. The system minimization was followed by a production step of 200 ns that was simulated by the NTP ensemble at 310 K and 1 atm. For each system, three replicas with different random seeds were performed. MD simulation results were analyzed through the Simulation interaction diagram (SID) tool provided by Schrödinger.

2.2. *Chemistry*

2.2.1. Synthetic Procedures

All the compounds evaluated in the biological assays (Table S2) were synthesized as previously reported by us, with the exception of compound **T187** which has never been reported so far and, therefore, is described in this paper, and compounds **T124, T126,** and **T128** that were synthesized as reported in the literature [32–35].

Compound **T187** was synthesized as reported in Scheme 1, starting from a one-pot Gewald reaction entailing the reaction of cycloheptanone with cyanoacetamide in the presence of sulphur and morpholine in EtOH at reflux, to give the intermediate 1 [36]. Successive amidation of compound **1** [36] with 2,4-difluorobenzoic acid was performed in CH_2Cl_2 at 50 °C, in the presence of the Mukaiyama reagent (2-chloro-1-methylpyridinium iodide, CMPI), 4-dimethylaminopyridine (DMAP), and triethylamine (TEA), furnishing compound **T187**.

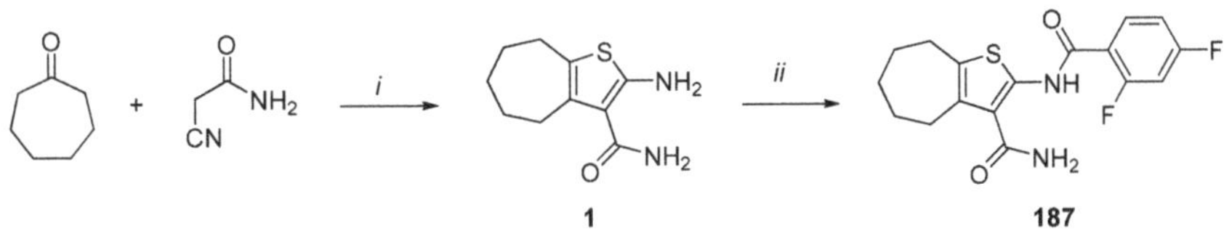

Scheme 1. Reagents and conditions: (i) S8, morpholine, EtOH, reflux; (ii) 2,4-difluorobenzoic acid, CMPI, DMAP, TEA, CH_2Cl_2, 50 °C.

2.2.2. Experimental Section

The commercially available starting materials, reagents, and solvents were used as supplied. All reactions were routinely monitored by TLC on silica gel 60F254 (Merck) and visualized by using UV or iodine. Flash column chromatography was performed on Merck silica gel 60 (mesh 230–400). After extraction, the organic solutions were dried over anhydrous Na_2SO_4, filtered, and concentrated with a Büchi rotary evaporator at reduced pressure. Yields are of the purified product and were not optimized. The purities of compounds **T124**, **T126**, and **T128** (96.5%, 100%, and 100%, respectively) were determined by HPLC analysis using a Jasco LC-4000 instrument equipped with a DAD Jasco MD-4015 detector from 200 to 600 nm. The purity was revealed at 344 nm using an XTerra MS C18 Column, 5 μm, 4.6 mm × 150 mm with a flow rate: 0.4 mL/min; an acquisition time: 10 min; an isocratic: acetonitrile containing 0.1% of formic acid (85%) and water (15%); a column temperature: 25 °C; and an injection volume: 20 μL. The peak retention time is given in minutes and the Chromatograms were analyzed by Chromatography Data System software ChromNAV version number 2.03.06 [Build 4] (2003 JASCO Corporation, Tokyo, Japan). The purity of compound **T187** (95.09%) was determined by LC/MS using an Agilent 1290 Infinity System machine equipped with a DAD detector from 190 to 640 nm. The purity was revealed at 254 nm using a Phenomenex AERIS Widepore C4, 4.6 mm, 100 mm (6.6 lm) with a flow rate: 0.6 mL/min; an acquisition time: 10 min; a gradient: acetonitrile in water containing 0.1% of formic acid (0 100% in 10 min); a column temperature: 40 °C; and an injection volume: 2 μL. The peak retention time is given in minutes. Accurate mass measurements were performed with HRMS (Agilent Q-TOF 6540) and the monitored *m/z* values ranged from 100 *m/z* to 3000 *m/z* in positive polarity. The Mass spectrometry was equipped with an ESI source. The gas temperature, gas flow, nebulizer, sheath gas, and sheath gas flow were set at 350 °C, 9 L/min, 35 psig, 400 °C, and 9 L/min, respectively. The capillary voltage was set at 4000 V and the fragmentor at 120 V. The 1H NMR and 13C NMR spectra were recorded on a Bruker Avance DRX-400 MHz using a residual solvent such as dimethylsulfoxide (δ = 2.48) as an internal standard. The chemical shifts were recorded in ppm (δ) and the spectral data are consistent with the assigned structures. The spin multiplicities are indicated by the: d (doublet), t (triplet), and m (multiplet).

2-(2,4-Difluorobenzamido)-5,6,7,8-tetrahydro-4*H*-cyclohepta[*b*]thiophene-3-carboxamide (**T187**). 2,4-Difluorobenzoic acid (0.09 g, 0.61 mmol), CMPI (0.14 g, 0.57 mmol), DMAP (0.03 g, 0.24 mmol), and TEA (0.14 g, 1.4 mmol) were added to a solution of compound **1 [36]** (0.10 g, 0.47 mmol) in dry CH_2Cl_2 (5 mL). The reaction mixture was maintained at 40 °C overnight and then, after colling, it was evaporated to dryness. The obtained residue was suspended in a 1 M HCl solution, stirred for 10 min, and then filtered. The solid was purified by flash chromatography eluting with $CHCl_3$/MeOH (98:2), to give **T187** (0.07 g, 42%). ^{1}H-NMR (400 MHz, DMSO-d_6) δ 1.55–1.65 (m, 4H, cycloeptane CH_2), 1.76–1.80, 2.70–2.75, 2.78–2.80 (m, each 2H, cycloheptane CH_2), 7.28–7.32 (m, 1H, aromatic CH),

7.47–7.70 (m, 3H, aromatic CH and NH_2), 8.00–8.10 (m, 1H, aromatic CH), 11.66–1168 (d, J = 8.5 Hz, 1H, NH); ^{13}C-NMR (101 MHz, DMSO-d_6) δ 27.7, 28.1, 28.7, 28.8, 32.1, 105.6 (t, J_{C-F} = 27.3 Hz), 113.3 (d, J_{C-F} = 20.2 Hz), 117.6 (d, J_{C-F} = 9.1 Hz), 122.1, 131.6, 133.8 (d, J_{C-F} = 11.1 Hz), 135.6, 137.2, 158.7, 159.6, 160.8 (dd, J_{C-F} = 14.1 and 253.5 Hz), 165.0 (dd, J_{C-F} = 13.1 and 266.6 Hz), 167.9. HRMS: m/z calcd for $C_{17}H_{16}F_2N_2O_2S$ 351.0089 $(M + H)^+$, found 351.0978 $(M + H)^+$. HPLC, ret. time: 5.633 min, peak area: 95.09%.

2.3. *Enzymatic Assays*

The studied compounds, including the standard inhibitor Staurosporine (Thermo-Fisher , Madrid, Spain, Invitrogen BP2541), were incubated with 3 nM AKT1 (#Aps 01-101; Kinase Logistics, Meyn, Germany) and 1.5 μM Profiler Pro peptide 6 (#760350; Perkin Elmer, Tres Cantos, Madrid, Spain) in the presence of different concentrations of Adenosine 5′triphosphate disodium salt, ATP (#A2383; Sigma Aldrich, Madrid, Spain) in a 384-well Microplate (#781076; Greiner, San Sebastián de los Reyes, Madrid, Spain). The reaction rate was calculated from kinetic measurements carried out measured in a Caliper EzReader LabChip 3000 (Caliper, Hopkinton, MA, USA) reader with samples taken each 10 min for a total time of 60 min. The percentage of substrate conversion rate was plotted versus the ATP concentration. The data was analyzed by non-linear fitting with GraphPad Prism by employing the Michaelis–Menten equation. The buffer assay used in the inhibition hAKT1 assay is: Hepes 100 mM, DMSO 4%, Brij 0.003%, Tween 20 0.004%, $MgCl_2$ 10 mM, with a pH = 7.5.

The K_i values were calculated by employing the GraphPad Prism equations.

2.4. *Cell-Based Assays*

2.4.1. Cells and Compound Treatments

The human AML cell lines, OCI-AML3 (carrying the *NPM1* gene mutation A and the *DNMT3A*^{R882C} mutation), IMS-M2 (carrying the *NPM1* gene mutation A), OCI-AML2 (with the wild-type *NPM1* and *DNMT3A*^{R635W} mutation), and SKM-1 and MOLM-13 (with the wild-type *NPM1*) were previously reported [37,38]. The cell lines were purchased from the Deutsche Sammelung von Mikroorganismen und Zellkulturen (DSMZ) (Braunschweig, Germany) cell bank and maintained according to the vendors' instructions. Briefly, cultured cells were split every 3 days and maintained in an exponential growth phase. The cell lines were tested for mycoplasma using the Universal Mycoplasma Detection Kit (ATCC). The OCI-AML3 were grown in α-MEM supplemented with 20% fetal bovine serum, 100 U/mL penicillin, and 100 μg/mL streptomycin (P/S), and all other cell lines were cultured in RPMI and supplemented with a 10–20% fetal bovine serum and 100 U/mL penicillin and 100 μg/mL streptomycin. The cells were kept at 37 °C in a 5% CO_2 incubator. Prior to screening, the appropriate cell concentration for all cell lines was determined. For the cell lines' treatments, cells were maintained as follows: OCI-AML3 at 0.25 × 10^6/mL, OCI-AML2 and IMS-M2 at 0.35 × 10^6/mL, and SKM-1 and MOLM-13 at 0.5 × 10^6/mL. The compounds were re-suspended at 10 mM stock in DMSO. Further dilutions were performed in phosphate-buffer-saline (PBS) for pharmacological experiments. The cells were exposed to compounds at different concentrations for the indicated time points and sampled as described below. Briefly, the cells (4 × 10^5/mL) were seeded in 384-well plate and treated for 48 h with various concentrations (0.001 μM to 100 μM) of either one of the compounds, **T101**, **T126**, and **T159**, or **NSC348884**, as a positive control [39]. Each plate included DMSO at 0.1% as a negative control. The metabolic cell activity was measured by using the CellTiter Blue proliferation agent (Roche) as described in the manufacturer's procedure. Briefly, CellTiter Blue solution (Resazaurin) was added to each well and the plates were incubated at 37 °C with 5% CO_2 for 4 h. During the incubation time, the resazurin is reduced to fluorescent resorufin in metabolically and proliferating living cells. The fluorescence intensity was measured using the automated Tecan plate reader (Tecan Group Ltd., Männedorf, Switzerland) at an excitation/emission wavelength of 531/572 nm. The fluorescence readouts were reported as relative fluorescence units (RFU) and read on a luminescence plate reader (Infinite 200 Pro, Tecan, CA, USA). The dose response curves (maximal inhibitory concentration (IC_{50})) were obtained by non-linear regression analysis using the GraphPad Prism 5.0 program. The concentration is shown as a logarithmic function. All the experiments were performed at least in triplicate, with technical repetition. The results are expressed as mean $\pm$ standard error of the mean (SEM). For the Colony Forming Unit (CFU) assay, 10×10^6 CD34+ human hematopoietic progenitors were received from unique healthy donors, upon signed informed consent, and were cultured in SFEM II (Serum-free medium) and P/S at a density of 1×10^6/mL with cytokines that are important for cell division and self-renewal: fms-like tyrosine kinase 3 (FLT3)-ligand, FLT3L (100 ng/mL), Stem Cell Factor, SCF (100 ng/mL) and Trombopoietin, and TPO (50 ng/mL) for 48 h before plating them on a Methocult medium.

2.4.2. Colony Forming Unit (CFU) Assay

For the CFU assay, 500 CD34+ human cells were plated in a MethoCult™ enriched medium (see Table S4) on a 6-well plate (untreated cells versus **T126** at 1, 2.5, and 5 μM, each condition in triplicate). The plates were then incubated for 14 days at 37 °C (5% CO_2), and the images were captured and the colonies counted by STEMvision™. STEMvision™ is a bench-top instrument and computer system designed specifically for the automated imaging and counting of hematopoietic colonies in the CFU assay. This system has been optimized for use with MethoCult™ media and meniscus-free SmartDish™ culture ware. The use of this standardized platform significantly improves the accuracy and reproducibility of the human CFU assay. Mobilized peripheral blood (MPB) analysis packages have been used to accurately count colonies.

2.4.3. Cell Lysates Preparation and Western Blot Analysis

Immunoblotting was conducted as described previously [37]. Briefly, fresh cell pellets ($0.2–0.5 \times 10^6$/test for AML cell lines) were dissolved directly in 30–60 μL of Laemmli sample buffer 1× (1.5 M Tris-HCl pH 6.8, glycerol, β-mercaptoethanol, SDS, bromophenol blue) and boiled at 95 °C for 5 min. The proteins were separated by SDS-polyacrylamide gel electrophoresis (SDS-PAGE) on Precast 4–15% gradient gels (Biorad, Hercules, CA, USA) transferred onto nitrocellulose membranes (GE Healthcare, Piscataway, NJ, USA) and probed with specific primary antibodies followed by horseradish peroxidase-conjugated secondary antibodies (GE Healthcare Lifesciences). The polypeptides were visualized using enhanced chemiluminescence (Luminata Crescendo, Merck Millipore, Burlington, MA, USA) according to the manufacturer's instructions. The bands were visualized using the Biorad Chemidoc and the images were processed using the image lab software (Biorad) and Adobe Photoshop (Ps CC 2019). β-actin expression levels were used as a control for protein loading, as indicated.

2.4.4. Cell Growth and Apoptosis Evaluation

The viable cell count was evaluated daily using the Invitrogen Countess automated cell counter (Invitrogen, Carlsbad, CA, USA) by trypan blue exclusion, and individual growth curves were generated from these data.

The cell apoptosis was assessed using Annexin V APC- or FITC-conjugated antibodies and counterstained with Propidium Iodide (PI) or 7AAD for the detection of necrotic cells (Becton Dickinson, Franklin Lakes, NJ, USA), according to the manufacturer's protocol. The data acquisition and analysis were done with a FACS Canto II flow cytometer and data were acquired and analyzed using Diva software (Becton Dickinson, Franklin Lakes, NJ, USA).

2.4.5. Antibodies

The primary antibodies used for western blot analyses are the following: mouse monoclonal anti-human β-actin (Clone AC-15; dilution 1:5000) from Sigma Aldrich (St. Louis, MO, USA); rabbit monoclonal anti-human Cleaved poly (ADP-ribose) polymerase (c-PARP) (Clone Asp 214; dilution 1:1000); rabbit monoclonal phospho-Akt

(Ser473) (clone D9E, dilution 1.500); rabbit polyclonal phospho-mTOR (Ser2448) (dilution 1:500); rabbit polyclonal phospho-4E-BP1 (Ser65) (dilution 1:500); rabbit monoclonal Akt1 (clone D9R8K, dilution 1:500); rabbit polyclonal 4E-BP1 (dilution 1:500); and rabbit polyclonal mTOR (dilution 1:500), all from Cell Signaling (Cell Signaling Technology, Inc., Danvers, MA, USA).

2.4.6. Statistical Analysis

All the experiments were performed at least in triplicate, as indicated, with technical repetition when possible. The results are expressed as mean ± standard error of the mean (SEM). A one- or two-tailed paired Student *t*-test with a normal-based 95% CI was applied for statistical analysis, as indicated. Statistical significance was defined as $p < 0.05$.

2.5. Validation Assays to Confirm T126 as a True Hit

2.5.1. Intrinsic Fluorescence Assay

Compound **T126** was serially diluted in 384-well black plates with Hepes 100 mM, DMSO 4%, Brij 0.003%, Tween 20 0.004%, $MgCl_2$ 10 mM, at a pH = 7.5 and incubated for 2 h at 37 °C. After the incubation time the intensity of fluorescence (λexc: 485 nm; λemm: 535 nm) was detected in a Tecan M1000 Pro.

2.5.2. Solubility Screen Assay to Determine Chemical Aggregation

The stock solutions (10–2 M) of the compound **T126** were diluted to a decreased molarity, from 100 μM to 0.0001 μM, in 384-well transparent plates (Greiner 781801) with 1% DMSO: 99% buffer Hepes 100 mM, DMSO 4%, Brij 0.003%, Tween 20 0.004%, $MgCl_2$ 10 mM, at a pH = 7.5. Incubation occurred at 37 °C and the results were read after 2 h in a NEPHELOstar Plus (BMG LABTECH, Ortenberg, Germany). The results were adjusted to a segmented regression to obtain the maximum concentration in which the compounds were soluble.

2.5.3. UV-VIS Titration Assay with an Increasing Amount of $MgCl_2$

The potential ability of **T126** to chelate Mg^{2+} ions was evaluated by spectrophotometric methods using a T70+ UV-Vis Spectrometer (PG Instruments, Leicestershire, UK) and a quartz cuvette with a 1 cm optical path. $MgCl_2$. $5H_2O$ 1M solution was prepared from powder (VWR International, Milan, Italy). The compound **T126** was dissolved in DMSO to a final concentration of 10 mM and diluted with a milliQ water/DMSO mixture to a final concentration of 100 μM (4% DMSO). The obtained solution was placed in a cuvette and the UV-Vis spectra was recorded between 200 and 500 nm using water milliQ (with 4% DMSO) as reference. Thereafter, small volumes of $MgCl_2$ aqueous solution were added both in the sample and in the reference cuvettes to obtain a series of solutions containing increasing concentrations of $MgCl_2$ (1.25, 2.5, 5 and 10 mM), carefully pipetted for mixing, and the UV-Vis spectra were repeated. Each experiment was performed in triplicate.

3. Results and Discussion

3.1. Computer-Aided Identification of Novel AKT1 Inhibitors

The idea of exploring AKT1 as a pharmacological target arose from serendipity during our efforts to discover new bioactive compounds against other protein kinases that were already the objects of our studies [40,41].

Among the investigated compounds, the in-house cyclohepta[*b*]thiophen-3-carboxyamide derivative **T187** (Figure 1) turned out to be inactive against the desired target (i.e., p38 MAPK), showing instead an unexpected inhibitory activity towards the catalytic domain of the AKT1 protein, with a percentage of enzymatic inhibition at 10 μM equal to 56%. This unforeseen result prompted us to better characterize the small molecule, which showed an IC_{50} = 11.4 ± 2.8 μM and a K_i = 4.19 ± 1.36 μM. Interestingly, a data literature search using SciFinder [42] pointed out that neither compound **T187** nor its chemical scaffold *N*-(5,6,7,8-tetrahydro-4*H*-cyclohepta[*b*]thiophen-2-yl)benzamide had ever been investigated in the context of AKT1 inhibition.

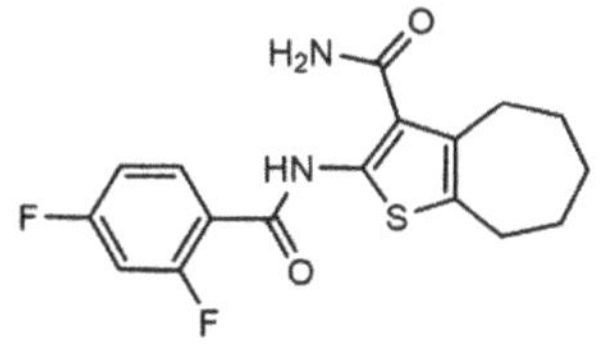

T187
IC_{50}: 11.4 ± 2.8 μM
K_i: 4.19 ± 1.36 μM
LE: 0.28

Figure 1. Chemical structure and biological data for **T187**. IC_{50} represents the compound concentration necessary to reduce by 50% the ADP formation in the ADP-Glo Assay; K_i is the AKT1 inhibition constant; Ligand efficiency (LE) refers to the potency of a compound averaged by its non-hydrogen atom count (LE = 1.4 pIC_{50}/N, where N is the number of heavy atoms). LE is expressed as units of kcal mol^{-1} N^{-1}.

Intrigued by these results and stimulated by our interest and expertise in the field of kinases [43–45] and AML [46–48], we decided to take advantage of the serendipitous discovery of **T187** to rationally search for other AKT1 inhibitors within our in-house compound library (herein after called UNIPG-lib). Specifically, the UNIPG-lib is a chemical collection of ~3000 small molecules, including both target compounds and intermediates, synthesized and/or collected by the research team in the context of several drug discovery projects.

For our AKT1-targeted program, we carried out both ligand- and structure-based modeling studies to aid the selection of **T187**-like compounds for biological testing. First, the BioSolveIT Feature Trees (FTrees) [22] software tool was used to perform a fast 2D-similarity screening of the UNIPG-lib using **T187** as a query molecule. This ligand-based approach allowed us to select 247 available neighbors to the initial compound with a similarity threshold of at least 0.65. At this stage, to further rationalize the identification of potential AKT1 inhibitors, the collected compounds were submitted to molecular docking simulations to explore their binding potential to the investigated kinase. Within this aim, the RCSB PDB [23] was consulted to retrieve structural information on AKT1 and its ATP-site binders, highlighting that 28 crystal structures had been disclosed prior to December 2021. In particular, the most interesting co-crystallized ligands were present in the PDB structures 4EKL [25] and 4GV1 [24], showing the clinical trial inhibitors **ipatasertib** (GDC-0068) and **capivasertib** (AZD5363), respectively [20]. These small molecules shared a double interaction with the hinge region (residues Glu228 and Ala230) that anchored the ligand to the ATP-binding site (Figure S1). With such valuable structural information available, we decided to explore the ATP-binding pocket in the rational search of new potential AKT1 inhibitors. Among the two previously mentioned structures, 4GV1 showed the most favorable resolution and was thus selected as a protein model for molecular docking experiments of the previously cited 247 in-house compounds. Each molecule was docked using Glide [30] in the extra precision (XP) mode and the generated poses were then rescored with the MM-GBSA method within Prime [28]. The two co-crystallized advanced inhibitors were added to the ligand set for comparison, and the obtained binding poses were evaluated mainly based on the predicted intermolecular interactions, paying particular attention to the contacts that involved the key residues Ala230 and Glu228. The best docking conformations of **ipatasertib** and **capivasertib** agreed well with their experimental poses and in both cases the key interactions with the hinge residues were correctly reproduced (Figure S2).

Subsequently, the docking-based approach allowed the selection of 12 virtual hits as the most promising in-house AKT1 small-molecule binders (Tables S1 and S2), which were then evaluated for the anti-AKT1 activity.

In the employed biochemical assay, a compound was considered as a 'primary active' only when the inhibition of the AKT1 phosphorylation activity (hereinafter called %inh) was reduced by at least of 60%, employing a 10 μM compound concentration. Among the tested virtual hits, five compounds were thus selected as primary actives and evaluated in a dose–response assay to calculate the 50% inhibitory concentration (IC_{50}) value (Figure 2 and Table S2).

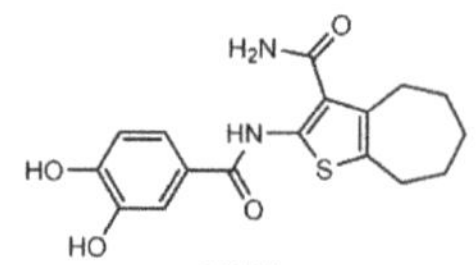

T159
IC_{50} : 2.12 ± 0.26 μM
K_i: 1.91 ± 0.17 μM
LE: 0.32

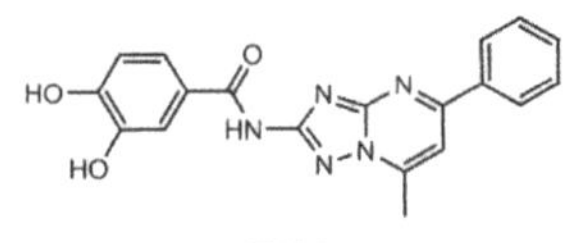

T101
IC_{50} : 1.62 ± 0.15 μM
K_i: 0.24 ± 0.55 μM
LE: 0.29

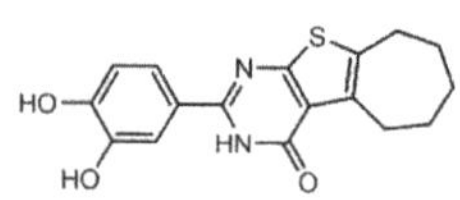

T124
IC_{50} : 30.9 ± 5.3 μM
K_i: ND
LE: 0.27

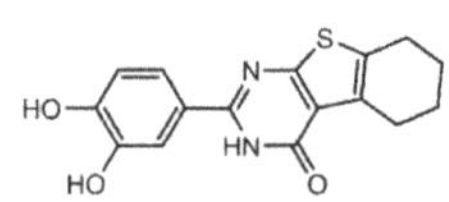

T126
IC_{50} : 1.99 ± 0.11 μM
K_i: 0.41 ± 0.08 μM
LE: 0.35

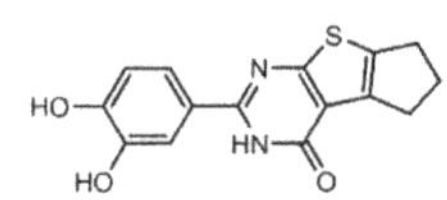

T128
IC_{50} : 26.6 ± 4.8 μM
K_i: ND
LE: 0.30

Figure 2. Chemical structure and biological data for the experimentally validated in-house compounds as AKT1 inhibitors. ND = not determined.

The [1,2,4]triazolo[1,5-*a*]pyrimidine derivative **T101** [35] and the cyclohepta[*b*]thiophene-3-carboxamide **T159** [32] were previously reported as allosteric inhibitors of HIV-1 reverse transcriptase-associated ribonuclease H. Analogously, compounds **T124, T126,** and **T128,** showing a common [4,5]thieno[2,3-*d*]pyrimidin-4(3*H*)-one moiety, were described as HIV-1 reverse transcriptase-associated ribonuclease H inhibitors by Masaoka et al. [34], and re-synthesized by us to serve as internal reference compounds.

Among them, compounds **T101, T126,** and **T159** exhibited an AKT1 inhibitory potency higher than **T187**, displaying IC_{50} values of 1.62 ± 0.15 μM, 1.99 ± 0.11 μM, and 2.12 ± 0.26 μM, respectively.

Notably, these three molecules turned out to be efficient inhibitors, as highlighted by the corresponding experimental values of LE. Particularly, **T159** and **T126** had an LE value ≥ 0.3 (Figure 2), which is the generally accepted lower limit of efficiency in a typical drug discovery program. It is worth noting that the 5,6,7,8-tetrahydrobenzo[4,5]thieno[2,3-*d*]pyrimidin-4(3*H*)-one derivative **T126** showed a LE value comparable to that of the clinical trial inhibitors **capivasertib** and **ipatasertib** (Figure S1). In addition, preliminary structure-activity relationships (SAR) analysis around this in-house compound can be deduced by examining the collected biological results (Table S2). First, the replacement of the amide nitrogen atom in the tricyclic core with an oxygen atom seemed to be detrimental for kinase inhibition, as highlighted by comparing **T126** (%inh = 87%, IC_{50} = 1.99 ± 0.11 μM) and **T128** (%inh = 78%, IC_{50} = 26.6 ± 4.8 μM) with **T125** (%inh = 9%) and **T127** (%inh = 43%), respectively. This observation underlined that, although the in silico approach was successful in identifying new AKT1 inhibitors, the different impact on the biological activity of the pyrimidin-4(3*H*)-one moiety (**T126** and **T128**) compared to that of the oxazin-4-one (**T125** and **T127**) was not predicted correctly because similar docking poses and ligand-target interactions were proposed for the two series of compounds (Figure S3).

Therefore, we carried out molecular dynamics (MD) simulations to elucidate the molecular basis of the SAR data by overcoming the limitations of ordinary molecular docking studies, where the protein is typically treated as a rigid entity and no water molecules are included in the target preparation. The docking-generated AKT1–**T126** and AKT1-**T125** complexes were thus used as starting structures for three 200 ns MD-simulation replicas using Desmond software [31], and the analysis of the produced trajectories provided valuable clues to explain the SAR observation.

Specifically, the **T126** orientation generated from the docking study was well preserved during the entire execution time, with the overall ligand stability having been reached in the last 100 ns of the MD simulations (Figure S4). Interestingly, the nitrogen of the endocyclic amide established well-conserved, water-mediated interactions with the residues Thr291 and Asp292 (Table S3 and Figure 3).

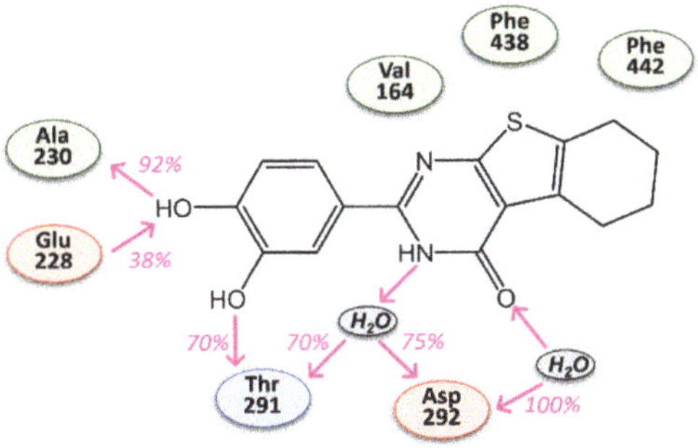

Figure 3. A schematic overview of the MD-derived intermolecular interactions established by **T126** within the AKT1 ATP-binding pocket. The key contacts are represented with their occurrence calculated as the average value of the three replicas in the time window 100–200 ns. The protein residues are color-coded as follows: red, charged (negative); green, hydrophobic; cyan, polar.

Conversely, the three generated replicas for the inactive analog **T125** (Figure S4) did not detect common stable ligand-protein contacts, with the oxygen atom of the oxazin-4-one moiety never having been involved in significantly conserved intermolecular interactions (Table S3).

As a second SAR insight, the size of the aliphatic ring of the tricyclic core appeared to be crucial to gain high inhibitory potency against AKT1. Indeed, the 6-membered ring of **T126** turned out to be the best substitution with respect to the 5-membered (**T128**) and 7-membered (**T124**, %inh = 60%, IC_{50} = 30.9 μM) systems.

In addition, it is well known that the level of interest in kinase targets has created a crowded chemical space landscape, rendering the discovery of novel chemical entities as kinase inhibitors challenging. Against this backdrop, compounds **T126, T159,** and **T101** were used as inputs for a structure search in SciFinder. To the best of our knowledge prior to December 2021, none of the three explored compounds had been reported in the literature, including publications and patents, as inhibitor of AKT1 or other kinases, as well as no information on the effect of these compounds on AML has emerged from our bibliographic survey.

3.2. *Evaluation of the Identified AKT1 Inhibitors on AML Cell Lines*

Encouraged by the variety of information gathered so far, we evaluated these compounds in cellular assays of AML models. Specifically, a cell viability assay was initially performed to determine the cellular IC_{50} of **T101, T126,** and **T159** on different human AML cell lines, namely, OCI-AML3, IMS-M2, OCI-AML2, MOLM-13, and SKM-1, which carry some of the most recurrent AML mutations [37,38]. We measured the ability of living cells to convert a redox dye (resazurin) into a fluorescent end-product (resorufin). A dose response curve was generated for each cell line using GraphPad Prism and the IC_{50} was established. While no interesting (IC_{50} higher than 30 μM) inhibitory activity was detected for **T101** and **T159** in all the explored cell lines, we obtained promising results for **T126** (Figure 4).

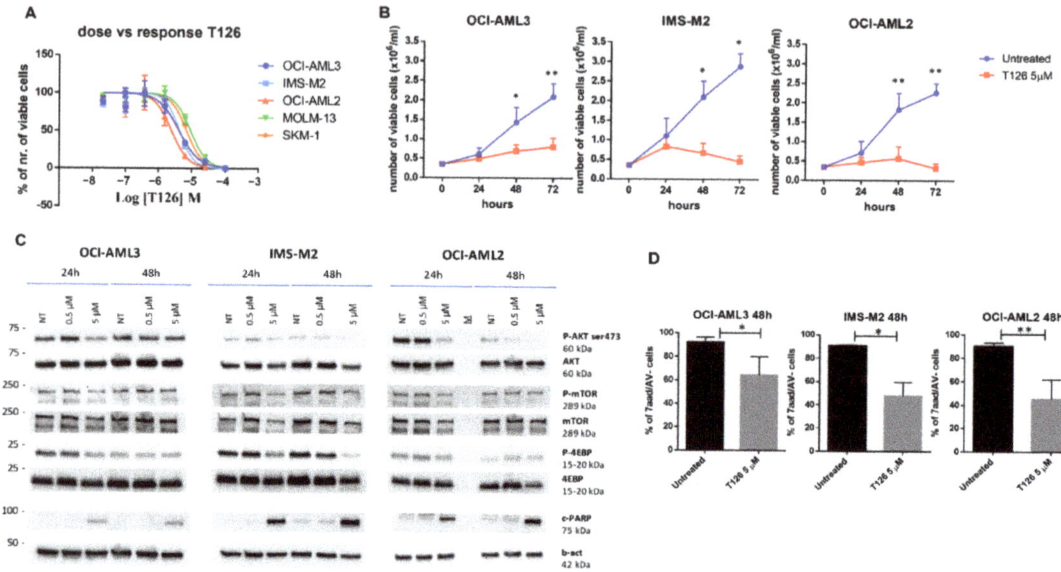

Figure 4. **T126** induces growth arrest and apoptosis in AML cells. (**A**) **T126** dose-response curves as generated by CellTiter Blue assay for the human AML cell lines OCI-AML3, IMS-M2, OCI-AML2, MOLM-13, and SKM-1. (**B**) **T126** induced a significant cell growth inhibition on OCI-AML3, IMS-M2, and OCI-AML2, at 48 and 72 h from the start of the treatment at a 5 μM dose (student paired *t*-test, 48 h $p = 0.02$ and 72 h 0.006, 48 h $p = 0.04$ and 72 h $p = 0.01$, and 48 h $p = 0.006$ and 72 h $p = 0.004$, respectively). (**C**) The effects of **T126** at a 0.5 and 5 μM dose on phosphorylation of AKT1 and downstream signaling molecules, namely mTOR and 4E-BP1, at 24 and 48 h, in OCI-AML3, IMS-M2, and OCI-AML2 cells. Activation of the apoptotic signaling molecule cleaved-PARP (c-PARP) under the same experimental conditions. β-actin (b-act) expression levels were used as a control for protein loading. The representative results of at least three independent experiments are shown. All the original blots are reported in Figure S8. (**D**) **T126** at a 5 μM dose induced a significant level of apoptosis (% of Annexin V negative/7aad negative cells) in OCI-AML3, IMS-M2, and OCI-AML2 cells at 48 h (Student paired *t*-test, where * and ** stand for $p = 0.01$ and $p = 0.005$, respectively).

Indeed, **T126** IC_{50} values were 4.2, 4.3, 2.4, 9.2, and 6.9 μM for OCI-AML3, IMS-M2, OCI-AML2, MOLM-13, and SKM-1 cell lines, respectively (Figure 4A).

It is noteworthy that the low-micromolar inhibitory activities observed in the cell-based assays were similar to those obtained for the reference compound **NSC348884** [39] used as positive control (Figure S5), and were in keeping with the low-micromolar IC_{50} of **T126** on the isolated AKT1 protein (i.e., 1.99 μM).

We then investigated the effect of **T126** on OCI-AML3, IMS-M2, and OCI-AML2 cell growth (0–72 h), using a concentration of 5 μM, and found that **T126** was able to induce a significant growth inhibition at 48 and 72 h in all the three of the AML cell lines (Figure 4B).

To determine the effect of **T126** on the AKT1 intracellular mediated signaling pathway and cell death, we exposed our cell lines to an increasing concentration of **T126** and checked whether this small molecule acted upon their functional proteomic profiles. The effect of **T126** on the AKT signaling pathway was assessed by western blot analysis, focusing on the key targets. Upon increasing concentrations of **T126**, we observed at various levels in the different AML cell lines a decreased phosphorylation of AKT1 (phospho-AKT, at site S473) as an expression of the AKT1 activity inhibition [17], and of the downstream signaling molecule mTOR (phospho-mTOR, at site S2448). As a representative target of mTOR activity, we also analyzed the E4-BP1 protein and we noticed a reduction of its phosphorylation levels at site S65 (Figure 4C). As a marker of cell apoptosis, the apoptosis-related protein cleaved-PARP was detected upon treatment with **T126** 5 μM at 48 h (Figure 4C). The effect of **T126** on cell apoptosis was also measured by flow cytometry using Annexin V/7AAD probes for the identification of apoptotic cells. Notably, in accordance with western blot analysis, at the concentration of 5 μM, **T126** induced a significant level of apoptosis at 48 h in OCI-AML3, IMS-M2, and OCI-AML2 (Figure 4D), confirming in this setting, its anti-leukemic effect.

Finally, we performed a colony forming unit assay (CFU) using normal donor CD34+ hematopoietic stem/progenitor cells (HSPC) which were exposed to escalating doses of **T126** (1, 2.5 and 5 μM) and plated in a Methocult medium for 14 days. Remarkably, at 1 and 2.5 μM, compound **T126** did not inhibit the colony forming ability of human CD34+ HSPC, suggesting that there was no toxicity at these lower doses. Importantly, whilst a slight inhibition emerged at the maximum dose of 5 μM, the clonogenic efficiency was not blocked, but only decreased by about 20% as compared to the untreated control (Figure S6A). The obtained results were quite encouraging considering that the inhibition of p-AKT was already observed at an early time point (4 h) at the concentrations of 1 and 2.5 μM in the OCI-AML3 cell line (Figure S6B), suggesting that the compound could also be used at lower doses in additional biological experiments. These results confirm the pivotal role of the AKT pathway in the maintenance of cell growth and survival of AML cells, as previously reported [4,5]. Moreover, AKT pathway activation has also been implicated as a mechanism of resistance to therapy, including novel agents [18,49], pointing to AKT1 as an attractive target for AML, more specifically by using AKT1 inhibitors in combination with other drugs. Considering this, it is worth noting that the three AKT isoforms are highly conserved and share about 80% of the amino acid sequence identity [50]. As a consequence, the majority of currently developed AKT inhibitors are pan-AKT inhibitors, and this concept is evident when considering the two AKT inhibitors mentioned earlier and undergoing clinical trials (i.e., **capivasertib** IC_{50}-AKT1 = 3 nM, IC_{50}-AKT2 = 8 nM, IC_{50}-AKT3 = 8 nM, **ipatasertib** IC_{50}-AKT1 = 5 nM, IC_{50}-AKT2 = 18 nM, and IC_{50}-AKT3 = 8 nM) [51]. Based on this, we are aware that **T126** may potentially also inhibit AKT2 and AKT3. However, in the AML cell lines used in this study, AKT1 is the most expressed isoform among the AKT family members. Indeed, the analyses of RNA seq data from these cell lines (https://cellmodelpassports.sanger.ac.uk/passports, accessed on 14 October 2022) clearly indicate that AKT1 shows a higher degree of expression compared to the other isoforms. In the light of this information and considering that the work herein reported falls within the early-stage drug discovery process, we have focused our efforts exclusively on AKT1 to identify a hit compound. Nevertheless, during the next planned hit-to-lead steps, biological characterization studies involving the other AKT family members will be also carried out.

3.3. Validation of T126 as a True Hit

To further validate **T126** as a promising hit, focused studies were conducted to rule out the possibility that the inhibitory potency detected in the AKT1 assays could be related to an artifact. Indeed, this small molecule possessed a catechol moiety that seemed to be important for AKT1 binding by establishing polar interactions (Figure 3). Although the presence of the catechol by itself did not guarantee a satisfactory potency against AKT1 as the moderately active (**T124** and **T128**) and inactive (**T125**, **T127**, and **T129**) derivatives shared this feature as well, it is well-known that the mentioned chemical group has been counted as one of the Pan-Assay Interference Compounds (PAINS) motifs [52,53].

Various mechanisms of assay interference or promiscuous behaviors have been reported as responsible for PAINS activity, including compound fluorescence effect, metal chelation, and chemical aggregation [52–54].

In the AKT1 inhibition assay, the enzymatic activity was evaluated by measuring the phosphorylation of a fluorescently-labelled peptide induced by the kinase. Indeed, the detection of the substrate peptide and the product phosphorylated peptide after a microfluidic mobility shift was performed by measuring the fluorescence given to the substrate and the product of the kinase reaction. Based on the methodology employed for quantifying the AKT1 activity, it was not expected that the intrinsic fluorescence of the compounds could interfere with the assay. Nevertheless, we determined the intrinsic fluorescence of **T126**, which emerged as not fluorescent at the excitation/emission wavelength (485–535 nm) used in the assay.

Second, since the AKT1 inhibition assay required $MgCl_2$, the Mg-complexation ability has been investigated for the **T126** derivative to rule out any interference. Specifically, UV-vis spectra were recorded for the compound alone and in the presence of increasing concentrations of $MgCl_2$. The results showed no shift in the maximum of absorbance (hypsochromic effect), nor any presence of an isosbestic point (Figure S7) up to 10 mM concentration of $MgCl_2$ (as used in the biochemical assay), thus suggesting the inability of **T126** to chelate Mg^{2+} ions.

Finally, to discard the hypothesis of chemical aggregation, the presence of aggregates was evaluated in the same buffer employed in the AKT1 activity assay by a NEPHELOstar Plus (BMG LABTECH). Compound **T126** did not aggregate at concentrations up to 100 μM, thus rejecting that the AKT1 inhibition effect may be due to aggregation.

In conclusion, the 5,6,7,8-tetrahydrobenzo[4,5]thieno[2,3-*d*]pyrimidin-4(3*H*)-one derivative **T126** showed a clear anti-AKT1 effect and the growth inhibition and induction of apoptosis in AML cells at low micromolar concentrations, making it an appealing candidate for further testing in other biological assays. Furthermore, the low molecular weight (MW = 314,37) and promising LE value of **T126** render this small molecule an attractive starting point for future medicinal chemistry efforts directed at hit-to-lead optimization.

Supplementary Materials: The following supporting information can be downloaded at: https://www.mdpi.com/article/10.3390/pharmaceutics14112295/s1. Figure S1: 2D representation of the protein-ligand interactions of **capivasertib** (A) and **ipatasertib** (B) with AKT1, Figure S2: Superimposition of the experimental and docking poses for **capivasertib** and **ipatasertib**, Figure S3: Superimposition of the pyrimidin-4(3*H*)-one derivative **T126** and the oxazin-4-one derivative **T125** binding modes, Figure S4: Ligand RMSD computed for the three MD replicas on the **T126**-AKT1 and **T125**-AKT1 complexes, Figure S5: Dose-response curves for the reference compound **NSC348884**, Figure S6: (A) CFU assay on normal donor CD34+ hematopoietic stem/progenitor cells (HSPC). (B) Effects of **T126** at 0.5, 1.5 and 2.5 μM dose on phosphorylation of AKT at 4 hours (4 h) in OCI-AML3, Figure S7: UV-vis spectra of compound **T126** measured alone and in presence of increasing concentrations of $MgCl_2$, Figure S8: Uncropped full-length pictures of western blotting membranes, Table S1: Summary of the docking results, Table S2: Chemical structure and biological activity for the selected in-house virtual hits against AKT1, Table S3: Protein-ligand interaction occurrences observed in the three MD replicas for **T125**-AKT1 and **T126**-AKT1 complexes, Table S4: Composition of Enriched Methocult medium.

Author Contributions: Conceptualization, A.A. and M.L.B.; methodology, investigation, and validation: (i) computational studies, A.A., D.P. and M.L.B. (ii) chemistry, solubility screen, and UV-VIS titration assays, M.C.P., O.T. and S.M. (iii) enzymatic and intrinsic fluorescence assays, J.B. and M.I.L. (iv) cell-based assays, F.M., M.M. and M.P.M.; writing—original draft preparation, A.A., F.M., D.P., S.M. and M.L.B.; writing—review and editing, all authors; supervision, M.L.B. All authors have read and agreed to the published version of the manuscript.

Funding: Part of this research was funded by the ERC Cons Grant 2016, grant number 725725 (to M.P.M.).

Institutional Review Board Statement: Not applicable.

Informed Consent Statement: Not applicable.

Data Availability Statement: Data are contained within the article or Supplementary Materials.

Acknowledgments: Fondazione Umberto Veronesi—Post-doctoral Fellowships 2021 (to A.A.) and 2022 (to M.C.P.).

Conflicts of Interest: The authors declare no conflict of interest. M.P.M. declares an honoraria from Rasna Therapeutics, Inc. for scientific advisor activities. M.P.M. also declares an honoraria/consultancy at the scientific advisory board for Abbvie, Amgen, Celgene, Janssen, Novartis, Pfizer, Jazz Pharmaceuticals. M.L.B. is a co-founder and shareholder of the innovative start up Sibylla Biotech SRL (https://www.sibyllabiotech.it).

Abbreviations

AML, acute myeloid leukemia; ATP, adenosine triphosphate; 4E-BP1, eukaryotic translation initiation factor 4E (eIF4E)-binding protein 1; FLT3, fms-like tyrosine kinase 3; LE, ligand efficiency; mTOR, mammalian target of rapamycin; p70S6K, p70 ribosomal S6 kinase; PARP, poly (ADP-ribose) polymerase; CFU, colony forming unit; HSPC, hematopoietic stem/progenitor cells; PAINS, Pan-Assay Interference Compounds; KRAS, Kirsten rat sarcoma viral oncogene; NRAS, neuroblastoma ras viral oncogene homolog; SAR, structure-activity relationship; XPO1, exportin 1.

References

1. Scheid, M.P.; Woodgett, J.R. PKB/AKT: Functional insights from genetic models. *Nat. Rev. Mol. Cell Biol.* **2001**, *2*, 760–768. [CrossRef] [PubMed]
2. Vanhaesebroeck, B.; Stephens, L.; Hawkins, P. PI3K signalling: The path to discovery and understanding. *Nat. Rev. Mol. Cell Biol.* **2012**, *13*, 195–203. [CrossRef] [PubMed]
3. Altomare, D.A.; Testa, J.R. Perturbations of the AKT signaling pathway in human cancer. *Oncogene* **2005**, *24*, 7455–7464. [CrossRef] [PubMed]
4. Nepstad, I.; Hatfield, K.J.; Gronningsaeter, I.S.; Reikvam, H. The PI3K-Akt-mTOR Signaling Pathway in Human Acute Myeloid Leukemia (AML) Cells. *Int. J. Mol. Sci.* **2020**, *21*, 2907. [CrossRef] [PubMed]
5. Sabnis, H.; Bradley, H.L.; Bunting, S.T.; Cooper, T.M.; Bunting, K.D. Capillary nano-immunoassay for Akt 1/2/3 and 4EBP1 phosphorylation in acute myeloid leukemia. *J. Transl. Med.* **2014**, *12*, 166. [CrossRef] [PubMed]
6. The Cancer Genome Atlas Research Network. Genomic and epigenomic landscapes of adult de novo acute myeloid leukemia. *N. Eng. J. Med.* **2013**, *368*, 2059–2074. [CrossRef]
7. Tamburini, J.; Elie, C.; Bardet, V.; Chapuis, N.; Park, S.; Broet, P.; Cornillet-Lefebvre, P.; Lioure, B.; Ugo, V.; Blanchet, O.; et al. Constitutive phosphoinositide 3-kinase/Akt activation represents a favorable prognostic factor in de novo acute myelogenous leukemia patients. *Blood* **2007**, *110*, 1025–1028. [CrossRef]
8. Recher, C.; Dos Santos, C.; Demur, C.; Payrastre, B. mTOR, a new therapeutic target in acute myeloid leukemia. *Cell Cycle* **2005**, *4*, 1540–1549. [CrossRef]
9. Chow, S.; Minden, M.D.; Hedley, D.W. Constitutive phosphorylation of the S6 ribosomal protein via mTOR and ERK signaling in the peripheral blasts of acute leukemia patients. *Exp. Hematol.* **2006**, *34*, 1183–1191. [CrossRef]
10. Fransecky, L.; Mochmann, L.H.; Baldus, C.D. Outlook on PI3K/AKT/mTOR inhibition in acute leukemia. *Mol. Cell* **2015**, *3*, 2. [CrossRef]
11. Gallay, N.; Dos Santos, C.; Cuzin, L.; Bousquet, M.; Simmonet Gouy, V.; Chaussade, C.; Attal, M.; Payrastre, B.; Demur, C.; Recher, C. The level of AKT phosphorylation on threonine 308 but not on serine 473 is associated with high-risk cytogenetics and predicts poor overall survival in acute myeloid leukaemia. *Leukemia* **2009**, *23*, 1029–1038. [CrossRef]
12. Tazzari, P.L.; Cappellini, A.; Grafone, T.; Mantovani, I.; Ricci, F.; Billi, A.M.; Ottaviani, E.; Conte, R.; Martinelli, G.; Martelli, A.M. Detection of serine 473 phosphorylated Akt in acute myeloid leukaemia blasts by flow cytometry. *Br. J. Haematol.* **2004**, *126*, 675–681. [CrossRef] [PubMed]
13. Min, Y.H.; Eom, J.I.; Cheong, J.W.; Maeng, H.O.; Kim, J.Y.; Jeung, H.K.; Lee, S.T.; Lee, M.H.; Hahn, J.S.; Ko, Y.W. Constitutive phosphorylation of Akt/PKB protein in acute myeloid leukemia: Its significance as a prognostic variable. *Leukemia* **2003**, *17*, 995–997. [CrossRef] [PubMed]
14. Kornblau, S.M.; Womble, M.; Qiu, Y.H.; Jackson, C.E.; Chen, W.; Konopleva, M.; Estey, E.H.; Andreeff, M. Simultaneous activation of multiple signal transduction pathways confers poor prognosis in acute myelogenous leukemia. *Blood* **2006**, *108*, 2358–2365. [CrossRef] [PubMed]
15. Bhutani, J.; Sheikh, A.; Niazi, A.K. Akt inhibitors: Mechanism of action and implications for anticancer therapeutics. *Infect. Agent. Cancer* **2013**, *8*, 49. [CrossRef] [PubMed]
16. Huang, W.C.; Hung, M.C. Induction of Akt activity by chemotherapy confers acquired resistance. *J. Med. Assoc.* **2009**, *108*, 180–194. [CrossRef]
17. Martelli, A.M.; Evangelisti, C.; Chappell, W.; Abrams, S.L.; Basecke, J.; Stivala, F.; Donia, M.; Fagone, P.; Nicoletti, F.; Libra, M.; et al. Targeting the translational apparatus to improve leukemia therapy: Roles of the PI3K/PTEN/Akt/mTOR pathway. *Leukemia* **2011**, *25*, 1064–1079. [CrossRef]

18. Lin, K.H.; Rutter, J.C.; Xie, A.; Killarney, S.T.; Vaganay, C.; Benaksas, C.; Ling, F.; Sodaro, G.; Meslin, P.A.; Bassil, C.F.; et al. P2RY2-AKT activation is a therapeutically actionable consequence of XPO1 inhibition in acute myeloid leukemia. *Nat. Cancer* **2022**, *3*, 837–851. [CrossRef]
19. Coleman, N.; Moyers, J.T.; Harbery, A.; Vivanco, I.; Yap, T.A. Clinical Development of AKT Inhibitors and Associated Predictive Biomarkers to Guide Patient Treatment in Cancer Medicine. *Pharmgenomics. Pers. Med.* **2021**, *14*, 1517–1535. [CrossRef]
20. Lazaro, G.; Kostaras, E.; Vivanco, I. Inhibitors in AKTion: ATP-competitive vs allosteric. *Biochem. Soc. Trans.* **2020**, *48*, 933–943. [CrossRef]
21. *Schrödinger Release 2019-2: LigPrep*; Schrödinger, LLC: New York, NY, USA, 2019.
22. Rarey, M.; Dixon, J.S. Feature trees: A new molecular similarity measure based on tree matching. *J. Comput. Aided Mol. Des.* **1998**, *12*, 471–490. [CrossRef] [PubMed]
23. Berman, H.M.; Battistuz, T.; Bhat, T.N.; Bluhm, W.F.; Bourne, P.E.; Burkhardt, K.; Feng, Z.; Gilliland, G.L.; Iype, L.; Jain, S.; et al. The Protein Data Bank. *Acta Cryst. D Biol. Cryst.* **2002**, *58*, 899–907. [CrossRef] [PubMed]
24. Addie, M.; Ballard, P.; Buttar, D.; Crafter, C.; Currie, G.; Davies, B.R.; Debreczeni, J.; Dry, H.; Dudley, P.; Greenwood, R.; et al. Discovery of 4-amino-N-[(1S)-1-(4-chlorophenyl)-3-hydroxypropyl]-1-(7H-pyrrolo[2,3-d]pyrimidin-4-yl)piperidine-4-carboxamide (AZD5363), an orally bioavailable, potent inhibitor of Akt kinases. *J. Med. Chem.* **2013**, *56*, 2059–2073. [CrossRef] [PubMed]
25. Lin, K.; Lin, J.; Wu, W.I.; Ballard, J.; Lee, B.B.; Gloor, S.L.; Vigers, G.P.; Morales, T.H.; Friedman, L.S.; Skelton, N.; et al. An ATP-site on-off switch that restricts phosphatase accessibility of Akt. *Sci. Signal* **2012**, *5*, ra37. [CrossRef] [PubMed]
26. Sastry, G.M.; Adzhigirey, M.; Day, T.; Annabhimoju, R.; Sherman, W. Protein and ligand preparation: Parameters, protocols, and influence on virtual screening enrichments. *J. Comput. Aided Mol. Des.* **2013**, *27*, 221–234. [CrossRef]
27. Greenwood, J.R.; Calkins, D.; Sullivan, A.P.; Shelley, J.C. Towards the comprehensive, rapid, and accurate prediction of the favorable tautomeric states of drug-like molecules in aqueous solution. *J. Comput. Aided Mol. Des.* **2010**, *24*, 591–604. [CrossRef]
28. Jacobson, M.P.; Pincus, D.L.; Rapp, C.S.; Day, T.J.F.; Honig, B.; Shaw, D.E.; Friesner, R.A. A Hierarchical Approach to All-Atom Protein Loop Prediction. *Proteins Struct. Funct. Bioinform.* **2004**, *55*, 351–367. [CrossRef]
29. Halgren, T.A.; Murphy, R.B.; Friesner, R.A.; Beard, H.S.; Frye, L.L.; Pollard, W.T.; Banks, J.L. Glide: A New Approach for Rapid, Accurate Docking and Scoring. 2. Enrichment Factors in Database Screening. *J. Med. Chem.* **2004**, *47*, 1750–1759. [CrossRef]
30. Friesner, R.A.; Murphy, R.B.; Repasky, M.P.; Frye, L.L.; Greenwood, J.R.; Halgren, T.A.; Sanschagrin, P.C.; Mainz, D.T. Extra precision glide: Docking and scoring incorporating a model of hydrophobic enclosure for protein-ligand complexes. *J. Med. Chem.* **2006**, *49*, 6177–6196. [CrossRef]
31. Bowers, K.J.; Chow, E.; Xu, H.; Dror, R.O.; Eastwood, M.P.; Gregersen, B.A.; Klepeis, J.L.; Kolossvary, I.; Moraes, M.A.; Sacerdoti, F.D.; et al. Scalable Algorithms for Molecular Dynamics Simulations on Commodity Clusters. In Proceedings of the ACM/IEEE Conference on Supercomputing (SC06), Tampa, FL, USA, 11–17 November 2006.
32. Massari, S.; Corona, A.; Distinto, S.; Desantis, J.; Caredda, A.; Sabatini, S.; Manfroni, G.; Felicetti, T.; Cecchetti, V.; Pannecouque, C.; et al. From cycloheptathiophene-3-carboxamide to oxazinone-based derivatives as allosteric HIV-1 ribonuclease H inhibitors. *J. Enzym. Inhib. Med. Chem.* **2019**, *34*, 55–74. [CrossRef]
33. Desantis, J.; Nannetti, G.; Massari, S.; Barreca, M.L.; Manfroni, G.; Cecchetti, V.; Palu, G.; Goracci, L.; Loregian, A.; Tabarrini, O. Exploring the cycloheptathiophene-3-carboxamide scaffold to disrupt the interactions of the influenza polymerase subunits and obtain potent anti-influenza activity. *Eur. J. Med. Chem.* **2017**, *138*, 128–139. [CrossRef] [PubMed]
34. Masaoka, T.; Chung, S.; Caboni, P.; Rausch, J.W.; Wilson, J.A.; Taskent-Sezgin, H.; Beutler, J.A.; Tocco, G.; Le Grice, S.F. Exploiting drug-resistant enzymes as tools to identify thienopyrimidinone inhibitors of human immunodeficiency virus reverse transcriptase-associated ribonuclease H. *J. Med. Chem.* **2013**, *56*, 5436–5445. [CrossRef] [PubMed]
35. Desantis, J.; Massari, S.; Corona, A.; Astolfi, A.; Sabatini, S.; Manfroni, G.; Palazzotti, D.; Cecchetti, V.; Pannecouque, C.; Tramontano, E.; et al. 1,2,4-Triazolo[1,5-a]pyrimidines as a Novel Class of Inhibitors of the HIV-1 Reverse Transcriptase-Associated Ribonuclease H Activity. *Molecules* **2020**, *25*, 1183. [CrossRef] [PubMed]
36. Wang, Y.D.; Johnson, S.; Powell, D.; McGinnis, J.P.; Miranda, M.; Rabindran, S.K. Inhibition of tumor cell proliferation by thieno[2,3-d]pyrimidin-4(1H)-one-based analogs. *Bioorg. Med. Chem. Lett.* **2005**, *15*, 3763–3766. [CrossRef] [PubMed]
37. Quentmeier, H.; Martelli, M.P.; Dirks, W.G.; Bolli, N.; Liso, A.; Macleod, R.A.; Nicoletti, I.; Mannucci, R.; Pucciarini, A.; Bigerna, B.; et al. Cell line OCI/AML3 bears exon-12 NPM gene mutation-A and cytoplasmic expression of nucleophosmin. *Leukemia* **2005**, *19*, 1760–1767. [CrossRef] [PubMed]
38. Chi, H.T.; Vu, H.A.; Iwasaki, R.; Nagamura, F.; Tojo, A.; Watanabe, T.; Sato, Y. Detection of exon 12 type A mutation of NPM1 gene in IMS-M2 cell line. *Leuk Res.* **2010**, *34*, 261–262. [CrossRef]
39. Balusu, R.; Fiskus, W.; Rao, R.; Chong, D.G.; Nalluri, S.; Mudunuru, U.; Ma, H.; Chen, L.; Venkannagari, S.; Ha, K.; et al. Targeting levels or oligomerization of nucleophosmin 1 induces differentiation and loss of survival of human AML cells with mutant NPM1. *Blood* **2011**, *118*, 3096–3106. [CrossRef]
40. Sancineto, L.; Iraci, N.; Massari, S.; Attanasio, V.; Corazza, G.; Barreca, M.L.; Sabatini, S.; Manfroni, G.; Avanzi, N.R.; Cecchetti, V.; et al. Computer-aided design, synthesis and validation of 2-phenylquinazolinone fragments as CDK9 inhibitors with anti-HIV-1 Tat-mediated transcription activity. *ChemMedChem* **2013**, *8*, 1941–1953. [CrossRef]
41. Astolfi, A.; Kudolo, M.; Brea, J.; Manni, G.; Manfroni, G.; Palazzotti, D.; Sabatini, S.; Cecchetti, F.; Felicetti, T.; Cannalire, R.; et al. Discovery of potent p38alpha MAPK inhibitors through a funnel like workflow combining in silico screening and in vitro validation. *Eur. J. Med. Chem.* **2019**, *182*, 111624. [CrossRef]

42. Available online: https://scifinder.cas.org (accessed on 15 February 2022).
43. Sabatini, S.; Manfroni, G.; Barreca, M.L.; Bauer, S.M.; Gargaro, M.; Cannalire, R.; Astolfi, A.; Brea, J.; Vacca, C.; Pirro, M.; et al. The Pyrazolobenzothiazine Core as a New Chemotype of p38 Alpha Mitogen-Activated Protein Kinase Inhibitors. *Chem. Biol. Drug Des.* **2015**, *86*, 531–545. [CrossRef]
44. Astolfi, A.; Iraci, N.; Sabatini, S.; Barreca, M.L.; Cecchetti, V. p38alpha MAPK and Type I Inhibitors: Binding Site Analysis and Use of Target Ensembles in Virtual Screening. *Molecules* **2015**, *20*, 15842–15861. [CrossRef] [PubMed]
45. Astolfi, A.; Iraci, N.; Manfroni, G.; Barreca, M.L.; Cecchetti, V. A Comprehensive Structural Overview of p38alpha MAPK in Complex with Type I Inhibitors. *ChemMedChem* **2015**, *10*, 957–969. [CrossRef] [PubMed]
46. Martelli, M.P.; Rossi, R.; Venanzi, A.; Meggendorfer, M.; Perriello, V.M.; Martino, G.; Spinelli, O.; Ciurnelli, R.; Varasano, E.; Brunetti, L.; et al. Novel NPM1 exon 5 mutations and gene fusions leading to aberrant cytoplasmic nucleophosmin in AML. *Blood* **2021**, *138*, 2696–2701. [CrossRef] [PubMed]
47. Gionfriddo, I.; Brunetti, L.; Mezzasoma, F.; Milano, F.; Cardinali, V.; Ranieri, R.; Venanzi, A.; Pierangeli, S.; Vetro, C.; Spinozzi, G.; et al. Dactinomycin induces complete remission associated with nucleolar stress response in relapsed/refractory NPM1-mutated AML. *Leukemia* **2021**, *35*, 2552–2562. [CrossRef]
48. Falini, B.; Brunetti, L.; Sportoletti, P.; Martelli, M.P. NPM1-mutated acute myeloid leukemia: From bench to bedside. *Blood* **2020**, *136*, 1707–1721. [CrossRef]
49. Grandage, V.L.; Gale, R.E.; Linch, D.C.; Khwaja, A. PI3-kinase/Akt is constitutively active in primary acute myeloid leukaemia cells and regulates survival and chemoresistance via NF-kappaB, Mapkinase and p53 pathways. *Leukemia* **2005**, *19*, 586–594. [CrossRef]
50. Toker, A.; Marmiroli, S. Signaling specificity in the Akt pathway in biology and disease. *Adv. Biol. Regul.* **2014**, *55*, 28–38. [CrossRef]
51. Martorana, F.; Motta, G.; Pavone, G.; Motta, L.; Stella, S.; Vitale, S.R.; Manzella, L.; Vigneri, P. AKT Inhibitors: New Weapons in the Fight Against Breast Cancer? *Front. Pharm.* **2021**, *12*, 662232. [CrossRef]
52. Baell, J.B.; Holloway, G.A. New substructure filters for removal of pan assay interference compounds (PAINS) from screening libraries and for their exclusion in bioassays. *J. Med. Chem.* **2010**, *53*, 2719–2740. [CrossRef]
53. Baell, J.; Walters, M.A. Chemistry: Chemical con artists foil drug discovery. *Nature* **2014**, *513*, 481–483. [CrossRef]
54. Baell, J.B. Feeling Nature's PAINS: Natural Products, Natural Product Drugs, and Pan Assay Interference Compounds (PAINS). *J. Nat. Prod.* **2016**, *79*, 616–628. [CrossRef] [PubMed]

pharmaceutics

MDPI

Article

A ROR1 Small Molecule Inhibitor (KAN0441571C) Induced Significant Apoptosis of Mantle Cell Lymphoma (MCL) Cells

Amineh Ghaderi [1], Wen Zhong [1,†], Mohammad Ali Okhovat [1,†], Johanna Aschan [1], Ann Svensson [1], Birgitta Sander [2], Johan Schultz [3], Thomas Olin [3], Anders Österborg [1,4], Mohammad Hojjat-Farsangi [1,*] and Håkan Mellstedt [1]

1 Department of Oncology-Pathology, BioClinicum, Karolinska University Hospital Solna and Karolinska Institutet, 171 64 Stockholm, Sweden
2 Department of Laboratory Medicine, Division of Pathology, Karolinska Institutet, 171 77 Stockholm, Sweden
3 Kancera AB, Nanna Svartz Väg 4, 171 65 Solna, Sweden
4 Department of Hematology, Karolinska University Hospital Solna, 171 77 Stockholm, Sweden
* Correspondence: mohammad.hojat-farsangi@ki.se; Tel.: +46-735-234-706
† These authors contributed equally to this work.

Abstract: The receptor tyrosine kinase orphan receptor 1 (ROR1) is absent in most normal adult tissues but overexpressed in various malignancies and is of importance for tumor cell survival, proliferation, and metastasis. In this study, we evaluated the apoptotic effects of a novel small molecule inhibitor of ROR1 (KAN0441571C) as well as venetoclax (BCL-2 inhibitor), bendamustine, idelalisib (PI3Kδ inhibitor), everolimus (mTOR inhibitor), and ibrutinib (BTK inhibitor) alone or in combination in human MCL primary cells and cell lines. ROR1 expression was evaluated by flow cytometry and Western blot (WB). Cytotoxicity was analyzed by MTT and apoptosis by Annexin V/PI staining as well as signaling and apoptotic proteins (WB). ROR1 was expressed both in patient-derived MCL cells and human MCL cell lines. KAN0441571C alone induced significant time- and dose-dependent apoptosis of MCL cells. Apoptosis was accompanied by decreased expression of MCL-1 and BCL-2 and cleavage of PARP and caspase 3. ROR1 was dephosphorylated as well as ROR1-associated signaling pathway molecules, including the non-canonical WNT signaling pathway (PI3Kδ/AKT/mTOR). The combination of KAN0441571C and ibrutinib, venetoclax, idelalisib, everolimus, or bendamustine had a synergistic apoptotic effect and significantly prevented phosphorylation of ROR1-associated signaling molecules as compared to KAN0441571C alone. Our results suggest that targeting ROR1 by a small molecule inhibitor, KAN0441571C, should be further evaluated particularly in combination with other targeting drugs as a new therapeutic approach for MCL.

Keywords: MCL; ROR1; small molecules; apoptosis; targeted therapy

Citation: Ghaderi, A.; Zhong, W.; Okhovat, M.A.; Aschan, J.; Svensson, A.; Sander, B.; Schultz, J.; Olin, T.; Österborg, A.; Hojjat-Farsangi, M.; et al. A ROR1 Small Molecule Inhibitor (KAN0441571C) Induced Significant Apoptosis of Mantle Cell Lymphoma (MCL) Cells. *Pharmaceutics* **2022**, *14*, 2238. https://doi.org/10.3390/pharmaceutics14102238

Academic Editor: Gabriele Grassi

Received: 22 August 2022
Accepted: 17 October 2022
Published: 20 October 2022

1. Introduction

Receptor tyrosine kinases (RTKs) are therapeutic targets in cancer. The RTK ROR1 plays a crucial role in the development of different tissues and organs during embryogenesis [1]. ROR1 is essentially not expressed in normal adult postpartum tissues but overexpressed in several malignancies [2]. Mutations, translocations, deletions, and overexpression of RTKs have been identified. Overexpression of ROR1 in patients was first described in chronic lymphocytic leukemia (CLL) by applying gene expression profiling [3]. ROR1 has also been shown to be overexpressed in several other hematological malignancies, including mantle cell lymphoma (MCL), as well as in solid tumors [4–12]. Small molecules and monoclonal antibodies have been used to target dysregulated RTKs in malignancies in a therapeutic approach [13,14].

In neoplastic cells, ROR1 is of importance for survival, proliferation, migration, and metastasis [15]. ROR1 mediates chemotactic and proliferative signals induced by the binding of WNT5a to ROR1 [16], resulting in the activation of guanine exchange factors and the phosphoinositol-3 kinase (PI3K)/JNK signaling pathway [17]. Activation of the canonical and non-canonical WNT signaling pathways correlated with activated ROR1 [18]. Moreover, high expression of ROR1 was associated with disease progression and short survival in several malignancies [15,19,20].

Hematological malignancies account for around 6.5% of cancers around the world [21]. Various lymphomas, including MCL, mostly spread to the bone marrow, and different RTKs, such as focal adhesion kinase (FAK), are involved in tumor–stromal cells interaction (in the tumor microenvironment) to enhance tumor cell survival and drug resistance [22]. MCL is a rare but aggressive B-cell lymphoma characterized by the chromosomal translocation (11; 14) (q13; q32) and constitutive overexpression of cyclin D1 [23,24] contributing to the uncontrolled growth of MCL cells [25]. Despite the introduction of novel drugs such as BTK inhibitors and new therapeutic principles, including CAR-T, drug resistance and relapses occur, and MCL remains incurable [26]. Thus, there is a need for novel therapeutic approaches with other mechanisms of action (MOAs) than those in clinical use. ROR1 is an important molecule for the malignant phenotype, e.g., tumor cell proliferation, survival, epithelial–mesenchymal transition (EMT), migration, and metastasis [15,26,27], and been suggested to be an interesting target molecule for precision cancer medicine (PCM).

In the present study, the expression of ROR1 in MCL cells was characterized and the apoptotic effects of a novel ROR1 inhibitor, KAN0441571C, were evaluated in in vitro and ex vivo preclinical models as well as in combination with therapeutics with other MOAs of clinical relevance (ibrutinib, idelalisib, bendamustine, everolimus, and venetoclax) that are currently used for the treatment of patients with MCL [28] or have shown promising results in clinical trials. The results indicate that a small molecule ROR1 inhibitor might be a promising new drug candidate that warrants further evaluation in preclinical models and clinical trials in difficult-to-treat MCL patients.

2. Materials and Methods

2.1. Patients

Patients were diagnosed and treated according to the Swedish MCL guidelines at the Department of Hematology, Karolinska University Hospital, Stockholm, Sweden. The study was authorized by the National Ethics Authority (www.etikprovningsmyndigheten.se (accessed on 20 December 2016)). A written consent form was collected from patients before blood sampling. Eleven patients with MCL were included, consisting of two females and nine males with a median age of 70 years (range 62–84). Five had early-stage untreated MCL and six had active MCL, and all but two were treatment-naïve. The selection was based on the availability of stored blood MCL cells (liquid nitrogen).

2.2. Cell Lines

Five human MCL cell lines, Granta-519 (ACC 342), Mino (ACC 687), JeKo1 (ACC 553), JVM-2 (ACC 12), and Z-138 (CRL-3001), obtained from the American Type Culture Collection (ATCC), were used. All cell lines were tested for mycoplasma contamination. The cell lines were cultured in RPMI-1640 medium (Sigma, St. Louis, MO, USA) with 10% fetal bovine serum (FBS) (Gibco, Life Technologies, Karlsruhe, Germany), 100 ug/mL penicillin/streptomycin (Biochrom KG, Berlin, Germany), and 2% glutamine (Biochrom KG) in humidified air at 37 °C with 5% CO_2.

2.3. ROR1 Small Molecule Tyrosine Kinase Inhibitor (KAN0441571C)

The development of the first small molecule ROR1 tyrosine kinase inhibitor (KAN0439834) was reported in 2018 [29]. The second generation of a ROR1 small molecule inhibitor, KAN0441571C, with a higher cytotoxic potency against tumor cells in vitro, as well as longer halftime (T1/2) in a mouse model (11 h) compared to KAN0439834 (2 h), was recently described [30,31] and was applied in the present study.

2.4. Cell Surface Marker Analysis (Flowcytometry)

Analysis of surface ROR1 expression was performed using an anti-ROR1 monoclonal antibody (Miltenyi Biotec, Bergisch Gladbach, Germany) [27]. Cells were analyzed by a FACS Canto II flow cytometer (BD Biosciences, San Jose, CA, USA). The FlowJo software program (Tree Star Inc., Ashland, OR, USA) was applied for analysis of cells.

2.5. Annexin V/PI Apoptosis Assay

Human MCL cell lines were plated in 24-well plates at a concentration of 5×10^4 cells per well in RPMI-1640 medium containing 10% FBS and incubated with KAN0441571C or venetoclax (BCL-2 inhibitor), ibrutinib (BTK inhibitor), idelalisib (PI3k inhibitor), everolimus (mTOR inhibitor), and bendamustine (an alkylating agent) as single drugs or in combination with KAN0441571C. After 24 h of incubation, the cells were centrifuged and resuspended in 100 µL of Annexin V binding buffer (BD Biosciences) containing 1 µL fluorescein isothiocyanate (FITC)-conjugated Annexin V and 1 µL propidium iodide (PI) (BD Biosciences) and incubated for 15 min at RT.

Frozen primary MCL cells were thawed at 37 °C for 1–2 min and resuspended in 10 mL of RPMI-1640 medium with 10% FBS. Cell viability was analyzed by EVE™ Automatic Cell Counter NanoEnTek (VWR, Atlanta, GA, USA) according to the manufacturer's description. Cells were cultured in RPMI-1640 medium supplemented with 10% FBS and 20% conditioned media (CM). CM was harvested from 70% confluent HS-5 cells (CRL-11882, ATCC) grown in RPMI-1640 medium with 10% FBS for 72 h. The supernatant was collected, filtered, and used fresh. Then, 2×10^5 MCL cells were seeded in 96-well plates and incubated at 37 °C in 5% CO_2 for 24 h with and without drugs. Cells were stained with Annexin V/PI gated for $CD45^+/ROR1^+/CD3^-$ cells, as exemplified in Figure S1. Cells were analyzed with NovoCyte Advanteon flow cytometry (Agilent, Santa Clara, CA, USA), and data were collected with the NovoExpress® 1.4.1 software program (Agilent, Santa Clara, CA, USA).

2.6. MTT Cytotoxicity (or % Inhibition of Cell Proliferation or Cell Growth) Assay

MCL cell lines (10^4 cells/well) were incubated in quadruplicates in 200 µL of RPMI-1640 containing 10% FBS in 96-well plates for 24–72 h at 37 °C with 5% CO_2. Then, 20 µL MTT solution (5 mg/mL) was added to each well and incubated for another 4 h at 37 °C. Stopping solution (solubilizing reagent) (10% sodium dodecyl sulphate (SDS) in 0.01M HCL) was added to a final volume of 100 µL/well. After overnight incubation at 37 °C, the plates were read in a microplate reader at 570 nm.

2.7. Drug Combination Analyses

Zero Interaction Potency (ZIP) is a reference model [32] for evaluating drug combinations by comparing the deviation in potency of the dose–response curves between the individual agents and the combinations. A ZIP delta score (Δ) for each pair of drugs was used to quantify the deviation by taking the average of all delta scores over their dose combinations. Summary synergy scores are defined as the average excess response of drug interactions, e.g., a score of 20 corresponds to 20% of response above the expectation, i.e., synergism. A score close to 0 gives low confidence for synergism or antagonism. A score less than −10 indicates that the relationship between two drugs is possibly antagonistic; from −10 to 10, the relationship between two drugs might be additive; and if the score is >10, the interaction between two drugs is likely to be synergistic [32].

2.8. Western Blot (WB) Analysis

Western blot was carried out as previously described [26]. MCL cells were lysed on ice for 30 min in lysis buffer containing 1% Triton X-100, 150 mM NaCl, 5 mM EDTA, and 50 mM Tris-HCl with 1% protease inhibitor cocktail (Sigma-Aldrich, St. Louis, MO, USA) and phosphatase inhibitors (Roche Ltd., Basel, Switzerland) and centrifuged at 14,000 rpm. Protein concentration was measured by Pierce™ BCA Protein Assay Kit (Thermo Fisher

Scientific, Waltham, MA, USA). Then, 10–20 μg protein was loaded into each well for WB analysis. The following antibodies were used for the detection of total (t) and phosphorylated (p) proteins: ROR1 (R & D Systems, Minneapolis, MN, USA), own produced polyclonal rabbit anti-p-ROR1 (Tyr 641 and 646, Ser 652) [26], AKT and p-AKT (Ser 473), mammalian target of rapamycin (mTOR) and p-mTOR (Ser 2448), phosphoinositide 3-kinase (PI3K) p110δ, cleaved poly ADP ribose polymerase (PARP), myeloid cell leukemia-1 B-cell lymphoma (BCL)-2, cleaved caspase 3 (Cell Signaling Technology, Danvers, MA, USA), and p-PI3K p110δ (Tyr 585) (Santa Cruz Biotechnology, Santa Cruz, CA, USA). β-actin (Sigma-Aldrich, St. Louis, MO, USA) was used as a loading control. Densitometry measurements of proteins were calculated by Image J1.44p software (National Institute of Health, Bethesda, MD, USA).

2.9. Statistical Analysis

Statistical analysis was performed using GraphPad Prism software (GraphPad Software, Inc., La Jolla, CA, USA). EC_{50} values were calculated from the dose–response curve by non-linear curve fitting (HillSlope). The Mann–Whitney U test was applied for comparison of non-Gaussian distributed data. p-values < 0.05 were considered significant. Asterisks represent p-values of * 0.01 to 0.05, ** 0.001 to < 0.01, *** < 0.001, and **** < 0.0001.

3. Results

3.1. Surface ROR1 Was Heterogeneously Expressed on MCL Cell Lines and Primary MCL Cells

The frequency of surface ROR1-positive cells of JeKo-1, Mino, Z138, and Granta-519 cell lines was 100%, 99%, 71%, and 21%, respectively, while JVM-2 did not express surface ROR1 (Figure S2A). However, in WB, JVM-2 showed intense expression of ROR1 (Figure S2B). This indicates that JVM-2 expressed ROR1 as a splice variant lacking the extracellular part. Surface ROR1 was also analyzed in MCL cells from peripheral blood of leukemia patients (n = 11). MCL cells were identified as $CD19^+/CD5^+$ cells. The frequency of blood $CD19^+/CD5^+$ cells was 80 ± 4% (mean ± SEM), and 68 ± 10% of the blood MCL cells exhibited surface ROR1.

3.2. KAN0441571C Induced Significant Cell Death of MCL Cell Lines

KAN0441571C induced a time- and dose-dependent cell death (MTT) of all MCL lines but to varying degrees. Seventy-two hours of incubation were required to achieve the optimal cytotoxic effect. JeKo1 seemed to be the most sensitive cell line and Z138 the least sensitive (Figure 1A). Next, apoptosis analysis (Annexin V/PI staining) was performed after incubation with KAN0441571C, venetoclax, or bendamustine (Figures 1B and S6D). JeKo1 was again the most sensitive cell line to ROR1 inhibition, reaching almost 100% killing at 250 nM, while venetoclax and bendamustine were less effective. A similar pattern was seen for Mino, Z138, and JVM-2, while Granta-519 was the most sensitive to venetoclax. The data support the notion that ROR1 inhibition by a small molecule inhibitor seems to effectively induce death of MCL cells. We analyzed protein phosphorylation and found, as expected, that KAN0441571C prevented phosphorylation of ROR1 in a dose-dependent manner in all MCL cell lines (Figure S2B). Previous studies of ROR1 inhibitors in tumor cells from other hematological malignancies have indicated the involvement of both WNT non-canonical (PI3K/AKT/mTOR) and WNT canonical (beta-catenin) signaling pathways, as well as downregulation of the intrinsic apoptotic pathway, including cleavage of caspase 3 [12,27,28]. When primary MCL cells were analyzed, we also found that in addition to the prevention of ROR1 phosphorylation, PI3K, AKT, and mTOR were also subsequently dephosphorylated (Figure 2).

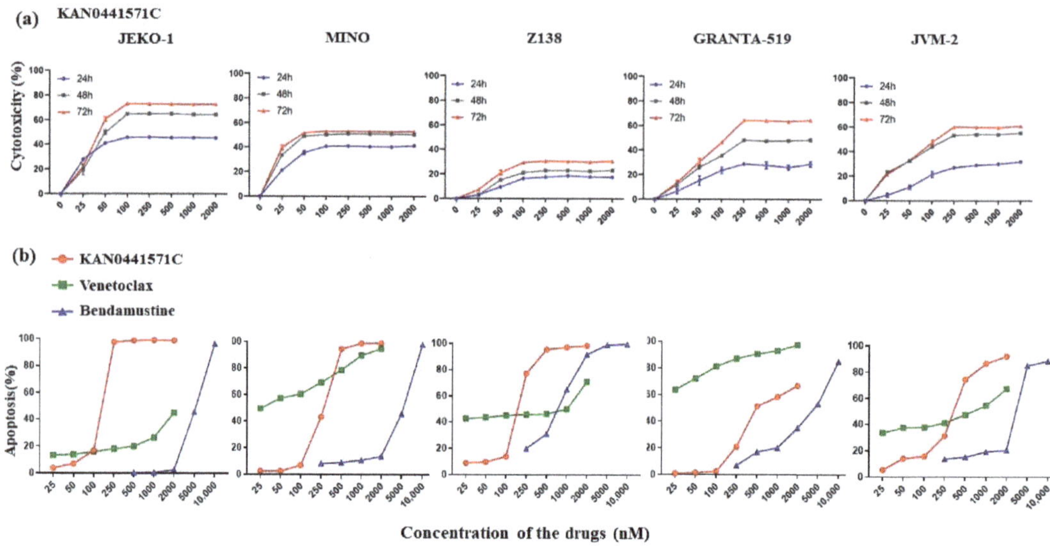

Figure 1. Effects of KAN0441571C, venetoclax, and bendamustine on MCL cell lines. (**a**) Cytotoxicity (MTT) of 5 ROR1$^+$ MCL cell lines incubated with the ROR1 inhibitor KAN0441571C for various time periods. (**b**) Apoptosis (Annexin V/PI) of MCL cell lines incubated with KAN0441571C, venetoclax, or bendamustine alone for 48 h. Data are shown as mean ± SEM of triplicate samples.

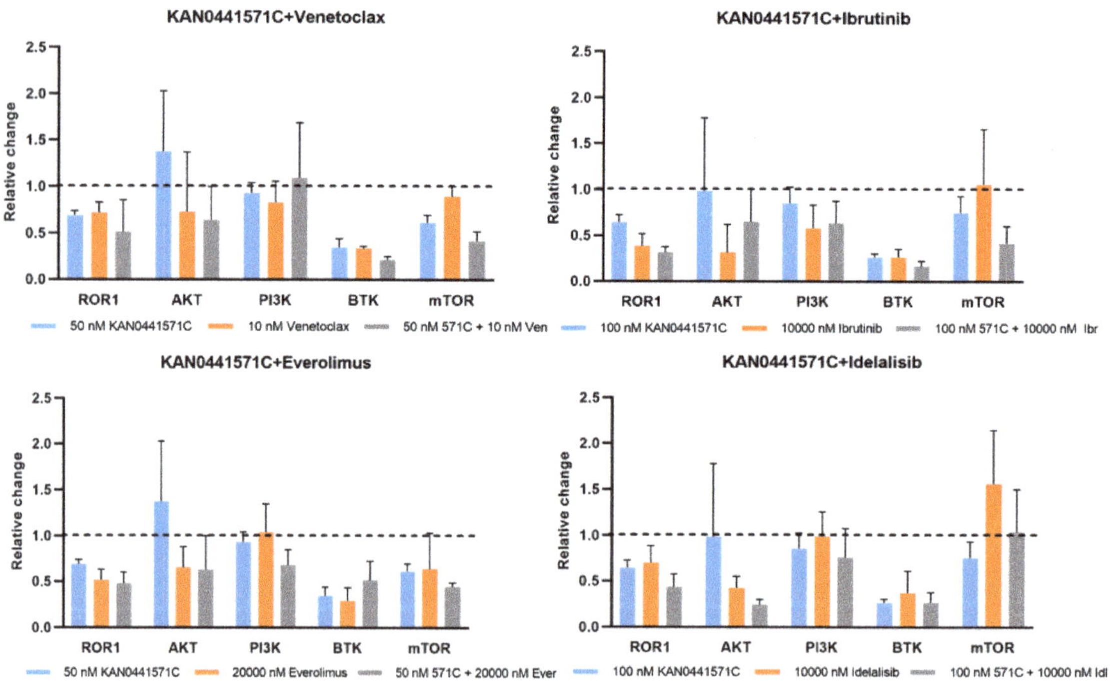

Figure 2. Effects of KAN0441571C, venetoclax, ibrutinib, and idelalisib as well as the combinations on intracellular signaling molecules. Relative change in primary MCL cells in the levels of phosphorylated

ROR1, AKT, PI3K, BTK, and mTOR after single-drug incubation with KAN0441571C (571C) and venetoclax (Ven), ibrutinib (Ibr), everolimus (Ever), and idelalisib (Idel), respectively, and in combination, as indicated at the bottom of the figure. The dotted lines indicate the level of the phosphorylated protein in untreated cells. The figure shows mean ± SEM of 3 patients.

3.3. *Effects of KAN0441571C in Combination with Other Drugs*

First, we evaluated the apoptotic effect of single-drug KAN0441571C, venetoclax, ibrutinib, idelalisib, everolimus, or bendamustine on patient-derived blood MCL cells (n = 11). Dose–response curves for each individual drug, including EC_{50} values, are shown in Figure 3. Venetoclax and KAN0441571C were the most effective single agents among the tested drugs. Next, MCL cell lines were incubated with the EC_{50} concentration of KAN0441571C and equimolar concentrations of venetoclax, ibrutinib, or bendamustine, alone and in combination (Figures S3 and S6). In single-drug experiments, KAN0441571C, venetoclax, and ibrutinib were similarly effective in inducing cell death of the four different MCL cell lines, while bendamustine seemed to be inferior in two of five cell lines (Mino and Granta-519). In drug combination experiments, KAN0441571C plus venetoclax or KAN0441571C plus ibrutinib showed a statistically significant increased cytotoxic effect, indicating synergism for three cell lines (not for Granta-519) compared to each drug alone, while the combination of KAN0441571C with bendamustine showed synergism in all cell lines except for Granta-519.

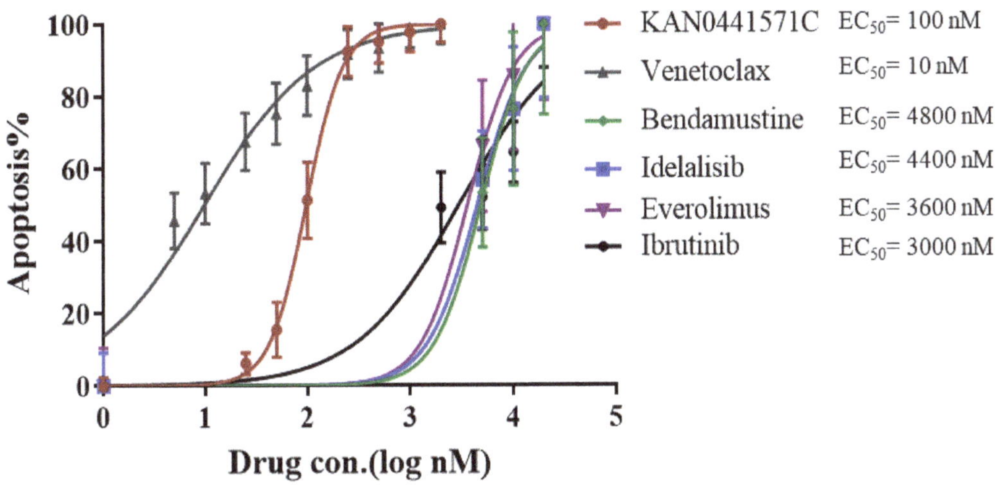

Figure 3. Dose–response curves and EC_{50} values for each individual drug. Apoptosis (Annexin V/PI) (mean ± SEM) of MCL cells from patients (n = 11) incubated in vitro with KAN0441571C, venetoclax, bendamustine, idelalisib, everolimus, and ibrutinib for 24 h. EC_{50} values for each drug are shown. R-square values are 0.119, 0.6161, 0.2826, 0.3542, 0.3077, and 0.5438 for KAN0441571C, venetoclax, bendamustine, idelalisib, everolimus, and ibrutinib, respectively.

We then repeated the drug combination experiments using patient-derived MCL cells. The results are shown in Figure 4. There was a significant increase in apoptosis (Annexin V/PI) of MCL cells when KAN0441571C was incubated in combination with either venetoclax, ibrutinib, idelalisib, everolimus, or bendamustine as compared to each drug alone. Furthermore, when drugs were combined, low doses of each drug seemed to be as effective as high doses.

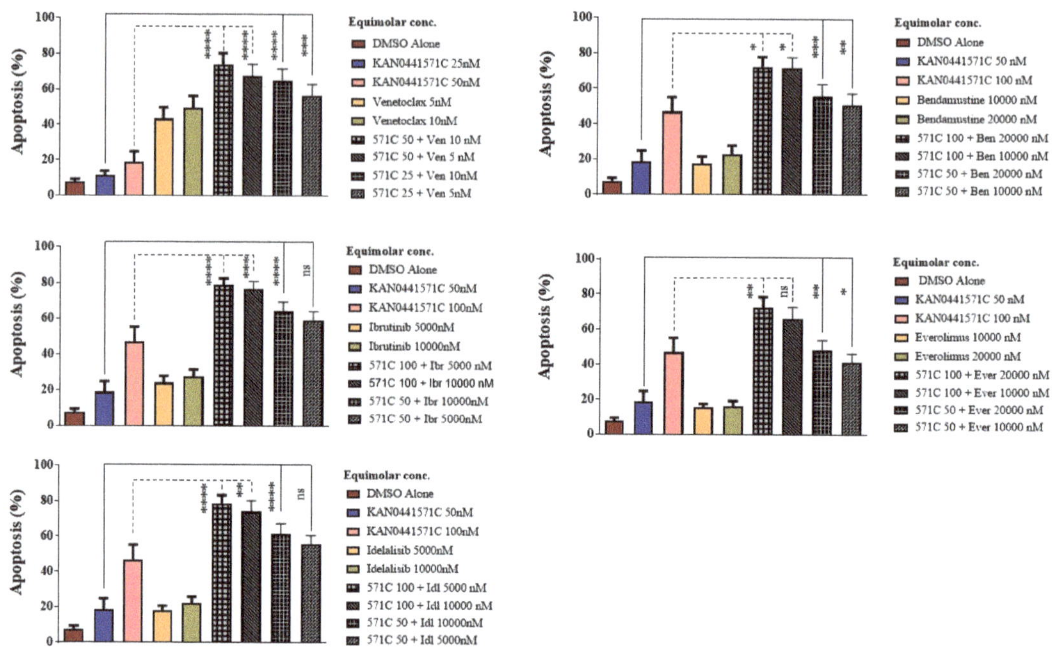

Figure 4. Induction of apoptosis by inhibitors alone or in combination on primary MCL cells. Apoptosis (Annexin V/PI) (mean ± SEM) of primary MCL cells (n = 11) incubated in vitro with equimolar concentrations of KAN0441571C (571C), ibrutinib (Ibr), venetoclax (Ven), idelalisib (Idl), everolimus (Ever), or bendamustine (Ben) alone and in combination. Statistical significances are shown at the top: * $p < 0.05$, ** $p < 0.01$, *** $p < 0.001$, **** $p < 0.0001$; ns: not significant (Mann–Whitney test was used for comparisons between different groups).

A summary of apoptosis ranking for all drug combinations in primary MCL cells is displayed in Figure 5. The highest synergism was seen between a low dose (50 nM) of KAN0441571C and a low dose (5000 nM) of idelalisib. Moreover, a low dose (50 nM) of KAN0441571C and a low dose (5 nM) of venetoclax showed significant synergism. A high dose (100 nM) of KAN0441571C and everolimus (20,000 nM) induced a significant synergistic effect, as well. To confirm apoptosis, we analyzed the expression of BCL-2 and MCL-1, which were markedly downregulated by the combination of drugs as compared to each drug alone, while cleaved caspase 3 and cleaved PARP were upregulated (Figures 6 and S4).

As shown above, KAN0441571C alone dephosphorylated ROR1, followed by the inactivation of AKT, BTK, and mTOR, which are involved in the ROR1 signaling pathway (Figures S2B and 2). When KAN0441571C was combined with venetoclax, ibrutinib, everolimus, or idelalisib, phosphorylation of ROR1, AKT, BTK, and mTOR was decreased as compared to single drugs (Figure 2). The most effective combination seemed to be KAN0441571C with venetoclax or ibrutinib. Interestingly, phospho-PI3K was least affected by all drugs and combinations including idelalisib (Figure 2). Overall, the pattern of phosphorylation of target proteins was less clear as compared to that of apoptotic proteins, probably reflecting the complex network of intracellular signaling.

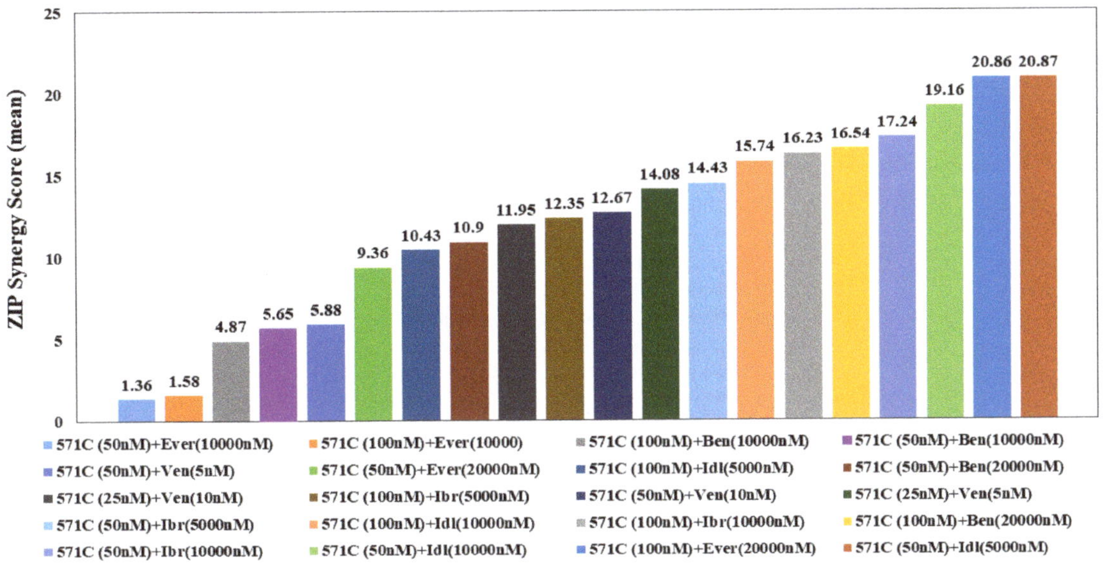

Figure 5. Ranking (ZIP synergy score) for apoptosis induced by drug combinations in primary MCL cells. ZIP synergy score (mean) for apoptosis (Annexin V/PI) of drug combinations using MCL cells derived from patients (n = 11) incubated for 24 h by indicated drugs and conc. The ZIP score captures the effect of drug combinations by comparing the change in the potency of the dose–response curves between individual drugs and their combination (https://synergyfinder.org/ (accessed on 21 August 2021)). A score between 1 and 10 indicates an additive effect and >10 synergism. KAN0441571C (571C), everolimus (Ever), ibrutinib (Ibr), idelalisib (Idl), bendamustine (Bend), venetoclax (Ven).

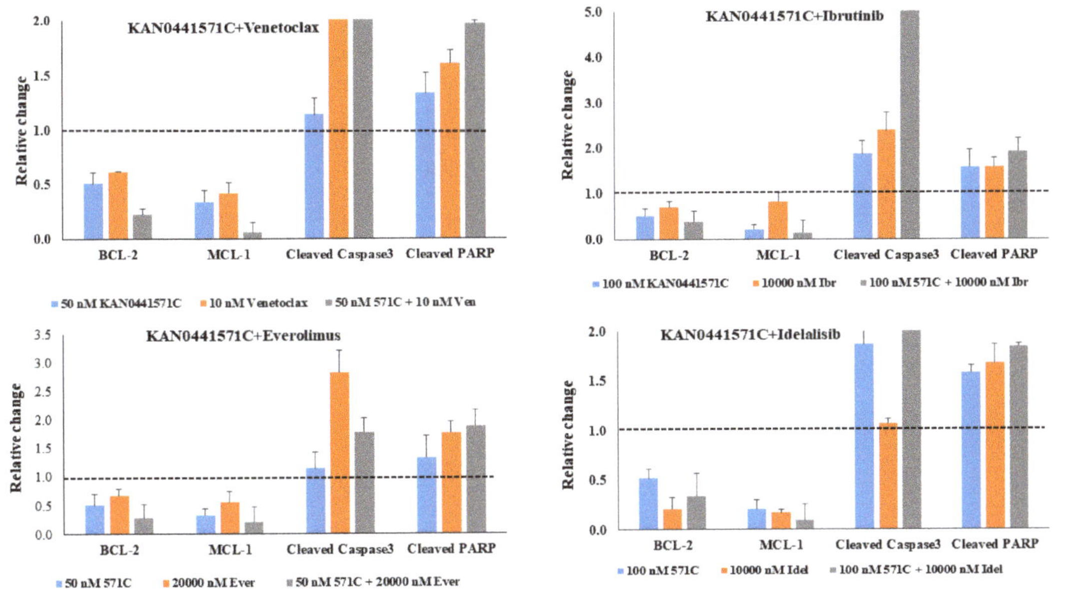

Figure 6. Induction of apoptosis following treatment of MCL primary cells with KAN0441571C, venetoclax, ibrutinib, everolimus, or idelalisib. Relative change in MCL cells from patients of BCL-2,

MCL-1, cleaved caspase 3, and cleaved PARP after incubation with single-agent KAN0441571C (571C), venetoclax (Ven), ibrutinib (Ibr), everolimus (Ever), or idelalisib (Idel) alone and in combination, as indicated at the bottom of the figure. The dotted lines indicate the level of untreated cells. The figure shows the results of one representative patient out of three.

4. Discussion

The introduction of new drugs has improved the prognosis for patients with MCL. However, there is an urgent medical need for novel treatment options to increase the response rate and depth of remission and a drug with other MOAs than those currently used at clinical relapse and resistance. ROR1 has been proposed to be a druggable target [26,31] and is reported to be heterogeneously expressed in MCL [33]. In this study, we confirmed and extended the expression pattern of ROR1 in MCL cell lines and primary MCL cells. There are also splice variants of ROR1. A truncated ROR1 lacking the extracellular domain with an intact intracellular region has been described [2]. Forouzesh et al. [34] also reported a splice variant which only had the extracellular region. These results might explain the findings in JeKo1 and Mino cell lines comparing surface staining and Western blot. Different isotypes of ROR1 have also been reported with molecular weights from 100 to 130 kDa, which may represent different degrees of glycosylation [6,34].

KAN0441571C is the second generation of a small molecule ROR1 tyrosine kinase inhibitor [30]. The drug induced apoptosis of ROR1$^+$ MCL cell lines as well as primary MCL cells. The ROR1 inhibitor prevented phosphorylation of ROR1. Molecules of the non-canonical WNT signaling pathway were inactivated, which is in line with previous reports showing that ROR1 signals through at least the β-catenin-independent (non-canonical) WNT pathway [35–37]. Involvement of the PI3K/AKT/mTOR pathway in MCL is also supported by a previous report showing that inhibition of mTOR in MCL cells induced cell cycle arrest and apoptosis [38]. Furthermore, AKT activates mTOR through the interaction of several signaling molecules [35], including the downstream target CREB. AKT inactivation is of importance for apoptosis induced by many targeting drugs [39]. The ERK pathway, which is frequently mutated in malignancies [40], was also inactivated by ROR1 inhibition, which is in agreement with a report showing that ERK was dephosphorylated followed by treatment with KAN0441571C in small-cell lung cancer cells [41]. Moreover, ROR1 interacts with SRC, a key regulator of cancer cells. KAN0441571C prevented phosphorylation of SRC, which should also be of significance for induction of tumor cell death.

Intracellular signaling is a complex network of interactions between proteins. Malignant cells are masters to overcome signal suppression through upregulation of other signaling proteins. In MCL, various signaling pathways seemed to be used, which may vary between patients. A drug which can act on several key oncogenic signaling molecules should be of interest in a therapeutic attempt. KAN0441571C not only prevented phosphorylation of ROR1 but also inactivation of the PI3K/AKT/mTOR molecules, an important axis in tumorigenesis and a significant therapeutic target [42].

The combination of KAN0441571C with other agents had a synergistic apoptotic effect. Increased tumor cell death induced by KAN0441571C in combination with other drugs was supported by enhanced downregulation of, e.g., the anti-apoptotic MCL-1 and BCL-2 proteins, which are highly expressed in MCL cells [34]. The BCL-2 family proteins are crucial regulators of the mitochondrial apoptotic pathway. Genetic aberrations of these genes correlated with lymphomagenesis and chemotherapy resistance [33,43]. Altering the balance between anti-apoptotic and pro-apoptotic BCL-2 proteins may lead to the evasion of apoptosis and the extension of the survival of malignant cells. Targeting MCL-1 and BCL-2 might be a rewarding therapeutic approach in MCL.

Drugs which are in clinical use for MCL such as ibrutinib, venetoclax, idelalisib, everolimus, and bendamustine showed significant killing of MCL cells. The ROR1 inhibitor KAN0441571C seemed to act synergistically with these drugs, enhancing tumor cell death. Of specific interest was that low doses of KAN0441571C and low doses of venetoclax, ibrutinib, and idelalisib had a significant synergistic apoptotic effect. The increased apoptotic effect of KAN0441571C in combination with other targeting drugs might be due to enhanced inactivation of various signaling molecules associated with ROR1 signaling.

This is the first study on the effects of a small molecule ROR1 inhibitor alone on MCL tumor cell survival and signaling. The data also indicated that combining the ROR1 inhibitor with drugs targeting other dysregulated pathways contributes to an increased anti-tumor effect. Previous studies in diffuse large B-cell lymphoma (DLBCL) and small-cell lung cancer (SCLC) have indicated that combining ROR1 inhibitors with venetoclax [30,41,44] or with erlotinib and chemotherapeutics in pancreatic carcinoma [5] acted synergistically to induce tumor cell death. This report also adds PI3K and BTK inhibitors to the group of combinatorial partners for ROR1 inhibitors. Further pre-clinical and clinical studies are warranted to evaluate this new treatment concept with respect to side effects and anti-tumor activity. As the tumor microenvironment is of importance for tumor progression [45], models where a tumor microenvironment is incorporated should be included. There is a substantial amount of data indicating the importance of the microenvironment in MCL [26]. We have previously shown that when CLL cells were cultured in the presence of stromal cells, low concentrations of the ROR1 inhibitor were not as effective in killing leukemic cells as if cultured without stromal cells. When a high concentration of the inhibitor was used, stromal cells had no preventive effect [29].

5. Conclusions

ROR1 is an oncogenic RTK involved in the survival of tumor cells of various origins. Due to its unique expression in tumors, ROR1 has been recognized as an interesting target for cancer treatment. ROR1 small molecule inhibitor in hematologic malignancies, including MCL, induced significant apoptosis of tumor cells in vitro as well as the inhibition of several important signaling pathways for tumor cell survival. In combination with other small molecules targeting MCL tumor cells, a significantly increased tumor cell death was seen.

Supplementary Materials: The following supporting information can be downloaded at: https://www.mdpi.com/article/10.3390/pharmaceutics14102238/s1, Figure S1: Dose-dependent apoptosis effects of KAN0441571C. Analysis of apoptosis in primary MCL cells from a patient using Annexin V/PI staining in $CD45^+/ROR1^+/CD3^-$ gated cells after incubation for 24 h with KAN0441571C (25–2000 nM); Figure S2: Analysis of ROR1 expression in MCL cell lines. Analysis of ROR1 surface expression in 5 MCL cell lines (A). Western blot analysis of total (t) ROR1 and phosphorylated (p) ROR1 proteins in 5 MCL cell lines before and after incubation (4 h) with the ROR1 inhibitor, KAN044157C (250 and 500 nM) (B). pROR1/ROR1 density has been shown under each blot; Figure S3: MTT assay on 5 MCL cell lines. Cytotoxicity (MTT) in 5 MCL cell lines incubated for 72 h with EC_{50} concentration of KAN0441571C and equimolar concentration of venetoclax, ibrutinib, and bendamustine. p-values comparing cytotoxicity of KAN0441571C alone against the combination of KAN0441571C with other drugs are shown at the top. ** $p < 0.01$, **** $p < 0.0001$; ns: not significant. The Mann–Whitney test was used for comparisons; Figure S4: Western blots analysis of apoptotic proteins in patient-derived MCL cells. Western blots for apoptotic proteins in patient-derived MCL cells after incubation with KAN0441571C, venetoclax, ibrutinib, or everolimus alone and in combination, as indicated at the top. The molecular weight of protein is shown for each blot (A). Relative density of each protein band to beta actin (B); Figure S5: Effects of KAN0441571C, venetoclax, ibrutinib, idelalisib, or everolimus alone and in combination on ROR1 and intracellular signaling molecules. Western blots of total and phosphorylated proteins after incubation with KAN0441571C, venetoclax, ibrutinib, idelalisib, or everolimus alone and in combination, as indicated at the top (A). Densitometry analysis of pROR1/ROR1, pAKT/AKT, pPI3K/PI3K, and pBTK/BTK (B); Figure S6: Induction of apoptosis (flowcytometry Annexin/PI) in primary MCL cells. Representative figures of patients are shown in (A) (2013#020), (B) (2010#040), and (C) (2013#058) using a combination of KAN0441571C (25 nM) and venetoclax (5 nM). Untreated or DMSO (10 µL/mL) were used as controls (24 h). Induction of apoptosis (48 h) in the MCL cell line MINO (D) using various concentrations of KAN0441571C (25–2000 nM) and DMSO (10 µL/mL) as a negative control is also shown.

Author Contributions: A.G., M.A.O., W.Z., J.A. and A.S. performed experiments, analyzed data, and finalized the manuscript. J.S. and T.O. produced the small molecule and finalized the manuscript. B.S. provided clinical material, analyzed data, and finalized the manuscript. A.Ö., H.M. and M.H.-F. designed the study, performed experiments, supervised the study, and wrote the manuscript. All authors have read and agreed to the published version of the manuscript.

Funding: This research was funded by the Åke Olsson Foundation for Hematology Research (grant numbers 2017-00436, 2019-00396 to M.H.-F.), the Cancer and Allergy Foundation (grant numbers 2016/5, 2017/63, 2018/63, 2019/189, 2020/258 to A.Ö. and 2017/64, 2018/140, 2019/194, 2020/260 to H.M.), the Cancer Society in Stockholm (grant numbers 151313, 184203 to A.Ö. and 144142, 164122 to H.M.), the Swedish Cancer Society (grant numbers CAN 2015/408, CAN 2018/499 to A.Ö.), AFA Insurance (grant number 130054 to A.Ö.), the Karolinska Institutet Foundation (A.Ö., H.M.), and Region Stockholm (grant numbers 20150070, 20180030 to A.Ö.).

Institutional Review Board Statement: The study was conducted in accordance with Good Clinical Practice Guidelines and the Declaration of Helsinki. All patients or their legally acceptable representative, or both (if possible), provided written informed consent. The study was authorized by the National Ethics Authority of Sweden (www.etikprovningsmyndigheten.se (accessed on 20 December 2016)) (approval code: 2016/2506-32, date of approval: 20 December 2016).

Informed Consent Statement: Informed consent was obtained from all subjects involved in the study.

Data Availability Statement: Not applicable.

Acknowledgments: For excellent secretarial assistance, we thank Leila Relander.

Conflicts of Interest: J.S. is an employee and shareholder of Kancera AB. T.O. is a founder, employee, and shareholder of Kancera AB. H.M. is a founder and shareholder of Kancera AB and received research grants and honorarium from Kancera AB. The remaining authors declare that they have no conflict of interest.

References

1. Minami, Y.; Oishi, I.; Endo, M.; Nishita, M. Ror-family receptor tyrosine kinases in noncanonical Wnt signaling: Their implications in developmental morphogenesis and human diseases. *Dev. Dyn.* **2010**, *239*, 1–15. [CrossRef]
2. Reddy, U.R.; Phatak, S.; Pleasure, D. Human neural tissues express a truncated Ror1 receptor tyrosine kinase, lacking both extracellular and transmembrane domains. *Oncogene* **1996**, *13*, 1555–1559. [PubMed]
3. Klein, U.; Tu, Y.; Stolovitzky, G.A.; Mattioli, M.; Cattoretti, G.; Husson, H.; Freedman, A.; Inghirami, G.; Cro, L.; Baldini, L.; et al. Gene expression profiling of B cell chronic lymphocytic leukemia reveals a homogeneous phenotype related to memory B cells. *J. Exp. Med.* **2001**, *194*, 1625–1638. [CrossRef] [PubMed]
4. Baskar, S.; Kwong, K.Y.; Hofer, T.; Levy, J.M.; Kennedy, M.G.; Lee, E.; Staudt, L.M.; Wilson, W.H.; Wiestner, A.; Rader, C. Unique cell surface expression of receptor tyrosine kinase ROR1 in human B-cell chronic lymphocytic leukemia. *Clin. Cancer Res.* **2008**, *14*, 396–404. [CrossRef] [PubMed]
5. Daneshmanesh, A.H.; Hojjat-Farsangi, M.; Ghaderi, A.; Moshfegh, A.; Hansson, L.; Schultz, J.; Vågberg, J.; Byström, S.; Olsson, E.; Olin, T.; et al. A receptor tyrosine kinase ROR1 inhibitor (KAN0439834) induced significant apoptosis of pancreatic cells which was enhanced by erlotinib and ibrutinib. *PLoS ONE* **2018**, *13*, e0198038. [CrossRef] [PubMed]
6. Daneshmanesh, A.H.; Mikaelsson, E.; Jeddi-Tehrani, M.; Bayat, A.A.; Ghods, R.; Ostadkarampour, M.; Akhondi, M.; Lagercrantz, S.; Larsson, C.; Osterborg, A.; et al. Ror1, a cell surface receptor tyrosine kinase is expressed in chronic lymphocytic leukemia and may serve as a putative target for therapy. *Int. J. Cancer* **2008**, *123*, 1190–1195. [CrossRef]
7. Karvonen, H.; Niininen, W.; Murumägi, A.; Ungureanu, D. Targeting ROR1 identifies new treatment strategies in hematological cancers. *Biochem. Soc. Trans.* **2017**, *45*, 457–464. [CrossRef]
8. O'Connell, M.P.; Marchbank, K.; Webster, M.R.; Valiga, A.A.; Kaur, A.; Vultur, A.; Li, L.; Herlyn, M.; Villanueva, J.; Liu, Q.; et al. Hypoxia induces phenotypic plasticity and therapy resistance in melanoma via the tyrosine kinase receptors ROR1 and ROR2. *Cancer Discov.* **2013**, *3*, 1378–1393. [CrossRef]
9. Shabani, M.; Asgarian-Omran, H.; Jeddi-Tehrani, M.; Vossough, P.; Faranoush, M.; Sharifian, R.A.; Toughe, G.R.; Kordmahin, M.; Khoshnoodi, J.; Roohi, A.; et al. Overexpression of orphan receptor tyrosine kinase Ror1 as a putative tumor-associated antigen in Iranian patients with acute lymphoblastic leukemia. *Tumour Biol.* **2007**, *28*, 318–326. [CrossRef]
10. Shabani, M.; Asgarian-Omran, H.; Vossough, P.; Sharifian, R.A.; Faranoush, M.; Ghragozlou, S.; Khoshnoodi, J.; Roohi, A.; Jeddi-Tehrani, M.; Mellstedt, H.; et al. Expression profile of orphan receptor tyrosine kinase (ROR1) and Wilms' tumor gene 1 (WT1) in different subsets of B-cell acute lymphoblastic leukemia. *Leuk. Lymphoma* **2008**, *49*, 1360–1367. [CrossRef]

11. Yamaguchi, T.; Yanagisawa, K.; Sugiyama, R.; Hosono, Y.; Shimada, Y.; Arima, C.; Kato, S.; Tomida, S.; Suzuki, M.; Osada, H.; et al. NKX2-1/TITF1/TTF-1-Induced ROR1 is required to sustain EGFR survival signaling in lung adenocarcinoma. *Cancer Cell* **2012**, *21*, 348–361. [CrossRef] [PubMed]
12. Zhang, S.; Chen, L.; Cui, B.; Chuang, H.Y.; Yu, J.; Wang-Rodriguez, J.; Tang, L.; Chen, G.; Basak, G.W.; Kipps, T.J. ROR1 is expressed in human breast cancer and associated with enhanced tumor-cell growth. *PLoS ONE* **2012**, *7*, e31127. [CrossRef] [PubMed]
13. Hojjat-Farsangi, M. Small-molecule inhibitors of the receptor tyrosine kinases: Promising tools for targeted cancer therapies. *Int. J. Mol. Sci.* **2014**, *15*, 13768–13801. [CrossRef]
14. Yamaoka, T.; Kusumoto, S.; Ando, K.; Ohba, M.; Ohmori, T. Receptor Tyrosine Kinase-Targeted Cancer Therapy. *Int. J. Mol. Sci.* **2018**, *19*, 3491. [CrossRef]
15. Hojjat-Farsangi, M.; Moshfegh, A.; Daneshmanesh, A.H.; Khan, A.S.; Mikaelsson, E.; Osterborg, A.; Mellstedt, H. The receptor tyrosine kinase ROR1—An oncofetal antigen for targeted cancer therapy. *Semin. Cancer Biol.* **2014**, *29*, 21–31. [CrossRef] [PubMed]
16. Vaisitti, T.; Arruga, F.; Vitale, N.; Lee, T.T.; Ko, M.; Chadburn, A.; Braggio, E.; Di Napoli, A.; Iannello, A.; Allan, J.N.; et al. ROR1 targeting with the antibody-drug conjugate VLS-101 is effective in Richter syndrome patient-derived xenograft mouse models. *Blood* **2021**, *137*, 3365–3377. [CrossRef] [PubMed]
17. Yoda, A.; Oishi, I.; Minami, Y. Expression and function of the Ror-family receptor tyrosine kinases during development: Lessons from genetic analyses of nematodes, mice, and humans. *J. Recept. Signal Transduct. Res.* **2003**, *23*, 1–15. [CrossRef]
18. Karvonen, H.; Perttilä, R.; Niininen, W.; Hautanen, V.; Barker, H.; Murumägi, A.; Heckman, C.A.; Ungureanu, D. Wnt5a and ROR1 activate non-canonical Wnt signaling via RhoA in TCF3-PBX1 acute lymphoblastic leukemia and highlight new treatment strategies via Bcl-2 co-targeting. *Oncogene* **2019**, *38*, 3288–3300. [CrossRef]
19. Cui, B.; Ghia, E.M.; Chen, L.; Rassenti, L.Z.; DeBoever, C.; Widhopf, G.F., 2nd; Yu, J.; Neuberg, D.S.; Wierda, W.G.; Rai, K.R.; et al. High-level ROR1 associates with accelerated disease progression in chronic lymphocytic leukemia. *Blood* **2016**, *128*, 2931–2940. [CrossRef]
20. Mao, Y.; Xu, L.; Wang, J.; Zhang, L.; Hou, N.; Xu, J.; Wang, L.; Yang, S.; Chen, Y.; Xiong, L.; et al. ROR1 associates unfavorable prognosis and promotes lymphoma growth in DLBCL by affecting PI3K/Akt/mTOR signaling pathway. *Biofactors* **2019**, *45*, 416–426. [CrossRef]
21. Tietsche de Moraes Hungria, V.; Chiattone, C.; Pavlovsky, M.; Abenoza, L.M.; Agreda, G.P.; Armenta, J.; Arrais, C.; Avendano Flores, O.; Barroso, F.; Basquiera, A.L.; et al. Epidemiology of Hematologic Malignancies in Real-World Settings: Findings from the Hemato-Oncology Latin America Observational Registry Study. *J. Glob. Oncol.* **2019**, *5*, 1–19. [CrossRef] [PubMed]
22. Rudelius, M.; Rosenfeldt, M.T.; Leich, E.; Rauert-Wunderlich, H.; Solimando, A.G.; Beilhack, A.; Ott, G.; Rosenwald, A. Inhibition of focal adhesion kinase overcomes resistance of mantle cell lymphoma to ibrutinib in the bone marrow microenvironment. *Haematologica* **2018**, *103*, 116–125. [CrossRef]
23. Cheah, C.Y.; Seymour, J.F.; Wang, M.L. Mantle Cell Lymphoma. *J. Clin. Oncol.* **2016**, *34*, 1256–1269. [CrossRef]
24. Dreyling, M.; Aurer, I.; Cortelazzo, S.; Hermine, O.; Hess, G.; Jerkeman, M.; Le Gouill, S.; Ribrag, V.; Trněný, M.; Visco, C.; et al. Treatment for patients with relapsed/refractory mantle cell lymphoma: European-based recommendations. *Leuk. Lymphoma* **2018**, *59*, 1814–1828. [CrossRef]
25. Chen, R.; Chubb, S.; Cheng, T.; Hawtin, R.E.; Gandhi, V.; Plunkett, W. Responses in mantle cell lymphoma cells to SNS-032 depend on the biological context of each cell line. *Cancer Res.* **2010**, *70*, 6587–6597. [CrossRef]
26. Jain, P.; Wang, M.L. Mantle cell lymphoma in 2022—A comprehensive update on molecular pathogenesis, risk stratification, clinical approach, and current and novel treatments. *Am. J. Hematol.* **2022**, *97*, 638–656. [CrossRef]
27. Zhao, Y.; Zhang, D.; Guo, Y.; Lu, B.; Zhao, Z.J.; Xu, X.; Chen, Y. Tyrosine Kinase ROR1 as a Target for Anti-Cancer Therapies. *Front. Oncol.* **2021**, *11*, 680834. [CrossRef]
28. Wang, M.L.; Jurczak, W.; Jerkeman, M.; Trotman, J.; Zinzani, P.L.; Belada, D.; Boccomini, C.; Flinn, I.W.; Giri, P.; Goy, A.; et al. Ibrutinib plus Bendamustine and Rituximab in Untreated Mantle-Cell Lymphoma. *N. Engl. J. Med.* **2022**, *386*, 2482–2494. [CrossRef]
29. Hojjat-Farsangi, M.; Daneshmanesh, A.H.; Khan, A.S.; Shetye, J.; Mozaffari, F.; Kharaziha, P.; Rathje, L.S.; Kokhaei, P.; Hansson, L.; Vågberg, J.; et al. First-in-class oral small molecule inhibitor of the tyrosine kinase ROR1 (KAN0439834) induced significant apoptosis of chronic lymphocytic leukemia cells. *Leukemia* **2018**, *32*, 2291–2295. [CrossRef]
30. Ghaderi, A.; Daneshmanesh, A.H.; Moshfegh, A.; Kokhaei, P.; Vågberg, J.; Schultz, J.; Olin, T.; Harrysson, S.; Smedby, K.E.; Drakos, E.; et al. ROR1 Is Expressed in Diffuse Large B-Cell Lymphoma (DLBCL) and a Small Molecule Inhibitor of ROR1 (KAN0441571C) Induced Apoptosis of Lymphoma Cells. *Biomedicines* **2020**, *8*, 170. [CrossRef]
31. Hojjat-Farsangi, M.; Moshfegh, A.; Schultz, J.; Norin, M.; Olin, T.; Osterborg, A.; Mellstedt, H. Targeting the Receptor Tyrosine Kinase ROR1 by Small Molecules. *Handb. Exp. Pharmacol.* **2021**, *269*, 75–99.
32. Yadav, B.; Wennerberg, K.; Aittokallio, T.; Tang, J. Searching for Drug Synergy in Complex Dose-Response Landscapes Using an Interaction Potency Model. *Comput. Struct. Biotechnol. J.* **2015**, *13*, 504–513. [CrossRef] [PubMed]
33. Karvonen, H.; Chiron, D.; Niininen, W.; Ek, S.; Jerkeman, M.; Moradi, E.; Nykter, M.; Heckman, C.A.; Kallioniemi, O.; Murumägi, A.; et al. Crosstalk between ROR1 and BCR pathways defines novel treatment strategies in mantle cell lymphoma. *Blood Adv.* **2017**, *1*, 2257–2268. [CrossRef] [PubMed]

34. Forouzesh, F.; Tabarian, S.S.; Emami, S.; Tehrani, M.J.; Hadavi, R.; Rabbani, H. Construction and Stable Expression of a Truncated Human Receptor Tyrosine Kinase Ror1 (Ror1-ECD). *Avicenna J. Med. Biotechnol.* **2012**, *4*, 41–45. [PubMed]
35. Daneshmanesh, A.H.; Hojjat-Farsangi, M.; Moshfegh, A.; Khan, A.S.; Mikaelsson, E.; Österborg, A.; Mellstedt, H. The PI3K/AKT/mTOR pathway is involved in direct apoptosis of CLL cells induced by ROR1 monoclonal antibodies. *Br. J. Haematol.* **2015**, *169*, 455–458. [CrossRef]
36. Green, J.L.; Kuntz, S.G.; Sternberg, P.W. Ror receptor tyrosine kinases: Orphans no more. *Trends Cell Biol.* **2008**, *18*, 536–544. [CrossRef]
37. Ho, H.Y.; Susman, M.W.; Bikoff, J.B.; Ryu, Y.K.; Jonas, A.M.; Hu, L.; Kuruvilla, R.; Greenberg, M.E. Wnt5a-Ror-Dishevelled signaling constitutes a core developmental pathway that controls tissue morphogenesis. *Proc. Natl. Acad. Sci. USA* **2012**, *109*, 4044–4051. [CrossRef]
38. Peponi, E.; Drakos, E.; Reyes, G.; Leventaki, V.; Rassidakis, G.Z.; Medeiros, L.J. Activation of mammalian target of rapamycin signaling promotes cell cycle progression and protects cells from apoptosis in mantle cell lymphoma. *Am. J. Pathol.* **2006**, *169*, 2171–2180. [CrossRef]
39. Rudelius, M.; Pittaluga, S.; Nishizuka, S.; Pham, T.H.; Fend, F.; Jaffe, E.S.; Quintanilla-Martinez, L.; Raffeld, M. Constitutive activation of Akt contributes to the pathogenesis and survival of mantle cell lymphoma. *Blood* **2006**, *108*, 1668–1676. [CrossRef]
40. McCubrey, J.A.; Steelman, L.S.; Chappell, W.H.; Abrams, S.L.; Wong, E.W.; Chang, F.; Lehmann, B.; Terrian, D.M.; Milella, M.; Tafuri, A.; et al. Roles of the Raf/MEK/ERK pathway in cell growth, malignant transformation and drug resistance. *Biochim. Biophys. Acta* **2007**, *1773*, 1263–1284. [CrossRef]
41. Wang, W.Z.; Shilo, K.; Amann, J.M.; Shulman, A.; Hojjat-Farsangi, M.; Mellstedt, H.; Schultz, J.; Croce, C.M.; Carbone, D.P. Predicting ROR1/BCL2 combination targeted therapy of small cell carcinoma of the lung. *Cell Death Dis.* **2021**, *12*, 577. [CrossRef] [PubMed]
42. Lien, E.C.; Dibble, C.C.; Toker, A. PI3K signaling in cancer: Beyond AKT. *Curr. Opin. Cell Biol.* **2017**, *45*, 62–71. [CrossRef] [PubMed]
43. Frenzel, A.; Grespi, F.; Chmelewskij, W.; Villunger, A. Bcl2 family proteins in carcinogenesis and the treatment of cancer. *Apoptosis* **2009**, *14*, 584–596. [CrossRef] [PubMed]
44. Ghaderi, A.; Okhovat, M.A.; Wikanthi, L.S.S.; Svensson, A.; Palma, M.; Schultz, J.; Olin, T.; Osterborg, A.; Mellstedt, H.; Hojjat-Farsangi, M. A ROR1 small molecule inhibitor (KAN0441571C) induced significant apoptosis of ibrutinib-resistant ROR1(+) CLL cells. *EJHaem* **2021**, *2*, 498–502. [CrossRef]
45. Hanahan, D. Hallmarks of Cancer: New Dimensions. *Cancer Discov.* **2022**, *12*, 31–46. [CrossRef]

pharmaceutics

Article

Optimization of a Novel Mandelamide-Derived Pyrrolopyrimidine Series of PERK Inhibitors

Michael E. Stokes [1,*,†], Matthew D. Surman [2,*,†], Veronica Calvo [1], David Surguladze [1], An-Hu Li [1], Jennifer Gasparek [3], Matthew Betzenhauser [3], Guangyu Zhu [3], Hongwen Du [4], Alan C. Rigby [1] and Mark J. Mulvihill [1]

1 HiberCell Inc., 619 West 54th Street, New York, NY 10019, USA
2 Curia, 26 Corporate Circle, Albany, NY 12203, USA
3 Curia, 1001 Main Street, Buffalo, NY 14203, USA
4 Pharmaron Beijing Co., Ltd., No. 6, TaiHe Road, BDA, Beijing 100176, China
* Correspondence: mstokes@hibercell.com (M.E.S.); matthew.surman@curiaglobal.com (M.D.S.)
† These authors contributed equally to this work.

Abstract: The protein kinase R (PKR)-like endoplasmic reticulum kinase (PERK) is one of three endoplasmic reticulum (ER) transmembrane sensors of the unfolded protein response (UPR) responsible for regulating protein synthesis and alleviating ER stress. PERK has been implicated in tumorigenesis, cancer cell survival as well metabolic diseases such as diabetes. The structure-based design and optimization of a novel mandelamide-derived pyrrolopyrimidine series of PERK inhibitors as described herein, resulted in the identification of compound **26**, a potent, selective, and orally bioavailable compound suitable for interrogating PERK pathway biology in vitro and in vivo, with pharmacokinetics suitable for once-a-day oral dosing in mice.

Keywords: ER stress; PERK; UPR; kinase; inhibitor; cancer; diabetes; small molecule; structure–activity-relationship (SAR)

Citation: Stokes, M.E.; Surman, M.D.; Calvo, V.; Surguladze, D.; Li, A.-H.; Gasparek, J.; Betzenhauser, M.; Zhu, G.; Du, H.; Rigby, A.C.; et al. Optimization of a Novel Mandelamide-Derived Pyrrolopyrimidine Series of PERK Inhibitors. *Pharmaceutics* **2022**, *14*, 2233. https://doi.org/10.3390/pharmaceutics14102233

Academic Editor: Francesca Musumeci

Received: 2 September 2022
Accepted: 17 October 2022
Published: 19 October 2022

1. Introduction

Cells must match the rates of protein synthesis with demand to avoid the onset of proteotoxic endoplasmic reticulum (ER) stress. In conditions of excess protein production, an accumulation of misfolded proteins in the ER lumen induces the unfolded protein response (UPR), which arrests global translation while enabling specific processes that restore protein homeostasis [1,2]. The UPR is mediated by three discrete signaling cascades, IRE1, ATF6, and EIF2AK3 (PERK) [3]. Upon activating the UPR, PERK phosphorylates eIF2a at Ser51, leading to widespread inhibition of protein translation with a few selected regulatory proteins translated to restore homeostasis and alleviate ER stress [1,3]. In cancer, acute activation of UPR enables tumors to overcome deleterious ER stress induced by rapid cell proliferation, nutrient deprivation, hypoxia, and drug treatments [4–6]. Pharmacological or genetic inhibition of PERK slows proliferation of tumor xenografts in mice [7,8], and small molecule PERK inhibitors have recently entered clinical trials for several oncology indications, including clear cell renal cell carcinoma (ccRCC) and multiple myeloma (NCT04834778; NCT05027594).

Acute ER stress and activation of UPR restores protein homeostasis and supports cell survival, whereas long-term ER stress shifts cells toward an apoptotic cell fate [9]. While intermittent ER stress occurs regularly in cell types that produce high abundance of protein, metabolic dysfunction can result in chronic ER stress that coincides with a number of metabolic and neurological diseases [9–12]. Type-1 Diabetes (T1D) is an example of a metabolic disease and autoimmune disorder characterized by PERK hyperactivation in insulin-producing β-cells of the pancreatic islets [11–13]. Immunohistochemistry (IHC) staining of pancreas sections from T1D patients revealed PERK pathway activation in insulin-secreting beta cells, relative to healthy control patient samples [13], and PERK activation precedes the destructive autoimmune attack of beta cells in animal models of T1D [12,14]. Together, these findings highlight a potential role for PERK in diabetes and possible therapeutic benefit of PERK inhibitors beyond oncology.

While some cancers and metabolic diseases are characterized by PERK hyperactivation, genetic ablation of PERK results in glucose dysregulation and pancreatitis, suggesting that some basal level of PERK activity is essential to pancreatic health [15,16]. It is possible that the attenuation of PERK activity, rather than complete suppression, could provide therapeutic benefit while supporting normal pancreatic function. To enable future testing of this hypothesis, we sought to create a series of PERK inhibitors with stable PK profiles that could be administered at low doses to attenuate PERK yet tuned through higher doses to sufficiently inhibit PERK for tumor inhibition.

Previously described medicinal chemistry efforts around PERK inhibitors have made use of a variety of core hinge-binding regions, including aminopyridines, pyrrolopyrimidines, quinazolines, and quinolines [7,8,17]. Pyrrolopyrimidine cores have been developed and approved as JAK kinase inhibitors [18,19], and modifications to the core scaffold can have profound effects on stability and selectivity [17]. Here, we optimized the pyrrolopyrimidine core and leveraged extensive SAR information from our recently reported aminopyridine series of inhibitors to guide chemistry efforts aimed at optimizing the in vivo properties of pyrrolopyrimidine PERK inhibitors [7].

Our efforts were focused on developing potent and selective PERK inhibitors with suitable pharmacokinetic profiles to enable once or twice per day oral dosing in mouse models with a flat steady state exposure profile. The goal was to avoid large differences between C_{max} to C_{12h} or C_{24h} in order to achieve a constant and sustained pharmacodynamic response. This paper focuses on the structure-based design, structure–activity relationship (SAR), absorption, distribution, metabolism and excretion (ADME), and pharmacokinetic/pharmacodynamic (PK/PD) properties of a novel class of mandelamide-derived pyrrolopyrimidine PERK inhibitors leading to the identification of Compound **26,** with the desired flat steady state exposure profile. Compound **26** was characterized in a mouse pancreas assay and has been selected for further evaluation in animal models, which will be reported in due course.

2. Materials and Methods

2.1. Biochemical Assays

The inhibitory potencies of **26** inhibition against PKR, HRI, PERK and GCN2 were evaluated in cell-free biochemical assays following manufacturer's instructions (Carna; Table 1). In brief, **26** was dissolved in DMSO and serially diluted 3-fold to generate concentrations ranging from 10 μM to 0.2 nM in a 384-well plate using a TECAN EVO200 automated liquid handler. 10 nL stock solution was transferred into each well of a 384well assay plate by an Echo550 and an equivalent volume of DMSO was added to the control wells. To each assay well, recombinant protein was suspended in 20 μL reaction buffer (listed below) and incubated at 25 °C for 1.5 h while shaking at 1250 rpm. Following incubation, 10 μL of a FRET dilution buffer including 20 mM EDTA and 4 nM Tb-anti-peIF2a [pSer51] (Invitrogen, Cat# PV4815) was added to each well and incubated at 25 °C for 2 h while shaking at 600 rpm in the dark. Analysis was performed using an Envision Microplate Reader (PerkinElmer).

Table 1. Reagents and conditions used in biochemical characterization of ISR kinases.

Protein	Carna Cat#	Assay Buffer
GCN2	05-153	GCN2 (1 nM); DTT (2 mM); ATP (150 μM); GFP-eIF2a (100 nM; Invitrogen Cat# PV4809)
HRI	05-154	HRI (0.12 nM); DTT (2 mM); ATP (30 μM); GFP-eIF2a (100 nM; Invitrogen Cat# PV4809)
PERK	05-155	PERK (0.09 nM); DTT (2 mM); ATP (8 μM); GFP-eIF2a (100 nM; Invitrogen Cat# PV4809)
PKR	05-156	PKR (0.11 nM); DTT (2 mM); ATP (6 μM); GFP-eIF2a (100 nM; Invitrogen Cat# PV4809)

2.2. PERK Crystallization and Structure Determination

Human PERK (575-1094 Δ670-874) was purified as described previously [7]. Purified PERK protein at 11 mg/mL was mixed with 10 mM compound **11**, **24**, or **26** (in DMSO) to a final protein-inhibitor molar ratio 1:2. The PERK-inhibitor mixture was incubated on ice for 2 h before crystallization. The crystals with compound **11** or **26** were grown at 20 °C in sitting drops by combining 2.0 μL PERK-inhibitor mixture, 2.0 μL reservoir solution (12–14% PEG3350, 4% tacsimate pH7.0), and 0.4 μL seed stock which was equilibrated over a 500 μL reservoir solution. The crystal with compound **24** was grown at 4 °C in a hanging drop by combining 1.5 μL PERK-inhibitor mixture, 1.5 μL reservoir solution (9% PEG3350, 180 mM Na/K tartrate, 100 mM HEPES pH 7.0) which was equilibrated over a 500 μL reservoir solution. The crystals were grown to 0.2–0.5 mm over a three-week period before harvesting for analysis. The crystal was transferred stepwise to a cryo-solution with the crystallization cocktail plus 20% glycerol before being frozen in liquid nitrogen. Diffraction data were collected at the IMCA-CAT beamline 17ID for compound **11** or GMCA-CAT beamline 23IDD for compound **24** or **26** at the Advanced Photon Source at Argonne National Laboratory using a Pilatus 6M detector. The diffraction images were processed with DIALS [20] and scaled with AIMLESS [21]. The structure was solved by molecular replacement using PDB 4X7J [17] as starting model by Phaser [22]. The structure was manually built using Coot [23] and subsequently refined using Refmac5 [24]. The crystallographic figures were generated by CCP4MG [25], and the statistics of data collection and refinement are summarized in Supplemental Table S1.

2.3. Pharmacokinetic and Pharmacodynamic Analysis in Plasma and Pancreas

For PK analysis of pyrrolopyrimidine analogs in mouse, rat, and dog plasma, the oral formulation was prepared as a solution containing 20% (*w*/*v*) Captisol in 25 mM NaH_2PO_4 buffer (pH 2). Compounds were administered orally by gavage at 10 mg/kg to female CD1 mice. After dosing, plasma samples were collected to characterize the PK profiles at 0.25, 0.5, 1, 2, 4, 8, 12 and 24 h post-dose. The plasma concentration of compound was determined by protein precipitation with acetonitrile and liquid chromatography with mass spectrometric detection (LC-MS/MS). Parameters were estimated using Phoenix (WinNonlin) pharmacokinetic software version 6.1.0 using a non-compartmental approach consistent with the oral route of administration. PK analysis of **26** in plasma and pancreas followed a similar analytical method as above, with minor modifications. Compound **26** was suspended in a vehicle consisting of 0.5% methylcellulose (400 cP) and 0.1% Tween80 in water and administered to BALB/c nude mice by oral gavage at 0.3, 1, 3, 10, 30 mg/kg. Plasma and pancreas were sampled from 5 mice per group following a single oral administration at 1, 4, 8, 12, 24 h post-dose.

Total protein was extracted from pancreas tissues sampled during PK studies described above. Frozen pancreas tissue was homogenized by polytron in lysis buffer consisting of 2x Laemmli SDS sample buffer (Novex), supplemented with 10% BME (Gibco), 1X benzonase (EMD Millipore Sigma), phosphatase inhibitors (Sigma) and Mini protease inhibitor tablet (Roche). Homogenate was incubated at room temperature for 10′, then boiled for 10′, followed by centrifugation for 10′ at max speed on a benchtop centrifuge. Protein analysis performed using JESS high-throughput protein analysis instrument (Biotechne) using antibodies for phosphoPERK developed internally and PERK (CST).

2.4. Biochemical and Cell-Based Characterization of Pyrrolopyrimidine Analogs

Methods used to evaluate biochemical PERK activity are described and published elsewhere [7]. In brief, biochemical PERK activity is evaluated using a LanthaScreenTM (PerkinElmer) TR-FRET assay to detect phosphorylation of a GFP-tagged eIF2a substrate (ThermoFisher). Excitation at 340 nM of a terbium chelate donor fluorophore in an eIF2a antibody results in energy transfer to GFP acceptor fluorophore upon eIF2a phosphorylation by PERK. Reaction buffer was composed of 50 mM HEPES (pH 7.4), 10 mM $MgCl_2$, 1 mM EGTA and 0.01% (*v*/*v*) BrijTM-35. Reactions were initiated by the addition of substrate and

ATP, followed by incubation for 1 h at RT prior to addition of EDTA and the anti-peIF2a antibody. FRET was measured by an EnVision Plate Reader (PerkinElmer). Data were plotted as percent inhibition along a 10-point 3-fold dilution series of inhibitors. IC_{50} values were calculated using 4-parameter logistical fitting in XLFit.

Methods used to evaluate in vitro cellular PERK inhibition are described and published elsewhere [7]. In brief, stableHEK293 cells were created using lentiviral particles harboring a GFP-eIF2a expression vector. Transfectants were selected by puromycin and enriched using fluorescent cell sorting based on GFP expression. HEK293-EGFP-eIF2a cells were plated at 5000 cells/well in 384-well assay plates and incubated overnight. The following day, pyrrolopyrimidine analogs were added to wells using an Echo acoustic dispenser and incubated for 30 min. ER stress was induced by the addition of tunicamycin to the wells at 1 mM for 2 h. Cells were lysed and peIF2a was evaluated using a TR-FRET system, as described above.

2.5. *In Vivo Tumor Xenograft Studies*

In vivo studies of the effect of pyrrolopyrimidines on 786-O xenograft inhibition follow a method published elsewhere [7]. Female BALB/c nude mice were inoculated subcutaneously with 786-O tumor cells (5×10^6) in 0.1 mL of PBS. Animals were randomized when tumors reached 100–200 mm^3 into treatment groups. Pyrrolopyrimidine PERK inhibitors **24**, **26** or **70** were dissolved in vehicle solution consisting of 20% (w/v) Captisol in 25 mM $NaPO_4$ buffer (pH2) and administered twice daily (BID) by oral gavage for 29 days. Tumor volumes were measured by caliper and the volume was expressed in mm^3 using the formula: $V = (L \times W \times W)/2$, where V, L, and W represent tumor volume, length, and width, respectively. All studies were conducted following an IACUC-approved protocol (AN-1903-05-1798). Experimental data management and reporting were in accordance with applicable Crown Bioscience's Guidelines and Standard Operating Procedures.

3. Results and Discussion

Small molecule PERK inhibitors across a variety of chemotypes have been previously described in the literature and captured in an earlier publication from our group around an aminopyridine series of PERK inhibition [7]. Herein, we report the synthesis and discovery of a series of novel mandelamide-derived pyrrolopyrimidines (I) that are highly potent, selective, and orally bioavailable PERK inhibitors. These medicinal chemistry efforts were iteratively supported by an X-ray crystallography structure-based approach to develop and refine SAR information. Collectively this approach permitted us to optimize the physiochemical and ADME/PK properties of this chemical series. Different R^1, R^2, R^3, Ar and L groups have been introduced into our PERK small molecule series to tune the potency, selectivity and drug-like properties, such as aqueous solubility and cellular permeability.

The general synthesis for our pyrrolopyrimidine PERK inhibitors is shown in Scheme 1. The terminal ring substitution with different R^2 groups was introduced through the corresponding aldehyde **1**. Cyanohydrin **2** formation with TMS-cyanide and zinc(II) iodide followed by acid hydrolysis of the nitrile in methanol provided methyl mandelic ester **3**. Hydrolysis of the ester with lithium hydroxide and acylation of the benzylic alcohol gave the protected mandelic acid **5**.

The proximal ring substitution with different R^3 groups was introduced through bromo aniline **6**. Palladium-mediated borylation of **6** to give the aniline boronate ester **7** was followed by T_3P coupling with acid **5** to provide amide **8**. Boronate ester **8** underwent Suzuki coupling with pyrrolopyrimidine bromide **9** to give racemic acetyl-protected product **10**. Deacylation under basic conditions followed by chiral separation of the enantiomers provided the final product **I**.

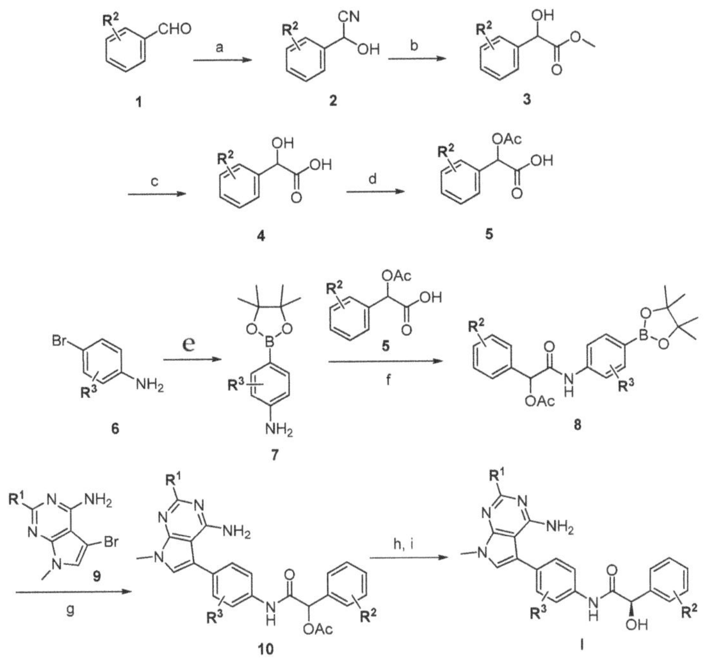

Scheme 1. General synthesis of PERK inhibitors. Reaction conditions: (a) TMS-CN (1.2 equiv.), ZnI_2 (0.1 equiv.), 0 °C; (b) 3M HCl in MeOH, 0 °C to rt, 60–80% over two steps; (c) $LiOH{\cdot}H_2O$ (1.2 equiv.), THF, H_2O, MeOH, rt, 50–85%; (d) AcCl (2 vol.), 0 °C to rt, 70–85%; (e) B_2pin_2 (1.2 equiv.), KOAc (3 equiv.), $PdCl_2$ $(PPh_3)_2$ (5 mol. %), NEt_3 (10 vol.), 90 °C, 50–70%; (f) **5** (1.2 equiv.), 50% propylphosphonic anhydride (T_3P) in EtOAc (1.5 equiv.), DIPEA (3 equiv.), CH_2Cl_2, 0 °C to rt, 50–80%; (g) XPhos-Pd-G_2 (10 mol. %), K_2CO_3 (3 equiv.), THF, H_2O, 80 °C, 25–70%; (h) K_2CO_3 (2 equiv.), MeOH, rt, 80–90%; (i) Chiral SFC separation.

The PERK activity of analog **I** was evaluated using cell-free biochemical assays. In these assays, PERK phosphorylation of eIF2a was assessed using a FRET-based labelled protein assay [7]. Compounds were then evaluated for cellular activity by quantifying inhibition of eIF2a phosphorylation in HEK293 cells treated with the ER inducer tunicamycin [7]. Together, these data read out the potency of a given analog against PERK as well as the cellular permeability to enable bioactivity.

Initial SAR studies in the pyrrolopyrimidine series focused on the linker (**L**) region between the central amide and terminal aryl ring (Table 2). Hydroxyl substitution on this group was found to increase potency by 2-3-fold (**13** vs. **14**) with preference for the (*R*)-over the (*S*)-enantiomer (**11** vs. **12**). Expansion or contraction of the length of the linker **L** was not tolerated, resulting in at least 20-fold losses in PERK biochemical potency and nearly 50- to 100-fold losses in cellular potency (**14** vs. **15** and **16**, respectively). Difluoro substitution of the linker also caused a substantial loss in potency, as did replacement of the CH(OH) group with an O, NH, or NCH_3 group (**13** vs. **17**, **18**, **19**, and **20**, respectively). Seven-fold losses in both biochemical and cellular potency were observed when the hydroxyl group was converted to the methyl ether (**13** vs. **21**). Even more substantial losses were associated with replacement of the hydroxyl with an amino group (**13** vs. **23**). Addition of a methyl group to the linker to form a tertiary alcohol also caused a substantial loss in potency (**13** vs. **22**). Based on these findings, it was concluded that (R)-CH(OH) is the preferred **L** group.

Table 2. In vitro data of PERK inhibitors.

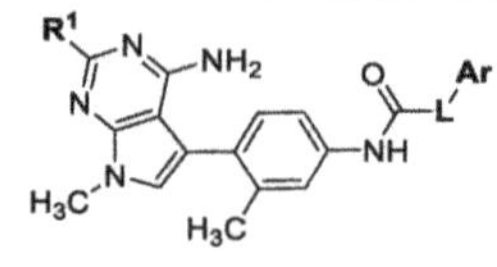

Cmpd	R^1	L	Ar	PERK IC_{50} (μM)	p-eIF2α IC_{50} (μM)
11	H	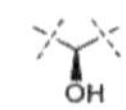	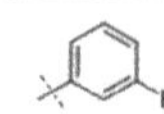	0.0022	0.0084
12	H	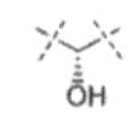	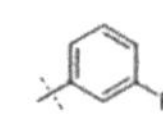	0.0078	0.057
13	H	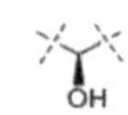	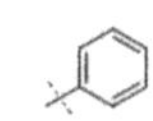	0.0018	0.023
14	H		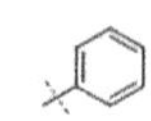	0.0052	0.041
15	H	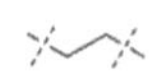	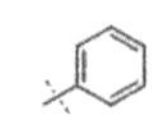	0.126	1.93
16	H	- - -	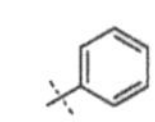	0.168	3.96
17	H		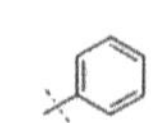	0.3	0.532
18	H		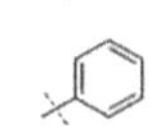	3.98	>10.0
19	H	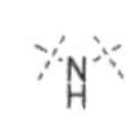	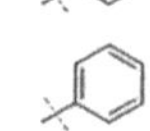	0.021	0.211
20	H	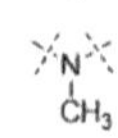	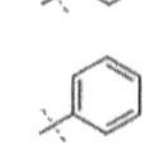	0.108	1.75
21	H		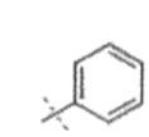	0.014	0.174
22	H	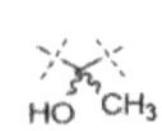	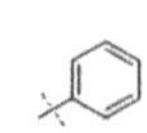	0.056	0.184
23	H	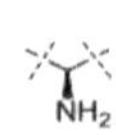	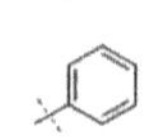	0.033	0.581
24	Me	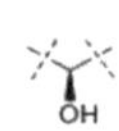	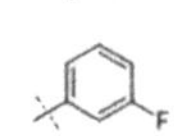	0.02	0.051
25	Me	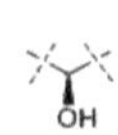	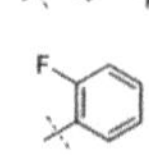	0.102	0.303
26	Me	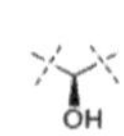	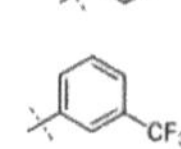	0.0028	0.015
27	Me	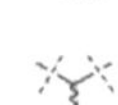	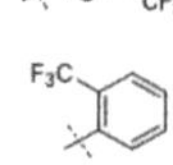	0.343	2.05
28	Me		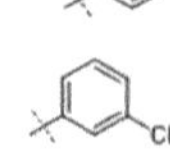	0.0018	0.016
29	Me	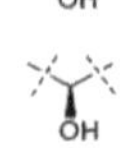	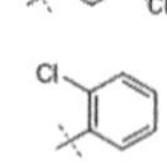	0.107	0.713

Table 2. *Cont.*

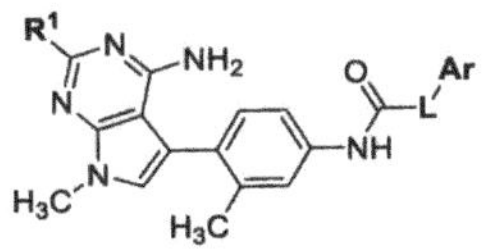

Cmpd	R^1	L	Ar	PERK IC_{50} (μM)	p-eIF2α IC_{50} (μM)
30	Me	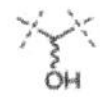	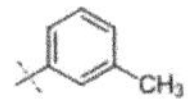	0.014	0.083
31	Me	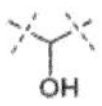	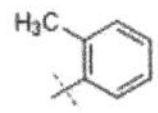	0.026	0.122
32	Me	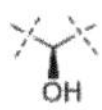	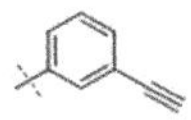	0.003	0.0078
33	Me	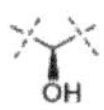	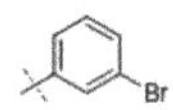	0.003	0.014
34	Me	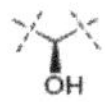	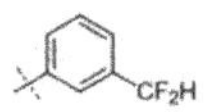	0.0059	0.021
35	Me	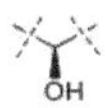	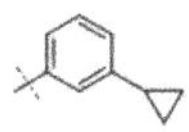	0.0059	0.017
36	Me	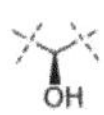	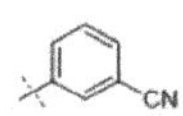	0.008	0.325
37	Me	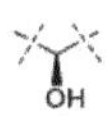	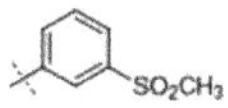	>1	>10
38	Me	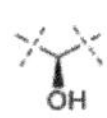	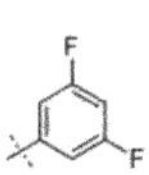	0.0019	0.014
39	Me	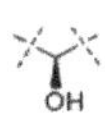	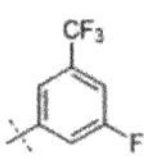	0.0026	0.011
40	Me	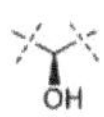	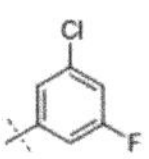	0.0029	0.0085
41	Me	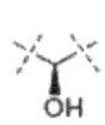	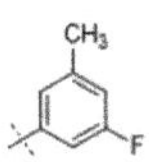	0.0014	0.01
42	Me	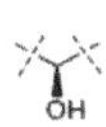	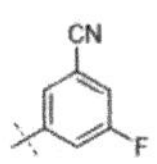	0.0042	0.032
43	Me	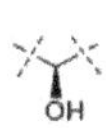	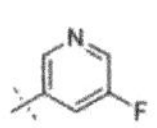	0.052	0.562
44	Me	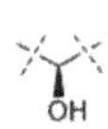	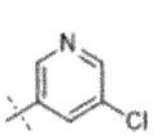	0.017	0.129
45	Me	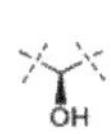	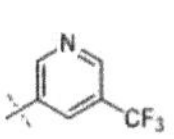	0.019	0.165

Previously reported SAR of PERK inhibitors have revealed modifications to the hinge binding region that have profound effects on binding affinity to PERK as well as selectivity versus the kinome [7,17]. Structure based design modelling was used to focus efforts on optimizing the pyrrolopyrimidine hinge-binding region of the molecule. Inhibitor **11** was co-crystallized with PERK into a trigonal space group$P3_221$ with comparable cell parameters as PDB 4X7O. Compound **11** formed three hydrogen bonds directly with the carbonyl of Q889, the amide of C891, and the carbonyl of F956 from PERK, respectively (Figure 1A). In addition, compound **11** forms two hydrogen bonds with a water molecule, and this molecule hydrogen bonds to the amide of V652. In the ATP binding hinge region, substitution at the C2-position (R1) of the pyrrolopyrimidine ring appeared feasible, filling a small pocket near C891. A methyl group located at the R1 position of compound **11** afforded analog **24** whose co-crystal structure with PERK confirmed the occupation of the small pocket near C891 along with the close contact of the C2-methyl group with the carbonyl of C891 with a distance of 3.14 Å (Figure 1B). Although a weak CH···O hydrogen bond is feasible, this contact appears to be more repulsive. As a result, a small rotation of the main chain amino acid Q889 and L890 of PERK, along with a minor shift of the pyrrolopyrimidine ring of inhibitor **24** are observed compared to those in the structure of **11** (Figure 1C). These shifts maintain the hydrogen bonds with similar distances. Due to a lower resolution (2.92Å) of the compound **24** structure (Table S1), the aforementioned water molecule in the compound **11** structure was not observed. This water molecule is assumed to locate at the same position, as all hydrogen bond doner and acceptor atoms are still presented at similar positions in the compound **24** structure. These observations indicate that the reduction of potency of **24** is likely caused by repulsion from the methyl group rather than the disruption of the hydrogen bonding pattern.

While having a minor effect on PERK potency, the methyl substituent on the pyrrolopyrimidine ring had a profound effect on the kinase selectivity. Compounds **11** and **24** were screened in the DiscoverX scanMAX^{SM} panel of 468 kinases [26]. The resulting TREEspot^{TM} interaction maps (Figure 2) revealed that the methyl substituent in **24** led to superior selectivity versus **11**. The S(35), S(10), and S(1) selectivity scores at a concentration of 1 μM were 0.005, 0, and 0 for **24** versus 0.072, 0.035, and 0 for **11**. Given the superior selectivity of **24** but slight loss in potency versus **11**, further optimization of the series was carried out with the 2-methyl substitution on the pyrrolopyrimidine ring to further improve potency and drug-like properties.

Exploration of the terminal aryl ring revealed several trends (Table 2): (1) Substitution at the 3-position was preferred to substitution at the 2-position of the ring (compare **24** vs. **25**, **26** vs. **27**, **28** vs. **29**, and **30** vs. **31**). (2) Lipophilic substituents, such as CF_3, Cl, Br, CCH, CF_2H, and *c*-Pr provided enhanced biochemical and cellular potency compared with F substitution (**24** vs. **26**, **28**, **33**, **32**, **34**, and **35**) leading to the recovery of the potency lost with the incorporation of the selectivity-enhancing 2-methyl pyrrolopyrimidine ring (compare **11** vs. **24** and **11** vs. **26**, **28**, **33**, and **32**). (3) Polar substituents were not well tolerated and resulted in substantially reduced cellular potency, as in the case of CN derivative **36**, or in complete loss of activity (biochemical and cellular activity at the concentrations tested), as in the case of sulfone analog **37**. (4) 3,5-Di-substitution with lipophilic groups provided further enhancements in potency (**38**, **39**, **40**, and **41**). Interestingly, addition of a fluorine substituent to CN on the terminal phenyl ring largely reversed the loss of cellular potency due to the CN group (**36** vs. **42**). (5) Replacement of the terminal phenyl ring with heterocycles, such as F-, Cl-, or CF_3-substituted pyridines, resulted in 8- to 11-fold losses in cellular potency (compare **24** vs. **43**, **28** vs. **44**, and **26** vs. **45**).

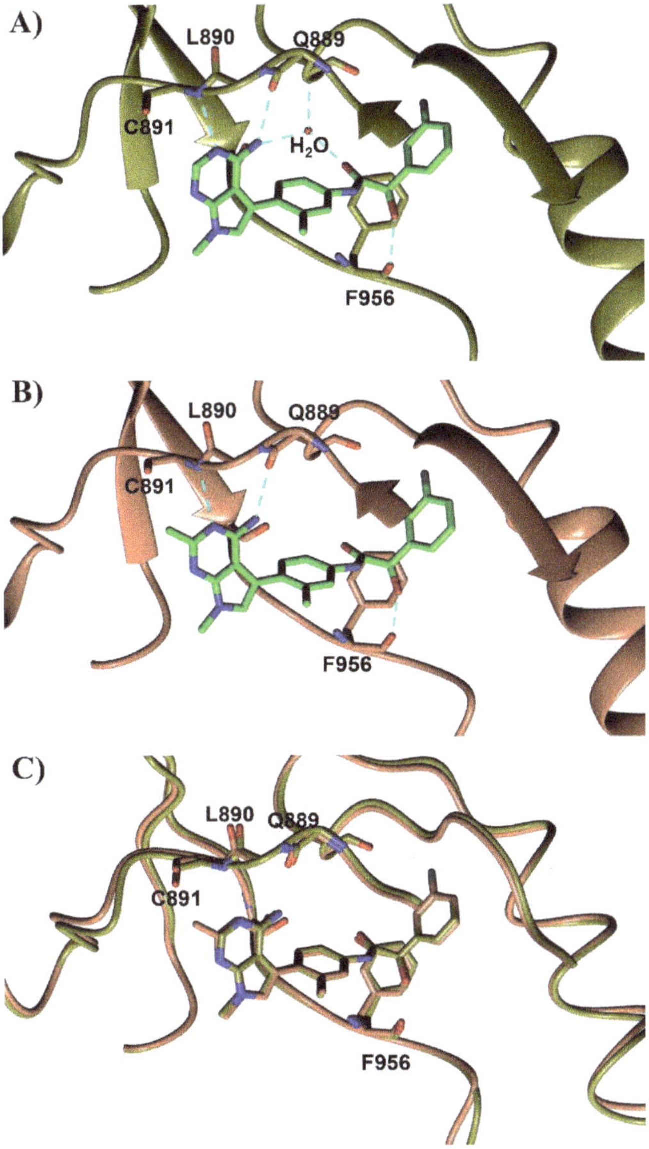

Figure 1. Crystal structures of PERK with selected inhibitors **11** (**A**) and **24** (**B**). Overlay of the crystal structures of **11** and **24** (**C**). The same orientation is presented for all panels. The PERK backbone is shown as a ribbon or solid strand in yellow for **11** and coral for **24**. The side chains are depicted as cylinders in the same color as the main chain for carbon, blue for nitrogen, and red for oxygen atoms. Inhibitors are shown as cylinders with same color scheme except green for carbon in panels (**A**,**B**) and grey for fluorine. Hydrogen bonds are shown as cyan dashed lines.

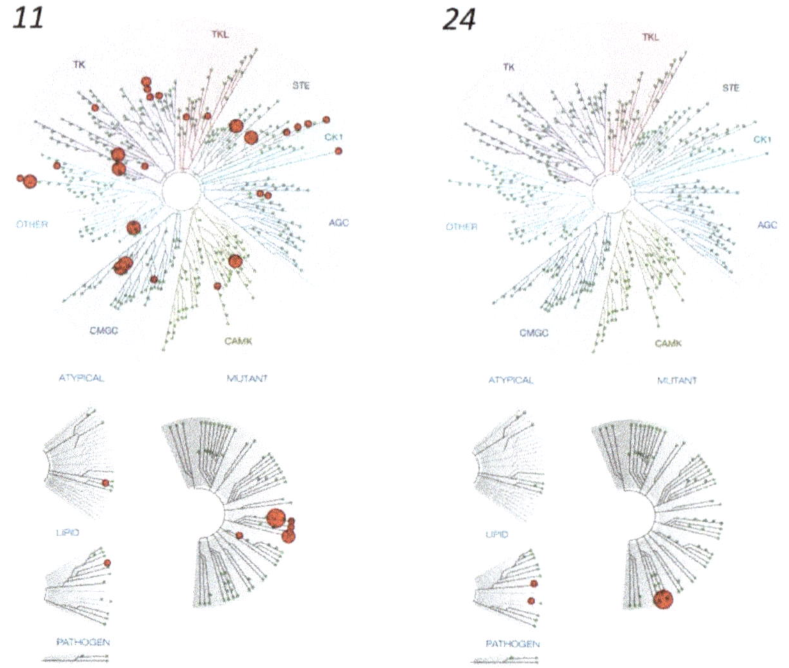

Figure 2. The TREEspot™ interaction map of **11** and **24** assayed at 1000 nM in the 468-kinase scanMAXSM panel. The panel includes 403 unique wild-type kinases and 59 mutant isoforms. Note that PERK is not included in the TREEspot assay but was evaluated in separate biochemical assays. Complete tabulated results presented in Supplemental Table S2.

Next, the SAR of the proximal aryl ring was examined with new analogs incorporating the best substituted phenyl groups from the survey of the terminal aryl ring (3-F-Ph; 3-CF_3-Ph; 3-Cl-Ph; 3-CH_3-Ph; 3-Et-Ph; 3-F,5-CF_3-Ph; and 3-F,5-CH_3-Ph) was (Table 3). Replacement of the proximal ring methyl substituent with a F, Cl, or Et group was found to be well tolerated (**24** vs. **46**, **53**, and **59**). These changes to the proximal ring could be combined with modifications to the terminal ring to provide analogs with <5 nM biochemical and <15 nM cellular potency (**47**, **48**, **52**, **57**). Modest losses in cellular potency (2.5- to 4-fold) were observed for H and CF_3 methyl replacements on the proximal ring (**24** vs. **62** and **58**). More substantial reductions in potency (8- to 36-fold) occurred with CF_3O and CH_3O replacements (**24** vs. **61** and **60**). Addition of a fluorine substituent to the proximal ring of **24** resulted in the general maintenance of cellular potency with a 10-fold range in biochemical potency, depending upon the F regioisomer (**24** vs. **65**, **66**, and **67**). 3,5-Dimethyl proximal ring substitution (**68**) provided comparable in vitro potency to the 3-fluoro, 5-methyl analog **67**. Movement of the fluorine substituent from the 3-position of the proximal ring (**46**) to the 2-position (**69**) had little impact on the biochemical or cellular potency of the compounds. Addition of a second fluorine to the proximal ring of **46** similarly had little impact on the cellular potency of the compounds, although some variation in the biochemical data was observed (**46** vs. **70** and **71**). Replacement of the proximal phenyl ring of **24** with a pyridine ring resulted in modest (2- to 3-fold) losses in cellular potency (**24** vs. **72** and **74**). Some of the lost potency could be restored by replacing the terminal ring F with CF_3 (**73** vs. **72** and **75** vs. **74**). Replacement of the proximal phenyl ring of **24** with a pyrimidine ring was not tolerated, resulting in a 30-fold loss in biochemical potency and complete loss in cellular activity (**24** vs. **76**).

Table 3. In vitro data of PERK inhibitors with proximal and terminal ring modifications.

Cmpd No.	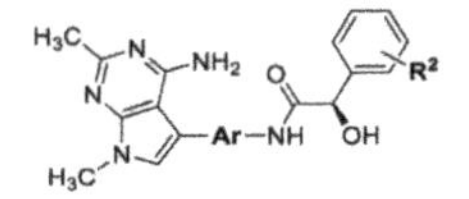 Ar	R²	PERK IC_{50} (μM)	p-eIF2α IC_{50} (μM)
24	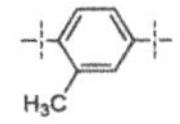	3-F	0.020	0.051
46	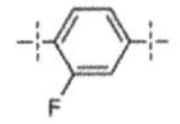	3-F	0.0022	0.041
47	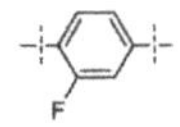	3-CF_3	0.0051	0.011
48	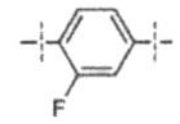	3-Cl	0.0010	0.010
49	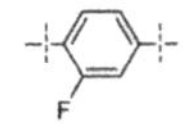	3-Me	0.0011	0.017
50	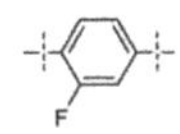	3-Et	0.0009	0.025
51	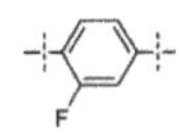	3-F, 5-CF_3	0.0027	0.062
52	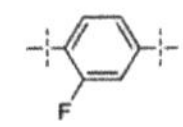	3-F, 5-Me	0.0011	0.011
53	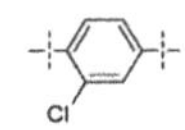	3-F	0.017	0.056
54	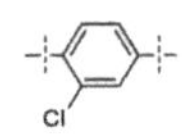	3-CF_3	0.0027	0.021
55	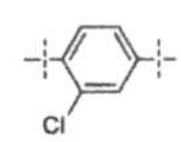	3-Cl	0.0016	0.017
56	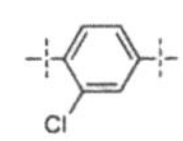	3-Me	0.0041	0.018
57	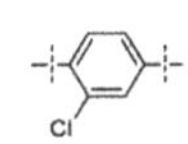	3-Et	0.059	0.014
58	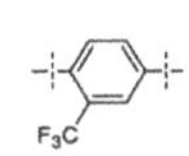	3-F	0.0085	0.204
59	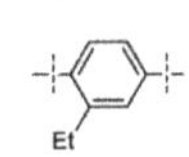	3-F	0.0040	0.041

Table 3. *Cont.*

Cmpd No.	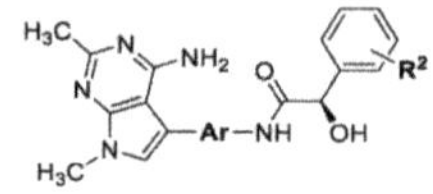		PERK IC_{50} (μM)	p-eIF2α IC_{50} (μM)
	Ar	**R^2**		
60	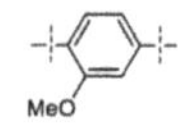	3-F	0.236	1.85
61	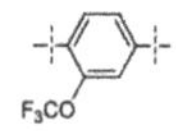	3-F	0.023	0.435
62	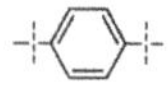	3-F	0.0020	0.126
63	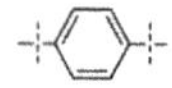	3-CF_3	0.0028	0.013
64	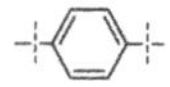	3-Br	0.0015	0.014
65	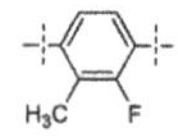	3-F	0.0019	0.030
66	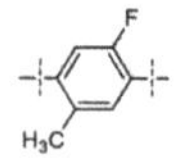	3-F	0.011	0.042
67~	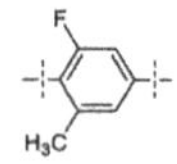	3-F	0.022	0.078
68~	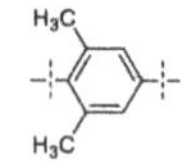	3-F	0.013	0.082
69	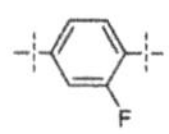	3-F	0.0010	0.026
70	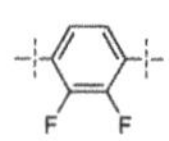	3-F	0.0015	0.024
71~	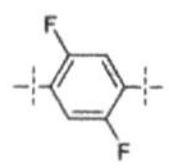	3-F	0.0010	0.034
72	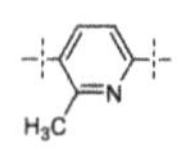	3-F	0.017	0.105
73	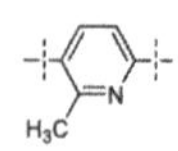	3-CF_3	0.013	0.051
74	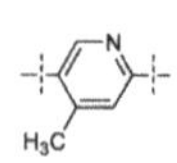	3-F	0.0083	0.146

Table 3. *Cont.*

Cmpd No.	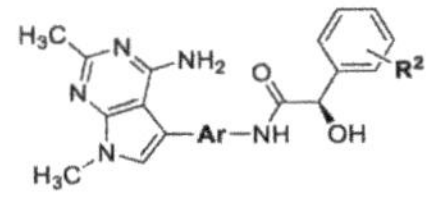		PERK IC_{50} (μM)	p-eIF2α IC_{50} (μM)
	Ar	**R^2**		
75	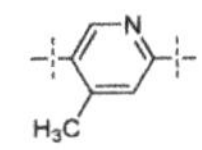	3-CF_3	0.0068	0.095
76	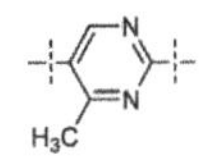	3-F	0.647	>10

~Racemic mixture of alcohols.

The SAR campaign successfully identified a number of PERK inhibitors that demonstrated biochemical and cellular potency. These promising compounds were evaluated in in vitro ADME assays (aqueous solubility, Caco-2 permeability, plasma protein binding, and hepatocyte stability) to determine their drug-like properties (Table 4). In general, the compounds showed moderate kinetic aqueous solubility in PBS at pH 7.4, ranging from 10–66 μM. The Caco-2 permeability was found to be high for all compounds ($P_{app} \geq 9 \times 10^{-6}$ cm/s), predicting good oral absorption. Additionally, with the exception of **24** (efflux ratio of 5.2), the compounds showed little to no efflux. Plasma protein binding was found to be high across all species tested, a characteristic common for many kinase inhibitors.

Table 4. Solubility, permeability, plasma protein binding and hepatocyte clearance properties of selected PERK inhibitors.

Cmpd No.	Kinetic Aqueous Solubility (μM)	Caco-2		Plasma Protein Binding				Hepatocyte Clearance ($t_{1/2}$ min)			
		P_{app} (A→B) (10^{-6} cm/s)	Efflux Ratio	Human	Rat	Mouse	Dog	Human	Rat	Mouse	Dog
26	19	21	2.2	99.8 ± 0.0	99.8 ± 0.0	99.9 ± 0.0	99.5 ± 0.0	194	166	118	130
28	35	23	1.9	99.8 ± 0.1	99.7 ± 0.0	>99.9	98.3 ± 0.3	778	27	62	355
65	58	26	1.6	99.6 ± 0.1	98.9 ± 0.0	99.3 ± 0.2	98.7 ± 0.1	390	22	64	567
24	76	9	5.2	99.2 ± 0.2	99.1 ± 0.2	99.4 ± 0.0	97.0 ± 0.4	733	23	31	>1000
41	24	20	2.5	99.6 ± 0.2	99.5 ± 0.2	99.9 ± 0.0	98.8 ± 0.3	68	16	24	232
39	10	16	2.0	99.9 ± 0.0	99.9 ± 0.0	>99.9	>99.9	277	252	320	223
38	19	19	1.9	99.0 ± 0.1	99.1 ± 0.3	99.2 ± 0.3	98.4 ± 0.1	897	121	805	460
48	14	24	1.6	99.9 ± 0.0	99.6 ± 0.0	99.8 ± 0.0	99.6 ± 0.1	556	25	53	215
66	61	24	1.5	99.3 ± 0.2	98.6 ± 0.1	98.8 ± 0.2	98.2 ± 0.2	358	13	167	39
70	66	29	1.7	99.4 ± 0.0	99.3 ± 0.1	99.7 ± 0.1	99.4 ± 0.1	360	15	45	268

In vitro metabolic stability was assessed in human, rat, mouse, and dog hepatocytes. In general, the compounds were found to be relatively stable in human and dog hepatocytes and exhibited long half-lives. For several compounds (**28**, **65**, **24**, **41**, **66**, and **70**), stability was reduced in rodent hepatocytes, particularly rat. Compounds with alkyl groups (i.e., Me) on the terminal phenyl ring tended to have lower stability (**41** vs. **39**). Several compounds (**26**, **39**, and **38**) maintained good metabolic stability across all species tested.

To connect the in vitro ADME results with the in vivo setting, select compounds (**26**, **28**, **65**, **24**, **41**, **39**, **38**, **48**, **66**, and **70**) were examined in mouse PK studies (Table 5). All of the compounds showed robust oral exposure with C_{max} values for most compounds >8000 ng/mL and $AUC_{0–last}$ values >25,000 h·ng/mL. The compounds also showed good in vivo half-lives of 2–3.5 h and excellent bioavailability (≥70%) apart from compound **41**. Compound **26** stood out with a particularly attractive pharmacokinetics profile with C_{max} of 17,200 ng/mL, $AUC_{0–last}$ of 126,019 h·ng/mL, half-life of 3.5 h, and bioavailability of 82%, a significant improvement over our earlier analog **24** and was therefore chosen as one of the key leads in this series PERK inhibitors.

Table 5. In vivo pharmacokinetics (PK) properties of selected PERK inhibitors in mice. Compounds administered at 10 mg/kg by oral gavage (PO).

Compound	C_{max} (ng/mL)	$AUC_{0–last}$ (h·ng/mL)	$t_{1/2}$ (h)	F (%)
26	17,200	126,019	3.5	82
28	14,067	65,764	2.4	84
65	8150	40,105	2.2	129
24	13,100	37,495	2.1	77
41	12,467	26,010	2.7	40
39	14,300	86,444	2.9	78
38	8597	28,932	2.0	94
48	3017	10,889	2.1	70
66	9610	39,835	2.0	126
70	9723	32,258	2.0	117

Inhibitor **26** was co-crystallized with PERK into a trigonal space group$P3_221$ with comparable cell parameters as previously noted with compounds **11** and **24**. Replacement of the 3-fluorine on the terminal phenyl group in **11** and **24** and with a larger CF_3 group in **26** induces a flip in the phenyl ring since the enzyme position containing the fluorine cannot accommodate the bulkier CF_3 group (Figures 1 and 3). The fluorine atom of **24** forms four contacts with a distance of 3.8 Å or less with residues within the back pocket region of PERK, whereas the CF_3 group of **26** forms ten contacts with PERK. These extra contacts may stabilize the binding of **26** to PERK and explain the increased potency of **26** compared to **24**.

Compound **26** was found to have excellent selectivity in the DiscoverX kinase panel (Figure 4; Supplemental Table S2) comparable to that of compound **24**. The S(35), S(10), and S(1) selectivity scores at a concentration of 1 μM were 0.01, 0.002, and 0.002. Importantly, **26** demonstrated potent selectivity for PERK over the other three highlyconserved ISR kinases (GCN2, HRI, and PKR; Supplemental Table S3). In biochemical assays, the IC_{50} of **26** against PERK was 4.7 nM, compared to IC_{50} values greater than 10 μM for each of the other ISR kinases. The primary off-target activity of **26** was associated with FLT3 kinase and its mutant isoforms. Radiometric kinase assays were used to validate the kinome panel and determine the relative IC_{50} value of **26** against wild-type FLT3 and a mutant isoform (D835Y). When evaluated individually using concentrations of **26** up to 10 μM, the relative IC_{50} against wild-type FLT3 was >1 μM, and >10 μM against the mutant variant FLT3 (D385Y), indicating weak to no activity against FLT3wt or the mutant form (Supplemental Table S3).

To understand the effect of **26** in target organs, the PK/PD relationship of **26** was investigated in mouse pancreas. PK analysis of plasma and pancreas samples from mice treated with **26** revealed dose-proportionate exposure at doses ranging from 0.3 to 30 mg/kg following a single administration by oral gavage (Supplemental Figure S2). Pancreas tissue was sampled to evaluate the effect of **26** on PERK autophosphorylation in pancreas, following oral administration of **26** at 0.3, 1, 3, 10, and 30 mg/kg. At 1 h following administration, phosphoPERK (T980), relative to total PERK protein (pPERK/PERK) decreased in a dose-dependent manner, reaching approximately 80% inhibition at 30 mg/kg (Figure 5A). When PERK inhibition was evaluated across time in mouse pancreas, a dose-dependent inhibition was observed which was relatively stable across a 12 h period following administration of **26** (Figure 5B). This observation was consistent with the dose-proportionate exposure and low clearance rate of **26** observed in both mouse plasma and pancreas (Supplemental Figure S2). To confirm that the PK profile of **26** in other species was suitable to enable further development, plasma concentration of **26** was quantified in mouse, rat and dog following a single oral administration at 10 mg/kg. Consistent with observations in mice, rat and dog both shared similar clearance rates as mouse, albeit with different exposure levels following administration (Figure 5C).

A) B)

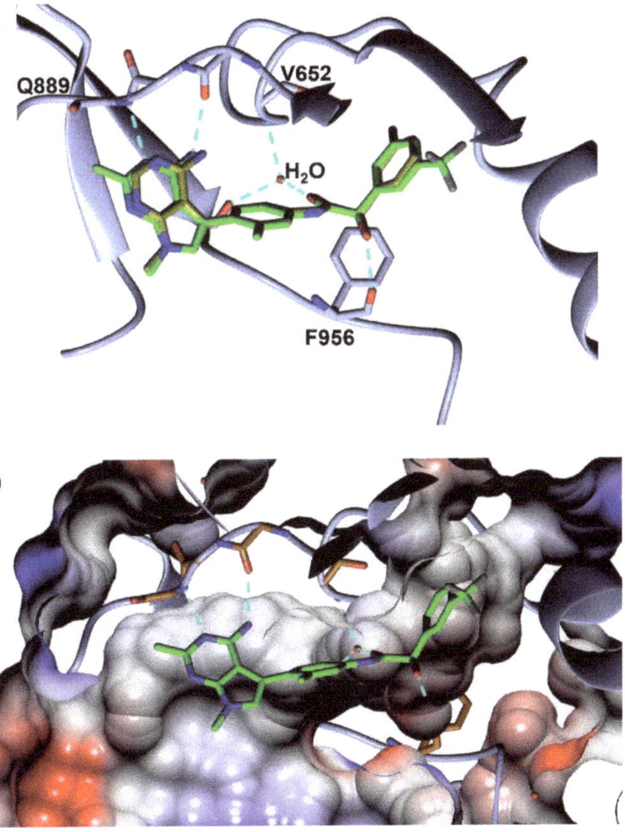

Figure 3. Crystal structure of compound **26** bound to the PERK kinase domain. (**A**) View of **26** in the PERK active site. PERK backbone is shown as a blue ribbon, and the side chains are shown as cylinder in brown for carbon, blue for nitrogen, and red for oxygen atoms, respectively. Compound **26** is shown as cylinder with same color scheme except green for carbon and grey for fluorine. Compound **11** is shown in a similar manner to compound **26**, except the color gold is used for carbon atoms. Hydrogen bonds for **26** are shown as cyan dashed lines. The phenyl ring is flipped due to a replacement of the 3-fluorine in **11** to CF_3 in **26**. (**B**) View of **26** in the PERK active site with surface rendering. The color intensity corresponds to the calculated electrostatic potential, from −20 kT/e (red) to +20 kT/e (blue).

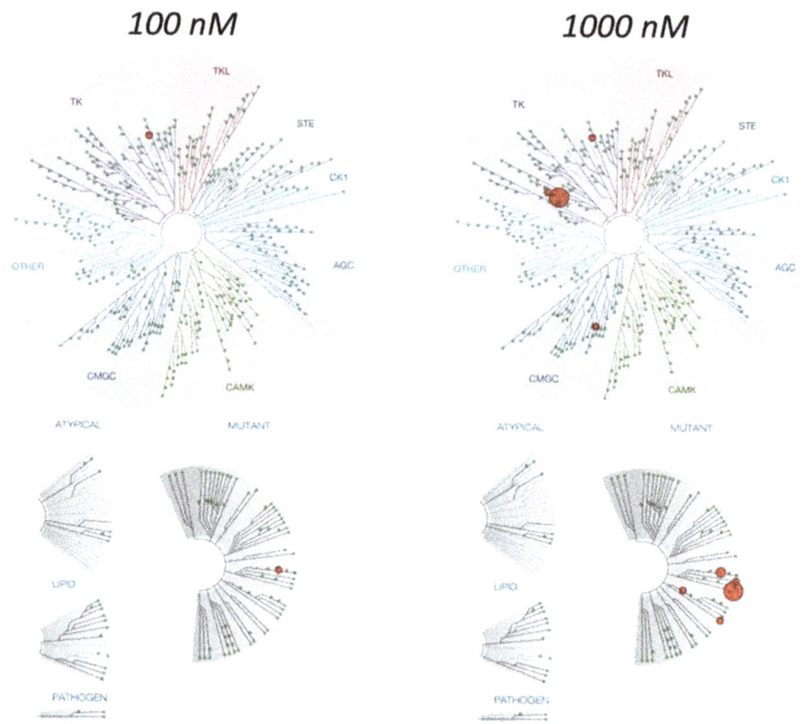

Figure 4. The TREEspot^{TM} interaction map of compound **26** in the 468-kinase scanMAXSM panel at 100 and 1000 nM. The panel includes 403 unique wild-type kinases and 59 mutant isoforms. Note that PERK is not included in the TREEspot assay. Complete tabulated results presented in Supplemental Table S2.

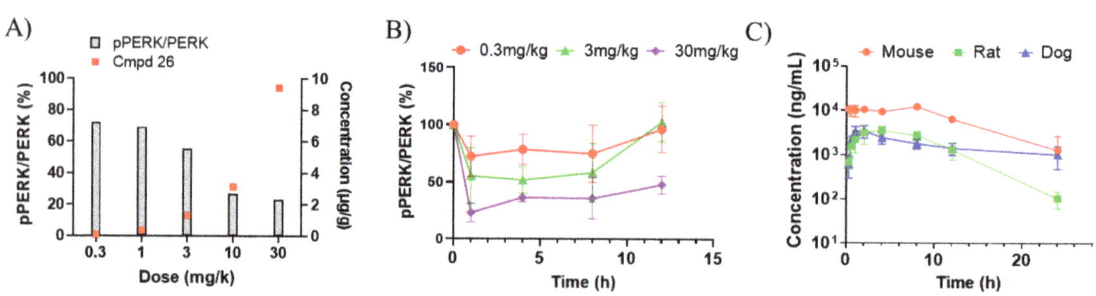

Figure 5. In vivo PK/PD analysis of **26**. (**A**) Inhibition of pPERK relative to total PERK protein (pPERK/PERK) in pancreas following oral administration of **26**. Abundance of **26** quantified by LC-MS/MS from matching pancreas samples at Cmax (1 h post administration). (**B**) Inhibition of pPERK/PERK in pancreas following oral administration of **26** at 0.3, 3, or 30 mg/kg. Pancreas samples taken at 1, 4, 8, 12 h following oral administration of **26**. Data represent mean ± SD of four mice. (**C**) Plasma PK analysis of **26** in mouse, rat and dog. **26** administered orally at 10 mg/kg, and sampled across a 24-h period; **26** quantified by LC-MS/MS.

As the focus of this program was to develop a novel, highly potent, selective and orally bioavailable PERK inhibitor to probe the role of PERK biology in animal models, compounds **24**, **26** and **70** were tested for in vivo efficacy in a clear cell renal cell carcinoma (ccRCC) xenograft model, 786-O (Figure 6). The 786-O model is driven by the Von Hippel-Lindau (VHL) mutation, which leads to PERK pathway activation and is therefore a suitable model to assess the biological activity of these PERK inhibitors [17]. The compounds were dosed at 30 mg/kg, twice per day (BID) for 29 days and were well tolerated as determined by body weight measurement (Supplemental Figure S3). Treatment with pyrrolopyrimidine PERK inhibitors resulted in significant tumor growth inhibition that was in line with our previously reported aminopyridine series of PERK inhibitors [7].

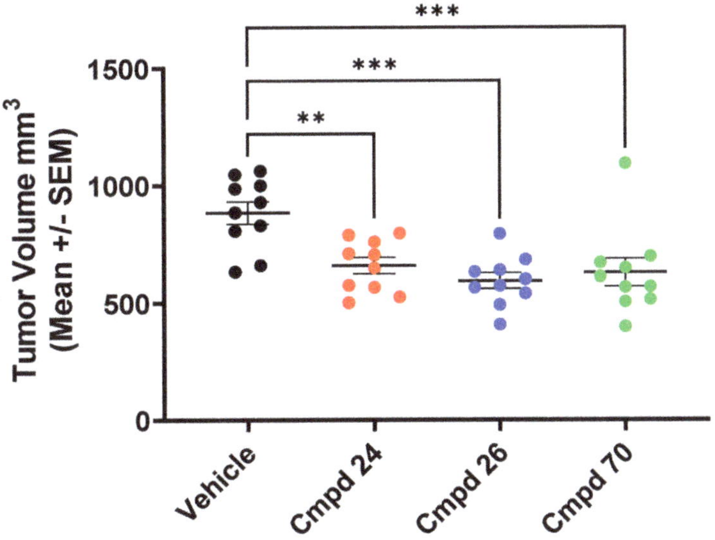

Figure 6. PERK inhibitors **24**, **26** and **70** inhibit tumor growth of 786-O RCC xenografts. Animals were treated at 30 mg/kg BID dosing for 29 days. Statistical analysis of group means by one-way ANOVA (** $p < 0.01$, *** $p < 0.001$).

In summary, we have developed a novel class of potent pyrrolopyrimidine PERK inhibitors with excellent ADME and PK/PD properties. Robust kinase selectivity was achieved with substitution on the pyrrolopyrimidine hinge binder region driving kinase selectivity, with a methyl substitution at the 2-position of the ring resulting in exquisite selectivity for PERK over other kinases. Further optimization of the 2-methyl substituted analogs led to identification of lead molecule **26** that retained excellent potency and selectivity and had good in vitro metabolic stability across all species tested. When **26** was evaluated in vivo, high AUC_{last} values and a flat, sustained plasma exposure profile were observed in mouse plasma and pancreas tissue. A single administration of **26** resulted in stable PERK inhibition across a 12 h period in mouse pancreas. These characteristics led to the selection of **26** for further development, and evaluation in animal models is ongoing.

Supplementary Materials: The following supporting information can be downloaded at: https://www.mdpi.com/article/10.3390/pharmaceutics14102233/s1, Supplemental Table S1: Crystallography data collection and refinement statistics; Supplemental Table S2:scanMAXSM Kinome Assay Results; Supplemental Table S3: Potency of Cmpd **26** against four ISR kinases and two FLT3 isoforms; Supplemental Figure S1: Compound **26** is selective against cell lines driven by FLT3-ITD; Supplemental Figure S2: In vivo PK in plasma and pancreas; Supplemental Figure S3: Compound **26** slows growth of 786-O RCC tumor xenografts; Supplemental Scheme S1: Synthesis of (*R*)-*N*-(4-(4-amino-7-methyl-7*H*-pyrrolo[2,3-*d*]pyrimidin-5-yl)-3-methylphenyl)-2-(3-fluorophenyl)-2-hydroxy acetamide (**11**) and (*S*)-*N*-(4-(4-amino-7-methyl-7*H*-pyrrolo[2,3-*d*]pyrimidin-5-yl)-3-methylphenyl) -2-(3-fluorophenyl) -2-hydroxyacetamide (**12**); Supplemental Scheme S2: Synthesis of (*S*)-*N*-(4-(4-amino-2,7-dimethyl-7*H*-pyrrolo[2,3-*d*]pyrimidin-5-yl)-2,3-difluoro phenyl)-2-(3-fluorophenyl)-2-hydroxyacetamide (**96**) and (*R*)-*N*-(4-(4-amino-2,7-dimethyl-7*H*-pyrrolo[2,3-*d*]pyrimidin-5-yl)-2,3-difluorophenyl)-2-(3-fluoro phenyl)-2-hydroxyacetamide (**70**); Supplemental Scheme S3: Synthesis of (*S*)-*N*-(4-(4-amino-2,7-dimethyl-7*H*-pyrrolo[2,3-*d*]pyrimidin-5-yl)-3-fluorophenyl)-2-hydroxy-2-(3-(trifluoromethyl)phenyl)acetamide (**106**) and (*R*)-*N*-(4-(4-amino-2,7-dimethyl-7*H*-pyrrolo[2,3-*d*]pyrimidin-5-yl)-3-fluorophenyl)-2-hydroxy-2-(3-(trifluoromethyl) phenyl)acetamide (**47**). References [7,17,20–25,27] are cited in the supplementary materials.

Author Contributions: Conceptualization, M.J.M., A.C.R., M.E.S. and M.D.S.; methodology, M.J.M., M.E.S., J.G., M.B., V.C., D.S., A.-H.L. and M.D.S.; formal analysis, M.D.S., M.E.S., D.S. and V.C.; investigation, M.E.S., V.C., G.Z., M.B. and J.G.; resources, A.C.R. and M.J.M.; data curation, G.Z. and M.D.S.; writing—original draft preparation, M.J.M., M.E.S., M.D.S. and V.C.; writing—review and editing, M.E.S., M.J.M. and M.D.S.; visualization, H.D. and G.Z.; supervision, A.C.R., M.J.M., M.E.S., J.G., M.B. and M.D.S.; funding acquisition, A.C.R. and M.J.M. All authors have read and agreed to the published version of the manuscript.

Funding: All laboratory work was funded by HiberCell Inc.

Institutional Review Board Statement: All the procedures related to animal handling, care, and the treatment in this study were performed according to guidelines approved by the Institutional Animal Care and Use Committee (IACUC) of Pharmaron following the guidance of the Association for Assessment and Accreditation of Laboratory Animal Care (AAALAC). All animal studies were approved and carried out in accordance with Pharmaron's Institutional Animal Care and Use Committee, protocol ON-TOL-09012021 approved on 26 March 2021.

Data Availability Statement: Coordinates and structural factors have been deposited in the PDB under codes 8EQ9 (Compound **11**), 8EQD (Compound **24**), and 8EQE (Compound **26**).

Acknowledgments: We acknowledge the support of Advanced Photon Source beam lines IMCA-CAT and GMCA-CAT. The use of the IMCA-CAT beamline 17-ID at APS was supported by the companies of the Industrial Macromolecular Crystallography Association through a contract with the Hauptman-Woodward Medical Research Institute. GM/CA@APS has been funded in whole or in part with Federal funds from the NCI, National Institutes of Health, Grant ACB-12002 and NIGMS, National Institutes of Health Grant AGM-12006. This research used resources of the Advanced Photon Source, a United States Department of Energy (DOE) Office of Science User Facility operated for the DOE Office of Science by Argonne National Laboratory under Contract DE-AC02-06CH11357. We also thank Arie Zask for reviewing the manuscript and sharing thoughtful editorial feedback. HiberCell would like to thank and acknowledge the chemists at Curia, Hyderabad: Srikanth Malibhatla, Madhavarao Bandaru, Prabhakara Reddy Goluguri, Satyanarayana Yatam, Ravi Adarasandi, Madhusudhan Velmala, Durga Rama Prasad Singireddi, Mahendar Velpuri, Bhaskar Reddy Nareddy, Visweswara Sastry, Chiranjeevi Mandati, Kumaraswamy Battula, Rajashekar Reddy Chittem, Sujatha Kanuganti, Shapi Siddiqui, and Subir Sadhukhan. The authors would also like to thank technical assistance by Elena Chad, Jennifer Wolfley, Richard Miller, Kirsten Feldman, Brent Weins, and Mary Koszelak-Rosenblum.

Conflicts of Interest: At the time of execution, M.E.S., V.C., D.S., A.-H.L., A.C.R. and M.J.M. were employed by HiberCell Inc. As such, HiberCell was involved in all aspects of this work, including conceptualization and methodology, investigation, data analysis, resources, and writing of this manuscript. Curia Global is a contract research organization engaged by HiberCell Inc to support chemical synthesis and characterization of novel molecules. At the time of experimentation, M.D.S., M.B., J.G. and G.Z. were employed by Curia Global, which was involved in conceptualization, methodology, resources, data analysis and manuscript composition. Pharmaron Co. Ltd. is a contract research organization engaged by HiberCell to support chemical characterization of novel compounds. At the time of experimentation, H.D. was employed by Pharmaron, which was involved in data visualization and manuscript composition.

References

1. Hetz, C.; Zhang, K.; Kaufman, R.J. Mechanisms, regulation and functions of the unfolded protein response. *Nat. Rev. Mol. Cell Biol.* **2020**, *21*, 421–438. [CrossRef] [PubMed]
2. Walter, P.; Ron, D. The Unfolded Protein Response: From Stress Pathway to Homeostatic Regulation. *Science* **2011**, *334*, 1081–1086. [CrossRef]
3. Hetz, C. The unfolded protein response: Controlling cell fate decisions under ER stress and beyond. *Nat. Rev. Mol. Cell Biol.* **2012**, *13*, 89–102. [CrossRef] [PubMed]
4. Mann, M.J.; Hendershot, L.M. UPR activation alters chemosensitivity of tumor cells. *Cancer Biol. Ther.* **2006**, *5*, 736–740. [CrossRef] [PubMed]
5. Ma, Y.; Hendershot, L.M. The role of the unfolded protein response in tumour development: Friend or foe? *Nat. Rev. Cancer* **2004**, *4*, 966–977. [CrossRef] [PubMed]
6. Avril, T.; Vauléon, E.; Chevet, E. Endoplasmic reticulum stress signaling and chemotherapy resistance in solid cancers. *Oncogenesis* **2017**, *6*, e373. [CrossRef] [PubMed]
7. Calvo, V.; Surguladze, D.; Li, A.-H.; Surman, M.D.; Malibhatla, S.; Bandaru, M.; Jonnalagadda, S.K.; Adarasandi, R.; Velmala, M.; Singireddi, D.R.P.; et al. Discovery of 2-amino-3-amido-5-aryl-pyridines as highly potent, orally bioavailable, and efficacious PERK kinase inhibitors. *Bioorg. Med. Chem. Lett.* **2021**, *43*, 128058. [CrossRef] [PubMed]
8. Atkins, C.; Liu, Q.; Minthorn, E.; Zhang, S.-Y.; Figueroa, D.J.; Moss, K.; Stanley, T.B.; Sanders, B.; Goetz, A.; Gaul, N.; et al. Characterization of a Novel PERK Kinase Inhibitor with Antitumor and Antiangiogenic Activity. *Cancer Res.* **2013**, *73*, 1993–2002. [CrossRef] [PubMed]
9. Oakes, S.A.; Papa, F.R. The Role of Endoplasmic Reticulum Stress in Human Pathology. *Annu. Rev. Pathol. Mech. Dis.* **2015**, *10*, 173–194. [CrossRef]
10. Sharma, R.B.; Landa-Galván, H.V.; Alonso, L.C. Living Dangerously: Protective and Harmful ER Stress Responses in Pancreatic β-Cells. *Diabetes* **2021**, *70*, 2431–2443. [CrossRef] [PubMed]
11. Brozzi, F.; Eizirik, D.L. ER stress and the decline and fall of pancreatic beta cells in type 1 diabetes. *Upsala J. Med. Sci.* **2016**, *121*, 133–139. [CrossRef] [PubMed]
12. Marré, M.L.; James, E.A.; Piganelli, J.D. β cell ER stress and the implications for immunogenicity in type 1 diabetes. *Front. Cell Dev. Biol.* **2015**, *3*, 67. [CrossRef] [PubMed]
13. Marhfour, I.; Lopez, X.M.; Lefkaditis, D.; Salmon, I.; Allagnat, F.; Richardson, S.J.; Morgan, N.G.; Eizirik, D.L. Expression of endoplasmic reticulum stress markers in the islets of patients with type 1 diabetes. *Diabetologia* **2012**, *55*, 2417–2420. [CrossRef] [PubMed]
14. Tersey, S.A.; Nishiki, Y.; Templin, A.T.; Cabrera, S.M.; Stull, N.D.; Colvin, S.C.; Evans-Molina, C.; Rickus, J.L.; Maier, B.; Mirmira, R.G. Islet β-Cell Endoplasmic Reticulum Stress Precedes the Onset of Type 1 Diabetes in the Nonobese Diabetic Mouse Model. *Diabetes* **2012**, *61*, 818–827. [CrossRef] [PubMed]
15. Zhang, P.; McGrath, B.; Li, S.; Frank, A.; Zambito, F.; Reinert, J.; Gannon, M.; Ma, K.; McNaughton, K.; Cavener, D.R. The PERK Eukaryotic Initiation Factor 2α Kinase Is Required for the Development of the Skeletal System, Postnatal Growth, and the Function and Viability of the Pancreas. *Mol. Cell. Biol.* **2002**, *22*, 3864–3874. [CrossRef] [PubMed]
16. Harding, H.P.; Zeng, H.; Zhang, Y.; Jungries, R.; Chung, P.; Plesken, H.; Sabatini, D.D.; Ron, D. Diabetes Mellitus and Exocrine Pancreatic Dysfunction in Perk−/− Mice Reveals a Role for Translational Control in Secretory Cell Survival. *Mol. Cell* **2001**, *7*, 1153–1163. [CrossRef]
17. Smith, A.L.; Andrews, K.L.; Beckmann, H.; Bellon, S.F.; Beltran, P.J.; Booker, S.; Chen, H.; Chung, Y.-A.; D'Angelo, N.D.; Dao, J.; et al. Discovery of 1*H*-Pyrazol-3(2*H*)-ones as Potent and Selective Inhibitors of Protein Kinase R-like Endoplasmic Reticulum Kinase (PERK). *J. Med. Chem.* **2015**, *58*, 1426–1441. [CrossRef] [PubMed]
18. Ayala-Aguilera, C.C.; Valero, T.; Lorente-Macías, Á; Baillache, D.J.; Croke, S.; Unciti-Broceta, A. Small Molecule Kinase Inhibitor Drugs (1995–2021): Medical Indication, Pharmacology, and Synthesis. *J. Med. Chem.* **2021**, *65*, 1047–1131. [CrossRef]
19. Xing, L.; Klug-Mcleod, J.; Rai, B.; Lunney, E.A. Kinase hinge binding scaffolds and their hydrogen bond patterns. *Bioorg. Med. Chem.* **2015**, *23*, 6520–6527. [CrossRef] [PubMed]

20. Winter, G.; Waterman, D.G.; Parkhurst, J.M.; Brewster, A.S.; Gildea, R.J.; Gerstel, M.; Fuentes-Montero, L.; Vollmar, M.; Michels-Clark, T.; Young, I.D.; et al. *DIALS*: Implementation and evaluation of a new integration package. *Acta Crystallogr. Sect. D Struct. Biol.* **2018**, *74 Pt 2*, 85–97. [CrossRef]
21. Evans, P.R.; Murshudov, G.N. How good are my data and what is the resolution? *Acta Crystallogr. Sect. D Biol. Crystallogr.* **2013**, *69 Pt 7*, 1204–1214. [CrossRef] [PubMed]
22. McCoy, A.J.; Grosse-Kunstleve, R.W.; Adams, P.D.; Winn, M.D.; Storoni, L.C.; Read, R.J. Phaser crystallographic software. *J. Appl. Crystallogr.* **2007**, *40 Pt 4*, 658–674. [CrossRef] [PubMed]
23. Emsley, P.; Lohkamp, B.; Scott, W.G.; Cowtan, K. Features and development of Coot. *Acta Crystallogr. Sect. D Biol. Crystallogr.* **2010**, *66 Pt 4*, 486–501. [CrossRef]
24. Murshudov, G.N.; Skubák, P.; Lebedev, A.A.; Pannu, N.S.; Steiner, R.A.; Nicholls, R.A.; Winn, M.D.; Long, F.; Vagin, A.A. REFMAC5 for the refinement of macromolecular crystal structures. *Acta Crystallogr. Sect. D Biol. Crystallogr.* **2011**, *67 Pt 4*, 355–367. [CrossRef]
25. McNicholas, S.; Potterton, E.; Wilson, K.S.; Noble, M.E.M. Presenting your structures: The*CCP*4mgmolecular-graphics software. *Acta Crystallogr. Sect. D Biol. Crystallogr.* **2011**, *67 Pt 4*, 386–394. [CrossRef] [PubMed]
26. Fabian, M.A.; Biggs, W.H., III; Treiber, D.K.; Atteridge, C.E.; Azimioara, M.D.; Benedetti, M.G.; Carter, T.A.; Ciceri, P.; Edeen, P.T.; Floyd, M.; et al. A small molecule–kinase interaction map for clinical kinase inhibitors. *Nat. Biotechnol.* **2005**, *23*, 329–336. [CrossRef]
27. Williams, C.J.; Headd, J.J.; Moriarty, N.W.; Prisant, M.G.; Videau, L.L.; Deis, L.N.; Verma, V.; Keedy, D.A.; Hintze, B.J.; Chen, V.B.; et al. MolProbity: More and better reference data for improved all-atom structure validation. *Protein Sci.* **2018**, *27*, 293–315. [CrossRef] [PubMed]

pharmaceutics

Review

Heterocyclic Compounds as Hsp90 Inhibitors: A Perspective on Anticancer Applications

Mina Ardestani [1], Zahra Khorsandi [1], Fariba Keshavarzipour [1], Siavash Iravani [2], Hojjat Sadeghi-Aliabadi [1,*] and Rajender S. Varma [3,*]

1 Pharmaceutical Sciences Research Center, School of Pharmacy and Pharmaceutical Sciences, Isfahan University of Medical Sciences, Isfahan 81746-73461, Iran
2 School of Pharmacy and Pharmaceutical Sciences, Isfahan University of Medical Sciences, Isfahan 81746-73461, Iran
3 Regional Centre of Advanced Technologies and Materials, Czech Advanced Technology and Research Institute, Palacký University in Olomouc, Šlechtitelů 27, 783 71 Olomouc, Czech Republic
* Correspondence: sadeghi@pharm.mui.ac.ir (H.S.-A.); varma.rajender@epa.gov (R.S.V.)

Abstract: Heat shock proteins (Hsps) have garnered special attention in cancer therapy as molecular chaperones with regulatory/mediatory effects on folding, maintenance/stability, maturation, and conformation of proteins as well as their effects on prevention of protein aggregation. Hsp90 ensures the stability of various client proteins needed for the growth of cells or the survival of tumor cells; therefore, they are overexpressed in tumor cells and play key roles in carcinogenesis. Accordingly, Hsp90 inhibitors are recognized as attractive therapeutic agents for investigations pertaining to tumor suppression. Natural Hsp90 inhibitors comprising geldanamycin (GM), reclaimed analogs of GM including 17-AAG and DMAG, and radicicol, a natural macrocyclic antifungal, are among the first potent Hsp90 inhibitors. Herein, recently synthesized heterocyclic compounds recognized as potent Hsp90 inhibitors are reviewed along with the anticancer effects of heterocyclic compounds, comprising purine, pyrazole, triazine, quinolines, coumarin, and isoxazoles molecules.

Keywords: Hsp90 inhibitor; co-chaperone; heterocycle molecules; anticancer agents; heat shock proteins

Citation: Ardestani, M.; Khorsandi, Z.; Keshavarzipour, F.; Iravani, S.; Sadeghi-Aliabadi, H.; Varma, R.S. Heterocyclic Compounds as Hsp90 Inhibitors: A Perspective on Anticancer Applications. *Pharmaceutics* **2022**, *14*, 2220. https://doi.org/10.3390/pharmaceutics14102220

Academic Editor: Ja-Hyoung Ryu

Received: 6 September 2022
Accepted: 17 October 2022
Published: 18 October 2022

1. Introduction

Cancer refers to a variety of diseases caused by escalating uninhibited cell proliferation, which may spread beyond tissue range. Cancer is initiated with the deformation of a normal cell caused by several hereditary factors, lifestyle, immune system defects, environmental/occupational factors, possible toxicity from medications, including the aging process, causing damage or genetic mutations [1–4]. The mechanisms of control, proliferation, and differentiation of cells are disrupted [5–9], which leads to deviation of normal cells from their regular growth path [10–12]. Tumor cells may acquire autonomy in two ways, namely activation of a growth-promoting oncogene or inactivation of a growth-inhibiting gene [13–15]. Cancer is one of the major death reasons which ranked after cardiovascular diseases; therefore, anticancer explorations have attracted much attention, especially through the evaluation of the deactivation of various proteins, such as tubulin, aromatase, and heat shock proteins [16–24].

Heat shock proteins (Hsps) inhibitors are one of the most eminent active anticancer agents with appropriate effects on several strange signaling paths in tumors; notably, these inhibitors can help to surmount several notorious problems regarding resistance cancers. Hsps act as proteins for vital cellular activities, including protein accumulation, secretion, and regulation of gene expression through direct correlation with transcription factors; the cellular expression is increased due to various stressors. Hsps are classified by their molecular weight into the Hsp110, Hsp90, Hsp70, Hsp60, Hsp40, and Hsp27; most of them being generally characterized as ATP-dependent [25–27]. Additionally, Hsp90 can regulate the activity and stability of different client proteins with a wide range of sizes and functions as these client proteins have critical roles in proliferation, survival, protein misfolding, aggregation, and apoptosis. There are four various homologs of Hsp90: cytosolic Hsp90 (including Hsp90α, Hsp90β), TRAP1 (tumor necrosis factor receptor-associated protein 1) in mitochondrial, GRP94 (94 kDa glucose-regulated protein) in the endoplasmic reticulum (ER), and Hsp90C in chloroplasts [27].

The crystal formation of Hsp90 was first defined in 1996 as a homodimer with a three-part monomer. The N-terminal domain (~35-kDa) is fabricated from layers of a/b sandwich structures formed in the pocket, acting as a binding site for adenine nucleotides. The requirement of ATP in Hsp90 is associated with auto-phosphorylation with the N-terminal folding pattern; the superfamily of ATPase has shown similar activity to Hsp90 by type II topoisomerases and MutL (Mutator L) [28,29].

A client protein binding site and nuclear localization signal, Hsp90 middle domain (~35 kDa), entails precise identification of client proteins and molecular regulator chaperones to activate the appropriate substrate, such as ATP hydrolysis [30]. The C-terminal domain (~12 kDa) is the site of dimerization close to the ATP binding site and a pentapeptide domain (Met-Glu-Glu-Val-Asp or MEEVD), as well as the binding site of co-chaperones of Hsp90 consisting of Sti15 and Hop [31]; the Hsp90 ATPase cycle is depicted in Figure 1.

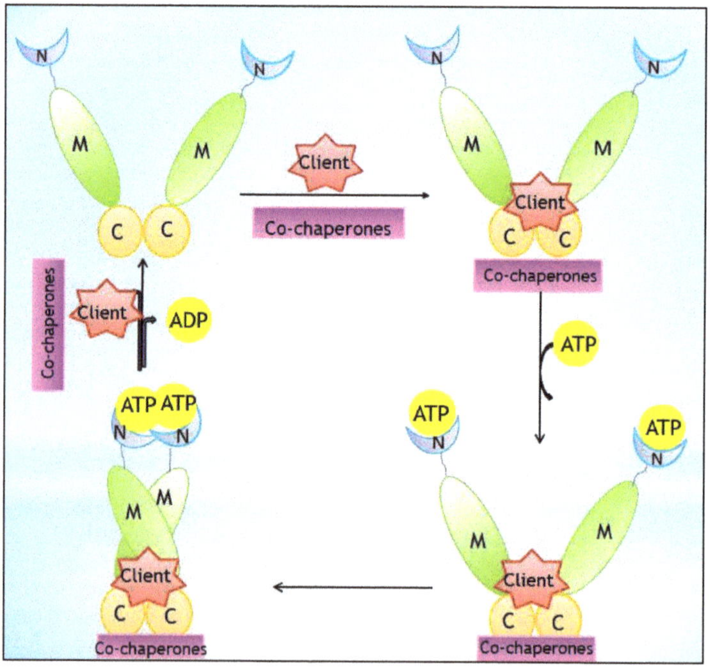

Figure 1. ATPase cycle in Hsp90.

Hsp90 consists of a chaperoning subsidiary with the assistance of co-chaperones and ATP. Initially, the client protein attaches to the M domain with the co-chaperones source; afterward, the ATP binds to the N domain to dimer the Hsp90, leading to the production of "closed form" protein. The client protein is reformed, while the required energy is supplied from the bond division process, and ATP is hydrolyzed to ADP and unfastened phosphate. The release of client proteins and ADP from the complex occurs after covering the chaperoning function; ATP binds to the N terminal via a standard chaperoning cycle and then the client protein binds to Hsp90 by assisting co-chaperones [32–34].

Due to the unique effectiveness of Hsp90 inhibitors in cancer therapy, researchers have focused on them in recent decades [35–47]. Hsp90 is responsible for the conformational maturation of 500 client protein substrates embracing transcription factors, receptors, kinases, or oncoproteins, which might be overexpressed and/or mutated in most cancers [48]. Consequently, the inhibition of Hsp90 is contemplated an attractive cancer-treating strategy due to its impacts on oncoprotein and pathways, concurrently [48,49]. Herein, recent advances in potential inhibitory effects of lately synthesized heterocyclic compounds, such as purine, pyrazole, triazine, quinolines, coumarin, and isoxazoles against Hsp90, are deliberated, focusing on their anticancer and antitumor applications.

2. Hsp90 Inhibitors

2.1. Natural Inhibitors

The first reported natural Hsp90 inhibitor has been a macrocyclic product called geldanamycin (**1**) (GM), extracted from the culture of *Streptomyces hygroscopicus* in 1970 which is known as an antibiotic compound (Scheme 1) [50]. Due to its unacceptable toxicological properties, the additional development of this compound was averted. However, the Hsp90 inhibitory activity of several geldanamycin analogs, namely, 17-dimethylaminoethylamino-17-demethoxygeldanamycin (17-DMAG), and 17-allylamino-17-demethoxygeldanamycin (17-AAG) are studied, but their clinical trials have been interrupted due to their poor solubility, complex formulation, and hepatotoxicity [51–56].

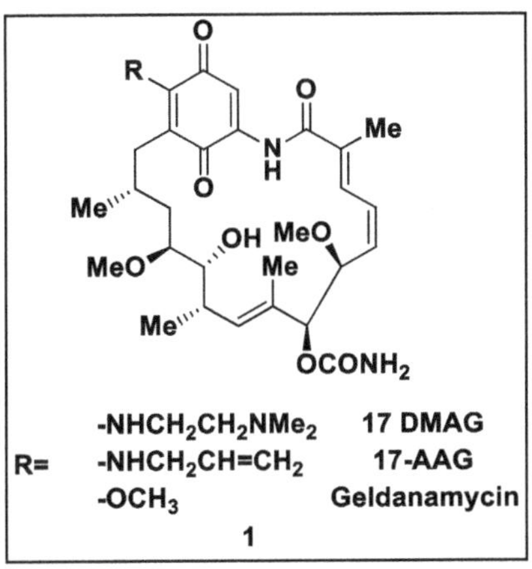

Scheme 1. Chemical structure of geldanamycin and its analogs.

One of the developed GM analogs is herbimycin A in which the methoxy substitution is located at positions C-11 and C-15, and the methoxy group is not present at position C-17; GM and herbimycin A (**2**) can induce Hsp70 expression in fibroblasts [57]. Macbecin I (**3**) as a GM analog has been developed with a similar structure to herbimycin A, except that the methoxy group at C-6 is replaced by a methyl group. Macbecin II (**4**) is a hydroquinone analog of macbecin I with less separation tendency due to its inferior structural instability [58,59]. The chemical structures of Herbimycin A and macbecin (I and II) are depicted in Scheme 2.

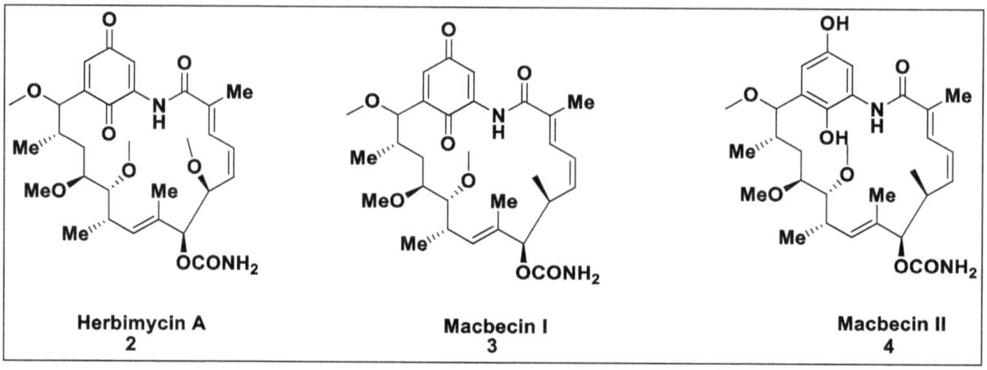

Scheme 2. Chemical structures of herbimycin A, macbecin I, and macbecin II.

IPI-504 (retaspimycin hydrochloride) (**5**) and IPI-493 (17-AG) (**6**) (Scheme 3) are more soluble analogs of GM and both of them inhibit Hsp90 activity through the degradation of client proteins. Several clinical trials (phase I and II) have been disclosed on different types of cancers including multiple myeloma, metastatic gastrointestinal stromal tumor (GIST), refractory non-small cell lung cancer (NSCLC), and chronic myelogenous leukemia (CML) [60]. IPI-493, equivalent to 17-AAG, obstructs the growth of SKBr3 breast cancerous cells, which may be derived from a similar major metabolite. IPI-493 exhibited highly effective results in GIST xenografts carrying heterogeneous KIT mutations in a preclinical investigation; these findings led to the start of its phase I clinical trial studies [61–63].

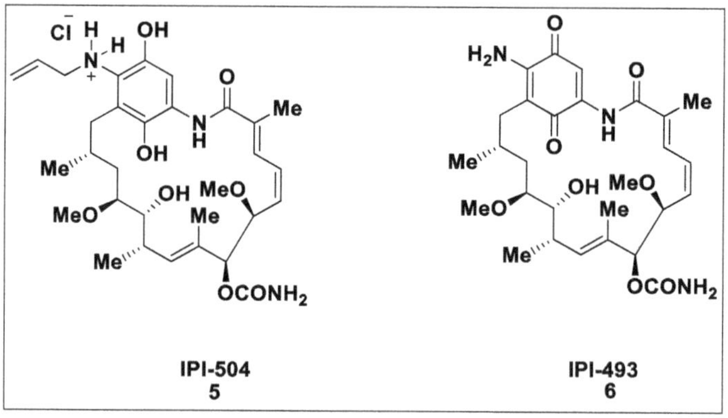

Scheme 3. Chemical structures of IPI-504 and IPI-493.

Benzoquinone ansamycins are also known as Hsp90 inhibitors that act as anti-malarial, antiviral, and anti-surra agents, and their efficacy in the treatment of cardiac arrest, stroke, and Alzheimer's has been explored [64]. One of the natural Hsp90 inhibitors is Radicicol (RD) (7), which has been isolated from Monosporium bonorden, it can link to the N-terminal ATP binding site. RD is a potential cell growth inhibitor, but its unstable metabolite in the body leads to inactivity; therefore, elaborative studies have been focused on the modification of its structures to obtain analogs with better stability [65].

KF25706 (**8**) and KF2711(**9**) have been introduced as more soluble analogs of RD with significant inhibition of the Hsp90 activity. As depicted in their chemical structure (Scheme 4), the carbonyl group in RD has been replaced with the oxime group, from the point of view of chemistry; this replacement can reduce its Michael acceptor electrophilicity thus improving the stability. However, their anticancer investigations have been limited to in vivo and animal model studies [65,66].

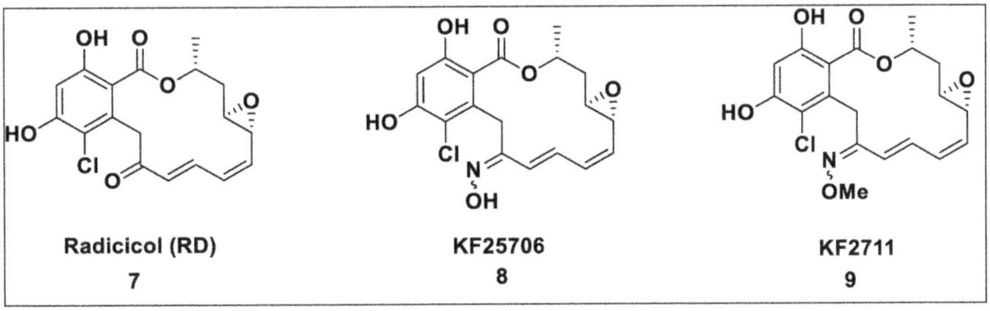

Scheme 4. Chemical structures of Radicicol, KF25706, and KF2711.

2.2. *Synthetic Hsp90 Inhibitors*

2.2.1. Purine-Based Structures

Naturally-derived Hsp90 inhibitors may suffer from some disadvantages, including low solubility and restricted activity. To overcome these limitations, the usage of synthesized Hsp90 inhibitors has garnered much attention; for instance, purine-scaffold-based compounds were reported as effective Hsp90 inhibitors with good solubility and an acceptable level of cell permeability [67]. Purine is an aromatic heterocycle compound consisting of two fused rings. Purine derivatives have exhibited significant pharmaceutical activities such as anticancer, anti-HIV-1, and antimicrobial properties [68–72]. Purine and pyrimidine-based entities are essential natural heterocyclic compounds that have a critical role in several metabolic and cellular conversion processes in deoxyguanosine monophosphate (DGMP) nucleotide (AMP). Moreover, the chemical structure of several biologically important molecules, such as ATP, GTP, cyclic AMP, NADH, and 3′-phosphoadenosine-5′-phosphosulfate (PAPS) contain fused purine ring. In chemical terms, purine is a nitrogen-rich heterocyclic compound comprising two rings of pyrimidine and imidazole, and because of the lack of natural sources, they ought to be prepared through synthetic organic reactions [73–75].

The activity of a purine-based molecule as a synthesized Hsp90 inhibitor (PU_3) has been investigated by assessing its interaction in Hsp90 K ADP/ADP binding site. As a theoretical outcome, this molecule covered all the necessary interactions with hosting protein; the Lys112 interacted with methoxy groups, and the hydrophobic pocket was occupied (Figure 2) [76]. The purine derivatives PU_3 and Pu24FCl (Scheme 5) have been introduced as potential small-molecule Hsp90 inhibitors through binding to the Hsp90 ADP/ATP site. The trimethoxy phenyl functional group of Pu24FCl makes it suitable to bind into the phosphate region of the host protein. Bao et al. reported a purine-based Hsp90 inhibitor with potential oral activity, named CUDC-305 (**12**), it is a unique compound due to its: ability to extremely privilege into the tissue of the brain, prolonged duration in intracranial tumors in animal models, and function in intracranial glioblastoma models; these attractive features may be raised from its fascinating high lipophilicity (clog P of 4.0) (Scheme 5) [77].

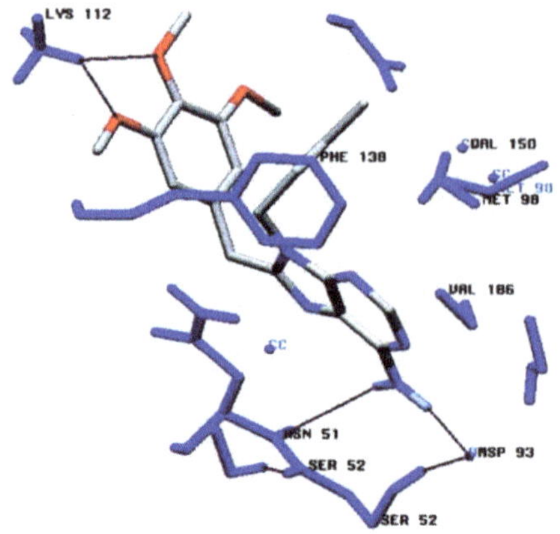

Figure 2. Molecular docking analysis of PU_3 in interaction with Hsp90.

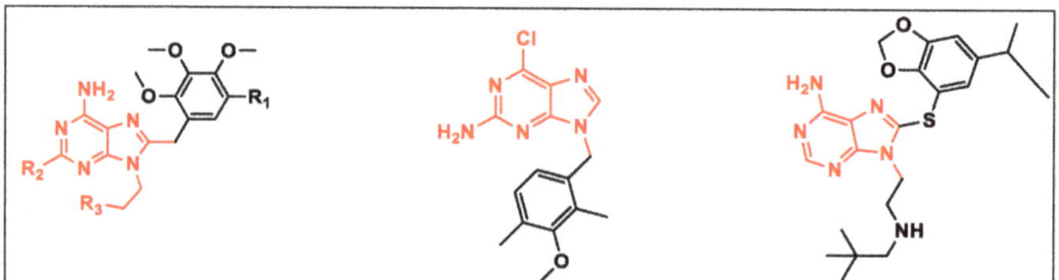

Scheme 5. Chemical structures of PU3, PU24FCl, BIIB021, and CUDC-305.

A family of 8-arylsulfonyl analogs of PU_3 (**10**) have been synthesized to evaluate the influence of the aryl part on inhibitory function and are being introduced as a purine-based Hsp90 inhibitor with selective activities which successfully enters clinical trials [78–81]. BIIB021 (**11**) has been the first fully synthetic Hsp90 inhibitor that moved in clinical trials endowed with its special properties that facilitated its formulation and bioavailability improvement; it binds to Hsp90 with high affinity and inhibits tumor growth. In phase I clinical trials, BIIB021 exhibited well-tolerated and good antitumor activity [82]. Therefore, this most developed purine-based compound, entered phase II clinical trials for GIST treatment. Pharmacokinetic parameters for BIIB021 600 mg, the mean C_{max} was 1.5 μmol and the mean AUC was 2.9 μmol h; a $C_{max} > 1.5$ μmol being associated with a decrease in standardized uptake value (SUV_{max}) [83].

2.2.2. Coumarin-Based Structures

Coumarin and its derivatives are important heterocyclic molecules endowed with various biological activities, such as platelet aggregation inhibition, antibacterial effects, and anticancer activity [84–88]. Novobiocin (**13**) is one of the first established organic compounds with a coumarin core that functioned as an Hsp90 inhibitor; it is a natural product with significant antibacterial DNA gyrase activity. However, its low efficiency in degrading Hsp90 clients (IC_{50} = ~700 μM) has discontinued more evaluations; therefore, research has mainly focused on structure-activity relationship (SAR) studies to identify other coumarin-based compounds with stronger inhibitor activity [89]. Modification of the 3-position of coumarin, from amide in novobiocin to urea, creates a new link to the hosting protein. The chemical structure of modified coumarin (**14**) is depicted in Scheme 6.

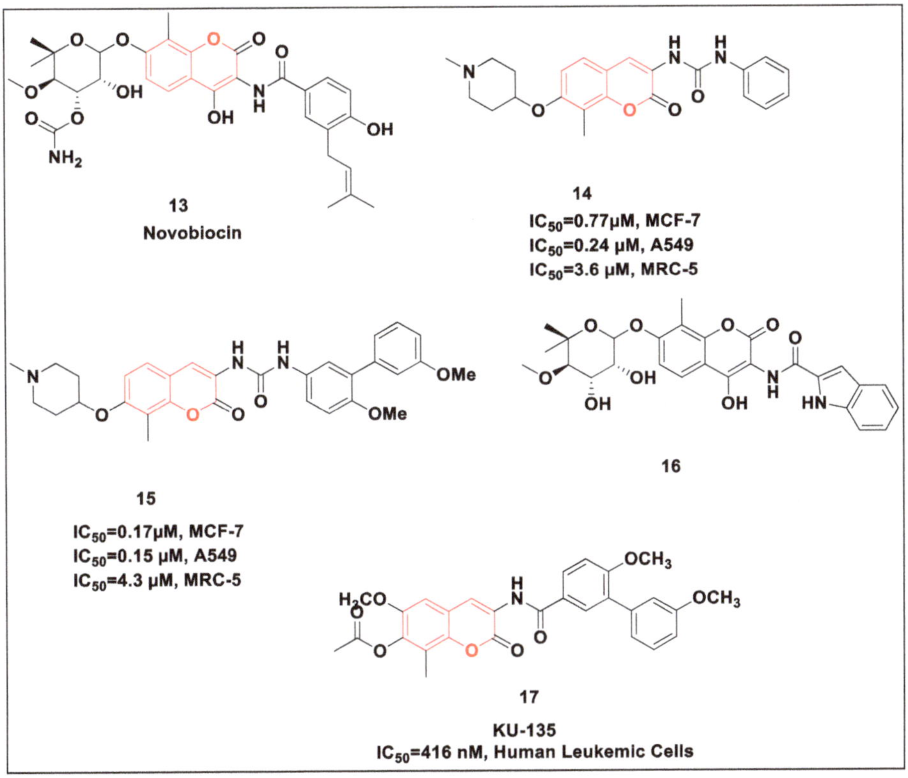

Scheme 6. Hsp90 inhibitors containing coumarin motif.

Among coumarin derivatives, compound **15** (Scheme 6) did not show any Hsp90 inhibitory activities, but additional investigations revealed its ability to interrupt MAPK signaling pathway by inhibiting the level of p-ERK and p-MEK; this function would be useful in anticancer activities [90]. Blagg et al. illustrated that the 4-hydroxyl and the 3′-carbamate functional groups of novobiocin have proven to be detrimental as their essential components for Hsp90 inhibitory activity by SAR studies. Therefore, indole moiety replacement in compound **16** significantly increased its activity (Scheme 6), more than 500 times that of novobiocin [91]. Furthermore, Shelton et al. synthesized KU135 as an Hsp90 inhibitor agent with anti-proliferative activity. The results showed that this compound can degrade Hsp90 client proteins through signaling pathways, a process that has stronger anti-proliferative effects than the N-terminal Hsp90 inhibitor 17-AAG. Notably, the explorations on this Hsp90 inhibitor demonstrated that this compound could inhibit G2/M cell cycle and have mitochondria-mediated apoptosis effects [92].

Garg et al, produced different analogs of ring-bound novobiocin (Scheme 7) wherein SAR and computational studies illustrated that when lactam was in α position, it produced more effective analogs than sugars. Activity of these derivatives was assessed as anti-proliferative agents against SKBr3 and MCF-7 cell lines. Among these derivatives, the cyclohexylamine analog demonstrated the best inhibitory effect with IC_{50} compared to bicycloalkyl and tricyclic amino analogs (0.35 μM against MCF-7 and 0.2 μM against SKBr3 cells) [93].

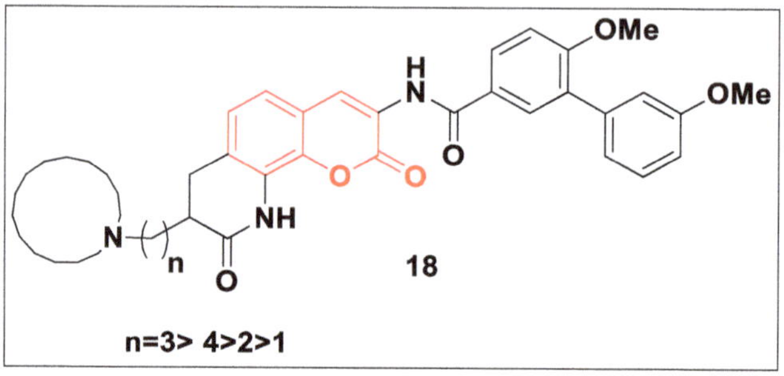

Scheme 7. Chemical structure of ring-bound novobiocin analogs synthesized and evaluated by Garg et al. [93].

Wei et al. reported some assorted coumarin compounds comprising the pyrazoline functional group, (compound **19** (a–f), Scheme 8) and evaluated their anticancer activity through several biological assays. Based on docking study results, all of these compounds are located in the active site of the N-terminus of Hsp90. Among these six derivatives, the structure of 19a exhibited higher binding energy and Hsp90 inhibitory function (IC_{50} = ~4.7 µM). All these derivatives reduced the viability of A549 lung cancerous cells, without any necrosis induction on them as they stimulated the apoptosis with blocking effects on the autophagic flux of HCP1 in A549 lung cancerous cells [94].

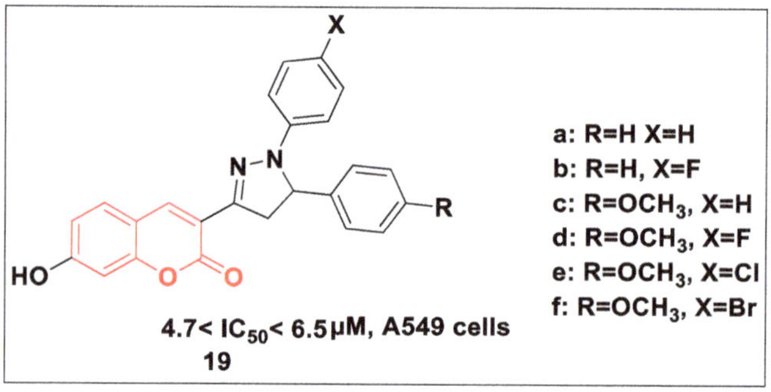

Scheme 8. Chemical structures of pyrazoline containing coumarin compounds, reported by Wei et al. [94].

2.2.3. Quinolone-Based Structures

Quinolines are one of the most important nitrogenous heterocyclic compounds that have been extensively examined due to their widespread pharmacology appliances, such as anti-malarial, antitumor, anti-parasitic, antibacterial, anti-asthma, antidiabetic, anti-inflammatory, antiplatelet, and antihypertensive activities [95–102]. Streptonigrin (isolated from *Streptomyces flocculus*) and lavendamycin have already been known as antimicrobial and antitumor compounds with quinoline skeleton (Scheme 9); they create efficient interactions with targets to act as a cancer chemotherapy agent and as Hsp90 inhibitors [103–105].

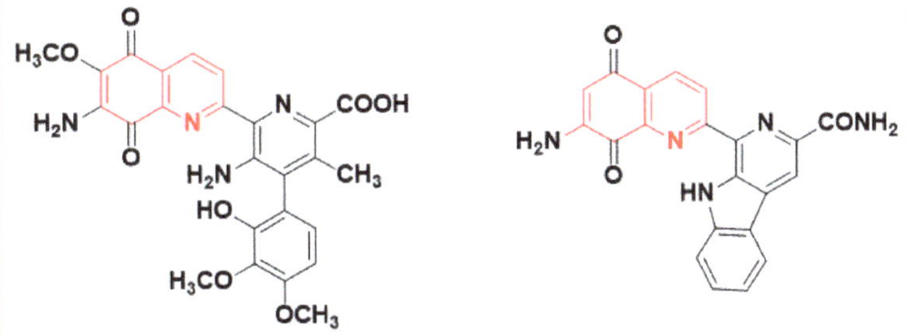

Scheme 9. Chemical structures of quinolones-based HSP90 inhibitors: Streptonigrin and Lavendamycin.

The chemical structures of synthesized quinolones-core organic compounds with Hsp90 inhibitory potentials properties have been investigated; their chemical structure is depicted in Scheme 10, named 22 to 32. Ganesh and co-workers reported the modest Hsp90 inhibitor activity of compound **22** (Scheme 10). Several quinoline-based organic compounds were synthesized and their activities were evaluated in micromolar concentrations using cell-based Western blot (WB) and fluorescent polarization (FP) techniques [106].

Studies of SAR, optimization of structures, and re-synthesis of several aminoquinoline compounds indicated their low Hsp90 inhibitor activity. However, the synthesized compound **23** (Scheme 10) exhibited high activity, with IC_{50} of ~0.73 μM and 1 μM in the low micro-molar range obtained in FP and WB assays, respectively. These compounds, with their simple chemical structures and facile synthesis pathways, are recognized as a series of new Hsp90 inhibitors [106]. Audisio and co-workers synthesized a novel array of 3-(*N*-substituted) aminoquinolin-2(1H)-one derivative and evaluated their anticancer activity using cell proliferation and flow cytometry, including biological assays.

Among these synthetic derivatives, compounds **24** and **25** (Scheme 10) offered the most effective inhibitory activity against various genes, such as Raf-1, HER2, CDK4, and estrogen receptors. It was indicated that compound **24** could stimulate apoptosis in MCF-7 breast cancer cells by activating caspases and subsequent division of poly (ADP-ribose) polymerase (PARP) and inhibiting the growth of all tumor cell-independent cell lines with growth inhibition of 50% (GI_{50}) values in the range of 2 to 32 μM. The examined cell lines included MCF-7, T47D, IRGOV-1, Ishikawa, HT-29, Caco-2, and MDA-MB-231. In addition to these properties, only compound **24** was identified as mediated cell death inductor in a p23-independent procedure; the p23 was a small and important co-chaperone for the Hsp90 chaperoning pathway [107].

A series of 2-aroylquinoline-5,8-diones have been synthesized and their potential biological activities have been evaluated by Nepali and coworkers [108]. Among these Hsp90 inhibitors, compounds **26** and **27** could inhibit the growth of cancerous cells (IC_{50} = ~0.14 (compound **26**) and 0.27 μM (compound **27**). Moreover, compound **27** displayed an IC_{50} of ~5.9 μM to inhibit tubulin polymerization as it persuaded the degradation of Hsp70 and Akt protein through WB analysis. Different quinoline analogs have been reported as Hsp90 inhibitors with a cytotoxic function against cancerous cell lines. Accordingly, cytotoxic derivatives encompassing alcohol functional groups exhibited significant activity against MCF-7 cells. Among these evaluated compounds with anti-proliferative activity, compound **28** distinguished itself as the most effective analog in the degradation of Her2 protein, a client protein of Hsp90. The possible state of interaction between compound **28** and the N-terminal ATP binding pocket of Hsp90 was demonstrated by molecular modeling studies [109].

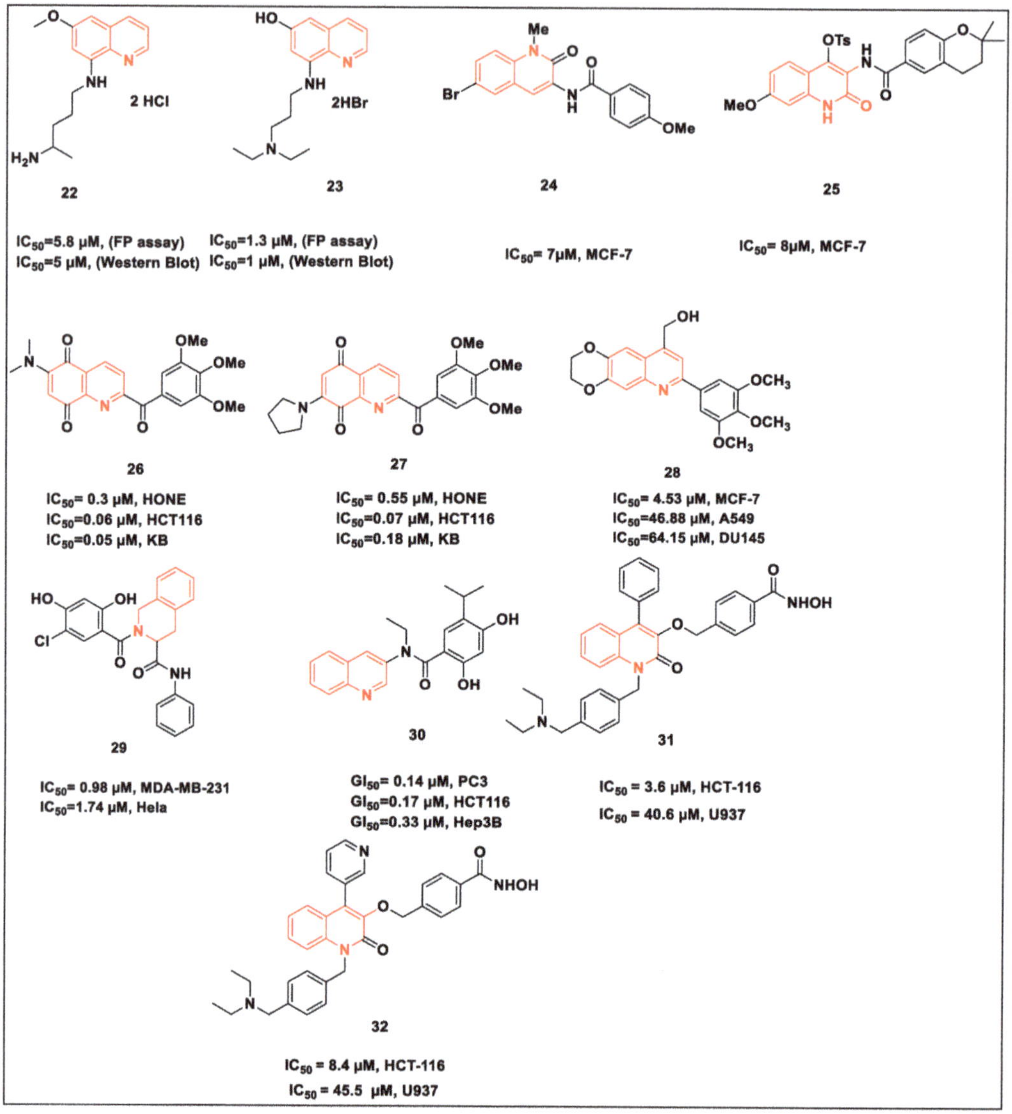

Scheme 10. Hsp90 inhibitors containing quinoline motif.

Liang et al. synthesized two new series of compounds of *N*-(5-chloro-2,4-dihydroxybenzoyl)-1,2,3,4-tetrahydroisoquinoline-3-carboxamides as Hsp90 inhibitors, which illustrated acceptable anti-proliferative activities against MDA-MB-231 and HeLa cell lines. Compound **29** unveiled strong cytotoxicity with inhibitory effects against the molecular proliferation of MDA-MB-231 (IC_{50} = ~0.98 μM) and HeLa (IC_{50} = ~1.74 μM). Moreover, the effective intracellular interaction of compound **29** with Hsp90α in 293T cells was confirmed through isothermal dose-response fingerprint curves. The induction activity of compound **29** in the degradation of CDK4, Her2, Cdc-2, and C-RAF Hsp90 client proteins was assessed by WB evaluation on MDA-MB-231 breast cancer cell lines [110].

Molecular docking and dynamic (MD) analyses on the complex of compound **29** and Hsp90 displayed its effective binding to Hsp90 through the interaction of its benzyl amino moiety with the residue Phe138, leading to the formation of a Π-stacking interaction [110]. Nepali et al. reported different fused quinoline-resorcinol compounds with powerful inhibitory activity against Hsp90. Through MD analysis of synthesized compounds in interactions with the Hsp90 chaperone protein receptor (Figure 3), compound **30** was determined as the ideal candidate. It interacted with the amino acid residues of the Hsp90 chaperone protein, resorcinol ring of compound **30**, 1,3-dihydroxybenzene, link to chaperone function through amide bond formation with 1,3-dihydroxybenzene. In vitro studies indicated the effective cell growth inhibitory effect of compound **30**, as one of the most active entities through in silico studies, against HCT-116 (colon), Hep3B (liver), and PC-3 (bone metastasis) cell lines [111].

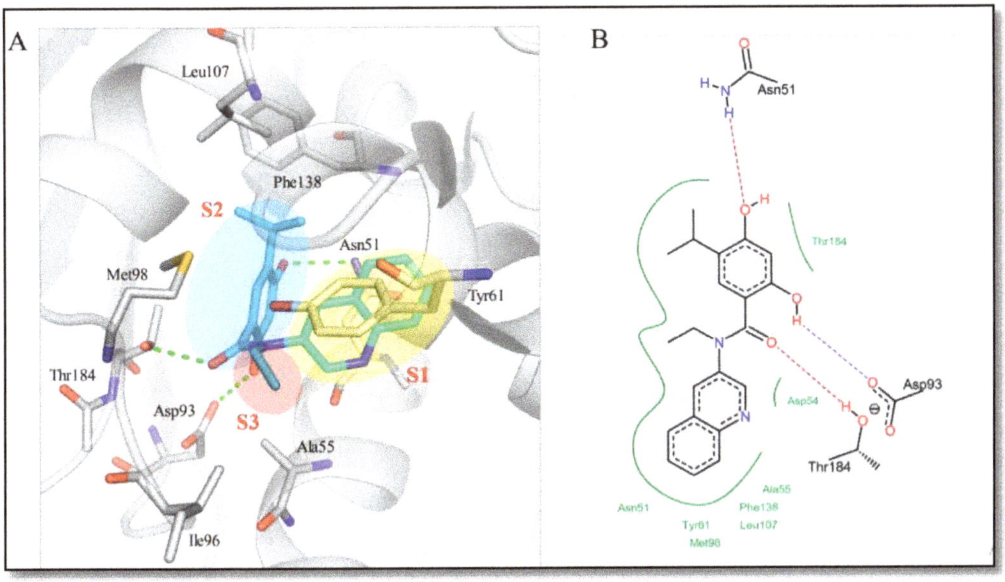

Figure 3. MD analysis of compound **30** in interaction with Hsp90. (**A**) compound **30** (blue) is anchored within the Hsp90 (gray) binding site. The three distinct sections of MPTG0G256 that are located in the S1, S2, and S3 sites, are colored as yellow, blue, and red, respectively. Interacting residues are shown as sticks and labeled as indicated. Hydrogen bonds are indicated by dotted green lines. (**B**) 2D representation of compound **30** docked in Hsp90. Green lines show regions of hydrophobic interactions. Interacting residues are labeled as indicated. Adapted from Ref. [111] with permission. Copyright 2019 Elsevier.

Relitti and co-workers synthesized various quinolone-based organic compounds as histone deacetylase 6 (HDAC6), wherein compounds **31** and **32** were introduced as the most promising active molecules against HDAC6. In addition, they displayed potent activity in cellular studies with development of inhibition against human cancer cell lines, HCT-116, and histiocytic lymphoma (U9347). It was an effective compound against tumor cells via apoptosis induction [112].

2.2.4. Triazine and Triazolothione-Based Structures

Triazine is an organic heterocyclic compound with a 6-membered ring containing three carbon atoms and three adjacent nitrogen atoms. Pharmacology studies have revealed that triazine derivatives have significant activity as antimicrobial, antituberculosis, anticancer, antiviral, and antimalarial agent [113–117]. BX-2919 was discovered by Feldman et al.

as an Hsp90 inhibitor during HTS, through a high-throughput screening assay that was performed to identify Hsp90 inhibitors that compete with GA binding. The resorcinol **33a** analog (Scheme 11) was identified as one of Hsp90 inhibitors optimized by applying parallel paths on its aryl rings. After assessing the strength of a large number of aryl compounds, ethyl carbamate 33c was found as one of the best candidates. Among the resorcinol ring alternatives, the ethyl analog (**33d**) increased relative strength; finally, the isopropyl analog (**33e**) was the optimal alternative in this situation. A combination of optimal alternatives with the BX-2819 was examined. This compound could bind strongly to Hsp90 (IC_{50} = ~41 nM) to inhibit GM-Bodipy binding, which was less than both above-mentioned compounds 17AAG (IC_{50} = ~350 nM) or radicicol (IC_{50} = ~87 nM) [118].

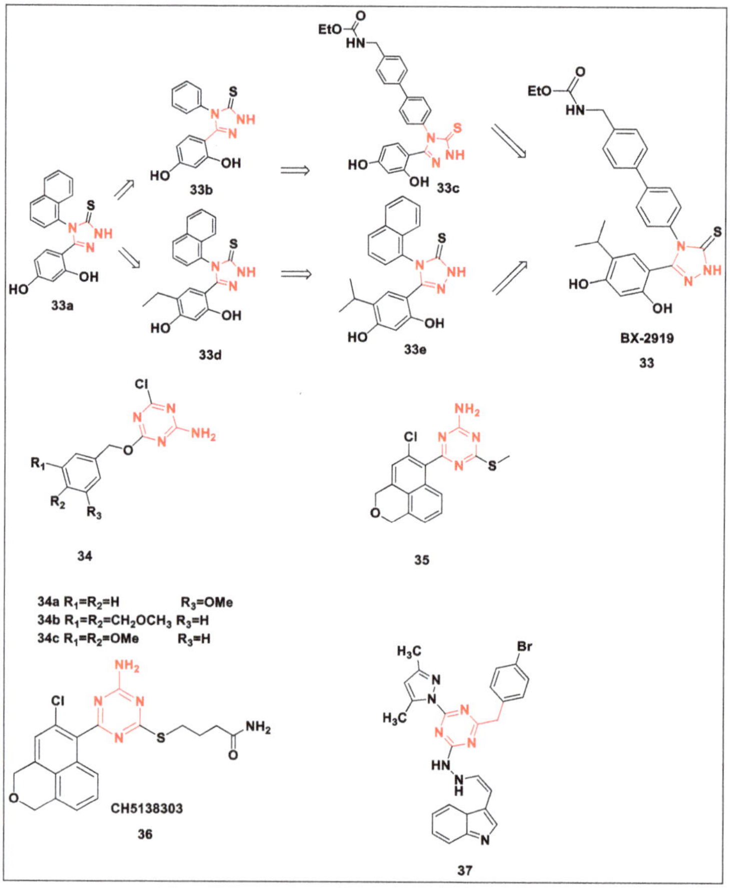

Scheme 11. Hsp90 inhibitors containing triazine and triazolothione motif.

Seo et al. reported a set of 2-amino-1,3,5 triazines containing a tricyclic part as Hsp90 inhibitors with high activity against gefitinib-resistant H1975 cells (Scheme 11) [119]. Several derivatives of compound **34** (Scheme 11) were synthesized, most of which entailed cell multiplication in a highly dose-dependent behavior with regularity effects on cell proliferation. Among them, compound **34a** demonstrated the highest activity as an inhibitor of cell proliferation and the product containing 2,6 dimethyl phenyl was able to reach the hydrophobic part of Hsp90 by van der Waals interactions, which led to anti-proliferative activity against H1975. Miura et al. synthesized various tricyclic molecules, including 2-amino-1 and 3,5-triazines as Hsp90 inhibitors with suitable anti-proliferative function against HCT-116 (IC_{50} = ~0.46 mM). Among these synthesized compounds, hybrid CH5015765 (35, Scheme 11) significantly improved the binding affinity (Kd = 3.4 nm) [120].

The structural properties are involved in the possible interactions with Hsp90 at the active region, thus the fabricated tricyclic molecules should be precisely analyzed. Consequently, it was indicated that the hydrogen bonds can be formed between the amino group on the two positions of the triazine core and the carboxylic group of Asp93. A hydrophobic interaction can be formed between the methyl thio group at the 4-position and Ile96/Met98. Additionally, a hydrophobic portion can interact with the side chains of Leu107, Phe138, and Val150. Finally, the ether oxygen may bind to the remnant Phe138 and Asn51 [121]. Compound **36** (CH5138303) showed potent inhibitory activity against HCT-116, in vitro. Its efficacy and chemical properties led to reducing the phosphorylation and protein level of several Hsp90 client proteins. By further improving the chemical structure of tricyclic molecules, a category of triazines derivatives could be produced via the functionalization of the sulfur atom in compound **36** with hydrogen bond acceptor functional groups. They are primed to interact with the Lys58, which concerned in several hydrogen bonds with geldanamycin as a natural and specific inhibitor of Hsp90. In general, the derivative of compound **36** exhibited significant antiproliferative activity, in vitro and in vivo (Kd = 0.48 nM) [122].

The antitumor activity of compound **37** as an Hsp90 inhibitor agent was reported by Zhao et al. [123]. Their interaction with Hsp90 was studied by surface plasmon resonance (SPR) assay; consequently, the results indicated the binding of compound **37** into the N-terminal of the protein with a very extraordinary binding mode with respect to the old inhibitors of Hsp90. Its anti-proliferative function was evaluated in vitro and in vivo. Accordingly, the results obtained from in vitro assessments demonstrated that these compounds could inhibit the molecular proliferation of BT-474 (IC_{50} = ~8.9 mM), SK-BR-3 (IC_{50} = ~7.1 mM), A549 (IC_{50} = ~7.5 mM), K562 (IC_{50} = ~8.6 mM), and HCT-116 (IC_{50} = ~6.7 mM). After in vivo assessments, it was revealed that these compounds could obstruct the increase in proliferation of tumors that increased with a dose-dependent behavior in the BT-474 [123].

2.2.5. Isoxazole-Based Structures

One of the prominent heterocyclic organic compounds comprises the isoxazole ring, a 5-membered ring, with adjacent oxygen and nitrogen atoms. Various chemical structures/compounds encompassing the isoxazole moiety have shown attractive biomedical and pharmaceutical potentials, especially for cancer therapy [124–130]. Medicinal chemistry studies have revealed that the inclusion of isoxazole in the chemical structure of Hsp90 inhibitors could lead to improved efficacy, decreased toxicity, and enhanced pharmacokinetics profiles. Aromatase is an enzyme for the conversion of androgenic hormone into estrogen; thus, considering the high-level expression of aromatase in breast tissue, intense generation of estrogen can cause breast cancer. Isoxazole derivatives could stop the conversion of androgen into estrogen by inhibiting the aromatase enzyme, thus serving as anticancer drugs [131].

Apoptosis is an extremely regulated process enabling cells to destroy themselves and kill unwanted cells; this has a crucial role in homeostasis development. Isoxazoles as an inducer of apoptosis causes the elimination of or discontinues the progression of cancer [132]. Enzymes of protein tyrosine phosphatases are involved in mechanisms of cellular signaling by regulating the levels of phosphorylation in a clear tyrosine residue in proteins or peptides, controlling various functions of cells, including metabolism, migration, survival, adhesion, and multiplication. Mutations in proteins can cause cancers. In this context, isoxazole derivatives can attach to the ATP-binding region of the protein kinase and hinder the signaling cascade, resulting in cell cycle arrest and apoptosis [133].

Many preclinical studies have indicated that Hsp90 inhibition is associated with antineoplastic effects; these properties motivated researchers to find novel small-molecule containing isoxazoles as Hsp90 inhibitors. Their capability to bind with the NH_2-terminal nucleotide-binding region of human Hsp90 was found extremely strong along with inhibitory effects on the multiplication of growth cells through inducing the arrest of G1-G2 and apoptosis. Pharmacological compounds containing isoxazole are active against different examined cell lines of tumors, especially breast cancer, which stimulated numerous efforts to design and synthesize innovative isoxazole derivatives [134].

Eccles et al. synthesized organic compounds containing isoxazole, such as VER-50589 (Scheme 12). Their Hsp90 inhibitory activity was assessed through in vivo studies, wherein these small synthetic molecules exhibited strong inhibitory activity. The results were found comparable with the activity of the clinically approved drug 17-AAG, an analog of geldanamycin; additionally, their lipophilicity renders them capable of easily penetrating the cell membrane. The encouraging outcomes illustrated the value of a structure-based design and optimization process on the way to achieve a potential clinical candidate. Prodigious findings motivated more research on the optimization of their potency, pharmacokinetic, and pharmacodynamic properties via structural modifications, which led to the discovery of VER52296 (NVPAUY922) (Scheme 12). The anticancer activity of VER52296 was investigated against cell lines, namely, HCT-116, Hun7, and SW620, by applying 3-(4,5-dimethylthiazol-2-yl)-2,5-diphenyltetrazolium bromide tetrazolium (MTT) assay. The obtained results demonstrated high cytotoxic activity with IC_{50} values lower than 20 μM (IC_{50} = ~21 nM and Kd = 1.7 nM), besides the low glucuronidation metabolite levels. Thus, VER52296 was introduced as a potential anticancer drug, and it is under evaluation in a second phase clinical trial [135].

A set of Hsp90 inhibitors comprising *N*-(isoxazol-5-yl) amides structures presented unique pharmacokinetic characteristics as well as high antitumor activities (both in vitro and in vivo). Compounds **39a**, **39b**, and **39c** exhibited the GI_{50} lower than 0.1 μM; thus, studies have been performed on their activity against the various cell lines. They displayed potential cytotoxicity with common GI_{50} which indicated their extensive antitumor activities [135]. Chen et al. reported various small molecules comprising isoxazole rings and identified them by X-ray crystallographic techniques. Their Hsp90 inhibitory effects were evaluated, and compound **40** exhibited potent inhibitory effects on Hsp90 at cellular/molecular levels; moreover, in vivo assessments showed its remarkable capability for inhibiting the growth of tumors. The ratio between treated and control (T/C) value of compound **41** was equal to 18.35% at 50 mg/kg, which was found to be nearly twice as strong as NVP-AUY 922 as a traditional Hsp90 (T/C values at 50 mg/kg reach 34.06%) [136].

Sun et al. reported a set of isoxazole scaffolds associated with alkynes to study their inhibitory activity on the human Hsp90 protein as well as their anti-proliferative function against the examined cell lines of tumors. During conservation, it was found that the substitution of alkynes at C-4, in compound **42**, could induce nice cation–π bond interaction with the Lys58 residue of the Hsp90 protein, along with the interaction of the resorcinol hydroxyl and C-3 amide groups. Based on the results, a set of 4-alkynylisoxazole structures were synthesized with 3,5-substitutes and their binding affinity to Hsp90 protein was also indicated. In addition, their anti-proliferative effects were evaluated against five cancerous cell lines, such as A549, human basal alveolar epithelial cell adenocarcinoma, K562, human immortal myeloma leukemia line, MCF-7, and the cell line of most breast cancers. Promising results were obtained, wherein compound **42** demonstrated an IC_{50} value of ~0.066 μM [137].

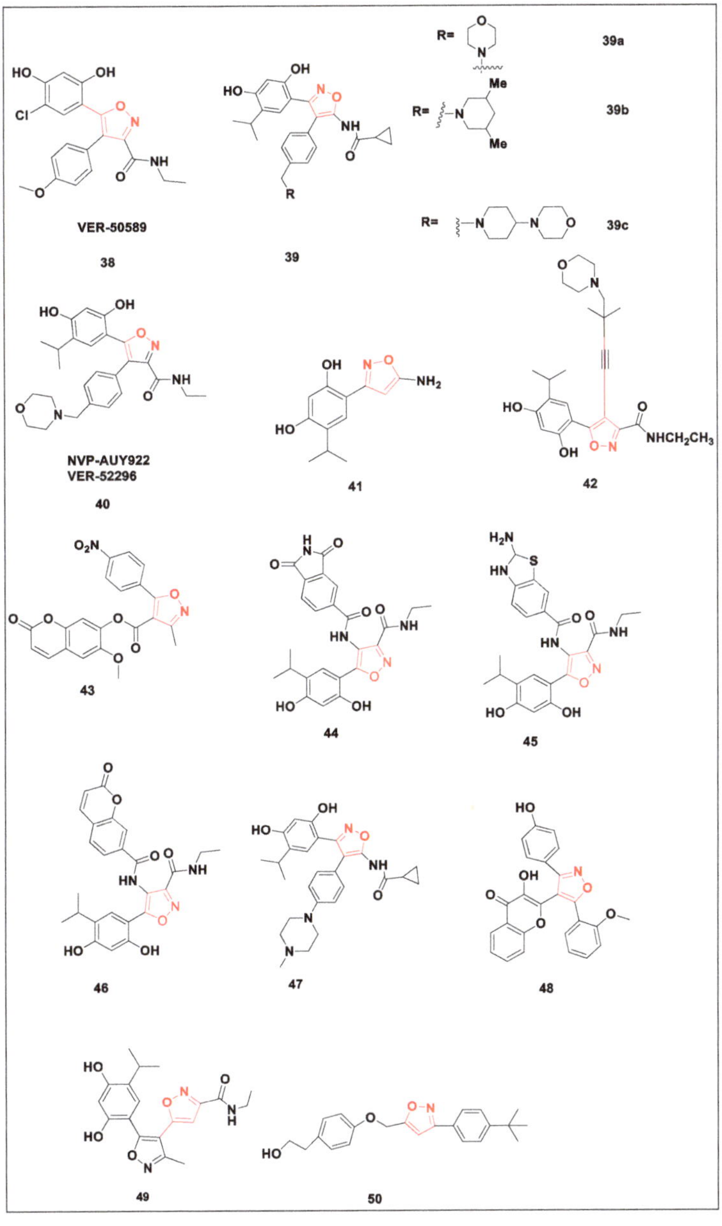

Scheme 12. Hsp90 inhibitors containing isoxazole motif.

Shi et al. synthesized a set of scopoletin-isoxazole and scopoletin-pyrazole hybrids and their anticancer effects were evaluated against SW620, Hun7, and HCT-116. Scopoletin-isoxazole derivatives illustrated higher cytotoxic activity in comparison with scopoletin-pyrazole hybrids ($IC_{50} \leq 20$ μM). Impressively, compound **43** showed anti-proliferative effects similar to sunitinib, the FAD-approved drug, with IC_{50} = ~8.76–9.83 μM against SW620, Hun7, and HCT-116, and it disclosed low toxicity on normal HFL-1 cells (IC_{50} = ~90.9 μM) [138]. Sadeghi et al. studied the potential Hsp90 inhibitor activity of 50 organic molecules, including 3,4-isoxazolediamides functional groups; they discovered that three compounds, **44**, **45**, and **46**, could be considered as promising target molecules for further evaluations [139]. They also introduced compound **47** as an efficient Hsp90 inhibitor after the assessment of several isoxazoles-based compounds [140]. Additionally, several chromone-linked isoxazole compounds were synthesized, and their ER-α and their anti-proliferative activity against MCF-7 were evaluated. The replacement of isoxazole's positions of 3 and 5 with phenyl ring and 3-hydroxychromone increased their anticancer activity. Compound **48** displayed cytotoxicity against ERα-positive (MCF-7) with IC_{50} of 32 $\mu g\ mL^{-1}$ by linking with a key residue of Glu353 and Arg394. These residues were stabilized by forming H-bonds of 4-hydroxyl phenyl in 4HT (4-hydroxytamoxifen) through molecular docking studies investigations [141].

Jung et al. reported a variety of isoxazole-containing compounds as Hsp90 inhibitors. Their capability of client protein degradation as well as ATPase performance was analyzed through in vitro cytotoxic assessments against cancerous cell lines. The targeted compounds were assessed for their possible antitumor effects in tumor xenograft models in vivo. Molecular modeling studies demonstrated the binding mode inside the inhibitors and N-terminal ATP binding pocket. To evaluate the antitumor effects of isoxazole-containing compounds, their cytotoxicity was analyzed against A2780 (ovarian cancer) and HCT-116 (colon cancer). Consequently, the IC_{50} of compound **49** for ATPase, Her2, A2780, and HCT-116 were found to be 0.284, 0.028, 0.043, and 0.014 μg, respectively [142]. Aissa and co-workers reported the synthesis of 3, 5-disubstituted isoxazoles and 1,4-disubstituted triazoles and evaluated their possible anticancer/hemolytic activities. The anticancer derivatives of isoxazole exhibited an active apoptotic trend in the glioma U251-MG and T98G. These derivatives illustrated significant anti-proliferative function against most human glioblastoma cancerous cell lines (U87) in a dose-dependent manner wherein compound **50** (IC_{50} 15.2 $\pm$ 1.0 $\mu g\ mL^{-1}$) exhibited stronger anticancer activity. Notably, the hemolytic activity of compound **53** caused hemolysis up to 40% at a concentration of 400 $\mu g\ mL^{-1}$. These results revealed that isoxazole derivatives did not damage red blood cells and release hemoglobin, which strengthened the potential of these compounds as innovatively synthesized anticancer prototypes with low toxicity [143].

2.2.6. Pyrazole-Based Structure

Pyrazole is a 5-membered ring heterocycle compound consisting of three carbon atoms and two adjacent nitrogen atoms. Pyrazole derivatives have significant activities, such as high efficacy as partial agonists in G protein-coupled and cannabinoid receptors [144–147]. Synthetic small pyrazole-fused heterocycle molecules, PF-04929113 and PF-04929113, were developed as inhibitors of Hsp90, but their restrictions/disadvantages, such as poor solubility and cytotoxicity prompted a search for the structural improvement of pyrazole-based compounds (Scheme 13) [148]. Bai et al. synthesized various fluorescent pyrazoline coumarin derivatives and investigated their anticancer activity against lung cancer cell lines in vitro. As a result, compound **52** had the strongest growth inhibition (IC50 = ~7.9 mM), with a strong fluorescence strip; subsequently, it was introduced as a potential and promising fluorescent Hsp90 inhibitor [149].

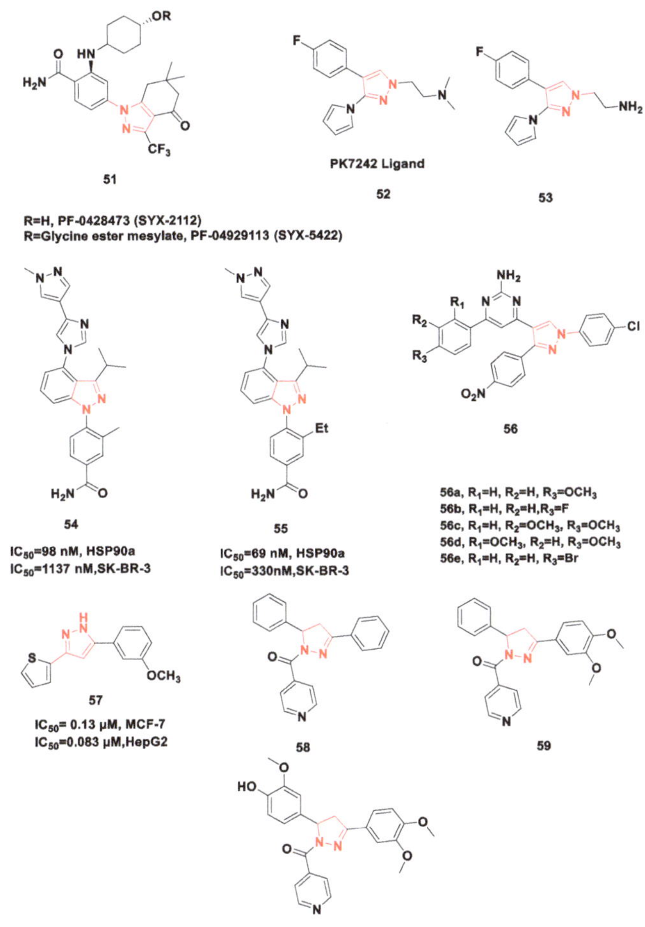

Scheme 13. Hsp90 inhibitors containing pyrazole motif.

Sadeghi et al. predicted dual agents as mutant P53 activators and Hsp90 inhibitors by using docking and MD analysis; according to their published results, compound **53** was proposed as a dual-mode agent, P53 activator, and Hsp90 inhibitor [150]. Uno et al. reported various pyrazolo [3,4-b]pyridine derivatives as potential Hsp90α and Hsp90β inhibitory agents with oral availability properties. Compound **55** exhibited the same binding state of Hsp90 as an analog of pyrazole compound **54** via X-ray crystallography. Oral administration of compound **54** demonstrated potent antitumor effects on human lung cancer (NCI-H1975) in a xenograft mouse model [151]. Mettu et al. innovatively synthesized a variety of pyrazolyl 2-aminopyrimidine derivatives and evaluated their Hsp90 inhibitory and anticancer activity. Among them, compound **56e** established the highest binding affinity to Hsp90 (20 nM) with anti-proliferative function against MCF7 (IC50 = ~2.4 μM), MDA-MB-231 (IC50 = ~0.8 μM), and HCT-116 (IC50 = ~4.8 μM), in vitro. According to Western blotting (WB) analysis, two compounds (**56b**, **56e**) generated dose-dependent degradation of two client proteins (pHER2 and pERK1/2) [152]. Molecular docking studies demonstrated that the compounds **56b–56e** were significant Hsp90 inhibitors. These studies revealed that the para substitution on pyrazole rings A and B, especially with the p-nitro group on ring B and 2 amino groups on the pyrimidine ring, had unique consequences on the development of new Hsp90 inhibitors. The binding potential of the disubstituted pyrazolyl pyrimidine scaffold and pocket of Hsp90 had effective interactions with Thr184, Asn51, and AsH93 (a protonated form of Asp93), Asp54, and Lys58 [152]. According to the results reported by Mettu et al. compound **56e** was the most active compound; the apoptosis potential of this compound was recognized by the Annexin V assay, where it could induce mitochondrial stress that increases membrane permeability, causing induction of apoptosis in MCF-7 cells. This was demonstrated by increasing the generation of J-monomer by applying JC-1 stain. Furthermore, a change in mitochondrial membrane potential could inevitably give rise to ROS strains as evidenced by DCFDA staining. In addition, this compound could stop the subG1 phase cell cycle. Thus, compound **56e** could serve as a potential compound with anticancer activity [153].

Mohamady et al. synthesized and evaluated possible biological activities of 3,5-diarylpyrazoles derivatives against HepG2 and MCF-7 cancerous cells. Among them, compound **57** exhibited the highest cytotoxic activity on HepG2 (IC_{50} = ~0.083 μM) and MCF7 (IC_{50} = ~0.13 μM) cells, with efficient anti-proliferation effectiveness. Compound **57** induced the inhibition of tumor cell proliferation by stimulating G2 phase inhibition, obstructing client proteins, such as c-Raf, EGFR, Akt, and c-Met and increasing the levels of Hsp70 [154]. Kadasi et al. reported the preparation of a novel series of *N*-pyridoyl-2-pyrazolines and evaluated their Hsp90 inhibitory activities for possible anticancer effects. The docking simulations revealed the binding potential of the synthesized compounds in the N-terminal ATP in Hsp90; moreover, the optimized compounds had significant interactions with Asp93 and Thr184. Consequently, compounds **58**, **59**, and **60** demonstrated an effective cytotoxic activity against MDA-MB-468 with IC_{50} of 1.60, 2.8, and 12 μM, respectively. Compound **58** exhibited efficient inhibitory effect as well (IC_{50} = ~7.7 μM) against A375 as human melanoma cells [155].

3. Conclusions and Future Outlook

Several investigations have focused on the design of inhibitors with heterocyclic structures to bring down the increased level of Hsp90 proteins in cancer or tumor cells, thus providing promising treatment strategies for cancer therapy with higher targeting and efficiency/efficacy. Hsp90 proteins have crucial roles in regulatory activity: folding, maintenance, function, and stability of various vital proteins, especially in client proteins. The inhibition of Hsp90s overexpressed in tumor cells can be considered an effective treatment strategy for various cancers. Numerous natural products and rationally designed synthetic Hsp90 inhibitors encompassing purine, pyrazole, triazine, quinolines, coumarin, and isoxazole structures can be viable candidates for further anticancer evaluations as deliberated. However, more systematic analyses and explorations are still needed for clinical applications and formulations of these compounds Hsp90 proteins have serious roles with regulatory activity in folding, maintenance, function, and stability of various vital proteins, especially, client proteins, which make them attractive candidates for cancer therapy targets. The inhibition of these Hsp90s overexpressed in tumor cells can be considered as an effective treatment strategy for various cancers and malignancies.

In past years, several investigations have focused on heterocyclic structures, which are capable of inhibiting Hsp90 proteins, which are overexpressed in cancer or tumor cells, providing promising treatment strategies for cancer therapy with high targeting and efficiency. These Hsp90 inhibitors are attractive candidates due to their inhibitory effect against several cellular signaling pathways simultaneously in cancer cells or tumors. Natural compounds (**1–9**) were proposed as Hsp90 inhibitors, and some compounds were selected for animal and in-vivo studies. Compounds based on the purine structure (**10–12**) were synthesized which are claimed to bind to the Hsp90 inhibitors strength and inhibit tumor growth, these compounds were known as good candidates for clinical phase studies. Coumarin and its derivatives (**13–19**) have different biological activities such as platelet aggregation inhibition, antibacterial effects, and anticancer activity, these scaffolds also act as Hsp90 inhibitors. Modification of 4-hydroxyl and the 3′-carbamate functional groups of novobiocin in compound **16** led to an increase in its activity more than 500 times from novobiocin.

Compounds aroylquinolone-based compounds were also categorized as Hsp90 inhibitors that prevent significant growth of cancerous cells. For example, compound **27** displayed an IC50 of ~5.9 μM to inhibit tubulin polymerization and degradation of Hsp70 and Akt protein through WB analysis. Compound **35**, hybrid CH5015765, containing the 2-amino-1,3,5-triazines was recognized as an Hsp90 inhibitor with good anti-proliferative activity against HCT-116 and excellent binding affinity (Kd = 3.4 nm). Triazines were also known as Hsp90 inhibitors with high activity and good anti-proliferative activity against HCT-116. In general, the triazine derivatives showed significant anti-proliferative activity in vitro and in vivo. Medicinal chemistry studies showed that the inclusion of isoxazole in the chemical structure of Hsp90 inhibitors can lead to improved efficacy. Isoxazole derivatives (**38–53**) can block the signaling cascade by binding to the ATP of the kinase protein and cause cell cycle arrest and apoptosis, investigations exposed that the isoxazole derivatives did not damage red blood cells, which strengthened the potential of these compounds as synthesized anticancer prototypes with low toxicity. Pyrazole derivatives (**54–64**) have remarkable activities, such as high efficiency as partial agonists at G protein and cannabinoid receptors, which make these properties useful inhibitors. This combination inhibits the proliferation of tumor cells by stimulating G2 phase inhibition. However, more systematic analyses are still needed for clinical applications and formulations of these compounds.

Author Contributions: M.A., Z.K. and F.K.: contributed equally to conceptualization and writing—review; S.I., H.S.-A. and R.S.V.: conceptualization, writing—review, and editing. All authors have read and agreed to the published version of the manuscript.

Funding: This work was supported by Projects 298121, 140192 and 988767 of the Pharmaceutical Sciences Research Center; National Institute for Medical Research Development.

Institutional Review Board Statement: Not applicable.

Informed Consent Statement: Not applicable.

Data Availability Statement: Not applicable.

Acknowledgments: The authors are thankful to National Institute for Medical Research Development for providing necessary research facilities. NIMAD (Project No. 988767) and the Pharmaceutical Sciences Research Center (Projects No. 298121, No. 140192).

Conflicts of Interest: There are no conflict of interest.

References

1. Mahboubi-Rabbani, M.; Zarghi, A. Lipoxygenase Inhibitors as Cancer Chemopreventives: Discovery, Recent Developments and Future Perspectives. *Curr. Med. Chem.* **2021**, *28*, 1143–1175. [CrossRef] [PubMed]
2. Schopf, F.H.; Biebl, M.M.; Buchner, J. The Hsp90 chaperone machinery. *Nat. Rev. Mol. Cell Biol.* **2017**, *18*, 345–360. [CrossRef] [PubMed]
3. Mahalingam, D.; Swords, R.; Carew, J.S.; Nawrocki, S.T.; Bhalla, K.; Giles, F.J. Targeting HSP90 for cancer therapy. *Br. J. Cancer* **2009**, *100*, 1523–1529. [CrossRef] [PubMed]
4. Ischia, J.; So, A.I. The role of heat shock proteins in bladder cancer. *Nat. Rev. Urol.* **2013**, *10*, 386–395. [CrossRef]
5. Davenport, J.; Manjarrez, J.R.; Peterson, L.; Krumm, B.; Blagg, B.S.J.; Matts, R.L. Gambogic Acid, a Natural Product Inhibitor of Hsp90. *J. Nat. Prod.* **2011**, *74*, 1085–1092. [CrossRef] [PubMed]
6. Bhat, R.; Tummalapalli, S.R.; Rotella, D.P.J. Progress in the Discovery and Development of Heat Shock Protein 90 (Hsp90) Inhibitors: Miniperspective. *Med. Chem.* **2014**, *57*, 8718–8728. [CrossRef] [PubMed]
7. Jaeger, A.M.; Whitesell, L. Hsp90: Enabler of Cancer Adaptation. *Annu. Rev. Cancer Biol.* **2019**, *3*, 275–297. [CrossRef]
8. Gupta, S.D.; Bommaka, M.K.; Banerjee, A. Inhibiting protein-protein interactions of Hsp90 as a novel approach for targeting cancer. *Eur. J. Med. Chem.* **2019**, *178*, 48–63. [CrossRef]
9. Wang, L.; Zhang, Q.; You, Q. Targeting the HSP90–CDC37–kinase chaperone cycle: A promising therapeutic strategy for cancer. *Med. Res. Rev.* **2022**, *42*, 156–182. [CrossRef]
10. Li, L.; Chen, N.-N.; You, Q.-D.; Xu, X.-L. An updated patent review of anticancer Hsp90 inhibitors (2013–present). *Expert Opin. Ther. Patents* **2021**, *31*, 67–80. [CrossRef]

11. Niu, M.; Song, S.; Su, Z.; Wei, L.; Li, L.; Pu, W.; Zhao, C.; Ding, Y.; Wang, J.; Cao, W.; et al. Inhibition of heat shock protein (HSP) 90 reverses signal transducer and activator of transcription (STAT) 3-mediated muscle wasting in cancer cachexia mice. *Br. J. Pharmacol.* **2021**, *178*, 4485–4500. [CrossRef] [PubMed]
12. Jafari, A.; Rezaei-Tavirani, M.; Farhadihosseinabadi, B.; Taranejoo, S.; Zali, H. HSP90 and Co-chaperones: Impact on Tumor Progression and Prospects for Molecular-Targeted Cancer Therapy. *Cancer Investig.* **2020**, *38*, 310–328. [CrossRef] [PubMed]
13. Sanchez, J.; Carter, T.R.; Cohen, M.S.; Blagg, B.S.J. Old and New Approaches to Target the Hsp90 Chaperone. *Curr. Cancer Drug Targets* **2020**, *20*, 253–270. [CrossRef] [PubMed]
14. Trepel, J.; Mollapour, M.; Giaccone, G.; Neckers, L. Targeting the dynamic HSP90 complex in cancer. *Nat. Rev. Cancer* **2010**, *10*, 537–549. [CrossRef]
15. Nazaripour, E.; Mousazadeh, F.; Moghadam, M.D.; Najafi, K.; Borhani, F.; Sarani, M.; Ghasemi, M.; Rahdar, A.; Iravani, S.; Khatami, M. Biosynthesis of lead oxide and cerium oxide nanoparticles and their cytotoxic activities against colon cancer cell line. *Inorg. Chem. Commun.* **2021**, *131*, 108800. [CrossRef]
16. Iravani, S.; Varma, R.S. MXenes for Cancer Therapy and Diagnosis: Recent Advances and Current Challenges. *ACS Biomater. Sci. Eng.* **2021**, *7*, 1900–1913. [CrossRef]
17. Delfi, M.; Sartorius, R.; Ashrafizadeh, M.; Sharifi, E.; Zhang, Y.; De Berardinis, P.; Zarrabi, A.; Varma, R.S.; Tay, F.R.; Smith, B.R.; et al. Self-assembled peptide and protein nanostructures for anti-cancer therapy: Targeted delivery, stimuli-responsive devices and immunotherapy. *Nano Today* **2021**, *38*, 101119. [CrossRef]
18. Alavi, M.; Varma, R.S. Overview of Novel Strategies for the Delivery of Anthracyclines to Cancer Cells by Liposomal and Polymeric Nano Formulations. *Int. J. Biol. Macromol.* **2020**, *164*, 2197–2203. [CrossRef]
19. Makvandi, P.; Baghbantaraghdari, Z.; Zhou, W.; Zhang, Y.; Manchanda, R.; Agarwal, T.; Wu, A.; Maiti, T.K.; Varma, R.S.; Smith, B.R. Gum polysaccharide/nanometal hybrid biocomposites in cancer diagnosis and therapy. *Biotechnol. Adv.* **2021**, *48*, 107711. [CrossRef]
20. Hajipour, A.R.; Khorsandi, Z.; Mortazavi, M.; Farrokhpour, H. Green, efficient and large-scale synthesis of benzimidazoles, benzoxazoles and benzothiazoles derivatives using ligand-free cobalt-nanoparticles: As potential anti-estrogen breast cancer agents, and study of their interactions with estrogen receptor by molecular docking. *RSC Adv.* **2015**, *5*, 107822–107828. [CrossRef]
21. Khorsandi, Z.; Hajipour, A.R.; Sarfjoo, M.R.; Varma, R.S. A Pd/Cu-Free magnetic cobalt catalyst for C–N cross coupling reactions: Synthesis of abemaciclib and fedratinib. *Green Chem.* **2021**, *23*, 5222–5229. [CrossRef]
22. Khorsandi, Z.; Keshavarzipour, F.; Varma, R.S.; Hajipour, A.R.; Sadeghi-Aliabadi, H. Sustainable synthesis of potential antitumor new derivatives of Abemaciclib and Fedratinib via C-N cross coupling reactions using Pd/Cu-free Co-catalyst. *Mol. Catal.* **2021**, *517*, 112011. [CrossRef]
23. Whitesell, L.; Lindquist, S.L. HSP90 and the chaperoning of cancer. *Nat. Rev. Cancer.* **2005**, *5*, 761–772. [CrossRef] [PubMed]
24. Jego, G.; Hazoumé, A.; Seigneuric, R.; Garrido, C. Targeting heat shock proteins in cancer. *Cancer Lett.* **2013**, *332*, 275–285. [CrossRef]
25. Calderwood, S.K.; Khaleque, M.A.; Sawyer, D.B.; Ciocca, D.R. Heat Shock Proteins in Cancer: Chaperones of Tum Origenesis. *J. Pet. Sci. Eng.* **2006**, *31*, 164–172.
26. Patel, H.J.; Patel, P.D.; Ochiana, S.O.; Yan, P.; Sun, W.; Patel, M.R.; Shah, S.K.; Tramentozzi, E.; Brooks, J.; Bolaender, A.J. Structure–Activity Relationship in a Purine-Scaffold Compound Series with Selectivity for the Endoplasmic Reticulum Hsp90 ParalogGrp94. *Med. Chem.* **2015**, *58*, 3922–3943. [CrossRef]
27. Johnson, J.L. Evolution and function of diverse Hsp90 homologs and cochaperone proteins. *Biochim. Biophys. Acta Mol. Cell Res.* **2012**, *1823*, 607–613. [CrossRef]
28. Csermely, P.; Miyata, Y.; Schnaider, T.; Yahara, I. Autophosphorylation of grp94 (Endoplasmin). *J. Biol. Chem.* **1995**, *270*, 6381–6388. [CrossRef]
29. Mishra, P.; Bolon, D.N. Designed Hsp90 Heterodimers Reveal an Asymmetric ATPase-Driven Mechanism In Vivo. *Mol. Cell* **2014**, *53*, 344–350. [CrossRef]
30. Meyer, P.; Prodromou, C.; Hu, B.; Vaughan, C.; Roe, S.M.; Panaretou, B.; Piper, P.W.; Pearl, L.H. Structural and Functional Analysis of the Middle Segment of Hsp90: Implications for ATP Hydrolysis and Client Protein and Cochaperone Interactions. *Mol. Cell* **2003**, *11*, 647–658. [CrossRef]
31. Gupta, S.D. Hsp90 Flexibility and Development of its Inhibitors for the Treatment of Cancer. *Curr. Chem. Biol.* **2018**, *12*, 53–64. [CrossRef]
32. Gupta, S.D. Novel Anti-Cancer Drugs Based on Hsp90 Inhibitory Mechanisms: A Recent Report. In *Medicinal Chemistry with Pharma-Ceutical Product Development*; Apple Academic Press: New York, NY, USA, 2019; pp. 57–104.
33. Gupta, S.D.; Pan, C.H. Recent update on discovery and development of Hsp90 inhibitors as senolytic agents. *Int. J. Biol. Macromol.* **2020**, *161*, 1086–1098. [CrossRef] [PubMed]
34. Hanahan, D.; Weinberg, R.A. The Hallmarks of Cancer. *Cell* **2000**, *100*, 57–70. [CrossRef]
35. Hanahan, D.; Weinberg, R.A. Hallmarks of cancer: The next generation. *Cell* **2011**, *144*, 646–674. [CrossRef]
36. Vartholomaiou, E.; Echeverría, P.C.; Picard, D. Unusual Suspects in the Twilight Zone Between the Hsp90 Interactome and Carcinogenesis. *Adv. Cancer Res.* **2016**, *129*, 1–30. [CrossRef]

37. Hayat, U.; Elliott, G.T.; Olszanski, A.J.; Altieri, D.C. Feasibility and safety of targeting mitochondria for cancer therapy–preclinical characterization of gamitrinib, a first-in-class, mitochondriaL-targeted small molecule Hsp90 inhibitor. *Cancer Biol. Ther.* **2022**, *23*, 117–126. [CrossRef]
38. Biondini, M.; Kiepas, A.; El-Houjeiri, L.; Annis, M.G.; Hsu, B.E.; Fortier, A.-M.; Morin, G.; Martina, J.A.; Sirois, I.; Aguilar-Mahecha, A.; et al. HSP90 inhibitors induce GPNMB cell-surface expression by modulating lysosomal positioning and sensitize breast cancer cells to glembatumumab vedotin. *Oncogene* **2022**, *41*, 1701–1717. [CrossRef]
39. Epp-Ducharme, B.; Dunne, M.; Fan, L.; Evans, J.C.; Ahmed, L.; Bannigan, P.; Allen, C. Heat-activated nanomedicine formulation improves the anticancer potential of the HSP90 inhibitor luminespib in vitro. *Sci. Rep.* **2021**, *11*, 11103. [CrossRef]
40. Wu, Y.-W.; Chao, M.-W.; Tu, H.-J.; Chen, L.-C.; Hsu, K.-C.; Liou, J.-P.; Yang, C.-R.; Yen, S.-C.; HuangFu, W.-C.; Pan, S.-L. A novel dual HDAC and HSP90 inhibitor, MPT0G449, downregulates oncogenic pathways in human acute leukemia in vitro and in vivo. *Oncogenesis* **2021**, *10*, 39. [CrossRef]
41. Mshaik, R.; Simonet, J.; Georgievski, A.; Jamal, L.; Bechoua, S.; Ballerini, P.; Bellaye, P.-S.; Mlamla, Z.; de Barros, J.-P.P.; Geissler, A.; et al. HSP90 inhibitor NVP-BEP800 affects stability of SRC kinases and growth of T-cell and B-cell acute lymphoblastic leukemias. *Blood Cancer J.* **2021**, *11*, 61. [CrossRef]
42. Abdelmoaty, A.A.A.; Zhang, P.; Lin, W.; Fan, Y.-J.; Ye, S.-N.; Xu, J.-H. C0818, a novel curcumin derivative, induces ROS-dependent cytotoxicity in human hepatocellular carcinoma cells in vitro via disruption of Hsp90 function. *Acta Pharmacol. Sin.* **2022**, *43*, 446–456. [CrossRef] [PubMed]
43. Zhao, S.; Zhou, L.; Dicker, D.T.; Lev, A.; Zhang, S.; Ross, E.; El-Deiry, W.S. Anti-Cancer Efficacy Including Rb-Deficient Tumors and VHL-Independent HIF1α Proteasomal Destabilization by Dual Targeting of CDK1 or CDK4/6 and Hsp90. *Sci. Rep.* **2021**, *11*, 20871. [CrossRef] [PubMed]
44. Chen, X.-L.; Liu, P.; Zhu, W.-L.; Lou, L.-G. DCZ5248, a novel dual inhibitor of Hsp90 and autophagy, exerts antitumor activity against colon cancer. *Acta Pharmacol. Sin.* **2021**, *42*, 132–141. [CrossRef] [PubMed]
45. Cheng, C.-J.; Liu, K.-X.; Zhang, M.; Shen, F.-K.; Ye, L.-L.; Wu, W.-B.; Hou, X.-T.; Hao, E.-W.; Hou, Y.-Y.; Bai, G. Okicamelliaside targets the N-terminal chaperone pocket of HSP90 disrupts the chaperone protein interaction of HSP90-CDC37 and exerts antitumor activity. *Acta Pharmacol. Sin.* **2021**, *43*, 1046–1058. [CrossRef] [PubMed]
46. Konstantinopoulos, P.A.; Cheng, S.-C.; Supko, J.G.; Polak, M.; Wahner-Hendrickson, A.E.; Ivy, S.P.; Bowes, B.; Sawyer, H.; Basada, P.; Hayes, M.; et al. Combined PARP and HSP90 inhibition: Preclinical and Phase 1 evaluation in patients with advanced solid tumours. *Br. J. Cancer* **2021**, *126*, 1027–1036. [CrossRef]
47. Magwenyane, A.M.; Lawal, M.M.; Amoako, D.G.; Somboro, A.M.; Agoni, C.; Khan, R.B.; Mhlongo, N.N.; Kumalo, H.M. Exploring the inhibitory mechanism of resorcinylic isoxazole amine NVP-AUY922 towards the discovery of potential heat shock protein 90 (Hsp90) inhibitors. *Sci. Afr.* **2022**, *15*, e01107. [CrossRef]
48. Serwetnyk, M.A.; Blagg, S.J. The Disruption of Protein−Protein Interactions with Co-chaperones and Client Substrates as a Strategy Towards Hsp90 Inhibition. *Acta. Pharm. Sin. B* **2021**, *11*, 1446–1468. [CrossRef]
49. Roe, S.M.; Prodromou, C.; O'Brien, R.; Ladbury, J.E.; Piper, P.W.; Pearl, L.H. Structural Basis for Inhibition of the Hsp90 Molecular Chaperone by the Antitumor Antibiotics Radicicol and Geldanamycin. *J. Med. Chem.* **1999**, *42*, 260–266. [CrossRef]
50. Supko, J.G.; Hickman, R.L.; Grever, M.R.; Malspeis, L. Preclinical Pharmacologic Evaluation of Geldanamycin as an Antitumor Agent. *Cancer Chemother. Pharmacol.* **1995**, *36*, 305–315. [CrossRef]
51. Banerji, U.; O'Donnell, A.; Scurr, M.; Benson, C.; Hanwell, J.; Clark, S.; Raynaud, F.; Turner, A.; Walton, M.; Workman, P.; et al. Phase I Trial of the Heat Shock Protein 90 (Hsp90) Inhibitor 17-Allylamino 17-Demethoxygeldanamycin (17AAG). Pharmacokinetic (PK) Profile and Pharmacodynamic (PD) Endpoints. *Proc. Am. Soc. Clin. Oncol.* **2001**, *20*, 326.
52. Egorin, M.J.; Zuhowski, E.G.; Rosen, D.M.; Sentz, D.L.; Covey, J.M.; Eiseman, J.L. Plasma pharmacokinetics and tissue distribution of 17-(allylamino)-17-demethoxygeldanamycin (NSC 330507) in CD2F1 mice1. *Cancer Chemother. Pharmacol.* **2001**, *47*, 291–302. [CrossRef] [PubMed]
53. Egorin, M.J.; Lagattuta, T.F.; Hamburger, D.R.; Covey, J.M.; White, K.D.; Musser, S.M.; Eiseman, J.L. Pharmacokinetics, tissue distribution, and metabolism of 17-(dimethylaminoethylamino)-17-demethoxygeldanamycin (NSC 707545) in CD2F1 mice and Fischer 344 rats. *Cancer Chemother. Pharmacol.* **2002**, *49*, 7–19. [CrossRef] [PubMed]
54. Munster, P.N.; Tong, L.; Schwartz, L.; Larson, S.; Kenneson, K.; De La Cruz, A.; Rosen, N.; Scher, H. Phase I Trial of 17(Allylamino)-17-demethoxygeldanamycin (17AAG) in Patients (Pts) with Advanced Solid Malignancies. *Proc. Am. Soc. Clin. Oncol.* **2001**, *20*, 83–91.
55. Soga, S.; Sharma, S.V.; Shiotsu, Y.; Shimizu, M.; Tahara, H.; Yamaguchi, K.; Ikuina, Y.; Murakata, C.; Tamaoki, T.; Kurebayashi, J.; et al. Stereospecific antitumor activity of radicicol oxime derivatives. *Cancer Chemother. Pharmacol.* **2001**, *48*, 435–445. [CrossRef] [PubMed]
56. Johnson, V.A.; Singh, E.K.; Nazarova, L.A.; Alexander, L.D.; McAlpine, S.R. Macrocyclic inhibitors of hsp90. *Curr. Top. Med. Chem.* **2010**, *10*, 1380–1402. [CrossRef] [PubMed]
57. Porter, J.R.; Ge, J.; Lee, J.; Normant, E.; West, K. Ansamycin Inhibitors of Hsp90: Nature's Prototype for Anti-chaperone Therapy. *Curr. Top. Med.* **2009**, *9*, 1386–1418. [CrossRef]
58. Tanida, S.; Hasegawa, T.; Higashide, E.; Macbecins, I.; Macbecins, I.I. New Antitumor Antibiotics. I. Producing Organism, Fermentation and Antimicrobial Activities. *J. Antibiot.* **1980**, *33*, 199–204. [CrossRef]

59. Tian, Z.-Q.; Liu, Y.; Zhang, D.; Wang, Z.; Dong, S.D.; Carreras, C.W.; Zhou, Y.; Rastelli, G.; Santi, D.V.; Myles, D.C. Synthesis and biological activities of novel 17-aminogeldanamycin derivatives. *Bioorg. Med. Chem.* **2004**, *12*, 5317–5329. [CrossRef]
60. Wagner, A.J.; Chugh, R.; Rosen, L.S.; Morgan, J.A.; George, S.; Gordon, M.; Dunbar, J.; Normant, E.; Grayzel, D.; Demetri, G.D. A Phase I Study of the HSP90 Inhibitor Retaspimycin Hydrochloride (IPI-504) in Patients with Gastrointestinal Stromal Tumors or Soft-Tissue Sarcomas. *Clin. Cancer Res.* **2013**, *19*, 6020–6029. [CrossRef]
61. Floris, G.; Debiec-Rychter, M.; Wozniak, A.; Stefan, C.; Normant, E.; Faa, G.; Machiels, K.; Vanleeuw, U.; Sciot, R.; Schöffski, P. The Heat Shock Protein 90 Inhibitor IPI-504 Induces KIT Degradation, Tumor Shrinkage, and Cell Proliferation Arrest in Xenograft Models of Gastrointestinal Stromal Tumors. *Mol. Cancer Ther.* **2011**, *10*, 1897–1908. [CrossRef]
62. Floris, G.; Sciot, R.; Wozniak, A.; Van Looy, T.; Wellens, J.; Faa, G.; Normant, E.; Debiec-Rychter, M.; Schöffski, P. The Novel Hsp90 inhibitor, IPI-493, is Highly Effective in Human Gastrostrointestinal Stromal Tumor Xenografts Carrying Heterogeneous KIT Mutations. *Clin. Cancer Res.* **2011**, *17*, 5604–5614. [CrossRef] [PubMed]
63. Whitesell, L.; Mimnaugh, E.G.; De Costa, B.; Myers, C.E.; Neckers, L.M. Inhibition of heat shock protein HSP90-pp60v-src heteroprotein complex formation by benzoquinone ansamycins: Essential role for stress proteins in oncogenic transformation. *Proc. Natl. Acad. Sci. USA* **1994**, *91*, 8324–8328. [CrossRef] [PubMed]
64. Stebbins, C.E.; Russo, A.A.; Schneider, C.; Rosen, N.; Hartl, F.; Pavletich, N.P. Crystal Structure of an Hsp90–Geldanamycin Complex: Targeting of a Protein Chaperone by an Antitumor Agent. *Cell* **1997**, *89*, 239–250. [CrossRef]
65. Soga, S.; Shiotsu, Y.; Akinaga, S.; Sharma, S. Development of Radicicol Analogues. *Curr. Cancer Drug Targets* **2003**, *3*, 359–369. [CrossRef] [PubMed]
66. Miyata, Y. Hsp90 Inhibitor Geldanamycin and Its Derivatives as Novel Cancer Chemotherapeutic Agents. *Curr. Pharm. Des.* **2005**, *11*, 1131–1138. [CrossRef]
67. Chiosis, G.; Lucas, B.; Huezo, H.; Solit, D.; Basso, A.; Rosen, N. Development of Purine-Scaffold Small Molecule Inhibitors of Hsp90. *Curr. Cancer Drug Targets* **2003**, *3*, 371–376. [CrossRef]
68. Conejo-García, A.; García-Rubiño, M.E.; Marchal, J.A.; Núñez, M.C.; Ramírez, A.; Cimino, S.; García, M.; Aránega, A.; Gallo, M.A.; Campos, J.M. Synthesis and anticancer activity of (RS)-9-(2,3-dihydro-1,4-benzoxaheteroin-2-ylmethyl)-9H-purines. *Eur. J. Med. Chem.* **2011**, *46*, 3795–3801. [CrossRef]
69. Pelliccia, S.; Amato, J.; Capasso, D.; Di Gaetano, S.; Massarotti, A.; Piccolo, M.; Irace, C.; Tron, G.C.; Pagano, B.; Randazzo, A.J. Bio-Inspired Dual-Selective BCL-2/c-MYC G-Quadruplex Binders: Design, Synthesis, and Anticancer Activity of Drug-like Imidazo [2,1-i] purine Derivatives. *Med. Chem.* **2019**, *63*, 2035–2050. [CrossRef]
70. Ashour, F.A.; Rida, S.M.; El-Hawash, S.A.; ElSemary, M.M.; Badr, M.H. Synthesis, Anticancer, Anti-HIV-1, and Antimicrobial Activity of Some Tricyclic Triazino and Triazolo [4,3-e] purine Derivatives. *Cancer Causes Control* **2012**, *21*, 1107–1119. [CrossRef]
71. Kinali-Demirci, S.; İdil, Ö.; Dişli, A. Synthesis of Some Novel Purine Derivatives Incorporating Tetrazole Ring and Investigation of their Antimicrobial Activity and DNA Interactions. *Cancer Cell Int. J. Mol. Histol.* **2015**, *24*, 1218–1225. [CrossRef]
72. Abdallah, A.E.; Elgemeie, G.H. Development; Therapy, Design, Synthesis, Docking, and Antimicrobial Evaluation of Some Novel Pyrazolo [1,5-a] Pyrimidines and their Corresponding Cycloalkane Ring-fused Derivatives as Purine Analogs. *Drug Des. Devel. Ther.* **2018**, *12*, 1785–1798. [CrossRef] [PubMed]
73. Rosemeyer, H. The Chemodiversity of Purine as a Constituent of Natural Products. *Chem. Biodivers.* **2004**, *1*, 361–401. [CrossRef] [PubMed]
74. Kaneko, K.; Aoyagi, Y.; Fukuuchi, T.; Inazawa, K.; Yamaoka, N. Total Purine and Purine Base Content of Common Foodstuffs for Facilitating Nutritional Therapy for Gout and Hyperuricemia. *Biol. Pharm. Bull.* **2014**, *37*, 709–721. [CrossRef] [PubMed]
75. Yin, J.; Ren, W.; Huang, X.; Deng, J.; Li, T.; Yin, Y. Potential Mechanisms Connecting Purine Metabolism and Cancer Therapy. *Front. Immunol.* **2018**, *9*, 1697. [CrossRef] [PubMed]
76. Chiosis, G.; Timaul, M.N.; Lucas, B.; Munster, P.N.; Zheng, F.F.; Sepp-Lorenzino, L.; Rosen, N. A small molecule designed to bind to the adenine nucleotide pocket of Hsp90 causes Her2 degradation and the growth arrest and differentiation of breast cancer cells. *Chem. Biol.* **2001**, *8*, 289–299. [CrossRef]
77. Bao, R.; Lai, C.-J.; Qu, H.; Wang, D.; Yin, L.; Zifcak, B.; Atoyan, R.; Wang, J.; Samson, M.; Forrester, J.; et al. CUDC-305, a Novel Synthetic HSP90 Inhibitor with Unique Pharmacologic Properties for Cancer Therapy. *Clin. Cancer Res.* **2009**, *15*, 4046–4057. [CrossRef]
78. Zhang, H.; Neely, L.; Lundgren, K.; Yang, Y.-C.; Lough, R.; Timple, N.; Burrows, F. BIIB021, a synthetic HSP90 inhibitor, has broad application against tumors with acquired multidrug resistance. *Int. J. Cancer* **2010**, *126*, 1226–1234. [CrossRef]
79. Yin, X.; Zhang, H.; Lundgren, K.; Wilson, L.; Burrows, F.; Shores, C.G. BIIB021, a novel HSP90 inhibitor, sensitizes head and neck squamous cell carcinoma to radiotherapy. *Int. J. Cancer* **2010**, *126*, 1216–1225. [CrossRef]
80. Chaurasiya, A.; Wahan, S.K.; Sahu, C.; Chawla, P.A. An insight into the rational design of recent purine-based scaffolds in targeting various cancer pathways. *J. Mol. Struct.* **2022**, 134308. [CrossRef]
81. Wang, X.-T.; Bao, C.-H.; Jia, Y.-B.; Wang, N.; Ma, W.; Liu, F.; Wang, C.; Wang, J.-B.; Song, Q.-X.; Cheng, Y.-F. BIIB021, a novel Hsp90 inhibitor, sensitizes esophageal squamous cell carcinoma to radiation. *Biochem. Biophys. Res. Commun.* **2014**, *452*, 945–950. [CrossRef]
82. ElFiky, A.; Saif, M.W.; Beeram, M.; Brien, S.O.; Lammanna, N.; Castro, J.E.; Woodworth, J.; Perea, R.; Storgard, C.; Von Hoff, D.D. BIIB021, an oral, synthetic non-ansamycin Hsp90 inhibitor: Phase I experience. *J. Clin. Oncol.* **2008**, *26*, 2503. [CrossRef]

83. Dickson, M.A.; Okuno, S.H.; Keohan, M.L.; Maki, R.G.; D'Adamo, D.R.; Akhurst, T.J.; Antonescu, C.R.; Schwartz, G.K. Phase II study of the HSP90-inhibitor BIIB021 in gastrointestinal stromal tumors. *Ann. Oncol.* **2013**, *24*, 252–257. [CrossRef] [PubMed]
84. Emami, S.; Dadashpour, S. Current developments of coumarin-based anti-cancer agents in medicinal chemistry. *Eur. J. Med. Chem.* **2015**, *102*, 611–630. [CrossRef] [PubMed]
85. Zhang, L.; Xu, Z. Coumarin-containing hybrids and their anticancer activities. *Eur. J. Med. Chem.* **2019**, *181*, 111587. [CrossRef] [PubMed]
86. Al-Warhi, T.; Sabt, A.; Elkaeed, E.B.; Eldehna, W.M. Recent advancements of coumarin-based anticancer agents: An up-to-date review. *Bioorg. Chem.* **2020**, *103*, 104163. [CrossRef] [PubMed]
87. Song, X.; Fan, J.; Liu, L.; Liu, X.; Gao, F. Coumarin derivatives with anticancer activities: An update. *Arch. Pharm.* **2020**, *353*, 2000025. [CrossRef]
88. Qin, H.-L.; Zhang, Z.-W.; Ravindar, L.; Rakesh, K. Antibacterial activities with the structure-activity relationship of coumarin derivatives. *Eur. J. Med. Chem.* **2020**, *207*, 112832. [CrossRef]
89. Keshavarzipour, F.; Tavakol, H. The synthesis of coumarin derivatives using choline chloride/zinc chloride as a deep eutectic solvent. *J. Iran. Chem. Soc.* **2016**, *13*, 149–153. [CrossRef]
90. Ahmed, E.Y.; Elserwy, W.S.; El-Mansy, M.F.; Serry, A.M.; Salem, A.M.; Abdou, A.M.; Abdelrahman, B.A.; Elsayed, K.H.; Elaziz, M.R.A. Angiokinase inhibition of VEGFR-2, PDGFR and FGFR and cell growth inhibition in lung cancer: Design, synthesis, biological evaluation and molecular docking of novel azaheterocyclic coumarin derivatives. *Bioorg. Med. Chem. Lett.* **2021**, *48*, 128258. [CrossRef]
91. Burlison, J.A.; Neckers, L.; Smith, A.B.; Maxwell, A.A.; Blagg, B.S.J. Novobiocin: Redesigning a DNA Gyrase Inhibitor for Selective Inhibition of Hsp90. *J. Am. Chem. Soc.* **2006**, *128*, 15529–15536. [CrossRef]
92. Shelton, S.N.; Shawgo, M.E.; Matthews, S.B.; Lu, Y.; Donnelly, A.C.; Szabla, K.; Tanol, M.; Vielhauer, G.A.; Rajewski, R.A.; Matts, R.L.; et al. KU135, a Novel Novobiocin-Derived C-Terminal Inhibitor of the 90-kDa Heat Shock Protein, Exerts Potent Antiproliferative Effects in Human Leukemic Cells. *Mol. Pharmacol.* **2009**, *76*, 1314–1322. [CrossRef] [PubMed]
93. Garg, G.; Zhao, H.; Blagg, B.S.J. Design, Synthesis, and Biological Evaluation of Ring-Constrained Novobiocin Analogues as Hsp90 C-Terminal Inhibitors. *ACS Med. Chem. Lett.* **2015**, *6*, 204–209. [CrossRef] [PubMed]
94. Wei, Q.; Ning, J.-Y.; Dai, X.; Gao, Y.-D.; Su, L.; Zhao, B.-X.; Miao, J.-Y. Discovery of novel HSP90 inhibitors that induced apoptosis and impaired autophagic flux in A549 lung cancer cells. *Eur. J. Med. Chem.* **2018**, *145*, 551–558. [CrossRef] [PubMed]
95. Kouznetsov, V.V.; Méndez, L.Y.V.; Galvis, C.E.P.; Villamizar, M.C.O. The direct C–H alkenylation of quinoline *N*-oxides as a suitable strategy for the synthesis of promising antiparasitic drugs. *New J. Chem.* **2020**, *44*, 12–19. [CrossRef]
96. Keshavarzipour, F.; Tavakol, H. Zinc Cation Supported on Carrageenan Magnetic Nanoparticles: A Novel, Green and Efficient Catalytic System for One-pot Three-Component Synthesis of Quinoline Derivatives. *Appl. Organomet. Chem.* **2017**, *31*, 3682–3692. [CrossRef]
97. Insuasty, D.; García, S.; Abonia, R.; Insuasty, B.; Quiroga, J.; Nogueras, M.; Laali, K.K. Design, Synthesis, and Molecular Docking Study of Novel Quinoline-Based Bis-Chalcones as Potential Antitumor Agents. *Arch. Pharm.* **2021**, *354*, 2100094–2100103. [CrossRef]
98. Jin, G.; Li, Z.; Xiao, F.; Qi, X.; Sun, X. Optimization of activity localization of quinoline derivatives: Design, synthesis, and dual evaluation of biological activity for potential antitumor and antibacterial agents. *Bioorg. Chem.* **2020**, *99*, 103837. [CrossRef]
99. Mishra, S.; Salahuddin, R.K.; Majumder, A.; Kumar, A.; Singh, C.; Tiglani, D. Updates on Synthesis and Biological Activities of Quinoline Derivatives: A Review. *Int. J. Pharm. Sci. Rev. Res.* **2021**, *13*, 3941–3960.
100. Goyal, S.; Binnington, B.; McCarthy, S.D.; Desmaële, D.; Férrié, L.; Figadère, B.; Loiseau, P.M.; Branch, D.R. Inhibition of in vitro Ebola infection by anti-parasitic quinoline derivatives. *F1000Research* **2020**, *9*, 268. [CrossRef]
101. Verma, C.; Quraishi, M.; Ebenso, E.E. Quinoline and its derivatives as corrosion inhibitors: A review. *Surfaces Interfaces* **2020**, *21*, 100634. [CrossRef]
102. Kayamba, F.; Malimabe, T.; Ademola, I.K.; Pooe, O.J.; Kushwaha, N.D.; Mahlalela, M.; van Zyl, R.L.; Gordon, M.; Mudau, P.T.; Zininga, T.; et al. Design and synthesis of quinoline-pyrimidine inspired hybrids as potential plasmodial inhibitors. *Eur. J. Med. Chem.* **2021**, *217*, 113330. [CrossRef] [PubMed]
103. Rao, K.V.; Cullen, W.P. Streptonigrin, an Antitumor Substance. I. Isolation and Characterization. *Antibiot. Annu.* **1959**, *7*, 950–953.
104. Chirigos, M.A.; Pearson, J.W.; Papas, T.S.; Woods, W.A.; Wood, H.B., Jr.; Spahn, G. Effect of Three Strains of BeG Against a Murine Leukemia After Drug Therapy. *Cancer Chemother. Rep.* **1973**, *57*, 305–309.
105. Balitz, D.M.; Bush, J.A.; Bradner, W.T.; Doyle, T.W.; O'Herron, F.A.; Nettleton, D.E. Isolation of lavendamycin. A new antibiotic from Streptomyces lavendulae. *J. Antibiot.* **1982**, *35*, 259–265. [CrossRef] [PubMed]
106. Ganesh, T.; Min, J.; Thepchatri, P.; Du, Y.; Li, L.; Lewis, I.; Wilson, L.; Fu, H.; Chiosis, G.; Dingledine, R.; et al. Discovery of aminoquinolines as a new class of potent inhibitors of heat shock protein 90 (Hsp90): Synthesis, biology, and molecular modeling. *Bioorg. Med. Chem.* **2008**, *16*, 6903–6910. [CrossRef] [PubMed]
107. Audisio, D.; Messaoudi, S.; Cegielkowski, L.; Peyrat, J.-F.; Brion, J.-D.; Methy-Gonnot, D.; Radanyi, C.; Renoir, J.-M.; Alami, M. Discovery and Biological Activity of 6BrCaQ as an Inhibitor of the Hsp90 Protein Folding Machinery. *Chem. Med. Chem.* **2011**, *6*, 804–815. [CrossRef]

108. Nepali, K.; Kumar, S.; Huang, H.-L.; Kuo, F.-C.; Lee, C.-H.; Kuo, C.-C.; Yeh, T.-K.; Li, Y.-H.; Chang, J.-Y.; Liou, J.-P.; et al. 2-Aroylquinoline-5,8-diones as potent anticancer agents displaying tubulin and heat shock protein 90 (HSP90) inhibition. *Org. Biomol. Chem.* **2015**, *14*, 716–723. [CrossRef]
109. Malayeri, S.O.; Abnous, K.; Arab, A.; Akaberi, M.; Mehri, S.; Zarghi, A.; Ghodsi, R. Design, Synthesis and Biological Evaluation of 7-(Aryl)-2,3-dihydro-[1,4] dioxino [2,3-g] Quinoline Derivatives as Potential Hsp90 Inhibitors and Anticancer Agents. *Bioorg. Med. Chem.* **2017**, *25*, 1294–1302. [CrossRef]
110. Liang, C.; Hao, H.; Wu, X.; Li, Z.; Zhu, J.; Lu, C.; Shen, Y. Design and synthesis of N-(5-chloro-2,4-dihydroxybenzoyl)-(R)-1,2,3,4-tetrahydroisoquinoline-3-carboxamides as novel Hsp90 inhibitors. *Eur. J. Med. Chem.* **2016**, *121*, 272–282. [CrossRef]
111. Nepali, K.; Lin, M.-H.; Chao, M.-W.; Peng, S.-J.; Hsu, K.-C.; Lin, T.E.; Chen, M.-C.; Lai, M.-J.; Pan, S.-L.; Liou, J.-P. Amide-tethered quinoline-resorcinol conjugates as a new class of HSP90 inhibitors suppressing the growth of prostate cancer cells. *Bioorg. Chem.* **2019**, *91*, 103119. [CrossRef]
112. Relitti, N.; Saraswati, A.P.; Chemi, G.; Brindisi, M.; Brogi, S.; Herp, D.; Schmidtkunz, K.; Saccoccia, F.; Ruberti, G.; Ulivieri, C.; et al. Novel quinolone-based potent and selective HDAC6 inhibitors: Synthesis, molecular modeling studies and biological investigation. *Eur. J. Med. Chem.* **2021**, *212*, 112998. [CrossRef] [PubMed]
113. Liu, H.; Long, S.; Rakesh, K.P.; Zha, G.F. Structure-Activity Relationships (SAR) of Triazine Derivatives: Promising Antimicrobial Agents. *Eur. J. Med. Chem.* **2020**, *185*, 111804. [CrossRef] [PubMed]
114. Gavade, S.N.; Markad, V.L.; Kodam, K.M.; Shingare, M.S.; Mane, D.V. Synthesis and Biological Evaluation of Novel 2, 4, 6-Triazine Derivatives as Antimicrobial Agents. *Bioorg. Med. Chem. Lett.* **2012**, *22*, 5075–5077. [CrossRef]
115. Cascioferro, S.; Parrino, B.; Spanò, V.; Carbone, A.; Montalbano, A.; Barraja, P.; Diana, P.; Cirrincione, G. An overview on the recent developments of 1,2,4-triazine derivatives as anticancer compounds. *Eur. J. Med. Chem.* **2017**, *142*, 328–375. [CrossRef] [PubMed]
116. Singla, P.; Luxami, V.; Paul, K. Triazine as a Promising Scaffold for its Versatile Biological Behavior. *Eur. J. Med. Chem.* **2015**, *102*, 39–57. [CrossRef]
117. Patel, P.K.; Patel, R.V.; Mahajan, D.H.; Parikh, P.A.; Mehta, G.N.; Pannecouque, C.; Chikhalia, K.H. Different Heterocycles Functionalized s-Triazine Analogues: Design, Synthesis and In Vitro Antimicrobial, Antituberculosis, and Anti-HIV Assessment. *J. Heterocycl. Chem.* **2014**, *51*, 1641–1658. [CrossRef]
118. Feldman, R.I.; Mintzer, B.; Zhu, D.; Wu, J.M.; Biroc, S.L.; Yuan, S.; Emayan, K.; Chang, Z.; Chen, D.; Arnaiz, D.O.; et al. Potent Triazolothione Inhibitor of Heat-Shock Protein-90. *Oncol. Lett.* **2009**, *74*, 43–50. [CrossRef]
119. Lee, T.; Seo, Y.H. Targeting the hydrophobic region of Hsp90's ATP binding pocket with novel 1,3,5-triazines. *Bioorg. Med. Chem. Lett.* **2013**, *23*, 6427–6431. [CrossRef]
120. Miura, T.; Fukami, T.A.; Hasegawa, K.; Ono, N.; Suda, A.; Shindo, H.; Yoon, D.-O.; Kim, S.-J.; Na, Y.-J.; Aoki, Y.; et al. Lead generation of heat shock protein 90 inhibitors by a combination of fragment-based approach, virtual screening, and structure-based drug design. *Bioorg. Med. Chem. Lett.* **2011**, *21*, 5778–5783. [CrossRef]
121. Suda, A.; Koyano, H.; Hayase, T.; Hada, K.; Kawasaki, K.-I.; Komiyama, S.; Hasegawa, K.; Fukami, T.A.; Sato, S.; Miura, T.; et al. Design and synthesis of novel macrocyclic 2-amino-6-arylpyrimidine Hsp90 inhibitors. *Bioorg. Med. Chem. Lett.* **2012**, *22*, 1136–1141. [CrossRef]
122. Suda, A.; Kawasaki, K.-I.; Komiyama, S.; Isshiki, Y.; Yoon, D.-O.; Kim, S.-J.; Na, Y.-J.; Hasegawa, K.; Fukami, T.A.; Sato, S.; et al. Design and synthesis of 2-amino-6-(1H,3H-benzo[de]isochromen-6-yl)-1,3,5-triazines as novel Hsp90 inhibitors. *Bioorg. Med. Chem.* **2014**, *22*, 892–905. [CrossRef] [PubMed]
123. Zhao, Z.; Zhu, J.; Quan, H.; Wang, G.; Li, B.; Zhu, W.; Xie, C.; Lou, L. X66, a novel N-terminal heat shock protein 90 inhibitor, exerts antitumor effects without induction of heat shock response. *Oncotarget* **2016**, *7*, 29648–29663. [CrossRef] [PubMed]
124. Ali, I.; Lone, M.; Al-Othman, Z.; Al-Warthan, A.; Sanagi, M. Heterocyclic Scaffolds: Centrality in Anticancer Drug Development. *Curr. Drug Targets* **2015**, *16*, 711–734. [CrossRef]
125. Çalışkan, B.; Sinoplu, E.; İbiş, K.; Akhan Güzelcan, E.; Atalay, R.Ç.; Banoglu, E. Synthesis and Cellular Bioactivities of Novel Isoxazole Derivatives Incorporating an Arylpiperazine Moiety as Anticancer Agents. *J. Enzym. Inhib. Med. Chem.* **2018**, *33*, 1352–1361. [CrossRef] [PubMed]
126. Shaik, A.; Bhandare, R.R.; Palleapati, K.; Nissankararao, S.; Kancharlapalli, V.; Shaik, S. Antimicrobial, Antioxidant, and Anticancer Activities of Some Novel Isoxazole Ring Containing Chalcone and Dihydropyrazole Derivatives. *Molecules* **2020**, *25*, 1047. [CrossRef] [PubMed]
127. Barmade, M.A.; Murumkar, P.R.; Sharma, M.K.; Yadav, M.R. Medicinal Chemistry Perspective of Fused Isoxazole Derivatives. *Curr. Top. Med. Chem.* **2016**, *16*, 2863–2883. [CrossRef] [PubMed]
128. Galenko, A.V.; Khlebnikov, A.F.; Novikov, M.S.; Pakalnis, V.V.; Rostovskii, N.V. Recent advances in isoxazole chemistry. *Russ. Chem. Rev.* **2015**, *84*, 335–377. [CrossRef]
129. Sysak, A.; Obmińska-Mrukowicz, B. Isoxazole ring as a useful scaffold in a search for new therapeutic agents. *Eur. J. Med. Chem.* **2017**, *137*, 292–309. [CrossRef]
130. Arya, G.C.; Kaur, K.; Jaitak, V. Isoxazole derivatives as anticancer agent: A review on synthetic strategies, mechanism of action and SAR studies. *Eur. J. Med. Chem.* **2021**, *221*, 113511. [CrossRef]
131. Fernald, K.; Kurokawa, M. Evading apoptosis in cancer. *Trends Cell Biol.* **2013**, *23*, 620–633. [CrossRef]

132. Tonks, N.K. Protein Tyrosine Phosphatases: From Genes, to Function, to Disease. *Nat. Rev. Mol. Cell Biol.* **2006**, *7*, 833–846. [CrossRef] [PubMed]
133. Taldone, T.; Gozman, A.; Maharaj, R.; Chiosis, G. Targeting Hsp90: Small-molecule inhibitors and their clinical development. *Curr. Opin. Pharmacol.* **2008**, *8*, 370–374. [CrossRef] [PubMed]
134. Sharp, S.Y.; Prodromou, C.; Boxall, K.; Powers, M.V.; Holmes, J.L.; Box, G.; Workman, P. Inhibition of The Heat Shock Protein 90 Molecular Chaperone in vitro and in vivo by Novel, Synthetic, Potent Resorcinylic Pyrazole/isoxazole Amide Analogues. *Mol. Cancer Ther.* **2007**, *6*, 1198–1211. [CrossRef] [PubMed]
135. Eccles, S.A.; Massey, A.; Raynaud, F.I.; Sharp, S.Y.; Box, G.; Valenti, M.; Patterson, L.; de Haven Brandon, A.; Gowan, S.; Boxall, F.; et al. NVP-AUY922: A Novel Heat Shock Protein 90 Inhibitor Active against Xenograft Tumor Growth, Angiogenesis, and Metastasis. *Cancer Res.* **2008**, *68*, 2850–2860. [CrossRef] [PubMed]
136. Chen, D.; Shen, A.; Li, J.; Shi, F.; Chen, W.; Ren, J.; Liu, H.; Xu, Y.; Wang, X.; Yang, X.; et al. Discovery of potent N-(isoxazol-5-yl)amides as HSP90 inhibitors. *Eur. J. Med. Chem.* **2014**, *87*, 765–781. [CrossRef] [PubMed]
137. Sun, J.; Lin, C.; Qin, X.; Dong, X.; Tu, Z.; Tang, F.; Chen, C.; Zhang, J. Synthesis and biological evaluation of 3,5-disubstituted-4-alkynylisoxozales as a novel class of HSP90 inhibitors. *Bioorg. Med. Chem. Lett.* **2015**, *25*, 3129–3134. [CrossRef]
138. Shi, W.; Hu, J.; Bao, N.; Li, D.; Chen, L.; Sun, J. Design, synthesis and cytotoxic activities of scopoletin-isoxazole and scopoletin-pyrazole hybrids. *Bioorg. Med. Chem. Lett.* **2017**, *27*, 147–151. [CrossRef]
139. Abbasi, M.; Sadeghi-Aliabadi, H.; Amanlou, M. Prediction of New Hsp90 Inhibitors Based on 3, 4-Isoxazolediamide Scaffold Using QSAR Study, Molecular Docking and Molecular Dynamic Simulation. *Daru J. Pharm. Sci.* **2017**, *25*, 17. [CrossRef]
140. Abbasi, M.; Sadeghi-Aliabadi, H.; Amanlou, M. 3D-QSAR, Molecular Docking, and Molecular Dynamic Simulations for Prediction of New Hsp90 Inhibitors Based on Isoxazole Scaffold. *J. Biomol. Struct. Dyn.* **2018**, *36*, 1463–1478. [CrossRef]
141. Kaushik, S.; Sanawar, R.; Lekshmi, A.; Chandrasekhar, L.; Nair, M.; Bhatnagar, S.; Santhoshkumar, T.R. ER alpha selective chromone, isoxazolylchromones, induces ROS-mediated cell death without autophagy. *Chem. Biol. Drug Des.* **2019**, 1352–1367.
142. Jung, J.; Kwon, J.; Hong, S.; Moon, A.-N.; Jeong, J.; Kwon, S.; Kim, J.-A.; Lee, M.; Lee, H.; Lee, J.H.; et al. Discovery of novel heat shock protein (Hsp90) inhibitors based on luminespib with potent antitumor activity. *Bioorg. Med. Chem. Lett.* **2020**, *30*, 127165. [CrossRef] [PubMed]
143. Aissa, I.; Abdelkafi-Koubaa, Z.; Chouaïb, K.; Jalouli, M.; Assel, A.; Romdhane, A.; Harrath, A.H.; Marrakchi, N.; Ben Jannet, H. Glioblastoma-specific anticancer activity of newly synthetized 3,5-disubstituted isoxazole and 1,4-disubstituted triazole-linked tyrosol conjugates. *Bioorg. Chem.* **2021**, *114*, 105071. [CrossRef] [PubMed]
144. Chaudhary, M.; Kumar, N.; Baldi, A.; Chandra, R.; Babu, M.A.; Madan, J. Chloro and bromo-pyrazole curcumin Knoevenagel condensates augmented anticancer activity against human cervical cancer cells: Design, synthesis, in silico docking and in vitro cytotoxicity analysis. *J. Biomol. Struct. Dyn.* **2020**, *38*, 200–218. [CrossRef] [PubMed]
145. Rai, G.; Urban, D.J.; Mott, B.T.; Hu, X.; Yang, S.-M.; Benavides, G.A.; Johnson, M.S.; Squadrito, G.L.; Brimacombe, K.R.; Lee, T.D.; et al. Pyrazole-Based Lactate Dehydrogenase Inhibitors with Optimized Cell Activity and Pharmacokinetic Properties. *J. Med. Chem.* **2020**, *63*, 10984–11011. [CrossRef]
146. Azimi, F.; Azizian, H.; Najafi, M.; Hassanzadeh, F.; Sadeghi-Aliabadi, H.; Ghasemi, J.B.; Mahdavi, M. Design and Synthesis of Novel Quinazolinone-Pyrazole Derivatives as Potential α-Glucosidase Inhibitors: Structure-activity Relationship, Molecular Modeling and Kinetic Study. *Bioorg. Chem.* **2021**, *114*, 105127–105134. [CrossRef]
147. Abbasi, M.; Amanlou, M.; Aghaei, M.; Bakherad, M.; Doosti, R.; Sadeghi-Aliabadi, H. New heat shock protein (Hsp90) inhibitors, designed by pharmacophore modeling and virtual screening: Synthesis, biological evaluation and molecular dynamics studies. *J. Biomol. Struct. Dyn.* **2020**, *38*, 3462–3473. [CrossRef]
148. Bennani, F.E.; Doudach, L.; Cherrah, Y.; Ramli, Y.; Karrouchi, K.; Ansar, M.; Faouzi, M.E.A. Overview of recent developments of pyrazole derivatives as an anticancer agent in different cell line. *Bioorg. Chem.* **2020**, *97*, 103470. [CrossRef]
149. Bai, S.-Y.; Dai, X.; Zhao, B.-X.; Miao, J.-Y. Discovery of a novel fluorescent HSP90 inhibitor and its anti-lung cancer effect. *RSC Adv.* **2014**, *4*, 19887–19890. [CrossRef]
150. Abbasi, M.; Sadeghi-Aliabadi, H.; Hassanzadeh, F.; Amanlou, M. Prediction of dual agents as an activator of mutant p53 and inhibitor of Hsp90 by docking, molecular dynamic simulation and virtual screening. *J. Mol. Graph. Model.* **2015**, *61*, 186–195. [CrossRef]
151. Uno, T.; Kawai, Y.; Yamashita, S.; Oshiumi, H.; Yoshimura, C.; Mizutani, T.; Kitade, M. Discovery of 3-Ethyl-4-(3-isopropyl-4-(4-(1-methyl-1 H-pyrazol-4-yl)-1 H-imidazol-1-yl)-1 H-pyrazolo [3,4-b] pyridin-1-yl) benzamide (TAS-116) as a Potent, Selective, and Orally Available Hsp90 Inhibitor. *J. Med. Chem.* **2018**, *62*, 531–551. [CrossRef]
152. Mettu, A.; Talla, V.; Bajaj, D.M.; Subhashini, N.J. Design, Synthesis, Molecular Docking Studies of Novel Pyrazolyl 2-Aminopyrimidine Derivatives as Hsp90 Inhibitors. *Arch. Pharm.* **2019**, *352*, 1900063. [CrossRef] [PubMed]
153. Mettu, A.; Talla, V.; Naikal, S.J.P. Novel anticancer Hsp90 inhibitor disubstituted pyrazolyl 2-aminopyrimidine compound 7t induces cell cycle arrest and apoptosis via mitochondrial pathway in MCF-7 cells. *Bioorg. Med. Chem. Lett.* **2020**, *30*, 127470. [CrossRef] [PubMed]
154. Mohamady, S.; Ismail, M.; Mogheith, S.M.; Attia, Y.M.; Taylor, S.D. Discovery of 5-aryl-3-thiophen-2-yl-1H-pyrazoles as a new class of Hsp90 inhibitors in hepatocellular carcinoma. *Bioorg. Chem.* **2020**, *94*, 103433. [CrossRef] [PubMed]
155. Kadasi, S.; Yerroju, R.; Gaddam, S.; Pullanagiri, N.; Chary, M.; Pingili, D.; Raghavendra, N.M. Discovery of N-Pyridoyl-Δ2-pyrazolines as Hsp90 Inhibitors. *Arch. Der Pharm.* **2020**, *353*, 190019. [CrossRef]

pharmaceutics

Article

Bruton's Tyrosine Kinase Inhibitor Zanubrutinib Effectively Modulates Cancer Resistance by Inhibiting Anthracycline Metabolism and Efflux

Lucie Čermáková [1], Jakub Hofman [2], Lenka Laštovičková [1], Lucie Havlíčková [1], Ivona Špringrová [1], Eva Novotná [1] and Vladimír Wsól [1,*]

1 Department of Biochemical Sciences, Faculty of Pharmacy, Charles University, Akademika Heyrovskeho 1203, 50005 Hradec Kralove, Czech Republic
2 Department of Pharmacology and Toxicology, Faculty of Pharmacy, Charles University, Akademika Heyrovskeho 1203, 50005 Hradec Kralove, Czech Republic
* Correspondence: wsol@faf.cuni.cz

Abstract: Zanubrutinib (ZAN) is a Bruton's tyrosine kinase inhibitor recently approved for the treatment of some non-Hodgkin lymphomas. In clinical trials, ZAN is often combined with standard anthracycline (ANT) chemotherapy. Although ANTs are generally effective, drug resistance is a crucial obstacle that leads to treatment discontinuation. This study showed that ZAN counteracts ANT resistance by targeting aldo-keto reductase 1C3 (AKR1C3) and ATP-binding cassette (ABC) transporters. AKR1C3 catalyses the transformation of ANTs to less potent hydroxy-metabolites, whereas transporters decrease the ANT-effective concentrations by pumping them out of the cancer cells. In our experiments, ZAN inhibited the AKR1C3-mediated inactivation of daunorubicin (DAUN) at both the recombinant and cellular levels. In the drug combination experiments, ZAN synergistically sensitised AKR1C3-expressing HCT116 and A549 cells to DAUN treatment. Gene induction studies further confirmed that ZAN did not increase the intracellular level of AKR1C3 mRNA; thus, the drug combination effect is not abolished by enzyme induction. Finally, in accumulation assays, ZAN was found to interfere with the DAUN efflux mediated by the ABCB1, ABCG2, and ABCC1 transporters, which might further contribute to the reversal of ANT resistance. In summary, our data provide the rationale for ZAN inclusion in ANT-based therapy and suggest its potential for the treatment of tumours expressing AKR1C3 and/or the above-mentioned ABC transporters.

Keywords: zanubrutinib; anthracycline; drug resistance; aldo-keto reductase 1C3; ABC drug efflux transporter

Citation: Čermáková, L.; Hofman, J.; Laštovičková, L.; Havlíčková, L.; Špringrová, I.; Novotná, E.; Wsól, V. Bruton's Tyrosine Kinase Inhibitor Zanubrutinib Effectively Modulates Cancer Resistance by Inhibiting Anthracycline Metabolism and Efflux. *Pharmaceutics* **2022**, *14*, 1994. https://doi.org/10.3390/pharmaceutics14101994

Academic Editor: Francesca Musumeci

Received: 5 August 2022
Accepted: 14 September 2022
Published: 21 September 2022

1. Introduction

In recent decades, an understanding of the processes related to the pathogenesis of cancer has given rise to targeted therapy. Anticancer drugs target a wide range of cellular structures and signalling pathway components involved in cancer cell growth and survival. One such important target is Bruton's tyrosine kinase (BTK), belonging to non-receptor tyrosine kinases of the TEC family [1,2]. BTK has been recognised as a key player in the B-cell receptor signalling pathway, the dysregulation of which is implicated in B-cell malignancies [3,4]. Abundant expression of BTK has been identified in B-cell chronic lymphocytic leukaemia (CLL) and B-cell lymphomas, making it a promising target for the development of new small-molecule inhibitors [5–7].

The first BTK inhibitor that was given accelerated approval by the U.S. Food and Drug Administration (FDA) was ibrutinib (IBR) in 2013, followed by acalabrutinib (ACA) in 2017 [8,9]. The incorporation of IBR into cancer therapy is a milestone that has changed the fate of many patients with CLL, Waldenström's macroglobulinaemia, and some non-Hodgkin lymphomas [10–14]. IBR and ACA are strong BTK inhibitors that covalently and irreversibly interact with Cys-481 in the ATP-binding domain of BTK. The inhibitory activity of IBR has also been demonstrated in other kinases containing homologous cysteine residues such as Itk and Tec [6,15,16]. The combined inhibition of different kinases may contribute to some adverse effects of IBR, such as bleeding and cardiovascular complications [17,18]. Thus, although IBR and ACA are the most effective agents for treating several B-cell malignancies, treatment-limiting adverse effects and emerging cases of cancer resistance [19,20] have encouraged the search for new BTK inhibitors. Various compounds (e.g., vecabrutinib, fenebrutinib, and tirabrutinib) have been identified as BTK inhibitors and have entered clinical trials in recent years [21,22]. However, in addition to IBR and ACA, only zanubrutinib (ZAN) has been approved by the FDA [23–25].

In clinical trials, IBR, ACA, and ZAN have been tested alone and in combination with anthracycline (ANT)-containing standard chemotherapy [26–29]. ANTs, even in the era of targeted therapy, represent the gold standard for the treatment of haematological malignancies. Daunorubicin (DAUN) and doxorubicin (DOX) are routinely used to treat acute myeloid leukaemia (AML), acute lymphoblastic leukaemia, and aggressive lymphomas. Therefore, emerging ANT resistance is a serious complication that is often responsible for the failure of cancer therapy. The decreased cancer sensitivity to ANTs may be due to several mechanisms. Metabolic drug inactivation by anthracycline reductases and increased efflux mediated by ATP-binding cassette (ABC) transporters are known to contribute significantly to this phenomenon [30–33].

Recently, we reported that IBR and ACA interact with aldo-keto reductase 1C3 (AKR1C3) [34], an enzyme that is overexpressed in several malignancies [35–38]. AKR1C3 catalyses the two-electron reduction of DAUN and DOX, converting them to alcohol metabolites that possess lower anticancer activities than their parent drugs [32,39,40]. In our previous experiments, IBR and ACA inhibited the AKR1C3-mediated inactivation of DAUN, leading to an increase in its effectiveness against cancer cells [34]. Furthermore, IBR sensitises cancer cells to DOX by inhibiting the ABCC1 transporter [41]. Knowing these off-target activities of IBR and ACA and considering their structural similarities within BTK inhibitors, we aimed to investigate whether there might be a rationale for combining the novel second-generation BTK inhibitor ZAN and ANTs for therapeutic purposes. The assumption that targeting AKR1C3 and/or ABC transporters may improve the clinical outcome of ANTs [42,43] allowed us to investigate whether ZAN interacts with these resistance mediators and counteracts their activities. First, we performed inhibitory studies on recombinant enzymes at the cellular level. Next, drug combination studies were performed to examine whether ZAN acts as a dual-activity resistance modulator. As AKR1C3 may influence the resistance phenotype of cancer cells, we performed induction studies focusing on the expression of this enzyme.

2. Materials and Methods

2.1. Material

ZAN was purchased from MedChemExpress LLC (Monmouth Junction, NJ, USA). Nicotinamide adenine dinucleotide phosphate (NADP+), glucose-6-phosphate, glucose-6-phosphate dehydrogenase (Roche), fetal bovine serum (FBS), and high-performance liquid chromatography (HPLC) grade solvents were obtained from Merck (Prague, Czech Republic). Phenazine methosulfate (PMS) and 2,3-Bis-(2-Methoxy-4-Nitro-5-Sulfophenyl)-2H-Tetrazolium-5-Carboxanilide (XTT) tetrazolium salt were purchased from BioTech (Prague, Czech Republic). DAUN was supplied by SelleckChem (Houston, TX, USA). Daunorubicinol (DAUN-OL), a C13-hydroxy-metabolite of DAUN, was obtained from Toronto Research Chemicals (Toronto, ON, Canada). JetPrime transfection reagent (Polyplus Transfection, Illkirch, France) was provided by VWR International Ltd. (Radnor, PA, USA). Cell culture reagents were obtained from Lonza (Walkersville, MD, USA) and Merck (St. Louis, MO, USA). TRI reagent solution was purchased from Molecular Research Center Inc. (Cincinnati, OH, USA). The RNA extraction kit was supplied by Zymo Research (Irvine, CA, USA). AKR1C3-specific primers (Forward primer: 5′-GCCAGGTGAGGAACTTTCAC-

3′, Reverse primer: 5′-AGTTTGACACCCCAATGGAC-3′) and oligo-dT were purchased from Generi Biotech (Hradec Králové, Czech Republic). ProtoScript II buffer, dithiothreitol (DTT), dNTP mix, and ProtoScript II Reverse Transcriptase were obtained from New England BioLabs (Ipswich, MA, USA). The qPCR SG mix was acquired from the Institute of Applied Biotechnologies (Prague, Czech Republic).

2.2. Cell Culture

Human colorectal carcinoma HCT116 cell line and human bone marrow AML KG1α cells were purchased from the European Collection of Authenticated Cell Cultures (Salisbury, UK). Human liver carcinoma HepG2 and human non-small cell lung carcinoma A549 cells were obtained from the American Type Culture Collection (Manassas, VA, USA). HCT116, HepG2, and A549 cells were cultured in Dulbecco's modified Eagle's medium (DMEM) with 10% FBS. KG1α leukaemia cells were maintained in Iscove's modified Dulbecco's medium (IDMEM) supplemented with 20% FBS and 2 mM L-glutamine. All cells were cultured under standard conditions (37 °C, 5% CO_2) and examined for mycoplasma infections. Routine cultivation and all experiments were performed in an antibiotic-free medium. Cell passages of 10–25 were used in the experiments. ZAN was dissolved in dimethyl sulfoxide (DMSO), and the concentration of DMSO in the experiments did not exceed 0.5% (*v/v*). The effect of DMSO on cell viability was observed to be negligible in the validation experiments. Furthermore, possible interfering effects on other biological processes were eliminated by using of vehicle controls.

2.3. Cloning, Overexpression, and Purification of Recombinant Enzymes

As described previously, human recombinant AKR1C3 was prepared in *E. coli* BL21(DE3) [44] and purified using NGC chromatography system equipped with a 1 ml HisTrap FF column (GE Healthcare Life Sciences, Marlborough, MA, USA) [45].

2.4. Inhibitory Assays

To determine the value of half-maximal inhibitory concentration (IC_{50}), human recombinant AKR1C3 (1.5 µg) was incubated with ZAN (0.01–100 µM) or vehicle DMSO. The reaction was carried out in 0.1 M sodium phosphate buffer (pH 7.4) containing NADP+ regeneration system (2.54 mM NADP, 19.72 mM glucose-6-phosphate, 10 mM $MgCl_2$, and 0.88 U glucose-6-phosphate-dehydrogenase), as described previously [45]. After 10 min of pre-incubation, the reactions were started with a DAUN (500 µM) addition, and the solution was further incubated for 30 min. Methanol (300 µL) was used to stop the reactions, and the samples were cooled on ice and allowed to precipitate for 1 h at 4 °C. After centrifugation (9050$\times$ *g*, 10 min), the solution was filtered through a 0.2 µm PTFE membrane (Whatman, GE Healthcare, Uppsala, Sweden). The amount of DAUN-OL was analysed using ultra-high-performance liquid chromatography (UHPLC). To determine the mode of interaction between AKR1C3 and ZAN, AKR1C3 was incubated with DAUN (200–2000 µM) and ZAN (1, 5, 10 µM) or DMSO, and the Michaelis–Menten kinetic parameters were calculated and transformed into a Lineweaver–Burk plot using GraphPad Prism 9.3.1 (GraphPad Software, Inc., La Jolla, CA, USA).

2.5. Inhibition of AKR1C3 in Transiently Transfected HCT116 Cells

Vector coding for AKR1C3 (pCI_AKR1C3) and pCI empty vector (EV) was generated, purified from *E. coli* HB101, and used to transfect HCT116 cells, as described previously [43,46]. In brief, HCT116 cells (approximately 30 $\times$ 10^4 cells/well) were seeded into 24-well plates and cultured for 24 h under standard conditions (5% CO_2, 37 °C). The transfection mixture (0.25 µg plasmid and 0.75 µL jetPRIME® Transfection Reagent in jetPRIME buffer) was prepared according to the manufacturer's protocol (Polyplus Transfection, Illkirch, France). After 24 h, the culture medium was refreshed, the transfection mixture (37.5 µL) was added dropwise to each well, and the cells were incubated (37 °C, 5% CO_2) for 24 h before being used for subsequent experiments. The functional expression of AKR1C3 was verified, as previously described [43].

To study the inhibition of AKR1C3 in HCT116-transfected cells, the culture medium was harvested and replaced with fresh medium containing DAUN (1 μM) and ZAN (1, 5, 10, and 50 μM) or DMSO (vehicle). HCT116 cells harbouring EV (HCT116-EV) or pCI_AKR1C3 (HCT116-AKR1C3) were incubated with DAUN and ZAN, or its vehicle DMSO, for 2 and 4 h (37 °C, 5% CO_2). At specified intervals, the culture medium was collected and cells were lysed with 200 μL of lysis buffer (25 mM Tris, 150 mM NaCl, and 1% Triton X-100, pH 7.8) for 15 min at room temperature (25 °C). The medium was mixed with the cell lysate, DAUN and DAUN-OL were extracted twice with 1 mL of ethyl acetate, and the organic phases were evaporated under vacuum. These residues were then dissolved in the mobile phase and analysed using UHPLC.

2.6. UHPLC

The amount of DAUN-OL was determined using UHPLC with fluorescence detection. Briefly, an Agilent 1290 Series UHPLC chromatographic system equipped with a Zorbax C18 Eclipse Plus (2.1 × 50 mm, 1.8 μm) column and a 1290 Infinity inline filter (Agilent, Santa Clara, CA, USA) was used. The mobile phase was a mixture of 0.1% formic acid in water and acetonitrile at a ratio of 74:26 (*v/v*). Isocratic elution and a flow rate of 0.7 mL/min were used. The excitation and emission wavelengths of the detector were 480 and 560 nm, respectively.

2.7. Drug Combination Assays

Prior to drug combination studies, HCT116 cells were transfected using the same procedure as described above, but optimised for the 96-well plate. In detail, HCT116 cells (approximately 8×10^3 cells/well) were seeded in 96-well plates and cultured for 24 h (5% CO_2, 37 °C). The next day, the cells were transfected with pCI_AKR1C3 or EV, according to the manufacturer's protocol (Polyplus Transfection, Illkirch, France). In each well of the 96-well plate, a mixture of 0.10 μg plasmid and 0.20 μL jetPRIME® Transfection Reagent in jetPRIME buffer was added and incubated for 10 min at room temperature (25 °C). The polyplexes (5 μL) were then added dropwise to HCT116 cells cultured in 100 μL of fresh DMEM supplemented with 10% FBS and further incubated for 24 h (5% CO_2, 37 °C). A549 cells (approximately 5×10^3 cells/well) were seeded into 96-well plates, cultured under standard conditions (37 °C, 5% CO_2) for 24 h, and then directly used for the experiments. To determine whether ZAN counteracts AKR1C3-mediated DAUN resistance, the culture medium was changed to a fresh medium containing DAUN (0.1–5 μM for HCT116-AKR1C3 and HCT116-EV; 0.01–1 μM for A549) and ZAN (5 and 10 μM for HCT116-AKR1C3 and HCT116-EV; 1 and 5 μM for A549), or DMSO (a vehicle). The cells were incubated for 72 h under standard conditions (37 °C, 5% CO_2) before their viability assessment by XTT assay. Apart from these combinations, the cytotoxicity of ZAN itself (0.1–50 μM) was also determined for both cell lines. The Chou–Talalay method was used to quantify combination effects [47,48].

2.8. XTT Cell Proliferation Assay

An XTT solution consisting of XTT powder (1 mg/mL) and phenazine methosulfate (3 mg/mL) in phosphate-buffered saline (PBS) was added to HCT116, A549, HepG2, and KG1α cells and incubated for 1.5 h (HCT116, A549, HepG2) or 4 h (KG1α). Absorbance was measured at 450 nm using a microplate reader (Infinite M200; Tecan, Salzburg, Austria). This method was used to verify the lack of ZAN toxicity in cellular inhibitory assays, drug combination studies, and induction experiments.

2.9. Accumulation Studies in KG1α and A549 Cells

To study the effect of ZAN on DAUN accumulation, KG1α (approximately 50×10^4 cells/well) and A549 cells (approximately 35×10^4 cells/well) were seeded into 24- and 12-well plates, respectively. After 24 h, ZAN (1, 5, 10, and 25 μM), LY335979 (2 μM), Ko143 (2 μM), and MK571 (25 μM) (selective inhibitors of the ABC transporters ABCB1, ABCG2, and ABCC1, respectively) were added to the cells and incubated for 15 min. Then,

2 and 1 μM DAUN were pipetted into A549 and KG1α cell cultures, respectively. Following 1 h of incubation, the cells were washed twice with cold PBS and used for further analysis of DAUN accumulation using a Sony SA3800 Spectral Cell Analyser (Sony Biotechnology, San Jose, CA, USA) flow cytometer with a laser set to 488 nm. The SA3800 software (Sony Biotechnology, San Jose, CA, USA) was used to evaluate the data.

2.10. *Determination of ZAN Cytotoxicity in HepG2 and KG1α Cells*

Prior to the induction studies, the cytotoxicity of ZAN alone was tested to determine a relatively non-toxic concentration suitable for subsequent experiments. HepG2 (approximately 18 × 10^3 cells/well) and KG1α (approximately 25 × 10^3 cells/well) cells were seeded into 96-well plates, and either on the same day (KG1α), or after a 24 h-incubation (HepG2 cells), were treated with ZAN (0.1–50 μM) or DMSO (vehicle). After incubation for 48 h under standard conditions (37 °C, 5% CO_2), the effect of ZAN on cancer cell viability was measured using the XTT assay.

2.11. *Induction Studies in HepG2 and KG1α Cells*

Induction studies were performed as previously described [46]. HepG2 (approximately 20 × 10^4 cells/well) and KG1α (approximately 35 × 10^4 cells/well) cells were seeded into 48-well plates. While the KG1α cells were immediately treated with ZAN (0.5 μM) or DMSO (a vehicle), the HepG2 cells were first cultured for 24 h and then treated with 0.5 μM ZAN or a vehicle control. After 24 and 48 h of incubation, the medium was removed, the cells were lysed in TriReagent, and the total RNA was isolated using the Zymo Research Direct-zol™ RNA Miniprep kit and transcribed into cDNA, according to the manufacturer's instructions (ProtoScript® II Reverse Transcriptase, New England Biolabs, Ipswich, MA, USA). AKR1C3 mRNA levels were quantified using quantitative real-time PCR (qRT-PCR). The PCR contained 1x concentrated XCEED qPCR SG Mix, AKR1C3-specific primers (1 μM), and 20 ng of cDNA. qPCR was performed using QuantStudio 6flex (Applied Biosystems by Life Technologies, Carlsbad, CA, USA) set to the following conditions: initial denaturation at 95 °C for 10 min, followed by 40 cycles at 95 °C for 15 s and 65 °C for 1 min. Absolute quantification of AKR1C3 expression was achieved by comparing samples with the concomitantly amplified AKR1C3 standard, which was generated as described previously [43].

2.12. *Statistical Analysis*

Statistical analysis of the data was performed using GraphPad Prism software version 9.3.1 (GraphPad Software Inc., La Jolla, CA, USA) using ordinary one-way analysis of variance (ANOVA) followed by Dunnett's post hoc test or unpaired t-test with Welch's correction, as specified in the respective figure legends. Statistical significance was set at $p < 0.05$. In the case of drug combination studies, the combination indices (CI) were used to distinguish between synergism (CI < 0.9), additivity (CI = 0.9–1.1), and antagonism (CI > 1.1), which were calculated using CompuSyn software (version 1.0; ComboSyn Inc., Paramus, NJ, USA).

3. Results

3.1. *ZAN Inhibits the Recombinant AKR1C3-Mediated DAUN Reduction In Vitro*

First, we investigated whether ZAN could inhibit ANT metabolism using the recombinant enzyme. AKR1C3 was incubated with the substrate DAUN, with or without ZAN. As can be seen from the half-maximal inhibitory concentration (IC_{50}) (Figure 1A), ZAN interacted with AKR1C3 and potently inhibited the reduction of DAUN to C13-hydroxy-metabolite daunorubicinol (DAUN-OL) in a dose-dependent manner. Furthermore, Michaelis–Menten kinetics and the Lineweaver–Burk double reciprocal plot provided evidence of a mixed-type mode of inhibition (inhibition constant (Ki) = 2.9 μM, α > 1.0) (Figure 1B,C).

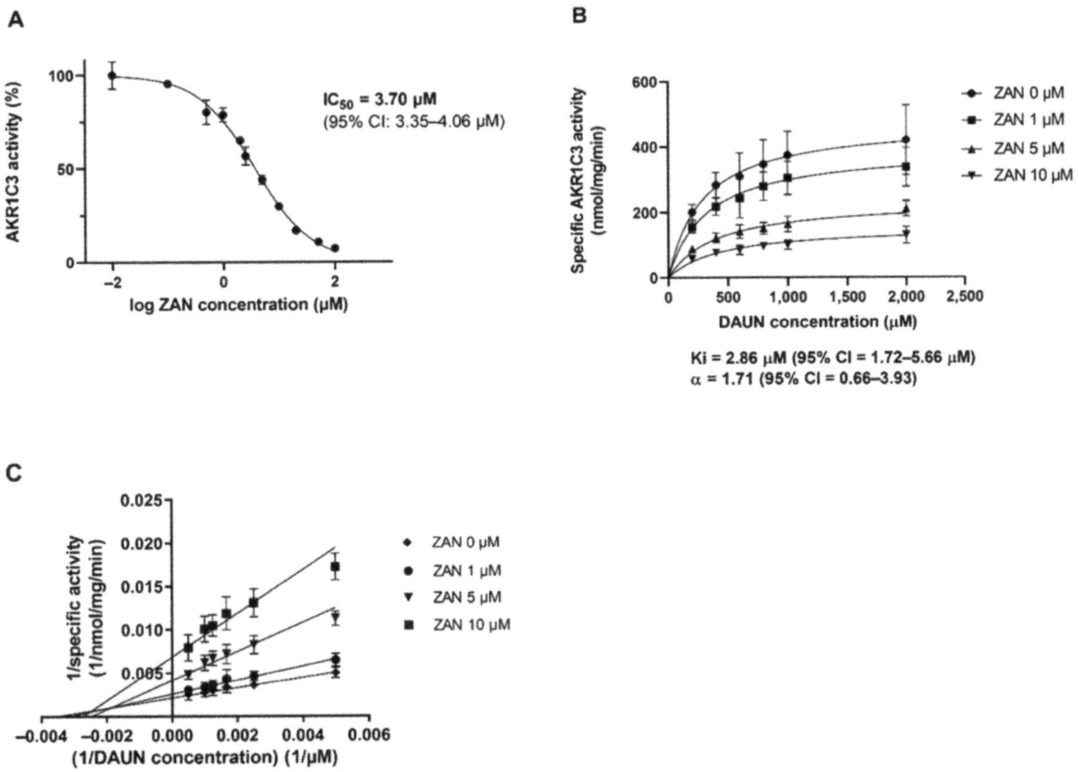

Figure 1. Zanubrutinib (ZAN) inhibits daunorubicin (DAUN) reduction catalysed by human recombinant AKR1C3. (**A**) IC_{50} of AKR1C3 inhibition. (**B**) Determination of the inhibition constant (Ki). (**C**) Mode of inhibition graphically represented by the Lineweaver–Burk plot. All data are expressed as the mean ± standard deviation (SD) from at least three independent experiments. 95% CI = 95% confidence interval.

3.2. Effect of ZAN on AKR1C3-Mediated DAUN Metabolism in Transfected HCT116 Cells

Next, we investigated the effect of ZAN on AKR1C3-mediated DAUN metabolism in intact cells. HCT116 cells were chosen because of their negligible endogenous expression of AKR1C3 and high transfectability [43]. HCT116 cells were transfected with pCI_AKR1C3 or an EV. Transfected cells were treated with DAUN (1 μM) and ZAN (or vehicle control), which were used at concentrations that were non-toxic to HCT116 cells, at a reaction interval of 2 and 4 h (data not shown). Figure 2 shows that ZAN, at all concentrations tested, significantly inhibited the conversion of DAUN to DAUN-OL in HCT116-AKR1C3 cells, proving that ZAN can influence the intracellular activity of the enzyme.

3.3. ZAN Overcomes DAUN Resistance in A549 and Transfected HCT116 Cells

After confirming that ZAN inhibits AKR1C3 intracellularly, we analysed whether a decrease in DAUN metabolism modulates ANT resistance in HCT116 cells. HCT116-AKR1C3 and HCT116-EV cells were treated with various combinations of DAUN and ZAN. In HCT116 cells expressing AKR1C3, the value of IC_{50} for DAUN alone was almost two times higher than the value of IC_{50} obtained from experiments with HCT116-EV cells. These data demonstrate that the expression of AKR1C3 protects cancer cells from DAUN toxicity. Importantly, as can be seen from the shifts in IC_{50} values, the DAUN resistance of HCT116-AKR1C3 cells was reversed when ZAN (5 or 10 μM) was added to the culture

(Figure 3A). No such significant changes in IC_{50} were found in experiments with HCT116-EV cells (Figure 3B), confirming the critical role of the ZAN-AKR1C3 interaction in the DAUN resistance reversal effect. Combination effects were precisely quantified using the Chou–Talalay method [47,48]. Combination index (CI) vs. fraction of cells affected (F_A) plots were created. Cells that survived the treatment were considered unaffected. The values of CI for DAUN with ZAN (5 or 10 μM) resulted in synergism (CI = 0.3–0.9) in HCT116-AKR1C3 cells. The lowest value of CI was detected when 10 μM ZAN was combined with 0.25 μM DAUN (CI = 0.33 ± 0.11). In contrast, additivity to antagonism was detected when the examined drug combinations were tested in HCT116-EV cells (CI = 0.9–10) (Figure 3C,D).

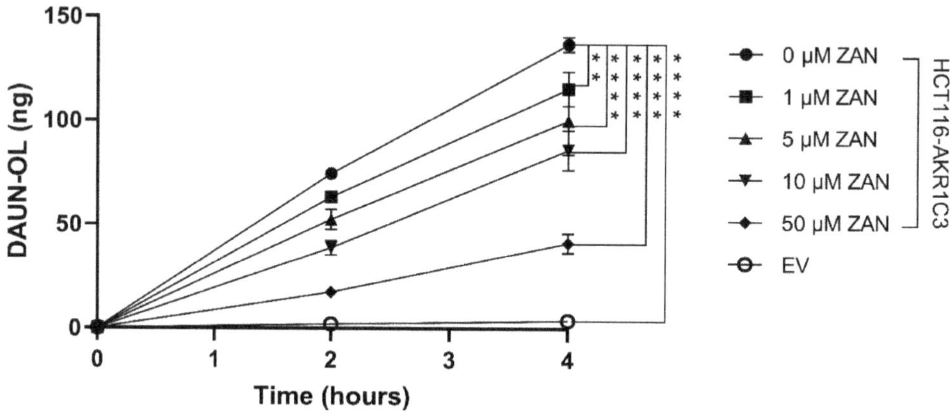

Figure 2. Zanubrutinib (ZAN) inhibits AKR1C3-mediated daunorubicin (DAUN) metabolism in intact cells. Transfected HCT116 cells were incubated with DAUN (1 μM) in combination with ZAN (1, 5, 10 and 50 μM) or its vehicle dimethyl sulfoxide (DMSO) for 2 and 4 h. DAUN-OL was then extracted and its amount was analysed by ultra-high-performance liquid chromatography (UHPLC). Data were evaluated with one-way analysis of variance (ANOVA), followed by Dunnett's post hoc test. ** $p < 0.01$ and **** $p < 0.0001$ compared to DMSO vehicle control. The values are expressed as mean ± standard deviation (SD) from three independent experiments.

After confirming that the inhibition of AKR1C3 remains beyond the synergy between DAUN and ZAN in transfected cells, combination studies were performed with the lung adenocarcinoma cell line A549, which naturally expresses considerable amounts of AKR1C3 [43]. Similar to HCT116-AKR1C3 cells, significant shifts in IC_{50} values were detected when A549 cells were treated with a combination of DAUN and ZAN (1 and 5 μM) (Figure 4A). Using the Chou–Talalay method, a synergistic effect was detected over the entire range $F_{A,}$ with the lowest CI value detected for the combination of 5 μM ZAN + 0.75 μM DAUN (CI = 0.06 ± 0.05) (Figure 4B).

3.4. ZAN Increases DAUN Accumulation by Inhibiting the Efflux Activity of ABC Transporters

As mentioned in the Introduction, not only AKR1C3, but also ABC transporters are IBR off-targets [41,49]. ABC transporters actively pump ANTs out of cancer cells, decreasing their effective concentration and contributing to drug resistance [50]. The fact that IBR re-sensitises drug-resistant ABCC1-overexpressing leukaemia cells to DOX [41] inspired us to investigate whether a similar interaction could be observed with ZAN. A549 and KG1α leukaemia cells were selected for our experiments because of their high ABC transporter expression. Using selective modulators LY335979 (inhibitor of ABCB1), Ko143 (ABCG2 inhibitor), and MK571 (ABCC1 inhibitor), along with substrate DAUN, we demonstrated the functional presence of the ABCB1 transporter in KG1α and the existence of ABCG2 and ABCC1 in the A549 cell line (Figure 5A,B). Furthermore, both the cell lines were incubated with a combination of DAUN and ZAN. The obtained data showed that ZAN at 10 and 25 μM inhibited DAUN efflux in A549 cells, while 5, 10, and 25 μM ZAN interacted with the ABCB1 transporter expressed in KG1α cells (Figure 5). Importantly, our results with model inhibitors correlate well with previous reports on the expressions and/or functional activities of examined transporters in selected cellular models [51–54].

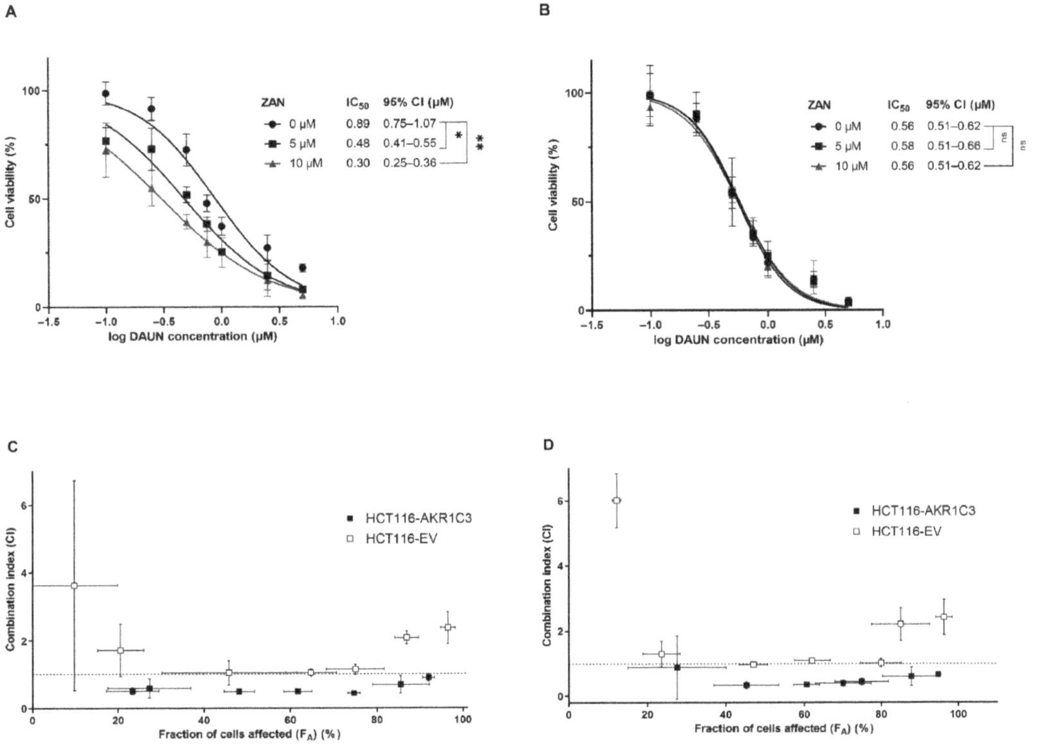

Figure 3. Zanubrutinib (ZAN) counteracts AKR1C3-mediated daunorubicin (DAUN) resistance in cancer cells. HCT116-AKR1C3 (**A**) and HCT116-EV (**B**) cells were treated with dimethyl sulfoxide (DMSO) (a vehicle) or ZAN (5 and 10 μM) and increasing concentrations of DAUN (0.1–5 μM). After 72 h of incubation, cell viability was evaluated using the XTT test. GraphPad Prism 9.3.1. was used for one-way analysis of variance (ANOVA), followed by Dunnett's post hoc test. ns = non-significant, * $p < 0.05$, and ** $p < 0.01$ compared to DMSO vehicle control. 95% CI = 95% confidence interval. Chou–Talalay analysis was conducted to create combination index (CI) vs. fraction of cells affected (F_A) plots for 5 (**C**) and 10 μM ZAN (**D**) in order to distinguish between synergism (<0.9), additivity (0.9–1.1), and antagonism (>1.1). The dotted line was positioned at CI = 1 for better orientation. Data are presented as mean ± standard deviation (SD) from three independent experiments.

3.5. ZAN Does Not Affect AKR1C3 Expression

In addition to inhibition, changes in AKR1C3 expression may affect the pharmacological fate of ANTs. Therefore, in the last part of our study, we investigated how ZAN affects the intracellular levels of AKR1C3 mRNA. Along with the KG1α leukaemia cells, the HepG2 hepatocarcinoma cell line was used as a liver model to evaluate whether ZAN has the potential to influence whole-body DAUN pharmacokinetics. First, we tested the effects of ZAN on the viability of HepG2 and KG1α cells to determine the concentration with tolerable cytotoxicity. Based on these experiments (Figure 6A,B), a pharmacologically relevant [55] and negligibly cytotoxic concentration of 0.5 μM ZAN was used for induction studies. The qRT-PCR results showed that ZAN did not cause any significant changes in

AKR1C3 mRNA expression in either the KG1α or HepG2 cells (Figure 6C,D). These results confirmed that exposure to ZAN does not strengthen the resistance phenotype inside the cancer cells or force DAUN elimination from the body.

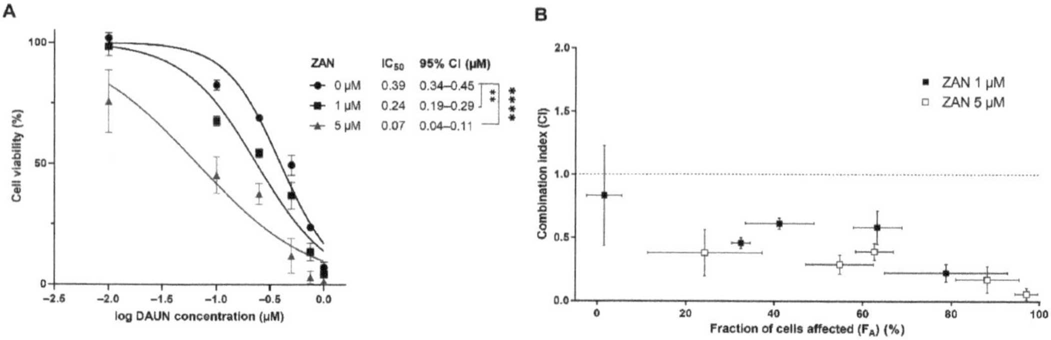

Figure 4. Zanubrutinib (ZAN) counteracts daunorubicin (DAUN) resistance in A549 cells. (**A**) A549 cells were treated with dimethyl sulfoxide (DMSO) (a vehicle) or ZAN (1 and 5 μM) and increasing concentrations of DAUN (0.01–1 μM), incubated for 72 h, and then cell viability was evaluated using the XTT test. GraphPad Prism 9.3.1. was used for one-way analysis of variance (ANOVA), followed by Dunnett's post hoc test. ** $p < 0.01$ and **** $p < 0.0001$ in regards to DMSO vehicle control. 95% CI = 95% confidence interval. (**B**) Chou–Talalay analysis was performed to create the combination index (CI) vs. fraction of cells affected (F_A) plot for 1 and 5 μM ZAN and to distinguish synergism (<0.9) from additivity (0.9–1.1) and antagonism (>1.1). The dotted line was positioned at CI = 1 for better orientation. Data are presented as mean ± standard deviation (SD) from three independent experiments.

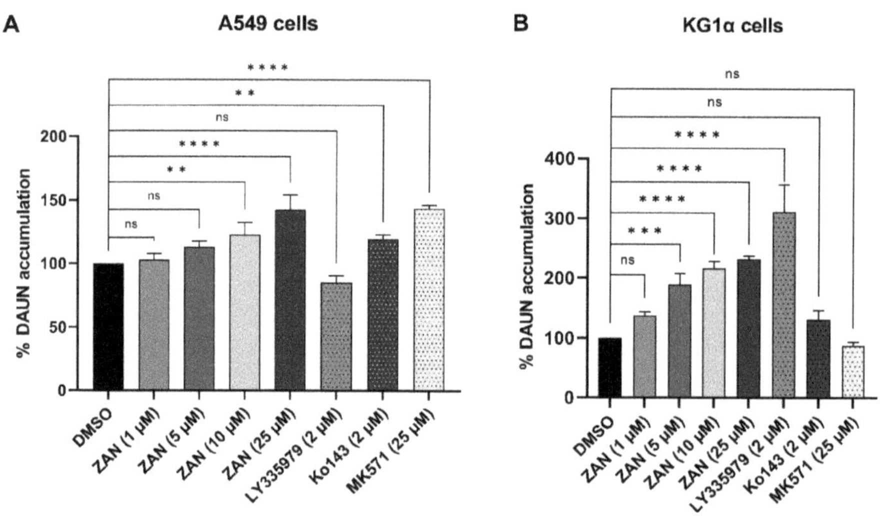

Figure 5. Flow cytometry analysis of daunorubicin (DAUN) accumulation in A549 (**A**) and KG1α cells (**B**). A549 and KG1α cells were preincubated with zanubrutinib (ZAN) (1, 5, 10, and 25 μM) or selective inhibitors of ABC transporters: LY335979 (inhibitor of ABCB1), Ko143 (inhibitor of ABCG2), or MK571 (inhibitor of ABCC1). After preincubation, 2 μM (A549) or 1 μM DAUN (KG1α) were added to the cells. The bar graphs represent the DAUN accumulation in % relative to the dimethyl sulfoxide (DMSO) vehicle control. Data were analysed by one-way analysis of variance (ANOVA), followed by Dunnett's post hoc test; ns = non-significant, ** $p < 0.01$, *** $p < 0.001$, and **** $p < 0.0001$. Data are presented as mean ± standard deviation (SD) from three independent experiments.

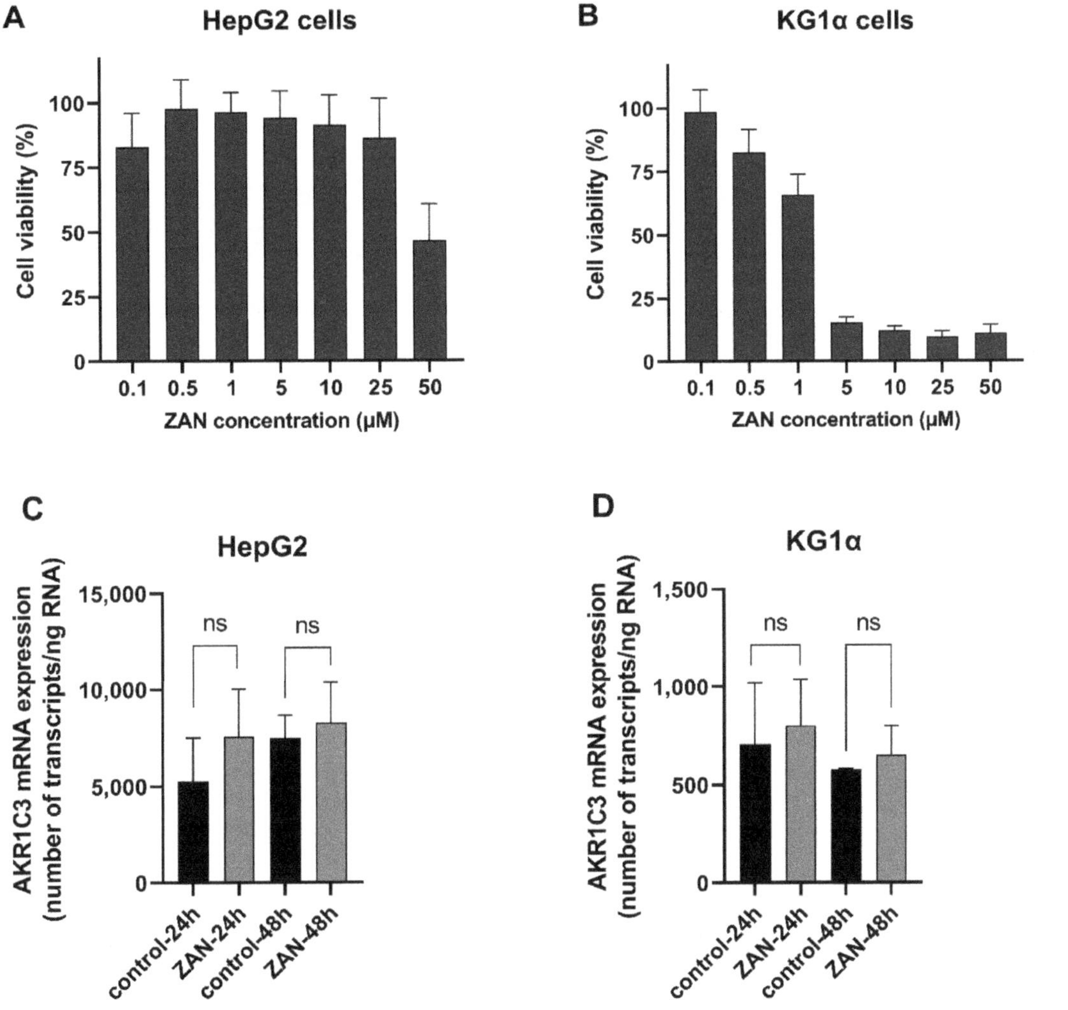

Figure 6. Effect of zanubrutinib (ZAN) on AKR1C3 mRNA expression in HepG2 and KG1α cell lines. To find a non-toxic ZAN concentration, HepG2 (**A**) and KG1α (**B**) cells were treated with ZAN (0.1–50 μM) or dimethyl sulfoxide (DMSO) (a vehicle) for 48 h, and the cell viability was evaluated by the XTT test. Furthermore, determination of AKR1C3 mRNA following ZAN (0.5 μM) exposure was performed in HepG2 (**C**) and KG1α (**D**) cells. qRT-PCR was used to monitor changes in AKR1C3 mRNA expression following a 24 and 48 h incubation. The Student's unpaired t-test with Welch's correction was used to assess statistical significance; ns = non-significant. The data represent mean ± standard deviation (SD) from at least three independent experiments.

4. Discussion

The FDA approved ZAN as a second-line therapy for relapsed/refractory marginal zone lymphoma (MZL) and mantle cell lymphoma (MCL), and as a first-line treatment for Waldenström's macroglobulinaemia [23–25]. Currently, ZAN, either alone or in combination with other drugs, is being studied in 66 clinical trials, of which 48 are currently recruiting, or have not started recruiting yet [56]. In addition to FDA-approved indications, the effectiveness of ZAN has been tested in other B-cell malignancies, such as CLL/small lymphocytic lymphoma (SLL), diffuse large B-cell lymphoma (DLBCL), CNS lymphoma, and follicular lymphoma [57–61]. CLL/SLL and indolent lymphomas are typically characterised by slow growth rates that allow a "watch and wait" strategy, if no symptoms are present [62–65]. However, genetic alterations leading to blastic transformations, such as Richter's syndrome in B-CLL, can cause serious complications. They drive the aggression of cancer cells, and the disease progression requires immediate treatment [66,67]. Although rituximab + cyclophosphamide + ANT + vincristine + prednisone (R-CHOP) and other ANT-containing chemotherapy regimens are generally effective in such cases, some patients do not respond to the therapy. Patients who are naturally insensitive or resistant to drug therapy during treatment require new strategies to overcome resistance issues and prevent disease relapse [68].

Several factors that impair the effectiveness of ANT treatment in cancer cells have been identified. Metabolic inactivation and drug efflux, mediated by ABC transporters, are among the most important factors. AKR1C3 reduces DAUN and DOX to less potent C13-hydroxy-metabolites, whereas ABC transporters pump ANTs out of cancer cells, decreasing their effective concentration [30–33]. In our experiments, ZAN inhibited AKR1C3 at the level of a recombinant enzyme in AKR1C3-overexpressing intact cells. Synergism between DAUN and ZAN was detected in HCT116-AKR1C3 cells when ZAN was tested at concentrations of 5 and 10 μM. In A549 cells, a similar effect was observed with ZAN, even at a concentration of 1 μM, which can be explained by the dual inhibition of both AKR1C3 and ABC transporters. In a phase 1 study, a therapeutic dose of 320 mg resulted in a ZAN plasma Cmax of 1.4 μM [55]. Therefore, it is presumed that the effects observed in our in vitro experiments could also be manifested in patients with cancer. Importantly, no upregulation of AKR1C3 mRNA was detected upon ZAN treatment of HepG2 or KG1α cells, suggesting that ZAN does not affect systemic DAUN metabolism or influence the resistance of cancer cells to DAUN by promoting changes in AKR1C3 expression.

There are several lines of evidence regarding the role of AKR1C3 and ABC transporters in aggressive lymphomas and their sensitivity to ANT therapy, identifying them as potential drug targets. First, AKR1C3 expression was detected in samples obtained from patients with DLBCL treated with anthracycline-based therapy. Second, the AKR1C3 single nucleotide polymorphism may influence the survival outcomes of patients receiving the CHOP regimen [69]. Finally, the role of ABC transporters in cancer responses to ANT-containing chemotherapy has been suggested in DLBCL [70], MCL [71], and Burkitt lymphoma [72]. In addition to the primary indication of BTK inhibitors in CLL and lymphomas, their efficacy has recently been demonstrated in AML and solid tumours [73–76]. Rushworth et al. (2014) discovered that BTK is phosphorylated (p-BTK), and is thus constitutively active in the majority of samples obtained from patients with AML. Importantly, ibrutinib was found to enhance the cytotoxic effect of DAUN in high p-BTK-expressing AML cells and in the U937 cell line, but not in CD34+ non-malignant cells [75]. However, the mechanism of DAUN-IBR synergism may potentially be BTK-independent [77]. Our current results suggest that inhibition of DAUN metabolism and/or its cellular efflux may help to explain the ANT-sensitising effects of the BTK inhibitor ZAN. Importantly, AKR1C3 is not only an ANT-reductase, but is also involved in the production of sex hormones and pro-proliferative prostaglandins [78–80]. The association between AKR1C3 expression and poor patient prognosis has been proven many times in different solid tumours (e.g., breast, prostate, and liver) [35,81–83]. Several studies have further pointed to a potential link between AKR1C3 and leukaemogenesis [78,80,84]. Moreover, maternal and offspring polymorphisms of AKR1C3 were found to be associated with an increased risk of the development of childhood leukaemia [85], contributing to the FDA's list of AKR1C3 as a relevant target in paediatric oncology [86,87]. Therefore, by targeting AKR1C3, ZAN may affect both leukaemia- and hormone-dependent, as well as independent, cancers. This further supports the rationale for combining ZAN with ANT therapy in cancers with a high expression of AKR1C3 and/or ABC transporters. If confirmed in vivo, this information may be translated into an effective therapeutic strategy beneficial for a significant number of patients.

Author Contributions: Conceptualization, V.W., J.H. and E.N.; data curation, L.Č., L.L., I.Š. and L.H.; funding acquisition, V.W. and J.H.; investigation, L.Č., L.L. and L.H.; methodology, E.N. and J.H.; project administration, V.W.; supervision, V.W. and J.H.; writing—original draft, L.Č. and E.N.; writing—review and editing, J.H., L.L. and V.W. All authors have read and agreed to the published version of the manuscript.

Funding: This work was supported by the InoMed (project no. CZ.02.1.01/0.0/0.0/18_069/0010046), co-funded by the European Union, by the Czech Science Foundation (project no. 20-20414Y), by the Grant Agency of Charles University (project no. 315221), and finally by the Charles University (SVV 260 550).

Institutional Review Board Statement: Not applicable.

Informed Consent Statement: Not applicable.

Data Availability Statement: Not applicable.

Conflicts of Interest: The authors declare no conflict of interest.

References

1. Uckun, F.M.; Tibbles, H.E.; Vassilev, A.O. Bruton's tyrosine kinase as a new therapeutic target. *Anticancer Agents Med. Chem.* **2007**, *7*, 624–632. [CrossRef] [PubMed]
2. Mano, H. Tec family of protein-tyrosine kinases: An overview of their structure and function. *Cytokine Growth Factor Rev.* **1999**, *10*, 267–280. [CrossRef]
3. Niemann, C.U.; Wiestner, A. B-cell receptor signaling as a driver of lymphoma development and evolution. *Semin. Cancer Biol.* **2013**, *23*, 410–421. [CrossRef] [PubMed]
4. Buggy, J.J.; Elias, L. Bruton tyrosine kinase (BTK) and its role in B-cell malignancy. *Int. Rev. Immunol.* **2012**, *31*, 119–132. [CrossRef] [PubMed]
5. Davis, R.E.; Ngo, V.N.; Lenz, G.; Tolar, P.; Young, R.M.; Romesser, P.B.; Kohlhammer, H.; Lamy, L.; Zhao, H.; Yang, Y.; et al. Chronic active B-cell-receptor signalling in diffuse large B-cell lymphoma. *Nature* **2010**, *463*, 88–92. [CrossRef]
6. Honigberg, L.A.; Smith, A.M.; Sirisawad, M.; Verner, E.; Loury, D.; Chang, B.; Li, S.; Pan, Z.; Thamm, D.H.; Miller, R.A.; et al. The Bruton tyrosine kinase inhibitor PCI-32765 blocks B-cell activation and is efficacious in models of autoimmune disease and B-cell malignancy. *Proc. Natl. Acad. Sci. USA* **2010**, *107*, 13075–13080. [CrossRef]
7. Herman, S.E.; Gordon, A.L.; Hertlein, E.; Ramanunni, A.; Zhang, X.; Jaglowski, S.; Flynn, J.; Jones, J.; Blum, K.A.; Buggy, J.J.; et al. Bruton tyrosine kinase represents a promising therapeutic target for treatment of chronic lymphocytic leukemia and is effectively targeted by PCI-32765. *Blood* **2011**, *117*, 6287–6296. [CrossRef]
8. FDA (US Food and Drug Administration). FDA Approves Imbruvica. Available online: https://www.drugs.com/newdrugs/fda-approves-imbruvica-ibrutinib-mantle-cell-lymphoma-3960.html (accessed on 8 July 2022).
9. FDA (US Food and Drug Administration). FDA Approves New Treatment for Adults with Mantle Cell Lymphoma. Available online: https://www.fda.gov/news-events/press-announcements/fda-approves-new-treatment-adults-mantle-cell-lymphoma (accessed on 8 July 2022).
10. Byrd, J.C.; Furman, R.R.; Coutre, S.E.; Burger, J.A.; Blum, K.A.; Coleman, M.; Wierda, W.G.; Jones, J.A.; Zhao, W.; Heerema, N.A.; et al. Three-year follow-up of treatment-naive and previously treated patients with CLL and SLL receiving single-agent ibrutinib. *Blood* **2015**, *125*, 2497–2506. [CrossRef]
11. Castillo, J.J.; Meid, K.; Gustine, J.N.; Leventoff, C.; White, T.; Flynn, C.A.; Sarosiek, S.; Demos, M.G.; Guerrera, M.L.; Kofides, A.; et al. Long-term follow-up of ibrutinib monotherapy in treatment-naive patients with Waldenstrom macroglobulinemia. *Leukemia* **2022**, *36*, 532–539. [CrossRef]
12. Rusconi, C.; Cheah, C.Y.; Eyre, T.A.; Tucker, D.L.; Klener, P.; Gine, E.; Crucitti, L.; Muzi, C.; Iadecola, S.; Infante, G.; et al. Ibrutinib improves survival compared to chemotherapy in mantle cell lymphoma with central nervous system relapse. *Blood* 2022, in press. [CrossRef]
13. Noy, A.; de Vos, S.; Coleman, M.; Martin, P.; Flowers, C.R.; Thieblemont, C.; Morschhauser, F.; Collins, G.P.; Ma, S.; Peles, S.; et al. Durable ibrutinib responses in relapsed/refractory marginal zone lymphoma: Long-term follow-up and biomarker analysis. *Blood Adv.* **2020**, *4*, 5773–5784. [CrossRef] [PubMed]
14. Byrd, J.C.; Hillmen, P.; O'Brien, S.; Barrientos, J.C.; Reddy, N.M.; Coutre, S.; Tam, C.S.; Mulligan, S.P.; Jaeger, U.; Barr, P.M.; et al. Long-term follow-up of the RESONATE phase 3 trial of ibrutinib vs ofatumumab. *Blood* **2019**, *133*, 2031–2042. [CrossRef] [PubMed]
15. Pan, Z.; Scheerens, H.; Li, S.J.; Schultz, B.E.; Sprengeler, P.A.; Burrill, L.C.; Mendonca, R.V.; Sweeney, M.D.; Scott, K.C.; Grothaus, P.G.; et al. Discovery of selective irreversible inhibitors for Bruton's tyrosine kinase. *ChemMedChem* **2007**, *2*, 58–61. [CrossRef] [PubMed]

16. Barf, T.; Covey, T.; Izumi, R.; van de Kar, B.; Gulrajani, M.; van Lith, B.; van Hoek, M.; de Zwart, E.; Mittag, D.; Demont, D.; et al. Acalabrutinib (ACP-196): A Covalent Bruton Tyrosine Kinase Inhibitor with a Differentiated Selectivity and In Vivo Potency Profile. *J. Pharmacol. Exp. Ther.* **2017**, *363*, 240–252. [CrossRef]
17. Von Hundelshausen, P.; Siess, W. Bleeding by Bruton Tyrosine Kinase-Inhibitors: Dependency on Drug Type and Disease. *Cancers* **2021**, *13*, 1103. [CrossRef]
18. Ahn, I.E. Cardiovascular adverse events of ibrutinib. *Blood* **2019**, *134*, 1881–1882. [CrossRef]
19. Woyach, J.A.; Furman, R.R.; Liu, T.M.; Ozer, H.G.; Zapatka, M.; Ruppert, A.S.; Xue, L.; Li, D.H.; Steggerda, S.M.; Versele, M.; et al. Resistance mechanisms for the Bruton's tyrosine kinase inhibitor ibrutinib. *N. Engl. J. Med.* **2014**, *370*, 2286–2294. [CrossRef]
20. Burger, J.A.; Landau, D.A.; Taylor-Weiner, A.; Bozic, I.; Zhang, H.; Sarosiek, K.; Wang, L.; Stewart, C.; Fan, J.; Hoellenriegel, J.; et al. Clonal evolution in patients with chronic lymphocytic leukaemia developing resistance to BTK inhibition. *Nat. Commun.* **2016**, *7*, 11589. [CrossRef]
21. Bond, D.A.; Woyach, J.A. Targeting BTK in CLL: Beyond Ibrutinib. *Curr. Hematol. Malig. Rep.* **2019**, *14*, 197–205. [CrossRef]
22. Byrd, J.C.; Smith, S.; Wagner-Johnston, N.; Sharman, J.; Chen, A.I.; Advani, R.; Augustson, B.; Marlton, P.; Renee Commerford, S.; Okrah, K.; et al. First-in-human phase 1 study of the BTK inhibitor GDC-0853 in relapsed or refractory B-cell NHL and CLL. *Oncotarget* **2018**, *9*, 13023–13035. [CrossRef]
23. FDA (US Food and Drug Administration). FDA Approves Therapy to Treat Patients with Relapsed and Refractory Mantle Cell Lymphoma Supported by Clinical Trial Results Showing High Response Rate of Tumor Shrinkage. Available online: https://www.fda.gov/news-events/press-announcements/fda-approves-therapy-treat-patients-relapsed-and-refractory-mantle-cell-lymphoma-supported-clinical (accessed on 8 July 2022).
24. FDA (US Food and Drug Administration). FDA Approves Zanubrutinib for Waldenström's Macroglobulinemia. Available online: https://www.fda.gov/drugs/resources-information-approved-drugs/fda-approves-zanubrutinib-waldenstroms-macroglobulinemia (accessed on 8 July 2022).
25. FDA (US Food and Drug Administration). FDA Grants Accelerated Approval to Zanubrutinib for Marginal Zone Lymphoma. Available online: https://www.fda.gov/drugs/resources-information-approved-drugs/fda-grants-accelerated-approval-zanubrutinib-marginal-zone-lymphoma (accessed on 8 July 2022).
26. Younes, A.; Thieblemont, C.; Morschhauser, F.; Flinn, I.; Friedberg, J.W.; Amorim, S.; Hivert, B.; Westin, J.; Vermeulen, J.; Bandyopadhyay, N.; et al. Combination of ibrutinib with rituximab, cyclophosphamide, doxorubicin, vincristine, and prednisone (R-CHOP) for treatment-naive patients with CD20-positive B-cell non-Hodgkin lymphoma: A non-randomised, phase 1b study. *Lancet Oncol.* **2014**, *15*, 1019–1026. [CrossRef]
27. Younes, A.; Sehn, L.H.; Johnson, P.; Zinzani, P.L.; Hong, X.; Zhu, J.; Patti, C.; Belada, D.; Samoilova, O.; Suh, C.; et al. Randomized Phase III Trial of Ibrutinib and Rituximab Plus Cyclophosphamide, Doxorubicin, Vincristine, and Prednisone in Non-Germinal Center B-Cell Diffuse Large B-Cell Lymphoma. *J. Clin. Oncol.* **2019**, *37*, 1285–1295. [CrossRef] [PubMed]
28. Davies, A.; Caddy, J.; Mercer, K.; Saunders, G.N.; Stanton, L.; Collins, G.P.; Cummin, T.E.C.; Schuh, A.; Ardeshna, K.M.; McMillan, A.; et al. Acalabrutinib in Combination with Rituximab, Cyclophosphamide, Doxorubicin, Vincristine and Prednisolone (R-CHOP) As First Line Therapy for Patients with Diffuse Large B-Cell Lymphoma (DLBCL): The Accept Phase Ib/II Single Arm Study. *Blood* **2020**, *136*, 38–39. [CrossRef]
29. Zhu, H.; Sha, Y.; Wu, W.; Chen, R.; Yang, Y.; Qiu, J.; Mi, H.; Peng, C.; Ding, C.; Wang, Z.; et al. Zanubrutinib, Lenalidomide Plus R-CHOP (ZR2-CHOP) as the Treatment for Diffused Large B-Cell Lymphoma (DLBCL). *Hematol. Oncol.* **2021**, *39*, 425–427. [CrossRef]
30. Xiao, H.; Zheng, Y.; Ma, L.; Tian, L.; Sun, Q. Clinically-Relevant ABC Transporter for Anti-Cancer Drug Resistance. *Front. Pharmacol.* **2021**, *12*, 648407. [CrossRef] [PubMed]
31. Bains, O.S.; Grigliatti, T.A.; Reid, R.E.; Riggs, K.W. Naturally occurring variants of human aldo-keto reductases with reduced in vitro metabolism of daunorubicin and doxorubicin. *J. Pharmacol. Exp. Ther.* **2010**, *335*, 533–545. [CrossRef]
32. Bains, O.S.; Szeitz, A.; Lubieniecka, J.M.; Cragg, G.E.; Grigliatti, T.A.; Riggs, K.W.; Reid, R.E. A correlation between cytotoxicity and reductase-mediated metabolism in cell lines treated with doxorubicin and daunorubicin. *J. Pharmacol. Exp. Ther.* **2013**, *347*, 375–387. [CrossRef]
33. Penning, T.M.; Jonnalagadda, S.; Trippier, P.C.; Rizner, T.L. Aldo-Keto Reductases and Cancer Drug Resistance. *Pharmacol. Rev.* **2021**, *73*, 1150–1171. [CrossRef]
34. Morell, A.; Cermakova, L.; Novotna, E.; Lastovickova, L.; Haddad, M.; Haddad, A.; Portillo, R.; Wsol, V. Bruton's Tyrosine Kinase Inhibitors Ibrutinib and Acalabrutinib Counteract Anthracycline Resistance in Cancer Cells Expressing AKR1C3. *Cancers* **2020**, *12*, 3731. [CrossRef]
35. Wang, B.; Gu, Y.; Hui, K.; Huang, J.; Xu, S.; Wu, S.; Li, L.; Fan, J.; Wang, X.; Hsieh, J.T.; et al. AKR1C3, a crucial androgenic enzyme in prostate cancer, promotes epithelial-mesenchymal transition and metastasis through activating ERK signaling. *Urol. Oncol.* **2018**, *36*, 472.e411–472.e420. [CrossRef]
36. Yamashita, N.; Kanno, Y.; Saito, N.; Terai, K.; Sanada, N.; Kizu, R.; Hiruta, N.; Park, Y.; Bujo, H.; Nemoto, K. Aryl hydrocarbon receptor counteracts pharmacological efficacy of doxorubicin via enhanced AKR1C3 expression in triple negative breast cancer cells. *Biochem. Biophys. Res. Commun.* **2019**, *516*, 693–698. [CrossRef] [PubMed]

37. Miller, V.L.; Lin, H.K.; Murugan, P.; Fan, M.; Penning, T.M.; Brame, L.S.; Yang, Q.; Fung, K.M. Aldo-keto reductase family 1 member C3 (AKR1C3) is expressed in adenocarcinoma and squamous cell carcinoma but not small cell carcinoma. *Int. J. Clin. Exp. Pathol.* **2012**, *5*, 278–289. [PubMed]
38. Bortolozzi, R.; Bresolin, S.; Rampazzo, E.; Paganin, M.; Maule, F.; Mariotto, E.; Boso, D.; Minuzzo, S.; Agnusdei, V.; Viola, G.; et al. AKR1C enzymes sustain therapy resistance in paediatric T-ALL. *Br. J. Cancer* **2018**, *118*, 985–994. [CrossRef] [PubMed]
39. Bernardini, N.; Giannessi, F.; Bianchi, F.; Dolfi, A.; Lupetti, M.; Zaccaro, L.; Malvaldi, G.; Del Tacca, M. Comparative activity of doxorubicin and its major metabolite, doxorubicinol, on fibroblasts: A morphofunctional study. *Exp. Mol. Pathol.* **1991**, *55*, 238–250. [CrossRef]
40. Heibein, A.D.; Guo, B.; Sprowl, J.A.; Maclean, D.A.; Parissenti, A.M. Role of aldo-keto reductases and other doxorubicin pharmacokinetic genes in doxorubicin resistance, DNA binding, and subcellular localization. *BMC Cancer* **2012**, *12*, 381. [CrossRef]
41. Zhang, H.; Patel, A.; Ma, S.L.; Li, X.J.; Zhang, Y.K.; Yang, P.Q.; Kathawala, R.J.; Wang, Y.J.; Anreddy, N.; Fu, L.W.; et al. In vitro, in vivo and ex vivo characterization of ibrutinib: A potent inhibitor of the efflux function of the transporter MRP1. *Br. J. Pharmacol.* **2014**, *171*, 5845–5857. [CrossRef]
42. Kosztyu, P.; Bukvova, R.; Dolezel, P.; Mlejnek, P. Resistance to daunorubicin, imatinib, or nilotinib depends on expression levels of ABCB1 and ABCG2 in human leukemia cells. *Chem. Biol. Interact.* **2014**, *219*, 203–210. [CrossRef]
43. Hofman, J.; Malcekova, B.; Skarka, A.; Novotna, E.; Wsol, V. Anthracycline resistance mediated by reductive metabolism in cancer cells: The role of aldo-keto reductase 1C3. *Toxicol. Appl. Pharmacol.* **2014**, *278*, 238–248. [CrossRef]
44. Skarydova, L.; Tomanova, R.; Havlikova, L.; Stambergova, H.; Solich, P.; Wsol, V. Deeper insight into the reducing biotransformation of bupropion in the human liver. *Drug Metab. Pharmacokinet.* **2014**, *29*, 177–184. [CrossRef]
45. Bukum, N.; Novotna, E.; Morell, A.; Zelazkova, J.; Lastovickova, L.; Cermakova, L.; Portillo, R.; Solich, P.; Wsol, V. Inhibition of AKR1B10-mediated metabolism of daunorubicin as a novel off-target effect for the Bcr-Abl tyrosine kinase inhibitor dasatinib. *Biochem. Pharmacol.* **2021**, *192*, 114710. [CrossRef]
46. Tavares, T.S.; Hofman, J.; Lekesova, A.; Zelazkova, J.; Wsol, V. Olaparib Synergizes the Anticancer Activity of Daunorubicin via Interaction with AKR1C3. *Cancers* **2020**, *12*, 3127. [CrossRef] [PubMed]
47. Chou, T.C. Drug combination studies and their synergy quantification using the Chou-Talalay method. *Cancer Res.* **2010**, *70*, 440–446. [CrossRef] [PubMed]
48. Chou, T.C. Theoretical basis, experimental design, and computerized simulation of synergism and antagonism in drug combination studies. *Pharmacol. Rev.* **2006**, *58*, 621–681. [CrossRef] [PubMed]
49. Zhang, H.; Patel, A.; Wang, Y.J.; Zhang, Y.K.; Kathawala, R.J.; Qiu, L.H.; Patel, B.A.; Huang, L.H.; Shukla, S.; Yang, D.H.; et al. The BTK Inhibitor Ibrutinib (PCI-32765) Overcomes Paclitaxel Resistance in ABCB1- and ABCC10-Overexpressing Cells and Tumors. *Mol. Cancer Ther.* **2017**, *16*, 1021–1030. [CrossRef] [PubMed]
50. Juan-Carlos, P.M.; Perla-Lidia, P.P.; Stephanie-Talia, M.M.; Monica-Griselda, A.M.; Luz-Maria, T.E. ABC transporter superfamily. An updated overview, relevance in cancer multidrug resistance and perspectives with personalized medicine. *Mol. Biol. Rep.* **2021**, *48*, 1883–1901. [CrossRef] [PubMed]
51. Shi, B.; Xu, F.F.; Xiang, C.P.; Jia, R.; Yan, C.H.; Ma, S.Q.; Wang, N.; Wang, A.J.; Fan, P. Effect of sodium butyrate on ABC transporters in lung cancer A549 and colorectal cancer HCT116 cells. *Oncol. Lett.* **2020**, *20*, 148. [CrossRef]
52. Lainey, E.; Sebert, M.; Thepot, S.; Scoazec, M.; Bouteloup, C.; Leroy, C.; De Botton, S.; Galluzzi, L.; Fenaux, P.; Kroemer, G. Erlotinib antagonizes ABC transporters in acute myeloid leukemia. *Cell Cycle* **2012**, *11*, 4079–4092. [CrossRef]
53. Grundy, M.; Seedhouse, C.; Russell, N.H.; Pallis, M. P-glycoprotein and breast cancer resistance protein in acute myeloid leukaemia cells treated with the aurora-B kinase inhibitor barasertib-hQPA. *BMC Cancer* **2011**, *11*, 254. [CrossRef]
54. Hauswald, S.; Duque-Afonso, J.; Wagner, M.M.; Schertl, F.M.; Lubbert, M.; Peschel, C.; Keller, U.; Licht, T. Histone deacetylase inhibitors induce a very broad, pleiotropic anticancer drug resistance phenotype in acute myeloid leukemia cells by modulation of multiple ABC transporter genes. *Clin. Cancer Res.* **2009**, *15*, 3705–3715. [CrossRef]
55. Tam, C.S.; Trotman, J.; Opat, S.; Burger, J.A.; Cull, G.; Gottlieb, D.; Harrup, R.; Johnston, P.B.; Marlton, P.; Munoz, J.; et al. Phase 1 study of the selective BTK inhibitor zanubrutinib in B-cell malignancies and safety and efficacy evaluation in CLL. *Blood* **2019**, *134*, 851–859. [CrossRef]
56. ClinicalTrials.gov. Available online: https://clinicaltrials.gov/ct2/home (accessed on 21 July 2022).
57. Hillmen, P.; Brown, J.R.; Eichhorst, B.F.; Lamanna, N.; O'Brien, S.M.; Qiu, L.; Salmi, T.; Hilger, J.; Wu, K.; Cohen, A.; et al. ALPINE: Zanubrutinib versus ibrutinib in relapsed/refractory chronic lymphocytic leukemia/small lymphocytic lymphoma. *Future Oncol.* **2020**, *16*, 517–523. [CrossRef] [PubMed]
58. Tam, C.S.; Giannopoulos, K.; Jurczak, W.; Šimkovič, M.; Shadman, M.; Österborg, A.; Laurenti, L.; Walker, P.; Opat, S.; Chan, H.; et al. SEQUOIA: Results of a Phase 3 Randomized Study of Zanubrutinib versus Bendamustine + Rituximab (BR) in Patients with Treatment-Naïve (TN) Chronic Lymphocytic Leukemia/Small Lymphocytic Lymphoma (CLL/SLL). *Blood* **2021**, *138*, 396. [CrossRef]
59. Yang, H.; Xiang, B.; Song, Y.; Zhang, H.; Zhao, W.; Zou, D.; Lv, F.; Guo, W.; Liu, A.; Li, C.; et al. Zanubrutinib monotherapy for relapsed or refractory non-germinal center diffuse large B-cell lymphoma. *Blood Adv.* **2022**, *6*, 1629–1636. [CrossRef] [PubMed]
60. Phillips, T.; Chan, H.; Tam, C.S.; Tedeschi, A.; Johnston, P.; Oh, S.Y.; Opat, S.; Eom, H.S.; Allewelt, H.; Stern, J.C.; et al. Zanubrutinib monotherapy in relapsed/refractory indolent non-Hodgkin lymphoma. *Blood Adv.* **2022**, *6*, 3472–3479. [CrossRef]

61. Zhang, Y.; Li, Y.; Zhuang, Z.; Wang, W.; Wei, C.; Zhao, D.; Zhou, D.; Zhang, W. Preliminary Evaluation of Zanubrutinib-Containing Regimens in DLBCL and the Cerebrospinal Fluid Distribution of Zanubrutinib: A 13-Case Series. *Front. Oncol.* **2021**, *11*, 760405. [CrossRef]
62. Armitage, J.O.; Longo, D.L. Is watch and wait still acceptable for patients with low-grade follicular lymphoma? *Blood* **2016**, *127*, 2804–2808. [CrossRef]
63. Brieghel, C.; Galle, V.; Agius, R.; da Cunha-Bang, C.; Andersen, M.A.; Vlummens, P.; Mattsson, M.; Rosenquist, R.; Smedby, K.E.; Herling, C.D.; et al. Identifying patients with chronic lymphocytic leukemia without need of treatment: End of endless watch and wait? *Eur. J. Haematol.* **2022**, *108*, 369–378. [CrossRef]
64. Sindel, A.; Al-Juhaishi, T.; Yazbeck, V. Marginal Zone Lymphoma: State-of-the-Art Treatment. *Curr. Treat. Options Oncol.* **2019**, *20*, 90. [CrossRef]
65. Durot, E.; Delmer, A. Watch and wait in Waldenstrom macroglobulinaemia: Looking for who to watch carefully and who can wait without worrying. Is it that simple? *Br. J. Haematol.* **2021**, *195*, 155–157. [CrossRef]
66. Rossi, D.; Spina, V.; Gaidano, G. Biology and treatment of Richter syndrome. *Blood* **2018**, *131*, 2761–2772. [CrossRef]
67. Gonzalez-Rincon, J.; Mendez, M.; Gomez, S.; Garcia, J.F.; Martin, P.; Bellas, C.; Pedrosa, L.; Rodriguez-Pinilla, S.M.; Camacho, F.I.; Quero, C.; et al. Unraveling transformation of follicular lymphoma to diffuse large B-cell lymphoma. *PLoS ONE* **2019**, *14*, e0212813. [CrossRef] [PubMed]
68. Coiffier, B.; Sarkozy, C. Diffuse large B-cell lymphoma: R-CHOP failure-what to do? *Hematol. Am. Soc. Hematol. Educ. Program* **2016**, *2016*, 366–378. [CrossRef] [PubMed]
69. Gustafson, H.L.; Yao, S.; Goldman, B.H.; Lee, K.; Spier, C.M.; LeBlanc, M.L.; Rimsza, L.M.; Cerhan, J.R.; Habermann, T.M.; Link, B.K.; et al. Genetic polymorphisms in oxidative stress-related genes are associated with outcomes following treatment for aggressive B-cell non-Hodgkin lymphoma. *Am. J. Hematol.* **2014**, *89*, 639–645. [CrossRef] [PubMed]
70. Liu, D.; Wu, N.; Sun, H.; Dong, M.; Guo, T.; Chi, P.; Li, G.; Sun, D.; Jin, Y. ABCG2 and NCF4 polymorphisms are associated with clinical outcomes in diffuse large B-cell lymphoma patients treated with R-CHOP. *Oncotarget* **2017**, *8*, 58292–58303. [CrossRef] [PubMed]
71. Takimoto-Shimomura, T.; Nagoshi, H.; Maegawa, S.; Fujibayashi, Y.; Tsukamoto, T.; Matsumura-Kimoto, Y.; Mizuno, Y.; Chinen, Y.; Mizutani, S.; Shimura, Y.; et al. Establishment and Characteristics of a Novel Mantle Cell Lymphoma-derived Cell Line and a Bendamustine-resistant Subline. *Cancer Genom. Proteom.* **2018**, *15*, 213–223. [CrossRef]
72. Tabata, M.; Tsubaki, M.; Takeda, T.; Tateishi, K.; Tsurushima, K.; Imano, M.; Satou, T.; Ishizaka, T.; Nishida, S. Dasatinib reverses drug resistance by downregulating MDR1 and Survivin in Burkitt lymphoma cells. *BMC Complement Med. Ther.* **2020**, *20*, 84. [CrossRef]
73. Molina-Cerrillo, J.; Alonso-Gordoa, T.; Gajate, P.; Grande, E. Bruton's tyrosine kinase (BTK) as a promising target in solid tumors. *Cancer Treat. Rev.* **2017**, *58*, 41–50. [CrossRef]
74. Szklener, K.; Michalski, A.; Zak, K.; Piwonski, M.; Mandziuk, S. Ibrutinib in the Treatment of Solid Tumors: Current State of Knowledge and Future Directions. *Cells* **2022**, *11*, 1338. [CrossRef]
75. Rushworth, S.A.; Murray, M.Y.; Zaitseva, L.; Bowles, K.M.; MacEwan, D.J. Identification of Bruton's tyrosine kinase as a therapeutic target in acute myeloid leukemia. *Blood* **2014**, *123*, 1229–1238. [CrossRef]
76. Yeh, C.T.; Chen, T.T.; Satriyo, P.B.; Wang, C.H.; Wu, A.T.H.; Chao, T.Y.; Lee, K.Y.; Hsiao, M.; Wang, L.S.; Kuo, K.T. Bruton's tyrosine kinase (BTK) mediates resistance to EGFR inhibition in non-small-cell lung carcinoma. *Oncogenesis* **2021**, *10*, 56. [CrossRef]
77. Rotin, L.E.; Gronda, M.; Hurren, R.; Wang, X.; Minden, M.D.; Slassi, M.; Schimmer, A.D. Investigating the synergistic mechanism between ibrutinib and daunorubicin in acute myeloid leukemia cells. *Leuk. Lymphoma* **2016**, *57*, 2432–2436. [CrossRef]
78. Birtwistle, J.; Hayden, R.E.; Khanim, F.L.; Green, R.M.; Pearce, C.; Davies, N.J.; Wake, N.; Schrewe, H.; Ride, J.P.; Chipman, J.K.; et al. The aldo-keto reductase AKR1C3 contributes to 7,12-dimethylbenz(a)anthracene-3,4-dihydrodiol mediated oxidative DNA damage in myeloid cells: Implications for leukemogenesis. *Mutat. Res.* **2009**, *662*, 67–74. [CrossRef]
79. Byrns, M.C.; Duan, L.; Lee, S.H.; Blair, I.A.; Penning, T.M. Aldo-keto reductase 1C3 expression in MCF-7 cells reveals roles in steroid hormone and prostaglandin metabolism that may explain its over-expression in breast cancer. *J. Steroid Biochem. Mol. Biol.* **2010**, *118*, 177–187. [CrossRef]
80. Desmond, J.C.; Mountford, J.C.; Drayson, M.T.; Walker, E.A.; Hewison, M.; Ride, J.P.; Luong, Q.T.; Hayden, R.E.; Vanin, E.F.; Bunce, C.M. The aldo-keto reductase AKR1C3 is a novel suppressor of cell differentiation that provides a plausible target for the non-cyclooxygenase-dependent antineoplastic actions of nonsteroidal anti-inflammatory drugs. *Cancer Res.* **2003**, *63*, 505–512.
81. Oduwole, O.O.; Li, Y.; Isomaa, V.V.; Mantyniemi, A.; Pulkka, A.E.; Soini, Y.; Vihko, P.T. 17beta-hydroxysteroid dehydrogenase type 1 is an independent prognostic marker in breast cancer. *Cancer Res.* **2004**, *64*, 7604–7609. [CrossRef]
82. Zhu, P.; Feng, R.; Lu, X.; Liao, Y.; Du, Z.; Zhai, W.; Chen, K. Diagnostic and prognostic values of AKR1C3 and AKR1D1 in hepatocellular carcinoma. *Aging* **2021**, *13*, 4138–4156. [CrossRef] [PubMed]
83. Zhao, S.F.; Wang, S.G.; Zhao, Z.Y.; Li, W.L. AKR1C1-3, notably AKR1C3, are distinct biomarkers for liver cancer diagnosis and prognosis: Database mining in malignancies. *Oncol. Lett.* **2019**, *18*, 4515–4522. [CrossRef]
84. Mahadevan, D.; DiMento, J.; Croce, K.D.; Riley, C.; George, B.; Fuchs, D.; Mathews, T.; Wilson, C.; Lobell, M. Transcriptosome and serum cytokine profiling of an atypical case of myelodysplastic syndrome with progression to acute myelogenous leukemia. *Am. J. Hematol.* **2006**, *81*, 779–786. [CrossRef]

85. Liu, C.Y.; Hsu, Y.H.; Pan, P.C.; Wu, M.T.; Ho, C.K.; Su, L.; Xu, X.; Li, Y.; Christiani, D.C.; Kaohsiung Leukemia Research, G. Maternal and offspring genetic variants of AKR1C3 and the risk of childhood leukemia. *Carcinogenesis* **2008**, *29*, 984–990. [CrossRef]
86. FDA (US Food and Drug Administration). U.S. Food & Drug Administration: List of the Relevant Molecular Targets in Paediatric Oncology. Available online: https://www.fda.gov/about-fda/oncology-center-excellence/pediatric-oncology (accessed on 16 July 2022).
87. FDA (US Food and Drug Administration). U.S. Food & Drug Administration: List of the Relevant Molecular Targets in Paediatric Oncology. Available online: https://www.fda.gov/media/120332/download (accessed on 16 July 2022).

pharmaceutics MDPI

Article

Vitamin E TPGS-Poloxamer Nanoparticles Entrapping a Novel PI3Kα Inhibitor Potentiate Its Activity against Breast Cancer Cell Lines

Suhair Sunoqrot *, Sundos Aliyeh, Samah Abusulieh and Dima Sabbah

Department of Pharmacy, Faculty of Pharmacy, Al-Zaytoonah University of Jordan, Amman 11733, Jordan
* Correspondence: suhair.sunoqrot@zuj.edu.jo; Tel.: +962-64291511 (ext. 197)

Abstract: N-(2-fluorphenyl)-6-chloro-4-hydroxy-2-quinolone-3-carboxamide (R19) is a newly synthesized phosphatidylinositol 3-kinase alpha (PI3Kα) inhibitor with promising activity against cancer cells. The purpose of this study was to develop a polymeric nanoparticle (NP) formulation for R19 to address its poor aqueous solubility and to facilitate its future administration in preclinical and clinical settings. NPs were prepared by nanoprecipitation using two polymers: D-α-tocopheryl polyethylene glycol 1000 succinate (vitamin E TPGS) and the poloxamer Pluronic P123 in different ratios. Physicochemical characterization of the NPs revealed them to be around 100 nm in size with high monodispersity, a spherical morphology, and an almost neutral surface charge. The NPs achieved ~60% drug loading efficiency and sustained release of R19 for up to 96 h, with excellent colloidal stability in serum-containing cell culture media. NPs containing TPGS enhanced R19's potency against MCF-7 and MDA-MB-231 breast cancer cells in vitro, with half-maximal inhibitory concentrations (IC_{50}) ranging between 1.8 and 4.3 μM compared to free R19, which had an IC_{50} of 14.7–17.0 μM. The NPs also demonstrated low cytotoxicity against human dermal fibroblasts and more significant induction of apoptosis compared to the free drug, which was correlated with their cellular uptake efficiency. Our findings present a biocompatible NP formulation for the delivery of a cancer-targeted PI3Kα inhibitor, R19, which can further enhance its potency for the treatment of breast cancer and potentially other cancer types.

Keywords: breast cancer; nanomedicine; PI3Kα; poloxamers; polymeric nanoparticles; TPGS

Citation: Sunoqrot, S.; Aliyeh, S.; Abusulieh, S.; Sabbah, D. Vitamin E TPGS-Poloxamer Nanoparticles Entrapping a Novel PI3Kα Inhibitor Potentiate Its Activity against Breast Cancer Cell Lines. *Pharmaceutics* **2022**, *14*, 1977. https://doi.org/10.3390/pharmaceutics14091977

Academic Editor: Francesca Musumeci

Received: 4 August 2022
Accepted: 13 September 2022
Published: 19 September 2022

1. Introduction

Despite advances in early detection and medical treatments, cancer remains one of the leading causes of death globally [1]. Cancer treatment regimens typically include surgery, radiotherapy, and chemotherapy [2]. Chemotherapy involves the use of cytotoxic agents, which are accompanied by immediate and chronic side effects that can significantly impact cancer patients' quality of life. The side effects are mainly attributed to the systemic distribution of the cytotoxic agent and a lack of selectivity in killing normal and cancer cells alike [3]. For this reason, significant efforts have been dedicated to the discovery of targeted therapies that can reduce the systemic side effects of chemotherapy.

Protein kinases are important enzymes that modify other proteins by phosphorylation, which consequently regulates the function of these proteins and their downstream signaling pathways. Several kinases have been found to be dysregulated in various cancers, which has inspired the development of kinase inhibitors as cancer-targeted therapies [4]. The phosphatidylinositol-3-kinases (PI3Ks) are a family of lipid kinases that are important for the regulation of cell metabolism, proliferation, migration, and survival [5]. PI3Ks are part of a signaling pathway that includes Akt and the mammalian target of rapamycin (mTOR). The PI3K/Akt/mTOR signaling pathway regulates the growth of both normal and cancer cells and has been shown to be hyperactivated in cancer cells [6–8]. Three classes of PI3Ks are known; class I, II, and III. Class I can be divided into Class I_A and

Class I_B. Class I_A is the class of concern since it is directly linked to cancer progression. Class I_A PI3K consists of α, β, and δ isoforms, which are encoded by the genes PIK3CA, PIK3CB, and PIK3CD, respectively [5,9]. Several cancers have demonstrated abnormal PI3Kα/Akt/mTOR signaling pathway activation, in which PIK3CA gene expression has been implicated [9].

N-phenyl-6-chloro-4-hydroxy-2-quinolone-3-carboxamides (Figure 1) are a group of newly developed anticancer agents with inhibitory activity against PI3Kα. A series of different analogues were recently synthesized to achieve the maximum inhibition of cancer cells with the lowest possible toxicity on normal cells. N-(2-fluorphenyl)-6-chloro-4-hydroxy-2-quinolone-3-carboxamide (R19) was found to be an effective inhibitory agent against human colon carcinoma (HCT-116) and human epithelial colorectal adenocarcinoma (Caco-2) cell lines, with half-maximal inhibitory concentrations (IC_{50}) of 5.3 and 17.0 μM, respectively. Moreover, R19 was tested for its activity on the PI3K/Akt/mTOR signaling pathway, where gene expression analysis revealed a significant reduction in the expression of PI3K and Akt genes and a significant increase in the pro-apoptotic gene Bcl-2-associated death promoter (BAD) [9]. Despite its promising anticancer potency, R19 and its newly developed analogs suffer from poor aqueous solubility, which will be a limiting factor in the clinical translation of this class of compounds.

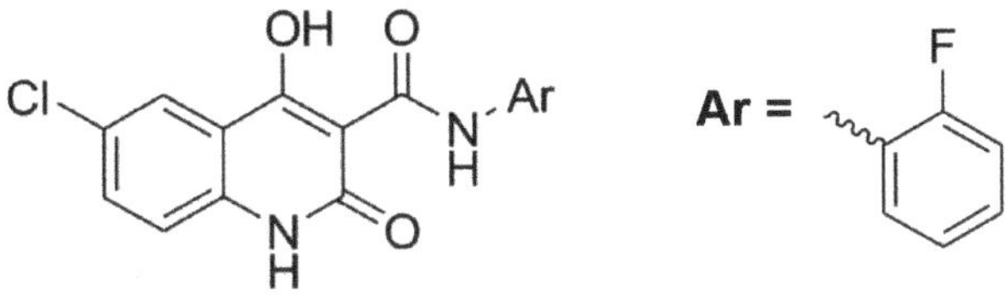

Figure 1. Chemical structure of the N-phenyl-6-chloro-4-hydroxy-2-quinolone-3-carboxamide series and R19 [9].

Nanotechnology has revolutionized the way that we treat diseases, especially cancer. Nanosized drug carriers or nanoparticles (NPs) can improve the solubility, permeability, absorption, delivery, and targeting of drug molecules to the site of action [10–12]. Polymeric NPs are a particularly attractive delivery approach due to their highly tunable properties, which can be leveraged to achieve better solubility, bioavailability, controlled release, and targeting of drug candidates by passive and active targeting mechanisms [13–16].

Poloxamers are amphiphilic triblock copolymers composed of poly(ethylene oxide) (PEO) and poly(propylene oxide) (PPO) (PEO-PPO-PEO) [17,18]. They are biocompatible polymers that are used in various applications, especially as nanocarriers for hydrophobic drugs [19]. Poloxamers have also been combined with other polymers to improve drug loading, impart NP stability, and enhance drug solubility [18]. On the other hand, D-α-tocopheryl polyethylene glycol 1000 succinate (vitamin E TPGS or TPGS) is a water-soluble form of vitamin E that is used in drug delivery to prolong the circulation half-life, enhance the bioavailability, improve the cytotoxicity, and synergize with other anticancer drugs [20–23]. TPGS is used as a nanocarrier for hydrophobic and poorly water-soluble drugs due to its amphiphilic nature [24,25].

In this study, we hypothesized that two biocompatible polymers, TPGS and a poloxamer (Pluronic P123), can be used as nanocarriers for R19 encapsulation and delivery, combining the benefits of both TPGS and poloxamers. To verify our hypothesis, R19 NPs at different ratios of TPGS and Pluronic P123 were prepared and characterized for their physicochemical properties. The NPs were also tested on breast cancer cell lines to evaluate their bioactivity and biocompatibility.

2. Materials and Methods

2.1. Materials

N-(2-fluorphenyl)-6-chloro-4-hydroxy-2-quinolone-3-carboxamide (R19) was synthesized and characterized as previously described [9]. Pluronic P123 (P123), TPGS, and Nile red (NR) were purchased from Sigma-Aldrich (St. Louis, MO, USA). Phosphate buffered saline solution (PBS, 10×, pH 7.4), potassium bromide (KBr), and dimethyl sulfoxide (DMSO) were procured from Fisher Chemicals (Pittsburgh, PA, USA). Spectra/Por dialysis membranes with 3500 and 12,000–14,000 Da molecular weight cut-offs (MWCO) were obtained from Repligen (Waltham, MA, USA). Ultrapure water was prepared using a Millipore Direct-Q 5UV system (EMD Millipore, Burlington, MA, USA).

2.2. Preparation of R19-Loaded NPs

R19-loaded NPs were produced via the nanoprecipitation technique, as previously described [26]. Different NP formulations were prepared by varying the P123: TPGS ratio (Table 1), and each batch was prepared in triplicate. Stock solutions of each polymer (P123 and TPGS) were first prepared at 20 mg/mL in DMSO. Each batch consisted of a 20 mg polymer or polymer mixture, and the amount of drug was fixed at 2 mg. The organic phase was prepared by mixing the appropriate volumes of polymer solutions in a microtube (total volume 1 mL), to which 2 mg of R19 powder was added, and the mixture was vortexed until the drug was completely dissolved. The resultant organic phase was then added dropwise to 10 mL of ultrapure water under continuous stirring. After 30 min of stirring, the NP dispersion was transferred to a dialysis membrane (3500 MWCO) and was dialyzed against 2 L of deionized water under gentle stirring for 24 h. After 24 h, the dialysis bag was emptied into a vial, and the samples were either used directly or were stored at 4 °C for further characterization. Blank NPs were prepared as described above but without the addition of R19. NR-labeled NPs were prepared by replacing R19 with NR (0.2 mg per NP formulation), and the NPs were purified by dialysis, as described above. A schematic of the NP preparation process is presented in Figure 2.

Table 1. Composition of R19-loaded NPs.

NP	R19 (mg)	Pluronic P123 (mg)	TPGS (mg)
100% TPGS	2	0	20
25% P123/75% TPGS	2	5	15
50% P123/50% TPGS	2	10	10
75% P123/25% TPGS	2	15	5
100% P123	2	20	0

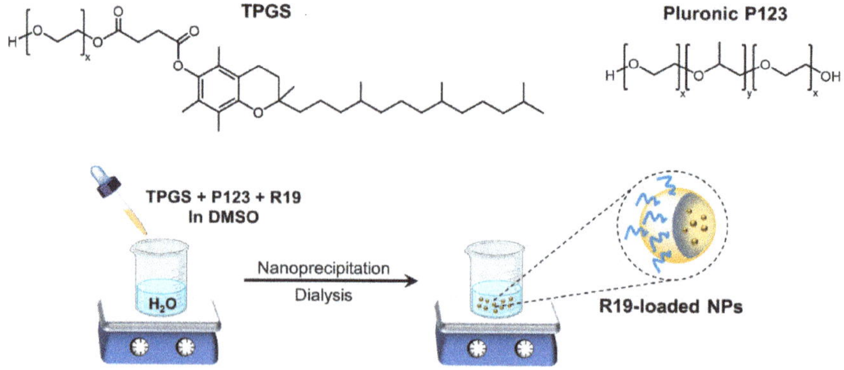

Figure 2. Structures of TPGS and P123 and a schematic of R19 NP preparation by nanoprecipitation.

2.3. Particle Size and Zeta Potential Measurements

Dynamic light scattering (DLS) was used for the measurement of the particle size and polydispersity index (PDI) of the NPs. Electrophoretic light scattering was used for zeta potential measurements. Samples were analyzed using a Nicomp® Nano Z3000 instrument (Entegris, Billerica, MA, USA). For the analysis, fresh NPs (200 μL) were diluted at a ratio of 1:1 with ultrapure water to obtain a scattering intensity between 200 and 300 kHz. Results were expressed as the mean ± standard deviation (SD), and each formulation was prepared in triplicate.

2.4. Drug Loading Efficiency Determination

The amount of R19 loaded in the NPs and the drug loading efficiency were determined by UV-Vis (UV-1800 spectrophotometer, Shimadzu, Kyoto, Japan). Fresh NPs (100 μL) were diluted 10× in DMSO, and their UV absorbance was measured at 309 nm (λ_{max} for R19). The concentration of loaded R19 was then calculated based on a calibration curve of drug absorbance at 309 nm versus concentration in DMSO. The concentration was multiplied by the total volume of each formulation to obtain the total amount of loaded drug; then, the drug loading efficiency (DL%) was calculated according to Equation (1):

$$\text{Drug loading efficiency (DL\%)} = (\text{Actual amount of drug loaded})/(\text{Initial amount of drug added}) \times 100\% \quad (1)$$

NR-labeled NPs were also diluted 10× in DMSO, and the absorbance was measured at 552 nm. A calibration curve of NR in DMSO was prepared at 552 nm and was used to determine the NR concentration in the labeled NPs.

2.5. Transmission Electron Microscopy (TEM) Imaging

R19-loaded NPs were visualized by transmission electron microscopy (TEM). Samples were prepared and were imaged as previously described [27].

2.6. Characterization by Fourier Transform-Infrared (FTIR) Spectroscopy

For analysis by FTIR, fresh R19-loaded NPs were lyophilized (FreeZone 4.5 L, Labconco, Kansas City, MO, USA) and transformed into KBr pellets. The FTIR spectra of R19, Pluronic P123, TPGS, and the R19-loaded NPs were acquired using an IR Affinity-1 spectrometer (Shimadzu, Kyoto, Japan) within a wavenumber range of 3600–500 cm^{-1}.

2.7. Differential Scanning Calorimetry (DSC)

The thermal transitions of the R19-loaded NPs were examined using a Netzsch DSC 204 F1 instrument (Selb, Germany). Approximately 10 mg of lyophilized NPs, individual polymers, and R19 were used for the analysis. The samples' thermal behavior was investigated by heating from ambient temperature to 250 °C at a rate of 10 °C/min under a nitrogen atmosphere.

2.8. Colloidal Stability of R19-Loaded NPs

The stability of the prepared NPs was evaluated upon incubation in cell culture media for 24 h at 37 °C by monitoring the changes in particle size and in the PDI. In total, 200 μL of each sample were diluted with 200 μL of cell culture medium supplemented with 10% fetal bovine serum (FBS), and the samples were incubated for 24 h with shaking at 100 rpm and 37 °C (Biosan ES-20 Orbital Shaker-Incubator, Riga, Latvia). Particle size and PDI of the NPs were measured before and after 24 h incubation by DLS. The experiment was performed in triplicate, and the results were reported as the mean ± SD.

2.9. In Vitro Release of R19 from R19-Loaded NPs

The in vitro release of R19 was evaluated using the dialysis method. Briefly, 1-mL aliquots of fresh NPs were transferred to a dialysis bag (12,000–14,000 MWCO) in triplicate and were inserted into vials containing 30 mL of phosphate-buffered saline (PBS, pH 7.4) supplemented with 0.5% *w/v* Tween 80 to maintain sink conditions [25]. The samples were incubated under shaking at 100 rpm and 37 °C (Biosan ES-20 Orbital Shaker-Incubator). At the designated sampling time points, 10 mL of each release medium was withdrawn and replaced with 10 mL of fresh buffer. At the 24, 48, 72, and 96 h time points, 10 mL samples were withdrawn, and the remaining release medium was replaced with 30 mL of fresh buffer. The collected samples were lyophilized, redissolved in 1 mL of DMSO, and centrifuged for 5 min at 4000 rpm to precipitate the undissolved buffer salts. The amount of R19 released was then measured by analyzing the UV absorbance of the supernatant based on the calibration curve of R19 in DMSO. Results were expressed as the average cumulative release (%) versus time (h). The release profiles were compared by calculating the similarity factor (f_2) between each of the two formulations according to Equation (2) [28]:

$$f_2 = 50 \times \log[(1 + 1/n \sum(R_t - T_t)^2)^{-0.5} \times 100] \quad (2)$$

where n is the number of sampling time points, and R_t and T_t are the cumulative amounts released from every two formulations at each time point t.

2.10. Cell Viability Assays

Cell viability assays were conducted on the MCF-7 and MDA-MB-231 breast cancer cells as well as on human dermal fibroblasts (HDFs) as normal cells. Cell lines were obtained from the American Type Culture Collection (ATCC, Manassas, VA, USA). MCF-7 cells were grown in Minimum Essential Medium (MEM; Gibco, Thermo Fisher Scientific, Waltham, MA, USA). MDA-MB-231 cells were grown in Roswell Park Memorial Institute (RPMI) 1640 medium (Euroclone SpA, Milan, Italy). HDFs were grown in Dulbecco's Modified Eagle's Medium (DMEM; Euroclone SpA). All media were supplemented with 10% heat-inactivated fetal bovine serum (FBS; Capricorn Scientific, Ebsdorfergrund, Germany) and penicillin–streptomycin (100 U/mL–100 μg/mL; Euroclone SpA). Cells were maintained at 37 °C in a humidified 5% CO_2 incubator. For the viability assays, cells were seeded in 96-well plates at a density of 10,000 cells/well (MCF-7 and MDA-MB-231) or 7000 cells/well (HDFs) and were incubated overnight. The next day, cells (n = 5 per treatment group) were treated with R19 (from a 20 mg/mL stock solution in DMSO) and R19-loaded NPs (freshly prepared and dispersed in water) at concentrations between 0 and 300 μM of R19 diluted in complete culture medium for 48 h. Another group of cells was treated with blank NP formulations at concentrations equivalent to the concentrations used for R19-loaded NPs. At the end of the incubation period, the medium was removed, and 100 μL of fresh medium containing 0.5 mg/mL of 3-(4,5-dimethylthiazol-2-yl)-2,5-diphenyltetrazolium bromide (MTT, Wheeling, IL, USA) was added to each well. Following incubation for 2 h, the medium was removed, and 100 μL of DMSO was added to each well to dissolve the formazan crystals. The optical density of each well was measured at 540 nm using a BioTek Synergy HTX Multi-Mode Reader. Cell viability was expressed as the % relative to the untreated cells. IC_{50} values were determined by a non-linear regression analysis of cell viability versus concentration curves using GraphPad Prism software version 7.00 (San Diego, CA, USA).

2.11. Mitochondrial Membrane Potential Assay

The mitochondrial membrane potential of MCF-7 and MDA-MB-231 cells treated with R19 and the R19-loaded NPs was measured using the JC-1 probe (Abcam, Cambridge, UK) [29]. Briefly, cells were seeded in 96-well plates at a density of 10,000 cells/well overnight. The next day, cells were treated with 20 μM of free R19 or an equivalent concentration of R19-loaded NPs for 24 h in triplicate (100 μL per well). After 24 h, JC-1 was diluted in PBS to a concentration of 20 μM, and 100 μL of the dye solution was added to each well. The cells were incubated with JC-1 for 10 min at 37 °C, and the fluorescence intensity of the wells was then measured using a BioTek Synergy HTX Multi-Mode Reader at 485 nm/528 nm excitation/emission wavelengths for the detection of the green JC-1 monomers and at 540 nm/620 nm excitation/emission wavelengths for the detection of the red JC-1 aggregates. The results were expressed as the ratio of green/red fluorescence after subtracting the background fluorescence of the unstained cells.

2.12. Cellular Uptake of NR-Labeled NPs

Cellular uptake assays were conducted in MCF-7 and MDA-MB-231 cells using NR-labeled NPs. Briefly, cells were seeded in 24-well plates at a density of 50,000 cells/well. The next day, cells were treated with NR dissolved in complete culture medium (MEM for MCF-7 cells or RPMI 1640 for MDA-MB-231 cells) at a concentration of 1 μg/mL or at an equivalent concentration of NR NPs for 1 h ($n = 3$). Then, the medium was removed, the cells were washed with PBS twice, and 500 μL of PBS was added to each well. NR fluorescence was quantified using a BioTek Synergy HTX Multi-Mode Reader at 540 nm/620 nm excitation/emission wavelengths. Untreated cells were used as background controls, and cell-associated fluorescence was normalized relative to the untreated cells.

2.13. Statistical Analysis

All data are presented as mean ± SD of at least three different trials. Differences between sample means were compared using one- or two-way analysis of variance (ANOVA) followed by Tukey's or Sidak's multiple comparison test, respectively, where a p-value of 0.05 was considered statistically significant. All analyses were performed in GraphPad Prism version 7.00.

3. Results and Discussion

3.1. Preparation and Characterization of R19-Loaded NPs

Here, we investigated the feasibility of incorporating R19 in a polymeric NP formulation to ensure its solubility, stability, and potential targeting ability when applied in vivo due to the enhanced permeability and retention (EPR) effect [15,30]. NPs were prepared by nanoprecipitation, as it is a simple and convenient method for entrapping hydrophobic drugs in polymeric NPs (Figure 2) [31]. R19 was formulated using two types of biocompatible polymers to evaluate the loading capacity of the NPs for the drug, the size of the NPs produced, and the effectiveness and safety toward cancer and normal cells.

Several formulations were prepared using different ratios of P123: TPGS (Table 1), and the characteristics of the NPs in terms of particle size, PDI, and DL% are presented in Table 2. The R19-loaded NPs were initially prepared using 100% P123. The average DL% was 59%, and the average particle size of the NPs was 100 nm, with a PDI of 0.17. Then, TPGS was used in combination with P123 to evaluate its effect on the DL% and particle size. Different ratios of both polymers were prepared: 100% TPGS (0% P123), 25% P123/75% TPGS, 50% P123/50% TPGS, and 75% P123/25% TPGS. Interestingly, as shown in Table 2, the different NP formulations showed similar average particle sizes ranging between 92 and 112 nm. With respect to the DL%, it ranged between 57 and 61%. Overall, the different ratios of P123 with TPGS did not reveal a significant difference with regard to the DL% nor the particle size of the NPs obtained. Likewise, the PDI of the formulations was relatively small and ranged between 0.14 and 0.20. PDI is a measure of the uniformity of the particle size distribution [32]. The PDI results revealed a narrow particle size distribution in all of the NP formulations as the values were close to 0, reflecting the excellent monodispersity of the NPs. Furthermore, the NPs were characterized by a spherical morphology when observed under TEM (Figure 3).

Zeta potential is used as an indicator of a colloidal dispersion's stability that is based on the surface charge of the particles in contact with water [33]. The zeta potential values for all of the NP formulations ranged between −30 mV and +30 mV, indicating that the NPs carried an almost neutral surface charge. Since this may affect their aggregation tendency, particularly when in contact with biological fluids, this led us to investigate the stability of the R19-loaded NPs in cell culture media, as described later.

Table 2. Characteristics of R19-loaded NPs reported as the mean ± SD of at least three different batches of each NP.

NP	Particle Size * (nm)	PDI	Zeta Potential (mV)	DL%
100% TPGS	108 ± 8	0.20 ± 0.06	−1.7 ± 5.0	60 ± 6
25% P123/75% TPGS	92 ± 7	0.20 ± 0.03	−6.6 ± 19.6	57 ± 7
50% P123/50% TPGS	111 ± 8	0.14 ± 0.01	−5.4 ± 15.3	59 ± 10
75% P123/25% TPGS	112 ± 8	0.15 ± 0.05	16.0 ± 2.9	59 ± 6
100% P123	100 ± 8	0.17 ± 0.02	−10.5 ± 17.1	59 ± 7

* Intensity-weighted.

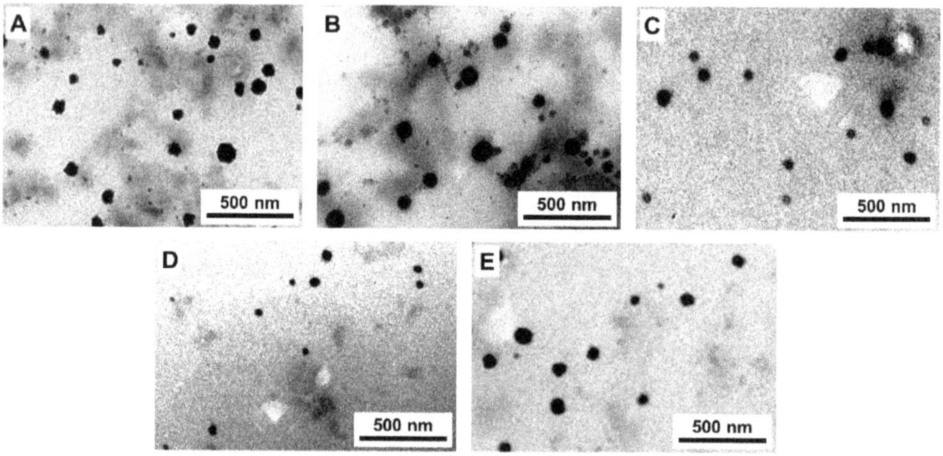

Figure 3. TEM images of R19-loaded NPs prepared using (**A**) 100% TPGS, (**B**) 25% P123/75% TPGS, (**C**) 50% P123/50% TPGS, (**D**) 75% P123/25% TPGS, and (**E**) 100% P123.

The five NP formulations were characterized by FTIR spectroscopy and were compared to free R19 and the individual polymers P123 and TPGS (Figure 4). TPGS displayed characteristic bands at 1750 cm^{-1} and at 2900–3000 cm^{-1}, corresponding to the O−C=O stretching of its ester groups and C−H stretching, respectively [34]. The O−C=O stretching band appeared in the R19 NP formulation prepared with 100% TPGS and in the other formulations containing TPGS but with lower intensities due to the lower concentration. Moreover, this band disappeared in the formulations containing 100% P123. Pluronic P123 is a polyether characterized by a broad C−O−C stretching band between 1000 and 1100 cm^{-1} [35]. P123 also showed a C−H stretching band at 2900–3000 cm^{-1}, which is related to asymmetric C−H stretching in the methyl groups in PPO in addition to the band at 1300–1400 cm^{-1} [36]. R19 NPs exhibited the same bands at 1300–1400 cm^{-1}. R19 and TPGS showed similar bands that overlapped with P123. In addition to a broad -OH stretching band between 2600 and 3200 cm^{-1}, R19 displayed a sharp band at 1660 cm^{-1} corresponding to the amide C=O stretching. This band was present in the NP formulations but had a weaker intensity, most likely due to the low % of R19 in the formulations (around 6% *w*/*w*). Overall, no new peaks were formed, which indicated no drug–excipient interaction.

The thermal transitions of the NP formulations were examined by DSC. As shown in Figure 5A, R19 displayed broad endothermic melting peaks between 80 and 100 °C and between 150 and 270 °C, indicating the nearly amorphous nature of the compound. Neat TPGS (Figure 5B) exhibited a sharp endothermic melting peak at around 45.9 °C, which is close to the literature value [37]. The R19-loaded NPs prepared using 100% TPGS (Figure 5C) showed a reduction in the TPGS melting peak to 40.9 °C, suggesting an interaction between the drug and the polymer, most likely representing van der Waals and potentially H-bonding interactions. As P123 was added to the formulation, the melting peak of TPGS was slightly lowered to 38.2–39.5 °C in the R19 NPs composed of 25% P123/75% TPGS (Figure 5D) and 50% P123/50% TPGS (Figure 5E), reflecting additional polymer–polymer and polymer–drug interactions. Notably, the peaks that were originally present in R19 disappeared in all of the NP formulations, suggesting that it is contained in the NPs in an amorphous state. The neat P123 and NPs with 75% and 100% P123 could not be analyzed due to the liquification of the polymer upon heating and the technical limitations of the DSC instrument. However, according to the literature, P123 is known for its thermal stability, with a weak endothermic peak appearing at 75 °C corresponding to the melting of PEO chains and a strong endothermic peak at around 400 °C due to polymer decomposition [38].

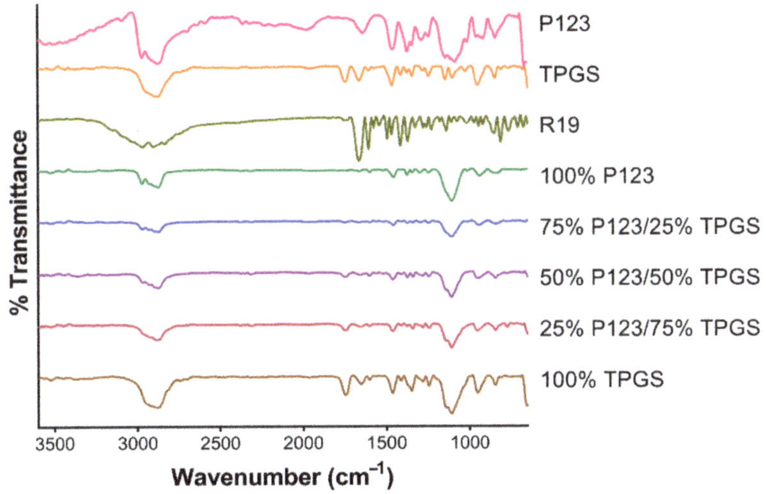

Figure 4. FTIR spectra of R19, P123, TPGS, and R19-loaded NPs.

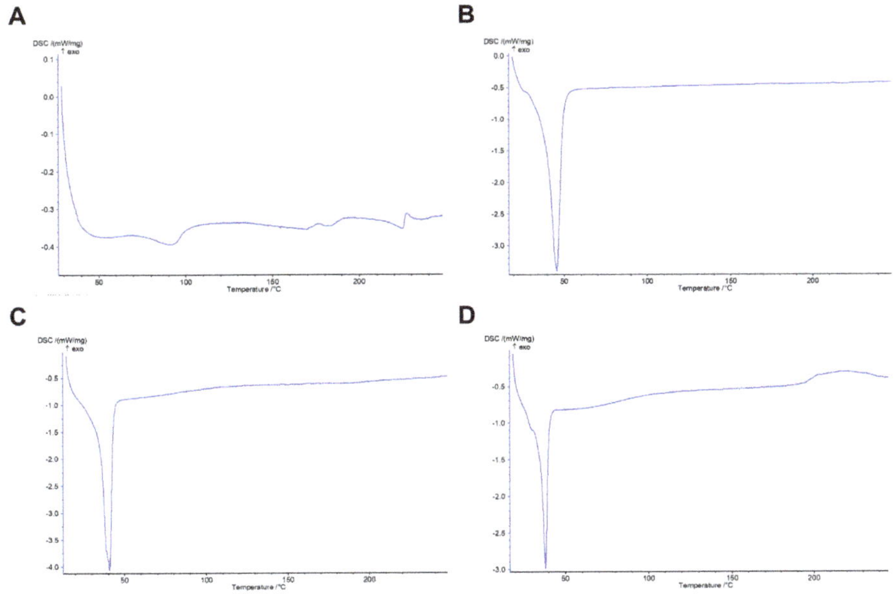

Figure 5. *Cont.*

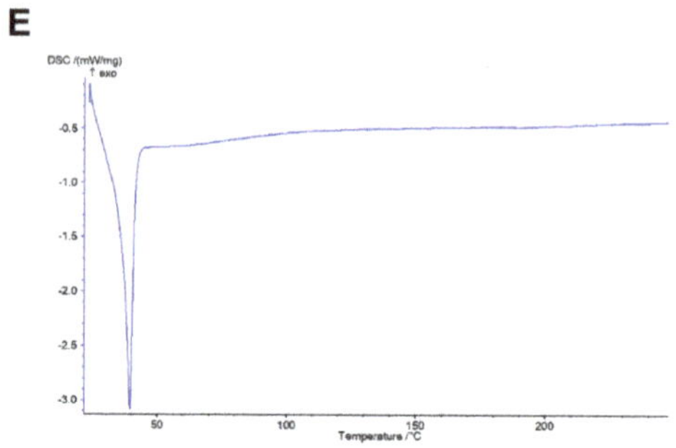

Figure 5. DSC thermograms of (**A**) R19, (**B**) neat TPGS, and R19-loaded NPs composed of (**C**) 100% TPGS, (**D**) 25% P123/75% TPGS, and (**E**) 50% P123/50% TPGS.

3.2. Stability of R19-Loaded NPs

As mentioned earlier, the R19-loaded NPs may have a tendency toward aggregation due to their low surface charge. It is challenging to maintain a stabilized particle size distribution of the NPs in a colloidal dispersion, particularly in biological fluids. Since the NPs were going to be evaluated in in vitro cell culture, we tested the colloidal stability of the NPs in a serum-supplemented cell culture medium for 24 h. Figure 6 represents the particle sizes of the samples before and after 24 h of incubation at 37 °C in the cell culture medium. Interestingly, all of the NPs, except for those prepared using 100% TPGS, showed no significant changes in the particle size nor in the PDI upon incubation. On the other hand, the NPs composed of 100% TPGS underwent a significant size increase ($p < 0.05$) from 88 nm to 108 nm that was accompanied by a marked increase in the PDI from 0.21 to 0.38 ($p < 0.01$). These results strongly indicate that the presence of P123, even at small percentages, can make the NPs more stable against aggregation when applied in biological systems.

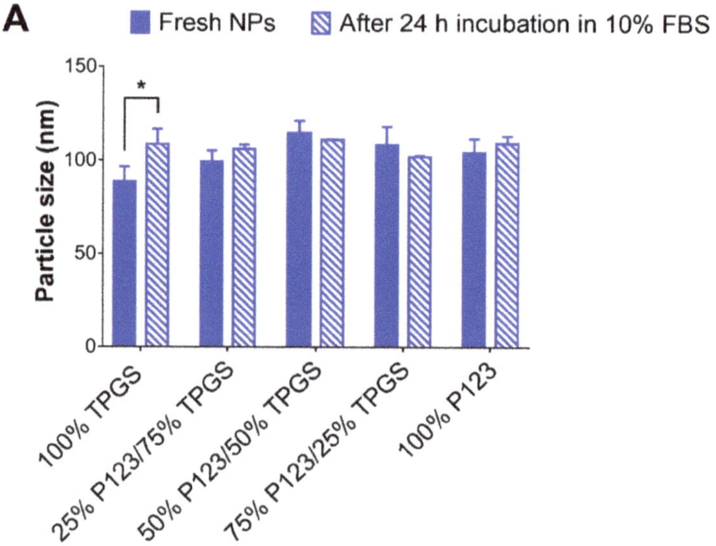

Figure 6. *Cont.*

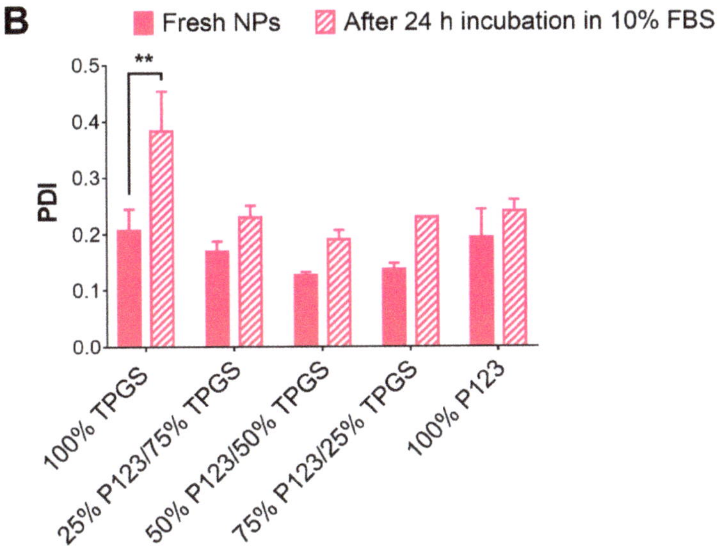

Figure 6. Change in (**A**) particle size and (**B**) PDI of R19-loaded NPs after 24 h incubation in cell culture medium supplemented with 10% FBS at 37 °C. Results are presented as the mean ± SD (n = 3). * $p < 0.05$ and ** $p < 0.01$ based on two-way ANOVA followed by Sidak's multiple comparison test.

3.3. *In Vitro Release of R19 from R19-Loaded NPs*

The different formulations of the R19-loaded NPs were subjected to release testing in PBS buffer with a pH of 7.4 at 37 °C to mimic physiological conditions. As shown in Figure 7, the cumulative drug release from all of the formulations was almost superimposable. All of the profiles were characterized by a biphasic release pattern, with a relatively fast release phase within the first 8 h followed by a more sustained release phase up to 96 h. The first phase may be related to the drug release from the surface of the NPs, whereas the second phase may be attributed to the slow diffusion of the drug from inside the NPs. Within the first 8 h of incubation, the cumulative drug release across all of the formulations was between 40 and 49%. After 24 h, the cumulative drug release increased more gradually to 50–58%. The same trend was observed at 48 and 72 h. After 96 h, the cumulative drug release reached 72−79% among the various formulations. Our results are in line with previous work on similar polymeric systems. For example, TPGS-formulated liposomes or NPs have been studied to determine in vitro release and showed a relatively slow release and a cumulative release of 82% after 48 h [39,40]. As for P123-based nanoformulations, the release of different drugs was affected by the percentage of P123 and ranged between 50 and 80% [41]. On the other hand, formulations with P123/TPGS mixtures showed a slow in vitro release that ranged between 40 and 50%, which was believed to be caused by the micelles formed from P123/TPGS and the preference of the hydrophobic drug to remain in the hydrophobic core of the micelles due to its low solubility in the release medium [42]. The similarity factor (f_2) was calculated between the various NP formulations to examine the differences between the release profiles. As shown in Table 3, f_2 was greater than 50 in all of the formulations, confirming their similar release behavior [28].

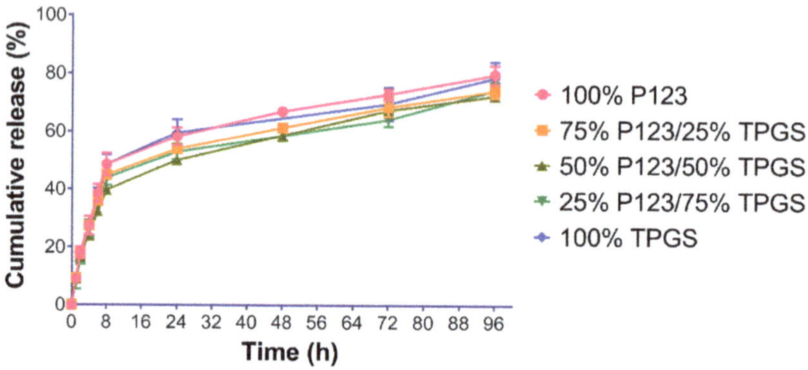

Figure 7. In vitro release of R19 from R19-loaded NPs in PBS pH 7.4 at 37 °C, demonstrating sustained drug release up to 96 h. Results are expressed as the mean cumulative release % ± SD (n = 3) plotted against time (h).

Table 3. Similarity factor (f_2) results for R19-loaded NPs.

Sample 1	Sample 2	f_2
100% P123	75% P123/25% TPGS	71
	50% P123/50% TPGS	60
	25% P123/75% TPGS	66
	100% TPGS	88
75% P123/25% TPGS	50% P123/50% TPGS	75
	25% P123/75% TPGS	84
	100% TPGS	75
50% P123/50% TPGS	25% P123/75% TPGS	78
	100% TPGS	62
25% P123/75% TPGS	100% TPGS	69

3.4. Cell Viability Assays

The anticancer activity of R19 and R19-loaded NPs was evaluated against two breast cancer cell lines: MCF-7 and MDA-MB-231, because breast cancer is the most common malignancy affecting women and accounts for the highest percentage of cancer-related deaths [43]. Cells were treated with increasing concentrations of R19 and the NPs for 48 h, followed by the MTT assay. As shown in Figures 8A and 9A, all of the treatments caused a dose-dependent inhibition of cell growth, but with varying degrees. The dose–response curves were fitted by non-linear regression analysis to obtain the IC_{50} values, which correspond to the treatments' potency. As shown in Table 4, R19 displayed similar potency against both cell lines, with IC_{50} values of 14.7 and 17.0 µM in the MCF-7 and MDA-MB-231 cells, respectively. These values were similar to the R19 potency in colorectal cancer cells [9]. Since the PI3K/Akt/mTOR pathway is dysregulated in these cancer cell lines, it was expected that R19 would show similar inhibitory activity. The potent anticancer effect of R19 in breast cancer cell lines strongly support its application as a potential targeted therapy in hormone receptor-positive and triple-negative breast cancer.

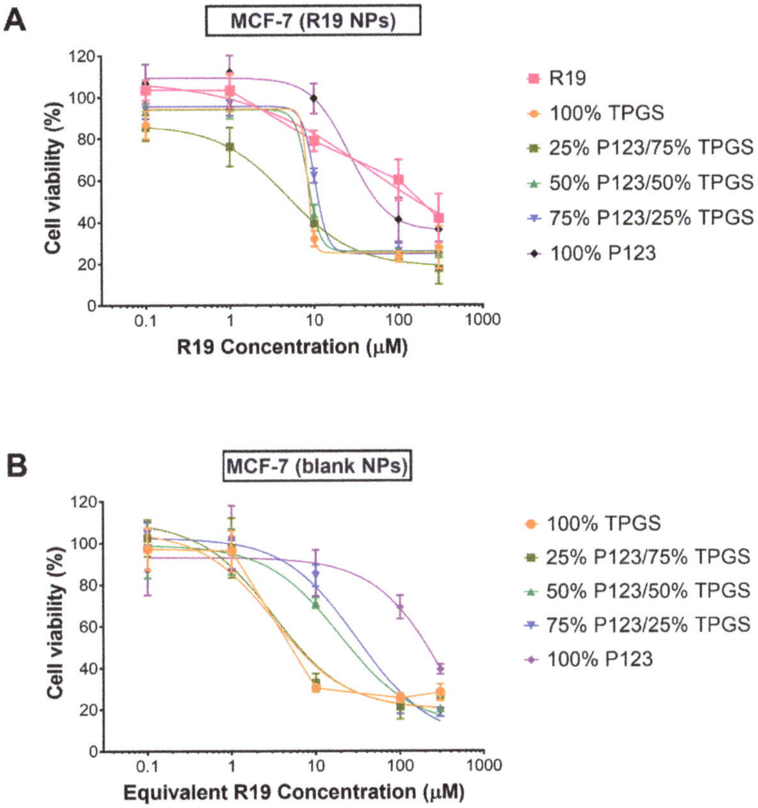

Figure 8. Percentage viability of MCF-7 cells treated with (**A**) R19 and R19-loaded NPs and (**B**) blank NPs for 48 h (n = 5).

As for the R19 NP formulations, lower IC_{50} values were obtained compared to the free drug in both cell lines, with the exception of 100% P123 NPs. The highest potency was observed for the R19 NPs prepared using 100% TPGS and 25% P123/75% TPGS in both the MCF-7 and MDA-MB-231 cells, with almost a three-fold reduction in the IC_{50} values in MCF-7 cells and a two- to nine-fold reduction in MDA-MB-231 cells. Generally, the IC_{50} value increased as the % of TPGS decreased, and the NPs with 100% P123 displayed the highest IC_{50} values and the poorest fit to the dose–response curve ($R^2 < 0.9$). The enhancement of anticancer potency in the NP formulations can likely be attributed to the polymeric carriers, particularly TPGS. TPGS has been reported to inhibit cancer cell proliferation through cell cycle arrest and the induction of apoptosis [44,45]. In addition, TPGS has been shown to overcome cancer drug resistance by downregulating the expression of P-glycoprotein efflux pumps [46,47]. These findings were confirmed by testing equivalent concentrations of the NPs without the drug under the same conditions. As shown in Table 4, blank NPs containing more than 50% TPGS were almost equally potent as the drug-loaded formulations. However, as the % of P123 increased, a reduction in potency was observed, similar to the drug-loaded NPs. These results provide a promising approach for the application of TPGS-containing R19 NPs, not only in breast cancer but also in multidrug-resistant cancer types.

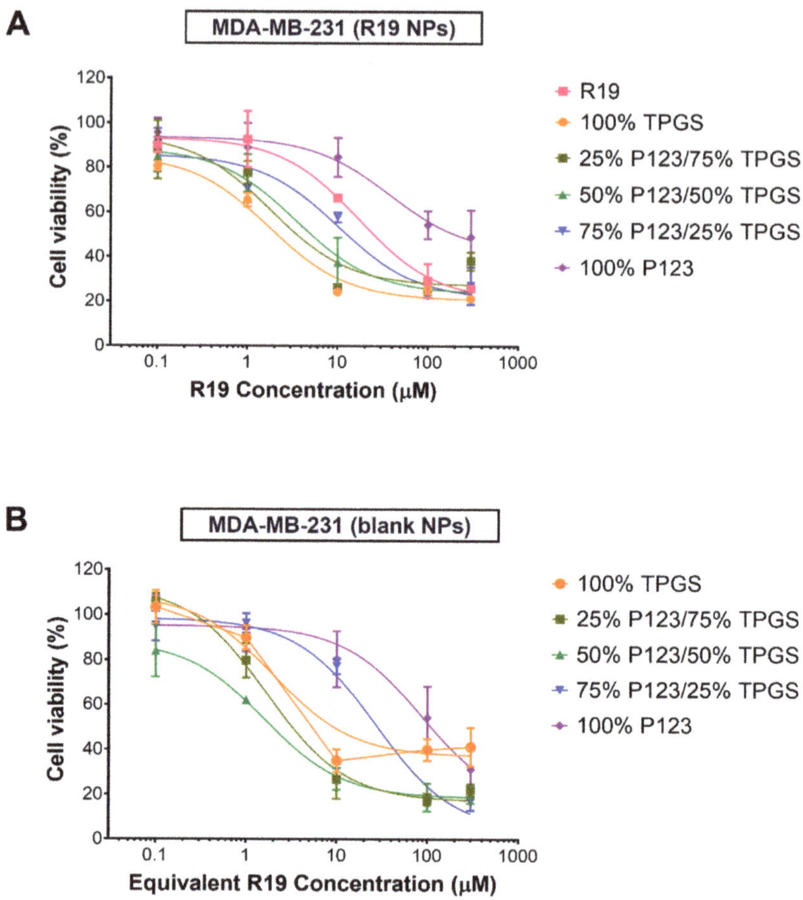

Figure 9. Percentage viability of MDA-MB-231 cells treated with (**A**) R19 and R19-loaded NPs and (**B**) blank NPs for 48 h ($n = 5$).

Table 4. IC_{50} values of R19 and R19-loaded NPs and equivalent concentrations of blank NPs in MCF-7 and MDA-MB-231 cells after 48 h of incubation.

Treatment	IC_{50} (μM; Mean ± SEM *)	
	MCF-7	MDA-MB-231
R19	14.7 ± 5.3	17.0 ± 4.2
100% TPGS	4.3 ± 1.9	1.8 ± 0.4
25% P123/75% TPGS	4.7 ± 1.0	1.8 ± 0.7
50% P123/50% TPGS	4.9 ± 1.1	3.5 ± 1.0
75% P123/25% TPGS	12.3 ± 2.4	10.2 ± 3.2
100% P123	45.4 ± 18.6	37.5 ± 24.3
Blank 100% TPGS	3.2 ± 1.1	2.0 ± 0.8
Blank 25% P123/75% TPGS	3.1 ± 1.0	1.7 ± 0.3
Blank 50% P123/50% TPGS	19.5 ± 4.9	1.6 ± 0.4
Blank 75% P123/25% TPGS	31.0 ± 7.2	26.8 ± 5.8
Blank 100% P123	415.1 ± 448.8	87.1 ± 52.0

* SEM: standard error.

Having shown excellent activity against cancer cells, equivalent concentrations of R19-loaded NPs were also tested in HDF to determine their selectivity and biocompatibility.

As depicted in Figure 10, the NPs exhibited more than 60% cell viability at the highest concentration tested (300 μM), whereas the free drug caused an almost 50% reduction in cell viability at the same concentration. Moreover, the NPs prepared using 100% TPGS and 50% P123/50% TPGS maintained a cell viability of ~100% at the highest concentration, indicating their superior biocompatibility. As all of the groups resulted in more than 50% cell viability, the IC_{50} values could not be accurately determined. Altogether, the results clearly demonstrate the selectivity of R19 in killing cancer cells, which was enhanced upon incorporation into the NPs. Moreover, the low IC_{50} values obtained for the TPGS-containing NPs in the MDA-MB-231 cells, which represent the more aggressive type of breast cancer, represent a highly promising result for the potential clinical application of the NP formulation as a breast cancer nanomedicine.

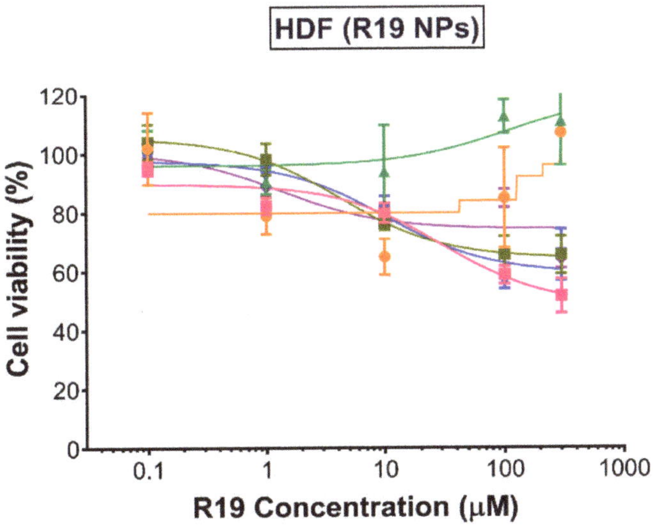

Figure 10. Percentage viability of HDF treated with R19 and R19-loaded NPs for 48 h (n = 5).

3.5. Mitochondrial Membrane Potential Assay of R19-Treated Cells

Mitochondrial depolarization is one of the early signs of apoptosis and can be conveniently detected by JC-1. Under normal conditions, JC-1 exists in an aggregated state within the mitochondrial matrix and emits red fluorescence. As the mitochondrial transmembrane potential dissipates in cells undergoing apoptosis, JC-1 is effluxed to the cytoplasm, where it dissociates into monomers that emit green fluorescence. Therefore, the change in the ratio of green/red fluorescence may be used as an indicator of apoptosis induction [48]. As seen in Figure 11A, MCF-7 cells treated with free R19 and R19 NPs composed of 75% P123/25% TPGS and 100% P123 did not show a significant difference in the JC-1 monomer/aggregate ratio compared to the control (untreated cells). Likewise, the mitochondrial membrane potential of MDA-MB-231 cells treated with R19 and R19 NPs composed of 100% P123 was similar to that of the control (Figure 11B). On the other hand, cells treated with the R19 NPs containing TPGS displayed a significantly higher JC-1 monomer/aggregate ratio, signifying that these formulations were able to induce a greater degree of apoptosis compared to the free drug and R19 NPs composed entirely of P123.

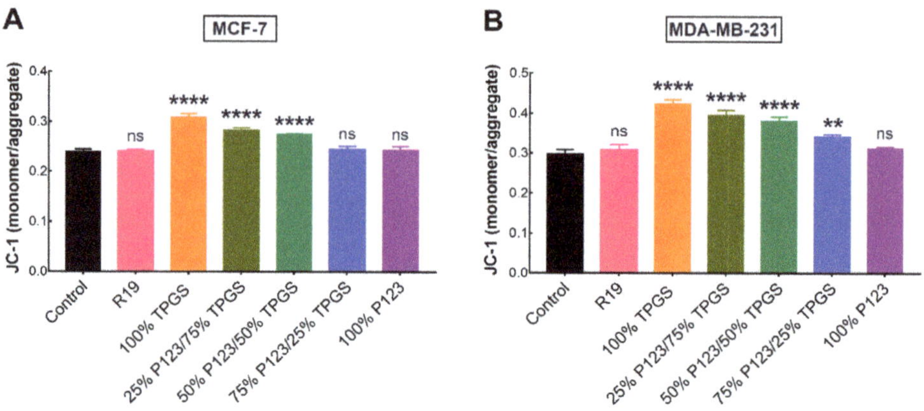

Figure 11. Change in mitochondrial membrane potential of (**A**) MCF-7 and (**B**) MDA-MB-231 cells after 24 h of treatment with 20 μM of R19 and R19-loaded NPs. Results are expressed as the ratio of the green/red fluorescence signals of the JC-1 monomers and aggregates, respectively (mean ± SD; $n = 3$). ** $p < 0.01$, **** $p < 0.0001$, and ns: not significantly different, compared to the control (untreated cells), based on one-way ANOVA followed by Tukey's multiple comparisons test.

3.6. Cellular Uptake of NR-Labeled NPs

The observed enhancement in the bioactivity of R19 upon incorporation into the NP formulations may in part be attributed to an enhancement in cellular uptake. To test this hypothesis, R19 was replaced with NR, and the NR-labeled NPs were incubated with the cells. As depicted in Figure 12A, in MCF-7 cells, the NPs composed of high percentages of TPGS (100%, 75%, and 50%) were associated with the highest increase in intracellular fluorescence compared to the free dye, denoting enhanced cellular uptake. As for MDA-MB-231 cells (Figure 12B), all of the NPs exhibited a significant increase in intracellular fluorescence compared to the free dye regardless of their composition. Interestingly, the TPGS-containing NPs achieved a four- to five-fold increase in fluorescence compared to the control in MCF-7 cells, whereas in the MDA-MB-231 cells, the NPs averaged around a 2.7-fold increase in fluorescence. These results are most likely attributed to differences in the membrane permeability between the two cell lines. Nonetheless, the results strongly support the ability of the NP formulations, particularly those containing TPGS, to enhance the cellular uptake of hydrophobic molecules such as R19, consistent with the cell viability and JC-1 assays.

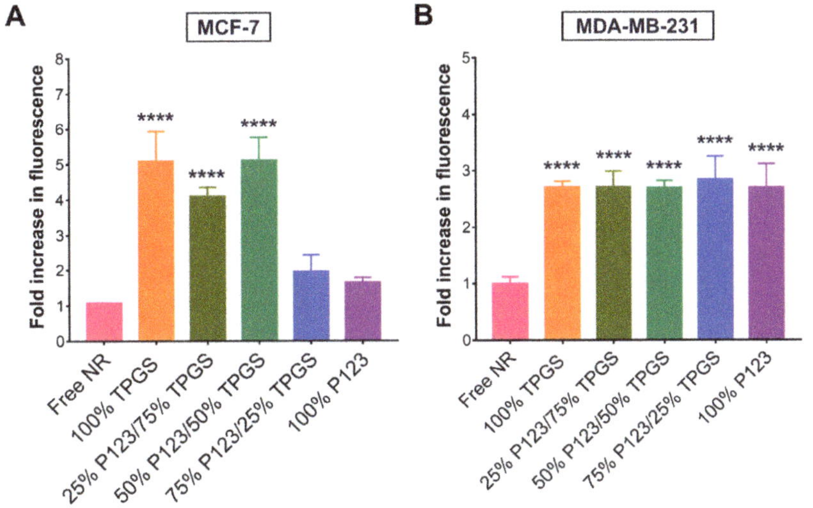

Figure 12. Normalized fluorescence intensities (mean ± SD; $n = 3$) of (**A**) MCF-7 and (**B**) MDA-MB-231 cells treated with free NR or NR-labeled NPs for 1 h. **** $p < 0.0001$ based on one-way ANOVA followed by Tukey's multiple comparison test.

4. Conclusions

R19 is a new chemical entity that has been found to be an effective anticancer agent. In this work, R19 was successfully formulated in NPs composed of different combinations of TPGS and Pluronic P123. The NPs had particle sizes of approximately 100 nm and high monodispersity. Additionally, the DL% reached about 60% in the different polymer combinations and all of the NPs sustained the release of the drug up to 96 h. The NPs were highly stable in serum-supplemented cell culture medium, with the exception of the formulations composed of 100% TPGS, which exhibited a tendency for aggregation. Cell viability assays in MCF-7 and MDA-MB-231 breast cancer cell lines revealed an enhanced potency of R19 when it was incorporated into TPGS-containing NPs with reduced cytotoxicity against HDF, most likely by enhancing its cellular uptake and promoting apoptosis. The observed cancer cell selectivity and high biocompatibility of the NP formulations emphasize the positive attributes of the designed polymeric NPs as a promising delivery approach for R19 and its analogs, bringing them one step closer to clinical translation.

Author Contributions: Conceptualization, S.S. and D.S.; methodology, S.S.; software, S.S.; validation, S.A. (Sundos Aliyeh), S.A. (Samah Abusulieh) and S.S.; formal analysis, S.A. (Sundos Aliyeh) and S.S.; investigation, S.A. (Sundos Aliyeh), S.A. (Samah Abusulieh) and S.S.; resources, S.S.; data curation, S.A. (Sundos Aliyeh), S.A. (Samah Abusulieh) and S.S.; writing—original draft preparation, S.A. (Sundos Aliyeh); writing—review and editing, D.S. and S.S.; visualization, S.S.; supervision, D.S. and S.S.; project administration, S.S.; funding acquisition, S.S. All authors have read and agreed to the published version of the manuscript.

Funding: This research was funded by Al-Zaytoonah University of Jordan, grants number 11/18/2018-2019 and 14/08/2021-2022. The APC was funded by the authors.

Institutional Review Board Statement: Not applicable.

Informed Consent Statement: Not applicable.

Data Availability Statement: Not applicable.

Acknowledgments: The authors thank Faten Alregeb (University of Jordan) for the assistance with the DSC analysis.

Conflicts of Interest: The authors declare no conflict of interest. The funders had no role in the design of the study; in the collection, analyses, or interpretation of data; in the writing of the manuscript; or in the decision to publish the results.

References

1. World Health Organization (WHO). Cancer Fact Sheet. Available online: https://www.who.int/news-room/fact-sheets/detail/cancer (accessed on 14 July 2022).
2. Rallis, K.S.; Lai Yau, T.H.; Sideris, M. Chemoradiotherapy in cancer treatment: Rationale and clinical applications. *Anticancer Res.* **2021**, *41*, 1. [CrossRef] [PubMed]
3. Schirrmacher, V. From chemotherapy to biological therapy: A review of novel concepts to reduce the side effects of systemic cancer treatment (Review). *Int. J. Oncol.* **2019**, *54*, 407–419.
4. Kannaiyan, R.; Mahadevan, D. A comprehensive review of protein kinase inhibitors for cancer therapy. *Expert Rev. Anticancer Ther.* **2018**, *18*, 1249–1270. [CrossRef] [PubMed]
5. Andrs, M.; Korabecny, J.; Jun, D.; Hodny, Z.; Bartek, J.; Kuca, K. Phosphatidylinositol 3-inase (PI3K) and phosphatidylinositol 3-kinase-related kinase (PIKK) inhibitors: Importance of the morpholine ring. *J. Med. Chem.* **2015**, *58*, 41–71. [CrossRef] [PubMed]
6. Zhang, Y.; Kwok-Shing Ng, P.; Kucherlapati, M.; Chen, F.; Liu, Y.; Tsang, Y.H.; de Velasco, G.; Jeong, K.J.; Akbani, R.; Hadjipanayis, A.; et al. A Pan-cancer proteogenomic atlas of PI3K/Akt/mTOR pathway alterations. *Cancer Cell* **2017**, *31*, 820–832.e823. [CrossRef]
7. Yu, J.S.; Cui, W. Proliferation, survival and metabolism: The role of PI3K/Akt/mTOR signalling in pluripotency and cell fate determination. *Development* **2016**, *143*, 3050–3060. [CrossRef]
8. Hamadneh, L.; Bahader, M.; Abuarqoub, R.; AlWahsh, M.; Alhusban, A.; Hikmat, S. PI3K/AKT and MAPK1 molecular changes preceding matrix metallopeptidases overexpression during tamoxifen-resistance development are correlated to poor prognosis in breast cancer patients. *Breast Cancer* **2021**, *28*, 1358–1366. [CrossRef]
9. Sabbah, D.A.; Haroon, R.A.; Bardaweel, S.K.; Hajjo, R.; Sweidan, K. N-phenyl-6-chloro-4-hydroxy-2-quinolone-3-carboxamides: Molecular docking, synthesis, and biological investigation as anticancer agents. *Molecules* **2020**, *26*, 73. [CrossRef]
10. Zhong, Y.; Su, T.; Shi, Q.; Feng, Y.; Tao, Z.; Huang, Q.; Li, L.; Hu, L.; Li, S.; Tan, H.; et al. Co-administration of iRGD enhances tumor-targeted delivery and anti-tumor effects of paclitaxel-loaded PLGA nanoparticles for colorectal cancer treatment. *Int. J. Nanomed.* **2019**, *14*, 8543–8560. [CrossRef]
11. Jin, C.; Wang, K.; Oppong-Gyebi, A.; Hu, J. Application of nanotechnology in cancer diagnosis and therapy—A mini-review. *Int. J. Med. Sci.* **2020**, *17*, 2964–2973. [CrossRef]
12. Sunoqrot, S.; Hamed, R.; Abdel-Halim, H.; Tarawneh, O. Synergistic interplay of medicinal chemistry and formulation strategies in nanotechnology—From drug discovery to nanocarrier design and development. *Curr. Top. Med. Chem.* **2017**, *17*, 1451–1468. [CrossRef] [PubMed]
13. Gagliardi, A.; Giuliano, E.; Venkateswararao, E.; Fresta, M.; Bulotta, S.; Awasthi, V.; Cosco, D. Biodegradable polymeric nanoparticles for drug delivery to solid tumors. *Front. Pharmacol.* **2021**, *12*, 601626. [CrossRef] [PubMed]
14. Kamaly, N.; Yameen, B.; Wu, J.; Farokhzad, O.C. Degradable controlled-release polymers and polymeric nanoparticles: Mechanisms of controlling drug release. *Chem. Rev.* **2016**, *116*, 2602–2663. [CrossRef] [PubMed]
15. Nakamura, H.; Jun, F.; Maeda, H. Development of next-generation macromolecular drugs based on the EPR effect: Challenges and pitfalls. *Expert Opin. Drug Deliv.* **2015**, *12*, 53–64. [CrossRef]
16. Sunoqrot, S.; Alsadi, A.; Tarawneh, O.; Hamed, R. Polymer type and molecular weight dictate the encapsulation efficiency and release of Quercetin from polymeric micelles. *Colloid Polym. Sci.* **2017**, *295*, 2051–2059. [CrossRef]
17. Parmar, A.; Singh, K.; Bahadur, A.; Marangoni, G.; Bahadur, P. Interaction and solubilization of some phenolic antioxidants in Pluronic® micelles. *Colloids Surf. B Biointerfaces* **2011**, *86*, 319–326. [CrossRef] [PubMed]
18. Zhao, L.; Du, J.; Duan, Y.; Zang, Y.n.; Zhang, H.; Yang, C.; Cao, F.; Zhai, G. Curcumin loaded mixed micelles composed of Pluronic P123 and F68: Preparation, optimization and in vitro characterization. *Colloids Surf. B Biointerfaces* **2012**, *97*, 101–108. [CrossRef]
19. Nguyen, V.T.; Nguyen, T.H.; Dang, L.H.; Vu-Quang, H.; Tran, N.Q. Folate-conjugated chitosan-pluronic P123 nanogels: Synthesis and characterizations towards dual drug delivery. *J. Nanomater.* **2019**, *2019*, 1067821. [CrossRef]
20. Gan, H.; Chen, L.; Sui, X.; Wu, B.; Zou, S.; Li, A.; Zhang, Y.; Liu, X.; Wang, D.; Cai, S.; et al. Enhanced delivery of sorafenib with anti-GPC3 antibody-conjugated TPGS-b-PCL/Pluronic P123 polymeric nanoparticles for targeted therapy of hepatocellular carcinoma. *Mater. Sci. Eng. C* **2018**, *91*, 395–403. [CrossRef]
21. Zou, T.; Gu, L. TPGS emulsified zein nanoparticles enhanced oral bioavailability of daidzin: In vitro characteristics and in vivo performance. *Mol. Pharm.* **2013**, *10*, 2062–2070. [CrossRef]
22. Gaonkar, R.; Ganguly, S.; Dewanjee, S.; Sinha, S.; Gupta, A.; Ganguly, S.; Chattopadhyay, S.; Debnath, C. Garcinol loaded vitamin E TPGS emulsified PLGA nanoparticles: Preparation, physicochemical characterization, in vitro and in vivo studies. *Sci. Rep.* **2017**, *7*, 530–544. [CrossRef] [PubMed]
23. Zhu, D.; Tao, W.; Zhang, H.; Liu, G.; Wang, T.; Zhang, L.; Zeng, X.; Mei, L. Docetaxel (DTX)-loaded polydopamine-modified TPGS-PLA nanoparticles as a targeted drug delivery system for the treatment of liver cancer. *Acta Biomater.* **2016**, *30*, 144–154. [CrossRef] [PubMed]

24. Zhang, J.; Tao, W.; Chen, Y.; Chang, D.; Wang, T.; Zhang, X.; Mei, L.; Zeng, X.; Huang, L. Doxorubicin-loaded star-shaped copolymer PLGA-vitamin E TPGS nanoparticles for lung cancer therapy. *J. Mater. Sci. Mater. Med.* **2015**, *26*, 165. [CrossRef] [PubMed]
25. Mu, L.; Feng, S.S. PLGA/TPGS nanoparticles for controlled release of paclitaxel: Effects of the emulsifier and drug loading ratio. *Pharm. Res.* **2003**, *20*, 1864–1872. [CrossRef]
26. Sunoqrot, S.; Orainee, B.; Alqudah, D.A.; Daoud, F.; Alshaer, W. Curcumin-tannic acid-poloxamer nanoassemblies enhance curcumin's uptake and bioactivity against cancer cells in vitro. *Int. J. Pharm.* **2021**, *610*, 121255. [CrossRef]
27. Sunoqrot, S.; Al-Shalabi, E.; Al-Bakri, A.G.; Zalloum, H.; Abu-Irmaileh, B.; Ibrahim, L.H.; Zeno, H. Coffee bean polyphenols can form biocompatible template-free antioxidant nanoparticles with various sizes and distinct colors. *ACS Omega* **2021**, *6*, 2767–2776. [CrossRef]
28. Duan, J.Z.; Riviere, K.; Marroum, P. In vivo bioequivalence and in vitro similarity factor (f2) for dissolution profile comparisons of extended release formulations: How and when do they match? *Pharm. Res.* **2011**, *28*, 1144–1156. [CrossRef]
29. Sivandzade, F.; Bhalerao, A.; Cucullo, L. Analysis of the Mitochondrial Membrane Potential Using the Cationic JC-1 Dye as a Sensitive Fluorescent Probe. *Bio-Protocol* **2019**, *9*, e3128. [CrossRef]
30. Kalyane, D.; Raval, N.; Maheshwari, R.; Tambe, V.; Kalia, K.; Tekade, R.K. Employment of enhanced permeability and retention effect (EPR): Nanoparticle-based precision tools for targeting of therapeutic and diagnostic agent in cancer. *Mater. Sci. Eng. C* **2019**, *98*, 1252–1276. [CrossRef]
31. Martínez Rivas, C.J.; Tarhini, M.; Badri, W.; Miladi, K.; Greige-Gerges, H.; Nazari, Q.A.; Galindo Rodríguez, S.A.; Román, R.Á.; Fessi, H.; Elaissari, A. Nanoprecipitation process: From encapsulation to drug delivery. *Int. J. Pharm.* **2017**, *532*, 66–81. [CrossRef]
32. Danaei, M.; Dehghankhold, M.; Ataei, S.; Hasanzadeh Davarani, F.; Javanmard, R.; Dokhani, A.; Khorasani, S.; Mozafari, M.R. Impact of particle size and polydispersity index on the clinical applications of lipidic nanocarrier systems. *Pharmaceutics* **2018**, *10*, 57. [CrossRef] [PubMed]
33. Ferraris, S.; Cazzola, M.; Peretti, V.; Stella, B.; Spriano, S. Zeta potential measurements on solid surfaces for in vitro biomaterials testing: Surface charge, reactivity upon contact with fluids and protein absorption. *Front. Bioeng. Biotechnol.* **2018**, *6*, 60. [CrossRef] [PubMed]
34. Khare, V.; Sakarchi, W.A.; Gupta, P.N.; Curtis, A.D.M.; Hoskins, C. Synthesis and characterization of TPGS–gemcitabine prodrug micelles for pancreatic cancer therapy. *RSC Adv.* **2016**, *6*, 60126–60137. [CrossRef]
35. Dehvari, K.; Lin, K.-S.; Hammouda, B. Small-angle neutron scattering studies of microenvironmental and structural changes of Pluronic micelles upon encapsulation of paclitaxel. *J. Taiwan Inst. Chem. Eng.* **2017**, *71*, 405–413. [CrossRef]
36. Su, Y.-L.; Wang, J.; Liu, H.-Z. FTIR spectroscopic study on effects of temperature and polymer composition on the structural properties of PEO−PPO−PEO block copolymer micelles. *Langmuir* **2002**, *18*, 5370–5374. [CrossRef]
37. Mu, L.; Feng, S.S. A novel controlled release formulation for the anticancer drug paclitaxel (Taxol®): PLGA nanoparticles containing vitamin E TPGS. *J. Control. Release* **2003**, *86*, 33–48. [CrossRef]
38. Zhang, G.; Chen, X.; Zhao, Y.; Ma, F.; Jing, B.; Qiu, H. Lyotropic liquid-crystalline phases formed by pluronic p123 in ethylammonium nitrate. *J. Phys. Chem. B* **2008**, *112*, 6578–6584. [CrossRef]
39. Mu, L.; Feng, S.S. Vitamin E TPGS used as emulsifier in the solvent evaporation/extraction technique for fabrication of polymeric nanospheres for controlled release of paclitaxel (Taxol®). *J. Control. Release* **2002**, *80*, 129–144. [CrossRef]
40. Farooq, M.A.; Xinyu, H.; Jabeen, A.; Ahsan, A.; Seidu, T.A.; Kutoka, P.T.; Wang, B. Enhanced cellular uptake and cytotoxicity of vorinostat through encapsulation in TPGS-modified liposomes. *Colloids Surf. B Biointerfaces* **2021**, *199*, 111523. [CrossRef]
41. Jindal, N.; Mehta, S.K. Nevirapine loaded Poloxamer 407/Pluronic P123 mixed micelles: Optimization of formulation and in vitro evaluation. *Colloids Surf. B Biointerfaces* **2015**, *129*, 100–106. [CrossRef]
42. Duan, Y.; Cai, X.; Du, H.; Zhai, G. Novel in situ gel systems based on P123/TPGS mixed micelles and gellan gum for ophthalmic delivery of curcumin. *Colloids Surf. B Biointerfaces* **2015**, *128*, 322–330. [CrossRef]
43. Lei, S.; Zheng, R.; Zhang, S.; Wang, S.; Chen, R.; Sun, K.; Zeng, H.; Zhou, J.; Wei, W. Global patterns of breast cancer incidence and mortality: A population-based cancer registry data analysis from 2000 to 2020. *Cancer Commun.* **2021**, *41*, 1183–1194. [CrossRef]
44. Neophytou, C.M.; Constantinou, C.; Papageorgis, P.; Constantinou, A.I. D-alpha-tocopheryl polyethylene glycol succinate (TPGS) induces cell cycle arrest and apoptosis selectively in Survivin-overexpressing breast cancer cells. *Biochem. Pharmacol.* **2014**, *89*, 31–42. [CrossRef]
45. Ruiz-Moreno, C.; Jimenez-Del-Rio, M.; Sierra-Garcia, L.; Lopez-Osorio, B.; Velez-Pardo, C. Vitamin E synthetic derivate-TPGS-selectively induces apoptosis in Jurkat T cells via oxidative stress signaling pathways: Implications for acute lymphoblastic leukemia. *Apoptosis* **2016**, *21*, 1019–1032. [CrossRef]
46. Guan, Y.-Y.; Zeng, S.-Q.; Qin, Y.; Mu, Y.; Liu, H. Vitamin E-tocopheryl polyethylene glycol succinate decorated drug delivery system with synergistic antitumor effects to reverse drug resistance and immunosuppression. *Colloids Surf. A Physicochem. Eng. Asp.* **2021**, *628*, 127387. [CrossRef]
47. Yan, H.; Du, X.; Wang, R.; Zhai, G. Progress in the study of D-α-tocopherol polyethylene glycol 1000 succinate (TPGS) reversing multidrug resistance. *Colloids Surf. B Biointerfaces* **2021**, *205*, 111914. [CrossRef]
48. Tusskorn, O.; Khunluck, T.; Prawan, A.; Senggunprai, L.; Kukongviriyapan, V. Mitochondrial division inhibitor-1 potentiates cisplatin-induced apoptosis via the mitochondrial death pathway in cholangiocarcinoma cells. *Biomed. Pharmacother.* **2019**, *111*, 109–118. [CrossRef]

pharmaceutics

MDPI

Article

Fragment-Based and Structural Investigation for Discovery of JNK3 Inhibitors

Men Thi Hoai Duong and Hee-Chul Ahn *

College of Pharmacy, Dongguk University-Seoul, Goyang, Gyeonggi 10326, Korea
* Correspondence: hcahn@dongguk.edu; Tel.: +82-31-961-5221

Abstract: The c-Jun N-terminal kinases (JNKs) are members of the mitogen-activated protein kinase (MAPK) family and are related to cell proliferation, gene expression, and cell death. JNK isoform 3 (JNK3) is an important therapeutic target in varieties of pathological conditions including cancers and neuronal death. There is no approved drug targeting JNKs. To discover chemical inhibitors of JNK3, virtual fragment screening, the saturation transfer difference (STD) NMR, in vitro kinase assay, and X-ray crystallography were employed. A total of 27 fragments from the virtually selected 494 compounds were identified as initial hits via STD NMR and some compounds showed the inhibition of the activity of JNK3 in vitro. The structures of JNK3 with a fragment and a potent inhibitor were determined by X-ray crystallography. The fragment and inhibitor shared a common JNK3-binding feature. The result shows that fragment screening by NMR spectroscopy is a very efficient method to screen JNK3 binders and the structure of JNK3-inhibitor complex can be used to design and develop more potent inhibitors.

Keywords: JNK3; fragment; saturation transfer difference NMR; X-ray crystallography

Citation: Duong, M.T.H.; Ahn, H.-C. Fragment-Based and Structural Investigation for Discovery of JNK3 Inhibitors. *Pharmaceutics* **2022**, *14*, 1900. https://doi.org/10.3390/pharmaceutics14091900

Academic Editor: Francesca Musumeci

Received: 29 July 2022
Accepted: 5 September 2022
Published: 8 September 2022

1. Introduction

JNKs are the serine/threonine kinases and one of the members of the mitogen-activated protein kinases [1,2]. JNK phosphorylates several substrates, such as c-Jun and activating transcription factor 2 in response to environmental stress and pro-inflammatory cytokines [2]. JNKs are involved in varieties of physiological processes including neuronal function, immune activity, and embryonic development [3,4]. Thus, it is implicated that the JNK pathway is related to various pathological conditions, including neurodegenerative diseases, cancer, and inflammation [5–7]. Since the JNK signaling pathway is involved in numerous inflammatory diseases, the inhibition of JNK signaling could decrease the expression of pro-inflammatory cytokines [8,9]. In cancer, some isoforms of JNKs are pro-oncogenic [10], whereas others act as tumor suppressors [11,12]. These findings imply the necessity of isoform-specific JNK inhibitors in the development of the cancer therapeutics.

Particularly, JNK3 is almost exclusively expressed in the brain, with very low levels in the kidneys and the testis, while JNK1 and JNK2 are widely expressed in a variety of tissues [3]. The level of the phosphorylated JNKs increased in postmortem brain tissue samples of Alzheimer's disease (AD) patients [13] and was connected to the rate of cognitive decline [14]. Several studies demonstrated that JNKs are activated in Parkinson's disease (PD) mouse models [15–17] and dopaminergic neurons were protected from apoptosis due to the inhibition of JNK by its specific inhibitor, SP600125 [18]. This evidence supports JNK3 as a promising target for the treatment of neurodegenerative diseases.

There have been great efforts to design and develop pan- and isoform-specific JNK inhibitors. Particularly, the structure determination of the JNK-inhibitor complex aided the pursuit of the development of novel and potent JNK inhibitors. Recently, we reviewed most of the available structures of JNK-inhibitor complexes [19]. Several hinge-binding scaffolds, such as aminopyrimidine, (iso)quinoline, quinazoline, and other scaffolds similar to purine, were frequently utilized for the development of ATP-competitive kinase inhibitors and 63 ATP competitive kinase inhibitors were FDA-approved till in the year 2021 [20,21]. Even though great efforts were paid to develop JNK inhibitors, there are no clinically available drug molecules as JNK therapeutics until now.

Fragment-based drug discovery (FBDD) has been an important paradigm for new drug discovery and is actively used in many pharmaceutical companies [22,23]. FBDD starts with the fragment screening in which weak binders to a target protein are selected and followed by creating lead compounds based on the fragment hits. Therefore, the screening methods generally rely on the biophysical methods to observe the weak interaction between the fragments and proteins. Those include the fluorescence-based method, NMR, surface plasmon resonance (SPR), isothermal titration calorimetry (ITC), mass spectroscopy, and X-ray crystallography. FBDD can be applied to the discovery of new drugs against most of the therapeutic targets, regardless of the types and characteristics of the proteins. Therefore, by using this method, several drugs have already entered the market, and many drug candidates are in clinical trials [24–26]. In discovering new kinase inhibitors, the fragment-based approaches have provided powerful insights into the selectivity enhancement and the lead optimization [27,28].

Compared with other methods, the NMR method has many advantages. The NMR spectroscopy is versatile and includes various methods of measurements to identify the weak binders to proteins. For example, STD NMR [29–31], relaxation filter experiments [32,33], and waterLOGSY [34] are commonly used for a ligand-based approach. A protein-based method is also utilized to identify protein binders, where the chemical shifts of the backbond -HNs of a target protein in two-dimensional (2D) ^{1}H-^{15}N heteronuclear single quantum correlation (HSQC) spectrum are monitored in the absence and presence of fragments. If the backbone chemical shift assignments are completed, the ligand binding "hot spot" can be identified without the knowledge of the detailed three-dimensional (3D) structure of the target. This provides an important information to create more potent compounds based on the fragments [22,23]. Even though the NMR methods are some of the most popular tools in FBDD campaign, it is not a stand-alone method and should be used complementarily with other methods to reduce the false positives, such as with X-ray crystallography for the identification of the exact binding mode and the assessment of structural novelty of the binding fragment. In this work, we present a collection of a fragment library by virtual screening, STD NMR spectroscopy, in vitro kinase assay, and the 3D structures of JNK3 in complex with fragments and an inhibitor, cyclopropyl[(3R)-3-({4-[6-hydroxy-2-(naphthalen-2-yl)-1H-benzimidazol-1-yl]pyrimidin-2-yl}amino)piperidine-1-yl]methanone, which previously showed a strong binding affinity to JNK3 and presented a potency of the cell protective effect in neuronal cell apoptosis [35]. In the structural investigation, we found a common JNK3-binding feature of the fragment and inhibitor, which occurred in the solvent-exposed region near the ATP binding site. We suppose this interaction potentiates the affinity of a chemical binder to JNK3 and could be utilized to design novel JNK3 inhibitors.

2. Materials and Methods

2.1. Fragment Library

The collection of compounds for the fragment library was conducted by virtual screening with the known structure of JNK3 (PDB ID, 1JNK) [36]. The bound molecules such as AMP-PNP, magnesium ions, and water were removed for the docking simulation. Docking simulations were performed on the ATP binding site of JNK3 by using the software Autodock vina [37] with the standard protocol provided by developer. The chemicals for the docking simulation were collected virtually from the various vendors' websites (Asinex, Winston-Salem, NC, USA; Maybridge, Waltham, NA, USA; ChemiDive, San Diego, CA, USA) with the molecular weight being less than 300 Da. A total of 15,000 compounds were screened, and top-scored molecules were chosen. Of those commercially available,

494 compounds were purchased from Asinex. The information of the fragments is provided in the Supplementary Files S1 (list) and S2 (structure data file, sdf).

2.2. Preparation of JNK3

The gene encoding the catalytic domain of human JNK3 which spans the amino acid residues from Ser40 to was amplified by PCR and inserted into the pET-15b expression vector between NcoI and BamHI restriction sites. The plasmid was transformed into the *Escherichia coli* strain BL21 (DE3) (Merck KGaA, Darmstadt, Germany). Cells were grown at 37 °C utill OD_{600} reached to ~0.6–0.7 and induced with a final concentration of 0.5 mM isopropyl-β-D-thiogalactopyranoside (IPTG) for 16–20 h at 18 °C. Cells were harvested by centrifugation 4000 rpm at 4 °C. The cell pellet was re-suspended in 20 mL lysis buffer (20 mM Hepes, pH = 7.0, 20 mM NaCl, 10% (*v*/*v*) glycerol, 2 mM dithiothreitol (DTT)). Cell membranes were disrupted by sonication on ice. The lysate was clarified by centrifugation at 18,000 rpm for 40 min. The supernatant was used for the following steps of purification.

Ion exchange chromatography using SP-Sepharose FF 16/10 (GE Healthcare, Chicago, IL, USA) and size-exclusion chromatography using HiLoad 16/600 Superdex 75 prep-grade (GE Healthcare) were applied sequentially to purify JNK3. The purity of JNK3 was more than 95% judged by SDS-PAGE. The purified JNK3 was collected and concentrated using an Amicon (Merck KGaA, Darmstadt, Germany) centrifugal filter with MWCO 10,000 to 10 mg/mL. The protein was frozen under liquid nitrogen and finally stored at −80 °C.

2.3. NMR Experiments

All NMR data were measured at 25 °C on Agilent DD2 600 MHz NMR spectrometer with TR^{TM} probe (Agilent, Santa Clara, CA, USA). The 494 compounds were respectively dissolved in $_{d6}$-DMSO for the preparation of stock solutions. The stock was diluted into $_{d11}$-Tris-HCl buffer (in D_2O), pH 7.5 with the final concentration of 320 μM in 500 μL. The simple ^{1}H NMR spectrum of each compound was recorded with the sweep widths, 8370.5 Hz, and the time-domain points, 32K, for all experiments. Eight compounds were mixed, and the mixture was put to a single NMR tube, followed by the ^{1}H NMR measurement. The final concentration of a single compound was 160 μM.

The purified JNK3 was added to an NMR tube containing 8 compounds to the final concentration of 4 μM. In this way, a total of 62 NMR tubes containing 8 compounds (in the last tube, 6 compounds) and JNK were prepared. ^{1}H and STD NMR spectra were recorded, respectively. For STD NMR, the reference spectrum was acquired with the off-resonance saturation at 30 ppm, and the saturation spectrum was acquired with the on-resonance saturation at 0.5 ppm. The duration of the saturations was 3 s for both on- and off-resonance experiments. All data were processed and analyzed with the software Mnova (Mestrelab Research, Santiago de Compostela, Spain).

2.4. Crystallization and Data Collection

The recombinant human JNK3 (10 mg/mL) solution containing 0.02% n-octyl-glucopyranoside was mixed with ATP to a final concentration of 2 mM and crystallized as described previously [38] with some modification. The JNK3-AMP crystal was obtained by vapor diffusion in hanging drops at 4 °C with a reservoir solution containing 16% PEG MME 550, 10% ethylene glycol, 0.1 M HEPES pH 7.25, and 10 mM TCEP. JNK3-fragment (or inhibitor) crystals were obtained by micro-seeding the JNK3-AMP crystal under the addition of the fragment (or inhibitor) to the crystallization solution. The inhibitor was cyclopropyl[(3R)-3-({4-[6-hydroxy-2-(naphthalen-2-yl)-1H-benzimidazol-1-yl]pyrimidin-2-yl}amino)piperidine-1-yl]methanone (inhibitor, here after), which bound very strongly to JNK3 (K_d, 46 nM) [35]. The diffraction data were collected at the 7A and 5C beamline of Pohang Light Source, Korea. The raw data were processed with HKL2000 [39]. The crystals belonged to space group of P212121, and the asymmetric unit contained one monomer (Table 1).

2.5. Structure Determination and Refinement

The structures were solved by the molecular replacement method using the program PHASER [40] with the template model PDB ID 3OY1, one of the JNK3 structure in complex with a selective inhibitor [41]. A manual model was built using the program COOT [42], and the resulting models were refined with the program REFMAC5 [43]. The quality of the refined models was evaluated by the program MolProbity [44]. Data collection and refinement statistics are summarized in Table 1. All figures representing the structure were generated by the *PyMOL* molecular–graphics program (The PyMOL Molecular Graphics System, Schrödinger, LLC., New York, NY, USA) [45] and by UCSF Chimera program [46]. The atomic coordinate and structure factors have been deposited in the Protein Data Bank (http://www.rcsb.org) with accession number of 4KKH and 7YL1 for fragment- and inhibitor-bound JNK3, respectively.

Table 1. Data collection and refinement statistics.

Crystals	3A8	Inhibitor
PDB ID	7YL1	4KKH
Data collection		
Space group	P212121	P212121
Cell dimensions		
a, b, c (Å)	50.15, 71.98, 107.16	52.32, 71.50, 107.08
α, β, γ (°)	90, 90, 90	90, 90, 90
Wavelength (Å)	0.97951	0.97951
Resolution (Å) *	59.75–2.48 (2.55–2.49)	50–2.0 (2.03–2.00)
Redundancy *	13.3 (12.5)	13.5 (12.8)
Completeness (%) *	99.1 (99.4)	100 (100)
I/σ_I *	33.6 (3.7)	58.9 (9.4)
R_{merge} (%) [a,*]	9.0 (35.3)	9.0 (44.2)
Refinement		
Resolution (Å)	36.61–2.49	37.45–2.0
No. of reflections	13,406	26,471
R_{work}/R_{free} (%) [b]	20.4/29.4	22.9/27.7
No. of atoms	2925	2942
Average B-factors	Protein/water/3A8 43.139/42.040/48.722	Protein/water/inhibitor 43.774/47.582/35.827
R.m.s. deviations		
Bond lengths (Å)	0.007	0.007
Bond angle (°)	1.157	1.109
Ramachandran plot		
Residues in favored/allowed/disallowed regions (%) [c]	93.9/5.5/0.6	97.4/2.6/0/0

* Values in the parentheses refer to the highest resolution shells. [a] $R_{merge} = \Sigma_i\Sigma_i \mid I(h)_i - \langle I(h) \rangle \mid / \Sigma_h\Sigma_i I(h)_i$, where $I(h)$ is the intensity of reflection h, Σ_h is the sum over all reflections, and Σ_i is the sum over i measurements of reflection h. [b] $R_{work} = \Sigma \mid \mid F_{obs} \mid - \mid F_{calc} \mid \mid / \Sigma \mid F_{obs} \mid$, 5% of the data was set aside for R_{free} calculation. [c] Statistics according to MolProbity Ramachandran analysis [44].

2.6. In Vitro Kinase Assay

The in vitro kinase assay was carried out with ADP-Glo^{TM} Kinase assay kit (Promega, Madison, WI). The assay was prepared with the final concentration of 2 ng of active JNK3, 50 μM ATP, 100 μM fragments, 1X reaction buffer, and 0.04 μg/μL p38 substrate. The experiments were set up with three control experiments including without ATP, without active JNK3, and without compound.

The mixture was incubated for 1 h at room temperature. The incubated solution was then supplied with 50 μL ADP-Glo reagent, centrifuged with 6000 rpm at 26 °C for 30 s, and incubated for 40 min at room temperature. The samples were transferred to a 96-white plate and added kinase-detection reagent with the ratio of 1:1. After incubating for 40 min at room temperature, the luminescence signals were recorded using the GloMax®-Multi Microplate Reader (Promega, Madison, WI, USA). The kinase assays were duplicated for each fragment.

3. Results

3.1. Virtual Screening

To build a fragment library to screen potential JNK3 inhibitors, about 15,000 compounds with molecular weights less than 300 Da from the vendors Asinex, Maybridge, and ChemiDive were collected virtually. We performed structure-based virtual screening using the 3D structure of JNK3 (PDB ID, 3DA6) retrieved from Protein Data Bank (PDB) (http://www.rcsb.org). AutoDock Vina [37] was utilized for the molecular docking. The docking site of fragments was set at the ATP-binding site of JNK3. From the simulation, 494 compounds presenting docking scores less than −7 kcal/mol were chosen. The structure data file of 494 compounds is available in the Supplementary File S2.

3.2. Fragment Screening

First, the individual ^{1}H NMR spectra of 494 compounds were measured and used as 'reference spectra' of the fragment library. Since we performed the NMR experiments in an aqueous buffer, the NMR spectra of almost 100 compounds were not able to be measured because of the poor solubility of those chemicals. To reduce the time of NMR measurements, we added eight compounds in a single NMR tube and recorded ^{1}H NMR spectrum. Since the number of compounds in the library was 494, the number of 8 chemical mixtures in total were 62, where 61 contained 8 compounds, and the last tube contained 6 compounds.

If there is no reaction or interaction between compounds in a single NMR tube, the NMR spectrum of the mixture is a simple sum of the spectra of each compound. In Figure 1, it is clear that the spectra of 8 compound mixtures is exactly the same of the overlay of individual spectrum of 8 compounds. When JNK3 was added to the 8 compound mixtures, the NMR spectrum was crowded in the region 2–4 ppm since the added protein solution contained some additives such as buffer ingredients, Hepes, and DTT; however most of the compound signals were clearly shown. In the STD NMR spectrum, we found several NMR signals which were the signals of possible JNK3-binding molecules. In this way, we found that compound #3 would have the possibility to bind to JNK3 (bottom of Figure 1).

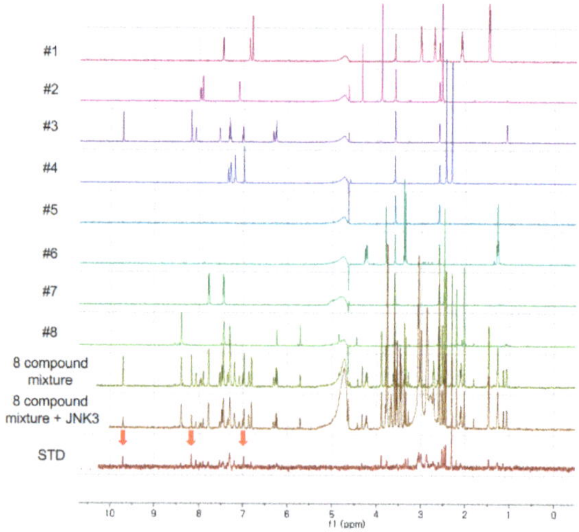

Figure 1. Reference NMR spectra of fragments and STD NMR spectrum. The spectra from #1 to #8 were the individual spectrum of fragment in a single NMR tube, respectively. STD NMR spectrum showed that compound #3 might bind to JNK3 (indicated by red arrow).

Through STD NMR experiments, we chose 27 compounds which might bind to JNK3; thus, the hit rate was about 6.9% when we excluded the insoluble compounds. Some of the selected compounds shared a common chemical structure which is similar to purine (Table 2). Among them, aliphatic chains at R_1 position and sulfur at R_2 are in common.

Table 2. Compounds sharing a common structure.

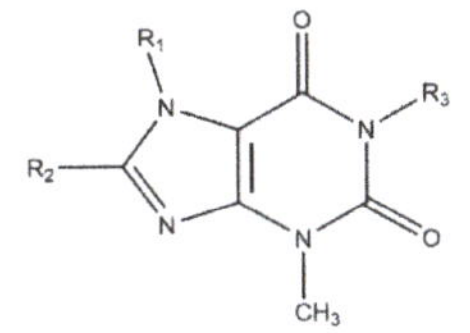

Code	R_1	R_2	R_3
1A4	$-CH_3$	$-SCH_2COOH$	-H
2H4	$-CH_2-CH=CH_2$	$-SCH_2COOH$	-H
1F9	$-CH_3$	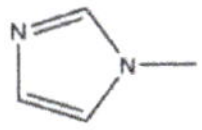	$-CH_3$
2D5	$-CH_2-CH(CH_3)_2$	$-NH-NH_2$	-H
2D8	$-CH_2-CH=CH-CH_3$	-SH	-H
3A1	$-CH_3$	-SH	-H
3A8	$-CH_2-CH_2-CH_2-CH_2-CH_2-CH_3$	-SH	-H
3B11	$-CH_2-CO-CH_3$	-H	$-CH_3$
3B7	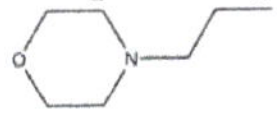	-H	-H
1C9	$-CH(CH_3)_2$	$-S-CH_2-CO-NH_2$	-H
2A4	$-CH_2-CH_2-CH_3$	$-S-CH_2-CO-NH_2$	-H

3.3. In Vitro Kinase Assay

The in vitro kinase assay was performed in the presence of the selected compounds to check whether those compounds were able to inhibit the catalytic activity of JNK3. For the fragments 2D5, 2D8, 3A8, 3B7, and 3B11, it was not possible to measure the inhibitory effects, possibly because of the solubility issue of those compounds in the reaction condition. The positive control without the treatment of any fragments showed significant luminescence since in ADP-Glo^{TM} assay ADP converted from ATP by active JNK3 was monitored by luminescence. Thus, if any fragmented molecules inhibited the enzymatic activity of JNK3, the production of ADP would be reduced, resulting in the decrease of luminescence signal. In Figure 2, the summary of the inhibitory effects of selected 22 fragments was provided. Some of those did not show any inhibitory effect against JNK3; however, several fragments including 1B7 showed about 40–50% inhibition of JNK3 activity at the fragment concentration of 100 μM.

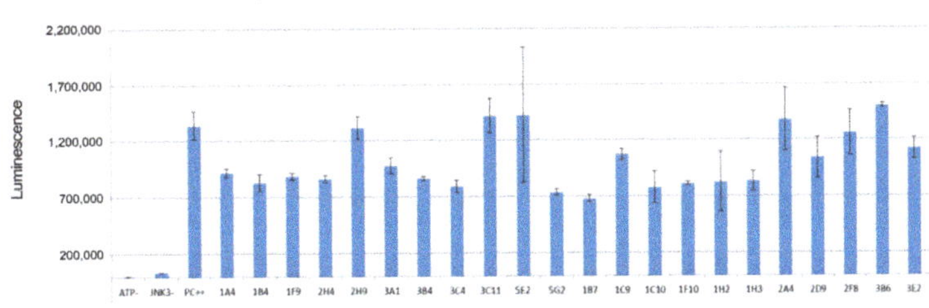

Figure 2. In vitro kinase assay and the inhibitory effects of selected fragments. PC++ stands for the positive control without any fragment molecules. The experiments were duplicated, and the errors were indicated.

3.4. Overall Structure of JNK3 in Complex with Fragment and Inhibitor

We tried the crystallization of JNK3 with all the selected fragments by using the micro-seeding or fragment-soaking methods, but a limited number of crystals was acquired. Of those, only the structure with a fragment 3A8 (7-hexyl-3-methyl-8-sulfanylidene-9H-purine-2,6-dione) was determined, since other complex crystals gave rise to poor diffraction. Another crystal structure of human JNK3 in a complex with a potent inhibitor, cyclopropyl[(3R)-3-({4-[6-hydroxy-2-(naphthalen-2-yl)-1H-benzimidazol-1-yl]pyrimidin-2-yl}amino) piperidin-1-yl]methanone was also determined by molecular replacement. The Ramachandran plot of 3A8 bound JNK3 shows that 93.9% of non-glycine and non-proline residues are in the most favored regions; 5.5% of residues are in the additional allowed regions; and two residues, Arg212 and Asp381, in the disallowed regions. For the inhibitor-bound JNK3, those values were 97.4, 2.6, and 0%, respectively. The data and refinement statistics are summarized in Table 2. The structures of JNK3 were almost identical to the previously reported structures of JNK3. The root mean square (r. m. s.) deviation between the 3A8 bound JNK3 and AMP-PNP-bound JNK3 (PDB ID, 1JNK) was 0.392 Å for 343 Cα atom pairs and the r. m. s. deviation between the inhibitor-bound forms and AMP-PNP-bound JNK3 was 0.495 Å for 340 Cα atom pairs. The r. m. s. deviation of the two structures determined in this work was 0.533 Å for 345 Cα atom pairs (Figure 3). Very subtle differences were found in the glycine-rich region (Gly71-Ser-Gly-Ala-Gln-Gly-Ile-Val78) and the phosphorylation lip (Ser217-Thr226). In the inhibitor-bound structure, the glycine-rich loop and the phosphorylation lip moved little bit toward ATP-binding site.

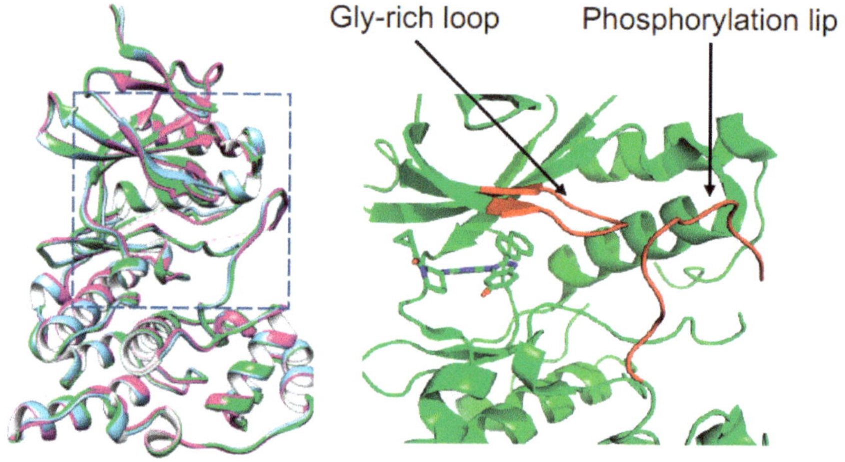

Figure 3. Overall structure of apo-(purple), fragment-bound (cyan), and inhibitor-bound (green) JNK3. The structural difference occurred at glycine-rich region and phosphorylation lip, which were enlarged on right and indicated. In the overlayed structures, the bound molecules were removed for the clarity of structural comparison.

3.5. Structure of JNK3 in Complex with Fragment 3A8

The ATP-binding site of JNK3 is located between N-terminal lobe (residues 45–149) and C-terminal lobe (residues 150–211 and 217–374) [36]. The fragment 3A8 bound to the ATP binding site of JNK3. Since 3A8 contained a similar structure to purine, the binding mode of 3A8 was thought to be similar to AMP-PNP. The 3A8 was positioned at the adenine binding site of AMP-PMP, but the orientations of the pyrine moiety of 3A8 and the adenine of AMP-PNP were different by almost 90° degrees (Figure 4A,B). 3A8 binding to JNK3 was completely opposite to the docking pose. In virtual screening, the lowest energy docking pose of 3A8 to JNK3 scored −7.0 kcal/mol and the hexyl chain penetrated deeply into the hydrophobic region of the ATP binding site (Figure 4C). The hexyl chain of 3A8 in JNK3 crystal structure protruded from the ATP binding site toward the surface of the protein. The hydrophobic interaction between the hexyl group of 3A8 with residues in β-strands above the ATP binding site might contribute to the binding of fragment to enzyme.

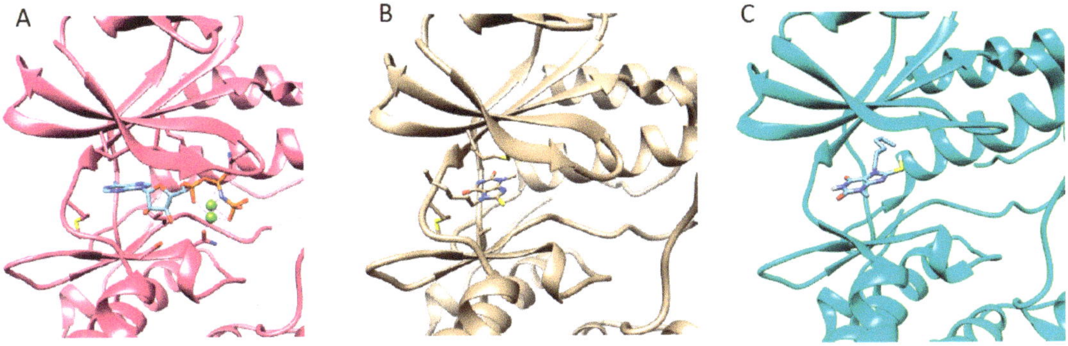

Figure 4. Structure of JNK3 in complex with AMP-PNP ((**A**), PDB ID, 1JNK) and fragment 3A8 (**B**), and docking simulated structure of JNK3 with 3A8 (**C**). Small molecules were presented by stick. In AMP-PNP bound structure, two magnesium ions were involved in the complex.

3.6. Structure JNK3 in Complex with a Potent Inhibitor

In the previous study, 1-Heteroaryl-2-aryl-1H-benzimidazole derivatives showed the inhibitory effect against JNK3 [35]. Cyclopropyl[(3R)-3-({4-[6-hydroxy-2-(naphthalen-2-yl)-1H-benzimidazol-1-yl]pyrimidin-2-yl}amino)piperidin-1-yl]methanone was one of the derivatives of those inhibitors with a high binding affinity (K_d, 46 nM) and inhibitory potency (IC_{50}, 27.7 nM) to JNK3 (Figure S1). In this work, we solved the crystal structure of JNK3 in complex with this inhibitor. The binding of inhibitor to JNK3 is illustrated in Figure 5A.

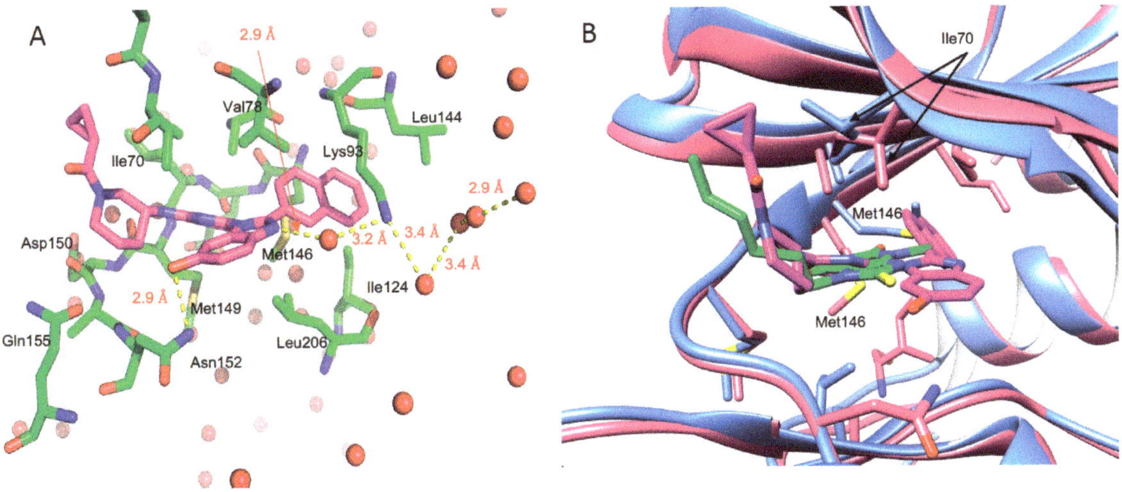

Figure 5. Structure of JNK3 in complexed with cyclopro-pyl[(3R)-3-({4-[6-hydroxy-2-(naphthalen-2-yl)-1H-benzimidazol-1-yl]pyrimidin-2-yl}amino)piperidin-1-yl]methanone (**A**). The residues involved in the interaction were indicated. Water molecules were indicated as red dots and hydrogen bonds as yellow broken lines. The lengths of hydrogen bonds are labeled in red. (**B**) Overlaid structure of JNK3 with 3A8 and inhibitor. Ile70 were indicated by arrow. Different positioning of Met146 in both structures are also indicated. JNK3 were colored blue for 3A8 bound form and pink for inhibitor bound form, respectively.

The inhibitor resided on the ATP binding site of JNK3, and the naphthalene moiety penetrated deeply into the interior of ATP binding pocket, where ATP or AMP-PNP did not interact. The naphthalene ring moiety interacted with Met 146. A naphthalene ring-binding induced the movement of the side chain of Met146 (gatekeeper in JNK3), which had a different orientation in AMP-PNP bound JNK3 structures. The naphthalene ring moiety possessed hydrophobic interaction with Lys93, Met146, Ile124, Val78, Ala91, and Leu144. Met149 formed a hydrogen bond with a 2-aminopyrimidine moiety of the inhibitor, and Ile70 interacted with piperidine and 1-cyclopropyl moieties of inhibitor. The side chain of Ile70 was rotated for stable interaction with 1-cyclopropyl. Leu206, Asn152, Val78, and Ile70 interacted with the 6-Hydroxy-1H-benzimidazole moiety of inhibitor. The side chain of Asn152 formed a direct hydrogen bond with the hydroxyl group of the 6-Hydroxy-1H-benzimidazole moiety that formed water-mediated hydrogen bond with Lys93. Several water molecules were coordinated in the vicinity of Lys93, which interacted with the naphthalene ring moiety.

When we compare the structures of JNK3 with a fragment and inhibitor, it was noticeable that the binding of 3A8 to JNK3 very much resembled the binding to the inhibitor, even though two chemicals did not share structural similarities, and the bicyclic rings of the two molecules were different. In the 3A8-bound structure, the 6-membered ring was outward from ATP binding site; however, in the inhibitor-bound one, 5-membered ring was outward and 6-membered ring inward. The common feature was the orientation and interaction of the hexyl group of 3A8 and the 1-cyclopropyl moiety of the inhibitor. Those two moieties protruded from the ATP binding site and presented hydrophobic interactions with the β-strand region of JNK3, where Ile 90 played a role. The importance of the hydrophobic interaction involving the 1-cyclopropyl moiety of the inhibitor on the binding and inhibitory activity to JNK3 are discussed below.

4. Discussion

We have conducted the in silico, in vitro, and structural investigations for the discovery of JNK3 inhibitors by adapting the fragment-based approach. 494-membered fragment library was constructed, and the library of NMR spectra was also built, even though about 20% of compounds did not give rise to the spectrum, because of the solubility problem. The hit rate from the screening from the STD NMR experiments was 6.9%, which is a relatively high value in a drug discovery campaign. In a drug repurposing project, we have shown that the longitudinal (T_1) and transverse (T_2) relaxation experiments and the STD NMR spectroscopy detected the JNK3 binding of azelastine, which has an original indication for anti-allergy and anti-inflammation effects and proven that azelastine binds to ATP binding pocket of JNK3 via competition STD experiment with AMP-PNP [47]. In this regard, we have shown that NMR spectroscopy is very efficient in drug discovery process, especially in initial fragment screening.

The kinase assay presented that most of the selected compounds had the inhibitory activity against JNK3, even though the effects were not potent. Since fragments are small and have simple structures, it is very hard to find potent molecules from a fragment-based approach. In this work, we just checked the inhibitory effects of the screened fragments to see whether those could inhibit the activity of JNK3 or not. The versatility of the fragment-based approach in drug discovery would be whether the screened fragment compounds could be used for designing and creating potent inhibitors. In this context, the structural information between fragments and the target proteins are very valuable. We have determined a structure of JNK3 in complex with one of the screened fragments as well as with a potent inhibitor. The fragment 3A8 had limited water solubility, thus, it was impossible to run the kinase assay. However, in the crystallization, certain additives such as 0.02% n-octyl-glucopyranoside might enhance the solubility of 3A8 in an aqueous solution. The 3A8 shared structural similarities to the adenine of ATP; however, the binding pattern was completely different from the adenine moiety of ATP. The hydrophobic hexyl group of 3A8 protruded from the ATP binding site and gave rise to additional interactions with the glycine-rich region of JNK3. We focused on the hydrophobic interaction to discover an isoform-specific JNK3 inhibitors. Every kinase shares a common ATP binding mode, which hampers the discovery of the specific kinase inhibitor, especially ATP-competitive inhibitors. For the discovery of potent and selective ATP-competitive JNK inhibitors, targeting hydrophobic interactions in combination with the interaction on ATP-binding sites was suggested [38]. Those interactions were ones with solvent-exposed hydrophobic regions, including a glycine-rich loop and another with deep hydrophobic region where the naphthalene moiety of inhibitor in this work penetrated. In the previous work, among 1-Heteroaryl-2-aryl-1H-benzimidazole derivatives, the inhibitor cyclopropyl[(3R)-3-({4-[6-hydroxy-2-(naphthalen-2-yl)-1H-benzimidazol-1-yl]pyrimidin-2-yl}amino)piperidin-1-yl]methanone showed the strongest binding affinity to JNK3. The binding affinity of this inhibitor was 60-fold stronger than the derivatives without 1-cyclopropyl moiety [35]. We performed the measurement of the inhibitory activity of the 1-Heteroaryl-2-aryl-1H-benzimidazole derivatives (Figure S1). The compound with the 1-cyclopropyl moiety (inhibitor in this work) showed the IC_{50} of 27 nM, whereas the compound without the moiety showed the IC_{50} of 1.41 mM, which coincides very well with previous binding experiments. When the naphthalene moiety was replaced by the dichlorophenyl moiety, the IC_{50} increased 178-fold. We confirmed experimentally that the importance of targeting hydrophobic interactions in discovering JNK inhibitors. We suppose that those interactions would be related to the selectivity of JNK inhibitors. In Figure S2, the inhibitory activity of cyclopropyl[(3R)-3-({4-[6-hydroxy-2-(naphthalen-2-yl)-1H-benzimidazol-1-yl]pyrimidin-2-yl}amino)piperidin-1-yl]methanone against several MAPKs is shown. The inhibitor in this work had potency against JNK3, JNK1, p38α, and p38β, and showed none or little inhibitory effects on p38γ, and p38δ., ERK, and GSK. We suppose that the hydrophobic interaction mentioned above would provide the selectivity. Thus, we suggest that keeping this hydrophobic interaction is very important for the discovery of novel and potent JNK3 inhibitors.

5. Conclusions

We presented the fragment-based approach in discovering JNK3 kinase with the structural investigation. We showed that virtual screening and in vitro assay supported by the NMR-based fragment screening were very efficient in early drug-discovery projects. In the structural investigation, we have identified the hydrophobic interaction between a fragment and the enzyme and presented the importance of this interaction in the potency and selectivity of the JNK3 inhibitors. We believe that this work provides valuable insight in designing lead compounds based on the fragment in the discovery of kinase inhibitors.

Supplementary Materials: The following supporting information can be downloaded at: https://www.mdpi.com/article/10.3390/pharmaceutics14091900/s1, Figure S1: Structures and inhibitory activities of 1-Heteroaryl-2-aryl-1H-benzimidazole derivatives; Figure S2: Inhibitory activity of cyclopropyl[(3R)-3-({4-[6-hydroxy-2-(naphthalen-2-yl)-1H-benzimidazol-1-yl]pyrimidin-2-yl}amino)piperidin-1-yl]methanone against various kinases; File S1: list_fragments.xlsx; File S2: SDF_fragments.sdf.

Author Contributions: Conceptualization, H.-C.A.; methodology, M.T.H.D.; software, M.T.H.D.; validation, H.-C.A.; formal analysis, M.T.H.D.; resources, H.-C.A.; writing—original draft preparation, M.T.H.D.; writing—review and editing, H.-C.A.; visualization, M.T.H.D.; supervision, H.-C.A.; project administration, H.-C.A.; funding acquisition, H.-C.A. All authors have read and agreed to the published version of the manuscript.

Funding: This work was supported by the National Research Foundation of Korea (NRF), grants funded by the Korean Government (MSIT) (NRF-2018R1D1A1B07050975 and 2021R1F1A1063558 to H.-C.A.), Korea.

Institutional Review Board Statement: Not applicable.

Informed Consent Statement: Not applicable.

Data Availability Statement: Not applicable.

Acknowledgments: We thank the staff at the Pohang Light Source (beamlines 5C and 7A) for their assistance with synchrotron data collection.

Conflicts of Interest: The authors declare no conflict of interest.

References

1. Johnson, G.L.; Lapadat, R. Mitogen-activated protein kinase pathways mediated by ERK, JNK, and p38 protein kinases. *Science* **2002**, *298*, 1911–1912. [CrossRef]
2. Ip, Y.T.; Davis, R.J. Signal transduction by the c-Jun N-terminal kinase (JNK)—From inflammation to development. *Curr. Opin. Cell Biol.* **1998**, *10*, 205–219. [CrossRef]
3. Mohit, A.A.; Martin, J.H.; Miller, C.A. p493F12 kinase: A novel MAP kinase expressed in a subset of neurons in the human nervous system. *Neuron* **1995**, *14*, 67–78. [CrossRef]
4. Chang, L.; Karin, M. Mammalian MAP kinase signalling cascades. *Nature* **2001**, *410*, 37–40. [CrossRef] [PubMed]
5. Zeke, A.; Misheva, M.; Remenyi, A.; Bogoyevitch, M.A. JNK Signaling: Regulation and Functions Based on Complex Protein-Protein Partnerships. *Microbiol. Mol. Biol. Rev.* **2016**, *80*, 793–835. [CrossRef] [PubMed]
6. Tournier, C.; Dong, C.; Turner, T.K.; Jones, S.N.; Flavell, R.A.; Davis, R.J. MKK7 is an essential component of the JNK signal transduction pathway activated by proinflammatory cytokines. *Genes Dev.* **2001**, *15*, 1419–1426. [CrossRef]
7. Graczyk, P.P. JNK inhibitors as anti-inflammatory and neuroprotective agents. *Future Med. Chem.* **2013**, *5*, 539–551. [CrossRef]
8. Assi, K.; Pillai, R.; Gomez-Munoz, A.; Owen, D.; Salh, B. The specific JNK inhibitor SP600125 targets tumour necrosis factor-alpha production and epithelial cell apoptosis in acute murine colitis. *Immunology* **2006**, *118*, 112–121. [CrossRef]
9. Hefetz-Sela, S.; Stein, I.; Klieger, Y.; Porat, R.; Sade-Feldman, M.; Zreik, F.; Nagler, A.; Pappo, O.; Quagliata, L.; Dazert, E.; et al. Acquisition of an immunosuppressive protumorigenic macrophage phenotype depending on c-Jun phosphorylation. *Proc. Natl. Acad. Sci. USA* **2014**, *111*, 17582–17587. [CrossRef]
10. Das, M.; Garlick, D.S.; Greiner, D.L.; Davis, R.J. The role of JNK in the development of hepatocellular carcinoma. *Genes Dev.* **2011**, *25*, 634–645. [CrossRef]
11. Shibata, W.; Maeda, S.; Hikiba, Y.; Yanai, A.; Sakamoto, K.; Nakagawa, H.; Ogura, K.; Karin, M.; Omata, M. c-Jun NH2-terminal kinase 1 is a critical regulator for the development of gastric cancer in mice. *Cancer Res.* **2008**, *68*, 5031–5039. [CrossRef] [PubMed]
12. Han, M.S.; Barrett, T.; Brehm, M.A.; Davis, R.J. Inflammation Mediated by JNK in Myeloid Cells Promotes the Development of Hepatitis and Hepatocellular Carcinoma. *Cell Rep.* **2016**, *15*, 19–26. [CrossRef] [PubMed]
13. Zhu, X.; Raina, A.K.; Rottkamp, C.A.; Aliev, G.; Perry, G.; Boux, H.; Smith, M.A. Activation and redistribution of c-jun N-terminal kinase/stress activated protein kinase in degenerating neurons in Alzheimer's disease. *J. Neurochem.* **2001**, *76*, 435–441. [CrossRef] [PubMed]
14. Gourmaud, S.; Paquet, C.; Dumurgier, J.; Pace, C.; Bouras, C.; Gray, F.; Laplanche, J.-L.; Meurs, E.F.; Mouton-Liger, F.; Hugon, J. Increased levels of cerebrospinal fluid JNK3 associated with amyloid pathology: Links to cognitive decline. *J. Psychiatry Neurosci.* **2015**, *40*, 151–161. [CrossRef] [PubMed]
15. Pan, J.; Li, H.; Zhang, B.; Xiong, R.; Zhang, Y.; Kang, W.-Y.; Chen, W.; Zhao, Z.-B.; Chen, S.D. Small peptide inhibitor of JNK3 protects dopaminergic neurons from MPTP induced injury via inhibiting the ASK1-JNK3 signaling pathway. *PLoS ONE* **2015**, *10*, e0119204. [CrossRef]
16. Badshah, H.; Ali, T.; Rehman, S.-U.; Amin, F.-U.; Ullah, F.; Kim, T.H.; Kim, M.O. Protective Effect of Lupeol Against Lipopolysaccharide-Induced Neuroinflammation via the p38/c-Jun N-Terminal Kinase Pathway in the Adult Mouse Brain. *J. Neuroimmune Pharmacol.* **2016**, *11*, 48–60. [CrossRef]
17. Zhang, S.; Gui, X.-H.; Huang, L.-P.; Deng, M.-Z.; Fang, R.-M.; Ke, X.-H.; He, Y.-P.; Li, L.; Fang, Y.-Q. Neuroprotective Effects of beta-Asarone Against 6-Hydroxy Dopamine-Induced Parkinsonism via JNK/Bcl-2/Beclin-1 Pathway. *Mol. Neurobiol.* **2016**, *53*, 83–94. [CrossRef]
18. Wang, W.; Ma, C.; Mao, Z.; Li, M. JNK inhibition as a potential strategy in treating Parkinson's disease. *Drug News Perspect.* **2004**, *17*, 646–654. [CrossRef]
19. Duong, M.T.H.; Lee, J.-H.; Ahn, H.-C. C-Jun N-terminal kinase inhibitors: Structural insight into kinase-inhibitor complexes. *Comput. Struct. Biotechnol. J.* **2020**, *18*, 1440–1457. [CrossRef]
20. Attwood, M.M.; Fabbro, D.; Sokolov, A.V.; Knapp, S.; Schioth, H.B. Trends in kinase drug discovery: Targets, indications and inhibitor design. *Nat. Rev. Drug Discov.* **2021**, *20*, 839–861. [CrossRef]
21. Roskoski, R., Jr. Properties of FDA-approved small molecule protein kinase inhibitors: A 2021 update. *Pharmacol. Res.* **2021**, *165*, 105463. [CrossRef] [PubMed]
22. Hajduk, P.J.; Greer, J. A decade of fragment-based drug design: Strategic advances and lessons learned. *Nat. Rev. Drug Discov.* **2007**, *6*, 211–219. [CrossRef] [PubMed]
23. Erlanson, D.A.; Fesik, S.W.; Hubbard, R.E.; Jahnke, W.; Jhoti, H. Twenty years on: The impact of fragments on drug discovery. *Nat. Rev. Drug Discov.* **2016**, *15*, 605–619. [CrossRef] [PubMed]
24. Wang, P.-F.; Qiu, H.-Y.; Wang, Z.-F.; Zhang, Y.-J.; Wang, Z.-C.; Li, D.-D.; Zhu, H.-L. Identification of novel B-Raf(V600E) inhibitors employing FBDD strategy. *Biochem. Pharmacol.* **2017**, *132*, 63–76. [CrossRef] [PubMed]

25. Lamoree, B.; Hubbard, R.E. Current perspectives in fragment-based lead discovery (FBLD). *Essays Biochem.* **2017**, *61*, 453–464. [CrossRef]
26. Irie, T.; Sawa, M. 7-Azaindole: A Versatile Scaffold for Developing Kinase Inhibitors. *Chem. Pharm. Bull.* **2018**, *66*, 29–36. [CrossRef] [PubMed]
27. Mortenson, P.N.; Berdini, V.; O'Reilly, M. Fragment-based approaches to the discovery of kinase inhibitors. *Methods Enzym.* **2014**, *548*, 69–92. [CrossRef]
28. Zhang, J.; Yang, P.L.; Gray, N.S. Targeting cancer with small molecule kinase inhibitors. *Nat. Rev. Cancer* **2009**, *9*, 28–39. [CrossRef]
29. Meyer, B.; Peters, T. NMR spectroscopy techniques for screening and identifying ligand binding to protein receptors. *Angew. Chem. Int. Ed. Engl.* **2003**, *42*, 864–890. [CrossRef]
30. Mayer, M.; Meyer, B. Group epitope mapping by saturation transfer difference NMR to identify segments of a ligand in direct contact with a protein receptor. *J. Am. Chem. Soc.* **2001**, *123*, 6108–6117. [CrossRef]
31. Streiff, J.H.; Juranic, N.O.; Macura, S.I.; Warner, D.O.; Jones, K.A.; Perkins, W.J. Saturation transfer difference nuclear magnetic resonance spectroscopy as a method for screening proteins for anesthetic binding. *Mol. Pharmacol.* **2004**, *66*, 929–935. [CrossRef] [PubMed]
32. Peng, C.; Frommlet, A.; Perez, M.; Cobas, C.; Blechschmidt, A.; Dominguez, S.; Lingel, A. Fast and Efficient Fragment-Based Lead Generation by Fully Automated Processing and Analysis of Ligand-Observed NMR Binding Data. *J. Med. Chem.* **2016**, *59*, 3303–3310. [CrossRef]
33. Johnson, J.A.; Olson, N.M.; Tooker, M.J.; Bur, S.K.; Pomerantz, W.C.K. Combined Protein- and Ligand-Observed NMR Workflow to Screen Fragment Cocktails against Multiple Proteins: A Case Study Using Bromodomains. *Molecules* **2020**, *25*, 3949. [CrossRef] [PubMed]
34. Dalvit, C.; Fogliatto, G.; Stewart, A.; Veronesi, M.; Stockman, B. WaterLOGSY as a method for primary NMR screening: Practical aspects and range of applicability. *J. Biomol. NMR* **2001**, *21*, 349–359. [CrossRef]
35. Kim, M.-H.; Lee, J.; Jung, K.; Kim, M.; Park, Y.-J.; Ahn, H.; Kwon, Y.H.; Hah, J.-M. Syntheses and biological evaluation of 1-heteroaryl-2-aryl-1H-benzimidazole derivatives as c-Jun N-terminal kinase inhibitors with neuroprotective effects. *Bioorganic Med. Chem.* **2013**, *21*, 2271–2285. [CrossRef] [PubMed]
36. Xie, X.; Gu, Y.; Fox, T.; Coll, J.T.; Fleming, M.A.; Markland, W.; Caron, P.R.; Wilson, K.P.; Su, M.S. Crystal structure of JNK3: A kinase implicated in neuronal apoptosis. *Structure* **1998**, *6*, 983–991. [CrossRef]
37. Trott, O.; Olson, A.J. AutoDock Vina: Improving the speed and accuracy of docking with a new scoring function, efficient optimization, and multithreading. *J. Comput. Chem.* **2010**, *31*, 455–461. [CrossRef]
38. Scapin, G.; Patel, S.B.; Lisnock, J.; Becker, J.W.; LoGrasso, P.V. The structure of JNK3 in complex with small molecule inhibitors: Structural basis for potency and selectivity. *Chem. Biol.* **2003**, *10*, 705–712. [CrossRef]
39. Otwinowski, Z.; Minor, W. Processing of X-ray diffraction data collected in oscillation mode. *Methods Enzymol.* **1997**, *276*, 307–326.
40. McCoy, A.J.; Grosse-Kunstleve, R.W.; Adams, P.D.; Winn, M.D.; Storoni, L.C.; Read, R.J. Phaser crystallographic software. *J. Appl. Crystallogr.* **2007**, *40*, 658–674. [CrossRef]
41. Probst, G.D.; Bowers, S.; Sealy, J.M.; Truong, A.P.; Hom, R.K.; Galemmo, R.A., Jr.; Konradi, A.W.; Sham, H.L.; Quincy, D.A.; Pan, H.; et al. Highly selective c-Jun N-terminal kinase (JNK) 2 and 3 inhibitors with in vitro CNS-like pharmacokinetic properties prevent neurodegeneration. *Bioorg. Med. Chem. Lett.* **2011**, *21*, 315–319. [CrossRef] [PubMed]
42. Emsley, P.; Cowtan, K. Coot: Model-building tools for molecular graphics. *Acta Cryst. D* **2004**, *60*, 2126–2132. [CrossRef] [PubMed]
43. Murshudov, G.N.; Skubak, P.; Lebedev, A.A.; Pannu, N.S.; Steiner, R.A.; Nicholls, R.A.; Winn, M.D.; Long, F.; Vagin, A.A. REFMAC5 for the refinement of macromolecular crystal structures. *Acta Cryst. D* **2011**, *67*, 355–367. [CrossRef] [PubMed]
44. Davis, I.W.; Leaver-Fay, A.; Chen, V.B.; Block, J.N.; Kapral, G.J.; Wang, X.; Murray, L.W.; Arendall, W.B., 3rd; Snoeyink, J.; Richardson, J.S.; et al. MolProbity: All-atom contacts and structure validation for proteins and nucleic acids. *Nucleic Acids Res.* **2007**, *35*, W375–W383. [CrossRef] [PubMed]
45. Schrödinger, L.; DeLano, W. PyMOL. Available online: http://www.pymol.org/pymol (accessed on 28 July 2022).
46. Pettersen, E.F.; Goddard, T.D.; Huang, C.C.; Couch, G.S.; Greenblatt, D.M.; Meng, E.C.; Ferrin, T.E. UCSF Chimera—A visualization system for exploratory research and analysis. *J. Comput. Chem.* **2004**, *25*, 1605–1612. [CrossRef]
47. Nguyen, P.L.; Bui, B.P.; Duong, M.T.H.; Lee, K.; Ahn, H.-C.; Cho, J. Suppression of LPS-Induced Inflammation and Cell Migration by Azelastine through Inhibition of JNK/NF-kappaB Pathway in BV2 Microglial Cells. *Int. J. Mol. Sci.* **2021**, *22*, 9061. [CrossRef]

pharmaceutics

MDPI

Article

Kinase Inhibitor Screening Displayed *ALK* as a Possible Therapeutic Biomarker for Gastric Cancer

Felipe Pantoja Mesquita [1], Pedro Filho Noronha Souza [1,2], Emerson Lucena da Silva [1], Luina Benevides Lima [1], Lais Lacerda Brasil de Oliveira [1], Caroline Aquino Moreira-Nunes [1,3,*], William J. Zuercher [4], Rommel Mario Rodríguez Burbano [3,5,6], Maria Elisabete Amaral de Moraes [1] and Raquel Carvalho Montenegro [1,*]

1 Department of Medicine, Pharmacogenetics Laboratory, Drug Research and Development Center (NPDM), Federal University of Ceará, Fortaleza 60430-275, CE, Brazil
2 Department of Biochemistry and Molecular Biology, Federal University of Ceará, Fortaleza 60430-275, CE, Brazil
3 Department of Biological Sciences, Oncology Research Center, Federal University of Pará, Belém 66073-005, PA, Brazil
4 Division of Chemical Biology and Medicinal Chemistry, University of North Carolina at Chapel Hill, Chapel Hill, NC 27599, USA
5 Laboratory of Human Cytogenetics, Institute of Biological Sciences, Federal University of Pará, Belém 66073-005, PA, Brazil
6 Molecular Biology Laboratory, Ophir Loyola Hospital, Belém 66073-005, PA, Brazil
* Correspondence: carolfam@gmail.com (C.A.M.-N.); rcm.montenegro@gmail.com (R.C.M.)

Citation: Mesquita, F.P.; Souza, P.F.N.; da Silva, E.L.; Lima, L.B.; de Oliveira, L.L.B.; Moreira-Nunes, C.A.; Zuercher, W.J.; Burbano, R.M.R.; de Moraes, M.E.A.; Montenegro, R.C. Kinase Inhibitor Screening Displayed *ALK* as a Possible Therapeutic Biomarker for Gastric Cancer. *Pharmaceutics* **2022**, *14*, 1841. https://doi.org/10.3390/pharmaceutics14091841

Academic Editor: Murali Mohan Yallapu

Received: 20 July 2022
Accepted: 30 August 2022
Published: 1 September 2022

Abstract: Despite advances in cancer chemotherapy, gastric cancer (GC) continues to have high recurrence rates and poor prognosis with limited treatment options. Understanding the etiology of GC and developing more effective, less harmful therapeutic approaches are vital and urgent. Therefore, this work describes a novel kinase target in malignant gastric cells as a potential therapeutic strategy. Our results demonstrate that among 147 kinase inhibitors (KI), only three molecules were significantly cytotoxic for the AGP-01 cell line. Hence, these three molecules were further characterized in their cellular mode of action. There was significant cell cycle impairment due to the expression modulation of genes such as *TP53*, *CDKN1A*, *CDC25A*, *MYC*, and *CDK2* with subsequent induction of apoptosis. In fact, the Gene Ontology analysis revealed a significant enrichment of pathways related to cell cycle regulation (GO:1902749 and GO:1903047). Moreover, the three selected KIs significantly reduced cell migration and Vimentin mRNA expression after treatment. Surprisingly, the three KIs share the same target, *ALK* and *INSR*, but only the *ALK* gene was found to have a high expression level in the gastric cancer cell line. Additionally, lower survival rates were observed for patients with high *ALK* expression in TCGA-STAD analysis. In summary, we hypothesize that *ALK* gene overexpression can be a promising biomarker for prognosis and therapeutic management of gastric adenocarcinoma.

Keywords: gastric cancer; kinase inhibitor; *ALK*; targeted therapy

1. Introduction

Gastric cancer (GC) is the second major cause of cancer-related deaths worldwide and the third leading cause in industrialized nations [1,2]. Despite improvements in the management of GC patients with distant metastases, significant recurrence rates and a poor prognosis persist, with limited treatment choices and a median survival around one year [3,4]. Another problem is that GC is a very complex illness with a complicated etiology involving complex host genetic and environmental variables [5,6].

Until now, only a few targeted therapeutic agents, such as trastuzumab and ramucirumab, have been approved by the US Food and Drug Administration for gastric cancer patients identified with the respective genetic defects [7]. However, only 7–34% of GC patients have the altered molecular target for these agents and therefore most GC patients must continue to depend on cytotoxic chemotherapy and/or surgical resection as the current standard of care [8]. As a result, a greater number of patients evolve with poor prognostic and low survival rate because of therapeutic failure [9,10]. Therefore, understanding the etiology of GC and the development of more effective, less hazardous therapy techniques are critical and urgent.

In this sense, protein kinases are central actors in most of signal transduction pathways. Because their function is frequently disrupted in signal transduction networks leading to illnesses like cancer and inflammation, several of the 518 protein kinases encoded by the human genome have emerged as potential therapeutic targets [11]. Nowadays, there are around 250 kinase inhibitors (KIs) in clinical studies, with 37 of them being authorized for human usage [12]. Thus, it is critical to search new compounds' targets in depth in order to comprehend their molecular and cellular modes of action (MoA).

Thus, several kinase activity and binding screens have been published in the past few years. Among them, Elkins and co-workers reported a comprehensive characterization of 367 tool compounds screened against 224 recombinant kinases [13]. Despite the considerable value of these studies, the cellular or clinical evaluation of KIs has not yet been systematically analyzed. Therefore, the published kinase inhibitors library (PKIS) can be a useful tool for target screening in different cancer models. Taking all together, the present study aimed to evaluate possible novel kinase targets in malignant gastric cells in order to identify a new therapeutic strategy based on kinase inhibition using PKIs library.

2. Materials and Method

2.1. Cell Culture

Leal and co-workers (2009) developed a cell line from a patient with metastasis of gastric adenocarcinoma (AGP-01) as an alternative model for the anticancer drug screening model. A normal gastric mucosa cell line MNP-01 was used as normal cellular control [14,15]. Cells were cultured in sterilized flasks containing filtered Dulbecco's modified Eagle's medium (DMEM; Gibco®, New York, NY, USA) supplemented with 10% fetal bovine serum (Gibco®), 100 U/mL penicillin, and 100 µg/mL streptomycin. The culture flasks were maintained in a 5% CO_2 air-humidified atmosphere at 37 °C.

2.2. Alamar Blue Method

The growth inhibitory activity of 147 kinase inhibitors (KI) against the AGP-01 cell line was first evaluated at 10 µM concentration (72 h of treatment) using the Alamar Blue method. The growth inhibition (%) was calculated as comparing both KIs based on the cellular viability of treated and KI-untreated AGP-01 cells. Then, relevant KIs that obtained >80% of growth inhibition were selected for IC_{50} determination by a concentration-response curve. KI screening was carried out on 384-well plate (Corning®, Glendale, AZ, USA). Firstly, 300 cells per well were seeded and incubated overnight at 37 °C and 5% CO_2. Then, KIs were diluted with DMEM medium and subsequently added in the plate to achieve final desired concentration (10 µM). After 72 h of incubation, Alamar Blue reagent was added to measure cell viability. Fluorescence intensity (560/590 nm) was measured using Beckman Coulter Microplate Reader DTX 880. Cells treated with 0.1% dimethyl sulfoxide (DMSO) were used as negative control and doxorubicin (Doxo, Milan, Italy) was used as positive control.

2.3. Cell Cycle Progression

AGP-01 cells were seeded at 2×10^3 cells per well in a 96-well plate and treated with a single concentration (1 µM) of each KI selected in the screening. After 72 h of exposure, cells were harvested, washed with PBS 1X, and fixed with ice-cold 80% ethanol for 1 h. Then, cells were incubated with RNAse A (200 µg/mL) and propidium iodide (50 µg/mL) for 30 min. DNA content was analyzed using flow cytometry (BD FACSVerse™, BD Biosciences, East Rutherfor, NJ, USA). In total, 10,000 events were acquired, and data were analyzed using FlowJo® software (Version 10.5.3, FlowJo LLC, Ashland, Wilmington, DE, USA).

2.4. Caspase 3/7 Activity

The caspase-3 and caspase-7 activities were evaluated using the CellEvent® Caspase-3/7 kit according to the manufacturer's protocol (Invitrogen™, Waltham, MA, USA). AGP-01 cells were plated (3×10^3 cells/well) in a 96-well plate and treated with KIs (1 μM) for 72 h. Then, cells were incubated in 5 μM of CellEvent® reagent for 30 min and evaluated in Cytation™ with absorption and fluorescence emission spectra of 511/533 nm. Apoptotic cells were labelled with bright fluorescent green nuclei.

2.5. Wound Healing Assay

The migratory potential of AGP-01 cells were evaluated by the wound healing assay as previously described by Mesquita et al. (2021) [16]. Cells were seeded (2×10^4 cells/well) in a 96-well plate and scratches were made in each well with a sterile pipette tip. Then, cells were treated with KIs (1 μM) for 24 h. Pictures were taken at time 0 and 24 h from wells, and the scratch area was measured using ImageJ software (National Institutes of Health, Rockville, MD, USA).

2.6. Gene Expression by Quantitative Real-Time PCR

Quantitative real-time PCR (qPCR) analyses were performed with QuantStudio5 Real-Time PCR system (Applied Biosystems®, Waltham, MA, USA) in a 96-well PCR plate using 3 μL of cDNA obtained after reverse-transcription of total RNA (control and treated sample) using High-Capacity cDNA Reverse Transcription Kit (Applied Biosystems™), 1 μL of each primer/probe in forward and reverse, 12.5 μL of PowerUp SYBR® Master Mix (Life Technologies, Carlsbad, CA, USA), and 8.5 μL of ultra-pure sterile water. The thermocycling was applied according to the manufacturer's protocol of PowerUp SYBR® Master Mix. Each assay was performed at least three times according to Minimum Information for Publication of Quantitative Real-Time PCR Experiments Guidelines [17]. The gene expression levels were based on relative analyses and calculated using the $2^{-\Delta\Delta CT}$ (delta-delta threshold cycle) method [18].

2.7. PPI Network and GENE Ontology Enrichment

The Protein–Protein interaction (PPI) Network and Gene Ontology (GO) enrichment analyses of the differentially expressed genes were constructed using the string (https://string-db.org/ (accessed on 9 March 2022) freely available and yfile plug-in in Cytoscape software to analyze the interactions between them. False discovery rate was setup in <0.005.

2.8. Survival Analysis

The GEPIA database (http://gepia.cancer-pku.cn/ (accessed on 22 March 2022) was used to perform in silico correlation between gene expression and survival analyses [19]. The Cancer Genome Atlas-Stomach Adenocarcinoma (TCGA-STAD) is one of the cancer transcriptome data sets available through the GEPIA server. The database was used to compare the expression profiles of the *ALK* gene in normal and tumor tissue from gastric adenocarcinoma patients, categorizing them as high or low expression, which was then correlated with patient survival probability.

2.9. Statistical Analysis

All tests were performed in three independent experiments in triplicate and shown as a mean ± standard deviation (SD). The treated groups (KIs) were compared to the untreated group (DMSO) by Analysis of Variance (ANOVA) followed by Bonferroni's posttest or by the *t*-test, considering significant differences with an interval of confidence of 95% ($p < 0.05$). GraphPad Prism 5.01 (GraphPad Software, San Diego, CA, USA) was used for data analysis and graph design.

3. Results

3.1. Screening of KIs Compounds against Metastatic Gastric Cancer Cell Line

AGP-01 cells were treated with a single concentration (10 μM) of 147 PKIS compounds to search for the most potent inhibitor based on cell growth inhibition (Figure 1). Among all, only six compounds suppressed the cell growth equally or even higher than the positive control doxorubicin (87% of growth inhibition): GW580509X, GSK1713088A, GSK1751853A, GSK2186269A, GSK1173862A, GSK2220400A (Figure 1B). Then, the non-linear regression was performed to assess the IC_{50} values of these inhibitors (Figure 1C), which confirmed that only three compounds had highly cytotoxic potential, in the micromolar range, GSK2186269A, GSK1173862A, and GSK2220400A (IC_{50} < 2.2 μM). Therefore, these three compounds were selected to proceed with further analyses.

A

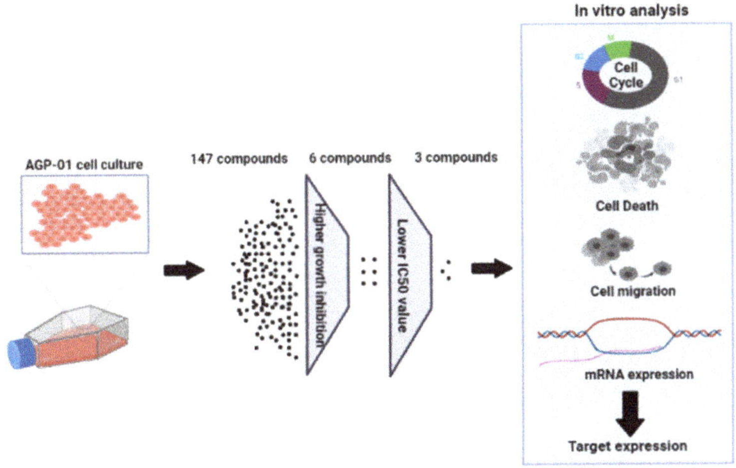

B

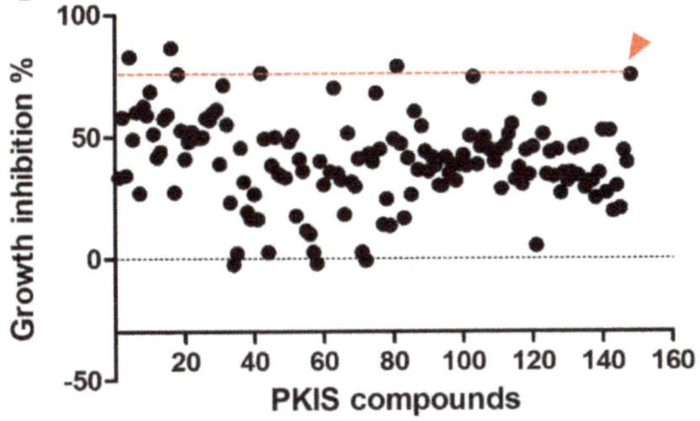

C

ID Compound	IC50 μM (CI95%)
GW580509X	4.06 (2.02 – 8.16)
GSK1713088A	7.71 (4.54 – 13.11)
GSK1751853A	5.40 (2.16 – 13.52)
GSK2186269A	0.93 (0.53 – 1.63)
GSK1173862A	0.99 (0.45 – 2.15)
GSK2220400A	2.21 (1.22 – 3.99)
Doxorubicin	0.78 (0.45 – 1.33)

Figure 1. Cytotoxicity of 147 published kinase inhibitor set (PKIS) against metastatic gastric cancer cells (AGP-01). (**A**) Experimental design. After exposure to 147 published kinase inhibitors set (PKIS), only three kinase inhibitors were selected for further analyses based on their highly cytotoxic effect (lowest IC_{50} values). (**B**) Growth inhibition normalized by the untreated control (DMSO) of 147 kinase inhibitors after 72 h of treatment. The red head arrow represents the positive control doxorubicin. (**C**) IC_{50} values obtained from non-linear regression of curve concentration-response for the kinase with IC_{50} values higher than doxorubicin.

3.2. Kinase Inhibitors Act as Anti-Cancer Agents by Regulating Cell Cycle and Mesenchymal Genes

The anti-proliferative effects of GSK2186269A, GSK1173862A and GSK2220400A were then evaluated by the cell cycle interaction on flow cytometry. GSK2220400A (1 μM) induced a significant G0/G1-phase arrest compared to the untreated control, while the

GSK2186269A (1 μM) and GSK1173862A (1 μM) compounds triggered extensive nuclei fragmentation, considered a sub-G1 phase alteration (Figure 2A,B). Furthermore, the three compounds significantly increased caspase 3 and 7 activity, suggesting the induction of apoptosis (Figure 2C). Another cancer hallmark significantly inhibited by the selected KIs was the cell migration. All three inhibitors (1 μM) were capable of preventing the wound closure by the AGP-01 cells (Figure 2D,E).

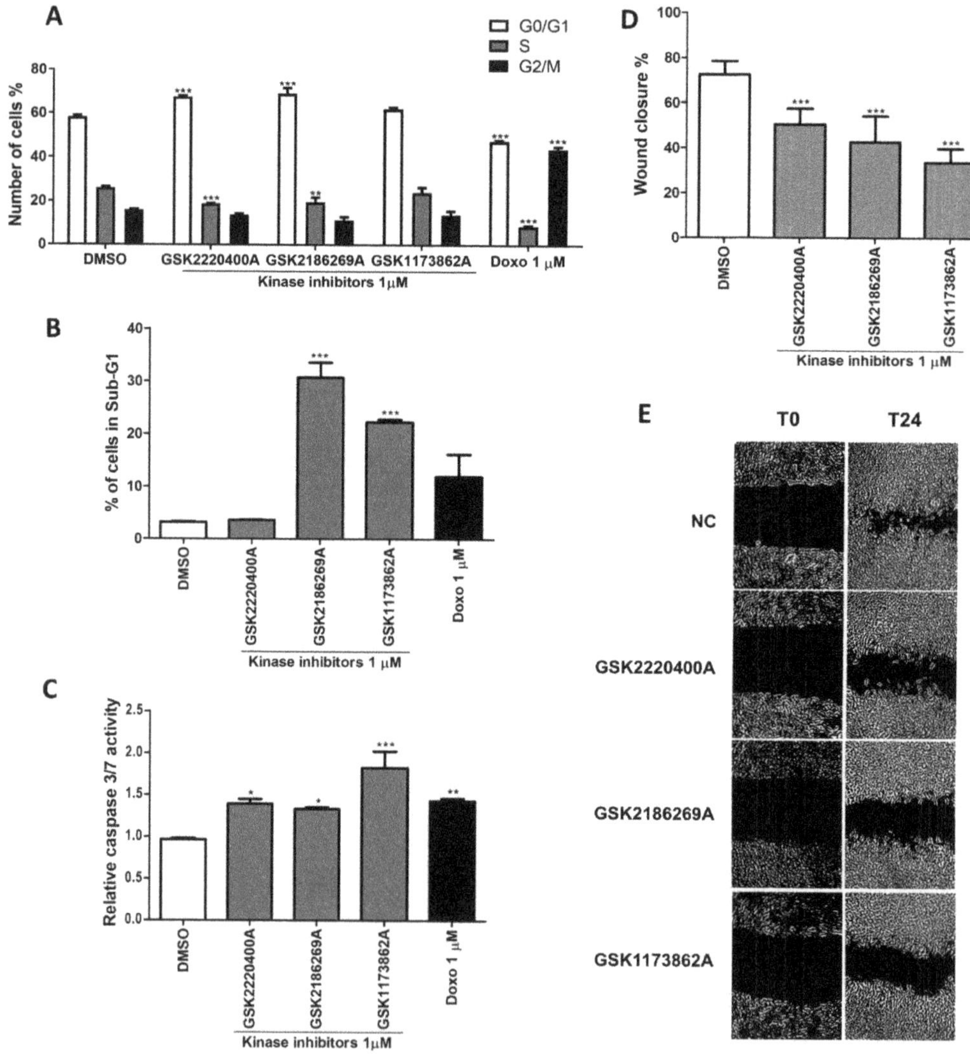

Figure 2. Anti-cancer effects of the three selected kinase inhibitors. (**A**) Cell cycle phase distribution of metastatic gastric cancer cells after 72 h of kinase inhibitors treatment. (**B**) Percentage of cells in sub-G1 phase after kinase inhibitors treatment. (**C**) Caspase 3 and 7 activities after 72 h of 1 μM kinase inhibitors exposure against metastatic gastric cancer cells. (**D**,**E**) Inhibitory effect on cell migration after 24 h of kinase inhibitors treatment. Mean ± standard deviation. Significant differences compared to untreated control: * $p < 0.05$, ** $p < 0.01$, *** $p < 0.001$.

Furthermore, after KI exposure, an alteration in gene expression involved in the cell cycle regulation, cell death, and migration was observed. In AGP-01 cells, the treatment with KIs led to the suppression of *CDK2*, *c-MYC*, *CDC25A*, and *VIM* expression while increasing *CDK1NA* expression (Figure 3A). On the other hand, GSK2220400A did not influence the expression of *CDC25A*, but it was the only drug that increased *TP53* levels. The analysis of PPI networks produced by those were involved in processes such as *cell cycle*, *mitotic processes*, *UV and radiation response*, *DNA damage process*, and *DNA metabolic process* with an FDR < 0.05 (Figure 3B,C).

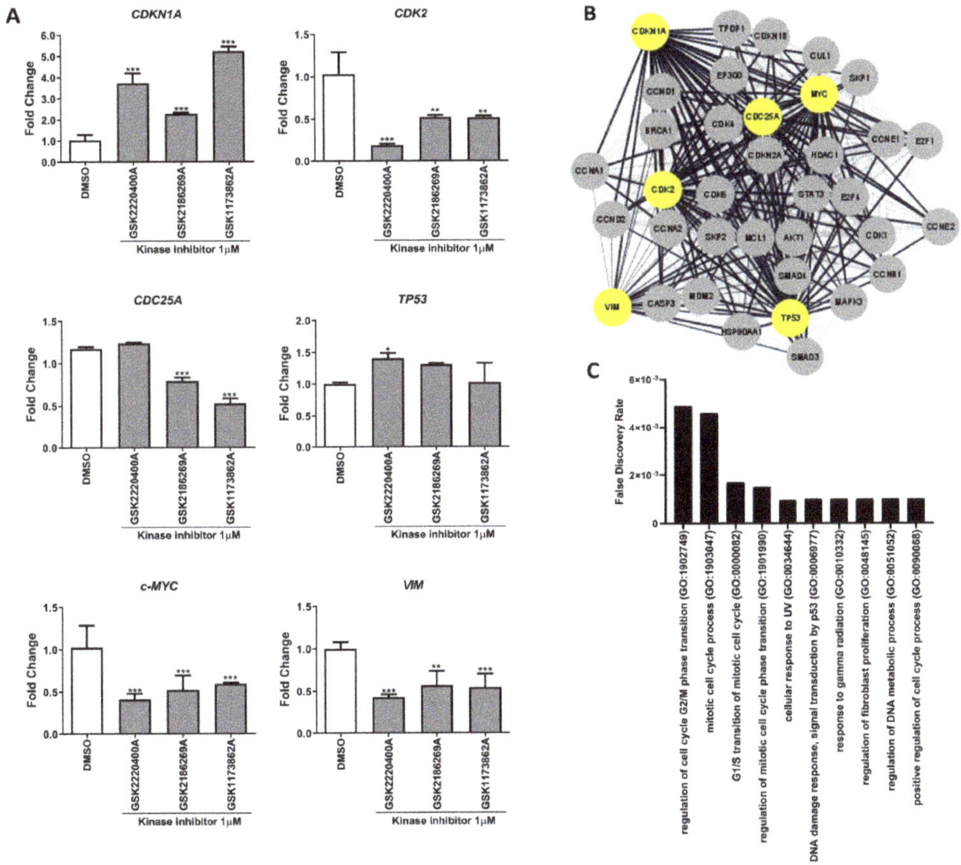

Figure 3. Gene expression pattern of oncogenes and tumor suppressors genes in metastatic gastric cancer cells after kinase inhibitors treatment (1 μM). (**A**) Mean ± standard deviation of mRNA expression fold change of *CDK1NA*, *CDK2*, *CDC25A*, *TP53*, *c-MYC*, and *VIM* after KI exposure. (**B**) Protein–protein interaction network analysis of the modulated genes after kinase inhibitor treatment. Analyzed genes by RT-qPCR after treatment are represented with yellow color. (**C**) False Discovery Rate (FDR) of functional analysis with GO biological processes gene set. The top 10 GO processes are sorted by the FDR. Significant differences compared to the untreated control: * $p < 0.05$, ** $p < 0.01$, *** $p < 0.001$.

3.3. *Kinase Target Expression in the Metastatic Gastric Cancer Cells*

Gene expression of AGP-01 cells and non-neoplastic gastric mucosa MNP-01 (normal cell) revealed that the *ALK* gene is overexpressed in the gastric cancer cells when compared to the normal cell line. As show in Figure 4A, among the three kinase identified as the target for the three studied compounds [13], only the *ALK* gene was overexpressed in the gastric cancer cells compared to the normal cell line. Moreover, we performed the GEPIA analysis to determine the impact of high expression on patients' survival. Gastric adenocarcinoma patients with higher expression of *ALK* gene had a lower survival probability compared to the low expression of *ALK* group ($p = 0.0045$).

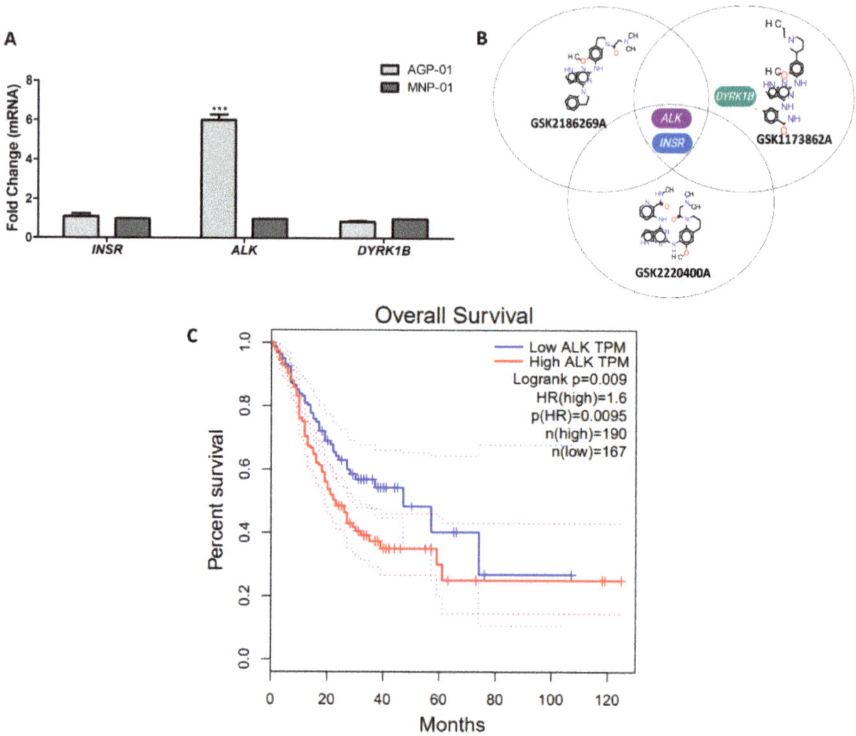

Figure 4. Overexpression of the *ALK* gene in the metastatic gastric cancer cells AGP-01. (**A**) Gene expression of the three-target kinase for the three selected inhibitors were assessed by RT-qPCR. The comparative expression between AGP-01 cells (metastatic gastric cancer cell) and MNP-01 cells (non-malignant gastric mucosa) are shown as mean ± standard deviation (SD). (**B**) Diagram showing the three molecules selected in this study and their targets. (**C**) GEPIA server analysis for *ALK* expression on Cancer Genome Atlas Stomach Adenocarcinoma (TCGA-STAD) comparing tumor samples and normal samples. Red and blue dotted lines represent high and low *ALK* expression, respectively. Significant differences compared to untreated control: *** $p < 0.001$.

4. Discussion

ALK is considered to play a key role in the nervous system's development and function, where it regulates the basic principles of cell proliferation, survival, and differentiation in response to extracellular stimuli [20,21]. Besides chromosome rearrangements, gene overexpression has been described as the relevant abnormal alterations in the *ALK* gene in neuroblastoma, lung, and oesophageal cancer [22–24]. Chen et al. (2012) also showed *ALK* overexpression in non-small cell lung cancer detected by immunohistochemical techniques [25]. Another recent evidence demonstrated that the ovarian high-grade serous carcinoma (HGSC) had significantly higher cytoplasmic *ALK* expression without chromosomal rearrangement or gene alterations compared to non-HGSC ovarian carcinomas [26]. However, *ALK* overexpression in gastric adenocarcinoma has not been well investigated.

To our knowledge, Fan and co-workers (2020) described the first case of gastrointestinal stromal tumor (GIST) with *ALK* overexpression; however, GIST are a different cancer identity when compared to gastric adenocarcinoma [27].

Our study aimed to screen 147 different kinase inhibitors identified and characterized by Elkins et al. (2016) in a metastatic gastric adenocarcinoma model in order to identify a possible kinase target with therapeutic relevance. After the initial screening, only six compounds suppressed the cell growth as much as the positive control and only three compounds (GSK2186269A, GSK1173862A, and GSK2220400A) continued to be investigated regarding their mode of action and to confirm their anti-cancer properties. The results were clear in demonstrating the cell cycle blockage, apoptosis induction, and the inhibitory effect on migration as well as in gene expression modulation. In fact, based on the gene expression of *CDKN1A*, *CDK2*, *CDC25A*, *TP53*, and *MYC*, the GO biological analysis revealed that the regulation of the cell cycle (GO:1902749, GO:1903047, GO:0000082, GO:1901990) and DNA damage response (GO:0034644, GO:0006977, GO:0010332) were induced after treatment with the three kinase inhibitors, which resulted in the eventual cell death [28–31]. The *VIM* gene suppression also revealed that the kinase inhibitors probably induced mesenchymal-epithelium transition which corroborated with the inhibition on cell migration [32]. A single study by [27] showed a decrease in cell viability and cell cycle progression, as well as tumor growth in a xenograft model of gastric cancer after *ALK* inhibition [33]. Therefore, our study corroborates with the possibility of the use of *ALK* target therapy against gastric cancer.

Evaluating the three compounds with high cytotoxic potency, compounds GSK2186269A, GSK2220400A, and GSK1173862A share the same targets [13], Anaplastic Lymphoma Receptor Tyrosine Kinase (*ALK*) and Insulin Receptor (*INSR*). Additionally, the GSK1173862A molecule also interacts with Dual Specificity Tyrosine Phosphorylation Regulated Kinase 1B (*DYRK1B*). These findings suggest that the gastric cancer cells probably overexpress these genes and therefore could be a relevant therapeutic target for gastric cancer management. Moreover, we performed the gene expression analysis of these targets on gastric cancer cells (AGP-01) and, interestingly, only the *ALK* gene showed a high expression level in gastric cancer cells when compared to the non-malignant cell line (sixfold changes). Looking for another evidence that *ALK* can be a biomarker for gastric cancer, we identified a lower survival rate for patients with gastric cancer with high levels of *ALK* expression in TCGA-STAD analysis [19]. Our hypothesis is that *ALK* gene overexpression can be a therapeutic target and/or a prognosis biomarker for patients with high-grade gastric adenocarcinoma. However, further studies are needed to confirm this hypothesis.

To our knowledge, only one study evaluated the *ALK* expression and this was in four Asian gastric cancer cell lines and a single Asian cohort of gastric cancer patients. *ALK* gene amplification and protein overexpression were not seen in any of these samples [34]. It is worth it to notice that gastric carcinogenesis differs between Asian and Caucasian tumors [35–38]. A few studies have described other *ALK* molecular alterations in gastric tumors which do not include gene overexpression. For example, a study showed 2.3% positive cases of *ALK* translocation by FISH using the standard criteria of at least 15% positive cells for the break-apart signal in signet ring cell carcinoma of the gastrointestinal tract [39].

Interestingly, a novel form of *ALK* gene fusion was identified, being the first gastric adenocarcinoma case with *RAB10-ALK* fusion [40]. On the contrary, another study did not find *ALK* fusions by FISH and gene sequencing methods in gastric adenocarcinoma [41]. Therefore, it seems to have a relevant difference among the population tested or genetic alteration found within *ALK* gene in those studies. Indeed, the genetic background can affect the molecular analysis of the cell line or cohorts [42]. Small-molecule *ALK* tyrosine kinase inhibitors are very effective against a group of cancers defined by chromosomal rearrangements involving the anaplastic lymphoma kinase (*ALK*) gene. The first- and second-generation of *ALK* inhibitors was designed to act by inhibiting the *ALK* chromosomal rearrangements, and currently, treatment resistance and recurring illness have gained importance [43]. Therefore, there is still a need for the discovery of new *ALK* gene-related alterations and the development of new inhibitors. There is ample evidence that *ALK* overexpression may be a promising therapeutic biomarker for other populations in different continents, which could benefit gastric cancer patients.

5. Conclusions

In conclusion, in vitro screening of a kinase inhibitor library reveals three potential kinase inhibitors against gastric cancer cell model. These three molecules showed significant anticancer activity by inhibiting cell proliferation and migration, as well as provoking cell death. Interestingly, the three molecules share the same targets, *ALK* and *INSR*, in which only the *ALK* gene is overexpressed in the gastric cancer cell line. Looking for a possible impact on clinical aspects, we identified that higher *ALK* expression reduces patients' survival rate in TCGA-STAD data. We hypothesize that *ALK* gene overexpression can be a promising biomarker for gastric adenocarcinoma. However, further studies, in particular clinical studies, need to be conducted to prove this hypothesis.

Author Contributions: Invitation received, C.A.M.-N.; Conceptualization, F.P.M. and R.C.M.; Provision of data and subsequent analysis and interpretation, F.P.M., P.F.N.S., E.L.d.S., L.L.B.d.O., L.B.L., C.A.M.-N. and R.C.M.; Data Validation, F.P.M., E.L.d.S. and W.J.Z.; Writing—original draft preparation, F.P.M., P.F.N.S. and R.C.M.; Writing—review and editing, F.P.M., P.F.N.S. and C.A.M.-N., R.M.R.B. and R.C.M.; Supervision, M.E.A.d.M. and R.C.M.; Funding acquisition, R.C.M. All authors have read and agreed to the published version of the manuscript.

Funding: This study was supported by Brazilian funding agencies: Coordination for the Improvement of Higher Education Personnel (CAPES; to F.P.M.), National Council of Technological and Scientific Development (CNPq grant number 404213/2021-9 to C.A.M.-N.; CNPQ/MCTI/FNDCT no. 18/2021—no. 403493/2021-8 to R.C.M. and R.M.R.B., and Productivity in Research PQ to M.E.A.d.M., R.M.R.B. and R.C.M.) We also would like to thank PROPESP/UFPA for publication payment.

Institutional Review Board Statement: Not applicable.

Informed Consent Statement: Not applicable.

Acknowledgments: We are grateful to the Brazilian Agencies CNPq, CAPES, and FUNCAP, along with the Federal University of Ceará and the Structural Genomics Consortium for fellowships and research grants. We also thank the staff of the Multi-User Facility of the Drug Research and Development Center of Federal University of Ceará for technical support.

Conflicts of Interest: The authors declare no conflict of interest. The funders had no role in the design of the study; in the collection, analyses, or data interpretation; in the writing of the manuscript, or in the decision to publish the results.

References

1. Sung, H.; Ferlay, J.; Siegel, R.L.; Laversanne, M.; Soerjomataram, I.; Jemal, A.; Bray, F. Global Cancer Statistics 2020: GLOBOCAN Estimates of Incidence and Mortality Worldwide for 36 Cancers in 185 Countries. *CA Cancer J. Clin.* **2021**, *71*, 209–249. [CrossRef] [PubMed]
2. Bray, F.; Ferlay, J.; Soerjomataram, I.; Siegel, R.L.; Torre, L.A.; Jemal, A. Global Cancer Statistics 2018: GLOBOCAN Estimates of Incidence and Mortality Worldwide for 36 Cancers in 185 Countries. *CA Cancer J. Clin.* **2018**, *68*, 394–424. [CrossRef]
3. Lee, J.; Bass, A.J.; Ajani, J.A. Gastric Adenocarcinoma: An Update on Genomics, Immune System Modulations, and Targeted Therapy. *Am. Soc. Clin. Oncol. Educ. book. Am. Soc. Clin. Oncol. Annu. Meet.* **2016**, *35*, 104–111. [CrossRef]
4. Eusebi, L.H.; Telese, A.; Marasco, G.; Bazzoli, F.; Zagari, R.M. Gastric Cancer Prevention Strategies: A Global Perspective. *J. Gastroenterol. Hepatol.* **2020**, *35*, 1495–1502. [CrossRef] [PubMed]
5. Yin, J.; Wu, X.; Li, S.; Li, C.; Guo, Z. Impact of Environmental Factors on Gastric Cancer: A Review of the Scientific Evidence, Human Prevention and Adaptation. *J. Environ. Sci.* **2020**, *89*, 65–79. [CrossRef]
6. Bass, A.J.; Thorsson, V.; Shmulevich, I.; Reynolds, S.M.; Miller, M.; Bernard, B.; Hinoue, T.; Laird, P.W.; Curtis, C.; Shen, H.; et al. Comprehensive Molecular Characterization of Gastric Adenocarcinoma. *Nature* **2014**, *513*, 202–209.
7. Jácome, A.A.; Coutinho, A.K.; Lima, E.M.; Andrade, A.C.; Dos Santos, J.S. Personalized Medicine in Gastric Cancer: Where Are We and Where Are We Going? *World J. Gastroenterol.* **2016**, *22*, 1160–1171. [CrossRef] [PubMed]
8. Catalano, V.; Labianca, R.; Beretta, G.D.; Gatta, G.; de Braud, F.; Van Cutsem, E. Gastric Cancer. *Crit. Rev. Oncol. Hematol.* **2009**, *71*, 127–164. [CrossRef]

9. Bang, Y.J.; Van Cutsem, E.; Feyereislova, A.; Chung, H.C.; Shen, L.; Sawaki, A.; Lordick, F.; Ohtsu, A.; Omuro, Y.; Satoh, T.; et al. Trastuzumab in Combination with Chemotherapy versus Chemotherapy Alone for Treatment of HER2-Positive Advanced Gastric or Gastro-Oesophageal Junction Cancer (ToGA): A Phase 3, Open-Label, Randomised Controlled Trial. *Lancet* **2010**, *376*, 687–697. [CrossRef]
10. Wagner, A.D.; Syn, N.L.; Moehler, M.; Grothe, W.; Yong, W.P.; Tai, B.C.; Ho, J.; Unverzagt, S. Chemotherapy for Advanced Gastric Cancer. *Cochrane Database Syst. Rev.* **2017**, *8*. [CrossRef] [PubMed]
11. Bhullar, K.S.; Lagarón, N.O.; McGowan, E.M.; Parmar, I.; Jha, A.; Hubbard, B.P.; Rupasinghe, H.P.V. Kinase-Targeted Cancer Therapies: Progress, Challenges and Future Directions. *Mol. Cancer* **2018**, *17*, 48. [CrossRef]
12. Wu, P.; Nielsen, T.E.; Clausen, M.H. FDA-Approved Small-Molecule Kinase Inhibitors. *Trends Pharmacol. Sci.* **2015**, *36*, 422–439. [CrossRef]
13. Elkins, J.M.; Fedele, V.; Szklarz, M.; Abdul Azeez, K.R.; Salah, E.; Mikolajczyk, J.; Romanov, S.; Sepetov, N.; Huang, X.P.; Roth, B.L.; et al. Comprehensive Characterization of the Published Kinase Inhibitor Set. *Nat. Biotechnol.* **2016**, *34*, 95–103. [CrossRef] [PubMed]
14. Leal, M.F.; do Nascimento, J.L.M.; da Silva, C.E.A.; Lamarão, M.F.V.; Calcagno, D.Q.; Khayat, A.S.; Assumpção, P.P.; Cabral, I.R.; Smith, M.D.A.C.; Burbano, R.R. Establishment and Conventional Cytogenetic Characterization of Three Gastric Cancer Cell Lines. *Cancer Genet. Cytogenet.* **2009**, *195*, 85–91. [CrossRef]
15. Leal, M.F.; Ribeiro, H.F.; Rey, J.A.; Pinto, G.R.; Smith, M.C.; Moreira-Nunes, C.A.; Assumpção, P.P.; Lamarão, L.M.; Calcagno, D.Q.; Montenegro, R.C.; et al. YWHAE Silencing Induces Cell Proliferation, Invasion and Migration through the up-Regulation of CDC25B and MYC in Gastric Cancer Cells: New Insights about YWHAE Role in the Tumor Development and Metastasis Process. *Oncotarget* **2016**, *7*, 85393–85410. [CrossRef] [PubMed]
16. Mesquita, F.P.; Silva, E.L.; Souza, P.F.N.; Lima, L.B.; Amaral, J.L.; Zuercher, W.; Albuquerque, L.M.; Rabenhorst, S.H.B.; Moreira-Nunes, C.A.; Moraes, M.E.; et al. Kinase Inhibitor Screening Reveals Aurora-a Kinase Is a Potential Therapeutic and Prognostic Biomarker of Gastric Cancer. *J. Cell. Biochem.* **2021**, *122*, 1376–1388. [CrossRef]
17. Bustin, S.A.; Benes, V.; Garson, J.A.; Hellemans, J.; Huggett, J.; Kubista, M.; Mueller, R.; Nolan, T.; Pfaffl, M.W.; Shipley, G.L.; et al. The MIQE Guidelines: Minimum Information for Publication of Quantitative Real-Time PCR Experiments. *Clin. Chem.* **2009**, *55*, 611–622. [CrossRef]
18. Livak, K.J.; Schmittgen, T.D. Analysis of Relative Gene Expression Data Using Real-Time Quantitative PCR and the 2(-Delta Delta C(T)) Method. *Methods* **2001**, *25*, 402–408. [CrossRef]
19. Tang, Z.; Li, C.; Kang, B.; Gao, G.; Li, C.; Zhang, Z. GEPIA: A Web Server for Cancer and Normal Gene Expression Profiling and Interactive Analyses. *Nucleic Acids Res.* **2017**, *45*, W98–W102. [CrossRef]
20. Holla, V.R.; Elamin, Y.Y.; Bailey, A.M.; Johnson, A.M.; Litzenburger, B.C.; Khotskaya, Y.B.; Sanchez, N.S.; Zeng, J.; Shufean, M.A.; Shaw, K.R.; et al. ALK: A Tyrosine Kinase Target for Cancer Therapy. *Cold Spring Harb. Mol. Case Stud.* **2017**, *3*, a001115. [CrossRef]
21. Hallberg, B.; Palmer, R.H. The Role of the ALK Receptor in Cancer Biology. *Ann. Oncol. Off. J. Eur. Soc. Med. Oncol.* **2016**, *27* (Suppl. 3), iii4–iii15. [CrossRef] [PubMed]
22. Schönherr, C.; Ruuth, K.; Yamazaki, Y.; Eriksson, T.; Christensen, J.; Palmer, R.H.; Hallberg, B. Activating ALK Mutations Found in Neuroblastoma Are Inhibited by Crizotinib and NVP-TAE684. *Biochem. J.* **2011**, *440*, 405–413. [CrossRef]
23. Murugan, A.K.; Xing, M.M. Anaplastic Thyroid Cancers Harbor Novel Oncogenic Mutations of the ALK Gene. *Cancer Res.* **2011**, *71*, 4403–4411. [CrossRef] [PubMed]
24. Schoppmann, S.F.; Streubel, B.; Birner, P. Amplification but Not Translocation of Anaplastic Lymphoma Kinase Is a Frequent Event in Oesophageal Cancer. *Eur. J. Cancer* **2013**, *49*, 1876–1881. [CrossRef] [PubMed]
25. Huang, S.-F.; Chen, T.-D.; Chang, I.-C.; Chen, Y.-T.; Liu, H.-P.; Wu, Y.-C.; Wang, C.-L.; Chen, Y.-R. Correlation of Anaplastic Lymphoma Kinase Overexpression and the EML4-ALK Fusion Gene in Non-Small Cell Lung Cancer by Immunohistochemical Study. *Chang Gung Med. J.* **2012**, *35*, 309–317. [CrossRef]
26. Matsumoto, T.; Oda, Y.; Hasegawa, Y.; Hashimura, M.; Oguri, Y.; Inoue, H.; Yokoi, A.; Tochimoto, M.; Nakagawa, M.; Jiang, Z.; et al. Anaplastic Lymphoma Kinase Overexpression Is Associated with Aggressive Phenotypic Characteristics of Ovarian High-Grade Serous Carcinoma. *Am. J. Pathol.* **2021**, *191*, 1837–1850. [CrossRef]
27. Fan, J.; Yang, M.; Huang, B.; Wang, Z.; Luo, D.; Zhang, J.; Zhang, P.; Shi, H.; Li, Y.; Nie, X. ALK Expressed in a Gastrointestinal Stromal Tumor Harboring PDGFRA p. D842V Mutation:A Case Report. *Diagn. Pathol.* **2020**, *15*, 8. [CrossRef]
28. Suski, J.M.; Braun, M.; Strmiska, V.; Sicinski, P. Targeting Cell-Cycle Machinery in Cancer. *Cancer Cell* **2021**, *39*, 759–778. [CrossRef]
29. Strasser, A.; Vaux, D.L. Cell Death in the Origin and Treatment of Cancer. *Mol. Cell* **2020**, *78*, 1045–1054. [CrossRef]
30. Razaghi, A.; Heimann, K.; Schaeffer, P.M.; Gibson, S.B. Negative Regulators of Cell Death Pathways in Cancer: Perspective on Biomarkers and Targeted Therapies. *Apoptosis* **2018**, *23*, 93–112. [CrossRef]
31. Carneiro, B.A.; El-Deiry, W.S. Targeting Apoptosis in Cancer Therapy. *Nat. Rev. Clin. Oncol.* **2020**, *17*, 395–417. [CrossRef] [PubMed]
32. Debnath, P.; Huirem, R.S.; Dutta, P.; Palchaudhuri, S. Epithelial-Mesenchymal Transition and Its Transcription Factors. *Biosci. Rep.* **2022**, *42*, BSR20211754. [CrossRef]
33. Chen, X.; Ji, J.; Chen, W.; Lian, W.; Chen, R.; Yang, J.; Zhang, Q.; Weng, Q.; Khan, Z.; Hu, J.; et al. (S)-Crizotinib Reduces Gastric Cancer Growth through Oxidative DNA Damage and Triggers pro-Survival Akt Signal. *Cell Death Dis.* **2018**, *9*, 660.

34. Yang, Y.; Wu, N.; Shen, J.; Teixido, C.; Sun, X.; Lin, Z.; Qian, X.; Zou, Z.; Guan, W.; Yu, L.; et al. MET Overexpression and Amplification Define a Distinct Molecular Subgroup for Targeted Therapies in Gastric Cancer. *Gastric Cancer* **2016**, *19*, 778–788. [CrossRef]
35. Montenegro, R.C.; Clark, P.G.K.; Howarth, A.; Wan, X.; Ceroni, A.; Siejka, P.; Nunez-Alonso, G.A.; Monteiro, O.; Rogers, C.; Gamble, V.; et al. BET Inhibition as a New Strategy for the Treatment of Gastric Cancer. *Oncotarget* **2016**, *7*, 43997. [CrossRef] [PubMed]
36. Wang, J.; Sun, Y.; Bertagnolli, M.M. Comparison of Gastric Cancer Survival between Caucasian and Asian Patients Treated in the United States: Results from the Surveillance Epidemiology and End Results (SEER) Database. *Ann. Surg. Oncol.* **2015**, *22*, 2965–2971. [CrossRef] [PubMed]
37. Jin, H.; Pinheiro, P.S.; Callahan, K.E.; Altekruse, S.F. Examining the Gastric Cancer Survival Gap between Asians and Whites in the United States. *Gastric Cancer* **2017**, *20*, 573–582. [CrossRef] [PubMed]
38. Trumbull, D.; Lemini, R.; Attwood, K.; Kukar, M.; Gabriel, E. Gastric Cancer Disparities Among Asian American Subpopulations. *Anticancer Res.* **2020**, *40*, 6381–6385. [CrossRef]
39. Alese, O.B.; Rayes, B.F.; Zhang, G.; Khuri, F.R.; Sica, G.; Adsay, N.V.; Alexis, D.; Rosa, F.G.; Varella Garcia, M.; Chen, Z.; et al. Anaplastic Lymphoma Kinase (ALK) Gene Alteration in Signet Ring Cell Carcinoma of the Gastrointestinal Tract. *Ther. Adv. Med. Oncol.* **2015**, *7*, 56–62. [CrossRef]
40. Wen, Z.; Xiong, D.; Zhang, S.; Liu, J.; Li, B.; Li, R.; Zhang, H. Case Report: RAB10-ALK: A Novel ALK Fusion in a Patient With Gastric Cancer. *Front. Oncol.* **2021**, *11*, 645370. [CrossRef]
41. Ying, J.; Lin, C.; Wu, J.; Guo, L.; Qiu, T.; Ling, Y.; Shan, L.; Zhou, H.; Zhao, D.; Wang, J.; et al. Anaplastic Lymphoma Kinase Rearrangement in Digestive Tract Cancer: Implication for Targeted Therapy in Chinese Population. *PLoS ONE* **2015**, *10*, e0144731.
42. Turajlic, S.; Sottoriva, A.; Graham, T.; Swanton, C. Resolving Genetic Heterogeneity in Cancer. *Nat. Rev. Genet.* **2019**, *20*, 404–416. [CrossRef] [PubMed]
43. Shaw, A.T.; Bauer, T.M.; de Marinis, F.; Felip, E.; Goto, Y.; Liu, G.; Mazieres, J.; Kim, D.-W.; Mok, T.; Polli, A.; et al. First-Line Lorlatinib or Crizotinib in Advanced ALK -Positive Lung Cancer. *N. Engl. J. Med.* **2020**, *383*, 2018–2029. [CrossRef] [PubMed]

pharmaceutics

MDPI

Review

Kinase Inhibition in Multiple Myeloma: Current Scenario and Clinical Perspectives

Igor Valentim Barreto [1,†], Caio Bezerra Machado [1,†], Davi Benevides Almeida [2,†], Flávia Melo Cunha de Pinho Pessoa [1], Renan Brito Gadelha [1], Laudreísa da Costa Pantoja [3], Deivide de Sousa Oliveira [4], Rodrigo Monteiro Ribeiro [4], Germison Silva Lopes [5], Manoel Odorico de Moraes Filho [1], Maria Elisabete Amaral de Moraes [1], André Salim Khayat [3], Edivaldo Herculano Correa de Oliveira [6,7] and Caroline Aquino Moreira-Nunes [1,3,8,*]

1 Pharmacogenetics Laboratory, Department of Medicine, Drug Research and Development Center (NPDM), Federal University of Ceará, Fortaleza 60430-275, CE, Brazil
2 Unichristus University Center, Faculty of Biomedicine, Fortaleza 60430-275, CE, Brazil
3 Department of Biological Sciences, Oncology Research Center, Federal University of Pará, Belém 66073-005, PA, Brazil
4 Department of Hematology, Fortaleza General Hospital (HGF), Fortaleza 60150-160, CE, Brazil
5 Department of Hematology, César Cals General Hospital, Fortaleza 60015-152, CE, Brazil
6 Faculty of Natural Sciences, Institute of Exact and Natural Sciences, Federal University of Pará (UFPA), Rua Augusto Correa, 01, Belém 66075-990, PA, Brazil
7 Laboratory of Cytogenomics and Environmental Mutagenesis, Environment Section (SAMAM), Evandro Chagas Institute (IEC), BR 316, KM 7, s/n, Levilândia, Ananindeua 67030-000, PA, Brazil
8 Northeast Biotechnology Network (RENORBIO), Itaperi Campus, Ceará State University, Fortaleza 60740-903, CE, Brazil
* Correspondence: carolfam@gmail.com
† These authors contributed equally to this work.

Citation: Barreto, I.V.; Machado, C.B.; Almeida, D.B.; Pessoa, F.M.C.d.P.; Gadelha, R.B.; Pantoja, L.d.C.; Oliveira, D.d.S.; Ribeiro, R.M.; Lopes, G.S.; de Moraes Filho, M.O.; et al. Kinase Inhibition in Multiple Myeloma: Current Scenario and Clinical Perspectives. *Pharmaceutics* **2022**, *14*, 1784. https://doi.org/10.3390/pharmaceutics14091784

Academic Editor: Ross McKinnon

Received: 21 July 2022
Accepted: 23 August 2022
Published: 25 August 2022

Abstract: Multiple myeloma (MM) is a blood cell neoplasm characterized by excessive production of malignant monoclonal plasma cells (activated B lymphocytes) by the bone marrow, which end up synthesizing antibodies or antibody fragments, called M proteins, in excess. The accumulation of this production, both cells themselves and of the immunoglobulins, causes a series of problems for the patient, of a systemic and local nature, such as blood hyperviscosity, renal failure, anemia, bone lesions, and infections due to compromised immunity. MM is the third most common hematological neoplasm, constituting 1% of all cancer cases, and is a disease that is difficult to treat, still being considered an incurable disease. The treatments currently available cannot cure the patient, but only extend their lifespan, and the main and most effective alternative is autologous hematopoietic stem cell transplantation, but not every patient is eligible, often due to age and pre-existing comorbidities. In this context, the search for new therapies that can bring better results to patients is of utmost importance. Protein tyrosine kinases (PTKs) are involved in several biological processes, such as cell growth regulation and proliferation, thus, mutations that affect their functionality can have a great impact on crucial molecular pathways in the cells, leading to tumorigenesis. In the past couple of decades, the use of small-molecule inhibitors, which include tyrosine kinase inhibitors (TKIs), has been a hallmark in the treatment of hematological malignancies, and MM patients may also benefit from TKI-based treatment strategies. In this review, we seek to understand the applicability of TKIs used in MM clinical trials in the last 10 years.

Keywords: multiple myeloma; tyrosine kinase; treatment; ibrutinib

1. Introduction

Multiple myeloma (MM) is a systemic hematological neoplasm in which there is an abnormal proliferation of malignant monoclonal plasmocytes that release antibodies or antibody fragments, called M proteins, in excess [1–5]. This feature is responsible for a range of symptoms of MM such as blood hyperviscosity and damage to the renal tubules. Along with this accumulation, the interaction of malignant plasma cells with other cells in the bone marrow (BM) causes several problems for patients such as anemia, destructive bone lesions, and infections due to compromised immunity [5–7].

MM is the third most reported neoplasm representing approximately 1% of all cancers and approximately 10% of all hematological malignancies, with risk factors for MM development involving gender, age, family history of malignancies, and ethnicity [1,8,9]. MM is more commonly reported in men than in women with a ratio of 1.5 to 1 and being twice as common in African Americans compared to Caucasians [2,10,11]. The median age at diagnosis tends to be around 65 years [2,12].

MM emerges from molecular changes caused by DNA damage and failures in DNA repair mechanisms [13]. As a genetically complex disease, MM development is a process composed of several stages, with initial mutations appearing in hematopoietic stem cells (HSC) of the bone marrow (BM), the site most affected by the disease [14]. With the onset of malignant transformations, patients enter a pre-malignant stage called smoldering multiple myeloma (SMM) or monoclonal gammopathy of undetermined significance (MGUS) that occurs due to genetic events such as chromosomal translocations involving immunoglobulin heavy chain *(IgH)* genes and aneuploidy [15]. Secondary genetic events, such as copy number abnormalities and acquired mutations, are linked to tumor progression [16–19]. With the accumulation of mutations that guarantee competitive advantages, HSCs evolve into malignant cells and begin to proliferate causing accumulation of malignant plasma cells in the BM [19]. This proliferation is sustained also due to the release of cytokines, such as interleukin 6 (IL-6), carried out by BM stromal cells [5,20,21]. The genetic alterations found in the MGUS stage are involved in tumor development, while the events present in the MM stages that were not found in MGUS are responsible for tumor progression [12,17].

MGUS is an asymptomatic pre-malignant phase, preceding most cases of MM and being present in approximately 3–4% of the population over 50 years of age. Of the total number of MGUS cases, only 1% per year has a chance of progression to MM [22–25]. Patients with translocation t(4;14), del(17p), gain(1q), and trisomies have a higher risk of progression from the SMM stage to MM, reaching 10% per year. This being also a risk factor for progression from MGUS to MM [26–28].

Regarding genetic alterations, MM is known to present a great heterogeneity; however, mutations that could be considered precursors of the disease are now well known, and these genetic alterations can be used as a prognostic factor. An example of this would be mutations in the *IgH* gene locus, responsible for producing the heavy chains of immunoglobulins, which are considered to be early precursor mutations [3,29,30]. Chromosomal translocations involving *IgH* and other genes such as *Nuclear Receptor Binding SET Domain Protein* (*NSD2*), *Fibroblast Growth Factor Receptor 3* (*FGFR3*), *Cyclin D3* (*CCND3*), *Cyclin D1* (*CCND1*), *MAF bZIP Transcription Factor* (*MAF*), and *MAF bZIP Transcription Factor B* (*MAFB*), resulting in the translocations t(4;14), t(6;14), t(11;14), t(14;16), and t(14;20), respectively, are also an important clinical finding, as they deregulate checkpoints of the cell cycle due to increased gene transcription under the activity of *IgH* transcription enhancer (Figure 1) [14,31–34].

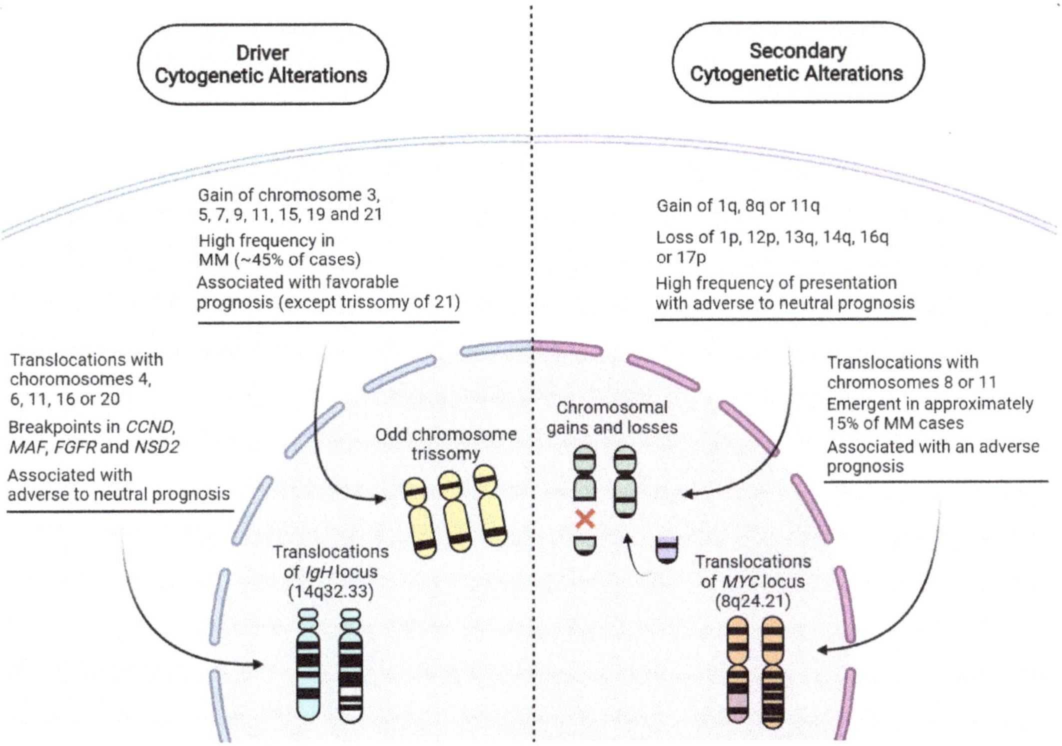

Figure 1. Cytogenetic alterations associated with multiple myeloma (MM). In pre-malignant settings, driver alterations correlate with deregulation of oncogene activity and carcinogenesis onset. At later disease stages, secondary alterations emerge due to genomic instability in malignant clones, and MM cells from the same patient may even harbor different secondary alterations, following the concepts of linear and branching evolution. Created with BioRender.com.

2. Current Clinical MM Treatment Options

Overall survival of MM cases has more than doubled in recent decades due to the introduction of new combinations of chemotherapy, small molecule inhibitors, and the use of monoclonal antibodies [35–38].

Hematopoietic stem cell transplantation (HSCT) is still the most recommended treatment for MM, being the first choice in most cases, although not all patients are eligible. Non-eligibility can happen for a variety of reasons, including age, which is an important cut-off point for inclusion criteria, pre-existing comorbidities, and performance score, which is used to predict poor outcomes in patients with MM [39–41].

When eligible for HSCT, patients are submitted to one of two induction regimens: VTD, which includes bortezomib, thalidomide, and dexamethasone, or VRD, which includes bortezomib, lenalidomide, and dexamethasone, being the most adopted pre-transplant induction regimens available [42]. The role of induction chemotherapy is to reduce the neoplastic burden at the patient's BM in order to increase response rates and effectiveness of an autologous transplantation graft [43–46]. Satisfactory response rates can be seen in young, transplant-eligible patients receiving high-dose melphalan therapy with autologous stem cell transplantation achieving a >60% effective response [47].

Most patients afflicted with MM end up relapsing and those who relapse and are not eligible for a new HSCT require a triple therapy regimen that varies from case to case.

Although highly cytotoxic, triple therapy regimens should be continued until the toxicity outweighs the benefits or until patients are eligible for autologous HSCT [35,38]

On the other hand, the use of proteasome inhibitors revolutionized the management of hematological malignancies emerging as one of the most important agents for the treatment of MM [48]. Tumor cells are proteasome-dependent to eliminate excess proteins that arise due to the continuous production of monoclonal immunoglobulin chains. Proteasome hyperactivity in MM results in the degradation of important proteins such as the nuclear factor kappa B (NFκB), the enzymatic complex of inhibitors of nuclear factor kappa B (IκB), tumor protein p53 (TP53) suppressor proteins, among other proteins responsible for the cell cycle [49,50]. Its inhibition leads to cellular stress induced by high protein load in the endoplasmic reticulum due to the accumulation of intracellular proteins, leading to cell death in MM cells [51,52]. Three agents in this class are approved by the Food and Drug Administration (FDA) for use, bortezomib, carfilzomib, and ixazomib [53–57].

Approved in 2003, bortezomib was the first proteasome inhibitor to be used for the treatment of relapsed and refractory multiple myeloma. Reversibly binding with high affinity to the 20S proteolytic core withing the 26S proteasome without inhibiting other types of proteases commonly present in the human body, bortezomib inhibits the ubiquitin–proteasome pathway, triggering a series of events such as induction of apoptosis, cell cycle inhibition, angiogenesis and adhesion, and cell proliferation. Despite being highly potent and effective, bortezomib has a limitation during treatment: the dose that will be used is limited by the toxicity of the drug that is often associated with peripheral neuropathy [52,55,56,58,59].

Carfilzomib is a second-generation proteasome inhibitor approved in 2012 for use in monotherapy or doublet or triplet combination regimens, especially for patients with relapsed or refractory multiple myeloma. Unlike bortezomib, which binds reversibly, carfilzomib binds irreversibly and highly selectively to the 20S proteasome, precisely in the chymotrypsin-like β5 subunit. The inhibition of this subunit is enough to cause apoptosis due to the accumulation of proteasome substrates inside the cell, which could explain the high sensitivity of MM cells to this drug, use of carfilzomib being possible in a monotherapy regimen aiming to reduce adverse effects. In this context, the drug has lower neurotoxicity compared to bortezomib, but there is a higher possibility of causing hypertension, congestive heart failure, and coronary artery disease, not being recommended for patients with heart diseases [54–56].

Ixazomib was approved in 2015 for use in combination with lenalidomide and dexamethasone (Rd) in patients who have received at least one therapy regimen previously. It is the first proteasome inhibitor that can be administered orally. Ixazomib is a potent and selective inhibitor of 20S proteasome, also binding in the chymotrypsin-like β5 subunit, as carfilzomib, but reversibly, as bortezomib. In addition, at high concentrations, the drug can also bind in two other subunits of this proteasome: β1 caspase-like and β2 trypsin-like, increasing the selectivity of ixazomib by proteasome 20S. Ixazomib has no cardiotoxic effects, but neurotoxicity is present, although at a lower level when compared to bortezomib. The most reported adverse effects are thrombocytopenia, skin rash, and GI symptoms—diarrhea, nausea, and vomiting [54,55,60–62].

Another important therapy in the treatment of MM is the use of immunomodulators (IMiDs), such as thalidomide, pomalidomide, and lenalidomide. In general, immunomodulators act by inhibiting cell growth by inducing apoptosis in MM cells through the inhibition of interferon regulatory factor 4 (IRF4), thus, affecting expression of the *MYC* proto-oncogene (MYC). The overexpression of these two factors is linked to oncogenesis in several types of cancer, including MM, as they are involved in the processes of regulation of cell growth and metabolism, differentiation, apoptosis, angiogenesis, DNA repair, protein translation, and hematopoietic cell formation. In MM, they act by regulating the immune response and the development of immune cells [63–66].

In addition, IMiDs act favoring the production of interleukin-2 (IL-2) and interferon-gamma (IFN-γ), activating T lymphocytes and natural-killer (NK) cells, on the other hand also inhibiting the production of the tumor necrosis (TNF)-α. The discovery of the inhibitory effects of thalidomide on tumor progression led to the development of two other analogues, lenalidomide and pomalidomide, which were authorized and released for use in MM after clinical trials in 2006 and 2013 [64,67–69].

Daratumumab is a new kind of drug that has been used since 2015 for the treatment of MM and it has been promoting an incredible improvement in the efficiency of the treatment [70]. Daratumumab is an anti-CD38 monoclonal antibody, which is a pleiotropic glycoprotein highly expressed on plasma cells and MM cells, acting as a transmembrane receptor on these cells [70,71]. The role of CD38 as a receptor involves signaling for cell activation and proliferation and inducing cell adhesion processes [72], which may explain its high expression in MM tumor cells. Therefore, the daratumumab's monoclonal antibodies directly target and destroy tumor cells due to several mechanisms, including antibody-dependent cell-mediated cytotoxicity, complement-dependent cytotoxicity, antibody-dependent phagocytosis, and immune cell depletion or inhibition of immunosuppressive cells, constituting a specific type of immunotherapy for patients with multiple myeloma [46,73–76]. Initially, it was only used as a monotherapy regimen for patients with relapsed or refractory MM, but its low toxicity allowed it to be added to other drugs in several different combinations, both in triple and quadruple therapies [70,71].

Emerging therapies with better outcomes involve immunotherapy and personalized medicine based on the molecular characteristics of the patient's tumor [33,71,77]. Specific antibodies against B-cell maturation antigen (BCMA) have shown promise in the treatment of MM, since it is an antigen whose expression is higher in myeloma cells than in healthy plasma cells, having an essential role in the process of maturation and differentiation of B-cells, being only expressed on antibody-producing B lymphocytes [78,79]. Antibodies with dual specificity are also under development, the central idea of which is to bind a T cell to a tumor cell, with the antibody acting as a binding bridge, and thus induce the destruction of the tumor cell by the T lymphocyte connected to it [71,77].

Due to the evolution of molecular profiling techniques, it is possible to identify several genetic and molecular abnormalities that constitute the neoplastic clones of a specific patient. The knowledge of the existing mutations helps guide the best available treatment based on the genetic characteristics of each patient, in addition to assessing the patient's prognosis, thus, determining whether a more aggressive drug therapy is necessary or if the patient is in a situation of good prognosis and may be eligible for autologous HSCT [31,33,80].

With the understanding that MM is a disease still treated with a non-curative approach, the constant development of new therapeutic strategies is one of the main goals in oncologic investigations and routine clinical practice. In the past couple of decades, the use of small-molecule inhibitors, which include tyrosine kinase inhibitors (TKI), has been a hallmark in the treatment of hematological malignancies, and MM patients may also benefit from TKI-based treatment strategies [31,32].

3. Tyrosine Kinase Inhibitors in MM

Protein tyrosine kinases (PTKs) are part of a large, multigene family and their main functions are to coordinate cellular behavior, regulate mitosis, differentiation, apoptosis, and a series of physiological and biochemical processes [81–83]. Structurally, PTKs can be divided into receptor PTKs (RTK), acting as receptors for external signals of growth and survival factors and phosphorylating other protein residues in the intracellular compartment, and non-receptor PTKs (NRTK), which are cytoplasmic or nuclear proteins that act as second messengers. Examples of both classes include insulin-like growth factor 1 receptor (IGFR), mast/stem cell growth factor receptor Kit (KIT), hepatocyte growth factor receptor (MET), fibroblast growth factor receptor (FGFR3), vascular endothelial growth factor receptor (VEGFR), and platelet derived growth factor receptor (PDGFR), as RKTs, and Bruton's tyrosine kinase (BTK), Janus kinase (JAK), SRC proto-oncogene (SRC), ABL proto-oncogene (ABL), and FA complementation group (FAC), as NRTKs [81,84–87].

Several groups of diseases present alterations linked to PTK, as their abnormal expression is linked to disorders in the regulation of cell proliferation, leading to the process of tumorigenesis, and their overexpression is also related to invasion and metastasis, tumor neovascularization, and resistance to chemotherapy [81,85,86,88,89].

There are currently 71 TKIs approved by the FDA for the treatment of neoplasms (http://www.brimr.org/PKI/PKIs.htm, accessed on 21 July 2022). Acquired resistance remains a problem in cancer-targeted therapies as a variety of resistance mechanisms are described in TKI treatment protocols, such as amplification of target receptor expression, mutations in tyrosine kinase inhibitor binding receptors, overactivation of alternative cell survival pathways, and activation of downstream signaling effectors linked to cell proliferation [82,90]

Table 1 is composed of a series of clinical trials over the last 10 years using TKIs as monotherapy or in combination with other cytotoxic agents to treat patients afflicted with refractory multiple myeloma and their results with degrees of efficacy.

Table 1. TKIs used in clinical trials to treat refractory multiple myeloma in the last 10 years.

Clinical Study Phase	Targeted Kinase	Kinase Inhibitor	Associated Treatment	Clinical Outcome	Adverse Events	References
I	BTK	Ibrutinib	Carfilzomib/Dexmethasone	ORR of 67% and a PFS of 7.2 months.	Hypertension, anemia, pneumonia, fatigue, diarrhea, and thrombocytopenia.	[91]
I/IIb	BTK	Ibrutinib	Carfilzomib/Dexamethasone	Acceptable safety profiles with PFS and ORR of 7.7 months and 71%, respectively. The average one-year OS rate was 77%.	Thrombocytopenia, anemia, diarrhea, fatigue, nausea, and hypertension.	[92]
II	BTK	Ibrutinib	Bortezomib/Dexamethasone	The drug combination initially increased the levels of infections and risk minimization measures were necessary. Clinical response was observed in 57% of patients with a duration of 9.5 months.	Thrombocytopenia, diarrhea, anemia, asthenia, and pneumonia.	[93]
II	BTK	Ibrutinib	Dexamethasone	The highest CBR was achieved in the combination of ibrutinib 840 mg with dexamethasone 40 mg. With CBR of 28%, ORR of 5%, sustained SD of 23%, and median PFS of 4.6 months.	Diarrhea, fatigue, nausea, anemia, and thrombocytopenia.	[94]
I	JAK	Ruxolitinib	Lenalidomide and metilprednisolone	The drug showed the ability to abrogate resistance to lenalidomide. Featuring CBR of 46% and ORR of 38%.	Anemia, thrombocytopenia and lymphopenia, sepsis, and pneumonia.	[95]
Ib	HGF and MET	Cabozantinib	NR	The drug alone has no significant activity in patients with refractory MM. The study was interrupted, and the rates were not calculated.	Grade 2 congestive heart failure and grade 3 APN. The remaining AEs were related to intestinal events.	[96]
Ib	JAK1	INCB052793	NR	No significant responses were observed. ORR of 24% and OS of 6.7 months.	Thrombocytopenia, anemia, fatigue, nausea, and vomiting.	[97]
II	c-MET	Tivantinib	NR	In isolation, the drug did not present a satisfactory response in refractory patients. SD of 36% and PD of 63% were obtained.	Neutropenia, hypertension, syncope, infection, and pain.	[98]
II	MEK and AKT	Trametinib and Afuresertib	NR	MTDs were found, being concentrations that are below the monotherapy concentration of each drug. However, these doses were considered subtherapeutic.	Diarrhea, acneiform dermatitis, maculopapular rash, fatigue, dry skin, nausea, dyspnea, and vomiting.	[99]

Table 1. *Cont.*

Clinical Study Phase	Targeted Kinase	Kinase Inhibitor	Associated Treatment	Clinical Outcome	Adverse Events	References
II	VEGF	Sorafenib	NR	Only one patient completed the 13 cycles of treatment and achieved PR, another 7 patients remained in PD.	Fatigue, nausea, hypertension, dermal toxicity, hematologic toxicity, and heart attack.	[100]
II	FGFR3	Dovitinib	NR	The SD rate in t(4;14)-positive patients was higher, being 61.5%, compared with 34.6% rates for those translocation-negative	Diarrhea, nausea, vomiting, and fatigue.	[101]

Legend: NR: not reported; SD: stable disease; PR: partial response; ORR: overall response rate; PD: patients showed progression; OS: median overall survival; PFS: progression-free survival; CBR: clinical benefit rate; MTD: maximum tolerated dose.

Of the 11 articles described in Table 1, 36.4% (4) addressed the treatment with ibrutinib as a major option for MM. The other studies addressed treatments with other kinase inhibitors, such as Ruxolitinib, Cabozantinib, INCB052793, Tivantinib, Trametinib, Afuresertibe, Sorafenib, and Dovitinib. In total, 36.4% (4) of the articles described in the table are phase I clinical trials, while the other 63.6% (7) are phase II clinical trials [92–101].

4. Ibrutinib: A BTK Inhibition Approach

BTK inhibitors (BTKi) are one of the most popular and advanced approaches to targeting the BCR pathway. In addition to having revolutionized the treatment of B-lymphocyte malignancy, they also have a high level of efficacy in relation to chronic lymphocytic leukemia (CLL) patients, especially those with high-risk mutations [102,103]. Among these inhibitors, Ibrutinib stands out, already showing robust and durable efficacy in the treatment of refractory CLL and being one of the first inhibitors of the pathway to be approved by the FDA [104,105].

Ibrutinib is an FDA-approved drug for the treatment of B-cell malignancies [106], it works by irreversibly binding BTK through a covalent bond with a cysteine residue at position 481 (C481) [107]. It has demonstrated clinical responses mainly related to refractory CLL and mantle cell lymphoma (MCL), but it also has approval for use in cases of Waldenström's macroglobulinemia, small lymphocytic lymphoma, and marginal zone lymphoma. Chronic inhibition of BTK has been shown to be so effective in terms of its anticancer activity that its drugs are being widely tested in hematologic malignancies and solid malignancies [108,109]. Since overexpression of BTK is present in 85% of MM cases, ibrutinib appears as a promising therapy for MM patients, and the roles of BTK in the development of bone resorption by osteoclasts, as well as in cell migration, are characteristics that support BTK's research in the context of MM development [108,110–113].

BTK belongs to the Tec tyrosine kinase family that is involved in the B-cell antigen receptor (BCR) signaling pathway, being related to the survival, proliferation, and progression of malignancies in these cells. In MM, BTK is related to drug resistance, bone disease, and increased cell proliferation [114]. BTK activity, together with tyrosine phosphorylation, triggers the action of protein kinase B (AKT) which in turn will mediate transcription factors for proliferation, differentiation, and signaling cascades for survival—RAS/RAF/MEK/ERK and PI3K/AKT/mTOR (Figure 2) [110,114–117].

In the four studies presented in Table 1 in which ibrutinib was used, satisfactory results were reported, reaching conclusions where the clinical response to the use of ibrutinib encourages its use. This can also be seen in the statistical data of these articles, such as ORR above 60% [91,92], clinical response of 57% [93], and CBR of 28% [94]. The number of patients presented in the studies totaled 268, with an average age above 60 years. The preferred dose of ibrutinib was 840 mg, which is the limit dose for patients with MM, in addition to being a higher dose than that used in the treatment of other diseases such as lymphoma and chronic lymphocytic leukemia (CLL), which already have approved doses of 560 and 420 mg, respectively. It is worth mentioning that patients with CLL do not usually use doses of 840 mg due to the risk of discontinuation caused by adverse effects (AEs) [118].

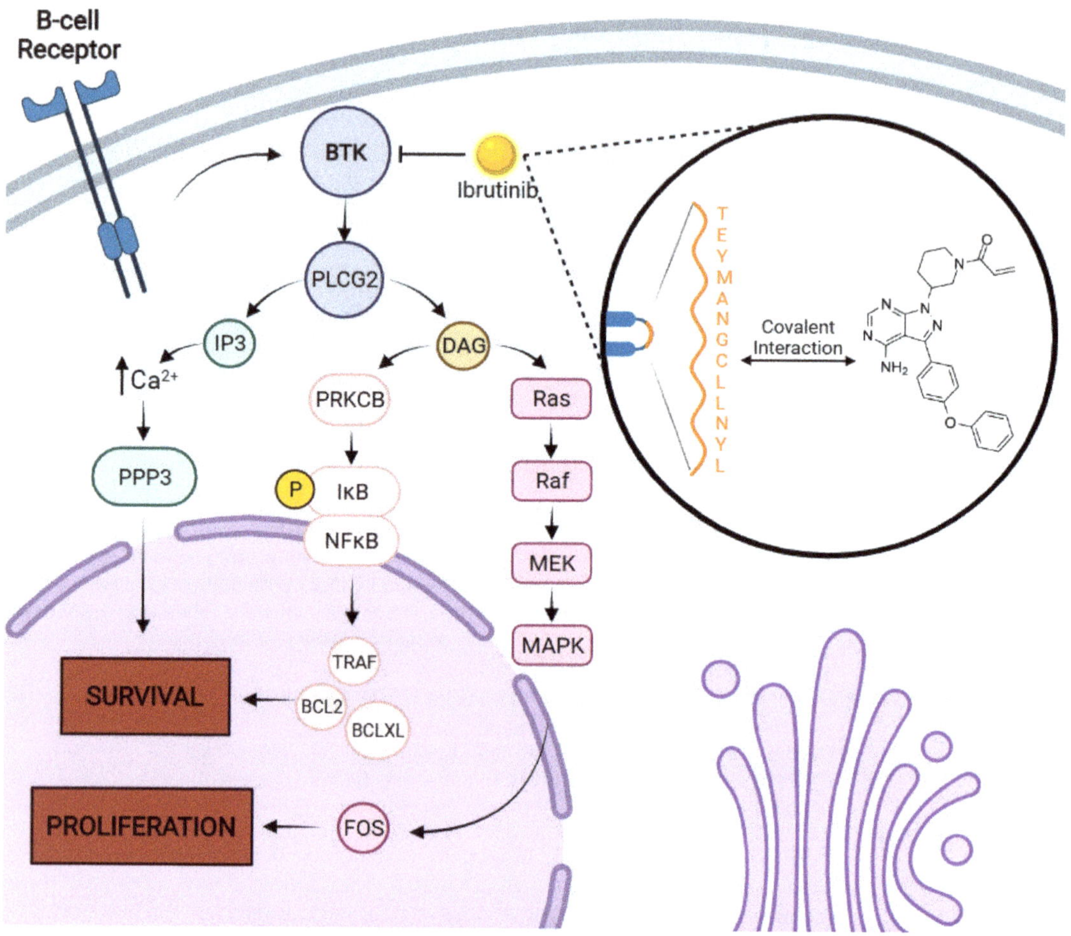

Figure 2. Bruton's tyrosine kinase (BTK) survival and proliferation pathways and interaction with ibrutinib. In multiple myeloma (MM), B-cell receptors of malignant plasma cells signal for BTK cascade to begin, which is highly overexpressed and induces survival and proliferation of neoplastic clones. Downstream activity happens mainly through phospholipases which catalyze formation of IP3 (inositol 1-4-5 trisphosphate) and DAG (diacylglycerol), being effectors of intracellular calcium signaling pathways, in the case of IP3, and NFκB and MAPK signaling pathways, in the case of DAG. Inhibition of BTK by ibrutinib, however, can drastically disrupt the metabolism of MM cells and happens through the covalent and irreversible interaction of ibrutinib with the cysteine residue on position 481 of the BTK active domain. Created with BioRender.com.

All four studies [91–94] utilized drugs which are standards for MM therapies in combination with ibrutinib, varying between carfilzomib and bortozomib, but the association of ibrutinib with dexamethasone was unanimous among all groups. The association of the BTK inhibitor with other drugs seeks to potentiate the therapeutic action and increase rates such as overall response rates (ORR) and progression-free survival (PFS) [94]. This association is also observed in cases of CLL, since BTKis in monotherapy are not enough to obtain profound responses and are therefore used in combination with other drugs to increase efficiency and not lose their effects due to resistance mechanisms [103,105]. In the study carried out by Richardson et al. [94], the combination of ibrutinib with dexamethasone demonstrated better results when compared to ibrutinib monotherapy. Chari et al. [91] observed that the combination of ibrutinib with carfilzomib and dexamethasone showed a promising response.

In studies using ibrutinib, among the AEs suffered by patients, the presence of anemia, diarrhea, and thrombocytopenia was constant [91–94]. AEs related to BTKi inhibitors are still being described, however, cardiac effects are the main concern due to the risk of combining these with hemorrhagic effects [108]. In in vitro studies, ibrutinib has already been demonstrated to interact with collagen-dependent platelet activation and von Willebrand factor, and it has been linked to an increased incidence of ventricular arrhythmia, hypertension, and neutropenia [110,111,118,119].

Dickerson et al. [54] observe that 78.3% of the patients who were using ibrutinib developed or worsened hypertension in an average period of 30 months. Although the causal relationship between ibrutinib and cardiotoxicity has not yet been fully elucidated, there are some theories, one of which would be the ability of ibrutinib to bind with other Tec kinases [120]. Despite cases of cardiotoxicity such as atrial fibrillation (AF), as long as the patient is benefiting from the therapy, they may continue with the treatment, with follow-up and medication to control the possible AEs. Next-generation BTK inhibitors, with less capacity to cause cardiotoxicity, are being studied and considered for MM treatment [120–122].

5. Clinical Perspectives with Other TKIs

In general, the use of tyrosine kinase inhibitors alone does not present a satisfactory response in patients with refractory MM [123]. The use of sorafenib alone in 7 out of 11 patients did not stop progression of MM, presenting a PFS of 2.6 months [100], while the isolated use of tivantinib was shown to be well tolerated, but in a clinical trial with 11 patients the agent was only able to stabilize 4 of 11 (36%) patients with progressive myeloma [124].

Other studies show that isolated administration of trametinib and afuresertib presented a significant clinical improvement in patients with MM, but when they are associated with other medications patients have even better responses [125,126]. Since trametinib is a drug also used to treat melanoma, its mechanism is based on MEK inhibition [127] and afuresertib is an AKT inhibitor that can also be used in the treatment of ovarian cancer [128]. However, the association between these two kinase inhibitors for the treatment of refractory MM patients performed in the study by Tolcher et al. [99] did not show such promising results, because despite having obtained maximum tolerated dose (MTD) values, these were compatible with subtherapeutic doses, thus, rending the maintenance of the drugs in clinically significant concentrations for a long period of time impossible due to the AEs.

In the study carried out by Scheid et al. [101], the activity of dovitinib, which acts on the fibroblast growth factor receptor 3 (*FGFR3*), was evaluated in MM patients. *FGFR3* is an RTK of the FGFR family that is responsible for cell growth, differentiation, and migration in a wide variety of cell types and is present in MM and a variety of cancers. Upon ligand stimulation, a dimerization of the receptor occurs followed by transphosphorylation of tyrosine residues in the intracellular domain signaling mainly through extracellular signal-regulated kinase (ERK) 1 and 2 pathways, PI3K and PLC [129–134].

Better results from dovitinib were observed in patients with MM with the presence of t (4;14). This translocation is associated with a worse prognosis, causing the overexpression of the *MMSET* (multiple myeloma SET domain protein) and *FGFR3* genes [135,136]. Thus, the interaction between genetic abnormalities can be further explored, bringing new insights into personalized therapy.

It was possible to perceive a wide variety of occurrences in the AEs manifested by the patients participating in the analyzed studies. In general, the most recurrent AEs were diarrhea, nausea, fatigue, dermal toxicity, anemia, leukopenia, thrombocytopenia, infections, hypertension, and congestive heart failure [95–100].

Due to many clinical difficulties in the treatment of refractory MM, it is important to point out that the search for newer effective therapies extends also to the promising use of TKIs, which demonstrated the positive ability to prevent the progression of the disease in several studies described here. However, it is worth emphasizing the importance of follow-up studies and clinical trials with different drug combinations to obtain better clinical results and avoiding AEs [137].

A new alternative to this would be other BTK inhibitors that are already approved by the FDA and even those in the study phase. Ibrutinib, despite its satisfactory results, presents the problem, already discussed, regarding the AE profile and its susceptibility to resistance pathways. However, since its mechanism of action demonstrates benefits, drugs that act in a similar way appear as a good proposal for future research and treatment. An example would be zanubrutinib and acalabrutinib, which are already approved by the FDA, and despite having irreversible links in their sites of action, are able to be more selective and show fewer problems related to platelet dysfunction and bleeding. Another proposal that has been showing good prospects is the non-covalent, reversible BTK inhibitors, that, among their advantages, have a greater selectivity for the site of action and are also effective in patients with resistance to ibrutinib [138–141].

6. Conclusions

In this review, we observed that among the TKIs tested in the last 10 years for the treatment of refractory MM, Ibrutinib was the most used and presented better clinical results, manly when administered in association with other drugs to avoid the emergence of resistance mechanisms that have already been found in other hematological neoplasms. Although AEs emerging from TKI's clinical administration is a major problem, when properly addressed and managed, treatment-emergent AEs are not considered serious, and the patient benefit versus risk ratio must be measured and taken into account individually from case to case.

Author Contributions: Invitation received, C.B.M. and C.A.M.-N.; conceptualization, I.V.B., C.B.M., F.M.C.d.P.P. and C.A.M.-N.; provision of data and sub-sequent analysis and interpretation, I.V.B., C.B.M., D.B.A., F.M.C.d.P.P., R.B.G., L.d.C.P., D.d.S.O., R.M.R., G.S.L., M.O.d.M.F., M.E.A.d.M., E.H.C.d.O. and A.S.K.; writing—original draft preparation, I.V.B., C.B.M., D.B.A., F.M.C.d.P.P., R.B.G. and C.A.M.-N.; writing—review and editing, I.V.B., C.B.M., D.B.A. and C.A.M.-N.; funding acquisition, A.S.K., E.H.C.d.O. and C.A.M.-N. All authors have read and agreed to the published version of the manuscript.

Funding: This study was supported by Brazilian funding agencies: Coordination for the Improvement of Higher Education Personnel (CAPES; to C.B.M), National Council of Technological and Scientific Development (CNPq grant number 404213/2021-9 to C.A.M.-N.; and Productivity in Research PQ scholarships to M.O.d.M.F., M.E.A.d.M., R.C.M., E.H.C.d.O. and A.S.K.), Cearense Foundation of Scientific and Technological Support (FUNCAP grant number P20-0171-00078.01.00/20 to F.M.C.d.P.P., M.O.d.M.F. and to C.A.M.-N.) and we also thank PROPESP/UFPA for publication payment.

Conflicts of Interest: The authors declare no conflict of interest. The funders had no role in the design of the study; in the collection, analyses, or data interpretation; in the writing of the manuscript, or in the decision to publish the results.

References

1. Tsang, M.; Le, M.; Ghazawi, F.M.; Cyr, J.; Alakel, A.; Rahme, E.; Lagacé, F.; Netchiporouk, E.; Moreau, L.; Zubarev, A.; et al. Multiple Myeloma Epidemiology and Patient Geographic Distribution in Canada: A Population Study. *Cancer* **2019**, *125*, 2435–2444. [CrossRef] [PubMed]
2. Padala, S.A.; Barsouk, A.; Barsouk, A.; Rawla, P.; Vakiti, A.; Kolhe, R.; Kota, V.; Ajebo, G.H. Epidemiology, Staging, and Management of Multiple Myeloma. *Med. Sci.* **2021**, *9*, 3. [CrossRef] [PubMed]
3. Bird, S.; Cairns, D.; Menzies, T.; Boyd, K.; Davies, F.; Cook, G.; Drayson, M.; Gregory, W.; Jenner, M.; Jones, J.; et al. Sex Differences in Multiple Myeloma Biology but Not Clinical Outcomes: Results from 3894 Patients in the Myeloma XI Trial. *Clin. Lymphoma Myeloma Leuk.* **2021**, *21*, 667–675. [CrossRef] [PubMed]

4. Marinac, C.R.; Ghobrial, I.M.; Birmann, B.M.; Soiffer, J.; Rebbeck, T.R. Dissecting Racial Disparities in Multiple Myeloma. *Blood Cancer J.* **2020**, *10*, 19. [CrossRef]
5. Brigle, K.; Rogers, B. Pathobiology and Diagnosis of Multiple Myeloma. *Semin. Oncol. Nurs.* **2017**, *33*, 225–236. [CrossRef]
6. Cowan, A.J.; Green, D.J.; Kwok, M.; Lee, S.; Coffey, D.G.; Holmberg, L.A.; Tuazon, S.; Gopal, A.K.; Libby, E.N. Diagnosis and Management of Multiple Myeloma: A Review. *JAMA J. Am. Med. Assoc.* **2022**, *327*, 464–477. [CrossRef]
7. Firth, J. Haematology: Multiple Myeloma. *Clin. Med.* **2019**, *19*, 58–60. [CrossRef]
8. Joshi, H.; Lin, S.; Fei, K.; Renteria, A.S.; Jacobs, H.; Mazumdar, M.; Jagannath, S.; Bickell, N.A. Multiple Myeloma, Race, Insurance and Treatment. *Cancer Epidemiol.* **2021**, *73*, 101974. [CrossRef]
9. Rajkumar, S.V. Multiple Myeloma: 2022 Update on Diagnosis, Risk Stratification, and Management. *Am. J. Hematol.* **2022**, *97*, 1086–1107. [CrossRef]
10. Landgren, O.; Weiss, B.M. Patterns of Monoclonal Gammopathy of Undetermined Significance and Multiple Myeloma in Various Ethnic/Racial Groups: Support for Genetic Factors in Pathogenesis. *Leukemia* **2009**, *23*, 1691–1697. [CrossRef]
11. Bunce, C.M.; Drayson, M.T. Dissecting Racial Disparities in Multiple Myeloma—Clues from Differential Immunoglobulin Levels. *Blood Cancer J.* **2020**, *10*, 44. [CrossRef] [PubMed]
12. Joshua, D.E.; Bryant, C.; Dix, C.; Gibson, J.; Ho, J. Biology and Therapy of Multiple Myeloma. *Med. J. Aust.* **2019**, *210*, 375–380. [CrossRef] [PubMed]
13. Huang, R.; Zhou, P.K. DNA Damage Repair: Historical Perspectives, Mechanistic Pathways and Clinical Translation for Targeted Cancer Therapy. *Signal Transduct. Target. Ther.* **2021**, *6*, 254. [CrossRef] [PubMed]
14. Castaneda, O.; Baz, R. Multiple Myeloma Genomics—A Concise Review. *Acta Med. Acad.* **2019**, *48*, 57–67. [CrossRef] [PubMed]
15. Wallington-Beddoe, C.T.; Mynott, R.L. Prognostic and Predictive Biomarker Developments in Multiple Myeloma. *J. Hematol. Oncol.* **2021**, *14*, 151. [CrossRef] [PubMed]
16. Kumar, S.K.; Rajkumar, V.; Kyle, R.A.; van Duin, M.; Sonneveld, P.; Mateos, M.-V.; Gay, F.; Anderson, K.C. Multiple Myeloma. *Nat. Rev. Dis. Primers* **2017**, *3*, 17046. [CrossRef]
17. Manier, S.; Salem, K.Z.; Park, J.; Landau, D.A.; Getz, G.; Ghobrial, I.M. Genomic Complexity of Multiple Myeloma and Its Clinical Implications. *Nat. Rev. Clin. Oncol.* **2017**, *14*, 100–113. [CrossRef]
18. Zingone, A.; Kuehl, W.M. Pathogenesis of Monoclonal Gammopathy of Undetermined Significance and Progression to Multiple Myeloma. *Semin. Hematol.* **2011**, *48*, 4–12. [CrossRef]
19. Furukawa, Y.; Kikuchi, J. Molecular Basis of Clonal Evolution in Multiple Myeloma. *Int. J. Hematol.* **2020**, *111*, 496–511. [CrossRef]
20. Barwick, B.G.; Gupta, V.A.; Vertino, P.M.; Boise, L.H. Cell of Origin and Genetic Alterations in the Pathogenesis of Multiple Myeloma. *Front. Immunol.* **2019**, *10*, 1121. [CrossRef]
21. Maura, F.; Rustad, E.H.; Boyle, E.M.; Morgan, G.J. Reconstructing the Evolutionary History of Multiple Myeloma. *Best Pr. Res. Clin. Haematol.* **2020**, *33*, 101145. [CrossRef] [PubMed]
22. Dispenzieri, A.; Katzmann, J.A.; Kyle, R.A.; Larson, D.R.; Melton, L.J.; Colby, C.L.; Therneau, T.M.; Clark, R.; Kumar, S.K.; Bradwell, A.; et al. Prevalence and Risk of Progression of Light-Chain Monoclonal Gammopathy of Undetermined Significance: A Retrospective Population-Based Cohort Study. *Lancet* **2010**, *375*, 1721–1728. [CrossRef]
23. Kyle, R.A.; Therneau, T.M.; Rajkumar, S.V.; Larson, D.R.; Plevak, M.F.; Offord, J.R.; Dispenzieri, A.; Katzmann, J.A.; Melton, L.J. Prevalence of Monoclonal Gammopathy of Undetermined Significance. *N. Engl. J. Med.* **2006**, *354*, 1362–1369. [CrossRef] [PubMed]
24. Kyle, R.A.; Therneau, T.M.; Rajkumar, S.V.; Offord, J.R.; Larson, D.R.; Plevak, M.F.; Melton, L.J. A Long-Term Study of Prognosis in Monoclonal Gammopathy of Undetermined Significance. *N. Engl. J. Med.* **2002**, *346*, 564–569. [CrossRef]
25. Greenberg, A.J.; Vachon, C.M.; Rajkumar, S. v Disparities in the Prevalence, Pathogenesis and Progression of Monoclonal Gammopathy of Undetermined Significance and Multiple Myeloma between Blacks and Whites. *Leukemia* **2012**, *26*, 609–614. [CrossRef]
26. Lakshman, A.; Paul, S.; Rajkumar, S.V.; Ketterling, R.P.; Greipp, P.T.; Dispenzieri, A.; Gertz, M.A.; Buadi, F.K.; Lacy, M.Q.; Dingli, D.; et al. Prognostic Significance of Interphase FISH in Monoclonal Gammopathy of Undetermined Significance. *Leukemia* **2018**, *32*, 1811–1815. [CrossRef]
27. Neben, K.; Jauch, A.; Hielscher, T.; Hillengass, J.; Lehners, N.; Seckinger, A.; Granzow, M.; Raab, M.S.; Ho, A.D.; Goldschmidt, H.; et al. Progression in Smoldering Myeloma Is Independently Determined by the Chromosomal Abnormalities Del(17p), t(4;14), Gain 1q, Hyperdiploidy, and Tumor Load. *J. Clin. Oncol.* **2013**, *31*, 4325–4332. [CrossRef]
28. Rajkumar, S.V.; Gupta, V.; Fonseca, R.; Dispenzieri, A.; Gonsalves, W.I.; Larson, D.; Ketterling, R.P.; Lust, J.A.; Kyle, R.A.; Kumar, S.K. Impact of Primary Molecular Cytogenetic Abnormalities and Risk of Progression in Smoldering Multiple Myeloma. *Leukemia* **2013**, *27*, 1738–1744. [CrossRef]
29. Avet-Loiseau, H.; Attal, M.; Moreau, P.; Charbonnel, C.; Garban, F.; Hulin, C.; Leyvraz, S.; Michallet, M.; Yakoub-Agha, I.; Garderet, L.; et al. Genetic Abnormalities and Survival in Multiple Myeloma: The Experience of the Intergroupe Francophone Du Myélome. *Blood* **2007**, *109*, 3489–3495. [CrossRef]
30. Chretien, M.-L.; Corre, J.; Lauwers-Cances, V.; Magrangeas, F.; Cleynen, A.; Yon, E.; Hulin, C.; Leleu, X.; Orsini-Piocelle, F.; Blade, J.-S.; et al. Understanding the Role of Hyperdiploidy in Myeloma Prognosis: Which Trisomies Really Matter? *Blood* **2015**, *126*, 2713–2719. [CrossRef]

31. Willenbacher, W.; Seeber, A.; Steiner, N.; Willenbacher, E.; Gatalica, Z.; Swensen, J.; Kimbrough, J.; Vranic, S. Towards Molecular Profiling in Multiple Myeloma: A Literature Review and Early Indications of Its Efficacy for Informing Treatment Strategies. *Int. J. Mol. Sci.* **2018**, *19*, 2087. [CrossRef] [PubMed]
32. Ziogas, D.C.; Dimopoulos, M.A.; Kastritis, E. Prognostic Factors for Multiple Myeloma in the Era of Novel Therapies. *Expert Rev. Hematol.* **2018**, *11*, 863–879. [CrossRef] [PubMed]
33. Pawlyn, C.; Davies, F.E. Toward Personalized Treatment in Multiple Myeloma Based on Molecular Characteristics. *Blood* **2019**, *133*, 660–675. [CrossRef]
34. Hanamura, I. [Advances in Multiple Myeloma Molecular Biology Research]. *[Rinsho Ketsueki] Jpn. J. Clin. Hematol.* **2019**, *60*, 1236–1242.
35. Rajkumar, S.V. Multiple Myeloma: 2020 Update on Diagnosis, Risk-stratification and Management. *Am. J. Hematol.* **2020**, *95*, 548–567. [CrossRef] [PubMed]
36. Usmani, S.Z.; Hoering, A.; Cavo, M.; Miguel, J.S.; Goldschimdt, H.; Hajek, R.; Turesson, I.; Lahuerta, J.J.; Attal, M.; Barlogie, B.; et al. Clinical Predictors of Long-Term Survival in Newly Diagnosed Transplant Eligible Multiple Myeloma—An IMWG Research Project. *Blood Cancer J.* **2018**, *8*, 123. [CrossRef] [PubMed]
37. Child, J.A.; Morgan, G.J.; Davies, F.E.; Owen, R.G.; Bell, S.E.; Hawkins, K.; Brown, J.; Drayson, M.T.; Selby, P.J. High-Dose Chemotherapy with Hematopoietic Stem-Cell Rescue for Multiple Myeloma. *N. Engl. J. Med.* **2003**, *348*, 1875–1883. [CrossRef]
38. Rajkumar, S.V. Multiple Myeloma: Every Year a New Standard? *Hematol. Oncol.* **2019**, *37*, 62–65. [CrossRef]
39. Gonsalves, W.I.; Buadi, F.K.; Ailawadhi, S.; Bergsagel, P.L.; Chanan Khan, A.A.; Dingli, D.; Dispenzieri, A.; Fonseca, R.; Hayman, S.R.; Kapoor, P.; et al. Utilization of Hematopoietic Stem Cell Transplantation for the Treatment of Multiple Myeloma: A Mayo Stratification of Myeloma and Risk-Adapted Therapy (MSMART) Consensus Statement. *Bone Marrow Transplant.* **2019**, *54*, 353–367. [CrossRef]
40. Saad, A.; Mahindra, A.; Zhang, M.-J.; Zhong, X.; Costa, L.J.; Dispenzieri, A.; Drobyski, W.R.; Freytes, C.O.; Gale, R.P.; Gasparetto, C.J.; et al. Hematopoietic Cell Transplant Comorbidity Index Is Predictive of Survival after Autologous Hematopoietic Cell Transplantation in Multiple Myeloma. *Biol. Blood Marrow Transplant.* **2014**, *20*, 402–408.e1. [CrossRef]
41. Rajkumar, S.V.; Sonneveld, P. Front-Line Treatment in Younger Patients with Multiple Myeloma. *Semin. Hematol.* **2009**, *46*, 118–126. [CrossRef]
42. Hus, I.; Mańko, J.; Jawniak, D.; Jurczyszyn, A.; Charliński, G.; Poniewierska-Jasak, K.; Usnarska-Zubkiewicz, L.; Sawicki, M.; Druzd-Sitek, A.; Świderska, A.; et al. High Efficacy and Safety of VTD as an Induction Protocol in Patients with Newly Diagnosed Multiple Myeloma Eligible for High Dose Therapy and Autologous Stem Cell Transplantation: A Report of the Polish Myeloma Study Group. *Oncol. Lett.* **2019**, *18*, 5811–5820. [CrossRef]
43. Attal, M.; Lauwers-Cances, V.; Hulin, C.; Leleu, X.; Caillot, D.; Escoffre, M.; Arnulf, B.; Macro, M.; Belhadj, K.; Garderet, L.; et al. Lenalidomide, Bortezomib, and Dexamethasone with Transplantation for Myeloma. *N. Engl. J. Med.* **2017**, *376*, 1311–1320. [CrossRef] [PubMed]
44. al Hamed, R.; Bazarbachi, A.H.; Malard, F.; Harousseau, J.-L.; Mohty, M. Current Status of Autologous Stem Cell Transplantation for Multiple Myeloma. *Blood Cancer J.* **2019**, *9*, 44. [CrossRef] [PubMed]
45. Cavo, M.; Tacchetti, P.; Patriarca, F.; Petrucci, M.T.; Pantani, L.; Galli, M.; di Raimondo, F.; Crippa, C.; Zamagni, E.; Palumbo, A.; et al. Bortezomib with Thalidomide plus Dexamethasone Compared with Thalidomide plus Dexamethasone as Induction Therapy before, and Consolidation Therapy after, Double Autologous Stem-Cell Transplantation in Newly Diagnosed Multiple Myeloma: A Randomised Phase 3 Study. *Lancet* **2010**, *376*, 2075–2085. [CrossRef]
46. Ghandili, S.; Weisel, K.C.; Bokemeyer, C.; Leypoldt, L.B. Current Treatment Approaches to Newly Diagnosed Multiple Myeloma. *Oncol. Res. Treat.* **2021**, *44*, 690–699. [CrossRef] [PubMed]
47. Attal, M.; Harousseau, J.-L.; Stoppa, A.-M.; Sotto, J.-J.; Fuzibet, J.-G.; Rossi, J.-F.; Casassus, P.; Maisonneuve, H.; Facon, T.; Ifrah, N.; et al. A Prospective, Randomized Trial of Autologous Bone Marrow Transplantation and Chemotherapy in Multiple Myeloma. *N. Engl. J. Med.* **1996**, *335*, 91–97. [CrossRef] [PubMed]
48. Ito, S. Proteasome Inhibitors for the Treatment of Multiple Myeloma. *Cancers* **2020**, *12*, 265. [CrossRef]
49. Hideshima, T.; Mitsiades, C.; Akiyama, M.; Hayashi, T.; Chauhan, D.; Richardson, P.; Schlossman, R.; Podar, K.; Munshi, N.C.; Mitsiades, N.; et al. Molecular Mechanisms Mediating Antimyeloma Activity of Proteasome Inhibitor PS-341. *Blood* **2003**, *101*, 1530–1534. [CrossRef]
50. Besse, A.; Besse, L.; Kraus, M.; Mendez-Lopez, M.; Bader, J.; Xin, B.T.; de Bruin, G.; Maurits, E.; Overkleeft, H.S.; Driessen, C. Proteasome Inhibition in Multiple Myeloma: Head-to-Head Comparison of Currently Available Proteasome Inhibitors. *Cell Chem. Biol.* **2019**, *26*, 340–351.e3. [CrossRef]
51. Niewerth, D.; Jansen, G.; Assaraf, Y.G.; Zweegman, S.; Kaspers, G.J.L.; Cloos, J. Molecular Basis of Resistance to Proteasome Inhibitors in Hematological Malignancies. *Drug Resist. Updates* **2015**, *18*, 18–35. [CrossRef] [PubMed]
52. Adams, J. The Proteasome: A Suitable Antineoplastic Target. *Nat. Rev. Cancer* **2004**, *4*, 349–360. [CrossRef] [PubMed]
53. Nunes, A.T.; Annunziata, C.M. Proteasome Inhibitors: Structure and Function. *Semin. Oncol.* **2017**, *44*, 377–380. [CrossRef] [PubMed]
54. Gandolfi, S.; Laubach, J.P.; Hideshima, T.; Chauhan, D.; Anderson, K.C.; Richardson, P.G. The Proteasome and Proteasome Inhibitors in Multiple Myeloma. *Cancer Metastasis Rev.* **2017**, *36*, 561–584. [CrossRef]

55. Groen, K.; van de Donk, N.W.C.J.; Stege, C.A.M.; Zweegman, S.; Nijhof, I.S. Carfilzomib for Relapsed and Refractory Multiple Myeloma. *Cancer Manag. Res.* **2019**, *11*, 2663–2675. [CrossRef]
56. Tan, C.R.C.; Abdul-Majeed, S.; Cael, B.; Barta, S.K. Clinical Pharmacokinetics and Pharmacodynamics of Bortezomib. *Clin. Pharmacokinet.* **2018**, *58*, 157–168. [CrossRef]
57. Zinatizadeh, M.R.; Schock, B.; Chalbatani, G.M.; Zarandi, P.K.; Jalali, S.A.; Miri, S.R. The Nuclear Factor Kappa B (NF-KB) Signaling in Cancer Development and Immune Diseases. *Genes Dis.* **2021**, *8*, 287–297. [CrossRef]
58. Chauhan, D.; Bianchi, G.; Anderson, K.C. Targeting the UPS as Therapy in Multiple Myeloma. *BMC Biochem.* **2008**, *9*, S1. [CrossRef]
59. Richardson, P.G.; Sonneveld, P.; Schuster, M.W.; Irwin, D.; Stadtmauer, E.A.; Facon, T.; Harousseau, J.-L.; Ben-Yehuda, D.; Lonial, S.; Goldschmidt, H.; et al. Bortezomib or High-Dose Dexamethasone for Relapsed Multiple Myeloma. *N. Engl. J. Med.* **2005**, *352*, 2487–2498. [CrossRef]
60. Richardson, P.G.; Zweegman, S.; O'Donnell, E.K.; Laubach, J.P.; Raje, N.; Voorhees, P.; Ferrari, R.H.; Skacel, T.; Kumar, S.K.; Lonial, S. Ixazomib for the Treatment of Multiple Myeloma. *Expert Opin. Pharmacother.* **2018**, *19*, 1949–1968. [CrossRef]
61. Touzeau, C.; Moreau, P. Ixazomib in the Management of Relapsed Multiple Myeloma. *Future Oncol.* **2018**, *14*, 2013–2020. [CrossRef]
62. Zanwar, S.; Abeykoon, J.P.; Kapoor, P. Ixazomib: A Novel Drug for Multiple Myeloma. *Expert Rev. Hematol.* **2018**, *11*, 761–771. [CrossRef] [PubMed]
63. Lopez-Girona, A.; Heintel, D.; Zhang, L.-H.; Mendy, D.; Gaidarova, S.; Brady, H.; Bartlett, J.B.; Schafer, P.H.; Schreder, M.; Bolomsky, A.; et al. Lenalidomide Downregulates the Cell Survival Factor, Interferon Regulatory Factor-4, Providing a Potential Mechanistic Link for Predicting Response. *Br. J. Haematol.* **2011**, *154*, 325–336. [CrossRef] [PubMed]
64. Martinez-Høyer, S.; Karsan, A. Mechanisms of Lenalidomide Sensitivity and Resistance. *Exp. Hematol.* **2020**, *91*, 22–31. [CrossRef] [PubMed]
65. Agnarelli, A.; Chevassut, T.; Mancini, E.J. IRF4 in Multiple Myeloma—Biology, Disease and Therapeutic Target. *Leuk. Res.* **2018**, *72*, 52–58. [CrossRef] [PubMed]
66. Duffy, M.J.; O'Grady, S.; Tang, M.; Crown, J. MYC as a Target for Cancer Treatment. *Cancer Treat. Rev.* **2021**, *94*, 102154. [CrossRef] [PubMed]
67. Sampaio, E.P.; Sarno, E.N.; Galilly, R.; Cohn, Z.A.; Kaplan, G. Thalidomide Selectively Inhibits Tumor Necrosis Factor Alpha Production by Stimulated Human Monocytes. *J. Exp. Med.* **1991**, *173*, 699–703. [CrossRef]
68. Davies, F.E.; Raje, N.; Hideshima, T.; Lentzsch, S.; Young, G.; Tai, Y.-T.; Lin, B.; Podar, K.; Gupta, D.; Chauhan, D.; et al. Thalidomide and Immunomodulatory Derivatives Augment Natural Killer Cell Cytotoxicity in Multiple Myeloma. *Blood* **2001**, *98*, 210–216. [CrossRef]
69. Abe, Y.; Ishida, T. Immunomodulatory Drugs in the Treatment of Multiple Myeloma. *Jpn. J. Clin. Oncol.* **2019**, *49*, 695–702. [CrossRef]
70. Goldsmith, S.R.; Foley, N.; Schroeder, M.A. Daratumumab for the Treatment of Multiple Myeloma. *Drugs Today* **2021**, *57*, 591. [CrossRef]
71. Hosoya, H.; Sidana, S. Antibody-Based Treatment Approaches in Multiple Myeloma. *Curr. Hematol. Malign. Rep.* **2021**, *16*, 183–191. [CrossRef] [PubMed]
72. Nooka, A.K.; Kaufman, J.L.; Hofmeister, C.C.; Joseph, N.S.; Heffner, T.L.; Gupta, V.A.; Sullivan, H.C.; Neish, A.S.; Dhodapkar, M.V.; Lonial, S. Daratumumab in Multiple Myeloma. *Cancer* **2019**, *125*, 2364–2382. [CrossRef] [PubMed]
73. Kastritis, E.; Palladini, G.; Minnema, M.C.; Wechalekar, A.D.; Jaccard, A.; Lee, H.C.; Sanchorawala, V.; Gibbs, S.; Mollee, P.; Venner, C.P.; et al. Daratumumab-Based Treatment for Immunoglobulin Light-Chain Amyloidosis. *N. Engl. J. Med.* **2021**, *385*, 46–58. [CrossRef] [PubMed]
74. Petrucci, M.T.; Vozella, F. The Anti-CD38 Antibody Therapy in Multiple Myeloma. *Cells* **2019**, *8*, 1629. [CrossRef]
75. Vozella, F.; Fazio, F.; Lapietra, G.; Petrucci, M.T.; Martinelli, G.; Cerchione, C. Monoclonal Antibodies in Multiple Myeloma. *Panminerva Med.* **2021**, *63*, 21–27. [CrossRef] [PubMed]
76. D'Agostino, M.; Boccadoro, M.; Smith, E.L. Novel Immunotherapies for Multiple Myeloma. *Curr. Hematol. Malig. Rep.* **2017**, *12*, 344–357. [CrossRef]
77. Minnie, S.A.; Hill, G.R. Immunotherapy of Multiple Myeloma. *J. Clin. Investig.* **2020**, *130*, 1565–1575. [CrossRef]
78. Tan, C.R.; Shah, U.A. Targeting BCMA in Multiple Myeloma. *Curr. Hematol. Malig. Rep.* **2021**, *16*, 367–383. [CrossRef]
79. Mateos, M.-V.; Masszi, T.; Grzasko, N.; Hansson, M.; Sandhu, I.; Pour, L.; Viterbo, L.; Jackson, S.R.; Stoppa, A.-M.; Gimsing, P.; et al. Impact of Prior Therapy on the Efficacy and Safety of Oral Ixazomib-Lenalidomide-Dexamethasone vs. Placebo-Lenalidomide-Dexamethasone in Patients with Relapsed/Refractory Multiple Myeloma in TOURMALINE-MM1. *Haematologica* **2017**, *102*, 1767–1775. [CrossRef]
80. Venezian Povoa, L.; Ribeiro, C.H.C.; Silva, I.T. da Machine Learning Predicts Treatment Sensitivity in Multiple Myeloma Based on Molecular and Clinical Information Coupled with Drug Response. *PLoS ONE* **2021**, *16*, e0254596. [CrossRef]
81. Lemmon, M.A.; Schlessinger, J. Cell Signaling by Receptor Tyrosine Kinases. *Cell* **2010**, *141*, 1117–1134. [CrossRef] [PubMed]
82. Esteban-Villarrubia, J.; Soto-Castillo, J.J.; Pozas, J.; Román-Gil, M.S.; Orejana-Martín, I.; Torres-Jiménez, J.; Carrato, A.; Alonso-Gordoa, T.; Molina-Cerrillo, J. Tyrosine Kinase Receptors in Oncology. *Int. J. Mol. Sci.* **2020**, *21*, 8529. [CrossRef] [PubMed]

83. Tibes, R.; Trent, J.; Kurzrock, R. Tyrosine Kinase Inhibitors and the Dawn of Molecular Cancer Therapeutics. *Annu. Rev. Pharmacol. Toxicol.* **2005**, *45*, 357–384. [CrossRef] [PubMed]
84. Krause, D.S.; van Etten, R.A. Tyrosine Kinases as Targets for Cancer Therapy. *N. Engl. J. Med.* **2005**, *353*, 172–187. [CrossRef] [PubMed]
85. Wang, Z.; Cole, P.A. Catalytic Mechanisms and Regulation of Protein Kinases. In *Methods in Enzymology*; Elsevier: Amsterdam, The Netherlands, 2014; Volume 548.
86. Regad, T. Targeting RTK Signaling Pathways in Cancer. *Cancers* **2015**, *7*, 1758–1784. [CrossRef]
87. Lind, J.; Czernilofsky, F.; Vallet, S.; Podar, K. Emerging Protein Kinase Inhibitors for the Treatment of Multiple Myeloma. *Expert Opin. Emerg. Drugs* **2019**, *24*, 133–152. [CrossRef]
88. Knösel, T.; Kampmann, E.; Kirchner, T.; Altendorf-Hofmann, A. Tyrosinkinasen in Weichgewebstumoren. *Pathologe* **2014**, *35*, 198–201. [CrossRef]
89. Drake, J.M.; Lee, J.K.; Witte, O.N. Clinical Targeting of Mutated and Wild-Type Protein Tyrosine Kinases in Cancer. *Mol. Cell. Biol.* **2014**, *34*, 1722–1732. [CrossRef]
90. Jiao, Q.; Bi, L.; Ren, Y.; Song, S.; Wang, Q.; Wang, Y.S. Advances in Studies of Tyrosine Kinase Inhibitors and Their Acquired Resistance. *Mol. Cancer* **2018**, *17*, 1–12. [CrossRef]
91. Chari, A.; Larson, S.; Holkova, B.; Cornell, R.F.; Gasparetto, C.; Karanes, C.; Matous, J.V.; Niesvizky, R.; Valent, J.; Lunning, M.; et al. Phase 1 Trial of Ibrutinib and Carfilzomib Combination Therapy for Relapsed or Relapsed and Refractory Multiple Myeloma. *Leuk. Lymphoma* **2018**, *59*, 2588–2594. [CrossRef]
92. Chari, A.; Cornell, R.F.; Gasparetto, C.; Karanes, C.; Matous, J.V.; Niesvizky, R.; Lunning, M.; Usmani, S.Z.; Anderson, L.D.; Chhabra, S.; et al. Final Analysis of a Phase 1/2b Study of Ibrutinib Combined with Carfilzomib/Dexamethasone in Patients with Relapsed/Refractory Multiple Myeloma. *Hematol. Oncol.* **2020**, *38*, 353–362. [CrossRef]
93. Hajek, R.; Pour, L.; Ozcan, M.; Martin Sánchez, J.; García Sanz, R.; Anagnostopoulos, A.; Oriol, A.; Cascavilla, N.; Terjung, A.; Lee, Y.; et al. A Phase 2 Study of Ibrutinib in Combination with Bortezomib and Dexamethasone in Patients with Relapsed/Refractory Multiple Myeloma. *Eur. J. Haematol.* **2020**, *104*, 435–442. [CrossRef]
94. Richardson, P.G.; Bensinger, W.I.; Huff, C.A.; Costello, C.L.; Lendvai, N.; Berdeja, J.G.; Anderson, L.D.; Siegel, D.S.; Lebovic, D.; Jagannath, S.; et al. Ibrutinib Alone or with Dexamethasone for Relapsed or Relapsed and Refractory Multiple Myeloma: Phase 2 Trial Results. *Br. J. Haematol.* **2018**, *180*, 821–830. [CrossRef]
95. Berenson, J.R.; To, J.; Spektor, T.M.; Martinez, D.; Turner, C.; Sanchez, A.; Ghermezi, M.; Eades, B.M.; Swift, R.A.; Schwartz, G.; et al. A Phase I Study of Ruxolitinib, Lenalidomide, and Steroids for Patients with Relapsed/Refractory Multiple Myeloma. *Clin. Cancer Res.* **2020**, *26*, 2346–2353. [CrossRef] [PubMed]
96. Lendvai, N.; Yee, A.J.; Tsakos, I.; Alexander, A.; Devlin, S.M.; Hassoun, H.; Korde, N.; Lesokhin, A.M.; Landau, H.; Mailankody, S.; et al. Phase IB Study of Cabozantinib in Patients with Relapsed and/or Refractory Multiple Myeloma. *Blood* **2016**, *127*, 2355–2356. [CrossRef] [PubMed]
97. Zeidan, A.M.; Cook, R.J.; Bordoni, R.; Berenson, J.R.; Edenfield, W.J.; Mohan, S.; Zhou, G.; Asatiani, E.; Srinivas, N.; Savona, M.R. A Phase 1/2 Study of the Oral Janus Kinase 1 Inhibitors INCB052793 and Itacitinib Alone or in Combination with Standard Therapies for Advanced Hematologic Malignancies. *Clin. Lymphoma Myeloma Leuk.* **2022**, *22*, 523–534. [CrossRef] [PubMed]
98. Baljevic, M.; Zaman, S.; Baladandayuthapani, V.; Lin, Y.H.; de Partovi, C.M.; Berkova, Z.; Amini, B.; Thomas, S.K.; Shah, J.J.; Weber, D.M.; et al. Phase II Study of the C-MET Inhibitor Tivantinib (ARQ 197) in Patients with Relapsed or Relapsed/Refractory Multiple Myeloma. *Ann. Hematol.* **2017**, *96*, 977–985. [CrossRef]
99. Tolcher, A.W.; Patnaik, A.; Papadopoulos, K.P.; Rasco, D.W.; Becerra, C.R.; Allred, A.J.; Orford, K.; Aktan, G.; Ferron-Brady, G.; Ibrahim, N.; et al. Phase I Study of the MEK Inhibitor Trametinib in Combination with the AKT Inhibitor Afuresertib in Patients with Solid Tumors and Multiple Myeloma. *Cancer Chemother. Pharmacol.* **2015**, *75*, 183–189. [CrossRef]
100. Yordanova, A.; Hose, D.; Neben, K.; Witzens-Harig, M.; Gütgemann, I.; Raab, M.S.; Moehler, T.; Goldschmidt, H.; Schmidt-Wolf, I.G. Sorafenib in Patients with Refractory or Recurrent Multiple Myeloma. *Hematol. Oncol.* **2013**, *31*, 197–200. [CrossRef]
101. Scheid, C.; Reece, D.; Beksac, M.; Spencer, A.; Callander, N.; Sonneveld, P.; Kalimi, G.; Cai, C.; Shi, M.; Scott, J.W.; et al. Phase 2 Study of Dovitinib in Patients with Relapsed or Refractory Multiple Myeloma with or without t(4;14) Translocation. *Eur. J. Haematol.* **2015**, *95*, 316–324. [CrossRef]
102. Estupiñán, H.Y.; Bouderlique, T.; He, C.; Berglöf, A.; Gupta, D.; Saher, O.; Daza Cruz, M.Á.; Peña-Perez, L.; Yu, L.; Zain, R.; et al. Novel Mouse Model Resistant to Irreversible BTK Inhibitors: A Tool Identifying New Therapeutic Targets and Side Effects. *Blood Adv.* **2020**, *4*, 2439–2450. [CrossRef] [PubMed]
103. Ahn, I.E.; Brown, J.R. Targeting Bruton's Tyrosine Kinase in CLL. *Front. Immunol.* **2021**, *12*, 687458. [CrossRef] [PubMed]
104. Munir, T.; Brown, J.R.; O'Brien, S.; Barrientos, J.C.; Barr, P.M.; Reddy, N.M.; Coutre, S.; Tam, C.S.; Mulligan, S.P.; Jaeger, U.; et al. Final Analysis from RESONATE: Up to Six Years of Follow-up on Ibrutinib in Patients with Previously Treated Chronic Lymphocytic Leukemia or Small Lymphocytic Lymphoma. *Am. J. Hematol.* **2019**, *94*, 1353–1363. [CrossRef]
105. Lipsky, A.; Lamanna, N. Managing Toxicities of Bruton Tyrosine Kinase Inhibitors. *Hematology* **2020**, *2020*, 336–345. [CrossRef] [PubMed]
106. Pan, Z.; Scheerens, H.; Li, S.-J.; Schultz, B.E.; Sprengeler, P.A.; Burrill, L.C.; Mendonca, R.V.; Sweeney, M.D.; Scott, K.C.K.; Grothaus, P.G.; et al. Discovery of Selective Irreversible Inhibitors for Bruton's Tyrosine Kinase. *ChemMedChem* **2007**, *2*, 58–61. [CrossRef]

107. Bender, A.T.; Gardberg, A.; Pereira, A.; Johnson, T.; Wu, Y.; Grenningloh, R.; Head, J.; Morandi, F.; Haselmayer, P.; Liu-Bujalski, L. Ability of Bruton's Tyrosine Kinase Inhibitors to Sequester Y551 and Prevent Phosphorylation Determines Potency for Inhibition of Fc Receptor but Not B-Cell Receptor Signaling. *Mol. Pharmacol.* **2017**, *91*, 208–219. [CrossRef]
108. Pal Singh, S.; Dammeijer, F.; Hendriks, R.W. Role of Bruton's Tyrosine Kinase in B Cells and Malignancies. *Mol. Cancer* **2018**, *17*, 57, Erratum in *Mol. Cancer* **2019**, *18*, 79. [CrossRef]
109. Burger, J.A. Bruton Tyrosine Kinase Inhibitors. *Cancer J.* **2019**, *25*, 386–393. [CrossRef]
110. von Suskil, M.; Sultana, K.N.; Elbezanti, W.O.; Al-Odat, O.S.; Chitren, R.; Tiwari, A.K.; Challagundla, K.B.; Srivastava, S.K.; Jonnalagadda, S.C.; Budak-Alpdogan, T.; et al. Bruton's Tyrosine Kinase Targeting in Multiple Myeloma. *Int. J. Mol. Sci.* **2021**, *22*, 5707. [CrossRef]
111. Lampson, B.L.; Yu, L.; Glynn, R.J.; Barrientos, J.C.; Jacobsen, E.D.; Banerji, V.; Jones, J.A.; Walewska, R.; Savage, K.J.; Michaud, G.F.; et al. Ventricular Arrhythmias and Sudden Death in Patients Taking Ibrutinib. *Blood* **2017**, *129*, 2581–2584. [CrossRef]
112. Edwards, C.M. BTK Inhibition in Myeloma: Targeting the Seed and the Soil. *Blood* **2012**, *120*, 1757–1759. [CrossRef] [PubMed]
113. Tai, Y.T.; Chang, B.Y.; Kong, S.Y.; Fulciniti, M.; Yang, G.; Calle, Y.; Hu, Y.; Lin, J.; Zhao, J.J.; Cagnetta, A.; et al. Bruton Tyrosine Kinase Inhibition Is a Novel Therapeutic Strategy Targeting Tumor in the Bone Marrow Microenvironment in Multiple Myeloma. *Blood* **2012**, *120*, 1877–1887. [CrossRef] [PubMed]
114. Tai, Y.T.; Anderson, K.C. Bruton's Tyrosine Kinase: Oncotarget in Myeloma. *Oncotarget* **2012**, *3*, 913–914. [CrossRef] [PubMed]
115. Liu, Y.; Dong, Y.; Jiang, Q.L.; Zhang, B.; Hu, A.M. Bruton's Tyrosine Kinase: Potential Target in Human Multiple Myeloma. *Leuk. Lymphoma* **2014**, *55*, 177–181. [CrossRef]
116. Campbell, R.; Chong, G.; Hawkes, E.A. Novel Indications for Bruton's Tyrosine Kinase Inhibitors, beyond Hematological Malignancies. *J. Clin. Med.* **2018**, *7*, 62. [CrossRef]
117. Mohamed, A.J.; Yu, L.; Bäckesjö, C.M.; Vargas, L.; Faryal, R.; Aints, A.; Christensson, B.; Berglöf, A.; Vihinen, M.; Nore, B.F.; et al. Bruton's Tyrosine Kinase (Btk): Function, Regulation, and Transformation with Special Emphasis on the PH Domain. *Immunol. Rev.* **2009**, *228*, 58–73. [CrossRef]
118. de Weerdt, I.; Koopmans, S.M.; Kater, A.P.; van Gelder, M. Incidence and Management of Toxicity Associated with Ibrutinib and Idelalisib: A Practical Approach. *Haematologica* **2017**, *102*, 1629–1639. [CrossRef]
119. Tang, C.P.S.; McMullen, J.; Tam, C. Cardiac Side Effects of Bruton Tyrosine Kinase (BTK) Inhibitors. *Leukemia and Lymphoma* **2018**, *59*. [CrossRef]
120. Ahn, I.E. Cardiovascular Adverse Events of Ibrutinib. *Blood* **2019**, *134*, 1881–1882. [CrossRef]
121. Dickerson, T.; Wiczer, T.; Waller, A.; Philippon, J.; Porter, K.; Haddad, D.; Guha, A.; Rogers, K.A.; Bhat, S.; Byrd, J.C.; et al. Hypertension and Incident Cardiovascular Events Following Ibrutinib Initiation. *Blood* **2019**, *134*, 1919–1928. [CrossRef]
122. Castillo, J.J.; Meid, K.; Gustine, J.N.; Leventoff, C.; White, T.; Flynn, C.A.; Sarosiek, S.; Demos, M.G.; Guerrera, M.L.; Kofides, A.; et al. Long-Term Follow-up of Ibrutinib Monotherapy in Treatment-Naive Patients with Waldenstrom Macroglobulinemia. *Leukemia* **2022**, *36*, 532–539. [CrossRef] [PubMed]
123. Srkalovic, G.; Hussein, M.A.; Hoering, A.; Zonder, J.A.; Popplewell, L.L.; Trivedi, H.; Mazzoni, S.; Sexton, R.; Orlowski, R.Z.; Barlogie, B. A Phase II Trial of BAY 43-9006 (Sorafenib) (NSC-724772) in Patients with Relapsing and Resistant Multiple Myeloma: SWOG S0434. *Cancer Med.* **2014**, *3*, 1275–1283. [CrossRef] [PubMed]
124. Orlowski, R.Z.; Zaman, S.; Thomas, S.K.; Alexanian, R.; Shah, J.J.; Weber, D.M.; Wang, M.; Holloway, A.N.; Baladandayuthapani, V.; Lin, H.Y.; et al. Phase II Study of The C-MET Inhibitor ARQ 197 (Tivantinib) In Patients With Relapsed Or Relapsed/Refractory Multiple Myeloma (RRMM). *Blood* **2013**, *122*, 1953. [CrossRef]
125. Spencer, A.; Yoon, S.S.; Harrison, S.J.; Morris, S.R.; Smith, D.A.; Brigandi, R.A.; Gauvin, J.; Kumar, R.; Opalinska, J.B.; Chen, C. The Novel AKT Inhibitor Afuresertib Shows Favorable Safety, Pharmacokinetics, and Clinical Activity in Multiple Myeloma. *Blood* **2014**, *124*, 2190–2195. [CrossRef] [PubMed]
126. Trudel, S.; Bahlis, N.J.; Venner, C.P.; Hay, A.E.; Kis, O.; Chow, S.; Li, Z.H.; Wei, E.N.; Wang, L.; Tran, C.; et al. Biomarker Driven Phase II Clinical Trial of Trametinib in Relapsed/Refractory Multiple Myeloma with Sequential Addition of the AKT Inhibitor, GSK2141795 at Time of Disease Progression to Overcome Treatment Failure: A Trial of the Princess Margaret Phase II Consortium. *Blood* **2016**, *128*, 4526. [CrossRef]
127. Hoffner, B.; Benchich, K. Trametinib: A Targeted Therapy in Metastatic Melanoma. *J. Adv. Pract. Oncol.* **2018**, *9*, 741. [CrossRef]
128. Blagden, S.P.; Hamilton, A.L.; Mileshkin, L.; Wong, S.; Michael, A.; Hall, M.; Goh, J.C.; Lisyanskaya, A.S.; DeSilvio, M.; Frangou, E.; et al. Phase IB Dose Escalation and Expansion Study of Akt Inhibitor Afuresertib with Carboplatin and Paclitaxel in Recurrent Platinum-Resistant Ovarian Cancer. *Clin. Cancer Res.* **2019**, *25*, 1472–1478. [CrossRef]
129. Moreau, P.; Attal, M.; Garban, F.; Hulin, C.; Facon, T.; Marit, G.; Michallet, M.; Doyen, C.; Leyvraz, S.; Mohty, M.; et al. Heterogeneity of t(4;14) in Multiple Myeloma. Long-Term Follow-up of 100 Cases Treated with Tandem Transplantation in IFM99 Trials. *Leukemia* **2007**, *21*, 2020–2024. [CrossRef]
130. Hart, K.C.; Robertson, S.C.; Donoghue, D.J. Identification of Tyrosine Residues in Constitutively Activated Fibroblast Growth Factor Receptor 3 Involved in Mitogenesis, Stat Activation, and Phosphatidylinositol 3-Kinase Activation. *Mol. Biol. Cell* **2001**, *12*, 931–942. [CrossRef] [PubMed]
131. Kanai, M.; Göke, M.; Tsunekawa, S.; Podolsky, D.K. Signal Transduction Pathway of Human Fibroblast Growth Factor Receptor 3. *J. Biol. Chem.* **1997**, *272*, 6621–6628. [CrossRef] [PubMed]

132. Xin, X.; Abrams, T.J.; Hollenbach, P.W.; Rendahl, K.G.; Tang, Y.; Oei, Y.A.; Embry, M.G.; Swinarski, D.E.; Garrett, E.N.; Pryer, N.K.; et al. CHIR-258 Is Efficacious in a Newly Developed Fibroblast Growth Factor Receptor 3-Expressing Orthotopic Multiple Myeloma Model in Mice. *Clin. Cancer Res.* **2006**, *12*, 4908–4915. [CrossRef] [PubMed]
133. Chell, V.; Balmanno, K.; Little, A.S.; Wilson, M.; Andrews, S.; Blockley, L.; Hampson, M.; Gavine, P.R.; Cook, S.J. Tumour Cell Responses to New Fibroblast Growth Factor Receptor Tyrosine Kinase Inhibitors and Identification of a Gatekeeper Mutation in FGFR3 as a Mechanism of Acquired Resistance. *Oncogene* **2013**, *32*, 3059–3070. [CrossRef] [PubMed]
134. Chesi, M.; Brents, L.A.; Ely, S.A.; Bais, C.; Robbiani, D.F.; Mesri, E.A.; Kuehl, W.M.; Bergsagel, P.L. Activated Fibroblast Growth Factor Receptor 3 Is an Oncogene That Contributes to Tumor Progression in Multiple Myeloma. *Blood* **2001**, *97*, 729–736. [CrossRef] [PubMed]
135. Chesi, M.; Bergsagel, P.L. Advances in the Pathogenesis and Diagnosis of Multiple Myeloma. *Int. J. Lab. Hematol.* **2015**, *37*, 108–114. [CrossRef]
136. Cardona-Benavides, I.J.; de Ramón, C.; Gutiérrez, N.C. Genetic Abnormalities in Multiple Myeloma: Prognostic and Therapeutic Implications. *Cells* **2021**, *10*, 336. [CrossRef]
137. Richter, J.; Ramasamy, K.; Rasche, L.; Bladé, J.; Zweegman, S.; Davies, F.; Dimopoulos, M. Management of Patients with Difficult-to-Treat Multiple Myeloma. *Futur. Oncol.* **2021**, *17*, 2089–2105. [CrossRef]
138. Palma, M.; Mulder, T.A.; Österborg, A. BTK Inhibitors in Chronic Lymphocytic Leukemia: Biological Activity and Immune Effects. *Front. Immunol.* **2021**, *12*, 686768. [CrossRef]
139. Brullo, C.; Villa, C.; Tasso, B.; Russo, E.; Spallarossa, A. Btk Inhibitors: A Medicinal Chemistry and Drug Delivery Perspective. *Int. J. Mol. Sci.* **2021**, *22*, 7641. [CrossRef]
140. Tasso, B.; Spallarossa, A.; Russo, E.; Brullo, C. The Development of Btk Inhibitors: A Five-Year Update. *Molecules* **2021**, *26*, 7411. [CrossRef]
141. Lewis, K.L.; Cheah, C.Y. Non-Covalent Btk Inhibitors—The New Btkids on the Block for b-Cell Malignancies. *J. Pers. Med.* **2021**, *11*, 764. [CrossRef]

pharmaceutics MDPI

Article

Anti-Survival Effect of SI306 and Its Derivatives on Human Glioblastoma Cells

Lorenzo Monteleone [1,†], Barbara Marengo [1,2,†], Francesca Musumeci [3,†], Giancarlo Grossi [3], Anna Carbone [3], Giulia E. Valenti [1], Cinzia Domenicotti [1,2,*] and Silvia Schenone [3]

1 Department of Experimental Medicine (DIMES), General Pathology Section, University of Genoa, 16132 Genoa, Italy; lorenzomonteleone94@gmail.com (L.M.); barbara.marengo@unige.it (B.M.); giuliaelda.valenti@edu.unige.it (G.E.V.)
2 Inter-University Center for the Promotion of the 3Rs Principles in Teaching & Research (Centro 3R), 56122 Pisa, Italy
3 Department of Pharmacy, University of Genoa, 16132 Genoa, Italy; francesca.musumeci@unige.it (F.M.); grossi@difar.unige.it (G.G.); carbone@difar.unige.it (A.C.); schenone@difar.unige.it (S.S.)
* Correspondence: cinzia.domenicotti@unige.it; Tel.: +39-010-353-8830
† These authors contributed equally to this work.

Abstract: Glioblastoma (GBM) is the most common adult brain tumor and, although many efforts have been made to find valid therapies, the onset of resistance is the main cause of recurrence. Therefore, it is crucial to identify and target the molecular mediators responsible for GBM malignancy. In this context, the use of Src inhibitors such as SI306 (C1) and its prodrug (C2) showed promising results, suggesting that SI306 could be the lead compound useful to derivate new anti-GBM drugs. Therefore, a new prodrug of SI306 (C3) was synthesized and tested on CAS-1 and U87 human GBM cells by comparing its effect to that of C1 and C2. All compounds were more effective on CAS-1 than U87 cells, while C2 was the most active on both cell lines. Moreover, the anti-survival effect was associated with a reduction in the expression of epidermal growth factor receptor (EGFR)^{WT} and EGFR-vIII in U87 and CAS-1 cells, respectively. Collectively, our findings demonstrate that all tested compounds are able to counteract GBM survival, further supporting the role of SI306 as progenitor of promising new drugs to treat malignant GBM.

Keywords: glioblastoma; Src; pyrazolo [3,4-*d*]pyrimidine scaffold; EGFR

Citation: Monteleone, L.; Marengo, B.; Musumeci, F.; Grossi, G.; Carbone, A.; Valenti, G.E.; Domenicotti, C.; Schenone, S. Anti-Survival Effect of SI306 and Its Derivatives on Human Glioblastoma Cells. *Pharmaceutics* **2022**, *14*, 1399. https://doi.org/10.3390/pharmaceutics14071399

Academic Editor: Gabriele Grassi

Received: 6 June 2022
Accepted: 30 June 2022
Published: 1 July 2022

1. Introduction

Glioblastomas (GBM) are malignant and aggressive astrocytic tumors, classified grade IV according to the World Health Organization [1]. They are the most common adult brain tumors, with an annual incidence of about 1/33.330 and with a current median survival of 15 months [2].

Current therapy includes surgery, radiotherapy, and chemotherapy but, unfortunately, in many cases such approaches are ineffective and facilitate the onset of relapse and the development of therapy resistance [3]. Therefore, there is a great need to identify new treatments capable of improving the prognosis of patients. In recent years, several genetic alterations and pathways responsible for GBM malignancy were identified and indicated as targets for new therapies [4–6]. In this context, Src-Family Kinases (SFKs), having a key role in the development, tumorigenicity, invasion, and progression of GBM [7], have received particular attention. In this regard, our research group has recently developed new Src inhibitors endowed with a pyrazolo[3,4-*d*]pyrimidine scaffold [8–13]. Among this series of compounds, SI306 (compound **1**, C1, Figure 1) showed significant efficacy against U87 human GBM cells [14–16]. Based on these successful results, we decided to study this compound further by synthesizing prodrugs potentially endowed with improved pharmacokinetic (PK) properties and testing them on U87 and CAS-1 GBM cells in order to evaluate their efficacy.

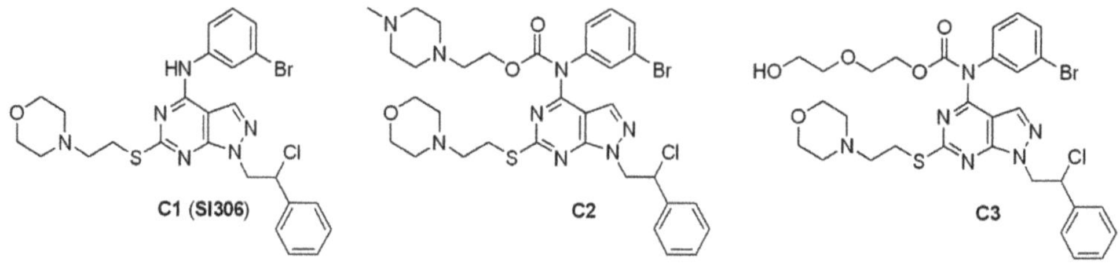

Figure 1. Structures of C1, C2, and C3.

In particular, we selected compound **2** (C2, Figure 1), an SI306 prodrug that already showed an improved activity in biological assays [13,16], and the ethylene glycol derivative **3** (C3, Figure 1) as a potential valuable prodrug of a new synthesis. Indeed, a careful analysis of the literature highlighted that introducing a poly(ethylene glycol) (PEG) chain can successfully modulate drug delivery [17]. As different studies pointed out that the activity of many small molecules is improved by PEGylation [18], we synthesized compound **3** (Figure 1), which bears two ethylene glycol units on the carbamate moiety. We chose this short PEG chain driven by the rationale to balance a feasible synthesis with the chance to afford a compound characterized by improved ADME (Absorption, Distribution, Metabolism, Excretion) properties. Compound **3**, and previous analog compounds **1** and **2**, were tested on CAS-1 and U87 cells, and their activities were compared.

The obtained results showed that all tested compounds exhibited a more significant anti-survival effect on CAS-1 than on U87 cells and that C2 was the most active on both cancer cell lines.

Given that Src has been demonstrated to interact and activate epidermal growth factor receptor (EGFR) [19,20], an oncogene frequently amplified and mutated in GBM [21–24], the action of SI306 (C1) and its derivatives (C2 and C3) on the expression of both wild type (wt) and mutated (vIII) EGFR was evaluated. In this regard, our results showed that CAS-1 cells express EGFR-vIII while U87 cells express $EGFR^{WT}$. Moreover, 48 h exposure to the highest dose (10 μM) of all inhibitors markedly reduced the expression of $EGFR^{WT}$ and EGFR-vIII in U87 and CAS-1 cells, respectively.

Collectively, our findings demonstrate that SI306 and its prodrugs are able to counteract GBM survival by inhibiting Src, as expected. Moreover, for the first time to our knowledge, we have shown that this event is accompanied by a reduction in $EGFR^{WT}$ and EGFR-vIII expression.

2. Materials and Methods

2.1. Chemistry

All commercially available chemicals were used as purchased from Sigma-Aldrich (St. Louis, MO, USA). DCM was dried over molecular sieves (3 Å, 10% *m*/*v*). TLC was carried out using Merck TLC silica gel 60 F254. Chromatographic purifications were performed on columns packed with Merck silica gel 60, 23−400 mesh, for flash technique. ^{1}H-NMR and ^{13}C-NMR spectra were recorded on a JEOL JNM ECZ-400S/L1 FT (400 MHz). Chemical shifts were reported relative to tetramethylsilane at 0.00 ppm. Elemental analysis for C, H, N, and S was determined using Thermo Scientific Flash 2000 and results were within ±0.4% of the theoretical value. All target compounds possessed a purity of ≥95% verified by elemental analysis. Compounds **1** and **2** were previously synthesized by us [13,25,26].

2.2. Synthesis of 2-(2-hydroxyethoxy)ethyl (3-bromophenyl)(1-(2-chloro-2-phenylethyl)-6-((2-morpholinoethyl)thio)-1H-pyrazolo[3,4-d]pyrimidin-4-yl)carbamate (C3)

A solution of triphosgene (0.45 mmol, 133 mg) in anhydrous CH_2Cl_2 (8 mL) was added dropwise to a mixture of $NaHCO_3$ (2.25 mmol, 189 mg) and N-(3-bromophenyl)-1-(2-chloro-2-phenylethyl)-6-((2-morpholinoethyl)thio)-1H-pyrazolo[3,4-d]pyrimidin-4-amine 1

(0.45 mmol, 258 mg) in anhydrous CH_2Cl_2 (8 mL) precooled at 0 °C in an ice bath. After 30 min, the reaction was allowed to warm to room temperature and stirred for 5 h. Then, a solution of ethylene glycol (20.00 mmol, 1.9 mL) in anhydrous CH_2Cl_2 (8 mL) was added and the mixture stirred at room temperature for 16 h. Cold water (25 mL) was added, and the two phases were separated. The aqueous phase was extracted twice with CH_2Cl_2 (2 × 25 mL). The organic phase was dried over anhydrous Na_2SO_4 and evaporated. The crude oil was purified through two flash chromatographies on silica gel, eluting first with THF/EP 1:1 and then with ethyl acetate/acetone (1:1) affording the desired C3 as a pure light amber oil. Yield 54%. ^{1}H-NMR ($CDCl_3$): δ 2.48–2.57 (m, 6H, $2CH_2N$ morph + NCH_2CH_2), 298–3.13 (m, 2H, CH_2S), 3.47 (t, *J* = 8 Hz, 2H, $HOCH_2CH_2O$), 3.64–3.72 (m, 8H, $2CH_2O$ morph + $HOCH_2$ + COO-CH_2CH_2), 4.41 (t, J = 4 Hz, 2H, COO-CH_2), 4.68–4.72 and 492–4.98 (2m, 2H, CH_2N pyraz), 5.48–5.52 (m, 1H, CHCl), 7.16–7.18 (m, 1H Ar), 7.25–7.34 (m, 4H Ar), 7.41–7.43 (m, 3H Ar), 7.49–7.51 (m, 1H Ar), and 7.98 (s, 1H, H-3). ^{13}C-NMR ($CDCl_3$): δ NMR (101 MHz,) δ = 31.87, 53.45, 53.93, 54.32, 57.66, 60.29, 61.70, 66.58, 66.82, 68.74, 72.62, 76.85, 77.16, 77.48, 103.66, 122.36, 127.44, 127.56, 128.88, 129.17, 130.49, 131.33, 131.93, 135.39, 137.88, 140.98, 153.45, 154.27, 156.07, and 168.36. Anal. calcd. for $C_{30}H_{34}N_6O_5SClBr$: C 51.03, H 4.85, N 11.90, and S 4.54; found: C 50.77, H 5.19, N 11.57, and S 4.09.

2.3. Cell Cultures

U87 and CAS-1 human GBM cell lines were obtained from Ospedale Policlinico San Martino (Genova, Italy). Cells were periodically tested for mycoplasma contamination (Mycoplasma Reagent Set, Aurogene s.p.a, Pavia, Italy). Cells were maintained in DMEM low glucose medium (Euroclone SpA, Pavia, Italy) supplemented with 10% fetal bovine serum (FBS; Euroclone), 2 mM of glutamine (Euroclone).

2.4. Treatments

U87 and CAS-1 cells were treated with C1, C2, and C3 for 24, 48, and 72 h with increasing concentrations (1–10 μM) of each compound. The stock solutions of the three compounds were prepared in DMSO (Sigma-Aldrich Co., Saint Louis, MO, USA) and pilot experiments demonstrated that final DMSO doses did not affect any cell response analyzed.

2.5. Cell Viability Assay

Cell viability was determined by using the CellTiter 96® AQueous One Solution Cell Proliferation Assay (Promega, Madison, WI, USA), as previously described [27]. Briefly, cells (9 × 10^4 cells/well) were seeded in 96-well plates (Corning Incorporate, Corning, NY, USA) and then treated. Next, the cells were incubated with MTT solution according to manufacturer's instructions and the absorbance at 570/630 nm was recorded using a microplate reader (EL-808, BIO_TEK Instruments Inc., Winooski, VT, USA). The relative IC50 values for were calculated by non-linear regression analysis using GraphPad Prism 6.0 software (GraphPad Software, La Jolla, CA, USA).

2.6. Blot Analyses

Immunoblots were performed according to standard methods [28] using rabbit antibody anti-human anti-Src (#2109), anti-Phospho-Src (Tyr527, #2105) corresponding in humans to tyrosine 530 residue, anti-Phospho-Src (Tyr416, #2101) corresponding in humans to tyrosine 419 residue, anti PARP (#9542), anti-EGFR (#4267), and anti-EGFR-vIII (#2232; Cell Signaling Technology Inc., Danvers, MA, USA, Upstate, Lake Placid, NY, USA). Anti-rabbit secondary antibody coupled with horseradish peroxidase (Cell Signaling Technologies, Danvers, MA, USA) was utilized.

2.7. Statistical Analyses

Data are expressed as means ± S.E.M. Statistical significance of differences was determined by one-way analysis of variances (ANOVA) followed by Tukey's test. $p < 0.05$ was considered statistically significant.

3. Results

3.1. Chemistry

Synthesis of C1 and C2 was already reported by us [13,25,26]. C3 was synthesized starting from SI306 with a one-pot two-step procedure. First, trisphosgene was added to a pre-cooled solution of SI306, then ethylene glycol was added, and the reaction was stirred at room temperature affording C3 with good yield (Figure 2).

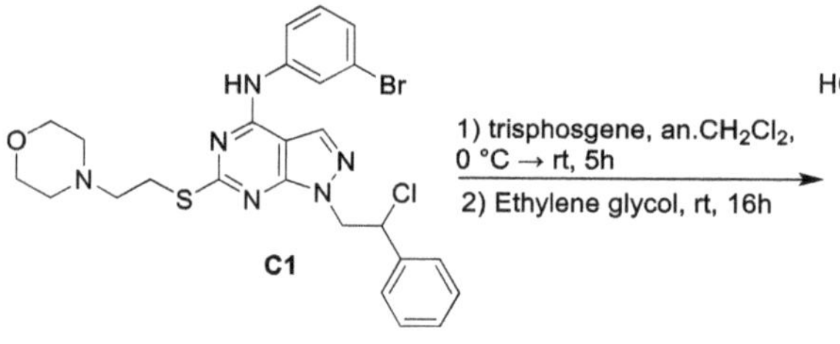

Figure 2. Synthesis of C3.

3.2. U87 Cells Are Less Sensitive Than CAS1 to the Effect of SI306- and SI306-Derived Drugs and Show a Major Susceptibility to Compound **2**

In order to compare the effects of C2 and C3 with those of SI306, two human GBM cell lines (U87 and CAS-1) were exposed to increasing concentrations of compounds for 24–72 h.

The analysis of cell viability showed that all compounds, at the same doses and time exposure, induced a major anti-survival effect on CAS-1 cells. In fact, as reported in Table 1, the IC_{50} of all three compounds was lower in CAS-1 than in U87 cells. Moreover, C2 was the most effective on both cell lines, while C3 exerted a lower activity (Table 1).

Table 1. IC_{50} evaluated in CAS-1 and U87 cell lines treated with increasing doses (1–10 μM) of C1 (SI306), C2, and C3 for 24, 48, and 72 h. The data are the means ± S.E.M. of five independent experiments. Statistical significance of differences was determined by ANOVA followed by Tukey's test. ** $p < 0.01$ vs. C1; *** $p < 0.001$ vs. C1; **** $p < 0.0001$ vs. C1; °°° $p < 0.001$ vs. C2; and °°°° $p < 0.0001$ vs. C2.

	CAS-1			U87		
Compound	**24 h**	**48 h**	**72 h**	**24 h**	**48 h**	**72 h**
C1	11.74 ± 0.34	3.88 ± 0.15	3.03 ± 0.1	52.46 ± 1.24	18.63 ± 0.32	5.28 ± 0.35
C2	6.05 ± 0.15 ***	3.66 ± 0.28 ***	3.13 ± 0.16 ***	46.46 ± 0.8 ***	9.24 ± 0.14 ****	3.77 ± 0.15 **
C3	13.92 ± 0.42 °°°°	6.97 ± 0.27 ***/°°°	4.47 ± 0.2 ***/°°°	54.51 ± 1.16 °°°°	28.36 ± 0.2 ****/°°°°	5.79 ± 0.11 °°°

In detail, SI306 induced a concentration-dependent decrease in the viability of CAS-1 cells, reaching a 40% reduction at 10 μM after 24 h-exposure (Figure 3a, left panel). Instead, no effects on cell viability of U87 were induced by 24 h-treatment with SI306 (Figure 3a, right panel). Analyzing the viability at 48 h, the highest doses of SI306 (8 μM and 10 μM) reduced the viability of CAS1 and U87 by 85% and 25%, respectively (Figure 3b,c, respectively). Moreover, after 72 h, the highest doses decreased the viability of U87 cells by over 70% (Figure 3c), while no further changes were observed in CAS-1 in respect to those observed after 48 h.

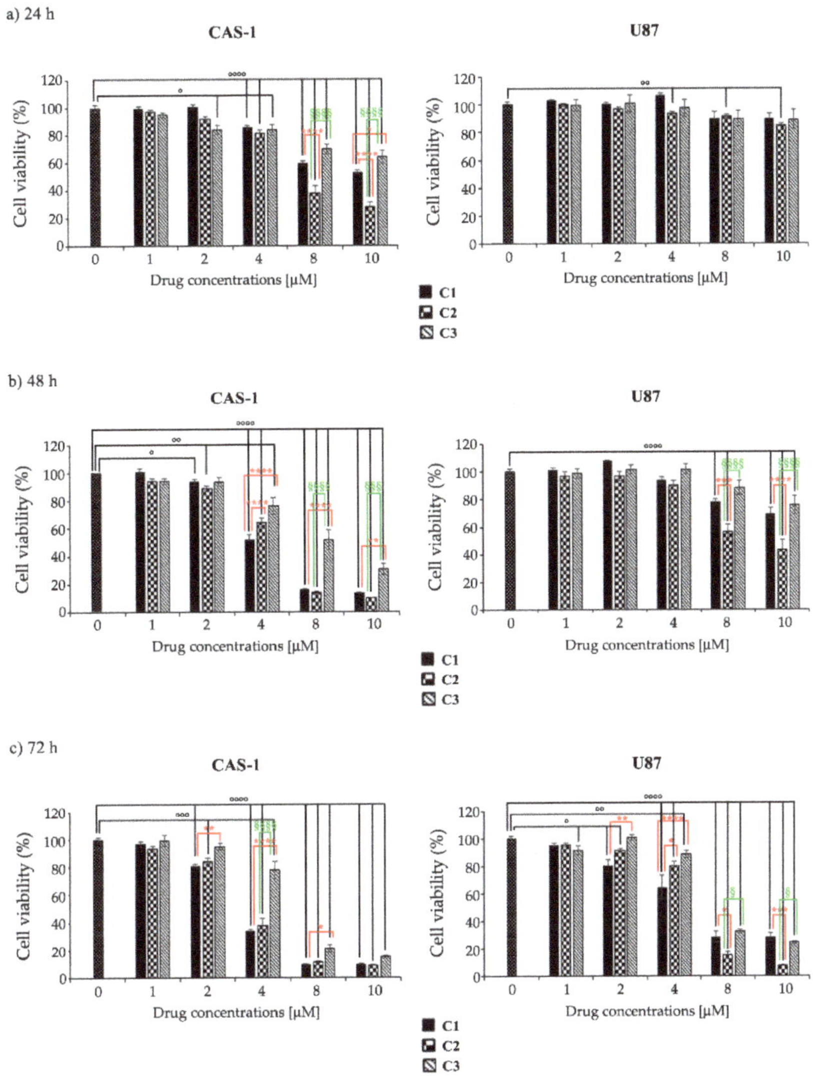

Figure 3. U87 cells are less sensitive than CAS1 to the anti-survival effect of SI306- and SI306-derived drugs. CAS-1 (left panels) and U87 (right panels) cells were exposed to increasing concentrations (1–10 μM) of compound **1** (C1, SI306), compound **2** (C2), and compound **3** (C3) for 24 h (**a**), 48 h (**b**), and 72 h (**c**). Cell viability was evaluated by MTS assay. Histograms summarize quantitative data of the means ± S.E.M. of five independent experiments. Statistical significance of differences was determined by ANOVA followed by Tukey's test. ° $p < 0.05$ vs. untreated cells; °° $p < 0.01$ vs. untreated cells; °°° $p < 0.001$ vs. untreated cells; °°°° $p < 0.0001$ vs. untreated cells; * $p < 0.05$ vs. compound **1**; ** $p < 0.01$ vs. compound **1**; *** $p < 0.001$ vs. compound **1**; **** $p < 0.0001$ vs. compound **1**; § $p < 0.05$ vs. compound **2**; §§§ $p < 0.001$ vs. compound **2**; and §§§§ $p < 0.0001$ vs. compound **2**.

With regard to SI306-derived drugs, in both cell lines it has been observed that C2 had a major anti-survival action than SI306 (Figure 3), whilst C3 was less active than SI306 and C2 (Figure 3). In detail, after 24 h-treatment, the highest doses (8 μM and 10 μM) of C2 were able to reduce cell viability more effectively than SI306 (Figure 3b, right panel).

Instead, after 48 and 72 h, the highest doses of C2 and SI306 analogously decreased CAS1 cell viability by 85% and 92%, respectively (Figure 3b,c, respectively, left panels). The viability of U87 cells was significantly impaired after 48 h and, in particular, C2 was more active than SI306 and C3. Interestingly, after 72 h, the highest doses (8 μM and 10 μM) of C2 reduced U87 cell viability by about 90% (Figure 3, right panels).

Moreover, C3 at these highest doses (8 μM and 10 μM) exerted an anti-survival effect on U87 cells comparable to that of SI306, while it was found less effective on CAS1 (Figure 3).

In order to shed light on the kind of cell death, the levels of full length (116 kDa) and cleaved (89 kDa) PARP, a known marker of apoptosis [29], were evaluated. As reported in Figure 4, C1, at 10 μM, and C2 and C3, at 4 and 10 μM, induced apoptosis of U87 cells after 24 and 48 h. The same treatments did not induce PARP cleavage in CAS-1 cells (Figure 4).

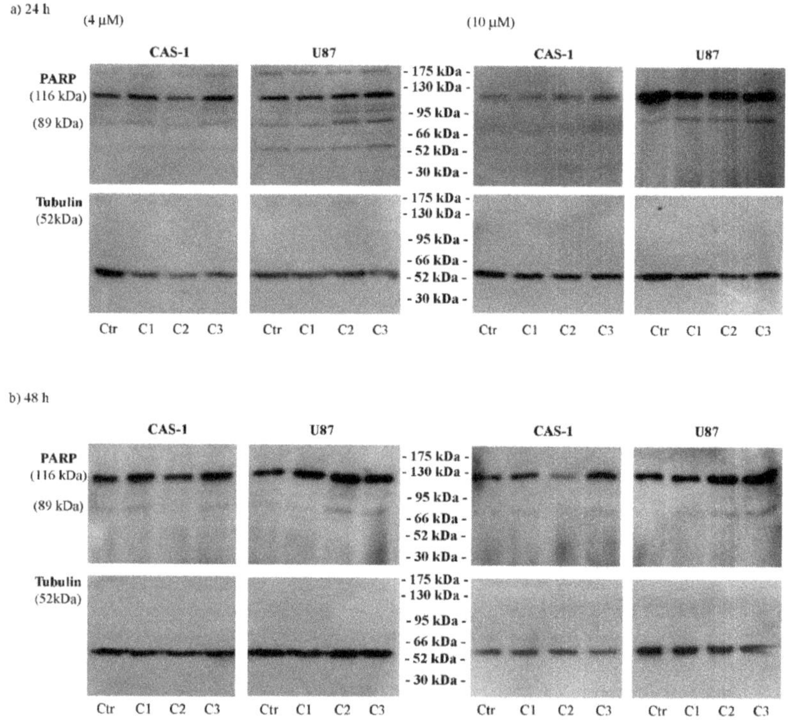

Figure 4. C2 and C3 induce apoptotic death of U87 cells. The expression levels of full length (116 kDa) and cleaved (89 kDa) PARP were evaluated in CAS-1 (left panels) and U87 (right panels) cells exposed to 4 μM and 10 μM of compound **1** (SI306, C1), compound **2** (C2), and compound **3** (C3) for 24 h (**a**) and 48 h (**b**). Immunoblots shown are representative of three independent experiments. Tubulin is the internal loading control.

3.3. SI306 Derivatives Inhibit Src Activity in U87 Cells

Considering that SI306 [25,30] is a Src inhibitor [8,10,11,14–16], the inhibitory ability of its derivatives was tested by evaluating the expression levels of Src phosphorylated at Tyr530, a marker of Src inactivation [31], and at Tyr416 (Tyr419 in humans), a marker of Src activation [32]. As shown in Figure 5, in U87 cells, C2 (10 μM) increased the p-Tyr530/Src ratio by 15% and by 20% at 24 and 48 h, respectively (Figure 5, right panels), while C3 (10 μM) induced a 15% increase after 24 h. The analysis of the expression of Src phosphorylated at Tyr416 (Tyr419) showed that 24 h-treatment of U87 cells with all compounds at 4 μM reduced the p-Src/Src ratio by 45% (Figure 5a, left panel). After 48 h, only C1 and C3 reduced the ratio by 20% (Figure 5b, left panel). No significant changes were observed in CAS-1 cells exposed to 4 μM C1, C2, or C3 (Figure 5a,b, left panels). However, 10 μM C1 or C3 reduced the p-Src/Src ratio by 55% in CAS-1 cells treated for 24 h, while only C3 reduced the ratio in U87 cells by 35% (Figure 5a, right panel). After 48 h, C1 and C2 also decreased the p-Src/Src ratio by 30% in U87 cells, while no significant changes were found in CAS-1 cells (Figure 5b, right panel).

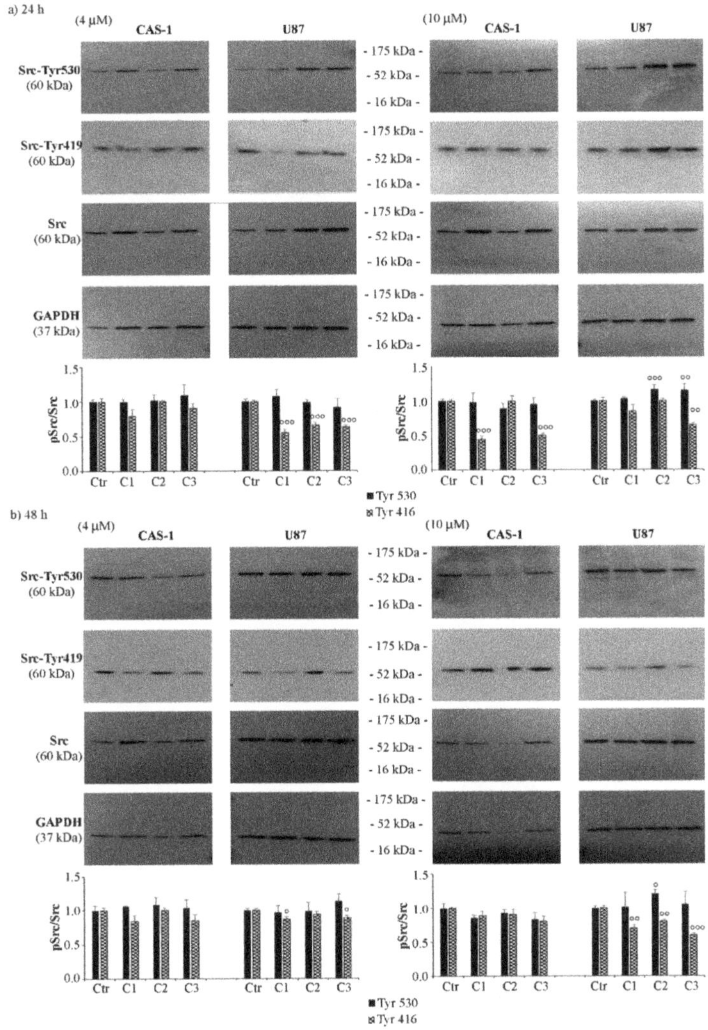

Figure 5. SI306 derivatives inhibit Src activity in U87 cells. Src inactivation was evaluated by immunoblot analyses of p-Src (Tyr530 and Tyr419) and total Src. The expression levels of p-Src and Src were evaluated in CAS-1 (left panels) and U87 (right panels) cells exposed to 4 μM and 10 μM of compound **1** (SI306, C1), compound **2** (C2), and compound **3** (C3) for 24 h (**a**) and 48 h (**b**). Immunoblots shown are representative of three independent experiments. GAPDH is the internal loading control. Histograms summarize quantitative data of means ± S.E.M. of three independent experiments. Statistical significance of differences was determined by ANOVA followed by Tukey's test. Data are expressed as a ratio of the levels of p-Src to the levels of total Src. ° $p < 0.05$ vs. untreated cells; °° $p < 0.01$ vs. untreated cells; and °°° $p < 0.001$ vs. untreated cells.

3.4. CAS-1 Cells Express EGFR-vIII While U87 Cells Express EGFR^{WT} and the Exposure to SI306 or Its Derivatives Has a Different Impact on Their Expression Levels

As it has been demonstrated that progression of GBM is related to Src up-regulation [33] and it is also a consequence of EGFR mutations [34], the expression of both wild type EGFR (EGFR^{WT}) and EGFR-vIII, the most common mutant found in GBM, was investigated. Firstly, as shown in Figure 6, EGFR^{WT} was found in U87 while EGFR-vIII was detected in CAS-1 cells.

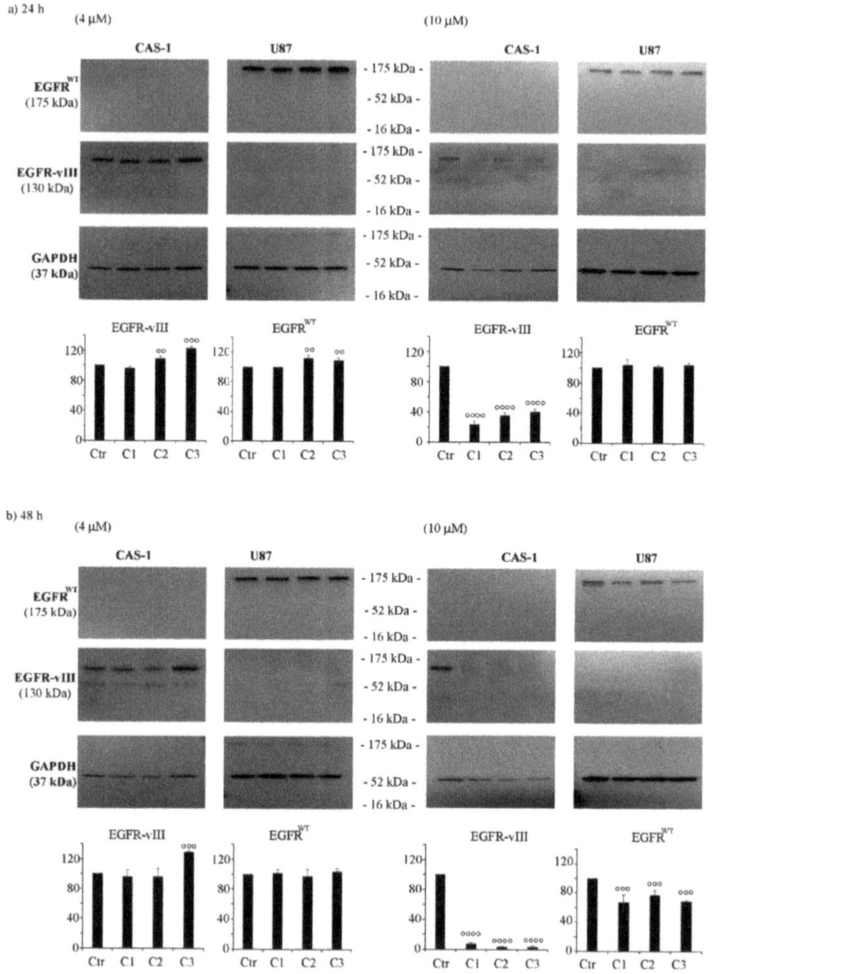

Figure 6. CAS-1 and U87 cells show a different type of EGFR, and treatment with SI306 or its derivatives differently alters their expression levels. The expression of EGFR^{WT} or EGFR-vIII was evaluated by immunoblot analyses performed on protein extracts obtained by CAS-1 (left panels) and U87 (right panels) cells exposed to 4 µM and 10 µM of compound **1** (SI306, C1), compound **2** (C2), and compound **3** (C3) for 24 h (**a**) and 48 h (**b**). Immunoblots shown are representative of three independent experiments. GAPDH is the internal loading control. Histograms summarize quantitative data of means ± S.E.M. of three independent experiments. Statistical significance of differences was determined by ANOVA followed by Tukey's test. °° $p < 0.01$ vs. untreated cells; °°° $p < 0.001$ vs. untreated cells; °°°° $p < 0.0001$ vs. untreated cells.

Notably, 24 h-treatment with 4 μM of C2 or C3 increased the expression of $EGFR^{WT}$ in U87 cells by 10%, while it enhanced $EGFR^{-}$vIII levels in CAS-1 by about 15%, in respect to control cells (Figure 6a, left panels). However, 48 hour-exposure to C3 induced a 30% increase in EGFR-vIII protein levels in CAS-1 cells (Figure 6b, left panel), while 24 h-treatment with 4 μM of SI306 did not significantly change the expression of either $EGFR^{WT}$ or EGFR-vIII in ether cell population (Figure 6a,b, respectively, left panels).

Surprisingly, all compounds at the dose of 10 μM reduced the expression of EGFR-vIII in CAS-1 cells by 70% and 95% after 24 and 48 h, respectively (Figure 6b, right panels), and a 30–35% decrease in $EGFR^{WT}$ protein levels was observed in U87 cells only after 48 h of treatment (Figure 6b, right panels).

4. Discussion

Despite ongoing efforts to find valid therapies, GBM remains a highly aggressive tumor with low recovery expectations. In fact, today, the standard cure for GBM is surgery, followed by radiotherapy and/or chemotherapy [2]. However, such approaches are not currently curative, and the onset of resistance is the main cause of recurrence. Therefore, it is crucial to identify the molecular mediators responsible for therapeutic resistance in order to develop targeted strategies capable of overcoming this obstacle. In this regard, several studies have attributed a crucial role to the Src family, both in GBM carcinogenesis and in the mechanisms involved in therapeutic failure [14,35,36]. This evidence suggests that targeting Src could be a valuable approach to improve the life expectancy of GBM patients. Most of the Src inhibitors available today are non-specific and offer cross-reactivity [37]. Moreover, the particular location of GBM makes necessary to develop drugs able to cross the Blood–brain barrier (BBB), and to over-express ATP-binding cassette (ABC) transporters, preventing the drug supply and accumulation in the brain. Many FDA-approved drugs are directed towards these transporters [7] but, although they induce an adequate intake of the drug to the brain, are not effective for GBM treatment [38–40].

Therefore, based on these considerations, an effective therapy must hit GBM acting on several fronts and target the molecular pathways involved in the resistance. In recent years, our research team developed an extensive library of pirazolo[3,4-*d*]pyrimidines able to cross the BBB [9,10]. Among these compounds, SI306 (namely C1 in the paper), by inhibiting Src [41,42] and the ABC transporter Pgp [9,10], has demonstrated a powerful antitumor action in vitro and in vivo, both as monotherapy and combined therapy [8,10,13,14]. Similar results were obtained with the prodrug of SI306 (namely C2) [10,13]. Starting from these promising results, a new SI306 derivative (namely C3) was designed and synthesized in order to extend our knowledge on the effect of a different polar chain on the carbamate moiety. C3 was decorated with an ethylene glycol unit in place of the *N*-methyl piperazine fragment present in C2. Both prodrugs were tested on two human GBM cell lines and their effects compared with those of the parent drug SI306. The obtained results herein demonstrate that the anti-survival action of SI306 and its derivatives is both time- and concentration-dependent. In fact, the lowest doses (2 and 4 μM) of all compounds were able to reduce cell survival after 48 h while the highest ones (8 and 10 μM) decreased cell survival already at 24 h with a progressive reduction at 48 and 72 h.

However, although an MTT test revealed that U87 cells are less sensitive than CAS-1, PARP cleavage, as a recognized marker of apoptosis [43], was detected only in U87 treated cells, suggesting that in CAS-1 cells other death pathways occur and further investigation is needed. Interestingly, the pro-apoptotic effect of these compounds on U87 cells is accompanied by Src inactivation measured by evaluating the phosphorylation of Tyr530 and 416 (Tyr419 in humans) [31,32].

With regard to CAS-1 cells, the reduction in cell viability is not related to Src activity and might be due to other mechanisms that deserve further investigation. However, the data demonstrates that Src phosphorylation status does not predict the drug sensitivity in agreement with a recent study reporting that patients with different GBM subtypes expressed similar levels of the unphosphorylated and phosphorylated Src [44]. In the present study, the different sensitivity of two GBM cells to the tested compounds could be related to their pharmacokinetic profile, but also to the fact that CAS-1 cells express EGFR-vIII while U87 cells express $EGFR^{WT}$. This result is in contradiction with a previous data reporting that U87 cells express EGFR-vIII [45] and we believe that this discrepancy could be due to the different specificity of the antibody used in Western blotting analysis.

However, to investigate if all tested compounds are effective on GBM with a different status of EGFR could have important therapeutic implications. In fact, considering that patients with EGFR-vIII have a reduced survival rate [46] and that EGFR-vIII is not expressed in healthy tissues, EGFR-vIII could be considered an excellent target for therapy [47]. To date, drugs developed to treat patients with EGFR-vIII or EGFR have proven to be ineffective due to the inability of such drugs to overcome the BBB, to tumor heterogeneity, and to the activation of compensatory pathways such as Src [21–24]. Although further investigations are needed, our studies demonstrate that SI306 and its derivatives impact GBM cell survival and that this event is accompanied by a reduction in the expression of $EGFR^{WT}$ and EGFR-vIII in U87 and Cas-1 cells, respectively. These results lead us to believe that these compounds could represent a promising alternative approach to treat GBM.

Author Contributions: Conceptualization, B.M., F.M., S.S. and C.D.; methodology, B.M., L.M. and G.G.; validation, B.M. and C.D.; visualization, B.M., L.M., F.M. and C.D.; investigation, B.M., L.M., G.G. and G.E.V.; formal analysis, B.M., L.M. and G.E.V.; writing—original draft preparation, B.M., L.M. and G.E.V.; writing—review and editing, B.M., F.M., A.C., S.S. and C.D.; resources, S.S. and C.D., supervision, S.S. and C.D.; project administration, S.S. and C.D.; and funding acquisition, S.S. All authors have read and agreed to the published version of the manuscript.

Funding: This research was funded by the AIRC Foundation for Cancer Research in Italy (Fondazione AIRC per la Ricerca sul Cancro, AIRC), Grant IG-2019, project code 23725, PI Silvia Schenone.

Informed Consent Statement: Not applicable.

Acknowledgments: We would like to thank Giuseppe Catalano (DIMES-University of Genoa) for his technical assistance.

Conflicts of Interest: The authors declare no conflict of interest.

References

1. Louis, D.N.; Perry, A.; Wesseling, P.; Brat, D.J.; Cree, I.A.; Figarella-Branger, D.; Hawkins, C.; Ng, H.K.; Pfister, S.M.; Reifenberger, G.; et al. The 2021 WHO Classification of Tumors of the Central Nervous System: A summary. *Neuro-Oncology* **2021**, *23*, 1231–1251. [CrossRef] [PubMed]
2. Tan, A.C.; Ashley, D.M.; López, G.Y.; Malinzak, M.; Friedman, H.S.; Khasraw, M. Management of glioblastoma: State of the art and future directions. *CA Cancer J. Clin.* **2020**, *70*, 299–312. [CrossRef] [PubMed]
3. Olar, A.; Aldape, K.D. Using the molecular classification of glioblastoma to inform personalized treatment. *J. Pathol.* **2014**, *232*, 165–177. [CrossRef] [PubMed]
4. Martins, T.A.; Schmassmann, P.; Shekarian, T.; Boulay, J.L.; Ritz, M.F.; Zanganeh, S.; Vom Berg, J.; Hutter, G. Microglia-Centered Combinatorial Strategies Against Glioblastoma. *Front. Immunol.* **2020**, *11*, 571951. [CrossRef]
5. Majc, B.; Novak, M.; Kopitar-Jerala, N.; Jewett, A.; Breznik, B. Immunotherapy of Glioblastoma: Current Strategies and Challenges in Tumor Model Development. *Cells* **2021**, *10*, 265. [CrossRef] [PubMed]
6. Liu, E.K.; Sulman, E.P.; Wen, P.Y.; Kurz, S.C. Novel Therapies for Glioblastoma. *Curr. Neurol. Neurosci. Rep.* **2020**, *20*, 19. [CrossRef] [PubMed]
7. Cirotti, C.; Contadini, C.; Barilà, D. SRC Kinase in Glioblastoma News from an Old Acquaintance. *Cancers* **2020**, *12*, 1558. [CrossRef] [PubMed]
8. Calgani, A.; Vignaroli, G.; Zamperini, C.; Coniglio, F.; Festuccia, C.; Di Cesare, E.; Gravina, G.L.; Mattei, C.; Vitale, F.; Schenone, S.; et al. Suppression of SRC Signaling Is Effective in Reducing Synergy between Glioblastoma and Stromal Cells. *Mol. Cancer Ther.* **2016**, *15*, 1535–1544. [CrossRef] [PubMed]
9. Ceccherini, E.; Indovina, P.; Zamperini, C.; Dreassi, E.; Casini, N.; Cutaia, O.; Forte, I.M.; Pentimalli, F.; Esposito, L.; Polito, M.S.; et al. SRC family kinase inhibition through a new pyrazolo[3,4-*d*]pyrimidine derivative as a feasible approach for glioblastoma treatment. *J. Cell Biochem.* **2015**, *116*, 856–863. [CrossRef] [PubMed]
10. Fallacara, A.L.; Zamperini, C.; Podolski-Renić, A.; Dinić, J.; Stanković, T.; Stepanović, M.; Mancini, A.; Rango, E.; Iovenitti, G.; Molinari, A.; et al. A New Strategy for Glioblastoma Treatment: In Vitro and In Vivo Preclinical Characterization of Si306, a Pyrazolo[3,4-*d*]Pyrimidine Dual Src/P-Glycoprotein Inhibitor. *Cancers* **2019**, *11*, 848. [CrossRef] [PubMed]

11. Greco, C.; Taresco, V.; Pearce, A.K.; Vasey, C.E.; Smith, S.; Rahman, R.; Alexander, C.; Cavanagh, R.J.; Musumeci, F.; Schenone, S. Development of Pyrazolo[3,4-*d*]pyrimidine Kinase Inhibitors as Potential Clinical Candidates for Glioblastoma Multiforme. *ACS Med. Chem. Lett.* **2020**, *11*, 657–663. [CrossRef] [PubMed]
12. Musumeci, F.; Fallacara, A.L.; Brullo, C.; Grossi, G.; Botta, L.; Calandro, P.; Chiariello, M.; Kissova, M.; Crespan, E.; Maga, G.; et al. Identification of new pyrrolo[2,3-*d*]pyrimidines as Src tyrosine kinase inhibitors in vitro active against Glioblastoma. *Eur. J. Med. Chem.* **2017**, *127*, 369–378. [CrossRef] [PubMed]
13. Vignaroli, G.; Iovenitti, G.; Zamperini, C.; Coniglio, F.; Calandro, P.; Molinari, A.; Fallacara, A.L.; Sartucci, A.; Calgani, A.; Colecchia, D.; et al. Prodrugs of Pyrazolo[3,4-*d*]pyrimidines: From Library Synthesis to Evaluation as Potential Anticancer Agents in an Orthotopic Glioblastoma Model. *J. Med. Chem.* **2017**, *60*, 6305–6320. [CrossRef] [PubMed]
14. Cammarata, F.P.; Torrisi, F.; Forte, G.I.; Minafra, L.; Bravatà, V.; Pisciotta, P.; Savoca, G.; Calvaruso, M.; Petringa, G.; Cirrone, G.A.P.; et al. Proton Therapy and Src Family Kinase Inhibitor Combined Treatments on U87 Human Glioblastoma Multiforme Cell Line. *Int. J. Mol. Sci.* **2019**, *20*, 4745. [CrossRef]
15. Kostić, A.; Jovanović Stojanov, S.; Podolski-Renić, A.; Nešović, M.; Dragoj, M.; Nikolić, I.; Tasić, G.; Schenone, S.; Pešić, M.; Dinić, J. Pyrazolo[3,4-*d*]pyrimidine Tyrosine Kinase Inhibitors Induce Oxidative Stress in Patient-Derived Glioblastoma Cells. *Brain Sci* **2021**, *11*, 884. [CrossRef] [PubMed]
16. Nešović, M.; Divac Rankov, A.; Podolski-Renić, A.; Nikolić, I.; Tasić, G.; Mancini, A.; Schenone, S.; Pešić, M.; Dinić, J. Src Inhibitors Pyrazolo[3,4-*d*]pyrimidines, Si306 and Pro-Si306, Inhibit Focal Adhesion Kinase and Suppress Human Glioblastoma Invasion In Vitro and In Vivo. *Cancers* **2020**, *12*, 1570. [CrossRef]
17. Greenwald, R.B.; Choe, Y.H.; McGuire, J.; Conover, C.D. Effective drug delivery by PEGylated drug conjugates. *Adv. Drug Deliv. Rev.* **2003**, *55*, 217–250. [CrossRef]
18. Wu, T.; Chen, K.; He, S.; Liu, X.; Zheng, X.; Jiang, Z.X. Drug Development through Modification of Small Molecular Drugs with Monodisperse Poly(ethylene glycol)s. *Org. Process Res. Dev.* **2020**, *24*, 1364–1372. [CrossRef]
19. Cvrljevic, A.N.; Akhavan, D.; Wu, M.; Martinello, P.; Furnari, F.B.; Johnston, A.J.; Guo, D.; Pike, L.; Cavenee, W.K.; Scott, A.M.; et al. Activation of Src induces mitochondrial localisation of de2-7EGFR (EGFRvIII) in glioma cells: Implications for glucose metabolism. *J. Cell Sci.* **2011**, *124*, 2938–2988. [CrossRef]
20. Eskilsson, E.; Rosland, G.V.; Talasila, K.M.; Knappskog, S.; Keunen, O.; Sottoriva, A.; Foerster, S.; Solecki, G.; Taxt, T.; Jirik, R.; et al. EGFRvIII mutations can emerge as late and heterogenous events in glioblastoma development and promote angiogenesis through Src activation. *Neuro-Oncology* **2016**, *18*, 1644–1655. [CrossRef] [PubMed]
21. An, Z.; Aksoy, O.; Zheng, T.; Fan, Q.W.; Weiss, W.A. Epidermal growth factor receptor and EGFRvIII in glioblastoma: Signaling pathways and targeted therapies. *Oncogene* **2018**, *37*, 1561–1575. [CrossRef] [PubMed]
22. Banisadr, A.; Eick, M.; Beri, P.; Parisian, A.D.; Yeoman, B.; Placone, J.K.; Engler, A.J.; Furnari, F. EGFRvIII uses intrinsic and extrinsic mechanisms to reduce glioma adhesion and increase migration. *J. Cell Sci.* **2020**, *133*, jcs247189. [CrossRef] [PubMed]
23. Rutkowska, A.; Stoczyńska-Fidelus, E.; Janik, K.; Włodarczyk, A.; Rieske, P. $EGFR^{vIII}$: An Oncogene with Ambiguous Role. *J. Oncol.* **2019**, *2019*, 1092587. [CrossRef] [PubMed]
24. Sigismund, S.; Avanzato, D.; Lanzetti, L. Emerging functions of the EGFR in cancer. *Mol. Oncol.* **2018**, *12*, 3–20. [CrossRef] [PubMed]
25. Tintori, C.; Fallacara, A.L.; Radi, M.; Zamperini, C.; Dreassi, E.; Crespan, E.; Maga, G.; Schenone, S.; Musumeci, F.; Brullo, C.; et al. Combining X-ray crystallography and molecular modeling toward the optimization of pyrazolo[3,4-*d*]pyrimidines as potent c-Src inhibitors active in vivo against neuroblastoma. *J. Med. Chem.* **2015**, *58*, 347–361. [CrossRef] [PubMed]
26. Vignaroli, G.; Zamperini, C.; Dreassi, E.; Radi, M.; Angelucci, A.; Sanità, P.; Crespan, E.; Kissova, M.; Maga, G.; Schenone, S.; et al. Pyrazolo[3,4-*d*]pyrimidine prodrugs: Strategic optimization of the aqueous solubility of dual Src/Abl inhibitors. *ACS Med. Chem. Lett.* **2013**, *4*, 622–626. [CrossRef] [PubMed]
27. Marengo, B.; De Ciucis, C.; Ricciarelli, R.; Passalacqua, M.; Nitti, M.; Zingg, J.M.; Marinari, U.M.; Pronzato, M.A.; Domenicotti, C. PKCδ sensitizes neuroblastoma cells to L-buthionine-sulfoximine and etoposide inducing reactive oxygen species overproduction and DNA damage. *PLoS ONE* **2011**, *6*, e14661. [CrossRef] [PubMed]
28. Monteleone, L.; Speciale, A.; Valenti, G.E.; Traverso, N.; Ravera, S.; Garbarino, O.; Leardi, R.; Farinini, E.; Roveri, A.; Ursini, F.; et al. PKCα Inhibition as a Strategy to Sensitize Neuroblastoma Stem Cells to Etoposide by Stimulating Ferroptosis. *Antioxidants* **2021**, *10*, 691. [CrossRef] [PubMed]
29. Boulares, A.H.; Yakovlev, A.G.; Ivanova, V.; Stoica, B.A.; Wang, G.; Iyer, S.; Smulson, M. Role of poly(ADP-ribose) polymerase (PARP) cleavage in apoptosis. Caspase 3-resistant PARP mutant increases rates of apoptosis in transfected cells. *J. Biol. Chem.* **1999**, *274*, 22932–22940. [CrossRef]
30. Tintori, C.; Laurenzana, I.; La Rocca, F.; Falchi, F.; Carraro, F.; Ruiz, A.; Esté, J.A.; Kissova, M.; Crespan, E.; Maga, G.; et al. Identification of Hck inhibitors as hits for the development of antileukemia and anti-HIV agents. *ChemMedChem* **2013**, *8*, 1353–1360. [CrossRef] [PubMed]
31. Ayrapetov, M.K.; Wang, Y.H.; Lin, X.; Gu, X.; Parang, K.; Sun, G. Conformational basis for SH2-Tyr(P)527 binding in Src inactivation. *J. Biol. Chem.* **2006**, *281*, 23776–23784. [CrossRef] [PubMed]
32. Boczek, E.E.; Luo, Q.; Dehling, M.; Röpke, M.; Mader, S.L.; Seidl, A.; Kaila, V.R.I.; Buchner, J. Autophosphorylation activates c-Src kinase through global structural rearrangements. *J. Biol. Chem.* **2019**, *294*, 13186–13197. [CrossRef]

33. Lewis-Tuffin, L.J.; Feathers, R.; Hari, P.; Durand, N.; Li, Z.; Rodriguez, F.J.; Bakken, K.; Carlson, B.; Schroeder, M.; Sarkaria, J.N.; et al. Src family kinases differentially influence glioma growth and motility. *Mol. Oncol.* **2015**, *9*, 1783–1798. [CrossRef] [PubMed]
34. Frederick, L.; Wang, X.Y.; Eley, G.; James, C.D. Diversity and frequency of epidermal growth factor receptor mutations in human glioblastomas. *Cancer Res.* **2000**, *60*, 1383–1387. [PubMed]
35. Stettner, M.R.; Wang, W.; Nabors, L.B.; Bharara, S.; Flynn, D.C.; Grammer, J.R.; Gillespie, G.Y.; Gladson, C.L. Lyn kinase activity is the predominant cellular SRC kinase activity in glioblastoma tumor cells. *Cancer Res.* **2005**, *65*, 5535–5543. [CrossRef]
36. Du, J.; Bernasconi, P.; Clauser, K.R.; Mani, D.R.; Finn, S.P.; Beroukhim, R.; Burns, M.; Julian, B.; Peng, X.P.; Hieronymus, H.; et al. Bead-based profiling of tyrosine kinase phosphorylation identifies SRC as a potential target for glioblastoma therapy. *Nat. Biotechnol.* **2009**, *27*, 77–83. [CrossRef]
37. Knight, Z.A.; Lin, H.; Shokat, K.M. Targeting the cancer kinome through polypharmacology. *Nat. Rev. Cancer* **2010**, *10*, 130–137. [CrossRef]
38. Agarwal, S.; Mittapalli, R.K.; Zellmer, D.M.; Gallardo, J.L.; Donelson, R.; Seiler, C.; Decker, S.A.; Santacruz, K.S.; Pokorny, J.L.; Sarkaria, J.N.; et al. Active efflux of Dasatinib from the brain limits efficacy against murine glioblastoma: Broad implications for the clinical use of molecularly targeted agents. *Mol. Cancer Ther.* **2012**, *11*, 2183–2192. [CrossRef]
39. Lassman, A.B.; Pugh, S.L.; Gilbert, M.R.; Aldape, K.D.; Geinoz, S.; Beumer, J.H.; Christner, S.M.; Komaki, R.; DeAngelis, L.M.; Gaur, R.; et al. Phase 2 trial of dasatinib in target-selected patients with recurrent glioblastoma (RTOG 0627). *Neuro-Oncology* **2015**, *17*, 992–998. [CrossRef]
40. Galanis, E.; Anderson, S.K.; Twohy, E.L.; Carrero, X.W.; Dixon, J.G.; Tran, D.D.; Jeyapalan, S.A.; Anderson, D.M.; Kaufmann, T.J.; Feathers, R.W.; et al. A phase 1 and randomized, placebo-controlled phase 2 trial of bevacizumab plus dasatinib in patients with recurrent glioblastoma: Alliance/North Central Cancer Treatment Group N0872. *Cancer* **2019**, *125*, 3790–3800. [CrossRef]
41. Molinari, A.; Fallacara, A.L.; Di Maria, S.; Zamperini, C.; Poggialini, F.; Musumeci, F.; Schenone, S.; Angelucci, A.; Colapietro, A.; Crespan, E.; et al. Efficient optimization of pyrazolo[3,4-d]pyrimidines derivatives as c-Src kinase inhibitors in neuroblastoma treatment. *Bioorg. Med. Chem. Lett.* **2018**, *28*, 3454–3457. [CrossRef] [PubMed]
42. Radi, M.; Tintori, C.; Musumeci, F.; Brullo, C.; Zamperini, C.; Dreassi, E.; Fallacara, A.L.; Vignaroli, G.; Crespan, E.; Zanoli, S.; et al. Design, synthesis, and biological evaluation of pyrazolo[3,4-*d*]pyrimidines active in vivo on the Bcr-Abl T315I mutant. *J. Med. Chem.* **2013**, *56*, 5382–5394. [CrossRef] [PubMed]
43. Colla, R.; Izzotti, A.; De Ciucis, C.; Fenoglio, D.; Ravera, S.; Speciale, A.; Ricciarelli, R.; Furfaro, A.L.; Pulliero, A.; Passalacqua, M.; et al. Glutathione-mediated antioxidant response and aerobic metabolism: Two crucial factors involved in determining the multi-drug resistance of high-risk neuroblastoma. *Oncotarget* **2016**, *7*, 70715–70737. [CrossRef] [PubMed]
44. Alhalabi, O.T.; Fletcher, M.N.C.; Hielscher, T.; Kessler, T.; Lokumcu, T.; Baumgartner, U.; Wittmann, E.; Schlue, S.; Rahman, M.G.; Hai, L.; et al. A novel patient stratification strategy to enhance the therapeutic efficacy of dasatinib in glioblastoma. *Neuro-Oncology* **2022**, *24*, 39–51. [CrossRef]
45. Nishikawa, R.; Ji, X.D.; Harmon, R.C.; Lazar, C.S.; Gill, G.N.; Cavenee, W.K.; Huang, H.J. A mutant epidermal growth factor receptor common in human glioma confers enhanced tumorigenicity. *Proc. Natl. Acad. Sci. USA* **1994**, *91*, 7727–7731. [CrossRef]
46. Shinojima, N.; Tada, K.; Shiraishi, S.; Kamiryo, T.; Kochi, M.; Nakamura, H.; Makino, K.; Saya, H.; Hirano, H.; Kuratsu, J.; et al. Prognostic value of epidermal growth factor receptor in patients with glioblastoma multiforme. *Cancer Res.* **2003**, *63*, 6962–6970.
47. Congdon, K.L.; Gedeon, P.C.; Suryadevara, C.M.; Caruso, H.G.; Cooper, L.J.; Heimberger, A.B.; Sampson, J.H. Epidermal growth factor receptor and variant III targeted immunotherapy. *Neuro-Oncology* **2014**, *16* (Suppl. 8), viii20-5. [CrossRef]

pharmaceutics

MDPI

Article

Choline Kinase α Inhibitors MN58b and RSM932A Enhances the Antitumor Response to Cisplatin in Lung Tumor Cells

Juan Carlos Lacal [1,2,*], Rosario Perona [1,2], Javier de Castro [2] and Arancha Cebrián [1,3]

1 Instituto de Investigaciones Biomédicas, CSIC/UAM, 28029 Madrid, Spain; rperona@iib.uam.es (R.P.); arancha.cebrian@quironsalud.es (A.C.)
2 Instituto de Investigación Sanitaria Hospital La Paz, IDIPAZ, 28046 Madrid, Spain; javier.decastro@salud.madrid.org
3 Instituto de Investigación Sanitara Fundación Jiménez Díaz, 28040 Madrid, Spain
* Correspondence: jclacal@iib.uam.es; Tel.: +34-914-975-438

Abstract: Lung cancer is one of the main causes of death in developed countries, and non-small cell lung cancer (NSCLC) is the most frequent type (80% of patients). In advanced NSCLC, platinum-based chemotherapy is the frontline palliative treatment, but less than 5% of patients achieve prolonged survival. Immunotherapy has recently been proposed as the standard of care (SoC) as either monotherapy or in combination with chemotherapy for advanced NSCLC. The levels of expression of PD-L1 are the only predictive biomarkers for patient assessment. Although around 30% of patients receiving immunotherapy achieve 5-year survival, a significant number does not benefit from this novel therapeutic approach. Therefore, there is a need for novel strategies to improve clinical outcomes. The expression level of choline kinase α (ChoKα) is increased in a large number of human tumors, including NSCLC tumors, and constitutes an independent prognostic factor for early-stage NSCLC patients. Thus, ChoKα has been postulated as a new target drug in cancer therapy. The combination of cisplatin with novel targeted drugs such as choline kinase inhibitors may improve both the survival rates and the quality of life of NSCLC patients and may serve as the basis for the development of new therapeutic approaches. To that aim, we developed several in vitro and in vivo approaches to assess the antitumor activity of a novel combination regimen using cisplatin and ChoKα inhibitors. Our results suggest that a proper combination of specific inhibitors of the NSCLC prognostic factor ChoKα and platinum-based conventional chemotherapy might constitute a new, efficient treatment approach for NSCLC patients. This novel approach may help reduce the toxicity profile associated with cisplatin since, despite the advances in NSCLC management in recent years, the overall 5-year survival rate is still poor.

Keywords: choline kinase; lipid metabolism; cisplatin; lung tumors; combinatorial chemotherapy

Citation: Lacal, J.C.; Perona, R.; de Castro, J.; Cebrián, A. Choline Kinase α Inhibitors MN58b and RSM932A Enhances the Antitumor Response to Cisplatin in Lung Tumor Cells. *Pharmaceutics* **2022**, *14*, 1143. https://doi.org/10.3390/pharmaceutics14061143

Academic Editor: Francesca Musumeci

Received: 7 April 2022
Accepted: 24 May 2022
Published: 27 May 2022

Publisher's Note: MDPI stays neutral with regard to jurisdictional claims in published maps and institutional affiliations.

1. Introduction

Over 2.2 million cases of lung cancer are diagnosed each year worldwide, constituting the leading cause of cancer-related mortality, with 1.8 million deaths in 2020 (Globocan, 2020; see https://gco.iarc.fr/today/online-analysis-table?v=2020&mode=cancer&mode_population=continents&population=900&populations=900&key=asr&sex=0&cancer=39&type=1&statistic=5&prevalence=0&population_group=0&ages_group%5B%5D=0&ages_group%5B%5D=17&group_cancer=1&include_nmsc=0&include_nmsc_other=1, accessed on 23 May 2022). Over 80% of all lung cancers belong to the "non-small cell" histological subgroup, known as non-small cell lung cancer (NSCLC). A total of 20–25% of NSCLC patients present with stage III, and 35–40% are stage IV. For this advanced disease, firstline treatment options have evolved in recent years with immunotherapy in patients with high levels of PD-L1. This approach has rendered an impressive 30% survival rate at 5 years; however, the remaining 70% experience no benefit, and the mainstay still consists of platinum-based chemotherapy doublets. Furthermore, concurrent chemoradiotherapy is still the therapeutic approach used in patients with unresectable stage III and good performance status. However, a wide range of side effects are associated with this regimen, including nausea, vomiting, and myelosuppression, as well as oto-, neuro-, and nephro-toxicity, accounting for a strong decrease in quality of life [1,2].

In order to reduce the toxicity profile of cisplatinum-based regimens, the use of new generations of platinum drugs, including carboplatin and oxaliplatin, has been introduced in the backbone of systemic chemotherapy in NSCLC, but adverse events still display a relevant impact on the quality of life of these patients. Subsequently, the introduction of third-generation cytotoxic drugs to platinum agents (known as doublets), including paclitaxel, docetaxel, gemcitabine, pemetrexed, and vinorelbine, has yielded higher response rates, maintaining the use of platinum doublets as standard therapy for NSCLC treatment [3,4].

More recently, new targeted, personalized therapies based on identified oncogenic drivers usually alone, but also combined with platinum doublets, with the introduction of patient selection procedures, have been explored with the main goals of improving clinical outcomes and minimizing side effects of treatments within drug optimization. Among these, interferences with EGFR, ALK, ROS 1, BRAF, NTRK, RET, and MET signaling have provided improved outcome in these patients [5]. Finally, the advent of immunotherapy based on specific antibodies against the programmed death (PD-1) receptor, the programmed death-ligand 1 (PD-L1), and the cytotoxic T-lymphocyte–associated protein 4 receptor (CTLA-4) has had a great impact on survival for some NSCLC patients. In metastatic tumors, recent clinical trials favored the use of these immune checkpoint inhibitors (ICIs) as first- or second-line treatments alone or in combination with chemotherapy [6]. The levels of expression of PD-L1 in tumor cells are the reference biomarkers for patient assessment of this approach. However, despite these advances, a significant number of patients cannot benefit from these therapies. Therefore, there is still a need for testing new targets and new combinatorial regimens.

Choline kinase alpha (ChoKα), the first enzyme of the Kennedy pathway for the biosynthesis of the major phospholipid of the plasma membrane, phosphatidylcholine (PC), has been described as a novel oncogene [7] whose gene expression levels constitute a new prognostic factor in NSCLC patients [8]. Consequently, ChoKα-specific inhibitors have been designed and showed strong in vitro antiproliferative activity against human lung-cancer-derived models, as well as efficient antitumor activity in vivo in nude mice against different human xenografts [9]. These results were further confirmed using siRNA-driven interference molecules [10–12]. The development of RSM-932A/TCD-717, a highly specific ChoKα inhibitor [13] allowed entrance into a Phase I clinical trial as a new "first in class" targeted therapy (http://clinicaltrials.gov/ct2/show/NCT01215864; accessed on 6 April 2022). Here, we present in vitro and in vivo approaches to assess the antitumor activity of a novel combination regimen using cisplatin and ChoKα inhibitors.

2. Materials and Methods

2.1. Patients

Specimens of lung cancer tissue from 63 randomly selected patients who underwent surgical resection for NSCLC between 2001 and 2004 and who were followed up on by the Medical Oncology division at La Paz Hospital in Madrid were used for this analysis. No adjuvant therapy was administrated to these patients. The institutional review board of the hospital approved the study, and written informed consent was obtained from all the patients.

2.2. Primary Cultures of NSCLC Tumors

The resected tissues from the NSCLC patients were enzymatically and chemically dissociated (Cell dissociation sieve-tissue grinder Kit, SIGMA-Aldrich, Madrid, Spain), and the obtained cells were seeded into 96-well plates (BD, Falcon, Bioscience, San Jose, CA, USA), as previously described [14]. The cells were treated with increasing concentrations (0, 0.5, 1, 5, 10, and 20 μM) of cDDP, taxol, vinolrelbine, gemcitabine, or MN58b for 10 days in DMEM:F12HAM (Ref:D8437, SIGMA-Aldrich) supplemented with 10% fetal bovine serum (FBS) (Life Technologies, Grand Island, NY, USA). The final persistent population in each well was quantified with an Alamar blue assay, as previously described [14]. Alamar blue was added directly into the culture media at a final concentration of 10%, and the plate was returned to the incubator. The optical density of the plate was measured at 570 and 600 nm with a standard spectrophotometer at 3 h after adding Alamar blue. The cell viability was calculated according to the manufacturer's protocol (Biosource Europe, Nivelles, Belgium).

2.3. Cell Lines and Chemicals

All the cell lines used in this study were maintained under standard conditions of temperature (37 °C), humidity (95%), and carbon dioxide (5%). Epithelial non-small lung cancer H460 cells (generated from a pleural effusion of a patient with large cell lung cancer) and H1299 cells (derived from an NSCLC carcinoma) were maintained in RPMI supplemented with 10% fetal bovine serum (FBS) (Life Technologies, Grand Island, NY, USA). Cell lines resistant to the ChoKα-specific inhibitors MN58b or RSM-932A/TCD-717 were generated by prolonged continuous exposure to increasing concentrations of each drug and were maintained in RPMI supplemented with 10% FBS. A parallel control (H460 stock) of the cell line in the absence of the compounds was kept in culture for the same time. In-house-generated MN58b and RSM-932A/TCD-717 were dissolved in PBS and DMSO: PBS 2:1, respectively. Cisplatin was obtained from Ferrer Farma (EGF speciality).

2.4. Cell Proliferation Assays and Combined Index Evaluation

An amount of 6000 cells/well was seeded into 96-well flat-bottomed plates (BD, Falcon, Bioscience, San Jose, CA, USA) and incubated for 24 h under standard conditions. The cells were then treated with different concentrations of ChoKα inhibitors and cisplatin either concomitantly or in a sequential manner, with cisplatin for 5 h followed by a ChoKα inhibitor (MN58b or RSM-932A/TCD-717) for the next 40 h. Quantification of the number of cells remaining in each well was carried out using the MTT (3-(4,5-dimethylthiazol-2-yl)-2,5-diphenyltetrazolium bromide) method. The absorbance was read at 560 nm with a VersaMax Microplate Reader (Molecular Devices, Sunnyvale, CA, USA). Drug interaction between the ChoKα inhibitors and cisplatin was assessed using a combination index (CI) [15] where CI < 1, CI = 1, and CI > 1 indicated synergistic, additive, and antagonistic effects, respectively. The CI value was calculated as $CI = D_1/Df_1 + D_2/Df_2$, where Df_1 and Df_2 are the concentrations of MN58b or RSM-932A/TCD-717 and cisplatin, respectively, required to inhibit cell growth by 50%, and D_1 and D_2 are the drugs' concentrations in the combination treatment that also inhibited cell growth by 50%. The data analysis was performed with Calcusyn software (Biosoft, Oxford, UK).

2.5. Flow Cytometric Assay

The cell cycle distribution was determined by DNA content analysis after propidium iodide staining. The cells were treated with ChoKα inhibitors and cisplatin alone or in combination for 45 h. The cells were then harvested, stained, and incubated with 50 µL detergent and 1 mL IP (50 µg/mL) containing RNase (20 µg/mL). The DNA content of approximately 4×10^5 stained cells was analyzed using a Coulter XL-MZL flow cytometer. The fractions of cells in apoptosis, G0-G1, S, and G2-M phases were analyzed with DNA program software.

2.6. Western Blot Analysis

The cells were incubated with ChoKα inhibitors and cisplatin alone or in combination. Equal amounts of protein (30 µg) were loaded into 15% SDS-PAGE acrylamide gels, and the resolved proteins were transferred onto nitrocellulose membranes. The following antibodies were used: anti-p-JNK (Promega, v7932), anti-p-p38 (Cell Signaling 9211S), and anti-P38 (Santa Cruz sc-535), with anti-α-tubulin (Sigma T9026) as a load control.

2.7. *In Vivo Antitumoral Assays*

Female athymic BALB/C nude mice were supplied by Jackson. The animal care and surgery protocols were approved by the Animal Care Committee. The mice were maintained with autoclaved water and sterile food ad libitum. The mice were inoculated subcutaneously with injections of 1×10^6 H460 cells in the flank of each mouse mixed 1:1 with matrigel (354234, BD Bioscience, Haryana, India). The tumor sizes were determined using micrometer calipers, and when the size of the tumors was approximately 0.1 cm^3, the mice were divided into 8 different groups: control (vehicle); 2 mg MN58b/kg for 3 days/week; 1 mg cisplatin/kg for 2 days/week; 2 mg RSM-932A/TCD-717/kg for 2 days/week; and MN58b-cisplatin or RSM-932A/TCD-717-cisplatin combinations following both a sequential and concomitant schedule. MN58b and cisplatin were dissolved in PBS and injected i.p. in the amount of 0.1 mL/mouse. RSM-932A/TCD-717 was dissolved in DMSO:PBS 2:1 and diluted with PBS to give appropriate concentrations. The tumor sizes were measured twice a week at their greatest lengths and widths, and the volumes were calculated as (tumor width^2 $\times$ tumor length)/2.

2.8. *Statistical Analysis*

The correlations in responsiveness to the different treatments in the tumor samples of patients with NSCLC were performed following Pearson's correlations (*r*) to measure the linear correlation between two different treatments. This is achieved by determining the ratio between their covariance and the product of their standard deviations according to the following formula:

$$r_{x,y} = \mathrm{COV}_{(x,y)} / \sigma_x \sigma_y$$

where COV is the covariance, σ_x is the standard deviation of x, and σ_y is the standard deviation of y. This rendered a normalized measurement of the covariance, such that the result always had a value between −1 and +1. A Pearson's correlation of ($r = -1$) meant a negative perfect correlation, while ($r = 1$) was a perfect positive alignment. A value of 0 implied that there was no linear dependency between the variables. The values for *r* represented comparisons between pairs of the drugs used in the study and are reflected in Figure 1. The dispersion graphics are also reported.

Comparisons in the tumor volumes between the untreated and treated groups were performed using a nonparametric Mann–Whitney test. Two-sided *p*-values less than 0.05 were considered statistically significant. All the calculations were performed using SPSS software, version 20.0 (SPSS Inc., Chicago, IL, USA).

3. Results

3.1. *Cisplatin-Intrinsic-Resistant NSCLC Tumors Are Sensitive to ChoKα Inhibition*

The primary cultures from 63 resected NSCLC tumors were established and cultivated for 10 days, during which they were treated with increasing concentrations of the ChoKα-specific inhibitor MN58b, cisplatin (cDDP), taxol, vinorelbine, or gemcitabine. Considering resistance when nearly 100% of the cells remained alive for the maximum concentration of the drug at day 10, different resistance rates to the treatments were found. As shown in Table 1, more than 50% of the samples were resistant independently to the treatment, in keeping with the lack of response to chemotherapy observed in the clinic.

A correlation analysis of the responsiveness of these tumors to the different treatments was then performed (Figure 1). Pearson's correlation indicated significant cross-resistance among cisplatin, taxol, vinorelbine, and gemcitabine. By contrast, MN58b showed no significant cross-resistance to any of the chemotherapeutic drugs tested, consistent with its different mechanism of action [16,17].

Table 1. Responsiveness of NSCLC primary cultures to different antitumor agents. Resected tissues from NSCLC patients were dissociated to obtain primary cultures that were treated for 10 days with different concentrations of the indicated drugs. Cell viability was determined by Alamar blue. The number of total samples examined in each case and the number of samples sensitive or resistant to each treatment are reported with their proportions (%).

Drug	NSCLC Samples	Sensitive (%)	Resistant (%)
MN58b	63	35 (55.6)	28 (44.4)
cDDP	62	31 (50.0)	31 (50.0)
Taxol	62	27 (43.5)	35 (56.5)
Vinorelbine	39	15 (38.5)	24 (61.5)
Gemcitabine	52	18 (34.6)	34 (65.4)

		MN58b	**cDDP**	**Taxol**	**Vinorelbine**	**Gemcitabine**
MN58b	*r*	1				
	s					
	n	63				
cDDP	*r*	−0.096	1			
	s	0.46				
	n	62	62			
Taxol	*r*	−0.09	0.319(*)	1		
	s	0.488	0.012			
	n	62	62	62		
Vinorelbine	*r*	−0.124	0.469(**)	0.421(**)	1	
	s	0.45	0.003	0.008		
	n	39	39	39	39	
Gemcitabine	*r*	0.133	0.074	0.364(**)	0.174	1
	s	0.346	0.603	0.008	0.304	
	n	52	52	52	37	52

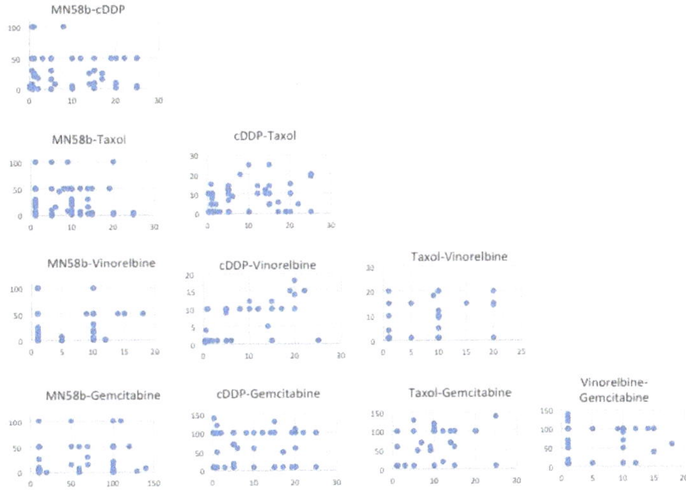

Figure 1. Pearson's correlation analysis of intrinsic resistance of primary tumors of patients with NSCLC to different antitumoral agents. Upper panel: the statistical analysis was performed as indicated in Material and Methods for the response rate of the samples to the different treatments. MN58b was the only drug whose response did not correlate in responsiveness to any other agent. *r*: Pearson's correlations coefficient; s: significant correlation, bilateral (* $p \leq 0.05$; ** $p \leq 0.01$); n: number of primary tumors analyzed. Lower panel: dispersion graphs for each pair of drugs.

To further confirm the absence of cross-resistance between ChoKα inhibitors and cisplatin, resistant H460 cells were generated by exposure to increasing doses of MN58b or RSM-932A/TCD717. The cell lines acquired drug resistance over a long-term period of 9–10 months, after which their sensitivity to both ChoKα inhibitors and cisplatin was determined (Table 2). The fold resistance was from 7–8 for RSM-932A/TCD-717 and from 49–73 for MN58b. A control cell line (H460 stock) was maintained in culture for the same time without treatment and was included as a comparison with the parental cell line. As a consequence of the similar mechanisms of action of the two ChoKα inhibitors, resistant cells to either MN58b (H460 MN58R) or to RSM-932A/TCD-717 (H460 TCD717R) showed an important cross-resistance between them. By contrast and consistent with a different mechanism of resistance acquisition, a significant lack of cross-resistance was found with cisplatin in both cases. In fact, cells resistant to the ChoKα inhibitors were slightly more sensitive to cisplatin.

Table 2. Absence of cross-resistance between cisplatin and ChoKα inhibitors. NSCLC H460 cells and H460 cells made resistant to ChoKα inhibitors MN58b (H460 MN58R) or RSM-932A/TCD-717 (H460 TCD717R) were treated with different concentrations of MN58b, RSM-932A/TCD-717, and cisplatin. The concentration where 50% of the cell proliferation was inhibited (IC_{50}) was determined by MTT method. Data represent the mean ± SD of 3 to 9 independent experiments, each performed in quadruplicate. The fold induction of resistance is shown within parentheses.

Cell Line	IC_{50} MN58b (μM)	IC_{50} TCD-717 (μM)	IC_{50} cDDP (μM)
H460	0.28 ± 0.12	1.11 ± 0.4	16.60 ± 1.8
H460 Stoc	0.39 ± 0.19	1.32 ± 0.43	14.95 ± 2.6
H460 MN58R	19.2 ± 2.6 (49)	9.5 ± 1.19 (7)	6.60 ± 1.2 (0.4)
H460 TCD717R	28.8 ± 10.5 (73)	10.8 ± 2.26 (8)	4.98 ± 0.9 (0.3)

3.2. *Synergism of Cisplatin and ChoKα Inhibitors against NSCLC Cells*

The results shown above suggest the putative effectiveness of combined therapy between cisplatin and ChoKα inhibitors. In order to address this issue, the cytotoxic effects of combinations of cisplatin and the ChoKα inhibitors MN58b or RSM-932A/TCD-717 were determined in the human NSCLC H460 cell line (Figure 2A,B). To this end, cell viability was determined after treatment with different drug concentrations and schedules. Strong synergism (combination index, CI ≤ 0.4) was observed when the cells were treated following a sequential treatment initiated with cisplatin for a short period of time (5 h), followed by treatment with RSM-932A/TCD-717 (Figure 2A) or MN58b (Figure 2B) for longer times. Synergism was observed in the different timepoints analyzed, treating with ChoKα inhibitors from 24 to 72 h. A moderate synergistic effect was also found with both agents following a concomitant treatment (CI = 0.5–0.8) (data not shown). These results were confirmed using a second NSCLC cell line, H1299 (Figure 2C,D).

With the aim of further validating these results, we investigated the effect of this combination on cell cycle distribution using flow cytometry analysis. As shown in Figure 2, the combination of these agents displayed a stronger antiproliferative effect than when treating with these drugs separately. Furthermore, the promotion of cell death mediated by the combination was observed even under conditions when these agents separately displayed a minimum effect.

With the aim of elucidating the mechanism of this synergism, the putative modulation mediated by ChoKα inhibitors of JNK and p38 known to be involved in cell death induced by cisplatin [18] was investigated. As shown in Figure 3A, a two-fold activation of JNK was promoted by cisplatin after 12 h of treatment. ChoKα inhibitors induced a significant greater increase of activation of this pathway (3–4 fold). However, the major effect was observed after the combination of cisplatin and RSM-932A/TCD-717, increasing basal levels of p-JNK up to seven-fold induction. On the other hand, the p38 pathway was significantly more affected by cisplatin than by the ChoKα inhibitors (Figure 3B), also with greater effects after combination. These results suggested that the p38 pathway is mainly activated by cisplatin, whereas JNK is mostly modulated by the ChoKα inhibitors, suggesting that the synergistic effect may be a consequence of a complementary increase in both pathways of the combinatory regimen.

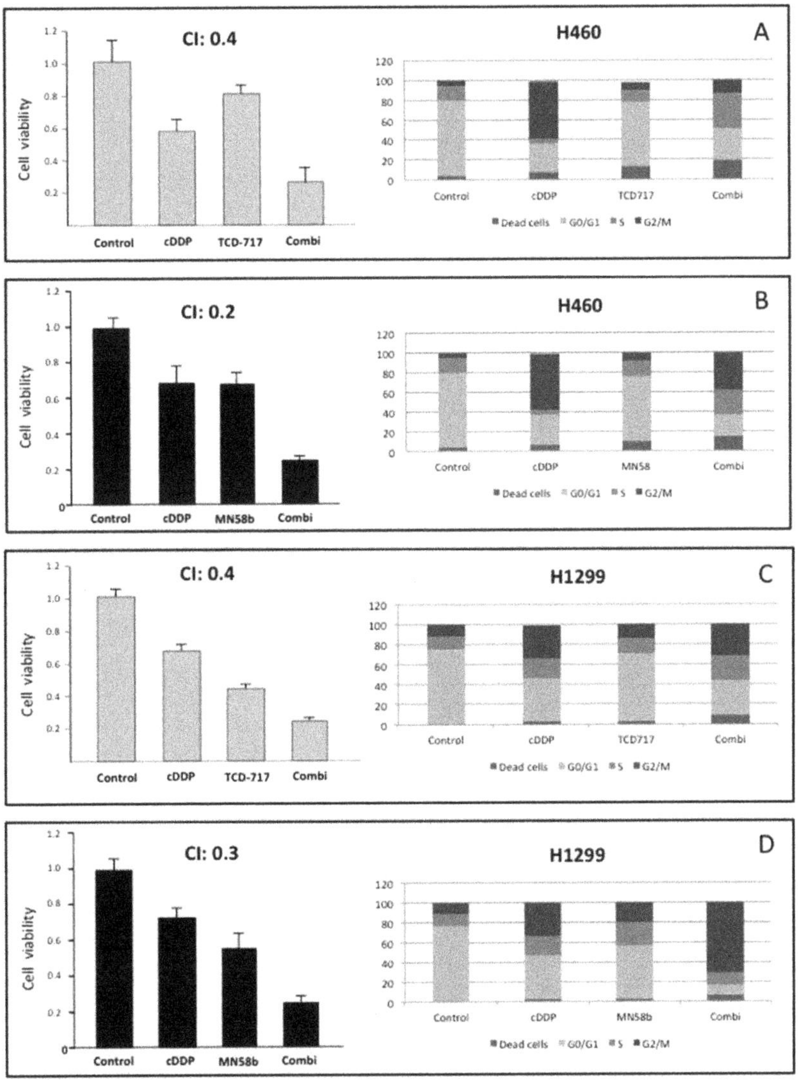

Figure 2. Synergistic effect of ChoKα inhibitors and cisplatin in H460 and H1299 cells. (**A**,**B**) H460 cells were exposed to cisplatin for 5 h. Then, cells were treated with RSM-932A/TCD-717 (**A**) or MN58b (**B**) for 40 h. Cytotoxicity was evaluated by MTT assay (left panels). Combinations of the two drugs was also tested using flow cytometry analysis (right panels). (**C**,**D**) H1299 cells were treated as indicated for H460 cells with either RSM932A/TCD-717 (**C**) or MN58b (**D**). Represented CI value in each case is the mean of three independent experiments for each concentration in quadruplicate. As shown for both drugs, a CI < 1 indicated a strong synergistic effect. Furthermore, combination of the two drugs increased cell death compared to the two drugs alone, as determined by flow cytometry analysis. As shown, similar results were obtained with both cell lines.

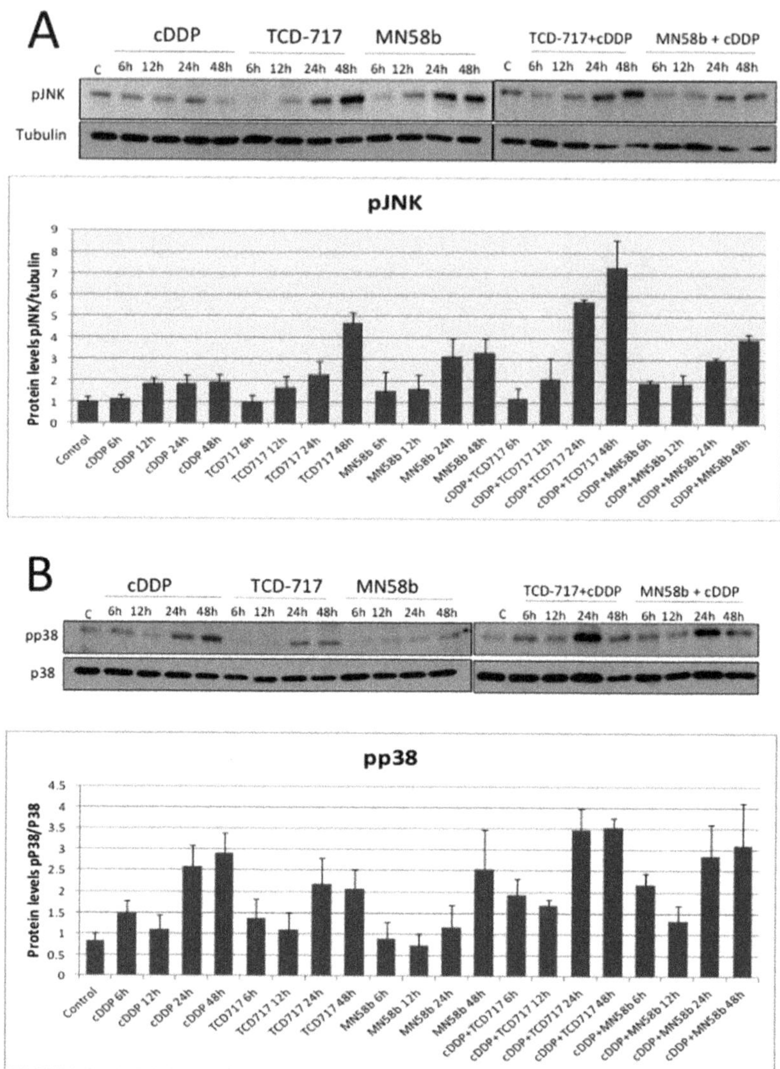

Figure 3. MAPK pathway activation in response to treatment with cisplatin combined with ChoKα inhibitors. H460 human NSCLC cells were exposed to cisplatin, ChoKα inhibitors, or both treatments following a sequential schedule and showing a synergistic effect (cisplatin followed by ChoKα inhibitors). p-JNK (**A**) and p-p38 (**B**) levels were analyzed by Western blotting at different timepoints. Tubulin or total p38 protein were used to normalize. Results are the means of three independent experiments. Western blots are representative of one single experiment.

3.3. ChoKα Inhibitors Potentiate the Antitumoral Efficacy of Cisplatin In Vivo against NSCLC Xenografts, Reducing the Toxicity Associated with Cisplatin Treatment

Cisplatin is still used in firstline chemotherapeutic regimens for NSCLC patients. As shown above, synergistic effects were observed when this agent was combined with ChoKα inhibitors in vitro in NSCLC cells. In order to further evaluate the potential therapeutic effect of these combinations, a series of experiments was carried out in vivo using human NSCLC H460 xenografts. Mice were inoculated subcutaneously with an injection of

1×10^6 H460 cells in the flank mixed with Matrigel, as described in Material and Methods. When tumor volumes reached 0.1 cm^3, the animals were randomly distributed in groups ($n = 8$), and treatments were assayed according to the in vitro regimens. The schedules of the treatments were determined when the maximum efficacy was reached following sequential treatment regimes initiated by cisplatin. A schematic representation of the followed treatments is shown in Table 3. Treatment of the NSCLC H460 tumors with ChoKα inhibitors in combination with cisplatin resulted in a significant reduction in tumor growth compared to these drugs alone (Figure 4).

Table 3. Synergistic antitumoral effect of cisplatin and ChoKα inhibitors in vivo. The antitumoral effect of the combination of cisplatin and MN58b or RSM932A/TCD-717 was evaluated in vivo using NSCLC xenografts. H460 cells were injected in nude mice, and when tumor volumes reached 0.1 cm^3, animals were randomly distributed in groups ($n = 8$) to start treatments according to the schedules described in Material and Methods (indicated as the days treatment was performed for each drug or vehicle (DMSO)). Tumor growth inhibition, mean body weight, and statistical significance are shown. More similar efficacy was observed with the combinations than with much higher concentrations of cisplatin alone, significantly reducing toxicity. * statistical significance $p < 0.05$.

MN58b	**Drug**	**Schedule (Days)**	**Mean Body Weigth (Day 22)**	**Tumor Gowth Inhibition (Day 22)**	***p***
Control-1	Vehicle	1,3,4,5,8,10,11,12,15,17,18,19	23.4		
cDDP	cDDP (1 mg/kg)	1,4,8,11,15,18	23	33%	0.3
MN58b	MN58b (2 mg/kg)	1,3,5,8,10,12,15,17,19	24.4	35%	0.09
Sequential	cDDP (1 mg/kg)	1,4	23.9	67%	0.017 *
	MN58b (2 mg/kg)	8,10,12,15,17,19			
Concomitant	cDDP (1 mg/kg)	1,4,8,11,15,18	22.5	66%	0.160 *
	MN58b (2 mg/kg)	1,3,5,8,10,12,15,17,19			
Control cDDP	cDDP (4 mg/kg)	1,3,5,8,10,12,15,17,19	17.4 *	69%	0.014 *
TCD-717	**Drug**	**Schedule (Days)**	**Mean Body Weigth (Day 22)**	**Tumor Gowth Inhibition (Day 22)**	***p***
Control-2	Vehicle	1,4,8,11,15,18	20.6		
cDDP	cDDP (1 mg/kg)	1,4,8,11,15,18	20.9	24%	0.4
TCD-717	TCD717 (2 mg/kg)	1,4,8,11,15,18	19.9	31%	0.16
Sequential	cDDP (1 mg/kg)	1,4	21.2	51%	0.03 *
	TCD717 (2 mg/kg)	8,11,15,18			
Concomitant	cDDP (1 mg/kg)	1,4,8,11,15,18	20.1	55%	0.04 *
	TCD717 (2 mg/kg)	1,4,8,11,15,18			

With this combination therapy, no external signs of toxicity were observed during the treatment, as indicated by the maintenance of body weight. The concomitant treatment of MN58b reached the same efficacy as that of the sequential group, but much higher doses of cytotoxic agents were required. By contrast, to obtain similar effectiveness in the inhibition of tumor growth (65%) using cisplatin as a single agent, its dosage had to be increased from 1 to 4 mg/kg, conditions in which clear signs of toxicity were observed, including a significant reduction in body weight (Table 3). Similar results were obtained when the RSM-932A/TCD-717 inhibitor was used (Table 3).

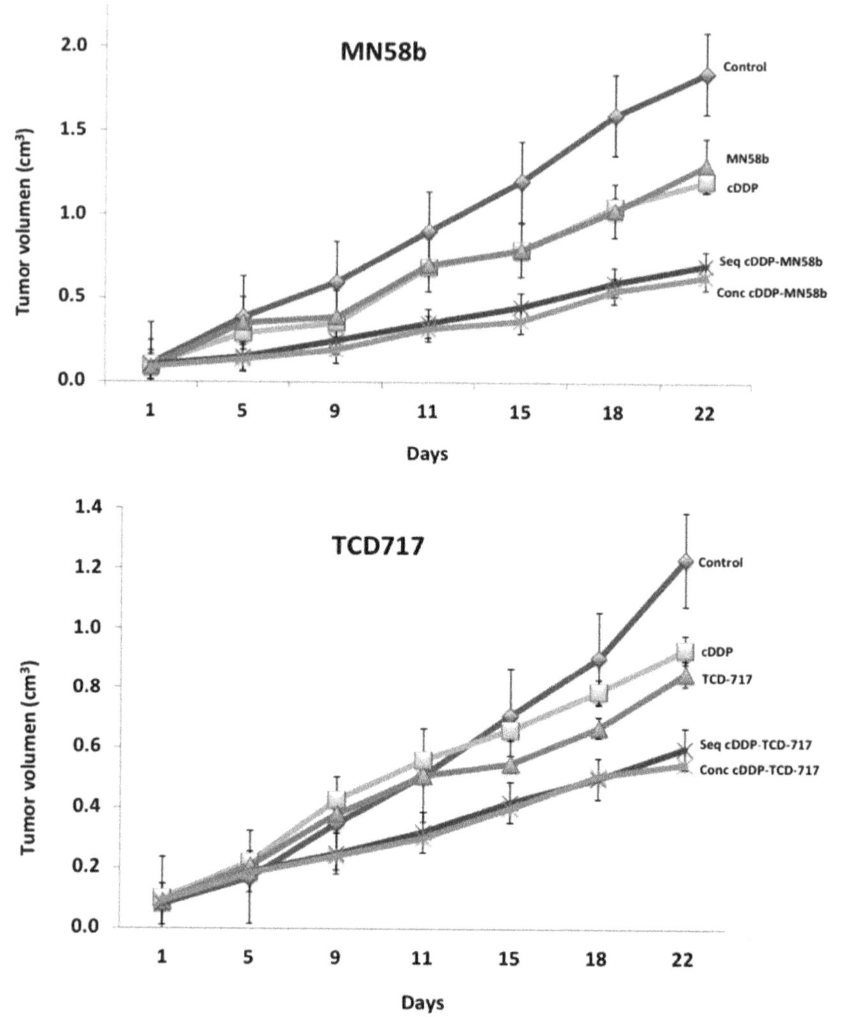

Figure 4. Tumor growth inhibition on H460 xenografts by combination of MN58b or RSM-932A/TCD-717 and cisplatin. H460 cells were injected in nude mice, and when tumor volumes reached 0.1 cm^3, animals were randomly distributed in groups (n = 8) to start treatments. Mice were treated with the indicated ChoKα and cisplatin (cDPP) either alone or following sequential or concomitant schedules. A significant synergistic effect was observed with the combined therapy in both cases: MN58b (upper) and RSM-932A/TCD-717 (lower).

4. Discussion

NSCLC accounts for 80–85% of all bronchogenic malignancies. Included in this category are squamous cell carcinoma (25%), adenocarcinoma (40%), and large cell carcinoma (10%). Extensive research has been performed in the last decade using newly developed drugs as an important attempt to improve clinical outcomes, aiming at achieving higher responses and longer survival rates. In this sense, although the introduction of new, active agents, including paclitaxel, docetaxel, gemcitabine, pemetrexed, and vinorelbine, was promising, the finding of strong cross-resistances is a serious drawback [4]. Furthermore, cancer embraces different diseases, and specific drugs and regimens are required for each specific tumor type. Currently, among the different drugs approved by the FDA as anti-cancer agents, only a few are related to NSCLC. Bevacizumab, a monoclonal antibody that binds to the vascular endothelial growth factor (VEGF), blocking interaction with its receptors on endothelial cells, prevents the initiation of new blood vessel growth. Some studies have demonstrated that the addition of bevacizumab to platinum-based chemotherapy in addition to gemcitabine or paclitaxel improves chemotherapy response, progression-free survival, and overall survival. The use of the tyrosine kinase inhibitor gefitinib first and, more recently, osimertinib proved to be efficient as a first therapy for NSCLC habouring EGFR mutations [5].

Efforts have been dedicated in recent years to immunotherapy based on specific antibodies against PD-1, its ligand (PD-L1), and CTLA-4. This effort has translated into improved survival for NSCLC patients. Both first- and second-line treatments have been intensively explored, either alone or in combination with diverse chemotherapy treatments. As a result of this search for improved protocols and regimens, the use of these ICIs along with chemotherapy seems to be most promising and has translated into improved survival for NSCLC patients [6]. However, for 70% of these NSCLC patients, the OS is still in very poor figures and makes it necessary to investigate further alternative targets and new combinatorial regimens. Keeping with this, the use of drugs with different mechanisms of action may be useful for overcoming drug resistance to individual treatments. Here, we reported the successful combination of platinum-based chemotherapy with a novel targeted therapy approach as an innovative strategy against solid tumors, such as inhibition of an enzyme that is critical for phospholipid metabolism.

Both primary and acquired resistance is one of the major problems in cancer treatment. The mechanism of action for cisplatin consists of the interaction with N7 sites of purine bases to form inter- and intrastrand crosslinks. The main mechanisms of resistance to cisplatin are reduction in the accumulation of cisplatin by changing uptake or efflux; inactivation of the drug by glutathione, metallothionein, or other sulphur-containig molecules; increased repair of adducts; and increased cisplatin adduct tolerance [19–23]. In contrast, the mechanism of action for ChoKα has been related to the metabolic disbalance produced by the inhibition of phosphatdylcholine synthesis. The inhibition of this enzyme by pharmacological or siRNA approaches produces a loss of mitochondrial potential and cytochrome c release, increased ceramide production, ER stress, unfold protein response (UPR), and ROS homeostasis via glutathione levels. Additional effects due to mitochondria function render a reduction in citrate synthase expression and AMPK activation. Finally, this cascade of events induces an increase in glucose and acetate uptake to overcome the metabolic stress. All these events induce apoptosis or necrosis, specifically in cancer cells (reviewed in [9]).

Resistance to ChoKα has been related to an increase in acid ceramidase, which drains the elevated levels of ceramides and allows cancer cells to keep proliferating [16]. This effect was found in primary NSCLC cultures and in ChoKα-resistant H460 cells, and it was reverted by acid ceramidase inhibitors [24]. However further studies are needed to define precisely a complete set of biomarkers to define ChoKα inhibitor resistance. This will allow efficient selection for NSCLC patients who are potentially resistant to the proposed combination.

In this study, we demonstrated that neither cisplatin nor other drugs usually taking part in platinum-based chemotherapy of NSCLC patients displayed cross-resistance with ChoKα inhibitors. These results may be explained by the differential mechanisms of action of these novel drugs, whose activities rely on inhibiting ChoKα, the first enzyme of the biosynthesis of the major phospholipid component of cell membranes [16,17], or interfering with the replication of DNA [20]. This effect is reflected in the responsiveness of primary tumors from NSCLC patients, and also in tumor-derived cell lines [24]. Therefore, when cells are promoted to division, they are simultaneously blocked at two different phases of the cell cycle, increasing the possibilities to overcome resistance to any of these agents when used alone. This is further supported by the evidence presented in our study where the p38 and the JNK pathways were differentially activated by cisplatin and ChoKα inhibitors, while the combination of both drugs triggered a complementary increase in both pathways.

We also demonstrated here, both in vitro and in vivo, that the use of this combined therapy achieved another important aim of a combination regimen: obtaining similar efficacy using lower concentrations of the chemotherapeutic agents and, therefore, reducing their associated toxicities. ChoKα inhibitors have entered Phase I clinical trials, in part based on the involvement of this enzyme and the effect of its inhibitors on NSCLC. In this sense, ChoKα gene expression levels are independent prognosis factors for early-stage NSCLC patients [8], suggesting an easy molecular method for identifying patients that could benefit from this treatment. Therefore, since platinum-based chemotherapy continues to constitute a key tool in the management of NSCLC and ChoKα inhibition constitutes a "first in class" targeted therapy for NSCLC patients, our study provides the basis for a promising new alternative of combination treatment therapy for patients with this disease.

5. Conclusions

Cancer is a multifactorial disease with a large diversity of clinical manifestations. Lung cancer represents one the most devastating types of cancer, being responsible for over 1.8 million deaths every year worldwide. NSCLC represents over 80% of all lung cancers. Significant improvements in clinical management have been achieved in the last decade with the introduction of third-generation cytotoxic drugs, targeted personalized therapies, and immune checkpoint inhibitors (ICIs). However, the clinical outcome of a 5-year survival rate is still poor. New therapeutic approaches are urgently needed for these patients. We provided evidence that ChoKα is a bona fide new target for NSCLC patients and that its inhibitors are potent therapeutic agents that can be efficiently combined with cisplatin-based chemotherapeutic regimens. A limitation of our study is that we presented this effect in two established NSCLC cell lines, H460 (derived from a large cell lung cancer) and H1299 (derived from a carcinoma), which represent only 50% of the NSCLC types. This novel strategy requires confirmation in appropriate Phase II clinical trials.

Author Contributions: Conceptualization, J.C.L.; methodology, J.C.L., R.P., J.d.C. and A.C.; resources, J.C.L., R.P. and J.d.C.; writing—original draft preparation, J.C.L.; writing—review and editing, J.C.L., R.P., J.d.C. and A.C.; supervision, J.C.L.; funding acquisition, J.C.L. and R.P. All authors have read and agreed to the published version of the manuscript.

Funding: This research was funded by MCIN/AEI/10.13039/501100011033, Grant PID2020-116165RB-C21 and Fondo de Investigaciones Sanitarias, Instituto de Salud Carlos III, Spain, Grant PI20-00335 supported by FEDER funds.

Institutional Review Board Statement: The study was conducted in accordance with the Declaration of Helsinki and approved by the Institutional Ethics Committee of the La Paz Hospital (HULP: PI-1409) on 10 December 2012. The animal study protocol was approved by the Institutional Ethics Committee of CSIC (CEEA-CNB: 080025) on 18 February 2008.

Informed Consent Statement: Informed consent was obtained from all subjects involved in the study.

Acknowledgments: We acknowledge Ana Ramírez de Molina for discussion of the research design, as well as Ana de la Cueva, Rosario Machado-Pinilla, and Vanessa Rodríguez-Fanjul for their excellent technical contributions.

Conflicts of Interest: The authors declare no conflict of interest. The funders had no role in the design of the study; in the collection, analysis, or interpretation of data; in the writing of the manuscript; or in the decision to publish the results.

References

1. Chaitanya Thandra, K.; Barsouk, A.; Saginala, K.; Sukumar Aluru, J.; Barsouk, A. Epidemiology of Lung Cancer. *Contemp. Oncol.* **2021**, *25*, 45–52. [CrossRef]
2. Griesinger, F.; Korol, E.E.; Kayaniyil, S.; Varol, N.; Ebner, T.; Goring, S.M. Efficacy and safety of first-line carboplatin-versus cisplatin-based chemotherapy for non-small cell lung cancer: A meta-analysis. *Lung Cancer* **2019**, *135*, 196–204. [CrossRef] [PubMed]

3. Schiller, J.H.; Harrington, D.; Belani, C.; Langer, C.; Sandler, A.; Krook, J.; Zhu, J.; Johnson, D.H. Comparison of Four Chemotherapy Regimens for Advanced Non–Small-Cell Lung Cancer. *N. Engl. J. Med.* **2002**, *346*, 92–98. [CrossRef] [PubMed]
4. Baxevanos, P.; Mountzios, G. Novel chemotherapy regimens for advanced lung cancer: Have we reached a plateau? *Ann. Transl. Med.* **2018**, *6*, 139. [CrossRef] [PubMed]
5. Remon, J.; Hendriks, L.E.L. Targeted therapies for unresectable stage III non-small cell lung cancer. *Mediastinum* **2021**, *5*, 22. [CrossRef]
6. Reck, M.; Remon, J.; Hellmann, M.D. First-Line Immunotherapy for Non–Small-Cell Lung Cancer. *J. Clin. Oncol.* **2022**, *40*, 586–597. [CrossRef]
7. Ramirez de Molina, A.; Gallego-Ortega, D.; Sarmentero, J.; Bañez-Coronel, M.; Martin-Cantalejo, Y.; Lacal, J.C. Choline kinase is a novel oncogene that potentiates RhoA-induced carcinogenesis. *Cancer Res.* **2005**, *65*, 5647–5653. [CrossRef]
8. Ramirez de Molina, A.; Sarmentero-Estrada, J.; Belda-Iniesta, C.; Taron, M.; Ramirez de Molina, V.; Cejas, P.; Skrzypski, M.; Gallego-Ortega, D.; de Castro, J.; Casado, E.; et al. Expression of choline kinase alpha to predict outcome in patients with early-stage non-small-cell lung cancer: A retrospective study. *Lancet Oncol.* **2007**, *8*, 889–897. [CrossRef]
9. Lacal, J.; Zimmerman, T.; Campos, J. Choline Kinase: An Unexpected Journey for a Precision Medicine Strategy in Human Diseases. *Pharmaceutics* **2021**, *13*, 788. [CrossRef]
10. Banez-Coronel, M.; de Molina, A.; Rodriguez-Gonzalez, A.; Sarmentero, J.; Ramos, M.; Garcia-Cabezas, M.; Garcia-Oroz, L.; Lacal, J. Choline Kinase Alpha Depletion Selectively Kills Tumoral Cells. *Curr. Cancer Drug Targets* **2008**, *8*, 709–719. [CrossRef]
11. Glunde, K.; Raman, V.; Mori, N.; Bhujwalla, Z.M. RNA interference-mediated choline kinase suppression in breast cancer cells induces differentiation and reduces proliferation. *Cancer Res.* **2005**, *65*, 11034–11043. [CrossRef]
12. Yalcin, A.; Clem, B.; Makoni, S.; Clem, A.; Nelson, K.; Thornburg, J.; Siow, D.; Lane, A.N.; E Brock, S.; Goswami, U.; et al. Selective inhibition of choline kinase simultaneously attenuates MAPK and PI3K/AKT signaling. *Oncogene* **2009**, *29*, 139–149. [CrossRef]
13. Lacal, J.C.; Campos, J. Preclinical Characterization of RSM-932A, a Novel Anticancer Drug Targeting the Human Choline Kinase Alpha, an Enzyme Involved in Increased Lipid Metabolism of Cancer Cells. *Mol. Cancer Ther.* **2015**, *14*, 31–39. [CrossRef]
14. De Cáceres, I.I.; Cortés-Sempere, M.; Moratilla, C.; Machado-Pinilla, R.; Rodriguez-Fanjul, V.; Manguan-García, C.; Cejas, P.; Lopez-Rios, F.; Paz-Ares, L.; de CastroCarpeño, J.; et al. IGFBP-3 hypermethylation-derived deficiency mediates cisplatin resistance in non-small-cell lung cancer. *Oncogene* **2010**, *29*, 1681–1690. [CrossRef]
15. Chou, T.C.; Talalay, P. Quantitative analysis of dose-effect relationships: The combined effects of multiple drugs or enzyme inhibitors. *Adv. Enzym. Regul.* **1984**, *22*, 27–55. [CrossRef]
16. Rodríguez-González, A.; de Molina, A.R.; Fernández, F.; Lacal, J.C. Choline kinase inhibition induces the increase in ceramides resulting in a highly specific and selective cytotoxic antitumoral strategy as a potential mechanism of action. *Oncogene* **2004**, *23*, 8247–8259. [CrossRef]
17. Rodríguez-González, A.; de Molina, R.; Fernández, F.; Ramos, M.A.; Núñez, M.D.C.; Campos, J.; Lacal, J.C. Inhibition of choline kinase as a specific cytotoxic strategy in oncogene-transformed cells. *Oncogene* **2003**, *22*, 8803–8812. [CrossRef]
18. Sánchez-Pérez, I.; Perona, R. Lack of c-Jun activity increases survival to cisplatin. *FEBS Lett.* **1999**, *453*, 151–158. [CrossRef]
19. Kelland, L. The resurgence of platinum-based cancer chemotherapy. *Nat. Rev. Cancer* **2007**, *7*, 573–584. [CrossRef]
20. Lai, G.-M.; Ozols, R.F.; Young, R.C.; Hamilton, T.C. Effect of Glutathione on DNA Repair in Cisplatin-Resistant Human Ovarian Cancer Cell Lines. *JNCI J. Natl. Cancer Inst.* **1989**, *81*, 535–539. [CrossRef]
21. Zeng-Rong, N.; Paterson, J.; Alpert, L.; Tsao, M.; Viallet, J.; Alaoui-Jamali, M.A. Elevated DNA repair capacity is associated with intrinsic resistance of lung cancer to chemotherapy. *Cancer Res.* **1995**, *55*, 4760–4764.
22. Komatsu, M.; Sumizawa, T.; Mutoh, M.; Chen, Z.S.; Terada, K.; Furukawa, T.; Yang, X.L.; Gao, H.; Miura, N.; Sugiyama, T.; et al. Copper-transporting P-type adenosine triphosphatase (ATP7B) is associated with cisplatin resistance. *Cancer Res.* **2000**, *60*, 1312–1316.
23. Mamenta, E.L.; Poma, E.E.; Kaufmann, W.K.; Delmastro, D.A.; Grady, H.L.; Chaney, S.G. Enhanced replicative bypass of platinum-DNA adducts in cisplatin-resistant human ovarian carcinoma cell lines. *Cancer Res.* **1994**, *54*, 3500–3505.
24. Ramírez de Molina, A.; de la Cueva, A.; Machado-Pinilla, R.; Rodríguez-Fanjul, V.; Gómez del Pulgar, T.; Cebrián, A.; Perona, R.; Lacal, J.C. Acid ceramidase as a chemotherapeutic target to overcome resistance to the antitumoral effect of choline kinase α inhibition. *Curr. Cancer Drug Targets* **2012**, *12*, 617–624. [CrossRef]

pharmaceutics

MDPI

Review

A Comprehensive Overview of Globally Approved JAK Inhibitors

Ahmed M. Shawky [1], Faisal A. Almalki [2,*], Ashraf N. Abdalla [3,4], Ahmed H. Abdelazeem [5,6] and Ahmed M. Gouda [5,*]

1 Science and Technology Unit (STU), Umm Al-Qura University, Makkah 21955, Saudi Arabia; amesmail@uqu.edu.sa
2 Department of Pharmaceutical Chemistry, Faculty of Pharmacy, Umm Al-Qura University, Makkah 21955, Saudi Arabia
3 Department of Pharmacology and Toxicology, Faculty of Pharmacy, Umm Al-Qura University, Makkah 21955, Saudi Arabia; anabdrabo@uqu.edu.sa
4 Department of Pharmacology and Toxicology, Medicinal and Aromatic Plants Research Institute, National Center for Research, Khartoum 2404, Sudan
5 Department of Medicinal Chemistry, Faculty of Pharmacy, Beni-Suef University, Beni-Suef 62514, Egypt; ahmed.haasn77@yahoo.com
6 Department of Pharmacy, College of Pharmacy, Riyadh Elm University, Riyadh 11681, Saudi Arabia
* Correspondence: famalki@uqu.edu.sa (F.A.A.); ahmed.gouda@pharm.bsu.edu.eg (A.M.G.)

Citation: Shawky, A.M.; Almalki, F.A.; Abdalla, A.N.; Abdelazeem, A.H.; Gouda, A.M. A Comprehensive Overview of Globally Approved JAK Inhibitors. *Pharmaceutics* **2022**, *14*, 1001. https://doi.org/10.3390/pharmaceutics14051001

Academic Editor: Francesca Musumeci

Received: 1 April 2022
Accepted: 28 April 2022
Published: 6 May 2022

Abstract: Janus kinase (JAK) is a family of cytoplasmic non-receptor tyrosine kinases that includes four members, namely JAK1, JAK2, JAK3, and TYK2. The JAKs transduce cytokine signaling through the JAK-STAT pathway, which regulates the transcription of several genes involved in inflammatory, immune, and cancer conditions. Targeting the JAK family kinases with small-molecule inhibitors has proved to be effective in the treatment of different types of diseases. In the current review, eleven of the JAK inhibitors that received approval for clinical use have been discussed. These drugs are abrocitinib, baricitinib, delgocitinib, fedratinib, filgotinib, oclacitinib, pacritinib, peficitinib, ruxolitinib, tofacitinib, and upadacitinib. The aim of the current review was to provide an integrated overview of the chemical and pharmacological data of the globally approved JAK inhibitors. The synthetic routes of the eleven drugs were described. In addition, their inhibitory activities against different kinases and their pharmacological uses have also been explained. Moreover, their crystal structures with different kinases were summarized, with a primary focus on their binding modes and interactions. The proposed metabolic pathways and metabolites of these drugs were also illustrated. To sum up, the data in the current review could help in the design of new JAK inhibitors with potential therapeutic benefits in inflammatory and autoimmune diseases.

Keywords: JAK; synthesis; kinase inhibitory activity; pharmacological uses; binding mode/interactions

1. Introduction

Janus kinases (JAKs) are intracellular, non-receptor tyrosine kinases [1]. The JAK family consists of four members, including JAK1, JAK2, JAK3, and TY2K. Since the first discovery of JAKs by Wilks thirty years ago [2], great efforts have been made to understand their structure and functions. The four JAKs play an essential role in the transduction of the cytokine-mediated signals, which takes place through the JAK-signal transducers and activators of the transcription (STAT) pathway [3]. Four members of the JAK family have emerged as potential drug targets in different types of diseases [4].

In the current review, we aimed to provide an integrated overview of the chemical and pharmacological data of the globally approved JAK inhibitors that were approved for the treatment of inflammatory, autoimmune, and myeloproliferative diseases. Highlighting the differences in chemical structure, binding interactions, kinase inhibitory activities, pharmacological activities, and metabolic pathways of these drugs could help in the design of new, more potent, and safer JAK inhibitors.

1.1. Structure of JAKs

The structure of JAKs (Figure 1) consists of 7 domains, JH1-JH7 [5]. The four JAKs have the same domains with an overall similarity of 48% [6]. The first domain is JH1, which exists at the C-terminal. This domain is also called the kinase domain because it is responsible for the enzymatic activity of the kinase. The second domain, JH2, is the pseudo-kinase domain, which lacks the tyrosine kinase activity [7]. However, JH2 plays an important role in the regulation of kinase activity [8]. JAKs also include two domains, JH3-JH4, which share homology with the Src-homology-2 (SH2) domain, while the fourth region of JAKs is the FERM domain, which exists at the N-terminal and plays a role in the binding of JAKs with cytokine receptors [9].

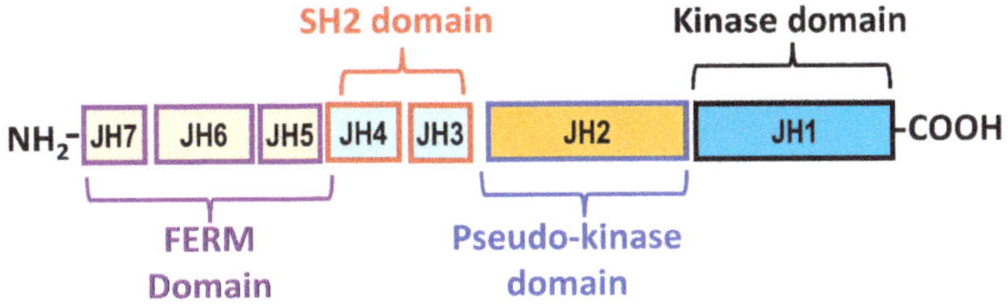

Figure 1. Domain structure of JAKs.

1.2. JAK-STAT Pathway

Several ligands such as cytokines and growth factors have been reported to activate the JAK-STAT pathway [10]. Following the activation of JAKs, phosphorylation and dimerization of STATs take place (Figure 2). The phosphorylated STATs enter the nucleus, where they initiate a suitable transcriptional response in the genes that regulate immunity, inflammation, and hematopoiesis [10–12].

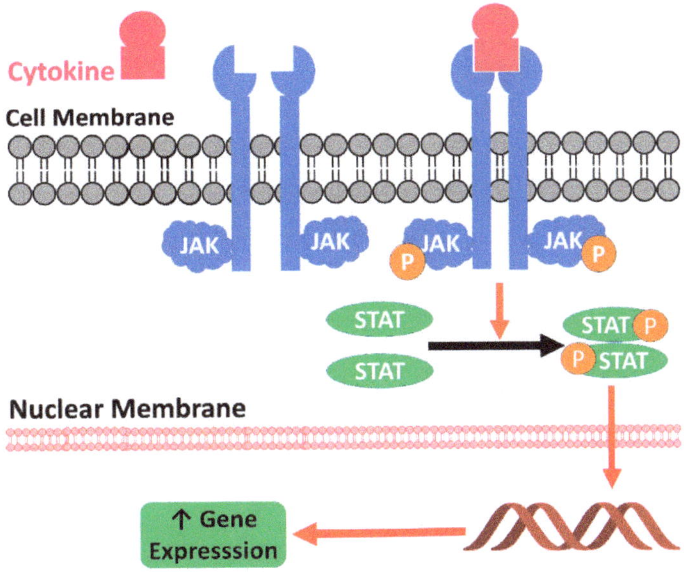

Figure 2. JAK-STAT pathway.

1.3. Therapeutic Potential of JAK Inhibition

The JAK-STAT pathway is responsible for the transduction of cytokines and growth factor signals that play a crucial role in inflammation and autoimmune diseases [13]. Among these kinases, JAK1 seems to play a crucial role in pruritic dermatitis [14], allergic rhinitis [15], asthma [16], and inflammatory bowel disease [17]. Targeting JAK1 with small-molecule inhibitors has proved efficacy in the treatment of these diseases [16,18,19]. Moreover, several small molecules with JAK1 and JAK2 inhibitory activity have also provided therapeutic benefits in the treatment of rheumatoid arthritis, psoriasis, and pruritis [20]. In addition, several JAK3 selective inhibitors have been evaluated for their efficacy in the treatment of rheumatoid arthritis [20]. In addition, JAK1/TYK2 dual inhibitors have also been developed with potential therapeutic options in inflammatory diseases [21]. Furthermore, TYK2 selective inhibitors may also be useful in the treatment of autoimmune diseases [20,22].

On the other hand, excessive activation of JAKs has also been reported in different types of cancer [23]. The JAK/STAT3 pathway plays an important role in the proliferation and angiogenesis of solid tumors [23]. In 2005, the discovery of JAK2 mutation (JAK2V617F) in myeloproliferative neoplasms has attracted much attention [24,25]. This discovery led to a better understanding of these diseases. In addition, JAK2V617F has emerged as a potential therapeutic target for myeloproliferative neoplasms [26,27]. In addition, the JAK1/JAK2 inhibitor, ruxolitinib, has also been approved for the treatment of myelofibrosis and polycythemia vera [28,29]. Moreover, the dual inhibition of JAK2 and FLT3 could also provide therapeutic option in the treatment of acute myelogenous leukemia and myeloproliferative neoplasms [30–32].

1.4. Classification of JAK Inhibitors

JAK inhibitors can be divided into two generations [33]. The first-generation includes small molecules such as baricitinib and tofacitinib, which act as non-selective inhibitors of JAKs. On the other hand, second-generation drugs such as filgotinib and upadacitinib have selective inhibitory activity against JAKs [33]. This difference in the selectivity of the two generations is associated with some differences in their safety and efficacy.

One the other hand, JAK inhibitors may also be classified based on their binding mode and the type of interactions with the amino acids in JAKs into reversible (competitive) and irreversible (covalent) inhibitors (Figure 3).

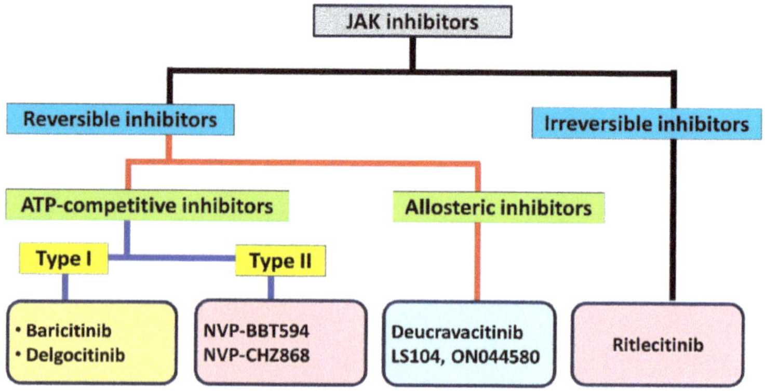

Figure 3. Classification of JAK inhibitors with representative examples for each type.

1.4.1. Reversible JAK Inhibitors

Competitive JAK inhibitors form reversible (non-covalent) binding interactions with the amino acids in the four JAKs. The binding interactions formed by this type of JAK inhibitors include hydrogen bonds and hydrophobic interactions. The class of reversible JAK inhibitors can also be classified into two sub-classes.

ATP-Competitive Inhibitors

The mechanism of action of these inhibitors depends on their competition with ATP for the catalytic ATP-binding site in JAKs [27,34]. These inhibitors may also be classified based on the conformation of the kinase domain to which they bind:

- Type I JAK Inhibitors

These inhibitors bind to the ATP-binding site of the JAKs under the active conformation of the kinase domain [27,34]. This includes clinically approved drugs such as filgotinib, which acts and is classified as a selective JAK1, while fedratinib exhibits the selective inhibition of JAK2 [20,35]. On the other hand, tofacitinib and peficitinib act by blocking multiple JAKs [20,36]. The ability of type I JAK inhibitors to bind to multiple kinases and act as non-selective inhibitors could be due to the highly conserved structure of the ATP-binding site in the four JAKs [18].

- Type II JAK Inhibitors

Type II JAK inhibitors also bind to the ATP-binding site of the kinase domain in the inactive conformation of JAKs [27,34]. NVP-BBT594 and NVP-CHZ868 (Figure 4) are representative examples of type II inhibitor, which target JAK2 [37,38].

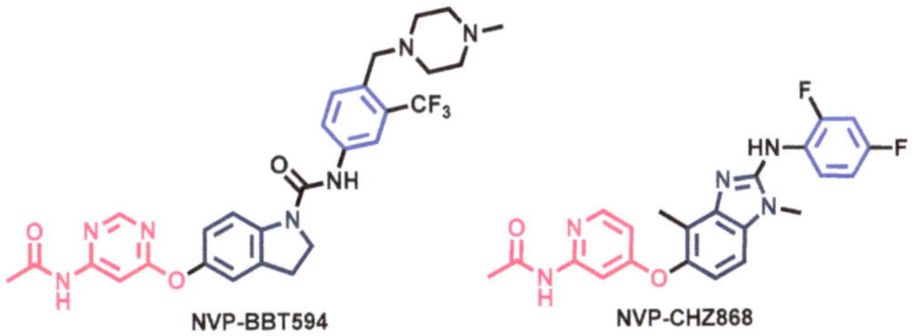

Figure 4. JAK2 inhibitors, NVP-BBT594 and NVP-CHZ868.

- Allosteric JAK Inhibitors

The allosteric JAK inhibitors (Figure 5) include small molecule inhibitors that bind to a site other than the ATP-binding site in JAKs [27,34]. Among these inhibitors, deucravacitinib (BMS-986165) act as a selective allosteric inhibitor of TYK2 [39]. In addition, LS104, and ON044580 are examples of JAK2 allosteric inhibitors [34,40–42].

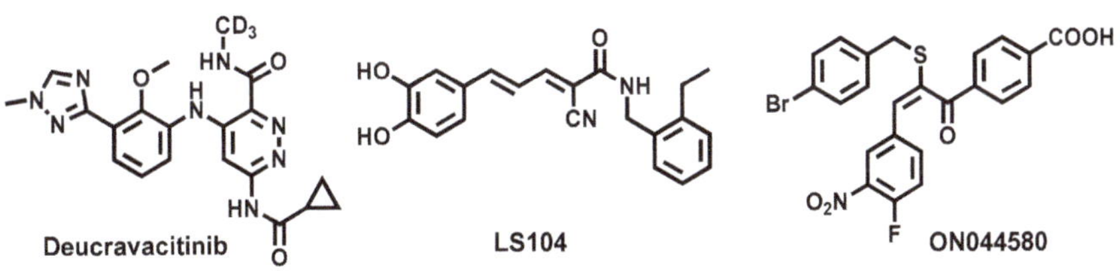

Figure 5. Allosteric JAKs inhibitors.

1.4.2. Irreversible JAK3 Inhibitors

This class of irreversible JAK inhibitors (Figure 6) that target JAK3 was also reported [43–45]. The mechanism of action of these inhibitors depends on the covalent interaction with the unique Cys909 residue in JAK3 [43]. The chemical structure of these inhibitors has a covalent-bond forming group such as acrylamide and α-cyanoacrylamide, which can bind covalently with Cys909 residue (Figure 6).

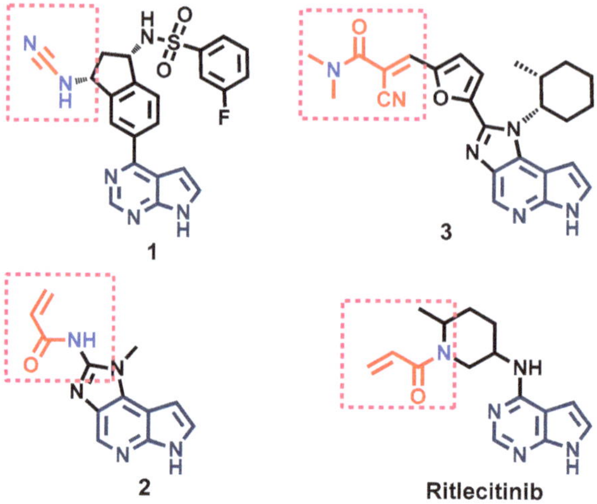

Figure 6. JAK3 covalent inhibitors.

Compound **1** was reported among a series of cyanamide-based JAK3 covalent inhibitors [43]. Moreover, compounds **2** (IC_{50} = 0.003 μM) and **3** (IC_{50} = 154 pM) were also reported with potent and selective inhibitory activity against JAK3 [44,45].

Ritlecitinib (Figure 6) is an irreversible inhibitor of JAK3, which is currently under evaluation in clinical trials in humans [43,46]. The inhibitory activity of ritlecitinib was mediated by a covalent interaction with Cys909 residue in JAK3 [43]. The binding mode and interaction of ritlecitinib in JAK3 (pdb: 5TOZ) are visualized in Figure 7. The crystal structure was downloaded from the protein data bank (https://www.rcsb.org/, accessed on 11 November, 2021). In this work, the crystal structures of JAK kinases were visualized using the Discovery Studio Visualizer [47]. The crystals were prepared by removing water molecules following the previous reports [48,49]. Moreover, 2/3D binding modes of the JAK inhibitors into the co-crystallized kinases were visualized according to the previous report [50].

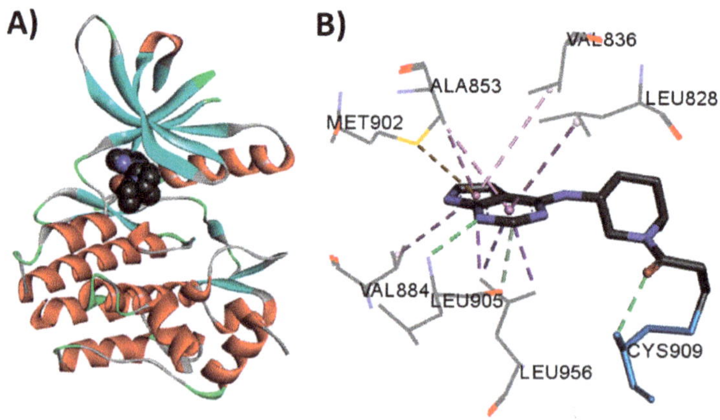

Figure 7. JAK3 (pdb: 5TOZ) bound to ritlecitinib: (**A**) 3D representation of JAK3 bound to ritlecitinib (shown as CPK); (**B**) 3D binding mode of ritlecitinib showing one covalent bond with Cys909, hydrogen bonds (shown as green-dotted lines), and multiple hydrophobic interactions, this figure was generated using Discovery Studio Visualizer (V16.1.0.15350).

1.5. Mutation of JAKs

Besides the role of wild-type of JAKs in autoimmune and inflammatory diseases, JAK mutants also play crucial roles in myeloproliferative, lymphomas, and leukemias diseases [6]. JAK1 mutation has been found in acute lymphoblastic leukemia [51]. In addition, JAK2 mutation (JAK2V617F) has been associated with the incidence of myeloproliferative neoplasms [24,52]. The occurrence of JAK2 mutation was reported in nearly all patients with polycythaemia vera (PV) [53].

Moreover, the expression of the JAK3 mutant was associated with the induction of leukemia in mice [54]. JAK3 mutation was also reported in small percent of patient with T-cell acute lymphoblastic leukemia. It was also associated with the occurrence of severe combined immunodeficiency [55].

On the other hand, TYK2 mutations were also associated with immunodeficiency with T-cell lymphopenia [56]. It was also associated with increased susceptibility to bacterial and/or viral infections. The deficiency of TYK2 was associated with recurrent respiratory infection [57].

1.6. Crystal Structure of JAKs

The advances in X-ray crystallography have provided a lot of data about the structure and function of different members of the JAK family. This data has helped in the design of selective JAK inhibitors [58]. Many of the crystal structures of different JAKs are available from the protein data bank (https://www.rcsb.org/). Visualization of the binding modes and interactions of the different small molecules inhibitors into the four JAKs was done using the Discovery Studio Visualizer [47].

The binding mode and interactions of KEV, a pyrazolopyridine inhibitor with JAK1 kinase (pdb: 6N7A) [19], were visualized in Figure 8. KEV displays hydrogen bond interactions with Glu957 and Leu959 in the hinge region of JAK1.

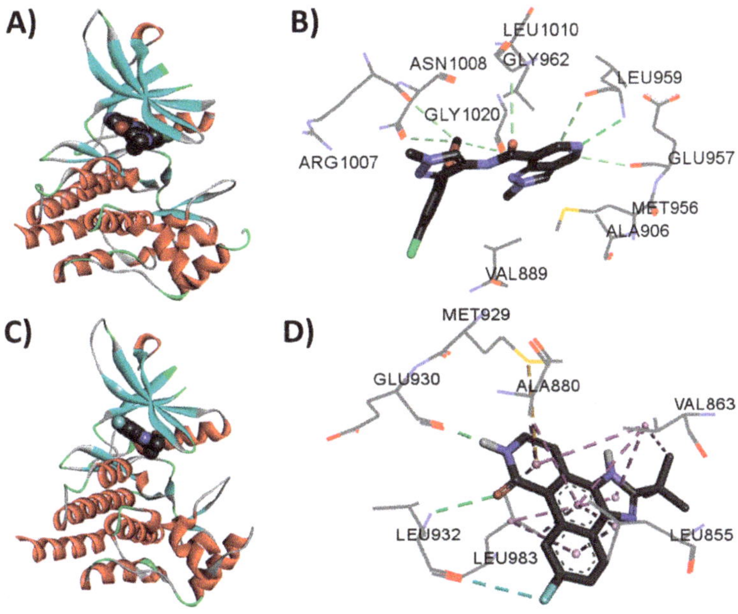

Figure 8. JAK1 (pdb: 6N7A) and JAK2 (pdb: 2B7A) kinases bound to small-molecule inhibitors: (**A**) 3D representation of JAK1 bound to KEV (shown as CPK); (**B**) 3D binding mode of KEV into JAK1 showing the hydrogen bond interactions; (**C**) 3D representation of JAK2 bound to IZA (shown as CPK); (**D**) 3D binding mode of IZA into JAK2, hydrogen bonds are shown as green-dotted lines, this figure was generated using Discovery Studio Visualizer (V16.1.0.15350).

In addition, many small molecules of divers chemical nature were reported as co-crystallized ligands with JAK2 [59,60]. Among these ligands, IZA is an isoquinoline derivative with potent and pan-JAK inhibitory activity. The binding interactions of IZA show two conventional hydrogen bonds with Leu932 and Glu930 in the hinge region in JAK2 (Figure 8).

The crystal structure of JAK3 (pdb: 3PJC) in complex with PJC, a pyrazolopyridine inhibitor [61] was visualized in Figure 9. PJC shows hydrogen bond interactions with Leu905 and Glu903. On the other hand, the crystal structure of TYK2 kinase (pdb: 6VNX) in complex with R4V, a pyrimidine inhibitor [62] was also visualized in Figure 9. R4V shows one conventional hydrogen bond with Asp1041.

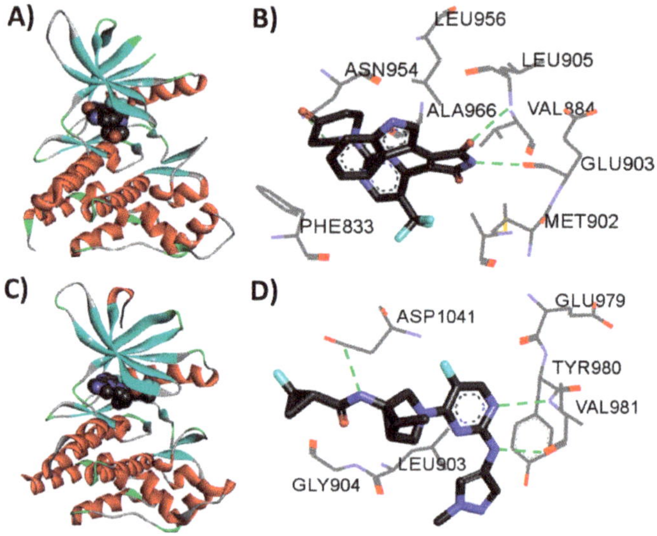

Figure 9. JAK3 (pdb: 3PJC) and TYK2 (pdb: 6VNX) kinases bound to small-molecule inhibitors: (**A**) 3D representation of JAK3 bound to PJC (shown as CPK); (**B**) 3D binding mode of PJC into JAK3; (**C**) 3D representation of TYK2 bound to R4V (shown as CPK); (**D**) 3D binding mode of R4V into TYK2, this figure was generated using Discovery Studio Visualizer (V16.1.0.15350).

1.7. *JAK Inhibitors Approved for Clinical Use*

In 1996, Meydan et al. reported AG-490 (Figure 10), a JAK2 inhibitor with antileukemic activity [63]. Following this discovery, great efforts have been made in the last two decades to develop potent and selective JAK inhibitors.

In 2011, the FDA-approved ruxolitinib (Figure 10) became the first JAK inhibitor [64]. One year later, tofacitinib was also approved for the treatment of rheumatoid arthritis [64]. In 2017, baricitinib was approved by the FDA for rheumatoid arthritis [65]. However, in 2019, three JAK inhibitors were approved for clinical use [66–68]. These drugs include fedratinib and upadacitinib, which were approved by the FDA, whereas peficitinib had already been approved in Japan for rheumatoid arthritis. In 2020, delgocitinib and filgotinib were also approved in Japan for the treatment of atopic dermatitis and rheumatoid arthritis, respectively [69,70].

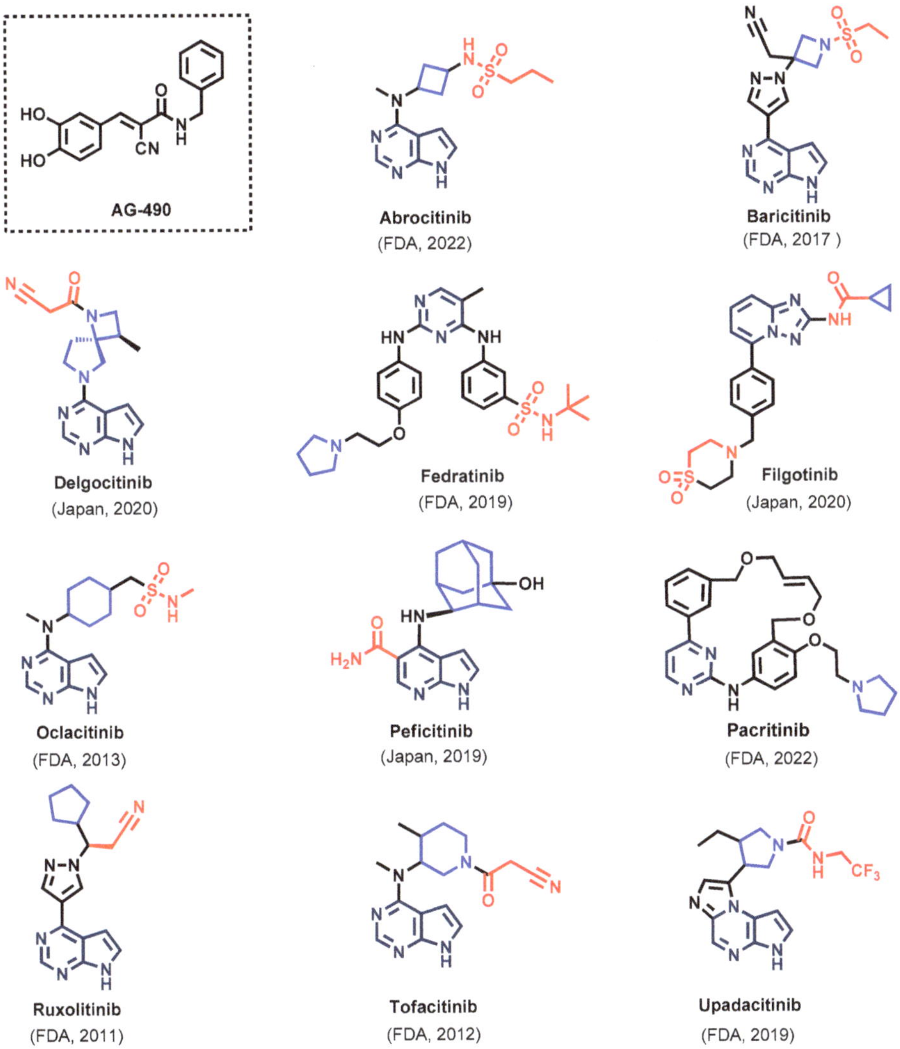

Figure 10. Chemical structure of AG-490 and the globally proven JAK inhibitors with the dates of their first approval.

1.7.1. Abrocitinib

Approval History

Abrocitinib (Figure 11) is a JAK1 kinase inhibitor [71,72]. It was approved by the FDA in January 2020 for the treatment of refractory, moderate-to-severe atopic dermatitis [72].

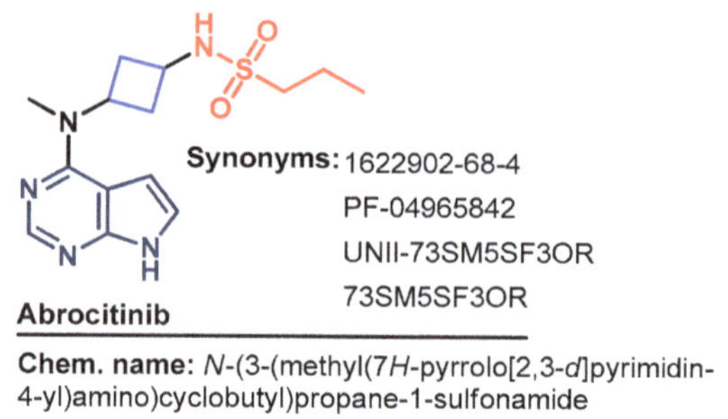

Figure 11. Chemical structure/name/synonyms of abrocitinib.

Synthesis

Vazquez et al. [59] have reported the synthesis of abrocitinib from cyclobutyl carbamate **4**, Scheme 1. The ketone group in **4** was condensed with methylamine and reduced with sodium borohydride to give **5**. The coupling of **6** with the *cis*-isomer of **5** afforded **7**, which underwent the acid-catalyzed hydrolysis of the carbamate group to give **8**. The base-catalyzed sulfonylation of **8** with propane-1-sulfonyl chloride afforded **9**, which underwent deprotection to remove the tosyl moiety and give abrocitinib.

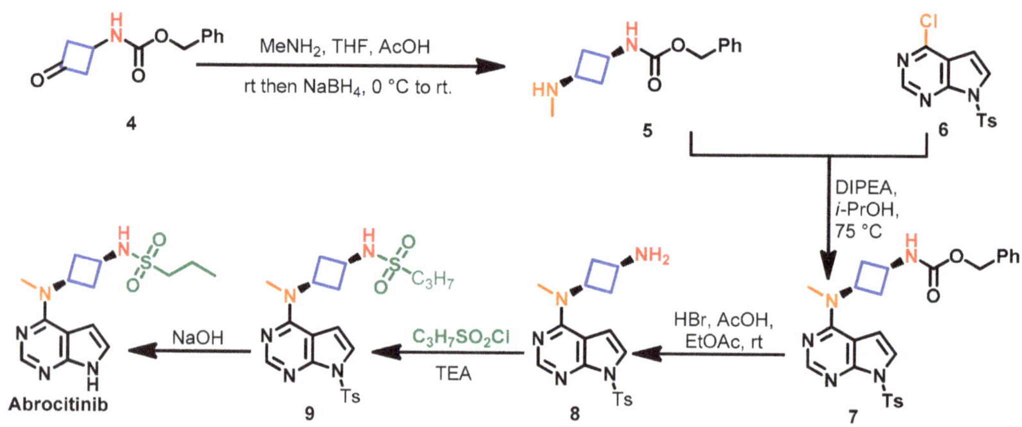

Scheme 1. Synthesis of abrocitinib (route 1).

Connor et al. [71] have also developed a commercial route for the synthesis of abrocitinib. The synthesis takes place through a nitrene-type rearrangement, Scheme 2. Compound **10** underwent biocatalytic reductive amination using the wild-type *Sp*RedAm enzyme from *S. purpureus*, which gave the *cis*-amino ester compound **11** in 74% yield. The reaction of **11** and **12** afforded the isopropyl ester **13**. The reaction of **13** with hydroxylamine hydrochloride afforded the hydroxamic acid derivative **14**, which underwent Lossen rearrangement on the reaction with 1,1-carbonyldiimidazole (CDI) to give **15**. The sulfonylation of **15** was achieved using the triazole derivative **16** to avoid the sulfonylation of the pyrrole nitrogen.

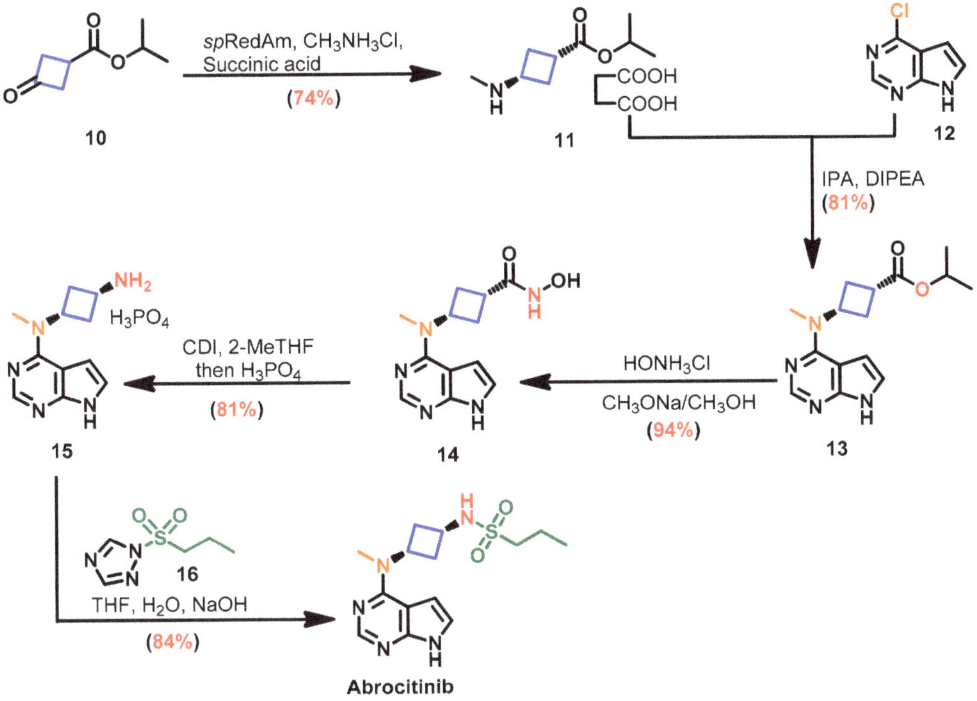

Scheme 2. Synthesis of abrocitinib (route 2).

Target Kinases

The inhibitory activity of abrocitinib against JAK kinases was evaluated by Vazquez et al. [59]. The results revealed inhibitory activity against JAK1 and JAK2 at IC_{50} values of 0.029 and 0.803 μM, respectively (Figure 12).

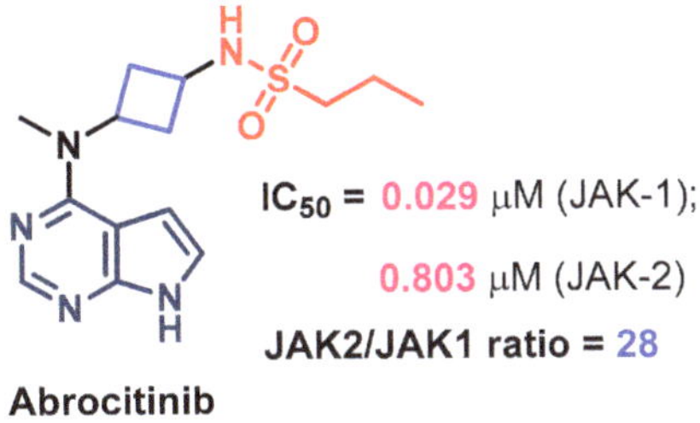

Figure 12. JAKs inhibitory activities of abrocitinib.

Crystal Structures

Abrocitinib exists as a co-crystallized ligand in two crystal structures, including its crystal structure with JAK1 (pdb: 6BBU) and JAK2 (pdb: 6BBV) [59].

The binding mode of abrocitinib in the active site of JAK1 is illustrated in Figure 13. Abrocitinib shows three conventional hydrogen bonds with Glu957, Leu959, and Asn1008. In addition, abrocitinib shows three carbon–hydrogen bonds with LEU881 and Leu959. Several hydrophobic interactions were also observed between abrocitinib and amino acids in JAK1.

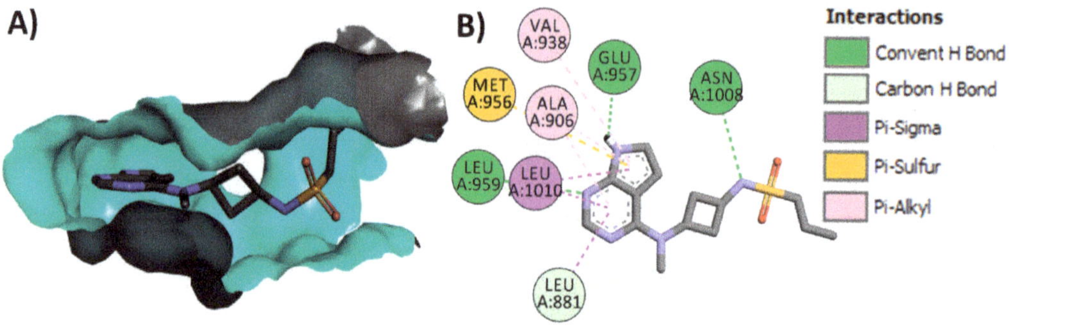

Figure 13. Binding modes of abrocitinib (shown as sticks) into JAK1 (pdb: 6BBU): (**A**) 3D binding mode, receptor shown as a hydrogen bond surface; (**B**) 2D binding mode showing different types of binding interactions with amino acids in JAK1; this figure was generated using Discovery Studio Visualizer (V16.1.0.15350).

On the other hand, abrocitinib exhibited four conventional hydrogen bonding interactions with Glu930, Leu932, Arg980, and Asn981, Figure 14. The binding interactions of abrocitinib also included one carbon–hydrogen bond with Leu932 and several hydrophobic interactions, with Leu855, Val863, Ala880, Val911, and Leu983.

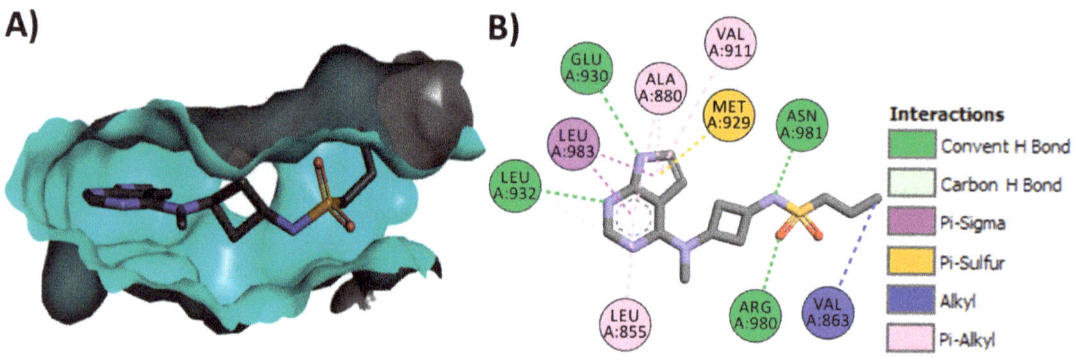

Figure 14. Binding modes of abrocitinib (shown as sticks) into JAK2 (pdb: 6BBV): (**A**) 3D binding mode, receptor shown as a hydrogen bond surface; (**B**) 2D binding mode showing different types of binding interactions with amino acids in JAK2; this figure was generated using Discovery Studio Visualizer (V16.1.0.15350).

Pharmacological Activities and Uses

Abrocitinib combined with topical therapy displayed higher effectiveness in the treatment of atopic dermatitis compared to the placebo [73]. In another clinical trial (NCT03720470), abrocitinib showed higher therapeutic benefits regarding the symptoms of atopic dermatitis compared to the placebo [74].

The abrocitinib-induced response in patients with moderate-to-severe atopic dermatitis was also evaluated by Blauvelt et al. [75]. The results revealed the effectiveness of the induction treatment with abrocitinib, as most responders did not flare.

Metabolism

The metabolic study of abrocitinib in humans revealed the formation of several oxidative metabolites, in addition to the parent drug (26%) [76,77]. The results of the in vitro metabolic study of abrocitinib revealed that it undergoes metabolism by several CYP450 enzymes, where CYP2C19 and CYP2C9 were the major metabolizing enzymes. Among these metabolites, the 3-hydroxypropyl M1 (PF-06471658) and the 2-hydroxypropyl metabolite M2 (PF-07055087) were retrieved in 11% and 12%, respectively, Figure 15. Both M1 and M2 exhibited JAK inhibitory activity similar to the parent drug with higher selectivity toward JAK1 over the other JAKs. On the other hand, a third oxidative metabolite M4 (PF-07054874) that lacks JAK inhibitory activity was also isolated [77].

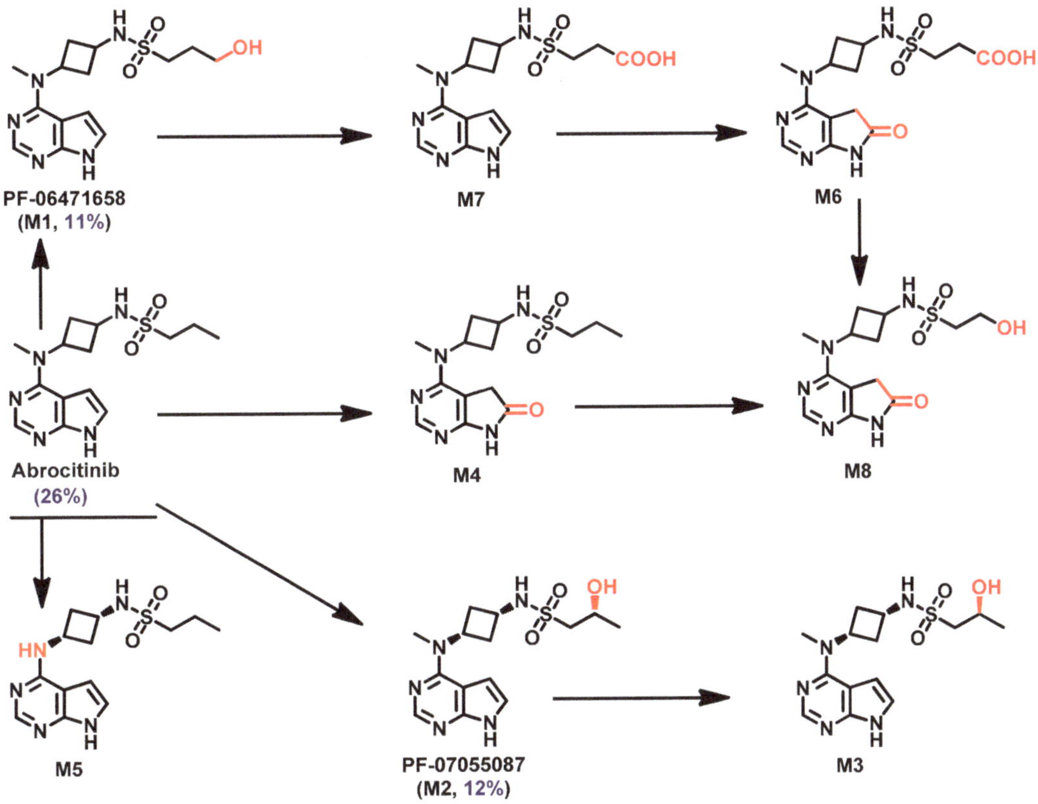

Figure 15. Proposed metabolic pathways and metabolites of abrocitinib.

1.7.2. Baricitinib

Approval History

Baricitinib (Figure 16) is an orally active small-molecule inhibitor of JAK1/2. It was approved by the European Medicine Agency (EMA) in 2017 for the treatment of rheumatoid arthritis [65]. In June, 2018, baricitinib was approved by the FDA for the treatment of moderate-to-severe rheumatoid arthritis in adults [64]. Recently, the FDA issued an emergency use authorization for the combination of baricitinib and remdesivir to treat hospitalized patients with COVID-19 [78].

Synthesis

The original synthesis of baricitinib was reported by Rodgers et al. [79]. In the first step, 3-oxoazetidine-1-carboxylate **18** underwent a Horner–Emmons reaction with diethyl (cyanomethyl)phosphonate **17** to give **19**, Scheme 3. Treatment of **19** with hydrochloric acid resulted in the removal of the protective group and produced **20**. The reaction of compound **20** with ethane sulfonyl chloride **21** afforded **22**.

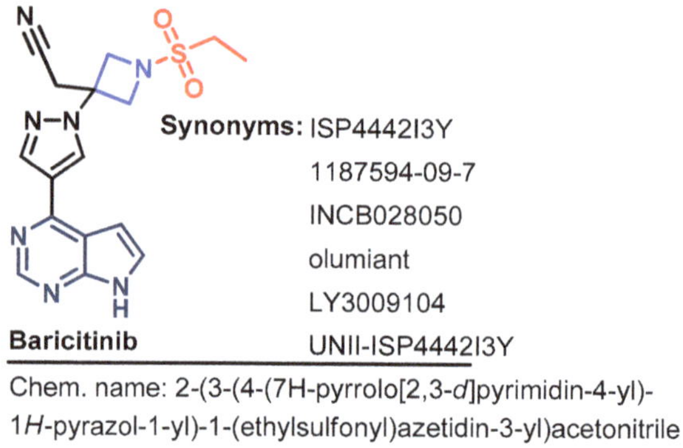

Figure 16. Chemical structure/name/synonyms of baricitinib.

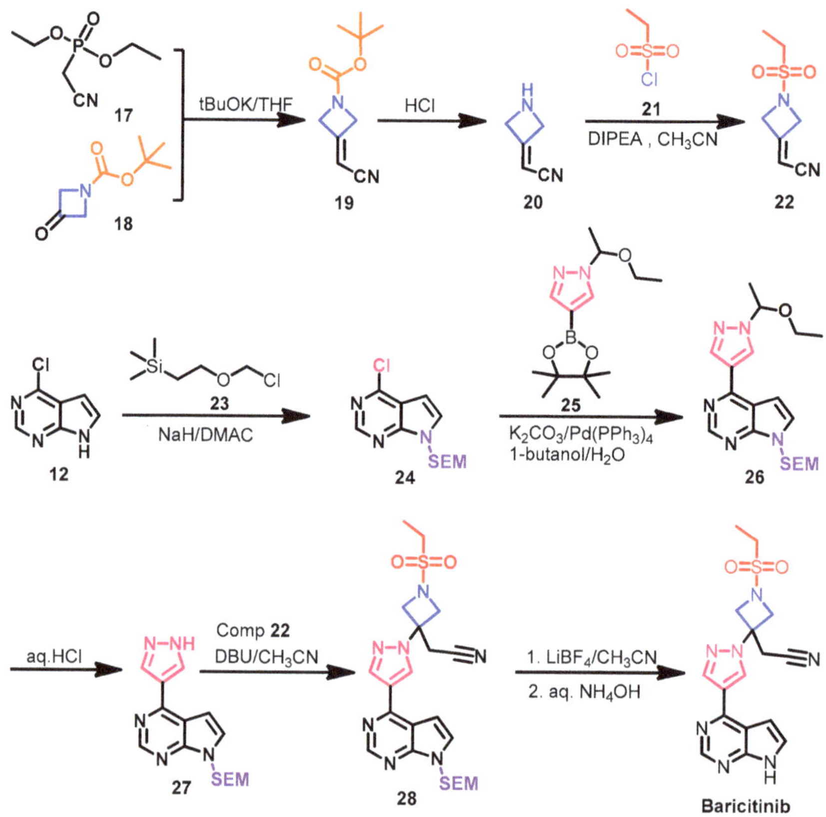

Scheme 3. Synthesis of baricitinib (route 1).

On the other hand, protection of the pyrrole nitrogen in **12** was achieved on the reaction of 12 with (2-(chloromethoxy)ethyl)trimethylsilane (SEM) 23, which gave **24**. Compound **24** underwent Suzuki coupling with 4-(1,3,2-dioxaborolan-2-yl)-1*H*-pyrazole **25** to give **26**, Scheme 3. The acid hydrolysis of **26** afforded **27**, which was then reacted with **22** to give compound **28**. Compound **28** was treated with lithium tetrafluoroborate followed by ammonia solution to give baricitinib.

Xu et al. [80] have reported an efficient five-step synthetic route of baricitinib (Scheme 4) with an overall yield of 49% compared to the eight-step method reported by Rodgers et al. [79]. Both methods share the first 3 steps. In the fourth step of Xu et al.'s method, a nucleophilic addition of **29** to the double bond in **22** afforded compound **30**. Suzuki coupling of **30** with compound **12** gave baricitinib. This method may be suitable for industrial synthesis due to its low cost and simple operating requirements [80].

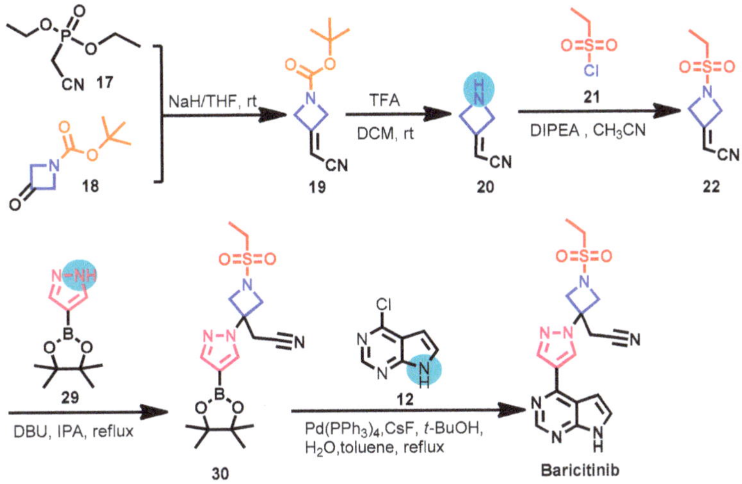

Scheme 4. Synthesis of baricitinib (route 2).

Furthermore, several other methods have been reported, describing alternative synthetic routes of baricitinib using the key intermediates **12** and **22** [81,82]. Among these methods, Cui et al. [82] reported a green and facile synthesis of the azetidine intermediate **22** using inexpensive and environmentally friendly starting materials.

Target Kinases

The inhibitory activity of baricitinib against JAKs was evaluated by Clark et al. [83]. These results revealed high inhibitory activity against JAK1/2 (Figure 17). However, baricitinib also exhibited moderate inhibitory activity against TYK2 (IC_{50} = 61 nM).

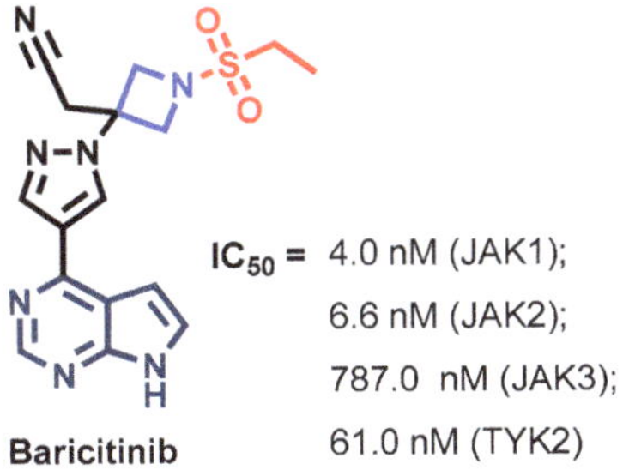

Figure 17. JAKs inhibitory activities of baricitinib.

Crystal Structures

Baricitinib is found in three crystal structures in the protein data bank. The first is the crystal structure of baricitinib with BMP-2-inducible kinase (pdb: 4W9X) [84]. In addition baricitinib also exists as a co-crystallized ligand in two crystal structures of JAK2 [85,86].

The bone morphogenic proteins (BMPs) are involved in skeletal morphogenesis [87]. The binding mode of baricitinib int BMP-2-inducible kinase is visualized in Figure 18. The binding interactions of baricitinib into BMP-2-inducible kinase include 3 conventional hydrogen bonds with Ser63, Cys133 and Gln137. Baricitinib also shows one carbon hydrogen bond with Asp198 and one pi-sulfur interaction with Met130. Several hydrophobic interactions could also be observed between baricitinib and the hydrophobic residues in the kinase.

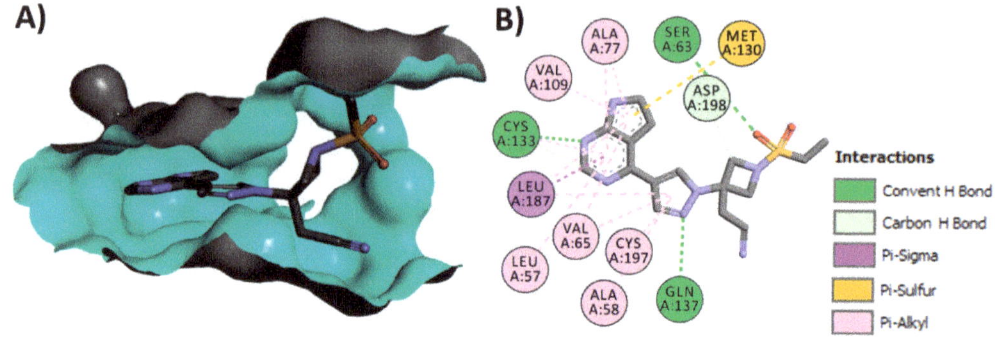

Figure 18. Binding modes of baricitinib (shown as sticks) into BMP-2-inducible kinase (pdb: 4W9X): (**A**) 3D binding mode, receptor shown as a hydrogen bond surface; (**B**) 2D binding mode showing different types of binding interactions with amino acids in BMP-2-inducible kinase; this figure was generated using Discovery Studio Visualizer (V16.1.0.15350).

Visualization of the binding mode/interactions of baricitinib into JAK2 JH1 (pdb: 6VN8) shows one conventional hydrogen bond with Leu932 [85]. Baricitinib also shows three carbon hydrogen bonds with Ser862, Lys882, and Asp994 (Figure 19). Baricitinib also shows one pi-sulfur interaction with Met929 and several hydrophobic interactions of the pi-sigma and pi-alkyl types.

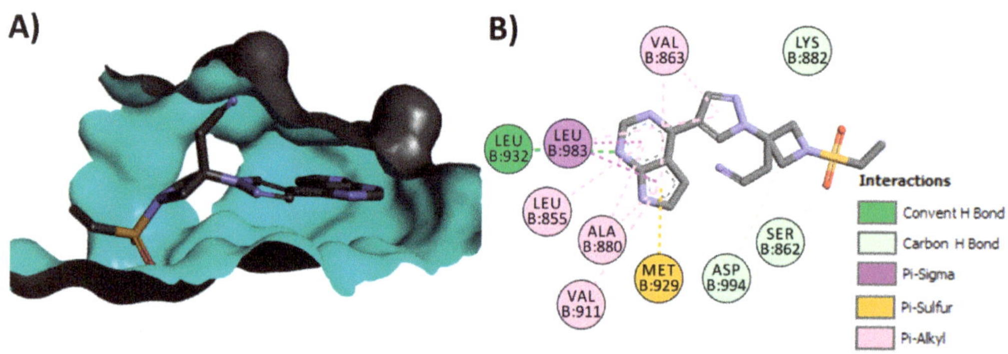

Figure 19. Binding modes of baricitinib (shown as sticks) into JAK2 JH1 (pdb: 6VN8): (**A**) 3D binding mode, receptor shown as a hydrogen bond surface; (**B**) 2D binding mode showing different types of interactions with JAK2 JH1; the figure was generated using Discovery Studio Visualizer (V16.1.0.15350).

In addition, the binding mode of baricitinib into human JAK2 JH1 was also reported in another crystal structure (pdb: 6WTO) [86]. Visualization of the binding mode/interactions (Figure 20) shows identical hydrogen bonds and hydrophobic interactions with those of the above crystal (pdb: 6VN8), Figure 19.

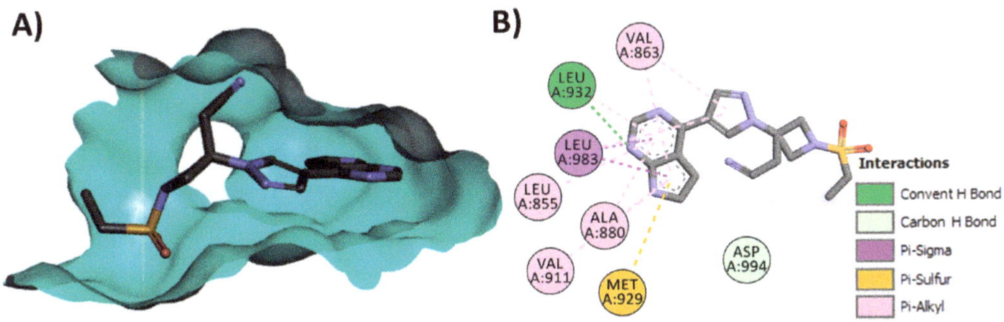

Figure 20. Binding modes of baricitinib (shown as sticks) into human JAK2 JH1 (pdb: 6WTO): (**A**) 3D binding mode, receptor shown as a hydrogen bond surface; (**B**) 2D binding mode showing different types of interactions with JAK2 JH1; the figure was generated using Discovery Studio Visualizer (V16.1.0.15350).

Pharmacological Activities and Uses

Baricitinib was the second JAK inhibitor to be approved for rheumatoid arthritis after the approval of ruxolitinib. It showed rapid and long-lasting therapeutic benefits in the treatment of rheumatoid arthritis [88].

Baricitinib also improved the signs and symptoms of systemic lupus erythematosus in a double-blind study [89]. Moreover, baricitinib decreased inflammation/pruritus in atopic dermatitis patients when used with topical corticosteroids [90].

On the other hand, The inhibition of JAK1/2 by baricitinib leads to the inhibition of the JAK-STAT signaling pathway and subsequent inhibition of the production of the pro-inflammatory cytokines that could be useful for patients with COVID-19 [91]. The results of the clinical trial of baricitinib in combination with remdesivir (NCT04401579) revealed superior activity in reducing the recovery period compared to remdesivir alone [92]. Similar results were also obtained when using a combination of dexamethasone with baricitinib/remdesivir [93]. Finally, the combination of baricitinib and remdesivir received an Emergency Use Authorization (EUA) from the FDA to treat hospitalized patients with COVID-19.

Metabolism

The in vivo metabolism of baricitinib was performed in mice, rats, and dogs using [^{14}C]-baricitinib [94,95]. The metabolic pathways include mono- and bis-oxidation, oxidative ring opening, and oxidative decarboxylation. The in vitro metabolism of baricitinib incubated with human liver enzymes was also very limited, with the formation of four metabolites.

Baricitinib undergoes minor in vivo metabolic transformation mediated by the CYP3A4 enzyme [94,95]. Nearly 6–10% of baricitinib undergoes metabolism, producing four oxidative metabolites (Figure 21). Accordingly, it is excreted mainly as an unchanged drug in urine (69%) and feces (15%), while no circulating metabolites were identified in human plasma. The proposed metabolite M3 was formed through oxidative ring opening of the pyrrole ring, which undergoes oxidative decarboxylation giving M12. On the other hand, mono- or bis-oxidation of baricitinib affords the oxidative metabolites M10 and M22, respectively. The metabolic profile of baricitinib in animals was similar to that observed in humans.

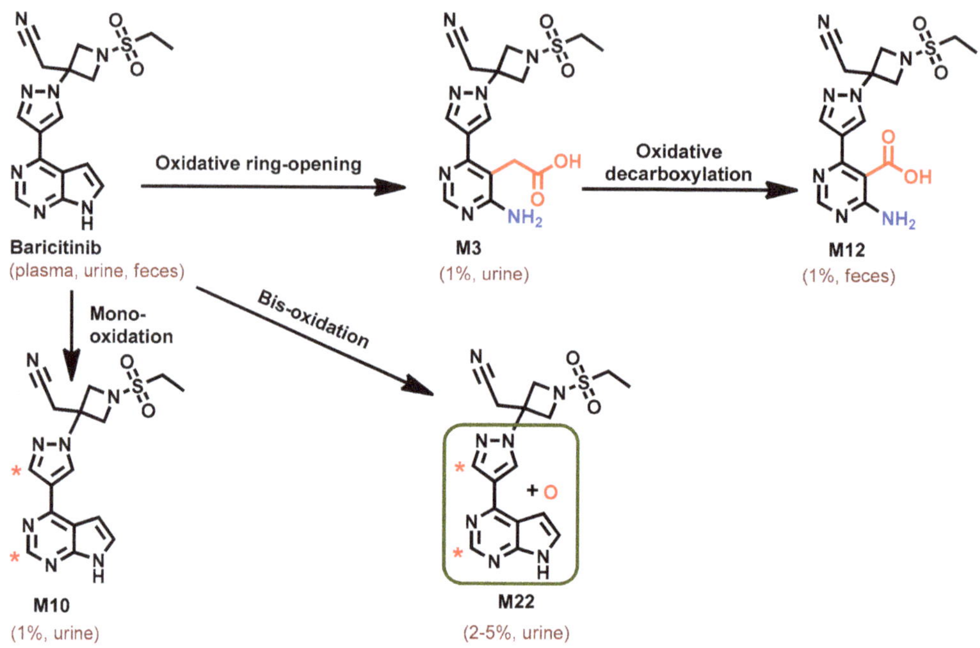

Figure 21. Proposed metabolic pathways and major metabolites of baricitinib in human, asterisks indicate the sites of oxidation.

1.7.3. Delgocitinib

Approval History

Delgocitinib (Figure 22) is a pan-JAK inhibitor that was approved for atopic dermatitis in 2020 in Japan [69]. Delgocitinib demonstrated inhibitory activity against the four JAKs [96].

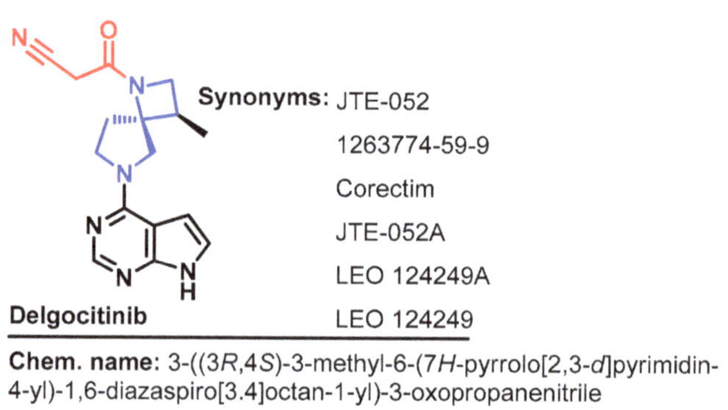

Figure 22. Chemical structure/name/synonyms of delgocitinib.

Synthesis

Noji et al. have reported the original stereoselective synthesis of delgocitinib [96]. In the first step, compound **31** was alkylated with *t*-butyl bromoacetate to give **32**, Scheme 5. The chlorination of **32** resulted in the formation of the reactive aziridinium intermediate **34**, which gave **35** on heating. Compound **35** underwent intramolecular cyclization to afford **36** as a single enantiomer. The protecting group in **36** was replaced by *t*-butyloxycarbonyl protecting (Boc) in **37**. Compound **37** underwent α-alkylation to give **38**, which underwent ozonolysis and reductive amination to give **39**. Removal of the Boc group and intramolecular cyclization of **39** afforded **40**, which was reacted with di-*tert*-butyl dicarbonate to protect the azetidine nitrogen, resulting in **41**. The debenzylation of **41** followed by reaction with 4-chloro-7*H*-pyrrolo[2,3-*d*]pyrimidine **12** afforded **43**, which underwent deprotection by the removal of the Boc group, followed by cyanoacetylation, using **45** to give delgocitinib.

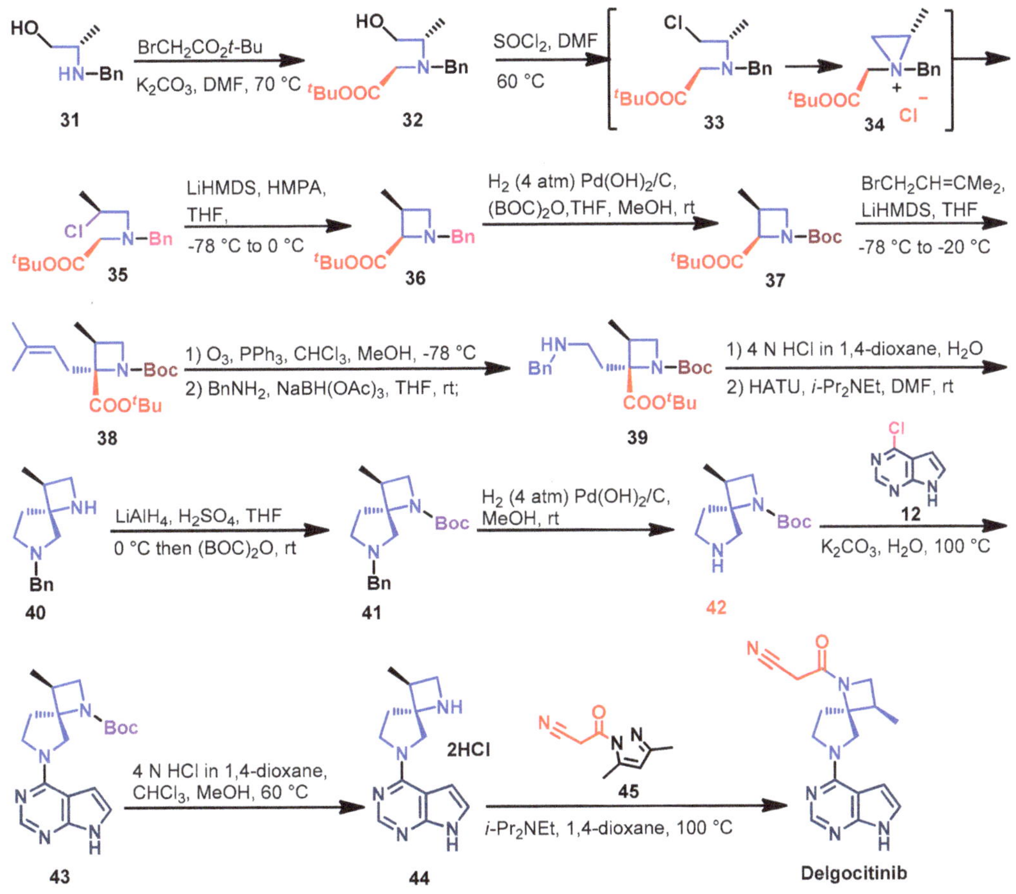

Scheme 5. Synthesis of delgocitinib (route 1).

In addition, Takiguchi et al. have also reported the stereocontrolled synthesis of delgocitinib [97]. The synthesis was started with 3-bromodihydrofuran-2(3*H*)-one **46**, which was converted to **49** in a three-step synthesis, Scheme 6. Compound **49** underwent intramolecular cyclization catalyzed by Cs_2CO_3 to afford a diastereomixture of **50** and **51** in 98% and 2% yield, respectively. Compound **50** was isolated chromatographically and reacted with potassium phthalimide, followed by esterification with ethyl iodide to give the intermediate **52**. Dephthaloylation of compound **52** gave the dilactam derivative **53**. The LiAlH4/TMSCl-catalyzed reduction of **53** afforded **54**, which was isolated as a stereochemically pure succinate salt in 86% yield. The reaction of **12** with **54** afforded compound **44**, which was reacted with the cyanoacetyl pyrazole derivative **45** to give delgocitinib.

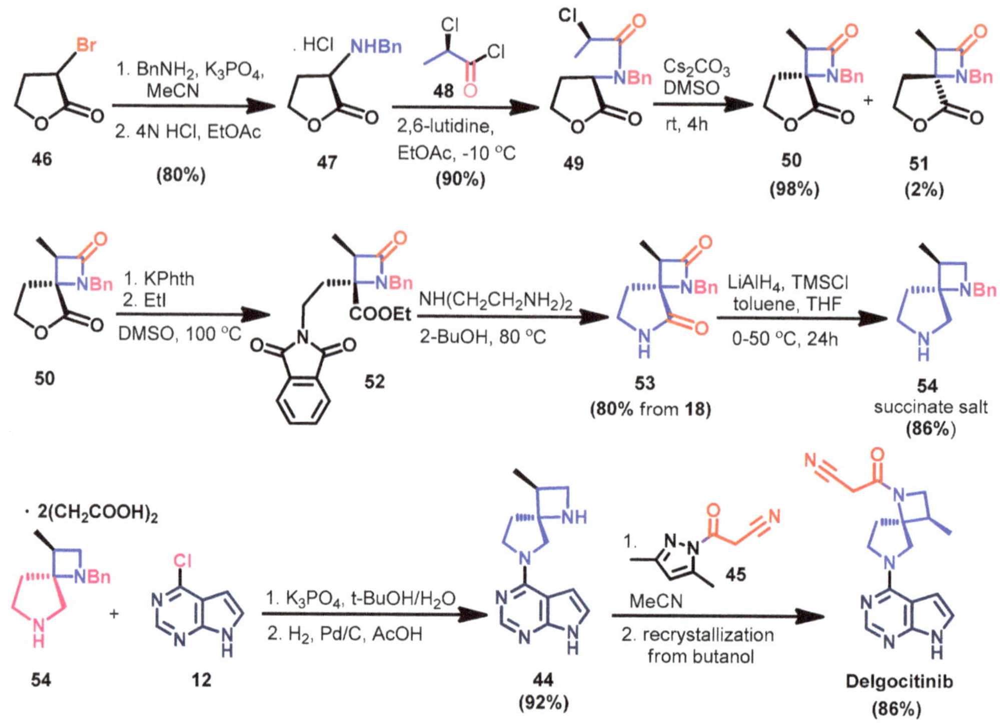

Scheme 6. Synthesis of delgocitinib (route 2).

Target Kinases

Noji et al. [96] have investigated the kinase inhibitory activity of delgocitinib. The results (Figure 23) show potent activities against the four JAKs (IC_{50} = 2.6–58 nM). Delgocitinib also exhibited inhibitory activity against lymphocyte-specific protein tyrosine kinase (LCK) at IC_{50} value of 5.8 μM.

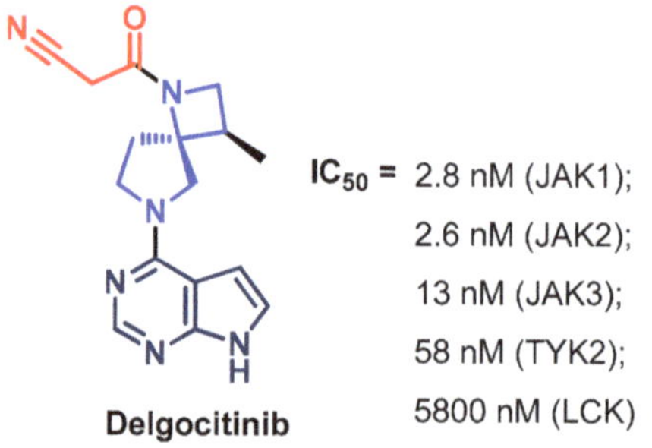

Figure 23. Kinase inhibitory activity of delgocitinib.

Crystal Structures

One crystal structure of delgocitinib in complex with JAK3 (pdb: 7C3N) is available on the protein data bank [96]. The binding mode of delgocitinib into JAK3 is visualized in Figure 24. Delgocitinib shows one conventional hydrogen bond with Leu905 and an electrostatic interaction (attractive charge) with Asp967.

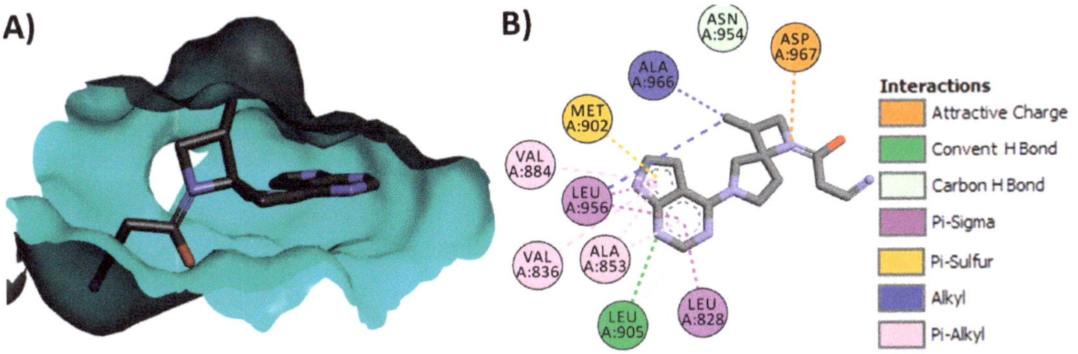

Figure 24. Binding modes of delgocitinib (shown as sticks) into JAK3 (pdb: 7C3N): (**A**) 3D binding mode, receptor shown as a hydrogen bond surface; (**B**) 2D binding mode showing different types of interactions with JAK3; this figure was generated using Discovery Studio Visualizer (V16.1.0.15350).

Pharmacological Activities and Uses

Delgocitinib displayed therapeutic efficacy in the treatment of patients with atopic dermatitis [98]. Delgocitinib was approved in Japan for atopic dermatitis [69].

Due to its pan-JAK inhibitory activity, the pharmacological effects of delgocitinib depend on the competitive inhibition of the four JAKs, which plays an important role in the pathophysiology of chronic inflammatory skin diseases [69]. Delgocitinib also inhibited the activation of inflammatory cells such as mast cells and T cells [99].

Delgocitinib was also evaluated for the treatment of psoriasis and chronic hand eczema (CHE) [69]. After 8 weeks of topical treatment, delgocitinib showed clearance of CHE in a significant number of patients compared to the vehicle [100]. On the other hand, the clinical trial of delgocitinib for discoid lupus erythematosus (NCT03958955) was terminated [101].

Metabolism

The pharmacokinetic study of delgocitinib (JTE-052) revealed that it is excreted mainly as the parent drug in urine, which indicates its stability to metabolism [102]. Delgocitinib also has higher stability values in human cells compared to animal cells [96,103]. The in vitro metabolic study performed using mouse, rat, rabbit, and dog hepatocytes revealed the recovery of the parent drug in 95.5–97.7%. These results indicate that less than 5% of the total dose was metabolized. The metabolic study was performed using animal microsomes and revealed the formation of three metabolites, M1 (oxidative metabolites of the side chain), M2, and M3 (oxidative metabolites of pyrrolopyrimidine), Figure 25.

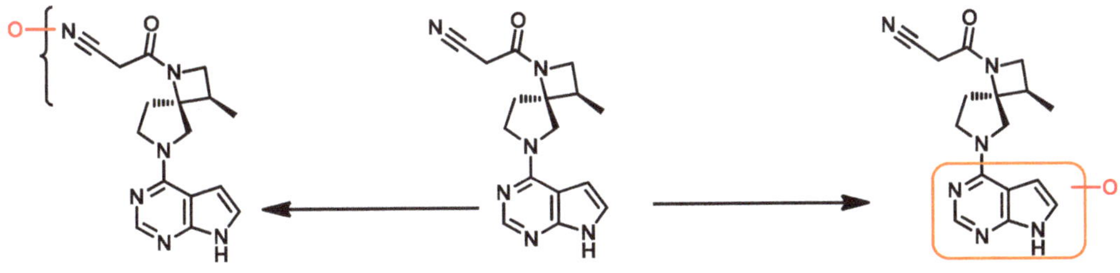

Figure 25. Proposed metabolic pathways and major metabolites of delgocitinib based on the data of the deliberation results [103].

1.7.4. Fedratinib

Approval History

Fedratinib (Figure 26) is a competitive inhibitor of JAK2, BRD4, and FMS-like tyrosine kinase 3 (FLT3) [35]. Fedratinib was approved by the FDA in August 2019 to treat patients with intermediate-2 or high-risk primary or secondary myelofibrosis [66].

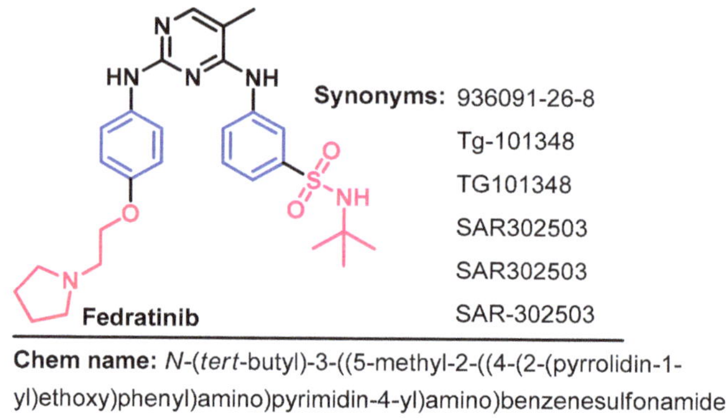

Figure 26. Chemical structure/name/synonyms of fedratinib.

Synthesis

The synthesis of fedratinib could be achieved by the reaction of the intermediates **58** and **61** [104]. Firstly, compound **58** was obtained from the reaction of *p*-nitrophenol **55** and 1-(2-chloroethyl)pyrrolidine **56** in the presence of $CsCO_3$ according to the previous report [105]. The reduction of the nitro group in **57** afforded the amino ether **58**, Scheme 7.

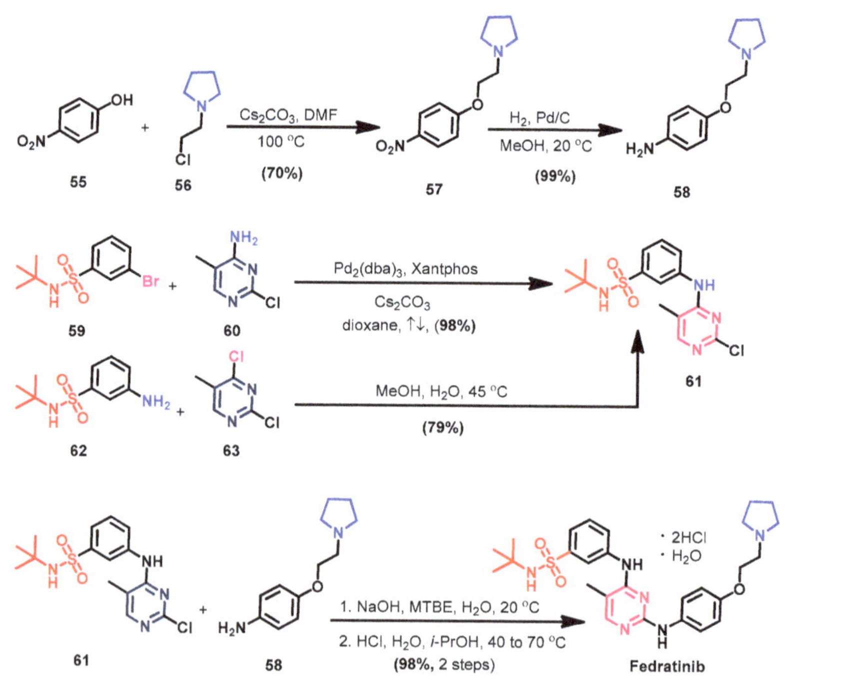

Scheme 7. Synthesis of fedratinib.

On the other hand, compound **61** was obtained in 98% yield from the Pd-catalyzed coupling of **59** and **60** [104,106,107]. Compound **61** could also be prepared in 79% yield from the reaction of **62** with **63** in methanol. The reaction of **58** and **61** in the presence of sodium hydroxide followed by treatment with acidic isopropyl alcohol gave fedratinib, Scheme 7.

Target Kinases

Wernig et al. investigated the kinase inhibitory activity of fedratinib [35]. The results revealed the highest inhibitory activity against JAK2 and JAK2V617F at IC_{50} 3 nM against the two kinases, Figure 27. On the other hand, fedratinib exhibited weak inhibitory activity against JAK3. Fedratinib also showed inhibitory activity against FLT3 and RET kinases with IC_{50} values of 15 nM and 48 nM, respectively. In addition, fedratinib exhibited potent bromodomain inhibitory activity at IC_{50} value in the nanomolar range [108].

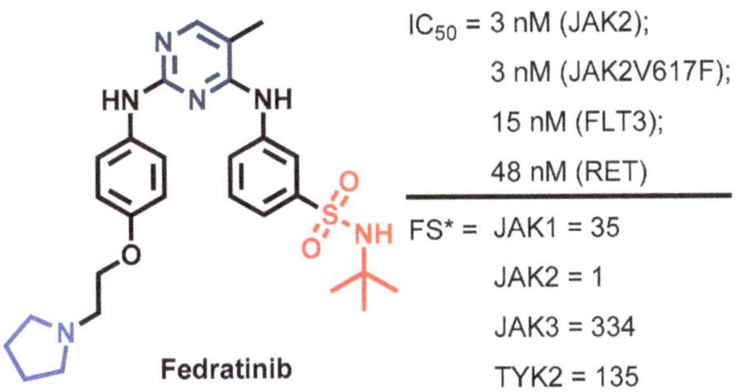

Figure 27. Kinase inhibitory activity of fedratinib (* fold selectivity compared with JAK2).

Crystal Structures

Three crystal structures of fedratinib are available on the protein data bank. Two of these crystals are bromodomain of human dual kinase-bromodomain (BRD4) in complex with fedratinib (pdb: 4OGJ and 4PS5) [108,109], while the third crystal is a complex of fedratinib with JAK2 JH1 (pdb: 6VNE) [85].

The binding mode of fedratinib into human bromodomain BRD4 was visualized in Figure 28. Fedratinib shows two conventional hydrogen bonds with Asn140 and multiple hydrophobic interactions with BRD4.

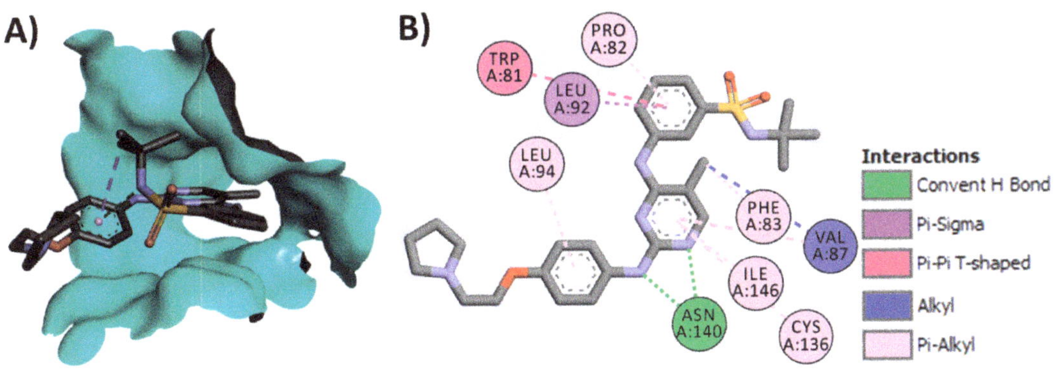

Figure 28. Binding modes of fedratinib (shown as sticks) into human bromodomain BRD4 (pdb: 4OGJ): (**A**) 3D binding mode, receptor shown as a hydrogen bond surface; (**B**) 2D binding mode showing different types of interactions with BRD4; the figure was generated using Discovery Studio Visualizer (V16.1.0.15350).

On the other hand, fedratinib displays two conventional hydrogen bond and Asn140 into human bromodomain BRD4 (pdb: 4PS5). However, fedratinib also shows an additional hydrogen bond with Pro82 (Figure 29).

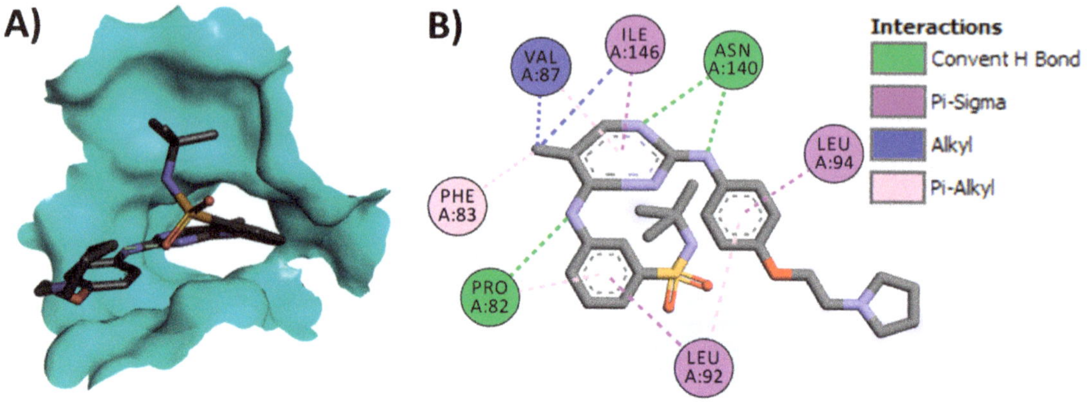

Figure 29. Binding modes of fedratinib (shown as sticks) into human bromodomain BRD4 (pdb: 4PS5): (**A**) 3D binding mode, receptor shown as a hydrogen bond surface; (**B**) 2D binding mode showing different types of interactions with BRD4; the figure was generated using Discovery Studio Visualizer (V16.1.0.15350).

In addition, the binding mode of fedratinib into JAK2 JH1 was visualized in Figure 30. Fedratinib exhibits two conventional hydrogen bonds with Leu932 in JAK2 JH1. Moreover, two carbon hydrogen bonds could be observed with Leu855 and Glu930.

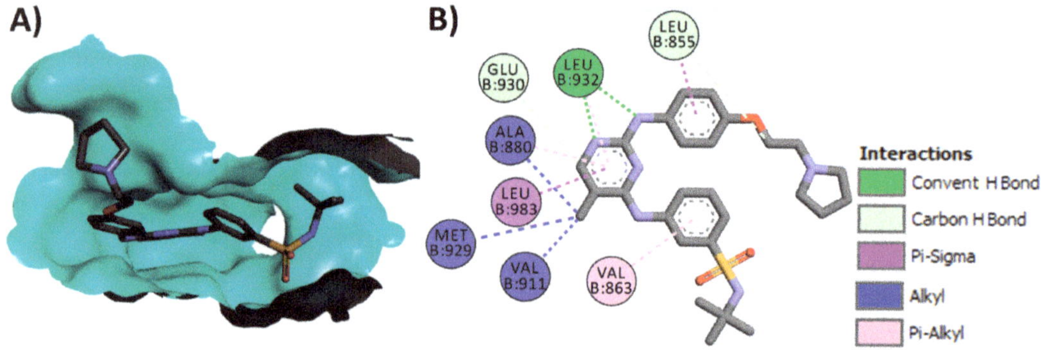

Figure 30. Binding modes of fedratinib (shown as sticks) into JAK2 JH1 (pdb: 6VNE): (**A**) 3D binding mode, receptor shown as a hydrogen bond surface; (**B**) 2D binding mode showing different types of interactions with JAK2 JH1; the figure was generated using Discovery Studio Visualizer.

Pharmacological Activities and Uses

Fedratinib has equal inhibitory activity against the wild-type JAK2 and its mutated form, JAK2V617F [110]. It was approved for the treatment of myelofibrosis in adult patients [66].

Moreover, fedratinib displayed inhibitory activity against other kinases such as FLT3 and RET kinases (Figure 27). The overexpression of FLT3 has been reported in acute leukaemia [111]. In addition, the overexpression of BRD proteins has been reported in different types of cancers [112]. Accordingly, the inhibition of BRD4 by fedratinib could contribute to its pharmacological activity [110].

Metabolism

Ogasawara et al. evaluated the metabolic transformation of [^{14}C]-labeled fedratinib in healthy subjects [113]. The results revealed that nearly 80% of the radioactivity in plasma was due to the parent compound. Among the identified metabolites, SAR317981, a metabolite obtained through oxidation of the pyrrolidine ring, was retrieved in 9.4%, Figure 31. In addition, the *N*-butyric acid derivative (SAR318031) was retrieved at 5.7%. Fedratinib and its metabolites are eliminated mainly through feces (50–70%) and partly in urine (~3%). The amino ethanoic acid derivatives (M17 and M21b) were also detected among 19 different metabolites identified in feces.

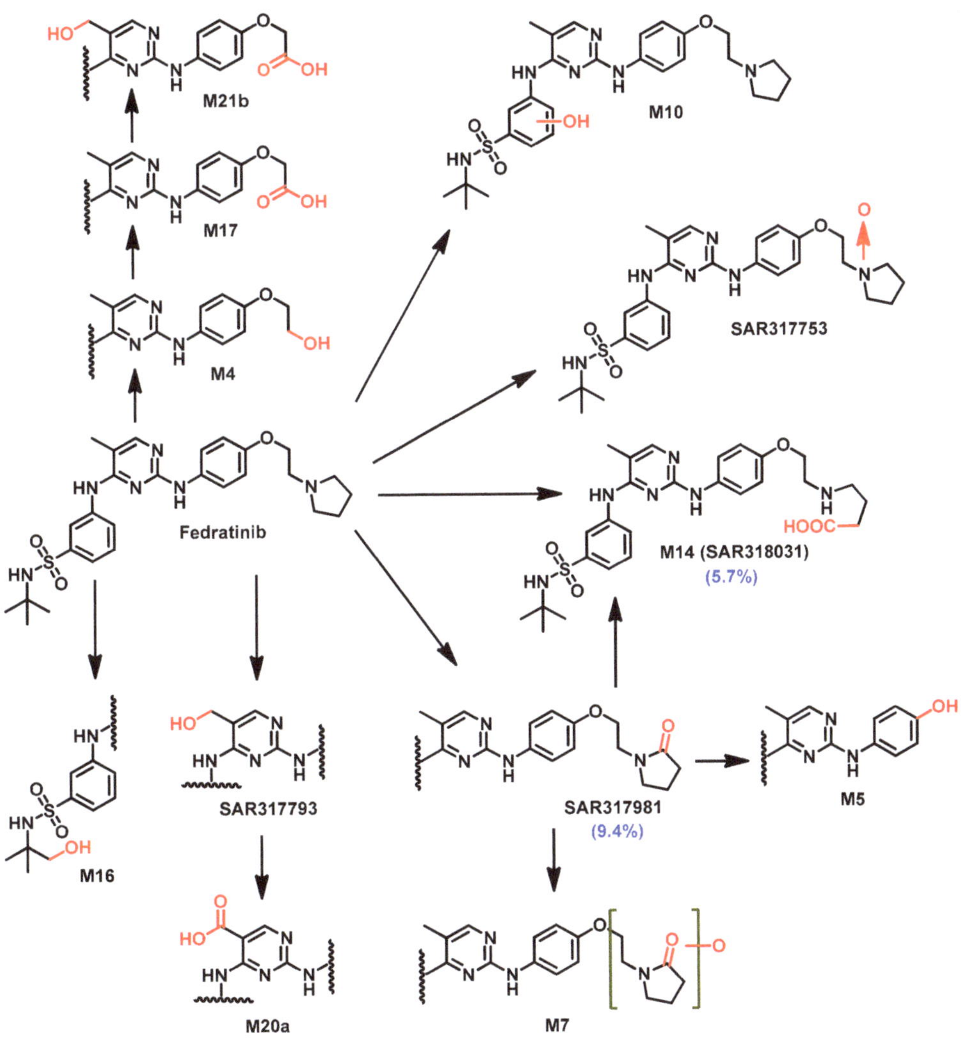

Figure 31. Proposed metabolic pathway and metabolites of fedratinib in humans.

1.7.5. Filgotinib

Approval History

Filgotinib (Figure 32) is classified as a selective, ATP-competitive inhibitor of JAK1 [70]. Filgotinib was approved by the European Medicines Agency (EMA) in September 2020, for adult patients with moderate to severely active rheumatoid arthritis [70]. Filgotinib was also approved for the treatment of rheumatoid arthritis in Japan [70].

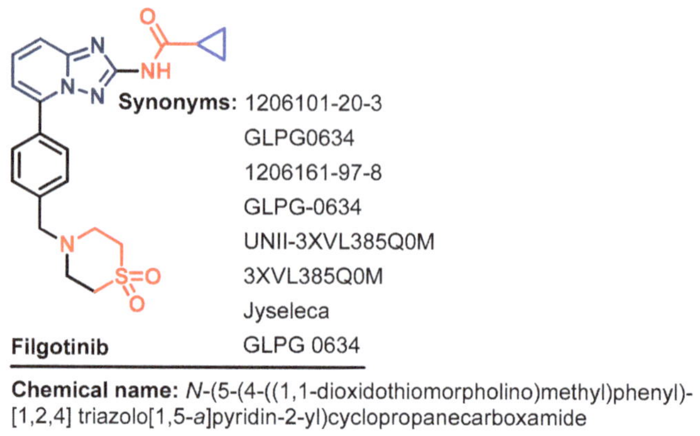

Figure 32. Chemical structure/name/synonyms of filgotinib.

Synthesis

The synthesis of filgotinib (Scheme 8) was achieved in a five-step synthesis [114,115]. Initially, compound **64** was reacted with thiomorpholine 1,1-dioxide **65** to give **66**. On the other hand, the triazolopyridine **70** was prepared from compound **67** in a two-step synthesis. The first step involved the reaction of **67** with ethoxycarbonyl isothiocyanate **68** to give **69**, which underwent intramolecular cyclization with the hydrolysis of the carbamate group to give compound **70**. Acylation of the amino group in **70** with cyclopropanecarbonyl chloride **71** afforded **72**. Finally, filgotinib was obtained from the reaction between **66** and **72**.

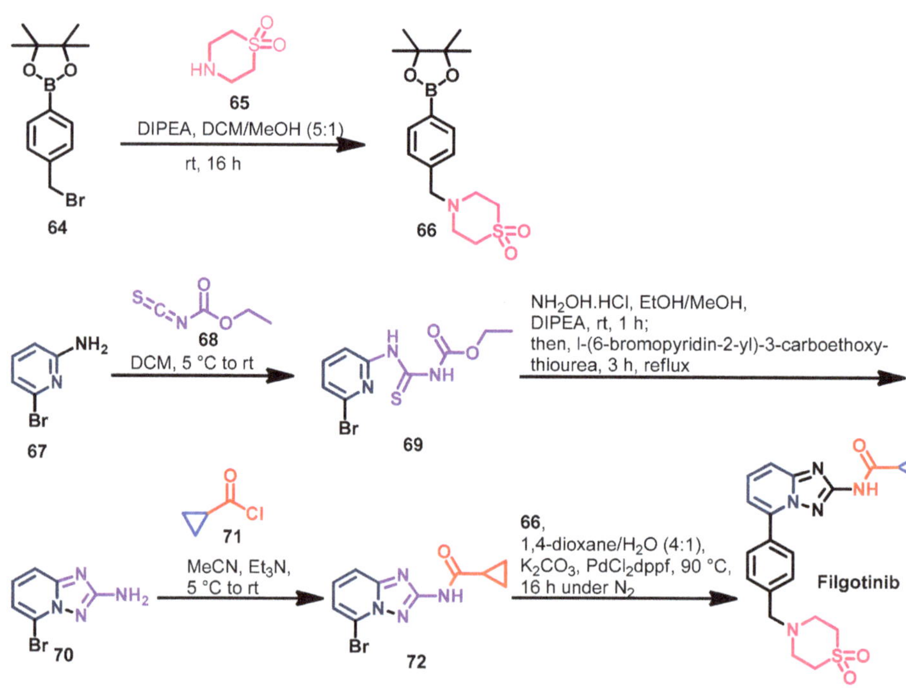

Scheme 8. Synthesis of filgotinib.

Target Kinases

The inhibitory activity of filgotinib was evaluated using recombinant JAKs and whole blood assay [116,117]. The results revealed the highest inhibitory activity against JAK1 (IC_{50} = 10 nM), Figure 33. In addition, filgotinib exhibited 2.8-, 81-, and 116-fold selectivity for JAK1 over JAK2, JAK3, and TYK2, respectively.

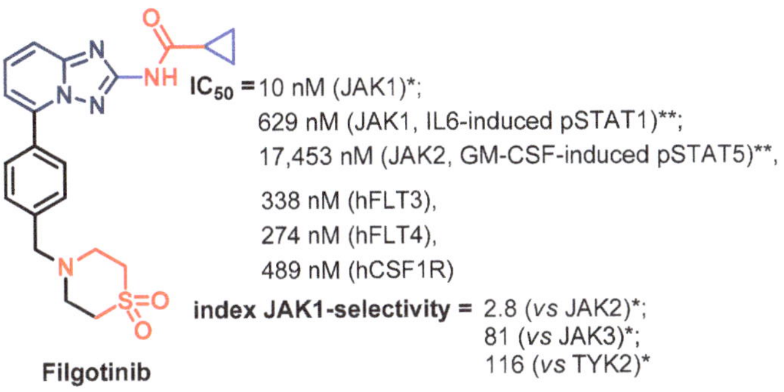

Figure 33. Kinase inhibitory activity of filgotinib, * indicates the IC_{50} and index selectivity values determined using recombinant JAKs, ** indicates the IC_{50} values determined using a human whole blood assay [116,117].

Filgotinib was also tested against a panel of 170 kinases [117]. The results revealed weak inhibitory activities against hFLT3, hFLT4 and hCSF1R (Figure 33).

Crystal Structures

Filgotinib exists in two crystal structures with JAKs protein in the protein data bank. The first crystal (pdb: 4P7E) [117] was refined at a resolution of 2.40 Å, while the second (pdb: 5UT5) [118] was refined at 1.90 Å resolution. The binding mode/interactions of filgotinib into the ATP-binding site of JAK1 (pdb: 4P7E) shows two conventional hydrogen bonds with Leu932 and one carbon hydrogen bond Lys882, Figure 34.

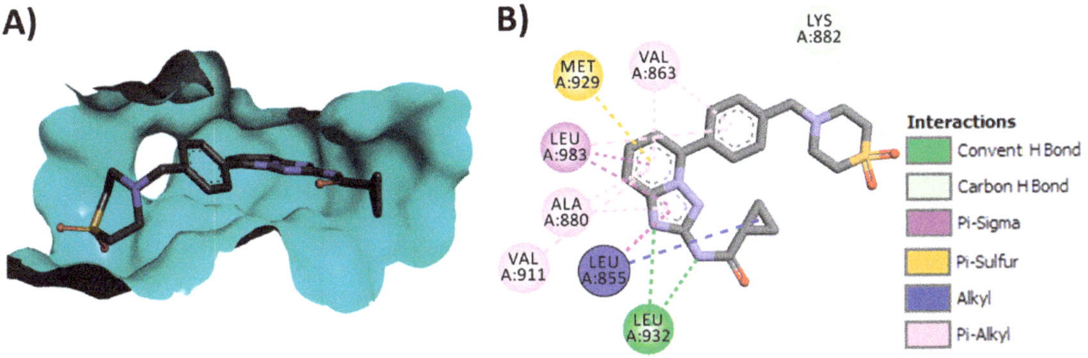

Figure 34. Binding modes of filgotinib (shown as sticks) into JAK1 (pdb: 4P7E): (**A**) 3D binding mode, receptor shown as a hydrogen bond surface; (**B**) 2D binding mode showing different types of interactions with JAK1; the figure was generated using Discovery Studio Visualizer (V16.1.0.15350).

On the other hand, the binding mode of filgotinib into the JH2 domain of JAK2 was visualized in Figure 35. Filgotinib shows two conventional hydrogen bonds with Val629 and two carbon hydrogen bonds with Gln553 and Ser633.

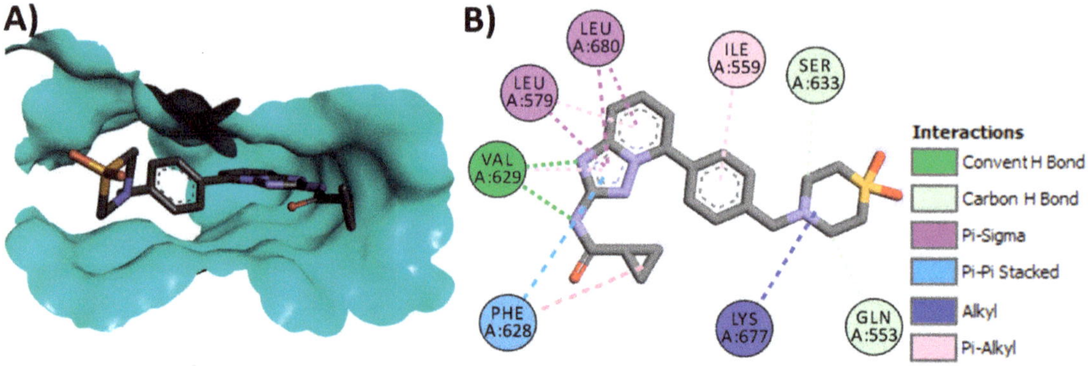

Figure 35. Binding modes of filgotinib (shown as sticks) into JAK2 JH2 (pdb: 5UT5): (**A**) 3D binding mode, receptor shown as a hydrogen bond surface; (**B**) 2D binding mode showing different types of interactions with JAK2 JH2; the figure was generated using Discovery Studio Visualizer (V16.1.0.15350).

Pharmacological Activities and Uses

Filgotinib was first investigated for the treatment of rheumatoid arthritis, where the clinical results proved that rheumatoid arthritis can be treated with a selective inhibitor of JAK1 [119]. Later, filgotinib showed promising efficacy in the treatment of rheumatoid arthritis in two randomized phase IIa trials of filgotinib that were performed by Vanhoutte et al. [120]. In addition, filgotinib showed a rapid improvement in the signs and symptoms of rheumatoid arthritis [121]. In 2020, filgotinib was approved for the treatment of moderate-to-severe rheumatoid arthritis. It acts as a selective inhibitor of JAK1, which leads to the prevention of STAT phosphorylation and activation [3].

Filgotinib was also evaluated in a phase-II study for the treatment of Crohn's disease [122]. The results obtained from this study were also promising for patients with moderate-to-severe Crohn's disease. Moreover, the clinical remission induced by filgotinib in patients with active Crohn's disease was associated with an acceptable safety profile [123].

Moreover, the efficacy of filgotinib in the treatment of ulcerative colitis was also evaluated in a clinical trial (NCT02914522) [124]. The results revealed the efficacy of filgotinib in the treatment of ulcerative colitis.

In addition, the safety and efficacy of filgotinib were also evaluated in a clinical trial (NCT03101670) for the treatment of active psoriatic arthritis [125].

Metabolism

Namour et al. [126] investigated the pharmacokinetics of filgotinib and its active metabolite in healthy male volunteers. The results revealed the formation of an active metabolite, aminotriazolopyridine metabolite (Figure 36), which contributes to the inhibitory activity against JAKs [119,127]. In addition, the active metabolite of filgotinib undergoes further metabolism by glucuronidation of the free amino group to give a glucuronide metabolite.

The aminotriazolopyridine active metabolite of filgotinib (Figure 36) exhibited an inhibitory activity against human JAK1 (IC_{50} = 307 nM) using human recombinant JAK1 [117,127]. On the other hand, the inhibitory activity of this metabolite against JAK1 was observed at IC_{50} of 11.9 μM in a whole blood assay (IL6-induced pSTAT1).

In another study, Namour et al. identified the aminotriazolopyridine active metabolite of filgotinib in three animals (mouse, monkey, and dog) and humans [128]. The study revealed a significant interaction between filgotinib or its active metabolite with CYP450, glucuronosyltransferases, or drug transporters. This study suggests that filgotinib can be combined with the other drugs used for rheumatoid arthritis.

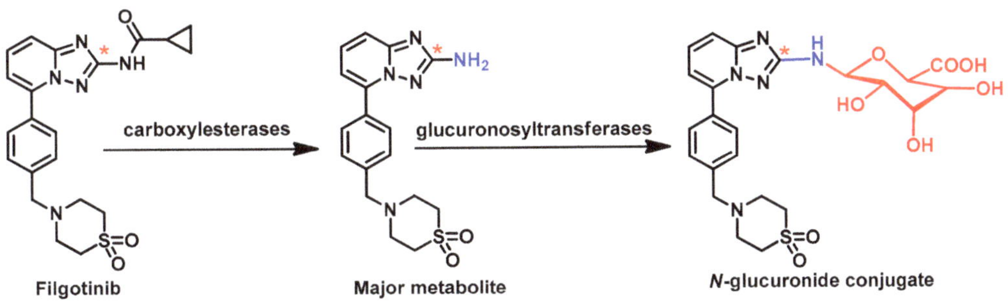

Figure 36. Proposed metabolic pathway of filgotinib.

1.7.6. Oclacitinib

Approval History

In 2013, oclacitinib (Figure 37) was approved by the FDA to treat dogs with atopic dermatitis and pruritus associated with allergic dermatitis [129].

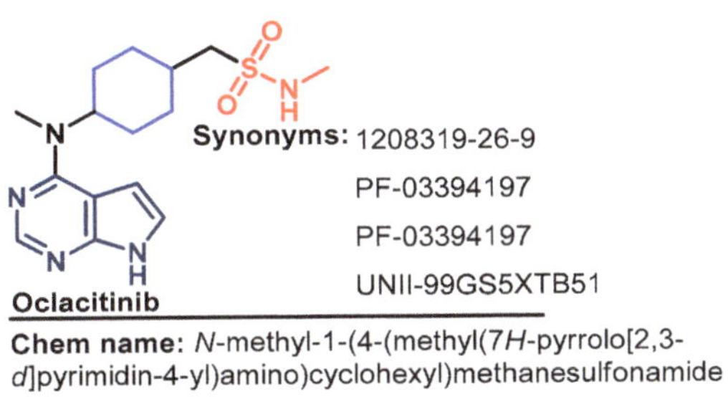

Figure 37. Chemical structure/name/synonyms of oclacitinib.

Synthesis

The preparation of oclacitinib was patented by Berlinski et al. [130]. The synthetic route depends on the reaction of the commercially available 4-chloro-7*H*-pyrrolo[2,3-*d*]pyrimidine **12** with tosyl chloride to give compound **6**, Scheme 9.

In addition, compound **6** was reacted with **73** to afford **74**, which underwent sulfonylation with methanesulfonyl chloride **75** to give **76**, Scheme 9. The reaction of **76** with sodium sulfite gave **77**, which was treated with thionyl chloride to give **78**. Finally, the reaction of compound **78** with methylamine gave **79**, which was treated with lithium hydroxide to liberate oclacitinib as a free base.

Target Kinases

Gonzales et al. evaluated the inhibitory activity of oclacitinib against the four JAKs using isolated enzymes [129]. The results revealed inhibitory activity against the four kinases at IC_{50} values in the range of 10–99 nM, Figure 38. The highest inhibitory activity of oclacitinib was observed against JAK1 (IC_{50} = 10 nM) with 1.8- and 9.9-fold selectivity toward JAK1 compared to JAK2 and JAK3, respectively.

Oclacitinib was also evaluated for its inhibitory activity against a panel of 38 non-JAK kinases [129]. The results revealed 32%–42% inhibition in the activity of aurora-related kinase 1 (ARK1), serine/threonine protein kinase MARK, high-affinity nerve growth factor receptor (TRK-A), and vascular endothelial growth factor receptor 2 (VEGFR-2/FLK1), Figure 38.

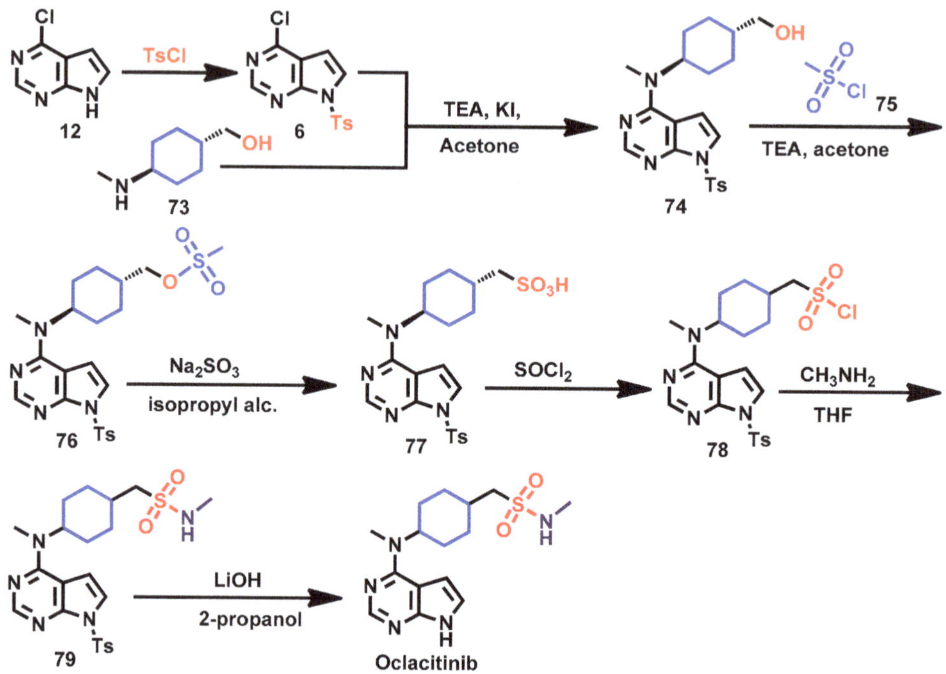

Scheme 9. Synthesis of oclacitinib.

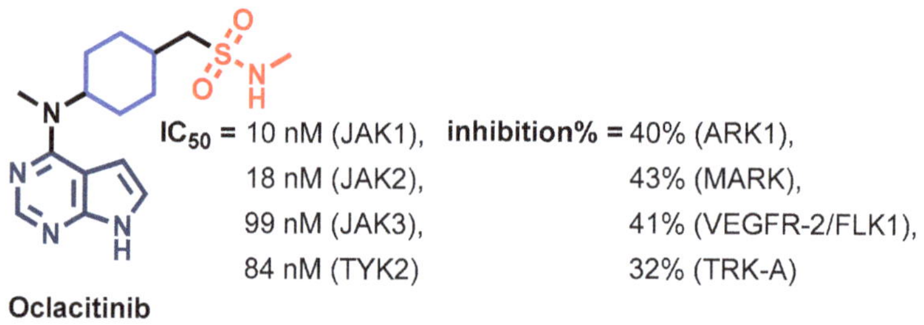

Figure 38. Kinase inhibitory activity of oclacitinib.

Crystal Structures

Oclacitinib has not yet been reported in a crystal structure with any of its target kinases.

Pharmacological Activities and Uses

Gonzales et al. evaluated the effect of oclacitinib on the function of different cytokines in canine and human cell model systems [129]. The results revealed the inhibition of IL-2, IL-4, IL-6, IL-13, and IL-31, which play an essential role in inflammation, allergy, and pruritus at IC_{50} values in the range of 36–249 nm. In 2013, oclacitinib was approved by the FDA to treat dogs ≥12 months old with atopic dermatitis and pruritus associated with allergic dermatitis [129]. Moreover, Haugh et al. also reported the first use of oral oclacitinib in the treatment of a man with atopic dermatitis [131].

Oclacitinib also showed safe and effective control in the treatment of dogs with pruritus associated with allergic dermatitis [132].

Rynhoud et al. investigated the association between the use of oclacitinib and antibacterial therapy [133]. The results suggested a reduction of the antibacterial use when combined with oclacitinib, compared to other antipruritic agents.

Banovic et al. [134] have reported an immunosuppressive effect of oclacitinib in dogs when used at a dose higher than that used in the treatment of allergic pruritus. However, an increase in CD4+ lymphocyte populations in dogs was observed after long-term treatment with oclacitinib [135].

Metabolism

Oclacitinib maleate has rapid oral absorption with 89% absolute bioavailability and a low clearance rate in dogs [136]. It undergoes metabolism into several metabolites. Among these metabolites, one major oxidative metabolite was detected in plasma and urine [137]. However, oclacitinib exhibited very weak inhibitory activity against Cyp450 enzymes, which indicates a low potential for drug–drug interaction.

1.7.7. Pacritinib

Approval History

Pacritinib (Figure 39), a JAK2/FLT3 inhibitor, was approved by the FDA for the treatment of myelofibrosis in adult patients with thrombocytopenia [138].

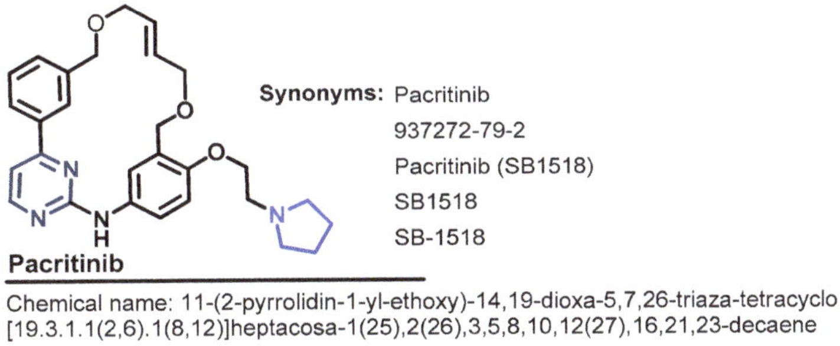

Figure 39. Chemical structure, name, and synonyms of pacritinib.

Synthesis

The synthesis of pacritinib was reported by William et al. [139]. It depends on the reaction of the intermediate compounds **83** and **88**, Scheme 10.

At the first step (Scheme 10), compound **80** was reacted with **81** to give **82**, which underwent a base-catalyzed alkylation using allyl bromide to give **83** [139].

On the other hand, 2-hydroxy-5-nitrobenzaldehyde **84** was reacted with dichloroethane, followed by a reduction in the resulting product **85** to give the corresponding alcohol **86** [139]. Compound **86** was alkylated using allyl bromide to give **87**, which saw a decrease in the nitro group to give **87**. The reaction of **88** with **83** afforded **89**, which was reacted with pyrrole to give pacritinib, Scheme 10.

Target Kinases

The kinase inhibitory activity of pacritinib was evaluated by William et al. [139]. The results revealed inhibitory activity against JAK2 and JAK2^{V617F}, at IC_{50} values of 22 and 19 nM, respectively, Figure 40.

Pacritinib also showed weaker inhibitory activity against the other types of JAK kinases. It also showed potent inhibitory activity against FLT3 (IC_{50} = 22 nM) [139]. In addition, only weak inhibitory activity was observed against CDK2, Figure 40.

Pacritinib also underwent extensive evaluation of its inhibitory activity against 439 recombinant kinases [140]. Besides the inhibition of JAK2, JAK2V617F, and FLT3, pacritinib also showed inhibitory activity against colony-stimulating factor 1 receptor, and interleukin-1 receptor-associated kinase 1 at IC_{50} < 50 nM.

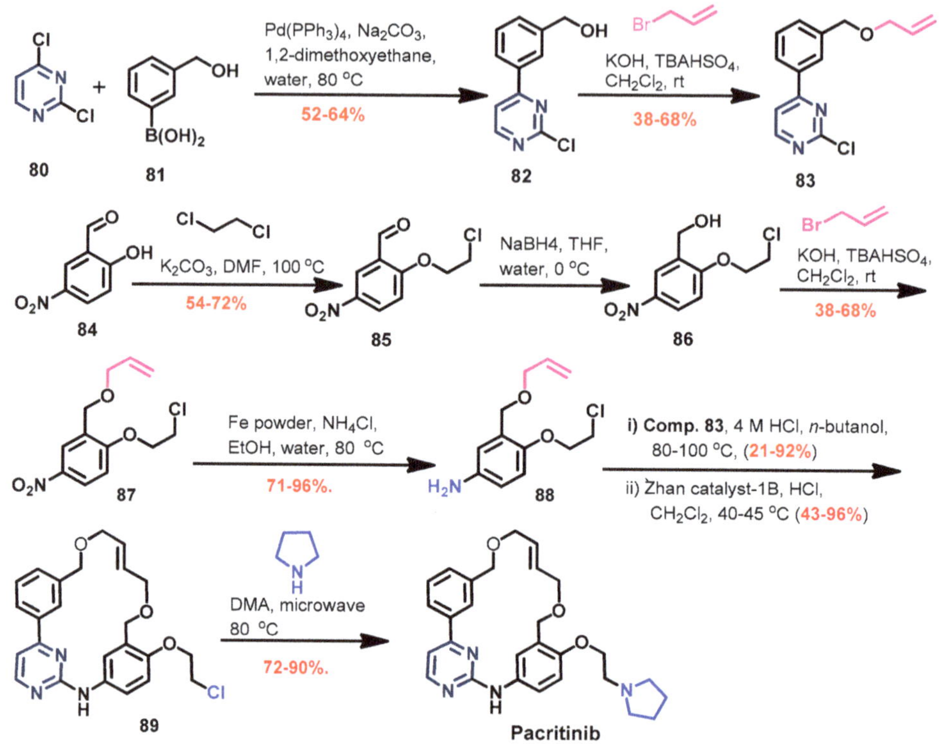

Scheme 10. Synthesis of pacritinib.

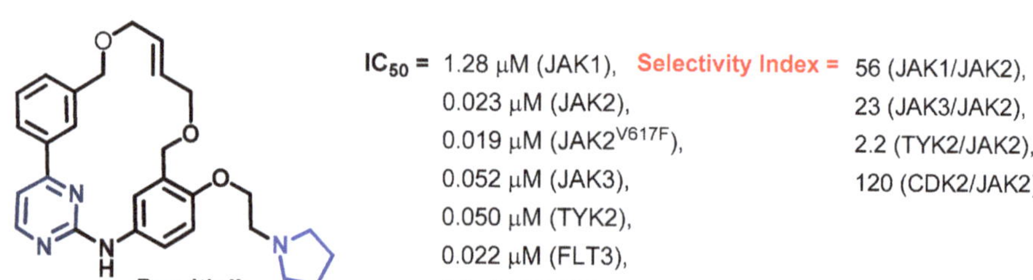

Figure 40. Kinase inhibitory activities of pacritinib.

Crystal Structures

Pacritinib exists in only one crystal structure in the protein data bank. This crystal includes pacritinib in complex with human quinone reductase 2 (NQO2) (pdb: 5LBZ) [141]. Pacritinib shows only multiple hydrophobic interactions of the pi-pi stacked and pi-pi T-shaped types with Phe126 and Phe178 in this reductase enzyme, Figure 41. However, no crystal structure with any of the target JAKs has been reported yet.

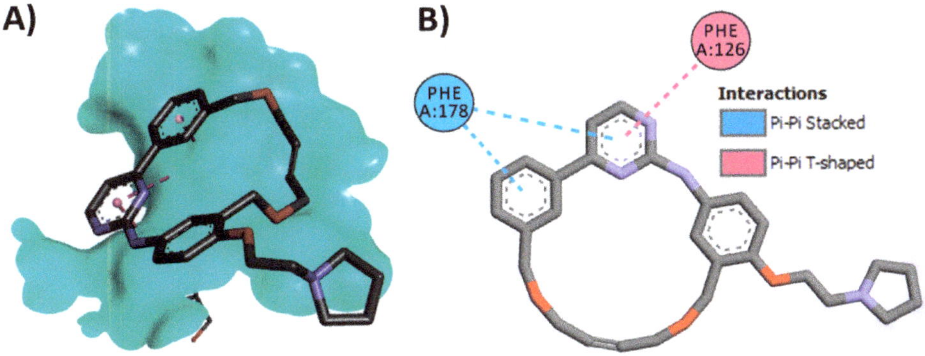

Figure 41. Binding modes of pacritinib (shown as sticks) into human quinone reductase 2 (NQO2) (pdb: 5LBZ): (**A**) 3D binding mode, receptor shown as a hydrogen bond surface; (**B**) 2D binding mode showing different types of binding interactions; this figure was generated using Discovery Studio Visualizer (V16.1.0.15350).

Pharmacological Activities and Uses

Pacritinib acts as an inhibitor of both JAK2 and FLT3 that could be used to overcome the resistance problems in patients with acute myeloid leukaemia (AML) [30]. The dual activity against the two kinases could enable pacritinib to overcome the resistance mediated by the upregulation of JAK2 in FLT3-TKI-resistant AML cells [30].

Pacritinib was later tested for its activity in the treatment of myelofibrosis in several clinical trials [142,143]. The efficacy of pacritinib was evaluated against the best available therapy in patients with myelofibrosis (NCT01773187) [142]. The results revealed a significant reduction in the volume of the spleen (SVR) in patients receiving pacritinib therapy, indicating that it could be used in the treatment of myelofibrosis.

In addition, pacritinib also showed higher efficacy in reducing splenomegaly than the best available therapy in patients with myelofibrosis and thrombocytopenia (NCT02055781) [143].

Pacritinib also showed potential anti-leukemic activity when combined with chemotherapeutic agents [144,145].

The efficacy of pacritinib in the treatment of patients with glioblastoma multiforme was also reported when combined with temozolomide [146].

Metabolism

The preliminary results of the metabolic study of pacritinib revealed that CYP3A4 is the main metabolizing enzyme [139]. However, no inhibitory activity was observed against the other isoenzymes.

On the other hand, the metabolism of pacritinib was investigated in vitro/in vivo by Jayaraman et al. [147]. The results revealed the formation of four metabolites by liver microsomes in both humans and mice, Figure 42.

The detected metabolites include two oxidation metabolites formed by oxidation of the pyrrole ring (M1) and pyrrole nitrogen (M3) [147]. In addition, the third metabolite (M2) formed by *O*-dealkylation of the pyrrole-bearing side chain, while the fourth one (M4) was formed by reduction of the double bond, Figure 42.

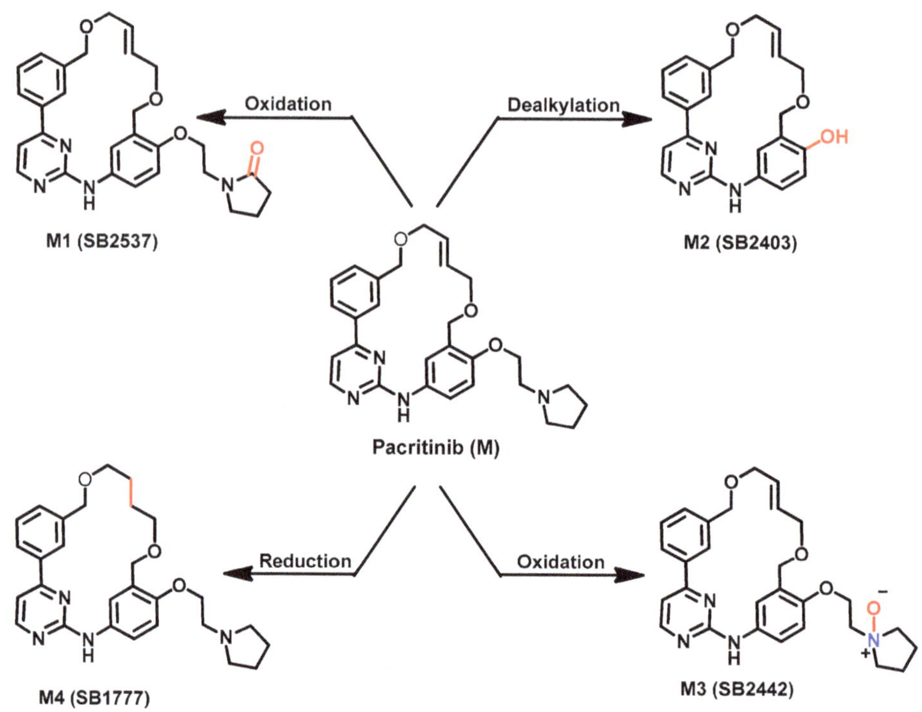

Figure 42. Proposed metabolic pathways and metabolites of pacritinib in human and mouse plasma.

1.7.8. Peficitinib

Approval History

Peficitinib (Figure 43) is a pan-JAK inhibitor. It was approved in 2019 in Japan for the treatment of rheumatoid arthritis [68].

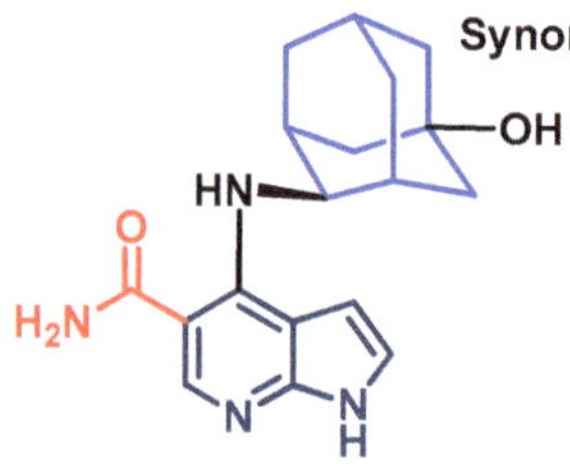

Peficitinib

Synonyms: ASP015K
944118-01-8
UNII-HPH1166CKX
ASP-015K
HPH1166CKX
JNJ-54781532
GTPL8315

Chem. name: 4-(((1*R*,2*s*,3*S*,5*s*,7*s*)-5-hydroxyadamantan-2-yl)amino)-1*H*-pyrrolo[2,3-*b*]pyridine-5-carboxamide

Figure 43. Chemical structure/name/synonyms of peficitinib.

Synthesis

Peficitinib was obtained from pyrrolo[2,3-*b*]pyridine **7** in a seven-step synthesis [104,148]. In the first step, *N*-protected pyrrolo[2,3-*b*]pyridine **90** was obtained from the reaction of **12** with triisopropylsilyl chloride (TIPSCl). Compound **91** was obtained from the reaction of **90** with *sec*-BuLi and ethyl chloroformate followed by treatment with *tetra*-butylammonium fluoride (TBAF) to remove the TIPS group, Scheme 11.

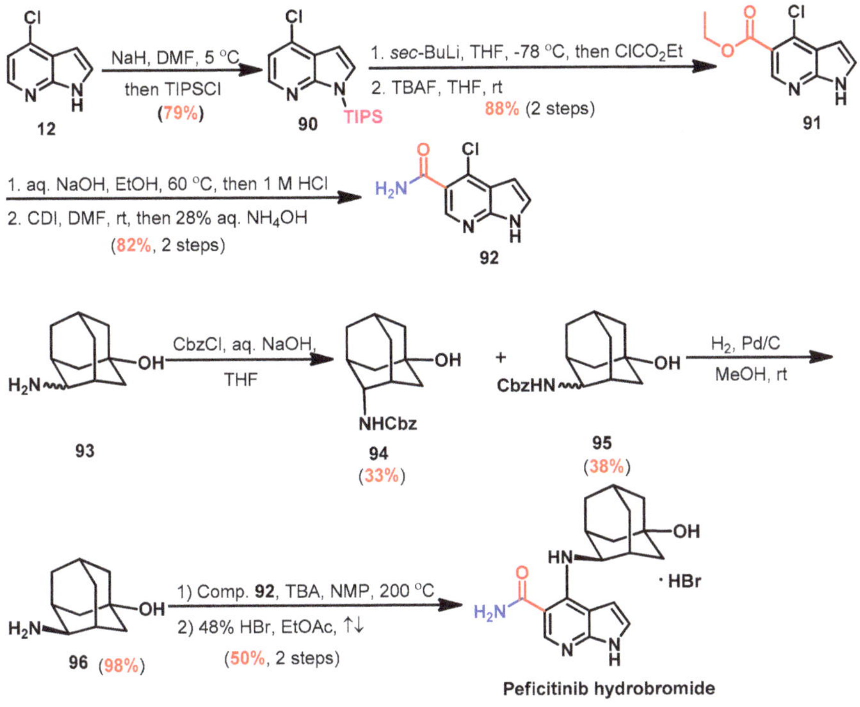

Scheme 11. Synthesis of peficitinib.

Conversion of the ethyl ester group in **91** to the corresponding amide group in **92** was achieved in two steps (Scheme 11). The first step proceeded through hydrolysis of the ester group in **91** to liberate the corresponding carboxylic acid derivative, which was reacted with ammonium hydroxide in the presence of CDI to give the amide derivative **92**.

On the other hand, the diastereomeric mixture of **93** was reacted with benzyl chloroformate followed by chromatographic separation to isolate the *trans*-isomer **95**. Pd/C-catalyzed hydrogenation of **95** yielded *trans*-4-aminoadamantan-1-ol **96**. The reaction between **92** and **96** afforded peficitinib as a free base. Treatment of peficitinib with hydrobromic acid afforded peficitinib hydrobromic acid salt, Scheme 11.

Target Kinases

The kinase inhibitory activity of peficitinib was evaluated against the four JAKs [149,150]. The results (Figure 44) revealed inhibitory activity against the four kinases with IC_{50} values in the range of 0.70–5.0 nM [149]. These results indicate that the highest inhibitory activity of peficitinib was against JAK3.

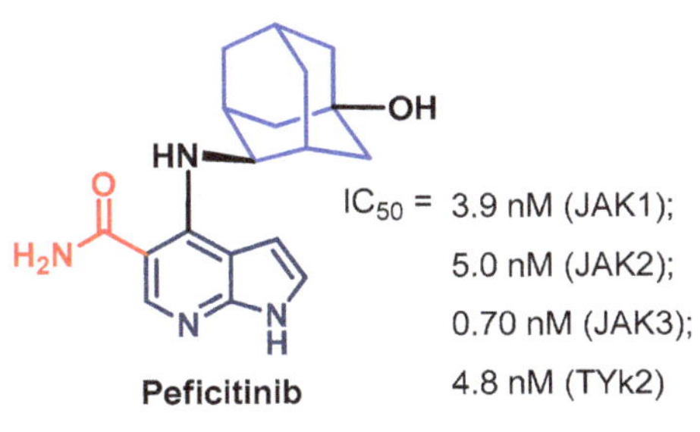

Figure 44. Kinase inhibitory activities of peficitinib.

Crystal Structures

Peficitinib is available as a co-crystallized ligand in four crystal structures with JAKs in the protein data bank. These crystals include the crystal structure of peficitinib bound to JAK1 (pdb: 6AAH) [151], JAK2 (pdb: 6AAJ) [151], JAK3 (pdb: 6AAK) [151], and TYK2 (6AAM) [151]. All at 1.83–2.67 Å resolution. The binding mode and interactions of peficitinib with JAK1 (pdb: 6AAH) are presented in Figure 45. Peficitinib shows three conventional hydrogen bonds with Leu959 and Asn1008.

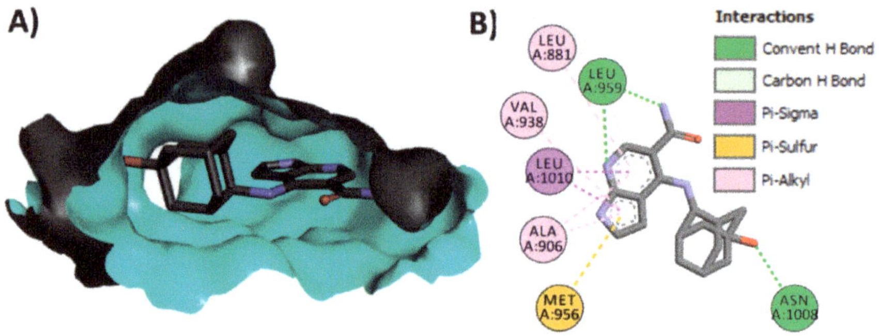

Figure 45. Binding modes of peficitinib (shown as sticks) into JAK1 (pdb: 6AAH): (**A**) 3D binding mode, receptor shown as a hydrogen bond surface; (**B**) 2D binding mode showing different types of interactions with JAK1; the figure was generated using Discovery Studio Visualizer (V16.1.0.15350).

The binding interactions of peficitinib in JAK2 (pdb: 6AAJ) show two conventional hydrogen bonds with Glu930 and Leu932, Figure 46. In addition, peficitinib forms one carbon hydrogen bond with Glu930.

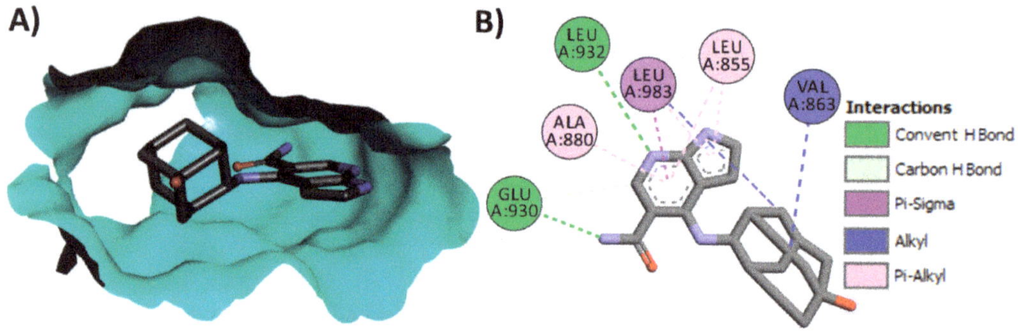

Figure 46. Binding modes of peficitinib (shown as sticks) into JAK2 (pdb: 6AAJ): (**A**) 3D binding mode, receptor shown as a hydrogen bond surface; (**B**) 2D binding mode showing different types of interactions with JAK2; hydrogen atoms were omitted for clarity; this figure was generated using Discovery Studio Visualizer (V16.1.0.15350).

The binding mode and interactions of peficitinib are presented in Figure 47. Peficitinib shows two conventional hydrogen bonds with Glu903 and Leu905, as well as one carbon–hydrogen bond with Glu903.

On the other hand, the binding orientation, and interactions of peficitinib with TYK2 are presented in Figure 48. Peficitinib shows four conventional hydrogen bonds: Val981, Glu979, Asn1028, and Asp1041.

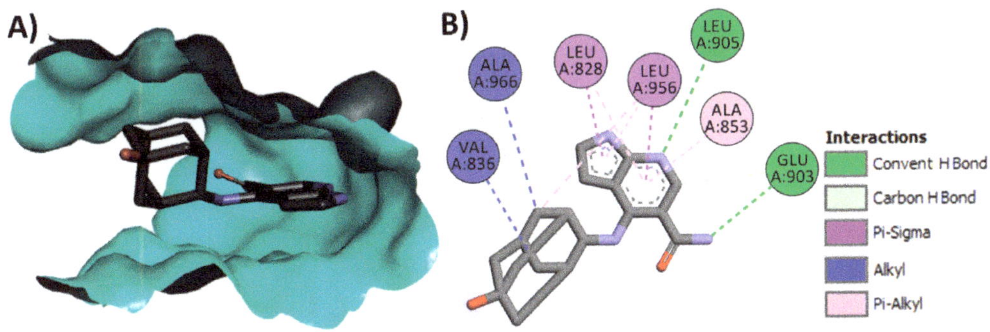

Figure 47. Binding modes of peficitinib (shown as sticks) into JAK3 (pdb: 6AAK): (**A**) 3D binding mode, receptor shown as a hydrogen bond surface; (**B**) 2D binding mode showing different types of interactions with JAK3; hydrogen atoms were omitted for clarity; this figure was generated using Discovery Studio Visualizer (V16.1.0.15350).

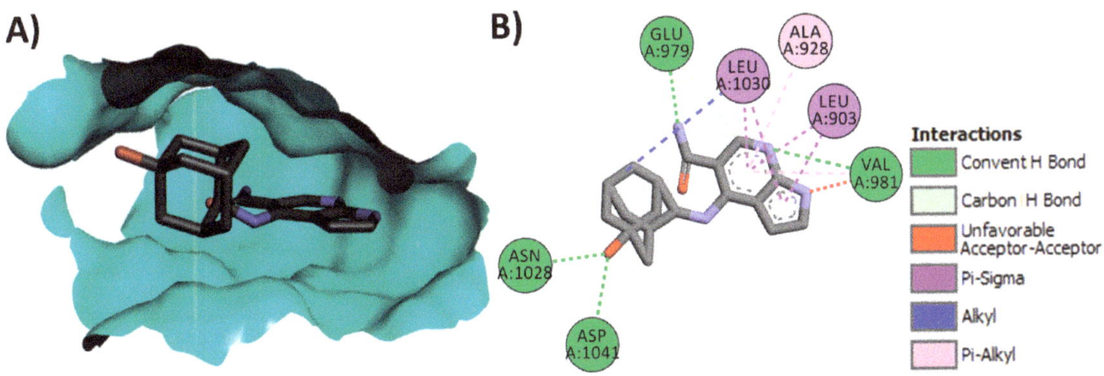

Figure 48. Binding modes of peficitinib (shown as sticks) into TYK2 (pdb: 6AAM): (**A**) 3D binding mode, receptor shown as a hydrogen bond surface; (**B**) 2D binding mode showing different types of interactions with TYK2; hydrogen atoms were omitted for clarity; this figure was generated using Discovery Studio Visualizer (V16.1.0.15350).

Pharmacological Activities and Uses

Cytokines play an important role in pain, inflammatory reactions, and nerve sensitization [152]. The pan-JAK inhibitor, peficitinib, was evaluated for the treatment of rheumatoid arthritis, which is characterized by joint destruction and inflammation [153]. Peficitinib suppressed bone destruction and paw swelling in rats with adjuvant-induced arthritis [149].

In a clinical trial (NCT01565655), peficitinib exhibited a dose-dependent ACR20 response rate when given orally to patients with moderate-to-severe rheumatoid arthritis [154]. Peficitinib also demonstrated clinical efficacy and prevention of joint destruction in Asian patients who displayed an inadequate response to conventional DMARDs [155]. In addition, no drop in the effectiveness of peficitinib was observed after long-term use [156]. However, the use of peficitinib was also associated with a risk of herpes zoster infection, similar to other JAK inhibitors [157].

Metabolism

Oda et al. investigated the metabolism of peficitinib in healthy male subjects using [^{14}C]-labelled peficitinib [158]. The results revealed the formation of a sulfate-conjugated metabolite (M2). The sulfate conjugate and the parent drug constituted the major components in plasma and urine, Figure 49. On the other hand, peficitinib was metabolized to give the *N*-methylated derivative (M4), which was identified in feces.

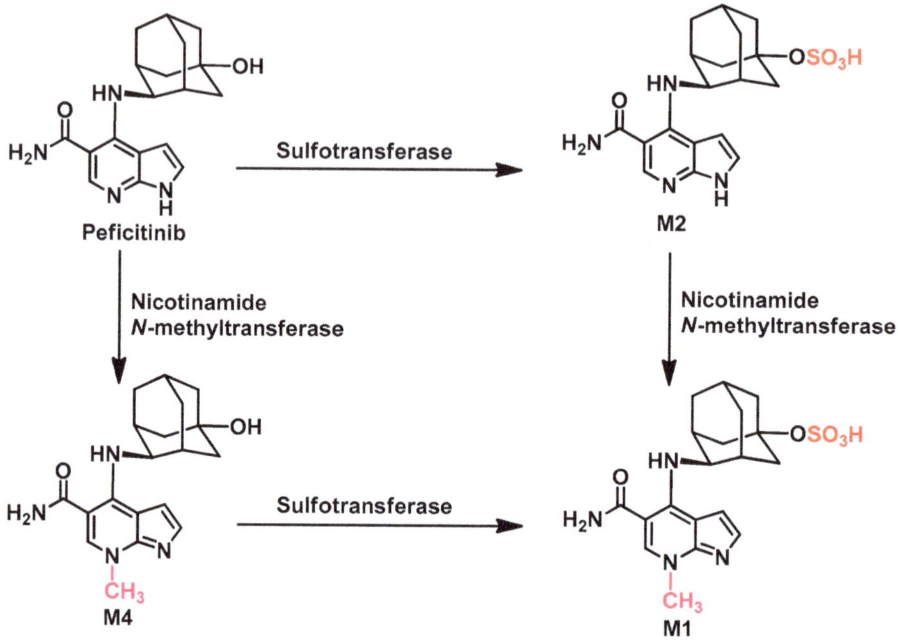

Figure 49. Proposed metabolic pathways and metabolites of peficitinib.

Miyatake et al. investigated the effect of hepatic impairment on the pharmacokinetic profile of peficitinib [159]. The results of this study suggest a decrease in the dose of peficitinib in patients with moderate hepatic impairment.

1.7.9. Ruxolitinib

Approval History

Ruxolitinib (Figure 50) was approved by the FDA in November 2011 to treat myelofibrosis [28]. The FDA also approved ruxolitinib in December 2014 for the treatment of patients with polycythemia vera [29]. In addition, ruxolitinib received approval in 2019 and 2021 for the treatment of acute and chronic graft-versus-host disease, respectively [160].

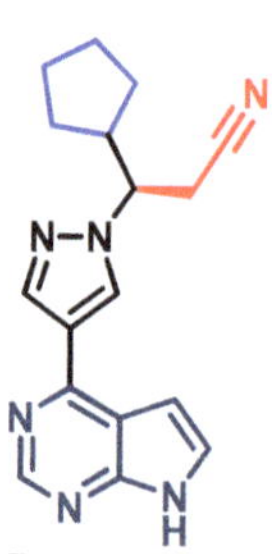

Ruxolitinib

Synonyms: 941678-49-5
INCB018424
UNII-82S8X8XX8H
82S8X8XX8H
INC424
CHEBI:66919
(R)-ruxolitinib

Chem. name: (*R*)-3-(4-(7*H*-pyrrolo[2,3-*d*]pyrimidin-4-yl)-1*H*-pyrazol-1-yl)-3-cyclopentylpropanenitrile

Figure 50. Chemical structure/name/synonyms of ruxolitinib.

Synthesis

The synthesis of ruxolitinib (Scheme 12) was achieved from compound **12** in a five-step synthesis [115,161]. Compound **12** was first reacted with 2-(trimethylsilyl)ethoxyethyl chloride (SEM-Cl) to give **24**.

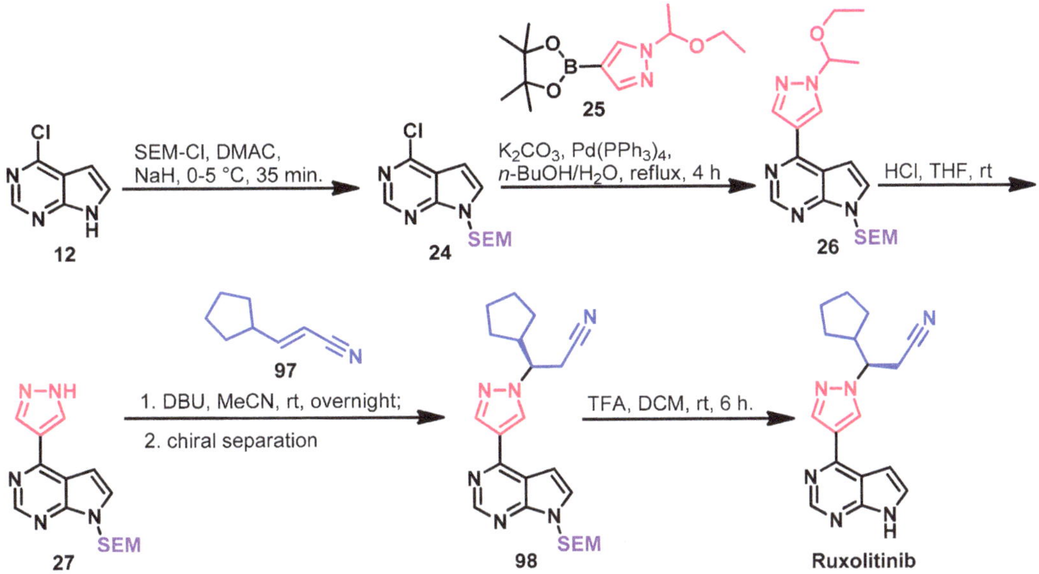

Scheme 12. Synthesis of ruxolitinib (route 1).

In addition, compound **27** was prepared from compound **24** via two steps, Scheme 12. The first step is a Suzuki coupling of **24** and **25** to give **26**. In the second step, compound **26** underwent acid-catalyzed hydrolysis to give **27**. The reaction of **27** with 3-cyclopentylacrylonitrile **97** afforded a mixture of *R*- and *S*-enantiomers **98**, which underwent chiral separation. The *R*-enantiomer of **98** was treated with trifluoroacetic acid to give ruxolitinib.

Haydl et al. [162] reported a regio- and enantioselective synthesis of *N*-substituted pyrazoles that can be used in the synthesis of ruxolitinib, Scheme 13. Compound **101** was prepared from the reaction of cyclopentyl magnesium bromide **99** and 3-bromoprop-1-yne **100**.

Asymmetric addition reaction of the substituted pyrazole **102** to the cyclohexylallene **101** afforded **103**; Scheme 13. Hydroboration of **103** using 9-borabicyclo(3.3.1)nonane followed by the Swern oxidation of the alcoholic group afforded **105**. Compound **105** was then reacted with hydroxylamine in the presence of iodine to give the corresponding nitrile **106**, which was then reacted with bis(pinacolato)diboron to give **107**. The Suzuki coupling of compounds **12** and **107** afforded (*R*)-ruxolitinib.

Lin et al. [163] have also reported an enantioselective synthesis of ruxolitinib (INCB018424). This synthesis depends on the addition of a substituted pyrazole to (*E*)-3-cyclopentylacry laldehyde, which was catalyzed by diarylprolinol silyl ether.

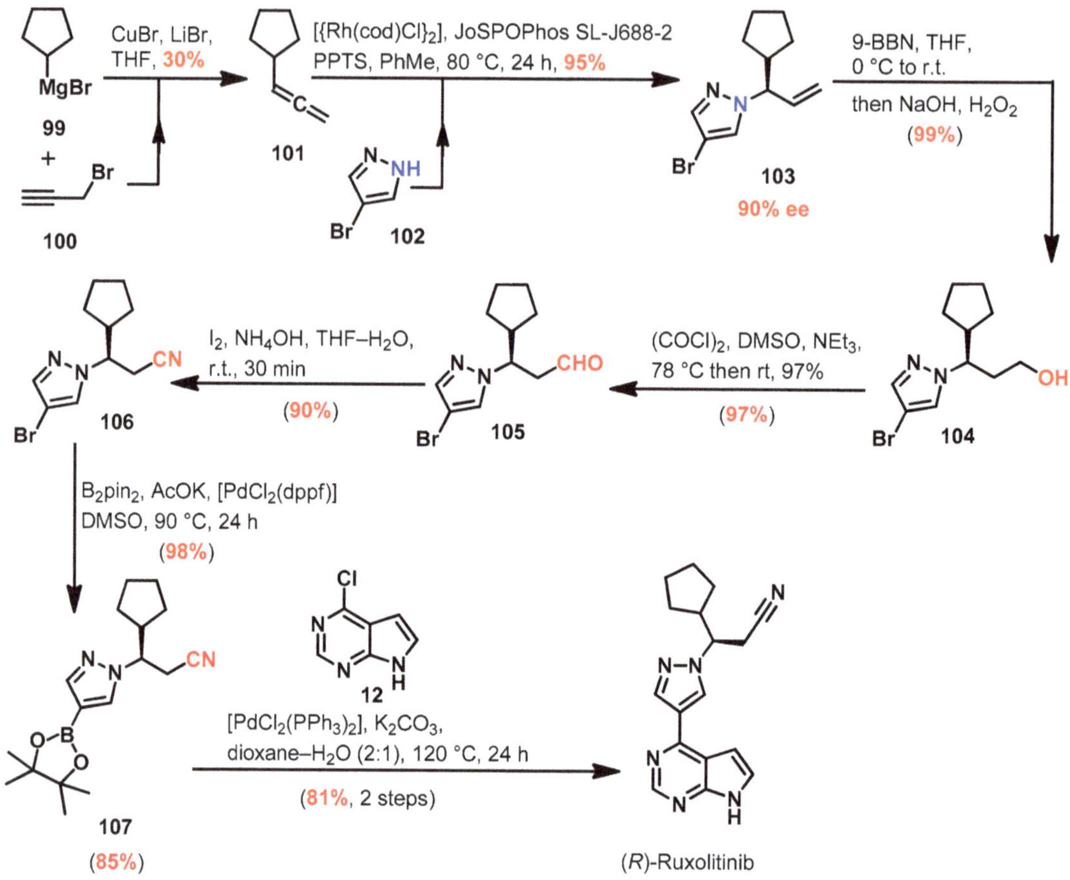

Scheme 13. Synthesis of ruxolitinib (route 2).

Target Kinases

Clark et al. investigated the inhibitory activity of ruxolitinib against JAKs [83]. The results (Figure 51) revealed the highest inhibitory activity against JAK1 and JAK2 with IC_{50} values of 6.4 and 8.8 nM, respectively.

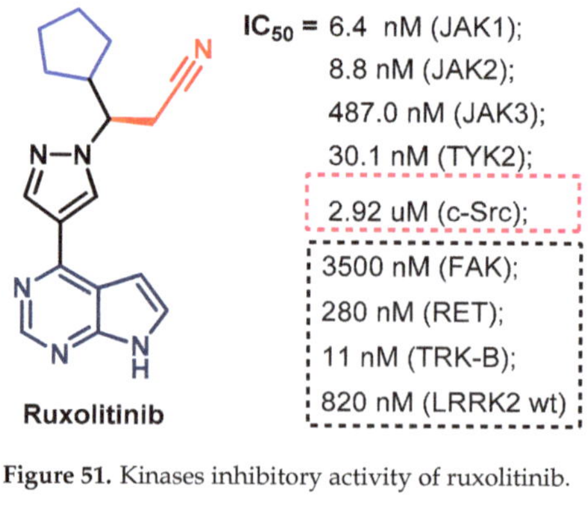

Figure 51. Kinases inhibitory activity of ruxolitinib.

The kinase inhibitory activity of ruxolitinib was also evaluated against 368 kinases by Zhou et al. [164]. At 1 μM, ruxolitinib displayed 96%-100% inhibition in the activity of the four JAKs. Ruxolitinib also showed inhibitory activity against other kinases such as FAC, RET, TRK-B, and LRRK2 (wild-type), Figure 51. Ruxolitinib also exhibited inhibitory activity against c-Src kinase at IC_{50} value of 2.92 nM [165].

In another study performed by Quintás-Cardama et al. [166], ruxolitinib displayed inhibitory activity against the four JAKs at IC_{50} in the range of 2.8–428 nM. The results of this study revealed the inhibitory activity of ruxolitinib against JAK1 (IC_{50} = 3.3 nM) and JAK2 (IC_{50} = 2.8 nM). Ruxolitinib also exhibited moderate inhibitory activity against TYK2 (IC_{50} = 19 nM). In addition, high efficacy of ruxolitinib was also observed in tumour cells with JAK2V617F mutation.

Ruxolitinib was also evaluated for its inhibitory activity against CHK2 and c-Met kinases, where the results revealed IC_{50} values exceeding 1000 nM [166].

Crystal Structures

Ruxolitinib exists as a co-crystallized ligand in five crystal structures in the protein data bank. Three of these crystals include complexes of ruxolitinib with the JH1 domain of JAK2 (pdb: 6VGL, 6WTN, and 6VNK) [85,86]. Ruxolitinib also exists as a co-crystallized ligand with c-Src (pdb: 4U5J) [165] and DRLK1 (pdb: 7F3G). The binding mode and interactions of ruxolitinib into JAK2 JH1 (pdb: 6VGL) are visualized in Figure 52. Ruxolitinib shows one conventional hydrogen bond with Leu932 and two carbon hydrogen bonds with Leu855 and Gly856.

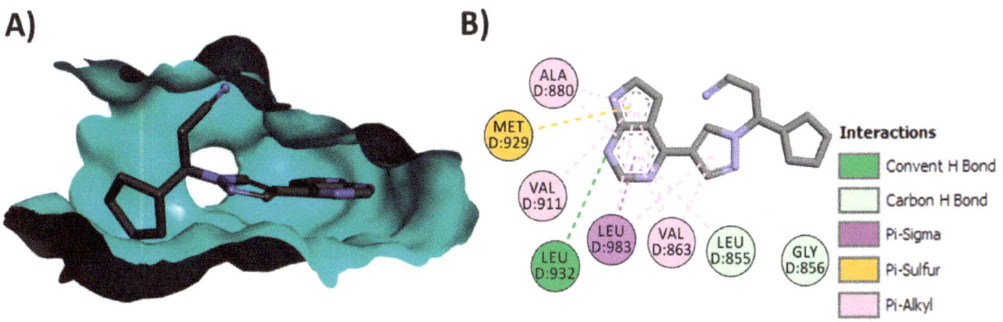

Figure 52. Binding modes of ruxolitinib (shown as sticks) into JAK2 JH1 (pdb: 6VGL): (**A**) 3D binding mode, receptor shown as a hydrogen bond surface; (**B**) 2D binding mode showing different types of interactions with JAK2 JH1; hydrogen atoms were omitted for clarity; this figure was generated using Discovery Studio Visualizer (V16.1.0.15350).

Ruxolitinib also exists as a co-crystallized ligand into the JH1 domain of human JAK2 (pdb: 6WTN) at 1.83 Å resolution. Visualization of the binding mode and interactions of ruxolitinib shows one conventional and one carbon hydrogen bond with Leu932, Figure 53.

In addition, ruxolitinib was also isolated as a co-crystallized ligand with JAK2 JH1 (pdb: 6VNK) at 2.00 Å resolution [85]. The binding interactions also shows one conventional hydrogen bond with Leu932 and one carbon hydrogen bond with Gly856, Figure 54.

The Src and JAK family kinases share ~34% sequence identity in the kinase domain [165]. Ruxolitinib was also refined in a crystal structure with c-Src (pdb: 4U5J) [165]. The binding interactions of ruxolitinib with c-Src show one conventional hydrogen bond with Met341 and one carbon–hydrogen bond with Glu339, Figure 55.

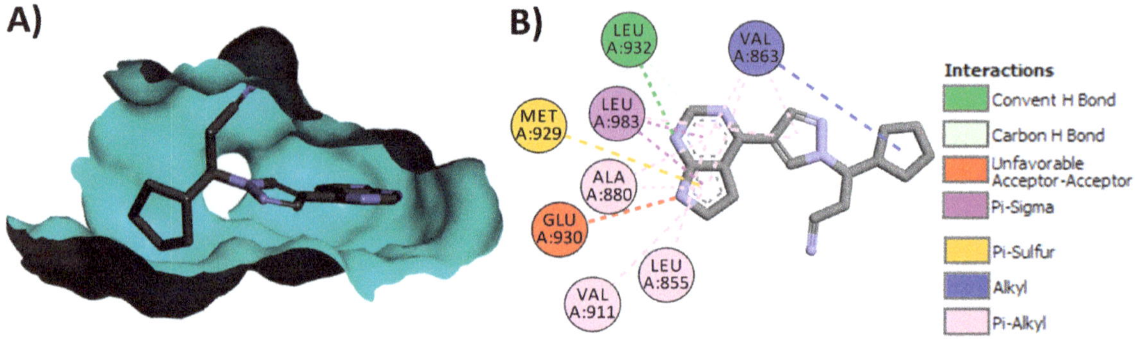

Figure 53. Binding modes of ruxolitinib (shown as sticks) into human JAK2 JH1 (pdb: 6WTN): (**A**) 3D binding mode, receptor shown as a hydrogen bond surface; (**B**) 2D binding mode showing different types of interactions with JAK2 JH1; hydrogen atoms were omitted for clarity; this figure was generated using Discovery Studio Visualizer (V16.1.0.15350).

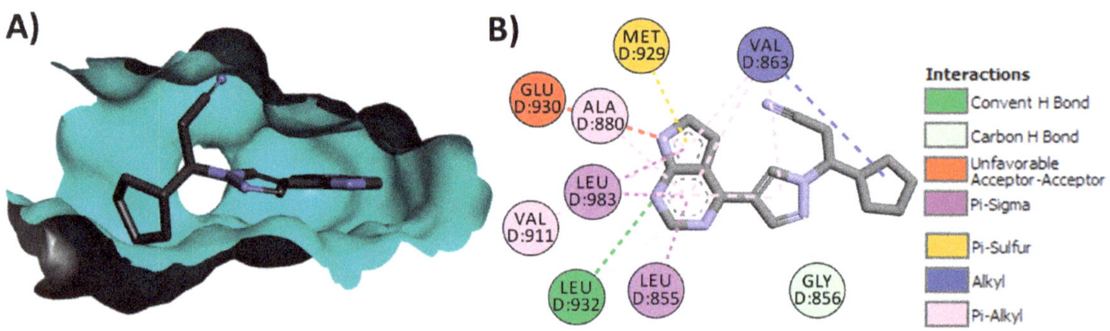

Figure 54. Binding modes of ruxolitinib (shown as sticks) into JAK2 JH1 (pdb: 6VNK): (**A**) 3D binding mode, receptor shown as a hydrogen bond surface; (**B**) 2D binding mode showing different types of interactions with JAK2 JH1; hydrogen atoms were omitted for clarity; this figure was generated using Discovery Studio Visualizer (V16.1.0.15350).

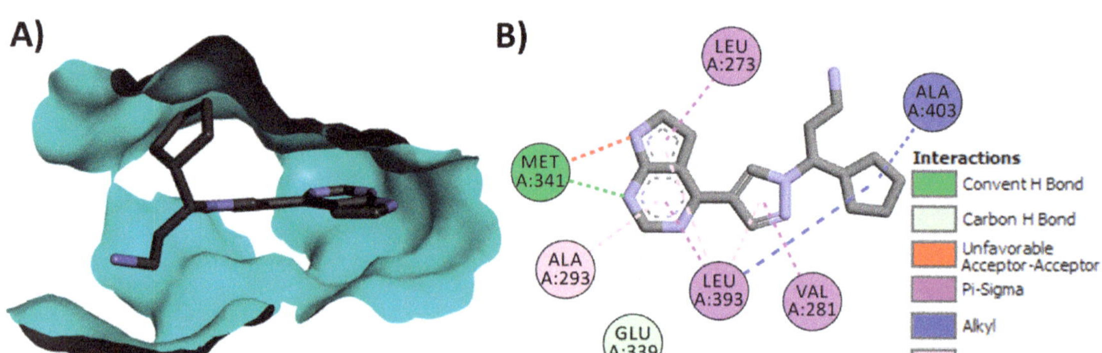

Figure 55. Binding modes of ruxolitinib (shown as sticks) into c-Src (pdb: 4U5J): (**A**) 3D binding mode, receptor shown as a hydrogen bond surface; (**B**) 2D binding mode showing different types of interactions with c-Src; hydrogen atoms were omitted for clarity; this figure was generated using Discovery Studio Visualizer (V16.1.0.15350).

Ruxolitinib also exists as a co-crystallized ligand with DCLK1 kinase (pdb: 7F3G). Visualization of the binding mode/interactions of ruxolitinib into DCLK1 shows one conventional hydrogen bond with Val468, Figure 56.

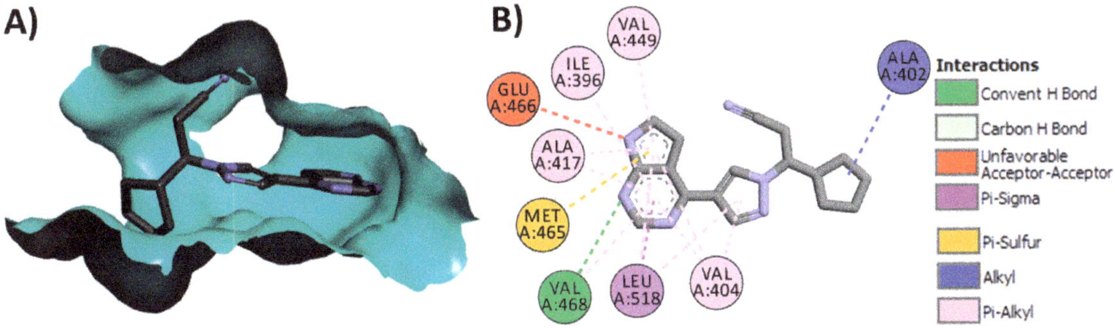

Figure 56. Binding modes of ruxolitinib (shown as sticks) into the DCLK1 kinase domain (pdb: 7F3G): (**A**) 3D binding mode, receptor shown as a hydrogen bond surface; (**B**) 2D binding mode showing different types of interactions with DCLK1 kinase domain; hydrogen atoms were omitted for clarity; this figure was generated using Discovery Studio Visualizer (V16.1.0.15350).

Pharmacological Activities and Uses

Ruxolitinib is a JAK1/2 inhibitor which showed the strong inhibition of JAK2V617F-positive Ba/F3 cells [167]. The efficacy of ruxolitinib in the treatment of myelofibrosis was evaluated in several clinical trials [167,168]. The results of the clinical trial (NCT00952289) revealed significant therapeutic outcomes compared to placebo [168].

In addition to its inhibitory activity against JAKs, ruxolitinib also targets other kinases such as CHK2 and c-Met [166]. Furthermore, ruxolitinib exhibited antiproliferative activity against JAK2V617F+ Ba/F3 cells at IC_{50} 127 nM [166]. The combination of ruxolitinib and ERBB1/2/4 inhibitors also displayed synergistic anticancer activity against lung, breast, and ovarian cancer cells [169].

In 2019, Kim et al. [170] investigated the efficacy of ruxolitinib cream in the treatment of atopic dermatitis in a phase 2 study. A fast improvement in the symptoms of atopic dermatitis was observed, which persisted for up to 12 weeks without significant site reactions. Papp et al. also reported similar findings in another study [171]. Ruxolitinib cream was later approved by the FDA for the treatment of atopic dermatitis [172].

Furthermore, ruxolitinib has also displayed therapeutic benefits in the treatment of acute or chronic graft-versus-host disease (a/cGVHD) [173,174].

Metabolism

Shilling et al. [175] investigated the metabolic profile of ruxolitinib in healthy human subjects using [^{14}C]-labeled ruxolitinib. The results revealed that the parent drug constitutes the major circulating component in plasma (58–74%). Ruxolitinib underwent extensive metabolism, mainly through oxidative pathways that occurred preferentially at the 2- or 3-position of the cyclopentyl ring and resulted in a series of hydroxy/oxo-metabolites, Figure 57. In addition, *O*-glucuronide conjugates (M28 and M51) were also identified. Among these metabolites, the 2-hydroxycyclopentyl derivative of ruxolitinib (M18) was the major one.

Shi et al. [176] evaluated the impact of the CYP3A4 inhibitor and inducer on the pharmacokinetics of ruxolitinib. The results of this study revealed an increase of the plasma concentration of ruxolitinib by the CYP3A4 inhibitors, ketoconazole, and erythromycin. On the other hand, a decrease in total ruxolitinib was observed on co-administration with the CYP3A4 inducer rifampin.

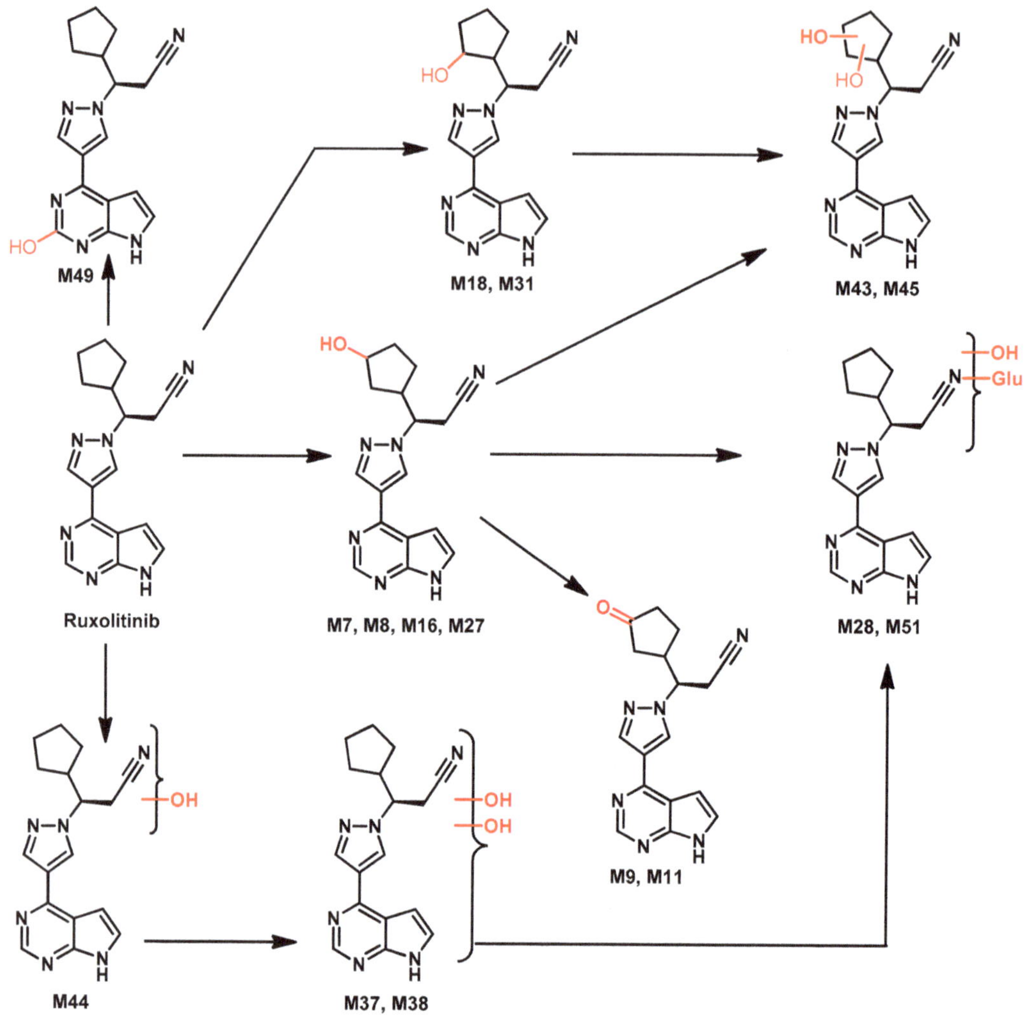

Figure 57. Proposed metabolic pathways and metabolites of ruxolitinib in humans.

1.7.10. Tofacitinib

Approval History

Tofacitinib (Figure 58) is a JAK inhibitor that was approved by the FDA for rheumatoid arthritis in 2012 [64,177,178]. It was also approved for the treatment of psoriatic arthritis and ulcerative colitis in 2017 and 2018, respectively [115]. In addition, tofacitinib has received FDA approval for the treatment of juvenile idiopathic arthritis in 2020 [179]. In December 2021, tofacitinib was approved for the treatment of active ankylosing spondylitis [180].

Synthesis

Tofacitinib was obtained from the reaction of compound **12** and (3*R*,4*R*)-1-benzyl-*N*,4-dimethylpiperidin-3-amine **112** [115,181]. Preparation of the substituted piperidine **112** could be prepared from different starting materials using diverse reaction conditions [64,182]. Among these materials, the 4-methylpyridin-3-amine **108** was reacted with dimethyl carbonate to give **109**, Scheme 14. Rhodium-catalyzed hydrogenation of **109** gave **110**, which underwent reductive amination to give **111** as a racemic mixture. The (3*R*,4*R*)-1-benzyl-*N*,4-

dimethylpiperidin-3-amine enantiomer of **111** was resolved using L-di-*p*-toluoyl-tartaric acid (L-DTTA) to give **112**.

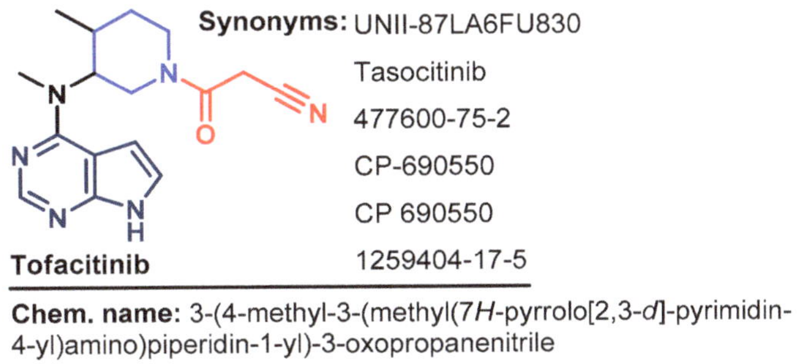

Figure 58. Chemical structure/name/synonyms of tofacitinib.

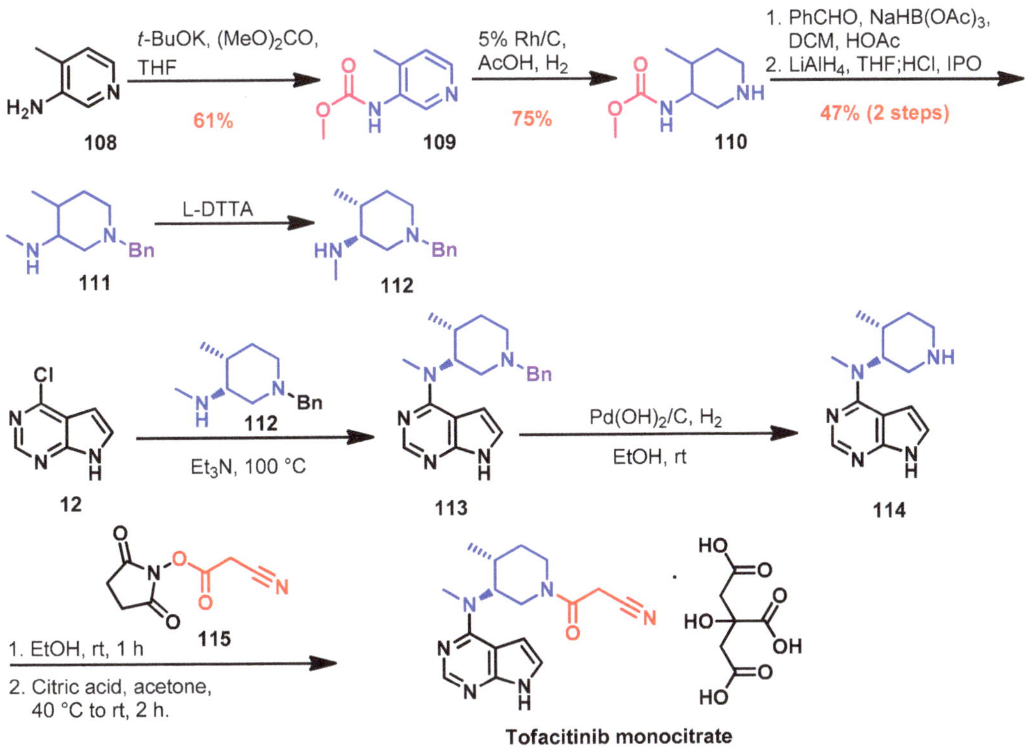

Scheme 14. Synthesis of tofacitinib citrate.

The reaction of compound **12** with the piperidine derivative **112** afforded compound **113**, Scheme 14. Removal of the benzyl group in **113** was achieved using Pd/C-catalyzed hydrogenation, which gave **114**. Tofacitinib was then obtained from the reaction of **114** with cyanoacetic acid 2,5-dioxopyrrolidin-1-yl ester **115**, while treatment of tofacitinib with citric acid afforded the citrate salt.

Target Kinases

The kinase inhibitory activity of tofacitinib was evaluated in several studies [83,96]. The results (Figure 59) revealed the inhibitory activity of tofacitinib against the four JAKs at IC_{50} values in the range of 1.1–42 nM, where the highest inhibitory activity was against JAK3 [96]. In addition, tofacitinib inhibited JAK1 and JAK2 at IC_{50} values of 2.9 nM and 1.2 nM, respectively.

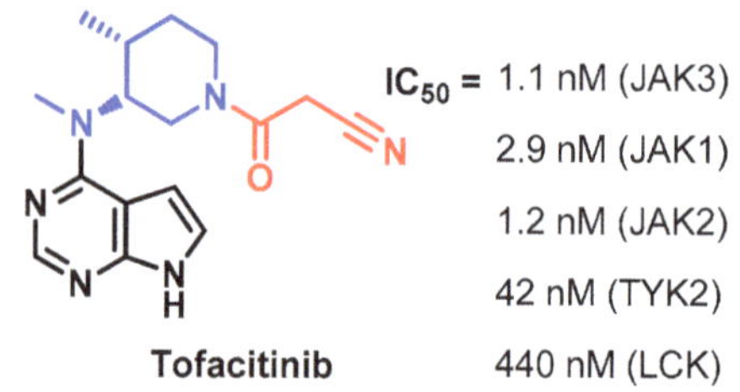

Figure 59. Kinase inhibitory activity of tofacitinib.

On the other hand, the inhibitory activity of tofacitinib against JAKs was also evaluated by Clark et al. [83]. The results of that study revealed its inhibitory activity against JAK3 at IC_{50} of 55 nM. In addition, tofacitinib inhibited the enzymatic activity of JAK1 and JAK2 at IC_{50} values of 15.1 nM and 77.4 nM, respectively.

Crystal Structures

Tofacitinib exists as a co-crystallized ligand in five crystal structures. These crystals include tofacitinib in complex with JAK1 (pdb: 3EYG) [183], JAK2 (pdb: 3FUP) [183], JAK3 (pdb: 3LXK) [184], TYK2 (pdb: 3LXN) [184], and PRK1 (pdb: 4OTI) [185]. The binding interactions of tofacitinib into JAK1 (pdb: 3EYG) are depicted in Figure 60. Tofacitinib shows two conventional hydrogen bonds with Gly884 and Leu959.

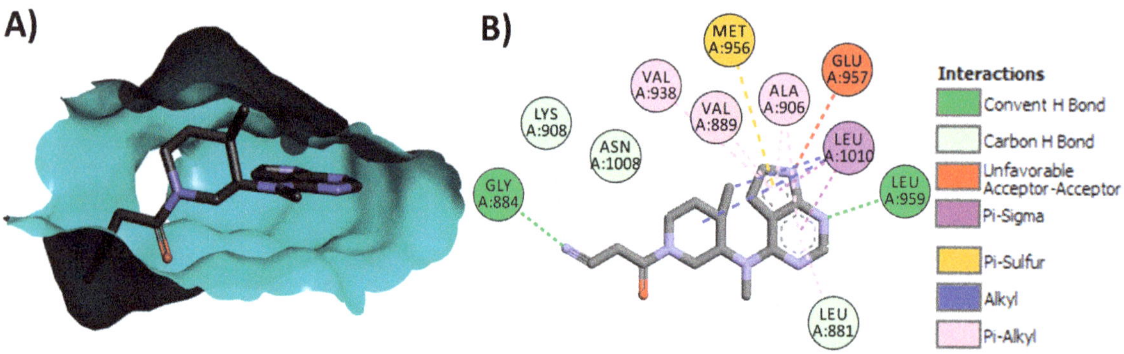

Figure 60. Binding modes of tofacitinib (shown as sticks) into JAK1 (pdb: 3EYG): (**A**) 3D binding mode, receptor shown as a hydrogen bond surface; (**B**) 2D binding mode showing different types of interactions with JAK1; hydrogen atoms were omitted for clarity; this figure was generated using Discovery Studio Visualizer (V16.1.0.15350).

The binding mode and interactions of tofacitinib with JAK2 (pdb: 3FUP) are depicted in Figure 61. Tofacitinib shows two conventional hydrogen bonds with Gly858 and Leu932. In addition, tofacitinib forms three carbon hydrogen bonds with Leu855, Arg980, and Asn981.

In addition, the binding mode, and interactions of tofacitinib with JAK3 (pdb: 3LXK) are visualized in Figure 62. Tofacitinib shows one conventional hydrogen bond with Leu905. In addition, tofacitinib displays four carbon hydrogen bonds with Leu828, Lys855, Arg953, and Asn954 amino acids in JAK3.

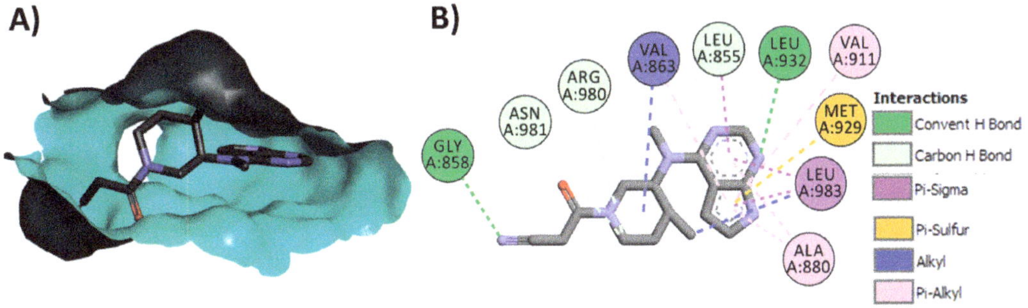

Figure 61. Binding modes of tofacitinib (shown as sticks) into JAK2 (pdb: 3FUP): (**A**) 3D binding mode, receptor shown as a hydrogen bond surface; (**B**) 2D binding mode showing different types of interactions with JAK2; hydrogen atoms were omitted for clarity; this figure was generated using Discovery Studio Visualizer (V16.1.0.15350).

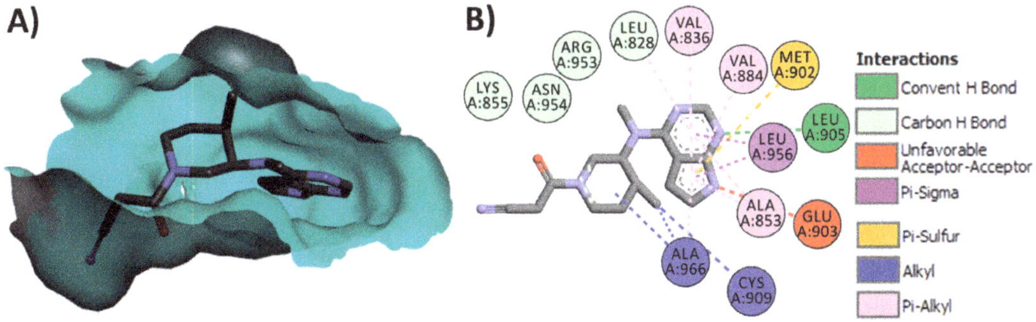

Figure 62. Binding modes of tofacitinib (shown as sticks) into JAK3 (pdb: 3LXK): (**A**) 3D binding mode, receptor depicted as a hydrogen bond surface; (**B**) 2D binding mode showing different types of interactions with JAK3; hydrogen atoms were omitted for clarity; this figure was generated using Discovery Studio Visualizer (V16.1.0.15350).

On the other hand, the orientation of tofacitinib into TYK2 (pdb: 3LXN) is visualized in Figure 63. Tofacitinib shows two conventional hydrogen bonds with Gly906 and Val981.

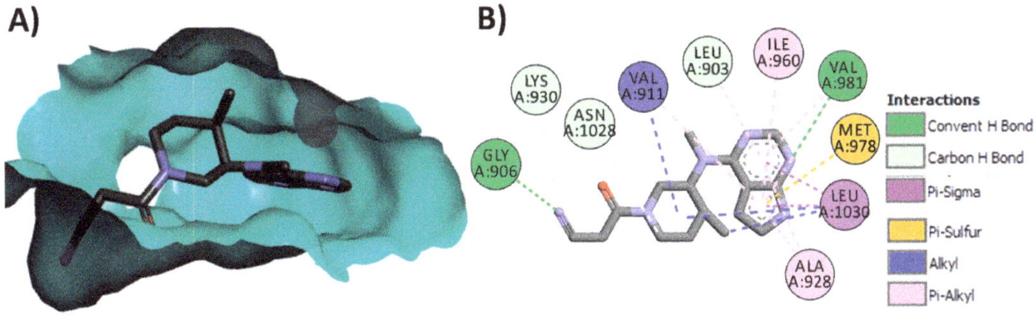

Figure 63. Binding modes of tofacitinib (shown as sticks) into TYK2 (pdb: 3LXN): (**A**) 3D binding mode, receptor shown as a hydrogen bond surface; (**B**) 2D binding mode showing different types of interactions with TYK2; hydrogen atoms were omitted for clarity; this figure was generated using Discovery Studio Visualizer (V16.1.0.15350).

In addition, the binding mode/interactions of tofacitinib into PRK1 (pdb: 4OTI) are visualized in Figure 64. Tofacitinib shows one conventional hydrogen bond with Ser704 and one carbon hydrogen bond with Leu627.

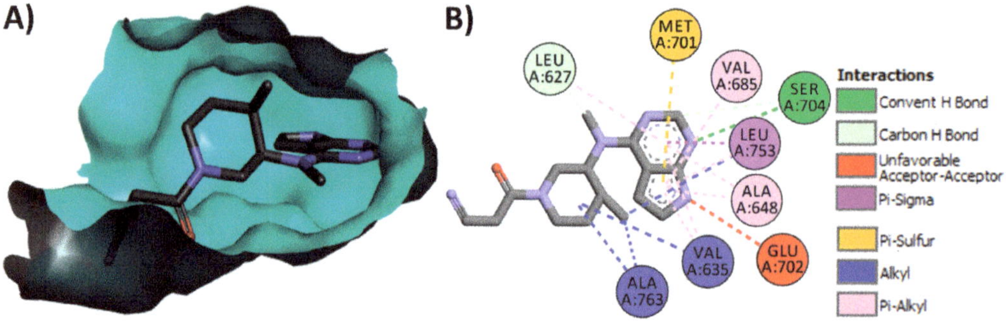

Figure 64. Binding modes of tofacitinib (shown as sticks) into the PRK1 catalytic domain (pdb: 4OTI): (**A**) 3D binding mode, receptor shown as a hydrogen bond surface; (**B**) 2D binding mode showing different types of interactions with PRK1 catalytic domain; hydrogen atoms were omitted for clarity; this figure was generated using Discovery Studio Visualizer (V16.1.0.15350).

Pharmacological Activities and Uses

Several clinical trials were performed to evaluate the efficacy of tofacitinib in the treatment of rheumatoid arthritis. In 2008, the results of a clinical trial (NCT00814307) of tofacitinib revealed improvement in the signs and symptoms of rheumatoid arthritis [186].

In another clinical trial (NCT00853385), tofacitinib showed similar efficacy to adalimumab in patients with rheumatoid arthritis [187]. In patients receiving methotrexate, tofacitinib also stopped the progression of structural damage [188]. By the end of 2012, tofacitinib received the first approval for the treatment of rheumatoid arthritis.

Since the first approval of tofacitinib in 2012 for rheumatoid arthritis, several studies have been performed to evaluate its efficacy in several types of inflammatory and immune diseases. In a clinical trial (NCT01882439), tofacitinib reduced active psoriatic arthritis in patients who had an inadequate response to TNF inhibitors [189]. Mease et al. also evaluated the efficacy of tofacitinib in patients with psoriatic arthritis who had an inadequate response to DMARDs [190].

In another clinical trial (NCT00787202), tofacitinib was investigated for the treatment of patients with severely active ulcerative colitis [191]. The results showed that clinical response and remission were more expected in the treated patients than in those receiving placebo. In addition, Huang et al. also reported an improvement in arthritis in a 13-year-old girl, with complete remission within three months [192]. Tofacitinib was approved by the FDA for the treatment of active psoriatic arthritis, ulcerative colitis, and juvenile idiopathic arthritis [177,178].

On the other hand, tofacitinib displayed inhibitory activity against LCK, which could also contribute to its pharmacological activities [64].

Metabolism

The pharmacokinetics parameters of tofacitinib were evaluated by Dowty et al. [193] using [^{14}C]-labeled tofacitinib in healthy males. The results showed rapid absorption with the parent drug forming ~70% of the circulating activity in plasma.

Guo et al. investigated the metabolism of tofacitinib in vitro using a recombinant CYP3A4 enzyme [194]. Tofacitinib was incubated with mixed male human liver microsomes (HLMs) or individual human recombinant P450 enzymes. The results revealed the formation of a tofacitinib epoxide metabolite which was trapped by *N*-acetyl-L-cysteine (NAC), Figure 65. On the other hand, the metabolism of tofacitinib afforded also an α-keto-aldehyde metabolite, which was catalyzed by CYP3A4. This metabolite was trapped with *N*α-acetyl-L-lysine.

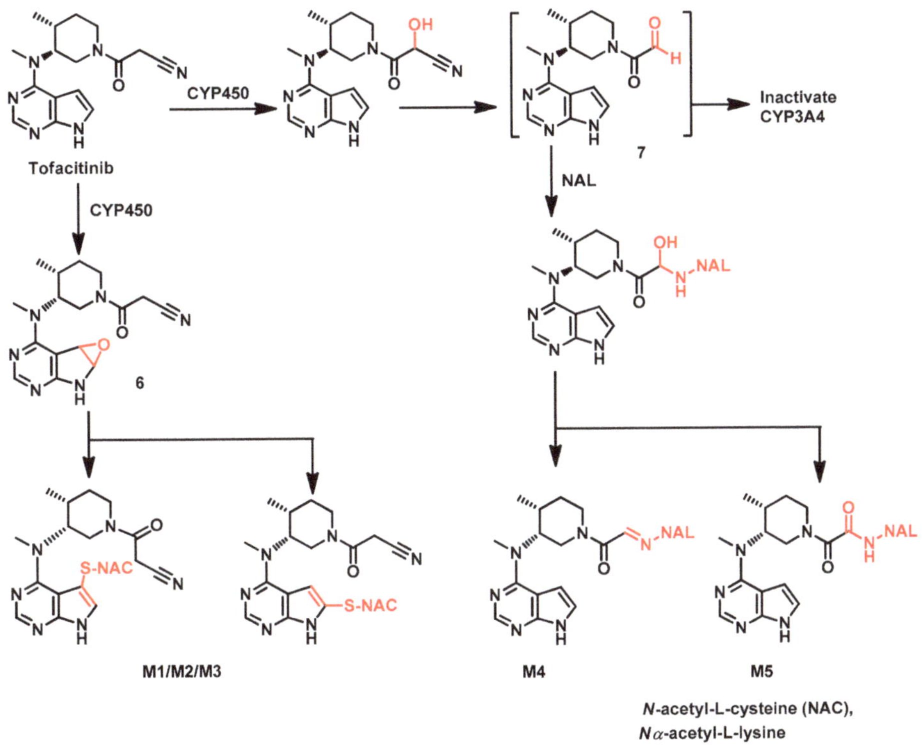

Figure 65. Proposed metabolic pathways and metabolites of tofacitinib.

1.7.11. Upadacitinib

Approval History

Upadacitinib (Figure 66) is a JAK1 inhibitor that was approved to treat rheumatoid arthritis by the FDA in August 2019 [67]. It was also approved for the treatment of patients with psoriatic arthritis [195]. In 2022, upadacitinib was approved for treatment of atopic dermatitis [196]. In addition, it was approved in March 2022 to treat patients with moderate-to-severe active ulcerative colitis [197].

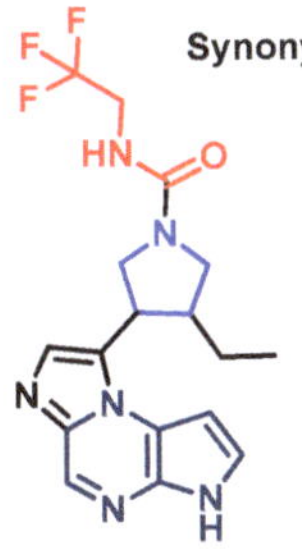

Synonyms: 1310726-60-3
ABT-494
4RA0KN46E0
MFCD30502663
GTPL9246
CHEMBL3622821
BCP19011
EX-A1628
BDBM50503287

Upadacitinib

Chem. name: 3-ethyl-4-(3*H*-imidazo[1,2-*a*]pyrrolo[2,3-*e*]pyrazin-8-yl)-*N*-(2,2,2-trifluoroethyl)pyrrolidine-1-carboxamide

Figure 66. Chemical structure/name/synonyms of upadacitinib.

Synthesis

Upadacitinib was synthesized from the coupling of compounds **121** and **128** [104,198]. Compound **121** was obtained **116** in a three steps synthesis. In the first step, 3,5-dibromopy razin-2-amine **116** was reacted with ethynyltrimethylsilane **117** to afford **118**, which was reacted with *p*-toluenesulfonyl chloride to give **119**. Palladium-catalyzed amination of **119** with ethyl carbamate **120** afforded **121**. Scheme 15.

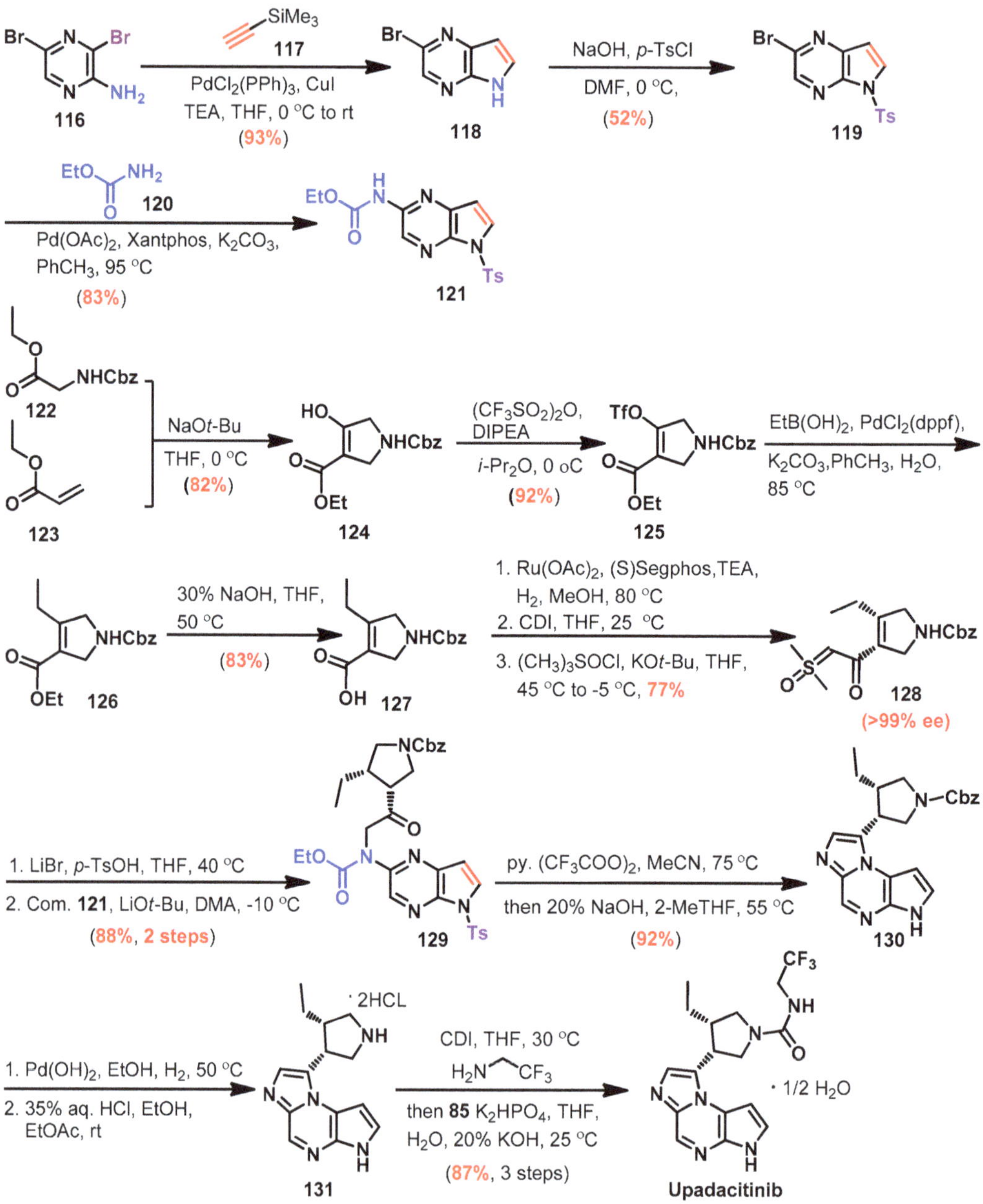

Scheme 15. Synthesis of upadacitinib.

To prepare compound **128**, the ethyl acrylate **123** was first reacted with compound **122** to give **124**, which was then reacted with triflic anhydride to give **125**, Scheme 15. The reaction of **125** with ethyl boronic acid gave **126**, which underwent alkaline hydrolysis to give **127**. Compound **128** was obtained from **127** in a three-step synthesis, which included ruthenium-catalyzed hydrogenation, reaction with carbonyldiimidazole (CDI), and reaction with trimethyl sulfoxonium chloride.

The coupling of **121** and **128** afforded **129**, which underwent intramolecular cyclization to give the tricyclic imidazo[1,2-*a*]pyrrolo[2,3-*e*]pyrazine **130**, Scheme 15. Palladium-catalyzed hydrogenation of **130** was performed to remove the protecting group, followed by treatment with hydrochloric acid to give the salt **131**. The reaction of **131** with CDI and trifluoroethylamine afforded upadacitinib.

Target Kinases

The JAKs inhibitory activity of upadacitinib was evaluated by Parmentier et al. [199]. The results revealed the highest inhibitory activity against JAK1 (IC_{50} = 47 nM), Figure 67. The study also revealed inhibitory activity for upadacitinib against JAK2 at IC_{50} 120 nM, which indicates 2.5-fold lower inhibitory activity compared to JAK1.

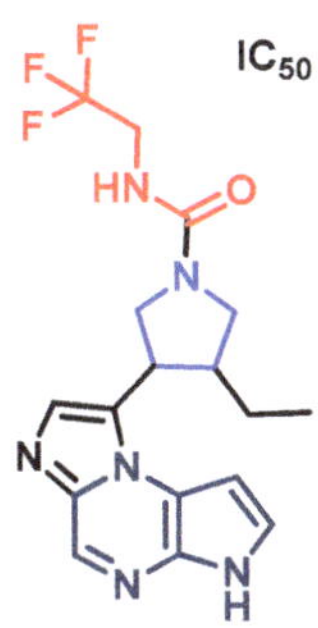

Upadacitinib

IC_{50} =	47 nM (JAK1);	**FS*** = 1;
	120 nM (JAK2);	2.5;
	2304 nM (JAK3);	49;
	4690 nM (TYK2);	100
	0.92 μM (Rock1);	
	0.43 μM (Rock2);	
	4.5 μM (Aurora 2);	
	2.3 μM (CAMK1D);	
	2.4 μM (RET)	

Figure 67. Kinase inhibitory activity of upadacitinib (* FS, fold selectivity compared to JAK1).

In addition, the results of the kinase inhibitory assay of upadacitinib against 70 kinases revealed weak inhibitory activity against kinases other than JAKs [199].

Crystal Structures

Upadacitinib has not yet been reported in a crystal structure with any of its target kinases.

Pharmacological Activities and Uses

Upadacitinib is a second-generation selective JAK inhibitor that was evaluated for the treatment of different types of inflammatory and immune diseases [200–203]. Evaluation of the efficacy of upadacitinib in the treatment of rheumatoid arthritis revealed a fast and favorable efficacy profile [202]. In addition, Smolen et al. [204] evaluated the efficacy of upadacitinib monotherapy in rheumatoid arthritis. The results also revealed significant therapeutic outcomes compared to methotrexate. The medical use of upadacitinib, either alone or in combination therapy for the treatment of rheumatoid arthritis, was associated with lower direct medical costs [203]. In August 2019, upadacitinib was approved for the treatment of moderate-to-severe rheumatoid arthritis.

Moreover, several clinical trials were also performed to evaluate the efficacy of upadacitinib in the treatment of psoriatic arthritis. In a clinical trial (NCT03104400.) for treatment of psoriatic arthritis, upadacitinib produced a rapid and sustained improvement in patient outcomes [201]. Upadacitinib was also evaluated in a 24-week, phase 3 trial to treat psoriatic arthritis [205]. The results revealed a significantly higher number of patients with ACR20 compared to the placebo. In addition, upadacitinib at a daily dose of 30 mg showed superior results to those of adalimumab. Furthermore, no significant safety signals were observed when upadacitinib was evaluated in patients with psoriatic arthritis. Burmester et al. [206] also evaluated the safety of upadacitinib in patients with psoriatic arthritis for up to 3 years, where the results revealed a safety profile similar to that observed in rheumatoid arthritis. Currently, upadacitinib has been approved by the FDA and EMA for the treatment of patients with active psoriatic arthritis [195,207].

Furthermore, upadacitinib was also studied in adult patients with atopic dermatitis, where the results showed superior efficacy compared to the human monoclonal antibody dupilumab [208]. In January 2022, upadacitinib was also approved by the FDA to treat refractory, moderate to severe atopic dermatitis in children aged $\geq$ 12 years.

Metabolism

The results of the in vitro metabolic study of upadacitinib suggested that it undergoes metabolism by cytochrome P450 [209]. On the other hand, the contribution of CYP2D6 in the metabolism of upadacitinib was very minor [209]. Coadministration of upadacitinib with the CYP3A4 inhibitor, ketoconazole, resulted in a weak effect on its concentration [210]. However, coadministration of upadacitinib with rifampin, a broad CYP inducer, resulted in a decrease in upadacitinib concentration by ~50% [211].

In conclusion, the target kinases of the eleven JAK inhibitors and the approval data including the approval date and the disease for which these inhibitors were approved are presented in Table 1.

Table 1. The globally approved JAK inhibitors, their target kinase, clinical trials, and the approval data.

Drug	Target JAKs	Disease	Approval Date	Reference	Clinical Trials
Abrocitinib	JAK1, JAK2	Atopic dermatitis	2022	[72]	NCT03627767, NCT03720470
Baricitinib	JAK1, JAK2	Rheumatoid arthritis	2017 (EMA) 2018 (FDA)	[64,65]	NCT02265705, NCT01710358
		COVID-19	2020 (EUA)	[78]	NCT04421027, NCT04401579
Delgocitinib	Nonselective	Atopic dermatitis	2020 (Japan)	[69]	NCT03826901, NCT03725722
Fedratinib	JAK2, JAK2V617F	Myelofibrosis	2019 (FDA)	[66]	NCT00724334, NCT00631462 NCT01437787
Filgotinib	JAK1	Rheumatoid arthritis	2020 (EMA)	[70]	NCT02873936, NCT02886728
Oclacitinib	JAK1	Canine allergic dermatitis	2013	[129]	NA *
Pacritinib	JAK2, JAK2V617F	Myelofibrosis	2022 (FDA)	[138]	NCT04884191
Peficitinib	Pan-JAK inhibitor	Rheumatoid arthritis	2019 (Japan)	[68]	NCT01565655, NCT02308163
Ruxolitinib	JAK1, JAK2 JAK2V617F	Myelofibrosis	2011	[28]	NCT00952289
		Polycythemia vera	2014	[29]	NCT02038036
		Acute and chronic graft-versus-host disease	2019, 2021	[160]	NCT03112603 NCT03147742

Table 1. *Cont.*

Drug	Target JAKs	Disease	Approval Date	Reference	Clinical Trials
Tofacitinib	JAK1, JAK2, JAK3	Rheumatoid arthritis	2012 (FDA)	[64]	NCT02187055
		Psoriatic arthritis	2017	[115]	NCT01877668
		Ulcerative colitis	2018	[115]	NCT03281304
		Juvenile idiopathic arthritis	2020	[179]	NCT02592434
		Ankylosing spondylitis	2021	[180]	NCT03502616
Upadacitinib	JAK1	Rheumatoid arthritis	2019	[67]	NCT02706847
		Psoriatic arthritis	2021	[195]	NCT03104400
		Atopic dermatitis	2022	[196]	NCT03738397
		Ulcerative colitis	2022	[197]	NCT02819635

* NA, not applicable.

2. Conclusions

In the current review, eleven of the JAK inhibitors that received approval for the treatment of inflammatory, autoimmune, and myeloproliferative neoplasms were discussed. These drugs are abrocitinib, baricitinib, delgocitinib, fedratinib, filgotinib, oclacitinib, pacritinib, peficitinib, ruxolitinib, tofacitinib, and upadacitinib. The synthetic routes of these drugs, including the original and/or alternative pathways, were described. The crystal structures of these drugs in various kinases were also listed. Their binding modes and interactions were visualized, where two key hydrogen-bonding interactions were observed with Leu959 and Leu932 in JAK1 and JAK2, respectively. Furthermore, the kinase inhibitory activities and pharmacological uses of the eleven drugs were also summarized. Based on their inhibitory activity against the target kinases, these drugs could be classified as either selective or nonselective JAK inhibitors. Among these drugs, several JAK1 inhibitors have been approved for the treatment of inflammatory and autoimmune conditions. On the other hand, the drugs approved for treatment of myeloproliferative neoplasms target JAK2 and its mutant form (JAK2V617F). In addition, the metabolic studies of the eleven drugs revealed the formation of several oxidation metabolites, which were mediated by CYP450 enzymes. On the other hand, few conjugation metabolites such as glucuronide acid and sulfate conjugates were detected among the metabolites of filgotinib and peficitinib, respectively. To sum up, the data in this review may assist in the design of new JAK inhibitors with potential therapeutic benefits.

3. Perspective

The high efficacy of the clinically approved JAK inhibitors in the treatment of inflammatory and autoimmune diseases has attracted much attention. However, most of them are non-selective inhibitors, which may account for some of their adverse effects, such as anemia, thrombocytopenia, upper respiratory tract infection, and herpes infection [212–216]. Accordingly, design and development of new potent, selective, and more safe JAK inhibitors could provide a solution to these adverse effects [217].

Currently, several nonselective JAK inhibitors, including brepocitinib, cerdulatinib, gusacitinib, and momelotinib (Figure 68), are being investigated for their efficacy in inflammatory and cancer diseases [218–221]. In addition, the efficacy and safety of several selective JAK inhibitors are being investigated in clinical trials. Among these inhibitors, the selective JAK1 inhibitor, itacitinib showed promising efficacy when evaluated in patients with aGVHD [222].

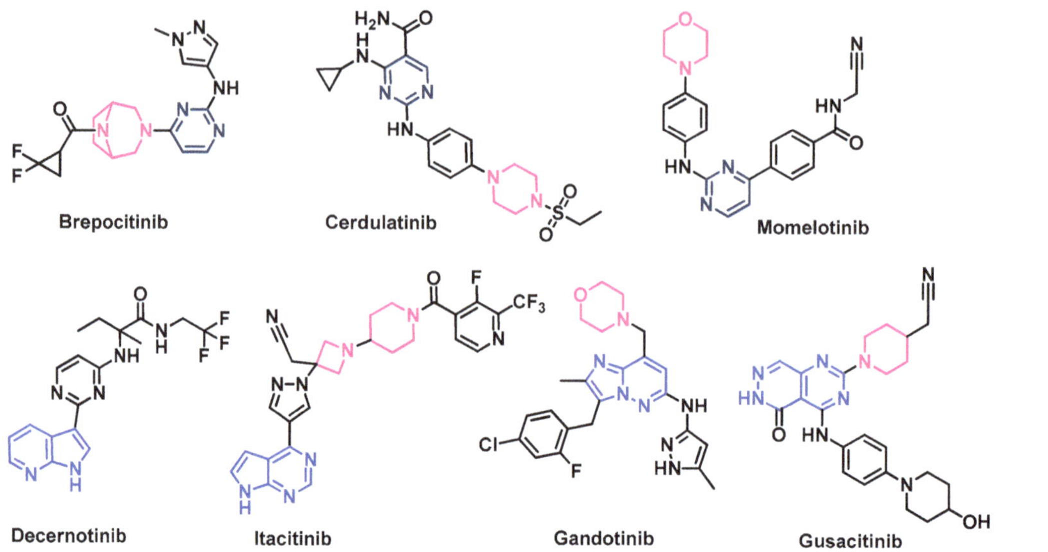

Figure 68. JAK inhibitors under development.

The JAK2 inhibitor, gandotinib (Figure 68), showed high potency toward the JAK2V617F mutation and showed promising potential in the treatment of myeloproliferative disorders [223]. In addition, JAK2/FLT3 dual inhibitors could provide better therapeutic option for acute myeloid leukemia [30,224]. Currently, it is seeking FDA approval for the treatment of myelofibrosis. On the other hand, the JAK3 inhibitor, decernotinib also showed a high potential activity in the treatment of rheumatoid arthritis [225].

Several JAK inhibitors have been reported to bind reversibly to allosteric sites in JAKs [34,42]. Among these inhibitors, deucravacitinib and LS104 are being investigated in clinical trials. This type of JAK inhibitor could provide advantages over the currently used ATP-competitive inhibitors [27]. However, none of these allosteric inhibitors has been approved for clinical use. Future research in this area could lead to approval of the allosteric inhibitors for clinical use.

Development of irreversible JAK3 inhibitors that can bind covalently with the unique Cys909 reside has attracted a great attention in the last few years [43]. Several small molecules have displayed potent and selective inhibition of JAK activity [43,44,226]. The design of new JAK3 covalent inhibitors could be supported by the success of ritlecitinib in reaching the clinical trial (NCT04517864). In addition, advances in X-ray and covalent docking may also play a crucial role in the design of this type of JAK inhibitors.

Recently, JAK inhibitors showed promising potential in the treatment of COVID-19-related cytokine storm [227]. However, among several drug combinations evaluated in the treatment of COVID-19 [228,229], the combination of baricitinib plus remdesivir has received EUA by the FDA [92,230,231]. Moreover, the clinical trials of tofacitinib (NCT04469114) and nezulcitinib (NCT04402866) are being performed in patients with COVID-19 related lung problems [36,232]. Although these results also support the investigation of other JAK inhibitors in the treatment of COVID-19 related problems, the adverse effects of these inhibitors on the immune responses must be evaluated [233].

Author Contributions: Conceptualization, A.M.S., F.A.A., A.N.A., A.H.A. and A.M.G.; methodology, F.A.A., A.N.A., A.M.S., A.H.A. and A.M.G.; software, A.M.S., F.A.A., A.N.A., A.H.A. and A.M.G.; validation, F.A.A., A.N.A., A.M.S., A.H.A. and A.M.G.; formal analysis, A.M.S., F.A.A., A.N.A. and A.M.G.; investigation, F.A.A., A.N.A., A.M.S., A.H.A. and A.M.G.; resources, A.M.S., F.A.A., A.N.A. and A.M.G.; data curation, F.A.A., A.N.A., A.M.S. and A.M.G.; writing—original draft preparation,

A.M.S., F.A.A., A.N.A., A.H.A. and A.M.G.; writing—review and editing, F.A.A., A.N.A., A.M.S. and A.M.G.; visualization, F.A.A., A.N.A., A.M.S., A.H.A. and A.M.G.; supervision, A.M.S., F.A.A., A.N.A. and A.M.G.; Project administration, A.M.S.; funding acquisition, A.M.S. All authors have read and agreed to the published version of the manuscript.

Funding: The authors would like to thank the Deanship of Scientific Research at Umm Al-Qura University for supporting this work by Grant Code: (22UQU4331174DSR08).

Institutional Review Board Statement: Not applicable.

Informed Consent Statement: Not applicable.

Conflicts of Interest: Authors declared that there is no conflict of interest and have approved the article.

References

1. Yamaoka, K.; Saharinen, P.; Pesu, M.; Holt, V.E.T., 3rd; Silvennoinen, O.; O'Shea, J.J. The Janus kinases (Jaks). *Genome Biol.* **2004**, *5*, 253. [CrossRef] [PubMed]
2. Wilks, A.F. Two putative protein-tyrosine kinases identified by application of the polymerase chain reaction. *Proc. Natl. Acad. Sci. USA* **1989**, *86*, 1603–1607. [CrossRef] [PubMed]
3. Banerjee, S.; Biehl, A.; Gadina, M.; Hasni, S.; Schwartz, D.M. JAK-STAT signaling as a target for inflammatory and autoimmune diseases: Current and future prospects. *Drugs* **2017**, *77*, 521–546. [CrossRef]
4. Aittomäki, S.; Pesu, M. Therapeutic targeting of the Jak/STAT pathway. *Basic Clin. Pharmacol. Toxicol.* **2014**, *114*, 18–23. [CrossRef] [PubMed]
5. Abroun, S.; Saki, N.; Ahmadvand, M.; Asghari, F.; Salari, F.; Rahim, F. STATs: An old story, yet mesmerizing. *Cell J.* **2015**, *17*, 395–411. [CrossRef]
6. Liau, N.P.D.; Laktyushin, A.; Morris, R.; Sandow, J.J.; Nicola, N.A.; Kershaw, N.J.; Babon, J.J. Enzymatic characterization of wild-type and mutant janus kinase 1. *Cancers* **2019**, *11*, 1701. [CrossRef]
7. Min, X.; Ungureanu, D.; Maxwell, S.; Hammarén, H.; Thibault, S.; Hillert, E.-K.; Ayres, M.; Greenfield, B.; Eksterowicz, J.; Gabel, C.; et al. Structural and functional characterization of the JH2 pseudokinase domain of JAK family tyrosine kinase 2 (TYK2). *J. Biol. Chem.* **2015**, *290*, 27261–27270. [CrossRef]
8. Garrido-Trigo, A.; Salas, A. Molecular structure and function of Janus kinases: Implications for the development of inhibitors. *J. Crohns. Colitis* **2020**, *14*, S713–S724. [CrossRef]
9. Ferrao, R.; Lupardus, P.J. The Janus kinase (JAK) FERM and SH2 domains: Bringing specificity to JAK-receptor interactions. *Front. Endocrinol.* **2017**, *8*, 71. [CrossRef]
10. Kiu, H.; Nicholson, S.E. Biology and significance of the JAK/STAT signalling pathways. *Growth Factors* **2012**, *30*, 88–106. [CrossRef]
11. Bryan, M.C.; Rajapaksa, N.S. Kinase inhibitors for the treatment of immunological disorders: Recent advances. *J. Med. Chem.* **2018**, *61*, 9030–9058. [CrossRef] [PubMed]
12. Xu, P.; Shen, P.; Yu, B.; Xu, X.; Ge, R.; Cheng, X.; Chen, Q.; Bian, J.; Li, Z.; Wang, J. Janus kinases (JAKs): The efficient therapeutic targets for autoimmune diseases and myeloproliferative disorders. *Eur. J. Med. Chem.* **2020**, *192*, 112155. [CrossRef] [PubMed]
13. Hammarén, H.M.; Virtanen, A.T.; Raivola, J.; Silvennoinen, O. The regulation of JAKs in cytokine signaling and its breakdown in disease. *Cytokine* **2019**, *118*, 48–63. [CrossRef]
14. Yasuda, T.; Fukada, T.; Nishida, K.; Nakayama, M.; Matsuda, M.; Miura, I.; Dainichi, T.; Fukuda, S.; Kabashima, K.; Nakaoka, S.; et al. Hyperactivation of JAK1 tyrosine kinase induces stepwise, progressive pruritic dermatitis. *J. Clin. Investig.* **2016**, *126*, 2064–2076. [CrossRef] [PubMed]
15. Shen, Y.; Liu, Y.; Ke, X.; Kang, H.-Y.; Hu, G.-H.; Hong, S.-L. Association between JAK1 gene polymorphisms and susceptibility to allergic rhinitis. *Asian Pac. J. Allergy Immunol.* **2016**, *34*, 124–129. [CrossRef] [PubMed]
16. Wills-Karp, M.; Luyimbazi, J.; Xu, X.; Schofield, B.; Neben, T.Y.; Karp, C.L.; Donaldson, D.D. Interleukin-13: Central mediator of allergic asthma. *Science* **1998**, *282*, 2258–2261. [CrossRef] [PubMed]
17. Harris, C.; Cummings, J.R.F. JAK1 inhibition and inflammatory bowel disease. *Rheumatology* **2021**, *60*, ii45–ii51. [CrossRef] [PubMed]
18. Virtanen, A.T.; Haikarainen, T.; Raivola, J.; Silvennoinen, O. Selective JAKinibs: Prospects in inflammatory and autoimmune diseases. *BioDrugs* **2019**, *33*, 15–32. [CrossRef]
19. Zak, M.; Hanan, E.J.; Lupardus, P.; Brown, D.G.; Robinson, C.; Siu, M.; Lyssikatos, J.P.; Romero, F.A.; Zhao, G.; Kellar, T.; et al. Discovery of a class of highly potent Janus Kinase 1/2 (JAK1/2) inhibitors demonstrating effective cell-based blockade of IL-13 signaling. *Bioorg. Med. Chem. Lett.* **2019**, *29*, 1522–1531. [CrossRef]
20. Schwartz, D.M.; Kanno, Y.; Villarino, A.; Ward, M.; Gadina, M.; O'Shea, J.J. JAK inhibition as a therapeutic strategy for immune and inflammatory diseases. *Nat. Rev. Drug Discov.* **2017**, *16*, 843–862. [CrossRef]
21. Menet, C.J. A dual inhibition, a better solution: Development of a JAK1/TYK2 inhibitor. *J. Med. Chem.* **2018**, *61*, 8594–8596. [CrossRef] [PubMed]

22. Wrobleski, S.T.; Moslin, R.; Lin, S.; Zhang, Y.; Spergel, S.; Kempson, J.; Tokarski, J.S.; Strnad, J.; Zupa-Fernandez, A.; Cheng, L.; et al. Highly selective inhibition of tyrosine kinase 2 (TYK2) for the treatment of autoimmune diseases: Discovery of the allosteric inhibitor BMS-986165. *J. Med. Chem.* **2019**, *62*, 8973–8995. [CrossRef] [PubMed]
23. Buchert, M.; Burns, C.J.; Ernst, M. Targeting JAK kinase in solid tumors: Emerging opportunities and challenges. *Oncogene* **2016**, *35*, 939–951. [CrossRef] [PubMed]
24. Kilpivaara, O.; Levine, R.L. JAK2 and MPL mutations in myeloproliferative neoplasms: Discovery and science. *Leukemia* **2008**, *22*, 1813–1817. [CrossRef] [PubMed]
25. Treliński, J.; Robak, T. JAK inhibitors: Pharmacology and clinical activity in chronic myeloprolipherative neoplasms. *Curr. Med. Chem.* **2013**, *20*, 1147–1161. [CrossRef] [PubMed]
26. Nielsen, C.; Birgens, H.S.; Nordestgaard, B.G.; Kjaer, L.; Bojesen, S.E. The JAK2 V617F somatic mutation, mortality and cancer risk in the general population. *Haematologica* **2011**, *96*, 450–453. [CrossRef]
27. Leroy, E.; Constantinescu, S.N. Rethinking JAK2 inhibition: Towards novel strategies of more specific and versatile Janus kinase inhibition. *Leukemia* **2017**, *31*, 1023–1038. [CrossRef]
28. Mascarenhas, J.; Hoffman, R. Ruxolitinib: The first FDA approved therapy for the treatment of myelofibrosis. *Clin. Cancer Res. Off. J. Am. Assoc. Cancer Res.* **2012**, *18*, 3008–3014. [CrossRef]
29. Raedler, L.A. Jakafi (Ruxolitinib): First FDA-approved medication for the treatment of patients with polycythemia vera. *Am. Health Drug Benefits* **2015**, *8*, 75–79.
30. Hart, S.; Goh, K.C.; Novotny-Diermayr, V.; Tan, Y.C.; Madan, B.; Amalini, C.; Ong, L.C.; Kheng, B.; Cheong, A.; Zhou, J.; et al. Pacritinib (SB1518), a JAK2/FLT3 inhibitor for the treatment of acute myeloid leukemia. *Blood Cancer J.* **2011**, *1*, e44. [CrossRef]
31. Yang, T.; Hu, M.; Qi, W.; Yang, Z.; Tang, M.; He, J.; Chen, Y.; Bai, P.; Yuan, X.; Zhang, C.; et al. Discovery of potent and orally effective dual Janus kinase 2/FLT3 inhibitors for the treatment of acute myelogenous leukemia and myeloproliferative neoplasms. *J. Med. Chem.* **2019**, *62*, 10305–10320. [CrossRef] [PubMed]
32. Hu, X.; Li, J.; Fu, M.; Zhao, X.; Wang, W. The JAK/STAT signaling pathway: From bench to clinic. *Signal Transduct. Target. Ther.* **2021**, *6*, 402. [CrossRef] [PubMed]
33. Angelini, J.; Talotta, R.; Roncato, R.; Fornasier, G.; Barbiero, G.; Dal Cin, L.; Brancati, S.; Scaglione, F. JAK-Inhibitors for the treatment of rheumatoid arthritis: A focus on the present and an outlook on the future. *Biomolecules* **2020**, *10*, 1002. [CrossRef] [PubMed]
34. Vainchenker, W.; Leroy, E.; Gilles, L.; Marty, C.; Plo, I.; Constantinescu, S.N. JAK inhibitors for the treatment of myeloproliferative neoplasms and other disorders. *F1000Research* **2018**, *7*, 82. [CrossRef] [PubMed]
35. Wernig, G.; Kharas, M.G.; Okabe, R.; Moore, S.A.; Leeman, D.S.; Cullen, D.E.; Gozo, M.; McDowell, E.P.; Levine, R.L.; Doukas, J.; et al. Efficacy of TG101348, a selective JAK2 inhibitor, in treatment of a murine model of JAK2V617F-induced polycythemia vera. *Cancer Cell* **2008**, *13*, 311–320. [CrossRef]
36. Singh, D.; Bogus, M.; Moskalenko, V.; Lord, R.; Moran, E.J.; Crater, G.D.; Bourdet, D.L.; Pfeifer, N.D.; Woo, J.; Kaufman, E.; et al. A phase 2 multiple ascending dose study of the inhaled pan-JAK inhibitor nezulcitinib (TD-0903) in severe COVID-19. *Eur. Respir. J.* **2021**, *58*, 2100673. [CrossRef]
37. Andraos, R.; Qian, Z.; Bonenfant, D.; Rubert, J.; Vangrevelinghe, E.; Scheufler, C.; Marque, F.; Régnier, C.H.; De Pover, A.; Ryckelynck, H.; et al. Modulation of activation-loop phosphorylation by JAK inhibitors is binding mode dependent. *Cancer Discov.* **2012**, *2*, 512–523. [CrossRef]
38. Wu, S.-C.; Li, L.S.; Kopp, N.; Montero, J.; Chapuy, B.; Yoda, A.; Christie, A.L.; Liu, H.; Christodoulou, A.; van Bodegom, D.; et al. Activity of the type II JAK2 inhibitor CHZ868 in B cell acute lymphoblastic leukemia. *Cancer Cell* **2015**, *28*, 29–41. [CrossRef]
39. Burke, J.R.; Cheng, L.; Gillooly, K.M.; Strnad, J.; Zupa-Fernandez, A.; Catlett, I.M.; Zhang, Y.; Heimrich, E.M.; McIntyre, K.W.; Cunningham, M.D.; et al. Autoimmune pathways in mice and humans are blocked by pharmacological stabilization of the TYK2 pseudokinase domain. *Sci. Transl. Med.* **2019**, *11*, eaaw1736. [CrossRef]
40. Lipka, D.B.; Hoffmann, L.S.; Heidel, F.; Markova, B.; Blum, M.-C.; Breitenbuecher, F.; Kasper, S.; Kindler, T.; Levine, R.L.; Huber, C.; et al. LS104, a non-ATP-competitive small-molecule inhibitor of JAK2, is potently inducing apoptosis in JAK2V617F-positive cells. *Mol. Cancer Ther.* **2008**, *7*, 1176–1184. [CrossRef]
41. Jatiani, S.S.; Cosenza, S.C.; Reddy, M.V.R.; Ha, J.H.; Baker, S.J.; Samanta, A.K.; Olnes, M.J.; Pfannes, L.; Sloand, E.M.; Arlinghaus, R.B.; et al. A non-ATP-competitive dual inhibitor of JAK2 and BCR-ABL kinases: Elucidation of a novel therapeutic spectrum based on substrate competitive inhibition. *Genes Cancer* **2010**, *1*, 331–345. [CrossRef] [PubMed]
42. Alexander, M.; Luo, Y.; Raimondi, G.; O'Shea, J.J.; Gadina, M. Jakinibs of all trades: Inhibiting cytokine signaling in immune-mediated pathologies. *Pharmaceuticals* **2022**, *15*, 48. [CrossRef] [PubMed]
43. Casimiro-Garcia, A.; Trujillo, J.I.; Vajdos, F.; Juba, B.; Banker, M.E.; Aulabaugh, A.; Balbo, P.; Bauman, J.; Chrencik, J.; Coe, J.W.; et al. Identification of cyanamide-based Janus kinase 3 (JAK3) covalent inhibitors. *J. Med. Chem.* **2018**, *61*, 10665–10699. [CrossRef] [PubMed]
44. Goedken, E.R.; Argiriadi, M.A.; Banach, D.L.; Fiamengo, B.A.; Foley, S.E.; Frank, K.E.; George, J.S.; Harris, C.M.; Hobson, A.D.; Ihle, D.C.; et al. Tricyclic covalent inhibitors selectively target Jak3 through an active site thiol. *J. Biol. Chem.* **2015**, *290*, 4573–4589. [CrossRef] [PubMed]

45. Forster, M.; Chaikuad, A.; Bauer, S.M.; Holstein, J.; Robers, M.B.; Corona, C.R.; Gehringer, M.; Pfaffenrot, E.; Ghoreschi, K.; Knapp, S.; et al. Selective JAK3 Inhibitors with a covalent reversible binding mode targeting a new induced fit binding pocket. *Cell Chem. Biol.* **2016**, *23*, 1335–1340. [CrossRef]
46. Telliez, J.-B.; Dowty, M.E.; Wang, L.; Jussif, J.; Lin, T.; Li, L.; Moy, E.; Balbo, P.; Li, W.; Zhao, Y.; et al. Discovery of a JAK3-selective inhibitor: Functional differentiation of JAK3-selective inhibition over pan-JAK or JAK1-selective inhibition. *ACS Chem. Biol.* **2016**, *11*, 3442–3451. [CrossRef]
47. Dassault Systems BIOVIA. *Discovery Studio Visualizer*, version 16.1.0.15350; Dassault Systems: San Diego, CA, USA, 2016.
48. Almalki, F.A.; Shawky, A.M.; Abdalla, A.N.; Gouda, A.M. Icotinib, almonertinib, and olmutinib: A 2D similarity/docking-based study to predict the potential binding modes and interactions into EGFR. *Molecules* **2021**, *26*, 6423. [CrossRef]
49. Abourehab, M.A.S.; Alqahtani, A.M.; Almalki, F.A.; Zaher, D.M.; Abdalla, A.N.; Gouda, A.M.; Beshr, E.A.M. Pyrrolizine/indolizine-NSAID hybrids: Design, synthesis, biological evaluation, and molecular docking studies. *Molecules* **2021**, *26*, 6582. [CrossRef]
50. Abourehab, M.A.S.; Alqahtani, A.M.; Youssif, B.G.M.; Gouda, A.M. Globally approved EGFR inhibitors: Insights into their syntheses, target kinases, biological activities, receptor interactions, and metabolism. *Molecules* **2021**, *26*, 6677. [CrossRef]
51. Arulogun, S.O.; Choong, H.-L.; Taylor, D.; Ambrosoli, P.; Magor, G.; Irving, I.M.; Keng, T.-B.; Perkins, A.C. JAK1 somatic mutation in a myeloproliferative neoplasm. *Haematologica* **2017**, *102*, e324–e327. [CrossRef]
52. Musumeci, F.; Greco, C.; Giacchello, I.; Fallacara, A.L.; Ibrahim, M.M.; Grossi, G.; Brullo, C.; Schenone, S. An update on JAK inhibitors. *Curr. Med. Chem.* **2019**, *26*, 1806–1832. [CrossRef] [PubMed]
53. Levine, R.L.; Pardanani, A.; Tefferi, A.; Gilliland, D.G. Role of JAK2 in the pathogenesis and therapy of myeloproliferative disorders. *Nat. Rev. Cancer* **2007**, *7*, 673–683. [CrossRef] [PubMed]
54. Degryse, S.; Bornschein, S.; de Bock, C.E.; Leroy, E.; Vanden Bempt, M.; Demeyer, S.; Jacobs, K.; Geerdens, E.; Gielen, O.; Soulier, J.; et al. Mutant JAK3 signaling is increased by loss of wild-type JAK3 or by acquisition of secondary JAK3 mutations in T-ALL. *Blood* **2018**, *131*, 421–425. [CrossRef] [PubMed]
55. O'Shea, J.J.; Husa, M.; Li, D.; Hofmann, S.R.; Watford, W.; Roberts, J.L.; Buckley, R.H.; Changelian, P.; Candotti, F. JAK3 and the pathogenesis of severe combined immunodeficiency. *Mol. Immunol.* **2004**, *41*, 727–737. [CrossRef]
56. Nemoto, M.; Hattori, H.; Maeda, N.; Akita, N.; Muramatsu, H.; Moritani, S.; Kawasaki, T.; Maejima, M.; Ode, H.; Hachiya, A.; et al. Compound heterozygous TYK2 mutations underlie primary immunodeficiency with T-cell lymphopenia. *Sci. Rep.* **2018**, *8*, 6956. [CrossRef]
57. Wu, P.; Chen, S.; Wu, B.; Chen, J.; Lv, G. A TYK2 gene mutation c.2395G>A leads to TYK2 deficiency: A case report and literature review. *Front. Pediatr.* **2020**, *8*, 253. [CrossRef]
58. Menet, C.J.; van Rompaey, L.; Geney, R. Advances in the discovery of selective JAK inhibitors. *Prog. Med. Chem.* **2013**, *52*, 153–223. [CrossRef]
59. Vazquez, M.L.; Kaila, N.; Strohbach, J.W.; Trzupek, J.D.; Brown, M.F.; Flanagan, M.E.; Mitton-Fry, M.J.; Johnson, T.A.; TenBrink, R.E.; Arnold, E.P.; et al. Identification of N-{cis-3-[Methyl(7H-pyrrolo[2,3-d]pyrimidin-4-yl)amino]cyclobutyl}propane-1-sulfonamide (PF-04965842): A selective JAK1 clinical candidate for the treatment of autoimmune diseases. *J. Med. Chem.* **2018**, *61*, 1130–1152. [CrossRef]
60. Lucet, I.S.; Fantino, E.; Styles, M.; Bamert, R.; Patel, O.; Broughton, S.E.; Walter, M.; Burns, C.J.; Treutlein, H.; Wilks, A.F.; et al. The structural basis of Janus kinase 2 inhibition by a potent and specific pan-Janus kinase inhibitor. *Blood* **2006**, *107*, 176–183. [CrossRef]
61. Thoma, G.; Nuninger, F.; Falchetto, R.; Hermes, E.; Tavares, G.A.; Vangrevelinghe, E.; Zerwes, H.-G. Identification of a potent Janus kinase 3 inhibitor with high selectivity within the Janus kinase family. *J. Med. Chem.* **2011**, *54*, 284–288. [CrossRef]
62. Fensome, A.; Ambler, C.M.; Arnold, E.; Banker, M.E.; Clark, J.D.; Dowty, M.E.; Efremov, I.V.; Flick, A.; Gerstenberger, B.S.; Gifford, R.S.; et al. Design and optimization of a series of 4-(3-azabicyclo[3.1.0]hexan-3-yl)pyrimidin-2-amines: Dual inhibitors of TYK2 and JAK1. *Bioorg. Med. Chem.* **2020**, *28*, 115481. [CrossRef] [PubMed]
63. Meydan, N.; Grunberger, T.; Dadi, H.; Shahar, M.; Arpaia, E.; Lapidot, Z.; Leeder, J.S.; Freedman, M.; Cohen, A.; Gazit, A.; et al. Inhibition of acute lymphoblastic leukaemia by a JAK-2 inhibitor. *Nature* **1996**, *379*, 645–648. [CrossRef] [PubMed]
64. Coricello, A.; Mesiti, F.; Lupia, A.; Maruca, A.; Alcaro, S. Inside perspective of the synthetic and computational toolbox of JAK inhibitors: Recent updates. *Molecules* **2020**, *25*, 3321. [CrossRef] [PubMed]
65. Markham, A. Baricitinib: First global approval. *Drugs* **2017**, *77*, 697–704. [CrossRef]
66. Blair, H.A. Fedratinib: First approval. *Drugs* **2019**, *79*, 1719–1725. [CrossRef]
67. Duggan, S.; Keam, S.J. Upadacitinib: First approval. *Drugs* **2019**, *79*, 1819–1828. [CrossRef]
68. Markham, A.; Keam, S.J. Peficitinib: First global approval. *Drugs* **2019**, *79*, 887–891. [CrossRef]
69. Dhillon, S. Delgocitinib: First approval. *Drugs* **2020**, *80*, 609–615. [CrossRef]
70. Dhillon, S.; Keam, S.J. Filgotinib: First approval. *Drugs* **2020**, *80*, 1987–1997. [CrossRef]
71. Connor, C.G.; DeForest, J.C.; Dietrich, P.; Do, N.M.; Doyle, K.M.; Eisenbeis, S.; Greenberg, E.; Griffin, S.H.; Jones, B.P.; Jones, K.N.; et al. Development of a nitrene-type rearrangement for the commercial route of the JAK1 inhibitor abrocitinib. *Org. Process Res. Dev.* **2021**, *25*, 608–615. [CrossRef]
72. Deeks, E.D.; Duggan, S. Abrocitinib: First approval. *Drugs* **2021**, *81*, 2149–2157. [CrossRef] [PubMed]

73. Eichenfield, L.F.; Flohr, C.; Sidbury, R.; Siegfried, E.; Szalai, Z.; Galus, R.; Yao, Z.; Takahashi, H.; Barbarot, S.; Feeney, C.; et al. Efficacy and safety of abrocitinib in combination with topical therapy in adolescents with moderate-to-severe atopic dermatitis: The JADE TEEN randomized clinical trial. *JAMA Dermatol.* **2021**, *157*, 1165–1173. [CrossRef] [PubMed]
74. Bieber, T.; Simpson, E.L.; Silverberg, J.I.; Thaçi, D.; Paul, C.; Pink, A.E.; Kataoka, Y.; Chu, C.-Y.; DiBonaventura, M.; Rojo, R.; et al. Abrocitinib versus placebo or dupilumab for atopic dermatitis. *N. Engl. J. Med.* **2021**, *384*, 1101–1112. [CrossRef]
75. Blauvelt, A.; Silverberg, J.I.; Lynde, C.W.; Bieber, T.; Eisman, S.; Zdybski, J.; Gubelin, W.; Simpson, E.L.; Valenzuela, F.; Criado, P.R.; et al. Abrocitinib induction, randomized withdrawal, and retreatment in patients with moderate-to-severe atopic dermatitis: Results from the JAK1 Atopic dermatitis efficacy and safety (JADE) REGIMEN phase 3 trial. *J. Am. Acad. Dermatol.* **2022**, *86*, 104–112. [CrossRef]
76. Wang, E.Q.; Le, V.; O'Gorman, M.; Tripathy, S.; Dowty, M.E.; Wang, L.; Malhotra, B.K. Effects of hepatic impairment on the pharmacokinetics of abrocitinib and its metabolites. *J. Clin. Pharmacol.* **2021**, *61*, 1311–1323. [CrossRef] [PubMed]
77. European Medicines Agency EMA/MB/69923/2010, Abrocitinib Assessment Report. Available online: https://www.ema.europa.eu/en/documents/assessment-report/cibinqo-epar-public-assessment-report_en.pdf (accessed on 14 February 2022).
78. FDA Fact Sheet for Healthcare Providers Emergency use Authorization (EUA) of Baricitinib. Available online: https://www.fda.gov/media/143823/download (accessed on 20 April 2022).
79. Rodgers, J.D.; Shepard, S. Azetidine and Cyclobutane Derivatives as JAK Inhibitors. US Patent 8,158,616, 17 April 2012.
80. Xu, J.; Cai, J.; Chen, J.; Zong, X.; Wu, X.; Ji, M.; Wang, P. An efficient synthesis of baricitinib. *J. Chem. Res.* **2016**, *40*, 205–208. [CrossRef]
81. Azad, M.A.K.; Pandey, G.; Singh, K.; Prasad, M. Process for the Preparation of Baricitinib and an Intermediate Thereof. U.S. Patent US10526350B2, 7 January 2020.
82. Cui, X.; Du, J.; Jia, Z.; Wang, X.; Jia, H. A green and facile synthesis of an industrially important quaternary heterocyclic intermediates for baricitinib. *BMC Chem.* **2019**, *13*, 123. [CrossRef] [PubMed]
83. Clark, J.D.; Flanagan, M.E.; Telliez, J.-B. Discovery and development of Janus kinase (JAK) inhibitors for inflammatory diseases. *J. Med. Chem.* **2014**, *57*, 5023–5038. [CrossRef]
84. Sorrell, F.J.; Szklarz, M.; Abdul Azeez, K.R.; Elkins, J.M.; Knapp, S. Family-wide structural analysis of human numb-associated protein kinases. *Structure* **2016**, *24*, 401–411. [CrossRef]
85. Davis, R.R.; Li, B.; Yun, S.Y.; Chan, A.; Nareddy, P.; Gunawan, S.; Ayaz, M.; Lawrence, H.R.; Reuther, G.W.; Lawrence, N.J.; et al. Structural insights into JAK2 inhibition by ruxolitinib, fedratinib, and derivatives thereof. *J. Med. Chem.* **2021**, *64*, 2228–2241. [CrossRef]
86. Chang, Y.; Min, J.; Jarusiewicz, J.; Actis, M.; Bradford, S.Y.-C.; Mayasundari, A.; Yang, L.; Chepyala, D.; Alcock, L.J.; Roberts, K.G.; et al. Degradation of Janus kinases in CRLF2-rearranged acute lymphoblastic leukemia. *Blood.* **2021**, *138*, 2313–2326. [CrossRef] [PubMed]
87. Wan, M.; Cao, X. BMP signaling in skeletal development. *Biochem. Biophys. Res. Commun.* **2005**, *328*, 651–657. [CrossRef] [PubMed]
88. Al-Salama, Z.T.; Scott, L.J. Baricitinib: A review in rheumatoid arthritis. *Drugs* **2018**, *78*, 761–772. [CrossRef] [PubMed]
89. Wallace, D.J.; Furie, R.A.; Tanaka, Y.; Kalunian, K.C.; Mosca, M.; Petri, M.A.; Dörner, T.; Cardiel, M.H.; Bruce, I.N.; Gomez, E.; et al. Baricitinib for systemic lupus erythematosus: A double-blind, randomised, placebo-controlled, phase 2 trial. *Lancet* **2018**, *392*, 222–231. [CrossRef]
90. Guttman-Yassky, E.; Silverberg, J.I.; Nemoto, O.; Forman, S.B.; Wilke, A.; Prescilla, R.; de la Peña, A.; Nunes, F.P.; Janes, J.; Gamalo, M.; et al. Baricitinib in adult patients with moderate-to-severe atopic dermatitis: A phase 2 parallel, double-blinded, randomized placebo-controlled multiple-dose study. *J. Am. Acad. Dermatol.* **2019**, *80*, 913–921.e9. [CrossRef]
91. Hasan, M.J.; Rabbani, R.; Anam, A.M.; Huq, S.M.R. Additional baricitinib loading dose improves clinical outcome in COVID-19. *Open Med.* **2021**, *16*, 41–46. [CrossRef]
92. Kalil, A.C.; Patterson, T.F.; Mehta, A.K.; Tomashek, K.M.; Wolfe, C.R.; Ghazaryan, V.; Marconi, V.C.; Ruiz-Palacios, G.M.; Hsieh, L.; Kline, S.; et al. Baricitinib plus remdesivir for hospitalized adults with COVID-19. *N. Engl. J. Med.* **2020**, *384*, 795–807. [CrossRef]
93. Izumo, T.; Kuse, N.; Awano, N.; Tone, M.; Sakamoto, K.; Takada, K.; Muto, Y.; Fujimoto, K.; Saiki, A.; Ito, Y.; et al. Clinical impact of combination therapy with baricitinib, remdesivir, and dexamethasone in patients with severe COVID-19. *Respir. Investig.* **2021**, *59*, 799–803. [CrossRef]
94. Eli Lilly & Co. Baricitinib. U.S. Food and Drug Administration Website. Available online: https://www.accessdata.fda.gov/drugsatfda_docs/nda/2018/207924Orig1s000PharmR.pdf (accessed on 16 February 2022).
95. Eli Lilly Australia Pty Ltd. Olumiant Australian Public Assessment Report for Baricitinib. Available online: https://www.tga.gov.au/sites/default/files/auspar-baricitinib-190321.pdf (accessed on 16 February 2022).
96. Noji, S.; Hara, Y.; Miura, T.; Yamanaka, H.; Maeda, K.; Hori, A.; Yamamoto, H.; Obika, S.; Inoue, M.; Hase, Y.; et al. Discovery of a Janus kinase inhibitor bearing a highly three-dimensional spiro scaffold: JTE-052 (delgocitinib) as a new dermatological agent to treat inflammatory skin disorders. *J. Med. Chem.* **2020**, *63*, 7163–7185. [CrossRef]
97. Takiguchi, H.; Higashi, A.; Watanabe, T.; Takeichi, T.; Shimazaki, T.; Inaba, T. Stereocontrolled synthesis of delgocitinib, a JAK inhibitor for the treatment of atopic dermatitis. *Org. Process Res. Dev.* **2021**, *25*, 342–348. [CrossRef]

98. Nakagawa, H.; Nemoto, O.; Igarashi, A.; Saeki, H.; Kaino, H.; Nagata, T. Delgocitinib ointment, a topical Janus kinase inhibitor, in adult patients with moderate to severe atopic dermatitis: A phase 3, randomized, double-blind, vehicle-controlled study and an open-label, long-term extension study. *J. Am. Acad. Dermatol.* **2020**, *82*, 823–831. [CrossRef] [PubMed]
99. Tanimoto, A.; Ogawa, Y.; Oki, C.; Kimoto, Y.; Nozawa, K.; Amano, W.; Noji, S.; Shiozaki, M.; Matsuo, A.; Shinozaki, Y.; et al. Pharmacological properties of JTE-052: A novel potent JAK inhibitor that suppresses various inflammatory responses in vitro and in vivo. *Inflamm. Res.* **2015**, *64*, 41–51. [CrossRef] [PubMed]
100. Worm, M.; Bauer, A.; Elsner, P.; Mahler, V.; Molin, S.; Nielsen, T.S.S. Efficacy and safety of topical delgocitinib in patients with chronic hand eczema: Data from a randomized, double-blind, vehicle-controlled phase IIa study. *Br. J. Dermatol.* **2020**, *182*, 1103–1110. [CrossRef] [PubMed]
101. Xie, Z.; Yang, X.; Duan, Y.; Han, J.; Liao, C. Small-molecule kinase inhibitors for the treatment of nononcologic diseases. *J. Med. Chem.* **2021**, *64*, 1283–1345. [CrossRef]
102. Nakagawa, H.; Nemoto, O.; Yamada, H.; Nagata, T.; Ninomiya, N. Phase 1 studies to assess the safety, tolerability and pharmacokinetics of JTE-052 (a novel Janus kinase inhibitor) ointment in Japanese healthy volunteers and patients with atopic dermatitis. *J. Dermatol.* **2018**, *45*, 701–709. [CrossRef]
103. Japan Tobacco Inc. Report on the Deliberation Results, Corectim Ointment 0.5% (Delgocitinib). Available online: https://www.pmda.go.jp/files/000240793.pdf (accessed on 16 February 2022).
104. Flick, A.C.; Leverett, C.A.; Ding, H.X.; McInturff, E.; Fink, S.J.; Mahapatra, S.; Carney, D.W.; Lindsey, E.A.; DeForest, J.C.; France, S.P.; et al. Synthetic approaches to the new drugs approved during 2019. *J. Med. Chem.* **2021**, *64*, 3604–3657. [CrossRef]
105. Wang, Y.; Li, H. Diphenylaminopyrimidine Compound for Inhibiting Kinase Activity. U.S. Patent US20200071303A1, 5 March 2020.
106. Noronha, G.; Mak, C.C.; Cao, J.; Renick, J.; McPherson, A.; Zeng, B.; Pathak, V.P.; Lohse, D.L.; Hood, J.D.; Soll, R.M. Preparation of Pyrimidine Derivatives as JAK Kinases Inhibitors. U.S. Patent US20110212077A1, 1 September 2011.
107. Tefferi, A. N-tert-Butyl-3-[(5-methyl-2-{[4-(2-pyrrolidin-1- ylethoxy)phenyl]amino}pyrimidin-4-yl)amino]benzenesulfonamide for treating myelofibrosis. WO2012060847A1, 10 May 2012.
108. Ciceri, P.; Müller, S.; O'Mahony, A.; Fedorov, O.; Filippakopoulos, P.; Hunt, J.P.; Lasater, E.A.; Pallares, G.; Picaud, S.; Wells, C.; et al. Dual kinase-bromodomain inhibitors for rationally designed polypharmacology. *Nat. Chem. Biol.* **2014**, *10*, 305–312. [CrossRef]
109. Ember, S.W.J.; Zhu, J.-Y.; Olesen, S.H.; Martin, M.P.; Becker, A.; Berndt, N.; Georg, G.I.; Schönbrunn, E. Acetyl-lysine binding site of bromodomain-containing protein 4 (BRD4) interacts with diverse kinase inhibitors. *ACS Chem. Biol.* **2014**, *9*, 1160–1171. [CrossRef]
110. Talpaz, M.; Kiladjian, J.-J. Fedratinib, a newly approved treatment for patients with myeloproliferative neoplasm-associated myelofibrosis. *Leukemia* **2021**, *35*, 1–17. [CrossRef]
111. Poubel, C.P.; Mansur, M.B.; Boroni, M.; Emerenciano, M. FLT3 overexpression in acute leukaemias: New insights into the search for molecular mechanisms. *Biochim. Biophys. Acta Rev. Cancer* **2019**, *1872*, 80–88. [CrossRef]
112. Muller, S.; Filippakopoulos, P.; Knapp, S. Bromodomains as therapeutic targets. *Expert Rev. Mol. Med.* **2011**, *13*, e29. [CrossRef] [PubMed]
113. Ogasawara, K.; Xu, C.; Kanamaluru, V.; Siebers, N.; Surapaneni, S.; Ridoux, L.; Palmisano, M.; Krishna, G. Excretion balance and pharmacokinetics following a single oral dose of [14C]-fedratinib in healthy subjects. *Cancer Chemother. Pharmacol.* **2020**, *86*, 307–314. [CrossRef] [PubMed]
114. Menet, C.J.M.; van Rompaey, L.J.C.; Robert, S.; Blanc, J.; Jouannigot, N.; Hodges, A.J.; Smits, K.K. Novel Compounds Useful for the Treatment of Degenerative and Inflammatory Diseases. WO2010010190A1, 28 January 2021.
115. Ayala-Aguilera, C.C.; Valero, T.; Lorente-Macías, Á.; Baillache, D.J.; Croke, S.; Unciti-Broceta, A. Small molecule kinase inhibitor drugs (1995–2021): Medical indication, pharmacology, and synthesis. *J. Med. Chem.* **2021**, *65*, 1047–1131. [CrossRef] [PubMed]
116. Van Rompaey, L.; Galien, R.; van der Aar, E.M.; Clement-Lacroix, P.; Nelles, L.; Smets, B.; Lepescheux, L.; Christophe, T.; Conrath, K.; Vandeghinste, N.; et al. Preclinical characterization of GLPG0634, a selective inhibitor of JAK1, for the treatment of inflammatory diseases. *J. Immunol.* **2013**, *191*, 3568–3577. [CrossRef] [PubMed]
117. Menet, C.J.; Fletcher, S.R.; Van Lommen, G.; Geney, R.; Blanc, J.; Smits, K.; Jouannigot, N.; Deprez, P.; van der Aar, E.M.; Clement-Lacroix, P.; et al. Triazolopyridines as selective JAK1 inhibitors: From hit identification to GLPG0634. *J. Med. Chem.* **2014**, *57*, 9323–9342. [CrossRef] [PubMed]
118. Newton, A.S.; Deiana, L.; Puleo, D.E.; Cisneros, J.A.; Cutrona, K.J.; Schlessinger, J.; Jorgensen, W.L. JAK2 JH2 Fluorescence polarization assay and crystal structures for complexes with three small molecules. *ACS Med. Chem. Lett.* **2017**, *8*, 614–617. [CrossRef]
119. Norman, P. Selective JAK inhibitors in development for rheumatoid arthritis. *Expert Opin. Investig. Drugs* **2014**, *23*, 1067–1077. [CrossRef] [PubMed]
120. Vanhoutte, F.; Mazur, M.; Voloshyn, O.; Stanislavchuk, M.; Van der Aa, A.; Namour, F.; Galien, R.; Meuleners, L.; van 't Klooster, G. Efficacy, safety, pharmacokinetics, and pharmacodynamics of filgotinib, a selective JAK-1 inhibitor, after short-term treatment of rheumatoid arthritis: Results of two randomized phase IIa trials. *Arthritis Rheumatol.* **2017**, *69*, 1949–1959. [CrossRef] [PubMed]

121. Westhovens, R.; Taylor, P.C.; Alten, R.; Pavlova, D.; Enríquez-Sosa, F.; Mazur, M.; Greenwald, M.; Van der Aa, A.; Vanhoutte, F.; Tasset, C.; et al. Filgotinib (GLPG0634/GS-6034), an oral JAK1 selective inhibitor, is effective in combination with methotrexate (MTX) in patients with active rheumatoid arthritis and insufficient response to MTX: Results from a randomised, dose-finding study (DARWIN 1). *Ann. Rheum. Dis.* **2017**, *76*, 998–1008. [CrossRef] [PubMed]
122. Labetoulle, R.; Paul, S.; Roblin, X. Filgotinib for the treatment of Crohn's disease. *Expert Opin. Investig. Drugs* **2018**, *27*, 295–300. [CrossRef] [PubMed]
123. Vermeire, S.; Schreiber, S.; Petryka, R.; Kuehbacher, T.; Hebuterne, X.; Roblin, X.; Klopocka, M.; Goldis, A.; Wisniewska-Jarosinska, M.; Baranovsky, A.; et al. Clinical remission in patients with moderate-to-severe Crohn's disease treated with filgotinib (the FITZROY study): Results from a phase 2, double-blind, randomised, placebo-controlled trial. *Lancet* **2017**, *389*, 266–275. [CrossRef]
124. Feagan, B.G.; Danese, S.; Loftus, E.V.J.; Vermeire, S.; Schreiber, S.; Ritter, T.; Fogel, R.; Mehta, R.; Nijhawan, S.; Kempiński, R.; et al. Filgotinib as induction and maintenance therapy for ulcerative colitis (SELECTION): A phase 2b/3 double-blind, randomised, placebo-controlled trial. *Lancet* **2021**, *397*, 2372–2384. [CrossRef]
125. Mease, P.; Coates, L.C.; Helliwell, P.S.; Stanislavchuk, M.; Rychlewska-Hanczewska, A.; Dudek, A.; Abi-Saab, W.; Tasset, C.; Meuleners, L.; Harrison, P.; et al. Efficacy and safety of filgotinib, a selective Janus kinase 1 inhibitor, in patients with active psoriatic arthritis (EQUATOR): Results from a randomised, placebo-controlled, phase 2 trial. *Lancet* **2018**, *392*, 2367–2377. [CrossRef]
126. Namour, F.; Diderichsen, P.M.; Cox, E.; Vayssière, B.; Van der Aa, A.; Tasset, C.; Van't Klooster, G. Pharmacokinetics and pharmacokinetic/pharmacodynamic modeling of filgotinib (GLPG0634), a selective JAK1 inhibitor, in support of phase IIB dose selection. *Clin. Pharmacokinet.* **2015**, *54*, 859–874. [CrossRef] [PubMed]
127. Van't Klooster, G.A.E.; Brys, R.C.X.; Van Rompaey, L.J.C.; Namour, F.S. Aminotriazolopyridine for Use in the Treatment of Inflammation, and Pharmaceutical Compositions Thereof. WO2013189771A1, 27 December 2013.
128. Namour, F.; Desrivot, J.; Van der Aa, A.; Harrison, P.; Tasset, C.; van't Klooster, G. Clinical confirmation that the selective JAK1 inhibitor filgotinib (GLPG0634) has a low liability for drug-drug interactions. *Drug Metab. Lett.* **2016**, *10*, 38–48. [CrossRef] [PubMed]
129. Gonzales, A.J.; Bowman, J.W.; Fici, G.J.; Zhang, M.; Mann, D.W.; Mitton-Fry, M. Oclacitinib (APOQUEL®) is a novel Janus kinase inhibitor with activity against cytokines involved in allergy. *J. Vet. Pharmacol. Ther.* **2014**, *37*, 317–324. [CrossRef] [PubMed]
130. Berlinski, P.J.; Birchmeier, M.J.; Bowman, J.W.; Gonzales, A.J.; Kamerling, S.G.; Mann, D.W.; Mitton-Fry, M.J. Pyrrolo[2,3-d]Pyrimidine Compounds. WO2010020905A1, 25 February 2010.
131. Haugh, I.M.; Watson, I.T.; Alan Menter, M. Successful treatment of atopic dermatitis with the JAK1 inhibitor oclacitinib. *Bayl. Univ. Med. Cent. Proc.* **2018**, *31*, 524–525. [CrossRef] [PubMed]
132. Cosgrove, S.B.; Wren, J.A.; Cleaver, D.M.; Martin, D.D.; Walsh, K.F.; Harfst, J.A.; Follis, S.L.; King, V.L.; Boucher, J.F.; Stegemann, M.R. Efficacy and safety of oclacitinib for the control of pruritus and associated skin lesions in dogs with canine allergic dermatitis. *Vet. Dermatol.* **2013**, *24*, 479-e114. [CrossRef] [PubMed]
133. Rynhoud, H.; Gibson, J.S.; Meler, E.; Soares Magalhães, R.J. The association between the use of oclacitinib and antibacterial therapy in dogs with allergic dermatitis: A retrospective case-control study. *Front. Vet. Sci.* **2021**, *8*, 631443. [CrossRef] [PubMed]
134. Banovic, F.; Tarigo, J.; Gordon, H.; Barber, J.P.; Gogal, R.M.J. Immunomodulatory in vitro effects of oclacitinib on canine T-cell proliferation and cytokine production. *Vet. Dermatol.* **2019**, *30*, 17-e6. [CrossRef] [PubMed]
135. De Caro Martins, G.; da Costa-Val, A.P.; Coura, F.M.; Diamantino, G.M.L.; Nogueira, M.M.; de Oliveira Melo-Junior, O.A.; Giunchetti, R.C.; da Silveira-Lemos, D.; Melo, M.M. Immunomodulatory effect of long-term oclacitinib maleate therapy in dogs with atopic dermatitis. *Vet. Dermatol.* **2021**, *33*, 142-e40. [CrossRef] [PubMed]
136. Collard, W.T.; Hummel, B.D.; Fielder, A.F.; King, V.L.; Boucher, J.F.; Mullins, M.A.; Malpas, P.B.; Stegemann, M.R. The pharmacokinetics of oclacitinib maleate, a Janus kinase inhibitor, in the dog. *J. Vet. Pharmacol. Ther.* **2014**, *37*, 279–285. [CrossRef] [PubMed]
137. Zoetisus.com APOQUEL®(Oclacitinib Tablet): Fast-Acting and Safe Itch Relief for Dogs. Available online: https://www.zoetisus.com/products/dogs/apoquel/assets/downloadable-resources/1413113_m03r_apl_infosheet_fda_labelupdate-_1__new2019.pdf (accessed on 14 February 2022).
138. FDA Approves Drug for Adults with Rare Form of Bone Marrow Disorder. Available online: https://www.fda.gov/drugs/news-events-human-drugs/fda-approves-drug-adults-rare-form-bone-marrow-disorder (accessed on 17 April 2022).
139. William, A.D.; Lee, A.C.-H.; Blanchard, S.; Poulsen, A.; Teo, E.L.; Nagaraj, H.; Tan, E.; Chen, D.; Williams, M.; Sun, E.T.; et al. Discovery of the macrocycle 11-(2-pyrrolidin-1-yl-ethoxy)-14,19-dioxa-5,7,26-triaza-tetracyclo[19.3.1.1(2,6).1(8,12)]heptacosa-1(25),2(26),3,5,8,10,12(27),16,21,23-decaene (SB1518), a potent Janus kinase 2/fms-like tyrosine kinase-3 (JAK2/FLT3) inhibitor. *J. Med. Chem.* **2011**, *54*, 4638–4658. [CrossRef] [PubMed]
140. Singer, J.W.; Al-Fayoumi, S.; Ma, H.; Komrokji, R.S.; Mesa, R.; Verstovsek, S. Comprehensive kinase profile of pacritinib, a nonmyelosuppressive Janus kinase 2 inhibitor. *J. Exp. Pharmacol.* **2016**, *8*, 11–19. [CrossRef]
141. Klaeger, S.; Heinzlmeir, S.; Wilhelm, M.; Polzer, H.; Vick, B.; Koenig, P.-A.; Reinecke, M.; Ruprecht, B.; Petzoldt, S.; Meng, C.; et al. The target landscape of clinical kinase drugs. *Science* **2017**, *358*, eaan4368. [CrossRef] [PubMed]
142. Mesa, R.A.; Vannucchi, A.M.; Mead, A.; Egyed, M.; Szoke, A.; Suvorov, A.; Jakucs, J.; Perkins, A.; Prasad, R.; Mayer, J.; et al. Pacritinib versus best available therapy for the treatment of myelofibrosis irrespective of baseline cytopenias (PERSIST-1): An international, randomised, phase 3 trial. *Lancet. Haematol.* **2017**, *4*, e225–e236. [CrossRef]

143. Mascarenhas, J.; Hoffman, R.; Talpaz, M.; Gerds, A.T.; Stein, B.; Gupta, V.; Szoke, A.; Drummond, M.; Pristupa, A.; Granston, T.; et al. Pacritinib vs best available therapy, including ruxolitinib, in patients with myelofibrosis: A randomized clinical trial. *JAMA Oncol.* **2018**, *4*, 652–659. [CrossRef]
144. Hosseini, M.M.; Kurtz, S.E.; Abdelhamed, S.; Mahmood, S.; Davare, M.A.; Kaempf, A.; Elferich, J.; McDermott, J.E.; Liu, T.; Payne, S.H.; et al. Inhibition of interleukin-1 receptor-associated kinase-1 is a therapeutic strategy for acute myeloid leukemia subtypes. *Leukemia* **2018**, *32*, 2374–2387. [CrossRef]
145. Jeon, J.Y.; Zhao, Q.; Buelow, D.R.; Phelps, M.; Walker, A.R.; Mims, A.S.; Vasu, S.; Behbehani, G.; Blachly, J.; Blum, W.; et al. Preclinical activity and a pilot phase I study of pacritinib, an oral JAK2/FLT3 inhibitor, and chemotherapy in FLT3-ITD-positive AML. *Investig. New Drugs* **2020**, *38*, 340–349. [CrossRef]
146. Jensen, K.V.; Cseh, O.; Aman, A.; Weiss, S.; Luchman, H.A. The JAK2/STAT3 inhibitor pacritinib effectively inhibits patient-derived GBM brain tumor initiating cells in vitro and when used in combination with temozolomide increases survival in an orthotopic xenograft model. *PLoS ONE* **2017**, *12*, e0189670. [CrossRef]
147. Jayaraman, R.; Pasha, M.K.; Williams, A.; Goh, K.C.; Ethirajulu, K. Metabolism and disposition of pacritinib (SB1518), an orally active Janus kinase 2 inhibitor in preclinical species and humans. *Drug Metab. Lett.* **2015**, *9*, 28–47. [CrossRef] [PubMed]
148. Inoue, T.; Tanaka, A.; Nakai, K.; Sasaki, H.; Takahashi, F.; Shirakami, S.; Hatanaka, K.; Nakajima, Y.; Mukoyoshi, K.; Hamaguchi, H.; et al. Heterocyclic Janus Kinase 3 Inhibitors. WO2007077949A1, 12 July 2007.
149. Ito, M.; Yamazaki, S.; Yamagami, K.; Kuno, M.; Morita, Y.; Okuma, K.; Nakamura, K.; Chida, N.; Inami, M.; Inoue, T.; et al. A novel JAK inhibitor, peficitinib, demonstrates potent efficacy in a rat adjuvant-induced arthritis model. *J. Pharmacol. Sci.* **2017**, *133*, 25–33. [CrossRef] [PubMed]
150. Takeuchi, T.; Tanaka, Y.; Iwasaki, M.; Ishikura, H.; Saeki, S.; Kaneko, Y. Efficacy and safety of the oral Janus kinase inhibitor peficitinib (ASP015K) monotherapy in patients with moderate to severe rheumatoid arthritis in Japan: A 12-week, randomised, double-blind, placebo-controlled phase IIb study. *Ann. Rheum. Dis.* **2016**, *75*, 1057–1064. [CrossRef] [PubMed]
151. Hamaguchi, H.; Amano, Y.; Moritomo, A.; Shirakami, S.; Nakajima, Y.; Nakai, K.; Nomura, N.; Ito, M.; Higashi, Y.; Inoue, T. Discovery and structural characterization of peficitinib (ASP015K) as a novel and potent JAK inhibitor. *Bioorg. Med. Chem.* **2018**, *26*, 4971–4983. [CrossRef] [PubMed]
152. Zhang, J.-M.; An, J. Cytokines, inflammation, and pain. *Int. Anesthesiol. Clin.* **2007**, *45*, 27–37. [CrossRef]
153. Takeuchi, T.; Tanaka, Y.; Tanaka, S.; Kawakami, A.; Song, Y.-W.; Chen, Y.-H.; Rokuda, M.; Izutsu, H.; Ushijima, S.; Kaneko, Y. Safety and effectiveness of peficitinib (ASP015K) in patients with rheumatoid arthritis: Final results (32 months of mean peficitinib treatment) from a long-term, open-label extension study in Japan, Korea, and Taiwan. *Rheumatol. Ther.* **2021**, *8*, 425–442. [CrossRef]
154. Genovese, M.C.; Greenwald, M.; Codding, C.; Zubrzycka-Sienkiewicz, A.; Kivitz, A.J.; Wang, A.; Shay, K.; Wang, X.; Garg, J.P.; Cardiel, M.H. Peficitinib, a JAK inhibitor, in combination with limited conventional synthetic disease-modifying antirheumatic drugs in the treatment of moderate-to-severe rheumatoid arthritis. *Arthritis Rheumatol.* **2017**, *69*, 932–942. [CrossRef]
155. Tanaka, Y.; Izutsu, H. Peficitinib for the treatment of rheumatoid arthritis: An overview from clinical trials. *Expert Opin. Pharmacother.* **2020**, *21*, 1015–1025. [CrossRef]
156. Takeuchi, T.; Tanaka, Y.; Tanaka, S.; Kawakami, A.; Song, Y.-W.; Chen, Y.-H.; Rokuda, M.; Izutsu, H.; Ushijima, S.; Kaneko, Y.; et al. Safety and effectiveness of peficitinib (ASP015K) in patients with rheumatoid arthritis: Interim data (22.7 months mean peficitinib treatment) from a long-term, open-label extension study in Japan, Korea, and Taiwan. *Arthritis Res. Ther.* **2020**, *22*, 47. [CrossRef]
157. Kaneko, Y. Efficacy and safety of peficitinib in rheumatoid arthritis. *Mod. Rheumatol.* **2020**, *30*, 773–778. [CrossRef]
158. Oda, K.; Cao, Y.J.; Sawamoto, T.; Nakada, N.; Fisniku, O.; Nagasaka, Y.; Sohda, K.-Y. Human mass balance, metabolite profile and identification of metabolic enzymes of [^{14}C]ASP015K, a novel oral janus kinase inhibitor. *Xenobiotica* **2015**, *45*, 887–902. [CrossRef] [PubMed]
159. Miyatake, D.; Shibata, T.; Toyoshima, J.; Kaneko, Y.; Oda, K.; Nishimura, T.; Katashima, M.; Sakaki, M.; Inoue, K.; Ito, T.; et al. Pharmacokinetics and safety of a single oral dose of peficitinib (ASP015K) in Japanese subjects with normal and impaired hepatic function. *Clin. Pharmacol. Drug Dev.* **2020**, *9*, 699–708. [CrossRef] [PubMed]
160. Yang, W.; Zhu, G.; Qin, M.; Li, Z.; Wang, B.; Yang, J.; Wang, T. The effectiveness of ruxolitinib for acute/chronic graft-versus-host disease in children: A retrospective study. *Drug Des. Devel. Ther.* **2021**, *15*, 743–752. [CrossRef] [PubMed]
161. Rodgers, D.J.; Shepard, S. Heteroaryl Substituted Pyrrolo[2,3-b]pyridines and Pyrrolo[2,3-b]pyrimidines as Janus Kinase Inhibitors. U.S. Patent US7598257B2, 16 October 2009.
162. Haydl, A.M.; Xu, K.; Breit, B. Regio- and enantioselective synthesis of N-substituted pyrazoles by rhodium-catalyzed asymmetric addition to allenes. *Angew. Chemie Int. Ed.* **2015**, *54*, 7149–7153. [CrossRef] [PubMed]
163. Lin, Q.; Meloni, D.; Pan, Y.; Xia, M.; Rodgers, J.; Shepard, S.; Li, M.; Galya, L.; Metcalf, B.; Yue, T.-Y.; et al. Enantioselective synthesis of Janus kinase inhibitor INCB018424 via an organocatalytic Aza-Michael reaction. *Org. Lett.* **2009**, *11*, 1999–2002. [CrossRef] [PubMed]
164. Zhou, T.; Georgeon, S.; Moser, R.; Moore, D.J.; Caflisch, A.; Hantschel, O. Specificity and mechanism-of-action of the JAK2 tyrosine kinase inhibitors ruxolitinib and SAR302503 (TG101348). *Leukemia* **2014**, *28*, 404–407. [CrossRef]
165. Duan, Y.; Chen, L.; Chen, Y.; Fan, X. c-Src binds to the cancer drug Ruxolitinib with an active conformation. *PLoS ONE* **2014**, *9*, e106225. [CrossRef]

166. Quintás-Cardama, A.; Vaddi, K.; Liu, P.; Manshouri, T.; Li, J.; Scherle, P.A.; Caulder, E.; Wen, X.; Li, Y.; Waeltz, P.; et al. Preclinical characterization of the selective JAK1/2 inhibitor INCB018424: Therapeutic implications for the treatment of myeloproliferative neoplasms. *Blood* **2010**, *115*, 3109–3117. [CrossRef]
167. Arana Yi, C.; Tam, C.S.; Verstovsek, S. Efficacy and safety of ruxolitinib in the treatment of patients with myelofibrosis. *Future Oncol.* **2015**, *11*, 719–733. [CrossRef]
168. Verstovsek, S.; Mesa, R.A.; Gotlib, J.; Levy, R.S.; Gupta, V.; DiPersio, J.F.; Catalano, J.V.; Deininger, M.; Miller, C.; Silver, R.T.; et al. A double-blind, placebo-controlled trial of ruxolitinib for myelofibrosis. *N. Engl. J. Med.* **2012**, *366*, 799–807. [CrossRef]
169. Tavallai, M.; Booth, L.; Roberts, J.L.; Poklepovic, A.; Dent, P. Rationally repurposing ruxolitinib (Jakafi (®)) as a solid tumor therapeutic. *Front. Oncol.* **2016**, *6*, 142. [CrossRef]
170. Kim, B.S.; Howell, M.D.; Sun, K.; Papp, K.; Nasir, A.; Kuligowski, M.E. Treatment of atopic dermatitis with ruxolitinib cream (JAK1/JAK2 inhibitor) or triamcinolone cream. *J. Allergy Clin. Immunol.* **2020**, *145*, 572–582. [CrossRef] [PubMed]
171. Papp, K.; Szepietowski, J.C.; Kircik, L.; Toth, D.; Eichenfield, L.F.; Leung, D.Y.M.; Forman, S.B.; Venturanza, M.E.; Sun, K.; Kuligowski, M.E.; et al. Efficacy and safety of ruxolitinib cream for the treatment of atopic dermatitis: Results from 2 phase 3, randomized, double-blind studies. *J. Am. Acad. Dermatol.* **2021**, *85*, 863–872. [CrossRef]
172. Witkoff, B.; Logas, C.M.; Glick, B.P.; Del Rosso, J.Q. JAK inhibitors in the treatment of atopic dermatitis. *Dermatol. Rev.* **2022**, *3*, 20–28. [CrossRef]
173. Sarmiento Maldonado, M.; Ramírez Villanueva, P.; Bertín Cortes-Monroy, P.; Jara Arias, V.; Soto Donoso, K.; Uribe Gonzalez, P.; Ocqueteau Tachini, M.; Perez-Simón, J.A. Compassionate use of ruxolitinib in acute and chronic graft versus host disease refractory both to corticosteroids and extracorporeal photopheresis. *Exp. Hematol. Oncol.* **2017**, *6*, 32. [CrossRef]
174. Przepiorka, D.; Luo, L.; Subramaniam, S.; Qiu, J.; Gudi, R.; Cunningham, L.C.; Nie, L.; Leong, R.; Ma, L.; Sheth, C.; et al. FDA approval summary: Ruxolitinib for treatment of steroid-refractory acute graft-versus-host disease. *Oncologist* **2020**, *25*, e328–e334. [CrossRef] [PubMed]
175. Shilling, A.D.; Nedza, F.M.; Emm, T.; Diamond, S.; McKeever, E.; Punwani, N.; Williams, W.; Arvanitis, A.; Galya, L.G.; Li, M.; et al. Metabolism, excretion, and pharmacokinetics of [14C]INCB018424, a selective Janus tyrosine kinase 1/2 inhibitor, in humans. *Drug Metab. Dispos.* **2010**, *38*, 2023–2031. [CrossRef]
176. Shi, J.G.; Chen, X.; Emm, T.; Scherle, P.A.; McGee, R.F.; Lo, Y.; Landman, R.R.; McKeever, E.G.J.; Punwani, N.G.; Williams, W.V.; et al. The effect of CYP3A4 inhibition or induction on the pharmacokinetics and pharmacodynamics of orally administered ruxolitinib (INCB018424 phosphate) in healthy volunteers. *J. Clin. Pharmacol.* **2012**, *52*, 809–818. [CrossRef]
177. Aschenbrenner, D.S. Tofacitinib trial prompts FDA review of adverse effects. *Am. J. Nurs.* **2019**, *119*, 25. [CrossRef]
178. Berbert Ferreira, S.; Berbert Ferreira, R.; Neves Neto, A.C.; Assef, S.M.C.; Scheinberg, M. Topical tofacitinib: A Janus kinase inhibitor for the treatment of vitiligo in an adolescent patient. *Case Rep. Dermatol.* **2021**, *13*, 190–194. [CrossRef]
179. Kostik, M.M.; Raupov, R.K.; Suspitsin, E.N.; Isupova, E.A.; Gaidar, E.V.; Gabrusskaya, T.V.; Kaneva, M.A.; Snegireva, L.S.; Likhacheva, T.S.; Miulkidzhan, R.S.; et al. The safety and efficacy of tofacitinib in 24 cases of pediatric rheumatic diseases: Single centre experience. *Front. Pediatr.* **2022**, *10*, 820586. [CrossRef] [PubMed]
180. Mohanakrishnan, R.; Beier, S.; Deodhar, A. Tofacitinib for the treatment of active ankylosing spondylitis in adults. *Expert Rev. Clin. Immunol.* **2022**, *18*, 273–280. [CrossRef] [PubMed]
181. Blumenkopf, T.A.; Flanagan, M.E.; Munchhof, M.J. Pyrrolo[2,3-d]pyrimidine Compounds. U.S. Patent US6956041B2, 18 October 2005.
182. Cai, W.; Colony, J.L.; Frost, H.; Hudspeth, J.P.; Kendall, P.M.; Krishnan, A.M.; Makowski, T.; Mazur, D.J.; Phillips, J.; Ripin, D.H.B.; et al. Investigation of practical routes for the kilogram-scale production of cis-3-methylamino-4-methylpiperidines. *Org. Process Res. Dev.* **2005**, *9*, 51–56. [CrossRef]
183. Williams, N.K.; Bamert, R.S.; Patel, O.; Wang, C.; Walden, P.M.; Wilks, A.F.; Fantino, E.; Rossjohn, J.; Lucet, I.S. Dissecting specificity in the Janus kinases: The structures of JAK-specific inhibitors complexed to the JAK1 and JAK2 protein tyrosine kinase domains. *J. Mol. Biol.* **2009**, *387*, 219–232. [CrossRef] [PubMed]
184. Chrencik, J.E.; Patny, A.; Leung, I.K.; Korniski, B.; Emmons, T.L.; Hall, T.; Weinberg, R.A.; Gormley, J.A.; Williams, J.M.; Day, J.E.; et al. Structural and thermodynamic characterization of the TYK2 and JAK3 kinase domains in complex with CP-690550 and CMP-6. *J. Mol. Biol.* **2010**, *400*, 413–433. [CrossRef]
185. Chamberlain, P.; Delker, S.; Pagarigan, B.; Mahmoudi, A.; Jackson, P.; Abbasian, M.; Muir, J.; Raheja, N.; Cathers, B. Crystal structures of PRK1 in complex with the clinical compounds lestaurtinib and tofacitinib reveal ligand induced conformational changes. *PLoS ONE* **2014**, *9*, e103638. [CrossRef]
186. Fleischmann, R.; Kremer, J.; Cush, J.; Schulze-Koops, H.; Connell, C.A.; Bradley, J.D.; Gruben, D.; Wallenstein, G.V.; Zwillich, S.H.; Kanik, K.S. Placebo-controlled trial of tofacitinib monotherapy in rheumatoid arthritis. *N. Engl. J. Med.* **2012**, *367*, 495–507. [CrossRef]
187. Van Vollenhoven, R.F.; Fleischmann, R.; Cohen, S.; Lee, E.B.; García Meijide, J.A.; Wagner, S.; Forejtova, S.; Zwillich, S.H.; Gruben, D.; Koncz, T.; et al. Tofacitinib or Adalimumab versus Placebo in Rheumatoid Arthritis. *N. Engl. J. Med.* **2012**, *367*, 508–519. [CrossRef] [PubMed]
188. Van der Heijde, D.; Tanaka, Y.; Fleischmann, R.; Keystone, E.; Kremer, J.; Zerbini, C.; Cardiel, M.H.; Cohen, S.; Nash, P.; Song, Y.-W.; et al. Tofacitinib (CP-690,550) in patients with rheumatoid arthritis receiving methotrexate: Twelve-month data from a twenty-four–month phase III randomized radiographic study. *Arthritis Rheum.* **2013**, *65*, 559–570. [CrossRef]

189. Gladman, D.; Rigby, W.; Azevedo, V.F.; Behrens, F.; Blanco, R.; Kaszuba, A.; Kudlacz, E.; Wang, C.; Menon, S.; Hendrikx, T.; et al. Tofacitinib for psoriatic arthritis in patients with an inadequate response to TNF inhibitors. *N. Engl. J. Med.* **2017**, *377*, 1525–1536. [CrossRef]
190. Mease, P.; Hall, S.; FitzGerald, O.; van der Heijde, D.; Merola, J.F.; Avila-Zapata, F.; Cieślak, D.; Graham, D.; Wang, C.; Menon, S.; et al. Tofacitinib or adalimumab versus placebo for psoriatic arthritis. *N. Engl. J. Med.* **2017**, *377*, 1537–1550. [CrossRef] [PubMed]
191. Sandborn, W.J.; Ghosh, S.; Panes, J.; Vranic, I.; Su, C.; Rousell, S.; Niezychowski, W. Tofacitinib, an oral Janus kinase inhibitor, in active ulcerative colitis. *N. Engl. J. Med.* **2012**, *367*, 616–624. [CrossRef] [PubMed]
192. Huang, Z.; Lee, P.Y.; Yao, X.; Zheng, S.; Li, T. Tofacitinib treatment of refractory systemic juvenile idiopathic arthritis. *Pediatrics* **2019**, *143*, e20182845. [CrossRef] [PubMed]
193. Dowty, M.E.; Lin, J.; Ryder, T.F.; Wang, W.; Walker, G.S.; Vaz, A.; Chan, G.L.; Krishnaswami, S.; Prakash, C. The pharmacokinetics, metabolism, and clearance mechanisms of tofacitinib, a Janus kinase inhibitor, in humans. *Drug Metab. Dispos.* **2014**, *42*, 759–773. [CrossRef]
194. Guo, X.; Li, W.; Li, Q.; Chen, Y.; Zhao, G.; Peng, Y.; Zheng, J. Tofacitinib is a mechanism-based inactivator of cytochrome P450 3A4. *Chem. Res. Toxicol.* **2019**, *32*, 1791–1800. [CrossRef]
195. Muensterman, E.; Engelhardt, B.; Gopalakrishnan, S.; Anderson, J.K.; Mohamed, M.-E.F. Upadacitinib pharmacokinetics and exposure-response analyses of efficacy and safety in psoriatic arthritis patients—Analyses of phase III clinical trials. *Clin. Transl. Sci.* **2022**, *15*, 267–278. [CrossRef]
196. AbbVie U.S. FDA Approves RINVOQ®(Upadacitinib) to Treat Adults and Children 12 Years and Older with Refractory, Moderate to Severe Atopic Dermatitis. Available online: https://news.abbvie.com/news/press-releases/us-fda-approves-rinvoq-upadacitinib-to-treat-adults-and-children-12-years-and-older-with-refractory-moderate-to-severe-atopic-dermatitis.htm#:~{}:text=14%2C2022%2FPRNewswire%2F (accessed on 20 April 2022).
197. AbbVie RINVOQ®(Upadacitinib) Receives FDA Approval for the Treatment of Adults with Moderately to Severely Active Ulcerative Colitis. Available online: https://news.abbvie.com/news/press-releases/rinvoq-upadacitinib-receives-fda-approval-for-treatment-adults-with-moderately-to-severely-active-ulcerative-colitis.htm (accessed on 19 April 2022).
198. Allian, A.; Jayanth, J.; Mohamed, M.-E.; Mulhern, M.; Nordstroem, L.F.; Othman, A.; Rozema, M.; Bhagavatula, L.; Marroum, P.J.; Mayer, P.T. Processes for the Preparation of (3S,4R)-3-Ethyl-4-(3H-Imidazo[1,2-a]Pyrrolo[2,3-e]-Pyrazin-8-yl)-N-(2,2,2-Trifluoroethyl)Pyrrolidine-1-Carboxamide and Solid State Forms Thereof. WO2017066775A1, 20 April 2017.
199. Parmentier, J.M.; Voss, J.; Graff, C.; Schwartz, A.; Argiriadi, M.; Friedman, M.; Camp, H.S.; Padley, R.J.; George, J.S.; Hyland, D.; et al. In vitro and in vivo characterization of the JAK1 selectivity of upadacitinib (ABT-494). *BMC Rheumatol.* **2018**, *2*, 23. [CrossRef]
200. Padda, I.S.; Bhatt, R.; Parmar, M. *Upadacitinib*; StatPearls Publishing: Treasure Island, FL, USA, 2021.
201. Strand, V.; Mease, P.J.; Soriano, E.R.; Kishimoto, M.; Salvarani, C.; Saffore, C.D.; Zueger, P.; McDearmon-Blondell, E.; Kato, K.; Gladman, D.D. Improvement in patient-reported outcomes in patients with psoriatic arthritis treated with upadacitinib versus placebo or adalimumab: Results from SELECT-PsA 1. *Rheumatol. Ther.* **2021**, *8*, 1789–1808. [CrossRef]
202. Serhal, L.; Edwards, C.J. Upadacitinib for the treatment of rheumatoid arthritis. *Expert Rev. Clin. Immunol.* **2019**, *15*, 13–25. [CrossRef]
203. Bergman, M.; Tundia, N.; Yang, M.; Orvis, E.; Clewell, J.; Bensimon, A. Economic benefit from improvements in quality of life with upadacitinib: Comparisons with tofacitinib and methotrexate in patients with rheumatoid arthritis. *Adv. Ther.* **2021**, *38*, 5649–5661. [CrossRef]
204. Smolen, J.S.; Pangan, A.L.; Emery, P.; Rigby, W.; Tanaka, Y.; Vargas, J.I.; Zhang, Y.; Damjanov, N.; Friedman, A.; Othman, A.A.; et al. Upadacitinib as monotherapy in patients with active rheumatoid arthritis and inadequate response to methotrexate (select-monotherapy): A randomised, placebo-controlled, double-blind phase 3 study. *Lancet* **2019**, *393*, 2303–2311. [CrossRef]
205. McInnes, I.B.; Anderson, J.K.; Magrey, M.; Merola, J.F.; Liu, Y.; Kishimoto, M.; Jeka, S.; Pacheco-Tena, C.; Wang, X.; Chen, L.; et al. Trial of upadacitinib and adalimumab for psoriatic arthritis. *N. Engl. J. Med.* **2021**, *384*, 1227–1239. [CrossRef]
206. Burmester, G.R.; Winthrop, K.; Blanco, R.; Nash, P.; Goupille, P.; Azevedo, V.F.; Salvarani, C.; Rubbert-Roth, A.; Lesser, E.; Lippe, R.; et al. Safety profile of upadacitinib up to 3 years in psoriatic arthritis: An integrated analysis of two pivotal phase 3 trials. *Rheumatol. Ther.* **2021**, *9*, 521–539. [CrossRef] [PubMed]
207. Funk, P.J.; Perche, P.O.; Singh, R.; Kelly, K.A.; Feldman, S.R. Comparing available JAK inhibitors for treating patients with psoriasis. *Expert Rev. Clin. Immunol.* **2022**, *18*, 281–294. [CrossRef]
208. Blauvelt, A.; Teixeira, H.D.; Simpson, E.L.; Costanzo, A.; De Bruin-Weller, M.; Barbarot, S.; Prajapati, V.H.; Lio, P.; Hu, X.; Wu, T.; et al. Efficacy and safety of upadacitinib vs dupilumab in adults with moderate-to-severe atopic dermatitis: A randomized clinical trial. *JAMA Dermatol.* **2021**, *157*, 1047–1055. [CrossRef] [PubMed]
209. Mohamed, M.-E.F.; Camp, H.S.; Jiang, P.; Padley, R.J.; Asatryan, A.; Othman, A.A. Pharmacokinetics, safety and tolerability of ABT-494, a novel selective JAK 1 inhibitor, in healthy volunteers and subjects with rheumatoid arthritis. *Clin. Pharmacokinet.* **2016**, *55*, 1547–1558. [CrossRef] [PubMed]
210. Mohamed, M.-E.F.; Jungerwirth, S.; Asatryan, A.; Jiang, P.; Othman, A.A. Assessment of effect of CYP3A inhibition, CYP induction, OATP1B inhibition, and high-fat meal on pharmacokinetics of the JAK1 inhibitor upadacitinib. *Br. J. Clin. Pharmacol.* **2017**, *83*, 2242–2248. [CrossRef]

211. Mohamed, M.-E.F.; Klünder, B.; Othman, A.A. Clinical pharmacokinetics of upadacitinib: Review of data relevant to the rheumatoid arthritis indication. *Clin. Pharmacokinet.* **2020**, *59*, 531–544. [CrossRef]
212. Vainchenker, W.; Favale, F. Myelofibrosis, JAK2 inhibitors and erythropoiesis. *Leukemia* **2013**, *27*, 1219–1223. [CrossRef]
213. Wood, H.; Chandler, A.; Nezamololama, N.; Papp, K.; Gooderham, M.J. Safety of Janus kinase (JAK) inhibitors in the short-term treatment of atopic dermatitis. *Int. J. Dermatol.* **2021**. [CrossRef]
214. Minshawi, F.; Lanvermann, S.; McKenzie, E.; Jeffery, R.; Couper, K.; Papoutsopoulou, S.; Roers, A.; Muller, W. The generation of an engineered interleukin-10 protein with improved stability and biological function. *Front. Immunol.* **2020**, *11*, 1794. [CrossRef] [PubMed]
215. Almoallim, H.M.; Alharbi, L.A. Rheumatoid arthritis in Saudi Arabia. *Saudi Med. J.* **2014**, *35*, 1442–1454. [PubMed]
216. Hussain, W.; Janoudi, N.; Noorwali, A.; Omran, N.; Baamer, M.; Assiry, E.H.; Alrayes, H.; Alosaimi, H.; Ibrahim, A.; Gohary, S.; et al. Effect of adalimumab on work ability assessed in rheumatoid arthritis disease patients in Saudi Arabia (AWARDS). *Open Rheumatol. J.* **2015**, *9*, 46–50. [CrossRef]
217. O'Shea, J.J.; Gadina, M. Selective Janus kinase inhibitors come of age. *Nat. Rev. Rheumatol.* **2019**, *15*, 74–75. [CrossRef] [PubMed]
218. Pardanani, A.; Lasho, T.; Smith, G.; Burns, C.J.; Fantino, E.; Tefferi, A. CYT387, a selective JAK1/JAK2 inhibitor: In vitro assessment of kinase selectivity and preclinical studies using cell lines and primary cells from polycythemia vera patients. *Leukemia* **2009**, *23*, 1441–1445. [CrossRef]
219. Sideris, N.; Vakirlis, E.; Tsentemeidou, A.; Kourouklidou, A.; Ioannides, D.; Sotiriou, E. Under development JAK inhibitors for dermatologic diseases. *Mediterr. J. Rheumatol.* **2020**, *31*, 137–144. [CrossRef]
220. Ma, J.; Xing, W.; Coffey, G.; Dresser, K.; Lu, K.; Guo, A.; Raca, G.; Pandey, A.; Conley, P.; Yu, H.; et al. Cerdulatinib, a novel dual SYK/JAK kinase inhibitor, has broad anti-tumor activity in both ABC and GCB types of diffuse large B cell lymphoma. *Oncotarget* **2015**, *6*, 43881–43896. [CrossRef]
221. Nogueira, M.; Puig, L.; Torres, T. JAK inhibitors for treatment of psoriasis: Focus on selective TYK2 inhibitors. *Drugs* **2020**, *80*, 341–352. [CrossRef]
222. Schroeder, M.A.; Khoury, H.J.; Jagasia, M.; Ali, H.; Schiller, G.J.; Staser, K.; Choi, J.; Gehrs, L.; Arbushites, M.C.; Yan, Y.; et al. A phase 1 trial of itacitinib, a selective JAK1 inhibitor, in patients with acute graft-versus-host disease. *Blood Adv.* **2020**, *4*, 1656–1669. [CrossRef]
223. Verstovsek, S.; Mesa, R.A.; Salama, M.E.; Li, L.; Pitou, C.; Nunes, F.P.; Price, G.L.; Giles, J.L.; D'Souza, D.N.; Walgren, R.A.; et al. A phase 1 study of the Janus kinase 2 (JAK2)(V617F) inhibitor, gandotinib (LY2784544), in patients with primary myelofibrosis, polycythemia vera, and essential thrombocythemia. *Leuk. Res.* **2017**, *61*, 89–95. [CrossRef]
224. Verstovsek, S.; Odenike, O.; Singer, J.W.; Granston, T.; Al-Fayoumi, S.; Deeg, H.J. Phase 1/2 study of pacritinib, a next generation JAK2/FLT3 inhibitor, in myelofibrosis or other myeloid malignancies. *J. Hematol. Oncol.* **2016**, *9*, 137. [CrossRef] [PubMed]
225. Genovese, M.C.; van Vollenhoven, R.F.; Pacheco-Tena, C.; Zhang, Y.; Kinnman, N. VX-509 (decernotinib), an oral selective JAK-3 inhibitor, in combination with methotrexate in patients with rheumatoid arthritis. *Arthritis Rheumatol.* **2016**, *68*, 46–55. [CrossRef] [PubMed]
226. Tan, L.; Akahane, K.; McNally, R.; Reyskens, K.M.S.E.; Ficarro, S.B.; Liu, S.; Herter-Sprie, G.S.; Koyama, S.; Pattison, M.J.; Labella, K.; et al. Development of selective covalent Janus kinase 3 inhibitors. *J. Med. Chem.* **2015**, *58*, 6589–6606. [CrossRef] [PubMed]
227. Tanaka, Y.; Luo, Y.; O'Shea, J.J.; Nakayamada, S. Janus kinase-targeting therapies in rheumatology: A mechanisms-based approach. *Nat. Rev. Rheumatol.* **2022**, *18*, 133–145. [CrossRef]
228. Jin, W.; Stokes, J.M.; Eastman, R.T.; Itkin, Z.; Zakharov, A.V.; Collins, J.J.; Jaakkola, T.S.; Barzilay, R. Deep learning identifies synergistic drug combinations for treating COVID-19. *Proc. Natl. Acad. Sci. USA* **2021**, *118*, e2105070118. [CrossRef] [PubMed]
229. Sayed Ahmed, H.A.; Merrell, E.; Ismail, M.; Joudeh, A.I.; Riley, J.B.; Shawkat, A.; Habeb, H.; Darling, E.; Goweda, R.A.; Shehata, M.H.; et al. Rationales and uncertainties for aspirin use in COVID-19: A narrative review. *Fam. Med. Community Health* **2021**, *9*, e000741. [CrossRef]
230. Marconi, V.C.; Ramanan, A.V.; de Bono, S.; Kartman, C.E.; Krishnan, V.; Liao, R.; Piruzeli, M.L.B.; Goldman, J.D.; Alatorre-Alexander, J.; de Cassia Pellegrini, R.; et al. Efficacy and safety of baricitinib for the treatment of hospitalised adults with COVID-19 (COV-BARRIER): A randomised, double-blind, parallel-group, placebo-controlled phase 3 trial. *Lancet Respir. Med.* **2021**, *9*, 1407–1418. [CrossRef]
231. Kmietowicz, Z. COVID-19: WHO recommends baricitinib and sotrovimab to treat patients. *BMJ* **2022**, *376*, o97. [CrossRef]
232. Guimarães, P.O.; Quirk, D.; Furtado, R.H.; Maia, L.N.; Saraiva, J.F.; Antunes, M.O.; Kalil Filho, R.; Junior, V.M.; Soeiro, A.M.; Tognon, A.P.; et al. Tofacitinib in patients hospitalized with COVID-19 pneumonia. *N. Engl. J. Med.* **2021**, *385*, 406–415. [CrossRef] [PubMed]
233. Seror, R.; Camus, M.; Salmon, J.-H.; Roux, C.; Dernis, E.; Basch, A.; Germain, V.; Leske, C.; Brousseau, S.; Truchetet, M.-E.; et al. Do JAK inhibitors affect immune response to COVID-19 vaccination? Data from the MAJIK-SFR Registry. *Lancet. Rheumatol.* **2022**, *4*, e8–e11. [CrossRef]

MDPI
St. Alban-Anlage 66
4052 Basel
Switzerland
www.mdpi.com

Pharmaceutics Editorial Office
E-mail: pharmaceutics@mdpi.com
www.mdpi.com/journal/pharmaceutics

www.ingramcontent.com/pod-product-compliance
Lightning Source LLC
Chambersburg PA
CBHW061502220926
51850CB00004B/147
9783725812905